HANDBOOK OF DEVELOPMENTAL NEUROTOXICOLOGY

HANDBOOK OF DEVELOPMENTAL NEUROTOXICOLOGY

DR. WILLIAM SLIKKER, JR.
Division of Neurotoxicology
National Center for Toxicological Research
Food and Drug Administration
Jefferson, Arkansas 72079

DR. LOUIS W. CHANG
Departments of Pathology, Pharmacology, and Toxicology
College of Medicine
University of Arkansas for Medical Sciences
Little Rock, Arkansas 72205

Academic Press
San Diego London Boston New York Sydney Tokyo Toronto

Front cover photograph: Images © 1998 Photodisc, Inc.

This book is printed on acid-free paper. ∞

Academic Press
a division of Harcourt Brace & Company
525 B Street, Suite 1900, San Diego, California 92101-4495, USA
http://www.apnet.com

Academic Press Limited
24-28 Oval Road, London NW1 7DX, UK
http://www.hbuk.co.uk/ap/

Library of Congress Catalog Card Number: 98-85622

International Standard Book Number: 0-12-648860-6

PRINTED IN THE UNITED STATES OF AMERICA
98 99 00 01 02 03 MM 9 8 7 6 5 4 3 2 1

In memory of our mentors, colleagues, and friends, Keith F. Killam, Jr., Ph.D., neuropharmacologist/toxicologist, and Donald E. Hill, M.D., pediatrician/perinatologist. Their courageous pursuit of knowledge provided the framework and inspiration for this comprehensive review of developmental neurotoxicology.

To our families; their constant support and encouragement helped us overcome the many challenges we confronted during the preparation of this volume.

W.S., Jr.
L.W.C.

CONTENTS

Part IV
Nutrient and Chemical Disposition

Part V
Behavioral Assessment

Part VI
Clinical Assessment and Epidemiology

Part VII
Specific Neurotoxic Syndromes

Part VIII
Risk Assessment

CONTRIBUTORS

Numbers in parentheses indicate the pages on which the authors' contributions begins.

Adams, Jane (631) Department of Psychology, University of Massachusetts, Boston, Massachusetts

Ali, Syed F. (353) Neurochemistry Laboratory, Division of Neurotoxicology, National Center for Toxicological Research/FDA, Jefferson, Arkansas

Andersen, Melvin (709) ICF Kaiser, Inc., Research Triangle Park, North Carolina

Aschner, Michael (339) Department of Physiology and Pharmacology, Wake Forest University School of Medicine, Winston-Salem, North Carolina

Audesirk, Gerald (61) Department of Biology, University of Colorado, Denver, Colorado

Audesirk, Teresa (61) Department of Biology, University of Colorado, Denver, Colorado

Bennett, Gregory D. (189) Departments of Veterinary Anatomy and Public Health, Texas A&M University, College Station, Texas

Bogle, Margaret L. (371) Delta Nutrition Intervention Research Initiative, U.S. Department of Agriculture/ARS, Little Rock, Arkansas

Broening, Harry W. (257) Exxon Biomedical Sciences, Inc., East Millstone, New Jersey

Casaccia-Bonnefil, Patrizia (141) Molecular Neurobiology Program, Skirball Institute, New York University Medical Center, New York, New York

Catalano, Susan M. (3) Howard Hughes Medical Institute, Department of Molecular and Cell Biology, University of California, Berkeley, California

Chang, Louis W. (507) Departments of Pathology, Pharmacology, and Toxicology, College of Medicine, University of Arkansas for Medical Sciences, Little Rock, Arkansas

Chao, Moses V. (141) Molecular Neurobiology Program, Skirball Institute, New York University Medical Center, New York, New York

Chen, Hao (321) Department of Pharmacology, School of Medicine, University of Washington, Seattle, Washington

Coles, Claire D. (455) Department of Psychiatry and Behavioral Sciences, Marcus Institute at Emory University, and Department of Pediatrics, Emory University School of Medicine, Atlanta, Georgia

Costa, Lucio G. (275) Department of Environmental Health, University of Washington, Seattle, Washington, and Toxicology Units, Fondazione S. Maugeri, Pavia, Italy

Day, Nancy L. (487) Western Psychiatric Institute and Clinic, University of Pittsburgh School of Medicine, Pittsburgh, Pennsylvania

Desaiah, Durisala (559) University of Mississippi Medical Center, Jackson, Mississippi

Finnell, Richard H. (189) Departments of Veterinary Anatomy and Public Health, Texas A&M University, College Station, Texas

Fried, Peter A. (469) Department of Psychology, Carleton University, Ottawa, Ontario, Canada

Friedman, J. M. (497) Department of Medical Genetics, University of British Columbia, Vancouver, British Columbia, Canada

Gaylor, David W. (727) Division of Neurotoxicology, National Center for Toxicological Research, Food and Drug Administration, Jefferson, Arkansas

Gladstone, Jonathan (567) The Motherisk Program, Division of Clinical Pharmacology and Toxicology, Hospital for Sick Children, Faculty of Medicine, University of Toronto, Toronto, Ontario, Canada

Greene, Robert M. (119) Department of Biological and Biophysical Sciences, University of Louisville School of Dentistry, Louisville, Kentucky

Grunwald, Gerald B. (43) Department of Pathology, Anatomy, and Cell Biology, Thomas Jefferson University, Philadelphia, Pennsylvania

Guilarte, Tomás R. (285) Department of Environmental Health Sciences, Johns Hopkins University, School of Hygiene and Public Health, Baltimore, Maryland

Guo, Grace Liejun (507) Department of Pharmacology and Toxicology, University of Kansas Medical Center, Kansas City, Kansas

Hansen, Deborah K. (643) Division of Reproductive and Developmental Toxicology, Department of Health and Human Services, Food and Drug Administration, National Center for Toxicological Research, Jefferson, Arkansas

Harry, G. Jean (87) National Toxicology Program, National Institute of Environmental Health Sciences, Research Triangle Park, North Carolina

Hastings, Lloyd (517) Department of Environmental Health, College of Medicine, University of Cincinnati, Cincinnati, Ohio

Hendrickx, Andrew G. (225) California Regional Primate Research Center, University of California, Davis, California

Holson, R. Robert (631, 643) Division of Reproductive and Developmental Toxicology, Department of Health and Human Services, Food and Drug Administration, National Center for Toxicological Research, Jefferson, Arkansas

Hussain, Saber (353) Neurochemistry Laboratory, Division of Neurotoxicology, National Center for Toxicological Research/FDA, Jefferson, Arkansas

Jensen, Karl F. (3) Neurotoxicology Division, National Health and Environmental Research Laboratory, U.S. Environmental Protection Agency, Research Triangle Park, North Carolina

Jett, David A. (257) Department of Environmental Health Science, Johns Hopkins University, Baltimore, Maryland

Juchau, Mont R. (321) Department of Pharmacology, School of Medicine, University of Washington, Seattle, Washington

Kable, Julie A. (455) Marcus Institute at Emory University, Department of Pediatrics, Emory University School of Medicine, Atlanta, Georgia

Kimmel, Carole A. (675) National Center for Environmental Assessment, U.S. Environmental Protection Agency, Washington, DC

Knudsen, Thomas B. (209) Department of Pathology, Anatomy, and Cell Biology, Jefferson Medical College, Philadelphia, Pennsylvania

Kong, Haeyoung (141) Molecular Neurobiology Program, Skirball Institute, New York University Medical Center, New York, New York

Koren, Gideon (567) The Motherisk Program, Division of Clinical Pharmacology and Toxicology, Department of Pediatrics, and The Research Institute, Hospital for Sick Children, University of Toronto, Toronto, Ontario, Canada

Krishnan, Kannan (709) Group of Research on Human Toxicology, Faculty of Medicine, University of Montreal, Montreal, Quebec, Canada

Lauder, Jean M. (153) Department of Cell Biology and Anatomy, University of North Carolina School of Medicine, Chapel Hill, North Carolina

Levin, Edward D. (587) Departments of Psychiatry and Pharmacology, Duke University Medical Center, Durham, North Carolina

Liu, Jiangping (153) Department of Cell Biology and Anatomy, University of North Carolina School of Medicine, Chapel Hill, North Carolina

Meyer, Jerrold S. (403) Department of Psychology, Neuroscience and Behavior Program, University of Massachusetts, Amherst, Massachusetts

Miller, Marian L. (517) Department of Environmental Health, College of Medicine, University of Cincinnati, Cincinnati, Ohio

Mirkes, Philip E. (159) Department of Pediatrics, University of Washington, Seattle, Washington

Nugent, Paul (119) Department of Biological and Biophysical Sciences, University of Louisville School of Dentistry, Louisville, Kentucky

Nulman, Irena (567) The Motherisk Program, Division of Clinical Pharmacology and Toxicology, Department of Pediatrics, Hospital for Sick Children and University of Toronto, Toronto, Ontario, Canada

O'Flaherty, Ellen J. (307) Retired, University of Cincinnati Department of Environmental Health, Cincinnati, Ohio

O'Hayon, Bonnie (567) The Motherisk Program, Division of Clinical Pharmacology and Toxicology, Hospital for Sick Children, Faculty of Medicine, University of Toronto, Toronto, Ontario, Canada

Paule, Merle G. (427, 617) Behavioral Toxicology Laboratory, Division of Neurotoxicology, National Center for Toxicological Research, Jefferson, Arkansas

Peterson, Pamela E. (225) California Regional Primate Research Center, University of California, Davis, California

Pisano, M. Michele (119) Department of Biological and Biophysical Sciences, University of Louisville School of Dentistry, Louisville, Kentucky

Polifka, Janine E. (383) Department of Pediatrics, University of Washington, Seattle, Washington

Potchinsky, Merle (119) Department of Pathology, Anatomy, and Cell Biology, Daniel Baugh Institute, Jefferson Medical College, Thomas Jefferson University, Philadelphia, Pennsylvania

Rice, Deborah C. (539) Toxicology Research Division, Bureau of Chemical Safety, Food Directorate, Health Protection Branch, Health Canada, Ottawa, Ontario, Canada

Richardson, Gale A. (487) Western Psychiatric Institute and Clinic, University of Pittsburgh School of Medicine, Pittsburgh, Pennsylvania

Rodier, Patricia M. (661) Department of Obstetrics and Gynecology, University of Rochester, Rochester, New York

Schardein, James L. (687) WIL Research Laboratories, Inc., Ashland, Ohio

Shield, Margaret A. (159) Department of Pediatrics, University of Washington, Seattle, Washington

Slikker, William Jr. (245, 727) Division of Neurotoxicology, National Center for Toxicological Research, Food and Drug Administration, Jefferson, Arkansas

Slotkin, Theodore A. (587) Departments of Psychiatry and Pharmacology, Duke University Medical Center, Durham, North Carolina

Toews, Arrel D. (87) Department of Biochemistry and Biophysics and Neuroscience Center and Department of Biology, University of North Carolina, Chapel Hill, North Carolina

van Baar, Anneloes (439) Academic Medical Center, Department of Neonatology, Amsterdam, The Netherlands, and Saint Joseph Hospital, Department of Psychosocial Care, Veldhoven, The Netherlands

Weston, Wayde (119) Department of Pathology, Anatomy, and Cell Biology, Daniel Baugh Institute, Jefferson Medical College, Thomas Jefferson University, Philadelphia, Pennsylvania

Wubah, Judith A. (209) Department of Pathology, Anatomy, and Cell Biology, Jefferson Medical College, Philadelphia, Pennsylvania

FOREWORD

The *Handbook of Developmental Neurotoxicology* heralds a remarkable advance in the field of toxicology that will likely have a substantial impact on policy as it unites developmental neuroscience with the principles of neurotoxicology. Although the demonstration of gross morphologic defects provides clear evidence of neurotoxicologic effects, the critical issue that has bedeviled the field of developmental neurotoxicology is the identification of threshold effects. Thus, we must assume that doses of neurotoxins below those causing evident structural damage still disrupt important but subtle aspects of synaptic circuitry and function.

A creative attempt to address this problem was the conceptualization of "behavioral teratology" to identify behavioral surrogates of neural toxicity in the developing brain. This strategy was based on the inference that toxins might disrupt brain maturation without causing gross lesions and that these subtle disruptions could become apparent from the behavioral deviance of the treated subjects compared to controls. It was hoped that a better understanding of the brain mechanisms responsible for the specific aberrant behaviors might disclose neuronal systems at risk. Nevertheless, in the absence of a verifiable "lesion," how could one infer that a behavioral difference was pathologic and not simply statistical? Indeed, some have argued that the "redundancy" and plasticity of the developing brain could prove to be self-correcting.

In fact, the problem of behavioral teratology shares important similarities with the challenge of understanding the causes of severe mental disorders. In the absence of objective evidence of structural changes in the brain, severely disabling disorders such as schizophrenia were ascribed in the past to pathologic mother–infant interactions, the consequences of an oppressive society, creative adaptations to life circumstances, or abnormal brain function. The advances in our understanding of corticolimbic synaptic circuitry coupled with informed postmortem studies and the application of sophisticated brain imaging technology have resulted in compelling evidence that schizophrenia is a neural developmental disorder that often results from an interaction between heritable vulnerabilities and perinatal insults.

As laid out in the *Handbook* and verified by experience, a thorough understanding of the principles that direct brain development at the molecular, cellular, and systems levels provides the greatest promise for identifying mechanisms of action for toxins that adversely affect the developing brain. The *Handbook* provides compelling case studies drawing on the rapidly advancing field of developmental neuroscience that have transformed our understanding of the developmental neurotoxic effects of an expanding family of agents. The important message to be derived from this evolving knowledge base is that developmental neurotoxicology cannot succeed by working in isolation but only through continuous dialogue with developmental neuroscientists. Developmental neuroscientists, in fact, are opportunistic and exploit selective agents to alter signal transduction, cell–cell interaction, and structural components of cells to understand how these processes determine brain maturation. Thus, toxins become "tools," and "tools" are elucidated as toxins.

An important advance in developmental neurosciences that has undercut one pathologic criterion for identifying lesions has been elucidation of programmed cell death, or apoptosis. Although apoptosis has long been known to be an important mechanism for sculpting the developing nervous system by eliminating excess or redundant neurons, its role in neuropathologic processes such as stroke, Alzheimer's disease, and Huntington's disease is only now coming to the foreground. A notable characteristic of apoptosis is the negligible inflammatory response to neuronal loss that it evokes; in contrast, necrosis produces microglial proliferation that results in a "scar," a footprint of prior neuronal death. If the neuropathologic criterion for neurotoxin-induced neuronal degeneration requires, as traditionally has been the case, gliosis, then apoptosis represents a "silent" assassin that leaves no clues. Indeed, the developing nervous system is most vulnerable to apoptotic neuronal elimination.

The salience of apoptosis has been demonstrated through the role of glutamate, the brain's major excitatory neurotransmitter, in regulating neuronal differentiation. The evidence is accumulating that use-dependent activation of the NMDA subtype of glutamate receptors not only has direct trophic effects on neurons but also enhances their response to endogenous growth factors. These insights have been particularly helpful in elucidating the developmental neurotoxic effects of agents that directly or indirectly inhibit NMDA receptors, especially the commonly abused substance ethanol. *In vitro* studies indicate that ethanol's inhibition of NMDA receptors promotes apoptotic neuronal degeneration, consistent with the hypoplasia observed in certain brain regions as a consequence of fetal ethanol exposure. The importance of these mechanistic studies is that they argue against a nonspecific effect of poor diet in causing the effects of fetal alcohol syndrome on the brain and support a direct neurotoxic effect of ethanol. The obvious policy implication is that dietary supplements alone will not protect against the neurotoxic effects of ethanol on the fetal brain.

Another unfolding story that merits mention is the role of retinoic acid in regulating spatially and temporally specific developmental events in the brain. One of the conundrums of developmental neurotoxicology is relating toxin exposure to region-specific sequelae in the brain. Recent studies now demonstrate that retinoic acid dehydrogenases that convert the precursor to the active signaling agent are transiently expressed in discrete brain structures to cue differentiation. Understanding these spatial–temporal events is essential for clarifying how a toxin might produce quite focused adverse effects and is relevant to the expanding number of transiently expressed developmental cues.

Developmental neurotoxicology must also walk hand-in-hand with the Human Genome Project. Although population variance in sensitivity to carcinogens is now well appreciated, the same heritable variations in enzyme and receptor characteristics likely impact sensitivity to the neurodevelopmental consequences of toxins. With the ascendance of gene knock-out and transgenic strategies, it is now possible to explore the consequences of human gene variants on susceptibility to toxins in animal models. Such heritable variability likely contributes substantially to the developmental disorders for which no known etiology is obvious. The old paradigm of "nature versus nurture" or "gene versus environment" has been replaced by a more sophisticated appreciation of complex genetic endowment interacting with environment, including an expanding array of potential developmental neurotoxins.

In closing, the *Handbook* charts a new course. The future that it projects is not represented by toxicologists working in isolation, attempting to define the surrogates of developmental neurotoxicity in experimental animals. Rather, the future requires the embrace of developmental neuroscience and molecular genetics to define the mechanisms of risk and their temporal relationships in brain development to achieve findings that compellingly will shape policy and thereby reduce risk for future generations.

JOSEPH COYLE

PREFACE

The Congressional designation of the 1990s as the *Decade of the Brain* underscores the tremendous opportunities offered by the current and anticipated advances in brain research and the enormous cost of mental disorders to the national economy. In the United States, brain-related disorders account for more hospitalizations than any other major disease group, including cancer or cardiovascular diseases. One out of four Americans will suffer from a brain-related disorder at some point in life, and the cost to the national economy for treatment, rehabilitation, and related consequences is an estimated $400 billion each year. The discipline of neurotoxicology is devoted to developing a better understanding of the extent, causes, and underlying mechanisms of brain-related disorders.

Nowhere is the pain and suffering from brain-related disorders felt more than by the very young members of our society who must live with their disabilities for a lifetime. Of the 250,000 malformed or impaired children born each year in the United States, approximately half suffer from a nervous system or behavioral deficit. With 4 to 8% of children born in the United States exhibiting anatomical and/or functional deficits, and the occurrence of several tragic clinical syndromes resulting from developmental exposure to such agents as ethanol, lead, and methylmercury, there is good reason to focus attention on the principles of developmental neurotoxicology. Various animal models have been used to confirm the developmental neurotoxicity that results from exposure to these agents and, along with clinical evidence, have implicated several other chemical classes such as antimitotics, insecticides, polyhalogenated hydrocarbons, psychoactive drugs, solvents, and vitamins as specific agents with developmental neurotoxic potential.

Clinical case reports and other human studies primarily have been responsible for identifying approximately two dozen human teratogens. Of these, over half are known to affect the developing nervous system. In those cases where human data exist, comparisons may be made to data generated in animal models. There is extensive concordance between animal and human databases for effects or biomarkers for a variety of human developmental toxicants of several different chemical classes. There is ample evidence that effects seen in some animal models can be predictive of human outcome for every agent. There is also evidence that the selection of an animal model is important because the use of certain animals under certain study conditions results in a lack of concordance.

As for developmental toxicity in general, the nature and extent of neurotoxic effects are often dependent on the timing of exposure, and because stages of nervous system development can vary significantly between species in relation to the time of birth, variations in neurotoxic outcome across species are expected. Organogenesis, one of the critical periods of susceptibility when many organ systems are formed, varies from species to species in time from conception and in overall length. Differences also exist concerning the various stages of nervous system growth and development in different species. It is imperative, therefore, that the time and duration of exposure in the animal model be selected to match the window of exposure in the human situation.

The overall strategy to understand developmental neurotoxicity is based on two assumptions: (1) the developing nervous system may be more or less susceptible to neurotoxic insult than the adult depending on the stage of development; and (2) neuropathological, neurochemical, neurophysiological, and behavioral evaluations are necessary and complimentary approaches to determining the type and degree of nervous system toxicity. The research approach consists of the following four steps: (1) gather information from all available endpoints and use it to generate a developmental neurotoxicity profile; (2) correlate structural and chemical lesions with overt behavioral manifestations of neurotoxicity; (3) compare with the adult neurotoxicity profile and determine relative susceptibility of the developing organism; and (4) develop a pharmacokinetic/metabolic basis for interspecies extrapolation.

When provided the opportunity to edit a special volume of the *Handbook of Developmental Neurotoxicology* we accepted because of the tremendous importance of this subject to neurotoxicology. Indeed, developmental neurotoxicology is no longer "an appendix" of neurotoxicology; in fact, it has become an independent discipline of neurotoxicology. It is our belief that to understand developmental neurotoxicology, one must first be familiar with developmental neuroscience. Portions of this volume, therefore, are devoted to developmental neurobiology and neurochemistry as well as how the normal developmental processes can be influenced by various biological, chemical, and physical agents. A presentation of some of the most prominent neurotoxic syndromes, such as fetal Minamata disease, developmental lead poisoning, and fetal alcohol syndrome are provided to underscore the human health impact of developmental neurotoxicants. Other sections of the volume are presentations and discussions of various assessment approaches including behavioral and clinical processes and the risk assessment process itself. Although there are other books related to this subject, we believe that this volume represents the most comprehensive and in-depth coverage of the subject of developmental neurotoxicology.

William Slikker, Jr.
Louis W. Chang

PART I

Cellular and Molecular Morphogenesis of the Nervous System

Developmental neurotoxicology involves the study of adverse effects on the developing nervous system induced by biological, chemical, or physical agents. It is therefore important to understand the various stages and factors that are critical in the morphogenic development of the nervous system, such as neuronal differentiation, neuronal migration, cell–cell interactions, neuritic developments, synaptogenesis, and myelinogenesis. The cellular events in the maturation and cytoarchitectural development of the central nervous system (CNS) are underlied by a series of molecular elements and processes such as cytoskeletal elements, adhesion molecules, signal transductions, and so on. The role of these elements and processes in both normal and chemically intoxicated brains is presented and discussed in Part I.

The developing CNS is a constantly remodeling organ with active neuronal differentation, migration, synaptogenesis, and circuitry establishments. Chapter 1 by Karl F. Jensen introduces and discusses all these basic processes of morphogenesis of the brain. Factors such as hormonal and nutritional homeostasis, xenobiotic metabolism, and the development of the blood–brain barrier that influence the development of the brain are also presented. The normal development and function of the nervous system can only be achieved if all these critical processes and stages of development remain intact. Obviously, any disturbance or disruption of these processes by exogenous chemicals will result in pathologic and functional (neurobehavioral) changes of the brain.

The CNS is uniquely different from other organ systems by having a complex circuitry of communication network between various nervous cells, neuronal groups, and the glial elements. The abilities of developing nerve cells and glial elements to migrate to their proper and "predestined" positions, recognition of the target cells, and signaling are largely influenced and directed by cadherins, a family of calcium-dependent cell adhesion molecules. In Chapter 2, Gerald B. Grunwald provides a detailed presentation and discussion on the cadherin adhesion molecules and their critical roles in the development of both the CNS and the PNS. An informative account on the structural and functional diversity of the cadherin molecules, the regulations and expressions of these molecules in relation to neural development, nuclear and ganglionic organization, axonal growth, and fiber tract development is provided and discussed.

Another characteristic feature of nerve cells, different from other cell types, is the existence of neuronal processes or neurites for the establishment of the communicative network (circuitry) among the nerve cells. The neuritic development is critically controlled by the cytoskeletal components. The fundamental structures and roles of cytoskeleton in developing neurites (dendrites and axons) is presented and discussed in Chapter 3 by Gerald Audesirk and Teresa Audesirk. The regulation of cytoskeleton by intracellular calcium and by protein phosphorylation, the relationship between cytoskeleton and growth cone formation, and the roles of adhesion, attraction, and repulsion in pathfinding of the growth cones are also discussed in conjunction with neuritic initiation and elongation. Various neurotoxicants such as lead, mercury, and ethanol are known to affect neuritic development. These toxicants and their toxic actions on the developing neurites are provided to illustrate chemical impact on the developing nervous system.

One of the most fundamental functions of the nervous system is impulse transmission along the axons of the neurons. The myelin sheath surrounding the axons serves as an insulator and facilitates such impulse transmission. Development of the myelin is most active at an early neonatal age. In Chapter 4 by G. Jean Harry and Arrel D. Toews, the basic structures, morphogenesis, and chemistry of myelin are presented. Molecular aspects of myelin assembly, axon–glial interactions, myelin disorders, and dysmyelination–demyelination induced by toxic chemicals are also discussed.

It is obvious that a single neurotoxicant (e.g., lead, mercury) can exert its toxic influence and action on one or combinations of developmental parimeters (e.g., cadherin adhesion molecules, cytoskeleton, cell migration, neuritic development, synaptogenesis, myelination). Investigators must avoid the "tunnel vision" mentality, but approach the problem or issue with a broad view and the understanding that neural development involves multiple factors, processes and stages. Any given developmental neurotoxicant can affect multiple factors and events at the same time. Neurotoxicology is a complex issue, this complexity is even more so in the situation of developmental neurotoxicology.

Louis W. Chang

CHAPTER

1

Brain Morphogenesis and Developmental Neurotoxicology

KARL F. JENSEN
Neurotoxicology Division
National Health and Environmental Research Laboratory
U.S. Environmental Protection Agency
Research Triangle Park, North Carolina 27711

SUSAN M. CATALANO
Howard Hughes Medical Institute
Department of Molecular and Cell Biology
University of California
Berkeley, California 94720

*Abbreviations: APP, amyloid precursor protein; BMP, bone morphogenetic proteins; Ci, Cubitus Interuptus; Dix, distalless; DS, Down syndrome; Emx, empty spiracles; EN, engrailed; FGF, fibroblast growth factor; HASAS, hydrocephalus as a result of stenosis of the Aqueduct of Sylvius; Mad, Mothers against decaptaplegic; MAP2, microtubule-associated protein 2; MAPK, mitogen-activated protein kinase; MASA, mental retardation, aphasia, shuffling gaits and adducted thumbs; NMDA, *N*-methyl-*d*-aspartate; NRC, National Research Council; Otx, orthodenticle; PCB, polychlorinated biphenyls; PKC, protein kinase C; ptc, patched; Shh, Sonic hedgehog; smo, smothered; T3, triiodothyrione; T4, thyroxine; TGF-β, transforming growth factor β; VEGF, vascular endothelial growth factor; Wnt, wingless.

I. Introduction

A postulate of developmental neurotoxicology is that particular toxicants can cause more severe injury to the nervous system during development than at maturity. The outbreak of Minamata disease demonstrated the capacity of the archetypal developmental neurotoxicant methylmercury to devastate the developing human brain. However, as the National Research Council (NRC) has emphasized: "A major unanswered question—indeed, a central issue confronting neurotoxicology today—is whether the causal associations observed in epidemics of neurotoxic diseases reflect isolated events or are merely the most obvious examples of a widespread association between environmental chemicals and nervous system impairment" (p. 2) (NRC, 1992). This question poses at least two major challenges. One challenge, considered by many to be a classic objective of behavioral teratology, is to detect a significant diminution in neurologic function that can occur as a direct result of developmental exposure to a toxicant. A second challenge is to determine the extent to which developmental exposure to a toxicant can enhance the incidence or severity of neurologic diseases or disorders of diverse etiologies.

Advances in the genetics of neurologic disorders together with the identification of diverse molecular interactions critical to the development of the neural circuitry provide a foundation on which to address the second challenge. The tasks facing toxicologists in this regard include determining: (1) the vulnerability of these molecular interactions to disruption by toxicants; (2) the extent to which disruption of these molecular interactions results in aberrant neural circuitry; and (3) the relationship of such aberrant circuitry to developmental neurologic disorders.

The difficulty of these tasks is due in large part to two considerations. One consideration is related to our limited understanding of the morphologic basis of developmental neurobehavioral disorders. Our ignorance is exemplified at one extreme by debilitating disorders that do not appear to be associated with any detectable morphologic defect. At another extreme is the occurrence of dramatic congenital alterations in the brains of individuals that exhibit only minimal, if any, functional impairments. These extremes indicate that a morphologically "normal" brain is neither necessary nor sufficient for "normal" functional capacity.

One approach to addressing this paradox is to postulate two types of morphogenetic processes. One involves progressive "instructive elaboration" of neural circuitry by enhancing the acquisition and growth of neurons and their connections and a second type involves "selective consolidation" of neural circuitry by the loss of neurons or the regression of particular phenotypic characteristics expressed transiently during development (Cowan *et al.*, Easter *et al.*, 1985). To account for the paradox, these two types of processes should be differentially vulnerable to disruption. The roles of instructive and selective processes in normal brain development continue to be debated (Purves *et al.*, 1996; Changeux, 1997; Purves, 1997; Sporns, 1997), and it is important to emphasize that use of the terms "instructive" and "selective" in the present context is intentionally broad and no rigorous definitions are implied. Nonetheless, this distinction may be useful in attempting to relate vulnerability of morphogenetic processes to the formation of aberrant neural circuitry, and the extent to which morphologic and functional alterations may be associated with particular developmental neurobehavioral disorders.

A second consideration contributing to the difficulty of the tasks facing developmental neurotoxicologists is the enormous anatomic complexity of the nervous system. Each of our 10^{11} neurons has a unique integrative capacity derived from its individual morphology, neurochemical properties, and connections. This neuronal diversity is dependent on cellular interactions during development, and numerous theoretic frameworks have been proposed relating developmental processes to the complexity of the neural circuitry of the brain (Herrick, 1925; Weiss, 1939; Ramon y Cajal, 1960; Sperry, 1963; Gaze, 1970; Kuhlenbeck, 1973; Edelman, 1987; Jacobson, 1991). Although the capacity of neuronal precursors to interact at the molecular level is derived from their genetic potential, the structure of the developing brain is the matrix in which such interactions occur. Thus, the significance of toxicant-induced disruption of growth and differentiation at the molecular level can only be fully appreciated in the context of brain morphogenesis.

In the nanosecond time frame and submicron resolution of molecular events, the developing brain is an expanding universe in which explosive increases in cellular heterogeneity occur over hours, days, and weeks. The disparity between these temporal and spatial frames of reference presents a formidable obstacle to unraveling the molecular mechanisms that are potential targets of toxicants. One approach to addressing such disparities is to integrate molecular manipulations with descriptive morphology. Characterization of altered morphogenesis can provide a critical link between investigations of the actions of a toxicant at the molecular level and the localization of aberrant circuitry responsible for neurologic impairment.

Ramon y Cajal's (1906, 1984) application of cell theory to the nervous system and his postulate that neurons are "dynamically polarized" and conduct information in only one direction, are seminal to the study of localization of function. Sherrington (1906) coined the term *synapse* to describe the polarized connections between neurons, and morphologic characterization of these specialized connections were an early result of the application of electron microscopy to biologic tissues (De Robertis and Bennet, 1954; Palade and Palay, 1954). That the synapse is an elemental construct of neural circuitry is the foundation of two of the major premises of neuroscience: that regional patterns of synaptic connections subserve particular neurologic functions, and that the heterogeneity of these synaptic patterns is a morphologic correlate of the integrative capacity of the brain.

A major concern for developmental neurotoxicologists is whether an evolutionary increase in the integrative capacity of the brain engenders a corresponding increase in a susceptibility to environmental insult resulting from the vulnerability of developmental processes. If so, then environmental exposure of human populations to toxicants presents serious risk to public health. The answer to this question remains elusive. The objectives of this chapter are to provide an overview of developmental processes important for brain morphogenesis, to describe examples of developmental neurobehavioral disorders in which these processes are disrupted, and to discuss how the vulnerability of morphogenetic processes to environmental insult could influence the incidence and severity of neurobehavioral disorders.

II. Instructive and Selective Processes Shape Brain Morphogenesis

The heterogeneous patterns of synaptic connectivity within the brain arises from the vast number of neurons, the unique combination of morphologic and biochemical characteristics of each of these neurons and the multitude of their synaptic connections. The coordinated generation of this neuronal diversity is dependent on the appropriate regulation of the proliferation of neuronal precursors and the acquisition of the unique morphologic and biochemical characteristics by differentiating neurons. The prodigious increase in the number of neurons together with the extensive growth and specialization of their axons and dendrites is perhaps the most convincing evidence that processes subserving the instructive elaboration of neural circuitry dominate brain morphogenesis. However, the precision inherent in the patterns of connections within various sensory and motor pathways also indicates that processes of refinement and the selective consolidation of neural circuitry are also essential to the effective integration of connections during brain development. Similarities and differences in the molecular pathways that mediate these two types of processes can help reveal distinctive aspects of their vulnerabilities.

A. *Intercellular Signals Organize the Brain Primordium*

Morphogenesis of the nervous system begins when the mesoderm induces the formation of the neural plate on the dorsal surface of the embryo. In the absence of mesodermal influence, the dorsal ectoderm differentiates into epidermis under the intrinsic influence of members of the transforming growth factor-β (TGF-β) superfamily and, in particular, the bone morphogenetic proteins (BMPs). Neural induction occurs when the mesoderm secretes factors that block the binding of TGF-β and BMPs to the ectoderm. Candidate mesodermal factors participating in this blockade in the anterior ectoderm include chordin, noggin, and follistatin. Members of the fibroblast growth factor (FGF) family may be involved in this blockade within the posterior ectoderm (Holland and Graham, 1995; Valenzuela *et al.*, 1995; Bier, 1997; Graff, 1997; Hemmati-Brivanlou and Melton, 1997).

Following induction, the neural plate folds and curls up to form the neural tube, which subsequently develops three evaginations that form the rhombencephalic, mesecephalic, and prosocenphalic vesicles. At the time of closure, the neural tube begins to exhibit both an anteroposterior and a dorsoventral axis. The anteroposterior axis forms as the neural fold closes and becomes segmented along its posterior to anterior extent prior to the formation of the rhombencephalic vesicle. The patterning of these segments, the *rhombomeres,* occurs in response to several morphogens, including those involved in induction of the neural plate, and is regulated by transcription factors derived by several homeobox genes such as Hox, Pax, and Barx. The regulation of this segmentation process by morphogen-induced transcription factors appears to have been highly conserved in evolution (Bachiller *et al.,* 1994; Bartlett *et al.,* 1994; Chalepakis *et al.,* 1994; Coletta *et al.,* 1994; Graham *et al.,* 1994; Kessel, 1994; Keynes and Krumlauf, 1994; Lumsden and Krumlauf, 1996; Nonchev *et al.,* 1996; Grapin-Botton *et al.,* 1997; Jones *et al.,* 1997).

The dorsoventral axis of the neural tube appears to arise differentially in the rostral and caudal regions. In the caudal region, a dorsoventral axis also develops as the neural plate folds. In the chick embyro, *Pax3* and

Pax7 are initially expressed uniformly in the caudal neural plate. As the plate folds, *Pax3* and *Pax7* become repressed in the medial aspect of the plate and *Pax6* begins to be expressed. By the time the neural tube closes, cells of the ventral portion of the tube express only *Pax6*, whereas the expression of *Pax3* and *Pax7* are restricted to the dorsal region. The repression of *Pax3* and *Pax7* appears to be the result of the induction of a transcription factor homologous to Cubitus Interuptus (Ci), Gli, in response the morphogen Sonic hedgehog (Shh), a morphogen secreted from the notochord and floor plate. The development of the dorsal identity of the caudal neural tube is associated with the secretion of BMPs from the epidermal ectoderm. At more rostral levels, establishment of the dorsoventral axis involves the same family of morphogens but the process appears to be transposed, with the ventral source of BMP cooperating with Shh to induce ventralization (Holst *et al.,* 1994, 1997; Ericson *et al.,* 1995, 1997; Liem *et al.,* 1995; Marigo and Tabin, 1996; Tanabe and Jessell, 1996; Dale *et al.,* 1997; Hynes *et al.,* 1997b; Knecht and Harland, 1997; Muhr *et al.,* 1997; Placzek, 1997).

Whereas mesencephalic and prosecenphalic vesicles do not exhibit a pronounced segmentation paralleling that of the rhombencephalic vesicle, they do exhibit regional specialization. One of the earliest markers for mesencephalic polarity is the expression of morphogens that are the products of the *engrailed* (*EN*) gene family, induced by signals from the isthmus that forms the border between the rhombencephalic and mesencephalic vesicles.

A number of genes appear to impart regionalization within the prosecenphalic vesicle, most notably homologues of *Drosophilia* genes *distal-less* (*Dlx*), *empty spiracles* (*Emx*), *orthrodenticle* (*Otx*), and *wingless* (*Wnt*) gene families (Bonicinelli, 1994; Bang and Goulding, 1996; Lumsden and Krumlauf, 1996; Roelink, 1996; Tanabe and Jessell, 1996; Cadigan and Nusse, 1997; Fishell, 1997; Mastick *et al.,* 1997; Moon *et al.,* 1997; Rodriguez and Basler, 1997; Shimamura and Rubenstein, 1997).

Morphogen-mediated regional specification of the neural tube and subsequently formed trivesicular brain primordia involves integration of opposing influences of diverse receptor-activated intracellular pathways. For example, the opposing effects of the BMP and EGF families are mediated by the convergence of two distinct postranslational pathways on a family of tumor suppressor proteins, the Smads, that are homologous to *Drosophilia* Mothers against *decapentaplegic* (Mad). In one pathway, BMP receptor serine/threonine phosphorylation of particular sites on Smad subunits results in the formation of Smad complexes that accumulate in the nucelus and regulate transcription. Another pathway involves EGF-receptor tyrosine kinase activation of an Erk subfamily of the mitogen-activated protein kinases (MAPKs) via the Ras and PI3 kinase pathways. Erk phosphorylation of SMAD proteins prevents the complex formation and thereby inhibits subsequent nuclear accumulation and the accompanying transcriptional activity (Liu *et al.,* 1996; Baker and Harland, 1997; Hata *et al.,* 1997; Imamura *et al.,* 1997; Kretzschmar *et al.,* 1997; Persson *et al.,* 1997; Shi *et al.,* 1997; Suzuki *et al.,* 1997; Wilson *et al.,* 1997).

The influence of the hedgehog family in *Drosophila* has been proposed to be mediated via an equally complex but strikingly different set of intracellular signals. Hedgehog proteins bind to a receptor complex comprised of proteins from the Smothered (smo) and Patched (ptc) families. The binding of hedgehog reverses the normal inhibition of smo by ptc, allowing smo to activate a serine–threonine kinase, Fu. Fu, in turn activates Ci, which is normally inhibited by protein kinase A (PKA). The transcriptional activation of Ci results in of the formation of a large complex comprised of Fu, Ci, microtubules and a kinesin-like molecule, Cos2. It is not known whether the kinesin involvement in this pathway is related primarily to translocation of the complex to the nucleus or represents a substrate for divergent influences on the cytoskeleton (Macdonald *et al.,* 1995; Concordet *et al.,* 1996; Epstein *et al.,* 1996; Goodrich *et al.,* 1996; Hammerschmidt *et al.,* 1996; Altaba, 1997; Lee *et al.,* 1997).

Morphogen-induced transcription factors also regulate the expression of adhesion molecules, which can, in turn, modulate proliferative and histogenetic events through a variety of intracellular pathways. One example of a pathway by which cell adhesion molecules can influence proliferation is the cyclin-dependent pathway. This pathway is composed of a series of kinases that regulate passage through various stages of the cell cycle. The activity of these kinases is modulated via an anchorage-mediated process involving calcium-independent adhesion molecules of the integrin family (van den Heuvel and Harlow, 1993; Holst *et al.,* 1994, 1997; Copertino *et al.,* 1995; Cillo *et al.,* 1996; Lewis, 1996; Ross, 1996; Wang *et al.,* 1996; Assoian and Zhu, 1997; Assoian *et al.,* 1997; Ben-Zeev and Bershadsky, 1997; Edelman and Jones, 1997; Giancotti, 1997). Another pathway is the Notch receptor pathway, which can amplify small differences in gradients of morphogens and contact signals. The Notch pathway can confer selectivity on proliferative activities of the neuroepithelium by regulating asymmetric division or mediate the differential expression of distinctive phenotypic characteristics in adjacent cells by lateral inhibition (Artavanis-Tsakonas *et al.,* 1995; Bettenhausan *et al.,* 1995; Copertino *et al.,* 1995; Blaumueller and Artavanis-Tsakonas,

1997; Cohen *et al.*, 1997; de la Pompa *et al.*, 1997; Ma *et al.*, 1997; Robey, 1997).

The signaling processes involved in the transformation of the neural tube to a trivesicular brain primordium subserve not only stage-specific regional specification, but also confer particular cell lineages with the competence to respond to a variety of extracellular stimuli encountered at later stages. Such competence may be one of the molecular correlates of the classic concept of the commitment of a cell to a particular fate. Consistent with this conjecture are observations that alterations of the expression of the transcription factors and adhesion molecules involved in the early regional specialization also can influence phenotypic properties of neurons that emerge later in development (Pimenta *et al.*, 1995; Anderson *et al.*, 1997; Barondes *et al.*, 1997; Ding *et al.*, 1997; Eagleson *et al.*, 1997; Levitt *et al.*, 1997).

B. Time and Place of Origin of a Neuron in the Germinal Neuroepithelium Is Predictive of Its Adult Location

As the trivesicular brain primordium enlarges and increases in cell mass, regional differences in rate of proliferation together with histogenetic evagination results in the formation of consistently located enlargements that give rise to neurons destined for predictable locations in the adult brain (Rakic, 1974; Altman, 1992). The majority of neurons of the brain arise in the primary germinal matrix formed by the neuroepithelium that lines the ventricles. In the rat, the majority of neurons derived from the primary germinal neuroepithelium are formed between embryonic days 11 and 22. There are also three specialized secondary germinal matrices that give rise to the interneurons of the cerebellar cortex, hippocampus, and olfactory bulb. In the rat, neurons continue to be generated in these secondary germinal matrices until the third postnatal week (Bayer *et al.*, 1993).

The primary germinal matrix is initially a pseudostratified columnar epithelium exhibiting interkinetic nuclear migration, in which the nucleus of the germinal cell moves toward the apex during the S phase of the cell cycle, then returns toward the ventricular border to enter the M phase (Sauer, 1935; Sauer and Walker, 1959). Two kinds of cleavages are observed. Vertical cleavage is symmetric and results in two daughter cells having a parallel orientation, and both continuing to undergo subsequent divisions resulting in an expansion of the germinal matrix. Horizontal cleavage is asymmetric and results in one of the daughter cells occupying a basal position and returning to the germinal matrix while the more apical cell begins its migration from the germinal epithelium. Asymmetric, but not symmetric, cleavages are associated with a preferential accumulation of Notch in the apical regions of the cell with subsequent transfer of Notch to the apical daughter cell. Whether Notch is critical to the initiation of migration is not yet clear (Chenn and McConnell, 1995).

At the time cells begin leaving the ventricular neuroepithelium, the brain primordium is dominated by a radial organization. This radial organization is due at least in part to the distribution of radial glia that extend from the ventricle to the pial surface and provide a permissive substrate on which many neurons migrate (Rakic, 1971, 1972, 1974; Sidman and Rakic, 1973; Schmechel and Rakic, 1979; Anton *et al.*, 1996). The migratory trajectory of many developing neurons, however, is more complex than can be accounted for by radial dispersal (Rakic, 1995; Karten, 1997; Kuan *et al.*, 1997). Altman (1992) has identified four different patterns of neuronal migration. The simplest is when neurons take a radial path over a short distance and is exemplified by the early formed neurons of the diencephalon. A slightly more complicated migratory pattern is when neurons traverse paths with substantial curvatures, most notably in the trigeminal and facial nuclei of the brainstem.

A third type of migratory pattern is still more complex and occurs when the various neuronal populations comprising a structure have both converging and diverging migratory paths. This is a characteristic of regions in which some of the neurons are generated by secondary germinal matrices and is typified by the cerebellum system (Altman and Bayer, 1997). The neurons of the precerebellar nuclei of the brainstem, although generated in close proximity within the germinal neuroepithelium, take widely arcing and superficially situated migratory paths that tend to converge on their adult locations. The two major constituents of the cerebellar cortex, the granule and Purkinje cells, also have their origins in close proximity within the germinal epithelium, but the precursors of the granule cells initially migrate to the surface of the cerebellar anlage, where they disperse and continue to proliferate to form the transient secondary germinal matrix of the external germinal layer of the cerebellum. The migrating granule cells leaving this secondary germinal matrix traverse the primordial molecular layer and migrate through the Purkinje cell layer to occupy the internal granule cell layer. A key feature of this pattern is that the Purkinje and granule cells have opposing migratory trajectories that allow them to intermingle before their final disposition is achieved.

The fourth and most complex pattern of migration is cell sorting and lateral migration and describes the paths of the neuronal precursors of the neocortex (Bayer and Altman, 1991; Altman, 1992). One of the first events in the development of the neocortex is the formation of the primordial plexiform layer consisting of cells destined to occupy marginal and subplate layers (Marin-Padilla, 1971, 1978, 1983). As neuronal precursors destined for the neocortex leave the germinal epithelium they form the cortical plate, partitioning the primordial plexiform layer into deep and superficial zones that in turn form the marginal and subplate layers. These early formed layers have been implicated in the regulation of several aspects of neocortical development, including proliferation, migration, and axonal growth (Marin-Padilla, 1978; Lauder *et al.*, 1983; Lauder, 1988, 1993; Hellendall *et al.*, 1993; Anton *et al.*, 1997; Clark *et al.*, 1997; D'Arcangelo *et al.*, 1997; Del Rio *et al.*, 1997; Frotscher, 1997; Schiffmann *et al.*, 1997). The radial migration of neocortical neurons forms an inside-out pattern, with the earlier formed neurons occuping the deeper layers and later formed neurons migrating through them to form the more superficial layers. There is also a significant lateral component to migration, with neurons traversing significant distances to occupy regions such as the ventrolateral neocortex.

Complex migratory trajectories involving transduction of both diffusable and local signals may encompass a series of precisely timed and sequence-dependent molecular interactions providing a neuron with instructive signals essential to acquiring its distinctive morphology and connectivity. Evidence from mutants and developmental manipulations, however, indicates that the migratory signal may not play such an extensive role in inducing mature neuronal phenotype. A number of characteristics of the morphology and connectivity of a neuron can be retained when migration is disrupted (Jensen and Killackey, 1984a,b; McConnell, 1988, 1995a,b). Complex migratory pathways may therefore reflect adaptive strategies that accommodate constraints arising during the evolution of the mammalian brain, including the increased proliferation from a spatially restricted germinal matrix; the dispersion of neurons to distant locations, permitting increases in cell number and size; and the preservation of access to particular substrates for axonal pathfinding. In this view, migratory signals are not the direct inducers of specific morphologic phenotypes or pattern of connectivity, but provide the means to accommodate an expanding neuronal populations while maintaining the essential aspects of morphologic framework for the elaboration of specific patterns of connectivity.

A variety of receptors and transduction pathways involved in signaling between neurons, glia, and the extracellular matrix have been implicated in neuronal migration. These diverse pathways include those that mediate signals from adhesion molecules, growth factors, neurotransmitters, and calcium ions and converge on several elements of the cytoskeleton (Edelman *et al.*, 1985; Burgoyne and Cambray-Deakin, 1988; McConnell, 1988, 1995a,b; Burgoyne, 1991; Komuro and Rakic, 1992, 1995, 1996; Hirokawa, 1993; Rakic *et al.*, 1994, 1996; Carraway and Burden, 1995; O'Rourke *et al.*, 1995; Rakic and Komuro, 1995; Blackshear *et al.*, 1996, 1997; Lemke, 1996; Anton *et al.*, 1997; Carraway *et al.*, 1997; Chothia and Jones, 1997; Messersmith *et al.*, 1997; Meyer *et al.*, 1997; Rio *et al.*, 1997; Schmidt-Wolf *et al.*, 1997).

C. Axonal Pathfinding Involves Attractive and Repulsive Cues

Axonal growth begins as migrating neurons leave the germinal matrix. Both cell-surface and diffusable molecules influence the direction of axonal growth. Four mechanisms have been postulated; contact-mediated attraction, contact-mediated inhibition, hemoattraction, and chemorepulsion. The direction of axon growth at any one point in time appears to be the result of the balance of the combined influence of attractive and repulsive forces. Molecules mediating these influences include members of the laminin, netrin, and semaphorin families, as well as members of the immunoglobulin superfamily. Each of these families include membrane-bound and diffusable molecules (Snow *et al.*, 1990; Lemmon *et al.*, 1992; Keynes and Cook, 1995a,b; Burden-Gulley and Lemmon, 1996; Giger *et al.*, 1996; Goodman, 1996; Tessier-Lavigne and Goodman, 1996).

The motility of axonal growth cones arises, at least in part, from the coordinated regulation of the stabilizing of microtubules located more proximally to the cell the body and the extension of actin filaments into the distal aspect of filapodia. Extracellular guidance signals influence the direction of growth cone movements by stimulating actin fiber assembly through pathways that involve receptor and nonreceptor tyrosine kinases, calmodulin-dependent serine–threonine kinases, proline-directed kinases, and cAMP (Meiri *et al.*, 1986; Kater *et al.*, 1988; Goslin *et al.*, 1989; Cheng and Sahyoun, 1990; Matten *et al.*, 1990; Tucker, 1990; Meakin and Shooter, 1991; Lohof *et al.*, 1992; Niclas *et al.*, 1994; Zheng *et al.*, 1994, 1996a,b; Bloom and Endow, 1995; Brady and Sperry, 1995; Garrity and Zipursky, 1995; Langford, 1995; Lopez and Sheetz, 1995; Maccioni and Cambiazo, 1995; Mandelkow *et al.*, 1995; Tanaka and Sabry, 1995; VanBerkum and Goodman, 1995; Hirokawa, 1995; Mandell and Banker, 1996; Maness *et al.*, 1996; Vallee and Sheetz, 1996; Waters and Salmon, 1996; Wong *et al.*, 1996a,b; Benowitz and Routtenberg, 1997; Burden-Gulley *et al.*, 1997; Ming *et al.*, 1997; Zisch and Pasquale, 1997).

The association of these guidance molecules with the floor plate and midline have a prominent organizing influence on the formation of axonal pathways in the neural tube. Guidance molecules associated with regional boundaries within the trivesicular brain primordium also provide additional intermediate guidance for growing axons in the embryonic brain. Developmentally transient structures, such as the subplate of the developing neocortex and the perireticular nucleus, are also critical intermediate targets of growing axons. The interactions of axons with such transient structures can involve transient synaptic contacts, axonal sorting, and shifts in axonal responsiveness to particular guidance cues (Goodman and Shatz, 1993; Wilson *et al.,* 1993; Allendoerfer and Shatz, 1994; Colamarino and Tessier-Lavigne, 1995; Adams *et al.,* 1997; Fukuda *et al.,* 1997a,b; Shirasaki *et al.,* 1998).

The complexity related to the diversity of guidance molecules may impart accuracy and resolution to the pathfinding process, but it may also engender potential for variability and sensitivity to perturbation. The remarkable consistency in the organization of major axonal pathways, however, indicates that guidance signals are likely to have a robust and stabilizing influence on growing axons. This robust influence could arise from cooperative interactions between heterogeneous populations of guidance molecules that impart a greater stability than individual molecules alone. Another possibility is that combinations of signals converge on particular transduction pathways, resulting in an amplification of a response that overrides more variable signals. In any case, it appears guidance molecules are likely to have sufficient combinational possibilities to provide axons with the instructive information necessary to reach their eventual target.

D. Cell Death and Collateral Loss Refine Connectivity

There is a close correspondence between gradients in the time of origin of projection neurons and corresponding gradients of their targets (Bayer and Altman, 1991). For example, efferent projection neurons of both the spinal cord and neocortex originate in a rostral to caudal gradient, setting the stage for a coordinated maturation between source and target. Additional refinement of axonal projections, however, also occur by selective loss of axonal collaterals. In the case of corticospinal neurons, individual neurons initially project to several subcortical targets, including thalamus, brainstem, or spinal cord. Over the course of several postnatal days these projection neurons selectively lose one or more of these subcortically projecting axon collaterals (Stanfield *et al.,* 1982; Bates and Killackey, 1984; Stanfield, 1992; Koester and O'Leary, 1994). This selective consolidation may reflect competition at terminal sites or a matching process between afferent input and efferent projections of these neurons.

A variety of factors may influence such competitive or matching processes. The rate of descent of the axons of telencephalic neurons projecting to the spinal cord appears to vary to a greater extent than can be accounted for by differences in their time of origin, consistent with the notion that guidance molecules may differentially influence rates of axonal growth (Lakke, 1997). A competitive advantage may be acquired by the early arrival of an axon at a target. Alternatively, a comparative matching process may involve both retrograde signals from the target and anterograde signals from developing afferent innervation to the neuron of origin. Selective loss of axon collaterals also occurs in many sensory pathways such as the optic nerve (Fawcett *et al.,* 1984; Levitt *et al.,* 1984) and associational pathways such as the corpus callosum (O'Leary *et al.,* 1981; Easter *et al.,* 1985; LaMantia and Rakic, 1994; Innocenti, 1995). Patterns of connectivity are also modified by the selective loss of neurons via programmed cell death such as in the neocortex, where intracortical connections are refined by the loss of neurons in the upper layers (Finlay and Slattery, 1983; Windrem *et al.,* 1988). Both programmed cell death and the loss of axon collaterals play a significant role in the selective consolidation of circuitry throughout the brain (Oppenheim, 1989; Houenou *et al.,* 1991; Homma *et al.,* 1994). Whether collateral loss or programmed cell death predominates in the refinement process of a particular pathway may reflect several factors such as the number of collaterals neurons initially project, the nature and extent of competition at a target, the developmental pattern of afferent innervation of the neurons, and intrinsic patterns of activity of source and target.

Competition for target-derived trophic factors is likely to be the primary mechanisms for programmed cell death, or *apoptosis,* during normal brain development. Such competition may be limited primarily by the extent of terminal branching and contact sites for the uptake of trophic factor, rather than the production of trophic factor by the target (Oppenheim, 1981, 1989, 1997; Bothwell, 1995; Bredesen, 1995; Karavanov *et al.,* 1995; Brugger *et al.,* 1996). The intracellular pathways mediating apoptosis depend on cell type and initiating stimulus. Regulation of normal and induced apoptosis can involve a number of signaling pathways, such as receptor and nonreceptor tyrosine kinases, transcription factors of the BCL-2 and JUN families, p53 tumor suppresser proteins, CREB, the caspases, ICE protease, and the CaM, Ras, Raf, and MAPK pathways (Kameyama and Inouye, 1994; Wood and Youle, 1995; Ferrer, 1996; Ferrer *et al.,* 1996; Henderson, 1996; Kuida *et al.,* 1996; Martinou and Sadoul, 1996; Schwartz and Milligan, 1996; Means *et al.,* 1997; Merry and Krosmeyer, 1997).

E. Dendritic Form Is Shaped by Afferent Innervation

The distinctive integrative capacity of a neuron is largely dependent on the shape of its dendritic tree and the orientation of that tree to afferent axons (Johnston *et al.,* 1996). The growth of dendritic processes occurs after migration is completed and the axon has made initial contact with a target. The attainment of the mature form of the dendrite tree is a dynamic process in which branches are extended, and then some are retracted whereas others are stabilized and continue to grow and extend additional branches (Jacobson, 1991).

The influence of afferent axons on the patterning of dendritic morphology has been demonstrated by the selective elimination of neurons that innervate cerebellar Purkinje cells. Basket, stellate, and granule cells each contact various parts of the Purkinje cell dendritic tree. When basket cells, which normally contact the base of the dendritic tree, are eliminated, the Purkinje cells display numerous primary dendrites instead of just one. When stellate cells that normally innervate intermediate aspects of the dendrites are eliminated, the pattern of secondary branches is disrupted. When granule cells are eliminated, the planar organization of the tree and the distribution spines is disrupted (Altman, 1972).

The growth and retraction of dendrites is likely to involve both instructive and selective influences mediated by neurotrophins (Duffy and Rakic, 1983; Greenough and Chang, 1988; McAllister *et al.,* 1995). Key elements in the transduction of such signals may be serine–threonine kinases such as CaM kinase II, one of the most abundant proteins in the postsynaptic density (Kennedy, 1983; Kennedy *et al.,* 1983). CaM kinase II has been shown to stabilize growing dendrites, perhaps via the phosphorylation-dependent association of microtubule-associated protein 2 (MAP2) with microtubules, which has been postulated to impart stability to the microtubular array within growing dendrites (Garner *et al.,* 1988; Matus, 1988; Brugg and Matus, 1991; Ludin and Matus, 1993; Weisshaar and Matus, 1993; Lopez and Sheetz, 1995; Maccioni and Cambiazo, 1995; Ludin *et al.,* 1996; Mandell and Banker, 1996; Felgner *et al.,* 1997; Gang-Yi and Cline, 1998).

The dendritic spine is a specialization for the reception and integration of synaptic signals (Harris and Kater, 1994). Lund (1978) summarized the developmental acquisition of dendritic spines in four phases. In the first phase, the processes are short and have a varicose appearance, display barblike extensions along their length, and terminate in a growth cone. In the second stage, the processes attain a more even diameter in a proximal to distal direction, and the barblike extensions attain a finer hairlike appearance. In the third stage, the fine hairlike appendages attain the more classic morphology of mature spines. The presence of spines at this developmental stage occurs in all neurons, regardless of whether they display spines at maturity.

In addition to a role of CaM kinase II in the stabilization of dendritic processes, spine ontogenesis may also involve the Rac GTPase regulation of actin. In transgenic mice expressing constitutively active Rac1, the form of the Purkinje cell dendritic trees are normal, but the spines are finer and more numerous (Leopold and Logothetis, 1996), consistent with an immature appearance. The possibility exists that, in contrast to the importance of tubulin regulation in dendritic branching, actin regulation may be more critical to spine formation.

F. Activity Patterns Synaptic Connections

The role of sensory stimulation in the patterning of connections has been studied extensively in the developing visual system (Hubel and Wiesel, 1965; Wiesel and Hubel, 1965; Hubel *et al.,* 1977; LeVay *et al.,* 1978, 1980; Hubel, 1979; Antonini and Stryker, 1996; Campbell *et al.,* 1997; Crair *et al.,* 1997), and one focus of this work has been the patterning of the axonal terminal arbors (Shatz and Stryker, 1978; Sretavan *et al.,* 1988; Antonini and Stryker, 1996; Katz and Shatz, 1996; Shatz, 1996). In the retinogeniculate pathway, eye-specific patterns develop prior to the time at which photoreceptor outer segments are functional. Initially, retinal axons from each eye occupy overlapping territories within the lateral geniculate nucleus. The majority of geniculate neurons are responsive to stimulation of either optic nerve. Subsequently, the terminal arbors selectively retract certain branches while extending others such that axons from each eye become segregated into different layers. This process coincides with the majority of geniculate neurons becoming responsive to stimulation of only one of the optic nerves. The differential growth of retinogeniculate terminal arbors responsible for segregation is dependent on the spontaneous, spatially correlated activity of retinal ganglion cells (Shatz, 1996). A similar process takes place in the formation of ocular dominance columns in the visual cortex (Hubel *et al.,* 1977; Hubel, 1979; LeVay *et al.,* 1980; Herrmann and Shatz, 1995).

There are several candidate mechanisms by which activity influences the patterning of terminal axonal arbors. One mechanism involves the selective strengthening of Hebb type *N*-methyl-D-aspartate (NMDA) synapses as a result of the coordination of pre- and postsynaptic activity. Such strengthening stabilizes the synapse and promotes the growth of additional axonal branches. In contrast, discordant activity weakens synapses, resulting in the withdrawal of axonal branches. Several observations are consistent with this possibility including the regulation of the NMDA receptor subunit immunoreactivity in the developing visual cortex by retinal activity and disruption of ocular dominance column segregation by NMDA receptor blockade (Kleinschmidt *et al.,* 1987; Gu *et al.,* 1989; Catalano *et al.,* 1997). Another possible mechanism is that, when postsynaptic neurons are activated, they release a neurotrophic factor that can selectively enhance the growth of axons that have also been active (Purves, 1988; Katz and Shatz, 1996). This possibility is supported by the observation that selective blockade of the TrkB neurotrophin receptor in the lateral geniculate nucleus inhibits the formation of ocular dominance columns (Cabelli *et al.,* 1997).

The importance of these different mechanisms may vary depending on the architecture of the circuitry and the nature of information processed by a pathway. For example, in the somatosensory system of the rodent there is a representation of the body surface at each synaptic level from the periphery to the neocortex. In the face region of this representation there is a segmented pattern corresponding to the distribution of vibrissae on the face of the rat, with each segment representing an individual vibrissae. The segmented nature of the representation is dramatically altered by neonatal disruption of the innervation of the vibrissae (Woolsey and Van Der Loos, 1970; Van Der Loos and Woolsey, 1973; Belford and Killackey, 1980; Woolsey, 1984, 1996). As in the visual system, the formation of vibrissae-related patterns may involve NMDA receptors, as indicated by the lack of segmentation in the trigeminal pathway in NMDAR1 knockout mice (Li *et al.,* 1994). Also, like the visual system, the branching of individual thalamocortical axons corresponds to the segmented pattern. Unlike the visual system, however, the axons are not intermixed as they grow into the cortex and they do not extensively overlap prior to emergence of the segmented pattern. Furthermore, neonatal disruption of vibrissae innervation disrupts both the segmented pattern and alters the size and branching of terminal arbors. Such terminal arbors tend to be larger with fewer branches and the first signs of this altered morphology are apparent shortly after birth (Jensen and Killackey, 1987a,b; Catalano *et al.,* 1991, 1995, 1996; Killackey *et al.,* 1995). Thus, the pruning of terminal arbors does not make a significant contribution to the establishment of the somatotopic branching pattern of somatosensory axons. Instead, the patterning appears to be dominated by instructive elaboration of axonal processes under the influence of the sensory periphery.

G. Neurotransmitters and Hormones Regulate Morphogenetic Events

Neurotransmitters and their receptors, in particular acetylcholine, norepinephrine, serotonin, GABA, and glutamate, modulate a spectrum of morphogenetic events that include the regionalization of the brain primodium, axonal pathfinding, elaboration of terminal axonal arbors, and activity-dependent patterning of synaptic connections. They potentate the actions of morphogens, growth factors, and cell adhesion molecules during development. The expression of neurotransmitters and their receptors is associated with developmentally transient projections during times of intense elaboration and/or consolidation of synapses. The convergence of two or more types of neurotransmitter projections, such as the cholinergic and adrenergic projections to the neocortex, can enhance synaptic stabilization (Lauder and Krebs, 1976, 1978; Johnston *et al.,* 1979; Lauder *et al.,* 1981, 1985, 1988; Bear *et al.,* 1985; Bear and Singer, 1986; Konig *et al.,* 1988; Lauder, 1988, 1993; McDonald *et al.,* 1988; Mobley *et al.,* 1989; Lieth *et al.,* 1990; Hellendall *et al.,* 1993; Johnston, 1994, 1995; Bennett-Clarke *et al.,* 1995; Buznikov *et al.,* 1996; Hensch and Stryker, 1996; Romano *et al.,* 1996; Stoop and Poo, 1996; Berger-Sweeney and Hohmann, 1997; Colquhoun and Patrick, 1997; Reid *et al.,* 1997; Roerig and Katz, 1997; Roerig *et al.,* 1997; Wang and Dow, 1997).

The influences of growth, thyroid, and steroid hormones can have a profound effect on brain morphogenesis (Hashimoto *et al.,* 1989; Kawata, 1995; Stratakis and Chrousos, 1995; Noguchi, 1996; Spindler, 1997). One of the most dramatic effects is the developmental consequences of severe congenital thyroid abnormality, cretinism, which results in mental retardation and deafness (Balazs, 1971; Stein *et al.,* 1991; Brent, 1994; Bernal and Nunez, 1995; Forrest *et al.,* 1996a,b). The thyroid releases triiodothyrione (T3) and thyroxine (T4) in response to throtropin from the pituitary. T4 is the primary form in the circulation and is converted to T3 in target tissues by the iodothyronine deiodinases, selenium-dependent enzymes regulated by T3 at both the pre- and posttranslational levels. Thyroid hormone receptors, TR-alpha and TR-beta, are expressed by the thyroid and target tissues, and their expression is regulated by thyroid hormone (Arthur *et al.,* 1992; Berry and Larsen, 1992; Saunier *et al.,* 1993; Calomme *et al.,* 1995; Larsen and Berry, 1995; Biesiada *et al.,* 1996; Forrest *et al.,* 1996a,b; Kohrle, 1996; Sohmer and Freeman, 1996; Burmeister *et al.,* 1997; Farsetti *et al.,* 1997; Liang *et al.,* 1997). T3 can substantially increase the rate of intermediary metabolism and, in particular, sodium–potassium ATPase activity in a variety tissues during development (Schmitt and McDonough, 1988; McNabb, 1995). Even though there is significant transplacental transfer of maternal T4 during fetal development, the dramatic effects of altered thyroid function typically observed during the postnatal development of the cerebellum are not seen when maternal hypothyroidism is induced late in gestation (Morreale de Escobar *et al.*, 1993; Pickard *et al.*, 1993; Porterfield and Hendrich, 1993; Ekins *et al.*, 1994; Pasquini and Adamo, 1994; Sinha *et al.*, 1994; Oppenheimer and Schwartz, 1997; Schwartz *et al.*, 1997). The cerebellum exhibits higher levels of mRNA for TR-beta postnatally than in the adult. T3 can influence several signaling pathways in both neurons and glia that regulate proliferation, migration, axonal growth, apoptosis, dendritic differentiation, synaptogenesis, and myelination. Elements of morphogenetic signaling pathways that have been reported to be regulated by T3 include NT-3, BCL-2, p75-NGFR; G-proteins, GAP-43, protein kinase C (PKC), CaM kinase IV, adenylate cyclase, CREB, GFAP, actin, tubulin, and MBP (Umesono *et al.*, 1988; Gravel and Hawkes, 1990; Gravel *et al.*, 1990; Nunez *et al.*, 1991; Figueiredo *et al.*, 1993a,b; Lindholm *et al.*, 1993; Poddar and Sarkar, 1993; Saunier *et al.*, 1993; Sherer *et al.*, 1993; Andres-Barquin *et al.*, 1994; Besnard *et al.*, 1994; Das and Paul, 1994; Wong *et al.*, 1994; Muller *et al.*, 1995; Krebs and Honegger, 1996; Krebs *et al.*, 1996; Lakshmy and Srinivasarao, 1997; Leonard and Farwell, 1997; Lima *et al.*, 1997; Martinez-Galan *et al.*, 1997a,b; Pal *et al.*, 1997; Sarkar *et al.*, 1997).

H. Astrocytes Induce Expression of the Blood–Brain Barrier in Endothelial Cells

The brain is dependent on other organ systems of the body for oxygen, nutrients, and the elimination of metabolic by-products. The primary route for such transfers is the cerebrovasculature, and its development is a critical aspect of brain morphogenesis. The cerebrovasculature also participates in regulating pH and osmolarity of the cerebrospinal fluid, selectively distributing hormones, segregating components of the immune response, preventing pathogenic invasion, and metabolizing or excluding xenobiotics. Various aspects of these diverse functions are attributed to the structural and molecular properties of cerebral endothelial cells and constitute the blood–brain barrier (Lathja and Ford, 1968; Rapoport, 1976; Bradbury, 1979; Jacobs, 1982; Goldstein and Betz, 1986; Neuwelt, 1989; Minn *et al.*, 1991; Tatsuta *et al.*, 1994; Bellamy, 1996; Schinkel *et al.*, 1996). The cardinal feature of the blood–brain barrier is the continuity of a cellular layer between the blood and the interstitial space of the brain. The anatomic basis of this contiguity is the tight junctions between the endothelial cells that line the capillary of the brain (Koella and Sutin, 1967; Bradbury, 1979; Saunders, 1992; Staddon and Rubin, 1996). Other structural specializations include close apposition by astrocytic endfeet, sparsely distributed pericytes, and extensive association of microglia. Molecular specializations include endothelial expression of numerous transporters, and a variety of enzymes including those involved in xenobiotic metabolism (Minn *et al.*, 1991; Peters *et al.*, 1991; Banks *et al.*, 1992; Aschner and Gannon, 1994; Boado *et al.*, 1994; Lagrange *et al.*, 1993; Tatsuta *et al.*, 1994; Yamada *et al.*, 1994; Beaulieu *et al.*, 1995; Fischer *et al.*, 1995; Jette *et al.*, 1995a,b; Nag, 1995; Zlokovic, 1995; Drion *et al.*, 1996; Huwyler *et al.*, 1996; Lechardeur *et al.*, 1996; Arboix *et al.*, 1997; Desrayaud *et al.*, 1997; Kanai, 1997; Lankas *et al.*, 1997; Pardridge *et al.*, 1997).

The vasculature of several brain regions lack these specializations, presumably because of their role in chemoreception or neurohumoral regulation, and are called the *circumventricular organs.* These brain regions include area postremia, median eminence, neurohypophysis, subfornical organ, lamina terminalis, and choroid plexus. The vasculature of several other regions of the nervous system also exhibit greater permeability, most notably the arcute nucleus of the hypothalamus and the dorsal root ganglia (Koella and Sutin, 1967; Rapoport, 1976; Bradbury, 1979; Jacobs, 1982).

The cerebrovasculature also exhibits regional variations in density and complexity. For example, regions of the spinal cord can have less than 150 mm of capillaries per cubic millimeter of tissue, whereas regions of the hypothalamus have in excesss of 2,000 mm of capillaries per cubic millimeter of tissue (Zeman and Innes, 1963). Local differences in the vascular density of gray matter parallels synaptic density and may be one of the requisites for a link between blood flow, metabolism, and regional neural activity. Although the extent and potential mechanisms of such neurovascular coupling remains controversial, blood flow and vascular permeability can be modified by the cholinergic, adrenergic, and serotonergic innervation of blood vessels, as well as autoregulatory pathways within the endothelium that involve endothelial growth factors and adhesion molecules, their receptors, nitric oxide, and calcium and potassium ion channels (Pardridge *et al.*, 1985; Minn *et al.*, 1991; Moncada *et al.*, 1991; Janigro *et al.*, 1994; Lagrange *et al.*, 1994; Jaffrey and Snyder, 1995; Lincoln, 1995; Miyawaki *et al.*, 1995; Haller, 1997; Hanahan, 1997; Northington *et al.*, 1997; Villringer and Dirnagl, 1997).

Marin-Padilla (1988) has described in exquisite detail the process of cerebral vasculogenesis based on extensive light and electron microscopic observations, and his description is largely consistent with that of other investigators (Bar, 1980; Sturrock, 1981; Kniesel *et al.*, 1996). Cerebral capillaries grow by the local generation of new endothelial cells from those of preexisting endothelium. The newly formed endothelial cells extend filopodia and begin migrating away from parent vessel. Several migrating endothelial cells clusters to form capillary sprouts, begin to secrete a basal lamina, and form loose intercellular contacts. This process transforms extracellular space into accessory lumina, which enlarge and coalesce. Dynamic rearrangement of intercellular contacts between the endothelial cells results in these accessory lumina gaining continuity with the lumen of the parent vessel and ultimately forming an extension of the original lumen.

The development of the cerebrovasculature entails the growth of capillaries into three tissue compartments: the meningial compartment, the Virchow–Robin compartment, and the perivascular compartment. Capillaries become established within the meninges at the same time that these membranes segregate into their distinctive layers (dura, arachnoid, and pia). As migrating endothelial cells contact the pial surface, several of their filopodia pierce through both the endothelial and nervous system basal lamina to come into direct contact with the glial endfeet at the surface of the brain. The filopodia then grow between the glial endfeet, and the endothelial cell migrates along extending filopodia. Several of the migrating endothelial cells cluster to form a capillary sprout that continues to grow into the brain by addition of newly generated endothelial cells.

At the time endothelial cells penetrate the surface of the brain, the basal lamina of the endothelium and the neural tissue fuse to form the anlage of the Virchow–Robin compartment. This compartment forms as the space widens between the basal lamina of neural tissue and the basal lamina of the endothelium. The Virchow–Robin compartment may serve some functions analogous to the lymphatics of other tissues, and appears to be involved in a variety of pathogenic processes. The association of the Virchow–Robin compartment with the vessels that first penetrate the brain distinguishes them from the subsequently formed perivascular vessels that invade the parachemya to form dense anastomosing networks. The transition from penetrating vessels to the perivascular vessels entails the fusion of the neural and endothelial basal lamina into a single layer. Also at this point, the perivascular vessels become surrounded by their characteristic glial sheath. The perivascular network of anastamosing vessels within the brain paracheyma represents the most dynamic of the three compartments, and can be altered by a variety of developmental manipulations (Argandona and Lafuente, 1996) and continues to undergo substantial growth and remodeling during development and into adulthood.

Specialization characteristic of the blood–brain barrier are induced in endothelial cells by contact with astrocytes, and thus it is likely that they are primarily, if not exclusively, properties of the vessels of the perivascular compartment (Joo, 1987). As noted previously, the tight junctions between endothelial cells, the cardinal structural specializations of the blood–brain barrier, develop during the initial stages of capillary formation. This contrasts with the commonly cited notion that the barrier characteristics of the cerebrovasculature are acquired late in development. Saunders (1992) reviewed carefully the earlier studies typically cited as indicating that the blood–brain barrier of developing brain is more permeable than that of the adult. Saunders pointed out that in the adult, the tracers that have been used to assess the integrity of the blood–brain barrier are largely bound to plasma proteins in the blood, and there is relatively only low levels of unbound tracer typically available to reach the brain. Because developing animals have lower levels of plasma proteins per unit of body weight than adults, a dose comparable to that administered to adult animals would result in significantly greater levels of unbound tracer in the blood of developing animals. These higher levels of tracer could result in a toxic disruption of the intact, but possibly vulnerable, blood–brain barrier of the neonate.

Additional features of the blood–brain barrier, most notably transporters for glucose, amino acids, and the multidrug resistance transporter (p-glycoprotein) appear to have distinctive developmental profiles over more a protracted postnatal period (Dobbing, 1968; Pardridge *et al.,* 1988, 1997; Jones *et al.,* 1992; Laterra *et al.,* 1992; Cornford *et al.,* 1993; Dwyer and Pardridge, 1993; Vannucci *et al.,* 1993, 1994; Engelhardt *et al.,* 1994; Ibiwoye *et al.,* 1994; Sada *et al.,* 1994; Xu and Ling, 1994; Brett *et al.,* 1995; Bellamy, 1996; Bolz *et al.,* 1996; Cavalcante *et al.,* 1996; Kuwahara *et al.,* 1996; Stonestreet *et al.,* 1996; al-Sarraf *et al.,* 1997; Cassella *et al.,* 1997; Plateel *et al.,* 1997). Signaling pathways regulating vascular morphogenesis and the remodeling of the vasculature involve the paracrine actions of vascular endothelial growth factors (VEGFs/VPFs, PDGF, Tie2, Angl/2) and their tyrosine kinase receptors, as well as adhesion molecule signaling pathways (VE-cadherein, PECAM-1, beta-catenin) (Farrell and Risau, 1994; Achen *et al.,* 1995; Breier *et al.,* 1995; Flamme *et al.,* 1995; Kniesel *et al.,* 1996; Carmeliet *et al.,* 1997; Lampugnani *et al.,* 1997; Risau, 1997).

III. Developmental Neurobehavioral Disorders Result from the Disruption of Morphogenesis

Some of the signaling pathways involved in brain morphogenesis may be interdependent and organized in a temporal hierarchy such that early transformations are a requisite for the subsequent induction of events that occur later in development. Others may play a permissive role and modulate developmental interactions over extended periods of development. Although morphogenetic signaling pathways tend to be highly conserved in vertebrate evolution, the relatively protracted development of the mammalian brain may increase the potential for variations in interactions between signaling pathways. The extent to which such variations overlap in their specificity may afford a degree of protection from genetic and environmental perturbations. The limits of any such protection, however, are demonstrated by the existence of diverse and severe developmental neurobehavioral disorders. Identification of alterations in brain morphogenesis that occur in developmental neurobehavioral disorders is critical to defining the fundamental boundaries between the adaptive capacity of the developing brain and its susceptibility to disruption of normal development.

Developmental neurobehavioral disorders can be defined as pathologic alterations of the developing nervous system resulting from genetic defect, infection, or other environmental insult, and are characterized by a unique group of signs and symptoms. Diagnostic criteria can be derived from three overlapping domains: the morphologic, neurologic, and psychiatric. In the morphologic domain, disorders can be grouped as acquired lesions, malformations, and metabolic disorders in which the pattern of gross and microscopic alterations in the postmortem brain are the defining characteristics (Friede, 1989). In the neurologic domain, diagnosis is based on a neurologic examination, history, and special diagnostic tests (Willis and Grossman, 1977). The neurologic exam is conducted to determine the anatomic site of a lesion and to gain information on its pathologic condition. Together with the history, this information is used to select special diagnostic tests to aid in the differential diagnosis. In the psychiatric domain, syndromes are defined by the presence of several related behavioral abnormalities (American Psychiatric Association, 1994). Although there is a general concordance between these three domains, there are a number of difficulties in establishing clear parallels between the different diagnostic criteria for developmental neurobehavioral disorders. From a morphologic point of view, the most obvious is that there is seldom a unique or consistent structural alteration associated with a particular syndrome. Even when there is an association with a characteristic set of morphologic alterations, their developmental origins can be obscure. The high frequency of the coexistence of different disorders within an individual also contributes to the difficulty of selectively associating a particular set of morphologic alterations with a particular syndrome. Nonetheless, significant progress continues to be made in understanding the neurobiologic basis of these disorders. Two examples, mental retardation and schizophrenia, illustrate advances in identifying alterations in morphogenetic processes associated with developmental neurobehavioral disorders.

A. *Mental Retardation Results from the Disruption of Diverse Morphogenetic Signals*

Mental retardation is typically defined by three criteria: subaverage intellectual functioning as indicated by an IQ of 70–75 or less, impaired adaptive behavior, and evidence these impairments originated prior to the age of 18 years (Schaefer and Bodensteiner, 1992; King *et al.,* 1997). Based on these criteria, the prevalence of mental retardation is 2.5%, with estimates of diagnosis ranging from 1–3%. The most common causes of the severe form are chromosomal aberrations, accounting for 30% of patients with mental retardation. Of the chromosomal aberrations, Down syndrome (DS) accounts for 20% of patients with mental retardation, whereas other chromosomal aberrations, including fragile X syndrome, account for approximately 10%. Congenital anomalies, including nervous system malformations, occur in 14–20%, whereas hormonal and metabolic disorders contribute to 3–5%. Infections and teratogens such as ethanol account for 15–20%.

In spite of the widely cited nature of the previously mentioned causes of mental retardation, the association of a singular genetic or environmental cause with a specific proportion of cases of mental retardation is misleading. Interactions between genetic and environmental factors contribute significantly to the incidence and severity of mental retardation (Gibbons *et al.,* 1995; Liu and Elsner, 1995; Matilainen *et al.,*1995; Simonoff *et al.,* 1996; Curry *et al.,* 1997; Evrard *et al.,* 1977). Genetic alterations responsible for mental retardation can be influenced dramatically by environmental factors. For example, in phenylketonuria, in which genotypic variations are highly correlated with phenotypic enzyme activity (Burgard *et al.,* 1996; Eiken *et al.,* 1996; Eisensmith *et al.,* 1996) and blood levels of phenylalanine are predictive of the extent of neurologic impairment (Burgard *et al.,* 1997), perinatal dietary modification can virtually eliminate the associated mental retardation, although other types of neurologic and cognitive impairments can persist (Acosta, 1996; Azen *et al.,* 1996; Diamond and Herzberg, 1996; Guldberg *et al.,* 1996; Weglage *et al.,* 1996).

Environmental factors influencing the incidence and severity of mental retardation involve complex interactions. There is dramatic variability in estimates of the percent of congenital infections such as cytomegalavirus, influenza, or rubella that result in mental retardation (Conover and Roessmann, 1990; Bale and Murph, 1992; Raynor, 1993; Koedood *et al.,* 1995; Steinlin *et al.,* 1996). Such variability may be related to hormonal, nutritional, and genetic factors that influence the severity of the infection, modulate infection-induced injury to the nervous system, and influence the extent of persistent neurologic impairment (Connolly and Kvalsvig, 1993; McCann *et al.,* 1995; Arvin *et al.,* 1996; Bobryshev and Ashwell, 1996; An and Scaravilli, 1997; Brooke *et al.,* 1997; Goujon *et al.,* 1997; Johnson *et al.,* 1997; Lipton, 1997).

Given the spectrum of such interactions, the fact that a diversity of neuropathologic alterations are associated with mental retardation is not surprising. Nor is it surprising that no one alteration, or combination of alterations, are considered diagnostic of mental retardation. Estimates of the occurrence of detectable overt morphologic alterations in mental retardation range from 34% to 98%, and between 2% and 16% exhibit major anomalies (Schaefer and Bodensteiner, 1992). Alterations in the hippocampus and cerebellum are commonly reported in imaging studies (Pulsifer, 1996; Kumada *et al.,* 1997). Neocortical hypoplasia and ectopias can be observed on neuropathologic examination (Sidman and Rakic, 1973, 1975). A potentially revealing microscopic feature of brains of patients with mental retardation, with and without overt anatomic defects, is the alteration in dendritic spines. Such spines tend to be long, thin, and have abnormally large terminal heads (Marin-Padilla, 1972; Purpura, 1974, 1977). CaM kinase II has been hypothesized to play a role in determining morphologic characteristics of dendrites and their spines (Sahyoun *et al.,* 1985; Burgin *et al.,* 1990; Gang-Yi and Cline, 1998), and may therefore represent a point of convergence for diverse influences contributing to mental retardation.

Three advances in the characterization of mental retardation are particularly relevant to developmental neurotoxicology. One advance is in the understanding of factors contributing to precocious dementia in DS (Takashima, 1997). The gene for amyloid precursor protein (APP) is located on chromosome 21, and APP overexpression can be detected in the brains of DS patients as early as 21 weeks of gestation. This early expression of APP may be responsible for the subsequent accumulation of the Aβ42 form of β-amyloid that precedes the appearance of the amyloid plaques in the brains of DS patients. In addition to amyloid accumulation, cultured cortical neurons from fetal DS brain have been observed to have a fourfold increase in reactive oxygen species compared to those from age-matched controls. Because amyloid accumulation and free radical generation are mechanisms hypothesized to account for neurotoxicant-related injury, they constitute potential pathways by which the DS genotype could confer susceptibility to environmental toxicants.

Significant advances have also been made in the characterization of the fragile X syndrome (Chakrabarti and Davies, 1997). Morphologic hallmarks of the syndrome include increased cerebral and ventricular size as well as alterations in the hippocampus, basal ganglia, and cerebellum. Observations of brains stained with the Golgi method reveal an increased density of long, thin spines suggestive of an immature stage of development (Comery *et al.,* 1997). The fragile X syndrome is one of the several triplet expansion repeat mutations associated with several neurologic disorders (La Spada, 1997; Li and el-Mallakh, 1997; Mitas, 1997; Reddy and Housman, 1997). The mutation likely to be responsible for the fragile X syndrome has been localized to the 5′ untranslated region of exonl of the FMR1 gene and results in a developmental methylation of the promoter with transcriptional silencing of the gene. The product of the FMR1 gene, FMRP, is normally expressed ubiquitously, but is most abundant in the brain. Immunohistochemical and ultrastructural studies reveal a preferential distribution in cytoplasm and dendrites, and a distinct association with polysomes. In addition, it can be detected within nuclear pores, suggesting a role in cytoplasmic–nuclear shuttling (Feng *et al.,* 1997). This pattern of distribution is consistent with its involvement in RNA processing (Chakrabarti and Davies, 1997). One potential consequence of alterations in RNA processing is an increased frequency of transcript mutations. Altered RNA processing has also been implicated in the increased frequency of transcript mutations of APP and ubiquitin-B in the brains of Alzheimer's disease and DS patients (van Leeuwen *et al.,* 1998). Whether the absence of FMRP is also associated with an increase in transcript mutations remains to be determined. These observations raise the possibility that genetic factors for certain developmental neurobehavioral disorders may engender increased vulnerability to toxicants capable of disrupting transcription.

A third advance in understanding the neurobiologic basis of mental retardation has been the identification of several mutations of the L1 adhesion molecule in inherited syndromes that include Hydrocephalus As a result of Stenosis of the Aqueduct of Sylvius (HASAS syndrome), Mental retardation, Aphasia, Shuffling gaits and Adducted thumbs (MASA syndrome), and X-linked spastic paraplegia (Fransen *et al.,* 1995; Wong *et al.,* 1995; Kenwrick *et al.,* 1996; Uyemura *et al.,* 1996; Yamasaki *et al.,* 1997). In addition to hydrocephalus, frequent signs include adducted thumbs, an enlarged and sometimes asymmetric head, spasticity, and aphasia. L1 is a member of the immunoglobulin superfamily of cell adhesion molecules and is involved in both neuronal migration and the guidance of growing axons. Surface interactions of L1 may mediate the influence of growth factors such as FGF on axonal extension by activating several intracellular signaling cascades that involve both tyrosine–and serine–threonine kinases. Likely consequences of alterations in the function of L1 at the cellular level correspond to several of the morphologic alterations observed in these syndromes, the most striking of which are a diminution or absence of the corticospinal tract and corpus callosum.

The range of morphologic alterations associated with mental retardation is consistent with the involvement of a variety of mechanisms, with the most dramatic morphologic alterations associated with disruption of instructive processes during early morphogenesis. A more difficult question is whether the occurrence of mental retardation in the absence of overt morphologic alterations is related specifically to disruption of the selective consolidation of synaptic connections. The altered morphology of dendritic spines, observed in brains with and without overt morphologic malformations, is consistent with this possibility.

B. Schizophrenia Is Associated with Structural Alterations Indicative of Alterations in Development

In contrast to the wide range of morphologic alterations associated mental retardation, there are other neurobehavioral disorders in which historically there have been found no consistent evidence of gross morphologic defects in the brain. Schizophrenia has been an example of such a disorder. Identification of the disorder is based on three hallmark behavioral symptoms: delusions, hallucinations, and disorganization of thought processes. The etiology of schizophrenia is unknown. Familial and twin studies indicate a highly significant polygenetic component. Concordance rates of about 50% in identical twins and 20% in fraternal twins indicate that both genetic and environmental factors play a significant role in the emergence of the disorder, yet no single type of environmental influence has yet emerged to account for disconcordant monozygotic twins (Polymeropoulos *et al.,* 1994; Crow, 1995a,b 1997; Faraone *et al.,* 1995; Goodman, 1995; Mednick and Hollister, 1995; Tsuang and Faraone, 1995; Franzek and Beckmann, 1996; Mallet, 1996; Morris, 1996; Murphy and McGuffin, 1996; Murphy *et al.,* 1996; O'Donovan and Owen, 1996; Ross and Pearlson, 1996; Deakin *et al.,* 1997; Hafner and an der Heiden, 1997; Karayiorgou and Gogos, 1997; Mohler, 1997; Mowry *et al.,* 1997; Portin and Alanen, 1997a,b; Stefan and Murray, 1997; Weinberger, 1997).

Structural alterations that have been reported in the brains of schizophrenics are indicative of abnormal development (Waddington, 1993; Ebadi *et al.,* 1995; Goldman-Rakic, 1995; Jones, 1995, 1997; Selemon *et al.,* 1995; Chua and Murray, 1996; Frangou and Murray, 1996; Shioiri *et al.,* 1996; Wolf and Weinberger, 1996; Barondes *et al.,* 1997; Benes, 1997; Catts *et al.,* 1997; Deakin *et al.,* 1997; Deakin and Simpson, 1997; Goldman-Rakic and Selemon, 1997; Harrison, 1997; Lewis, 1997; Rapoport *et al.,* 1997; Seidman *et al.,* 1997; Weinberger, 1997). These alterations include ventricular enlargement, decrease in cortical and subcortical volume, increased cerebral symmetry, and alterations in cerebral architectonics. The prefrontal and occipital regions of the neocortex of schizophrenics exhibit increased density of neuronal cell bodies together with a decrease in cortical thickness. Such a combination of alterations is consistent with the atrophy of neuronal and glial processes within the cortical neuropil. The anomalous expression of NCAM and GAP-43 together with the finding that cerebral atrophy and ventricle enlargement continue to progress in adolescence indicate that it is likely that at least some aspects of these alterations are the result of dynamic processes that may continue into adolescence.

The patterns of morphologic alterations associated with schizophrenia are consistent with the possibility that a normal selective consolidation of neural circuitry responsible for the emergence of hemispheric asymmetry may be disrupted. The finding of a significantly greater number of neurons in the striatum of schizophrenics compared to age-matched controls (Beckmann and Lauer, 1997) is consistent with the possibility of a disruption in normally occurring programmed cell death. Such disruptions may be an antecedent to a later atrophy of cortical neurons during adolescence that trigger the emergence of symptoms. The plausibility of this or related scenarios is dependent on more direct identification of the specific developmental events that are disrupted in schizophrenia.

C. Predominance of Structural and Functional Alterations May Be a Reflection of the Relative Disruption of Instructive and Selective Morphogenetic Processes

The examples of mental retardation and schizophrenia illustrate the difficulties in associating specific morphologic alterations with particular developmental neurobehavioral disorders. Nonetheless, the consistent patterns of morphologic alterations are compatible with the possibility that particular syndromes may be related to the relative differences in the disruption of instructive and selective morphogenetic processes. Instructive processes are likely to have robust influences on the acquisition of gross morphologic features of the brain, whereas selective processes have a more profound effect on the accuracy and precision of connections. These differences can be demonstrated by comparing the effect of an increase in the length of the cell cycle with the effect of programmed cell death on the acqusition of neurons from a hypothetic germinal matrix (Fig. 1). In the context of the normal exponential rate of production of neurons, the decrease in the number of neurons resulting from a 75% loss of differentiating neurons appears small compared to the substantial reduction in the number of neurons resulting from a 50% increase in the length of the cell cycle. There are two important implications of this comparison. First, it is unlikely that a reduction in normally occurring cell loss could compensate for disruptions in proliferation. Second, instructive influences, such as those that regulate proliferation rate, can have profound effects on gross morphologic characteristics of the brain. In contrast, selective influences have profound effects on patterns of connectivity that occur in the absence of overt morphologic alterations. These two kinds of processes are presumably coordinated during normal development. The extent to which genetic or environmental factors disrupt such coordination, or have a greater selectivity towards one or the other type of process may account for relative prominence of morphologic or functional deficits (Fig. 2).

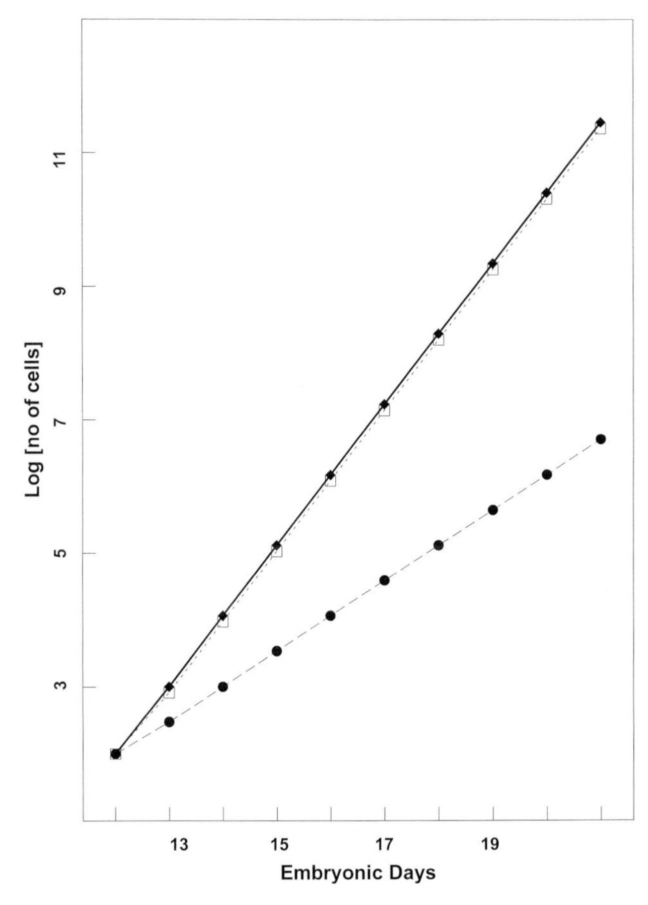

FIGURE 1 Comparison of the effect of cell cycle time and programmed cell death on number of neurons. The rate of production of neurons from three hypothetical populations of germinal cells over 10 days. Each population begins with 200 proliferating germinal cells. At the end of each cell cycle, one cell leaves the matrix to differentiate and one cell returns to continue to proliferate. The percent cell loss is the proportion of differentiating neurons that undergo programmed cell death at the end of each cell cycle. Population A (◆; —). Cell cycle = 3 hours, 0% cell loss. Population B (□; · · · ·). Cell cycle = 3 hours, 75% cell loss. Population C (●; - - - -). Cell cycle = 6 hours, 0% cell loss. The small difference between populations A and B as compared to their large difference with population C indicate that alterations in cell cycle time can have a far more dramatic influence on the total number of neurons than programmed cell death. Programmed cell death, however, may have a far more dramatic effect on the selective consolidation of connections between highly localized regions of the nervous system.

There are several important caveats to this speculative proposal of a dichotomy between instructive and selective processes. The role of the selective elimination of neurons and their connections in the development of the brain remains controversial (Purves *et al.*, 1996; Changeux, 1997; Purves, 1997; Sporns, 1997). In particular, the extent to which consolidation of connections can compensate for a disruption in instructive processes is not known. The role of the selective consolidation of neural connections may not be primarily in correcting errors or eliminating redundant connections. Developmentally transient connections may serve specific behaviors expressed exclusively during development. If the consolidation of neural connections is necessary for the orderly retention and recall of information in the adult, the developing organism may depend on qualitatively different kinds of processing requiring these transient connections. Such processing might be based on more distributed pathways and involve more global interactions such as those proposed by Pribram (1969), for which working models have been developed (Psaltis *et al.*, 1990; Shen *et al.*, 1997).

An alternative possibility is that developmentally transient connections mediate instructive influences derived from environmental stimulation or behavioral activation. In this case their preservation in response to injury may extend the time during which inductive interactions may take place. This latter possibility is consistent with the amearolative effects of environmental enrichment following perinatal insults (Greenough *et al.*, 1987; Hannigan, 1995).

The self-organizing interactions responsible for the plasticity of the developing brain remain elusive, and the instructive–selective dichotomy may be a small com-

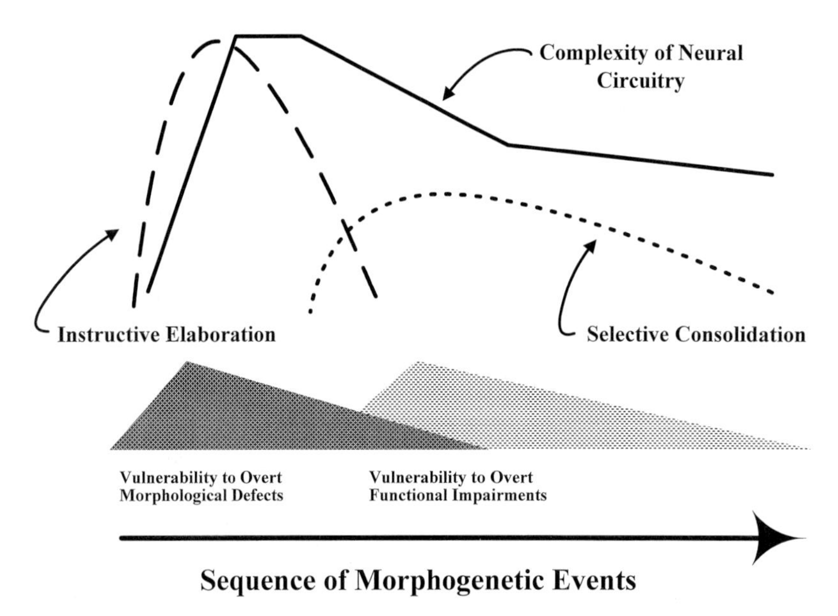

FIGURE 2 The interrelationship of disruptions of the progressive elaboration and selective consolidation of neural circuitry and the extent of morphological defects and functional impairments. The dahsed line indicates the magnitude of instructive influences on the growth of complexity of neural circuitry (solid line). The dotted line indicates the magnitude of the selective influences on the consolidation of neural circuitry. The darkly hatched area indicates the corresponding vulnerability to morphologic defects as a result of the disruption of instructive signals. The lighter hatched area indicates the corresponding vulnerability to functional impairments resulting from the disruption consolidate signals.

ponent of a far more complex and heterogeneous process. One indication of such complexity is that the classic observation that injury to the neocortex earlier in development results in less dramatic motor deficits than does injury at later times (Kennard, 1942) is likely to be limited to particular types of circuits or behaviors. Age-dependent variations in recovery can depend on the nature and location of the lesion and the task used to assess functional recovery (Nonneman and Isaacson, 1973).

Another limitation of this instructive–selective dichotomy relates to the presumption of differential vulnerability. There appears to substantial overlap in the signaling pathways mediating instructive and selective influences. Consequently, the extent to which they are temporally segregated may be a primary determinant of selective vulnerability. Distinguishing vulnerable periods for the two types of processes requires greater temporal resolution than that provided by the classic notion of the "brain growth spurt." Whereas the intensity of developmental events occurring during the "brain growth spurt" clearly engenders susceptibility to nutritional deprivation (Dobbing and Sands, 1979), insults that occur earlier or later in development can also result in significant alterations in neuronal phenotype, defects in brain morphology, or functional deficits that emerge later in life. Developmental insults may also predispose an individual to the cumulative effects of biologically significant events that contribute to a decline in cognitive function late in life (Houx and Jolles, 1992). Such temporal considerations are critical to determining the relative vulnerability of instructive and selective processes.

As investigations of the effects of toxicants on brain development become more sharply focused, the concept of a vulnerable period may be more appropriately described in terms of specific molecular interactions rather than the time period in which they occur. For example, injury to the brain caused by exogenous glutamate was first associated with exposure during a perinatal vulnerable period. The characterization of the developmental patterns of expression of glutamate receptors provided a more proximate determinant of localized patterns of neuronal damage, and further characterization of the role of such receptors in the injury resulted in the concept of "excitotoxicity," which has been hypothesized to play a role in a number of neurologic disorders (Zorumski and Olney, 1992).

IV. Convergence of Genetic and Environmental Influences May Increase the Severity and Incidence of Developmental Neurobehavioral Disorders

Epidemiologic studies of developmental neurobehavioral disorders often demonstrate significant genetic components to their etiologies, but rarely do genetic factors account for major sources of the variance of the affected population (Paneth, 1993; Cote and Vandenburg, 1994; Nass, 1994; Palomaki, 1996; Sham, 1996; Baraitser, 1997; Beitchman and Young, 1997; Hafner and an der Heiden, 1997; Roeleveld *et al.,* 1997). Epidemiologic studies also demonstrate a significant role for environmental toxicants in developmental neurobehavioral impairments and, as in the case of genetic determinants, markers of exposure tend to account for a small portion of the variance of measures of behavioral impairments (Bellinger *et al.,* 1987, 1992; Needleman and Bellinger, 1991; Riess and Needleman, 1992; Needleman, 1994, 1995; Winneke, 1996). These two observations are consistent with the suggestion that an interaction of genetic and environmental factors is a critical determinant in the incidence and severity of developmental neurobehavioral disorders. Two significant points at which the influences of genotype and exposure to toxicants may convergence are morphogenetic signaling pathways and the homeostatic processes that protect the developing brain from xenobiotic insult.

A. *Developmental Neurotoxicants Disrupt Morphogenetic Signals*

Developmental neurotoxicants are typically identified by the detection of morphologic or functional alterations observed at various times after exposure during development. The protracted period of brain morphogenesis makes it difficult to infer the primary targets of a toxicant based on persistent alterations. Consequently, the action of developmental neurotoxicants are typically characterized in terms of their effects on major classes of developmental events such as proliferation and migration, dendritic and axonal differentiation, cell death, collateral loss, synaptogenesis, and myelination (Rodier, 1976, 1977, 1986, 1988, 1994, 1995; Chang *et al.,* 1980a,b; Chang, 1990; Eriksson, 1997). Even though such characterization accentuates similarities in outcome rather than selectivity for targets, it represents an essential step in identifying potential signals that may be targets of neurotoxicants.

A number of considerations contribute to the complexity of attempting to relate the disruption of developmental events to specific perturbations in morphogenetic signals. A particular toxicant can disrupt a variety of morphogenetic signals. Varying degrees of evidence exist that developmental neurotoxicants such as metals, pesticides, polychlorinated biphenyls (PCBs), or solvents disrupt the function of adhesion molecules, cytoskeletal elements, ion channels, and other aspects of ion regulation. They may also disrupt intracellular the transduction pathways that regulate the transcription, translation, and postranslational modification of proteins. Toxicants may also disrupt developmental events by interfering with the morphogenetic roles of hormones, neurotransmitters, and their receptors. Some developmental neurotoxicants can increase oxidative stress, which may result in the aberrant stimulation of neuronal differentiation or programmed cell death. (Chang *et al.,* 1980; Parke, 1982; Byrne and Sepkovic, 1987; Cockerill *et al.,* 1987; Atchison, 1988; Byrne *et al.,* 1988; Wasteneys *et al.,* 1988; Bressler and Goldstein, 1991; Ecobichon, 1991; Levesque and Atchison, 1991; Chang, 1992; Nicotera *et al.,* 1992; Seegal and Shain, 1992; Sutoo, 1992; Tilson and Harry, 1992; von Euler, 1992; Zorumski and Olney, 1992; Peterson *et al.,* 1993; Atchison and Hare, 1994; Balduini *et al.,* 1994; Costa, 1994; Lagunowich *et al.,* 1994; O'Callaghan, 1994; Reuhl *et al.,* 1994; Kovacs *et al.,* 1995; Tan and Costa, 1995; Aschner, 1996; Corey *et al.,* 1996; Denny and Atchison, 1996; Hanas and Gunn, 1996; Holmuhamedov *et al.,* 1996; Miller, 1996; Seegal, 1996; Yang *et al.,* 1996; Yu *et al.,* 1996; Aschner, 1997; Conner, 1997; Desaulniers *et al.,* 1997; Graff *et al.,* 1997; Guerri and Renau-Piqueras, 1997; Schuur *et al.,* 1997; Song *et al.,* 1997; Windmill *et al.,* 1997; Yuan and Atchison, 1997).

The extent to which a developmental neurotoxicant may concomitantly disrupt more than one morphogenetic signaling pathway may be critical in determining the nature of alterations it can induce in brain morphogenesis. The disruption of developmental events may not result from a cascade of alterations emanating from the disruption of one particular morphogentic signal, but may be the consequence of a "web" of interactions resulting from alterations in several signals (Fig. 3). Identifying the critical features of such a web that determine that pattern of disruptions associated with exposure to a toxicant is a complex challenge. One determinant of the complexity of such a web is the developmental pattern of expression of potential targets at the time of exposure. Critical features of such a web may be dramatically different depending on the concentration of a toxicant at various targets. The maturational state of developing neurons may also contribute to the complexity of such interactions. For example, mercury has been hypothesized to interact with ion channels,

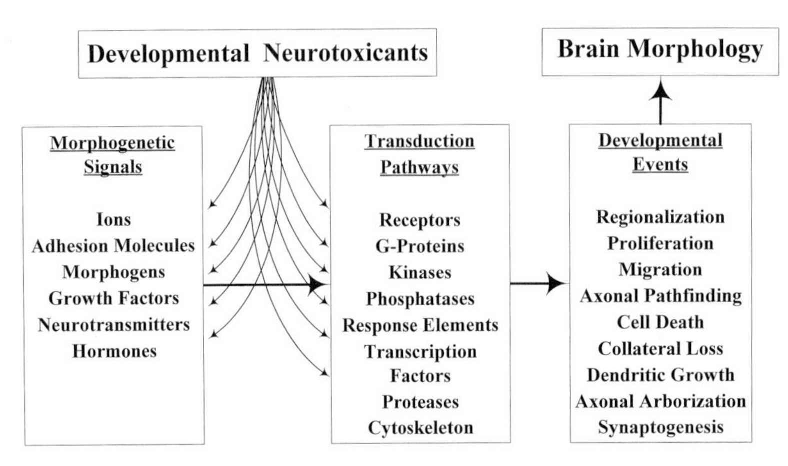

FIGURE 3 "Web" of interactions between developmental neurotoxicants and signaling pathways that regulate morphogenic events.

adhesion molecules, and cytoskeletal elements and alterations in these targets can directly or indirectly influence calcium distribution (Atchison *et al.,* 1986; Atchison, 1987, 1988; Shafer *et al.,* 1990; Levesque and Atchison, 1991; Russell *et al.,* 1992; Denny *et al.,* 1993; Hare *et al.,* 1993; Atchison and Hare, 1994; Denny and Atchison, 1994, 1996; Reuhl *et al.,* 1994; Sirois and Atchison, 1996). Because these targets may themselves be developmentally regulated by calcium-dependent mechanisms, the extent of disruption of signals would be dependent on the state of maturation of cellular calcium homeostasis.

B. Homeostatic Processes Protect the Developing Brain from Xenobiotic Insult

If the sole determinants of toxicant-induced neurologic impairments were the abundance of natural and human-made developmental neurotoxicants and the diversity of their potential targets in the developing brain, then it is very likely such impairments would be ubiquitous throughout the population. That this is not the case is due in large part to a variety of homeostatic mechanisms that protect the brain from xenobiotic insult. Processes that regulate absorption, metabolism, elimination, and distribution throughout the body play a central role in minimizing the probability of a toxicant reaching the developing brain (Khera and Clegg, 1969; Lake *et al.,* 1973; Benke and Murphy, 1975; Neubert, 1988; Birnbaum, 1991; Pang *et al.,* 1992; deBethizy and Hayes, 1994; Commandeur *et al.,* 1995; Leveque and Jehl, 1995; Davies *et al.,* 1996). The metabolism and elimination of xenobiotics by the placenta, liver, and kidney are, however, developmentally and hormonally regulated and the signaling pathways regulating the development of organs and tissues involved in the distribution and metabolism of xenobiotics may be disrupted by toxicants (Basu *et al.,*1971; Chakraborty *et al.,*1971; Chang *et al.,* 1972, 1980; Walker *et al.,* 1973; Ware *et al.,* 1974; Desnoyers and Chang, 1975; Ware *et al.,* 1975; Chang and Sprecher, 1976; Chang and Desnoyers, 1978; Symons *et al.,* 1982; Ebstein *et al.,* 1986; Simmons and Kasper, 1989; Goodlett and West, 1990; Goyer, 1990; Menez *et al.,* 1993; Simone *et al.,* 1994; Fleck and Braunlich, 1995; Lee *et al.,* 1995; Abe-Dohmae *et al.,* 1996; Chen *et al.,* 1996; Lash and Zalup, 1996; Prough *et al.,* 1996; Segall *et al.,* 1996; Shimada *et al.,* 1996; Taylor *et al.,* 1996; Wells and Winn, 1996; Wharton and Scott, 1996; Chung *et al.,* 1997; Dekant, 1997; Ladle *et al.,* 1997; Mitani *et al.,* 1997; Notenboom *et al.,* 1997; Rich and Boobis, 1997; Zangar *et al.,* 1997). The brain itself also has a specialized capacity to metabolize certain xenobiotics and the development of this capacity may be a determinant in susceptibility to toxicant insult (Lowndes *et al.,*1995; Ravindranath and Boyd, 1995; Bergh and Strobel, 1996; Strobel *et al.,* 1997). The development of the cerebrovasculature, in particular the blood–brain barrier and neurovascular coupling also influence the distribution of toxicants in the developing brain and their regulation may be altered by toxicants (Chang and Hartmann, 1972; Ware *et al.,* 1974; Domer and Wolf, 1980; Cornford *et al.,* 1985; Press, 1985; Bradbury and Deane, 1988; Moorhouse *et al.,* 1988; Aschner and Aschner, 1991; Petrali *et al.,* 1991; Aschner *et al.,* 1992; Ray *et al.,* 1992a,b; Bradbury and Deane, 1993; Vorbrobt and Trowbridge, 1993; Lowndes *et al.,* 1994; Vorbrodt *et al.,* 1994a,b; Ghersi-Egea *et al.,* 1995; Lowndes *et al.,* 1995; Ravindranath and Boyd, 1995; Romero *et al.,* 1995; Bergh and Strobel, 1996; Blake *et al.,* 1996; Duchini, 1996; Lephart, 1996; Negri-Cesi *et al.,* 1996; Kerper and Hinkle, 1997; Ladle *et al.,* 1997; Ray, 1997a,b; Rose *et al.,* 1997; Struzynska *et al.,* 1997). Taken together, these considerations are consistent with the notion that the capacity of a toxicant to disrupt brain morphogenesis is the result of its potential effects, not only on the brain, but also on diverse cell types, tissues, and organs that influence xenobiotic detoxification, elimination, activation, and distribution. An important determinant in the potency of a toxicant to alter brain morphogenesis may the extent to which it induces alterations in tissues and organs that influence the level of the toxicant, or its active metabolite, that reach potential targets within the developing brain.

C. Interactions between Genetic and Environmental Factors Influence Homeostatic and Morphogenetic Processes

The actions of developmental neurotoxicants on diverse organ systems parallels the complex symptomology of developmental neurobehaviorial disorders. In such disorders, disruption of morphogenesis is rarely confined to the nervous system. Other organ systems in which alteration often occur include the heart, kidney, immune system, as well as features of the head, face, and limbs (Daft *et al.,* 1986; Rosenberg and Harding, 1988; Gage and Sulik, 1991; Cohen and Sulik, 1992; Kotch and Sulik, 1992; Lorke, 1994; Baraitser, 1997). A developmentally related concordance between alterations in diverse organ systems does not, however, necessarily indicate that an associated neurologic impairment is the direct consequence of the disruption of a singular developmental event. Instead, some neurologic manifestations may be secondary to the alterations in other organ systems. For example, there is high frequency of combined autoimmune dysfunction and neurologic impairment in the BXSB strain of New Zealand Black mice resulting from a recessively inherited trait with incomplete penetrance. These display several impairments that parallel those observed in dyslexia (Geschwind, 1979; Galaburda *et al.,* 1985; Sherman *et al.,* 1990, 1992, 1994; Denenberg *et al.,* 1991; Schrott *et al.,* 1993; Boehm *et al.,* 1996a,b). One of the most striking parallels is that about half of the BXSB mice exhibit superficial ectopias in the neocortex, a pathologic alteration commonly found in the brains of patients with dyslexia. BXSB mice are impaired in tasks of avoidance learning and working memory, but exhibit superior reference memory. Alterations in some neurologic functions, such as avoidance learning and paw preference, appear to be a consequence of intrauterine immune dysfunction, whereas others such as discrimination learning and Morris water maze performance appear to be under more direct genetic control and are associated with the presence of ectopias. Early environment enrichment can eliminate the more genetically related functional deficits, but does not reduce impairments associated with immune dysfunction.

In addition to the influence of other organ systems on the expression of neurologic impairments, there are also a variety of genetic and environmental factors that can influence either distribution of toxicants to the developing brain or their capacity to disrupt morphogenic signals (Fig. 4). One of the most obvious genetic factors are genetic polymorphisms in enzymes responsible for xenobiotic metabolism. Such polymorphisms have been hypothesized to contribute to variability in the response of populations to toxicants (Hattis *et al.,* 1987, 1996; Kalow, 1992; Daly *et al.,* 1994; McFadden, 1996; Miller *et al.,* 1997). Genetic

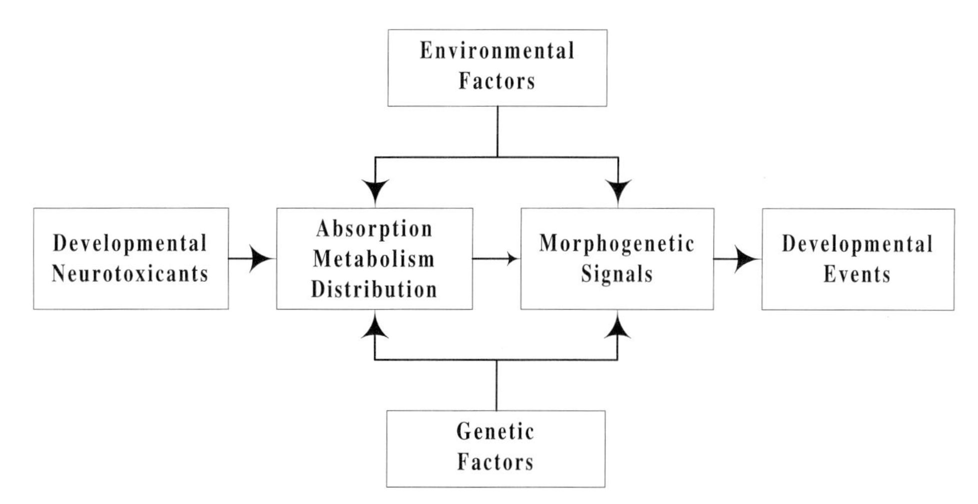

FIGURE 4 Schematic diagram of the genetic and environmental influences on processes that regulate the availability of the toxicants to the brain and on signals that regulate developmental events critical to brain morphogenesis.

mutations may also target critical elements of morphogenic signal pathways. As previously discussed, a number of syndromes are associated with mutations in the adhesion molecule L1. In addition, it has also been demonstrated that remarkably low levels of ethanol can disrupt L1-mediated neuronal adhesion (Wong *et al.*, 1995). The question remains, however, whether mutations in L1 can exacerbate morphologic defects that result from the effect of ethanol on L1.

Environmental factors can influence how toxicants effect the developing brain by altering morphogenetic signaling pathways or by altering the distribution of toxicants relative to their potential targets. Of all environmental factors that influence brain development, the most important from a perspective of human health is nutrition. In spite of the fundamental protection afforded the developing brain during nutritional deprivation, malnutrition can, depending on its severity and the developmental period in which it occurs, alter any aspect of brain morphogenesis (Morgane *et al.*, 1992). Specific nutritional deficiencies or supplimentations can intensify or ameliorate the effects of toxicants on specific developmental events as well as alter xenobiotic metabolism (Dickerson *et al.*, 1971; Basu *et al.*, 1973; Dyson and Jones, 1976; Chang *et al.*, 1978; Chang and Suber, 1982; Yip and Chang, 1982, 1983; Gilbert *et al.*, 1983; Komsta-Szumska *et al.*, 1983a,b; Sundstrom *et al.*, 1984, 1985; Aruoma, 1994; Manzo *et al.*, 1994; Parke and Ioannides, 1994; Butterworth, 1995; Almeida *et al.*, 1996; Segall *et al.*, 1996; Gluckman, 1997; Goyer, 1997; Gronskov *et al.*, 1997).

Stress can have a potent influence on the toxicant-induced disruptions of morphogenetic signaling pathways as well as influence xenobiotic metabolism. The pathways mediating corticosteroid influences on neurotoxic outcomes are complex and may be an important mechanism that contributes to interindividual variability in response to toxicants (Chang *et al.*, 1989; Miller, 1992, 1997; De Kloet *et al.*, 1996; Strobel *et al.*, 1997).

Congenital infection has long been known to be a risk factor for a variety of neurobehavioral disorders. Two general mechanisms have been proposed by which infectious agents can alter brain morphogenesis. One involves the direct invasion of the infectious organism into the developing brain, where secreted or membrane-bound toxins disrupt morphogenetic events. A second mechanisms involves the induction of an autoimmune response, presumably related to cross-reactivity between proteins of the infectious agent and those of the nervous system. Susceptibility to adverse neurologic outcome from either these mechanisms appears to depend on genetic and environmental factors in addition to the severity of infection. Because infection can alter xenobiotic metabolism and as well as damage the blood–brain barrier, it may also predispose the neonate to adverse consequences of toxicants at levels that it would normally be protected against (Bale and Murph, 1992; Raynor, 1993; Swedo *et al.*, 1994; March, 1995; Takei *et al.*, 1995; Annegers *et al.*, 1996; Morris, 1996; Pellegrini *et al.*, 1996; Tuomanen, 1996; Dammann and Leviton, 1997; Henderson *et al.*, 1997; Morgan, 1997).

This chapter provides a brief encapsulation of signaling pathways the regulate developmental events critical to brain morphogenesis. Select examples are used to illustrate how the genetic predisposition and environmental influences that may contribute to the emergence of developmental neurobehavioral disorders are likely to be similar to those that engender susceptibility to developmental neurotoxicants. To whatever extent these influences overlap and converge, they engender greater risk of developmental neurologic impairment (Fig. 5). Extremely demanding epidemiologic studies are required to determine if low level exposure to environmental pollutants contributes to the incidence and severity of the developmental neurobehavioral disorders. However, as illustrated with the examples of mental retardation and schizophrenia, investigations of basic developmental processes are critical to knowing where to look and what to look for.

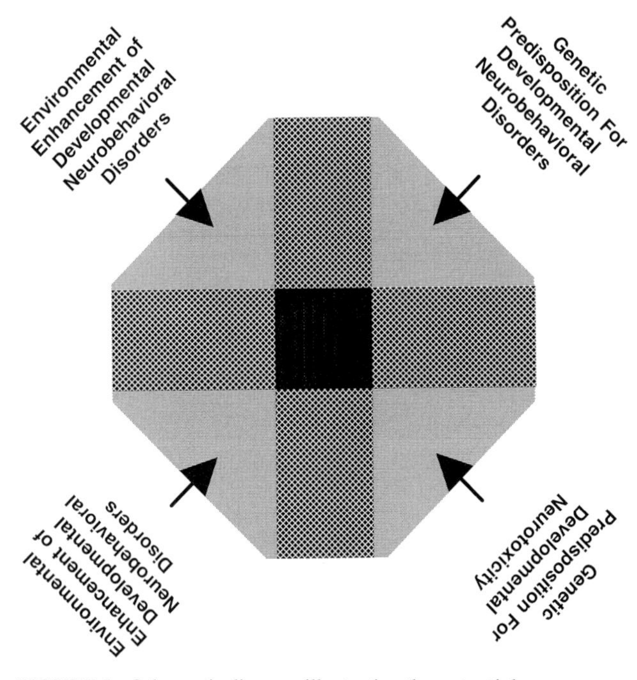

FIGURE 5 Schematic diagram illustrating the potential convergence of genetic and environmental determinants of susceptibility to the effects of developmental neurotoxicants and genetic and environmental determinants of susceptibility to the developmental neurobehavioral disorders. Because these determinants are not likely to be either coextensive or mutually exclusive, the areas of overlap are illustrated by the darker grays and represent the potential for greater susceptibility.

V. Summary

Identifying the molecular basis of the role toxicants may play in the pathogenesis of developmental neurobehavioral disorders is a complex challenge. The development of the brain involves both the instructive elaboration of neural circuitry and the selective consolidation of connections. The extent to which morphologic alterations in the brain correspond to behavioral impairment may be related to the relative disruption of these two types of processes. The developing brain may modify its developmental trajectory in response to insult in a manner that minimizes dramatic morphologic alterations, but significant functional impairments can persist.

Brain morphogenesis is regulated by a variety signals such as ions, growth factors, diffusable morphogens, adhesion molecules, neurotransmitters, hormones and neural activity. Diverse signal transduction pathways integrate these signals at transcriptional, translational, and postranslational levels. Interactions between genetic and environmental factors resulting in perturbations of these signals may contribute to the pathogenesis of developmental neurobehavioral disorders.

Little is known about the extent to which variations in signal transduction pathways afford protection from toxicants. A variety of toxicants may disrupt the same pathway, or conversely, a given toxicant may disrupt several different pathways. The effect of a toxicant on brain morphogenesis may be the result of the convergence of such actions on several tissues and organs. Environmental factors such as nutrition, traumatic injury, infection, or exposure to multiple toxicants can alter the distribution of toxicants and the interactions of toxicants with specific targets in the developing brain. Genetic variations occur in enzymes involved in the metabolism of toxicants and in critical elements of morphogenetic signaling pathways.

Epidemiologic approaches play a central role in determining the relative importance of genetic and environmental factors in the incidence of developmental neurobehavioral disorders. Experimental approaches can play a central role in refining how developmental neurobehavioral disorders are identified in the molecular and clinical arenas, and can investigate interactions between specific genetic and environmental factors. Identifying genetic and environmentally related alterations in morphogenetic signaling pathways is critical to this task and fundamental to protecting human health.

Acknowledgments

This manuscript has been reviewed by the NHEERL, U.S. EPA. Approval for publication does not indicate that the views expressed reflect Agency policy nor does mention of trade names constitute an endorsement for use. Dr. Karl F. Jensen wishes to thank Drs. Robert Isaacson, John March, Linda Ide, Mark Stanton, David Stewart for their valuable discussions. The authors also wish to express their heartfelt gratitude to Dr. Louis W. Chang for his long-standing encouragement and infinite patience, without which this chapter would not exist.

References

Abe-Dohmae, S., Takagi, Y., and Harada, N. (1996). Neurotransmitter-mediated regulation of brain aromatase: protein kinase C- and G-dependent induction. *J. Neurochem.* **67,** 2087–2095.

Achen, M. G., Clauss, M., Schnurch, H., and Risau, W. (1995). The non-receptor tyrosine kinase Lyn is localised in the developing murine blood–brain barrier. *Differentiation* **59,** 15–24.

Acosta, P. B. (1996). Nutrition studies in treated infants and children with phenylketonuria: vitamins, minerals, trace elements. *Eur. J. Pediatr.* 155 (Suppl 1), S136–S139.

Adams, N. C., Lozsadi, D. A., and Guillery, R. W. (1997). Complexities in the thalamocortical and corticothalamic pathways. *Eur. J. Neurosci.* **9,** 204–209.

Allendoerfer, K. L., and Shatz, C. J. (1994). The subplate, a transient neocortical structure: its role in the development of connections between thalamus and cortex. *Annu. Rev. Neurosci.* **17,** 185–218.

Almeida, S. S., Tonkiss, J., and Galler, J. R. (1996). Malnutrition and reactivity to drugs acting in the central nervous system. *Neurosci. Biobehav. Rev.* **20,** 389–402.

al-Sarraf, H., Preston, J. E., and Segal, M. B. (1997). Changes in the kinetics of the acidic amino acid brain and CSF uptake during development in the rat. *Dev. Brain Res.* **102,** 127–134.

Altaba, A. R. I. (1997). Catching a Gli-mpse of hedgehog. *Cell* **90,** 193–196.

Altman, J. (1972). Postnatal development of the cerebellar cortex in the rat II. Phases in the maturation of Purkinje cells and the molecular layer. *J. Comp. Neurol.* **145,** 399–464.

Altman, J. (1992). The early stages of nervous system development: neurogenesis and neuronal migration. In *Handbook of Chemical Neuroanatomy. Vol. 10. Ontogeny of Transmitters and Peptides in the CNS.* A. Bjorklund, T. Hokfelt, and M. Tohyama, Eds. Esivier, Amsterdam, pp. 1–31.

Altman, J. and Bayer, S. A. (1997). *Development of the Cerebellar System in Relation to Its Evolution, Structure and Functions.* CRC Press, Boca Raton.

American Psychiatric Association (APA). (1994). *Diagnostic and Statistical Manual of Mental Disorders IV.* APA, Washington DC.

An, S. F., and Scaravilli, F. (1997). Early HIV-1 infection of the central nervous system. *Arch. Anat. Cytol. Pathol.* **45,** 94–105.

Anderson, S., Qiu, M., Bulfone, A., Eisenstat, D. D., Meneses, J. J., Penderson, R., and Rubenstein, J. L. (1997). Mutations of the hemeobox genes Dlx-1 Dlx-2 disrupt the striatal subventricular zone and differentiation of late born striatal neurons. *Neuron* **19,** 27–37.

Andres-Barquin, P. J., Fages, C., Le Prince, G., Rolland, B., and Tardy, M. (1994). Thyroid hormones influence the astroglial plasticity: changes in the expression of glial fibrillary acidic protein (GFAP) and of its encoding message. *Neurochem. Res.* **19,** 65–69.

Annegers, J. F., Rocca, W. A., and Hauser, W. A. (1996). Causes of epilepsy: contributions of the Rochester epidemiology project. *Mayo Clin. Proc.* **71,** 570–575.

Anton, E. S., Cameron, R. S., and Rakic, P. (1996). Role of neuron–glial junctional domain proteins in the maintenance and termination of neuronal migration across the embryonic cerebral wall. *J. Neurosci.* **16,** 2283–2293.

Anton, E. S., Marchionni, M. A., Lee, K., and Rakic, P. (1997). Role of GGF/neuregulin signaling in interactions between migrating neuron and radial glia in the developing cerebral cortex. *Development* **124,** 3501–3510.

Antonini, A., and Stryker, M. P. (1996). Plasticity of geniculocortical afferents following brief or prolonged monocular occlusion in the cat. *J. Comp. Neurol.* **369,** 64–82.

Arboix, M., Paz, O. G., Colombo, T., and D'Incalci, M. (1997). Multidrug resistance-reversing agents increase vinblastine distribution in normal tissues expressing the P-glycoprotein but do not enhance drug penetration in brain and testis. *J. Pharmacol. Exp. Ther.* **281,** 1226–1230.

Argandona, E. G., and Lafuente, J. V. (1996). Effects of dark-rearing on the vascularization of the developmental rat visual cortex. *Brain Res.* **732,** 43–51.

Artavanis-Tsakonas, S., Matsuno, K., and Fortini, M. E. (1995). Notch signaling. *Science* **268,** 225–232.

Arthur, J. R., Nicol, F., and Beckett, G. J. (1992). The role of selenium in thyroid hormone metabolism and effects of selenium deficiency on thyroid hormone and iodine metabolism. *Biol. Trace Elem. Res.* **34,** 321–325.

Aruoma, O. I. (1994). Nutrition and health aspects of free radicals and antioxidants. *Food Chem. Toxicol.* **32,** 671–683.

Arvin, B., Neville, L. F., Barone, F. C., and Feuerstein, G. Z. (1996). The role of inflammation and cytokines in brain injury. *Neurosci. Biobehav. Rev.* **20,** 445–452.

Aschner, M. (1996). The functional significance of brain metallothioneins. *FASEB J.* **10,** 1129–1136.

Aschner, M. (1997). Manganese neurotoxicity and oxidative damage. In *Metals and Oxidative Damage in Neurological Disorders.* J. Conner, Ed. Plenum Press, New York, pp. 77–93.

Aschner, M., and Aschner, J. (1991). Manganese neurotoxicity: cellular effects and blood–brain transport. *Neurosci. Behav. Physiol.* **15,** 333–340.

Aschner, M., and Gannon, M. (1994). Manganese (Mn) transport across the rat blood–brain barrier: saturable and transferrin-dependent transport mechanisms. *Brain Res. Bull.* **33,** 345–349.

Aschner, M., Aschner, J., and Kimmelburg, H. K. (1992). Methylmercury neurotoxicity and its uptake across the blood–brain barrier. In *The Vulnerable Brain and Environmental Risk. Vol. 2 Toxins in Food.* R. L. Isaacson and K. F. Jensen Eds. Plenum Press, New York. pp. 3–18.

Assoian, R. K., and Zhu, X. (1997). Cell anchorage and the cytoskeleton as partners in growth factor dependent cell cycle progression. *Curr. Opin. Cell Biol.* **9,** 93–98.

Assoian, R. K., Zhu, X., Davey, G., and Bottazzi, M. (1997). Role of the extracellular matrix and cytoskeleton in the regulation of cyclins, cyclin dependent kinase inhibitors, and achorage-dependent growth. *Adv. Mol. Cell Biol.* **24,** 57–75.

Atchison, W. D. (1987). Effects of activation of sodium and calcium entry on spontaneous release of acetylcholine induced by methylmercury. *J. Pharmacol. Exp. Ther.* **241,** 131–139.

Atchison, W. D. (1988). Effects of neurotoxicants on synaptic transmission: lessons learned from electrophysiological studies. *Neurotoxicol. Teratol.* **10,** 393–416.

Atchison, W. D., and Hare, M. F. (1994). Mechanisms of methylmercury-induced neurotoxicity. *FASEB J.* **8,** 622–629.

Atchison, W. D. Joshi, U., and Thornburg, J. E. (1986). Irreversible suppression of calcium entry into nerve terminals by methylmercury. *J. Pharmacol. Exp. Ther.* **238,** 618–624.

Azen, C., Koch, R., Friedman, E., Wenz, E., and Fishler, K. (1996). Summary of findings from the United States Collaborative Study of children treated for phenylketonuria. *Eur. J. Pediatr.* **155** (Suppl 1), S29–32.

Bachiller, D., Macias, A., Duboule, D., and Morata, G. (1994). Conservation of a functional hierarchy between mammalian and insect Hox/HOM genes. *EMBO J.* **13,** 1930–1941.

Baker, J. C., and Harland, R. M. (1997). From receptor to nucleus: the Smad pathway. *Curr. Opin. Genet. Dev.* **7,** 467–473.

Balazs, R. (1971). Biochemical effects of thyroid hormones in the developing brain. *UCLA Forum. Med. Sci.* **14,** 273–320.

Balduini, W., Reno, F., Costa, L. G., and Cattabeni, F. (1994). Developmental neurotoxicity of ethanol: further evidence for an involvement of muscarinic receptor–stimulated phosphoinositide hydrolysis. *Eur. J. Pharmacol.* **266,** 283–289.

Bale, J. F. Jr., and Murph, J. R. (1992). Congenital infections and the nervous system. *Pediatr. Clin. North Am.* **39,** 669–690.

Bang, A. G., and Goulding, M. (1996). Regulation of vertebrate neural cell fate by transcription factors. *Curr. Opin. Neurobiol.* **6,** 25–32.

Banks, W. A., Audus, K. L., and Davis, T. P. (1992). Permeability of the blood–brain barrier to peptides: an approach to the development of therapeutically useful analogs. *Peptides* **13,** 1289–1394.

Bar, T. (1980). The vascular system of the cerebral cortex. *Adv. Anat. Embryol. Cell Biol.* **59,** 1–62.

Baraitser, M. (1997). *The Genetics of Neurological Disorders. Third Edition.* Oxford University Press, Oxford.

Barondes, S. H., Alberts, B. M., Andreasen, N. C., Bargmann, C., Benes, F., Goldman-Rakic, P. S., Gottesman, I., Heinemann, S. F., Jones, E. G., Kirschner, M., Lewis, D., Raff, M., Roses, A., Rubenstein, J., Snyder, S., Watson, S. J., Weinberger, D. R., and Yolken, R. H. (1997). Workshop on schizophrenia. *Proc. Natl. Acad. Sci. USA* **94,** 1612–1614.

Bartlett, P. F., Kilpatrick, T. J., Richards, L. J., Talman, P. S., and Murphy, M. (1994). Regulation of the early development of the nervous system by growth factors. *Pharmacol. Ther.* **64,** 371–393.

Basu, T. K., Dickerson, J. W., and Parke, D. V. (1971). Effect of development on the activity of microsomal drug-metabolizing enzymes in rat liver. *Biochem. J.* **124,** 19–24.

Basu, T. K., Dickerson, J. W., and Parke, D. V. (1973). Effect of underfeeding suckling rats on the activity of hepatic drug metabolizing enzymes. *Biol. Neonate.* **23,** 109–115.

Bates, C., and Killackey, H. P. (1984). The emergence of a discretely distributed pattern of corticospinal projection neurons. *Dev. Brain Res.* **13,** 265–273.

Bayer, S. A., and Altman, J. (1991). *Neocortical Development.* Raven Press, New York.

Bayer, S. A., Altman, J., Russo, R. J., and Zhang, X. (1993). Timetables of neurogenesis in the human brain based on experimentally determined patterns in the rat. *Neurotoxicology* **14,** 83–144.

Bear, M. F., and Singer, W. (1986). Modulation of visual cortical plasticity by acetylcholine and noradrenaline. *Nature* **320,** 172–176.

Bear, M. F., Carnes, K. M., and Ebner, F. F. (1985). Postnatal changes in the distribution of acetylcholinesterase in kitten striate cortex. *J. Comp. Neurol.* **237,** 519–532.

Beaulieu, E., Demeule, M., Pouliot, J. F., Averill-Bates, D. A., Murphy, G. F., and Beliveau, R. (1995). P-glycoprotein of blood brain barrier: cross-reactivity of Mab C219 with a 190 kDa protein in bovine and rat isolated brain capillaries. *Biochim. Biophys. Acta* **1233,** 27–32.

Beckmann, H., and Lauer, M. (1997). The human striatum in schizophrenia. II. Increased number of striatal neurons in schizophrenics. *Psychiatry Res.* **68,** 99–109.

Beitchman, J. H., and Young, A. R. (1997). Learning disorders with a special emphasis on reading disorders: a review of the past 10 years. *J. Am. Acad. Child Adolesc. Psychiatry* **36,** 1020–1032.

Belford, G., and Killackey, H. P. (1980). The sensitive period in the development of the trigeminal system of the neonatal rat. *J. Comp. Neurol.* **193,** 335–350.

Bellamy, W. T. (1996). P-glycoproteins and multidrug resistance. *Annu. Rev. Pharmacol. Toxicol.* **36,** 161–183.

Bellinger, D., Leviton, A., Waternaux, C., Needleman, H. L., and Rabinowitz, M. (1987). Longitudinal analyses of prenatal and postnatal lead exposure and early cognitive development. *N. Engl. J. Med.* **316,** 1037–1043.

Bellinger, D. C., Stiles, K. M., and Needleman, H. L. (1992). Low-level lead exposure, intelligence and academic achievement: a long-term follow-up study [see comments]. *Pediatrics* **90,** 855–861.

Benes, F. M. (1997). The role of stress and dopamine–GABA interactions in the vulnerability for schizophrenia. *J. Psychiatr. Res.* **31,** 257–275.

Benke, G. M., and Murphy, S. D. (1975). The influence of age on the toxicity and metabolism of methyl parathion and parathion in male and female rats. *Toxicol. Appl. Pharmacol.* **31,** 254–269.

Bennett-Clarke, C. A., Lane, R. D., and Rhoades, R. W. (1995). Fenfluramine depletes serotonin from the developing cortex and alters thalamocortical organization. *Brain Res.* **702,** 255–260.

Benowitz, L. I., and Routtenberg, A. (1997). GAP-43: an intrinsic determinant of neuronal development and plasticity. *Trends Neurosci.* **20,** 84–91.

Ben-Zeev, A., and Bershadsky, A. (1997). The role of the cytoskeleton in adhesion-mediated signaling and gene expression. *Adv. Mol. Cell Biol.* **24,** 125–163.

Berger-Sweeney, J., and Hohmann, C. F. (1997). Behavioral consequences of abnormal cortical development: insights into developmental disabilities. *Behav. Brain Res.* **86,** 121–142.

Bergh, A. F., and Strobel, H. W. (1996). Anatomical distribution of NADPH-cytochrome P450 reductase and cytochrome P4502D forms in rat brain: effects of xenobiotics and sex steroids. *Mol. Cell Biochem.* **162,** 31–41.

Bernal, J., and Nunez, J. (1995). Thyroid hormones and brain development. *Eur. J. Endocrinol.* **133,** 390–398.

Berry, M. J., and Larsen, P. R. (1992). The role of selenium in thyroid hormone action. *Endocr. Rev.* **13,** 207–219.

Besnard, F., Luo, M., Miehe, M., Dussault, J. H., Puymirat, J., and Sarlieve, L. L. (1994). Transient expression of 3,5,3′-triiodothyronine nuclear receptors in rat oligodendrocytes: *in vivo* and *in vitro* immunocytochemical studies. *J. Neurosci. Res.* **37,** 313–323.

Bettenhausen, B., Hrabe de Angelis, M., Simon, D., Guenet, J. L., and Gossler, A. (1995).Transient and restricted expression during mouse embryogenesis of Dll1, a murine gene closely related to *Drosophila* delta. *Development* **121,** 2407–2418.

Bier, E. (1997). Anti–neural-inhibition: a conserved mechanism for neural induction. *Cell* **89,** 681–684.

Biesiada, E., Adams, P. M., Shanklin, D. R., Bloom, G. S., and Stein, S. A. (1996). Biology of the congenitally hypothyroid hyt/hyt mouse. *Adv. Neuroimmunol.* **6,** 309–346.

Birnbaum, L. S. (1991). Pharmacokinetic basis of age-related changes in sensitivity to toxicants. *Ann. Rev. Pharmacol.* **31,** 101–128.

Blackshear, P. J., Lai, W. S., Tuttle, J. S., Stumpo, D. J., Kennington, E., Nairn, A. C., and Sulik, K. K. (1996). Developmental expression of MARCKS and protein kinase C in mice in relation to the exencephaly resulting from MARCKS deficiency. *Dev. Brain Res.* **96,** 62–75.

Blackshear, P. J., Silver, J., Nairn, A. C., Sulik, K. K., Squier, M. V., Stumpo, D. J., and Tuttle, J. S. (1997). Widespread neuronal ectopia associated with secondary defects in cerebrocortical chondroitin sulfate proteoglycans and basal lamina in MARCKS-deficient mice. *Exp. Neurol.* **145,** 46–61.

Blake, B. L., Philpot, R. M., Levi, P. E., and Hodgson, E. (1996). Xenobiotic biotransforming enzymes in the central nervous system: an isoform of flavin-containing monooxygenase (FMO4) is expressed in rabbit brain. *Chem. Biol. Interact.* **99,** 253–261.

Blaumueller, C. M., and Artavanis-Tsakonas, S. (1997). Comparative aspects of Notch signaling in lower and higher eukaryotes. *Perspect. Dev. Neurobiol.* **4,** 325–343.

Bloom, G. S., and Endow, S. A. (1995). Motor proteins 1: kinesins. *Protein Profile* **2,** 1105–1171.

Boado, R. J., Black, K. L., and Pardridge, W. M. (1994). Gene expression of GLUT3 and GLUT1 glucose transporters in human brain tumors. *Mol. Brain Res.* **27,** 51–57.

Bobryshev, Y. V., and Ashwell, K. W. (1996). Activation of microglia in haemorrhage microzones in human embryonic cortex. An ultrastructural description. *Pathol. Res. Pract.* **192,** 260–270.

Boehm, G. W., Sherman, G. F., Hoplight, B. J. II, Hyde, L. A., Waters, N. S., Bradway, D. M., Galaburda, A. M., and Denenberg, V. H. (1996). Learning and memory in the autoimmune BXSB mouse: effects of neocortical ectopias and environmental enrichment. *Brain Res.* **726,** 11–22.

Boehm, G. W., Sherman, G. F., Rosen, G. D., Galaburda, A. M., and Denenberg, V. H. (1996). Neocortical ectopias in BXSB mice: effects upon reference and working memory systems. *Cereb. Cortex.* **6,** 696–700.

Bolz, S., Farrell, C. L., Dietz, K., and Wolburg, H. (1996). Subcellular distribution of glucose transporter (GLUT-1) during development of the blood–brain barrier in rats. *Cell Tissue Res.* **284,** 355–365.

Boncinelli, E. (1994). Early CNS development: distal-less related genes and forebrain development. *Curr. Opin. Neurobiol.* **4,** 29–36.

Bothwell, M. (1995). Functional interactions of neurotrophins and neurotrophin receptors. *Annu. Rev. Neurosci.* **18,** 223–253.

Bradbury, M. (1979). *The Concept of a Blood–Brain Barrier.* Wiley, New York.

Bradbury, M. W., and Deane, R. (1988). Brain endothelium and interstitium as sites for effects of lead. *Ann. NY Acad. Sci.* **529,** 1–8.

Bradbury, M. W., and Deane, R. (1993). Permeability of the blood–brain barrier to lead. *Neurotoxicology* **14,** 131–136.

Brady, S. T., and Sperry, A. O. (1995). Biochemical and functional diversity of microtubule motors in the nervous system. *Curr. Opin. Neurobiol.* **5,** 551–558.

Bredesen, D. E. (1995). Neural apoptosis. *Ann. Neurol.* **38,** 839–851.

Breier, G., Clauss, M., and Risau, W. (1995). Coordinate expression of vascular endothelial growth factor receptor-1 (flt-1) and its ligand suggests a paracrine regulation of murine vascular development. *Dev. Dyn.* **204,** 228–239.

Brent, G. A. (1994). The molecular basis of thyroid hormone action. *N. Engl. J. Med.* **331,** 847–853.

Bressler, J. P., and Goldstein, G. W. (1991). Mechanisms of lead neurotoxicity. *Biochem. Pharmacol.* **41,** 479–484.

Brett, F. M., Mizisin, A. P., Powell, H. C., and Campbell, I. L. (1995). Evolution of neuropathologic abnormalities associated with blood–brain barrier breakdown in transgenic mice expressing interleukin-6 in astrocytes. *J. Neuropathol. Exp. Neurol.* **54,** 766–775.

Brooke, S., Chan, R., Howard, S., and Sapolsky, R. (1997). Endocrine modulation of the neurotoxicity of gp 120: implications for AIDS-related dementia complex. *Proc. Natl. Acad. Sci. U.S.A.* **94,** 9457–9462.

Brugg, B., and Matus, A. I. (1991). Phosphorylation determines the binding of microtubule-associated protein 2 (MAP2) to microtubules in living cells. *J. Cell Biol.* **114,** 735–743.

Brugger, P., Monsch, A. U., Salmon, D. P., and Butters, N. (1996). Random number generation in dementia of the Alzheimer type: a test of frontal executive functions. *Neuropsychologia* **34,** 97–103.

Burden-Gulley, S. M., and Lemmon, V. (1996). L1, N-cadherin, and laminin induce distinct distribution patterns of cytoskeletal elements in growth cones. *Cell Motil. Cytoskeleton* **35,** 1–23.

Burden-Gulley, S. M., Pendergast, M., and Lemmon, V. (1997). The role of cell adhesion molecule L1 in axonal extension, growth cone motility, and signal transduction. *Cell Tissue Res.* **290,** 415–422.

Burgard, P., Rey, F., Rupp, A., Abadie, V., and Rey, J. (1997). Neuropsychologic functions of early treated patients with phenylketonuria, on and off diet: results of a cross-national and cross-sectional study. *Pediatr. Res.* **41,** 368–374.

Burgard, P., Rupp, A., Konecki, D. S., Trefz, F. K., Schmidt, H., and Lichter-Konecki, U. (1996). Phenylalanine hydroxylase genotypes, predicted residual enzyme activity and phenotypic parameters of diagnosis and treatment of phenylketonuria. *Eur. J. Pediatr.* **155** (Suppl 1), S11–5.

Burgard, P., Schmidt, E., Rupp, A., Schneider, W., and Bremer, H. J. (1996). Intellectual development of the patients of the German Collaborative Study of children treated for phenylketonuria. *Eur. J. Pediatr.* **155** (Suppl 1), S33–8.

Burgin, K. E., Waxham, M. N., Rickling, S., Westgate, S. A., Mobley, W. C., and Kelly, P. T. (1990), *In situ* hybridizan histochemistry of Ca^{2+} Calmodulin dependent protein kinase in developing rat brain. *J. Neurosci.* **10,** 1788–1798.

Burgoyne, R. D. (1991). *The Neuronal Cytoskeleton.* Wiley-Liss, New York.

Burgoyne, R. D., and Cambray-Deakin, M. A. (1988). The cellular neurobiology of neuronal development: the cerebellar granule cell. *Brain Res. Rev.* **13,** 77–101.

Burmeister, L. A., Pachucki, J., and St. Germain, D. L. (1997). Thyroid hormones inhibit type 2 iodothyronine deiodinase in the rat cerebral cortex by both pre- and posttranslational mechanisms. *Endocrinology* **138,** 5231–5237.

Butterworth, R. F. (1995). Pathophysiology of alcoholic brain damage: synergistic effects of ethanol, thiamine deficiency and alcoholic liver disease. *Metab. Brain Dis.* **10,** 1–8.

Buznikov, G. A., Shmukler, Y. B., and Lauder, J. M. (1996). From oocyte to neuron: do neurotransmitters function in the same way throughout development? *Cell Mol. Neurobiol.* **16,** 537–559.

Byrne, J. J., and Sepkovic, D. W. (1987). Inhibition of monovalent cation transport across the cell membrane by polychlorinated biphenyl but not by polybrominated biphenyl. *Arch. Environ. Contam. Toxicol.* **16,** 573–577.

Byrne, J. J., Carbone, J. P., and Pepe, M. G. (1988). Suppression of serum adrenal cortex hormones by chronic low-dose polychlorobiphenyl or polybromobiphenyl treatments. *Arch. Environ. Contam. Toxicol.* **17,** 47–53.

Cabelli, R. J., Shelton, D. L., Segal, R. A., and Shatz, C. J. (1997). Blockade of endogenous ligands of trkB inhibits formation of ocular dominance columns. *Neuron* **19,** 63–76.

Cadigan, K. M., and Nusse, R. (1997). Wnt signaling: a common theme in animal development. *Genes Dev.* **11,** 3286–3305.

Calomme, M., Vanderpas, J., Francois, B., Van Caillie-Bertrand, M., Vanovervelt, N., Van Hoorebeke, C., and Vanden Berghe, D. (1995). Effects of selenium supplementation on thyroid hormone metabolism in phenylketonuria subjects on a phenylalanine restricted diet. *Biol. Trace Elem. Res.* **47,** 349–353.

Campbell, G., Ramoa, A. S., Stryker, M. P., and Shatz, C. J. (1997). Dendritic development of retinal ganglion cells after prenatal intracranial infusion of tetrodotoxin. *Vis. Neurosci.* **14,** 779–788.

Carmeliet, P., Moons, L., Dewerchin, M., Mackman, N., Luther, T., Breier, G., Ploplis, V., Muller, M., Nagy, A., Plow, E., Gerard, R., Edgington, T., Risau, W., and Collen, D. (1997). Insights in vessel development and vascular disorders using targeted inactivation and transfer of vascular endothelial growth factor, the tissue factor receptor, and the plasminogen system. *Ann. NY Acad. Sci.* **811,** 191–206.

Carraway, K. L. III, and Burden, S. J. (1995). Neuregulins and their receptors. *Curr. Opin. Neurobiol.* **5,** 606–612.

Carraway, K. L., 3rd, Weber, J. L., Unger, M. J., Ledesma, J., Yu, N., Gassmann, M., and Lai, C. (1997). Neuregulin-2, a new ligand of ErbB3/ErbB4-receptor tyrosine kinases. *Nature* **387,** 512–516.

Cassella, J. P., Lawrenson, J. G., and Firth, J. A. (1997). Development of endothelial paracellular clefts and their tight junctions in the pial microvessels of the rat. *J. Neurocytol.* **26,** 567–575.

Catalano, S. M., Chang, C. K., and Shatz, C. J. (1997). Activity-dependent regulation of NMDAR1 immunoreactivity in the developing visual cortex. *J. Neurosci.* **17,** 8376–8390.

Catalano, S. M., Robertson, R. T., and Killackey, H. P. (1991). Early ingrowth of thalamocortical afferents to the neocortex of the prenatal rat. *Proc. Natl. Acad. Sci. U.S.A.* **88,** 2999–3003.

Catalano, S. M., Robertson, R. T., and Killackey, H. P. (1995). Rapid alteration of thalamocortical axon morphology follows peripheral damage in the neonatal rat. *Proc. Natl. Acad. Sci. U.S.A.* **92,** 2549–2552.

Catalano, S. M., Robertson, R. T., and Killackey, H. P. (1996). Individual axon morphology and thalamocortical topography in developing rat somatosensory cortex. *J. Comp. Neurol.* **367,** 36–53.

Catts, S. V., Ward, P. B., Lloyd, A., Huang, X. F., Dixon, G., Chahl, L., Harper, C., and Wakefield, D. (1997). Molecular biological investigations into the role of the NMDA receptor in the pathophysiology of schizophrenia. *Aust. N.Z. J. Psychiatry* **31,** 17–26.

Cavalcante, L. A., Barradas, P. C., and Vieira, A. M. (1996). The regional distribution of neuronal glycogen in the opossum brain, with special reference to hypothalamic systems. *J. Neurocytol.* **25,** 455–463.

Chakrabarti, L., and Davies, K. E. (1997). Fragile X syndrome. *Curr. Opin. Neurol.* **10,** 142–147.

Chakraborty, J., Hopkins, R., and Parke, D. V. (1971). Biological oxygenation of drugs and steroids in the placenta. *Biochem. J.* **125,** 15P–16P.

Chalepakis, G., Jones, F. S., Edelman, G. M., and Gruss, P. (1994). Pax-3 contains domains for transcription activation and transcription inhibition. *Proc. Natl. Acad. Sci. U.S.A.* **91,** 12745–12749.

Chang, L. (1992). The concept of direct and indirect neurotoxicity and the concept of toxic metals/essential element interactions as a common biomechanism underlying metal toxicity. In *The Vulnerable Brain and Environmental Risks. Vol. 2. Toxins in Food.* R. L. Isaacson and K. F. Jensen, Eds. Plenum, New York, pp. 61–82.

Chang, L. W. (1983). Protective effects of selenium against methylmercury neurotoxicity: a morphological and biochemical study. *Exp. Pathol.* **23,** 143–156.

Chang, L. W. (1990). The neurotoxicology and pathology of organomercury, organolead, and organotin. *J. Toxicol. Sci.* **15** (Suppl 4), 125–151.

Chang, L. W., and Desnoyers, P. A. (1978). Methylmercury induced biochemical and microsomal changes in the rat liver. *J. Environ. Pathol. Toxicol.* **1,** 569–579.

Chang, L. W., and Hartmann, H. A. (1972). Blood–brain barrier dysfunction in experimental mercury intoxication. *Acta Neuropathol. (Berl)* **21,** 179–184.

Chang, L. W., and Sprecher, J. A. (1976). Pathological changes in the kidney after methyl cadmium intoxication. *Environ. Res.* **12,** 92–109.

Chang, L. W., and Suber, R. (1982). Protective effect of selenium on methylmercury toxicity: a possible mechanism. *Bull. Environ. Contam. Toxicol.* **29,** 285–289.

Chang, L. W., Desnoyers, P. A., and Hartmann, H. A. (1972). Quantitative cytochemical studies of RNA in experimental mercury poisoning. Changes in RNA content. *J. Neuropathol. Exp. Neurol.* **31,** 489–501.

Chang, L. W., Gilbert, M., and Sprecher, J. (1978). Modification of methylmercury neurotoxicity by vitamin E. *Environ. Res.* **17,** 356–366.

Chang, L. W., Hough, A. J., Bivins, F. G., and Cockerill, D. (1989). Effects of adrenalectomy and corticosterone on hippocampal lesions induced by trimethyltin. *Biomed. Environ. Sci.* **2,** 54–64.

Chang, L. W., Wade, P. R., Pounds, J. G., and Reuhl, K. R. (1980). Prenatal and neonatal toxicology and pathology of heavy metals. *Adv. Pharmacol. Chemother.* **17,** 195–231.

Chang, L. W., Wade, P. R., Reuhl, K. R., and Olson, M. J. (1980). Ultrastructural changes in renal proximal tubules after tetraethyllead intoxication. *Environ. Res.* **23,** 208–223.

Changeux, J. P. (1997). Variation and selection in neural function. *Trends Neurosci.* **20,** 291–293.

Chen, B. P., Wolfgang, C. D., and Hai, T. (1996). Analysis of ATF3, a transcription factor induced by physiological stresses and modulated by gadd153/Chop 10. *Mol. Cell Biol.* **16,** 1157–1168.

Cheng, N., and Sahyoun, N. (1990). Neuronal tyrosine phosphorylation in growth cone glycoproteins. *J. Biol. Chem.* **265,** 2417–2420.

Chenn, A., and McConnell, S. K. (1995). Cleavage orientation and the asymmetric inheritance of Notch1 immunoreactivity in mammalian neurogenesis. *Cell* **82,** 631–641.

Chothia, C., and Jones, E. Y. (1997). The molecular structure of cell adhesion molecules. *Annu. Rev. Biochem.* **66,** 823–862.

Chua, S. E., and Murray, R. M. (1996). The neurodevelopmental theory of schizophrenia: evidence concerning structure and neuropsychology. *Ann. Med.* **28,** 547–555.

Chung, B. C., Guo, I. C., and Chou, S. J. (1997). Transcriptional regulation of the CYP11A1 and ferredoxin genes. *Steroids* **62,** 37–42.

Cillo, C., Cantile, M., Mortarini, R., Barba, P., Parmiani, G., and Anichini, A. (1996). Differential patterns of HOX gene expression are associated with specific integrin and ICAM profiles in clonal populations isolated from a single human melanoma metastasis. *Int. J. Cancer* **66,** 692–697.

Clark, G. D., Mizuguchi, M., Antalffy, B., Barnes, J., and Armstrong, D. (1997). Predominant localization of the LIS family of gene products of Cajal–Retzius cells and ventricular neuroepithelium in the developing human cortex. *J. Neuropathol. Exp. Neurol.* **56,** 1044–1052.

Cockerill, D., Chang, L. W., Hough, A., and Bivins, F. (1987). Effects of trimethyltin on the mouse hippocampus and adrenal cortex. *J. Toxicol. Environ. Health* **22,** 149–161.

Cochen, B., Bashirullah, A., Dagnino, L., Campbell, C., Fisher, W. W., Leow, C. C., Whiting, E., Ryan, D., Zinyk, D., Boulianne, G., Hui, C. C., Gallie, B., Phillips, R. A., Lipshitz, H. D., and Egan, S. E. (1997). Fringe boundaries coincide with Notch-dependent patterning centres in mammals and alter Notch-dependent development in *Drosophila. Nat. Genet.* **16,** 283–288.

Cohen, M. M. Jr., and Sulik, K. K. (1992). Perspectives on holoprosencephaly: Part II. Central nervous system, craniofacial anatomy, syndrome commentary, diagnostic approach, and experimental studies. *J. Craniofac. Genet. Dev. Biol.* **12,** 196–244.

Colamarino, S. A., and Tessier-Lavigne, M. (1995). The role of the floor plate in axon guidance. *Annu. Rev. Neurosci.* **18,** 497–529.

Coletta, P. L., Shimeld, S. M., and Sharpe, P. T. (1994). The molecular anatomy of Hox gene expression. *J. Anat.* **184,** 15–22.

Colquhoun, L. M., and Patrick, J. W. (1997). Pharmacology of neuronal nicotinic acetylcholine receptor subtypes. *Adv. Pharmacol.* **39,** 191–220.

Comery, T. A., Harris, J. B., Willems, P. J., Oostra, B. A., Irwin, S. A., Weiler, I. J., and Greenough, W. T. (1997). Abnormal dendritic spines in fragile X knockout mice: maturation and pruning deficits. *Proc. Natl. Acad. Sci. U.S.A.* **94,** 5401–5404.

Commandeur, J. N., Stijntjes, G. J., and Vermeulen, N. P. (1995). Enzymes and transport systems involved in the formation and disposition of glutathione S-conjugates. Role in bioactivation and detoxication mechanisms of xenobiotics. *Pharmacol. Rev.* **47,** 271–330.

Concordet, J. P., Lewis, K. E., Moore, J. W., Goodrich, L. V., Johnson, R. L., Scott, M. P., and Ingham, P. W. (1996). Spatial regulation of a zebrafish patched homologue reflects the roles of sonic hedgehog and protein kinase A in neural tube and somite patterning. *Development* **122,** 2835–2846.

Conner, J. R. (1997). *Metals and Oxidative Damage in Neurological Disorders.* Plenum Press, New York.

Connolly, K. J., and Kvalsvig, J. D. (1993). Infection, nutrition and cognitive performance in children. *Parasitology* **107** (Suppl), S187–200.

Conover, P. T., and Roessmann, U. (1990). Malformational complex in an infant with intrauterine influenza viral infection. *Arch. Pathol. Lab. Med.* **114,** 535–538.

Copertino, D. W., Jenkinson, S., Jones, F. S., and Edelman, G. M. (1995). Structural and functional similarities between the promoters for mouse tenascin and chicken cytotactin. *Proc. Natl. Acad. Sci. U.S.A.* **92,** 2131–2135.

Corey, D. A., Juarez de Ku, L. M., Bingman, V. P., and Meserve, L. A. (1996). Effects of exposure to polychlorinated biphenyl (PCB) from conception on growth, and development of endocrine, neurochemical, and cognitive measures in 60 day old rats. *Growth Dev. Aging* **60,** 131–143.

Cornford, E. M., Diep, C. P., and Pardridge, W. M. (1985). Blood–brain barrier transport of valproic acid. *J. Neurochem.* **44,** 1541–1550.

Cornford, E. M., Hyman, S., and Pardridge, W. M. (1993). An electron microscopic immunogold analysis of developmental up-regulation of the blood–brain barrier GLUT1 glucose transporter. *J. Cereb. Blood Flow Metab.* **13,** 841–854.

Costa, L. G. (1994). Signal transduction mechanisms in developmental neurotoxicity: the phosphoinositide pathway. *Neurotoxicology* **15,** 19–27.

Cote, I. L., and Vandenburg, J. J. (1994). Overview of health effects and risk-assessment issues associated with air pollution. In *The Vulnerable Brain and Environmental Risks. Vol. 3 Toxins in Air and Water.* R. L. Isaacson and K. F. Jensen, Eds. Plenum Press, New York. pp. 231–245.

Cowan, W. M., Fawcett, J. W., O'Leary, D. D., and Stanfield, B. B. (1984). Regressive events in neurogenesis. *Science* **225,** 1258–1265.

Crair, M. C., Ruthazer, E. S., Gillespie, D. C., and Stryker, M. P. (1997). Relationship between the ocular dominance and orientation maps in visual cortex of monocularly deprived cats. *Neuron* **19,** 307–318.

Crow, T. J. (1995a). A continuum of psychosis, one human gene, and not much else—the case for homogeneity. *Schizophr. Res.* **17,** 135–145.

Crow, T. J. (1995b). A Darwinian approach to the origins of psychosis. *Br. J. Psychiatry* **167,** 12–25.

Crow, T. J. (1997). Schizophrenia as failure of hemispheric dominance for language. *Trends Neurosci.* **20,** 339–343.

Curry, C. J., Stevenson, R. E., Aughton, D., Byrne, J., Carey, J. C., Cassidy, S., Cunniff, C., Graham, J. M., Jr., Jones, M. C., Kaback, M. M., Moeschler, J., Schaefer, G. B., Schwartz, S., Tarleton, J., and Opitz, J. (1997). Evaluation of mental retardation: recommendations of a Consensus Conference: American College of Medical Genetics. *Am. J. Med. Genet.* **72,** 468–477.

Daft, P. A., Johnston, M. C., and Sulik, K. K. (1986). Abnormal heart and great vessel development following acute ethanol exposure in mice. *Teratology* **33,** 93–104.

Dale, J. K., Vesque, C., Lints, T. J., Sampath, T. K., Furley, A., Dodd, J., and Placzek, M. (1997). Cooperation of BMP7 and SHH in the induction of forebrain ventral midline cells by prechordal mesoderm. *Cell* **90,** 257–269.

Daly, A. K., Cholerton, S., Armstrong, M., and Idle, J. R. (1994). Genotyping for polymorphisms in xenobiotic metabolism as a predictor of disease susceptibility. *Environ. Health Perspect.* **102** (Suppl 9), 55–61.

Dammann, O., and Leviton, A. (1997). Maternal intrauterine infection, cytokines, and brain damage in the preterm newborn. *Pediatr. Res.* **42,** 1–8.

D'Arcangelo, G., Nakajima, K., Miyata, T., Ogawa, M., Mikoshiba, K., and Curran, T. (1997). Reelin is a secreted glycoprotein recognized by the CR-50 monoclonal antibody. *J. Neurosci.* **17,** 23–31.

Das, S., and Paul, S. (1994). Decrease in beta-adrenergic receptors of cerebral astrocytes in hypothyroid rat brain. *Life Sci.* **54,** 621–629.

Davies, H. G., Richter, R. J., Keifer, M., Broomfield, C. A., Sowalla, J., and Furlong, C. E. (1996). The effect of the human serum paraoxonase polymorphism is reversed with diazoxon, soman and sarin. *Nat. Genet.* **14,** 334–336.

Dawson, D. R., and Killackey, H. P. (1985). Distinguishing topography and somatotopy in the thalamocortical projections of the developing rat. *Dev. Brain Res.* **17,** 309–313.

De Kloet, E. R., Korte, S. M., Rots, N. Y., and Kruk, M. R. (1996). Stress hormones, genotype, and brain organization. Implications for aggression. *Ann. NY Acad. Sci.* **794,** 179–191.

de la Pompa, J. L., Wakeham, A., Correia, K. M., Samper, E., Brown, S., Aguilera, R. J., Nakano, T., Honjo, T., Mak, T. W., Rossant, J., and Conlon, R. A. (1997). Conservation of the Notch signaling pathway in mammalian neurogenesis. *Development* **124,** 1139–1148.

De Robertis, E., and Bennet, H. S. (1954). Submicroscopic vesicular component in the synapse. *Fed. Proc.* **13,** 35.

Deakin, F. W., Simpson, M. D., Slater, P., and Hellewell, J. S. (1997). Familial and developmental abnormalities of front lobe function and neurochemistry in schizophrenia. *J. Psychopharmacol. (Oxf).* **11,** 133–142.

Deakin, J. F., and Simpson, M. D. (1997). A two-process theory of schizophrenia: evidence from studies in post-mortem brain. *J. Psychiatr. Res.* **31,** 277–295.

deBethizy, J. D., and Hayes, J. R. (1994). Metabolism: a determinant of toxicity. In *Principles and Methods of Toxicology,* Third Edition. A. W. Hayes, Ed. Raven Press, New York, pp. 59–105.

Dekant, W. (1997). Glutathione-dependent bioactivation and renal toxicity of xenobiotics. *Recent. Results Cancer Res.* **143,** 77–87.

Del Rio, J. A., Heimrich, B., Borrell, V., Forster, E., Drakew, A., Alcantara, S., Nakajima, K., Miyata, T., Ogawa, M., Mikoshiba, K., Derer, P., Frotscher, M., and Soriano, E. (1997). A role for Cajal–Retzius cells and reelin in the development of hippocampal connections. *Nature* **385,** 70–74.

Denenberg, V. H., Mobraaten, L. E., Sherman, G. F., Morrison, L., Schrott, L. M., Waters, N. S., Rosen, G. D., Behan, P. O., and Galaburda, A. M. (1991). Effects of the autoimmune uterine/maternal environment upon cortical ectopias, behavior and autoimmunity. *Brain Res.* **563,** 114–122.

Denny, M. F., and Atchison, W. D. (1994). Methylmercury-induced elevations in intrasynaptosomal zinc concentrations: an 19F-NMR study. *J. Neurochem.* **63,** 383–386.

Denny, M. F., and Atchison, W. D. (1996). Mercurial-indiced alterations in neuronal divalent cation homeostasis. *Neurotoxicology* **17,** 47–61.

Denny, M. F., Hare, M. F., and Atchison, W. D. (1993). Methylmercury alters intrasynaptosomal concentrations of endogenous polyvalent cations. *Toxicol. Appl. Pharmacol.* **122,** 222–232.

Desaulniers, D., Poon, R., Phan, W., Leingartner, K., Foster, W. G., and Chu, I. (1997). Reproductive and thyroid hormone levels in rats following 90-day dietary exposure to PCB 28 (2,4,4′-trichlorobiphenyl) or PCB 77 (3,3′4,4′-tetrachlorobiphenyl). *Toxicol. Ind. Health* **13,** 627–638.

Desnoyers, P. A., and Chang, L. W. (1975). Ultrastructural changes in the liver after chronic exposure to methylmercury. *Environ. Res.* **10,** 59–75.

Desrayaud, S., Guntz, P., Scherrmann, J. M., and Lemaire, M. (1997). Effect of the P-glycoprotein inhibitor, SDZ PSC 833, on the blood and brain pharmacokinetics of colchicine. *Life Sci.* **61,** 153–163.

Diamond, A., and Herzberg, C. (1996). Impaired sensitivity to visual contrast in children treated early and continuously for phenylketonuria. *Brain* **119,** 523–538.

Dickerson, J. W., Basu, T. K., and Parke, D. V. (1971). Protein nutrition and drug-metabolizing enzymes in the liver of the growing rat. *Proc. Nutr. Soc.* **30,** 5A–6A.

Ding, M., Robel, L., James, A. J., Eisenstat, D. D., Leckman, J. F., Rubenstein, J. L., and Vaccarino, F. M. (1997). Dlx-2 homeobox gene controls neuronal differentiation in primary cultures of developing basal ganglia. *J. Mol. Neurosci.* **8,** 93–113.

Dobbing, J. (1968). The development of the blood–brain barrier. *Prog. Brain Res.* **29,** 417–427.

Dobbing, J., and Sands, J. (1979). Comparative aspects of the brain growth spurt. *Early Hum. Dev.* **3,** 79–83.

Domer, F. R., and Wolf, C. L. (1980). Effects of lead on movement of albumin into brain. *Res. Commun. Chem. Pathol. Pharmacol.* **29,** 381–384.

Drion, N., Lemaire, M., Lefauconnier, J. M., and Scherrmann, J. M. (1996). Role of P-glycoprotein in the blood–brain transport of colchicine and vinblastine. *J. Neurochem.* **67,** 1688–1693.

Duchini, A. (1996). The role of central nervous system endothelial cell activation in the pathogenesis of hepatic encephalopathy. *Med. Hypotheses.* **46,** 239–244.

Duffy, C. J., and Rakic, P. (1983). Differentiation of granule cell dendrites in the dentate gyrus of the rhesus monkey: a quantitative Golgi study. *J. Comp. Neurol.* **214,** 224–237.

Dwyer, K. J., and Pardridge, W. M. (1993). Developmental modulation of blood–brain barrier and choroid plexus GLUT1 glucose transporter messenger ribonucleic acid and immunoreactive protein in rabbits. *Endocrinology* **132,** 558–565.

Dyson, S. E., and Jones, D. G. (1976). Undernutrition and the developing nervous system. *Prog. Neurobiol.* **7,** 171–196.

Eagleson, K. L., Lillien, L., Chan, A. V., and Levitt, P. (1997). Mechanisms specifying area fate in cortex include cell-cycle–dependent decisions and the capacity of progenitors to express phenotype memory. *Development* **124,** 1623–1630.

Easter, S. S. Jr., Purves, D., Rakic, P., and Spitzer, N. C. (1985). The changing view of neural specificity. *Science* **230,** 507–511.

Ebadi, M., Iversen, P. L., Hao, R., Cerutis, D. R., Rojas, P., Happe, H. K., Murrin, L. C., and Pfeiffer, R. F. (1995). Expression and regulation of brain metallothionein. *Neurochem. Int.* **27,** 1–22.

Ebstein, R. P., Oppenheim, G., Ebstein, B. S., Amiri, Z., and Stessman, J. (1986). The cyclic AMP second messenger system in man: the effects of heredity, hormones, drugs, aluminum, age and disease on signal amplification. *Prog. Neuropsychopharmacol. Biol. Psychiatry* **10,** 323–353.

Ecobichon, D. J. (1991). Toxic effects of pesticides. In *Casarett and Doull's Toxicology, Fourth Edition.* M. O. Amdur, J. Doull, and C. D. Klaasen, Eds. Pergamon Press, New York.

Edelman, G. M. (1987). *Neural Darwinism.* Basic Books, New York.

Edelman, G. M., and Jones, F. S. (1997). Gene regulation of cell adhesion molecules in neural morphogenesis. *Acta Paediatr. Suppl.* **422,** 12–19.

Edelman, G. M., Gall, W. E., and Cowan, W. M. (1985). *Molecular Basis of Neural Development.* John Wiley & Sons, New York.

Eiken, H. G., Knappskog, P. M., Motzfeldt, K., Boman, H., and Apold, J. (1996). Phenylketonuria genotypes correlated to metabolic phenotype groups in Norway. *Eur. J. Pediatr.* **155,** 554–560.

Eisensmith, R. C., Martinez, D. R., Kuzmin, A. I., Goltsov, A. A., Brown, A., Singh, R., Elsas, L. J. II, and Woo, S. L. (1996). Molecular basis of phenylketonuria and a correlation between genotype and phenotype in a heterogeneous southeastern US population. *Pediatrics* **97,** 512–516.

Ekins, R. P., Sinha, A. K., Pickard, M. R., Evans, I. M., and al Yatama, F. (1994). Transport of thyroid hormones to target tissues. *Acta Med. Austriaca.* **21,** 26–34.

Engelhardt, B., Conley, F. K., and Butcher, E. C. (1994). Cell adhesion molecules on vessels during inflammation in the mouse central nervous system. *J. Neuroimmunol.* **51,** 199–208.

Epstein, D. J., Marti, E., Scott, M. P., and McMahon, A. P. (1996). Antagonizing cAMP-dependent protein kinase A in the dorsal CNS activates a conserved Sonic hedgehog signaling pathway. *Development* **122,** 2885–2894.

Ericson, J., Muhr, J., Placzek, M., Lints, T., Jessell, T. M., and Edlund, T. (1995). Sonic hedgehog induces the differentiation of ventral forebrain neurons: a common signal for ventral patterning within the neural tube. *Cell* **81,** 747–756.

Ericson, J., Rashbass, P., Schedl, A., Brenner-Morton, S., Kawakami, A., van Heyningen, V., Jessell, T. M., and Briscoe, J. (1997). Pax6 controls progenitor cell identity and neuronal fate in response to graded Shh signaling. *Cell* **90,** 169–180.

Eriksson, P. (1997). Developmental neurotoxicity of environmental agents in the neonate. *Neurotoxicology* **18,** 719–726.

Evrard, P., Marret, S., and Gressens, P. (1997). Environmental and genetic determinants of neural migration and postmigratory survival. Acta Paediatr. Suppl. **422,** 20–26.

Faraone, S. V., Seidman, L. J., Kremen, W. S., Pepple, J. R., Lyons, M. J., and Tsuang, M. T. (1995). Neuropsychological functioning among the nonpsychotic relatives of schizophrenic patients: a diagnostic efficiency analysis. *J. Abnorm. Psychol.* **104,** 286–304.

Farrell, C. L., and Risau, W. (1994). Normal and abnormal development of the blood–brain barrier. *Microsc. Res. Tech.* **27,** 495–506.

Farsetti, A., Lazar, J., Phyillaier, M., Lippoldt, R., Pontecorvi, A., and Nikodem, V. M. (1997). Active repression by thyroid hormone receptor splicing variant alpha$_2$ requires specific regulatory elements in the context of native triiodothyronine-regulated gene promoters. *Endocrinology* **138,** 4705–4712.

Fawcett, J. W., O'Leary, D. D., and Cowan, W. M. (1984). Activity and the control of ganglion cell death in the rat retina. *Proc. Natl. Acad. Sci. U.S.A.* **81,** 5589–5593.

Felgner, H., Frank, R., Biernat, J., Mandelkow, E. M., Mandelkow, E., Ludin, B., Matus, A. I., and Schliwa, M. (1997). Domains of neuronal microtubule-associated proteins and flexural rigidity of microtubules. *J. Cell Biol.* **138,** 1067–1075.

Feng, Y., Gutekunst, C. A., Eberhart, D. E., Yi, H., Warren, S. T., and Hersch, S. M. (1997). Fragile X mental retardation protein: nucleocytoplasmic shuttling and association with somatodendritic ribosomes. *J. Neurosci.* **17,** 1539–1547.

Ferrer, I. (1996). Cell death in the normal developing brain, and following ionizing radiation, methyl–azoxymethanol acetate, and hypoxia–ischaemia in the rat. *Neuropathol. Appl. Neurobiol.* **22,** 489–494.

Ferrer, I., Olive, M., Blanco, R., Cinos, C., and Planas, A. M. (1996). Selective c-Jun overexpression is associated with ionizing radiation-induced apoptosis in the developing cerebellum of the rat. *Brain Res. Mol. Brain Res.* **38,** 91–100.

Figueiredo, B. C., Almazan, G., Ma, Y., Tetzlaff, W., Miller, F. D., and Cuello, A. C. (1993a). Gene expression in the developing cerebellum during perinatal hypo- and hyperthyroidism. *Brain Res. Mol. Brain Res.* **17,** 258–268.

Figueiredo, B. C., Otten, U., Strauss, S., Volk, B., and Maysinger, D. (1993b). Effects of perinatal hypo- and hyperthyroidism on the levels of nerve growth factor and its low-affinity receptor in cerebellum. *Dev. Brain Res.* **72,** 237–244.

Finlay, B. L., and Slattery, M. (1983). Local differences in the amount of early cell death in neocortex predict adult local specializations. *Science* **219,** 1349–1351.

Fischer, S., Sharma, H. S., Karliczek, G. F., and Schaper, W. (1995). Expression of vascular permeability factor/vascular endothelial growth factor in pig cerebral microvascular endothelial cells and its upregulation by adenosine. *Mol. Brain Res.* **28,** 141–148.

Fishell, G. (1997). Regionalization in the mammalian telencephalon. *Curr. Opin. Neurobiol.* **7,** 62–69.

Flamme, I., Breier, G., and Risau, W. (1995). Vascular endothelial growth factor (VEGF) and VEGF receptor 2 (flk-1) are expressed during vasculogenesis and vascular differentiation in the quail embryo. *Dev. Biol.* **169,** 699–712.

Fleck, C., and Braunlich, H. (1995). Renal handling of drugs and amino acids after impairment of kidney or liver function—influences of maturity and protective treatment. *Pharmacol. Ther.* **67,** 53–77.

Forrest, D., Erway, L. C., Ng, L., Altschuler, R., and Curran, T. (1996a). Thyroid hormone receptor beta is essential for development of auditory function. *Nat. Genet.* **13,** 354–357.

Forrest, D., Golarai, G., Connor, J., and Curran, T. (1996b). Genetic analysis of thyroid hormone receptors in development and disease. *Recent. Prog. Horm. Res.* **51,** 1–22.

Forrest, D., Hanebuth, E., Smeyne, R. J., Everds, N., Stewart, C. L., Wehner, J. M., and Curran, T. (1996). Recessive resistance to thyroid hormone in mice lacking thyroid hormone receptor beta: evidence for tissue-specific modulation of receptor function. *EMBO J.* **15,** 3006–3015.

Frangou, S., and Murray, R. M. (1996). Imaging as a tool in exploring the neurodevelopment and genetics of schizophrenia. *Br. Med. Bull.* **52,** 587–596.

Fransen, E., Lemmon, V., Van Camp, G., Vits, L., Coucke, P., and Willems, P. J. (1995). CRASH syndrome: clinical spectrum of corpus callosum hypoplasia, retardation, adducted thumbs, spastic paraparesis and hydrocephalus due to mutations in one single gene, L1. *Eur. J. Hum. Genet.* **3,** 273–284.

Franzek, E., and Beckmann, H. (1996). Gene-environment interaction in schizophrenia: season-of-birth effect reveals etiologically different subgroups. *Psychopathology* **29,** 14–26.

Friede, R. L. (1989). *Developmental Neuropathology.* Springer-Verlag, Berlin.

Frotscher, M. (1997). Dual role of Cajal–Retzius cells and reelin in cortical development. *Cell Tissue Res.* **290,** 315–322.

Fukuda, T., Kawano, H., Ohyama, K., Li, H. P., Takeda, Y., Oohira, A., and Kawamura, K. (1997). Immunohistochemical localization of neurocan and L1 in the formation of thalamocortical pathway of developing rats. *J. Comp. Neurol.* **382,** 141–152.

Gage, J. C., and Sulik, K. K. (1991). Pathogenesis of ethanol-induced hydronephrosis and hydroureter as demonstrated following in vivo exposure of mouse embryos. *Teratology* **44,** 299–312.

Galaburda, A. M., Sherman, G. F., Rosen, G. D., Aboitiz, F., and Geschwind, N. (1985). Developmental dyslexia: four consecutive patients with cortical anomalies. *Ann. Neurol.* **18,** 222–233.

Gang-Yi, W., and Cline, H. T. (1998). Stabilization of dendritic arbor structure *in vivo* by CamKII. *Science* **279,** 222–226.

Garner, C. C., Tucker, R. P., and Matus, A. I. (1988). Selective localization of messenger RNA for cytoskeletal protein MAP2 in dendrites. *Nature* **336,** 674–677.

Garrity, P. A., and Zipursky, S. L. (1995). Neuronal target recognition. *Cell* **83,** 177–185.

Gaze, R. M. (1970). *Formation of Nerve Connections.* Academic Press, New York.

Geschwind, N. (1979). Specializations of the human brain. *Sci. Am.* **241,** 180–199.

Ghersi-Egea, J. F., Leininger-Muller, B., Cecchelli, R., and Fenstermacher, J. D. (1995). Blood–brain interfaces: relevance to cerebral drug metabolism. *Toxicol. Lett.* **82,** 645–653.

Giancotti, F. G. (1997). Integrin signaling: specificity and control of cell survival and cell cycle progression. *Curr. Opin. Cell Biol.* **9,** 691–700.

Gibbons, R. J., Picketts, D. J., and Higgs, D. R. (1995). Syndromal mental retardation due to mutations in a regulator of gene expression. *Hum. Mol. Genet.* **4,** 1705–1709.

Giger, R. J., Wolfer, D. P., De Wit, G. M., and Verhaagen, J. (1996). Anatomy of rat semaphorin III/collapsin-1 mRNA expression and relationship to developing nerve tracts during neuroembryogenesis. *J. Comp. Neurol.* **375,** 378–392.

Gilbert, M. M., Sprecher, J., Chang, L. W., and Meisner, L. F. (1983). Protective effect of vitamin E on genotoxicity of methylmercury. *J. Toxicol. Environ. Health* **12,** 767–773.

Gluckman, P. D. (1997). Endocrine and nutritional regulation of prenatal growth. *Acta Paediatr. Suppl.* **423,** 153–157.

Goldman-Rakic, P. S. (1995). More clues on "latent" schizophrenia point to developmental origins. *Am. J. Psychiatry* **152,** 1701–1703.

Goldman-Rakic, P. S., and Selemon, L. D. (1997). Functional and anatomical aspects of prefrontal pathology in schizophrenia. *Schizophr. Bull.* **23,** 437–458.

Goldstein, G. W., and Betz, A. L. (1986). The blood–brain barrier. *Sci. Am.* **255,** 74–83.

Goodlett, C. R., and West, J. R. (1990). Cortical vascular damage and astroglial reactions induced by binge exposure to alcohol during the brain growth spurt. *Teratology* **41,** 561.

Goodman, A. B. (1995). Chromosomal locations and modes of action of genes of the retinoid (vitamin A) system support their involvement in the etiology of schizophrenia. *Am. J. Med. Genet.* **60,** 335–348.

Goodman, C. S. (1996). Mechanisms and molecules that control growth cone guidance. *Annu. Rev. Neurosci.* **19,** 341–377.

Goodman, C. S., and Shatz, C. J. (1993). Developmental mechanisms that generate precise patterns of neuronal connectivity. *Cell* **72** (Suppl), 77–98.

Goodrich, L. V., Johnson, R. L., Milenkovic, L., McMahon, J. A., and Scott, M. P. (1996). Conservation of the hedgehog/patched signaling pathway from flies to mice: induction of a mouse patched gene by Hedgehog. *Genes Dev.* **10,** 301–312.

Goslin, K., Birgbauer, E., Banker, G., and Solomon, F. (1989). The role of cytoskeleton in organizing growth cones: a microfilament-associated growth cone component depends upon microtubules for its localization. *J. Cell Biol.* **109,** 1621–1631.

Goujon, E., Laye, S., Parnet, P., and Dantzer, R. (1997). Regulation of cytokine gene expression in the central nervous system by glucocorticoids: mechanisms and functional consequences. *Psychoneuroendocrinology* **22** (Suppl 1), S75–80.

Goyer, R. A. (1990). Transplacental transport of lead. *Environ. Health Perspect.* **89,** 101–105.

Goyer, R. A. (1997). Toxic and essential metal interactions. *Annu. Rev. Nutr.* **17,** 37–50.

Graff, J. M. (1997). Embryonic patterning: to BMP or not to BMP, that is the question. *Cell* **89,** 171.

Graff, R. D., Falconer, M. M., Brown, D. L., and Reuhl, K. R. (1997). Altered sensitivity of posttranslationally modified microtubules to methylmercury in differentiating embryonal carcinoma–derived neurons. *Toxicol. Appl. Pharmacol.* **144,** 215–224.

Graham, A., Francis-West, P., Brickell, P., and Lumsden, A. (1994). The signalling molecule BMP4 mediates apoptosis in the rhombencephalic neural crest. *Nature* **372,** 684–686.

Grapin-Botton, A., Bonnin, M. A., and Le Douarin, N. M. (1997). Hox gene induction in the neural tube depends on three parameters: competence, signal supply and paralogue group. *Development* **124,** 849–859.

Gravel, C., and Hawkes, R. (1990). Maturation of the corpus callosum of the rat: I. Influence of thyroid hormones on the topography of callosal projections. *J. Comp. Neurol.* **291,** 128–146.

Gravel, C., Sasseville, R., and Hawkes, R. (1990). Maturation of the corpus callosum of the rat: II. Influence of thyroid hormones on the number and maturation of axons. *J. Comp. Neurol.* **291,** 147–161.

Greenough, W. T., and Chang, F.-L. F. (1988). Dendritic pattern formation involves both oriented regression and oriented growth in the barrels of mouse somatosensory cortex. *Dev. Brain Res.* **43,** 148–152.

Greenough, W. T., Black, J. E., and Wallace, C. S. (1987). Experience and brain development. *Child Dev.* **58,** 539–559.

Gronskov, K., Hjalgrim, H., Bjerager, M. O., and Brondum-Nielsen, K. (1997). Deletion of all CGG repeats plus flanking sequences in FMR1 does not abolish gene expression. *Am. J. Hum. Genet.* **61,** 961–967.

Gu, Q. A., Bear, M. F., and Singer, W. (1989). Blockade of NMDA-receptors prevents ocularity changes in kitten visual cortex after reversed monocular deprivation. *Dev. Brain Res.* **47,** 281–288.

Guerri, C., and Renau-Piqueras, J. (1997). Alcohol, astroglia, and brain development. *Mol. Neurobiol.* **15,** 65–81.

Guldberg, P., Levy, H. L., Hanley, W. B., Koch, R., Matalon, R., Rouse, B. M., Trefz, F., de la Cruz, F., Henriksen, K. F., and Guttler, F. (1996). Phenylalanine hydroxylase gene mutations in the United States: report from the Maternal PKU Collaborative Study. *Am. J. Hum. Genet.* **59,** 84–94.

Hafner, H., and an der Heiden, W. (1997). Epidemiology of schizophrenia. *Can. J. Psychiatry* **42,** 139–151.

Haller, H. (1997). Endothelial function. General considerations. *Drugs* **53 Suppl 1:1–10,** 1–10.

Hammerschmidt, M., Bitgood, M. J., and McMahon, A. P. (1996). Protein kinase A is a common negative regulator of Hedgehog signaling in the vertebrate embryo. *Genes Dev.* **10,** 647–658.

Hanahan, D. (1997). Signaling vascular morphogenesis. *Science* **277,** 48–50.

Hanas, J. S., and Gunn, C. G. (1996). Inhibition of transcription factor IIIA-DNA interactions by xenobiotic metal ions. *Nucleic Acids Res.* **24,** 924–930.

Hannigan, J. H. (1995). Behavioral plasticity after teratogenic alcohol exposure as recovery of function. In *Neurobehavioral Plasticity.* N. Spear, L. P. Spear, and M. L. Woodruff, Eds. Lawrence Erlbaum Associates, Hillsday, NJ. pp. 283–309.

Hare, M. F., McGinnis, K. M., and Atchison, W. D. (1993). Methylmercury increases intracellular concentrations of Ca^{++} and heavy metals in NG108-15 cells. *J. Pharmacol. Exp. Ther.* **266,** 1626–1635.

Harris, K. M., and Kater, S. B. (1994). Dendritic spines: cellular specializations imparting both stability and flexibility to synaptic function. *Annu. Rev. Neurosci.* **17,** 341–371.

Harrison, P. J. (1997). Schizophrenia: a disorder of neurodevelopment? *Curr. Opin. Neurobiol.* **7,** 285–289.

Hashimoto, H., Marystone, J. F., Greenough, W. T., and Bohn, M. C. (1989). Neonatal adrenalectomy alters dendritic branching of hippocampal granule cells. *Exp. Neurol.* **104,** 62–67.

Hata, A., Lo, R. S., Wotton, D., Lagna, G., and Massague, J. (1997). Mutations increasing autoinhibition inactivate tumour suppressors Smad2 and Smad4. *Nature* **388,** 82–87.

Hattis, D. (1996). Human interindividual variability in susceptibility to toxic effects: from annoying detail to a central determinant of risk. *Toxicology* **111,** 5–14.

Hattis, D., Erdreich, L., and Ballew, M. (1987). Human variability in susceptibility to toxic chemicals—a preliminary analysis of pharmacokinetic data from normal volunteers. *Risk Analysis* **7,** 415–426.

Hellendall, R. P., Schambra, U. B., Liu, J. P., and Lauder, J. M. (1993). Prenatal expression of 5-HT1C and 5-HT2 receptors in the rat central nervous system. *Exp. Neurol.* **120,** 186–201.

Hemmati-Brivanlou, A., and Melton, D. (1997). Vertebrate neural induction. *Annu. Rev. Neurosci.* **20,** 43–60.

Henderson, B., Wilson, M., and Wren, B. (1997). Are bacterial exotoxins cytokine network regulators? *Trends Microbiol.* **5,** 454–458.

Henderson, C. E. (1996). Programmed cell death in the developing nervous system. *Neuron* **17,** 579–585.

Hensch, T. K., and Stryker, M. P. (1996). Ocular dominance plasticity under metabotropic glutamate receptor blockade. *Science* **272,** 554–557.

Herrick, C. J. (1925). Morphogenetic factors in the differentiation of the nervous system. *Physiol. Rev.* **5,** 112–130.

Herrmann, K., and Shatz, C. J. (1995). Blockade of action potential activity alters initial arborization of thalamic axons within cortical layer 4. *Proc. Natl. Acad. Sci. U.S.A.* **92,** 11244–11248.

Hirokawa, N. (1993). *Neuronal Cytoskeleton: Morphogenesis, Transport and Synaptic Transmission.* CRC Press, Boca Raton.

Hirokawa, N. (1996). The molecular mechanism of organelle transport along microtubules: the identification and characterization of KIFs (kinesin superfamily proteins). *Cell Struct. Funct.* **21,** 357–367.

Holland, P. W., and Graham, A. (1995). Evolution of regional identity in the vertebrate nervous system. *Perspect. Dev. Neurobiol.* **3,** 17–27.

Holmuhamedov, E. L., Kholmoukhamedova, G. L., and Baimuradov, T. B. (1996). Non-cholinergic toxicity of organophosphates in mammals: interaction of ethaphos with mitochondrial functions. *J. Appl. Toxicol.* **16,** 475–481.

Holst, B. D., Goomer, R. S., Wood, I. C., Edelman, G. M., and Jones, F. S. (1994). Binding and activation of the promoter for the neural cell adhesion molecule by Pax-8. *J. Biol. Chem.* **269,** 22245–22252.

Holst, B. D., Wang, Y., Jones, F. S., and Edelman, G. M. (1997). A binding site for Pax proteins regulates expression of the gene for the neural cell adhesion molecule in the embryonic spinal cord. *Proc. Natl. Acad. Sci. U.S.A.* **94,** 1465–1470.

Homma, S., Yaginuma, H., and Oppenheim, R. W. (1994). Programmed cell death during the earliest stages of spinal cord development in the chick embryo: a possible means of early phenotypic selection. *J. Comp. Neurol.* **345,** 377–395.

Houenou, L. J., Haverkamp, L. J., McManaman, J. L., and Oppenheim, R. W. (1991). The regulation of motoneuron survival and differentiation by putative muscle-derived neurotrophic agents: neuromuscular activity and innervation. *Development* **2** (Suppl), 149–155.

Houx, P. J., and Jolles, J. (1992). Vulnerability factors for age-related cognitive decline. In *The Vulnerable Brain and Environmental Risks. Vol. 3. Toxins in Air and Water.* R. L. Isaacson and K. F. Jensen, Eds. Plenum Press, New York, pp. 25–42.

Hubel, D. H. (1979). The visual cortex of normal and deprived monkeys. *Am. Sci.* **67,** 532–543.

Hubel, D. H., and Wiesel, T. N. (1965). Binocular interaction in striate cortex of kittens reared with artificial squint. *J. Neurophysiol.* **28,** 1041–1059.

Hubel, D. H., Wiesel, T. N., and LeVay, S. (1977). Plasticity of ocular dominance columns in monkey striate cortex. *Philos. Trans. R. Soc. Lond. B. Biol. Sci.* **278,** 377–409.

Huwyler, J., Drewe, J., Klusemann, C., and Fricker, G. (1996). Evidence for P-glycoprotein–modulated penetration of morphine-6-glucuronide into brain capillary endothelium. *Br. J. Pharmacol.* **118,** 1879–1885.

Hynes, M., Poulson, K., Tessier-Lavigne, M., and Rosenthal, A. (1997a). Control of neuronal diversity by the floor plate: contact-mediated induction of midbrain dopaminergic neurons. *Cell* **80,** 95–101.

Hynes, M., Stone, D. M., Dowd, M., Pitts-Meek, S., Goddard, A., Gurney, A., and Rosenthal, A. (1997b). Control of cell pattern in the neural tube by the zinc finger transcription factor and oncogene Gli-1. *Neuron* **19,** 15–26.

Ibiwoye, M. O., Sibbons, P. D., Howard, C. V., and van Velzen, D. (1994). Immunocytochemical study of a vascular barrier antigen in the developing rat brain. *J. Comp. Pathol.* **111,** 43–53.

Imamura, T., Takase, M., Nishihara, A., Oeda, E., Hanai, J., Kawabata, M., and Miyazono, K. (1997). Smad6 inhibits signalling by the TGF-beta superfamily. *Nature* **389,** 622–626.

Innocenti, G. M. (1995). Exuberant development of connections, and its possible permissive role in cortical evolution. *Trends Neurosci.* **18,** 397–402.

Jacobs, J. M. (1982). Vascular permeability and neurotoxicity. In *Nervous System Toxicology.* C. Mitchell, Ed. Raven Press, New York, pp. 285–298.

Jacobson, M. (1991). *Developmental Neurobiology, Third Edition.* Plenum Press, New York.

Jaffrey, S. R., and Snyder, S. H. (1995). Nitric oxide: a neural messenger. *Ann. Rev. Cell Dev. Biol.* **11,** 417–440.

Janigro, D., West, G. A., Nguyen, T. S., and Winn, H. R. (1994). Regulation of blood–brain barrier endothelial cells by nitric oxide. *Circ. Res.* **75,** 528–538.

Jensen, K. F., and Killackey, H. P. (1984). Subcortical projections from ectopic neocortical neurons. *Proc. Natl. Acad. Sci. U.S.A.* **81,** 964–968.

Jensen, K. F., and Killackey, H. P. (1987a). Terminal arbors of axons projecting to the somatosensory cortex of the adult rat. I. The normal morphology of specific thalamocortical afferents. *J. Neurosci.* **7,** 3529–3543.

Jensen, K. F., and Killackey, H. P. (1987b). Terminal arbors of axons projecting to the somatosensory cortex of the adult rat. II. The altered morphology of thalamocortical afferents following neonatal infraorbital nerve cut. *J. Neurosci.* **7,** 3544–3553.

Jette, L., Murphy, G. F., Leclerc, J. M., and Beliveau, R. (1995a). Interaction of drugs with P-glycoprotein in brain capillaries. *Biochem. Pharmacol.* **50,** 1701–1709.

Jette, L., Pouliot, J. F., Murphy, G. F., and Beliveau, R. (1995b). Isoform I (mdr3) is the major form of P-glycoprotein expressed in mouse brain capillaries. Evidence for cross-reactivity of antibody C219 with an unrelated protein. *Biochem. J.* **305,** 761–766.

Johnson, R. W., Arkins, S., Dantzer, R., and Kelley, K. W. (1997). Hormones, lymphohemopoietic cytokines and the neuroimmune axis. *Comp. Biochem. Physiol. A. Physiol.* **116,** 183–201.

Johnston, D., Magee, J. C., Colbert, C. M., and Cristie, B. R. (1996). Active properties of neuronal dendrites. *Annu. Rev. Neurosci.* **19,** 165–186.

Johnston, M. V. (1994). Developmental aspects of NMDA receptor agonists and antagonists in the central nervous system. *Psychopharmacol. Bull.* **30,** 567–575.

Johnston, M. V. (1995). Neurotransmitters and vulnerability of the developing brain. *Brain Dev.* **17,** 301–306.

Johnston, M. V., Grzanna, R., and Coyle, J. T. (1979). Methyloxazymethanol treatment of fetal rats results in abnormally dense noradrenergic innervation of neocortex. *Science* **203,** 369–371.

Jones, E. G. (1995). Cortical development and neuropathology in schizophrenia. *Ciba Found. Symp.* **193,** 277–295.

Jones, F. S., Kioussi, C., Copertino, D. W., Kallunki, P., Holst, B. D., and Edelman, G. M. (1997). Barx2, a new homeobox gene of the Bar class, is expressed in neural and craniofacial structures during development. *Proc. Natl. Acad. Sci. U.S.A.* **94,** 2632–2637.

Jones, H. C., Keep, R. F., and Butt, A. M. (1992). The development of ion regulation in the blood–brain barrier. *Prog. Brain Res.* **91,** 123–131.

Jones, P. (1997). The early origins of schizophrenia. *Br. Med. Bull.* **53,** 135–155.

Joo, F. (1987). Current aspects of the development of the blood–brain barrier. *Int. J. Dev. Neurosci.* **5,** 369–372.

Kalow, W. (1992). *Pharmacogenetics of Drug Metabolism.* Plenum Press, New York.

Kameyama, Y., and Inouye, M. (1994). Irradiation injury to the developing nervous system: mechanisms of neuronal injury. *Neurotoxicology* **15,** 75–80.

Kanai, Y. (1997). Family of neutral and acidic amino acid transporters: molecular biology, physiology and medical implications. *Curr. Opin. Cell Biol.* **9,** 565–572.

Karavanov, A., Sainio, K., Palgi, J., Saarma, M., Saxen, L., and Sariola, H. (1995). Neurotrophin 3 rescues neuronal precursors from apoptosis and promotes neuronal differentiation in the embryonic metanephric kidney. *Proc. Natl. Acad. Sci. U.S.A.* **92,** 11279–11283.

Karayiorgou, M., and Gogos, J. A. (1997). Dissecting the genetic complexity of schizophrenia. *Mol. Psychiatry* **2,** 211–223.

Karten, H. J. (1997). Evolutionary developmental biology meets the brain: the origins of the mammalian cortex. *Proc. Natl. Acad. Sci. U.S.A.* **94,** 2800–2804.

Kater, S. B., Mattson, M. P., Cohan, C. C., and Connor, J. (1988). Calcium regulation of the neuronal growth cone. *Trends Neurosci.* **11,** 315–322.

Katz, L. C., and Shatz, C. J. (1996). Synaptic activity and the construction of cortical circuits. *Science* **274,** 1133–1138.

Kawata, M. (1995). Roles of steroid hormones and their receptors in structural organization in the nervous system. *Neurosci. Res.* **24,** 1–46.

Kennard, M. A. (1942). Cortical reorganization of motor function. Studies on series of monkeys of various ages from infancy to maturity. *Arch. Neurol. Psychiatr.* **48,** 227–240.

Kennedy, M. B. (1983). Experimental approaches to understanding the role of protein phosphorylation in the regulation of neuronal function. *Annu. Rev. Neurosci.* **6,** 493–525.

Kennedy, M. B., Bennett, M. K., and Erondu, N. E. (1983). Biochemical and immunochemical evidence that the "major postsynaptic density protein" is a subunit of a calmodulin-dependent protein kinase. *Proc. Natl. Acad. Sci. U.S.A.* **80,** 7357–7361.

Kenwrick, S., Jouet, M., and Donnai, D. (1996). X linked hydrocephalus and MASA syndrome. *J. Med. Genet.* **33,** 59–65.

Kerper, L. E., and Hinkle, P. M. (1997). Lead uptake in brain capillary endothelial cells: activation by calcium store depletion. *Toxicol. Appl. Pharmacol.* **146,** 127–133.

Kessel, M. (1994). Hox genes and the identity of motor neurons in the hindbrain. *J. Physiol. Paris* **88,** 105–109.

Keynes, R., and Cook, G. M. (1995). Axon guidance molecules. *Cell* **83,** 161–169.

Keynes, R., and Krumlauf, R. (1994). Hox genes and regionalization of the nervous system. *Annu. Rev. Neurosci.* **17,** 109–132.

Khera, K. S., and Clegg, D. J. (1969). Perinatal toxicity to pesticides. *Can. Med. Assoc. J.* **100,** 167–172.

Killackey, H. P., Rhoades, R. W., and Bennett-Clarke, C. A. (1995). The formation of a cortical somatotopic map. *Trends Neurosci.* **18,** 402–407.

King, B. H., State, M. W., Shah, B., Davanzo, P., and Dykens, E. (1997). Mental retardation: a review of the past 10 years. Parts 1&2. *J. Am. Acad. Child Adolesc. Psychiatry* **36,** 1656–1671.

Kleinschmidt, A., Bear, M. F., and Singer, W. (1987). Blockade of "NMDA" receptors disrupts experience-dependent plasticity of kitten striate cortex. *Science* **238,** 355–358.

Knecht, A. K., and Harland, R. M. (1997). Mechanisms of dorsaventral patterning in noggin-induced neural tissue. *Development* **124,** 2477–2488.

Kniesel, U., Risau, W., and Wolburg, H. (1996). Development of blood–brain barrier tight junctions in the rat cortex. *Dev. Brain Res.* **96,** 229–240.

Koedood, M., Fichtel, A., Meier, P., and Mitchell, P. J. (1995). Human cytomegalovirus (HCMV) immediate-early enhancer/promoter specificity during embryogenesis defines target tissues of congenital HCMV infection. *J. Virol.* **69,** 2194–2207.

Koella, W. P., and Sutin, J. (1967). Extra-blood–brain barrier brain structures. *Int. Rev. Neurobiol.* **10,** 31–55.

Koester, S. E., and O'Leary, D. D. (1994). Development of projection neurons of the mammalian cerebral cortex. *Prog. Res.* **102,** 207–215.

Kohrle, J. (1996). Thyroid hormone deiodinases—a selenoenzyme family acting as gate keepers to thyroid hormone action. *Acta Med. Austriaca.* **23,** 17–30.

Komsta-Szumska, E., Reuhl, K. R., and Miller, D. R. (1983a). Effect of selenium on distribution, demethylation, and excretion of methylmercury by the guinea pig. *J. Toxicol. Environ. Health* **12,** 775–785.

Komsta-Szumska, E., Reuhl, K. R., and Miller, D. R. (1983b). The effect of methylmercury on the distribution and excretion of selenium by the guinea pig. *Arch. Toxicol.* **54,** 303–310.

Komuro, H., and Rakic, P. (1992). Selective role of N-type calcium channels in neuronal migration. *Science* **257,** 806–809.

Komuro, H., and Rakic, P. (1995). Dynamics of granule cell migration: a confocal microscopic study in acute cerebellar slice preparations. *J. Neurosci.* **15,** 1110–1120.

Komuro, H., and Rakic, P. (1996). Intracellular Ca^{2+} fluctuations modulate the rate of neuronal migration. *Neuron* **17,** 275–285.

Konig, N., Wilkie, M. B., and Lauder, J. M. (1988). Tyrosine hydroxylase and serotonin containing cells in embryonic rat rhombencephalon: a whole-mount immunocytochemical study. *J. Neurosci. Res.* **20,** 212–223.

Kotch, L. E., and Sulik, K. K. (1992). Experimental fetal alcohol syndrome: proposed pathogenic basis for a variety of associated facial and brain anomalies. *Am. J. Med. Genet.* **44,** 168–176.

Kovacs, K. A., Kavanagh, T. J., and Costa, L. G. (1995). Ethanol inhibits muscarinic receptor–stimulated phosphoinositide metabolism and calcium mobilization in rat primary cortical cultures. *Neurochem. Res.* **20,** 939–949.

Krebs, J., and Honegger, P. (1996). Calmodulin kinase IV: expression and function during rat brain development. *Biochim. Biophys. Acta.* **313,** 217–222.

Krebs, J., Means, R. L., and Honegger, P. (1996). Induction of calmodulin kinase IV by the thyroid hormone during the development of rat brain. *J. Biol. Chem.* **271,** 11055–11058.

Kretzschmar, M., Doody, J., and Massague, J. (1997). Opposing BMP and EGF signaling pathways converge on the TGF-beta family mediator Smad1. *Nature* **389,** 618–622.

Kretzschmar, M., Liu, F., Hata, A., Doody, J., and Massague, J. (1997). The TGF-beta family mediator Smad1 is phosphorylated directly and activated functionally by the BMP receptor kinase. *Genes Dev.* **11,** 984–995.

Kuan, C. Y., Elliott, E. A., Flavell, R. A., and Rakic, P. (1997). Restrictive clonal allocation in the chimeric mouse brain. *Proc. Natl. Acad. Sci. U.S.A.* **94,** 3374–3379.

Kuhlenbeck, H. (1973). *The Central Nervous System of Vertebrates.* Karger, Basel.

Kuida, K., Zheng, T. S., Na, S., Kuan, C., Yang, D., Karasuyama, H., Rakic, P., and Flavell, R. A. (1996). Decreased apoptosis in the brain and premature lethality in CPP32-deficient mice. *Nature* **384,** 368–372.

Kumada, S., Hayashi, M., Umitsu, R., Arai, N., Nagata, J., Kurata, K., and Morimatsu, Y. (1997). Neuropathology of the dentate nucleus in developmental disorders. *Acta Neuropathol. (Berl)* **94,** 36–41.

Kuwahara, S., Sada, Y., Moriki, T., Yamane, T., and Hara, H. (1996). Spatial and temporal expression of P-glycoprotein in the congenitally hydrocephalic HTX rat brain. *Pathol. Res. Pract.* **192,** 496–507.

La Spada, A. R. (1997). Trinucleotide repeat instability: genetic features and molecular mechanisms. *Brain Pathol.* **7,** 943–963.

Ladle, D. R., Jacobson, N. A., and Lephart, E. D. (1997). Hypothalamic aromatase cytochrome P450 and 5alpha-reductase enzyme activities in pregnant and female rats. *Life Sci.* **61,** 2017–2026.

Lagrange, P., Livertoux, M. H., Grassiot, M. C., and Minn, A. (1994). Superoxide anion production during monoelectronic reduction of xenobiotics by preparations of rat brain cortex, microvessels, and choroid plexus. *Free Radic. Biol. Med.* **17,** 355–359.

Lagunowich, L. A., Stein, A. P., and Reuhl, K. R. (1994). N-cadherin in normal and abnormal brain development. *Neurotoxicology* **15.** 123–132.

Lake, B. G., Hopkins, R., Chakraborty, J., Bridges, J. W., and Parke, D. V. (1973) The influence of some hepatic enzyme inducers and inhibitors on extrahepatic drug metabolism. *Drug Metab. Dispos.* **1,** 342–349.

Lakke, E. A. (1997). The projections to the spinal cord of the rat during development: a timetable of descent. *Adv. Anat. Embryol. Cell Biol.* **135,** 1–143.

Lakshmy, R., and Srinivasarao, P. (1997). Effect of thiocyanate on microtubule assembly in rat brain during postnatal development. *Int. J. Dev. Neurosci.* **15,** 87–94.

LaMantia, A. S., and Rakic, P. (1994). Axon overproduction and elimination in the anterior commissure of the developing rhesus monkey. *J. Comp. Neurol.* **340,** 328–336.

Lampugnani, M. G., Corada, M., Andriopoulou, P., Esser, S., Risau, W., and Dejana, E. (1997). Cell confluence regulates tyrosine phosphorylation of adherens junction components in endothelial cells. *J. Cell Sci.* **110,** 2065–2077.

Langford, G. M. (1995). Actin- and microtubule-dependent organelle motors: interrelationships between the two motility systems. *Curr. Opin. Cell Biol.* **7,** 82–88.

Lankas, G. R., Cartwright, M. E., and Umbenhauer, D. (1997). P-glycoprotein deficiency in a subpopulation of CF-1 mice enhances avermectin-induced neurotoxicity. *Toxicol. Appl. Pharmacol.* **143,** 357–365.

Larsen, P. R., and Berry, M. J. (1995). Nutritional and hormonal regulation of thyroid hormone deiodinases. *Annu. Rev. Nutr.* **15,** 323–352.

Lash, L. H., and Zalup, R. K. (1996). Alterations in renal cellular glutathione metabolism after *in vivo* administration of a subtoxic dose of mercuric chloride. *J. Biochem. Toxicol.* **11,** 1–12.

Laterra, J., Wolff, J. E., Guerin, C., and Goldstein, G. W. (1992). Formation and differentiation of brain capillaries. *NIDA Res. Monogr.* **120,** 73–86.

Lathja, A., and Ford, D. H. (1968). *Brain Barrier Systems.* Elsevier, Amsterdam.

Lauder, J. M. (1988). Neurotransmitters as morphogens. *Prog. Brain Res.* **73,** 365–387.

Lauder, J. M. (1993). Neurotransmitters as growth regulatory signals: role of receptors and second messengers. *Trends Neurosci.* **16,** 233–240.

Lauder, J. M., and Krebs, H. (1976). Effects of p-chlorophenylalanine on time of neuronal origin during embryogenesis in the rat. *Brain Res.* **107,** 638–644.

Lauder, J. M., and Krebs, H. (1978). Serotonin as a differentiation signal in early neurogenesis. *Dev. Neurosci.* **1,** 15–30.

Lauder, J. M., Tamir, H., and Sadler, T. W. (1988). Serotonin and morphogenesis. I. Sites of serotonin uptake and binding protein immunoreactivity in the midgestation mouse embryo. *Development* **102,** 709–720.

Lauder, J. M., Towle, A. C., Patrick, K., Henderson, P., and Krebs, H. (1985). Decreased serotonin content of embryonic raphe neurons following maternal administration of p-chlorophenylalanine: a quantitative immunocytochemical study. *Brain Res.* **352,** 107–114.

Lauder, J. M., Wallace, J. A., and Krebs, H. (1981). Roles for serotonin in neuroembryogenesis. *Adv. Exp. Med. Biol.* **133,** 477–506.

Lauder, J. M., Wallace, J. A., Wilkie, M. B., DiNome, A., and Krebs, H. (1983). Roles for serotonin in neurogenesis. *Monogr. Neural Sci.* **9,** 3–10.

Lechardeur, D., Phung-Ba, V., Wils, P., and Scherman, D. (1996). Detection of the multidrug resistance of P-glycoprotein in healthy tissues: the example of the blood–brain barrier. *Ann. Biol. Clin. (Paris)* **54,** 31–36.

Lee, J., Platt, K. A., Censullo, P., and Ruiz, I. A. (1997). Glil is a target of Sonic hedgehog that induces ventral neural tube development. *Development* **124,** 2537–2552.

Lee, M. Y., Smiley, S., Kadkhodayan, S., Hines, R. N., and Williams, D. E. (1995). Developmental regulation of flavin-containing monooxygenase (FMO) isoforms 1 and 2 in pregnant rabbit. *Chem. Biol. Interact.* **96,** 75–85.

Lemke, G. (1996). Neuregulins in development. *Mol. Cell Neurosci.* **7,** 247–262.

Lemmon, V., Burden, S. M., Payne, H. R., Elmslie, G. J., and Hlavin, M. L. (1992). Neurite growth on different substrates: permissive versus instructive influences and the role of adhesive strength. *J. Neurosci.* **12,** 818–826.

Leonard, J. L., and Farwell, A. P. (1997). Thyroid hormone–regulated actin polymerization in brain. *Thyroid* **7,** 147–151.

Leopold, D. A., and Logothetis, N. K. (1996). Activity changes in early visual cortex reflect monkeys' percepts during binocular rivalry. *Nature* **379,** 549–553.

Lephart, E. D. (1996). A review of brain aromatase cytochrome P450. *Brain Res. Rev.* **22,** 1–26.

LeVay, S., Stryker, M. P., and Shatz, C. J. (1978). Ocular dominance columns and their development in layer IV of the cat's visual cortex: A quantitive study. *J. Comp. Neurol.* **179,** 223–244.

LeVay, S., Wiesel, T. N., and Hubel, D. H. (1980). The development of ocular dominance columns in normal and visually deprived monkeys. *J. Comp. Neurol.* **191,** 1–51.

Leveque, D., and Jehl, F. (1995). P-glycoprotein and pharmacokinetics. *Anticancer Res.* **15,** 331–336.

Levesque, P. C., and Atchison, W. D. (1991). Disruption of brain mitochondrial calcium sequestration by methylmercury. *J. Pharmacol. Exp. Ther.* **256,** 236–242.

Levitt, P., Eagleson, K. L., Chan, A. V., Ferri, R. T., and Lillien, L. (1997). Signaling pathways that regulate specification of neurons in developing cerebral cortex. *Dev. Neurosci.* **19,** 6–8.

Levitt, P., Rakic, P., De Camilli, P., and Greengard, P. (1984). Emergence of cyclic guanosine 3′:5′-monophosphate-dependent protein kinase immunoreactivity in developing rhesus monkey cerebellum: correlative immunocytochemical and electron microscopic analysis. *J. Neurosci.* **4,** 2553–2564.

Lewis, D. A. (1997). Development of the prefrontal cortex during adolescence: insights into vulnerable neural circuits in schizophrenia. *Neuropsychopharmacology* **16,** 385–398.

Lewis, J. (1996). Neurogenic genes and vertebrate neurogenesis. *Curr. Opin. Neurobiol.* **6,** 3–10.

Li, R., and el-Mallakh, R. S. (1997). Triplet repeat gene sequences in neuropsychiatric diseases. *Harv. Rev. Psychiatry* **5,** 66–74.

Li, Y., Erzurumlu, R. S., Chen, C., Jhaveri, S., and Tonegawa, S. (1994). Whisker-related neuronal patterns fail to develop in the trigeminal brainstem nuclei of NMDAR1 knockout mice. *Cell* **76,** 427–437.

Liang, H., Juge-Aubry, C. E., O'Connell, M., and Burger, A. G. (1997). Organ-specific effects of 3,5,3′-triiodothyroacetic acid in rats. *Eur. J. Endocrinol.* **137,** 537–544.

Liem, K. F. Jr., Tremml, G., Roelink, H., and Jessell, T. M. (1995). Dorsal differentiation of neural plate cells induced by BMP-mediated signals from epidermal ectoderm. *Cell* **82,** 969–979.

Lieth, E., McClay, D. R., and Lauder, J. M. (1990). Neuronal–glial interactions: complexity of neurite outgrowth correlates with substrate adhesivity of serotonergic neurons. *Glia* **3,** 169–179.

Lima, F. R., Trentin, A. G., Rosenthal, D., Chagas, C., and Moura Neto, V. (1997). Thyroid hormone induces protein secretion and

morphological changes in astroglial cells with an increase in expression of glial fibrillary acidic protein. *J. Endocrinol.* **154,** 167–175.
Lincoln, J. (1995). Innervation of cerebral arteries by nerves containing 5-hydroxytryptamine and noradrenaline. *Pharmacol. Ther.* **68,** 473–501.
Lindholm, D., Castren, E., Tsoulfas, P., Kolbeck, R., Berzaghi, M. D., P., Leingartner, A., Heisenberg, C. P., Tesarollo, L., Parada, L. F., and Thoenen, H. (1993). Neurotrophin-3 induced by triiodothyronine in cerebellar granule cells promotes Purkinje cell differentiation. *J. Cell Biol.* **122,** 443–450.
Lipton, S. A. (1997). Neuropathogenesis of acquired immunodeficiency syndrome dementia. *Curr. Opin. Neurol.* **10,** 247–253.
Liu, F., Hata, A., Baker, J. C., Doody, J., Carcamo, J., Harland, R. M., and Massague, J. (1996). A human Mad protein acting as a BMP-regulated transcriptional activator. *Nature* **381,** 620–623.
Liu, G., and Elsner, J. (1995). Review of the multiple chemical exposure factors which may disturb human behavioral development. *Soz. Praventivmed.* **40,** 209–217.
Lohof, A. M., Quillan, M., Dan, Y., and Poo, M. M. (1992). Asymmetric modulation of cytosolic cAMP activity induces growth cone turning. *J. Neurosci.* **12,** 1253–1261.
Lopez, L. A., and Sheetz, M. P. (1995). A microtubule-associated protein (MAP2) kinase restores microtubule motility in embryonic brain. *J. Biol. Chem.* **270,** 12511–12517.
Lorke, D. E. (1994). Developmental characteristics of trisomy 19 mice. *Acta Anta. (Basel)* **150,** 159–169.
Lowndes, H. E., Beiswanger, C. M., Philbert, M. A., and Reuhl, K. R. (1994). Substrates for neural metabolism of xenobiotics in adult and developing brain. *Neurotoxicology* **15,** 61–73.
Lowndes, H. E., Philbert, M. A., Beiswanger, C. M., Kauffman, F. C., and Reuhl, K. R. (1995). Xenobiotic metabolism in the brain as mechanistic bases for neurotoxicity. In *Handbook of Neurotoxicology.* L. W. Chang and R. S. Dyer, Eds. Marcel Dekker, New York, pp. 1–27.
Ludin B., and Matus, A. I. (1993). The neuronal cytoskeleton and its role in axonal and dendritic plasticity. *Hippocampus* **3,** 61–71.
Ludin, B., Ashbridge, K., Funfschilling, U., and Matus, A. L. (1996). Functional analysis of the MAP2 repeat domain. *J. Cell Sci.* **109,** 91–99.
Lumsden, A., and Krumlauf, R. (1996). Patterning the vertebrate neuraxis. *Science* **274,** 1109–1115.
Lund, R. D. (1978). *Development and Plasticity of the Brain.* Oxford University Press, New York.
Ma, Q., Sommer, L., Cserjesi, P., and Anderson, D. J. (1997). Mashl and neurogenin1 expression patterns define complementary domains of neuroepithelium in the developing CNS and are correlated with regions expressing notch ligands. *J. Neurosci.* **17,** 3644–3652.
Maccioni, R. B., and Cambiazo, V. (1995). Role of microtubule-associated proteins in the control of microtubule assembly. *Physiol. Rev.* **75,** 835–864.
Macdonald, R., Barth, K. A. Xu, Q., Holder, N., Mikkola, I., and Wilson, S. W. (1995). Midline signaling is required for Pax gene regulation and patterning of the eyes. *Development* **121,** 3267–3278.
Mallet, J. (1996). The TiPS/TINS lecture. Catecholamines: from gene regulation to neuropsychiatric disorders. *Trends Biochem. Sci.* **17,** 129–135.
Mandelkow, E. M., Biernat, J., Drewes, G., Gustke, N., Trinczek, B., and Mandelkow, E. (1995). Tau domains, phosphorylation, and interactions with microtubules. *Neurobiol. Aging* **16,** 335–62; discussion 362–3.
Mandell, J. W., and Banker, G. A. (1996). Microtubule-associated proteins, phosphorylation gradients, and the establishment of neuronal polarity. *Perspect. Dev. Neurobiol.* **4,** 125–135.
Maness, P. F., Beggs, H. E., Klinz, S. G., and Morse, W. R. (1996). Selective neural cell adhesion molecule signaling by Src family tyrosine kinases and tyrosine phosphatases. *Perspect. Dev. Neurobiol.* **4,** 169–181.
Manzo, L., Locatelli, C., Candura, S. M., and Costa, L. G. (1994). Nutrition and alcohol neurotoxicity. *Neurotoxicology* **15,** 555–565.
March, J. S. (1995). *Anxiety Disorders in Children and Adolescents.* Guilford Press, New York.
Marigo, V., and Tabin, C. J. (1996). Regulation of patched by sonic hedgehog in the developing neural tube. *Proc. Natl. Acad. Sci. U.S.A.* **93,** 9346–9351.
Marin-Padilla, M. (1971). Early prenatal ontogenesis of the cerebral cortex (neocortex) of the cat (*Felis domestica*). A golgi study. I. The primordial neocortical organization. *Z. Anat. Entwicklungsgesch.* **134,** 117–145.
Marin-Padilla, M. (1972). Structural abnormalities of the cerebral cortex in human chromosomal aberrations: a Golgi study. *Brain Res.* **44,** 625–629.
Marin-Padilla, M. (1978). Dural original of the mammalian neocortex and evolution of the cortical plate. *Anta. Embryol.* **152,** 109–126.
Marin-Padilla, M. (1983). Structural organization of the human cerebral cortex prior to the appearance of the cortical plate. *Anat. Embryol.* **168,** 21–40.
Marin-Padilla, M. (1988). Embryonic vascularization of the mammalian cerebral cortex. In *Cerebral Cortex, Vol. 7. Development and Maturations of the Cerebral Cortex.* A. Peters and E. G. Jones Eds. Plenum Press, New York, pp. 479–510.
Martinez-Galan, J. R., Pedraza, P., Santacana, M., Escobar del Rey, F., Morreale de Escobar, G., and Ruiz-Marcos, A. (1997). Myelin basic protein immunoreactivity in the internal capsule of neonates from rats on a low iodine intake or on methylmercaptoimidazole (MMI). *Dev. Brain Res.* **101,** 249–256.
Martinez-Galan, J. R., Pedraza, P., Santacana, M., Rey, F. E., Escobar, G. M., and Ruiz-Marcos, A. (1997). Early effects of iodine deficiency on radial glial cells of the hippocampus of the rat fetus. A model of neurological cretinism. *J. Clin. Invest.* **99,** 2701–2709.
Martinou, J. C., and Sadoul, R. (1996). ICE-like proteases execute the neuronal death program. *Curr. Opin. Neurobiol.* **6,** 609–614.
Mastick, G. S., Fan, C., Tessier-Levagne, M., Serbedzija, G. N., McMahaon, A. P., and Easter, S. S. Jr. (1997). Early deletion of neuromeres in Wnt-1-/1-mutant mice: evaluation by morphological and molecular markers. *J. Comp. Neurol.* **374,** 246–258.
Matilainen, R., Airaksinen, E., Mononen, T., Launiala, K., and Kaariainen, R. (1995). A population-based study on the causes of mild and severe mental retardation. *Acta Paediatr.* **84,** 261–266.
Matten, W. T., Aubry, M., West, J., and Maness, P. F. (1990). Tubulin is phosphorylated at tyrosine by pp60c-src in nerve growth cone membranes. *J. Cell Biol.* **111,** 1959–1970.
Matus, A. I. (1988). Microtubule-associated proteins: their potential role in determining neuronal morphology. *Annu. Rev. Neurosci.* **11,** 29–44.
McAllister, A. K., Lo, D. C., and Katz, L. C. (1995). Neurotrophins regulate dendritic growth in developing visual cortex. *Neuron* **15,** 791–803.
McCann, S. M., Lyson, K., Karanth, S., Gimeno, M., Belova, N., Kamat, A., and Rettori, V. (1995). Mechanism of action of cytokines to induce the pattern of pituitary hormone secretion in infection. *Ann. NY Acad. Sci.* **771,** 386–395.
McConnell, S. K. (1988). Development and decision-making in the mammalian cerebral cortex. *Brain Res. Rev.* **13,** 1–23.
McConnell, S. K. (1995a). Constructing the cerebral cortex: neurogenesis and fate determination. *Neuron* **15,** 761–768.
McConnell, S. K. (1995b). Plasticity and commitment in the developing cerebral cortex. *Prog. Brain Res.* **105,** 129–143.

McConnell, S. K. (1995). Strategies for the generation of neuronal diversity in the developing central nervous system. *J. Neurosci.* **15,** 6987–6998.

McDonald, J. W., Silverstein, F. S., and Johnston, M. V. (1988). Neurotoxicity of *N*-methyl-D-aspartate is markedly enhanced in developing rat central nervous system. *Brain Res.* **459,** 200–203.

McFadden, S. A. (1996). Phenotypic variation in xenobiotic metabolism and adverse environmental response: focus on sulfur-dependent detoxification pathways. *Toxicology* **111,** 43–65.

McNabb, F. M. (1995). Thyroid hormones, their activation, degradation and effects on metabolism. *J. Nutr.* **125,** 1773S–1776S.

Meakin, S. O., and Shooter, E. M. (1991). Tyrosine kinase activity coupled to the high-affinity nerve growth factor-receptor complex. *Proc. Natl. Acad. Sci. U.S.A.* **88,** 5862–5866.

Means, A. R., Ribar, T. J., Kane, C. D., Hook, S. S., and Anderson, K. A. (1997). Regulation and properties of the rat Ca^{2+}/calmodulin-dependent protein kinase IV gene and its protein products. *Recent. Prog. Horm. Res.* **52,** 389–406.

Mednick, S. A., and Hollister, J. M. (1995). *Neural Development and Schizophrenia: Theory and Research.* Plenum Press, New York.

Meiri, K. F., Pfenninger, K. H., and Willard, M. B. (1986). Growth-associated protein, GAP-43, a polypeptide that is induced when neurons extend axons, is a component of growth cones and corresponds to pp46, a major polypeptide of a subcellular fraction enriched in growth cones. *Proc. Natl. Acad. Sci. U.S.A.* **83,** 3537–3541.

Menez, J. F., Machu, T. K., Song, B. J., Browning, M. D., and Deitrich, R. A. (1993). Phosphorylation of cytochrome P4502E1 (CYP2E1) by calmodulin dependent protein kinase, protein kinase C and cAMP dependent protein kinase. *Alcohol Alcohol* **28,** 445–451.

Merry, D. E., and Korsmeyer, S. J. (1997). Bcl-2 gene family in the nervous system. *Annu. Rev. Neurosci.* **20,** 245–267.

Messersmith, E. K., Feller, M. B., Zhang, H., and Shatz, C. J. (1997). Migration of neocortical neurons in the absence of functional NMDA receptors. *Mol. Cell Neurosci.* **9,** 347–357.

Meyer, D., Yamaai, T., Garrett, A., Riethmacher-Sonnenberg, E., Kane D., Theill, L. E., and Birchmeier, C. (1997). Isoform-specific expression and function of neuregulin. *Development* **124,** 3575–3586.

Miller, D. B. (1992). Caveats in hazard assessment: stress and neurotoxicity. In *The Vulnerable Brain and Environmental Risks. Vol. 1. Malnutrition and Hazard Assessment.* R. L. Isaacson, and K. F. Jensen, Eds. Plenum Press, New York. pp. 239–266.

Miller, M. S., McCarver, D. G., Bell, D. A., Eaton, D. L., and Goldstein, J. A. (1997). Genetic polymorphisms in human drug metabolic enzymes. *Fundam. Appl. Toxicol.* **40,** 1–14.

Miller, M. W. (1996). Mechanisms of ethanol induced neuronal death during development: from the molecule to behavior. *Alcohol Clin. Exp. Res.* **20,** 128A–132A.

Ming, G. L., Song, H. J., Berning, B., Holt, C. E., Tessier-Lavigne, M., and Poo, M. M. (1997). cAMP-dependent growth cone guidance by netrin-1. *Neuron* **19,** 1225–1235.

Minn, A., Ghersi-Egea, J., Perrin, R., Leininger, B., and Siest, G. (1991). Drug metabolizing enzyme in the brain and cerebral microvessels. *Brain Res. Rev.* **16,** 65–82.

Mitani, F., Mukai, K., Ogawa, T., Miyamoto, H., and Ishimura, Y. (1997). Expression of cytochrones P450aldo and P45011 beta in rat adrenal gland during late gestational and neonatal stages. *Steroids* **62,** 57–61.

Mitas, M. (1997). Trinucleotide repeats associated with human disease. *Nucleic Acids Res.* **25,** 2245–2254.

Miyawaki, T., Sohma, O., Mizuguchi, M., and Takashima, S. (1995). Development of endothelial nitric oxide synthase in endothelial cells in the human cerebrum. *Dev. Brain Res.* **89,** 161–166.

Mobley, W. C., Woo, J. E., Edwards, R. H., Riopelle, R. J., Longo, F. M., Weskamp, G., Otten, U., Valletta, J. S., and Johnston, M. V. (1989). Developmental regulation of nerve growth factor and its receptor in the rat caudate–putamen. *Neuron* **3,** 655–664.

Mohler, H. (1997). Genetic approaches to CNS disorders with particular reference to GABAA-receptor mutations. *J. Recept. Signal. Transduct. Res.* **17,** 1–10.

Moncada, S., Palmers, R. M., and Higgs, E. A. (1991). Nitric oxide: physiology, pathophysiology, and pharmacology. *Pharmacol. Rev.* **43,** 109–142.

Moon, R. T., Brown, J. D., and Torres, M. (1997). WNTs modulate cell fate behavior during vertebrate development. *Trends Genet.* **13,** 157–161.

Moorhouse, S. R., Carden, S., Drewitt, P. N., Eley, B. P., Hargreaves, R. J., and Pelling, D. (1988). The effect of chronic low level lead exposure on blood–brain barrier function in the developing rat. *Biochem. Pharmacol.* **37,** 4539–4547.

Morgan, E. T. (1997). Regulation of cytochromes P450 during inflammation and infection. *Drug Metab. Rev.* **29,** 1129–1188.

Morgane, P. J., Austin-LaFrance, R. J., Bronzino, J. D., Tonkiss, J., and Galler, J. R. (1992). Malnutrition and the developing nervous system. In *The Vulnerable Brain and Environmental Risks. Vol. 1. Malnutrition and Hazard Assessment.* R. L. Isaacson and K. F. Jensen, Eds. Plenum Press, New York. pp. 3–44.

Morreale de Escobar, G., Obregon, M. J., Calvo, R., and Escobar del Rey, F. (1993). Effects of iodine deficiency on thyroid hormone metabolism and the brain in fetal rats: the role of the maternal transfer of thyroxin. *Am. J. Clin. Nutr.* **57,** 280S–285S.

Morris, J. A. (1996). Schizophrenia, bacterial toxins and the genetics of redundancy. *Med. Hypotheses.* **46,** 362–366.

Mowry, B. J., Nancarrow, D. J., and Levinson, D. F. (1997). The molecular genetics of schizophrenia: an update. *Aust. NZ J. Psychiatry* **31,** 704–713.

Muhr, J., Jessell, T. M., and Edlund, T. (1997). Assignment of early caudal identity to neural plate cells by a signal from caudal paraxial mesoderm. *Neuron* **19,** 487–502.

Muller, Y., Rocchi, E., Lazaro, J. B., and Clos, J. (1995). Thyroid hormone promotes BCL-2 expression and prevents apoptosis of early differentiating cerebellar granule neurons. *Int. J. Dev. Neurosci.* **13,** 871–885.

Murphy, K. C., and McGuffin, P. (1996). The role of candidate genes in the etiology of schizophrenia. *Mol. Med.* **2,** 665–669.

Murphy, K. C., Cardno, A. G., and McGuffin, P. (1996). The molecular genetics of schizophrenia. *J. Mol. Neurosci.* **7,** 147–157.

Nag, S. (1995). Role of the endothelial cytoskeleton in blood–brain-barrier permeability to protein. *Acta Neuropathol. (Berl)* **90,** 454–460.

Nass, R. (1994). Advances in learning disabilities. *Curr. Opin. Neurol.* **7,** 179–186.

National Research Council (NRC). (1992). *Environmental Neurotoxicology.* National Academy Press, Washington, DC.

Needleman, H. L. (1994). Childhood lead poisoning. *Curr. Opin. Neurol.* **7,** 187–190.

Needleman, H. L. (1995). Behavioral toxicology. *Environ. Health Perspect.* **103,** (Suppl 6) 77–79.

Needleman, H. L., and Bellinger, D. (1991). The health effects of low level exposure to lead. *Annu. Rev. Public Health* **12,** 111–140.

Negri-Cesi, P., Poletti, A., and Celotti, F. (1996). Metabolism of steroids in the brain: a new insight into the role of 5alpha-reductase and aromatase in brain differentiation and functions. *J. Steroid Biochem. Mol. Biol.* **58,** 455–466.

Neubert, D. (1988). Significance of pharmacokinetic variables in reproductive and developmental toxicity. *Xenobiotica* **18,** 45–58.

Neuwelt, E. A. (1989). *Implication of the Blood–Brain Barrier and Its Manipulation.* Plenum Press, New York.

Niclas, J., Navone, F., Hom-Booher, N., and Vale, R. D. (1994). Cloning and localization of a conventional kinesin motor expressed exclusively in neurons. *Neuron* **12,** 1059–1072.

Nicotera, P., Bellomo, G., and Orrenius, S. (1992). Calcium-mediated mechansims in chemically induced cell death. *Annu. Rev. Pharmacol. Toxicol.* **32,** 449–470.

Noguchi, T. (1996). Effects of growth hormone on cerebral development: Morphological studies. *Horm. Res.* **45,** 5–17.

Nonchev, S., Maconochie, M., Vesque, C., Aparicio, S., Ariza-McNaughton, L., Manzanares, M., Maruthainar, K., Kuroiwa, A., Brenner, S., Charnay, P., and Krumlauf, R. (1996). The conserved role of Krox-20 in directing Hox gene expression during vertebrate hindbrain segmentation. *Proc. Natl. Acad. Sci. U.S.A.* **93,** 9339–9345.

Nonneman, A. J., and Isaacson, R. L. (1973). Task dependent recovery after early brain damage. *Behav. Biol.* **8,** 143–172.

Northington, F. J., Tobin, J. R., Harris, A. P., Traystman, R. J., and Koehler, R. C. (1997). Developmental and regional differences in nitric oxide synthase activity and blood flow in the sheep brain. *J. Cereb. Blood Flow Metab.* **17,** 109–115.

Notenboom, R. G. E., Moorman, A. F. M., and Lamers, W. H. (1997). Developmental appearance of ammonia-metabolizing enzymes in prenatal murine liver. *Microsc. Res. Technique* **39,** 413–423. (Abstract)

Nunez, J., Couchie, D., Aniello, F., and Bridoux, A. M. (1991). Regulation by thyroid hormone of microtubule assembly and neuronal differentiation. *Neurochem. Res.* **16,** 975–982.

O'Callaghan, J. P. (1994). A potential role for altered protein phosphorylation in the mediation of developmental neurotoxicity. *Neurotoxicology* **15,** 29–40.

O'Donovan, M. C., and Owen, M. J. (1996). The molecular genetics of schizophrenia. *Ann. Med.* **28,** 541–546.

O'Leary, D. D., Stanfield, B. B., and Cowan, W. M. (1981). Evidence that the early postnatal restriction of the cells of origin of the callosal projection is due to the elimination of axonal collaterals rather than to the death of neurons. *Brain Res.* **227,** 607–617.

Oppenheim, R. W. (1981). Neuronal cell death and some related regressive phenomena during neurogenesis: a selective historical review and a progress report. In *Studies in Developmental Neurobiology: Essays in Honor of Viktor Hamburger.* W. M. Cowan, Ed. Oxford University Press, New York, pp. 74–133.

Oppenheim, R. W. (1989). The neurotrophic theory and naturally occurring motoneuron death. *Trends Neurosci.* **12,** 252–255.

Oppenheim, R. W. (1997). Related mechanisms of action of growth factors and antioxidants in apoptosis: an overview. *Adv. Neurol.* **72,** 69–78.

Oppenheimer, J. H., and Schwartz, H. L. (1997). Molecular basis of thyroid hormone-dependent brain development. *Endocr. Rev.* **18,** 462–475.

O'Rourke, N. A., Sullivan, D. P., Kaznowski, C. E., Jacobs, A. A., and McConnell, S. K. (1995). Tangential migration of neurons in the developing cerebral cortex. *Development* **121,** 2165–2176.

Pal, U., Biswas, S. C., and Sarkar, P. K. (1997). Regulation of actin and its mRNA by thyroid hormones in cultures of fetal human brain during second trimester of gestation. *J. Neurochem.* **69,** 1170–1176.

Palade, G. E., and Palay, S. L. (1954). Electron microscopic observations of the interneuronal and neuromuscular synapses. *Anat. Rec.* **118,** 335–336.

Palomaki, G. E. (1996). Down's syndrome epidemiology and risk estimation. *Early Hum. Dev.* **47** (Suppl), S19–26.

Paneth, N. (1993). The causes of cerebral palsy. Recent evidence. *Clin. Invest. Med.* **16,** 95–102.

Pang, S. K., Xu, X., and St. Pierre, M. V. (1992). Determinants of metabolite disposition. *Annu. Rev. Pharmacol. Toxicol.* **32,** 623–669.

Pardridge, W. M., Eisenberg, J., Fierer, G., and Musto, N. A. (1988). Developmental changes in brain and serum binding of testosterone and in brain capillary uptake of testosterone-binding serum proteins in the rabbit. *Brain Res.* **466,** 245–253.

Pardridge, W. M., Golden, P. L., Kang, Y. S., and Bickel, U. (1997). Brain microvascular and astrocyte localization of P-glycoprotein. *J. Neurochem.* **68,** 1278–1285.

Pardridge, W. M., Yang, J., and Eisenberg, J. (1985). Blood–brain barrier protein phosphorylation and dephosphorylation. *J. Neurochem.* **45,** 1141–1147.

Parke, D. V. (1982). Mechanisms of chemical toxicity—a unifying hypothesis. *Regul. Toxicol. Pharmacol.* **2,** 267–286.

Parke, D. V., and Ioannides, C. (1994). The effects of nutrition on chemical toxicity. *Drug. Metab. Rev.* **26,** 739–765.

Pasquini, J. M., and Adamo, A. M. (1994). Thyroid hormones and the central nervous system. *Dev. Neurosci.* **16,** 1–8.

Pellegrini, M., O'Brien, T. J., Hoy, J., and Sedal, L. (1996). *Mycoplasma pneumoniae* infection associated with an acute brain stem syndrome. *Acta Neurol. Scand.* **93,** 203–206.

Persson, U., Souchelnytskyi, S., Franzen, P., Miyazono, K., ten Dijke, P., and Heldin, C. H. (1997). Transforming growth factor (TGF-beta)-specific signaling by chimeric TGF-beta type II receptor with intracellular domain of activin type IIB receptor. *J. Biol. Chem.* **272,** 21187–21194.

Peters, A., Palay, S. L., and Webster, H. (1991). *The Fine Structure of the Nervous System. Third Edition.* Oxford University Press, Oxford.

Peterson, R. E., Theobald, H. M., and Kimmel, G. L. (1993). Developmental and reproductive toxicity of dioxins and related compounds: cross-species comparisons. *CRC Crit. Rev. Toxicol.* **23,** 283–335.

Petrali, J. P., Maxwell, D. M., Lenz, D. E., and Mills, K. R. (1991). Effect of an anticholinesterase compound on the ultrastructure and function of the rat blood–brain barrier: a review and experiment. *J. Submicrosc. Cytol. Pathol.* **23,** 331–338.

Pickard, M. R., Sinha, A. K., Ogilvie, L., and Ekins, R. P. (1993). The influence of the maternal thyroid hormone environment during pregnancy on the ontogenesis of brain and placental ornithine decarboxylase activity in the rat. *J. Endocrinol.* **139,** 205–212.

Pimenta, A. F., Zhurkareva, V., Barbe, M. F., Reinoso, B. S., Grimley, C., Henzel, W., Fischer, I., and Levitt, P. (1995). The limbic system–associated membrane protein is an Ig superfamily member that mediates selective neuronal growth and axon targeting. *Neuron* **15,** 287–297.

Placzek, M. (1997). The role of the notocord and floor plate in inductive interactions. *Curr. Opin. Dev. Biol.* **5,** 499–506.

Plateel, M., Teissier, E., and Cecchelli, R. (1997). Hypoxia dramatically increases the nonspecific transport of blood-borne proteins to the brain. *J. Neurochem.* **68,** 874–877.

Poddar, R., and Sarkar, P. K. (1993). Delayed detyrosination of alpha-tubulin from parallel fibre axons and its correlation with impaired synaptogenesis in hypothyroid rat cerebellum. *Brain Res.* **614,** 233–240.

Polymeropoulos, M. H., Coon, H., Byerley, W., Gershon, E. S., Goldin, L., Crow, T. J., Rubenstein, J., Hoff, M., Holik, J., and Smith, A. M. (1994). Search for a schizophrenia susceptibility locus on human chromosome 22. *Am. J. Med. Genet.* **54,** 93–99.

Porterfield, S. P., and Hendrich, C. E. (1993). The role of thyroid hormones in prenatal and neonatal neurological development—current perspectives. *Endocr. Rev.* **14,** 94–106.

Portin, P., and Alanen, Y. O. (1997a). A critical review of genetic studies of schizophrenia. I. Epidemiological and brain studies. *Acta Psychiatr. Scand.* **95,** 1–5.

Portin, P., and Alanen, Y. O. (1997b). A critical review of genetic studies of schizophrenia. II. Molecular genetic studies. *Acta Psychiatr. Scand.* **95,** 73–80.

Press, M. F. (1985). Lead-induced permeability changes in immature vessels of the developing cerebellar mnicrocirculation. *Acta Neuropathol. (Berl)* **67,** 86–95.

Pribram, K. H. (1969). The neurophysiology of remembering. *Sci. Am.* **220,** 73–87.

Prough, R. A., Linder, M. W., Pinaire, J. A., Xiao, G. H., and Falkner, K. C. (1996). Hormonal regulation of hepatic enzymes involved in foreign compound metabolism. *FASEB J.* **10,** 1369–1377.

Psaltis, D., Brady, D., Xiang-Guang, G., and Lin, S. (1990). Holography in artificial neural networks. *Nature* **343,** 325–330.

Pulsifer, M. B. (1996). The neuropsychology of mental retardation. *J. Int. Neuropsychol. Soc.* **2,** 159–176.

Purpura, D. P. (1974). Dendritic spines "dysgenesis" and mental retardation. *Science* **186,** 1126–1128.

Purpura, D. P. (1977). Factors contributing to abnormal neuronal development in the cerebral cortex of the human infant. In *Brain: Fetal and Infant.* S. R. Berenberg, Ed. Martinus Nijhoff, The Hague. pp. 55–78.

Purves, D. (1988). *Body and Brain.* Harvard University Press, Cambridge.

Purves, D. (1997). Reply. *Trends Neurosci.* **20,** 294.

Purves, D., White, L. E., and Riddle, D. R. (1996). Is neural development Darwinian? *Trends Neurosci.* **19,** 460–464.

Rakic, P. (1971). Neuron–glia relationship during granule cell migration in developing cerebellar cortex. A Golgi and electronmicroscopic study in *Maccacus rhesus. J. Comp. Neurol.* **141,** 283–312.

Rakic, P. (1972). Mode of cell migration of the superficial layers of fetal monkey neocortex. *J. Comp. Neurol.* **145,** 61–84.

Rakic, P. (1974). Neurons in Rhesus monkey visual cortex: systematic relation between time of origin and eventual disposition. *Science* **183,** 425–427.

Riakic, P. (1975). Cell migration and neuronal ectopias in the brain. *Birth Defects* **11,** 95–129.

Rakic, P. (1995). Radial versus tangential migration of neuronal clones in the developing cerebral cortex [comment]. *Proc. Natl. Acad. Sci. U.S.A.* **92,** 11323–11327.

Rakic, P., and Komuro, H. (1995). The role of receptor/channel activity in neuronal cell migration. *J. Neurobiol.* **26,** 299–315.

Rakic, P., Cameron, R. S., and Komuro, H. (1994). Recognition, adhesion, transmembrane signaling and cell motility in guided neuronal migration. *Curr. Opin. Neurobiol.* **4,** 63–69.

Rakic, P., Knyihar-Csillik, E., and Csillik, B. (1996). Polarity of microtubule assemblies during neuronal cell migration. *Proc. Natl. Acad. Sci. U.S.A.* **93,** 9218–9222.

Ramon y Cajal S. (1906). The structure and connections of neurons. In *Nobel Lectures: Physiology and Medicine, 1901–1921.* (1987). Elsevier, Amsterdam, pp. 220–253.

Ramon y Cajal, S. (1960). *Studies on Vertebrate Neurogenesis.* Charles C. Thomas, Springfield, IL.

Ramon y Cajal, S. (1984). *The Neuron and the Glial Cell.* Charles C. Thomas, Springfield, IL.

Rapoport, J. L., Giedd, J., Kumura, S., Jacobsen, L., Smith, A., Lee, P., Nelson, J., and Hamburger, S. (1997). Childhood-onset schizophrenia. Progressive ventricular change during adolescence. *Arch. Gen. Psychiatry* **54,** 897–903.

Rapoport, S. I. (1976). *Blood–Brain Barrier in Physiology and Medicine.* Raven Press, New York.

Ravindranath, V., and Boyd, M. R. (1995). Xenobiotic metabolism in the brain. *Drug Metab. Rev.* **27,** 419–448.

Ray, D. E. (1997). Physiological factors predisposing to neurotoxicity. *Arch. Toxicol. Suppl.* **19,** 219–226.

Ray, D. E., Brown, A. W., Cavanagh, J. B., Nolan, C. C., Richards, H. K., and Wylie, S. P. (1992). Functional/metabolic modulation of the brain stem lesions caused by 1,3-dinitrobenzene in the rat. *Neurotoxicology* **13,** 379–388.

Raynor, B. D. (1993). Cytomegalovirus infection in pregnancy. *Semin. Perinatol.* **17,** 394–402.

Reedy, P. S., and Housman, D. E. (1997). The complex pathology of trinucleotide repeats. *Curr. Opin. Cell Biol.* **9,** 364–372.

Reid, S. N., Romano, C., Hughes, T., and Daw, N. W. (1997). Developmental and sensory-dependent changes of phosphoinositide-linked metabotropic glutamate receptors. *J. Comp. Neurol.* **389,** 577–583.

Reuhl, K. R. (1991). Delayed expression of neurotoxicity: the problem of silent damage. *Neurotoxicology* **12,** 341–346.

Reuhl, K. R., Lagunowich, L. A., and Brown, D. L. (1994). Cytoskeleton and cell adhesion molecules: critical targets of toxic agents. *Neurotoxicology* **15,** 133–145.

Rich, K. J., and A. R. Boobis (1997). Expression and inducibility of P450 enzymes during liver ontogeny. *Microsc. Res. Technique* **39,** 424–435. (Abstract)

Riess, J. A., and Needleman, H. L. (1992). Cognitive, neural, and behavioral effects of low-level lead exposure. In *The Vulnerable Brain and Environmental Risks. Vol. 2. Toxins in Food.* R. L. Isaacson and K. F. Jensen, Eds. Plenum Press, New York. pp. 111–126.

Rio, C., Rieff, H. I., Qi, P., and Corfas, G. (1997). Neuregulin and erbB receptors play a critical role in neuronal migration. *Neuron* **19,** 39–50.

Risau, W. (1997). Mechanisms of angiogenesis. *Nature* **386,** 671–674.

Robey, E. (1997). Notch in vertebrates. *Curr. Opin. Genet. Dev.* **7,** 551–557.

Rodier, P. M. (1976). Critical periods for behavioral anomalies in mice. *Environ. Health Perspect.* **18,** 79–83.

Rodier, P. M. (1977). Correlations between prenatally-induced alterations in CNS cell populations and postnatal function. *Teratology* **16,** 235–246.

Rodier, P. M. (1986). Time of exposure and time of testing in developmental neurotoxicology. *Neurotoxicology* **7,** 69–76.

Rodier, P. M. (1988). Structural–functional relationships in experimentally induced brain damage. *Prog. Brain Res.* **73,** 335–348.

Rodier, P. M. (1994). Vulnerable periods and processes during central nervous system development. *Environ. Health Perspect.* **102** (Suppl 2), 121–124.

Rodier, P. M. (1995). Developing brain as a target of toxicity. *Environ. Health Perspect.* **103** (Suppl 6), 73–76.

Rodriguez, I., and Basler, K. (1997). Control of compartmental affinity boundries by hedgehog. *Nature* ***389,*** 614–618.

Roeleveld, N., Zielhuis, G. A., and Gabreels, F. (1997). The prevalence of mental retardation: a critical review of recent literature. *Dev. Med. Child Neurol.* **39,** 125–132.

Roelink, H. (1996). Tripartite signaling of pattern: interactions between hedgehogs, BMPs, and Wnts in the control of vertebrate development. *Curr. Opin. Neurobiol.* **6,** 33–40.

Roerig, B., and Katz, L. C. (1997). Modulation of intrinsic circuits by serotonin 5-HT3 receptors in developing ferret visual cortex. *J. Neurosci.* **17,** 8324–8338.

Roerig, B., Nelson, D. A., and Katz, L. C. (1997). Fast synaptic signaling by nicotinic acetylcholine and serotonin 5-HT3 receptors in developing visual cortex. *J. Neurosci.* **17,** 8353–8362.

Romano, C., van den Pol, A. N., and O'Malley, K. L. (1996). Enhanced early developmental expression of the metabotropic glutamate receptor mGluR5 in rat brain; protein, mRNA splice variants, and regional distribution. *J. Comp. Neurol.* **367,** 403–412.

Romero, I. A., Lister, T., Richards, H. K., Seville, M. P., Wylie, S. P., and Ray, D. E. (1995). Early metabolic changes during m-Dinitrobenzene neurotoxicity and the possible role of oxidative stress. *Free Radic. Biol. Med.* **18,** 311–319.

Rose, K. A., Stapleton, G., Dott, K., Kieny, M. P., Best, R., Schwarz, M., Russell, D. W., Bjorkhem, I., Seckl, J., and Lathe, R. (1997). Cyp7b, a novel brain cytochrome P450, catalyzes the synthesis of neurosteroids 7alpha-hydroxy dehydroepiandrosterone and 7alpha-hydroxy pregnenolone. *Proc. Natl. Acad. Sci. U.S.A.* **94,** 4925–4930.

Rosenberg, R. N., and Harding, A. E. (1988). *The Molecular Biology of Neurological Disease.* Butterworths, London.

Ross, C. A., and Pearlson, G. D. (1996). Schizophrenia, the heteromodal association neocortex and development: potential for a neurogenetic approach. *Trends Neurosci.* **19,** 171–176.

Ross, M. E. (1996). Cell division and the nervous system: regulating the cycle from neural differentiation to death. *Trends Neurosci.* **19,** 62–68.

Russell, R. W., McGaugh, J. L., and Isaacson, R. L. (1992). Nitrates, nitrites, nitroso compounds: balancing benefits and risks. In *The Vulnerable Brain and Environmental Risks. Vol. 3. Toxins in Air and Water.* R. L. Isaacson and K. F. Jensen, Eds. Plenum, New York, pp. 323–340.

Sada, Y., Moriki, T., Kuwahara, S., Yamane, T., and Hara, H. (1994). Immunohistochemical study on blood–brain barrier in congenitally hydrocephalic HTX rat brain. *Zentralbl. Pathol.* **140,** 289–298.

Sahyoun, N., LeVine, H. I., Burgess, S. K., Blanchard, S., Chang, K.-J., and Cuatrecasas, P. (1985). Early postnatal development of calmodulin-dependent protein kinase II in rat brain. *Biochem. Biophys. Res. Comm.* **132,** 878–884.

Sarkar, S., Chaudhury, S., and Sarkar, P. K. (1997). Regulation of beta- and gamma-acin mRNA by thyroid hormone in the developing rat brain. *Neuroreport* **8,** 1267–1271.

Sauer, F. (1935). Mitosis in the neural tube. *J. Comp. Neurol.* **62,** 377–405.

Sauer, M., and Walker, B. (1959). Radioautographic study of interkinetic nuclear migration in the neural tube. *Proc. Soc. Exp. Biol. Med.* **101,** 557–560.

Saunders, N. R. (1992). Development of the blood–brain barrier to macromolecules. In *Barriers and Fluids of the Eye and Brain.* M. B. Segal, Ed. CRC Press, Boca Raton, pp. 128–158.

Saunier, B., Pierre, M., Jacquemin, C., and Courtin, F. (1993). Evidence for cAMP-independent thyrotropin effects on astroglial cells. *Eur. J. Biochem.* **218,** 1091–1094.

Schaefer, G. B., and Bodensteiner, J. B. (1992). Evaluation of the child with idiopathic mental retardation. *Pediatr. Clin. North Am.* **39,** 929–943.

Schiffmann, S. N., Bernier, B., and Goffinet, A. M. (1997). Reelin mRNA expression during mouse brain development. *Eur. J. Neurosci.* **9,** 1055–1071.

Schinkel, A. H., Wagenaar, E., Mol, C. A., and van Deemter, L. (1996). P-glycoprotein in the blood–brain barrier of mice influences the brain penetration and pharmacological activity of many drugs. *J. Clin. Invest.* **97,** 2517–2524.

Schmechel, D. E., and Rakic, P. (1979). A Golgi study of radia glial cells in developing monkey telencephalon: morphogenesis and transformation into astrocytes. *Anat. Embryol.* **156,** 115–152.

Schmidt-Wolf, G. D., Negrin, R. S., and Schmidt-Wolf, I. G. (1997). Activated T cells and cytokine-induced CD3+CD56+ killer cells. *Ann. Hematol.* **74,** 51–56.

Schmitt, C. A., and McDonough, A. A. (1988). Thyroid hormone regulates a and a+ isoforms of Na,K-ATPase during development in neonatal rat brain. *J. Biol. Chem.* **263,** 17643–17649.

Schrott, L. M., Waters, N. S., Boehm, G. W., Sherman, G. F., Morrison, L., Rosen, G. D., Behan, P. O., Galaburda, A. M., and Denenberg, V. H. (1993). Behavior, cortical ectopias, and autoimmunity in BXSB-Yaa and BXSB-Yaa+ mice. *Brain Behav. Immun.* **7,** 205–223.

Schuur, A. G., Boekhorst, F. M., Brouwer, A., and Visser, T. J. (1997). Extrathyroidal effects of 2,3,7,8-tetrachlorodibenzo-*p*-dioxin on thyroid hormone turnover in male Sprague–Dawley rats. *Endocrinology* **138,** 3727–3734.

Schwartz, H. L., Ross, M. E., and Oppenheimer, J. H. (1997). Lack of effect of thyroid hormone on late fetal rat brain development. *Endocrinology* **138,** 3119–3124.

Schwartz, L. M., and Milligan, C. E. (1996). Cold thoughts of death: the role of ICE proteases in neuronal cell death. *Trends Neurosci.* **19,** 555–562.

Seegal, R. F. (1996). Epidemiological and laboratory evidence of PCB-induced neurotoxicity. *Crit. Rev. Toxicol.* **26,** 709–737.

Seegal, R. F., and Shain, W. (1992). Neurotoxicity of polychlorinated biphenyls: The role of ortho-substituted congenets in altering neurochemical function. In *The Vulnerable Brain and Environmental Risks. Vol. 2. Toxins in Food.* R. L. Isaacson and K. F. Jensen, Eds. Plenum Press, New York. pp. 169–196.

Segall, M. A., French, T. A., and Weiner, N. (1996). Effect of neonatal thyroid hormone alterations in CNS ethanol sensitivity in adult LS and SS mice. *Alcohol* **13,** 559–567.

Seidman, L. J., Faraone, S. V., Goldstein, J. M., Goodman, J. M., Kremen, W. S., Matsuda, G., Hoge, E. A., Kennedy, D., Makris, N., Caviness, V. S., and Tsuang, M. T. (1997). Reduced subcortical brain volumes in nonpsychotic siblings of schizophrenic patients: a pilot magnetic resonance imaging study. *Am. J. Med. Genet.* **74,** 507–514.

Selemon, L. D., Rajkowska, G., and Goldman-Rakic, P. S. (1995). Abnormally high neuronal density in the schizophrenic cortex: A morphometric analysis of prefrontal area 9 and occipital area 17. *Arch. Gen. Psychiatry* **52,** 805–18.

Shafer, T. J., Contreras, M. L., and Atchison, W. D. (1990). Characterization of interactions of methylmercury with Ca^{2+} channels in synaptosomes and pheochromocytoma cells: radiotracer flux and binding studies. *Mol. Pharmacol.* **38,** 102–113.

Sham, P. (1996). Genetic epidemiology. *Br. Med. Bull.* **52,** 408–433.

Shatz, C. J. (1996). Emergence of order in visual system development. *J. Physiol. Paris.* **90,** 141–150.

Shatz, C. J., and Stryker, M. P. (1978). Ocular dominance in layer IV of the cat's visual cortex and the effects of monocular deprivation. *J. Physiol.* (*Lond*) **281,** 267–283.

Shen, X. A., Nguyen, A. D., Perry, J. W., Huestis, D. L., and Kachru, R. (1997). Time-domain holographic digital memory. *Science* **278,** 96–100.

Sherer, T. T., Sylvester, S. R., and Bull, R. J. (1993). Differential expression of c-erbA mRNAs in the developing cerebellum and cerebral cortex of the rat. *Biol. Neonate.* **63,** 26–34.

Sherman, G. F., Rosen, G. D., Stone, L. V., Press, D. M., and Galaburda, A. M. (1992). The organization of radial glial fibers in spontaneous neocortical ectopias of newborn New Zealand black mice. *Dev. Brain Res.* **67,** 279–283.

Sherman, G. F., Stone, J. S., Press, D. M., Rosen, G. D., and Galaburda, A. M. (1990). Abnormal architecture and connections disclosed by neurofilament staining in the cerebral cortex of autoimmune mice. *Brain Res.* **529,** 202–207.

Sherman, G. F., Stone, L. V., Denenerg, V. H., and Beier, D. R. (1994). A genetic analysis of neocortical ectopias in New Zealand black autoimmune mice. *Neuroreport* **5,** 721–724.

Sherrington, C. (1906). *The Integrative Action of the Nervous System.* Yale University Press, New Haven.

Shi, Y., Hata, A., Lo, R. S., Massague, J., and Pavletich, N. P. (1997). A structural basis for mutational inactivation of the tumour suppressor Smad4. *Nature* **388,** 87–93.

Shimada, T., Yamazaki, H., Mimura, M., Wakamiya, N., Ueng, Y. F., Guengerich, F. P., and Inui, Y. (1996). Characterization of microsomal cytochrome P450 enzymes involved in the oxidation of xenobiotic chemicals in human fetal liver and adult lungs. *Drug Metab. Dispos.* **24,** 515–522.

Shimamura, K., and Rubenstein, J. L. (1997). Inductive interactions direct early regionalization of the mouse forebrain. *Development* **124,** 2709–2718.

Shioiri, T., Oshitani, Y., Kato, T., Murashita, J., Hamakawa, H., Inubushi, T., Nagata, T., and Takahashi, S. (1996). Prevalence of cavum septum pellucidum detected by MRI in patients with bipolar disorder, major depression and schizophrenia. *Psychol. Med.* **26,** 431–434.

Shirasaki, R., Katsumata, R., and Murakami, F. (1998). Change in chemoattractant responsiveness of developing axons at an intermediate target. *Science* **279,** 105–107.

Sidman, R. L., and Rakic, P. (1973). Neuronal migration, with special reference to developing human brain: a review. *Brain Res.* **62,** 1–35.

Simmons, D. L., and Kasper, C. B. (1989). Quantitation of mRNAs specific for the mixed-function oxidase system in rat liver and extrahepatic tissues during development. *Arch. Biochem. Biophys.* **271,** 10–20.

Simone, C., Derewlany, L. O., Oskamp, M., Johnson, D., Knie, B., and Koren, G. (1994). Acetylcholinesterase and butyrylcholinesterase activity in the human term placenta: implications for fetal cocaine exposure. *J. Lab. Clin. Med.* **123,** 400–406.

Simonoff, E., Bolton, P., and Rutter, M. (1996). Mental retardation: genetic findings, clinical implications and research agenda. *J. Child Psychol. Psychiatry* **37,** 259–280.

Sinha, A. K., Pickard, M. R., Kim, K. D., Ahmed, M. T., al Yatama, F., Evans, I. M., and Elkins, R. P. (1994). Perturbation of thyroid hormone homeostasis in the adult and brain function. *Acta Med. Austriaca.* **21,** 35–43.

Sirois, J. E., and Atchison, W. D. (1996). Effects of mercurials on ligand- and voltage-gated ion channels: a review. *Neurotoxicology* **17,** 63–84.

Snow, D. M., Lemmon, V., Carrino, D. A., Caplan, A. I., and Silver, J. (1990). Sulfated proteoglycans in astroglial barriers inhibit neurite outgrowth *in vitro*. *Exp. Neurol.* **109,** 111–130.

Sohmer, H., and Freeman, S. (1996). The importance of thyroid hormone for auditory development in the fetus and neonate. *Audiol. Neurootol.* **1,** 137–147.

Song, X., Seidler, F. J., Saleh, J. L., Zhang, J., Padilla, S., and Slotkin, T. A. (1997). Cellular mechanisms for developmental toxicity of chlorpyrifos: targeting the adenylyl cyclase signaling cascade. *Toxicol. Appl. Pharmacol.* **145,** 158–174.

Sperry, R. W. (1963). Chemoaffinity in the orderly growth of nerve fiber patterns and connections. *Proc. Natl. Acad. Sci. U.S.A.* **50,** 703–710.

Spindler, K. D. (1997). Interactions between steroid hormones and the nervous system. *Neurotoxicology* **18,** 745–754.

Sporns, O. (1997). Variation and selection in neural function. *Trends Neurosci.* **20,** 290–291.

Sretavan, D. W., Shatz, C. J., and Stryker, M. P. (1988). Modification of retinal ganglion cell axon morphology by prenatal infusion of tetrodotoxin. *Nature* **336,** 468–471.

Staddon, J. M., and Rubin, L. L. (1996). Cell adhesion, cell junctions and the blood–brain barrier. *Curr. Opin. Neurobiol.* **6,** 622–627.

Stanfield, B. B. (1992). The development of the corticospinal projection. *Prog. Neurobiol.* **38,** 169–202.

Standfield, B. B., O'Leary, D. D., and Fricks, C. (1982). Selective collateral elimination in early postnatal development restricts cortical distribution of rat pyramidal tract neurones. *Nature* **298,** 371–373.

Stefan, M. D., and Murray, R. M. (1997). Schizophrenia: developmental disturbance of brain and mind? *Acta Paediatr. Suppl.* **422,** 112–116.

Stein, S. A., Adams, P. M., Shanklin, D. R., Mihailoff, G. A., and Palnitkar, M. B. (1991). Thyroid hormone control of brain and motor development: molecular, neuroanatomical, and behavioral studies. *Adv. Exp. Med. Biol.* **299,** 47–105.

Steinlin, M. I., Nadal, D., Eich, G. F., Martin, E., and Boltshauser, E. J. (1996). Late intrauterine cytomegalovirus infection: clinical and neuroimaging findings. *Pediatr. Neurol.* **15,** 249–253.

Stonestreet, B. S., Patlak, C. S., Pettigrew, K. D., Reilly, C. B., and Cserr, H. F. (1996). Ontogeny of blood–brain barrier function in ovine fetuses, lambs, and adults. *Am. J. Physiol.* **271,** R1594–601.

Stoop, R., and Poo, M. M. (1996). Synaptic modulation by neurotrophic factors: differential and synergistic effects of brain-derived neurotrophic factor and ciliary neurotrophic factor. *J. Neurosci.* **16,** 3256–3264.

Stratakis, C. A., and Chrousos, G. P. (1995). Neuroendocrinology and pathophysiology of the stress system. *Ann. N.Y. Acad. Sci.* **771,** 1–18.

Strobel, H. W., Geng, J., Kawashima, H., and Wang, H. (1997). Cytochrome P450-dependent diotransformation of drugs and other xenobiotic substrates in neural tissue. *Drug Metab. Rev.* **29,** 1079–1105.

Struzynska, L., Walski, M., Gadamski, R., Dabrowska-Bouta, B., and Rafalowska, U. (1997). Lead-induced abnormalities in blood-brain barrier permeability in experimental chronic toxicity. *Mol. Chem. Neuropathol.* **31,** 207–224.

Sturrock, R. R. (1981). A quantitative and morphological study of vascularisation of the developing mouse spinal cord. *J. Anat.* **132,** 203–221.

Sundstrom, R., Conradi, N. G., and Sourander, P. (1984). Vulnerability to lead in protein-deprived suckling rats. *Acta Neuropathol. (Berl)* **62,** 276–283.

Sundstrom, R., Muntzing, K., Kalimo, H., and Sourander, P. (1985). Changes in the integrity of the blood–brain barrier in suckling rats with low dose lead encephalopathy. *Acta Neuropathol. (Berl)* **68,** 1–9.

Sutoo, D. (1992). Disturbances of brain function by exogenous cadmium. In *The Vulnerable Brain and Environmental Risks. Vol. 1. Malnutrition and Hazard Assessment.* R. L. Isaacson, and K. F. Jensen, Eds. Plenum Press, New York, pp. 281–299.

Suzuki, A., Chang, C., Yingling, J. M., Wang, X. F., and Hemmati-Brivanlou, A. (1997). Smad5 induces ventral fates in *Xenopus* embryo. *Dev. Biol.* **184,** 402–405.

Swedo, S. E., Leonard, H. L., and Kiessling, L. S. (1994). Speculations on antineuronal antibody-mediated neuropsychiatric disorders of childhood. *Pediatrics* **93,** 323–326.

Swedo, S. E., Leonard, H. L., Mittleman, B. B., Allen, A. J., Rapoport, J. L., Dow, S. P., Kanter, M. E., Chapman, F., and Zabriskie, J. (1997). Identification of children with pediatric autoimmune neuropsychiatric disorders associated with streptococcal infections by a marker associated with rheumatic fever. *Am. J. Psychiatry* **154,** 110–112.

Symons, A. M., Turcan, R. G., and Parke, D. V. (1982). Hepatic microsomal drug metabolism in the pregnant rat. *Xenobiotica* **12,** 365–374.

Takashima, S. (1997). Down syndrome. *Curr. Opin. Neurol.* **10,** 148–152.

Takei, N., Murray, G., O'Callaghan, E., Sham, P. C., Glover, G., and Murray, R. M. (1995). Prenatal exposure to influenza epidemics and risk of mental retardation. *Eur. Arch. Psychiatry Clin. Neurosci.* **245,** 255–259.

Tan, X. X., and Costa, L. G. (1995). Postnatal development of muscarinic receptor–stimulated phosphoinositide metabolism in mouse cerebral cortex: sensitivity to ethanol. *Dev. Brain Res.* **86,** 348–353.

Tánabe, Y., and Jessell, T. M. (1996). Diversity and pattern in the developing spinal cord. *Science* **274,** 1115–1123.

Tanaka, E., and Sabry, J. (1995). Making the connection: cytoskeletal rearrangements during growth cone guidance. *Cell* **83,** 171–176.

Tatsuta, T., Naito, M., Mikami, K., and Tsuruo, T. (1994). Enhanced expression by the brain matrix of P-glycoprotein in brain capillary endothelial cells. *Cell Growth Differ.* **5,** 1145–1152.

Taylor, C. G., Nagy, L. E., and Bray, T. M. (1996). Nutritional and normonal regulation of glutathione homeostasis. *Curr. Top. Cell Regul.* **34,** 189–208.

Tessier-Lavigne, M., and Goodman, C. S. (1996). The molecular biology of axon guidance. *Science* **274,** 1123–1133.

Tilson, H. A., and Harry, G. J. (1992). Developmental neurotoxicology of polychlorinated biphenyls and related compounds. In *The Vulnerable Brain and Environmental Risks. Vol. 3. Toxins in Air and Water.* R. L. Isaacson and K. F. Jensen, Eds. Plenum, New York, pp. 267–280.

Tsuang, M. T., and Faraone, S. V. (1995). The case for heterogeneity in the etiology of schizophrenia. *Schizophr. Res.* **17,** 161–175.

Tucker, R. P. (1990). The roles of microtubule-associated proteins in brain morphogenesis: a review. *Brain Res. Rev.* **15,** 101–120.

Tuomanen, E. (1996). Entry of pathogens into the central nervous system. *FEMS Microbiol. Rev.* **18,** 289–299.

Umesono, K., Giguere, V., Glass, C. K., Rosenfeld, M. G., and Evans, R. M. (1988). Retinoic acid and thyroid hormone induce gene expression through a common responsive element. *Nature* **336,** 262–265.

Uyemura, K., Ason, H., Yazaki, T., and Takeda, Y. (1996). Cell-adhesion proteins of the immunoglobulin superfamily in the nervous system. *Essays Biochem.* **31,** 37–48.

Valenzuela, D. M., Economides, A. N., Rojas, E., Lamb, T. M., Nunez, L., Jones, P., Ip, N.Y., Espinosa, R. III, Brannan, C. I., Gilbert, D. J., Copeland, N. G., Jenkins, N. A., Le Beau, M. M., Harland, R. M., and Yancopoulos, G. D. (1995). Identification of mammalian noggin and its expression in the adult nervous system. *J. Neurosci.* **15,** 6077–6084.

Vallee, R. B., and Sheetz, M. P. (1996). Targeting of motor proteins. *Science* **271,** 1539–1544.

van den Heuvel, S., and Harlow, E. (1993). Dinstinct roles for cyclin-dependent kinases in cell cycle control. *Science* **262,** 2050–2054.

Van Der Loos, H., and Woolsey, T. A. (1973). Somatosensory cortex: structural alterations following early injury to sense organs. *Science* **179,** 395–398.

van Leeuwen, F. W., de Kleijn, P. V., van der Hurk, H. H., Neubauer, A., Sonnemans, M. A., Sluijs, J. A., Koycu, S., Ramdjielal, R. D., Salehi, A., Martens, G. J., Grosveld, F. G., Burbach, J. P., and Hol, E. M. (1998). Frameshift mutants of b-amyloid precursor protein and ubiguitin-B in Alzheimer's and Down patients. *Science* **279,** 242–247.

VanBerkum, M. F., and Goodman, C. S. (1995). Targeted disruption of Ca(2+)-calmodulin signaling in *Drosophila* growth cones leads to stalls in axon extension and errors in axon guidance. *Neuron* **14,** 43–56.

Vannucci, S. J., Seaman, L. B., Brucklacher, R. M., and Vannucci, R. C. (1994). Glucose transport in developing rat brain: glucose transporter proteins, rate constants and cerebral glucose utilization. *Mol. Cell Biochem.* **140,** 177–184.

Vannucci, S. J., Willing, L. B., and Vannucci, R. C. (1993). Developmental expression of glucose transporters, GLUT1 and GLUT3, in postnatal rat brain. *Adv. Exp. Med. Biol.* **331,** 3–7.

Villringer, A., and Dirnagl, U. (1997). *Optical Imaging of Brain Function and Metabolism 2. Physiological Basis and Comparison to Other Functional Neuroimaging Methods.* Plenum Press, New York.

von Euler, G. (1992). Toluene, and dopaminergic transmission. In *The Vulnerable Brain and Environmental Risks, Vol. 3. Toxins in Air and Water.* R. L. Isaacson and K. F. Jensen, Eds. Plenum, New York, pp. 301–322.

Vorbrodt, A. W., and Trowbridge, R. S. (1993). Aluminum-induced alteration of surface anionic sites in cultured brain microvascular endothelial cells. *Acta Neuropathol. (Berl)* **86,** 371–377.

Vorbrodt, A. W., Dobrogowska, D. H., and Lossinsky, A. S. (1994). Ultracytochemical studies of the effects of aluminum on the blood–brain barrier of mice. *J. Histochem. Cytochem.* **42,** 203–212.

Vorbrodt, A. W., Trowbridge, R. S., and Dobrogowska, D. H. (1994). Cytochemical study of the effect of aluminum on cultured brain microvascular endothelial cells. *Histochem. J.* **26,** 119–126.

Waddington, J. L. (1993). Schizophrenia: developmental neuroscience and pathology. *Lancet* **341,** 531–538.

Walker, R., Rahim, A., and Parke, D. V. (1973). The effects of antioxidants on enzyme induction in developing and ageing rats. *Proc. R. Soc. Med.* **66,** 780

Wang, W., and Dow, K. E. (1997). Differential regulation of neuronal proteoglycans by activation of excitatory amino acid receptors. *Neuroreport* **8,** 659–663.

Wang, Y., Jones, F. S., Krushel, L. A., and Edelman, G. M. (1996). Embryonic expression patterns of the neural cell adhesion molecule gene are regulated by homeodomain binding sites. *Proc. Natl. Acad. Sci. U.S.A.* **93,** 1892–1896.

Ware, R. A., Burkholder, P. M., and Chang, L. W. (1975). Ultrastructural changes in renal proximal tubules after chronic organic and inorganic mercury intoxication. *Environ. Res.* **10,** 121–140.

Ware, R. A., Chang, L. W., and Burkholder, P. M. (1974). An ultrastructural study on the blood–brain barrier dysfunction following mercury intoxication. *Acta Neuropathol. (Berl)* **30,** 211–224.

Wasteneys, G. O., Cadrin, M., Reuhl, K. R., and Brown, D. L. (1988). The effects of methylmercury on the cytoskeleton of murine embryonal carcinoma cells. *Cell Biol. Toxicol.* **4,** 41–60.

Waters, J. C., and Salmons, E. D. (1996). Cytoskeleton: a catastrophic kinesin. *Curr. Biol.* **6,** 361–363.

Weglage, J., Pietsch, M., Funders, B., Koch, H. G., and Ullrich, K. (1996). Deficits in selective and sustained attention processes in early treated children with phenylketonuria—result of impaired frontal lobe functions? *Eur. J. Pediatr.* **155,** 200–204.

Weinberger, D. R. (1997). The biological basis of schizophrenia: new directions. *J. Clin. Psychiatry* **58** (Suppl 10), 22–27.

Weiss, P. (1939). *Principles of Development.* Hebry Holt, New York.

Weisshaar, B., and Matus, A. I. (1993). Microtubule-associated protein 2 and the organization of cellular microtubules. *J. Neurocytol.* **22,** 727–734.

Wells, P. G., and Winn, L. M. (1996). Biochemical toxicology of chemical teratogenesis. *Crit. Rev. Biochem. Mol. Biol.* **31,** 1–40.

Wharton, B. A., and Scott, P. H. (1996). Distinctive aspects of metabolism and nutrition in infancy. *Clin. Biochem.* **29,** 419–428.

Weisel, T. N., and Hubel, D. H. (1965). Comparison of the effects of unilateral and bilateral eye closure on cortical unit response in kittens. *J. Neurophysiol.* **28,** 1029–1040.

Willis, J. W. D., and Grossman, R. G. (1977). *Medical Neurobiology Second Edition.* C.V. Mosby, St. Louis.

Wilson, P. A., Lagna, G., Suzuki, A., and Hemmati-Brivanlou, A. (1997). Concentration-dependent patterning of the *Xenopus* ectoderm by BMP4 and its signal transducer Smad1. *Development* **124,** 3177–3184.

Wilson, S. W., Placzek, M., and Furley, A. (1993). Border disputes: do boundaries play a role in growth-cone guidance. *Trends Neurosci.* **16,** 316–323.

Windmill, K. F., McKinnon, R. A., Zhu, X., Gaedigk, A., Grant, D. M., and McManus, M. E. (1997). The role of xenobiotic metabolizing enzymes in arylamine toxicity and carcinogenesis: functional and localization studies. *Mutat. Res.* **376,** 153–160.

Windrem, M. S., Jan de Beur, S., and Finlay, B. L. (1988). Control of cell number in the developing neocortex. II. Effects of corpus callosum section. *Dev. Brain Res.* **43,** 13–22.

Winneke, G. (1996). Inoranic lead as a developmental neurotoxicant: some basic issues and the Dusseldorf experience. *Neurotoxicology* **17,** 565–580.

Wolf, S. S., and Weinberger, D. R. (1996). Schizophrenia: a new frontier in developmental neurobiology. *Isr. J. Med. Sci.* **32,** 51–55.

Wong, C. C., Warsh, J. J., Sibony, D., and Li, P. P. (1994). Differential ontogenetic appearance and regulation of stimulatory G protein isoforms in rat cerebral cortex by thyroid hormone deficiency. *Dev. Brain Res.* **79,** 136–139.

Wong, E. V., Kenwick, S., Willems, P., and Lemmon, V. (1995). Mutations in the cell adhesion molecule L1 cause mental retardation. *Trends Neurosci.* **18,** 168–172.

Wong, E. V., Schaefer, A. W., Landreth, G., and Lemmon, V. (1996). Casein kinase II phosphorylates the neural cell adhesion molecule L1. *J. Neurochem.* **66,** 779–786.

Wong, E. V., Schaefer, A. W., Landreth, G., Lemmon, V. (1996). Involvement of p90rsk in neurite outgrowth mediated by the cell adhesion molecule L1. *J. Biol. Chem.* **271,** 18217–18223.

Wood, K. A., and Youle, R. J. (1995). The role of free radicals and p53 in neuron apoiptosis *in vivo. J. Neurosci.* **15,** 5851–5857.

Woolsey, T. A. (1984). The postnatal development and plasticity of the somatosensory system. In *Neuronal Growth and Plasticity.* M. Kuno, Ed. Japan Scientific Societies Press, Tokoyo, pp. 241–259.

Woolsey, T. A. (1996). Barrels: 25 years later. *Somatosens. Mot. Res.* **13,** 181–186.

Woolsey, T. A., and Van Der Loos, H. (1970). The structural organization of layer IV in the somatosensory region (SI) of the mouse cerebral cortex: the description of a cortical field composed of discrete cytoarchitectonic units. *Brain Res.* **17,** 205–242.

Xu, J., and Ling, E. A. (1994). Studies of the ultrastructure and permeability of the blood–brain barrier in the developing corpus callosum in postnatal rat brain using electron dense tracers. *J. Anat.* **184,** 227–237.

Yamada, T., Hosokawa, M., Satoh, T., Moroo, I., Takahashi, M., Akatsu, H., and Yamamoto, T. (1994). Immunohistochemistry with an antibody to human liver carboxylesterase in human brain tissues. *Brain Res.* **658,** 163–167.

Yamasaki, M., Thompson, P., and Lemmon, V. (1997). CRASH syndrome: mutations in L1CAM correlate with severity of the disease. *Neuropediatrics* **28,** 175–178.

Yang, X., Diehl, A. M., and Wand, G. S. (1996). Ethanol exposure alters the phosphorylation of cyclic AMP responsive element binding protein and cyclic AMP responsive element binding activity in rat cerebellum. *J. Pharmacol. Exp. Ther.* **278,** 338–346.

Yip, R. K., and Chang, L. W. (1982). Protective effects of vitamin E on methylmercury toxicity in the dorsal root ganglia. *Environ. Res.* **28,** 84–95.

Yu, D., Zhang, L., Eisele, J. L., Bertrand, D., Changeux, J. P., and Weight, F. F. (1996). Ethanol inhibition of nicotinic acetylcholine type alpha 7 receptors involves the amino-terminal domain of the receptor. *Mol. Pharmacol.* **50,** 1010–1016.

Yuan, Y., and Atchison, W. D. (1997). Action of methylmercury on GABA(A) receptor-mediated inhibitory synaptic transmission is primarily responsible for its early stimulatory effects on hippocampal CA1 excitatory synaptic transmission. *J. Pharmacol. Exp. Ther.* **282,** 64–73.

Zangar, R. C., Hernandez, M., and Novak, R. F. (1997). Posttranscriptional elevation of cytochrome P450 3A expression. *Biochem. Biophys. Res. Commun.* **231,** 203–205.

Zeman, W., and Innes, J. (1963). *Craigie's Neuroanatomy of the Rat.* Academic Press, New York.

Zheng, J. Q., Poo, M. M., and Connor, J. A. (1996a). Calcium and chemotropic turning of nerve growth cones. *Perspect. Dev. Neurobiol.* **4,** 205–213.

Zheng, J. Q., Wan, J. J., and Poo, M. M. (1996b). Essential role of filopodia in chemotropic turning of nerve growth cone induced by a glutamate gradient. *J. Neurosci.* **16,** 1140–1149.

Zheng, J. Q., Zheng, Z., and Poo, M. (1994). Long-range signaling in growing neurons after local elevation of cyclic AMP–dependent activity. *J. Cell Biol.* **127,** 1693–1701.

Zisch, A. H., and Pasquale, E. B. (1997). The Eph family: a multitude of receptors that mediate cell recognition signals. *Cell Tissue Res.* **290,** 217–226.

Zlokovic, B. V. (1995). Cerebrovascular permeability to peptides: manipulations of transport systems at the blood–brain barrier. *Pharm. Res.* **12,** 1395–1406.

Zorumski, C. F., and Olney, J. W. (1992). Acute and chronic neurodegenerative disorders produced by dietary excitotoxins. In *The Vulnerable Brain and Environmental Risks. Vol. 2. Toxins in Food.* R. L. Isaacson and K. F. Jensen, Eds. Plenum Press, New York, pp. 273–292.

CHAPTER

2

Cadherin Cell Adhesion Molecules in Normal and Abnormal Neural Development

GERALD B. GRUNWALD
Department of Pathology, Anatomy, and Cell Biology
Thomas Jefferson University
Philadelphia, Pennsylvania 19107

*Abbreviations: CNS, central nervous system; PCR, polymerase chain reaction.

I. Introduction

A. *The Cadherin Cell Adhesion Molecule Family and Neural Development*

The human brain is the most highly complex structure in the known universe. Its assembly is the result of the carefully orchestrated interactions of some 10^{11} neurons and 10 times as many glial cells, with perhaps on the order of 10^{14} neuronal interconnections forming its functional circuitry. Much has been learned about neural development from studies of animal model systems. The development of vertebrate nervous systems is characterized by the initial segregation of subpopulations of embryonic cells that, during the process of neurulation, form the neuroepithelial primordium of the central nervous system (CNS) as well as the migratory cells of the neural crest that will form the peripheral nervous system. The subsequent migrations and reorganizations of neurons and glia to form the characteristic arrangements of cellular cortices, nuclei, and ganglia, as well as the further elaboration by neurons within these tissues of nerve fiber tracts and synaptic circuits, are regulated by the actions of a variety of genetically programmed as well as epigenetically modulated processes. These developmental events may be subject to perturbation by physical and chemical environmental influences, with resultant subtle or gross abnormalities of structure or function recognizable as neuronal birth defects. A rational approach to the prevention and possible treatment of such neuroteratogenic insults ultimately requires a mechanistic understanding of normal neural developmental processes as well as the manner in which these processes are subject to environmental modification. Experimental analysis of nervous system development has revealed that one major class of mechanisms guiding neural development is through cell–cell and cell–matrix adhesion and signaling processes mediated by cell surface proteins called *cell adhesion molecules.* Three major types of such molecules have been identified and are grouped into the cadherin, integrin, and immunoglobulin families (Bixby, 1992; Buck, 1992; Grunwald, 1992; Hynes and Lander, 1992; Brummendorf and Rathjen, 1996; Gumbiner, 1996). This review is focused on progress in our understanding of the role of cadherins in normal and abnormal neural development.

B. Identification and Structural Classification of Neural Cadherins

Cadherins are cell surface transmembrane glycoproteins that mediate calcium-dependent homophilic intercellular adhesion and signaling interactions. Cadherins have been identified in a wide range of tissues and vertebrate species, including humans, as well as more recently in invertebrates. Identification of the cadherins and the elucidation of their fundamental structural and functional characteristics culminated over a century of research investigating the cellular basis of morphogenesis (see Grunwald, 1991, for a historic review). A number of additional recent reviews are available that discuss in detail aspects of cadherin structure and function from a variety of perspectives (Takeichi, 1988, 1990, 1991, 1992, 1995; Geiger and Ayalon, 1992; Kemler, 1992; Grunwald, 1993, 1996a,b; Gumbiner, 1996; Huber *et al.,* 1996; Marrs and Nelson, 1996). The cadherins represent an ancient, large, and diverse family of proteins that play a fundamental role both in guiding cell interactions during development and in the maintenance of adult tissues. A number of classification schemes for the diverse family of cadherins have been suggested (Pouliot, 1992; Tanihara *et al.,* 1994). Based on structural considerations the cadherins have been divided into three major groups, including the classic type I and type II cadherins as well as the nonclassic cadherins. Although cadherins contribute to the development and organization of many tissues, an especially large number are expressed in the nervous system (Table 1). The specific role of cadherins in aspects of normal and abnormal nervous system development has been the subject of several reviews (Ranscht, 1991; Redies, 1995; Redies and Takeichi, 1996; Grunwald, 1996b; Reuhl and Grunwald, 1997) and is discussed further later in this chapter.

The structure of a classic cadherin family member is illustrated in Fig. 1. Classic type I cadherins typically consist of extracellular, transmembrane, and cytoplasmic domains, forming transmembrane glycoproteins of about 130 kDa molecular weight. The extracellular domain is composed of five repeated subdomains that contain several conserved cadherin sequence motifs. Classic type II cadherins differ somewhat from the type I cadherins in the presence or pattern of certain cadherin sequence motifs. The nonclassic cadherins include a variety of more distantly related proteins, including the desmosomal cadherins and protocadherins. Several critical functional regions that have been identified in the extracellular domain of cadherins include calcium-binding regions and cell adhesion regions. The latter is especially important in determining the selectivity of the homophilic adhesive interactions that characterize cadherin function. Recent crystallographic studies of cadherin extracellular domains have begun to yield additional insights into the organization of these functionally critical portions of the molecules that seem to function as dimers, as shown in Fig. 1 (Overduin *et al.,* 1995; Shapiro *et al.,* 1995; Nagar *et al.,* 1996). The cytoplasmic domain of classic cadherins contains highly conserved sequences that are sites of interaction with a variety of cytoskeletal and signaling proteins (Geiger *et al.,* 1995; Yamada and Geiger, 1997). The cadherins had been shown to colocalize with actin, and a group of three proteins called α-, β-, and γ-catenin were originally identified in cadherin complexes providing cytoskeletal linkage (Ozawa and Kemler, 1992; Aberle *et al.,* 1996). A cadherin typically associates with either β- or γ-catenin, which then associates with α-catenin and α-actinin, and finally the actin cytoskeleton. α-Catenin also exists in a neural specific isoform (Hirano *et al.,* 1992; Uchida *et al.,* 1994). Additional proteins, such as $p120^{cas}$ (Reynolds *et al.,* 1994), have also been found in association with the cytoplasmic domain of cadherins, and several members of the growing list of proteins that form part of the cadherin complex are illustrated in Fig. 1. Optimal function of cadherins in the formation of cell junctions, and modulation of cadherin function, is dependent on complex formation with these proteins.

TABLE 1 Cadherins of the Nervous System

Classic Type I	
N-cadherin (gp130/4.8, A-CAM, NcalCAM)	Grunwald *et al.* (1982); Hatta and Takeichi (1986); Crittenden *et al.* (1987)
R-cadherin (cadherin-4)	Inuzuka *et al.* (1991a,b)
E-cadherin (L-CAM, uvomorulin)	Gallin *et al.* (1983): Shimamura and Takeichi (1992)
B-cadherin (P-cadherin, XB/U-cadherin)	Napolitano *et al.* (1991); Murphy-Erdosh *et al.* (1994)
M-cadherin	Rose *et al.* (1995)
T-cadherin (cadherin-13)	Ranscht and Dours-Zimmerman (1991)
PB-cadherin (OB-cadherin)	Sugimoto *et al.* (1996)
XmN-cadherin	Tashior *et al.* (1995)
Br-cadherin	Selig *et al.* (1995)
Classic Type II	
Cadherin-5 (VE-cadherin)	Suzuki *et al.* (1991); Tanihara *et al.* (1994)
Cadherin-6 (K-cadherin)	Suzuki *et al.* (1991); Nakagawa and Takeichi (1995)
Cadherin-7	Suzuki *et al.* (1991); Nakagawa and Takeichi (1995)
Cadherin-8	Tanihara *et al.* (1994); Korematsu and Redies (1997)
Cadherin-9	Suzuki *et al.* (1991)
Cadherin-10	Suzuki *et al.* (1991)
Cadherin-11 (OB-cadherin)	Suzuki *et al.* (1991); Kimura *et al.* (1995)
Cadherin-12	Suzuki *et al.* (1991)
F-cadherin	Espeseth *et al.* (1995)
Cadherin-14	Shibata *et al.* (1997)
DE-cadherin	Oda *et al.* (1994)
Nonclassic	
Protocadherins	Sano *et al.* (1993)
ret protooncogene	Pachnis *et al.* (1993)

II. Expression Patterns of Cadherins during Neural Development

A. Classic Type I Cadherins of the Nervous System

For many years the cadherin family consisted of three classic type I cadherins, E-, P-, and N-cadherin, which were identified in biochemical, immunologic, and functional studies. The cadherin family was so named on the realization that these proteins constituted a related group of calcium-dependent adhesion molecules (Yoshida-Noro *et al.,* 1984). N-cadherin (neural) was the first neural cadherin to be identified and functionally and structurally characterized (Grunwald *et al.,* 1982; Hatta and Takeichi, 1986; Crittenden *et al.,* 1987; Hatta *et al.,* 1988). N-cadherin has a broad distribution in the developing nervous system and is expressed from neurulation through synaptogenesis (Hatta and Takeichi, 1986; Hatta *et al.,* 1987; Matsunaga *et al.,* 1988b; Lagunowich and Grunwald, 1989; Miyatani *et al.,* 1989; Lagunowich *et al.,* 1992; Redies and Takeichi, 1993a,b). N-cadherin is first expressed during neurulation, is expressed at highest levels during further neural histogenesis, and is reduced in mature neural tissues, when other cadherins are up-regulated or expressed in more discrete patterns. Later in development N-cadherin remains expressed at high levels in choroid plexus and ependyma, and is also associated with synaptogenesis as discussed later in this chapter. E-cadherin as well as other adhesion molecules are down-regulated during this phase of neurulation, and as discussed later cadherin switching is an important aspect of the morphogenetic program of development. In an analysis of both chick and mouse embryos during neurulation, as the neural plate undergoes morphogenetic shape changes to form the neural tube, it was found that a switch occurs between expression of occludin, the principle transmembrane glycoprotein of tight junctions, and N-cadherin (Aaku-Saraste *et al.,* 1996). Interestingly, the submembrane protein ZO-1, which is normally found in association with occludin at tight junctions, remains colocalized with N-cadherin, raising the possibility of its association with the latter.

Although most early studies of N-cadherin were done in the chick, the availability of new probes has permitted analysis of other species. In the mouse, as in the chick, N-cadherin expression in the developing nervous system is widespread, for example not corresponding during early development in the mouse to the delineation of any neuromeres, as opposed to most other cadherins

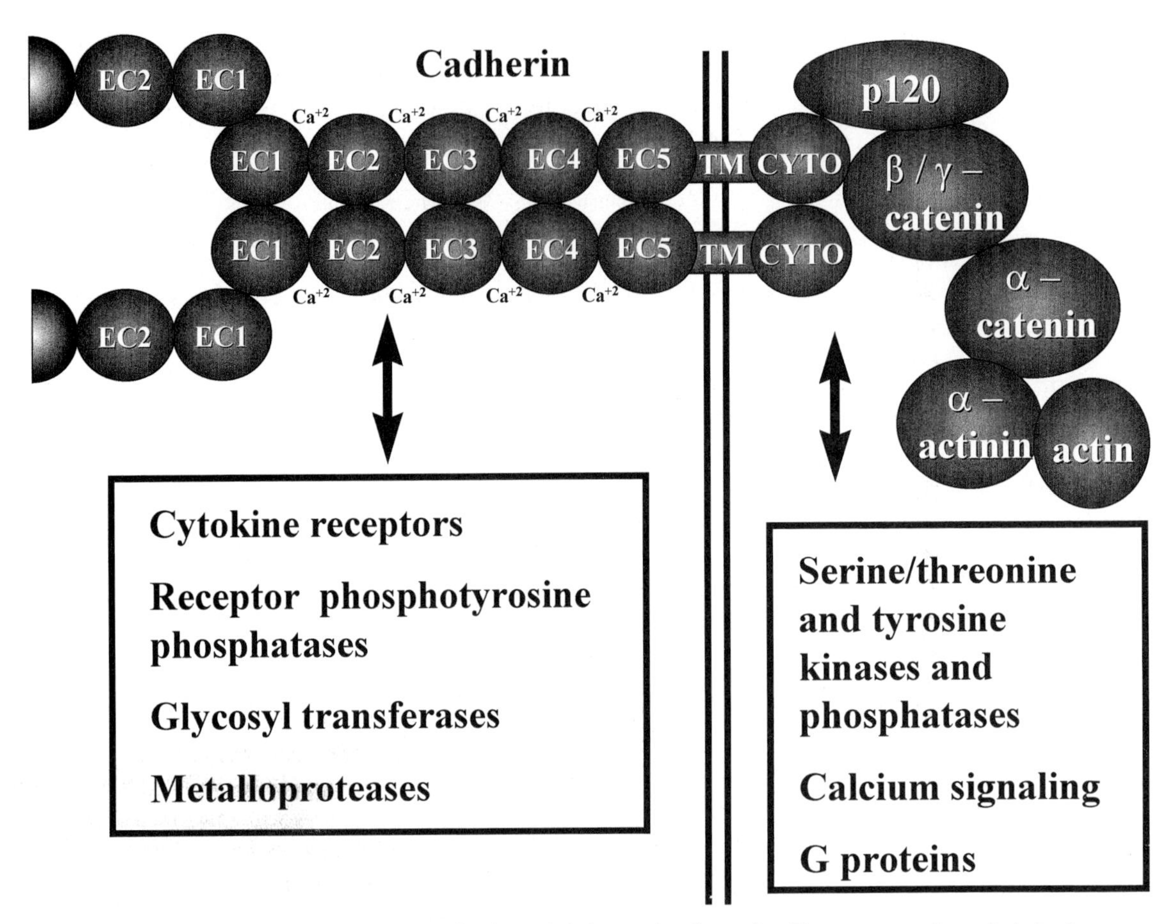

FIGURE 1 The structure of classic cadherins and their associated proteins. The structure of a typical classic cadherin is illustrated with the subdomains shown including the five extracellular repeats, EC1–EC5, the transmembrane region (TM), and the cytoplasmic domain (CYTO). Also indicated are the calcium-binding regions between the subdomains of the extracellular region. Structural studies suggest that cadherins may exist in the plane of the membrane as dimers, and that intercellular interactions between cadherins are mediated principally through the EC1 domains. The cytoplasmic domain interacts directly with either β- or γ-catenin, and also with p120. β- or γ-catenin in turn bind to α-catenin, which ultimately links the complex to the actin cytoskeleton via α-actinin. The stoichiometry of these associations is still under investigation. The figure also indicates that cadherins interact with a variety of extracellular signaling proteins and enzymes that modulate cadherin expression and function. Intracellular cadherin signaling also involves a variety of kinases and phosphatases that modulate cadherin and catenin function, as well as calcium and G protein–mediated pathways. See text for details.

that have been examined (Matsunami and Takeichi, 1995). cDNA library screening of Zebrafish led to the identification of the Zebrafish homolog of N-cadherin, which was shown to be structurally and functionally similar to its avian and mammalian homologs (Bitzur *et al.,* 1994). Expression patterns indicate a role in gastrulation movements of cells and a high level of expression in all developing nervous tissues, as in other species.

R-cadherin (retinal) was first identified in chick and mouse neural tissues and its structure is very similar to that of N-cadherin (Inuzuka *et al.,* 1991a,b; Matsunami and Takeichi, 1995). Its pattern of expression has been studied during chick embryo nervous system development (Redies *et al.,* 1992, 1993; Redies and Takeichi, 1993b; Ganzler and Redies, 1995; Arndt and Redies, 1996). R-cadherin is expressed later in neurogenesis than N-cadherin, and these two cadherins were shown to have generally complementary spatial and temporal patterns of expression. In a study that focused on the diencephalon during the first few days of chick embryo brain development, R-cadherin expression was observed initially in a pattern composed of various patchy and striped regions of the primitive brain that correspond somewhat to neuromeric divisions. Among the boundaries distinguished by R-cadherin-positive and -negative boundaries are the dorsal and ventral thalamus and the mesencephalon and metencephalon. Within the hypothalamus, the percentage of R-cadherin expressing cells increased with further distance from the ventricular surface, and these cells had morphologies consistent with cells migrating along glial processes. When R- and N-cadherin-expressing cells were compared, it was found that in several instances initially mixed populations became segregated into R- and N-cadherin positive zones, respectively. During later stages of brain development, R-cadherin was found in a number of gray and white matter regions. Although less ubiquitous than N-cadherin, R-cadherin nevertheless occurs in a wide range of neural structures. Indeed, a detailed neuroanatomic analysis in the developing chick embryo brain using immunocytochemical detection resulted in the identification of at least 20 distinct cortical and subcortical telencephalic regions, at least 20 diencephalic nuclei, a dozen distinct mesencephalic nuclei, the cerebellum, more than 20 rhombencephalic nuclei, several cranial nerves, and more than a dozen major central fiber tracts that express significant amounts of R-cadherin in all or a portion of their extent (Arndt and Redies, 1996). The detailed pattern was not as ubiquitous as that of N-cadherin, because R-cadherin showed clear distinctions between certain nuclear groups, in many cases distinguished subsets of neurons within nuclear groups, divided the cerebellum into characteristic parasagital striped regions, and delineated distinct fiber tracts that could in some cases be correlated with R-cadherin–positive targets. Although this pattern is clearly rather complex, and the function of R-cadherin within these distinct regions remains to be elucidated, the already demonstrated role of R-cadherin in cell adhesion and axon growth suggests that it makes a particularly important contribution to the establishment of those regions identified in this study.

To extend these studies into mammalian systems, generation of a new set of monoclonal antibodies specific for murine R- and N-cadherin permitted their analysis during development of the mouse nervous system (Matsunami and Takeichi, 1995). During early development of the neural tube, R-cadherin expression was first detected with a patchy expression pattern in the brain. However, the highest level of expression was found in the ventral thalamus, with significant expression also in the telencephalic cortex, with the highest level in the ventricular zone. Sharp boundaries were delineated between the dorsal and ventral thalamus, between the cortical and ganglionic regions of the telencephalon, and between the positive pretectum and negative mesencephalon. Interestingly, this pattern was relatively coextensive with that found for patterning genes Pax-6 and Dlx-2, suggesting a possible regulatory role for these factors. A particularly sharp and high level expression of R-cadherin was observed in the primitive trochlear nuclei. As development progressed, R-cadherin–positive regions became further subdivided into more complex organizations, although these were not further described. In general, the mammalian and avian patterns were similar.

E-cadherin (epithelial) was one of the initially identified classic cadherins but was not recognized as a neural cadherin for several years (Gallin *et al.*, 1983). However, more recent studies have demonstrated that E-cadherin is reexpressed in a limited fashion in specific localized regions of the developing brain and peripheral nervous system (Shimamura and Takeichi, 1992; Shimamura *et al.*, 1992). Regions of the CNS in which reexpression of E-cadherin could be detected included portions of the diencephalon (dorsal thalamus), and mesencephalon (pretectum), where sharp borders were formed (Matsunami and Takeichi, 1995).

A novel role for E-cadherin in the nervous system was revealed when it was demonstrated that this protein is expressed in a highly localized fashion among Schwann cells (Fannon *et al.*, 1995). E-cadherin was restricted to the paranodal regions, the Schmidt–Lanterman clefts, and the inner and outer mesaxons, but was not present in compact myelin. E-cadherin was shown to be associated with adherens junctions in these membrane regions, which link parts of the same Schwann cell to itself, helping to stabilize the myelinating loops. No staining was observed between adjacent Schwann cells or between Schwann cells and axons. These observations raise the possibility that E-cadherin plays a critical role in the establishment and maintenance of peripheral nerve myelination. E-cadherin is also expressed among satellite cells of the peripheral nervous system, as well as at Schwann cell–axon contacts among regenerating nerves following injury, indicating that such E-cadherin–mediated Schwann cell–axon interactions may be involved during myelination even if not observed in mature peripheral nerves (Hasegawa *et al.*, 1996).

B-cadherin (brain) was first identified in the chick embryo nervous system (Napolitano *et al.*, 1991; Murphy-Erdosh *et al.*, 1994). During neurulation, B-cadherin was detected in the neural tube, especially in the lumen and floor plate, showing a general anterior–posterior gradient of expression as well with no expression caudal to the rhombencephalon. However, no expression was seen later in neural development except in the circumventricular organs such as choroid plexus and in the neural-tube derived retinal pigment epithelium, all of which interestingly share barrier functions. It has been suggested on the basis of structural similarities and related expression patterns that avian B-cadherin is closely related to mammalian P-cadherin and *Xenopus* XB/U-cadherin (Redies and Muller,

1994). Thus, despite is name, B-cadherin is actually expressed very little by neurons.

The most unique of the classic cadherins, T-cadherin (truncated) was also first identified in the chick embryo nervous system (Ranscht and Dours-Zimmerman, 1991; Vestal and Ranscht, 1992). At 95 kDa molecular weight, T-cadherin is unusual for the classic cadherins in that it lacks a transmembrane and cytoplasmic domain, but rather is linked to the cell surface by a glycosyl phosphatidylinositol linkage, and also occurs in more than one RNA splice variant (Sacristan *et al.,* 1993). Despite the unusual linkage of T-cadherins, they can mediate calcium-dependent adhesions between cells. In spite of this adhesive function, the expression patterns of T-cadherin suggested it could play a role as a negative regulator of cell adhesion and neurite growth in establishing exclusion zones, which direct neural crest cell migration and limb innervation by motor neurons (Ranscht and Bronner-Fraser, 1991; Fredette and Ranscht, 1994). Experimental evidence in support of this is discussed later in this chapter.

Rat cDNA cloning led to the identification of a novel type I cadherin called PB-cadherin due to its preferential expression in the pituitary gland and brain (Sugimoto *et al.,* 1996). Within the brain the highest expression was observed in specific layers of the olfactory bulb and cerebellum, and its structure suggests it may be related to mouse OB-cadherin. Interestingly, two mRNA splice variants were identified that give rise to proteins with identical extracellular domains but distinct cytoplasmic domains, both of which were shown to be expressed as functional adhesion molecules. The classic cadherins generally do not exhibit splice variants, which in addition to T-cadherin have been described for the desmosomal members of the cadherin gene family, but has been described as well for OB-cadherin.

Identification of additional cadherins has continued in a number of vertebrate and invertebrate species that provide important model systems for experimental analysis of neural development. *Xenopus* XmN-cadherin is closely related but not identical to R- and N-cadherin (Tashiro *et al.,* 1995). Although maternal transcripts showed wide distribution, zygotic and adult expression was found to be mainly in neural tissues. Whereas little information about this cadherin is available, it appears to be structurally most highly related to R-cadherin. Another recently reported neural cadherin was given the name Br-cadherin because of its expression pattern, which was found to be specific to the brain and was upregulated late in development (Selig *et al.,* 1995, 1997). Although perhaps structurally related to K-cadherin and cadherin-12, little is known about this cadherin's structure or function. Interestingly, this cadherin was first identified genomically as a possible candidate gene for spinal muscular atrophy, when a curious repetitive distribution of a pseudogene apparently derived from Br-cadherin was observed in the region of this disease locus.

M-cadherin (muscle) was initially identified and so named due to its expression in muscle cells. However, it has been shown to have a unique distribution in cerebellar granule layer synaptic glomeruli of adult mice. These glomeruli are intricate synaptic complexes formed between mossy fibers and granule cell dendrites (Rose *et al.,* 1995). M-cadherin was localized here along with α-, β-, and γ-catenins but was not observed in other regions of the brain. In a subsequent study that examined M-cadherin expression during cerebellar development, M-cadherin was not found during early stages of development during cell migration and lamination, but only after the establishment of glomerular synapses (Bahjaoui-Bouhaddi *et al.,* 1997). Such expression was absent from mutant mice in which cerebellar granule cell development is disturbed.

B. Classic Type II Cadherins of the Nervous System

Several variant members of the cadherin family were originally identified by homology-based degenerate oligonucleotide reverse transcription polymerase chain reaction (PCR) amplification and cloning from rat and human brain tissues (Suzuki *et al.,* 1991; Tanihara *et al.,* 1994). As discussed previously, these cadherins, numbered 5–11, have come to be collectively referred to as classic type II cadherins due to their distinct structural organization relative to the type I cadherins. Until recently, only fragmentary information has been available on their structure and function.

Cadherin-6 is a type II cadherin that was originally identified by polymerase chain reaction amplification of brain cDNAs, but about which little is known. Cadherin-6 provides another example of a cadherin that displays a very dynamic pattern of expression during various stages of development, suggestive of a role in regionalization of the nervous system. A study of the expression of cadherin-6 indicated that it may play a role in the early establishment of segmental brain organization of both the CNS as well as neural crest derivatives (Inoue *et al.,* 1997). Cadherin-6 expression correlated with subdivision of the neural tube into distinct regions during neural tube closure. Expression was initiated in the neural plate in the presumptive hindbrain region extending caudally to the junction with the primitive spinal cord, with sharp borders seen at both ends. Expression was subsequently detected in the forebrain extending caudally to the junction with the midbrain, with the latter showing very localized expression only at the neural folds. Later in development, expression was more highly restricted to the neural crest as well as to a narrow band in the hindbrain that corresponded to rhombomere 6. Subsequent to neural tube closure, cells identifiable as neural crest cells remained positive for cadherin-6 as the cells migrated away from the neural tube and condensed to form cranial ganglia, although this expression was transient. Within the CNS, cadherin-6 continued to be expressed along the dorsal midline, the presumptive retina and optic stalk, and other parts of the diencephalon, midbrain, and hindbrain. A preliminary report states that later in development, cadherin-6 is associated with specific functional systems, such as those of the auditory pathway (Takeichi *et al.,* 1997).

The cloning of the chick homologs of cadherins 6 and 7 has shed additional light on their nervous system function. In developing chick embryos cadherin-6B and cadherin-7 were identified in the developing neural tube by PCR amplification of early chick embryo cDNAs and *in situ* hybridization using the resulting clones (Nakagawa and Takeichi, 1995). Chick cadherin-6B is similar in structure to mammalian cadherin-6 (also called K-cadherin) and is likely to be its interspecies homolog. Interestingly, cadherin-6B has a unique and highly specific distribution and is found in those regions of the neural fold that give rise to the neural crest. Later during neurulation, when neural crest cells begin their migration, cadherin-6B is down-regulated while cadherin-7 is up-regulated. Interestingly, only a subpopulation of the migrating neural crest cells expressed cadherin-7. These cells remained close to the neural tube in regions associated with the developing dorsal and ventral roots and spinal nerves, with a similar pattern found in association with cranial nerve roots. The pattern is consistent with expression by Schwann cells.

Although cadherin-8 was also originally identified by PCR of human cDNA (Tanihara *et al.,* 1994), its expression pattern was analyzed in detail in the developing mouse nervous system following cloning of the murine homolog (Korematsu and Redies, 1997). Cadherin-8 expression delineates the borders of several distinct regions of the primitive CNS including the ventral versus dorsal thalamus, and the caudato–pallial angle versus the intermediate zone of the lateral ganglionic eminence. Later in development, cadherin-8 expression distinguishes between different nuclear groups within the ventral thalamus and is expressed in the developing facial nucleus and pontine nuclei of the hindbrain.

Because cadherins have been generally identified as epithelial adhesion molecules, it was thus of interest that cadherin-11, a type II cadherin, has been associated principally with adhesion among mesenchymal cells (Kimura *et al.,* 1995). Interestingly, this cadherin has also been found to have a unique expression pattern during nervous system development (Kimura *et al.,* 1996). Within the nervous system expression was first detected in the neural folds and this expression persisted in the roof plate after neural tube closure, from the epithalamus of the diencephalon caudal throughout the spinal cord. Expression was also observed to demarcate divisions between the mesencephalon and metencephalon, and the basal and alar regions of the diencephalon. As brain segmentation continued, localized expression was observed in certain thalamic nuclei, in the mammillary bodies, and among telencephalic derivatives in the cortical plate as well as in forming putamen, septal, and striatal nuclei. Further developmental changes exhibited expression in association with formation of certain midbrain nuclei including the oculomotor, red, and substantia nigra. Additional nuclei expressing this marker were located more caudally and included the trigeminal, facial, hypoglossal, and inferior olive. In general, these patterns were grouped by the authors into three patterns: (1) distinguishing the ventricular versus parenchymal regions; (2) distinguishing boundaries between major subdivisions; and (3) distinguishing specific nuclei. A fourth pattern that emerges is that connecting neuronal groups into functional systems.

In *Xenopus* embryos a novel cadherin called F-cadherin was so named due its preferential expression at the flexures of the developing neural tube (Espeseth *et al.,* 1995). Another cadherin found in the developing nervous system was identified by PCR amplification of Zebrafish embryo neural tube cDNAs and was termed VN-cadherin due to its localization in the ventral nervous system (Franklin and Sargent, 1996). Expression was detected at sequential stages of development first in the neuroectoderm, then the neural keel, and later in two ventral neural tube regions lateral to the floor plate. Later stages of development found VN-cadherin localized to several discrete regions of the CNS. It is too early in the analysis of this cadherin to know much of its function in the developing CNS. However, among the known cadherins, VN-cadherin is most closely related to the type II cadherin-11 previously identified in mammalian nervous systems. Finally, a human neural cadherin designated cadherin-14 was identified by protein interaction cloning using β-catenin as the expression library probe (Shibata *et al.,* 1997). This type II cadherin is most closely related to cadherins 11 and 12, although it remains to be further studied.

The extent of conservation of cadherin structure and function between vertebrates and invertebrates is well illustrated by the DE-cadherin from *Drosophila* (Oda *et al.,* 1994). Studies have shown that DE-cadherin is required for cellular reorganizations that occur during development of the neuroectodermal epithelium of the *Drosophila* embryo (Tepass *et al.,* 1996). Although morphogenesis of the insect nervous system does not quite parallel that of the vertebrate nervous system, the general principles of cadherin action during *Drosophila* neurogenesis seem to hold insofar as DE-cadherin is required for stabilization of epithelia including the neuroectoderm, the protein is down-regulated during delamination of neural precursors that is similar to what occurs during vertebrate neural crest development, and interaction with catenins follows the classic cadherin pattern.

C. Nonclassic Cadherins of the Nervous System

A large group of cadherin gene superfamily members identified by PCR homology amplification and cloning was named the protocadherins due to their widespread expression in many species including human, mouse, *Xenopus, Drosophila,* and *Caenorhabditis elegans* (Sano *et al.,* 1993; Suzuki, 1996). As opposed to the classic cadherins, protocadherins have more than the typical five extracellular domain repeats and also have highly variable cytoplasmic domains. Among the protocadherins, Pcdh2 has been best characterized in the nervous system. Initial studies of the cerebellum showed it to be expressed on the cell bodies but not the axons of Purkinje cells (Obata *et al.,* 1995). Although protocadherins have been shown to have a weak adhesive function, apparently due to the absence of classic cadherin cytoplasmic domains, chimeras containing cytoplasmic sequences of E-cadherin showed much stronger adhesion, confirming the adhesive function of the protocadherin extracellular domain.

Other variant members of the cadherin family expressed in nervous tissue include the c-ret protooncogene, which consists of cadherin-like sequences in its extracellular domain and encodes a tyrosine kinase in its intracellular domain (Schneider, 1992; Iwamoto *et al.,* 1993; Pachnis *et al.,* 1993). There is as yet no evidence that the desmosomal members of the cadherin family, which differ from the classic cadherins mainly in the structure of their cytoplasmic domains, are expressed in neural tissues (Buxton and Magee, 1992; Garrod *et al.,* 1996).

III. *In Vitro* Studies of Neural Cadherin Function

A. Neural and Glial Cell Adhesion

A fundamental action of the cadherins is to mediate neuronal cell adhesion, and the proper expression and function of N-cadherin is critical for early neural development. *In vitro* neural cell and organ culture studies showed that antibodies directed against N-cadherin inhibit neural cell–cell adhesion and inhibit neural histogenesis (Grunwald *et al.,* 1982; Hatta and Takeichi, 1986; Matsunaga *et al.,* 1988b). More recently, the possible role of differential cadherin expression in the establishment of discrete regions of the developing CNS has been tested using *in vitro* aggregation studies of mouse embryonic brain cells (Matsunami and Takeichi, 1995). In order to test the possibility that differential R-cadherin expression could contribute to the establishment of distinct neuronal compartments, cells derived from brain tissues that were respectively positive or negative for R-cadherin expression were mixed in aggregation cultures. When the resulting aggregates that formed were stained to detect R-cadherin, it was found that the cells had sorted out within the aggregates into regions of R-cadherin positive and negative cells. This segregation was calcium dependent, indicating that it was cadherin dependent, because aggregates formed under conditions in which calcium-independent adhesion molecules such as N-CAM were active formed randomly mixed aggregates with respect to R-cadherin expression. This finding indicates that cell sorting is the predominant result of cadherin function even when other families of adhesion molecules are present. Similar results were found comparing E-cadherin positive and negative cells, and this sorting was inhibited in the presence of anti–E-cadherin antibodies.

In addition to neuronal adhesion, there is ample evidence that cadherins participate in neural–glial cell adhesion as well. Glial cells support neurite growth in an N-cadherin–dependent fashion (Bixby *et al.,* 1988; Neugebauer *et al.,* 1988; Drazba and Lemmon, 1990), and glial cells themselves use N-cadherin as an adhesive substrate (Payne and Lemmon, 1993). A study of N-cadherin expression in the normal and regenerating sciatic nerve of the chicken demonstrated N-cadherin expression on both axonal and glial cell surfaces (Shibuya *et al.,* 1995). Among unmyelinated axons, N-cadherin was localized at regions of fiber–Schwann cell contact. However, among myelinated axons, no N-cadherin was found at the neural–glial junction, but was found instead at the mesaxon, which is the region of the Schwann cell where self contacts are made between adjacent wrappings of the myelin sheath and where E-cadherin is also found. Sites of nerve injury induced by nerve ligation were found to contain a high level of N-cadherin proximal to the ligation site, which was determined to be associated with growth cones of regenerating axons. In a subsequent study that examined the localization of the neuronal isoform of α-catenin, αN-catenin, it was found that Schwann cells did not express this isoform, nor did myelinated axons, although expression was detected in unmyelinated axons (Shibuya *et al.,* 1996). Both axon types expressed this protein following injury. Curiously, most of the immunoreactivity was located in the cytoplasm and not at the plasma membrane. A study of oligodendrocytes identified N-cadherin expression at all stages of their development, including precursor cells as well as immature and mature oligodendrocytes (Payne *et al.,* 1996). Because N-cadherin had been shown to serve as a strong adhesive substrate for oligodendrocytes, but did not significantly promote the migration of their precursors, it was concluded that this protein is more likely involved in their interaction with neurons during myelination rather than in their migration during development.

Although cadherins were first identified and have been studied primarily in the context of embryonic development, it soon became apparent that many cadherins are also expressed in mature adult tissues. This implies that cadherins continue to exert an influence on the function and stability of these tissues, and in particular that misexpression or malfunction of cadherins could contribute to the onset or progression of various disease states. The most obvious such relationship could be in cancers and other proliferative disorders in which normal cell–cell interactions are disturbed. Indeed, among carcinomas loss of E-cadherin function has been correlated with enhanced invasiveness and malignancy (Takeichi, 1993). N-cadherin expression has been examined in studies by Schiffman and Grunwald (1992) and by Shinoura *et al.* (1995) of different human nervous systems tumor cells. A study comparing several lines of retinoblastoma cells demonstrated that they functionally expression calcium-dependent adhesions and also expressed N-cadherin (Schiffman and Grunwald, 1992). The three cell lines studied exhibited different degrees of calcium-dependent adhesion and distinct morphologies that correlated with the level of N-cadherin expression. N-cadherin and α-catenin expression were also analyzed in a second study among a number of astrocytomas and glioblastomas (Shinoura *et al.,* 1995). No significant differences in expression were found between expression of either protein and the adhesiveness, migration, or invasiveness of the tumors and cell lines studied. Thus the question remains whether N-cadherin and E-cadherin play similar metastasis-suppressing roles in neural or glial versus epithelial cancers.

B. Neurite Growth and Synaptogenesis

Growth cones at the tips of elongating nerve cell processes act as sensory transducing structures that play a critical role in determining the trajectories of fiber growth. Cadherins are present on growth cones and have been shown to actively influence signaling pathways in developing nerve terminals and to affect neurite growth and morphology. Cells expressing surface N-cadherin naturally or following transfection, or the purified protein itself, have been shown to support neurite outgrowth (Matsunaga *et al.,* 1988a; Neugebauer *et al.,* 1988; Bixby and Zhang, 1990; Drazba and Lemmon, 1990; Doherty *et al.,* 1991a,b; Paradies and Grunwald, 1993). The response of neurons to N-cadherin can be cell-type selective (Kjlavin *et al.,* 1994). Growth cones respond uniquely to N-cadherin as compared to other adhesion molecules (Bixby and Zhang, 1990; Payne *et al.,* 1992). As retinal ganglion cell growth cones were observed *in vitro* to cross from a laminin substrate to an N-cadherin substrate, or vice versa, a preference was observed for remaining on or moving onto N-cadherin in both the timing and direction of neurite growth (Burden-Gulley *et al.,* 1995). These transitions were also marked by significant changes in growth cone size and shape. In a subsequent study, these changes were shown to likely result from adhesion molecule-induced changes in cytoskeletal organization of microfilaments and microtubules (Burden-Gulley and Lemmon, 1996).

A study of the distribution of a large array of adhesion molecules during development of the chick optic tectum demonstrated that many such molecules, including N-cadherin, exhibit dynamic temporal and spatial distributions during retinotectal innervation (Yamagata *et al.,* 1995). Although at early developmental stages N-cadherin showed a wide distribution as found in earlier studies, during later stages N-cadherin was found to be concentrated in those tectal laminae receiving retinal input. Examination of the distribution of N-cadherin at the electron microscopic level showed it to be expressed at synaptic contacts. Interestingly, enucleation of embryos prevented this accumulation, suggesting that retinal ganglion cell axon innervation was responsible for this accumulation of N-cadherin in the tectum. In a subsequent study of visual system development, an *in vitro* retinal–tectal coculture system was used to examine the effects of anti–N-cadherin antibodies on innervation (Inoue and Sanes, 1997). Overall neurite growth as well as secondary branching within retinorecipient laminae was reduced, and the laminar distribution itself was selectively perturbed.

In addition to cadherin expression at synapses, the associated cytoskeletal catenins have also been shown to localize at these junctions (Uchida *et al.,* 1996). In a study of both mouse and chick brains, both αN-catenin and β-catenin were found to be localized in adherens junctions at synaptic contacts, but were excluded from the synaptic neurotransmitter release zones. The timing and distributions of expression suggested a widespread role in the establishment and stabilization of synapse in widely varied regions of the brain. When the distributions of N- and E-cadherin were compared at the light and electron microscopic level in the brain of the adult mouse, it was found that both could be identified at synapses in various regions of the brain (Fannon and Colman, 1996). These two cadherins were found to have a widespread punctate distribution, and were particularly noted to be associated with synaptic regions of the cerebellum and hippocampus. The distributions of E- and N-cadherin were generally mutually exclusive, which was interpreted to suggest that these and other cadherins could contribute to the formation, sorting, and stabilization of distinct synaptic contacts during development.

It has become increasingly clear that both positive attractive as well as negative repulsive forces play a role in neural interactions and axonal pathfinding during development. Although the vast majority of cadherin-mediated interactions fall into the positive pool, studies have confirmed the inference from expression patterns discussed earlier that at least one cadherin, the truncated, lipid-linked T-cadherin, can be the source of repulsive stimuli that may help guide nerves along their trajectories during embryonic development (Fredette *et al.,* 1996). This was demonstrated in studies where neurite growth was compared on a variety on cellular substrata composed of control cells or cell transfected to express T- or N-cadherin. Although the control cells were permissive for neurite outgrowth, and a similar pattern was seen on N-cadherin–expressing cells, the T-cadherin expressing cells were found to inhibit such growth significantly. A recombinantly expressed soluble form of the extracellular domain of T-cadherin was also found to be an effective inhibitor of neurite growth, presumably through binding to cell surface T-cadherin. Interestingly, the inhibitory effects on motor neurons showed an age-dependence, which paralleled changes in T-cadherin expression, which had previously been observed *in vivo* during chick embryo neurogenesis (Fredette and Ranscht, 1994). This specificity was also cell-type dependent, as in addition to inhibitory effects on motor axon elongation, such effects were also observed with neurons isolated from sympathetic and dorsal root ganglia, but had no effect on cells from ciliary ganglia or the dorsal spinal cord.

C. Cell Signaling and Regulation of Cadherin Expression and Function

A number of studies have investigated mechanisms by which the dynamic spatial and temporal patterns of cadherin expression are regulated during neural development. Expression of N-cadherin in developing neural tissues has been shown to be regulated by multiple genetic and epigenetic mechanisms, including modulation of mRNA levels, cytokine-mediated signaling, and cell surface proteolytic turnover. Although the genomic organization of N-cadherin has been analyzed, little is yet known about the complexities of its regulation at this level, but mRNA levels are known to be regulated during neural development (Walsh *et al.,* 1990; Miyatani *et al.,* 1992; Roark *et al.,* 1992). *In vivo* and *in vitro* studies have shown that N-cadherin is cleaved at the cell surface by a metalloprotease to yield a soluble fragment of N-cadherin called NCAD90. NCAD90 retains biological activity and can mediate adhesion and neurite growth if sequestered to the substrate (Paradies and Grunwald, 1993).

Mutations in Hoxa-1 expression led to changes in cadherin-6 expression, although it is not clear whether this is a direct effect (Inoue *et al.,* 1997). Expression of both E-cadherin and the neural isoform of α-catenin were found to be regulated in part by the signaling molecule Wnt-1 (Shimamura *et al.,* 1994). Studying expression of E-cadherin as well as the αE-, αN-, β-, and γ-catenins in normal and Wnt-deficient mice, it was found that among the catenins, the only correlation was a positive one between αN-catenin and Wnt-1 expression. However, the patterns of expression suggested negative control of E-cadherin expression by Wnt-1. This is of interest because vertebrate Wnt-1 is related to the *Drosophila* wingless (wg) gene, which is known to use the *Drosophila* armadillo protein as a downstream effector molecule, which is related to the vertebrate β-catenin protein, which itself has signaling functions (McRea *et al.,* 1993; Huber *et al.,* 1996).

The function of N-cadherin has been shown to depend in part on the action of a variety of neuronal second messenger signaling systems. These include serine/threonine and tyrosine kinases, calcium channels, and G-protein pathways (Bixby and Jhabvala, 1990; Doherty *et al.,* 1991a,b; Saffel *et al.,* 1992; Williams *et al.,* 1994a). These studies have demonstrated bidirectional signaling between N-cadherin at the cell surface and the intracellular milieu that seems especially important in regulation of neurite growth (Walsh and Doherty, 1996). Modulation of these pathways by various pharmacological agents has been found to affect neurite growth and, in turn, cadherin-mediated interactions also result in intracellular changes. Direct evidence for N-cadherin–induced calcium signaling in neurons was obtained using the NCAD90 soluble form of N-cadherin as a stimulus (Bixby *et al.,* 1994). When cultured chick embryo ciliary ganglion neurons were treated with NCAD90, this was found to induce an increase in intracellular calcium levels in cell bodies and growth cones, and opposite effects in glial cells. Pharmacological modulation indicated this signaling involved voltage-sensitive calcium channels. Aspects of cadherin-mediated signaling may also be mediated in part by interaction with the fibroblast growth factor receptor (Williams *et al.,* 1994b). Morphologic and biochemical responses of neurons to N-cadherin were shown to be similar to those induced by receptor activation, and a common sequence motif shared by these two proteins was identified. However, it remains to be seen whether this is a direct or an indirect effect.

N-cadherin is a phosphoprotein whose posttranslational modification is developmentally regulated (Lagunowich and Grunwald, 1991). In developing retinal neurons the turnover of N-cadherin to yield NCAD90 was found to be modulated by treatments that altered the state of N-cadherin phosphorylation (Lee *et al.,* 1997). Enhanced tyrosine phosphorylation was correlated with enhanced turnover of N-cadherin. N-cadherin was shown to be multiply phosphorylated on distinct peptides containing both serine and tyrosine residues, indicating it was the target of multiple kinase/phosphatase pathways. Signaling pathways that modulate N-cadherin were investigated in a study of N-cadherin expression among migrating neural crest cells (Monier-Gavelle and Duband, 1995). Although N-cadherin was found to be expressed *in vitro* to a greater extent than indicated by *in vivo* studies, the cultured neural crest cells did not form close contacts. This was found to be correlated with the lack of cytoskeletal association of the N-cadherin that was expressed. Inhibition of tyrosine or serine/threonine kinases generally resulted in an increase in cell–cell associations among the neural crest cells, although this could not be correlated with a change in overall phosphorylation of N-cadherin or catenins. However, an increase in the level of N-cadherin and its localization to cell–cell junctions was noted. In a subsequent study it was demonstrated that inhibition of integrin-mediated interactions resulted in an up-regulation of N-cadherin activity among migrating neural crest cells (Monier-Gavelle and Duband, 1997). Enhanced intracellular calcium levels were found to reduce N-cadherin–mediated adhesion, whereas increased serine/threonine phosphorylation of α- and β-catenins had the opposite effect. These results indicate that cross talk between distinct cell adhesion molecule systems mediated through a calcium- and kinase/phosphatase-dependent signaling pathway may be important in maintaining normal cell function.

The interaction of cadherins with complex signaling pathways is supported by a number of proteins that have been found to be associated with cadherins. Cadherins in brain are complexed with the receptor protein tyrosine phosphatase PTPμ (Brady-Kalnay *et al.,* 1995). Modulation of cadherin function as assessed both by cell adhesion and neurite growth is modulated by the interaction of the proteoglycan neurocan with a cell surface N-acetylgalactosaminyltransferase, which had itself been shown to exist previously in a complex with N-cadherin (Balsamo and Lilien, 1990; Balsamo *et al.,* 1995). This modulation involves signaling as detected by altered tyrosine phosphorylation of β-catenin, which is in turn regulated by an N-cadherin-associated tyrosine nonreceptor phosphatase PTP1B (Balsamo *et al.,* 1996). Neurocan may thus serve as the endogenous ligand for this enzyme as antibodies directed against the enzyme were previously shown to also inhibit N-cadherin function through uncoupling from the cytoskeleton (Balsamo *et al.,* 1991; Gaya-Gonzales *et al.* 1991).

IV. *In Vivo* Studies of Neural Cadherin Function

A. *Immunologic Perturbation Studies*

The fundamental role of cadherins in neural cell adhesion, migration, and neurite growth indicated by the previously mentioned *in vitro* studies has been supported by studies carried out in intact animal systems. Embryos microinjected with anti–N-cadherin antibodies results in abnormal neural tube development and neural crest cell migration, further supporting a role for N-cadherin in neurulation and nerve cell migration (Bronner-Fraser *et al.,* 1992). The adult songbird brain provides a unique model system for studies of this early phase of neural differentiation because generation, migration, and incorporation of neurons into neural circuits continues into adulthood. This model system was used advantageously to analyze the role of N-cadherin during neurogenesis in the avian higher vocal center (Barami *et al.,* 1994). N-cadherin was found to be localized to the subventricular zone and generally absent from the parenchyma. However, down-regulation of N-cadherin expression was found to be coincident with migration of neuronal precursors from the subependymal zone into the parenchyma. A role for down-regulation of N-cadherin in this process was tested by treatment of subventricular zone organ culture explants with function-blocking anti–N-cadherin antibodies that was found to increase the rate of neuronal migration. Subsequent interactions of migrating neurons and glial cells were found to be dependent on adhesions mediated by members of the immunoglobulin family.

Innervation patterns in both the central and peripheral nervous systems have been shown to involve cadherin function. A role for N-cadherin in establishment of the retinotectal projection was demonstrated by injection of function-blocking antibodies into the optic tract and optic tectum of *Xenopus* embryos (Stone and Sakaguchi, 1996). However, injection of anti–N-cadherin antibodies alone had little if any effect on development of the optic projection. Similarly, injection of antibodies against β1-integrins also had little effect. However, when both antibodies were co-injected, a profound effect was observed on the retinotectal pathway, with ectopic projections and defasciculation of axons. This synergistic effect may suggest either that redundancies exist or that cooperative interactions between these various adhesive systems exist *in vivo*. The function of N-cadherin in the projection of peripheral axons was studied by injection of function-blocking antibodies into the hindlimbs of chick embryos (Honig and Rutishauser, 1996). Although the gross pattern of innervation was normal, an effect was observed when the detailed pattern of sensory projections was analyzed. This appeared to result from an effect on fasciculation of processes, which reduced the ability of growing axons to follow pre-existing pathways along earlier generated peripheral nerves, which affected growth cone choices while navigating through the plexus on entering the limb bud.

B. Molecular Genetic Perturbation Studies

An approach to analysis of cadherin function during development has been the generation of mutant mice that lack expression of the N-cadherin gene (Radice *et al.*, 1997). These mice were generated by introduction of a mutant gene into embryonic stem cells followed by generation of a knockout line of mutant mice. The resulting mice die as embryos on about the tenth day of development, principally due to cardiac defects. Thus, these mutant mice are not directly useful for studies of neural development subsequent to this stage. Interestingly, however, these mutant mice have provided further insights into the role of N-cadherin in early neurogenesis. In the most severely affected mice, the neural tube was abnormal by day 9 of development and was described as exhibiting abnormal undulations. Although this study supports an important role for N-cadherin in early neural development, the embryonic lethality due to cardiac defects precludes analysis of later stages of neural development in this system. In a sense it is surprising that the neural tube, although abnormal, developed to the extent that it did, because perturbation of N-cadherin as previously discussed results in severe neural tube defects. It is possible that in the case of a genetic knockout the embryo has compensatory mechanisms such as up-regulation of alternative cadherins that can to some extent substitute for N-cadherin.

Transgenic approaches using ectopic expression of normal or mutant dominant–negative cadherins have been taken to circumvent the problems associated with total gene knockouts. As previously discussed, N-cadherin is a transmembrane protein whose critical extracellular adhesive domains and intracellular cytoskeletal binding domains have been identified through site-directed mutagenesis and expression of truncated, nonfunctional cadherins lacking these domains. Interestingly, *in vivo* expression of such truncated cadherins has shown that they can function as dominant–negative inhibitors of cadherin function. Depending on whether the extracellular or intracellular domain has been deleted, such mutant proteins can serve as either type-specific cadherin or pan-cadherin inhibitors. If the extracellular domain is deleted, the resulting protein when expressed in a cell competes for catenin binding and blocks the function of other endogenous cadherins. If the intracellular domain is deleted, the resulting protein binds transcellularly, and even perhaps in a *cis*-fashion, to homotypic cadherins, but cannot complete the formation of stable bonds because it lacks a catenin-binding domain, thus blocking the function of a specific cadherin. Both such mutational approaches, as well as ectopic expression of normal N-cadherin, have been used to analyze cadherin function *in vivo*. Microinjection of mRNA encoding normal N-cadherin into early *Xenopus* embryos resulted in ectopic expression of N-cadherin causing aberrant neural tube development (Detrick *et al.*, 1990; Fujimori *et al.*, 1990). In Zebrafish, ectopic expression of N-cadherin via mRNA microinjection caused aberrant cellular aggregations and malformed embryos (Bitzur *et al.*, 1994). That such defects result from ectopic expression of normal N-cadherin indicates that the E- and N-cadherin switching that occurs during neurulation, and the appropriate compartmentation of expression, is critical to early neural tube development. Exogenous expression of either type of dominant–negative N-cadherin mutant caused abnormal neurulation (Kintner, 1992; Levine *et al.*, 1994). Preliminary studies using a similar dominant–negative N-cadherin mutation expressed under the control of a retinal photoreceptor–specific promoter indicate that localized N-cadherin inhibition can selectively perturb development of specific retinal layers (Hickman *et al.*, 1997).

A requirement for cadherin function in the outgrowth of retinal ganglion cell axons and dendrites was demonstrated *in vivo* using dominant–negative cadherins expressed in *Xenopus* embryo retinas (Riehl *et al.*, 1996). These experiments were done via lipofection of ocular tissues, advantageously resulting in expression of the mutant truncated cadherin lacking the extracellular domain, and also engineered to carry an epitope tag permitting identification of transfected cells in a limited number of cells against a normal background. Although control tagged cells exhibited normal morphologies, those cells carrying the mutant cadherin showed a lack of process outgrowth. These studies were the first to demonstrate an *in vivo* role for cadherins in neuronal process outgrowth. The results furthermore indicated some degree of cell-type specificity, because retinal ganglion cells were perturbed to a greater extent than other types of neurons examined. The effect was also specific for cadherin-mediated process outgrowth, because the inhibition when tested *in vitro* was found on cadherin substrates but not on laminin. Interestingly, comparison of the inhibitory effects of several distinct mutations lacking different regions of the cytoplasmic domain indicated that the catenin-binding domain may not be the critical element in neuronal process outgrowth, but rather that a more membrane-proximal domain could play the critical role.

V. Cadherins as Targets of Developmental Neurotoxicants

The critical role of cadherins in multiple aspects of neural development provides fertile ground for potential neurotoxicant action. The pathways through which adhesion molecule function could indeed be subject to such effects have been reviewed (Reuhl and Grunwald, 1997). Such effects could be either direct, as in inhibition of normal cadherin function via structural alterations, or indirect through modulation of one or more portions of cadherin and associated protein signaling pathways. One possible mechanism of direct perturbation could be through displacement of calcium ions by heavy metals such as lead or cadmium. Heavy metal neurotoxicants could mediate indirect effects through alterations in calcium channel function, calcium or other second messenger signaling, kinase or phosphatase function, or other cadherin-associated metalloenzymes such as metalloproteases. As previously discussed, these all represent known regulatory and signaling pathways with which the cadherins interact. Indeed, several studies have indicated that these compounds can effect cadherin expression or adhesion (Prozialeck and Niewenhuis, 1991; Chen and Hales, 1994; Lagunowich *et al.,* 1994). In the study by Lagunowich and co-workers (1994), chick embryo retinal neuron adhesion was assayed in the presence and absence of lead acetate. It was observed that formation of typical cadherin-dependent histotypic aggregation of cells was inhibited in the presence of lead. However, the nature of this effect and whether it

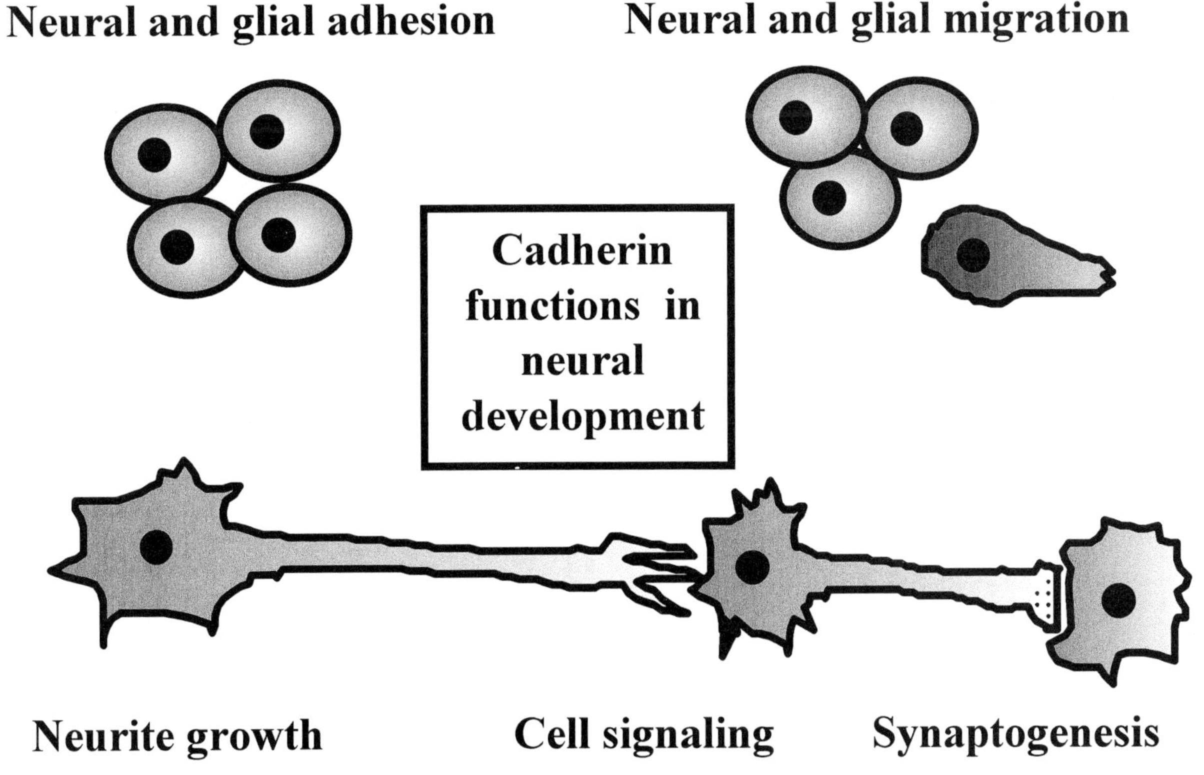

FIGURE 2 Functions of cadherins in neural development. Experimental evidence discussed in this review indicates that cadherins play important roles in a variety of events during neural development, including neural and glial cell adhesion and migration. Neurite growth and synaptogenesis also involve cadherins. Cadherin expression and function during these events is modulated by a number of cellular signaling pathways. See text for details.

resulted from direct or indirect perturbation of cadherin function remains to be elucidated.

VI. Conclusions and Future Directions

There now exists a wealth of data obtained from many *in vivo* and *in vitro* model systems that indicate that the cadherins play diverse and critical roles in all aspects of neural development, as summarized in Fig. 2. From the initial formation of the neural tube to the final establishment of synaptic circuitry, neural and glial adhesion, neurite growth, synapse formation, and associated recognition and signaling processes are controlled in part through the concerted action of many members of the cadherin family. Although much has been learned about these processes, there remain many important issues awaiting resolution. Why are there so many cadherins, and how do the type I and type II classic and nonclassic cadherins differ functionally between groups and within a group? What are the genetic and epigenetic mechanisms by which the dynamic spatial and temporal expression patterns of cadherins are regulated during development? Are one or more of the cadherins directly or indirectly subject to the actions of developmental neurotoxicants? If so, do resultant neuronal deficits arise from either the misexpression or malfunction of cadherins either directly or indirectly through the signaling pathways by which cadherin function is modulated? Future investigations into these issues should provide additional novel and important insights into the role of cadherins in both normal and abnormal nervous system development.

Acknowledgments

The author's work is supported by grants EY06658 and EY10965 from the National Institutes of Health.

References

Aaku-Saraste, E., Hellwig, A., and Huttner, W. B. (1996). Loss of occludin and functional tight junctions, but not ZO-1, during neural tube closure–remodeling of the neuroepithelium prior to neurogenesis. *Dev. Biol.* **180,** 664–679.

Aberle, H., Schwartz, H., and Kemler, R. (1996). Cadherin–catenin complex: protein interactions and their implications for cadherin function. *J. Cell. Biochem.* **61,** 514–523.

Arndt, K., and Redies, C. (1996). Restricted expression of R-cadherin by brain nuclei and neural circuits of the developing chicken brain. *J. Comp. Neurol.* **373,** 373–399.

Bahjaoui-Bouhaddi, M., Padilla, F., Nicolet, M., Cifuentes-Diaz, C., Fellman, D., and Mege, R. M. (1997). Localized deposition of M-cadherin in the glomeruli of the granular layer during postnatal development of the mouse cerebellum. *J. Comp. Neurol.* **378,** 180–195.

Balsamo, J., Ernst, H., Zanin, M. K. B., Hoffman, S., and Lilien, J. (1995). The interaction of the retina cell surface N-acetylgalactosminylphosphotransferase with an endogenous proteoglycan ligand results in inhibition of cadherin-mediated adhesion. *J. Cell Biol.* **129,** 1391–1401.

Balsamo, J., Leung, T., Ernst, H., Zanin, M. K. B., Hoffman, S., and Lilien, J. (1996). Regulated binding of a PTP1B-like phosphatase to N-cadherin: control of cadherin-mediated adhesion by dephosphorylation of β-catenin. *J. Cell Biol.* **134,** 801–813.

Balsamo, J., and Lilien, J. (1990). N-cadherin is stably associated with and is an acceptor for a cell surface N-acetylgalactosaminylphosphotransferase. *J. Biol. Chem.* **265,** 2923–2928.

Balsamo, J., Thiboldeaux, R., Swaminathan, N., and Lilien, J. (1991). Antibodies to the retina N-acetylgalactosaminyl–phosphotransferase modulate N-cadherin–mediated adhesion and uncouple the N-cadherin transferase complex from the actin-containing cytoskeleton. *J. Cell Biol.* **113,** 429–436.

Barami, K., Kirshcenbaum, N., Lemmon, V., and Goldman, S. A. (1994). N-cadherin and Ng-CAM/8D9 are involved serially in the migration of newly generated neurons into the adult songbird brain. *Neuron* **13,** 567–582.

Bitzur, S., Kam, S., and Geiger, B. (1994). Structure and distribution of N-cadherin in developing zebrafish embryos: morphogenetic effects of ectopic over-expression. *Dev. Dynamics* **201,** 121–136.

Bixby, J. L. (1992). Diversity of axonal growth–promoting receptors and regulation of their function. *Curr. Opin. Neurobiol.* **2,** 66–69.

Bixby, J. L., Grunwald, G. B., and Bookman, R. J. (1994). Ca^{2+} influx and neurite growth in response to purified N-cadherin and laminin. *J. Cell Biol.* **127,** 1461–1475.

Bixby, J. L., and Jhabvala, P. (1990). Extracellular matrix molecules and cell adhesion molecules induce neurites through different mechanisms. *J. Cell Biol.* **111,** 2725–2732.

Bixby, J. L., Lilien, J., and Reichardt, L. F. (1988). Identification of the major proteins that promote neuronal process outgrowth on Schwann cells *in vitro*. *J. Cell Biol.* **107,** 353–361.

Bixby, J. L., and Zhang, R. (1990). Purified N-cadherin is a potent substrate for the rapid induction of neurite outgrowth. *J. Cell Biol.* **110,** 1253–1260.

Brady-Kalnay, S. M., Rimm, D. L., and Tonks, N. K. (1995). Receptor protein tyrosine phosphatase PTPμ associates with cadherins and catenins *in vivo*. *J. Cell Biol.* **130,** 977–986.

Bronner-Fraser, M., Wolf, J. J., and Murray, B. A. (1992). Effects of antibodies against N-cadherin and N-CAM on the cranial neural crest and neural tube. *Dev. Biol.* **153,** 291–301.

Brummendorf, T., and Rathjen, F. G. (1996). Structure/function relationships of axon-associated adhesion molecules of the immunoglobulin family. *Curr. Opin. Neurobiol.* **6,** 584–593.

Buck, C. (1992). Immunoglobulin superfamily: structure, function and relationship to other receptor molecules. *Sem. Cell Biol.* **3,** 179–188.

Burden-Gulley, S. M., and Lemmon, V. (1996). L1, N-cadherin, and laminin induce distinct distribution patterns of cytoskeletal elements in growth cones. *Cell Motil. Cytoskel.* **35,** 1–23.

Burden-Gulley, S. M., Payne, H. R., and Lemmon, V. (1995). Growth cones are actively influenced by substrate-bound adhesion molecules. *J. Neurosci.* **15,** 4370–4381.

Buxton, R. S., and Magee, A. I. (1992). Structure and interactions of desmosomal and other cadherins. *Semin. Cell Biol.* **3,** 157–167.

Chen, B., and Hales, B. F. (1994). Cadmium-induced rat embryotoxicity *in vitro* is associated with an increased abundance of E-cadherin protein in the yolk sac. *Toxicol. Appli. Pharmacol.* **128,** 293–301.

Crittenden, S. L., Pratt, R. S., Cook, J. H., Balsamo, J., and Lilien, J. (1987). Immunologically unique and common domains within a family of proteins related to the retina $Ca2^+$-dependent cell adhesion molecule, NcalCAM. *Devel.* **101,** 729–740.

Detrick, R. J., Dickey, D., and Kintner, C. R. (1990). The effects of N-cadherin misexpression on morphogenesis in *Xenopus* embryos. *Neuron.* **4,** 493–506.

Doherty, P., Ashton, S. V., Moore, S. E., and Walsh, F. S. (1991a). Morphoregulatory activities of NCAM and N-cadherin can be accounted for by G protein–dependent activation of L- and N-type neuronal calcium channels. *Cell* **67,** 21–33.

Doherty, P., Rowett, L. H., Moore, S. E., Mann, D. A., and Walsh, F. S. (1991b). Neurite outgrowth in response to transfected N-CAM and N-cadherin reveals fundamental differences in neuronal responsiveness to CAMs. *Neuron* **6,** 247–258.

Drazba, J., and Lemmon, V. (1990). The role of cell adhesion molecules in neurite outgrowth on Muller cells. *Dev. Biol.* **138,** 82–93.

Espeseth, A., Johnson, E., and Kintner, C. (1995). *Xenopus* F-cadherin, a novel member of the cadherin family of cell adhesion molecules, is expressed at boundaries in the neural tube. *Mol. Cell. Neurosci.* **6,** 199–211.

Fannon, A. M., and Colman, D. R. (1996). A model for central synaptic junctional complex formation based on the differential adhesive specificities of the cadherins. *Neuron* **17,** 423–434.

Fannon, A. M., Sherman, D. L., Ilyina-Gragerova, G., Brophy, P. J., Friedrich V. L. Jr., and Colman, D. R. (1995). Novel E-cadherin–mediated adhesion in peripheral nerve: Schwann cell architecture is stabilized by autotypic adherens junctions. *J. Cell Biol.* **129,** 189–202.

Franklin, J. L., and Sargent, T. D. (1996). Ventral neural cadherin, a novel cadherin expressed in a subset of neural tissues in the zebrafish embryo. *Dev. Dynamics* **206,** 121–130.

Fredette, B. J., Miller, J., and Ranscht, B. (1996). Inhibition of motor axon growth by T-cadherin substrata. *Development* **122,** 3163–3173.

Fredette, B. J., and Ranscht, B. (1994). T-cadherin expression delineates specific regions of the developing motor axon–hindlimb projection pathway. *J. Neurosci.* **14,** 7331–7346.

Fujimori, T., Miyatani, S., and Takeichi, M. (1990). Ectopic expression of N-cadherin perturbs histogenesis in *Xenopus* embryos. *Development* **110,** 97–104.

Gallin, W. J., Edelman, G. M., and Cunningham, B. A. (1983). Characterization of L-CAM, a major cell adhesion molecule from embryonic liver cells. *Proc. Natl. Acad. Sci. USA* **80,** 1038–1042.

Ganzler, S. I. I., and Redies, C. (1995). R-cadherin expression during nucleus formation in chicken forebrain neuromeres. *J. Neurosci.* **15,** 4157–4172.

Garrod, D., Chidgey, M., and North, A. (1996). Desmosomes: differentiation, development, dynamics and disease. *Curr. Opin. Cell Biol.* **8,** 670–678.

Gaya-Gonzalez, L., Balsamo, J., Swaminathan, N., and Lilien, J. (1991). Antibodies to the retina N-acetylgalactosaminylphosphotransferase inhibit neurite outgrowth. *J. Neurosci. Res.* **29,** 474–480.

Geiger, B., and Ayalon, O. (1992). Cadherins. *Annu. Rev. Cell Biol.* **8,** 307–332.

Geiger, B., Yehuda-Levenberg, S., and Bershadsky, A. D. (1995). Molecular interactions in the submembrane plaque of cell–cell and cell–matrix adhesions. *Acta Anat.* **154,** 46–62.

Grunwald, G. B. (1991). The conceptual and experimental foundations of vertebrate embryonic cell adhesion research. In *A Conceptual History of Modern Embryology* (S. F. Gilbert, Ed.), Vol. 7 of *Developmental Biology: A Comprehensive Synthesis,* pp. 129–158. Plenum Press, New York.

Grunwald, G. B. (1992). Cell adhesion and recognition in development. In *Fundamentals of Medical Cell Biology* Vol. 7: *Developmental Biology* (E. E. Bittar, Ed.), pp. 103–132. JAI Press, Greenwich, Connecticut.

Grunwald, G. B. (1993). The structural and functional analysis of cadherin calcium-dependent cell adhesion molecules. *Curr. Opin. Cell Biol.* **5,** 797–805.

Grunwald, G. B. (1996a). Discovery and analysis of the classical cadherins. In *Advances in Molecular and Cell Biology.* Vol. 16. pp. 63–112. JAI Press, Greenwich, Connecticut.

Grunwald, G. B. (1996b). Cadherin cell adhesion molecules in retinal development and pathology. *Prog. Ret. Eye Res.* **15,** 363–392.

Grunwald, G. B., Pratt, R.S., and Lilien, J. (1982). Enzymatic dissection of embryonic cell adhesive mechanisms. III. Immunological identification of a component of the calcium-dependent adhesive system of embryonic chick neural retina cells. *J. Cell Sci.* **55,** 69–83.

Gumbiner, B. M. (1996). Cell adhesion: the molecular basis of tissue architecture and morphogenesis. *Cell* **84,** 345–357.

Hasegawa, H., Seto, A., Uchiyama, N., Kida, S., Yamashima, T., and Yamashita, J. (1996). Localization of E-cadherin in peripheral glia after nerve injury and repair. *J. Neuropath. Exp. Neurol.* **55,** 424–434.

Hatta, K., Nose, A., Nagafuchi, A., and Takeichi, M. (1988). Cloning and expression of cDNA encoding a neural calcium-dependent cell adhesion molecule: its identity in the cadherin gene family. *J. Cell Biol.* **106,** 873–881.

Hatta, K., Takagi, S., Fujisawa, H., and Takeichi, M. (1987). Spatial and temporal expression pattern of N-cadherin cell adhesion molecules correlated with morphogenetic processes of chicken embryos. *Dev. Biol.* **120,** 215–227.

Hatta, K., and Takeichi, M. (1986). Expression of N-cadherin adhesion molecules associated with early morphogenetic events in chick development. *Nature* **320,** 447–449.

Hickman, Y., Veneziale, R., Rock, M., Khillan, J., Shinohara, T., Menko, A. S., and Grunwald, G. B. (1997). *In vivo* analysis of N-cadherin cell adhesion molecule function in ocular development using transgenic mice expressing tissue-specific dominant–negative mutations. *Invest. Ophthalmol. Vis. Sci.* **38,** S32.

Hirano, S., Kimoto, N., Shimoyama, Y., Hirohashi, S., and Takeichi, M. (1992). Identification of a neural α-catenin as a key regulator of cadherin function and multicellular organization. *Cell* **70,** 293–301.

Honig, M. G., and Rutishauser, U. S. (1996). Changes in the segmental pattern of sensory neuron projections in the chick hindlimb under conditions of altered cell adhesion molecule function. *Dev. Biol.* **175,** 325–337.

Huber, O., Bierkamp, C., and Kemler, R. (1996). Cadherins and catenins in development. *Curr. Opin. Cell Biol.* **8,** 685–691.

Hynes, R. O., and Lander, A. D. (1992). Contact and adhesive specificities in the associations, migrations, and targeting of cells and axons. *Cell* **68,** 303–322.

Inoue, T., Chisaka, O., Matsunami, H., and Takeichi, M. (1997). Cadherin-6 expression transiently delineates specific rhombomeres, other neural tube subdivisions, and neural crest subpopulations in mouse embryos. *Dev. Biol.* **183,** 183–194.

Inoue, A., and Sanes, J. R. (1997). Lamina-specific connectivity in the brain: regulation by N-cadherin, neurotrophins and glycoconjugates. *Science* **276,** 1428–1431.

Inuzuka, H., Miyatani, S., and Takeichi, M. (1991a). R-cadherin: a novel $Ca2^+$-dependent cell–cell adhesion molecule expressed in the retina. *Neuron* **7,** 69–79.

Inuzuka, H., Redies, C., and Takeichi, M. (1991b). Differential expression of R- and N-cadherin in neural and mesodermal tissues during early chicken development. *Development* **113,** 959–967.

Iwamoto, T., Taniguchi, M., Asai, N., Ohkusu, K., Nakashima, I., and Takahashi, M. (1993). cDNA cloning of mouse ret proto-oncogene and its sequence similarity the cadherin superfamily. *Oncogene* **7,** 1331–1337.

Kemler, R. (1992). Classical cadherins. *Sem. Cell Biol.* **3,** 149–155.

Kimura, Y., Matsunami, H., Inoue, T., Shimamura, K., Uchida, N., Ueno, T., Miyazaki, T., and Takeichi, M. (1995). Cadherin-11 expressed in association with mesenchymal morphogenesis in the head, somite, and limb bud of early mouse embryos. *Dev. Biol.* **169,** 347–358.

Kimura, Y., Matsunami, H., and Takeichi, M. (1996). Expression of cadherin-11 delineates boundaries, neuromeres, and nuclei in the developing mouse brain. *Dev. Dynamics* **206,** 455–462.

Kintner, C. (1992). Regulation of embryonic cell adhesion by the cadherin cytoplasmic domain. *Cell* **69,** 225–236.

Kjlavin, I. J., Lagenaur, C., Bixby, J. L., and Reh, T. A. (1994). Cell adhesion molecules regulating neurite growth from amacrine and rod photoreceptor cells. *J. Neurosci.* **14,** 5035–5049.

Korematsu, K., and Redies, C. (1997). Restricted expression of cadherin-8 in segmental and functional subdivisions of the embryonic mouse brain. *Dev. Dynamics* **208,** 178–189.

Lagunowich, L. A., and Grunwald, G. B. (1989). Expression of calcium-dependent cell adhesion during ocular development: a biochemical, histochemical and functional analysis. *Dev. Biol.* **135,** 158–171.

Lagunowich, L. A., and Grunwald, G. B. (1991). Tissue and age-specificity of post-translational modifications of N-cadherin during chick embryo development. *Differentiation* **47,** 19–27.

Lagunowich, L. A., Schneider, J. C., Chasen, S., and Grunwald, G. B. (1992). Immunohistochemical and biochemical analysis of N-cadherin expression during CNS development. *J. Neurosci. Res.* **32,** 202–208.

Lagunowich, L. A., Stein, A. P., and Reuhl, K. R. (1994). N-cadherin in normal and abnormal brain development. *Neurotoxicol.* **15,** 123–132.

Lee, M. M., Fink, B. D., and Grunwald, G. B. (1997). Evidence that tyrosine phosphorylation regulates N-cadherin turnover during retinal development. *Dev. Genetics* **20,** 224–234.

Levine, E., Lee, C. H., Kintner, C., and Gumbiner, B. M. (1994). Selective disruption of E-cadherin function in early *Xenopus* embryos by a dominant negative mutant. *Development* **120,** 901–909.

Marrs, J. A., and Nelson, W. J. (1996). Cadherin cell adhesion molecules in differentiation and embryogenesis. *Int. Rev. Cytol.* **165,** 159–205.

Matsunaga, M., Hatta, K., Nagafuchi, A., and Takeichi, M. (1988a). Guidance of optic nerve fibers by N-cadherin adhesion molecules. *Nature* **334,** 62–64.

Matsunaga, M., Hatta, K., and Takeichi, M. (1988b). Role of N-cadherin cell adhesion molecules in the histogenesis of neural retina. *Neuron* **1,** 289–295.

Matsunami, H., and Takeichi, M. (1995). Fetal brain subdivisions defined by R- and E-cadherin expressions: evidence for the role of cadherin activity in region-specific, cell–cell adhesion. *Dev. Biol.* **172,** 466–478.

McRea, P. D., Brieher, W. M., and Gumbiner, B. M. (1993). Induction of a secondary body axis in *Xenopus* by antibodies to β-catenin. *J. Cell Biol.* **123,** 477–484.

Miyatani, S., Copeland, N. G., Gilbert, D. J., Jenkins, N. A., and Takeichi, M. (1992). Genomic structure and chromosomal mapping of the mouse N-cadherin gene. *Proc. Natl. Acad. Sci. USA* **89,** 8443–8447.

Miyatani, S., Shimamura, K., Hatta, M., Nagafuchi, A., Nose, A., Matsunaga, M, Hatta, K., and Takeichi, M. (1989). Neural cadherin: role in selective cell–cell adhesion. *Science* **245,** 631–635.

Monier-Gavelle, F., and Duband, J. L. (1995). Control of N-cadherin–mediated intercellular adhesion in migrating neural crest cells *in vitro*. *J. Cell Sci.* **108,** 3839–3853.

Monier-Gavelle, F., and Duband, J. L. (1997). Cross talk between adhesion molecules: control of N-cadherin activity by intracellular signals elicited by β1 and β3 integrins in migrating neural crest cells. *J. Cell Biol.* **137,** 1663–1681.

Murphy-Erdosh, C., Napolitano, E. W., and Reichardt, L. F. (1994). The expression of B-cadherin during embryonic chick development. *Dev. Biol.* **161,** 107–125.

Nagar, B., Overduin M., Ikura, M., and Rini, J. M. (1996). Structural basis of calcium-induced E-cadherin rigidification and dimerization. *Nature* **380,** 360–364.

Nakagawa, S., and Takeichi, M. (1995). Neural crest cell–cell adhesion controlled by sequential and subpopulation-specific expression of novel cadherins. *Development* **121,** 1321–1332.

Napolitano, E. W., Venstrom, K., Wheeler, E. F., and Reichardt, L. F. (1991). Molecular cloning and characterization of B-cadherin, a novel chick cadherin. *J. Cell Biol.* **113,** 893–905.

Neugebauer, K. M., Tomaselli, K. J., Lilien, J., and Reichardt, L. F. (1988). N-cadherin, NCAM, and integrins promote retinal neurite outgrowth on astrocytes *in vitro*. *J. Cell Biol.* **107,** 1177–1188.

Obata, S., Sago, H., Mori, N., Rochelle, J. M., Seldin, M. F., Davidson, M., St. John., T., Taketani, S., and Suzuki, S. T. (1995). Protocadherin Pcdh2 shows properties similar to, but distinct from, those of classical cadherins. *J. Cell Sci.* **108,** 3765–3773.

Oda, H., Uemura, T., Harada, Y., Iwai, Y., and Takeichi, M. (1994). A *Drosophila* homologue of cadherin associated with armadillo and essential for embryonic cell–cell adhesion. *Dev. Biol.* **165,** 716–726.

Overduin, M., Harvey, T. S., Bagby, S., Tong, K. I., Yau, P., Takeichi, M., and Ikura, M. (1995). Solution structure of the epithelial cadherin domain responsible for selective cell adhesion. *Science* **267,** 386–389.

Ozawa, M., and Kemler, R. (1992). Molecular organization of the uvomorulin–catenin complex. *J. Cell Biol.* **116,** 989–996.

Pachnis, V., Mankoo, B., and Constantini, F. (1993). Expression of the c-ret proto-oncogene during mouse embryogenesis. *Development* **119,** 1005–1017.

Paradies, N., and Grunwald, G. B. (1993). Purification and characterization of NCAD90, a soluble endogenous form of N-cadherin, which is generated by proteolysis during retinal development and retains adhesive and neurite-promoting function. *J. Neurosci. Res.* **36,** 33–45.

Payne, H. R., Burden, S. M., and Lemmon, V. (1992). Modulation of growth cone morphology by substrate-bound adhesion molecules. *Cell Motil. Cytoskel.* **21,** 65–73.

Payne, H. R., Hemperley, J. J., and Lemmon, V. (1996). N-cadherin expression and function in cultured oligodendrocytes. *Dev. Brain Res.* **97,** 9–15.

Payne, H. R., and Lemmon, V. (1993). Glial cells of the O-2A lineage bind preferentially to N-cadherin and develop distinct morphologies. *Dev. Biol.* **159,** 595–607.

Pouliot, Y. (1992). Phylogenetic analysis of the cadherin superfamily. *Bioessays* **14,** 743–748.

Prozialeck, W. C., and Niewenhuis, R. J. (1991). Cadmium (Cd+2) disrupts Ca+2-dependent cell–cell junctions and alters the pattern of E-cadherin expression in LLC-PK1 cells. *Biochem. Biophys. Res. Commun.* **181,** 1118–1124.

Radice, G. L., Rayburn, H., Matsunami, H., Knudsen, K. A., Takeichi, M., and Hynes, R. O. (1997). Developmental defects in mouse embryos lacking N-cadherin. *Dev. Biol.* **181,** 64–78.

Radice, G. L., Rayburn, H., Matsunami, H., Knudsen, K. A., Takeichi, M., and Hynes, R. O. (1997). Developmental defects in mouse embryos lacking N-cadherin. *Dev. Biol.* **181,** 64–78.

Ranscht, B. (1991). Cadherin cell adhesion molecules in vertebrate neural development. *Semin. Neurosci.* **3,** 285–296.

Ranscht, B., and Bronner-Fraser, M. (1991). T-cadherin expression alternates with migrating neural crest cells in the trunk of the avian embryo. *Development* **111,** 15–22.

Ranscht, B., and Dours-Zimmerman, M. T. (1991). T-cadherin, a novel cadherin cell adhesion molecule in the nervous system lacks the conserved cytoplasmic region. *Neuron* **7,** 391–402.

Redies, C. (1995). Cadherin expression in the developing vertebrate CNS: from neuromeres to brain nuclei and neural circuits. *Exp. Cell Res.* **220,** 243–256.

Redies, C., Engelhardt, K., and Takeichi, M. (1993). Differential expression of N- and R-cadherin in functional neuronal systems and other structures of the developing chicken brain. *J. Comp. Neurol.* **333,** 398–416.

Redies, C., Inuzuka, H., and Takeichi, M. (1992). Restricted expression of N- and R-cadherin on neurites of the developing chicken CNS. *J. Neurosci.* **12,** 3525–3534.

Redies, C., and Muller, H. A. J. (1994). Similarities in structure and expression between mouse P-cadherin, chicken B-cadherin and frog XB/U-cadherin. *Cell Ad. Commun.* **2,** 511–520.

Redies, C., and Takeichi, M. (1993a). Expression of N-cadherin mRNA during development of the mouse brain. *Dev. Dynamics* **197,** 26–39.

Redies, C., and Takeichi, M. (1993b). N- and R-cadherin expression in the optic nerve of the chicken embryo. *Glia* **8,** 161–171.

Redies, C., and Takeichi, M. (1996). Cadherins in the developing central nervous system: an adhesive code for segmental and functional subdivisions. *Dev. Biol.* **180,** 413–423.

Reuhl, K. R., and Grunwald, G. B. (1997). Cell adhesion molecules as targets of neurotoxicants. In *Comprehensive Toxicology, Vol. 11: Nervous System and Behavioral Toxicology* (H. E. Lowndes and K. R. Reuhl, Eds.), Elsevier, pp. 115–137.

Reynolds, A. B., Daniel, J., McRea, P. D., Wheelock, M. J., Wu, J., and Zhang, Z. (1994). Identification of a new catenin: the tyrosine kinase substrate p120^{cas} associates with E-cadherin complexes. *Mol. Cell Biol.* **14,** 8333–8342.

Riehl, R., Johnson, K., Bradley, R., Grunwald, G. B., Cornel, E., Lilienbaum, A., and Holt, C. (1996). Cadherin function is required for axon outgrowth in retinal ganglion cells *in vivo. Neuron* **17,** 837–848.

Roark, E. F., Paradies, N. E., Lagunowich, L. A., and Grunwald, G. B. (1992). Evidence for endogenous proteases, mRNA level and insulin as multiple mechanisms of N-cadherin down-regulation during retinal development. *Development* **114,** 973–984.

Rose, O., Grund, C., Reinhardt, S., Starzinski-Powitz, A., and Franke, W. W. (1995). *Contactus adherens,* a special type of plaque-bearing adhering junction containing M-cadherin, in the granule cell layer of the cerebellar glomerulus. *Proc. Natl. Acad. Sci. USA* **92,** 6022–6026.

Sacristan, M. P., Vestal, D. J., Dours-Zimmerman, M. T., and Ranscht, B. (1993). T-cadherin 2: molecular characterization, function in cell adhesion, and co-expression with T-cadherin and N-cadherin. *J. Neurosci. Res.* **34,** 664–680.

Saffel, J. L., Walsh, F. S., and Doherty, P. (1992). Direct activation of second messenger pathways mimics cell adhesion molecule–dependent neurite outgrowth. *J. Cell Biol.* **118,** 663–670.

Sano, K., Tanihara, H., Heimark, R. L., Obata, S., Davidson, M., St. John, T., Taketani, S., and Suzuki, S. (1993). Protocadherins: a large family of cadherin-related molecules in central nervous system. *EMBO J.* **12,** 2249–2256.

Schiffman, J. S., and Grunwald, G. B. (1992). Differential cell adhesion and expression of N-cadherin among retinoblastoma cell lines. *Invest. Ophthalmol. Vis. Sci.* **33,** 1568–1574.

Schneider, R. (1992). The human protooncogene ret: a communicative cadherin? *Trends Biochem. Sci.* **17,** 468–469.

Selig, S., Bruno, S., Wang, C. H., Vitale, E., Gilliam, T. C., and Kunkel, L. M. (1995). Expressed cadherin pseudogenes are localized to the critical region of the spinal muscular atrophy gene. *Proc. Natl. Acad. Sci. USA* **92,** 3702–3706.

Selig, S., Lidov, H. G. W., Bruno, S. A., Segal, M. M., and Kunkel, L. M. (1997). Molecular characterization of Br-cadherin, a developmentally regulated, brain-specific cadherin. *Proc. Natl. Acad. Sci. USA* **94,** 2398–2403.

Shapiro, L., Fannon, A. M., Kwong, P. D., Thompson, A., Lehmann, M. S., Grubel, G., Legrand, J. F., Als-Neilsen, J., Colman, D. R., and Hendrickson, W. A. (1995). Structural basis of cell–cell adhesion by cadherins. *Nature* **374,** 327–337.

Shibata, T., Shimoyama, Y., Gotoh, M., and Hirohashi, S. (1997). Identification of human cadherin-14, a novel neurally specific type II cadherin, by protein interaction cloning. *J. Biol. Chem.* **272,** 5236–5240.

Shibuya, Y., Mizoguchi, A., Takeichi, M., Shimada, K., and Ide, C. (1995). Localization of N-cadherin in the normal and regenerating nerve fibers of the chicken peripheral nervous system. *Neurosci.* **67,** 253–261.

Shibuya, Y., Yasuda, H., Tomatsuri, M., Mizoguchi, A., Takeichi, M., Shimada, K., and Ide, C. (1996). αN-catenin expression in the normal and regenerating chick sciatic nerve. *J. Neurocytol.* **25,** 615–624.

Shimamura, K., Hirano, S., McMahon, A. P., and Takeichi, M. (1994). Wnt-1-dependent regulation of local E-cadherin and αN-catenin expression in the embryonic mouse brain. *Development* **120,** 2225–2234.

Shimamura, K., and Takeichi, M. (1992). Local and transient expression of E-cadherin involved in mouse embryonic brain morphogenesis. *Development* **116,** 1011–1019.

Shimamura, K., Takahashi, T., and Takeichi, M. (1992). E-cadherin expression in a particular subset of sensory neurons. *Dev. Biol.* **152,** 242–254.

Shinoura, N., Paradies, N. E., Warnick, R. E., Chen, H., Larson, J. J., Tew, J. J., Simon, M., Lynch, R. A., Kanai, Y., Hirohashi, S., Hemperley, J. J., Menon, A. G., and Brackenbury, R. (1995). Expression of N-cadherin and α-catenin in astrocytomas and glioblastomas. *Brit. J. Cancer* **72,** 627–633.

Stone, K. E., and Sakaguchi, D. S. (1996). Perturbation of the developing *Xenopus* retinotectal projection following injections of antibodies against β1 integrin receptors and N-cadherin. *Dev. Biol.* **180,** 297–310.

Sugimoto, K., Honda, S., Yamamoto, T., Ueki, T., Monden, M., Kaji, A., Matsumoto, K., and Nakamura, T. (1996). Molecular cloning and characterization of a newly identified member of the cadherin family, PB-cadherin. *J. Biol. Chem.* **271,** 11548–11556.

Suzuki, S., Sano, K., and Tanihara, H. (1991). Diversity of the cadherin family: evidence for eight new cadherins in nervous tissue. *Cell Reg.* **2,** 261–270.

Suzuki, S. T. (1996). Protocadherins and diversity of the cadherin superfamily. *J. Cell Sci.* **109,** 2609–2611.

Takeichi, M. (1988). The cadherins: cell–cell adhesion molecules controlling animal morphogenesis. *Development* **102,** 639–655.

Takeichi, M. (1990). Cadherins: a molecular family important in selective cell–cell adhesion. *Annu. Rev. Biochem.* **59,** 237–252.

Takeichi, M. (1991). Cadherin cell adhesion receptors as a morphogenetic regulator. *Science* **251,** 1451–1455.

Takeichi, M. (1992). The cadherins: cell–cell adhesion molecules controlling animal morphogenesis. *Development* **102,** 639–655.

Takeichi, M. (1993). Cadherins in cancer: implications for invasion and metastasis. *Curr. Opin. Cell Biol.* **5,** 806–811.

Takeichi, M. (1995). Morphogenetic roles of classic cadherins. *Curr. Opin. Cell Biol.* **7,** 619–627.

Takeichi, M., Matsunami, H., Inoue, T., Kimura, Y., Suzuki, S., and Tanaka, T. (1997). Roles for cadherins in patterning of the developing brain. *Dev. Neurosci.* **19,** 86–87.

Tanihara, H., Kido, M., Obata, S., Heimark, R. L., Davidson, M., St. John, T., and Suzuki, S. (1994). Characterization of cadherin-4 and cadherin-5 reveals new aspects of cadherins. *J. Cell Sci.* **107,** 1697–1704.

Tashiro, K., Toi, O., Nakamura, H., Koga, C., Ito, Y., Hikasa, H., and Shikawa, K. (1995). Cloning and expression studies of cDNA for a novel *Xenopus* cadherin (XmN-cadherin), expressed maternally and later neural-specifically in embryogenesis. *Mech. Dev.* **54,** 161–171.

Tepass, U., Gruszynski-DeFeo, E., Haag, T. A., Omatyar, L., Torok, T., and Hartenstein, V. (1996). *shotgun* encodes *Drosophila* E-cadherin and is preferentially required during cell rearrangement in the neurectoderm and other morphogenetically active epithelia. *Genes Dev.* **10,** 672–685.

Uchida, N., Honjo, Y., Johnson, K. R., Wheelock, M. J., and Takeichi, M. (1996). The catenin/cadherin adhesion system is localized in synaptic junctions bordering transmitter release zones. *J. Cell Biol.* **135,** 767–779.

Uchida, N., Shimamura, K., Miyatani, S, Copeland, N. G., Gilbert, D. J., Jenkins, N. A., and Takeichi, M. (1994). Mouse αN-catenin: two isoforms, specific expression in the nervous system, and chromosomal localization of the gene. *Dev. Biol.* **163,** 75–85.

Vestal, D., and Ranscht, B. (1992). Glycosyl phosphatidylinositol-anchored T-cadherin mediates calcium-dependent, homophilic adhesion. *J. Cell Biol.* **119,** 451–461.

Walsh, F. S., Barton, C. H., Putt, W., Moore, S. E., Kelsell, D., Spurr, N., and Goodfellow, P. N. (1990). N-cadherin gene maps to human chromosome 18 and is not linked to the E-cadherin gene. *J. Neurochem.* **55,** 805–812.

Walsh, F., and Doherty, P. (1996). Cell adhesion molecules and neuronal regeneration. *Curr. Opin. Cell Biol.* **8,** 707–713.

Williams, E. J., Furness, J., Walsh, F. S., and Doherty, P. (1994b). Activation of the FGF receptor underlies neurite outgrowth stimulated by L1, N-CAM and N-cadherin. *Neuron* **13,** 583–594.

Williams, E. J., Walsh, F. S., and Doherty, P. (1994a). Tyrosine kinase inhibitors can differentially inhibit integrin-dependent and CAM-stimulated neurite outgrowth. *J. Cell Biol.* **124,** 1029–1037.

Yamada, K. M., and Geiger, B. (1997). Molecular interactions in cell adhesion complexes. *Curr. Op. Cell Biol.* **9,** 76–85.

Yamagata, M., Herman, J. P., and Sanes, J. R. (1995). Lamina-specific expression of adhesion molecules in developing chick optic tectum. *J. Neurosci.* **15,** 4556–4571.

Yoshida-Noro, C., Suzuki, M., and Takeichi, M. (1984). Molecular nature of the calcium-dependent cell–cell adhesion system in mouse teratocarcinoma and embryonic cells studied with a monoclonal antibody. *Dev. Biol.* **101,** 19–27.

CHAPTER

3

Neurite Development

GERALD AUDESIRK
TERESA AUDESIRK
Department of Biology
University of Colorado
Denver, Colorado 80217

*Abbreviations: CaATPase, calcium adenosine triphosphatase; cAMP (cyclic AMP), cyclic adenosine monophosphate; CNS, central nervous system; CSF, cerebrospinal fluid; GAP, growth associated protein; MAPs, microtubule-associated proteins; NMDA, *N*-methyl-D-aspartate; Pb, lead; PKA, protein kinase A; PKC, protein kinase C; PP2B, calcineurin.

The behavior of humans and other animals arises from a complex and highly specific set of interconnections among billions of neurons. Individual neurons put forth an array of neurites (axons and dendrites) that follow chemical guideposts toward appropriate target regions. Initiation of neurites, elongation, branching and elongation of the branches, and synapse formation with target cells connect most neurons with thousands or tens of thousands of other neurons during development. Many of these connections atrophy during development; some persist into adulthood; some are capable of remarkable plasticity, either strengthening or weakening as a result of experience. Specificity and functionality of behavior depends on specificity and functionality of neurite development and connections among neurons.

Many neurotoxicants, including as diverse substances as heavy metals, pesticides, and ethanol, can perturb neuronal development. In some cases, particularly with high doses, the neurotoxicant may cause cell death. In other cases, particularly with lower doses, neurons remain alive but grow abnormally. In this chapter, we review the cellular mechanisms underlying normal neurite development and then examine a few specific instances of toxicant-induced alterations of neurite development.

I. Neurite Development under Normal Conditions

The term *neurite* is often applied to any neuronal process and includes not only axons or dendrites, but also processes that cannot be assigned definitively to either category, such as the processes produced by some immortal cell lines. We use the term *neurite* when the literature does not clearly define the process under study, and when a mechanism or molecular event appears to apply equally to axons or dendrites. However, axons and dendrites have important differences—for example, in cytoskeletal proteins, other developmentally important proteins, and the distribution of certain organelles. Therefore, we specify the type of process where the literature provides specific data on axonal versus dendritic traits.

The cellular mechanisms and interactions among cells that control neurite development are incompletely understood, but are subjects of intense study, particularly in cultured neurons. Compared to the intact nervous system, cell cultures are simpler and easier to manipulate. One or a few defined cell types can be grown in culture dishes, the geometry is two-dimensional rather than three-dimensional, and, in some cases at least, the extracellular environment can be completely specified by the investigator. Cell signalling pathways can often be more easily manipulated *in vitro* as well. Therefore, most of what is known or suspected about the cellular and subcellular mechanisms of neurite development has been discovered using culture systems. However, many techniques commonly used in culture are also being applied *in vivo,* such as immunolocalization of cytoskeletal components and determination of their phosphorylation states. Although many of these *in vivo* studies are correlative (e.g., at what stages during development certain phosphorylated forms of microtubule-associated proteins appear), they provide crucial reference points for validation of *in vitro* studies. Mechanistic *in vivo* studies are also increasingly common. For example, experiments with knockout mutant mice, although often difficult to interpret, can provide mechanistic information about the role of individual proteins in regulating neurite development.

A. *Overview of Neurite Development*

The development of the mature neuronal form from a relatively undifferentiated neuroblast (*in vivo*) or from a process-free soma (*in vitro*) involves several steps, some of which may overlap temporally. The first step is *neurite initiation* from the soma. The nascent neurite is in reality a *growth cone,* a flattened, mobile structure that appears to sample the environment, responding to chemical, mechanical, and perhaps electrical cues that guide the direction and rate of growth of the neurite. During *neurite elongation,* a thin, rounded cable, the presumptive axon or dendrite, connects the growth cone with the soma. *Neurite branching* occurs when either the growth cone splits or new growth cones emerge from the side of the neurite shaft. Contact with inappropriate targets may be ignored by the growth cone or may even cause growth cone collapse and neurite retraction; contact with appropriate targets results in *synaptogenesis.* Synaptogenesis is covered in detail in Section III of this handbook; in this section, we focus on the earlier events in neurite development.

In many cultured neurons, cell adhesion to a suitable substrate in a suitable medium is followed by extension of lamellipodia (veils of plasma membrane) around the cell body. Some of these begin the process of neurite initiation; that is, they differentiate into growth cones and begin to extend from the soma (Dotti *et al.,* 1988). Tension, such as that exerted by an emerging growth cone, is an adequate stimulus for the initiation and elongation of a neurite (Bray, 1984; Zheng *et al.,* 1991). The traditional view of neurite elongation is that lamellipodia add membrane at their leading edge, presumably by transport of vesicles from the Golgi apparatus into the cone, where they fuse with the lamellipodial membrane (Purves and Lichtmann, 1985), but recent studies using cultured chick sensory neurons (Zheng *et al.,* 1991) and *Xenopus* spinal neurons (Popov *et al.,* 1993) have demonstrated that membrane may also be added near the soma and along the neurite shaft (for a review of the sites of membrane addition during elongation, see Futerman and Banker, 1996). As the growth cone advances, microtubules and neurofilaments "consolidate" the advance by polymerizing into relatively stable structures in the neurite shaft. Neurite branching has received relatively little study, although it has been suggested that phosphorylation of microtubule-associated proteins may promote branching (Friedrich and Aszodi, 1991). A review by Heidemann (1996) provides superb coverage of the mechanisms of neurite initiation, growth cone motility, and neurite elongation.

B. *The Neurite Cytoskeleton*

The cytoskeleton of a neurite consists largely of three types of proteins: microfilaments, neurofilaments, and microtubules (for reviews of the cytoskeletal elements in neurites and their role in neurite function, see Riederer, 1990; Cambray-Deakin, 1991).

Microfilaments are composed of actin polymers. Actin microfilaments are in a state of dynamic equilibrium, continuously adding actin monomers at the "plus" end and losing them at the "minus" end. The assembly of actin microfilaments is regulated by a variety of proteins, including profilins, villin, and gelsolin (Bamburg and Bernstein, 1991). Actin microfilaments are the most abundant cytoskeletal component of growth cones, particularly of filopodia. Interactions of microfilaments with nonmuscle myosin molecules cause the movements characteristic of growth cones (Evans and Bridgman, 1995; Lin *et al.,* 1996). Several proteins, including fimbrin, spectrin, talin, and vinculin, link actin filaments to the plasma membrane (Bamburg and Bernstein, 1991; Luna and Hitt, 1992; Hitt and Luna, 1994). Neurite growth may require actin–plasma membrane interactions, probably mediated by these proteins (Sihag *et al.,* 1996; Sydor *et al.,* 1996). Many of the proteins that modulate actin assembly, linkage to the plasma membrane, and movement within a growth cone are regulated by the intracellular Ca^{2+} concentration (Janmey, 1994).

Microtubules are polymers of tubulin. Like microfilaments, microtubules are constantly polymerizing at the "plus" end and depolymerizing at the "minus" end. They provide structural support and act as conveyers of organelles within neurons and other cells. A variety of microtubule-associated proteins (MAPs) are found in conjunction with microtubules. Some of these MAPs promote microtubule assembly and/or stabilize microtubules by forming cross-bridges between them (for reviews of microtubule and MAP function, see Gelfand and Bershadsky, 1991; Joshi and Baas, 1993; Hirokawa, 1994; and Matus, 1994).

Neurofilaments are intermediate filaments that are unique to neurons. They are composed of three subunits, usually designated NF-L, NF-M, and NF-H for low, medium and high molecular weight neurofilament subunits. Unlike microfilaments and microtubules, neurofilaments usually remain in a stable polymerized state. Neurofilaments are synthesized within the cell body and then transported into neurites. They are found distributed throughout the neuron singly or bound by cross-bridges into bundles, and oriented longitudinally. Neurofilaments presumably have a stabilizing structural role in neuronal cytoarchitecture, but their function is poorly understood (for reviews, see Shaw, 1991; Lee and Cleveland, 1996).

1. Cytoskeletal Components of Axons and Dendrites

Axons and dendrites share many similarities, but also differ in several important respects (for review, see Craig and Banker, 1994). Axons are usually longer than dendrites and elongate more rapidly. When the leading growth cone of an axon or axon branch contacts an appropriate target cell, it differentiates into a presynaptic terminal; with some exceptions, dendrites do not form presynaptic terminals. Axons and dendrites also differ in intracellular proteins and organelles. Neurofilaments are more abundant in axons than in dendrites. Microtubules, although found in both types of neurites, are uniformly oriented in axons, with the plus end directed toward the periphery, but have mixed orientation in dendrites (Baas *et al.,* 1988). Microtubule-associated proteins are differentially distributed, with MAP2 localized in dendrites and tau in axons (see review by Ginzburg, 1991). Dendrites, but not axons, contain pericentriolar material (Ferreira *et al.,* 1993), ribosomes, rough and smooth endoplasmic reticulum, and Golgi elements (Sargent, 1989). These structural differences between axons and dendrites suggest that neurotoxicants may have differential effects on these two types of neurites.

Observations of cultured neurons undergoing differentiation have provided information as to the timing and sequence of events that leads to the differentiation of neurites into axons and dendrites (Dotti *et al.,* 1988). Cultured hippocampal neurons initially extend several apparently identical processes. At this stage, all processes contain MAP2, which, in a fully differentiated cell, is confined to dendrites. After a few hours, one neurite, the presumptive axon, begins to elongate rapidly, but all neurites still express MAP2 until about 48 hrs in culture. Over the next few days, the axon becomes distinct from the dendrites, losing MAP2 and expressing a neurofilament subunit confined to axons (Pennypacker *et al.,* 1991). Microtubule polarity remains uniform in all neurites until 4–5 days in culture, when dendritic microtubules begin to become nonuniform (Baas *et al.,* 1989). After a week in culture, axons and dendrites are fully differentiated.

The importance of cytoskeletal elements and associated proteins in governing neurite development has been shown in studies in which the synthesis or activity of one or more of these proteins was inhibited or enhanced. For example, antisense oligonucleotides directed against MAP1B (also called MAP5 in some papers) mRNA block neurite outgrowth in PC12 cells (Brugg *et al.,* 1993). Similarly, antisense oligonucleotides directed against MAP2 mRNA inhibit neurite outgrowth in embryonal carcinoma cells treated with retinoic acid to induce neuronal differentiation (Dinsmore and Solomon, 1991). Inhibiting tau expression with antisense oligonucleotides inhibits axonal, but not dendritic, growth in cultured cerebellar macroneurons (Caceres *et al.,* 1992), and inhibits the development of axon-like processes produced by NB2a/d1 neuroblastoma cells in response to serum deprivation (Shea *et al.*, 1992). Transgenic mice overexpressing any single neurofilament subunit (NF-L, NF-M, or NF-H) show decreased axonal diameter, whereas increasing NF-L along with either NF-M or NF-H causes increased axonal diameter (Xu *et al.*, 1996).

2. Regulation of the Cytoskeleton by Intracellular Calcium

Changes in intracellular free calcium ion concentrations, $[Ca^{2+}]_i$, can affect the cytoskeleton, and consequently neurite growth, in many ways. For example, actin remodelling is necessary for the growth cone to guide neurite extension along appropriate pathways (Marsh and Letourneau, 1984). A Ca^{2+}-activated enzyme, gelsolin, severs actin filaments and produces a transition from a gel to a sol state that allows cytoplasmic streaming and movements of cells and growth cones. Activation of gelsolin by a brief rise in $[Ca^{2+}]_i$ greatly increases its affinity for actin, causing it to shut down new actin filament assembly and disrupt the existing actin network. This breakdown may be an important first step in the restructuring of the membrane cytoskeleton that is required for growth cone motility, neurite initiation, and other Ca^{2+}-influenced processes (see Forscher, 1989).

High, probably localized, $[Ca^{2+}]_i$ may stimulate calcium-activated proteases (calpains), which can cleave a large assortment of proteins, including neurofilaments (Johnson *et al.*, 1991), tubulin (Billger *et al.*, 1988), microtubule associate proteins (Billger *et al.*, 1988; Fischer *et al.*, 1991) and several protein kinases (Melloni and Pontremoli, 1989). Calpain activity appears to be regulated through a complex set of interactions among calpain, $[Ca^{2+}]_i$, and a natural, Ca^{2+}-dependent inhibitor protein, calpastatin (Melloni and Pontremoli, 1989).

Perhaps the most ubiquitous protein activated by Ca^{2+} is calmodulin, which, when bound to Ca^{2+}, regulates many other proteins. For example, calmodulin binds to the 200 kD neurofilament protein and inhibits its calpain-mediated hydrolysis (Johnson *et al.*, 1991). Ca^{2+}-calmodulin also binds to MAP2 and inhibits microtubule assembly (Wolff, 1988). Calmodulin may also modulate the attachment of the cytoskeleton to the plasma membrane (Liu and Storm, 1990). Ca^{2+}-calmodulin–dependent protein kinases (CaM kinases) phosphorylate many cytoskeletal components, including microtubules and MAPs, and this phosphorylation helps to control microtubule bundling, the interactions between microtubules, MAPs, and neurofilaments, and the stability of microtubules (for reviews see Matus, 1988a,b; Nixon and Sihag, 1991).

A minimum concentration of Ca^{2+} is required for activation of some isoforms of protein kinase C (PKC), which also phosphorylates cytoskeletal proteins (Matus, 1988a,b). Finally, elevated $[Ca^{2+}]_i$ activates calcineurin (PP2B), a calcium-dependent phosphatase. Calcineurin dephosphorylates many of the same proteins that are phosphorylated by CaM kinases or other protein kinases (Armstrong, 1989; Liu and Storm, 1990).

The importance of Ca^{2+} in controlling neurite growth has been demonstrated in many studies. For example, Ca^{2+} influx through voltage-sensitive calcium channels can stimulate neurite development in depolarized PC12 cells (Rusanescu *et al.*, 1995). Transmitter-induced growth cone turning in cultured *Xenopus* spinal neurons appears to require Ca^{2+} influx from the external medium, activating Ca^{2+}-calmodulin-dependent protein kinase II (Zheng *et al.*, 1994). In transgenic *Drosophila* expressing high levels of calmodulin-binding proteins (thereby inhibiting calmodulin function), axonal extension and guidance are both disrupted (VanBerkum and Goodman, 1995). Axon branch retraction is increased in *Xenopus* infected with viruses carrying constitutively active CaM kinase II (Zou and Cline, 1996). As these examples suggest, depending on the exact Ca^{2+} dependence of these regulatory proteins, their spatial distribution within a neuron, and the spatial distribution of $[Ca^{2+}]_i$, neurites may initiate, extend, or retract. We return to regulation by $[Ca^{2+}]_i$ as we discuss several specific aspects of neurite development.

3. Regulation of the Cytoskeleton by Protein Phosphorylation

Changes in the phosphorylation state of cytoskeletal molecules is receiving increasing attention as a mechanism whereby extracellular and intracellular signals are translated into changes in neuronal morphology (for reviews, see Nixon and Sihag, 1991; Avila *et al.*, 1994a; Doering, 1994). Although not all phosphorylation or dephosphorylation events are Ca^{2+} dependent, changes in $[Ca^{2+}]_i$ may regulate protein phosphorylation in several ways: (1) Ca^{2+} may activate protein kinases such as PKC and CaM kinases; (2) Ca^{2+} may activate protein phosphatases, specifically calcineurin; and (3) Ca^{2+} may regulate other cellular processes that modulate kinase activity, for example altering cyclic AMP–dependent protein kinase activity by its stimulatory effects on some isoforms of adenylyl cyclase and on phosphodiesterase.

Many cytoskeletal proteins are phosphorylated both *in vivo* and *in vitro* by a variety of protein kinases, including PKC, cyclic AMP–dependent protein kinase (PKA), CaM kinase II, and proline-directed kinases (e.g., MAP2: Vallee, 1980; Schulman, 1984; Steiner *et al.*, 1990; tau: Johnson, 1992; Scott *et al.*, 1993). Similarly, cytoskeletal proteins have been found to be substrates for dephosphorylation by protein phosphatases (e.g., MAP2: Murthy and Flavin, 1983; tau: Furiya *et al.*, 1993). The phosphorylation state of cytoskeletal proteins has functional effects. For example, phosphorylation of MAP2 by CaM kinase II inhibits the ability of MAP2 to cross-link actin filaments (Yamauchi and Fujisawa, 1988). Phosphorylation also reduces the binding of MAP2 (Brugg and Matus, 1991) and tau (Biernat *et al.,* 1993) to microtubules.

Many studies have shown that phosphorylation of cytoskeletal proteins, particularly microtubule-associated proteins, is correlated with neuronal development both in culture and *in vivo.* For example, in cultured chick spinal cord neurons and rat cerebral cortex neurons, tau phosphorylated at serine 202 is concentrated in the proximal parts of axons, close to the cell body, while tau that lacks phosphorylation at this residue is concentrated in the distal axon and growth cone (Rebhan *et al.,* 1995). In cultured rat hippocampal neurons, MAP1B phosphorylated by either proline-directed protein kinase or casein kinase II shows opposite distributions in growing axons; proline-directed kinase phosphorylation is concentrated in distal segments and growth cones, whereas casein kinase II phosphorylation is concentrated in proximal segments (Avila *et al.,* 1994b). Phosphorylation of many proteins, including MAP2, increases during neuronal development in cultured hippocampal neurons (Diez-Guerra and Avila, 1993). In rat brain, MAP1B is more highly phosphorylated in growing axons than in nongrowing axons (Harrison *et al.,* 1993). Similarly, tau phosphorylated at serine 202 is concentrated in growing regions of developing rat cortex, and phosphorylation decreases postnatally (Brion *et al.,* 1994).

Further support for a link between protein phosphorylation and neurite development comes from *in vitro* studies in which the phosphorylation state of neuronal proteins was experimentally manipulated with inhibitors and/or activators of protein kinases or phosphatases. For example, depending on the cell type and exact experimental conditions, various aspects of neurite development have been found to be stimulated by inhibition (Tsuda *et al.,* 1989; Felipo *et al.,* 1990; Leli *et al.,* 1993) or activation (Hsu *et al.,* 1989; Cambray-Deakin *et al.,* 1990) of PKC. Overexpression of CaM kinase II in neuro2a or NG108-15 neuroblastoma cells promotes neurite development (Goshima *et al.,* 1993), but inhibits neurite development stimulated by cyclic AMP in PC 12 cells (Tashima *et al.,* 1996). Phosphatase inhibitors such as okadaic acid inhibit neurite development in PC12 cells (Chiou and Westhead, 1992) and SH-SY5Y neuroblastoma cells (Leli *et al.,* 1993).

C. Growth Cones

Growth cones consist of lamellipodia spanning finger-like membrane extensions called *filopodia.* The growth cone adheres to the substrate (including other cells) by means of one or more of several cell adhesion molecules (the role of cell adhesion molecules in neurite development is covered in the chapter by Grunwald, and is not discussed here). As the growth cone advances, it exerts tension on the extending neurite (Letourneau, 1975; Lamoureux, *et al.,* 1989; Heidemann *et al.,* 1990). Growth cones appear to detect both physical and chemical environmental cues and follow them to appropriate targets for innervation (see reviews by Bray and Hollenbeck, 1988; Goodman, 1996). Appropriate target cells may provide "stop" signals that inhibit further neurite extension (Baird *et al.,* 1992).

1. The Cytoskeleton of Growth Cones

Growth cones contain a high concentration of filamentous actin. Many actin-binding proteins, including gelsolin (Tanaka *et al.,* 1993), α-actinin, and spectrin (Sobue and Kanda, 1989), are found in growth cones and regulate actin assembly and attachment to the plasma membrane. Nonmuscle myosin interacts with the actin filaments, apparently causing movement of actin within the growth cone (Lin *et al.,* 1996) and movement of organelles along the actin filaments (Evans and Bridgman, 1995). The importance of actin in controlling growth cone movement has been demonstrated using actin inhibitors. For example, when cultured dorsal root ganglion cells are treated with the actin-binding drug cytochalasin B, growth-cone advance is halted (Yamada *et al.,* 1971). Similarly, in cultured cerebellar granule cells, cytochalasin B induces growth cone collapse (Abosch and Lagenaur, 1993). Neurite elongation may persist, but even when neurites continue to elongate, they are more curved than control neurites on the same substrate (Marsh and Letourneau, 1984; Abosch and Lagenaur, 1993). These findings suggest that actin-mediated pull, rather than push from behind, mediates growth-cone advance and normal neurite elongation. Furthermore, actin filaments in growth cones undergo retrograde flow (Okabe and Hirokawa, 1991; Lin and Forscher, 1995), driven by interactions with nonmuscle myosin (Lin *et al.,* 1996). One model of growth-cone advance suggests that actin monomers are added to filaments at the leading edge of a growth cone; meanwhile, the filament flows back toward the neurite shaft, probably driven by myosin. Variable attachment of the flowing actin filament to the plasma membrane and the plasma membrane to the substrate causes the growth cone to advance in specific directions (Fig. 1; Lin and Forscher, 1995; Lin *et al.,* 1996).

Microtubules, the principal cytoskeletal component in most neurite shafts, also invade the growth cone. Although the evidence cited previously indicates that growth cone movements are actin-based, microtubule polymerization and bundling at the base of the growth cone also appears to be required for growth cone ad-

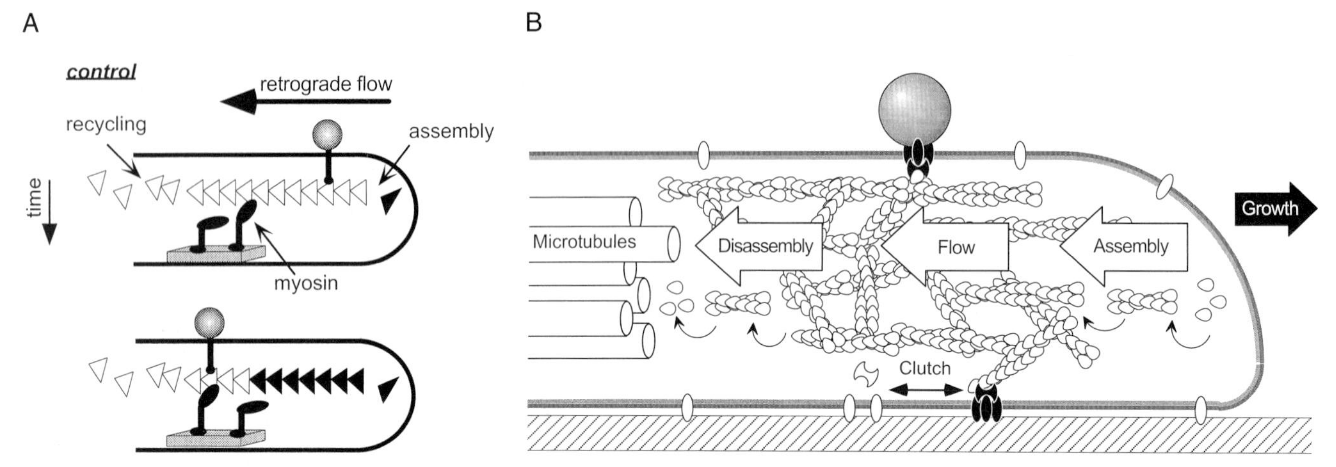

FIGURE 1 A model of growth cone advance. (A) Actin monomers at the leading edge of the growth cone assemble into actin filaments. Assembly may cause protrusion of the leading edge. Simultaneously, interactions between the actin filaments and myosin drive the filaments in a retrograde direction. (B) The combination of assembly of actin filaments at the leading edge, retrograde flow of filaments driven by myosin (myosin molecules not shown here) and variable attachment of filaments and/or myosin to the plasma membrane and substrate ("clutch") drives growth cone advance. (Part (A): reprinted from Lin, C. H., Espreafico, E. M., Mooseker, M. S., and Forscher, P. Myosin drives retrograde F-actin flow in neuronal growth cones. *Neuron* **16,** 769–782 (1996). Part (B): reprinted from Lin, C. H., and Forscher, P. Growth cone advance is inversely proportional to retrograde F-actin flow. *Neuron* **14,** 763–771 (1995). Both figures are copyrighted by Cell Press and are reprinted by the kind permission of P. Forscher and Cell Press.)

vance. For example, low doses of vinblastine, which inhibit microtubule dynamics, inhibit growth cone advance (Tanaka *et al.,* 1995). Apparently, the consolidation of the base of the growth cone into the elongating neurite shaft, with the resulting collapse of the relatively flat growth cone into a more cylindrical structure surrounding microtubule bundles in the shaft, is required if the growth cone is to advance further. Interactions between actin and microtubules in growth cone motility are explored by Challacombe *et al.* (1996).

2. Control of Growth Cone Movement by Intracellular Calcium and Protein Phosphorylation

There seems to be relatively good agreement that Ca^{2+} influx and/or fairly high $[Ca^{2+}]_i$ is required for growth-cone motility (e.g., Anglister *et al.,* 1982; Connor, 1986; Cohan *et al.,* 1987; Kater *et al.,* 1988). For example, Mattson and Kater (1987), using cultured *Helisoma* (pond snail) neurons, found that low concentrations of heavy metals, which partially inhibit Ca^{2+} influx through voltage-sensitive calcium channels, cause filopodia to retract and growth-cone motility to cease. At least moderate levels of Ca^{2+} influx may enhance growth-cone motility in N1E-115 neuroblastoma cells (Silver *et al.,* 1990). These investigators also found that calcium channels were clustered in "hot spots" in the growth cone but not the neurite shaft. Therefore, action potentials would result in localized Ca^{2+} influx and higher $[Ca^{2+}]_i$ levels in the growth cone than in the neurite shaft (Silver *et al.* 1990). At least in some cell types, elongation seems to be promoted by relatively low $[Ca^{2+}]_i$, so this distribution of calcium channels may be important both for growth cone motility and neurite elongation.

The intracellular molecular targets whereby $[Ca^{2+}]_i$ regulates growth cone movements are not completely defined. However, it is known that several of the actin-binding proteins in growth cones, including gelsolin (Tanaka *et al.,* 1993), α-actinin (Sobue and Kanda, 1989), and caldesmon (Kira *et al.,* 1995), are Ca^{2+}-sensitive. Furthermore, spectrin is a substrate for the Ca^{2+}-dependent protease, calpain. Through these proteins, changing Ca^{2+} concentrations in the growth cone may regulate assembly of actin filaments (gelsolin), attachment of the actin filaments to the plasma membrane (α-actinin and spectrin), and interactions between actin and myosin (caldesmon).

Growth cones contain a wide variety of protein kinases, including both serine–threonine kinases (e.g., PKC, CaM kinases: Katz *et al.,* 1985) and tyrosine kinases (e.g., Matten *et al.,* 1990; Wu and Goldberg, 1993), and protein phosphatases. Kinase activity appears to be required for growth-cone motility, at least in some cell types (Polak *et al.,* 1991; Jian *et al.,* 1994). Probably the best studied example of the effects of phosphorylation on growth-cone structure and motility is the growth-associated protein GAP-43. GAP-43, which is highly enriched in axonal growth cones, is involved in growth-cone spreading, filopodial induction, branching, and adhesion to the substrate (Zuber *et al.,* 1989; Aigner and Caroni, 1993, 1995; Strittmatter *et al.,* 1994). GAP-43 is a PKC substrate (for review, see Gispen *et al.,* 1992). Phosphorylation of GAP-43 appears to regulate growth-cone spreading and motility (Dent and Meiri, 1992; Widmer and Caroni, 1993).

Many growth-cone proteins, including GAP-43 (Meiri and Burdick, 1991) and most of the cytoskeletal proteins (see Section I.B.3., this chapter), are phosphatase substrates. Phosphatase activity may regulate growth-cone structure and motility, although in general the specific substrates involved are not known. For example, in growth cones of chick dorsal root ganglion neurons, focal inactivation of calcineurin causes retraction of filopodia and lamellipodia (Chang *et al.,* 1995).

3. Growth Cone Guidance and Pathfinding: The Roles of Adhesion, Attraction, and Repulsion

As the leading structure of a neurite, growth cones control the rate and direction of growth of the neurite. Growth cones accomplish this by adhesion to the substrate, turning in response to asymmetrical cues and advancing over the substrate (which applies tension to the neurite shaft, causing it to elongate). Adhesion molecules are discussed in Chapter 2 by Grunwald in this volume. Here, we merely mention that the growth-cone plasma membrane contains many receptors for adhesion molecules and that contact between growth cone and adhesion molecules trigger changes in the growth cone, including events such as Ca^{2+} influx and protein phosphorylation that alter cytoskeletal function and regulate the rate and direction of growth-cone advance (see reviews by Rosales *et al.,* 1995; Yamada and Miyamoto, 1995).

Growth cones turn in response to a wide variety of cues, including electric fields (e.g., Davenport and McCaig, 1993) and soluble chemicals (e.g., dibutryl cyclic AMP: Lohof *et al.,* 1992; neurotransmitters: Zheng *et al.,* 1994). Turning may be the result of attraction (e.g., toward neurotransmitters [Zheng *et al.,* 1994] or cyclic AMP [Lohof *et al.,* 1992]) or repulsion (e.g., away from semaphorins [Messersmith *et al.,* 1995]). Some molecules, such as netrin, may be attractants for neurite growth in some neuronal types (Kennedy *et al.,* 1994) and repellants for other neuronal types (Colamarino and Tessier-Lavigne, 1995; see review by Kolodkin, 1996).

Finally, some signals cause growth cones to stop moving and begin differentiation into presynaptic terminals. In most cases, these appear to be contact with an appropriate target cell. For example, neurites of neurons from the pontine nucleus (the source of cerebellar mossy fibers) stop growing when they contact cerebellar granule neurons in culture (Baird *et al.,* 1992). The nature of the "stop" signals and the subsequent processes of synaptogenesis after the growth cone halts are not well understood, particularly among mammalian neurons (see Haydon and Drapeau, 1995).

D. Neurite Initiation

Neurite initiation, which is essentially the emergence of a growth cone from a neuronal soma, appears to involve a complex interplay among the cytoskeleton, the plasma membrane, the intracellular free calcium ion concentration and the activities of protein kinases. The involvement of the cytoskeleton and the plasma membrane has been demonstrated in many studies. For example, antisense oligonucleotides to mRNA encoding the microtubule-associated protein MAP1B inhibit neurite initiation in PC12 cells (Brugg *et al.,* 1993) and antisense oligonucleotides to MAP2 inhibit initiation in cerebellar macroneurons (Caceres *et al.,* 1992). Membrane-associated proteins such as GAP-43 may also be important in initiation; antisense oligonucleotides to GAP-43 prevent initiation in PC12 cells (Jap Tjoen San *et al.,* 1992).

In cultured rat sympathetic neurons (Rogers and Hendry, 1990), chick neurons and N1E-115 neuroblastoma cells (Audesirk *et al.,* 1990), neurite initiation is reduced by inhibitors of L-type voltage-sensitive calcium channels. These inhibitors should reduce transmembrane Ca^{2+} influx and may reduce $[Ca^{2+}]_i$, at least locally, implicating intracellular Ca^{2+} as an important regulator of initiation. Changes in $[Ca^{2+}]_i$ caused by release of Ca^{2+} from intracellular stores may also modulate neurite initiation; for example, in rat dorsal root ganglion neurons, depletion of stores with thapsigargin, which presumably renders the stores incapable of releasing Ca^{2+} in response to a subsequent physiological stimulus, inhibits initiation (Lankford *et al.,* 1995).

Protein phosphorylation by various protein kinases, some of which may be modulated by $[Ca^{2+}]_i$, also affects neurite initiation, although the role of a given kinase appears to vary among cell types. For example, inhibiting PKC stimulates neurite initiation in N18TG2 neuroblastoma cells (Tsuda *et al.,* 1989), but stimulating PKC with phorbol esters enhances neurite initiation in chick dorsal root ganglia (Hsu *et al.,* 1989) and cerebellar granule cells (Cambray-Deakin *et al.,* 1990). In rat hippocampal neurons, inhibiting PKC with calphostin C or chelerythrine reduces neurite initiation, whereas stimulating PKC with phorbol esters has no effect (Cabell and Audesirk, 1993; 1996, unpublished results).

E. Neurite Elongation

In many cultured neurons, a soma initiates several morphologically and biochemically similar neurites. Soon, one neurite elongates more rapidly than the others and becomes the axon; the other neurites become dendrites. The "switch" that triggers transformation of an apparently generic neurite to an axon is not known. Most studies of neurite elongation have examined either axons or undefined "generic" neurites growing *in vitro.*

Depending on the cell type and/or culture conditions, elongation is modulated by many factors, including substrate (Wheeler and Brewer, 1994); protein kinase activity (Cabell and Audesirk, 1993); depolarization (Fields *et al.,* 1990), stimulation of the NMDA subtype of glutamate receptors (Mattson *et al.,* 1988; Brewer and Cotman, 1989), contact with other neurons (Kapfhammer and Raper, 1987), and Ca^{2+} influx (see following).

1. Cytoskeletal Mechanisms of Neurite Elongation

As a growth cone advances along a substrate, it exerts tension on its attached neurite shaft. This tension causes elongation of the neurite (Bray, 1984; Lamoureux *et al.,* 1989; Zheng, *et al.,* 1991). According to one model, tension applied to the microtubules in the neurite shaft stimulates addition of tubulin subunits to the microtubules, consolidating the advance begun by the growth cone (Heidemann and Buxbaum, 1994). Whether caused by tension or some other stimulus, microtubules must add tubulin as the neurite elongates. In cultured chick dorsal root ganglion cells, tubulin monomers are added to microtubules at the distal ends of the neurites (Lim, *et al.,* 1990). In cultured frog embryonic neurons, axonal elongation is associated with bundling of microtubules that extend into the growth cone (Tanaka and Kirschner, 1991). In addition to cytoskeletal components, new membrane must also be added, either in the growth cone or along the neurite shaft. In cultured neurons of the mollusk *Aplysia,* the growth cone lamellipodium spreads and then thickens as it is invaded by cytoplasm and organelles. As the old lamellipodium becomes an extension of the neurite shaft, new lamellipodial membrane elaborates distally (Aletta and Greene, 1988). New membrane in elongating *Xenopus* neurites is added at the cell body and along the neurite shaft (Popov, *et al.,* 1993). In cultured chick sensory neurons, when neurites are artificially stimulated to elongate by applying tension, new membrane is added along the neurite shaft (Zheng, *et al.,* 1991).

2. Modulation of Elongation by Intracellular Calcium

The role of Ca^{2+} in neurite elongation is poorly understood, perhaps due to variability among cell types. In some cell types, neurite elongation appears to be enhanced by low $[Ca^{2+}]_i$ and inhibited by Ca^{2+} influx. For example, the Ca^{2+} ionophore A23187 inhibits neurite elongation in rat hippocampal neurons (Mattson *et al.,* 1988) and *Helisoma* neurons (Mattson and Kater, 1987). Low levels of heavy metals (La^{3+}, Co^{2+}), which reduce Ca^{2+} influx through calcium channels, enhance neurite elongation in *Helisoma* neurons, but higher concentrations of heavy metals inhibit elongation (Mattson and Kater, 1987). These data suggest that low, but not zero, Ca^{2+} influx and/or low $[Ca^{2+}]_i$ enhance neurite elongation.

Intermediate levels of $[Ca^{2+}]_i$ favor neurite elongation in cultured chick dorsal root ganglion neurons. Neurite outgrowth is inhibited both by removing Ca^{2+} from the culture medium and by the addition of Ca^{2+} ionophores, suggesting that the permissive level of Ca^{2+} for neurite outgrowth in these cells is neither very high nor very low (Lankford and Letourneau, 1989).

Neurite elongation in some cell types is apparently enhanced by Ca^{2+} influx. Anglister *et al.* (1982) found that depolarization by elevated potassium or the Ca^{2+} ionophore A23187 promotes neurite elongation in N1E-115 neuroblastoma cells, suggesting that high Ca^{2+} influx stimulates elongation.

Neurite elongation in still other cell types may be relatively unaffected by changes in $[Ca^{2+}]_i$. For example, Garyantes and Regehr (1992) electrically stimulated cultured rat superior cervical ganglion neurons and monitored a dramatic increase in $[Ca^{2+}]_i$, but found no change in the rate of elongation. Neurites from *Xenopus* spinal neurons elongate in Ca^{2+}-free media with as much as 5 m*M* EGTA, suggesting that Ca^{2+} influx is not required at all in this system (Bixby and Spitzer, 1984).

This sampling of data suggests that although Ca^{2+} influx and $[Ca^{2+}]_i$ often play important roles in neurite elongation, their roles are complex and may be cell-type specific.

F. Neurite Branching

Neurite branching is perhaps the least understood aspect of neurite development. Branches may arise as bifurcations of a growth cone or as new growth cones emerging from a neurite shaft. Although mechanisms of neurite branching are generally unknown, branching is modulated by a large number of factors, including electrical activity (Schilling *et al.,* 1991), glutamate receptor stimulation (Brewer and Cotman, 1989), and protein kinase activity (Cabell and Audesirk, 1993; Zou and Cline, 1996).

1. Modulation of Branching by Protein Phosphorylation

Based on the structural changes that occur in MAP2 with increasing phosphorylation and the fact that MAP2 is largely confined to neuronal dendrites, Friedrich and Aszodi (1991) hypothesized that increasing MAP2 phosphorylation promotes dendritic branching. In support of this hypothesis, the highest concentration of MAP2 in rat cerebellar Purkinje neurons *in vivo* occurs at the point of maximum dendrite branching (Matus *et al.,* 1990). In cultured hippocampal neurons, phosphorylation of MAP2 (and many other proteins) increases over time in parallel with the development of dendrite branching (Diez-Guerra and Avila, 1993). More direct evidence for a role of protein phosphorylation in branching was obtained by Weeks *et al.* (1991), who found that increasing intracellular cAMP levels, which should increase protein phosphorylation by PKA, increases neurite branching in NG108-15 neuroblastoma cells cultured on a laminin substrate. In hippocampal neurons, treatments that increase protein phosphorylation, such as stimulation of PKA or PKC or inhibition of phosphatases, increase dendrite branching, whereas treatments that decrease protein phosphorylation, such as inhibition of PKA or CaM kinases, reduce dendrite branching (Audesirk *et al.,* 1997).

II. Toxicant Effects on Neurite Development

As this summary indicates, the control of neurite growth involves interactions among the plasma membrane, adhesion molecules and their receptors, the cytoskeleton, intracellular Ca^{2+} concentrations, and numerous second messenger systems, including protein kinases and phosphatases. Biochemical studies have shown that many neurotoxicants disturb one or more of these structures or molecules. Therefore, one would predict that some type of perturbation of neurite development would be a common consequence of neurotoxicant exposure. However, many neurotoxicants may affect several cellular events; for example, inorganic lead has been shown to stimulate PKC, activate calmodulin, inhibit Ca^{2+} influx through voltage-sensitive calcium channels and NMDA receptor/channels, and alter $[Ca^{2+}]_i$ (see Section III.A). Given this multiplicity of neurotoxicant actions and the complex control mechanisms underlying neurite development, the precise effects of a given neurotoxicant on neurite development are not easily predicted.

Neurotoxicant effects on neurite development have been studied both *in vivo* or *in vitro.* During *in vivo* studies, toxicants are usually administered to embryos *in utero* via the pregnant mother or postnatally via mother's milk and later in food or water. Some combination of anatomic, biochemical, and physiologic studies are then performed on the brains of the offspring at various postnatal ages. Most *in vivo* studies to date have been anatomic, providing evidence that a given toxicant causes abnormal neuronal development of certain brain regions. *In vitro* studies assess a variety of anatomic, biochemical, or physiologic parameters in neurons exposed to toxicants in cell culture. Most studies directed at determining the mechanisms whereby toxicants affect neurite development have been performed on cultured neurons.

A. *Targets for Toxicant Action*

We will focus here on only three of the many possible mechanisms whereby toxicants may alter neurite development: alterations of calcium-regulated processes, cytoskeletal structure and function, and protein phosphorylation. Although listed separately, there is broad overlap among these mechanisms of toxicant action; for example, increased $[Ca^{2+}]_i$ may stimulate CaM kinases that phosphorylate cytoskeletal elements.

1. Alteration of Calcium-Regulated Processes

The intracellular free Ca^{2+} ion concentration in neurons is about 50–100 n*M*, approximately 10,000-fold lower than in the extracellular fluid. $[Ca^{2+}]_i$ is regulated by sequestration and release of Ca^{2+} by organelles, binding of Ca^{2+} to intracellular molecules, Na^+–Ca^{2+} exchange and CaATPase extrusion pumps in the plasma membrane, and permeability of neurotransmitter-gated and voltage-sensitive calcium channels. This diversity of mechanisms reflects the multiplicity of roles of Ca^{2+} in a neuron. As described previously, $[Ca^{2+}]_i$ is important in regulating or modulating the initiation of neurites from the soma, the elongation of neurites, and the movement of growth cones. Interactions of $[Ca^{2+}]_i$ with second messengers, such as actin-binding proteins, calmodulin, protein kinases, and protein phosphatases, are important in mediating neuronal differentiation. Because they share certain chemical properties with calcium, a number of metal neurotoxicants affect Ca^{2+} homeostasis by altering Ca^{2+} influx through the plasma membrane, extrusion through the plasma membrane, or sequestration and/or release from intracellular stores. Metals may also bind to Ca^{2+}-regulated molecules such as calmodulin, inhibiting or activating them.

2. Alteration of Cytoskeletal Assembly and Function

Neurite development depends on intricate interactions among multiple cytoskeletal elements, including actin, neurofilaments, and microtubules, and their associated proteins. Several neurotoxicants disrupt the neuronal cytoskeleton, either indirectly by acting on intracellular messengers (including $[Ca^{2+}]_i$, Ca^{2+}-regulated molecules, and a number of kinases), or by direct effects on cytoskeletal elements or their associated proteins.

3. Alteration of Protein Phosphorylation

Several toxicants alter kinase and/or phosphatase activity, either directly or through changes in other intracellular molecules such as Ca^{2+}. As noted previously, extremely low concentrations of inorganic lead stimulate PKC. Manganese and nickel stimulate purified calcineurin *in vitro* (Li, 1984), although whether this happens at environmentally relevant concentrations in living cells has not been determined. Ethanol would be expected to alter $[Ca^{2+}]_i$ and PKC activity by a combination of inhibition of NMDA receptor function (during acute exposure), upregulation of NMDA receptor density (after chronic exposure), inhibition of phosphoinositide metabolism, and upregulation of PKC (see Section III.C.3).

III. Effects of Specific Neurotoxicants on Neurite Development

Many toxicants either are known to alter neurite development or, based on their cellular effects, have the potential to alter neurite development. In the remainder of this chapter, we discuss three specific developmental neurotoxicants: inorganic lead, methylmercury, and ethanol.

A. *Inorganic Lead*

Most heavy metals have no known role in the metabolism of living organisms. Even those that are essential, such as zinc or iron, are normally present in very low amounts in living organisms and/or are regulated very precisely by specialized molecules such as ferritin. The nonessential metals, with inorganic lead (Pb^{2+}) as perhaps the best example, were practically unavailable to living organisms throughout most of evolutionary history, because they were bound in rocks and seldom mobilized in significant amounts in forms that could be assimilated by living organisms, particularly animals. Therefore, it appears that there has been little or no selective pressure favoring the evolution of macromolecules that could bind essential metals, such as zinc or calcium, while excluding nonessential heavy metals. The result is significant susceptibility of modern organisms to metals that have been mobilized by industrial processes.

As a result of human activities, Pb^{2+} is now so ubiquitous that even the most remote human populations probably have body burdens of lead that are orders of magnitude greater than those of our prehistoric ancestors (Settle and Patterson, 1980). Although overt lead poisoning is rare today, subtle dysfunctions can be caused by very low level exposure, especially in children (for reviews, see Bellinger *et al.,* 1991; Schwartz, 1994; Rosen, 1995). Neurobehavioral effects of lead have been observed at blood lead concentrations of 10–15 μg/dl in children, and less than 20 μg/dl in rodents (Davis *et al.,* 1990).

1. *In Vivo* Effects of Inorganic Lead on Neurite Development

A large body of evidence suggests that chronic exposure to Pb^{2+} *in vivo* causes abnormal neuronal development. Exposure of fetal and neonatal rodents, usually rats, to Pb^{2+} causes a variety of changes in the fine structure of neurons and in their synaptic connections. Administration of Pb^{2+} to neonatal rats via milk of dams maintained on a diet of 4.0% $PbCO_3$ until postnatal day 25 caused changes in the morphology of pyramidal cells in the sensorimotor cerebral cortex. A reduction in the number of dendritic branches occurred at distances 80–100 μm away from the cell body (Petit and LeBoutillier, 1979). Lorton and Anderson (1986a) dosed rat pups with 600 mg/kg of lead acetate daily via a stomach tube for 4 days after birth. Blood lead averaged 526 μg/dl after 10 days. At 30 days of age, pyramidal cells in the motor cortex of Pb^{2+}-exposed rats had a significantly decreased length of branches of both apical and basal dendrites. Cerebellar Purkinje cells from these rats showed a 40% decrease in dendritic arborization (Lorton and Anderson 1986b). In hippocampal dentate gyrus cells, rats whose dams were fed a diet containing either 0.4% or 4.0% lead carbonate showed increased dendritic branching close to the cell body, but reduced branching at distances from the cell body of 160 μm and greater (Alfano and Petit, 1982). In 25-day-old rats fed by dams on a 4.0% $PbCO_3$ diet, both the length and width of the mossy fiber tract (composed of axons of dentate gyrus neurons) were significantly reduced (Alfano *et al.,* 1982). Kiraly and Jones (1982) investigated lead effects on the dendritic spine density of hippocampal pyramidal neurons. Newborn rats were fed by Pb^{2+}-exposed dams (given 1% lead acetate in drinking water) until day 25, then received drinking water with 1% lead acetate until day 56. At both 20 and 56 days of age, apical dendritic spine density was significantly reduced in the lead-exposed pups. *In vivo* Pb^{2+} exposure has also been reported to reduce synaptogenesis in rat cortex (Krigman *et al.,* 1974; Petit and LeBoutillier, 1979; Averill and Needle, 1980). In a study by Reuhl *et al.* (1989), monkeys were dosed with Pb^{2+} from birth until 6 years of age and their visual cortices examined. Monkeys receiving 2 mg Pb/kg/day had blood lead levels of about 50 μg/dl, whereas those receiving 25 μg/kg/day had levels of 20 μg/dl or lower. Dendritic arborization was decreased in pyramidal neurons of the visual cortex in the high-dose group compared to the low-dose group. Although the Pb^{2+} dosing regimes varied enormously among these studies, the predominant finding was a decrease in dendrite branching, dendritic spine number, or synapses, all of which would indicate a decrease in connectivity among neurons. In an elegant study by Cline *et al.* (1996), Pb^{2+}-dosed plastic blocks were implanted in the optic tectum of frog tadpoles. Pb^{2+} released from the blocks inhibited both branching and elongation at extracellular concentrations calculated to be as low as 1 pM (Fig. 2).

Somewhat different results were obtained when cortical pyramidal neurons were measured in guinea pig pups at 34 days of age whose dams had been dosed with several concentrations of lead acetate from gestational day 22 until birth (Legare *et al.*, 1993). The pups showed increased apical and basal dendrite length, increased numbers of apical dendrites per cell, and increased branching of basal dendrites. Although blood lead levels were not reported in these animals, extrapolation from an earlier study by these investigators (Sierra *et al.*, 1989) suggests that blood lead peaked in the pups at well below 100 μg/dl.

The differences among these studies are not readily reconciled. However, differences among species and different stages of development, routes of administration, and Pb^{2+} concentrations may produce different results. As we describe later, *in vitro* data both from intact cells and cell-free enzyme studies suggest that Pb^{2+} may have multimodal effects. Furthermore, as briefly described in Section I, normal neurite development may require optimal levels of intracellular Ca^{2+} concentrations, kinase activity, etc. Departures from these optimal levels may either enhance or reduce various aspects of neurite development, and, of course, either hypo- or hypertrophy of neurites may be detrimental to brain functioning.

2. *In Vitro* Effects of Inorganic Lead on Neurite Development

The concentration of lead in cerebrospinal fluid (CSF) or brain extracellular fluid in the experimental animals described previously is unknown. In human populations not known to be exposed to lead, Cavalleri *et al.* (1984) found the total lead concentration in CSF to be approximately 25 n*M* (mean total lead in whole blood approximately 21 μg/dl and total lead in plasma about 30 n*M*). Similiarly, Manton and Cook (1984) found CSF total lead to be less than 50 nM except in rare cases of overt plumbism. Compared to CSF, culture media normally contain higher protein and amino acid concentrations and have somewhat higher pH. Many proteins and some amino acids (e.g., cysteine) bind lead with fairly high affinity; furthermore, ionic lead is less soluble and less likely to be in free form at high pH. Therefore, most culture media would be expected to contain a higher proportion of bound lead and less free lead than CSF of the same total lead concentration. Simple physiologic salines, however, usually contain no protein or amino acids, and thus may have higher free lead concentrations than CSF of equivalent total lead concentrations. With these caveats, the concentration of lead in CSF nevertheless provides a starting point for evaluation of the likely environmental relevance of culture studies.

The effects of *in vitro* exposure of cultured neurons to Pb^{2+} vary considerably, depending on the species, cell type, and parameter of neurite development measured. The effects of Pb^{2+} exposure *in vitro* on neuronal differentiation have been studied in rat dorsal root ganglion cells or explants, IMR32 human neuroblastoma cells, embryonic chick brain neurons, embryonic rat cortical neurons, embryonic rat hippocampal neurons, N1E-115 neuroblastoma cells (derived from mouse peripheral nervous system), and B-50 neuroblastoma cells (derived from rat central nervous system [CNS]). In rat dorsal root ganglion explants, neurite elongation was inhibited by Pb^{2+} exposure for 2–3 days at concentrations of approximately 500 μM and higher (Windebank, 1986). Myelination, however, was almost completely prevented by concentrations of 1 μM or higher. Scott and Lew (1986) found that survival of rat dorsal root ganglion cells was reduced by 50% following 18 days of exposure to 35 μM Pb^{2+}. In IMR32 neuroblastoma cells, survival was reduced by 50% by exposure to 336 μM for 10 days, and 260 μM Pb^{2+} reduced neurite initiation by about 45%, although elongation remained almost normal (Gotti *et al.*, 1987). In chick neurons, Pb^{2+} exposure for 3–4 days (beginning with plating) inhibited neurite initiation at concentrations of 10 μM and higher, and enhanced neurite elongation at concentrations of 100 μM and higher (Audesirk *et al.*, 1989). Two day exposures (beginning with plating) to Pb^{2+} concentrations between 10 n*M* and 500 μM had very little effect on any parameter of development in B50 neuroblastoma cells (survival, initiation rate, number of neurites per cell, elongation of neurites; Audesirk *et al.*, 1991). For N1E-115 neuroblastoma cells, rat hippocampal neurons, and rat cortical neurons, there were complex, multimodal dose–response curves for several parameters of neuronal differentiation (Audesirk *et al.*, 1991; Kern *et al.*, 1993). Generally, in cortex and hippocampal neurons, mid-to-high nanomolar and mid-to-high micromolar Pb^{2+} concentrations inhibited initiation and enhanced branching, with low micromolar concentrations having little effect on any parameter of neurite development.

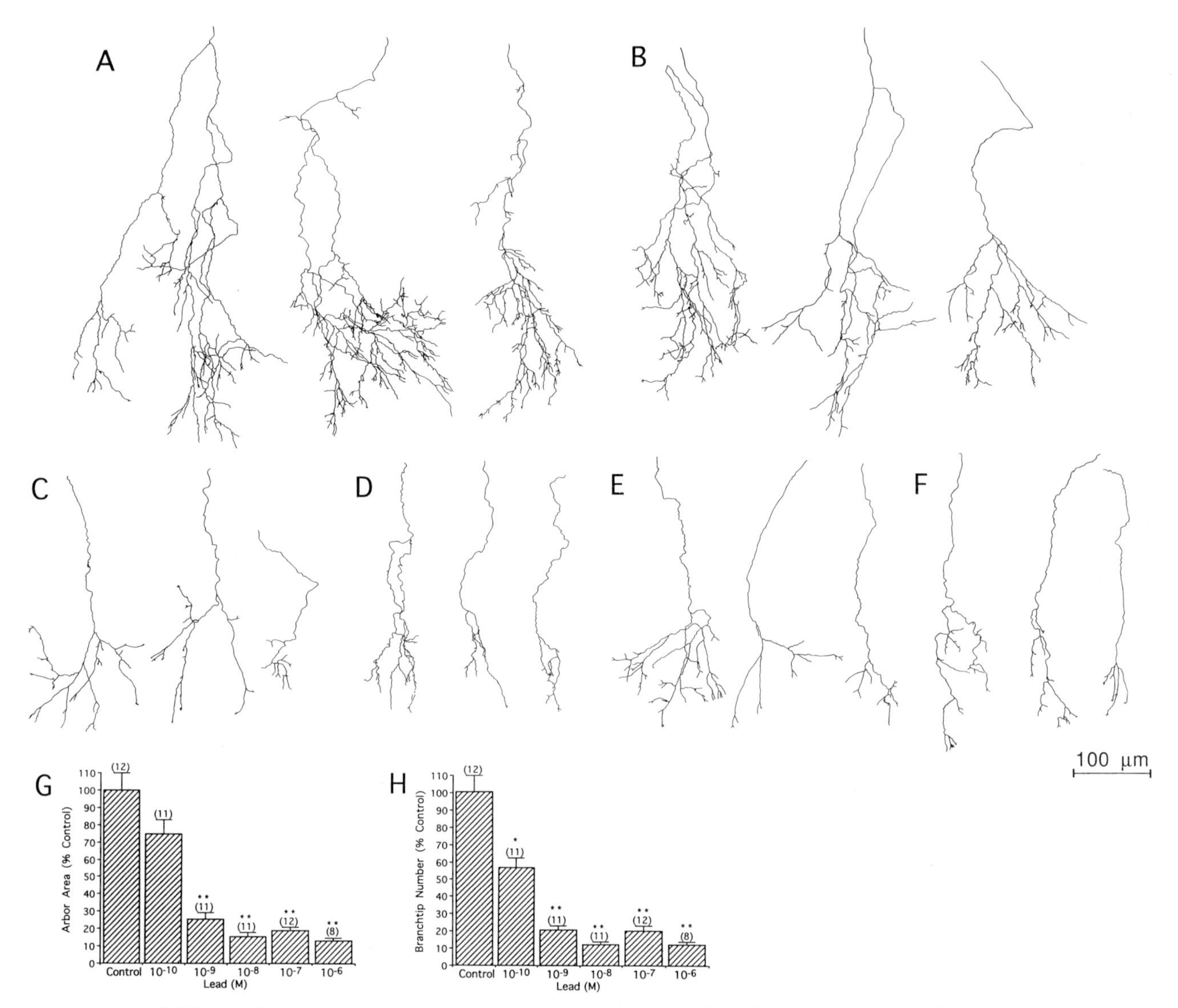

FIGURE 2 Camera lucida drawings of axon arbors of retinal ganglion cells in the optic tecta of *Xenopus.* Lead-impregnated plastic blocks were implanted into the optic tectum during development, producing estimated lead concentrations in the tectum of 10^{-10} (B) to 10^{-6} M (F). (A) Axon arbors from control tadpoles into which lead-free plastic blocks were implanted. Chelation therapy with 2,3-dimercaptosuccinic acid (DMSA) reversed the effects of lead (not shown). (Reprinted from Cline, H. T., Witte, S., and Jones, K. W. Low lead levels stunt neuronal growth in a reversible manner. *Proc. Natl. Acad. Sci. USA* **93,** 9915–9920, copyright 1996, National Academy of Sciences, U.S.A., and with the kind permission of H. T. Cline.)

It is difficult to summarize such disparate data, but a few generalizations may be made. First, immortal cell lines appear to be relatively insensitive to Pb^{2+} effects, as shown by the high concentrations needed to have any significant effects (Gotti *et al.,* 1987; Audesirk *et al.,* 1991). Second, neuronal survival is also relatively resistant to Pb^{2+}, even in primary neurons (Audesirk *et al.,* 1991; Kern *et al.,* 1993). Third, myelination, neurite initiation, and, perhaps, branching appear to be the most sensitive parameters of neuronal differentiation, with inhibition or enhancement occurring at concentrations of 1 μM or less (Windebank, 1986; Audesirk *et al.,* 1991; Kern *et al.,* 1993).

3. Cellular Mechanisms

Inorganic lead inhibits Ca^{2+} influx through voltage-sensitive calcium channels at low micromolar concentrations in a variety of cell types, including mouse N1E-115 neuroblastoma cells (Oortgiesen *et al.,* 1990; Audesirk and Audesirk, 1991), human neuroblastoma cells (Reuveny and Narahashi, 1991), rat dorsal root gangion cells (Evans *et al.,* 1991), bovine chromaffin cells (Sun and Suszkiw, 1995), and embryonic rat hippocampal neurons (Audesirk and Audesirk, 1993; Ujihara *et al.,* 1995). Pb^{2+} also inhibits current flow through some Ca^{2+}-permeable receptor-operated channels, notably icotinic acetylcholine receptor/channels (Oortgiesen *et al.,* 1990; Ishihara *et al.,* 1995) and NMDA receptor/channels (Alkondon *et al.,* 1990; Uteshev *et al.,* 1993).

Normal rates of Ca^{2+} influx through voltage-sensitive calcium channels and receptor-operated channels promote neurite development (see Section I), so levels of Pb^{2+} sufficient to block these calcium channels would be expected to alter neurite growth. However, the Pb^{2+} concentrations reported to block voltage-sensitive calcium channels and NMDA receptor-channels (generally in the micromolar range) are considerably higher than those found in CSF of Pb^{2+}-exposed humans (low to mid-nanomolar). Therefore, it would seem that blocking Ca^{2+} influx through these channels would be unlikely to be an important mechanism of developmental neurotoxicity of Pb^{2+} unless there are much higher localized extracellular Pb^{2+} concentrations in the brain, for example in synaptic clefts. Pb^{2+} may block certain types of nicotinic receptor/channels at much lower concentrations (Oortgiesen *et al.,* 1990).

Chronic Pb^{2+} exposure may also alter NMDA channel and/or receptor density. For example, chronic exposure *in vivo* has been found to increase (Brooks *et al.,* 1993; Guilarte *et al.,* 1993; Jett and Guilarte, 1995) or decrease (Guilarte and Miceli, 1992; Jett and Guilarte, 1995) NMDA receptor density, apparently depending on the brain region studied and the exposure regime (including animal age). A decrease in NMDA receptor/channel density would certainly be expected to alter neurite development. Alternatively, increased NMDA receptor/channel density may overcompensate for an ongoing partial block of channel function by Pb^{2+}, allowing increased Ca^{2+} influx, which would also alter neurite development. Much more research is needed to decide among these possibilities.

Other interactions of Pb^{2+} with Ca^{2+}-permeable channels may also be important. For example, Pb^{2+} ions readily permeate through calcium channels (Simons and Pocock, 1987; Tomsig and Suszkiw, 1991). Once inside a neuron, Pb^{2+} may influence a variety of Ca^{2+}- mediated intracellular processes (see reviews by Bressler and Goldstein, 1991; Goldstein, 1993; Simons, 1993). For example, intracellular Pb^{2+} may increase Ca^{2+} influx through calcium channels by inhibiting Ca^{2+}-dependent channel inactivation (Sun and Suszkiw, 1995). Calcium channels are also modulated by phosphorylation, with both inhibitory and excitatory effects reported, apparently dependent on the specific channel subtype and the phosphorylating kinase under study. Intracellular Pb^{2+}, by activating PKC, calmodulin (and hence CaM kinases), or other kinases, either directly or indirectly, may therefore have similarly complex effects on calcium channel function.

One difficulty in elucidating the mechanisms underlying intracellular effects of Pb^{2+} is the paucity of information on Pb^{2+} concentrations inside cells. Following short-term exposure to relatively high concentrations of Pb^{2+}, the intracellular free Pb^{2+} ion concentration has been reported to be in the range of tens of picomolar (Schanne *et al.,* 1989a; Tomsig and Suszkiw, 1991). In brain slices taken from rats chronically exposed to a high dose of lead acetate (100 mg/kg/day), Singh (1995) reported free intracellular lead Pb^{2+} concentrations of 178 p*M*.

Only a few biochemical studies of Pb^{2+} effects on intracellular proteins have used such low Pb^{2+} concentrations. For example, Pb^{2+} in picomolar concentrations may activate PKC (Markovac and Goldstein, 1988; Rosen *et al.,* 1993; Long *et al.,* 1994), whereas higher levels may block PKC activation (Speizer *et al.,* 1989; Rosen *et al.,* 1993). PKC, in turn, has diverse effects on differentiation and growth in several types of cultured neurons (see Section I). Pb^{2+} may also substitute for Ca^{2+} in the activation of calmodulin (Goldstein and Ar, 1983; Habermann *et al.,* 1983). Lead may also increase $[Ca^{2+}]_i$ (Schanne *et al.,* 1989a,b; Schanne *et al.,* 1990), and this increase may activate calmodulin (Goldstein, 1993); however, Singh (1995) found no effect of chronic Pb^{2+} exposure *in vivo* on $[Ca^{2+}]_i$, so the effects of Pb^{2+} on $[Ca^{2+}]_i$ remain uncertain. Ca^{2+}-calmodulin or Pb^{2+}-calmodulin in turn may regulate the activity of protein kinases, such as CaM kinase II, that modulate cytoskeletal polymerization, crosslinking, and attachment to membranes (see, for example, Matus, 1988b; Goedert *et al.,* 1991). Pb^{2+}-induced hyperphosphorylation may inhibit neurite initiation in cultured hippocampal neurons; inhibitors of calmodulin or several protein kinases completely reversed the inhibition of initiation by 100 n*M* Pb^{2+}, whereas stimulators of protein kinases and inhibitors of protein phosphatases mimic the inhibition by Pb^{2+} (Kern and Audesirk, 1995). In general, however, because the molecular mechanisms whereby the cytoskeleton and cytoskeleton-associated molecules generate normal neurite development are incompletely understood, exactly how the enhancement or inhibition of the function of these molecules by Pb^{2+} would disturb neurite development remains unclear.

Although there are very few studies available, it appears unlikely that Pb^{2+} has significant direct interactions with the cytoskeleton. For example, Roderer and Doenges (1983) found no effects on tubulin polymerization *in vitro* with Pb^{2+} at concentrations as high as

650 μM, a concentration several orders of magnitude greater than would ever occur inside a cell. However, because Pb^{2+} can substitute for Ca^{2+}, and because Ca^{2+} mediates so many of the intracellular processes governing cytoskeletal integrity (discussed previously), direct or indirect interference with the neuronal cytoskeleton may be an important, but as yet undocumented, mechanism underlying lead neurotoxicity.

Finally, Pb^{2+} also has substantial effects on glial cells *in vivo* (e.g., Cookman *et al.*, 1988; Selvin-Testa *et al.*, 1991; Buchheim *et al.*, Stoltenburg-Didinger *et al.*, 1996) and *in vitro* (e.g., Sobue and Pleasure, 1985; Cookman *et al.*, 1988; Rowles *et al.*, 1989; Stark *et al.*, 1992; Dave *et al.*, 1993; Opanashuk and Finkelstein, 1995; for reviews, see Tiffany-Castiglioni *et al.*, 1989, 1996). Pb^{2+} also affects cell adhesion molecules (Lagunowich *et al.*, 1993; Regan, 1993). Impacts on glia or cell adhesion would also be expected to alter neuronal differentiation. A discussion of these effects is beyond the scope of this chapter.

B. Methylmercury

Mercury enters the environment by weathering of geologic deposits and from industrial sources and agriculture. After entry into aquatic ecosystems, inorganic mercury is methylated into more toxic methylmercury, which bioaccumulates in fish and shellfish. Methylmercury has devastating effects on the developing fetal nervous system (for reviews see Choi, 1989; O'Kusky, 1992; Massaro, 1996). Although inorganic mercury has effects on microtubules, Ca^{2+} channels and Ca^{2+} homeostasis, and therefore would be expected to impair neurite development, it appears to be less toxic than methylmercury. This review focuses exclusively on methylmercury.

1. *In Vivo* Effects of Methylmercury on Neurite Development

At relatively high doses, methylmercury causes widespread cell loss in the brain, with some regional differences in sensitivity (Burbacher *et al.*, 1990; Choi, 1989; O'Klusky, 1992). There are relatively few detailed studies of neurite structure and development *in vivo*. Stoltenburg-Didinger and Markwort (1990) exposed pregnant rats to methylmercury in doses of 0.025 to 5 mg/kg/day on days 6–9 of gestation. At 250 days of age, pyramidal neurons in the somatosensory cortex of the offspring of all exposed groups showed normal dendritic arborization, but both apical and basilar dendrites from the 5 mg/kg group had abnormal spines. Although controls had stubby, mushroom-shaped spines, methylmercury-exposed neurons had spines that were longer, thinner, more distorted and more numerous. Because the methylmercury-exposed dendritic spines were similar to those found normally up until 30 days after birth, the authors suggested that this may be related to the developmental retardation seen in the offspring. Choi *et al.* (1981b) injected mice on postnatal days 3, 4, and 5 with 5 mg/kg methylmercuric chloride and examined the morphology of Purkinje cells in the cerebellum. They found a dramatic reduction in the dendritic arborization of Purkinje cells in the exposed mice.

2. *In Vitro* Effects of Methylmercury on Neurite Development

Methylmercury is highly cytotoxic to cultured neurons; for example, cerebellar granule neurons are killed at concentrations as low as 0.1 μM, probably by apoptosis (Kunimoto, 1994). There are few studies specifically of neurite development *in vitro*. Neurite outgrowth and filopodial activity rapidly ceased in organotypic cultures of human fetal cerebrum exposed to 20 μM methylmercuric chloride (Choi *et al.*, 1981a). Furthermore, the plasma membrane became separated from the neurites, especially in growth cones. Neurites degenerated and separated from the cell bodies within 4 hrs. Electron microscopic examination of cultures revealed neurotubular damage.

3. Cellular Mechanisms

Methylmercury has been reported to increase $[Ca^{2+}]_i$ and block Ca^{2+} channels. In rat brain synaptosomes, methylmercury increases $[Ca^{2+}]_i$ in a dose-dependent manner (Komulainen and Bondy, 1987; Denny *et al.*, 1993). Methylmercury also increases $[Ca^{2+}]_i$ in cultured cerebellar granule neurons (Sarafian, 1993) and NG108-15 neuroblastoma cells (Hare *et al.*, 1993). Although the specific mechanisms leading to increased $[Ca^{2+}]_i$ are not well understood, methylmercury may both increase release of Ca^{2+} from intracellular stores (Hare and Atchison, 1995) and cause increased influx from the extracellular fluid (Komulainen and Bondy, 1987; Denny *et al.*, 1993; Hare *et al.*, 1993).

Acute methylmercury exposure inhibits neuromuscular transmission (Atchison and Narahashi, 1982), blocks calcium channels in PC12 cells (Shafer and Atchison, 1991) and cerebellar granule neurons (Sirois and Atchison, 1996), and reduces ^{45}Ca uptake into synaptosomes (Shafer and Atchison, 1989), mostly in the low-to-mid micromolar concentration range, although inhibition of current flow through calcium channels in granule neurons occurs at submicromolar concentrations. This appears difficult to reconcile with the increased $[Ca^{2+}]_i$ noted previously that is dependent on extracellular Ca^{2+}; one possibility is that methylmercury causes Ca^{2+} influx through other channels, perhaps those responsible for refilling depleted intracellular Ca^{2+} stores.

Because neurite initiation, at least in some cell types, is modulated by Ca^{2+} influx through calcium channels and/or by Ca^{2+} release from intracellular stores, these data suggest that neurite production might be altered by methylmercury concentrations in the low micromolar or even submicromolar range.

Methylmercury has long been known to inhibit cytoskeletal assembly. Methylmercury disrupts microtubules and inhibits their polymerization *in vitro* at low micromolar levels (Abe *et al.,* 1975; Sager *et al.,* 1983; Miura *et al.* 1984; Vogel *et al.,* 1985). In cultured cells, methylmercury disrupts existing microtubules (fibroblasts: Sager *et al.* 1983, Graff *et al.,* 1993; mouse glioma cells: Miura *et al.,* 1984; PtK_2 kidney epithelial cells: Sager, 1988). It appears that methylmercury inhibits microtubule assembly or stability at lower concentrations in intact cells than *in vitro.* This indicates that interference with other factors that influence cytoskeletal structure or function is probably more important than direct effects on the cytoskeletal proteins themselves.

Exposing cerebellar granule cells in culture to methylmercury at concentrations from 200 n*M* to 3.0 μM stimulates phosphorylation of a variety of proteins, including cytoskeletal proteins, in a dose-dependent manner (Sarafian and Verity, 1990). Protein phosphorylation, in turn, influences the interactions between microtubules, MAPs, and neurofilaments, as well as the stability of microtubules (see Section I). Because methylmercury inhibits several protein kinases (Inoue *et al.,* 1988; Saijoh *et al.,* 1993), this increased protein phosphorylation may be mediated indirectly via a second messenger, such as Ca^{2+} or inositol phosphate. Methylmercury-induced increases in $[Ca^{2+}]_i$ could activate calmodulin and therefore CaM kinases, which phosphorylate many cytoskeletal components. In addition, methylmercury (0.5–10 μM) produces a dose-dependent increase in levels of inositol phosphate in cultured cerebellar granule cells (Sarafian, 1993). This suggests that methylmercury may activate phospholipase C and could stimulate PKC by production of diacylglycerol. PKC phosphorylates many cytoskeletal proteins.

Finally, injecting methylmercury into frog sciatic nerve impairs axonal transport (Abe *et al.,* 1975). Neurite development is dependent on axonal transport for the delivery of many materials to the growth cone and elongating neurite shaft. Therefore, inhibiting axonal transport would be expected to inhibit neurite development.

C. Ethanol

Ethanol is the most widely and voluntarily consumed neurotoxicant in most human societies. In many cultures, the vast majority of adults drink at least small amounts of ethanol, and many adults drink moderate to large quantities. Fetal alcohol syndrome has long been recognized as a sequel of moderate to heavy ethanol consumption by pregnant women (Jones and Smith, 1973; reviewed by Abel, 1984). Animal models of fetal alcohol syndrome resemble the human condition, including substantial loss of neurons in the hippocampus (Walker *et al.,* 1980) and cerebellum (Bauer-Moffett and Altman, 1977). Animal models also show that ethanol alters neurite development (reviewed by Jones, 1988; Pentney and Miller, 1992).

1. *In Vivo* Effects of Ethanol on Neurite Development

Numerous studies have examined the effects of *in vivo* ethanol exposure on neuronal morphology, particularly of the dendritic trees of neurons in the cerebellum, hippocampus, and neocortex. Interestingly, there are reports of both increases and decreases in dendritic development. For example, decreased neuronal development, most often decreased dendritic trees, has been observed in CA1 hippocampal pyramidal neurons (Davies and Smith, 1981; McMullen *et al.,* 1981), cerebellar granule neurons (Smith *et al.,* 1986), neocortical pyramidal neurons (Hammer and Scheibel, 1981), and substantia nigra pars compacta neurons (Fig. 3) (Shetty *et al.,* 1993). In apparent contrast, Cadete-Leite *et al.* (1988) found that neurons of the dentate gyrus of ethanol-exposed rats had significantly larger dendritic trees. Ethanol also caused dendritic hypertrophy in Purkinje neurons of the cerebellum (Pentney and Quackenbush, 1990; Pentney, 1995). Similarly conflicting data occur for dendritic spines, with both decreases (e.g., Stoltenburg-Didinger and Spohr, 1983; King *et al.,* 1988) and increases (Ferrer *et al.,* 1988) having been reported.

There are several confounding variables that make these results somewhat difficult to interpret. For example, in some (e.g., Davies and Smith, 1981; Smith *et al.,* 1986; Shetty *et al.,* 1993) but not all (e.g., Hammer and Scheibel, 1981; McMullen *et al.,* 1981; Cadete-Leite *et al.,* 1988) studies, the body weight of ethanol-exposed animals was significantly smaller than either the chow-fed or sucrose-fed controls, or both. Other differences among studies include the species and/or strain of animal (e.g., rats vs. mice; Sprague-Dawley vs. Fischer 344 rats), the ages of the animals, the durations and levels of ethanol exposure, and the timing of morphologic evaluation relative to ethanol exposure. Despite these uncertainties, the concensus seems to be that ethanol exposure during development generally causes a decrease in dendritic trees and dendritic spines. After withdrawal there may be "catch-up" growth, masking probable earlier decreases, or even hypertrophy, for example as fewer remaining neurons elaborate more extensive

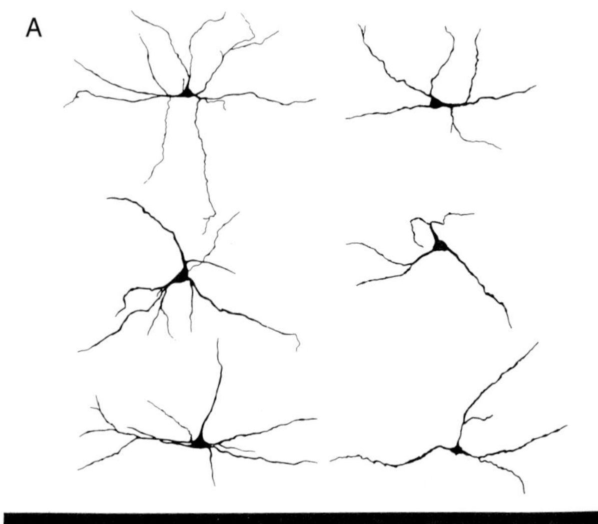

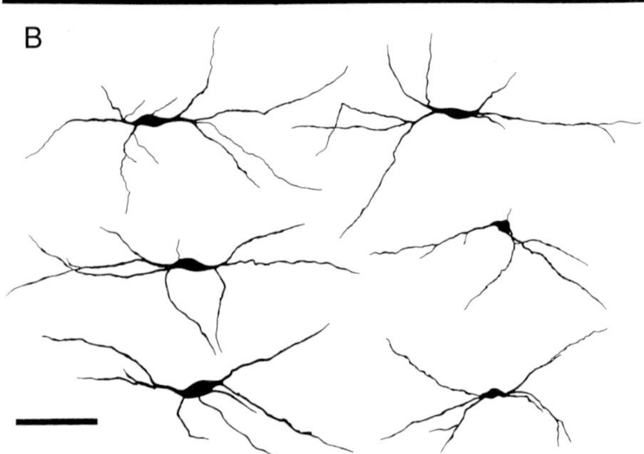

FIGURE 3 Camera lucide drawings of (A) Golgi-stained pyramidal neurons and (B) fusiform neurons of the substantia nigra pars compacta of the rat. In both parts of the figure, neurons in the left column are from control rat pups (P15) and neurons in the right column are from P15 rat pups exposed prenatally to ethanol. Dendrites of both types of neurons from the ethanol-exposed rats are less developed than dendrites of neurons from control rats. (Reprinted from Shetty, A. K., Burrows R. C., and Phillips, D. E. Alterations in neuronal development in the substantia nigra pars compacta following *in utero* ethanol exposure: immunohistochemical and Golgi studies. *Neuroscience* **52,** 311–322, copyright 1993, with kind permission from Elsevier Science Ltd., The Boulevard, Langford Lane, Kidlington OX5 1GB, U.K., and from A. K. Shetty and D. E. Phillips.)

dendritic fields per neuron, presumably compensating for the missing dendritic fields of neurons lost during development.

2. *In Vitro* Effects of Ethanol on Neurite Development

All possible effects of *in vitro* ethanol exposure on neurite development have been reported: increase, no effect, and decrease. In rat cerebellar macroneurons (75 m*M*; Zou *et al.,* 1993) and PC12 cells (25–150 m*M*; Messing *et al.,* 1991a; Roivainen *et al.,* 1993), ethanol exposure increases neurite initiation; elongation and branching also increase in cerebellar macroneurons (Fig. 4). In neuro-2A neuroblastoma cells, ethanol has no effect on initiation at concentrations up to about 250 m*M* (Leskawa *et al.,* 1995). In contrast, a wide range of ethanol concentrations inhibit neurite initiation in chick dorsal root ganglion cells (Dow and Riopelle, 1985; Heaton *et al.,* 1993), and rat septal and hippocampal neurons (Heaton *et al.,* 1994). In LA-N-5 human neuroblastoma cells (Saunders *et al.,* 1995) and rat hippocampal neurons (Matsuzawa *et al.,* 1996), ethanol exposure reduces neurite elongation. In many cases, ethanol also inhibits cell proliferation or enhances cell death, or both, but usually only at concentrations much higher than those required to alter neurite development.

In vitro ethanol exposure may also exert complex effects on neurite development. For example, ethanol (100 m*M*) inhibits neurite outgrowth toward appropriate target tissues in both chick embryo spinal cord explants and fetal rat hippocampal explants (Heaton *et al.,* 1995). If such effects occur *in vivo,* ethanol would be expected to profoundly alter pathfinding and connectivity patterns in the brain.

3. Cellular Mechanisms

In general, there have been relatively few studies specifically designed to elucidate cellular mechanisms whereby ethanol might alter neurite development. Research by Messing and colleagues in PC12 cells is perhaps the most thorough investigation of mechanisms of ethanol actions on neurite development. Ethanol, which enhances neurite initiation in this cell type (Messing *et al.* 1991a), upregulates the δ and ε isoforms of PKC (Roivainen *et al.,* 1993) and stimulates PKC-mediated protein phosphorylation (Messing *et al.,* 1991b). Stimulating PKC activity with low concentrations of phorbol esters at least qualitatively mimics the effects of ethanol in stimulating neurite initiation, whereas downregulating the β, δ, and ε isoforms of PKC with higher concentrations of phorbol esters completely prevents the stimulation of initiation by ethanol (Roivainen *et al.,* 1993). Finally, exposure to ethanol causes upregulation of L-type calcium channels, an effect that also appears to be mediated by PKC (Messing *et al.,* 1990). Given that PKC, protein phosphorylation, and Ca^{2+} influx through L-type calcium channels have all been implicated in regulation of neurite development, it is likely that the stimulation of neurite initiation by ethanol, at least in PC12 cells, may be caused by these cellular effects.

Other potential mechanisms of ethanol actions on neurite development include interference with cellular responses to neurotrophic substances such as nerve growth factor (Heaton *et al.,* 1993, 1994), reduced production of neurotrophic or neurotropic factors (Heaton

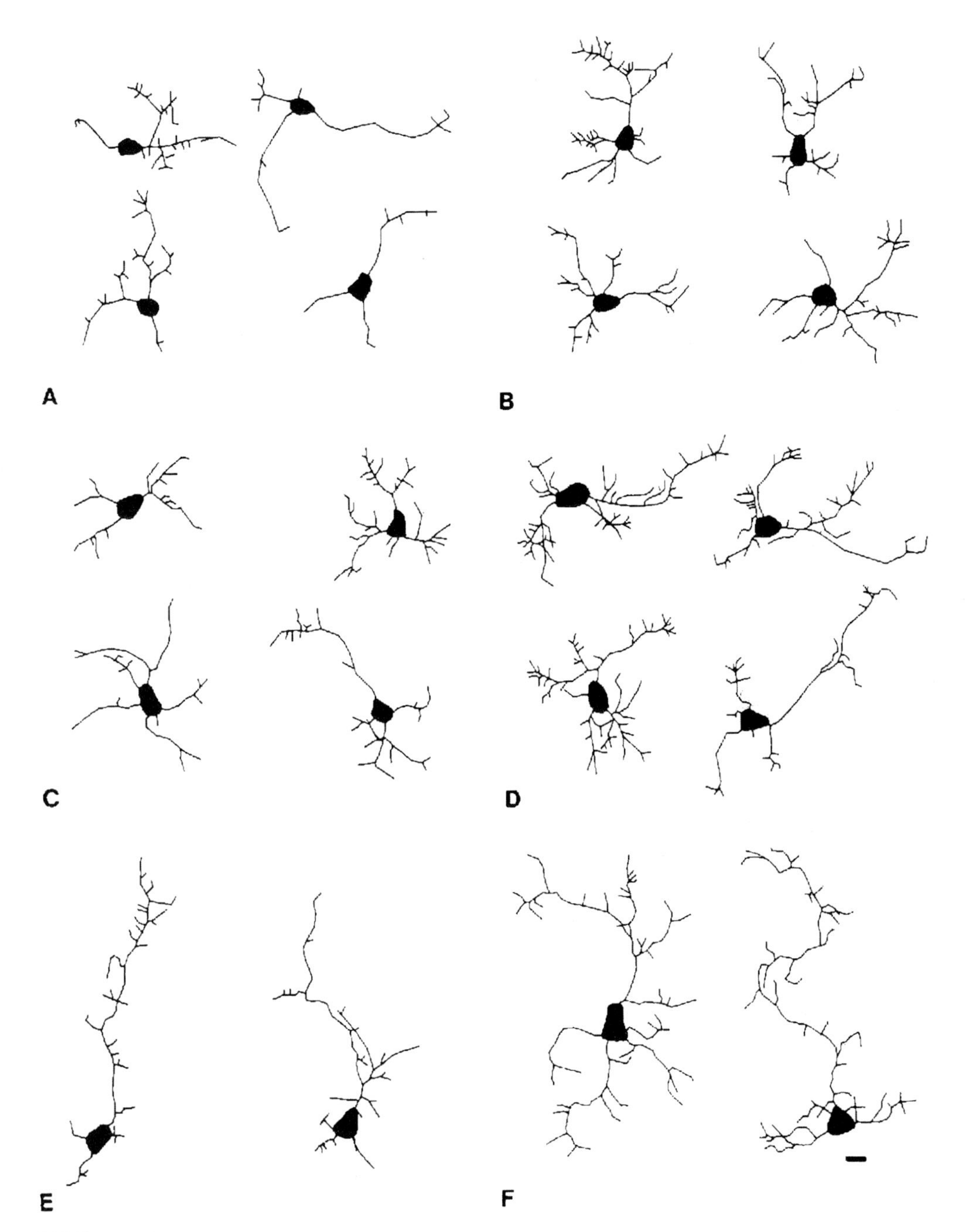

FIGURE 4 Computer-assisted drawings of cultured, MAP2-immunostained cerebellar macroneurons. Parts A, C, and E are drawing of neurons from control cultures after 8, 24, and 48 hours in culture, respectively. Parts B, D, and F are drawings of neurons from cultures exposed to 75 m*M* ethanol for the 8, 24, and 48 hours, respectively. Ethanol exposure enhanced both elongation and branching of neurites. (Reprinted from Zou, J.-Y., Rabin, R. A., and Pentney, R. J. Ethanol enhances neurite outgrowth in primary cultures of rat cerebellar macroneurons. *Dev. Brain Res.* **72,** 75–84, copyright 1993, with kind permission from Elsevier Science—NL, Sara Burgerhartstraat 25, 1055 KV Amsterdam, The Netherlands and from R. J. Pentney.)

et al., 1995), reduction of neural cell adhesion (Charness *et al.,* 1994), elevated intracellular Ca^{2+} concentrations (Zou *et al.,* 1995), reduced phosphoinositide metabolism (Kovacs *et al.,* 1995; Tan and Costa, 1995), decreased IP_3 receptors that govern Ca^{2+} release from intracellular stores (Saito *et al.,* 1996), and impaired function of glial cells (e.g., Miller and Potempa, 1990; Lokhorst and Druse, 1993). Finally, acute ethanol exposure *in vitro* reduces Ca^{2+} influx through both NMDA receptor/channels (Hoffmann *et al.,* 1989; Lovinger *et al.,* 1989) and L-type calcium channels (e.g., Twombly *et al.*, 1990; Mullikin-Kilpatrick and Treistman, 1993), whereas chronic exposure both *in vitro* and *in vivo* cause upregulation of both channel types (NMDA receptor/channels: Snell *et al.*, 1993; Trevisan *et al.*, 1994; L-type channels: Messing *et al.*, 1986; Guppy and Littleton, 1994; Little, 1995). All of these processes have been shown to be involved in neuronal differentiation; their alteration by ethanol would therefore be expected to interfere with normal neurite development.

IV. Summary

Neurite development is a series of complex processes whose control mechanisms are far from understood. A large number of cellular processes may modulate neurite development, including Ca^{2+} fluxes and concentrations, a variety of second messenger molecules such as cyclic AMP and IP_3, protein phosphorylation and dephosphorylation, cytoskeletal assembly and disassembly, and the actions of specific growth-associated proteins such as GAP-43. Many of these processes are in turn modulated by responses to growth factors, substrate attachment molecules, neurotransmitters, and electric fields. Insofar as neurotoxicants may alter one or more of these processes, toxicant exposure would be expected to alter neurite development.

However, although neurite development is clearly affected by *in vivo* exposure to a number of neurotoxicants, it remains unclear in most cases whether altered development is a direct effect on neurons and their processes, on glial cells, on general bodily development with secondarily deleterious effects on the nervous system, or some combination of these. In at least some cases, different studies report seemingly opposite effects. Differences in the *in vivo* and *in vitro* models used, differences among brain regions, great disparities in the concentrations of toxicants used in different laboratories, and variable methodologies and levels of analysis often lead to significant differences in effects reported even for single toxicants. Even in culture systems, effects do not appear to be consistent, for example as noted above with *in vitro* ethanol exposure.

Inadequate knowledge of the fundamental mechanisms of neurite development, both *in vivo* and *in vitro,* hampers investigations of effects and mechanisms of toxicant action. However, as the mechanisms of neurite development and the cellular targets of toxicant action become better understood, our understanding of how toxicants alter axonal and dendritic morphology and connectivity will advance rapidly. Because of the obvious importance of neurite development for proper wiring of the brain and, consequently, for normal behavior, we predict that the effects of toxicants on neurite development will become increasingly recognized as an important aspect of developmental neurotoxicity.

References

Abe, T., Haga, T., Kurokawa, M. (1975). Blockage of axoplasmic transport and depolymerisation of reasssembled microtubules by methyl mercury. *Brain Res.* **86,** 504–508.

Abel, E. L. (1984). *Fetal Alcohol Syndrome and Fetal Alcohol Effects.* Plenum Press, New York.

Abosch, A., and Lagenaur, C. (1993). Sensitivity of neurite outgrowth to microfilament disruption varies with adhesion molecule substrate. *J. Neurobiol.* **24,** 344–355.

Aigner, L., and Caroni, P. (1993). Depletion of 43-kD growth-associated protein in primary sensory neurons leads to diminshed formation and spreading of growth cones. *J. Cell Biol.* **123,** 417–429.

Aigner, L., and Caroni, P. (1995). Absence of persistent spreading, branching, and adhesion in GAP-43–depleted growth cones. *J. Cell Biol.* **128,** 647–660.

Aletta, J., and Greene, L. A. (1988). Growth cone configuration and advance: a time-lapse study using video-enhanced differential interference contrast microscopy. *J. Neurosci.* **8,** 1425–1435.

Alfano, D. P., LeBoutillier, J. C., and Petit, T. L. (1982). Hippocampal mossy fiber development in normal and postnatally lead-exposed rats. *Exp. Neurol.* **75,** 308–319.

Alfano, D. P. and Petit, T. L. (1982). Neonatal lead exposure alters the dendritic development of hippocampal dentate granule cells. *Exp. Neurol.* **75,** 275–288.

Alkondon, M., Costa, A. C. S., Radhakrishnan, V., Aronstam, R. S., and Albuquerque, E. X. (1990). Selective blockade of NMDA-activated channel currents may be implicated in learning deficits caused by lead. *FEBS Lett.* **261,** 124–130.

Anglister, L., Farber, I. C., Shahar, A., and Grinvald, A. (1982). Localization of voltage-sensitive calcium channels along developing neurites: their possible role in regulating neurite elongation. *Dev. Biol.* **94,** 351–365.

Armstrong, D. L., (1989). Calcium channel regulation by calcineurin, a Ca^{2+}-activated phosphatase in mammalian brain. *Trends Neurosci.* **12,** 117–122.

Atchison, W. D., and Narahashi, T. (1982). Methylmercury-induced depression of neuromuscular transmission in the rat. *Neurotoxicol.* **3,** 37–50.

Audesirk, G., and Audesirk, T. (1991). Effects of inorganic lead on voltage-sensitive calcium channels in N1E-115 neuroblastoma cells. *Neurotoxicol.* **12,** 519–528.

Audesirk, G., and Audesirk, T. (1993). The effects of inorganic lead on voltage-sensitive calcium channels differ among cell types and among channel subtypes. *Neurotoxicol.* **14,** 259–266.

Audesirk, G., Shugarts, D., Nelson, G., and Przekwas J. (1989). Organic and inorganic lead inhibit neurite growth in vertebrate and invertebrate neurons in culture. *In Vitro Cell Dev. Biol.* **25,** 1121–1128.

Audesirk, G., Audesirk, T., Ferguson, C., Lomme, M., Shugarts, D., Rosack, J., Caracciolo, P., Gisi, T., and Nichols, P. (1990). L-type calcium channels may regulate neurite initiation in cultured chick embryo brain neurons and N1E-115 neuroblastoma cells. *Dev. Brain Res.* **55,** 109–120.

Audesirk, T., Audesirk, G., Ferguson, C., and Shugarts, D. (1991). Effects of inorganic lead on the differentiation and growth of cultured hippocampal and neuroblastoma cells. *Neurotoxicol.* **12,** 529–538.

Audesirk, G., Cabell, L., and Kern, M. (1997). Modulation of neurite branching by protein phosphorylation in cultured rat hippocampal neurons. *Dev. Brain. Res.,* **102,** 247–260.

Averill, D. R., and Needleman, H. L. (1980). Neonatal lead exposure retards cortical synaptogenesis in the rat. In *Low Level Lead Exposure: The Clinical Implications of Current Research.* (H. L. Needleman, Ed.), pp 201–210. Raven Press, New York.

Avila, J., Dominguez, J., and Diaz-Nido, J. (1994a) Regulation of microtubule dynamics by microtubule-associated protein expression and phosphorylation during neuronal development. *Int. J. Dev. Biol.* **38,** 13–25.

Avila, J., Ulloa, L., Diez-Guerra, J., and Diaz-Nido, J. (1994b). Role of phosphorylated MAP1B in neuritogenesis. *Cell Biol. Int.* **18,** 309–314.

Baas, P. W., Deitch, J. S., Black, M. M., and Banker, G. A. (1988). Polarity orientation of microtubules in hippocampal neurons: uniformity in the axon and nonuniformity in the dendrite. *Proc. Nat. Acad. Sci. USA* **85,** 8335–8339.

Baas, P. W., Black, M. M., and Banker, G. A. (1989). Changes in microtubule polarity orientation during the development of hippocampal neurons in culture. *J. Cell Biol.* **109,** 3085–3094.

Baird, D. H., Hatten, M. E., and Mason, C. A. (1992). Cerebellar target neurons provide a stop signal for afferent neurite extension *in vitro. J. Neurosci.* **12,** 619–634.

Bamburg, J. R., and Bernstein, B. W. (1991). Actin and actin-binding proteins in neurons. *J. Neurosci. Res.* **30,** 121–160.

Bauer-Moffett, C., and Altman, J. (1977). The effect of ethanol chronically administered to preweanling rats on cerebellar development: a morphological study. *Brain Res.* **119,** 249–268.

Bellinger D. C., Sloman, J., Leviton, A., Rabinowitz, M., Needleman, H. L., and Waternaux, C. (1991). Low-level lead exposure and children's cognitive function in the preschool years. *Pediatrics* **87,** 219–227.

Biernat, J., Gustke, N., Drewes, G., Mandelkow, E.-M., and Mandelkow, E. (1993). Phosphorylation of ser^{262} strongly reduces binding of tau to microtubules: distinction between PHF-like immunoreactivity and microtubule binding. *Neuron* **11,** 153–163.

Billger, M., Wallin, M., and Karlsson, J. O. (1988). Proteolysis of tubulin and microtubule-associated proteins 1 and 2 by calpain I and II. Difference in sensitivity of assembled and disassembled microtubules. *Cell Calcium* **9,** 33–44.

Bixby, J. L., and Spitzer, N. C. (1984). Early differentiation of vertebrate spinal neurons in the absence of voltage-dependent Ca^{2+} and Na^{+} influx. *Dev. Biol.* **106,** 89–96.

Bray, D. (1984). Axonal growth in response to experimentally applied mechanical tension. *Dev. Biol.* **102,** 379–389.

Bray, D., and Hollenbeck, P. J. (1988). Growth cone guidance and motility. *Annu. Rev. Cell. Biol.* **4,** 43–61.

Bressler, J. P., and Goldstein, G. W. (1991). Mechanisms of lead toxicity. *Biochem. Pharmacol.* **41,** 479–484.

Brewer, G. J., and Cotman, C. W. (1989). NMDA receptor regulation of neuronal morphology in cultured hippocampal neurons. *Neurosci. Lett.* **99,** 268–273.

Brion, J.-P., Octave, J. N., and Couck, A. M. (1994). Distribution of the phosphorylated microtubule-associated protein tau in developing cortical neurons. *Neurosci.* **63,** 895–909.

Brooks, W. J., Petit, T. L., LeBoutillier, J. C., Nobrega, J. N., and Jarvis, M. F. (1993). Differential effects of early chronic lead exposure on postnatal rat brain NMDA, PCP, and adenosine A_1 receptors: an autoradiographic study. *Drug. Dev. Res.* **29,** 40–47.

Brugg, B., and Matus, A. (1991). Phosphorylation determines the binding of microtubule-associated protein 2 (MAP2) to microtubules in living cells. *J. Cell Biol.* **114,** 735–743.

Brugg, B., Reddy, D., and Matus, A. (1993). Attenuation of microtubule-associated protein 1B expresson by antisense oligodeoxynucleotides inhibits initiation of neurite outgrowth. *Neurosci.* **52,** 489–496.

Buccheim, K., Noack, S., Stoltenburg, G., Lilienthal, H., and Winneke, G. (1994). Developmental delay of astrocytes in hippocampus of rhesus monkeys reflects the effect of pre- and postnatal chronic low level lead exposure. *NeuroToxicol.* **15,** 665–670.

Burbacher, T. M., Rodier, P. M., and Weiss, B. (1990). Methylmercury developmental neurotoxicity: a comparison of effects in humans and animals. *Neurotoxicol. Teratol.* **12,** 191–202.

Cabell, L., and Audesirk, L. (1993). Effects of selective inhibition of protein kinase C, cyclic AMP-dependent protein kinase, and Ca^{2+}–calmodulin-dependent protein kinase on neurite development in cultured rat hippocampal neurons. *Int. J. Dev. Neurosci.* **11,** 357–368.

Caceres, A., Mautino, J., and Kosik, K. S. (1992) Suppression of MAP2 in cultured cerebellar macroneurons inhibits minor neurite formation. *Neuron* **9,** 607–618.

Cadete-Leite, A., Tavares, M. A., Uylings, H. B. M., and Paula-Barbosa, M. (1988). Granule cell loss and dendritic regrowth in the hippocampal dentate gyrus of the rat after chronic alcohol consumption. *Brain Res.* **473,** 1–14.

Cambray-Deakin, M. A. (1991). Cytoskeleton of the growing axon. *J. Neurosci. Res.* **30,** 233–255.

Cambray-Deakin, M. A., Adu, J., and Burgoyne, R. D. (1990). Neuritogenesis in cerebellar granule cells *in vitro*: a role for protein kinase C. *Dev. Brain. Res.* **53,** 40–46.

Cavalleri, A., Minoia, C., Ceroni, M., and Poloni, M. (1984). Lead in cerebrospinal fluid and its relationship to plasma lead in humans. *J. Appl. Toxicol.* **4,** 63–65.

Challacombe, J. F., Snow, D. M., and Letourneau, P. C. (1996). Role of the cytoskeleton in growth cone motility and axonal elongation. *Seminars Neurosci.* **8,** 67–80.

Charness, M. E., Safran, R. M., and Perides, G. (1994). Ethanol inhibits neural cell–cell adhesion. *J. Biol. Chem.* **269,** 9304–9309.

Chiou, J.-Y., and Westhead, E. W. (1992). Okadaic acid, a protein phosphatase inhibitor, inhibits nerve growth factor–directed neurite outgrowth in PC12 cells. *J. Neurochem.* **59,** 1963–1966.

Choi, B. H. (1989). The effects of methylmercury on the developing brain. *Prog. Neurobiol.* **32,** 447–470.

Choi, B. H., Cho, K. H., and Lapham, L. W. (1981a). Effects of methylmercury on human fetal neurons and astrocytes *in vitro*: a time-lapse cinematographic, phase and electron microscopic study. *Env. Res.* **24,** 61–74.

Choi, B. H., Kudo, M., and Lapham, L. W. (1981b). A Golgi and electron-micrographic study of cerebellum in methylmercury-poisoned neonatal mice. *Acta Neuropathol.* **54,** 233–237.

Cline, H. T., Witte, S., and Jones, K. W. (1996). Low lead levels stunt neuronal growth in a reversible manner. *Proc. Natl. Acad. Sci. USA* **93,** 9915–9920.

Cohan, C. S., Connor, J. A., and Kater, S. B. (1987). Electrically and chemically mediated increases in intracellular calcium in neuronal growth cones. *J. Neurosci.* **7,** 3588–3599.

Colamarino, S. A., and Tessier-Lavigne, M. (1995). The axonal chemoattractant netrin-1 is also a chemorepellent for trochlear motor axons. *Cell* **81,** 621–629.

Connor, J. A. (1986). Digital imaging of free calcium changes and spatial gradients in growing processes in single, mammalian central nervous system cells. *Proc. Natl. Acad. Sci. USA* **83,** 6179–6183.

Cookman, G. R., Hemmens, S. E., Keane, G. J., King, W. B., and Regan, C. M. (1988). Chronic low level lead exposure precociously induces rat glial development in vitro and in vivo. *Neurosci. Lett.* **86,** 33–37.

Craig, A. M., and Banker, G. (1994). Neuronal polarity. *Annu. Rev. Neurosci.* **17,** 267–310.

Dave, V., Vitarella, D., Aschner, J. L., Fletcher, P., Kimelberg, H. K., and Aschner, M. (1993). Lead increases inositol 1,4,5-trisphosphate levels but does not interfere with calcium transients in primary rat astrocytes. *Brain Res.* **618,** 9–18.

Davenport, R. W., and McCaig, C. D. (1993). Hippocampal growth cone responses to focally applied electric fields. *J. Neurobiol.* **24,** 89–100.

Davies, D. L., and Smith, D. E. (1981). A Golgi study of mouse hippocampal CA1 pyramidal neurons following perinatal ethanol exposure. *Neurosci. Lett.* **26,** 49–54.

Davis, J. M., Otto, D. A., Weil, D. E., and Grant, L. D. (1990). The comparative developmental neurotoxicity of lead in humans and animals. *Neurotoxicol. Teratol.* **12,** 215–229.

Denny, M. F., Hare, M. F., and Atchison, W. D. (1993). Methylmercury alters intrasynaptosomal concentrations of endogenous polyvalent cations. *Toxicol. Appl. Pharmacol.* **122,** 222–232.

Dent, E. W., and Meiri, K. F. (1992). GAP-43 phosphorylation is dynamically regulated in individual growth cones. *J. Neurobiol.* **23,** 1037–1053.

Diez-Guerra, F. J., and Avila, J. (1993). MAP2 phosphorylation parallels dendrite arborization in hippocampal neurones in culture. *NeuroReport* **4,** 412–419.

Dinsmore, J. H., and Solomon, F. (1991). Inhibition of MAP2 expression affects both morphological and cell division phenotypes of neuronal differentiation. *Cell* **64,** 817–826.

Doering, L. C. (1994). Probing modifications of the neuronal cytoskeleton. *Molec. Neurobiol.* **7,** 265–291.

Dotti, C. G., Sullivan, C. A., and Banker, G. A. (1988). The establishment of polarity by hippocampal neurons in culture. *J. Neurosci.* **8,** 1454–1468.

Dow, K. E., and Riopelle, R. J. (1985). Ethanol neurotoxicity: effects on neurite formation and neurotrophic factor production *in vitro.* *Science* **228,** 591–593.

Evans, L. L., and Bridgman, P. C. (1995). Particles move along actin filament bundles in nerve growth cones. *Proc. Natl. Acad. Sci USA* **92,** 10954–10958.

Evans, M. L., Busselberg, D., and Carpenter, D. O. (1991). Pb^{2+} blocks calcium currents of cultured dorsal root ganglion cells. *Neurosci. Lett.* **129,** 103–106.

Felipo, V., Minana, M.-D., and Grisolia, S. (1990). A specific inhibitor of protein kinase C induces differentiation of neuroblastoma cells. *J. Biol. Chem.* **265,** 9599–9601.

Ferreira, A., Palazzo, R. E., and Rebhun, L. I. (1993). Preferential dendritic localization of pericentriolar material in hippocampal pyramidal neurons in culture. *Cell Motil. Cytoskel.* **25,** 336–344.

Ferrer, I., Galofre, E., Fabregues, I., and Lopez-Tejero, D. (1988). Effects of chronic ethanol consumption beginning at adolescence: increased numbers of dendritic spines on cortical pyramidal cells in the adulthood. *Acta Neuropathol.* **78,** 528–532.

Fields, R. D., Neele, E. A., and Nelson, P. G. (1990). Effects of patterned electrical activity on neurite outgrowth from mouse sensory neurons. *J. Neurosci.* **10,** 2950–2964.

Fischer, I., Romano-Clarke, G., and Grynspan, F. (1991). Calpain-mediated proteolysis of microtubule associated proteins MAP1B and MAP2 in developing brain. *Neurochem. Res.* **16,** 891–898.

Forscher, P. (1989). Calcium and polyphosphoinositide control of cytoskeletal dynamics. *Trends Neurosci.* **12,** 468–474.

Friedrich, P., and Aszodi, A. (1991). MAP2: a sensitive cross-linker and adjustable spacer in dendritic architecture. *FEBS Lett.* **295,** 5–9.

Furiya, Y., Sahara, N., and Mori. H. (1993). Okadaic acid enhances abnormal phosphorylation on tau proteins. *Neurosci. Lett.* **156,** 67–69.

Futerman, A. H., and Banker, G. A. (1996). The economics of neurite outgrowth—the addition of new membrane to growing axons. *Trends Neurosci.* **19,** 144–149.

Garyantes, T. K., and Regehr, W. G. (1992). Electrical activity increases growth cone calcium but fails to inhibit neurite outgrowth from rat sympathetic neurons. *J. Neurosci.* **12,** 96–103.

Gelfand, V. I., and Bershadsky, A. D. (1991). Microtubule dynamics: mechanism, regulation, and function. *Annu. Rev. Cell Biol.* **7,** 93–116.

Ginzburg, I. (1991). Neuronal polarity: targeting of microtubule components into axons and dendrites. *Trends Biochem. Sci.* **16,** 257–261.

Gispen, W. H., Nielander, H. B., De Graan, P. N. E., Oestreicher, A. B., Schrama, L. H., and Schotman, P. (1992). Role of the growth-associated protein B-50/GAP-43 in neuronal plasticity. *Molec. Neurobiol.* **5,** 61–85.

Goedert, M., Crowther, R. A., and Garner, C. C. (1991). Molecular characterization of microtubule-associated proteins tau and MAP2. *Trends Neurosci.* **14,** 193–199.

Goldstein, G. W. (1993). Evidence that lead acts as a calcium substitute in second messenger metabolism. *Neurotoxicol.* **14,** 97–101.

Goldstein, G. W., and Ar, D. (1983). Lead activates calmodulin sensitive processes. *Life Sci.* **33,** 1001–1006.

Goodman, C. S. (1996). Mechanisms and molecules that control growth cone guidance. *Annu. Rev. Neurosci.* **19,** 341–377.

Goshima, Y., Ohsako, S., and Yamauchi, T. (1993). Overexpression of Ca^{2+}/calmodulin–dependent protein kinase II in Neuro2a and NG108-15 neuroblastoma cell lines promotes neurite outgrowth and growth cone motility. *J. Neurosci.* **13,** 559–567.

Gotti, C., Cabrini, D., Sher, E., and Clementi, F. (1987). Effects of long-term *in vitro* exposure to aluminum, cadmium, or lead on differentiation and cholinergic receptor expression in a human neuroblastoma cell line. *Cell Biol. Toxicol.* **3,** 431–440.

Graff, R. D., Philbert, M. A., Lowndes, H. E., and Reuhl, K. R. (1993). The effect of glutathione depletion on methylmercury–induced microtubule disassembly in cultured embryonal carcinoma cells. *Toxicol. Appl. Pharmacol.* **120,** 20–28.

Guilarte, T. R., and Miceli, R. C. (1992). Age-dependent effects of lead on [^{3}H]MK-801 binding to the NMDA receptor–gated ionophore: *in vitro* and *in vivo* studies. *Neurosci. Lett.* **148,** 27–30.

Guilarte, T. R., Miceli, R. C., Altmann, L., Weinsberg, F., Winneke, G., and Wiegand, H. (1993). Chronic prenatal and postnatal Pb^{2+} exposure increases [^{3}H]MK801 binding sites in adult rat forebrain. *Eur. J. Pharmacol.* **248,** 273–275.

Guppy, L. J., and Littleton, J. M. (1994). Binding characteristics of the calcium channel antagonist [^{3}H]-nitrendipine in tissues from ethanol-dependent rats. *Alcohol Alcoholism* **29,** 283–293.

Habermann, E., Crowell, K., and Janicki, P. (1983). Lead and other metals can substitute for Ca^{2+} in calmodulin. *Arch. Toxicol.* **54,** 61–70.

Hammer, R. P. Jr., and Scheibel, A. B. (1981). Morphologic evidence for a delay of neuronal maturation in fetal alcohol exposure. *Exp. Neurol.* **74,** 587–596.

Hare, M. F., and Atchison, W. D. (1995). Methylmercury mobilizes Ca^{2+} from intracellular stores sensitive to inositol 1,4,5,-trisphosphate in NG108–15 cells. *J. Pharmacol. Exp. Ther.* **272,** 1016–1023.

Hare, M. F., McGinnis, K. M., and Atchison, W. D. (1993). Methylmercury increases intracellular concentrations of Ca^{2+} and heavy metals in NG108–15 cells. *J. Pharmacol. Exp. Ther.* **266,** 1626–1635.

Harrison, L., Cheetham, M. E., and Calvert, R. A. (1993). Investigation of the changes in neuronal distribution and phosphorylation state of MAP1X during development. *Dev. Neurosci.* **15,** 68–76.

Haydon, P. G., and Drapeau, P. (1995). From contact to connection: early events during synaptogenesis. *Trends Neurosci.* **18,** 196–201.

Heaton, M. B., Paiva, M., Swanson, D. J., and Walker, D. W. (1993). Modulation of ethanol neurotoxicity by nerve growth factor. *Brain Res.* **620,** 78–85.

Heaton, M. B., Paiva, M., Swanson, D. J., and Walker, D. W. (1994). Responsiveness of cultured septal and hippocampal neurons to ethanol and neurotrophic substances. *J. Neurosci. Res.* **39,** 305–318.

Heaton, M. B., Carlin, M., Paiva, M., and Walker, D. W. (1995). Perturbation of target-directed neurite outgrowth in embryonic CNS co-cultures grown in the presence of ethanol. *Dev. Brain Res.* **89,** 270–280.

Heidemann, S. R. (1996). Cytoplasmic mechanisms of axonal and dendritic growth in neurons. *Int. Rev. Cytology* **165,** 235–296.

Heidemann, S. R., and Buxbaum, R. E. (1994). Mechanical tension as a regulator of axonal development. *Neurotoxicol.* **15,** 95–108.

Heidemann, S. R., Lamoureux, P., and Buxbaum, R. E. (1990). Growth cone behavior and production of traction force. *J. Cell. Biol.* **111,** 1949–1957.

Hirokawa, N. (1994). Microtubule organization and dynamics dependent on microtubule-associated proteins. *Curr. Opin. Cell Biol.* **6,** 74–81.

Hitt, A. L., and Luna, E. J. (1994). Membrane interactions with the actin cytoskeleton. *Curr. Opin. Cell Biol.* **6,** 120–130.

Hoffman, P. L., Rabe, C. S., Moses, F., and Tabakoff, B. (1989). *N*-methyl-D-aspartate receptors and ethanol: inhibition of calcium influx and cyclic GMP production. *J. Neurochem.* **52,** 1937–1940.

Hsu, L., Jeng, A. Y., and Chen, K. Y. (1989). Induction of neurite outgrowth from chick embryonic ganglia explants by activators of protein kinase C. *Neurosci. Lett.* **99,** 257–262.

Ishihara, K., Alkondon, M., Montes, J. G., and Albuquerque, E. X. (1995). Nicotinic responses in acutely dissociated rat hippocampal neurons and the selective blockade of fast-desensitizing nicotinic currents by lead. *J. Pharmacol. Exp. Ther.* **273,** 1471–1482.

Janmey, P. A. (1994). Phosphoinositides and calcium as regulators of cellular actin assembly and disassembly. *Annu. Rev. Physiol.* **56,** 169–191.

Jap Tjoen San, E. R. A., Schmidt-Michels, M., Oestreicher, A. B., Gispen, W. H., and Schotman, P. (1992) Inhibition of nerve growth factor–induced B-50/GAP-43 expression by antisense oligomers interferes with neurite outgrowth of PC12 cells. *Biochem. Biophys. Res. Comm.* **187,** 839–846.

Jett, D. A., and Guilarte, T. R. (1995). Developmental lead exposure alters *N*-methyl-D-aspartate and muscarinic cholinergic receptors in the rat hippocampus: an autoradiographic study. *Neurotoxicol.* **16,** 7–18.

Jian, X., Hidata, H., and Schmidt, J. T. (1994). Kinase requirement for retinal growth cone motility. *J. Neurobiol.* **25,** 1310–1328.

Johnson, G. V. W. (1992). Differential phosphorylation of tau by cyclic AMP–dependent protein kinase and Ca^{2+}/calmodulin–dependent protein kinase II: metabolic and functional consequences. *J. Neurochem.* **59,** 2056–2062.

Johnson, G. V. W., Greenwood, J. A., Costello, A. C., and Troncoso, J. C. (1991). The regulatory role of calmodulin in the proteolysis of individual neurofilament proteins by calpain. *Neurochem. Res.* **16,** 869–873.

Jones, D. G. (1988). Influence of ethanol on neuronal and synaptic maturation in the central nervous system—morphological investigations. *Prog. Neurobiol.* **31,** 171–197.

Jones, K. L., and Smith, D. W. (1973). Recognition of the fetal alcohol syndrome in early infancy. *Lancet* **2,** 999–1001.

Joshi, H. C., and Baas, P. W. (1993). A new perspective on microtubules and axon growth. *J. Cell Biol.* **121,** 1191–1196.

Kapfhammer, J. P., and Raper, J. A. (1987). Collapse of growth cone structure on contact with specific neurites in culture. *J. Neurosci.* **7,** 201–212.

Kater, S. B., Mattson, M. P., Cohon, C., and Connor, J. (1988). Calcium regulation of the neuronal growth cone. *Trends Neurosci.* **11,** 315–321.

Katz, F., Ellis, L., and Pfenninger, K. (1985). Nerve growth cones isolated from fetal rat brain. III. Calcium-dependent protein phosphorylation. *J. Neurosci.* **5,** 1402–1411.

Kennedy, T. E., Serafini, T., de la Torre, J. R., and Tessier-Lavigne, M. (1994). Netrins are diffusible chemotropic factors for commissural axons in the embryonic spinal cord. *Cell* **78,** 425–435.

Kern, M., and Audesirk, G. (1995). Inorganic lead may inhibit neurite development in cultured rat hippocampal neurons through hyperphosphorylation. *Toxicol. Appl. Pharmacol.* **124,** 111–123.

Kern, M., Audesirk, T., and Audesirk, G. (1993). Effects of inorganic lead on the differentiation and growth of cortical neurons in culture. *Neurotoxicol.* **14,** 319–328.

King, M. A., Hunter, B. E., and Walker, D. W. (1988). Alterations and recovery of dendritic spine density in rat hippocampus following long-term ethanol ingestion. *Brain Res.* **459,** 381–385.

Kira, M., Tanaka, J., and Sobue, K. (1995). Caldesmon and low Mr isoform of tropomyosin are localized in neuronal growth cones. *J. Neurosci. Res.* **40,** 294–305.

Kiraly, E., and Jones, D. G. (1982). Dendritic spine changes in rat hippocampal pyramidal cells after postnatal lead treatment: a Golgi study. *Exp. Neurol.* **77,** 236–239.

Kolodkin, A. L. (1996). Growth cones and the cues that repel them. *Trends Neurosci.* **19,** 507–513.

Komulainen, H., and Bondy, S. C. (1987). Increased free intrasynaptosomal Ca^{2+} by neurotoxic organometals: distinctive mechanisms. *Toxicol. Appl. Pharmacol.* **88,** 77–86.

Kovacs, K. A., Kavanagh, T. J., and Costa, L. G. (1995). Ethanol inhibits muscarinic receptor–stimulated phosphoinositide metabolism and calcium mobilization in rat primary cortical cultures. *Neurochem. Res.* **20,** 939–949.

Krigman, M. R., Druse, M. J., Traylor, T. D., Wilson, M. H., Newell, L. R., and Hogan, E. L. (1974). Lead encephalopathy in the developing rat: effect on cortical ontogenesis. *J. Neuropathol Exp. Neurol.* **33,** 671–687.

Kunimoto, M. (1994). Methylmercury induces apoptosis of rat cerebellar neurons in primary culture. *Biochem. Biophys. Res. Comm.* **204,** 310–317.

Lagunowich, L. A., Stein, A. P., and Reuhl, K. R. (1993). N-cadherin mediated aggregation and neurulation are disrupted by exposure to lead. *Toxicologist* **13,** 168.

Lamoureux, P., Buxbaum, R. E., Heidemann, S. R. (1989). Direct evidence that growth cones pull. *Nature* **340,** 159–162.

Lankford, K. L., and Letourneau, P. C. (1989). Evidence that calcium may control neurite outgrowth by regulating the stability of actin filaments. *J. Cell Biol.* **109,** 1229- 1243.

Lankford, K. L., Rand, M. N., Waxman, S. G., and Kocsis, J. D. (1995). Blocking Ca^{2+} mobilization with thapsigargin reduces neurite initiation in cultured adult rat DRG neurons. *Dev. Brain Res.* **84,** 151–163.

Lee, M. K., and Clevelend, D. W. (1996). Neuronal intermediate filaments. *Annu. Rev. Neurosci.* **19,** 187–217.

Legare, M. E., Castiglioni, A. J., Rowles, T. K., Calvin, J. A., Snyder-Armstead, C., and Tiffany-Castiglioni, E. (1993). Morphological alterations of neurons and astrocytes in guinea pigs exposed to low levels of inorganic lead. *Neurotoxicol.* **14,** 77–80.

Leli, U., Beermann, M. L., Hauser, G., and Shea, T. B. (1993). Phosphatase inhibitors prevent staurosporine-induced neurite outgrowth in human neuroblastoma cells. *Neurosci. Res. Comm.* **12,** 17–21.

Leskawa, K. C., Maddox, T., and Webster, K. A. (1995). Effects of ethanol on neuroblastoma cells in culture: role of gangliosides in neuritogenesis and substrate adhesion. *J. Neurosci. Res.* **42,** 377–384.

Letourneau, P. C. (1975). Cell-to-substratum adhesion and guidance of axonal elongation. *Dev. Biol.* **44,** 92–101.

Li, H.-C. (1984). Activation of brain calcineurin phosphatase towards nonprotein phosphoesters by Ca^{2+}, calmodulin, and Mg^{2+}. *J. Biol. Chem.* **259,** 8801–8807.

Lim, S. S., Edson, K. L., Letourneau, P. C., and Borisy, G. G. (1990). A test of microtubule translocation during neurite elongation. *J. Cell Biol.* **111,** 123–130.

Lin, C.-H. , Espreafico, E. M., Mooseker, M. S., and Forscher, P. (1996). Myosin drives retrograde F-actin flow in neuronal growth cones. *Neuron* **16,** 769–782.

Lin, C.-H., and Forscher, P. (1995). Growth cone advance is inversely proportional to retrograde F-actin flow. *Neuron* **14,** 763–771.

Little, H. J. (1995). The role of calcium channels in drug dependence. *Drug Alcohol Dep.* **38,** 173–194.

Liu, Y., and Storm, D. R. (1990). Regulation of free calmodulin levels by neuromodulin: neuron growth and regeneration. *Trends Pharmacol. Sci.* **11,** 107–111.

Lohof, A. M., Quillan, M., Dan, Y., and Poo, M.-M. (1992). Asymmetric modulation of cytosolic cAMP activity induces growth cone turning. *J. Neurosci.* **12,** 1253–1261.

Lokhorst, D. K., and Druse, M. J. (1993). Effects of ethanol on cultured fetal astroglia. *Alcohol Clin. Exp. Res.* **17,** 810–815.

Long, G. J., Rosen, J. F., and Schanne, F. A. X. (1994). Lead activation of protein kinase C from rat brain. Determination of free calcium, lead and zinc by ^{19}F NMR. *J. Biol. Chem.* **269,** 834–837.

Lorton, D., and Anderson, W. J. (1986a). Altered pyramidal cell dendritic development in the motor cortex of lead intoxicated neonatal rats. A Golgi study. *Neurobehav. Toxicol. Pharmacol.* **8,** 45–50

Lorton, D., and Anderson, W. J. (1986b). The effects of postnatal lead toxicity on the development of cerebellum in rats. *Neurobehav. Toxicol. Pharmacol.* **8,** 51–59.

Lovinger, D. M., White, G., and Weight, F. F. (1989). Ethanol inhibits NMDA-activated ion current in hippocampal neurons. *Science* **243,** 1721–1724.

Luna, E. J., and Hitt, A. L. (1992). Cytoskeleton–plasma membrane interactions. *Science* **258,** 955–964.

Manton, W. I., and Cook, J. D. (1984). High accuracy (stable isotope dilution) measurements of lead in serum and cerebrospinal fluid. *Brit. J. Ind. Med.* **41,** 313–319.

Markovac, J., and Goldstein, G. W. (1988). Picomolar concentrations of lead stimulate brain protein kinase C. *Nature* **334,** 71–73.

Marsh, L., and Letourneau, P. C. (1984). Growth of neurites without filopodial or lamellipodial activity in the presence of cytochalasin B. *J. Cell Biol.* **99,** 2041–2047.

Massaro, E. J. (1996). The developmental cytotoxicity of mercurials. In *Toxicology of Metals* (L. Chang, Ed.), pp. 1047–1081. CRC Press, Boca Raton, Florida.

Matsuzawa, M., Weight, F. F., Potember, R. S., and Liesi, P. (1996). Directional neurite outgrowth and axonal differentiation of embryonic hippocampal neurons are promoted by a neurite outgrowth domain of the B2-chain of laminin. *Int. J. Dev. Neurosci.* **14,** 283–295.

Matten, W. T., Aubry, M., West, J., and Maness, P. F. (1990). Tubulin is phosphorylated at tyrosine by $pp60^{c\text{-}src}$ in nerve growth cone membranes. *J. Cell Biol.* **111,** 1959–1970.

Mattson, M. P., and Kater, S. B. (1987). Calcium regulation of neurite elongation and growth cone motility. *J. Neurosci.* **7,** 4034–4043.

Mattson, M. P., Dou, P., and Kater, S. B. (1988). Outgrowth-regulating actions of glutamate in isolated hippocampal pyramidal neurons. *J. Neurosci.* **8,** 2087–2100.

Matus, A. (1988a). Neurofilament protein phosphorylation—where, when and why. *Trends Neurosci.* **11,** 291–292.

Matus, A. (1988b). Microtubule-associated proteins. Their potential role in determining neuronal morphology. *Annu. Rev. Neurosci.* **11,** 29–44.

Matus, A. (1994). Stiff microtubules and neuronal morphology. *Trends Neurosci.* **17,** 19–22.

Matus, A., Delhaye-Couchard, N., and Mariani, J. (1990). Microtubule-associated protein 2 (MAP2) in Purkinje cell dendrites: evidence that factors other than binding to microtubules are involved in determining its cytoplasmic distribution. *J. Comp. Neurol.* **297,** 435–440.

McMullen, P. A., Saint-Cyr, J. A., and Carlen, P. L. (1981). Morphological alterations in rat CA1 hippocampal pyramidal cell dendrites resulting from chronic ethanol consumption and withdrawal. *J. Comp. Neurol.* **225,** 111–118.

Meiri, K. F., and Burdick, D. (1991). Nerve growth factor stimulation of GAP-43 phosphorylation in intact isolated growth cones. *J. Neurosci.* **11,** 3155–3164.

Melloni, E., and Pontremoli, S. (1989). The calpains. *Trends Neurosci.* **12,** 438–444.

Messersmith, E. K., Leonardo, E. D., Shatz, C. J., Tessier-Lavinge, M., Goodman, C. S., and Kolodkin, A. L. (1995). Semaphorin III can function as a selective chemorepellent to pattern sensory projections in the spinal cord. *Neuron* **14,** 949–959.

Messing, R. O., Carpenter, C. L., Diamond, J., and Greenberg, D. A. (1986). Ethanol regulates calcium channels in clonal neural cells. *Proc. Natl. Acad. Sci. USA* **83,** 6213–6215.

Messing, R. O., Sneade, A. B., and Savidge, B. (1990). Protein kinase C participates in upregulation of dihydropyridine-sensitive calcium channels by ethanol. *J. Neurochem.* **55,** 1383–1389.

Messing, R. O., Henteleff, M., and Park, J. J. (1991a). Ethanol enhances growth factor–induced neurite formation in PC12 cells. *Brain Res.* **565,** 301–311.

Messing, R. O., Petersen, P. J., and Henrich, C. J. (1991b). Chronic ethanol exposure increases levels of protein kinase C δ and ε and protein kinase C–mediated phosphorylation in cultured neural cells. *J. Biol. Chem.* **266,** 23428–23432.

Miller, M. W., and Potempa, G. (1990). Numbers of neurons and glia in mature somatosensory cortex: effects of prenatal exposure to ethanol. *J. Comp. Neurol.* **293,** 92–102.

Miura, K., Inokawa, M., and Imura, N. (1984). Effects of methylmercury and some metal ions on microtubule networks in mouse glioma cells and *in vitro* tubulin polymerization. *Toxicol. Appl. Pharmacol.* **73,** 218–231.

Mullikin-Kilpatrick, D., and Treistman, S. N. (1993). Ethanol inhibition of L-type Ca^{2+} channels in PC12 cells: role of permeant ions. *Eur. J. Pharmacol.* **270,** 17–25.

Murthy, A. S. N., and Flavin, M. (1983). Microtubule assembly using the microtubule-associated protein MAP2 prepared in defined states of phosphorylation with protein kinase and phosphatase. *Eur. J. Biochem.* **137,** 37–46.

Nixon, R. A., and Sihag, R. K. (1991). Neurofilament phosphorylation: a new look at regulation and function. *Trends Neurosci.* **14,** 501–506.

Okabe, S., and Hirokawa, N. (1991). Actin dynamics in growth cones. *J. Neurosci.* **11,** 1918–1929.

O'Kusky, J. R. (1992). The neurotoxicity of methylmercury in the developing nervous system. In *The Vulnerable Brain and Environmental Risk, Volume 2: Toxins in Food* (R. L. Isaacson and K. F. Jensen, Eds.), pp. 19–34. Plenum Press, New York.

Oortgiesen, M., van Kleef, R. D. G. M., Bajnath, R. B., and Vijerberg, H. P. M. (1990). Nanomolar concentrations of lead selectively block nicotininc acetylcholine responses in mouse neuroblastoma cells. *Toxicol. Appl. Pharmacol.* **103,** 165–174.

Opanashuk, L. A., and Finkelstein, J. N. (1995). Induction of newly synthesized proteins in astroglial cells exposed to lead. *Toxicol. Appl. Pharmacol.* **131,** 21–30.

Pennypacker, K., Fischer, I., and Levitt, P. (1991). Early *in vitro* genesis and differentiation of axons and dendrites by hippocampal neurons analyzed quantitatively with neurofilament-H and microtubule-associated protein 2 antibodies. *Exp. Neurol.* **111,** 25–35.

Pentney, R. J. (1995). Measurements of dendritic path lengths provide evidence that ethanol-induced lengthening of terminal dendritic segments may result from dendritic regression. *Alcohol Alcoholism* **30,** 87–96.

Pentney, R. J., and Miller, M. W. (1992). Effects of ethanol on neuronal morphogenesis. In *Development of the Central Nervous System: Effects of Alcohol and Opiates* (M. W. Miller, Ed.), pp. 71–107. Wiley-Liss: New York.

Pentney, R. J., and Quackenbush, L. J. (1990). Dendritic hypertrophy in Purkinje neurons of old Fischer 344 rats after long-term ethanol treatment. *Alcohol Clin. Exp. Res.* **14,** 878–886.

Petit, T. L., and LeBoutillier, J. (1979). Effects of lead exposure during development on neocortical dendritic and synaptic structure. *Exp. Neurol.* **64,** 482–492.

Polak, K. A., Edelman, A. M., Wasley, J. W. F., and Cohan, C. S. (1991). A novel calmodulin antagonist, CGS 9343B, modulates calcium-dependent changes in neurite outgrowth and growth cone movements. *J. Neurosci.* **11,** 534–542.

Popov, S., Brown, A., and Poo, M. (1993). Forward plasma membrane flow in growing nerve processes. *Science* **259,** 244–246.

Purves, D., and Lichtman, J. W. (1985) *Principles of Neural Development.* Sinauer Associates, Sunderland, Massachusetts.

Rebhan, M., Vacun, G., and Rosner, H. (1995). Complementary distribution of tau proteins in different phosphorylation states within growing axons. *NeuroReport* **6,** 429–432.

Regan, C. M. (1993). Neural cell adhesion molecules, neuronal development and lead toxicity. *Neurotoxicol.* **14,** 69–74.

Reuhl, K. R., Rice, D. C., Gilbert, S. G., and Mallett, J. (1989). Effects of chronic lead exposure on monkey neuroanatomy: visual system. *Toxicol. Appl. Pharmacol.* **99,** 501–509.

Reuveny, E., and Narahashi, T. (1991). Potent blocking action of lead on voltage-activated calcium channels in human neuroblastoma cells SH-SY5Y. *Brain Res.* **545,** 312–314.

Riederer, B. M. (1990). Some aspects of the neuronal cytoskeleton in development. *Eur. J. Morphol.* **28,** 347–378.

Roderer, G., and Doenges, K. H. (1983). Influence of trimethyl lead and inorganic lead on the *in vitro* assembly of microtubules from mammalian brain. *Neurotoxicol.* **4,** 171–180.

Rogers, M., and Hendry, I. (1990). Involvement of dihydropyridine-sensitive calcium channels in nerve growth factor–dependent neurite outgrowth by sympathetic neurons. *J. Neurosci. Res.* **26,** 447–454.

Roivainen, R., McMahan, T., and Messing, R. O. (1993). Protein kinase C isozymes that mediate enhancement of neurite outgrowth by ethanol and phorbol esters in PC12 cells. *Brain Res.* **624,** 85–93.

Rosales, C., O'Brien, V., Kornberg, L., and Juliano, R. (1995). Signal transduction by cell adhesion receptors. *Biochim. Biophys. Acta* **1242,** 77–98.

Rosen, J. F. (1995). Adverse health effects of lead at low exposure levels: trends in management of childhood lead poisoning. *Toxicol.* **97,** 11–17.

Rosen, J. F., Schanne, F. A., and Long, G. J. (1993). Use of ^{19}F NMR to examine interactions between lead (Pb), calcium (Ca) and protein kinase C. *Toxicologist* **13,** 413.

Rowles, T. K., Womac, C., Bratton, G. R., and Tiffany-Castiglioni, E. (1989). Interaction of lead and zinc in cultured astroglia. *Metabolic Brain Disease* **4,** 187–201.

Rusanescu, B., Qi, H., Thomas, S. M., Brugge, J. S., and Halegoua, S. (1995). Calcium influx induces neurite growth through a Src-Ras signalling cassette. *Neuron* **15,** 1415–1425.

Sager, P. R. (1988). Selectivity of methyl mercury effects on cytoskeleton and mitotic progression in cultured cells. *Toxicol. Appl. Pharmacol.* **94,** 473–486.

Sager, P. R., Doherty, R. A., and Olmstead, J. B. (1983). Interaction of methylmercury with microtubules in cultured cells and *in vitro. Exp. Cell. Res.* **146,** 127–137.

Saijoh, K., Fukunaga, T., Katsuyama, H., Lee, M. J., and Sumino, K. (1993). Effects of methylmercury on protein kinase A and protein kinase C in the mouse brain. *Environ. Res.* **63,** 264–273.

Saito, H., Nishida, A., Shimizu, M., Motohashi, N., and Yamawaki, S. (1996). Effects of chronic ethanol treatment of inositol 1,4,5-trisphosphate receptors and inositol 1,3,4,5-tetrakisphosphate receptors in rat brain. *Neurospsychobiol.* **33,** 60–65.

Sarafian, T. A. (1993). Methyl mercury increases intracellular Ca^{2+} and inositol phosphate levels in cultured cerebellar granule neurons. *J. Neurochem.* **61,** 648–657.

Sarafian, T., and Verity, M. A. (1990). Methyl mercury stimulates protein ^{32}P phospholabeling in cerebellar granule cell culture. *J. Neurochem.* **55,** 913–921.

Saunders, D. W., Zajac, C. S., and Wappler, N. L. (1995). Alcohol inhibits neurite extension and increases N-myc and c-myc proteins. *Alcohol* **12,** 475–483.

Schanne, F. A. X., Dowd, T. L., Gupta, R. K., and Rosen, J. F. (1989a). Lead increases free Ca^{2+} concentration in cultured osteoblastic bone cells. Simultaneous detection intracellular free Pb^{2+} by ^{19}F-NMR. *Proc. Natl. Acad. Sci. USA* **86,** 5133–5135.

Schanne, F. A. X., Dowd, T. L., Gupta, R. K., and Rosen, J. F. (1989b). Effect of lead on intracellular free calcium ion concentration in a presynaptic neuronal model. ^{19}F-NMR study of NG108-15 cells. *Brain Res.* **503,** 308–311.

Schanne, F. A. X., Dowd, T. L., Gupta, R. K., and Rosen, J. F. (1990). Effect of lead on parathyroid hormone–induced responses in rat osteoblastic osteosarcoma cells (ROS 17/2.8) using ^{19}F-NMR. *Biochim. Biophys. Acta* **1054,** 250–255.

Schilling, K., Dickinson, M. H., Connor, J. A., and Morgan, J. I. (1991). Electrical activity in cerebellar cultures determines Purkinje cell dendritic growth patterns. *Neuron* **7,** 891–902.

Schulman, H. (1984). Phosphorylation of microtubule-associated proteins by a Ca^{2+}/calmodulin–dependent protein kinase. *J. Cell Biol.* **99,** 11–19.

Schwartz, J. (1994). Low-level lead exposure and children's IQ: a meta-analysis and search for a threshold. *Env. Res.* **65,** 42–55.

Scott, B., and Lew, J. (1986). Lead neurotoxicity: neuronal and non-neuronal cell survival in fetal and adult DRG cell cultures. *Neurotoxicol.* **7,** 57–68.

Scott, C. W., Spreen, R. C., Herman, J. L., Chow, F. P., Davison, M. D., Young, J., and Caputo, C. B. (1993). Phosphorylation of recombinant tau by cAMP-dependent protein kinase. *J. Biol. Chem.* **268,** 1166–1173.

Selvin-Testa, A., Lopez-Costa, J. J., Nessi De Avinon, A. C., and Pecci Saavedra, J. (1991). Astroglial altaerations in rat hippocampus during chronic lead exposure. *Glia* **4,** 384–392.

Settle, D. M., and Patterson, C. C. (1980). Lead in albacore: guide to lead pollution in Americans. *Science* **207,** 1167–1176.

Shafer, T. J., and Atchison, W. D. (1989). Block of ^{45}Ca uptake into synaptosomes by methylmercury: Ca^{2+}- and Na^{+}-dependence. *J. Pharmacol. Exp. Ther.* **248,** 696–702.

Shafer, T. J., and Atchison, W. D. (1991). Methylmercury blocks N- and L-type Ca^{2+} channels in nerve growth factor–differentiated pheochromocytoma (PC12) cells. *J. Pharmacol. Exp. Ther.* **258,** 149–157.

Shaw, G. (1991). Neurofilament proteins. *J. Neurosci. Res.* **30,** 185–214.

Shea, T. B., Beermann, M. L., Nixon, R. A., and Fischer, I. (1992). Microtubule-associated protein tau is required for axonal neurite elaboration by neuroblastoma cells. *J. Neurosci. Res.* **32,** 363–374.

Shetty, A. K., Burrows, R. C., and Phillips, D. E. (1993). Alterations in neuronal development in the substantia nigra pars compacta following *in utero* ethanol exposure: immunohistochemical and Golgi studies. *Neurosci.* **52,** 311–322.

Sierra, E. M., Rowles, T. K., Martin, J., Bratton, G. R., Womac, C., and Tiffany-Castiglioni, E. (1989). Low level lead neurotoxicity in a pregnant guinea pigs model: neuroglial enzyme activities and brain trace metal concentrations. *Toxicol.* **59,** 81–96.

Sihag, R. K., Shea, T. B., and Wang, F.-S. (1996). Spectrin-actin interaction is required for neurite extension in NB2a/dl neuroblastoma cells. *J. Neurosci. Res.* **44,** 430–437.

Silver, R. A., Lamb, A. G., Bolsover, S. R. (1990). Calcium hotspots caused by L-channel clustering promote morphological changes in neuronal growth cones. *Nature* **343,** 751–754.

Simons, T. J. B. (1993). Lead-calcium interacations as a basis for lead toxicity. *Neurotoxicol.* **14,** 77–85.

Simons, T. J. B., and Pocock, G. (1987). Lead enters bovine adrenal medullary cells through calcium channels. *J. Neurochem.* **56,** 568–574.

Singh, A. K. (1995). Neurotoxicity in rats chronically exposed to lead ingestion: measurement of intracellular concentrations of free calcium and lead ions in resting or depolarized brain slices. *Neuro-Toxicol.* **16,** 133–138.

Sirois, J. E., and Atchison, W. D. (1996). Methylmercury decreases whole cell barium current in cerebellar granule neurons. *Toxicologist* **30,** 26.

Smith, D. E., Foundas, A., and Canale, J. (1986). Effect of perinatally administered ethanol on the development of the cerebellar granule cell. *Exp. Neurol.* **92,** 491–501.

Snell, L. D., Tabakoff, B., and Hoffman, P. L. (1993). Radioligand binding to the *N*-methyl-D-aspartate receptor/ionophore complex: alterations by ethanol *in vitro* and by chronic in vivo ethanol ingestion. *Brain Res.* **602,** 91–98.

Sobue, G., and Pleasure, D. (1985). Experimental lead neuropathy: inorganic lead inhibits proliferation but not differentiation of Schwann cells. *Ann. Neurol.* **17,** 462–468.

Sobue, K., and Kanda, K. (1989). α-Actinins, calspectin (brain spectrin or fodrin), and actin participate in adhesion and movement of growth cones. *Neuron* **3,** 311–319.

Speizer, L. A., Watson, M. J., Kanter, J. R., and Brunton, L. L. (1989). Inhibition of phorbol ester binding and protein kinase C activity by heavy metals. *J. Biol. Chem.* **264,** 5581–5585.

Stark, M., Wolff, J. E. A., and Korbmacher, A. (1992). Modulation of glial cell differentiation by exposure to lead and cadmium. *Neurotoxicol. Teratol.* **14,** 247–252.

Steiner, B., Mandelkow, E.-M., Biernat, J., Gustke, N., Meyer, H. E., Schmidt, B., Mieskes, G., Soling, H. D., Drechsel, D., Kirshner, M. W., Goedert, M., and Mandelkow, E. (1990). Phosphorylation of microtubule-associated protein tau: identification of the site for Ca^{2+}-calmodulin dependent kinase and relationship with tau phosphorylation in Alzheimer tangles. *EMBO J.* **9,** 3539–3544.

Stoltenburg-Didinger, G., and Markwort, S. (1990). Prenatal methylmercury exposure results in dendritic spine dysgenesis in rats. *Neurotoxicol. Teratol.* **12,** 573–576.

Stoltenburg-Didinger, G., and Spohr, H. L. (1983). Fetal alcohol syndrome and mental retardation: spine distribution of pyramidal cells in prenatal alcohol-exposed rat cerebral cortex: a Golgi study. *Dev. Brain Res.* **11,** 119–123.

Strittmatter, S. M., Valenzuela, D., and Fishman, M. C. (1994). An amino-terminal domain of the growth-associated protein GAP-43 mediates its effects on filopodial formation and cell spreading. *J. Cell Science* **107,** 195–204.

Sun, L. R., and Suszkiw, J. B. (1995). Extracellular inhibition and intracellular enhancement of Ca^{2+} currents by Pb^{2+} in bovine adrenal chromaffin cells. *J. Neurophysiol.* **74,** 574–581.

Sydor, A. M., Su, A. L., Wang, F.-S., Xu, A., and Jay, D. G. (1996). Talin and vinculin play distinct roles in filopodial motility in the neuronal growth cone. *J. Cell Biol.* **134,** 1197–1207.

Tan, X.-X., and Costa, L. G. (1995). Postnatal development of muscarinic receptor-stimulated phosphoinositide metabolism in mouse cerebral cortex: sensitivity to ethanol. *Dev. Brain Res.* **86,** 348–353.

Tanaka, E., and Kirschner, M. W. (1991). Microtubule behavior in the growth cones of living neurons during elongation. *J. Cell Biol.* **115,** 345–363.

Tanaka, E., Ho, T., and Kirschner, M. W. (1995). The role of microtubule dynamics in growth cone motility and axonal growth. *J. Cell Biol.* **128,** 139–155.

Tanaka, J., Kira, M., and Sobue, K. (1993). Gelsolin is localized in neuronal growth cones. *Dev. Brain Res.* **76,** 268–271.

Tashima, K., Yamamoto, H., Setoyama, C., Ono, T., and Miyamoto, E. (1996). Overexpression of Ca^{2+}/calmodulin–dependent protein kinase II inhibits neurite outgrowth of PC12 cells. *J. Neurochem.* **66,** 57–64.

Tiffany-Castiglioni, E., Sierra, E. M., Wu, J-N., and Rowles, T. K. (1989). Lead toxicity in neuroglia. *Neurotoxicol.* **10,** 417–444.

Tiffany-Castiglioni, E., Legare, M. E., Schneider, L. A., Harris, E. D., Barhoumi, R., Zmudzki, J., Qian, Y., and Burghardt, R. C. (1996). Heavy metal effects on glia. *Methods Neurosci.* **30,** 135–166.

Tomsig, J. L., and Suszkiw, J. B. (1991). Permeation of Pb^{2+} through calcium channels: fura-2 measurements of voltage- and dihydropyridine-sensitive Pb^{2+} entry in isolated bovine chromaffin cells. *Biochim Biophys Acta* **1069,** 197–200.

Trevisan, L., Fitzgerald, L. W., Brose, N., Gasic, G. P., Heinemann, S. F., Duman, R. S., and Nestler, E. J. (1994). Chronic ingestion of ethanol up-regulates NMDAR1 receptor subunit immunoreactivity in rat hippocampus. *J. Neurochem.* **62,** 1635–1638.

Tsuda, M., Ono, K., Katayama, N., Yamagata, Y., Kikuchi, K., and Tsuchiya, T. (1989). Neurite outgrowth from mouse neuroblastoma and cerebellar cells induced by the protein kinase inhibitor H-7. *Neurosci. Lett.* **105,** 241–245.

Twombly, D. A., Herman, M. D., Kye, C. H., and Narahashi, T. (1990). Ethanol effects on two types of voltage-activated calcium channels. *J. Pharmacol. Exp. Ther.* **254,** 1029–1037.

Ujihara, H., Sasa, M., and Ban, T. (1995). Selective blockade of P-type calcium channels by lead in cultured hippocampal neurons. *Japan J. Pharmacol.* **67,** 267–269.

Uteshev, V., Busselberg, D., and Haas, H. L. (1993). Pb^{2+} modulates the NMDA receptor-channel–complex. *Naunyn-Schmiedeber's Arch. Pharmacol.* **347,** 209–213.

Vallee, R. (1980). Structure and phosphorylation of microtubule-associated protein 2 (MAP2). *Proc. Natl. Acad. Sci. USA* **77,** 3206–3210.

VanBerkum, M. F. A., and Goodman, C. S. (1995). Targeted disruption of Ca^{2+}- calmodulin signaling in *Drosophila* growth cones leads to stalls in axon extension and errors in axon guidance. *Neuron* **14,** 43–56.

Vogel, D. G., Margolis, R. L., and Mottet, N. K. (1985). The effects of methyl mercury binding to microtubules. *Toxicol. Appl. Pharmacol.* **80,** 473–486.

Walker, D. W., Barnes, D. E., Zornetzer, S. F., Hunter, B. E., and Kubanis, P. (1980). Neuronal loss in hippocampus induced by prolonged ethanol consumption in rats. *Science* **209,** 711–713.

Weeks, B. S., Papadopoulos, V., Dym, M., and Kleinman, H. K. (1991). cAMP promotes branching of laminin-induced neuronal processes. *J. Cell. Physiol.* **B147,** 62–67.

Wheeler, B. C., and Brewer, G. J. (1994). Selective hippocampal neuritogenesis: axon growth on laminin or pleiotrophin, dendrite growth on poly-D-lysine. *Soc. Neurosci. Abs.* **20,** 1292.

Widmer, F., and Caroni, P. (1993). Phosphorylation-site mutagenesis of the growth-associated protein GAP-43 modulates its effects on cell spreading and morphology. *J. Cell Biol.* **120,** 503–512.

Windebank, A. J. (1986). Specific inhibition of myelination by lead *in vitro;* comparison with arsenic, thallium, and mercury. *Exp. Neurol.* **94,** 203–212.

Wolff, J. (1988). Regulation of the *in vitro* assembly of microtubules by calcium and calmodulin. In *Structure and Function of the Cytoskeleton* (B. A. F. Rousset, Ed.), Vol. 171, pp. 477–480. Paris, Colloque INSERM/John Libbey Eurotext Ltd.

Wu, D.-Y., and Goldberg, D. J. (1993). Regulated tyrosine phosphorylation at the tips of growth cone filopodia. *J. Cell Biol.* **123,** 653–664.

Xu, Z., Marszalek, J. R., Lee, M. K., Wong, P. C., Folmer, J., Crawford, T. O., Hsieh, S.-T., Griffin, J. W., and Cleveland, D. W. (1996). Subunit composition of neurofilaments specifies axonal diameter. *J. Cell Biol.* **133,** 1061–1069.

Yamada, K. M., and Miyamoto, S. (1995). Integrin transmembrane signalling and cytoskeletal control. *Curr. Opin. Cell. Biol.* **7,** 681–689.

Yamada, K. M., Spooner, B. S., and Wessells, N. K. (1971). Ultrastructure and function of growth cones and axons of cultured nerve cells. *J. Cell Biol.* **49,** 614–635.

Yamauchi, T., and Fujisawa, H. (1988). Regulation of the interaction of actin filaments with microtubule-associated protein 2 by calmodulin-dependent protein kinase II. *Biochim. Biophys. Acta* **968,** 77–85.

Zheng, J., Lamoureux, P., Santiago, V., Dennerli, T., Buxbaum, R. E., and Heidemann, S. R. (1991). Tensile regulation of axonal elongation and initiation. *J. Neuroscience* **11,** 1117–1125.

Zheng, J., Felder, M., Connor, J. A., and Poo, M. (1994). Turning of nerve growth cones induced by neurotransmitters. *Nature* **368,** 140–144.

Zou, D.-J., and Cline, H. T. (1996). Expression of constitutively active CaMKII in target tissue modifies presynaptic axon arbor growth. *Neuron* **16,** 529–539.

Zou, J.-Y., Rabin, R. A., and Pentney, R. J. (1993). Ethanol enhances neurite outgrowth in primary cultures of rat cerebellar macroneurons. *Dev. Brain Res.* **72,** 75–84.

Zou, J.-Y., Cohan, C., Rabin, R. A., and Pentney, R. J. (1995). Continuous exposure of cultured rat cerebellar macroneurons to ethanol-depressed NMDA and KCl-stimulated elevations of intracellular calcium. *Alcohol Clin. Exp. Res.* **19,** 840–845.

Zuber, M. X., Goodman, D. W., Karns, L. R., and Fishman, M. C. (1989). The neuronal growth-associated protein GAP-43 induces filopodia in non-neuronal cells. *Science* **244,** 1193–1195.

CHAPTER

4

Myelination, Dysmyelination, and Demyelination

G. JEAN HARRY
National Toxicology Program
National Institute of Environmental Health Sciences
Research Triangle Park, North Carolina 27709

ARREL D. TOEWS
Department of Biochemistry and Biophysics and Neuroscience Center
Department of Biology
University of North Carolina
Chapel Hill, North Carolina 27599

*Abbreviations: ARIA, acetylcholine inducing activity; ATP, adenosine triphosphate; bFGF, basic fibroblast growth factor; CNP, 2′-3′-cyclicnucleotide-3′-phosphodiesterase; CNS, central nervous system; CPG, choline phosphoglycerides; EAE, experimental autoimmune encephalomyelitis; E-NCAM, embryonic neural cell adhesion molecule; EPG, ethanolamine phosphoglycerides; FABP, fatty-acid binding protein; Gal-C, galactocerebroside; GAP, growth associated protein; GGF, glial growth factor; HMG-CoA, hydroxymethylglutaryl-Coenzyme A; HNK-1, human natural killer cell–antigen 1; IGF-1, insulin-like growth factor-1; IHN, isonicotimic acid hydrazide; IPG, inositol phosphoglycerides; jp, jimpy mouse; jp-msd, myelin-synthesis deficient mouse; MAG, myelin-associated protein; MBP, myelin basic protein; MOG, myelin/oligodendrocyte glycoprotein; MyTI, myelin transcription factor; N-CAM, neural cell adhesion molecules; NDF, Neu-differentiation factor; NGF, nerve growth factor; NGF-R, nerve growth factor receptor; OL, oligodendroblast; Omgp, oligodendrocyte-myelin glycoprotein; P_0, peripheral myelin protein zero; PDGF, platelet-derived growth factor; PLP, proteolipid protein; PMD, Pelizaeus–Merzbacher disease; PMP-22, peripheral myelin protein, PNS, peripheral nervous system; RER, rough endoplasmic reticulum; rsh, rumpshaker mouse; SCIP, suppressed cAMP–inducible Pou-domain transcription factor; shi, shiverer mouse; shi-mid, myleindeficient mouse; SPG, serine phosphoglycerides; TET, triethyltin; TGF-β, transforming growth factor-β; Tr, trembler mutant mouse.

I. Introduction

Normal functioning of the nervous system involves the transmission, processing, and integration of information as nervous impulses. Impulse transmission along axons is greatly facilitated by the presence of myelin, the compact multilamellar extension of the plasma membrane of specialized glial cells that spirals around larger axons. In the central nervous system (CNS), oligodendroglial cells are responsible for the synthesis and maintenance of myelin, whereas Schwann cells subserve this role in the peripheral nervous system (PNS). Schwann cells produce a single segment of myelin (called an *internode*), whereas oligodendroglial cells furnish multiple myelin segments around different axons, although only one segment for a given axon. There are periodic interruptions along the axons between adjacent myelin internodes; termed *nodes of Ranvier,* these short intervals where axons are not enveloped by myelin are vital for normal nervous system function (see later this chapter).

Myelin is an electrical insulator, and the periodic interruptions at the nodes allow for rapid and efficient transmission of nervous impulses. In unmyelinated axons, impulse transmission involves a wave of membrane depolarization that moves down the axon in a continuous sequential manner. However, in myelinated axons, the myelin internodes function as high-resistance insulators, so the excitable axonal membrane, containing a high concentration of voltage-sensitive sodium channels, is exposed only at the nodes of Ranvier. Impulse conduction thus involves excitation at the nodes only, and the impulse jumps from node to node (saltatory conduction) (see Funch and Faber, 1984; Waxman *et al.,* 1989; and Morell *et al.,* 1994, for details). Saltatory conduction is much more rapid and requires much less energy for membrane repolarization than conduction in unmyelinated axons. Myelin thus greatly increases the efficiency of the nervous system, facilitating conduction while conserving metabolic energy and space. It is not difficult to imagine how even minor loss of myelin or perturbations in its structure and/or function could have deleterious effects on normal nervous system function.

II. Morphologic and Structural Aspects of Myelin Formation

Structural aspects of the process of myelination are most easily illustrated in the PNS. Each myelin-forming Schwann cell produces an elaborate specialized extension of its plasma membrane, which is wrapped spirally around a segment of one axon (Fig. 1). Schwann cells, the glial cells of the PNS, are derived from the portion of the neural epithelium that gives rise to the neural crest (Le Douarin, 1982). During development, Schwann cells invade the developing nerves, where they migrate along bundles of axons, proliferate (probably in response to an axonal mitogen; Webster and Favilla, 1984), and segregate the axons individually within invaginations on the surface of Schwann cells. As they cease migrating, they synthesize a basal lamina (Billings-Gagliardi *et al.,* 1974), composed of laminin, merosin, type IV collagen, fibronectin, nidogen/entactin, and heparan sulfate proteoglycan (Sanes and Cheney, 1982; Tohyama and Ide, 1984; Bannerman *et al.,* 1986; Leivo and Engvall, 1989; Sanes *et al.,* 1990). As the axons continue to enlarge, the larger axons become further segregated so that a single Schwann cell envelops a single axon. After the plasma membrane of the Schwann cell has completely enclosed the axon, the external surfaces of the plasma membrane fuse to form a structure known as the *mesaxon.* The mesaxon then elongates and spirals around the axon, eventually resulting in a "jelly-roll" structure consisting of double layers of the Schwann cell plasma membrane. Myelin internodes can be as much as 2 mm long and contain 5 mm of myelin spiral (Friede and Bischhausen, 1980). Myelin compaction occurs as cytoplasm is extruded and the cytoplasmic faces are condensed to produce the dark major period line visible in electron micrographs (Fig. 2). The juncture of what was originally the outer faces of the apposing plasma membranes form the lighter appearing intraperiod line. Mature myelin thus has a characteristic compact multilamellar structure, but cytoplasmic inclusions continu-

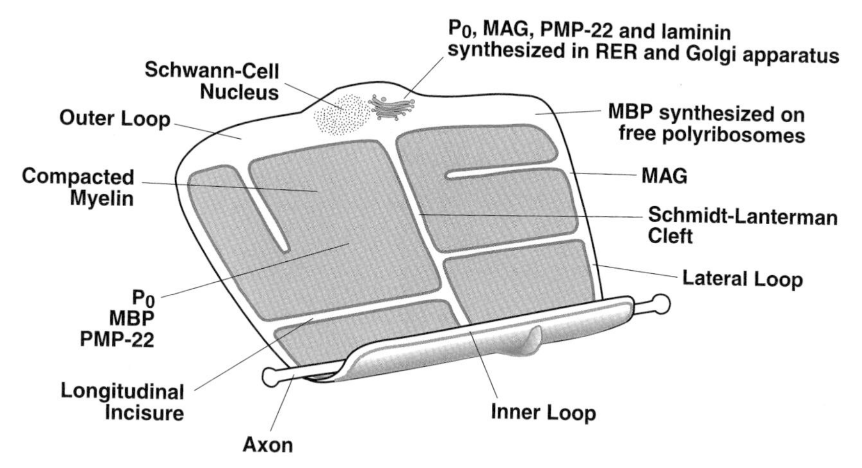

FIGURE 1 Diagram of a single myelinated internode of the PNS, as it would appear if the myelin sheath, with its associated Schwann cell, were unrolled from around the axon. Note the very large myelin membrane surface area, as well as the many cytoplasm-containing structures (lateral, inner, and outer loops; Schmidt–Lanterman clefts; and longitudinal incisures), all continuous with the cytoplasm of the Schwann cell perikarya. This cytoplasmic continuity is presumably necessary for metabolic maintenance of compact myelin. Sites of synthesis, as well as ultrastructural locations, of the major myelin proteins (discussed in a later section) are also shown. In mature compact myelin, the large trapezoidal sheet of myelin membrane is tightly spiraled around the axons, with apposing inner surfaces of the original plasma membrane fusing to give the electron-dense intraperiod lines and the outer surfaces of the original plasma membrane forming the lighter appearing intraperiod line (see also electron micrograph in Fig. 2). Abbreviations: RER, rough endoplasmic reticulum; MBP, myelin basic protein; MAG, myelin-associated glycoprotein; PMP-22, peripheral myelin protein-22.

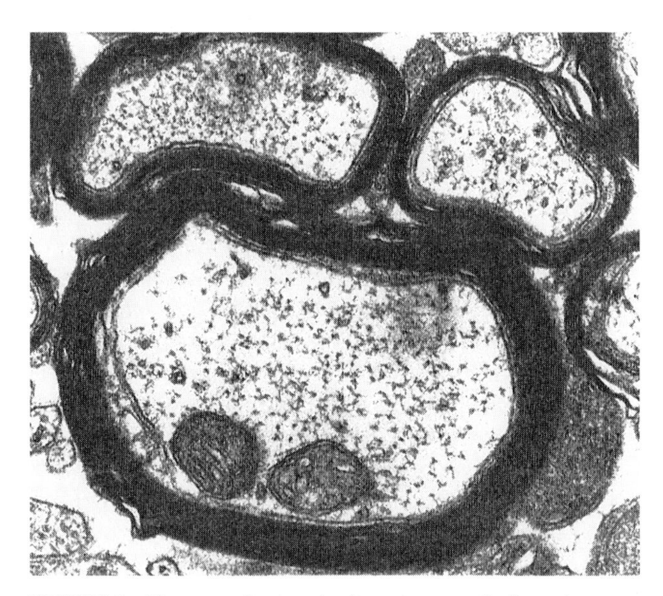

FIGURE 2 Electron micrograph of compact myelin from the mammalian CNS. Although there are minor ultrastructural differences, compact PNS myelin has a similar ultrastructural appearance. Note the alternating pattern of darker major dense lines and paler intraperiod lines, originally formed by fusion of apposing surfaces of the inner and outer leaflets, respectively, of the oligodendroglial plasma membrane. Cytoplasm-containing internal mesaxons can be seen on two of the myelinated axons.

ous with the perikarya cytoplasm of the Schwann cells are also present (Figs. 1 and 2). In addition, the myelin internode contains several ultrastructurally and biochemically distinct membrane domains, including the outer plasma membrane of the myelin-forming cell and the compact myelin itself, as well as the cytoplasm-containing Schmidt–Lanterman incisures, paranodal loops, nodal microvilli, and outer and inner mesaxons. The latter cytoplasm-containing structures provide connections with the perikaryal cytoplasm, and are vital for myelin maintenance.

The process of myelination in the CNS is similar, except that a single oligodendroglial cell extends a number of processes from its cell body; each process then envelopes and myelinates a single segment of a given axon (Fig. 3). Much of the local membrane assembly to give mature, compact myelin occurs within the oligodendroglial cytoplasmic processes (Waxman and Sims, 1984). The size of the fibers and the thickness of the sheaths are very different in the PNS and the CNS, but the overall surface area of myelin generated by an oligodendrocyte around multiple axons may be no larger than that formed by a Schwann cell around a single internode. The term *oligodendrocyte,* meaning "few processes," is actually somewhat of a misnomer,

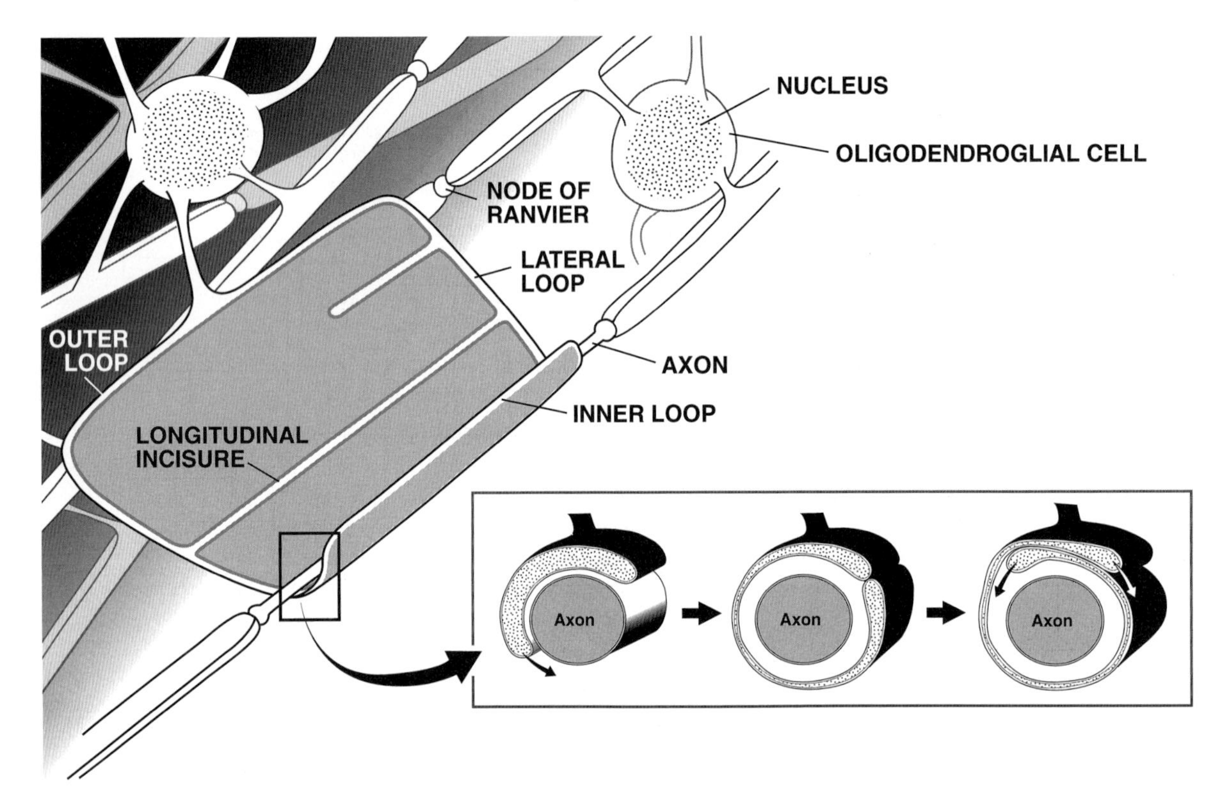

FIGURE 3 Diagram illustrating ultrastructural features of CNS myelin. Two myelin-forming oligodendroglial cells are shown, with each extending numerous processes to form myelin internodes around axons. Myelin from one internode has been unrolled from the axon to illustrate the large membrane surface area as well as various cytoplasm-containing structures (inner, outer, and lateral loops, longitudinal incisures) that are continuous with the perikaryal cytoplasm of the oligodendroglial cell. The nodes of Ranvier, short segments of axon between adjacent myelin internodes, are vital to normal nervous system function (see text). The expanded inset illustrates the process of formation of the initial wrap of newly forming myelin. An oligodendroglial cell process makes contact with the axon and eventually surrounds it. The apposing layers of membrane then fuse to form the mesaxon, which spirals around the axon a number of times. The process of myelination is completed when most cytoplasm is extruded from the uncompacted myelin to give mature, compact myelin (step not shown).

as a given oligodendrocyte may myelinate anywhere from less than five axons up to dozens of axons (Butt and Ransom, 1989; Bjartmar *et al.,* 1994). As in the PNS, the onset of myelination is preceded by proliferation of oligodendroglia. As development continues, both the diameter and length of the axons increase, and this is associated with a corresponding increase in nodal length as well as increases in myelin thickness. Thus, despite its compact highly ordered appearance, myelin continues to expand in all planes during growth and development.

In general, myelination follows the order of phylogenetic development, with the peripheral nerves myelinating first, then the spinal cord, and finally the brainstem, cerebellum, and cerebrum. There is, however, considerable overlap in this progression. In addition, each fiber tract may have its own spatiotemporal pattern of myelination, so that the degree of myelination may differ in different fiber tracts at a given developmental stage. For example, myelination in the spinal cord proceeds in a rostral–caudal gradient, whereas in the optic nerve, myelination progresses with a retinal to chiasmal gradient. Although not necessarily an absolute prerequisite for function, in general, fiber tracts are myelinated before they become fully functional.

Myelination is a major metabolic and structural event that occurs during a relatively brief but precisely defined period in the normal progression of events involved in nervous system development. In both the CNS and PNS, an enormous amount of myelin membrane is formed (some PNS axons may have as many as 100 layers) and this membrane must be maintained at a considerable distance from the supporting glial cell body. The surface area of myelin per adult oligodendrocyte in the rat brain has been calculated to be $1–20 \times 10^5$ μm^2, several orders of magnitude greater than the perikaryal membrane surface area of about 100 μm^2 (Pfeiffer *et al.,* 1993). Myelinating glial cells are thus maximally stressed in terms of their metabolic and synthetic capacity during this time, with each cell synthesizing myelin equivalent to up to 3 times the weight of its perikarya each day (Morell *et al.,* 1994). Because of these very high levels of synthetic and metabolic activity, these myelinating cells are especially vulnerable to nutritional deficits and/or to toxic insults or injuries during this period.

III. Axon–Glia Interactions during Development and Myelination

The tightly programmed sequence of events eventually resulting in the formation of mature compact myelin and the consequent initiation of impulse transmission is regulated by interactions between axons and glial cells at numerous stages (see Waxman and Black, 1995, for detailed discussion). During the early stages of myelination, the axon is loosely ensheathed by processes arising from immature, relatively undifferentiated glial cells. Loose glial ensheathment of axons in the optic nerve is seen in the rat beginning at postnatal day 6. This is followed by spiral wrapping of the axon by oligodendroglial processes that form compact myelin. In some tracts, the immature myelin sheath is initially close to the oligodendroglial cell body (Remahl and Hildebrand, 1990), but with maturation the sheath is displaced radially and often is connected to the cell body by only a thin cytoplasmic bridge. A single oligodendrocyte can myelinate axons of various diameters in their vicinity and can form myelin sheaths of different thicknesses around axons of differing diameter.

The development, maturation, and maintenance of the myelin sheath is dependent on both the normal functioning of the myelinating glial cells and the integrity of its relationship to the axon it ensheaths. During development, physical features of the myelin sheath, such as thickness and number of lamellae, are not preprogrammed within the myelin-forming cells, but rather depend on local regulation by the axon, with larger axons having thicker myelin sheaths (Waxman and Sims, 1984). Myelin internodal distance is also matched to fiber diameter (Hess and Young, 1952) and the internodal distance : diameter ratio is different for fibers in different tracts. There is some evidence that myelination is initiated when a developing axon reaches a "critical diameter." However, myelination occurs over a range of axonal diameters (Fraher, 1972) and at various times along a single axon (Waxman *et al.,* 1972; Waxman, 1985). The signal for initiation of myelination thus appears to be specific for particular axons, or for specific domains along axons.

The axonal membrane contains molecules that trigger mitogenesis in Schwann cells and oligodendrocytes (Salzer *et al.,* 1980a; 1980b; DeVries *et al.,* 1983; Chen and DeVries, 1989) and regulate the rate and degree of myelin formation (Black *et al.,* 1986; Waxman 1987a,b). Although some Schwann cells go on to myelinate axons, others only ensheath bundles of unmyelinated axons. These nonmyelinated Schwann cells express distinct molecular markers such as the low-affinity nerve growth factor receptor (NGF-R), neural cell adhesion molecules (N-CAM and L1), and growth associated protein-43 (GAP-43), but none of the myelin specific proteins (see following sections, as well as Mirsky and Jessen, 1990; Curtis *et al.,* 1992). Transplantation studies have shown that the axon determines the phenotype of the Schwann cell (Aguayo *et al.,* 1976; Weinberg and Spencer, 1976). Thus, there are a number of potentially distinct physical interactions between Schwann cells and axons as Schwann cells proliferate, migrate, ensheath axons, and form myelin sheaths during development of the PNS.

Although the axon influences and helps direct the formation of myelin, the myelin sheath also significantly defines features of the axon. The premyelinated axon is electrically excitable (Foster *et al.,* 1982; Waxman *et al.,* 1989) and the loss of excitability in the internodal axonal membrane that occurs with myelination involves an active suppression of Na^+ channels by the overlying oligodendrocyte or myelin sheath (Black *et al.,* 1985, 1986). Proliferation of oligodendrocyte precursors in the optic nerve is dependent on axonal electrical activity in that the blockage of optic nerve electrical activity by transection or exposure to tetrodotoxin resulted in a dramatic loss of oligodendrocyte precursor cells (Barres and Raff, 1993). Potassium channels may also participate in myelin formation; the potassium channel blocker, TEA^+ was shown to be effective in eliminating myelination in spinal cord explants while leaving axonal conduction and synapse formation intact.

IV. Myelin-Forming Cells and Their Ontogenic Development

The ontogenic development of the myelinating Schwann cell lineage has been relatively well-characterized, particularly in rodents. Most of our knowledge derives from cell-culture studies, but *in vivo* developmental studies in normal as well as in transgenic and gene knock-out mice have also proved useful. Understanding of the events and stages involved is of potential clinical relevance, not only with respect to various toxic neuropathies and to PNS nerve regeneration, but also because Schwann cell precursors may be attractive

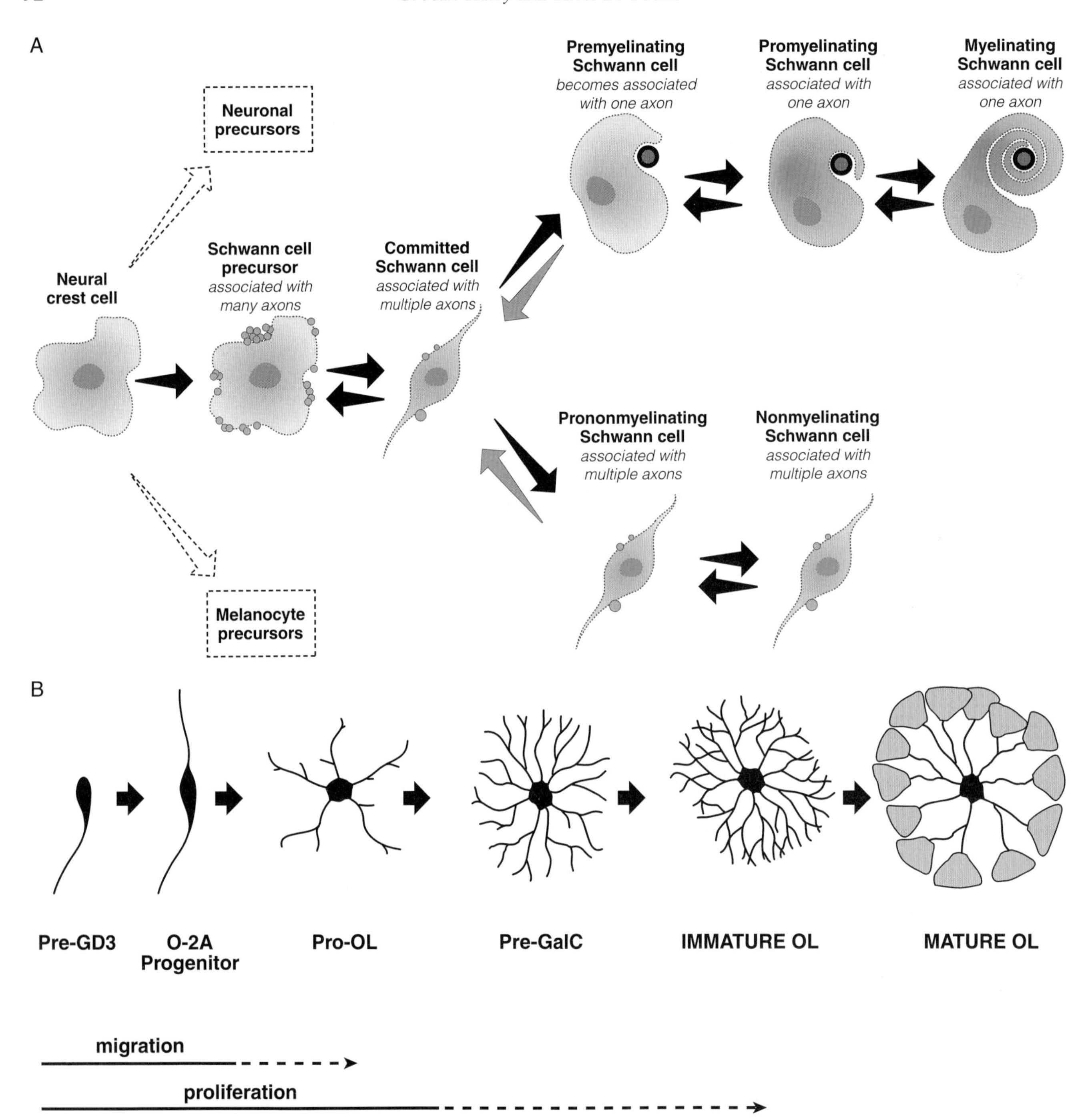

FIGURE 4 Ontogenic development of myelin-forming Schwann cells of the PNS and CNS. (A) Myelinating Schwann cells of the PNS originate from primitive neural crest cells, proliferative multipotential cells also capable of differentiating into neurons or melanocytes. The first step along the Schwann cell lineage gives the Schwann cell precursor, a proliferative cell that becomes associated with many axons and expresses the low-affinity nerve growth factor receptor (NGF-R), growth-associated protein 43 (GAP-43), and the neural cell adhesion molecules N-CAM and L1. The subsequent "committed" Schwann cell becomes associated with progressively fewer axons and expresses, in addition to the previously noted markers, S-100 protein (from this stage onward, all Schwann cells express S-100). Committed Schwann cells develop into either nonmyelinating Schwann cells, which remain associated with several axons and express galactocerebroside (GalC) in addition to the previous markers, or into myelinating Schwann cells. Myelinating Schwann cells progress through a proliferative "premyelinating" stage, characterized by transient expression of suppressed cAMP–inducible Pou-domain transcription factor (SCIP), followed by a "promyelinating" GalC-positive stage, becoming associated with a single axon in the process. The final differentiation into a mature myelinating Schwann cell involves

candidates for transplantation to facilitate remyelination and repair in injured CNS. A detailed discussion of this subject is beyond the scope of this chapter, but the reader is referred to several comprehensive reviews on this subject if more details are desired (Jessen and Mirsky, 1991; Gould *et al.,* 1992; Mirsky and Jessen, 1996; Zorick and Lemke, 1996). Schwann cells develop from neural crest cells and most of the key developmental stages in Schwann cell maturation seem to depend on axon-associated signals. Schwann cell precursors give rise to immature Schwann cells, which have a distinct phenotype (Fig. 4A). These cells then develop into either myelin-forming or nonmyelin-forming Schwann cells, under the control of reversible processes regulated by axon-associated signals. Even mature Schwann cells show a high degree of plasticity. Following a demyelinating insult, Schwann cells dedifferentiate into more primitive precursor cells, but generally do not die. When appropriate conditions present themselves (such as the regrowth of axons that occurs following a nerve-crush injury), these cells can proliferate, reestablish contact with axons, redifferentiate to the myelin-forming phenotype, and remyelinate the axons, thereby restoring normal function.

It is possible that primitive neural crest cells enter the Schwann cell lineage only when they first encounter axons in the developing nerves, and that a signal related to axonal contact guides them down the path to mature Schwann cells. Such signals may involve members of the Neu-differentiation factor (NDF) family. Members of the NDF growth factor family, including glial growth factor (GGF), heregulin, acetylcholine-inducing activity (ARIA), and neuregulin, are alternatively spliced products of a single gene, and these molecules are emerging as important regulators of Schwann cell lineage development (Dong *et al.,* 1995; Zorick and Lemke, 1996). Alternatively, it is possible that selected neural crest cells have already entered the Schwann cell lineage before they encounter axons (see Mirsky and Jessen, 1996, for discussion).

Investigation of which transcription factors regulate Schwann cell development is a current area of active study. Among factors likely to play significant roles are the zinc-finger transcription factors Krox-20 (Topilko, 1994), SCIP (see Zorick and Lemke, 1996), Pax3 (Kioussi *et al.,* 1995), and possibly c-jun (Stewart, 1995). A number of signalling molecules also regulate Schwann cell proliferation and myelination, including insulin-like growth factor-1 (IGF-1), which promotes expression of a myelinating phenotype in cell culture, and transforming growth factor βs (TGF-β), which inhibit myelin formation. The latter may be involved in generating the nonmyelinating Schwann cells, which ensheath smaller PNS axons.

The developmental lineage of the oligodendrocyte, the myelin-producing cell of the CNS, is also relatively well characterized, particularly in rodent *in vitro* systems (Fig. 4B). Oligodendrocytes originate as neuroectodermal cells of the subventricular zones and then migrate, proliferate, and further differentiate into mature, postmitotic myelin-forming oligodendroglial cells. Their development is discussed only briefly here, but a more comprehensive review is available (Warrington and Pfeiffer, 1992). Panels of cell- and stage-specific antibodies have proved especially useful in characterizing the sequential expression of various developmental markers, and this has allowed identification of distinct phenotypic stages, each characterized by its proliferative capacity, migratory ability, and distinct morphologies. Primitive precursor cells differentiate into proliferative, migratory bipolar O2A progenitor cells. These cells are bipotential, being capable of differentiating into either astrocytes or oligodendrocytes. The oligodendrocyte lineage progresses through several additional stages, including an immature oligodendrocyte expressing galactocerebroside, sulfatide, and cyclic nucleotide phospho-

downregulation of NGF-R, GAP-43, N-CAM, and L1 expression, with upregulation of expression of GalC and myelin proteins, and *in vivo,* the synthesis and elaboration of myelin. Because of the high degree of plasticity of Schwann cells, most of the developmental steps shown are reversible. (Modified from Mirsky and Jessen [1996] and Zorick and Lemke [1996]). (B) Myelinating oligodendroglial cells of the CNS originate from neuroectodermal cells of the subventricular zones of the developing brain. The earliest precursor cells recognized to date (Pre-GD3 stage) are proliferative, unipolar cells that express the embryonic neural cell adhesion molecule (E-NCAM). These cells develop into GD3 ganglioside-expressing proliferative bipolar cells, termed *O-2A progenitor cells* because they are capable (in culture, at least) of developing into either "type 2" astrocytes or oligodendrocytes. Development continues through a postmigratory but proliferative multipolar pro-oligodendroblast (Pro-OL) and a Pre-GalC stage, characterized by lack of expression of GalC. The onset of terminal oligodendroglial cell differentiation (immature OL stage) is identified by the surface appearance of a subset of "myelin components," consisting of the lipids GalC and sulfatide, as well as the enzyme cyclic nucleotide phosphodiesterase (CNP). Immature OLs then undergo final differentiation into mature oligodendrocytes (mature OL), characterized by regulated expression of myelin components such as MBP and PLP and by the synthesis and elaboration of sheets of myelin membrane.

diesterase (all myelin components), finally arriving several days later at the mature oligodendrocyte stage. Mature oligodendrocytes express all of the myelin-specific proteins and are capable of myelin synthesis *in vitro* in the absence of axons.

It is worth noting that oligodendroglial cells show some developmental plasticity, and this may be of clinical relevance with respect to CNS remyelination. In addition, a small population of oligodendrocyte progenitors persist in the adult rat brain (see Wolswijk and Noble, 1995, for details), and these constitute a potential source of myelin-forming cells for CNS remyelination. Immature, cycling oligodendroglial progenitor cells endogenous to adult white matter are capable of remyelinating CNS axons following lysolecithin-induced demyelination (Gensert and Goldman, 1997). Also, O2-A progenitor cells from mice subjected to coronavirus-induced demyelination show increased phenotypic plasticity and enhanced mitotic potential, properties that may be linked to the efficient remyelination that occurs following the demyelinating phase of this disease (Armstrong *et al.,* 1990). Manipulation of these progenitor cells by various factors (see later this chapter) thus may also be useful in promoting remyelination in a clinical context.

As is the case for Schwann cells in the PNS, development of oligodendrocytes is governed by a number of growth factors, including platelet-derived growth factor (PDGF), basic fibroblast growth factor (bFGF), IGF-1, TGF-β, and nerve growth factor (NGF), as well as several cytokines (see Pfeiffer *et al.,* 1993). Oligodendrocytes differ from Schwann cells in that they can be induced to produce myelin in culture in the absence of axons. This raises the question of the extent to which neurons and their axons influence oligodendrocytes with respect to development and myelination *in vivo.* It is, however, difficult to imagine the lack of significant interaction between these two cells in the developing nervous system, and in fact, many such interactions are known. Neurons produce many of the growth factors involved, and they are known to modulate steady-state levels of myelin components as well as mRNA levels for these components (Singh and Pfeiffer, 1985; Macklin *et al.,* 1986; Kidd *et al.,* 1990; Barres and Raff, 1993,).

As in the PNS, production of myelin by oligodendrocytes requires the coordinated synthesis of massive amounts of myelin components. The marked up-regulation of myelin-specific proteins, as well as of enzymes involved in synthesis of myelin lipids (see later this chapter), reflects corresponding increases in abundance of the respective mRNA transcripts, suggesting that most regulation of the program for myelination occurs at the level of transcription (see Hudson, 1990, for review). A possible candidate for the coordinate control of CNS myelination is myelin transcription factor 1 (MyTI), a zinc-finger DNA-binding protein first identified by its ability to recognize the myelin proteolipid protein (PLP) gene (Kim and Hudson, 1992). MyTI mRNA transcripts are most abundant in oligodendrocyte progenitor cells, suggesting that this factor acts at a very early stage in the regulation of transcription for myelinogenesis (Armstrong *et al.,* 1995).

V. Composition of Myelin

Myelin of both the CNS and PNS has a distinctive composition that differs somewhat from that of most cellular membranes. It is the major component of white matter of the CNS, accounting for about half the dry weight of this tissue, and is responsible for its glistening white appearance. The same is true for larger nerves of the PNS, such as the sciatic nerve. Myelin *in situ* has a water content of about 40%, and is characterized by its high lipid content (70–85% of its dry mass) and its correspondingly low protein content (15–30%). Most biologic membranes have a much higher protein : lipid ratio, usually somewhere around unity. The insulating properties of myelin, vital to its physiological function, are related to this high lipid content. The high lipid content of myelin also results in a buoyant density less than that of other biologic membranes, and advantage can be taken of this to isolate myelin with a high yield and a high degree of purity (Norton and Poduslo, 1973a).

The lipid content of CNS myelin (Table 1) is characterized by high levels of galactolipids (about 32% of lipid dry weight), including galactocerebroside (Gal-C) and its sulfated derivative, sulfatide, and cholesterol (about 26% of total lipid weight), with phospholipids accounting for most of the remainder (Norton and Cammer, 1984; DeWille and Horrocks, 1992; Morell *et al.,* 1994). Plasmalogens, phospholipids having a fatty aldehyde linked to the C1 of glycerol instead of a fatty acid, are especially prominent in myelin. Ethanolamine plasmalogens and phosphatidylcholine (lecithin) are the major phospholipid species. Gangliosides are also present in minor amounts. PNS myelin has a similar composition, although there are quantitative differences (see Smith, 1983). PNS myelin has less cerebroside and sulfatide and more sphingomyelin than CNS myelin. These differences are minor, however, relative to the larger differences in protein composition discussed later.

The protein composition of myelin is relatively simple in that a few major structural proteins account for the bulk of the total protein (Table 2). The PLP and the myelin basic proteins (MBP) together account for about 80% of the protein content of CNS myelin (Nor-

TABLE 1 Lipid Composition of Rat and Human CNS and PNS Myelin and Rat Whole Brain[a]

Component[b]	Human CNS myelin	Rat CNS myelin	Rat whole brain	Human PNS myelin[c]	Rat PNS myelin[c]
			(percent total dry weight)		
Protein	30.0	29.5	56.9	28.7	—
Lipid	70.0	70.5	37.0	71.3	—
			(percent total lipid weight)		
Cholesterol	27.7	27.3	23.0	23.0	27.2
Total galactolipid	27.5	31.5	21.3	22.1	21.5
Cerebroside	22.7	23.7	14.6	—	15.8
Sulfatide	3.8	7.1	4.8	—	5.7
Total phospholipid	43.1	44.0	57.6	54.9	50.6
EPG[d]	15.6	16.7	19.8	19.4	19.2
CPG	11.2	11.3	22.0	8.3	9.6
Sphingomyelin	7.9	3.2	3.8	17.7	7.0
SPG	4.8	7.0	7.2	9.4[f]	11.0
IPG	0.6	1.2	2.4	—[f]	2.5
Plasmalogens[e]	12.3	14.1	11.6	16.1	—

[a]See Norton and Cammer, 1984; Morell *et al.,* 1994; and Morell and Toews, 1996b, for references and additional details.

[b]Other lipids are also present in myelin, including gangliosides, galactosyl diglycerides, and fatty acid esters of cerebroside; although not shown in the table, polyphosphoinositides may account for up to 7% of total myelin lipid phosphorus (see Morell *et al.,* 1994).

[c]Calculated from data in Norton and Cammer (1984), using 800 and 750 as average molecular weights for phosphoglycerides and sphingomyelin, respectively.

[d]Abbreviations: EPG, ethanolamine phosphoglycerides; CPG, choline phosphoglycerides; SPG, serine phosphoglycerides; IPG, inositol phosphoglycerides.

[e]Primarily ethanolamine plasmalogens.

[f]Value includes both SPG and IPG.

ton and Cammer, 1984; Morell *et al.,* 1994). In contrast, P_0 protein, a protein not found in the CNS, accounts for more than half the total protein of PNS myelin.

TABLE 2 Protein Composition of CNS and PNS Myelin[a]

Component	CNS	PNS
	(percent total myelin protein)[c]	
PLP/DM20[b]	50	<0.01[d]
P_0	<0.01[d]	>50
MBP	30	18
CNP	4	0.4
MAG	1	0.1
P2	<1	1–15
PMP-22	<0.01[d]	5–10

[a]Only major proteins are shown; see text for discussion of other myelin proteins, and Morell *et al.* (1994) and Newman *et al.* (1995) for additional references and details.

[b]Abbreviations: PLP, proteolipid protein; MBP, myelin basic protein; CNP, cyclic nucleotide phosphodiesterase; MAG, myelin-associated glycoprotein; PMP-22, peripheral myelin protein-22.

[c]Composite values representative of adult mammalian myelin.

[d]Although mRNA for this protein has been detected, the protein itself, if present at all, is present in myelin at only low to undetectable levels.

MBP, P2-protein, and PMP-22 account for most of the remainder of PNS myelin proteins. PLP, the major protein component of CNS myelin, is present in PNS myelin at only very low levels, if at all (see later). In both CNS and PNS myelin, there are a number of other minor but integral protein components, and the list will continue to grow as research continues. These include structural proteins and proteins involved in cell–cell interactions, as well as a large number of enzymes, receptors, and second messenger–related proteins. All of these have vital roles in maintaining the complex structure of myelin and/or in its function. Characteristics of some myelin proteins, including selected aspects of their gene structure and expression, follows, but it is necessarily brief. For a more detailed discussion of individual myelin proteins, see Lemke (1988), Morell *et al.* (1994), Campagnoni (1995), and Newman *et al.* (1995).

It is worth noting at this point that the composition of myelin changes during development, with the first myelin deposited having a somewhat different composition than that present in adults (Norton and Cammer, 1984). In the rat brain, myelin galactolipids increase by about 50%, and phosphatidylcholine decreases by a similar amount. Similar changes have been noted in human myelin as well. Other minor changes in lipids and gangliosides also occur. The protein portion of myelin also changes somewhat during development; both MBP and PLP increase during development, whereas the amount of higher molecular weight proteins decreases.

A. *Myelin Basic Proteins (MBPs)*

Myelin basic proteins (MBPs) are highly basic proteins of related isoforms derived from alternative splicing of a single gene. The MBP gene consists of seven exons distributed over about 32 kb of chromosome 18 in the mouse (Roach *et al.,* 1985) and human (Sparkes *et al.,* 1987) and chromosome 1 in the rat (Koizumi *et al.,* 1991); at least six transcripts are expressed via alternative splicing of RNA (Table 3). The MBP gene is actually a "gene within a gene," being part of a much larger (approx. 105 kb in mice and 179 kb in humans) transcriptional unit, called the *Golli-mbp* gene (Campagnoni *et al.,* 1993; Pribyl *et al.,* 1993). Portions of the *Golli-mbp* gene are expressed outside the nervous system, including the immune system, although the exact function of these gene products remains unknown. This may be of relevance to clinical disorders related to autoimmunity against MBP.

Expression of the various MBP protein products is also very complex; in addition to alternative splicing of a number of exons, there is also considerable transcriptional and posttranscriptional control (see Campagnoni, 1995, for details). This complexity of gene expression is augmented by posttranslational protein modifications, including loss of C-terminal arginine, N-acylation, glycosylation, phosphorylation, methylation, deamination, and substitution of some arginine residues with citrulline (Toews and Morell, 1987; Smith, 1992; Morell *et al.,* 1994). MBPs are extrinsic membrane proteins localized to the cytoplasmic membrane surface (major dense line) of myelin of both the CNS and PNS. In the CNS this protein accounts for approximately 30% of the total myelin protein, whereas in the PNS it accounts for only 18% (Greenfield *et al.,* 1980). Myelin lipids can promote MBP self-association, suggesting it may exist as oligomers on the cytoplasmic surface of the myelin membrane (Smith, 1992). It has been suggested that stabilization and maintenance of the myelin structure may be due to specific associations between MBPs and sulfatides and gangliosides (Ong and Yu, 1984; Maggio and Yu, 1989, 1992; Mendz, 1992; Smith, 1992).

B. *Proteolipid Protein (PLP) and DM20*

Proteolipid protein (PLP) and DM20 integral membrane proteins with several transmembrane domains. PLP is the most abundant protein in CNS myelin (50%), and although mRNA for this protein is present in the PNS, the protein itself is present at only very low levels in PNS myelin (Lemke, 1992) and its function in PNS myelin is unknown (Puckett *et al.,* 1987; Kamholz *et al.,* 1992). A report (Garbern *et al.,* 1997) of a human PLP null mutation phenotype characterized by a demyelinating peripheral neuropathy suggests that PLP/DM20 is necessary for proper myelin formation in the PNS as well as the CNS. This report also demonstrates by immunoelectron microscopy the presence of PLP in compact PNS myelin (however, see also Puckett *et al.,* 1987). Like MBP, PLP is one of the products of alternative splicing of a single gene, having a molecular weight of approximately 25kDa. DM20, a second isoform that migrates as a 20kDa band on SDS gel electrophoresis, is identical to PLP except for the deletion of amino acid residues 116–150 (Macklin, 1992). The PLP–DM20 gene, located on chromosome X in the mouse, rat, and human, is approximately 17 kb in length and consists of seven exons. The alternative splicing of this gene to give PLP and/or DM20 is developmentally regulated, with the DM20 splice product predominating during early myelination. Some mutations of the PLP/DM20

TABLE 3 "Myelin" Mutant Mice

Myelin gene (chromosome)	Mouse mutant	Human disease equivalent	References[a]
MBP	*shiverer*		Roach *et al.,* 1985
(18)	*shiverer^{mld}*		Popko *et al.,* 1988
PLP	*jimpy*	Pelizaeus–Merzbacher Disease	Nave *et al.,* 1986
(X)	*jimpy^{msd}*		Gencic and Hudson, 1990
	rumpshaker		Schneider *et al.,* 1992
P_0	*P_0-deficient*	Charcot–Marie–Tooth disease, type 1B	
(1)			Giese *et al.,* 1992
PMP-22	*trembler*	Charcot–Marie–Tooth disease type 1A	Suter *et al.,* 1992a
(1)	*trembler^{j}*		Suter *et al.,* 1992b
??	*quaking^{viable}*		Ebersole *et al.,* 1992
(17)			

[a]See text for additional references and discussion.

gene (e.g., *jimpy* mice; see later this chapter) result in developmental abnormalities prior to any myelination, suggesting that gene may be involved in other functions besides myelination. In addition, expression is not confined to myelin-producing oligodendrocytes in the CNS. The role of protein products of this gene outside the nervous system remain unknown, however. Missense mutations of the PLP/DM20 gene give rise to a host of CNS pathologies, the most devastating being Pelizaeus–Merzbacher disease (PMD) (see Seitelberger, 1995). In some forms of PMD, there is complete deletion of the PLP/DM20 gene (Raskind *et al.,* 1991). The described single base pair deletion in humans leads to the absence of PLP/DM20 expression; this produces a disease similar to, but less severe than, classic PMD, but also involving a progressive demyelinating peripheral neuropathy (Garbern *et al.,* 1997).

A number of both positive and negative *cis*-acting elements, as well as some *trans*-acting factors, have been identified for this gene (see Campagnoni, 1995, for details). In its orientation in the myelin membrane, the extracellular domains of PLP may be instrumental in stabilizing the intraperiod line of myelin (Morell *et al.,* 1994). In addition, PLP may play an active role as an ion channel (Toews and Morell, 1987; Lees and Bizzozero, 1992). As noted previously, DM20 is generally a relatively minor product of the PLP gene but it shows a pattern of developmental regulation distinct from PLP. It is expressed earlier in development than PLP and is the major PLP gene product in the developing embryo (Ikenaka *et al.,* 1992; Macklin, 1992; Timsit *et al.,* 1992). Its presence in "premyelinating" glial cells and in cells outside the glial cell lineage suggest a possible functional role unrelated to myelination.

C. Myelin-associated Glycoprotein (MAG)

Myelin-associated glycoprotein (MAG) is the principal glycoprotein of central nervous system myelin (for review, see Milner *et al.,* 1990; Quarles *et al.,* 1992). MAG is heavily glycosylated and is specific to myelin sheaths, with an especially high concentration in the periaxonal regions of both CNS and PNS myelin. It is a member of the immunoglobulin gene superfamily (Sutcliffe *et al.,* 1983; Lai *et al.,* 1987; Salzer *et al.,* 1987). Examination of the extracellular, N-terminal domains suggest that MAG is most closely related to the cell adhesion molecules N-CAM, L1, and contactin. In the PNS, MAG immunostaining is seen in glial membranes of the Schmidt–Lantermann incisures, paranodal loops, and mesaxons (Trapp, 1990) and is distinctly absent from the compact myelin sheath. It is thought that MAG plays a major role in membrane–membrane interactions during myelin formation and maintenance (Salzer *et al.,* 1990; Quarles *et al.,* 1992; Morell *et al.,* 1994). It is presumed to be involved in the adhesion of the myelin sheath to the axonal plasmalemma and in membrane spacing (Trapp, 1990), and it has been implicated in various peripheral neuropathies (Mendell *et al.,* 1985; Tatum, 1993). MAG exists as two isoforms (L-MAG and S-MAG) that are derived by alternative splicing from a single gene. L-MAG is produced almost exclusively in the CNS and is the predominant variant during early development and active myelination (Campagnoni, 1988; Trapp, 1990). S-MAG is the major isoform in the adult CNS and in the PNS at all ages. It is thought that the differences in distribution within the CNS and PNS may be associated with either the phosphorylation or other posttranslational modifications (e.g., sulfation of oligosaccharide moieties, acylation of the transmembrane domain) altering interactions with the cytoskeleton (Trapp, 1990; Quarles *et al.,* 1992). Homotypic interaction may be operational in Schmidt–Lantermann incisures, paranodal loops, and mesaxon membranes in PNS myelin, whereas heterotypic interactions with axolemmal constituents may mediate glia–axon adhesion (Salzer *et al.,* 1990; Trapp, 1990; Quarles *et al.,* 1992).

D. 2′-3′-Cyclic Nucleotide-3′-Phosphodiesterase (CNP)

2′-3′-Cyclic nucleotide-′-phosphodiesterase (CNP) is localized within oligodendrocytes in the CNS and within Schwann cells in the PNS. One of the earliest markers for cells of the oligodendroglial lineage, CNP is an enzyme that hydrolyzes 2′,3′-cyclic nucleotides to form 2′-nucleotides exculsively. Because no physiologically relevant substrate molecules have been found in myelin, however, this enzymatic activity may actually be vestigial and unrelated to its function in myelin. The current view of CNP is that it may be a key component of an interactive protein network within oligodendroglial cells, possibly involved in extension of processes (e.g., see Braun *et al.,* 1990). It is isoprenylated, suggesting possible involvement in signal transduction processes (Braun *et al.,* 1991). Furthermore, the presence of potential nucleotide-binding domains on CNP (Sprinkle, 1989) suggest it might exert regulatory influence on various cellular processes such as growth and differentiation by serving as a link between extracellular signals and intracellular effector molecules. Its exact function within myelin and oligodendroglial cells, however, remains unknown.

E. Myelin/Oligodendrocyte Glycoprotein (MOG)

Mylein/Oligodendrocyte glycoprotein (MOG) is localized primarily at the external surfaces of the myelin sheath and oligodendrocytes. It is developmentally regulated, appearing with the onset of myelination as one of the last myelin protein genes expressed (Scolding *et al.,* 1989). Features of the protein structure suggest it may be a member of the immunoglobulin gene superfamily (Gardinier *et al.,* 1992). Because anti-MOG antibodies can cause demyelination *in vivo* (Schluesener *et al.,* 1987), it has received attention for a potential role in autoimmune-mediated demyelination such as experimental autoimmune encephalomyelitis (EAE) and multiple sclerosis (Gunn *et al.,* 1989; Bernard and Kerlero de Rosbo, 1991).

F. Oligodendrocyte–Myelin Glycoprotein (Omgp)

Oligodendrocyte–myelin glycoprotein (Omgp) is one of the minor protein components of myelin that appears in the CNS during the period of myelination. It is highly glycosylated and appears to be specific to oligodendrocytes and myelin membranes in the CNS (Mikol and Stefansson, 1988). The protein is anchored to the membrane through a glycosylphosphatidylinositol intermediate. A subpopulation of Omgp molecules contain the human natural killer cell antigen–1 (HNK-1) carbohydrate (Mikol *et al.,* 1990b). The presence of a tandem leucine repeat domain in the predicted polypeptide sequence and the apparent presence of Omgp at the paranodal regions of the myelin sheath have lead to speculation that its major role is as an adhesion molecule that mediates axon–glial cell interactions (Mikol *et al.,* 1990a).

G. P_0 Protein (Peripheral Myelin Protein Zero)

P_0 protein (peripheral myelin protein zero) accounts for more than 50% of the protein in peripheral nerve myelin (Ishaque *et al.,* 1980). This has lead to the suggestion that P_0 is the PNS equivalent of PLP in the CNS, although the properties of these two proteins are very different. P_0 is a transmembrane protein with a glycosylated extracellular domain, a single membrane spanning region, and a highly basic intracellular domain (Lemke,(Lemke and Axel, 1985). It is thought to play an important role in the compaction of myelin through the homotypic interaction of molecules on adjacent myelin lamellae (Lemke, 1988). Like MAG, it is a member of the immunoglobulin superfamily. Unlike many of the other myelin protein genes, expression of the P0 gene is highly restricted to Schwann cells. The P_0 gene contains six exons distributed over about 7 kb in both rats and mice (Lemke, 1988; You *et al.,* 1991). Based on transgenic experiments, elements regulating its expression appear to reside in first 1.1 kb of the 5′-flanking region (Messing *et al.,* 1992). Mutation of the P_0 gene in humans is associated with Charcot–Marie–Tooth disease, type 1B, an inherited peripheral neuropathy (Hayasaka *et al.,* 1993; Kulkens *et al.,* 1993).

H. Peripheral Nerve Protein P2

Peripheral nerve protein P2 is a basic protein distinct from MBP. Interest in P2 arose from its ability to induce experimental allergic neuritis, a demyelinating disease of the PNS (Kadlubowski and Hughes, 1980). It has sequence homology to cellular retinol and retinoic acid–binding proteins (Crabb and Saari,1981; Eriksson *et al.,* 1981) and fatty acid–binding proteins (FABPs) (Veerkamp *et al.,* 1991), and has high affinity for oleic acid, retinoic acid, and retinol (Uyemura *et al.,* 1984). The P2 protein gene belongs to an ancient family of FABPs that diverged into two major subfamilies (Medzihradszky *et al.,* 1992). P2 mRNA levels parallel myelination during development as well as the levels of microsomal enzymes involved in fatty acid elongation. This suggests the P2 protein may be involved in fatty acid elongation or in the transport of very long-chain fatty acids to myelin (Narayanan *et al.,* 1988).

I. Peripheral Myelin Protein-22 (PMP-22)

Peripheral myelin protein-22 (PMP-22) is a glycoprotein with an apparent molecular weight of about 22 kDa (Kitamura *et al.,* 1976; Smith and Perret, 1986). The rat and human genes have been cloned (Spreyer *et al.,* 1991; Welcher *et al.,* 1991; Hayasaka *et al.,* 1992); although these have a high homology to the growth arrest–specific mRNAs for gas 3 and PASII-glycoprotein, the function of PMP-22 in PNS myelin remains unknown. The PMP-22 gene maps to mouse chromosome 11 and a point mutation in this gene is apparently responsible for the autosomal dominant mutation in the Trembler (Tr) mutant mouse (Suter *et al.,* 1992a,b). In humans, the gene maps to chromosome 17; this gene is duplicated in patients with Charcot–Marie–Tooth disease, type 1A, and this alteration is presumably related to the pathology (Patel *et al.,* 1992; Valentijn *et al.,* 1992; see Campagnoni, 1995, and Newman *et al.,* 1995 for additional references and discussion). The rat PMP-22 gene, the expression of which is largely confined to the PNS, is developmentally regulated in Schwann cells, where its expression coincides with myelination (Spreyer *et al.,* 1991; Snipes *et al.,* 1992). mRNA expression is coordinately down-regulated along with other myelin proteins during tellurium-induced primary demyelination and during degeneration induced by nerve transection or crush; message levels are consequently up-regulated along with other myelin genes during remyelination if it occurs (Spreyer *et al.,* 1991; Toews *et al.,* 1997).

J. Enzymes Associated with Myelin

Although myelin initially was believed to be metabolically inert (due largely to the very slow metabolic turnover of some of its components) and to function exclusively as an electrical insulator, it is now known that the picture is considerably more complex and interesting. All protein and lipid components of myelin turn over with measurable turnover rates (see Benjamins and Smith, 1984; Morell *et al.,* 1994). Although some structural components of myelin are indeed relatively stable metabolically, with half-lives of several months, there is also a very rapid turnover of some myelin components. The phosphate groups modifying myelin basic protein in compact CNS myelin turn over with half-lives of minutes or less (DesJardins and Morell, 1983), and the phosphate groups on myelin polyphosphoinositides also show a very rapid turnover rate. At least 40 different enzyme activities have been documented in CNS myelin (Newman *et al.,* 1995). In addition to CNP described previously, these include enzymes related to second messenger signalling, as well as associated receptors and G-proteins (Larocca *et al.,* 1990). A number of enzymes involved in myelin lipid metabolism are also present, including those for phospholipid synthesis and catabolism. Noteworthy among the latter are phospholipase C activities for polyphosphoinositides, and these may have important roles in signal transduction mechanisms in myelin (Ledeen, 1992). Also of note are a cholesterol ester hydrolase, and UDP-galactose: ceramide galactosyltransferase, the terminal enzyme in biosynthesis of galactocerebroside, the most "myelin-specific" lipid. Various proteases, protein kinases and phosphatases, and transport-related enzymes are also present; of particular interest with respect to the transport-related enzymes are carbonic anhydrase (Cammer *et al.,* 1976) and Na^+,K^+-ATPase (Zimmerman and Cammer, 1982). These enzymes may be involved in controlling K^+ levels at nodes of Ranvier (Lees and Sapirstein, 1983) and/or in removal of carbonic acid from metabolically active axons.

K. Cytoskeletal Proteins in Myelin and Myelinating Glia

The cell surface area of myelin-forming cells is so large as to suggest the need for specialized structures and mechanisms for transporting components between the perikaryon and the remote extensions. There is indeed a great deal of protein and lipid transport and targeting within the myelinating glial cell, and cytoskeletal elements play important roles in these processes. P_0, MAG, and laminin (a secreted extracellular matrix component; Cornbrooks *et al.,* 1983) are synthesized and modified in the rough endoplasmic reticulum (RER) and Golgi membranes of the perinuclear cytoplasm and then sorted into different carrier vesicles upon exit from the *trans* Golgi network (Trapp *et al.,* 1993). These proteins must be transported over millimeter distances prior to insertion into their proper surface membrane locations. Various proteins enriched in compact myelin reach their proper destinations via different mechanisms. For example, P_0 reaches compact myelin by vesicular transport, whereas it is the mRNA for MBP that is translocated, with synthesis of MBP occurring close to the site of its insertion into the forming myelin (Colman *et al.,* 1982; Trapp *et al.,* 1987; Griffiths *et al.,* 1989).

As noted previously, the cytoskeleton plays a major role in transport and assembly of myelin. In myelinating Schwann cells, microfilaments are enriched beneath the membranes of the Schmidt–Lanterman incisures, the outer and inner mesaxons, and portions of the outermost compact myelin lamellae and Schwann cell plasma membrane (Trapp *et al.,* 1989; Zimmerman and Vogt, 1989; Kordeli *et al.,* 1990). It is thought that interactions between MAG and microfilaments play a role in membrane motility during myelination (Trapp and Quarles, 1982; Trapp *et al.,* 1984; Martini and Schachner, 1986; Salzer *et al.,* 1987). MAG colocalizes with microfilaments at membranes that move during internodal growth (Trapp *et al.,* 1989). A role for MAG in myelin wrapping and spacing is supported by studies showing precocious spiral wrapping by myelinating Schwann cells transfected with additional copies of MAG (Owens *et al.,* 1990), and impaired or prevented wrapping when MAG expression is reduced or eliminated in Schwann cells (Owens and Bunge, 1991; Mendell *et al.,* 1985). Microfilaments also help to define and maintain organelle-rich channel regions and organelle-free non-channel regions of the myelin internode. The channel regions are important to formation and maintenance of the myelin internode and to intracellular transport of myelin components. They include the external cytoplasmic channels, Schmidt–Lanterman incisures, para nodal loops, and periaxonal cytoplasm. Microfilaments associated with the abaxonal plasma membrane and the adaxonal periaxonal membrane may have multiple functions in dealing with the external environment, including endocytosis (Blok *et al.,* 1982), pinocytosis (Phaire-Washington *et al.,* 1980), and exocytosis (John *et al.,* 1983; Koffer *et al.,* 1990), as well as stress resistance.

Just as they do in axons, microtubules may also subserve a role in the directional movement of organelles within glial cells. Post-Golgi vesicles transported in this way could be involved in the growth, turnover, or modification of compact myelin at distant sites. Microtubules are the largest of the cytoskeletal filaments and provide a dynamic substrate for organelle trafficking and structural organization (for review see Dustin, 1984; Kirschner and Mitchison, 1986; Schroer and Sheetz, 1991). They are present in all major cytoplasmic compartments of the myelin internode, but are excluded from compact myelin (Peters *et al.,* 1991). In myelinating Schwann cells, microtubules are crucial to the transport of myelin proteins and organelles (Trapp *et al.,* 1995). This function is determined by the orientation and organization of microtubules, which in turn are influenced by axons (Kidd *et al.,* 1994). During the formation of the myelin sheath, contact with a myelin-inducing axon results in a more complex microtubule organization (Kidd *et al.,* 1994). As in axons, microtubules are inherently unstable and oscillate between phases of elongation and collapse (Dustin, 1984; Kirschner and Mitchison, 1986). The extent of depolymerization and repolymerization is determined by complex assembly/disassembly kinetics and can be influenced by modifications such as binding of MAPs (Sloboda *et al.,* 1976; Pryer *et al.,* 1992). Microtubule disassembly causes marked accumulation of P_0, MAG, and laminin in the perinuclear cytoplasm of myelinating Schwann cells (Trapp *et al.,* 1995). Because of this, chemically induced neurotoxicity involving microtubules may lead to alterations not only in axons, but also in myelin and myelinating cells as well.

In myelinating Schwann cells, the major intermediate filament is vimentin (Dahl *et al.,* 1982; Schachner *et al.,* 1984; Kobayashi and Suzuki, 1990). In most cells, intermediate filaments are considered to have a structural role in mechanically maintaining cell shape against external forces (Klymkowsky *et al.,* 1989; Skalli and Goldman, 1991) It has been proposed that vimentin intermediate filaments interact with microfilament-associated molecules and with microtubules in resisting stress (Wang *et al.,* 1993). It is thought that intermediate filaments play a role in the process of myelination in the PNS because myelinating Schwann cells contain abundant intermediate filaments and the content of these filaments between myelinating and nonmyelinating Schwann cells can vary substantially.

VI. Molecular Aspects of Myelin Synthesis and Assembly

Because the structure and composition of myelin is unique, its formation involves activation of a set of unique genes (see Lemke, 1988, for details). These genes include those related to induction of myelination (e.g., glial-specific receptors for differentiation signals), those involved in controlling and directing the initial deposition of myelin (e.g., axon–glial cell-adhesion molecules), and those involved in actual production of compact myelin (e.g., structural proteins of myelin). Genes for enzymes involved in synthesis of lipids enriched in myelin are also preferentially activated as well. The process of myelination is a highly regulated event that begins postnatally during the first few weeks of life in the rodent brain and within the third fetal trimester in the human spinal cord. In the CNS of rodents, maximum levels of synthesis of myelin components and actual accumulation of myelin occurs at about 3 weeks of age (Norton and Poduslo, 1973b; Norton and Cammer, 1984; Morell *et al.,* 1994), and although myelin accumulation continues for an extended time, the rate of synthesis declines considerably by about 6 weeks of age. This time course is similar to the profile of expression of myelin protein genes (see Campagnoni and Macklin, 1988). In the rodent PNS, myelination begins at about birth, peaks at about 2 weeks, and then decreases to a low basal level by the end of the first month (Webster, 1971). As is the case in the CNS, the pattern of myelin synthesis and accumulation is closely paralleled by expression of mRNA for PNS myelin protein components (Lemke and Axel, 1985; Stahl *et al.,* 1990) and for enzymes involved in synthesis of major myelin lipids (hydroxymethylglutaryl-Coenzyme A [HMG-CoA] reductase, the rate-limiting enzyme in cholesterol biosynthesis; and ceramide galactosyltransferase, the rate-limiting enzyme in cerebroside biosynthesis) (Lemke and Axel, 1985). Regulation of the expression of myelin genes occurs at a number of different levels including promotor choice, transcription, mRNA splicing and stability, translation, and posttranslational processing (for reviews see Campagnoni, 1988; Campagnoni and Macklin, 1988; Lemke, 1988; Nave and Milner, 1989; Ikenaka *et al.,* 1991; Mikoshiba *et al.,* 1991; Campagnoni, 1995).

The synthesis and assembly of myelin has been examined by measuring incorporation of radioactive precursors into myelin components both *in vivo* and in tissue slices, by measuring the *in vitro* activities of enzymes involved in synthesis of myelin components, by examining levels of expression of mRNA species for myelin-related genes, and by actual isolation and analysis of myelin (for reviews see Benjamins and Smith, 1984; DeWille and Horrocks, 1992; Morell *et al.,* 1994). After individual myelin components have been synthesized, they must be assembled to form mature compact myelin. Some components, such as PLP, which is synthesized on bound polyribosomes in the oligodendroglial perikaryon, show a time lag of about 45 min between their synthesis and their appearance in myelin, reflecting the time required for transport from their site of synthesis to the forming myelin. Other components, such as MBP, show only a short lag, in keeping with their synthesis of free polyribosomes in oligodendroglial processes, near the actual site of myelin assembly. In keeping with this difference, studies have shown that myelinating cells spatially segregate mRNA species for myelin-specific proteins (see Trapp *et al.,* 1987). mRNA for MBP is transported to near the sites of myelin assembly before the protein is synthesized (Trapp *et al.,* 1988; Gillespie *et al.,* 1990), whereas PLP mRNA is present in a perinuclear location. Individual lipids also show different kinetics of entry into myelin following synthesis, and some of this may be due to synthesis in, or movement through, different intracellular pools (Benjamins *et al.,* 1976; Benjamins and Iwata, 1979; for review, see Benjamins and Smith, 1984).

The mature myelin internode contains several ultrastructurally and biochemically distinct membrane domains that include the outer plasma membrane of the myelin-forming cell and its attached compact myelin, as well as the Schmidt–Lanterman incisures, paranodal loops, nodal microvilli, the periaxonal membrane, and the membranes of the outer and inner mesaxons (see Figs. 1 and 3). Some of these membrane domains are compositionally distinct, containing different structural proteins and differing in lipid composition as well. With respect to the PNS, P_0, MBP, and PMP-22 are enriched in compact PNS myelin (Trapp *et al.,* 1981; Omlin *et al.,* 1982), whereas PLP and MBP predominate in compact CNS myelin. The exact mechanisms by which individual components are targeted to their respective membrane domains and other molecular aspects of actual myelin membrane assembly are not well understood, but this continues to be an active area of investigation. It seems clear that cytoskeletal elements are closely involved in the intracellular sorting and transport of myelin components, as discussed in a previous section, and they presumably also function in the actual process of myelin assembly.

VII. Mutant Dysmyelinating Mouse Models

In the mouse, terminal differentiation of myelin-forming cells occurs mostly after birth, following establishment of the basic wiring of the nervous system. Many neurologic mutations result in dysmyelination, the inability of myelin-forming glial cells to assemble qualitatively and/or quantitatively normal myelin (see Quarles *et al.,* 1994; Nave, 1995). Because myelination is a postnatal event in rodents, this inability to assembe normal myelin is not immediately lethal, and the animals usually survive for at least several weeks. Mutations that affect myelin are characterized behaviorally by abnormalities such as shivering, ataxia, and frequently including seizures; these signs of abnormal nervous system function begin at about the time when myelin accumulation becomes significant, probably an indication of the importance of myelin for motor control and normal brain function. In general, myelin-deficient mice possess a mutant gene for some structural myelin protein (see Lemke, 1988). The major known "myelin-deficient" mutations in mice are described as follows, as are some related transgenic and "knockout" mouse models.

A. *Shiverer Mice*

The shiverer mouse (shi, mouse chromosome 18) was one of the first neurologic mouse mutants examined at the molecular–genetic level (Roach *et al.,* 1983). Affected homozygotes lack any detectable MBP and fail to make normal CNS myelin. The behavioral phenotype of this mouse is first observed within the second postnatal week, when a general body tremor, which becomes more pronounced with intentional movements, develops (Biddle *et al.,* 1973; Chernoff, 1981). The shivering behavior derives from a loss of spinal motor and reflex control and increases with age, often progressing to include seizures. The life span of the shiverer mouse is limited to approximately 6 months of age.

Histologic examination shows severe dysmyelination throughout the entire CNS, but with normal-appearing PNS myelin (Privat *et al.,* 1979; Kirschner and Ganser, 1980; Rosenbluth, 1980). At the ultrastructural level, the pattern of dysmyelination is dominated by a severe lack of myelin. The myelin-like structures that are occasionally present are loosely wrapped around the axon and the intracellular adhesion zone of the extended cell process that normally forms the "major dense line" of myelin cannot be discerned (Readhead and Hood, 1990). The lack of proper myelin sheath formation may be the result of a defect in myelin compaction or a related process, because oligodendroglia appear to be normally differentiated. Histologic evidence of dysmyelination is supported biochemically by a dramatic reduction of all major myelin proteins, and more specifically by the complete lack of MBP. This is the result of a large (20 kb) genomic deletion encompassing exons 3–7 of the MBP gene (or exons 7–11 of the larger golli-MBP gene) resulting in no coding capacity for any of the MBP isoforms (Roach *et al.,* 1985; Molineaux *et al.,* 1986).

The tremoring phenotype can be cured by a number of manipulations associated with restoration of MBP expression, indicating that the amount of MBP available to the oligodendrocytes is a rate-limiting step in the assembly of CNS myelin. Successful approaches include reintroduction of the entire wild-type MBP gene into the germ line of the shiverer mouse (Readhead *et al.*, 1987), increasing the transgene copy number (Popko *et al.*, 1987; Shine *et al.*, 1990), and the reintroduction of a MBP minigene encoding only the smallest (14 kDa) MBP isoform (Kimura *et al.*, 1989).

A shiverer-like phenotype can also be generated in normal mice by specifically down-regulating the amount of MBP mRNA available for protein synthesis via transgenic expression of the MBP gene in "anti-sense" orientation under the control of its cognate MBP promotor (Katsuki *et al.*, 1988). Similarly, the formation of antisense MBP mRNA is the presumed primary defect of the *myelin-deficient* (shi-mld) mouse mutant, an allele of the shiverer mutation on chromosome 18 (Doolittle and Schweikart, 1977; Popko *et al.*, 1988). The presence of this antisense RNA is thought to reduce and dysregulate the amount of normal MBP mRNA functionally available, thereby resulting in a level insufficient for normal myelin formation (Fremeau and Popko, 1990; Tosic *et al.*, 1990). The dysmyelination is less severe in the myelin deficient mouse when compared to the shiverer mouse. Isolated white matter tracts in CNS have a 3–4-fold increase in sodium channel density and it has been suggested that some myelin-associated molecule absent from the Shiverer white matter tracts could cause a down-regulation of either the synthesis or accumulation of sodium channels in myelinating axons (Noebels *et al.*, 1991). The lack of dysmyelination in the PNS is thought to be due to the substitution of P_0, the major integral protein of PNS myelin, for the structural function of MBP (Lemke and Axel, 1985).

B. Jimpy Mice—Dysmyelination and Glial Cell Death

The mammalian PLP gene is linked to the X chromosome and defects in this gene are associated with neurologic abnormalities in the mouse and with Pelizaeus–Merzbacher disease in humans. In the mouse, three mutations have been characterized: the jimpy (jp), myelin synthesis–deficient (jp msd), and rumpshaker (rsh). Each derives from a point mutation that alters the structure of the encoded protein. In the jimpy mouse, the mutation is a single nucleotide change in the PLP gene that inactivates the splice-acceptor site of intron 4. The last α-helical transmembrane domain is replaced by an aberrant carboxy terminus, and the resulting abnormally folded protein is degraded in the endoplasmic reticulum shortly after synthesis, failing to reach the Golgi apparatus for further processing and transport (Roussel *et al.*, 1987). The CNS is nearly completely devoid of myelin, with less than 1% of the axons ensheathed; PNS myelin, however, is ultrastructurally intact (Sidman 1964; Herschkowitz *et al.*, 1971). There are only a few layers of abnormally thin myelin around CNS myelin, consisting of either uncompacted membrane whorls or compacted myelin with abnormal ultrastructure (Duncan *et al.*, 1989). The major cause of the dysmyelination in the jimpy mouse seems to be a lack of differentiated oligodendrocytes. The proliferation rate of oligodendrocyte precursor cells is increased but an abnormally high rate of apoptotic cell death eliminates most of these maturating oligodendrocytes (Skoff, 1982; Knapp *et al.*, 1986; Barres *et al.*, 1992). Islands of myelinated fibers are formed by the few oligodendrocytes that escape degeneration and developmental arrest. The behavioral phenotype is evident at 2 weeks of age and consists of general body tremor and ataxia; the animals die with seizures and convulsions by about 4 weeks of age. Heterozygous jimpy females, which are mosaics with respect to the X-linked PLP gene, display normal behavior.

In the allelic mouse mutant, jimpy msd, there are similar ultrastructural alterations in myelin as seen in the jimpy mouse, but about twice as many glial cells escape premature degeneration (Billings-Gagliardi *et al.*, 1980). The rumpshaker mutant is the result of a novel mutation of the PLP gene (Schneider *et al.*, 1992) and displays a phenotype very different from the jimpy and the jimpy msd. Rumpshaker mice have more myelin than other dysmyelinated mutants and the degree of dysmyelination varies among CNS regions, with early myelinated regions appearing normal whereas late myelinating regions are severely hypomyelinated. The oligodendrocytes appear differentiated and most escape apoptotic cell death resulting in a normal complement of mature oligodendrocytes (Griffiths *et al.*, 1990). The rumpshaker mutation appears to allow the oligodendrocyte to survive but somehow interferes with its ability to normally deposit PLP in the myelin membrane (Schneider *et al.*, 1992). Although sparse, some myelin sheaths subsist in the rumpshaker mutants and these show selective immunostaining for DM20 (Schneider *et al.*, 1992). These findings suggest that DM20 may serve a critical purpose in glial cell development that is distinct from any function in myelin formation and maintenance.

C. P_0-Deficient Mice

P_0-deficient mice have been generated by homologous recombination of the P_0 gene in mouse embryonic stem cells with the cloned gene and subsequent generation of germline chimeric mice (Giese *et al.*, 1992). Animals lacking one functional copy of the P_0 gene are phenotypically normal, but the homozygotes develop a behavioral phenotype by the third week of life. These mice show a body tremor and dragging movements of the hindlimbs. There is no evidence of paralysis or seizures and the mutants have a normal life span. Histologically, the deficit is characterized by the inability of the Schwann cell to assemble a compacted multilamellar PNS myelin sheath. The high degree of variability in the pathology is thought to be due to the intervening actions of other proteins such as cell adhesion molecules (MAG; N-CAM), and perhaps other myelin proteins. Using promoter and regulatory regions of the P_0 gene in a fusion gene construct, Schwann cells were destroyed when they began to express P_0, after associating in a 1:1 ratio with axons (Messing *et al.,* 1992). The behavioral phenotype of the Schwann cell–ablated mice was similar to the phenotype displayed in the homogygous P_0 mutants. A proliferation of nonmyelin-forming Schwann cells was induced along with skeletal muscle atrophy.

D. Trembler Mice

The trembler mouse (Tr; mouse chromosome 11) contains a mutation of the PMP-22 gene, which results in PNS dysmyelination. The PMP-22 gene encodes an integral membrane protein specific to Schwann cells (22 kDa peripheral myelin protein) believed to be important for normal Schwann cell development (Spreyer *et al.,* 1991; Welcher *et al.,* 1991). Histologically, the majority of large caliber axons in the sciatic nerve are devoid of a myelin sheath and, if present at all, these are abnormally thin and relatively uncompacted (Henry and Sidman, 1988). The total number of Schwann cells is dramatically increased at the time of segregation of axons, and myelin deposition is arrested. In the absence of PNS myelin, the mice display a behavioral phenotype characterized by a coarse-action tremor that begins at the end of the second postnatal week and results in moderate quadriparesis and a waddling gate. Under controlled conditions, these animals can experience a normal life span.

E. Quaking Mice

The quaking mouse (qk; mouse chromosome 17) is the result of an autosomal recessive mutation (Sidman *et al.,* 1964). Homozygous mice carrying the viable quaking allele (qkv/qkv) show the typical motor coordination signs of dysmyelination in the absence of seizures and have a normal life span. The myelin deficiency, characterized by fewer than normal myelin lamellae, is predominately in the brain and spinal cord, although a lack of normal compaction and enlarged intraperiod lines of some myelinated fibers has been noted in the PNS as well (Trapp, 1988). Interestingly, the distribution of MAG is shifted from the innermost myelin layer facing the axon to throughout the compact myelin sheath.

F. Myelin Mutants in Other Species

Myelin mutants in a number of other species besides humans and mice have also been described (see Duncan, 1995, for detailed discussion). These include X-linked mutations in the dog (shaking pup, Griffiths *et al.,* 1981), pig (type A III hypomyelinogenesis congenita, Blakemore *et al.,* 1974), rat (myelin-deficient; Dentinger *et al.,* 1982; Jackson and Duncan, 1988), and rabbit (paralytic tremor, Taraszewska, 1988), as well as the autosomal recessive *taiep* mutant rat (Duncan *et al.,* 1992).

VIII. Other "Disorders" of Myelination

As noted previously, myelination is a critical process in the maturation of the nervous system. It involves the synthesis of an enormous amount of specialized membrane within a relatively short period of time. Much of the myelin in both the CNS and the PNS is formed during a relatively short "developmental window" (first few years in humans; first 30 days of age in rodents), and this period is preceded by a burst of myelinating-cell proliferation. During these time periods, a large portion of the nervous system's metabolic capacity is devoted to myelinogenesis. During these "vulnerable periods," the process of myelination is especially susceptible to perturbations such as toxic insults, nutritional deficiencies, genetic disorders of metabolism, viral infections, substances of abuse, and other environmental factors (for review, see Wiggins, 1986). Insults occurring during the period of proliferation of myelinating cells may be especially disruptive, as this may lead to an irreversible deficit of myelin-forming cells and consequent permanent hypomyelination. Perturbation of myelination at a later stage may result in a myelin deficit that can be reversed. Depending on the timing of the insult, a myelin deficiency can result from alterations related to several different developmental events, including failure of myelin-forming cells to proliferate, reduction of axonal development resulting in fewer and/or smaller axons to myelinate, and decreased formation of myelin at time of maximal synthesis.

The morphologic term *myelinopathy* describes damage to white matter or myelin, and disorders of myelin can be classified by a number of different factors. Such factors include the preferential effects on either the CNS or on the PNS or an involvement of both systems. In addition, effects on myelin can sometimes be delineated as the result of a primary effect on myelin itself or the myelinating glial cell. Myelin loss due to a primary insult to myelin or the myelinating cell are termed *primary demyelination.* There are a number of factors relevant to selective targeting of various toxic or metabolic insults to myelin (for discussion, see Morell *et al.,* 1994; Morell and Toews, 1996a). An intact axon is a prerequisite for maintenance of normal myelin; alterations in myelin due to an effect on the neuron or the underlying axon is termed *secondary demyelination.* Secondary demyelination is an inevitable consequence of serious damage to neurons supporting myelinating axons or to axonal transcetion or crush (Wallerian degeneration). However, the distinction between primary and secondary demyelination is often somewhat vague; the basis for this distinction usually involves morphologic evidence of the initial target site. The term *hypomyelination* is used to describe developmental alterations of myelination in which an insufficient amount of myelin accumulates. Hypomyleination can be the result of disease processes, undernutrition, or toxic insult. The term *dysmyelination,* when used in its strictest sense, refers to certain inborn errors of metabolism in which a block in the breakdown of a myelin lipid causes accumulation of myelin of an abnormal composition (which eventually leads to a collapse and degeneration of myelin), but it is also in wide use as a general descriptor of situations characterized by any abnormalities in myelin.

Some specific myelinopathies that are preferential to developing organisms are discussed later in this chapter. Additional toxicants have been demonstrated to disturb myelin in the adult animal and the morphologic descriptions and mechanistic processes involved have been previously discussed (Morell, 1994; Morell and Toews, 1997). These include tellurium, diphtheria toxin, 2′3′-dideoxycytidine, vigabatrin, carbon monoxide, triethyltin, lead, hexachlorophene, cuprizone, and isoniazid.

A. *Undernutrition*

In the human infant, several studies have provided evidence supporting the concept of a critical period from birth to about 2 years of age, during which time the nervous system is most vulnerable to malnutrition. The production of neurons is virtually completed by about the midpoint of gestation, but glial cell production continues through the end of gestation into the second postnatal year. The vulnerability of the developing nervous system to various factors is determined by the developmental stage of the cellular activities targeted by a specific insult. The effects of an agent or condition may vary depending on the agent, the time of insult during development, and the species under study (Dobbing, *et al.,* 1971). General factors such as undernutrition can have maximal effects on processes that are most active during what has been called the "brain growth spurt" (Dobbing and Sands, 1979). Depending on the developmental process ongoing at the time of exposure, alterations can be produced in either the number of neurons and extent of axonal arborizations, the number of glial cells, or the degree of myelination.

Myelination in both the CNS and PNS is sensitive to nutritional factors (see Wiggins, 1986; Blass, 1994). Brains of rats undernourished from birth contain a lower amount (20% deficit) of total lipids, cholesterol, and phospholipids and a 50% deficit in cerebrosides (Benton *et al.,* 1966). Following severe nutritional deprivation during lactation and post-weaning, total brain galactolipids, cholesterol, and lipid phosphorus showed a slower rate of accumulation (Krigman and Hogan, 1976). Lipid phosphorus and cholesterol levels recovered by adulthood, whereas galactolipids remained at a 60% decreased level. Myelin recovered from undernourished rats was normal in lipid composition but siginificantly reduced in total amount (Fishman *et al.,* 1971). A reduced proportion of basic and proteolipid protein was seen in myelin isolated from undernourished rats at postnatal days 15 and 20, but the composition was similar to normals by postnatal day 30. At all time points examined, myelin yield was 25% less than normal levels (Wiggins *et al.,* 1976). These studies suggested that undernutrition produced a delay in myelin maturation. Morphologic examination of animals undernourished from birth showed a decreased number of mature oligodendroglia and poorly stained myelin (Bass *et al.,* 1970; Krigman and Hogan, 1976). The number of myelin lamellae per axon and the number of myelin lamellae for a given axon diameter were both lower (Krigman and Hogan, 1976).

Some studies suggest that in some cases, myelination is able to catch-up and achieve normal levels once unrestricted feeding is initiated. Rats deprived by an increased litter size rapidly gained body and brain weight and normal brain lipid composition within 3 weeks after weaning to an unrestricted diet (Benton *et al.,* 1966). However, nutritional deprivation during the first 21 days of life resulted in reduced levels of total brain lipids, cerebrosides, cholesterol, and PLP, and this deficit persisted through 120 days of age (Bass *et al.,* 1970). Similar persistent myelin deficits were found in brains of rats subjected to either moderate or severe food deprivation during the first 30 days of life (Toews *et al.,* 1983). Although metabolic studies showed that after 6 days of free feeding following 20 days of postnatal starvation, incorporation of labeled precursors into myelin proteins was higher than in animals starved for the entire 26 days, and it was still depressed relative to controls (Wiggins *et al.,* 1976). Severe underfeeding in rats from 1 to 14 days of age resulted in a lasting significant deficit in myelin, even with rehabilitation (Wiggins and Fuller, 1978). Overall, these studies point to the possibility of irreversible deficits in myelin resulting from nutritional deficiencies during development. Additional studies suggest the most vulnerable period for myelin may be the time of oligodendroglial proliferation. Animals deprived during this period are left with a permanent deficit of myelin-forming cells, resulting in irreversible hypomyelination (Wiggins and Fuller, 1978). Apparently, once normal numbers of oligodendroglia have been formed, the process of myelin formation itself is somewhat more capable of nutritional rehabilitation.

Similar effects can be found with a specific nutritional manipulation of depleting protein in the diet. In rats subjected to a protein and calorie deficiency during gestation and lactation, glial numbers were greatly reduced, and by postnatal day 19 the majority of cells in the corpus callosum appeared to be glioblasts rather than differentiated oligodendroglia (Robain and Ponsot, 1978). An early postnatal protein deficiency resulted in reduced levels of brain myelin and an altered myelin composition in rats (Nakhasi *et al.,* 1975). In myelin from offspring of rats maintained on a 4% protein diet during lactation, an excess of high molecular weight proteins and a deficiency of PLP in heavy myelin was found at postnatal day 17, with normal protein composition seen at 53 days (Figlewicz *et al.,* 1978). The MAG persisted in its higher molecular weight form longer than normal, suggesting that protein deficiency results in a delay in development and maturation of the myelination process (Druse and Krett, 1979).

Animals raised on a fat-deficient diet are able to synthesize all fatty acids except the essential fatty acids (linoleic and linolenic acid families). Essential fatty acid deficiency induced prenatally in the mothers and postnatally in the offspring resulted in lower brain weights (White *et al.,* 1971), and a low level of galactolipid and PLP (McKenna and Campagnoni, 1979). In the optic nerve of essential fatty acid–deficient rats, vacuolation, intramyelinic splitting, and Wallerian degeneration were present (Trapp and Bernsohn, 1978). Several studies have demonstrated hypomyelination in the offspring of copper-deficient mothers (DiPaolo *et al.,* 1974; Prohaska and Wells, 1974). In the third generation of mice maintained on a copper-deficient diet, the offspring showed approximately 60% decrease in myelin yield with the major glycoprotein shifted to a higher molecular weight (Zimmerman *et al.,* 1976).

B. Thyroid Deficiency

Thyroid hormones influence the temporal onset of myelination and its compositional maturation. Neonatal thyroidectomy in rats results in a lasting reduction of total cerebroside in the brain and a 30% reduction in myelin yield (Balazs *et al.,* 1969). It is thought, however, that hypothyroidism does not exert a specific effect on myelin but rather delays myelin development and maturation (Dalal *et al.,* 1971; Walters and Morell, 1981). Whereas hypothyroidism resulted in a 1–2 day delay of myelinogenesis with prolonged immature myelin formation, it eventually attained a normal composition, although the myelin deficit persisted.

C. Inorganic Lead

A classic example of differential susceptibility of the developing organism to the effects of an environmental chemical is that of inorganic lead exposure. Children are more vulnerable to lead in terms of external exposure sources, internal levels of lead, and timing of exposures during development. At high exposure levels, lead induces encephalopathy in children and can be life threatening. Experimental animal studies have allowed examination of various specific target sites and processes of development susceptible to lead toxicity (Krigman *et al.,* 1980). During development, the process of CNS myelination shows an increased vulnerability to lead exposure. The amount of lead that accumulates in the brain of the developing animal during lactation can be as much as 4 times higher than brain levels in the lactating dam receiving lead in the drinking water. Under these conditions, myelin was significantly reduced; however, the relationship between the axon diameter and myelin lamellae remained normal, suggesting that the hypomyelination was the result of altered axonal growth (Krigman *et al.,* 1974). Direct administration of lead via to pups intubation from 2 to 30 days of age resulted in a reduction of myelin accumulation in the forebrain and optic nerve. These effects were not due to undernutrition, as the accumulation of brain myelin was decreased significantly relative to controls undergoing a similar degree of malnourishment (Toews *et al.,* 1980). In developing rats, there is a synergistic interaction between lead exposure and mild malnutrution induced by milk deprivation with respect to decreasing the normal developmental accumulation of myelin (Harry *et al.,* 1985). This interaction appeared to be more prevalent in females as compared to males. The decrement in myelin induced by development exposure to inorganic lead is a long-lasting effect that persists into adulthood (Toews *et al.,* 1980, 1983). Myelination is not necessarily the most sensitive target for lead as low doses sufficient to produce some microscopically discernible hemorrhagic encephalopathy in the cerebellum of young rats did not depress myelination (Sundstrom and Karlsson, 1987); this hemorrhagic encephalopathy may be related to concentration of lead in brain capillaries (Toews *et al.,* 1978).

D. Triethyltin (TET)

The basic CNS change induced by exposure to TET is a massive cerebral edema, restricted primarily to the white matter (Magee *et al.,* 1957; Torak *et al.,* 1970), with the formation of intramyelinic vacuoles (Jacobs *et al.,* 1977). The pathologic effect varies with the age of the animal (Suzuki, 1971). Young rats exposed to TET develop severe spongious white matter similar to that seen in the adult, but with the absence of major clinical signs seen in the adult (Suzuki, 1971; Blaker *et al.,* 1981). It is thought that the severe paralysis seen in the adult animal is due in part to the intracranial pressure developed during severe edema, whereas the open cranial sutures in the young rat may allow for edema in the absence of increased pressure. When newborn rats are exposed to TET, brains became swollen and petechial hemorrhages are observed, particularly in the cerebellum. Necrotic cells were found diffusely throughout the brain (Watanabe, 1977). When older (postnatal day 8) animals were exposed, both the hemorrhagic and necrotic changes occurred, but damage was also seen in the myelinated fibers of the brain stem and cerebellum. Although the morphologic alterations in myelin dissipate with time, biochemical evidence suggests that the amount of myelin produced is decreased and that this myelin deficit persists through adulthood (Blaker *et al.,* 1981; Toews *et al.,* 1983). Chronic exposure to TET from 2 to 30 days after birth decreases myelin yield and cerebroside content (55%) and 2′,3′-cyclic nucleotide 3′-phosphohydrolase activity (20%) (Blaker *et al.,* 1981). In studies using radioactive tracer, Smith (1973) demonstrated that it is the newly forming CNS myelin that is preferentially susceptible to degradation by TET. Interestingly, administration of TET to quaking mice did not produce intramyelinic edema (Nagara *et al.,* 1981).

E. Trialkyllead

When young animals are exposed to trialkyllead, the process of myelination is inhibited (Konat and Clausen, 1976). Unlike triethyltin, this impairment in myelinogenesis is not accompanied by edema of white matter. The impairment appears to be primarily in the deposition of myelin rather than in the program for myelination, as the protein composition of forebrain myelin isolated from triethyllead-intoxicated young rats was normal (Konat and Clausen, 1978). *In vitro* studies suggest that the alteration involves posttranslational processing and transport of integral membrane proteins, processes particularly important for myelin proteins during development (Konat and Clausen, 1980; Konat and Offner, 1982).

F. Hexachlorophene

Hexachlorophene (2,2′-methylene*bis*-3,4,6-trichlorophenol) is an antimicrobial agent that has been used previously in soaps and detergents, as well as in the bathing of newborn babies to prevent bacterial infections (Herter, 1959; Powell *et al.,* 1973; for review, see Towfighi, 1980). Both CNS and PNS myelin show a severe white matter edema following exposure, and young rats are more vulnerable than adults (Towfighi *et al.,* 1974). In young rats, edema of the myelin sheath becomes evident after postnatal day 15, probably because the myelin membrane provides a hydrophobic reservoir for accumulation of this toxic compound and thereby becomes a significant site for fluid accumulation (Nieminen *et al.,* 1973). Developmental exposure results in a decrease of the normal accumulation of myelin during development (Matthieu *et al.,* 1974). In 22-day-old rats nursed by mothers fed hexachlorophene, there was a decrease in myelin yield, yet the myelin composition remained normal. Abnormal "dissociated" myelin accounted for about 10% of the total myelin and contained the typical myelin constituents with the exception of MAG, which was absent (Matthieu *et al.,* 1974).

G. 6-Aminonicotinamide

Degeneration caused by this antimetabolite involves myelin, neurons, astrocytes, and oligodendroglia. In young animals injected with 6-aminonicotinamide, the PNS shows a selective swelling of Schwann cell cytoplasm at the inner surface of the myelin sheath. The nerve is compressed by the swelling and results in an overgrowth of the myelin sheath (Friede and Bischhausen, 1978).

H. Isonicotinic Acid Hydrazide (IHN)

Young ducklings fed a diet containing IHN developed a wobbling gait and head tremor after 2 weeks, progressing to ataxia and inability to stand (Lampert and Schochet, 1968). Examination of the CNS showed spongy degeneration of the myelin-containing white matter.

I. Cuprizone

Cuprizone (*bis*-cyclohexanoneoxalyhydrazone) is a copper chelator that results in CNS demyelination following dietary exposure to weanling mice. The loss of myelin can reach as much as 70% in white matter regions of the cerebrum (Carey and Freeman, 1983). Deficits in adenosine triphosphate (ATP) production secondary to reduced activity of cytochrome oxidase (a copper-requiring enzyme) may lead to alterations in energy-requiring ion transport mechanisms, but the underlying reason for targeting of this compound to CNS myelin is not clear. Interestingly, cuprizone inhibits carbonic anhydrase, an enzyme present in myelin, and this inhibition takes place well before any demyelination is observed (Komoly *et al.*, 1987). In experimental studies of cuprizone neurotoxicity, mRNA for MAG, a protein located at the myelin–axonal interface, is down-regulated during demyelination and returns to normal levels following cessation of exposure (Fujita *et al.*, 1990). The mRNA for this glycoprotein exists in two major splice variants that are both severely down-regulated. On recovery, one splice variant returns to normal levels whereas the other shows an accumulation above control levels. Prolonged exposure to cuprizone (9 weeks or longer in mice) results in irreversible demyelination (Tansey *et al.*, 1996), possibly due to death of oligodendrocytes and/or oligodendrogilal precursor cells.

J. Tellurium

Exposure of weanling rats to a diet containing tellurium (element 52) leads to a highly synchronous demyelinating peripheral neuropathy (Lampert and Garrett, 1971; Duckett *et al.*, 1979; Said and Duckett, 1981; Takahashi, 1981; Bouldin *et al.*, 1988). When tellurium exposure is discontinued, there is rapid and synchronous remyelination. Although tellurium toxicity in humans is rare, this model is of considerable interest as a system for studying the manner in which a specific metabolic insult can lead to demyelination. Because there is little or no associated axonal degeneration, it has also proved useful for examining events and processes related to PNS remyelination, independently of processes related to axonal regeneration.

Inclusion of 1–1.5% elemental tellurium in the diet of weanling rats leads to a primary segmental demyelination of about 20–25% of myelinating internodes in the sciatic nerve, but with sparing of axons (Bouldin *et al.*, 1988; Harry *et al.*, 1989). This demyelination results in a peripheral neuropathy characterized by hindlimb paresis and paralysis. Older rats are much more resistant to tellurium, and the CNS is generally not affected, although some pathologic alterations can be induced with prolonged exposure periods. The nature of the underlying metabolic insult has been delineated. Tellurium blocks cholesterol synthesis, specifically by inhibiting the enzyme squalene epoxidase, an obligate step in the cholesterol biosynthesis pathway (Harry *et al.*, 1989; Wagner-Recio *et al.*, 1991; Wagner *et al.*, 1995). Tellurite, a water-soluble oxidized metabolite of the administered insoluble element, is the active species *in vitro*, effective at micromolar concentrations in a cell-free system (Wagner *et al.*, 1995). The organotellurium compound dimethyltelluronium dichloride, $(CH_3)_2TeCl_2$, is also effective in inhibiting squalene epoxidase in cultured Schwann cells and in inducing demyelination when administered intraperitoneally (Goodrum, 1997). Presumably, the resulting cholesterol deficit in Schwann cells eventually leads to an inability to maintain preexisting myelin and to assemble new myelin; this in turn leads to the observed demyelination. Although the tellurium-induced inhibition of cholesterol biosynthesis is systemic, deleterious effects are confined largely to the sciatic nerve. In the liver, which supplies cholesterol for most body tissues, the resulting intracellular cholesterol deficit results in a marked upregulation of the cholesterol biosynthetic pathway (Toews *et al.*, 1991b; Wagner-Recio *et al.*, 1991; Wagner *et al.*, 1995), presumably via well-characterized feedback mechanisms (see Goldstein and Brown, 1990, for review). This allows normal levels of cholesterol synthesis in this tissue despite considerable inhibition of one of the steps in the synthesis pathway, and normal levels of lipoprotein-associated circulating cholesterol are maintained. However, unlike many other tissues, the sciatic nerve cannot use circulating cholesterol; all cholesterol required for myelin in the sciatic nerve must be synthesized locally (Jurevics and Morell, 1994). This fact, coupled with the great demand for cholesterol in the rapidly myelinating PNS at the time of tellurium exposure, may account for the specificity of toxicity observed.

Expression of mRNA for myelin proteins is markedly down-regulated during the demyelinating phase of tellurium neuropathy, as is gene expression for enzymes involved in synthesis of lipids enriched in myelin (Toews *et al.*, 1990, 1991a,b, 1997). The latter include HMG–CoA reductase, the rate-limiting enzyme in cholesterol biosynthesis. Although this enzyme is markedly up-regulated in the liver (as expected from the tellurium-induced intracellular sterol deficit), it is down-regulated in the sciatic nerve in concert with other myelin-related genes (Toews *et al.*, 1991b). Failure to up-regulate the cholesterol biosynthesis pathway in the sciatic nerve is, in fact, probably the major underlying reason for the preferential susceptibility of this tissue. The co-ordinate down-regulation for myelin-related genes seen following exposure to tellurium suggests that gene expression of all proteins involved in myelin synthesis and assembly may be under the co-ordinate control of the overall program for myelination (see Morell and Toews, 1996a; Toews *et al.*, 1997, for discussion). The coordinate down-regulation of myelin gene expression takes place in all myelinating Schwann cells and not just in those undergoing demyelination (Toews *et al.*, 1992). Thus, this down-regulation is not just a secondary response to injury but rather reflects the co-ordinate control of myelin gene expression. When tellurium exposure is discontinued, there is co-ordinate up-regulation of these messages during the remyelinating period.

Thus, tellurium toxicity specifically leads to PNS demyelination because (1) synthesis of cholesterol, a major myelin lipid, is severely inhibited; (2) unlike other tissues, peripheral nerve cannot up-regulate the synthesis of cholesterol in response to the tellurium-induced cholesterol deficit; (3) because the PNS is isolated from the circulation by barriers, it cannot use circulating cholesterol derived from the diet or from synthesis in the liver; and (4) there is a particularly high demand for cholesterol in the myelinating PNS at the time of tellurium exposure.

IX. Concluding Remarks

The process of myelination by oligodendroglia in the CNS and by Schwann cells in the PNS represents a complex series of metabolic and cell–biologic events involving intercellular recognition and interaction, adhesion, synthesis, sorting and assembly of specialized myelin membranes, compaction of myelin lamellae, and axonal (and possibly glial) ion channel reorganization. This entire process must be completed during an intense burst of metabolic activity at a specific predetermined interval during development, and failure to complete this program of myelination within the proper "developmental window" may have permanent deleterious effects. Because myelin-forming cells are operating at near their metabolic capacity, they are especially sensitive to toxic or other types of insults during this "vulnerable period" of nervous system development, and deficiencies in myelin, either qualitative or quantitative, may result. These myelin deficiencies can result from underlying alterations of various developmental events, including failure of myelin-forming cells to proliferate in normal numbers, reduction of axonal development, and/or decreased or altered formation of myelin at its normal time of maximal synthesis. The timing of any toxicant exposure or other insult can also differentially effect one or more of these events necessary for normal myelination. Although these processes are best examined in the developing nervous system, a clearer understanding of the biochemistry, molecular biology, and cell biology of these events is also of particular relevance with regards to remyelination in injured adult nervous tissue. Further delineation of the underlying nature of insults that result from toxic, genetic, nutritional, or other perturbations will also be useful in better understanding these vital processes.

References

Aguayo, A. J., Epps, J., Charron, J., and Bray, G. M. (1976). Multipotentiality of Schwann cells in cross-anastomosed and grafted unmyelinated nerves—quantitative microscopy and radioautography. *Brain Res.* **104,** 1–20.

Armstrong, R. A., Kim, J. G., and Hudson, L. D. (1995). Expression of Myelin Transcription Factor I (MyTI), a "zinc-finger" DNA-binding protein, in developing oligodendrocytes. *Glia* **14,** 303–321.

Armstrong, R. A., Friedrich, V. L. Jr, Holmes, K. V., and Dubois-Dalcq, M. (1990). *In vitro* analysis of the oligodendrocyte lineage in mice during demyelination and remyelination. *J. Cell Biol.* **111,** 1183–1195.

Balazs, R., Brooksbank, B. W. L., Davison, A. N., Eayrs, J. T., and Wilson, D. A. (1969). The effects of neonatal thyroidectomy on myelination in the rat brain. *Brain Res.* **15,** 219–232.

Bannerman, P. G. C., Mirsky, R., Jessen, K. R., Timpi, R., and Duance, V. C. (1986). Light microscopic immunolocalization of laminin, type IV collagen, nidogen, heparan sulfate proteoglycan and fibronectin in the enteric nervous system of rat and guinea pig. *J. Neurocytol.* **15,** 432–443.

Barres, B. A., Hart, I. K., Coles, H. S. R., Burne, J. F., Voyvodic, J. T., Richardson W. D., and Raff, M. C. (1992). Cell death and control of cell survival in the oligodendrocyte lineage. *Cell* **70,** 31–46.

Barres, B. A., and Raff, M. C. (1993). Proliferation of oligodendrocyte precursor cells depends on electrical activity in axons. *Nature* **361,** 258–263.

Bass, N. H., Netsky, M. G., and Young, E. (1970). Effect of neonatal malnutrition on developing cerebrum. II. Microchemical and histologic study of myelin formation in the rat. *Arch. Neurol.* **23,** 303–313.

Benjamins, J. A., and Iwata, R. (1979). Kinetics of entry of galactolipids and phospholipids into myelin. *J. Neurochem.* **32,** 921–926.

Benjamins, J. A., Miller, S., and Morell, P. (1976). Metabolic relationships between myelin subfractions: entry of galactolipids. *J. Neurochem.* **27,** 565–570.

Benjamins, J. A., and Smith, M. E. (1984). Metabolism of myelin. In: *Myelin* (Morell, P., Ed.), Plenum Press, New York, pp. 225–258.

Benton, J. W., Moser, H. W., Dodge, P. R., and Carr, S. (1996). Modification of the schedule of myelination in the rat by early nutritional deprivation. *Pediatrics* **38,** 801–807.

Bernard, C. C. A., and Kerlero, de Rosbo, N. (1991). Immunopathological recognition of autoantigens in multiple sclerosis. *Acta Neurol.* **13,** 170–178.

Biddle, F., March, E., and Miller, J. R. (1973). Research news. *Mouse News Lett.* **48,** 24.

Billings-Gagliardi, S., Adcock, L. H., and Wolf, M. (1980). Hypomyelinated mutant mice: description of jpmsd and comparison with jp and qk on their present genetic background. *Brain Res.* **194,** 325–338.

Bjartmar, C., Hildebrand, C., and Loinder, K. (1994). Morphological heterogeneity of rat oligodendrocytes: I. Electronmicroscopic studies on serial sections. *Glia* **11,** 235–244.

Black, J. A., Sims, T. J., Waxman, S. G., and Gilmore, S. A. (1985). Membrane ultrastructure of developing axons in glial cell deficient rat spinal cord. *J. Neurocytol.* **14,** 79–104.

Black, J. A., Waxman, S. G., Sims, T. J., and Gilmore, S. A. (1986). Effects of delayed myelination by oligodendrocytes and Schwann cells on the macromolecular structure of axonal membrane in rat spinal cord. *J. Neurocytol.* **15,** 745–762.

Blakemore, W. F., Harding, H. D., and Done, J. T. (1974). Ultrastructural observations on the spinal cord of a Landrace pig with congenital tremor type A III. *Res. Vet. Sci.* **17,** 174–178.

Blaker, W. D., Krigman, M. R., Thomas, D. J., Mushak, P. and Morell, P. (1981). Effect of triethyl tin on myelination in the developing rat. *J. Neurochem.* **36,** 44–52.

Blass, J. P. (1994). Vitamin and nutritional deficiencies. In: *Basic Neurochemistry* (5th ed.), (R. W. Albers, G. W. Siegel, P. Molinoff, and B. Agranoff, Eds.), Raven Press, pp. 749–760, New York.

Blok, J., Scheven, B. A. A., Mulder-Stapel, A. A., Gensel, L. A., and Daems, W. T. H. (1982). Endocytosis in adsorptive cells of cultured human small-intenstine tissue: effects of cytochalasin B and D. *Cell Tissue Res.* **222,** 113–126.

Bouldin, T. W., Samsa, G., Earnhardt, T., and Krigman, M. R. (1988). Schwann-cell vulnerability to demyelination is associated with internodal length in tellurium neuropathy. *J. Neuropathol. Exp. Neurol.* **47,** 41–47.

Braun, P. E., Babbrick, L. L., Edwards, A. M., and Bernier, L. (1990). 2′,3′-cyclic nucleotide 3′-phosphodiesterase has characteristics of cytoskeletal proteins: a hypothesis for its function. *Ann. NY Acad. Sci.* **605,** 55–65.

Braun, P. E., DeAngelis, D., Shtybel, W. W., and Bernier, L. (1991) Isoprenoid modification permits 2′,3′-cyclic nucleotide 3′-phosphodiesterase to bind to membranes. *J. Neurosci, Res.* **30,** 540–544.

Butt, A. M., and Ransom, B. R. (1989). Visualization of oligodendrocytes and astrocytes in the intact rat optic nerve by intracellular injection of lucifer yellow and horseradish peroxidase. *Glia* **2,** 470–475.

Cammer, W., Fredman, T., Rose, A. L., and Norton, W. T. (1976). Brain carbonic anhydrase: activity in isolated myelin and the effect of hexachlorophene. *J. Neurochem.* **27,** 165–171.

Campagnoni, A. T. (1988). Molecular biology of myelin proteins from the central nervous system. *J. Neurochem.* **51,** 1–14.

Campagnoni, A. T. (1995). Molecular biology of myelination In: *Neuroglia.* H. Kettenman and B. R. Ransom, Eds. pp. 555–570. New York, Oxford University Press.

Campagnoni, A. T., and Macklin, W. B. (1988). Cellular and molecular aspects of myelin protein gene expression. *Mol. Neurobiol.* **2,** 41–89.

Campagnoni, A. T., Pribyl, T. M., Campagnoni, C. W., Kampf, K., Amur-Umarjee, C. F., Handley, V. W., Newman, S. L., Garbay, B., and Kitamura, K. (1993). Structure and developmental regulation of Golli-mbp, a 105 kilobase gene that encompasses the myelin basic protein gene and is expressed in cells in the oligodendrocyte lineage in the brain. *J. Biol. Chem.* **268,** 4930–4938.

Carey, E. M., and Freeman, N. M. (1983). Biochemical changes in Cuprizone-induced spongiform encephalopathy. I. Changes in the activities of 2′,3′-cyclic nucleotide 3′-phosphohydrolase, oligodendroglial ceramide galactosyl transferase, and the hydrolysis of the alkenyl group of alkenyl, acyl-glycerophospholipids by plasmalogenase in different regions of the brain. *Neurochem. Res.* **8,** 1029–1044.

Chen, S. J., and DeVries, G. H. (1989). Mitogenic effect of axolemma-enriched fraction on cultured oligodendrocytes. *J. Neurochem.* **52,** 325–327.

Chernoff, G. F. (1988). Shiverer, an autosomal recessive mutant mouse with myelin-deficiency. *J. Hered.* **72,** 128.

Colman, D. R., Kreibich, G., Frey, A. B., and Sabatini, D. D. (1982). Synthesis and incorporation of myelin polypeptide into CNS myelin. *J. Cell Biol.* **95,** 598–608.

Cornbrooks, C. J., Carey, D. J., McDonald, J. A., Timpi, R., and Bunge, R. P. (1983). *In vivo* and *in vitro* observations on laminin production by Schwann cells. *Proc. Natl. Acad. Sci. USA* **80,** 3850–3854.

Crabb, J. W., and Saari, J. C. (1981). N-terminal sequence homology among retinoid-binding proteins from bovine retina. *FEBS Lett.* **130,** 15–18.

Dahl, D., Chi, N. H., Miles, L. E., Nguyen, B. T., and Bignami A. (1982). Glial fibrillary acidic (GFA) protein in Schwann cells: fact or artifact? *J. Histochem. Cytochem.* **30,** 912–918.

Dalal, K. B., Valcana, T., Timiras, P. S., and Einstein, E. R. (1971). Regulatory role of thyroxine on myelinogenesis in the developing rat. *Neurobiology* **1,** 211–224.

Dentinger, M. P., Barron, K. D., and Csiza, C. K. (1982). Ultrastructure of the central nervous system in a myelin deficient rat. *J. Neurocytol.* **11,** 671–691.

DesJardins, K. C., and Morell, P. (1983). The phosphate groups modifying myelin basic proteins are metabolically labile; the methyl groups are stable. *J. Cell Biol.* **97,** 438–446.

DeVries, G. H., Minier, L. N., and Lewis, B. (1983). Further studies on the mitogenic response of cultured Schwann cells to rat CNS axolemma-enriched fractions. *Dev. Brain Res.* **9,** 87–93.

DeWille, J. W., and Horrocks, L. A. (1992). Synthesis and turnover of myelin phospholipids and cholesterol. In: *Myelin: Biology and Chemistry.* (R. E. Martenson, Ed.), pp. 213–234. CRC Press, Boca Raton, Florida.

DiPaolo, R. V., Kanfer, J. N., and Newberne, P. M. (1974). Copper deficiency and the central nervous system: myelination in the rat—morphological and biochemical studies. *J. Neuropathol. Exp. Neurol.* **33,** 226–236.

Dobbing, J., Hopewell, J. W., and Lynch, A. (1971). Vulnerability of developing brain: VII. Permanent deficit of neurons in cerebral and cerebellar cortex following early mild undernutrition. *Exp. Neurol.* **32,** 439–447.

Dobbing, J., and Sands, J. (1979). Comparative aspects of the brain growth spurt. *Early Hum. Dev.* **3,** 79–83.

Doolittle, D. P., and Schweikart, K. M. (1977). Myelin deficient, a neurological mutation in the mouse. *J. Hered.* **68,** 331–332.

Dong, Z., Brennan, A., Liu, N., Yarden, Y., Lefkowitz, G., Mirsky, R., and Jessen, K. R. (1995). neu differentiation factor is a neuron–glia signal and regulates survival, proliferation, and maturation of rat Schwann cell precursors. *Neuron* **15,** 585–596.

Druse, M. J., and Krett, N. L. (1979). CNS myelin-associated glycoproteins in the offspring of protein deficient rats. *J. Neurochem.* **32,** 665–667.

Duckett, S., Said, G., Streletz, L. G., White, R. G., and Galle, P. (1979). Tellurium-induced neuropathy: correlative physiological, morphological and electron microprobe studies. *Neuropathol. Appl. Neurobiol.* **5,** 256–278.

Duncan, I. D. (1995). Inherited disorders of myelination of the central nervous system. In: *Neuroglia.* (H. Kettenmann, and B. R. Ransom, Eds.), pp. 990–1009, Oxford University Press, New York.

Duncan, L. D., Hammang, J. P., Goda, S., and Quarles, R. H. (1989). Myelination in the jimpy mouse in the absence of proteolipid protein. *Glia* **2,** 148–154.

Duncan, I. D., Lunn, K. F., Holmgren, B., Urba-Holmgren, R., and Brignolo-Homes, L. (1992). The taiep rat: a myelin mutant with an associated oligodendrocyte microtubular defect. *J. Neurocytol.* **21,** 870–884.

Dustin, P. (1984). *Microtubules* 2nd ed. Springer-Verlag, Berlin.

Ebersol, T., Tho, O., and Artzt, K. (1992). The proximal end of mouse chromosome 17: new molecular markers identify a deletion associated with quaking viable. *Genetics* **131,** 183–190.

Eriksson, U., Sundelin, J., Rask, L., and Peterson, P. A. (1981). The NH2-terminal amino acid sequence of cellular retinoic-acid binding protein from rat testis. *FEBS Lett.* **135,** 70–72.

Figlewicz, D. A., Hofteig, J. H., and Druse, M. J. (1978). Maternal deficiency of protein or protein and calories during lactation: effect upon CNS myelin subfraction formation in rat offspring. *Life Sci.* **23,** 2163.

Fishman, M. A., Madyastha, P., and Prensky, A. L. (1971). The effect of undernutrition on the development of myelin in the rat central nervous system. *Lipids* **6,** 458–465.

Fraher, J. P. (1972). A quantitative study of anterior root fibres during early myelination. *J. Anat.* **112,** 99–124.

Fremeau, R. T. Jr., and Popko, B. (1990). *In situ* analysis of myelin basic protein gene expression in myelin-deficient oligodendrocytes: antisense hmRNA and read through transcription. *EMBO J.* **9,** 3533–3538.

Friede, R. L., and Bischhausen, R. (1980). The precise geometry of large internodes. *J. Neurol. Sci.* **48,** 367–381.

Fujita, N., Ishiguro, H., Sato, S., Kurihara, T., Kuwano, R., Sakimura, K., Takahashi, Y., and Miyatake, T. (1990). Induction of myelin-associated mRNA in experimental remyelination. *Brain Res.* **513,** 152–155.

Funch, P. G., and Faber, D. S. (1984). Measurement of myelin sheath resistances: implications for axonal conduction and pathophysiology. *Science* **225,** 538–540.

Garbern, J. Y., Cambi, F., Tang, X.-M., Sima, A. A. F., Vallat, J. M., Bosch, E. P., Lewis, R., Shy, M., Sohi, J., Kraft, G., Chen, K. L., Joshi, I., Leonard, D. G. B., Johnson, W., Raskind, W., Dlouhy, S. R., Pratt, V., Hodes, M. E., Bird, T., and Kamholz, J. (1997). Proteolipid protein is necessary in peripheral as well as central myelin. *Neuron* **19,** 205–218.

Gardinier, M. V., Imaged, P., Linington, C., and Matthieu, J. M. (1992). Myelin–oligodendrocyte glycoprotein is a unique member of the immunoglobulin superfamily. *J. Neurosci. Res.* **33,** 177–187.

Gencic, S., and Hudson, L. (1990). Conservative amino acid substitution in the myelin proeolipid protein of jimpy msd mice. *J. Neurosci.* **10,** 117–124.

Gensert, J. M., and Goldman, J. E. (1997). Endogenous progenitors remyelinate demyelinated axons in the adult CNS. *Neuron,* **19,** 197–203.

Giese, K. P., Martin, R., Lemke, G., Soriano, P., and Schachner, M. (1992). Disruption of the PO gene in mice leads to abnormal expression of recognition molecules, and degeneration of myelin and axons. *Cell* **71,** 565–576.

Gillespie, C. S., Trapp, B. D., Colman, D. R., and Brophy, P. J. (1990). Distribution of myelin basic protein and P2 mRNAs in rabbit spinal cord oligodendrocytes. *J. Neurochem.* **54,** 1556–1561.

Goldstein, J. L., and Brown, M. S. (1990). Regulation of the mevalonate pathway. *Nature* **343,** 425–430.

Goodrum, J. F. (1997). Tellurium-induced demyelination is correlated with squalene epoxidase inhibition. (Abstr.) *J. Neurochem.* **69**(Suppl), S13C.

Gould, R. M., Jessen, K. R., Mirsky, R., and Tennekoon, G. (1992). The cell of Schwann: an update. In: *Myelin: Biology and Chemistry.* (Martenson R. E., Ed.) pp. 121–171, CRC Press, Boca Raton, Florida.

Greenfield, S., Brostoff, S. W., and Hogan, E. L. (1980). Characterization of the basic proteins from rodent peripheral nervous system myelin. *J. Neurochem.* **34,** 453–455.

Griffiths, I. R., Duncan, I. D., McCulloch, M., and Harvey, M. J. A. (1981). Shaking pups: a disorder of central myelination in the Spaniel dog. I. Clinical, genetic, and light microscopical observations. *J. Neurol. Sci.* **50,** 423–433.

Griffiths, I. R., Mitchell, L. S., McPhilemy, K., Morrisson, S., Kyriakides, E., and Barrie, J. A. (1989). Expression of myelin protein genes in Schwann cells. *J. Neurocytol.* **18,** 345–352.

Griffiths, I. R., Scott, I., McCulloch, M. C., Barrie, J. A., McPhilemy, K., and Cattanach, B. M. (1990). Rumpshaker mouse: a new X-linked mutation affecting myelination: Evidence for a defect in PLP expression. *J. Neurocytol.* **19,** 273–283.

Gunn, C., Suckling, A. J., and Linington, C. (1989). Identification of a common idiotype on myelin oligodendrocyte glycoprotein-specific autoantibodies in chronic relapsing experimental allergic encephalomyelitis. *J. Neuroimmunol.* **23,** 101–108.

Harry, G. J., Goodrum, J. F., Bouldin, T. W., Wagner-Recio, M., Toews, A. D., and Morell, P. (1989). Tellurium-induced neuropathy: metabolic alterations associated with demyelination and remyelination in rat sciatic nerve. *J. Neurochem.* **52,** 938–945.

Harry, G. J., Toews, A. D., Krigman, M. R., and Morell, P. (1985). The effect of lead toxicity and milk deprivation on myelination in the rat. *Toxicol. Appl. Pharmacol.* **77,** 458–464.

Hayasaka, K., Himoro, M., Nanao, K., Sato, W., Miura, M., Uyemura, K., Takahashi, E., and Takada, G. (1992). Isolation and sequence determination of cDNA encoding PMP-22 (PAS-II/SR13/GAS-3) of human peripheral myelin. *Biochem. Biophys. Res. Commun.* **186,** 827–831.

Hayasaka, K., Himoro, M., Sato, W., Takada, G., Uyemura, K., Shimizu, N., Bird, T. D., Conneally, M., and Chance, P. (1993). Charcot–Marie–Tooth neuropathy type 1B is associated with mutations of the myelin P_0 gene. *Nature Genet.* **5,** 31–34.

Henry, E. W., and Sidman, R. L. (1988). Long lives for homozygous trembler mutant mice despite virtual absence of peripheral nerve myelin. *Science* **241,** 344–346.

Herschkowitz, N., Vassella, F., and Bischoff, A. (1971). Myelin differences in the central and peripheral nervous system in the 'jimpy' mouse. *J. Neurochem.* **18,** 1361–1363.

Herter, W. B. (1959). Hexachlorophene poisoning. *Kaiser Found. Med. Bull.* **7,** 228–230.

Hess, A., and Young, J. Z. (1952). The nodes of Ranvier. *Proc. Roy. Soc. London. B.* **140,** 301–319.

Hudson, L. D. (1990). Molecular genetics of X-linked mutants. *Annu. NY Acad. Sci.* **605,** 155–165.

Ikenaka, K., Kagawa, T., and Mikoshiba, K. (1992). Selective expression of DM-20, an alternatively spliced myelin proteolipid protein gene product, in developing nervous system and in non-glial cells. *J. Neurochem.* **58,** 2248–2253.

Ishaque, A., Roomi, M. W., Szymanska, I., Kowalski, S., and Eylar, E. H. (1980). The PO glycoprotein of peripheral nerve myelin. *Can. J. Biochem.* **58,** 913–921.

Jackson, K. F., and Duncan, I. D. (1992). Cell kinetics and cell death in the optic nerve of the myelin deficient rat. *J. Neurocytol.* **17,** 657–670.

Jacobs, J. M., Cremer, J. E., and Cavanagh, J. B. (1977). Acute effects of triethyltin on the rat myelin sheath. *Neuropathol. Appl. Neurobiol.* **3,** 169–181.

Jessen, K. R., and Mirsky, R. (1991). Schwann cell precursors and their development. *Glia* **4,** 185–194.

John, S. M., Rathke, P. C., and Kern, H. P. (1983). Effect of cytochalasin D on secretion by rat pancreatic acini. *Cell Biol. Int. Rep.* **7,** 603–610.

Jurevics, H. A., and Morell, P. (1994). Sources of cholesterol for kidney and nerve during development. *J. Lipid Res.* **35,** 112–120.

Kadlubowski, M., and Hughes, R. A. (1980). The neuritogenicity and encephalitogenicity of P2 in the rat, guinea-pig and rabbit. *J. Neurol. Sci.* **48,** 171–178.

Kamholz, J., Sess, M., Scherer, S., Vogelbacker, H., Mokuno, K., Baron, P., Wrabetz, L., Shy, M., and Pleasure, D. (1992). Structure and expression of proteolipid protein in the peripheral nervous system. *J. Neurosci. Res.* **31,** 231–244.

Katsuki, M., Sato, M., Kimura, M., Yokoyama, M., Kobayashi, K., and Nomura, T. (1988). Conversion of normal behavior to shiverer by myelin basic protein antisense cDNA in transgenic mice. *Science* **241,** 593–595.

Kidd, G. J., Andrews, S. B., and Trapp, B. D., (1994). Organization of microtubules in myelinating Schwann cells. *J. Neurocytol.* **23,** 801–810.

Kim, J. G., and Hudson, L. D. (1992). Novel member of the zinc finger superfamily: a C_2-HC finger that recognizes a glia-specific gene. *Mol. Cell. Biol.* **12,** 5632–5639.

Kimura, M., Sato, M., Akatsuka, A., Nozawa-Kimura, S., Takahashi, R., Yokoyama, M., Nomura, T., and Katsuki, M. (1989). Restoration of myelin formation by a single type of myelin basic protein in transgenic shiverer mice. *Proc. Natl. Acad. Sci. USA* **86,** 5661–5665.

Kioussi, C., Gross, M. K., and Gruss, P. (1995). Pax3: a paired domain gene as a regulator in PNS myelination. *Neuron* **15,** 553–562.

Kirschner, D. A., and Ganser, A. L. (1980). Compact myelin exists in the absence of basic protein in the shiverer mutant mouse. *Nature* **283,** 207–210.

Kirschner, M., and Mitchison, T. (1986). Beyond self-assembly: from microtubules to morphogenesis. *Cell* **45,** 329–342.

Kitamura, K., Suzuki, M., and Uyemura, K. (1976). Purification and partial characterization of two glycoproteins in bovine peripheral nerve myelin membrane. *Biochem. Biophys. Acta* **455,** 806–816.

Klymkowsky, M. W., Bachant, J. B., and Domingo, A. (1989). Functions of intermediate filaments. *Cell Motil. Cytoskel.* **14,** 309–331.

Knapp, P. E., Skoff, R. P., and Redstone, D. W. (1986). Oligodendroglial cell death in jimpy mice: an explanation for the myelin deficit. *J. Neurosci.* **6,** 2813–2822.

Kobayashi, S., and Suzuki, K. (1990). Development of unmyelinated fibers in peripheral nerve: an immunohistochemical and electron-microscopic study. *Brain Dev.* **12,** 237–246.

Koffer, A., Tatham, P. E. R., and Gomperts, B. D. (1990). Changes in the state of actin during the exocytotic reaction of permeabilized rat mast cells. *J. Cell Biol.* **111,** 919–927.

Koizumi, T., Katsuki, M., Kimura, M., and Hayakawa, J. (1991). Localization of the gene encoding myelin basic protein to mouse chromosome 18E3-4 and rat chromosome 1 p11-p12. *Cytogenet. Cell Genet.* **56,** 199–201.

Komoly, S., Jeyasingham, M. D., Pratt, O. E., and Lantos, P. L. (1987). Decrease in oligodendrocyte carbonic anhydrase activity preceding myelin degeneration in cuprizone induced demyelination. *J. Neurol. Sci.* **79,** 141–148.

Konat, G., and Clausen, J. (1976). Triethyllead-induced hypomyelination in the developing rat forebrain. *Exp. Neurol.* **50,** 124–133.

Konat, G., and Clausen, J. (1978). Protein composition of forebrain myelin isolated from triethyllead-intoxicated young rats. *J. Neurochem.* **30,** 907–909.

Konat, G., and Clausen, J. (1980). Suppressive effect of triethyllead on entry of proteins into the CNS myelin sheath *in vitro*. *J. Neurochem.* **35,** 382–387.

Konat, G., and Offner, H. (1982). Effect of triethyllead on post-translational processing of myelin proteins. *Exp. Neurol.* **75,** 89–94.

Kordeli, E., Davis, J. D., Trapp, B. D., and Bennett, V. (1990). An isoform of ankyrin is localized at nodes of Ranvier in myelinated axons of central and peripheral nerves. *J. Cell Biol.* **110,** 1341–1352.

Krigman, M. R., Bouldin, T. W., and Mushak, P. (1980). Lead. In: *Experimental and Clinical Neurotoxicology* (P. S. Spencer, and H. H. Schaumburg, Eds.), pp. 490–507, Williams & Wilkins, Baltimore.

Krigman, M. R., Druse, M. J., Traylor, T. D., Wilson, M. H., Newell, L. R., and Hogan, E. L. (1974). Lead encephalopathy in the developing rat: effect upon myelination. *J. Neuropathol. Exp. Neurol.* **33,** 58–73.

Krigman, M. R., and Hogan, E. L. (1976). Undernutrition in the developing rat: effect upon myelination. *Brain Res.* **107,** 239–255.

Kulkens, T., Bolhuis, P. A., Wolterman, R. A., Kemp, S., Te Nijenhuis, S., Valentijn, L. J., Hensels, G. W., Jennekens, F. G. I., De Visser, M., Hoogendijk, J. E., and Baas, F. (1993). Deletion of the serine 34 codon from the major peripheral myelin protein P_0 gene in Charcot–Marie–Tooth disease type 1B. *Nature Genet.* **5,** 35–39.

Lai, C., Brow, M. A., Nave, K.-A., Noronha, A. B., Quarles, R. H., Bloom, F. E., Milner, R. J., and Sutcliffe, J. G. (1987). Two forms of 1B236/myelin–associated glycoprotein, a cell adhesion molecule for postnatal neural development, are produced by alternative splicing. *Proc. Natl. Acad. Sci. USA* **84,** 4337–4341.

Lampert, P. W., and Garrett, R. S. (1971). Mechanism of demyelination, in tellurium neuropathy: electron microscopic observations. *Lab. Invest.* **25,** 380–388.

Lampert, P. W., and Schochet, S. S. (1968). Demyelination and remyelination in lead neuropathy: electron microscopic studies. *J. Neuropathol. Exp. Neurol.* **27,** 527–545.

Larocca, J. N., Golly, F., Makman, M. H., Cervone, A., and Ledeen, R. W. (1990). Receptor activity and signal transduction in myelin. In *Cellular and Molecular Biology of Myelination* (G. Jeserich, H. H. Althaus, and T. V. Waehneldt, Eds.), pp. 405–416, Springer-Verlag, Berlin.

Ledeen, R. W. (1992). Enzymes and receptors of myelin. In: *Myelin: Biology and Chemistry* (R. E. Martenson, Ed.), pp. 531–570, CRC Press, Boca Raton, Florida.

Le Douarin, N. M. (1982). *The Neural Crest.* Cambridge University Press, Cambridge, England.

Lees, M. B., and Bizzozero, O. A. (1992). Structure and acylation of proteolipid protein. In: *Myelin: Biology and Chemistry* (R. E. Martenson, Ed.), pp. 237–255, CRC Press, Boca Raton, Florida.

Lees, M. B., and Sapirstein, V. S. (1983). Myelin-associated enzymes. In: *Handbook of Neurochemistry,* 2nd ed. (A. Lathja, Ed.), p. 435, Plenum Press, New York.

Leivo, I., and Engvall, E. (1989). Merosin, a protein specific for basement membranes of Schwann cells, striated muscle and trophoblast, is expressed late in nerve and muscle development. *Proc. Natl. Acad. Sci. USA* **1548.**

Lemke, G. (1988). Unwrapping the genes of myelin. *Neuron* **1,** 535–543.

Lemke, G., and Axel, R. (1985). Isolation and sequence of a cDNA encoding the major structural protein of peripheral myelin. *Cell* **40,** 501–508.

Macklin, W. B. (1992). The myelin proteolipid protein gene and its expression. In: *Myelin: Biology and Chemistry* (R. E. Martenson, Ed.), pp. 257–276, CRC Press, Boca Raton, Florida.

Macklin, W. B., Weill, C. L., and Deininger, P. L. (1986). Expression of myelin proteolipid and basic protein mRNAs in cultured cells. *J. Neurosci. Res.* **16,** 217–230.

Magee, P. N., Stoner, H. B., and Barnes, J. M. (1957). The experimental production of edema in the central nervous system of the rat by triethyltin compounds. *J. Pathol. Bacteriol.* **73,** 107–124.

Maggio, B., and Yu, R. K. (1989). Interaction and fusion of unilamellar vesicles containing cerebrosides and sulfatides induced by myelin basic protein. *Chem. Phys. Lipids* **51,** 127–136.

Maggio, B., and Yu, R. K. (1992). Modulation by glycosphingolipids of membrane-membrane interaction induced by myelin basic protein and mellitin. *Biochim. Biophys. Acta* **112,** 105–114.

Martini, R., and Schachner, M. (1986). Immunoelectron microscopic localization of neural cell adhesion molecules (L1, N-CAM, and MAG) and their shared carbohydrate epitope and myelin basic protein in developing sciatic nerve. *J. Cell Biol.* **103,** 2439–2448.

Matthieu, M.-M., Zimmerman, A. W., Webster, H., deF, Ulsamer, A. G., Brady, R. O., and Quarles, R. H. (1974). Hexachlorophene intoxication: characterization of myelin and myelin related fractions in the rat during early postnatal development. *Exp. Neurol.* **45,** 558–575.

Mendell, J. R., Sahenk, Z., Whitaker, J. N., Trapp, B. D., Yates, A. J., Griggs, R. C., and Quarles, R. H. (1985). Polyneuropathy and IgM monoclonal gammopathy: studies on the pathogenetic role of anti–myelin-associated glycoprotein antibody. *Ann. Neurol.* **17,** 243–254.

McKenna, M. C., and Campagononi, A. T. (1979). Effect of pre- and postnatal essential fatty acid deficiency on brain in development and myelination. *J. Nutrition* **109,** 1195.

Medzihradsky, K. F., Gibson, B. W., Kaur, S., Yu, Z. H., Medzihradszky, D., Burlingame, A. L., and Bass, N. M. (1992). The primary structure of fatty acid–binding protein from nurse shark liver: structural and evolutionary relationship to the mammalian fatty acid–binding protein family. *Eur. J. Biochem.* **203,** 327–339.

Mendz, G. L. (1992). Structure and molecular interactions of myelin basic protein and its antigenic peptides. In: *Myelin: Biology and Chemistry* (R. E. Martenson, Ed.), pp. 277–366, CRC Press, Boca Raton, Florida.

Messing, A., Behringer, R. R., Hammang, J. P., Palmiter, R. D., Brinster, R. L., and Lemke, G. (1992). P_0 promoter directs expression of reporter and toxin genes to Schwann cells of transgenic mice. *Neuron* **8,** 507–520.

Mikol, D. D., Alexakos, M. J., Bayley, C. A., Lemons, R. S., LeBeau, M. M., and Stefansson, K. (1990a). Structure and chromosomal localization of the gene for the oligodendrocyte–myelin glycoprotein. *J. Cell Biol.* **111,** 2673–2679.

Mikol, D. D., Gulcher, J. R., and Stefansson, K. (1990b). The oligodendrocyte–myelin glycoprotein belongs to a distinct family of proteins and contains the HNK-1 carbohydrate. *J. Cell Biol.* **110,** 471–479.

Mikol, D. D., Stefansson, K. (1988). A phosphatidylinositol-linked peanut agglutinin-binding glycoprotein in central nervous system myelin and on oligodendrocytes. *J. Cell Biol.* **106,** 1273–1279.

Mikoshiba, K., Okano, H., Tamura, T. A., and Ikenake, K. (1991). Structure and function of myelin protein genes. *Annu. Rev. Neurosci.* **14,** 201–217.

Milner, R. J., Lai, C., Nave, K.-A., Montag, D., Farber, L., and Sutcliffe, J. G. (1990). Organization of myelin protein genes: myelin-associated glycoprotein. *Ann. NY Acad. Sci.* **605,** 254–261.

Mirsky, R., and Jessen, K. R. (1990). Schwann cell development and the regulation of myelination. *Sem. Neurosci.* **2,** 423–435.

Mirsky, R., and Jessen, K. R. (1996). Schwann cell development, differentiation, and myelination. *Curr. Opin. Neurobiol.* **6,** 89–96.

Molineaux, S. M., Engh, H., DeFerra, F., Hudson, L., and Lazzarini, R. A. (1986). Recombination within the myelin basic protein gene created the dysmyelinating shiverer mouse mutation. *Proc. Natl. Acad. Sci. USA* **83,** 7542–7546.

Morell, P. (1994). Biochemical and molecular bases of myelinopathy. In: *Principles of Neurotoxicology* (L. W. Chang, Ed.), pp. 583–608, Marcel Dekker, Inc, New York.

Morell, P., Quarles, R. H., and Norton, W. T. (1994). Myelin formation, structure, and biochemistry. In: *Basic Neurochemistry,* (G. W. Siegel, P. Molinoff, B. Agranoff, Eds.), 5th ed, Raven Press, pp. 117–143, New York.

Morell, P., and Toews, A. D. (1996a). Schwann cells as targets for neurotoxicants. *Neurotoxicology* **17,** 685–696.

Morell, P., and Toews, A. D. (1996b). Biochemistry of lipids. In: *Neurodystrophies and Neurolipidoses, Handbook of Clinical Neurology,* (H. W. Moser, Ed.), Vol. 22, pp. 33–49, Elsevier Press, Amsterdam.

Morell, P., and Toews, A. D. (1997). Myelin and myelination as affected by toxicants. In: *Nervous System and Behavioral Toxicology* (H. E. Lowndes, and K. R. Reuhl, Eds.), Vol. 11 of *Comprehensive Toxicology,* pp. 201–215, Elsevier Science, Amsterdam.

Nave, K. A., Lai, C., Bloom, F. E., and Milner, R. J. (1986). Jumpy mutant mouse: a 74-base deletion in the mRNA for myelin proteolipid protein and evidence for a primary defect in RNA splicing. *Proc. Natl. Acad. Sci. USA* **83,** 9264–9268.

Nagara, H., Suzuki, K., Tiffany, C. W., and Suzuki, K. (1981). Triethyl tin does not induce intramyelinic vacuoles in the CNS of the quaking mouse. *Brain Res.* **225,** 413–420.

Nakhasi, H. L., Toews, A. D., and Horrocks, L. A. (1975). Effects of a postnatal protein deficiency on the content and composition of myelin from brains of weanling rats. *Brain Res.* **83,** 176–179.

Narayanan, V., Barbosa, E., Reed, R., and Tennekoon, G. I. (1988). Characterization of a cloned cDNA encoding rabbit myelin P2 protein. *J. Biol. Chem.* **263,** 8332–8337.

Nave, K.-A. (1995). Neurological mouse mutants: a molecular–genetic analysis of myelin proteins. In: *Neuroglia.* (H. Kettenmann, and B. R. Ransom, Eds.), pp. 571–586, Oxford University Press, New York.

Nave, K.-A., and Milner, R. J. (1989). Proteolipid proteins: structure and genetic expression in normal and myelin-deficient mutant mice. *Crit. Rev. Neurobiol.* **5,** 65–91.

Newman, S., Saito, M., and Yu, R. K. (1995). Biochemistry of myelin proteins and enzymes. In: *Neuroglia* (H. Kettenman and B. R. Ransom, Eds.), pp. 535–554, Oxford University Press, New York.

Nieminen, L., Bjondahl, K., and Mottonen, M. (1973). Effect of hexachlorphene on the rat brain during organogenesis. *Food Cosmet. Toxicol.* **11,** 635–639.

Noebels, J. L., Marcom, P. K., and Jalilian-Tehrani, M. (1991). Sodium channel density in hypomyelinated brain increased by myelin basic protein gene deletion. *Nature* **352,** 431–434.

Norton, W. T., and Cammer, W. (1984). Isolation and characterization of myelin. In *Myelin.* (P. Morell, Ed.), pp. 147–180, Plenum Press, New York.

Norton, W. T., and Poduslo, S. E. (1973a). Myelination in rat brain: method of myelin isolation. *J. Neurochem.* **21,** 749–757.

Norton, W. T., and Poduslo, S. E. (1973b). Myelination in rat brain: changes in myelin composition during brain maturation. *J. Neurochem.* **21,** 759–773.

Omlin, F. X., Webster, H. deF., Palkovits, C. G., and Cohen, S. R. (1982). Immunocytochemical localization of basic protein in major dense line regions of central and peripheral myelin. *J. Cell Biol.* **95,** 242–248.

Ong, R. L., and Yu, R. K. (1984). Interaction of ganglioside GM1 and myelin basic protein studied by ^{12}C and ^{1}H nuclear magnetic resonance. *J. Neurosci. Res.* **12,** 377–393.

Owens, C. G., Boyd, C. J., Bunge, R. P., and Salzer, J. L. (1990). Expression of recombinant myelin-associated glycoprotein in primary Schwann cells promotes the initial investment of axons by myelinating Schwann cells. *J. Cell Biol.* **111,** 1171–1182.

Owens, G. C., and Bunge, R. P. (1991). Schwann cells infected with a recombinant retrovirus expressing myelin-associated glycoprotein antisense RNA do not form myelin. *Neuron* **7,** 565–575.

Patel, P. I., Roa, B. B., Welcher, A. A., Schoener-Scott, R., Trask, B. J., Pentao, L., Jackson, Snipes, G., Garcia, C. A., Francke, U., Shooter, E. M., Lupski, J. R., and Suter, U. (1992). The gene for the peripheral myelin protein PMP-22 is a candidate for Charcot–Marie–Tooth disease type 1A. *Nature Genet.* **1,** 159–165.

Peters, A., Palay, S. L., and Webster, H. deF. (1991). *The Fine Structure of the Nervous System: Neurons and Their Supporting Cells.* 3rd ed. Oxford University Press, New York.

Pfeiffer, S. E., Warrington, A. E., and Bansal, R. (1993). The oligodendrocyte and its many cellular processes. *Trends Cell Biol.* **3,** 191–197.

Phaire-Washington, L., Wang, E., and Silverstein, S. C. (1980). Phorbol myristate acetate stimulates pinocytosis and membrane spreading in mouse peritoneal macrophages. *J. Cell Biol.* **86,** 634–640.

Popko, B., Puckett, C., and Hood, L. (1988). A novel mutation in myelin-deficient mice results in unstable myelin basic protein gene transcripts. *Neuron* **1,** 221–225.

Popko, B., Puckett, C., Lai, E., Shine, H. D., Readhead, C., Takahashi, N., Hunt, S. W., III, Sidman, R. L., and Hood, L. (1987). Myelin deficient mice: expression of myelin basic protein and generation of mice with varying levels of myelin. *Cell* **48,** 713–721.

Powell, H., Searner, O., Gluck, L., and Lampert, P. (1973). Hexachlorophene myelinopathy in premature infants. *J. Pediatr.* **82,** 976–981.

Pribyl, T. M., Campagnoni, C. W., Kampf, K., Kashima, T., Handley, V. W., McMahon, J., and Campagnoni, A. T. (1993). The human myelin basic protein gene is included within a 179-kilobase transcription unit: expression in the immune and centran nervous sustems. *Proc. Natl. Acad. Sci. USA* **90,** 10695–10699.

Privat, A., Jaque, C., Bourre, J. M., Dupouye, P., and Baumann, N. (1979). Absence of the major dense line in myelin of the mutant mouse 'shiverer.' *Neurosci. Lett.* **12,** 107–112.

Prohaska, J. R., and Wells, W. W. (1974). Copper deficiency in the developing rat brain: a possible model for Menkes steely hair disease. *J. Neurochem.* **25,** 91–98.

Pryer, N. K. Walker, R. A., Skeen, V. P., Bourns, B. D., Soboeiro, M. F., and Salmon, E. D. (1992). Brain microtubule-associated proteins modulate microtubule dynamic instability *in vitro:* real-time observations using video microscopy. *J. Cell Sci.* **103,** 965–976.

Puckett, C., Hudson, L., Ono, K., Friedrich, V., Benecke, J., Dubois-Dalcq, M., and Lazzarini, R. A. (1987). Myelin-specific proteolipid protein is expressed in myelinating Schwann cells but is not incorporated into myelin sheaths. *J. Neurosci. Res.* **18,** 511–518.

Quarles, R. H., Colman, D. R., Salzer, J. L., and Trapp, B. D. (1992). Myelin-associated glycoprotein: structure–function relationships and involvement in neurological diseases. In: *Myelin: Biology and Chemistry* (R. E. Martenson, Ed.), pp. 413–448, CRC Press, Boca Raton, Florida.

Quarles, R. H., Morell, P., and McFarlin, D. E. (1994). Diseases involving myelin. In: *Basic Neurochemistry* (5th ed. R. W. Albers, G. W., Siegel, P. Molinoff, and B. Agranoff, Eds.), pp. 771–792, Raven Press, New York.

Raskind, W. H., Williams, C. A., Hudson, L. D., and Bird, T. D. (1991). Complete deletion of the proteolipid protein gene (PLP) in a family with X-linked Pelizaeus–Merzbacher disease. *Am. J. Hum. Genet.* **49,** 1355–1360.

Readhead, C., and Hood, L. (1990). The dysmyelinating mouse mutations shiverer (shi) and myelin deficient (shi mld). *Behav. Genet.* **20,** 213–234.

Readhead, C., Popko, B., Takahashi, N., Shine, H. D., Saavedra, R. A., Sidman, R. L., and Hood, L. (1987). Expression of a myelin basic protein gene in transgenic shiverer mice: correction of the dysmyelinating phenotype. *Cell* **48,** 703–712.

Remahl, S., and Hildebrand, C. (1990). Relation between axons and oligodendroglial cells during initial myelination: the glial unit. *J. Neurocytol.* **19,** 313–328.

Roach, A., Boylan, K., Horvath, S., Prusiner, S. B., and Hood, L. E. (1983). Characterization of cloned cDNA representing rat myelin basic protein: absence of expression in brain of shiverer mutant mice. *Cell* **34,** 799–806.

Roach, A., Takahashi, N., Pravtcheva, D., Ruddle, E., and Hood, L. (1985). Chromosomal mapping of mouse myelin basic protein gene and structure and transcription of the partially deleted gene in shiverer mutant mice. *Cell* **42,** 149–155.

Robain, O., and Ponsot, G. (1978). Effects of undernutrition on glial maturation. *Brain Res.* **149,** 379–397.

Rosenbluth, J. (1980). Central myelin in the mouse mutant shiverer. *J. Comp. Neurol.* **194,** 639–648.

Roussel, G., Neskovic, N. M., Trifilieff, E., Artault, J.-C., and Nussbaum, J. L. (1987). Arrest of proteolipid transport through the Golgi apparatus in jimpy brain. *J. Neurocytol.* **16,** 195–204.

Said, G., and Duckett, S. (1981). Tellurium-induced myelinopathy in adult rats. *Muscle Nerve* **4,** 319–325.

Salzer, J. L., Bunge, R. P., and Glaser, L. (1980a). Studies of Schwann cell proliferation. III. Evidence for the surface localization of the neurite mitogen. *J. Cell Biol.* **84,** 767–778.

Salzer, J. L., Williams, A. K., Glasser, L., and Bunge, R. P. (1980b). Studies of Schwann cell proliferation. II. Characterization of the stimulation and specificity of the response to a neurite membrane fraction. *J. Cell Biol.* **84,** 753–766.

Salzer, J. L., Holmes, W. P., and Colman, D. R. (1987). The amino acid sequences of the myelin-associated glycoproteins: homology to the immunoglobulin gene superfamily. *J. Cell Biol.* **104,** 957–965.

Salzer, J. L., Pedraza, L., Brown, M., Struyk, A., Afar, D., and Bell, J. (1990). Structure and function of the myelin-associated glycoprotein. *Ann. NY Acad. Sci.* **605,** 302–312.

Sanes, J. R., and Cheney, J. M. (1982). Laminin, fibronectin, and collagen in synaptic and extrasynaptic portions of muscle fiber basement membrane. *J. Cell Biol.* **93,** 442–451.

Sanes, J. R., Engvall, E., Butkowsky, R., and Hunter, D. D. (1990). Molecular heterogeneity of basal laminae: isoforms of laminin and collagen IV at neuromuscular junction and elsewhere. *J. Cell Biol.* **111,** 1685–1699.

Schachner, M., Commer, I., and Lagenaur, C. (1984). Expression of glial antigens C1 and M1 in the peripheral nervous system during development and regeneration. *Dev. Brain Res.* **14,** 165–178.

Schluesener, H. J., Sobel, R. A., Linington, C., and Weiner, H. L. (1987). A monoclonal antibody against a myelin oligodendrocyte glycoprotein induces relapses and demyelination in central nervous system autoimmune disease. *J. Immunol.* **139,** 4016–4021.

Schneider, A., Montague, P., Griffiths, I. R., Fanarragam, Kennedy, P., Brophy, P., and Nave, K. A. (1992). Uncoupling of hypomyelination and glial cell death by a mutation in the proteolipid protein gene. *Nature* **358,** 758–761.

Schroer, T. A., and Sheetz, M. P. (1991). Functions of microtubule-based motors. *Ann. Rev. Physiol.* **53,** 629–652.

Scolding, N. J., Frith, S., Linington, C., Morgan, B. P., Campbell, A. K., and Compston, D. A. S. (1989). Myelin–oligodendrocyte glycoprotein (MOG) is a surface marker of oligodendrocyte maturation. *J. Neuroimmunol.* **22,** 169–176.

Seitelberger, F. (1995). Neuropathology and genetics of Pelizaeus–Merzbacher disease. *Brain Pathol.* **5,** 267–273.

Shine, H. D., Readhead, C., Popko, B., Hood, L., and Disman, R. L. (1990). Myelin basic protein myelinogenesis: morphometric analysis of normal, mutant and transgenic central nervous system. *Prog. Clin. Biol. Res.* **336,** 81–92.

Sidman, R. L. (1964). Mutant mice (quaking and jimpy) with deficient myelination in the central nervous system. *Science* **144,** 309–311.

Singh, H., and Pfeiffer, S. E. (1985). Myelin-associated galactolipids in primary cultures from dissociated fetal rat brain: biosynthesis, accumulation, and cell surface expression. *J. Neurochem.* **45,** 1371–1381.

Skalli, O., and Goldman, R. D. (1991). Recent insights into the assembly, dynamics, and function of intermediate filament networks. *Cell Motil. Cytoskel.* **19,** 67–79.

Skoff, R. P. (1982). Increased proliferation of oligodendrocyte in the hypomyelinated mouse mutant-jimpy. *Brain Res.* **248,** 19–31.

Sloboda, R. D., Dentler, W. L., and Rosenbaum, J. L. (1976). Microtubule-associated proteins and the stimulation of tubulin assembly *in vitro. Biochemistry* **15,** 4497–4505.

Smith, M. E. (1973). Studies on the mechanism of demyelination: triethyl tin–induced demyelination. *J. Neurochem.* **21,** 357–372.

Smith, M. E. (1983). Peripheral nervous system myelin properties and metabolism. In: *Handbook of Neurochemistry,* 2nd ed. (A. Lajtha, Ed.) pp. 201–223, Plenum Press, New York.

Smith, M. E., and Perret, V. (1986). Immunological non-identity of 19K protein and P_0 in peripheral nervous system myelin. *J. Neurochem.* **47,** 924–929.

Smith, R. (1992). The basic protein of CNS myelin: its structure and ligand binding. *J. Neurochem.* **59,** 1589–1608.

Snipes, G. J., Suter, U., Welcher, A. A., and Shooter, E. M. (1992). Characterization of a novel peripheral nervous system myelin protein (PMP-22/SR13). *J. Cell Biol.* **117,** 225–238.

Sparkes, R. S., Mohandas, T., Heinzmann, C., Roth, H. J., Klisak, I., and Campagnoni, A. T. (1987). Assignment of the myelin basic protein gene to human chromosome 18q22-qter. *Hum. Genet.* **75,** 147–150.

Spreyer, P., Kuhn, G., Hanemann, C. O., Gillen, C., Schaal, H., Kuhn, R., Lemke, G., and Muller, H. W. (1991). Axon-regulated expression of a Schwann cell transcript that is homologous to a growth arrest-specific gene. *EMBO J.* **10,** 3661–3668.

Sprinkle, T. J. (1989). 2′,3′-Cyclic nucleotide 3′-phosphodiesterase, an oligodendrocyte–Schwann cell and myelin-associated enzyme of the nervous system. *Crit. Rev. Neurobiol.* **4,** 235–301.

Stahl, N., Harry, J., and Popko, B. (1990). Quantitative analysis of myelin protein gene expression during development in the rat sciatic nerve. *Mol. Brain Res.* **8,** 209–212.

Stewart, H. J. S. (1995). Expression of c-Jun, Jun B, Jun D, and cAMP response element binding protein by Schwann cells and their precursors *in vivo* and *in vitro Eur. J. Neurosci.* **7,** 1366–1375.

Sundstrom, R., and Karlsson, B. (1987). Myelin basic protein in brains of rat with low dose lead encephalopathy. *Arch. Toxicol.* **59,** 341–345.

Sutcliffe, J. G., Milner, R. J., Shinnick, T. M., and Bloom, F. E. (1983). Identifying the protein products of brain-specific genes with antibodies to chemically synthesized peptides. *Cell* **33,** 671–672.

Suter, U., Moskow, J. J., Welcher, A. A., Sniper, G. J., Kosaras, B., Sidman, R. L., Buchberg, A. M., and Shooter, E. M. (1992a). A leucine-to-proline mutation in the putative first transmembrane domain of the 22-kDa peripheral myelin in the trembler-J mouse. *Proc. Natl. Acad. Sci. USA* **89,** 4382–4386.

Suter, U., Welcher, A. A., Ozcelik, T., Snipes, G. J., Kosaras, B., Francke, U., Billings-Gagliardi, S., Sidman, R. L., and Shooter, E. M. (1992b). Trembler mouse carrier a point mutation in a myelin gene. *Nature* **356,** 241–244.

Suzuki, K. (1971). Some new observations in triethyl-tin of rats. *Exp. Neurol.* **31,** 207–213.

Takahashi, T. (1981). Experimental study on segmental demyelination in tellurum neuropathy. *Hokkaido Igaku Zasshi* **856,** 105–131.

Tansey, F. A., Zhang, H., and Cammer, W. (1996). Expression of carbonic anhydrase II mRNA and protein in oligodendrocytes during toxic demyelination in the young adult mouse. *Neurochem. Res.* **21,** 411–416.

Taraszewska, A. (1988). Ultrastructure of axons in disturbed CNS myelination in pt rabbit. *Neuropatol. Pol.* **26,** 385–402.

Tatum, A. H. (1993). Experimental paraprotein neuropathy, demyelination by passive transfer of IgM anti–myelin-associated glycoprotein. *Ann Neurol.* **33,** 502–506.

Timsit, S., Sinoway, M. P., Levy, L., Allinquant, B., Stempak, J., Staugaitis, S. M., and Colman, D. R. (1992). The DM20 protein of myelin: intracellular and surface expression patterns in transfectants. *J. Neurochem.* **58,** 1936–1942.

Toews, A. D., Blaker, W. D., Thomas, D. J., Gaynor, J. J. Krigman, M. R., Mushak, P., and Morell, P. (1983). Myelin deficity produced by early postnatal exposure to inorganic lead or triethyltin are persistent. *J. Neurochem.* **41,** 816–822.

Toews, A. D., Eckermann, C. E., Lee, S. Y., and Morell, P. (1991a). Primary demyelination induced by exposure to tellurium alters mRNA levels for nerve growth factor receptor, SCIP, 2′,3′-cyclic nucleotide 3′-phosphodiesterase, and myelin proteolipid protein in rat sciatic nerve. *Mol. Brain Res.* **11,** 321–325.

Toews, A. D., Goodrum, J. F., Lee, S. Y., Eckermann, E., and Morell, P. (1991b). Tellurium-induced alterations in HMG-CoA reductase gene expression and enzyme activity: differential effects in sciatic nerve and liver suggests tissue-specific regulation of cholesterol synthesis. *J. Neurochem.* **57,** 1902–1906.

Toews, A. D., Griffiths, I. R., Kyriakides, E., Goodrum, J. F., Eckermann, C. E., Morell, P., and Thomson, C. E. (1992). Primary demyelination induced by exposure to tellurium alters Schwann-cell gene expression: a model for intracellular targeting of NGF-receptor. *J. Neurosci.* **12,** 3676–3687.

Toews, A. D., Hostettler, J., Barrett, C., and Morell, P. (1997). Alterations in gene expression associated with primary demyelination and remyelination in the peripheral nervous system. *Neurochem. Res.* **22,** 1271–1280.

Toews, A. D., Kolber, A., Hayward, J., Krigman, M. R., and Morell, P. (1978). Experimental lead encephalopathy in the suckling rat: concentration of lead in cellular fractions enriched in brain capillaries. *Brain Res.* **147,** 131–138.

Toews, A. D., Krigman, M. R., Thomas, D. J., and Morell, P. (1980). Effect of inorganic lead exposure on myelination in the rat. *Neurochem. Res.* **5,** 605–616.

Toews, A. D., Lee, S. Y., Popko, B., and Morell, P. (1990). Tellurium-induced neuropathy: a model for reversible reductions in myelin protein gene expression. *J. Neurosci. Res.* **26,** 501–507.

Toews, A. D., and Morell, P. (1987). Posttranslational modification of myelin proteins. In: *A Multidisciplinary Approach to Myelin Diseases* (G. Serlupi-Crescenzi, Ed.), pp. 59–75, Plenum Press, New York.

Tohyama, K., and Ide, C. (1984). The localization of laminin and fibronectin on the Schwann cell basal lamina. *Arch. Histol. Japan* **47,** 519–532.

Topilko, P., Schneider-Manoury, S., Levi, G., Baron-Van Evercooren, A., Chennoufi, A. B. Y., Seitanidou, T., Babinet, C., and Charney, P. (1994). Krox-20 controls myelination in the peripheral nervous system. *Nature* **371,** 796–799.

Torack, R., Gordon, J., and Prokop, J. (1990). Pathobiology of acute triethyltin intoxication. *Int. Rev. Neurobiol.* **12,** 45–86.

Tosic, M., Roach, A., deRivaz, J. C., Dolivo, M., and Matthieu, J. M. (1990). Post-transcriptional events are responsible for low expression of myelin basic protein in myelin-deficient mice: role of natural antisense RNA. *EMBO J.* **9,** 401–406.

Towfighi, J. (1980). Hexachlorophene. In: *Experimental and Clinical Neurotoxicology* (P. S. Spencer, and H. H. Schaumburg, Eds.), pp. 440–455, Williams & Williams, Baltimore.

Towfighi, J., Gonatas, N. K., and McCree, L. (1974). Hexachlorophene-induced changes in central and peripheral myelinated axons of developing and adult rats. *Lab. Invest.* **31,** 712–721.

Trapp, B. (1988). Distribution of the myelin-associated glycoprotein and P_0 protein during myelin compaction in quaking mouse peripheral nerve. *J. Cell Biol.* **107,** 675–685.

Trapp, B. D. (1990). The myelin-associated glycoprotein: location and potential functions. In: *Myelination and Dysmyelination* (I. D. Duncan, R. P. Skoff, and D. R. Colman, Eds.), pp. 29–43. New York Academy of Science, New York.

Trapp, B. D., and Bernsohn, J. (1978). Essential fatty acid deficiencies and CNS myelin. *J. Neurol. Sci.* **37,** 249.

Trapp, B. D., and Quarles, R. H. (1982). Presence of the myelin-associated glycoprotein correlates with alterations in the periodicity of peripheral myelin. *J. Cell Biol.* **92,** 877–882.

Trapp, B. D., Quarles, R. H., and Griffin, J. W. (1984). Myelin-associated glycoprotein and myelinating Schwann cell–axon interaction in chronic β, β'-iminodipropionitrile neuropathy. *J. Cell Biol.* **98,** 1272–1278.

Trapp, B. D., Bernier, L., Andrews, S. B., and Colman, D. R. (1988). Cellular and subcellular distribution of 2′,3′-cyclic nucleotide 3′-phosphodiesterase and its mRNA in the rat central nervous system. *J. Neurochem.* **51,** 859–868.

Trapp, B. D., Andrews, S. B., Wong, A., O'Connel, M., and Griffin, J. W. (1989). Co-localization of the myelin-associated glycoprotein and the microfilament components f-actin and spectrin in Schwann cells of myelinated fibers. *J. Neurocytol.* **18,** 47–60.

Trapp, B. D., Itoyama, Y., Sternberger, N. H., Quarles, R. H., and Webster, H. deF. (1981). Immunocytochemical localization of P_0 protein in Golgi complex membranes and myelin of developing rat Schwann cells. *J. Cell Biol.* **90,** 1–6.

Trapp, B. D., Moench, T., Puller, M., Barbosa, E., Tennekoon, G., and Griffin, J. (1987). Spatial segregation of mRNA encoding myelin-specific proteins. *Proc. Natl. Acad. Sci. USA* **84,** 7773–7777.

Trapp, B. D., Kidde, G. J., Hauer, P., Mulrenin, E., Haney, C. A., and Andrews, S. B. (1995). Polarization of myelinating Schwann cell surface membranes: role of microtubules and the trans-Golgi network. *J. Neurosci.* **15,** 1797–1807.

Uyemura, K., Yoshimura, K., Suzuki, M., and Kitamura, K. (1984). Lipid binding activities of the P2 protein in peripheral nerve myelin. *Neurochem. Res.* **9,** 1509–1514.

Valentijn, L. J., Baas, F., Wolterman, R. A., Hoogendijk, J. E., Van Den Bosch, N., Zorn, I., Gabreels-Festen, A., DE Visser, M., and Bolhuis, P. A. (1992). Identical point mutation of PMP-22 in trembler-J mouse and Charcot–Marie–Tooth disease type 1A. *Nature Genet.* **2,** 288–291.

Veerkamp, J. H., Peeters, R. A., and Maatman, J. H. (1991). Structural and functional features of different types of cytoplasmic fatty acid–binding proteins. *Biochim. Biophys. Acta* **1081,** 1–24.

Wagner, M., Toews, A. D., and Morell, P. (1995). Tellurite specifically affects squalene epoxidase: investigations examining the mechanism of tellurium-induced neuropathy. *J. Neurochem.* **64,** 2169–2176.

Wagner-Recio, M., Toews, A. D., and Morell, P. (1991). Tellurium blocks cholesterol synthesis by inhibiting squalene metabolism: preferential vulnerability to this metabolic block leads to peripheral nervous system demyelination. *J. Neurochem.* **57,** 1891–1901.

Walters, S. N., and Morell, P. (1981). Effects of altered thyroid states on myelogenesis. *J. Neurochem.* **36,** 1792–1801.

Wang, N., Butler, J. P., and Ingber, D. E. (1993). Mechano-transduction across the cell surface and through the cytoskeleton. *Science* **260,** 1124–1127.

Warrington, A. E., and Pfeiffer, S. E. (1992). Proliferation and differentiation of O4+ oligodendrocytes in postnatal rat cerebellum: analysis in unfixed tissue slices using anti-glycolipid antibodies. *J. Neurosci. Res.* **33,** 338–353.

Watanabe, I. (1977). Effect of triethyltin on the developing brain of the mouse. In: *Neurotoxicology* (L. Roizin, H. Shiraki, and N. Grcevic, Eds.), pp. 317–326, Raven Press, New York.

Waxman, S. G. (1985). Structure and function of the myelinated fiber. In: *Handbook of Clinical Neurology, The Demyelinating Diseases* (J. C. Koetsier, Ed.), Vol. 3, pp. 1–28, Elsevier, Amsterdam.

Waxman, S. G. (1987a). Molecular organization of the cell membrane in normal and pathological axons: relation to glial contact. In: *Glial–Neuronal Communication in Development and Regeneration* (H. Althaus, and W. Seifert, Eds.), pp. 711–736, Springer-Verlag, Berlin.

Waxman, S. G. (1987b). Rules governing membrane reorganization and axon–glial interactions during the development of myelinated fibers. In: *Progress in Brain Research, Neural Regeneration* (E. J. Seil, E. Herbert, and B. Carlson, Eds.), Vol. 71, pp. 121–142, Raven Press, New York.

Waxman, S. G., and Black, J. A. (1995). Axoglial interactions at the cellular and molecular levels in central nervous system myelinated fibers. In: *Neuroglia.* (H. Kettenmann, and B. R. Ransom, Eds.), pp. 587–610, Oxford University Press, New York.

Waxman, S. G., and Sims, T. J. (1984). Specificity in central myelination: evidence for local regulation of myelin thickness. *Brain Res.* **292,** 179–185.

Waxman, S. G., Pappas, G. D., and Bennett, M. V. L. (1972). Morphological correlates of functional differentiation of nodes of Ranvier along single fibers in the neurogenic electric organ of the knife fish Sternarchus. *J. Cell Biol.* **53,** 210–224.

Waxman, S. G., Black, J. A., Kocsis, J. D., and Ritchie, J. M. (1989). Low density of sodium channels supports action potential conduction in axons of neonatal rat optic nerve. *Proc. Natl. Acad. Sci. USA* **86,** 1406–1410.

Webster, H. and de F. (1971). The geometry of peripheral myelin sheaths during their formation and growth in rat sciatic nerves. *J. Cell Biol.* **48,** 348–367.

Webster, H. de F., and Favilla, J. T. (1984). Development of peripheral nerve fibers. In: *Peripheral Neuropathy* (P. J. Dyck, P. K. Thomas, E. H. Lambert, and R. Bunge, Eds.), pp. 329–359, WB Saunders, Philadelphia.

Weinberg, H. J., and Spencer, P. S. (1976). Studies on the control of myelinogenesis. II. Evidence for neuronal regulation of myelin production. *Brain Res.* **113,** 363–378.

Welcher, A. A., Suter, U., De Leon, M., Snipes, G. J., and Shooter, E. M. (1991). A myelin protein is encoded by the homologue of a growth arrest–specific gene. *Proc. Natl. Acad. Sci. USA* **88,** 7195–7199.

White, H. B. Jr., Galli, C., and Paoletti, R. (1971). Brain recovery from essential fatty acid deficiency in developing rats. *J. Neurochem.* **18,** 869–882.

Wiggins, R. C. (1986). Myelination: a critical stage in development. *Neurotoxicology* **7,** 103–120.

Wiggins, R. C., and Fuller, G. N. (1978). Early postnatal starvation causes lasting brain hypomyelination. *J. Neurochem.* **30,** 1231–1237.

Wiggins, R. C., Miller, S. L., Benjamins, J. A., Krigman, M. R., and Morell, P. (1976). Myelin synthesis during postnatal nutritional deprivation and subsequent rehabilitation. *Brain Res.* **107,** 257–273.

Wolswijk, G., and Noble, M. (1995). *In vitro* studies of the development, maintenance and regeneration of the oligodendrocyte–type-2 astrocyte (O-2A) lineage in the adult central nervous system. In: *Neuroglia* (H. Kettenman, and B. R. Ransom, Eds.), pp. 149–161, Oxford University Press, New York.

You, K. H., Hsieh, C. L., Hayes, C., Stahl, N., Francke, U., and Popko, B. (1991). DNA sequence, genomic organization, and chromosomal localization of the mouse peripheral myelin protein zero gene: identification of polymorphic alleles. *Genomics* **9,** 751–757.

Zimmerman, A. W., Matthieu, J. M., Quarles, R. H., Brady, R. O., and Hsu, J. M. (1976). Hypomyelination in copper-deficient rats. *Arch. Neurol.* **33,** 111–119.

Zimmerman, T. R. Jr., and Cammer, W. (1982). ATPase activities in myelin and oligodendrocytes isolated from the brains of developing rats and from bovine brain white matter. *J. Neurosci. Res.* **8,** 73–81.

Zimmerman, H., and Vogt, M. (1989). Membrane proteins of synaptic vesicles and cytoskeletal specializations at the node of Ranvier in electric ray and rat. *Cell Tissue Res.* **258,** 617–629.

Zorick, T. S., and Lemke, G. (1996). Schwann cell differentiation. *Curr. Opin. Cell. Biol.* **8,** 870–876.

PART II

Developmental Biology/Toxicology

For some time now investigators in the fields of reproductive toxicology and teratology have been searching for mechanisms of normal and abnormal development. With the advent of molecular biologic techniques this searching has adopted a new level of sophistication, illustrated elegantly in the reports in this part. From neurotransmitters to growth factors, from cell signaling to cell response, the level of our understanding is growing exponentially. How do cells respond differently to the same signal molecule? How does the same signaling pathway elicit two or more different cellular responses? How is cell death determined, and how can the same factor stimulate cell proliferation on the one hand and cell death on the other? These and other critical questions are addressed in these reports, as are models for further exploration.

Chapters 5, 6, and 7 offer excellent reviews of the role of signal molecules and their effector pathways. In Chapter 5, Robert M. Greene, Wade Weston, Paul Nugent, Merle Potchinsky, and M. Michelle Pisano provide an excellent overview of the role of protein kinases and their ligands in craniofacial development. In Chapter 6, Moses V. Chao, Haeyoung Kong, Sung Ok Yoon, Bruce Carter, and Patrizia Casaccia-Bennefil explain how neurotrophins can elicit a cell death response and conversely inhibit cell suicide pathways. Controls for these events reside in factors such as the strength and duration of signaling through phosphorylation and dephosphorylation events and the types and numbers of receptors present—that is the ratio of the two classes of neurotrophin receptors. This report represents an excellent review of the divergent effects of neurotrophins and incorporates many of the phenomena described in the other chapters in this section.

In Chapter 7, Jean M. Lauder and Jiangping Liu remind us that other types of signaling molecules exist, including neurotransmitters, and that environmental agents such as pesticides can easily disrupt these signals, resulting in serious developmental defects. Thus, this report reminds us that environmental toxicants act on cellular pathways at specific sites and that, as our level of understanding of these events increases, our ability to assess these effects at the biochemical and molecular levels will also increase, in turn increasing our ability to classify toxic agents and predict developmental outcomes following exposures.

Because many insults to the central nervous system produce cell death by apoptosis, which results in developmental abnormalities, Chapter 8 by Philip E. Mirkes and Margaret A. Shield on apoptosis is particularly appropriate. In fact, it is one of the best overviews on this complex topic that I have ever read. With respect to toxic insults and their effects on genetic regulation of morphogenesis, Chapter 9 by Gregory D. Bennett and Richard H. Finnell reminds us that timing is everything. However, they have expanded the windows for insults to the central nervous system and propose that even insults occurring prior to implantation may have adverse effects. Chapter 10, by Thomas B. Knudsen and Judith A. Wubak, explores the role of p53 and transgenic mice as models for investigating the phenomenon of apoptosis in development and, again, explains the complexities of a dichotomous response to a signal, where cells may undergo differentiation or apoptosis.

Together, these reports clearly indicate that we are beginning the true era of developmental toxicology. Investigators will combine the best of developmental biology, embryology, toxicology, and teratology to derive a fundamental understanding of normal and abnormal development. It is an exciting time, but a complex one that will probably require interactions among investigators from all of the aforementioned disciplines.

Thomas Sadler

CHAPTER

5

Signal Transduction Pathways as Targets for Induced Embryotoxicity

ROBERT M. GREENE
PAUL NUGENT
M. MICHELE PISANO
Department of Biological and Biophysical Sciences
University of Louisville
School of Dentistry
Louisville, Kentucky 40292

WAYDE WESTON
MERLE POTCHINSKY
Department of Pathology, Anatomy, and Cell Biology
Daniel Baugh Institute
Jefferson Medical College
Thomas Jefferson University
Philadelphia, Pennsylvania 19107

*Abbreviations: AKAP, A kinase anchoring proteins; cAMP, cyclic adenosine monophosphate (cyclic AMP); cdk, cyclin dependent kinase; CKDI, cdk inhibitory protein; CNS, central nervous system; CRABP, cytoplasmic cellular retinoic acid-binding proteins; CREB, CRE-binding protein; CRE, cAMP-response element; CREM, cAMP-responsive element modulator; DAG, diacylglycerol; EGF, epidermal growth factor; ERK1, extracellular signal regulated kinase 1; ERK2, extracellular signal regulated kinase 2; IP_3, inositol triphosphate; LTBP, latent-transforming growth factor-β; LTGF-β, latent transforming growth factor-β; Mad, mothers against dpp; MAPK, mitogen-activated protein kinase; MEE, medial edge epithelial; MEK, MAPkinase/ERK kinase; MEPM, murine embryonic palate mesenchymal; PC, phospatidylcholine; PGE_2, prostaglandin E_2; PGI_2, prostaglandin I_2; PI, phosphoinositide; PKA, protein kinase A; PKC, protein kinase C; RA, retinoic acid; RAR, retinoic acid receptors; RAR-β, retinoic acid receptor beta; Rb, retinoblastoma; RXR, retinoid X receptors; Sma, structurally related Mad; β-LAP, latency-associated peptide; TSG, tumor suppressor gene.

In the 1930s, Ross Harrison discussed developmental processes in terms of the cell. This important step forward in the history of embryology continues to provide the intellectual framework on which the majority of current developmental studies are based. Normal embryonic development involves coordinated regulation of a mutiplicity of cellular processes such as proliferation, migration, secretion, transformation, and cell–cell and cell–matrix interactions. What these processes have in common is dependence on the ability of individual cells to transmit multiple messages from the cell surface to the nucleus and to translate those messages into the correct genomic response. This transfer of information is mediated by multiple parallel and interacting signal transduction pathways—the original information superhighway. The dialog of these cellular signaling cascades may be perturbed by either teratogens or mutant genes with resultant congenital malformations.

Because it is one of the most rapidly growing and developing areas in the embryo, the craniofacial region is highly susceptible to malformations. Craniofacial malformations occur with a frequency of 1 in 600 live births annually in the United States. Of these, 65% manifest clefts of the lip and/or palate. The embryonic craniofacial region has provided several excellent model systems that have allowed examination of multiple developmental/cellular phenomena known to play critical roles in embryonic development. The embryonic orofacial region, specifically the developing secondary palate, has provided an excellent paradigm with which to study the means by which cellular signaling regulates embryonic growth and differentiation. This chapter examines interacting intracellular signal transduction pathways and how they regulate gene expression with subsequent alterations in cell proliferation. The developing orofacial region is profiled as a model with which the role of these pathways in normal and abnormal development can be assessed.

I. Cell Proliferation in Developing Orofacial Tissue

The developing primary and secondary palate have proved to be valuable model systems for gaining insight into cellular aspects of growth and tissue differentiation, as well as into mechanisms of reproductive toxicity. The midfacial region in mammalian and avian embryos develops largely from the first pharyngeal arch, which contains mesenchymal cells derived from the cranial neural crest. The bilateral maxillary processes enlarge and fuse with the nasal processes, thereby forming the primary palate, which includes the entire upper lip. Later in embryonic development, the secondary palate originates as bilateral extensions from the oral aspect of the maxillary processes. In mammals, these extensions, the palatal processes, make contact, fuse with one another, and give rise to the secondary palate or roof of the oral cavity.

Normal growth of the entire craniofacial region is critically important for proper development of the palate in both the human (Trenouth, 1984; Sandham, 1985) and laboratory animals (Diewert, 1982; Johnston and Nash, 1982). The importance of proper quantitative and spatiotemporal growth in the orofacial region is supported by the finding that experimental elimination of mesencephalic neural crest cells results in insufficient growth of the mesenchyme in the maxillary processes with resultant cleft lip and/or cleft palate. Extensive changes in craniofacial dimensions and spatial relations occur during development of the orofacial region. Differential growth of craniofacial structures such as the first branchial arch, the palatal processes, the tongue, Meckel's cartilage, and the mandible appears to be an important factor in normal facial ontogenesis. Enlargement of the embryonic facial processes and development of the primary and secondary palates are accompanied by distinct patterns and rates of proliferation (Nanda and Romeo, 1975; Minkoff, 1980). Moreover, various cleft palate–producing teratogens have been shown to inhibit orofacial mesenchymal cell proliferation (Kochhar, 1968; Salomon and Pratt, 1978; Tassinari *et al.,* 1981).

Studies have provided fascinating insights regarding control of cellular growth and differentiation of embryonic craniofacial tissue, which relies on exquisitely orchestrated interactions between several signal transduction pathways (see later this chapter). Thus, these signaling pathways represent targets for exogenous teratogens and mutant gene products and define potential mechanisms for induced embryotoxicity.

II. Cyclic Nucleotide–Mediated Signal Transduction

A. *Transmembrane Signaling*

The search for molecules regulating growth and differentiation during metazoan ontogenesis has led to the realization that the developing embryo is exposed to a wide variety of molecular signals. Control mechanisms governing development of the palate clearly involve interactions between various extracellular hormones, neurotransmitters, growth factors, prostaglandins, and a number of intracellular second messenger systems.

Because mesenchyme of the embryonic facial prominences is of neural crest origin, the ability of mesenchymal cells of the developing palate to synthesize adrenergic catecholamines *in situ* comes as little surprise. Indeed, the levels of the primary catecholamines (dopamine, norepinephrine, epinephrine) in embryonic palatal tissue have been determined and quantitative alterations noted over the course of palatogenesis (Zimmerman *et al.,* 1981; Pisano and Greene, 1984). Functional β_2-adrenergic receptors have been characterized in embryonic palatal tissue during the period of palatal cellular differentiation (Fig. 1) (Garbarino and Greene, 1984). Both the β-adrenergic catecholamine, isoproterenol, and prostaglandin E_2 inhibit the growth of murine embryonic palate mesenchymal (MEPM) cells by delaying the onset of DNA synthesis through a cAMP-mediated mechanism (Pisano *et al.,* 1986). It is reasonable therefore to hypothesize that endogenous growth-modulating factors such as catecholamines, prostaglandins, serotonin, glucocorticoids, and several growth factors may contribute to differential palatal growth rates by virtue of their ability to modulate the initiation of DNA synthesis.

Most of the differentiative processes that occur during palate development *in vivo* also occur *in vitro,* suggesting that factors regulating tissue growth and differentiation are locally derived. Embryonic palate mesenchymal cells possess several phospholipases that hydrolyze membrane phospholipids to yield free fatty acids used in the synthesis of locally active prostaglandins (George and Chepenik, 1985) such as prostaglandin E_2 (PGE_2) and prostaglandin I_2 (PGI_2) (Chepenik and Greene, 1981; Alam *et al.,* 1982). Prostaglandin E_2 receptor sites of high affinity and low capacity, functionally coupled to adenylate cyclase, have been demonstrated in the murine embryonic palate during the period of maximal palatal growth and tissue differentiation (Jones and Greene, 1986). Prostaglandin binding to these receptors results in stimulation of palatal adenylate cyclase activity (Palmer *et al.,* 1980) and increased intracellular levels of cAMP (Greene *et al.,* 1981). These data suggest that developing palatal tissue is exposed to several catecholamines and prostaglandins, contains functional receptors for both of these classes of biologically active compounds, and is capable of responding

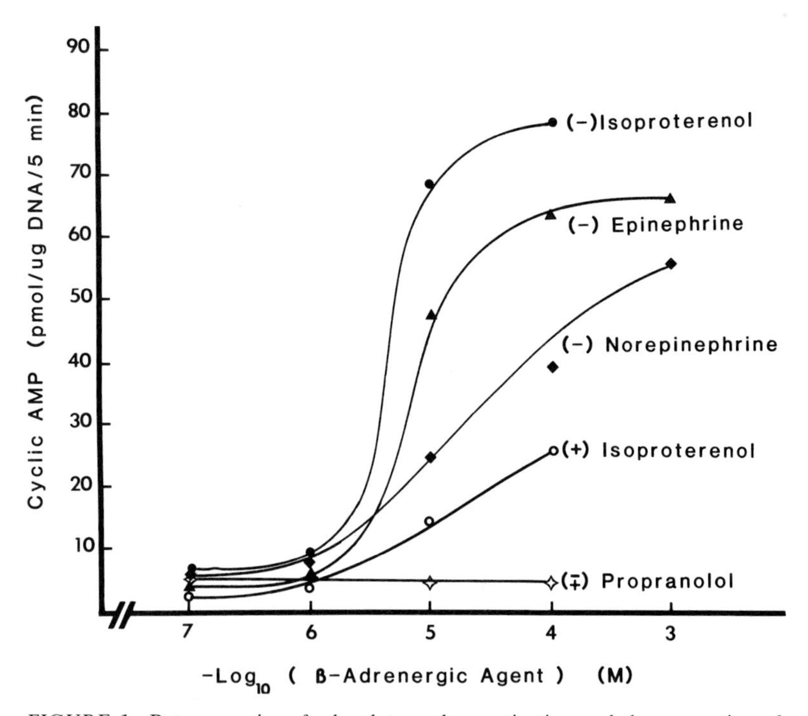

FIGURE 1 Potency series of adenylate cyclase activation and the generation of cAMP by β-adrenergic catecholamines. Fetal palate mesenchymal cells were grown to subconfluency and incubated in serum-free conditions with increasing concentrations of various catecholamines. Levels of cAMP were determined by radioimmunoassay and expressed as pmol cAMP/μg DNA/5 min. Basal activity averaged 3.5 pmol cAMP/μg DNA. (*From Garbarino and Greene (1984) Biochem. Biophys. Res. Comm.* ***119,*** *193–202.*)

to stimulation by these molecules with increases in intracellular levels of cAMP.

Tight control is maintained over cAMP levels via receptor-mediated transmembrane signaling systems comprised of the "effector" enzyme adenylate cyclase and the "transducing" guanine nucleotide-binding G protein (Codina *et al.*, 1984). An interesting correlation can be drawn between the effects of ethanol exposure on the developing embryo and G-protein mediated transmembrane signaling. The toxic effects of prenatal alcohol exposure are well recognized as the *fetal alcohol syndrome* and include prenatal and perinatal mortality, intrauterine growth retardation, and the production of various congenital malformations (Abel and Sokol, 1986). Craniofacial malformations are consistently produced by prenatal ethanol exposure and consist of central nervous system (CNS), ocular, and facial anomalies. Clefts of the lip and palate, agnathia, micrognathia, and facial hypoplasia have all been observed in various laboratory animal models (Blakely, 1988). Although the mechanisms of ethanol-induced craniofacial teratogenicity remain to be clarified, it is interesting to note that studies using whole embryo cultures indicate that ethanol can induce malformations by acting directly on the embryo (Prescott, 1982).

Ethanol has profound effects on G-protein mediated activation of adenylate cyclase, stimulating the production of cAMP in response to acute exposure and reducing receptor-stimulated cAMP production in response to chronic exposure (Charness *et al.*, 1988; Mochly-Rosen *et al.*, 1988; Hoffman and Tabakoff, 1990). This inhibitory response is particularly pronounced in lymphocytes taken from alcoholic subjects (Nagy *et al.*, 1988). Although chronic exposure of MEPM cells to ethanol also resulted in desensitization of adenylate cyclase to hormone stimulation (Weston and Greene, 1991), as well as growth inhibition (Weston *et al.*, 1994), it did so without altering expression of either Gsα or Giα (Weston *et al.*, 1996). These glimpses offer rationale for the pursuit of studies directed at clarifying cellular mechanisms of ethanol-induced craniofacial teratogenicity and invite speculation that ethanol may perturb normal craniofacial ontogenetic processes at the level of receptor–adenylate cyclase coupling.

In eukaryotes, cAMP has been shown to be a versatile regulator of cellular function. Normal growth and differentiation of embryonic palatal tissue depends on regulated levels of intracellular cAMP. Maximal levels of adenylate cyclase, which catalyze the synthesis of cAMP from ATP, occur during differentiation of palatal medial edge epithelium (Waterman *et al.*, 1976), a tissue in which immunohistochemical staining for cAMP is particularly intense during its differentiation (Greene *et al.*, 1980). Intracellular levels of cAMP (Greene and Pratt, 1979; Olson and Massaro, 1980) and the activity of its intracellular receptor, cAMP-dependent protein kinase (PKA) (Linask and Greene, 1989b), increase transiently prior to fusion of the palatal processes. The importance of cAMP is highlighted by the observation that alterations in palatal cAMP levels are seen in response to a number of teratogenic agents and mutant genes known to affect palate development adversely (Erickson *et al.*, 1979; Olson and Massaro, 1980; Harper *et al.*, 1981).

B. Intracellular PKA-Mediated Signaling

Receptor-mediated generation of the second messenger molecule cAMP and the subsequent activation of cAMP-dependent protein kinases (PKA) (EC 2.7.1.37) is a widely used mechanism of signal transduction in mammalian cells (Kuo and Greengard, 1969). This class of enzymes, one of the best characterized classes of kinases, is composed of a regulatory subunit dimer and two catalytic subunits. Cyclic AMP activates the tetrameric holoenzyme by combining with the regulatory subunits to form a dimeric noncovalent cAMP-regulatory subunit complex and two free catalytic subunits that initiate cellular responses by catalyzing the phosphorylation of specific serine and threonine residues in cellular proteins (Bouvier *et al.*, 1987). cDNA cloning and DNA sequencing studies have demonstrated a number of PKA regulatory and catalytic subunit isoforms representing different gene products. Four different regulatory subunits, RIα (Lee *et al.*, 1983), RIβ (Clegg *et al.*, 1988), RIIα (Scott *et al.*, 1987), RIIβ (Jahnsen *et al.*, 1986), and three different catalytic subunits, Cα, Cβ (Chrivia *et al.*, 1988), and Cγ (Beebe *et al.*, 1990), for PKA have now been identified at the level of the gene, mRNA, and protein.

The cellular events of tissue differentiation and morphogenesis have been shown in a number of systems to be associated with specific changes in PKA levels or PKA-binding activities (Smales and Biddulph, 1985; Hedin *et al.*, 1987). However, with at least seven genes encoding for components of the PKA holoenzyme, regulation of cAMP-mediated developmental processes is likely to depend less on levels of total kinase activity than on patterns of expression of different subunits of the holoenzyme, tissue distribution of active enzyme, subcellular location of regulatory subunits, and positive and negative transcriptional control. Indeed, cAMP itself may be responsible for developmental alterations in PKA in the embryonic palate in that PKA exhibits adaptational changes in response to conditions of increased intracellular levels of cAMP (Greene *et al.*, 1995). In addition, PKA isozyme protein and mRNA profiles as well as cellular and subcellular distribution (Linask and Greene, 1989a,b; Linask *et al.*, 1991; Greene *et al.,* 1995) demonstrate clear patterns of temporal alterations in embryonic palatal tissue during the critical period of murine palatal ontogeny. Whereas the type II PKA isozyme predominates early during palatal ontogeny, during the period characterized by tissue growth, the predominant class of PKA present during later stages (during reorientation of the embryonic palatal processes) is the type I isozyme. This suggests that the two isozymes may have different biologic functions and that the type of kinase that is activated may confer specificity to the second messenger response. Thus, precise tissue and cellular compartmentalization of kinase isozymes may be a contributing factor in defining hormonal specificity and responsiveness during development of the secondary palate.

C. Signal Compartmentalization

Paradoxically, diverse hormonal signals acting via a common second messenger, cAMP, promote PKA-mediated phosphorylation of *distinct* substrate proteins. This triggers diverse physiologic processes in the same cell. Because PKA is a multifunctional enzyme with broad substrate specificity, its compartmentalization is likely to be a key regulatory event in controlling intracellular sites of hormone action. Support for such intracellular partitioning comes from compartmentalized elevations of cAMP (Adams *et al.,* 1991; Barsony and Marks, 1991) and differential localization of PKA (Nigg *et al.,* 1985). Moreover, both RI isoforms are soluble and cytoplasmic, whereas up to 75% (depending on the tissue source) of the RII isoforms are particulate and attached to specific subcellular structures (Rubin *et al.,* 1979) through a site in the first 50 amino acids of either RIIα or RIIβ that is not present in the RI isoform (Luo *et al.,* 1990). Indeed, during development of embryonic palatal tissue, RI and RII exhibit distinct subcellular compartmentalization (Linask and Greene, 1989a). Moreover, the embryonic palate exhibits distinct spatial distribution of the type II regulatory subunit of PKA (Linask and Greene, 1989b) and developmental regulation of RII mRNA expression and kinase activity (Greene *et al.,* 1995). This indicates the likelihood of targeting of PKAs to specific cellular addresses.

Subcellular attachment is directed by a family of A-kinase anchoring proteins (AKAPs), which tether the type II PKA to specific subcellular sites (Hirsch *et al.,* 1992; Rios *et al.,* 1992; Keryer *et al.,* 1993) presumably colocalizing the kinase close to—and ensuring rapid and preferential phosphorylation of—physiologic substrates. Each AKAP contains a carboxyl terminal targeting domain that defines the cellular address of each PKA/AKAP complex. Indeed, immunocytochemical analyses show site specific localization of RII with specific AKAPs (Coghlan *et al.,* 1993). Nearly 30 RII-binding proteins, selectively expressed and ranging in size from 34 to 300 kD, have been detected (Carr and Scott, 1992). This suggests that tissue-specific expression of AKAPs may represent a mechanism to explain tissue-specific cAMP responsiveness.

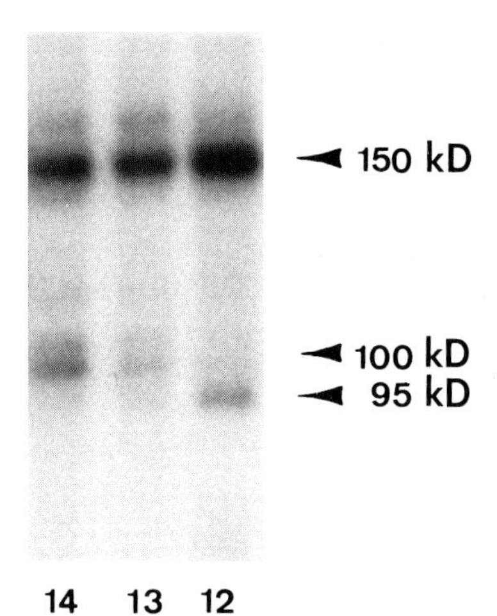

FIGURE 2 AKAP expression by embryonic palatal tissue. A modified Western blot procedure, using radiolabeled RII as a probe, was employed to detect AKAPs in embryonic palatal protein extracts from gestational days 12, 13, and 14. Three proteins of 150, 100, and 95 kD were identified. Whereas the 150 kD protein appeared to be ubiquitously expressed, the 100 and 95 kD proteins exhibited changing temporal patterns of expression. (*Done collaboratively with T. Klauk and J. Scott, Oregon Health Sci. Univ., Portland, OR*)

Preliminary studies using a modified Western blot procedure wherein radiolabeled RII was used as a probe (Carr and Scott, 1992) indicate the presence in the embryonic palate of three distinct RII-binding proteins (AKAPs), two of which exhibit distinct temporal expression profiles (Fig. 2). A standard Western blot (Fig. 3) confirmed the presence of both AKAP 95 and AKAP 100. AKAP-100 is a cytoplasmic, organelle-binding anchoring protein with wide tissue distribution (McCart-

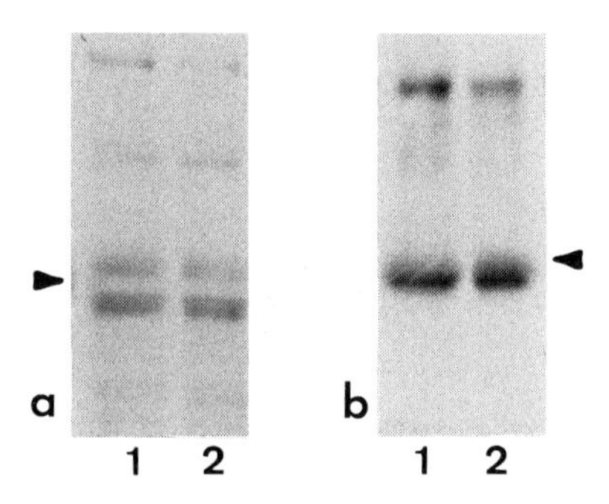

FIGURE 3 Western blot demonstrating the presence of AKAP 95 in nuclear extracts (Fig. 3a, arrow) and AKAP 100 in cytosolic extracts (Fig. 3b, arrow) derived from embryonic palatal tissue on day 13 (lane 1) and day 14 (lane 2) of gestation.

ney *et al.,* 1995), and AKAP 95 contains a DNA-binding domain and is localized to the nucleus (Coghlan *et al.,* 1994), thus providing the first evidence of nuclear anchoring sites for RII. AKAP95 could thus play a role in targeting type II PKA for cAMP-responsive nuclear events. Immunostaining of embryonic palatal tissue confirmed cytoplasmic distribution of AKAP100 and nuclear localization of AKAP 95 (Fig. 4).

D. Transcription Factor–Mediated Regulation of Gene Expression

Cyclic AMP-directed transcriptional regulation is mediated by a family of transcription factors—*trans*-acting DNA-binding proteins—that recognize a conserved *cis*-acting cAMP-response element (CRE) identified in the promoter of several cAMP-regulated genes (Gonzalez *et al.,* 1991). In mammalian cells, including those of the embryonic palate (Weston and Greene, 1995), the linkage between cAMP, PKA, and activation of specific genes is mediated via the phosphorylation of one or more members of this family of nuclear proteins. PKA-mediated phosphorylation of the nuclear CRE-binding protein (CREB) at serine 133 is required for transcriptional activation (Gonzalez and Montminy, 1989). Indeed, a CREB has been identified and characterized in embryonic palatal tissue and it is regulated at the level of ser 133 phosphorylation during development of the palate (Fig. 5) (Weston and Greene, 1995).

The ability to repress the cAMP response pathway appears to be an important mechanism of negative control of gene expression. A murine gene encoding a protein identified as cAMP-responsive element modulator (CREM), highly homologous to CREB and exhibiting cell- and tissue-specific expression, has been characterized (Foulkes *et al.,* 1991). Like CREB, this protein is transcriptionally activated by PKA-mediated phosphorylation (Brindle *et al.,* 1993; deGroot *et al.,* 1993; Masquilier *et al.,* 1994) and exhibits the same specificity of binding to CRE sequences as CREB. Human homologues have also recently been identified (Fujimoto *et al.,* 1994). CREM mRNA—like CREB—is alternatively spliced to produce functionally different proteins (Laoide *et al.,* 1993). The translational products of the CREM gene are unusual in that, due to alternative splicing, they consist of activators or inhibitors of CRE-mediated transcription. CREMα, β, and γ isoforms act as down-regulators of cAMP-induced transcription, presumably via competitive binding to CRE, thus suppressing CREB-stimulated transcription (Foulkes *et al.,* 1991) or by complexing with CREB into a transcriptionally inactive heterodimer unable to recognize and bind to CREs (Laoide *et al.,* 1993). Such a mechanism has also been suggested for another specific repressor of CRE-dependent transcription—the cloned CRE-binding protein CREB-2 (Karpinski *et al.,* 1992). Immunostaining reveals the presence of CREB-2 in embryonic palatal tissue *in vivo* and *in vitro* (Fig. 6). Thus, the potential for transcriptional competition between CREB, CREM,

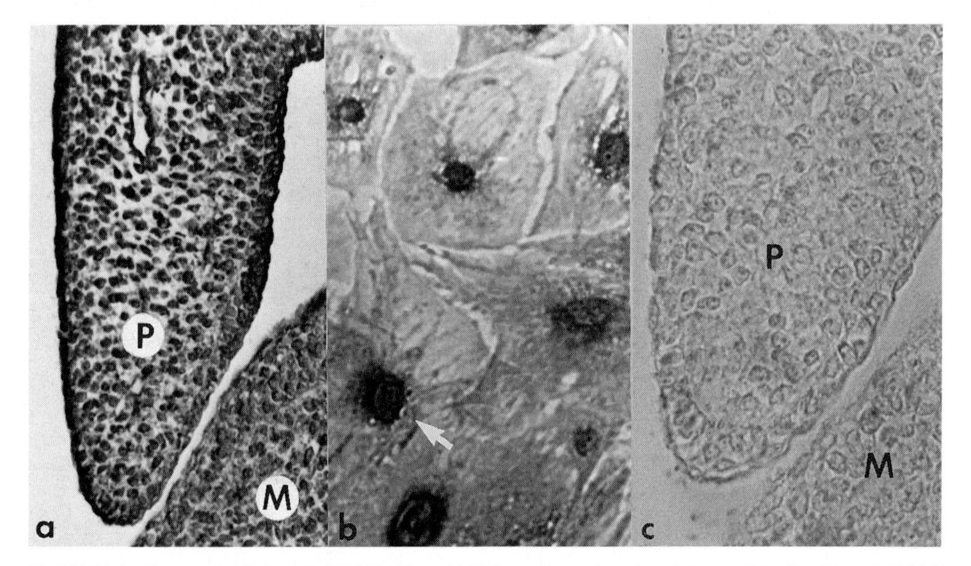

FIGURE 4 Immunostaining of AKAPs 95 and 100 in embryonic palatal tissue. Distribution of AKAP 95 in gestational day 13 murine embryonic palate *in vivo* (A). Palatal tissue (P) vertically oriented in the oral cavity. Maxilla (M). Immunostaining of AKAP 95 in MEPM cells *in vitro* (B). Note the nuclear distribution of this anchoring protein (arrow). (C) A more diffuse cytoplasmic distribution of AKAP 100 in gestational day 13 murine embryonic palate. Palatal tissue (P) vertically oriented in the oral cavity. Maxilla (M).

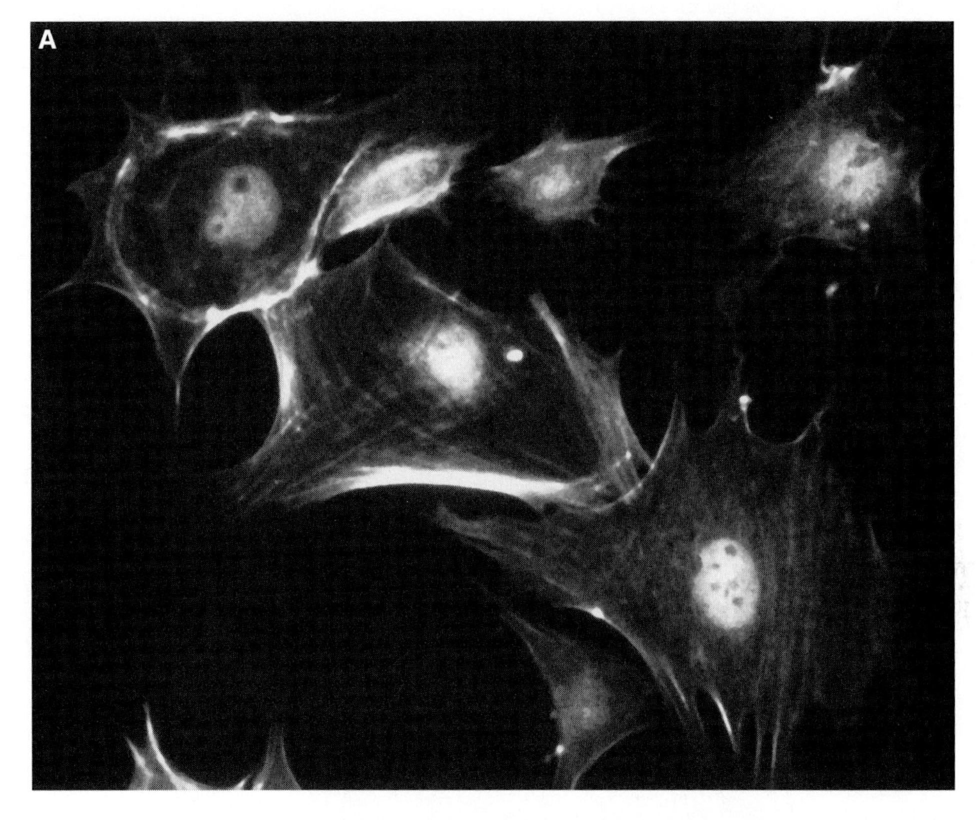

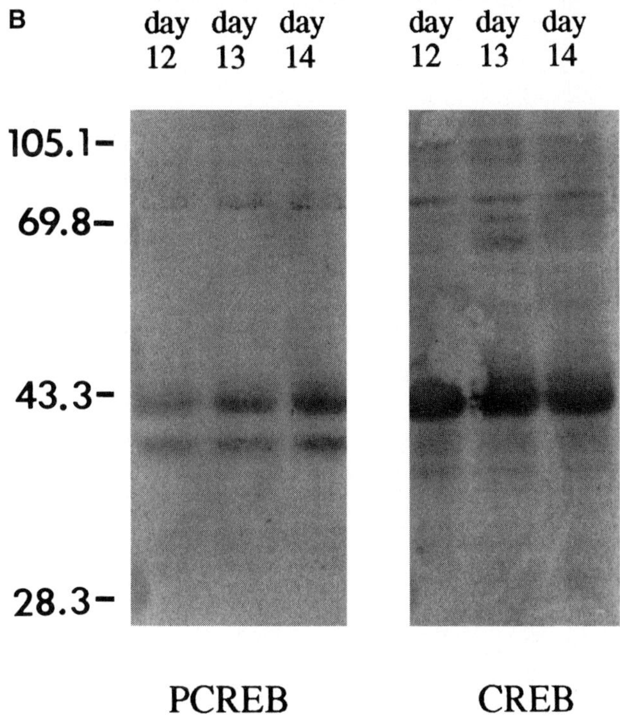

FIGURE 5 Immunostained nuclear localization of CREB in MEPM cells (A). Rhodamine–phalloidin was used to highlight cell membranes and cytoskeletal elements. (B) Western blot of CREB and ser 133 phospho–CREB expression in embryonic palatal tissue on gestational days 12–14. Staining of phospho–CREB demonstrates temporal alterations with maximal CREB ser 133 phosphorylation–and presumably CREB-mediated transcriptional activity–on day 14 of gestation. (*From Weston and Greene* (*1995*) *J. Cellular Physiol.* ***164:*** *277–285*)

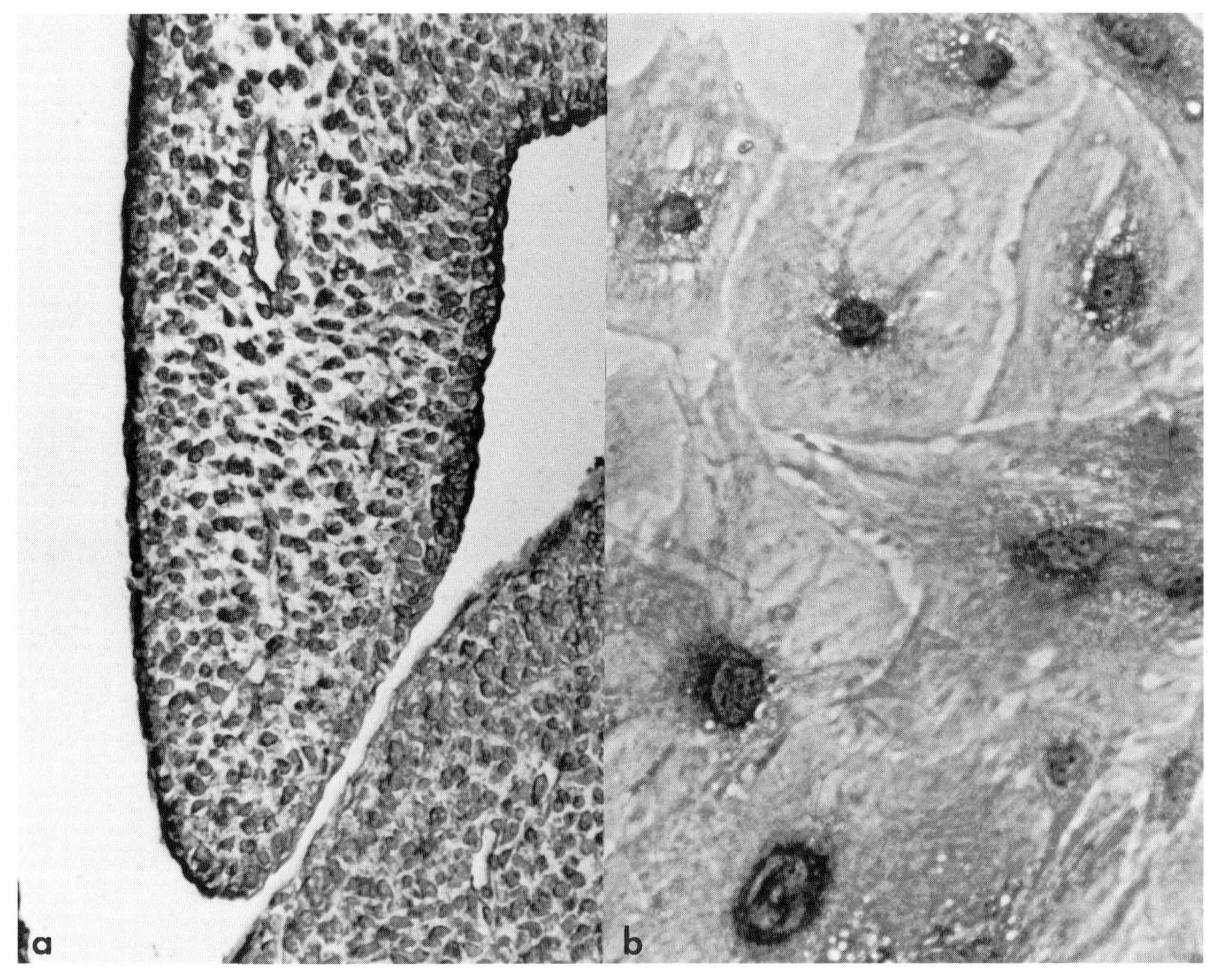

FIGURE 6 Immunohistochemical localization of the CREB-2 transcription factor in murine embryonic secondary palatal tissue (A) and murine embryonic palate mesenchymal cells in culture (B).

and CREB-2 makes CREM and CREB-2 important new members in this family of transcription factors.

III. Protein Kinase C

A. Activation and Crosstalk

Receptor-mediated hydrolysis of inositol phospholipids is a common means for the transduction of extracellular signals into the cell. Protein kinase Cs (PKCs) represent a family of phospholipid and Ca^{2+}-dependent ser/threo kinases, each of which contains a regulatory (lipid binding) and catalytic (phosphotransferase) domain. Endogenous phosphoinositide (PI) or phosphatidylcholine (PC) hydrolysis results in the intracellular release of two second messengers, inositol triphosphate (IP_3) and diacylglycerol (DAG). Inositol triphosphate mediates Ca^{2+} mobilization from intracellular stores (Berridge and Irvine, 1984), whereas DAG activates PKC (Nishizuka, 1984). Both have a transient existence and are rapidly degraded, but not before eliciting dramatic effects on cellular metabolism. Cellular responses may be immediate, such as secretion of cellular constituents, or long-term, such as gene expression and cell proliferation. The multiplicity of these responses has been reviewed by Nishizuka (1986).

Hydrolysis of inositol phospholipids may also be a critical metabolic process in assuring proper growth and differentiation of palatal tissue. Support for such a hypothesis comes from evidence that phosphatidylinositol metabolism in embryonic palate mesenchymal cells can be modulated by corticoids (Grove *et al.,* 1986) and that some prostaglandins, important factors in palatal tissue differentiation also stimulate phosphatidylinositol turnover (Macphee *et al.,* 1984). Epidermal growth factor

(EGF), which is capable of modulating various aspects of palate development including cell proliferation and extracellular matrix synthesis, has also been shown to modulate PI turnover in embryonic palate mesenchymal cells (Chepenik and Haystead, 1989).

A major function of PKC appears to be down-regulation of a number of cell surface receptors. Indeed, some of the transmembrane signaling pathways thought to be operative during palate development may be regulated by PKC because, at least in other systems, both the EGF receptor (Moon *et al.*, 1984) and the β-adrenergic receptor (Sibley *et al.*, 1984) are down-regulated by this enzyme. Whether these developmentally important receptors are modulated by PKC in embryonic palatal tissue is an intriguing possibility that requires investigation.

PKC consists of a large family of proteins representing at least 12 isozymes representing individual gene products. The distribution of several PKC protein isoforms in embryonic palatal tissue exhibits distinct spatiotemporal patterns of expression (Fig. 7) as does expression of their mRNAs (Fig. 8). Isozymes may possess different biologic functions and the type of PKC isozyme that is activated may confer specificity to the signal transduction response (Pears, 1995). Although PKC and cAMP-dependent protein kinase transmit information along different intracellular signal pathways, similar cellular responses are often evoked (Berridge, 1987). Moreover, these two enzymes often phosphorylate the same proteins (Kishimoto *et al.*, 1985). Whether these two pathways, both operative in cells of the embryonic palate, cooperate positively to amplify cellular responses is an intriguing possibility.

In some systems PKC activation results in inhibition of receptor- or forskolin-mediated cAMP formation (Gusovsky and Gutkind, 1991). For example, the β_2-adrenoreceptor, linked to cAMP-regulated proliferation of embryonic palate mesenchymal cells (Garbarino and Greene, 1984), can be phosphorylated on its cytoplasmic domain by PKC (Sibly *et al.*, 1984) resulting in uncoupling of G_s from the β_2-adrenoreceptor (Garte and Belman, 1980) and subsequent desensitization of β-adrenergic stimulation of adenylate cyclase (Nambi *et al.*, 1985). In addition, adenylate cyclase activity is often augmented subsequent to phosphorylation by PKC (Aasheim *et al.*, 1989). Evidence indicates that whether the cAMP generating system is inhibited or augmented may depend on which PKC isozymes the particular cell type possesses (Gusovsky and Gutkind, 1991). PKC can thus either down- or up-regulate the cAMP pathway at the level of the plasma membrane.

B. Early Response Genes

Activation of PKCs mediates the induction of a set of nuclear "early response" genes (Herschman, 1991). These genes, named due to the rapidity with which they are activated and their independence from any intervening protein synthesis (Cantley *et al.*, 1991), are typified by members of the fos, myc, and jun proto-oncogene families (Adamson, 1987). They act as general transcription factors and thus regulate the expression of selected "secondary response" target genes (Herschman, 1991).

The evolutionarily well-conserved nature of the proto-oncogenes initially suggested that their gene products might play a critical role in governing the growth and differentiation of normal cells. This notion has been strengthened by the identification of the proto-oncogene protein products as growth factors, growth factor receptors, GTP-binding proteins, protein kinases, and DNA-binding proteins that serve as cellular transcription factors (Schönthal, 1990; Cantley *et al.*, 1991), all components of signal transduction pathways.

The PKA and PKC signal transduction pathways thus represent integrated signaling avenues that provide multiple levels of regulation and remarkable sensitivity in responding to developmental effectors in embryonic palatal tissue. It has become increasingly apparent that "cross talk" between these pathways may be more the rule than the exception, providing the cell with greater sensitivity in discerning the multiple extracellular signals impinging on it and providing a means for amplification of short-term signals. Such interactions render cells more responsive to endogenous modulators of proliferation and differentiation and afford exquisite control of gene expression.

IV. Transforming Growth Factor-β

A. TGF-β and Tissue Differentiation

The TGF-βs represents a family of 25kD, structurally related, dimeric polypeptide growth factors whose multifunctional nature has shown them to have profound effects on growth, differentiation, and embryogenesis (Massagúe, 1990; Attisano *et al.*, 1994). The TGF-β proteins are synthesized and secreted as high molecular weight complexes that are unable to bind TGF-β receptors, and are therefore biologically inactive or "latent" (Flaumenhaft *et al.*, 1993). Latent TGF-β (LTGF-β) is composed of three subunits: the mature, biologically active TGF-β, a propeptide or latency-associated peptide (β-LAP), and the latent–TGF-β binding protein (LTBP) (Miazono and Heldin, 1991). Embryonic palate cells (Nugent *et al.*, 1997) and embryonic cranial neural crest cells, from which MEPM cells are derived (Brauer and Yee, 1993), release primarily LTGF-β *in vitro*.

Active TGF-β and the β-LAP are transcribed from the same gene and synthesized as a 390-amino acid

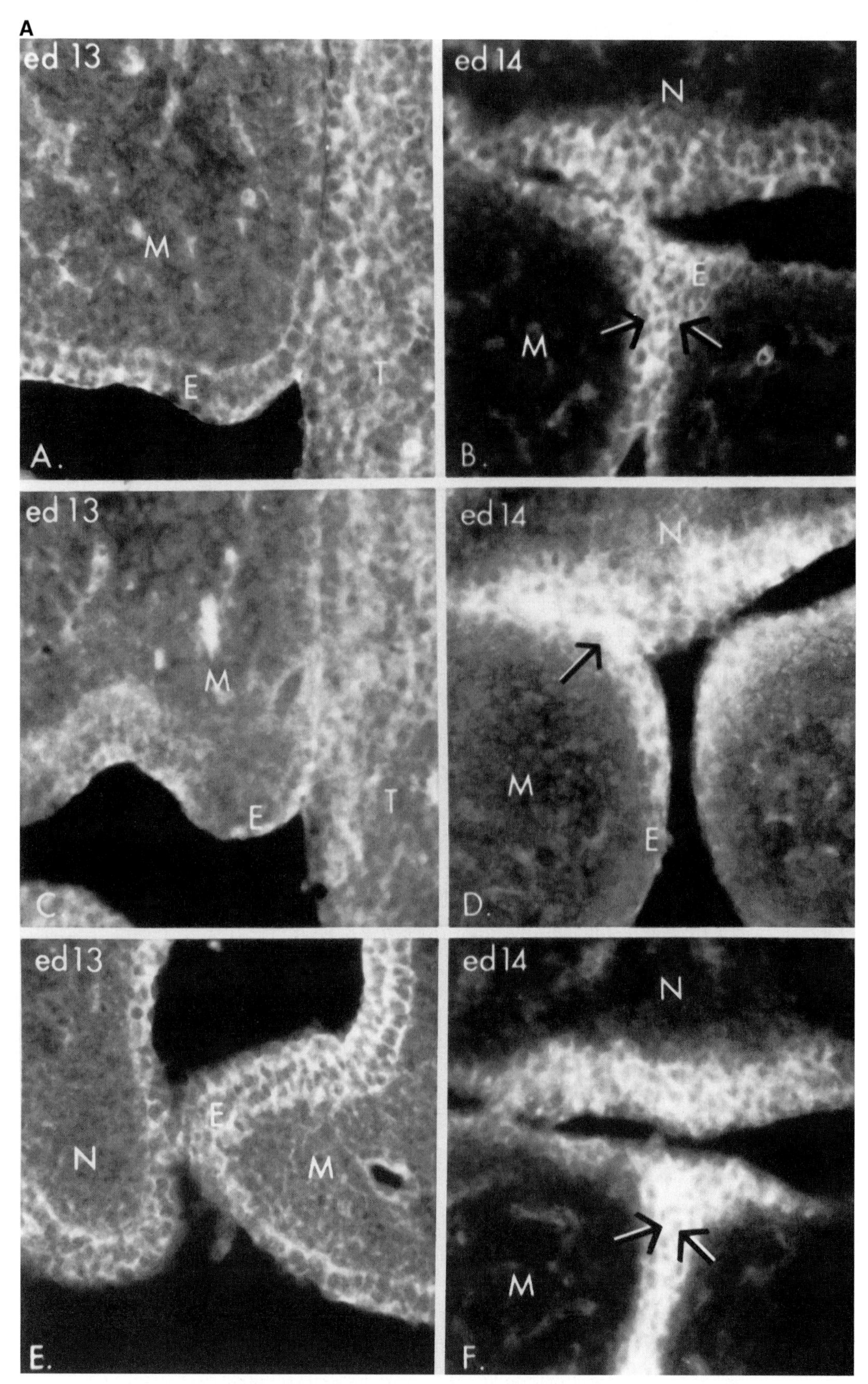

FIGURE 7 Immunohistochemical distribution of protein kinase C (PKC) in murine embryonic palatal tissue. Coronal sections of murine embryonic orofacial tissue on embryonic days (ed) 13 and 14 of gestation immunostained for PKC isozymes α (A & B), β (C & D), and γ (E & F), δ (G & H), ε (I & J) and ζ (K & L). (M) Palate mesenchyme; (E) palate epithelium; (T) tongue; (N) nasal septum. Control immunostaining in the absence of PKC antibodies (M). (*Adapted from Pisano et al., (1998) In Vitro and Devel Biol.; submitted for publication*).

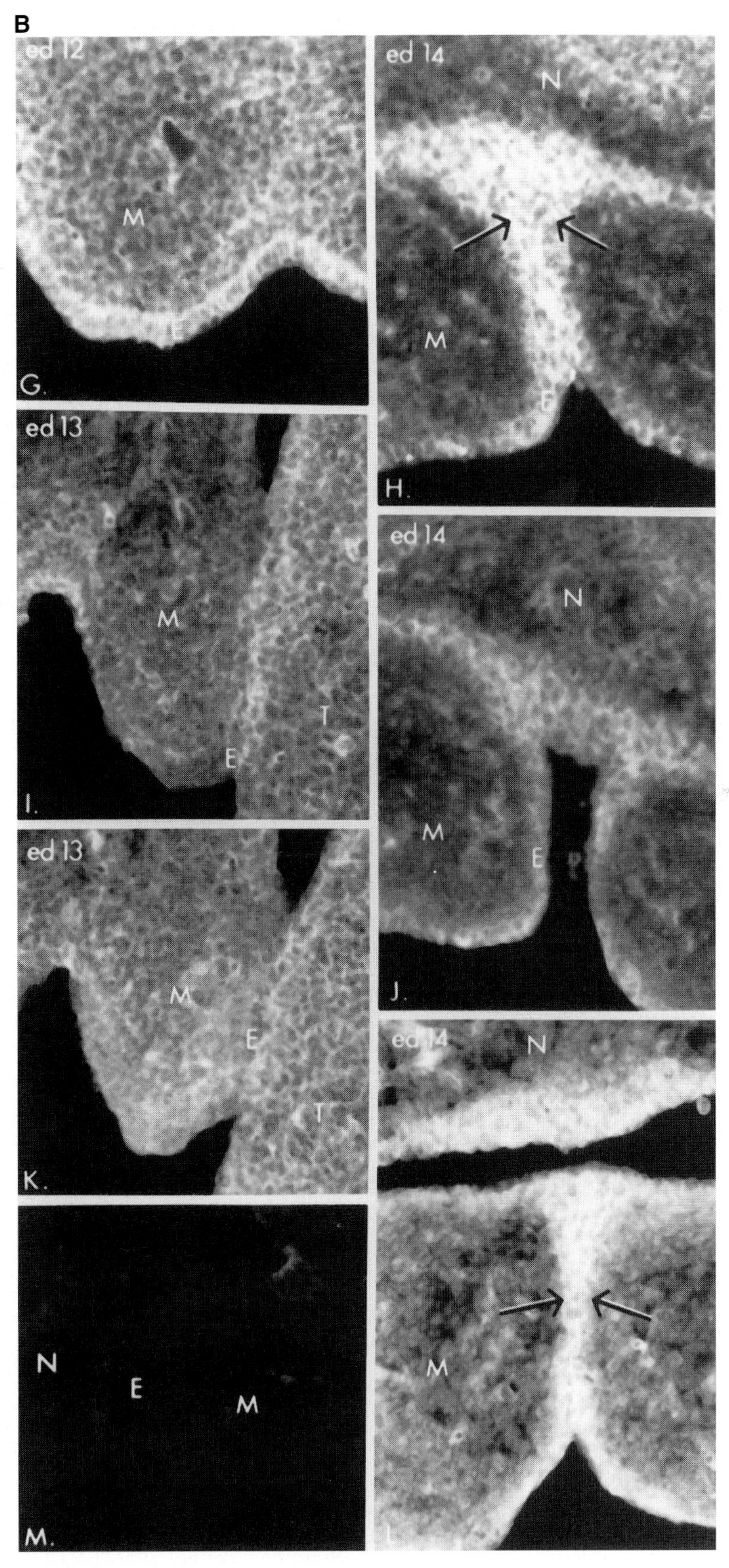

FIGURE 7 (Continued)

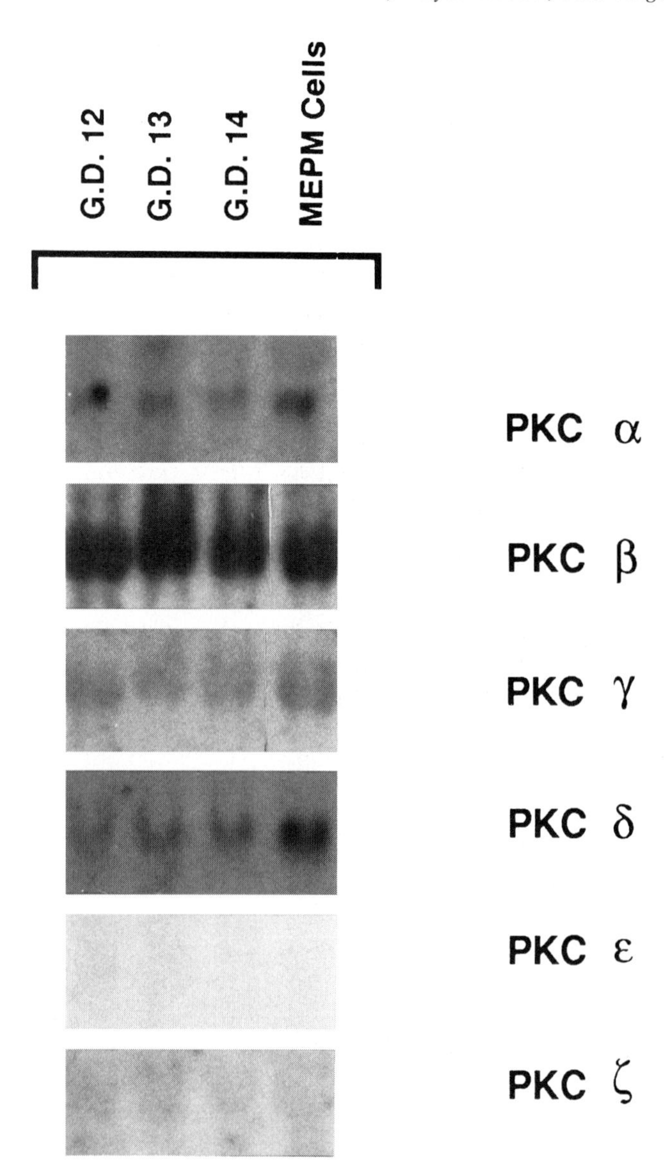

FIGURE 8 Northern blots representing steady state levels of PKC isoform mRNAs extracted from murine embryonic palate mesenchymal (MEPM) cells *in vitro* and from murine embryonic palatal tissue on gestational days (GD) 12, 13, and 14 . Transcript sizes: PKCα, 8.5 Kb; PKCβ, 9.5 Kb; PKCγ, 3.4 Kb; PKCδ, 3.1 Kb PKCε, 7.4 Kb; PKCζ, 2.3 Kb. Note that although these genes do not appear to exhibit temporal changes in expression during development of the palate, they are themselves differentially expressed with PKCβ mRNA being in far greater abundance than any other isoform, and PKCε being least abundant. In addition, steady state levels of PKCα, β, γ, and δ isoforms appear in greater abundance in MEPM cells than *in vivo*. (*Adapted from Pisano et al.,* (*1998*) *In Vitro and Devel Biol.; submitted for publication*).

preproprotein which is cleaved intracellularly (Gentry *et al.,* 1988). Active TGF-β1 is a homodimer of 112 amino acids derived from the carboxy terminus of the preproprotein. Three different isoforms (TGF-β1, -β2, and -β3) that exert quantitatively and qualitatively different effects have been identified in mammalian cells. The expression of these members of the TGF-β family of genes in embryonic craniofacial tissue is strictly regulated both spatially and temporally (Heine *et al.,* 1987; Fitzpatrick *et al.,* 1990; Pelton *et al.,* 1990; Gehris *et al.,* 1991, 1994), suggesting distinct functional roles for these molecules in orofacial development. Moreover, in the developing palate, the expression of the TGF-βs may be modulated via both autoregulatory pathways, other growth factor signaling peptides or the cAMP cascade (Gehris *et al.,* 1994). In addition, glucocorticoids modulate TGF-β signaling pathways by downregulating TGF-β expression and altering the availability of mature TGF-β necessary to exert its biological effects in the developing palate (Potchinsky *et al.,* 1996).

TGF-β has been shown to play a central role in tissue differentiation in the embryonic palate. A role for TGF-β in embryonic palate epithelial differentiation is supported by (1) localization of TGF-β isoforms (Gehris *et al.,* 1991) and their mRNA transcripts (Pelton *et al.,* 1990) to palatal epithelium during the gestational period of epithelial differentiation; (2) the ability of exogenously administered TGF-β to precociously induce (Gehris and Greene, 1992) and TGF-β3 antisense oligonucleotides to prevent (Brunet *et al.,* 1993) medial edge epithelial differentiation; and (3) the failure of palate epithelial fusion resulting in cleft palate in TGF-β3 null mutant mice (Kaartinen *et al.,* 1995; Proetzel *et al.,* 1995). In addition, exogenously added TGF-β1 and -β2 affect both the cell cycle time of MEPM cells (Linask *et al.,* 1991) and their synthesis of ECM components (D'Angelo and Greene, 1991; D'Angelo *et al.,* 1994). Thus, in addition to a clearly established role for TGF-β in palatal epithelial differentiation, palate mesenchymal cells also represent an important biologic target tissue for TGF-β.

Linkage between the cAMP cascade and TGF-β signaling has been demonstrated in embryonic palatal tissue. The activation of signal transduction pathways—such as the cAMP cascade—by developmental stimuli results in alterations in specific gene expression (discussed earlier). The demonstration of a cAMP-responsive element in the promoter of both the TGF-β2 and TGF-β3 genes (Lafyatis *et al.,* 1990) adds support to the notion that cAMP may mediate some of its effects via activation of TGF-β. Indeed, elevation of intracellular levels of cAMP in MEPM cells *in vitro* resulted in a transient increase in TGF-β3 gene expression (Gehris *et al.,* 1994). Recently it was demonstrated that selective inhibition of cyclic AMP–dependent protein kinase (PKA) prevented terminal differentiation of palatal medial edge epithelial (MEE) cells and exogenous TGF-β was able to abrogate this inhibition of MEE differentiation (Fig. 9) (Greene *et al.,* 1997). These data indicate that the cAMP/PKA signal transduction pathway plays a critical role in modulating embryonic palate MEE differentiation and that the TGF-β and cAMP signal transduction pathways may converge distal to PKA to regulate this differentiation.

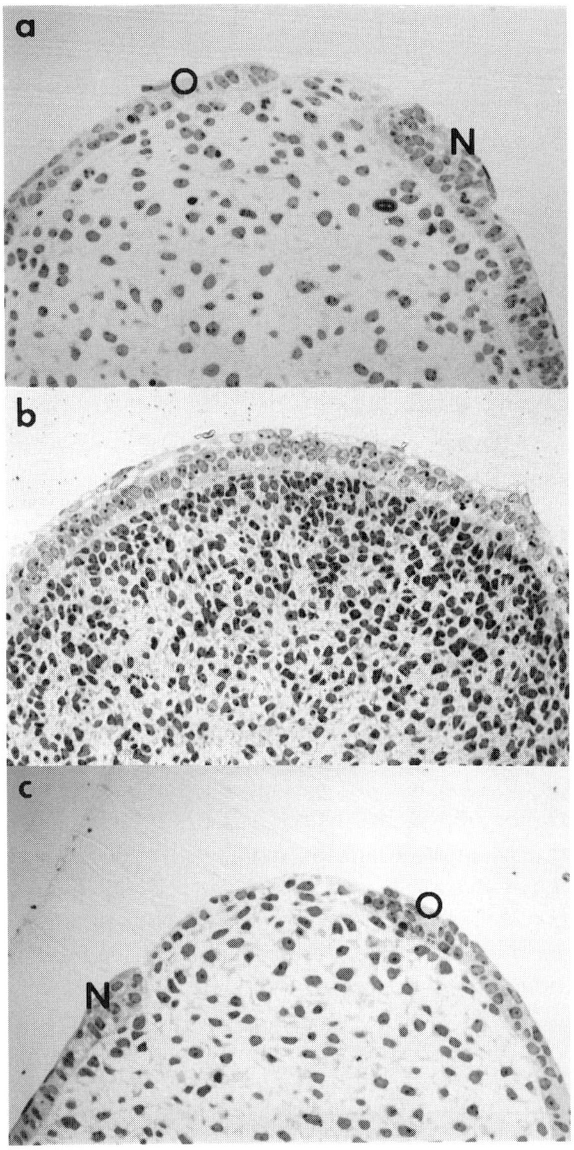

FIGURE 9 Histologic sections of the medial edge epithelial (MEE) region of gestational day 12 murine embryonic palates organ cultured on the surface of hydrated collagen gels for 48 hr. (A) Control conditions in which medial edge epithelial cells have transdifferentiated into mesenchymal cells and a denuded area is seen between the remaining oral (O) and nasal (N) epithelium. Incubation in the presence of the PKA specific inhibitor Rp-cAMP (10^{-5}M) (B) prevented terminal differentiation of palatal MEE cells. These data demonstrate that the cAMP/PKA signal transduction pathway plays a critical role in regulating MEE differentiation. Simultaneous treatment with either TGF-β1, TGF-β2, or TGF-β3 (10 ng/ml) was able to abrogate the Rp-cAMP–induced inhibition of MEE differentiation (C). (*Adapted from Greene et al., (1998) Functional analysis of protein kinase A in embryonic palatal tissue. In Vitro and Develop. Biol.; submitted for publication*).

B. TGF-β and Regulation of Proliferation

Cell proliferation is regulated largely by a balance between the activities of growth-promoting proto-oncogenes and growth-constraining tumor suppresser genes (TSGs) (Cross and Dexter, 1991; Baserga *et al.,* 1993). One of the most thoroughly characterized tumor suppressers is the retinoblastoma (Rb) gene product (Levine, 1993), which is present in MEPM cells (Fig. 10). Rb proteins play a crucial role in restraining not only neoplasia but also in regulating non-neoplastic cell proliferation through control of cell cycle traverse (Hollingsworth *et al.,* 1993). The Rb family of tumor suppressers consists of three distinct gene products (pRb, p130(Rb2), p107) (Whyte, 1995). The ability of the Rb proteins to bind transcription factors (such as the E2Fs) defines the means by which they regulate gene expression. Rb-E2F interactions are regulated by cyclin dependent kinase (cdk)-mediated phosphorylation (Akiyama *et al.,* 1992). Evidence indicates that the *hypo*-phosphorylated forms of Rb and Rb2 bind and sequester E2F transcription factors and prevent interaction with promoters of genes necessary for cell cycle progression (Claudio *et al.,* 1996).

The TGF-βs arrest the growth of cells in the G1-phase of the cell cycle and prevent progression of cells into S-phase DNA synthesis. The growth-suppressive capabilities of the TGF-βs appear to be mediated by alterations in the steady-state levels of Rb mRNA and/or protein (Polyak, 1996) and/or cdk-mediated Rb protein phosphorylation (Landesman *et al.,* 1992; Belbrahem *et al.,* 1996). Indeed, TGF-β treatment of MEPM cells results in an increase in the *hypo*-phosphorylated, growth suppressive form of Rb (Fig. 11). The discovery of novel cdk inhibitory proteins (CDKIs), which bind to and inactivate the kinases (Harper and Elledge, 1996), has added an additional level of regulation to cell growth control. The expression of some of the CDKIs are upregulated by TGF-β, thus inactivating cdks and leading to cell growth arrest (Li *et al.,* 1995).

Signaling pathways used by EGF, cAMP, and TGF-β have all been demonstrated to have the capability of regulating the proliferative capacity of embryonic palatal tissue. Precise control of proliferation may in fact be exerted through crosstalk among these signaling pathways. Both TGF-β and cAMP inhibited the proliferative response of cells to treatment with EGF, whereas H89, a serine/threonine protein kinase inhibitor with selectivity towards cAMP-dependent protein kinase, increased the cells' proliferative response to EGF

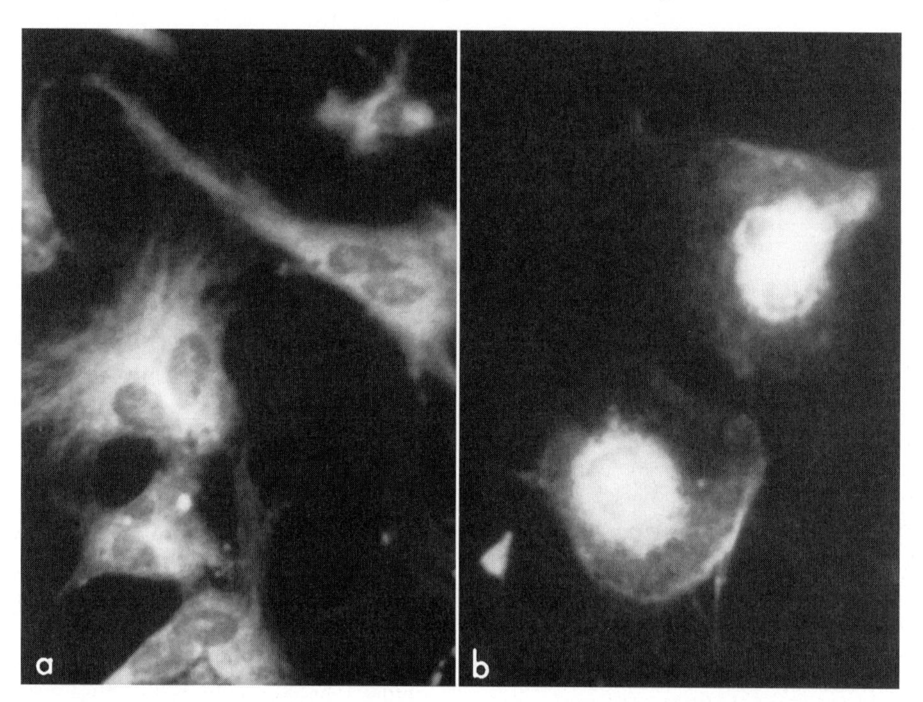

FIGURE 10 Immunolocalization of Rb protein in primary cultures of MEPM cells. Examination of pRb expression in exponentially growing cells (A) reveals intense perinuclear and cytoplasmic immunostaining while immunostaining in proliferatively quiescent cells is predominantly nuclear (B). (*Adapted from Pisano et al., Modifications of cell cycle controlling nuclear proteins by TGFβ in embryonic palatal cells. In preparation*)

(Weston *et al.,* 1997). These data suggest that both the TGF-β– and cAMP-mediated signaling pathways may be involved in modulation of the effects of EGF on palate cell growth.

C. Crosstalk between the TGF-β and the Retinoid Signaling Pathways

Retinoic acid (RA) is a potent cleft palate–inducing teratogen in some strains of mice and has been implicated in human teratogenesis. The effects of RA on gene expression follow mainly from its translocation to the nucleus and activation of specific elements within the promoter/enhancer of its target genes (Giguere, 1994). Both translocation and promoter activation are mediated by two classes of protein that specifically bind RA: the nuclear retinoic acid receptors (RARs) and retinoid X receptors (RXRs), and the cytoplasmic cellular retinoic acid–binding proteins (CRABPs).

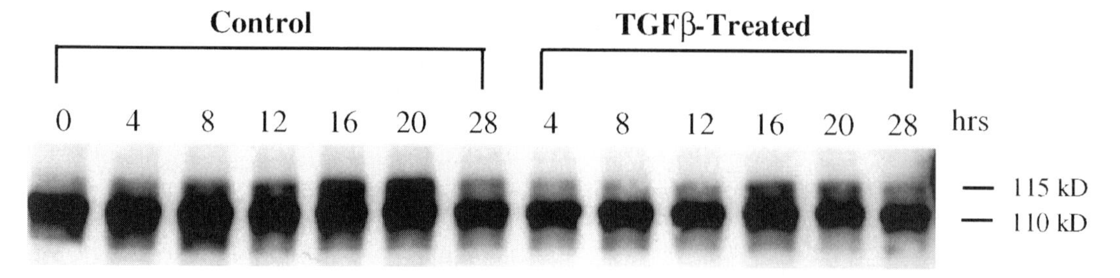

FIGURE 11 Western blot of Rb expression in synchronized and released MEPM cells cultured for 28 hr in the presence or absence of 1ng/ml TGF-β1. Maximal S-phase DNA synthesis in these cells occurs at approximately 16–20 hr and TGF-β arrests cell growth (by approximately 50%) in G1. Note that cells exhibit various phosphoforms of the Rb protein that migrate between 110 and 115 kD. In synchronized, proliferatively quiescent cells (0 hr) the smaller, faster migrating/*hypo*phosphorylated forms of Rb predominate. As cells progress into S-phase (4–20 hr) increasing amounts of the larger, slower migrating/*hyper*phosphorylated forms become more evident. Thus, under conditions of growth inhibition (0 hr control; TGFβ-treatment) Rb is in the *hypo*phosphorylated, growth inhibitory form. (*Adapted from Pisano et al., Modifications of cell cycle controlling nuclear proteins by TGFβ in embryonic palatal cells. In preparation*)

Reciprocal interactions between growth and differentiation factors provide a complexity of signaling pathways necessary to orchestrate the ordered manifestation of embryonic tissue development. Interactions between the retinoic acid and TGF-β signal transduction pathways have been demonstrated in several systems, including the developing palate (Sporn and Roberts, 1991; Roberts *et al.,* 1992; Nugent and Greene, 1994). Treatment of pregnant mice with RA during early embryonic development elicits changes in TGF-β protein expression in the embryo (Mamhood *et al.,* 1992). Cells derived from the mammalian developing palate express CRABP-I and CRABP-II, both of which may be regulated by RA and TGF-β (Nugent and Greene, 1994). In addition, TGF-β, including the endogenous form(s), can modulate the expression of the nuclear retinoic acid receptor–beta (RAR-β) (Nugent *et al.,* 1995). Conversely, TGF-β isoform expression in embryonic palatal tissue may be regulated by RA (Nugent and Greene, 1994; Nugent *et al.,* 1997). These results provide evidence for complex interactions between TGF-β and RA in the regulation of gene expression in embryonic palatal cells and suggest a role for endogenous TGF-β in the regulation of expression of gene encoding elements of the RA signal transduction pathway.

D. A Mad Way for TGF-β to Signal

Recently, a family of very highly conserved proteins that are essential components of TGF-β signaling pathways has been identified (Derynck and Zhang, 1996; Massaqué, 1996; Massaqué *et al.,* 1997). The first family member to be cloned was a *Drosophila* gene called *Mothers against dpp* (Sekelsky *et al.,* 1995), hence the acronym "Mad." Parallel genetic studies of *Caenorhabditis elegans* development identified structurally related proteins referred to as "Sma" (Savage *et al.,* 1996). Several homologs have been identified in *C. elegans, Xenopus,* mouse, and human (Derynck and Zhang, 1996). Five vertebrate Mad genes have been identified in different laboratories resulting in a bewildering diversity of names. Recent unification of the nomenclature refers to these latter genes and their products as *Smad1* through *Smad5* (Derynck *et al.,* 1996), a merger of "Sma" and "Mad." The murine homolog, called *Mad-related* (Madr) (Baker and Harland, 1996) is now referred to as *Smad2.*

Whereas Smad1 is thought to mediate BMP signaling (Hoodless *et al.,* 1996), Smad2 mediates signaling in response to TGF-β (Macias-Silva *et al.,* 1996). In response to members of the TGF-β family of ligands, Smad2 associates with, and is phosphorylated by, the type I TGF-β receptor (Macias-Silva *et al.,* 1996) and accumulates in the nucleus (Hoodless *et al.,* 1996), where it displays transcriptional activity (Liu *et al.,* 1996). Smad2 is regulated specifically by TGF-β (Eppert *et al.,* 1996) and transmits signals directly from the receptor to the nucleus.

The demonstration of the ability of *Xenopus* XMAD2 to combine with a specific transcription factor to generate a site-specific regulatory element binding complex (Chen *et al.,* 1996) has defined a potential mechanism by which TGF-β signaling may regulate transcriptional responses. TGF-β treatment of palate mesenchymal cells increases phosphorylation of the CREB transcription factor on a residue (ser133) known to induce transcriptional activation (Potchinsky *et al.,* 1997). It is thus reasonable to speculate that Smad2 mediates the transduction of a TGF-β signal via interaction with specific nuclear transcription factors.

E. Does TGF-β Signal via Mitogen Activated Protein Kinases (MAPK)?

The myriad effects of TGF-β support the notion that signal transduction from plasma membrane to the nucleus involves multiple pathways. In multicellular organisms, individual cells respond to signals in their environment by relaying information from the cell surface to the nucleus via signal transduction pathways that usually involve cascades of phosphorylation/dephosphorylation reactions.

Among the candidates for intracellular molecules that may act downstream of TGF-β receptors are members of the mitogen-activated protein kinase (MAPK) family (Yamaguchi *et al.,* 1995). The MAPKs comprise a family of highly conserved eukaryotic enzymes that act as key signal transduction vehicles. MAPKs are activated via upstream cascades of kinases, which, in turn, can be triggered by a wide diversity of signaling molecules including protein tyrosine kinase receptors (Hill and Treisman, 1995). Once activated, MAPKs translocate to the nucleus and effect downstream ser/threo phosphorylation of regulatory molecules such as transcription factors and components of the cell-cycle machinery thereby effecting changes in gene expression and proliferative activity (Marshall, 1994).

Two related mammalian MAPKs, $p44^{mapk}$ and $p42^{mapk}$, also called *extracellular signal-regulated kinase 1* (ERK1) and ERK2, have been cloned and found to be ubiquitously expressed and activated by many growth factors and cytokines (Boulton *et al.,* 1991). These are phosphorylated (activated) by MAP kinase/ERK kinase (MEK) (Gomez and Cohen, 1991), which is in turn phosphorylated (activated) primarily by the ser/threo protein kinase Raf-1 (Howe *et al.,* 1992) after having been recruited to the plasma membrane by Ras (Fabian *et al.,* 1994). The bucket brigade of protein kinases upstream of MAPK is thus represented by: Ras—Raf-1—MEK—MAPK. At least three distinct MAPK cas-

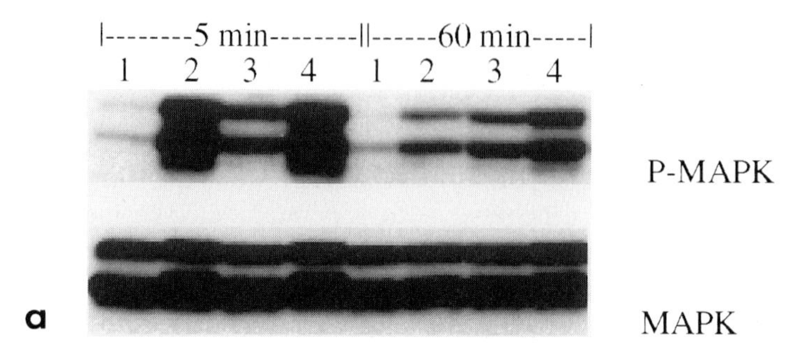

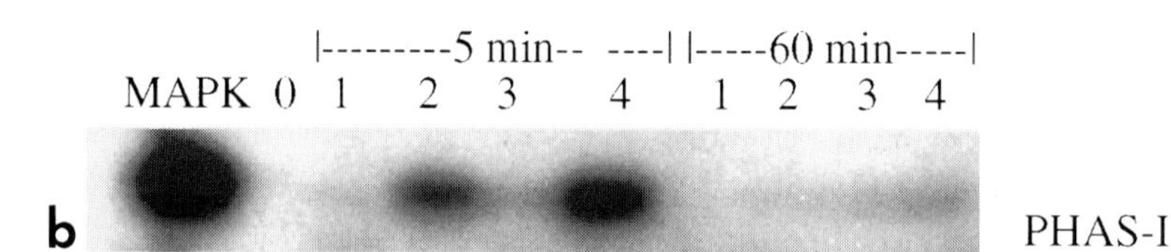

FIGURE 12 Effects of EGF and phorbol myristate acetate (PMA) on induction of p42/p44 MAPK tyrosine phosphorylation (A) and MAPK activity (B). MEPM cells were treated with vehicle control (Lane 1), 20 ng/ml EGF (Lane 2), 1 μM PMA (Lane 3), or EGF and PMA (Lane 4) for the indicated times. (A) Protein was extracted, separated by 10% SDS–PAGE and blots probed with antiphosphoMAPK or anti-erk-1, which recognizes p42/44 MAPK. Top: Tyrosine phosphorylation of p42/44 MAPK. Bottom: total MAPK. (B) Protein was extracted and MAPK activity was determined by measuring incorporation of γ^{32}ATP into PHAS-I (phosphorylated heat- and acid-stable protein regulated by insulin), a highly specific substrate for MAPK. Radioactive bands represent ^{32}P incorporated by cellular MAPKs into p42/44 MAPK-specific 21 kD PHAS-I substrate. The lane labeled "MAPK" represents the activity of purified p42/44 MAPK; the lane labeled "0" represents incorporation in the absence of protein. Note that both EGF and PMA are effective in stimulating MAPK activity. (*Adapted from Potchinsky, Lloyd, Weston, and Greene. (1998). Selective modulation of MAP kinase in embryonic palate cells. I. Cellular Physiol.; in press.*)

cades, functioning as separate modules, are known to exist in mammalian cells. The terminal MAPK components are grouped as ERKs (extracellular signal related kinases), SAPK/JNKs (stress activated protein kinase/JUN N-terminal kinase) and p38MAPK. Whereas JNKs and P38 are activated by many of the same stimuli, they are activated by distinct MAPK kinases.

Several lines of evidence support the notion that TGF-β signaling may be mediated in part through the MAPK cascade: TGF-β has been shown to rapidly activate Ras (Hartsough *et al.,* 1996) and p44^{mapk} (Hartsough and Mulder, 1995); TGF-β activates SAPK/JNK (Atfi *et al.,* 1997); TGF-β inhibits bFGF-induced MAPK activity (Berrou *et al.,* 1996); a recently identified kinase, termed *TAK1* for TGF-β activated kinase-1 (Yamaguchi *et al.,* 1995), has a kinase domain with extensive sequence identity with c-Raf and MEKK, which act as MAPK kinase kinases in the Ras/ERK and SAPK/JNK MAPK pathways, respectively. In palatal tissue, TGF-β–mediated signaling can modulate the effects of EGF on palate cell growth (Weston *et al.,* 1997). The MAPK cascade is a signal transduction pathway that mediates cellular responsiveness to EGF. Hence, it is likely that a MAPK lies downstream of TGF-β receptors in much the same way that the RAS/ERK MAPK pathway lies downstream of tyrosine kinase receptors such as EGF. In embryonic palatal tissue, MAPK tyrosine-185 phosphorylation and activity may be stimulated via multiple pathways (Fig. 12) (Potchinsky *et al.,* 1997). Moreover, other signaling pathways known to play significant roles in differentiation of palatal tissue converge with the MAPK cascade and may use this pathway in the regulation of alternative cellular processes.

V. Concluding Remarks

Science is an inferential exercise, not a simple formulary of facts. The wealth of studies on palatal ontogeny have allowed inferences that in turn have provided illumination for the enlargement of our conceptual understanding of orofacial ontogeny. The developing mammalian palate offers a valuable paradigm in that examinations of cellular mechanisms involved in signal transduction have proved instructive with regard to craniofacial development, and have also offered insight into the fundamental basis of ontogeny.

Complex interactions between different signal transduction pathways provide sensitive mechanisms by which embryonic cells may respond to multiple and subtle developmental cues. Our scientific understanding relating to signal transduction in the embryonic palate has been presented as a construct of what we currently understand to be the facts. Molecular analysis of gene function in the embryo is currently providing greater depth of understanding of mechanisms underlying embryonic development. Such approaches, using the developing craniofacial region as a model system, promise the exploration of exciting new territories of thought.

References

Aasheim, L., Kleine, L., and Franks, D. (1989). Activation of protein kinase C sensitizes the cyclic AMP signalling system of T51B rat liver cells. *Cell. Signal.* **1,** 617–625.

Abel, E., and Sokol, R., (1986). Maternal and fetal characteristics affecting alcohol's teratogenicity. *Neurobehav. Toxicol. Teratol.* **8,** 329–334.

Adams, S., Harootunian, A., Beuchler, Y., Taylor, S., and Tsien, R. (1991). Fluorescence ratio imaging of cAMP in single cells. *Nature* **349,** 694–697.

Adamson, E. (1987) Oncogenes in development. *Development* **99,** 449–471.

Akiyama, T., Ohuchi, T., Sumida, S., Matsumoto, K., and Toyoshima, K. (1992). Phosphorylation of the retinoblastoma protein by cdk2. *Proc. Natl. Acad. Sci. USA* **89,** 7900–7904.

Alam, I., Capitanio, A., Smith, J., Chepenik, K., and Greene, R. (1982). Radioimmunologic identification of prostaglandins produced by serum-stimulated mouse embryo palate mesenchyme cells. *Biochim. Biophys. Acta.* **712,** 408–411.

Atfi, A., Djelloul, S., and Chastre, E. (1997). Evidence for a role of rho-like GTPases and stress-activated protein kinase/c-Jun N-terminal kinase (SAPK/JNK) in transforming growth factorβ-mediated signaling. *J. Biol. Chem.* **272,** 1429–1432.

Attisano, L., Wrana, J., Lopez-Casillas, F., and Massague, J. (1994). TGF-beta receptors and actions. *Biochim. Biophys. Acta* **1222,** 71–80.

Baker, J., and Harland, R. (1996). A novel mesoderm inducer, Madr2, functions in the activin signal transduction pathway. *Genes Dev.* **10,** 1880–1889.

Barsony, J., and Marks, S. (1991). Immunocytology on microwave-fixed cells reveals rapid and agonist-specific changes in subcellular accumulation patterns for cAMP or cGMP. *Proc. Natl. Acad. Sci. USA* **87,** 1188–1192.

Baserga, R., Porcu, P., and Sell, C. (1993). Oncogenes, growth factors and control of the cell cycle. *Cancer Surv.* **16,** 201–213.

Beebe, S., Øyen, O., Sandberg, M., Frøysa, A., Hansson, V., and Jahnsen, T. (1990). Molecular cloning of a tissue-specific protein kinase (Cγ) from human testis representing a third isoform for the catalytic subunit of cAMP-dependent protein kinase. *Molec. Endoc.* **4,** 465–475.

Belbrahem, A., Godden-Kent, D., and Mittnacht, P. (1996). Regulation and activity of the retinoblastoma protein family in growth factor–deprived and TGFβ-treated keratinocytes. *Exper. Cell Res.* **225,** 286–293.

Berridge, M., (1987). Inositol triphosphate and diacylglycerol: two interacting second messengers. *Ann. Rev. Biochem.* **56,** 159–193.

Berridge, M., and Irvine, R., (1984). Inositol triphosphate, a novel second messenger in cellular signal transduction. *Nature* **312,** 315–321.

Berrou, E., Fontenay, M., Quarck, R. *et al.* (1996). Transforming growth factor β1 inhibits mitogen-activated protein kinase induced by basic fibroblast growth factor in smooth muscle cells. *Biochem J.* **316,** 167–173.

Blakely, P. (1988). Experimental teratology of ethanol. In *Issues and Reviews in Teratology.* H. Kalter, Ed., Vol. 4, Plenum Publ. Co. pp. 237–241, New York.

Boulton, T., Nye, S., Robbins, D., *et al.* (1991). ERK's: a family of protein-serine/threonine kinases that are activated and tyrosine phosphorylated in response to insulin and NGF. *Cell* **65,** 663–675.

Bouvier, M., Leeb-Lundberg, L., Benovic, J., Caron, M., and Lefkowitz, R. (1987). Regulation of adrenergic receeptor function by phosphorylation. *J. Biol. Chem.* **262,** 3106–3113.

Brauer, P., and Yee, J. (1993). Cranial neural crest cells synthesize and secrete a latent form of transforming growth factor β that can be activated by neural crest cell proteolysis. *Develop. Biol.* **155,** 281–285.

Brindle, P., Linke, S., and Montminy, M., (1993). Protein-kinase-A–dependent activator in transcription factor CREB reveals new role for CREM repressors. *Nature* **364,** 821–824.

Brunet, C., Sharpe, P., and Ferguson, M. (1993). Inhibition of TGFβ3 (but not TGFβ1 or TGFβ2) actively prevents normal mouse embryonic palate fusion. *Int. J. Develop. Biol.* **39,** 345–355.

Cantley, L., Auger, K., Carpenter, C., Duckworth, B., Graziani, A., Kapeller, R., and Soltoff S. (1991). Oncogenes and signal transduction. *Cell* **64,** 281–302.

Carr, D., and Scott, J. (1992). Blotting and band-shifting: techniques for studying protein–protein interactions. *TIBS* **17,** 246–149.

Charness, M., Querimit, L., and Henteleff, A. (1988). Ethanol differentially regulates G proteins in neural cells. *Biochem. Biophys. Res. Comm.* **155,** 138–143.

Chen, X., Rubock, M., and Whitman, M. (1996). A transcriptional partner for MAD proteins in TGFβ signalling. *Nature* **383,** 691–696.

Chepenik, K., and Greene, R., (1981). Prostaglandin synthesis by primary cultures of mouse embryo palate mesenchyme cells. *Biochem. Biophys. Res. Comm.* **100,** 951–958.

Chepenik, K., and Haystead, A. (1989). Epidermal growth factor alters metabolism of inositol lipids and activity of protein kinase C in mouse embryo palate mesenchyme cells. *J. Craniofacial Genet. Dev. Biol.* **9,** 285–301.

Chrivia, J., Uhler, M., and McKnight, G. (1988). Characterization of genomic clones coding for the Cα and Cβ subunits of mouse cAMP–dependent protein kinase. *J. Biol. Chem,* **263,** 5739–5744.

Claudio, P., DeLuca, A., Howard, C., Baldi, A., Firpo, E., Koff, A., Paggi, M., and Giordano, A. (1996). Functional analysis of pRb2/p130 interaction with cyclins. *Cancer Res.* **56,** 2003–2008.

Clegg, C., Cadd, G., and McKnight, G. (1988). Genetic characterization of a brain specific form of the type-I regulatory subunit of cAMP-dependent protein kinase. *Proc. Natl. Acad. Sci. USA* **85,** 3703–3707.

Codina, J., Hildebrandt, J., Sunyer, T., Sekura, R., Manclark, C., Iyengar, R., and Birnbaumer, L. (1984). Mechanisms in the vectorial receptor–adenylate cyclase signal transduction. *Adv. Cyc. Nucleo. Prot. Phos. Res.* **17,** 111–125.

Coghlan, V., Bergeson, S., Langeberg, L., Nilaver, G., and Scott, J. (1993). A-Kinase anchoring proteins: a key to selective activation of cAMP-response events? *Molec. Cell Biochem.* **127,** 309–319.

Coghlan, V., Langeberg, L., Fernandez, A., Lamb, N., and Scott, J. (1994). Cloning and characterization of AKAP95, a nuclear protein that associates with the regulatory subunit of type II cAMP-dependent protein kinase. *J. Biol. Chem.* **269,** 7658–7665.

Cross, M., and Dexter, T. (1991). Growth factors in development, transformation, and tumorigenesis. *Cell* **64,** 271–280.

D'Angelo, M., Chen, J.-M., Ugen, K., and Greene, R. (1994). TGFβ1 regulation of collagen metabolism by embryonic palate mesenchymal cells. *J. Exp. Zool.* **270,** 189–201.

D'Angelo, M., and Greene, R. (1991). Transforming growth factor-β modulation of glycosaminoglycan production by mesenchymal cells of the developing murine secondary palate. *Develop. Biol.* **145,** 374–378.

deGroot, R., Hertog, J., Vandenheede, J., Goris, J., and Sassone-Corsi, P. (1993). Multiple and cooperative phosphorylation events regulate the CREM activator function. *EMBO J.* **12,** 3903–3911.

Derynck, R., Gelbart, W., Harland, R., *et al.* (1996). Nomenclature: vertebrate mediators of the TGFβ family signals. *Cell* **87,** 173.

Derynck, R., and Zhang, Y., (1996). Intracellular signalling: the MAD way to do it. *Curr. Biol.* **6,** 1226–1229.

Diewert, V. (1982). A comparative study of craniofacial growth during secondary palate development in four strains of mice. *J. Craniofacial Genet. Develop. Biol.* **2,** 247–263.

Eppert, K., Scherer, S., Ozcelik, H., *et al,* (1996). MADR2 maps to 18q21 and encodes a TGFβ MAD-related protein that is functionally mutated in colorectal carcinoma. *Cell* **86,** 543–552.

Erickson, R., Butley, M., and Sing, C., (1979). H-2 and non–H-2 determined strain variation in palatal shelf and tongue adenosine 3′:5′ cyclic monophosphate. *J. Immunogenet.* **6,** 253–262.

Fabian, J., Vojtek, A., Cooper, J., *et al.* (1994). A single amino acid change in Raf-1 inhibits Ras binding and alters Raf-1 function. *Proc. Natl. Acad. Sci. USA* **91,** 5982–5986.

Fitzpatrick, D., Denhez, F., Kondaiah, P., and Akhurst, R. (1990). Differential expression of TGF beta isoforms in murine palatogenesis. *Development* **109,** 585–595.

Flaumenhaft, R., Kojima, S., Abe, M., and Rifkin, D. (1993). Activation of latent transforming growth factor β. *Adv. Pharmacol.* **24,** 51–76.

Foulkes, N., Borrelli, E., and Sassone-Corsi, P. (1991). CREM gene: use of alternative DNA-binding domains generates multiple antagonists of cAMP-induced transcription. *Cell* **64,** 739–749.

Fujimoto, T., Fujisawa, J., and Yoshida, M. (1994). Novel isoforms of human cyclic AMP-response element modulator (hCREM) mRNA. *J. Biochem.* **115,** 298–303.

Garbarino, M., and Greene, R. (1984). Identification of adenylate cyclase–coupled β adrenergic receptors in the developing mammalian palate. *Biochem. Biophys. Res. Comm.* **119,** 193–202.

Garte, S., and Belman, S. (1980). Tumour promoter uncouples β-adrenergic receptor from adenylate cyclase in mouse epidermis. *Nature* **284,** 171–173.

Gehris, A., D'Angelo, M., and Greene, R. (1991). Immunodetection of the transforming growth factor-β1 and β2 in the developing palate. *Int. J. Develop. Biol.* **35,** 17–24.

Gehris, A., and Greene, R. (1992). Regulation of murine embryonic epithelial cell differentiation by transforming growth factors β. *Differentiation* **49,** 167–173.

Gehris, A., Pisano, M., Nugent, P., and Greene, R. (1994). Regulation of TGFβ gene expression in embryonic palatal tissue. *In Vitro Develop. Biol.* **30A,** 671–679.

Gentry, L., Lioubin, M., Purchio, A., and Marquardi, H. (1988). Molecular events in the processing of recombinant type I pre–pro-transforming growth factor beta to the mature polypeptide. *Molec. Cell. Biol.* **8,** 4162–4168.

George, M., and Chepenik, K. (1985). Phospholipase A activities in embryonic palate mesenchyme cells *in vitro. Biochim. Biophys. Acta* **836,** 45–55.

Giguere, V. (1994). Retinoic acid receptors and cellular retinoid binding proteins: complex interplay in retinoid signaling. *Endo. Rev.* **15,** 61–79.

Gomez, N., and Cohen, P. (1991). Dissection of the protin kinase cascade by which nerve growth factor activates MAP kinase. *Nature* **353,** 170–173.

Gonzalez, G., Menzel, P., Leonard, J., Fischer, W., and Montminy, M. (1991). Characterization of motifs which are critical for activity of the cyclic AMP-responsive transcription factor CREB. *Molec. Cell. Biol.* **11,** 1306–1312.

Gonzalez, G., and Montminy, M. (1989). Cyclic AMP stimulates somatostatin gene transcription by phosphorylation of CREB at serine 133. *Cell* **59,** 675–680.

Greene, R., Lloyd, M., and Nicolau, K. (1981). Agonist-specific desensitization of prostaglandin stimulated cAMP accumulation in palate mesenchymal cells. *J. Craniofacial Gen. Devel. Biol.* **1,** 261–272.

Greene, R., Lloyd, M., and Pisano, M. (1998). Functional analysis of cAMP-dependent protein kinase in embryonic palatal tissue. *In Vitro Develop. Biol.* (submitted for publication).

Greene, R., Lloyd, M., Uberti, M., Nugent, P., and Pisano, M. (1995). Patterns of cyclic AMP–dependent protein kinase gene expression in developing orofacial tissue. *J. Cell. Physiol.* **163,** 431–440.

Greene, R., and Pratt, R. (1979). Correlation between cAMP levels and cytochemical localization of adenylate cyclase during development of the secondary palate. *J. Histochem. Cytochem.* **27,** 924–931.

Greene, R., Shanfeld, J., Davidovitch, Z., and Pratt, R. (1980). Immunohistochemical localization of cyclic AMP in the developing rodent secondary palate. *J. Embryol. Exp. Morph.* **60,** 271–281.

Grove, R., Willis, W., and Pratt, R. (1986). Studies on phosphatidylinositol metabolism and dexamethasone inhibition of proliferation of human palatal mesenchyme cells. *J. Craniofacial Gen. Devel. Biol.* **2,** 285–292.

Gusovsky, F., and Gutkind, J. (1991). Selective effects of activation of protein kinase C isozymes on cyclic AMP accumulation. *Molec. Pharm.* **39,**124–129.

Harper, K., Burns, R., and Erickson, R. (1981). Genetic aspects of the effects of methylmercury in mice: the incidence of cleft palate and concentrations of adenosine 3′:5′ cyclic monophosphate in tongue and palatal shelf. *Teratology* **23,** 397–401.

Harper, J., and Elledge, S. (1996). Cdk inhibitors in development and cancer. *Curr. Op. Gen. Dev.* **6,** 56–64.

Hartsough, M., Frey, R., Zipfel, P., Buard, A., Cook, S., McCormick, F., and Mulder, K. (1996). Altered transforming growth factor signaling in epithelial cells when ras activation is blocked. *J. Biol. Chem.* **271,** 22368–22375.

Hartsough, M., and Mulder, K. (1995). Transforming growth factor beta activation of p44mapk in proliferating cultures of epithelial cells. *J. Biol. Chem.* **270,** 7117–7124.

Hedin, L., McKnight, G., Lifka, J., Durika, J., and Richards, J. (1987). Tissue distribution and hormonal regulation of messenger ribonucleic acid for regulatory and catalytic subunits of adenosine 3′,5′-monophosphate-dependent protein kinases during ovarian follicular development and luteinization in the rat. *Endocrinology* **120,** 1928–1935.

Heine, U., Munoz, E., Flanders, K., Ellingsworth, L., Lain, H., Thompson, N., Roberts, A., and Sporn, M. (1987). Role of transforming growth factor–beta in the developing mouse embryo. *J. Cell Biol.* **105,** 2861–2876.

Herschman, H. (1991). Primary response genes induced by growth factors and tumor promoters. *Ann. Rev. Biochem.* **60,** 281–319.

Hill, C., and Treisman, R. (1995). Transcriptional regulation by extracellular signals: mechanisms and specificity. *Cell* **80,** 199–211.

Hirsch, A., Glantz, S., Li, Y., You, Y., and Rubin, C. (1992). Cloning and expression of an intron-less gene for AKAP 75, an anchor protein for the regulatory subunit of cAMP-dependent protein kinase IIb. *J. Biol. Chem.* **267,** 2131–2134.

Hoffman, P., and Tabakoff, B. (1990). Ethanol and guanine nucleotide binding proteins: a selective interaction. *FASEB J.* **4,** 2612–2622.

Hollingsworth, R., Hensey, C., and Lee, W-H. (1993). Retinoblastoma protein and the cell cycle. *Curr. Op. Gen. Dev.* **3,** 55–62.

Hoodless, P., Haerry, T., Abdollah, S., *et al.* (1996). MADR1, a MAD-related protein that functions in BMP2 signalling pathways. *Cell* **85,** 489–500.

Howe, L., Leevers, S., Gomez, N., *et al.* (1992). Activation of the MAP kinase pathway by the protein kinase raf. *Cell* **71,** 335–342.

Jahnsen, T., Hedin, L., Kidd, V., Lohmann, S., Walter, U., Durica, J., Schultz, T., Schlitz, E., Browner, M., Goldman, D., Ratoosh, S., and Richards, J. (1986). Molecular cloning, cDNA structure and regulation of the regulatory subunit (RII_{51}) of type II cAMP dependent protein kinase from rat granulosa cells. *J. Biol. Chem.* **261,** 12352–12361.

Johnston, L., and Nash, D. (1992). Sagittal growth trends of the development of cleft palate in mice homozygous for the "paddle" gene. *J. Craniof. Gen. Dev. Biol.* **2,** 265–275.

Jones, J., and Greene, R. (1986). Identification of prostaglandin E_2 receptor sites in the embryonic murine palate. *Prostag. Leuko. Med.* **22,** 139–151.

Kaartinen, V., Voncken, J., Shuler, C., Warburton, D., Heisterkamp, N., and Groffen, J. (1995). Abnormal lung development and cleft palate in mice lacking TGFβ-3 indicates defects of epithelial–mesenchymal interaction. *Nat. Gen.* **11,** 415–421.

Karpinski, B., Morle, G., Huggenvik, J., Uhler, M., and Leiden, J. (1992). Molecular cloning of human CREB-2: an ATF/CREB transcription factor that can negatively regulate transcription from the cAMP response element. *Proc. Natl. Acad. Sci. U.S.A.* **89,** 4820–4824.

Keryer, G., Rios, R., Landmark, B., Shalhegg, B., Lohmann, S., and Bornens, M. (1993). A high affinity binding protein for the regulatory subunit of the cAMP-dependent protein kinase II in the centrosome of human cells. *Exp. Cell Res.* **204,** 230–240.

Kishimoto, A., Nishiyama, K., Nakanishi, H., Uratsuji, Y., Nomura, H., *et al.* (1985). Studies on the phosphorylation of myelin basic protein by protein kinase C and adenosine-3′,5′-monophosphate-dependent protein kinase. *J. Biol. Chem.* **260,** 12492–12499.

Kochhar, D. (1968). Studies on vitamin A–induced teratogenesis: effects on embryonic mesenchyme and epithelium, and on incorporation of ^{3}H-thymidine. *Teratology* **1,** 299–310.

Kuo, J., and Greengard, P. (1969). Cyclic nucleotide dependent protein kinases. IV. Widespread occurence of adenosine 3′:5′-monophosphate-dependent protein kinase in various tissues and phyla of the animal kingdom. *Proc. Natl. Acad. Sci. USA* **64,** 1349–1355.

Lafyatis, R., Lechleider, R., Seong-Jin, K., *et al.* (1990). Structural and functional characterization of the transforming growth factor β3 promotor. *J. Biol. Chem.* **265,** 19128–19136.

Landesman, Y., Pagano, M., Draetta, G., Rotter, V., Fusenig, N., and Kimchi, A. (1992). Modifications of cell cycle controlling nuclear proteins by transforming growth factor β in the HaCaT keratinocyte cell line. *Oncogene* **7,** 1661–1665.

Laoide, B., Foulkes, N., Schlotter, F., and Sassone-Corsi, P. (1993). The functional versatility of CREM is determined by its modular structure. *EMBO J.* **12,** 1179–1191.

Lee, D., Carmichael, D., Krebs, E., and McNight, G. (1983). Isolation of a cDNA clone for the type I regulatory subunit of bovine cAMP-dependent protein kinase. *Proc. Natl. Acad. Sci. USA* **80,** 3608–3612.

Levine, A. (1993). The tumor suppressor genes. *Annu. Rev. Biochem.* **62,** 623–651.

Li, J., Nichols, M., Chandrasekharan, S., and Xiong, Y. (1995). Transforming growth factor β activates the promoter of cyclin-dependent kinase inhibitor $p15^{ink4b}$ through an Sp1 consensus site. *J. Biol. Chem.* **270,** 26750–26753.

Linask, K., D'Angelo, M., Gehris, A., and Greene, R. (1991). Transforming growth factor-β receptor profiles of human and murine embryonic palate mesenchymal cells. *Exp. Cell Res.* **192,** 1–9.

Linask, K., and Greene, R. (1989a). Subcellular compartmentalization of cAMP-dependent protein kinase regulatory subunits during palatal ontogeny. *Life Sci.* **45,** 1863–1868.

Linask, K., and Greene, R. (1989b). Ontogenetic analysis of embryonic palatal type I and type II cAMP-dependent protein kinase isozymes. *Cell Diff. Dev.* **28,** 189–202.

Liu, F., Hata, A., Baker, J., *et al.* (1996). A human Mad protein acting as a BMP-regulated transcriptional activator. *Nature* **381,** 620–623.

Luo, Z., Shafit-Zagardo, B., and Erlichman, J. (1990). Identification of the MAP2- and P75-binding domain in the regulatory subunit of type II cAMP-dependent protein kinase. *J. Biol. Chem.* **265,** 21804–21810.

Macias-Silva, M., Abdollah, S., Hoodless, P., *et al.* (1996). MADR2 is a substrate of the TGFβ receptor and its phosphorylation is required for nuclear accumulation and signaling. *Cell* **87,** 1215–1224.

Macphee, C., Drummond, A., Otto, A., and DeAsua, L. (1984). Prostaglandin $F_{2\alpha}$ stimulates phosphatidylinositol turnover and increases the cellular content of 1,2-diacylglycerol in confluent resting Swiss 3T3 cells. *J. Cell. Physiol.* **119,** 35–40.

Mamhood, R., Flanders, K., and Morris-Kay, G. (1992). Interactions between retinoids and TGF βs in mouse morphogenesis. *Development* **115,** 67–74.

Marshall, C. (1994). MAP kinase kinase kinase, MAP kinase kinase and MAP kinase. *Curr. Op. Gen. Dev.* **4,** 82–89.

Masquilier, D., Foulkes, N., Mattei, M., and Sassone-Corsi, P. (1994). Human CREM gene: evolutionary conservation, chromosomal localization, and inducibility of the transcript. *Cell Growth Diff.* **4,** 931–937.

Massagué, J. (1990). The transforming growth factor-β family. *Ann. Rev. Cell Biol.* **6,** 597–641.

Massagué J. (1996). TGFβ signaling: receptors, transducers, and Mad proteins. *Cell* **85,** 947–950.

Massagué, J., Hata, A., and Liu, F. (1997). TGFβ signalling through the Smad pathway. *Trends Cell Biol.* **7,** 187–192.

McCartney, S., Little, B., Langeberg, L., and Scott, J. (1995). Cloning and characterization of A-kinase anchor protein 100 (AKAP100). *J. Biol. Chem.* **270,** 9327–9333

Minkoff, R. (1980). Regional variation of cell proliferation withing the facial processes of the chick embryo: a study of the role of 'merging' during development. *J. Embryol. Exp. Morph.* **57,** 37–49.

Miyazono, K., and Heldin, C.-H. (1991). Latent forms of TGFβ: molecular structure and mechanisms of activation. *Ciba Found. Sym.* **157,** 81–92.

Mochly-Rosen, D., Chang, F.-H., Cheevers, L., Kim, M., Diamond, I., and Gordon, A. (1988). Chronic ethanol causes heterologous desensitization of receptors by reducing a G_s messenger RNA. *Nature* **333,** 848–850.

Moon, S., Palfrey, H., and King, A. (1984). Phorbol esters potentiate tyrosine phosphorylation of epidermal growth factor receptors in A431 mmembranes by a calcium-independent mechanism. *Proc. Natl. Acad. Sci. USA* **81,** 2298–2302.

Nagy, L., Diamond, I., and Gordon, A. (1988). Cultured lymphocytes from alcoholic subjects have altered cAMP signal transduction. *Proc. Natl. Acad. Sci. USA* **85,** 6973–6976.

Nambi, P., Peters, J., Sibley, D., and Lefkowitz, R. (1985). Desensitization of the turkey erythrocyte S-adrenergic receptor in a cell-free system. *J. Biol. Chem.* **260,** 2165–2171.

Nanda, R., and Romeo, D. (1975). Differential cell proliferation of embryonic rat palatal processes as determined by incorporation of tritiated thymidine. *Cleft Pal. J.* **12,** 436–443.

Nigg, E., Schafer, G., Hilz, H., and Eppenberger, H. (1985). Cyclic AMP dependent protein kinase type II is associated with the golgi complex and with centrosomes. *Cell* **41,** 1039–1051.

Nishizuka, Y. (1984). Turnover of inositol phospholipids and signal transduction. *Science* **225,** 1365–1370.

Nishizuka, Y. (1986). Studies and perspectives of protein kinase C. *Science* **233,** 305–312.

Nugent, P., and Greene, R. (1994). Interactions between the TGFβ and retinoic acid signal transduction pathways in embryonic palatal cells. *Differentiation* **58,** 149–155.

Nugent, P., Lafferty, C., and Greene, R. (1998). Differential expression and biological activity of retinoic acid–induced TGFβ isoforms in embryonic palate mesenchymal cells. *J. Cell. Physiol.* (submitted for publication).

Nugent, P., Lafferty, C., Potchinsky, M., and Greene, R. (1995). TGFβ modulates the expression of retinoic acid–induced RARβ in primary cultures of embryonic palate cells. *Exp. Cell Res.* **220,** 495–500.

Olson, F., and Massaro. E. (1980). Developmental pattern of cAMP, adenyl cyclase, and cAMP phosphodiesterase in the palate, lung, and liver of the fetal mouse: alterations resulting from exposure to methylmercury at levels inhibiting palate closure. *Teratology* **22,** 155–166.

Palmer, G., Palmer, J., Waterman, R., and Palmer, S. (1980). *In vitro* activation of adenylate cyclase by norepinephrine, parathyroid hormone, calcitonin, and prostaglandins in the developing maxillary process and palatal shelf of the golden hamster. *Ped. Pharm.* **1,** 45–54.

Pears, C. (1995). Structure and function of the protein kinase C gene family. *J. Biosci.* **20,** 311–332.

Pelton, R., Hogan, B., Miller, D., and Moses, H. (1990). Differential expression of genes encoding TGFs β1, β2, and β3 during murine palate formation. *Dev. Biol.* **141,** 456–460.

Pisano, M., and Greene, R. (1984). Quantitation of catecholamines in embryonic oro-facial tissue. *IRCS Med. Sci.* **13,** 900–901.

Pisano, M., Schneiderman, M., and Greene, R. (1986). Catecholamine modulation of embryonic palate mesenchymal cell DNA synthesis. *J. Cell. Physiol.* **126,** 84–92.

Polyak, K. (1996). Negative regulation of cell growth by TGFβ. *Biochim. Biophys. Acta* **1242,** 185–199.

Potchinsky, M., Nugent, P., and Greene, R. (1996). Effects of dexamethasone on the expression of TGFβ in mouse embryonic palate mesenchymal cells. *J. Cell. Physiol.* **166,** 380–386.

Potchinsky, M., Weston, W., and Greene, R. (1998). Selective modulation of MAP kinase in embryonic palatal cells. *J. Cell. Physiol.* (in press).

Potchinsky, M.,Weston, W., Lloyd, M., and Greene, R. (1997). TGF-β signaling in murine embryonic palate cells involves activation of the CREB transcription factor. *Exp. Cell Res.* **231,** 96–103.

Prescott, P. (1982). The effects of ethanol on rat embryos developing *in vitro. Biochem. Pharmacol.* **31,** 3641–3646.

Proetzel, G., Pawlowski, S., Wiles, M., Yin, M., Boivin, G., Howles, P., Ding, J., Ferguson, M., and Doetschman, T. (1995). Transforming growth factor-β3 is required for secondary palate fusion. *Nature Gen.* **11,** 409–414.

Rios, R., Celati, C., Lohmann, S., Bornens, M., and Keryer, G. (1992). Identifiaction of a high affinity binding protein for the regulatory subunit of RIIβ of cAMP-dependent protein kinase in golgi enriched membranes of human lymphoblasts. *EMBO J.* **11,** 1723–1731.

Roberts, A., Glick, A., and Sporn, M. (1992). Interrelationships between two families of multifunctional effectors: retinoids and TGF-β. In *Retinoids in Normal Development and Teratogenesis.* G. Morris-Kay, Ed., Oxford University Press: Oxford, pp. 137–148.

Rubin, C., Rangal-Aldao, R., Sarkar, D., Erlichman, J., and Fleischer, N. (1979). Characterization and comparison of membrane-associated and cytosolic cAMP-dependent protein kinases. *J. Biol. Chem.* **254,** 3797–3805.

Salomon, D., and Pratt, R. (1978). Inhibition of growth *in vitro* by glucocorticoids in mouse embryonic facial mesenchyme cells. *J. Cell. Physiol.* **97,** 315–328.

Sandham, A. (1985). Embryonic facial vertical dimension and its relationship to palatal shelf elevation. *Early Hum. Dev.* **12,** 241–245.

Savage, C., Das, P., Finelli, A., Townsend, S., Sun, C., Baird, S., and Padgett, R. (1996). *Caenorhabdtis elegans* genes sma-2, sma-3 and sma-4 define a conserved family of transforming growth factor β pathway components. *Proc. Natl. Acad. Sci. USA* **93,** 790–794.

Schönthal, A. (1990). Nuclear protooncogene products: fine-tuned components of signal transduction pathways. *Cell. Signal* **2,** 215–225.

Scott, J., Glaccum, M., Zoller, M., Duhler, M., Helfman, D., McKnight, G., and Krebs, E. (1987). The molecular cloning of a type II regulatory subunit of the cAMP-dependent protein kinase from rat skeletal muscle and mouse brain. *Proc. Natl. Acad. Sci. USA* **84,** 5192–5196.

Sekelsky, J., Newfeld, S., Raftery, L., Chartoff, E., and Gelbart, W. (1995). Genetic characterization and cloning of Mother's against dpp, a gene required for decapentaplegic function in *Drosophila melanogaster. Genetics* **139,** 1347–1358.

Sibley, D., Nambi, P., Peters, J., and Lefkowitz, R. (1984). Phorbol diesters promote β-adrenergic receptor phosphorylation and adenylate cyclase desensitization in duck erythrocytes. *Biochem. Biophys. Res. Comm.* **121,** 973–979.

Smales, W., and Biddulph, D. (1985). Limb development in chick embryos: cyclic AMP-dependent protein kinase activity, cyclic AMP, and prostaglandin concentrations during cytodifferentiation and morphogenesis. *J. Cell. Physiol.* **122,** 259–265.

Sporn, M., and Roberts, A. (1991). Interactions of retinoids and transforming growth factor-β in regulation of cell differentiation and proliferation. *Mol. Endo.* **5,** 3–7.

Tassinari, M., Lorente, C., and Keith, D. (1981). Effect of prenatal phenytoin exposure on tissue protein and DNA levels in the rat. *J. Cranio. Gen. Dev. Biol.* **1,** 315–331.

Trenouth, M. (1984). Shape changes during human fetal craniofacial growth. *J. Anat.* **139,** 639–651.

Waterman, R., Palmer, G., Palmer, J., and Palmer, S. (1976). Catecholamine-sensitive adenylate cyclase in the developing golden hamster palate. *Anat. Rec.* **185,** 125–138.

Weston, W., and Greene, R. (1991). Effects of ethanol on cAMP production in murine embryonic palate mesenchymal cells. *Life Sci.* **49,** 489–494.

Weston, W., and Greene, R. (1995). Developmental changes in phosphorylation of the transcription factor CREB in the embryonic murine palate. *J. Cell. Physiol.* **164,** 277–285.

Weston, W., Greene, R., Uberti, M., and Pisano, M. (1994). Ethanol effects on embryonic craniofacial growth and development: implications for study of the Fetal Alcohol Syndrome. *Alcoholism: Clin. Exp. Res.* **18,** 177–182.

Weston, W., Potchinsky, M., Lafferty, C., Ma, L., and Greene, R. (1998). Crosstalk between signaling pathways in murine embryonic palate cells: effect of TGFβ and cAMP on EGF-induced DNA synthesis. *In Vitro Dev. Biol.* (in press).

Weston, W., Shah-Quazi, K., Lafferty, C., Pisano, M., and Greene, R. (1996). Ethanol does not affect expression of Gsα and Giα in murine embryonic palate mesenchymal cells. *Res. Comm. Alcohol Sub. Abuse* **17,** 207–218.

Whyte, P. (1995). The retinoblastoma protein and its relatives. *Semin. Can. Biol.* **6,** 83–90.

Yamaguchi, Y., Shirakabe, K., Shibava, H., *et al.* (1995). Identification of a member of the MAPKKK family as a potential mediator of TGFβ signal transduction. *Science* **270,** 2008–2011.

Zimmerman, E., Wee, E., Phillips, N., and Roberts, N. (1981). Presence of serotonin in the palate just prior to shelf elevation. *J. Embryol. Exp. Morph.* **64,** 233–250.

CHAPTER

6

Trophic Nerve Growth Factors

HAEYOUNG KONG
PATRIZIA CASACCIA-BONNEFIL
MOSES V. CHAO
Molecular Neurobiology Program
Skirball Institute
New York University Medical Center
New York, New York 10016

The formation of the vertebrate nervous system is characterized by considerable programmed cell death, which determines cell number and appropriate target innervation during development. Neurotrophic factors, such as nerve growth factor (NGF), support the survival of selective populations of neurons during different developmental time periods. It has been established that competition between neurons for limiting amounts of neurotrophin molecules produced by target cells accounts for cell survival. Two predictions emanate from this hypothesis. First, the efficacy of neurotrophic action depends on the concentration of trophic factors. Second, the binding and activation of receptors expressed in responsive cell populations dictate neuronal responsiveness.

The neurotrophins represent a family of survival and differentiation factors that exert profound effects in both the central nervous system (CNS) and the peripheral nervous system (PNS) (Barde, 1989). Five neurotrophins—NGF, brain derived neurotrophic factor (BDNF), neurotrophin-3 (NT-3), neurotrophin-4/5 (NT-4/5), and neurotrophin-6 (NT-6)—associate as noncovalent homodimers in their biologically active

*Abbreviations: BDNF, brain derived neutotrophic factor; CNS, central nervous system; CNTF, ciliary neurotrophic factor; FADD, fas-associated death domain protein; FGF, fibroblast growth factor; FLICE, FADD-like ICE; GDNF, glial derived neurotrophic factor; hCG, human choriaonic gonadotrophin; IGF-1, insulin-like growth factor; JNK, c-jun amino-terminal kinase; LIF, leukemia inhibitory factor; MACH, MORT-1–associated CED-3 homolog; NGF, nerve growth factor; NT-3, neurotrophin-3; NT-4/5, neurotrophin-4/5; NT-6, neurotrophin-6; O-2A, progenitor oligodendrocytes; p75, $p75^{NTR}$; PDGF, platelet-derived growth factor; PNS, peripheral nervous system; SHC, Src homology collagen; TGF-β, transforming growth factor-β; TNF, tumor necrosis factor; TRADD, TNF-R1–associated death domain protein; TRAF, tumor necrosis factor (TNF)-associated factor; TUNEL, terminal deoxynucleotidyl transferase-mediated biotinylated UTP nick end-labeling.

form (Bradshaw *et al.,* 1993; Gotz *et al.,* 1994). These target-derived trophic factors are active on distinct sets of embryonic neurons whose dependence is in many cases restricted in duration during development (Davies, 1994). The neurotrophins have been proposed as therapeutic agents for the treatment of neurodegenerative disorders and nerve injury, either individually or in combination with other trophic factors such as ciliary neurotrophic factor (CNTF) or insulin-like growth factor (IGF-1) or fibroblast growth factor (FGF) (Lindsay *et al.,* 1994; Nishi, 1994), although recent clinical efforts have met with disappointing results. Improvements in overcoming difficulties of delivery and pharmacokinetics in the CNS will provide more impetus for the application of neurotrophins for neurodegenerative diseases. Clearly, an understanding of the mechanisms of neurotrophin action will provide new insights into the use of these proteins therapeutically.

Responsiveness of neurons to a given neurotrophin is governed by the expression of two classes of cell surface receptors (Chao, 1992b). For NGF, these are $p75^{NTR}$ (p75) and TrkA receptor tyrosine kinase. Three vertebrate trk receptor genes have been isolated, including numerous variants of trk structure (Barbacid, 1994). The related TrkB receptor tyrosine kinase binds both BDNF and NT-4/5 (Klein *et al.,* 1992), the most closely related neurotrophins from phylogenetic analysis, whereas TrkC receptor binds only NT-3 (Lamballe *et al.,* 1991). NT-3 and NT-4/5 can also bind to TrkA receptors (Berkemeier *et al.,* 1991; Ip *et al.,* 1993). The p75 receptor binds to all neurotrophins with similar affinity. Hence, the neurotrophins can engage two separate receptor moieties, which can act independently or may interact with Trk receptors.

In this chapter, the functional properties of neurotrophins and their receptors are discussed, with special emphasis on mechanisms determined from receptor binding and signal transduction, gene targeting, and structure–function experiments. The mechanism of action of neurotrophins has continued to provide a challenging and formidible problem in signal transduction. In addition to promoting cell differentiation and survival, NGF can paradoxically be an inducer of cell death in rare instances. Several mechanisms are proposed to explain how NGF might act as a trophic factor and as a cell killer. The survival and cell death properties of the receptors are dependent on the relative ratio of receptors and the persistent nature of the signaling events.

I. Neurotrophin Structure

Neurotrophins are produced as precursor proteins, which are cleaved at dibasic amino acids to form a mature form of 118–120 amino acids (Angeletti and Bradshaw, 1971; Leibrock *et al.,* 1989; Hohn *et al.,* 1990; Maisonpierre *et al.,* 1990; Berkemeier *et al.,* 1991; Hallbook *et al.,* 1991; Gotz *et al.,* 1994). The X-ray crystal structure of NGF has been solved (McDonald *et al.,* 1991) and provides a structural framework for this family. Of the two dozen accumulated neurotrophin sequences, only 28 residues are invariant (Bradshaw *et al.,* 1994). The preservation of these features indicates that the neurotrophins adopt similar conformations to that of NGF.

The dimeric NGF possesses a novel tertiary fold that results in a flat asymmetric molecule with dimensions of 60 Å by 25 Å by 30 Å (McDonald *et al.,* 1991). Each NGF subunit is characterized by two pairs of antiparallel β strands that contribute to the molecule's flat, elongated shape. These β strands are connected at one end of the neurotrophin by three short loops. These loops are known to be highly flexible and represent the regions in the neurotrophin structure where many amino acid differences exist between the neurotrophins.

The three disulfide bridges in each neurotrophin are clustered at the opposite end of molecule and provide rigidity to the structure. The topologic arrangement of the disulfides is quite unusual. Two of the disulfide bridges and their connecting residues form a ring structure through which the third disulfide bridge passes to form a tightly packed cystine knot motif (McDonald and Hendrickson, 1993). This cystine knot allows the two pairs of β strands from each neurotrophin to pack against each other, generating an extensive subunit interface. The interface has a largely hydrophobic character composed primarily of aromatic residues, consistent with the tight association constant (10^{13} M) measured for NGF. The NGF tertiary fold and "cystine knot" motif have been identified in structures of transforming growth factor-β (TGF-β), platelet-derived growth factor (PDGF), and in human chorionic gonadotrophin (hCG) (McDonald and Hendrickson, 1993). Members of this diverse structural superfamily of ligands typically form homo- or heterodimeric species.

The structural features of the neurotrophin family, in particular the dimer interface, are highly conserved, as evidenced in the ability of these members to form heterodimers (Radziewjewski and Robinson, 1993; Arakawa *et al.,* 1994; Jungbluth *et al.,* 1994). These heterodimeric proteins give functional activity in many cases, indicating there is overall compatibility of these structures.

II. Neurotrophin Receptors

A schematic drawing representing the structural features of Trk tyrosine kinases and the p75 neurotrophin receptor is displayed in Fig. 1. The Trk subfamily of receptor tyrosine kinases is distinguished by immunoglobulin-C2 domains and repeats rich in leucine and cysteine residues in the extracellular domain, and a consensus tyrosine kinase domain with a small interruption and a short cytoplasmic tail. The p75 receptor contains four negatively charged cysteine-rich extracellular repeats, and a unique cytoplasmic domain that is highly conserved among species. There are no sequence similarities between the two receptors in either ligand binding or cytoplasmic domains.

After binding neurotrophins, the ligand-receptor complex is internalized and retrogradely transported in the axon to the cell body (Distefano *et al.*, 1992). Both Trk and p75 receptors play a role in the retrograde transport of neurotrophins. For example, in the isthmo-optic nucleus, BDNF is taken up at the axon terminal and transported to the cell body by both p75 and TrkB receptors (von Bartheld *et al.*, 1995).

Each receptor type undergoes ligand-induced dimerization (Grob *et al.*, 1985; Jing *et al.*, 1992) that activates multiple signal transduction pathways. Neurotrophin binding to Trk family members produces biologic responses through activation of the tyrosine kinase domain resulting in a rapid increase in the phosphorylation of selected effector enzymes, such as phospholipase C-γ and phosphotidylinositol 3′-kinase. Increased ras activity, a common signal from all tyrosine kinase receptors, results from the stimulation of guanine nucleotide exchange factors coupled to SHC adaptor proteins, which interact directly with Trk after ligand binding (Kaplan and Stephens, 1994). The p75 receptor signals via pathways involved with activation of sphingomyelinase activities (Dobrowsky *et al.*, 1994, 1995), NFκB (Carter *et al.*, 1996) and c-jun kinase (Casaccia-Bonnefil *et al.*, 1996) and modulates trk activity (Barker and Shooter, 1994; Hantzopoulos *et al.*, 1994; Verdi *et al.*, 1994).

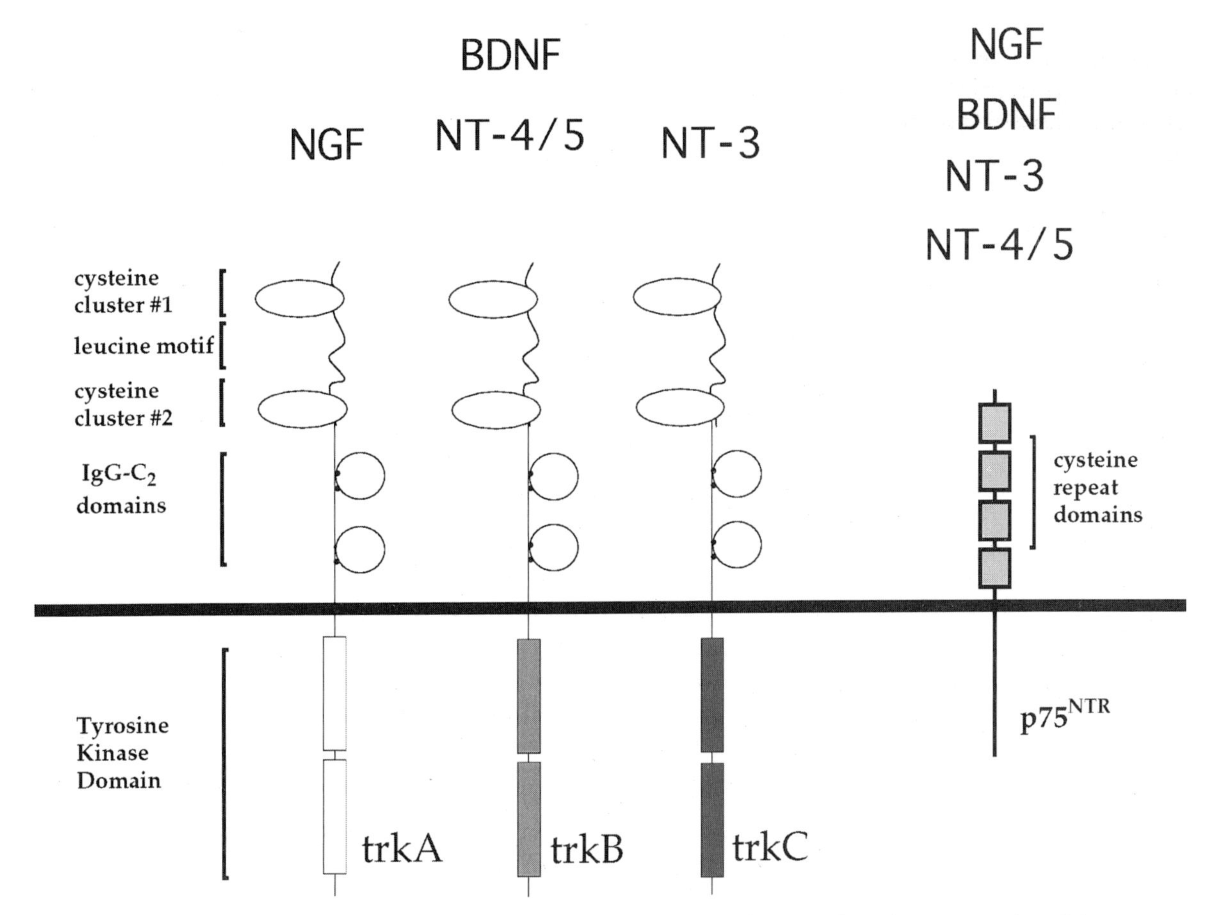

FIGURE 1 Transmembrane receptors for neurotrophin family members. A schematic representation of the Trk receptor tyrosine kinase family (TrkA, TrkB, and TrkC) and the p75 neurotorphin receptor are shown with their perferred ligands indicated at the top of the figure. Signal transduction by the neurotrophins use these two receptor classes. Whereas the Trk receptors form a subfamily of tyrosine kinase receptors, the p75 neurotrophin receptor belongs to the TNF superfamily of receptors.

The substrates for neurotrophin Trk receptors—phospholipase C-γ, PI-3 kinase, and SHC and Grb2 adaptor proteins—are used by many tyrosine kinase receptors. This raises the question of how these phosphorylation events lead to different biologic outcomes (Chao, 1992a). There are several possibilities. First, the strength and duration of the receptor autophosphorylation events may determine downstream signaling. Second, differential signaling may be controlled by specific dephosphorylation events. Third, there may be unique second messengers or substrates that determine the specific nature of the response.

For neurotrophic factors, the duration of signaling has provided an important criteria for differential signaling. In PC12 cells, NGF induces a prolonged activation of ras and MAP kinase activity, lasting for several hours, whereas EGF-mediated MAP kinase activation is transient in nature (Qiu and Green, 1991; Qiu and Green, 1992). The time course of downstream signaling is therefore one of the major differences that accounts for the differentiation program elicited by NGF versus the action of other mitogenic growth factors, such as EGF. However, the strength of receptor action is alone not sufficient, because ligand-dependent autophosphorylation of the EGF receptor is actually higher than for the TrkA NGF receptor in PC12 cells (Berg *et al.,* 1992), so that the level of receptor activation cannot solely account for the difference in trophic factor action. Other novel receptor substrates, such as docking or adaptor proteins (Kouhara *et al.,* 1997) may be recruited by a subset of tyrosine kinase receptors.

III. Patterns of Expression

The widely accepted view of neurotrophin synthesis is that limiting amounts are produced by nonneuronal target tissues in the periphery. In the CNS, neurotrophins are produced not only by astroyctes, oligodendrocytes, and microglia, but also by neuronal cells. Hence, local autocrine production and paracrine mechanisms of neurotrophin factor can take place. Activity-dependent regulation of neurotrophin synthesis has consequences in formation of the visual system and neurotransmitter release (Thoenen, 1995).

The neuronal cell populations responsive to NGF are restricted, in contrast to the more numerous neuronal populations that are dependent on BDNF, NT-3, and NT-4/5. During development, TrkA expression is limited to sympathetic and a subset of sensory neurons in the PNS and cholinergic neurons of the basal forebrain, whereas more extensive expression is found for TrkB and TrkC receptors in the CNS (Barbacid, 1994; Snider, 1994). The p75 neurotrophin receptor has a much wider distribution and is expressed on numerous cell types, including Schwann cells, motor neurons, meningeal cells, dental pulp cells, hair follicle cells, and cerebellar Purkinje cells. The widespread pattern of p75 expression is consistent with its role as a potential receptor for BNDF, NT-3, or NT-4/5, in addition to NGF. The p75 neurotrophin receptor often is up-regulated in response to injury and inflammation. The enhanced expression of this receptor on glial cells during inflammation, infection, or traumatic lesion could be responsible for greater sensitivity of these cells to growth factors and cytokines released at the site of injury.

The majority of NGF responsive cells, including sensory, sympathetic, and basal forebrain cholinergic neurons, express both p75 and TrkA. By contrast, cells expressing TrkB or TrkC may or may not express significant levels of p75. Although p75 and Trk receptors are coexpressed in many cells, independent expression of p75 and of individual Trk family members and their isoforms is also observed.

IV. Consequences of Neurotrophin Gene Ablation

Targeted mouse mutations in the genes encoding the neurotrophins and their receptors have been generated by homologous recombination. Although germ-line mutation of many vertebrate genetic loci do not necessarily lead to a detectable phenotype, inactivation of the three mammalian *trk* genes and their ligands results in considerable neuronal cell loss in mutant mice (Table 1). These significant cell losses provide a striking verification of the biologic effects of each member of the NT family, the distribution of neurotrophin expression, and the binding specificities of each neurotrophin.

Several general conclusions can be made regarding null mice, in which the expression of neurotrophins or their receptors has been eliminated. First, there is a striking similarity in the phenotypes displayed by NGF, BDNF, and NT-3 mutant mice, and homozygous mutations in TrkA, TrkB, and TrkC, respectively (Table 1). The defects in each case correlate with the cellular populations known to express each neurotrophin and its cognate receptors (Barbacid, 1994; Snider, 1994). This conclusion has been verified by evidence that the small diameter nociceptive neurons are largely lost in the NGF and TrkA mutant mice, and that the large sensory neurons responsible for proprioception are largely absent in mice with either a disrupted NT-3 or a TrkC gene. Consistent with the ablation experiments, NT-3 is synthesized by muscle spindles and is retrogradely transported to large proprioceptive neurons.

TABLE 1 Targeted Mutations of Neurotrophins and Their Receptors[a]

Gene	Phenotype	Reference
NGF	SCG, DRG, TG loss	Crowley *et al.*, 1994
BDNF	DRG, nodose, TG, vestibular loss	Ernfors *et al.*, 1994, 1995; Jones *et al.*, 1994
NT-3	DRG, nodose, TG, cochlear loss	Ernfors *et al.*, 1994, 1995; Tessarollo *et al.*, 1994
NT-4/5	DRG, nodose/petrosal loss	Conover *et al.*, 1995; Erickson *et al.*, 1996
TrkA	SCG, DRG, TG loss	Smeyne *et al.*, 1994
TrkB	DRG, nodose, TG, vestibular loss	Klein *et al.*, 1993
TrkC	DRG, nodose, TG loss	Klein *et al.*, 1994
p75	DRG loss	Lee *et al.*, 1992
	SCG innervation to pineal	Lee *et al.*, 1994
	Increase in cholinergic neurons	Van der Zee *et al.*, 1996

[a]The most prominent phenotypes are listed that reflect defects detected in mice carrying homozygous mutations in each gene. References for the initial characterization for each mouse mutation are provided.

DRG, dorsal root ganglia; SCG, sympathetic cervical ganglia; TG, trigeminal ganglia.

The pattern and degree of neuronal cell death indicate that each neurotrophin acts on distinct neuronal populations and that survival of each population may depend on more than one neurotrophin. For instance, development of the inner ear relies on BDNF and NT-3 (Ernfors *et al.,* 1995). A substantial loss of of cochlear neurons is observed in mice lacking NT-3 or TrkC, whereas vestibular neurons are more sensitive to a lack of BDNF. When both neurotrophins are absent, all cochlear neurons are missing. Therefore, BDNF and NT-3 display survival effects that are additive. Although some classes of neurons exhibit a correspondence between Trk receptor expression and neurotrophin responsiveness, other subsets of neurons can express multiple Trk receptors, or may exhibit different patterns of Trk receptor expression at distinctive developmental stages.

Another striking conclusion is that survival of cells in the CNS in null mutant mice appear to be relatively unaffected compared to the PNS. Although neurotrophins are able to influence the survival of CNS neurons and can protect against excitotoxity, ischemia, and nerve lesion, preliminary analysis indicates the same extent of neurodegeneration observed in the PNS was not observed in the CNS of each mutant animal. The absence of overt cell loss in the CNS of neurotrophin-deficient animals is most likely attributable either to the action of other growth factors that can substitute for the lack of neurotrophin signaling or to other adaptive mechanisms. In addition, more local versus target-derived, cell–cell interactions may exist to promote cell survival in the CNS. These cellular interactions may involve regulation of neurotrophin availability through changes in hormonal levels, neurotransmitter release, or by other growth factors (Lindholm *et al.,* 1994). Further analysis of these null mutations and crosses between different mice will likely reveal additional deficits in neuronal morphology and physiology.

Finally, targeted disruptions of the trk receptor genes lead to a slightly more severe phenotype than ligand mutations, measured by the extent of cell death and the target populations that are affected. This difference in phenotype suggests that individual neurotrophin ligands may be more dispensable than individual receptors and underscores the ability of multiple trophic factors to interact with a single receptor. Another level of regulation is provided by truncated forms of the TrkB and TrkC receptors, which are involved in signal transduction, as suggested by *in vitro* experiments (Baxter *et al.,* 1997).

Neurotrophins do not display the same effects with each *trk* receptor or with the p75 neurotrophin receptor. For example, targeted mutations in BDNF and TrkB result in lethal phenotypes, whereas mice deficient in NT-4/5 are viable (Conover *et al.,* 1995). This difference is attributable to a selective loss of chemosensory input in the BDNF-null mice, leading to a fatal respiratory defect (Erickson *et al.,* 1996).

Another example is the differential effects of the disruption of the murine p75 gene, which results in loss of sensory neurons and a lack in sympathetic innervation in selective targets, such as the pineal gland (Lee *et al.,* 1994). Death of sensory neurons deficient in p75 is due to the higher concentration of NGF necessary to maintain cell survival with TrkA alone (Davies *et al.,* 1993). A lack of p75 expression leads to a greater requirement for NGF by TrkA. However, a change in the dose response for sensory neuron survival was not detected for BDNF and NT-3 (Davies *et al.,* 1993). The behavior of p75-deficient neurons toward the neurotrophins can be attributed to the subtle differences in binding properties of each neurotrophin for p75, largely governed by their electrostatic surfaces. As a consequence, the rates of dissociation are markedly different, with NGF $>$ NT-3 $>$ BDNF (Rodriguez-Tebar *et al.,* 1992).

The neurotrophin sequence responsible for p75 binding has been localized. Mutation of three lysine residues (K32, K34, and K95) to alanine within NGF results in a complete loss of binding to p75, but maintains binding to *trkA,* and produced a survival response (Ibanez *et al.,* 1992). The three lysine residues in NGF are not conserved among all neurotrophins. Modeling of each neurotrophin protein indicates the presence of a positively charged surface in a topologically similar region (Ryden *et al.,* 1994). Small local structural differences could account for the distinct association and dissociation rates exhibited by NGF, BDNF, and NT-3 for p75 (Rodriguez-Tebar *et al.,* 1992).

Although p75 is regarded as a common neurotrophin binding protein, interactions of each neurotrophin with p75 are different and may lead to alternative binding specificities with *trk* family members. For example, neurons in the chick isthmo-optic nucleus normally depend on BDNF for cell survival. However, NGF can cause neuronal cell death (see later this chapter). This unexpected effect requires an interaction of NGF with p75, which is thought to interfere with BDNF/TrkB-mediated cell survival (von Bartheld *et al.,* 1994). Differences in neurotrophin action may be also attributed to the differential transport of NGF, BDNF, and NT-4/5 in sensory neurons (Curtis *et al.,* 1995).

V. Interactions of *trk* and p75 Receptors

The neurotrophins are unusual among polypeptide growth factors in that two different transmembrane proteins exist for each neurotrophin. In addition to *trk* family members, the p75 receptor is a member of the family of receptors represented by tumor necrosis factor (TNF) and the Fas receptor (Smith *et al.,* 1994). The p75 receptor is the founding member of the TNF receptor superfamily. The TNF family of receptors is defined by unique cysteine-rich domains in the extracellular region. Other members include the lymphoid cell-specific receptor CD30, CD40, and CD27 and the most recently discovered members DR-3, DR-4, and GITR (Smith *et al.,* 1994; Brojatsch *et al.,* 1996; Chinnaiyan *et al.,* 1996; Montgomery *et al.,* 1996; Pan *et al.,* 1997). All of these transmembrane proteins share in their extracellular domains a cysteine motif that spans 40 amino acids and is repeated two to six times.

An important criteria for the neurotrophin family of receptors is the relative ratio of p75 and *trk* receptors, which may dictate the degree of responsiveness to individual neurotrophins (Benedetti *et al.,* 1993; Lee *et al.,* 1994; Verdi *et al.,* 1994) and equilibrium binding properties. The binding properties exhibited by neurotrophins for *trk* family tyrosine kinase receptors are likely to result in different consequences in the presence of p75 or other Trk isoforms.

From Scatchard analyses, two classes of NGF binding sites exist on the surface of responsive neurons (Sutter *et al.,* 1979). These sites differ 100-fold in equilibrium binding constants, which can be further distinguished by the rates of ligand association and dissociation. The proteins responsible for the high affinity NGF binding site have been the subject of considerable debate, because p75 and TrkA each exhibit predominately low affinity binding and a small percentage of high affinity sites has been detected for TrkA (Jing *et al.,* 1992). The contribution of each receptor type to the high affinity NGF binding site has been clarified by kinetic analysis (Mahadeo *et al.,* 1994). Whereas p75 displays fast rates of association and dissociation with NGF, TrkA interacts with much slower on- and off-rates. Due to its unusually slow on-rate, NGF binding to TrkA results in a K_d of 10^{-9} to 10^{-10} M. A similar K_d has also been determined for BDNF binding to *trkB* (Dechant *et al.,* 1993) . These affinities are distinct from the high affinity binding site, $K_d = 10^{-11}$ *M,* measured in sensory neurons (Sutter *et al.,* 1979; Dechant *et al.,* 1993). It is therefore not correct to refer to p75 and Trk receptors as low and high affinity receptors, because p75 receptors have been found to exist in a high affinity state (Dechant and Barde, 1997) and previous studies found that TrkA and TrkB display predominantly low affinity values (Hempstead *et al.,* 1991; Dechant *et al.,* 1993).

When TrkA and p75 receptors are coexpressed, the on-rate is accelerated 25-fold, creating a new kinetic site whose features are consistent with the high affinity NGF binding site ($K_d = 10^{-11}$ *M*). This site requires an excess ratio of p75 to TrkA (Chao, 1994). Hence, one function of p75 is to increase the binding affinity of NGF for TrkA.

In addition to binding, signal transduction by TrkA can be influenced by p75 (Berg *et al.,* 1991; Barker and Shooter, 1994; Hantzopoulos *et al.,* 1994; Verdi *et al.,* 1994). Cell culture experiments indicate that p75 is capable of enhancing TrkA autophosphorylation (Barker and Shooter, 1994; Verdi *et al.,* 1994). A potential function of the p75 receptor may be to increase the effective concentration of neurotrophin at the cell surface in order to enhance TrkA binding (Barker and Shooter, 1994). This is consistent with the limited amounts of neurotrophin available to competing neurons during development. Another model is that an altered conformation of *trk* may be formed in the presence of p75 that facilitates ligand binding and subsequent signaling functions (Mahadeo *et al.,* 1994).

In support of this model is the intrinsic flexibility of the loops and the N- and C-termini of NGF. These domains of NGF are important for TrkA interaction (McDonald *et al.,* 1991; Holland *et al.,* 1994) and may induce conformational changes on formation of a productive NGF–*trk* complex. A common binding motif has not been identified for the *trk* receptor–neurotrophin binding site. This is interesting in light of the high conservation found for species variants of certain neurotrophins, such as NT-3. Ligand specificity could be determined directly by enhanced binding to TrkA, or indirectly, by preventing binding to other trks (Ilag *et al.*, 1994).

Neurotrophin binding to cognate Trk receptors requires the two extracellular immunoglobulin-like domains, which are probably also necessary for dimerization (Perez *et al.*, 1995; Urfer *et al.*, 1995). The use of an immunoglobulin-like domain for neurotrophin binding is a recurrent theme for ligand-receptor signaling by receptor tyrosine kinase, such as FGF and c-kit receptors. Besides the IgC2-like domains, the region between the transmembrane and the IgG domain nearest to the membrane may be also critical to binding of neurotrophins (McDonald and Meakin, 1996). The localization of neurotrophin binding sites on their receptors may allow for the generation of small peptide agonists. From these binding properties, it is plausible that homo- and heterodimeric complexes may be formed by neurotrophin factors or by different receptor types.

Direct interactions between p75 and Trk receptors have been difficult to document biochemically. However, immunoprecipitation experiments carried out in crosslinked spinal cord and brain tissues with ^{125}I-NGF suggest that an association between the TrkA and p75 may take place (Huber and Chao, 1995). Photobleaching experiments following a fluorescently tagged p75 receptor have also revealed a potential physical interaction with *TrkA* (Wolf *et al.*, 1995).

VI. Cell Death Signaling by Neurotrophins

An unexpected property that has emerged for NGF is that it can promote apoptotic cell death in addition to its survival function. For example, treatment of medulloblastoma cells with NGF resulted in apoptosis that was directly inhibited by anti-NGF antibodies (Muragaki *et al.*, 1997). Furthermore, this action was dependent on TrkA autophosphorylation by NGF. Although the reason that accounts for this killing activity through TrkA is not known, one explanation is that an alteration in cell cycle control has subverted these medulloblastoma cells to this fate.

The similarity in the intracellular domains of the p75 with other family members, such as the Fas antigen and the p55 TNF receptor (Liepinsh *et al.*, 1997), also suggested that p75 might function as a cell death molecule. The Fas and TNF receptor share significant homology within their intracellular domains—an 80 amino acid region called the *death domain,* which has been shown to be required for the apoptosis-promoting activities of these receptors. The death domain is a novel protein–protein association motif found in many pro-apoptotic molecules (Feinstein *et al.*, 1995).

The first report of a cell death activity for p75 was made in immortalized cerebellar neuronal cell lines, which undergo a faster rate of cell death after transfection of p75 (Rabizadeh *et al.*, 1993). Binding of NGF and agonist antibodies rescue cells from apoptotic cell death. Moreover, sensory neuronal survival experiments indicate that expression of p75 has a negative effect on cell survival, particularly during the postnatal periods (Barrett and Bartlett, 1994). Although these effects are presumed to occur independently of TrkA expression, the effects of coexpression have not been examined. For example, earlier in development, sensory neurons depend on the coexpression of trk and p75 receptors (Barrett and Bartlett, 1994). These results indicated that p75 can signal through a death pathway, similar to its family members Fas and TNF, but that the receptor is capable of supporting cell survival.

Other examples implicating a cell death function for p75 have been documented (Table 2). The number of dying cholinergic neurons in the basal forebrain is higher in wild-type animals compared to p75-deficient animals, suggesting that the expression of p75 is detrimental to viability in this cell population. This property was traced to basal forebrain cells that express p75 and are deficient in TrkA (Van der Zee *et al.*, 1996), but may also be strain dependent (Peterson *et al.*, 1997). Differentiated PC12 cells undergoing NGF deprivation seem to die more rapidly in cells expressing high levels of p75. These combined studies have suggested that p75 promotes cell death in the absence of NGF binding.

In contrast to a ligand-independent mechanism, cell death induced by binding of NGF to p75 also has been observed. A clear example of the *in vivo* consequences of p75-mediated cell death was detected in developing chick retina (Frade *et al.*, 1996). Neuronal precursor cells at embryonic day 5 undergo apoptotic cell death, which can be blocked effectively by antibodies to either p75 or NGF. This indicates that endogenous NGF causes the death of retinal neurons at an early developmental age, which is diametrically opposed to its well established survival function. Similarly, NGF

TABLE 2 Examples of Neurotrophin-Related Apoptosis

Cell type	Reference
Immortalized neural cell lines	Rabizadeh *et al.*, 1993
Postnatal sensory neurons	Barrett and Bartlett, 1994
Chick retina precursors	Frade *et al.*, 1996
Basal forebrain cholinergic neurons	Van der Zee *et al.*, 1996
PC12 cells after NGF withdrawal	Barrett and Georgiou, 1996
Mature oligodendrocytes	Casaccia-Bonnefil *et al.*, 1996
Medulloblastoma	Muragaki *et al.*, 1997

treatment of fully differentiated oligodendrocytes results in cell death through binding to p75 receptors (Casaccia-Bonnefil *et al.,* 1996). Apoptosis is detected by fluorescent TUNEL labeling and increased DNA fragmentation. Interestingly, the effect of NGF on oligodendrocyte cultures is only observed during terminal differentiation and requires high concentrations (100 ng/ml) and cannot be reproduced by similar concentrations of BDNF or NT-3. Progenitor oligodendrocytes (O-2A) do not express p75 and are not susceptible to death by neurotrophins. Instead, these cells can also use neurotrophins as trophic factors (Cohen *et al.,* 1996).

Although the majority of oligodendrocytes do not express p75 *in vivo* in physiological conditions, the oligodendrocyte cultures provide an experimental system to study signal transduction events following NGF treatment. Glial cells are highly susceptible to injury and inflammation. Recent evidence indicates that p75-positive oligodendrocytes can be detected in white matter plaques from cases of multiple sclerosis (Dowling *et al.,* 1997). Indeed, oligodendrocytes can express Trk receptors, depending on the growth conditions of the cells. When cells are grown with mitogenic growth factors such as FGF or PDGF, neurotrophins respond with increased survival to neurotrophin factors (Cohen *et al.,* 1996), such as NT-3.

In the absence of TrkA expression, binding of NGF to p75 receptors can increase intracellular ceramide levels through increased sphingomyelin hydrolysis (Dobrowsky *et al.,* 1994). When the levels of intracellular ceramide are measured in differentiated oligodendrocytes only expressing p75 receptors, NGF leads to an elevation of ceramide levels that is sustained for a period of several hours instead of minutes. This time course is reminiscent of the activation of ras and MAP kinase activities through TrkA. A sustained versus transient duration in signaling may determine which pathway is dominant. Interestingly, other neurotrophins such as BDNF and NT-3, which also bind to p75, do not induce long-term ceramide production in mature oligodendrocytes. Moreover, cell death in oligodendrocytes through increased intracellular ceramide levels can also be mimicked by the application of exogenous ceramide analogs or bacterial sphingomyelinase or by inhibitors of the glucosylceramide synthase. These data suggest that the high sustained ceramide level elicited by the p75 receptor may contribute to NGF-dependent death.

Another criteria for the dose-dependent cell death of primary cortical oligodendrocytes is the activation of c-jun amino-terminal kinase (JNK). The JNK enzymatic activity, a downstream effector of the stress-activated protein kinase family, has been implicated in cell death progression in PC12 cells (Xia *et al.,* 1996). In addition, ceramide is a potential activator of JNK. Significantly, incubation of mature and not progenitor oligodendrocytes with NGF led to an induction of JNK. The activation of JNK activity by NGF was similar in magnitude to the effects by ceramide and was not observed in cells whose viability was not affected by NGF, such as NIH3T3 fibroblasts expressing p75. The stress-activated protein kinase JNK can therefore be regulated by p75 in older oligodendrocytes.

When both NGF receptors are coexpressed, as in the case of PC12 cells, TrkA exerts a suppressive effect on p75, as assayed by ceramide production (Dobrowsky *et al.,* 1995). In support of this model, introduction of TrkA receptors into oligodendrocytes results in a reduction of ceramide production, but leaves NFκB activation through p75 unaffected (Yoon *et al.,* 1998). JNK activities also are suppressed by coexpression of TrkA, whereas MAP kinase activity is increased by NGF binding to TrkA. From these experiments, TrkA signaling can selectively down-regulate certain responses elicited from p75.

Although p75 can modulate high affinity binding by NGF and TrkA signaling, p75 can also display independent signaling properties through ceramide production (Dobrowsky *et al.,* 1994, 1995) and increased NFκB activity. The activation of JNK by NGF binding to the p75 receptor and the sustained level of ceramide has been observed only in limited cases, such as end-stage oligodendrocytes. Strikingly, other cell types that express p75, such as 3T3 cells, respond to all neurotrophins with a transient increase in ceramide levels (Dobrowsky *et al.,* 1995), but do not undergo cell death after neurotrophin treatment.

Indeed, Schwann cells and melanoma cells that express high levels of p75 do not undergo neurotrophin-dependent cell death. These data indicate that p75 produces a dichotomy of effects, some of which are extremely cell specific. Thus, neurotrophin treatment of p75–3T3 cells results in a transient elevation of ceramide levels, but a persistent, long-lasting increase in ceramide and JNK activity levels is observed in mature oligodendrocytes. Similar to findings in Schwann cells, NFκB activation by p75 appears to be exclusive to NGF, as BDNF and NT-3 are ineffective (Carter *et al.,* 1996). These differences in ligand specificity and time courses may result in separate signaling cascades and different biologic responses, depending on the cell type and differentiative stage. A critical requirement for the death pathway is the terminal nature of cell differentiation. From these observations, the mere expression of p75 is not sufficient for apoptosis.

There are several explanations that may account for the mechanism of cell death initiated by neurotrophins. Because the majority of cases involve elevated levels of p75 receptor expression, the responses may involve the same signaling intermediates used by TNF and

Fas receptors. This includes adaptor proteins Fas-associated death domain protein (FADD) and TNF-R1–associated death domain protein (TRADD) and downstream effectors, such as caspase-8, the FADD-like ICE (FLICE)/MORT-1–associated CED-3 homolog (MACH) interacting enzyme (Cleveland and Ihle, 1995). It remains to be seen whether p75 overexpression might co-op the use of these proteins or other interacting proteins of different specificity.

Also, a new family of cytokine associated signaling intermediates, the TNF-associated factors (TRAFs), have been described. Originally discovered as molecules associated with the p75 TNF receptor (Rothe *et al.*, 1994), it is now apparent the TRAF proteins represent a large family of molecules that carry potential signaling capabilities through NFκB activation. Moreover, these molecules have been found to be associated with receptors required for different biologic activities. These receptors include CD30, CD40, and lymphotoxin β (Rothe *et al.*, 1995; Gedrich *et al.*, 1996; Nakano *et al.*, 1996) and the p55 TNF receptor (Hsu *et al.*, 1996). It seems likely that TRAF or other proteins are used by p75, because neurotrophins share similar NFκB responses as IL-1 and the TNF family.

VII. Summary

The decision of the cells to undergo cell division as opposed to terminal differentiation and cell suicide depends on a delicate balance between many opposing biochemical interactions. Neurotrophins are secreted proteins that can influence this balance by signaling growth arrest through different pathways and intermediates. On one level, these decisions are dictated by gene regulatory events, which determine the levels of trophic factors and their receptors; essential enzymatic functions, such as phospholipase C-γ and MAP kinases; and mitochondrial proteins, such as Bcl-2 family members, which confer susceptibility to apoptosis. Protein–protein interactions together with protein phosphorylation serve as the primary regulatory mechanisms for determining cell survival. Cellular responses depend on the strength of the binding interactions from ligand-receptor interactions, transcription factor binding to target DNA sequences, and key enzymatic complexes, such as cyclin/CDK proteins.

Trophic factors exemplified by NGF and its family members, CNTF and glial derived neurotrophic factor (GDNF), all use increased tyrosine phosphorylation of cellular substrates to mediate neuronal cell survival. Actions of the NGF family of neurotrophins are not only dictated by *ras* activation through the *trk* family of receptor tyrosine kinases, but also a survival pathway defined by phosphatidylinositol-3-kinase activity (Yao and Cooper, 1995), which gives rise to phosphoinositide intermediates that activate the serine/threonine kinase Akt/PKB (Dudek *et al.*, 1997). Induction of the serine–threonine kinase Akt is critical for cell survival as well as cell proliferation. Hence, for many trophic factors, multiple subunits constitute a functional transmembrane complex that activates *ras*-dependent and *ras*-independent intracellular signaling.

Neurotrophins also mediate other critical functions in the nervous system, including neurotransmitter release, formation of ocular dominance columns in the visual system, and increased synaptic efficacy, as measured by long-term potentiation (Thoenen, 1995). This chapter indicates that neurotrophins have the capability to kill cells through a proapoptotic mechanism in addition to their ability to prevent a cell suicide pathway.

There are other examples of growth factors and cytokines that carry apoptotic as well as cell survival functions. Bone morphogenetic proteins such as BMP4 can influence early developmental events by inducing apoptosis (Graham *et al.*, 1994) and leukemia inhibitory factor (LIF) can cause sympathetic neuron cell death in culture (Kessler *et al.*, 1993). The prominent role of cytokines in inducing apoptosis is underscored by the actions of TNF family of ligands and receptors.

The NGF receptors provide a clear example of bidirectional crosstalk. In the presence of TrkA receptors, p75 can participate in the formation of high affinity binding sites and enhanced neurotrophin responsiveness leading to a survival signal. In the absence of TrkA receptors, p75 can generate a death signal only in specific cell populations. These activities include the induction of NFκB (Carter *et al.*, 1996); the hydrolysis of sphingomyelin to ceramide (Dobrowsky *et al.*, 1995); and the proapoptotic functions attributed to p75. An important finding is that Trk tyrosine kinase action appears to negate the signaling properties of p75, particularly with respect to ceramide production (Dobrowsky *et al.*, 1995) and programmed cell death activities.

Neurotrophin receptors are generally drawn and viewed as isolated integral membrane proteins that span the lipid bilayer, with signal transduction proceeding in a linear stepwise fashion. There are now numerous examples that indicate that each receptor acts not only in a linear and independent manner, but can also influence the activity of other cell surface receptors, either directly or through signaling intermediates. Which step and which intermediates are used for crosstalk between the receptors is an ongoing question.

Biologically, cooperative actions of two receptors imply that the possibilities for extracellular signaling are expanded. Cell differentiation is determined by biochemical reactions representing the combined effects of many growth factors and cytokines. This diversity may reflect inherent differences in receptor structure and substrate specificities. Regulation of cellular differentiation and proliferation decisions is likely to be determined by the additive effects of multiple receptors and the duration of second messengers and phosphorylation events. Unlike studies carried out *in vitro* in which cell lines are treated with single factors, growth and survival of cells *in vivo* are subject to the simultaneous action of multiple polypeptide growth factors.

The goals are to define the molecular mechanisms that dictate the action of macromolecular receptor complexes during transmembrane signaling. This will likely address the long-standing question of how cells make decisions determining growth versus cell cycle arrest versus death. Growing evidence has established a TNF family of transmembrane receptors that serve as mediators of cell death. The pathways leading from death receptor activation and execution of an apoptotic program have quickly become understood on a molecular level. For neurotrophins, their primary function in sustaining the viability of neurons is counterbalanced by a receptor mechanism to eliminate cells by an apoptotic mechanism. It is conceivable that this bidirectional system may be used selectively during development and in neurodegenerative diseases.

References

Angeletti, R. H., and Bradshaw, R. A. (1971). The amino acid sequence of 2.5S mouse submaxillary gland nerve growth factor. *Proc. Natl. Acad. Sci.* **68,** 2417–2420.

Arakawa, T., Haniu, M., Narhi, L. O., Miller, J. A., Talvenheimo, J., Philo, J. S., Chute, H. T., Matheson, C., Carnahan, J., Louis, J.-C., Yan, Q., Welcher, A. A., and Rosenfeld, R. (1994). Formation of heterodimers from three neurotrophins, nerve growth factor, neurotrophin-3, and brain-derived neurotrophic factor. *J. Biol. Chem.* **269,** 27833–27839.

Barbacid, M. (1994). The trk family of neurotrophin receptors. *J. Neurobiol.* **25,** 1386–1403.

Barde, Y. A. (1989). Trophic factors and neuronal survival. *Neuron* **2,** 1525–1534.

Barker, P. A., and Shooter, E. M. (1994). Disruption of NGF binding to the low-affinity neurotrophin receptor p75 reduces NGF binding to trkA on PC12 cells. *Neuron* **13,** 203–215.

Barrett, G. L., and Bartlett, P. F. (1994). The p75 receptor mediates survival or death depending on the stage of sensory neuron development. *Proc. Natl. Acad. Sci. USA* **91,** 6501–6505.

Baxter, G. T., Radeke, M. J., Kuo, R. C., Makrides, V., Hinkle, B., Hoang, R., Medina-Selby, A., Coit, D., Valenzuela, P., and Feinstein, S. C. (1997). Signal transduction mediated by the truncated trkB receptor isoforms, trkB.T1 and trkB.T2. *J. Neurosci.* **17,** 2683–2690.

Benedetti, M., Levi, A., and Chao, M. V. (1993). Differential expression of nerve growth factor receptors leads to altered binding affinity and neurotrophin responsiveness. *Proc. Natl. Acad. Sci. USA* **90,** 7859–7863.

Berg, M. M., Sternberg, D. W., Hempstead, B. L., and Chao, M. V. (1991). The low affinity nerve growth factor (NGF) receptor mediates NGF-induced tyrosine phosphorylation. *Proc. Natl. Acad. Sci. USA* **88,** 857–866.

Berg, M. M., Sternberg, D., Parada, L. F., and Chao, M. V. (1992). K-252a inhibits NGF-induced trk proto-oncogene tyrosine phosphorylation and kinase activity. *J. Biol. Chem.* **267,** 13–16.

Berkemeier, L. R., Winslow, J. W., Kaplan, D. R., Nikolics, K., Goeddel, D. V., and Rosenthal, A. (1991). Neurotrophin-5: a novel neurotrophic factor that activates Trk and TrkB. *Neuron* **7,** 857–866.

Bradshaw, R. A., Blundell, T. L., Lapatto, R., McDonald, N. Q., and Murray-Rust, J. (1993). Nerve growth-factor revisited. *Trends Biochem. Sci.* **18,** 48–52.

Bradshaw, R. A., Murray-Rust, J., Ibanez, C. F., McDonald, N. Q., Lapatto, R., and Blundell, T. L. (1994). Nerve growth factor: structure/function relationships. *Protein Sci.* **3,** 1901–1913.

Brojatsch, J., Naughton, J., Rolls, M. M., Zingler, K., and Young, J. A. T. (1996). CAR1, a TNFR-related protein is a cellular receptor for cytopathic avian leukosis–sarcoma viruses and mediates apoptosis. *Cell* **87,** 845–855.

Carter, B. D., Kaltschmidt, C., Kaltschmidt, B., Offenhauser, N., Bohm-Matthaei, R., Baeuerle, P. A., and Barde, Y.-A. (1996). Selective activation of NK-kB by nerve growth factor through the neurotrophin receptor p75. *Science* **272,** 542–545.

Casaccia-Bonnefil, P., Carter, B. D., Dobrowsky, R. T., and Chao, M. V. (1996). Death of oligodendrocytes mediated by the interaction of nerve growth factor with its receptor p75. *Nature* **383,** 716–719.

Chao, M. V. (1992a). Growth factor signaling: where is the specificity? *Cell* **68,** 995–997.

Chao, M. V. (1992b). Neurotrophin receptors—a window into neuronal differentiation. *Neuron* **9,** 583–593.

Chao, M. V. (1994). The p75 neurotrophin receptor. *J. Neurobiol.* **25,** 1373–1385.

Chinnaiyan, A. M., O'Rourke, K., Yu, G.-L., Lyons, R. H., Garg, M., Duan, D. R., Xing, L., Gentz, R., Ni, J., and Dixit, V. M. (1996). Signal transduction by DR3, a death domain–containing receptor related to TNFR-1 and CD95. *Science* **274,** 990–992.

Cleveland, J. L., and Ihle, J. N. (1995). Contenders in FasL/TNF death signaling. *Cell* **81,** 479–482.

Cohen, R. I., Marmur, R., Norton, W. T., Mehler, M. F., and Kessler, J. A. (1996). Nerve growth and neurotrophin-3 differentially regulate the proliferation and survival of developing rat brain oligodendrocytes. *J. Neurosci.* **16,** 6433–6442.

Conover, J. C., Erikson, J. T., Katz, D. M., Bianchi, L. M., Poueymirou, W. T., McClain, J., Pan, L., Helgen, M., Ip, N. Y., Boland, P., Friedman, B., Wiegand, S., Vejsada, R., Kato, A. C., DeChiara, T. M., and Yancopoulos, G. (1995). Neuronal deficits, not involving motor neurons, in mice lacking BDNF and/or NT4. *Nature* **375,** 235–238.

Curtis, R., Adryan, K. M., Stark, J. L., Park, J. S., Compton, D. L., Weskamp, G., Huber, L. J., Chao, M. V., Jaenisch, R., Lee, K.-F., Lindsay, R. M., and DiStefano, P. S. (1995). Differential role of the low affinity neurotrophin receptor (p75) in retrograde axonal transport of neurotrophins. *Neuron* **14,** 1201–1211.

Davies, A., Lee, K.-F., and Jaenisch, R. (1993). p75 deficient trigeminal sensory neurons have an altered response to NGF but not to other neurotrophins. *Neuron* **11,** 1–20.

Davies, A. M. (1994). Neurotrophic factors—switching neurotrophin dependence. *Curr. Biol.* **4,** 273–276.

Dechant, G., Biffo, S., Okazawa, H., Kolbeck, R., Pottgiesser, J., and Barde, Y.-A. (1993). Expression and binding characteristics of the BDNF receptor chick trkB. *Development* **119,** 545–558.

Dechant, G., and Barde, Y.-A. (1997). The neurotrophin receptor p75 binds neurotrophin-3 on sympathetic neurons with high affinity and specificity. *J. Neurosci.* **17,** 5281–5287.

Distefano, P. S., Friedman, B., Radziejewski, C., Alexander, C., Boland, P., Schick, C. M., Lindsay, R. M., and Wiegand, S. J. (1992). The neurotrophins BDNF, NT-3, and NGF display distinct patterns of retrograde axonal-transport in peripheral and central neurons. *Neuron* **8,** 983–993.

Dobrowsky, R. T., Werner, M. H., Castellino, A. M., Chao, M. V., and Hannun, Y. A. (1994). Activation of the sphingomyelin cycle through the low-affinity neurotrophin receptor. *Science* **265,** 1596–1599.

Dobrowsky, R. T., Jenkins, G. M., and Hannun, Y. A. (1995). Neurotrophins induce sphingomyelin hydrolysis. *J. Biol. Chem.* **270,** 22135–22142.

Dowling, P., Raval, S., Husar, W., Casaccia-Bonnefil, P., Chao, M., Cook, S., and Blumberg, B. (1997). Expression of the p75 neurotrophin receptor in oligodendrocytes in multiple sclerosis. *Neurology* **48,** 60–83.

Dudek, H., Datta, S. R., Franke, T. F., Birnbaum, M. J., Yao, R., Cooper, G. M., Segal, R. A., Kaplan, D. R., and Greenberg, M. E. (1997). Regulation of neuronal survival by the serine-threonine protein kinase. *Akt. Sci.* **275,** 661–665.

Erickson, J. T., Conover, J. C., Borday, V., Champagnat, J., Barbacid, M., Yancopoulos, G., and Katz, D. M. (1996). Mice lacking brain-derived neurotrophic factor exhibit visceral sensory neuron losses distinct from mice lacking NT-4 and display a severe developmental deficit in control of breathing. *J. Neurosci.* **16,** 5361–5371.

Ernfors, P., van de Water, T., Loring, J., and Jenisch, R. (1995). Complementary roles of BDNF and NT-3 in vestibular and auditory development. *Neuron* **14,** 1153–1164.

Feinstein, E., Kimchi, A., Wallach, D., Boldin, M., and Varfolomeev, E. (1995). The death domain: a module shared by proteins with diverse cellular functions. *TIBS* **20,** 342–344.

Frade, J.-M., Rodriguez-Tebar, A., and Barde, Y.-A. (1996). Induction of cell death by endogenous nerve growth factor through its p75 receptor. *Nature* **383,** 166–168.

Gedrich, R. W., Gilfillan, M. C., Duckett, C. S., Van Dongen, J. L., and Thompson, C. B. (1996). CD30 contains two binding sites with different specificities for members of the tumor necrosis factor receptor-associated factor family of signal transducing proteins. *J. Biol. Chem.* **271,** 12852–12858.

Gotz, R., Koster, R., Winkler, C., Raulf, F., Lottspeich, F., Schartl, M., and Thoenen, H. (1994). Neurotrophin-6 is a new member of the nerve growth factor family. *Nature* **372,** 266–269.

Graham, A., Francis-West, P., Brickell, P., and Lumsden, A. (1994). The signaling molecule BMP4 mediates apoptosis in the rhombencephalic neural crest. *Nature* **372,** 684–686.

Grob, P. M., Ross, A. H., Koprowski, H., and Bothwell, M. (1985). Characterization of the human melanoma nerve growth factor receptor. *J. Biol. Chem.* **260,** 8044–8049.

Hallbook, F., Ibanez, C. F., and Persson, H. (1991). Evolutionary studies of the nerve growth factor family reveal a novel member abundantly expressed in Xenopus ovary. *Neuron* **6,** 845–858.

Hantzopoulos, P. A., Suri, C., Glass, D. J., Goldfarb, M. P., and Yancopoulos, G. D. (1994). The low affinity NGF receptor, p75, can collaborate with each of the Trks to potentiate functional responses to the neurotrophins. *Neuron* **13,** 187–207.

Hempstead, B. L., Martin-Zanca, D., Kaplan, D. R., Parada, L. F., and Chao, M. V. (1991). High-affinity NGF binding requires co-expression of the trk proto-oncogene ANd the low-affinity NGF receptor. *Nature* **350,** 678–683.

Hohn, A., Leibrock, J., Bailey, K., and Barde, Y. A. (1990). Identification and characterization of a novel member of the nerve growth factor/brain-derived neurotrophic factor family. *Nature* **344,** 339–341.

Holland, D. R., Cousens, L. S., Meng, W., and Matthews, B. W. (1994). Nerve growth factor in different crystal forms displays structural flexibility and reveals zinc binding sites. *J. Mol. Biol.* **239,** 385–400.

Hsu, H., Shu, H.-B., Pan, M.-P., and Goeddel, D. V. (1996). TRADD-TRAF2 and TRADD-FADD interactions define two distinct TNF receptor-1 signal transduction pathways. *Cell* **84,** 299–308.

Huber, L. J., and Chao, M. V. (1995). A potential interaction of p75 and trkA NGF receptors revealed by affinity crosslinking and immunoprecipitation. *J. Neurosci. Res.* **40,** 557–563.

Ibanez, C. F., Ebendal, T., Barbany, G., Murrayrust, J., Blundell, T. L., and Persson, H. (1992). Disruption of the low affinity receptor-binding site in NGF allows neuronal survival and differentiation by binding to the trk gene-product. *Cell* **69,** 329–341.

Ilag, L. L., Lonnerberg, P., Persson, H., and Ibanez, C. F. (1994). Role of variable beta-hairpin loop in determining biological specificities in neurotrophin family. *J. Biol. Chem.* **269,** 19941–19946.

Ip, N. Y., Stitt, T. N., Tapley, P., Klein, R., Glass, D. J., Fandl, J., Greene, L. A., Barbacid, M., and Yancopoulos, G. D. (1993). Similarities and differences in the way neurotrophins interact with the trk receptors in neuronal and nonneuronal cells. *Neuron* **10,** 137–149.

Jing, S. Q., Tapley, P., and Barbacid, M. (1992). Nerve growth-factor mediates signal transduction through trk homodimer receptors. *Neuron* **9,** 1067–1079.

Jungbluth, S., Bailey, K., and Barde, Y. A. (1994). Purification and characterization of a brain-derived neurotrophic factor neurotrophin-3 (bdnf/nt-3) heterodimer. *Eur. J. Biochem.* **221,** 677–685.

Kaplan, D. R., and Stephens, R. M. (1994). Neurotrophin signal transduction by the trk receptor. *J. Neurobiol.* **25,** 1404–1417.

Kessler, J. A., Ludlam, W. H., Freidin, M. M., Hall, D. H., Michaelson, M. D., Spray, D. C., Dougherty, M., and Batter, D. K. (1993). Cytokine-induced programmed death of cultured sympathetic neurons. *Neuron* **11,** 1123–1132.

Klein, R., Lamballe, F., Bryant, S., and Barbacid, M. (1992). The trkb tyrosine protein-kinase is a receptor for neurotrophin-4. *Neuron* **8,** 947–956.

Kouhara, H., Hadari, Y. R., Spivak-Kroizman, T., Schilling, J., Bar-Sagi, D., Lax, I., and Schlessinger, J. (1997). A lipid-anchored Grb2-binding protein that links FGF-receptor activation to the Ras/MAPK signaling pathway. *Cell* **89,** 693–702.

Lamballe, F., Klein, R., and Barbacid, M. (1991). trkC, a new member of the trk family of tyrosine protein kinases, is a receptor for neurotrophin-3. *Cell* **66,** 967–979.

Lee, K.-F., Bachman, K., Landis, S., and Jaenisch, R. (1994). Dependence on p75 for innervation of some sympathetic targets. *Science* **263,** 1447–1449.

Liepinsh, E., Ilag, L. L., Otting, G., and Ibanez, C. F. (1997). NMR structure of the death domain of the p75 neurotrophin receptor. *EMBO J.* **16,** 4999–5005.

Lindholm, D., Castren, E., Berzaghi, M., Blochl, A., and Thoenen, H. (1994). Activity-dependent and hormonal regulation of neurotrophin mRNA levels in the brain–implications for neuronal plasticity. *J. Neurobiol.* **25,** 1362–1372.

Lindsay, R. M., Wiegand, S. J., Altar, C. A., and Distefano, P. S. (1994). Neurotrophic factors—from molecule to man. *Trends Neurosci.* **17,** 182–190.

Mahadeo, D., Kaplan, L., Chao, M. V., and Hempstead, B. L. (1994). High affinity nerve growth factor binding displays a faster rate of association than p140(trk) binding—implications for multisubunit polypeptide receptors. *J. Biol. Chem.* **269,** 6884–6891.

Maisonpierre, P. C., Belluscio, L., Squinto, S., Ip, N. Y., Furth, M. E., Lindsay, R. M., and Yancopoulos, G. D. (1990). Neurotrophin-3:

a neurotrophic factor related to NGF and BDNF. *Science* **247,** 1446–1451.

McDonald, J. I., and Meakin, S. O. (1996). *Mol. Cell. Neurosci.* **7,** 371–390.

McDonald, N. Q., Lapatto, R., Murray-Rust, J., Gunning, J., Wlodawer, A., and Blundell, T. L. (1991). A new protein fold revealed by a 2.3A resolution crystal structure of nerve growth factor. *Nature* **354,** 411–414.

McDonald, N. Q., and Hendrickson, W. A. (1993). A new structural superfamily of growth factors defined by a cystine knot motif. *Cell* **73,** 421–424.

Montgomery, R. I., Warner, M. S., Lum, B. J., and Spear, P. G. (1996). Herpes simplex virus-1 entry into cells mediated by a novel member of the TNF/NGF receptor family. *Cell* **87,** 427–436.

Muragaki, Y., Chou, T. T., Kaplan, D. R., Trojanowski, J. Q., and Lee, V. M. Y. (1997). Nerve growth factor induces apoptosis in human medulloblastoma cell lines that express trkA receptors. *J. Neurosci.* **17,** 530–542.

Nakano, H., Oshima, H., Chung, W., Williams-Abbott, L., Ware, C. F., Yagita, H., and Okumura, K. (1996). TRAF5, an activator of NF-kB and putative signal transducer for the lymphotoxin-b receptor. *J. Biol. Chem.* **271,** 14661–14664.

Nishi, R. (1994). Neurotrophic factors: Two are better than one. *Science* **265,** 1052–1053.

Pan, G., O'Rourke, K., Chinnaiyan, A. M., Gentz, R., Ebner, R., Ni, J., and Dixit, V. M. (1997). The receptor for the cytotoxic ligand TRAIL. *Science* **276,** 111–113.

Perez, P., Coll, P. M., Hempstead, B. L., Martin-Zanca, D., and Chao, M. V. (1995). NGF binding to the trk tyrosine kinase receptor requires the extracellular immunoglobulin-like domains. *Mol. Cell. Neurosci.* **6,** 97–105.

Peterson, D. A., Leppert, J. T., Lee, K.-F., and Gage, F. H. (1997). Basal forebrain neuronal loss in mice lacking neurotrophin receptor p75. *Science* **277,** 837–838.

Qiu, M.-S., and Green, S. H. (1992). PC12 cell neuronal differentiation is associated with prolonged p21ras activity and consequent prolonged ERK activity. *Neuron* **9,** 705–717.

Qui, M.-S., and Green, S. H. (1991). NGF and EGF rapidly activate p21ras in PC12 cells by distinct, convergent pathways involving tyrosine phosphorylation. *Neuron* **7,** 937–946.

Rabizadeh, S., Oh, J., Zhong, L. T., Yang, J., Bitler, C. M., Butcher, L. L., and Bredesen, D. E. (1993). Induction of apoptosis by the low-affinity NGF receptor. *Science* **261,** 345–348.

Radziewjewski, C., and Robinson, R. C. (1993). Heterodimers of the neurotrophic factors—formation, isolation, and differential stability. *Biochemistry* **32,** 13350–13356.

Rodriguez-Tebar, A., Dechant, G., Gotz, R., and Barde, Y. A. (1992). Binding of neurotrophin-3 to its neuronal receptors and interactions with nerve growth-factor and brain-derived neurotrophic factor. *Embo. J.* **11,** 917–922.

Rothe, M., Wong, S. C., Henzel, W. J., and Goeddel, D. V. (1994). A novel family of putative signal transducers associated with the cytoplasmic domain of the 75 kDa tumor necrosis factor receptor. *Cell* **78,** 681–692.

Rothe, M., Sarma, V., Dixit, V. M., and Goeddel, D. V. (1995). TRAF2-mediated activation of NF-kB by TNF receptor 2 and CD40. *Science* **269,** 1424–1427.

Ryden, M., Murray-Rust, J., Glass, D., Ilag, L., Trupp, M., Yancopoulos, G. D., McDonald, N. Q., and Ibanez, C. F. (1994). Functional analysis of mutant neurotrophins deficient in low-affinity binding reveals a role for p75LNGFR in NT-4 signaling. *EMBO J.* **14,** 1979–1990.

Smith, C. A., Farrah, T., and Goodwin, R. G. (1994). The TNF receptor superfamily of cellular and viral proteins: Activation, costimulation and death. *Cell* **76,** 959–962.

Snider, W. D. (1994). Functions of the neurotrophins during nervous-system development—what the knockouts are teaching us. *Cell* **77,** 627–638.

Sutter, A., Riopelle, R. J., Harris-Warrick, R. M., and Shooter, E. M. (1979). NGF receptors: characterization of two distinct classes of binding sites on chick embryo snesory ganglia cells. *J. Biol. Chem.* **254,** 5972–5982.

Thoenen, H. (1995). Neurotrophins and neuronal plasticity. *Science* **270,** 593–598.

Urfer, R., Tsoulfas, P., O'Connell, L., Shelton, D. L., Parada, L. F., and Presta, L. G. (1995). An immmunoglobulin-like domain determines the specificity of neurotrophin receptors. *EMBO J.* **14,** 2795–2805.

Van der Zee, C. E. E. M., Ross, G. M., Riopelle, R. J., and Hagg, T. (1996). Survival of cholinergic forebrain neurons in developing p75NGFR-deficient mice. *Science* **274,** 1729–1732.

Verdi, J. M., Birren, S. J., Ibanez, C. F., Persson, H., Kaplan, D. R., Benedetti, M., Chao, M. V., and Anderson, D. J. (1994). p75(LNGFR) regulates trk signal transduction and NGF-induced neuronal differentiation in MAH cells. *Neuron* **12,** 733–745.

von Bartheld, C. S., Kinoshita, Y., Prevette, D., Yin, Q.-W., Oppenheim, R. W., and Bothwell, M. A. (1994). Positive and negative effects of neurotrophins on the isthmo-optic nucleus in chick embryos. *Neuron* **12,** 639–654.

von Bartheld, C. S., Williams, R., Lefcort, F., Clary, D. O., Reichardt, L. F., and Bothwell, M. (1995). Retrograde transport of neurotrophins from the eye to the brain in chick embryos: Roles of the p75 NTR and trkB receptors. *J. Neurosci.* **16,** 2995–3008.

Wolf, D. E., McKinnon, C. A., Daou, M.-C., Stephens, R. M., Kaplan, D. R., and Ross, A. H. (1995). Interaction with trkA immobilizes gp75 in the high affinity nerve growth factor receptor complex. *J. Biol. Chem.* **270,** 2133–2138.

Xia, Z., Dickens, M., Raingeaud, J., Davis, R. J., and Greenberg, M. E. (1996). Opposing effects of ERK and JNK-p38 MAP kinases on apoptosis induced by neurotrophic factor withdrawal. *Science* **270,** 1326–1331.

Yao, R., and Cooper, G. M. (1995). Requirement for phosphatidylinositol-3 kinase in the prevention of apoptosis by nerve growth factor. *Science* **267,** 2003–2006.

Yoon, S. O., Casaccia-Bonnefil, P., Carter, B., and Chao, M. V. (1998). Competitive signaling between TrkA and p75 nerve growth factor receptors determines cell survival. *J. Neurosci.* **18.**

CHAPTER

7

Neurotoxic and Neurotrophic Effects of GABAergic Agents on Developing Neurotransmitter Systems

JEAN M. LAUDER
JIANGPING LIU
Department of Cell Biology and Anatomy
University of North Carolina School of Medicine
Chapel Hill, North Carolina 27599

*Abbreviations: [^{35}S]TBPS, [^{35}S]TBPS*t*-butylbicyclophosphorothionate; 5-HT, serotonin; GABA, γ-aminobutyric acid; PTX, picrotoxin; RT–PCR, reverse transcriptase–polymerase chain reaction; TH, tyrosine hydroxylase; γ-HCH; hexachlorocyclohexane.

I. Developing Neurotransmitter Systems May Be Especially Vulnerable to Environmental Neurotoxins

Neurotransmitters appear to have evolved to their highly specialized roles in the vertebrate nervous system from more primitive functions in lower organisms in which they act as paracrine and autocrine growth regulatory signals. Many of the signal transduction mechanisms used by neurotransmitters in the mammalian nervous system are highly conserved across species (Venter *et al.,* 1988). The evolutionary history of neurotransmitters appears to be reiterated in embryos of many species in which these substances serve as developmental signals regulating basic cellular functions, such as cell proliferation, differentiation, migration, and morphogenesis, by receptor-mediated mechanisms (reviewed by Lauder, 1993, 1995; Buznikov *et al.,* 1996). These functions could make the developing nervous system especially vulnerable to environmental neurotoxins that target neurotransmitter receptors. This may be especially true of the organochlorine pesticides, which act as potent antagonists of $GABA_A$ receptors in both insects and vertebrates.

II. Organochlorine Pesticides Are Potent Antagonists of $GABA_A$ Receptors

In the insect nervous system, $GABA_A$ receptors are targeted by organochlorine pesticides, including dieldrin and lindane (Costa, 1988; Bloomquist, 1992). These neurotoxins block the inhibitory effects of GABA and produce hyperexcitation of the nervous system by preventing GABA-induced Cl^- influx (Eldefrawi and Eldefrawi, 1987; Ogata *et al.,* 1988; Narahashi *et al.,* 1992). They also bind to $GABA_A$ receptors on vertebrate neurons (Abalis *et al.,* 1986; Gant *et al.,* 1987; Bloomquist, 1992). Dieldrin is a member of the chlorinated hydrocarbon (cyclodiene) group of pesticides that includes aldrin, endrin, chlordane, and heptachlor. Lindane is the gamma isomer of hexachlorocyclohexane (γ-HCH). Dieldrin and lindane both bind to the picrotoxinin (picrotoxin, PTX) site on vertebrate and insect $GABA_A$ receptors (Nagata and Narahashi, 1994; Nagata *et al.,* 1994; Tokutomi *et al.,* 1994). Cyclodiene resistence in insects has been linked to a point mutation at this site on the $GABA_A$ receptor located within the wall of the Cl^- channel (reviewed by ffrench-Constant, 1994). This site is labelled by [^{35}S]*t*-butylbicyclophosphorothionate ([^{35}S]TBPS) (Pomes *et al.,* 1994). Dieldrin and lindane competitively inhibit [^{35}S]TBPS binding to $GABA_A$ receptors (Lawrence and Casida, 1984; Cole and Casida, 1986; Llorens *et al.,* 1990), and acute treatment of adult rats with lindane causes a rapid decrease in [^{35}S]TBPS binding (Sola *et al.,* 1993).

III. Organochlorine Pesticides Are Ubiquitous Environmental Neurotoxins

Dieldrin has been banned in the United States due to its documented toxicity but is still used as an agricultural insecticide in many developing countries (Mohamed, 1990). Lindane is one of the components of technical grade HCH that is used extensively in developing countries as both an agricultural and a household insecticide (Srivastava and Raizada, 1993; Deo and Karanth, 1994). Even in the United States, lindane is still a key component of shampoos used to treat head lice infestations, frequently in children (Brown *et al.,* 1995). It is becoming increasingly clear that organochlorines are accumulating in the environment (in water and foods) and building up in human tissues. Significant levels of these compounds have been measured in breast milk of women and in fatty tissues of men, women, and children around the world (Nair *et al.,* 1992; Stevens *et al.,* 1993; Deo and Karanth, 1994; Gold-Bouchot *et al.,*1995; Kuhnlein *et al.,* 1995; Vaz, 1995), suggesting that chronic low-dose exposure is widespread. Although the acute toxicity of high doses of organochlorines is well recognized, little is yet known about effects of chronic low dose exposure. Prenatal or perinatal exposure to these substances should be of special concern, because organochlorines readily pass the placenta, accumulate in uterine tissues and breast milk (Seiler *et al.,* 1994).

Until recently, little was known about effects on the developing brain, although a few studies had reported neurobehavioral changes in postnatal rats and children chronically exposed to organochlorines during gestation or infancy (Gray *et al.,* 1981; Castro and Palermo-Neto, 1988, 1989; Castro *et al.,* 1992; Hall, 1992). As discussed in more detail later, studies from the authors' laboratory have revealed evidence that dieldrin interferes with trophic actions of GABA on monoamine neurons in embryonic rat brainstem cultures while positively regulating development of GABA neurons. We have also found evidence that exposure to dieldrin or lindane *in utero* alters gene expression of $GABA_A$ receptor subunits and reduces [^{35}S]TBPS binding in fetal rat brainstem (Brannen *et al.,* 1997a,b). These findings raise the possibility that *in utero* exposure to organochlorine pesticides could alter functioning of monoaminergic and GABAergic systems and have behavioral consequences in offspring.

IV. Organochlorine Pesticides Block the Trophic Actions of GABA on Embryonic Monoamine Neurons, but Promote Development of GABA Neurons

GABA acts as a trophic signal for many types of developing central nervous system (CNS) neurons (reviewed by Wolff *et al.,* 1993; Belhage *et al.,* 1997; Lauder *et al.,* 1997). GABA is remarkable as the most precocious neurotransmitter known in the developing rat brain (Lauder *et al.,* 1986; Ma *et al.,* 1992; Lundgren *et al.,* 1995; Dupuy and Houser, 1996). In the rat embryo, GABAergic axons appear in the brainstem as early as embryonic day 12 (E12) and grow rapidly through regions where serotonin (5-HT) and noradrenergic neurons are beginning to appear (Lauder *et al.,* 1986). This has led us to speculate that development of brainstem monoamine neurons might be influenced by GABA released from developing GABAergic axons.

To test this hypothesis, the authors have used dissociated cell cultures from E14 rat brainstem, which contain developing 5-HT, noradrenaline (tyrosine hydroxylase; TH) and GABA neurons (Liu *et al.,* 1997a). These cultures express multiple $GABA_A$ receptor subunit mRNA transcripts, including high levels of $\beta3$ transcripts, and have functional $GABA_A/Cl^-$ channels. Treatment of these cultures with GABA and/or a $GABA_A$ antagonist (dieldrin or bicuculline) differentially regulates survival and growth of monoamine and GABA neurons. GABA exerts positive effects on monoamine neurons, which are countered by dieldrin or bicuculline, suggesting involvement of $GABA_A$ receptors. Dieldrin also strongly inhibits growth and survival of 5-HT neurons when given alone. In contrast, GABA neurons respond positively to these $GABA_A$ antagonists, which promote their growth and survival.

Taken together, results of these *in vitro* studies suggest that functional inhibition of $GABA_A$ receptors by environmental neurotoxins such as dieldrin could inhibit development of embryonic brainstem monoamine neurons, but stimulate development of GABA neurons. This could lead to imbalances in monoaminergic and GABAergic neurotransmission in the developing brain. Moreover, because GABAergic activity at $GABA_A$ receptors appears to regulate (modulate) developmental expression of these receptors (discussed later), *in utero* exposure to these neurotoxins could signifcantly alter functional development of the GABAergic system.

V. Organochlorine Pesticides Alter Developmental Expression of $GABA_A$ Receptors

Transient, region-specific patterns of $GABA_A$ receptor subunit expression in developing rat brain have been described using *in situ* hybridization or immunocytochemistry (Gambarana *et al.,* 1991; Laurie *et al.,* 1992; Poulter *et al.,* 1992, 1993; Zdilar *et al.,* 1992; Ma *et al.,* 1993; Fritschy *et al.,* 1994). Transient $GABA_A$ subunit expression patterns have also been found in developing primate brain (Hornung and Frtischy, 1996). These expression patterns appear to confer specific functional characteristics on developing $GABA_A$ receptors that may underlie the trophic actions of GABA (Schousboe and Redburn, 1995).

$GABA_A$ receptors develop in spatiotemporal synchrony with GABAergic pathways in the rat embryo (Schlumpf *et al.,* 1989, 1992; Cobas *et al.,* 1991). These receptors appear to be functional, as judged by electrophysiologic criteria in acutely dissociated cells (Fiszman *et al.,* 1990; Ma *et al.,* 1993; Walton *et al.,* 1993) and the ability of GABA to stimulate Cl^- influx in fetal synaptoneurosomes (Kellogg and Pleger, 1989). In the human embryo, $GABA_A$ receptors are detectable by [^{3}H]flunitrazepam binding at 7 weeks gestation, and increase sharply between 8–11 weeks gestation. GABA enhances ligand binding to these receptors, suggesting that they are functional (Hebebrand *et al.,* 1988).

Coordinate development of GABAergic axons and $GABA_A$ receptors has led to speculation that GABA may regulate (modulate) expression of its own receptors (Schousboe and Redburn,1995). Therefore, it is possible that enhancing or blocking the actions of GABA could alter $GABA_A$ receptor development. It is well known that prenatal exposure to ligands for dopamine or serotonin receptors significantly alters postnatal expression of these receptors (Whitaker-Azmitia *et al.,* 1987; Miller and Friedhoff, 1988; Lauder *et al.,* 1994). Similarly, prenatal exposure to drugs or neurotoxins that target $GABA_A$ receptors alters expression of $GABA_A$ receptors and produces long-lasting behavioral changes. Chronic administration of benzodiazepines (BZD) to pregnant rats produces behavioral and physiological abnormalities in offspring, especially following pharmacologic challenge, which suggest that alterations in $GABA_A$ receptor expression have occurred (Kellogg, 1988; Schlumpf *et al.,* 1989, 1992). Direct evidence that activity at developing $GABA_A$ receptors regulates developmental expression of $GABA_A$ receptor subunits and binding sites has been provided by both *in vitro* and *in vivo* studies.

A. *In Vitro Studies*

In cell cultures derived from embryonic rodent or chick brain, chronic exposure to $GABA_A$ agonists up- or down-regulates expression of particular $GABA_A$ subunit transcripts, whereas $GABA_A$ antagonists have opposite effects and block the effects of agonists (Montpied *et al.,* 1991; Hirouchi *et al.,* 1992).

The authors of this chapter determined whether exposure of E14 brainstem cultures to dieldrin alters developmental expression of $GABA_A$ receptor subunit mRNA transcripts (Liu *et al.,* 1997b). These cultures were treated for 48 hr with 10 μM dieldrin in serum-free medium. Effects on abundance of $\alpha1$, $\beta3$, $\gamma1$, $\gamma2S$, and $\gamma2L$ mRNA transcripts were absoutely quantified by competitive reverse transcriptase–polymerase chain reaction (RT–PCR) using subunit-selective internal standards (Grayson *et al.,* 1993). Dieldrin differentially regulated expression of these transcripts, such that $\beta3$ transcripts were significantly increased, whereas $\gamma2S$ and $\gamma2L$ transcripts were decreased. Dieldrin slightly increased expression of $\alpha1$ subunit transcripts, but this was not statistically significant, and did not alter expression of $\gamma1$ transcripts. These subunit selective effects could alter sensitivities of $GABA_A$ receptors expressed in these cultures to BZD and other ligands (see review by Möhler *et al.,* 1997).

These studies support the hypothesis that *in utero* exposure to organochlorine pesticides could alter developmental expression of $GABA_A$ receptor subunits and perhaps produce $GABA_A$ receptors with altered functional properties.

B. In Vivo Studies

One study found that treatment of pregnant rats from E12 through E17 with dieldrin, lindane, or bicuculline caused a significant reduction in near-maximal [^{35}S] TBPS binding in brainstem of E17 fetuses, but had no significant effects on the rest of brain when analyzed as a whole (Brannen *et al.,* 1997a,b). In a separate study, using the same experimental paradigm, it was found that these treatments decreased expression of multiple $GABA_A$ receptor subunit mRNA transcripts in fetal brainstem (Liu *et al.,* 1998).

VI. Summary and Conclusions

Taken together, the studies reviewed here provide evidence that *in utero* exposure to organochlorine pesticides acting as $GABA_A$ receptor antagonists can alter prenatal development of $GABA_A$ receptors. They may also block trophic effects of GABA on monoamine neurons while promoting development of GABA neurons. If such effects are long lasting, they could seriously impact development of functional GABAergic and monoaminergic brain circuitry and have significant behavioral consequences in offspring. These findings raise the possibility that exposure of pregnant women to organochlorine pesticides could pose a significant risk for brain development of the fetus.

Acknowledgments

This work was supported by grant ES07017 from NIEHS. The authors are grateful to Mary Beth Wilkie for editorial assistance.

References

Abalis, I. M., Eldefrawi, M. E., and Eldefrawi, A. T. (1986). Effects of insecticides on GABA-induced chloride influx into rat brain microsacs. *Toxicol. J. Environ. Health* **18,** 13–23.

Belhage, B., Hansen, G. H., Elster, L., and Schousboe, A. (1997). Effects of γ-aminobutyric acid (GABA) on synaptogenesis and synaptic function. *Persp. Dev. Neurobiol.* **5(1–2),** (In Press).

Bloomquist, J. R. (1992). Intrinsic lethality of chloride-channel–directed insecticides and convulsants in mammals. *Toxicol. Lett.* **60,** 289–298.

Brannen, K. C., Devaud, L. L., Liu, J., and Lauder, J. M. (1997a). The effects of prenatal exposure to environmental neurotoxicants on *tert*-butylbicyclophosphorothinate binding properties. *Fund. Appl. Toxicol.* **36,** 71.

Brannen, K. C., Devaud, L. L., Liu, J., and Lauder, J. M. (1997b). Prenatal exposure to neurotoxicants dieldrin or lindane alters *tert*-butylbicyclophosphorothionate binding to $GABA_A$ receptors in fetal rat brainstem. *Dev. Neurosci.* (In Press).

Brown, S., Becher, J., and Brady, W. (1995). Treatment of ectoparasitic infections: review of the English-language literature, 1982–1992. *Clin. Infect. Dis.* **20,** S104–S109.

Buznikov, G. A., Shmukler, Y. B., and Lauder, J. M. (1996). From oocyte to neuron: do neurotransmitters function the same way throughout development? *Cell. Mol. Neurobiol.* **16,** 533–559.

Castro, V. L., Bernardi, M. M., and Palermo-Neto, J. (1992). Evaluation of prenatal aldrin intoxication in rats. *Arch. Toxicol.* **66,** 149–152.

Castro, V. L., and Palermo-Neto, J. (1988). Alterations in the behavior of young and adult rats exposed to aldrin during lactation. *Braz. J. Med. Biol. Res.* **21,** 987–990.

Castro, V. L., and Palermo-Neto, J. (1989). Effect of a specific stress (aldrin) on rat progeny behavior. *Braz. J. Med. & Biol. Res.* **22,** 979–982.

Cobas, A., Fairen, A., Alvarez-Bolado, G., and Sanchez, M. P. (1991). Prenatal development of the intrinsic neurons of the rat neocortex: a comparative study of the distribution of the GABA-immunoreactive cells and the $GABA_A$ receptor. *Neuroscience* **40,** 375–397.

Cole, L. M., and Casida, J. E. (1986). Polychlorocycloalkane insecticide-induced convulsions in mice in relation to disruption of the GABA-regulated chloride ionophore. *Life Sci.* **39,** 1855–1862.

Costa, L. G. (1988). Interactions of neurotoxicants with neurotransmitter systems. *Toxicology* **49,** 359–366.

Deo, P. G., and Karanth, N. G. (1994). Biodegradation of hexachlorocyclohexane isomers in soil and food environment. *Crit. Rev. Microbiol.* **20,** 57–78.

Dupuy, S. T., and Houser, C. R. (1996). Prominent expression of two forms of glutamate decarboxylase in the embryonic and early postnatal rat hippocampal formation. *J. Neurosci.* **16,** 6919–6932.

Eldefrawi, A. T., and Eldefrawi, M. E. (1987). Receptors for γ-aminobutyric acid and voltage-dependent chloride channels as targets for drugs and toxicants. *FASEB J.* **1,** 262–271.

ffrench-Constant, R. H. (1994). The molecular and population genetics of cyclodiene insecticide resistance. *Insect Biochem. Mol. Biol.* **24,** 335–345.

Fiszman, M. L., Novotny, E. A., Lange, G. D., and Barker, J. L. (1990). Embryonic and early postnatal hippocampal cells respond to nanomolar concentrations of muscimol. *Dev. Brain Res.* **53,** 186–193.

Fritschy, J. M., Paysan, J., Enna, A., and Mohler, H. (1994). Switch in the expression of rat $GABA_A$-receptor subtypes during postnatal development: an immunohistochemical study. *J. Neurosci.* **14,** 5302–5324.

Gambarana, C., Beattie, C. E., Rodriguez, Z. R., and Siegel, R. E. (1991). Region-specific expression of messenger RNAs encoding $GABA_A$ receptor subunits in the developing rat brain. *Neuroscience* **45,** 423–432.

Gant, D. B., Eldefrawi, M. E., and Eldefrawi, A. T. (1987). Cyclodiene insecticides inhibit $GABA_A$ receptor-regulated chloride transport. *Toxicol. Appl. Pharmacol.* **88,** 313–321.

Gold-Bouchet, G., Silva-Herrera, T., and Zapata-Perez, O. (1995). Organochlorine pesticide residue concentrations in biota and sediments from Rio Palizada, Mexico. *Bull. Environ. Contam. Toxicol.* **54,** 554–561.

Gray, L. E. Jr., Kavlock, R. J., Chernoff, N., Gray, J. A., and McLamb, J. (1981). Perinatal toxicity of endrin in rodents. III. Alterations of behavioral ontogeny. *Toxicology* **21,** 187–202.

Grayson, D. R., Bovolin, P., and Santi, M. R. (1993). Absolute quantitation of γ-aminobutyric acid A receptor subunit mRNAs by competitive polymerase chain reaction. *Meth. Neurosci.* **12,** 191–208.

Hall, R. H. (1992). A new threat to public health: organochlorines and food. *Nutrit Health* **8,** 33–43.

Hebebrand, J., Hofmann, D., Reichelt, R., Scnarr, S., Knapp, M., Propping, P., and Födisch, H. J. (1988). Early ontogeny of the central benzodiazepine receptor in human embryos and fetuses. *Life Science* **43,** 2127–2136.

Hirouchi, M., Ohkuma, S., and Kuriyama, K. (1992). Muscimol-induced reduction of $GABA_A$ receptor α1-subunit mRNA in primary cultured cerebral cortical neurons. *Mol. Brain Res.* **15,** 327–331.

Hornung, J. P., and Fritschy, J.-M. (1996). Developmental profile of $GABA_A$-receptors in the marmoset monkey: expression of distinct subtypes in pre- and postnatal brain. *J. Comp. Neurol.* **367,** 413–430.

Kellogg, C. K. (1988). Benzodiazepines: influence on the developing brain. *Prog. Brain Res.* **73,** 207–228.

Kellogg, C. K., and Pleger, G. L. (1989). GABA-stimulated chlroide uptake and enhancement by diazepam in synaptoneurosomes from rat brain during prenatal and postnatal development. *Brain Res. Dev. Brain Res.* **49,** 87–95.

Kuhnlein, H. V., Receveur, O., Muir, D. C., Chan, H. M., and Soueida, R. (1995). Arctic indigenous women consume greater than acceptable levels of organochlorines. *J. Nutrit.* **125,** 2501–2510.

Lauder, J. M. (1993). Neurotransmitters as growth regulatory signals: role of receptors and second messengers. *Trends Neurosci.* **16,** 233–240.

Lauder, J. M. (1995). Ontogeny of Neurotransmitter Systems: Substrates for Developmental Disabilities? *Men. Retard. Devel. Disabil. Res. Rev.* **1,** 151–168.

Lauder, J. M., Han, V. K. M., Henderson, P., and Verdoorn, T. (1986). Prenatal ontogeny of the GABAergic systemin the rat brain: an immunocytochemical study. *Neuroscience* **19,** 465–493.

Lauder, J. M., Liu, J., Devaud, L., and Morrow, A. L. (1997). GABA as a trophic factor for developing monoamine neurons. *Perspect. Dev. Neurobiol.* **5(1–2),** (In press)

Lauder, J. M., Moiseiwitsch, J., Liu, J., and Wilkie, M. B. (1994). Serotonin in development and pathophysiology. *Brain Lesions in the Newborn, Alfred Benzon Symp.* **37,** 60–73.

Laurie, D. J., Wisden, W., and Seeburg, P. H. (1992). The distribution of thirteen $GABA_A$ receptor subunit mRNAs in the rat brain. III. Embryonic and postnatal development. *J. Neurosci.* **12,** 4151–4172.

Lawrence, L. J., and Casida, J. E. (1984). Interactions of lindane, toxaphene and cyclodienes with brain-specific *t*-butylbicyclophosphorothionate receptor. *Life Sciences* **35,** 171–178.

Liu, J., Brannen, K. C., Grayson, D. R., Morrow, A. L., Devaud, L. L., and Lauder, J. M. (1998). Prenatal exposure to the pesticide dieldrin or the GABA A receptor antagonist bicuculline differentially alters expression of GABA A receptor subunit mRNAs in fetal brainstem. *Dev. Neurol.* (In press).

Liu, J., Morrow, A. L., Devaud, L., Grayson, D. R., and Lauder, J. M. (1997a). $GABA_A$ receptors mediate trophic effects of GABA on embryonic brainstem monoamine neurons. *J. Neurosci.* **17,** 2420–2428.

Liu, J., Morrow, A. L., Devaud, L. L., Grayson, D. R., and Lauder, J. M. (1997b). Regulation of $GABA_A$ subunit mRNA expression by the pesticide dieldrin in embryonic brainstem cultures: a quantitative competitive RT-PCR study. *J. Neurosci. Res.* **49,** 645–653.

Llorens, J., Sunol, C., Tusell, J. M., and Rodriguez-Farre, E. (1990). Lindane inhibition of [^{35}S]TBPS binding to the $GABA_A$ receptor in rat brain. *Neurotox. Teratol.* **12,** 607–610.

Lundgren, P., Mattsson, M. O., Johansson, L., Ottersen, O. P., and Sellstrom, A. (1995). Morphological and GABA-immunoreactive development of the embryonic chick telencephalon. *Int. J. Dev. Neurosci.* **13,** 463–472.

Ma, W., Behar, T., and Barker, J. L. (1992). Transient expression of GABA immunoreactivity in the developing rat spinal cord. *J. Comp. Neurol.* **325,** 271–290.

Ma, W., Saunders, P. A., Somogyi, R., Poulter, M. O., and Barker, J. L. (1993). Ontogeny of $GABA_A$ receptor subunit mRNAs in rat spinal cord and dorsal root ganglia. *J. Comp. Neurol.* **338,** 37–359.

Miller, J. C., and Friedhoff, A. J. (1988). Prenatal neurotransmitter programming of postnatal receptor function. *Prog. Brain Res.* **73,** 509–522.

Mohamed, B. (1990). *Neurotoxicology: New Developments in Neuroscience.* Van Nostrand, New York, pp. 249–250.

Möhler, H., Benke, D., Benson, J., Lüscher, B., Rudolph, U., and Fritschy, J.-M. (1997). Diversity in structure, pharmacology, and regulation of $GABA_A$ receptors. In *The GABA Receptors.* S. J. Enna and N. G. Bowery, Eds., Humana Press, Totowa, New Jersey, pp. 11–36.

Montpied, P., Ginns, E. I., Martin, B. M, Roca, D., Farb, D. H., and Paul, S. M. (1991). γ-Aminobutyric acid (GABA) induces a receptor-mediated reduction in $GABA_A$ receptor alpha subunit messenger RNAs in embryonic chick neurons in culture. *J. Biol. Chem.* **266,** 6011–6014.

Nagata, K., Hamilton, B. J., Carter, D. B., and Narahashi, T. (1994). Selective effects of dieldrin on the $GABA_A$ receptor-channel complex subunits expressed in human embryonic kidney cells. *Brain Res.* **645,** 19–26.

Nagata, K., and Narahashi, T. (1994). Dual action of the cyclodiene insecticide dieldrin on the γ-aminobutyric acid receptor–chloride channel complex of rat dorsal root ganglion neurons. *J. Pharm. Exp. Ther.* **269,** 164–172.

Nair, A., Dureja, P., and Pillai, M. K. (1992). Aldrin and dieldrin residues in human fat, milk and blood serum collected from Delhi. *Hum. Exp. Toxicol.* **11,** 43–45.

Narahashi, T., Frey, J. M., Ginsburg, K. S., and Roy, M. L. (1992). Sodium and GABA-activated channels as the targets of pyrethroids and cyclodienes. *Toxicol. Lett.* **64–65,** 429–436.

Ogata, N., Vogel, S. M., and Narahashi, T. (1988). Lindane but not deltamethrin blocks a component of a GABA-activated chloride channel. *FASEB J.* **2,** 2895–2900.

Pomes, A., Frandsen, A. A., Sunol, C., Sanfeliu, E., Rodriguez-Farre, E., and Schousboe, A. (1994). Lindane cytotoxicity in cultured neocortical neurons is ameliorated by GABA and flunitrazepam. *J. Neurosci. Res.* **39,** 663–668.

Poulter, M. O., Barker, J. L., O'Carroll, A.-M., Lolait, S. J., and Mahan, L. C. (1992). Different and transient expression of $GABA_A$ receptor α–subunit mRNAs in the developing rat CNS. *J. Neurosci.* **12,** 2888–2900.

Poulter, M. O., Barker, J. L., O'Carroll, A.-M., Lolait, S. J., and Mahan, L. C. (1993). Co-existent expression of $GABA_A$ receptor β2, β3 and γ2 subunit messenger RNAs during embryogenesis and early postnatal development of the rat central nervous system. *Neuroscience* **53,** 1019–1033.

Schlumpf, M., Parmar, R., Schreiber, A., Ramseier, H. R., Butikofer, E., Abriel, H., Barth, M., Rhyner, T., and Lichtensteiger, W. (1992). Nervous and immune systems as targets for developmental effects of benzodiazepines. A review of recent studies. *Dev. Pharmacol. Ther.* **18,** 145–158.

Schlumpf, M., Ramseier, H., Abriel, H., Youmbi, M., Baumann, J. B., Lichtensteiger, W. (1989). Diazepam effects on the fetus. *Neurotoxicology* **10,** 501–516.

Schousboe, A., and Redburn, D. A. (1995). Modulatory actions of gamma aminobutyric acid (GABA) and GABA type A receptor subunit expression and function. *J. Neurosci. Res.* **41,** 1–7.

Seiler, P., Fischer, B., Lindenau, A., and Beier, H. M. (1994). Effects of persistent chlorinated hydrocarbons on fertility and embryonic development in the rabbit. *Human Reprod.* **9,** 1920–1926.

Sola, C., Martinez, E., Camon, L., Pazos, A., and Rodriguez-Farre, E. (1993). Lindane administration to the rat induces modifications in the regional cerebral binding of [^{3}H]muscimol, [^{3}H]flunitrazepam, and *t*-[^{35}S]butyl-bicyclophosphorothionate: an autoradiographic study. *J. Neuroschem.* **60,** 1821–1834.

Srivastava, M. K., and Raizada, R. B. (1993). Prenatal effects of technical hexachlorocyclohexane in mice. *J. Toxicol. Environ. Health* **40,** 105–115.

Stevens, M. F., Ebell, G. F., and Psaila-Savona, P. (1993). Organochlorine pesticides in Western Australian nursing mothers. *Med. J. Australia* **158,** 238–241.

Tokutomi, N., Ozoe, Y., Katayama, N., and Akaika, N. (1994). Effects of lindane (γ-BHC) and related convulsants on $GABA_A$ receptor-operated chloride channels in frog dorsal root ganglion neurons. *Brain Res.* **643,** 66–73.

Vaz, R. (1995). Average Swedish dietary intakes of organochlorine contaminants via foods of animal origin and their relation to levels in human milk, 1975-90. *Food Addit. Contam.* **12,** 543–558.

Venter, J. C., diPorzio, U., Robinson, D. A., Shreeve, S. M., Lai, J., Kerlavage, A. R., Fracek, S. P. Jr., Lentes, K. U. and Fraser, C. M. (1988). Evolution of neurotransmitter systems. *Prog. Neurobiol.* **30,** 105–169.

Walton, M. K., Schaffner, A. E., and Barker, J. L. (1993). Sodium channels, $GABA_A$ receptors, and glutamate receptors develop sequentially on embryonic rat spinal cord cells. *J. Neurosci.* **13,** 2068–2084.

Whitaker-Azmitia, P., Lauder, J., Shemmer, A., and Azmitia, E. (1987). Postnatal changes in serotonin receptors following prenatal alteration in serotonin levels: further evidence for functional fetal serotonin receptors. *Dev. Brain Res.* **33,** 285–289.

Wolff, J. R., Joo, F., and Kasa, P. (1993). Modulation by GABA of neuroplasticity in the central and peripheral nervous system. *Neurochem. Res.* **18,** 453–461.

Zdilar, D., Luntz-Leybman, V., Frostholm, A., and Rotter, A. (1992). Differential expression of $GABA_A$/benzodiazepine receptor beta 1, beta 2, and beta 3 subunit mRNAs in the developing mouse cerebellum. *J. Comp. Neurol.* **326,** 580–594.

CHAPTER

8

Apoptosis

MARGARET A. SHIELD
PHILIP E. MIRKES
Department of Pediatrics
University of Washington
Seattle, Washington 98195

*Abbreviations: AAF, acetylaminofluorene; ADP, adenosine phosphate; AIF, apoptosis inducing factor; bp, base pairs; CAPK, ceramide-activated protein kinase; CAPP, ceramide-activated phosphatase; CNS, central nervous system; CTL, cytotoxic T lymphocyte cells; DED, death-effector domain; DFF, DNA fragmentation factor; DNA-PK, DNA-dependent protein kinase; DRG, dorsal root ganglia; dUTP, deoxy uridine transplant; EGF, epidermal growth factor; ERK, extracellular signal-regulated kinase; FasL, Fas ligand; FGF, fibroblast growth factor; GFP, green fluorescent protein; GM-CSF, granulocyte-macrophage colony stimulating factor; HIV, human immunodeficiency virus; IAP, inhibitor of apoptosis protein; ICE, interleukin-1β converting enzyme; IL, interleukin; IL-1β, interleukin-1β; JNK, c-Jun N-terminal kinase; JNKK, JNK kinase; kDa, kilodaltons; MAPK, mitogen-activated protein kinase; MAPKKK, mitogen-activated protein kinase kinase kinase; MEKK, MAP kinase/ERK kinase kinase; MPT, mitochondrial permeability transition; NAIP, neuronal apoptosis inhibitor protein; NGF, nerve growth factor; NK, natural killer; PARP, poly (ADP-ribose) polymerase; PCD, programmed cell death; PDGF, platelet-derived growth factor; PKA, protein kinase A; PKC, protein kinase C; PKC-ζ, protein kinase C-ζ; PNS, peripheral nervous system; RA, retinoic acid; ROS, reactive oxygen species; SMA, spinal muscular atrophy; SMase, sphingomyelinase; SRE/BP, sterol regulatory element-binding protein; TdT, terminal deoxynucleotidyl transferase; TGF-β, transforming growth factor-β; TM, transmembrane; TNF, tumor necrosis factor; TRAIL, TNF-related apoptosis inducing ligand; TUNEL, TdT-mediated dUTP-biotin nick end labeling.

I. Introduction

The term *apoptosis* was first used by Kerr, Wyllie, and Currie (1972) to describe the distinct type of cell death observed by many different researchers during normal development and in response to various toxins. The word *apoptosis* was derived from the ancient Greek for "dropping off" of leaves from trees, signifying the natural role of cell suicide. Results of intense study of apoptotic cell death over the past decade have shown that apoptosis involves an orderly, reproducible series of molecular events controlled by specific genes that have been conserved throughout evolution. Any discussion of nervous system development requires consideration of the role of apoptosis because this cell death mechanism is critical to both the normal development of the mammalian nervous system and the damage produced by a wide range of toxins. The magnitude of recent research articles in this field is astounding; therefore, we apologize for not being able to cite all relevant articles.

II. Overview of Morphology and Molecular Events of Apoptosis

Apoptotic cell death can be divided into four sequential stages: initiation, decision to die, execution, and engulfment. A range of cellular stresses can initiate apoptosis by the activation of intracellular signaling pathways or the release of second messengers. Some examples of cell death triggers are ionizing radiation, chemotherapeutic drugs, hyperthermia, and growth factor deprivation. Molecules presented by other cells, such as Fas ligand (FasL) and tumor necrosis factor (TNF), can also induce apoptosis in cells bearing the appropriate receptors. A cell may receive a variety of signals simultaneously, including both proliferation and cell death signals. In response to these signals, a cell must assess its situation and decide whether to commit suicide by apoptosis. Key regulatory molecules at this stage are members of the Bcl-2 protein family (White, 1996; Yang and Korsmeyer, 1996), although the mechanism of their action is not fully understood. Once a cell has made the decision to die, molecules that act as executioners are activated. In all types of apoptosis studied, in a variety of multicellular organisms, the executioners include cytosolic proteases, the majority of which belong to a family called the *caspases* (Nicholson and Thornberry, 1997; Alnemri, 1997). These proteases attack specific cellular proteins resulting in the irreversible destruction of critical cellular processes and structures. In the final stage of apoptosis, the remnants of the dying cell are recognized, engulfed, and degraded, either by a neighboring cell or by a scavenging macrophage.

The morphologic changes associated with apoptosis are distinct and well characterized (Wyllie, 1987; Darzynkiewicz *et al.,* 1997). Cells undergoing apoptosis rapidly shrink and condense, pulling away from neighboring cells. As a result of the loss of cytoplasmic volume, the plasma membrane forms blebs and protrusions, giving the cell a blistered look. A rapid increase in intracellular calcium ion concentration is also commonly observed (Schwartzman and Cidlowski, 1993). The nuclear membrane rounds up and the chromatin condenses and aggregates into dense, crescent-like shapes near the nuclear membrane. Chromatin condensation is associated with double-stranded cleavage of DNA by one or more nuclear endonucleases (Wyllie *et al.,* Montague and Cidlowski, 1996) and degradation of the nuclear lamina by proteases (Lazebnik *et al.,* 1995).

DNA degradation during apoptosis occurs in a precise, reproducible pattern (Wyllie *et al.,* 1992; Montague and Cidlowski, 1996). Initially, chromosomal DNA is cleaved into large segments of about 50–300 kilobases (Oberhammer *et al.,* 1993). In most, but not all, instances of apoptosis, these domains are then further digested into smaller fragments by cleavage between nucleosomes. Separation of DNA from apoptotic nuclei by agarose gel electrophoresis yields a characteristic pattern of these mono- or oligonucleosomal fragments in multiples of 185–200 base pairs (bp), described as a *DNA ladder* (Tilly and Hsueh, 1993). The endonucleases responsible for this DNA fragmentation have not been convincingly identified, although several candidates have been suggested (Montague and Cidlowski, 1996). Some studies have suggested a role for DNase I or II, but these enzymes are not normally located in the nucleus. Perhaps a better candidate is an 18 kD Ca^{2+}/Mg^{2+}–dependent endonuclease called *NUC18* that was isolated from apoptotic nuclear extracts and found to be highly homologous to cyclophilin A. NUC18 may be responsible for cleavage of DNA into 50 kD fragments. In addition, a novel 95 kD Ca^{2+}/Mg^{2+}–dependent endonuclease that is active during apoptosis has been identified (Pandey *et al.,* 1997). A protein called *DNA fragmentation factor* (*DFF*) that appears to activate the apoptotic endonuclease(s) has also been isolated from HeLa cells (Liu *et al.,* 1997).

Changes in the mitochondria of apoptotic cells precede cellular condensation and nuclear disintegration and may be an essential early event in apoptosis (Petit *et al.,* 1996; Kroemer, 1997). The mitochondria exhibit a depolarization in membrane potential ($\Delta\psi_m$) (Marchetti *et al.,* 1996; Zamzami *et al.,* 1995b; Zamzami *et al.,* 1996), which appears to be due to the opening of mitochondrial permeability transition (MPT) pores (Zoratti and Szabo, 1995). Proteins released from apoptotic mitochondria into the cytoplasm either just before or just after the MPT are capable of inducing chromatin condensation and DNA fragmentation (Liu *et al.,* 1996; Susin *et al.,* 1996; Zamzami *et al.,* 1996).

The dying cell breaks up into several round membrane-enclosed pieces called *apoptotic bodies* that are phagocytosed and degraded by phagocytic cells (Savill *et al.,* 1993; Hart *et al.,* 1996). Epithelial cells, endothelial cells, and fibroblasts adjacent to sites of apoptosis may engulf apoptotic bodies. Several studies also demonstrate that "professional" macrophages are recruited to sites of cell death and are responsible for much of the phagocytosis of apoptotic bodies (Hopkinson-Woolley *et al.,* 1994; Camp and Martin, 1996). This engulfment prevents an inflammatory response from leakage of cellular debris into intercellular spaces. Apoptotic cells display a range of signals to attract phagocytes, including changes in cell surface molecules such as sugars, lipids, and proteins. Although phagocytic recognition of apoptotic cells is a critical process and an active field of research, details of this final event in apoptosis are beyond the scope of this review.

The cellular events of apoptosis contrast with those of necrosis, although both eventually result in the death of the cell (Darzynkiewicz *et al.,* 1997). Necrosis, or accidental cell death, is characterized by rapid, almost instantaneous death of a cell due to a catastrophic injury. Necrotic cells swell to a large volume, exhibiting a dramatic increase in mitochondrial volume. The plasma membrane is disrupted and cellular contents are released, typically producing an inflammatory response that damages neighboring cells. Degradation of DNA sometimes occurs during necrosis; however, the cleavage sites are random, resulting in a full range of fragment sizes. Instances of cell death can often be clearly distinguished as necrotic or apoptotic, but in some cases a dying cell exhibits characteristics of both processes.

The term *programmed cell death* (*PCD*) is commonly used to describe the demise of cells during normal development of an organism. In most but not all cases, PCD proceeds by the same stereotypical process as apoptosis; therefore, the terms apoptosis and PCD are often used interchangeably. An alternative practice is to define the term apoptosis as descriptive of one of the mechanisms of PCD. Throughout this chapter, we use the term apoptosis to describe cell death induced by abnormal extracellular stresses and the term PCD to indicate apoptotic cell death that is developmentally predestined and normally occurring.

III. Biologic Significance of Apoptosis

A. *Developmental Events Involving Apoptosis*

PCD occurs in a wide range of developing tissues (Jacobson *et al.,* 1997). Among the critical developmental roles of PCD are the sculpting of tissues, the removal of embryonic structures that are unnecessary in the adult, and the controlled removal of excess, nonfunctional, or misplaced cells. Formation of the digits during limb development demonstrates the importance of PCD in sculpting tissues. The distal portion of the limb bud differentiates into digits as a result of dramatic cell death in the interidgital mesoderm (Garcia-Martinez *et al.,* 1993; Hurle *et al.,* 1995). In this situation, the PCD is so rapid and localized that a dense region of apoptotic bodies is visible between the developing digits. Genetic defects or teratogenic exposures that result in the formation of webbed fingers or toes result from incomplete cell death in the interdigital region.

PCD plays a critical role in normal development of the vertebrate nervous system because the number of both neurons and oligodendrocytes formed at early embryonic stages is in great excess of the number required in the mature organism. Fifty percent or more of sensory neurons and motoneurons formed during development undergo PCD (Oppenheim, 1991; Johnson and Deckwerth, 1993; Raff *et al.,* 1993). This massive amount of cell death begins as axons connect to their target tissues during embyrogenesis and continues postnatally (Naruse and Keino, 1995). Most of the PCD occurring during nervous system development appears to be apoptotic cell death.

The current mechanistic model explaining this PCD, called the *neurotrophic theory,* is that the survival of neurons is dependent on specific neurotrophic factors that are secreted by the synaptic target cells (Oppenheim, 1991; Yuen *et al.,* 1996). Neurons that fail to reach appropriate targets also fail to receive the appropriate growth factor stimuli and subsequently die by apoptosis. Other factors related to proper matching of neuron to target, such as electrical activity, afferent stimulation, and cell–cell interactions, may also influence this PCD. Thus cell death matches the number of neurons to the number of target cells and guards against inappropriate connections. Oligodendrocytes are also initially in excess, but undergo PCD until their number matches that of the axons they myelinate. Although it is now well accepted that cells in the nervous system die in large numbers during development, the rationale for this developmental process is still not understood. Does an overproduction of neurons offer organisms a greater variety of options for innervation of target tissues? Or does the PCD of neurons correct a lack of regulation in the early stages of neurogenesis?

The critical neurotrophic factor for sensory and sympathetic neurons is nerve growth factor (NGF), although these neurons are also responsive to other neurotrophins (Yuen *et al.,* 1996). Exogenously supplemented NGF blocks neuronal PCD, whereas removal of NGF *in vivo* results in an increase in neuronal PCD (Johnson and Deckwerth, 1993). Cultured neuronal cells exhibit a similar requirement for NGF. Motoneurons do not respond to NGF, but are responsive to a larger array of survival factors including fibroblast growth factor (FGF), platelet-derived growth factor (PDGF), transforming growth factor-β (TGF-β), and insulin and insulin-like growth factors (Oppenheim, 1996; Yuen *et al.,* 1996).

PCD is also observed during the joining of epithelial sheets to form tubular structures in the embryo, such as during the closure of the neural tube (Naruse and Keino, 1995). Developing chick embryos exposed to peptide inhibitors of the apoptotic caspases exhibit less PCD in the developing neural tube and the neural tube fails to close (Weil *et al.,* 1997). These results suggest that apoptosis is required for correct formation of the neural tube.

B. Apoptosis and Disease

The important role of apoptosis in development and tissue homeostasis presages the dire consequences that occur when this process goes awry. Many disease states are now believed to involve the detrimental effects of too much or too little apoptosis (Thompson, 1995), prompting an intense interest in the design of drugs that can repress or stimulate apoptotic pathways.

Tumor cells have lost the ability to respond appropriately to extracellular signals, resulting in uncontrolled proliferation. This pathology may result in part from the inability of these cells to undergo apoptosis. Follicular lymphomas, for example, are associated with the hyperexpression of the antiapoptotic factor Bcl-2 (Hockenbery, 1994), resulting in their enhanced protection against cell death. A current model of carcinogenesis holds that a defect in apoptosis alone may not result in cancer, but when coupled with an alteration in a gene controlling cell division, a malignancy develops (Harrington *et al.,* 1994).

Apoptosis is an important cellular mechanism in the defense against viral infections (Ravzi and Welsh, 1995). Infected cells commit suicide or are killed by cytotoxic lymphocytes that initiate apoptosis in the infected cell. However, a number of viruses express proteins that block a cell's ability to commit suicide. Other viruses cause damage by triggering inappropriate apoptosis of healthy cells, for example, the destruction of immunocompetent lymphocytes by human immunodeficiency virus (HIV) type I.

During ischemic heart attacks and strokes, cells deprived of blood flow die immediately by necrosis, but over the next several days surrounding cells commit suicide by apoptosis after sensing the damage to their neighbors. Restoration of blood flow to the affected areas often produces even more apoptosis, presumably due to an increase in reactive oxygen species. Greater understanding of this apoptotic mechanism may lead to the design of drugs that can be administered after an ischemic attack to mitigate tissue damage by apoptosis.

Many neurodegenerative diseases, such as Parkinson's, Alzheimer's, and amyotrophic lateral sclerosis, in which specific populations of neurons are lost, may result from an increased tendency of the neurons to undergo apoptosis. The β-amyloid protein that accumulates in neuronal tissue of Alzheimer's patients can induce apoptosis (Cotman and Su, 1996). Spinal muscular atrophy, which results in a devastating loss of motoneurons, has been found to involve a defect in a gene encoding an antiapoptotic factor called *neuronal apoptotic inhibitory protein* (*NAIP*) (Liston *et al.,* 1996). Death of photoreceptor cells in retinitis pigmentosa has also been shown to be apoptotic with characteristic DNA laddering (Chang *et al.,* 1993; Porter-Cailiu *et al.,* 1994).

C. Apoptosis in Response to Environmental Stresses

A wide range of cellular stresses has been shown to induce apoptosis, including hyperthermia, radiation, genotoxic agents, chemotherapeutic drugs (e.g., DNA synthesis inhibitors, antimetabolites, alkylating agents), reactive oxygen species, steroid hormones, oxidant stresses (e.g., hydrogen peroxide, menadione), and growth factor withdrawal. Different cell types often exhibit different sensitivities to the same stress. However, some agents like staurosporine, a broad-specificity protein kinase inhibitor, can induce apoptosis in virtually every type of cell (Weil *et al.,* 1996). Whether a cell dies by apoptosis or necrosis in response to a cellular stress often depends on the magnitude of the exposure. If exposure to the apoptotic agent occurs during development, the resulting cell death may produce congenital malformations.

IV. Overview of Experimental Approaches in Detection of Apoptotic Cells

A. Dye Staining

The visualization of apoptotic cells in whole embryos can be achieved by staining with supravital dyes such as nile blue sulphate and acridine orange (Knudsen, 1997). These dyes are excellent for providing a global view of PCD or developmental toxicant-induced cell death. Alternatively, differentiation of live, dead, and apoptotic cells is possible by staining with membrane sensitive dyes such as propidium iodide, Hoechst 33342, Hoechst 33258, and carboxyfluorescein followed by flow cytometry (Elstein and Zucker, 1997). In addition, characteristic changes in the cell membrane associated with apoptosis can be detected by staining with the anticoagulant protein, Annexin V, a calcium-dependent phospholipid-binding protein that specifically recognizes phosphatidylserine (Koopman *et al.,* 1994) exposed on the surface of apoptotic cells. This technique is especially useful for separating cell populations by flow cytometry.

B. TUNEL Staining

Internucleosomal degradation of DNA by the apoptotic endonuclease(s) results in formation of free 3′-hydroxyl ends of DNA. Taking advantage of these free ends, Gavrieli *et al.* (1992) designed the TUNEL (TdT-mediated dUTP-biotin nick end labeling) method, in which terminal deoxynucleotidyl transferase (TdT) binds to the nicked ends and catalyzes synthesis of a deoxynucleotide (dUTP) hetropolymer. TUNEL is performed on fixed tissue sections and standard detection is by avidin–peroxidase staining or, in a modification of Gavrieli's original technique, by immunogold–silver labeling (Tornusciolo *et al.,* 1995).

C. DNA Fragmentation Assay

When analyzed by polyacylamide gel electrophoresis and ethidium bromide staining, the characteristic degradation of DNA during apoptosis presents a series of DNA fragments that are multiples of 185–200 bp, called a *DNA ladder* (Wyllie *et al.,* 1992). This technique was quickly applied to a variety of situations in which cell death was induced. In most, but not all cases, apoptosis induced by a variety of physical and chemical agents is associated with this specific pattern of DNA degradation. Tilly and Hsueh (1993) have refined this technique to make it more quantitative.

V. Molecular Events in Apoptosis

Characterization of the genes involved in PCD in the roundworm *Caenorhabditis elegans* provided the basis of current understanding of the mechanism of apoptosis and led to the discovery of evolutionarily conserved gene families in vertebrates. Of the 1090 somatic cells formed during development of an adult *C. elegans* hermaphrodite reproducibly, 131 die by PCD (Ellis *et al.,* 1991). Mutations in 14 different *C. elegans* genes were found to affect different aspects of PCD in the developing nematode, but only three of these genes—ced-3, ced-4, and ced-9—were shown to be involved in all instances of cell death (Steller, 1995; Yuan, 1996). Ced-3 and ced-4 are required for cell death to occur. Ced-9 acts as a survival factor capable of blocking the action of ced-3 and ced-4.

Discovery of these three critical cell death genes in the roundworm led to a search for their mammalian counterparts. Ced-9 is homologous to the Bcl-2 family of proteins that function as a checkpoint to determine whether apoptosis will occur. Ced-3 is related to the caspase family of cysteine proteases that is activated during apoptosis and proteolytically degrades specific cellular substrates. Identification of the ced-4 homolog proved to be the most challenging; however, biochemical fractionation of cellular extracts has yielded the likely human homolog, named *Apaf-1* (Hengartner, 1997; Vaux, 1997; Zou *et al.,* 1997). Zou *et al.* (1997) found that Apaf-1 is significantly homologous to ced-4 and is required for caspase activation during apoptosis. Ced-4 appears to be a central player in cell death by interacting with the other two gene products and linking their actions (Chinnaiyan *et al.,* 1997; Wu *et al.,* 1997). Ced-9 or its mammalian homologs can regulate the activity of ced-4, and ced-4 can activate ced-3 or the mammalian caspases. Cytochrome c can also bind to ced-4 and appears to be required for activation of caspases by ced-4.

As discovery of the mammalian homolog of ced-4 is so recent, many questions remain about its role in apoptosis in mammalian cells. The Bcl-2 family of apoptosis regulators and the caspase family of proteases, however, have been more extensively characterized and are described in detail in later sections in this chapter.

A. Molecular Signals/Events That Trigger Cell Death

Different cell death signals are likely to activate distinct intracellular signaling pathways or intermediates. Sensitivity to cell death signals varies between different cell types, therefore these signaling pathways may also exhibit cell-type differences. At some downstream point, however, these diverse pathways converge to produce the stereotypical morphology of apoptosis. The molecular events that occur between exposure of a cell to an apoptotic signal and the cell's decision to die remain the least understood aspects of apoptosis.

1. Kinase Signaling Pathways: ERK, JNK/SAPK, and p38 Cascades

A number of kinase signaling pathways leading to apoptosis have been identified; however, many questions still remain about how extracellular stresses activate these kinase cascades, the identity of downstream targets of these cascades and how they link to other molecules involved in apoptosis (such as the Bcl-2 family and the caspases), and how these signaling pathways interact or balance each other in the cell fate decision. Three parallel cascades of the mitogen-activated protein kinase (MAPK) family are known to be involved in cell fate decision: the ERK, JNK/SAPK, and p38 cascades (Fig. 1) (Bokemeyer *et al.,* 1996; Kyriakis and Avruch, 1996).

The ERK (or extracellular signal–regulated kinase) cascade is activated by growth factors and leads to cell growth, differentiation, and protection against apoptosis. ERK is also called MAPK, as it was the first member of the MAPK family to be cloned in mammalian cells. Growth factors, such as epidermal growth factor (EGF), PDGF, fibroblast growth factor (FGF), and NGF, bind to their extracellular receptors that activate adaptor proteins and intermediaries such as Ras and Raf. These events lead to phosphorylation and activation of ERK, which can be translocated into the nucleus. ERK may then phosphorylate a variety of cytoplasmic or nuclear targets. Suppression of the ERK cascade may be required for apoptosis to occur (Xia *et al.,* 1995).

C-Jun N-terminal kinase (JNK) or stress-activated protein kinase (SAPK) is a member of an ERK subfamily that has been implicated in apoptosis induced by withdrawal of NGF and by environmental stresses (Xia *et al.,* 1995; Verheij *et al.,* 1996; Bokemeyer *et al.,* 1996). The JNK/SAPK cascade is also activated by the cytokines TNF, Fas, and IL-1β (Goillat *et al.,* 1997; Bokemeyer *et al.,* 1996). The JNK/SAPK cascade is weakly activated by growth factors, but strongly activated by stress. JNK/SAPK activation can be mediated by a number of upstream kinases, including JNK kinase (JNKK) and MAP kinase/ERK kinase kinase (MEKK) (Fanger *et al.,* 1997, Bokemeyer *et al.,* 1996). Different extracellular stresses may activate JNK through different upstream mechanisms. Targets of the JNK/SAPK pathway include the transcription factors c-Jun, ELK-1, ATF-2, and AP-1. Transcriptional activation of c-Jun appears

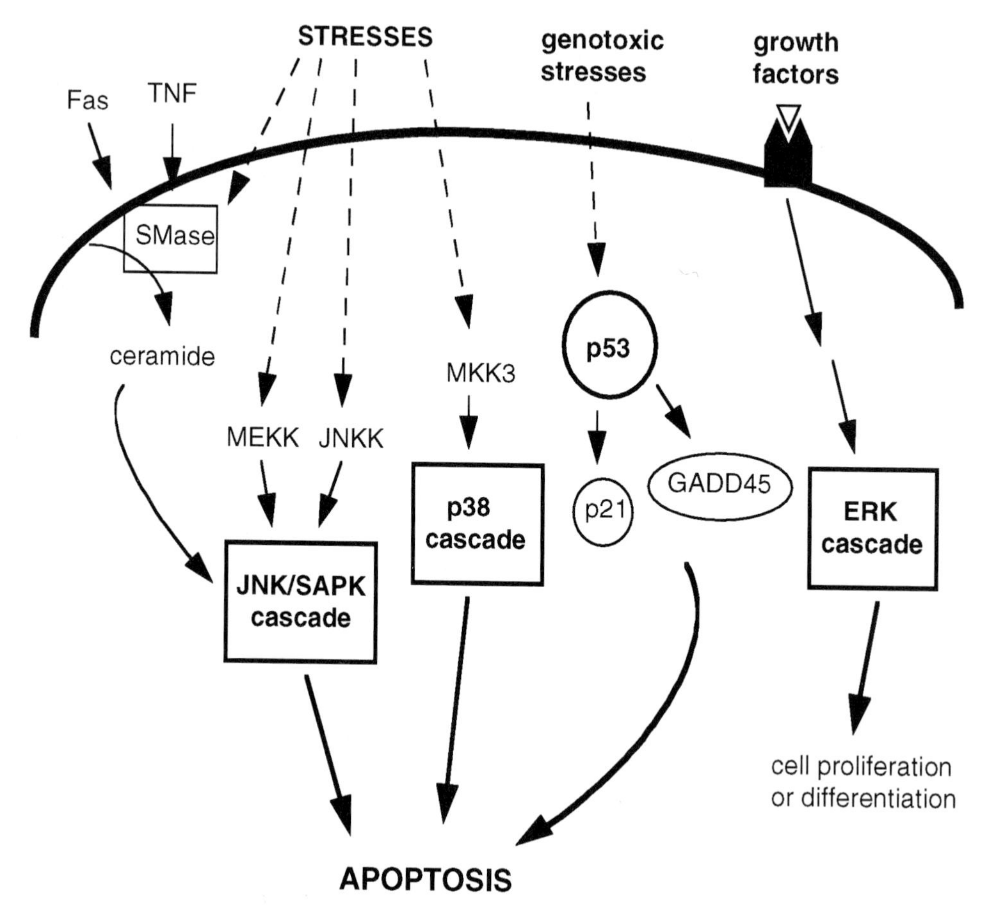

FIGURE 1 Signaling pathways involved in cell fate decisions. The JNK/SAPK and p38 kinase cascades are activated by environmental stresses. p53 is activated by genotoxic damage. The ERK kinase cascade is activated by growth factors. A balance between these signaling pathways results in either cell proliferation or cell death by apoptosis. SMase: sphingomyelinase.

to play a critical role in apoptosis of neuronal cells affected by neurotrophin withdrawal (Estus *et al.*, 1994).

p38 kinase, the mammalian homolog of the yeast HOG1 kinase (Kyriakis and Avruch, 1996), is also activated in response to extracellular stresses such as UV radiation, hyperosmolarity, cytokines, and bacterial lipopolysaccharide. A kinase called MKK3 activates p38, but not JNK/SAPK or ERK. The upstream activator of MKK3 is not yet known, and the downstream targets of the p38 pathway are also not well understood.

Apoptosis induced by withdrawal of NGF from neuronal cells demonstrates how a balance between activation of these signalling pathways may determine cell fate. When PC12 pheochromocytoma cells are deprived of NGF, apoptosis is induced. Under these conditions, both the JNK/SAPK and p38 pathways are activated and the ERK cascade is simultaneously inhibited (Xia *et al.*, 1995). If the ERK pathway is constitutively activated by transfection of dominant mutants, apoptosis is inhibited. These results suggest that the ERK cascade that promotes cell growth must be inactivated in order for the stress-activated JNK/SAPK and p38 pathways to induce apoptosis.

Fas and TNF also stimulate these MAPK signaling pathways. During Fas-induced apoptosis of a human neuroblastoma cell line, both the JNK/SAPK and ERK cascades were activated (Goillot *et al.*, 1997). However, in Jurkat cells, Fas activated the JNK/SAPK and p38 stress-activated cascades but not the ERK pathway (Juo *et al.*, 1997). Addition of inhibitors of caspases, the protease family activated during apoptosis, blocked activation of the p38 cascade, suggesting that activity of these proteases is critical for activation of p38.

2. Ceramide

A number of cellular stresses that induce apoptosis, such as growth factor withdrawal, TNF, Fas, hyperthermia and radiation have been shown to cause the intracellular release of the sphingolipid ceramide (Hannun, 1996; Pena *et al.*, 1997). Addition of exogenous ceramide to cultured cells results in either apoptosis or cell cycle arrest, depending on the cellular system used. Treatment of cultured oligodendrocytes with ceramide or sphingomyelinase, for example, results in the classic morphologic changes associated with apoptosis (Larocca *et al.*, 1997). These effects are specific to ceramide because other fatty acids or even dihydroceramide do not produce the same effect. Ceramide is known to dephosphorylate the cell cycle checkpoint factor Rb, suggesting this as a mechanism for its induction of cell cycle arrest. The mechanism by which ceramide induces apoptosis is not fully understood, but a number of studies suggest ceramide, or derivatives of ceramide, may be a second messenger for a variety of apoptotic stimuli.

Ceramide is generated by degradation of plasma membrane sphingomyelin by the action of acidic and neutral sphingomyelinases (Pena *et al.*, 1997) (Fig. 1). Cytoplasmic domains of the TNF receptor (including the death domain) have been shown to associate with these sphingomyelinases. Once in the cytosol, ceramide influences a number of cellular targets. Ceramide directly activates a serine/threonine kinase called ceramide-activated protein kinase (CAPK). CAPK is known to phosphorylate Raf-1, which in turn activates the ERK signaling cascade (Bokemeyer *et al.*, 1996). This pathway may be predominantly involved in the ceramide-mediated proliferative and inflammatory response induced by TNF. Ceramide also directly activates two other enzymes, ceramide-activated phosphatase (CAPP) and PKC-ζ, a PKC isoform, but the roles of these proteins are not known. Ceramide-induced apoptosis can also involve activation of the JNK/SAPK protein kinase cascade (Verheij *et al.*, 1996). Disruption of the JNK/SAPK pathway, by expression of dominant negative mutants for example, blocks ceramide or TNF-induced apoptosis. These results suggest that the JNK/SAPK cascade is critical for the apoptotic action of ceramide. A novel member of the mitogen-activated protein kinase kinase kinase (MAPKKK) family called TAK1 has been suggested as the protein that mediates ceramide's activation of the JNK/SAPK cascade (Shirakabe *et al.*, 1997). Apoptosis induced by ceramide eventually involves the action of the protease caspase-3 (CPP32) (Mizushima *et al.*, 1996).

3. p53

p53, a tumor-suppressor protein that activates or represses transcription of a number of genes, has been called the guardian of the genome. DNA damage induces an increase in p53 protein levels resulting in cell cycle arrest to allow time for DNA repair or apoptosis (Ko and Prives, 1996). p53-mediated apoptosis is induced by genotoxic agents, as well as viral infection, withdrawal of growth factors, and some more general stresses such as hyperthermia.

The mechanism of p53 activation in response to DNA damage is not entirely understood and is likely to involve multiple pathways given the diversity of agents that activate p53. p53 is known to activate a number of genes, including the p21/WAF1/CIP1 gene, which mediates G_1 cell cycle arrest and the GADD45 gene, which prevents cells from entering S phase (Ko and Prives, 1996). A link between the action of p53 and the Bcl-2 family of proteins that act as regulators of apoptosis has also been demonstrated. p53 up-regulates transcription of the Bax gene, which has p53-responsive elements in its promoter, and down-regulates the Bcl-2 gene (Miyashita *et al.*, 1994).

4. Fas-Mediated Apoptosis

When the membrane associated cytokine FasL binds to its receptor on a target cell, the receptor-bearing cell dies rapidly by apoptosis. The cellular events provoked by FasL are currently the best understood apoptotic signaling pathway (Nagata, 1997) (Fig. 2). Fas (also called Apo-1 or CD95), the receptor for FasL, is a 45–48 kD type I membrane protein belonging to the TNF/NGF receptor family (Nagata and Golstein, 1995). The functional form of FasL appears to be a trimer that induces formation of a Fas trimer, which in turn activates the *death domain* of Fas, an 80 amino acid cytoplasmic region that is required for its apoptotic activity (Itoh and Nagata, 1993). Apoptosis induced by the binding of TNF to the TNF receptor involves a similar series of molecular events as Fas-mediated apoptosis (Nagata, 1997).

In Fas-mediated apoptosis, a cytoplasmic death domain containing protein called FADD or MORT1 binds to the death domain(s) of the activated Fas receptor multimer. The N-terminal *death-effector domain* (DED) of FADD/MORT1 then interacts with a homologous DED in the N-terminus of the protease caspase-8 (also called FLICE or MACH) (Boldin *et al.,* 1995; Muzio *et al.,* 1996), resulting in activation of the caspase. Caspase-8 in turn activates other caspases that proteolyze cellular substrates during the final stages of the apoptotic pathway. Studies in mouse lymphoma cells treated with Fas antibodies demonstrate sequential activation of two sets of caspases (Enari *et al.,* 1996). *In vitro,* caspase-8

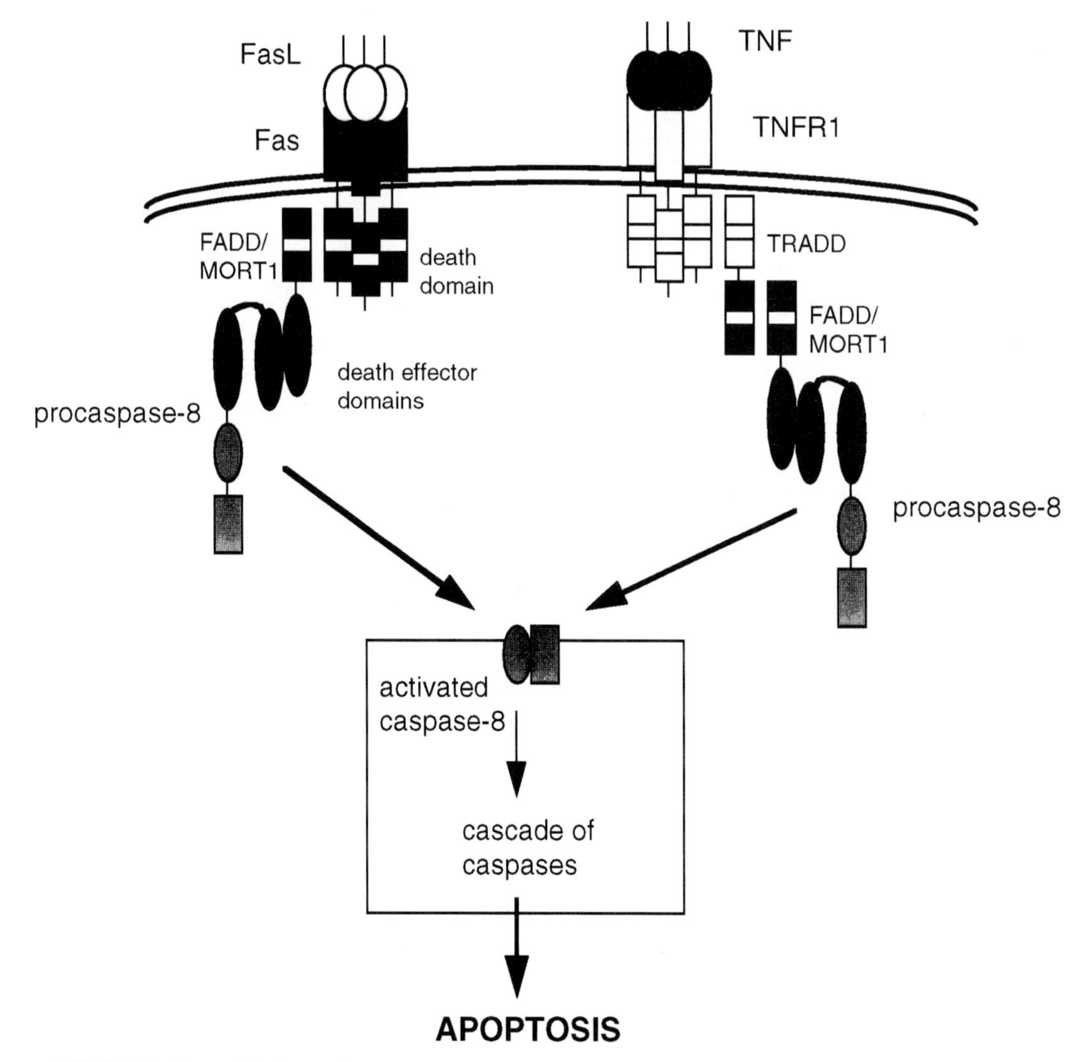

FIGURE 2 Fas- and TNF-mediated apoptosis. Binding of ligand to cell surface receptor results in recruitment of the death domain containing proteins FADD/MORT1 and TRADD to the receptor death domains. These molecules interact with procaspase-8 through homologous death effector domains (DEDs), resulting in activation of caspase-8. Caspase-8 then proteolytically activates downstream effector caspases that carry out the demise of the cell. (Adapted from Nagata, 1997.)

(FLICE/MACH) has been shown to proteolytically activate caspase-3 (CPP32), a key protease in apoptosis (Muzio *et al.,* 1997). Apoptosis by Fas triggering is blocked by inhibitors of the caspases (Armstrong *et al.,* 1996; Enari *et al.,* 1996). The cellular factors required for the Fas apoptotic pathway are constitutively expressed because enucleated cells undergo Fas-induced cell death (Schulze-Osthoff *et al.,* 1994).

Fas-induced apoptosis is critical to a number of important physiologic processes involving natural killer (NK) or cytotoxic T lymphocyte (CTL) cells that function to remove infected or defective cells (Nagata, 1997; Nagata and Golstein, 1995). When virally infected cells are recognized by a CTL, the activated T cell expresses FasL that stimulates Fas-mediated apoptosis of the infected cell. Activated T cells also trigger one another's suicide via the Fas pathway resulting in down-regulation of an immune response. Cancer cells often express FasL as protection against CTL or NK cells that express Fas. A number of anticancer agents (doxorubicin, cisplatin, VP16, and methotrexate) appear to mediate cell death by activating expression of Fas and FasL, leading to Fas-mediated autocrine or paracrine demise of the tumor cells (Fulda *et al.,* 1997; Friesen *et al.,* 1996). Resistance of some tumor cells to such anticancer agents may result from some block in this up-regulation of Fas and/or FasL (Friesen *et al.,* 1997). Upregulation of Fas expression in response to genotoxic agents may be mediated by p53 because p53 has been shown to increase expression of Fas in a number of cell lines (Owen-Schaub *et al.,* 1995). Because Fas-mediated apoptosis plays such a critical role in both normal and abnormal physiology, there is great interest in designing drugs that could block or stimulate the Fas apoptotic pathway.

Fas and FasL mRNA are found in a broad array of embryonic and adult murine tissues as detected by *in situ* hybridization (French *et al.,* 1996), although their role in nonlymphoid tissues is not well understood. FasL is expressed in the developing nervous system, brain, and spinal cord, as well as the submaxillary gland. Abundant expression of FasL in the developing brain may contribute to the immune-privileged status of this tissue by protecting neural tissue from inappropriate attack by Fas-bearing T cells (Griffith *et al.,* 1995). The embryonic expression patterns of Fas and FasL do not clearly correlate with regions of developmentally significant cell death.

Progress in understanding death factor mediated apoptotic signalling has been extremely rapid, and homologous factors and receptors are still being discovered. TRAIL (TNF-related apoptosis inducing ligand) or APO-2 ligand is another death factor homologous to TNF (Pitti *et al.,* 1996) that binds to the DR-4 receptor (Pan *et al.,* 1997). Two other members of the human TNFR family that possess a cytoplasmic death domain have been identified: DR-3 (Chinnaiyan *et al.,* 1996a; Kitson *et al.,* 1996) and APO-3 (Marsters *et al.,* 1996).

B. Cell-Death Regulators

1. The Bcl-2 Gene Family

In 1985, Bcl-2 was identified as a proto-oncogene associated with the majority of human t(14:18) follicular lymphomas (Hockenbery, 1994). In these cancer cells, a chromosomal translocation juxtaposes the Bcl-2 gene at 18q21 with the Ig heavy chain locus at 14q32, resulting in deregulated expression of the Bcl-2 gene product. Subsequent experiments showed that Bcl-2 overexpression in hematopoietic cells enhanced cell survival during cytokine withdrawal (Vaux *et al.,* 1988). The functional ability to block cell death distinguishes Bcl-2 from other known proto-oncogenes that act by stimulating cell proliferation. Mammalian Bcl-2 was subsequently found to be 23% identical to the *C. elegans* ced-9 gene (Hengartner and Horvitz, 1994) and capable of substituting for the action of ced-9 in *C. elegans* (Vaux *et al.,* 1992b).

Bcl-2 is capable of inhibiting apoptosis induced by a variety of stresses, such as withdrawal of growth factors, oxidative stress, hyperthermia, ionizing radiation, chemotherapeutic drugs, ceramide, and viral infection (Yang and Korsmeyer, 1996). However, Bcl-2 does not effectively block some other types of cell death, such as developmental selection of thymocytes (Sentman *et al.,* 1991), cytotoxic T cell-mediated cell death (Vaux *et al.,* 1992a), and withdrawal of ciliary neurotrophic factor from ciliary neurons (Allsopp *et al.,* 1993). Observations that Bcl-2 fails to block all types of apoptosis suggest that parallel pathways to induce cell death exist and have different sensitivities to regulation by the Bcl-2 family. Despite intense examination, the mechanism(s) of action of Bcl-2 is unknown and remains a central enigma in the complex process of apoptosis. However, Bcl-2 acts upstream of the caspases that carry out execution of the cell (Chinnaiyan *et al.,* 1996b; Shimizu *et al.,* 1996).

a. Characteristics of the Bcl-2 Protein Family At this time, 12 members of the Bcl-2 family have been identified in mammalian cells (White, 1996; Yang and Korsmeyer, 1996; Merry and Korsmeyer, 1997). The family contains both antagonists and agonists of apoptosis (Tables 1 and 2). Exogenous expression of apoptosis antagonists, such as Bcl-2 and Bcl-x_L, enhances the survival of cells exposed to an apoptotic signal. Conversely, expression of apoptosis agonists, such as Bax, Bad, and Bak, results in increased apoptosis. The terms *antagonist* and *agonist,* rather than survival promoter

TABLE 1 Apoptosis Antagonists of the Mammalian Bcl-2 Family

Factor	Physical properties[a]	Identified by	Dimerization partners[b]	TM domain	Knockout phenotype
Bcl-2	239 aa 25 kD	Characterization of the t(14:18) translocation in human B cell follicular lymphomas	Homodimerizes, Bax, Bad, Bak, Bik, Bcl-x_L, Bcl-x_s, Mcl-1, A1	√	Smaller at birth, hair hypopigmentation; neuronal development normal, but neuronal numbers decrease postnatally; die prematurely of polycystic kidney disease and lymphoid apoptosis; heterozygotes normal
Bcl-x_L	233 aa 29 kD	Screen of chick lymphoid library for Bcl-2 homologs	Bax, Bak, Bik, Bad, Bcl-x_s	√	Die at E13 due to massive CNS and hematopoietic cell death; heterozygotes normal
Bcl-w	193 aa 22 kD	Homology to BH1 and BH2 of Bcl-2	ND	√	ND
Mcl-1	350 aa 37 kD	Early induction in ML-1 cell differentiation	Bax	√	ND
A1	172 aa 16 kD	Homology to BH1 and BH2 of Bcl-2	Bcl-2, Bax	√	ND

[a]Molecular weights indicated are approximate as estimated by SDS-polyacrylamide gel electrophoresis.
[b]Dimerization demonstrated in the yeast two-hybrid assay (Sato *et al.*, 1994; Sedlak *et al.*, 1995; Farrow *et al.*, 1995).
TM: transmembrane; aa: amino acids; ND: no data.

or death promoter, are more appropriately used at this time because the mechanism(s) of action of these factors is not yet fully understood. The apoptosis antagonists (or survival factors) may be dominant mechanistically with the apoptosis agonists functioning solely as inhibitors of their action. Alternatively both groups of factors may possess distinct functional abilities. Proteins in the Bcl-2 family interact with each other via heterodimerization suggesting this as a means of regulation of each other's functions.

The greatest sequence homology between most members of the Bcl-2 family are found in two domains called

TABLE 2 Apoptosis Agonists of the Mammalian Bcl-2 Family

Factor	Physical properties[a]	Identified by	Dimerization partners[b]	TM domain	Knockout phenotype
Bax	192 aa 21 kD	Ability to bind to Bcl-2	Homodimerizes, Bcl-2, Bcl-x_L, A1, Mcl-1	√	Males are infertile; hyperplasia or hypoplasia depending on tissue type; hyperplasia in spleen, T and B cells; heterozygotes normal
Bad	204 aa 30 kD	Ability to bind to Bcl-2 in yeast two-hybrid assay	Bcl-2, Bcl-x_L,		ND
Bak	211 aa 23 kD	Homology to BH1 and BH2 of Bcl-2 and binding to E1B19K in yeast two-hybrid assay	Bcl-x_L	√	ND
Bcl-x_s	170 aa 18 kD	Screen of chick lymphoid library for Bcl-2 homologs	Bcl-2, Bcl-x_L		ND
Bik (Nbk)	160 aa 26 kD	Binding to Bcl-2 in yeast two-hybrid assay	Bcl-2, Bcl-x_L, Bcl-x_s, E1B19K, BHFR1	√	ND
Bid	195 aa 22 kD	Binding to Bcl-2 and Bax in screen of T-cell hybridoma library	Bcl-2, Bcl-x_L, Bax		ND

[a]Molecular weights indicated are approximate as estimated by SDS-polyacrylamide gel electrophoresis.
[b]Dimerization demonstrated in the yeast two-hybrid assay (Sato *et al.*, 1994; Sedlak *et al.*, 1995; Farrow *et al.*, 1995).
TM: transmembrane; aa: amino acids; ND: no data.

BH1 and BH2 that are important for homo- and heterodimerization between members of the Bcl-2 family (Yin *et al.,* 1994). In a broader definition, Bcl-2 family members contain at least one of four conserved domains, named BH1, BH2, BH3, and BH4 (Fig. 3). Sequence homology to at least one of the conserved protein domains of Bcl-2 or the ability to heterodimerize with Bcl-2 in screens such as the yeast two-hybrid assay has been used to identify novel members of the Bcl-2 family. The structural basis for functional differences between family members is not known.

The three-dimensional structure of Bcl-x_L shows that the BH1, BH2, and BH3 domains form a hydrophobic pocket that can interact with the BH3 domain of a dimerization partner such as Bax (Muchmore *et al.,* 1996). The BH4 region is an amphipathic helix that may stabilize the structure of the other three BH domains. None of the apoptosis agonists, with the exception of Bcl-x_s, contain the BH4 domain, but it is highly conserved between Bcl-2, Bcl-x_L, and Bcl-w (Gibson *et al.,* 1996). The BH4 region appears to be very important in the function of Bcl-2 because its deletion or substitution results in a loss or reversal of function of Bcl-2 (Borner *et al.,* 1994; Hanada *et al.,* 1995; Hunter *et al.,* 1996). The BH4 domain is not required for dimerization of the Bcl-2 family, but it may be involved in interactions with other proteins. The four BH domains form a compact globular shape, but the region between BH4 and BH3, which exhibits considerable sequence variability between family members, appears to exist as a loop without defined structure (Muchmore *et al.,* 1996). This flexible loop region is not required for the anitapoptotic function of Bcl-x_L, but may constitute a regulatory domain affected by posttranslational modification or involved in interaction with other proteins (Chang *et al.,* 1997).

The Bcl-2 protein contains a hydrophobic region at its carboxyl terminus that functions as a membrane-anchor sequence (Nguyen *et al.,* 1993), orienting the majority of the protein towards the cytoplasm (Chen-Levy and Cleary, 1990). Bcl-2 is associated with the mitochondrial membrane, the smooth endoplasmic reticulum, and the nuclear membrane (Hockenbery *et al.,* 1990; Krajewski *et al.,* 1993). The transmembrane (TM) domain is found in all of the apoptosis antagonists, but only some of the apoptosis agonists (Tables 1 and 2). A truncated Bcl-2 protein that lacks the TM domain has been shown to be less efficient at blocking apoptosis (Tanaka *et al.,* 1993; Hockenbery, 1994; Nguyen *et al.,* 1994), but conflicting results in which deletion of the TM domain has no effect have also been reported (Borner *et al.,* 1994). These differing results may arise from use of different cell lines in these studies. When the TM domain of Bcl-2 was altered to preferentially target all of the protein to the mitochondria, the endoplasmic reticulum, or the cytoplasm, only the mitochondrially targeted Bcl-2 was capable of blocking apoptosis induced by serum deprivation of kidney epithelial cells; however, both the mitochondrial and endoplasmic reticulum localized forms block serum deprivation induced apoptosis in Rat-1/myc fibroblasts (Zhu *et al.,* 1996).

Studies using green fluorescent protein (GFP) fused to Bcl-2 family proteins show that distribution of Bcl-2 is restricted to intracellular organelles (Wolter *et al.,* 1997). Bcl-x_L, which is structurally and functionally similar to Bcl-2, is largely distributed on organelle membranes, but is also found in a diffuse distribution in the cytosol. In contrast, the apoptosis agonist Bax is normally entirely in the cytosol, not associated with membranes. Most interestingly, when cells were exposed to an apoptotic signal, the GFP–Bax relocalized from the cytosol to the mitochondria (Wolter *et al.,* 1997). These results suggest the intracellular localization of Bax may be important for its ability to promote cell death.

Temporal and tissue specific expression patterns have not been fully characterized for all members of the Bcl-2 family, especially Mcl-1, Bcl-w, Bad, Bik, and Bid. Expression of several factors in the same cell type may provide functional redundancy or be required for the

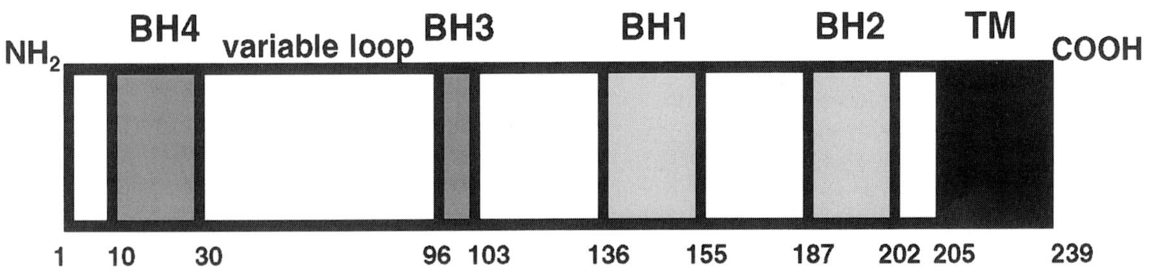

FIGURE 3 Human Bcl-2 protein domains. Numbers indicate approximate amino acid positions of the domain boundaries. Comparisons of sequence homology between Bcl-2 family members can be found in Merry and Korsmeyer (1997) and Gibson *et al.* (1996). TM: transmembrane domain.

unique properties of each factor. Temporal expression patterns of Bcl-2 family members often suggest their involvement in developmental PCD. For example, a number of antiapoptotic factors, Bcl-2, Bcl-x_L, and A1, are expressed in the developing limb, but not in the interdigital region that will undergo apoptosis to form the digits (Carrio *et al.,* 1996). Expression patterns of the Bcl-2 family is described briefly in the following section, and neural expression of the Bcl-2 family is also reviewed by Merry and Krosmeyer (1997).

b. Apoptosis Antagonists Bcl-x_L and Bcl-x_s are alternatively spliced products of the same gene (Boise *et al.,* 1993). Bcl-x_s is a much smaller protein that lacks the BH1 and BH2 domains and promotes apoptosis; however, *in vivo* expression of Bcl-x_s has not been detected in the mouse (Gonzalez-Garcia *et al.,* 1994). Bcl-x_L exhibits the greatest homology (56%) to Bcl-2 and is also similar in function, and cellular localization. These similarities might suggest that they are functionally redundant; however, they differ in expression patterns, especially in the developing and adult murine central nervous system (CNS) (Merry *et al.,* 1994; Gonzalez-Garcia *et al.,* 1995). Bcl-2 protein levels in the brain peak during embryonic development and decline postnatally, whereas Bcl-x_L levels are low in the brain during embryonic stages and increase postnatally. Bcl-2 expression remains high postnatally in the sympathetic neurons of the peripheral nervous system (PNS). Differences are also apparent in the phenotypes of mice in which the Bcl-2 or Bcl-x genes have been disrupted. Mice deficient for the Bcl-2 gene are fairly normal at birth but develop polycystic kidney disease, increased lymphoid apoptosis, gray hair, and die by 6 weeks of age (Veis *et al.,* 1993). In contrast, disruption of both alleles of the Bcl-x gene results in death by embryonic day 13 (Motoyama *et al.,* 1995). Bcl-x null mice exhibit excessive cell death in the developing brain, spinal cord, and hematopoietic cells. Characteristics of neuronal development in these knockout mice is described in greater detail later in this section.

Bcl-w is another family member that promotes cell survival and exhibits similar functional ability as Bcl-2 and Bcl-x_L (Gibson *et al.,* 1996). Bcl-w messenger RNA was detected in a wide range of adult murine tissues with the highest expression levels in brain, colon, and salivary gland.

Mcl-1 was identified as a gene up-regulated during phorbol ester-induced differentiation of the ML-1 human myeloid leukemia cell line. It was found to have homology to the BH1 and BH2 domains of Bcl-2 and act as an apoptosis antagonist (Kozopas *et al.,* 1993; Zhou *et al.,* 1997a). Mcl-1 is localized primarily to the mitochondrial and light membrane fraction, whereas Bcl-2 is found in the mitochondrial and nuclear membranes of ML-1 cells (T. Yang *et al.,* 1995). Mcl-1 exhibits widespread expression in human tissues in a pattern distinct from that of Bcl-2 (Krajewski *et al.,* 1995). It is expressed in sympathetic neurons and dorsal root ganglia, but is not detected, except at very low levels, in brain and spinal cord neurons.

A1 is another apoptosis antagonist that has homology to the BH1 and BH2 domains of Bcl-2 (Lin *et al.,* 1993, 1996). A1 was originally identified as a gene whose expression in hematopoietic cell lineages is induced by granulocyte-macrophage colony-stimulating factor (GM-CSF). In the mouse, A1 is expressed in the early embryo in brain, liver, and limbs (Carrio *et al.,* 1996). In adult tissues, expression is more widespread with the highest levels in thymus and spleen. A human protein named Bfl-1 that is highly homologous (72%) to A1 has also been identified and found to be overexpressed in normal bone marrow as well as some cancer cells (Choi *et al.,* 1995). Bfl-1 prevents apoptosis and may be involved in oncogenesis (D'Sa-Eipper *et al.,* 1996).

c. Apoptosis Agonists Bax was the first member of the Bcl-2 family found to promote apoptosis. It was identified as a protein that co-immunoprecipitates with Bcl-2 (Oltvai *et al.,* 1993). Overexpression of Bax results in increased cell death and can overwhelm the protective effect of Bcl-2. The Bax gene promoter contains several p53 binding sites and is transcriptionally activated by p53 (Miyashita *et al.,* 1994). However, thymocytes from Bax $-/-$ mice still undergo p53-dependent cell death in response to gamma radiation, indicating Bax is not required for p53-dependent apoptosis (Knudson *et al.,* 1995). Nonetheless, Bax appears to act as a tumor suppressor once it is activated by p53 (Yin *et al.,* 1997).

Bax mRNA is expressed in the developing nervous system at times when neuronal populations are undergoing PCD (Deckwerth *et al.,* 1996). Bax protein levels decrease during postnatal development of the cerebral cortex and cerebellum (Vekrellis *et al.,* 1997). During this same time period, levels of Bcl-x_L remain constant in these tissues. Bax protein is widely expressed in adult murine tissues (Krajewski *et al.,* 1994), especially at high levels in sympathetic neurons of the adult PNS, which also express Bcl-2. Neurons of the CNS also exhibit varying levels of immunohistochemical staining for Bax protein with higher levels of expression in the cortex and cerebellum, than in the brain stem and spinal cord. Astrocytes and oligodendrocytes did not exhibit positive staining for Bax. Mice in which the Bax gene has been knocked out exhibit hyperplasias in some lymphoid tissues, such as thymocytes and spleen, and males are infertile (Knudson *et al.,* 1995).

The apoptosis agonist Bad was isolated from a murine embryonic day 14.5 cDNA library by its ability to interact with other Bcl-2 family members in the yeast two-hybrid assay (Yang, E., *et al.*, 1995). Overexpression of Bad in FL5.12 cells does not increase cell death. Instead, Bad may promote apoptosis indirectly by tightly binding to Bcl-x_L and blocking it from interacting with Bax. Bad lacks a TM domain, but is found to be membrane associated due to its interactions with other proteins, potentially Bcl-x_L and Bcl-2, which are membrane associated. Bad expression overlaps with that of Bcl-x_L in many tissues (Yang, E., *et al.*, 1995) and Bad mRNA has been detected in the CNS and PNS (Merry and Korsmeyer, 1997).

Another homolog of Bcl-2, named Bak, was isolated by three groups at the same time using degenerate PCR primers to the BH1 and BH2 domains of Bcl-2 (Chittenden *et al.*, 1995; Kiefer *et al.*, 1995) or screening for factors that dimerize with the adenovirus E1B19 protein (Farrow *et al.*, 1995). Bak exhibits significant homology to both Bcl-2 and Bcl-x in the BH1 and BH2 domains and contains a hydrophobic C-terminal domain, suggesting it may be a membrane associated protein. In contrast to the protective effect of Bcl-2, cytokine-deprived FL5.12 cells and primary cultures of sympathetic neurons that overexpress Bak also die at an accelerated rate (Farrow *et al.*, 1995). Bak mRNA is expressed in a wide range of adult and fetal human tissues with notably high levels in the heart (Kiefer *et al.*, 1995). Analysis of Bak protein levels in the mouse also showed widespread but not ubiquitous expression, with moderate levels of expression in PNS neurons (Krajewski *et al.*, 1996).

Two recently identified death agonists—Bik and Bid—share sequence homology with Bcl-2 only in the BH3 domain and were identified by their ability to bind to Bcl-2. Bik also binds to Bcl-x_L and viral homologs of the Bcl-2 family, but not to Bax (Boyd *et al.*, 1995; Han *et al.*, 1996a). The tissue distribution of Bik is widespread in the adult. Bid is a soluble protein that binds to Bcl-2 and Bax (Wang, K., *et al.*, 1996). A model has been suggested by K. Wang *et al.* (1996) that Bid acts as a soluble ligand that can bind to membrane-associated Bcl-2, resulting in reduced function, or Bax, resulting in enhanced function.

d. Bcl-2 Family Function The mechanism(s) of action of Bcl-2 and its family members is unknown, but a number of observations about their physical properties that are likely to relate to function have been made.

1. Dimerization of Bcl-2 Family Members Most members of the Bcl-2 family can dimerize with themselves as well as heterodimerize with other family members (Tables 1 and 2). The BH1 and BH2 domains near the C-terminus of one protein interact with the BH3 domain near the N-terminus of a second protein. For example, the BH3 domain of Bax is essential for its formation of dimers with Bcl-2 (Zha *et al.*, 1996a). Some Bcl-2 family members may also be capable of forming higher order oligomers. These interactions have been characterized by the ability of these proteins to coimmunoprecipitate and by their ability to form active dimers in the yeast two-hybrid assay. Some Bcl-2 family members have not been detected as homodimers. Bad, for example, does not homodimerize (Yang, E., *et al.*, 1995), perhaps because it lacks a BH3 domain.

Dimerization between Bcl-2 family members appears to play a key role in their functional abilities. In FL5.12 mouse lymphoid cells, the ratio of Bax/Bax homodimers to Bax/Bcl-2 heterodimers appears to be a rheostat that regulates cell death (Oltvai *et al.*, 1993; Yang and Korsmeyer, 1996). An excess of Bax/Bax homodimers tips the scale towards cell death, whereas an excess of Bcl-2/Bax and Bcl-2 homodimers results in increased cell survival. This model suggests that the susceptibility of a tissue to a cell death signal may be due to its native ratio of Bax/Bax dimers to Bax/Bcl-2 dimers or Bcl-2/Bcl-2 dimers. Furthermore, alterations in this dimer ratio could change a cell's propensity to undergo apoptosis. However, the dimerization state of Bcl-2 family members has not been so extensively examined or correlated with sensitivity to cell death signals in developing tissues *in vivo*.

Mutations in the BH1 and BH2 regions of Bcl-2 and Bcl-x_L block their heterodimerization with Bax and block the survival function of these factors (Yin *et al.*, 1994). These results suggested that dimerization of the apoptosis antagonists is critical to the functional activity of Bcl-2 and Bcl-x_L. However, several studies have identified mutations in Bcl-x_L and/or Bcl-2 that reduce survival function (Cheng *et al.*, 1996) or augment survival function (Chang *et al.*, 1997) without affecting their ability to heterodimerize. Therefore, protein–protein interactions may not account for all the functional attributes of Bcl-2 family members.

2. Interactions of Bcl-2 Family with Nonfamily Members Bcl-2 family members also exhibit protein–protein interactions with nonfamily members, but these interactions are generally less well-characterized and their physiological relevance is not yet understood. Bad can bind to the cytosolic protein 14-3-3 (Zha *et al.*, 1996b). Bcl-2 binds to Raf-1 kinase (Wang, H., *et al.*, 1996), the NIP family of proteins (Boyd *et al.*, 1994), the human ras-related protein $p23^{R\text{-}ras}$ (Fernandez-Sarabia and Bischoff, 1993), and $p21^{ras}$ (Chen and Faller, 1996). Bcl-2 also binds to a protein called Bag-1 (Takayama *et al.*, 1995). Bag-1 is not an apoptosis antagonist by itself, but it enhances the antiapoptotic activity of Bcl-2, notably in the survival of hematopoietic cells from a death-inducing Fas signal.

3. Phosphorylation of Bcl-2 Family Members Post-translational modification, especially phosphorylation, is a common mechanism of regulating the activity of cellular proteins. Bcl-2 and Bad have been shown to be phosphorylated on serine residues (Haldar *et al.,* 1995; Zha *et al.,* 1996b; Gajewski and Thompson, 1996). The physiological consequences of phosphorylation of Bcl-2 are not yet clear due to conflicting results from several studies. In some cases, phosphorylation of Bcl-2 decreases its antiapoptotic ability (Haldar *et al.,* 1995). Other studies correlate phosphorylation of Bcl-2, induced by the presence of growth factors, with an increased antiapoptotic activity (May *et al.,* 1996; Chen and Faller, 1996). Whatever the physiological function of Bcl-2 phosphorylation, the regulation of activity of Bcl-2 family members by phosphorylation suggests a link to signaling pathways that involve kinases and phosphatases.

Cells treated with agents that damage the integrity of microtubules—such as taxol, okadaic acid, and vincristine—are associated with phosphorylation of Bcl-2 (Blagosklonny *et al.,* 1997; Haldar *et al.,* 1997). Other agents that induce apoptosis do not alter the phosphorylation state of Bcl-2, these agents include DNA damaging drugs and alkylating reagents. Severe microtubule damage threatens a cell's ability to undergo proper chromosomal segregation during cell division. Thus inactivation of Bcl-2 in such situations might enable the cell to undergo apoptosis rather than produce genetically defective progeny (Haldar *et al.,* 1997).

Studies characterizing the variable loop domains of Bcl-2 and Bcl-x_L support the hypothesis that phosphorylation inactivates these proteins. Bcl-2 and Bcl-x_L mutants lacking the loop domain are not phosphorylated and exhibit greater than normal antiapoptotic activity (Chang *et al.,* 1997). The loop contains serine and threonine residues that may be phosphorylation sites or the entire domain may be critical for interaction with a kinase.

Phosphorylation of the apoptosis agonist Bad correlates with a decrease in its apoptosis-promoting ability (Zha *et al.,* 1996b). Phospho–Bad is sequestered by the cytosolic phosphoserine binding protein 14-3-3 and cannot heterodimerize with Bcl-x_L. Mutagenesis of serine residues to abrogate phosphorylation increases the death-promoting ability of Bad. The serine–threonine kinase Akt is responsible for phosphorylation of Bad in response to stimulation by the cytokine IL-3 (del Peso *et al.,* 1997). In several cell lines and in *in vitro* studies, Akt phosphorylates Bad on the same residues (Ser^{112} and Ser^{136}) that are phosphorylated in response to IL-3 treatments *in vivo.*

To link regulation of Bcl-2 homologs to signaling pathways, the kinase(s) responsible for phosphorylation of Bcl-2 family members must be identified. Candidate kinases include Raf-1, protein kinase C (PKC), and protein kinase A (PKA)-related kinases. Bcl-2 associates with the protein kinase Raf-1 and targets it to the mitochondrial membrane (Wang, H., *et al.,* 1996). Raf-1, which is an effector of the Ras signaling pathway, does not directly phosphorylate Bcl-2 (Wang, H., *et al.,* 1994, 1996); however, Raf-1 is required for the taxol-stimulated phosphorylation of Bcl-2 and apoptosis in MCF-7 breast carcinoma cells (Blagosklonny *et al.,* 1996). Raf-1 can phosphorylate Bad *in vitro* and in 293 cells, but it cannot *in vitro* phosphorylate Bcl-2, Bcl-x, Mcl-1, Bax, or Bak (Wang, H., *et al.,* 1996). PKC can phosphorylate Bcl-2 and Bad on serine residues *in vitro* (May *et al.,* 1996; Zha *et al.,* 1996b). However, both Raf-1 and PKC phosphorylate Bad at different sites than are modified *in vivo* (Zha *et al.,* 1996b), suggesting they may not modify Bad *in vivo.* Heart muscle kinase, which is related to PKA, has been shown to phosphorylate Bad at the correct sites (Zha *et al.,* 1996b). Further studies are required to understand the mechanism and role of phosphorylation of Bcl-2 and Bad *in vivo.*

4. Membrane Pore Formation The three-dimensional crystal structure of Bcl-x_L is similar to that of diphtheria toxin and colicin, membrane associated proteins that form membrane-spanning pores (Muchmore *et al.,* 1996). *In vitro* studies using artificial phospholipid bilayers confirm that Bcl-x_L forms ion channels with a preference for cations (Minn *et al.,* 1997). It is not yet known whether a pore formed in the mitochondrial membrane by Bcl-2, Bcl-x_L, or Bax would span both the inner and outer membranes. However, the distribution of Bcl-2 on the mitochondria is patchy, with greatest concentrations at contact sites between the inner and outer mitochondrial membranes (Riparbelli *et al.,* 1995). Observations that Bax moves from the cytosol to the mitochondrial membrane in response to an apoptotic stimulus (Wolter *et al.,* 1997) suggest that Bax may promote apoptosis by forming a membrane pore or by blocking the action of Bcl-2 on the mitochondrial membrane. These observations are intriguing in the light of recent reports characterizing critical changes in mitochondrial function during apoptosis (see later section).

5. Cell Cycle Effects Some members of the Bcl-2 family have been shown to regulate entry into or exit from the cell cycle. Increases in Bcl-2 expression can delay cell proliferation (Mazel *et al.,* 1995; Borner, 1996) or increase cell cycle withdrawal into G_O (Vairo *et al.,*

1996). Overexpression of Bcl-2, Bcl-x_L, or the related adenovirus protein E1B19K delays the transition from G_O to S in T cells and fibroblasts (O'Reilly *et al.*, 1996). In contrast, Bax appears to stimulate entry into the cell cycle. T cells that overexpress Bax exhibit accelerated degradation of the cell cycle inhibitor $p27^{Kip1}$ and enter S phase more rapidly (Brady *et al.*, 1996). Understanding of the rationale for these effects of Bcl-2 family members on the cell cycle requires further examination. Results of these studies, however, suggest a link between cell death and cell proliferation (King and Cidlowski, 1995; Meikrantz and Schlegel, 1995).

6. Bcl-2 Family and Reactive Oxygen Species Reactive oxygen species (ROS), especially hydroxyl radicals, are known biologic toxins that can induce apoptosis (Jacobson, 1996). An increase in intracellular ROS levels is often observed during apoptosis, such as when NGF is withdrawn from cultured sympathetic neurons (Greenlund *et al.*, 1995), suggesting that ROS are effectors of the apoptotic pathway. ROS may be generated by the mitochondria during apoptosis as a result of uncoupling of the respiratory chain (Zamzami *et al.*, 1995a; Petit *et al.*, 1996). Intracellular localization of membrane associated Bcl-2 family members corresponds with this site of ROS generation. It has been proposed that Bcl-2 acts as an antioxidant to block cell death induced by oxidative stresses (Hockenbery *et al.*, 1993), or that Bcl-2 acts as a pro-oxidant that induces endogenous antioxidants to modulate ROS levels (Steinman, 1995). However, further studies showed that apoptosis can occur in cells cultured under extremely low oxygen conditions in which ROS would not be produced (Muschel *et al.*, 1995) and Bcl-2 and Bcl-x_L can still block apoptotic stimuli under these conditions (Jacobson and Raff, 1995; Shimizu *et al.*, 1995). Therefore, it now appears that ROS are capable of activating the apoptosis pathway but are not required for apoptosis.

e. Role of Bcl-2 Family in Neuronal Cell Death Results from analysis of Bcl-x knockout mice suggest that Bcl-x_L is a critical factor for normal neuronal development. Mice in which both copies of the Bcl-x gene were knocked out die at embryonic day 13 due to massive cell death in the CNS and in hematopoietic cells (Motoyama *et al.*, 1995). Abnormally large amounts of neuronal cell death were noted just before or during terminal differentiation of neurons in Bcl-x −/− mice. Exogenous expression of Bcl-x protects cultured sympathetic neurons from NGF withdrawal-induced apoptosis (Frankowski *et al.*, 1995; Gonzalez-Garcia *et al.*, 1995). Thus, Bcl-x_L appears to be required for neuronal survival during embryonic periods of PCD.

Despite the widespread embryonic expression of Bcl-2 in the developing nervous system (Merry *et al.*, 1994), Bcl-2–deficient mice appear to be grossly normal in development (Veis *et al.*, 1993), suggesting that Bcl-2 does not play a critical role in embryonic neuronal development. Further analysis of Bcl-2 −/− mice suggests that Bcl-2 is essential for postnatal neuronal survival (Michaelidis *et al.*, 1996). Bcl-2 −/− mice exhibit postnatal decreases in number of facial motoneurons compared to normal mice. Dorsal root ganglion sensory neurons and sympathetic neurons of the superior cervical ganglion also exhibited decreased numbers postnatally in Bcl-2 knockout mice. Axotomy of facial motoneurons, however, did not produce greater damage in Bcl-2–deficient mice, suggesting that Bcl-2 does not play a protective role in motoneuron survival following axotomy.

Overexpression of Bcl-2 protects CNS neurons from a range of damaging agents, including the oxygen radical generator adriamycin (Lawrence *et al.*, 1996). Overexpression of Bcl-2 also protects cultured primary neurons and neuronal cell lines from death when NGF is withdrawn (Garcia *et al.*, 1992; Allsopp *et al.*, 1993; Farlie *et al.*, 1995). However, Bcl-2 cannot rescue all cell types from neurotrophin withdrawal because Bcl-2 hyperexpression does not protect ciliary neurons from withdrawal of ciliary neurotrophic factor (Allsopp *et al.*, 1993).

Evidence from a number of studies suggests that Bax is required for neuronal cell death during normal development and in response to trophic factor withdrawal. Cultures of sympathetic neurons lacking the Bax gene exhibit dramatically increased survival in the absence of NGF, whereas sympathetic neurons overexpressing Bax exhibit increased apoptosis even in the presence of NGF (Deckwerth *et al.*, 1996; Vekrellis *et al.*, 1997). Similarly, neurons in Bax −/− mice survived after facial nerve axotomy, whereas neurons in normal mice die within a week (Deckwerth *et al.*, 1996).

2. Other Apoptosis Regulatory Factors

A number of viruses encode proteins that prevent removal of virally infected cells by apoptosis and are homologous in structure and/or function to apoptosis antagonists of the Bcl-2 family. The adenovirus E1B19K protein is functionally similar enough that Bcl-2 can substitute for it (Rao *et al.*, 1992). E1B19K also forms protein–protein interactions with Bcl-2 family members such as Bak and Bax (Farrow *et al.*, 1995; Han *et al.*, 1996b). Other viral Bcl-2 homologues include HMW5-HL of African swine fever virus (Afonso *et al.*, 1996), BHRF1 of Epstein–Barr virus (Henderson *et al.*, 1993), and KSbcl-2 from human herpesvirus 8 (Cheng *et al.*, 1997). At least one viral antiapoptotic factor that is not homologous to Bcl-2 has also been identified (Chiocca *et al.,* 1997).

Normally occurring PCD in *Drosophila* embryos is controlled by several neighboring genes: reaper (White *et al.,* 1994), hid (Grether *et al.,* 1995), and grim (P. Chen *et al.,* 1996). These gene products appear to act in parallel to trigger cell death (Hengartner, 1996). Reaper mRNA is expressed in cells a few hours before cell death in every case of PCD in the developing fruit fly and appears to initiate a common apoptosis pathway involving caspase-like proteases (Song *et al.,* 1997). No mammalian homologs of reaper, hid, or grim have been discovered; however, reaper has homology to the cytoplasmic death domain of Fas (Golstein *et al.,* 1995).

C. Cell-Death Mediators

Many of the characteristic morphologic changes of a cell undergoing apoptosis result from cleavage of specific cellular substrates by proteases. These proteases are expressed as inactive zymogens that are enzymatically cleaved to yield an active protease once a cell makes the decision to die. Targets of the proteins include critical repair enzymes and structural proteins. Protease activation must be under strict control to avoid accidental awakening of these killer enzymes; however, this process is not yet entirely understood. Much evidence now supports the concept that these cytosolic proteases are a critical component of the cell execution machinery in a wide range of cell types from a variety of organisms (Martin and Green, 1995; Weil *et al.,* 1996); therefore, every cell has the capacity to die by apoptosis at any time. Many unanswered questions remain, however, about the complex cast of cellular factors involved in this process and how cells regulate the activity of these proteases.

1. Caspases

A great deal of research has focused on identification and characterization of a family of cytosolic proteases involved in apoptosis called the *caspases.* The first of these cell death proteases to be identified in mammals was interleukin-1β converting enzyme (ICE), which proteolytically cleaves the inflammatory cytokine interleukin-1β (IL-1β) from an inactive 33 kD form to an active 17 kD protein (Thornberry and Molineaux, 1995). Early evidence that ICE was critical for PCD came from studies in which crmA, an inhibitor of ICE, blocked apoptosis in cultured neurons deprived of growth factors (Gagliardini *et al.,* 1994), indicating that apoptosis could not proceed without the action of ICE-like proteases. The *C. elegans* ced-3 gene is absolutely required for cell death and was shown to process significant sequence homology to mammalian ICE (Yuan *et al.,* 1993). Subsequent discoveries of additional ICE/ced-3–like proteases occurred rapidly. A caspase homologous to ced-3 has also been identified in *Drosophila* (Song *et al.,* 1997). ICE-like proteases have therefore been conserved throughout evolution from nematodes to humans. It is now well accepted that apoptosis in neuronal cells as well as many other cell types proceeds via the action of proteases belonging to the ICE/ced-3 family (Schwartz and Milligan, 1996).

Concurrent identification of ICE/ced-3 homologs in different species has resulted in multiple names for many of the proteases; therefore, a new naming system has been adopted using the root name "caspase" and a number to identify each enzyme (Alnemri *et al.,* 1996). The name caspase signifies cysteinyl aspartate-protease, in reference to their two common features: a cysteine residue in the active site and proteolytic cleavage after an aspartate residue in the protein substrate. There are 10 members in the mammalian caspase family (Alnemri, 1977) (Table 3); however, a potentially novel murine caspase has been identified (Van de Craen *et al.,* 1997). It is possible that additional homologs will be found. The 10 caspases can be further grouped into three subfamilies based on their phylogenetic relationships. *In vivo* expression patterns of the caspases have not been fully characterized; although expression of caspase-3 (CPP32) in adult human tissues have been described (Krajewska *et al.,* 1997). It remains to be seen whether the caspases function in any normal cellular processes or whether their activity is restricted to the mechanism of apoptosis.

The caspases are themselves activated posttranslationally from their zymogen (or procaspase) form by cleavage after specific aspartic acids. This processing removes a prodomain, which varies considerably in size between the different caspases, and divides the caspase into a large subunit and a small subunit. These subunits heterodimerize and associate to form an active protease tetramer (Cohen, 1997; Nicholson and Thornberry, 1997). Caspase activity has been examined *in vitro* with purified proteins or by use of caspase inhibitors in cell lines. Some caspases can activate other caspases, which then cleave cellular death substrates. Once activated, a caspase may also activate itself autocatalytically. The result is several waves of protease activation and the production of a self-amplifying proteolytic cascade that overwhelms the cell's ability to synthesize new molecules.

Analysis of caspase activity in many different types of cell death suggests a common theme in the order of caspase activation (Nicholson and Thornberry, 1997). Caspases 2, 3, and 7 usually function as effector caspases that are activated downstream in the caspase cascade and proteolytically destroy cellular targets. These effector caspases are activated by caspases that are more proximal to the cell death signal, such as caspases 8

TABLE 3 Mammalian Caspases

Subfamily[a]	Name	Alternative names	Substrates that are other caspases[b]	Substrate(s) that are cellular proteins/enzymes[b]
ICE	Caspase-1	ICE	pro-caspase-3 and 4	pro-IL-1β, PARP, U1-70K, actin
	Caspase-4	ICErel-II, Ich-2, TX, Mih1	pro-caspase-1	PARP, U1-70K
	Caspase-5	ICErel-III, TY	?	?
Nedd-2	Caspase-2	ICH-1, Nedd-2	?	PARP
	Caspase-9	ICE-LAP6, Mch6	?	PARP
ced-3	Caspase-3	CPP32, apopain, Yama	pro-caspase-6, 7, and 9	PARP, U1-70K, SRE/BP, DNA-PK, actin, D4-GDI
	Caspase-6	Mch2	pro-caspase-3	PARP, lamin A, U1-70K
	Caspase-7	Mch3/ICE-LAP3/CMH-1	pro-caspase-6	PARP, U1-70K
	Caspase-8	FLICE, MACH, Mch5	pro-caspase-3	?
	Caspase-10	Mch-4	pro-caspase-3 and 7	PARP, U1-70K

[a]Caspases are divided into subfamilies based on their phylogenetic relationships (Alnemri *et al.,* 1996).

[b]References for caspase substrates can be found in: Duan *et al.,* 1996; Fernandes-Alnemri *et al.,* 1996; Kayalar *et al.,* 1996; Na *et al.,* 1996; Patel *et al.,* 1996; Srinivasula *et al.,* 1996; Alnemri, 1997; Mashima *et al.,* 1997. Additional caspase substrates are catalogued in Nicholson and Thornberry (1997) and Cohen (1997). In many cases, substrates were identified in *in vitro* contexts, where concentration of the caspase may be in large excess; therefore, it is possible that the caspase may not cleave the substrate under *in vivo* conditions.

PARP: poly (ADP-ribose) polymerase; SRE/BP: sterol regulatory element–binding protein; DNA-PK: DNA-dependent protein kinase; U1-70K: U1-70 kDa small nuclear ribonucleoprotein; ?: substrates not yet identified.

(FLICE), 6 (Mch2), or 9. The activation of terminal effector caspases by more proximal caspases is especially clear in the Fas or TNF-mediated apoptosis pathways where caspase-8 (FLICE) is activated by death adapter molecules (FADD and TRADD) that interact directly with a transmembrane receptor (Fig. 2).

How the caspase cascade is initially activated in response to most apoptotic stimuli is not yet fully understood. However, caspase activation can be regulated by members of the Bcl-2 family (Chinnaiyan *et al.,* 1996b; Shimizu *et al.,* 1996). For example, overexpression of Bcl-2 blocks the activation of caspases in the GT1–7 neural cell line after exposure to a range of apoptotic stimuli (Srinivasan *et al.,* 1996) and apoptosis in neuronal cells induced by overexpression of Bax can be blocked by the addition of a caspase inhibitor (Vekrellis *et al.,* 1997). In most or perhaps all cells the caspases can be activated without synthesis of additional factors (Schulze-Osthoff *et al.,* 1994; Weil *et al.,* 1996).

The question of why so many different caspases exist has not been fully answered. The substrate specificities of different caspases is distinct in some cases, but also overlapping in many others. Expression of the caspases may vary in different cell types or distinct sets of caspases may be activated by different death triggers. Redundancy in caspase function may also be advantageous to ensure efficient and irreversible execution of the cell.

Although caspase-1 (ICE) was the first mammalian caspase to be identified, it is not essential for apoptosis in most situations, except perhaps in T-cell mediated cell death (Los *et al.,* 1995; Enari *et al.,* 1996). ICE knockout mice do not exhibit dramatic deficiencies in PCD and stress-induced apoptosis (Kuida *et al.,* 1995; Li *et al.,* 1995). This may result from functional redundancy of members of the protease family. Caspase-1 knockout mice are deficient in mature IL-1β suggesting a role for this caspase in cytokine maturation. Transgenic mice expressing a dominant negative mutant of caspase-1 in neuronal cells develop normally, but exhibit reduced apoptosis after induced ischemia (Friedlander *et al.,* 1997).

Caspase-3 (or CPP32) is the mammalian caspase with the greatest similarity to ced-3 in terms of sequence homology and substrate specificity (Xue and Horvitz, 1995). Caspase-3 is cleaved from an inactive 32 kD form to one of two active forms composed of heterodimeric subunits of either 20 kD and 11 kD or 17 kD and 12 kD (Fernandes-Alnermri *et al.,* 1994; Nicholson *et al.,* 1995). Several caspases, including ICE (caspase-1), FLICE (caspase-8), and Mch-4 (caspase-10), have been shown to activate caspase-3 (Boldin *et al.,* 1995; Tewari *et al.,* 1995; Fernandes-Alnemri *et al.,* 1996). CPP32 knockout mice die perinatally and exhibit a large excess of cells in the CNS (Kuida *et al.,* 1996), suggesting that the actions of caspase-3 are critical for normal neuronal development; however, other major organs appear to be unaffected in these mice. Caspase-3 is expressed in murine dorsal root ganglia (DRG) neurons and other sensory neurons during development and inhibitors of caspase-3 block apoptosis of DRG neurons during neurotrophin factor withdrawal (Mukasa *et al.,* 1997).

a. Caspase Substrates Thus far, the number of cellular substrates of the caspases appears to be fairly limited, suggesting that apoptotic proteolysis is not random (Table 3). *In vitro* substrate specificity of the caspases has been examined using synthetic peptides by Talanian *et al.* (1997). Death substrates include structural proteins and nuclear enzymes. It is not yet known how the cytosolic caspases enter the nucleus to attack their nuclear substrates.

Structural proteins targeted by the caspases include fodrin (Martin *et al.,* 1995), actin (Kayalar *et al.,* 1996; Mashimi *et al.,* 1997), and nuclear lamins (Lazebnik *et al.,* 1995). Disruption of the nuclear lamina may block structural interactions between chromosomal DNA and the nuclear envelope, thereby leading to breakdown of the nucleus and formation of apoptotic bodies. Inactivation of enzymes involved in DNA repair may accelerate degradation of nuclear material. An enzyme critical for recognition and repair of damaged DNA called poly (ADP-ribose) polymerase (PARP) is inactivated during apoptosis by caspase-3 (CPP32) (Nicholson *et al.,* 1995; Tewari *et al.,* 1995; Kaufmann *et al.,* 1993) as well as other caspases (Table 3). Cleavage of PARP, however, does not appear to be essential for downstream events in apoptosis because mice in which the PARP gene has been disrupted undergo apoptosis normally (Wang *et al.,* 1995). During apoptosis the catalytic subunit of DNA-dependent protein kinase (DNA-PK), which is involved in the repair of double-stranded DNA breaks, is degraded into three fragments (Song *et al.,* 1996). DNA-PK is a substrate for caspase-3 (CPP32) in Fas-mediated apoptosis (Casciola-Rosen *et al.,* 1996). Another substrate common to several of the caspases is the U1-70kDa small ribonucleoprotein, which is essential for mRNA processing (Sharp, 1994). A number of caspases are capable of degrading these nuclear enzymes; however, studies of the efficiency of substrate cleavage by different caspases shows that caspase-3 appears to be much more efficient at proteolysis of PARP, U1-70K, and DNA-PK than caspase-1 (Casciola-Rosen *et al.,* 1996). Caspase-3 has also been linked to the apoptotic pathway that results in DNA fragmentation. Caspase-3 activates a 45kD protein called *DNA fragmentation factor* (DFF) that in turn induces cleavage of nuclear DNA, presumably by activation of an endonuclease (Liu *et al.,* 1997).

b. Use of Caspase Activation and Substrate Cleavage in Studies of Apoptosis Activation of caspases and cleavage of death substrates is a reproducible event in apoptosis of a wide range of cell types in response to a wide range of cell death signals. Therefore, these molecular events are extremely useful in assays to detect apoptotic cell death and examine its kinetics. Cleavage of caspase-3 (CPP32) from its inactive 32 kD form to an active 17 kD fragment is readily detectable using specific antibodies to CPP32 in Western analysis (Erhardt and Cooper, 1996). Similarly the cleavage of PARP from its active 118 kD form to an inactive 85 kD fragment can easily be detected by Western blot (Erhardt and Cooper, 1996).

c. Peptide Caspase Inhibitors The caspases can be inhibited by peptides that are analogous to their native substrate recognition sequences. It is often difficult, however, to identify the exact caspase or caspases being inactivated by a particular agent because the specificity of these inhibitors is not entirely understood (Patel *et al.,* 1996). The modified tetrapeptide Ac-YVAD-cmk (acetyl-Tyr-Val-Ala-Asp-chloromethylketone) is a potent inhibitor of caspase-1 (ICE) and a weaker inhibitor of caspase-3 (CPP32) and potentially other caspases. Conversely, Ac-DEVD-CHO (acetyl-Asp-Gln-Val-Asp-aldehyde) is a potent inhibitor of caspase-3 and a weak inhibitor of caspase-1. zVAD-fmk (benzyloxycarbonyl-Val-Ala-Asp-fluoromethylketone) is a commonly used inhibitor that appears to affect all the caspases as well as cathepsin b and calpain (Armstrong *et al.,* 1996). Aldehyde, nitrile, and ketone derivatives of aspartate-containing peptides are reversible inhibitors of the caspases, whereas methylketone derivatives are irreversible inhibitors (Nicholson, 1996).

d. Naturally Occurring Caspase Inhibitors A number of viruses express proteins that inhibit the activity of caspases. These factors presumably act as part of the virus' defense against the host's apoptotic clearance of virally infected cells. The coxpox virus protein CrmA and the baculovirus p35 protein inhibit a number of caspases and can be used as exogenously expressed agents to block apoptosis (Bump *et al.,* 1995; Zhou *et al.,* 1997b). CrmA is a good inhibitor of caspases 1 and 8 (ICE and FLICE) and is a weaker inhibitor of caspase-3 (CPP32) and is effective in blocking apoptosis induced by some stimuli, but not others (Cohen, 1997).

Caspase inhibitors that are naturally expressed in mammalian cells have also been identified. Some of these proteins are members of the inhibitor of apoptosis protein (IAP) gene family, which also has viral members (Clem and Duckett, 1997). Initially, a neuronal apoptosis inhibitor protein (NAIP) was identified (Liston *et al.,* 1996). Partial deletion of the human NAIP gene is associated with SMA, a severe neuromuscular disorder resulting from excessive motoneuron apoptosis. When expressed in mammalian cells, NAIP blocks induction of apoptosis by menadione, serum deprivation, and staurosporine. Identification of NAIP led to discovery of several other human IAP genes whose molecular function has now been linked to the caspases. The X-linked IAP, c-IAP-1, and c-IAP-2 proteins have been shown to bind directly and inhibit caspases 3 and 7 (Deveraux *et al.,* 1997; Roy *et al.,* 1997). Although these IAPs inhibit the downstream effector caspases, they do not block the action of more proximal caspases such as caspases 8, 6, or 1. NAIP is structurally related to the other IAPs; however, it appears to act on other cellular targets because it does not bind to the caspases (Roy *et al.*, 1997).

Proteins such as the viral and cellular FLIPs (Irmler *et al.*, 1997; Thome *et al.*, 1997), I-FLICE (Hu *et al.*, 1997), Casper (Shu *et al.*, 1997), and CASH (Goltsev *et al.*, 1997) constitute another group of apoptosis inhibitors that act at the level of the caspases. These factors appear to block caspase activation by acting as dominant negative forms of the proximal caspases that are activated by the death effector proteins FADD/MORT1 and TRADD in the Fas and TNF-receptor cell death pathways (Fig. 2). The inhibitors are generally similar in structure to caspase-8 (FLICE) and caspase-10 (Mch-4), possessing two DEDs and a caspase-like domain. However, the caspase-like domain of the inhibitors lacks the critical cysteine residue in the active site and cannot form a functional protease. These inhibitors have been shown to bind to FADD/MORT1 and/or TRADD and to caspase-8 (FLICE) and/or caspase-10. Presumably, in cells exposed to Fas (or TNF), the inhibitor proteins are recruited to the death effector domain proteins instead of procaspase-8 and are cleaved into inactive subunits, thereby blocking activation of caspase-8. The inactive caspase-like inhibitor subunits may also heterodimerize with functional FLICE subunits, resulting in an inactive caspase tetramer.

2. Calpain

Calpain is a ubiquitously expressed cytosolic protease that has long been implicated in necrosis and has been found to be activated in some cases of apoptosis. Calpain is activated by elevated intracellular concentrations of calcium; therefore, it could be activated by the calcium increase that occurs during apoptosis. There are two isozymes of calpain, calpain I (or μ-calpain) and calpain II (or m-calpain), that have different calcium requirements. The substrate sequence recognized by calpain is distinct from that of the caspases, but calpain is also a cysteine protease.

Calpain is activated during apoptosis in a number of different cell types exposed to a variety of apoptotic signals; however, it may not be required for all incidences of apoptosis. Calpain I is activated during apoptosis in thymocytes treated with dexamethasone (Squier *et al.*, 1994). In cultured hepatocytes, inhibitors of either calpain I or II blocked apoptotic cell death induced by TGF-β (Gressner *et al.*, 1997). Both calpain and caspases are required for TNF-induced cell death in the U937 cell line, but calpain is not required for Fas-induced cell death in Jurkat cells (Vanags *et al.*, 1996). A number of cellular proteins and enzymes are substrates for calpain, including α-fodrin (Nath *et al.*, 1996) and p53 (Kubbutat and Vousden, 1997).

Cleavage of cellular proteins by calpain is involved in some neurodegenerative disorders, including brain ischemia (Bartus *et al.*, 1995) and Alzheimer's disease (Nixon *et al.*, 1994); however, the cell death involved in these processes involves both necrosis and apoptosis. Inhibitors of calpain partially block apoptosis of T cells induced by HIV infection (Sarin *et al.*, 1994). These observations have prevoked great interest in the design of therapeutic drugs that inhibit the action of calpain.

3. Granzyme B

Granzyme B belongs to a family of serine proteases expressed in the granules of CTL and NK cells (Shi *et al.*, 1992; Smyth *et al.*, 1996). CTL and NK cells destroy infected or malignant cells by binding to them and injecting the contents of their granules. The target cell then exhibits apoptotic characteristics such as activation of caspases and DNA fragmentation, as well as some characteristics of necrosis such as plasma membrane degradation. Other granzymes have also been implicated in apoptosis, but Granzyme B appears to be the most potent.

The mechanism of Granzyme B action involves another CTL and NK-specific protein called *perforin.* Granzyme B can enter the cytoplasm in the absence of perforin (Jans *et al.*, 1996), but addition of perforin results in translocation of Granzyme B into the nucleus and initiation of apoptosis (Shi *et al.*, 1997). Once in the nucleus, Granzyme B induces DNA fragmentation, perhaps by activating an endonuclease (Montague and Cidlowski, 1996) and cleaves PARP producing a 64 kD degradation product instead of the 85 kD fragment produced by caspase cleavage of PARP (Froelich *et al.*, 1996). Granzyme B can activate a number of the caspases (e.g., Darmon *et al.*, 1996; Duan *et al.*, 1996). Thus a protease delivered by CTL or NK cells may serve to activate the target cell's own cell death machinery.

D. Mitochondrial Events in Apoptosis

Cells exposed to a wide range of apoptosis-inducing signals exhibit dramatic mitochondrial changes, such as a reduction in membrane potential ($\Delta\psi_m$), that precede nuclear disintegration (Petit *et al.*, 1996, Kroemer, 1997). The mitochondrial proteins cytochrome c and apoptosis inducing factor (AIF) are released into the cytoplasm triggering nuclear degradation and activation of caspases (Liu *et al.*, 1996; Susin *et al.*, 1996; Kluck *et al.*, 1997; Yang *et al.*, 1997) (Fig. 4). The importance of mitochondrial events in apoptosis is supported by the observations that apoptosis in a cell-free system requires a mitochondrial-rich cellular fraction (Newmeyer *et al.*,

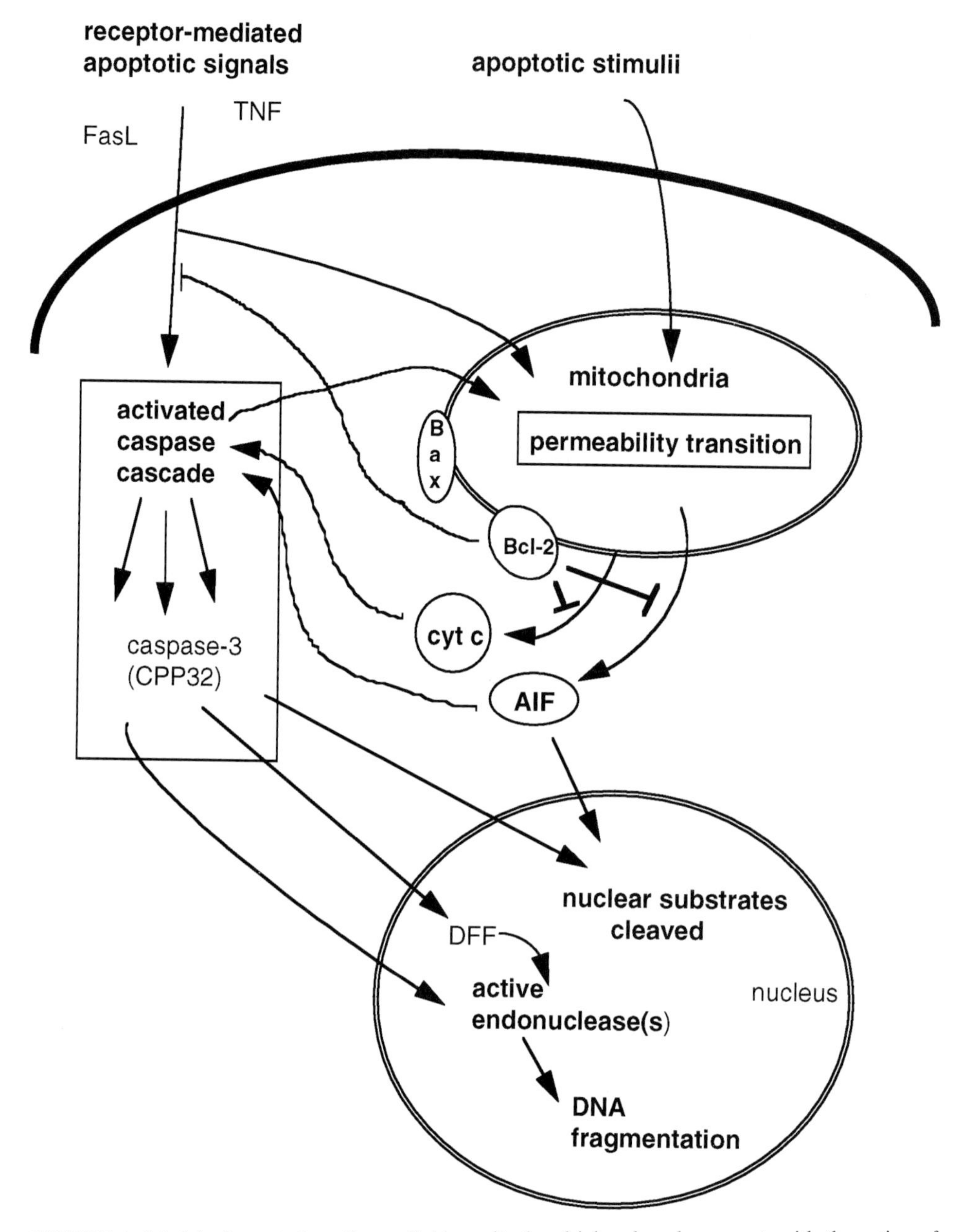

FIGURE 4 Model of apoptotic pathways linking mitochondrial and nuclear events with the action of the caspases and the Bcl-2 family.

1994), isolated mitochondria from apoptotic cells are capable of inducing DNA fragmentation in normal unstressed nuclei (Newmeyer *et al.,* 1994; Zamzami *et al.,* 1996), and anucleated cells can be induced to undergo apoptosis (Jacobson *et al.,* 1994; Schulze-Osthoff *et al.,* 1994; Nakajima *et al.,* 1995). These studies suggest that mitochondrial events are essential for the apoptotic process and nuclear events are not; however, the ability to undergo apoptosis requires only nuclear genes, not mitochondrial genes (Zamzami *et al.,* 1996).

The permeability transition pores that open during the MPT (Zorrati and Szabo, 1995) produce the change in membrane potential that commits a cell to the apoptotic pathway (Zamzami *et al.,* 1995b; Machetti *et al.,* 1996). A disruption in $\Delta\psi_m$ has been observed in a variety of cell types reacting to a range of cell death triggers (Petit *et al.,* 1996; Kroemer, 1997), including cultured sympathetic neurons deprived of NGF (Deckwerth and Johnson, 1993). A number of compounds that induce the MPT also induce apoptosis, whereas several inhibitors of the MPT block apoptosis. For example, bongkrekic acid, a specific inhibitor of the MPT, blocks all the typical cellular events of apoptosis (Marchetti *et al.*, 1996; Zamzami *et al.*, 1996).

In mammalian cell lines and *Xenopus* cell-free systems, release of cytochrome c from the mitochondria can be monitored kinetically after exposure to a death-inducing signal (Liu *et al.*, 1996; Kluck *et al.*, 1997; Yang *et al.*, 1997). Mitochondrial release of cytochrome c appears to occur before the MPT (Kluck *et al.*, 1997; Yang *et al.*, 1997). Cytochrome c is a 12 kD soluble protein that is normally located in the intermitochondrial membrane space, where it serves as a critical component of the electron transport chain. Some mitochondrial proteins, such as cytochrome oxidase, are not released during apoptosis, but other soluble mitochondrial proteins may also be involved in the MPT (Kroemer, 1997). The presence of cytochrome c in the cytosol results in the activation of caspase-3 (CPP32), which is critical for the proteolysis of nuclear proteins, but this occurs only when a cytosolic fraction and dATP are also present. The mechanism by which cytochrome c activates the caspases is unknown; however, cytochrome c has also been linked to the activity of Apaf-1, the mammalian homolog of ced-4 (Zou *et al.*, 1997).

AIF is a 50 kD mitochondrial protease that is released during the MPT (Susin *et al.*, 1996). Cytosolic AIF induces chromatin condensation and DNA fragmentation, but is not itself an endonuclease. AIF, activity can be inhibited by ZVAD-fmk but not by Ac-DEVD-CHO (an inhibitor of caspase-3) or Ac-YVAD-CHO (an inhibitor of caspase-1), indicating that it is a distinct protease and its activities are not dependent on caspase-3 or caspase-1. Although it induces nuclear condensation, AIF activity does not result in cleavage of PARP or lamin, suggesting it must attack other nuclear substrates. The AIF gene has not yet been identified, but AIF was purified from mouse livers and detected in other murine and human tissues. However, AIF may not be present or active in all cell types or all species. For example, AIF does not appear to be involved in the *Xenopus* cell-free system because inhibitors of caspase-3 effectively block all apoptosis (Yang *et al.*, 1997).

Further investigation is required to identify the molecular signals that trigger mitochondrial events directly. However, in some instances, activation of caspases precedes the mitochondrial changes during apoptosis. In Fas-mediated cell death, for example, activation of caspase-1 (ICE) occurs very rapidly, before detectable reduction in the mitochondrial transmembrane potential (Enari *et al.*, 1996). Therefore, observations that mitochondrial events precede activation of caspases do not preclude that, in some circumstances, activated caspases induce mitochondrial events.

Hyperexpression of either Bcl-2 and Bcl-x_L can block $\Delta\psi_m$ (Decaudin *et al.*, 1997) and release of cytochrome c (Kluck *et al.*, 1997; Yang *et al.*, 1997) or AIF (Susin *et al.*, 1996) from mitochondria in cells exposed to a death signal. Once AIF or cytochrome c is released into the cytoplasm, however, Bcl-2 cannot block its actions on the nucleus or activation of CPP32. These results suggest that Bcl-2 acts upstream of the signals that trigger mitochondrial changes, but cannot block the downstream events. The observations that Bcl-2 family members are capable of forming membrane spanning pores and that Bax moves from the cytosol to the mitochondria during apoptosis (Wolter *et al.*, 1997) offer an intriguing connection to their role in regulating mitochondrial events in apoptosis. Mitochondrial membrane pores formed by Bcl-2 homologs may directly regulate the MPT or the release of mitochondrial factors such as cytochrome c and AIF.

VI. Apoptosis in Abnormal Development

In the seventh edition of the *Catalog of Teratogenic Agents* (Shepard, 1992), approximately 1200 agents are listed that produce congenital anomalies in various experimental animals, primarily mice, rats, and rabbits. For the vast majority of these teratogens, little is known about the mechanisms by which they disrupt normal development leading to the anomalies observed at term. Despite the paucity of information concerning teratogenic mechanisms, numerous observations have been reported that many teratogens produce cell death in excess of the PCD occurring as a part of normal development. Although excessive cell death in tissues destined to give rise to malformations is now a well-documented phenomenon (Scott, 1977), less is known about the relationship between teratogen-induced cell death and the pathogenesis culminating in congenital anomalies. Another well-documented finding is that different cells (tissues) within the embryo respond to a teratogenic insult differently, that is, some cells die while other do not. Again, although that is a common observation, few experimental data are available that explain this differential sensitivity.

A. *Teratogens Whose Effects Are Mediated by Apoptosis*

Although excessive cell death is a common feature of teratogenesis, only recently have studies been reported in which this cell death has been characterized formally as apoptosis, that is, demonstration of hallmarks of apoptosis, such as internucleosomal DNA fragmentation and activation of caspases. In fact, many reports in the 1970s and 1980s characterized teratogen-induced cell death as necrosis (e.g., Scott, 1977; Mirkes *et al.,* 1983). These characterizations were made before it became clear that necrosis and apoptosis can be distinguished biochemically. Although to date only four teratogens have been formally shown to induce apoptosis, it is likely that most, if not all, teratogens induce an apoptotic form of cell death given the effects of cytotoxic agents in cultured cells (Schwartzman and Cidlowski, 1993).

1. Hyperthermia

Two groups (Mirkes, 1985a; Webster *et al.,* 1985) have reported that exposures to hyperthermia, either *in vitro* or *in vivo,* induce excessive cell death, particularly in the developing CNS but also in other tissues. None of these studies, however, formally characterized this hyperthermia-induced cell death as apoptosis. Using a gel electrophoresis–based assay for nucleosomal DNA fragmentation as a marker of apoptosis, it has been shown that hyperthermia induces apoptosis in day 10–11 rat embryos (Mirkes *et al.,* 1997). A significant increase in DNA fragmentation is observed as early as 2.5 hr after a 15 min exposure to 43°C. By TUNEL analysis, in which DNA fragmentation is localized within cells of different tissues, we also demonstrate that much of the hyperthermia-induced apoptosis occurs in the developing CNS.

2. Chemical Teratogens

Cyclophosphamide is a well-studied animal teratogen (Mirkes, 1985b) that is known to induce cell death. Two groups have now reported that cyclophosphamide-induced cell death is apoptotic in nature. Chen *et al.,* (1994) exposed rat embryos *in vitro* to the active metabolite of cyclophosphamide, phosphoramide mustard, and showed that this metabolite induced DNA fragmentation in the embryo but not in the surrounding yolk sac. In a subsequent study, Moallem and Hales (1995) reported that 4-hydroxycyclophosphamide, a preactivated analog of cyclophosphamide, induced DNA ladder formation in cultured rat embryo limb buds exposed *in vitro.* A study by Nomura *et al.* (1996) shows that cyclophosphamide administered to pregnant mice on day 11 of gestation induced phocomelia and apoptosis on the basis of TUNEL staining and DNA ladder formation.

Another alkylating agent, acetylaminofluorene (AAF), has been shown to disrupt development (Faustman-Watts *et al.,* 1983) and to induce excessive cell death characteristic of apoptosis (Thayer and Mirkes, 1995). In this study, cell death was characterized by DNA ladder formation in isolated DNA, TUNEL staining in histologic sections and by transmission electron microscopy.

Retinoic acid (RA) is a well-known animal and human teratogen (Shepard, 1992) and is known to induce cell death in tissues destined to develop abnormally (Sulik and Alles, 1991). Using the TUNEL assay for DNA fragmentation, Ahuja *et al.* (1997) showed that RA induces apoptosis in day 14.5 embryonic hindlimbs.

B. Susceptibility of Developing Tissues to Teratogens

Despite the fact that many, if not all, teratogens induce excessive cell death in tissues destined to develop abnormally, it is equally clear that within a sensitive tissue teratogens kill some cells and spare others. The mechanisms underlying this differential sensitivity are unknown, although there is increasing evidence that teratogens preferentially kill cells in areas of ongoing normal PCD (Alles and Sulik, 1989). Attempts to determine the mechanisms by which teratogens kill some cells but spare neighboring cells in a multicellular, multitissue embryo represent an almost insurmountable task. Therefore, we have focused on our finding that the cells of the developing heart and yolk sac are resistant and cells in the developing CNS are sensitive to the cytoxic potential of a variety of teratogens. The resistance of heart and yolk sac cells and the concomitant sensitivity of CNS cells have been demonstrated for the following teratogens: hyperthermia (Mirkes, 1985a), cyclophosphamide (Mirkes, 1985b; Chen *et al.,* 1994), 2′-deoxyadenosine (Gao *et al.,* 1994), cysteine proteinase inhibitors that disrupt yolk sac proteolysis (Ambroso and Harris, 1994), and N-acetoxy-2-acetlyaminofluorene (Thayer and Mirkes, 1995).

As outlined earlier in this review, the activation of caspases represents a common event in the apoptotic pathway initiated by a wide variety of cytotoxic agents. In addition, the activation of specific caspases, for example, caspase-3 (CPP32), is now thought to represent an irreversible commitment to cell suicide. Results from our laboratory show that this protease is activated in cells of embryonic head (containing the sensitive neuroepithelial cells), but is not activated in cells of the heart after embryos are exposed to hyperthermia, cyclophosphamide, and arsenite (Mirkes and Little, 1997). As expected from previous TUNEL data, we also showed by gel electroporesis that DNA fragmentation occurs in cells of the head but not of the heart. Thus, some mechanism in heart cells allows them to escape the irreversible decision to commit suicide.

C. Future Directions in Understanding Regulation of Apoptosis in Embryonic Tissues

As outlined previously, PCD is an essential part of normal development and teratogen-induced cell death plays a role in abnormal development. Thus, a comprehensive understanding of both normal and abnormal development requires a detailed understanding of the apoptotic pathway from initiation through commitment to execution. Currently, we know the following about apoptosis induced by developmental toxicants. First, a variety of developmental toxicants can induce apoptosis in embryos; however, almost nothing is known concerning the way in which teratogens engage the apoptosis pathway. Thus, elucidating the cellular factors that interact with teratogens should be a high priority. Second, we know that a few teratogens induce DNA fragmentation and activate at least one key cysteine protease as part of the teratogenic process; however, the role of additional factors in the embryonic cell death pathway are largely unknown. Thus, another high priority should be to elucidate the molecular players in what are likely to be complex, interacting pathways to cell death. Finally, we know that teratogen-induced cell death preferentially occurs in areas of normal PCD and that the sensitivity of embryonic cells to a teratogen varies both within and between tissues of the embryo. Thus, a third high priority area of research should focus on the molecular mechanisms underlying this differential sensitivity. It is hoped that information from these priority areas of research will provide not only a detailed map of the apoptosis pathway in the mammalian embryo, but also insights into ways in which the apoptotic pathway can be pharmacologically modulated in an embryo/fetus exposed to a potential teratogen.

References

Afonso, C. L., Neilan, J. G., Kutish, G. F., and Rock, D. L. (1996). An African swine fever virus Bcl-2 homolog, 5-HL, suppresses apoptotic cell death. *J. Virol.* **70,** 4858–4863.

Ahuja, H. S., James, W., and Zakeri, Z. (1997). Rescue of the limb deformity in hammertoe mutant mice by retinoic acid-induced cell death. *Dev. Dynamics* **208,** 466–481.

Alles, A. J., and Sulik, K. K. (1989). Retinoic acid-induced limb defects: pertubation of zones of programmed cell death as a pathogenetic mechanism. *Teratology* **40,** 163–171.

Allsopp, T. E., Wyatt, S., Paterson, H. F., and Davies, A. M. (1993). The proto-oncogene bcl-2 can selectively rescue neurotrophic factor-dependent neurons from apoptosis. *Cell* **73,** 295–307.

Alnemri, E. S. (1997) Mammalian cell death proteases: a family of highly conserved aspartate-specific cysteine proteases. *J. Cell. Biochem.* **64,** 33–42.

Alnemri, E. S., Livingston, D. J., Nicholson, D. W., Salvesen, G., Thornberry, N. A., Wong, W. W., and Yuan, J. (1996). Human ICE/CED-3 protease nomenclature. *Cell* **87,** 171.

Ambroso, J. L., and Harris, C. (1994). *In vitro* embryotoxicity of the cysteine proteinase inhibitors benzyloxycarbonyl-phenylalanine-alanine-diazomethane (Z-Phe-Ala-CHN_2), and benzyloxycarbonyl-phenylalanine-phenylalanine-diazomethane (Z-Phe-Phe-CHN_2). *Teratology* **50,** 214–228.

Armstrong, R. C., Aja, T., Xiang, J., Gaur, S., Krebs, J. F., Hoang, K., Bai, X., Korsmeyer, S. J., Karanewsky, D. S., Fritz, L. C., and Tomaselli, K. J. (1996). Fas-induced activation of the cell death-related protease CPP32 is inhibited by Bcl-2 and ICE family protease inhibitors. *J. Biol. Chem.* **271,** 16850–16855.

Bartus, R. T., Elliott, P. J., Hayward, N. J., Dean, R. L., Harbeson, S., Straub, J. A., Li, Z., and Powers, J. C. (1995). Calpain as a novel target for treating acute neurodegenerative disorders. *Neurol. Res.* **17,** 249–258.

Blagosklonny, M. V., Schulte, T., Nguyen, P., Trepel, J., and Neckers, L. M. (1996). Taxol-induced apoptosis and phosphoylation of Bcl-2 protein involves c-Raf-1 and represents a novel c-Raf-1 signal transduction pathway. *Cancer Res.* **56,** 1851–1854.

Blagosklonny, M. V., Giannakakou, P., El-Deiry, W. S., Kingston, D. G. I., Higgs, P. I., Neckers, L., and Fojo, T. (1997). Raf-1/bcl-2 phosphorylation: a step from micotubule damage to cell death. *Cancer Res.* **57,** 130–135.

Boise, L. H., Gonzalez-Garcia, M., Postema, C. E., Ding, L., Lindsten, T., Turka, L. A., Mao, X., Nunez, G., and Thompson, C. B. (1993). bcl-x, a bcl-2-related gene that functions as a dominant regulator of apoptotic cell death. *Cell* **74,** 597–608.

Bokemeyer, D., Sorokin, A., and Dunn, M. J. (1996). Multiple intracellular MAP kinase signaling cascades. *Kidney Intl.* **49,** 1187–1198.

Boldin, M. P., Varfolomeev, E. E., Pancer, Z., Mett, I. L., Camonis, J. H., and Wallach, D. (1995). A novel protein that interacts with the death domain of Fas/APO1 contains a sequence motif related to the death domain. *J. Biol. Chem.* **270,** 7795–7798.

Borner, C. (1996). Diminished cell proliferation associated with the death-protective activity of Bcl-2. *J. Biol. Chem.* **271,** 12695–12698.

Borner, C., Martinou, I., Mattmann, C., Irmler, M., Schaerer, E., Martinou, J.-C., and Tschopp, J. (1994). The protein bcl-2α does not require membrane attachment, but two conserved domains to suppress apoptosis. *J. Cell Biol.* **126,** 1059–1068.

Boyd, J. M., Malstrom, S., Subramanian, T., Venkatesh, L. K., Schaeper, U., Elangovan, B., D'Sa-Eipper, C., and Chinnadurai, G. (1994). Adenovirus E1B 19 kDa and Bcl-2 proteins interact with a common set of cellular proteins. *Cell* **79,** 341–351.

Boyd, J. M., Gallo, G. J., Elangovan, B., Houghton, A. B., Malstrom, S., Avery, B. J., Ebb, R. G., Subramanian, T., Chittenden, T., Lutz, R. J., and Chinnadurai, G. (1995). Bik, a novel death inducing protein shares a distinct sequence motif with Bcl-2 family proteins and interacts with viral and cellular survival-promoting proteins. *Oncogene* **11,** 1921–1928.

Brady, H. J. M., Gil-Gomez, G., Kirberg, J., and Berns, A. J. M. (1996). Baxα perturbs T cell development and affects cell cycle entry of T cells. *EMBO J.* **15,** 6991–7001.

Bump, N. J., Hackett, M., Hugunin, M., Seshagiri, S., Brady, K., Chen, P., Ferenz, C., Franklin, S., Ghayur, T., Li, P., Licari, P., Mankovich, J., Shi, L., Greenberg, A. H., Miller, L. K., and Wong, W. W. (1995). Inhibition of ICE family proteases by baculovirus antiapoptotic protein p35. *Science* **269,** 1885–1888.

Camp, V., and Martin, P. (1996). The role of macrophages in clearing programmed cell death in the developing kidney. *Anat. Embryol. (Berl.)* **194,** 341–348.

Carrio, R., Lopez-Hoyos, M., Jimeno, J., Benedict, M. A., Merino, R., Benito, A., Fernandez-Luna, J. L., Nunez, G., Garcia-Porrero, J. A., and Merino, J. (1996). A1 demonstrates restricted tissue distribution during embryonic development and functions to protect against cell death. *Am. J. Path.* **149,** 2133–2142.

Casciola-Rosen, L., Nicholson, D. W., Chong, T., Rowan, K. R., Thornberry, N. A., Miller, D. K., and Rosen, A. (1996). Apopain/CPP32 cleaves proteins that are essential for cellular repair: a fundamental principle of apoptotic death. *J. Exp. Med.* **183,** 1957–1964.

Chang, B. S., Minn, A. J., Muchmore, S. W., Fesik, S. W., and Thompson, C. B. (1997). Identification of a novel regulatory domain in Bcl-x_L and Bcl-2. *EMBO J.* **16,** 968–977.

Chang, G.-Q., Hao, Y., and Wong, F. (1993). Apoptosis: final common pathway of photoreceptor death in rd, rds, and rhodopsin mutant mice. *Neuron* **11,** 595–605.

Chen, B., Cyr, D. G., and Hales, B. F. (1994). Role of apoptosis in mediating phosphoramide mustard-induced rat embryo malformations *in vitro. Teratology* **50,** 1–12.

Chen, C.-Y., and Faller, D. V. (1996). Phosphorylation of Bcl-2 protein and association with $p21^{Ras}$ in Ras-induced apoptosis. *J. Biol. Chem.* **271,** 2376–2379.

Chen, P., Nordstrom, W., Gish, B., and Abrams, J. M. (1996). grim, a novel cell death gene in Drosophila. *Genes Dev.* **10,** 1773–1782.

Chen-Levy, Z., and Cleary, M. L. (1990). Membrane topology of the Bcl-2 proto-oncogenic protein demonstrated *in vitro. J. Biol. Chem.* **265,** 4929–4933.

Cheng, E. H.-Y., Levine, B., Boise, L. H., Thompson, C. B., and Hardwick, J. M. (1996). Bax-independent inhibition of apoptosis by Bcl-x_L. *Nature* **379,** 554–556.

Cheng, E. H., Nicholas, J., Bellows, D. S., Hayward, G. S., Guo, H. G., Reitz, M. S., and Hardwick, J. M. (1997). A Bcl-2 homolog encoded by Kaposi sarcoma-associated virus, human herpesvirus 8, inhibits apoptosis but does not heterodimerize with Bax or Bak. *Proc. Natl. Acad. Sci.* (*USA*) **94,** 690–694.

Chinnaiyan, A. M., O'Rourke, K., Yu, G.-L., Lyons, R. H., Gang, M., Duan, D. R., Xing, L., Gentz, R., Ni, J., and Dixit, V. M. (1996a). Signal transduction by DR-3, a death domain-containing receptor related to TNFR-1 and CD95. *Science* **274,** 990–992.

Chinnaiyan, A. M., Orth, K., O'Rourke, K., Duan, H., Poirier, G. G., and Dixit, V. M. (1996b). Molecular ordering of the cell death pathway. *J. Biol. Chem.* **271,** 4573–4576.

Chinnaiyan, A. M., O'Rourke, K., Lane, B. R., and Dixit, V. M. (1997). Interaction of CED-4 with CED-3 and CED-9: a molecular framework for cell death. *Science* **275,** 1122–1126.

Chiocca, S., Baker, A., and Cotten, M. (1997). Identification of a novel antiapoptotic protein, GAM-1, encoded by the CELO adenovirus. *J. Virol.* **71,** 3168–3177.

Chittenden, T., Harrington, E. A., O'Connor, R., Flemington, C., Lutz, R. J., Evan, G. I., and Guild, B. C. (1995). Induction of apoptosis by the Bcl-2 homologue Bak. *Nature* **374,** 733–736.

Choi, S. S., Park, I. C., Yun, J. W., Sung, Y. C., Hong, S. I., and Shin, H. S. (1995). A novel Bcl-2 related gene, Bfl-1, is overexpressed in stomach cancer and preferentially expressed in bone marrow. *Oncogene* **11,** 1693–1698.

Clem, R. J., and Duckett, C. S. (1997). The iap genes: unique arbitrators of cell death. *Trends Cell Biol.* **7,** 337–339.

Cohen, G. M. (1997). Caspases: the executioners of apoptosis. *Biochem. J.* **326,** 1–16.

Cotman, C. W., and Su, J. H. (1996). Mechanisms of neuronal death in Alzheimer's disease. *Brain Pathol.* **6,** 493–506.

Darmon, A. J., Ley, T. J., Nicholson, D. W., and Bleackley, R. C. (1996). Cleavage of CPP32 by granzyme B represents a critical role for granzyme B in the induction of target cell DNA fragmentation. *J. Biol. Chem.* **271,** 21709–21712.

Darzynkiewicz, Z., Juan, G., Li, X., Gorczyca, W., Murakami, T., and Traganos, F. (1997). Cytometry in cell necrobiology: analysis of apoptosis and accidental cell death (necrosis). *Cytomet.* **27,** 1–20.

Decaudin, D., Geley, S., Hirsch, T., Castedo, M., Marchetti, P., Macho, A., Kofler, R., and Kroemer, G. (1997). Bcl-2 and Bcl-x_L antagonize the mitochondrial dysfunction preceding nuclear apoptosis induced by chemotherapeutic agents. *Cancer Res.* **57,** 62–67.

Deckwerth, T. L., and Johnson, E. M. Jr. (1973). Temporal analysis of events associated with programmed cell death (apoptosis) of sympathetic neurons deprived of nerve growth factor. *J. Cell Biol.* **123,** 1207–1222.

Deckwerth, T. L., Elliott, J. L., Knudson, C. M., Johnson, E. M. Jr., Snider, W. D., and Korsmeyer, S. J. (1996). BAX is required for neuronal death after trophic factor deprivation and during development. *Neuron* **17,** 401–411.

del Peso, L., Gonzalez-Garcia, M., Page, C. Herrera, R., and Nunez, G. (1997). Interleukin-3-induced phosphorylation of BAD through the protein kinase Akt. *Science* **278,** 687–689.

Deveraux, Q. L., Takahashi, R., Salvesen, G. S., and Reed, J. C. (1997). X-linked IAP is a direct inhibitor of cell death proteases. *Nature* **388,** 300–304.

D'Sa-Eipper, C., Subramanian, T., and Chinnadurai, G. (1996). *bfl-1,* a *bcl-2* homologue, suppresses p53-induced apoptosis and exhibits potent cooperative transforming activity. *Cancer Res.* **56,** 3879–3882.

Duan, H. Orth, K., Chinnaiyan, A. M., Poirier, G. G., Froelich, C. J., He, W. W., and Dixit, V. M. (1996). ICE-LAP6, a novel member of the ICE/Ced-3 gene family, is activated by the cytotoxic T cell protease granzyme B. *J. Biol. Chem.* **271,** 16720–16724.

Ellis, R. E., Yuan, J., and Horvitz, H. R. (1991). Mechanisms and functions of cell death. *Annu. Rev. Cell Biol.* **7,** 663–698.

Elstein, K. H., and Zucker, R. M. (1997). Applications of Flow Cytometry in Developmental and Reproduction Toxicology. In *Molecular and Cellular Methods in Development Toxicology* G. P. Daston, Ed., CRC Press, Inc., Boca Raton, FL. pp. 195–230.

Enari, M., Talanian, R. V., Wong, W. W., and Nagata, S. (1996). Sequential of ICE-like and CPP32-like proteases during Fas-mediated apoptosis. *Nature* **380,** 723–726.

Erhardt, P., and Cooper, G. M. (1996). Activation of the CPP32 apoptotic protease by distinct signaling pathways with differential sensitivity to Bcl-x_L. *J. Biol. Chem.* **271,** 17601–17604.

Estus, S., Zaks, W. J., Freeman, R. S., Gruda, M., Bravo, R., and Johnson, E. M. (1994). Altered gene expression in neurons during programmed cell death: identification of c-*jun* as necessary for neuronal apoptosis. *J. Cell Biol.* **127,** 1717–1727.

Fanger, G. R., Gerwins, P., Widmann, C., Jarper, M. B., and Johnson, G. L. (1997). MEKKs, GCKs, MLKs, PAKs, TAKs, and Tpls: upstream regulators of the c-Jun amino-terminal kinases? *Curr. Opin. Genet. Devel.* **7,** 67–74.

Farlie, P. G., Dringen, R., Rees, S. M., Kannourakis, G., and Bernard, O. (1995). *bcl-2* transgene expression can protect neurons against developmental and induced cell death. *Proc. Natl. Acad. Sci.* (*USA*) **92,** 4397–4401.

Farrow, S. N., White, J. H. M., Martinou, I., Raven, T., Pun, K.-T., Grinham, C. J., Martinou, J.-C., and Brown R. (1995). Cloning of a *bcl-2* homologue by interaction with adenovirus E1B 19K. *Nature* **374,** 731–733.

Faustman-Watts, E., Greenaway, J. C., Namkung, M. J., Fantel, A. G., and Juchau, M. R. (1983). Teratogenicity *in vitro* of 2-acetylaminofluorene: role of biotransformation in the rat. *Teratology* **27,** 19–28.

Fernandes-Alnemri, T., Litwack, G., and Alnemri, E. S. (1994). CPP32, a novel human apoptotic protein with homology to *Caenorhabditis elegans* cell death protein Ced-3 and mammalian interleukin-1 β-converting enzyme. *J. Biol. Chem.* **269,** 30761–30764.

Fernandes-Alnemri, T., Armstrong, R. C., Krebs, J., Srinivasula, S. M., Wang, L., Bullrich, F., Fritz, L. C., Trapani, J. A., Tomaselli, K. J., Litwack, G., and Alnemri, E. S. (1996). *In vitro* activation of CPP32 and Mch3 by Mch 4, a novel human apoptotic cysteine protease containing two FADD-like domains. *Proc. Natl. Acad. Sci.* (*USA*) **93,** 7464–7469.

Fernandez-Sarabia, M. J., and Bischoff, J. R. (1993). Bcl-2 associates with the ras-related protein R-ras p23. *Nature* **366,** 274–275.

Frankowski, H., Missotten, M., Fernandez, P.-A., Martinou, I., Michel, P., Sadoul, R., and Martinou, J.-C. (1995). Function and expression of the Bcl-x gene in the developing and adult nervous system. *NeuroReport* **6,** 1917–1921.

French, L. E., Hahne, M., Viard, I., Radlgruber, G., Zanone, R., Becker, K., Muller, C., and Tchsopp, J. (1996). Fas and Fas ligand in embryos and adult mice: ligand expression in several immune-priviledged tissues and coexpression in adult tissues characterized by apoptotic cell turnover. *J. Cell Biol.* **133,** 335–343.

Friedlander, R. M., Gagliardini, V., Hara, H., Fink, K. B., Li, W., MacDonald, G., Fishman, M. C., Greenberg, A. H., Moskowitz, M. A., and Yuan, J. (1997). Expression of a dominant negative mutant of interleukin-1β converting enzyme in transgenic mice prevents neuronal cell death induced by trophic factor withdrawal and ischemic brain injury. *J. Exp. Med.* **185,** 933–940.

Friesen, C., Herr, I., Krammer, P. H., and Debatin, K.-M. (1996). Involvement of the CD95 (APO-1/Fas) receptor/ligand system in drug-induced apoptosis in leukemia cells. *Nat. Med.* **2,** 574–577.

Friesen, C., Fulda, S., and Debatin, K.-M. (1997). Deficient activation of the CD95 (APO-1/Fas) system in drug-resistant cells. *Leukemia* **11,** 1833–1841.

Froelich, C. J., Hanna, W. L., Poirier, G. G., Duriez, P. J., D'Amours, D., Salvesen, G. S., Alnemri, E. S., Earnshaw, W. C., and Shah, G. M. (1996). Granzyme B/perforin-mediated apoptosis of Jurkat cells results in cleavage of poly(ADP-ribose) polymerase to the 89-kDa apoptotic fragment and less abundant 64-kD fragment. *Biochem. Biophys. Res. Comm.* **227,** 658–665.

Fulda, S., Sieverts, H., Friesen, C., Herr, I., and Debatin, K.-M. (1997). The CD95 (APO-1/Fas) system mediates drug-induced apoptosis in neuroblastoma cells. *Cancer Res.* **57,** 3823–3829.

Gagliardini, V., Fernandez, P.-A., Lee, R. K., Drexler, H. C. A., Rotello, R. J., Fishman, M. C., and Yuan, J. (1994). Prevention of vertebrate neuronal death by the crmA gene. *Science* **263,** 826–828.

Gajewski, T. F., and Thompson, C. B. (1996). Apoptosis meets signal transduction: elimination of a BAD influence. *Cell* **87,** 589–592.

Gao, X., Blackburn, M. R., and Knudsen, T. B. (1994). Activation of apoptosis in early mouse embryos by 2′-deoxyadenosine exposure. *Teratology* **49,** 1–12.

Garcia, I., Marinous, I., Tsujimoto, Y., and Martinou, J.-C. (1992). Prevention of programmed cell death of sympathetic neurons by the bcl-2 proto-oncogene. *Science* **258,** 302–304.

Garcia-Martinez, V., Macias, D., Ganan, Y., Garcia-Lobo, J. M., Francia, M. V., Fernandez-Teran, M. A., and Hurle, J. M. (1993). Internucleosomal DNA fragmentation and programmed cell death (apoptosis) in the interdigitial tissue of the embryo chick leg bud. *J. Cell. Sci.* **106,** 201–208.

Gavrieli, Y., Sherman, Y., and Ben-Sasson, S. A. (1992). Identification of programmed cell death in situ via specific labeling of nuclear DNA fragmentation. *J. Cell Biol.* **119,** 493–501.

Gibson, L., Holmgreen, S. P., Huang, D. C. S., Bernard, O., Copeland, N. G., Jenkins, N. A., Sutherland, G. R. Baker, E., Adams, J. M., and Cory, S. (1996). *bcl-w,* a novel member of the *bcl-2* family promotes cell survival. *Oncogene* **13,** 665–675.

Goillot, E., Raingeaud, J., Range, A., Tepper, R. I., Davis, R. J., Harlow, E., and Sanchez, I. (1997). Mitogen-activated protein kinase-mediated Fas apoptotic signaling pathway. *Proc. Natl. Acad. Sci. (USA)* **94,** 3302–3307.

Golstein, P., Marguet, D., Depraetere, V. (1995). Fas bridging cell death and cytotoxicity: the reaper connection. *Imm. Rev.* **146,** 45–56.

Goltsev, Y. V., Kovalenko, A. V., Arnold, E., Varfolomeev, E. E., Brodianskii, V. M., and Wallach, D. (1997). CASH, a novel caspase homologue with death effector domains. *J. Biol. Chem.* **272,** 19641–19644.

Gonzalez-Garcia, M., Perez-Ballestero, R., Ding, L., Duan, L., Boise, L. H., Thompson, C. B., and Nunez, G. (1994). bcl-x_L is the major bcl-x mRNA expressed during muring development and its product localizes to mitochondria. *Devel.* **120,** 3033–3042.

Gonzalez-Garcia, M., Garcia, I., Ding, L., O'Shea, S., Boise, L. H., Thompson, C. B., and Nunez, G. (1995). bcl-x is expressed in embryonic and postnatal neural tissues and functions to prevent neuronal cell death. *Proc. Natl. Acad. Sci. (USA)* **92,** 4304–4308.

Greenlund, L. J. S., Deckwerth, T. L., and Johnson, E. M. Jr. (1995). Superoxide dismutase delays neuronal apoptosis: a role of reactive oxygen species in programmed neuronal death. *Neuron* **14,** 303–315.

Gressner, A. M., Lahme, B., and Roth, S. (1997). Attenuation of TGF-β-induced apoptosis in primary cultures of hepatocytes by calpain inhibitors. *Biochem. Biophys. Res. Commun.* **231,** 457–462.

Grether, M. E., Abrams, J. M., Agapite, J., White, K., and Steller, H. (1995). The head involution defective gene of *Drosophila melanogaster* functions in programmed cell death. *Genes Dev.* **9,** 1694–1708.

Griffith, T. S., Brunner, T., Fletcher, S. M., Green, D. R., and Ferguson, T. A. (1995). Fas ligand-induced apoptosis as a mechanism of immune privilege. *Science* **270,** 1189–1192.

Haldar, S., Jena, N., and Croce, C. M. (1995). Inactivation of Bcl-2 by phosphorylation. *Proc. Natl. Acad. Sci. (USA)* **92,** 4507–4511.

Haldar, S., Basu, A., and Croce, C. M. (1997). Bcl-2 is the guardian of microtubule integrity. *Cancer Res.* **57,** 229–233.

Han, J., Sabbatini, P., and White, E. (1996a). Induction of apoptosis by human Nbk/Bik, a BH3-containing protein that interacts with E1B 19K. *Mol. Cell. Biol.* **16,** 5857–5864.

Han, J., Sabbatini, P., Perez, D., Rao, L., Modha, D., and White, E. (1996b). The E1B 19K protein blocks apoptosis by interacting with and inhibiting the p53-inducible and death-promoting Bax protein. *Genes Dev.* **10,** 461–477.

Hanada, M., Aim'e-Semp'e, C., Sato, T., and Reed, J. C. (1995). Structure-function analysis of Bcl-2 protein. Identification of conserved domains important for homodimerization with Bcl-2 and heterodimerization with Bax. *J. Biol. Chem.* **270,** 11962–11969.

Hannun, Y. A. (1996). Functions of ceramide in coordinating cellular responses to stress. *Science* **274,** 1855–1859.

Harrington, E. A., Fanidi, A., and Evan, G. I. (1994). Oncogenes and cell death. *Curr. Opin. Genet. Dev.* **4,** 120–129.

Hart, S. P., Haslett, C., and Dransfield, I. (1996). Recognition of apoptotic cells by phagocytes. *Experientia* **52,** 950–956.

Henderson, S., Huen, D., Rowe, M., Dawson, C., Johnson, G., and Rickinson, A. (1993). Epstein-Barr virus-coded BHRF1 protein, a viral homologue of Bcl-2, protects human B cells from programmed cell death. *Proc. Natl. Acad. Sci. (USA)* **90,** 8479–8483.

Hengartner, M. O. (1996). Programmed cell death in invertebrates. *Curr. Opin. Genet. Dev.* **6,** 34–38.

Hengartner, M. O. (1997). Ced-4 is a stranger no more. *Nature* **388,** 714–715.

Hengartner, M. O., and Horvitz, H. R. (1994). C. elegans cell survival gene ced-9 encodes a functional homolog of the mammalian proto-oncogene bcl-2. *Cell* **76,** 665–676.

Hockenbery, D., Nunez, G., Milliman, C., Schreiber, R. D., and Korsmeyer, S. J. (1990). Bcl-2 is an inner mitochondrial membrane protein that blocks programmed cell death. *Nature* **348,** 334–336.

Hockenbery, D. M. (1994). bcl-2 in cancer, development, and apoptosis. *J. Cell Science (Suppl.)* **18,** 51–55.

Hockenbery, D. M., Oltvai, Z. N., Yin, X.-M., Milliman, C. L., and Korsmeyer, S. J. (1993). bcl-2 functions in an antioxidant pathway to prevent apoptosis. *Cell* **75,** 241–251.

Hopkinson-Woolley, J., Hughes, D., Gordon, S., and Martin, P. (1994). Macrophage recruitment during limb development and wound healing in the embryonic and foetal mouse. *J. Cell Sci.* **107,** 1159–1167.

Hu, S., Vincenz, C., Ni, J., Gentz, R., and Dixit, V. M. (1997). I-FLICE, a novel inhibitor of tumor necrosis factor receptor-1 and CD-95–induced apoptosis. *J. Biol. Chem.* **272,** 17255–17257.

Hunter, J. J., Bond, B. L., and Parslow, T. G. (1996). Functional dissection of the human Bcl2 protein: sequence requirements for inhibition of apoptosis. *Mol. Cell. Biol.* **16,** 877–883.

Hurle, J. M., Ros, M. A., Garcia-Martinez, V., Macias, D., and Ganan, Y. (1995). Cell death in the embryonic developing limb. *Scanning Microsc.* **9,** 519–534.

Irmler, M., Thome, M., Hahne, M., Schneider, P., Hofmann, K., Steiner, V., Bodmer, J.-L., Schroter, M., Burns, K., Mattmann, C., Rimoldi, D., French, L. E., and Tschopp, J. (1997). Inhibition of death receptor signals by cellular FLIP. *Nature* **388,** 190–195.

Itoh, N., and Nagata, S. (1993). A novel protein domain required for apoptosis: mutational analysis of human Fas antigen. *J. Biol. Chem.* **268,** 10932–10937.

Jacobson, M. D. (1996). Reactive oxygen species and programmed cell death. *Trends Biochem. Sci.* **21,** 83–86.

Jacobson, M. D., Burne, J. F., and Raff, M. C. (1994). Programmed cell death and Bcl-2 protection in the absence of a nucleus. *EMBO J.* **13,** 1899–1910.

Jacobson, M. D., and Raff, M. C. (1995). Programmed cell death and Bcl-2 protection in very low oxygen. *Nature* **374,** 814–816.

Jacobson, M. D., Weil, M., and Raff, M. C. (1997). Programmed cell death in animal development. *Cell* **88,** 347–354.

Jans, D. A., Jans, P., Briggs, L. J., Sutton, V., and Trapani, J. A. (1996). Nuclear transport of perforin *in vivo* and cytosolic factors *in vitro*. *J. Biol. Chem.* **271,** 30781–30789.

Johnson, E. M., and Deckwerth, T. L. (1993). Molecular mechanisms of developmental neuronal death. *Annu. Rev. Neurosci.* **16,** 31–46.

Juo, P., Kuo, C. J., Reynolds, S. E., Konz, R. F., Raingeaud, J., Davis, R. J., Biemann, H.-P., and Blenis, J. (1997). Fas activation of the p38 mitogen-activated protein kinase signalling pathway requires ICE/CED-3 family proteases. *Mol. Cell. Biol.* **17,** 24–35.

Kaufmann, S. H., Desnoyers, S., Ottaviano, Y., Davidson, N. E., and Poirier, G. G. (1993). Specific proteolytic cleavage of poly(ADP-ribose) polymerase: an early marker of chemotherapy-induced apoptosis. *Cancer Res.* **53,** 3976–3985.

Kayalar, C., Ord, T., Testa, M. P., Zhong, L. T., and Bredesen, D. E. (1996). Cleavage of actin by interleukin-1β converting enzyme to reverse DNase I inhibition. *Proc. Natl. Acad. Sci.* (*USA*) **93,** 2234–2238.

Kerr, J. F., Wyllie, A. H., and Currie, A. R. (1972). Apoptosis: a basic biological phenomenon with wide ranging implications in tissue kinetics. *Br. J. Cancer* **26,** 239–257.

Kiefer, M. C., Brauer, M. J., Powers, V. C., Wu, J. J., Umansky, S. R., Tomei, L. D., and Barr, P. J. (1995). Modulation of apoptosis by widely distributed Bcl-2 homologue Bak. *Nature* **374,** 736–739.

King, K. L., and Cidlowski, J. A. (1995). Cell cycle and apoptosis: common pathways to life and death. *J. Cell. Biochem.* **58,** 175–180.

Kitson, J., Raven, T., Jiang, Y.-P., Goeddel, D. V., Giles, K. M., Pun, K.-T., Grinham, C. J., and Brown, R., and Farrow, S. N. (1996). A death-domain-containing receptor that mediates apoptosis. *Nature* **384,** 372–375.

Kluck, R. M., Bossy-Wetzel, E., Green, D. R., and Newmeyer, D. D. (1997). The release of cytochrome c from mitochondria: a primary site of Bcl-2 regulation of apoptosis. *Science* **275,** 1132–1136.

Knudsen, T. B., (1997). Cellular Techniques for Teratological Cell Death. In *Molecular and Cellular Methods in Developmental Toxicology*. G. P. Daston, Ed., CRC Press, Inc., Boca Raton, FL. pp. 183–193.

Knudson, C. M., Tung, K. S. K., Tourtellotte, W. G., Brown, G. A. J., and Korsmeyer, S. J. (1995). Bax-deficient mice with lymphoid hyperplasia and male germ cell death. *Science* **270,** 96–99.

Ko, L. J., and Prives, C. (1996). p53: puzzle and paradigm. *Genes Dev.* **10,** 1054–1072.

Koopman, G., Reutelingsperger, C. P. M., Kuijten, G. A. M., Keehnen, R. M. J., Pals, S. T. and van Oers, M. H. J. (1994). Annexin V for flow cytometric detection of phosphatidylserine expression on B cells undergoing apoptosis. *Blood* **84,** 1415–1420.

Kozopas, K. M., Yang, T., Buchan, H. L., Zhou, P., and Craig, R. W. (1993). MCL1, a gene expressed in programmed myeloid cell differentiation, has sequence similarity to BCL2. *Proc. Natl. Acad. Sci.* (*USA*) **90,** 3516–3520.

Krajewska, M., Wang, H.-G., Krajewski, S., Zapata, J. M., Shabaik, A., Gascoyne, R., and Reed, J. C. (1997). Immunohistochemistry analysis of *in vivo* patterns of expression of CPP32 (Caspase-3), a cell death protease. *Cancer Res.* **57,** 1605–1613.

Krajewski, S., Tanaka, S., Takayama, S., Schibler, M. J., Fenton, W., and Reed, J. C. (1993). Investigation of the subcellular distribution of the bcl-2 oncoprotein: residence in the nuclear envelope, endoplasmic reticulum, and outer mitochondrial membranes. *Cancer Res.* **53,** 4701–4714.

Krajewski, S., Krajewska, M., Shabaik, A., Miyashita, T., Wang, H.-G., and Reed, J. C. (1994). Immunohistochemical determination of *in vivo* distribution of Bax, a dominant inhibitor of Bcl-2. *Am. J. Path.* **145,** 1323–1336.

Krajewski, S., Bodrug, S., Krajewska, M., Shabaik, A., Gascoyne, R., Berean, K., and Reed, J. C. (1995). Immunohistochemical analysis of Mcl-1 protein in human tissues. Differential regulation of Mcl-1 and Bcl-2 protein production suggests a unique role for Mcl-1 in control of programmed cell death *in vivo*. *Am. J. Path.* **146,** 1309–1319.

Krajewski, S., Krajewska, M., and Reed, J. C. (1996). Immunohistochemical analysis of *in vivo* patterns of *Bak* expression, a proapoptotic member of the Bcl-2 protein family. *Cancer Res.* **56,** 2849–2855.

Kroemer, G. (1997). Mitochondrial implication in apoptosis. Towards an endosymbiont hypothesis of apoptosis evolution. *Cell Death Diff.* **4,** 443–456.

Kubbutat, M. H., and Vousden, K. H. (1997). Proteolytic cleavage of human p53 by calpain: a potential regulator of protein stability. *Mol. Cell. Biol.* **17,** 460–468.

Kuida, K., Lippke, J. A., Ku, G., Harding, M. W., Livingston, D. J., Su, M. S., and Flavell, R. A. (1995). Altered cytokine export and apoptosis in mice deficient in interleukin-1β converting enzyme. *Science* **267,** 2000–2003.

Kuida, K., Zheng, T. S., Na, S., Kuan, C.-Y., Yang, D., Karasuyama, H., Rakic, P., Flavell, R. A. (1996). Decreased apoptosis in the brain and premature lethality in CPP32-deficient mice. *Nature* **384,** 368–372.

Kyriakis, J. M., and Avruch, J. (1996). Protein kinase cascades activated by stress and inflammatory cytokines. *BioEssays* **18,** 567–577.

Larocca, J. N., Farooq, M., Norton, W. T. (1997). Induction of oligodendrocyte apoptosis by C-2-ceramide. *Neurochem. Res.* **22,** 529–534.

Lawrence, M. S., Ho, D. Y., Sun, G. H., Steinberg, G. K., and Sapolsky, R. M. (1996). Overexpression of Bcl-2 with herpes simplex virus vectors protects CNS neurons against neurological insults *in vitro* and *in vivo*. *J. Neurosci.* **16,** 486–496.

Lazebnik, Y. A., Takahashi, A., Moir, R. D., Goldman, R. D., Poirier, G. G., Kaufmann, S. H., Earnshaw, W. C. (1995). Studies of the lamin proteinase reveal multiple parallel biochemical pathways during apoptotic execution. *Proc. Natl. Acad. Sci.* (*USA*) **92,** 9042–9046.

Li, P., Allen, H., Bannerjee, S., Franklin, S., Herzog, L., Johnston, C., McDowell, J., Paskind, M., Rodman, L., Salfeld, J., Towne, E., Tracey, D., Wardwell, S., Wei F-Y., Wong, W., Kamen, R., and Seshadri, T. (1995). Mice deficient in IL-1β-converting enzyme are defective in production of mature IL-1β and resistant to endotoxic shock. *Cell* **80,** 401–411.

Lin, E. Y., Orlofsky, A., Berger, M. S., and Prystowsky, M. B. (1993). Characterization of A1, a novel hemopoietic-specific early-response gene with sequence similarity to bcl-2. *J. Immunol.* **151,** 1979–1988.

Lin, E. Y., Orlofsky, A., Wang, H.-G., Reed, J. C., and Prystowsky, M. B. (1996). A1, a*Bcl-2* family member, prolongs cell survival and permits myeloid differentiation. *Blood* **87,** 983–992.

Liston, P., Roy, N., Tamai, K., Lefebre, C., Baird, S., Cherton-Horvat, G., Farahani, R., McLean, M., Ikeda, J.-E., MacKenzie, A., and Korneluk, R. G. (1996). Suppression of apoptosis in mammalian cells by NAIP and a related family of IAP genes. *Nature* **379,** 349–353.

Liu, X., Kim, C. N., Yang, J., Jemmerson, R., and Wang, X. (1996). Induction of apoptotic program in cell-free extracts: requirement for dATP and cytochrome c. *Cell* **86,** 147–157.

Liu, X. S., Zou, H., Slaughter, C., and Wang, X. D. (1997). DFF, a heterodimeric protein that functions downstream of caspase-3 to trigger DNA fragmentation during apoptosis. *Cell* **89,** 175–184.

Los, M., Van de Craen, M., Penning, L. C., Schenk, H., Westendorp, M., Baeuerle, P. A., Droge, W., Krammer, P. H., Fiers, W., Schulze-Osthoff, K. (1995). Requirement of an ICE/CED-3 protease for Fas/APO-1-mediated apoptosis. *Nature* **375,** 81–83.

Marchetti, P., Castedo, M., Susin, S. A., Zamzami, N., Hirsch, T., Macho, A., Haeffner, A., Hirsch, F., Geuskens, M., and Kroemer, G. (1996). Mitochondrial permeability transition is a central coordinating event of apoptosis. *J. Exp. Med.* **184,** 1155–1160.

Marsters, S. A., Sheridan, J. P., Donahue, C. J., Pitti, R. M., Gray, C. L., Goddard, A. D., Bauer, K. D., and Ashkenazi, A. (1996). Apo-3, a new member of the tumor necrosis factor receptor family, contains a death domain and activates apoptosis and NF-κB. *Curr. Biol.* **12,** 1669–1676.

Martin, S. J., O'Brien, G. A., Nishioka, W. K., McGahon, A. J., Mahboubi, A., Saido, T. C., Green, D. R. (1995). Proteolysis of fodrin (non-erythroid spectrin) during apoptosis. *J. Biol. Chem.* **270,** 6425–6428.

Mashimi, T., Naito, M., Noguchi, K., Miller, D. K., Nicholson, D. W., and Tsuruo, T. (1997). Actin cleavage by CPP-32/apopain during the development of apoptosis. *Oncogene* **14,** 1007–1012.

May, W. S., Tyler, P. G., Ito, T., Armstrong, D. K., Qatsha, K. A., and Davidson, N. E. (1996). Interleukin-3 and bryostatin-1 mediated hyperphosphorylation of Bcl-2α in association with suppression of apoptosis. *J. Biol. Chem.* **269,** 26865–26870.

Mazel, S., Burtrum, D., and Petrie, H. T. (1995). Regulation of cell division cycle progression by bcl-2 expression: a potential mechanism for inhibition of programmed cell death. *J. Exp Med.* **183,** 2219–2226.

Meikrantz, W., and Schlegel, R. (1995). Apoptosis and the cell cycle. *J. Cell. Biochem.* **58,** 160–174.

Merry, D. E., and Korsmeyer, S. J. (1997). Bcl-2 gene family in the nervous system. *Annu. Rev. Neurosci.* **20,** 245–267.

Merry, D. E., Veis, D. J., Hickey, W. F., and Korsmeyer, S. J. (1994). bcl-2 protein expression is widespread in the developing nervous system and retained in the adult PNS. *Devel.* **120,** 301–311.

Michaelidis, T. M., Sendtner, M., Cooper, J. D., Airaksinen, M. S., Holtmann, B., Meyer, M., and Thoenen, H. (1996). Inactivation of bcl-2 results in progressive degeneration of motoneurons, sympathetic and sensory neurons during early postnatal development. *Neuron* **17,** 75–89.

Minn, A. J., Velez, P., Schendel, S. L., Liang, H., Muchmore, S. W., Fesik, S. W., Fill, M., and Thompson, C. B. (1997). Bcl-x_L forms an ion channel in synthetic lipid membranes. *Nature* **385,** 353–357.

Mirkes, P. E., Greenaway, J. C., and Shepard, T. H. (1983). A kinetic analysis of rat embryo response to cyclophosphamide exposure *in vitro. Teratology* **28,** 249–256.

Mirkes, P. E. (1985a). Effects of acute exposures to elevated temperatures on rat embryo growth and development in vitro. *Teratology* **32,** 259–266.

Mirkes, P. E. (1985b). Cyclophosphamide teratogenesis: a review. *Teratog. Carcinog. Mutagen.* **5,** 75–88.

Mirkes, P. E., and Little, S. A. (1997). Teratogen-induced cell death in postimplantation mouse embryos: differential tissue sensitivity and hallmarks of apoptosis. *Cell Death Differen.* (In Press).

Mirkes, P. E., Huynh, L. T., and Little, S. A. (1997). Activation of CPP32 and internucleosomal DNA fragmentation and inactivation of Poly (ADP-ribose) polymerase (PARP) during hyperthermia and sodium arsenite–induced apoptosis in early postimplantation mouse embryos. *Teratol. Soc. Abstracts* **55,** 44.

Miyashita, T., Krajewski, S., Krajewska, M., Wang, H.-G., Lin, H. K., Liebermann, D. A., Hoffman, B., and Reed, J. C. (1994). Tumor suppressor p53 is a regulator of bcl-2 and bax gene expression in vitro and in vivo. *Oncogene* **9,** 1799–1805.

Mizushima, N., Koike, R., Kohsaka, H., Kushi, Y., Handa, S., Yagita, H., and Miyasaka, N. (1996). Ceramide induces apoptosis via CPP32 activation. *FEBS Letts.* **395,** 267–271.

Moallem, S. A., and Hales, B. F. (1995). Induction of apoptosis and cathepsin D in limbs exposed in vitro to an activated analog of cyclophosphamide. *Teratology* **52,** 3–14.

Montague, J. W., and Cidlowski, J. A. (1996). Cellular catabolism in apoptosis: DNA degradation and endonuclease activation. *Experientia* **52,** 957–962.

Motoyama, N., Wang, F., Roth, K. A., Sawa, H., Nakayama, K.-I., Nakayama, K., Negishi, I., Senju, S., Zhang, Q., Fujii, S., Loh, D. Y. (1995). Massive cell death of immature hematopoietic cells and neurons in bcl-x-deficient mice. *Science* **267,** 1506–1510.

Muchmore, S. W., Sattler, M., Liang, H., Meadows, R. P., Harlan, J. E., Yoon, H. S., Nettesheim, D., Chang, B. S., Thompson, C. B., Wong, S.-L., Ng, S.-C., and Fesik, S. W. (1996). X-ray and NMR structure of human Bcl-x_L, an inhibitor of programmed cell death. *Nature* **381,** 335–341.

Mukasa, T., Urase, K., Momoi, M. Y., Kimura, I., and Momoi, T. (1997). Specific expression of CPP32 in sensory neurons of mouse embryos and activation of CPP32 in the apoptosis induced by a withdrawal of NGF. *Biochem. Biophys. Res. Comm.* **231,** 770–774.

Muschel, R. J., Bernhard, E. J., Garza, L., McKenna, W. G., and Koch, C. J. (1995). Induction of apoptosis at different oxygen tensions: evidence that oxygen radicals do not mediate apoptotic signaling. *Cancer Res.* **55,** 995–998.

Muzio, M., Chinnaiyan, A. M., Kischkel, F. C., O'Rourke, K., Shevchenko, A., Ni, J., Seaffiche, C., Bretz, J. D., Zhang, M., Gentz, R., Mann, M., Krammer, P. H., Peter, M. E., and Dixit, V. M. (1996). FLICE, a novel FADD-homologous ICE/CED-3-like protease, is recruited to the CD95 (Fas/APO-1) death-inducing signaling complex. *Cell* **85,** 817–827.

Muzio, M., Salvesen, G. S., and Dixit, V. M. (1997). FLICE induced apoptosis in a cell-free system. Cleavage of caspase zymogens. *J. Biol. Chem.* **272,** 2952–2956.

Na, S. Q., Chuang, T. H., Cunningham, A., Turi, T. G., Hanke, J. H., Bokoch, G. M., and Danley, D. E. (1996). D4-GDI, a substrate of CPP32, is proteolyzed during Fas-induced apoptosis. *J. Biol. Chem.* **271,** 11209–11213.

Nagata, S. (1997). Apoptosis by death factor. *Cell* **88,** 355–365.

Nagata, S., and Golstein, P. (1995). The Fas death factor. *Science* **267,** 1449–1456.

Nakajima, H., Golstein, P., and Henkart, P. A. (1995). The target cell nucleus is not required for cell-mediated granzyme- or Fas-based cytotoxicity. *J. Exp. Med.* **181,** 1905–1909.

Naruse, I., and Keino, H. (1995). Apoptosis in the developing CNS. *Prog. Neurobiol.* **47,** 135–155.

Nath, R., Raser, K. J., Stafford, D., Hajimohammadreza, I., Posner, A., Allen, H., Talanian, R. V., Yuen, P.-W., Gilbertsen, R. B., and Wang, K. K. W. (1996). Non-erythroid α-spectrin breakdown by calpain and interleukin 1β-converting-enzyme-like protease(s) in apoptotic cells: contributory roles of both protease families in neuronal apoptosis. *Biochem. J.* **319,** 683–690.

Newmeyer, D. D., Farschon, D. M., and Reed, J. C. (1994). Cell-free apoptosis in Xenopus egg extracts: inhibition by Bcl-2 and requirement for an organelle fraction enriched in mitochondria. *Cell* **79,** 353–364.

Nguyen, M., Millar, D. G., Yong, V. W., Korsmeyer, S. J., and Shore, G. C. (1993). Targeting of bcl-2 to the mitochondrial outer membrane by a COOH-terminal signal anchor sequence. *J. Biol. Chem.* **268,** 25265–25268.

Nguyen, M., Branton, P. E., Walthon, P. A., Oltvai, Z. N., Korsmeyer, S. J., and Shore, G. C. (1994). Role of membrane anchor domain of bcl-2 in suppression of apoptosis caused by E1B-defective adenovirus. *J. Biol. Chem.* **269,** 16521–16524.

Nicholson, D. W. (1996). ICE/CED3-like proteases as therapeutic targets for the control of inappropriate apoptosis. *Nature Biotech.* **14,** 297–301.

Nicholson, D. W., Ali, A., Thornberry, N. A., Vaillancourt, J. P., Ding, C. K., Gallant, M., Gareau, Y., Griffin, P. R., Labelle, M., Lazebnik, Y. A., Munday, N. A., Raju, A. M., Smulson, M. E., Yamin, T.-T., Yu, V. L., and Miller, D. K. (1995). Identification and inhibition of the ICE/CED-3 protease necessary for mammalian apoptosis. *Nature* **376,** 37–43.

Nicholson, D. W., and Thornberry, N. A. (1997). Caspase: killer proteases. *Trends Biochem. Sci.* **22,** 299–306.

Nixon, R. A., Saito, K. I., Grynspan, F., Griffin, W. R., Katayama, S., Honda, T., Mohan, P. S., Shea, T. B., and Beermann, M. (1994). Calcium-activated neutral proteinase (calpain) system in aging and Alzheimer's disease. *Annal. N.Y. Acad. Sci.* **747,** 77–91.

Nomura, M., Suzuki, M., Suzuki, Y., Ikeda, H., Tamura, J., Koike, M., Jie, T., and Itoh, G. (1996). Cyclophosphamide-induced apoptosis induces phocomelia in the mouse. *Arch. Toxicol.* **70,** 672–677.

Oberhammer, F., Wilson, J. W., Dive, C., Morris, I. D., Hickman, J. A., Wakeling, A. E., Walker, P. R., and Sikorska, M. (1993). Apoptotic death in epithelial cells: cleavage of DNA to 300 and/or 50 kb fragments prior to or in the absence of internucleosomal fragmentation. *EMBO J.* **12,** 3679–3684.

Oltvai, Z. N., Milliman, C. L., and Korsmeyer, S. J. (1993). Bcl-2 heterodimerizes in vivo with a conserved homolog, Bax, that accelerates programmed cell death. *Cell* **74,** 609–619.

Oppenheim, R. W. (1991). Cell death during development of the nervous system. *Annu. Rev. Neurosci.* **14,** 453–501.

Oppenheim, R. W. (1996). Neurotrophic survival molecules for motoneurons: an embarassment of riches. *Neuron* **17,** 195–197.

O'Reilly, L. A., Huang, D. C. S., and Strasser, A. (1996). The cell death inhibitor Bcl-2 and its homologues influence control of cell cycle entry. *EMBO J.* **15,** 6979–6990.

Owen-Schaub, L. B., Zhang, W., Cusack, J. C., Angelo, L. S., Santee, S. M., Fujiwara, T., Roth, J. A., Deisseroth, A. B., Zhang, W. W., Kruzel, E., and Radinsky, R. (1995). Wild-type human p53 and a temperature-sensitive mutant induce Fas/APO-1 expression. *Mol. Cell. Biol.* **15,** 3032–3040.

Pan, G., O'Rourke, K., Chinnaiyan, A. M., Gentz, R., Ebner, R., Ni, J., and Dixit, V. M. (1997). The receptor for the cytotoxic ligand TRAIL. *Science* **276,** 111–113.

Pandey, S., Walker, P. R., and Sikorska, M. (1997). Identification of a novel 97 kD protein capable of internucleosomal DNA cleavage. *Biochem.* **36,** 711–720.

Patel, T., Gores, G. J., and Kaufmann, S. H. (1996). The role of proteases during apoptosis. *FASEB J.* **10,** 587–597.

Pena, L. A., Fuks, Z., and Kolesnick, R. (1997). Stress-induced apoptosis and the sphingomyelin pathway. *Biochem. Pharm.* **53,** 615–621.

Petit, P. X., Susin, S.-A., Zamzami, N., Mignotte, B., and Kroemer, G. (1996). Mitochondria and programmed cell death: back to the future. *FEBS Letts* **396,** 7–13.

Pitti, R. M., Marsters, S. A., Ruppert, S., Donahue, C. J., Moore, A., and Ashkenazi, A. (1996). Induction of apoptosis by Apo-2 ligand, a new member of the tumor necrosis factor cytokine family. *J. Biol. Chem.* **271,** 12687–12690.

Porter-Cailliau, C., Sung, C.-H., Nathans, J., and Adler, R. (1994). Apoptotic photoreceptor cell death in mouse models of retinitis pigmentosa. *Proc. Natl. Acad. Sci.* (*USA*) **91,** 974–978.

Raff, M. C., Barres, B. A., Burne, J. F., Coles, H. S., Ishizaki, Y., and Jacobson, M. D. (1993). Programmed cell death and the control of cell survival: lessons from the nervous system. *Science* **262,** 695–700.

Rao, L., Debbas, M., Sabbatini, P., Hockenbery, D., Korsmeyer, S., and White, E. (1992). The adenovirus E1A proteins induce apoptosis, which is inhibited by the E1B 19-kDa and Bcl-2 proteins. *Proc. Natl. Acad. Sci.* (*USA*) **89,** 7742–7746.

Ravzi, E. S., and Welsh, R. M. (1995). Apoptosis in viral infections. *Adv. Virus Res.* **45,** 1–60.

Riparbelli, M. G., Callaini, G., Tripodi, S. A., Cintorino, S. A., Tosi, P., and Dallai, R. (1995). Localization of the Bcl-2 protein to the outer mitochondrial membrane by electron microscopy. *Exp. Cell Res.* **221,** 363–369.

Roy, N., Deveraux, Q. L., Takahashi, R., Salvesen, G. S., and Reed, J. C. (1997). The c-IAP-1 and c-IAP-2 proteins are direct inhibitors of specific caspases. *EMBO J.* **16,** 6914–6925.

Sarin, M. A., Clerici, M., Blatt, S. P., Hendrix, C. W., Shearer, G. M., and Henkart, P. A. (1994). Inhibition of activation-induced programmed cell death and restoration of defective immune responses of HIV+ donors by cysteine protease inhibitors. *J. Immunol.* **153,** 862–872.

Sato, T., Hanada, M., Bodrug, S., Irie, S., Iwama, N., Boise, L. H., Thompson, C. B., Golemis, E., Fong, L., Wang, H.-G., and Reed, J. C. (1994). Interactions among members of the Bcl-2 protein family analyzed with a yeast two-hybrid system. *Proc. Natl. Acad. Sci.* (*USA*) **91,** 9238–9242.

Savill, J., Fadok, V., Henson, P., and Haslett, C. (1993). Phagocytic recognition of cells undergoing apoptosis. *Immunol. Today* **14,** 131–136.

Schulze-Osthoff, K., Walczak, H., Droege, W., and Krammer, P. H. (1994). Cell nucleus and DNA fragmentation not required for apoptosis. *J. Cell Biol.* **127,** 15–20.

Schwartz, L. M., and Milligan, C. E. (1996). Cold thoughts of death: the role of ICE proteases in neuronal cell death. *Trends Neurosci.* **19,** 555–562.

Schwartzman, R. A., and Cidlowski, J. A. (1993). Apoptosis: The biochemistry and molecular biology of programmed cell death. *Endocrine Rev.* **14,** 133–151.

Scott, W. J. (1977). Cell Death and Reduced Proliferative Rate. In *Handbook of Teratology.* J. G. Wilson and F. C. Fraser, Eds. Plenum Press, New York. vol. 2, pp. 81–98.

Sedlak, T. W., Oltvai, Z. N., Yang, E., Wang, K., Boise, L. H., Thompson, C. B., and Korsmeyer, S. J. (1995). Multiple Bcl-2 family members demonstrate selective dimerizations with Bax. *Proc. Natl. Acad. Sci.* (*USA*) **92,** 7834–7838.

Sentman, C. L., Shutter, J. R., Hockenbery, D., Kanagawa, O., and Korsmeyer, S. J. (1991). bcl-2 inhibits multiple forms of apoptosis but not negative selection in thymocytes. *Cell* **67,** 879–888.

Sharp, P. A. (1994). Split genes and RNA splicing. *Cell* **77,** 805–815.

Shepard, T. H. (1992). *Catalog of Teratogenic Agents,* 7th ed. Johns Hopkins University Press, Baltimore.

Shi, L., Kraut, R. P., Aebersold, R., Greenberg, A. H. (1992). A natural killer cell granule protein that induces DNA fragmentation and apoptosis. *J. Exp. Med.* **175,** 553–566.

Shi, L., Mai, S., Israels, S., Browne, K., Trapani, J. A., and Greenberg, A. H. (1997). Granzyme B (GraB) autonomously crosses the cell membrane and perforin initiates apoptosis and GraB nuclear localization. *J. Exp. Med.* **185,** 855–866.

Shimizu, S., Eguchi, Y., Kosaka, H., Kamiike, W., Matsuda, H., and Tsujimoto, Y. (1995). Prevention of hypoxia-induced cell death by Bcl-2 and Bcl-x_L. *Nature* **374,** 811–813.

Shimizu, S., Eguchi, Y., Kamiike, W., Matsuda, H., and Tsujimoto, Y. (1996). Bcl-2 expression prevents activation of the ICE protease cascade. *Oncogene* **12,** 2251–2257.

Shirakabe, K., Yamaguchi, K., Shibuya, H., Irie, K., Matsuda, S., Moriguchi, T., Gotoh, Y., Matsumoto, K., and Nishida, E. (1997). TAK1 mediates the ceramide signaling to stress-activated protein kinase/c-Jun N-terminal kinase. *J. Biol. Chem.* **13,** 8141–8144.

Shu, H. B., Halpin, D. R., and Goeddel, D. V. (1997). Casper is a FADD- and caspase-related inducer of apoptosis. *Immunity* **6,** 751–763.

Smyth, M. J., O'Connor, M. D., and Trapani, J. A. (1996). Granzymes: a variety of serine protease specificities encoded by genetically distinct subfamilies. *J. Leukocyte Biol.* **60,** 555–562.

Song, Q., Lees-Miller, S. P., Kumar, S., Zhang, Z., Chan, D. W., Smith, G. C., Jackson, S. P., Alnemri, E. S., Litwack, G., Khanna, K. K., and Lavin, M. F. (1996). DNA-dependent protein kinase catalytic subunit: a target for an ICE-like protease in apoptosis. *EMBO J.* **15,** 3238–3246.

Song, Z., McCall, K., and Steller, H. (1997). DCP-1, a Drosophila cell death protease essential for development. *Science* **275,** 536–540.

Squier, M. K. T., Miller, A. C. K., Malkinson, A. M., and Cohen, J. J. (1994). Calpain activation in apoptosis. *J. Cell. Phys.* **159,** 229–237.

Srinivasan, A., Foster, L. M., Testa, M.-P., Ord, T., Keane, R. W., Bredesen, D. E., and Kayalar, C. (1996). Bcl-2 expression in neural cells blocks activation of ICE/CED-3 family proteases during apoptosis. *J. Neurosci.* **16,** 5654–5660.

Srinivasula, S. M., Fernandes-Alnemri, T., Zangrilli, J., Robertson, N., Armstrong, R. C., Wang, L., Trapani, J. A., Tomaselli, K. J., Litwack, G., and Alnemri, E. S. (1996). The Ced-3/interleukin-1β converting enzyme-like homolog Mch6 and the lamin-cleaving enzyme Mch2α are substrates for the apoptotic mediator CPP32. *J. Biol. Chem.* **271,** 27099–27106.

Steinman, H. M. (1995). The Bcl-2 oncoprotein functions as a prooxidant. *J. Biol. Chem.* **270,** 3487–3490.

Steller, H. (1995). Mechanisms and genes of cellular suicide. *Science* **267,** 1445–1449.

Sulik, K. K., and Alles, A. J. (1991). Teratogenicity of the retinoids. In *Retinoids: 10 Years On* (J. H. Sauiat, Ed.) Karger, Basel. pp. 282–295.

Susin, S. A., Zamzami, N., Castedo, M., Hirsch, T., Marchetti, P., Macho, A., Daugas, E., Geuskins, M., and Kroemer, G. (1996). Bcl-2 inhibits the mitochondrial release of an apoptogenic protease. *J. Exp. Med.* **184,** 1331–1341.

Takayama, S., Sato, T., Krajewski, S., Kochel, K., Irie, S., Millan, J., and Reed, J. C. (1995). Cloning and functional analysis of BAG-1: a novel Bcl-2-binding protein with anti-cell death activity. *Cell* **80,** 279–284.

Talanian, R. V., Quinlan, C., Trautz, S., Hackett, M. C., Mankovich, J. A., Banach, D., Ghayur, T., Brady, K. D., and Wong, W. W. (1997). Substrate specificities of caspase family proteases. *J. Biol. Chem.* **272,** 9677–9682.

Tanaka, S., Saito, K., and Reed, J. (1993). Structure-function analysis of the apoptosis-suppressing bcl-2 oncoproteins: substitution of a heterologous transmembrane domain to portions of the Bcl-2 β protein restores function as a regulator of cell survival. *J. Biol. Chem.* **268,** 10920–10926.

Tewari, M., Quan, L. T., O'Rourke, K., Desnoyers, S., Zeng, Z., Beidler, D. R., Poirier, G. G., Salvesen, G. S., Dixit, V. M. (1995). Yama/CPP32β, a mammalian homolog of ced-3, is a crmA-inhibitable protease that cleaves the death substrate poly (ADP-ribose) polymerase. *Cell* **81,** 801–809.

Thayer, J. M., and Mirkes, P. E. (1995). Programmed cell death and N-acetoxy-2acetylaminofluorene-induced apoptosis in the rat embryo. *Teratology* **51,** 418–429.

Thome, M., Schneider, P., Hofmann, K., Fickenscher, H., Meinl, E., Neipel, F., Mattmann, C., Burns, K., Bodmer, J.-L., Schroter, M., Scaffidi, C., Krammer, P. H., Peter, M. E., and Tschopp, J. (1997). Viral FLICE-inhibitory proteins (FLIPs) prevent apoptosis induced by death receptors. *Nature* **386,** 517–521.

Thompson, C. B. (1995). Apoptosis in the pathogenesis and treatment of disease. *Science* **267,** 1456–1462.

Thornberry, N. A., and Molineaux, S. M. (1995). Interleukin-1β converting enzyme: a novel cysteine protease required for IL-1β production and implicated in programmed cell death. *Protein Sci.* **4,** 3–12.

Tilly, J. L., and Hsueh, A. J. (1993). Microscale autoradiographic method for the qualitative and quantitative analysis of apoptotic DNA fragmentation. *J. Cell Physiol.* **154,** 519–526.

Tornusciolo, D. R. Z., Schmidt, R. E., and Roth, K. A. (1995). Simultaneous deltection of TDT-mediated dUTP biotin nick end-labeling (TUNEL)-positive cells and multiple immunohistochemical markers in single tissue sections. *Biotechniques* **19,** 800–805.

Vairo, G., Innes, K. M., and Adams, J. M. (1996). Bcl-2 has a cell cycle inhibitory function separable from its enhancement of cell survival. *Oncogene* **13,** 1511–1519.

Vanags, D. M., Porn-Ares, M. I., Coppola, S., Burgess, D. H., and Orrenius, S. (1996). Protease involvement in fodrin cleavage and phosphatidylserine exposure in apoptosis. *J. Biol. Chem.* **271,** 31075–31085.

Van de Craen, M., Vandenabeele, P., Declerq, W., Van den Brande, I., Van-Loo, G., Molemans, F., Schotte, P., Van Criekinge, W., Beyaert, R., and Fiers, W. (1997). Characterization of seven murine caspase family members. *FEBS Letts.* **403,** 61–69.

Vaux, D. L., Cory, S., and Adams, J. M. (1988). Bcl-2 gene promotes haemopoietic cell survival and cooperates with c-myc to immortalize pre-B cells. *Nature* **335,** 440–442.

Vaux, D. L., Aguila, H. L., and Weissman, I. L. (1992a). Bcl-2 prevents death of factor-deprived cells but fails to prevent apoptosis in targets of cell mediated killing. *Int. Immunol.* **4,** 821–824.

Vaux, D. L., Weissman, I. L., and Kim, S. K. (1992b). Prevention of programmed cell death in *Caenorhabditis elegans* by human bcl-2. *Science* **258,** 1955–1957.

Vaux, D. L. (1997). Ced-4—the third horseman of apoptosis. *Cell* **90,** 405–413.

Veis, D. J., Sorenson, C. M., Shutter, J. R., and Korsmeyer, S. J. (1993). Bcl-2 deficient mice demonstrate fulminant lymphoid apoptosis, polycystic kidneys, and hypopigmented hair. *Cell* **75,** 229–240.

Vekrellis, K., McCarthy, M. J., Watson, A., Whitfield, J., Rubin, L. L., and Ham, J. (1997). Bax promotes neuronal cell death and is downregulated during the development of the nervous system. *Devel.* **124,** 1239–1249.

Verheij, M., Bose, R., Lin, X. H., Yao, B., Jarvis, W. D., Grant, S., Birrer, M. J., Szabo, E., Zon, L. I., Kyriakis, J. M., Haimovitz-Friedman, A., Fuks, Z., and Kolesnick, R. N. (1996). Requirement for ceramide-initiated SAPK/JNK signalling in stress-induced apoptosis. *Nature* **380,** 75–79.

Wang, H.-G., Miyashita, T., Takayama, S., Sato, T., Torigoe, T., Krajewski, S., Tanaka, S., Hovery, L. III, Troppmair, J., Rapp, U., and Reed, J. (1994). Apoptosis regulation by interaction of bcl-2 protein and Raf-1 kinase. *Oncogene* **9,** 2751–2756.

Wang, H.-G., Rapp, U. R., and Reed, J. C. (1996). Bcl-2 targets the protein kinase Raf-1 to mitochondria. *Cell* **87,** 629–638.

Wang, K., Yin, X.-M., Chao, D. T., Milliman, C. L., and Korsmeyer, S. J. (1996). BID: a novel BH3 domain-only death agonist. *Genes Dev.* **10,** 2859–2869.

Wang, Z. Q., Auer, B., Stingl, L., Berghammer, H., Haidacher, D., Schweiger, M., and Wagner, E. F. (1995). Mice lacking ADPRT and poly(ADP-ribosyl)ation develop normally but are susceptible to skin disease. *Genes Dev.* **9,** 509–520.

Webster, W. S., Germain, M.-A., and Edwards, M. J. (1985). The induction of microphthalmia, encephalocele, and other head defects following hyperthermia during the gastrulation process in the rat. *Teratology* **31,** 73–82.

Weil, M., Jacobson, M. D., Coles, H. S. R., Davies, T. J., Gardner, R. L., Raff, K. D., and Raff, M. C. (1996). Constitutive expression of the machinery for programmed cell death. *J. Cell. Biol.* **133,** 1053–1059.

Weil, M., Jacobson, M. D., and Raff, M. C. (1997). Is programmed cell death required for neural tube closure? *Curr. Biol.* **7,** 281–284.

White, E. (1996). Life, death, and the pursuit of apoptosis. *Genes Dev.* **10,** 1–15.

White, K., Grether, M. E., Abrams, J. M., Young, L., Farrell, K., and Steller, H. (1994). Genetic control of programmed cell death in Drosophila. *Science* **264,** 677–683.

Wolter, K. G., Hsu, Y.-T., Smith, C. L., Nechushtan, A., Xi, X.-G., and Youle, R. J. (1997). Movement of Bax from the cytosol to mitochondria during apoptosis. *J. Cell Biol.* **139,** 1281–1292.

Wu, D., Wallen, H. D., and Nunez, G. (1997). Interaction and regulation of subcellular localization of ced-4 by ced-9. *Science* **275,** 1126–1129.

Wyllie, A. H. (1987). Apoptosis: cell death in tissue regulation. *J. Pathology* **153,** 313–316.

Wyllie, A. H., Arends, M. J., Morris, R. G., Walker, S. W., and Evan, G. (1992). The apoptosis endonuclease and its regulation. *Immunology* **4,** 389–397.

Xia, Z., Dickens, M., Raingeaud, J., Davis, R. J., and Greenberg, M. E. (1995). Opposing effects of ERK and JNK-p38 MAP kinases on apoptosis. *Science* **270,** 1326–1331.

Xue, D., and Horvitz, R. H. (1995). Inhibition of the *Caenorhabditis elegans* cell-death protease CED-3 by a CED-3 cleavage site in baculovirus p35 protein. *Nature* **377,** 248–251.

Yang, E., and Korsmeyer, S. J. (1996). Molecular thanatopsis: a discourse on the BCL2 family and cell death. *Blood* **88,** 386–401.

Yang, E., Zha, J., Jockel, J., Boise, L. H., Thompson, C. B., and Korsmeyer, S. J. (1995). Bad, a heterodimeric partner for Bcl-x_L and Bcl-2, displaces Bax and promotes cell death. *Cell* **80,** 285–291.

Yang, J., Liu, X., Bhalla, K., Kim, C. N., Ibrado, A. M., Cai, J., Peng, T.-I., Jones, D. P., and Wang, X. (1997). Prevention of apoptosis by bcl-2: release of cytochrome c from mitochondria blocked. *Science* **275,** 1129–1132.

Yang, T., Kozopas, K. M., and Craig, R. W. (1995). The intracellular distribution and pattern of expression of Mcl-1 overlap with, but are not identical to, those of Bcl-2. *J. Cell Biol.* **128,** 1173–1184.

Yin, C., Knudson, C. M., Korsmeyer, S. J., and Van Dyke, T. (1997). Bax suppresses tumorigenesis and stimulates apoptosis *in vivo. Nature* **385,** 637–640.

Yin, X.-M., Oltvai, Z. N., and Korsmeyer, S. J. (1994). BH1 and BH2 domains of Bcl-2 are required for inhibition of apoptosis and heterodimerization with Bax. *Nature* **369,** 321–323.

Yuan, J. (1996). Evolutionary conservation of a genetic pathway of programmed cell death. *J. Cell. Biochem.* **60,** 4–11.

Yuan, J., Shaham, S., Ledoux, S., Ellis, H. M., and Horvitz, H. R. (1993). The *C. elegans* cell death gene ced-3 encodes a protein similar to mammalian interleukin-1 β-converting enzyme. *Cell* **75,** 641–652.

Yuen, E. C., Howe, C. L., Li, Y., Holtzman, D. M., and Mobley, W. C. (1996). Nerve growth factor and the neurotrophic theory hypothesis. *Brain Devel.* **18,** 362–368.

Zamzami, N., Marchetti, P., Castedo, M., Decaudin, D., Macho, A., Hirsch, T., Susin, S. A., Petit, P. X., Mignotte, B., and Kroemer, G. (1995a). Sequential reduction of mitochondrial transmembrane potential and generation of reactive oxygen species in early programmed cell death. *J. Exp. Med.* **182,** 367–377.

Zamzami, N., Marchetti, P., Castedo, M., Zanin, C., Vayssiere, J.-L., Petit, P. X., and Kroemer, G. (1995b). Reduction in mitochondrial potential constitutes an early irreversible step of programmed lymphocyte death *in vivo. J. Exp. Med.* **181,** 1661–1672.

Zamzami, N., Susin, S. A., Marchetti, P., Hirsch, T., Gomez-Monterrey, I., Castedo, M., and Kroemer, G. (1996). Mitochondrial control of nuclear apoptosis. *J. Exp. Med.* **183,** 1533–1544.

Zha, J., Aim'e-Semp'e, C., Sato, T., and Reed, J. C. (1996a). Proapoptotic protein Bax heterodimerizes with Bcl-2 and homodimerizes with Bax via a novel domain (BH3) distinct from BH1 and BH2. *J. Biol. Chem.* **271,** 7440–7444.

Zha, J., Harada, H., Yang, E., Jockel, J., and Korsmeyer, S. J. (1996b). Serine phosphorylation of death agonist BAD in response to survival factor results in binding to 14-3-3 not BCL-X_L. *Cell* **87,** 619–628.

Zhou, P., Qian, L., Kozopas, K. M., and Craig, R. W. (1997a). Mcl-1, a Bcl-2 family member, delays the death of hematopoietic cells under a variety of apoptosis-inducing conditions. *Blood* **89,** 630–643.

Zhou, Q., Snipas, S., Orth, K., Muzio, M., Dixit, V. M., and Salvesen, G. S. (1997b). Target protease specificity of the viral serpin CrmA. Analysis of five caspases. *J. Biol. Chem.* **272,** 7797–8000.

Zhu, W., Cowie, A., Wasfy, G. W., Penn, L. Z., Leber, B., and Andrews, D. W. (1996). Bcl-2 mutants with restricted subcellular location reveal spatially distinct pathways for apoptosis in different cell types *EMBO J.* **15,** 4130–4141.

Zoratti, M., and Szabo, I. (1995). The mitochondrial permeability transition. *Biochim. Biophys. Acta* **1241,** 139–176.

Zou, H., Henzel, J., Liu, X., Lutschg, A., and Wang, X. (1997). Apaf-1, a human protein homologous to *C. elegans* CED-4, participates in cytochrome c–dependent activation of caspase-3. *Cell* **90,** 405–413.

CHAPTER

9

Periods of Susceptibility to Induced Malformations of the Developing Mammalian Brain

GREGORY D. BENNETT
RICHARD H. FINNELL
Departments of Veterinary Anatomy and Public Health
Texas A&M University
College Station, Texas 77843

Periods of susceptibility to teratogen-induced congenital defects have historically been partitioned into three distinct categories: the all-or-nothing period, the period when structural malformations are induced, and the period of functional deficits. Depending on the time during gestation when the insult occurred, one could anticipate certain outcomes based on one of these three categories. Environmental insults during the early stages of development are thought to result in either sufficient damage so as to be lethal or to have no lasting effect. A classic example that illustrates this all-or-nothing effect of a teratogen is ionizing radiation, in which the most common developmental effect observed following exposure during the very early stages of development is death (Russell, 1950; Russell and Russell, 1954). Because the embryonic cells at these very early stages are totipotent, teratogenic insults either lead to embryonic death, if the dosage is sufficiently high, or exert minimal damage, which does not adversely effect the embryo or derail its developmental path.

*Abbreviations: CAM, cell adhesion molecules; *Cd*, crooked; CNS, central nervous system; *ct*, curlytail; ECM, extracellular matrix; FAS, fetal alcohol syndrome; *gas-5*, growth arrest specific-5; GD, gestational day; GFAP, glial fibrillary acidic protein; MAG, myelin associated glycoproteins; MNU, methylnitrosurea; NCAM, neural cell adhesion molecules; NTC, neural tube closure; NTD, neural tube defects; PKC, protein kinase C; PLP, proteolipid proteins; RA, retinoic acid; *Rf*, rib fusion; *Sp*, splotch; ST, straight tails; VPA, valproic acid.

As gestation progresses and the embryo develops from the early stages of being a blastocyst through organogenesis, teratogenic insults disrupt the formation of the embryonic structures, leading to the development of congenital malformations. Depending on the time and nature of these insults, different structural anomalies can be observed. The teratology literature is replete with examples of either physical or chemical insults producing various structural malformations. At the very early phases of organogenesis, as the neural tube is beginning to form, there are a myriad of agents as diverse as hyperthermia, arsenic, carbamazepine, valproic acid, and retinoic acid that can disrupt these processes and ultimately lead to a variety of malformations, not the least of which are neural tube defects (NTDs) (Kochhar, 1967; Steele *et al.,* 1983; Finnell *et al.,* 1986; 1988; Tibbles and Wiley, 1988; Wlodarczyk *et al.,* 1996a,b). As gestation progresses and limb buds begin to emerge, many of the same agents that induce NTDs also induce limb dysgenesis (Cusic and Dagg, 1985; Sulik and Dehart, 1988; Collins *et al.,* 1990; Kimmel *et al.,* 1993). During the later phases of organogenesis, as the craniofacial structures are developing, a variety of agents can disrupt essential cellular migratory events, with a result being a cleft lip and/or palate (Sulik *et al.,* 1988). It is clear from these observations that teratogen-induced malformations are not necessarily agent dependent—rather it is largely the timing of the teratogenic insult that determines which of the developing structures and/or organs are going to be adversely affected.

During the final stage of gestation, which is usually referred to as the *histogenesis* or *maturation phase,* teratogenic agents are believed to produce functional deficits. A classic example of the repercussions of exposure to a teratogenic agent during these later stages of gestation is ethanol (Jones *et al.,* 1973; West *et al.,* 1986). Prenatal exposure to this well-established teratogen is associated with intrauterine growth retardation, microcephaly, craniofacial abnormalities, and mental deficiencies. The neural defects are thought to be an accumulative result of reduced cell numbers, as evidence by decreases in transcriptional and translational activity, as well as delays in myelination, astrocyte development, and neuronal migration in the brain of ethanol-exposed individuals (Clarren and Smith, 1978; Colangelo and Jones, 1982; Miller, 1986).

The theme of developmental timing is recurrent to all the issues involved with periods of teratogenic susceptibility. This is true for all of the embryo's organs and major structures. The question that needs to be addressed is: What is/are the period(s) of susceptibility for the developing nervous system? The answer is far more complex than it may initially appear. This is due to the fact that the nervous system is one of the first systems to begin developing in the embryo, and that maturation in humans is not completed until several years after birth. Therefore, its susceptible period includes the complete continuum of gestation. There are several reasons why this simplistic view has now become antiquated. First, the system is innately plastic, with a highly coordinated network of cellular crosstalk and compensatory mechanisms between different components of the developing neuronal systems. Second, not all of the defects of the developing nervous system are as obvious and disfiguring as are anencephaly or craniorachischisis. Last, our ability to dissect and examine the developing nervous system at more sophisticated biochemical and molecular levels has expanded in recent years, revealing that teratogenic insults occurring during the early stages of neurulation can lead to very subtle yet profound deficiencies in structure and/or functions of the developing nervous system.

In this review the concepts pertaining to the periods of susceptibility of the nervous system are discussed. Initially, the embryology of the neural tube is presented to provide the foundation from which to launch this discussion into more speculative territory, such as neural defects arising from insults occurring during the very early stages of gestation, or the expression of specific genes during neurulation that may lead to an NTD, ultimately arriving at a better understanding of how and why these malformations occur.

I. The Embryology of Neural Tube Closure

There are two widely accepted versions of how neural tube closure (NTC) occurs in mammals. In the first version, NTC is thought to be comprised of two stages. The first stage, or *primary neurulation,* consists of the formation of the neural plate followed by the elevation, apposition, and fusion of the neural folds, which forms the majority of the neural tube (Campbell *et al.,* 1986). This phase in NTC is followed by secondary neurulation, which occurs below the level of the 32–34 somite or the future 3–5 sacral vertebrae in humans, where the canalization of a solid core of mesenchymal cells develops into the secondary neural tube without the previous formation of neural folds, this joins together with the primary neural tube to complete NTC (Lemire, 1969; Schoenwolf, 1984; Seller, 1987; Copp *et al.,* 1990). More recently, it has been proposed that NTC results from a discontinuous fusion of the neural folds that ultimately results in the formation of a single neural tube (MacDonald *et al.,* 1989; Golden and Chernoff, 1993; Van Allen *et al.,* 1993).

In several different inbred mouse strains, NTC has been observed to be initiated at four distinct sites (Geelen and Langman, 1977; Kaufman, 1979; MacDonald *et al.,* 1989; Sakai, 1989; Juriloff *et al.,* 1991; Golden and Chernoff, 1993; Gunn *et al.,* 1993). The anterior neuropore is formed by the fusion of closures II–IV, whereas the fusion of closure I is responsible for the formation of the posterior neuropore. Initially, contact and fusion of the neural folds (closure I), occurs in the midcervical region at the level of somites 2–4, which become the boundary between the spinal cord and the myelencephalon of the hindbrain. Closure I proceeds bidirectionally, where caudally it forms the thoracic portion of the neural tube, or the presumptive spinal cord, and cranially it proceeds to a point just rostral to the otic pits on the inferior aspect of the hindbrain (Golden and Chernoff, 1993). NTC proceeds with closure II, which begins at the junction between the prosencephalon and mesencephalon. This bidirectional closure creates two cranial neuropores, with one in the fore- and one in the midbrain regions of the developing neural tube (Van Allen *et al.,* 1993). Closure II progresses caudally to fuse the neural folds in the entire mesencephalon, terminating at the cranial-most portion of the hindbrain. Closure III, which develops concurrently with closure II, begins at the stomodeum and progresses caudally to close off the prosencephalon and the anterior neuropore by meeting closure II. The initiation of closure IV takes place at the caudal part of the hindbrain, and, like closure III, is unidirectional, proceeding cranially until it meets closure II, thereby completing NTC in the anterior portion of the embryo. However, unlike the preceding three sites of NTC initiation, fusion of closure IV occurs by the growth of a membrane that covers the hindbrain, rather than actual fusion of the neural folds (Geelen and Langman, 1977; Golden and Chernoff, 1993).

The discontinuous model of NTC has the distinct advantage over previous models in that it is better at explaining the morphologic etiology of the spectrum of NTDs observed in human fetuses (Busam *et al.,* 1993; Van Allen *et al.,* 1993). The failure of neural fold fusion within the anatomically restricted boundaries described by this model can be used morphologically to explain the majority of observed clinical cases of NTDs (Van Allen *et al.,* 1993). Prevailing opinion now holds that anencephaly is the result of a failure of closure II (meroacranium type) or closures II and IV (holoacranium type). Spina bifida cystica results from a failure in the cranial and/or caudal portions of closure I. Craniorachischisis results from failure of closures II–IV, whereas the isolated failure of closure III results in embryos with midfacial clefts. Frontal and parietal encephaloceles result from a failure of fusion at the junctions between the cranial closures. Therefore, frontal encephaloceles are a failure of closures II and III to unite, whereas parietal defects are an absence of closures II and IV to connect (Van Allen *et al.,* 1993). Last, occipital encephaloceles are believed to occur when closures I and IV fail to completely fuse.

Further evidence in support of the discontinuous closure model comes from human fetuses and neonates with multiple open neural tube lesions. Busam and colleagues (1993) described a fetus with two open NTDs in the cranial region of the skull. These were midline defects that were separated by a bony cutaneous bridge. However, this case is not an isolated report, for Lemire and co-workers (1978) have described several infants with anencephaly and a lumbosacral myelomeningocele. Although the multiple closure site hypothesis has not been universally accepted for humans, fetuses such as these bearing multiple neural tube lesions are difficult, if not impossible, to explain using the continuous closure model.

Based on existing models of NTC, three general hypotheses have been proposed to explain the pathogenesis of NTDs. The first of these theories involves the decreased rate, or lack of cellular proliferation, that leads to the failure of the neural folds to elevate. The fundamental role of cell proliferation in mammalian neurulation has been known since 1874 (His, 1874), and factors that interfere with this process often result in an NTD. Jelinek and Friebova (1966) observed a rapid proliferation of neuroepithelial cells during NTC in the chick embryo, and hypothesized that as the cell number increases in the neural plate, which is maintained at a constant width due to tethering to the neuroepithelial basement membrane, the plate is forced to fold with continued elevation driven by continued cell proliferation. It has been demonstrated in the murine mutants Splotch, Looptail, and curlytail that there is an association between a lengthening of the cell cycle in the neuroepithelium and a reduction in cellular proliferation in regions that failed to achieved neural fold fusion (Wilson, 1974; Wilson and Center, 1974; Copp *et al.,* 1988, 1990). Additional evidence from teratogen-induced NTDs supports this hypothesis, as Walsh and Morris (1989) reported that cultured rat embryos exposed to elevated temperatures increased the time for the neuroepithelial cells to progress through both G1/S and G2/M interfaces, resulting in a pronounced decrease in cellular proliferation in effected embryos.

The second general hypothesis proposes that NTC is dependent on specific changes in the cytoarchitecture and shape of the developing neuroectoderm (Langman and Welch, 1967; Bannigan and Cottell, 1984). Studies have been conducted to discern the intracellular cues that shape the neuroepithelial cells during the process of neural fold fusion. It has been suggested that microfilament bundles at the cell apex provide the driving force that results in the transition of the columnar into wedge-shaped epithelium during the folding of the neural tube (Lee and Nagele, 1986). Contraction of basal actin microfilament bundles in the neuroepithelium plays a role in maintaining the biconvex shape of the early neural folds, whereas the contraction of apical actin microfilament bundles is essential for altering the cell shape and subsequent elevation of the neural folds (Sadler *et al.*, 1982, 1986; Gunn *et al.*, 1993). The contractility of these microfilament bundles is likely being modulated by intracellular calcium ions (Moran and Rice, 1976; Lee and Nagele, 1978, 1979). Thus, in some instances, failure of NTC may be due to an alteration in the availability of intracellular ions. A failure in the contraction of these filament bundles is believed to result in misshapen neuroepithelial cells, that prevents the normal alteration in cellular shape required for proper elevation and neural fold fusion. This appears to be the problem in the murine SELH embryos that spontaneously develop NTDs (Gunn *et al.*, 1993).

Finally, the third hypothesis suggests that development of the vasculature supplying the neural folds is the critical factor, whereby a vascular disruption leads to inadequate perfusion of the neural folds and a delay in their normal development eventually resulting in an NTD (Vogel, 1961; Stevenson *et al.*, 1987; Sulik and Sadler, 1993). This hypothesis has recently become a focal point of renewed interest for several reasons. First, it has been recognized that cell proliferation may be arrested by a vascular disruption. It is known, in general, that the internal environment of cells needs to be slightly alkaline for proliferation to occur. However, if the vascular supply is disrupted, cellular waste products are not removed resulting in a cellular acidosis, which could potentially delay proliferation at a critical developmental time point. Second, a vascular disruption may produce an indirect effect when the neural epithelia is reperfused, which may result in the generation of oxygen radicals leading to oxidative stress, the subsequent destruction of the neural tissue. A third possibility involves a link between vascular disruption, folic acid, and their etiologic roles in the development of NTDs.

Regardless of the exact mechanism of NTC, it is clear that the process involves a highly orchestrated series of events that becomes increasingly complex as one goes from the morphologic to the biochemical to the molecular queues that govern neurulation, and subsequent neural development. The obvious consequences of abnormalities in these processes are NTDs, and there are numerous examples of how different compounds can interfere with normal neural development. What is not obvious, is the direct mechanism by which these compounds exert their actions on the developing neuroepithelia and supportive tissues.

II. Environmental Exposure during Early Gestation

The early stages of development have historically been considered by teratologist as "all or none" or a resistant (Austin, 1973; Tuchmann-Du Plessis, 1975) period of susceptibility to the induction of congenital defects. This is primarily due to the totipotent nature of the embryonic cells at these stages of development (Langman, 1981). Although embryonic lethality remains the primary adverse consequence when embryos are exposed to toxic agents during the early stages of gestation, a growing number of studies have begun to identify specific malformations, especially NTDs associated with such exposures. One possibility is that these agents or conditions produce a generalized disruption in the timing of embryonic events leading to a developmental delay, and ultimately to an NTD. It is also possible that these agents are acting in a more specific manner on the developing neuroepithelium or supporting tissues to induce NTDs. As we obtain a greater appreciation and understanding, as well as the technical ability to investigate the developmental processes that govern the ontogeny of the nervous system, we will be better poised to resolve these and other controversies. Historically, the bias has been towards believing that these agents or conditions produce a generalized developmental delay, with little or no specific effects. Although this may well be true, investigations have described specific effects on the embryo when they are exposed during the preimplantation stages. For example, a human maternal condition that has been associated with a high prevalence rate of NTDs is diabetes. Because this disease is continuous from its onset throughout life, it is impossible to accurately define when, during gestation, this disease causes deleterious effects on the development of the embryo. However, using animal models, such as nonobese diabetic mice, or streptozotocin-induced diabetic rats, it has been possible to demonstrate that if the diabetic conditions are limited to the preimplantation stages, NTDs are still observed (Eriksson *et al.*, 1983; Otani *et al.*, 1991). Suggesting that, under certain conditions, NTDs may be induced earlier than previously thought, which means well before the onset of neurulation.

Teratogens have also been shown to induce NTDs when exposure was limited to the preimplantation stages, which defies conventional wisdom. Morris (1972) produced central nervous system (CNS) malformations in approximately 6% of the murine embryos exposed to large doses of vitamin A that were administered prior to implantation. Similarly, retinoic acid administered 60 hr postcopulation to murine embryos induced malformations, including exencephaly, in 59% of the offspring (Pillans *et al.,* 1988). However, it could be argued that these compounds have extremely long half-lives, and the effects observed are actually in response to the toxicologic effects occurring later in gestation. There are other agents with shorter half-lives that have also reproducibly induced NTDs in experimental animals during the preimplantation period. Methylnitrosurea (MNU) administered to pregnant mice on gestation day 2 or 3 produced a dose-dependent increase in the incidence of abnormal fetuses, compared to controls. Although cleft palate was the most commonly observed defect, 26% of the exposed fetuses had exencephaly (Nagao and Fujikawa, 1990). Whereas MNU would be considered to induce generalized effects, the fact that these results were reproducible in a dose-dependent manner argues in favor of this agent disrupting specific events or processes in the developing embryo.

An intriguing observation has been reported on the reproductive toxicity of the estrogenic compound clomiphere citrate. When this pseudoestrogen was administered to female mice 2 days prior to copulation, there was a significant increase in the number of exencephalic fetuses (Dziadek, 1993). Treatments with this compound during the pre- or postimplantation period, however, failed to induce any congenital defects. Because this compound does not have a prolonged half-life, Dziadek concluded that clomiphere citrate indirectly affected the embryonic development by disrupting the normal maternal "support of pregnancy," which retarded embryonic growth, and ultimately lead to the failure of the neural tube to close. Exactly how this might occur remains to be determined; however, it is interesting to consider just how exposure of a female mouse 2 days before conception, which is some 10 days before neurulation, would have any direct effect on NTC. Even more surprising is that fact that there is limited human clinical data that support this relationship between the preconception use of this drug and the occurrence of NTDs (Milunsky *et al.,* 1989; Cuckle and Wald, 1989; Vollset, 1990).

Due to their obvious limitations, none of the aforementioned studies can definitively prove that these insults during the early stages of gestation directly leads to an NTD or to any effects other than death. It may be that the neural defects seen in response to exposure to these compounds were just a consequence of embryonic cells being predetermined at an earlier time than was previously believed. Alternatively, the maternal environment may play a greater role in the normal development of mammalian embryos than was previously appreciated. Given these data, it is important to begin looking at this period of gestation as a time when teratogenic exposure may be problematic. Further research, using more sophisticated techniques, probing into the molecular and biochemical events at these early stages of development, will provide much needed insight.

III. Teratogenic Exposures during Organogenesis

There are a number of agents that cause NTDs when administered either prior to or during the initial stages of neurulation. Only three of these agents—arsenic, hyperthermia, and valproic acid—are discussed in this chapter. As structurally, physically, and biologically diverse as these three compounds are, all have been associated to some extent with NTDs in both humans and experimental animal models.

A. *Arsenic*

The continuing controversy surrounding the teratogenic potential of arsenic in humans is most likely due to the fact that, as with most, if not all, human teratogens, only a limited number of studies have been conducted on large populations of pregnant women exposed to this metal. In several human studies, prenatal exposure to arsenic has been linked with multiple adverse reproductive outcomes including abortion, stillbirth, developmental impairment, congenital malformation, and early neonatal death (Beckman, 1978; Zierler *et al.,* 1988; Aschengrau *et al.,* 1989; Bolliger *et al.,* 1992; Borzsonyi *et al.,* 1992). Among the specific congenital defects that have been observed, NTDs, including both anencephaly and spina bifida, have been frequently reported (for review see Shalat *et al.,* 1996). There are many potential explanations as to why the nervous system might be particularly sensitive to arsenic. This metalloid specifically targets sulfhydrile groups, which have the potential of disrupting ATP production, as well as other cellular processes. Disruption of energy production could affect the embryo in a variety of ways, from a reduction in the proliferative rate of the neuroepithelium or supportive tissues, to a delay in cellular events, such as cytoskeletal changes, which could suspend NTC and ultimately lead to an NTD.

Whereas the effects of arsenic on human development remain controversial, the situation in experimental animal systems is far better understood. Arsenic is known to readily cross the placental barrier and to selectively accumulate in the fetal neuroepithelium during early embryogenesis (Hanlon and Ferm, 1977; Lindgren *et al.,* 1984). A range of fetal malformations have been observed in experimental animals following embryonic arsenic exposure, although NTDs appear as a predominant and consistent outcome in multiple mammalian species. This is especially true when exposure to arsenic occurs just prior to the initiation of neurulation (Beaudoin, 1974; Willhite and Ferm, 1984; Carpenter, 1987; Hood *et al.,* 1988; Wlodarczyk *et al.,* 1996a). In all of these studies, direct damage to the developing neuroectodermal and mesoderm cells were observed secondary to the arsenic exposure.

Although it is generally agreed that organic arsenic compounds are less toxic and, therefore, less teratogenic than are the inorganic derivatives, these organic compounds have a decided teratogenic potential. Harrison and colleagues (1980) observed significant teratogenicity when sodium dimethyl arsenate was administered to pregnant mice on gestational day (GD) 9 or 10. This treatment resulted in an increased frequency of resorptions, fetal death, and NTDs. Similar congenital defects were observed following the culture of whole murine embryos in the presence of arsenate (Chaineau *et al.,* 1990; Mirkes and Cornel, 1992). Clearly, arsenic compounds are not just toxic to adult organisms, but they have the capacity to alter normal developmental processes following *in utero* exposure in experimental animal models.

In an attempt to better understand how this environmental contaminant with a broad spectrum of toxicity could specifically increase the risk for NTDs, an animal model has been used to examine the impact of *in utero* arsenic exposure on neuroepithelial cell gene expression (Wlodarczyk *et al.,* 1996a,b). Arsenate administered twice during early neurulation induced exencephaly, specifically a failure of closure II, in approximately 95% of the exposed embryos. This treatment also induced a decrease in the expression of *HOX3.1,* as well as an overexpression of the *PAX-3* and *EMX-1* genes in the neural tubes of these embryos at the time when closure II should have occurred (Wlodarczyk *et al.,* 1996a). Given that these genes are involved in embryonic pattern formation, it is easy to imagine how their aberrant expression could lead to NTDs. In fact, several other investigations that have reported an association between altered homeobox gene expression and abnormal neural development (Crossin *et al.,* 1985; Kinter and Melton, 1987; Akam, 1989; Prieto *et al.,* 1989; Edelman and Crossin, 1991; Chalepakis *et al.,* 1993; Edelman and Jones, 1995).

Among the many components that are present within the neuroepithelium and its surrounding tissues that are critical to cellular migration and cytoarchitectural changes are the cell adhesion molecules (CAM). Neural-CAM (NCAM) expression has been shown to be regulated by the products of the *HOX* and *PAX* genes, which bind to their distinct and separate regions within the NCAM promoter (Akam, 1989; Chalepakis *et al.,* 1993). These transmembrane glycoproteins have a fundamental role in vertebrate pattern formation, and their expression is highly restricted within the developing neural tube (Crossin *et al.,* 1985; Kinter and Melton, 1987; Prieto *et al.,* 1989; Edelman and Crossin, 1991; Edelman and Jones, 1995). The expression patterns of these molecules are rigorously maintained in order to provide local signals to ensure proper cell migration and cytoarchitecture changes that need to occur for normal NTC (Thiery *et al.,* 1982; Edelman, 1984; Edelman and Jones, 1995). The combined data from these studies suggests the arsenic-induced disruption of *PAX-3* and *HOX3.1* expression leads to further transcriptional alterations that may be responsible for the observed NTDs.

Two other genes that are affected by an acute arsenic exposure on GD 7.5 and 8.5 were *bcl-2* and *p53.* Arsenic exposure *in utero* results in the overexpression of both of these genes in the developing neural tube during the critical stages of closure II (Wlodarczyk *et al.,* 1996b). These two genes appear to act antagonistically, where the gene product of *bcl-2* functions as an antioxidant by eliminating oxygen-free radicals and inhibiting programmed cell death, whereas the *p53* gene is a transcription regulator of cellular proliferation by either directly arresting proliferation, or inducing apoptosis (Tsuijimoto, 1989; Hartwell, 1992; Korsmeyer, 1992; Shaw *et al.,* 1992; Vogelstein and Kinzler, 1992; Miyashita and Reed, 1993; Larsen, 1994). In transgenic mice, the overexpression of *bcl-2* results in an accumulation of cells due to the evasion of normal apoptotic signals, whereas overexpression of *p53* produces excessive cell death (McDonnell *et al.,* 1989; Hoever *et al.,* 1994). Because *in utero* arsenic exposure caused both of these genes to be overexpressed and their relative expression remained constant throughout NTC, it is not surprising that it induced a disruption in the proliferative rate of the neuroepithelium (Wlodarczyk *et al.,* 1996b). Although the exact mechanism of how these defects are induced by arsenic remains to be elucidated, clearly this neural tube teratogen disrupts molecular events within the developing neural tube that could readily explain the defects resulting from exposure to this metalloid.

B. Hyperthermia

The adverse effects of maternal hyperthermia on embryonic development have been known since the mid-1960s (Edwards, 1986). However, not until the late 1970s did the medical community recognize its potential reproductive hazards. This recognition was largely the result of a number of anecdotal clinical reports, as well as retrospective epidemiologic studies (Edwards, 1986). Although it can be argued that humans, being homoethermic, are able to tolerate great variations in environmental temperatures, the literature suggests that even small elevations, 1.5–2.5°C, in the maternal core temperature are sufficient to disrupt the normal morphology of the developing mammalian brain (Edwards, 1986; Edwards *et al.,* 1986). These slight temperature elevations are not uncommon events and can occur as a result of febrile illness or certain occupational or recreational (sauna and hot tubs) exposures. Anencephaly, meningomyeloceles, and encephaloceles have all been associated with an elevated maternal core temperature during the first 6 weeks of gestation (Chance and Smith, 1978; Fraser and Skleton, 1978; Halperin and Wilroy, 1978; Miller *et al.,* 1978; Layde *et al.,* 1980; Pleet *et al.,* 1981; Shiota and Nishimura, 1982). Regardless of its source, hyperthermia is strongly associated with NTDs, especially if the maternal temperature reaches 38.9°C, and more commonly when it exceeds 40°C (Chance and Smith, 1978; Miller *et al.,* 1978).

Raising the maternal core temperature 3–5°C during the initial stages of neurulation has proved to be teratogenic to many laboratory rodent species. Edwards (1968) was the first to show this relationship by exposing pregnant guinea pigs to a hyperthermic insults just prior to the onset of neurulation. These treatments produced a spectrum of defects including microcephaly, hydranencephaly, defects of the abdominal wall, amyoplasia, and hypoplasic digits, but the most common defects observed were those involving the brain and skeleton. NTDs were the most common defects observed in hamsters that were exposed *in utero* to elevated temperatures (Kilham and Ferm, 1976). Interestingly, NTDs have been induced regardless of the method or route of hyperthermic insult. Using microwaves, pregnant rats were subjected to 27.12 MHz of radiation on GD 9 that elevated the core temperature 5°C and induced congenital malformations that were primarily restricted to the anterior neural tube (Lary *et al.,* 1982). These studies demonstrate that regardless of the etiology, elevating the maternal core temperature 3–5°C induces malformations that are primarily confined to the developing neural tube.

In terms of the mechanism by which elevated temperatures induce congenital malformations, it has been suggested that they occur as a direct result of maternal toxicity. To test this hypothesis, investigators used whole embryo culture, subjecting rat embryos at GD 10 to a mild (40–41°C) hyperthermic insult. This treatment resulted in head and somite malformations (Cockroft and New, 1975). More than half of the embryos that were exposed to this treatment were microcephalic, and the somite development was so disrupted that they could not be discerned from the surrounding tissues. These studies demonstrated that the defects do not arise from some generalized disruption in maternal health; rather, they are the result of a specific underlying mechanism.

To further this idea that hyperthermia acts in a specific way, a dose–response relationship has been demonstrated for NTDs. In a study by Germain and colleagues (1985), the core temperature of pregnant rats were elevated to 43.5°C for only a brief period (a spike exposure), and then returned to normal basal temperature. This treatment resulted in encephaloceles in nearly half of the surviving fetuses. When the core temperature was raised to 43°C, only 7% of the fetuses had microophthalmia. However, when the temperature was maintained at 43°C for 2–5 min, 24% and 97%, respectively, of the surviving fetuses had either encephaloceles or cleft faces (Germain *et al.,* 1985). Raising the core temperature to 42.5°C proved to be teratogenic only when this temperature was maintained for 5 min or longer. Temperatures of 42°C or less were rarely teratogenic in this study, which clearly illustrates a dose and time dependency to heat-induced NTDs. It also demonstrates just how sensitive the developing neural tissue is to elevated temperatures.

Another variable that has been shown to be important in hyperthermia-induced NTDs is the genetic background of the individual. Exposing pregnant dams from several murine strains to a 43°C heatshock at the onset of NTC induced varying rates of exencephaly that was strain dependent. When treating the pregnant dams on GD 8.5, which proved to be the peak period of sensitivity, there was a hierarchy of susceptibility ranging from the completely resistant strain (DBA/2J), to the highly susceptible SWV/Fnn strain, in which 44.3% of the exposed fetuses were exencephalic (Finnell *et al.,* 1986). This variation in susceptibility is suggestive of a strong genetic component in the induction of these defects. The association between genetic and hyperthermia-induced NTDs was further supported in a reciprocal cross study between strains of variable susceptibility (Finnell *et al.,* 1986). The high susceptibility of the SWV strain was lost in the F_1 hybrids outcrossed to less susceptible, more resistant strains. The data from these studies are suggestive of only a few genes being responsible for susceptibility of the SWV strain, and that it is the genotype of the embryo, not the mother, that is the determining factor in hyperthermia-induced NTDs.

To further define the genes that might confer susceptibility to hyperthermia-induced NTDs in the SWV embryos, it has been demonstrated that the *growth arrest specific-5* (*gas-5*) gene is expressed in neurulating SWV embryos but not in embryos from the more heat-resistant LMBc strain (Vacha *et al.,* 1997). It is not surprising that this gene would be differentially expressed in the susceptible embryos, for it has been associated with both growth arrest and ribsomal assembly, two cellular processes that have been reported to be disrupted by hyperthermia (Peterson, 1990; Walsh *et al.*, 1993). Although the gene expression of gas-5 was not found to be affected by hyperthermia in this investigation, its localized strain-specific expression may be symptomatic of an altered proliferative rate in the neural folds of this strain, leading to an enhanced susceptibility to NTDs. The putative role of gas-5 along with its observed inhibition of rRNA methylation following hyperthermia in Chinese hamster ovary cells (Bouche *et al.*, 1981) supports the association between the expression of this gene and hyperthermia-induced cellular damage. Additional studies are needed, however, in order to determine whether *gas-5* is a direct cause or a secondary phenomenon of compromised cellular metabolism in the developing neural epithelia.

Experiments on the effects of maternal hyperthermia on the developing neural tissue have also been performed on nonhuman primates. If the maternal core temperature of pregnant Bonnet monkeys was elevated to 42.9°C on GDs 24–27, it resulted in fetuses with midfacial hypoplasia (Hendrickx *et al.*, 1979). In addition, when pregnant females were heatshocked once on GD 26, this treatment induced major malformations of the skeleton, heart, and brain. More specifically, the brain possessed only rudimentary structures. As with other species, the induction of these neural defects was limited to the initial stages of neurulation, for treatment outside of this gestational window failed to induce neural abnormalities (Hendrickx *et al.*, 1979).

These experiments demonstrate that hyperthermia has a very dramatic effect on the developing organisms. In almost every case, if the pregnant female is exposed to a hyperthermic insult early in gestation, NTDs occur. This phenomenon is independent of species, but is dependent on the temperature, duration, and genotype of the individual embryo.

C. Valproic Acid

In humans, valproic acid (VPA) has been associated with a distinct pattern of malformations that include neural, craniofacial, cardiovascular, and skeletal defects (Bjerkedal *et al.*, 1982; Jäger-Roman *et al.*, 1986; Lindhout and Schmidt, 1986). Arguably, the most devastating of these defects are the ones involving the developing nervous system. It has been estimated that 1–2% of the infants exposed to VPA *in utero* during the first trimester will develop an NTD, specifically spina bifida, which is 10–20 times the occurrence rate of this defect within the general population (Bjerkedal *et al.*, 1982; MMWR, 1982).

Humans are not unique in their response to VPA, for this drug has also been shown to cause NTDs in laboratory animals (Brown *et al.*, 1980; Diaz and Shields, 1981; Paulson *et al.*, 1985; Finnell *et al.*, 1988). NTDs, specifically exencephaly, are the most commonly observed defects when VPA is administered to pregnant mice during the initial stages of NTC (Kao *et al.*, 1981; Nau *et al.*, 1981). Similar results have also been reported for both rat and mouse embryos exposed to VPA in culture (Kao *et al.*, 1981; Bruckner *et al.*, 1983). Nau and Loscher (1984) administered VPA on GD 8, which did not produce any overt maternal toxicity, yet induced exencephaly in 30% of the exposed fetuses. Although the defects could occur at any place on the cranium, the most common place was in the forebrain region (Nau and Loscher, 1984). This same observation has been made in the several other strains of mice (Finnell *et al.*, 1988; Wlodarczyk *et al.*, 1996a). Interestingly, the same genetic susceptibility to induced NTDs that was observed from a hyperthermic insult was also observed following a VPA treatment (Finnell and Chernoff, 1987; Finnell *et al.*, 1988; Naruse *et al.*, 1988). When embryos from the highly susceptible SWV murine strain were treated with VPA on GD 8.5 and the embryos are collected at selected time points posttreatment, it was possible to examine the alternation in the pattern and progression through NTC. Unlike control embryos, which undergo NTC at four distinct sites in a discontinuous manner, the treated embryo fails to complete closure at the prosencephalon–mesencephalon border, which results in an exencephalic fetus (Kaufman, 1979; MacDonald *et al.*, 1989; Finnell, 1991). Embryos that have been subjected to subteratogenic doses of VPA are able to close their neural tube by compensating for any minor failure in closure II by the overextension of closure III caudally, and closure IV rostrally (Finnell, 1991). The mechanism by which VPA produces these defects remains controversial; however, it has been demonstrated that VPA accumulates or concentrates in the neuroepithelium (Dencker *et al.*, 1990; Turner *et al.*, 1990). It has also been reported that VPA exposure *in utero* produces a thinning of the forebrain neuroepithelium, with blood collecting in the lumen of the neural tube from vessels in which the endothelial lining was ruptured. The embryonic murine basal lamina, which has been shown to be important in the morphologic shaping of the neural tube and in establishing polarity in the neuroepithelium, was found to be irregular and discontinuous following exposure to teratogenic concentrations of VPA (O'Shea and Liu, 1987; Turner *et al.*, 1990). Although VPA exposure has been associated with NTDs in both humans and laboratory animals, and despite numerous investigations, a clear understanding of the mechanism(s) by which VPA initiates the cascade of molecular and biochemical events that ultimately lead to aberrant neurulation remains unknown.

IV. Types of Neural Defects Induced during Organogenesis

As diverse as the aforementioned compounds are, they all induce the same neural lesion in laboratory animals. Previously it was stated that the neural tube closes at four discrete locations; however, these sites possess differential susceptibilities to teratogenic insults. In categorizing the locations of isolated NTDs based on the literature, there is no evidence for teratogenic agents' interfering with the anterior portion of closure I, for defects at the base of the neck and most cranial portions of this site have not been reported. Although the posterior segment of closure I has been shown to be susceptible to perturbations resulting in spina bifida, this has only been demonstrated for a relatively few teratogenic agents. In general, closure I is thought to be relatively resistant to induced NTDs in laboratory animals. The majority of the anterior defects induced by teratogenic agents involve closure II, which is initiated at the prosencephalon–mesencephalon border, 6–12 hr after closure I, depending on the mouse strain. Failure of closure II produced fetuses with exencephaly, the murine equivalent of anencephaly in humans. Failure of closure III is also seen, but in a small fraction of the teratogen-exposed and affected embryos, indicated phenotypically by fetuses with cleft face. Finally, isolated defects involving closure IV have not been reported. Because agents that are clearly divergent in their chemical and/or physical properties appear to cause the same type of defects, there must be something unique to closures II and III that make them particularly liable to NTC defects.

Greater research efforts are necessary in order to understand the mechanisms involved in the induction of posterior NTDs. Whether these defects arise in a similar fashion as do anterior defects remains unresolved, in large part because there is not an adequate or appropriate animal model with which to study teratogen-induced spina bifida. Although posterior NTDs have been induced by a few teratogenic agents (alloxan, phthalate esters, salicylates, and tryptan blue), laboratory animals are quite resistant to the induction of these types of NTDs (Murakami *et al.,* 1954; Barber and Greer, 1964; Trasler, 1965; Horii *et al.,* 1966; Shiota and Nishimura, 1982; Yasuda *et al.,* 1986; Tibbles and Wiley, 1988; Alles and Sulik, 1990). Two clinically important drugs (VPA and retinoic acid) have been shown to induce both anterior and posterior defects in mice (Ehlers *et al.,* 1992a,b). Administering VPA (300–500 mg/kg) three times, 6 hr apart, on GD 9, Ehlers and co-workers (1992a) observed fetuses with spina bifida occulta in a dose-dependent manner (50–95%), whereas exencephaly was observed in 2–4% of the fetuses. This treatment regimen also induced the more serious spina bifida aperta, although this was only observed at the higher doses (450 and 500 mg/kg) and occurred in a very low frequency (4–6%). Retinoic acid (RA) has also been shown to produce spina bifida aperta, and at a greater frequency than does VPA (Ehlers *et al.,* 1992b). Administering 12.5 mg/kg of all *trans*-retinoic acid three times, 6 hr apart, on GD 8 induced spina bifida aperta in 22–63% of the exposed fetuses. The response frequency of NTDs depended on precisely when on GD 8 the treatments were initiated. The most sensitive, or peak, time point for RA to induce these defects was shown to be GD 8.5. As expected, these treatments also produced exencephaly in a number of the exposed fetuses. Not only do the peak sensitivity time points for inducing these defects differ between VPA and RA, but the histopathology of the observed malformations are also distinct. VPA treatment produced a posterior lesion characterized by necrosis and disorganization of the spinal column, with an abundance of cellular debris, blood cells, macrophages, and meninges in the spinal canal at the site of the lesion (Ehlers *et al.,* 1992b). In contrast, RA did not disrupt the morphology of the spinal cord, for it retained its gray/white matter and dorsal/ventral horn organization. In response to a teratogenic dose of RA, the neural canal did not exist, and there was only a thin layer of ependymal cells that lay on the surface of the spinal cord (Ehlers *et al.,* 1992b). Whereas some studies provide data in support of the idea that spina bifida is caused by a relatively uniform set of cellular effects (Copp *et al.,* 1990), the results from the VPA and RA studies suggest that these teratogenic agents induce spina bifida aperta via different cellular mechanisms.

Teratogenic exposures during organogenesis are thought to induce the development of structural defects or malformations. However, some reports have suggested that more subtle abnormalities can also be produced at this gestational time point. Autism, a behavioral disorder in which the individual has impaired social and speech development as well as limited activities and interests, has been suggested to be the result of a specific and subtle defect that occurs during the latter stages of NTC (Rodier *et al.,* 1996). Although the symptoms of this disorder do not suggest a specific region of brain as being the focal point in the development of this neurologic disorder, many hypotheses have been proposed for its etiology. Unfortunately, the actual morphologic lesion(s) that leads to autism remains unknown. It has been reported that one of the effects caused by thalidomide, besides the obvious limb defects that have been associated with this notorious teratogen, is autism. Miller and Stromland (1993) reported that thalidomide exposure during a very narrow window of neural tube development lead to a surprising increase in the number of autistic individuals. Interestingly, if exposure to thalidomide occurred outside of this time frame, autism was not observed (Miller, 1991). Because the teratogenic effects of thalidomide are well known, as are the precise timing of these defects, the observed exposure and the eventual behavioral outcomes has stimulated several new hypotheses. One promising hypothesis that may explain the morphologic events that eventually lead to autism is that thalidomide interferes with the pattern formation of the rhombomeres, which give rise to the brainstem and cranial nerve motor nuclei that form during the later stages of NTC (Rodier *et al.*, 1996). Indirect evidence in support of this hypothesis has been presented, using both the anatomic pathology observed in autistic individuals and a rodent model that used VPA, a drug that can induce a very similar behavioral disorder in humans (Ardinger *et al.*, 1988; Collins *et al.*, 1991; Ehlers *et al.*, 1992a) and in rats (Vorhees, 1987). From this investigation, it was shown that the autistic individuals had an almost complete absence of the facial nucleus and superior olive, as well as a shortening of the brainstem between the trapezoid body and the inferior olive (Rodier *et al.*, 1996). In the rat, 350 mg/kg of VPA administered on GD 11.5, 12.0, or 12.5 reduced the number of neurons from the V and XII motor nuclei, as well as the VI and III cranial nerve nuclei. This treatment appeared to be quite specific, as several other nuclei (dorsal nucleus of the vagus, the locus ceruleus and the mesencephalic nucleus of the trigeminal nerves) that were forming at the same time were not adversely affected by the VPA treatment (Rodier *et al.*, 1996). Whereas these experiments cannot causally connect the observed neurologic damage with autism or the VPA-induced behavioral defects, it is an interesting association that warrants further investigation. It appears that thalidomide and VPA induce similar damage in a very distinct set of developing neural processes that lead to nearly identical behavioral endpoints. This happens in the absence of gross structural defects or disruption in the subsequent development of the brain. Illustrating that subtle neural damage, even during the earliest stages of neural development, can have very specific yet profound effects without derailing the overall process of NTC.

V. Postneurulation Development

Once the human embryo's neural tube has completely formed and fused along its entire length, the rudimentary brain enters into a period characterized by rapid growth and differentiation. This includes a series of complex subdivisions, flexures, and cellular differentiation that is not complete until well into the first few years of neonatal life. Although this has become a fascinating and growing area of research, it is not the aim of this discussion to delve into the intricate embryology of these processes. Instead, a description and explanation of the types of the neural defects that are observed when the developing neural tissue is exposed to a teratogen during the later stages of development is pursued, using ethanol and methylmercury as examples.

A. *Ethanol*

In utero exposure to ethanol is probably the most common teratogen in Western society (Department of Health and Human Services, 1993). The incidence of the fetal alcohol syndrome (FAS) has been estimated to be 1–2% of the general population of the United States, but among alcoholic women this figure jumps to between 2.5 and 40% (Hoyseth and Jones, 1989; Webster, 1989). Despite public awareness, FAS continues to be a major societal problem (Fitzgerald, 1988). Ethanol exposures during the later stages of gestation, when synaptogenesis, neuronal proliferation, dendritic arborization, protein synthesis, and increased brain weight occur, cause the most debilitating effects. The disruption of these critical events leads to CNS dysfunctions and neurologic abnormalities (Clarren and Smith, 1978; Dobbing and Sands, 1979; Meyer *et al.*, 1990). Despite the fact that nutrition with pre- and postnatal care can remedy the growth and craniofacial displasia associated with FAS, the mental deficiencies persist into adulthood (Spohr *et al.*, 1993). Therefore, FAS is not just a childhood disorder, but rather a life-long impairment (Streissguth *et al.*, 1991; Spohr *et al.*, 1993). As in humans, exposure of the fetus/neonate to ethanol during the brain's growth spurt causes the most damaging effects in experimental animals (Dobbing and Sands, 1979). Although the adverse affects of ethanol on the developing CNS are numerous, it has been reported that most of these can be attributed to a delay or disruption in the development of the fetal hippocampus (West *et al.*, 1986; Reynolds and Brien, 1995). Biochemical evidence that supports this hypothesis has shown that ethanol alters a number of stimulatory and inhibitory neurons, as well as disrupting the cAMP signal transduction system of the developing hippocampus (Farr *et al.*, 1988; Ledig *et al.*, 1988; Pennington, 1990). Other regions of the developing brain (cerebral cortex, pons, cerebellum, olfactory bulb, and amygdala) have also been reported as being similarly effected by ethanol (Fabregues *et al.*, 1985; Druse *et al.*, 1990; Meyer *et al.*, 1990; Abdollah and Brien, 1995). At the molecular level, ethanol has been shown to produce a generalized decrease in the transcriptional activity of the developing nervous system. A 2-week prenatal exposure produced a 22% reduction of total RNA in the brain of newborn rat pups (Rawat, 1980). This subchronic ethanol treatment also reduced the DNA synthesis in the brain of the exposed pups by 45% (Rawat, 1980). The reduction in DNA content has been suggested to be a reflection of the ethanol-induced decrease in mitosis or cell death (Miller, 1986).

There is limited data on the affects of *in utero* ethanol exposures on specific molecular events in the developing embryo. Exposing pregnant rats to ethanol vapors from GD 7–21 produced a delay in the expression pattern of genes for the myelin proteolipid proteins (PLP) and myelin associated glycoproteins (MAG) as well as tubulin (Milner *et al.,* 1987). Aberrant expression of these myelin genes could explain reports that have demonstrated a delay in myelination in ethanol-exposed fetal rats (Jacobsen *et al.,* 1979). Glial fibrillary acidic protein (GFAP) and somatostatin expression have also been shown to be altered by ethanol (Naus and Bechberger, 1991). Alterations in these genes could account for the delay in the differentiation of somatostatin neurons and the disruption of migration of cortical neurons observed following ethanol exposure (Miller, 1986; Naus and Bechberger, 1991). Therefore, although this teratogen continues to be a major social problem, it appears that most of its embryonic/fetal effects occur during the later stages of brain development. Ethanol does not appear to have a specific site of action, as it affects several regions within the developing CNS.

B. Methylmercury

Most of what is known about the neurologic effects of methylmercury stem from accidental human exposures in Japan and Iran, as well as experimental work conducted on nonhuman primates (Mottet *et al.,* 1985). This organometal readily passes through the placenta and has a high specificity for producing toxicity in neural tissue (Marsh *et al.,* 1980; Hackett and Kellman, 1983; Clarkson, 1987). Pathologic changes within the brain include neuronal lysis, phagocytosis, gliosis, as well as cerebral and cerebellar necrosis. These changes are most obvious in the deep sulic, and may be due to a vascular deficiency (Mottet *et al.,* 1985). In humans, fetal methylmercury exposure results in a pattern of dysmorphogenesis known as *fetal minamata disease,* which is characterized by cerebral palsy, blindness, deafness, and microcephaly (Bakir *et al.,* 1973; Mottet *et al.,* 1985). The most severely affected individuals are those are exposed during neuronal migration, which occurs during the later stages of development (Choi *et al.,* 1978; Sager *et al.,* 1982; Rodier *et al.,* 1984). Although severe neurologic effects were observed when the maternal body burden was between 165–320 ppm, moderate effects were seen at significantly lower concentrations (37 ppm) (Choi *et al.,* 1981). In fact, when nonhuman primates were exposed to as little as 0.7–0.9 ppm of methylmercury, this treatment produced developmental delays and learning deficits (Newland *et al.,* 1994). Thus, even small doses of this oganometal may lead to profound neurologic deficiencies.

Laboratory rodents have also been shown to be susceptible to this teratogen, for when neonatal rats pups were exposed to methylmercury for 5 consecutive days during the time when the brain is undergoing rapid growth, the pups had impaired motor performance, reduced brain weight, and degeneration of the nerve cells in the inner granular layer of the cerebellum (Naruse *et al.,* 1993; Roca *et al.,* 1993). The exact biochemical or molecular events that lead to these neurologic deficits is unknown, but it has been demonstrated that methylmercury blocks the action potential of motor neurons in culture, suggesting that this neurotoxicant induced changes in the lipid membrane of these nerves that altered their ion permeability (Mitolo-Chieppa, 1981). There is further evidence that suggests that proteins that contain sulfhydride groups are more susceptible to methylmercury's adverse effects, for reducing agents protected against, or at least delayed, the toxic impact of methylmercury on neuronal cells *in vitro* (Marquis, 1978). Therefore, as with ethanol, there is sufficient appreciation of the pathology induced by methylmercury, and the period during gestation when it is most toxic. Nonetheless, the underlying mechanisms of methylmercury toxicity remain controversial. Determining which effects and which specific regions of the brain are responsible for the observed neurologic abnormalities must be of paramount importance if we are to fully understand the extent of functional teratogens.

VI. Genetic Mutations

Once it has been established that significant congenital defects can occur throughout all of gestation and the types of abnormalities have been cataloged, it is important to attempt to understand how these defects may arise. In spite of the serious consequences associated with NTDs, efforts to prevent these congenital defects have been limited by our lack of understanding of the major risk factors involved and their interrelationship with underlying developmental and genetic mechanisms. Therefore, murine genetic mutants that give rise to spontaneous NTDs have been exploited in an attempt to understand some of the molecular and biochemical parameters that are involved in the development of these malformations. To date, one of the often studied model systems is the murine mutant *curlytail* (*ct*). This mutation arose in 1950 in a GFF stock and was subsequently transferred onto a CBA/Gr-*ct* background (Seller and Adinolfi, 1981). The inheritance pattern of the curlytail gene best fits the model of a recessive gene with incomplete penetrance (Embury *et al.,* 1979). The mating of apparently homozygous animals results in the production of two distinct classes of pups. The group have obvious tail malformations, which range from complete twists to slight bends, and may be associated with NTDs, including exencephaly, lumbosacral myeloschisis, and spina bifida (Embury *et al.,* 1979; Copp *et al.,* 1982). The littermates with normal or straight tails (ST) have no neural abnormalities. Thus, animals with the same genotype can express an abnormal phenotype or be perfectly normal. Clearly, the NTD phenotype in the *ct* mice is modified by the genetic background and exogenous environmental influences to which the embryos are exposed.

What makes the *ct* mouse mutant such an attractive model system to study the morphogenetic and molecular mechanisms of NTC defects is their many similarities to human NTDs. That is, there is a preponderance of affected female pups compared to affected males, there is an elevation in the amniotic fluid levels of alpha fetoprotein in affected fetuses, and other complicating factors such as hydrocephaly and polyhydramnios are observed in fetuses with NTDs in both species (Embury *et al.,* 1979; Seller and Adinolfi, 1981). Most importantly, the *ct* NTDs resemble human NTDs morphologically, with the majority of the lesions located in the lumbosacral region. Mechanistically, the spinal lesions appear to be the result of a delay in the closure of the posterior neuropore. This was determined by morphologic assessment of CT and ST embryos of a specific developmental stage that was standardized by the somite number. In *ct/ct* homozygotes with more than 35 somite pairs, between 50% and 64% had tail flexion defects and NTDs (Copp *et al.,* 1982). It was evident that the somite number was not depressed in affected fetuses, indicating that although the posterior neuropore was open and its closure had been delayed, there was not an overall delay in fetal development. Copp and colleagues (1982) suggested that the delayed closure of the posterior neuropore in the curlytail mutants in the presence of normal development of other posterior structures, imposes stress on the developing trunk and tail regions and that these stresses ultimately are responsible for the observed congenital defects.

Subsequent studies on curlytail embryos grown *in vitro* revealed that the posterior neuropore could be classified into five categories according to the cranial extent of the posterior neuropore, with category 1 being most closed and category 5 having the greatest degree of neuropore opening (Copp, 1985). Those embryos with the greatest cranial extension of the posterior neuropore were associated with delayed NTC and the development of spinal NTDs (Copp, 1985). To determine the underlying causes of the delay in posterior neuropore closure, Copp and colleagues (1988a) examined the rate of cellular proliferation using mitotic index, length of S-phase, and tritiated thymidine pulse chase studies in both normal and abnormal curly tail embryos. Among the abnormal embryos there was a reduced rate of cell proliferation in the gut endoderm and notochord when compared with normal littermates. There was also some indication of a lengthening of the cell-cycle in the surface ectoderm and mesoderm of the affected fetuses (Copp *et al.,* 1988).

It was further suggested that the cell-type–specific abnormality associated with the reduced cellular proliferation in the curlytail mutants could be the result of a molecular defect that is expressed in cells of the gut endoderm and notochord, or in adjacent cell types and in adjoining extracellular spaces. With respect to the latter, it has also been demonstrated that in comparison to normal embryos (Copp and Bernfield, 1988a), the levels of the extracellular matrix (ECM) protein hyaluronate are significantly depressed in the posterior neuropore region of curlytail embryos with spinal defects (Copp and Bernfield, 1988b). The failure of this ECM protein to accumulate specifically effects basement membranes developing beneath the neuroepithelium and around the notochord. These proteins are known to play a significant role in promoting normal cellular proliferation, and their absence in the mutants could be a contributing factor to the delayed closure of the posterior neuropore and the subsequent expression of a spinal NTD (Copp *et al.,* 1988). It is also possible that the absence of sufficient hyaluronate could preclude normal interaction with specific cell-surface receptors that are necessary for promoting cytoskeletal-induced shape changes in neuroepithelial cells that are also essential for normal NTC (Karfunkel, 1974; Copp and Bernfield, 1988b). Whether it is a specific problem in cellular proliferation or a failure to undergo normal cellular shaping and rearrangements, the posterior NTDs observed in the mutant curlytail mouse represents an excellent animal model with which to elucidate the complex processes involved in the development of posterior NTDs.

The Splotch (*Sp*) mutation on murine chromosome 1 has also been used heavily by investigators as a genetic model for both anterior and posterior NTDs (Trasler and Morriss-Kay, 1991). Embryos carrying the mutant

Splotch allele have been shown to have delayed NTC as well as problems associated with neural crest cell migration (Dempsey and Trasler, 1983; Moase and Trasler, 1992). Splotch homozygous (*Sp/Sp*) fetuses die at approximately GD 13, with more than 50% of the fetuses having cranioschisis, overgrowth of neural tissue, and a reduction in the spinal ganglion (Dempsey and Trasler, 1983). Crosses between heterozygous *Sp*/+ animals resulted in the anticipated 25% incidence of homozygous embryos whose failure of NTC lead to spina bifida with or without exencephaly (Auerbach, 1954; Dickie, 1964). When Splotch embryos were examined using immunohistochemical techniques prior to the development of the posterior NTD, it was clear that the temporal and spatial expression of selected extracellular matrix proteins in the notochord, mesoderm, and neuroepithelium were markedly different from that pattern observed in the wild-type embryos who completed NTC normally. Specifically, the chondroitin sulphate and heparin sulphate proteoglycans, compounds that have been shown to play essential roles during NTC (Solursh and Morriss, 1977; Morriss-Kay *et al.,* 1986; Copp and Bernfield, 1988a,b), appear more intensely in the neuroepithelial basement membrane at the 5-, 10-, and 15-somite stages of affected Splotch embryos, rather than at the 16- and 19-somite stages, which is the pattern observed in the wild-type controls (Trasler and Morriss-Kay, 1991). The altered spatial and temporal profiles of these proteins in the mutant embryos, when compared to the controls, suggests a possible interaction between the *Sp* gene locus and neural crest cell adhesion molecules (Trasler and Morriss-Kay, 1991). Interestingly, the *Sp* mutant mice also had increased amounts of hyaluronic acid at the same developmental stages as did the control embryos (McLone and Knepper, 1986), which contrasts markedly to the situation in the *ct* embryos (Copp and Bernfield, 1988b). Although the Splotch mutation differs in certain developmental aspects from the curly tail mouse, it nonetheless represents a valued genetic model with which to examine the molecular mechanisms underlying NTC defects.

The SELH stock has a high incidence of spontaneous NTDs (Juriloff *et al.,* 1989; MacDonald *et al.,* 1989; Tom *et al.,* 1991; Gunn *et al.,* 1995). Although this mutant has only recently been described, it possesses traits that make it an attractive animal model for studying anterior NTDs. Currently, most of the murine mutants used to study NTDs are single gene mutations with very high penetrance, such as rib fusion (*Rf*) and crooked (*Cd*) (Cole and Trasler, 1980), Splotch (Dempsey and Trasler, 1983), or Looptail (Wilson, 1974; Wilson and Center, 1974; Wilson and Finta, 1980). However, the SELH stock is more reflective of what has been observed in the human population, where the majority of human NTDs are multifactorial in nature, with a low penetrance and an unclear transmission pattern. Approximately 17% of the SELH embryos are exencephalic, due to a failure of closure II (Juriloff *et al.,* 1989; MacDonald *et al.,* 1989; Tom *et al.,* 1991; Gunn *et al.,* 1993). In fact, none of the SELH embryos have a normal closure pattern, for closure II never occurs in these embryos. In the embryos that are not exencephalic, the lack of closure II is compensated for by the overextension of closures III and IV. Why one embryo can complete neural tube closure and its littermate cannot remains unexplained. It has been shown that it is not due to some maternal effects, but rather an accumulative effect of a few modifying genes (Juriloff *et al.,* 1989; Gunn *et al.,* 1992). Histologic examination of exencephalic SELH embryos has shown that the mesenchyme at the site of closure II was collapsed and the neuroepithelium rippled. There were no differences in the cell shapes or numbers compared to embryos that successfully closed their neural tubes (MacDonald *et al.,* 1989). This mutant stock may ultimately become a highly useful tool in unraveling the complex molecular and biochemical events that lead to NTDs.

VII. Specific Genes Involved in NTDs

Although NTDs clearly have a strong genetic component to their etiology and numerous genes have either been localized or in some cases shown to be involved in normal neurulation specific genes that can be identified as being responsible for the induction or occurrence of NTDs remain relatively few in number. The complex nature of neurulation and the plasticity of the developing embryo suggest that more than one gene is involved in the development of these defects. One tool that has begun to make significant strides in elucidating potential genes that are important to the process of NTC is the development of transgenic mice. Using this technique, the disruption of several genes has proved to result in embryos with NTDs. Not surprising is the diversity of these genes, which further illustrates the complexity of the developing nervous system.

One gene that may play a role in the development of NTDs is cartilage homeoprotein 1 (*Cart-1*). Originally localized to chondrocytes, cartilage, and testes in the rat fetus, it was predicted to play an important role in the development of cartilage (Zhao *et al.,* 1996). However, *Cart-1* null embryos were found to have anencephaly and severe craniofacial defects. Interestingly, no cartilage defects were observed in these embryos. In the exencephalic embryos, the mesenchymal tissue between the forebrain neuroepithelium and the surface ectoderm was absent. *In situ* hybridization studies have confirmed the location of *Cart-1* to the forebrain mesenchyme (Zhao *et al.,* 1996). Although the neural fold elevation was not apparently affected in *Cart-1* deficient mice, the resulting defects appear to be due to improper neural bending and fusion. These studies demonstrate the important role of the underlying mesenchyme in proper neural fold fusion (Zhao *et al.,* 1996). One intriguing issue concerning this murine mutant is that the normal phenotype can be rescued by the administration of folic acid. Only 19% of the *Cart-1* deficient embryos from dams receiving supplemental folic acid were exencephalic. All of the *Cart-1* deficient mice died within 48 hr of birth, indicative of the presence of other associated developmental defects (Zhao *et al.,* 1996). Although it is currently unclear as to how folic acid helps in the process of NTC, further clarification and characterization of this mutant should provide insight into the relationship between folic acid and NTDs in humans.

Another gene whose deletion appears to interfere with normal neural tube closure is *hairy* and *enhancer of split homologue-1* (*HES-1*). This gene is the mammalian homolog of two *Drosophila* genes (*h* and *E(spl)*) that have been show to be negative regulators of neurogenesis (Ishibashi *et al.,* 1995). If *HES-1* is overexpressed in neural precursors of the ventricular zone following retroviral transfection, normal cell migration and differentiation failed to occur. This suggests that the *HES-1* gene product is acting much like its *Drosophila* homologs. *HES-1* appears to repress transcription by binding to a conserved N-box on different genes, specifically helix-loop-helix activators (Sasai *et al.,* 1992; Ishibashi *et al.,* 1995). It is the absence of this repression in the transgenic mice that alters the timing of neurogenesis and results in the observed anencephaly, reduced telecephalic vesicles, and midfacial clefts (Ishibashi *et al.,* 1995). When *HES-1* is inactivated, it disrupts the regulation of the differentiation process, which alters the rate of proliferation of these cells and ultimately the timing of NTC, with the result being anencephaly. Although the exact role of *HES-1* in NTD development or pathogenesis has yet to be determined, it is a useful model for determining how timing or delays in NTC might lead to cranial defects, as well as understanding the multifactorial nature of these defects.

A third gene whose function has been linked to normal NTC is the macrophage myristoylated alanine rich C kinase (*MacMARCK*) gene. Because the product of this gene has the potential to serve as a substrate for protein kinase C (PKC), a transgenic mouse was developed to assess *MacMARCK* involvement in early development (Blackshear, 1993). In the homozygous nullizygous embryos there is a high rate (60%) of exencephaly. These embryos lack brain structures as well as the overlying skull and skin (Chen *et al.,* 1996). Although the neuroepithelial layer and neural patterning appeared normal, the neural folds in the forebrain regions failed to fuse. The proposed mechanism of how a deletion of *MacMARCKS* leads to an NTD is believed to involve a misregulation of actin-based cell-shape changes that need to occur in the lateral region of the neural tube for normal closure to occur. This hypothesis is supported by the findings that the expression of *MacMARCKS* is limited to the lateral edges of the neural tube and surrounding cranial mesenchyme. Furthermore, normal embryos that have been treated with cytochalasin B, a cytoskeleton disrupter, have a similar morphology to the *MacMARCKS* nullizygous embryos (Hartwig *et al.,* 1992). Although the exact nature of how this genetic deletion leads to exencephaly remains unknown, this model is valuable in understanding how cellular cytoarchitecture plays a role in both normal and abnormal neurulation.

The *PAX* genes encode a family of transcription factors that are expressed in restricted regions of the developing embryo. Although several of these genes are believed to be involved in the ontogeny of the nervous system (Goulding and Paquett, 1994), *PAX-3* is probably the most studied of these genes. *PAX-3* expression has been localized to the dorsal neural groove, dorsal neuroepithelium, and neural tube (Goulding *et al.,* 1991). Because the expression of this gene has only been detected in embryos and never in adult tissues, it has been considered to play an vital role in morphogenesis. Its involvement has been demonstrated and characterized in the murine mutant Splotch. Nine different alleles of Splotch have been identified, and all involve a mutation in the *PAX-3* gene that leads to an aberrant protein. Homozygous Splotch embryos have a myriad of defects including exencephaly, spina bifida, and cardiac and muscle defects and die *in utero* by GD 14 (Goulding *et al.,* 1993; Underhill *et al.,* 1995). Although the *PAX-3* gene has been directly implicated as the cause of the Splotch phenotype, how this gene deletion produces embryos with NTDs remains largely unexplained (Epstein *et al.,* 1991). One possible explanation is that *PAX-3* regulates the anterior–posterior patterning of the neural tube. Using transgenic mice in which *PAX-3* expression was under the control of the *HOXb-4* promoter, it was shown that while *PAX-3* did not govern the anterior–posterior patterning, it was involved with the ventralization of the neural tube (Tremblay *et al.,* 1996). The exact nature of this regulation and how it is involved in neurulation is slowly being defined.

Although this review was not meant to be an index of all the transgenic mice that have reported NTDs as a phenotype, it does illustrate just how transgenic mice can be useful in understanding the impact of a deletion of one gene that could potentially be involved in inducing these malformations. Granted, the major limitation of this technique is that only a single gene is inactivated, and the primary defect may be due to a gene that is either down stream or regulated by the inactivated gene.

VIII. Conclusions

In summary, the developing nervous system is vulnerable to any agent, either physical or chemical, that interferes with any of the processes involved in its development. Because the nervous system has such a long ontogeny, the tissues involved are so complex, and the interconnections of the nervous system, it is uniquely susceptible to teratogenic agents. It should now be clear that agents can differentially affect these tissues depending on when during gestation the insult occurred. This may be the result of a failure of normal embryonic homeostatic properties having a chance to develop prior to the teratogenic insult. In categorizing neuroteratogens, the ones that are most familiar (arsenic, RA, VPA) are those that cause structural defects. When these are administered to pregnant laboratory animals just prior to neurulation, these agents can produce major malformations such as exencephaly and/or spina bifida. These types of agents will always be useful as tools that can be used to unravel the biochemical and molecular events that govern how the neural tube closes and how the rudimentary nervous system evolves into a highly complex adult brain. However, the more subtle changes that disrupt synaptogenesis, histogenesis, and/or the cell–cell communication of the developing brain leading to functional deficits may prove to be more of a challenge as well as being more problematic for society. Certainly structural defects, such as spina bifida, have a high societal price, but compared to functional (memory, learning, and social) deficits, their significance pales. We have only recently begun to identify agents that cause these more subtle neural injuries. Using various models together with more sophisticated neurobehavorial assessment techniques, it is our hope that the etiology of both minor and major structural defects of the nervous system will be elucidated and ultimately prevented.

Acknowledgments

This work was supported in part by grant number ES07165 and ES35396 from the National Institute of Environmental Health Sciences, NIH, as well as DE11303, from the National Institutes of Health. Its content is solely the responsibility of the authors and do not necessarily represent the official views of the NIEHS or the NIH. The authors would like to express their appreciation to Ms. Jennifer Wormuth, whose clerical and editorial assistance on this project was invaluable.

References

Abdollah, S., and Brien, J. F. (1995). Effect of chronic maternal ethanol administration on glutamate and *N*-methyl-D-asparate binding sites in the hippocampus of the near term fetal guinea pig. *Alcohol* **12,** 377–382.

Akam, M. (1989). Hoc and HOM: homologous gene clusters in insects and vertebrates. *Cell* **57,** 347–349.

Alles, A. J., and Sulik, K. K. (1990). Retinoic acid induced spina bifida: evidence for a pathogenetic mechanism. *Development* **108,** 73–81.

Ardinger, H. H., Atkin, J. F., Blackston, R. D., Elsas, L. J., Clarren, S. K., Livingstone, S., Flannery, D. B., Pellock, J. M., Harrod, M. J., Lammer, E. J., Majewski, F., Schinzel, A., Toriello, H. V., and Hanson, J. W. (1988). Verification of the fetal valproate syndrome phenotype. *Am. J. Med. Genet.* **29,** 171–185.

Aschengrau, A., Zierler, S., and Cohen, A. (1989). Quality of community drinking water and the occurrence of spontaneous abortion. *Arch. Environ. Health* **44,** 283–290.

Auerbach, R. (1954). Analysis of the developmental effects of a lethal mutation in the house mouse. *J. Exp. Zool.* **127,** 305–329.

Austin, C. R. (1973). Embryo transfer and sensitivity to teratogenesis. *Nature* **244,** 333–334.

Bakir, F., Damluji, S. F., Amin-Zaki, L., Murtadha, M., Khalida, A., al-Rawi, N. Y., Tikriti, S., Dahahir, H. I., Clarkson, T. W., Smith, J. C., and Doherty, R. A. (1973). Methylmercury poisoning in Iraq. *Science* **181,** 230–241.

Bannigan, J., and Cottell, D. (1984). Ethanol teratogenicity in mice: an electron microscopic study. *Teratology* **30,** 281–290.

Barber, A. N., and Greer, J. C. (1964). Studies on the teratogenic properties of trypan blue and its components in mice. *J. Embryol. Exp. Morphol.* **12,** 1–14.

Beaudoin, A. R. (1974). Teratogenicity of sodium arsenate in rats. *Teratology* **10,** 153–157.

Beckman, L. (1978). The Ronnskar smelter—occupational and environmental effects in and around a polluting industry in Northern Sweden. *Ambio.* **7,** 226–231.

Bjerkedal, T., Cziezel, A., Goujard, J., Kallen, B., Mastroiacova, P., Nevin, N., Oakley, G. Jr., and Robert, E. (1982). Valproic acid and spina bifida. *Lancet* **2,** 1096.

Blackshear, P. J. (1993). The MARCKS family of cellular protein kinase C substrates. *J. Biol. Chem.* **268,** 1501–1504.

Bollinger, C. T., van Zijl, P., and Louw, J. A. (1992). Multiple organ failure with the adult respiratory distress syndrome in homicidal arsenic poisoning. *Respiration* **59,** 57–61.

Borzsonyi, M., Bereczky, A., Rudnai, P., Csanady, M., and Horvath, A. (1992). Epidemiological studies on human subjects exposed to arsenic in drinking water in Southeast Hungary. *Arch. Toxicol.* **66,** 77–78.

Bouche, G., Raynal, F., Amalric, F., and Zalta, J. P. (1981). Unusual processing of nucleolar RNA synthesized during a heat shock in CHO cells. *Mol. Biol. Rep.* **7,** 253–258.

Brown, N. A., Kao, J., and Fabro, S. (1980). Teratogenic potential of valproic acid. *Lancet* **1,** 660–661.

Bruckner, A., Lee, Y. J., O'Shea, K. S., and Henneberry, R. C. (1983). Teratogenic effects of valproic acid and diphenylhydantoin on mouse embryos in culture. *Teratology* **27,** 29–42.

Busam, K. J., Roberts, D. J., and Golden, J. A. (1993). Clinical teratology counseling and consultation case report: two distinct anterior neural tube defects in a human fetus: evidence for an intermittent pattern of neural tube closure. *Teratology* **48,** 399–403.

Campbell, L. R., Dayton, D. H., and Sohal, G. S. (1986). Neural tube defects: a review of human and animal studies on the etiology of neural tube defects. *Teratology* **34,** 171–187.

Carpenter, S. J. (1987). Developmental analysis of cephalic axial dysraphic disorders in arsenic-treated hamster embryos. *Anat. Embryol.* **176,** 345–365.

Chaineau, E., Binet, S., Pol D., Chatellier, G., and Meininger, V. (1990). Embryotoxic effects of sodium arsenite and sodium arsenate on mouse embryos in culture. *Teratology* **41,** 105–112.

Chalepakis, G., Stoykova, A., Wijnholds, J., Temblay, P., and Gruss, P. (1993). Pax: gene regulators in the developing nervous system. *J. Neurobiol.* **24,** 1367–1384.

Chance, P. F., and Smith, D. W. (1978). Hyperthermia and the neural tube. *Lancet* **2,** 560–561.

Chen, J. S., Chang, S. A., Duncan, H. J., Okano, A., Fishell, G., and Aderem, A. (1996). Disruption of the *MacMARCKS* gene prevents cranial neural tube closure and results in anencephaly. *Proc. Natl. Acad. Sci. USA* **93,** 6275–6279.

Choi, B. H., Cho, K. H., and Lapham, L. W. (1981). Effects of methylmercury on human fetal neurons and astrocytes *in vitro:* a time-lapse cinematographic, phase and electron microscopic study. *Environ. Res.* **24,** 61–74.

Choi, B. H., Lapham, L. W., Amin-Zaki, L., and Saleem, T. (1978). Abnormal neuronal migration, deranged cerebral cortical organization and diffuse white matter astrocytosis of human fetal brain: a major effect of methylmercury *in utero. J. Neuropathol. Exp. Neurol.* **37,** 719–733.

Clarkson, T. W. (1987). Metal toxicity in the central nervous system. *Environ. Health. Perspect.* **75,** 59–64.

Clarren, S. K., and Smith, D. W. (1978). The fetal alcohol syndrome. *N. Engl. J. Med.* **298,** 1063–1067.

Cockroft, D. L., and New, D. A. (1975). Effects of hyperthermia on rat embryos in culture. *Nature* **258,** 604–606.

Colangelo, W., and Jones, D. G. (1982). The fetal alcohol syndrome: a review and assessment of the syndrome and its neurological sequelae. *Prog. Neurobiol.* **19,** 271–314.

Cole, W. A., and Trasler, D. G. (1980). Gene–teratogen interaction in insulin-induced mouse exencephaly. *Teratology* **22,** 125–139.

Collins, M. D., Fradkin, R., and Scott, W. J. (1990). Induction of postaxial forelimb ectrodactyly with anticonvulsant agents in A/J mice. *Teratology* **41,** 61–70.

Collins, M. D., Walling K. M., Resnick E., and Scott, W. J., Jr. (1991). The effect of administration time on malformations induced by three anticonvulsant agents in C57BL/6J mice with emphasis on forelimb ectrodactyly. *Teratology* **44,** 617–627.

Copp, A. J. (1985). Relationship between timing of posterior neuropore closure and development of spinal neural tube defects in mutant (curly tail) and normal mouse embryos in culture. *J. Embryol. Exp. Morphol.* **88,** 39–54.

Copp, A. J., and Bernfield, M. (1988a). Glycosaminoglycans vary in accumulation among the neuraxis during spinal neurulation in the mouse embryo. *Dev. Biol.* **130,** 573–582.

Copp, A. J., and Bernfield, M. (1988b). Accumulation of basement membrane–associated hyaluronate is reduced in the posterior neuropore region of mutant (curly tail) mouse embryos developing spinal neural tube defects. *Dev. Biol.* **130,** 583–590.

Copp, A. J., Brook, F. A., Estibeiro, J. P., Shum, A. S., and Cockroft, D. L. (1990). The embryonic development of mammalian neural tube defects. *Prog. Neurobiol.* **35,** 363–403.

Copp, A. J., Brook, F. A., and Roberts, H. J. (1988). A cell-type–specific-abnormality of cell proliferation in mutant (curly tail) mouse embryos developing spinal neural tube defects. *Development* **104,** 285–295.

Copp, A. J., Seller, M. J., and Polani, P. E. (1982). Neural tube development in mutant (curly tail) and normal mouse embryos: the timing of posterior neuropore closure in vivo and in vitro. *J. Embryol. Exp. Morphol.* **69,** 151–167.

Crossin, K. L., Chuong, C. M., and Edelman, G. M. (1985). Expression sequences of cell adhesion molecules. *Proc. Natl. Acad. Sci. USA* **82,** 6942–6946.

Cuckle, H., and Wald, N. (1989). Ovulation induction and neural tube defects. *Lancet* **2,** 1281.

Cusic, A. M., and Dagg, C. P. (1985). Spontaneous and retinoic acid induced postaxial polydactyly in mice. *Teratology* **31,** 49–59.

Dempsey, E. E., and Trasler, D. G. (1983). Early morphological abnormalities in splotch mouse embryos and predisposition to gene- and retinoic acid–induced neural tube defects. *Teratology* **28,** 461–472.

Dencker, L., Nau, H., and D'Argy, R. (1990). Marked accumulation of valproic acid in embryonic neuroepithelium of the mouse during early embryogenesis. *Teratology* **41,** 699–706.

Department of Health and Human Services, U.S. (1993). Eighth special report to the US Congress on alcohol and health. Department of Health and Human Services, Washington DC, pp. 203–232.

Diaz, J., and Shields, W. D. (1981). Effects of dipropylacetate on brain development. *Ann. Neurol.* **10,** 465–468.

Dickie, M. M. (1964). New splotch alleles in the mouse. *J. Hered.* **55,** 97–101.

Dobbing, J., and Sands, J. (1979). Comparative aspects of the brain growth spurt. *Early Hum. Dev.* **3,** 79–83.

Druse, M. J., Tajuddin, N., Kuo, A., and Connerty, M. (1990). Effects of *in utero* ethanol exposure on the developing dopaminergic system in rats. *J. Neurosci. Res.* **27,** 233–240.

Dziadek, M. (1993). Preovulatory administration of clomiphene citrate to mice causes growth retardation and neural tube defects (exencephaly) by an indirect maternal effect. *Teratology* **47,** 263–273.

Edelman, G. M. (1984). Expression of cell adhesion molecules during embryogenesis and regeneration. *Exp. Cell Res.* **161,** 1–16.

Edelman, G. M., and Crossin, K. L. (1991). Cell adhesion molecules: implications for a molecular histology. *Ann. Rev. Biochem.* **60,** 155–190.

Edelman, G. M., and Jones, F. S. (1995). Developmental control of N-CAM expression by Hox and Pax gene products. *Philos. Trans. Roy. Soc. Lond. Ser. B* **349,** 305–312.

Edwards, M. J. (1968). Congenital malformations in the rat following induced hyperthermia during gestation. *Teratology* **1,** 173–177.

Edwards, M. J. (1986). Hyperthermia as a teratogen: a review of experimental studies and their clinical significance. *Teratog. Carcinog. Mutagen.* **6,** 563–582.

Edwards, M. J., Walsh, D. A., Webster, W. S., and Lipson, A. H. (1986). Hyperthermia: is it a "direct" embryonic teratogen? *Teratology* **33,** 375–378.

Ehlers, K., Sturje, H., Merker, H.-J., and Nau, H. (1992a). Valproic acid induced spina bifida: a mouse model. *Teratology* **45,** 145–154.

Ehlers, K., Sturje, H., Merker, H.-J., and Nau, H. (1992b). Spina bifida aperta induced by valproic acid and by all *trans* retinoic acid in the mouse: distinct differences in morphology and periods of sensitivity. *Teratology* **46,** 117–130.

Embury, S., Seller, M. J., Adinolfi, M., and Polani, P. E. (1979). Neural tube defects in curly-tail mice. I. Incidence, expression and similarity to the human condition. *Proc. Roy. Soc. Lond. Ser. B* **206,** 85–94.

Epstein, D. J., Vekemans, M., and Gros, P. (1991). Splotch (Sp2H), a mutation affecting the development of the mouse neural tube, shows a deletion within the paired homeodomain of Pax-3. *Cell* **67,** 767–774.

Eriksson, U. J., Dahlstrom, E., and Hellerstrom, C. (1983). Diabetes in pregnancy. *Diabetes* **32,** 1141–1145.

Fabregues, I., Ferrer, I., Gairi, J. M., Cahuana, A., and Giner, P. (1985). Effects of prenatal exposure to ethanol on the maturation of the pyramidal neurons in the cerebral cortex of the guinea pig: quantitative Golgi study. *Neuropathol. Appl. Neurobiol.* **11,** 291–298.

Farr, K. L., Montano, C. Y., Paxton, L. L., and Savage, D. D. (1988). Prenatal ethanol exposure decreases hippocampal [3H]-glutamate binding in 45 day old rats. *Alcohol* **5,** 125–133.

Finnell, R. H. (1991). Genetic differences in susceptibility to anticonvulsant drug–induced developmental defects. *Pharmacol. Toxicol.* **69,** 1–5.

Finnell, R. H., Bennett, G. D., Karras, S. B., and Mohl, V. K. (1988). Common hierarchies of susceptibility to the induction of neural tube defects by valproic acid and its 4-propyl-4-pentenoic acid metabolite. *Teratology* **38,** 313–320.

Finnell, R. H., and Chernoff, G. F. (1987). Gene–teratogen interactions: an approach to understanding the metabolic basis of birth defects. In *Drug Disposition in Teratogenesis.* Nau, H., and Scott, W. J., Eds., Vol. II. CRC Press, Boca Raton, Florida, pp. 97–112.

Fitzgerald, P. (1988) FAS persists despite broad public awareness. *Mich. Med.* **87,** 262–264.

Fraser, F. C., and Skelton, J. (1978). Possible teratogenicity of maternal fever. *Lancet* **2,** 634.

Geelen, J. A., and Langman, J. (1977). Closure of the neural tube in the cephalic region of the mouse embryo. *Anat. Rec.* **189,** 635–640.

Germain, M. A., Webster, W. S., and Edwards, M. J. (1985). Hyperthermia as a teratogen: parameters determining hyperthermia-induced head defects in the rat. *Teratology* **31,** 265–272.

Golden, J. A., and Chernoff, G. F. (1993). Intermittent pattern of neural tube closure in two strains of mice. *Teratology* **47,** 73–80.

Goulding, M. D., Chalepakis, G., Deutsch, U. Erselius, J. R., and Gruss, P. (1991). Pax-3, a novel murine DNA binding protein expressed during early neurogenesis. *EMBO J.* **10,** 1135–1147.

Goulding, M., and Paquette, A. (1994). Pax genes and neural tube defects in the mouse. *Ciba Found. Symp.* **181,** 103–113.

Goulding, M., Sterrer, S., Fleming, J., Balling, R., Nadeau, J., Moore, K. J., Brown, S. D., Steelm, K. P., and Gruss, P. (1993). Analysis of the *pax 3* gene in the mouse mutant *splotch. Genomics* **17,** 355–363.

Gunn, T. M., Juriloff, D. M., and Harris, M. J. (1992). Further genetic studies of the cause of exencephaly in SELH mice. *Teratology* **45,** 679–686.

Gunn, T. M., Juriloff, D. M., and Harris, M. J. (1995). Genetically determined absence of an initiation site of cranial neural tube closure is causally related to exencephaly in SELH/Bc mouse embryos. *Teratology* **52,** 101–108.

Gunn, T. M., Juriloff, D. M., Vogel, W., Harris, M. J., and Miller, J. E. (1993). Histological study of the cranial neural folds of mice genetically liable to exencephaly. *Teratology* **48,** 459–471.

Hackett, P. L., and Kelman, B. J. (1983). Availability of toxic trace metals to the conceptus. *Sci. Total Environ.* **28,** 433–442.

Halperin, L. R., and Wilroy, R. S. Jr. (1978). Maternal hyperthermia and neural-tube defects. *Lancet* **2,** 212–213.

Hanlon, D. P., and Ferm, V. H. (1977). Placental permeability of arsenate ion during early embryogenesis in the hamster. *Experientia* **33,** 1221–1222.

Harrison, W. P., Frazier, J. C., Mazzanti, E. M., and Good, R. D. (1980). Teratogenicity of disodium methanearsenate and sodium dimethylarsinate (sodium cacodylate) in mice. *Teratology* **21,** 43A.

Hartwell, L. (1992). Defects in a cell cycle checkpoint may be responsible for the genomic instability of cancer cells. *Cell* **71,** 543–546.

Hartwig, J. H., Thelen, M., Rosen, A., Janmey, P. A., Nairn, A. C., and Aderern, A. (1992). MARCKS is an actin filament crosslinking protein regulated by protein kinase C and calcium–calmodulin. *Nature* **356,** 618–622.

Hendrickx, A. G., Stone, G. W., Henrickson, R. V., and Matayoshi, K. (1979). Teratogenic effects of hyperthermia in the bonnet monkey. *Teratology* **19,** 177–182.

His, W. (1874) Unsere Korperform und des Physiologische Problem ihrer Entstehung Leipzig. Cited by Holtfreter, J. (1943). *J. Exp. Zool.* **94,** 261–318.

Hoever, M., Clement, J. H., Wedlich, D., Montenarh, M., and Knochel W. (1994). Overexpression of wild type *p53* interferes with normal development in *Xenopus laevis* embryos. *Oncogene* **9,** 109–120.

Hood, R. D., Vegel, G. C., Zaworotko, M. J., Tatum, F. M., and Meeks, R. G. (1988). Uptake, distribution, and metabolism of trivalent arsenic in the pregnant mouse. *J. Toxicol. Env. Health* **25,** 423–434.

Horii, K., Watanabe, G., and Ingalls, H. (1966). Experimental diabetes in pregnant mice. Prevention of congenital malformations in offspring by insulin. *Diabetes* **15,** 194–204.

Hoyseth, K. S., and Jones, P. J. (1989). Mini-review: ethanol induced teratogenesis: characterization, mechanism and diagnostic approaches. *Life Sci.* **44,** 643–649.

Ishibashi, M., Ang, S., Shiota, K., Nakanishi, S., Kageyama, R., and Guillemot, F. (1995). Targeted disruption of mammalian hairy and enhancer of split homolog-1 (HES-1) leads to up-regulation of neural helix-loop-helix factors, premature neurogenesis, and severe neural tube defects. *Genes Dev.* **9,** 3136–3148.

Jacobsen, S., Rich, J., and Tovsky, N. J. (1979). Delayed myelination and lamination in the cerebral cortex of the albino rat as a result of fetal alcohol syndrome. *Curr. Alcohol* **6,** 123–133.

Jäger-Roman, E., Deichl, A., Jakob, S., Hartmann, A. M., Koch, S., Rating, D., Steldinger, R., Nau, H., and Helge, H. (1986). Fetal growth, major malformations, and minor anomalies in infants born to women receiving valproic acid. *J. Pediatr.* **108,** 997–1004.

Jelinek, R., and Friebova, Z. (1966). Influence of mitotic activity on neurulation movements. *Nature* **209,** 822–823.

Jones, K. L., Smith, D. W., Ulleland, C. N., and Streissguth, A. P. (1973). Patterns of malformations in offspring of chronic alcoholic mothers. *Lancet* **1,** 1267–1271.

Juriloff, D. M., Harris, M. J., Tom, C., and MacDonald, K. B. (1991). Normal mouse strains differ in the site of initiation of closure of the cranial neural tube. *Teratology* **44,** 224–233.

Juriloff, D. M., MacDonald, K. B., and Harris, M. J. (1989). Genetic analysis of the cause of exencephaly in the SELH/Bc mouse stock. *Teratology* **40,** 395–405.

Kao, J., Brown, N. A., Schmidt, B., Goulding, E. H., and Fabro, S. (1981). Teratogenicity of valproic acid: *in vivo* and *in vitro* systems. *Teratog. Carcinog. Mutagen* **1,** 367–382.

Karfunkel, P. (1974). The mechanisms of neural tube formulation. *Int. Rev. Cytol.* **38,** 245–271.

Kaufman, M. H. (1979). Cephalic neurulation and optic vessel formation in the early mouse embryo. *Am. J. Anat.* **155,** 425–444.

Kilham, L., and Ferm, V. H. (1976). Exencephaly in fetal hamsters following exposure to hyperthermia. *Teratology* **14,** 323–326.

Kimmel, G. L., Cuff, J. M., Kimmel, C. A., Heredia, D. J., Tutor, N., and Silverman, P. M. (1993). Embryonic development *in vitro* following short-duration exposure to heat. *Teratology* **47,** 243–251.

Kinter, C. R., and Melton, D. M. (1987). Expression of *Xenopus* N-CAM RNA is an early response of ectoderm to induction. *Development* **99,** 311–325.

Kochhar, D. M. (1967). Teratogenic activity of retinoic acid. *Acta Pathol. Microbiol. Immunol. Scand.* **70,** 398–404.

Korsmeyer, S. J. (1992). Bcl-2 initiates a new category of oncogenes; regulators in human B lymphoblastoid cell line. *Oncogene* **4,** 1331–1336.

Ladye, P. M., Edmonds, L. D., and Erickson, J. D. (1980). Maternal fever and neural tube defects. *Teratology* **21,** 105–108.

Langman, J. (1981). *Medical Embryology,* 4th Ed. Williams & Wilkins Co., Baltimore.

Langman, J., and Welch, G. (1967). Excess vitamin A on development of the cerebral cortex. *J. Comp. Neurol.* **128,** 1–15.

Larsen, C. J. (1994). The BCL2 gene prototype of a gene family that controls programmed cell death (apoptosis). *Ann. Genet.* **37,** 121–134.

Lary, J. M., Conover, D. L., Foley, E. D., and Hanser, P. L. (1982). Teratogenic effects of 27.12MHz radiofrequency radiation in rats. *Teratology* **26,** 299–309.

Ledig, M., Ciesielski, L., Simler, S., Lorentz, J. G., and Mandel, P. (1988). Effect of pre- and postnatal alcohol consumption on GABA levels of various brain regions in the rat offspring. *Alcohol Alcoholism* **23,** 63–67.

Lee, H., and Nagele, R. G. (1978). Inhibition of neural tube closure by ionophore A23187 in chick embryos. *Experientia* **34,** 518–520.

Lee, H., and Nagele, R. G. (1979). Neural tube closure defects caused by papaverine in explanted early chick embryos. *Teratology* **20,** 321–332.

Lee, H., and Nagele, R. G. (1986). Toxic and teratologic effects of verapamil on early chick embryos: evidence for the involvement of calcium in neural tube closure. *Teratology* **33,** 203–211.

Lemire, R. J. (1969). Variations in development of the caudal neural tube in human embryos. *Teratology* **2,** 361–370.

Lemire, R. J., Beckwith, J. B., and Warkary, J. (1978). *Anencephaly.* Raven Press, New York.

Lindgren, A., Danielsson, B. R., Dencker, L. and Vahter, M. (1984) Embryotoxicity of arsenite and arsenate: distribution in pregnant mice and monkeys and effects on embryonic cells *in vitro. Acta Pharmacol. Toxicol.* **54,** 311–320.

Lindhout, D., and Schmidt, D. (1986). *In-utero* exposure to valproate and neural tube defects. *Lancet* **1,** 1392–1393.

MacDonald, K. B., Juriloff, D. M., and Harris, M. J. (1989). Developmental study of neural tube closure in a mouse stock with high incidence of exencephaly. *Teratology* **39,** 195–213.

Marquis, J. K. (1978). Analysis of the nerve-blocking action of mercurochrome, a fluorescent thiol reagent. *Neuropharmacology* **17,** 631–635.

Marsh, D. O., Myers, G. J., Clarkson, T. W., Amin-Zaki, L., Tikriti, S., and Majeed, M. A. (1980). Fetal methylmercury poisoning: clinical and toxicological data on twenty nine cases. *Ann. Neurol.* **7,** 348.

McDonnell, T. J., Deane, N., Platt, F. M., Nunez, G., Jaeger, U., McKearn, J. P., and Korsmeyer, S. J. (1989). bcl-2 immunogloblin transgenic mice demonstrate extended B cell survival and follicular lymphoproliferation. *Cell* **57,** 79–88.

McLone, D. G., and Knepper, P. A. (1986). Role of complex carbohydrates and neurulation. *Pediatr. Neurosci.* **12,** 2–9.

Meyer, L. S., Kotch, L. E., and Riley, E. P. (1990). Neonatal ethanol exposure: functional alterations associated with cerebellar growth retardation. *Neurotoxicol. Teratol* **12,** 15–22.

Miller, M. T. (1991). Thalidomide embryopathy: a model for the study of congenital incomitant horizontal strabismus. *Trans. Am. Ophthalmol. Soc.* **89,** 623–674.

Miller, M. T., and Stromland, S. (1993) Thalidomide embryopathy: an insight into autism? *Teratology* **47,** 387–388.

Miller, M. W. (1986). Effects of alcohol on the generation and migration of cerebral cortical neurons. *Science* **233,** 1308–1311.

Miller, P., Smith, D. W., and Shepard, T. H. (1978). Maternal hyperthermia as a possible cause of anencephaly. *Lancet* **1,** 519–521.

Milner, R. J., Randolph, L., Bahr, D., Cappello, M., Lenoir, D., Miller, F., and Bloom, F. E. (1987). Molecular biological approaches to the brain and their application to the study of alcoholism. In *Genetics and Alcoholism.* H. W. Goedde and D. P. Agaral, Eds. Alan R. Liss, Inc., New York, pp. 291–302.

Milunsky, A., Jink, H., and Jick, S. S. (1989). Multivitamin/folic acid supplementation in early pregnancy reduces the prevalence of neural tube defects. *JAMA* **262,** 2847–2852.

Mirkes, P. E., and Cornel, L. (1992). A comparison of sodium arsenite- and hyperthermia-induced stress responses and abnormal development in cultured postimplantation rat embryos. *Teratology* **46,** 251–259.

Mitolo-Chieppa, D. (1981). Effects of methylmercury on the electric activity of the node of Ranvier. *Clin. Toxicol.* **18,** 1319–1325.

Miyashita, T., and Reed, J. C. (1993). Bcl-2 oncoprotein blocks chemotherapy-induced apoptosis in a human leukemia cell line. *Blood* **87,** 151–157.

MMWR. (1982). Valproic acid and spina bifida: a preliminary report—France. *Morbidity and Mortality Weekly Report* **31,** 565–566.

Moase, C. E., and Trasler, D. G. (1992). Splotch locus mouse mutants: models for neural tube defects and Waardenburg syndrome type I in humans. *J. Med. Genet.* **29,** 145–151.

Moran, D., and Rice, R. W. (1976). Action of papaverine and ionophore A23187 on neuralation. *Nature* **261,** 497–499.

Morris, G. M. (1972). Morphogenesis of the malformations induced in rat embryos by maternal hypervitaminosis A. *J. Anat.* **113,** 241–250.

Morriss-Kay, G. M., Tuckett, F., and Solursh M. (1986) The effects of *Streptomyces* hyaluronidase on tissue organization and cell cycle time in rat embryos. *J. Embryol. Exp. Morphol.* **98,** 59–70.

Mottet, N. K., Shaw, C. M., and Burbacher, T. M. (1985). Health risks from increases in methylmercury exposure. *Environ. Health. Perspect.* **63,** 133–40.

Murakami, U., Kameyama, Y., and Kato, T. (1954). Basic processes seen in disturbances of early development of the central nervous system. *Nagoya J. Med. Sci.* **17,** 74–84.

Nagao, T., and Fujikawa, K. (1990). Genotoxic potency in mouse spermatogonial stem cells of triethylenemelamine, mitomycin C, ethylnitrosourea, procarbazine, and propyl methanesulfonate as measured by F1 congenital defects. *Mutation Res.* **229,** 123–128.

Naruse, I., Arakawa, H. and, Fukui, Y. (1993). Effects of methylmercury on the brain of infant rats reared artificially. *Tokushima J. Exp. Med.* **40,** 69–74.

Naruse, I., Collins, M. D., and Scott, W. D. Jr. (1988). Strain differences in the teratogenicity induced by sodium valproate in cultured mouse embryos. *Teratology* **38,** 87–96.

Nau, H., and Loscher, W. (1984). Valproic acid and metabolites: pharmacological and toxicological studies. *Epilepsia* **25** (suppl 1), S14–S22.

Nau, H., Zierer, R., Spielmann, H., Neubert, D., and Gansau, C. (1981). A new model for embryotoxicity testing: teratogenicity and pharmacokinetics of valproic acid following constant-rate administration in the mouse using human therapeutic drug and metabolite concentrations. *Life Sci.* **29,** 2803–2814.

Naus, C. C. G., and Bechberger, J. F. (1991). Effects of prenatal ethanol exposure on postnatal neural gene expression in the rat. *Develop. Genet.* **12,** 293–298.

Newland, M. C., Yezhou, S., Logdberg, B., and Berlin, M. (1994). Prolonged behavioral effects of *in utero* exposure to lead or methyl mercury: recuded sensitivity to changes in reinforcement contingencies during behavioral transitions and in steady state. *Toxicol. Appl. Pharmacol.* **126,** 6–15.

O'Shea, K. S., and Liu, L. H. (1987). Basal lamina and extracellular matrix alterations in the caudal neural tune of the delayed Splotch embryo. *Brain Res.* **465,** 11–20.

Otani, H., Tanaka, O., Tatewaki, R., Naora, H., and Yoneyama, T. (1991). Diabetic environment and genetic predisposition as causes of congenital malformations in NOD mouse embryos. *Diabetes* **40,** 1245–1250.

Paulson, R. B., Sucheston, M. E., Hayes, T. G., Paulson, G. W. (1985). Teratogenic effects of valproate in the CD-1 mouse fetus. *Arch. Neurol.* **42,** 980–983.

Pennington, S. N. (1990). Molecular changes associated with ethanol induced growth suppression in the chick embryo. *Alcohol Clin. Exp. Res.* **14,** 832–837.

Peterson, N. S. (1990). Effects of heat and chemical stress on development. *Adv. Genet.* **28,** 275–295.

Pillans, P. I., Folb, P. I., and Ponzi, S. F. (1988). The effects of *in vivo* administration of teratogenic doses of vitamin A during the preimplantation period in the mouse. *Teratology* **37,** 7–11.

Pleet, H., Graham, J. M. Jr., and Smith, D. W. (1981). Central nervous system and facial defects associated with maternal hyperthermia at four to 14 weeks' gestation. *Pediatrics* **67,** 785–789.

Prieto, A. L., Crossin, K. L., Cunningham, B. A., and Edelman, G. M. (1989). Localization of mRNA for neural cell adhesion molecule (N-CAM) polypeptides in neural and nonneural tissues by *in situ* hybridization. *Proc. Natl. Acad. Sci. USA* **86,** 9579–9583.

Rawat, A. K. (1980). Biochemical aspect of neuroteratogenic effects of alcohol. *Neurobehav. Toxicol.* **2,** 259–265.

Reynolds, J. D., and Brien, J. F. (1995). Ethanol neurobehavioural teratogenesis and the role of L-glutamate in the fetal hippocampus. *Can. J. Physiol. Pharmacol.* **73,** 1209–1223.

Roca, J. B., Freitas, A. J., Marques, M. B., Pereira, M. E., Emanuelli, T., and Souza, D.O. (1993). Effects of methylmercury exposure during the second stage of rapid postnatal brain growth on negative geotaxis and on delta-aminolevulinate dehydratase of suckling rats. *Braz. J. Med. Biol. Res.* **26,** 1077–1083.

Rodier, P. M., Aschner, M., and Sager, P. R. (1984). Mitotic arrest in the developing CNS after prenatal exposure to methylmercury. *Neurobehav. Toxicol. Teratol.* **5,** 379–385.

Rodier, P. M., Ingram, J. I., Tisdale, B., Nelson, S., and Romano, J. (1996). Embryological origin for autism: developmental anomalies of the cranial nerve motor nuclei. *J. Comp. Neurol.* **370,** 247–261.

Russell, L. B. (1950). X-ray induced developmental abnormalities in the mouse and their use in the analysis of embryological patterns. I. External and gross visceral changes. *J. Exp. Zool.* **114,** 545–601.

Russell, L. B., and Russell, W. L. (1954). An analysis of the changing radiation response of the developing mouse embryo. *J. Cell Comp. Physiol.* **43,** 103–147.

Sadler, T. W., Burridge, K., and Yonker, T. (1986). A potential role for spectrin during neurulation. *J. Embryol. Exp. Morphol.* **94,** 73–82.

Sadler, T. W., Greenberg, D., Coughlin, P., and Lessard, J. L. (1982). Actin distribution patterns in the mouse neural tube during neurulation. *Science* **215,** 172–174.

Sager, P. R., Doherty, R. A., and Rodier, P. M. (1982). Effects of methylmercury on developing cerebellar cortex. *Exper. Neurol.* **77,** 179–193.

Sakai, Y. (1989). Neurulation in the mouse: manner and timing of neural tube closure. *Anat. Rec.* **223,** 194–203.

Sasai, Y., Kageyama, R., Tagawa, Y., Shigemoto, R., and Nakanishi, S. (1992). Two mammalian helix-loop-helix factors structurally related to *Drosophila* hairy and enhancer of split. *Genes. Dev.* **6,** 2620–2634.

Schoenwolf, G. C. (1984). Histological and ultrastructural studies of secondary neurulation of mouse embryos. *Am. J. Anat.* **169,** 361–374.

Seller, M. J. (1987). Neural tube defects and sex ratios. *Am. J. Med. Genet.* **26,** 699–707.

Seller, M. J., and Adinolfi, M. (1981). The curly-tail mouse: an experimental model for human neural tube defects. *Life Sci.* **29,** 1607–1615.

Shalat, S. L., Walker, D. B., and Finnell, R. H. (1996). Role of arsenic as a reproductive toxin with particular attention to neural tube defects. *J. Toxicol. Environ. Health* **48,** 253–272.

Shaw, P., Bovey, R., Tardy, S., Sahli, R., Sordat, B., and Costa, J. (1992). Induction of apoptosis by wild type p53 in a human colon tumor-derived cell line. *Proc. Natl. Acad. Sci. USA* **89,** 4495–4499.

Shiota, K., and Nishimura, H. (1982). Teratogenicity of di(2-ethylhexyl)phthalate (DEHP) and di-*n*-butylphthalate (DBHP) in mice. *Environ. Health Perspect.* **45,** 65–70.

Solursh, M., and Morriss, G. M. (1977). Glycosaminoglycan synthesis in rat embryos during the formation of the primary mesenchyme and neural folds. *Dev. Biol.* **57,** 75–86.

Spohr, H. L., Williams, J., and Steinhausen, H. C. (1993). Prenatal alcohol exposure and long term developmental consequences. *Lancet* **341,** 907–910.

Steele, C. E., Trasler, D. G., and New, D. A. T. (1983). An *in vivo/vitro* evaluation of the teratogenic action of excess vitamin A. *Teratology* **28,** 209–214.

Stevenson, R. E., Kelly, J. C., Aylsworth, A. S., and Phelan, M. C. (1987). Vascular basis for neural tube defects. *Proceedings of the Greenwood Genetic Center* **6,** 109–111.

Streissguth, A. P., Aase, J. M., Clarren, S. K., Randels, S. P., LaDue, R. A., and Smith, D. F. (1991). Fetal alcohol syndrome in adolescents and adults. *JAMA* **265,** 1961–1967.

Sulik, K. K., Cook, C. S., and Webster, W. S. (1988). Teratogens and craniofacial malformations: relationships to cell death. *Development* **103**(suppl), 213–232.

Sulik, K. K., and Dehart, D. B. (1988). Retinoic acid induced limb malformations resulting from apical ectodermal ridge cell death. *Teratology* **37,** 527–537.

Sulik, K. K., and Sadler, T. W. (1993). Postulated mechanisms underlying the development of neural tube defects: insights from *in vitro* and *in vivo* studies. *Ann. NY Acad. Sci.* **678,** 8–21.

Thiery, J. P., Duband, J. L., and Rutishauser, U. (1982). Cell adhesion molecules in early chick embryogenesis. *Proc. Natl. Acad. Sci. USA* **79,** 6737–6741.

Tibbles, L., and Wiley, M. J. (1988). A comparative study of the effects of retinoic acid given during the critical period for inducing spina bifida in mice and hamsters. *Teratology* **37,** 113–125.

Tom, C., Juriloff, D. M., and Harris, M. J. (1991). Studies of the effect of retinoic acid on anterior neural tube closure in mice genetically liable to exencephaly. *Teratology* **43,** 27–40.

Trasler, D. G. (1965). Aspirin induced cleft lip and other malformations in mice. *Lancet* **1,** 606–607.

Trasler, D. G., and Morriss-Kay, G. (1991). Immunohistochemical localization of chrondroitin and heparin sulfate proteoglycans in pre-spina bifida splotch embryos. *Tetratology* **44,** 571–579.

Tremblay, P., Pituello, F., and Gruss, P. (1996). Inhibition of floor plate differentiation by *Pax 3:* evidence from ectopic expression in transgenic mice. *Development* **122,** 2555–2567.

Tsuijimoto, Y. (1989). Stress-resistance conferred by high levels of bcl-2 protein in human B lymphoblastoid cell line. *Oncogene* **4,** 1331–1336.

Tuchmann-Du Plessis, H. (1975). Drug effects on the fetus. In *Monographs on Drugs.* H. Tuchmann-Du Plessis, Ed. Vol. II. Adis Press, London, p. 40.

Turner, S., Sucheston M. E., De Phillip, R. M., and Paulson, R. B. (1990). Teratogenic effects on the neuroepithelium of the CD-1 mouse embryo exposed *in utero* to sodium valproate. *Teratology* **41,** 421–442.

Underhill, D. A., Vogan, K. J., and Gros, P. (1995). Analysis of the mouse splotch-delayed mutation indicates that *Pax-3* paired domain can influence homeodomain DNA binding activity. *Proc. Natl. Acad. Sci. USA* **92,** 3692–3696.

Vacha, S. J., Bennett, G. D., Mackler, S. A., Koebbe, M. J., and Finnell, R. H. (1997). Identification of a growth arrest specific (gas-5) gene by differential display as a candidate gene for determining susceptibility to hyperthermia-induced exencephaly in mice. *Dev. Genet.* **21,** 212–222.

Van Allen, M. I., Kalousek, D. K., Chernoff, G. F., Juriloff, D., Harris, M., McGillivray, B. C., Yong, S. L., Langlois, S., MacLeod, P. M., and Chitayat, D. (1993). Evidence for multisite closure of the neural tube in humans. *Am. J. Med. Genet.* **47,** 723–743.

Vogel, F. S. (1961). The anatomic character of the vascular anomalies associated with anencephaly: with consideration of the role of abnormal angiogenesis in the pathogenesis of the cerebral malformations. *Am. J. Pathol.* **39,** 163–169.

Vogelstein, B., and Kinzler, K. W. (1992). p53 function and dysfunction. *Cell* **70,** 523–526.

Vollset, S. E. (1990). Ovulation induction and neural tube defects. *Lancet* **335,** 178.

Vorhees, C. V. (1987). Behavioral teratogenicity of valproic acid: selective effects on behavior after prenatal exposure to rats. *Psychopharmacology* **92,** 173–179.

Walsh, D., Li, K., Wass, J., Dolnikov, A., Zeng, F., Zhe, L., and Edwards, M. (1993). Heat-shock gene expression and cell cycle changes during mammalian embryonic development. *Dev. Genet.* **14,** 127–136.

Walsh, D. A., and Morris, V. B. (1989). Heat shock affects cell cycling in the neural plate of cultured rat embryos: a flow cytometric study. *Teratology* **40,** 583–592.

Webster, W. S. (1989). Alcohol as a teratogen: a teratological perspective of the fetal alcohol syndrome. In *Human Metabolism of Alcohol* K. E. Crow and R. D. Batt, Eds. Vol. I. CRC Press, Boca Raton, Florida, pp. 135–155.

West, J. R., Hamre, K. M., and Cassell, M. D. (1986). Effects of ethanol exposure during the third trimester equivalent on neuron number in rat hippocampus and dentate gyrus. *Alcohol Clin. Exp. Res.* **10,** 190–197.

Willhite, C. C., and Ferm, V. H. (1984). Prenatal and developmental toxicology of arsenicals. *Adv. Exp. Med. Biol.* **177,** 205–228.

Wilson, D. B. (1974). Proliferation in the neural tube of the splotch (Sp) mutant mouse. *J. Comp. Neurol.* **154,** 249–256.

Wilson, D. B., and Center, E. M. (1974). Neural cell cycle in the Looptail mutant of the mouse (Lp). *J. Embryol. Exp. Morphol.* **32,** 697–705.

Wilson, D. B., and Finta, L. A. (1980). Early development of the brain and spinal cord in dysraphic mice: a transmission electron microscopic study. *J Comp Neurol* **190,** 363–371.

Wlodarczyk, B., Bennett, G. D., Calvin, J. A., Craig, J. C., and Finnell, R. H. (1996a). Arsenic-induced alterations in embryonic transcription factor gene expression: implications for abnormal neural development. *Dev. Genet.* **18,** 306–315.

Wlodarczyk, B., Bennett, G. D., Calvin, J. A., and Finnell, R.H. (1996b). Arsenic-induced neural tube defects in mice: alterations in cell cycle gene expression. *Reprod. Toxicol.* **10,** 447–454.

Yasuda, Y., Okamoto, M., Konishi, H., Matsuo, T., Kihara, T., and Tanimura, T. (1986). Developmental anomalies induced by all *trans*-retinoic acid in fetal mice: I. Macroscopic findings. *Teratology* **34,** 37–49.

Zhao, Q., Behringer, R. R., and de Crombrugghe, B. (1996). Prenatal folic acid treatment suppresses acrania and meroanencephaly in mice mutant for the Cart1 homeobox gene. *Nature Gen.* **13,** 275–283.

Zierler, S., Theodore, M., Cohen, A., and Rothman, K. J. (1988). Chemical quality of maternal drinking water and congenital heart disease. *Int. J. Epidemiol.* **17,** 589–594.

CHAPTER

10

Transgenic Animal Models

Functional Analysis of Developmental Toxicity as Illustrated with the p53 Suppressor Model

THOMAS B. KNUDSEN
JUDITH A. WUBAH
Department of Pathology, Anatomy, and Cell Biology
Jefferson Medical College
Philadelphia, Pennsylvania 19107

*Abbreviations: 2CdA, 2-chloro-2′-deoxyadenosine; CAT, chloramphenicol acetyl transferase; Gy, grays; HLH, helix-loop-helix; IRBP, interstitial retinol-binding protein; kDa, kilodaltons; *lacZ*, β-galactosidase; MMTV LTR, mouse mammary tumor virus long terminal repeat; mtDNA, mitochondrial DNA; NGF, nerve growth factor; pc, postcoitum; *Sina*, *Seven-in-Absentia;* TCDD, 2,3,7,8-tetrachlorodibenzo-*p*-dioxin; XPC, *Xeroderma pigmentosum.*

I. Introduction

The potential adverse effects that result from exposure to environmental factors prior to conception or birth is an important public health issue (Sever *et al.*, 1993). To identify causes and preventive strategies of chemically induced malformations in humans, research must elucidate the mechanisms by which the environment can affect genome-directed events in the embryo. In particular, studies are needed to characterize the nature of the cell signals evoked by developmental toxicants, the critical environmentally responsive genes that transduce cellular changes in the embryo, and the biophysical mechanisms by which these cell-signaling pathways interact with normal pathways in development.

Cell-signaling pathways are woven into complex webs. Through functional redundancy, a cell may select between alternative pathways that compensate for the loss or gain of another gene's function. This makes it hard to predict, from knowledge of gene expression patterns alone, the impact that a specific genetic change will have on development. Transgenic animal models provide a powerful approach to dissect these mechanisms, permitting introduction or ablation of a specific gene function and the subsequent analysis of phenotype under conditions that maintain *in vivo* selective pressures of cell growth and development (Goldsworthy *et al.*, 1994). Their application to developmental neurotoxicology provides a conceptual framework onto which other working models can be devised for experimentation into causes and mechanisms of developmental toxicity in general and neurotoxicology in particular.

This chapter illustrates various principles in the application of transgenic animal models to problems in developmental toxicology. Specifically, the chapter focuses on lessons learned from *in vivo* analysis of the tumor suppressor gene *p53* during embryonic growth and differentiation. The *p53* gene encodes an environmentally responsive DNA-binding protein that invokes critical cellular responses to appropriate inductive signals. One of the most pervasive concepts pertaining to this pathway is "guardian of the genome" (Lane, 1992), which ensures that DNA replication, cell differentiation, and perhaps other genome-directed activities are completed with high fidelity.

Recently, *p53* has been nicknamed "guardian of the babies" in reference to an embryogenic role (Hall and Lane, 1997). This could have widespread implications on public health, given that 120,000–160,000 babies (3–4%) in the United States are born each year with a major birth defect and that causes remain unknown for the majority of cases (Sever *et al.*, 1993). Because this chapter focuses on the embryogenic functions of p53, and in particular how transgenic animal models have contributed to this research, the reader is referred to more comprehensive reviews on tumor suppressor *p53* (Donehower and Bradley, 1993; Hooper, 1994; Rotter *et al.*, 1994; Hall *et al.*, 1996; Harris, 1996; Ko and Prives, 1996; Hansen and Oren, 1997; Hall and Lane, 1997).

II. Overview of the *p53* Pathway

A. *Structure and Function of p53*

The *p53* gene contains 11 exons spanning about 20 kilobases of the genome. It resides on the short arm of human chromosome 17 (17p13) and mouse chromosome 11 (Donehower and Bradley, 1993). The gene encodes a 53 kDa protein that oligomerizes to a tetramer. The canonical *p53* homolog has been found in and among the genomes of fish, amphibians, avians, and mammals, but not of yeast, insects, or nematodes (Soussi and May, 1996). This ancestral pattern suggests that p53 activity is adapted for cell-signaling pathways in the *Vertebrata*. Loss of p53 activity leads to tumor formation as indicated by the detection of mutant p53 proteins in more than half of human tumor cases (Hollstein *et al.*, 1991) and of strong cancer predisposition in *p53*-deficient nullizygous mice (Donehower *et al.*, 1992). The vast majority of *p53*'s oncogenic mutations occur in four conserved DNA-binding regions in the hydrophobic core (Fig. 1), strongly implicating the importance of genome-based effects of p53 protein activation in suppressor functions (Cho *et al.*, 1994; Harris, 1996).

Under normal conditions the wild-type p53 protein resides in the cytoplasm. With a cellular half-life on the order of 20 min (Rogel *et al.*, 1985), this latent form of the protein is inactive with respect to sequence-specific DNA-binding activity (Ko and Prives *et al.*, 1996). The active conformation is induced by intracellular signals that lead to phosphorylation, protein stabilization, and nuclear translocation (Fig. 2). The consequence of nuclear p53 accumulation is sequence-specific DNA binding and transactivation of dozens of genes that contain a consensus p53 response element (Bourdon *et al.*, 1997). Among the most potent inductive stimulus of nuclear p53 protein accumulation is DNA damage, caused by either irradiation of cells (Kastan *et al.*, 1991; Kuerbitz *et al.*, 1992, Merritt *et al.*, 1994) or by exposure to genotoxic chemicals (Fritsche *et al.*, 1993; Lowe *et al.*, 1993b; Nelson and Kastan, 1994). Sublethal metabolic imbalances are also sufficient to induce p53 protein activity and cellular changes. This is demonstrated by studies with hypoxia (Graeber *et al.*, 1994, 1996) or ribonucleotide imbalance (Linke *et al.*, 1996; Liu *et al.*, 1998) as stress inducers. The biochemical nature of "sensor" signals evoked by these diverse inductive stimuli are not known.

Two cellular effects are classically associated with p53 protein induction: arrest of cell cycle progression and programmed cell death (apoptosis) (Kastan *et al.*, 1991; Clarke *et al.*, 1993; Debbas and White, 1993; El-Diery *et al.*, 1993; Harper *et al.*, 1993; Lowe *et al.*, 1993a; Yonish-Rouach *et al.*, 1993; Merritt *et al.*, 1994; Sabbatini *et al.*, 1995). Although the precise mechanisms for these cellular changes are not known, each effect maps to different DNA-binding regions of the p53 protein (Hansen and Braithwaite, 1996) and is elicited by different quantitative levels of p53 protein induction. For example, as a "regulator" of the cellular life cycle, p53 invokes behaviors that are graded by the level of intracellular p53 protein activity: low-grade p53 induction preferentially affects cell growth and differentiation, whereas high-grade p53 induction can initiate an apoptotic trigger (Chen *et al.* 1996; Eizenberg *et al.*, 1996). Thus, the biologic consequences of acute (stress-activated) and chronic (baseline) p53 responses depend on the nature and extent of the activating stimulus.

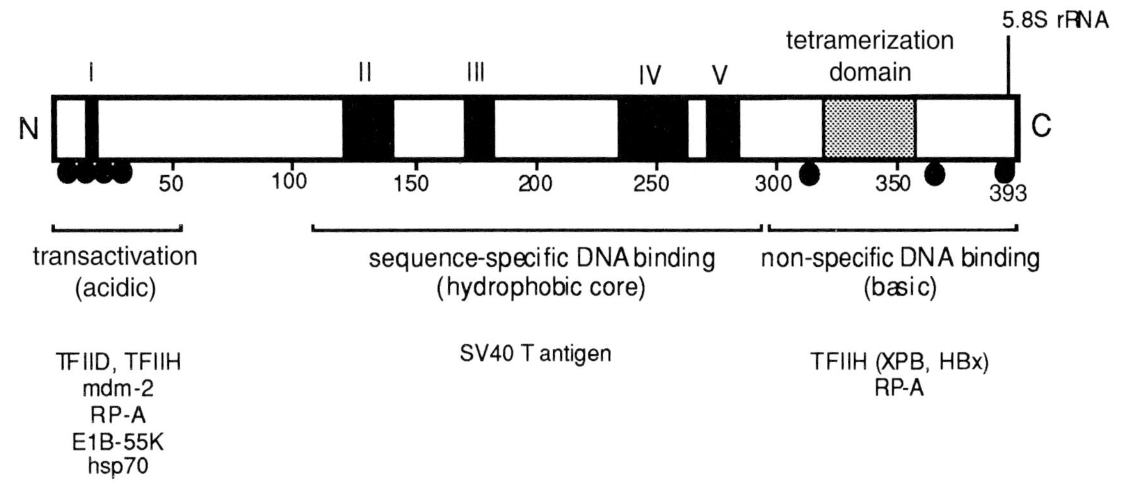

FIGURE 1 Human *p53* protein map. Three regions of the protein are indicated: N-terminal acidic domain, containing regions for protein–protein interations with replication protein A (RP-A) and heat shock protein 70 (hsp70); hydrophobic core domain involved in sequence-specific DNA binding; and C-terminal basic domain containing nuclear localization sequences, p53–p53 oligomerization domain, and an attachment site for ribosomal 5.8S rRNA. Potential phosphorylation sites are indicated by small spheres; boxes I through V indicate five highly conserved regions involved in DNA binding.

B. Role of p53 in Cell Differentiation and Apoptosis

The presence of a helix-loop-helix (HLH) binding site in the *p53* gene promoter operationally defines this suppressor gene as a target for the family of basic HLH transcription factors that regulate specific cell differentiation pathways (Ronen *et al.,* 1991). Differential expression of *p53* transcripts is therefore "hard-wired" into the ancestry of specific cell lineages in vertebrate embryos. *A priori,* status of the p53 response is likely to be subjected to programmed (genetic) as well as inducible (environmental) signals during embryo development.

Some of the earliest data concerning differential regulation of wild-type *p53* gene expression and protein subcellular distribution were obtained with teratocarcinoma cells. These pleuripotent cell lines, which resemble the inner cell mass of preimplantation mammalian embryos, accumulated p53 to levels characteristic of some tumors bearing tonically activated, and perhaps dysfunctional, mutant p53 proteins (Linzer and Levine, 1979; Oren *et al.,* 1982). In teratocarcinoma cells, however, the protein was wild type in nature and easily activated in cells induced to apoptosis with genotoxic agents or induced to differentiate with physiologic ligands (Lutzker and Levine, 1996). Presumably, the dichotomous effects of p53 on teratocarcinoma cell differentiation and apoptosis reflect a graded stimulus–response mechanism (low-grade differentiation, high-grade apoptosis).

There is also evidence for dichotomous effects of p53 during the differentiation or apoptosis of more ad-

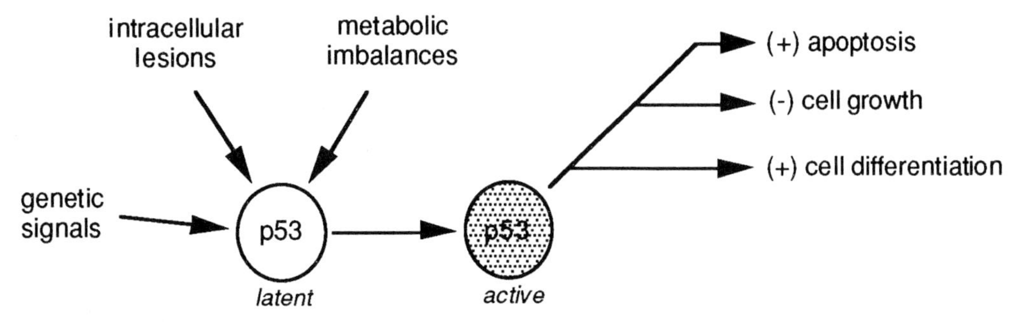

FIGURE 2 Cellular *p53* response. Latent p53 protein in the cytoplasm is activated by programmed (genetic) signals, as well as intracellular alterations caused by physical and chemical agents or intracellular metabolic imbalances. Low-grade p53 protein induction facilitates cell differentiation, whereas increasingly higher levels induce arrest of cell cycle advancement and apoptosis.

vanced cell types (Shaulsky *et al.*, 1991; Rotter *et al.*, 1994). In the developing nervous system, *p53* expression correlated precisely with differentiation of rat hippocampal neurons (Eizenberg *et al.*, 1996). When cultured hippocampal neurons were analyzed for p53 protein immunostaining, the subcellular distribution of p53 was found to change depending on the stage of differentiation. Neuronal p53 was first detected in the nucleus of postmitotic neurons. It later shifted to the cytoplasm as these cells matured. Nerve growth factor (NGF) stimulation of PC12 pheochromocytoma cells resulted in a similar change. The p53 was cytoplasmic in uncommitted mitotic PC12 progenitors, nuclear in cells committing to neuronal differentiation, and cytoplasmic in differentiated neurons. This general scheme implicated a role for p53 in the nucleus during early neuronal differentiation. To test whether neuron differentiation was dependent on p53 function, PC12 cells were infected with a recombinant retrovirus encoding a C-terminal p53 miniprotein (p53DD), a deletion mutant protein that acts as a transdominant-negative suppressor of wild-type p53 function. Neuroblasts expressing p53DD did not advance in differentiation when stimulated with NGF. These results suggest that p53 is a negative regulator of neuroblast cell growth and a positive regulator of neuroblast differentiation (Eizenberg *et al.*, 1996).

Within the context of developmental toxicity, a transitory increase in *p53* gene expression has been observed in neural tubes of early mouse embryos treated with arsenic (Wlodarczyk *et al.*, 1996). Although data on the status of the p53 protein was not given, that observation implies cell cycle perturbations associated with developmental neurotoxicity of environmental agents to which the human population is chronically exposed.

p53, in addition to facilitating neuronal differentiation, can initiate apoptosis in postmitotic neurons (Davies and Rosenthal, 1994; Sakhi *et al.*, 1994; Xiang *et al.*, 1996; Jordan *et al.*, 1997). Infecting hippocampal neurons with adenovirus expressing wild-type *p53* induced apoptosis spontaneously (Jordan, *et al.*, 1997). Hippocampal neurons were also induced to apoptosis with the excitotoxic agent kainic acid. This apoptosis was p53-dependent because neurons from *p53*-deficient nullizygous mice were refractory (Xiang *et al.*, 1996).

The precise mechanisms by which p53 initiates an apoptotic trigger are not known. Transcription-dependent mechanisms are partly responsible because wild-type p53 protein may directly transactivate genes such as *bax,* which are known to directly regulate the apoptosis effector (Miyashita and Reed, 1995; Sabbatini *et al.*, 1995). Other candidate genes have been identified through screening for differential gene expression in cells undergoing p53-dependent apoptosis (Vayssiere *et al.*, 1994; Amson *et al.*, 1996; Polyak *et al.*, 1997). One candidate that may be relevant for developmental neurotoxicity is the mammalian homolog to *Seven-in-Absentia* (*Sina*), a gene required for differentiation of photoreceptors in the developing compound eye of *Drosophila.* One of the *Sina* homologs in mice, *Siah1b,* was cloned by differential display of newly expressed genes during p53-dependent apoptosis (Amson *et al.*, 1996). *Siah1b* is a developmentally regulated gene associated with patterning of anterior neural structures in the mouse embryo (Della *et al.*, 1993). Although neural defects are associated with mutations or deletions in the *Sina*-regions of human and murine chromosomes (Holloway *et al.*, 1997), it remains to be determined whether *Siah1b* is a downstream effector of p53-dependent events (apoptosis) during developmental neurotoxicity.

Hypotheses concerning the mechanism by which p53 triggers apoptosis must take into consideration the fact that mitochondria, through release of cytochrome c, are key organelle transducers of the apoptotic effector mechanism (Newmeyer *et al.*, 1994; Vayssiere *et al.*, 1994; Petit *et al.*, 1995; Liu *et al.*, 1996; Zamzami *et al.*, 1996; Kluck *et al.*, 1997; Yang *et al.*, 1997). Although mitochondrial DNA (mtDNA) coded subunits of the electron transport chain are not required for apoptosis (Jacobson *et al.*, 1993; Marchetti *et al.*, 1996), the evidence suggests that at least part of the control of p53-dependent apoptosis may be exerted at the level of oxidative metabolism (Castedo *et al.*, 1995). For example, transient stimulation of mtDNA gene expression and oxygen consumption were observed during the committment of transformed cell lines to p53-dependent apoptosis (Vayssiere *et al.*, 1994). Transactivation of cellular oxidoreductases (Polyak *et al.*, 1997) and the subsequent generation of reactive oxygen intermediates (Johnson, *et al.*, 1996) were causally linked to p53-induced apoptosis. Taken together, these findings suggest that, during apoptosis, p53 may function in two ways. One is to stimulate uncoupled redox pressure (Johnson *et al.*, 1996), and a second is to promote the expression of genes (i.e., *bcl-*x_L) that may channel the release of cytochrome c from the mitochondrion (Vander Heiden *et al.*, 1997). To our knowledge, these hypotheses have not been tested during developmental neurotoxicity.

C. Developmental Expression of p53 in the Embryo

Evidence supporting the association between p53 and embryos date to initial discovery of this protein. Several groups studying cell transformation by the SV40 DNA tumor virus discovered a 54-55kDa host cell phosphoprotein that specifically bound the viral large T-antigen

(Lane and Crawford, 1979). The phosphoprotein, now known as p53, was present at high levels in uninfected murine embryonal carcinoma cells (Linzer and Levine, 1979) and in primary cell cultures from midgestational (days 12–14) mouse embryos (Mora *et al.,* 1980).

The first study to report on the expression pattern of *p53* during mammalian embryo development was published by Oren's laboratory (Rogel *et al.,* 1985). Using northern analysis, these investigators compared relative expression between various normal adult tissues and tissues at various stages of embryo development. Substantial amounts of *p53* mRNA were detected in embryonic stem cells and in whole embryos up to day 11 of gestation; however, a marked drop was detected in the fetus between days 13 and 16 of gestation (Rogel *et al.,* 1985). The sharp decline in steady-state *p53* mRNA was accompanied by down-regulation of p53 protein levels and was observed during fetal development of chickens also, suggesting a general pattern in vertebrate development (Louis *et al.,* 1988). Tissue distribution of *p53* transcripts in developing embryos was determined by *in situ* hybridization during mouse gestation between days 8–18 post coitum (Schmid *et al.,* 1991). Strong expression was found in essentially all tissues during neurulation (days 8–10). With advancement into organogenesis and histogenesis, *p53* expression became increasingly heterogeneous. In the brain, *p53* expression correlated more precisely with patterns of cell differentiation than cell proliferation (Schmid *et al.,* 1991). High-level expression of wild-type *p53* is thus observed in many early embryonic structures. The pattern of *p53* transcript accumulation within neural tissues reflects a potential role during early neuronal commitment, as suggested in studies described in the previous section of the chapter (Eizenberg *et al.,* 1996).

High-level expression of *p53* has also been detected in early *Xenopus* embryos (Tchang *et al.,* 1993; Hoever *et al.,* 1994). The accumulation of *p53* mRNA was most pronounced in growing oocytes. Maternally synthesized transcripts declined in abundance between fertilization and midblastula transition. At that stage, which marks the onset of zygotic gene expression, *p53* transcripts accumulated in most tissues of the neurula and tadpole (Tchang *et al.,* 1993; Hoever *et al.,* 1994). However, in contrast to dramatic shifts in *p53* mRNA accumulation, a relatively steady level of p53 protein was maintained in the cytoplasm between oocyte–tadpole stages. Furthermore, most of the p53 protein was synthesized during oocyte growth because cyclohexamide, a transcriptional inhibitor, abolished p53 protein in the growing oocyte but not the neurula (Tchang *et al.,* 1993). The general pattern that emerges from this and other studies is that the p53 protein of early embryos, as exemplified in *Xenopus* (Hoever *et al.,* 1994) and mammalian teratocarcinoma cell lines (Lutzker and Levine, 1996), is latent in the cytoplasm until called into play by programmed (genetic) or inducible (environmental) signals.

III. Transgenic Models to Dissect p53 Function in the Embryo

Analysis of *p53* expression patterns converge on a possible role for this tumor suppressor gene during early vertebrate development. High levels of *p53* mRNA and/or protein were detected in amphibian, avian, and mammalian embryos during cleavage, gastrulation, and neurulation. Ubiquitous expression in early embryos gives way to restricted expression during differentiation, but does not reflect a precise correlation with cell growth parameters. The biologic function of wild-type *p53* that is expressed during normal embryogenesis remains to be determined, and in this resides the potential for information gained through the use of transgenic approaches. Developmental studies using *p53* transgenic models is discussed in the remainder of this chapter and summarized in Table 1.

TABLE 1 Transgenic Animal Studies on *p53* in Development

Model	Refs.
Genetic overexpression of *p53*	
mRNA microinjection, *Xenopus* blastomeres	[1]
Transgenic mice, tissue-specific defects	[2] [3]
Functional inactivation of *p53*	
p53 knockout mice, neural tube defects	[4] [5]
Transgenic inactivation with papillomavirus	[6] [7]
Tissue-specific dominant negative	[8]
Developmental toxicity studies in *p53*-null mice	
Benzo[*a*]pyrene	[9]
Irradiation	[10]
2-Chloro-2′-deoxyadenosine	[11]
Promoter–reporter transgenics	
p53-CAT mice	[12]
p21-lacZ mice	[13]
mdm2-lacZ mice	[14]
Double mutants	
p53 (−/−), *Rb* (−/−)	[15]
p53 (−/−), *mdm2* (−/−)	[16] [17]
p53 (−/−), DNA repair genes	[18] [19]

Refs: [1] Hoever *et al.,* 1994; [2] Nakamura *et al.,* 1995; [3] Bowman *et al.,* 1996; [4] Sah *et al.,* 1995; [5] Armstrong *et al.,* 1995; [6] Howes *et al.,* 1994; [7] Pan and Griep, 1994; [8] Godley *et al.,* 1996; [9] Nicol *et al.,* 1995; [10] Norimura *et al.,* 1996; [11] Wubah *et al.,* 1996; [12] Almon *et al.,* 1993; [13] MacCallum *et al.,* 1996; [14] Gottlieb *et al.,* 1997; [15] Morgenbesser *et al.,* 1994; [16] de Oca Luna *et al.,* 1995; [17] Jones *et al.,* 1995; [18] Lim and Hasty, 1996; [19] Cheo *et al.,* 1996.

A. Microinjection of p53 mRNA

Because *p53* is highly conserved among vertebrates, clues to embryologic function can be gained from studies of model nonmammalian species that are amenable to direct experimental manipulation. For example, the large size of *Xenopus* blastomeres renders them amenable to direct microinjection of mRNA to test for developmental effects resulting from overexpression of *p53* (Hoever *et al.,* 1994). *Xenopus p53* mRNA was transcribed in a cell-free system and microinjected into single blastomeres of the *Xenopus* embryo at the two-cell or four-cell stage of development. In most cases the resulting embryos displayed severe disorganization of the neuraxis and eye reduction defects. Pathogenesis was attributed to irregular cleavage patterns induced by overexpression of p53 protein (Hoever *et al.,* 1994).

B. p53-Deficient Nullizygous (Knockout) Mice

Given the strong expression of *p53* in early embryos and potent embryogenic activity during neuraxis formation in *Xenopus,* a logical deduction was that p53 protein levels must be properly controlled during early development. This notion was consistent with early failures to generate transgenic animals carrying mutant p53 proteins. For example, Lavigueur *et al.* (1989) were unable to obtain transgenic mice carrying a Δp44 deletion mutant of *p53,* raising the possibility that overexpression of some mutant p53 proteins, in particular the 44kDa mutant protein lacking domains encoded in exon 2, was incompatible with early mouse development.

Viability of early embryos is not fully dependent on p53 function, however. Gene targeting in murine embryonic stem cells resulted in viable mice carrying germline null mutations in *p53* (Donehower *et al.,* 1992; Gondo *et al.,* 1994; Jacks *et al.,* 1994; Purdie *et al.,* 1994). Not surprisingly, these *p53*-deficient nullizygous mice displayed strong cancer predisposition. The unexpected survival of mouse embryos lacking *p53* was initially interpreted as indicating that *p53* plays a dispensable role in development. It has since been realized that the viability of *p53*(−/−) fetuses is not indicative of an indispensable role. In fact, a subset of *p53*-deficient mouse embryos, between 8–23% of them depending on the background strain, die from severe anterior neural tube defects (Armstrong *et al.,* 1995; Sah *et al.,* 1995). These altered phenotypes manifest as early as day 8–9 postcoitum (pc) (Ibrahim *et al.,* 1998). Although the reasons for incomplete penetrance of exencephaly among *p53*(−/−) fetuses are not known, developmental defects were strongly biased toward females (Sah *et al.,* 1995). They also became more frequent when *p53*(−/−) fetuses were sired by a sublethally γ-irradiated *p53*(−/−) male (Armstrong *et al.,* 1995). This suppressor model suggests that *p53*-deficiency predisposes developmental defects from genomic instability.

Because suppressor activity of p53 depends on DNA binding (Ko and Prives, 1996), a salient question in developmental biology pertains to the nature and function of genes potentially influenced by p53 in early embryos. Using the technique of mRNA differential display, Ibrahim *et al.* (1998) screened normal *p53*(+/+) and *p53*(−/−) embryos for differences in gene expression at the early head-fold stage (Ibrahim *et al.,* 1998). The results of this screen revealed deficiency of mitochondrial 16S ribosomal RNA. Subsequent analysis confirmed reduced 16S rRNA expression among p53(−/−) embryos relative to representative nuclear (COIV, β-actin) and mitochondrial (ND4L, COIII) transcripts, and also confirmed that 16S rRNA transcripts accumulated in p53(+/+) embryos during neurulation. Embryos lacking p53 showed weakened staining for cytochrome c oxidase and less ATP than normal counterparts. These findings suggest that p53 promotes a developmental transition of the embryo from an anaerobic (glycolytic) to an aerobic (oxidative) metabolism (Ibrahim *et al.,* 1998).

For some time it has been known that mammalian embryos are initially adapted for anaerobic (glycolytic) bioenergy production and then switch to aerobic (oxidative) respiration during neurulation and organogenesis (Mackler *et al.,* 1971). The pattern of glucose metabolism remains essentially anaerobic between implantation and the early somite stage as indicated by rates of lactate production (3–7 nmol μg^{-1} protein h^{-1}) that rival a rapidly growing tumor (Clough and Whittingham, 1983; Hunter and Tugman, 1995). As development continues, the rates of glucose utilization and lactate production drop in concert with increasing oxygen utilization (Tanimura and Shepard, 1970; Morriss and New, 1979; Miki *et al.,* 1988). Because the shift to oxidative metabolism is nearly completed by the 35 somite pair stage of development, the transitional period is congruent with neurulation and early organogenesis. If mitochondrial gene expression reflects the capacity for oxidative metabolism, then the alterations in *p53*-deficient embryos may reflect the ability of p53 to promote oxidative metabolism.

As the potential exists for systemic metabolic alterations in embryonic *p53* gene knockout models, analysis of developmental effects can benefit from a controlled way to specifically induce *p53* loss in somatic cells during specific stages. Tissue-specific inactivation of p53 has been accomplished by use of the p53DD transdominant-negative miniprotein, which retains the ability to oligomerize with endogenous p53 protein but blocks sequence-specific DNA binding (Bowman *et al.,* 1996). Transgenic mice expressing p53DD miniprotein under control of a brain-specific gene promoter displayed accelerated formation of brain tumors. Such animal models can become important adjuncts to traditional gene knockout models for developmental neurotoxicology.

C. Developmental Toxicology of p53-Null Mutant Mice

Because many natural and anthropogenic agents that damage DNA also induce birth defects, the embryonic *p53* response may be a critical determinant of developmental toxicity. Several studies have reported altered teratogenicity of mouse embryos having different *p53* genotypes (Nicol *et al.,* 1995; Norimura *et al.,* 1996; Wubah *et al.,* 1996). The results are consistent with the interpretation that embryogenic function of *p53* is a critical determinant of developmental toxicity; however, the data are still divided in terms of whether wild-type *p53* is a suppressor or promoter of embryotoxicity.

The first of these studies used benzo[*a*]pyrene as a model chemical agent (Nicol *et al.,* 1995). Pregnant *p53*(+/−) and *p53*(+/+) females, all of which were bred to *p53*(+/−) males, were primed with a nonteratogenic dose of 2,3,7,8-tetrachlorodibenzo-*p*-dioxin (TCDD) to induce CYP1A1, the major cytochrome P450 isoform in benzo[*a*]pyrene biotransformation. The pregnant females were then injected intraperitoneally with 200 mg Kg^{-1} benzo[*a*]pyrene on day 10 pc, which is a weakly teratogenic regimen, and the litters were evaluated at term. Both maternal genotypes were found to be equally CYP1A1-inducible, indicating that embryos in *p53*(+/+) × *p53*(+/−) and *p53*(+/−) × *p53*(+/−) litters probably received equal exposure to bioactivate metabolites. There were significantly higher incidences of developmental abnormalities reported for the fetuses from *p53*(+/−) females versus those from *p53*(+/+) females. In these respective groups of treated litters, fetal resorption incidences were 44% of total implants in *p53*(+/−) females versus 23% in *p53*(+/+) females, the incidences of collective teratologic defects were 47% versus 22%, and postnatal mortality rates were 74% versus 19%. These results were interpreted as indicating that p53 suppressed teratologic end-points, and that loss of one copy of the maternal *p53* gene resulted in enhanced susceptibility to benzo[*a*]pyrene-induced embryotoxicity (Nicol *et al.,* 1995). The direct correlation between *p53*-deficiency and developmental toxicity led to the hypothesis that wild-type p53 functions as a teratologic suppressor (Nicol *et al.,* 1995). However, an important issue not addressed in that particular study is whether the critical effect of *p53* deficiency was linked to the maternal *p53* genotype or to the zygotic *p53* genotype. Because data was not given for the distribution of severe developmental defects across embryonic *p53* genotypes, the correlation between *p53* gene function and teratologic suppressor activity did not indicate a direct mechanism.

Norimura *et al.* 1996 evaluated the impact of embryonic *p53* genotype on radiation-induced developmental toxicity. Mice were rendered *p53*-deficient by double-targeting, in which two steps of homologous recombination were used to replace the coding sequences of *p53* with those of *lacZ* (Gondo *et al.,* 1994). Embryos having different *p53* genotypes, *p53*(+/+), *p53*(+/−), and *p53*(−/−), were transferred to the uterus of pseudopregnant wild-type females and the dams were irradiated with 1 or 2 grays (Gy) of x-rays on day 3.5 or 9.5 of gestation. Incidences of early and late conceptal deaths, as well as fetal malformations, were proved to be p53-dependent. Malformation rates were higher in *p53*(−/−) fetuses versus *p53*(+/+) fetuses, and the reciprocal relationship was evident with respect to intrauterine deaths. For example, irradiation of *p53*(+/+) embryos during organogenesis resulted in an increase of resorption incidence from 24% (unirradiated) to 60% (irradiated) and respective malformation incidences from 14% to 20%. Malformation incidences in *p53*(−/−) embryos were increased from 29% (unirradiated) to 70% (irradiated) without an effect on resorption incidences. Although these data were regarded as supporting the teratologic suppressor model, malformations were expressed as a percentage of total implantations rather than survivors. When intrauterine deaths in the *p53*(+/+) group and spontaneous malformations in the *p53*(−/−) group are taken into consideration, the impact of *p53* was not evident: 32%, 33%, and 28% malformations among the *p53*(+/+), *p53*(+/−), and *p53*(−/−) fetuses, respectively. Therefore, a more precise correlation resided in the fact that irradiated embryos were selectively resorbed based on the level of *p53* gene activity. This suggests that embryogenic function of p53 prevented severely malformed embryos from surviving to term and indicates direct coupling of p53-dependent events with intrauterine death, a phenomenon more appropriately called *teratothanasia.*

The cellular basis for p53-dependent teratogenicity is likely to involve apoptosis. Excessive cell death is a common denominator in many examples of developmental toxicity, and the form of cell death frequently encountered in these cases is apoptosis (reviewed by Knudsen, 1996). To investigate the relationship between p53-dependent apoptosis and teratogenesis, mouse embryos of different *p53* genotypes were treated with 2-chloro-2′-deoxyadenosine (2-CdA). This compound is a potent analog of 2′-deoxyadenosine, a metabolic toxin that induces "whole-body" apoptosis in day 8 mouse embryos (Gao *et al.,* 1994). Exposure of day 8 mouse embryos to 2-CdA rapidly induced p53 protein and triggered apoptosis in some (head-fold) but not other (heart) developing structures. Induced cell death was *p53* gene-dose dependent as shown by the intermediate sensitivity of early head-fold stage embryos bearing only a single effective *p53* allele, and the lack of sensitivity of *p53*(−/−) mutants (Wubah *et al.*, 1996). Like early *Xenopus* embryos (Hoever *et al.*, 1994), the eye was most sensitive to overactivity of p53. Mouse embryos exposed to 2-CdA on day 8 of gestation displayed microphthalmia and anophthalmia in direct proportion to the embryonic *p53* genotype. Eye reduction defects were evident in 73.3% cases from treated *p53*(+/+) fetuses, 52.5% from *p53*(+/−), and 2.2% from *p53*(−/−) fetuses. Statistical analysis indicated that the interaction between prototype teratogen (2-CdA) and genotype (*p53*) was highly significant ($P \leq 0.001$) for both apoptosis on day 8 and eye reduction defects on day 17 (Wubah *et al.*, 1996).

In summary, studies with *p53* knockout mice have begun to establish a critical involvement of the embryonic p53 pathway in developmental toxicity. Physiologic processes influenced by p53 appear to be gene-dose dependent, because responses among heterozygotes fall somewhere between those of wild-type and nullizygous embryos. Fetal outcome may ultimately depend on the intracellular level of p53 protein activity. Whereas a low-grade response might suppress developmental defects, a high-grade activation may induce abnormalities depending on the extent of cellular changes (apoptosis) or developmental plasticity of the target structure.

D. Promoter–Reporter Transgenic Mice

Insight into the functional responses to p53 in mammalian systems has been gained from the analysis of transgenic mice in which a reporter gene, such as bacterial chloramphenicol acetyl transferase (CAT) or β-galactosidase (*lacZ*), is introduced into the genome under control of homologous *p53* promoter sequences. Enzymatic expression patterns in the *p53* promoter–reporter transgenic mice can then be followed during *in vivo* growth and development of the animal to yield clues on tissue specificity of *p53* expression.

Among the first studies, transgenic mouse lines were created in which a CAT activity was expressed under control of the murine *p53* promoter (Almon *et al.*, 1993). Analysis of the tissue distribution of CAT activity in several of the *p53* promoter–CAT strains revealed strong expression of the transgene in the testes. *In situ* hybridization revealed a cyclical pattern of *p53*–CAT transcripts in primary spermatocytes, indicating that expression of the *p53* gene may be up-regulated at certain stages of meiosis. Subsequent staining for p53 protein in nontransgenic control mice confirmed the cyclical accumulation of p53 in tetraploid primary spermatocytes of the meiotic pachytene phase (Schwartz *et al.*, 1993). Pachytene encompasses pairing of chromosomes and recombination and repair of DNA, suggesting that p53 controls some aspects of the meiotic cycle to permit DNA shuffling and repair.

Transgenic *p53* promoter–CAT mice displayed irregular seminiferous tubules (Almon *et al.*, 1993; Schwartz *et al.*, 1993). This phenotype was diagnosed as giant-cell degenerative syndrome and it involved transformation of a subset of primary spermatocytes into aberrant, multinucleated giant cells due to DNA replication without meiosis (Rotter *et al.*, 1993). The primary spermatocytes also displayed less endogenous *p53* mRNA and protein expression than nontransgenic control mice. The "squelching," or down-regulation of endogenous *p53* by the *p53*–CAT transgene, was attributed to the molar excess of transgene *p53* promoter sequences in competition with endogenous *p53* promoter sequences for binding to the same transcriptional activators (Rotter *et al.*, 1993). Thus, any hybrid transgene approach carries the potential for interference with expression and function of endogenous regulators and must be interpreted with caution.

Transgenic animal models have been generated to study the spatiotemporal transactivation of p53 responsive genes in developing and stressed embryonic tissues. In these models, a bacterial reporter gene (*lacZ*) was expressed under control of a consensus p53 response element (MacCallum *et al.*, 1996; Gottlieb *et al.*, 1997). In the first such study, a *p21* promoter element was selected by virtue of strong direct transactivation by p53. Pregnant transgenic *p21* promoter–*lacZ* mice were γ-irradiated (3–5 Gy) at day 8–10 pc and embryos were evaluated 3 hr later by staining for β-galactosidase activity and p53 protein. Presence of multiple copies (50–100) of the transgene rendered the assay of transcriptional activity of p53 *in vivo* quite sensitive, and apart from syncytiotrophoblast cells that expressed β-galactosidase activity without irradiation, the data showed a strong induction, by irradiation, of p53 activity in the neuraxis and developing eye. The p53 response was more restricted than the normal tissue distribution of *p53* mRNA (Schmid *et al.*, 1991). This indicated that radiation-induced p53 involved stabilization of latent p53 protein rather than new transcription (MacCallum *et al.*, 1996). Thus, *p53* response–reporter transgenic mice provide a model for *in situ* analysis of transcription dependent (and independent) p53 activities during developmental toxicity, with the drawback that the particular *p53* response element used to drive β-galactosidase may not be the only genomic cognate for p53.

A second example of a transgenic reporter strain carries *lacZ* under control of a p53-responsive promoter derived from intronic sequences of the murine *mdm2* oncogene. "Blue" staining embryos were produced when the β-galactosidase reporter was induced 3 hr after 5 Gy γ-irradiation, whereas the unirradiated embryos showed low background staining. Most tissues in an irradiated day 8 embryo were blue in accordance with p53 protein induction; however, a response was conspicuously absent in the primitive heart (Gottlieb *et al.,* 1997). This tissue specificity matched the lack of p53 protein induction and apoptosis in the primitive heart following 2-CdA treatment (Wubah *et al.,* 1996). Furthermore, when irradiation was performed in animals in which the *p53*-responsive promoter-reporter was bred onto *p53*(+/+), *p53*(+/−), and *p53*(−/−) backgrounds, the optimal response to DNA damage was found to require two functional *p53* alleles. The lack of a response in the *p53* nullizygous embryos and an intermediate effect in heterozygotes (Gottlieb *et al.,* 1997) also matched the intermediate responses of *p53*(+/−) fetuses to eye reduction defects induced by 2-CdA (Wubah *et al.,* 1996). These findings indicate that *p53* haploinsufficiency determines an intermediate response for both p53-dependent transactivation as well as developmental defects.

E. Double Mutants

Animals lacking *p53* have been used to study developmental interactions between different genetic loci. For example, mouse embryos lacking the tumor suppressor *retinoblastoma* (*Rb-1*) die between days 13–15 pc due to defects in erythropoiesis and neurogenesis, whereas heterozygotes survive with unusual lens tumors. In *Rb*(−/−) embryos, lens fiber cell proliferation was unchecked between days 12.5 and 14.5 pc, and massive programmed cell death (apoptosis) appeared at the time of normal fiber differentiation on day 14.5 (Morgenbesser *et al.,* 1994). The apoptosis was p53-dependent, because it was greatly reduced in embryos doubly null for *p53* and *Rb*. However, the rescue from excessive apoptosis did not correlate with changes in lens cell growth or ocular defects, and the double knockout embryos still died (Morgenbesser *et al.,* 1994). These experiments demonstrate in principle that oncogenic growth signals associated with the functional inactivation of one cell regulatory pathway may cooperatively induce autonomous changes coupled to the p53 stimulus–response (Symonds *et al.,* 1994).

The more intriguing finding was that p53-dependent apoptosis appeared in the rudimentary lens of *Rb*(−/−) embryos at a stage in development when secondary lens fibers would normally differentiate. These data suggest that p53-dependent apoptosis and cell differentiation were coupled to the same trigger. A critical question thus pertains to the nature of the apoptotic signals invoked in *Rb*(−/−) lenses when differentiation was supposed to have occurred. Because the functional inactivation of these tumor suppressors alters embryonic growth and development, other studies were needed to resolve between local and systemic signaling events. One approach was to perturb functions of either tumor suppressor by use of lens-specific gene promoters, such as αA crystallin (Pan and Griep, 1994; Nakamura *et al.,* 1995) or interstitial retinol-binding protein (IRBP) (Howes *et al.,* 1994). Transgenic mice were generated in which the papillomavirus oncogenes E6 and E7, which inactivate p53 and Rb, respectively, were expressed under control of the lens-specific αA-crystallin promoter. Overexpression of E7 affected lens fiber differentiation and triggered spatially inappropriate cell death, as expected, leading to microphthalmia. Overexpression of E6 reduced the amount of programmed cell death normally observed in the lens, leading to congenital cataracts. Combinatorial expression of E6 and E7 rescued microphthalmia, presumably because p53-dependent apoptosis was disengaged, leading to lens tumor formation (Pan and Griep, 1994). Essentially the same kinds of findings were observed in transgenic mice carrying papillomavirus E7 expression under direct control of the IRBP, which becomes active prior to the terminal differentiation of retinal photoreceptors. Retinas of these mice degenerate due to excessive cell death of the photoreceptors at a stage in development when photoreceptors would normally have undergone terminal differentiation (Howes *et al.,* 1994). On a *p53*-deficient gene background, retinal tumors resulted instead of apoptosis. Again, the peak cell death in the IRBP-E7 retina coincided with the terminal differentiation of photoreceptors, suggesting that a signal is given at the time of differentiation to invoke apoptosis in response to an oncogenic growth imbalance, and that signal is delivered through p53 (Howes *et al.,* 1994). Thus, imbalances in the regulation of endogenous *p53* and *Rb* gene function can significantly affect developmental outcome, resulting in malformations or neoplasms depending on an apoptotic signal.

There are other examples of developmental abnormalities that depend on the *p53* genotype (de Oca Luna *et al.,* 1995; Jones *et al.,* 1995; Cheo *et al.,* 1996; Lim and Hasty, 1996). For example, the MDM2 zinc-finger oncoprotein binds and inactivates p53 specifically. Mouse embryos nullizygous for *mdm2* died between implantation and day 5.5 pc (de Oca Luna *et al.,* 1995; Jones *et al.,* 1995). The failure of these *mdm2*(−/−) embryos to survive was entirely dependent on *p53* gene function because double-nullizygous mutant mice, lacking both *mdm2* and *p53* functions, were viable and normal. Because MDM2 protein is a negative regulator of p53 function *in vitro,* these studies provided evidence for the interaction between MDM2 and p53 *in vivo,* and also indicated that *mdm2*-null mutant embryos died because p53 function was not down-regulated (de Oca Luna *et al.,* 1995; Jones *et al.,* 1995).

The *p53* pathway also seems to modulate severe developmental defects associated with mutations in DNA repair systems (Cheo *et al.,* 1996; Lim and Hasty, 1996). Murine *rad51,* a homolog of the yeast recombinational repair gene *ScRad51,* plays a role in the repair of double-strand breaks in DNA during meiotic recombination. Targeting of the *rad51* gene in mice resulted in mutant embryos that arrested shortly after implantation, probably at the egg cylinder stage (Lim and Hasty, 1996). There was evidence of reduced cell growth and excessive apoptosis among *rad51(−/−)* blastocysts grown in culture; furthermore, when these mice were bred to a *p53(−/−)* gene background, the embryos survived normally to the head-fold stage (day 8.5). This suggested that damage caused by the *rad51* mutation alone must be at levels low enough to permit DNA replication and cell function up to the head-fold stage (i.e., day 8.5 pc).

However, targeting of the nucleotide excision–repair gene *Xeroderma pigmentosum* (*xpc*) resulted in neural tube defects similar in phenotype as those associated with *p53(−/−)* fetuses (Cheo *et al.,* 1996). Although the incidences of these defects were 54% in *XPC-p53* double-mutant females (compared with 19% reported by Sah *et al.,* 1995, for *p53(−/−)* females), no data was reported on neural tube defects in *xpc(−/−)* mutant embryos. Thus, a postulated synergistic interaction between XPC and p53 awaits definitive proof.

F. Overexpression of p53 in Transgenic Mice

Studies have shown that improper activation of p53 in the developing lens is sufficient to induce excessive apoptosis and lens malformations (Nakamura *et al.,* 1995). Transgenic mice expressing the wild-type human *p53* gene under control of the lens-specific αA-crystallin promoter developed microphthalmia as a result of excessive apoptosis in the fetal lens; a normal lens phenotype was restored in double transgenic mice that carry both wild-type and mutant human *p53* alleles, indicating that the improper balance of p53 can induce cell death abnormalities and that presence of the mutant protein can exert a dominant–negative influence (Nakamura *et al.,* 1995).

Another model involved transgenic mice expressing the wild-type murine *p53* gene under the control of the mouse mammary tumor virus long terminal repeat (MMTV LTR). These animals were predisposed to congenital defects of the kidney (Godley *et al.,* 1996). Excessive cell death (apoptosis) was observed in the undifferentiated renal mesenchyme sometime around day 17.5 pc. The loss of uncommitted renal mesenchyme led to fewer glomeruli in transgenic newborns, with compensatory hypertrophy of existing glomeruli and eventual renal failure. This disease resembled a rare human condition known as *oligomeganephronia.* Excessive apoptosis in this context was secondary to improper function of the embyronic ureteric buds between days 14 and 18. Thus, abnormally high activity of wild-type p53 altered cellular differentiation and caused secondary apoptosis in undifferentiated organ rudiments.

IV. Summary

Tumor suppressor p53 is an environmentally responsive DNA-binding protein that plays a critical role in spontaneous and induced neural malformations. Embryonic p53 is subject to cell-specific regulation by programmed (genetic) and inducible (environmental) signals. Genetic manipulation in transgenic embryo models has provided insights into the role of p53 protein in the normal and stressed embryo and demonstrated that proper regulation of p53 is essential for normal morphogenesis. Too little p53 renders the embryo susceptible to neural defects associated with genomic instability or chromosomal damage, whereas too much p53 may invoke cell death abnormalities, particularly in the eye. Future research must be directed at understanding the regulation and downstream consequences of the *p53* pathway in embryo development and its relevancy to the pathogenesis of human birth defects.

Acknowledgments

The authors are indebted to the National Institute of Child Health and Human Development (grant RO1 HD30302) and to the Environmental Protection Agency (EPA-CR 824 445-01) for support of this work (which does not reflect US EPA policy). Ms. Wubah is a predoctoral fellow from the National Institute of Environmental Health Sciences (training grant T32-ES07282).

References

Almon, E., Goldfinger, N., Kapon, A., Schwartz, D., Levine, A. J., and Rotter, V. (1993). Testicular tissue-specific expression of the *p53* suppressor gene. *Devel. Biol.* **156,** 107–116.

Amson, R. B., Nemani, M., Roberch, J.-P., Israeli, D., Bougueleret, L., LeGall, I., Medhioub, M., Linares-Cruz, G., Lethrosne, F., Pasturaud, P., Piouffre, L., Prieur, S., Susini, L., Alvaro, V., Millasseau, P., Guidicelli, C., Bui, H., Massart, C., Cazes, L., Dufour, F., Bruzzoni-Giovanelli, H., Owadi, H., Hennion, C., Charpak, G., Dausset, J., Calvo, F., Oren, M., Cohen, D., and Telerman, A. (1996). Isolation of 10 differentially expressed cDNAs in *p53*-induced apoptosis: activation of the vertebrate homologue of the *Drosophila* seven in absentia gene. *Proc. Natl. Acad. Sci. USA* **93,** 3953–3957.

Armstrong, J. F., Kaufman, M. A., Harrison, D. J., and Clarke, A. R. (1995). High-frequency developmental abnormalities in *p53*-deficient mice. *Curr. Biol.* **5,** 931–936.

Bourdon, J.-C., Deguin-Chambon, V., Lelong, J.-C., Dessen, P., May, P., Debuire, B., and May, E. (1997). Further characterization of the *p53* responsive element–identification of new candidate genes for *trans*-activation by *p53. Oncogene* **14,** 85–94.

Bowman, T., Symonds, H., Gu, L., Yin, C., Oren, M., and Van Dyke, T. (1996). Tissue-specific inactivation of *p53* tumor suppression in the mouse. *Genes Devel.* **10,** 826–835.

Castedo, M., Macho, A., Zamzami, N., Hirsch, T., Marchetti, P., Uriel, J., and Kroemer, G. (1995). Mitochondrial perturbations define lymphocytes undergoing apoptotic depletion *in vivo. Eur. J. Immunol.* **25,** 3277–3284.

Chen, X., Ko, L. J., and Prives, C. (1996). *p53* levels, functional domains, and DNA damage determine the extent of the apoptotic response of tumor cells. *Genes Devel.* **10,** 2438–2451.

Cheo, D. L., Meira, L. B., Hammer, R. E., Burns, D. K., Doughty, A. T. B., and Friedberg, E. C. (1996). Synergistic interactions between *XPC* and *p53* mutations in double-mutant mice: neural tube abnormalities and accelerated UV radiation-induced skin cancer. *Curr. Biol.* **6,** 1691–1694.

Cho, Y., Gorina, S., Jeffrey, P. D., and Pavletich, N. P. (1994). Crystal structure of a *p53* tumor suppressor–DNA complex: understanding tumorigenic mutations. *Science* **265,** 346–355.

Clarke, A. R., Purdie, C. A., Harrison, D. J., Morris, R. G., Bird, C. C., Hooper, M. L., and Wyllie, A. H. (1993). Thymocyte apoptosis induced by *p53*-dependent and independent pathways. *Nature* **362,** 849–852.

Clough, J., and Whittingham, D. G. (1983). Metabolism of [^{14}C]glucose by postimplantation mouse embryos *in vitro. J. Embryol. Exp. Morph.* **74,** 133–142.

Davies, A. M., and Rosenthal, A. (1994). Neurons from mouse embryos with a null mutation in the tumor suppressor gene *p53* undergo normal cell death in the absence of neurotrophins. *Neurosci. Lett.* **182,** 112–114.

Debbas, M., and White., E. (1993). Wild-type *p53* mediates apoptosis by E1A, which is inhibited by E1B. *Genes Dev.* **7,** 546–554.

Della, N. G., Senior, P. V., and Bowtell, D. D. L. (1993). Isolation and characterization of murine homologues of the *Drosophila seven in absentia* gene (*sina*). *Development* **117,** 1333–1343.

de Oca Luna, R. M., Wagner, D. S., and Lozano, G. (1995). Rescue of early embryonic lethality in *mdm2*-deficient mice by deletion of *p53. Nature* **378,** 203–206.

Donehower, L. A., and Bradley, A. (1993). The tumor suppressor *p53. Biochem. Biophys. Acta* **1155,** 181–205.

Donehower, L. A., Harvey, M., Slagle, B. L., McArthur, M. J., Montgomery, C. A. Jr., Butel, J. S., and Bradley, A. (1992). Mice deficient for *p53* are developmentally normal but susceptible to spontaneous tumours. *Nature* **356,** 215–221.

Eizenberg, O., Faber-Elman, A., Gottlieb, E., Oren, M., Rotter, V., and Schwartz, M. (1996). *p53* plays a regulatory role in differentiation and apoptosis of central nervous system–associated cells. *Mol. Cell. Biol.* **16,** 5178–5185.

El-Deiry, W. S., Tokino, T., Velculescu, V. E., Levy, D. B., Parsons, R., Trent, J. M., Lin, D., Mercer, W. E., Kinzler, K. W., and Vogelstein, B. (1993). WAF1, a potential mediator of *p53* tumor suppressor. *Cell* **75,** 817–825.

Fritsche, M., Haessler, C., and Brandner, G. (1993). Induction of nuclear accumulation of the tumor-suppressor protein *p53* by DNA-damaging agents. *Oncogene* **8,** 307–317.

Gao, X., Blackburn, M. R., and Knudsen, T. B. (1994). Activation of apoptosis in early mouse embryos by 2′-deoxyadenosine exposure. *Teratology* **49,** 1–12.

Godley, L. A., Kopp, J. B., Eckhaus, M., Paglino, J. J., Owens, J., and Varmus, H. E. (1996). Wild-type *p53* transgenic mice exhibit alteerd differentiation of the ureteric bud and possess small kidneys. *Genes Devel.* **10,** 836–850.

Goldsworthy, T. L., Recio, L., Brown, K., Donehower, L. A., Mirsalis, J. C., Tennant, R. W., and Purchase, I. F. H. (1994). Transgenic animals in toxicology. *Fund. Appl. Toxicol.* **22,** 8–19.

Gondo, Y., Nakamura, K., Nakao, K., Sasaoka, T., Ito, K-I., Kimura, M., and Katsuki, M. (1994). Gene replacement of the *p53* gene with the *lacZ* gene in mouse embryonic stem cells and mice by using two steps of homologous recombination. *Biochem. Biophys. Res. Commun.* **202,** 830–837.

Gottlieb, E., Haffner, R., King, A., Asher, G., Gruss, P., Lonai, P., and Oren, M. (1997). Transgenic mouse model for studying the transcriptional activity of the *p53* protein: age- and tissue-dependent changes in radiation-induced activation during embryogenesis. *EMBO J.* **16,** 1381–1390.

Graeber, T. G., Osmanian, C., Jacks, T., Housman, D. E., Koch, C. J., Lowe, S. W., and Giaccia, A. J. (1996). Hypoxia-mediated selection of cells with diminished apoptotic potential in solid tumors. *Nature* **379,** 88–91.

Graeber, T. G., Peterson, M., Tsai, M., Monica, K., Fornace, A. J. Jr., and Giaccia, A. J. (1994). Hypoxia induces accumulation of *p53* protein, but activation of a G1-phase checkpoint by low oxygen conditions is independent of *p53* status. *Mol. Cell Biol.* **14,** 6264–6277.

Hall, P. A., and Lane, D. P. (1997). *p53*—a developing role? *Curr. Biol.* **7,** R144–R147.

Hall, P. A., Meek, D., and Lane, D. P. (1996). *p53*—integrating the complexity. *J. Pathol.* **180,** 1–5.

Hansen, R., and Oren, M. (1997). *p53:* from inductive signal to cellular effect. *Curr. Opin. Genet. Devel.* **7,** 46–51.

Hansen, R. S., and Braithwaite, A. W. (1996). The growth-inhibitory function of *p53* is separable from transactivation, apoptosis and suppression of transformation by E1a and Ras. *Oncogene* **13,** 995–1007.

Harper, J. W., Adami, G. R., Wei, N., Keyomarsi, K., and Elledge, S. J. (1993). The p21 Cdk-interacting protein Cip1 is a potent inhibitor of G1 cyclin-dependent kinases. *Cell* **75,** 805–816.

Harris, C. C. (1996). *p53* tumor suppressor gene: at the crossroads of molecular carcinogenesis, molecular epidemiology, and cancer risk assessment. *Environ. Health. Persp.* **104** (suppl 3), 435–439.

Hoever, M., Clement, J. H., Wedlich, D., Montenarh, M., and Knochel, W. (1994). Overexpression of wild-type *p53* interferes with normal development in *Xenopus laevis* embryos. *Oncogene* **9,** 109–120.

Holloway, A. J., Della, N. G., Fletcher, C. F., Largespada, D. A., Copeland, N. G., Jenkins, N. A., and Bowtell, D. D. L. (1997). Chromosomal mapping of five highly conserved murine homologues of the *Drosophila* RING finger gene *Seven-in-absentia. Genomics* **41,** 160–168.

Hollstein, M., Sidransky, D., Vogelstein, B., and Harris, C. C. (1991). *p53* mutations in human cancers. *Science* **253,** 49–53.

Hooper, M. L. (1994). The role of the *p53* and *Rb-1* genes in cancer, development and apoptosis. *J. Cell Sci.* **suppl. 18,** 13–17.

Howes, K. A., Ransom, N., Papermaster, D. S., Lasudry, J. G. H., Albert, D. M., and Windle, J. J. (1994). Apoptosis or retinoblastoma: alternative fates of photoreceptors expressing the HPV-16 E7 gene in the presence or absence of *p53. Genes Dev.* **8,** 1300–1310.

Hunter, E. S. III, and Tugman, J. A. (1995) Inhibitors of glycolytic metabolism affect neurulation-staged mouse conceptuses *in vitro. Teratology* **52,** 317–323.

Ibrahim, M. M., Razmara, M., Nguyen, D., Donahue, R. J., Wubah, J. A., and Knudsen, T. B. (1998). Altered expression of mitochondrial 16S ribosomal RNA in *p53*-deficient mouse embryos revealed by differential display. *Biochim. Biophys. Acta* (submitted).

Jacks, T., Remington, L., Williams, B. O., Scmitt, E. M., Halachmi, S., Bronson, R. T., and Weinberg, R. A. (1994). Tumor spectrum analysis in *p53*-mutant mice. *Curr. Biol.* **4,** 1–7.

Jacobson, M. D., Burne, J. F., King, M. P., Miyashita, T., Reed, J. C., and Raff, M. C. (1993). Bcl-2 blocks apoptosis in cells lacking mitochondrial DNA. *Nature* **361,** 365–369.

Johnson, T. M., Yu, Z.-X., Ferrans, V. J., Lowenstein, R. A., and Finkel, T. (1996). Reactive oxygen species are downstream mediators of *p53*-dependent apoptosis. *Proc. Natl. Acad. Sci. USA* **93,** 11848–11852.

Jones, S. N., Roe, A. E., Donoehower, L. A., and Bradley, A. (1995). Rescue of early embryonic lethality in *Mdm2*-deficient mice by absence of *p53. Nature* **378,** 206–208.

Jordan, J., Galindo, M. F., Prehn, J. H. M., Weichselbaum, R. R., Beckett, M., Ghadge, G. D., Roos, R. P., Leiden, J. M., and Miller, R. J. (1997). *p53* expression induces apoptosis in hippocampal pyramidal neuron cultures. *J. Neurosci.* **17,** 1397–1405.

Kastan, M. B., Onyekwere, O., Sidransky, D., Vogelstein, B., and Craig, R. W. (1991). Participation of *p53* protein in the cellular response to DNA damage. *Canc. Res.* **51,** 6304–6311.

Kluck, R. M., Bossy-Wetzel, E., Green, D. R., and Newmeyer, D. D. (1997). The release of cytochrome c from mitochondria: a primary site for Bcl-2 regulation of apoptosis. *Science* **275,** 1132–1136.

Knudsen, T. B. (1996). Cell death. In *Drug Toxicity in Embryonic Development I. Advances in Understanding Mechanisms of Birth Defects: Morphogenesis and Processes at Risk.* R. J. Kavlock and G. P. Daston, Eds. Springer, New York, pp 211–244.

Ko, L. J., and Prives, C. (1996). *p53:* puzzle and paradigm. *Genes Dev.* **10,** 1054–1072.

Kuerbitz, S. J., Plunkett, B. S., Walsh, W. V., and Kastan, M. B. (1992). *p53* is a cell cycle checkpoint determinant following irradiation. *Proc. Natl. Acad. Sci. USA* **89,** 7491–7495.

Lane, D. P. (1992). *p53,* guardian of the genome. *Nature* **358,** 15–16.

Lane, D. P., and Crawford, L. V. (1979). T antigen is bound to a host protein in SV40-transformed cells. *Nature* **278,** 261–263.

Lavigueur, A., Maltby, V., Mock, D., Rossant, J., Pawson, T., and Bernstein, A. (1989). High incidence of lung, bone, and lymphoid tumors in transgenic mice overexpression mutant alleles of the *p53* oncogene. *Mol. Cell. Biol.* **9,** 3982–3991.

Lim, D-S., and Hasty, P. (1996). A mutation in mouse *rad51* results in an early embryonic lethal that is suppressed by a mutation in *p53. Mol. Cell. Biol.* **16,** 7133–7143.

Linke, S. P., Clarkin, K. C., Di Leonardo, A., Tsou, A., and Wahl, G. M. (1996). A reversible, *p53*-dependent Go/G1 cell cycle arrest induced by ribonucleotide depletion in the absence of detectable DNA damage. *Genes Dev.* **10,** 934–947.

Linzer, D. I., and Levine, A. J. (1979). Characterization of a 54K dalton cellular SV40 tumor antigen present in SV40-transformed cells and uninfected embryonal carcinoma cells. *Cell* **17,** 43–52.

Liu, Y., Bohn, S. A., and Sherley, J. L. (1998). Inosine-5′-monophosphate dehydrogenase is a rate-determining factor for p53-dependent growth regulation. *Mol. Biol. Cell* **9,** 15–28.

Liu, X., Kim, C. N., Yang, J., Jemmerson, R., and Wang, X. (1996). Induction of apoptotic program in cell-free extracts: requirement for dATP and cytochrome c. *Cell* **86,** 147–157.

Louis, J. M., McFarland, V. W., May, P., and Mora, P. T. (1988). The phosphoprotein *p53* is down-regulated post-transcriptionally during embryogenesis in vertebrates. *Biochim Biophys Acta,* **950,** 395–402.

Lowe, S. W., Ruley, H. E., Jacks, T., and Housman, D. E. (1993a). *p53*-dependent apoptosis modulates the cytotoxicity of anticancer agents. *Cell* **74,** 957–967.

Lowe, S. W., Schmitt, E. M., Smith, S. W., Osborne, B. A., and Jacks, T. (1993b) *p53* is required for radiation-induced apoptosis in mouse thymocytes. *Nature* **363,** 847–849.

Lutzker, S. G., and Levine, A. J. (1996). A functionally inactive *p53* protein in teratocarcinoma cells is activated by either DNA damage or cellular differentiation. *Nature Med.* **2,** 804–810.

MacCallum, D. E., Hupp, T. R., Midgley, C. A., Stuart, D., Campell, S. J., Harper, A., Walsh, F. S., Wright, E. G., Balmain, A., Lane, D. P., and Hall, P. A. (1996). The *p53* response to ionising radiation in adult and developing murine tissues. *Oncogene* **13,** 2575–2587.

Mackler, B., Grace, R., and Duncan, H. M. (1971). Studies of mitochondrial development during embryogenesis in the rat. *Arch. Biochem. Biophys.* **144,** 603–610.

Marchetti, P., Susin, S. A., Decaudin, D., Gamen, S., Castedo, M., Hirsch,T., Zamzami, N., Naval, J., Senik, A., and Kroemer, G. (1996). Apoptosis-associated derangement of mitochondrial function in cells lacking mitochondrial DNA. *Cancer Res.* **56,** 2033–2038.

Merritt, A. J., Potten, C. S., Kemp, C. J., Hickman, J. A., Balmain, A., Lane, D. P., and Hall, P. A. (1994). The role of *p53* in spontaneous and radiation-induced apoptosis in the gastrointestinal tract of normal and *p53*-deficient mice. *Cancer Res.* 614–617.

Miki, A., Fujimoto, E., Ohsaki, T., and Mizoguti, H. (1988). Effects of oxygen concentration on embryonic development in rats: a light and electron microscopic study using whole-embryo culture techniques. *Anat. Embryol.* **178,** 337–343.

Miyashita, T., and Reed, J. C. (1995). Tumor suppressor *p53* is a direct transcriptional activator of the human *bax* gene. *Cell* **80,** 293–299.

Mora, P. T., Chandrasekaran, K., and McFarland, V. W. (1980). An embryo protein induced by SV40 virus transformation of mouse cells. *Nature* **288,** 722–724.

Morgenbesser, S. D., Williams, B. O., Jacks, T., and DePinho, R. A. (1994). *p53*-dependent apoptosis produced by Rb-deficiency in the developing mouse lens. *Nature* **371,** 72–74.

Morriss, G. M., and New, D. A. T. (1979). Effect of oxygen concentration on morphogenesis of cranial neural folds and neural crest in cultured rat embryos. *J. Embryol. Exp. Morphol.* **54,** 17–35.

Nakamura, T., Pichel, J. G., Williams-Simons, L., and Westphal, H. (1995). An apoptotic defect in lens differentiation caused by human *p53* is rescued by a mutant allele. *Proc. Natl. Acad. Sci. USA* **92,** 6142–6146.

Nelson, W. G., and Kastan, M. B. (1994). DNA strand breaks: the DNA template alterations that trigger *p53*-dependent DNA damage response pathways. *Molec. Cell Biol.* **14,** 1815–1823.

Newmeyer, D. D., Farschon, D. M., and Reed, J. C. (1994). Cell-free apoptosis in *Xenopus* extracts. *Cell* **79,** 353–364.

Nicol, C. J., Harrison, M. L., Laposa, R. R., Gimelshtein, I. L., and Wells, P. G. (1995). A teratological suppressor role for *p53* in benzo[*a*]pyrene-treated transgenic *p53*-deficient mice. *Nature Gen.* **10,** 181–187.

Norimura, T., Nomoto, S., Katsuki, M., Gondo, Y., and Kondo, S. (1996). *p53*-dependent apoptosis suppresses radiation-induced teratogenesis. *Nature Med.* **2,** 577–580.

Oren, M., Reich, N. C., and Levine, A. J. (1982). Regulation of the cellular *p53* tumour antigen in teratocarcinoma cells and their differentiated progeny. *Mol. Cell. Biol.,* **2,** 443–449.

Pan, H., and Griep, A. E. (1994). Altered cell cycle regulation in the lens of HPV-16 E6 or E7 transgenic mice: implications for tumor suppressor gene function in development. *Genes Dev.* **8,** 1285–1299.

Petit, P. X., Lecoeur, H., Zorn, E., Dauget, C., Mignotte, B., and Gougeon, M-L. (1995). Alterations in mitochondrial structure and function are early events of dexamethasone-induced thymocyte apoptosis. *J. Cell Biol.* **130,** 157–167.

Polyak, K., Xia, Y., Zweler, J. L. Z., Kinzler, K. W., and Vogelstein, B. (1997). A model for p53-induced apoptosis. *Nature* **389,** 300–305.

Purdie, C. A., Harrison, D. J., Peter, A., Dobbie, L., White, S., Howie, S. E. M., Salter, D. M., Bird, C. C., Wyllie, A. H., Hooper, M. L., and Clarke, A. R. (1994). Tumor incidence, spectrum and ploidy in mice with a large deletion in the *p53* gene. *Oncogene* **9,** 603–609.

Rogel, A., Popliker, M., Webb, C. G., and Oren, M. (1985). *p53* cellular tumor antigen: analysis of mRNA levels in normal adult tissues, embryos, and tumors. *Molec. Cell. Biol.* **5,** 2851–2855.

Ronen, D., Rotter, V., and Reisman, D. (1991). Expression from murine *p53* promoter is mediated by factor-binding to a down stream helix-loop-helix recognition motif. *Proc. Natl. Acad. Sci. USA* **88,** 4128–4132.

Rotter, V., Aloni-Grinstein, R., Schwartz, D., Elkind, N. B., Simons, A., Wolkowicz, R., Lavigne, M., Beserman, P., Kapon, A., and Goldfinger, N. (1994). Does wild-type *p53* play a role in normal cell differentiation? *Sem. Canc. Biol.,* **5,** 229–236.

Rotter, V., Schwartz, D., Almon, E., Goldfinger, N., Kapon, A., Meshorer, A., Donehower, L. A., and Levine, A. J. (1993). Mice with reduced levels of *p53* protein exhibit the testicular giant-cell degenerative syndrome. *Proc. Natl. Acad. Sci. USA* **90,** 9075–9079.

Sabbatini, P., Lin, J., Levine, A. J., and White, E. (1995). Essential role for *p53*-mediated transcription in E1A-induced apoptosis. *Genes Dev.,* **9,** 2184–2192.

Sah, V. P., Attardi, L. D., Mulligan, G. J., Williams, B. O., Bronson, R. T., and Jacks, T. (1995). A subset of *p53*-deficient embryos exhibit exencephaly. *Nature Gen.* **10,** 175–180.

Sakhi, S., Bruce, A., Sun, N., Tocco, G., Baudry, M., and Schreiber, S. S. (1994). *p53* induction is associated with neuronal damage in the central nervous system. *Proc. Natl. Acad. Sci. USA* **91,** 7525–7529.

Schmid P., Lorenz, A., Hameister, H., and Montenarh, M. (1991). Expression of *p53* during mouse embryogenesis. *Development* **113,** 857–865.

Schwartz, D., Goldfinger, N., and Rotter, V. (1993). Expression of *p53* protein in spermatogenesis is confined to the tetraploid pachytene primary spermatocytes. *Oncogene,* **8,** 1487–1494.

Sever, L., Lynberg, M. C., and Edmonds, L. D. (1993). The impact of congenital malformations on public health. *Teratology* **48,** 547–549.

Shaulsky, S., Goldfinger, N., Peled, A., and Rotter, V. (1991). Involvement of wild type *p53* in pre-B cell differentiation *in vitro. Proc. Natl. Acad. Sci. USA* **88,** 8982–8986.

Soussi, T., and May, P. (1996). Structural aspects of the *p53* protein in relation to gene evolution: a second look. *J. Mol. Biol.,* **260,** 623–637.

Symonds, H., Krall, L., Remington, L., Robles, M. S., Lowe, S., Jacks, T., and Van Dyke, T. (1994). *p53*-dependent apoptosis suppresses tumor growth and progression *in vivo. Cell* **78,** 703–711.

Tanimura, T., and Shepard, T. H. (1970). Glucose metabolism by rat embryos *in vitro. Proc. Soc. Exp. Biol. Med.* **135,** 51–54.

Tchang, F., Gusse, M., Soussi, T., and Mechali, M. (1993). Stabilization and expression of high levels of *p53* during early development in *Xenopus laevis. Develop. Biol.,* **159,** 163–172.

Vander Heiden, M. G., Chandel, N. S., Williamson, E. K., Schumacker, P. T., and Thompson, C. B. (1997). *Bcl-*x_L regulates the membrane potential and volume homeostasis of mitochondria. *Cell* **91,** 627–637.

Vayssiere, J-L., Petit, P. X., Risler, Y., and Mignotte, B. (1994). Commitment to apoptosis is associated with changes in mitochondrial biogenesis and activity in cell lines conditionally immortalized with simian virus 40. *Proc. Natl. Acad. Sci. USA* **91,** 11752–11756.

Wlodarczyk, B. J., Bennett, G. D., Calvin, J. A., and Finnell, R. H. (1996). Arsenic-induced neural tube defects in mice: alterations in cell cycle gene expression. *Reprod. Toxicol.* **10,** 447–454.

Wubah, J., Ibrahim, M. M., Gao, X., Ngyuen, D., Pisano, M. M., and Knudsen, T. B. (1996). Teratogen-induced eye defects mediated by *p53*-dependent apoptosis. *Curr. Biol.* **6,** 60–69.

Xiang, H., Hochman, D. W., Saya, H., Fujiwara, T., Schwartzkroin, P. A., and Morrison, R. S. (1996). Evidence for *p53*-mediated modulation of neuronal viability. *J. Neurosci.* **16,** 6753–6765.

Yang, J., Liu, X., Bhalla, K., Kim, C. M., Ibrado, A. M., Cai, J., Peng, T.-I., Jones, D. P., and Wang, X. (1997). Prevention of apoptosis by Bcl-2: release of cytochrome c from mitochondria blocked. *Science* **275,** 1129–1132.

Yonish-Rouach, E., Resnitzky, D., Lotem, J., Sachs, L., Kimchi, A., and Oren, M. (1993). Wild-type *p53* induces apoptosis of myeloid leukaemic cells that is inhibited by interleukin-6. *Nature* **352,** 345–347.

Zamzami, A., Susin, S. A., Marchetti, P., Hirsch, T., Gomez-Monterrey, I., Castedo, M., and Kroemer, G. (1996). Mitochondrial control of nuclear apoptosis. *J. Exp. Med.* **183,** 1533–1544.

PART III

Synaptogenesis and Neurotransmission

The advent of computers and their link to information systems has revolutionized the way that humans communicate and thus enrich their lives. In an analogous fashion, synaptogenesis and neurotransmission provide the network by which neurons can tap into their own information highway to acquire multiple bits of information. This information is processed and analyzed to elicit a change in the neuron to maintain viability in a complex and dynamic environment. The inability of developing neural networks to form adequate connections or compensate for internal or external challenges can result in profound deficits and lead to abnormal function or demise. A fundamental aspect of developmental neuroscience and developmental neurotoxicology is in the understanding of defined genetic programs and modulatory factors that influence the formation of neuronal networks.

The section on synaptogenesis and neurotransmission encompasses the developmental aspects of the brain at the cellular and molecular levels. Effects of abnormal development resulting from exposure to xenobiotics are discussed. Chapter 11, by Hendrickx and Peterson, describes the migration of neural crest cells, genetic and epigenetic factors influencing their migration, and aberrations or disease that results from abnormal development. This is followed by chapters discussing the ontogeny of monoaminergic (Broening and Slikker, Chapter 12) and cholinergic (Jett, Chapter 13) innervation of the brain and their respective phenotypic markers. Chapter 15 on amino acid neurotransmitters (by Guilarte) focuses on *N*-methyl-D-aspartate receptor ontogeny, a glutamatergic receptor subtype that plays an important role in brain development and synaptic plasticity. Neuronal receptors play an essential role in capturing external signals and the information is processed with high fidelity by second-messenger systems. This allows the developing neuron to adapt to a changing environment by altering the expression of its genes. Chapter 13, by Costa, reviews the sequence of biochemical events that represent signal transduction in the ontogeny of the nervous system as it relates to mechanisms of developmental neurotoxicology.

The multitude of potential targets and the complexity of programmed cellular and molecular events in synaptogenesis and chemical neurotransmission makes these processes susceptible to environmental influences such as xenobiotics. The collaborative efforts of toxicologists, neuroscientists, and molecular biologists have the potential to define xenobiotic's toxic effects and strategies to alleviate their impact on human health.

Tomás R. Guilarte

CHAPTER

11

Neural Crest Cell Migration

ANDREW G. HENDRICKX
PAMELA E. PETERSON
California Regional Primate Research Center
University of California
Davis, California 95616

I. Introduction

The neural crest is a specialized cell population that originates at the apex of each neural fold, at the junction of the neuroectoderm and surface ectoderm (Selleck *et al.*, 1993). As the neural folds fuse to form the neural tube, neural crest cells detach by means of an epithelial–mesenchymal transformation and migrate through mesenchyme along specific pathways characteristic of their axial level of origin (Le Douarin, 1982; Newgreen and Erickson, 1986; Bronner-Fraser, 1993a). The last step in this process is the cessation of movement when neural crest cells settle in their final position and differentiate into a diverse array of cells and tissues (Carlson, 1994). Although the majority of trunk crest derivatives are neural (e.g., spinal ganglia), cranial crest derivatives in-

*Abbreviations: A-CAM, adherens cell adhesion molecules; BDNF, brain derived growth factor; bFGF, basic fibroblast growth factor; BM, basement membrane; CAM, cell adhesion molecules; CHARGE, coloboma of the eye, heart disease, choanal (nasal) atresia, retarded growth and development, genital hypoplasia, and ear anomalies; CNS, central nervous system; CRABP, cellular retinoic aced binding protein; DiI, 1,1-dioctadecyl-3,3,3′,3′-tetramethylindocarbocyanine percholate; ECM, extracellular matrix; EGF, epidermal growth factor; FGF, fibroblast growth factors; HGF, hepatocyte growth factor; L-CAM, liver cell adhesion molecules; MEN, multiple endocrine neoplasia; N-CAM, neural cell adhesion molecules; NF1, neurofibromatosis type 1 (von Recklinghausen's disease); NF2, neurofibromatosis type 2; NGF, nerve growth factor; NT-3, neurotrophin-3; PDGF, platelet-derived growth factor; PNS, peripheral nervous system; RAS, retinoic acid syndrome; SF scatter factor; TGF-β, transforming growth factor-β; WGA, wheat germ agglutinin.

clude both neural (e.g., cranial ganglia) as well as mesodermal (e.g., connective tissue, cartilage, and bone) tissues (Table 1, Fig. 1).

Much of the information available regarding neural crest migration and differentiation comes from studies in the amphibian, chick, and rodent embryo (Le Douarin, 1982; Hall and Horstadius, 1988; Morriss-Kay *et al.,* 1993; Bronner-Fraser, 1994). In all these species, neural crest morphogenesis is a multistep process that is tightly regulated both spatially and temporally by both internal factors and external signals. Whereas many common mechanisms of crest emigration exist between species, as well as between axial levels, there are differences in the time course of events that are species-specific and characteristic of cranial versus trunk neural crest (Duband *et al.,* 1995). In this chapter, we present an overview of these events with a focus on the crest cell contribution to neural development. No emphasis is placed on the subject of neural crest ontogeny (lineage analysis) because this extensive topic has been amply reviewed elsewhere (Dupin *et al.,* 1993; Le Douarin and Ziller, 1993; Murphy and Bartlett, 1993).

II. Neural Crest Migration

A. *Techniques*

Advances in our understanding of neural crest morphogenesis have largely been due to the development of innovative technologies to probe specific aspects of this complex process. Standard morphologic approaches, including special histologic stains (Nichols, 1981), transmission electron microscopy (Nichols, 1987), and scanning electron microscopy (Tan and Morriss-Kay, 1985) have elucidated cellular features as well as early phases of crest emigration. Characterization of migration pathways in both the head and trunk has been accomplished with various cell marking and grafting techniques using tritiated thymidine (Weston, 1963; Noden, 1975), chick-quail chimeras (Le Douarin, 1973),

TABLE 1 Major Derivatives of the Neural Crest

	Trunk crest	Cranial crest
Nervous system		
Sensory neurons	Spinal ganglia	Ganglia of trigeminal nerve (V) Ganglia of facial nerve (VII) Ganglia of glossopharyngeal nerve (IX) Ganglia of vagus nerve (X)
Sympathetic neurons	Superior cervical Paravertebral chain Celiac, mesenteric, adrenal, and retroaortic complexes	
Parasympathetic neurons	Pelvic plexus Visceral	Ciliary, ethmoidal, sphenopalatine, submandibular, otic, lingual, visceral
Nonneural support cells	Satellite cells of sensory ganglia Schwann cells of peripheral nerves Enteric glial cells	Satellite cells of sensory ganglia Schwann cells of peripheral nerves Leptomeninges of prosencephalon and part of mesenceophalon
Pigment cells	Melanocytes	Melanocytes
Endocrine/paraendocrine cells	Adrenal medulla Neurosecretory cells of heart and lungs	Carotid body (type I cells) Parafolicular cells (thyroid)
Skeletal/connective tissue		
Cartilage		Visceral arch and chondrocranium
Bones		Nasal, orbital, palatal, maxillary Parts of cranial vault, sphenoid complex, otic capsule
Connective tissue		Dermis and fat of skin Cornea of eye Dental papilla (odontoblasts) Stroma of thyroid, parathyroid, thymus, salivary, lacrimal Conotruncal region of heart Cardiac semilunar valves Walls of aorta and aortic arch–derived arteries
Muscle		Ciliary muscles Dermal smooth muscles Vascular smooth muscle

From the literature.

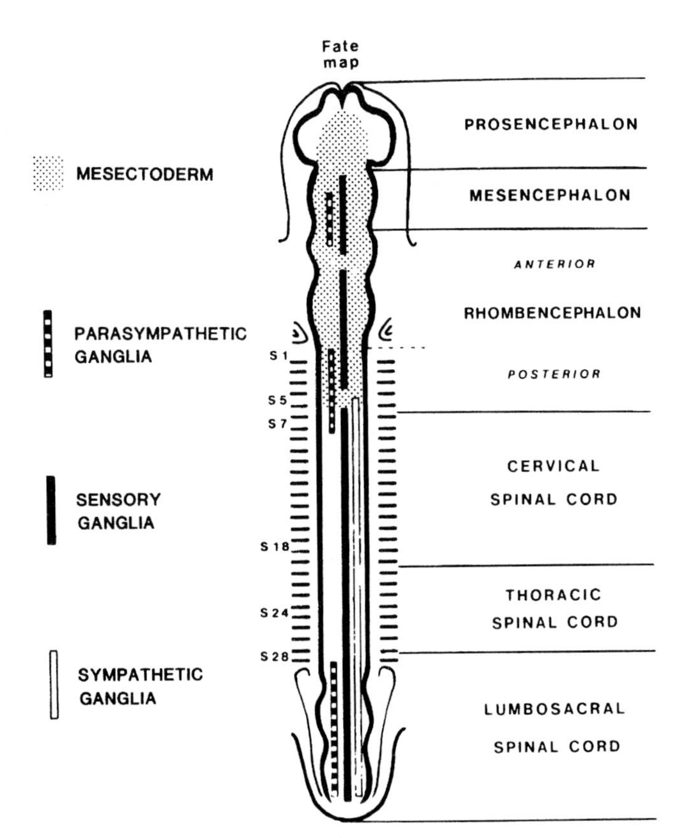

FIGURE 1 Fate map of the neural crest demonstrating the presumptive territories of the mesenchymal (mesectoderm) and neural (sensory and autonomic ganglia) derivatives in the avian embryo (established by quail-chick isotopic transplantation studies). (Reprinted with permission from N. M. Le Douarin, C. Ziller, and G. F. Couly. (1993). Patterning of neural crest derivatives in the avain embryo: *in vivo* and *in vitro* studies. *Dev. Biol.* 159, 24–29. Academic Press, San Diego.)

and wheat germ agglutinin (WGA)-gold conjugate (Chan and Tam, 1988). Rapid progress in this area has also been made possible by immunoctyochemistry using antibodies to label subpopulations of crest cells during specific phases of their development. NC-1 and HNK-1 antibodies, which recognize a surface carbohydrate on neural crest cells, are useful tools in studying crest migration in chick and rodent embryos (Tucker *et al.*, 1984; Vincent and Thiery, 1984). The use of antibodies to neural cell adhesion molecule (NCAM) has facilitated neural crest studies in the nonhuman primate (Peterson *et al.*, 1996). Cellular retinoic acid binding protein (CRABP) and neurofilament antibodies have also been used successfully to track neural crest cell pathways and their destinations (Maden *et al.*, 1992; Sechrist *et al.*, 1993).

These largely static techniques have been supplemented by more dynamic methodologies. The spatial and temporal aspects of crest migration have been studied using *in vitro* culture (Selleck and Bronner-Fraser, 1995), *exo utero* techniques (Serbedzija *et al.*, 1992), and intravital microscopy (Birgbauer *et al.*, 1995) combined with cell markers. One of the most useful of these methods involves labeling premigratory crest cells with the vital dye, DiI (1,1-dioctadecyl-3,3,3',3'-tetramethylindocarbocyanine perchlorate) (Serbedzija *et al.*, 1989). This technique has been particularly valuable in clarifying the varied pathways and destinations of cranial neural crest cells in the chick and rodent (Bronner-Fraser, 1993b; Osumi-Yamashita *et al.*, 1994). Finally, inhibition studies using antibodies (Bronner-Fraser *et al.*, 1992) and ablation techniques (Scherson *et al.*, 1993) have advanced our understanding of the regulatory and functional aspects of neural crest development.

Together, the use of new molecular technologies in conjunction with morphologic studies provides the basis for the information on neural crest development and maldevelopment presented in the following sections.

B. Cranial Crest

Cranial neural crest cells make a substantial contribution to normal craniofacial development (see Table 1). The pathways of migration, diverse derivatives, and regulatory factors are unique relative to trunk neural crest cells. Moreover, there are region-specific characteristics for neural crest emanating from the forebrain, midbrain, and hindbrain. Finally, despite apparent conservation of cranial crest development among species, there are species-specific differences that are particularly noteworthy from an evolutionary viewpoint.

One such difference between mammalian and avian embryos involves the initial phase of crest emigration, that is, epithelial–mesenchymal transformation. Cranial crest cell emigration in the chick embryo is an intrinsic part of neurulation. Crest cells are excluded from the neural epithelium concomitant with apposition and fusion of the neural folds (deVirgilio *et al.*, 1967; Anderson and Meier, 1981; Tosney, 1982). Crest cells accumulate and are observed as bulges in the dorsal neural tube, which build up as dorsolateral masses prior to emigration (Fig. 2A). In contrast, cranial crest emigration in the mouse and rat is not strictly correlated with the stage of neural tube closure; thus, crest cells migrate in a nonlinear sequence from open or fused neural folds in different brain regions (Fig. 2B) (Tan and Morriss-Kay, 1985; Nichols, 1987). The formation of crest cells before, during, and after neural tube closure has also been observed in human and nonhuman primate embryos (Müller and O'Rahilly, 1985, 1986, 1987; Peterson *et al.*, 1996). In all these mammalian species, cranial crest emigration is initiated when the neural folds are widely separated and continues as the folds appose each other and eventually fuse.

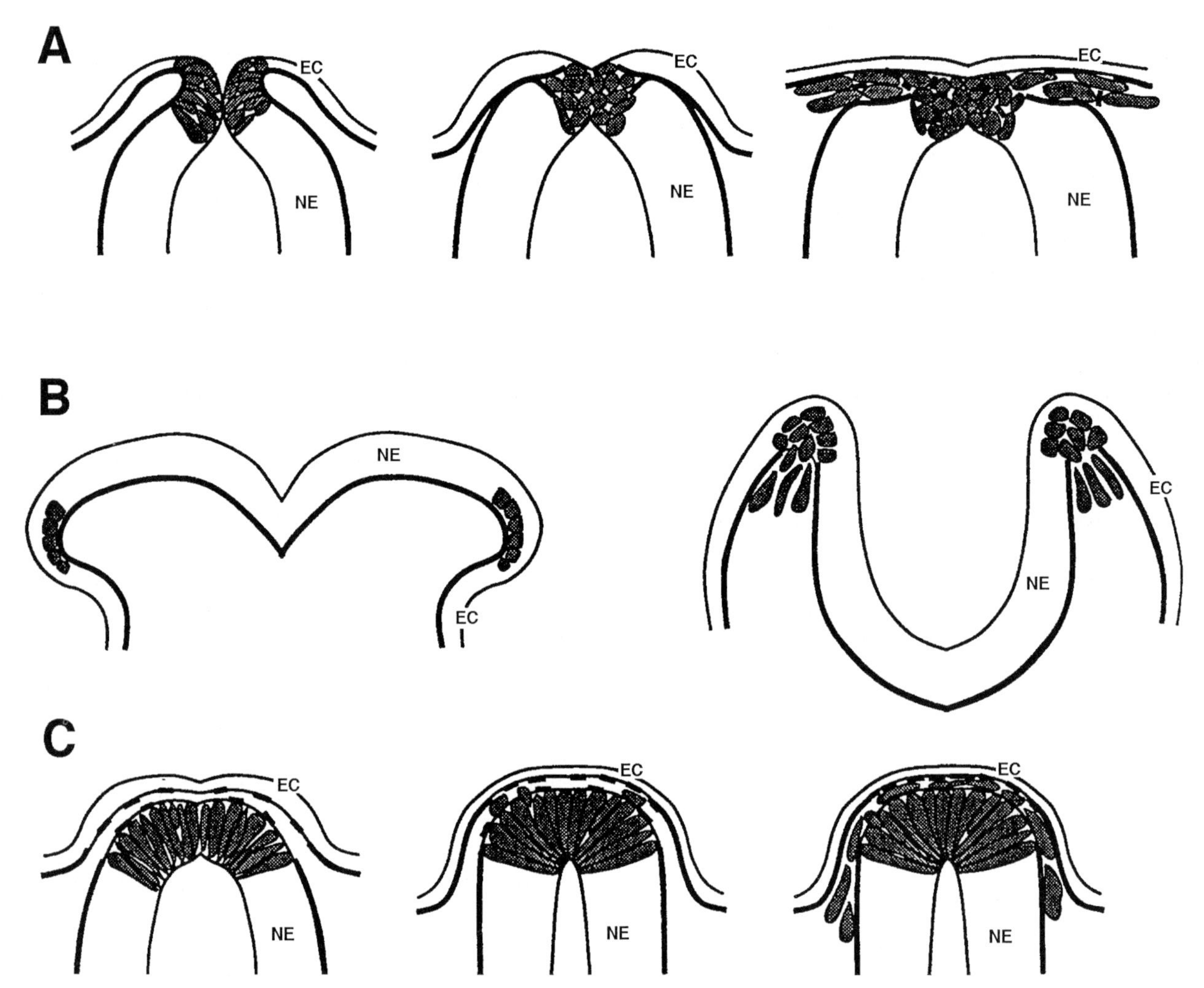

FIGURE 2 Schematic representation of the major modes of neural crest cell emigration from the neural epithelium that differ with the axial level in the avian and mammalian embyro. (A) Cranial region in the chick embyro, where crest emigration coincides with neural tube closure. (B) Cranial region of the mouse embyro, where crest emigration occurs before as well as after neural tube closure. (C) Trunk region in the chick and mouse embryo, where crest emigration occurs immediately after closure of the neural tube prior to repair of the neuroepithelial basement membranes. Thick lines = basement membranes; NE = neural epithelium; EC = epidermal ectoderm. (Adapted from J. L. Duband, F. Monier, Delannet, and D. Newgreen (1995). [Epithelium–mesenchyme transition during neural crest development. *Acta Anat.* **154,** 63–78. With permission from S. Karger AG, Basel].)

One consistent feature of the emigration process in all species studied is the dissolution and subsequent repair of the neuroepithelial basement membrane (BM) in areas of crest migration (Tosney, 1982; Innes, 1985; Müller and O'Rahilly, 1985; Tuckett and Morriss-Kay, 1986; Peterson *et al.,* 1996). Although the exact mechanism of BM degradation has not been established, it has been suggested that this step may provide the stimulus for crest cell emigration in the cranial region (Erickson and Perris, 1993).

Neural crest cells arise from all three major brain regions, that is, prosencephalon (forebrain), mesencephalon (midbrain), and rhombencephalon (hindbrain). The bulk of research has been directed towards hindbrain crest development, which exhibits many similarities across mammalian and avian species (Müller and O'Rahilly, 1985, 1986, 1987; Nichols, 1987; Lumsden *et al.,* 1991; Serbedzija *et al.,* 1992; Peterson *et al.,* 1996). Hindbrain crest cells emigrate in three broad streams from the second, fourth, and sixth rhombomeres, which are overt swellings of the hindbrain neuroepithelium during early development in all vertebrates. The first of these streams populates the trigeminal ganglion and mandibular arch; the second populates the hyoid arch and the facioacoustic ganglion; and the third stream populates the third and fourth arches and the glossopharyngeal–vagal ganglia (Fig. 3). In general, the earliest crest cells that emigrate from each region of the hindbrain move to the most ventral sites in the developing embryo to contribute to the mesenchymal progenitors of the facial skeleton. The last crest cells to exit from these areas seed the cranial sensory ganglia associated with the pharyngeal nerves (Fig. 4). In addition to the neural crest contribution to the cranial ganglia, there is also a significant cellular contribution to the sensory neurons associated with these ganglia that are made by the head placodes (Noden, 1993).

The mechanisms responsible for the segmentation of hindbrain crest have been well-studied in the chick embryo. Initial investigations attributed this phenomenon to apoptotic depletion of crest cells in rhombomeres 3 and 5, which ensures that streams of crest cells populating the arches are kept apart during migration by crest-free areas along the neuraxis (Lumsden *et al.,* 1991; Graham *et al.,* 1993). These studies suggest that there

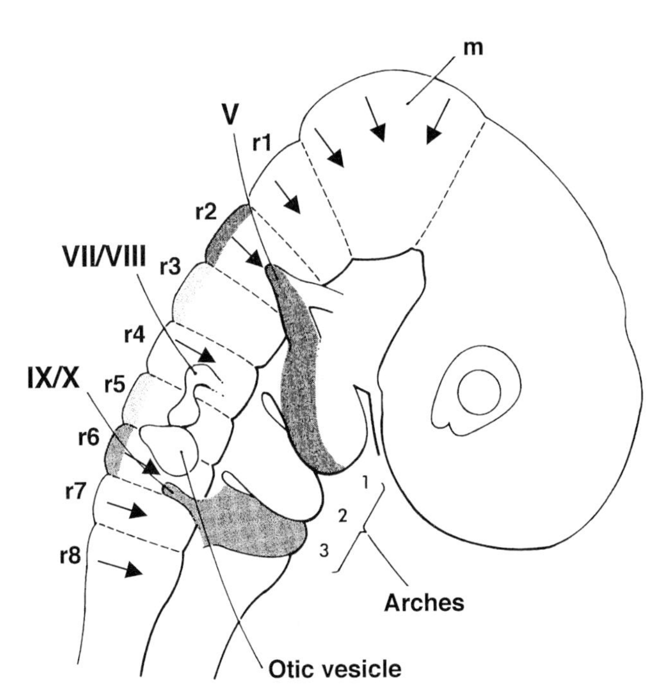

FIGURE 3 Diagram showing the sites of origin and migration paths of rhombencephalic (r1–r8) and mesencephalic (m) neural crest cells in the chick embryo. Separate streams of migrating cells emerge from m, r1, r2, r4, and r6, which enter the first, second, and third pharyngeal arches, respectively. These crest streams also contribute to cranial nerves V, VII/VIII, and IX/X, respectively. The intervening rhombomeres, r3 and r5, do not release crest cell directly into the periphery. Crest cells from the caudal hindbrain (r7 and r8) undergo extensive migrations to populate the cardiac and enteric ganglia. (Adapted with permission from A. Lumsden and A. Graham (1996). [Death in the neural crest: implications for pattern formation. *Semin. Cell Dev. Biol.* **7,** 169–174. Academic Press, San Diego].)

is regionalization of neural crest cells along the neural axis, which dictates a segmented migration pattern. However, the use of DiI labeling combined with intravital microscopy has shown that small numbers of surviving crest cells from these odd-numbered rhombomeres migrate within the neuroepithelium to exit at the level of adjacent even-numbered rhombomeres. Thus, rhombomere 3 contributes neural crest cells to both the first and second arches; rhombomere 5 contributes to the second and third arches (Sechrist *et al.,* 1993; Birgbauer *et al.,* 1995). These findings suggest that the segmental migration pattern of hindbrain neural crest may be due, in part, to inhibitory or attractive signals from the environment. Two structures that have been implicated as impediments to migration from the third and fifth rhombomeres are the ectoderm of the first pharyngeal groove and the otic vesicle, respectively (Sechrist *et al.,* 1993). Thus, the segmentation nature of crest formation from the hindbrain likely results from an interplay between intrinsic (e.g., Hox genes) and extrinsic (pathway) influences (Sechrist *et al.,* 1994).

Neural crest cells originating in the midbrain migrate primarily as a broad, unsegmented sheet of cells under the surface ectoderm (Fig. 4) to contribute to derivatives ranging from the periocular skeleton, the connective tissue of the eye (sclera, choroid, cornea, and iris), and the membrane bones of the face (maxillary, prefrontal, nasal, and temporal), to the ciliary ganglion, part of the trigeminal ganglion, and Schwann cells (Le Douarin, 1982).

In contrast to the well-studied contribution of the hindbrain and midbrain to the formation of neural crest cells, relatively little information is available regarding crest cells originating from the forebrain. Several studies in rodents have substantiated the presence of neural crest cells exiting various regions of the forebrain that contribute to the mesenchyme of the periocular region and the frontonasal mass (Fig. 4) (Nichols, 1981; Serbedzija *et al.,* 1992; Morriss-Kay *et al.,* 1993; Osumi-Yamashita *et al.,* 1994). In contrast, neural crest cells are generated only from the posterior portion of the forebrain in the chick embryo (Bronner-Fraser, 1994). The generation of crest cells from the optic vesicle has also been addressed in mammalian species, including the human and nonhuman primate, but not avian species (Bartelmez and Blount, 1954; Bartelmez, 1960; Blankenship *et al.,* 1996).

The cranial crest cells juxtaposed to somites 1 through 5 populate the third, fourth, and sixth pharyngeal arches. They provide mesenchymal support for the organs derived from the corresponding pouches including the parathyroids and thymus, as well as the nonendothelial layers of the arteries of the arches. Cells from this area also migrate ventrally and caudally to participate in

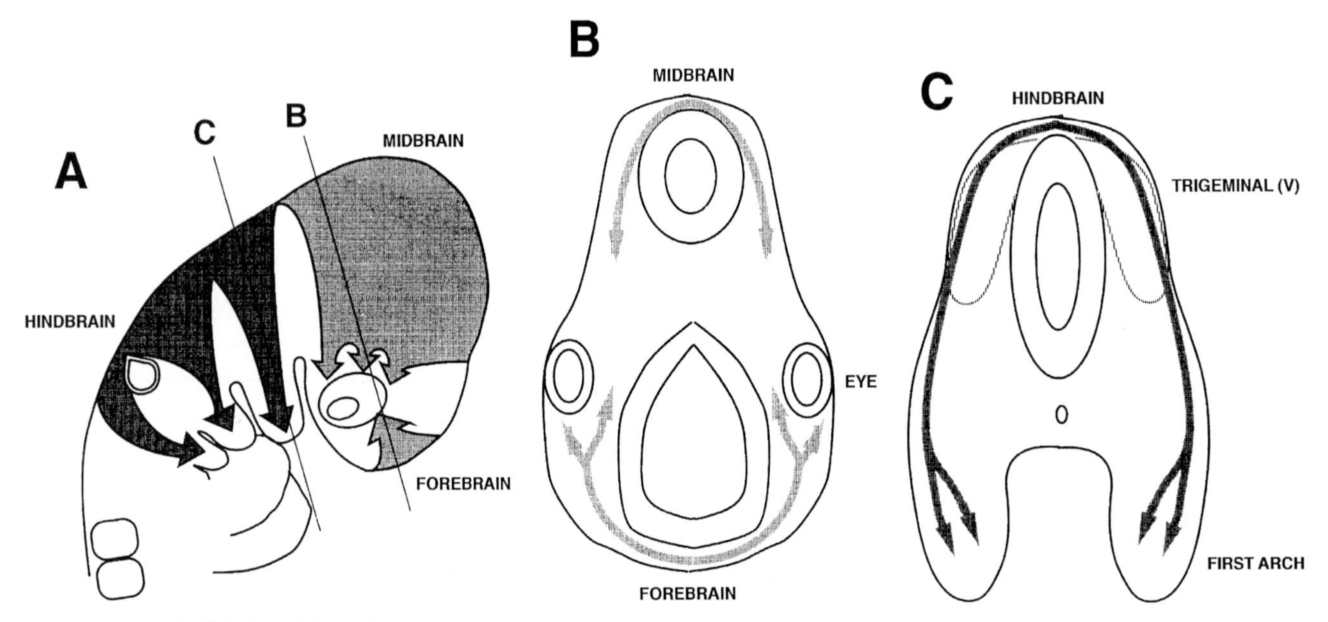

FIGURE 4 Schematic representation showing the pathways of neural crest cell migration in the mouse. (A) There are three regionally distinct pathways of migration seen in this whole embyro view. Neural crest cells from the level of the forebrain migrate ventrally in a sheet extending from the dorsal portion of the neural tube to the level of the eye. Neural crest cells from the level of the midbrain migrate ventrolaterally toward the maxillary process and the eye. At the level of the hindbrain, neural crest cells migrate ventrolaterally in three streams, from the dorsal portion of the neural tube to the distal portion of the first, second, and third pharyngeal arches. (B) A cross-sectional view through the level of the eye indicates that neural crest cells from both the midbrian and the forebrain migrate through the mesoderm. In addition, neural crest cells from the forebrain migrate between the eye and the forebrain. (C) A cross-sectional view through the level of the first pharyngeal arch shows that neural crest cells migrate along a subectodermal pathway that extends from the hindbrain into the first arch. (Adapted with permission from The Company of Biologists, Ltd., Cambridge. G. N. Serbedjiza, M. Bronner-Fraser, and S. E. Fraser. (1992). Vital dye analysis of cranial neural crest cell migration in the mouse embryo. *Development* **116,** 297–307.)

cardiac septation and to contribute to the enteric nervous system (Le Douarin, 1982).

C. Trunk Crest

The trunk neural crest extends from the area adjacent to the sixth somite to the lumbosacral region and develops in a craniocaudal sequence (Bronner-Fraser, 1994; Carlson, 1994). Trunk crest cells emigrate from the neural tube shortly after the neural folds have closed (Fig. 2C) and migrate into prescribed areas along two major pathways: (1) a dorsal or dorsolateral pathway between the surface ectoderm and the somites, and (2) a ventral or ventromedial pathway through the somites (Fig. 5). Crest cells taking the dorsal route migrate in an unsegmented fashion and eventually invade the surface ectoderm, synthesize melanin, and become melanocytes, which provide pigmentation to the skin. Crest cells taking the ventral route migrate in a segmental manner and become organized into dorsal root ganglia, which contain the sensory neurons that conduct impulses to the spinal cord from end organs in the viscera, body wall, and extremities. A subset of cells in the ventral group bypass the somites and colonize the region near the dorsal aorta. These cells contribute to the formation of the adrenal medulla and elements of the sympathetic nervous system (Fig. 6) (Le Douarin, 1982; Carlson, 1994). In the thoracic region, the neural crest cells that migrate adjacent to the aorta form a bilateral chain of sympathetic ganglia that are interconnected by longitudinal nerve fibers and are situated on each side of the vertebral column. Neuroblasts migrate from their original position in the thorax into the cervical and lumbar regions, which represent the full extension of the sympathetic chains. Initially the ganglia are segmentally arranged but the segmentation is gradually obscured by their fusion, especially in the cervical region. Migration of the neuroblasts near the aorta (preaortic ganglia) form the celiac and mesenteric ganglia; others migrate to the heart, lungs, and gastrointestinal tract and form sympathetic organ plexuses (Fig. 6) (Sadler, 1995).

To date, studies in chick and rodent embryos indicate similar mechanisms of emigration, migration pathways, and nervous system derivatives of trunk neural crest

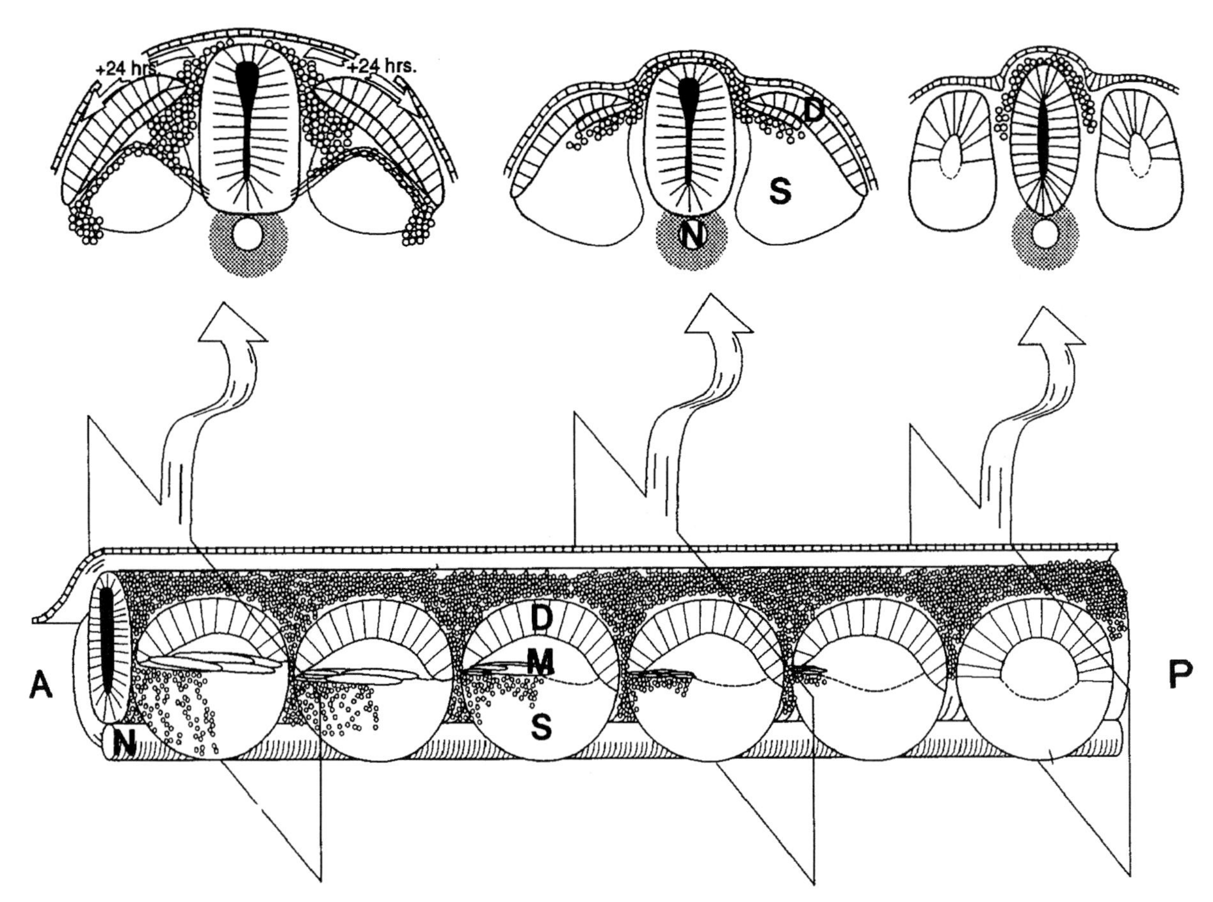

FIGURE 5 Summary of the migratory pathways followed by trunk neural crest cells. Bottom: A sagittal section through the somites indicating the simultaneous development of the myotome and neural crest cell invasion into the anterior somite. Top: Cross sections from different axial levels show the progressive spread of neural crest cells initially along the myotome and later filling the dorsal sclerotome. They avoid the more ventral sclerotome adjacent to the notochord. A = anterior; P = posterior; D = dermatome; M = myotome; S = sclerotome; N = notochord. (Reprinted with permission from Academic Press, San Diego. C. A. Erickson and R. Perris (1993). The role of cell–cell and cell–matrix interactions in the morphogenesis of the neural crest. *Dev. Biol.* **159,** 60–74.)

cells in avian and mammalian species (Morriss-Kay *et al.,* 1993). In both species, the lack of a continuous BM in the dorsal neural tube is a necessary condition for emigration (Erickson and Perris, 1993). However, because the BMs of the recently fused neural tube are discontinuous prior to and during the time of crest emigration, this condition is viewed as permissive, rather that a trigger for, migration in the trunk region. This is in contrast to the cranial region, where some neural crest cells begin their migration prior to the fusion of the neural folds and must degrade or penetrate the BM prior to departure (Tosney, 1982; Innes, 1985).

Segmentation of trunk neural crest cells is largely responsible for the segmental patterning of the peripheral nervous system (PNS) and its associated dorsal root ganglia, sensory ganglia, and spinal nerves (Fig. 7). This early segmentation process has been studied most extensively in the chick embryo using a variety of experimental approaches (Erickson and Perris, 1993; Bronner-Fraser, 1995). Collectively, these studies have shown that neural crest cell migration in this region is influenced primarily by the environment through which these cells migrate, rather than by craniocaudal compartmentalization of the trunk neuroepithelium (Le Douarin *et al.,* 1993). Trunk neural crest cells preferentially migrate through the rostral half of each somite, which is permissive to crest migration, rather than the caudal somite, which is nonpermissive or inhibitory to migration (Fig. 5). Experimental rostro–caudal rotation of the segmental plate mesenchyme results in a reversal of the normal migration route and suggests that the cues for migration are inherent in the somitic mesoderm prior to their differentiation from the segmental plate (Bronner-Fraser and Stern, 1991). Possible inhibitory

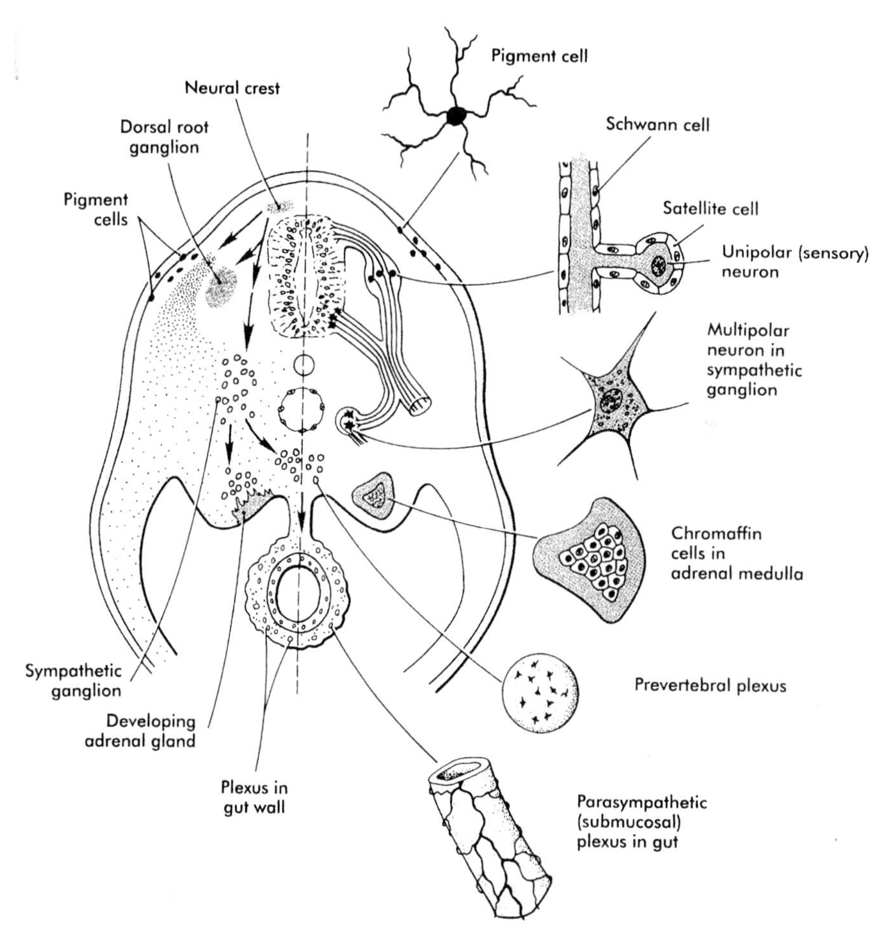

FIGURE 6 Major neural crest migratory pathways and derivatives in the trunk. Left: Pathways in the early embryo. Right: Derivatives of the trunk neural crest. (Reprinted with permission from Mosby-Year Book, Inc. St. Louis. B. M. Carlson (1994). *Human Embryology and Developmental Biology.* Mosby-Year Book, St. Louis. p. 245.)

molecules in the caudal somite include T-cadherin, chondroitin sulfate proteoglycans, and molecules recognized by peanut lectins. Possible permissive molecules in the rostral somite are butyrlcholinesterase and tenascin. Inhibitory cues from the notochord may also influence the migratory pathways of trunk neural crest cells (see Bronner-Fraser, 1995). Both *in vitro* and *in vivo* studies suggest that the notochord produces an inhibitory substance, chrondroitin sulfate proteoglycan, which has a negative regulatory effect on neural crest migration (Newgreen *et al.,* 1986; Bronner-Fraser, 1993 a,b).

III. Modulating Factors

A. *Genetic*

Our understanding of the molecular basis of neural crest specification along the anterior–posterior axis has come from analyses of the expression and function of homeobox genes, which are homologous to the homeotic genes of *Drosophila* (Hunt *et al.,* 1991). The most prominent of this group are the vertebrate *Hox* genes, which are organized in four clusters located on four different chromosomes, with all genes within each cluster having the same orientation relative to the 5′-3′ direction of transcription. With only a few exceptions, the genes located at the 3′ end of the clusters are expressed first in embryogenesis in the more anterior region of the embryo; the genes in the 5′ region are expressed in progressively more posterior regions as development proceeds (Hunt and Krumlauf, 1992). These genes, which have been identified in the human and mouse, encode transcription factors that control the expression of other genes (Murphy and Bartlett, 1993).

Although the expression of *Hox* genes occurs along the entire rostro–caudal axis of the developing embryo, particular attention has been directed to their expression pattern in the hindbrain region (Bronner-Fraser, 1994). As indicated in Fig. 8, there are overlapping expression patterns of different *Hox* genes that have their rostral borders of expression at rhombomere borders,

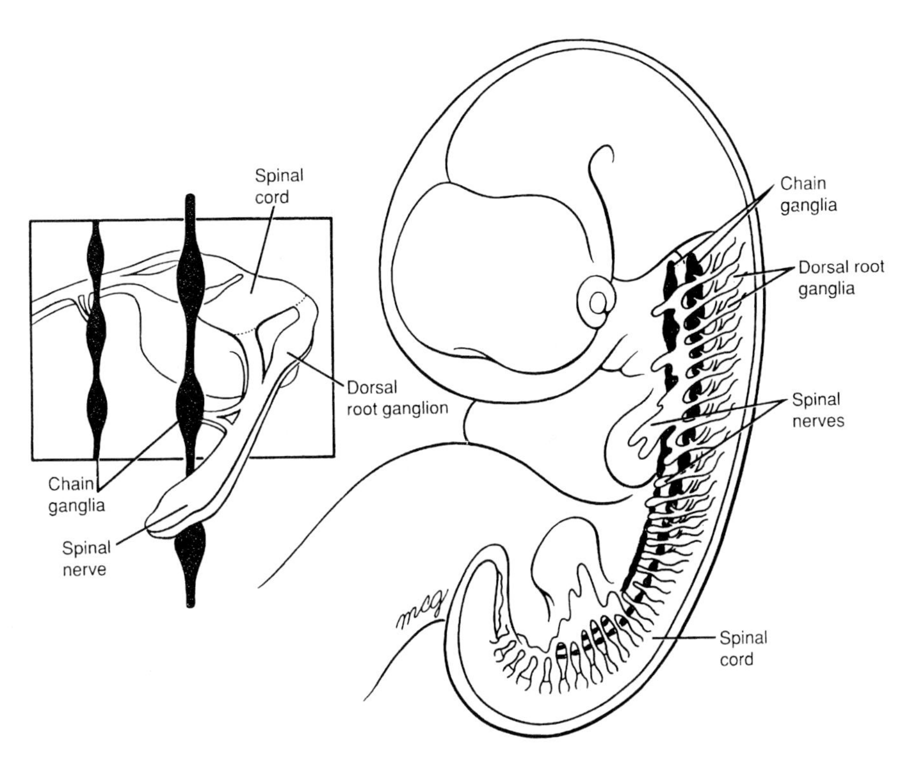

FIGURE 7 Schematic indicating that neural crest cells produce chain ganglia and dorsal root ganglia at almost every spinal segment. (Reprinted with permission from Churchill Livingston, New York. W. J. Larsen (1993). *Human Embryology*. Churchill Livingstone, Inc. p. 81.)

in a rostral to caudal order that corresponds with their 3′ to 5′ sequence along the chromosome. This figure also illustrates the expression of the zinc finger gene *Krox-20,* which is expressed in alternating domains in the neural plate that later corresponds to rhombomeres 3 and 5. *Krox-20* acts as a transcriptional regulator of *HoxB2* expression (Sham *et al.,* 1993). There is evidence that the pattern of gene expression in the developing hindbrain is replicated in the neural crest cells that migrate from each rhombomeric segment (Hunt *et al.,* 1991). This "code" of gene expression is proposed to underlie the patterning of pharyngeal arches and the cranial ganglia, the derivatives of the hindbrain neural crest cells (Bronner-Fraser, 1995). Targeted mutations of some of the genes comprising this "code" lead to perturbations in the development of these craniofacial structures (Le Douarin *et al.,* 1994).

Additional homeobox genes purported to play a role in patterning of the cranial neural crest and their mesectodermal derivatives are *Msx1* and *Msx2,* which belong to the *msh* gene family in *Drosophila* (Takahashi and Le Douarin, 1990). The *Pax* gene family, in particular *Pax-3,* also appears to play a significant role in early specification during neurulation. *Pax-3* is expressed just prior to neural tube closure in the dorsal aspect of the neuroectoderm along the length of the neural tube as well as in neural crest derived spinal ganglia, dorsal root ganglia, somitic mesoderm, and some craniofacial neural crest derivatives (Goulding *et al.,* 1991). It has been proposed that *Pax-3* regulates the expression of proteins essential for the migration of neural crest cells, such as extracellular matrix (ECM) and cell adhesion molecules (Moase and Trasler, 1991). Other molecules have been reported to show regional patterns of expression in the neural epithelium of the midbrain and forebrain, including the transcription factors Otx-2, Pax-6, and En-2, and the *Wnt* genes, which are homologs of the *Drosophila wingless* gene (Parr *et al.,* 1993; Bally-Cuif and Wassef, 1994).

A number of regulatory genes code for transcription factors required for autonomic neuronal development (Groves and Anderson, 1996). One of these, MASH-1, is expressed in neural crest cells as they arrive at sites of peripheral neurogenesis (Lo *et al.,* 1991). The expression of MASH-1 is restricted to precursor cells of the three autonomic lineages—sympathetic, parasympathetic, and enteric neurons—as well as the neuroendocrine chromaffin cells of the adrenal medulla. Gene knockout experiments demonstrate its necessity for the differentiation of many neurons in all three branches of the autonomic neuronal lineage. Several lines of circumstantial evidence suggest that a second factor, Phox2, may play a role in the determination of neurotransmitter phenotype in both the PNS and central ner-

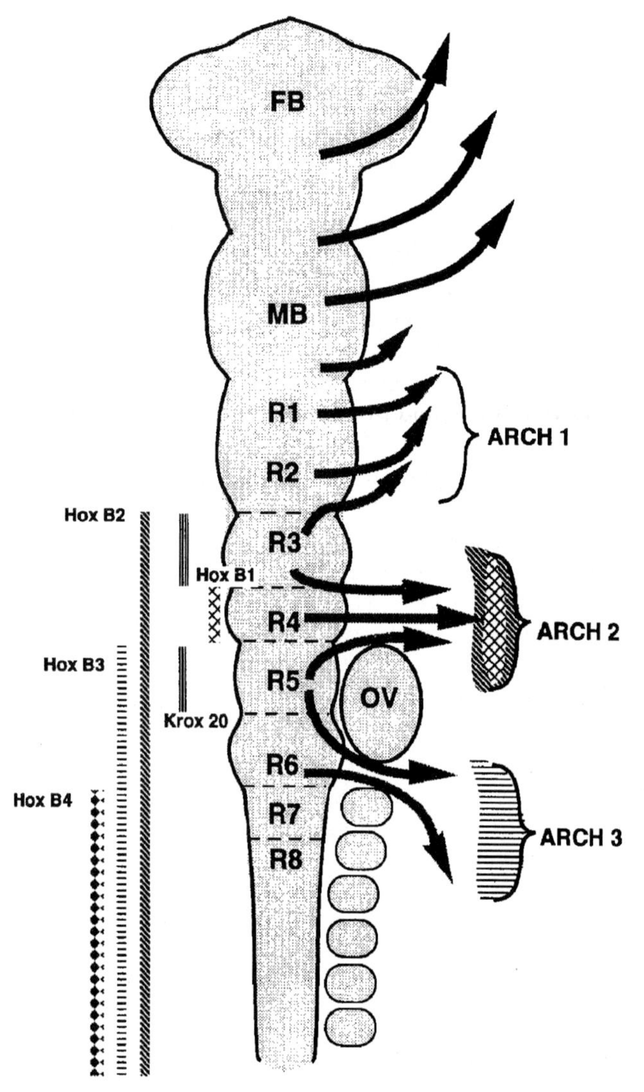

FIGURE 8 Schematic diagram illustrating the rhombomeres (R), pharyngeal arches (1, 2, and 3), and the expression patterns of known genes (*Hox* β1, β2, β3, and β4, and *Krox* 20). Arrows indicate the directions of migrating nerual crest cells arising from the rhombomeraes FB = forebrain; MB = midbrain. (Reprinted with permission from Lancaster Press, Pennsylvania. M. Bronner-Fraser (1994). Neural crest cell formation and migration in the developing embryo. *FASEB J.* **8,** 699–706.)

vous system (CNS). In addition, GATA-2 and GATA-3, which are zinc finger–containing genes, have been shown to be expressed at particular times during the maturation of subsets of autonomic neuronal precursors (Groves and Anderson, 1996).

There is a growing body of literature indicating that the vitamin A metabolite, retinoic acid, is involved in patterning of the developing CNS and the neural crest via its effects on the expression of *Hox* genes (Kessel and Gruss, 1991; Marshall *et al.,* 1996). The induction of homeotic transformations of vertebrae in addition to alterations in hindbrain and pharyngeal arch patterning by teratogenic doses of retinoic acid is consistent with this morphogenetic role (see retinoic acid syndrome in Sec. IV.A). Additional evidence for a regulatory role of retinoic acid in the developing nervous system and the neural crest include the induction of neurite outgrowth and stimulation of neural crest cell proliferation by retinoic acid in culture as well as the presence of retinoic binding proteins and receptors in early stages of development of the CNS (Maden and Holder, 1992; Morriss-Kay, 1992; Ito and Morita, 1995).

B. Epigenetic

1. Growth Factors

Growth factors are among the diverse molecules that are present in the extracellular environment that have been shown to play a critical role in neural crest development (Duband *et al.,* 1995). Although considerable data have accumulated on their role in governing the phenotypic expression of neural crest cells (Stemple and Anderson, 1993; Murphy and Bartlett, 1993), less attention has been directed at the involvement of growth factors in the initial phases of neural crest formation, that is, epithelial–mesenchymal transformation. The major growth factors that have been implicated in both of these processes are summarized in Table 2 and briefly summarized below.

Transforming growth factor-β (TGF-β) family members are active participants in controlling neural crest cell emigration from the neural tube and are involved

TABLE 2 Growth Factors Involved in Neural Crest Development

Activity	Growth factors
Emigration of crest cells from neural tube	TGF-β (transforming growth factor)
	FGF (fibroblast growth factor)
	Wnt gene products
	PDGF (platelet-derived growth factor)
	EGF (epidermal growth factor)
	SF/HGF (scatter factor/hepatocyte growth factor)
Proliferation of primary neural crest cells	FGF and bFGF
	NT-3 (neurotrophin-3)
Differentiation of neural crest cells	
Sympathoadrenal lineage	FGF and bFGF
	NGF (nerve growth factor)
Sensory lineage	BDNF (brain-derived neurotrophic factor)
	NT-3
	bFGF

From the literature.

in several key biologic events, including cell dispersion as well as regulation of synthesis of various extracellular matrix molecules and their integrin receptors. *In vitro* studies have demonstrated that TGF-β1 and TGF-β2 stimulate premature emigration of avian neural crest cells from premigratory neural tube explants. These growth factors also increase cellular substratum–adhesion properties but fail to regulate locomotory and intercellular adhesion properties of neural crest cells, indicating that they do not control the entire emigration process (Duband *et al.,* 1995). It has been proposed that TGF-β may stimulate neural crest cells to emigrate in response to a modification of expression pattern or binding properties of their cell surface integrins. At present, it is unknown which of the TGF-β family members play a role in controlling cell migration *in vivo,* although different types of TGF-β are present in the neural tube and in neural crest cells at the time of their emigration (Millan *et al.,* 1991).

Fibroblast growth factors (FGF) are also considered to be active participants in controlling emigration of neural crest cells due primarily to their presence in the neural tube during early development (Baird, 1994). In addition, *Wnt* gene products, a family of putative growth factors, are implicated in stabilizing intercellular junctions at the dorsal midline during closure of the neural tube as well as after neural crest cell emigration (Duband *et al.,* 1995). It has also been speculated that the neural ectoderm might trigger neural crest emigration through platelet-derived growth factor (PDGF), which is expressed in the ectoderm as well as in neural crest cells both prior to and during migration (Ho *et al.,* 1994). Additional growth factors that might be involved in controlling some of the events that accompany neural crest emigration out of the neural tube include epidermal growth factor (EGF) and hepatocyte growth factor or scatter factor (SF/HGF), whose expression patterns in early embryogenesis are consistent with a role in epithelial–mesenchymal transformation (Duband *et al.,* 1995).

Two growth factors that appear to play a role in proliferation of neural crest cells are FGF and neurotrophin-3 (NT-3), a member of a family of nerve growth factor–like peptides. Both *in vivo* and *in vitro* studies have demonstrated that FGF and basic FGF (bFGF) enhance survival of cells of neural crest origin and act as proliferation factors for primary neural crest cells (Kalcheim, 1989; Murphy *et al.,* 1991). NT-3 also stimulates neural crest cell numbers as well as incorporation of 3H-thymidine by crest cells in culture (Kalcheim *et al.,* 1992).

Several growth factors play a prominent role in the differentiation of neural crest cells into a variety of phenyotypes (Table 2). FGF and nerve growth factor (NGF) influence the differentiation of neural crest cells of sympathoadrenal lineage; that is, cell types that include the sympathetic neuron and the adrenal chromaffin cell (Murphy and Bartlett, 1993). NGF is also involved in the development of sensory neurons, where it likely acts as a target-derived survival factor during the period of natural neuron death similar to that which occurs during differentiation of sympathetic neurons. Studies by Kalcheim (1996) have investigated the role of neurotrophic factors in regulating the ontogeny of dorsal root ganglia. Both *in vivo* and *in vitro* approaches demonstrated that brain derived neurotrophic factor (BDNF), NT-3, and bFGF affected differentiation and survival of a subpopulation of crest cells destined to become sensory neurons. These studies emphasize that the mechanisms involved in regulation of these processes are complex and include the following features: (1) a single growth factor may elicit variable responses on progenitor cells during discrete phases of their differentiation; and (2) several growth factors may have converging activities on a subset of neural crest cells, inferring redundancy in their regulation; and (3) the rostral portion of each somite that promotes neural crest migration also regulates the number of neural crest cells by modulating the activity of these growth factors. These studies have also demonstrated that cooperation between the mesodermal cells and the neural tube is necessary for modulating cell number in the nascent sensory ganglia (Kalcheim, 1996).

2. Cell Adhesion Molecules

Changes in cell–cell adhesiveness has been proposed as a major mechanism involved in the initiation of crest cell emigration from the neural tube (Erickson, 1993; Newgreen and Tan, 1993). Before migration begins, presumptive crest cells are integrated into the ectodermal epithelium via prominent junctional cell adhesion complexes. The major cell adhesion molecules (CAMs) that play a role in this process include neural CAM (N-CAM), adherens CAM (A-CAM or N-cadherin), and liver CAM (L-CAM or E-cadherin). These molecules are cell-surface glycoproteins that mediate strong homophilic adhesion and are enriched at the adherens junctions between cells.

Initial studies in the chick regarding the involvement of all three CAMs in crest morphogenesis emphasized the loss of such molecules at the onset of migration and their reappearance at the end of migration (Thiery *et al.,* 1982; Duband *et al.,* 1988). Further studies in this model (Akitaya and Bronner-Fraser, 1992; Bronner-Fraser *et al.,* 1992) indicated that the sequence of this pattern of regulation may not be consistent for all CAMs. Although N-cadherin and N-CAM were both present in high levels on premigratory neural crest cells,

N-cadherin decreased prior to the onset of migration and was reexpressed on neural crest cells at the final stages of their migration and prior to formation of the dorsal root and sympathetic ganglia. In contrast, N-CAM remained on the surface of early migrating neural crest cells and was gradually down-regulated with migration, then gradually up-regulated several stages after ganglion formation. Perturbation studies using antibodies to N-CAM and N-cadherin demonstrated anomalous development of both the cranial neural tube and emigrating neural crest in the chick (Bronner-Fraser *et al.,* 1992).

Together, these findings are consistent with migration being cued by a generalized lowering of cell–cell adhesion between individual crest cells as well as between crest and their neural tube neighbors. The possible mechanisms responsible for this lowering include; (1) rapid down-regulation of adhesion molecules (Crossin *et al.,* 1990); (2) protease digestion of the adhesion molecules to allow detachment (Volk *et al.,* 1990); or (3) biochemical modification of the adhesion ligands that might alter affinity, such as phosphorylation (Volberg *et al.,* 1991). Although N-cadherin may be important for the maintenance of epithelial sheets and integrity of the neural tube as well as the initiation of emigration, N-CAM expression may be more important for regulating the proper migration of neural crest cells outside the neural tube. In this regard, it has been postulated that N-CAM may modulate cell–substrate interactions by influencing the configuration of the ECM. Evidence has been presented that indicates N-CAM binds to several different ECM ligands, including heparan sulfate proteoglycan, collagens I–VI and IX, and heparin (Cole *et al.,* 1986; Cole and Burg, 1989; Probstmeier *et al.,* 1989).

Work involving the expression of novel cadherins (e.g., c-cad6B and c-cad7) has expanded the function of CAMs to include maintenance of a loose cellular association between crest cells, which facilitates their oriented migration and homing to specific locations (Nakagawa and Takeichi, 1995). Moreover, different adhesion molecules may mediate intercellular communication in different subpopulations of migrating neural crest cells (Kimura *et al.,* 1995; Nakagawa and Takeichi, 1995).

All of these studies suggest that there must be a critical balance between cell–cell and cell–substrate adhesion that accounts for both the initiation of neural crest cell migration and its termination. This requirement for cell–matrix interactions for proper emigration of neural crest cells is summarized in the next section.

3. Extracellular Matrix and Integrins

There is abundant *in vivo* and *in vitro* evidence, primarily using the avian model, which supports a role for cell–matrix interactions in neural crest migration (Lallier and Bronner-Fraser, 1990; McCarthy and Hay, 1991). The ECM constituents identified to date as promoters of neural crest cell migration include laminin (Krotoski *et al.,* 1986), fibronectin (Newgreen and Thiery, 1980), tenascin/cytotactin (Perris *et al.,* 1991), proteoglycans (Perris *et al.,* 1991), hyaluronin (Pintar, 1978), and various collagens (e.g., I, IV, VI) (Duband and Thiery, 1987). Whereas some of these matrix molecules (e.g., laminin, collagen IV) comprise the BMs that are disorganized at the time of neural crest emigration, others constitute the migratory substratum for these cells following their exit from the neural tube. *In vitro* studies using cultured neural crest cells have demonstrated that several ECM molecules (e.g., fibronectin; laminin; collagens types I, IV, and VI) provide a substratum that is highly permissive for locomotion. Immunolocalization of ECM constituents (e.g., laminin, fibronectin, tenascin/cytotactin, collagens) in the matrix covering the dorsolateral neural tube surface or along neural crest pathways provide *in vivo* support for the involvement of these molecules in the motility of neural crest cells. In addition to promoting motility, some ECM components may also have an inhibitory effect on crest migration. The most prominent example of this is chondroitin sulfate, which is present in areas devoid of neural crest cells, such as the perinotochordal region as well as the caudal sclerotome (Bronner-Fraser, 1993 a,b).

Perturbation experiments carried out in the chick using antibodies against laminin, fibronectin, and tenascin cause abnormalities in cranial neural crest migration and neural tube development (Bronner-Fraser, 1993 a,b). The effects of these antibodies are similar to those observed with N-CAM and N-cadherin antibodies on cranial development (Bronner-Fraser *et al.,* 1992). This observation has led to the proposal that there may be a critical balance between cell–cell and cell–substrate adhesion necessary for proper initiation of neural crest cell migration. Thus, neural crest cell interactions may shift from favoring cell–cell adhesion prior to emigration to favoring cell–substrate adhesion during the migratory phase. After the migratory phase, cell adhesion molecules are reexpressed, consistent with the idea that neural crest cells then interact with each other rather than with ECM molecules. Interestingly, similar antibodies to ECM molecules do not adversely effect trunk crest development, suggesting that different guidance mechanisms may be involved at different axial levels. Cell–matrix interactions may play a predominant role in the cranial region, whereas tissue interactions (i.e., somites, notochord) may dominate in the trunk region.

Although a favorable substratum is a necessary precondition for migration, changes in the ECM are probably not the trigger for the initiation of migration. Most of the matrix molecules to which the crest cells are known to adhere are already present in the extracellular environment outside the neural tube prior to the time of emigration (Erickson, 1993). Alternatively, it is possible that changes in the affinity of neural crest cells to the matrix due to the expression of specific cell–matrix receptors, the integrins, may correlate with the onset of migration.

Integrins are recognized as a family of heterodimeric transmembrane receptors composed of noncovalently linked α and β glycoprotein subunits (Hynes, 1992). The $\beta 1$ integrins, the best characterized of this family, predominantly recognize and bind to specific ECM ligands, such as fibronectin and laminin, through their extracellular domain, thereby mediating cell–substrate adhesion. It has been well established that $\beta 1$ integrins play a functional role in the migration of cranial neural crest cells. There is immunocytochemical evidence of these receptors in tissue sections and immunoprecipitates of neural crest cells (Erickson and Perris, 1993). Moreover, *in vitro* application of antibodies to $\beta 1$ integrins disrupts neural crest cell adhesion to fibronectin and laminin substrates, and *in situ* microinjection of antibodies that bind and functionally inactivate the $\beta 1$ subunit causes major defects in neural crest development as well as neural tube anomalies (Bronner-Fraser, 1993 a,b). These studies suggest that the β receptor complex is important in the normal development of the cranial neural crest and neural tube.

Much less is known about the role played by the associated α-subunits on neural crest development. However, studies using antisense oligonucleotides against various integrin α-subunits suggest that there are three or more functionally distinct α-subunits that pair with the β-subunit on avian cranial neural crest cells. Moreover, the disruption of normal crest migration following injection of these oligonucleotides into the cranial mesenchyme suggest that the integrin α-subunits are also required for cranial crest cell development (Kil, *et al.*, 1996). Cranial and trunk neural crest cells appear to possess biochemically distinct integrins as indicated by cell attachment assays that quantitate the interactions of crest cells with ECM molecules (Lallier *et al.*, 1992). These studies demonstrate that the surface properties of trunk and cranial neural crest cells are distinct, and that these cells differ in their mechanisms of adhesion to the ECM.

IV. Neurocristopathies

Any aberration or disease arising from abnormal development of the neural crest may be considered a neurocristopathy. Due to the complexity of neural crest migration and differentiation into many diverse structures, it is not surprising that a variety of conditions may result from perturbation of these processes. These have commonly been subdivided into two main categories, malformations and tumors, in neural crest derivatives (Jones, 1990). The malformations are primarily a result of abnormal neural crest cell migration, which leads to a deficiency of tissue necessary for normal development of a given structure (Table 3). The tumors result primarily from errors in crest proliferation and differentiation (Table 4). Some of these defects involve only a single component of the neural crest (simple); others affect multiple components (complex) and are recognized as syndromes. Representative neurocristopathies in both categories are described in the following sections.

A. *Malformations*

There are four well-defined human malformation syndromes that have primary cranial neural crest involvement (Table 3). They are the retinoic acid syndrome (RAS), DiGeorge syndrome, hemifacial microsomia, and Treacher–Collins syndrome. All of these syndromes have common features that are cited later in this chapter and covered in more detail by Johnston and Bronsky (1995). RAS in humans has become one of the most commonly described malformation syndromes since it was first reported by Lammer and colleagues (1985). The primary adverse effects following exposure to 13-*cis* retinoic acid (Accutane, isotretinoin) during early pregnancy include malformations of the craniofacial region, including the external and middle ear and mandible, as well as the conotruncal region, cerebellum, and thymus. Since the mid-1980s, a number of relevant animal models have been developed that have increased our understanding of the cellular mechanisms and pathogenesis of retinoid embryopathy. Evidence ob-

TABLE 3 Major Neurocristopathies—Malformations

Cranial neural crest	Retinoic acid syndrome (RAS)
	DiGeorge syndrome
	Hemifacial microsomia
	Treacher–Collins syndrome
	Goldenhar syndrome
	Cardiac aortiocpulmonary septation defects
	Anterior chamber defects of eye
	Cleft lip and/or cleft palate
	Frontonasal dysplasia
Trunk neural crest	Hirschprung's disease
Cranial and trunk crest	CHARGE association
	Waardenburg syndrome

From the literature.

TABLE 4 Major Neurocristopathies—Tumors

Type	Tumor /disease	Affected cells
Simple	Pigmented nevus	Melanocyte
	Malignant melanoma	Melanocyte
	Ganglicytoma	Ganglion cell
	Ganglioneuroblastoma	Ganglion cell
	Neuroblastoma	Ganglion cell
	Paraganglioma	Paraganglion cell
	Schwannoma	Schwann cell
	Neurofibroma	Schwann cell
	Neurofibrosarcoma	Schwann cell
	Meningioma	Meningocyte
	Medullary thyroid carcinoma	Thyroid C-cell
	Carotid body tumor	Carotid body type I cell
	Pheochromocytoma	Adrenal medulla chromaffin cells
	Carcinoid tumors	Digestive tract enterochromaffin cells
	Ewing sarcoma	Undetermined
	Primitive neuroectodermal tumor	Undetermined
Complex	Neurofibromatosis type 1 (NF1)	Schwann cell, melanocyte, adrenal medulla
	Neurofibromatosis type 2 (NF2)	Schwann cell, meningocyte
	Multiple endocrine neoplasia type 2A (MEN 2A)	Adrenal medulla, thyroid C-cell
	Multiple endocrine neoplasia type 2B (MEN 2B)	Adrenal medulla, thyroid C-cell, ganglion cell, Schwann cell
	Dysplastic nevus syndrome	Melanocyte

From the literature.

tained from conventional experimental teratology as well as pathogenesis studies in mice, rats, hamsters, and the cynomolgus macaque support the original findings in the human that most of the malformed structures that comprise RAS are related to aberrant development of cranial neural crest cells (Webster *et al.*, 1986; Willhite *et al.*, 1986; Sulik *et al.*, 1988; Hart *et al.*, 1990; Hummler *et al.*, 1990; Makori *et al.*, 1997). These studies have also demonstrated that 13-*cis* retinoic acid and its isomer, all-*trans* retinoic acid, appear to have a similar adverse effect on cranial neural crest cells (Johnston and Bronsky, 1995).

Recent studies in the macaque embryo with 13-*cis* RA indicate a reduction and/or delay in the emigration of neural crest cells that exit the pre- and postotic regions of the hindbrain following exposure during gastrulation and early neurulation (Makori *et al.*, 1997). These findings are similar to earlier studies using all-*trans* RA in chick and mouse embryos that demonstrated altered patterns of migration of neural crest cells into the first three pharyngeal arches (Morriss-Kay and Mahmood, 1992; Moro Balbas *et al.* 1993). In addition to changes in neural crest cell migration, studies in the frog (Krumlauf *et al.*, 1991; Papalopulu *et al.*, 1991), zebrafish (Holder and Hill, 1991), rodent (Morriss-Kay *et al.*, 1991; Marshall *et al.*, 1992; Lee *et al.*, 1995), and macaque (Makori *et al.*, 1998) have demonstrated a consistent alteration in rhombomeric segmentation of the developing hindbrain subsequent to early exposure to RA. The prevailing hypothesis from these studies is that compaction of rhombomeres and a concomitant rostral shift of the otic vesicle alters early migratory patterns of neural crest cells, which results in abnormal development of their derivatives, including craniofacial structures and cranial ganglia (Morriss-Kay, 1993; Leonard *et al.*, 1995). There is a growing body of evidence suggesting that the deleterious effects of RA are mediated by changes in homeobox gene expression patterns in the developing brain (Wood *et al.*, 1994; Simeone *et al.*, 1995; Gale *et al.*, 1996).

DiGeorge syndrome is virtually identical to RAS except for the additional feature of a shortened upper lip. This syndrome is often associated with specific deletions in chromosome 22 (Scambler, 1994). In addition to a genetic component, studies in mice provide indirect evidence for a relationship between alcohol consumption and DiGeorge syndrome. Ethanol administration to pregnant mice prior to and during the period of cranial neural crest migration is lethal to premigratory as well as migrating neural crest cells and causes a spectrum of abnormalities similar to DiGeorge syndrome (Sulik *et al.*, 1986; Kotch and Sulik, 1992). Hemifacial microsomia is also characterized by a similar spectrum of craniofacial anomalies observed in DiGeorge syndrome and

RAS with the absence of thymus defects and presence of associated vertebral malformations (Johnston and Bronsky, 1995). The strongest evidence that supports a link between cranial neural crest aberrations and this syndrome is the presence of conotruncal and craniofacial defects. Treacher–Collins syndrome also involves defects of the craniofacial region; however, there is little or no involvement of the cardiovascular system. The pathogenesis of this condition has been linked to ganglionic placodal cell death leading to secondary defects in cranial neural crest derivatives (Webster *et al.*, 1986).

Several other syndromes involve aberrant trunk neural crest development with or without cranial crest perturbation. These include the CHARGE association, Waardenburg syndrome, and Hirschprung's disease. The CHARGE acronym characterizes a constellation of disorders stemming from aberrant cranial and trunk neural crest development. The main features of this condition are Coloboma of the eye, Heart disease, choanal (nasal) Atresia, Retarded growth and development, Genital hypoplasia, and Ear anomalies (Carlson, 1994). Facial palsy, cleft palate, and dysphagia may also be present. Waardenburg syndrome involves a combination of pigmentation defects, including white forelock and white eyelashes in addition to cochlear deafness, cleft palate, and ocular hypertelorism. This autosomal dominant condition involves disorders of both cranial and trunk neural crest cells (Jones, 1990).

Hirschprung's disease is characterized by a dilated segment of the proximal colon and absence of peristalsis in the segment distal to the dilation. Infants with this disease display delay in the passage of meconium, constipation, vomiting, and abdominal distention. Hirschprung's disease is often associated with other defects of neural crest development, such as Waardenburg type II syndrome, which involves deafness and facial clefts in addition to megacolon. Megacolon associated with Hirschprung's disease is believed to result from a failure of trunk neural crest cells to populate the bowel wall properly. Studies have been done to elucidate this process in the lethal spotted mutant mouse (*ls/ls*), which develops a congenital megacolon similar to Hirschprung's disease (Gershon, 1987). These studies have shown that neural crest cells migrate successfully to the affected bowel segment but fail to penetrate the bowel wall and form enteric ganglia. This inability to penetrate the bowel wall was attributed to a thickening of the smooth muscle BMs due to excessive production of several extracellular matrix molecules (Larsen, 1993).

B. Tumors

Neural crest–derived tumors (Table 4) were initially categorized by Bolande (1974) as simple or complex syndromes. The simple forms of neurocristopathy include tumors derived from neural crest cells of a particular type (e.g., melanocyte) and forming a specific tumor (e.g., malignant melanoma). The complex or systemic forms of neurocristopathy are hereditary disorders involving a combination of neural crest–derived cells; some of these are described in more detail following. Work directed at the gene alterations in both simple and systemic neurocristopathies in animal models has advanced our understanding of the genetic mechanisms of cancers (see Nakamura, 1995).

Neurofibromatosis type 1 (NF1, or von Rechlinghausen's disease), the most representative of the systemic neurocristopathies, is an autosomal dominant disorder characterized by alterations in Schwann cells, melanocytes, and the adrenal medulla (Riccardi and Eichner, 1992). Neurofibroma, one of the hallmarks of this disease, is a benign peripheral nerve sheath tumor with heterogeneous cellular components involving Schwann, fibroblastic, perineal, and mast cells as well as vascular endothelia and, possibly, axons. Other features include flat skin pigmented macula, cafe au lait spots (light brown pigmented lesions), and freckling. NF1 is regarded as one of the most cancer-prone syndromes. Neurofibrosarcoma (malignant peripheral nerve sheath tumor) is the most frequent of NF1-related malignant neoplasms. There is also an increase in the incidence of pheochromocytoma (tumor of chromaffin tissue of the adrenal medulla) in patients with NF1.

Neurofibromatosis type 2 (NF2), another systemic neurocristopathy, is an autosomal dominant disorder involving alterations of Schwann cells and meningocytes. NF2 is characterized by the presence of peripheral nerve tumors called *schwannomas,* which consist of monotonous proliferation of well-differentiated Schwann cells (Martuza and Eldridge, 1988). There is often bilateral involvement of the acoustic nerves and a lesser involvement of meningiomas (benign tumor of the meninges) and ependymomas (benign tumors composed of differentiated ependymal cells) in patients with NF2.

Systemic neurocristopathies involving tumors of the adrenal medulla and thyroid C-cell include multiple endocrine neoplasia (MEN) type 2A and 2B. MEN-2A is an autosomal dominant disorder that is characterized by medullary thyroid C-cell hyperplasia, pheochromocytoma, and, occasionally, parathyroid adenoma (Sizemore, 1987). MEN-2B also involves mucosal ganglioneuromas and skeletal anomalies (Hofstra *et al.*, 1994).

References

Akitaya, T., and Bronner-Fraser, M. (1992). Expression of cell adhesion molecules during initiation and cessation of neural crest cell migration. *Dev. Dyn.* **194,** 12–29.

Anderson, C. B., and Meier, S. (1981). The influence of the metameric pattern in the mesoderm on migration of cranial neural crest cells in the chick embryo. *Dev. Biol.* **85,** 385–402.

Baird, A. (1994). Fibroblast growth factors: activities and significance of nonneurotrophin neurotrophic growth factors. *Curr. Opin. Neurobiol.* **4,** 78–86.

Bally-Cuif, L., and Wassef, M. (1994). Ectopic induction and reorganization of *Wnt-1* expression in quail-chick chimeras. *Development* **120,** 3379–3394.

Bartelmez, G. W. (1960). Neural crest from the forebrain in mammals. *Anat. Rec.* **138,** 269–281.

Bartelmez, G. W., and Blount, M. P. (1954). The formation of neural crest from the primary optic vesicle in man. *Contrib. Embryol.* **35,** 55–71.

Birgbauer, E., Sechrist, J., Bronner-Fraser, M., and Fraser, S. (1995). Rhombomeric origin and rostrocaudal reassortment of neural crest cells revealed by intravital microscopy. *Development* **121,** 935–945.

Blankenship, T. N., Peterson, P. E., and Hendrickx, A. G. (1996). Emigration of neural crest cells from macaque optic vesicles is correlated with discontinuities in its basement membrane. *J. Anat.* **188,** 473–483.

Bolande, R. P. (1974). The neurocristopathies. A unifying concept of disease arising in neural crest maldevelopment. *Hum. Pathol.* **5,** 409–429.

Bronner-Fraser, M. (1993a). Mechanisms of neural crest cell migration. *Bioessays* **15,** 221–230.

Bronner-Fraser, M. (1993b). Environmental influences on neural crest cell migration. *J. Neurobiol.* **24,** 233–247.

Bronner-Fraser, M. (1994). Neural crest cell formation and migration in the developing embryo. *FASEB J.* **8,** 699–706.

Bronner-Fraser, M. (1995). Patterning of the vertebrate neural crest. *Perspect. Dev. Neurobiol.* **3,** 53–62.

Bronner-Fraser, M., and Stern, C. (1991). Effects of mesodermal tissues on avian neural crest cell migration. *Dev. Biol.* **143,** 213–217.

Bronner-Fraser, M., Wolf, J. J., and Murray, B. A. (1992). Effects of antibodies against N-cadherin and N-CAM on the cranial neural crest and neural tube. *Dev. Biol.* **153,** 291–301.

Carlson, B. M. (1994). *Human Embryology and Developmental Biology,* Mosby-Year Book, Inc., St. Louis.

Chan, W. Y., and Tam, P. P. L. (1988). A morphological and experimental study of the mesencephalic neural crest cells in the mouse embryo using wheat germ agglutinin–gold conjugate as the cell marker. *Development* **102,** 427–442.

Cole, G. J., and Burg, M. (1989). Characterization of a heparin sulfate proteoglycan that copurifies with the neural cell adhesion molecule. *Exp. Cell. Res.* **182,** 44–60.

Cole, G. J., Loewy, A., and Glaser, L. (1986). Neuronal cell–cell adhesion depends on interactions of N-CAM with heparin-like molecules. *Nature* **320,** 445–447.

Crossin, K. L., Prieto, A. L., Hoffman, S., Jones, F. S., and Friedlander, D. R. (1990). Expression of adhesion molecules and the establishment of boundaries during embryonic and neural development. *Exp. Neurol.* **109,** 6–18.

deVirgilio, G., Lavenda, N., and Worden, J. L. (1967). Sequence of events in neural tube closure and the formation of neural crest in the chick embryo. *Acta Anat.* **68,** 127–146.

Duband, J.-L., Monier, F., Delannet, M., and Newgreen, D. (1995). Epithelium–mesenchyme transition during neural crest development. *Acta Anat.* **154,** 63–78.

Duband, J.-L., and Thiery, J.-P. (1987). Distribution of laminin and collagens during avian neural crest development. *Development* **101,** 461–478.

Duband, J.-L., Volberg, T., Sabanay, I., Thiery, J. P., and Geiger, B. (1988). Spatial and temporal distribution of the adherens-junction-associated adhesion molecule A-CAM during avian embryogenesis. *Development* **103,** 325–344.

Dupin, E., Deville, F. S.-S.-C., Nataf, V., and Le Douarin, N. M. (1993). The ontogeny of the neural crest. *C.R. Acad. Sci. Paris, Life Sciences* **316,** 1072–1081.

Erickson, C. A. (1993). Morphogenesis of the avian trunk neural crest: use of morphological techniques in elucidating the process. *Micro. Res. Tech.* **26,** 329–351.

Erickson, C. A., and Perris, R. (1993). The role of cell–cell and cell–matrix interactions in the morphogenesis of the neural crest. *Dev. Biol.* **159,** 60–74.

Gale, E., Prince, V., Lumsden, A., Clarke, J., Holder, N., and Maden, M. (1996). Late effects of retinoic acid on neural crest and aspects of rhombomere identity. *Development* **122,** 783–793.

Gershon, M. D. (1987). Phenotypic expression by neural-crest derived precursors of enteric neurons and glia. In *Developmental and Evolutionary Aspects of the Neural Crest.* (P. F. A. Maderson, Ed.), John Wiley, New York. pp. 181–211.

Goulding, M. D., Chalapakis, G., Deutsch, U., Erselius, J., and Gruss, P. (1991). *Pax-3,* a novel murine DNA binding protein expressed during early neurogenesis. *EMBO J.* **10,** 1135–1147.

Graham, A., Heyman, I., and Lumsden, A. (1993). Even-numbered rhombomeres control the apoptotic elimination of neural crest cells from odd-numbered rhombomeres in the chick hindbrain. *Development* **119,** 233–245.

Groves, A. K., and Anderson, D. J. (1996). Role of environmental signals and transcriptional regulators in neural crest development. *Dev. Genet.* **18,** 64–72.

Hall, B. K., and Horstadius, S. (1988). *The Neural Crest.* Oxford Science Publication, London.

Hart, R. C., McCue, P. A., Ragland, W. L., Winn, K. J., and Unger, E. R. (1990). Avian model for 13-*cis* retinoic acid embryopathy: demonstration of neural crest related defects. *Teratology* **41,** 463–472.

Ho, L., Symes, K., Yordan, C., Gudas, L. J., and Mercola, M. (1994). Localization of PDGF-α and PDGF-α mRNA in *Xenopus* embryos suggests signaling from neural ectoderm and pharyngeal endoderm to neural crest cells. *Mech. Ageing Dev.* **48,** 165–174.

Hofstra, R. M. W., Landsvater, R. M., Ceccherini, I., Stulp, R. P., Stelwagen, T., Luo, Y., Pasisi, B., Hoppener, J. W. M., van Amstel, H. K. P., Romeo, G., Lips, C. J. M., and Buys, C. H. C. M. (1994). A mutation if the *RET* proto-oncogene associated with multiple endocrine neoplasia type 2B and sporadic medullary thyroid carcinoma. *Nature* **367,** 375–376.

Holder, N., and Hill, J. (1991). Retinoic acid modifies development of the midbrain–hindbrain border and affects cranial ganglion formation in zebrafish embryos. *Development* **113,** 1159–1170.

Hummler, H., Korte, R., and Hendrickx, A. G. (1990). Induction of malformations in the cynomolgus monkey with 13-*cis* retinoic acid. *Teratology* **42,** 263–272.

Hunt, P., Gulisano, M, Cook, M., Sham, M.-H., Faiella, A. Wilkinson, D., Boncinelli, E., and Krumlauf, R. (1991). A distinct *Hox* code for the branchial region of the vertebrate head. *Nature* **353,** 861–864.

Hunt, P., and Krumlauf, R. (1992). Hox codes and positional specification in vertebrate embryonic axes. *Annu. Rev. Cell Biol.* **8,** 227–256.

Hynes, R. O. (1992). Integrins: versatility, modulation, and signaling in cell adhesion. *Cell* **69,** 11–25.

Innes, P. B. (1985). The ultrastructure of early cephalic neural crest cell migration in the mouse. *Anat. Embryol.* **172,** 33–38.

Ito, K., and Morita, T. (1995). Role of retinoic acid in mouse neural crest cell development *in vitro. Dev. Dyn.* **204,** 211–218.

Johnston, M. C., and Bronsky, P. T. (1995). Prenatal craniofacial development: new insights on normal and abnormal mechanisms. *Crit. Rev. Oral. Biol. Med.* **6,** 25–79.

Jones, M. C. (1990). The neurocristopathies: reinterpretation based upon the mechanism of abnormal morphogenesis. *Cleft Pal J.* **27,** 136–140.

Kalcheim, C. (1989). Basic fibroblast growth factor stimulates survival of nonneuronal cells developing from trunk neural crest. *Dev. Biol.* **134,** 1–10.

Kalcheim, C. (1996). The role of neurotrophins in development of neural-crest cells that become sensory ganglia. *Phil. Trans. Roy. Soc. Lond. B.* **351,** 375–381.

Kalcheim, C., Carmeli, C., and Rosenthal, A. (1992). Neurotrophin 3 is a mitogen for cultured neural crest cells. *Proc. Natl. Acad. Sci. USA* **89,** 1661–1665.

Kessel, M., and Gruss, P. (1991). Homeotic transformations of murine vertebrae and concomitant alteration of *Hox* codes induced by retinoic acid. *Cell* **67,** 89–104.

Kil, S. H., Lallier, T., and Bronner-Fraser, M. (1996). Inhibition of cranial neural crest adhesion *in vitro* and migration *in vivo* using integrin antisense oligonucleotides. *Dev. Biol.* **179,** 91–101.

Kimura, Y., Matsunami, H., Inoue, T., Shimamura, K., Uchida, N., Ueno, T., Miyazaki, T., and Takeichi, M. (1995). Cadherin-11 expressed in association with mesenchymal morphogenesis in the head, somite, and limb bud of early mouse embryos. *Dev. Biol.* **169,** 347–358.

Kotch, L. E., and Sulik, K. K. (1992). Experimental fetal alcohol syndrome: proposed pathogenic basis for a variety of associated facial and brain anomalies. *Am. J. Med. Genet.* **44,** 168–176.

Krotoski, D., Domingo, C., and Bronner-Fraser, M. (1986). Distribution of a putative cell surface receptor for fibronectin and laminin in the avian embryo. *J. Cell Biol.* **103,** 1061–1072.

Krumlauf, R., Papalopulu, N., Clarke, J. D., and Holder, N. (1991). Retinoid acid in the *Xenopus* hindbrain. *Sem. Dev. Biol.* **2,** 181–188.

Lallier, T., and Bronner-Fraser, M. (1990). The role of the extracellular matrix in neural crest cell migration. *Sem. Dev. Biol.* **1,** 35–44.

Lallier, T., Leblanc, G., Artinger, K. B., and Bronner-Fraser, M. (1992). Cranial and trunk neural crest cells use different mechanisms for attachment to extracellular matrices. *Development* **116,** 531–541.

Lammer, E. J., Chen, D. T., Hoar, R. M., Agnish, N. D., Benke, P. J., Braun, J. T., Curry, C. J., Fernhoff, P. M., Grix, A. W., Lott, I. T., Richard, J. M., and Sun, S. C. (1985). Retinoic acid embryopathy. *N. Engl. J. Med.* **313,** 837–841.

Larsen, W. J. (1993). *Human Embryology.* Churchill Livingstone, Inc., New York.

Le Douarin, N. (1973). A biological cell labeling technique and its use in experimental embryology. *Dev. Biol.* **30,** 217–222.

Le Douarin, N. (1982). *The Neural Crest.* Cambridge University Press, Cambridge.

Le Douarin, N. M., Dupin, E., and Ziller, C. (1994). Genetic and epigenetic control in neural crest development. *Cur. Opin. Gen. Dev.* **4,** 685–695.

Le Douarin, N. M., and Ziller, C. (1993). Plasticity in neural crest cell differentiation. *Curr. Opin. Cell Biol.* **5,** 1036–1043.

Le Douarin, N. M., Ziller, C., and Couly, G. F. (1993). Patterning of neural crest derivatives in the avian embryo: *in vivo* and *in vitro* studies. *Dev. Biol.* **159,** 24–49.

Lee, Y. M., Osumi-Yamashita, N., Ninomiya, Y., Moon, C. K., Eriksson, U., and Eto, K. (1995). Retinoic acid stage-dependently alters the migration pattern and identity of hindbrain neural crest cells. *Development* **121,** 825–837.

Leonard, L., Horton, C., Maden, M., and Pizzey, J. A. (1995). Anteriorization of CRABP-I expression by retinoic acid in the developing mouse central nervous system and its relationship to teratogenesis. *Dev. Biol.* **168,** 514–528.

Lo, L.-C., Johnson, J. E., Wuenschell, C. W., Saito, T., and Anderson, D. J. (1991). Mammalian *achaete-scute* homolog 1 is transiently expressed by spatially-restricted subsets of early neuroepithelial and neural crest cells. *Genes Dev.* **5,** 1524–1537.

Lumsden, A., and Graham, A. (1996). Death in the neural crest: implications for pattern formation. *Semin. Cell Dev. Biol.* **7,** 169–174.

Lumsden, A., Sprawson, N., and Graham, A. (1991). Segmental origin and migration of neural crest cells in the hindbrain region of the chick embryo. *Development* **113,** 1281–1291.

Maden, M., and Holder, N. (1992). Retinoic acid and development of the central nervous system. *Bioessays* **14,** 431–438.

Maden, M. Horton, C., Graham, A., Leonard, L., Pizzey, J., Siegenthaler, G., Lumsden, A., and Eriksson, U. (1992). Domains of cellular retinoic acid–binding protein I (CRABP I) expression in the hindbrain and neural crest of the mouse embryo. *Mech. Dev.* **37,** 13–23.

Makori, N., Peterson, P. E., Blankenship, T. N., Dillard-Telm, L., Hummler, H., and Hendrickx, A. G. (1998). Effects of 13-*cis* retinoic acid on hindbrain and craniofacial morphogenesis in long-tailed macaques. *J. Med. Primatol.* (In Press).

Marshall, H., Morrison, A., Studer, M., Pöpperl, H., and Krumlauf, R. (1996). Retinoids and *Hox* genes. *FASEB J.* **10,** 969–978.

Marshall, H., Nochev, S., Sham, M. H., Muchamore, I., Lumsden, A., and Krumlauf, R. (1992). Retinoic acid alters hindbrain *Hox* code and induces transformation of rhombomeres 2/3 into a 4/5 identity. *Nature* **360,** 737–741.

Martuza, R. L., and Eldridge, R. N. (1988). Neurofibromatosis-2 (bilateral acoustic neurofibromatosis). *N. Engl. J. Med.* **318,** 684–688.

McCarthy, R. A., and Hay, E. D. (1991). Collagen I, laminin, and tenascin: ultrastructure and correlation with avian neural crest formation. *Int. J. Dev. Biol.* **35,** 437–452.

Millan, F. A., Denhez, F., Kondaiah, P., and Akhurst, R. J. (1991). Embryonic gene expression patterns of TGF $\beta 1$, $\beta 2$, and $\beta 3$ suggest different developmental functions *in vivo. Development* **111,** 131–144.

Moase, C. D., and Trasler, D. G. (1991). N-CAM alterations in *splotch* neural tube defect mouse embryos. *Development* **113,** 1049–1058.

Moro Balbás, J. A., Gato, A., Alonso, R. M. I., Pastor, J. F., Represa, J. J., and Barbosa, E. (1993). Retinoic acid induces changes in the rhombencephalic neural crest cell migration and extracellular matrix composition in chick embryos. *Teratology* **48,** 197–206.

Morriss-Kay, G. (1992). Retinoic acid and development. *Pathobiology* **60,** 264-270.

Morriss-Kay, G. (1993). Retinoic acid and craniofacial development: molecules and morphogenesis. *Bioessays* **15,** 9–15.

Morriss-Kay, G., and Mahmood, R. (1992). Morphogenesis-related changes in extracellular matrix induced by retinoic acid. In *Retinoids in Normal Development and Teratogenesis.* Morriss-Kay, G., Ed. Oxford Science Publications, New York, pp. 163–180.

Morriss-Kay, G. M., Murphy, P., Hill, R., and Davidson, D. (1991). Effects of retinoic acid on expression of *Hox-2.9* and *Krox-20* and on morphological segmentation of the hindbrain of mouse embryos. *EMBO J.* **10,** 2985–2995.

Morriss-Kay, G., Ruberte, E., and Fukiishi, Y. (1993). Mammalian neural crest and neural crest derivatives. *Ann. Anat.* **175,** 501–507.

Müller, F., and O'Rahilly, R. (1985). The first appearance of the neural tube and optic primordium in the human embryo at stage 10. *Anat. Embryol.* **172,** 157–169.

Müller, F., and O'Rahilly, R. (1986). The development of the human brain and the closure of the rostral neuropore at stage 11. *Anat. Embryol.* **175,** 205–222.

Müller, F., and O'Rahilly, R. (1987). The development of the human brain, the closure of the caudal neuropore, and the beginning of secondary neurulation at stage 12. *Anat. Embryol.* **176,** 413–430.

Murphy, M., and Bartlett, P. F. (1993). Molecular regulation of neural crest development. *Molec. Neurobiol.* **7,** 111–135.

Murphy, M., Bernard, O., Reid, K., and Bartlett, P. F. (1991). Cell lines derived from mouse neural crest are representative of cells at various stages of differentiation. *J. Neurobiol.* **22,** 522–535.

Nakagawa, S., and Takeichi, M. (1995). Neural crest cell–cell adhesion controlled by sequential and subpopulation-specific expression of novel cadherins. *Development* **121,** 1321–1332.

Nakamura, T. (1995). Genetic markers and animal models of neurocristopathy. *Histo. Histopathol.* **10,** 747–759.

Newgreen, D. F., and Erickson, C. A. (1986). The migration of neural crest cells. In *International Review of Cytology.* G. H. Bourne, K. W. Jeon, and M. Friedlander, Eds. Vol. 103. Academic Press, New York, pp. 89–146.

Newgreen, D. F., Scheel, M., and Kastner, V. (1986). Morphogenesis of sclerotome and neural crest in avian embryos: *in vivo* and *in vitro* studies on the role of notochordal extracellular matrix. *Cell Tissue Res.* **244,** 299–313.

Newgreen, D. F., and Tan, S. S. (1993). Adhesion molecules in neural crest development. *Pharmac. Ther.* **60,** 517–537.

Newgreen, D. F., and Thiery, J.-P. (1980). Fibronectin in early avian embryos: synthesis and distribution along the migration pathways of neural crest cells. *Cell Tissue Res.* **211,** 269–291.

Nichols, D. H. (1981). Neural crest formation in the head of the mouse embryo as observed using a new histological technique. *J. Embryol. Exp. Morphol.* **64,** 105–120.

Nichols, D. H. (1987). Ultrastructure of neural crest formation in the midbrain/rostral hindbrain and preotic hindbrain regions of the mouse embryo. *Am. J. Anat.* **179,** 143–154.

Noden, D. M. (1975). An analysis of the migratory behavior of avian cephalic neural crest cells. *Dev. Biol.* **42,** 106–130.

Noden, D. M. (1993). Spatial integration among cells forming the cranial peripheral nervous system. *J. Neurobiol.* **24,** 248–261.

Osumi-Yamashita, N., Ninomiya, Y., Doi, H., and Eto, K. (1994). The contribution of both forebrain and midbrain crest cells to the mesenchyme in the frontonasal mass of mouse embryos. *Dev. Biol.* **164,** 409–419.

Papalopulu, N., Clarke, J. D. W., Bradley, L., Wilkinson, D., Krumlauf, R., and Holder, N. (1991). Retinoic acid causes abnormal development and segmental patterning of the anterior hindbrain in *Xenopus* embryos. *Development* **113,** 1145–1158.

Parr, B. A., Shea, M. J., Vassileva, G., and McMahon, A. P. (1993). Mouse *Wnt* genes exhibit discrete domains of expression in the early embryonic CNS and limb buds. *Development* **119,** 247–261.

Perris, R., Krotoski, D., Domingo, C., Lallier, T., Sorrell, J. M., and Bronner-Fraser, M. (1991). Spatial and temporal changes in the distribution of proteoglycans during avian neural crest development. *Development* **111,** 583–599.

Peterson, P. E., Blankenship, T. N., Wilson, D. B., and Hendrickx, A. G. (1996). Analysis of hindbrain neural crest migration in the long-tailed monkey (*Macaca fascicularis*). *Anat. Embryol.* **194,** 235–246.

Pintar, J. (1978). Distribution and synthesis of glycosaminoglycans during quail neural crest morphogenesis. *Dev. Biol.* **67,** 444–464.

Probstmeier, R., Kühn, K, and Schachner, M. (1989). Binding properties of the neural cell adhesion molecule to different components of the extracellular matrix. *J. Neurochem.* **53,** 1794–1801.

Riccardi, V. M., and Eichner, J. E. (1992). *Neurofibromatosis: Phenotype, Natural History and Pathogenesis,* 2nd ed. Johns Hopkins University Press, Baltimore.

Sadler, T. W. (1995). *Langman's Medical Embryology,* 7th ed. Williams & Wilkins, Baltimore.

Scambler, P. J. (1994). DiGeorge syndrome and related birth defects. *Sem. Dev. Biol.* **5,** 303–310.

Scherson, T., Serbedzija, G., Fraser, S., and Bronner-Fraser, M. (1993). Regulative capacity of the cranial neural tube to form neural crest. *Development* **118,** 1049–1061.

Sechrist, J., Scherson, T., and Bronner-Fraser, M. (1994). Rhombomere rotation reveals that multiple mechanisms contribute to the segmental pattern of hindbrain neural crest migration. *Development* **120,** 1777–1790.

Sechrist, J., Serbedzija, G. N., Scherson, T., Fraser, S. E., and Bronner-Fraser, M. (1993). Segmental migration of the hindbrain neural crest does not arise from its segmental generation. *Development* **118,** 691–703.

Selleck, M. A. J., and Bronner-Fraser, M. (1995). Origins of the avian neural crest: the role of neural plate–epidermal interactions. *Development* **121,** 525–538.

Selleck, M. A. J., Scherson, T. Y., and Bronner-Fraser, M. (1993). Origins of neural crest cell diversity. *Dev. Biol.* **159,** 1–11.

Serbedzija, G. N., Bronner-Fraser, M., and Fraser, S. E. (1989). A vital dye analysis of the timing and pathways of neural crest cell migration. *Development* **106,** 809–819.

Serbedzija, G. N., Bronner-Fraser, M., and Fraser, S. E. (1992). Vital dye analysis of cranial neural crest cell migration in the mouse embryo. *Development* **116,** 297–307.

Sham, M. H., Vesque, C., Nonchev, S., Marshall, H., Frain, M., Das Gupta, R., Whiting, J., Wilkinson, D., Charnay, P., and Krumlauf, R. (1993). The zinc finger gene *Krox-20* regulates *Hox-B2* during hindbrain segmentation. *Cell* **72,** 183–196.

Simeone, A., Avantaggiato, V., Moroni, M. C., Mavilio, F., Arra, C., Cotelli, F., Nigro, V., and Acampora, D. (1995). Retinoic acid induces stage-specific antero-posterior transformation of rostral central nervous system. *Mech. Dev.* **51,** 83–98.

Sizemore, G. W. (1987). Medullary carcinoma of the thyroid gland. *Semin. Oncol.* **14,** 306–314.

Stemple, D. L., and Anderson, D. J. (1993). Lineage diversification of the neural crest: *in vitro* investigations. *Dev. Biol.* **159,** 12–23.

Sulik, K. K., Cook, C. S., and Webster, W. S. (1988). Teratogens and craniofacial malformations: relationships to cell death. *Development* **103** (Suppl.), 213–232.

Sulik, K. K., Johnston, M. C., Daft, P. A., and Russell, W. E. (1986). Fetal alcohol syndrome and DiGeorge anomaly: critical ethanol exposure periods for craniofacial malformations as illustrated in an animal model. *Am. J. Med. Genet.* **2,** 191–194.

Takahashi, Y., and Le Douarin, N. M. (1990). cDNA cloning of a quail homeobox gene and its expression in neural crest–derived mesenchyme and lateral plate mesoderm. *Proc. Natl. Acad. Sci. USA.* **87,** 7482–7486.

Tan, S. S., and Morriss-Kay, G. (1985). The development and distribution of the cranial neural crest in the rat embryo. *Cell Tissue Res.* **240,** 403–416.

Thiery, J.-P., Duband, J.-L., Rutishauser, U., and Edelman, G. M. (1982). Cell adhesion molecules in early chicken embryogenesis. *Proc. Natl. Acad. Sci. USA* **79,** 6737–6741.

Tosney, K. W. (1982). The segregation and early migration of cranial neural crest cells in the avian embryo. *Dev. Biol.* **89,** 13–24.

Tucker, G. C., Aoyama, H., Lipinski, M., Tursz, T., and Thiery, J.-P. (1984). Identical reactivity of monoclonal antibodies HNK-1 and NC-1: conservation in vertebrates on cells derived from the neural primordium and on some leukocytes. *Cell. Differ.* **14,** 223–230.

Tuckett, F., and Morriss-Kay, G. M. (1986). The distribution of fibronectin, laminin and entactin in the neurulating rat embryo studied by indirect immunofluorescence. *J. Embryol. Exp. Morphol.* **94,** 95–112.

Vincent, M., and Thiery, J.-P. (1984). A cell surface marker for neural crest and placodal cells: further evolution in peripheral and central nervous system. *Dev. Biol.* **103,** 468–481.

Volberg, T., Geiger, B., Dror, R., and Zick, Y. (1991). Modulation of intercellular adherens–type junctions and tyrosine phosphorylation of their components in RSV-transformed cultured chick lens cells. *Cell Regul.* **2,** 105–120.

Volk, T., Volberg, T., Sabanay, I., and Geiger, B. (1990). Cleavage of A-CAM by endogenous proteinases in cultured lens cells and in developing chick embryos. *Dev. Biol,* **139,** 314–326.

Webster, W. S., Johnston, M. C., Lammer, E. J., and Sulik, K. K. (1986). Isotretinoin embryopathy and the cranial neural crest: an *in vivo* and *in vitro* study. *J. Craniofac. Genet. Dev. Biol.* **6,** 211–222.

Weston, J. A. (1963). A radioautographic analysis of the migration and localization of trunk neural crest cells in the chick. *Dev. Biol.* **6,** 279–310.

Willhite, C., Hill, R. M., and Irving, D. W. (1986). Isotretinoin-induced craniofacial malformations in humans and hamsters. *J. Craniofac. Genet. Dev. Biol.* **2** (Suppl.), 193–209.

Wood, H., Pall, G., and Morriss-Kay, G. (1994). Exposure to retinoic acid before or after the onset of somitogenesis reveals separate effects on rhombomeric segmentation and 3′ *HoxB* gene expression domains. *Development* **120,** 2279–2285.

CHAPTER 12

Ontogeny of Neurotransmitters
Monoamines

HARRY W. BROENING
Exxon Biomedical Sciences, Inc.
East Millstone, New Jersey 08875

WILLIAM SLIKKER, JR.
Division of Neurotoxicology
National Center for Toxicological Research
Food and Drug Administration
Jefferson, Arkansas 72079

* Abbreviations: 5-HT, serotonin; CNS, central nervous system; DA, dopamine; DβH, dopamine-β-hydroxylase; ED, embryonic day; MA, methamphetamine; MDMA, 3,4-methylenedioxymethamphetamine; MPTP, *N*-methyl-4-phenyl-1,2,3,6-tetrahydropyridine; NE, norepinephrine; PN, postnatal day; TH, tyrosine hydroxylase; TPH, tryptophan hydroxylase.

I. Introduction

The monoaminergic neurotransmitter systems that innervate the mammalian central nervous system (CNS) include the catecholamine neurotransmitters dopamine and norepinephrine, and the indoleamine serotonin. A variety of drugs and neurotoxicants interact with these neurotransmitter systems. Among the more notable examples of these are the stimulant drugs of abuse, methamphetamine and 3,4-methylenedioxymethamphetamine, and the neurotoxicant *N*-methyl-4-phenyl-1,2,3,6-tetrahydropyridine (MPTP). Exposures to drugs or neurotoxicants may occur at any period throughout the life of an individual. Substantial effort has been expended in studying the developmental toxicology of stimulant drugs of abuse with a particular emphasis on their adverse effects to the developing CNS. The expression of the neurologic substrates affected by these drugs changes tremendously during development. Thus, it is essential to understand the developmental state of a neurotransmitter system when exposure to a neurotoxicant occurs in order to understand what the likely adverse effects of neurotoxicant exposure will be. This treatise examines the current knowledge regarding the development of dopaminergic, noradrenergic, and serotonergic innervation to the rodent CNS. Where possible, examples are provided to illustrate the differences that can arise from neurotoxicant exposure during different periods of development.

The development of monoaminergic neurotransmitter systems in the CNS can be subdivided into roughly three overlapping periods: (1) neurogenesis; (2) initial axon elongation and the development of axon pathways; and (3) terminal field innervation and synaptogenesis. This review explores the ontogeny of the central monoaminergic neurotransmitter systems primarily from the perspective of synaptogenesis. The rationale for focusing on synaptogenesis is that synaptogenesis represents the developmental period when many of the substrates for drug and neurotoxicant interactions are expressed. These substrates include, but are not limited to, monoamine reuptake transporters, neurotransmitter receptors, and enzymes responsible for neurotransmitter-biosynthesis. Thus, synaptogenesis represents adevelopmental period that is potentially vulnerable to disruption by agents that interact with these neuronal substrates. Thus, for the purposes of this review, the assumption is that the development of indices of monoaminergic synaptic function (receptor density, neurotransmitter content, etc.) provide insight into the progress of monoaminergic synaptogenesis in the CNS.

II. Innervation of Terminal Fields

Several common features of development are observed among the monoaminergic neurotransmitter systems. First, the perikarya of the monoaminergic neurotransmitter systems undergo histogenesis very early in CNS development. Neurons expressing dopamine, norepinephrine, or serotonin can be visualized by immunohistochemical or fluorescence histochemical methods by embryonic day (ED) 12–13 in the brainstem of the rat (Olson and Seiger, 1972; Seiger and Olson, 1973; Lauder and Bloom, 1974; Levitt and Moore, 1978; Specht *et al.,* 1981a; Lauder *et al.,* 1982; Lidov and Molliver, 1982a,b; Wallace and Lauder, 1983; and Voorn *et al.,* 1988). The development of the monoaminergic brainstem nuclei proceeds rapidly such that their organization resembles the adult distribution at parturition. Second, immediately following histogenesis, the monoaminergic neurons begin to send axonal processes into the developing forebrain. Axons begin arising from monoaminergic perikarya at ED 14–15 (Specht *et al.,* 1981a; Lauder *et al.,* 1982; Lidov and Molliver, 1982a; Wallace and Lauder, 1983; and Voorn *et al.,* 1988). It is from this point forward that the description of the forebrain innervation by each of the individual monoaminergic neurotransmitter systems proceeds.

A. *Ontogeny of Dopaminergic Innervation*

In the adult rat, the mesencephalic dopaminergic nuclei give rise to an ascending projection that can be divided into two subsystems: the nigrostriatal dopaminergic system and the mesolimbic dopaminergic system. The nigrostriatal system arises mainly from the substantia nigra, whereas the mesolimbic system arises primarily from the ventral tegmental area. These dopaminergic projections serve to innervate many areas of the forebrain, including prefrontal cortex, caudate nucleus, and nucleus accumbens. Dopamine-synthesizing neurons are also found in the hypothalamus and infundibular nucleus; however, this section focuses only on the ontogeny of innervation derived from the dopaminergic neurons in the mesencephalon.

Fibers from the dopaminergic neurons in the mesencephalon pass through the diencephalon at ED 14 to reach the telecephalon (Voorn *et al.,* 1988). These fibers follow the developing medial forebrain bundle to form the nascent nigrostriatal pathway (Specht *et al.,* 1981a,b). From this point, fibers derived from the dopaminergic neurons in the developing substantia nigra pars compacta extend into the ventral–lateral region of the primordium of the striatum. Dopaminergic axons can be seen in the developing striatum, including the primordium of the nucleus accumbens by ED 15–16 (Voorn *et al.,* 1988). By ED 19, dopaminergic axons have established pathways to most regions of the dorsal and ventral striatum. At this age, a patch-matrix type morphology can be observed in the developing caudate–putamen (Specht *et al.,* 1981b; Voorn *et al.,* 1988).

The dopaminergic terminal fields in the striatum develop primarily after parturition. Dopaminergic fibers in the developing striatum exhibit a morphologic change from thick, corkscrew-shaped fibers to thin fibers with varicosities. Fibers with varicosities can be visualized in the striatum at postnatal day (PN) 2, but do not become prevalent until PN 4 (Voorn *et al.,* 1988). The density of these varicose fibers gradually increases through the second and third postnatal week. By PN 21, the density of dopaminergic fibers in the striatum resembles that in the adult rat (Voorn *et al.,* 1988).

Dopaminergic fibers arising from the ventral tegmental area can be visualized in the neocortical anlage by ED 15 and in the subplate of the prefrontal cortex and the septum at ED 17 in the rat (Verney *et al.,* 1982; Kalsbeek *et al.,* 1988; Voorn *et al.,* 1988). Dopaminergic fibers reach the cingulate cortex by ED 20 (Verney *et al.,* 1982). Two bundles of dopaminergic fibers can be identified at this age: a lateral bundle passing ventrally to and through the striatum that diverges fanwise towards the intermediate zone of the frontal cortex; and a medial bundle ascending dorsally in the intermediate zone of the medial wall of the frontal cortex (Verney *et al.,* 1982; Kalsbeek *et al.,* 1988). Dopaminergic fibers do not penetrate the subplate until just before birth; somewhat later than the appearance of dopaminergic fibers in the developing striatum (Verney *et al.,* 1982; Kalsbeek *et al.,* 1988). By PN 2, dopaminergic fibers are seen in the area of the marginal zone that becomes layer I of the neocortex. The axons change from thick, straight fibers to thin, varicose fibers beginning at PN 4, suggesting the development of terminal fields (Kalsbeek *et al.,* 1988). By PN 6, some portions of the prefrontal cortex begin to take on characteristics of the mature innervation pattern. The topologic pattern of dopaminergic innervation to the neocortex resembles the adult pattern by PN 12. However, terminal field density continues to develop through PN 35, when innervation to this cortical region achieves the adult morphology (Kalsbeek *et al.,* 1988).

B. Ontogeny of Noradrenergic Innervation

Neurons that synthesize norepinephrine are found only in the pontine and medullary tegmental regions of the brainstem. The primary aggregation of noradrenergic neurons is the locus coeruleus. Although the locus coeruleus contains only about half of the noradrenergic neurons in the brain, it is quantitatively the most important noradrenergic center. The locus coeruleus sends two ascending fiber systems into the forebrain: the dorsal noradrenergic bundle and the rostral limb of the dorsal periventricular pathway. These afferents from the locus coeruleus innervate forebrain regions including cerebral cortex, hippocampus, and amygdala. The locus coeruleus also provides noradrenergic innervation to the cerebellum.

Axons arise from the noradrenergic perikarya in the developing locus coeruleus at ED 14 (Specht *et al.,* 1981a). By ED 15, these axons extend into the ventral mesencephalon and the dorsal pons to form the nascent ventral and dorsal noradrenergic bundles, respectively (Specht *et al.,* 1981a). Noradrenergic axons simultaneously innervate the medial and lateral cortex at this age; however, the dorsal cortex is not innervated until ED 19 (Levitt and Moore, 1979; Verney *et al.,* 1984; Berger and Verney, 1984). Noradrenergic fibers enter the neocortex primarily in the ventral–rostral regions of the cortical plate. Innervation of the neocortex then proceeds in a ventral-to-dorsal and rostral-to-caudal direction. By ED 18, noradrenergic fibers are first detected in the hippocampus. Noradrenergic axons course laterally under the anterior commisure at ED 20 and reach the marginal and intermediate zones of the lateral frontal cortex. Medially, noradrenergic axons course through the dorsal diagonal band and rostrally to it via the developing medial forebrain bundle (Levitt and Moore, 1979; Verney *et al.,* 1984). On reaching the corpus callosum, these axons diverge into two axon bundles: one running above the corpus callosum entering the cingulate cortex, and one running below the corpus callosum and entering the septum (Berger *et al.,* 1983; Verney *et al.,* 1984). The morphology of the noradrenergic axons begins to change from thick, straight fibers to thin, varicose fibers by ED 20 (Berger and Verney, 1984; Verney *et al.,* 1984). Noradrenergic fibers have innervated all layers of the neocortex by PN 7 (Lidov *et al.,* 1978; Levitt and Moore, 1979; Verney *et al.,* 1982; Berger *et al.,* 1983). Noradrenergic innervation to the neocortex begins to resemble the adult in density and distribution by PN 14 (Levitt and Moore, 1979; Berger *et al.,* 1983). Thus, noradrenergic innervation to the forebrain appears to mature at an earlier age than the dopaminergic and serotonergic innervation to the forebrain.

C. Ontogeny of Serotonergic Innervation

Serotonin synthesizing neurons are found in the mesencephalon, pons, and medulla oblongata. These neurons are distributed mainly within the raphe nuclei. The raphe nuclei that provide the majority of the serotonergic innervation to the forebrain are the dorsal raphe, the median raphe, and the caudal linear raphe nucleus (nucleus centralis superior). These nuclei give rise to the dorsal and ventral ascending serotonergic pathways. The cerebellum receives input from the cerebellar serotonergic pathways. Cerebellar afferents arise from all raphe nuclei, but primarily from the raphe pontis and raphe obscurus nuclei. Virtually the entire CNS receives serotonergic input.

Serotonergic neurons begin to sprout axons by ED 15 (Lidov and Molliver, 1982a,b; Wallace and Lauder, 1983). These axons grow rapidly, and by ED 16, axons from the primordial raphe nuclei condense into a bundle in the rostral pons and ascend through the tegmentum to pass over the vertex of the mesencephalic flexure (Lauder *et al.,* 1982; Lidov and Molliver, 1982a,b; Wallace and Lauder, 1983). Serotonergic axons run through the ventral portion of the midbrain ascending as far forward as the border between the diencephalon and the telencephalon. By ED 17 serotonergic axons enter the basal forebrain, with some fibers reaching as far forward as the septum and the frontal pole of the neocortex (Lauder *et al.,* 1982; Lidov and Molliver, 1982a;

Wallace and Lauder, 1983). Serotonergic axons have established pathways to all major divisions of the forebrain in the rat by ED 19 (Wallace and Lauder, 1983). Serotonergic fibers enter the cortical plate by ED 18, forming a superficial and a deep network of fibers spreading laterally and dorsally along the cortical surfaces (Lauder *et al.,* 1982). Innervation proceeds in a rostral-to-caudal direction such that occipital cortex is the last to be innervated. Serotonergic axons gradually arborize, sending fibers into all cortical layers.

Axon pathways increase in density, and terminal fields begin to appear by ED 21 in the rat (Lidov and Molliver, 1982a). Terminal innervation is limited mainly to subcortical regions at this stage, with most areas of the brainstem being densely innervated. Serotonergic axons are first noted to be present in the cerebellum on ED 21 (Lidov and Molliver, 1982a). Terminal field innervation continues into the postnatal developmental period and is the main feature of postnatal serotonergic development. By PN 3, elaboration of the serotonergic neuropil is underway in most cortical regions and in the hippocampus (Lidov and Molliver, 1982a; Nyakas *et al.,* 1994; Dori *et al.,* 1996). At this developmental stage, innervation in most brainstem regions is quite dense, and patterns of innervation have begun to resemble the adult.

By PN 6, the pons and medulla have essentially achieved an adult density of serotonergic innervation (Lidov and Molliver, 1982a). At this developmental stage serotonergic innervation of the cerebellum is underway, but there are still very few terminal arborizations present. Innervation of the thalamus and hypothalamus has reached adult densities in some of the nuclei within these regions. The serotonergic terminal field of the striatum is still sparse at this age. In the cerebral cortex, a gradient of terminal field innervation can be observed with fibers being dense at the rostral end of the cortex but few terminals can be ascertained in the more caudal regions (Lidov and Molliver, 1982a). By PN 10, when the six layers of the neocortex are readily apparent, serotonergic fibers can be found in all layers (Lidov and Molliver, 1982a; Dori *et al.,* 1996).

There are moderate to high levels of serotonergic innervation to be found throughout the entire cerebral cortex by PN 14 in the rat (Lidov and Molliver, 1982a; Nyakas *et al.,* 1994; Dori *et al.,* 1996). Low densities of serotonergic terminal arborizations are present in striatum and moderate levels of innervation are observed in the hippocampus. In the cerebellum, moderate levels of serotonergic innervation are observed in the internal granular layer with low levels of innervation being found in the molecular and Purkinje cell layers. Serotonergic nerve terminal densities resembles adult levels in all regions of the cerebral cortex, hippocampus, septal regions, and in the brainstem by PN 21 (Lidov and Molliver, 1982a; Dori *et al.,* 1996; Dinopoulos *et al.,* 1997). Only moderate levels of innervation are observed in the striatum and cerebellum at this age.

III. Neurochemical Synaptogenesis

Several aspects of monoaminergic neurotransmission lend themselves to biochemical characterization. The neurochemical processes that take place at the synapse may be characterized as being presynaptic or postsynaptic in nature. For the monoaminergic neurotransmitter systems, indices that denote presynaptic development include neurotransmitter content, neurotransmitter synthesis, and high affinity reuptake. Postsynaptic development may be followed in large part by the ontogeny of receptor density. As the ontogeny of these indices of monoaminergic function during development has often been taken as evidence for synaptogenesis within the monoaminergic neurotransmitter systems, it is necessary to study these biomarkers of monoaminergic function in the rat to provide some picture of the functional development of the central monoaminergic neurotransmitter systems.

A. *Neurotransmitter Content*

1. Dopamine

The concentrations of the monoaminergic neurotransmitters in the CNS are low at birth, ranging from 8% to 12% of the adult concentrations in whole brain. In general, concentrations of the monoamines attain adult values by the third or fourth postnatal week. In whole brain, dopamine (DA) content ranges from 5% to 15% of the adult values at birth (Breese and Traylor, 1972; Herregodts *et al.,* 1990). Whole brain DA content develops rapidly such that by 2 weeks of age, approximately 50% of the adult value of DA content is present. By 4 weeks of age, DA content ranges from 80% to 100% of adult values.

The striatum represents a region of the brain that receives the most dense dopaminergic innervation. DA content in the striatum ranges from 11% to 13% of the adult value at birth (Porcher and Heller, 1972; Coyle and Campochiaro, 1976; Giorgi *et al.,* 1987). Striatal DA content develops somewhat more slowly than what would be predicted from the ontogeny whole brain DA content. DA content in the striatum is only 28% of the adult value at 2 weeks of age and 50% by 3 weeks of age (Porcher and Heller, 1972; Coyle and Campochiaro, 1976; Kohno *et al.,* 1982; Giorgi *et al.,* 1987). Striatal DA content does not begin to approximate adult concentrations until about 50–60 days of age (Porcher and Heller, 1972; Giorgi *et al.,* 1987).

2. Norepinephrine

The development of norepinephrine (NE) concentrations in the CNS appears to precede that of the other monoamines. At birth, NE content in the rat ranges from 18% to 29% of the adult value in whole brain (Breese and Traylor, 1972; Coyle and Axelrod, 1972a; Herregodts *et al.,* 1990). By 2 weeks of age, whole brain concentrations of NE range from 55% to 85% of the adult value.

The development of NE content in the telencephalon has been the subject of several investigations. NE content in frontal cortex is 15–30% of the adult values at birth (Levitt and Moore, 1979; Kohno *et al.,* 1982). NE content in frontal cortex does not appear to develop as rapidly as does whole brain NE concentrations. NE concentrations in frontal cortex range from 22% to 38% of the adult values by 1 week of age, and only 37% to 55% at 3 weeks (Porcher and Heller, 1972; Levitt and Moore, 1979; Johnston and Coyle, 1980; Kohno *et al.,* 1982). At birth, NE content in hippocampus is approximately 8% of the adult value. The development of NE content in hippocampus is similar to that in frontal cortex. By 3 weeks of age, NE concentrations in hippocampus are 32% of the adult value (Kohno *et al.,* 1982). NE content in frontal cortex and hippocampus does not approximate adult values until 35–45 days of age.

3. Serotonin

Whole brain content of serotonin (5-HT) is also low at birth, ranging from 12% to 15% of the adult values (Bennett and Giarman, 1965; Herregodts *et al.,* 1990). 5-HT content in whole brain develops rapidly, such that concentrations range from 65% to 75% of the adult values by 2 weeks of age. Whole brain 5-HT content approximates adult values by 3 weeks of age. The development of 5-HT content in frontal cortex follows a similar time course to that seen in whole brain. Concentrations of 5-HT in frontal cortex are 18% of the adult value at birth and reach 60% by 3 weeks of age (Huether *et al.,* 1992).

B. Biosynthetic Enzymes

1. Tyrosine Hydroxylase

Whole brain tyrosine hydroxylase (TH), the rate limiting enzyme for DA biosynthesis, is low at birth in relation to the adult level of activity. Whole brain TH activity at birth ranges from 29% to 44% of the adult level of activity (Breese and Traylor, 1972; Coyle and Axelrod, 1972b). Thus, the development of TH activity in whole brain appears to precede the development of DA and NE content during the first postnatal week. Whole brain TH activity reaches 60–75% of the adult value by 2 weeks of age and approximates the adult value by 4 weeks.

Striatal TH activity is only 10% of the adult value at birth (Coyle and Axelrod, 1972b; Coyle and Campochiaro, 1976). This contrasts with the greater overall development seen in whole brain at this age. Thus, like DA content in the striatum, the ontogeny of striatal TH activity lags somewhat behind that seen in whole brain. Striatal TH activity is only 20% of the adult value at 1 week of age, and 75% at 4 weeks of age (Coyle and Axelrod, 1972b; Porcher and Heller, 1972; Coyle and Campochiaro, 1976).

TH activity in the cortex is relatively high at birth in comparison to other forebrain structures, ranging from 17% to 34% of the adult activity (Coyle and Axelrod, 1972b; Johnston and Coyle, 1980). This is probably due to the early development of dopaminergic and noradrenergic innervation to the frontal and lateral cortex. Despite this early innervation, cortical TH activity develops slowly after birth such that by 2 weeks of age, activity is only about 40–50% of the adult value (Coyle and Axelrod, 1972b; Porcher and Heller, 1972; Johnston and Coyle, 1980). Cortical TH activity does not begin to approximate adult values until sometime after 4–5 weeks of age.

2. Dopamine-β-Hydroxylase

Dopamine-β-hydroxylase (DβH) mediates the bioconversion of DA to NE. The ontogeny of this enzyme exhibits a similar time course to the development of NE content. Whole brain DβH activity is 14% of the adult activity at ED 17, the age when noradrenergic fibers begin to innervate the neocortex (Coyle and Axelrod, 1972a). Whole brain DβH activity reaches 29% of the adult value at birth. After parturition, the development of DβH activity slows somewhat, so that 42% of the adult value is achieved by 1 week of age and 71% by the end of the fourth postnatal week (Coyle and Axelrod, 1972a).

3. Tryptophan Hydroxylase

The development of tryptophan hydroxylase (TPH) activity, the rate-limiting enzyme for 5-HT biosynthesis, has been investigated in cortex and brainstem. In cortex, TPH activity is low at birth, not exceeding 5% of the adult value (Deguchi and Barchas, 1972). As with 5-HT content, TPH activity develops rapidly such that activity is approximately 30% of the adult value by 2 weeks of age and 60% by the end of the third postnatal week (Deguchi and Barchas, 1972; Huether *et al.,* 1992). Cortical TPH activity approximates the adult value by 4 weeks of age.

TPH activity in the brainstem develops in advance of that in cortex. Brainstem TPH activity is 27% of the adult value at birth and reaches 62% by 2 weeks of age (Deguchi and Barchas, 1972). TPH activity in the brainstem approximates the adult value by the end of the third postnatal week.

C. Reuptake Systems

The high affinity uptake of monoaminergic neurotransmitters is primarily a function of the presynaptic monoaminergic terminal. Thus, this index of monoaminergic function may be the best neurochemical descriptor of monoaminergic innervation to the CNS. This may be especially so in that the development of high affinity reuptake is observed to precede the development of the enzymes responsible for monoamine synthesis.

1. High-Affinity DA Uptake

The development of high-affinity DA uptake has been investigated by two different methods: (1) the specific uptake of tritiated DA ([^{3}H]DA); and (2) receptor binding of DA reuptake inhibitors to the DA reuptake transporter. Each of these methodologies produces a somewhat different profile of the development of the DA reuptake transporter.

Studies using [^{3}H]DA uptake to determine the development of the DA uptake transporter indicate that the activity of this transporter is low at birth in the rat. Striatal [^{3}H]DA uptake is only about 10% of the adult value at parturition (Coyle and Campochiaro, 1976; Kirksey and Slotkin, 1979). High-affinity [^{3}H]DA uptake develops rapidly in the striatum such that by 2 weeks of age, 60% of the adult value has been achieved. By the end of the third postnatal week, high-affinity [^{3}H]DA uptake is 75% of the adult value.

Investigating the development of the high-affinity DA uptake transporter with DA uptake inhibitors such as [^{125}I]RTI-55 reveals a different developmental profile. By this method, striatal DA uptake transporter densities are reported to be approximately 30% of the adult value at PN 0–1 (Coulter *et al.,* 1996; Jung and Bennett, 1996). The subsequent development of DA uptake transporters in the striatum is reported to achieve adult densities by 7–10 days of age. Autoradiographic studies indicates that the development of the DA reuptake transporter in striatum proceeds in a ventral–medial to dorsal–lateral direction. Although some regions of the striatum achieve adult densities of the DA reuptake transporter early in postnatal development, other regions of the striatum, namely the posterior regions, do not reach adult densities until the end of the third postnatal week (Coulter *et al.,* 1996).

A similar phenomenon exists regarding the development of DA uptake transporter densities in cortex. The uptake of [^{3}H]DA reveals that cortical DA reuptake is 12% of the adult value at birth. The development of [^{3}H]DA uptake proceeds such that, by 2 weeks of age, 50% of the adult value has been reached (Kirksey and Slotkin, 1979). [^{3}H]DA uptake approximates the adult value by 6 weeks of age. Conversely, [^{125}I]RTI-55 autoradiography reveals that, at birth, 40% of the adult density of DA uptake transporters has been achieved in the prefrontal cortex (Coulter *et al.,* 1996). The differences obtained by these two methods, [^{3}H]DA uptake versus [^{125}I]RTI-55 autoradiography, may merely be due to the greater anatomic specificity that can be achieved with autoradiographic methods. However, the possibility exists that, in an *in vitro* functional assay, [^{3}H]DA uptake may be measuring a qualitatively different aspect of the ontogeny of DA reuptake transporters in comparison to [^{125}I]RTI-55 autoradiography. It is unclear at present why these different methodologies should provide such disparate results in regards to the development of high-affinity DA uptake transporters.

2. High-Affinity NE Uptake

The development of the high-affinity NE uptake transporter has been studied using [^{3}H]NE uptake. Whole brain NE uptake ranges from 13% to 30% of the adult value at birth (Coyle and Axelrod, 1971; Kirksey *et al.,* 1978). [^{3}H]NE uptake in whole brain develops rapidly such that 70–100% of the adult value has been achieved by the end of the second postnatal week. Cortical [^{3}H]NE uptake develops early during postnatal life in the rat. This is consistent with the early innervation of the cortex by noradrenergic fibers. Cortical [^{3}H]NE uptake ranges from 13% to 15% of the adult value at birth (Levitt and Moore, 1979). By the end of the third postnatal week, [^{3}H]NE uptake approximates the adult value. There may be a transient overexpression of [^{3}H]NE uptake in frontal cortex as uptake in this brain region has been reported to exceed 200% of the adult value at 2 weeks of age (Levitt and Moore, 1979). [^{3}H]NE uptake then drops to adult values in frontal cortex by the end of the fourth postnatal week.

3. High-Affinity 5-HT Uptake

The development of the high affinity 5-HT reuptake transporter has been investigated using [^{3}H]5-HT uptake and [^{3}H]paroxetine binding. Cortical [^{3}H]5-HT uptake ranges from 19% to 22% of the adult value at parturition (Kirksey and Slotkin, 1979; Huether *et al.,* 1992). Cortical [^{3}H]5-HT uptake achieves 42–54% of the adult value by 2 weeks of age. By 4–5 weeks of age, [^{3}H]5-HT uptake in the cortex approximates adult values. Investigation of the ontogeny of the high affinity

5-HT uptake transporter by [^{3}H]paroxetine binding reveals a somewhat more advanced developmental time course versus that seen with studies using [^{3}H]5-HT uptake. Cortical [^{3}H]paroxetine binding to the 5-HT uptake transporter is reported to be 39% of the adult values by the end of the first postnatal week and approximates the adult value by 2 weeks of age (Pranzatelli and Martens, 1992).

D. Monoaminergic Receptors

1. Dopaminergic Receptors

The ontogeny of dopaminergic receptors has been investigated by a variety of methods including receptor binding assays, autoradiography, and immunohistochemistry. The majority of reports that investigate the ontogeny of dopaminergic receptors focus exclusively on the striatum. For the most part, the following synopsis of dopaminergic receptor development focuses on the ontogeny of these receptors in the striatum.

In the striatum of the rat, D_1 dopaminergic receptor density is approximately 10% of the adult value at birth (Giorgi *et al.,* 1987; Murrin and Zeng, 1990; Rao *et al.,* 1991; Jung and Bennett, 1996). Most investigators report that striatal D_1 receptors develop primarily during the first and second postnatal week. D_1 receptor density in the striatum ranges from 22% to 41% of the adult value by the end of the first postnatal week (Giorgi *et al.,* 1987; Gelbard *et al.,* 1989; Rao *et al.,* 1991; Jung and Bennett, 1996). By the end of the second postnatal week, D_1 receptor density begins to approximate the adult value (Giorgi *et al.,* 1987; Gelbard *et al.,* 1989; Broaddus and Bennett, 1990; Leslie *et al.,* 1991; Rao *et al.,* 1991; Schambra *et al.,* 1994; Jung and Bennett, 1996). Immunohistochemical studies using antibodies directed against the D_1 receptor confirm that these receptors attain the adult distribution and density early in postnatal life (Caille *et al.,* 1995). Several investigators observe that D_1 receptors are overexpressed in the caudate–putamen during development. In this area of the striatum, D_1 receptor density is reported to range from 110% to 200% of the adult value at 3 weeks of age (Giorgi *et al.,* 1987; Gelbard *et al.,* 1989; Murrin and Zeng, 1990; Schambra *et al.,* 1994; Teicher *et al.,* 1995).

Evidence also exists that D_1 receptors in the prefrontal cortex achieve the adult distribution and density early in postnatal life (Leslie *et al.,* 1991). D_1 receptor density in this cortical region ranges from 50% to 59% of the adult value at 1 week after birth. D_1 receptors in the prefrontal cortex also exhibit ontologic overexpression. D_1 receptor density is reported to exceed 200% of the adult value at 2 weeks after birth (Leslie *et al.,* 1991).

Striatal D_2 dopaminergic receptors exhibit a similar developmental time course to that exhibited by the D_1 receptor. D_2 receptor density ranges from 5% to 18% of the adult value at birth (Nomura *et al.,* 1982; Murrin and Zeng, 1986; Broaddus and Bennett, 1990; Rao *et al.,* 1991; Jung and Bennett, 1996). Striatal D_2 receptor density reaches 25–45% of the adult value by the end of the first postnatal week and begins to approximate adult values by the end of the second postnatal week (Nomura *et al.,* 1982; Bruinink *et al.,* 1983; Murrin and Zeng, 1986; Rao *et al.,* 1991; Schambra *et al.,* 1994; Jung and Bennett, 1996). As with the D_1 receptors, several investigators report that striatal D_2 receptors exhibit developmental overexpression in the caudate–putamen. D_2 receptor density in the caudate–putamen has been reported to exceed 200% of the adult value by 3 weeks of age (Gelbard *et al.,* 1989; Teicher *et al.,* 1995).

The ontogeny of striatal D_3 dopaminergic receptors has also been investigated in the rat. D_3 receptor density in the nucleus accumbens ranges from 5% to 13% of the adult value at parturition (Demotes-Mainard *et al.,* 1996; Stanwood *et al.,* 1997). In contrast to the D_1 and D_2 receptors, D_3 receptors exhibit no evidence of ontologic overexpression and develop over a longer period of postnatal life. D_3 receptor density in the nucleus accumbens reaches 45% of the adult value at 1 week of age and does not begin to approximate adult values until after 3 weeks of age (Demotes-Mainard *et al.,* 1996).

2. Noradrenergic Receptors

Receptor binding studies using [^{3}H]prazosin to label α_1-adrenergic receptors in the rat cerebral cortex demonstrate a progressive increase in receptor density during postnatal development (Schoepp and Rutledge, 1985; Slotkin *et al.,* 1990). At 1 week of age, α_1-receptor density has achieved approximately 25% of the adult value, and 54% by 2 weeks. α_1-receptor density in the cortex begins to approximate adult values by 3 weeks of age (Schoepp and Rutledge, 1985; Slotkin *et al.,* 1990).

Total β-adrenergic receptor density in whole brain is reported to be low at birth, reaching only 14% of the adult value (Erdtsieck-Ernste *et al.,* 1991). Adult values of whole brain β-adrenergic receptor density are achieved by 3 weeks of age. A similar pattern of β-adrenergic receptor development is observed in cortex. Total β-adrenergic receptor density is 32% of the adult value after 1 postnatal week and 72% at 2 postnatal weeks. Cortical β-adrenergic receptor density reaches adult values by 3 weeks of age (Harden *et al.,* 1977; Pittman *et al.,* 1980; McDonald *et al.,* 1982; Keshles and Levitzki, 1984; Bartolome *et al.,* 1987; Slotkin *et al.,* 1990; Kudlacz *et al.,* 1991).

In the cortex of the adult rat, β_1-receptors comprise approximately 80% of the total β-adrenergic receptors present. This proportion appears conserved somewhat during the postnatal ontogeny of β-adrenergic receptor density except for the perinatal period, when β_2 receptors contribute a greater percentage to the total (Pittman *et al.*, 1980; Erdtsieck-Ernste *et al.*, 1991). In cortex, β_1-receptor density achieves 24% of the adult value by 1 week of age and 65% of the adult value at 2 weeks of age (Pittman *et al.*, 1980; Lorton *et al.*, 1988). Cortical β_1-receptor density approximates the adult value by 3 weeks of age in the rat. The development of cortical β_2-receptor density follows a similar developmental time course as that observed for the β_1-receptor. β_2-receptor density is approximately 42% of the adult value at 1 week of age and achieves 76% of the adult value by 2 weeks of age (Pittman *et al.*, 1980; Lorton *et al.*, 1988).

β_2-receptors comprise approximately 90% of the total β-receptor density in the cerebellum in the adult rat. Cerebellar β_2-receptor density is approximately 33% of the adult value at 1 week of age and 71% at 3 weeks of age (Pittman *et al.*, 1980; Lorton *et al.*, 1988; Kudlacz *et al.*, 1991). The developmental time course of β_2-receptor density in the cerbellum begins to approximate the adult value by 6 weeks of age.

3. Serotonergic Receptors

Whole brain 5-HT_1 serotonergic receptor density is approximately 45% of the adult value at birth in the rat (Zilles *et al.*, 1985). By 2 weeks of age, 5-HT_1 receptor density reaches 68% of the adult value (Zilles *et al.*, 1985). The ontogeny of 5-HT_1 receptor density in frontal cortex and hippocampus follows a different developmental time course to that observed in whole brain. 5-HT_1 receptor density in frontal cortex is low at birth, achieving only 24% of the adult value at 1 day of age (Zilles *et al.*, 1985). By 10 days of age, 5-HT_1 receptor density is approximately 50% of the adult value. 5-HT_1 receptor density in frontal cortex approximates the adult value by 3 weeks of age in the rat. In hippocampus, 5-HT_1 receptor density is approximately 26% of the adult value by 5 days of age. 5-HT_1 receptor density reaches 50% of the adult value by 3 weeks of age and begins to approximate adult values by 4 weeks of age (Zilles *et al.*, 1985).

Several subtypes of the 5-HT_1 receptor have been identified. These subtypes currently include the 5-HT_{1A}, 5-HT_{1B}, and 5-HT_{1D} receptors (Hoyer *et al.*, 1994). The ontogeny of 5-HT_{1B} receptor density in the rat brain has been investigated using [I^{125}]iodocyanopindolol (Pranzatelli and Galvan, 1994). In cortex, 5-HT_{1B} receptor density is 31% of the adult value at 5 days of age and achieves 65% of the adult value at 3 weeks of age. In hippocampus, 5-HT_{1B} receptor density is 22% of the adult value at 5 days of age and reaches 78% of the adult value at 3 weeks of age. In striatum, 5-HT_{1B} receptor density is 31% of the adult value at 5 days of age and achieves 85% of the adult value at 3 weeks of age. By 4 weeks of age, 5-HT_{1B} receptor density approximates the adult receptor densities found in cortex, hippocampus, and striatum (Pranzatelli and Galvan, 1994).

Whole brain 5-HT_{2A} serotonergic receptor density is low in the perinatal period, achieving only 17% of the adult value 2 days after birth (Bruinink *et al.*, 1983; Roth *et al.*, 1991). 5-HT_{2A} receptor density reaches 76% of the adult value at 2 weeks of age and achieves the receptor density observed in the mature rat after 4 weeks of age. Investigations of the ontogeny of 5-HT_2 receptors using antibodies generated against this receptor confirm these findings (Morilak and Ciaranello, 1993).

5-HT_{2C} serotonergic receptors, formerly known as 5-HT_{1C} receptors (Hoyer *et al.*, 1994), are present in most regions of the brain. Whole brain 5-HT_{2C} receptor density reaches adult values early in postnatal development in the rat, achieving adult values by PN 5 (Roth *et al.*, 1991). This is reflected in the early maturation of 5-HT_{2C} receptor density in the diencephalon and brainstem (Pranzatelli, 1993). However, 5-HT_{2C} receptor density does not reach adult values until much later in postnatal development in cortex. 5-HT_{2C} receptor density in cortex is approximately 24% of the adult value at PN 1 in the rat, and reaches 37% of the adult value in cortex by PN 10 (Pranzatelli, 1993). 5-HT_{2C} receptor density achieves 76% of the adult value by 2 weeks of age and approximates the adult value by 3 weeks of age.

IV. Summary and Conclusions

A. *General Observations*

This review of the available information concerning the development of the monoaminergic innervation to the CNS of the rat has confirmed the observations made by others (Coyle, 1977; Lidov and Molliver, 1982a; Foote and Morrison, 1987) in that the development of monoaminergic innervation to the rat CNS can be divided into three distinct phases: (1) the histogenesis of the monoaminergic nuclei in the brainstem, (2) the establishment of axonal pathways to specific areas of the CNS that are to receive monoaminergic innervation, and (3) the elaboration of terminal fields. The first two phases occur primarily during prenatal development in the rat, whereas the third phase occurs primarily during postnatal development.

In general, the development of neurochemical indices of monoaminergic function follows closely with the demonstration of neuropil elaboration. Whereas the ontogeny of these neurochemical indices begins prenatally in the rat, the data presented in this review is consistent in showing that the majority of the development to the adult values occurs postnatally. It seems reasonable to suggest that as the monoaminergic neurotransmitter systems mature they exert differing functional influences on the CNS, based in part on their degree of maturation. Thus, the postnatal period of development in the rat offers up a unique experimental model for elucidating how drugs of abuse may interact with synaptogenetic events to disrupt the normal development of the CNS.

B. Interactions between Age of Exposure and Outcome for Stimulant Drugs of Abuse

Administration of neurotoxicants during different developmental phases may be expected to produce differing phenomena based on both the ages at exposure and the duration of exposure. One example of this is the neurotoxicity produced by administration of the neurotoxic amphetamine derivatives methamphetamine (MA) or 3,4-methylenedioxymethamphetamine (MDMA). In adult rats, MA is neurotoxic to the dopaminergic and serotonergic neurotransmitter systems, whereas MDMA primarily affects the serotonergic system. Administration of these compounds to experimental animals results in a reproducible profile of neurotoxicity that consists of persistent reductions in DA and 5-HT content in the forebrain. This phenomenon is accompanied by the loss of TH and TPH activity and diminished numbers of DA and 5-HT reuptake sites (Seiden *et al.,* 1976; Hotchkiss and Gibb, 1980; Ricaurte *et al.,* 1980; Wagner *et al.,* 1980; Bowyer *et al.,* 1994; O'Callaghan and Miller, 1994). Immunohistochemical examination reveals that MA reduces TH immunoreactivity in the striatum and other dopaminergic terminal fields in the forebrain (Trulson *et al.,* 1985; Hess *et al.,* 1990; Pu and Vorhees, 1993; Pu *et al.,* 1994). TPH activity is also lost following MA and MDMA administration. Astrogliosis and argyrophilia, indices of neuronal damage, are present in areas corresponding to the loss of these indices of monoaminergic function after MA and MDMA treatment, thereby supporting the conclusion these amphetamine derivatives injure monoaminergic terminals (Ricaurte *et al.,* 1982; Hess *et al.,* 1990; Pu and Vorhees, 1993; Bowyer *et al.,* 1994; Miller and O'Callaghan, 1994; O'Callaghan and Miller, 1994; Pu *et al.,* 1994).

Administration of MA or MDMA to the developing rat does not produce the pattern of neurotoxicity observed in the adult. Long-term reductions in forebrain monoamine content are not observed following MA or MDMA administration to rats before 4 weeks of age (Lucot *et al.,* 1982; Pu and Vorhees, 1993; Broening *et al.,* 1994, 1995; Cappon *et al.,* 1997). Furthermore, loss of 5-HT reuptake sites or TH immunoreactivity is not observed, nor is astrogliosis. Taken together, these findings indicate that developing rats are resistant to the "adult-type" neurotoxicity observed following the administration of these compounds to mature rats.

However, these findings should not be interpreted such that one would assume the developing rat does not experience adverse effects from exposure to MA or MDMA. Instead, the developing rat exhibits a different neurotoxic phenomena altogether. MA or MDMA exposure during postnatal development in the rat results in learning and memory deficits when the rats are tested as adults (Vorhees *et al.,* 1994a,b; Broening *et al.,* 1997a,b). These deficits have been noted in both egocentric and allocentric tasks of learning and memory. Furthermore, the expression of learning and memory deficits appears to be associated with neurotoxicant exposure during a specific developmental period—PN 10–20. That these learning and memory deficits are expressed in the absence of the "adult-type" neurotoxic response is even more remarkable. Indeed, functional deficits are rarely observed in the mature rats that exhibit the "adult-type" neurotoxic response (Dornan *et al.,* 1991; Ricaurte *et al.,* 1993, 1994; Robinson *et al.,* 1993).

The disparities that exist between the responses of adult rats versus developing rats following administration of MA or MDMA may be attributable in large part to the maturational status of the CNS at the time drug exposure takes place. Furthermore, the same characteristics that afford the developing CNS resistance to the "adult-type" neurotoxicity may underlie the functional deficits observed following neonatal MA or MDMA exposure. Further research will undoubtedly elucidate the underlying neurochemical and neuroanatomic changes that account for the cognitive deficits observed following developmental administration of MA or MDMA.

References

Bartolome, J. V., Kavlock, R. J., Cowdery, T., Orband-Miller, L., and Slotkin, T. A. (1987). Development of adrenergic receptor binding sites in brain regions of the neonatal rat: effects of prenatal or postnatal exposure to methylmercury. *Neurotoxicology.* **8,** 1–13.

Bennett, D. S., and Giarman, N. J. (1965). Schedule of appearance of 5-hydroxytryptamine (serotonin) and associated enzymes in the developing rat brain. *J. Neurochem.* **12,** 911–918.

Berger, B., and Verney, C. (1984). Development of catecholamine innervation in rat neocortex: morphological features. In *Monoamine Innervation of the Cerebral Cortex.* L. Descarries, T. R. Reader, and H. H. Jasper, Eds. Alan R. Liss, New York, pp. 95–121.

Berger, B., Verney, C., Gay, M., and Vigny, A. (1983). Immunocytochemical characterization of the dopaminergic and noradrenergic innervation of the rat neocortex during ontogeny. *Prog. Brain Res.* **58,** 263–267.

Bowyer, J. F., Davies, D. L., Schmued, L., Broening, H. W., Newport, G. D., Slikker, W. Jr., and Holson, R. R. (1994). Further studies of the role of hyperthermia in methamphetamine neurotoxicity. *J. Pharmacol. Exp. Ther.* **268,** 1571–1580.

Breese, G. R. and Traylor, T. D. (1972). Developmental characteristics of brain catecholamines and tyrosine hydroxylase in the rat: effects of 6-hydroxydopamine. *Br. J. Pharmacol.* **44,** 210–222.

Broaddus, W. C. and Bennett, J. P. Jr. (1990). Postnatal development of striatal dopamine function. I. An examination of D1 and D2 receptors, adenylate cyclase regulation and presynaptic dopamine markers. *Dev. Brain Res.* **52,** 265–271.

Broening, H. W., Bacon, L., and Slikker, W. Jr. (1994). Age modulates the long-term but not the acute effects of the serotonergic neurotoxicant 3,4-methylenedioxymethamphetamine. *J. Pharmacol. Exp. Ther.* **271,** 285–293.

Broening, H. W., Bowyer, J. F., and Slikker, W. Jr. (1995). Age dependent sensitivity of rats to the long-term effects of the serotonergic neurotoxicant (±)-3,4-methylenedioxymethamphetamine (MDMA) correlates with the magnitude of the MDMA-induced thermal response. *J. Pharmacol. Exp. Ther.* **275,** 325–333.

Broening, H. W., Morford, L., Inman, S., Moran, M. S., St. Clair, C., Fukumura, M., and Vorhees, C. V. (1997a). Developmental exposure to 3,4-methylenedioxymethamphetamine (MDMA) disrupts learning in the Morris water maze in rats. *Soc. Neurosci. Abs.* **23,** 935.13 (Abstract).

Broening, H. W., St. Clair, C., Morford, L., Inman, S., Moran, M. S., Fukumura, M., and Vorhees, C. V. (1997b). Developmental exposure to 3,4-methylenedioxymethamphetamine (MDMA) in rats produces persistent deficits in Cincinnati water maze performance. *Neurotoxicol. Teratol.* **19,** 246. (Abstract).

Bruinink, A., Lichtensteiger, W., and Schlumpf, M. (1983). Pre- and postnatal ontogeny and characterization of dopaminergic D_2, serotonergic S_2 and spirodecanone binding sites in rat forebrain. *J. Neurochem.* **40,** 1227–1236.

Caille, I., Dumartin, B., Le Moine, C., Begueret, J., and Bloch, B. (1995). Ontogeny of the D1 dopamine receptor in the rat striatonigral system: an immunohistochemical study. *Eur. J. Neurosci.* **7,** 714–722.

Cappon, G. D., Morford, L. L., and Vorhees, C. V. (1997). Ontogeny of methamphetamine-induced neurotoxicity and associated hyperthermic response. *Dev. Brain Res.* **103,** 155–162.

Coulter, C. L., Happe, H. K., and Murrin, L. C. (1996). Postnatal development of the dopamine transporter: a quantitative autoradiographic study. *Dev. Brain Res.* **92,** 172–181.

Coyle, J. T. (1977). Biochemical aspects of neurotransmission in the developing brain. *Int. Rev. Neurobiol.* **20,** 65–103.

Coyle, J. T., and Axelrod, J. (1971). Development of the uptake and storage of *l*-[3H]norepinephrine in the rat brain. *J. Neurochem.* **18,** 2061–2075.

Coyle, J. T., and Axelrod, J. (1972a). Dopamine-β-hydroxylase in the rat brain: developmental characteristics. *J. Neurochem.* **19,** 449–459.

Coyle, J. T. and Axelrod, J. (1972b). Tyrosine hydroxylase in rat brain: developmental characteristics. *J. Neurochem.* **19,** 1117–1123.

Coyle, J. T., and Campochiaro, P. (1976). Ontogenesis of dopaminergic-cholinergic interactions in the rat striatum: a neurochemical study. *J. Neurochem.* **27,** 673–678.

Deguchi, T. and Barchas, J. (1972). Regional distribution and developmental change of tryptophan hydroxylase activity in rat brain. *J. Neurochem.* **19,** 927–929.

Demotes-Mainard, J., Henry, C., Jeantet, Y., Arsaut, J., and Arnauld, E. (1996). Postnatal ontogeny of dopamine D3 receptors in the mouse brain: autoradiographic evidence for a transient cortical expression. *Dev. Brain Res.* **94,** 166–174.

Dinopoulos, A., Dori, I., and Parnavelas, J. G. (1997). The serotonin innervation of the basal forebrain shows a transient phase during development. *Dev. Brain Res.* **99,** 38–52.

Dori, I., Dinopoulos, A., Blue, M. E., and Parnavelas, J. G. (1996). Regional differences in the ontogeny of the serotonergic projection to the cerebral cortex. *Exp. Neurol.* **138,** 1–14.

Dornan, W. A., Katz, J. L., and Ricaurte, G. A. (1991). The effects of repeated administration of MDMA on the expression of sexual behavior in the male rat. *Pharmacol. Biochem. Behav.* **39,** 813–816.

Erdtsieck-Ernste, B. H. W., Feenstra, M. G. P., and Boer, G. J. (1991). Pre- and postnatal developmental changes of adrenoreceptor subtypes in rat brain. *J. Neurochem.* **57,** 897–903.

Foote, S. L. and Morrison, J. H. (1987). Development of the noradrenergic, serotonergic, and dopaminergic innervation of neocortex. In *Current Topics in Developmental Biology,* Vol. 21, Academic Press, pp. 391–423.

Gelbard, H. A., Teicher, M. H., Faedda, G., and Baldessarini, R. J. (1989). Postnatal development of dopamine D_1 and D_2 receptor sites in rat striatum. *Brain Res.* **49,** 123–130.

Giorgi, O., De Montis, G., Porceddu, M. L., Mele, S., Calderini, G., Toffano, G., and Biggio, G. (1987). Developmental and age-related changes in D1-dopamine receptors and dopamine content in the rat striatum. *Dev. Brain Res.* **35,** 283–290.

Harden, T. K., Wolfe, B. B., Sporn, J. R., Perkins, J. P., and Molinoff, P. B. (1977). Ontogeny of β-adrenergic receptors in rat cerebral cortex. *Brain Res.* **125,** 99–108.

Herregodts, P., Velkeniers, B., Ebinger, G., Michotte, Y., Vanhaelst, L., and Hooghe-Peters, E. (1990). Development of monoamine neurotransmitters in fetal and postnatal rat brain: analysis by HPLC with electrochemical detection. *J. Neurochem.* **55,** 774–779.

Hess, A., Desiderio, C., and McAuliffe, W. G. (1990). Acute neuropathological changes in the caudate nucleus caused by MPTP and methamphetamine: immunohistochemical studies. *J. Neurocytol.* **19,** 338–342.

Hotchkiss, A. J., and Gibb, J. W. (1980). Long-term effects of multiple doses of methamphetamine on tryptophan hydroxylase and tyrosine hydroxylase activity in rat brain. *J. Pharmacol. Exp. Ther.* **214,** 257–262.

Hoyer, D., Clarke, D. E., Fozard, J. R., Hartig, P. R., Martin, G. R., Mylecharane, E. J., Saxena, P. R., and Humphrey, P. P. A. (1994). International union of pharmacology classification of receptors for 5-hydroxytryptamine (serotonin). *Pharmacol. Rev.* **46,** 157–203.

Huether, G., Thomke, F., and Adler, L. (1992). Administration of tryptophan enriched diets to pregnant rats retards the development of the serotonergic system in their offspring. *Dev. Brain Res.* **68,** 175–181.

Johnston, M. V. and Coyle, J. T. (1980). Ontogeny of neurochemical markers for noradrenergic, GABAergic, and cholinergic neurons in neocortex lesioned with methylazoxymethanol acetate. *J. Neurochem.* **34,** 1429–1441.

Jung, A. B. and Bennett, J. P. (1996). Development of striatal dopaminergic function. I. Pre- and postnatal development of mRNAs and binding sites for striatal D1 (D1a) and D2 (D2a) receptors. *Dev. Brain Res.* **94,** 109–120.

Kalsbeek, A., Voorn, P., Buijs, R. M., Pool, C. W., and Uylings, H. B. M. (1988). Development of the dopaminergic innervation in the prefrontal cortex of the rat. *J. Comp. Neurol.* **269,** 58–72.

Keshles, O. and Levitzki, A. (1984). The ontogenesis of β-adrenergic receptors and of adenylate cyclase in the developing rat brain. *Biochem. Pharmacol.* **33,** 3231–3233.

Kirksey, D. F., Seidler, F. J., and Slotkin, T. A. (1978). Ontogeny of (−)-[^{3}H]norepinephrine uptake properties of synaptic storage vesicles of rat brain. *Brain Res.* **150,** 367–375.

Kirksey, D. F. and Slotkin, T. A. (1979). Concomitant development of [^{3}H]-dopamine and [^{3}H]-5-hydroxytryptamine uptake systems in rat brain regions. *Br. J. Pharmacol.* **67,** 387–391.

Kohno, Y., Tanaka, M., Nakagawa, R., Ida, Y., Iimori, K., and Nagasaki, N. (1982). Postnatal development of noradrenaline and 3-methoxy-4-hydroxyphenylethyleneglycol sulphate levels in rat brain regions. *J. Neurochem.* **39,** 878–881.

Kudlacz, E. M., Spencer, J. R., and Slotkin, T. A. (1991). Postnatal alterations in β-adrenergic receptor binding in rat brain regions after continuous prenatal exposure to propranolol via maternal infusion. *Res. Commun. Chem. Pathol. Pharmacol.* **71,** 153–161.

Lauder, J. M. and Bloom, F. E. (1974). Ontogeny of monoamine neurons in the locus coeruleus, raphe nuclei and substantia nigra of the rat. *J. Comp. Neurol.* **155,** 469–482.

Lauder, J. M., Wallace, J. A., Krebs, H., Petrusz, P., and McCarthy, K. (1982). *In vivo* and *in vitro* development of serotonergic neurons. *Brain Res. Bull.* **9,** 605–625.

Leslie, C. A., Robertson, M. W., Cutler, A. J., and Bennett, J. P. (1991). Postnatal development of D_1 dopamine receptors in the medial prefrontal cortex, striatum and nucleus accumbens of normal and neonatal 6-hydroxydopamine treated rats: a quantitative autoradiographic study. *Dev. Brain Res.* **62,** 109–114.

Levitt, P. and Moore, R. Y. (1978). Developmental organization of raphe serotonin neuron groups in the rat. *Anat. Embryol.* **154,** 241–251.

Levitt, P. and Moore, R. Y. (1979). Development of the noradrenergic innervation of neocortex. *Brain Res.* **162,** 243–259.

Lidov, H. G. W., and Molliver, M. E. (1982a). An immunohistochemical study of serotonin neuron development in the rat: ascending pathways and terminal fields. *Brain Res. Bull.* **8,** 389–430.

Lidov, H. G. W. and Molliver, M. E. (1982b). Immunohistochemical study of the development of serotonergic neurons in the rat CNS. *Brain Res. Bull.* **9,** 559–604.

Lidov, H. G. W., Molliver, M. E., and Zecevic, N. R. (1978). Characterization of the monoaminergic innervation of immature rat neocortex: a histofluorescence analysis. *J. Comp. Neurol.* **181,** 663–680.

Lorton, D., Bartolome, J., Slotkin, T. A., and Davis, J. N. (1988). Development of brain beta-adrenergic receptors after neonatal 6-hydroxydopamine treatment. *Brain Res. Bull.* **21,** 591–600.

Lucot, J. B., Wagner, G. C., Schuster, C. R., and Seiden, L. S. (1982). Decreased sensitivity of rat pups to long-lasting dopamine and serotonin depletions produced by methylamphetamine. *Brain Res.* **247,** 181–183.

McDonald, J. K., Petrovic, S. L., McCann, S. M., and Parnavelas, J. G. (1982). The development of beta-adrenergic receptors in the visual cortex of the rat. *Neuroscience* **7,** 2649–2655.

Miller, D. B. and O'Callaghan, J. P. (1994). Environment-, drug- and stress-induced alterations in body temperature affect the neurotoxicity of substituted amphetamines in the C57BL/6J mouse. *J. Pharmacol. Exp. Ther.* **270,** 752–760.

Morilak, D. A. and Ciaranello, R. D. (1993). Ontogeny of 5-hydroxytryptamine2 receptor immunoreactivity in the developing rat brain. *Neuroscience* **55,** 869–880.

Murrin, L. C. and Zeng, W. (1986). Postnatal ontogeny of dopamine D2 receptors in rat striatum. *Biochem. Pharmacol.* **35,** 1159–1162.

Murrin, L. C. and Zeng, W. Y. (1990). Ontogeny of dopamine D1 receptors in rat forebrain: a quantitative autoradiographic study. *Dev. Brain Res.* **57,** 7–13.

Nomura, Y., Oki, K., and Segawa, T. (1982). Ontogenic development of the striatal [^{3}H]spiperone binding: regulation by sodium and guanine nucleotide in rats. *J. Neurochem.* **38,** 902–908.

Nyakas, C., Buwalda, B., Kramers, R. J. K., Traber, J., and Luiten, P. G. (1994). Postnatal development of hippocampal and neocortical cholinergic and serotonergic innervation in rat: effects of nitrite-induced prenatal hypoxia and nimodipine treatment. *Neuroscience* **59,** 541–559.

O'Callaghan, J. P. and Miller, D. B. (1994). Neurotoxicity profiles of substituted amphetamines in the C57BL/6J mouse. *J. Pharmacol. Exp. Ther.* **270,** 741–751.

Olson, L. and Seiger, Å. (1972). Early prenatal ontogeny of central monoamine neurons in the rat: fluorescence histochemical observations. *Z. Anat. Entwickl.-Gesch.* **137,** 301–316.

Pittman, R. N., Minneman, K. P., and Molinoff, P. B. (1980). Ontogeny of β1- and β2-adrenergic receptors in rat cerebellum and cerebral cortex. *Brain Res.* **188,** 357–368.

Porcher, W. and Heller, A. (1972). Regional development of catecholamine biosynthesis in rat brain. *J. Neurochem.* **19,** 1917–1930.

Pranzatelli, M. R. (1993). Regional differences in the ontogeny of 5-hydroxytryptamine-1C binding sites in rat brain and spinal cord. *Neurosci. Lett.* **149,** 9–11.

Pranzatelli, M. R. and Galvan, I. (1994). Ontogeny of [125I]iodocyanopindolol-labeled 5-hydroxytryptamine1B-binding sites in the rat CNS. *Neurosci. Lett.* **167,** 166–170.

Pranzatelli, M. R. and Martens, J. M. (1992). Plasticity and ontogeny of the central 5-HT transporter: effect of neonatal 5,7-dihydroxytryptamine lesions in the rat. *Dev. Brain Res.* **70,** 191–195.

Pu, C., Fisher, J. E., Cappon, G. D., and Vorhees, C. V. (1994). The effects of amfonelic acid, a dopamine uptake inhibitor, on methamphetamine-induced dopaminergic terminal degeneration and astrocytic response in rat striatum. *Brain Res.* **649,** 217–224.

Pu, C. and Vorhees, C. V. (1993). Developmental dissociation of methamphetamine-induced depletion of dopaminergic terminals and astrocyte reaction in the rat striatum. *Dev. Brain Res.* **72,** 325–328.

Rao, P. A., Molinoff, P. B., and Joyce, J. N. (1991). Ontogeny of dopamine D1 and D2 receptor subtypes in rat basal ganglia: a quantitative autoradiographic study. *Dev. Brain Res.* **60,** 161–177.

Ricaurte, G. A., Guillery, R. W., Seiden, L. S., Schuster, C. R., and Moore, R. Y. (1982). Dopamine nerve terminal degeneration produced by high doses of methylamphetamine in the rat brain. *Brain Res.* **235,** 93–103.

Ricaurte, G. A., Markowska, A. L., Wenk, G. L., Hatzidimitriou, G., Wlos, J., and Olton, D. S. (1993). 3,4-Methylenedioxymethamphetamine, serotonin and memory. *J. Pharmacol. Exp. Ther.* **266,** 1097–1105.

Ricaurte, G. A., Sabol, K. E., and Seiden, L. S. (1994). Functional consequences of neurotoxic amphetamine exposure. In *Amphetamine and Its Analogs: Pharmacology, Toxicology and Abuse.* (A. K. Cho and D. S. Segal, Eds. Academic Press, San Diego, pp. 297–313.

Ricaurte, G. A., Schuster, C. R., and Seiden, L. S. (1980). Long-term effects of repeated methylamphetamine administration on dopamine and serotonin neurons in the rat brain: a regional study. *Brain Res.* **193,** 153–163.

Robinson, T. E., Castañeda, E., and Whishaw, I. Q. (1993). Effects of cortical serotonin depletion induced by 3,4-methylenedioxymethamphetamine (MDMA) on behavior, before and after additional cholinergic blockade. *Neuropsychopharmacology* **8,** 77–85.

Roth, B. L., Hamblin, M. W., and Ciaranello, R. D. (1991). Developmental regulation of 5-HT2 and 5-HT1c mRNA and receptor levels. *Dev. Brain Res.* **58,** 51–58.

Schambra, U. B., Duncan, G. E., Breese, G. R., Fornaretto, M. G., Caron, M. G., and Fremeau, R. T. (1994). Ontogeny of D_{1A} and D_2 dopamine receptor subtypes in rat brain using *in situ* hybridization and receptor binding. *Neuroscience* **62,** 65–85.

Schoepp, D. D. and Rutledge, C. O. (1985). Comparison of postnatal changes in alpha1-adrenoceptor binding and adrenergic stimulation of phosphoinositide hydrolysis in rat cerebral cortex. *Biochem. Pharmacol.* **34,** 2705–2711.

Seiden, L. S., Fischman, M. W., and Schuster, C. R. (1976). Long-term methamphetamine induced changes in brain catecholamines in tolerant rhesus monkeys. *Drug Alcohol Depend.* **1,** 215–219.

Seiger, Å, and Olson, L. (1973). Late prenatal ontogeny of central monoamines in the rat: fluorescence histochemical observations. *Z. Anat. Entwickl.-Gesch.* **140,** 281–318.

Slotkin, T. A., Kudlacz, E. M., Lappi, S. E., Tayyeb, M. I., and Seidler, F. J. (1990). Fetal terbutaline exposure causes selective postnatal increases in cerebellar α-adrenergic receptor binding. *Life Sci.* **47,** 2051–2057.

Specht, L. A., Pickel, V. M., Joh, T. H., and Reis, D. J. (1981a). Light-microscopic immunocytochemical localization of tyrosine hydroxylase in prenatal rat brain. I. Early ontogeny. *J. Comp. Neurol.* **199,** 233–253.

Specht, L. A., Pickel, V. M., Joh, T. H., and Reis, D. J. (1981b). Light-microscopic immunocytochemical localization of tyrosine hydroxylase in prenatal rat brain. II. Late ontogeny. *J. Comp. Neurol.* **199,** 255–276.

Stanwood, G. D., McElligot, S., Lu, L. H., and McGonigle, P. (1997). Ontogeny of dopamine D3 receptors in the nucleus accumbens of the rat. *Neurosci. Lett.* **223,** 13–16.

Teicher, M. H., Andersen, S. L., and Hostetter, J. C. (1995). Evidence for dopamine receptor pruning between adolescence and adulthood in striatum but not nucleus accumbens. *Dev. Brain Res.* **89,** 167–172.

Trulson, M. E., Cannon, M. S., Faegg, T. S., and Raese, J. D. (1985). Effects of chronic methamphetamine on the nigral–striatal dopamine system in rat brain: tyrosine hydroxylase immunochemistry and quantitative light microscopic studies. *Brain Res. Bull.* **15,** 569–577.

Verney, C., Berger, B., Adrien, J., Vigny, A., and Gay, M. (1982). Development of the dopaminergic innovation of the rat cerebral cortex. A light microscopic immunocytochemical study using anti-tyrosine hydroxylase antibodies. *Dev. Brain Res.* **5,** 41–52.

Verney, C., Berger, B., Baulac, M., Helle, K. B., and Alverez, C. (1984). Dopamine-β-hydroxylase-like immunoreactivity in the fetal cerebral cortex of the rat: noradrenergic ascending pathways and terminal fields. *Int. J. Dev. Neurosci.* **2,** 491–503.

Voorn, P., Kalsbeek, A., Jorritsma-Byham, B., and Groenewegen, H. J. (1988). The pre- and postnatal development of the dopaminergic cell groups in the ventral mesencephalon and the dopaminergic innervation of the striatum of the rat. *Neuroscience* **25,** 857–887.

Vorhees, C. V., Ahrens, K. G., Acuff-Smith, K. D., Schilling, M. A., and Fisher, J. E. (1994a). Methamphetamine exposure during early postnatal development in rats. 1. Acoustic startle augmentation and spatial learning deficits. *Psychopharmacology* **114,** 392–401.

Vorhees, C. V., Ahrens, K. G., Acuff-Smith, K. D., Schilling, M. A., and Fisher, J. E. (1994b). Methamphetamine exposure during early postnatal development in rats. 2. Hypoactivity and altered responses to pharmacological challenge. *Psychopharmacology* **114,** 402–408.

Wagner, G. C., Ricaurte, G. A., Seiden, L. S., Schuster, C. R., Miller, R. J., and Westley, J. (1980). Long-lasting depletions of striatal dopamine and loss of dopamine uptake sites following repeated administration of methamphetamine. *Brain Res.* **181,** 151–160.

Wallace, J. A., and Lauder, J. M. (1983). Development of the serotonergic system in the rat embryo: an immunocytochemical study. *Brain Res. Bull.* **10,** 459–479.

Zilles, K., Schleicher, A., Glaser, T., Traber, J., and Rath, M. (1985). The ontogenetic development of serotonin (5-HT1) receptors in various cortical regions of the rat brain. *Anat. Embryol.* **172,** 255–264.

CHAPTER

13

Central Cholinergic Neurobiology

DAVID A. JETT
Department of Environmental Health Science
Johns Hopkins University
Baltimore, Maryland 21205

I. Introduction

The process of chemical neurotransmission was first described by the classic experiments of Otto Loewi with acetylcholine (ACh) in frog heart in the early 1920s (see Loewi, 1960). It is thus surprising that specific methods for identifying cholinergic neurons in the brain have only recently been perfected. Consequently, our knowledge of the anatomic distribution of other types of neurons like those in the dopaminergic and adrenergic systems has ironically preceded that for the first described neurotransmitter. ACh is very old phylogenetically and

*Abbreviations: α–BGT, α–bungarotoxin; ACh, acetylcholine; AChE, acetylcholinesterase; ATP, adenosine triphosphate; ATPase, adenosine triphosphatase; BChE, butyrylcholinesterase; cAMP (cyclic AMP)—cyclic adenosine monophosphate; ChAT, choline acetyltransferase; CNS, central nervous system; GABA, γ-aminobutyric acid; GD, gestational day; GDP, guanosine diphophate; GTP, guanosine triphosphate; HC-3, hemicholinium-3; IC50, concentration at which inhibition is half maximal; Km, concentration at which the reaction rate is half maximal; LTP, long-term potentiation; MeHg, methylmercury; NGF, neurotrophic growth factor; NMDA, *N*-methyl-D-aspartate; OP, organophosphate; OPIDN, organophosphate-induced delayed neuropathy; PCB, polychlorinated biphenyls; PKC, protein kinase C; PN, postnatal day; PNS, peripheral nervous system; QNB, quinuclidinyl benzilate; T_4, thyroxine.

is found in bacteria, fungus, and lower invertebrates, where it has no known function. The function of ACh within the brain is excitatory. The cholinergic system plays more of a modulatory than a direct role in neurotransmission, influencing the excitatory and inhibitory actions of other neurotransmitters. Depending on the location of cholinergic terminals, and on the neurons with which they form synaptic connections; their influence on neurotransmission may be one of enhancement (e.g., presynaptic stimulation of glutamate release) or of suppression [e.g., stimulation of γ-aminobutyric acid (GABA) neurons]. This observation along with the widely disperse and diffuse nature of most cholinergic neurons as compared to, for example, well-defined striatal dopaminergic neurons, is indicative of the high degree of complexity in understanding the contribution of altered cholinergic activity in models of disease and neurotoxicology.

Development and maintenance of the cholinergic system is extremely important in a number of neurologic diseases. For example, the autoimmune disease myasthenia gravis causes changes in the steady-state levels and activity of skeletal muscle nicotinic ACh receptors, and degeneration of striatal cholinergic interneurons is a hallmark for Huntington's disease. In patients with Alzheimer's disease, a reduction in the ACh synthesizing enzyme, choline acetyltransferase (ChAT) is thought to be a result of degeneration and/or dysfunction of cholinergic nerve terminals in the cerebral cortex projected from the basal forebrain (Perry and Perry, 1996). The cholinergic system is also targeted by many neurotoxicants, most notably the organophosphate and carbamate insecticides that inhibit acetylcholinesterase (AChE), and those that alter cell-surface ACh receptor proteins such as lead (Pb). The developing brain is particularly vulnerable to neurotoxicants that have an impact on the cholinergic system because of the early appearance of ACh and cholinergic receptors, and their fundamental role in cell proliferation and differentiation.

Neurobiologists and toxicologists have acquired a great deal of knowledge on the effects of chemical and other perturbations on the development of the brain. The discussion in this chapter samples this knowledge from the perspective of the cholinergic system within the central nervous system (CNS). It is acknowledged that there are other developmental neurotoxicants that exhibit cholinergic toxicity, both within the CNS and peripheral nervous system (PNS). This chapter presents a general overview of the brain cholinergic system and a discussion of important neurotoxicants that alter its development.

II. The Cholinergic System

A. Central Cholinergic Anatomy

Cholinergic neurons in the CNS are of two types: interneurons or projection neurons. These neurons are widely dispersed throughout the brain, but five primary cholinergic systems have been identified (Martínez-Murillo and Rodrigo, 1995, review): (1) neostriatum, (2) cerebral cortex and hippocampus, (3) magnocellular basal nucleus, (4) pontomesencephalic tegmentum, and (5) cranial nerve motor nuclei and nerves of the spinal chord. Intrinsic cholinergic neurons are found in the cerebral cortex, striatum, hippocampus, nucleus accumbens, and other areas where they exist primarily as interneurons or in proximity with nonneuronal tissue. Cholinergic interneurons are often associated with the dopaminergic system, as is the case in the striatum. Some of these intrinsic neurons can form local circuits in the same regions that receive extrinsic cholinergic input. For example, in the hippocampus, interneurons are present along with major innervation from the medial septum and the magnocellular basal nucleus. Cholinergic projection neurons arising from the basal forebrain and projecting to the cerebral cortex are extensive and play a primary regulatory role in this structure. Other cholinergic neurons project to the diagonal band, prefrontal cortex, and thalamus. ChAT activity is the highest in the striatum, interpeduncular nucleus, and habenula, suggesting that these structures are major sites of cholinergic neurotransmission. Although the density of certain cholinergic neuronal groups may vary among species, the general organization of the cholinergic system is highly conserved.

The cellular morphology of cholinergic neurons is also diverse and varies from region to region within the brain. Four major morphologic groups have been suggested by Sofroniew *et al.* (1985): (1) very large motor (25–45 μm), (2) large forebrain (18–25 μm), (3) medium (14–20 μm), and (4) small (8–16 μm). These are generally distributed by region. For example, the cerebral cortex and hippocampus possess neurons that fall into the small-size class. Cell soma may be a variety of shapes and have long or short processes and sparse or extensive dendtritic arborization. One feature of cholinergic neurons that stands out among others is that these neurons more frequently than not make synaptic contact with noncholinergic structures. This again suggests a modulatory functional role for cholinergic neurons in the CNS.

B. The Chemistry of Acetylcholine

1. Synthesis of ACh

The synthesis of ACh occurs in the cytoplasm of cholinergic nerve terminals and is packaged in presynaptic vesicles (Fig. 1). The reaction following is catalyzed by the enzyme ChAT.

$$\text{Acetyl CoA} + \text{choline} \rightarrow \text{ACh} + \text{CoA}$$

The sources for acetyl CoA are different among species and tissue types. In mammalian brain tissue preparations, glucose (glycolysis) and citrate (citrate synthetase or citrate lyase reactions) are the primary sources, whereas in lobster giant axons, *Torpedo,* and corneal epithelium, acetate (acetatethiokinase reaction) is a more important source (Cooper *et al.,* 1996). Choline pools are maintained by exogenous sources and recycling after hydrolysis of ACh. Phosphatidylcholine is a membrane reservoir of choline for use in ACh synthesis, and it is possible that an alteration in the catabolism of this phospholipid may be involved in Alzheimer's disease (Wurtman, 1992).

Recent evidence suggests there is a solubilized form of ChAT found in the cytoplasm, and another associated with the cellular and possibly vesicular membranes (Cooper *et al.,* 1996). The latter is found at lower concentrations but with higher specific activity, and this membrane-associated form may be the active form *in vivo.* The concentration at which the reaction rate is half maximal (K_m) for ChAT purified from rat brain is 7.5×10^{-4} *M* for choline and 1.0×10^{-5} *M* for acetyl CoA. Kinetic analysis of choline uptake and the acetylation reaction indicates that the two reactions are tightly linked. To date there has not been much success in developing specific agents that modify ChAT activity.

2. Uptake, Storage, and Release of ACh

The packaging of synthesized ACh into synaptic vesicles occurs via proton exchange driven by glycosylated adenosine triphosphatase (ATPase) pumps (Fig. 1). Vesamicol inhibits the ACh transporter by binding to a site on the inner vesicular membrane surface and is another marker for cholinergic neurons. Adenosine triphosphate (ATP) is stored in vesicles along with ACh, and ATP levels fall during prolonged nerve stimulation. This may be due to the reported corelease of ATP with ACh (Unsworth and Johnson, 1990). The functional significance of this ATP is unclear. The concentration of ACh in vesicles is in the m*M* range, representing 2000 (mammalian) to 200,000 (*Torpedo*) molecules of ACh per vesicle. The classic model of ACh release describes a process initiated by depolarization of the nerve terminal and the entry of Ca^{2+} whereby vesicles fuse with the presynaptic membrane and their contents (ACh and ATP) are released by exocytosis (Fig. 1). The model was challenged several years ago by the finding that only newly synthesized ACh is released from vesicles, suggesting that preformed ACh, presumably stored in vesicles, was not (Adam-Vizi, 1995, review). The use of ACh analogs, vesamicol, and the discovery of vesicle heterogeneity has led to studies that support the original vesicular release model. Quantal ACh release is due to leakage of cytoplasmic ACh across the presynaptic membrane. The precise mechanism of vesicular release of ACh has not been determined, but several new proteins involved in the exocytotic process are being studied, for example, synaptotagmins, synaptophysins, synaptobrevins, and synapsin I. These proteins may be important in the developmental neurotoxicity of agents such as Pb that are known to affect ACh release.

3. Degradation and Uptake of Choline

ACh released from cholinergic nerve terminals binds to specific ACh receptors located on the post- and presynaptic membranes (Fig. 1). Negative feedback of ACh occurs on its release, and it is mediated through presynaptic receptors. Termination of neurotransmission is accomplished by the rapid catabolism of ACh by AChE. As much as 50% of the choline released by the hydrolysis of ACh is taken up into the presynaptic terminal by a sodium-dependent high affinity transport system and used for further ACh synthesis. The remaining choline may be reincorporated into membrane phospholipids or catabolized. Choline is rate limiting in the synthesis of ACh, and high-affinity uptake is critical to the regulation of ACh in nerve cells. Furthermore, the high-affinity choline uptake site appears to be a marker for cholinergic nerve terminals exclusively, and can be identified with the inhibitor hemicholinium-3 (HC-3). This is important because, as noted earlier, noncholinergic neurons may stain positive for other cholinergic markers such as AChE.

The hydrolytic cholinesterases are classified by their substrate selectivity. With few exceptions, the "true" cholinesterase, or AChE, exhibits fastest catalytic activity with ACh, whereas "pseudo" cholinesterase, or butyrylcholinesterase (BChE), is more selective for other esters such as butyrylcholine and propionylcholine. The concentrations of these two genetically distinct enzymes vary with tissue/organ type and species, and they are found throughout the body in such tissues as liver, lung, plasma, and erythrocytes. In the CNS, neurons are AChE-enriched and glia contain mostly BChE. The functional significance of BChE is unclear. AChE has one of the fastest turnover rates of any known enzyme, hydrolyzing 5000 molecules of ACh per molecule of AChE per second. The basic structure of the enzyme is designed to attract the positive charge in ACh to an anionic site at the active center of the molecule, and a nearby esteratic site that binds the carbonyl moiety on ACh. However, the crystal structure of AChE has revealed that other sites may in fact be more important in the reaction (Sussman *et al.,* 1991). A great deal has been learned about

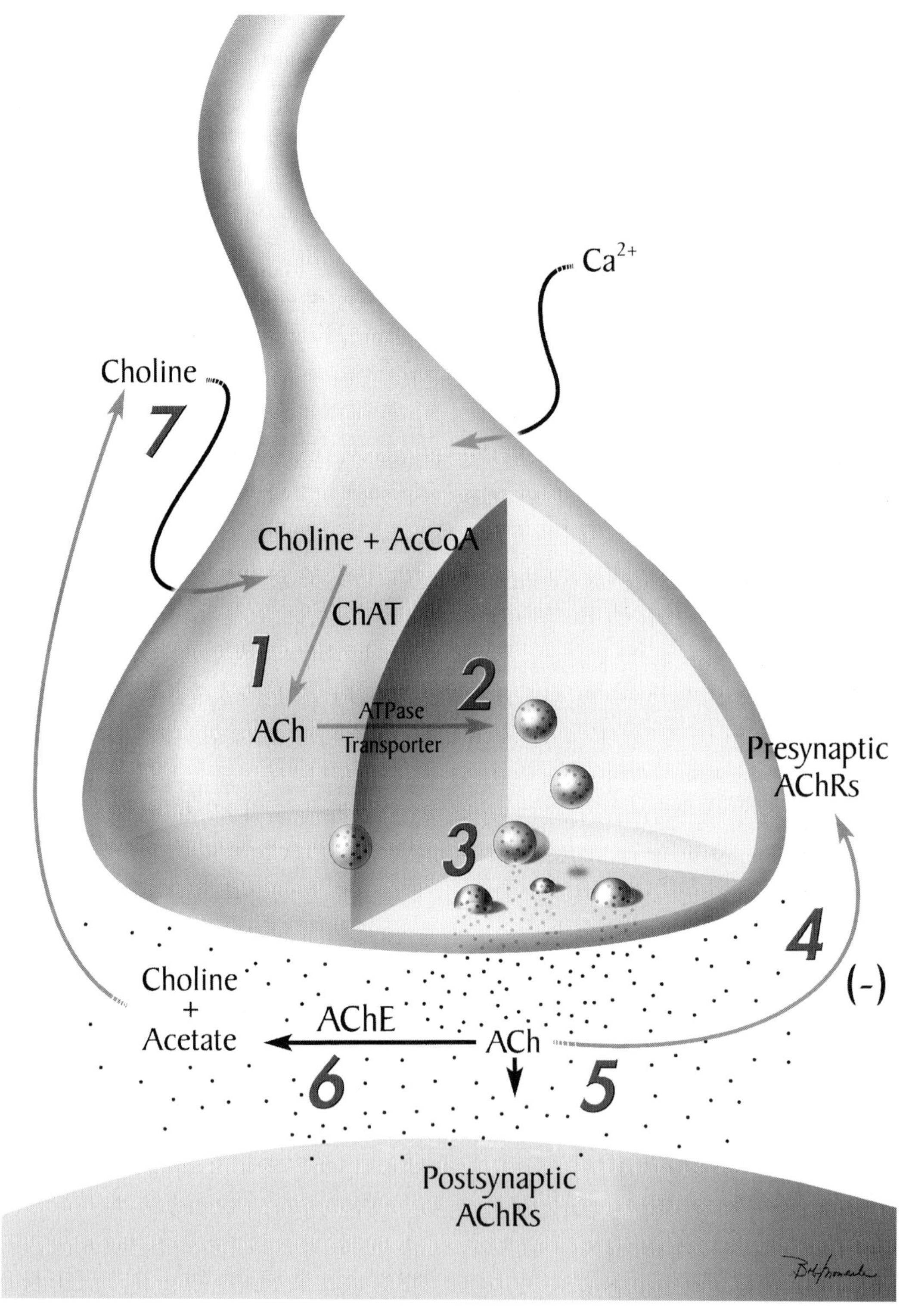

FIGURE 1 Synthesis, release, and metabolism of ACh at the cholinergic nerve terminal. (1) Synthesis of ACh by choline acetyltransferase (ChAT). (2) Packaging of ACh into vesicles by ATPase proton exchange transporter. (3) Exocytotic release of ACh from vesicles after depolarization and entry of Ca^{2+}. (4) Binding of ACh to presynaptic muscarinic autoreceptors. (5) Binding of ACh to postsynaptic muscarinic and nicotinic receptors. (6) Hydrolysis of ACh by acetylcholinesterase (AChE). (7) Sodium-dependent high-affinity uptake of choline released from the AChE reaction.

the molecule biology and genetics of AChE. Excellent reviews may be found elsewhere (Massoulie *et al.*, 1993; Fernandez *et al.*, 1996), and this molecule is particularly relevant to the model of organophosphate neurotoxicity discussed later in this chapter.

C. Muscarinic Cholinergic Receptors

Cell surface receptors found on cholinergic neurons are activated by muscarine or nicotine and hence are classified as muscarinic or nicotinic receptors, respectively. These structurally unrelated molecules belong to different superfamilies of genes and differ in their location, specificity for agonists and antagonists, and cellular responses mediated by their activation. Muscarinic receptors are the predominant cholinergic receptor in the CNS and they are abundant in smooth muscle, heart, and exocrine glands. Muscarinic receptors are activated by ACh and blocked by atropine and quinuclidinyl benzilate (QNB). Activation by ACh of muscarinic receptors is the first step in a G-protein–coupled signal transduction pathway that ultimately leads to cellular responses such as the modulation of K^+, Ca^{2+}, or Cl^- channel permeability, the regulation of cell growth, and the switching on of transcriptional factors in the nucleus (Baumgold, 1995, review). Muscarinic receptors exhibit slow responses of 100–250 ms due to the obligatory biochemical cascade required for coupling to an effector system. These receptors have been shown to be involved in neurologic disorders such as Alzheimer's disease (Perry and Perry, 1996), models of neuronal plasticity such as long-term potentiation (LTP) (Boddeke and Boeijinga, 1995, review) and the neurotoxicity of compounds such as organophosphate insecticides (e.g., Jett *et al.*, 1991).

The early observation that muscarinic receptors exhibited variable affinities for different antagonists suggested there might be subtypes within this class, and today there have been five distinct receptors cloned and characterized in mammals (Table 1). The m1 subtype predominates in the cortex and hippocampus, m2 in the cerebellum and thalamus, m3 at moderate density throughout, m4 in the striatum, and the m5 at very low density in the hippocampus and brainstem (Schliebs and Roßner, 1995, review). The agonist binding site on the muscarinic receptor is situated in a narrow cleft close to the membrane surface created by a circular arrangement of seven transmembrane helices. ACh as well as exogenous muscarinic agonists are attracted to this site by ionic forces between an aspartate residue on the protein and a positively charged ammonium or amino group on the ligand. Unfortunately for pharmacologists seeking to develop subtype-specific ligands for muscarinic receptors, this aspartate residue is highly conserved among the subtypes, and selective antagonists (Table 1) do not show more than a fivefold difference in affinities.

The coupling of muscarinic receptors to specific G-proteins occurs after a conformational change in the protein is induced by binding of ACh to the agonist site. The receptor then becomes associated with the heteromeric G-protein that releases guanosine diphosphate (GDP) from the binding pocket that is replaced with guanosine triphosphate (GTP) from the cytosol. The α-subunit dissociates from the G-protein complex and activates an effector molecule such as adenylyl cyclase. The dissociated α-subunit may also interact directly with ion channels to alter ion conductances. The G-protein is then returned to its resting state by the GTPase-mediated hydrolysis of GTP. The subtypes of muscarinic receptor show preferential coupling to different G-protein effector systems. The m1, m3, and m5 receptors couple to $G_{q/11}$, phospholipase Cβ, and the hydrolysis of phosphoinositides; the m2 and m4 receptors couple to $G_{i/o}$, adenylyl cyclase, and the inhibition of cyclic adenosine monophosphate (cAMP) synthesis (Table 1). Other biochemical effects of muscarinic receptor activation include the stimulation of adenylyl cyclase and phospholipases A2 and D (Baumgold, 1995, review).

D. Nicotinic Cholinergic Receptors

Nicotinic receptor–ion channel complexes are found in the PNS and the CNS. The possibility of neuronal nicotinic receptors that were structurally related to and contained α-subunits similar to the receptors found at the neuromuscular junction was first confirmed in 1986 by Boulter *et al.* The structural and functional heterogeneity of brain nicotinic receptors has led to the identification of several subtypes depending on various combinations of α and β subunits that make up a pentameric channel complex. The stoichiometry, subunit composition, and functional diversity of neuronal nicotinic receptors is still being defined (Lindstrom *et al.*, 1995; Albuquerque *et al.*, 1997, reviews). There have been eight α ($\alpha2$–$\alpha9$) and three β ($\beta2$–$\beta4$) subunits identified in the CNS, and novel subtypes have been reported (Pugh *et al.*, 1995). The α and β subunits associate to form heteromers or homomers with distinct pharmacologic and functional properties. In the CNS, there appears to be three major types of receptor complexes based on their electrophysiologic properties (Albuquerque *et al.*, 1995, review) and abundance (Role and Berg, 1996, review). The first is composed of $\alpha4$ and $\beta2$ subunits and is most abundant in the CNS, found primarily in the thalamus, midbrain, and brainstem (Table 1). They bind nicotine with high affinity and exhibit slow-inactivating or type II currents. The second type is homomeric and composed of $\alpha7$ subunits. This subtype

TABLE 1 Muscarinic and Nicotinic Cholinergic Receptors in the Rat CNS

Receptor type	Distribution	Effector	Selective ligands
Muscarinic			
m1	Cortex, hippocampus	PI hydrolysis	Pirenzepine; (+)telenzepine
m2	Cerebellum; thalamus	cAMP inhibition; slow activation of ion channels	AF-DX-116; methoctramine; himbacine; gallamine
m3	Low levels throughout CNS	PI hydrolysis	HHSiD;*p*-fluoro-HHSiD; 4-DAMP
m4	Striatum	cAMP inhibition	—
m5	Very low levels in brainstem and hippocampus	PI hydrolysis	—
Nicotinic			
$\alpha 4\beta 2$	Thalamus; other midbrain and brainstem nuclei	Type II currents; Na^+, K^+, Ca^{2+}	High-affinity nicotine; cytisine; epibatidine
$\alpha 7$ homomeric	Cortex, hippocampus, hypothalmus, inferior colliculus	Type IA currents; Na^+, K^+, Ca^{2+}; higher for Ca^{2+}	α-bungarotoxin
$\alpha 3$, $\beta 4$, (?)	Sparse autonomic neurons	Type III currents; Na^+, K^+, Ca^{2+}	

has widespread distribution and colocalizes with α-bungarotoxin (α-BGT) binding sites in the rat CNS (Del Toro *et al.,* 1994). Fast-inactivating type 1A currents are associated with this receptor–ion channel, and they are highly permeable to Ca^{2+}. The last type has a much more sparse distribution, primarily on autonomic neurons, and is composed of $\alpha 3$, $\beta 4$, and possibly other subunit types to form a complex heteromer. They subserve slow-inactivating type III currents and are sensitive to mecamylamine (Albuquerque *et al.,* 1995). The distribution of these subtypes overlaps; in fact, more than one subtype may be present on a single neuron (e.g., Connolly *et al.,* 1995). The mRNA for $\alpha 6$ was found to be associated with catecholaminergic nuclei such as the substantia nigra and locus coeruleus (Lenovere *et al.,* 1997).

The possible functions of nicotinic receptors in the CNS are described in several reviews (e.g., Albuquerque *et al.,* 1995, 1997; Role and Berg, 1996). This is a very active area of research, due in part to the discovery that nicotinic receptors in the brain are likely to be involved in neurologic disease, neurotoxicity, and in neuronal plasticity. It appears that a presynaptic modulatory role for central nicotinic receptors is more prominent than, but does not exclude, direct postsynaptic nicotinic neurotransmission. Nicotinic receptors are found on the presynaptic terminals of neurons, where they modulate the release of norepinephrine, dopamine, GABA, serotonin, glutamate, and ACh. The enhancement of fast excitatory neurotransmission is thought to be a primary role for presynaptic nicotinic receptors in the CNS (McGehee *et al.,* 1995). Activation of these receptors increases the probability of release of neurotransmitters by increasing presynaptic intracellular Ca^{2+}, and evidence from several types of studies suggests that some $\alpha 7$-bearing receptors, which are extremely permeable to Ca^{2+}, are presynaptic (Role and Berg, 1996; review).

Behavioral manifestations of the actions of nicotinic agonists and antagonists are well documented (Clarke *et al.,* 1995, review). There have been many reports on the role of nicotinic receptors in learning and memory processes and on their importance in synaptic plasticity (e.g., Amador and Dani, 1995; Picciotto *et al.,* 1995). Albuquerque *et al.* (1997) suggest a model for LTP involving the strengthening of the NMDA-dependent form by nicotinic pre- and postsynaptic receptors. These and other studies suggest that nicotinic receptor agonists may be useful in improving the dementias associated with aging and Alzheimer's disease. Furthermore, the nicotinic receptor is likely to be extremely important in the toxicity of chemical agents that alter cognitive function.

III. Cholinergic Neurobiology and CNS Development

Neuronal diversity and cytoarchitecture within the developing CNS is achieved by a complex series of cellular proliferations and differentiations guided by specific neurochemical cues. The early appearance of ACh and associated metabolic, receptor, and transport proteins suggests it may serve a basic functional role in cell growth and differentiation. Likewise, the early differentiation of cholinergic neurons may serve as a template for the subsequent development of other cell types.

A. *Cholinergic Markers in the Developing CNS*

The developmental profile of several cholinergic markers has been determined by many laboratories in a variety of tissues and species. The developmental profile of most cholinergic markers varies among different brain regions reflecting temporal differences in developmental activity. An illustration of the relative abundance of these markers generalized for brain regions where they are most abundant is presented in Fig. 2. It should be reiterated that there are exceptions to the developmental profile of the cholinergic markers depicted in Fig. 2, depending on the species and brain region examined.

Histochemical staining for AChE has revealed that this enzyme increases at a fairly constant rate from late embryonic development to adulthood, and a somewhat parallel pattern is observed for ACh levels (Fig. 2). There has been a report of AChE activity in the rat dorsal thalamus in early embryonic development, suggesting that it may function in events that occur before the establishment and fine-tuning of neuronal connections (Schlaggar *et al.,* 1993). The appearance of AChE may be very transient, as exemplified by the peak in the high-intensity AChE activity type of pyramidal neurons at postnatal day (PN) 8–10, and their disappearance after PN21 in the cerebral cortex (Geula *et al.,* 1995). The vesicular transport protein identified with [^{3}H]AH5183 also exhibits transient expression, peaking during postnatal development and declining during adulthood (Fig. 2) (Aubert *et al.,* 1996). A peak during the postnatal period is also observed for ChAT activity, and is indicative of a period of heightened synaptogenesis before pruning back to final adult cytoarchitecture. An interesting observation made by Zahalka *et al.* (1993a) was the precipitous drop in the ratio of HC-3 binding to ChAT activity in rat brain after embryonic development (Fig. 2). They ascribe this to asynchronous rises in cholinergic tone and the density of cholinergic nerve terminals, and/or the expression of HC-3 sites on noncholinergic neurons. Aubert *et al.* (1996) proposed that specialization of cholinergic nerve terminals, accompanied by increases in ACh synthesis, regulation of ACh release by m2 autoreceptors, and increased high affinity uptake, occurs after the embryonic period during which the trophic functions of ACh are more important.

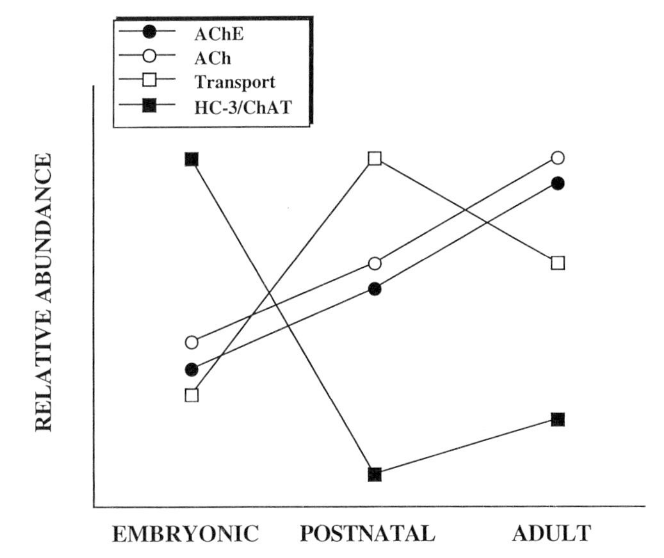

FIGURE 2 Relative abundance of cholinergic markers in the rodent forebrain during embryonic (GD1 to birth), postnatal (postnatal day 1 to weaning), and adult (>56 days) periods. AChE: acetylcholinesterase activity or staining; ACh: acetylcholine levels; Transport: [^{3}H]AH5183 binding; HC-3/ChAT: ratio of [^{3}H]hemicholinium-3 binding to choline acetyltransferase (ChAT) activity. [Derived from: Coyle and Yamamura (1976); Schlagger *et al.* (1993); Zahalka *et al.* (1993a); Aubert *et al.* (1996).]

B. Cholinergic Receptors in the Developing CNS

The cellular localization of muscarinic receptor immunoreactivity in parietal cortex of rats indicated an early appearance during the postnatal period and suggests their involvement in cell differentiation and process extension (Buwalda *et al.,* 1995). Growth cone membranes isolated from neonatal rat forebrain express muscarinic receptors (Saito *et al.,* 1991), and the muscarinic receptor–mediated phosphorylation of growth and cone B50/GAP43 protein is believed to be an important step in neuronal guidance and extension (VanHooff *et al.,* 1989). Similar results suggesting a functional role for muscarinic receptors in growing neurons were found in PC12 M1 cells (Pinkas-Kramarski *et al.,* 1992). A finding using muscarinic antibodies to study their cellular localization in mouse brain was a transient appearance of muscarinic receptors in several regions and brain capillaries during early development, supporting a functional role in the CNS and brain vascularization during early critical time periods (Hohmann *et al.,* 1995).

A clearer picture of the ontogeny of muscarinic receptor subtypes is emerging, primarily because of the development of subtype-specific antisera and oligonucleotide probes. The expression of the subtypes varies with tissue; however, all of the subtypes studied occur in the embryonic mammalian brain and increase through development into adulthood (Fig. 3). Antisera developed to recognize unique epitopes of the known subtypes (m1–m5) were used to determine the ontogenic profile of these receptors in rat forebrain (Wall *et al.,* 1992). It was found that relatively low levels of m3 receptors (11% of [^{3}H]QNB binding), and very low levels of m5 ($\leq$1% of [^{3}H]QNB binding) receptors were present in adults. Only subtle differences in the developmental profile of the different subtypes were reported (Wall *et al.,* 1992). In another study, immunoreactivity of m1 receptors in mouse brain was detected at PN5, and their developmental pattern was found to mature at PN60 (Hohmann *et al.,* 1995). In contrast, a gradual increase in levels of m2 immunore-

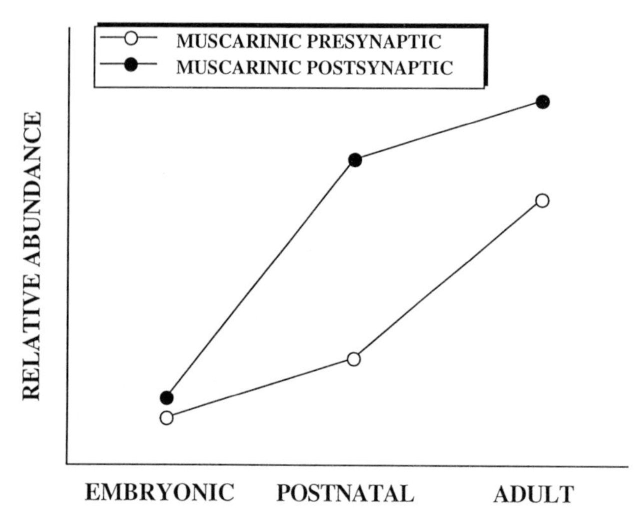

FIGURE 3 Relative abundance of pre- and postsynaptic muscarinic cholinergic receptors in the rodent forebrain during embryonic (GD1 to birth), postnatal (postnatal day 1 to weaning), and adult (>56 days) periods. [Derived from: Wall *et al.* (1992): Costa (1993); Hohman *et al.* (1995); Aubert *et al.* (1996).]

activity during the early to late postnatal period was observed in this study. A general pattern emerges from these and other studies (Costa, 1993; Aubert *et al.*, 1996) on the ontogeny of muscarinic subtypes: the m1 and m3 subtypes, generally found on the postsynaptic membrane increase rapidly during the postnatal period, whereas presynaptic m2 and m4-receptors increase at a somewhat slower rate (Fig. 3).

Nicotinic receptor subunit mRNA appears as early as gestational day (GD) 11 in rat brain (Zoli *et al.*, 1995). The $\alpha4$ mRNA in the cat cortex and hippocampus is present at pre- and postnatal time periods before cholinergic innervation has occurred (Ostermann *et al.*, 1995). In fact, nicotinic mRNA appears at the time of the last mitosis of progenitor neuroepithelial cells, which is the earliest among all the receptors studied. The intriguing possibility that these early nicotinic receptors play a role in guidance of developing cholinergic and noncholinergic neurons is supported by these findings. Generally, transcript levels for the prominent subunits ($\alpha3$, $\alpha4$, $\beta2$) increase from caudal to rostral CNS regions as embryonic development proceeds. The expression is transient and region-specific, depending on the subunit. For example, the $\alpha3$ mRNA decreases beginning at GD15–17 to nearly undetectable levels in most adult brain structures except for the medial habenula (Zoli *et al.*, 1995). Similarly, $\beta4$-mRNA disappears after GD13 in the brain except for the habenula and other structures. The $\alpha4$ and $\beta2$ subunits, clearly the two subunits that are most prominent in the adult brain, have widespread and heterogeneous distribution in the embryonic brain, and their expression is more stable than other subunits during embryonic development. There have been no reports on the changes of $\alpha4\beta2$ expression from gestational to late postnatal development (i.e., in the same study); however, Cimino *et al.* (1995) have shown that transcripts for these two subunits may be regulated somewhat differently during the postnatal period. They also found that near adult levels were already present at birth (Fig. 4), and a transient decrease in mRNA occurred at PN7–14.

The mRNA for the $\alpha7$ subunit has also been observed in the developing thalamus and cortex of rats as early as GD15 (Broide *et al.*, 1995). The expression is transient, increasing markedly in the late prenatal and early postnatal periods, and a remarkable 57% decrease from levels found at PN7 are observed in the adult (Fig. 4). The development of the $\alpha7$ subunit has been described in the cerebellum (Del Toro *et al.*, 1997). These receptors increased from PN3–15 and paralleled synaptogenesis within specific cerebellar nuclei. A similar pattern (i.e., parallel to synaptogenesis) was also observed when [^{125}I]BGT and $\alpha7$ immunoreactivity were examined in hippocampal cell culture (Samual *et al.*, 1997).

Several reports present evidence that nicotinic receptors are involved in CNS development (Zoli *et al.*, 1995; Role and Berg, 1996, reviews). Nicotinic antagonists prevent ACh-induced inhibition of neurite outgrowth in rat retinal ganglion cells (Lipton *et al.*, 1988), and nicotine regulates neurite retraction and outgrowth (Chan and Quik, 1993; Pugh and Berg, 1994). Also, in an elegant experiment using isolated embryonic spinal neurons in *Xenopus* cell cultures, Zheng *et al.* (1994)

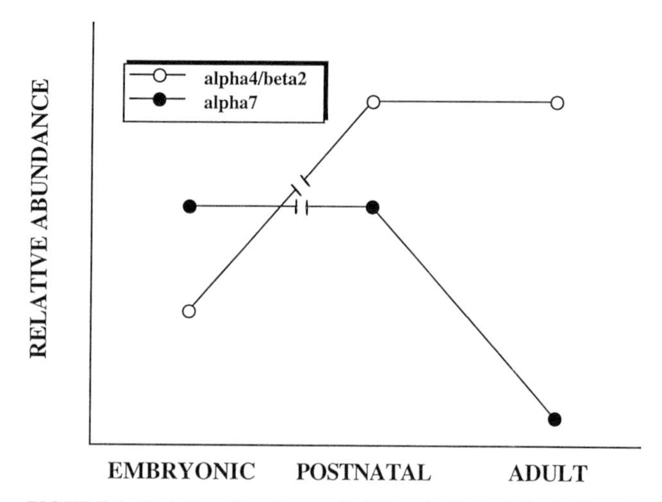

FIGURE 4 Relative abundance of $\alpha4\beta2$ and $\alpha7$ nicotinic cholinergic receptors in the rodent forebrain during embryonic (GD1 to birth), postnatal (postnatal day 1 to weaning), and adult (>56 days) periods. Note the curves are discontinuous from embryonic to postnatal periods, indicating no data on direct comparisons of these two time periods in the same study. [Derived from Broide *et al.* (1995); Cimino *et al.* (1995); Ostermann *et al.* (1995); Zoli *et al.* (1995); Aubert *et al.* (1996).]

demonstrated the importance of the activation of nicotinic receptors in the turning of nerve growth cones. Transcriptional events that occur very early in progenitor cells are extremely important for cell differentiation and specialization. Retinal ganglion cell differentiation was induced by the activation of the promoter region of the $\beta3$ nicotinic subunit (Matter *et al.*, 1995). In addition, α-BGT–sensitive sites have been shown to modulate the transcription of neurotrophic growth factor (NGF) mRNA (Freedman *et al.*, 1993). These observations, along with the coincidental transient peak in $\alpha7$-bearing receptors during a critical period of heightened thalamocortical development (Broide *et al.*, 1995), suggest $\alpha7$ receptors are involved in CNS development. The high Ca^{2+} permeability of $\alpha7$ receptors, and hence the likely influence on Ca^{2+}-mediated signal transduction and early gene expression, is consistent with this hypothesis.

The characterization of changes in cholinergic markers and receptors during development is of particular toxicologic relevance because there may be critical time windows during which the brain is most vulnerable to neurotoxicants. Moreover, the subtype distribution or subunit composition of receptors in the embryonic brain may switch as the organism matures to adulthood. This has been observed for nicotinic receptors in the developing neuromuscular junction (Duclert *et al.*, 1993) and N-methyl-D-aspartate (NMDA) receptors in the rat brain (Monyer *et al.*, 1994). The difference in receptor composition and potential functionality may explain why the neurotoxicity of many compounds is more pronounced in the developing CNS.

IV. Neurotoxicants That Affect the Developing Cholinergic System

A. Lead

1. Lead and Cognitive Development

There has been increasing concern for the effects of low-level exposure to lead (Pb) on cognitive function. This is due to earlier evidence suggesting that chronic low-level Pb exposure impairs cognitive development and is associated with decrements of intelligence measures in children (Bellinger *et al.*, 1987; McMichael *et al.*, 1988; Needleman *et al.*, 1990). These observations have been substantiated by studies using several animal models (Rice, 1993; Cory-Slechta, 1995, reviews), and the behavioral effects of Pb appear to be more pronounced in children and immature animals. It has been shown, for example, that rats exposed to Pb chronically during the developmental period and beyond were more impaired in a spatial memory task when tested at PN21 than as adults (Jett *et al.*, 1997a). Untreated adults, however were also impaired by direct injection of Pb into the dorsal hippocampus (Jett *et al.*, 1997b). These and other studies suggest that Pb may alter cognitive function irrespective of age and exposure history. The effect of Pb on cognitive development may have a morphologic basis. Developmental Pb exposure alters hippocampal morphology in the rat (Petit and LeBoutillier, 1979; Alfano and Petit, 1982; Kiraly and Jones, 1982), and many of the behavioral deficits associated with hippocampal lesions parallel those observed in Pb-treated rats and monkeys (Davis *et al.*, 1990; Ferguson and Bowman, 1990). There is evidence that other components of the septohippocampal pathway may also be important in Pb neurotoxicity (Bielarczyk *et al.*, 1994). The anatomic substrates for Pb-induced cognitive dysfunction may be located in several brain areas because learning and memory processes are not restricted to limbic system structures.

2. Effects of Pb on Cholinergic Function

Previous research since the mid-1970s on the effect of Pb exposure on cholinergic activity has been reviewed elsewhere (e.g., Goldberg, 1979; Winder and Kitchen, 1984; Cory-Slechta, 1995). Much of the historic data are contradictory except for general cholinergic hypofunction due to an inhibition by Pb of ACh turnover, and release of ACh from presynaptic vesicles. Pb *in vitro* augments spontaneous release, and inhibits evoked ACh release, presumably by its action at Ca^{2+} binding sites and ion channels (Shao and Suszkiw, 1991). It has been shown that cholinergic neuroanatomic plasticity in the hippocampus is compromised in rats exposed to Pb during development (Alfano *et al.*, 1983). In addition, several studies have implicated the cholinergic septohippocampal pathway in the developmental toxicity of Pb. Developmental exposure to Pb in the drinking water caused 30–40% reductions in ChAT activity in the septum and hippocampus of 7- and 28-day-old rats with blood Pb levels of only 20 μg/dL (Bielarczyk *et al.*, 1994), and Na-dependent high-affinity choline uptake was inhibited and the number of uptake sites were reduced by *in vitro* and *in vivo* Pb exposure (Silbergeld and Goldberg, 1975; Tian *et al.*, 1995). It has been demonstrated that the effects of Pb on the development of the septohippocampal pathway may persist into adulthood, and result in alterations of cholinergic and catecholaminergic systems similar to those observed after fimbria–fornix transection (Bielarczyk *et al.*, 1996).

3. Effects of Pb on Muscarinic Receptors

Several laboratories have shown that chronic *in vivo* exposure to Pb during development caused region-specific decreases in the number of muscarinic receptors in rat brain (Costa and Fox, 1983; Rossouw *et al.*, 1987;

Widmer *et al.,* 1992; Bielarczyk *et al.,* 1994; Jett and Guilarte, 1995). Changes in these receptors may be restricted to one or two subtypes and not be detected by nonselective methods of identification. This has been a crucial limitation of previous research because there is evidence that the ontogeny of muscarinic receptors varies among the different subtypes (Wall *et al.,* 1992), and certain subtypes may be more involved than others in the behavioral effects of Pb (Cory-Slechta and Pokora, 1995) and in aged cognitively impaired rats (Quirion *et al.,* 1995). One possible explanation for the effect of Pb on muscarinic receptors is by direct interaction. This effect was shown *in vitro* by Aronstam and Eldefrawi (1979), but only at higher concentrations of Pb. Studies with lower Pb concentrations indicated that Pb had no effect on muscarinic receptor binding *in vitro* (Costa and Fox, 1983; Jett and Guilarte, 1995). Several key elements within signal transduction pathways linked to muscarinic and other neurotransmitter receptors are affected by Pb. These include adenylyl cyclase activity (Nathanson and Bloom, 1975), phosphoinositide turnover (Dave *et al.,* 1993; Bressler *et al.,* 1994), and protein kinase C (PKC) (Markovac and Goldstein, 1988), but these effects have not been related directly to muscarinic receptor function. The effect of Pb on muscarinic receptors may be indirect, acting on the regulation of cell-surface receptors by feedback mechanisms within these pathways (e.g., receptor phosphorylation), or by altering the density and distribution of muscarinic subtypes during CNS development.

4. Effects of Pb on Nicotinic Receptors

There have been many reports on the role of nicotinic receptors in learning and memory processes (e.g., Picciotto *et al.,* 1995). The nicotinic receptor is one of the most sensitive neuronal targets to the effects of Pb at the molecular level. In N1E-115 neuroblastoma cells, 1 n*M* to 3 μM Pb reduced nicotinic ion conductances by 26–90% (Oortgiesen *et al.,* 1993). The mechanism of the inhibitory effect of Pb has been characterized in hippocampal cell culture as noncompetitive, voltage-independent, and primarily acting on the fast-desensitizing nicotinic current associated with $\alpha 7$ subunit-bearing receptors (Ishihara *et al.,* 1995). These effects occurred with an IC50 (concentration at which inhibition is half maximal) of 3 μM, and Pb was a less potent inhibitor of the slowly desensitizing currents associated with other nicotinic subtypes. The $\alpha 7$ bearing subtype is believed to be the presynaptic receptor that is extremely permeable to Ca^{2+} (McGehee *et al.,* 1995) and may serve to modulate synaptic plasticity (Amador and Dani, 1995). This Ca^{2+} permeability, involvement in plasticity, and sensitivity to Pb is strikingly similar to the NMDA glutamate receptor (see Chapter 15). Thus, the inhibitory action of Pb at presynaptic nicotinic receptors may contribute to impairment of behavioral processes believed to be mediated by postsynaptic cholinergic and glutamatergic receptors. The sensitivity of the $\alpha 7$ subtype to Pb may not be unique. In an attempt to investigate the subtype-specific effects of Pb on nicotinic receptors, $\alpha 3/\beta 2$ and $\alpha 3/\beta 4$ subtypes were expressed in *Xenopus* oocytes and Pb was found to potentiate and block ion conductances to different degrees depending on the subtype (Oortgiesen *et al.,* 1995). Preliminary data from our laboratory indicates that developmental exposure of rats to Pb alters the number of nicotinic receptor binding sites depending on the subunits (greater effect on $\alpha 7$-bearing subunits), age, and brain region. These studies suggest that nicotinic receptors may be involved in Pb-induced learning impairment, but research is needed on the effects of Pb *in vivo* and the behavioral consequences of changes in receptor density and function.

Organometals may pose a serious threat to the developing CNS because of their increased lipophilicity and potential effects on critical membrane components such as neurotransmitter receptors. For instance, tri-*n*-butyl Pb inhibited muscarinic receptor binding to rat brain membranes to a greater extent than Pb acetate (Bondy and Agrawal, 1980). Triethyl Pb, the primary metabolite of the antiknock gasoline additive tetraethyl Pb, altered several cholinergic parameters *in vitro* including decreases in evoked release of ACh, high-affinity choline uptake, and muscarinic receptor binding (Hoshi *et al.,* 1991). It is not known the effects of *in vivo* exposure to organoleads on the developing cholinergic system.

B. Mercury

A more extensively studied organometal is methylmercury (MeHg). The primary symptoms of MeHg poisoning consist of sensory and peripheral motor disturbances due to an unclear mechanism. The developing brain is particularly vulnerable to this mercurial because of its high lipophilicity and the affinity of Hg for cysteine residues on biologically important molecules. The cytotoxicity of developing neurons and glia in cell culture indicate that long-term exposure to MeHg or Hg may have an impact on CNS development (Monnettschudi *et al.,* 1996). MeHg exerts some of its toxic effects by stimulating spontaneous ACh release. It has been shown that this effect is dependent on nerve terminal mitochondrial Ca^{2+} sequestration and depression of high affinity choline uptake (Levesque and Atchison, 1991; Levesque *et al.,* 1992). The developmental neurotoxicity of MeHg is indicated by studies that show behavioral changes in rats exposed *in utero* (e.g., Fredriksson *et al.,* 1993). Little information on the effects of MeHg on

ACh release and ACh levels in the developing CNS exists, however, it is reasonable to assume that many of the important trophic and chemotaxic functions of ACh could be altered by perinatal exposure to MeHg.

Concern for the effects of inorganic mercury is due to the high levels contained in dental amalgam and the exposure of humans to mercury vapor. Studies indicate that both MeHg (Lärkfors *et al.*, 1991) and mercury vapor (Söderström *et al.*, 1995) may alter developing cholinergic neurons by reducing NGF at their target sites within the brain. Scattered reports of other metals interfering with the developing brain have not received much attention after initial observations. For example, changes in ChAT activity in the offspring of female mice treated with aluminum sulfate suggests that the cholinergic system may be impacted (Clayton *et al.*, 1992).

C. Organophosphorous Insecticides

1. Organophosphate Neurotoxicity

Organophosphate (OP) insecticides are a large class of compounds used in agriculture and in the home. Some were originally formulated as chemical warfare nerve gases and may have been used in Operation Desert Storm. These compounds were developed because of their ability to phosphorylate and deactivate the cholinesterase enzymes. OPs and other anticholinesterases acylate the enzyme at the esteratic site, as does ACh. The OPs phosphorylate the hydroxyl group of a serine residue on AChE in the CNS. In mammalian systems, their acute toxicity is derived from this anticholinesterase activity and subsequent parasympathomimetic effects. Acute poisonings are treated with atropine to inhibit central muscarinic effects, and an oxime reactivator, usually 2-PAM, to recover AChE activity. Chronic exposures may result in two distinct syndromes (Ecobichon, 1996): an OP-induced delayed neuropathy (OPIDN), or one that results in symptoms that are intermediate between OPIDN and acute effects. The development of tolerance to OPs has been well characterized and the mechanism involves the down-regulation of cholinergic receptors after repeated sublethal exposures (Costa and Murphy, 1982; Hoskins and Ho, 1992; review; Jett *et al.*, 1993; 1994). There has been concern for the effects of chronic low-level exposure in humans, especially as humans may be exposed to low levels of these compounds in food and in occupational environments. Some of the neuropsychological effects of low-level exposure appear to involve the cognitive processes that underlie memory (Annau, 1992, review). Many people exposed to these compounds complain of memory loss even in the absence of overt toxicity or significant cholinesterase inhibition. A poor correlation between cholinesterase inhibition and toxicity of many important OP insecticides has prompted research on alternative mechanisms of their actions (Eldefrawi *et al.*, 1992).

2. Effects of Organophosphates on CNS Development

Some OPs are teratogenic, causing various abnormalities such as growth retardation, micromelia, and axial skeletal problems in chicks, and somewhat less severe deformities in rodents (Kitos and Suntornwat, 1992, review). Spyker and Avery (1977) provided some of the earliest evidence that prenatal exposure to OPs causes behavioral effects in rat pups. Potential effects on the developing CNS were confirmed by animal studies that reported effects such as anencephaly, cerebellar hypoplasia, stunted growth of various brain structures, altered cholinergic and glutamatergic neurochemistry, and regional histopathology in the brains of neonatal mammals (Deacon *et al.*, 1980; Michalek *et al.*, 1985; Berge *et al.*, 1986; Hanafy *et al.*, 1986). Anecdotal reports of poisoning by OPs in humans also suggest that OPs may alter CNS development (Romero *et al.*, 1989). The insecticide methylparathion altered locomotor activity, operant behavior, and brain AChE in rat pups when administered to dams at 1.0 mg/kg on GD6–20 (Gupta *et al.*, 1985). Similarly, daily injections of 1.3 or 1.9 mg/kg parathion in rat pups during PN5–20 did not cause neuromuscular effects, but reduced the number of muscarinic receptors and AChE activity with a concomitant reduction in performance in tests of spatial memory (Stamper *et al.*, 1988). This study was later repeated by Veronesi and Pope (1990), using a slightly lower dose of parathion during PN5–20, a critical time period for development of the rat hippocampus. Morphologic as well as biochemical endpoints assessed in this study revealed significant reductions of AChE activity (73%) and muscarinic receptor binding (36%) at PN12, and cytopathologic lesions in the hippocampus confined to the dentate, CA4 and CA3 subfields.

Interest in the OP insecticide chlorpyrifos and its metabolite chlorpyrifos-oxon is largely due to its widespread agricultural and home use, long-acting anticholinesterase activity, and the several-fold greater degree of sensitivity of neonatal animals to the general toxicity and neurotoxicity relative to adults (Pope and Chakraborti, 1992; Chakraborti *et al.*, 1993; Whitney *et al.*, 1995). Chanda and Pope (1996) exposed rats to chlorpyrifos by injecting the dams daily from GD12–19 and found reductions in muscarinic and nicotinic receptor binding, AChE activity, and righting reflex and cliff avoidance behaviors in neonatal pups. It was observed that repeated exposures to low doses caused more neurotoxicity than was observed in their earlier studies using a single high dose. The mechanism of the developmental neurotoxicity of chlorpyrifos does not appear to be entirely attributable to differences in sensitivity of AChE to OPs in the neonate, and has thus prompted a study by Whitney *et al.* (1995) exploring other potential cellular mechanisms involved. Two findings were reported from these studies. First, significant reductions (15%) in DNA synthesis, as measured by [^{3}H]thymidine incorporation, were observed in PN1 rat cerebellum, forebrain, and brainstem 4 hr after receiving 2 mg/kg chlorpyrifos. At PN8, naive rats were injected with 11 mg/kg chlorpyrifos and DNA synthesis was also inhibited but with much more regional selectivity, reflecting the greater degree of regional cholinergic innervation and cholinergic receptors. The authors also found a similar decrease in [^{3}H]leucine incorporation indicating a depression in protein synthesis. These effects were not region specific and did not occur at PN8, however the reductions were greater in magnitude than those observed with DNA synthesis. Second, they suggest the effects were mediated by nicotinic receptors known to be involved in the regulation of DNA synthesis (McFarland *et al.,* 1991). The authors provide evidence for the specificity of these effects and suggest that chlorpyrifos targets regions within the CNS that are actively undergoing cell division (Whitney *et al.,* 1995).

D. Other Insecticides

Pyrethroid insecticides, modeled after natural pyrethrins, are also used to a large extent in agriculture and for domestic purposes. There are two classes of these compounds, Type I and Type II, based on different symptomology. These compounds cause tremor, choreoathetosis, and salivation. Their mechanism of action is through the prolongation of sodium channels in their depolarized state, leading to repetitive firing of action potentials (Narahashi, 1985). Pyrethroids, especially Type II compounds, also inhibit $GABA_A$ receptor function, thereby adding to CNS excitation (Eldefrawi and Eldefrawi, 1989). There have been several reports on the developmental neurotoxicity of pyrethroids in relation to the cholinergic system. Eriksson and Nordberg (1990) have observed changes in muscarinic and nicotinic receptor binding in neonatal mice after repeated administration of deltamethrin and bioallethrin for 7 days by gavage. The magnitude and direction of these changes depended on the brain region (cortex or hippocampus), pyrethroid compound, and binding sites on the receptor. Another study has related changes in muscarinic receptors and AChE activity to a delay in the functional maturation of the cholinergic system after gestational or lactational exposure to pyrethroids (Malaviya *et al.,* 1993). There is some indication that muscarinic receptors are involved in the behavioral teratogenicity observed from exposure to pyrethroid insecticides (Ericksson and Fredericksson, 1991).

E. Polychlorinated Biphenyls

The developmental neurotoxicity of polychlorinated biphenyls (PCBs) is a current public health concern because of their fairly ubiquitous disposition and bioconcentration in our environment, and the human toxicologic data indicating that PCBs selectively accumulate in the brain (Ness *et al.,* 1994) and target the developing CNS (e.g., Jacobson and Jacobson, 1993). PCBs are behavioral neurotoxicants that delay cognitive development during infancy. These effects may be apparent several years after initial exposures (Chen and Hsu, 1994), and the behavioral neurotoxicity associated with developmental exposure to PCBs has been demonstrated in several different species (Tilson *et al.,* 1990, review). The mechanism by which PCBs alter CNS development is still unclear. Persistent changes (observed as late as PN90) in neuronal and glial markers in the brains of rats exposed to Aroclor 1254 have been observed (Morse *et al.,* 1996a), suggesting that neuronal damage and reactive gliosis may occur during gestational exposure. PCBs are also reported to have weak estrogenic activity (Jansen *et al.,* 1993), and thus the potential for altering cell proliferation and differentiation during development. Several researchers have postulated that the developmental effects may be mediated by changes in various neurotransmitters including dopamine, norepinephrine, and serotonin (Seegal and Shain, 1992, review).

Of particular interest to this discussion of cholinergic neurobiology is the relationship between the cholinergic system and thyroid hormones in the developing CNS. ChAT activity in the basal forebrain and hippocampus is regulated by thyroxine (T_4) in the developing CNS (Patel *et al.,* 1988). These primary cholinergic brain areas are essential for the development of cognitive processes such as learning and memory. Hypothyroidism and hypothyroxinemia are hallmarks of PCB toxicity, and may in part explain the effects of PCBs on the immature brain (Morse *et al.,* 1996b). Juárez de Ku *et al.* (1994) have shown that maternal exposure of PCBs to rat pups reduced the circulating levels of T_4 and ChAT activity, and ChAT levels were restored to control levels by injecting T_4, but not triiodothyronine (T_3). The reduction of muscarinic receptors labeled with [^{3}H]QNB (Eriksson, 1988) also implicates the cholinergic system in the developmental neurotoxicity of PCBs.

F. Nicotine

It is generally accepted that nicotine is neurotoxic to the developing fetus, causing growth retardation, pregnancy complications, and even elevated perinatal mortality rates (e.g., Butler and Goldstein, 1973). Given the importance of ACh and neuronal receptors that have high affinity for ACh and nicotine in the developing CNS, it is not surprising that nicotine exposure from maternal smoking has been linked to neurologic dysfunction in children (e.g., Naeye and Peters, 1984; Levin *et al.,* 1996). Several hypotheses have been advanced to explain the toxic actions of nicotine on the fetus. These include an ischemic–hypoxic response involving the placenta and umbilical chord (Sastry, 1991), effects on maternal nutrition (Davies *et al.,* 1976), and direct teratogenicity by nicotine (Slotkin *et al.,* 1991). It is important to note, however, that most studies observed significant neurochemical effects at low doses of nicotine that do not cause reductions in fetal growth, possibly ruling out nonspecific effects associated with maternal toxicity. Nicotinic receptors are up-regulated by chronic nicotine exposure, an unusual paradoxical response to chronic receptor stimulation (Wonnacott, 1990, review). The mechanism by which these increases occur may not be the same for immature (including prenatal) and adult rodents. For example, Van De Kamp and Collins (1994) suggest that simple up-regulation occurs in the adult, whereas a teratologic effect may occur during development.

Slotkin and colleagues (McFarland *et al.,* 1991) have conducted a series of experiments using the regional and temporal selectivity of cholinergic activity in the developing CNS to study the effects and underlying mechanisms of prenatal nicotine exposure. One key finding from this group was that nicotine, as seen with the OP chlorpyrifos, inhibited DNA synthesis in fetal brain. These effects occurred with temporal and regional selectivity that reflects the ontogeny of nicotinic receptors and the cholinergic growth spurt during the early postnatal period. These and other data from this laboratory suggest that nicotine induces the premature termination of cell proliferation and initiation of cell differentiation, thus altering the normal progression of events during CNS development (McFarland *et al.,* 1991; Slotkin *et al.,* 1991; Zahalka *et al.,* 1992).

Different cholinergic receptor types with varied sensitivities to nicotine may be involved in the toxicity during CNS development. The greater effect of nicotine on high affinity [^{3}H]nicotine versus [^{125}I]BGT binding during development (Van De Kamp and Collins, 1994) suggests that α7-bearing nicotinic receptors may be less responsive to nicotine. Muscarinic m1 receptors identified by [^{3}H]pirenzepine binding in the striatum, which are most likely found on the postsynaptic membrane, were reduced in rats exposed to nicotine gestationally (Zahalka *et al.,* 1993b). This effect was transient, but functional uncoupling of receptors to G-proteins persisted into adulthood. The ontogeny of brain muscarinic receptors was delayed substantially by nicotine in the same rats in which nicotinic receptors were increased (Zhu *et al.,* 1996). In addition, multiple receptor systems with convergent signal transduction pathways may be involved in the cholinergic effects of nicotine. Adenylyl cyclase activity was altered by prenatal nicotine exposure (Slotkin *et al.,* 1992), and Seidler *et al.,* (1994) demonstrated that nicotine evokes the release of norepinephrine and dopamine during development in a pattern that parallels the regional ontogeny of nicotinic receptors in rat brain. This secondary effect of nicotine on other neurotransmitter systems has been observed previously by these (Navarro *et al.,* 1988) and other researchers (Peters *et al.,* 1979; Von Ziegler *et al.,* 1991). These results suggest that there are presynaptic and postsynaptic effects from developmental nicotine exposure mediated by several cholinergic receptor types.

V. Summary

Cholinergic neurons exhibit a widespread anatomic distribution and a diverse cellular morphology. These neurons are present in several subsystems in the brain where they exist as intrinsic interneurons or projection neurons arising from the basal forebrain. The cellular proliferation and differentiation of cholinergic neurons occur with temporal and regional selectivity during pre- and postnatal development. Synthesis and metabolism of ACh involves several enzymes and transporter proteins that are targeted by many developmental neurotoxicants. Exposure to environmental Pb is believed to cause cognitive dysfunction in children. The effects of Pb on ACh release and cholinergic receptors suggest that the impact of Pb on the developing cholinergic system may underlie some of these neurobehavioral effects. Organometals such as triethyl-Pb and MeHg have been shown to alter cholinergic function and may be of special concern because of their lipophilic properties and affinity for the developing CNS. Some of the anticholinesterase OP insecticides are neurobehavioral teratogens and are more toxic to the developing fetus than to adults. Changes in ACh levels by OP anticholinesterase activity may induce compensatory changes in ACh receptors during development, in addition to direct actions on receptors and/or signal transduction by the OPs. Abnormal expression of cholinergic receptors and markers after developmental exposure to OPs may also reflect an altered number of cholinergic neurons resulting from the nicotinic receptor–mediated inhibition of DNA synthesis and cellular proliferation during the cholinergic growth spurt of early postnatal life. Pyrethroid insecticides produce cholinergic effects in the developing brain, some of which are indicated by changes in AChE and muscarinic receptors. PCBs target the developing CNS, where they consistently produce neurobehavioral effects that may persist into adulthood. Several neurotransmitter and hormone systems may be involved in the developmental neurotoxicity of PCBs. One possible mechanism for the effects of PCBs is the deregulation of ChAT activity by PCB-induced hypothyroidism. Finally, the neurotoxic effects of maternal low-level nicotine exposure to the developing brain are due in part to changes in the modulatory function of nicotinic receptors on early DNA synthesis and catecholamine release, in conjunction with possible postsynaptic effects by nicotine.

Acknowledgment

My appreciation to Dr. A. T. Eldefrawi for reviewing an early draft of this manuscript.

References

Adam-Vizi, V. (1995). Compartmentalization and release from neurons in the CNS. In *CNS Neurotransmitters and Neuromodulators: Acetylcholine.* T. W. Stone, Ed. CRC Press, Boca, Raton, pp. 157–170.

Albuquerque, E. X., Alkondon, M., Pereira, F. R., Castro, N. G., Schrattenholz, A., Barbosa, C., Bonfante-Cabarcas, R., Arcava, Y., Eisenberg, H. M., and Maelicke, A. (1997). Properties of neuronal nicotinic acetylcholine receptors: pharmacological characterization and modulation of synaptic function. *J. Pharmacol. Exp. Ther.* **280,** 1117–1136.

Albuquerque, E. X., Pereira, F. R., Castro, N. G., Alkondon, M., Reinhardt, S., Schröder, H., and Maelicke, A. (1995). Nicotinic receptor function in the mammalian central nervous system. *NY Acad. Sci.* **757,** 48–72.

Alfano, D. P., and Petit, T. L. (1982). Neonatal lead exposure alters the dendritic development of hippocampal dentate granule cells. *Exp. Neurol.* **75,** 275–288.

Alfano, D. P., Petit, T. L., and LeBoutillier, J. C. (1983). Developmental plasticity of the hippocampal–cholinergic system in normal and early lead exposed rats. *Dev. Brain Res.* **10,** 117–124.

Amador, M., and Dani, J. A. (1995). Mechanisms for modulation of nicotinic acetylcholine receptors that can influence synaptic transmission. *J. Neurosci.* **15,** 4525–4532.

Annau, Z. (1992). Neurobehavioral effects of organophosphorous compounds. In *Organophosphates. Chemistry, Fate, and Effects.* J. E. Chambers and P. E. Levi, Eds. Academic Press, New York, 419–432.

Aronstam, R. S., and Eldefrawi, M. E. (1979). Transition and heavy metal inhibition of ligand binding to muscarinic acetylcholine receptors from rat brain. *Toxicol. Appl. Pharmacol.* **48,** 489–496.

Aubert, I., Cécyre, D., Gauthier, S., and Quirion, R. (1996). Comparative ontogenic profile of cholinergic markers, including nicotinic and muscarinic receptors, in the rat brain. *J. Comp. Neurol.* **369,** 369–355.

Baumgold, J. (1995). Neurochemical transduction processes associated with neuronal muscarinic receptors. In *CNS neurotransmitters and neuromodulators: acetylcholine.* T. W. Stone, Ed. CRC Press, Boca Raton, pp. 149–156.

Bellinger, D., Leviton, A., Waternaux, C., Needleman, H., and Rabinowitz, M. (1987). Longitudinal analyses of prenatal and postnatal lead exposure and early cognitive development. *N. Engl. J. Med.* **316,** 1037–1043.

Berge, G. N., Nafstad, I., and Fonnum, F. (1986). Prenatal effects of trichlorfon on the guinea pig brain. *Arch. Toxicol.* **59,** 30–35.

Bielarczyk, H., Tian, X., and Suszkiw, J. B. (1996). Cholinergic denervation-like changes in rat hippocampus following developmental lead exposure. *Brain Res.* **708,** 108–115.

Bielarczyk, H., Tomsig, J. L., and Suszkiw, J. B. (1994). Perinatal low-level lead exposure and the septo–hippocampal cholinergic system: selective reduction of muscarinic receptors and choline acetyltransferase in the rat septum. *Brain Res.* **643,** 211–217.

Boddeke, H. W. G. M., and Boeijinga, P. H. (1995). Muscarinic acetylcholine receptors and long-term potentiation of synaptic transmission. In *CNS Neurotransmitters and Neuromodulators: Acetylcholine.* T. W. Stone, Ed. CRC Press, Boca Raton, pp. 171–184.

Bondy, S. C., and Agrawal, A. K. (1980). The inhibition of cerebral high affinity receptor sites by lead and mercury compounds. *Arch. Toxicol.* **46,** 249–256.

Boulter, J., Evans, K., Goldman, D., Martin, G., Treco, D., Heinemann, S., and Patrick, J. (1986). Isolation of cDNA clone for a possible neural nicotinic acetylcholine receptor α-subunit. *Nature* **319,** 368–374.

Bressler, J., Forman, S., and Goldstein, G. W. (1994). Phospholipid metabolism in neural endothelial cells after exposure to lead *in vitro. Toxicol. Appl. Pharmacol.* **126,** 352–360.

Broide, R. S., O'Connor, L. T., Smith, M. A., Smith, A. M., and Leslie, F. M. (1995). Developmental expression of α_7 neuronal nicotinic receptor messenger RNA in rat sensory cortex and thalamus. *Neuroscience* **67,** 83–94.

Butler, N. R., and Goldstein, H. (1973). Smoking in pregnancy and subsequent child development. *Br. Med. J.* **4,** 573–575.

Buwalda, B., de Groote, L., Vandfer Zee, E. A., Matsuyama, T., and Luiten, P. G. M. (1995). Immunocytochemical demonstration of developmental distribution of muscarinic acetylcholine receptors in rat parietal cortex. *Dev. Brain Res.* **84,** 185–191.

Chakraborti, T. K., Farrar, J. D., and Pope, C. N. (1993). Comparative neurochemical and neurobehavioral effects of repeated chlorpyrifos exposures in young and adult rats. *Pharmacol. Biochem. Behav.* **46,** 219–224.

Chan, J., and Quik, M. (1993). A role for the nicotinic alpha-bungarotoxin receptor in neurite outgrowth in PC12 cells. *Neuroscience* **56,** 441–451.

Chanda, S. M., and Pope, C. N. (1996). Neurochemical and neurobehavioral effects of repeated gestational exposure to chlorphyrifos in maternal and developing rats. *Pharmacol. Biochem. Behav.* **53,** 771–776.

Chen, Y.-C., and Hsu, C.-C. (1994). Effects of prenatal exposure to PCBs on the neurological function of children: a neuropsychological and neurophysiological study. *Dev. Med. Child Neurol.* **36,** 312–320.

Cimino, M., Marini, P., Colombo, S., Andena, M., Cattabeni, F., Fornasari, D., and Clementi, F. (1995). Expression of neuronal acetylcholine nicotinic receptor $\alpha 4$ and $\beta 2$ subunits during postnatal development of the rat brain. *J. Neural Transm.* **100,** 77–92.

Clarke, P. B. S., Quik, M., Adkofer, F., and Tharau, K. (1995). *Advances in Pharmacological Sciences. Vol. 2: Effects of Nicotine on Biological Systems.* Verlag Press, Basal, Switzerland.

Clayton, R. M., Sedowofia, S. K., Rankin, J. M., and Manning, A. (1992). Long-term effects of aluminum on the fetal mouse brain. *Life Sci.* **51,** 1921–1928.

Connolly, J. G., Gibb, A. J., and Colquhoun, D. (1995). Heterogeneity of neuronal nicotinic acetylcholine receptors in thin slices of rat medial habenula. *J. Physiol.* **484,** 87–105.

Cooper, J. R, Bloom, F. E., Roth, R. H. (1996). *The Biochemical Basis of Neuropharmacology.* Oxford University Press, New York.

Cory-Slechta, D. A. (1995). Relationship between lead-induced learning impairments and changes in dopaminergic, cholinergic, and glutamatergic neurotransmitter system functions. *Ann. Rev. Pharmacol. Toxicol.* **35,** 391–415.

Cory-Slechta, D. A., and Pokora, M. J. (1995). Lead-induced changes in muscarinic cholinergic sensitivity. *Neurotoxicology* **16,** 337–348.

Costa, L. G. (1993). Muscarinic receptors and the developing nervous system. In *Receptors in the Developing Nervous System. Vol. 2: Neurotransmitters* I. S. Zagon and P. J. McLaughlin, Eds. Chapman and Hall, New York, pp. 21–42.

Costa, L. G., and Fox, D. A. (1983). A selective decrease of cholinergic muscarinic receptors in the visual cortex of adult rats following developmental lead exposure. *Brain Res.* **276,** 259–266.

Costa, L. G., and Murphy, S. D. (1982). Passive avoidance retention in mice tolerant to the organophosphorus insecticide disulfoton. *Toxicol. Appl. Pharmacol.* **65,** 451–458.

Coyle, J. T., and Yamamura, H. I. (1976). Neurochemical aspects of the ontogenesis of cholinergic neurons in the rat brain. *Brain Res.* **118,** 429–440.

Dave, V., Vitarella, D., Aschner, J., Fletcher, P., Kimelberg, H. K., and Aschner, M. (1993). Lead increases inositol 1,4,5-triphosphate levels but does not interfere with calcium transients in primary rat astrocytes. *Brain Res.* **618,** 9–18.

Davies, D. P., Gray, O. P., Ellwood, P. C., and Abernathy, M. (1976). Cigarette smoking in pregnancy: association with maternal weight gain and fetal growth. *Lancet* **1,** 385–387.

Davis, J. M., Otto, D. A., Weil, D. E., and Grant, L. D. (1990). The comparative developmental neurotoxicity of lead in humans and animals. *Neurotoxicol. Teratol.* **12,** 215–229.

Deacon, M. M., Murray, J. S., Pilny, M. K., Rao, K. S., Dittenber, D. A., Hanley, T. R., and John, J. A. (1980). Embryotoxicity and fetotoxicity of orally administered chlorpyrifos in mice. *Toxcol. Appl. Pharmcol.* **54,** 31–40.

Del Toro, E. D., Juiz, J. M., Peng, X., Lindstrom, J., and Criado, M. (1994). Immunocytochemical localization of the $\alpha 7$ subunit of the nicotinic acetylcholine receptor in the rat central nervous system. *J. Comp. Neurol.* **349,** 325–342.

Del Toro, E. D., Juiz, J. M., Smillie, F. I., Lindstrom, J., and Criado, M. (1997). Expression of alpha 7 neuronal nicotinic receptors during postnatal development of the rat cerebellum. *Dev. Brain Res.* **98,** 125–133.

Duclert, A., Savatier, N., and Changeux, J.-P. (1993). An 83-nucleotide promoter of the acetylcholine receptor ε-subunit confers preferential synaptic expression in mouse muscle. *Proc. Natl. Acad. Sci. USA* **90,** 3043–3047.

Ecobichon, D. J. (1996). Toxic effects of pesticides. In *Casarett and Doull's Toxicology: The Basic Science of Posisons.* C. D. Klaassen, Ed. McGraw-Hill, New York, pp. 643–690.

Eldefrawi, M. E., and Eldefrawi, A. T. (1989). Actions of insecticides on neurotransmitter receptors. *Comm. Toxicol.* **3,** 381–403.

Eldefrawi, A. T., Jett, D. A., and Eldefrawi, M. E. (1992). Direct actions of organophosphate anticholinesterases on muscarinic receptors. In *Organophosphates. Chemistry, Fate, and Effects.* J. E. Chambers and P. E. Levin, Eds. Academic Press, New York, pp. 419–432.

Eriksson, P., (1988). Effects of 3,3′,4,4′-tetrachlorobiphenyl in the brain of the neonatal mouse. *Toxicology* **49,** 43–48.

Eriksson, P., and Fredriksson, A. (1991). Neurotoxic effects of two different pyrethroids, bioallethrin and deltamethrin, on immature and adult mice: changes in behavioral muscarinic receptor variables. *Toxicol. Appl. Pharmacol.* **108,** 75–85.

Eriksson, P., and Norberg, A. (1990). Effects of two pyrethroids, bioallethrin and deltamethrin, on subpopulations of muscarinic and nicotinic receptors in the neonatal mouse brain. *Toxicol. Appl. Pharmacol.* **102,** 456–463.

Ferguson, S. A., and Bowman, R. E. (1990). Effects of post-natal lead exposure on open field behavior in monkeys. *Neurotoxicol. Teratol.* **12,** 91–97.

Fernandez, H. L., Moreno, R. D., and Inestrosa, N. C. (1996). Tetrameric (G4) acetylcholinesterase: structure, localization, and physiological regulation. *J. Neurochem.* **66,** 1335–1346.

Fredriksson, A., Teiling, G. A., Bergman, K., Oskarsson, A., Ohlin, B., Danielsson, B., and Archer, T. (1993). Effects of maternal dietary supplementation with selenite on the postnatal development of rat offspring exposed to methyl mercury *in utero. Pharmacol. Toxicol.* **72,** 377–382.

Freedman, R., Wetmore, C., Stromberg, I., Leonard, S., and Olson, L. (1993). α-bungarotoxin binding to hippocampal neurons: immunocytochemical characterization and effects on growth factor expression. *J. Neurosci.* **13,** 1965–1975.

Geula, C., Mesulam, M. M., Kuo, C. C., and Tokuno, H. (1995). Postnatal development of acetylcholinesterase-rich neurons in the rat brain: permanent and transient patterns. *Exp. Neurol.* **134,** 157–78.

Goldberg, A. M. (1979). The effects of inorganic lead on cholinergic transmission. *Prog. Brain Res.* **49,** 465–470.

Gupta, R. C., Rech, R. H., Lovell, K. L., Welsch, F., and Thornburg, J. E. (1985). Brain cholinergic, behavioral, and morphologic development in rats exposed *in utero* to methyl parathion. *Toxicol. Appl. Pharmacol.* **77,** 405–413.

Hanafy, M. S., Atta, A. H., and Hashim, M. M. (1986). Studies on the teratogenic effects of Tamaron (an organophosphorous pesticide). *Vet. Med. J.* **34,** 357–363.

Hohamann, C. F., Potter, E., and Levey, A. I. (1995). Development of muscarinic receptor subtypes in the forebrain of the mouse. *J. Comp. Neurol.* **358,** 88–101.

Hoshi, F., Kobayashi, H., Yuyama, A., and Matsusaka, N. (1991). Effects of triethyl lead on various cholinergic parameters in the rat brain. *Japan J. Pharmacol.* **55,** 27–33.

Hoskins, B., and Ho, I. K. (1992). Tolerance to organophosphorous cholinesterase inhibitors. In *Organophosphates. Chemistry, Fate, and Effects.* J. E. Chambers and P. E. Levi, Eds. Academic Press, New York, pp. 285–299.

Ishihara, K., Alkondon, M., Montes, J. G., and Albuquerque, E. X. (1995). Nicotinic responses in acutely dissociated rat hippocampal neurons and the selective blockade of fast-desensitizing nicotinic currents. *J. Pharmacol. Exp. Ther.* **273,** 1471–1482.

Jacobson, J. L., and Jacobson, S. W. (1993). A 4-year follow-up study of children born to consumers of Lake Michigan fish. *J. Great Lakes Res.* **19,** 776–783.

Jansen, H. T., Cooke, P. S., Porcelli, J., Liu, T.-C., and Hansen, L. G. (1993). Estrogenic and anti-estrogenic actions of PCBs in the female rat: *in vitro* and *in vivo* studies. *Reprod. Toxicol.* **7,** 237–248.

Jett, D. A., Abdallah, E. A. M., El-Fakahany, E. E., Eldefrawi, M. E., and Eldefrawi, A. T. (1991). The high affinity activation by paraoxon of a muscarinic receptor subtype in rat brain striatum. *Pest. Biochem. Physiol.* **39,** 149–157.

Jett, D. A., Fernando, J. C., Eldefrawi, M. E., and Eldefrawi, A. T. (1994). Differential regulation of muscarinic receptor subtypes in rat brain regions by repeated injections of parathion. *Toxicol. Lett.* **73,** 33–41.

Jett, D. A., and Guilarte, T. R. (1995). Developmental lead exposure alters *N*-methyl-D-aspartate and muscarinic cholinergic receptors in the rat hippocampus: an autoradiographic study. *Neurotoxicology* **16,** 7–18.

Jett, D. A., Hill, E. F., Fernando, J. C., Eldefrawi, M. E., and Eldefrawi, A. T. (1993). Down-regulation of muscarinic receptors and the m3 subtype in white-footed mice by dietary exposure to parathion. *J. Toxicol. Environ. Health* **39,** 395–415.

Jett, D. A., Kuhlmann, A. C., Farmer, S. J., and Guilarte, T. R. (1997a). Age-dependent effects of developmental exposure to lead on spatial learning and memory in the Morris water maze. *Pharmacol. Biochem. Behav.* **57,** 271–279.

Jett, D. A., Kuhlmann, A. C., and Guilarte, T. R. (1997b). Intrahippocampal injection of Pb alters performances of rats in the Morris water maze. Pharmacol. Biochem. Behav. **57,** 263–269.

Juárez de Ku, L. M., Sharma-Stokkermans, M., and Meserve, L. A. (1994). Thyroxine normalizes polychlorinated biphenyl (PCB) dose-related depression of choline acetyltransferase (ChAT) activity in hippocampus and basal forebrain of 15-day-old rats. *Toxicology* **94,** 19–30.

Kiraly, E., and Jones, D. G. (1982). Dendritic spine changes in rat hippocampal pyramidal cells after postnatal lead treatment: a golgi study. *Exp. Neurol.* **77,** 236–239.

Kitos, P. A., and Suntorwat, O. (1992). Teratogenic effects of organophosphorous compounds. In *Organophosphates. Chemistry, Fate, and Effects.* J. E. Chambers and P. E. Levi, Eds. Academic Press, New York, pp. 419–432.

Lärkfors, L., Oskarsson, A., Sundberg, J., and Ebendal, T. (1991). Methylmercury induced alterations in the nerve growth factor level in the developing brain. *Dev. Brain Res.* **62,** 287–291.

Lenovere, N., Zoli, M., and Changeux, J. P. (1997). Neuronal nicotinic receptor alpha-6 subunit mRNA is selectively concentrated in catecholaminergic nuclei of the rat brain. *Eur. J. Neurosci.* **8,** 2428–2439.

Levesque, P. C., and Atchison, W. D. (1991). Disruption of brain mitochondrial calcium sequestration by methylmercury. *J. Pharmacol. Exp. Ther.* **256,** 236–242.

Levesque, P. C., Hare, M. F., and Atchison, W. D. (1992). Inhibition of mitochondrial Ca^{2+} release diminishes the effectiveness of methyl mercury to release acetylcholine from synaptosomes. *Toxicol. Appl. Pharmacol.* **115,** 11–20.

Levin, E. D., Wilkerson, A., Jones, J. P., Christopher, N. C.,and Briggs, S. J. (1996). Prenatal nicotine effects on memory in rats—pharmacological and behavioral challenges. *Dev. Brain Res.* **97,** 207–215.

Lindstrom, J., Anand, R., Peng, X., Gerzanich, V., Wanf, F., and Li, Y. (1995). Neuronal nicotinic receptor subtypes. *NY Acad. Sci.* **757,** 100–116.

Lipton, S. A., Frosch, M. P., Phillips, M. D., Tauck, D. L., and Aizenman, E. (1988). Nicotinic antagonists enhance process outgrowth by rat retinal ganglion cells in culture. *Science* **239,** 1293–1296.

Loewi, O. (1960). An autobiographical sketch. *Perspect. Biol. Med.* **4,** 3–25.

Malaviya, M., Hussain, R., Seth, P. K., and Husain, R. (1993). Perinatal effects of two pyrethroid insecticides on brain neurotransmitter function in the neonatal rat. *Vet. Hum. Toxicol.* **35,** 119–122.

Markovac, J., and Goldstein, G. W. (1988). Picomolar concentrations of lead stimulate brain protein kinase C. *Nature* **334,** 71–73.

Martínez-Murillo, R., and Rodrigo, J. (1995). The localization of cholinergic neurons and markers within the CNS. In *CNS Neurotransmitters and Neuromodulators: Acetylcholine.* T. W. Stone, Ed. CRC Press, Boca Raton, pp. 1–38.

Massoulie, J., Sussman, J., Bon, S., and Silman, I. (1993). Structure and functions of acetylcholinesterase and butyrylcholinesterase. *Prog. Brain Res.* **98,** 139–146.

Matter, J.-M., Matter-Sadzinski, L., and Ballivet, M. (1995). Activity of the $\beta 3$ nicotinic receptor promoter is a marker of neuron fate determination during retina development. *J. Neurosci.* **15,** 5919–5928.

McFarland, B., Seidler, F. J., and Slotkin, T. A. (1991). Inhibition of DNA synthesis in neonatal rat brain regions caused by acute nicotine administration. *Dev. Brain Res.* **58,** 223–229.

McGehee, D. S., Heath, M. J. S., Gelber, S., Devay, P., and Role, L. W. (1995). Nicotine enhancement of fast excitatory synaptic transmission in CNS by presynaptic receptors. *Science* **269,** 1692–1696.

McMichael, A. J., Baghurst, P. A., Wigg, N. R., Vimpani, G. V., Robertson, E. F., and Roberts, R. J. (1988). Port Pirie cohort study: environmental exposure to lead and children's abilities at the age of four years. *N. Engl. J. Med.* **319,** 468–475.

Michalek, H., Pintor, A., Fortuna, S., and Bisso, G. M. (1985). Effects of diisopropylfluorophosphate on brain cholinergic systems of rats at early develomental stages. *Fund. Appl. Toxicol.* **5,** S204–S212.

Monnettschudi, F., Zurich, M. G., and Honegger, P. (1996). Comparisons of the developmental effects of two mercury compounds on glial cells and neurons in aggregate cultures of rat telencephalon. *Brain Res.* **741,** 52–59.

Monyer, H., Burnashev, N., Laurfie, D. J., Sakmann, B., and Seeberg, P. H. (1994). Developmental and regional expression in the rat brain and functional properties of four NMDA receptors. *Neuron* **12,** 529–540.

Morse, D. C., Plug, A., Wesseling, W., van den Berg, K. J., and Brouwer, A. (1996a). Persistent alterations in regional brain glial fibrillary acidic protein and synaptophysin levels following pre- and postnatal polychlorinated biphenyl exposure. *Toxicol. Appl. Pharmacol.* **139,** 252–261.

Morse, D. C., Wehler, E. K., Wesseling, W., Koeman, J. H., and Brouwer, A. (1996b). Alterations in rat brain thyroid hormone status following pre- and postnatal exposure to polychlorinated biphenyls (Aroclor 1254). *Toxicol. Appl. Pharmacol.* **136,** 269–279.

Naeye, R. L., and Peters, E. C. (1984). Mental development of children whose mothers smoked during pregnancy. *Obstet. Gynecol.* **64,** 601–607.

Narahashi, T. (1985). Nerve membrane ionic channels as the primary target of pyrethroids. *Neurotoxicology* **6,** 3–22.

Nathanson, J. A., and Bloom, F. E. (1975). Lead induced inhibition of brain adenyl cyclase. *Nature* **255,** 419–420.

Navarro, H. A., Seidler, F. J., Whitmore, W. L., and Slotkin, T. A. (1988). Prenatal exposure to nicotine via maternal infusions: effects on development of catecholamine systems. *J. Pharmacol. Exp. Ther.* **244,** 940–944.

Needleman, H. L., Schell, A., Bellinger, D., Leviton, A., and Allred, E. N. (1990). The long-term effects of exposure to low doses of lead in childhood. *N. Engl. J. Med.* **322,** 83–88.

Ness, D. K., Schantz, S. L., and Hansen, L. G. (1994). PCB congeners in the rat brain: selective accumulation and lack of regionalization. *J. Toxicol. Environ. Health* **43,** 453–468.

Oortgiesen, M., Leinders, T., van Kleef, R. G. D. M., and Vijverberg, H. P. M. (1993). Differential neurotoxicological effects of lead on voltage-dependent and receptor-operated ion channels. *Neurotoxicology* **14,** 87–96.

Oortgiesen, M., Zwart, R., van Kleef, R. G. D. M., and Vijverberg, H. P. M. (1995). Subunit-dependent action of lead on neuronal nicotinic acetylcholine receptors expressed in *Xenopus* oocytes. *Clin. Exper. Pharmacol. Physiol.* **22,** 364–365.

Ostermann, C.-H., Grunwald, J., Wevers, A., Lorke, D. E., Reinhardt, S., Maelicke, A., and Schroder, H. (1995). Cellular expression of $\alpha 4$ subunit mRNA of the nicotinic acetylcholine receptor in the developing rat telencephalon. *Neurosci. Lett.* **192,** 21–24.

Patel, A. J., Hayashi, M., and Hunt, A. (1988). Role of thyroid hormone and nerve growth factor in the development of choline acetyltransferase and other cell-specific marker enzymes in the basal forebrain of the rat. *J. Neurochem.* **50,** 803–811.

Perry, E. K., and Perry, R. H. (1996). Neurochemical pathology of cholinergic systems in the human brain. In *Muscarinic Agonists and the Treatment of Alzheimer's Disease.* A. Fisher, Ed. Chapman & Hall, New York, pp. 1–19.

Peters, D. A. V., Taub, H., and Tang, S. (1979). Postnatal effects of maternal nicotine exposure. *Neurobehav. Toxicol.* **1,** 221–225.

Petit, T. L., and LeBoutillier, J. C. (1979). Effects of lead exposure during development on neocortical dendritic and synaptic structure. *Exp. Neurol.* **64,** 482–492.

Picciotto, M. R., Zoli, M., Lena, C., Bessis, A., Lallemand, Y., LeNovere, N., Vincent, P., Pich, E. M., Brulet, P., and Changeux, J.-P. (1995). Abnormal avoidance learning in mice lacking functional high-affinity nicotine receptor in the brain. *Nature* **374,** 65–67.

Pinkas-Kramarski, R., Stein, R., Lindenboin, L., and Sokolovski, M. (1992). Growth factor–like effects mediated by muscarinic receptors in PC12M1 cells. *J. Neurochem.* **59,** 2158–2163.

Pope, C. N., and Chakraborti, T. K. (1992). Dose-related inhibition of brain and plasma cholinesterase in neonatal and adult rats following sublethal organophosphate exposures. *Toxicology* **73,** 35–43.

Pugh, P. C., and Berg, D. K. (1994). Neuronal acetylcholine receptors that bind α-bungarotoxin mediate neurite retraction in a calcium-dependent manner. *J. Neurosci.* **14,** 889–896.

Pugh, P. C., Corriveau, R. A., Conroy, W. G., and Berg, D. K. (1995). A novel population of neuronal acetylcholine receptors among those binding α-bungarotoxin. *Mol. Pharmacol.* **47,** 717–725.

Quirion, R., Wilson, A., Rowe, W., Aubert, I., Richard, J., Doods, H., Parent, A., and Meaney, M. J. (1995). Facilitation of acetylcholine release and cognitive performance by an M2-muscarinic receptor antagonist in aged memory-impaired rats. *J. Neurosci.* **15,** 1455–1462.

Rice, D. C. (1993). Lead-induced changes in learning: evidence for behavioral mechanisms from experimental animal studies. *Neurotoxicology* **14,** 167–178.

Role, L. W., and Berg, D. K. (1996). Nicotinic receptors in the development and modulation of CNS synapses. *Neuron* **16,** 1077–1085.

Romero, P., Barnett, P. G., and Midtling, J. E. (1989). Congenital anomalies associated with maternal exposure to oxydemeton methyl. *Environ. Res.* **50,** 256–261.

Rossouw, J., Offermeier, J., and van Rooyen, J. M. (1987). Apparent central neurotransmitter receptor changes induced by low lead exposure during different developmental phases in the rat. *Toxicol. Appl. Pharmacol.* **91,** 132–139.

Saito, S., Komiya, Y., and Igarashi, M. (1991). Muscarinic acetylcholine receptors are expressed and enriched in growth cone membranes isolated from fetal and neonatal rat forebrain: pharmacological demonstration and characterization. *Neurosci.* **45,** 735–745.

Samuel, N., Wonnacott, S., Lindstron, J., and Futerman, A. H. (1997). Parallel increases in [alpha-I-125]bungarotoxin binding and alpha-7 nicotinic subunit immunoreactivity during the development of rat hippocampal neurons in culture. *Neurosci. Lett.* **222,** 179–182.

Sastry, B. V. R. (1991). Placental toxicology: tobacco smoke, abused drugs, multiple chemical interactions, and placental function. *Reprod. Fertil. Dev.* **3,** 355–372.

Schlagger, B. L., DeCarlos, J. A., and O'Leary, D. D. (1993). Acetylcholinesterase as an early marker of the differentiation of dorsal thalamus in embryonic rats. *Dev. Brain Res.* **75,** 19–30.

Schliebs, R., and Roßner, S. (1995). Distribution of muscarinic acetylcholine receptors in the CNS. In *CNS Neurotransmitters and Neuromodulators: Acetylcholine.* T. W. Stone, Ed. CRC Press, Boca Raton, pp. 67–84.

Seegal, R. F., and Shain, W. (1992). Neurotoxicity of polychlorinated biphenyls: the role of *ortho*-substituted congeners in altering neurochemical function. In *The Vulnerable Brain and Environmental Risks. Vol. 2: Toxins in Food.* R. L. Isaacson and K. F. Jensen, Eds. Plenum Press, New York, pp 112–134.

Seidler, F. J., Albright, E. S., Lappi, S. E., and Slotkin, T. A. (1994). In search of a mechanism for receptor-mediated neurobehavioral teratogenesis by nicotine: catecholamine release by nicotine in immature rat brain regions. *Dev. Brain Res.* **82,** 1–8.

Shao, Z., and Suszkiw, J. B. (1991). Ca^{2+}-surrogate action of Pb^{2+} on acetylcholine release from rat brain synaptosomes. *J. Neurochem.* **56,** 568–574.

Silbergeld, E. K., and Goldberg, A. M. (1975). Pharmacological and neurochemical investigations of lead-induced hyperactivity. *Neuropharmacology* **14,** 431–444.

Slotkin, T. A. (1992). Prenatal exposure to nicotine: what can we learn from animal models? In *Maternal Substance Abuse and the Developing Nervous System.* I. S. Zagon and T. A. Slotkin, Eds. Academic Press, San Diego, pp. 97–124.

Slotkin, T. A., Lappi, S. E., Tayyeb, M. I., and Seidler, F. J. (1991). Chronic prenatal nicotine exposure sensitizes rat brain to acute postnatal nicotine challenge as assessed with ornithine decarboxylase. *Life Sci.* **49,** 665–670.

Slotkin, T. A., McCook, E. C., Lappi, S. E., and Seidler, F. J. (1992). Altered development of basal and forskolin-stimulated adenylate cyclase activity in brain regions of rats exposed to nicotine prenatally. *Dev. Brain Res.* **68,** 233–239.

Söderstrom, S., Fredriksson, A., Dencker, L., and Ebendal, T. (1995). The effect of mercury vapour on cholinergic neurons in the fetal brain: studies on the expression of nerve growth factor and its low- and high-affinity receptors. *Dev. Brain Res.* **85,** 96–108.

Sofroniew, M. V., Campbell, P. E., Cuello, A. C., and Eckenstein, F. (1985). Central cholinergic neurons visualized by immunohistochemical detection of choline acetyltransferase. In *The Rat Nervous System.* G. Paxinos, Ed. Vol. 1. Academic Press, Sydney, pp. 471–495.

Spyker, J. M., and Avery, D. L. (1977). Neurobehavioral effects of prenatal exposure to the organophosphate diazinon in mice. *J. Toxicol. Environ. Health* **3,** 989–1002.

Stamper, C. R., Balduini, W., Murphy, S. D., and Costa, L. G. (1988). Behavioral and biochemical effects of postnatal parathion exposure in the rat. *Neurotoxicol Teratol.* **10,** 261–266.

Sussman, J. L., Harel, F., Frolow, F., Oefner, C., Goldman, A., Toker, L., and Silman, I. (1991). Atomic structure of acetylcholinesterase from *Torpedo californica:* a prototypic acetylcholine-binding protein. *Science* **253,** 872–879.

Tian, X., Bourjeily, N., Bielarczyk, H., and Suszkiw, J. B. (1995). Reduced densities of sodium-dependent [^{3}H]hemicholinium-3 binding sites in hippocampus of developmental rats following perinatal low-level lead exposure. *Dev. Brain Res.* **86,** 268–274.

Tilson, H. A., Jacobson, J. L., and Rogan, W. J. (1990). Polychlorinated biphenyls and the developing nervous system: cross-species comparisons. *Neurotoxicol. Teratol.* **12,** 239–248.

Unsworth, C. D., and Johnson, R. G. (1990). Acetylcholine and ATP are coreleased from the electromotor nerve terminals of *Narcine brasiliensis* by an exocytotic mechanism. *Proc. Natl. Acad. Sci. USA* **87,** 553–60.

Van De Kamp, J. L., and Collins, A. C. (1994). Prenatal nicotine alters nicotinic receptor development in the mouse brain. *Pharmacol. Biochem. Behav.* **47,** 889–900.

Van Hoof, C. O. M., DeGraan, P. N. E., Oestreicher, A. B., and Gispen, W. H. (1989). Muscarinic receptor activation stimulates B-50/GAP-43 phosphorylation in isolated nerve growth cones. *J. Neurosci.* **9,** 3753–3759.

Veronesi, B., and Pope, C. (1990). The neurotoxicity of parathion-induced acetylcholinesterase inhibition in neonatal rats. *Neurotoxicology* **11,** 609–626.

Von Ziegler, N. I., Schumpf, M., and Lichtensteiger, W. (1991). Prenatal nicotine exposure selectively affects perinatal forebrain aromatase activity and fetal adrenal function in male rats. *Dev. Brain Res.* **62,** 23–31.

Wall, S. J., Yasuda, R. P., Li, M., Ceisla, W., and Wolfe, B. B. (1992). The ontogeny of m1–m5 muscarinic receptor subtypes in rat forebrain. *Dev. Brain Res.* **66,** 181–185.

Whitney, K. D., Seidler, F. J., and Slotkin, T. A. (1995). Developmental neurotoxicity of chlorpyrifos: cellular mechanisms. *Toxicol. Appl. Pharmacol.* **134,** 53–62.

Widmer, H. R., Vedder, H., Schlumpf, M., and Lichtensteiger, W. (1992). Concurrent changes in regional cholinergic parameters and nest odor preference in the early postnatal rat after lead exposure. *Neurotoxicology* **13,** 615–624.

Winder, C., and Kitchen, I. (1984). Lead neurotoxicity: a review of the biochemical, neurochemical and drug induced behavioral evidence. *Prog. Neurobiol.* **22,** 59–87.

Wonnacott, S. (1990). The paradox of nicotinic acetylcholine receptor upregulation. *Trends Pharmacol. Sci.* **11,** 216–219.

Wurtman, R. J. (1992). Choline metabolism as a basis for the selective vulnerability of cholinergic neurons. *Science* **14,** 117–119.

Zahalka, E. A., Seidler, F. J., Lappi, S. E., McCook, E. C., Yanai, J., and Slotkin, T. A. (1992). Deficits in development of central cholinergic pathways caused by fetal nicotine exposure: differential effects on choline acetyltransferase activity and [^{3}H]hemicholinium-3 binding. *Neurotoxicol. Teratol.* **14,** 375–382.

Zahalka, E. A., Seidler, F. J., Lappi, S. E., Yanai, J., and Slotkin, T. A. (1993a). Differential development of cholinergic nerve terminal markers in rat brain regions: implications for nerve terminal density, impulse activity and specific gene expression. *Brain Res.* **601,** 221–229.

Zahalka, E. A., Seidler, F. J., Yanai, J., and Slotkin, T. A. (1993b). Fetal nicotine exposure alters ontogeny of M_1-receptors and their link to G-proteins. *Neurotoxicol. Teratol.* **15,** 107–115.

Zheng, J. Q., Felder, M., Connor, J. A., and Poo, M. (1994). Turning of nerve growth cones induced by neurotransmitters. *Nature* **368,** 140–144.

Zhu, J., Takita, M., Konishi, Y., Sudo, M., and Muramatsu, I. (1996). Chronic nicotine treatment delays the developmental increase in brain muscarinic receptors in rat neonate. *Brain Res.* **732,** 257–260.

Zoli, M., LeNovere, N., Hill, J. A., and Changeux, J.-P. (1995). Developmental regulation of nicotinic ACh receptor subunit mRNAs in the rat central and peripheral nervous systems. *J. Neurosci.* **15,** 1912–1939.

CHAPTER

14

Ontogeny of Second Messenger Systems

LUCIO G. COSTA
Department of Environmental Health
University of Washington
Seattle, Washington 98105
and
Toxicology Units
Fondazione S. Maugeri
Pavia, Italy

I. Introduction

Neurotransmitters, growth factors, and hormones are considered "first messengers" in the transfer of information from one cell to another. On binding to specific receptors on the cell membrane, these compounds activate a cascade of events, usually referred to as *second messenger systems, cell signalling,* or *signal-transduction systems.* These systems are the subject of much research in fields such as neurobiology, immunology, and carcinogenesis, as it is apparent that these intracellular path-

*Abbreviations: AA, arachidonic acid; AP-1, activator protein-1; ATP, adenosine triphosphate; cAMP (cyclic AMP), cyclic adenosine monophosphate; CNS, central nervous system; DAG, diacylglycerol; ERK, extracellular signal kinase; FAS, fetal alcohol syndrome; GDP, guanosine diphosphate; GFAP, glial fibrillary acidic protein; GMP, guanosine monophosphate; GTP, guanosine triphosphate; ICE, interleukin 1β converting enzyme; IGF-I, insulin-like growth factor-I; IP, inositol phosphate; JAK-STAT, Janus Kinase-Signal transducers and activation factors; JNK, *cjun* N-terminal kinase; MAPK, mitogen-activated protein kinase; MBP, myelin basic protein; MEK, MAP kinase kinase; NFκB, nuclear factor κB; NMDA, *N*-methyl-D-aspartate; NO, nitric oxide; NOS, nitric oxide synthase; PA, phosphatidic acid; PARS, poly (ADP-ribose) synthetase; PDGF, platelet-derived growth factor; PKA, protein kinase A; PKC, protein kinase C; PKG, protein kinase G; SH-2, Src homology-2; TGF-β, transforming growth factor-β.

ways regulate most of the physiologic and pathologic events in the cell. Within the neurosciences, second messenger systems are being studied to elucidate their role in cognitive processes and other behaviors, such as in neuronal degeneration and neuropsychiatric disorders, to name only a few. The role of cell signaling in neurotoxic processes is also being increasingly investigated (Costa, 1990, 1994a, 1997).

Growth factors, neurotransmitters, hormones, and cytokines are thought to play multiple important roles in various stages of brain development. The ontogeny of their receptors and of enzymes involved in their synthesis and degradation has been studied for since the late 1970s, whereas only more recently has the attention focused on the signal-transduction systems activated by these compounds. These studies have revealed a complex pattern of ontogeny of receptor-activated second messenger systems, which do not always follow the development of the receptor themselves, indicating that at different developmental stages, these intracellular pathways can be differentially activated. These findings have led to the hypothesis that the particular expression or activity of a second messenger system may play a role in certain stages of brain development by regulating the proliferation and maturation of nerve cells (both neurons and glial cells). As it is thought that many injuries to the developing nervous system arise from interference with developmental processes, rather than destruction of tissue (Rodier, 1990), the interaction of toxicants with the biochemical events that follow receptor activation may represent an important aspect of developmental neurotoxicity (Costa, 1994b). There are several mechanisms by which receptors can transfer a message from the cell membrane to the cytosol and the nucleus; they consist initially in the opening of ion channels, activation of specific enzymes, or a combination of the two. These initial events are followed by a cascade of downstream events that amplify and/or modulate the response. It is not in the scope of this chapter to review all second messenger pathways and their ontogenetic aspects, but rather to focus on a selected few that have been the subject of extensive investigations and whose potential role in developmental neurotoxicity is, or appears to be, more relevant. A few considerations should be made beforehand. First, although the different second messenger systems are discussed as separate entities, all these pathways are highly interactive (a process also referred to as *crosstalk*) and they can often control and modulate each other. Second, totally unrelated compounds acting on different receptors can produce identical effects if they activate the same intracellular pathways. Third, the receptor–effector coupling should not be seen as a static concept, as the expression of intracellular pathways varies with the cell type, the developmental stage, and the pathophysiologic status of the cell.

In this chapter, some second messenger systems and their developmental aspects are summarized and some examples of their possible roles in brain development and their disruption by developmental neurotoxicants are provided. Additional discussion on these topics can be found in other chapters of this section on synaptogenesis and neurotransmission, as well as in the section on specific neurotoxic syndromes.

II. Second Messenger Systems: General and Ontogenetic Aspects

A. *G-Protein Coupled Receptors*

Several receptors in both neuronal and glial cells are coupled to specific enzymes through G-proteins, which are heterotrimetric guanosine triphosphate (GTP)-binding proteins composed of α-, β-, and γ-subunits. G-Protein–coupled receptors, when activated, promote the binding of GTP to the α-subunit, thereby causing its dissociation from the other subunits. Both the α-subunit and the β-γ complex regulate the activities of specific affected proteins, such as ion channels, phospholipases, and adenylate cyclase. Heterotrimeric G-proteins include G_i/G_o, $G_{q/11}$, G_s, and G_{12} (Post and Brown, 1996). Several G-protein–coupled, seven transmembrane spawning receptors exist, such as the glutamate metabotropic receptors, the cholinergic muscarinic receptors, or the β-adrenergic receptors.

1. Phospholipases

Agonists of some subtypes of muscarinic and excitatory amino acid receptors or other receptors (e.g., thrombin) activate a phosphoinositidase (phospholipase C) that hydrolyzes phosphoinositide 4,5-bisphosphate to form inositol 1,4,5-trisphosphate (IP_3) and diacylgycerol (DAG) (Martin, 1991; Exton, 1996). IP_3 binds to specific receptors on the endoplasmic reticulum (Joseph, 1996) and causes mobilization of calcium ions in the cytosol before being dephosphorylated by phosphatases to IP_2, IP, and inositol. IP_3 can also be phosphorylated by a 3-kinase to form IP_4, which may also play a role in modulating intracellular calcium levels (Berridge and Irvine, 1989; Fisher *et al.*, 1992). Other inositol phosphates have also been identified (e.g., IP_5, IP_6, cyclic IP_4), but their precise role is still not clear. Thus, a main consequence of the activation of the phosphoinositide pathway is a change in the intracellular concentration of calcium deriving from IP_3-, IP_4-, and Ca^{2+}-sensitive stores, or entering the cell through receptor-operated calcium channels. Intracellular calcium can bind to proteins such as calmodulin that do not possess enzymatic activity but that are capable of activating proteins (e.g., the $Ca2^+$, calmodulin-activated protein kinases), or to other proteins such as protein kinase C (PKC) that possess enzymatic activity that can be activated by calcium (Rasmussen *et al.,* 1990). Whereas a transient increase of intracellular calcium is necessary for the normal physiologic functions of the cell, a sustained increase can produce toxicity by activating phospholipases, proteases, and endonucleases and playing a role in both necrotic and apoptotic processes (Nicotera *et al.,* 1992; Harman and Maxwell, 1995).

In brain slices, agonist-stimulated hydrolysis of membrane phosphoinositides increases in parallel with the maturation of receptors for several neurotransmitter systems (e.g., norepinephrine, histamine). However, for other receptor systems, the ability of agonists (e.g., acetylcholine, glutamate) to stimulate phosphoinositide metabolism is enhanced in the immature brain, although the density of receptors is low and increases with age (Palmer *et al.,* 1990; Balduini *et al.,* 1991a; Tandon *et al.,* 1993). This may indicate that these two neurotransmitters, which are present at relatively high levels in the developing brain, are involved in various ontogenetic processes.

Several agents that stimulate the metabolism of phosphoinositides also induce hydrolysis of phosphatidylcholine. This membrane lipid, which accounts for almost 40% of the total membrane phospholipid content, can be hydrolyzed by both phospholipase C and phospholipase D. The former generates DAG and phosphocholine, whereas the latter leads to the formation of phosphatidic acid (PA) and choline (Billah and Anthes, 1990; Shukla and Halenda, 1991; Liscovitch, 1996). PA can be converted to DAG, which provides sustained activation of PKC while having effects on its own, such as playing a role in calcium entry (English, 1996). The ontogenetic aspects of phosphatidylcholine hydrolysis have not been studied in detail. Of interest, however, is the finding that muscarinic receptor–stimulated phospholipase D activity is seen in cortical slices from immature rats, but not in adult rats, suggesting the presence of an enhanced coupling of this system in early stages of brain development (Costa *et al.,* 1995).

A calcium-activated phospholipase, phospholipase A_2, is capable of generating arachidonic acid (AA) from membrane phospholipids. AA can directly activate PKC (Khan *et al.,* 1995) and is then metabolized by cyclooxygenase, lipoxygenase, and cytochrome P450 to an array of diverse compounds such as prostaglandins, tromboxanes, or leukotrienes, which have significant intracellular activities (Piomelli, 1993; Mukherjee *et al.,* 1994), and may be associated with neurotoxic injuries (Bonventre, 1996). High levels of phospholipase activity are present in growth cones, suggesting that this enzyme, in addition to its detrimental effects, may also play a role in synaptogenesis (Negre-Aminou and Pfenninger, 1993).

2. Adenylate Cyclase

Activation of a number of receptors coupled through the GTP-binding protein G_s to the enzyme adenylate cyclase causes the conversion of adenosine triphosphate (ATP) to cyclic adenosine monophosphate (AMP). Before being inactivated by phosphodiesterases, cyclic AMP can activate a cyclic AMP–dependent protein kinase A (PKA), that in turn can phosphorylate a number of substrates (Robison *et al.,* 1971). Certain receptors are negatively coupled to adenylate cyclase (through a G-protein known as G_i) and their activation leads to a decrease in the intracellular levels of cyclic AMP (Kendall-Harden, 1989).

The concentration of cyclic AMP increases with age in the rat brain, whereas basal adenylate cyclase activity does not show any major age-related differences (Keshles and Levitzki, 1984). However, agonist-stimulated adenylate cyclase increases significantly with age, both in *ex vivo* preparations, as well as in neurons in primary cultures (Ma *et al.,* 1991; Keshles and Levitzki, 1984). Thus, there appears to be a good correlation between the development of the receptor proteins and this second messenger system. Activation of the adenylate cyclase can influence cell replication or differentiation; a possible involvement of this pathway in the developmental neurotoxicity of chlorpyrifos has been suggested (Song *et al.,* 1997).

B. Tyrosine Kinases

The receptors for several growth and differentiation factors are transmembrane tyrosine–specific kinases, which, when activated, dimerize and autophosphorylate to initiate several intracellular signal transduction pathways (Malarkey *et al.,* 1995). Receptor tyrosine phosphorylation promotes the interaction with a number of key signalling enzymes/proteins that are defined by specific regions within the target molecules, known as SH-2 (Src homology-2) domain. Among proteins that contain the SH-2 domain are phospholipase C-γ, Ras GAP, or Grb 2; the last does not have intrinsic catalytic activity but functions as an apparent "molecular adaptor" to couple receptor tyrosine kinases to other signaling proteins that lack the SH-2 domain. As a result of these interactions, several downstream pathways are activated. For example, tyrosine phosphorylated–phospholipase C-γ can hydrolyze inositol lipids, thus representing another way of producing IP_3 and DAG. An interaction of Grb 2 with the activated receptor results in the recruitment of Sos, a nucleotide exchange protein, to the membrane; Sos then activates membrane-associated Ras by converting it from its guanosine diphosphate (GDP)-bound to its GTP-bound form (Bhat, 1995). Activated Ras in turn initiates the cascade involving activation of mitogen-activated protein kinases (MAPKs). It should be noted that nonreceptor tyrosine kinases also exist, such as those associated with cell adhesion molecules (Richardson and Parsons, 1995), which may play significant roles in brain development.

C. Downstream Signaling

1. Protein Kinases A and C

PKC is a family of protein kinases, usually subdivided in three groups: the classic proteins consisting of the α-, β-, and γ-isoforms, which are stimulated by calcium, DAG, and phosphatidylserine; the new isoforms (δ, ε, η, and μ), which are not regulated by calcium; and the atypical enzymes (ζ and λ), which are not regulated by diacylglycerol (Newton, 1995; Nishizuka, 1995; Liu, 1996). The N-terminal regulatory region and the C-terminal catalytic region of this enzymes, which consist of a single polypeptide, have been for the most part characterized (Newton, 1995). Different domains contain the binding sites for calcium ions, DAG, and phorbol esters. PKC is translocated from the cytosol to the membrane as part of its activation, and typically phosphorylates serine or threonine residues on a wide array of proteins. Substrates of PKC include receptors and G proteins, enzymes, cytoskeletal proteins, proto-oncogene products, nuclear proteins, and various other proteins, including ion channels (Liu, 1996). Ontogenetic studies on PKC have indicated that the α- and γ-subtypes are highly expressed in the immature rat brain (postnatal weeks 1 and 2), suggesting a role in various aspects of brain development (Hashimoto *et al.,* 1988; Shearman *et al.,* 1991).

Cyclic AMP activates another serine/threonine protein kinase, PKA, by binding cooperatively to two sites on each of its regulatory subunits and releasing the active catalytic subunits, which phosphorylate a large number of substrates (Krebs, 1989; Walaas and Greengard, 1991). PKA displays greater specificity than PKC and has more stringent requirements in terms of phosphorylation sites and stereospecificity.

2. MAP Kinases

MAPKs play a central role in intracellular signalling and can be activated by a variety of extracellular stimuli, including those mediated by receptor tyrosine kinases and G-protein–coupled receptors (Ferrell, 1996). Mammalian MAPKs include, for example, extracellular signal regulated kinase (ERKs) or *c-jun* N-terminal kinases (JNK). A common pathway starting from the receptor tyrosine kinase involves a small GTP-binding protein ($p21^{ras}$) that activates Raf-1, which then activates, by phosphorylation, a MAP kinase kinase (MEK) that in turn phosphorylates and activates MAPK (Gomez and Cohen, 1991). Evidence of opposite controls of MAPK activation by both PKC and cyclic AMP have also been provided (Kurino *et al.,* 1996; Van Biesen *et al.,* 1996). Following activation, MAPK translocates to the nucleus and initiates activation of nuclear oncogenes such as *c-jun,* whereas in the cytosol MAPK can phosphorylate various proteins such as cytosolic phospholipase A_2 (Malarkey *et al.,* 1995). MAPKs play a most relevant role in mitogenesis and are thus involved in various aspects of brain development.

D. Nitric Oxide

The ability of several neurotransmitters to stimulate the synthesis of cyclic guanosine monophosphate (GMP) from GTP has been known for some time, but has received new attention as it was discovered that the enzyme responsible for such conversion, guanylate cyclase, is a major target of nitric oxide (NO). NO is formed in neurons and glial cells as a result of the metabolism of L-arginine to L-citrulline, catalyzed by a family of enzymes known as NO synthases (Zhang and Snyder, 1995; Dawson and Dawson, 1996). NO is a free radical gas that easily diffuses across the cell membrane and has been shown to be involved in a number of central nervous system (CNS) effects, including neurotransmitter release, long-term potentiation, synaptic plasticity, nociception, and cerebrovascular functions. NO appears to possess both neuroprotective and neurotoxic properties, depending on its oxidative state (Paakkari and Lindsberg, 1995; Varner and Beckman, 1995; Rubbo *et al.,* 1996). NO^+ is thought to be neuroprotective through nitrosylation of free sulfhydryl groups in the redox modulatory site of the *N*-methyl-D-aspartate (NMDA) receptor. However, NO· can react with superoxide anions (O_2—·) to form peroxynitrite ($ONOO^-$). Peroxynitrite is a potent oxidant that can also decompose to the hydroxyl- and nitrogen-free radicals and is relevant in ischemic damage (Varner and Beckman, 1995; Rubbo *et al.,* 1996). The neurotoxic effects of NO and peroxynitrite can be mediated by different mechanisms (Dawson and Dawson, 1996). They can damage DNA, and this activates poly (ADP-ribose) synthetase (PARS) with the consumption of ATP, leading to depletion of the cell's energy stores, ultimately causing cell death.

Constitutive NO synthase (NOS) is expressed only in a selected number of neurons (1–2%) in areas such as the hippocampus and the cerebral cortex, although it is present in almost all granule cells, but not in the

Purkinje cells, in the cerebellum (Zhang and Snyder, 1995). An inducible form of NOS can be found in glial cells. The role of NO in brain development is not clear. Based on the observation in adult animals and/or in cells in culture, it has been hypothesized that NO may exert a trophic action for developing neurons. Alternatively, it may help to elicit the programmed cell death in which a good percentage of mammalian neurons die before maturation (Zhang and Snyder, 1995). It should also not be forgotten that, in addition to the many actions of NO that are not mediated by cyclic GMP, other effects on calcium signalling, cyclic nucleotides, and phosphatases appear to be mediated by cyclic GMP–dependent protein kinase (PKG) (Wang and Robinson, 1997).

E. Signalling to the Nucleus

The activation of transcription factors which control the expression of specific genes is the culminating event of the pathways that transfer signals from the cell membrane to the nucleus (Bhat, 1995). In some cases, there is activation of transducing kinases that translocate to the nucleus to change their DNA-binding activities or transactivation functions. Transcriptional control of *c-fos* is an example of such mechanism. On an increase in *c-fos* transcription, the Fos protein interacts with another nuclear proto-oncogene, *c-jun,* to form a heterodimeric complex (AP-1, activator protein-1) that binds to specific binding sites of target genes (Morgan and Curran, 1991). In other situations, there is a direct phosphorylation of a latent transcription factor and subsequent translocation of this to the nucleus; an example is represented by the so-called JAK-STAT (Janus Kinase–Signal transducers and activation factors) system, where a nonreceptor type tyrosine kinase (JAK), when activated, phosphorylates cytoplasmic proteins acting as transcription factors (STAT), which migrate to the nucleus (Finbloom and Larner, 1995; Schindler and Darnell, 1995). In yet other situations, phosphorylation may induce the release of a transcription factor from a cytoplasmic anchor and subsequent nuclear translocation, as is the case of nuclear factor κB (NFκB) (Thanos and Maniatis, 1995).

III. Some Roles of the Second Messenger System in Brain Development

As neurotransmitters, growth factors, and cytokines exert multiple important roles in various aspects of nervous system development and they transmit their signals to the nucleus through second messengers, it is evident how the latter are involved in the development of neurons and glial cells. Here, only two aspects of this complex process are briefly considered: the proliferation of glial cells and the modulation of neuronal programmed cell death.

A. Proliferation and Maturation of Glial Cells

One of the important events during brain development is the proliferation and maturation of glial cells (astrocytes, oligodendrocytes, and microglia). Second messenger systems are involved in the effects of neurotransmitters, cytokines, and growth factors on glial cells (Bhat, 1995). The development of oligodendrocytes is profoundly influenced by molecules, such as platelet-derived growth factor (PDGF) or insulin-like growth factor-I (IGF-I), which promote proliferation/differentiation of oligodendrocyte precursors, with PKC playing a most relevant role (Bhat, 1995). During development, the responsiveness of oligodendrocytes to purinergic and cholinergic receptor agonists decreases as a function of age, whereas the ability of bradikynin or histamine to increase intracellular calcium remains relatively stable (He and McCarthy, 1994). It has been suggested that the effect of these agonists, which would also lead to PKC activation, may have a role in the development of these cells. Indeed, PKC is responsible for phosphorylating myelin basic protein (MBP) in oligodendroglia (Vartanian *et al.,* 1986), it enhances the morphologic differentiation of oligodendrocytes, and is believed to play a primary role in myelination (Yong *et al.,* 1988). Both carbachol and glutamate can induce *c-fos* expression in oligodendrocyte progenitors; however, whereas activation of muscarinic receptors leads to an increase in cell proliferation (possibly mediated by PKC ξ), activation of non-NMDA glutamate receptors reduces the proliferation of these cells (Liu and Almazan, 1995; Cohen *et al.,* 1996).

A large number of growth factors, neurotransmitters, and cytokines are increasingly recognized as being important for the proliferation and maturation of astrocytes (Bhat, 1995; Post and Brown, 1996). For example, IGF-1, adenosine, thrombin, IL-2, and acetylcholine are all mitogens in astrocytes (Perraud *et al.,* 1987; Giulian *et al.,* 1988; Rathbone *et al.,* 1992; Tranque *et al.,* 1992; Guizzetti *et al.,* 1996). However, compounds such as glutamate, serotonin, and transforming growth factor-β (TGF-β) inhibit astrocyte proliferation (Nicoletti *et al.,* 1990; Lindholm *et al.,* 1992; Guizzetti *et al.,* 1996). A number of intracellular pathways are involved in the mitogenic effects in astrocytes; these include phosphoinositide metabolism and calcium mobilization (Stanimirovic *et al.,* 1995), MAPK (Lazarini *et al.,* 1996), PKC (Tranque *et al.,* 1992), and phospholipase D (Gonzalez *et al.,* 1996).

Activation of the cyclic AMP pathway tends to suppress growth of astrocytes, possibly by inhibition of the MAPK (Kurino *et al.*, 1996), and to promote their differentiation. Both PKC and cyclic AMP pathways have been implicated in the regulation of glial fibrillary acidic protein (GFAP) expression (Shafit-Zagardo *et al.*, 1988), which appears to be intimately associated with astrocyte morphogenesis.

B. Modulation of Neuronal Apoptosis

Many different cells of the nervous system undergo programmed cell death, or apoptosis, at different stages of development. Signals that can trigger apoptosis are diverse and include trophic deprivation, neuronal activity, and also death-inducing molecules (Henderson, 1996). A large number of agents, including cytokines, growth factors, and neurotransmitters, can also have survival-promoting effects on developing neurons. Cell-surface receptor-mediated mechanisms that control apoptosis often act through a signal transduction system that involves stimulation of the receptor, activation of protein kinases/phosphatases cascades, and the release of additional messengers to up-regulate or suppress the transcription of specific genes (Hale *et al.*, 1996). Protein tyrosine kinases, PKC, cyclic AMP, and the Ras/Raf/MAPK pathways play roles in apoptosis. For example, activation of muscarinic receptors and excitatory amino acid metabotropic receptors, leading to phospholipid hydrolysis and activation of PKC, has been shown to protect cerebellar granule cells against apoptotic death (Copani *et al.*, 1996; Yan *et al.*, 1995). Similar effects were also observed with NMDA, IGF-I, and brain-derived neurotrophic factor (D'Mello *et al.*, 1993; Yan *et al.*, 1994; Kubo *et al.*, 1995), suggesting a concerted action of calcium and PKC, among others. Muscarinic agonists have been shown, in SH-SY5Y neuroblastoma cells, to increase the level of Bcl-2 protein (which inhibits apoptosis) by a mechanism involving the activation of PKC (Itano *et al.*, 1996). However, in the same cells, cyclic AMP and PKA had an opposite effect on Bcl-2 protein levels (Itano *et al.*, 1996). The Ras proteins, which link receptor and nonreceptor tyrosine kinases to downstream serine/threonine kinases, including the MAP kinases, also appear to play a role in the inhibition of apoptosis (Hale *et al.*, 1996). An increasing role in mediating apoptosis induced by various mechanisms, appears to be also played by the family of interleukin 1β converting enzyme (ICE)-like proteases (Hale *et al.*, 1996). Clearly, it is not in the intent of this section to review the current knowledge on the role of signal transduction in apoptotic death, which is currently the subject of extensive investigations. Rather, it wants simply to point out how intracellular messengers have a key role in promoting and inhibiting this process. As several developmental neurotoxicants are capable of interacting with signal transduction systems, it is apparent that these biochemical interactions may be involved in their effects on the developing nervous system, when neuronal death is one of the ensuing effects of exposure. This neurotoxicity may be related either to a direct positive effect, or to the inhibition of a trophic, protective action of endogenous compounds, both leading to excess neuronal death.

IV. Developmental Neurotoxicants and Second Messenger Systems

Studies on second messenger systems as potential targets for developmental neurotoxicants are still limited and fragmented. Interactions of various chemicals with selected steps of second messenger systems have been reviewed elsewhere (Costa, 1994a, 1997). For the most part, it is difficult to relate these separate effects to general or specific mechanisms of developmental neurotoxicity, and there is certainly the need of more hypothesis-driven comprehensive studies to be carried out. Later in this chapter, data on two known human developmental neurotoxicants, ethanol and lead, are briefly reviewed to illustrate examples of the potential importance of second messenger systems in their effects.

A. Ethanol

Intake of alcohol during pregnancy is detrimental to fetal development. Offspring of alcoholics often present with fetal alcohol syndrome (FAS), the major features of which include growth deficiency, particularly facial features and CNS dysfunctions (Streissguth *et al.*, 1980). The latter, including mental retardation and microcephaly, appear to be long lasting if not irreversible (Streissguth *et al.*, 1991). Despite much research, the precise mechanism(s) underlying the neurotoxic effects of ethanol in the developing brain have not been elucidated. A series of mechanisms that have been proposed relate to the ability of alcohol to inhibit receptor-activated second messenger pathways, in particular the hydrolysis of membrane phospholipids. A line of research has focused on the cholinergic muscarinic receptors–coupled phosphoinositide metabolism; this signal transduction system has been shown to be inhibited by ethanol in a brain region-, neurotransmitter-, and age-specific manner (Balduini and Costa, 1989; Balduini *et al.*, 1991b, 1994), and this biochemical effect shows a good correlation with the ensuing microencephaly (Reno' *et al.*, 1994). By activating m_3 muscarinic receptors, which are coupled to phospholipid hydrolysis, acetylcholine has been shown to stimulate the proliferation of rat cortical astrocytes (Guizzetti *et al.,* 1996). This has led to the hypothesis that ethanol-induced microencephaly may be related to inhibition of the mitogenic effect of acetylcholine in glial cells. Ethanol is indeed a potent inhibitor (IC_{50} = 10 m*M*) of acetylcholine-induced astrocyte proliferation, and also inhibits acetylcholine-induced phospholipid metabolism and calcium mobilization in glial cells (Catlin and Costa, 1996; Guizzetti and Costa, 1996; and Costa, 1997, unpublished observations). Similar observations have also been made with IGF-I; this compound is also a potent mitogen for glial cells, and its action is inhibited by low millimolar concentrations of ethanol (Resnicoff *et al.,* 1994). Whereas in the case of acetylcholine the precise molecular target of ethanol has not been identified yet, in the case of IGF-I it has been proposed that ethanol's inhibition of stimulated proliferation is due to inhibition of the autophosphorylation of the IGF-I receptor (Resnicoff *et al.,* 1994). These interferences of ethanol with neurotransmitter- or growth factor–stimulated glial cell proliferation may be at the basis of the observed microencephaly and may also be indirectly responsible for abnormal neuronal development.

It is also becoming apparent that developmental exposure to ethanol can cause neuronal death, particularly of granule cells and Purkinje cells in the cerebellum (Pierce *et al.,* 1989). *In vitro* experiments have shown that ethanol can act directly on cerebellar granule cells and produce neuronal cell death (Pantazis *et al.,* 1993) and can also promote apoptosis by inhibiting the trophic effect of NMDA and of carbachol (Castoldi *et al.,* 1996; Bhave and Hoffman, 1997). These effects of ethanol would contribute to the observed neuronal loss and underlie some of the behavioral and cognitive abnormalities observed in FAS.

B. Lead

Subchronic or chronic exposure to lead is highly neurotoxic, particularly to the developing CNS. In children, lead blood levels as low as 0.5–10 μM may affect CNS development, leading to mental retardation, impaired visual motor coordination, and permanent cognitive deficits (Goldstein, 1992). As for ethanol, the precise mechanisms of lead neurotoxicity have not been fully elucidated. Lead is known to alter calcium-mediated processes and may also mimic calcium's effects (Bressler and Goldstein, 1991). Lead has a higher affinity than calcium for calmodulin and can activate some calmodulin-dependent processes. For example, inorganic lead inhibits neurite development in cultured rat hippocampal neurons at concentrations as low as 100 nM, by inappropriately stimulating protein phosphorylation by calcium calmodulin-dependent protein kinase II (Kern and Audesirk, 1995).

Lead can also substitute calcium in stimulating PKC activity (Markovac and Goldstein, 1988a,b; Long *et al.,* 1994). In isolated immature rat microvessels, lead causes activation of PKC in a concentration range of 0.1–10 μM, and also causes translocation of PKC from the cytosol to the particulate fraction. More surprisingly, lead has been shown to activate PKC, partially purified from rat brain, at concentrations as low as 10^{-15} *M,* being several orders of magnitude more potent than calcium itself. The reason for the apparent discrepancy between the potency of lead in the brain and in isolated microvessels is not clear, but may be linked to the relative distribution of PKC isozymes. It should be noted, however, that an inhibitory effect of lead on PKC activity has also been reported (Murakami *et al.,* 1993). As long-term potentiation, a possible equivalent of memory storage, is blocked by inhibitors of PKC, a lead-induced decrease in PKC activity (via inhibition or down-regulation due to sustained stimulation) may relate to learning and memory deficits caused by this compound. Activation of PKC by lead has also been suggested to be involved in its inhibition of astroglia-induced microvessel formation *in vitro* (Laterra *et al.,* 1992). Lead has also been shown to amplify glutamate-induced oxidative stress, possibly through a PKC-dependent mechanism (Naarala *et al.,* 1995). All or some of these interactions of lead with PKC may contribute to the neurotoxicity of lead in the developing nervous system. Indeed, developmental exposure to lead has been shown to alter the binding of phorbol esters (Farmer and Guillarte, 1994).

It has also been observed that lead promotes apoptosis of cerebellar granule cells *in vitro* (Oberto *et al.,* 1996). Although the involvement of PKC was not investigated (but may be inferred by its role in the apoptotic process), it has been shown that the effect of lead could be antagonized by a calcium channel agonist. Facilitation of the pathophysiologic process of suicidal death during neuronal selection in the CNS may thus represent another relevant mechanism of lead developmental neurotoxicity.

Lead has also been shown to alter the metabolism of cyclic GMP, and this biochemical effect has been associated with lead-induced rod-mediated alterations in visual functions (Fox *et al.,* 1994). Low concentrations of lead ($<$100 n*M*) can inhibit NO synthase (NOS) in rat cerebellum (Quinn and Harris, 1995). However, *in vivo* exposure to lead, resulting in lead blood levels as high as 49 μg/dl, did not cause any change in NOS synthase activity in rat cerebral cortex (Kala *et al.,* 1996). Thus, the potential role of NO in the developmental neurotoxicity of lead remains to be clarified.

V. Conclusions

The mechanisms by which extracellular signals transfer information from the cell surface to the nucleus are being recognized as key events in cell biology. The importance of neurotransmitters, cytokines, and growth factors in influencing and directing the development of the nervous system is also increasingly evident. Second messengers provide the link between these signaling molecules and changes in gene expression that are ultimately responsible for the proliferation, differentiation, and maturation of nerve cells. Several studies have evidenced that some cell signalling pathways may be overexpressed or overactive during certain stages of nervous system development, thus enhancing the action of extracellular signals. Furthermore, these pathways are highly interactive and can modulate each other in a complex pattern. The increasing understanding of these signal transduction systems and of their roles in specific ontogenetic stages suggests that certain developmental neurotoxicants may exert their toxicity by interfering with these processes. Building on the knowledge of the toxic manifestations of a developmental neurotoxicant and of the role of certain second messenger pathways, it is thus possible to formulate testable hypotheses on potential mechanisms of neurotoxic damage.

Acknowledgments

Research by the author was supported by grants from NIAAA (08154) and NIEHS (07033) and by the Fondazione S. Maugeri, Pavia, Italy.

References

Balduini, W., Candura, S. M., and Costa, L. G. (1991a). Regional development of carbachol-, glutamate-, norepinephrine-, and serotonin-stimulated phosphoinositide metabolism in rat brain. *Dev. Brain Res.* **62,** 115-120.

Balduini, W., Candura, S. M., Manzo, L., Cattabeni, F., and Costa, L. G. (1991b). Time-, concentration-, and age-dependent inhibition of muscarinic receptor-stimulated phosphoinositide metabolism by ethanol in the developing rat brain. *Neurochem. Res.* **16,** 1235–1240.

Balduini, W., and Costa, L. G. (1989). Effects of ethanol on muscarinic receptor–stimulated phosphoinositide metabolism during brain development. *J. Pharmacol. Exp. Ther.* **250,** 541–547.

Balduini, W., Reno', F., Costa, L. G., and Cattabeni, F. (1994). Developmental neurotoxicity of ethanol: further evidence for an involvement of muscarinic receptor–stimulated phosphoinositide metabolism. *Eur. J. Pharmacol. Mol. Pharmacol. Sect.* **266,** 283–289.

Berridge, M. J., and Irvine, R. F. (1989). Inositol phosphates and cell signalling. *Nature* **341,** 197–205.

Bhat, N. R. (1995). Signal transduction mechanisms in glial cells. *Dev. Neurosci.* **17,** 267–284.

Bhave, S. V., and Hoffman, P. L. (1997). Ethanol promotes apoptosis in cerebellar granule cells by inhibiting the trophic effect of NMDA. *J. Neurochem.* **68,** 578–586.

Billah, M. M., and Anthes, J. C. (1990). The regulation and cellular functions of phosphatidylcholine hydrolysis. *Biochem. J.* **269,** 281–291.

Bonventre, J. V. (1996). Roles of phospholipases A_2 in brain cell and tissue injury associated with ischemia and excitotoxicity. *J. Lipid Med. Cell. Signal.* **14,** 15–23.

Bressler, J. P., and Goldstein, G. W. (1991). Mechanisms of lead neurotoxicity. *Biochem. Pharmacol.* **41,** 479–484.

Castoldi, A. F., Li, B., Manzo, L., and Costa, L. G. (1996). Ethanol inhibits the neuroprotective action of NMDA in cerebellar granule cells. *Alcoh. Clin. Exp. Res.* **20,** (Suppl.), 8A (Abstr.).

Catlin, M. C., and Costa, L. G. (1996). The effect of alcohol on intracellular calcium mobilization in glial cells. *Toxicologist* **16,** 186.

Cohen, R. J., Molina-Holgado, E., and Almazan, G. (1996). Carbachol stimulates *c-fos* expression and proliferation in oligodendrocyte progenitors. *Mol. Brain Res.* **43,** 193–201.

Copani, A., Bruno, V. M. G., Barresi, V., Battaglia, G., Condorelli, D. F., and Nicoletti, F. (1996). Activation of metabotropic glutamate receptors prevents neuronal apoptosis in culture. *J. Neurochem.* **64,** 101–108.

Costa, L. G. (1990). The phosphoinositide/protein kinase C system as a potential target for neurotoxicity. *Pharmacol. Res.* **22,** 393–408.

Costa, L. G. (1994a). Cell signalling and neurotoxic events. In *Principles of Neurotoxicology.* L. W. Chang, Ed., Marcel Dekker, New York, pp. 475–493.

Costa, L. G. (1994b). Second messenger systems in developmental neurotoxicity. In *Developmental Neurotoxicology.* G. J. Harry, Ed., CRC Press, Boca Raton, pp. 77–101.

Costa, L. G. (1997). Role of cell signalling in neurotoxicity. In *Comprehensive Toxicology. Vol. 11. Nervous system and Behavioral Toxicology.* H. E. Lowndes and K. R. Reuhl, Eds., Elsevier, New York, pp. 99–113.

Costa, L. G., Balduini, W., and Reno', F. (1995). Muscarinic receptor stimulation of phospholipase D activity in the developing brain. *Neurosci. Res. Comm.* **17,** 169–176.

Dawson, V. L., and Dawson, T. M. (1996). Nitric oxide actions in neurochemistry. *Neurochem. Int.* **29,** 97–110.

D'Mello, R., Galli, C., Ciotti, T., and Calissano, P. (1993). Induction of apoptosis in cerebellar granule neurons by low potassium: inhibition by insulin-like growth factor-I and cAMP. *Proc. Natl. Acad. Sci. USA* **90,** 10989–10993.

English, D. (1996). Phosphatidic acid: a lipid messenger involved in intracellular and extracellular signalling. *Cell Signal.* **8,** 341–347.

Exton, J. H. (1996). Regulation of phosphoinositide phospholipases by hormones, neurotransmitters and other agonists linked to G proteins. *Annu. Rev. Pharmacol. Toxicol.* **36,** 481–509.

Farmer, S. J., and Guilarte, T. R. (1994). Quantitative autoradiography of ^{3}H-PdBu binding to hippocampal membrane–bound PKC in lead-exposed rats. *Toxicologist* **14,** 143 (Abstr.).

Ferrell, J. E. (1996). MAP kinases in mitogenesis and development. *Curr. Topics Dev. Biol.* **33,** 1–60.

Finbloom, D. S., and Larner, A. C. (1995). Regulation of the JAK/STAT signalling pathway. *Cell Signal.* **7,** 739–745.

Fisher, S. K., Heacock, A. M., and Agranoff, B. W. (1992). Inositol lipid and signal transduction in the nervous system: an update. *J. Neurochem.* **58,** 18–38.

Fox, D. A., Srivastava, D., and Hurwitz, R. L. (1994). Lead-induced alterations in rod-mediated signal functions and cGMP metabolism: new insights. *NeuroToxicology* **15,** 503–512.

Giulian, D., Young, D. G., Woodward, J., Brown, D. C., and Lachman, L. B. (1988). Interleukin-1 is an astroglial growth factor in the developing brain. *J. Neurosci.* **8,** 709–714.

Goldstein, G. W. (1992). Neurologic concepts of lead posoning in children. *Ped. Ann.* **21,** 384–388.

Gomez, N., and Cohen, P. (1991). Dissection of the protein kinase cascade by which nerve growth factor activates MAP kinases. *Nature* **353,** 170–173.

Gonzalez, R., Löffenholz, K., and Klein, J. (1996). Adrenergic activation of phospholipase D in primary rat astrocytes. *Neurosci. Lett.* **219,** 53–56.

Guizzetti, M., and Costa, L. G. (1996). Inhibition of muscarinic receptor–stimulated glial cell proliferation by ethanol. *J. Neurochem.* **67,** 2236–2245.

Guizzetti, M., Costa, P., Peters, J., and Costa, L. G. (1996). Acetylcholine as a mitogen: muscarinic receptor–mediated proliferation of rat astrocytes and human astrocytoma cells. *Eur. J. Pharmacol.* **297,** 265–273.

Hale, A. J., Smith, C. A., Sutherland, L. C., Stoneman, V. E. A., Longthorne, V. L., Culhane, A. C., and Williams, G. T. (1996). Apoptosis: molecular recognition of cell death. *Eur. J. Biochem.* **236,** 1–26.

Harman, A. W., and Maxwell, M. J. (1995). An evaluation of the role of calcium in cell injury. *Annu. Rev. Pharmacol. Toxicol.* **35,** 129–144.

Hashimoto, T., Katsuhiko, A., Sawamura, S., Kikkawa, U., Saito, N., Tanaka, C., and Nishizuka, Y. (1988). Postnatal development of a brain-specific subspecies of protein kinase C in rat. *J. Neurosci.* **8,** 1678–1683.

He, M., and McCarthy, K. D. (1994). Oligodendroglial signal transduction systems are developmentally regulated. *J. Neurochem.* **63,** 501–508.

Henderson, C. E. (1996). Programmed cell death in the developing nervous system. *Neuron* **17,** 579–585.

Itano, Y., Ito, A., Uehara, T., and Nomura, Y. (1996). Regulation of Bcl-2 protein expression in human neuroblastoma SH-SY5Y cells: positive and negative effects of protein kinase C and A, respectively. *J. Neurochem.* **67,** 131–137.

Joseph, S. K. (1996). The inositol triphosphate receptor family. *Cell Signal.* **8,** 1–7.

Kala, S. V., Hawkins, D. H., and Jodhav, A. L. (1996). Low-level subchronic exposure to lead does not affect nitric oxide synthase activity in rat frontal cortex. *Toxic Subst. Mech.* **15,** 333–341.

Kendall-Harden, T. (1989). Muscarinic cholinergic receptors-mediated regulation of cyclic AMP metabolism. In *The Muscarinic Receptors.* J. H. Brown, Ed., Humana Press, Clifton, NJ, pp. 221-258.

Kern, M., and Audesirk, G. (1995). Inorganic lead may inhibit neurite development in cultured rat hippocampal neurons through hyperphosphorylation. *Toxicol. Appl. Pharmacol.* **134,** 111–123.

Keshles, O., and Levitzki, A. (1984). The ontogeny of beta-adrenergic receptors and of adenylate cyclase in the developing rat brain. *Biochem. Pharmacol.* **33,** 3231–3233.

Khan, W. A., Blobe, G. C., and Hannun, Y. A. (1995). Arachidonic acid and free fatty acids as second messengers and the role of protein kinase C. *Cell Signal.* **3,** 171–184.

Krebs, E. G. (1989). Role of cyclic AMP–dependent protein kinase in signal transduction. *JAMA* **262,** 1815–1818.

Kubo, T., Nonomura, T., Enokido, Y., and Hatanaka, H. (1995). Brain-derived neurotrophic factor (BDNF) can prevent apoptosis of rat cerebellar granule neurons in culture. *Dev. Brain Res.* **85,** 249–258.

Kurino, M., Fukunaka, K., Ushio, Y., and Miyamoto, E. (1996). Cyclic AMP inhibits activation of mitogen-activated protein kinase and cell proliferation in response to growth factors in cultured rat cortical astrocytes. *J. Neurochem.* **67,** 2246–2255.

Laterra, J., Bressler, J. P., Indurti, R. R., Belloni-Ulivi, L., and Goldstein, G. W. (1992). Inhibition of astroglia-induced endothelial differentiation by inorganic lead: a role for protein kinase C. *Proc. Natl. Acad. Sci. USA* **89,** 10748–10752.

Lazarini, F., Strosberg, A. D., Courand, P. O., and Cazaubon, S. M. (1996). Coupling of ET_B endothelin receptor to mitogen-activated protein kinase stimulation and DNA synthesis in primary cultures of rat astrocytes. *J. Neurochem.* **66,** 459–465.

Lindholm, D., Castren, E., Kiefer R., Zefra, F., and Thoenen, H. (1992). Transforming growth factor-$\beta 1$ in the rat brain: increase after injury and inhibition of astrocyte proliferation. *J. Cell Biol.* **117,** 395–400.

Liscovitch, M. (1996). Phospholipase D: role in signal transduction and membrane traffic. *J. Lipid Med. Cell Signal.* **14,** 215–221.

Liu, H. N., and Almazan, G. (1995). Glutamate induces *c-fos* proto-oncogene expression and inhibits proliferation in oligodendrocyte progenitors: receptor characterization. *Eur. J. Neurosci.* **7,** 2355–2363.

Liu, J.-P. (1996). Protein kinase C and its substrates. *Mol. Cell Endocrinol.* **116,** 1–29.

Long, G. J., Rosen, J. F., and Schanne, F. A. X. (1994). Lead activation of protein kinase C from rat brain. Determination of free calcium, lead and zinc by ^{19}F NMR. *J. Biol. Chem.* **269,** 834–837.

Ma, F. H., Okhuma, S., Kishi, M., and Kuriyama, K. (1991). Ontogeny of beta-adrenergic receptor-mediated cyclic AMP generating system in primary cultured neurons. *Int. J. Dev. Neurosci.* **9,** 347–356.

Malarkey, K., Belham, C. M., Paul, A., Graham, A., McLees, A., Scott, P. H., and Plevin, R. (1995). The regulation of tyrosine kinase signalling pathways by growth factor and G-protein–coupled receptors. *Biochem. J.* **309,** 361–375.

Markovac, J., and Goldstein, G. W. (1988a). Lead activates protein kinase C in immature rat microvessels. *Toxicol. Appl. Pharmacol.* **96,** 14–23.

Markovac, J., and Goldstein, G. W. (1988b). Picomolar concentrations of lead stimulate brain protein kinase C. *Nature* **334,** 71–73.

Martin, T. F. (1991). Receptor regulation of phosphoinositidase C. *Pharmacol. Ther.* **49,** 329–345.

Morgan, J. I., and Curran, T. (1991). Stimulus-transcription coupling in the nervous system: involvement of the inducible proto-oncogenes *fos* and *jun*. *Annu. Rev. Neurosci.* **14,** 421–451.

Mukherjee, A. B, Miele, L., and Pattabiraman, N. (1994). Phospholipase A_2 enzymes: regulation and physiological role. *Biochem. Pharmacol.* **48,** 1–10.

Murakami, K., Feng, G., and Chen, S. G. (1993). Inhibition of brain protein kinase C subtypes by lead. *J. Pharmacol. Exp. Ther.* **264,** 757–761.

Naarala, J. T., Loikkanen, J. J., Ruotsalainen, M. H., and Savolainen, K. M. (1995). Lead amplifies glutamate-induced oxidative stress. *Free Rad. Biol. Med.* **19,** 689–693.

Negre-Aminou, P., and Pfenninger, K. H. (1993). Arachidonic acid turnover and phospholipase A_2 activity in neuronal growth cones. *J. Neurochem.* **60,** 1122–1136.

Newton, A. C. (1995). Protein kinase C: structure, function and regulation. *J. Biol. Chem.* **270,** 28495–28498.

Nicoletti, F., Magri, G., Ingras, F., Bruno, V., Catania, M. V., Dell'Albani, P., Condorelli, D. F., and Avola, R. (1990). Excitatory aminoacids stimulate inositol phospholipid hydrolysis and reduce proliferation in cultured astrocytes. *J. Neurochem.* **54,** 771–777.

Nicotera, P. Bellomo, G., and Orrenius, S. (1992). Calcium-mediated mechanisms in chemically-induced cell death. *Annu. Rev. Pharmacol. Toxicol.* **32,** 449–470.

Nishizuka, Y. (1995). Protein kinase C and lipid signalling for sustained cellular responses. *FASEB J.* **9,** 484–496.

Oberto, A., Marks, N., Evans, H. L., and Guidotti, A. (1996). Lead (Pb^{++}) promotes apoptosis in newborn rat cerebellar neurons: pathological implications. *J. Pharmacol. Exp. Ther.* **279,** 435–442.

Paakkari, I., and Lindsberg, P. (1995). Nitric oxide in the central nervous system. *Ann. Med.* **27,** 369–377.

Palmer, E., Nangel-Taylor, K., Krause, J. D., Roxas, A., and Cotman, C. W. (1990). Changes in excitatory amino acid modulation of phosphoinositide metabolism during development. *Dev. Brain Res.* **51,** 132–134.

Pantazis, N. J., Dohrman, D. P., Goodlett, C. R., Cook R. T., and West, J. R. (1993). Vulnerability of cerebellar granule cells to alcohol-induced cell death diminishes with time in culture. *Alcohol Clin. Exp. Res.* **17,** 1014–1021.

Perraud, F., Besnard, F., Sensenbrenner, M., and Labourdette, G. (1987). Thrombin is a potent mitogen for rat astroblasts but not for oligodendroblasts and neuroblasts in primary culture. *Int. J. Dev. Neurosci.* **5,** 181–188.

Pierce, D. R., Goodlett, C. R., and West, J. R. (1989). Differential neuronal loss following postnatal alcohol exposure. *Teratology* **40,** 113–126.

Piomelli, D. (1993). Arachidonic acid in cell signalling. *Curr. Op. Cell Biol.* **5,** 274–280.

Post, G. R., and Brown, J. H. (1996). G Protein-coupled receptors and signalling pathways regulating growth responses. *FASEB J.* **10,** 741–749.

Quinn, M. R., and Harris, C. L. (1995). Lead inhibits Ca^{2+}-stimulated nitric oxide synthase activity from rat cerebellum. *Neurosci. Lett.* **196,** 65–68.

Rasmussen, H., Barrett, P., Smallwood, J., Bollag, W., and Isales, C. (1990). Calcium ion as intracellular messenger and cellular toxin. *Env. Health Persp.* **84,** 17–25.

Rathbone, M. P., Middlemiss, P. J., Kim, J. K., Gysbers, J. W., DeForge, S. P., Smith, R. W., and Hughes, D. W. (1992). Adenosine and its nucleotides stimulate proliferation of chick astrocytes and human astrocytoma cells. *Neurosci. Res.* **13,** 1–17.

Reno', F., Tan, X. X., Balduini, W., and Costa, L. G. (1994). Administration of ethanol during the rat's brain growth spurt causes dose-dependent microencephaly and inhibition of muscarinic receptor–stimulated phosphoinositide metabolism. *Res. Comm. Alcoh. Subst. Abuse* **15,** 141–150.

Resnicoff, M., Rubini, M., Baserga, R., and Rubin, R. (1994). Ethanol inhibits insulin-like growth factor-1–mediated signalling and proliferation of C6 rat glioblastoma cells. *Lab. Invest.* **71,** 657–662.

Richardson, A., and Parsons, J. T. (1995). Signal transduction through integrins: a central role for focal adhesion kinase? *Bioessays* **17,** 229–236.

Robison, G. A., Butcher, R. W., and Sutherland, E. W. (1971). *Cyclic AMP.* Academic Press, New York, p. 531.

Rodier, P. M. (1990). Developmental neurotoxicology. *Toxicol. Pathol.* **18,** 89–95.

Rubbo, H., Darley-Usmar, V., and Freeman, B. A. (1996). Nitric oxide regulation of tissue free radical injury. *Chem. Res. Toxicol.* **9,** 809–820.

Schindler, C., and Darnell, J. E. (1995). Transcriptional responses to polypeptide ligands: the JAK-STAT pathway. *Annu. Rev. Biochem.* **64,** 621–651.

Shafit-Zagardo, B., Kume-Iwaki, A., and Goldman, J. E. (1988). Astrocytes regulate GFAP mRNA levels by cyclic AMP and protein kinase C–dependent mechanisms. *Glia* **1,** 346–354.

Shearman, M. S., Shinomura, T., Oda, T., and Nishizuka, Y. (1991). Synaptosomal protein kinase C subspecies: A. Dynamic changes in the hippocampus and cerebellar cortex concomitant with synaptogenesis. *J. Neurochem.* **56,** 1255–1262.

Shukla, S. D., and Halenda, S. P. (1991). Phospholipase D in cell signalling and its relationship to phospholipase C. *Life Sci.* **48,** 851–866.

Song, X., Seidler, F. J., Slotkin, T. A. (1997). Cellular mechanisms for developmental toxicity of chlorpyrifos: targeting the adenylyl cyclase signalling cascade. *Toxicologist* **36,** 100 (Abstr.).

Stanimirovic, D. B., Bell, R., Mealing, G., Morley, P., and Durkin, J. P. (1995). The role of intracellular calcium and protein kinase C in endothelin-stimulated proliferation of rat type I astrocytes. *Glia* **15,** 119–130.

Streissguth, A. P., Aase, J. M., Clarren, S. K., Randels, S. P., LaDue, R. A., and Smith, D. F. (1991). Fetal alcohol syndrome in adolescents and adults. *J. Am. Med. Assoc.* **265,** 1961–1967.

Streissguth, A. P., Landesman-Dwyer, S., Martin, J. C., and Smith, D. F. (1980). Teratogenic effects of alcohol in humans and laboratory animals. *Science* **209,** 353–361.

Tandon, P., Pope, C., Padilla, S., Tilson, H. A., and Harry, G. J. (1993). Developmental changes in carbachol-stimulated inositolphosphate release in pigmented rat retina. *Curr. Eye Res.* **12,** 439–449.

Thanos, D., and Maniatis, T. (1995). NF-κB: a lesson in family values. *Cell* **80,** 529–532.

Tranque, P. A., Calle, R., Naftolin, F., and Robbins, R. (1992). Involvement of protein kinase C in the mitogenic effect of insulin-like-growth–factor-I on rat astrocytes. *Endocrinology* **131,** 1948–1954.

van Biesen, T., Hawes, B. E., Raymond, J. R., Luttrell, L. M., Koch, W. J., and Lefkowitz, R. J. (1996). G_o-protein α-subunits activate mitogen-activated protein kinase via a novel protein kinase C–dependent mechanism. *J. Biol. Chem.* **271,** 1266–1269.

Varner, P. D., and Beckman, J. S. (1995). Nitric oxide in neuronal injury and degeneration. In *Nitric Oxide in the Nervous System.* Academic Press, San Diego, pp. 191–206.

Vartanian, T., Szuchet, S., Dawson, G., and Campagnoni, A. T. (1986). Oligodendrocyte adhesion activates protein kinase C–mediated phosphorylation of myelin basic protein. *Science* **234,** 1396–1398.

Walaas, S. I., and Grengaard, P. (1991). Protein phosphorylation and neuronal function. *Pharmacol. Rev.* **43,** 299–349.

Wang, X., and Robinson, P. J. (1997). Cyclic GMP–dependent protein kinase and cellular signalling in the nervous system. *J. Neurochem.* **68,** 443-456.

Yan, G. M., Lin, S. Z., Irwin, R. P., and Paul, S. M. (1995). Activation of muscarinic cholinergic receptors blocks apoptosis of cultured cerebellar granule neurons. *Mol. Pharmacol.* **47,** 248–257.

Yan, G. M., Ni, B., Weller, M., Wood, K. A., and Paul, S. M. (1994). Depolarization or glutamate receptor activation blocks apoptotic cell death of cultured cerebellar granule neurons. *Brain Res.* **656,** 43–51.

Yong, V. W., Sekiguchi, S., Kim, M. W., and Kim, S. U. (1988). Phorbol ester enhances morphological differentiation of oligodendrocytes in culture. *J. Neurosci. Res.* **19,** 187–194.

Zhang, J., and Snyder, S. H. (1995). Nitric oxide in the nervous system. *Annu. Rev. Pharmacol. Toxicol.* **35,** 213–233.

CHAPTER

15

The N-*Methyl*-D-*Aspartate Receptor*

Physiology and Neurotoxicology in the Developing Brain

TOMÁS R. GUILARTE
Department of Environmental Health Sciences
Johns Hopkins University
School of Hygiene and Public Health
Baltimore, Maryland 21205

I. Introduction

Glutamate and aspartate are the major excitatory amino acid neurotransmitters in the vertebrate central nervous system (CNS). Their physiologic actions are mediated by an interaction with receptor proteins, which are classified into two broad categories: (1) ionotropic and (2) metabotropic receptors. Ionotropic receptors are ligand-gated ion channels that allow the passage of cations such as sodium, potassium, and calcium. Metabotropic receptors are G-protein linked and modulate second messenger systems such as the activation of inositol phospholipid hydrolysis and inhibition of cyclic adenosine monophosphate (AMP) formation. The ionotropic excitatory amino acid receptors have been classified into three families based on their selective activation by the exogenous agonists *N*-methyl-D-aspartate (NMDA); AMPA (alpha-amino-3-hydroxy-5-methyl-4-isoxazolepropionic acid, also known as quisqualate receptors and kainate receptors. The AMPA and kainate receptors are referred to as *non-NMDA* receptors. This

*Abbreviations: AMPA, alpha-amino-3-hydroxy-5-methyl-4-isoxazoleproopionic acid; Bmax, maximal number of binding sites; CNS, central nervous system; DEPC, diethylpyrocarbonate; DTNB, 5-5-dithio-*bis*-nitrobenzoic acid; DTT, dithiothreitol; EPSP, excitatory postsynaptic potentials; GABA, γ-aminobutyric acid; Kd, affinity; LTP, long-term potentiation; MK-801, {(+)−5-methyl-10,11-dihydro-5*H*-dibenzo[a,d]cyclohepten-5,10-amine maleate}; NMDA, *N*-methyl-D-aspartate; NMDAR2, *N*-methyl-D-aspartate receptor 2; NMDAR1, *N*-methyl-D-aspartate receptor 1; NR1, NMDAR1; NR2, NMDAR2; Pb, lead; PbAc, lead acetate; PCP, phencyclidine; PKC, protein kinase C; PMA, phorbol 12-myristate 13-acetate; PN, postnatal [day]; ppm, parts per million; PS, population spikes.

is partly because the NMDA receptor has been the most widely studied and a great deal of knowledge has been gained on its role in physiologic and pathologic conditions. For each of these receptors there are a number of agonist and antagonist ligands that have been extremely valuable in understanding their physiology and pharmacology. Historically, the classification of excitatory amino acid receptors into these broad categories has been based on their pharmacology. However, since the advent of cloning technology and the isolation of the genes encoding these receptors, it is now well recognized that there is a great deal of heterogeneity within a receptor family. This molecular heterogeneity results in receptor subtypes with distinct pharmacology. There is a vast amount of information for each of the families of excitatory amino acid receptors and their role in the developing brain. This chapter is focused on the NMDA receptor complex and its role in the developmental neurotoxicity of the heavy metal, lead (Pb).

II. The NMDA Receptor Complex

A. *Molecular Biology*

The genes encoding the NMDA receptor have been cloned and two subunit families, namely the NMDAR1 (NR1) and NMDAR2 (NR2), have been characterized in both the rat (Moriyoshi *et al.,* 1991; Monyer *et al.,* 1992) and mouse (Kutsuwada *et al.,* 1992; Yamakazi *et al.,* 1992) brains. Currently, there is one known member of the NR1 gene family that encodes a single polypeptide. Expression of NR1 cDNA in *Xenopus* oocytes and electrophysiologic analysis indicate that the NR1 subunit can form homomeric receptors that express calcium permeability, voltage-dependent block by magnesium, glycine enhancement, and an agonist and antagonist pharmacologic profile similar to native receptors (Moriyoshi *et al.,* 1991; Nakanishi *et al.,* 1992). However, the amplitude of the currents associated with NR1 homomeric receptors is significantly smaller than that obtained with native receptors.

The NR2 subunit family consists of four closely related genes, the NR2A-D (Ikeda *et al.,* 1992; Kutsuwada *et al.,* 1992; Meguro *et al.,* 1992; Nakanishi *et al.,* 1992; Ishii *et al.,* 1993). The expression of each individual NR2 subunit in oocytes does not form functional receptors, but coexpression of NR1 with one or more of the NR2 subunits greatly enhances the amplitude of NMDA responses relative to NR1 homomeric receptors. NR1/NR2 heteromeric receptors have electrophysiologic and pharmacologic characteristics that more closely resemble native receptors (Ikeda *et al.,* 1992; Meguro *et al.,* 1992; Ishii *et al.,* 1993). Based on these findings, the NR1 but not the NR2 subunit appears to be necessary for the expression of functional receptors. The physiologic significance of the NR1 subunit has also been demonstrated in gene targeting studies in which disruption of the NR1 gene abolishes NMDA receptor currents and is lethal in neonatal rats (Forrest *et al.,* 1994). Native NMDA receptors are likely to be assemblies of NR1 with one or more NR2 subunits. The stoichiometry of native receptors is not presently known, but they may be pentameric structures similar to other ligand-gated ion channels such as the nicotinic acetylcholine receptor.

Molecular isoforms of the NR1 subunit have been described based on alternative RNA splicing (Nakanishi *et al.,* 1992). Differential splicing of three exons produces eight different variants. The eight variants are the result of alternative splicing of one cassette in the N-terminal region (presence or absence of 5′ insertion), and the individual or combined deletion of two cassettes in the C-terminal region (Sugihara *et al.,* 1992; Hollmann *et al.,* 1993; Zukin and Bennett, 1995). Hollmann *et al.* (1993) have designated the eight splice variants as NR1-1a-NR1-4a and NR1-1b-NR1-4b. Other approaches in the designation of the splice variants have been described (Zukin and Bennett, 1995). NR1 splice variants with the "a" and "b" designations refer to the absence or presence, respectively, of the N-terminal insertion cassette that is encoded by exon 5. The number designation indicates the individual or combined deletion of the two C-terminal cassettes encoded by exons 21 and 22. The splice variant with the number 1 is no deletion, 2 is deletion of cassette 1, 3 is deletion of cassette 2, 4 is deletion of both cassettes. Splice variant NR1-1a refers to the absence of the N-terminal cassette insertion and no deletion of the C-terminal cassettes. The splice variants of the NR1 subunit (Nakanishi *et al.,* 1992; Sugihara *et al.,* 1992; Hollmann *et al.,* 1993) as well as the NR2 subunits impart unique physiologic and pharmacologic properties to the NMDA receptor complex formed (Williams, 1993; McBain and Mayer, 1994; Kohr *et al.,* 1994; Laurie and Seeberg, 1994a). As a result, a diversity of NMDA receptor assemblies are possible based on the subunit combination and stoichiometry.

1. Developmental and Regional Brain Expression of NMDA Receptor Subunits

The use of subunit specific oligonucleotide probes and *in situ* hybridization techniques have advanced the understanding of the expression of NMDA receptor subunits in the brain. Subunit mRNA expression is dependent on the cell type, brain region, and developmental age of the animal. The NR1 subunit is ubiquitously distributed in the brain, whereas NR2 subunits have distinct regional specificity (Monyer *et al.,* 1992,

Monyer *et al.,* 1994; Zhong *et al.,* 1995; Portera-Cailliau *et al.,* 1996) (Fig. 1).

A detailed analysis of the regional and developmental expression of NR1 subunit splice variants has been described (Laurie and Seeberg, 1994b; Laurie *et al.,* 1995). The oligonucleotide probes used for *in situ* hybridization of NR1 splice variants are targeted against specific sequences of the original NR1 clone at the splice sites. Each probe is specific for its own splice variant and does not distinguish between other splice sites. For example, the NR1-a probe recognizes the 5′ splice region of NR1-1a through -4a. The other probes NR1-1, NR1-2, NR1-3, and NR1-4 recognize the full length mRNA (NR1-1) and deletions 1 (NR1-2) or 2 (NR1-3) or both (NR1-4), but do not distinguish between the presence or absence of the N-terminal cassette.

In general, it appears that there is widespread and diverse distribution of all spice variants with the exception of the NR1-3, which has low expression in most brain structures. The NR1-a and NR1-2 have overlapping distribution and little or no expression of the NR1-b and NR1-4 is present in some brain structures such as the corpus striatum (Laurie and Seeberg, 1994b; Laurie *et al.,* 1995). Most splice variants are prominently expressed in the pyramidal and granule cell layers of the hippocampus. Developmentally in rodents, most splice variants are expressed at low levels at birth and increase postnatally to reach maximal levels at approximately 3 weeks of age.

The expression of NR2 subunit mRNA is more regionally defined than the NR1 subunit mRNA (Fig. 1). In the adult brain, NR2A and NR2B have widespread and somewhat similar distribution although the level of expression may differ in some brain structures. There is high expression of these subunits in the pyramidal and granule cell layers of the hippocampus, and moder-

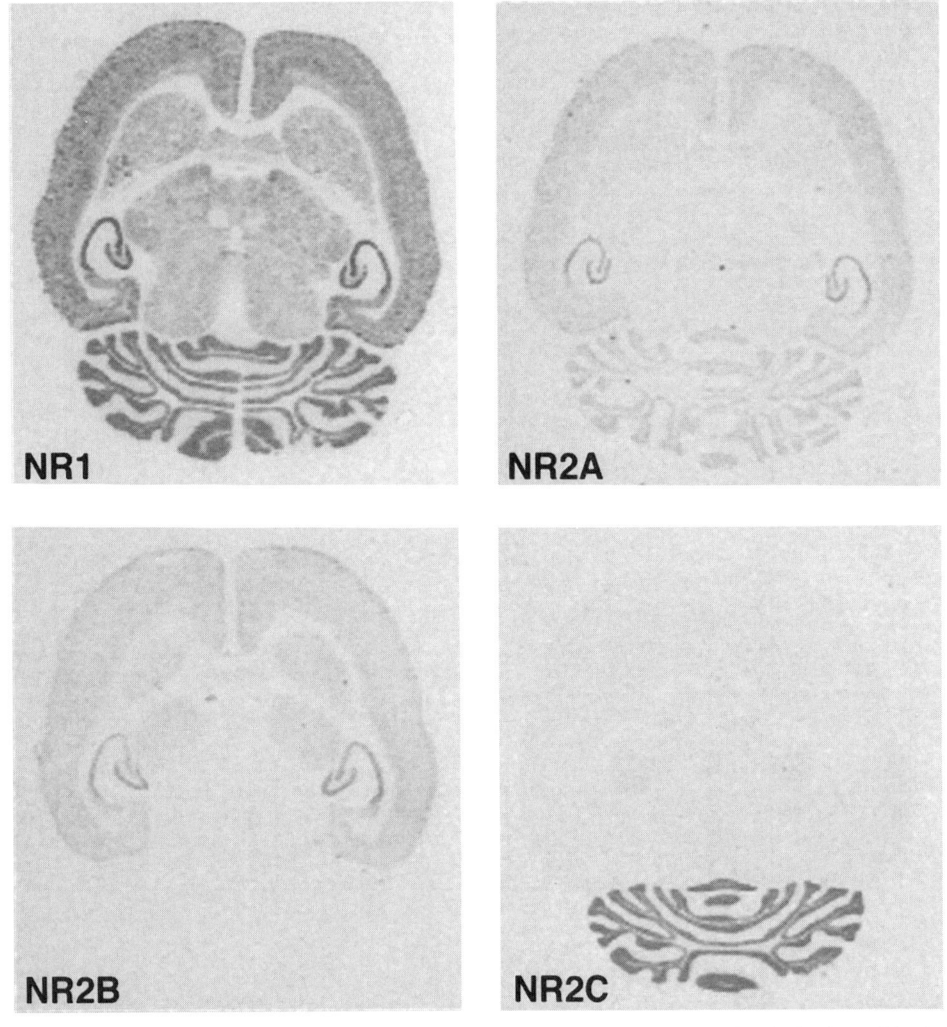

FIGURE 1 *In situ* hybridization of NMDA receptor subunit mRNA expression in horizontal brain sections from a 28-day-old rat.

ate levels in the cerebral cortex and corpus striatum. The most prominent difference between NR2A and NR2B transcripts is in the adult cerebellum, where NR2A mRNA is prominently expressed while NR2B is lacking (Fig. 1). Transcripts encoding for the NR2C and NR2D subunits have a more restricted distribution. NR2C transcripts are prominently and almost exclusively detected in the adult cerebellum, primarily localized in cerebellar granule cells (Monyer *et al.,* 1992). NR2D is predominantly found in brain structures such as thalamic nuclei, anteroventral nucleus, medial geniculate, and periaquaductal grey (Wenzel *et al.,* 1996).

Developmentally, there are significant changes in the expression of individual NR2 subunit mRNA in different brain structure. *In situ* hybridization studies have shown that in the cerebellum, NR2 subunit mNRAs undergo an age-dependent change in their expression. In the first 2 weeks after birth, the predominant mRNA is for the NR2B subunit, presumably forming NR1/NR2B receptors. However, the expression of the NR2B transcript disappears in the cerebellum by 28 days of age (Akazawa *et al.,* 1994; Monyer *et al.,* 1994; Riva *et al.,* 1994; Zhong *et al.,* 1995). Conversely, in the first 2 weeks after birth the expression of mRNAs for the NR2A and NR2C subunits is very weak or undetectable and their expression increases during the next 2 weeks to reach adult levels by 21–28 days of age (Akazawa *et al.,* 1994; Monyer *et al.,* 1994; Riva *et al.,* 1994; Zhong *et al.,* 1995). This change in NMDA receptor mRNA expression has been shown to be correlated with functional changes in receptor channel kinetics (Farrant *et al.,* 1994). The hippocampus, the brain region with the highest level of NMDA receptors, has the highest level of NR1, NR2A, and NR2B subunit mRNA. The mRNA for the NR2B subunit is highest at birth, and it remains fairly constant with a small decline in postnatal levels. However, the mRNA for the NR1 and NR2A subunits is low at birth and increase as a function of age to reach maximal levels at approximately 3–4 weeks of age (Fig. 2). The expression of NR2D transcripts is most prominent at birth and decreases postnatally (Wenzel *et al.,* 1996).

Based on the emerging evidence of the relative abundance, regional and anatomic distribution of NR1 splice variants, and NR2 transcripts, there is the potential for a wide variety of NMDA receptors with distinct pharmacologic and physiologic properties that may greatly influence the expression of synaptic plasticity and brain vulnerability to neurotoxic insult during development and in adults.

B. Regulatory Sites

The NMDA receptor is an oligomeric complex that forms a functional cation channel with high permeability for calcium ions (Fig. 3) (MacDermott *et al.,* 1986;

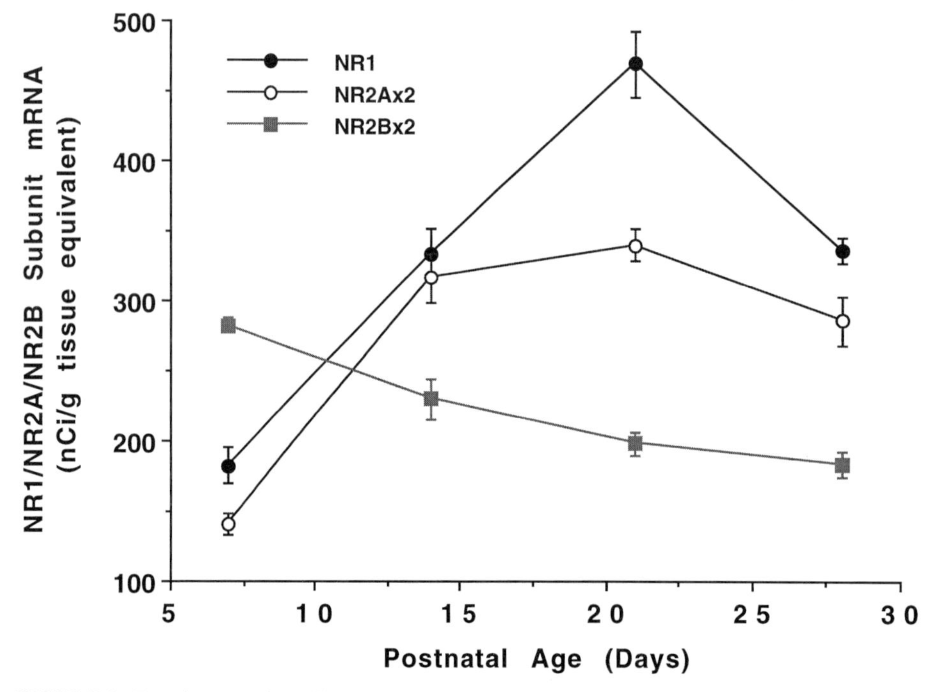

FIGURE 2 Developmental profile of NR1, NR2A, and NR2B mRNA in the CA1 subfield of the rat hippocampus. NR2A and NR2B values are twice their original in order to fit in the graph. Each value is the mean ± sem of 4–5 determinations.

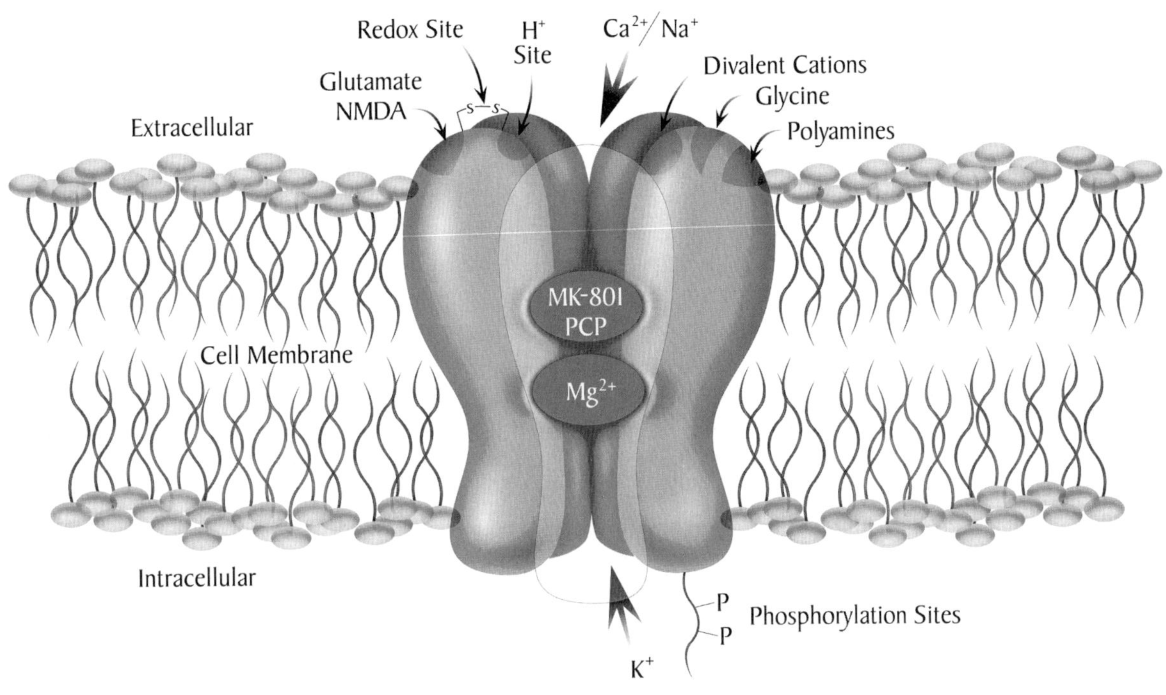

FIGURE 3 Representation of the NMDA receptor complex and modulatory sites.

Mayer and Westbrook, 1987). This receptor complex is allosterically modulated by a number of endogenous ligands, including the coagonists glutamate and glycine, divalent cations, polyamines, as well as having proton sensitive, redox, and phosphorylation sites (see McBain and Mayer, 1994, and Collingridge and Lester, 1989 for reviews).

1. Agonist Binding Sites

The amino acids glutamate and aspartate are thought to be the endogenous ligands at the NMDA recognition site. Other endogenous ligands have also been shown to have agonist properties at the glutamate binding site. These include L-cysteine sulphinate, L-homocysteate, and quinolinic acid. The NMDA receptor has a recognition site for glycine that enhances NMDA-associated currents. The first recognition that glycine plays a role in excitatory neurotransmission in addition to its well-documented role in inhibitory neurotransmission was from glycine receptor autoradiographic studies of the brain. Glycine receptors that mediate inhibitory neurotransmission and are labeled with the glycine receptor antagonist [^{3}H]-strychnine had a different regional brain distribution than those labeled with [^{3}H]-glycine (Bristow *et al.,* 1986; Probst *et al.,* 1986). [^{3}H]-strychnine labels glycine receptors in the lower brain axis (hind brain, brainstem, and spinal cord) with little or no labeling present in forebrain structures. However, glycine receptors labeled with [^{3}H]-glycine are widely distributed in the brain (Bristow *et al.,* 1986). The basis for the difference in brain labeling patterns was not apparent until the work of Johnson and Ascher (1987), which showed that NMDA receptor currents were enhanced by glycine. The distribution of glycine receptors in forebrain structures visualized using [^{3}H]-glycine receptor autoradiography was consistant with the distribution of NMDA receptors (Cotman *et al.,* 1987) but not of other excitatory amino acid receptors in the brain.

The glycine site associated with NMDA receptors is not sensitive to inhibition by strychnine and is referred to as a strychnine-insensitive glycine site. Glycine is not just a modulator of the NMDA receptor but the receptor has an absolute requirement for glycine (Kleckner and Dingledine, 1988; Dalkara *et al.,* 1992). A number of antagonists that interact at the glutamate and glycine binding sites of the NMDA receptor have been described.

2. Modulatory Domains Within the NMDA Receptor

The NMDA receptor channel has binding sites for allosteric modulators such as magnesium (described later under the divalent cation section). There is also a class of drugs that are classified as NMDA receptor open channel blockers that bind to and inhibit channel conductance. The most prominent among these are the dissociative anesthetic phencyclidine (PCP) and the anticonvulsant MK-801 {(+)-5-methyl-10,11-dihydro-5*H*-dibenzo[a,d]cyclohepten-5,10-imine maleate}, also called dizocilpine. These drugs are noncompetitive antagonists of the NMDA receptor because they bind at a different site than the glutamate or NMDA recognition site. They are use-dependent channel blockers because the NMDA receptor channel must be in the open state in order for the drug to enter the channel and bind to inhibit channel function (Wong *et al.,* 1986; Huettner and Bean, 1988; Javitt and Zukin, 1989; Rothman *et al.,* 1989). This unique property has been useful in developing a biochemical assay to measure the functional state of the receptor. The tritiated form of MK-801 has been used extensively to determine channel function and modulation in radioligand binding assays. MK-801 has a higher affinity and selectivity for NMDA receptors than PCP because the latter is also known to bind with high affinity to sigma opioid receptors (Wong *et al.,* 1986; Rothman *et al.,* 1989). [^{3}H]-MK-801 binding assays have been used in a wide variety of studies to determine NMDA receptor channel function, modulation, regional brain distribution, and pharmacology.

3. Divalent Cations

Divalent cations play an important role in modulating NMDA receptor channel function. Magnesium and zinc were originally demonstrated to inhibit NMDA receptor currents by binding at distinctly different sites on the receptor complex. Under normal physiologic conditions, the NMDA receptor channel is under a voltage-dependent magnesium block at a site located within the pore (Ault *et al.,* 1980; Nowak *et al.,* 1984). The NMDA receptor is unique among ligand-gated ion channels in that it acts as a coincident detector. In order for channel opening to occur, it requires both the binding of the coagonists glutamate and glycine in addition to a depolarizing pulse. If these conditions are present in a temporally defined fashion, there is a release of the voltage-dependent magnesium block of the channel allowing the passage of sodium and calcium into the cell.

A second magnesium binding site has been described by Wang and McDonald (1995) that distinctly differs in function and physical location from the magnesium binding site within the channel. While studying the effects of magnesium on NMDA receptor currents in cultured fetal mouse and acutely isolated neonatal rat hippocampal neurons, these investigators found that magnesium enhanced NMDA-activated currents at positive membrane potentials (Wang and McDonald, 1995). This effect of magnesium was different from the well-characterized voltage-dependent block of the channel pore at negative membrane potentials. Further investigations showed that the magnesium binding site mediating the enhancement of NMDA currents was not dependent on membrane potential and was physically located outside of the pore. It was also found that the mechanism by which magnesium enhances NMDA receptor currents is by increasing the affinity of the NMDA receptor for glycine (Wang and McDonald, 1995). This only occurs when the glycine site is not saturated. The magnesium-induced increase in the affinity of the strychnine-insensitive glycine site is due to a decrease in the apparent dissociation constant for glycine (Wang and McDonald, 1995).

Zinc is another divalent cation that was shown in early studies to be a potent inhibitor of NMDA receptor currents (Peters *et al.,* 1987; Mayer *et al.,* 1989; Legendre and Westbrook, 1990). The nature of the inhibitory effects of zinc was to reduce the frequency of channel opening and its effect was voltage-independent suggestive of a site outside of the channel pore (Mayer *et al.,* 1989; Legendre and Westbrook, 1990). Electrophysiologic studies by Christine and Choi (1990) subsequently showed that there were two putative zinc binding sites. Zinc, at concentrations of 1–10 μM, produced a reduction in channel opening that was voltage-independent indicative of a zinc binding site outside of the membrane field. At higher concentrations (10–100 μM), however, zinc produced an apparent reduction in single channel amplitude that was voltage dependent. Amplitude distribution analysis indicated that this site was within the channel pore and it may interfere directly with the passage of ions through the channel. Radioligand binding studies have also described two zinc binding sites (Guilarte *et al.,* 1995). One of these sites may be modulating the strychnine-insensitive glycine site as zinc has been shown to noncompetitively inhibit glycine binding (Yeh *et al.,* 1990).

Besides the well-recognized inhibitory effects of zinc on NMDA receptor currents, studies using NMDA receptor subunit mRNA expression in oocytes have shown that zinc at submicromolar concentrations can stimulate the current of homomeric NMDA receptors formed from certain splice variants of the NR1 subunit (Hollmann *et al.,* 1993). The potentiating effect of zinc was only present in receptor expressing the NR1-1a (absence of 5′ insertion cassette) type of splice variant but not the NR1-1b type. NR1/NR2 heteromeric NMDA receptors are not potentiated but are inhibited by zinc. The physiologic significance of the enhancing effect of zinc at homomeric NMDA receptors is not presently known.

Calcium is another important divalent cation with multiple sites of interaction that regulate the activity of the NMDA receptor complex. There is an extracellular calcium binding site at which low millimolar extracellular calcium concentrations enhances NMDA responses (Gu and Huang, 1994). Like magnesium, the effect of calcium only occurs at subsaturating glycine concentrations and is the result of a calcium-mediated increase in the affinity of the receptor for glycine. This extracellular calcium binding site is different from the site responsible for the decrease in NMDA receptor channel conductance known to occur at high extracellular calcium concentrations (Gu and Huang, 1994). These findings have been confirmed by other investigators (Paoletti *et al.,* 1995; Premkumar and Auerbach, 1996). Consistent with the electrophysiologic studies (Gu and Huang, 1994; Paoletti *et al.,* 1995; Premkumar and Auerbach, 1996) indicating an enhancing effect of low concentrations of extracellular calcium on NMDA receptor currents, radioligand binding studies have also shown an enhancing effect of calcium on [^{3}H]-MK-801 and [^{3}H]-glycine binding to the NMDA receptor channel and the strychnine-insensitive glycine site, respectively (Marvizon and Skolnick, 1988; Rajdev and Reynolds, 1992; Enomoto *et al.,* 1992). These radioligand binding studies, however, did not provide a mechanism by which calcium enhances ligand binding. The effects of low and high extracellular calcium concentrations on the NMDA receptor complex appear to be voltage-independent, placing both of these sites outside the electrostatic field of the pore (Sharma and Stevens, 1996). Premkumar and Auerbach (1996) have reported that NMDA receptors have an external divalent cation binding site that is near the mouth of the pore. When calcium is bound to this site, the sodium conductance is reduced but not abolished, perhaps by an electrostatic mechanism. This external binding site is an important feature of the calcium selectivity of the receptor because it determines the fraction of current that is carried by calcium (Premkumar and Auerbach, 1996).

Calcium also modulates NMDA receptor function through intracellular mechanisms (Legendre *et al.,* 1993). It is known that activation of the NMDA receptor leads to an increase in intracellular calcium levels with a subsequent inactivation of receptor activity through feedback inhibition via an intracellular calcium-sensitive protein or process. Evidence indicates that the calcium-dependent inactivation of NMDA receptors may be subunit specific. Receptor assemblies containing the NR2A and NR2D subunits express inactivation, whereas those with the NR2B and NR2C transcript do not (Krupp *et al.,* 1996). Ehlers *et al.* (1966) have suggested that calcium influx from NMDA receptor activation binds the regulatory protein calmodulin that binds directly the carboxy terminus of the NR1 subunit to inactivate the receptor. Another protein, calcineurin, a calcium/calmodulin–dependent phosphatase, also inactivates NMDA receptors (Liberman and Mody, 1994; Tong *et al.,* 1995). Thus, it is apparent that there are multiple extracellular and intracellular calcium binding domains that can modulate the functional state of the NMDA receptor.

4. Polyamines

The effects of polyamines on the NMDA receptor complex were first identified by Ransom and Stec (1988). Williams *et al.* (1989) in a search for endogenous modulators of the NMDA receptor isolated an active fraction from bovine brain that enhanced [^{3}H]-MK-801 binding to the NMDA receptor channel. This active fraction was identified to contain spermine and spermidine. Using radioligand binding techniques these investigators were able to show that polyamines enhanced the binding of [^{3}H]-MK-801 to neuronal membranes above the level of binding measured in the presence of saturating concentrations of glutamate and glycine. This finding suggested that the polyamines were interacting at a distinctly different site than the glutamate and glycine sites.

The effects of polyamines described in the radioligand binding studies were confirmed by electrophysiologic studies in cultured neurons where low concentrations of spermine enhanced NMDA currents with no effect on AMPA or kainate associated currents (Williams *et al.,* 1991). The stimulatory effects of polyamines appear to be glycine-independent because polyamines enhance NMDA receptor binding in the presence of saturating concentrations of glycine. However, a glycine-dependent stimulation has also been identified because spermine has been shown to induce an increase in the affinity of the receptor for glycine (Sacaan and Johnson, 1989; Ransom, 1991; Williams *et al.,* 1991). Radioligand and electrophysiologic studies have also identified an inhibitory effect of polyamines on NMDA receptor currents at higher concentrations than those which enhance receptor function (Williams *et al.,* 1991). The inhibitory effects appear to be voltage dependent because they are more pronounced at hyperpolarized membrane potentials and may be the result of a blockade of the channel (Williams *et al.,* 1991).

Since the cloning of the genes for the NMDA receptor, studies in expression systems have identified that the sensitivity of the NMDA receptor for polyamines is mediated by the NR2 subunits (Williams *et al.,* 1994). Heteromeric NMDA receptors comprised of NR1/NR2A subunits are inhibited by spermine, whereas NR1/NR2B receptors are stimulated. Spermine had no effect on NMDA receptors expressing the NR1/NR2C subunit. These findings help explain the apparently conflicting enhancing and inhibitory effects of polyamines on the NMDA receptor. Ifenprodil, a polyamine site antagonist, also has differential affinity for NMDA receptors based on the NR2 subunit expressed. Ifenprodil inhibits NR1/NR2B receptors with high affinity (IC_{50} = 0.34 μM), but it has a much lower affinity (IC_{50} = 146 μM) for NR1/NR2A receptors (Williams, 1993).

Furthermore, the inhibitory effect of ifenprodil at the low affinity site in NR1/NR2A receptors is not glycine dependent, whereas the inhibitory effect at NR1/NR2B receptors (high affinity site) can be reduced by increasing the glycine concentration. Therefore, part of the inhibitory effect of this polyamine site antagonist may be mediated by altering the receptor affinity for glycine.

5. Proton Sensitive and Redox Sites

Proton inhibition of NMDA receptor currents (Traynelis and Cull-Candy, 1990) and reducing agent enhancement of currents (Aizenman *et al.*, 1989) have been described. Proton inhibition of NMDA receptor currents occurs when the pH is reduced from 7.6 to 6.8. Potentiation of currents are measured when the pH is increased from 7.6 to 8.4 (Traynelis and Cull-Candy, 1990). This pH effect is within the physiologic range that may be present under pathologic conditions such as ischemia and seizures. Studies have shown that NMDA receptors assemblies lacking the 5′ insertion cassette (NR1-a splice variants) are more sensitive to proton inhibition than variants containing the insertion (NR1-b). The proton inhibition of NMDA-receptor currents is voltage-insensitive and is not due to a fast channel block or an effect on the glutamate or glycine modulation.

The disulfide reducing agent dithiothreitol (DTT) has been shown to potentiate NMDA-induced currents, whereas the sulfhydryl oxidant 5-5-dithio-bis-nitrobenzoic acid (DTNB) reduces currents (Aizenman *et al.*, 1989, 1990). Glutathione (Sucher and Lipton, 1991), oxygen free radicals (Aizenman *et al.*, 1990), and pyrroloquinoline quinone appear to modulate the redox site (Aizenman *et al.*, 1992). The functional significance of these regulatory sites has been demonstrated by the protective action of glutathione and ascorbic acid from NMDA-receptor mediated neuronal injury (Majewska and Bell, 1990; Levy *et al.*, 1991).

6. Phosphorylation Sites

Phosphorylation of neurotransmitter receptors is one way of posttranslationally modifying receptor function. NMDA receptor activation is known to increase intracellular calcium levels leading to the activation of a number of protein kinases [e.g., protein kinase C (PKC), tyrosine kinases, calcium-calmodulin kinase II] that can influence the phosphorylation state of the receptor and thus its activity. Chen and Huang (1992) have demonstrated that in trigeminal neuron preparations, intracellularly applied PKC potentiates NMDA responses by reducing the voltage-dependent block of magnesium. However, in rat hippocampal neurons, PKC activation has been reported to suppress NMDA-receptor mediated responses (Markram and Segal, 1992). The latter finding appears contradictory to the notion that stimulation of PKC activity increases NMDA receptor currents and the induction of long-term potentiation in the hippocampus (Ben-Ari *et al.*, 1992).

NMDA receptor phosphorylation has been studied in expression systems to determine the role of NMDA receptor subunits. Treatment with the PKC activator phorbol 12-myristate 13-acetate (PMA) increases responses in homomeric NMDA receptor assemblies containing the NR1-4b variant by 20-fold but only by fourfold in assemblies containing the NR1-1a variant (Durand *et al.*, 1992). The difference in PKC modulation between these two splice variants was attributed to the presence of the N-terminal insertion cassette encoded by exon 5. Other studies have suggested that the presence of exon 5 and the absence of the C-terminal cassettes encoded by exons 21 and 22 are the variants that are greatly enhanced by PKC activation. Tingley *et al.* (1993) have shown that PKC phosporylates serine residues in the NR1 subunit of the NMDA receptor on a single exon in the C-terminal domain and that alternative splicing regulates phosphorylation. However, other investigators have shown PKC activation of heteromeric NMDA receptors in variants in which the PKC concensus phosphorylation sites of the C-terminal region has been deleted (Sigel *et al.*, 1994).

Protein–tyrosine kinases have also been shown to phosphorylate the NMDA receptor and enhance NMDA-mediated currents in spinal dorsal horn neurons (Wang and Salter, 1994). Lau and Huganir (1995) studied the phosphorylation of tyrosine residues in NMDA receptor subunits from synaptic membranes of rat cerebral cortex. They showed that the NR2A and NR2B but not the NR1 subunit were phosphorylated on tyrosine residues. Taken together, these studies show that NMDA receptor phosphorylation occurs at different sites and are mediated by different kinases. This mechanism of regulating NMDA receptor function may be important in neuronal development and plasticity.

C. Brain Distribution and Allosteric Modulation

A number of methods have been used to measure the level of NMDA receptors in the vertebrate brain. Receptor binding assays or receptor autoradiography using a variety of ligands targeted to different sites in the NMDA receptor complex have been most commonly used to measure receptor levels. Electrophysiologic approaches have also been used in tissue slices to determine the modulation of the receptor complex by agonists and antagonists.

1. Radioligand Binding and Receptor Autoradiography

The most commonly used radioligands have been those targeted to the glutamate and glycine sites and the noncompetitive channel blocker MK-801. Using these radioligands, a number of studies have determined the neuroanatomic and developmental expression of NMDA receptors. Figure 4 shows a typical autoradiographic image of the distribution of [^{3}H]-MK-801 labeled NMDA receptors in the rat brain. The highest density of NMDA receptors is present in the CA1 region of the hippocampus (Fig. 4). Within the CA1 region, the stratum radiatum has a higher density than stratum oriens, followed by the stratum lacunosum moleculare and the stratum pyramidale (Sakurai *et al.,* 1991; Jett and Guilarte, 1995). Thus, in the hippocampus, NMDA receptors are primarily localized in the dendritic fields of the pyramidal cell layer with some expression in the soma of the cell. High density of NMDA receptors are also present in the molecular layer of the dentate gyrus. A similar neuroanatomic distribution as in the CA1 region is present in the CA3 region but with intermediate receptor levels. Relative to the hippocampus, intermediate levels of [^{3}H]-MK-801 labeled NMDA receptors are present in the cerebral cortex. Frontoparietal, cingulate, and entorhinal cortex displayed a layered distribution of receptors with higher levels in the outer cortical layers. Low to intermediate levels of NMDA receptors are present in structures associated with the basal ganglia and thalamic nuclei (Sakurai *et al.,* 1991). Little or no binding is found in the cerebellum and brainstem. The anatomic distribution of NMDA receptors using NMDA-sensitive [^{3}H]-glutamate and strychnine-insensitive [^{3}H]-glycine binding are very similar to the distribution obtained with [^{3}H]-MK-801 (Cotman *et al.,* 1987).

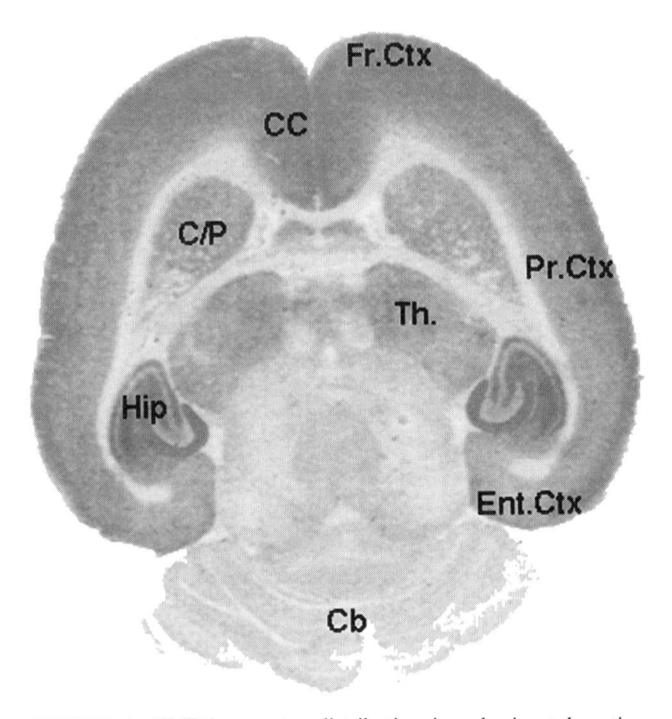

FIGURE 4 NMDA receptor distribution in a horizontal section from a 28-day-old rat brain as labeled by [^{3}H]-MK-801. Abbreviations: CC: cingulate cortex; Fr.Ctx: frontal cortex; Pr.Ctx: parietal cortex; Ent.Ctx: entorhinal cortex; C/P: caudate/putamen; Th.: thalamus; Hip: hippocampus; Cb: cerebellum.

NMDA receptor levels appear to be transiently increased during development. In the hippocampus and cerebral cortex, NMDA-sensitive [^{3}H]-glutamate and [^{3}H]-MK-801 binding sites are low at birth and increase postnatally to reach a maximal level at 2–3 weeks of age (Fig. 5). This is followed by a small but significant decline to adult levels (Tremblay *et al.,* 1988; Insel *et al.,* 1990; McDonald *et al.,* 1990a; Jett and Guilarte, 1995). Maximal levels of binding have small temporal variations dependent on the cortical layer or hippocampal region examined (Insel *et al.,* 1990; McDonald *et al.,* 1990). The development pattern of NMDA receptor overexpression in the hippocampus is consistent with the development of anatomic and physiologic processes that may require the activation of NMDA receptors. For example, cellular differentiation and synaptogenesis in the rat hippocampus occurs postnatally to a great extent. The growth of pyramidal dendrites in the stratum radiatum of the CA1 region of the hippocampus and its innervation by Schaffer collateral/comissural projections to this area are occurring rapidly during the second postnatal week of life and follow a temporal pattern similar to the development of NMDA receptors (Pokorny and Yamamoto, 1981). In this same region of the hippocampus the induction of long-term potentiation (LTP), a cellular model of learning and memory that is dependent on NMDA receptor activation, reaches a maximal magnitude of induction at 14–15 days of age, followed by a decline to adult levels (Harris and Teyler, 1984). Thus, the developmental expression of NMDA receptors may be associated and influence a number of physiologic processes such as synaptogenesis and synaptic plasticity.

2. Electrophysiologic Studies

Electrophysiologic studies have provided valuable information about the functional state of the NMDA receptor and changes in modulatory domains during development. Hamon and Heinemann (1988) have examined the developmental sensitivity to excitatory amino acids in the CA1 region of the hippocampus using brain slices. They found that NMDA receptors in the stratum radiatum of the CA1 region are most sensitive to NMDA application during the second week of life. The stratum radiatum is the hippocampal region where Schaffer collateral excitatory afferents form synapses at the level of the apical dendrites of CA1 pyramidal cells. The increased sensitivity to NMDA during the second week of life is consistent with the transient increase in

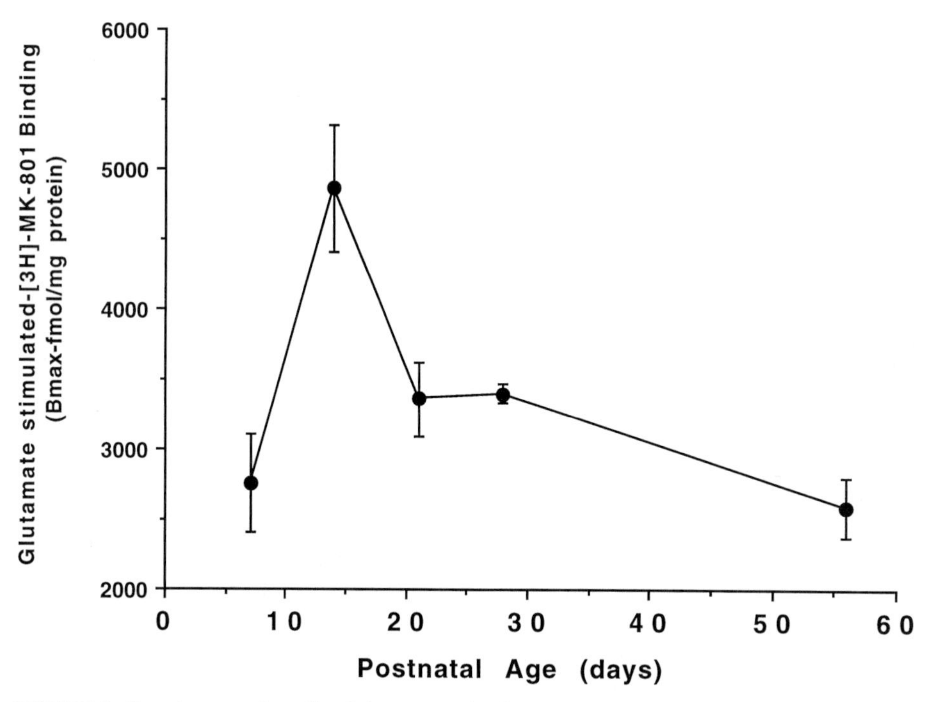

FIGURE 5 Developmental profile of glutamate-stimulated NMDA receptor levels in cerebral cortex membrane preparations. Each value is the mean ± sem of 5 determinations.

the population of NMDA receptors at this age and the maximal magnitude of LTP induction (Harris and Teyler, 1984).

The sensitivity of the NMDA receptor to inhibition by magnesium during development has also been studied in hippocampal slices. Bowe and Nadler (1990) found that in the CA1 region of the hippocampus there was an age-dependent increase in the potency of magnesium to inhibit NMDA receptor responses. The addition of 1 m*M* magnesium to the superfusion medium decreased NMDA receptor responses to a lesser degree in hippocampal slices from 10–15-day-old rats than from adults. These findings suggest that during the second week of life, a critical period of synapse formation, NMDA receptors in the CA1 region of the hippocampus have a reduced sensitivity to inhibition by magnesium and this may facilitate their participation in development processes occurring at this age. These findings are consistent with those of Morisett *et al.* (1990), who demonstrated a reduced sensitivity to magnesium inhibition of NMDA-mediated excitatory postsynaptic potentials in the CA1 region of hippocampal slices from immature rats relative to adults. It is likely that the changes in the sensitivity of NMDA receptors to inhibition by magnesium are mediated by developmental changes in NMDA receptor subunit mRNA expression.

D. Role of NMDA Receptors in Brain Development and Synaptic Plasticity

1. Neurotrophic Function

A number of studies have suggested that during early brain development the activation of NMDA receptors may play an important role in neuronal migration, elaboration of dendrites, synaptogenesis, and cell survival. Komuro and Ravik (1993) have demonstrated the involvement of NMDA receptors in regulating granule cell migration in the cerebellum. Enhancement of NMDA receptor activity by the removal of magnesium or by glycine application increased the rate of cell migration, whereas receptor inhibition, by using specific antagonists, decreased the rate of migration. NMDA receptor activation has also been shown to play an important role in the elimination of synapses in the developing cerebellum (Rabacchi *et al.,* 1992). Purkinje cells exhibit a transient sensitivity to NMDA-mediated responses during synapse elimination in the developing cerebellum resulting from their innervation by climbing fibers that are excitatory in nature. The presence of the NMDA receptor antagonist D,L-APV during this period prevented the regression of climbing fiber/Purkinje cell synapses, whereas the inactive isomer did not. In saline-treated animals or in animals treated with the inactive isomer (L-APV), 10–15% of Purkinje cells were multiply innervated by climbing fibers. In those treated with the active isomer, 50% of cells were multiply innervated indicating the role of NMDA receptor activity in synapse elimination. Cerebellar cells in culture have also been shown to be altered by NMDA receptor activation. Treatments that decrease NMDA receptor activity in granule cells (e.g., application of a receptor antagonist or the enzymatic removal of glutamate) inhibit granule cell neurite outgrowth and differentiation (Pearce *et al.*, 1987; Moran and Patel, 1989). Similar findings have been described using hippocampal neurons (Brewer and Cotman, 1989).

NMDA receptors expressed in cortical neurons prior to the formation of functional synapses are activated by NMDA and can be blocked by magnesium in a voltage-dependent manner (LoTurco *et al.*, 1991). The expression of NMDA receptors within the cortical plate coincides with the end of migration and the beginning of dendritic arborization and formation of synapses suggesting that they may be involved in neuronal development and synaptogenesis. Consistent with this notion, Brooks *et al.* (1991) have demonstrated rapid increases in synaptic number and thickening length in the occipital cortex of developing rats following the administration of NMDA. Taken together, these and other studies provide strong evidence that the activation of NMDA receptors may have a neurotrophic function during early development.

2. Synaptic Plasticity: Long-Term Potentiation

Synaptic plasticity has been thought to be a translation of experiences into an enduring physical change in the brain (Hebb, 1949). The use-dependent nature of the NMDA receptor makes it an appropriate candidate for the encoding of experiences in the mammalian brain. Activation of NMDA receptors has been shown to be essential for some forms of synaptic plasticity such as LTP in the hippocampus (Harris *et al.*, 1984; Collingridge and Singer, 1990; Madison *et al.*, 1991; Izquierdo, 1993). LTP has been advanced as a cellular model of learning and memory because it represents an enduring form of synaptic plasticity (Harris *et al.*, 1984; Collingridge and Singer, 1990; Madison *et al.*, 1991; Izquierdo, 1993). LTP has been associated with changes in dendritic spine morphology (Desmond and Levy, 1986). The entry of calcium through the NMDA receptor is responsible for a series of postsynaptic signalling cascades essential to the potentiation of the synaptic response. These include the activation of calcium-sensitive kinases followed by the transcription of early genes and protein synthesis that are essential for the maintenance phase of LTP (Izquierdo, 1993; Reymann, 1993).

The link between NMDA receptor activation in LTP and learning and memory has been examined using neurochemical, electrophysiologic and behavioral approaches. NMDA receptor antagonists such as AP-5, applied before a high-frequency stimulation necessary to induce LTP, prevents LTP formation in the CA1 region of the hippocampus (Morris *et al.*, 1986) but does not affect previously established LTP. Behavioral studies have shown that inactivation of the NMDA receptor selectively decreases LTP in the hippocampus and impairs NMDA receptor–mediated spatial learning in rats (Morris *et al.*, 1986; Upchurch and Wehner, 1990; Venable and Kelly, 1990). Gene targeting studies have shown that mice lacking the NR2A subunit of the NMDA receptor express impairments in NMDA receptor currents, hippocampal LTP, and spatial learning (Sakimura *et al.*, 1995). The most definitive studies to date in correlating NMDA receptor function, LTP in the hippocampus, and spatial learning has been described by a series of studies using brain-region-specific knockout mice. Tsien *et al.* (1996a) have developed a technique to produce knockout mice with a deletion of the NR1 gene selective for CA1 pyramidal cells of the hippocampus. Mice lacking NMDA receptors in the CA1 region have been found to lack NMDA receptor–mediated excitatory postsynaptic currents and LTP (Tsien *et al.*, 1996b). These mice are impaired in a spatial learning task (Morris water maze) and have impaired hippocampal representation of space (McHugh *et al.*, 1996; Tsien *et al.*, 1996b). These studies support the hypothesis that NMDA receptor–mediated synaptic plasticity is the cellular basis for some forms of learning and memory in the mammalian brain.

E. NMDA Receptors: Molecular Targets in Developmental Lead (Pb) Neurotoxicity

Children and experimental animals exposed to Pb during development express a wide range of CNS deficits (Needleman *et al.*, 1979; Cory-Slechta *et al.*, 1985; Rice, 1985; Bellinger *et al.*, 1991; Needleman, 1993). Since the early 1970s, efforts to phase out major sources of Pb in the environment (e.g., leaded gasoline and paint) have been effective in reducing exposure of the general population to this neurotoxicant. However, there is continuing concern on the part of government agencies, the scientific community, and the public regarding the detrimental effects of chronic low-level Pb exposure in children. These concerns are based on studies indicating that blood Pb levels as low as 7–10 μg/dl result in behavioral and cognitive impairments in children (Mushak *et al.*, 1989; Winneke *et al.*, 1994). This major public health problem is not going to disappear readily because environmental contamination by this very potent and ubiquitous neurotoxicant will continue to exist for many decades. This is partly a result of a high percentage of housing containing leaded paint that still exists in many cities and towns throughout the United States.

In 1991, the Centers for Disease Control set a blood Pb of 10 μg/dl as the maximal acceptable level (United States Department of Health and Human Services, 1991). It has been estimated that 17% of all children living in the United States have blood Pb levels exceeding 15 μg/dl (United States Department of Health and Human Services, 1988; Mushak *et al.,* 1989). Cases of Pb poisoning requiring clinical intervention with chelation therapy (blood Pb levels greater than 45 μg/dl) continue to be referred to treatment centers on a weekly basis. More importantly, thousands of children in the United States have blood Pb levels in the 15–30 μg/dl range for which no therapy or pharmacologic intervention currently exists.

1. Susceptibility of the Developing CNS to Pb Neurotoxicity

The susceptibility of the developing CNS to the neurotoxic actions of Pb is well documented. Experimental evidence suggests that the susceptibility of the developing organism is due, at least in part, to the fact that in children, the gastrointestinal tract can absorb a greater percentage of the ingested Pb than in adults (Ziegler *et al.,* 1978). Also, the blood–brain barrier, which when fully developed excludes many toxic substances from reaching the brain, is more likely to be compromised during childhood, allowing a greater percentage of the absorbed Pb to enter the brain (Goldstein *et al.,* 1974). Thus, part of the susceptibility of the developing brain to Pb-induced neurotoxicity may be due to a higher level of Pb accumulation. However, emerging experimental evidence indicates that the ontogeny of molecular targets in the developing brain are also important determinants of susceptibility to Pb neurotoxicity (Omelchenko *et al.,* 1996; Guilarte, 1997). Specifically, it has been suggested that the developmental expression of NMDA receptor subunits result in populations of receptors expressing differential sensitivity to inhibition by Pb (Omelchenko *et al.,* 1996; Guilarte, 1997). Furthermore, the vulnerability of different brain structures to Pb neurotoxicity may also be determined by the NMDA receptor subtypes expressed in that brain region (Omelchenko *et al.,* 1996; Guilarte, 1997).

2. Morphologic and Neurochemical Aspects of Pb Neurotoxicity

A great deal of work has been done to determine the morphologic, neurochemical, and behavioral correlates of chronic low-level developmental Pb neurotoxicity. These studies for the most part have examined individual end-points such as morphology or neurochemistry, but have seldom attempted to integrate them in the same or in similarly exposed animals. Morphologically, dendrites and dentritic spines are the "receptive" fields for the acquisition and integration of information to the neuron. Pb-induced alterations in dendritic morphology could result in a reduced capacity of the neuron to receive and process information, thus altering its capacity to "learn." The hippocampus and cerebral cortex undergo a great deal of change during development and are involved in cognition (Izquierdo, 1993). The morphology of neurons in these brain structures have been shown to be significantly altered by Pb. These changes include a reduction in the overall length of the dendritic field of pyramidal and granule cells (Petit and Leboutellier, 1979; Alfano and Petit, 1982; Kiraly and Jones, 1982; Petit *et al.,* 1983; Petit, 1986) and a reduction in the number of dendritic branches and spines (Petit and Leboutellier, 1979; Alfano and Petit, 1982; Kiraly and Jones, 1982; Petit *et al.,* 1983; Petit, 1986). These morphologic changes are present at moderate levels of Pb exposure and resemble changes observed in developing animals following the administration of NMDA receptor antagonists (Simon *et al.,* 1992; Fachinette *et al.,* 1993, 1994; Kalb, 1994). Pb is a potent inhibitor of the NMDA receptor (Alkondon *et al.,* 1990; Guilarte and Miceli, 1992), and NMDA receptor activation during development has been shown to be essential for neuronal development and differentiation (Moran and Patel, 1989; Brewer and Cotman, 1989). Thus, it is possible that NMDA receptor inhibition by Pb may account for some of the morphologic changes observed in developing animals.

The basis of Pb-induced alterations of specific neurotransmitter systems remains a very active area of investigation. The neuronal systems studied in Pb neurotoxicity include the cholinergic, dopaminergic, and γ-amino-butyric acid (GABA)ergic systems, among others. Despite these efforts, no definitive causal relationships to cognitive impairment have been made. In this respect, the glutamatergic system has been implicated as an important component of cellular processes associated with learning and memory. Emerging knowledge of the neurochemical basis of learning and memory has shown that NMDA, AMPA, and metabotropic excitatory amino acid receptors play an important role in use-dependent synaptic plasticity such as LTP (Collingridge and Singer, 1990; Izquierdo, 1993; Reymann, 1993). Of these excitatory amino acid receptors, the NMDA receptor has been shown to be potently and selectively inhibited by Pb (Alkondon *et al.,* 1990; Guilarte and Miceli, 1992). As a result, the NMDA receptor complex has been the focus of intense investigation in an attempt to understand the biochemical basis of the cognitive deficits known to occur in Pb-exposed children and in developing experimental animals.

3. NMDA Receptor Function in Pb Neurotoxicity

The first evidence that Pb had a direct inhibitory effect on the NMDA receptor was provided in the early 1990s. The laboratory of E. X. Albuquerque at the University of Maryland School of Medicine and our own laboratory were the first to independently describe the inhibitory action of Pb on the NMDA receptor using electrophysiologic and biochemical approaches, respectively (Alkondon *et al.*, 1990; Guilarte and Miceli, 1992). These early studies showed that Pb was a potent and selective inhibitor of the NMDA receptor and that the inhibitory effects of Pb were age- and brain-region specific (Alkondon *et al.*, 1990; Guilarte and Miceli, 1992; Ujihara and Albuquerque, 1992; Ishihara *et al.*, 1995; Guilarte, 1997). These findings have been confirmed by other investigators (Uteshev *et al.*, 1993; Schulte *et al.*, 1995).

4. Where does Pb Interact on the NMDA Receptor to Inhibit its Function?

It is now well documented that Pb is a potent inhibitor of the NMDA receptor. However, its site(s) of action is not well characterized. *In vitro* studies aimed at identifying the site(s) of action of Pb on the NMDA receptor have suggested that it may be interacting at a zinc allosteric site(s) (Guilarte *et al.*, 1994, 1995). This hypothesis is based on a number of similarities between the effects of Pb and zinc on the NMDA receptor. For example, both of these divalent cations (1) antagonize NMDA receptor activation in a noncompetitive fashion; (2) decrease the dissociation rate of [^{3}H]-MK-801 binding to neuronal membranes; (3) bind to sites that are sensitive to covalent modification of histidine residues by diethylpyrocarbonate (DEPC) treatment; and (4) decrease the potency of the other to inhibit [^{3}H]-MK-801 binding to the NMDA receptor (Guilarte *et al.*, 1995). The apparent competitive interaction of Pb at a zinc modulatory site has been confirmed by other investigators (Schulte *et al.*, 1995). Subsequent studies have also shown other similarities between Pb and zinc interactions with the NMDA receptor. Pb and zinc inhibit [^{3}H]-MK-801 binding to the NMDA receptor by interacting at high and low affinity sites (Guilarte *et al.*, 1995; Guilarte, 1997). The high and low affinity sites associated with NMDA receptor inhibition by Pb were shown to have developmental and regional specificity and may be associated with the expression of different NMDA receptor subunits (Omelchenko *et al.*, 1996; Guilarte, 1997). Another similarly between these two divalent cations is their ability to enhance NMDA receptor currents at nanomolar concentrations under certain experimental conditions. Zinc has been shown to enhance NMDA-induced currents in homomeric NMDA receptors containing certain splice variants of the NR1 subunit (Hollmann, 1993). Using the oocyte expression system and electrophysiological analysis, Omelchenko *et al.* (1996) have shown that NMDA receptor currents from NR1/NR2A/NR2B assemblies are enhanced by nanomolar concentrations of Pb but are inhibited by higher concentrations. The Pb-induced enhancement of NMDA receptor currents at nanomolar concentrations has also been shown in cultured hippocampal neurons, and the effect appears to be dependent on the glycine concentration (Marchioro *et al.*, 1996). The physiologic significance of the enhancing effects of zinc and Pb are not currently known. However, findings on the characterization of an extracellular divalent cation site on the NMDA receptor may provide valuable insights on the site(s) of Pb interaction.

Gu and Huang (1994) and Wang and McDonald (1995) have reported that there are extracellular binding sites for calcium and magnesium on the NMDA receptor complex where these divalent cations enhance NMDA-induced currents only when the strychnine-insensitive glycine site is not saturated. It has been suggested that a single site may be mediating the calcium and magnesium effects (Wang and McDonald, 1995; Paoletti *et al.*, 1995). Calcium and magnesium interaction at this site(s) appears to increase the affinity of the NMDA receptor for glycine (Gu and Huang, 1994; Paoletti *et al.*, 1995; Wang and McDonald, 1995). Therefore, they appear to act as agonists at a divalent cation binding site that modulates the affinity of the receptor for glycine. Zinc and Pb may be partial agonists at this site because they both typically inhibit NMDA receptor channel currents at micromolar concentrations but under certain conditions may cause an enhancement of conductance at submicromolar concentrations (Marchioro *et al.*, 1996). In addition, the magnitude of the zinc and Pb enhancement is not as large as calcium and magnesium. At saturating glycine concentrations and in the presence of micromolar Pb concentrations, Pb does not enhance but inhibits NMDA receptor responses (Marchioro *et al.*, 1996).

Yeh *et al.* (1990) have reported that zinc inhibits NMDA receptors by a noncompetitive antagonism of glycine binding at the strychnine-insensitive glycine site. It appears that the described calcium and magnesium extracellular binding sites may be functionally related to the Pb and zinc sites in that they all modulate the strychnine-insensitive glycine site on the NMDA receptor complex. It is possible that calcium and magnesium may be acting as agonists at a divalent cation site that allosterically modulates the glycine site and enhances NMDA receptor responses by increasing receptor affinity for glycine. However, zinc and Pb may be acting as antagonists at this divalent cation site or an overlapping site to inhibit or negatively modulate NMDA receptor function by decreasing the affinity of the receptor for glycine. A schematic of this hypothetical and complex interaction is provided in Fig. 6. Consistent with this hypothesis, two recent studies have shown that Pb competitively antagonizes the calcium enhancement of the NMDA receptor (Marchioro *et al.,* 1996; Hashemzadeh-Gargari and Guilarte, 1997).

5. *In Vivo* Effects of Developmental Lead Exposure on NMDA Receptor Function

In vivo studies have confirmed the *in vitro* findings on the direct interation of Pb with the NMDA receptor, because changes in NMDA receptor levels and modulation are present in animals exposed to Pb during development (Guilarte and Miceli, 1992; Brooks *et al.,* 1993; Guilarte *et al.,* 1993, 1994; Jett and Guilarte, 1995). Guilarte and Miceli (1992) described the first study that *in vivo* exposure to 750 ppm lead acetate (PbAc) in the diet has on NMDA receptor function in the rat brain. In this study, [^{3}H]-MK-801 binding to cortical membrane preparations was measured from rats exposed to Pb during development or rats exposed to Pb as adults. The results showed that there was a significant decrease in the apparent number of [^{3}H]-MK-801 binding sites (Bmax) and an increase in affinity (Kd) in cortical preparations from immature rat brain (postnatal day [PN 14]) but not in the brains of rats exposed to Pb as adults (Guilarte and Miceli, 1992). These findings suggested that the developmental period of Pb exposure influences the types of changes observed in NMDA receptor function.

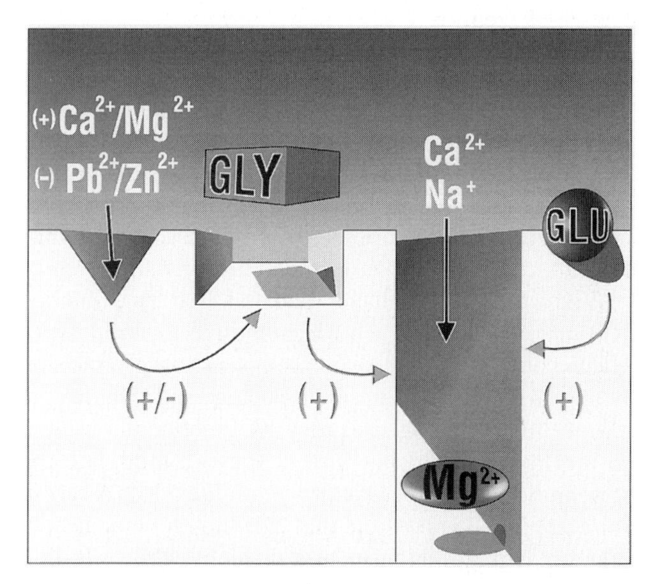

FIGURE 6 Graphic representation of divalent cation modulation of the strychnine-insensitive glycine site of the NMDA receptor.

A developmental profile of the effects of *in vivo* chronic Pb exposure on hippocampal NMDA receptors has been described using [^{3}H]-MK-801 quantitative autoradiography (Jett and Guilarte, 1995). This study found that there was a dose-dependent increase (19–49%) in the levels of [^{3}H]-MK-801 in virtually all hippocampal subfields and in the entorhinal cortex of PN 14 rats chronically exposed to Pb (Jett and Guilarte, 1995). The concentrations of Pb in the hippocampus of PN 14 rats exposed to 250 and 750 ppm lead acetate in the diet were: 0.33 ± 0.09 ($n = 3$) and 0.34 ± 0.04 ($n = 3$) μg/g wet weight, respectively, whereas those in control animals were 0.09 ± 0.03 μg/g ($n = 4$). In PN 28 and PN 56 rats, hippocampal Pb concentrations increased as a function of age, but the changes in [^{3}H]-MK-801 binding to the NMDA receptor complex were less dramatic (Jett and Guilarte, 1995). At PN 28, there were small but significant region-specific decreases (14–18%; CA1 subfield) or increases (12–15%; entorhinal cortex) in [^{3}H]-MK-801 binding, whereas in PN 56 hippocampus, no significant changes in [^{3}H]-MK-801 binding were present for any of the Pb-treated groups despite high levels of Pb measured at this age (Jett and Guilarte, 1995). A potential explanation for the increased level of [^{3}H]-MK-801 binding present in PN 14 hippocampus and entorhinal cortex of Pb-treated rats is that there is an increase in the number of NMDA receptors resulting from a compensatory response to the direct inhibitory effects of Pb *in situ.* Previous studies have suggested that a response to inhibition of NMDA receptor–mediated neurotransmission is to alter postsynaptic receptor levels (McDonald *et al.,* 1990b; Williams *et al.,* 1992). The lack of a change in the levels of [^{3}H]-MK-801 binding in PN 56 rat is difficult to interpret because high levels of Pb were present at this age (Jett and Guilarte, 1995).

Another study showed that a chronic Pb exposure (750 ppm PbAc in the diet) to the dams from preconception to the weaning of the pups and continued feeding the same diet to the offspring postweaning resulted in a 31% increase in the apparent number (Bmax) of [^{3}H]-MK-801 binding sites in forebrain membrane preparations of Pb-exposed rats at PN 70-210 (Guilarte *et al.,* 1993). The level of Pb in the blood measured in this study was 13.9 ± 2.8 μg/dl compared to 1.16 ± 0.28 μg/dl in controls. The blood Pb levels in the treated rats are similar to those measured in children environmentally exposed to Pb. Therefore, changes in NMDA receptor levels *in vivo* are present at blood Pb levels that are environmentally relevant.

Other laboratories have also provided evidence of alterations in NMDA receptor binding parameters as a result of chronic developmental Pb exposure. Brooks *et al.* (1993) used an experimental protocol in which maternal exposure to Pb was initiated at birth in the drinking water (4% $PbCO_3$). Because this is a very high level of Pb exposure that results in body weight differences between control and Pb-exposed pumps, they culled the litter size to five for the Pb-exposed group and ten for the control group. This procedure corrects for the differences in body weight of the pups. In this study, two different ligands that label different sites of the NMDA receptor were used. These were [^{3}H]-CGP 39653, which is an NMDA receptor site antagonist, and [^{3}H]-TCP, a ligand that presumably labels the same site as [^{3}H]-MK-801 inside of the NMDA receptor channel. Their findings indicate that, in rats that were exposed to Pb from PN 1 to PN 25, there was a significant increase (18–53%) in the levels of [^{3}H] CGP 39653 binding in several hippocampal subfields at PN 25. No significant increases in [^{3}H]-CGP 39653 binding were present in other brain regions measured. In the same study, the binding of [^{3}H]-TCP showed essentially no change in any brain region from Pb-treated rats relative to controls with the exception of a 17% increase in the CA3-radiatum of the hippocampus. No blood or brain concentrations were provided in this study. A study by Schulte *et al.* (1995) examining the *in vivo* effect of developmental Pb exposure in mice showed no significant differences in the level of [^{3}H]-MK-801 binding to forebrain membrane preparations. However, they found a 13–15% increase in binding in membrane preparations from mice exposed to 100 ppm Pb as adults. Based on these studies there is accumulating experimental evidence indicating that *in vivo* exposure to Pb during development produces changes in NMDA receptor levels measured using ligand binding and autoradiographic techniques. The magnitude and direction of the change may be influenced by the Pb dose, the period of Pb exposure, and the brain region analyzed.

Supporting evidence that changes in NMDA receptor levels and modulation following developmental Pb exposure have functional consequences is provided by a number of reports. Petit *et al.* (1992) have examined the effect of developmental Pb exposure on the seizure activity of rats injected with NMDA. Rat pups at birth were exposed via the dam's milk to three different levels of Pb. Four groups of dams were exposed to Pb in the diet that contained either 0.0, 0.05, 0.4, or 4% $PbCO_3$ by weight. The exposure protocol was similar to the one described by Brooks *et al.* (1993) in that the Pb exposure was initiated at birth. Blood Pb levels were measured in the rat pups at PN 25 and the values were 6 μg/dL for the control group and 43, 317, and 1323 μg/dL for the 0.05, 0.4, and 4% PbAc groups, respectively. Rat pups at PN 15 or PN 25 were injected with NMDA at doses ranging from 5–40 mg/kg (15 days of age) or 5 to 100 mg/kg (25 days of age) and the seizure activity monitored. In control rats, the injection of NMDA produced slowly developing seizures that required a higher dose of NMDA for their expression as the animals aged. However, in rat pups whose mothers were exposed to the 0.4 and 4% Pb-containing diets, there was an increased sensitivity to NMDA injections. In these groups of rats, the onset of seizure activity and death due to status epilepticus was present at lower doses of NMDA relative to rat pups in the control diet. However, the group of rats exposed to the lowest level of Pb in the diet displayed an attenuated response to NMDA relative to controls (Petit *et al.,* 1992). It is possible that this difference in seizure activity may be associated with changes in the levels of NMDA receptors in the brain of the Pb-exposed rats or in the expression of a population of NMDA receptors with either increased or decreased affinity for NMDA. An alternative explanation is that the differences in the sensitivity to NMDA injections may be related to the ability of NMDA to enter the brain. It is possible that in the two groups with the highest exposure to Pb, the blood–brain barrier is compromised, thus increasing the entry of NMDA into the brain and producing a much faster response to the induction of seizure activity at a lower dose of NMDA.

The most significant findings on the functional consequences of developmental Pb exposure are those describing an impairment in hippocampal LTP. Altmann *et al.* (1991) were the first to describe the effects of Pb on hippocampal LTP by acutely perfusing slices with PbAc-containing solutions (10–20 μM) and recording of excitatory postsynaptic potentials (EPSP) and population spikes (PS). This study indicated two different effects of *in vitro* Pb perfusion on LTP. In half of the slices, high frequency stimulation of the Schaffer collateral pathway and recording at the CA1 region of the hippocampus resulted in a significant impairment in the PS amplitude as a percent of the pretrain baseline value. In a second group of Pb-perfused slices, there was a gradual increase in the PS amplitude following the high frequency stimulation reaching a control value at approximately 80 min posttrain. EPSP slope as a percent of pretrain values was also plotted and expressed a similar pattern for Pb effects. In some slices Pb perfusion caused a gradual increase in EPSP slope, whereas in other slices a rapid decay of the EPSP slope occurred after the first 10 min of high frequency stimulation. These findings indicate that *in vitro* Pb interferes with both the induction and maintenance phases of CA1 hippocampal LTP. Because the induction of CA1 hippocampal LTP is dependent on the activation of NMDA receptors (Collingridge and Singer, 1990), these findings are consistent with the direct inhibitory actions of Pb on the NMDA receptor.

The same group of investigators examined the effects of *in vivo* Pb exposure during different periods of development on hippocampal LTP (Altmann *et al.,* 1993). In this study, the Pb exposure occurred *in vivo* and LTP was recorded in slices *ex vivo.* They found that continuous chronic exposure to environmentally relevant levels of Pb during gestation, lactation, and postweaning resulted in a significant impairment in CA1 hippocampal LTP in adult animals (70–120 days of age). Small but significant changes in hippocampal LTP were present in animals exposed to Pb during gestation and lactation up to 16 days of age and examined as adults (maternal group). No significant change in LTP was measured if the Pb exposure was initiated postweaning, indicative of the vulnerability of the immature brain to Pb-induced impairment in LTP. In another study, chronic developmental Pb exposure resulted in impaired CA1 hippocampal LTP induction in slices from postnatal day 16–20 but not in slices from younger rats (Altmann *et al.,* 1994).

Lasley *et al.* (1993) using a protocol in which the Pb exposure was initiated at birth with *in vivo* LTP recordings in the dentate gyrus of adult rats confirmed the findings of Altmann and colleagues (1993, 1994). They showed an essential lack of LTP induction in perforant path–dentate gyrus synapses of adult rats chronically exposed to environmentally relevant levels of Pb from birth. Gilbert *et al.* (1996) have shown that the lack of LTP was due to a higher threshold of induction in the Pb-exposed animals. LTP in this experimental paradigm is also NMDA receptor mediated. The importance of the NMDA receptor in the Pb-induced impairment in hippocampal LTP has been demonstrated by Gutowski *et al.* (1997), who have shown that LTP in the CA1 but not in the CA3 region of the hippocampus is impaired in rats exposed to Pb during development. Because the induction of LTP in the CA1 but not the CA3 region depends on NMDA receptor activation (Harris and Cotman, 1986; Ito *et al.,* 1991), this finding supports the hypothesis that the NMDA receptor plays a key role in the cognitive impairment associated with developmental Pb neurotoxicity.

Acknowledgments

The work presented in this chapter was supported by grant number ES06189 from the National Institute of Environmental Health Sciences. The author is greatful to Mr. Anthony Kuhlmann for reviewing the manuscript.

References

Aizeman, E., Frosch, M. P., and Lipton, S. A. (1989). Selective modulation of NMDA responses by reduction and oxidation. *Neuron* **2,** 1257–1263.

Aizenman, E., Hartnett, K. A., and Reynolds, I. J. (1990). Oxygen free radicals regulate NMDA receptor function via a redox modulatory site. *Neuron* **5,** 841–846.

Aizenman, E., Hartnett, K. A., Zhong, C., Gallop, P. M., and Rosenberg, P. A. (1992). Interaction of the putative essential nutrient pyrroloquinoline quinone with the *N*-methyl-D-aspartate receptor redox modulatory site. *J. Neurosci.* **12,** 2362–2369.

Akazawa, C., Shigemoto, R., Bessho, Y., Nakahishi, S., and Mizuno, N. (1994). Differential expression of five *N*-methyl-D-aspartate receptor subunit mRNAs in the cerebellum of developing and adult rats. *J. Comp. Neurol.* **347,** 150–160.

Alfano, D. P., and Petit, T. L. (1982). Neonatal lead exposure alters the dendritic development of hippocampal dentate granule cells. *Exp. Neurol.* **75,** 275–288.

Alkondon, M., Costa, A. C. S., Radhakrishnanm, V., Aronstam, R. S., and Albuquerque, E. X. (1990). Selective blockade of NMDA-activated channel currents may be implicated in learning deficits caused by lead. *FEBS Lett.* **261,** 124–130.

Altmann, L., Gutowski, M., and Wiegand, H. (1994). Effects of maternal lead exposure on functional plasticity in the visual cortex and hippocampus of immature rats. *Dev. Brain Res.* **81,** 50–56.

Altmann, L., Sveinsson, K., and Wiegand, H. (1991). Long-term potentiation in rat hippocampal slices is impaired following acute lead perfusion. *Neurosci. Lett.* **128,** 109–112.

Altmann, L., Weinsberg, F., Sveinsson, K., Lilienthal, H., Wiegand, H., Winneke, G. (1993). Impairment of long-term potentiation and learning following chronic lead exposure. *Toxicol. Lett.* **66,** 105–112.

Ault, B., Evans, R. H., Francis, A. A., Oakes D. J., and Watkins, J. C. (1980). Selective depression of excitatory amino acid induced depolarization by magnesium ions in isolated spinal cord preparations. *J. Physiol.* **307,** 413–428.

Bellinger, D., Sloman, J., Leviton, A., Rabinowitz, M., Needleman, H. L., and Waternaux, C. (1991). Low-level lead exposure and children's cognitive function in the preschool years. *Pediatrics* **87,** 219–277.

Ben-Ari, Y., Aniksztejn, L., and Bregestovski, P. (1992). Protein kinase C modulation of NMDA currents: an important link for LTP induction *TINS* **15,** 333–338.

Bow, M. A., and Nadler, J. V. (1990). Developmental increase in the sensitivity to magnesium of NMDA receptors on CA1 hippocampal pyramidal cells. *Dev. Brain Res.* **56,** 55–61.

Brewer, G. J., and Cotman, C. W. (1989). NMDA receptor regulation of neuronal morphology in cultured hippocampal neurons. *Neurosci. Lett.* **99,** 268–273.

Bristow, D. R., Bowery, N. G., and Woodruff, G. N. (1986). Light microscopic autoradiographic localization of [^{3}H]-glycine and [^{3}H]-strychnine binding sites in rat brain. *Eur. J. Pharmacol.* **126,** 303–307.

Brooks, W. J., Petit, T. L., LeBoutillier, J. C., and Lo, R. (1991). Rapid alterations of synaptic number and postsynaptic thickening length by NMDA: an electron microscopic study in the occipital cortex of postnatal rats. *Synapse* **8,** 41–48.

Brooks, W. J., Petit, T. L., Leboutillier, J. C., Nobrega, J. N., and Jarvis, M. F. (1993). Differential effects of early chronic lead exposure on postnatal rat brain NMDA, PCP, and adenosine A1 receptors: an autoradiographic study. *Drug Dev. Res.* **29,** 40–47.

Chen, L., and Huang, L. Y. M. (1992). Protein kinase C reduces Mg^{2+} block of NMDA-receptor channels as a mechanism of modulation. *Nature* **356,** 521–523.

Christine, C. W., and Choi, D. W. (1990). Effect of zinc on NMDA receptor-mediated channel currents in cortical neurons. *J. Neurosci.* **10,** 108–116.

Collingridge, G. L., and Lester, R. A. (1989). Excitatory amino acid receptors in the vertebrate central nervous system. *Pharmacol. Rev.* **40,** 143–195.

Collingridge, G. L., and Singer, W. (1990). Excitatory amino acid receptors and synaptic plasticity. *TIPS* **11,** 290–296.

Cory-Slechta, D. A., Weiss, B., and Cox, C. (1985). Performance and exposure indices of rats exposed to low concentrations of lead. *Toxicol. Appl. Pharmacol.* **78,** 291–299.

Cotman, C. W., Monaghan, D. T., Ottersen, O. P., and Storm-Mathisen, J. (1987). Anatomical organization of excitatory amino acid receptors and their pathways. *TINS* **10,** 273–280.

Dalkara, T., Erdemli, G., Barun, S., and Onur, R. (1992). Glycine is required for NMDA receptor activation: electrophysiological evidence from intact rat hippocampus. *Brain Res.* **576,** 197–202.

Desmond, N. L., and Levy, W. B. (1986). Changes in the numerical density of synaptic contacts associated with long-term potentiation in the hippocampal dentate gyrus. *J. Comp. Neurol.* **253,** 466–475.

Durand, G. M., Gregor, P., Zheng, X., Bennett, M. V. L., Uhl, G. R., and Zukin, R. S. (1992). Cloning of an apparent splice variant of the rat *N*-methyl-D-aspartate receptor NMDAR1 with altered sensitivity to polyamines and protein kinase C. *Proc. Nat. Acad. Sci.* **89,** 9359–9363.

Ehlers, M. D., Zhang, S., Bernhardt, J. P., and Huganir, R. L. (1996). Inactivation of NMDA receptors by direct interaction of calmodulin with the NR1 subunit. *Cell* **84,** 745–755.

Enomoto, R., Ogita, K., Han, D., and Yoneda, Y. (1992). Differential modulation by divalent cations of [^{3}H]-MK-801 binding in brain synaptic membranes. *J. Neurochem.* **59,** 473–481.

Facchinetti, F., Ciani, E., Dall'Olio, R., Virgili, M., Contestabile, A., and Fonnum, F. (1993). Structural, neurochemical and behavioral consequences of neonatal blockade of NMDA receptor through chronic treatment with CGP 39551 or MK-801. *Dev. Brain Res.* **74,** 219–224.

Facchinette, F., Dall'Olio, R., Sparapani, M., Virgili, M., and Contestabile, A. (1994). Long lasting effects of chronic neonatal blockade of *N*-methyl-D-aspartate receptor through the competitive antagonist CGP 39551 in rats. *Neurosci.* **60,** 343–353.

Farrant, M., Feldmeyer, D., Takahashi, T., and Cull-Candy, S. G. (1994). NMDA receptor channel diversity in the developing cerebellum. *Nature* **368,** 335–339.

Forrest, D., Yuzaki, M., Sores, H. D., Ng. L., Luk, D. C., Sheng, M., Stewart, C. L., Morgan, J. I., Connor, J. A., and Curran, T. (1994). Targeted disruption of NMDA receptor 1 gene abolishes NMDA response and results in neonatal death. *Neuron* **113,** 325–338.

Gilbert, M. E., Mack, C. M., and Lasley, S. M. (1996). Chronic developmental lead exposure increases the threshold for long-term potentiation in rat dentate gyrus *in vivo. Brain Res.* **736,** 118–124.

Goldstein, G. W., Asbury, A. K., and Diamond, I. (1974). Uptake of lead and reaction of brain capillaries. *Arch. Neurol.* **31,** 382–389.

Gu, Y., and Huang, L. Y. M. (1994). Modulation of glycine affinity for NMDA receptors by extracellular Ca^{2+} in trigeminal neurons. *J. Neurosci.* **14,** 4561–4570.

Guilarte, T. R. (1997). Pb^{2+} inhibits NMDA receptor function at high and low affinity sites: developmental and regional brain expression. *NeuroToxicology* **18,** 43–52.

Guilarte, T. R., and Miceli, R. C. (1992). Age-dependent effects of lead on [^{3}H]-MK-801 binding to the NMDA receptor-gated ionophore: *in vitro* and *in vivo* studies. *Neurosci. Lett.* **148,** 27–30.

Guilarte, T. R., Miceli, R. C., Altmann, L., Weinsburg, F., Winneke, G., and Weigand, H. (1993). Chronic prenatal and postnatal lead exposure increases [^{3}H]-MK-801 binding sites in adult rat forebrain *Eur. J. Pharmacol.* **248,** 273–275.

Guilarte, T. R., Miceli, R. C., and Jett, D. A. (1994). Neurochemical aspects of hippocampal and cortical lead neurotoxicity. *NeuroToxicology* **15,** 459–466.

Guilarte, T. R., Miceli, R. C., and Jett, D. A. (1995). Biochemical evidence of an interaction of lead at the zinc allosteric sites of the NMDA receptor complex: effects of neuronal development. *NeuroToxicology* **16,** 63–72.

Gutowski, M., Altmann, L., Sveinsson, K., and Wiegand, H. (1997). Postnatal development of synaptic plasticity in the CA3 hippocampal region of control and lead-exposed Wistar rats. *Dev. Brain Res.* **98,** 82–90.

Hamon, B., and Heinemann, U. (1988). Developmental changes in neuronal sensitivity to excitatory amino acids in area CA1 of the rat hippocampus. *Dev. Brain Res.* **38,** 286–290.

Harris, E. W., and Cotman, C. W. (1986). Long-term potentiation of guinea pig mossy fiber responses is not blocked by *N*-methyl-D-aspartate antagonists. *Neurosci. Lett.* **70,** 132–137.

Harris, E. W., Ganong, A. H., and Cotman, C. W. (1984). Long-term potentiation in the hippocampus involves activation of *N*-methyl-D-aspartate receptors. *Brain Res.* **323,** 132–137.

Harris, K. M., and Teyler, T. J. (1984). Developmental onset of long-term potentiation in area CA1 of the rat hippocampus. *J. Physiol.* **346,** 27–48.

Hashemzadeh-Gargari, H., and Guilarte, T. R. (1997). Lead inhibits calcium-mediated increases in [^{3}H]-MK-801 binding to the NMDA receptor in rat brain membranes. *The Toxicologist* (*Fund. Appl. Tox.*) **36**(1), 67.

Hebb, D. O. (1949). *The Organization of Behavior.* John Wiley & Sons, New York.

Hollmann, M., Boulter, J., Maron, C., Beasley, L., Sullivan, J., Pecht, G., and Heinemann, S. (1993). Zinc potentiates agonist-induced currents at certain splice variants of the NMDA receptor. *Neuron* **10,** 943–954.

Huettner, J. E., and Bean, B. P. (1988). Block of *N*-methyl-D-aspartate-activated current by the anticonvulsant MK-801: selective binding to open channels. *Proc. Nat. Acad. Sci.* **85,** 1307–1311.

Ikeda, K., Nagasawa, H., Mori, H., Araki, K., Sakimura, K., Watanabe, M., Inoue, Y., and Mishina, M. (1992). Cloning and expression of the epsilon-4 subunit of the NMDA receptor channel. *FEBS Lett.* **313,** 34–38.

Insel, T. R., Miller, L. P., and Gelhard, R. E. (1990). The ontogeny of excitatory amino acid receptors in rat forebrain-I. *N*-methyl-D-aspartate and quisqualate receptors. *Neurosci.* **35,** 31–43.

Ishihara, K., Alkondon, M., Montes, J. G., and Albuquerque E. X. (1995). Ontogenically related properties of *N*-methyl-D-aspartate receptors in rat hippocampal neurons and the age-specific sensitivity of developing neurons to lead. *J. Pharmacol. Exp. Ther.* **273,** 1459–1470.

Ishii, T., Moriyoshi, K., Sugihara, H., Sakurada, K., Kadotani, H., Yokoi, M., Akazawa, C., Shigemoto, R., Mizuno, N., Masu, M., and Nakanishi, S. (1993). Molecular characterization of the family of the *N*-methyl-D-aspartate receptor subunits. *J. Biol. Chem.* **268,** 2836–2843.

Ito, I., and Sugiyama, H. (1991). Roles of glutamate receptors in long-term potentiation at hippocampal mossy fiber synapses. *NeuroReport* **2,** 333–336.

Izquierdo, I. (1993). Long-term potentiation and the mechanism of memory. *Drug Dev. Res.* **30,** 1–17.

Izquierdo, I., Medina, J. H., Bianchin, M., Walz, R., Zanatta, M. S., Da Silva, R. C., Bueno e Silva, M., Ruschel, A. C., and Paczko, N. (1993). Memory processing by the limbic system: role of specific neurotransmitter systems. *Behav. Brain Res.* **58,** 91–98.

Javitt, D. C., and Zukin, S. R. (1989). Interaction of [^{3}H]-MK-801 with multiple states of the *N*-methyl-D-aspartate receptor complex of rat brain. *Proc. Nat Acad. Sci.* **86,** 740–744.

Jett, D. A., and Guilarte, T. R., (1995). Developmental Pb exposure alters *N*-methyl-D-aspartate and muscarinic cholinergic receptors in the rat hippocampus: an autoradiographic study. *Neurotoxicology* **16,** 7–18.

Jett, D. A., Kuhlmann, A. C., and Guilarte, T. R. (1997). Intrahippocampal administration of lead impairs performance of rats in the Morris Water Maze. *Pharmacol. Biochem. Beh.* **57,** 263–269.

Johnson, J. W., and Ascher, P. (1987). Glycine potentiates the NMDA response in cultured mouse brain neurons. *Nature* **325,** 529–531.

Kalb, R. G. (1994). Regulation of motor neuron dendrite outgrowth by NMDA receptor activation. *Development* **120,** 3063–3071.

Kiraly, E., and Jones, D. G. (1982). Dendritic spine changes in rat hippocampal pyramidal cells after postnatal lead exposure: a golgi study. *Exp. Neurol.* **77,** 236–239.

Kleckner, N. W., and Dingledine, R. (1988). Requirement of glycine in activation of NMDA receptors expressed in *Xenopus* oocytes. *Science* **241,** 835–837.

Kohr, G., Eckardt, S., Luddens, H., Monyer, H., and Seeberg, P. H. (1994). NMDA receptor channels: subunit specific potentiation by reducing agents. *Neuron* **12,** 1031–1040.

Komuro, H., and Rakic, P. (1993). Modulation of neuronal migration by NMDA receptors. *Science* **260,** 95–97.

Kurpp, J. J., Vissel, B., Heinemann, S. F., and Westbrook, G. L. (1996). Calcium-dependent inactivation of recombinant *N*-methyl-D-aspartate receptors is NR2 subunit specific. *Mol. Pharm.* **50,** 1680–1688.

Kutsuwada, T., Kashiwabushi, N., Mori, H., Sakimura, K., Kushiya, E., Araki, K., Meguro, H., Masak, H., Kumanishi, T., Arakawa, M., and Mishina, M. (1992). Molecular diversity of the NMDA receptor channel. *Nature* **358,** 36–41.

Lasley, S. M., Polan-Curtain, J., and Armstrong, D. L. (1993). Chronic exposure to environmental levels of lead impairs *in vivo* induction of long-term potentiation in rat hippocampal dentate. *Brain Res.* **614,** 347–351.

Lau, L. F., and Huganir, R. L. (1995). Differential tyrosine phosphorylation of *N*-methyl-D-aspartate receptor subunits. *J. Biol. Chem.* **270,** 20036–20041.

Laurie, D. J., Putzke, J., Zieglgänsberger, W., Seeberg, P. H., and Tölle, T. R. (1995). The distribution of splice variants of the NMDAR1 subunit mRNA in adult rat brain. *Mol. Brain. Res.* **32,** 94–108.

Laurie, D. J., and Seeberg, P. H. (1994a). Ligand affinities at recombinant *N*-methyl-D-aspartate receptors depend on subunit composition. *Eur. J. Pharm.* (*Mol. Pharm. Sec.*) **268,** 335–345.

Laurie, D. J., and Seeberg, P. H. (1994b). Regional and developmental heterogeneity in splicing of the rat brain NMDAR1 mRNA. *J. Neurosci.* **14,** 3180–3194.

Legendre, P., Rosenmund, C., and Westbrook, G. L. (1993). Inactivation of NMDA channels in cultured hippocampal neurons by intracellular calcium. *J. Neurosci.* **13,** 674–684.

Legendre, P., and Westbrook, G. L. (1990). The inhibition of single *N*-methyl-D-aspartate activated channels by zinc ions on cultured rat neurones. *J. Physiol.* **429,** 429–449.

Levy, D. I., Sucher, N. J., and Lipton, S. A. (1991). Glutathione prevents *N*-methyl-D-aspartate receptor–mediated neurotoxicity. *NeuroReport* **2,** 345–347.

Liberman, D. N., and Mody, I. (1994). Regulation of NMDA channel function by endogenous Ca^{2+}-dependent phosphatase. *Nature* **369,** 235–239.

LoTurco, J. L., Blanton, M. G., and Kriegstein, A. R. (1991). Initial expression and endogenous activation of NMDA channels in early neocortical development. *J. Neurosci.* **11,** 792–799.

MacDermott, A. B., Mayer, M. L., and Westbrook, G. L. (1986). NMDA receptor activation increases calcium concentration in cultured spinal cord neurones. *Nature* **321,** 519–522.

Madison, D. V., Malenka, R. C., and Nicoll, R. A. (1991). Mechanisms underlying long-term potentiation of synaptic transmission. *Ann. Rev. Neurosci.* **14,** 379–397.

Majewska, M. D., and Bell, J. A. (1990). Ascorbic acid protects neurons from injury induced by glutamate and NMDA. *NeuroReport* **1,** 194–196.

Marchioro, M., Swanson, K. L., Aracava, Y., and Albuquerque, E. X. (1996). Glycine- and calcium-dependent effects of lead on *N*-methyl-D-aspartate receptor function in rat hippocampal neurons. *J. Pharm. Exp. Ther.* **279,** 143–153.

Markram, H., and Segal, M. (1992). Activation of protein kinase C suppresses responses to NMDA in rat CA1 hippocampal neurones. *J. Physiol.* **457,** 491–501.

Marvizon, J. C. G., and Skolnick, P. (1988). [^{3}H]-glycine binding is modulated by Mg^{2+} and other ligands of the NMDA receptor–cation channel complex. *Eur. J. Pharm.* **151,** 157–158.

Mayer, M. L., Vyklicky, L., and Westbrook, G. L. (1989). Modulation of excitatory amino acid receptors by group IIB metal cations in cultured mouse hippocampal neurones. *J. Physiol.* **415,** 329–350.

Mayer, M. L., and Westbrook, G. L. (1987). Permeation and block of *N*-methyl-D-aspartate receptor channels by divalent cations in mouse cultured central neurons. *J. Physiol.* **394,** 501–527.

McBain, C. J., and Mayer, M. L. (1994). *N*-methyl-D-aspartate receptor structure and function. *Physiol. Rev.* **74,** 723–760.

McDonald, J. W., Johnston, M. V., and Young, A. B. (1990a) Differential ontogenic development of three receptors comprising the NMDA receptor/channel complex in the rat hippocampus. *Exp. Neurol.* **110,** 237–247.

McDonald, J. W., Silverstein, F. S., and Johnston, M. V. (1990b). MK-801 pretreatment enhances *N*-methyl-D-aspartate mediated brain injury and increases brain *N*-methyl-D-aspartate recognition site binding in rats. *Neurosci.* **38,** 103–113.

McHugh, T. J., Blum, K. I., Tsien, J. Z., Tonegawa, S., and Wilson, M. A. (1996). Impaired hippocampal representation of space in CA1-specific NMDAR1 knockout mice. *Cell* **87,** 1339–1349.

Meguro, H., Hori, H., Araki, K., Kushiya, E., Kutsuwada, T., Yamakazi, M., Kumanishi, T., Arakawa, M., Sakimura, K., and Mishina, M. (1992). Functional characterization of a heteromeric NMDA receptor channel expressed from cloned cDNAs. *Nature* **357,** 70–74.

Monyer, H., Sprengel, R., Shoepfer, R., Herb, A., Higuchi, M., Lomeli, H., Burnashev, N., Sakmann, B., and Seeberg, P. H. (1992). Heteromeric NMDA receptors: molecular and functional distinction of subtypes. *Science* **256,** 1217–1221.

Monyer, H., Burnashev, N., Laurie, D. J., Sakmann, B., Seeberg, P. H. (1994). Developmental and regional expression in the rat brain and functional properties of four NMDA receptors. *Neuron* **12,** 529–540.

Moran, J., and Patel, A. J. (1989). Stimulation of the *N*-methyl-D-aspartate receptor promotes the biochemical differentiation of cerebellar granule neurons and not astrocytes. *Brain Res.* **486,** 15–25.

Moriyoshi, K., Masu, M., Ishii, T., Shigemoto, R., Mizuno, N., and Nakanishi, S. (1991). Molecular cloning and characterization of the rat NMDA receptor. *Nature* **354,** 31–37.

Morris, R. G. M., Anderson, E., Lynch, G. S., and Baudry, M. (1986). Selective impairment of learning and blockade of long-term potentiation by an *N*-methyl-D-aspartate receptor antagonist, AP5. *Nature* **319,** 774–776.

Morrisett, R. A., Mott, D. D., Lewis, D. V., Wilson, W. A., and Swartzwelder, H. S. (1990). Reduced sensitivity of the *N*-methyl-D-aspartate component of synaptic transmission to magnesium in hippocampal slices from immature rats. *Dev. Brain Res.* **56,** 257–262.

Mushak, P., Davis, J. M., Crocetti, A. F., and Grant, L. D. (1989). Prenatal and postnatal effects of low level lead exposure: integrated summary of a report to the US Congress on childhood lead poisoning. *Env. Res.* **50,** 11–36.

Nakanishi, N., Axel, R., and Shneider, N. A. (1992). Alternative splicing generates functionally distinct *N*-methyl-D-aspartate receptors. *Proc. Nat. Acad. Sci.* **89,** 8552–8556.

Needleman, H. L. (1993). The current status of childhood low level lead toxicity. *NeuroToxicology* **14,** 161–166.

Needleman, H. L., Gunnoe, C., Leviton, A., Reed, R., and Peresie, H. (1979). Deficits in psychologic and classroom performance of children with elevated dentine lead levels. *N. Engl. J. Med.* **300,** 689–695.

Nowak, L., Bregestovski, P., Ascher, P., Herbet, A., and Prochiantz, A. (1984). Magnesium gates glutamate-activated channels in mouse central neurons. *Nature* **307,** 462–465.

Omelchenko, I. A., Nelson, C. S., Marino, J. L., and Allen, C. N. (1996). The sensitivity of *N*-methyl-D-aspartate receptors to lead is dependent on the receptor subunit composition. *J. Pharm. Exp. Ther.* **278,** 15–20.

Paoletti, P., Neyton, J., and Ascher, P. (1995). Glycine-independent and subunit-specific potentiation of NMDA responses by extracellular Mg^{2+} *Neuron* **15,** 1109–1120.

Pearce, I. A., Cambray-Deakin, M. A., and Burgoyne, R. D. (1987). Glutamate acting on NMDA receptors stimulate neurite outgrowth from cerebellar granule cells. *FEBS Lett.* **223,** 143–147.

Peters, S., Koh, J., and Choi, D. W. (1987). Zinc selectively blocks the action of *N*-methyl-D-aspartate on cortical neurons. *Science* **236,** 589–593.

Petit, T. L. (1986). Developmental effects of lead: its mechanisms in intellectual functioning and neural plasticity. *NeuroToxicology.* **7,** 483–496.

Petit, T. L., Alfano, D. P., and LeBoutillier, J. C. (1983). Early lead exposure and the hippocampus: a review and recent advances. *NeuroToxicology.* **4,** 79–94.

Petit, T. L., and LeBoutellier, J. C. (1979). Effects of lead exposure during development on neocortical dendritic and synaptic structure. *Exp. Neurol.* **64,** 482–492.

Petit, T. L., LeBoutillier, J. C., and Brooks, W. J. (1992). Altered sensitivity to NMDA following developmental lead exposure in rats. *Physiol. Beh.* **52,** 687–693.

Pokorny, J., and Yamamoto, T. (1981). Postnatal ontogenesis of hippocampal CA1 area in rats. I. Development of dendritic arborisation in pyramidal neurons. *Brain Res. Bull.* **7,** 113–120.

Portera-Cailliau, C., Price, D. L., and Martin, L. J. (1996). *N*-methyl-D-aspartate receptor proteins NR2A and NR2B are differentially distributed in the developing rat central nervous system by subunit-specific antibodies. *J. Neurochem.* **66,** 692–700.

Premkumar, L. S., and Auerback, A. (1996). Identification of a high affinity divalent cation binding site near the entrance of the NMDA receptor channel. *Neuron* **16,** 869–890.

Probst, A., Cortés, R., and Palacios, J. M. (1986). The distribution of glycine receptors in the human brain. A light microscopic autoradiographic study using [^{3}H]-strychnine. *Neurosci.* **17,** 11–35.

Rabacchi, S., Bailly, Y., Delhaye-Bouchaud, N., and Mariani, J. (1992). Involvement of the *N*-methyl-D-aspartate (NMDA) receptor in synapse elimination during cerebellar development. *Science* **256,** 1823–1825.

Rajdev, S., and Reynolds, I. J. (1992). Effects of monovalent and divalent cations on 3-(+)[^{125}I]Iododizocilpine binding to the *N*-methyl-D-aspartate receptor of rat brain membranes. *J. Neurochem.* **58,** 1469–1476.

Ransom, R. W. (1991). Polyamine and ifenprodil interactions with the NMDA receptor's glycine site. *Eur. J. Pharm.* (*Mol. Pharm. Sec.*) **208,** 67–71.

Ransom, R. W., and Stec, N. L. (1988). Cooperative modulation of [^{3}H]-MK-801 binding to the *N*-methyl-D-aspartate receptor–ion channel complex by 1-glutamate, glycine, and polyamines. *J. Neurochem.* **51,** 830–836, 1988.

Reymann, K. G. (1993). Mechanisms underlying synaptic long-term potentiation in the hippocampus: focus on postsynaptic glutamate receptors and protein kinases. *Funct. Neurol.* (suppl.to N5) **8,** 7–32.

Rice, D. C. (1985). Chronic low-lead exposure from birth produces deficits in discrimination reversal in monkeys. *Toxicol. Appl. Pharm.* **75,** 201–210.

Riva, M. A., Tascedda, F., Molteni, R., and Racagni, G. (1994). Regulation of NMDA receptor subunit mRNA expression in the rat brain during postnatal development. *Mol. Brain Res.* **25,** 209–216.

Rothman, R. B., Reid, A. A., Monn, J. A., Jacobson, A. E., and Rice, K. C. (1989). The psychotomimetic drug phencyclidine labels two high affinity binding sites in guinea pig brain: evidence for *N*-methyl-D-aspartate–coupled and dopamine reuptake carrier–associated phencyclidine binding sites. *Mol. Pharm.* **36,** 887–896.

Sacaan, A. I., and Johnson, K. M. (1989). Spermine enhances binding to the glycine site associated with the *N*-methyl-D-aspartate receptor complex. *Mol. Pharm.* **36,** 836–839.

Sakimura, K., Kutsuwada, T., Ito, I., Manabe, T., Akayama, C., Kushiya, E., Yagi, T., Alzawa, S., Inoue, Y., Sugiyama, H., and Mishina, M. (1995). Reduced hippocampal LTP and spatial learning in mice lacking the NMDA receptor epsilon1 subunit. *Nature* **373,** 151–155.

Sakurai, S. Y., Cha, J. H. J., Penney, J. B., and Young, A. B. (1991). Regional distribution and properties of [^{3}H]-MK-801 binding sites determined by quantitative autoradiography in rat brain. *Neurosci.* **40,** 533–543.

Schulte, S., Muller, W. E., and Friedberg, K. D. (1995). *In vitro* and *in vivo* effects of lead on specific [^{3}H]-MK-801 binding to NMDA receptors in the brain of mice. *NeuroToxicology* **16,** 309–318.

Sharma, G., and Stevens, C. F. (1996). Interactions between two divalent ion binding sites in *N*-methyl-D-aspartate receptor channels. *Proc. Nat. Acad. Sci.* **93,** 14170–14175.

Sigel, E., Baur, R., and Malherbe, P. (1994). Protein kinase C transiently activates heteromeric *N*-methyl-D-aspartate receptor channels independent of the phosphorylatable C-terminal splice domain and of consensus phosphorylation sites. *J. Biol. Chem.* **269,** 8204–8208.

Simon, D. K., Prusky, G. L., O'Leary, D. D. M., and Constantine-Paton, M. (1992). *N*-methyl-D-aspartate receptor antagonists disrupt the formation of a mammalian neural map. *Proc. Nat. Acad. Sci.* **89,** 10593–10597.

Sucher, N. J., and Lipton, S. A. (1991). Redox modulatory site of the NMDA receptor–channel complex: regulation by oxidized glutathione. *J. Neurosci. Res.* **30,** 582–591.

Sugihara, H., Moriyoshi, K., Ishii, T., Masu, M., and Nakanishi, S. (1992). Structures and properties of seven isoforms of the NMDA receptor generated by alternative splicing. *Bioch. Bioph. Res. Comm.* **185,** 826–832.

Tingley, W. G., Roche, K. W., Thompson, A. K., and Huganir, R. L. (1993). Regulation of NMDA receptor phosphorylation by alternative splicing of the C-terminal domain. *Nature* **364,** 70–73.

Tong, G., Shepard, D., and Jahr, C. E. (1995). Synaptic desensitization of NMDA receptors by calcineurin. *Science* **267,** 1510–1512.

Traynelis, S. F., and Cull-Candy, S. G. (1990). Proton inhibition of *N*-methyl-D-aspartate receptors in cerebellar neurons. *Nature* **345,** 347–350.

Tremblay, E., Rosin, M. P., Represa, A., Charriaut-Marlangue, C., and Ben-Ari, Y. (1988). Transient increased density of NMDA binding sites in the developing rat hippocampus. *Brain Res.* **461,** 393–396.

Tsien, J. Z., Chen, D. F., Gerber, D., Tom, C., Mercer, E. H., Anderson, D. J., Mayford, M., Kandel, E. R., and Tonegawa, S. (1996a). Subregion- and cell type–restricted gene knockout in mouse brain. *Cell* **87,** 1317–1326.

Tsien, J. Z., Huerta, P. T., and Tonegawa, S. (1996b). The essential role of hippocampal CA1 NMDA receptor-dependent synaptic plasticity in spatial learning. *Cell* **87,** 1327–1338.

Ujihara, H., and Albuquerque, E. X. (1992). Developmental change of the inhibition by lead of NMDA-activated currents in cultured hippocampal neurons. *J. Pharmacol. Exp. Ther.* **263,** 868–875.

United States Department of Health and Human Services. Agency for Toxic Substances and Disease Registry (1988). The nature and extent of lead poisoning in children in the United States: a report to Congress. US Department of Health and Human Services, Public Health Service, Atlanta.

United States Department of Health and Human Services. Centers for Disease Control (1991). Preventing lead poisoning in young children.

Upchurch, M., and Whener, J. M. (1990). Effects of *N*-methyl-D-aspartate antagonism on spatial learning in mice. *Psychopharmacol.* **100,** 209–214.

Uteshev, V., Busselberg, D., and Haas, H. L. (1993). Pb modulates the NMDA receptor complex. *Naunyn-Schmiedeberg's Arch. Pharmacol.* **347,** 209–213.

Venable, N., and Kelly, P. H. (1990). Effects of NMDA receptor antagonists on passive avoidance learning and retrival in rats and mice. *Psychopharmacol.* **100,** 215–221.

Wang, L. Y., and MacDonald, J. F. (1995). Modulation by magnesium of the affinity of NMDA receptors for glycine in murine hippocampal neurones. *J. Physiol.* **486,** 83–95.

Wang, Y. T., and Slater, M. W. (1994). Regulation of NMDA receptors by tyrosine kinases and phosphatases. *Nature* **369,** 233–235.

Wenzel, A., Villa, H., Mohler, H., and Benke, D. (1996). Developmental and regional expression of NMDA receptor subtypes containing the NR2D subunit in rat brain. *J. Neurochem.* **66,** 1240–1248.

Williams, K. (1993). Ifenprodil descriminates subtypes of the *N*-methyl-D-aspartate receptor: selectivity and mechanisms at recombinant heteromeric receptors. *Mol. Pharm.* **44,** 851–859.

Williams, K., Dichter, M. A., and Molinoff, P. B. (1992). Up-regulation of *N*-methyl-D-aspartate receptors in cultured cortical neurons after exposure to antagonists. *Mol. Pharm.* **42,** 147–151.

Williams, K., Romano, C., Dichter, M. A., and Molinoff, P. B. (1991). Modulation of the NMDA receptor by polyamines. *Life Sci.* **48,** 469–498.

Williams, K., Romano, C., and Molinoff, P. B. (1989). Effects of polyamines on the binding of [^{3}H]-MK-801 to the *N*-methyl-D-aspartate receptor: pharmacological evidence for the existence of a polyamine recognition site. *Mol. Pharm.* **36,** 575–581.

Williams, K., Zappia, A. M., Pritchett, D. B., Shen, Y. M., and Molinoff, P. B. (1994). Sensitivity of the *N*-methyl-D-aspartate receptor to polyamines is controlled by NR2 subunits. *Mol. Pharm.* **45,** 803–809.

Winneke, G., Altmann, L., Kramer, U., Turfeld, M., Behler, R., Gutsmuths, F. J., and Mangold, M. (1994). Neurobehavioral and neurophysiological observations in six year old children with low lead levels in East and West Germany. *Neurotoxicology* **15,** 705–714.

Wong, E. H. F., Kemp, J. A., Priestley, T., Knight, A. R., and Woodruff, G. N. (1986). The anticonvulsant MK-801 is a potent *N*-methyl-D-aspartate antagonist. *Proc. Nat. Acad. Sci.* **83,** 7104–7108.

Yamakazi, M., Hori, H., Araki, K., Mori, J., and Mishina, M. (1992) Cloning, expression and modulation of a mouse NMDA receptor subunit. *FEBS Lett.* **300,** 39–45.

Yeh, G. C., Bonhaus, D. W., and McNamara, J. O. (1990). Evidence that zinc inhibits *N*-methyl-D-aspartate receptor–gated ion channel by noncompetitive antagonism of glycine binding. *Mol. Pharm.* **38,** 14–19.

Zhong, J., Carroza, D. P., Williams, K., Pritchett, D. B., and Molinoff, P. B., (1995) and Expression of mRNA encoding subunits of the NMDA receptor in developing rat brain. *J. Neurochem.* **64,** 531–539.

Ziegler, E. E., Edwards, B. B., Jensen, R., Mahaffey, K. R., and Fomon, S. J. (1978). Absorption and retention of lead by infants. *Pediatr. Res.* **12,** 29–34.

Zukin, R. S., and Bennett, M. V. L. (1995). Alternatively spliced isoforms of the NMDAR1 receptor subunit. *TINS* **18,** 305–313.

PART IV

Nutrient and Chemical Disposition

In regard to developmental neurotoxicity, toxic agent delivery and nutritional status both have a major influence on outcome. If exposure occurs *in utero,* the placenta and extraembryonic membranes are the conduit for toxicant delivery and nutrient transfer. Postnatally, toxicant delivery is more direct, but just as during prenatal development, the blood–brain barrier plays a major role in determining nervous system exposure. The general topic of placental metabolism and transfer as it relates to developmental toxicity has been reviewed by Slikker and Miller (1994). As discussed, the placenta and extraembryonic membranes secure the embryo and fetus to the maternal source, the endronetrium. Another activity of these tissues is the transport of nutrients, waste products, and, of course, potential toxicants. As reviewed by Slikker (1994), the journey of an agent to the developing nervous system target begins with maternal considerations such as maternal distribution, metabolism, and elimination. The placenta may act to differentially transfer, metabolize, or store a toxicant. Fetal distribution, metabolism, and elimination may further modify an agent en route to the developing nervous system. Finally, the developing blood–brain barrier with its transport, metabolic, and membrane barrier characteristics has a decisive role to play. In the following chapters, several key aspects of both toxicant delivery and nutrition are considered in regards to developmental neurotoxicity.

Tissue dosimetry, as it relates to dose–response and developmental age, is the focus of Chapter 16 on physiologically based kinetic (PBK) models by Ellen J. O'Flahery. Because PBK models allow for prediction of dose to fetal neural tissues, they are powerful tools to predict and compare developmental neurotoxic risk across species. Likewise, similar models focused on early childhood growth can provide predictive power for postnatal development. The metabolic capability of the placenta is concisely addressed in Chapter 17 by Mont Jauchau and Hao Cheno. The various isoforms of cytochrome P-450 and related drug metabolizing enzymes are examined in relationship to developmental neurotoxicity. Knowledge of these enzymes and their relative abundance in the placenta provides for predictive capability in terms of which chemicals are metabolically activated or detoxified during placental passage to the developing conceptus.

The role of the blood–brain barrier in terms of its physiologic and functional characteristics is covered by Michael Aschner in Chapter 18. The blood–brain barrier is a specialized structure that serves to regulate central nervous system homeostasis and may also function to transport selected metals and other potential toxicants into the brain. The metabolic potential of the developing nervous system plays a role in the activation and deactivation of endogenous metabolic intermediates as well as exogenous chemicals. In Chapter 19, Saber Hussain and Syed F. Ali focus on the enzymes essential for controlling the oxidative stress level within the developing animal. The relative activities of sulfur containing scavenging molecules and various catalase and dismutase enzymes are determined in the developing animal and compared to activities in the adult.

Nutrient status is an important consideration in developmental neurotoxicologic outcome. In Chapter 20, Margaret L. Bogle addresses the balance of desired nutrient intake and overexposure to potentially toxic micronutrients that may well be essential at lower doses. Postnatal exposure to potentially toxic agents via lactation is the focus of Chapter 21 by Janine E. Polifka, who describes the maternal nutrient source of lactation not only in terms of its value to normal development, but also as a vehicle for toxicant exposure to the developing nervous system.

William Slikker, Jr.

References

Slikker, W. Jr. 1994. Placental transfer and pharmacokinetics of developmental neurotoxicants. In L. W. Chang, Ed. *Principles of Neurotoxicology,* Marcel Dekker, Inc., New York, pp. 659–680.

Slikker, W. Jr., and Miller, R. K. 1994. Placental metabolism and transfer: role in developmental toxicology. In C. A. Kimmel and J. Buelke-Sam, Eds. *Developmental Toxicology,* Raven Press, New York, pp. 245–283.

CHAPTER

16

PB/PK Models

ELLEN J. O'FLAHERTY
Retired
University of Cincinnati Department of Environmental Health
Cincinnati, Ohio 45267

I. Introduction

Tissue dosimetry is an emerging focus of dose–response characterization for risk assessment. From reliance on administered dose or on concentrations in foods, water, or air, risk assessors have progressed to the recognition that the amount, concentration, or rate of presentation of the proximal toxicant in target tissues is the dose metric of choice, and is the only meaningful dose metric whenever disposition or mechanism of action is not linearly related to exposure. An advance that has facilitated the linkage of external exposure with tissue dose is the development of physiologically based kinetic (PBK) models. This group of models relies on a set of differential rate equations to describe and predict the time course of uptake and loss of a toxicant in

*Abbreviations: PBK, physiologically based kinetic (models); pKa, the negative logarithm of the acid dissociation constant Ka.

body tissues. PBK models are designed as simplified but realistic descriptions of the body, the organs and tissues that compose it, and the blood flow rates to these organs and tissues. In this sense, the models may be thought of as generic, at least within a species. However, a PBK model must also be tailored to the toxicant of concern. Tissues of specific interest, either because they are target tissues or because they are tissues of entry or of elimination, are usually included explicitly in a model. Other tissues are grouped, most commonly in accordance with the flow rates of perfusing blood, because flow-limited distribution is assumed for many chemicals. PBK models are capable of incorporating the mechanism of tissue uptake (whether flow-limited or diffusion-limited), binding in blood plasma and/or tissues, other tissue accumulation and storage mechanisms, and the magnitude and localization of metabolism and excretion processes. In principle, they are readily convertible across species by means of allometric equations relating organ volumes, blood flow rates, and physiologic functions to body size, although allometric conversions are default procedures and may not be appropriate in some cases, particularly for conversion across species of parameters defining metabolism.

To date, the bulk of published PBK models have not been greatly concerned with changes in physiologic state during the time course of the simulation. Many controlled kinetic studies in animals use adult animals and are studies of toxicants whose half-lives are short relative to growth rates or other predictable physiologic changes. The volatile halogenated hydrocarbons are examples of such compounds. However, whole-body half-life can be comparable to life span for some toxicants, such as the bone volume–seeking element lead or the highly lipophilic and poorly metabolized dioxins. In these situations, anatomic and physiologic changes should be taken into account in development of a PBK model. This goal can be achieved by defining volumes and flow rates as functions of age or body weight.

The actions of developmental toxicants are expressed within a framework of rapid and profound physiologic change during pregnancy, infancy, and early childhood. Accordingly, and even for short–half-life chemicals, successful dose–response models for developmental toxicants must account for the physiologic changes that characterize pregnancy and early childhood growth and development. PBK models offer sufficient anatomic and physiologic realism and flexibility to recommend them as a technique for linking exposure to tissue dose, as well as for linking tissue dose to early biologic effects.

II. Physiologic Changes and Kinetics

Physiologic changes may have a marked impact on kinetics during pregnancy. Cummings (1983) compiled

TABLE 1 Changes in Tissue Weights, Blood Flows, and Physiologic Functions during Human Pregnancy

	Direction of change	Kinetic parameters potentially affected
Weights/volumes		
Liver	⇑	Distribution
Uterus	⇑	Distribution
Placenta	⇑	Distribution
Mammary glands	⇑	Distribution
Maternal body fat	⇑	Distribution
Maternal blood	⇑	Distribution
Total maternal body water	⇑	Distribution
Maternal plasma proteins	⇓	Distribution/elimination
Fetus	⇑	Distribution
Total fetal body water (as fraction of body weight)	⇓	Distribution in fetus
Blood flows		
Cardiac output	⇑	Distribution Pulmonary absorption/excretion
Renal plasma flow	⇑	Excretion
Glomerular filtration rate	⇑	Excretion
Physiologic functions		
Tidal volume (lung)	⇑	Pulmonary absorption/excretion
Minute volume (lung)	⇑	Pulmonary absorption/excretion
GI tract residence time	⇑	GI tract absorption/metabolism/excretion
Metabolism	⇑ or ⇓	Elimination

pharmacokinetic data from a number of sources for 20 drugs studied in pregnant and nonpregnant women, and found that half-life could be increased, decreased, or unchanged in pregnancy, whereas clearance and apparent volume of distribution tended to be either increased or unchanged. It was concluded that the complexity of possible interactions generated by the multifactorial nature of the changes taking place in both maternal physiology and drug metabolism during pregnancy rendered predictivity uncertain. Many of these changes, however, are known and their magnitudes can be specified. The expected effects on kinetics of known physiologic alterations during human pregnancy have been reviewed, summarized, and discussed (Krauer and Krauer, 1977; Sonawane and Yaffe, 1986; Mattison, 1986, 1990; Mattison *et al.*, 1991). At least some of these changes can be dealt with explicitly by their inclusion in PBK models. Many of the maternal changes in organ weights, blood flows, and physiologic functions during pregnancy are fairly well defined. Some of the more important ones are listed in Table 1, together with the particular aspect of kinetics (absorption, distribution, metabolism, or excretion) that they affect. Figure 1 illustrates a specific example of the use of a logistic function to describe the weight of the human uterus during gestation. As a component of a PBK model, this expression defines the constantly changing contribution of the uterus to the equally dynamic total potential volume of distribution. Any of the maternal changes of pregnancy can be incorporated directly into PBK models by means of such expressions.

Fetal growth and fetal physiologic and biochemical development are even more dramatic. Enough is known about fetal body weight, organ weights, and blood flows to generate reasonably reliable descriptions of their dependence on fetal age. The growth of the human fetus, with data taken from several sources, is shown in Fig. 2. This composite data set is described by a logistic

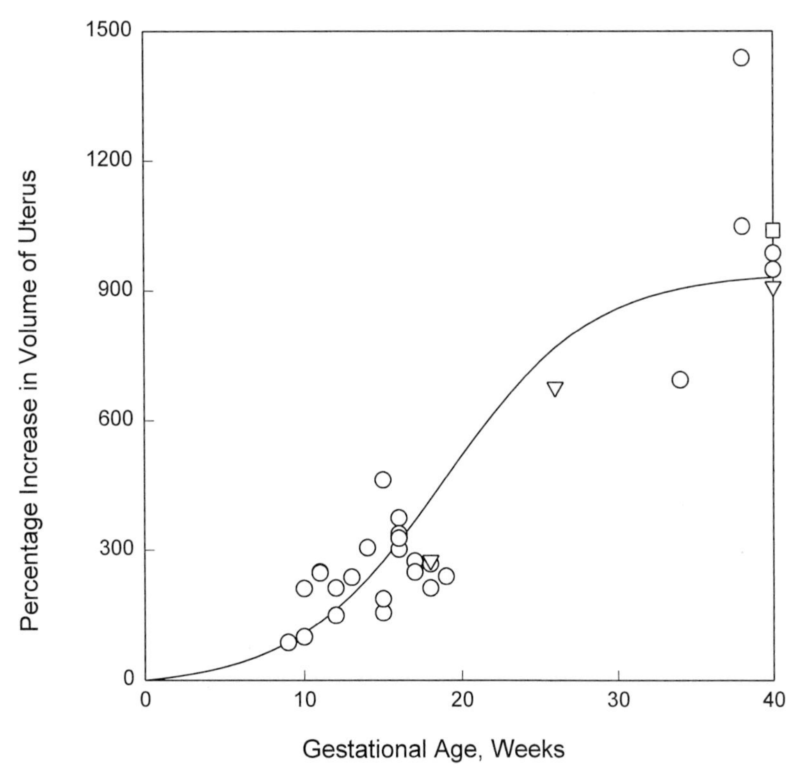

FIGURE 1 The increase in weight of the human uterus during pregnancy. The expression for uterine weight is

$$\mathrm{U0} + \frac{.76 \times (\mathrm{EXP}(\mathrm{AGEPRE} \times 11) - 1}{\mathrm{EXP}(4) + \mathrm{EXP}(\mathrm{AGEPRE} \times 11)}$$

where U0 is the weight of the nonpregnant uterus (kg) 0.14% of body weight) and AGEPRE is gestational age (yr). Data are from Hytten and Cheyne, 1969 (○); Morrione and Seifter, 1962 (▽); and Woessner and Brewer, 1963 (□).

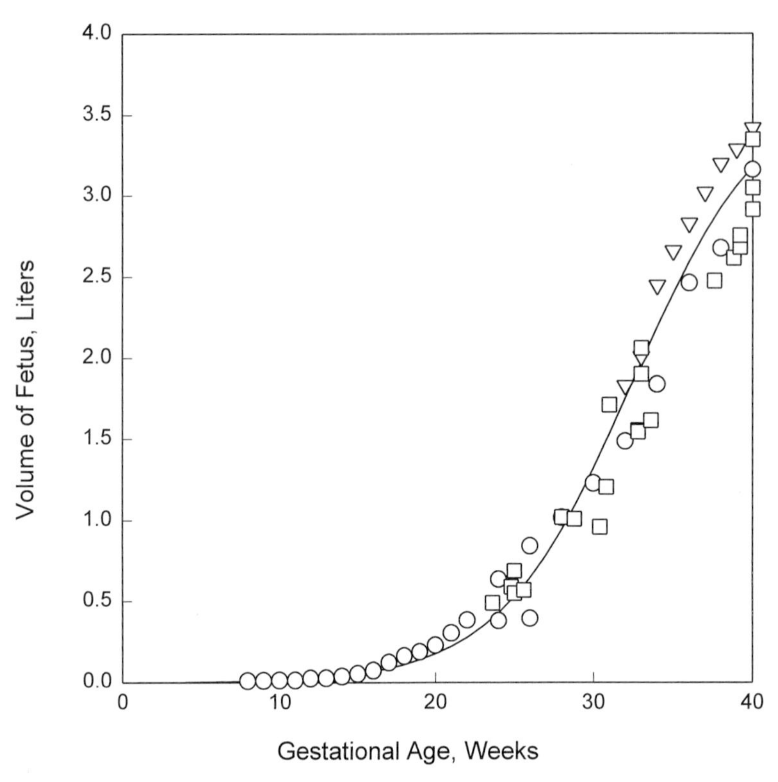

FIGURE 2 Human fetal weight as a function of gestational age. The expression plotted is

$$\mathrm{FW} = 1.055 \times \frac{3.5 \times \mathrm{EXP}(\mathrm{AGEPRE} \times 12.5) - 1}{\mathrm{EXP}(7.8) + \mathrm{EXP}(\mathrm{AGEPRE} \times 12.5)}$$

where FW is fetal weight (kg) and AGEPRE is gestational age (yr). Data are from Singer *et al.*, 1991 (○); Thomson *et al.*, 1968 (▽); and Ziegler *et al.*, 1976 (□).

expression with gestational age as the independent variable. Other investigators have used other expressions for the same purpose; for example, Luecke *et al.* (1994) have used a Gompertz function.

The growth of individual fetal organs is not necessarily proportional to total fetal body weight. Whereas fetal liver weight is an approximately constant fraction of fetal body weight throughout pregnancy, fetal bone is laid down largely during the last trimester. Figure 3 shows the weight of fetal marrow–free dry bone as a function of gestational age. It is described by an expoential expression based on fetal body weight to the 1.19 power. In the period during which fetal body weight doubles prior to birth, fetal marrow–free dry bone weight increases by a factor of 2.3.

Not only volumes but blood flow rates also may be described mathematically as functions of gestational age or of fetal body weight. Figure 4 illustrates fetal cardiac output as a function of gestational age, generated by an exponential expression with gestational age as the independent variable. In most cases, data sets such as the one in Fig. 4 can be fit satisfactorily by several different equations with different independent variables. To some extent, the choice of expressions is at the modeler's discretion. Those in common use are discussed in O'Flaherty (1994).

The rapid physical and biochemical growth of early childhood may also affect kinetics. Classic kinetics concepts have been applied to describe the effects of differences in metabolic capacity on disposition of therapeutic drugs in infants and adults. Little is known very precisely about the marked changes in body fluid volumes, metabolic capacities, and gastrointestinal absorption efficiency that take place shortly after birth (Sonawane and Yaffe, 1986), but most anatomic and physiologic features of growth during childhood can be modeled based on known relationships. Figure 5 shows glomerular filtration rate as a function of age. This relationship has been fit by two distinct allometric line segments—one for ages less than 1 year and one for ages greater

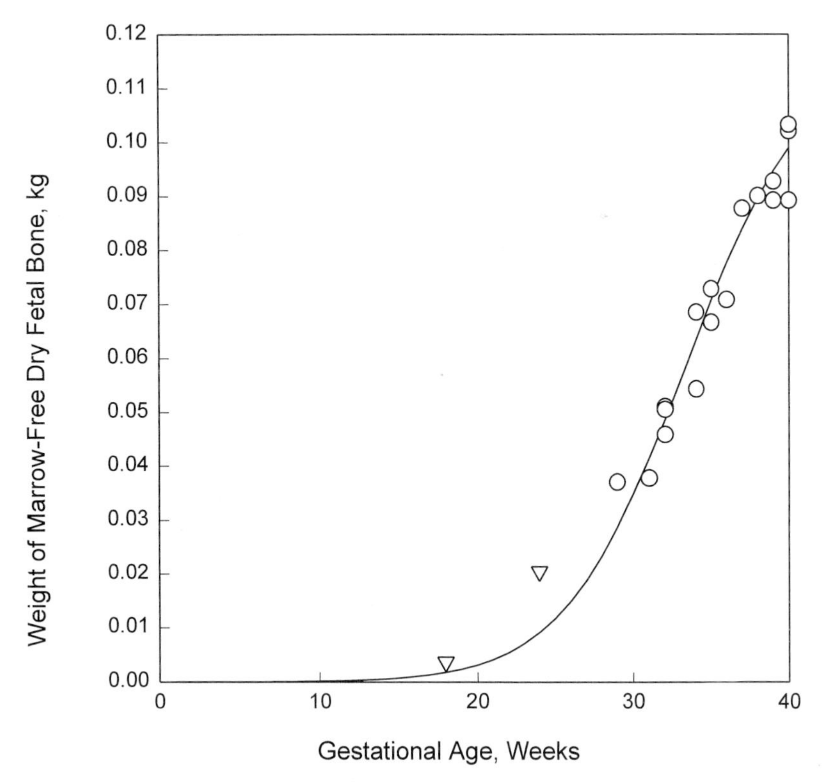

FIGURE 3 The dependence of marrow-free dry bone weight in the human fetus on gestational age. The expression plotted is

$$\text{WMFDB} = 0.022 \times (\text{FW}^{1.19})$$

where WMFDB is the weight of marrow-free dry bone (kg) and FW is fetal weight (kg). Data are from Hudson, 1965 (○) and Trotter and Peterson, 1969 (▽).

than 1 year. Incorporation of this expression into a PBK model allows the renal clearance of a toxicant, to the extent to which it is dependent on glomerular filtration rate, to be simulated from birth to adulthood.

III. Physiologically Based Models of Pregnancy

A. *What They Are*

PBK models of pregnancy incorporate the continuous changes in maternal and fetal tissue weights and blood flows and maternal pulmonary and renal function that accompany gestation. Depending on the absorption, distribution, and elimination characteristics of the toxicant, other features of gestation, such as prolonged gastrointestinal residence time or changes in metabolic function (Table 1), may also be incorporated into the PBK model.

B. *Examples of Physiologically Based Pregnancy Models*

Several PBK models of pregnancy have been published. Some have been used to assist with data organization and interpretation, others to make projections about kinetic behavior, and some simply to collect and codify data relating to growth and physiologic change during pregnancy.

1. Rodent Pregnancy

Early published PBK models dealt in general with pregnancy in rodents, the species in which developmental toxicity has been most intensively investigated in the laboratory. Especially rich sources of anatomic and physiologic data are available to support development of PBK models of pregnancy in the rat and, to a lesser extent, in the mouse (Sikov and Thomas, 1970; Goedbloed, 1972, 1976, 1977; Buelke-Sam *et al.*, 1982a,b).

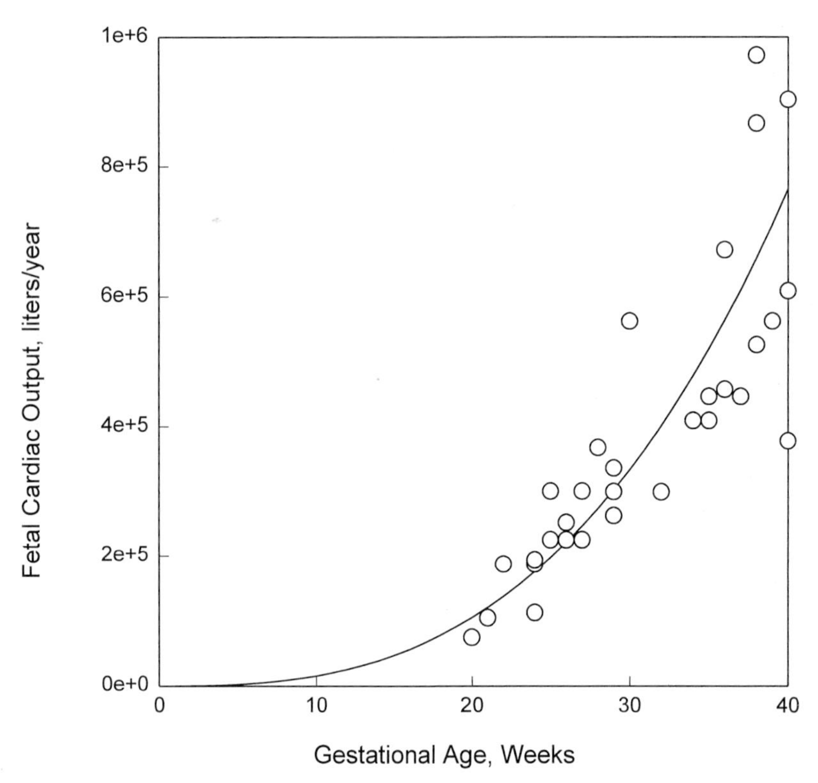

FIGURE 4 Human fetal cardiac output as a function of gestational age. The expression plotted is

$$\text{FCO} = 54 \times ((0.8 \times (\text{AGEPRE} \times 38 + 1))^3)$$

where FCO is fetal cardiac output (L/yr) and AGEPRE is gestational age (yr). Data are from Kenny *et al.,* 1968.

Olanoff and Anderson (1980) developed a PBK model that covered days 11–21 of rat gestation. The model included the growth of the fetus and of fetal organs and chorioallantoic placenta, and the increase in maternal blood volume. It accommodated variation in litter size. Transfers into maternal tissues and across the placenta were flow limited. Gabrielsson and Paalzow (1983) published a rat gestation model that considered "pregnancy" a set of conditions, the conditions being those of days 18–20 of gestation. Growth was not explicitly included in the model, and the model was insensitive to litter size. Transfer across the placenta was diffusion limited. Once the model had been titrated to the experimentally determined disposition of morphine in the maternal–fetal unit, it was used to explore the predicted effects on morphine disposition of changes in partitioning into maternal tissue (muscle) and in blood flow to the placenta. Because transfer across the placenta had been modeled as diffusion limited, changes in placental blood flow were predicted to have no effect except at very low flow rates, whereas increased partitioning of morphine into maternal muscle was associated with predicted increases in both maternal and fetal half-lives of morphine. A comparable model was devised for theophylline (Gabrielsson *et al.,* 1984). After it had been calibrated using experimental rat data and scaled to the human, it was used to simulate the effect of changes in pulmonary extraction ratio on the profile of theophylline concentrations in maternal blood.

The basic model structure was extended to methadone (Gabrielsson *et al.,* 1985; Gabrielsson and Groth, 1988). Again, the model was for days 18–20 of gestation in the rat. Tissues explicitly included were lung (which has high methadone concentrations), brain (the site of pharmacologic action), muscle (poorly perfused tissue and the major deposit of mathadone in the rat), intestine (the second largest deposit of methadone), liver and kidney (organs of elimination), and fetus (the site of embryotoxic effects). The model was calibrated to experimental data from intravenous infusions of methadone in the pregnant rat, scaled to the pregnant human, and then used to explore the simulated disposition of methadone under different physiologic conditions in both the rat and the human. This model was expanded

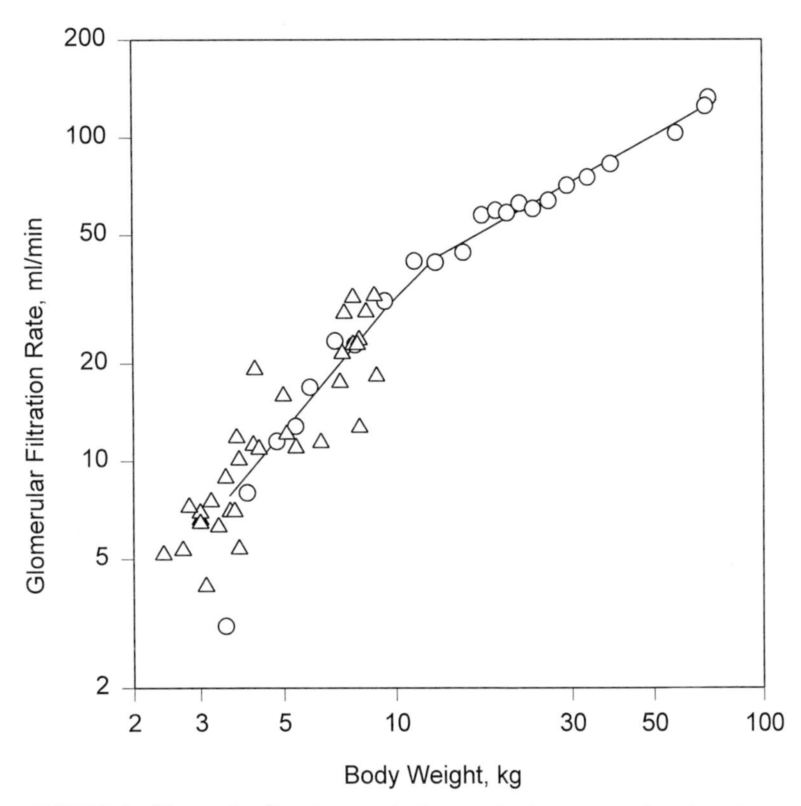

FIGURE 5 Glomerular filtration rate in the growing human as a function of age. Data are from multiple sources as tabulated in Johnson *et al.*, 1978 (○) and from Rubin *et al.*, 1949 (△). The line segments are allometric fits to data for ages less than or greater than 1 year.

to include additional tissues, and a sensitivity analysis was performed on the expanded model for both the rat and the human (Gabrielsson and Groth, 1988).

Fisher *et al.* (1989) designed a model that covered days 13–21 of rat pregnancy. It was written for a constant litter size but included growth of the fetus, chorioallantoic placenta, mammary glands, and maternal fat. Transfer across the placenta was diffusion limited. This model was calibrated to experimental data for the disposition of trichloroethylene and its metabolite, trichloroacetic acid, in the rat after oral (both gavage and drinking water) and inhalation exposure. Inclusion of trichloroacetic acid in the model required the development of a separate but linked model, which was not a PBK model, for metabolite formation and disposition. Metabolite formation was modeled as a capacity-limited transformation. The model was used to simulate fetal exposure after different routes of administration and to compare it with the corresponding maternal exposure. Fetal exposure was shown to be significant by all administration routes.

O'Flaherty *et al.* (1992) published gestation models for both the mouse and the rat. Both models covered days 0–22 of gestation, were sensitive to litter size and to the average weight of the pups at birth, and took into account growth of maternal tissues (uterus, mammary glands, and fat) as well as of the fetus. Both yolk sac and chorioallantoic placentation were explicitly included. Transfer across the placenta was flow limited. Inclusion of tissue and fluid pHs allowed the disposition of weak acids and bases to be simulated as functions of pH and pKa (the negative logarithm of the acid dissociation constant Ka); partition coefficients were not used in this model version. The model was applied successfully (O'Flaherty *et al.*, 1992) to simulate the disposition of the weak acid 5,5′-dimethyl-oxazolidine-2,4-dione in maternal plasma and muscle, in whole embryo and embryonic plasma in rats on day 13 of gestation, and in mice on days 10 and 11 of gestation, when the pHs of fetal tissues are changing relative to the pH of maternal plasma. It was only partially successful when it was used to predict the disposition of another weak acid, 2-methoxyacetic acid, the teratogenic metabolite of 2-methoxyethanol (O'Flaherty *et al.*, 1995). Specifically, although it performed well at low doses of 2-methoxyacetic acid, it consistently underestimated concentrations in maternal and embryonic tissues at intermediate and higher doses. A dose dependence of 2-methoxyacetic acid disposition, which suggested the possibility of capacity-limited elimination, was demonstrated.

This basic physiologic model was modified by eliminating the dependence of disposition on pH and pKa and introducing partition coefficients as determinants of tissue uptake (Clarke *et al.,* 1993). In this form, the model was used to simulate the absorption and disposition of 2-methoxyethanol and its oxidation to 2-methoxyacetic acid. The oxidation step was described by a Michaelis–Menten expression with initial values of the Michaelis constant and maximum rate obtained from published *in vitro* data. Maternal disposition of 2-methoxyacetic acid was described by a classic one-compartment linear kinetic model, whereas fetal disposition was handled with partition coefficients and a diffusion-limited placental transfer. The modified model was able to fit maternal and fetal 2-methoxyethanol and 2-methoxyacetic acid concentrations in the CD-1 mouse on gestation day 11, with 2-methoxyethanol given intravenously, subcutaneously, or by gavage. However, it too tended to underpredict embryo concentrations under some conditions (Terry *et al.,* 1995). Consequently, further exploration of this model (Terry *et al.,* 1996) was directed at examining the hypotheses that accumulation of 2-methoxyacetic acid in the embryo is determined by blood flow, ion trapping (pH/pKa considerations), active transport between placenta and embryo, or capacity-limited reversible binding to hypothetic embryonic binding sites. Blood flow–limited kinetics with partitioning into the fetus were determined to be satisfactory on day 8 of gestation, but invocation of active transport and/or reversible binding was necessary in order to reconcile model predictions to concentration data from gestation days 11 and 13 in the CD-1 mouse. Whereas the proposed possible mechanisms are empirical and operationally defined, this example illustrates how the application of physiologically based models can address alternative hypotheses and can reduce experimental work by identifying those studies most likely to yield relevant results.

2. Nonhuman Primate Pregnancy

The rhesus monkey has for years been the nonhuman primate of choice for studies of the effects of teratologic and other toxicants (Wilson *et al.,* 1977; Patterson *et al.,* 1997). Development of physiologically based models of nonhuman primate pregnancy should be facilitated by the availability of a rich fund of physiologic information concerning pregnancy in the rhesus monkey (Bourne, 1975a,b). A preliminary physiologically based model of pregnancy in the rhesus monkey (O'Flaherty *et al.,* 1991) took into account certain physiologic changes in the dam, such as expansion of plasma volume, as well as growth of the fetus, uterus, and placenta. This model has, however, not been developed further. There is a need for a more complete and well-calibrated model of rhesus monkey pregnancy to assist with interpretation of maternal and fetal disposition data during gestation.

Study of lead disposition in cynomolgus monkeys during pregnancy (Inskip *et al.,* 1996; Franklin *et al.,* 1997) has dictated the development of a physiologically based model of cynomolgus monkey pregnancy. In order to distinguish between lead originating from maternal stores—primarily bone (endogenous lead)—and lead originating from exogenous sources—primarily the diet—isotope mixes enriched in one or another of the three stable lead isotopes were administered sequentially in capsule form to mature female cynomolgus monkeys with histories of continuous constant high lead exposure from infancy. The relative amounts of the different isotope mixes in maternal blood indicate the relative contributions of endogenous and exogenous lead sources to maternal blood lead. Possible increases in both bone resorption rate and fractional absorption of lead from the gastrointestinal tract during later pregnancy were examined. It was found that a decrease in maternal bone lead mobilization during the first trimester was followed by a substantial increase in bone lead mobilization during the last trimester, consistent with what little is known of bone turnover during human pregnancy (Heaney and Skillman, 1971; Purdie *et al.,* 1988).

A physiologically based model of cynomolgus monkey pregnancy calibrated to these data is under development. It is superimposed on a physiologically based model of growth and mature bone turnover in nonpregnant cynomolgus monkeys (O'Flaherty *et al.,* 1997). Increases during pregnancy in the volumes of maternal blood, uterus, mammary glands, and carcass fat, as well as the growth of the placenta, fetus, and fetal bone and the development of fetal cardiac output are included in the model, along with increases in maternal cardiac output and shifts in the distribution of blood flow among maternal tissues. Transfers are blood flow–limited except for placental transfer, which is modeled as diffusion limited, and bone uptake and loss, which are determined by bone turnover and long-term movement of lead within bone. Trabecular and cortical bone, with characteristic different turnover rates, make up total bone in the model.

3. Human Pregnancy

Interestingly, the two major physiologically based models of human pregnancy are based on opposite philosophies of model development. The Luecke *et al.*

(1994) model includes 27 maternal tissues and 16 fetal tissues, one of which (bone) in each case is largely unperfused, although a perfusable maternal bone marrow/cartilage/bone subcompartment is included. The model originated in a detailed analysis of human embryonic and fetal growth (Wosilait *et al.,* 1992) that led to use of a Gompertz growth equation. Fetal tissue and organ weights are allometrically related to fetal body weight (Luecke *et al.,* 1995). Changes in maternal tissues (uterus, mammary glands, extracellular water, adipose tissue, and blood plasma) and changes in plasma flow rate and tissue perfusion rates are also allometrically related to fetal body weight (Luecke *et al.,* 1994). Tissue transfers can be blood flow–limited, affected to some degree by diffusional resistance, or fully diffusion limited. This model is versatile in that it allows exposure via ingestion, inhalation, dermal contact, and intravenous injection or infusion, accommodates chemicals with a variety of physicochemical and disposition characteristics including capacity-limited binding to plasma and tissue binding sites, and accommodates exposure to two chemicals simultaneously by the same or different routes. Although it has been exercised to simulate interactions of bromosulfophthalein with warfarin disposition in mother and fetus, it has not yet been tested against a range of data from experimental studies.

The Andriot and O'Flaherty (1997) model, in contrast, includes only seven maternal and three fetal tissues. It was developed specifically to simulate the disposition of lead during human pregnancy, but it is generalizable to other agents, including lipophilic chemicals, because it contains a maternal fat compartment. Because the model is directed at simulating the behavior of bone-seeking elements, bone is an important model component in both mother and fetus. Bone is not subdivided into cortical and trabecular bone in the current version of this model, but changes in bone formation and resorption rates can be incorporated to duplicate the limited information about bone turnover in human pregnancy (Heaney and Skillman, 1971; Purdie *et al.,* 1988). Maternal and fetal cardiac outputs and their distribution, and increases in the volumes of uterus, mammary glands, fat, blood plasma, red cells, placenta, fetus, and fetal tissues during pregnancy are based on information taken from the literature. This model has been exercised to predict the behavior of maternal blood lead concentration throughout pregnancy (Andriot and O'Flaherty, 1997), and the results have been compared with actual blood lead concentrations from a longitudinal study in pregnant women (Rothenberg *et al.,* 1994). However, it too has not yet been validated against a broader range of experimental data.

A possibility that has not been taken into account in either of these models is that the chemical may alter some of the parameters of the system. For example, it has been shown that caffeine in typical doses can alter maternal placental blood flow (Kirkinen *et al.,* 1983). Changes of this kind would generate kinetic nonlinearities. It is also possible that such changes, rather than actual transfer of the toxicant into the fetus, could be responsible for fetal toxicity in some instances. Such effects can be incorporated into physiologically based models, but it is likely that this will be done on a chemical-by-chemical basis.

C. Using a Physiologically Based Model to Predict Dose to Fetal Neural Tissues

A physiologically based model of the developing rabbit brain (Kim *et al.,* 1996) has been designed to simulate the disposition of neurotoxicants in the rabbit near term. This model is an extension of another physiologically based model (Kim *et al.,* 1995) that was focussed on the structure of the mature rabbit brain. Maternal brain subcompartments in both the original and expanded models are the hypothalamus, caudate nucleus, hippocampus, forebrain, brainstem, and cerebellum. Chemical passes from the blood plasma into the brain subcompartments in sequence; each compartment drains into the cerebrospinal fluid. Otherwise, the maternal body is approximated by two compartments. The fetal body is considered one compartment in addition to the brain and amniotic fluid. Placental transfer is flow limited. An interesting feature of this model is that maternal and fetal brain uptake are modeled as membrane limited by the blood–brain barrier, whereas clearance from the choroid plexus back to the blood plasma is modeled as capacity limited. Maternal renal clearance is also modeled as capacity limited.

This model was exercised to simulate the 2-hr time course of fetal brain concentrations of 2,4-dichlorophenoxyacetic acid (2,4-D) in rabbits at term after three different intravenous doses (Kim *et al.,* 1996). For this purpose, partition coefficients were obtained *in vitro* and the kinetic constants for clearance from the choroid plexus and the maternal kidney were obtained from earlier studies. Rate constants for transfer between fetal brain plasma and fetal brain tissue, fetal brain tissue and fetal cerebrospinal fluid, fetal body and amniotic fluid, and amniotic fluid and placenta were estimated by optimization to concentration data from Sandberg *et al.* (1996).

The ability of this model to simulate 2,4-D concentrations in fetal brain at the three doses is shown in Fig. 6. Although this exercise contains elements of model calibration (because of the optimization of key parameters), it demonstrates that the model is sufficiently flexible to reproduce the observed concentration curves

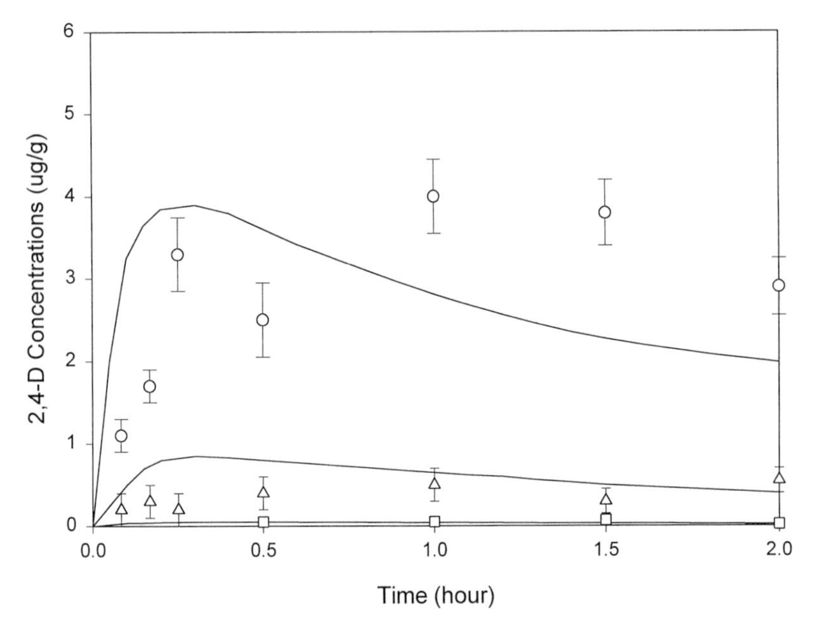

FIGURE 6 Concentration–time profiles of predicted and observed fetal brain concentrations of 2,4-D in the rabbit after single intravenous maternal doses of 1 mg/kg (□), 10 mg/kg (△), or 40 mg/kg (○). Data and simulations are from Kim *et al.,* 1996.

and suggests that the model has captured the principal features of organic acid disposition in the brain and central nervous system. Its future expansion to cover the entire period of fetal brain development will expand its range of potential applications.

IV. Physiologically Based Models of Early Childhood Growth

A. *What They Are*

Analogous to PBK models of pregnancy, these models incorporate the rapid changes in body and organ weights and in physiologic functions that characterize infancy and early childhood. They may also include the growth spurt and maturation associated with puberty.

B. *Example of a Physiologically Based Model of Early Childhood Growth*

A physiologically based model of human growth encompassing growth from birth to adulthood has been developed and calibrated (O'Flaherty, 1993). The expression for growth incorporates five parameters and describes the growth spurt of puberty as well as early childhood growth. It is sufficiently flexible that it can duplicate growth curves for individuals or for racially or ethnically distinct population subgroups. Its use to generate growth curves at the 95% upper and lower limits of a standardized growth chart is illustrated in the next section.

Tissue weights, cardiac output, respiratory rate, and glomerular filtration rate are scaled allometrically in this model in accordance with published measurements in children of different ages (Johnson *et al.,* 1978). An example of glomerular filtration rate as a function of age was given in Fig. 5. Bone mass is also linked to body weight. Bone formation rate as a function of age (O'Flaherty, 1995) is based on published studies of the kinetics of calcium tracers, and bone resorption rate is determined by bone mass and formation rate taken together.

C. *Using a Physiologically Based Growth Model to Predict the Effect of Growth Rate on Tissue Lead Levels*

A link between prenatal and early postnatal lead exposure and growth rate in young children is well established (Shukla *et al.,* 1989). Higher blood lead concentrations are associated with lower growth rates. It is tempting to assume that lead exposure is causal in this relationship. However, when the PBK model of lead kinetics in children (O'Flaherty, 1995) is exercised for hypothetic children whose growth rates fall on the upper 95% and lower 95% limits of a standardized growth chart (Fig. 7), it can be demonstrated that the reverse causality is equally plausible physiologically. The effect of the two extremes of growth rate (the upper 95% and lower 95% growth curves in Fig. 7) is illustrated in Fig. 8 (blood lead) and Fig. 9 (bone lead). The rapidly growing child is predicted to have lower blood and bone lead concen-

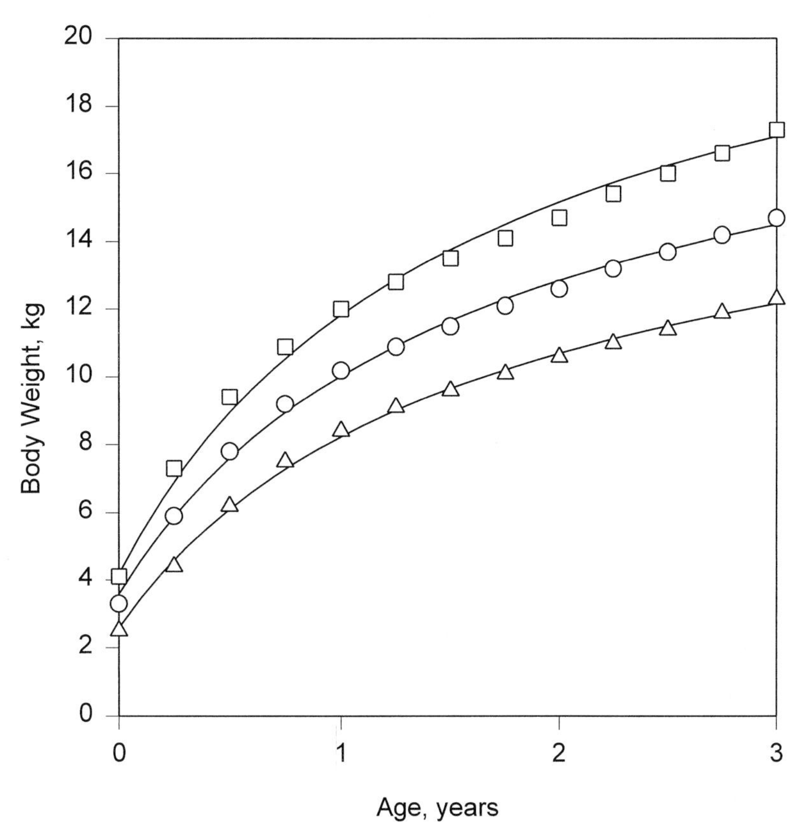

FIGURE 7 Modeled body weight curves, together with body weights at 3-month intervals taken from a standard National Center for Health Statistics growth chart for boys from birth to 3 years old (Hamill *et al.,* 1979). Fifth percentile (△); fiftieth percentile (○); ninety-fifth percentile (□).

trations than the slowly growing child. The origins of this model-predicted relationship are complex, but its principal determinant is the volumes of the tissues into which lead is distributed. Lead balance (the difference between rates of uptake from gastrointestinal tract and lung and rates of excretion from kidney and liver) is age-dependent but not strongly growth rate–dependent in the model, at least up to about age 1.5 years. Consequently, the total body burden of lead is predicted to be essentially the same in the growth scenarios modeled until about age 1.5 years, after which the more rapidly growing child begins to accumulate more lead than the slowly growing child with the same exposure. The greater rate of bone formation in the rapidly growing child generates a slowly increasing greater bone lead burden, which is the major component of the greater body burden. Expressed as tissue concentrations, however, this increasing body lead burden does not keep pace with growth rate, so that predicted concentrations of lead in both bone and soft tissues continue to be lower in the rapidly growing child than in the slowly growing child with the same lead exposure.

This application of the model suggests a different view of the posited relationship between blood lead and growth rate, and draws attention to the links among growth rate, blood lead concentration, and bone lead stores. It illustrates the importance of introducing physiologic changes (in this instance, differences in growth rates) into PBK models for agents with long half-lives, and also illustrates the ability of PBK models to clarify issues associated with the interpretation of experimental observations. It is particularly important to understand the association between growth and tissue lead concentrations because lead exposure during gestation and early childhood has been consistently linked with behavioral deficits (Dietrich *et al.,* 1987; Bellinger *et al.,* 1991) whose persistence into adolescence has been proposed (Needleman *et al.,* 1996).

V. New Directions in Physiologically Based Model Development

In the report of a symposium whose purpose was to examine the risk assessment process for neurotoxicants (Slikker *et al.,* 1996), Andersen set forth the prospects for application of PBK models in risk assessment for neurotoxicants, concluding that the increased emphasis on use of appropriate measures of tissue dose in risk assessment encourages the application of PBK models

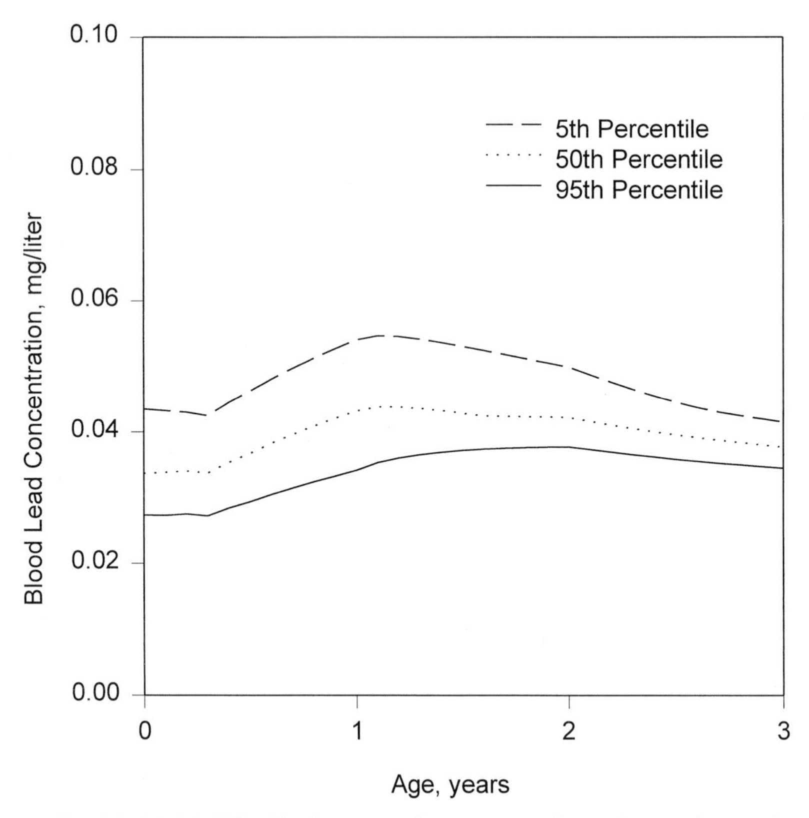

FIGURE 8 Modeled blood lead concentrations corresponding to the growth curves in Fig. 7. Fifth percentile (— —); fiftieth percentile (· · · · · · ·); ninety-fifth percentile (——).

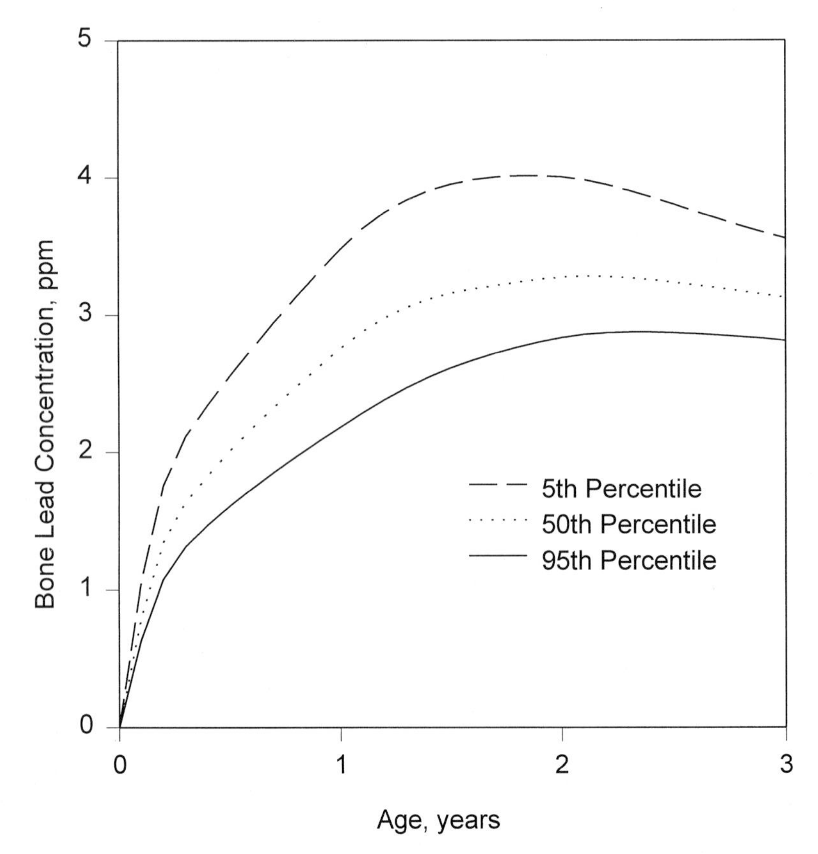

FIGURE 9 Modeled bone lead concentrations corresponding to the growth curve in Fig. 7. Fifth percentile (— —); fiftieth percentile (· · · · · · ·); ninety-fifth percentile (——).

to tissue dosimetry estimations and projections, as well as to dynamic models incorporating the mode of action of the neurotoxicant and linking tissue dose to early biologic effects. Such models have made some headway in risk assessment for carcinogens, but they are just beginning to be applied to risk assessment for noncarcinogens, including neurotoxicants. Their development and applications should increase dramatically in the near future.

References

Andriot, M. D., and O'Flaherty, E. J. (1998). A physiologically based model for bone-seeking elements in human pregnancy: Kinetics of lead disposition. Submitted.

Bellinger, D., Sloman, J., Leviton, A., Rabinowitz, M., Needleman, H. L., and Waternaux, C. (1991). Low-level lead exposure and children's cognitive function in the preschool years. *Pediatrics* **87,** 219–227.

Bourne, G. H., ed. (1975a). *The Rhesus Monkey, Volume I: Anatomy and Physiology.* Academic Press, New York, NY.

Bourne, G. H., ed. (1975b). *The Rhesus Monkey, Volume II: Management, Reproduction, and Pathology.* Academic Press, New York, NY.

Buelke-Sam, J., Nelson, C. J., Byrd, R. A., and Holson, J. F. (1982a). Blood flow during pregnancy in the rat: I. Flow patterns to maternal organs. *Teratology* **26,** 269–277.

Buelke-Sam, J., Holson, J. F., and Nelson, C. J. (1982b). Blood flow during pregnancy in the rat: II. Dynamics of and litter variability in uterine flow. *Teratology* **26,** 279–288.

Clarke, D. O., Elswick, B. A., Welsch, F., and Conolly, R. B. (1993). Pharmacokinetics of 2-methoxyethanol and 2-methoxyacetic acid in the pregnant mouse: a physiologically based mathematical model. *Toxicol. Appl. Pharmacol.* **121,** 239–252.

Cummings, A. J. (1983). A Survey of pharmacokinetic data from pregnant women. *Clin. Pharmacokinet.* **8,** 344–354.

Dietrich, K. N., Krafft, K. M., Bornschein, R. L., Hammond, P. B., Berger, O., Succop, P. A., and Bier, M. (1987). Low-level fetal lead exposure effect on neurobehavioral development in early infancy. *Pediatrics* **80,** 721–730.

Fisher, J. W., Whittaker, T. A., Taylor, D. H., Clewell, H. J., III, and Andersen, M. E. (1989). Physiologically based pharmacokinetic modeling of the pregnant rat: a multiroute exposure model for trichlorethylene and its metabolite, trichloroacetic acid. *Toxicol. Appl. Pharmacol.* **99,** 395–414.

Franklin, C. A., Inskip, M. J., Baccanale, C. L., Edwards, C. M., Manton, W. I., Edwards, E., and O'Flaherty, E. J. (1997). Use of sequentially administered stable lead isotopes to investigate changes in blood lead during pregnancy in a non-human primate (*Macaca fascicularis*). *Fundam. Appl. Toxicol.* **39,** 109–119.

Gabrielsson, J. L. and Groth, T. (1988). An extended physiological pharmacokinetic model of methadone disposition in the rat: validation and sensitivity analysis. *J. Pharmacokinet. Biopharmaceut.* **16,** 183–201.

Gabrielsson, J. L., Johansson, P., Bondesson, U., and Paalzow, L. K. (1985). Analysis of methadone disposition in the pregnant rat by means of a physiological flow model. *J. Pharmacokinet. Biopharmaceut.* **13,** 355–372.

Gabrielsson, J. L., and Paalzow, L. K. (1983). A physiological pharmacokinetic model for morphine disposition in the pregnant rat. *J. Pharmacokinet. Biopharmaceut.* **11,** 147–163.

Gabrielsson, J. L., Paalzow, L. K., and Nordström, L. (1984). A physiologically based pharmacokinetic model for theophylline disposition in the pregnant and nonpregnant rat. *J. Pharmacokinet. Biopharmaceut.* **12,** 149–165.

Goedbloed, J. F. (1972). The embryonic and postnatal growth of rat and mouse. I. The embryonic and early postnatal growth of the whole embryo. A model with exponential growth and sudden changes in growth rate. *Acta Anat.* **82,** 305–336.

Goedbloed, J. F. (1976). Embryonic and postnatal growth of rat and mouse. IV. Prenatal growth of organs and tissues: age determination, and general growth pattern. *Acta Anat.* **95,** 8–33.

Goedbloed, J. F. (1977). Embryonic and postnatal growth of rat and mouse. V. Prenatal growth of organs and tissues, general principles: allometric growth, absence of growth, and the genetic regulation of the growth process. *Acta Anat.* **98,** 162–182.

Hamill, P. V. V., Drizd, T. A., Johnson, C. L., Reed, R. B., Roche, A. F., and Moore, W. M. (1979). Physical growth: National Center for Health Statistics percentiles. *Am. J. Clin. Nutr.* **32,** 607–629.

Heaney, R. and Skillman, T. (1971). Calcium metabolism in normal human pregnancy. *J. Clin. Endocrinol. Metab.* **33,** 661–670.

Hudson, G. (1965). Bone-marrow volume in the human foetus and newborn. *Br. J. Haematol.* **II,** 446–452.

Hytten, F. E. and Cheyne, G. A. (1969). The size and composition of the human pregnant uterus. *J. Obstet. Gynaecol.* **65,** 400–403.

Inskip, M. J., Franklin, C. A., O'Flaherty, E. J., Manton, W. I., Edwards, C. M., and Baccanale, C. L. (1996). Blood lead in pregnancy: Stable isotopes reveal changes in both bone lead and current oral lead contribution in a non-human primate (*Macaca fascicularis*). *Fundam. Appl. Toxicol.* **30,** *The Toxicologist* (*Suppl*), **12** (abstr).

Johnson, T. R., Moore, W. M., and Jeffries, J. E. (Eds.) (1978). *Children Are Different: Developmental Physiology.* Ross Laboratories, Columbus, OH.

Kenny, J. F., Plappert, T., Doubilet, P., Saltzman, D. H., Cartier, M., Zollars, L., Leatherman, G. F., and St. John Sutton, M. G. (1986). Changes in intracardiac blood flow velocities and right and left ventricular stroke volumes with gestational age in the normal human fetus: a prospective Doppler echocardiographic study. *Circulation* **74,** 1208–1216.

Kim, C. S., Binienda, Z., and Sandberg, J. A. (1996). Construction of a physiologically based pharmacokinetic model for 2,4-dichlorophenoxyacetic acid dosimetry in the developing rabbit brain. *Toxicol. Appl. Pharmacol.* **136,** 250–259.

Kim, C. S., Slikker, W. Jr., Binienda, Z., Gargas, M. L., and Andersen, M. E. (1995). Development of a physiologically based pharmacokinetic model for 2,4-dichlorophenoxyacetic acid dosimetry in discrete areas of the rabbit brain. *Neurotoxicol. Teratol.* **17,** 111–120.

Kirkinen, P., Jouppila, P., Koivula, A., Vuori, J., and Puukka, M. (1983). The effect of caffeine on placental and fetal blood flow in human pregnancy. *Am. J. Obstet. Gynecol.* **147,** 939–942.

Krauer, B. and Krauer, F. (1977). Drug kinetics in pregnancy. *Clin. Pharmacokinet.* **2,** 167–181.

Luecke, R. H., Wosilait, W. D., Pearce, B. A., and Young, J. F. (1994). A physiologically based pharmacokinetic computer model for human pregnancy. *Teratology* **49,** 90–103.

Luecke, R. H., Wosilait, W. D., and Young, J. F. (1995). Mathematical representation of organ growth in the human embryo/fetus. *Int. J. Bio-Medical Computing* **39,** 337–347.

Mattison, D. R. (1986). Physiologic variations in pharmacokinetics during pregnancy. In *Drug and Chemical Action in Pregnancy: Pharmacologic and Toxicologic Principles.* S. Fabro and A. R. Scialli, Eds., Marcel Dekker, New York, pp. 37–102.

Mattison, D. R. (1990). Transdermal drug absorption during pregnancy. *Clin. Obstet. Gynecol.* **33,** 718–727.

Mattison, D. R., Blann, E., and Malek, A. (1991). Physiological alterations during pregnancy: impact on toxicokinetics. *Fund. Appl. Toxicol.* **16,** 215–218.

Morrione, T. G., and Seifter, S. (1962). Alteration in the collagen content of the human uterus during pregnancy and post partum involution. *J. Exp. Med.* **115,** 357–365.

Needleman, H. L., Riess, J. A., Tobin, M. J., Biesecker, G. E., and Greenhouse, J. B. (1996). Bone lead levels and delinquent behavior. *J. Am. Med. Assoc.* **275,** 363–369.

O'Flaherty, E. J. (1993). Physiologically based models for bone-seeking elements. IV. Kinetics of lead disposition in humans. *Toxicol. Appl. Pharmacol.* **118,** 16–29.

O'Flaherty, E. J. (1994). Physiologically-based pharmacokinetic models in developmental toxicology. *Risk Analysis* **14,** 605–611.

O'Flaherty, E. J. (1995). Physiologically based models for bone-seeking elements. V. Lead absorption and disposition in childhood. *Toxicol. Appl. Pharmacol.* **131,** 297–308.

O'Flaherty, E. J., Inskip, M. J., Franklin, C. A., Durbin, P. W., Manton, W. I., and Baccanale, C. L. (1998). Evaluation and modification of a physiologically-based model of lead kinetics using data from a sequential isotope study in cynomolgus monkeys. *Toxicol. Appl. Pharmacol.* **149,** 1–16.

O'Flaherty, E. J., Nau, H., McCandless, D., Beliles, R. P., Schreiner, C. M., and Scott, W. J., Jr. (1995). Physiologically based pharmacokinetics of methoxyacetic acid: dose–effect considerations in C57BL/6 mice. *Teratology* **52,** 78–89.

O'Flaherty, E. J., Scott, W., and Beliles, R. (1991). A physiologically-based model of gestation in the monkey: framework for simulation of kinetics of reproductive toxicants. *Teratology* **43,** 436 (abstr).

O'Flaherty, E. J., Scott, W., Schreiner, C., and Beliles, R. P. (1992). A physiologically based kinetic model of rat and mouse gestation: disposition of a weak acid. *Toxicol. Appl. Pharmacol.* **112,** 245–256.

Olanoff, L. S. and Anderson, J. M. (1980). Controlled release of tetracycline—III: a physiological pharmacokinetic model of the pregnant rat. *J. Pharmacokinet. Biopharmaceut.* **8,** 599–620.

Patterson, T. A., Binienda, Z. K., Lipe, G. W., Gillam, M. P., Slikker, W. J. Jr., and Sandberg, J. A. (1997). Transplacental pharmacokinetics and fetal distribution of azidothymidine, its glucuronide, and phosphorylated metabolites in late-term rhesus macaques after maternal infusion. *Drug Metab. Disp.* **25,** 453–459.

Purdie, D. W., Aaron, J. E., and Selby, P. L. (1988). Bone histology and mineral homeostasis in human pregnancy. *Brit. J. Obstet. Gynecol* **95,** 849–854.

Rothenberg, S. J., Karchmer, S., Schnaas, L., Perroni, E., Zea, F., and Alba, J. F. (1994). Changes in serial blood lead levels during pregnancy. *Environ. Health Perspect.* **102,** 876–880.

Rubin, M. I., Bruck, E., and Rapoport, M. (1949). Maturation of renal function in childhood: Clearance studies. *J. Clin. Invest.* **28,** 1144–1162.

Sandberg, J. A., Duhart, H. M., Lipe, G., Binienda, Z., and Slikker, W., Jr. (1996). Distribution of 2,4-dichlorophenoxyacetic acid (2,4-D) in maternal and fetal rabbits. *J. Toxicol. Environ. Health* **49,** 497–509.

Shukla, R., Bornschein, R. L., Dietrich, K. N., Buncher, C. R., Berger, O. G., Hammond, P. B., and Succop. P. A. (1989). Effects of fetal and infant lead exposure on growth in stature. *Pediatrics* **84,** 604–612.

Sikov, M. R., and Thomas, J. M. (1970). Prenatal growth of the rat. *Growth* **34,** 1–14.

Singer, D. B., Sung, C. J., and Wigglesworth, J. S. (1991). Fetal growth and maturation: with standards for body and organ development. In *Textbook of Fetal and Perinatal Pathology.* J. S. Wigglesworth and D. B. Singer, Eds. Blackwell Scientific Publications, Oxford, pp. 11–47.

Slikker, W. Jr., Crump, K. S., Andersen, M. E., and Bellinger, D. (1996). Biologically based, quantitative risk assessment of neurotoxicants. *Fund. Appl. Toxicol.* **29,** 18–30.

Sonawane, B. R. and Yaffe, S. J. (1986). Physiologic disposition of drugs in the fetus and newborn. In *Drug and Chemical Action in Pregnancy: Pharmacologic and Toxicologic Principles.* S. Fabro and A. R. Scialli, Eds. Marcel Dekker, New York, pp. 103–121.

Terry, K. K., Elswick, B. A., Welsch, F., and Conolly, R. B. (1995). Development of a physiologically based pharmacokinetic model describing 2-methoxyacetic acid disposition in the pregnant mouse. *Toxicol. Appl. Pharmacol.* **132,** 103–114.

Terry, K. K., Elswick, B. A., Welsch, F., and Conolly, R. B. (1996). A physiologically based pharmacokinetic model describing 2-methoxyacetic acid disposition in the pregnant mouse. *Occup. Hyg.* **2,** 57–65.

Thomson, A. M., Billewicz, W. Z., and Hytten, F. E. (1968). The assessment of fetal growth. *J. Obstet, Gynaecol.* **75,** 903–916.

Trotter, M. and Peterson, R. R. (1969). Weight of bone in the fetus during the last half of pregnancy. *Clin. Orthoped. Rel. Res.* **65,** 46–50.

Wilson, J. G., Ritter, E. J., Scott, W. J., and Fradkin, R. (1977). Comparative distribution and embryotoxicity of acetylsalicylic acid in pregnant rats and rhesus monkeys. *Toxicol. Appl. Pharmacol.* **41,** 67–78.

Woessner, J. F., and Brewer, T. H. (1963). Formation and breakdown of collagen and elastin in the human uterus during pregnancy and post-partum involution. *Biochem. J.* **89,** 75–82.

Wosilait, W. D., Luecke, R. H., and Young, J. F. (1992). A mathematical analysis of human embryonic and fetal growth data. *Growth Dev. Aging* **56,** 249–257.

Ziegler, E. E., O'Donnell, A. M., Nelson, S. E., and Fomon, S. J. (1976). Body composition of the reference fetus. *Growth* **40,** 329–341.

CHAPTER

17

Developmental Enzymology
Xenobiotic Biotransformation

MONT R. JUCHAU
HAO CHEN
Department of Pharmacology
School of Medicine
University of Washington
Seattle, Washington 98195

*Abbreviations: AKR1, aldehyde reductases; AKR2, aldose reductases; AKR3, ketone reductases; CNS, central nervous system; FAD, flavin-adenine dinucleotide; FAE, fetal alcohol effects; FAS, fetal alcohol syndrome; FMN, flavin mononucleotide; FMO, flavin monooxygenase; GSH, glutathione stimulating hormone; mEHb, microsomal epoxide hydrolase; mEHch, microsomal cholesterol oxide hydroxylase; NADPH, nicotinamide-adenine dinucleotide phosphate (reduced); PAPS, 3′-phosphoadenosine-5′-phosphosulfate; RT-PCR, reverse transcriptase-polymerase chain reaction; sEH, soluble epoxide hydrolase; UDPGT, uridine diphosphate glucuronosyltransferase.

I. Introduction

A. *Definitions and Scope*

This chapter discusses developmental aspects of the biotransformation and bioactivation of foreign organic chemicals (xenobiotics) as related to the capacity of such chemicals to elicit persistent, detrimental morphologic and/or functional effects on the developing nervous system. Emphasis on xenobiotics is related to the fact that nearly all chemicals currently classified as "established" human teratogens are xenobiotics (Schardein, 1993; Shepard, 1995; Juchau, 1997). Exceptions to this generalization are retinoids and steroids, which, interestingly, are also substrates for many of the same enzymes that catalyze the biotransformation of xenobiotics. A *xenobiotic* is defined as a chemical that is not used by the reference organism as a nutrient chemical, is not essential to the reference organism for maintenance of normal physiologic/biochemical function and homeostasis, and does not constitute a part of the conventional array of chemicals synthesized from nutrient chemicals by the reference organism in normal intermediary metabolism.

Primary focus in this chapter is on prenatal xenobiotic biotransformation/bioactivation in humans, rats, mice, and rabbits. It is now well recognized that profound species differences in xenobiotic biotransformation exist and that there is no known experimental animal that provides a suitable model for the human species. Rats, mice, and rabbits were selected for major focus because they are the experimental animals most commonly employed in studies of developmental toxicology and also because they are the species for which by far the most information pertaining to developmental aspects of xenobiotic biotransformation is available. All four species are placental mammals and the discussion is thus centered around the persistent, transplacental toxic effects of xenobiotic chemicals on conceptuses developing *in utero* and of the relationships of xenobiotic biotransformation/bioactivation in these conceptuses to the toxic effects elicited. Persistent transplacental toxic effects include genotoxicity (mutagenesis, clastogenesis, and aneugenesis), carcinogenesis, and cytotoxicity resulting in persistent morphologic and/or functional abnormalities in the offspring (birth defects or teratogenesis). Nonpersistent, transient conceptotoxic/neurotoxic effects are of much lesser interest and importance and is not considered here.

A very large number and huge variety of persistent morphologic and functional developmental defects (*terata*) are possible and clearly include persistent abnormalities of the nervous system. Effects on the central nervous system (CNS) appear to be of greatest current concern with considerable interest in the long-term effects of prenatal chemical insult on mental and behavioral functions. Thus, bioactivation of xenobiotic chemicals in cells/tissues of the prenatal CNS relates closely to this concern and emphasis is placed on this aspect, but primarily from a speculative point of view because so little pertinent information has been published to date.

The enzymes involved in xenobiotic bioactivation include virtually all enzymes classically regarded (e.g., Lamble, 1983; Gibson and Skett, 1986) as xenobiotic-biotransforming enzymes—that, is those oxidoreductases, hydrolases, and transferases that catalyze the oxidation, reduction, hydrolysis, and conjugation of foreign organic chemicals as substrates. Even those enzymes that were once classified as functional only in terms of inactivation (e.g., glucuronosyl transferases, glutathione transferases) are now well known to catalyze several examples of bioactivating reactions. Thus, consideration is given to each of these classes of enzymes in terms of their developmentally regulated expression. Lyases, isomerases, and ligases are not discussed, even though they occasionally participate as catalysts in xenobiotic biotransformation reactions. Emphasis is given to P450–dependent bioactivation in embryonic tissues for at least three reasons: (1) P450 heme–thiolate proteins are the enzymes most heavily implicated in bioactivation reactions (Nebert *et al.*, 1993; Parkinson, 1996); (2) these are the enzymes that have been most heavily investigated prenatally; thus much more information is available for discussion; and (3) the embryonic period is commonly regarded as (in general terms) the period of highest sensitivity of the conceptus to the toxic effects of chemicals. For the reader who desires recent background information on xenobiotic biotransformation and the enzymes that catalyze these reactions, the review of Parkinson (1996) is highly recommended.

Definition of a number of other terms may also be useful. The term *conceptus* is a nonspecific reference to the embryo or fetus with associated membranes of nonmaternal origin. *Embryo* refers to the conceptus without its associated membranes during the period extending from the time of fertilization to the end of organogenesis, usually marked by complete closure of the palatal folds. *Fetus* refers to the conceptus without its associated membranes during the period extending from the end of organogenesis until parturition. After parturition, the organism is referred to as a *neonate.* The period of *organogenesis* extends from shortly after implantation until the end of the embryonic period. These and other related terms have been discussed more extensively in other references (Schardein, 1993; Shepard, 1995; Juchau, 1997).

B. Brief History of Developmental Xenobiotic Biotransformation

A generally increased sensitivity during the neonatal period of many experimental animals and humans to the effects of exposures to drugs and other xenobiotics has long been recognized. Neonates typically [although certainly not always—Done (1964) has listed several notable exceptions] exhibit markedly increased intensity and duration of action responses after exposure to such chemicals even when the exposures are corrected to account for body weight and/or surface area. Recognition that a lack of xenobiotic biotransforming capacity during early development could account for a significant part of this marked sensitivity (Done, 1964, 1966) experienced a dramatic increase during the decades of the 1960s and 1970s. Several reviews of early research pertaining to prenatal biotransformation of xenobiotics appeared in the early 1970s (Pelkonen and Karki, 1973; Rane *et al.,* 1973; Juchau, 1975; Pelkonen, 1977). A highly comprehensive review of research in this area was published by Waddell and Marlow in 1976, with a subsequent comprehensive review in 1980 (Juchau *et al.,* 1980). This increased recognition of the importance of prenatal xenobiotic biotransformation coincided with the pronounced increase in research interest in xenobiotic biotransformation *per se,* with the concern surrounding the thalidomide disaster of 1958–1962, and also with the discovery and subsequent investigations of the P450 superfamily.

Early experiments, which were primarily investigations of the late fetal and early neonatal (perinatal) period, relied almost entirely on studies of rates of enzyme-catalyzed, xenobiotic-biotransforming reactions *in vitro* with cell-free hepatic preparations of common laboratory animals (rats, mice, rabbits, guinea pigs) as enzyme sources. Jondorf *et al.* (1958) and Fouts and Adamson (1959) are normally given credit for discovery of low levels of microsomal oxidative drug biotransformation. This discovery was almost immediately preceded by findings that prenatal glucuronidation of xenobiotic substrates also appeared to occur at very low rates (Brown and Suelzer, 1958). Investigators were initially unaware of the huge multiplicity of enzymic isoforms now known to participate in the catalysis of xenobiotic-biotransforming reactions. Consequently, many confusing and sometimes apparently conflicting results were reported. Nevertheless, it soon became apparent that, in very general terms, rates of xenobiotic biotransforming reactions and levels of enzymes catalyzing such reactions were relatively low in prenatal and neonatal hepatic tissues. Prenatal induction was demonstrable near term with phenobarbital (Hart *et al.,* 1962) and during earlier periods of gestation with polynuclear aromatic hydrocarbons (Bogdan and Juchau, 1970) but generally with decreased effectiveness relative to postnatal induction. Based primarily on experimental results obtained with rodents and rabbits, textbook and monograph discussions dealing with neonatal and prenatal xenobiotic biotransformation during the 1960s and 1970s (extending even into the 1980s in some cases) commonly indicated that expressions of enzymes catalyzing xenobiotic biotransformation during prenatal and early neonatal life were virtually negligible. It is now clearly recognized that there are numerous important exceptions to this early generalization.

A number of investigations in the 1970s began to initiate a change in perception of prenatal drug biotransformation. Studies with human tissues, in particular, indicated that a number of xenobiotic substrates underwent relatively rapid monooxygenation when primate hepatic or adrenal tissues were used as enzyme sources (Yaffe *et al.,* 1970; Juchau *et al.,* 1973; Yaffe and Juchau, 1974). Those investigations were subsequently confirmed and, since that time, studies in several laboratories have provided strong evidence for the expression of multiple active forms of P450 heme–thiolate proteins in the tissues of prenatal vertebrates, with much higher than originally expected levels in the hepatic tissues of prenatal human and subhuman primates. Another significant factor responsible for increased recognition of the importance of prenatal xenobiotic biotransformation was a growing awareness of the concept of bioactivation and of the importance of the generation of reactive intermediates in target cells and tissues, including those of the conceptus (Juchau, 1975, 1976, 1978, Berry *et al.,* 1976, 1977a,b). This awareness heightened interest in studies of even very low levels of monooxygenase activities if the reactions occurred in target tissues. Clearly, the conceptus is a very important target as was strongly emphasized by the thalidomide catastrophe. High levels of P450 and xenobiotic monooxygenation in the human fetal adrenal gland (Juchau and Pedersen, 1973; Zachariah and Juchau, 1975; Jones *et al.,* 1977) also appeared to contribute significantly to the increased interest. Investigations of bioactivation of prodysmorphogens in cultured whole embryo systems has stimulated considerable interest. A number of reviews describing many of the newer findings in this area have appeared in the more recent literature (Pelkonen, 1985; Juchau, 1987, 1989, 1990; Juchau *et al.,* 1987, 1991, 1992, 1997; Anderson *et al.,* 1989; Nebert, 1989; Krauer and Dayer, 1991; Raucy and Carpenter, 1993; Miller *et al.,* 1996; Wells *et al.,* 1996; Wells and Winn, 1996, 1997). With the exception of those of Wells *et al.* (1996) and Wells and Winn (1996, 1997), these reviews focus almost entirely on P450-dependent systems. The cited exceptions deal extensively with published data concerning peroxidase-catalyzed generation of free radical intermediates and biologic defense systems against free radical–mediated damage.

II. Oxidative Enzymes

A. *P450-Dependent Systems*

Applications of highly sensitive and specific techniques of modern molecular biology coupled with immunologic techniques and earlier developed biochemical/analytic techniques have combined to help provide a much clearer picture of the developmental expression of individual P450 isoforms in various species, and thereby of developmental aspects of P450 dependent biotransformation/bioactivation *per se.* It should be emphasized, however, that enormous gaps of knowledge still remain to be filled in this area. Until quite recently, many of those gaps were not even identifiable. Major gaps include the following: (1) A huge number of currently identified individual P450 isoforms still have not been investigated in terms of prenatal expression; (2) for those isoforms that have been investigated, we still do not have a clear understanding pertaining to quantitative levels of expression as a function of gestational age; (3) knowledge of distribution within and among various prenatal tissues/cell populations of most P450 isoforms is extremely meager at present; (4) knowledge of species differences in expression of individual P450 isoforms/orthologs during development is likewise very meager; (5) at least one P450 isoform (P4503A7) is expressed uniquely in the prenatal hepatic tissues of primates—it is unknown whether there other examples of prenatal-specific isoforms; (6) what specific factors control the expression of individual P450 isoforms in prenatal tissues? The experimental tools needed to fill these and other identifiable gaps are available and need only to be applied.

P450 isoforms contributing significantly to the catalysis of xenobiotic oxidative biotransformation and bioactivation belong almost exclusively to families 1, 2, and 3 of the P450 superfamily (rabbit P4504B1 is an example of a rare exception). An update of the nomenclature and functions of these families has appeared in the literature (Nelson *et al.,* 1996). A brief summary of the available information concerning the prenatal expression of P450 isoforms within families 1, 2, and 3 are presented here. A more extensive review of this aspect has been published by Juchau *et al.* (1997).

1. Family 1

At present, only two subfamilies (1A and 1B) and three individual isoforms (1A1, 1A2, and 1B1) have been identified within P450 family 1. Xenobiotic substrates for isoforms of this family are typically planar, aromatic, polynuclear chemicals, many of which are potent promutagens and procarcinogens. Endogenous substrates include estrogens, retinoids, and bilirubin. Constitutive expression (including prenatal constitutive expression) of 1A1 appears to be minimal or negligible, at least in the tissues investigated of the species under consideration. 1A1 is, however, highly inducible by dioxin/methylcholanthrene–type inducing agents and there is ample evidence for transplacental induction of 1A1 in both hepatic and extrahepatic prenatal tissues of common laboratory species. Dioxin induction of both 1A1 and 1B1 were demonstrated in explanted human embryonic hepatic tissues (Huang and Juchau, 1997). Responsiveness of prenatal tissues to inducers appears to increase as a function of gestational age, with marked increases near term, although induction of 1A1 appears to occur throughout gestation.

P4501A2 is expressed constitutively in adult hepatic tissues but appears neither to be constitutively expressed nor subject to induction in prenatal tissues, except that induction can be demonstrated in fetal hepatic tissues just prior to parturition. P4501B1 appears to be constitutively expressed in prenatal hepatic and extrahepatic tissues (including cephalic tissues) although rigorous investigations pertaining to quantitative levels of expression of enzymically functional hemoprotein appear not to have been reported. 1B1 is also inducible and is likely to be transplacentally induced by the same inducing agents that are effective inducers of 1A1 and 1A2, although experiments designed to demonstrate this phenomenon likewise have not yet been reported. In some tissues (e.g., adult rat adrenals), P4501B1 can be upregulated by cAMP but the capacity of cAMP to upregulate 1B1 in prenatal conceptal tissues also appears not yet to have been reported.

2. Family 2

This is an extremely large family with numerous subfamilies and, with some exceptions, several isoforms within each subfamily, especially subfamily 2C. Family 2 isoforms reported as detectable in human prenatal tissues include 2D6 (investigated extensively in terms of polymorphisms in adult liver), 2C8 and 2E1, an ethanol-inducible isoform that catalyzes preferentially the monooxygenation of lower molecular weight foreign organic chemicals, including ethanol. Investigations of prenatal expression of the 2D6 and 2E1 isoforms in common experimental animals suggest that neither is expressed significantly prior to parturition, at least in hepatic tissues. For human 2D6, Treluyer *et al.* (1991) reported ability to detect 2D6 message and functional protein at relatively late stages of human gestation (mostly after the 24th week) but at extremely low levels and reportedly not detectable at all in 70% of the fetal liver samples studied. In the 30% that were detectable, the dominant source of hepatic tissue was from spontaneously aborted fetuses rather than from therapeutic abortions. Subsequent investigations have tended to confirm those observations (Maenpaa *et al.,* 1993; Hakkola *et al.,* 1994). The latter authors also reported reverse transcriptase–polymerase chain reaction (RT–PCR) detection of P4502C8 but not 2A6/7, 2B6/7, 2C9, 2C18, 2C19, 2E1, 2F1, or 4B1. Interestingly, they reported inability to detect 1A1. Reasons for discrepant results relevant to 1A1 and 2E1 remain to be clarified. Further studies of 2C8 will be of interest relative to prenatal retinoid biotransformation as well as to xenobiotic biotransformation.

Reported observations of the expression of P4502E1 in human prenatal tissues (Carpenter *et al.,* 1996; Juchau and Yang, 1996; Juchau *et al.,* 1996, 1997) is of high interest because of (among other things) the potential implications for the toxic transplacental effects of ethyl alcohol. In human prenatal hepatic tissues, 2E1 appears to be expressed primarily only after about the 16th week of gestation, but at surprisingly high levels in view of the lack of expression in prenatal rodent hepatic tissues. In terms of developmental neurotoxicology, it was of great interest to us to detect expression of 2E1 mRNA in human prenatal cephalic tissues with RT–PCR, a highly sensitive detection technique. Subsequently, we demonstrated detectability with less sensitive but more readily quantifiable RNase protection assays and also found evidence for significant levels of functional 2E1 enzyme in embryonic and fetal brain tissues using P450-dependent conversion of *p*-nitrophenol to *p*-nitrocatechol as a probe (Boutelet-Bochan *et al.,* 1997). Although levels in homogenate preparations of prenatal cephalic tissues were very low, the possibility that certain specific cephalic cells/cell types may express 2E1 at relatively high levels remains a distinct possibility and should now be rigorously investigated.

It seems possible that prenatal cephalic 2E1 could be involved in the transplacental neurotoxic effects of ethyl alcohol via several mechanisms: First, 2E1 catalyzes the conversion of ethanol to acetaldehyde, a highly reactive, toxic metabolite. Second, 2E1 can catalyze the conversion of ethanol to a hydroxyethyl radical, likewise a highly toxic metabolite. Third, during 2E1 catalysis of ethanol oxidation, relatively large quantities of superoxide anion are generated and, under appropriate conditions, superoxide anions can undergo conversion to highly reactive/toxic hydroxyl (OH·) radicals as well as other reactive oxygen species. Fourth, generation of radicals during 2E1 catalysis of ethanol oxidation can initiate lipid peroxidation, a self-sustaining process that can lead to extensive damage to lipid membranes as well as generation of lipid hydroperoxides, which are also toxic. Fifth, 2E1 is capable of catalyzing the conversion of lipid hydroperoxides to highly reactive aldehydes such as 4-hydroxynonenal and malondialdehyde. Such aldehydes are not only directly and potently cytotoxic, but can also inhibit the biosynthesis of developmentally important retinoids (Chen and Juchau, 1996, 1998). Finally, it is established that 2E1, covalently bound to self-generated electrophilic (haptenic) metabolites, can behave as an antigen. Thus, subsequent damage from antigen–antibody reactions is not unexpected. Therefore, the expression of 2E1 in human prenatal cephalic tissues could feasibly contribute to the prenatal neurotoxic effects of ethanol via several mechanisms and it seems of importance to determine the extent to which each of these mechanisms may actually contribute in a significant fashion to fetal alcohol syndrome (FAS) and/or fetal alcohol effects (FAE).

3. Family 3

Only one subfamily (3A) has been identified for family 3 as of this writing. However, numerous 3A isoforms have been identified, and one of these (3A7) is of special interest because it is now known to be expressed at relatively high levels in human prenatal hepatic tissues, even during the period of organogenesis. Information concerning this human isoform and its surprisingly high prenatal expression and activity in hepatic tissues has been reviewed by Kitada and Kamataki (1994). Of additional special interest is that 3A7 appears to be expressed almost exclusively in prenatal hepatic tissues of primate species (exceptions have been noted by various investigators, particularly in adult tumor tissues). A closely related isoform, 3A4, is expressed extensively in adult human liver, but apparently not extensively prenatally, although Hakkola *et al.* (1994) have reported detection of 3A4 in human fetal liver with RT-PCR. Another 3A isoform, P4503A5, is variably and, as yet, unpredictably expressed as a minor isoform in both human fetal and adult tissues. With the possible exception of the sheep fetus (Ring *et al.,* 1995), corresponding or orthologous 3A isoforms appear not to be expressed significantly in a constitutive fashion in prenatal hepatic tissues of the more common experimental animals. 3A isoforms are, however, transplacentally inducible by glucocorticoids and antiglucocorticoids (pregnenolone 16α-carbonitrile) in rodent hepatic tissues during the later stages of gestation (Hulla and Juchau, 1989a,b). 3A7 has been shown to be capable of catalyzing the bioactivation of a variety of xenobiotic substrates (Kitada and Kamataki, 1994) and is expressed at relatively high levels even during organogenesis in hepatic but apparently not several extrahepatic tissues (Yang *et al.,*

1994). These observations generate a clear concern in terms of the transplacental toxic effects (including possible neurotoxic effects) of xenobiotics on human conceptuses. In general terms, most 3A isoforms appear to exhibit very broad substrate specificities, and 3A7 does not appear to be an exception to this rule. Thus, because nonprimate species do not appear to constitutively express 3A isoforms/orthologs prenatally, the 3A7 isoform may well be a major factor in the species differences in sensitivity to the transplacental effects of xenobiotic chemicals.

B. Flavin Monooxygenase-Dependent Systems

The flavin monooxygenases (FMOs) have been studied far less intensively than the P450 heme–thiolate proteins, particularly with respect to developmental aspects. For many years, the FMO isoform isolated from hog liver was thought to be the primary representative of a single-enzyme system located primarily in hepatic tissues of several species, including humans. Recent molecular analyses, however, have revealed the existence of at least 5 subfamilies of the enzyme, termed *FMO1–FMO5,* which share about 55% homology. These exhibit striking species and tissue distributions and also vary in terms of substrate specificities. Substrates for FMOs typically are heterocycles containing softly nucleophilic nitrogen, sulfur, or phosphorous atoms. Unlike P450s, FMOs do not ordinarily attack carbon atoms, although very softly nuclephilic carbon atoms may be attacked. FMO-dependent metabolisms of thioamides, thiols, and thiocarbamates have been associated with toxicity.

A very striking example of developmental regulation was provided recently by Dolphin *et al.* (1996), who reported with RNase protection assays that FMO1 mRNA was extensively expressed in human fetal liver and kidney (gestational ages not specified) but only minimally expressed in human adult liver and kidney. In contrast, FMO3 was expressed extensively in adult human liver and kidney but only minimally expressed in human fetal liver and kidney. Thus, a situation interestingly analogous to that with P4503A4/3A7 appears to exist for FMO3/FMO1 in terms of developmental expression in human hepatic tissues. Relatively low expression of FMO1 was also reported in the human fetal brain and was undetectable in adult human brain. The extent to which the apparently low expression of FMO1 in human fetal brain tissues may be of importance for prenatally elicited neurotoxicity remains an intriguing project for investigation. Sadeque *et al.* (1992) and Rettie *et al.* (1995) also reported in earlier studies that sulfoxidation of methyl *p*-tolyl sulfide (an FMO substrate) in cell-free preparations of human fetal livers proceeded at rates comparable to those observed in corresponding preparations of adult human livers and kidneys, noting that the fetal sulfoxidase activity appeared to be catalyzed by a separate FMO isoform. It now appears that FMO3 is the predominant isoform in human adult liver, whereas FMO1 is predominant in human fetal liver and that FMO1 may have been the major catalyst for N-oxidation of N,N-dimethylaniline observed in human fetal liver preparations much earlier by Rane (1973).

Other information pertaining to the prenatal expression of FMO isoforms appears not to be readily available. It is clear that an enormous amount remains to be learned concerning the developmental expression of FMO isoforms. Especially prominent in this regard is the need for more precise information concerning the level of expression of each isoform in various prenatal tissues as a function of the stage of gestation. Because of already noted profound species differences, human tissues would seem to be of prime consideration but information from the more commonly employed experimental animals also is of considerable value and there appears to be very little data relevant to the prenatal expression of FMOs in experimental animals at present.

C. Peroxidases

Investigations in the laboratories of Dr. Peter G. Wells and Arun P. Kulkarni have greatly increased interest in the expression of conceptal peroxidase enzymes during prenatal development. Because of the very low prenatal expression of several P450 isoforms (with exceptions as noted previously), the generally higher prenatal expression of peroxidase enzymes could play a more important role in bioactivation of xenobiotics in conceptal tissues. Several reviews of this aspect have appeared (Wells *et al.,* 1996; Wells and Winn, 1996, 1997). Greatest interest appears to have been in the potential roles of prostaglandin H synthase and lipoxygenases as catalysts for xenobiotic bioactivation, but it is clearly conceivable that numerous conceptal peroxidases could participate significantly in bioactivation reactions in conceptal cells/tissues. Kulkarni's laboratory has been especially active in investigations of the developmental expression and potential importance of lipoxygenases in catalysis of xenobiotic oxidations (e.g., Roy *et al.,* 1993).

Peroxidases are enzymes that, as the name implies, require a peroxide or similar oxidizing agent for the oxidation reaction. This is frequently hydrogen peroxide but may also be any one of a number of other peroxides, including organic hydroperoxides such as lipid hydroperoxides. Classic peroxidases are hemoproteins specific for a peroxide but that use a wide range of substrates as donors. Coenzyme peroxidases, however, are not hemoproteins, and some are flavoproteins. Flavoprotein peroxidoses are more similar to xanthine oxidase. Products of peroxidation reactions include the reduced peroxide (e.g., hydrogen peroxide is reduced to water) and oxidized substrate. Oxidation via removal of a single electron results in the generation of a free radical species that is a common oxidation product of peroxidase-catalyzed reactions. Thus, in order to act as a good substrate for such a peroxidase-catalyzed oxidation, the chemical should be a reasonably good electron donor. Several xenobiotic molecules can be oxidized in this fashion. Peroxidases may also catalyze incorporation of a peroxide oxygen atom into the substrate molecule. Generation of reactive epoxides can occur via this mechanism and several examples are now known.

Peroxidases such as prostaglandin H synthase and lipoxygenases are attractive as potential conceptal bioactivating enzyme catalysts because they are present with high content and activity in both human and rodent conceptuses even at very early stages of gestation (Hume, 1993; Wells *et al.*, 1996; Wells and Winn, 1996, 1997). Possibly the quantitatively most important reactive intermediates generated in peroxidase-catalyzed reactions are free radicals. Once generated, free radicals are capable of producing cell and tissue damage via a variety of mechanisms. They may interact directly with critical protein or nucleic acid macromolecules to oxidize these molecules or form covalent bonds; they may interact with lipid molecules to initiate self-sustatining lipid peroxidation reactions with consequent damage to cellular or subcellular membranes; or, if redox potentials are appropriate, they may interact with molecular oxygen to generate superoxide anion (peroxidase-generated radicals rarely donate electrons to O_2). In turn, under appropriate conditions, superoxide anion can be converted to a number of reactive oxygen species, the most important of which is probably the hydroxyl radical (OH·). Reactive oxygen species are known to be capable of producing extensive cell/tissue damage. Wells and Winn (1996) have summarized the evidence for involvement of peroxidases in the bioactivation of a number of conceptotoxic agents. Prostaglandin H synthase has been implicated in the bioactivations of phenytoin, mephenytoin, nirvanol, trimethadione, dimethadione, thalidomide, benzo(a)pyrene, 13-*cis*-retinoic acid (isotretinoin), diethylstilbestrol, cyclophosphamide, acetaminophen, 2-naphthylamine, and 2-acetylaminofluorene. Lipoxygenases have been mentioned in connection with bioactivation of phenytoin and cyclophosphamide. Thyroid peroxidase and myeloperoxidase were listed in connection with phenytoin bioactivation and horseradish peroxidase in connection with bioactivations of phenytoin, 2-naphthylamine and 2-acetylaminofluorene. The extent to which conceptal peroxidases contribute to bioactivation of the listed chemicals in terms of the mechanisms whereby such chemicals produce persistent conceptotoxic effects remains to be determined but appears to be a very promising area of investigation. Wells and Winn (1996, 1997) and Wells *et al.* (1996) have summarized the most recent evidence for the involvement of peroxidase-dependent mechanisms of teratogenicity for the previously mentioned chemicals. At present, such evidence is based primarily upon the interactive effects of supplemented chemicals (e.g., acetylsalicylic acid, eicosatetraynoic acid, arachidonic acid, spin trapping agents, etc.) that would be expected to either inhibit or diminish the generation of radical species. This is clearly an area of research that is deserving of much more attention than it has received in the past. A systematic analysis of the expression of various peroxidase enzymes in conceptal tissues at various stages of gestation and characterization of these enzymes in terms of tissue distribution, subcellular localization, regulation of expression, etc. would fill highly important gaps in our knowledge of this area. Critical analyses of the participation of specific conceptal peroxidases in the conceptotoxic effects of various chemicals will also require much additional attention.

D. Dehydrogenases

Reactions catalyzed by dehydrogenase enzymes (E.C. 1.1.1.–1.1.2) are often readily reversible (but not always; for instance, aldehyde dehydrogenation is usually irreversible) in contrast to reactions catalyzed by P450 hemeproteins, flavin monooxygenases, and peroxidases. Thus, dehydrogenase enzymes do catalyze reactions regarded as oxidation reactions as well as reactions regarded as reduction reactions (Section III). For oxidative dehydrogenation, an endogenous oxidant (usually NAD or NADP) accepts hydrogens and/or electrons from the molecule undergoing oxidation. Oxidative conversion of ethyl alcohol to acetaldehyde, catalyzed by alcohol dehydrogenase, is probably the most commonly familiar of these reactions. Alcohol dehydrogenases (E.C. 1.1.1.1.) are probably the dehydrogenases most heavily involved in xenobiotic biotransformation catalysis, but it should be noted that literally hundreds of dehydrogenase enzymes occur in nature and that very few have been investigated systematically in terms of their capacity to catalyze oxidative dehydrogenation of xenobiotics. Several aldehyde dehydrogenases and dihydrodiol dehydrogenases also are now known to be involved in xenobiotic biotransformation (Parkinson, 1996).

In very general terms, it is widely believed that the alcohol and aldehyde dehydrogenases significantly involved in xenobiotic oxidation are expressed at very low levels in prenatal tissues, based primarily on measurements of enzymatic activities in tissue-free preparations. Only relatively recently has it been recognized that alcohol and aldehyde dehydrogenases are actually relatively large families of enzymes rather than single entities, and it is therefore not unexpected that extensive knowledge of the characteristics, regulation, and so on, of the various individual isoforms is, for the most part, very rudimentary at present. This includes knowledge of developmental regulation, which appears to have received very little investigative attention, possibly because of the widespread notion that prenatal expression of dehydrogenases involved in xenobiotic biotransformation is minimal or negligible. Developmental aspects of dehydrogenase-dependent xenobiotic oxidation is an area in urgent need of a comprehensive and critical reevaluation/review and, to this author's knowledge, has not received attention in review form since Juchau *et al.,* in 1980.

Recognition of both ethanol and retinoids as highly important teratogenic agents capable of eliciting persistent morphologic and functional birth defects in humans, and interesting investigations in the laboratory of Greg Duester may combine to focus considerably more attention on the prenatal expression of dehydrogenase enzymes. Ethanol undergoes alcohol dehydrogenase-catalyzed oxidation to acetaldehyde, a highly toxic metabolite with considerable genotoxic, cytotoxic, and teratogenic (Campbell and Fantel, 1983) potential. Retinol (vitamin A_1), also undergoes alcohol dehydrogenase-catalyzed conversions from the inactive alcohol to the biologically highly potent all-*trans* retinoic acid. Both acetaldehyde and retinoic acid are heavily implicated in mechanisms of teratogenicity (Soprano and Soprano, 1995; Juchau *et al.,* 1997) and may be regarded as products of bioactivation. Duester (1991 and Deltour *et al.* (1996) have proposed that the teratogenicity of ethanol may be due to competitive inhibition of dehydrogenase-catalyzed conversion of retinol to retinoic acid, resulting in a deficiency of the morphogenic acid. Although the hypothesis appears to lack merit in terms of dysmorphogenic effects elicited by ethanol in rodent embryos (Chen *et al.,* 1995, 1996a,b; Chen and Juchau, 1997, 1998), it has not been disproved for humans and is deserving of further investigations. The investigations performed pertaining to ethanol and retinoids and their conceptotoxic interactions have pointed to the urgent need for a far better knowledge of the developmental regulation of individual dehydrogenase isoforms functional in the oxidation of embryo/fetotoxic agents, especially during prenatal development.

E. *Other Oxidative Enzymes*

Other oxidative enzymes with documented capacity to catalyze xenobiotic oxidation reactions include monoamine oxidases, diamine oxidases, dioxygenases, polyamine oxidases, molybdenum hydroxylases (xanthine oxidase and aldehyde oxidase), and probably others. Although certain xenobiotics act as substrates for these enzymes, they appear to function primarily in catalyses involving endobiotic substrates, and may therefore be expected to be expressed at relatively high levels during prenatal life. In terms of systematic investigations of the actual xenobiotic-biotransforming activities of these enzymes as a function of prenatal development, however, few solid statements can be made. Because of the dearth of information relative to prenatal expression as well as because of the relative lesser perceived importance of these enzymes in xenobiotic biotransformation, they are not discussed further here. Their possible participation as determinants of conceptotoxic effects, however, should be borne in mind.

III. Reductive Enzymes

A. *P450 Reductase*

This membrane-bound (endoplasmic reticulum) enzyme is also known as NADPH-dependent P450 reductase, P450 oxidoreductase, and (less frequently) as microsomal NADPH cytochrome c reductase. It is a 78 kDa flavoprotein containing 1 mole each of FAD and FMN. It functions in the transfer of electrons from NADPH (NADH is a much less efficient electron donor) first to the FAD moiety, subsequently to the FMN moiety, and finally to an electron acceptor such as P450 or cytochrome c. A very wide variety and large number of xenobiotics can also act as electron acceptors, the most notable of which are quinones, aromatic nitro compounds, heterocyclic nitro compounds and compounds with azo linkages. Nitro group and azo linkage reduction were some of the earliest xenobiotic reduction reactions investigated and are important historically. A common misconception is that P450 reductase is exclusively responsible for catalyses of these reduction reactions; this in spite of the fact that virtually any flavoprotein can act as a catalyst for such reactions and that even free FAD, FMN, or even riboflavin will act as a catalyst (Juchau *et al.,* 1970; Juchau, 1973). Hence, reduction of quinones, nitro groups, and azo linkages would be expected to occur extensively in virtually all tissues and at virtually all stages of gestation, and indeed, this has been shown to be the case (Juchau *et al.,* 1980). P450 reductase-catalyzed reduction occurs, as would be expected in flavin-catalyzed reduction, one electron at a time, and initial products are thus free radicals, capable in many cases of donating electrons to molecular oxygen. Superoxide anions and other reactive oxygen species can thus be generated from redox cycling with resultant toxicity.

P450 reductase serves as catalyst for transfer of electrons from the primary initial electron donor (NADPH) to substrate-bound P450 for all known mammalian P450s, and thus its distribution is also very widespread. As yet, no systematic investigations of developmentally regulated prenatal expression appear to have been undertaken, although it probably can be assumed that the gene coding for this flavoprotein would be significantly expressed at all stages of gestation. Nevertheless, it would also be helpful to have a more quantitative idea regarding its developmental expression.

B. Quinone Reductase

The official EC name for this enzyme is NAD(P)H quinone oxidoreductase, but it is probably more commonly known as DT-diaphorase. Two human forms have been cloned, one of which (NQO_1) is a dioxin-inducible form, and the other (NQO_2) is constitutive. These are 55 kDa, FAD containing, cytosolic enzymes that exist as homodimers. Importantly from a toxicologic perspective, they catalyze an obligatory 2-electron reduction with quinones as the most commonly recognized electron acceptor substrates. Thus, reduction of quinones by P450 reductase can lead to generation of redox cycling semiquinones and generation of reactive oxygen species whereas reduction of the same chemicals catalyzed by quinone reductase leads to direct generation of inactive hydroquinones without formation of semiquinone intermediates. Clearly, prenatal expression of this enzyme is expectedly of potentially high importance as a determinant of the prenatal toxic effects of electron accepting xenobiotic chemicals. Yet, again, no systematic investigations of the prenatal expression of the enzyme appear to have been undertaken. The fact that the enzyme can be overexpressed in tumor cells suggests that it may be expressed at high levels in embryonic/fetal cells, but there is as yet no documentation for this speculation. There seems to be, in fact, little information overall for the developmental expression of this enzyme in spite of the clear evidence for a highly important role in developmental toxicology.

C. Carbonyl Reductases

Carbonyl (C=O) groups such as aldehydes or ketones may undergo reductive biotransformation to the corresponding alcohols in reactions catalyzed by dehydrogenase enzymes or by various enzymes designated as carbonyl reductases (Weiner and Flynn, 1989). Carbonyl reductases include cytosolic aldehyde and ketone reductases which constitute a very poorly characterized group of NADPH-dependent enzymes with wide substrate specificity. Aldehyde reductases, aldose reductases, and ketone reductases, have been designated as AKR1, AKR2, and AKR3, respectively, but taxonomy in general is in a virtually total state of disarray. Thus, it is not surprising that little is known concerning the prenatal developmental expression of members of this group of enzymes. It should be emphasized that a very large number of oxidoreductase enzymes is capable of catalyzing such xenobiotic reduction reactions and that these reactions can sometimes also occur at very significant rates, even without the aid of enzyme catalysts. Therefore, the task of providing definitive information with respect to developmental regulation of enzymes importantly involved in xenobiotic reduction is indeed a daunting endeavor, and it may not be realistic to expect rapid progress in this area in the near future. However, the tissue distribution of these enzymes appears to be very broad/diverse, occurring in blood, liver, kidney, brain, and other tissues. It might, therefore, be expected that these enzymes would also be significantly expressed in prenatal tissues at several stages of gestation. Definitive studies designed to confirm or refute such a supposition, however, remain to be accomplished.

D. Other Reductases

Reductions of aromatic and heterocyclic nitro groups, azo linkages, quinones, ketones, and aldehydes appear to be catalyzed primarily by the enzymes discussed previously. In addition, numerous other groups present in xenobiotic chemicals may also undergo non-enzymic or enzyme-catalyzed reductions. These include epoxides, disulfides, sulfoxides, N-oxides, alkenes, alkynes, hydroxylamines, hydroxamic acids, halogens (reductive dehalogenation), and various other groups with redox potentials suitable for electron acceptance (e.g., halothane). Dehydroxylation reactions (rare) can also be catalyzed by oxidoreductase enzymes. The exact nature and relative contributions of enzymes involved in catalyses of these reactions is, for the most part, unknown. Thus, it is not possible at this time to discuss developmental regulation of other reductase enzymes except in speculative terms.

In terms of involvement in potentially conceptotoxic reduction reactions, the greatest concerns would be for catalysis of the generation of organic radicals (acceptance of a single electron by any organic molecule generates an organic radical species) and for the redox cycling of such organic radicals. Semiquinone and nitroanion radicals are prime examples. Redox cycling can participate in elicitation of toxic effects via several mechanisms including generation of reactive oxygen species depletion of molecular oxygen (important in tissues with relatively low O_2 tension); depletion of cellular reducing equivalents such as NADH, NADPH, GSH, and so on; generation of thiyl radicals; and initiation of lipid peroxidation. Clearly, reduction of xenobiotic chemicals in conceptal tissues/cells is of high potential importance toxicologically. Nevertheless, very few concrete examples of practical importance to conceptotoxicity *in vivo* can be pointed to at present. This may be because of highly efficient defense/repair mechanisms in prenatal tissues. A lesser expression of enzymes capable of catalyzing single electron reduction reactions may also be a factor, although there are few published studies that would support that speculation. It is obvious that much further study will be required to make definitive statements in this poorly developed but potentially important area of research.

IV. Hydrolytic Enzymes

Of enzymes catalyzing the three types of phase I xenobiotic biotransforming reactions (oxidations, reductions, hydrolyses), the least studied in terms of developmental regulation appear to be those catalyzing hydrolyses (hydrolases). This is in spite of the fact that hydrolytic enzymes are very ubiquitous, and therefore, at least some are likely to be expressed in prenatal tissues at relatively high levels. The more common xenobiotic substrates include carboxylic acid esters, amides, sulfate esters, glucuronide esters, and epoxides, and a very large number of xenobiotics exist as or are metabolically converted to chemicals with such groups. Early literature dealing with developmental aspects of xenobiotic hydrolysis was reviewed by Juchau *et al.* (1980) and by Klinger (1987). A brief review was provided by Juchau (1990). Early studies consisted primarily of measurements of enzymic activities with cell–free preparations *in vitro* and early investigations belied a lack of appreciation for the numerous proteins now known to be capable of catalyzing the hydrolytic reactions under study.

A. *Epoxide Hydrolases*

Epoxide hydrolases catalyze the *trans*-addition of water to alkene epoxides and arene oxides that are generated as the result of P450-dependent or peroxidase-dependent oxidations. Three immunologically distinct forms of epoxide hydrolase are recognized: (1) a microsomal epoxide hydrolase (mEHb) concentrated in the endoplasmic reticulum of adult liver cells but present in many other tissues as well; (2) a microsomal cholesterol oxide hydrolase (mEHch) that is selective for catalyzing the hydration of cholesterol (and a few related steroid) 5,6-oxides; (3) a soluble epoxide hydrolase originally termed *cytosolic* but now designated *soluble* because it also appears to be expressed in the peroxisomal matrix of hepatic cells. A principal function of mEHb appears to be catalysis of the conversion of electrophilic xenobiotic epoxides to vicinal *trans*-dihydrodiol metabolites that are normally much less reactive. As contrasted with soluble epoxide hydrolase (sEH), mEHb exhibits a substrate preference for hydrophobic epoxides without extensive substitution and displays a broad substrate specificity. It is inducible by phenobarbital, antioxidants, nitrosamines, and imidazole antimycotics. sEH catalyzes the hydration of a variety of aliphatic and extensively substituted arene oxides that include xenobiotics and is inducible by peroxisome proliferators such as clofibrate. mEHch does not appear to catalyze hydration of xenobiotic molecules significantly and likewise does not appear to be inducible by xenobiotic inducing agents. Although principally involved in catalyses of the inactivation of reactive electrophilic epoxides, it is established that both mEHb and sEH catalyze reactions that convert protoxins to more proximate toxic chemicals and are thus involved in bioactivation.

In rats, ontogenic development of mEHb was observed beginning at postnatal day 6 and continuing for approximately 9 weeks (Falany *et al.*, 1987). At postnatal day 6 activities were very low, but increased continually until approximately the postnatal day 42 and afterward began to decrease as a function of increasing age. Klinger (1987) also reported detection of the enzyme during prenatal life. Rouet *et al.* (1984) reported that relatively high levels of epoxide hydrolase activity were observable in the brain, lung, and liver tissues of perinatal mice. Benzo(a)pyrene 4,5-oxide was used as substrate and measurements were made at gestational days 17 and 19 as well as at postnatal days 4, 6, 10, and 20. Induction with 5,6-benzoflavone was also apparent pre- and postnatally, especially in the lung and liver. Human fetal livers at gestational ages ranging between 17 and 27 weeks were assayed both immunologically and enzymically with benzo(a)pyrene 4,5-epoxide as substrate (Cresteil *et al.*, 1985), and an excellent correlation between the two parameters was found ($r = 0.939$). Epoxide hydrolase activities in fetuses varied from approximately 1 to 3 nmol/mg/min as compared with 7 to 10 nmol/mg/min for corresponding preparations from four adult human hepatic samples. Earlier investigations (Juchau and Namkung, 1974) had also indicated that human fetuses possessed relatively high levels of mEHb. Taken together, the data suggest that human mEHB appears to be expressed at significant levels prenatally, whereas in common experimental animals it may not be expressed significantly until after parturition. More careful studies are needed to determine if this interpretation is accurate. Very little information concerning the developmental expression of sEH is currently available, and comments concerning this aspect appear not to be merited at present.

B. Esterases

A huge number of enzymes has exhibited the capacity to catalyze the hydrolysis of carboxylic acid esters and other esters. Classification has been difficult due to overlapping substrate specificities and other factors. They are frequently classified as A-esterases (including arylesterases with *p*-nitrophenyl acetate as a classic substrate, phosphotriester hydrolases with paraoxon as a classic substrate, and diisopropyl fluorophosphatases) and as B-esterases (including acetylcholinesterase with acetylcholine as a classic substrate, pseudocholinesterase with butyrylcholine as a classic substrate, and carboxylesterases). Most xenobiotic esters undergo hydrolytic biotransformation catalyzed by the B-esterases, often loosely termed the *carboxylesterases.* Multiple forms of hepatic microsomal carboxylesterases exist, but these kinds of enzymes are ubiquitous. Pseudocholinesterase is predominantly present in the plasma and functions to catalyze the hydrolysis of a very large number of xenobiotic substrates. Interestingly, genetic variants displaying deficiencies of pseudocholinesterase do not exhibit pathologic symptoms unless challenged with toxic xenobiotics such as succinylcholine. This suggests that evolution of this enzyme has been directed toward hydrolysis of xenobiotic substrates. Unfortunately, little information is available concerning the prenatal expression of these enzymes, and discussions must await the generation of such data.

C. Glucuronidases and Sulfatases

These hydrolytic enzymes are of particular interest because they catalyze hydrolysis of biotransformationally generated xenobiotic glucuronides and sulfates (discussed in Section V). Xenobiotic glucuronides and sulfate conjugates can be actively secreted by hepatic cells into the bile and, on reaching the gastrointestinal tract in the biliary fluid, can then undergo hydrolysis to the original unconjugated chemicals. These hydrolyses are catalyzed by β-glucuronidases and sulfatases present in the gastrointestinal tract. Because the unconjugated forms are far more lipophilic than the conjugates, they are then reabsorbed and redelivered to the liver via the circulation in the well-known process of enterohepatic circulation. The extent to which these hydrolase enzymes develop prenatally and perhaps play a role in the prenatal toxicity of xenobiotic chemicals appears to be entirely unknown at present.

D. Other Hydrolases

Just as numerous chemical groups are subject to reduction in biologic systems (as discussed previously), so are several chemical groups subject to biotransformation via hydrolyses. These include those discussed in Sections IV.A–C as well as amides, hydroxamic acids, hydrazides, carbamates, and nitriles. Certain ring structures (e.g., phenytoin and other hydantoin rings) may undergo hydrolytic ring scission, and various halogenated compounds (e.g., methylene chloride) undergo hydrolytic dehalogenation. Specific hydrolases that predominate in the catalysis of these hydrolytic reactions, however, are, for the most part, poorly defined and consequently, once more, statements concerning developmental aspects of regulatory control of such reactions are not practical at present.

V. Transferases

Transferases are enzymes that catalyze the conjugation of acceptor substrates to endogenous chemical groups including glucuronide, sulfate, phosphate, acetate, reduced glutathione, amino acids (primarily glycine and glutamine for xenobiotic conjugation) and methyl groups. Except for glutathione conjugation, these are energy-expensive, ATP-consuming reactions and, except for methylation (and sometimes acetylation), result in conjugates with far higher water solubility and polarity than the parent acceptor substrate molecules. Although such reactions are predominantly inactivating (i.e., conjugates are frequently far less biologically active than parent acceptor substrate molecules), all are also known to be capable of catalyzing bioactivation reactions as well—with the possible exception of methylation—and the literature is filled with examples of such transferase-mediated bioactivation reactions. Conjugation reactions involving xenobiotic acceptor substrates are commonly referred to as *phase II reactions* because an oxidation, reduction, or hydrolytic reaction (phase I) often precedes conjugation and generates the required acceptor moiety (hydroxyl, amino, imino, carboxyl, sulfhydryl, etc., groups).

A. Glucuronosyl Transferases

From a quantitative point of view, glucuronidation is the most important of the phase II (conjugation) reactions in humans and most other vertebrate species (the cat family is an exception). Historically, glucuronidation was the first xenobiotic-biotransforming reaction to be recognized as prenatally deficient (Brown and Suelzer, 1958), and it is currently a widespread and well-accepted belief that prenatal glucuronidations of xenbiotic acceptor molecules proceed at extremely low to negligible rates in humans as well as in most, if not all, experimental animals. It is now obvious that such a generalization is in need of refinement, although it must be conceded that high rates of glucuronidation in prenatal, perinatal, or neonatal tissues have never been confirmed in the open literature.

Burchell *et al.* (1989) have provided a review of developmental aspects of glucuronosyltransferase enzymes. Since publication of that review, the multiplicity and nature of these enzymes has become much better understood and developmental aspects should now be considered in light of this increased understanding. It is now appreciated that glucuronosyltransferases constitute a very large superfamily of enzymes, with at least 19 distinct human isozymes identified (reviewed by Burchell *et al.,* 1995). Many of these have now been cloned and expressed and each appears to have an identified ortholog in rats. It seems probable that many more will yet be identified. These have been divided into 2 gene families—Family 1 (UGT1) with at least 12 distinct forms, and Family 2 (UGT2A and UGT2B). UGT1*02 and UGT2B8 were designated as key isoforms in the glucuronidation of a wide range of xenobiotic substrates containing phenolic groups, but several other isoforms also appear to be extensively involved in xenobiotic biotransformation. Murine UGT1*06 is an established enzyme product of the dioxin-inducible Ah gene battery that also includes dioxin-inducible P450s 1A1 and 1A2, NAD(P)H quinone oxidoreductase, an aldehyde dehydrogenase, and a glutathione transferase subunit.

Because of the tools provided by modern molecular biology, this is now a very rapidly expanding area of research, but many pressing questions remaining to be answered. These questions include aspects of developmental regulation and expression, and future research will probably include investigations of quantitative aspects of the developmental expression of individual UGT isoforms. An example is provided by Coughtrie *et al.* (1988), who reported, on the basis of immunoblotting experiments, that microsomes from human fetal livers contain only one major immunoreactive protein (corresponding to 5-hydroxytryptamine and/or morphine glucuronidating activity). At 3 months postpartum, the liver exhibited the entire pattern of isozymes displayed by the adult human liver with the activity being 30% of adults. Another question pertaining to developmental aspects of glucuronosyltransferase enzymes is related to tissue/cellular distribution. One article (Hume *et al.,* 1996) pointed out that the human fetal kidney exhibited glucuronidation of estriol and *p*-nitrophenol as early as 10 weeks in gestation, and activity increased as a function of gestational age. In adult kidneys, uridine diphosphate-glucuronosyl transferases (UDPGTs) appeared, on the basis of immunohistochemical studies, to be localized in the proximal tubules; in human embryonic and early fetal development, UDPGT was reportedly localized in derivatives of the ureteric bud such as the ureter, pelvis, calyces, and collecting ducts. Differences in distribution and intracellular localization could well play an important role in chemical defenses against potential conceptotoxins, and further studies of this nature are strongly indicated.

The existence of multiple forms of UDPGTs was actually first suggested on the basis of ontogenic development and responses to inducers in rats. Rat liver enzymes were placed in four groups: 1. A group that exhibits a peak in expression 1–5 days prior to parturition and is inducible by dioxin-type inducing agents. Substrates for this group are planar and often aromatic in nature. 2. A group exhibiting an expression peak 1–5 days after parturition and inducible by phenobarbital-type inducing agents. Substrates for this group tend to be bulky and nonaromatic. 3. A group exhibiting an expression peak around the time of puberty and inducible by dexamethasone-type inducing agents. Substrates for this group include the cardiac glycosides and 4. A group exhibiting an expression peak around the time of puberty but inducible by clofibrate-type inducing agents (peroxisome proliferators). Bilirubin is a good substrate for group four. This classification system has some practical value, but it now should be investigated to determine which specific isoforms develop with similar patterns. Clearly, a great deal remains to be learned concerning the developmental expression of this important class of enzymes.

B. Sulfotransferases

A lack of data concerning developmental aspects of xenobiotic sulfation preclude confident generalizations at this time. Results of past studies have provided somewhat puzzling and often apparently conflicting pictures of the status in this research area. This subject was reviewed earlier by Pacifici *et al.* (1988) and by Juchau (1990). As with other xenobiotic-biotransforming enzymes, most of the earlier data consisted of measurements of enzymic activities *in vitro* using tissue homogenate/homogenate subfractions as enzyme sources. On the basis of those studies as well as various later ones (e.g., Gilissen *et al.,* 1994), it is evident that at least certain xenobiotics can undergo relatively rapid sulfation *in vitro* when cell-free preparations of prenatal liver, adrenal, or other tissues are used as enzyme sources. Such observations are certainly of importance because of the involvement of sulfation reactions in numerous examples of genotoxic bioactivation sequences. However, it is now recognized that for sulfation *in vivo,* the sulfotransferase(s) enzyme(s) catalyzing the reaction frequently do not limit and may not even be important determinants of the sulfation rate. Rather, quantities of inorganic sulfate (derived from low quantities of free cysteine), required for the biosynthesis of 3′-phosphoadenosine-5′-phosphosulfate (PAPS), are often rate limiting. Thus, evaluations of the developmental expression of enzyme levels assume lesser importance for sulfation than for glucuronidation, for example (but also for virtually all other xenobiotic biotransformations), for which enzyme levels almost invariably are the prime determinants of reaction rates. In terms of prenatal sulfation of xenobiotics, the extent to which sulfotransferase versus inorganic sulfate/PAPS levels limit reaction rates appears to be unknown at present. This would seem to merit a very high priority.

In cases for which levels of sulfotransferase enzyme(s) may be an important determinant of xenobiotic sulfation reaction rates in prenatal tissues, the large multiplicity of sulfotransferase enzymes must also be recognized. In humans, for example, at least five distinct forms exist and these remain poorly characterized in terms of substrate specificities, modes of regulation, kinetic parameters, and so on. It would appear that certain sulfotransferases are expressed at high levels prenatally, whereas others are expressed minimally. Definition is needed in this area. Profound species and tissue/cell differences also seem likely to exist prenatally, but rigorous studies of these aspects likewise appear not to have been published. Thus, for prenatal sulfation of xenobiotic substrates, it is tempting to argue that the time is ripe for initiation of an entirely fresh set of investigations with experiments designed to provide meaningful data. Hopefully this will occur in the near future. From the perspective of developmental neurotoxicology, observations of expression of aryl sulfotransferases in adult human brain tissues (Zhu *et al.,* 1993) would be hoped to stimulate investigations of sulfation in prenatal cephalic tissues.

C. Glutathione Transferases

The subject of prenatal glutathione transferase expression was discussed briefly in terms of pre-1980 studies in an early review by Juchau *et al.* (1980) and in later reviews by Klinger (1987) and by Juchau (1990). In each of these reviews, the observation was made that the available data suggested much more active glutathione conjugation in human prenatal tissues than in the prenatal tissues of experimental animals (rodents). This potentially important observation appears now to have been further supported in subsequent studies. The very large multiplicity of species and tissue-specific reduced glutathione (GSH)-S-transferase isoforms again presents a major challenge in terms of investigations of developmental regulation, however, and conclusions along these lines seem best made conservatively. Until quite recently, information pertaining to prenatal glutathione conjugation was almost entirely dependent on measurements of conjugating activities *in vitro* with various substrates. Such investigations are often difficult to interpret. It is now anticipated that future research with the tools of modern molecular biology will soon provide a much clearer picture.

Glutathione transferases catalyze reactions in which GSH reacts with electrophilic moieties present in endobiotic as well as xenobiotic chemicals (often generated in bioactivation reactions). They serve a highly important defense function against the deleterious effects of xenobiotic electrophiles that are capable of forming covalent bonds with nucleic acids and proteins. Interestingly, GSH transferases are also capable themselves of catalyzing numerous bioactivating reactions (for several examples, see Parkinson, 1996), and at least some of these transferases also exhibit significant peroxidase and isomerase activities in addition to their transferase functions.

Even though levels of GSH transferases appear to be generally very low in rodent conceptal tissues, localized activities in particular rodent conceptal tissues (including the visceral yolk sac and in various tissues within the embryo proper) may be sufficient to provide significant protection from the toxic effects of chemical electrophiles. In murine conceptal tissues, Dillio *et al.* (1995) reported detection of three classes of transferases (alpha, mu, and pi) as early as gestational day 9. Activities in embryonic tissues were reported as approximately half those observed in visceral yolk sac tissues and about 4% of maternal hepatic activities. Alpha and pi classes appeared to increase and mu to decrease as a function of gestational age. Investigations in rat conceptal tissues (Serafini *et al.,* 1991; Hales and Huang, 1994) yielded similar observations except that embryonic and yolk sac activities were roughly equal and only about 1% of that in adult rat liver. In humans, at least five fetal glutathione transferase isozymes have been reported (Guthenberg *et al.,* 1986; Kashiwada *et al.,* 1991; Mitra *et al.,* 1992) and activities reported have been comparable to those observed in adult tissues for at least some substrates (Grover and Sims, 1964; Chasseaud, 1973; Warholm *et al.,* 1980; Strange *et al.,* 1989). In contrast to rodent patterns, the human conceptal alpha and mu classes were reportedly increased at birth, while the pi class was decreased. It would appear that differences between humans and rodents in terms of prenatal expressions of glutathione transferases are profound and could possibly be a major determinant in species-specific conceptotoxicity. It will be of great interest to observe further progress in this important facet of research on developmental xenobiotic biotransformation.

D. Acyltransferases

Xenobiotic conjugation reactions involving acylation are of two categories. These are acetylation, for which acetyl coenzyme A is the reactive acetyl donor, and amino acid conjugation, for which the xenobiotic substrate undergoes activation prior to conjugation with an endogenous amino acid. Glycine is the amino acid most heavily involved and is almost exclusively involved for reactions occurring in rodents and rabbits. In humans and subhuman primates, glutamine is also frequently involved.

1. Acetyltransferases

N-Acetyltransferases (E.C. 2.3.1.5) catalyze the acetylation of arylamines encountered from medicinal, environmental, occupational, and dietary exposures. Humans exhibit genetic polymorphisms in hepatic N-acetyltransferase capacity yielding rapid, intermediate, and slow acetylator phenotypes. Because biotransformation of various arylamines is necessary for their mutagenicity and carcinogenicity, it has been suggested that acetylator phenotype can be a determining factor in predisposition of individuals to certain cancers related to arylamine exposures. Two human N-acetyltransferase genes have been cloned, sequenced, and expressed. These have been designated by Vatsis *et al.* (1995) as *NAT1**, previously thought to be monomorphically expressed, and *NAT2**, long recognized as polymorphically expressed. Similar genes have been identified in experimental animals. However, the developmental regulation of these genes has received very little attention and our knowledge of the prenatal expression of these genes in human, rodent, or rabbit prenatal tissues is extremely scant. Further discussion of these aspects must await future research.

2. Amino Acid Transferases

Although conjugation of xenobiotic molecules with a very large number of different endogenous amino acids is possible (Parkinson, 1996), by far the most important amino acids from a quantitative perspective are glycine and glutamine. Conjugation with glutamine appears to be limited in large measure to humans and subhuman primates. These reactions are catalyzed by acyl-CoA : amino acid N-acyltransferases and are important for conjugation of chemicals with carboxylic acid groups. At low substrate concentrations, the level of the acyltransferase represents the rate limiting factor, but at high concentrations, concentrations of coenzyme A and/or free amino acid may limit the rate. Investigations of the regulation of rates of these reactions as a function of development have been extremely scarce and our knowledge concerning prenatal amino acid conjugation of xenobiotics is very limited. As with prenatal acetylation, discussion of developmental aspects is probably not justified at this point in time.

E. Methyl Transferases

These enzymes are important for the biotransformation of catechols and phenols and, to a lesser degree, for various other chemicals. Because of the well-known conceptotoxicity of methyl mercury, it is of particular interest to this discussion to note that methylation of inorganic mercury to methyl mercury is an established biotransformation reaction, although the identity of a specific methyltransferase(s) involved in catalysis of the reaction is unknown at present. Arsenic and selenium also undergo methylation in biologic systems (Parkinson, 1996). Interest in methylation of catechols appears to have increased somewhat because of reports of catalysis of conversions of estrogenic chemicals to catechols by P4501B1. However, no systematic investigations of the developmental regulation of methyltransferase-catalyzed xenobiotic conjugations appear to have been undertaken. It seems likely, in view of extensive participation of methyltransferase enzymes in metabolism of endogenous substrates, that such enzymes might also be expressed extensively during prenatal development. This speculation, however, remains to be documented.

F. Other Transferases

Transferase enzymes are extremely numerous and a great many transferases capable of catalyzing conjugations of xenobiotic molecules other than those discussed previously have been noted in the literature. These include conjugations with glucose, N-acetylglucosamine, numerous other amino acids, phosphate, and several others. Such enzymes and such reactions, however, are normally regarded as very minor in mammalian species and, as expected, very little is known concerning developmental aspects.

VI. Summary and Conclusions

Although significant progress has been made in investigations of the developmental enzymology of several enzymes that catalyze xenobiotic-biotransformation reactions, progress in studies of many have been low to negligible, and it is clear that much better definition is sorely needed. The most significant progress has been in terms of P450 heme-thiolate proteins, peroxidases, and glutathione transferases, but even with these, progress has not been highly impressive. Some new findings with respect to flavin monooxygenases, dehydrogenases, and glucuronosyl transferases have also been made. For most hydrolases, reductases, sulfotransferases, acyltransferases, and methyltransferases, however, progress has been minimal. Interesting observations have been made that may well be of importance to the prenatally developing nervous system and it will be highly interesting to follow further research in those areas. Of particular interest are observations of the expression of various P450 isoforms in prenatal human cephalic tissues.

References

Anderson, L. M., Jones, A. B., Miller, M. S., and Chauhan, D. P. (1989). Metabolism of transplacental carcinogens. In *Transplacental and Multigenerational Carcinogenesis* N. P. Napalkov, J. M. Rice, L. Tomatis, and H. Yamasaki, Eds. IARC Scientific Publications No. 96.

Berry, D. L., Slaga, T. J., Wilson, N. M., Zachariah, P. K., Namkung, M. J., Bracken, W. R., and Juchau, M. R. (1977a). Transplacental induction of mixed function oxygenases in extrahepatic tissues by 2,3,7,8-tetrachlorodibenzo-*p*-dioxin (TCDD). *Biochem. Pharmacol.* **26,** 1383–1388.

Berry, D. L., Zachariah, P. K., Namkung, M. J., and Juchau, M. R. (1976). Transplacental induction of carcinogen hydroxylating systems with 2,3,7,8,-tetrachlorodibenzo-*p*-dioxin (TCDD). *Toxicol. Appl. Pharmacol.* **36,** 569–584.

Berry, D. L., Zachariah, P. K., Slaga, T. J., and Juchau, M. R. (1977b). Analysis of the biotransformation of benzo(a)pyrene in human fetal and placental tissues with high pressure liquid chromatography. *Europ. J. Cancer* **13,** 667–675.

Bogdan, D. P., and Juchau, M. R. (1970). Characteristics of induced benzpyrene hydroxylase in the rat foeto-placental unit. *Europ. J. Pharmacol.* **10,** 119–126.

Boutelet-Bochan, H., Huang, Y., and Juchau, M. R. (1997). Expression of *CYP2E1* during embryogenesis and fetogenesis in human cephalic tissues: implications for the fetal alcohol syndrome. *Biochem. Biophys. Res. Commun.* **238,** 443–447.

Brown, A. K., and Suelzer, W. W. (1958). Studies on the neonatal development of the glucuronide conjugating system. *J. Clin. Invest.* **37,** 332–340.

Burchell, B., Brierly, C. H., and Rance, D. (1995). Minireview: specificity of human UDP-glucuronosyltransferases and xenobiotic glucuronidation. *Life Sci.* **57,** 1819–1831.

Burchell, B., Coughtrie, M., Jackson, M., Harding, D., Fournel-Gigleux, S., Leakey, J., and Hume, R. (1989). Development of human liver UDP-glucuronosyltransferases. *Dev. Pharmacol. Ther.* **13,** 70–77.

Campbell, M. A., and Fantel, A. G. (1983). Teratogenicity of acetaldehyde *in vitro:* relevance to the fetal alcohol syndrome. *Life Sci.* **32,** 2641–2647.

Carpenter, S. P., Lasker, J. M., and Raucy, J. L. (1996). Expression, induction and catalytic activity of the ethanol-inducible cytochrome P450 (CYP2E1) in human fetal liver. *Mol. Pharmacol.* **49,** 260–268.

Chasseaud, L. F. (1973). Distribution of enzymes that catalyze reactions of glutathione with alpha, beta unsaturated compounds. *Biochem. J.* **131,** *765–772.*

Chen, H., and Juchau, M. R. (1996). Inhibition of embryonic retinoic acid synthesis by aldehydes generated from lipid peroxidation and reversal of inhibition by reduced glutathione and glutathione-S-transferases. *The Toxicologist* **36,** 353.

Chen, H., and Juchau, M. R. (1997). Biotransformation of all-*trans*-retinal, 13-*cis*-retinal and 9-*cis* retinal catalyzed by conceptal cytosol and microsomes. *Biochem. Pharmacol.* **53,** 877–886.

Chen, H., and Juchau, M. R. (1998). Inhibition of embryonic retinoic acid synthesis by aldehydic products of lipid peroxidation and reversal of inhibition by reduced glutathione and glutathione-S-transferases. *Free Rad. Biol. Med.* **24,** 408–417.

Chen, H., Namkung, M. J., and Juchau, M. R. (1995), Biotransformation of all-*trans*-retinol and all-*trans*-retinal to all-*trans*-retinoic acid in rat conceptal homogenates. *Biochem. Pharmacol.* **50,** 1257–1265.

Chen, H., Yang, H. L., Namkung, M. J., and Juchau, M. R. (1996a). Interactive dysmorphogenic effects of all-*trans*-retinol and ethanol on cultured whole rat embryos during organogenesis. *Teratology* **54,** 12–19.

Chen, H., Namkung, M. J., and Juchau, M. R. (1996b) The effects of ethanol on biotransformation of all-*trans*-retinol to all-*trans*-retinoic acid in rat conceptal cytosol. *Alcoholism: Clin. Exp. Res.* **20,** 942–948.

Coughtrie, M. W. H., Burchell, B., Leakey, J. E. A., and Hume, R. (1988). The inadequacy of perinatal glucuronidation: immunoblot analyses of the developmental expression of individual UDP-glucuronosyltransferase isozymes in rat and human liver microsomes. *Molec. Pharmacol.* **34,** 729–735.

Cresteil, T., Beaune, P., Kremers, P., Celier, C., Guengerich, P., and LeRoux, J. (1985). Immunoquantification of epoxide hydrolase and cytochrome P450 isozymes in fetal and adult human liver microsomes. *Eur. J. Biochem.* **151,** 345–354.

Deltour, H. L., Ang, G., and Duester, G. (1996). Ethanol inhibition of retinoic acid synthesis as a potential mechanism for fetal alcohol syndrome. *FASEB J.* **10,** 1050–1057.

Dillio, C., Tiboni, G. M., Sacchetta, P., Angelucci, S., Bucciarelli, T., Bellati, U., and Aceto, A. (1995). Time-dependent and tissue specific variations of glutathione transferase activity during gestation in the mouse. *Mech. Ageing Dev.* **78,** 47–54.

Dolphin, C. T., Cullingford, T. E., Shephard, E. A., Smith, R. L., and Phillips, I. R. (1996). Differential developmental and tissue-specific regulation of expression of the genes encoding three members of the flavin-containing monooxygenase family of man: FMO1, FMO3 and FMO4. *Eur. J. Biochem.* **235,** 683–689.

Done, A. K. (1964). Developmental pharmacology. *Clin. Pharmacol. Therap.* **5,** 432–479.

Done, A. K. (1966). Perinatal pharmacology. *Ann. Rev. Pharmacol.* **6,** 189–208.

Duester, G. (1991). A hypothetical mechanism for fetal alcohol syndrome involving ethanol inhibition of retinoic acid synthesis at the alcohol dehydrogenase step. *Alcohol Clin. Exp. Res.* **15,** 568–572.

Falany, C. N., McQuiddy, P., and Kasper, C. B. (1987). Structure and organization of the microsomal xenobiotic epoxide hydrolase gene. *J. Biol. Chem.* **262,** 5924–5929.

Fouts, J. R., and Adamson, R. H. (1959). Drug metabolism in the newborn rabbit. *Science* **129,** 897–898.

Gibson, G. G., and Skett, P. (1986). *Introduction to Drug Metabolism.* Chapman and Hall, London.

Gilissen, R. A. H. J., Hume, R., Meerman, J. H. N., and Coughtrie, M. W. H. (1994). Sulphation of N-hydroxy-4-amino-biphenyl and N-hydroxy-4-acetylaminobiphenyl by human foetal and neonatal sulphotransferase. *Biochem. Pharmacol.* **48,** 837–841.

Grover, P. L., and Sims, P. (1964). Conjugation with glutathione: distribution of glutathione S-aryltransferase in vertebrate species. *Biochem. J.* **90,** 603–611.

Guthenberg, C., Warholm, M., Rane, A., and Mannervik, B. (1986). Two distinct forms of glutathione transferase from human fetal liver. *Biochem. J.* **235,** 741–749.

Hakkola, J., Pasanen, M., Purkunen, R., Saarikoski, S., Pelkonen, O., Maenpaa, J., Rane, A., and Raunio, H. (1994). Expression of xenobiotic-metabolizing cytochrome P450 forms in human adult and fetal liver. *Biochem. Pharmacol.* **48,** 59–64.

Hales, B. F., and Huang, C. (1994). Regulation of the Yp subunit of glutathione S-transferase P in rat embryos and yolk sacs during organogenesis. *Biochem. Pharmacol.* **47,** 2029–2036.

Hart, L. G., Adamson, R. H., and Fouts, J. R. (1962). Stimulation of hepatic microsomal drug metabolism in the newborn and fetal rabbit. *J. Pharmacol Exp. Therap.* **137,** 103–106.

Huang, Y., and Juchau, M. R. (1997). *CYP1B1* and Ah receptor gene expression in tissues from human embryos and fetuses and *CYP1B1* inducibility by 2,3,7,8-tetrachlorodibenzo-*p*-dioxin during human organogenesis. Manuscript submitted.

Hulla, J. E., and Juchau, M. R. (1989a). Developmental aspects of P450IIIA: prenatal activity and inducibility. *Drug Metab. Rev.* **20,** 765–781.

Hulla, J. E., and Juchau, M. R. (1989b). Occurrence and inducibility of cytochrome P450IIIA in maternal and fetal rats during prenatal development. *Biochemistry* **28,** 4871–4880.

Hume, R. (1993). Proc. 7th Annual Workshop on Glucuronidation and the UDP-Glucuronosyltransferases, p. 7. Pitlochry, Scotland.

Hume, R., Burchell, A., Allan, B. B., Wolf, C. R., Kelly, R. W., Hallas, A., and Burchell, B. (1996). The ontogeny of key endoplasmic reticulum proteins in human embryonic and fetal red blood cells. *Blood* **87,** 762–770.

Jondorf, W. K., Maickel, R. P., and Brodie, B. B. (1958). Inability of newborn mice and guinea pigs to metabolize drugs. *Biochem. Pharmacol.* **1,** 352–354.

Jones, A. H., Fantel, A. G., Kocan, R. M., and Juchau, M. R. (1977). Bioactivation of procarcinogens to mutagens in human fetal and placental tissues. *Life Sci.* **21,** 1831–1837.

Juchau, M. R. (1973). Placental metabolism in relation to toxicology. *CRC Crit. Rev. Toxicol.* **2,** 125–159.

Juchau, M. R. (1975). Metabolic capabilities of the human fetus: drug biotransformation. *Int. J. Addict. Dis.* **2,** 37–43.

Juchau, M. R. (1987). Bioactivation of chemical teratogens by cultured whole embryos. In *Pharmacokinetics in Teratogenesis.* H. Nau and W. J. Scott, Eds. Vol. II. CRC Press, Boca Raton, Florida.

Juchau, M. R. (1989). Bioactivation in chemical teratogenesis. *Ann. Rev. Pharmacol. Toxicol.* **29,** 165–187.

Juchau, M. R. (1990). Developmental drug metabolism. In *Comprehensive Medicinal Chemistry,* Vol. 5. C. Hansch, Ed. Pergamon Press, New York, pp. 219–236.

Juchau, M.R. (1997). Chemical teratogenesis in humans: biochemical and molecular mechanisms. *Progr. Drug Res.* **49,** 25–92.

Juchau, M. R., Berry, D. L., Zachariah, P. K., Namkung, M. J., and Slaga, T. J. (1976). Prenatal biotransformation of benzo(a)pyrene and N-2-fluorenylacetamide in human and subhuman primates. In *Carcinogenesis: Polynuclear Aromatic Hydrocarbons.* R. Freudenthal and P. W. Jones, Eds. Vol. I. Raven Press, New York.

Juchau, M. R., Boutelet-Bochan, H., and Huang, Y. (1998). Cytochrome P450-dependent biotransformation of xenobiotics in human and rodent embryonic tissues. *Drug Metab. Rev.,* in press.

Juchau, M. R., Chao, S. T., and Omiecinski, C. J. (1980). Drug metabolism by the human fetus. *Pharmacokinetics* **5,** 320–340.

Juchau, M. R., Harris, C., Beyer, B. K., and Fantel, A. G. (1987). Reactive intermediates in chemical teratogenesis. In *Approaches to Elucidate Mechanisms in Teratogenesis.* F. Welsch, Ed. Hemisphere Press, New York, pp. 23–26.

Juchau, M. R., Huang, Y., Boutelet-Bochan, H., and Yang, Y. H. (1996). Expression of P450 cytochromes in human and rodent embryonic tissues during the period of organogenesis: implications for teratogenesis. Abstracts of the Eleventh International Symposium on Microsomes and Drug Oxidations, p. 246.

Juchau, M. R., Lee, Q. P., and Fantel, A. G. (1992). Xenobiotic biotransformation/bioactivation in organogenesis-stage conceptal tissues: implications for embryotoxicity and teratogenesis. *Drug Metab. Rev.* **14,** 195–238.

Juchau, M. R., Lee, Q. P., Louviaux, G. L., Symms, K. G., Krasner, J., and Yaffe, S. J. (1973). Oxidation and reduction of foreign compounds in tissues of the human placenta and fetus. In *Fetal Pharmacology.* L. Boreus, Ed. Raven Press, New York.

Juchau, M. R., and Namkung, M. J. (1974). Studies on the biotransformation of naphthalene-1,2-oxide in human and placental tissues of humans and monkeys. *Drug Metab. Disp.* **2,** 380–388.

Juchau, M. R., Namkung, M. J., Jones, A. H., and DiGiovanni, J. (1978). Biotransformation and bioactivation of 7,12-dimethylbenz(a)anthracene in human fetal and placental tissues. *Drug Metab. Disp.* **6,** 273–281.

Juchau, M. R., and Pedersen, M. G. (1973). Drug biotransformation reactions in the human fetal adrenal gland. *Life Sci. II. Biochem. Gen. Molec. Biol.* **12,** 193–204.

Juchau, M. R., and Yang, H. L. (1996). Cytochrome P450-dependent monooxygenation of embryotoxic/teratogenic chemicals in human embryonic tissues. *Fund. Appl. Toxicol.* **34,** 166–168.

Kashiwada, M., Kitada,M., Shimada, T., Itahashi, K., Sato, K., and Kamataki, T. (1991). Purification and characterization of acidic forms of glutathione S-transferase in human fetal livers: high similarity to placental form. *J. Biochem.* **110,** 743–750.

Kitada, M., and Kamataki, T. (1994). Cytochrome P450 in human fetal liver: significance and fetal-specific expression. *Drug Metab. Rev.* **26,** 305–323.

Klinger, W. (1987). Developmental aspects of enzyme induction and inhibition. *Pharmacol. Ther.* **33,** 55–88.

Krauer, B., and Dayer, P. (1991). Fetal drug metabolism and its possible clinical implications. *Clin. Pharmacokinet.* **21,** 70–80.

Lamble, J. W., ed. (1983). *Drug Metabolism and Distribution.* Elsevier Press, Amsterdam.

Maenpaa, J., Rane, A., Raunio, H., Honkakoski, P., and Pelkonen, O. (1993). Cytochrome P450 isoforms in human fetal tissues related to phenobarbital-inducible forms in the mouse. *Biochem. Pharmacol.* **45,** 899–907.

Miller, M. S., Juchau, M. R., Guengerich, F. P., Nebert, D. W., and Raucy, J. L. (1996). Drug metabolic enzymes in developmental toxicology. *Fund. Appl. Toxicol.* **34,** 165–176.

Mitra, A., Hilbelink, D. R., Dwornik, J. J., and Kulkarni, A. (1992). A novel model to assess developmental toxicity of dihaloalkanes in humans: biactivation of 1,2-dibromoethane by the isozymes of human fetal liver glutathione S-transferase. *Teratogen. Carcinogen. Mutagen.* **12,** 113–119.

Nebert, D. W. (1989). The Ah locus: genetic differences in toxicity, cancer, mutation and birth defects. *Crit. Rev. Toxicol.* **20,** 153–174.

Nebert, D. W., Puga, A., and Vasiliou, V. (1993). Role of the Ah receptor and the dioxin-inducible [Ah] gene battery in toxicity, cancer and in signal transduction. *Ann. NY Acad. Sci.* **685,** 624–640.

Nelson, D. R., Koymans, L., Kamataki, T., Stegeman, J. J., Feyereisen, R., Waxman, D. J., Waterman, M. R., Gotoh, O., Coon, M. J., Estabrook, R. W., Gunsalus, I. C., and Nebert, D. W. (1996). P450 superfamily: update on new sequences, gene mapping, accession numbers and nomenclature. *Pharmacogenetics* **6,** 1–42.

Pacifici, G. M., Franchi, M., Colizzi, C., Giuliani, L,, and Rane, A. (1988). Sulphotransferase in humans: development and tissue distribution. *Pharmacology* **36,** 411–419.

Parkinson, A. (1996). Biotransformation of xenobiotics. In *Casarett and Doull's Toxicology: The Basic Science of Poisons.* 5th ed. C. D. Klaassen, Ed. McGraw-Hill, New York, pp. 113–186.

Pelkonen, O. (1977). Transplacental transfer of foreign compounds and their metabolism by the fetus. *Progr. Drug Metab.* **2,** 119–161.

Pelkonen, O. (1985). Fetoplacental biochemistry—xenobiotic metabolism and pharmacokinetics. In *Occupational Hazards and Reproduction.* K. Hemminki, M. Sorsa, and H. Vainio, Eds. Hemisphere Press, Washington, DC, pp. 239–264.

Pelkonen, O., and Karki, N. T. (1973). Drug metabolism in human fetal tissues. *Life Sci.* **13,** 1163–1180.

Rane, A. (1973). N-Oxidation of a tertiary amine (N,N-dimethylaniline) by human fetal microsomes. *Clin. Pharmacol. Therap.* **15,** 256–261.

Rane, A., Sjoqvist, F., and Orrenius, S. (1973). Drugs and fetal metabolism. *Clin. Pharmacol. Therap.* **14,** 666–672.

Raucy, J. L., and Carpenter, S. J. (1993). The expression of xenobiotic metabolizing cytochromes P450 in fetal tissues. *J. Pharmacol. Toxicol. Methods* **29,** 121–128.

Rettie, A. E., Meier, G. P., and Sadeque, A. J. M. (1995). Prochiral sulfides as *in vitro* probes for multiple forms of the flavin-containing monooxygenase. *Chem.-Biol. Interact.* **96,** 3–15.

Ring, J. A., Ghabrial, H., Ching, M. S., Shulkes, A., Smallwood, R. A., and Morgan, D. J. (1995). Fetal hepatic propranolol metabolism: studies in the isolated, perfused fetal sheep liver. *Drug Metab. Disp.* **23,** 190–196.

Rouet, P., Dansette, P., and Frayssinet, C. (1984). Ontogeny of benzo (a)pyrene hydroxylase, epoxide hydrolase and glutathione-S-transferase in the brain, lung and liver of C57Bl/6 mice. *Dev. Pharmacol. Therap.* **7,** 245–254.

Roy, S. K., Mitra, A. K., Hilbelink, D. R., Dwornik, J. J., and Kulkarni, A. P. (1993). Lipoxygenase activity in rat embryos and its potential for xenobiotic oxidation. *Biol. Neonate* **63,** 297–302.

Sadeque, A. J. M., Eddy, A. C., Meier, G. P., and Rettie, A. E. (1992). Stereoselective sulfoxidation of human flavin-containing monooxygenases: evidence for catalytic diversity between hepatic, renal and fetal forms. *Drug Metab. Disp.* **20,** 832–839.

Schardein, J. L. (1993). *Chemically Induced Birth Defects.* 2nd Ed. Marcel Dekker, Inc., New York.

Serafini, M. T., Arola, L. I., and Romeu, A. (1991). Glutathione and related enzyme activity in the 11-day rat embryo, placenta and perinatal rat liver. *Biol. Neonate* **60,** 236–243.

Shepard, T. H. (1995). *Catalog of Teratogenic Agents.* 8th Ed. Johns Hopkins University Press, Baltimore.

Soprano, D. R., and Soprano, K. J. (1996). Retinoids as teratogens. *Ann. Rev. Nutr.* **15,** 111–132.

Strange, R. C., Howie, A. F., Hume, R., Matharoo, B., Bell, J., Hiley, C., Jones, P., and Beckett, G. J. (1989). The developmental expression of alpha-, mu- and pi-class glutathione S-transferase in human liver. *Biochim. Biophys. Acta* **993,** 186–193.

Treluyer, J.-M., Jacqz-Aigrain, E., Alvarez, F., and Cresteil, T. (1991). Expression of CYP2D6 in developing human liver. *Europ. J. Biochem.* **202,** 583–588.

Vatsis, K. P., Weber, W. W., Bell, D. A., Dupret, J., Price Evans, D. A., Grant, D. M., Hein, D. W., Lin,, H. J., Meyer, U. A., Relling, M. V., Sim, E., Suzuki, T., and Yamazoe, Y. (1995). Nomenclature for N-acetyltransferases. *Pharmacogenetics* **5,** 1–17.

Waddell, W. J., and Marlowe, G. C. (1976). Disposition of drugs in the fetus. In *Perinatal Pharmacology and Therapeutics.* B. L. Mirkin, Ed. Academic Press, New York.

Warholm, M., Guthenberg, C., Mannervik, B., von Bahr, C., and Glaumann, H. (1980). Identification of a new glutathione S-transferase in human liver. *Acta Chem. Scand.* **234,** 607–615.

Weiner, H., and Flynn, T. G., eds. (1989). *Enzymology and Molecular Biology of Carbonyl Metabolism 2: Aldehyde Dehydrogenase, Alcohol Dehydrogenase, and Aldo-Keto Reductase.* Alan R. Liss, New York.

Wells, P. G., and Winn, L. M. (1996). Biochemical toxicology of chemical teratogenesis. *Crit. Rev. Biochem. Mol. Biol.* **31,** 1–40.

Wells, P. G., and Winn, L. M. (1997). The role of biotransformation in developmental toxicology. In *Comprehensive Toxicology, Section 3: Developmental Toxicology.* C. Harris, Ed. Pergamon Elsevier Science Ltd., pp. 306–324.

Wells, P. G., Kim, P. M., Nicol, C. J., Parman, T., and Winn, L. M. (1996). Reactive intermediates. In *Handbook of Experimental Pharmacology and Drug Toxicity in Embryonic Development.* R. J. Kavlock and G. P. Daston, Eds. Springer Verlag, Heidelberg.

Yaffe, S. J., and Juchau, M. R. (1974). Perinatal pharmacology. *Ann. Rev. Pharmacol.* **14,** 219–238.

Yaffe, S. J., Rane, A., Sjoqvist, F., Boreus, O., and Orrenius, S. (1970). The presence of a monooxygenase system in human fetal liver microsomes. *Life Sci.* **9,** 1189–1200.

Yang, H. L., Lee, Q. P., Rettie, A. E., and Juchan, M. R. (1994). Functional CYP3A isoforms in human embryonic tissues: expression during organogenesis. *Molec. Pharmacol.* **46,** 922–929.

Zachariah, P. K., and Juchau, M. R. (1975). Spectral characteristics of human fetal adrenal microsomes. *Life Sci.* **16,** 55–63.

Zhu, X., Veronese, M. E., Bernard, C. C. A., Sansom, L. N., and McManus, M. E. (1993). Identification of two human brain aryl sulfotransferase cDNAs. *Biochem. Biophys. Res. Commun.* **195,** 120–127.

CHAPTER

18

Blood–Brain Barrier: Physiological and Functional Considerations

MICHAEL ASCHNER
Department of Physiology and Pharmacology
Wake Forest University School of Medicine
Winston-Salem, North Carolina 27157

*Abbreviations: AMPA, alpha-amino-3-hydroxy-5-methyl-isoxazole; BBB, blood–brain barrier; Bi, bismuth; CNS, central nervous system; CSF, cerebrospinal fluid; E, embryonic day; Fe, iron; γ-GT, gamma-glutamyl transpeptidase; GABA, γ-aminobutyric acid; GSH, glutathione stimulating hormone; Hg, mercury; HRP, horseradish peroxidase; HSV-IgG, herpes simplex virus IgG; MeHg, methylmercury; Mn, manganese; N-CAM, N-cell adhesion molecule; Pb, lead; SITS, 4′-isothiocyanostilbene-2,2′-disulphonic acid; VEGF, vascular endothelial cell growth factor; Zn, zinc.

I. Introduction

The concept of a blood–brain barrier (BBB) arose late in the 19th century, when the German bacteriologist, Paul Ehrlich (1906) noted that, when injected intravenously, certain dyes stained all the organs with the exception of the central nervous system (CNS). Ehrlich's interpretation of these results was that the CNS had low affinity for the dyes, an assertion that was subsequently proved erroneous (Goldmann, 1913). When injected directly into the cerebrospinal fluid (CSF), trypan blue readily stained the brain parenchyma but did not enter the vasculature, leading Goldmann (Ehrlich's student at the time) to conclude that a physical barrier separated the CNS from the systemic circulation. The term *blood–brain barrier* (*Bluthirnschranke*) was originally coined in 1900 by Lewandowsky, who showed the exclusion of Prussian blue from the CNS following its injection into the vascular bed.

The BBB is a specialized structure responsible for the maintenance of the neuronal microenvironment, playing a pivotal role in tissue homeostasis, fibrinolysis and coagulation, vasotonus regulation, the vascularization of normal and neoplastic tissues, as well as blood cell activation and migration during physiologic and pathologic processes (reviewed by Risau, 1995) (Table 1). Regulation of blood–tissue exchange is accomplished by individual endothelial cells that are continuously linked by tight junctions (referred to as *zonulae occludentes*), thus actively isolating the brain from the blood and negating oncotic and osmotic forces that govern blood–tissue exchange in peripheral tissues. A pivotal function of the endothelial cells is to regulate the selective transport and metabolism of substances from blood or brain. Because of the existence of tight junctions between adjacent endothelial cells, nonspecific paracellular ionic leakage across the BBB is thought to be minimal. It should be pointed out, however, that not all of the CNS vasculature conforms to the previously mentioned morphologic description. Structural attributes of endothelial cells in nonbarrier areas (i.e., areas that lack the BBB) have been examined most systematically in the circumventricular organs (Gross, 1992). In contrast to the *zonulae occludentes* junctions of tight barrier areas, endothelial cells in the circumventricular organs exhibit *maculae occludentes* junctions that only partially occlude the gaps between adjacent endothelial cells. Hence, the barrier is not as tight or "seamless," and diffusion across the capillaries is more prevalent. For the anatomy and physiology of CNS regions lacking a BBB, refer to Jacobs (1994). From the toxicologic point of view, those areas which lack a true BBB represent potential sites for the accumulation of neurotoxins, as their passage into the brain parenchyma is likely to be less restrictive.

A number of factors determine transport across the BBB (Rapoport, 1976). For most solutes and macromolecules, permeability is largely dependent on their lipophilicity. Certain molecules needed for brain metabolism, however, penetrate the BBB more readily than one would predict based on their lipid solubility alone (Oldendorf, 1970). Hydrophilic solutes and macromolecules are believed to cross the barrier through specific carrier mechanisms or facilitated diffusion. Some of these carriers are symmetrically distributed both on the luminal and abluminal membranes of the endothelial cells, whereas others have an asymmetric distribution (Goldstein and Betz, 1986). For example, the carriers for the essential neutral amino acids, which are required in the brain for neurotransmitter synthesis, are localized on both luminal and abluminal membranes. In contrast, the carrier for the amino acid glycine appears to be located only on the abluminal membrane. This asymmetric distribution functions to remove glycine from the CNS and to keep its concentration in the brain low. Similarly, the abluminal membrane contains more of the enzyme ATPase than does the luminal membrane. This enzyme forms the basis of a pump that simultaneously transports Na^+ out of the endothelium into the brain, and K^+ out of the brain into the endothelium. Like glycine, K^+ has a potent effect on the transmission of nerve impulses and neuron firing, and this asymmetric distribution functions to maintain low K^+ concentration in the extracellular fluid. For transport systems essential for nutrient and other substrates, refer to Fig. 1.

TABLE 1 Some of the Functions Ascribed to the Blood–Brain Barrier

1. Fibrinolysis and coagulation
2. Vasotonus regulation
3. Vascularization of normal and neoplastic tissues
4. Blood cell activation and migration during physiological and pathologic processes
5. Maintenance of the neuronal microenvironment and CNS homeostasis

Despite the inherent specialization of the BBB and its "built-in" safeguards, this highly restrictive barrier is incapable of preventing the exchange of toxins from the blood to the brain when their transport is governed by the same physiologic properties that govern the exchange of nutrients, therapeutic agents, or hormones. Furthermore, because the BBB in fetal and early neonatal life is not fully developed (see later this chapter), it allows for the diffusion of blood-borne macromolecules and toxins that are excluded from the mature CNS. Hence, one can surmise that fetus and neonate are at a greater risk for brain injury from both endogenous (bilirubin) and exogenous (metals, pesticides, etc.) sources.

This review commences with a brief discussion on the development of the BBB and physiologic considerations, specifically those addressing the source of barrier-inducing agents and its maintenance. It concludes with a brief survey of transport mechanisms for a number of metals, emphasizing the fact that properties of the BBB that normally serve to regulate the CNS microenvironment may also function to transport toxins into the CNS. The review is not intended to exhaustively address morphologic, evolutionary, and embryologic considerations of the BBB. For excellent review on these topics the reader is referred to Bradbury (1979), Neuwlet (1989), and Abbott (1991). Nor does the review address the effects of toxins on barrier function *per se,* because increased vascular permeability associated with neurotoxic agents does not appear to represent a primary target of damage. For those effects, the reader is referred to reviews by Jacobs (1994) and Romero *et al.* (1996).

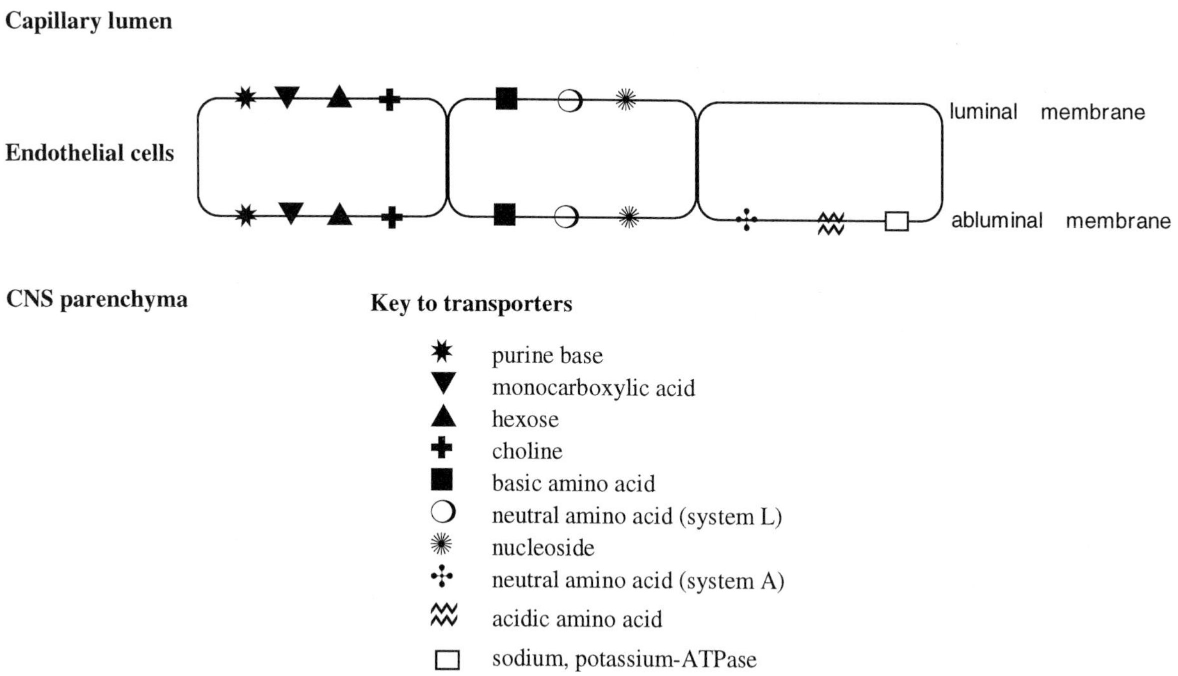

Capillary lumen

Endothelial cells

luminal membrane

abluminal membrane

CNS parenchyma

Key to transporters

- purine base
- monocarboxylic acid
- hexose
- choline
- basic amino acid
- neutral amino acid (system L)
- nucleoside
- neutral amino acid (system A)
- acidic amino acid
- sodium, potassium-ATPase

FIGURE 1 Transporters of the BBB. Additional transporters for peptides and vitamins are also known to exist, but there is scant information on their membrane localization. For further details refer to the text. [Adapted from Romero, I. A., Abbott, N. J., and Bradbury, M. W. B. (1996). The blood–brain barrier in normal CNS and in metal-induced neurotoxicity. In *Toxicology of Metals.* L. W. Chang, Ed. CRC Press, Boca Raton, FL, pp. 561–585.]

II. Blood–Brain Barrier: Developmental and Physiological Considerations

The development of the unique properties of the brain microvasculature is a consequence of tissue-specific interactions between endothelial cells of extraneural origin and developing CNS cells. Invading capillaries express BBB features in response to inductive signals from the surrounding neural tissue (Farrell and Risau, 1994; Risau, 1995). As pointed out by Risau (1995), "signal recognition, transduction, and processing are complex events which are dependent on the status of the endothelial cell, and is a consequence of inductive and permissive interactions of a pluripotent cell with soluble and insoluble signaling molecules during both embryonic and postnatal development" (p. 926). For biologic mechanisms involved in the differentiation of endothelial cells from the mesoderm and their subsequent functional heterogeneity in different organs and tissues under physiologic as well as pathologic conditions, the reader is referred to reviews by Risau (1995) and Farrell and Risau (1994).

In the mammalian CNS, brain capillaries develop from solid cords of endothelial cells. These cords develop a slitlike lumen, which increases its caliber progressively (reviewed by Jacobson, 1978). As the endothelial cell layer thickens, new cords of endothelial cells that sprout from more mature capillaries invade the cell layer. These newly formed cords are separated from juxtaposed neurons by a basement membrane and they become ensheathed progressively by resident astrocytes. The most rapid capillary sprouting corresponds to the period of glial cell proliferation and neuronal dendritic development and arborization. As suggested by Jacobson (1978), "it is reasonable to assume that the proliferation of the capillary endothelial cells may be stimulated by the rapid growth of the brain, and it certain that vascularization develops to satisfy the metabolic requirements of the rapidly growing central nervous system" (p. 93). It is thought that the density of brain capillaries in various CNS regions is directly correlated with the metabolic activity and oxygen consumption inherent to each specific region. Maximal growth of the capillaries into the developing brain corresponds to the time period in which the CNS is most sensitive to malnutrition. It is yet to be established, however, whether vascular insufficiency plays a causative role in the retardation in CNS development due to malnutrition (Rabin *et al.,* 1994). Furthermore, the effects of chemicals on the development of the blood supply to the CNS have not been adequately characterized. However, because collateral blood supply does not develop in animals deprived of an optimal blood supply during the critical developmental stages, it is reasonable to assume that where chemical-induced deficiencies in blood supply exist, CNS development is compromised.

Unlike the mammalian brain, in which the CNS is separated from the blood by the endothelium, the epithelial cells in the choroid plexus, and the arachnoid membrane (Abbott, 1991), in the invertebrate CNS the barrier is predominantly formed by glial cells (see later this chapter). As mentioned previously, of particular importance to the development of the BBB are the astrocytes, which are widely thought to provide trophic factor for barrier induction. It was thought at one time that the astrocytic foot-processes actually formed the restrictive barrier, because this was the most obvious distinguishing feature between brain capillaries (nonfenestrated) and all other capillaries in the periphery. However, electron microscope studies in the 1950s using electron-dense markers showed that the barrier to the diffusion of these markers resided in tight occluding junctions (*zonulae occludentes*) between the endothelial cells, and that there was free passage of such markers between the astrocytic end-feet (reviewed by Goldstein and Betz, 1986). This portion of the review focuses on advances in astrocytic functions, followed by an overview of data that both support and refute a possible role for astrocytes in BBB induction and maintenance.

A. A Survey of Astrocytic Functions in CNS Homeostasis

The role of astrocytes in CNS homeostasis has come to light since the late 1960s. Disposing of the old dogma that astrocytes are merely cytoskeletal support cells for neurons (*glia* in Greek means "glue") and "enlarged watery structures" with a seemingly absent extracellular space, Kuffler *et al.* (1966) showed that amphibian glial cells have a normal high intracellular K^+ with a membrane potential equal to the Nernst potential for K^+ (-80 to -90 mV). Later work on "electrically silent" cells in the mammalian CNS identified putative glial cells (astrocytes presumably) that had the same characteristic; namely, a nonexcitable cell with large negative membrane potential that is apparently sensitive only to changes in extracellular K^+ concentrations $[K^+]_o$. This led to one of the earliest functions proposed for astrocytes, the control of extracellular K^+ (referred to as *K^+ spatial buffering*) by locally removing K^+ released from active neurons (Kuffler *et al.*, 1966).

In addition to their role in K^+ spatial buffering, astrocytes perform a variety of other functions in the CNS of developing and adult organisms. Because review of these astrocytic functions is beyond the scope of this manuscript, the reader is referred to a number of comprehensive reviews on this topic (Kimelberg and Norenberg, 1989; Abbott, 1991; Murphy, 1993; Kimelberg and Aschner, 1994; Aschner and Kimelberg, 1996; Kettenmann and Ransom, 1996). A limited synopsis of astrocytic functions in CNS homeostasis is provided:

1. Reaction to and repair of neural damage induced by physical trauma or chemical insult by astrocytes has been reviewed by Norenberg (1996).

2. Uptake of transmitters by astrocytes has been reviewed by Kimelberg and Aschner (1994). There are very powerful uptake systems for a number of amino acid transmitters, such as glutamate, glycine, taurine, and γ-aminobutyric acid (GABA). These systems are Na^+-dependent and can also be electrogenic. There is considerable evidence for extremely active electrogenic uptake of glutamate in cultured and acutely isolated astrocytes and by autoradiographic and immunocytochemical evidence for electrogenic glutamate uptake *in situ.* Molecular biology studies have identified at least three members of a family of glutamate transporters, two of which are found exclusively in astrocytes (Kanner, 1993).

Uptake of a number of monoamine transmitters has been reported in cultured astrocytes, and astrocytes acutely isolated from the brain (reviewed by Kimelberg and Aschner, 1994). These systems appear to resemble the systems found in nerve terminals, being both Na^+-dependent and inhibitable by a variety of clinically relevant antidepressants, such as fluoxetine (Prozac) for serotonin. Astrocytic uptake systems for adenosine (Matz and Hertz, 1990), taurine (Shain and Martin, 1990), and histamine (Huszti, 1990) have also been described.

3. Astrocytic regulation of glycogen metabolism has been reviewed by Phelps (1972, 1975) and by Watanabe and Passoneau (1974).

4. Interactions between astrocytes and developing neurons are of utmost importance in the development of the CNS. In the immature CNS, neuronal cell body migration and axonal outgrowth occur on radial glia, which later lose their longitudinal orientation and are thought to form mature astrocytes (Rakic, 1971). Neuronal migration is the basis of CNS pattern formation and layering, and is characterized by astrocyte-guided translocation of nerve cells from the site of cell division (subventricular zone) to their final destination. The translocation of neuronal cell bodies requires the coordinated temporal and spatial expression of different adhesion molecules, [e.g., N-cell adhesion molecule (N-CAM), astrotactin, L1]. Adhesion molecules promote growth cone motility along astrocytic surfaces and also

provide the neurite with directional cues and other pertinent information regarding the surrounding microenvironment (reviewed by Hatten, 1990; Rakic, 1990).

5. An almost complete set of CNS neurotransmitter receptors have been found on astrocytes *in vitro,* and a number of these, specifically the adrenergic receptors and the ionotropic kainic acid/alpha-amino-3-hydroxy-5-methyl-isoxazole (AMPA) receptors have also been identified on astrocytes *in vivo.* Thus, these cells are poised to respond to the very same transmitters that at one time were thought to be exclusively located on post- or presynaptic neuronal membranes. The perisynaptic location of many astrocyte processes that form glial nets around neurons (Bruckner *et al.,* 1993) put these receptors and the uptake systems mentioned previously in direct apposition to their sites of release, namely presynaptic bouttons on the neuronal soma or dendrites.

B. Astrocytes and BBB-Induction: Pros and Cons

As early as the turn of the 20th century it was recognized by Lugaro (1907) that astrocytic endings surrounded the blood capillaries of the brain, both in the gray and white matter; these endings were termed *end-feet.* Other astrocytic processes were found to terminate around synapses, whereas others were shown to extend to the axonal nodes of Ranvier (reviewed by Kimelberg and Aschner, 1994). As suggested previously, these morphologic specializations led workers to speculate erroneously that astrocytes formed the physical basis of the restrictive BBB. More recent studies with endothelial cell and astrocyte cocultures refuted this speculation (see later this chapter) but maintained that astrocytes are responsible for the induction of tight gap junctions between the endothelial cells of the BBB (Abbott, 1991). This would be characterized by a high electrical resistance (around 2000 ohm. cm^2), indicative of a low conductance to even small ions (Risau and Wolburg, 1990). Shivers *et al.* (1988) reported that culturing of bovine aorta or pulmonary artery endothelial cells in media conditioned with primary astrocyte cultures showed ultrastructural features indicative of synthesis and plasma membrane insertion of tight junction components into the plasma membrane. However, no actual assemblies of tight junctions were seen. Growth of cultures of bovine brain capillary endothelial cells on one side of a filter with primary astrocyte cultures on the other, has been reported to lead to a transfilter resistance of 661 $\pm$ 48 ohms. cm^2 (Dehouck *et al.,*1990). However, these authors also reported that the endothelial cells grown alone had a resistance of 416 $\pm$ 58 ohm. cm^2. Because the astrocyte cultures formed multilayers of overlapping cellular sheets, it may well be that the small increase in resistance, especially taking into account the variability of each measurement, was due to the additive effect of the resistance of the astrocyte monolayer, thus adding two resistances in series. The resistance of an astrocyte multilayer of a similar thickness without the endothelial cells was not reported. In the coculture study, induction of the BBB-specific enzyme gamma-glutamyl transpeptidase (γ-GT) was found in brain endothelial cells when astrocytes and brain capillary endothelium, but not aortic artery endothelial cells, were cocultured (Dehouck *et al.,* 1990).

An interesting feature of the BBB is the extremely high density of intermembraneous particle assemblies in astrocytic membranes facing the blood capillaries (Landis and Reese, 1982). The exact nature of these systems remains unknown, but as discussed previously, it is thought that they might be K^+ channels because a high K^+ conductance has been found on these astrocyte membranes, which are thought to be involved in the spatial buffering concept or siphoning of K^+ from the neuropil into the blood (Newman, 1986). More recent work, however, has indicated that the astrocytic end-feet processes may play an important role in the induction of the BBB. Transplantation experiments showed that the formation of the BBB depended largely on the CNS environment because it formed in capillaries in systemic tissue transplanted into the CNS (Stewart and Wiley, 1981). Janzer and Raff (1987) demonstrated that injection of primary astrocyte cultures into the anterior eye chamber or chorioallantoic membrane of the chick induced a permeability barrier in the endothelial cells of the capillaries of these tissues that would otherwise lack such a barrier. Another line of evidence in support of the role of astrocytes in BBB induction is derived from studies by Tao-Cheng *et al.* (1987). When endothelial cells were cultured alone, their tight junctions appeared fragmentary. When cocultured with astrocytes, the tight junctions length, breadth, and the complexity between the endothelial cells were increased, closely resembling their counterparts *in vivo.* Interestingly, when astrocytes were substituted with other cell types, such as fibroblasts, the tight junctions remained fragmentary, giving rise to the idea that astrocytes could induce tight junctions in CNS capillaries. There appear to be several other lines of evidence for astrocytic modulation of the functional attributes of CNS capillaries:

1. Type I astrocytes injected into the anterior eye chamber of the rat or onto the chick chorioallantoic membrane were also able to induce a host-derived angiogenesis and some BBB properties in endothelial cells of nonneural origin (Janzer, 1993).

2. The activity of γ-GT, a specific marker of endothelial cells of the CNS endothelium, was abolished in the

absence of astrocytes in culture conditions; γ-GT was reexpressed, however, when C6 glioma cells were added to the culture (DeBault and Cancilla, 1980).

3. Addition of astrocytes to endothelial cell cultures also increased the incorporation of neutral amino acids by the endothelial cells (Cancilla and DeBault, 1983).

4. Expression of the barrier-specific marker, the GLUT-1 isoform of the glucose transporter (see later this chapter) is markedly down-regulated in cultured bovine brain capillary endothelial cells, presumably due to the absence of brain-derived or astrocyte trophic factors in the tissue culture medium (Boado *et al.,* 1994).

5. The fact that the differentiation and angiogenesis of the endothelial cells of the BBB is regulated by astrocytes is also supported by the observation that vascular endothelial cell growth factor (VEGF) expression is induced and strongly up-regulated in human malignant glioblastoma cells (Risau, 1994). VEGF is an angiogenic growth factor the expression of which appears to parallel embryonic brain angiogenesis. VEGF expression is highest in the embryonic brain during the angiogenic period, and is lowest in the adult brain when angiogenesis is turned off.

Despite the previously listed evidence, it remains unsettled as to whether astrocytes form and maintain the BBB. Although it is clear that induction of the BBB occurs at certain stages of development and there is circumstantial evidence that astrocytes may be responsible for it, the situation is still by no means clear (Brightman, 1991). A discussion of this issue can be found in a book by Abbott (1991). In this volume, Brightman (1991) concludes that "the precise role of perivascular astrocytes in the induction and maintenance of brain endothelium as a structural and functional barrier has yet to be fully elaborated" (p. 346). Reasons for doubt for the inductive effects of astrocytes on the BBB include the following:

1. The cerebral capillaries of a number of elasmobranchs are ensheathed by astrocytes; however, their endothelial cells do not express tight junctions, but rather exhibit open pores that are permeable to large molecules, including horseradish peroxidase (HRP) (Brightman *et al.,* 1971; Bundgaard and Cserr, 1981). As suggested by Brightman (1991), "either these particular astrocytes ensure that the endothelial junctions remain open or the junctional configuration is an inherent one that is not determined by the astrocytic investment" (p. 346).

2. Astrocytes are also found in close association with pituicytes, yet the endothelium in the neural lobe is of the fenestrated phenotype, and it is largely permeable to dyes such as HRP.

3. Without any astrocytes present, cloned endothelial cells *in vitro* can establish a functionally restrictive barrier with a relatively high resistance, as high as 700–800 ohm. cm^2. However, it does appear that morphologic differentiation and induction of specific BBB proteins can be induced by primary astrocyte cultures in endothelial cells *in vitro* (Dehouck *et al.,* 1990; Lobrinus *et al.,* 1992; Tagami *et al.,* 1992).

4. Astrocytes fail to form richly vascularized grafts when injected into the anterior chamber and chorioallantoic membrane of the eye (Holash *et al.,* 1993). These authors suggest that iridial vessels associated with astrocyte grafts do not change their ultrastructure to resemble brain capillaries, and that the grafting of the astrocytes to the chorioallantoic membrane leads an extensive inflammatory response, which, in turn, leads to poor delivery of tracers to graft vasculature as well as altering vessel permeability. Astrocyte graft vasculature also fails to express high levels of the GLUT-1 isoform of the glucose transporter even after treatment with anti-inflammatory agents (Holash *et al.,* 1993). Hence the authors question the general utility of the anterior chamber and chorioallantoic membrane for studying BBB induction.

5. Last but not least, as pointed out by Abbott (1991), the evolutionarily earlier barriers are formed between the glial cells as seen in the cephalopod mollusks and not between the endothelial cells. Abbott (1991) suggests that, in vertebrates, during evolution, the barrier has likely shifted from glial cells to endothelial cells, "perhaps to allow greater complexity and control of the CNS interstitial environment by the glial cells, superimposed upon a barrier which prevented interference by large and rapid changes in the blood" (p. 389).

C. Functional Development of the BBB

The problem of functional development of the BBB has been a subject of intensive debate for many years (Skultetyova *et al.,* 1993). Most studies support the notion of an immature BBB in late gestation fetuses and neonates. For example, in the optic tectum of the chick, the BBB permeability of neural vessels to Evans blue and HRP decreased progressively during development, and in 30-day-old chickens the nervous substrate is free of both tracers (Ribatti *et al.,* 1993). In the piglet, the passage of unbound bilirubin across the BBB declined with age (Lee *et al.,* 1989, 1995); both bilirubin concentrations and permeability across the BBB were higher in the 2-day-old than in the 2-week-old piglets. In the rat, a functional BBB is also not fully developed until postnatal day 24 (Schulze and Firth, 1992). Shortly after birth, the concentrations of exogenously administered albumin in the hypothalamus, hippocampus, cortex, striatum, brainstem, and cerebellum were demonstrated to be significantly higher compared to those of adult rats (Skultetyova *et al.*, 1993). Both the CNS mannitol space and ^{14}C-glycine transfer rate were also higher in 1-week-old neonatal rats compared with adults (Preston *et al.*, 1995). Developmental changes in BBB permeability to urea and K^+ were also noted between embryonic day 21 (E21) and adulthood in the rat (Keep *et al.*, 1995). Specifically, a marked perinatal decline in K^+ permeability was followed by a more gradual postnatal fall. The changes in BBB permeability coincided with changes in the rate of brain growth and the associated rate of brain K^+ accumulation. Keep and colleagues (1995) have suggested that, "as the K^+ permeability properties of the adult BBB would provide insufficient K^+ influx to meet the requirement associated with fetal brain growth, it is suggested that need for K^+ may be the reason for the greater BBB permeability early in development" (p. 439). Decreased permeability to macromolecules is likely to reflect a decrease in the BBB paracellular leak as a consequence of the morphologic maturation of the capillary endothelium. The immaturity of the developing BBB is also extended by observations on kinetic studies of virus-specific IgG subclasses among newborns and their mothers (Osuga *et al.*, 1992). These studies suggest that the BBB against herpes simplex virus IgG activities (HSV-IgG) is insufficient in newborns and greatly reduced compared to their dams.

In the rat capillaries, fenestrations, which were frequent at E11, disappeared at E13 in intraparenchymal vessels and at E17 in pial vessels (Stewart and Hayakawa, 1994). Although interendothelial junctions were not seen after birth, they were readily identified in fetal brain, suggestive of paracellular channels prior to birth. Furthermore, between E17 and young adulthood, the maturation of BBB interendothelial clefts was accompanied by the establishment of a characteristic ratio of *narrow zone* (complex tight junctions) to *wide zone* (15–20 nm), the latter disappearing progressively with age (Schulze and Firth, 1992). In the mouse, capillaries exhibited fenestrations on E9, and pericyte-like cells were found joined to the vessel walls. Around E10, the endothelial cells lost their pericyte-like cells as well as their fenestrations and exhibited numerous, partly extended junctional complexes, which appeared tight in some, but not all, areas (Bauer *et al.*, 1993). Taken together, these studies suggest that both the functional and the morphologic properties of the BBB develop progressively from the onset of intraneural vascularization. However, the morphologic characteristics of the BBB do not fully develop until the neonatal period.

Another marker of BBB maturity, the GLUT-1 isoform of the glucose transporter has been examined extensively. Glucose, an essential substrate for brain oxidative metabolism, was transported across the BBB by GLUT-1, a facilitative glucose transporter. Bolz *et al.* (1996) have quantified, by means of electron microscopy, the subcellular distribution of the GLUT-1 isoform of the developing microvessels of the brain of embryonic rats from E13 to E19 and in adult rats. Staining for GLUT-1 appeared weak at E13, and increased in density through adulthood, representing an increase in the absolute amount of transporter per vessel. In the rabbit, quantitative electron microscopic immunogold analyses of GLUT-1–immunoreactive sites per micrometer of capillary membrane indicated that GLUT-1 density increased with age, and correlated with *in vivo* measurements of V_{max} (Cornford *et al.*, 1994). Data from human postmortem tissue suggest, however, that the level of GLUT-1 in microvascular endothelial cells in both preterm (24–33 weeks) and term (38–40 weeks) neonates appears to be comparable to that of the adult (Mantych *et al.*, 1993). Work by Bauer *et al.* (1995) also supports GLUT-1 expression early during development in the mouse. GLUT-1 immunoreactivity was visible in intraneural capillaries as early as E11, corresponding to the course of intraneural neovascularization. Although somewhat unsettled, like many other markers of BBB maturity, the developmental up-regulation of the GLUT-1 transporter appears to occur at the BBB, and the modulation of the subcellular distribution of the transporter can be correlated with other observed changes in the microvessels as they develop the BBB phenotype (Cornford *et al.*, 1994).

III. Transport of Metals across the Blood–Brain Barrier

Advances in scientific methodologies have spurred interest in the mechanisms by which both essential and nonessential macromolecules are transported into the CNS. This portion of the chapter summarizes potential transport mechanisms for manganese (Mn), an essential metal, as well as lead (Pb) and mercury (Hg), both of which are nonessential metals. Although it is expected that Mn will readily traverse the BBB due to its essentiality (reviewed by Wedler, 1993), it is rather surprising that Pb and Hg enter the CNS quite freely. A cursory overview of the pharmacokinetics of these metals confirms that Mn, Pb, and Hg are potentially transported on endogenous carriers, which otherwise function in the transport of essential macromolecules.

A. *Manganese (Mn)*

Although only a small portion of systemic Mn is transported into the CNS, it has a defined role both in

CNS homeostasis and pathology. The mechanism of Mn transport into rat brain has been established by measuring the initial rate of Mn^{2+} uptake, demonstrating saturation kinetics (Aschner and Aschner, 1990; Murphy *et al.,* 1991; Ingersoll *et al.,* 1995).

Mn binds readily to transferrin, the protein carrier of iron (Fe) in plasma. When complexed with transferrin, Mn is exclusively present in the trivalent oxidation state, with 2 metal ions tightly bound to each transferrin molecule. A bicarbonate-binding site is activated for each bound Mn ion, a step requiring carbonic anhydrase—the enzyme converting CO_2 and H_2O to bicarbonate. In plasma, oxidizing agents such as ceruloplasmin may play a role in the oxidation of Mn^{2+} to Mn^{3+} (Gibbons *et al.,* 1976).

The role of transferrin in Mn transport across the BBB is favored by several lines of evidence. In the rat brain, the return of Mn to control levels is significantly slower when Mn is administered beyond days 18–20 of life (Rhenberg *et al.,* 1981), a developmental period that coincides with the appearance of immunocytochemical localization of transferrin (Connor and Fine, 1987), and carbonic anhydrase (Sapirstein, 1983) in the brain. There is additional theoretic and experimental evidence that transferrin, the principal Fe-carrying protein of the plasma, functions prominently in Mn transport across the BBB. In the absence of Fe, the binding sites of transferrin can accommodate a number of other metals including gallium, copper, chromium, cobalt, vanadium, aluminum, terbium, and plutonium, raising the possibility that transferrin functions *in vivo* as a transport agent for many of these metals. Fe^{3+} is taken up by cells after transferrin binds to a specific cell surface receptor, and the transferrin receptor complex is internalized (Karin and Mintz, 1981). Transferrin receptors have been demonstrated on numerous cell types, including CNS capillaries. Transferrin has also been shown to enter brain endothelial cells via receptor-mediated endocytosis and to subsequently enter the brain (Fishman *et al.,* 1985). At normal plasma Fe concentrations (0.9–2.8 μg/ml), normal iron binding capacity (2.5–4 μg/ml), and at normal transferrin concentration in plasma, 3 mg/ml, with 2 metal-ion–binding sites per molecule, of which only 30% are occupied by Fe^{3+}, transferrin has available 50 μmole of unoccupied Mn^{3+} binding sites per liter. Accordingly, analogous to the situation with aluminum, Mn may not need to displace Fe to bind to transferrin.

Examination of the distribution of transferrin receptors in relationship to Mn accumulation in the brain is intriguing. Mn concentrations are highest in the pallidum, thalamic nuclei, and substantia nigra. The areas with the highest transferrin receptor distribution are not identical to those which concentrate Mn. However, Mn-accumulating areas are efferent to areas of high transferrin receptor density, suggesting that perhaps these sites may accumulate Mn through neuronal transport. For example, the Mn rich areas of the ventral pallidum, globus pallidus, and substantia nigra receive input from the nucleus accumbens and the caudate putamen—two areas abundantly rich in transferrin receptors. Indeed, studies by Sloot and Gramsbergen (1994) confirmed axonal transport of Mn, and that Mn taken up by nerve cells in the striatum are subsequently transported into the substantia nigra by way of anterograde axonal transport. The presence of transferrin–receptors in the neuropil may provide the anatomic substrate for the "internalization" of Mn^{3+}–transferrin conjugates into neurons. A study by Suarez and Eriksson (1993) corroborated the internalization of a Mn–transferrin complex in SHSY5Y neuroblastoma cells. The transferrin receptor on SHSY5Y cells can bind and internalize a Mn–transferrin complex as efficiently as an Fe–transferrin complex. Accordingly, and analogous to the transport of Fe (Roberts *et al.,* 1993), the evidence supports an Mn transport model in which Mn-loaded transferrin is taken up by receptor-mediated endocytosis at the luminal membrane of brain capillaries. The Mn then dissociates from the transferrin complex, presumably in endosomal compartments, and is transcytosed by unknown mechanisms while the transferrin is retroendocytosed.

Most recently, the existence of transferrin receptor–mediated transport mechanisms on the capillaries of the CNS has been exploited for the delivery to the brain of nonlipophilic therapeutic compounds, especially proteins. In primate CNS, levels of recombinant human soluble CD4 (rsCD4), a potential anti-HIV therapeutic agent, were increased fivefold when the protein was administrated intravenously in the form of an antitransferrin receptor antibody–rsCD4 conjugate (Walus *et al.,* 1996).

B. Lead (Pb)

Lead (Pb) is among the best documented of the anthropogenic neurotoxins. In children, Pb neurotoxicity is commonly associated with learning deficit disorders and hyperactivity [blood Pb levels of 10–25 μg/ml (0.48–1.2 μM)]. The etiology of these cognitive disorders is unclear, and surprisingly very little is known about the transport of Pb from blood to brain.

Simons' (1986) studies on the transport of Pb into ghost erythrocytes postulate two pathways for its transport. The first pathway is dependent on HCO_3^-, because it is blocked by inhibitors of the anion transport exchange system, while the second pathway is independent of HCO_3^- concentration. Elevated concentrations of HCO_3^- in the media favor the anion-exchange pathway, while in the presence of perchlorate (ClO_4^-) (and in the absence of HCO_3^-) this transport pathway is minimized (Simons, 1986). Pb uptake depends on the presence of a second anion, and in the presence of HCO_3^- the rate is stimulated as follows: $ClO_4^- < NO_3^- < CH_3CO_2^-$, $F^- < Cl^- < Br^- < I^-$ (Simons, 1984, 1986). Transport of other cations by the anion-exchange transport system is well established (i.e., Li^+ and Na^+). In both cases, the uptake is inhibited by 4′-isothiocyanostilbene-2,2′-disulphonic acid (SITS) (Cabantchik *et al.,* 1978). The transport kinetics of Pb in ghost erythrocytes are consistent with the hypothesis that before transport of Pb can occur, a $PbCO_3$ complex must form, because the rate of Pb uptake is directly proportional to the concentrations of both Pb^{2+} and HCO_3^- and inversely proportional to the concentration of H^+. Furthermore, the temperature dependence of Pb uptake into ghost erythrocytes is similar to that of the HCO_3^-/Cl^- exchange. Accordingly, it was postulated that the transport of Pb may occur either via the exchange of $PbCO_3$ with an anion, or via exchange of an anion-ternary complex of $PbCO_3$ with another anion (Simons, 1986). Given the requirement for HCO_3^- for Pb^{2+} transport, the exchange of $PbCO_3$ by itself with an anion is favored over the anion-ternary complex model (Simons, 1986).

A second mechanism in adrenal medullary cells (Simons and Pocock, 1987; Tomsig and Suszkiw, 1990, 1991) postulates that Pb uptake occurs via Ca^{2+} channels. K^+ and veratridine stimulate the uptake of both Ca^{2+} and Pb^{2+}. Ca^{2+} acts as a competitive inhibitor for Pb^{2+} uptake with the K_i for the inhibitory effect of Ca^{2+} on Pb^{2+} uptake exhibiting a similar value to the K_m for calcium uptake. The entry of Pb into chromaffin cells consists of both voltage-independent and voltage-dependent (K^+-stimulated) components. The Ca^{2+} channel blockers, D-600, and nifedipine block Pb^{2+} uptake, whereas Bay K 8644, a Ca^{2+} channel agonist, stimulates the uptake of Pb, suggesting the involvement of the L-type Ca^{2+} channels (Simons and Pocock, 1987; Tomsig and Suszkiw, 1990, 1991). In contrast to ghost erythrocytes, SITS is ineffective in blocking Pb^{2+} and Ca^{2+} uptake in chromaffin medullary cells (Simons and Pocock, 1987).

A review by Bradbury and Deane (1993) examines the kinetics and mechanisms of Pb transport into the CNS. Initial studies with ^{203}Pb continuously infused intravenously into adult rats suggested that ^{203}Pb uptake into different brain regions was linear with time up to 4 hr after infusion. In the absence of organic ligands for Pb, the metal readily entered the CNS. However, presence of albumin, L-cysteine, or EDTA during the vascular perfusion completely abolished the measurable uptake of ^{203}Pb. Experiments designed to address the potential role of the anion exchanger or calcium channels (see earlier this chapter) provided negative results, suggesting that they may not be operative in the transport of Pb across the BBB. These authors (Bradbury and Deane, 1993) suggest, however, based on the effects of K^+ depolarization and of varying pH, that Pb may passively enter the CNS in the form of $PbOH^+$. It is clear that additional studies will have to be undertaken before the transport of Pb into the CNS is fully characterized.

C. Mercury (Hg)

Methylmercury (MeHg) poisonings are among the best documented chemical poisoning disasters in history. Because methylation of inorganic mercury species to MeHg by microorganisms is known to take place in waterways, resulting in its accumulation in the food chain, any source of environmental mercury represents a potential source for MeHg poisoning. Anthropogenic sources culminating in the acidification of freshwater streams and lakes in North America and the impoundment of water for large hydroelectric schemes have led to increases in MeHg concentrations in fish, posing increasingly greater risks to human populations. The most sensitive population to MeHg-induced neurotoxicity are the developing fetus and the neonate, both because MeHg is more readily transported across the immature BBB as well as because of its inhibitory effects on cell division (Sager *et al.,* 1984).

Hughes (1957) was the first to draw attention to the remarkable affinity of MeHg for the anionic form of -SH groups and to suggest that the principal chemical reaction of MeHg is with thiols; variations in the distribution and effects of MeHg seem principally to be dependent on this reaction. The affinity of MeHg for the anionic form of -SH groups is extremely high, with log K, where K is the association constant, on the order of 15–23; its affinity constants for oxygen-, chloride-, or nitrogen-containing ligands such as carboxyl or amino groups are about 10 orders of magnitude lower (Carty and Malone, 1979). Indeed, wherever a MeHg compound has been identified in biologic media, it has been complexed to -SH-containing ligands. Complexes with cysteine and glutathione stimulating hormone (GSH) have been identified in blood (Naganuma and Imura, 1979), and complexes with GSH have been identified in brain (Thomas and Smith, 1979), and the only therapeutic agents that are effective in reducing MeHg body–burden are those containing -SH groups. Mechanisms of membrane transport of MeHg invoke thiol-containing amino acids (Hirayama, 1980, 1985; Aschner and Clarkson, 1988; Aschner *et al.,* 1990; Kerper *et al.,* 1992). Aschner *et al.* (1990) demonstrated that MeHg conjugated to cysteine is transported across the BBB via the neutral amino acid transport L-system. Aschner and

Clarkson (1988) addressed the structural similarity of the L-cysteine–MeHg conjugate with the structure of methionine, and suggested that due the broad specificity of the L-system, it should transport cysteine-MeHg conjugates efficiently across the BBB. These studies were subsequently corroborated in *in vitro* suspensions of brain capillaries (Aschner and Clarkson, 1989), as well as in an *in vitro* model of the BBB. Uptake of MeHg–L-cysteine conjugates across the BBB followed Michaelis–Menten kinetics; it was stereoselective and Na^+-independent, and it was inhibited by the system L substrates, L-leucine, 2-amino-2-norbornanecarboxylic acid, and L-methionine (5 m*M*). This is consistent with transport of MeHg–L-cysteine by the L amino acid carrier (Aschner and Clarkson, 1988, 1989; Mokrzan *et al.*, 1995). From a therapeutic standpoint, these data suggest that the careful choice of thiol complexing agent may afford potential means for minimizing the transport of MeHg across the BBB. For a comprehensive review on molecular and ionic mimicry of Hg and additional toxic metals, both at the level of the BBB as well as other tissues, the reader is referred to Clarkson (1993).

L-cysteine was also postulated to play a role in the transport of bismuth (Bi) from blood to brain (Krari *et al.*, 1995). The concentration of Bi in the CNS was increased when it was administered by the intraperitoneal route to mice simultaneously as a Bi–L-cysteine complex. The authors further suggest that the increased toxicity associated with the Bi–L-cysteine complex versus Bi administered alone results from the stimulation of peroxidation by Bi and L-cysteine, as already observed for Fe and L-cysteine.

The potential role of histidine in the transport of Zn^{2+} was addressed by Buxani-Rice *et al.* (1994). ^{65}Zn transport into different regions of rat brain was measured during short vascular perfusion of one cerebral artery with either bovine serum albumin or histidine. The presence of L-histidine was shown to increase the uptake of ^{65}Zn into the CNS; this uptake was not affected by either L-arginine or L-phenylanine. The authors suggest that the enhancement due to histidine is attributable to diffusion of $ZnHis^+$ across unstirred layers "ferrying" Zn to and from transport sites.

IV. Summary

Concern about acute cytotoxicity has dominated discussions about the vulnerability of the BBB with little emphasis on the potential of an early adverse effect on subsequent barrier function. It is becoming apparent, however, that exposure to a neurotoxic agent early during development may give rise to a long-lasting dysfunctional barrier. For example, neonatal administration of peptides and opiates can affect later transport of peptides across BBB (Harrison *et al.*, 1993). Furthermore, a variety of changes in BBB transport processes have also been identified in the aging CNS (reviewed by Mooradian, 1994). These include diminished hexose and butyrate transport, choline transport, and triiodothyronine transport. In contrast, neutral and basic amino acids transport across the BBB does not appear to change as a function of age, or may actually decrease. A study by Tang and Melethil (1995) suggests that aging decreases the ability of the BBB to transport the neutral amino acid tryptophan. Mechanisms associated with the previously mentioned changes include hemodynamic alterations in the cerebral circulation of aged rats (namely increased occurrence of arteriovenous shunting), alterations in protein composition, increased accumulation of lipid peroxidation byproducts, and increased membrane fluidity (Mooradian, 1994).

The functional and structural changes of the BBB as it passes through different developmental stages should be kept in mind because they are likely to explain some of the fundamental differences in the effects of toxicants on the developing, mature, and aging brain. Because the transport of macromolecules into the CNS may be the limiting step in eliciting CNS damage, it would seem profitable to direct future studies towards BBB integrity, both in the young and aging brain as well as treatment modalities that can protect the BBB after injury.

References

Abbott, N. J. (1991). *Annal. NY Acad. Sci.* **633.**

Aschner, M., and Aschner, J. L. (1990). Manganese uptake across the blood–brain barrier: relationship to iron homeostasis. *Brain Res. Bull.* **24,** 857–860.

Aschner, M., and Clarkson, T. W. (1988). Uptake of methylmercury in the rat brain: effects of amino acid. *Brain Res.* **462,** 31–39.

Aschner, M., and Clarkson, T. W. (1989). Methylmercury uptake across bovine brain capillary endothelial cells in vitro: the role of amino acid. *Pharmacol. Toxicol.* **64,** 293–297.

Aschner, M., Eberle, N., Goderie, S., and Kimelberg, H. K. (1990). Methylmercury uptake in rat primary astrocyte cultures: the role of the neutral amino acid transport system. *Brain Res.* **521,** 221–228.

Aschner, M., and Kimelberg, H. K. (1996). Astrocytes: potential modulators of heavy metal–induced neurotoxicity. In: *Toxicology of Metals,* L. W. Chang, ed. CRC Press, Inc., Boca Raton, FL, pp. 587–608.

Bauer, H. C., Bauer, H., Lametschwandtner, A., Amberger, A., Ruiz, P., and Steiner, M. (1993). Neovascularization and the appearance of morphological characteristics of the blood–brain barrier in the embryonic mouse central nervous system. *Dev. Brain Res.* **75,** 269–278.

Bauer, H., Sonnleitner, U., Lametschwandtner, A., Steiner, M., Adam, H., and Bauer, H. C. (1995). Ontogenic expression of the erythroid-type glucose transporter (Glut 1) in the telencephalon of the mouse: correlation to the tightening of the blood-brain barrier. *Dev. Brain Res.* **86,** 317–325.

Boado, R. J., Wang, L., and Pardridge, W. M. (1994). Enhanced expression of the blood–brain barrier GLUT1 glucose transporter gene by brain-derived factors. *Mol. Brain Res.* **22,** 259–267.

Bolz, S., Farrell, C. L., Dietz, K., and Wolburg, H. (1996). Subcellular distribution of glucose transporter (GLUT-1) during development of the blood–brain barrier in rats. *Cell Tis. Res.* **284,** 355–65.

Bradbury, M. W. B. (Ed.) (1979). In *The Concept of a Blood–Brain Barrier.* Wiley. Chichester.

Bradbury, M. W. B., and Deane, R. (1993). Permeability of the blood–brain barrier to lead. *Neurotoxicology* **14,** 131–136.

Brightman, M. W. (1991). Implications of astroglia in the blood–brain barrier. *Ann. NY Acad. Sci.* **633,** 343–347.

Brightman, M. W., Reese, T. S., Olson, Y., and Klatzo, I. (1971). Morphological aspects of the blood–brain barrier to peroxidase in elasmobranchs. *Progr. Neuropathol.* **1,** 146–161.

Bruckner, G., Brauer, K., Hartig, W., Wolff, J. R., Rickmann, M. J., Derouiche, A., Delpech, B., Girard, N., Oertel, W. H., and Reichenbach, A. (1993). Perineuronal nets provide a polyanionic glia–associated form of microenvironment around certain neurons in many parts of the rat brain. *Glia* **8,** 183–200.

Bundgaard, M., and Cserr, H. F. (1981). A glial blood–brain barrier in elasmobranchs. *Brain Res.* **226,** 61–73.

Buxani-Rice, S., Ueda, F., and Bradbury, M. W. (1994). Transport of zinc-65 at the blood–brain barrier during short cerebrovascular perfusion in the rat: its enhancement by histidine. *J. Neurochem.* **62,** 665–72.

Cabantchik, Z. I., Knauf, P. A., and Rothstein, A. (1978). The anion transport system of the red blood cell. The role of membrane protein evaluated by the use of probes. *Biochim. Biophys. Acta* **515,** 239–302.

Cancilla, P. A., and DeBault, L. E. (1983). Neutral amino acid transport properties of cerebral endothelial cells *in vitro. J. Neuropathol. Exp. Neurol.* **42,** 191–199.

Carty, A. J., and Malone, S. F. (1979). The chemistry of mercury in biological systems. In *The Biogeochemistry of Mercury in the Environment.* Nrigau, J. O., Ed. Elsevier/North Holland Biomedical Press, Amsterdam, pp. 433–479.

Clarkson, T. W. (1993). Molecular and ionic mimicry of toxic metals. *Annu. Rev. Pharmacol. Toxicol.* **32,** 545–571.

Connor, J. R., and Fine, R. E. (1987). Development of transferrin-positive oligodendrocytes in the rat central nervous system. *J. Neurosc. Res.* **17,** 51–59.

Cornford, E. M., Hyman, S., and Landaw, E. M. (1994). Developmental modulation of blood–brain-barrier glucose transport in the rabbit. *Brain Res.* **663,** 7–18.

DeBault, L. E., and Cancilla, P. A. (1980). Gamma-glutamyl transpeptidase in isolated brain endothelial cells: induction by glial cells *in vitro. Science* **207,** 653–655.

Dehouck, M.-P., Meresse, S., Delorme, P., Fruchart, J-C., and Cecchelli, R. (1990). An easier, reproducible, and mass-production method to study the blood–brain barrier *in vitro. J. Neurochem.* **54,** 1798–1801.

Ehrlich, P. (1906). Über die Beziehungen von Chemische Constitution, Vertheilung, und Pharmakologisher Wirkung. Collected studies in *Immunity Repr. and Trans.* Wiley, New York, pp. 567–595.

Farrell, C. L., and Risau, W. (1994). Normal and abnormal development of the blood–brain barrier. *Micros. Res. Tech.* **27,** 495–506.

Fishman, J. B., Handrahan, J. B., Rubir, J. B., Connor, J. R., and Fine, R. E. (1985). Receptor-mediated trancytosis of transferrin across the blood–brain barrier. *J. Cell. Biol.* **101,** 423A.

Gibbons, R. A., Dixon, S. N., Hallis, K., Russell, A. M., Sansom, B. F., and Symonds, H. W. (1976). Manganese metabolism in cows and goats. *Biochim. Biophys. Acta* **444,** 1–10.

Goldmann, E. E. (1913). Vitalfärbung am Zentralnervensystem. *Abh. Preuss. Akad. Wiss. Phys.-Math.* **1,** 1–60.

Goldstein, G. A., and Betz, A. L. (1986). The blood–brain barrier. *Sci. Am.* **255,** 74–83.

Gross, P. M. (1992). Circumventricular organ capillaries. *Prog. Brain Res.* **91,** 219–234.

Harrison, L. M., Zadina, J. E., Banks, W. A., and Kastin, A. J. (1993). Effects of neonatal treatment with Tyr-MIF-1, morphiceptin, and morphine on development, tail flick, and blood–brain barrier transport. *Dev. Brain Res.* **75,** 207–12.

Hatten, M. E. (1990). Riding the glial monorail: a common mechanism for glial-guided neuronal migration in different regions of the developing mammalian brain. *Trends Neurosci.* **13,** 179.

Hirayama, K. (1980). Effects of amino acids on brain uptake of methyl mercury. *Toxicol. Appl. Pharmacol.* **55,** 318–323.

Hirayama, K. (1985). Effects of combined administration of thiol compounds and methylmercury chloride on mercury distribution in rats. *Biochem. Pharmacol.* **34,** 2030–2032.

Holash, H. A., Noden, D. M., and Stewart, P. A. (1993). Re-evaluating the role of astrocytes in blood–brain barrier induction. *Develop. Dynam.* **197,** 14–25.

Hughes, W. H. (1957). A physicochemical rationale for the biological activity of mercury and its compounds. *Ann. NY Acad. Sci.* **65,** 454–460.

Huszti, Z., Rimanoczy, A., Juhasz, A., Magyar, K. (1990). Uptake: metabolism, and release of h-histamine by glial cells in primary cultures of chicken cerebral hemisphere. *Glia* **3,** 159–168.

Ingersoll, R. T., Montgomery, E. B., and Aposhian, H. V. (1995). Central nervous system toxicity of manganese. I. Inhibition of spontaneous motor activity in rats after intrathecal administration of manganese chloride. *Toxicol. Appl. Pharmacol.* **27,** 106–113.

Jacobs, J. N. (1994). Blood–brain and blood–nerve barriers and their relationships. In *Principles of Neurotoxicology.* Chang, L. W., Ed. Marcel Dekker, Inc., New York, pp. 35–68.

Jacobson, M. (1978). Histogenesis and morphogenesis of the central nervous system. In *Developmental Neurobiology.* Jacobson, M., Ed. Plenum Press, New York, pp. 57–114.

Janzer, R. C. (1993). The blood–brain barrier: cellular basis. *J. Inherit. Metab. Dis.* **16,** 639–47.

Janzer, R. C., and Raff, M. C. (1987). Astrocytes induce blood–brain properties in endothelial cells. *Nature* **325,** 253–257.

Kanner, B. I. (1993). Glutamate transporters from brain: a novel neurotransmitter transporter family. *FEBS Lett.* **325,** 95–99.

Karin, M., and Mintz, B. (1981). Receptor-mediated endocytosis of transferrin in developmentally totipotent mouse teratocarcinoma stem cells. *J. Biol. Chem.* **256,** 3245–3252.

Keep, R. F., Ennis, S. R., Beer, M. E., and Betz, A. L. (1995). Developmental changes in blood–brain barrier potassium permeability in the rat: relation to brain growth. *J. Physiol.* **488,** 439–448.

Kerper, L., Ballatori, N., and Clarkson, T. W. (1992). Methylmercury transport across the blood–brain barrier by an amino acid. *Am. J. Physiol.* **31,** R761–R765.

Kettenmann, H. B., Ransom, B. (1996). In *Astrocytes.* 1996.

Kimelberg, H. K., Aschner, M. (1994). Astrocytes and their functions, past and present. In *National Institute on Alcohol Abuse and Alcoholism Research Monograph, Alcohol and Glial Cells.* NIH Publication No. 94-3742, Bethesda, MD, Monograph 27, pp. 1–40.

Kimelberg, H. K., and Norenberg, M. D. (1989). Astrocytes. *Sci. Amer.* **260,** 66–76.

Krari, N., Mauras, Y., and Allain, P. (1995). Enhancement of bismuth toxicity by L-cysteine. *Res. Comm. Mol. Pathol. Pharmacol.* **89,** 357–364.

Kuffler, S. W., Nicholls, J. G., and Orkand, R. K. (1966). Physiological properties of glial cells in the CNS system of amphibia. *J. Neurophysiol.* **29,** 768–787.

Landis, D., and Reese, T. S. (1982). Regional organization of astrocytic membranes in cerebellar cortex. *Neuroscience* **7,** 937–950.

Lee, C., Oh, W., Stonestreet, B. S., and Cashore, W. J. (1989). Permeability of the blood–brain barrier for ^{125}I-albumin–bound bilirubin in newborn piglets. *Ped. Res.* **25,** 452–456.

Lee, C., Stonestreet, B. S., Oh. W., Outerbridge, E. W., and Cashore, W. J. (1995). Postnatal maturation of the blood–brain barrier for unbound bilirubin in newborn piglets. *Brain Res.* **689,** 233–238.

Lewandowsky, M. (1900). Zur Lehre der Cerebrospinalflussigkeit. *Z. Klin. Med.* **40,** 480–494.

Lobrinus, J. A., Juillerat-Jeanneret, L., Darekar, P., Schlosshauer, B., and Janzer, R. C. (1992). Induction of the blood–brain barrier specific HT7 and neurothelin epitopes in endothelial cells of the chick chorioallantoic vessels by a soluble factor derived from astrocytes. *Dev. Brain Res.* **70,** 207–211.

Lugaro E. (1907). Sulle Funzioni Della Nevroglia. *Riv. D. Pat. Nerv. Ment.* **12,** 22–233.

Mantych, G, J., Sotelo-Avila, C., and Devaskar, S. U. (1993). The blood–brain barrier glucose transporter is conserved in preterm and term newborn infants. *J. Clin. Endocrinol. Metab.* **77,** 46–49.

Matz, H., Hertz, L. (1990). Effects of adenosine deaminase inhibition on active uptake and metabolism of adenosine in astrocytes in primary cultures. *Brain Res.* **515,** 168–172.

Mokrzan, E. M., Kerper, L. E., Ballatori, N., and Clarkson, T. W. (1995). Methylmercury–thiol uptake into cultured brain capillary endothelial cells on amino acid system L. *J. Pharmacol. Exp. Therap.* **27,** 1277–1284.

Mooradian, A. D. (1994). Potential mechanisms of the age-related changes in the blood–brain barrier. *Neurobiol. Aging.* **15,** 751–755.

Murphy S, Ed. 1993. In *Astrocytes: Pharmacology and Function.* Academic Press, New York.

Murphy, V. A., Wadhwani, K. C., Smith, Q. R. and Rapoport, S. I. (1991). Saturable transport of manganese (II) across the rat blood–brain barrier. *J. Neurochem.* **57,** 948–954.

Naganuma, A., and Imura, N. (1977). Methylmercury binds to a low molecular weight substance in rabbit and human erythrocytes. *Toxicol. Appl. Pharmacol.* **47,** 613–616.

Neuwlet, E. A., Ed. (1989). In *Implications of the Blood–Barrier and its Manipulation.* Plenum Publishing Co., New York.

Newman, E. (1986). High potassium conductance in astrocyte endfeet. *Science* **233,** 453–454.

Norenberg, M. D. (1996). Reactive astrocytosis. In *The Role of Glia in Neurotoxicity.* Aschner, M., and Kimelberg, H. K., Eds. CRC Press, Boca Raton, FL, pp. 93–107.

Oldendorf, W. H. (1970). Measurement of brain uptake of radiolabeled substances using a tritiated water internal standard. *Brain Res.* **24,** 372–376.

Osuga, T., Morishima, T., Hanada, N., Nishikawa, K., Isobe, K., and Watanabe, K. (1992). Transfer of specific IgG and IgG subclasses to herpes simplex virus across the blood–brain barrier and placenta in preterm and term newborns. *Acta Paed.* **81,** 792–796.

Phelps, C. H. (1972). Barbiturate-induced glycogen accumulation in brain. An electron microscope study. *Brain Res.* **39,** 225.

Phelps, C. H. (1975). An ultrastructural study of methionine sulfoximine-induced glycogen accumulation in astrocytes of the mouse cerebral cortex. *J. Neurocytol.* **4,** 479.

Preston, J. E., al-Sarraf, H., and Segal, M. B. (1995). Permeability of the developing blood–brain barrier to 14C-mannitol using the rat *in situ* brain perfusion technique. *Dev. Brain Res.* **87,** 69–76.

Rabin, O., Lefauconnier, J. M., Chanez, C., Bernard, G., and Bourre, J. M. (1994). Developmental effects of intrauterine growth retardation on cerebral amino acid transport. *Ped. Res.* **35,** 640–648.

Rakic, P. (1971). Neuron–glia relationship during granule cell migration in developing cerebellar cortex: a Golgi and electron microscopic study in *Macaques rhesus. J. Comp. Neurol.* **141,** 283.

Rakic, P. (1990). Principles of neural migration. *Experientia* **46,** 882–891.

Rapoport, S. I., Ed. (1976). Permeability and osmotic properties of the blood–brain barrier. In *Blood–brain barrier in physiology and medicine.* Raven Press, New York, pp. 87–127.

Rhenberg, G. L., Hein, F. J., Carter, D. S., Linko, S. R., and Laskey, W. J. (1981). Chronic ingestion of Mn_3O_4 by young rats: tissue accumulation, distribution and depletion. *J. Exp. Toxicol. Environ. Health* **7,** 263–272.

Ribatti, D., Nico, B., and Bertossi, M. (1993). The development of the blood–brain barrier in the chick. Studies with Evans blue and horseradish peroxidase. *Anatom. Anzeiger.* **175,** 85–88.

Risau, W. (1994). Molecular biology of blood–brain barrier ontogenesis and function. *Acta Neurochir.* (Suppl.) **60,** 109–112.

Risau, W. (1995). Differentiation of endothelium. *FASEB J.* **9,** 926–933.

Risau, W., and Wolfburg, H. (1990). Development of the blood–brain barrier. *Trend. Neurol. Sci.* **13,** 174–178.

Roberts, R. L., Fine, R. E., and Sandra A. (1993). Receptor-mediated endocytosis of transferrin at the blood–brain barrier. *J. Cell Sci.* **104,** 521–532.

Romero, I. A., Abbott, N. J., and Bradbury, M. W. B. (1996). The blood–brain barrier in normal CNS and in metal-induced neurotoxicity. In *Toxicology of Metals.* Chang, L. W., Ed. CRC Press, Boca Raton, FL, pp. 561–585.

Sager, P. R., Aschner, M., and Rodier, P. M. (1984). Persistent differential alterations in developing cerebellar cortex of male and female mice after methylmercury exposure. *Dev. Brain Res.* **12,** 1–11.

Sapirstein, V. S. (1983). Development of membrane-bound carbonic anhydrase and glial fibrillary acidic protein in normal and quaking mice. *Dev. Brain Res.* **6,** 13–19.

Schulze, C., and Firth, J. A. (1992). Interendothelial junctions during blood–brain barrier development in the rat: morphological changes at the level of individual tight junctional contacts. *Dev. Brain Res.* **69,** 85–95.

Shain, M. W., Martin, D. L. (1990). Uptake and release of taurine—an overview. In *Taurine, Functional Neurochemistry, Physiology and Cardiology* Pasantes-Morales, H., Martin, D. L., Shain, W., and del Rio, R., Eds. Wiley-Liss, Inc., New York, pp. 243–252.

Shivers, R. R., Arthur, F. E., and Bowman, P. D. (1988). Induction of gap junctions and brain endothelial-like tight junctions in cultured bovine endothelial cells: local control of cell specialization. *J. Submicrosc. Cytol. Pathol.* **20,** 1–14.

Simons, T. J. B. (1984). Active transport of lead by human red blood cells. *FEBS Lett.* **172,** 250–264.

Simons, T. J. B. (1986). The role of anion transport in the passive movement of lead across the human red cell membrane. *J. Physiol.* **378,** 287–312.

Simons, T. J. B., and, Pocock, G. (1987). Lead enters bovine adrenal medullary cells through calcium channels. *J. Neurochem.* **48,** 383–389.

Skultetyova, I., Tokarev, D., I., and Jezova, D. (1993). Albumin content in the developing rat brain in relation to the blood–brain barrier. *Endoc. Reg.* **27,** 209–13.

Sloot, W. N., and Gramsbergen, J. B. (1994). Axonal transport of manganese and its relevance to selective neurotoxicity in the rat basal ganglia. *Brain Res.* **657,** 124–132.

Stewart, P. A., and Hayakawa, K. (1994). Early ultrastructural changes in blood–brain barrier vessels of the rat embryo. *Dev. Brain Res.* **78,** 25–34.

Stewart, P. A. and Wiley, M. J. (1981). Developing nervous tissue induces formation of blood–brain barrier characteristics in invading endothelial cells: a study using quail-chick transplantation chimeras. *Develop. Biol.* **84,** 183–193.

Suarez, N., and Eriksson, H. (1993). Receptor-mediated endocytosis of a manganese complex of transferrin into neuroblastoma (SHSY5Y) cells in culture. *J. Neurochem.* **61,** 127–31.

Tagami, M., Yamagata, K., Fujino, H., Kubota, H., Nara, Y., and Yamori, Y. (1992). Morphological differentiation of endothelial cells co-cultured with astrocytes on type-1 or type-IV collagen. *Cell Tissue Res.* **268,** 225–232.

Tang, J. P., and Melethil, S. (1995). Effect of aging on the kinetics of blood–brain barrier uptake of tryptophan in rats. *Pharmacol. Res.* **12,** 1085–1091.

Tao-Cheng, J. H., Nagy, Z., and Brightman, M. W. (1987). Tight junctions of brain endothelium *in vitro* are enhanced by astrocytes. *J. Neurosci.* **7,** 3293–3299.

Thomas, D. J., and Smith, C. J. (1979). Effects of coadministered low molecular weight thiol compounds on short term distribution of methylmercury in the rat. *Toxicol. Appl. Pharmacol.* **62,** 104–110.

Tomsig, J. L., and Suszkiw, J. B. (1990). Pb^{2+}-induced secretion from bovine chromaffin cells: fura-2 as a probe for Pb^{2+}. *Am. J. Physiol.* **259,** C762–C768.

Tomsig, J. L., and Suszkiw, J. B. (1991). Permeation of Pb^{2+} through calcium channels: fura-2 measurements of voltage- and dihydropyridine-sensitive Pb^{2+} entry in isolated bovine chromaffin cells. *Biochim. Biophys. Acta* **1069,** 197–200.

Walus, L. R., Pardridge, W. M., Starzyk, R. M., and Friden, P. M. (1996). Enhanced uptake of $rsCD_4$ across the rodent and primate blood–brain barrier after conjugation to anti-transferrin receptor antibodies. *J. Pharmacol. Exp. Therap.* **277,** 1067–1075.

Watanabe, H., and Passonneau, J. V. (1974). The effect of trauma on cerebral glycogen and related metabolites and enzymes. *Brain Res* **66,** 147.

Wedler, F. C. (1993). Biological significance of manganese in mammalian systems. In *Progress in Medicinal Chemistry.* Ellis, G. P., and Luscombe, D. K., Eds. Vol. 30. Elsevier Science Publishers, Amsterdam, The Netherlands, pp. 89–133.

CHAPTER

19

Antioxidant Enzymes
Developmental Profiles and Their Role in Metal-Induced Oxidative Stress

SABER HUSSAIN
SYED F. ALI
Neurochemistry Laboratory
Division of Neurotoxicology
National Center for Toxicological Research/FDA
Jefferson, Arkansas 72079

*Abbreviations: AD, Alzheimer's disease; AIDS, acquired immunodeficiency syndrome; ALS, amyotrophic lateral sclerosis; ANOVA, analysis of variance; BS, brainstem; CAT, catalase; CE, cerebellum; CN, caudate nucleus; CNS, central nervous system; Cu,Zn-SOD, copper-zinc superoxide dismutase; CX, cerebral cortex; DTNB, 5,5-dithiobis-2nitrobenzoic acid; FC, frontal cortex; FDA, (U.S.) Food and Drug Administration; GPx, glutathione peroxidase; GSH, glutathione stimulating hormone; GST, glutathione transferase; HD, Huntington's disease; HIP, hippocampus; MDA, methamphetamine, methylenedioxyamphetamine; MDMA, methylenedioxymethamphetamine; MeHg, methylmercury; Mn-SOD, manganese-containing superoxide dismutase; PD, Parkinson's disease; PUFA, polyunsaturated fatty acids; ROS, reactive oxygen species; SOD, superoxide dismutase.

I. Introduction

Reactive oxygen species (ROS) such as superoxide ($O_2^{-}\cdot$) and hydroxyl ($OH^{-}\cdot$) radicals formed in the process of biologic oxidation during normal cellular metabolism are considered to be involved in many important biologic reactions. Superoxide anion radical ($O_2^{-}\cdot$) is produced by the addition of one electron to molecular oxygen. The superoxide radical itself undergoes a dismutation reaction in which one superoxide anion radical acts on another to produce hydrogen peroxide (H_2O_2). Although hydrogen peroxide is not by definition a free radical, the addition of an electron leads to the formation of the highly reactive hydroxyl radical. The hydroxyl ($OH^{-}\cdot$) radical is extremely reactive and interacts with membrane lipids leading to lipid peroxidation. In addition, ROS interact with proteins by oxidizing their amino acid residues. DNA degradation and destruction of endothelial cells have also been attributed to oxygen-derived ROS toxicity (Brown *et al.*, 1995; Vile and Tyrrell, 1995). Furthermore, ROS have been postulated to produce deleterious effects involved in various diseases such as cancer, acquired immunodeficiency disease syndrome (AIDS), chemical-induced liver damage, rheumatoid arthritis, and several autoimmune diseases (Cerutti, 1985; Halliwell and Cross, 1991; Jaeschke, 1995; Gibson *et al.,* 1996). Oxygen free radicals are also thought to be involved in neurodegenerative diseases such as Parkinson's and Alzheimer's disease (Coyle and Puttfarcken, 1993; Gotz *et al.,* 1994; Schapira, 1995).

The selective vulnerability of neuronal systems is a remarkable characteristic of age-related degenerative disorders of the brain including Parkinson's disease (PD), Huntington's disease (HD), and amyotrophic lateral sclerosis (ALS) (Coyle and Puttfarcken, 1993). The attractive feature of the oxidative stress hypothesis is that it can account for cumulative damage associated with the delayed onset and progressive nature of these conditions. The brain consumes a disproportionate amount of the body's oxygen, because it derives its energy almost exclusively from oxidative metabolism of the mitochondrial respiratory chain (Coyle and Puttfarcken, 1993).

It is known that free radicals are continuously generated under a variety of physiologic conditions (Fig. 1). In order to quench these toxic free radicals, all aerobic cells possess mechanisms to mitigate their harmful effects. As a defense, cells contain antioxidant scavengers or specific enzymes to remove ROSs, that is, superoxide dismutase (SOD), catalase (CAT), and glutathione peroxidase (GPx) (Fig. 2).

SOD (E.C. 1.15.1.6) catalyzes dismutation of $O_2^{-\cdot}$ to H_2O_2, thereby reducing the risk of $OH^{-\cdot}$ formation, which otherwise disrupts the cellular function and integrity by reacting with proteins, deoxynucleic acids, and lipid membranes. Two major forms of SOD exist in all eukaryotic cells that is, copper–zinc superoxide dismutase (Cu,Zn-SOD) is found in the cytosol and manganese containing superoxide dismutase (Mn-SOD) is localized in mitochondrial matrix (Fridovich, 1989). CAT and GPx remove H_2O_2 from the intracellular environment by reducing it to H_2O and O_2. In addition to eliminating H_2O_2, GPx also participates in pathways responsible for detoxification of lipid peroxy radicals.

This chapter focuses on the antioxidant enzymes (SOD, CAT, and GPx) that are unique enzymes responsible for degradation of $O^{-\cdot}$ and H_2O_2. Attention is focused primarily on the ontogeny of antioxidant enzymes in the central nervous system (CNS). Specific examples of the interaction of these antioxidant enzymes with known neurotoxic effects are also provided. The brain contains large amounts of polyunsaturated fatty acids (PUFA) that are particularly vulnerable to free radical attack (Haliwell,1989). There are two main stages during the development of the CNS. The first stage is early development, a period during which the general shape of the adult brain is acquired, and the spongioblasts and neuroblasts, precursors of glia cells and neurons, multiply. The second stage coincides with rapid growth of the brain, the brain "growth spurt" (Davison and Dobbing, 1968), which, in many mammalian species, occurs during early postnatal life, a period when animals acquire many new motor and sensory faculties (Bolles and Woods, 1964) and increase spontaneous motor behavior (Campbell *et al.,* 1969). The brain growth spurt is also associated with numerous biochemical changes that transform the fetal neonatal brain into that of the mature adult (Eriksson *et al.,* 1997). Interference by ROS during this period can cause several pathologic changes in the development of the CNS. High levels of antioxidant enzymes during CNS development have

$O_2^{\cdot-}$ Superoxide anion radical
H_2O_2 Hydrogen peroxide
$OH^{\cdot-}$ Hydroxyl radical
$^1O_2$ Singlet oxygen

$O_2 + 1e^- \longrightarrow O_2^{\cdot-}$ **(Superoxide anion)**

$O_2 + 2e^- + 2H^+ \longrightarrow H_2O_2$ **(Hydrogen peroxide)**

$O_2 + 3e^- + 3H^+ \longrightarrow {}^{\cdot}OH + H_2O$ **(Hydroxyl free radical)**

$O_2 + 4e^- + 4H^+ \longrightarrow 2H_2O$

FIGURE 1 Formation of different types of reactive oxygen species.

Superoxide dismutase (SOD)
Catalase (CAT)
Glutathione peroxidase (GPx)

$$O_2^{\cdot-} + O_2^{\cdot-} + 2H^+ \xrightarrow{\text{SOD}} H_2O_2 + O_2$$

$$H_2O_2 \xrightarrow{\text{CAT}} 2H_2O + 1/2\ O_2$$

$$H_2O_2 + R(OH)_2 \xrightarrow{\text{Peroxidase}} H_2O + RO_2$$

$$H_2O_2 + 2GSH \xrightarrow{\text{GPx}} GSSG + 2H_2O$$

FIGURE 2 Reactive oxygen species scavenged by the presence of antioxidant enzymes.

been postulated to protect against oxidative damage. Neurodegenerative disease such as Alzheimer's disease (AD) may pose a threat at all stages of brain development, even though it is generally manifested in the aging population. However, SOD increase may pose a threat to the CNS by accelerating H_2O_2 formation. Therefore, an understanding of the regulation of these enzymes during development is important to elucidate the role of oxidative stress in neurotoxicity. In particular, the role of antioxidant enzymes in scavenging ROS in the developing CNS is the main topic of this chapter.

II. Antioxidant Enzymes During Development and Aging

Oxygen free radicals such as $O_2^{-\cdot}$ and $OH^\cdot$ formed in the process of biologic oxidation have been reported to play an important role during the developmental process (Yuan *et al.*, 1996; Zima *et al.*, 1996). The balance between the formation and inactivation of ROS may be altered during the perinatal period as a consequence of rapid changes in tissue oxygen concentration and the development of antioxidant defense enzyme activities (Zima *et al.*, 1996). An imbalance in the ratio of SOD to GPx and CAT results in the accumulation of H_2O_2, which may participate in the Fenton reaction, resulting in the formation of noxious hydroxyl radicals, which may react with DNA, proteins, and lipids. De Haan *et al.* (1994) proposed that it is the balance in the activity of SOD to GPx plus CAT ratio [SOD/(GPx plus CAT)] that is an important determinant of cellular development and aging. Fujitani *et al.* (1997) have reported that the development of bovine embryo to blastocyte stage was suppressed in the presence of ROS. It has been reported (Candlish *et al.*, 1995) that SOD and CAT activities increase whereas GPx activity decrease in erythrocytes from neonates over a range of gestational ages. Abnormal development of the brain was observed in rats exposed to ROS (Fantel *et al.*, 1995). Yuan *et al.* (1996) demonstrated different developmental patterns of antioxidant enzymes expression in guinea pig lung and liver during both the prenatal and postnatal period. They observed that Mn-SOD, CAT, and GPx activities increase during the late gestational period, whereas a steady increase of Cu,Zn-SOD and CAT following birth. Cell-specific developmental changes in the cellular distribution of glutathione stimulating hormone (GSH) and glutathione transferase (GSTs) in the murine nervous system have been reported by Beiswanger *et al.* (1995). De Haan *et al.* (1994) investigated the expression of Cu,Zn-SOD and GPx in organs of developing mouse embryos, fetuses, and neonates. Their study demonstrated that an increase in both SOD and GPx mRNA occurs in lungs and liver of the late-gestational mouse fetus. An increase in SOD expression at or around the time of birth was observed in the CNS, whereas the kidney exhibited an elevation in GPx mRNA levels. The liver is the organ with the highest levels of SOD and GPx mRNA in embryos and neonates (immediately after birth). All these studies demonstrate that antioxidant enzymes play a substantial role during development.

ROS have been postulated to be responsible for aging (Harman,1971). Although several theoretic mechanisms have been proposed to explain aging of the CNS, the biochemical basis of aging is still unclear. Antioxidant enzymes such as CAT, SOD, and GPx have been postulated to protect against biologic oxidative damage by scavenging reactive oxygen species (Harris, 1992). Several studies have been conducted comparing the levels of antioxidant enzymes throughout development (Mizuno and Ohta, 1986; Cand and Verdetti, 1989; Sohal *et al.,* 1990; Nistico *et al.,* 1992), however, none of the studies demonstrated a clear correlation between development and antioxidative enzymes. Therefore, it remains unclear as to the relationship between age and defense mechanisms against the occurrence of oxidative stress in the CNS. This chapter describes the selective changes in antioxidant enzymes in different regions of the mouse brain across various age groups in an attempt to clarify the roles of reactive oxygen species in the aging process.

Male C57BL/6N mice of 1, 6, 12, and 24 months of age were sacrificed by cervical dislocation, and their brains were removed and dissected into five different regions: caudate nucleus, frontal cortex, hippocampus, cerebellum, and brainstem. Subsequently, these brain regions were frozen on dry ice and stored at −70°.

The brain regions were homogenized in ice-cold sucrose (0.32*M;* 20% w/v) followed by sonication for 10 sec on ice. Supernatants were collected after centrifugation at 10,000 × g for 45 min at 4°. CAT activity was analyzed on the same day and the rest of the supernatant was stored at −70° for subsequent determination of SOD, GPx activities, and total GSH levels. CAT activity was determined according to the method of Beers and Sizer (1952), in which the disappearance of substrate (H_2O_2) was measured spectrophotometrically at 240 nm. Total SOD activity was assayed by the method of Pattichis *et al.* (1994), which is based on the inhibition of nitrite formation from hydroxylammonium in the presence of O_2^- generators. One unit of SOD is defined as that amount needed to produce 50% inhibition of the initial rate of nitrite formation. GPx activity was determined according to the method described by Flohe and Gunzler (1984), based on NADPH oxidation determined by spectrophotometer at 340 nm. Reduced GSH concentration was determined by using 5,5-dithiobis-2-nitrobenzoic acid (DTNB) as described by Ellman (1959). Protein concentrations were determined with the use of bicinchoninic acid protein assay kit available from Sigma Chemical Company.

CAT activity in different regions of the brain in young and adult animals is presented in Fig. 3. In the caudate

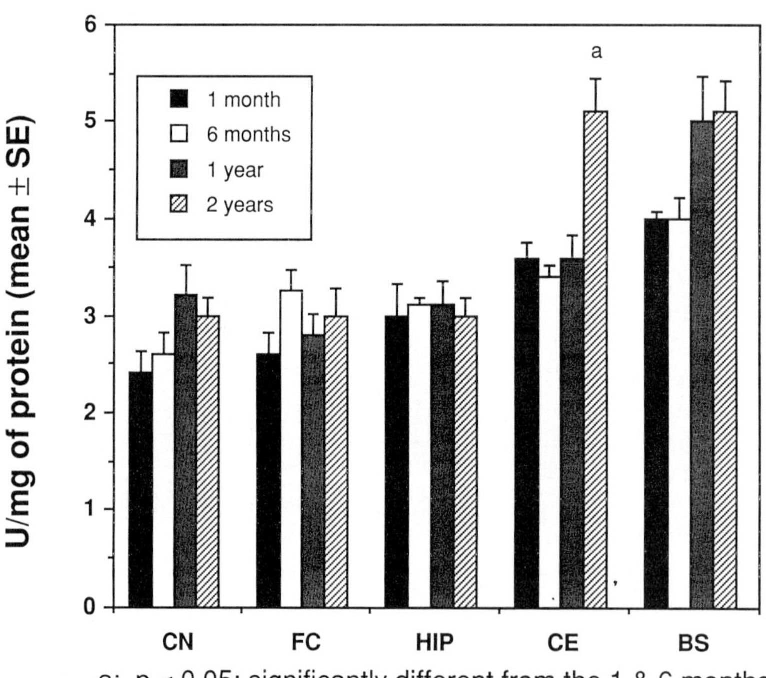

FIGURE 3 Age-related changes on catalase (CAT) activity in different brain regions: caudate nucleus (CN), frontal cortex (FC), hippocampus (HIP), cerebellum (CE), and brainstem (BS). The CAT activity is expressed as U/mg of protein; mean ± SEM for 8–10 animals per age group. Significant differences across the different age groups were evaluated by ANOVA, and *p* values are indicated on the top of the bars (adapted from Hussain *et al.,* 1995).

nucleus, the activity increased to 109%, 133%, and 125% in adult animals such as 6 months, 1- and 2-year-old animals, respectively compared to 1-month-old animals. There was no definitive pattern in the increase of CAT levels in frontal cortex, although the activity was higher in all age groups compared to 1-month-old animals. In the hippocampus, the activity remained constant in all age groups examined. In the cerebellum, the activity increased to 141% in 2-year-old animals, however, no change was found in 6-month- and 1-year-old animals. In brainstem the activity increased in 1- and 2-year-old animals, however, it remained unchanged in 6-month-compared to 1-month-old animals (Fig. 3).

SOD activity in developing animals at different age groups is illustrated in Fig. 4. SOD levels in caudate nucleus increased to 105%, 129%, and 148% in 6-month- and 1- and 2-year-old animals, respectively, compared to 1-month-old animals. Frontal cortex showed the same trend of increase in SOD activity in 6-month- (131%), 1-year- (170%) and 2-year-old animals (181%), compared to 1-month-old animals. In the hippocampus, the activity increased in 6-month- (125%) and 1-year-old animals; however, it decreased in 2-year-old animals (13%). In the cerebellum, the activity increased in 6-month- 1- and 2-year-old animals (122%, 117%, 128%, respectively) when compared to the 1-month old. SOD activity increased in brainstem to 109% at 6 months, 132% at 1 year, and 167% at 2 years. The data indicate that SOD activity significantly increases with age in caudate nucleus, frontal cortex, and brainstem (Fig. 4).

GPx activity in different age groups of animals is presented in Fig. 5. The GPx activity did not change at 1, 6, and 12 months in caudate nucleus; however, it was increased to 125% in 2-year-old animals. Frontal cortex showed elevated activity at 6 months (163%), followed by a threefold increase in 2-year-old as compared to 1-month-old animals. In 1-year-old animals, however, the activity did not change. Hippocampus showed an increase of GPx activity with increasing age. In the cerebellum and brainstem, the activity increased at 6 months but thereafter very little increase was observed up to 2 years of age (Fig. 5). GSH content in different age groups of animals is presented in Fig. 6. Although there was no linear relationship, GSH content increased in caudate nucleus, frontal cortex, hippocampus, cerebellum and brainstem with age.

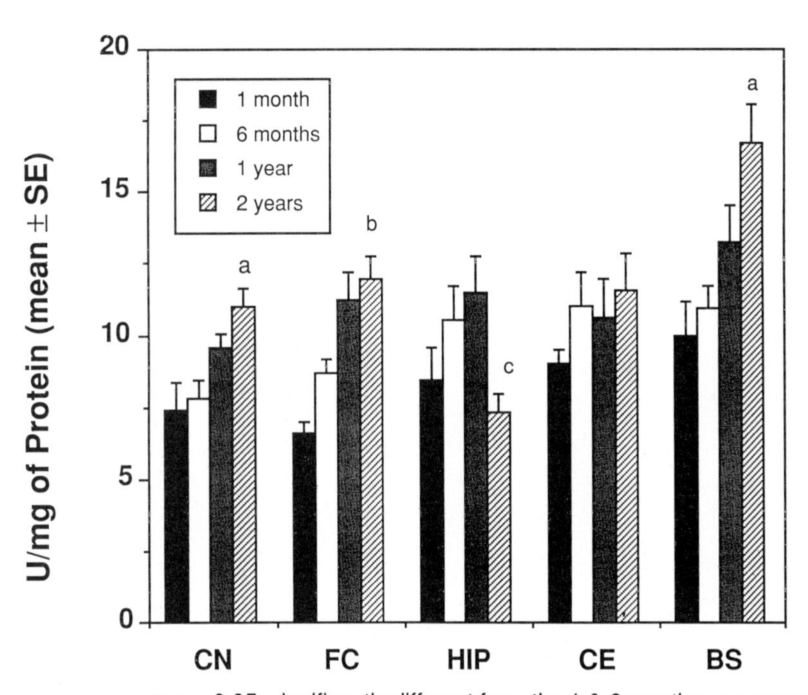

FIGURE 4 Age-related changes on superoxide dismutase (SOD) activity in different brain regions: caudate nucleus (CN), frontal cortex (FC), hippocampus (HIP), cerebellum (CE), and brainstem (BS). The SOD activity is expressed as U/mg of protein; mean ± SEM for 8–10 animals per age group. Significant differences across the different age groups were evaluated by ANOVA, and p values are indicated on the top of the bars (adapted from Hussain *et al.,* 1995).

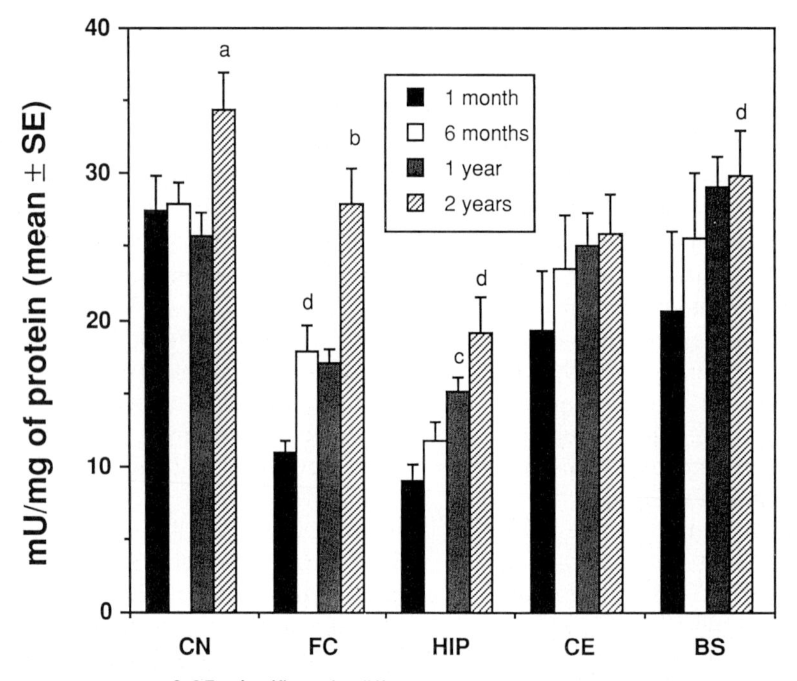

a: $p < 0.05$; significantly different from the 1year age group
b: $p < 0.05$; significantly different from the 1 & 6 months and 1 year age groups
c: $p < 0.05$; significantly different from the 1 & 6 months age groups
d: $p < 0.05$; significantly different from the 1 month age group

FIGURE 5 Age-related changes on glutathione peroxidase (GPx) activity in different regions: caudate nucleus (CN), frontal cortex (FC), hippocampus (HIP), cerebellum (CE), and brainstem (BS). The GPx activity is expressed as U/mg of protein; mean ± SEM for 8–10 animals per age group. Significant differences across the different age groups were evaluated by ANOVA, and *p* values are indicated on the top of the bars (adapted from Hussain *et al.*, 1995).

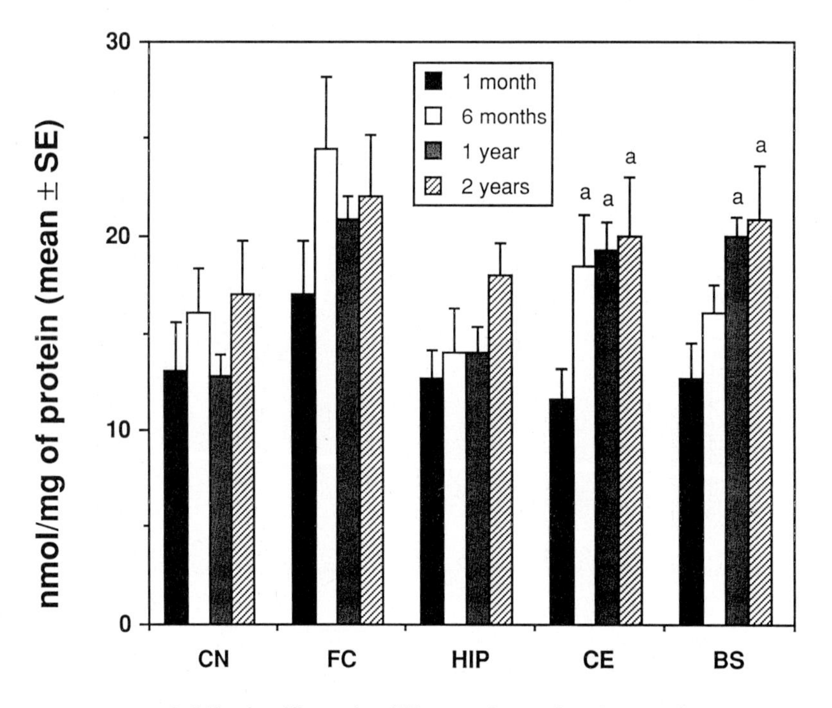

a: $p < 0.05$; significantly different from the 1 month age group

FIGURE 6 Age-related changes on glutathione (GSH) content in different regions: caudate nucleus (CN), frontal cortex (FC), hippocampus (HIP), cerebellum (CE), and brainstem (BS). The GSH content is expressed as nmol/mg of protein; mean ± SEM for 8–10 animals per age group. Significant differences across the different age groups were evaluated by ANOVA, and *p* values are indicated on the top of the bars (adapted from Hussain *et al.*, 1995).

The present study was conducted with the aim to understand the relationship between brain development and the enzymatic antioxidant defense system. With increasing age, there is an increase in the formation and release of ROS that induce membrane biochemical and functional alterations and DNA damage (Benzi and Moretti, 1995). Antioxidant enzymes such as SOD, GPx, and CAT play a major role in the intracellular defense against oxygen radical damage to aerobic cells (Harris, 1992). It has been proposed that the aging process of the brain is markedly influenced by oxidative and peroxidative mechanisms (Harman, 1983; Floyd *et al.,* 1984; Nohl, 1993). Therefore, ROS are increasingly discussed as important factors involved in the phenomenon of biologic aging. A comprehensive study performed on the brains of various vertebrate species showing different life energy potentials indicated that free radicals are important determinants of species-specific maximum life span (Del Maestro and McDonald, 1987). Results reported in the literature are contradictory and make it difficult to draw any conclusions about the relationship between antioxidative enzyme levels and age (Sagai and Ishinose, 1980; Ono and Okada, 1984; Ciriolo *et al.,* 1991; Barja *et al.,* 1994).

The present results show that the different brain regions have a different pattern of alteration of antioxidant enzyme activities with age. The activity of CAT did not show any significant change with increasing age except in the cerebellum of 2-year-old animals. Some investigators report the CAT activity decreased, whereas others reported an increase with age (Ono and Okada, 1984; Sagai and Ishinose, 1980; Barja *et al.,* 1994). However, due its low activity in the brain, CAT may appear to play a secondary role in H_2O_2 disposition in this organ compared with other organs (Benzi and Moretti, 1995). Increasing SOD activity was observed with age in rats or mice brain regions by a majority of the investigators (Cand and Verdetti, 1989; Semsei *et al.,* 1991; Nistico *et al.,* 1992; Sanitiago *et al.,* 1993; Barja *et al.,* 1994). The increased activity of SOD with age may indicate an augmented superoxide ion ($0_2^{-}\cdot$) flux with aging in certain brain regions (Coyle *et al.,* 1983). GPx activity was increased with age, however, with low activity in the hippocampus compared to other regions. This suggests that the hippocampus would be particularly vulnerable to oxidative stress with aging. Increased GPx activity with aging may be due to an increase in organic peroxide formation during aging. This finding supports the notion that PD, in which a specific degeneration of the dopaminergic nigrostriatal system occurs, may be related to a persistently high flux of H_2O_2 in nigrostriatal neurons as a consequence of oxidative deamination of dopamine by monoamine oxidase (Spina and Cohen, 1989). A small increase in glutathione content was found with age in all regions of the brain. Glutathione protects vulnerable cells from oxidative and other metabolite stress due to the metabolism of endogenous as well as exogenous substrates (Aebi and Suter, 1974). Therefore, the changes in GSH content may play a significant role in the aging process (Orlowski and Karkowsky, 1976).

SOD has been considered to be an important antioxidant enzyme that protects from the damage caused by superoxide radicals (Freeman and Crapo, 1982). For instance, significant increases of cytosolic Cu,Zn-SOD activity in transgenic mouse brain confers resistance against MPTP induced neurotoxicity (Przedborski *et al.,* 1992). Further, it was reported that transgenic mice over-expressing Cu,Zn-SOD genes showed resistance to a lethal dose of methamphetamine,methylene-dioxy-amphetamine (MDA) and methylenedioxymethamphetamine (MDMA), apparently by increasing SOD activity (Cadet *et al.,* 1994a,b). Increased resistance to paraquat-mediated cytotoxicity was observed in transfected cells that overproduce Cu,Zn-SOD (Stein *et al.,* 1986). All of these studies provide strong evidence for the significant protective role of SOD from oxidative stress–mediated intoxication of MPTP, methamphetamine, MDA, and MDMA. Because of its crucial role in the physiologic response to oxygen toxicity, it has been actively investigated as a potential therapeutic agent in pathologic conditions related to oxidative stress (Zweier *et al.,* 1987; Oda *et al.,* 1989). For instance, the Food and Drug Administration (FDA) designated SOD as an orphan drug for the prevention of bronchopulmonary dysplasia in premature infants resulting from pulmonary oxygen toxicity (Walther *et al.,* 1990).

The present results suggest an increased level of SOD may protect by scavenging superoxide radicals ($O_2^{-}\cdot$) generated during aging. Increases of brain GPx activity may remove H_2O_2 because CAT activity is reported to be very low in the brain compared to other organs (Hussain *et al.,* 1995). The results are consistent with the hypothesis that elevated SOD levels increase the ability of brain cells to prevent damage from acute oxidative stress during the aging process.

III. Effect of Selective Neurotoxicants on Antioxidant Enzymes

A. *Mercury (Hg)*

Mercury primarily occurs naturally in the environment, and occurs as a result of mining, smelting, and industrial discharge. Exposure to mercury may occur by ingestion, inhalation, and consumption via the food chain (Goyer, 1991). One of the most dramatic examples of neurotoxicity occurred with mercury and methylmercury (MeHg) and its association with the massive outbreak of poisoning in Japan (Minamata disease) and in Iraq (Bakir *et al.*, 1973; Takeuchi, 1977). Ganther (1978) hypothesized that MeHg generates free radicals *in vivo*, and thereby produces cytotoxicity. Ali *et al.* (1992) demonstrated that acute exposure to MeHg increases the formation of reactive oxygen species in the cerebellum. Le Bel *et al.* (1990, 1992) have reported that MeHg may exert its neurotoxicity through iron-mediated oxidative damage. Therefore, there is increasing evidence to suggest that oxidative damage may play a role in the neurotoxicity of MeHg (Ganther, 1978; Hussain *et al.*, 1997b).

Although there are several reports delineating the mechanisms of MeHg neurotoxicity, the mechanism of inorganic mercury neurotoxicity is poorly understood. Lund *et al.* (1993) demonstrated that the administration of mercuric chloride (1.5 or 2.25 mg $HgCl_2$/kg body weight) to rats results in increased hydrogen peroxide formation, glutathione depletion, and lipid peroxidation in kidney mitochondria. Zaman *et al.* (1994) reported that mercuric chloride acts as a prooxidant as it exerts oxidative stress. It was reported that $HgCl_2$ treatment (1 or 2 mg/kg body weight of rat) for 7 consecutive days significantly alters behavior and neurotransmitter systems in rats (Rodgers *et al.*, 1996a,b).

The alteration of antioxidant enzymes in response to mercury in rat renal mitochondria *in vitro* has been reported (Lund *et al.*, 1991). Shimojo and Arai (1994) found the induction of antioxidant enzymes due to exercise correlated with the distribution of mercury after mercury vapor exposure in mice. Because MeHg is known to produce oxygen free radicals in the CNS (Ali *et al.*, 1992), it is likely that it may exert an effect on antioxidant enzymes. Therefore, we hypothized that the mercuric chloride induces reactive oxygen species and alters antioxidant enzyme activities in different regions of the rat brain.

Acute exposure to mercury during the developmental process has been shown to induce behavioral and neurochemical alterations in rodents over the course of life. The developing CNS seems to be extremely susceptible to MeHg (Zanoli *et al.*, 1994). Administration of MeHg to pregnant rats during embryonic organogenesis induces depression of the spontaneous locomotor activity and learning and memory deficit in the offspring (Spyker *et al.*, 1972; Su and Okita, 1976; Shimai and Satoh, 1985). An increased accumulation of MeHg has been described to occur in rat fetus during the late period of gestation (Yang *et al.*, 1972). Exposure to MeHg (8 mg/kg) in the late stage of gestation (15th day) induces in the offspring transient effects on dopaminergic (Cagiano *et al.*, 1990) and GABA-benzodiazepine (Guidetti *et al.*, 1992) receptor systems and their pharmacological correlates. Soderstrom et al. (1995) reported that low levels of Hg exposure during development induce neuronal damage in rats. Prenatal MeHg exposure effects the development of cortical muscarinic receptors in rat offspring as reported by Zanoli *et al.* (1994). All these reports indicate that inorganic mercury and organic mercury are major concerns as environmental toxicants to the developing fetal CNS.

In vitro effects of $HgCl_2$ on ROS formation in synaptosomes of different regions of rat brain are presented in Fig. 7. A significant increase of ROS was observed following exposure to various (1–5 μM) concentrations of mercury in the caudate nucleus and cerebellum. In addition, the frontal cortex and hippocampus also showed augmented ROS formation following exposure to 2–5 μM mercury. The principle toxic effects of Hg (II) involve interactions with a large number of cellular processes, including the formation of complexes with free thiols and protein thiol groups, which may lead to oxidative stress (Ribarove and Benov, 1981; Stacey and Kappus, 1982). In addition, Hg (II) alters the structural integrity of the mitochondrial inner membrane, resulting in the loss of normal cation selectivity that permits it to participate effectively in oxidative metabolism (Weinberg *et al.*, 1982). It has been demonstrated that Hg (II) produces oxidative damage via H_2O_2 generation thereby leading to lipid perioxidation (Lund *et al.*, 1991; Stacey and Kappus, 1982; Karniski 1992). Karniski (1992) suggested that Hg (II) may also mediate electroneutral exchange of Cl^-/OH^- across lipid membranes, further contributing to the collapse of ionic and pH gradients. All these possible mechanisms of Hg (II) toxicity may lead to the formation of ROS, as found in the present investigation. Results from these and other studies have demonstrated that *in vivo* exposure to mercuric chloride produces alterations in antioxidant enzymes SOD and GPx activities in different regions of rat brain.

Total SOD, Cu,Zn-SOD, and Mn-SOD activities in different brain regions of mercury exposed rats (1–4 mg/kg body weight) are presented in Fig. 8. A significant dose-dependent reduction of total SOD and Cu,Zn-SOD activities were found in the cerebellum of mercury-treated rats. The Mn-SOD activity decreased significantly in the cerebellum at the high dose of mercury (4 mg/kg). Activities of total SOD, Cu,Zn-SOD,

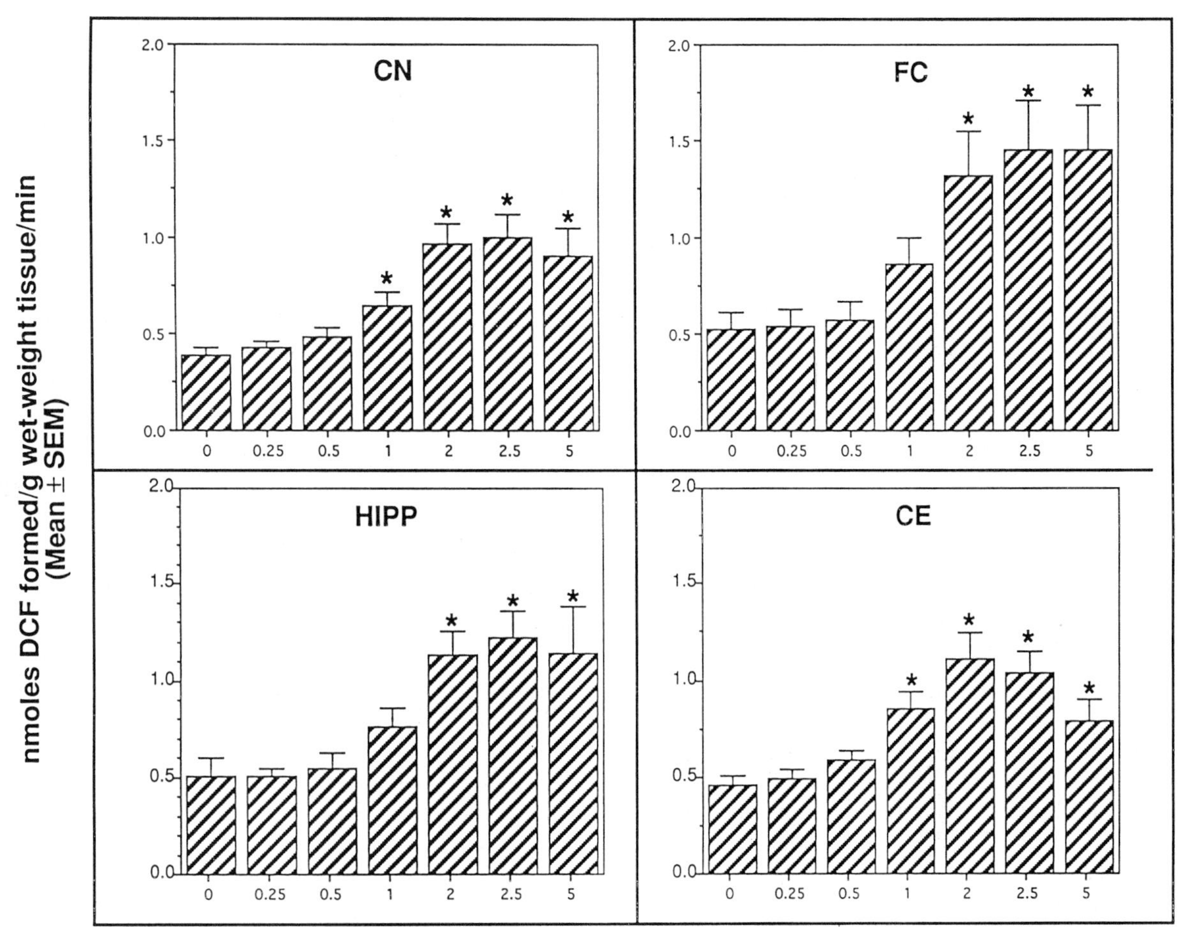

FIGURE 7 Reactive oxygen species (ROS) produced in different brain regions by mercury (II) *in vitro:* caudate nucleus (CN), frontal cortex (FC), hippocampus (HIPP), and cerebellum (CE). Control rat brains were quickly removed, dissected into different regions, and synaptosomal fractions (P2) were prepared for the ROS assay as described in material and methods. Control value represents zero concentration of $HgCl_2$. The formation of ROS was expressed in nmoles DCF formed/g wet weight of tissue/min. Mean ± SEM represents the values of 3–5 experiments. Significant differences across various treatments were evaluated by ANOVA, and p values (<0.05 significantly different from control) are indicated (*) on the top of each bar (adapted from Hussain *et al.,* 1997b).

and Mn-SOD did not vary significantly in cerebral cortex or brainstem.

Activity of GPx measured in different regions of the brains of mercury-treated (1–4 mg $HgCl_2$/kg body weight) rats are presented in Fig. 9. A significant decline in the activity of GPx was observed in the cerebellum at a 4 mg/kg dose of mercury, whereas in the brainstem an increased level of GPx activity was observed at 2 and 4 mg/kg mercury. Mercury did not alter GPx activity in the cerebral cortex (Fig. 9). GPx activity in the brainstem was increased at 2 and 4 mg/kg of mercury. This increase in GPx activity may be related to a persistently high flux of H_2O_2 due to mercury toxicity. Therefore, increased GPx activity in brainstem may remove H_2O_2 because CAT activity is reported to be very low in brain as compared to other organs (Sohal *et al.,* 1990). In summary, the results demonstrate that mercury induced ROS by altering antioxidant enzymes. Therefore, mercury poses a threat during the development of organisms.

B. Manganese (Mn)

Manganese (Mn) is an essential trace element and important to several key enzymes in living systems. Mn crosses the blood–brain barrier in both the adult and

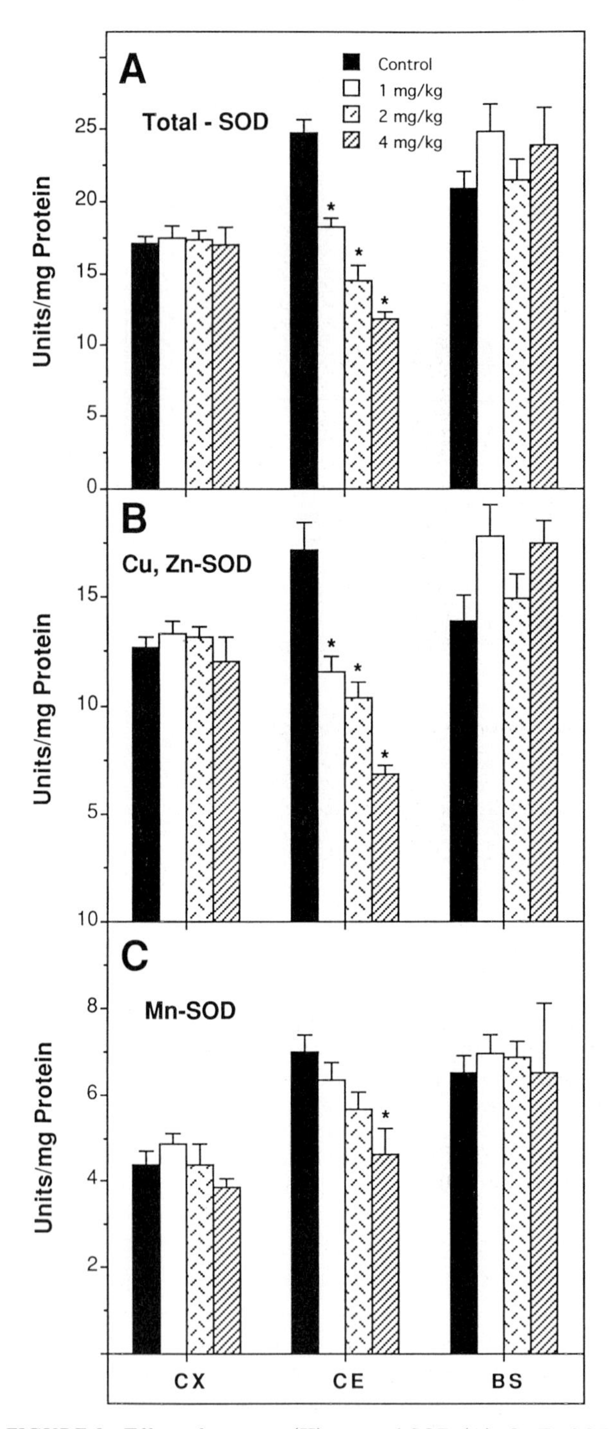

FIGURE 8 Effect of mercury (II) on total-SOD (A), Cu,Zn-SOD (B), and Mn-SOD (C) activities in different brain regions: cerebral cortex (CX), cerebellum (CE), and brainstem (BS). The activity of enzyme was measured as described in materials and methods. The SOD activity is expressed as units/mg of protein; mean ± SEM for 6–8 animals per group. Significant differences across the various treatments were evaluated by ANOVA, and *p* values (<0.05 significantly different from control) are indicated (*) on the top of each bar (adapted from Hussain *et al.,* 1997b).

the developing fetus. The brain normally contains a small amount of Mn (Cotzias *et al.,* 1968). In high concentrations, Mn causes an irreversible brain disorder with prominent psychologic and neurologic disturbances (Aschner and Aschner, 1992). Over time it has been hypothesized that Mn neurotoxicity is mediated by generation of ROS, which depends on the ability of Mn to undergo changes in its oxidative states (Ali *et al.,* 1995; Donaldson, 1987). A significant function of Mn is as a constituent of the antioxidant enzyme Mn-SOD, which catalyzes the dismutation of superoxide (O_2^-), the univalent reduction product of dioxygen (Coassin *et al.,* 1992; Singh *et al.,* 1992; Tampo and Yonaha, 1992). Mn-SOD is one of the cell's primary defenses against oxygen-derived free radicals and is vital for maintaining a healthy balance between oxidants and antioxidants (Flores *et al.,* 1993).

In vitro exposure to $MnCl_2$ (0.2–2.0 m*M*) produced a dose-dependent increase in the formation of ROS (Ali *et al.,* 1995). These results suggest that manganese-induced neurotoxicity may be mediated via generation of ROS. The formation of ROS was higher in younger animals as compared to older animals. This is in agreement with LaBel and Bondy (1991), who reported that, under basal conditions, there was an age-dependent decrease in the formation of ROS.

There are few reports of Mn toxicity during fetal development. Sanchez *et al.* (1993) demonstrated reduced body weight of mice during development after exposure to 8 or 16 mg/kg Mn (II). The CNS of neonatal rats was found to be more susceptible to Mn-induced neurotoxicity compared with the adult rat CNS (Kontur and Fechter, 1988).

The effect of $MnCl_2$ on the activities of SOD and GPx and its effect on GSH content were evaluated in different regions of rat brain after administration at 2.5 or 5 mg $MnCl_2$/kg. Total SOD, Cu,Zn-SOD, and Mn-SOD activities in different regions of the brains of control and Mn-treated rats are presented in Figs. 10 and 11. Total SOD activities did not vary significantly in any region of the brain examined. Mn-SOD activity was not altered in the caudate nucleus or frontal cortex, however, there was a significant increase of Mn-SOD activity in the hippocampus (Fig. 10). Mn-SOD activity in cerebellum and brainstem increased by 60% and 40%, respectively (Fig. 11). The Cu,Zn-SOD activity decreased in all brain regions; however, this decrease was not statistically significant (Figs. 10 and 11). The GPx activity in the brain regions of control and Mn-treated rats is presented in Fig. 12. The activity of GPx did not differ from control with either dose of $MnCl_2$. The effect of Mn on GSH content is illustrated in Fig. 13. Mn exposure reduced GSH content in cerebellum; however,

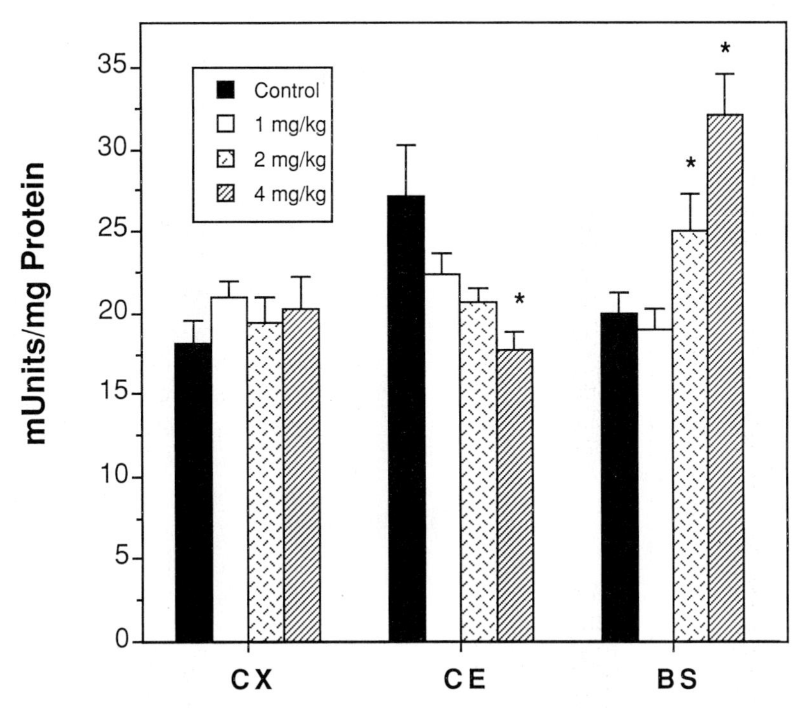

FIGURE 9 Effect of mercury (II) on glutathione peroxidase activity in different regions: cerebral cortex (CX), cerebellum (CE), and brainstem (BS). The activity of GPx was carried out as described in material and methods. The GPx activity is expressed as mUnits/mg of protein; mean ± SEM represents for 6–8 animals per group. Significant differences across the various treatments were evaluated by ANOVA, and p values (<0.05 significantly different from control) are indicated (*) on the top of each bar (adapted from Hussain *et al.,* 1997b).

no significant effect on GSH content was found in other brain regions.

There are reports that suggest that increased expression of Mn-SOD plays a central role by diminishing oxygen-mediated injuries and the cytotoxic effects of various toxicants and therapeutic agents (Hirose *et al.,* 1993; Cobbs *et al.,* 1996). Sprague-Dawley rats fed an Mn-deficient diet showed decreased Mn-SOD activity in kidney and heart (Thompson *et al.,* 1992). Liccione and Maines (1988) reported decreased levels of CAT, GSH, and GPx activities in rat striatum exposed to Mn. All these studies demonstrate conclusively the significant role of Mn-SOD in contending against oxidative stress. Studies also demonstrate that Mn-SOD activity increases due to Mn treatment (Hussain *et al.,* 1997a). In view of the biologic role of Mn, there are several possibilities that may explain the increasing levels of Mn-SOD activity. Because it is known that Mn is responsible for regulating Mn-SOD activity, this enzyme activity may increase due to the accumulation of excess Mn. It is also possible that Mn-SOD might have a critical role in Mn toxicity because either an Mn deficiency or an excess dose of Mn can alter Mn-SOD activity (Borello *et al.,* 1992; Davis and Greger, 1992; Hirose *et al.,* 1993; Cobbs *et al.,* 1996). Mn-SOD is primarily located in mitochondria, which play an important role in scavenging superoxide radicals (Weisiger and Fridowich,1973). An increased expression of Mn-SOD may serve to protect the mitochondria from the toxic effects of superoxide radicals that are generated due to Mn accumulation.

Studies also show that Mn-SOD is altered whereas other antioxidative enzymes such as Cu-Zn SOD and GPx remain unchanged (Hussain *et al.,* 1997a). This suggests that Mn may not effect cytosolic enzymes like Cu,Zn-SOD, but effects only Mn-SOD located in mitochondria. Warner *et al.* (1993) demonstrated that protection against paraquat-induced oxidative injury was directly related to increased Mn-SOD, occurring in the absence of changes in other antioxidant enzymes including CAT, Cu,Zn-SOD, and GSH associated with cellular antioxidant mechanisms. This suggests the possibility of mitochondria being a critical site for Mn toxicity.

IV. Summary and Conclusion

In summary, the neurotoxicity of mercury, MeHg, and Mn may be due to enhanced levels of ROS and

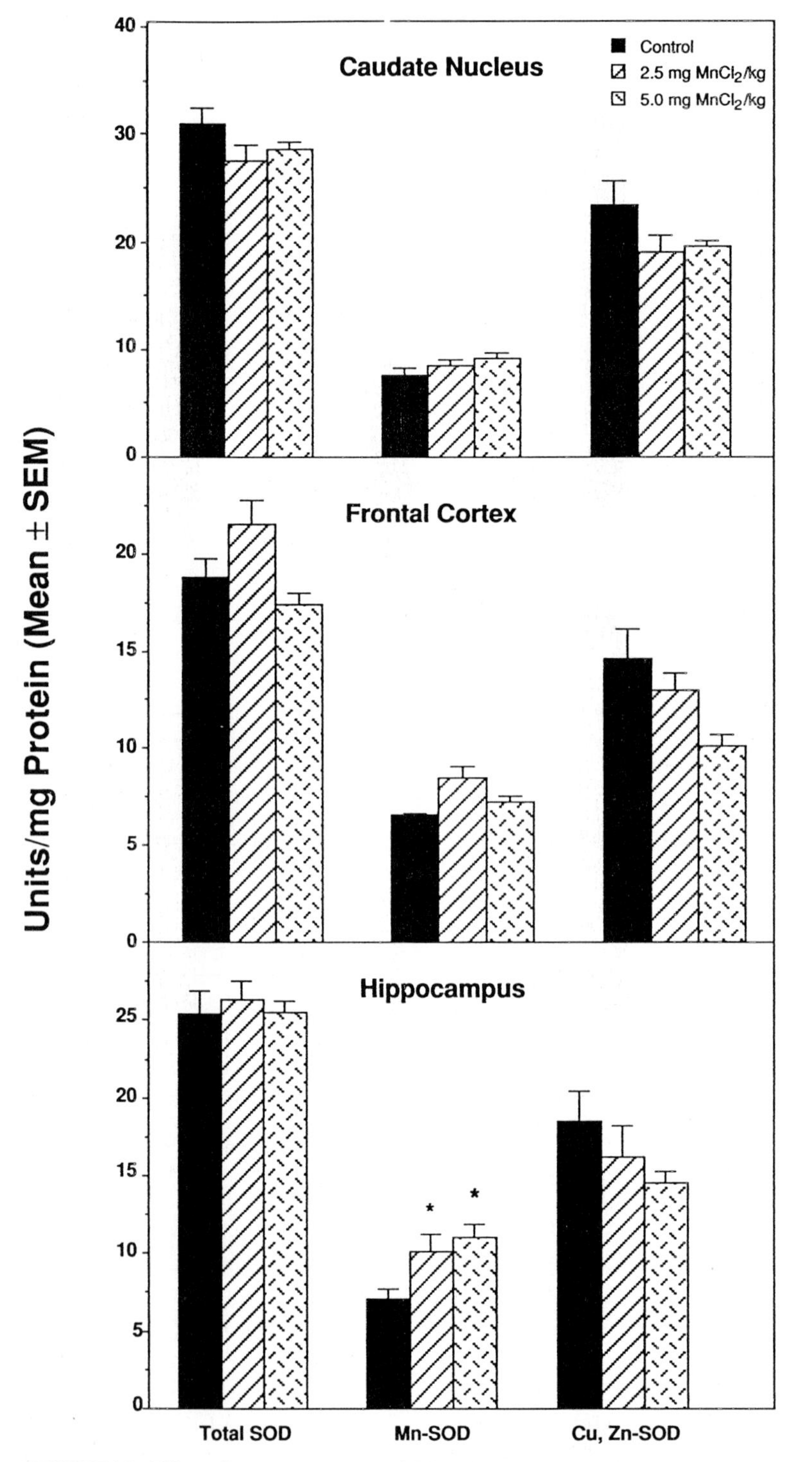

FIGURE 10 Effect of manganese on total-SOD, Cu,Zn-SOD, and Mn-SOD activities in different regions of rat brain. The SOD activity is expressed as units/mg of protein; mean ± SEM for 6–8 animals per group. Significant differences across the different groups were evaluated by ANOVA, and p values (<0.05) are indicated (*) on the top of the bars. Abbreviations: C-control; 2.5 mg $MnCl_2$,/kg, ip; 5 mg $MnCl_2$,/kg ip, (adapted from Hussain *et al.,* 1997a).

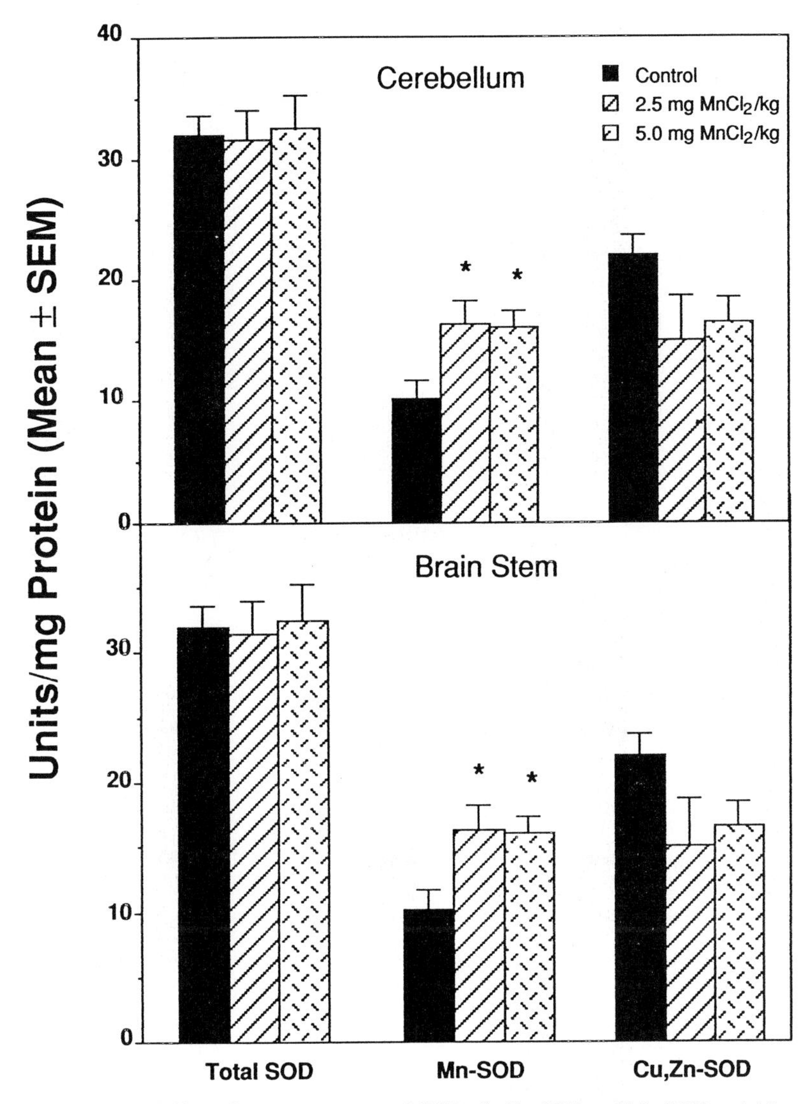

FIGURE 11 Effect of manganese on total-SOD, Cu,Zn-SOD and Mn-SOD activities in different regions of rat brain. The SOD activity is expressed as units/mg of protein; mean ± SEM for 6–8 animals per group. Significant differences across the different groups were evaluated by ANOVA, and p values (<0.05) are indicated (*) on the top of the bars (adapted from Hussain *et al.,* 1997a).

changes in the activities of antioxidant enzymes. The data suggest that the decreased activities of SOD in the cerebellum may decrease protection against mercury-induced oxidative stress. Therefore, cerebellum is more vulnerable following *in vivo* exposure to mercury. The findings with Mn indicate that the Mn-SOD was altered significantly due to Mn exposure. This increased level of Mn-SOD activity may contribute in detoxifying ROS generated by Mn exposure.

It has been proposed that the aging process of the brain is markedly influenced by oxidative and peroxidative mechanisms (Harman, 1983; Floyd *et al.,* 1984; Nohl, 1993). Therefore, free oxygen radicals are increasingly discussed as important factors involved in the phenomenon of biologic aging. Present study shows that although the activity of CAT did not change significantly, the activities of SOD and GPx increase with age. Present results suggest an increased level of SOD may protect by scavenging superoxide radicals ($O_2^{-\cdot}$) generated during aging. The results are consistent with the hypothesis that elevated SOD levels increase the ability of the brain cells (e.g., neurons, glia cells) to prevent damage from acute oxidative stress during the aging process. Increases of brain GPx activity may aid in the destruction of H_2O_2,

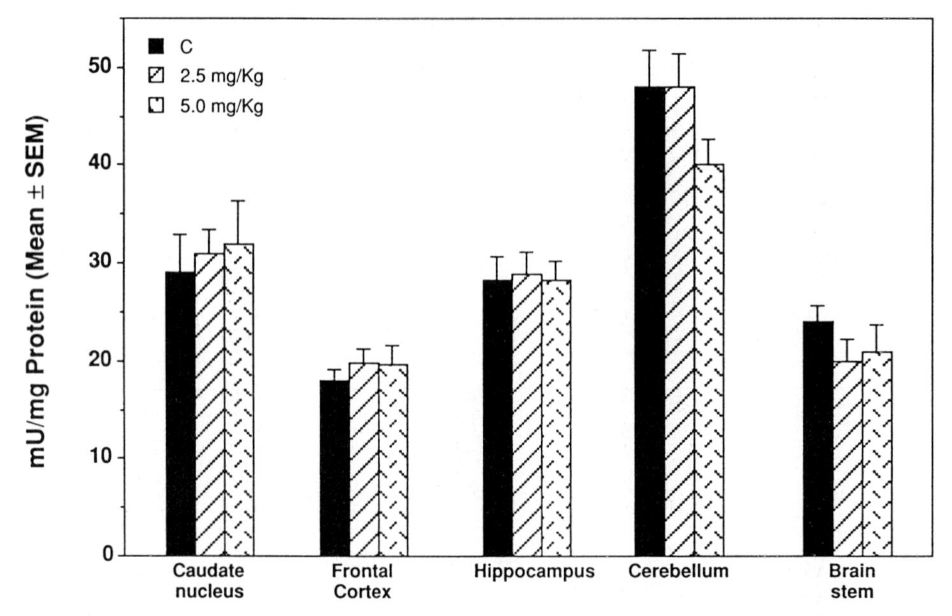

FIGURE 12 Effect of manganese on glutathione peroxidase activity in different regions of rat brain. The GPx activity is expressed as mU/mg of protein; mean ± SEM for 6–8 animals per group. Abbreviations: C, control; 2.5 mg $MnCl_2$/kg ip; 5 mg $MnCl_2$/kg ip (adapted from Hussain *et al.*, 1997a).

because CAT activity is reported to be very low in brain as compared to other organs (Hussain *et al.*, 1995). The increasing activity of antioxidant enzymes with age enhances the ability of the brain cells to prevent damage from acute oxidative stress during development and senescence.

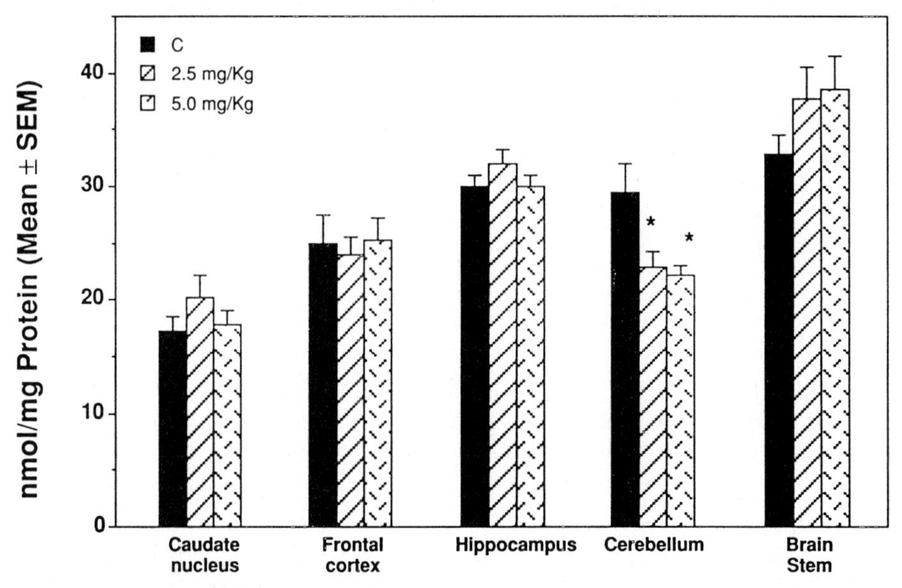

FIGURE 13 Effect of manganese on glutathione content in different regions rat brain. The GSH content was expressed as nmol/mg of protein; mean ± SEM for 6–8 animals per age group. Significant differences across the different groups were evaluated by ANOVA, and *p* values (<0.05) are indicated on the top of the bars. Abbreviations: C, control; 2.5 mg $MnCl_2$/kg ip; 5 mg $MnCl_2$/kg ip, (adapted from Hussain *et al.*, 1997a).

References

Ali, S. F., Duhart, H. M., Newport, G. D., Lipe G. W., and Slikker, W. Jr. (1995). Manganese-induced reactive oxygen species: comparison between Mn^{+2} and Mn^{+3}. *Neurodegeneration* **2,** 329–334.

Ali, S. F., LeBel, C. P., and Bondy, S. C. (1992). Reactive oxygen species formation as a biomarker of methyl mercury and trimethyltin neurotoxicity. *NeuroToxicology* **13,** 637–648.

Aebi, H., and Suter, H. (1974). Protective function of reduced glutathione (GSH) against the effect of prooxidative substances and of irradiation in the red cell. In *Glutathione.* Flohe L, Benohr HC, Sies H, Waller HD, Wendel A, Eds. Georg Thieme, Stuttgart, pp. 192–199.

Aschner, M., and Aschner, J. L., (1992). Manganese neurotoxicity. Cellular effects and blood–brain barrier transport. *Neurosci. Behav. Rev.* **15,** 333–340.

Bakir, F., Damluji, S. F., Amin-Zaki, L., Murtadha, M., Khalidi, A., Al-Rawi, N. Y., Tikriti, S., Dhahir, H. I., Clarkson, T.W ., Smith, J. C., and Doherty, R. A. (1973). Methylmercury poisoning in Iraq. *Science* **181,** 230–241.

Barja, G., Cadenas, S., Rojas, C., Lopes-Torres, M., and Perez-Campo, R. (1994). A decrease of free radical production near critical targets as a cause of maximum longevity in animals. *Comp. Biochem. Biophys.* **108,** 501–512

Beers, R. F. Jr. and Sizer I. W. (1952). A spectrophotometric method for measuring the break down of H_2O_2 by catalase. *J. Biol. Chem.* **195,** 133–140.

Beiswanger, C. M., Diegmann, M. H., Novak, R. F., Phillbert, M. A., Graessle, T. L., Reuhl, K. R., and Lowndes, H. E. (1995). Developmental changes in the cellular distribution of glutathione and glutathione-S-transferase in the murine nervous system. *NeuroToxicology* **16,** 425–440.

Benzi, G. and Moretti, A. (1995). Age- and peroxidative stress–related modifications of the cerebral enzymatic activities linked to mitochondria and glutathione system. *Free Rad. Biol. Med.* **19,** 77–101.

Bolles, R. G. and Woods, P. J. (1964). The ontogeny of behaviour in the albino rat. *Anim. Behav.* **12,** 427–441.

Borrello, S., DeLeo, M. E., and Galeotti, T. (1992). Transcriptional regulation of Mn-SOD by manganese in the liver of manganese-deficient mice and during rat development. *Biochem. Intern.* **28,** 595–601.

Brown R. K., Mcburney, A., Lunec, J., and Kelly F. J. (1995). Oxidative damage to DNA in patients with cystic fibrosis. *Free Rad. Biol. Med.* **18,** 801–806.

Cadet, J. L., Ladenheim, B., Baum, I., Carlson, E., and Epstein, C. (1994a). Cu,Zn-SOD transgenic mice show resistance to the lethal effects of methylenedioxyamphetamine (MDA) and of methylenedioxymethamphetamine (MDMA). *Brain Res.* **655,** 259–262.

Cadet, J. L., Sheng, P., Ali S., Rothman, R., Carlson, E., and Epstein, C. (1994b). Attenuation of methamphetamine-induced neurotoxicity in copper/zinc superoxide dismutase transgenic mice. *J. Neurochem.* **62,** 380–383.

Cagiano, R. M., De Salvia, M. A ., Renna, G., Tortella, E., Braghiroli, C,. Parenti, P., Zanoli, M., Barali, Z., Annau, Z., and Cuomo, V. (1990). Evidence that exposure to methyl mercury during gestation induces behavioural and neurochemical changes in offspring of rats. *Neurotoxicol. Teratol.* **8,** 23–28.

Campbell, B. A., Lytle, L. D., and Fibiger, H. C. (1969). Ontogeny of adrenergic arousal and cholinergic inhibitory mechanisms in the rat. *Science.* **166,** 635–637.

Cand F., and Verdetti J. (1989). Superoxide dismutase, glutathione peroxidase, catalase and lipid peroxidation in the major organs of the aged rats. *Free Radic. Biol. Med.* **7,** 59–63.

Candlish, J. K., Tho, L. L., and Lee, H. W. (1995). Erythrocyte enzymes decomposing reactive oxygen species and gestational age. *Early Hum. Dev.* **43,** 145-150.

Cerutti P. A. (1985). Prooxidant states and tumor promotion. *Science* **37,** 375–381.

Ciriolo, M. R, Fiskin, K., Martino, A. D., Corasaniti, M. T., Nistico, G., and Rotilio, G. (1991). Age-related changes in Cu-Zn SOD, Se-dependent and -independent GPx and catalase activities in specific areas of rat brain. *Mech. Aging Dev.* **61,** 287–297.

Coassin, M., Ursini, F., and Bindoli, A. (1992). Antioxidant effect of manganese. *Arch. Biochem. Biophys.* **299,** 330–333.

Cobbs, C. S., Levi, D. S., Aldape, K., and Israel, M.A. (1996). Manganese superoxide dismutase expression in human central nervous system tumors. *Cancer Res.* **6,** 3192–3195.

Cotzias, G. C., Horiuchi, K., Fuenzalida, S., and Mena, I. (1968). Chronic manganese poisoning. Clearance of tissue manganese concentrations with persistance of the neurological picture. *Neurology.* **18,** 376–82.

Coyle, J. T., Price D. L., and Delong M.R . (1983). Alzheimer's disease: a disorder of cortical cholinergic innervation. *Science* **219,** 1184–1189.

Coyle, J. T., Puttfarcken, P. (1993). Oxidative stress, glutamate, and neurodegenerative disorders. *Science* **263,** 689–695.

Davis, C. D., and Greger, J. L. (1992). Longitudinal changes of manganese-dependent superoxide dismutase and other indexes of manganese and iron status in women. *Am. J. Clin. Nutr.* **55,** 747–52.

Davison, A. N., and Dobbing, J. (1968). *Applied Neurochemistry.* Blackwell, Oxford, pp. 178–221; 253–316.

De Haan, J. B., Tymms, M. J., Cristiano, F., and Kola, I. (1994). Expression of Cu/Zn-superoxide dismuatse and glutathione peroxidase in organs of developing mouse embryo, fetuses and neonates. *Pediatric Res.* **35,** 188–196.

Del Maestro, R.F., and McDonald. (1987). Distribution of superoxide dismutase, glutathione peroxidase and catalase in developing rat brain. *.Mech. Ageing. Dev.* **41,** 29–38.

Donaldson, J. (1987). The physiopathologic significance of manganese in brain: its relation to schizophrenia and neurodegenerative disorders. *Neurotoxicology* **8,** 451–462.

Ellman, G. L. (1959). Tissue sulfhydryl groups. *Arch. Biochem. Biophys.* **82,** 70–77.

Eriksson, P., Ahlbom, J., and Fredrickson, A (1992). Exposure to DDT during a defined period in nonatal life induces permanent changes in brain muscarine receptors and behavioural in adult mice. *Brain Res.* **582,** 277–281.

Fantel, A. G., Person, R. E. Tumbic, R. W., Nguyen, T. D., Mackler, B. (1995). Studies of mitochondria in oxidative embryotoxicity. *Teratology* **52,** 190–195.

Flohe, L., and Gunzler, W. A. (1984.) Assay of glutathione peroxidase. In *Methods Enzymol.* Packer L., Ed. Vo.l 105. pp. 114–121.

Flores, S. C., Merecki, J. C., Harper, K. P., Bose, S. K., Nelson, S. K., and McCord, J. M. (1993). Tat protein of human immunodeficiency virus type 1 represses expression of manganese superoxide dismutase in HeLa cells. *Proc. Natl. Acad. Sci. U.S.A.* **90,** 7632–7636.

Floyd, R. A., Zaleska, M. M., and Harnon, J. (1984). Possible involvement of iron and oxygen free radicals in aspects of aging in brain. in Armstong D., Sohal R. S., Cutler R. G., and Slater T. F. Eds. *Free Radicals in Human Biology, Aging and Disease.* Raven Press, NY, pp. 143–161.

Freeman, B. A., and Crapo, J. D. (1982). Biology of disease. Free radicals and tissue injury. *Lab. Invest.* **47,** 412–426.

Fridovich I. (1989). Superoxide dismutase. *J. Biol. Chem.* **264,** 1776–1784.

Fujitani, Y., Kasai, K., Ohtani, S., Nishimura, K., Yamada, M., and Utsumi, K. (1997). Effect of oxygen concentration and free radicals on *in vitro* development of *in vitro*–produced bovine embryos. *J. Anim. Sci.* **75,** 483–489.

Ganther, H. E. (1978). Modification of methyl mercury toxicity and metabolism by selenium and vitamin E: possible mechanism. *Environ. Health Prospect.* **25,** 71–76.

Gibson, J. D., Pumford, N. R., Samokyszyn, V. M., and Hinson, J. A. (1996) Mechanism of acetaminophen-induced hepatotoxicity: Covalent binding versus oxidative stress. *Chem. Res. Toxicol.* **9,** 580–585.

Gotz, M. E., Kunig, G., Riederer, P., and Youdim, M. B. H. (1994). Oxidative stress: free radical production in neural degeneration. *Pharmacol. Therapeut.* **63,** 37–122.

Goyer, R. A. (1991). Toxic effects of metals. In Amdur, M. O., Doull, J., Klaassen, C. D. Eds. *Casarett and Doull's Toxicology: The Basic Science of Poisons.* 4th ed. Pergamon Press, New York, pp. 629–681.

Guidetti, P., Giacobazzi, P., Zanoli, P., and Baraldi, M. (1992). Prenatal exposure of rats to methylmercury: increased sensitivity of the GABA-benzodiazepine receptor function. In: *Metal Compounds in Environment and Life,* Vol. 4. E. Merian and W. Haerdi, Eds. Science Reviews, Wilmington, DE, pp. 365–371.

Halliwell, B. (1989). Free radicals, reactive oxygen species and human disease: a critical evaluation with special reference to atherosclerosis. [Review] *Brit. J. Exper. Path.* 70(6):737–557.

Halliwell, B., and Cross, C. E. (1991). Reactive oxygen species, antioxidants and acquired immunodeficiency syndrome. *Arch. Intern. Med.* **157,** 29–31.

Harris, E. D. (1992). Regulation of antioxidant enzymes. *FASEB J.* **6,** 2675–683.

Harman, D. (1971). Free radical theory of aging: effect of the amount and degree of unsaturation of dietary fat on mortality rates. *J. Gerontol.* **26,** 451–457.

Harman D. (1983). Free radical theory of aging: consequence of mitochondrial aging. *Age* **6,** 86–94.

Hirose, K., Longo, D. L., Oppenheim, J. J., and Matsushima, K. (1993). Overexpression of mitochondrial manganese super-oxide dismutase promotes the survival of tumor cells exposed to interleukin-1, tumor necrosis factor, selected anticancer drugs, and ionizing radiation. *FASEB J.* **7,** 361–8.

Hussain, S., Slikker, W. Jr., and Ali, S. F. (1995). Effect of aging on antioxidant enzymes, superoxide dismutase, catalase, glutathione peroxidase, glutathione in different regions of mouse brain. *Int. J. Dev. Neuroscience.* **13,** 811–817.

Hussain, S., Slikker, W. Jr., and Ali, S. F. (1997a). The effects of chronic exposure of manganese on antioxidant enzymes in different regions of rat brain. *Neurosci. Res. Commun.* **21,** 135–143.

Hussain, S., Slikker, W. Jr., and Ali, S. F. (1997b). Mercuric chloride induced formation of reactive oxygen species and its effect on antioxidant enzymes in different regions of rat brain. *J. Environ. Sci. Hlth.* **B32,** 395–409.

Jaeschke, H. (1995). Mechanisms of oxidant stress-induced acute tissue injury. *Proc. Soc. Exp. Biol. Med.* **209,** 104–111.

Karniski, L. P. (1992). Hg^{+2} and Cu^{+2} are ionophores mediating Cl^-/OH^- exchange in liposomes and rabbit renal brush border membranes. *J. Biol. Chem.* **267,** 19219–19225.

Kontur, P. J. and Fechter, L. D. (1988). Brain regional manganese levels and monoamine metabolism in manganese-treated neonatal rats. *Neurotoxicol. Teratol.* **10,** 295–303.

LaBel, C. P. and Bondy S. C. (1991). Persistent protein damage despite reduced oxygen radical formation in the aging rat brain. *Int. J. Devel. Neurosci.* **9,** 139–46.

LeBel, C. P., Ali, S. F., McKee, M., and Bondy, S. C. (1990). Organometals induced increases in oxygen radical activity: the potential of dichlorofluorescein diacetate as an index of neurotoxic damage. *Toxicol. Appl. Pharmacol.* **104,** 17–24.

LeBel, C. P., Ali, S. F., Bondy, S. C. (1992). Deforaxamine inhibits methylmercury-induced increases in reactive oxygen species formation in rat brain. *Toxicol. Appl. Pharmacol.* **112,** 161–165.

Liccione, J. J., and Maines, M. D. (1988). Selective vulnerability of glutathione metabolism and cellular defense mechanisms in rat striatum to manganese. *J. Pharmacol. Exp. Ther.* **247,** 156–61.

Lund, B. O., Miller, D. M., and Wood, J. S. (1991). Mercury-induced H_2O_2 production and lipid oxidation *in vitro* in rat kidney mitochondria. *Biochem. Pharmacol.* **42,** 181–187.

Lund, B. O., Miller, D. M., and Woods, J. S. (1993). Studies on Hg (II) induced H_2O_2 formation and oxidative stress *in vivo* and *in vitro* in rat kidney mitochondria. *Biochem. Pharmacol.* **45,** 2017–2024.

Mizuno, Y., and Ohta, K. (1986). Regional distribution of thiobarbituric acid reactive products, activity of enzymes regulating the metabolism of oxygen free radicals and some of the related enzymes in adult and aged rat brain. *J. Neurochem.* **46,** 1344–1346.

Nistico, G., Ciriolo, M. R., Fiskin, K., and Rotilio, G. (1992). NGF restores decrease in catalase activity and increases superoxide dismutase and glutathione peroxidase activity in the brain of aged rats. *Free Rad. Biol. Med.* **12,** 177–81.

Nohl, H. (1993). Involvement of free radicals in aging: a consequence or cause of senescence. *Brit. Med. Bull.* **49,** 653–667.

Oda, T., Akaike, T., Hamamoto, T., Suzuki, F., Hirano, T., and Maeda, H. (1989). Oxygen radicals in influenza-induced pathogenesis and treatment with pyran polymer-conjugated SOD. *Science* **244,** 974–976.

Ono, T., and Okada, S. (1984). Unique increase of superoxide dismutase levels in brains of long living mammals. *Exp. Gerontol.* **19,** 349–354.

Orlowski, M., and Karkowsky, A. (1976). Glutathione metabolism and some possible functions of glutathione in the nervous system. *Int. Rev. Neurobiol.* **19,** 75–121.

Pattichis, K., Louca, L. L., and Glover, V. (1994). Quantitation of soluble superoxide dismutase in rat striata, based on the inhibition of nitrite formation from hydroxylammonium chloride. *Anal. Biochem.* **221,** 428–431.

Przedborski, S., Jackson-Lewis, V., Kostic, V., Carlson, E., Epstein, C. J., and Cadet, J. L. (1992). Superoxide dismutase, catalase, and glutathione peroxidase activities in Cu,Zn-SOD transgenic mice. *J. Neurochem.* **58,** 1760–67.

Ribarov, S. R., and Benov, L. C. (1981). Relationship between the hemolytic action of heavy metals and lipid peroxidation. *Biochim. Biophys. Acta* **640,** 721–726.

Rodgers, D. A., Holson, R. R., Ferguson, S. A., Ali, S. F., and Soliman, M. R. I. (1996a). Effects of mercuric chloride exposure in rats Part I. Behavioral changes observed during open field, and complex maze evaluations. *Arch. Environ. Contam. Toxicol.* (In Press).

Rodgers, D. A., Soliman, M. R. I., and Ali, S. F. (1996b). Effects of mercuric chloride exposure in rats Part II. Alterations of neurotransmitter systems in various brain regions. *Arch. Environ. Contam. Toxicol.* (In Press).

Sagai, M. and Ishinose, T. (1980). Age related changes in lipid peroxidation as measured by ethylene, butane and pentane in respired gases of rats. *Life Sci.* **27,** 731–738.

Sanchez, D. J., Domingo, J. L., Llobet, J. M., and Keen, C. L. (1993). Maternal and developmental toxicity of manganese in the mouse. *Toxicol. Lett.* **69,** 45–52.

Santiago, L. A., Osato, J. A., Liu, J., and Mori, A. (1993). Age-related increase in superoxide dismutase activity and thiobarbituric acid-reactive substances: effect of bio-catalyzer in aged rat brain. *Neurochem. Res.* **18,** 711–718.

Schapira, A. H. V. (1995). Oxidative stress in Parkinson's disease. *Neuropathol. Appl. Neurobiol.* **21,** 3–9.

Semsei, I., Rao, G., and Richardson, A. (1991). Expression of superoxide dismutase and catalase in rat brain as a function of age. *Mech. Age. Dev.* **58,** 13–19.

Shimai, S., and Satoh, H. (1985). Behavioral teratology of methylmercury. *J. Toxicol. Sci.* **10,** 199–216.

Shimojo, N., and Arai, Y. (1994). Effects of exercise training on the distribution of metallic mercury in mice. *Hum. Exp. Toxicol.* **13,** 524–528.

Singh, R. K., Kooreman, K. M., Babbs, C. F., Fessler, J. F., Salaris, S. C., and Pham, J. (1992). Potential use of simple potential manganese salts as antioxidant drugs in horses. *Am. J. Vet. Res.* **53,** 1822–1829.

Soderstrom, S., Fredrickson, A,. Dencker, L., and Ebendal, T. (1995). The effect of mercury vapours on cholinergic neurons in the fetal brain: studies on the expression of nerve growth factor and its low and high-affinity receptors. *Brain Res. Dev. Brain. Res.* **85,** 96–108.

Sohal, R. S., Sohal, B. H., and Brunk, U. T. (1990). Relationship between antioxidant defenses and longevity in different mammalian species. *Mech. Aging Dev.* **53,** 217–227.

Spina, M. B., and Cohen, G. (1989). Dopamine turnover and glutathione oxidation: implication for Parkinson's disease. *Proc. Natl. Acad. Sci. U.S.A.* **86,** 1398–1400.

Spyker, J. M., Sparber, S. B., and Goldberg, A. M. (1972). Subtle consequences of methylmercury exposure: behavioral deviations in offspring of treated mothers. *Science* **62,** 621–623.

Stacey, N. H., and Kappus, H. (1982). Cellular toxicity and lipid peroxidation in response to mercury. *Toxicol. Appl. Pharmacol.* **63,** 29–35.

Stein, O. E., Bernstein, Y., and Groner, Y. (1986). Overproduction of human Cu/Zn-superoxide dismutase in transfected cells: extenuation of paraquat-mediated cytotoxicity and enhancement of lipid peroxidation. *EMBO Journal* **5,** 615–622.

Su, M., and Okita, G. T. (1976). Embryocidal and teratogenic effects of methylmercury in mice. *Toxicol. Appl. Pharmacol.* **38,** 195–205.

Takeuchi, T. (1977). Neuropathology of Minamata disease in Kumamato: especially at the chronic stages. Vol. I. Rozin, L., Shiraki, H., Grcevic, N., Eds. Raven Press, New York, pp 235–246.

Tampo, Y., and Yonaha, M. (1992). Antioxidant mechanism of Mn (II) in phospholipid peroxidation. *Free Radic. Biol. Med.* **13,** 115–120.

Thompson, K. H., Godin, D. V., and Lee, M. (1992). Tissue antioxidant status in streptozotocin-induced diabetes in rats. Effects of dietary manganese deficiency. *Biol. Trace. Element Res.* **35,** 213–224.

Vile, G. F. and Tyrrell, R. M. (1995). UVA radiation-induced oxidative damage to lipids and proteins *in vitro* and in human skin fibroblasts is dependent on iron and singlet oxygen. *Free Radic. Biol. Med.* **18,** 721–730.

Walther, F. J., Kuipers, I. M., Pavlova, Z., Willebrand, D., Abuchowski, A., and Viau, A. T. (1990). Mitigation of pulmonary oxygen toxicity in premature lambs with intravenous antioxidants. *Exp. Lung Res.* **16,** 177–189.

Warner, B. B., Papes, R., Heile, M., Spitz, D., and Wispe, J. (1993). Expression of human Mn-SOD in Chinese hamster ovary cells confers protection from oxidant injury. *Am. J. Physiol.* **264,** L598–L605.

Weinberg, J. M., Harding, P. G., and Humes, H. D., (1982). Mitochondrial bioenergetics during the initiation of mercuric chloride-induced renal injury. *J. Biol. Chem.* **257,** 60–67.

Weisiger, R. A., and Fridowich, I. (1973). Mitochondrial superoxide dismutase site of synthesis and intramitochondrial localization. *J. Biol. Chem.* **248,** 4793–4799.

Yang, M. G., Krawford, K. S., Gareia, J. D., Wang, J .H. C., and Lei, K. Y. (1972). Deposition of mercury in fetal and maternal brain. *Proc. Exp. Biol. Med.* **141,** 1004–1007.

Yuan, H. T., Bingle, C. D., and Kelly, F. J. (1996). Differential pattern of antioxidant enzyme mRNA expression in guinea pig lung and liver during development. *Biochim. Biophys. Acta* **1305,** 165–171.

Zaman, K., MacGill, R. S., Johnson, J. E., Ahmad, S., and Pardini, R. S. (1994). An insect model for assessing mercury toxicity: effect of mercury on antioxidant enzyme activities of the housefly (*Musca domestica*) and the cabbage looper moth (*Trichoplusia ni*). *Arch. Environ. Contam. Toxicol.* **26,** 114–118.

Zanoli, P., Truzzi, C., Veneri, C., Braghiroli, D., and Baraldi, M. (1994). Methylmercury during late gestation affects temporarily the development of cortical muscarinic receptors in rat offspring. *Pharmacol. Toxicol.* **75,** 261–264.

Zima, T., Stipek, S., Crkovska, J., Doudova, D., Mechurova, A., and Calda, P. (1996). Activity of antioxidant enzymes superoxide dismutase and glutathione peroxidase in fetal erythrocytes. *Prenatal Diagnosis* **16,** 1083–085.

Zweier, C. W., Flaherty, J. T. and Weisfeldt, M. L. (1987). Direct measurement of free radical generation following reperfusion of ischemic myocardium. *Proc. Natl. Acad. Sci. U.S.A.* **84,** 1404–1407.

CHAPTER

20

Food and Nutrient Exposure throughout the Life Span

How Does What We Eat Translate into Exposure, Deficiencies, and Toxicities?

MARGARET L. BOGLE
Delta Nutrition Intervention Research Initiative
U.S. Department of Agriculture/ARS
Little Rock, Arkansas 72204

I. Introduction

The problems encountered in attempting to evaluate or describe the effects of exposure to foods and nutrients throughout the human life span are the subjects of numerous textbooks. An attempt to discuss the effects relative to developmental neurotoxicology in a single chapter requires selectivity. Therefore, this chapter limits exposure primarily to that occurring with the consumption of food (orally), with a few references to gavage or tube feedings and intravenous infusion of some essential nutrients. This is not to discount the importance of exposure through inhaling or absorbing nutrients through the skin; but merely to indicate those are beyond the scope of this chapter. Naturally occurring toxic substances other than nutrients, either inherent in foods, as contaminants in or on foods, or as the results of food preparation, are not discussed in this chapter. The reader is referred to the comprehensive review by Pariza (1996). In addition, the references provided allow the reader to acquire more complete details about specific individual nutrients.

* Abbreviations: AI, adequate intake; DRI, dietary reference intake; EAR, estimated average requirement; EPA, Environmental Protection Agency; ESADDI, estimated safe and adequate daily dietary intake; LOAEL, lowest observed adverse effect level; NOAEL, no observed adverse effect level; NRC, National Research Council; RDA, recommended dietary allowance; RfD, oral reference dose; UL, tolerable upper intake level.

Many of the problems associated with food exposure are related to food composition, consumption behaviors of humans, and methods of data collection. Dietary exposure is very complex and varies considerably over time, both within individuals and among populations. The estimation of acute exposure is usually determined by recording foods and beverages an individual has consumed over the last 24 hr, commonly called a 24 hr *dietary recall* (Bingham, *et al.,* 1994; Thompson and Byers, 1994). This methodology is problematic for infants and children who require a proxy with knowledge of what has been consumed, when in fact the food or nutrient causing an acute reaction may have been ingested without the proxy's knowledge. The recall of foods eaten is difficult for adults in attempting to estimate portion sizes or amounts ingested, especially when acute problems are almost always dose related. Elderly persons, persons with cognitive problems, and persons consuming food away from home (no knowledge of food components or methods of preparation) also complicate the process of evaluating acute exposure (Bingham and Day, 1997).

Evaluating chronic exposure to foods and nutrients encompasses the problems mentioned previously and others as well. The within-subject or interindividual variability in food consumption from day to day complicate the collection of individual data and multiplies the effect of population data necessary for establishing cause and effect relationships of food and nutrient variables to outcomes. The lack of variability in consumption of foods in various population groups may make associations between outcomes (toxicity, developmental problems, disease) and food consumption insignificant as risk can only be inferred to the range of intake, which may limit the observations to markers of individual susceptibility (Beaton, 1994; Kaaks *et al.,* 1994).

The interrelationships between individual nutrients or food components within a diet also make it difficult if not impossible, to determine the effects of any one variable (nutrient). This is true for the macronutrients (Wacholder *et al.,* 1994; Beaton *et al.,* 1997), fat, protein, and carbohydrate, as well as for micronutrients, vitamins, and minerals (Mertz, 1991; Mertz, 1994). For example, the ingestion of Zinc supplements may interfere with the absorption of copper to the point of causing copper deficiency symptoms (Fiske *et al.,* 1994).

Closely related to nutrient interactions in foods is the bioavailability of specific nutrients within a given food (Bowman and Risher, 1994). Fiber content of foods may significantly alter the absorption of vitamins and minerals. Gallaher and Schneeman (1996) present a comprehensive review of the effects of dietary fiber in human nutrition and discuss in detail the role of fiber in the bioavailability of nutrients. Another interaction example is the increased absorption of nonheme iron from plant foods if eaten with a source of vitamin C at the same time.

The chemical form of the nutrient in food may also alter bioavailability. For instance, the heme forms of iron found in meats is two to three times more available than the nonheme form found in plants (Bjorn-Rasmussen *et al.,* 1974; Hallberg, 1981).

Bioavailability and the chemical form of the nutrient also alter the dose response of an individual as well as add to the problems of quantifying the dose from various foods. Just as Maines (1994) points out that there are many factors that affect the individual response to essential metals, these same factors affect the individual response to other essential nutrients; deficiency, toxicity, acute, chronic, mild, or lethal. These include physiologic functions, age, gender, disease state, diet, heredity, body composition, stress, medications, and nutrient supplements, all of which may be interacting simultaneously. At the cellular and organ level other factors effect the dose response: hormones, medications, drugs, transport receptors, oxidation, nutrient–nutrient interaction, cell type, and specific organ.

A final consideration in evaluating food exposure and dose response effect is the route of elimination from and/or storage of the nutrient within the body. Nutrients generally are eliminated from the body in urine or feces or through the skin (perspiration). Again, there are multiple factors that influence elimination: chemical form (water soluble vs. fat soluble), body temperature, season of the year, fluid consumption, and physiologic functioning of specific organs. In addition, fiber and fat in the diet, disease states, and altered enzyme systems alter elimination in fecal material. Many of these same factors also effect storage or accumulation of various nutrients within the body that may contribute to chronic exposure.

II. Current Nutrient Standards

Traditionally the nutrition community has relied on the National Research Council (NRC) for standards serving as goals for good nutrition. Since 1943, the NRC has provided at regular intervals recommended dietary allowances (RDA) of various essential nutrients for healthy people of different age groups (NRC, 1943). These recommendations and those in subsequent publications through the 10th edition (NRC, 1989) were meant for healthy people of various ages, based on existing scientific knowledge of physiologic requirements to prevent deficiency states of various nutrients. RDAs were established for essential nutrients only when the subcommittee agreed there was sufficient data to make reliable recommendations.

In the ninth edition (NRC, 1980), the subcommittee established estimated safe and adequate daily dietary intake (ESADDI) for nutrients for which the data were insufficient to establish RDAs. Three vitamins, six trace elements, and three electrolytes were included in the category (NRC, 1980). The tenth edition advanced vitamin K and selenium to RDA status, using new scientific data (NRC, 1989). The ESADDIs were to be used until adequate data were available to establish RDAs.

Because RDAs cannot be set for all known nutrients, the subcommittee has always based their recommendations on the concept that the RDAs would be met through the consumption of a diet of a variety of foods rather than through the use of dietary supplements (NRC, 1986). This followed the basic rationale of RDAs being meant to prevent deficiencies in healthy persons, not persons with disease or sensitivities not normally present in a healthy population. The RDAs are established from an estimate of the physiologic requirement (preventing deficiency state) for an absorbed nutrient. That value is then adjusted by the subcommittee to compensate for incomplete use, variation in requirements among individuals, and the bioavailability among various food sources. This provides for a safety factor such that each RDA exceeds the actual requirement for most individuals.

Although the RDAs are widely used and for a variety of purposes, it is important to remember that the amounts of nutrients are to be eaten as part of a normal diet, levels are not intended as minimals nor optimals, and are best applied to groups rather than individuals. Probable risks of deficiencies for individuals can be estimated by comparing individual intakes, averaged over a period of time, to the RDA levels of comparable age and gender.

Currently, the Standing Committee on the Scientific Evaluation of Dietary Reference Intakes (1997) has released the first in a series of reports that update and expand the RDAs. Three new categories of intakes have been added to the RDAs. Dietary reference intakes (DRIs) are reference values that can be used in planning and assessing diets for healthy populations and encompass the RDA, the estimated average requirement (EAR), the adequate intake (AI), and the tolerable upper intake level (UL) (Food and Nutrition Board, 1997).

The EAR is the intake value estimated to meet the requirement of that nutrient in 50% of the individuals in a given age and gender group. The EAR is based on a dietary intake averaged over time (usually 1 week) and incorporates the concept of reducing disease risk rather than just preventing nutrient deficiencies (as used in past reports). In addition, the EAR is now used in establishing the RDA for various nutrients, which is devised to meet the nutrient requirements of 97–98% of all healthy individuals.

The AI terminology is used when scientific data are not judged to be sufficient to calculate an EAR. The UL is used to indicate a "maximal level of nutrient intake that is unlikely to pose risks of adverse health effects to almost all individuals in the target group" (Food and Nutrition Board, 1997, p. 55). The UL is not meant to indicate health benefits, but provides a level that can be tolerated biologically. The UL is also based on total consumption—from food, supplements, and pharmacologic agents—and relates to consistent daily use over time.

It is important to note that the DRIs are based on scientific data from a variety of sources: clinical trials and studies (animals, as appropriate, and humans), food and supplement consumption studies—published in peer reviewed journals. The final scientific judgment of the subcommittee reflects considerable review and debate over the available data.

III. Oral Reference Dosages

For most of the essential nutrients for which recommendations are made, some information on toxicity is also provided when data are available. However, the role of hazard identification and risk of toxicity rests with the U.S. Environmental Protection Agency (EPA). Dourson (1994) has reviewed the terminology and methodology for establishing oral reference doses (RfD). The RfD is an estimate of daily exposure to humans that has no likely risk of negative (toxic) effects during the lifespan. The RfD is similar to the RDA in that both represent scientific estimates of exposure to nutrients that are protective of health (USEPA, 1993). It is only when these reference levels (RfD and RDA) overlap that the validity of both systems is called into question (Table 1). Bowman and Risher (1994) compared the methodology of each system relative to essen-

TABLE 1 Comparison of Selected Nutrients, 1989[a]

	RDA	ESADDI	RfD (current)[b]
Chromium (μg/day)		50–200	70,000
Manganese (mg/day)		2–5	10
Selenium (μg/day)	70, 55		350
Zinc (mg/day)	15, 12		21

[a]RDAs are for adults 19–51+ years (males, females).

[b]RfD established on body-weight per day basis—values shown are for hypothetical 70 kg adult.

tial nutrients. Table 2 provides some comparison of the similarities and differences of the two systems that are used in determining RDAs and RfDs. RDAs, ESADDI, and RfDs are intake levels intended to be compatible with normal physiologic function, protect health by preventing deficiencies, and protect from excessive intake or exposure.

IV. Effects of Deficiency and Excessive Amounts of Selected Nutrients

Table 3 provides summary information and reference detail for the effects of deficiency (consuming less than the required amount of nutrient) and excessive amounts (either on acute or chronic conditions) of selected nutrients. Particular emphasis is given to examples of developmental and neurologic problems related to deficiencies and excesses. It is interesting and of some significance to note that in some instances the symptoms of deficiency and excessive intakes may be very similar, which further complicates the assessment of cause and effect.

As indicated in Table 3, deficiencies and toxic effects are rare for most nutrients, but when those occasions occur for the few nutrients, the effects are devastating, including significant neurologic and developmental sequelae. Some consequences of deficiency and excessive intakes are lethal, resulting in death.

A few examples of the more critical consequences are discussed. In general, the water-soluble vitamins are not considered to pose any toxicity problems as excesses (amounts higher than what the body can use) are usually excreted in the urine. Even when supplemented in doses hundreds of times greater than the physiologic requirement, most of the water-soluble vitamins pose no threat of adverse effects. Certainly the amounts in food consumed in a varied diet are of little or no consequence. In other words, to produce toxic effects, the vitamin must be ingested in an extremely large dose. There are two notable exceptions: niacin and folic acid.

A. Niacin

Currently, some forms of this vitamin are being used to treat high levels of serum cholesterol at doses of greater than 1000 times the RDA (dose 1.5–3 g; RDA 13 mg–19 mg) (Canner, 1986). Side effects include flushing of the skin, liver and ocular abnormalities, and hyperuricemia and hyperglycemia. All of these effects can be reversed by decreasing the dose of niacin. If not reversed, the effects can have serious, long-term consequences that ultimately cannot be completely reversed. In the case of niacin, the deficiency effects, although extremely rare today, can have more serious consequences than the excessive amounts, particularly those effects relative to depression, memory loss, and mental disorientation. The niacin deficiency disease pellagra, which manifests all of the effects mentioned, is rare today because of the small amount of niacin required and because of the mandatory enrichment of grain products.

B. Folic Acid

This vitamin presents another example of the deficiency state producing much more serious developmental consequences than the known effects of excessive intake. Deficiency intakes of folic acid occurring prior to conception and continuing during pregnancy are known to significantly increase the risks of neural tube defects in the infants (Czeizel, 1995). A major effect of excessive intakes of folic acid is in the masking of vitamin B_{12} deficiencies, which, if untreated, can cause

TABLE 2 Differences/Similarities in Determining Standards

	RDA[a]	RfD[b]
Types of data used	Controlled studies	Controlled studies
Bioavailability of nutrient	Partial	100%
	Homeostatic regulatory mechanism	—
	Variation with food composition	—
	Synergy with other nutrients	—
	Antagonism with other nutrients	—
Nutritional assessment	Monitoring individuals	Food intake data
Route of exposure	Food (oral), water, only for fluoride	Food (oral), water, gavage, parenteral nutrition
Reference individuals	Multiple gender and age groups	1 reference (based on body weight)
Interindividual variability	Only healthy people	Based on most sensitive identified groups
Uncertainty factors	Increase levels	Decrease levels

[a]RDA—Recommended Dietary Allowances (NRC, 1989).
[b]RfD—Oral Reference Dose (USEPA, 1993).

TABLE 3 Selected Known Effects of Nutrients

	Functions	Deficiency	Excessive amounts
Vitamin A	Vision Embryonic development Cellular differentiation	Xerophthalmia Night blindness Bitot's spots Blindness Dry scaly skin Poor growth Reproductive problems Increased infections	Fetal resorption Birth defects Headaches Double vision Hair loss (alopecia) Bone abnormalities Liver damage Bone and joint pain
Vitamin D	Mediating calcium homeostasis	Loss of bone mass Risk of bone softening Rickets Defective bone growth	Kidney damage/stones Weakened muscles Excessive bleeding Calcium deposits in soft tissue Hypercalcemia Bone demineralization
Vitamin E	Prevents oxidative damage to cells Normal immune function Protects nervous system, skeletal muscle, and retina from oxidative damage	Rare Hemolytic anemia Infants—neurologic abnormalities	Rare Interfere with absorption of vitamins A and K Exaggerate effect of anticoagulants Decrease platelet adhesion Sepsis
Vitamin K	Synthesis of plasma clotting factors	Rare Defective blood coagulation Hemorrhagic disease of newborn Bruising	No known effects of natural vitamin Menadione (synthetic form) hemolytic anemia and liver toxicity
Vitamin C	Cofactor for enzyme formation Collagen synthesis Catecholamine synthesis Facilitates iron absorption	Fatigue Scurvy Follicular excess Hyperkeratosis Subcutaneous hemorrages Bleeding swollen gums Hypochondriasis Hematologic abnormalities Ecchymoses Petechiae	Rare Diarrhea Gastrointestinal problems Iron overabsorption Dependency Hyperoxaluria
Thiamin	Coenzyme reactions Ion activity in nerve tissue	Beriberi Fatigue Muscle weakness Convulsions Nerve damage Heart hypertrophy Tachycardia Polyneuritis	No known effects
Riboflavin	Precursor of coenzymes Drug metabolism Lipid metabolism Antioxidant	Eye disorders Brain dysfunction Dry, flaky skin Angular stomatitis Sore, red tongue Increased lipid peroxidation Impaired fat and protein metabolism Cheilosis	No confirmed effects
Niacin	Coenzyme system (electron acceptor or hydrogen donor) Reductive biosyntheses Nonredox reactions	Diarrhea Bright red tongue Mental disorientation Pigmented rash Pellagra Depression Memory loss	Flushed skin Liver damage Ocular abnormalities Hyperuricemia Hyperglycemia Decrease serum cholesterol levels

(*continues*)

TABLE 3 (Continued)

	Functions	Deficiency	Excessive amounts
Vitamin B6	Enzymatic reactions Gluconeogenesis Niacin formation Lipid metabolism Nervous system Immune system Hormone modulation	Convulsions Depression Nausea Greasy, flaky skin Glossitis Irritability Confusion	Neurotoxicity Photosensitivity
Folic acid	Coenzyme reactions Synthesis of adenine compounds Histidine metabolism Methionine synthesis Synthesis of nucleic acids	Elevated plasma homocysteine Abnormal cell division Macrocytic or Megaloblastic anemia Neural tube defects Impaired metabolism of histidine	Rare Mask B_{12} deficiency Interfere with convulsive disorder medications
Vitamin B_{12}	All DNA–synthesizing cells (hematopoietic and nervous systems) Metabolism of folic acid	Anemia Progressive demyelination of nerves Fatigue Elevated methylmalonic acid Sensitive skin Increased levels of homocysteine	No known effects
Biotin	Cofactor in intermediary metabolism (carboxylase) Glucose homeostasis	Rare Scaly skin rash Cerebral atrophy Depression Developmental delay Hypotonia Periorificial dermatitis Conjunctivitis Alopecia Ataxia	No known effects
Pantothenic acid	Fatty acid synthesis Membrane phospholipid synthesis Degradation of amino and fatty acids Amino acid synthesis Hormone synthesis Coenzyme A	Increased sensitivity to insulin Decreased antibody production Melalgea (numbness in toes and burning feet) Headache, fatigue Insomnia	No known effects
Calcium	Formation of bones and soft tissue Blood clotting Transmission of neuromuscular stimili	Bone density Muscle weakness	Kidney stones Poor kidney function Interferes with absorption of other nutrients Constipation Elevated blood pressure
Zinc	Cognition CNS activity Immune system development Maintenance of host defense Transcription Control of specific genes	Birth defects Reproductive failure Retarded growth (dwarfism) Loss of appetite/anorexia Retarded sexual development Decreased wound healing Skin lesions Reduced immune function Seizures	Gastrointestinal irritation Vomiting Decreased HDL Decreased immune function Interferes with copper absorption Sideroblastic anemia
Potassium	Energy management Membrane transport Transfer signals down nerves Contraction of muscles Hormone release	Weakness Fatigue Appetite loss Nausea Hypokalemia Metabolic alkalosis Hypochloremia	If not excreted may cause heart irregularities Hyperkalemia Cellular dehydration
Sodium	Blood pressure regulation	Nausea Dizziness	Fluid retention Swelling

TABLE 3 (Continued)

	Functions	Deficiency	Excessive amounts
		Muscle cramps Limited cardiovascular reserve Increased losses (unable to replace)	Increased blood pressure Hypernatremia
Phosphorus	Bone growth Hormone messengers Regulation of O_2 release from hemoglobin High energy bonds Buffer systems Enzyme regulation	Hypophosphatemia Bone loss Weakness Loss of appetite Rickets	Decreased calcium absorption Decreased renal excretion of calcium Increase of bone loss Tetany–hypocalcemia
Magnesium	Enzymatic reactions Beta oxidation of fatty acids Modulation of cardiac physiology	Hypomagnesemia Irregular heartbeat Hypokalemia Nausea Abnormal neuromuscular function Personality changes Convulsions	Decreased platelet adhesion Nausea/vomiting CNS depression Low blood pressure Coma Hyporeflexia Bradycardia Cardiac arrest
Chromium	Potentiates insulin action Growth Blood lipid profiles Regulation of cell growth Immune response	Increased blood sugar Impaired glucose tolerance Nerve damage Increased cell sensitivity to insulin Weight loss Peripheral neuropathy	Excess is stored in body Allergic contact dermatitis Rare
Copper	Many enzyme systems Antioxidant defense Maturation of collagen and elastin Wound healing Integrity of blood vessels Melanin formation Blood clotting Hormone formation and deactivation	Decreased catecholamine formation Hypochromic, microcytic anemia Cardiac hypertrophy Deficit in energy supply Fragility of cell membranes Shortened lifespan of erythrocytes Accumulation of lipid oxidation products Compromised iron metabolism	Rare Organ scar tissue Reduction in liver function Cell lysis Connective tissue deposition Liver cirrhosis
Fluoride	Prevention of tooth decay Stimulates bone formation Increases trabecular bone density Increases spinal bone density	Weakened tooth enamel Increased tooth decay	Tooth mottling/pitting Skeletal fluorosis Coma and paralysis Brittle bones Cardiac insufficiency Hypomineralization of teeth Excessive salivation Pulmonary distress
Iodine	Development Mediation of thyroid hormone	Reduced fertility Mental retardation Decreased thyroxin production Cretinism Cognitive and neuromuscular impairment Goiter Increased fetal and perinatal mortality	Rare Thyrotoxicosis Toxic nodular goiter Weight loss Tachycardia Muscle weakness
Iron	Hemoglobin formation Myoglobulin (muscle oxygen storage) Cytochromes (oxidative production of energy) Other enzymes	Negative pregnancy outcomes Decreased resistance to infection Anemia Increased absorption of lead Fatigue Impaired regulation of body temperature Reduction in work capacity Impaired psychomotor development Impaired intellectual performance	Metabolic acidosis Can be lethal in infants and children Hemorrhagic necrosis of GI tract Bloody diarrhea Coagulation defects Shock Hemolytic anemia Hemochromatosis

(continues)

TABLE 3 (Continued)

	Functions	Deficiency	Excessive amounts
Manganese	Enzyme activation Metaloenzyme formation	Rare Decreased insulin production Severe ataxia Abnormal lipid metabolism Congenital malformations	Psychiatric problems Severe CNS abnormalities Hyperirritability Hallucinations Incoordination Violent acts
Molybdenum	Enzyme cofactor Steroid binding of glucocorticoid receptor	Hypermethioninemia Hypouricemia Decreased sulfate excretion Coma	Growth depression Anemia Increased incidence of gout Elevated blood uric acid
Selenium	Protecting from lipid peroxidation Normal thyroid hormone formation and function Many enzymatic functions	Altered thyroid hormone metabolism Oxidative injury Increased injury from mercury Alterations in biotransformation enzymes Increase in plasma glutathione Keshan disease Muscle weakness Cardiomyopathy	Nausea, vomiting Abnormalities of nervous system Peripheral neuropathy Fatigue Irritability Loss of hair and nails Skin lesions Tooth decay

serious anemia and progressive demyelination of nerves (Selhub and Rosenberg, 1996).

Excessive intakes of the fat-soluble vitamins pose different problems than those of water-soluble vitamins and toxic doses can be acquired from foods. These effects can be both acute toxicity (if the dose is large enough) or chronic toxicity (because the vitamins are not eliminated and accumulate in the body). Again, occurrences of acute toxicity are rare and usually relate to overdoses of vitamin supplementation or the idea that "if a little is good, a whole lot is considerably better." However, chronic toxicity can occur with ingestion of only slightly elevated amounts over long periods of time. As with the water-soluble vitamins, the effects of deficiency are often more serious than the toxic effects of excessive ingestion.

C. Vitamin A

Reproductive problems (birth defects and fetal resorption) occur with both deficiencies and excessive amounts of the preformed vitamin A (Hofmann and Eichele, 1994). The ingestion of excessive amounts of carotenoids (precursor of vitamin A) does not result in toxicity even though some yellowing of the skin may occur in adults as well as infants (Micozzi *et al.,* 1988). Blindness is one of the most severe and nonreversible results of deficiency (Underwood, 1994). The dose of vitamin A required for producing birth defects is not well defined, but pregnant women and women of childbearing age should use vitamin A supplementation very sparingly. In fact, it is recommended they get the necessary vitamin A from carotenoids and preformed vitamin A in foodstuffs (Underwood, 1986, Teratology Society, 1987).

D. Vitamin K

The effects of vitamin K deficiency are rare, but can manifest as hemorrhagic disease of the newborn and defective coagulation of the blood (Lane and Hathaway, 1985). There are no known toxic effects of the naturally occurring vitamin K. However, excessive amounts of the synthetic form, menadione, have been shown to produce hemolytic anemia and liver damage (NRC, 1987).

Many of the essential minerals are not readily eliminated from the body and can also cause chronic toxicity if consumed in large amounts over time. The minerals are also much more susceptible to interaction with other minerals and elements of the diet than are the vitamins. As in the case of the vitamins, some mineral deficiencies have more serious developmental consequences than the ingestion of the excessive amounts.

E. Zinc

Deficiencies of zinc cause birth defects, dwarfism, and reproductive failure (Clegg *et al.,* 1989; Fosmire, 1990). The most serious consequence of excess is in the interference with copper absorption, causing a sideroblastic anemia that is usually reversed by decreasing the intake of zinc (Fiske, 1994).

F. Iron

A deficiency of iron is one of the most common nutritional disorders seen today (Yip and Dallman, 1996). It results in iron deficiency anemia, which can effect cognition, psychomotor, and intellectual performance (Earl and Woteki, 1993). Some of these effects are not completely corrected with iron supplementation, with adolescents and young adults continuing to show effects of iron deficiency they experienced as infants and children. Excessive doses of iron can be lethal, especially for infants and young children and usually occur when children accidentally take too much of an iron supplement (Yip, 1995). In addition, research is currently underway to determine if hemachromatosis (caused by excessive iron intake over time) has negative effects on coronary artery disease or the risk of developing heart diseases (Giles *et al.,* 1994).

V. Summary

This chapter provides an overview of the problems relating to the assessment of nutrient exposure in humans. An attempt is made to summarize the effects of deficiency and excessive intake of essential nutrients relative to developmental and toxicologic factors and physiologic requirements.

In addition, particular attention and detail are given to the roles of the National Research Council and the Food and Nutrition Board in providing dietary standards and recommendations to promote a healthy population by preventing nutrient deficiency diseases. The role of the USEPA is discussed with regard to providing oral reference dose (RfD) levels for some essential nutrients in order to protect individuals against toxicity or excessive exposure.

The definitions and references provided in this chapter should be helpful to both nutritionists and toxicologists in evaluating relative risks for individuals. There is a need for greater collaboration between the two disciplines to further the health and protection of the public.

References

Beaton, G. H. (1994). Approaches to analysis of dietary data: relationship between planned analyses and choice of methodology. *Am. J. Clin. Nutr.* **59** (suppl), 253S–261S.

Beaton, G. H., Burema, J., and Rittenbaugh, C. (1997). Errors in the interpretation of dietary assessments. *Am. J. Clin. Nutr.* **65** (suppl), 1100S–1107S.

Bingham, S. A., and Day, N. E. (1997). Using biochemical markers to assess the validity of prospective dietary assessment methods and the effect of energy adjustment. *Am. J. Clin. Nutr.* **65,** 1130S–1137S.

Bingham, S. A., Gill, C., Welch, A., *et al.* (1994). Comparison of dietary assessment methods in nutritional epidemiology: weighted records V 24hr recalls, food frequency questionnaires and estimated diet records. *Br. J. Nutr.* **72,** 619–643.

Bjorn-Rasmussen, E., Hallberg, L., Isakson, B., and Arvidsson, B. (1974). Food iron absorption in man: application of the two-pool extrinsic tag method to measure heme and non-heme iron absorption from the whole diet. *J. Clin. Invest.* **52,** 247–255.

Bowman, B. A., and Risher, J. F. (1994). Comparison of the methodological approaches used in the derivation of recommended dietary allowances and oral reference doses for nutritionally essential elements. In *Risk Assessment of Essential Elements.* W. Mertz, *et al.* Eds. ILSI Press, Washington, DC, pp. 63–73.

Canner, P. L., Berge, K. C., Wenger, N. K., *et al.* (1986). Fifteen year mortality in Coronary Drug Project patients: long-term benefit with niacin. *J. Am. Coll. Cardiol.* **8,** 1245–1255.

Clegg, M. S., Keen, C. C., and Hurley, L. S. (1989). Biochemical pathologies of zinc deficiency. In *Zinc in Human Biology.* C. F. Mills, Ed. Springer-Verlag, New York, pp. 129–145.

Czeizel, A. E. (1995). Folic acid in the prevention of neural tube defects. *J. Pediatr. Gastroenterol. Nutr.* **2,** 4–16.

Dourson, M. L. (1994). Methods for establishing oral reference doses. In *Risk Assessment of Essential Elements.* W. Mertz, *et al.* Eds. ILSI Press, Washington, DC, pp. 51–61.

Earl, R., and Woteki, C. Eds. (1994). Iron deficiency anemia: recommended guidelines for prevention, detection and management among U.S. children and women of childbearing age. National Academy Press, Washington, DC.

Fiske, D. N., McCoy, H. E., and Kitchens, C. S. (1994). Zinc-induced sideroblastic anemia: report of a case, review of the literature, and description of the hematologic syndrome. *Am. J. Dermatol.* **46,** 147–150.

Food and Nutrition Board, Institute of Medicine (1997). Dietary Reference Intakes: Calcium, Phosphorus, Magnesium, Vitamin D, and Fluoride. Standing Committee on the Scientific Evaluation of Dietary Reference Intakes. National Academy Press, Washington, DC, 350 pp.

Fosmier, G. J. (1990). Zinc toxicity. *Am. J. Clin. Nutr.* **51,** 225–227.

Gallaher, D. D., and Schneeman, B. O. (1996). Dietary fiber. In *Present Knowledge in Nutrition,* 7th ed. E. E. Ziegler and L. J. Filer, Eds. ILSI Press, Washington, DC., pp. 87–97.

Giles, W. H., Anda, R. F., Williamson, D. F., *et al.* (1994). Body iron stores and the risk of coronary heart disease. *N. Engl. J. Med.* **331,** 1159–1160.

Hallberg, L. (1981). Bioavailability of iron in man. *Annu. Rev. Nutr.* **1,** 123–147.

Hofmann, C., and Eichele, G. (1994). Retinoids in development. In *The Retinoids: Biology, Chemistry, and Medicine,* 2nd ed. M. B. Sporn, A. B. Roberts, and D. S. Goodman, Eds. Raven Press, New York, pp. 387–341.

Kaaks R., Plummer, M., Riboh, E., *et al.* (1994). Adjustment for bias due to errors in exposure assessments in multicenter cohort studies on diet and cancer: a calibration approach. *Am. J. Clin. Nutr.* **59** (suppl), 245S–250S.

Lane, P. A., and Hathaway, W. E. (1985). Vitamin K in infancy. *J. Pediatr.* **106,** 351–359.

Maines, M. D. (1994). Modulating factors that determine interindividual differences in response to metals. In *Risk Assessment of Essential Elements.* W. Mertz, *et al.* Eds. ILSI Press, Washington, DC, pp. 21–39.

Mertz, W. (1991). General considerations regarding requirements and toxicity of trace elements. In *Trace Elements in Nutrition of Children,* Vol. 2., R. K. Chandra, Ed. Raven Press, New York, pp. 1–13.
Mertz, W. (1994). Methodology for establishing recommended dietary allowances and estimated safe and adequate daily dietary intakes. In *Risk Assessment of Essentials Elements.* W. Mertz, *et al.* Eds. ILSI Press, Washington, DC. pp. 43–49.
Micozzi, M. S., Brown, E. D., Taylor, P. R., and Wolfe, E. (1988). Carotenodermia in men with elevated cartenoid intake from foods and B-carotene supplements. *Am. J. Clin. Nutr.* **48,** 1061–1064.
National Research Council. (1943). Recommended Dietary Allowances. Report of the Food and Nutrition Board, Reprint and Circular Series No. 115. National Research Council, Washington, DC, 6 pp.
National Research Council. (1980). Recommended Dietary Allowances, 9th revised ed. Report of the Committee on Dietary Allowances, Food and Nutrition Board, Division of Biological Sciences, Assembly of Life Sciences. National Academy of Sciences, Washington, DC, 185 pp.
National Research Council. (1986). Nutrient Adequacy: Assessment Using food Consumption Surveys. Report of the Subcommittee on Criteria for Dietary Evaluation, Food and Nutrition Board, Commission on Life Sciences. National Academy Press, Washington, DC, 146 pp.
National Research Council. (1987). Vitamin tolerances of animals. National Academy Press, Washington, DC.
National Research Council. (1989). Recommended Dietary Allowances, 10th Ed. Subcommittee on the Tenth Edition of the RDA's, Food and Nutrition Board, Commission on Life Sciences, National Academy Press, Washington, DC, 284 pp.
Pariza, M. W. (1996). Toxic substances in foods. In *Present Knowledge in Nutrition,* 7th Ed., E. E. Ziegler and Filer, L. J. Eds. ILSI Press, Washington, DC, pp. 563–573.
Selhub, J., and Rosenberg, I. H. (1996). Folic acid. In *Present Knowledge in Nutrition,* E. E. Ziegler and Filer, L. J. Eds, ILSI Press, Washington, DC, pp. 206–219.
Teratology Society (1987). Recommendations for vitamin A use during pregnancy (position paper). *Teratology* **35,** 269–275.
Thompson, F. E., and Byers, T. (1994). Dietary assessment resource manual. *J. Nutr.* **124,** (suppl), 2245S–2317S.
Underwood, B. A. (1986). The safe use of vitamin A by women during reproductive years. International Vitamin A Consultative Group, ILSI Nutrition Foundation, Washington, DC.
Underwood, B. A. (1994). Vitamin A in human nutrition: public health considerations. In *The Retinoids: Biology, Chemistry, and Medicine,* 2nd ed. M. B. Sporn, A. B. Roberts, and D. S. Goodman Eds. Raven Press, New York, pp. 211–227.
U.S. Environmental Protection Agency (U.S. EPA). (1993). Integrated Risk Information System (IRIS) (online). Office of Health and Environmental Assessment, Environmental Criteria and Assessment Office, Cincinnati, OH.
Wacholder, S., Schatzkin, A., Freedman, L. S., *et al.* (1994). Can energy adjustment separate the effects of energy from those of specific macronutrients? *Am. J. Epidemiol.* **140,** 848–855.
Yip, R. (1995). Toxicity of essential and beneficial metal ions—iron. In *Handbook on Metal Ligands Interactions of Biological Fluid.* G. Berthen Ed. Marcel Dekker, New York.
Yip, R., and Dallman, P. R. (1996). Iron. In *Present Knowledge in Nutrition,* 7th Ed., E. E. Ziegler and L. J. Filer Eds. ILSI Press, Washington, DC, pp. 277–292.

APPENDIX

Dietary requirement: Amount of the nutrient in the diet that must be consumed to meet the physiologic requirement.

Physiological requirement: The minimal intake to maintain all element-specific functions at near optimal levels.

AI — Adequate Intake: A goal for nutrient intake when sufficient scientific evidence is not available to estimate an average requirement.

RDA — Recommended Dietary Allowance: The level of intake of essential nutrients that, on the basis of scientific knowledge, are judged by the Food and Nutrition Board to be adequate to meet the known nutrient needs of practically all healthy persons.

DRI — Dietary Reference Intake: Reference values that can be used for planning and assessing diets for healthy populations and other purposes.

EAR — Estimated Average Requirement: The nutrient intake value that is estimated to meet the requirement defined by a specified indicator of adequacy in 50% of the individuals in a life-stage and gender group.

UL — Tolerable Upper Intake Level: The maximal level of nutrient intake that is unlikely to pose risks of adverse health effects to almost all individuals in the target group.

RfD — Oral Reference Dose: An estimate of a daily exposure to the human population, including sensitive subgroups, that is likely to be without an appreciable risk of deleterious effects during a lifetime.

ESADDI Estimated Safe and Adequate Daily Dietary Intake: A category established in 1989 for essential nutrients when data were sufficient to estimate a range of requirements, but insufficient for developing an RDA.

Acute toxicity: A toxic response, often immediate, induced by a single exposure.

Chronic toxicity: An effect that requires some time to develop (time may be lessened by increased concentration of dose exposure).

NOAEL No Observed Adverse Effect Level: An exposure level at which there are no statistically or biologically significant increases in the frequency or severity of adverse effects between the exposed population and an appropriate control (sometimes referred to as the highest level of exposure without adverse effect).

LOAEL Lowest Observed Adverse Effect Level: The lowest exposure level at which there are statistically or biologically significant increases in frequency or severity of adverse effects between the exposed population and its appropriate control group.

CHAPTER

21

Drug and Chemical Contaminants in Breast Milk

Effects on Neurodevelopment of the Nursing Infant

JANINE E. POLIFKA
Department of Pediatrics
University of Washington
Seattle, Washington 98195

The period of lactation in humans is often considered the fourth trimester of pregnancy because for 6 months following delivery, the mother continues to be the sole source of nutrition for her infant (Scialli, 1992). Indeed, breast milk has been referred to as *white blood* because of its similarities to the blood that perfuses the placenta in supplying nutrients and immunologic factors (Riordan and Auerbach, 1993). Because of its unique properties, breast milk is considered to be the ideal food to meet the needs of the rapidly growing infant for the first 6 months of life.

In addition to nutrients, breast milk contains minerals, hormones, growth factors, and immunologic factors that impart numerous health benefits to the nursing infant. For example, breast-fed infants have been found to get sick less often, have fewer respiratory and urinary tract infections, fewer episodes of middle ear infections,

---

*Abbreviations: AAP, American Academy of Pediatrics; ADI, allowable daily intake; CNS, central nervous system; DCM, concentration of the substance in breast milk (μg/mL); EPA, [U.S.] Environmental Protection Agency; GI, gastrointestinal; ID, estimated infant dose (mg/kg/d); NRC, National Research Council; PCB, polychlorinated biphenyls; PDI, Psychomotor Development Index; VM, volume of milk ingested by the infant per day on a weight-adjusted basis; WHO, World Health Organization.

and lower rates of mortality than formula-fed infants (Hanson and Bergstrom, 1990; Cunningham *et al.,* 1991; Paradise *et al.,* 1994).

Breast milk also contains various factors such as long-chain polyunsaturated fatty acids (e.g., arachidonic and docosahexaenoic acids) that have been shown to be important in retinal and brain development in both primates and humans (de Andraca and Uauy, 1995). During pregnancy, the placenta supplies the fetus with these essential fatty acids. After delivery, arachidonic and docosahexaenoic acids are transferred to the newborn through the breast milk (Lanting *et al.,* 1994). Few formulas contain these essential fatty acids. However, when docosahexaenoic acid was added to formula in one study, it was found to enhance the maturation of visual acuity in preterm infants (Uauy *et al.,* 1990).

The health benefits of breast feeding have a significant impact on infant mortality rates in underdeveloped countries where unsanitary conditions exist. In these countries, the majority of deaths related to formula feeding are due to diarrheal illnesses (Cunningham *et al.,* 1991). Increased use of breast feeding has been found to be one of the most important factors in reducing infant mortality caused by intestinal infections. In industrialized countries, where water is primarily piped and therefore more sanitary, breast feeding has less of an impact on infant mortality. This is because prepared formulas in these countries are more hygienic and nutritionally complete (Anonymous, 1994a). Nevertheless, even in industrialized countries, the mortality rates among breast-fed infants are significantly lower than those of formula-fed infants (Cunningham *et al.,* 1991). Recognition of these health benefits by policy makers is underscored by the implementation of national mandates in many countries, including the United States, to increase to 50% the proportion of women who continue to breastfeed at 6 months by the year 2000 (Lawrence, 1994).

However, there has been a growing concern among health care practitioners that the presence of drugs and chemicals in breast milk may jeopardize the normal development of the infant brain because exposure to these chemicals takes place during critical periods of brain growth and development. Many studies have found that drug use during the lactation period is high, with as many as 99% of all postpartum women receiving at least one drug during the first week after delivery (Taddio and Ito, 1994). The persistence of pesticides and other heavy metals in the food chain further increases the nursing infant's risk of exposure to potential neurotoxicants. For this reason, many health care professionals believe that women in developed countries who must take medication during the postpartum period or who are exposed to high levels of environmental toxicants should not expose their infants unnecessarily to these substances by breastfeeding. However, is there enough scientific evidence to justify this course of action?

Published information on the possible effects that breast milk contaminants may have on infant development is extremely dismal. Most of the reports that have been published are largely anecdotal. On the basis of these reports, it appears that the nursing infant is exposed to a small percentage (<5%) of the amount of substance that the mother is. However, for the nursing infant, whose brain is rapidly developing, this small amount may nevertheless be enough to alter development permanently. It is possible that the protective factors in breast milk may negate or attenuate any adverse effects associated with breast milk contaminants. However, few lactation studies have been undertaken to prove this. On the basis of such little information, how can mothers and clinicians be expected to make informed decisions?

The purpose of this chapter is to review what is known about the risks and benefits of breast feeding on infant neurodevelopment. Particular emphasis is placed on the potential risks of lactational exposure to drugs and chemicals that are known to have an affinity for the central nervous system (CNS). Because nutritional requirements as well as the capacity to absorb nutrients in the perinatal period of humans differ from rodents and other laboratory animals, only human studies are considered.

I. Benefits of Breast Feeding on Neurodevelopment of Nursing Infants

Many studies have found that breast milk confers an advantage to nursing infants in subsequent cognitive development and that this advantage becomes greater as the period of breast feeding increases. These studies have been critically reviewed by de Andraca and Uauy (1995). Most of the studies found that children who were breast fed had higher scores on tests of psychomotor development than children who were formula fed when tested during infancy and preschool. This was true even after adjustment for confounding factors, such as socioeconomic status, parental education, and maternal age. Breast-fed children continued to exhibit advantages over formula-fed children in cognitive and language skills at 8 and 15 years of age, but the difference became less marked as the children grew older (de Andraca and Uauy, 1995). Menkes (1977) compared 29 children who had learning disorders with 53 children who had other types of neurologic dysfunction and found a higher proportion of formula-fed infants among children with learning disorders than among children with other neurologic abnormalities. A small beneficial effect of breast feeding on the postnatal neurologic development of children was also found by Lanting *et al.* (1994).

The mechanism by which breast feeding benefits the neurodevelopment of children is not known. Methodologic limitations in studies on neurodevelopment limit the ability to make causal inferences on the basis of the results found. Neurodevelopment spans a long period of time (roughly 2 years) and therefore can be influenced by many intervening variables, such as the socioeconomic status, skills, attitudes, personalities, and educational levels of the parents as well as the quality of home environment and family size. In a study by Gale and Martyn (1996), it was found that adults who had been breast fed exclusively in infancy had slightly higher IQ scores than adults who had been formula fed or who had been fed by both methods. This advantage did not hold up, however, when other factors, such as use of a pacifier, number of older siblings, father's occupational class, and mother's age, were controlled for. Failure to control for these variables introduces biases in interpretation, usually in favor of breastfed infants (Anonymous, 1994a).

For obvious ethical reasons, women cannot be assigned randomly to breast feed or formula feed their infants. Therefore, breast-fed infants may differ from formula-fed infants in neurodevelopment for reasons other than exposure to a particular contaminant in their mothers' milk. For example, breast-fed infants tend to come from small, health-conscious families with relatively high standards of living (Rogan and Gladen, 1985; Cunningham *et al.,* 1991). The olfactory, tactile, and visual stimuli experienced by breast-fed infants may differ from those experienced by formula-fed infants, leading to more favorable transitions in subsequent development. Breast milk is initially low in fat, and therefore breast-fed infants spend more time feeding than formula-fed infants. Longer time at the breast presumably increases the period of maternal–infant interaction (de Andraca and Uauy, 1995). Mothers choose to formula feed for a variety of reasons, including depression, poor health, employment outside the home, and insufficient milk volume. Breast-fed infants, in general, may be healthier because infants who fail to gain weight or become ill are usually given supplements or are weaned. Any adequate study of the neurodevelopment of breast-fed infants must control for these many confounding variables.

Another methodologic limitation of breast feeding studies is that the duration and exclusivity of breast feeding cannot be controlled. In some hospitals, breast-fed neonates are routinely supplemented with other liquids (Anonymous, 1994a; Levitt *et al.,* 1996). Also, weaning practices vary considerably among women. These hospital and weaning practices tend to blur any distinctions that may exist between breast-fed and formula-fed infants. Finally, breast milk constituency varies among individuals. The experimental focus may be on one particular breast milk contaminant, but other putative contaminants present in the breast milk may also influence the outcome.

II. Exposure of the Nursing Infant to Drugs in the Breast Milk

Because of their low molecular weight and high lipid solubility, the majority of drugs and chemicals are excreted into breast milk. As a result, the nursing infant receives any drug or chemical that the mother is exposed to. The question is how much of the drug or chemical does the nursing infant ingest? It is not uncommon to come across the statement that a particular substance is "excreted in breast milk," as if to imply that the amount found in the breast milk is equivalent to the amount ingested by the mother. It is important to remember, however, that breast milk represents only one of many tissues or "compartments" in the body, and therefore the total amount ingested is diluted. The degree of dilution depends on the distribution space, or the number of "compartments" that the substance is distributed to. Substances with a high volume of distribution are generally found in small concentrations in plasma and breast milk (Atkinson *et al.,* 1988).

The rate of distribution into breast milk is effected by the physical characteristics of a substance, such as lipid solubility, degree of ionization, binding to plasma proteins, and molecular weight (Atkinson *et al.,* 1988). Lipid-soluble substances tend to be distributed in hydrophobic sites, such as adipose tissue and breast milk. Highly polar or charged substances tend to be distributed in body water, and weak bases, such as the beta-blockers atenolol and metoprolol, concentrate in milk (Scialli, 1992).

Molecules of substances that are highly bound to proteins in plasma are too large to diffuse into breast milk (Scialli, 1992). In this case, peak levels of a substance in plasma are much higher than those found in breast milk. In addition, distribution of substances to various body tissues is not always a one-way process. For example, substances that are rapidly eliminated from plasma via renal excretion or hepatic metabolism may diffuse back from breast milk to plasma if nursing does not occur for several hours (Wilson *et al.,* 1985).

Plasma levels of a substance are constantly changing as the substance is distributed throughout the body and eliminated. Consequently, single measures of a substance in milk or plasma do not accurately reflect the total amount that is transferred from the mother to the nursing infant. In addition, the amount that is actually delivered to the nursing infant is related to the concentration of the substance in the milk and the volume of breast milk that is ingested. The dose of a substance, then, can be estimated using the following formula:

$$ID = DCM \times VM$$

where ID is the estimated infant dose in mg/kg/d, DCM is the concentration of the substance in breast milk in μg/mL, and VM is the volume of milk ingested by the infant per day on a weight-adjusted basis. The volume of milk ingested by the nursing infant usually ranges between 600–1000 mL per day, depending on the age and size of the infant. In the case of medications, this dose can then be expressed as a proportion of the lowest therapeutic dose used in adults or children. For example, if the lowest therapeutic dose of a drug for adults is 320 mg/d and the concentration in the breast milk is 0.01 mg/mL, the dose received by a nursing infant weighing 4 kg and consuming 600 mL of milk per day would be:

$$0.01 \text{ mg/mL} \times 150 \text{ mL/kg/d} = 1.5 \text{ mg/kg/d}$$

Assuming the mother weighs 64 kg, the dose equivalent to a therapeutic dose in the infant would be:

$$\frac{320 \text{ mg/d}}{64 \text{ kg}} = 5 \text{ mg/kg/d}$$

The dose ingested by the infant would therefore be:

$$\frac{1.5 \text{ mg/kg/d}}{5 \text{ mg/kg/d}} = 0.3, \text{ or } 30\% \text{ of the lowest therapeutic dose for adults}$$

However, the amount of a substance ingested by the nursing infant is also affected by the bioavailability of a substance. For example, some drugs such as aminoglycoside antibiotics are poorly absorbed by the gastrointestinal (GI) tract. Thus, although drugs such as these may produce local effects on the GI system, such as alteration of bowel flora, they are not able to produce systemic effects in the nursing infant.

Chronic exposures to substances may cause the substance to accumulate in the nursing infant, particularly in preterm or newborn infants. During the first 2 weeks of life, hepatic metabolism is slow. Thus, although the amount of substance ingested by the nursing infant may be small, accumulation may occur as the substance is slowly and incompletely eliminated. After the initial 2-week period, the average infant is capable of metabolizing drugs at a rate that is similar to or faster than adults (Lawrence, 1994). The presence of active metabolites in milk must also be taken into consideration. Premature infants typically have low serum albumin. Therefore, substances such as sulfonamide antibiotics that are capable of displacing bilirubin from albumin have the *theoretic* potential for increasing circulating levels of bilirubin in premature infants at low levels (Pons *et al.*, 1994). Nevertheless, kernicterus resulting from bilirubin entering the CNS has not been reported in breast-fed infants.

III. Effects of Maternally Ingested Drugs on the Nursing Infant

Very few epidemiologic studies have been conducted to investigate the effects of drug exposure on the nursing infant. Most of the published data are based on case reports or small clinical series. This makes it difficult to know to what extent the findings in these reports can be generalized to all nursing infants. In the majority of reports, concentrations of a drug in maternal plasma and milk are often measured at a single point in time, thereby providing a crude and often misleading estimate of the amount of the drug that is actually available to the nursing infant. Concentrations in milk and plasma differ as a function of time because of several factors: (1) variability in maternal plasma levels relative to when the drug was taken and the milk sampled; (2) equilibrium of maternal plasma levels relative to whether the drug exposure is acute or chronic and the consequent changes in metabolism and excretion; and (3) variability in breast milk levels relative to the drug's volume of distribution, interval between infant feedings, and maturity of milk (Friedman and Polifka, 1996).

Without actual measurements of the drug in the infant's plasma it is difficult to get an accurate picture of how much of the drug that is excreted in the breast milk is actually absorbed by the infant. Generally, less than 5% of the maternal dose of a drug is ingested by the nursing infant. The infant doses provided in Tables 1–4 are calculated from concentrations measured in the breast milk. These doses are estimates only, and the effect of the ingested drug on the nursing infant may be modified by the infant's age, medical condition, ability to metabolize and excrete the drug, and supplementation with other foods. Finally, in many of the published reports measurements of drug concentrations are taken from lactating women who have chosen not to breast feed. As a result, very little information on symptoms in breast-fed infants is available.

A. *Antidepressants*

As many as 80% of women experience postpartum depression, with 20% of them requiring treatment with antidepressants or minor tranquilizers during the first few months following delivery (Pons *et al.,* 1994). Although many of the tricyclic antidepressants have been on the market since the mid-1960s, very little data has been published on the effects of antidepressants on the nursing infant. Table 1 summarizes the nursing infant's estimated daily intake of commonly used antidepressant medications based on breast milk concentrations reported in published case reports. The infant's daily intake is compared to the lowest therapeutic dose of the medication in children (when available) or in adults. It can be seen that for the majority of antidepressants studied, the amount of medication that is ingested by the nursing infant through the breast milk is very small. However, respiratory depression and sedation were observed in one infant whose mother was treated with doxepin three times a day while breast feeding. Serum levels of the metabolite of doxepin (*N*-desmethyldoxepin) in the nursed infant were found to be three times as high as the lowest therapeutic serum level of *N*-desmethyldoxepin in adults (0.02 μg/mL) (Matheson *et al.,* 1985). Serum levels of doxepin were below detection limits in another infant whose mother was treated with doxepin while breastfeeding (Kemp *et al.,* 1985). This infant's serum level of *N*-desmethyldoxepin (0.01 μg/mL) was close to the lowest therapeutic serum level of N-desmethyldoxepin in adults. No sedation was noted in the infant, who was also reported to have achieved motor developmental milestones at normal rates.

Serum levels of fluoxetine and its metabolite, norfluoxetine, were within the therapeutic range for adults in one infant (Lester *et al.,* 1993). Vomiting, excessive crying, watery stools, and decreased sleep were observed in this infant. These symptoms disappeared when the mother began formula feeding and reappeared when the infant was put to the breast again while the mother continued to take the medication.

Serum levels in one infant whose mother was treated with clomipramine while breast feeding were found to be between 7–60% of the lowest therapeutic serum levels in adults (0.15 μg/mL) (Schimmel *et al.,* 1991). In another study, neither clomipramine nor its metabolites could be detected in the sera of four nursing infants whose mothers were taking clomipramine (Wisner *et al.,* 1995). Detectable limits of amitriptyline and nortriptyline were not found in the sera of nursing infants whose mothers were treated with these drugs while breast feeding (Erickson *et al.,* 1979; Bader and Newman, 1980; Brixen-Rasmussen *et al.,* 1982; Wisner and Perel, 1991; Altshuler *et al.,* 1995); although very low concentrations of nortriptyline's metabolite, 10-hydroxynortriptyline, could be measured in two of the infants in one small series (Wisner and Perel, 1991). No adverse effects were observed in these two infants. The American Academy of Pediatrics (AAP) has published a list of 192 drugs that have been reported to be transferred in the breast milk and that may or may not have significant effects on the nursing infant (Committee on Drugs, American Academy of Pediatrics, 1994). The AAP rec-

TABLE 1 Estimated Daily Intakes of Commonly Used Antidepressants by the Nursing Infant

Drug	Number of subjects	Breast milk concentrations (μg/mL)	Estimated daily intake by infant (mg/kg/d)[a]	Percent of lowest therapeutic dose[b,c]	References
Amitriptyline	4	0.02–0.15	0.004–0.02	0.3–1.8[b]	Bader and Newman, 1980; Brixen-Rasmussen *et al.,* 1982; Pittard and O'Neal, 1986; Breyer-Pfaff *et al.,* 1995.
	6	Not detectable	—	—	Eschenhof and Rieder, 1969.
Clomipramine	1	0.22–0.62	0.03–0.09	2.5–7.5[b]	Schimmell *et al.,* 1991.
Doxepin	2	<0.02–0.22[d]	<0.003–0.03	<1–21[b]	Kemp *et al.,* 1985; Matheson *et al.,* 1985.
Fluoxetine	13	0.03–0.38[d]	0.003–0.06	<1–18[b]	Isenberg, 1990; Burch and Wells, 1992; Lester *et al.,* 1993; Taddlo *et al.,* 1994.
Imipramine	1	0.021–0.06[d]	0.004–0.01	<1[c]	Sovner and Orsulak, 1979.
Nortriptyline	1	0.09–0.4	0.01–0.06	1–6[c]	Matheson and Skjaeraasen, 1988.
Paroxetine	1	0.01	0.001	<1[b]	Spigset *et al.,* 1996.
Sertraline	1	0.01–0.04	0.001–0.01	<1[b]	Altshuler *et al.,* 1995.

[a]Based on assumption that the nursing infant consumes 600 mL breast milk per day and weighs 4 kg.
[b]In adults (USP DI, 1997).
[c]In children (USP DI, 1997).
[d]For both parent compound and active metabolite.

ommends that antidepressants be used with caution in breast feeding mothers because these drugs could "conceivably alter short-term and long-term central nervous system function" in the nursing infant (Committee on Drugs, American Academy of Pediatrics, 1994, p. 139).

B. Antipsychotics

Table 2 summarizes the nursing infant's estimated daily intakes of the antipsychotics, chlorpromazine, haloperidol, and lithium. Based on published data from seven lactating women, the amount of chlorpromazine or haloperidol that the nursing infant would be expected to ingest is less than 12% of the lowest therapeutic pediatric dose. The AAP recommends that chlorpromazine and haloperidol be used with caution in breast-feeding mothers because these drugs could "conceivably alter short-term and long-term central nervous system function" in the nursing infant (Committee on Drugs, American Academy of Pediatrics, 1994, p. 139). No adverse effects were observed in exposed infants, although drowsiness and lethargy were noted in one infant who was nursed (Wiles *et al.,* 1978).

Lithium has been found to be excreted in higher concentrations in breast milk than chlorpromazine and haloperidol (Schou and Amdisen, 1973; Sykes *et al.,* 1976). Serum levels measured in seven of the infants whose mothers were treated with lithium were found to range between 3.7–22.5 μg/mL. The upper limit of this range of concentrations is within the range of therapeutic serum concentrations of lithium in adults (15–37 μg/mL) (Schou and Amdisen, 1973; Sykes *et al.,* 1976). Because high serum concentrations have been measured in some nursed infants, the AAP considers lithium to be contraindicated in breast-feeding women (Committee on Drugs, American Academy of Pediatrics, 1994).

C. Anticonvulsants

With the exception of phenobarbital and ethosuximide, the anticonvulsants listed in Table 3 are excreted in small concentrations in the breast milk. Therefore, the nursing infant's daily intake of these medications has been estimated to be small. Infant serum concentrations of clonazepam and phenytoin have been found to be low (<5% of the lowest therapeutic serum level in adults). However, infant serum concentrations of carbamazepine were reported to be between 25–45% of the lowest therapeutic serum level in adults. Likewise, serum levels of diazepam in nursing infants have been found to be high, varying between less than 0.01–0.2 μg/mL. The maximum concentration observed in infant serum is within the range of therapeutic serum concentrations found in adults (0.07–2.8 μg/mL). In addition, infant serum concentrations of desmethyldiazepam were found to be similar to those of diazepam.

Phenobarbital and ethosuximide are excreted in breast milk in high concentrations (see Table 3). The amount of phenobarbital or ethosuximide that the nursing infant would be expected to ingest is between 2–163% or 13–58%, respectively, of the lowest therapeutic pediatric dose. Serum levels of ethosuximide in the nursed infants of treated women ranged between 26–100% of the lowest therapeutic serum concentration of ethosuximide in adults.

Sedation occurred in one infant shortly after the mother took diazepam (Wesson *et al.,* 1985), and mild jaundice was reported in three other infants (Cole and Hailey, 1975). Lethargy and weight loss that resolved after discontinuation of breast feeding were described in a nursing infant whose mother had taken diazepam (Patrick *et al.,* 1972). An infant whose mother took phenobarbital chronically for a seizure disorder exhibited withdrawal symptoms when breast feeding was abruptly

TABLE 2 Estimated Daily Intakes of Commonly Used Antipsychotics by the Nursing Infant

Drug	Number of subjects	Breast milk concentrations (μg/mL)	Estimated daily intake by infant (mg/kg/d)[a]	Percent of lowest therapeutic pediatric dose[b]	References
Chlorpromazine	5	0.007–0.29	0.001–0.04	0.03–1.3	Blacker *et al.,* 1962; Wiles *et al.,* 1978.
Haloperidol	2	0.002–0.02	0.0003–0.003	<1–12	Stewart *et al.,* 1980; Whalley *et al.,* 1981.
Lithium	9	4.5–28	0.7–4.5	5–30	Schou and Amdisen, 1973; Sykes *et al.,* 1976.

[a]Based on assumption that the nursing infant consumes 600 mL breast milk per day and weighs 4 kg.
[b](USP DI, 1997).

TABLE 3 Estimated Daily Intakes of Commonly Used Anticonvulsants by the Nursing Infant

Drug	Number of subjects	Breast milk concentrations (μg/mL)	Estimated daily intake by infant (mg/kg/d)[a]	Percent of lowest therapeutic dose[b,c]	References
Carbamazepine	54	1.3–4.8	0.2–0.7	2–7.2[c]	Pynnonen and Sillanpaa, 1975; Pynnonen *et al.*, 1977; Kaneko *et al.*, 1979; Kok *et al.*, 1982; Kuhnz *et al.*, 1983; Froescher *et al.*, 1984; Merlob *et al.*, 1992.
Clonazepam	1	0.011–0.013	0.002	16[b]	Fisher *et al.*, 1985.
Diazepam	18	0.01–0.31	0.003–0.07	<1–6[c]	Erkkola and Kanto, 1972; Cole and Hailey, 1975; Brandt, 1976; Wesson *et al.*, 1985; Dusci *et al.*, 1990.
Ethosuximide	21	18–77	2.7–11.5	13–58[c]	Koup *et al.*, 1978; Kaneko *et al.*, 1979; Rane and Tunell, 1981; Kuhnz *et al.*, 1984.
Phenobarbital	8	0.5–33	0.07–4.9	2.3–163[c]	Kaneko *et al.*, 1979.
Phenytoin	?	0.26–4	0.04–0.6	1–15[c]	Kaneko *et al.*, 1979, 1981; Chaplin *et al.*, 1982; Steen *et al.*, 1982.
Valproic acid	31	0.17–7.2	0.026–1.1	<1–7[c]	Espir *et al.*, 1976; Alexander, 1979; Dickinson *et al.*, 1979; Nau *et al.*, 1981; Bardy *et al.*, 1982; von Unruh *et al.*, 1984.

[a]Based on assumption that the nursing infant consumes 600 mL breast milk per day and weighs 4 kg.
[b]In adults (USP DI, 1997).
[c]In children (USP DI, 1997).

terminated at 7 months (Knott *et al.*, 1987). The infant's withdrawal symptoms resolved following treatment with phenobarbital and gradual withdrawal of the drug over a period of 9 months.

Although valproic acid has been found to be excreted in breast milk in low concentrations, serum levels of valproic acid (0.43–64.9 μg/mL) in seven of the infants studied fell within the therapeutic range of valproic acid serum levels (30–100 μg/mL) for adults.

The AAP considers carbamazepine, ethosuximide, phenytoin, and valproic acid to be compatible with breast feeding. Diazepam is listed as a drug that could potentially affect the developing CNS, and phenobarbital is listed as a drug that should be given to nursing mothers with caution (Committee on Drugs, American Academy of Pediatrics, 1994).

D. Recreational Drugs

Published breast milk concentrations of some commonly abused drugs and the estimated daily intakes by nursing infants can be found in Table 4. Although alcohol is absorbed in small amounts by the nursing infant, sedation is possible if the mother drinks a large amount. Psychomotor development, as assessed by the Psychomotor Development Index (PDI), was found to be significantly lower in nursing infants whose mothers consumed at least one alcoholic drink per day than those whose mothers consumed less (Little *et al.*, 1989, 1990). This effect persisted even after prenatal exposure to alcohol was controlled for, and was enhanced when nursing infants that had received some supplementation with formula were excluded from the analysis.

No information is available on the effects of maternal amphetamine abuse on the nursing infant, but adverse effects were not observed in one nursed infant whose mother was treated with dextroamphetamine while she was breastfeeding (Steiner *et al.*, 1984). Neurobehavioral development of the infant was said to be normal at 2 years of age.

Cocaine has been reported to be excreted in low concentrations (Chasnoff *et al.*, 1987), but actual values have not been published. Cocaine toxicity (vomiting, diarrhea, seizures, irritability, tachycardia, tachypnea, and tremulousness) have been described in two infants breast fed by mothers who used cocaine either intranasally or topically prior to breast feeding (Chasnoff *et al.*, 1987; Chaney *et al.*, 1988).

Marijuana and its metabolites, 11-hydroxy-tetrahydrocannabinol and 9-carboxy-tetrahydrocannabinol, are excreted in breast milk in small amounts. No adverse effects were observed in two infants whose mothers smoked between one and seven marijuana pipes per day (Perez-Reyes and Wall, 1982). Nursing infants of

TABLE 4 Estimated Daily Intakes of Recreational Drugs by the Nursing Infant

Drug	Number of subjects	Breast milk concentrations (μg/mL)	Estimated daily intake by infant (mg/kg/d)[a]	Relative dose[b,c,d,e]	References
Alcohol	11	35–410	5.2–61.5	4–11	Flores-Huerta *et al.,* 1992.
Amphetamine	1	0.055–0.138	0.008–0.02	10–25[b]	Steiner *et al.,* 1984.
Marijuana	2	0.06–0.34	0.01–0.05	3–17[c]	Perez-Reyes and Wall, 1982.
Nicotine	?	0.0005–0.2	0.0003–0.03	1–100[d] 1–150[e]	Hardee *et al.,* 1983; Trundle and Skellern, 1983; Luck and Nau, 1984; Woodward *et al.,* 1986; Piazza *et al.,* 1987; Dahlstrom *et al.,* 1990.

[a]Based on assumption that the nursing infant consumes 600 mL breast milk per day and weighs 4 kg.
[b]Percent of lowest therapeutic dose of amphetamine in adults.
[c]Percent of lowest pharmacologically active dose of marijuana in children (0.3 mg/kg/d).
[d]Percent of lowest adult dose of nicotine chewing gum.
[e]Percent of amount of nicotine obtained from smoking one cigarette (0.02 mg/kg).

women who smoked marijuana daily were found to have significantly decreased motor development at 1 year of age compared to nursed infants of women who did not smoke marijuana in one study (Astley and Little, 1990), but not in another (Tennes *et al.,* 1985). No effect on growth or other aspects of mental development was observed in the nursed infants of women who smoked marijuana daily.

Large amounts of nicotine and its metabolite, cotinine, can be ingested by nursing infants of women who smoke (see Table 4). Although mothers who smoke have been found to choose formula feeding over breast feeding more frequently than women who do not smoke, formula-fed infants whose mothers smoked during and after pregnancy had increased levels of nicotine and cotinine in their urine (Dahlstrom *et al.,* 1990). However, breast-fed infants had much higher levels.

IV. Effects of Maternal Exposure to Environmental Contaminants on the Nursing Infant

A. *Polychlorinated Biphenyls (PCBs)*

Polychlorinated biphenyls (PCBs) are components of a class of halogenated aromatic hydrocarbons that were first manufactured in the United States in 1930. PCBs were used extensively in the electrical industry, in die and machine cutting oils, as heat-exchange fluids in the preparation of edible oils, as well as in materials such as paints, plastics, and sealants (Hutzinger *et al.,* 1974). PCBs were widely distributed throughout the world until their use became severely restricted in the 1970s. Currrent use of PCBs is limited to closed systems in the electrical industry, such as transformers and condensers (Dewailly *et al.,* 1996).

PCB congeners are highly lipophilic and hydrophobic. Lipophilic substances are metabolized by enzyme systems in the liver that reduce the parent compounds to more water-soluble compounds that can be eliminated by the kidneys. However, PCBs contain halogen atoms, which inhibit oxidative metabolism and reduce the rate of enzymatic degradation (Sim and McNeil, 1992). As a result, PCB compounds tend to persist in the body. The half-life of some PCB congeners can be as long as 8.4 years (Sim and McNeil, 1992). Consumption of fat-containing foods, particularly fish, is one of the more prevalent sources of human PCB exposure (Albers *et al.,* 1996; Dewailly *et al.,* 1996).

Because of widespread industrial use and their biochemical stability, PCBs continue to persist and bioconcentrate in the food chain. Although PCBs have low acute toxicity, there is a concern that these substances may have subtle, long-term effects on humans. Dermatologic, immuno- and heptatotoxic, neurobehavioral, and reproductive effects as well as alterations in thyroid hormone status associated with exposure to PCBs have been described in laboratory and human studies (Gellert and Wilson, 1979; Kimbrough and Jensen, 1989; Gallo *et al.,* 1991; Ahlborg *et al.,* 1992; Seegal and Shain, 1992; Koopman-Esseboom *et al.,* 1994; Tryphonas, 1995).

Teratogenic effects have been observed among infants of women poisoned with PCBs in two major industrial accidents occurring in Japan in 1968 and Taiwan in 1978, known as the Yusho and YuCheng incidents. In these incidents, approximately 4000 people were poisoned when they consumed rice oil contaminated with large amounts of PCBs and related compounds (Higuchi, 1976; Wong and Hwang, 1981; Seegal and Shain,

1992). Infants of women who had been poisoned during pregnancy tended to be small in size and to be born prematurely (Funatsu *et al.,* 1972). Higher rates of mortality, lower body weights, and dermal evidence of poisoning occurred in infants whose mothers were poisoned with PCB-contaminated oil during pregnancy and breast feeding than in infants whose mothers were not poisoned (Yamaguchi *et al.,* 1971). Follow-up studies showed that exposed children continued to be small in size and exhibited dermal, nail, and gingival abnormalities; apathy; and hypotonia. These children also suffered from respiratory infections, such as pneumonia and bronchitis, more often than nonexposed children. Exposed children also were delayed in psychomotor development and performed significantly lower on standardized intelligence tests than nonexposed children (Harada, 1976; Rogan *et al.,* 1988). In these two incidents, the PCBs were contaminated with dibenzofurans, which are highly toxic and possibly responsible for some of the developmental effects observed.

Higher incidences of abnormalities in growth, reflex and tone, psychomotor development, and short-term memory function have been found among infants and children of women with high background levels of PCBs at term (Jacobson *et al.,* 1984, 1985, 1990a,b; Rogan *et al.,* 1986a; Gladen *et al.,* 1988; Jacobson and Jacobson, 1988; Koopman-Esseboom *et al.,* 1996). However, follow-up studies of these children show that the neurodevelopmental deficits seen at earlier ages do not persist (Gladen and Rogan, 1991; Koopman *et al.,* 1996).

Persistent lipophilic substances such as PCBs accumulate in fatty tissues of the body. Breast milk, which contains 3% fat, is a major route of excretion for these chemicals; consequently, the maternal body burden of PCBs is reduced during lactation (Yakushiji *et al.,* 1984; Duarte-Davidson and Jones, 1994; Dahl *et al.,* 1995). The average concentration of PCBs in breast milk has been found to decrease with duration of breastfeeding and number of breastfeeding episodes, but not with parity *per se* (Jensen, 1983; Rogan *et al.,* 1986b; Dewailly *et al.,* 1996). Dewailly *et al.* (1996) found statistically significant differences in breast milk concentrations of PCBs between primiparous and multiparous women who had breastfed in previous pregnancies, but not between primiparous and multiparous women who had never breast-fed their infants.

Transfer of PCBs through the breast milk lowers the body burden of the mother and can substantially increase the body burden of PCBs in their infants. Absorption of PCBs in the digestive tract of the nursing infant has been found to be greater than 90% for most of the congeners analyzed (Kodama and Ota, 1980; McLachlan, 1993; Dahl *et al.,* 1995). In one study (Yakushiji *et al.,* 1984), infants whose mothers had high blood levels of PCBs and who breast fed for more than 3 months were found to have blood PCB levels that were as high or higher than that of the mother. In other studies, the infant's daily or weekly intake of PCBs through breast milk was estimated to exceed the safe level established by regulatory agencies for adults (Rogan *et al.,* 1985; Johansen *et al.,* 1994; Becher *et al.,* 1995; Quinsey *et al.,* 1995). Although regulatory levels pertain to adult exposures over a lifetime, it is possible that the breast-fed infant may be at a special risk for toxic exposures to PCB.

The possible adverse effects of lactational exposure to PCBs on the cognitive development of nursing infants have been evaluated in several large cohorts of women who were exposed to high levels of PCBs in their diet (Jacobson *et al.,* 1990a,b; Koopman-Esseboom *et al.,* 1996) or occupation (Yakushiji *et al.,* 1984), or who had no special exposures (Rogan and Gladen, 1985, 1991; Rogan *et al.,* 1987; Gladen *et al.,* 1988) throughout pregnancy. Breast-fed infants were compared to formula-fed infants in these cohorts and followed up developmentally from birth to 5 years of age. No significant influence of lactational exposure to PCBs on body weight and height, psychomotor, or mental development could be found in either of these studies. In one study, however, breast milk levels of PCB and breast-feeding duration were associated with decreased spontaneous activity in 4-year-old children in a dose-dependent fashion, with the largest effect being seen in a subgroup of five children whose mothers had above-average levels of PCB in their breast milk and who had breast fed for at least 1 year (Jacobson *et al.,* 1990a). However, the authors concluded that this effect on activity was probably due to confounding effects associated with transplacental transfer of high levels of PCBs, because the mothers of these children also had the highest blood PCB levels and body burden levels (Jacobson, 1992).

On the basis of these studies it can be concluded that, although much larger amounts of PCBs are transferred through the breast milk than through the placenta, the greatest risks to the developing child are associated with exposure during pregnancy and not during breast feeding. Therefore, breast feeding is not contraindicated in the presence of detectable levels of PCBs in the breast milk. The AAP recommends that women not be discouraged from breast feeding because of exposure to PCBs (Anonymous, 1994b). Furthermore, the AAP does not recommend that breast milk be tested for PCBs because current laboratories in North America do not have standard procedures to determine PCB levels and because no "normal" or "abnormal" values have been established for clinical interpretation (Anonymous, 1994b). Because of the enormous variability found in breast milk samples over the course of lactation, the potential for misinterpretation of test results is high.

B. Lead

Lead is another environmental pollutant that is ubiquitous in the environment. Prior to 1970, leaded gasoline accounted for the majority of ambient lead levels with an estimated 500 million pounds being released into the environment from this source alone in 1968 (Schardein, 1993). Other sources of lead include water, soil, cans, paints, plaster, pipes, solder, newsprint, and ceramic glazes. Since federal guidelines were passed in the 1970s to remove 99.8% of lead from gasoline as well as to remove lead from soldered cans, blood lead levels of people between the ages of 1 and 74 years in the United States were found to have declined as much as 78% between the period of 1976 and 1991 (Pirkle *et al.,* 1994). Nevertheless, lead continues to be present in low levels from many other sources. For example, despite regulations in the United States to limit the amount of lead in household paints to 0.06%, millions of homes are known to still contain leaded paint (Cory-Slechta, 1996). Lead continues to be a major public health problem in many third-world countries where leaded gasolines, water, plumbing, and household supplies remain in wide-spread use (Cory-Slechta, 1996).

Approximately 2% of the total body burden of lead is found in the red blood cells and another 3% in the soft tissues of the brain, kidneys, and bone marrow. The remaining 95% is stored in bone and teeth (Tellier and Aronson, 1994). The major target organs of lead toxicity are the brain, hematopoetic and GI systems, and kidneys. The half-life of lead is 20–30 years in bone, 30–40 days in blood and soft tissue, and 13 weeks in breast milk (Tellier and Aronson, 1994). Lead is a well-known neurotoxicant in children when blood lead levels rise above 80 μg/dL (Tellier and Aronson, 1994). Typical manifestations of lead toxicity are anemia, ataxia, coma, and seizures (Anonymous, 1996; Dabney, 1997). Chronic low-level lead exposures in children have been associated with learning disabilities and lower scores on standardized intelligence tests (Bellinger *et al.,* 1984; Silva *et al.,* 1986; Schwartz and Otto, 1987; Needleman *et al.,* 1990; Winneke *et al.,* 1990; Hammond and Dietrich, 1991; Baghurst *et al.,* 1992; Bellinger, 1994). After reviewing the literature on the neurobehavioral effects of chronic low-level lead exposures in children, the United States Environmental Protection Agency (EPA) concluded that blood lead levels between 50–75 μg/dL are associated with a 5-point reduction in IQ, even when controlling for potential confounding factors (US EPA, 1986).

As stated previously, 95% of an adult's body burden of lead is stored in the bone. Thus, physiologic conditions that increase bone turnover and remodeling, such as osteoporosis, thyrotoxicosis, pregnancy, and lactation, can mobilize lead from bone and divert it to other tissues and organs. There is concern, then, that breast milk may serve as an additional source of lead for the developing infant. Although many studies have been published that report breast milk concentrations of women residing in different countries, none of these studies report on the effects of exposure to lead through breast milk on the nursing infant. Breast milk concentrations of lead vary widely among women in the general population, with levels ranging as high as 0.35 μg/mL for women living near smelters in Mexico City (Namihira *et al.,* 1993). In the United States, the highest breast milk concentration reported was 0.07 μg/mL during the period of 1971–1972 (Lamm *et al.,* 1973). Mean breast milk levels of lead in the general population have demonstrated a gradual decline from 0.02 μg/mL to 0.001 μg/mL (Tellier and Aronson, 1994). Using the figure of 0.001 μg/mL, it can be estimated that an exclusively breast-fed infant consuming 600 mL of breast milk per day could ingest as much as 0.6 μg of lead. Because only about 42% of lead is actually absorbed by an infant following ingestion (Ziegler *et al.,* 1978), the amount of lead absorbed by the nursing infant would be approximately 0.25 μg per day. This is lower than FDA estimates of an infant's total daily lead intake from food, water, and other beverages (Wolff, 1983) and substantially lower than the maximal daily intakes of 300 μg that were proposed by the National Research Council (NRC, 1980) and by the World Health Organization (WHO, 1977) for infants. Prepared formulas have been found to contain amounts of lead that approximate those found in breast milk and may even be several times higher than those in breast milk if the formula requires dilution with water or is not packaged in lead-free cans (Dabeka and McKenzie, 1988).

Although longitudinal studies on the neurodevelopment of infants exposed to lead via breast milk have not been published, the available data on breast milk concentrations do not indicate that this type of exposure is likely to have any adverse effects on subsequent neurodevelopment. Cord blood lead levels are generally two to five times higher than breast milk levels (Wolff, 1993). Therefore, prenatal exposure as well as postnatal exposure to lead from other sources (e.g., ingestion of leaded paint and contaminated water) are more likely to pose a risk to the developing infant than exposure to lead via breast milk.

C. Mercury

Mercury is present in the environment in two different forms: inorganic (or elemental) and organic. Inorganic mercury is used in dental amalgam, in making thermometers, switches, and fluorescent lamps, and in various industrial processes. Mercury compounds are sometimes used in paints, pesticides, and medicines. Exposure to organic mercury (methylmercury) is primarily through the consumption of mercury-contaminated food. Predatory fish such as freshwater pike and perch, ocean tuna, swordfish, and shark contain higher levels of mercury than other types of fish (World Health Organization, 1990).

Liquid inorganic mercury is poorly absorbed following ingestion, although mercury vapor is readily absorbed systemically through the lungs, intact skin, and GI tract. Following absorption, mercury passes through the bloodstream and is distributed in various organs, principally the brain and kidneys (Dabney, 1997). The half-life for mercury is long: 64 days for the kidney, 54 days for the whole body, and approximately 1 year for the brain (Dabney, 1997). The turnover rate of mercury in the brain is comparatively slower than that of other organs (Neathery and Miller, 1975). Consequently, high levels of mercury may accumulate in the brain as a result of chronic exposure to mercury. Indeed, retention of mercury in the brain has been found as long as 10 years after exposure to the mercury ceased (Dabney, 1997).

In contrast to inorganic mercury, organic mercury is easily absorbed in the GI tract (>90%) and penetrates the blood–brain barrier (Aschner and Aschner, 1990). Once in the body, organic mercury is slowly converted to inorganic mercury (Oskarsson *et al.,* 1995). Mercury is a well-known neurotoxicant. Neuropsychologic effects of mercury toxicity include increased irritability, insomnia, apathy, impaired concentration and memory, and changes in personality (Dabney, 1997). Neurobehavioral deficits, such as impaired memory and motor coordination, were found to persist for 18 years following the last exposure to high levels of mercury vapor in one controlled study of former mercury miners (Dabney, 1997).

Sensitivity of the developing brain to mercury has been well documented in two famous outbreaks of mercury poisoning. In one outbreak, the children of women in Minamata, Japan, were born with cerebral palsy and microcephaly after their mothers had eaten fish contaminated with high levels of mercury from an industrial spill while they were pregnant (Murakami, 1972). The mothers showed no or few signs of poisoning. In the other outbreak, children of Iraqi women who consumed homemade bread contaminated with a methylmercury fungicide throughout pregnancy were also born with cerebral palsy (Amin-Zaki *et al.,* 1974, 1979). Out of 15 nursing infants whose mothers consumed contaminated bread only after delivery, six were found to have neurologic and psychomotor deficits (Amin-Zaki *et al.,* 1981). A peak value of 200 ng/g was reported in the breast milk of Iraqi women following consumption of methylmercury-contaminated bread (Bakir *et al.,* 1973).

Diets consisting of seafood high in mercury have been found to be associated with high levels of mercury in breast milk (Skerfving, 1988; Grandjean *et al.,* 1994, 1995). Levels of mercury in breast milk are approximately 25–30% of levels found in blood (Skerfving, 1988; Oskarsson *et al.,* 1996). Most of the total mercury in breast milk is in the form of inorganic mercury (40–80%), whereas only 20% of the total mercury in blood is inorganic (Bakir *et al.,* 1973; Skerfving, 1988; Oskarsson *et al.,* 1996). Total mercury levels in blood and milk were found to be correlated with the number of amalgam fillings in Swedish women in one study (Oskarsson *et al.,* 1996). The authors of this study estimated that total mercury concentrations increased 0.1 ng/g in blood and 0.05 ng/g in milk for each filling. However, no relation to the number of fillings in the mother's teeth and milk mercury levels was found in another study (Grandjean *et al.,* 1995). In Table 5, the breast milk levels of mercury found in some published studies are shown along with the estimated daily intake by a nursing infant weighing 4 kg and consuming 600 mL of milk per day. It can be seen that maximum levels of mercury in breast milk may result in daily intakes by nursing infants that exceed the highest tolerable daily intake of 0.5 μg/kg proposed for adults (WHO/FAO, 1989). Whether breast milk concentrations of mercury as shown in Table 5 could have any possible adverse effect on the neurodevelopment of children is not known. Only one study investigating the effects of exposure to mercury via the breast milk on the development of infants has been published (Grandjean *et al.,* 1995). In this study, three developmental milestones (sitting, creeping, and standing) were examined between 5 and 12 months of age in 583 infants of women living in fishing communities in the Faroe Islands in the North Atlantic ocean. The Faroese diet consists primarily of seafood, including pilot whale meat and blubber, which contain high concentrations of mercury. The authors found that the age at which an infant reached a developmental milestone was not associated with prenatal exposure to mercury (as determined by mercury concentrations in umbilical cord blood and maternal hair). However, they did find that earlier milestone development was observed in infants that had significantly higher mercury concentrations in the hair. In addition, for all three milestones, those infants that nursed for 3 months or more reached the milestone earlier than those infants that nursed for less than 3

TABLE 5 Estimated Daily Intakes of Mercury by the Nursing Infant

Authors	Number of subjects	Source of mercury	Concentrations	Estimated daily intake by infant (mg/kg/d)[a,b]	Percent of highest tolerable daily intake for adults[c]	Comments
Grandjean *et al.*, 1995	88	Whale	<0.001–0.009 μg/mL	0.15–1.3[a]	30–260	Milk mercury concentrations were significantly associated with cord blood concentrations; milk mercury levels were related to number of pilot whale dinners, but not with number of fish dinners during pregnancy; milk mercury showed no relation to number of fillings in mother's teeth, fat content of milk, maternal age, or parity.
Skerfving, 1988	15	Offshore and coastal fish	0.2–6.3 ng/g (total mercury)	0.03–0.94[b]	6–188	Methylmercury fraction in milk averaged 20% of total mercury
			0.2–1.2 ng/g (methylmercury)	0.03–0.18[b]	6–36	
Oskarsson *et al.*, 1996	30	Fish	0.1–2.0 ng/g (total mercury)	0.01–0.30[b]	2–60	Milk levels were 27% of levels found in blood; 51% of the total mercury in milk was present in the inorganic form; only 26% of total mercury in blood was present as inorganic mercury; total mercury levels in blood and milk were correlated with the number of amalgam fillings; total mercury concentration in blood increased 0.1 ng/g and in milk 0.05 ng/g for each filling.
Schramel *et al.*, 1988	15	No special exposure	Mean = 2.0 ± 0.7 ng/mL	0.3[a]	60	Colostrum had a significantly higher mercury concentration than mature milk; 10 out of 15 women had mercury milk concentrations <1 ng/mL.
Negretti de Brätter, 1987	112	No special exposure	0.3–9.7 μg/kg	0.04–1.5[b]	8–300	Mercury level of milk decreases from an average of 2.6 μg/kg in colostrum to an average of 1.5 μg/kg in mature milk.

[a]Based on assumption that nursing infant consumes 600 mL breast milk per day and weighs 4 kg.
[b]Based on assumption that nursing infant consumes 600 gm breast milk per day and weighs 4 kg.
[c]0.5 μg/kg/d has been proposed as the highest tolerable daily intake of mercury for adults (WHO/FAO, 1989).

months or did not nurse at all. Developmental milestone data in this study were obtained by district health nurses who visited the family and interviewed the mothers several times during the months following parturition. Therefore, the reliability of these data is questionable. In addition, developmental milestone data are poor predictors of subsequent cognitive development, and therefore further research is needed to confirm the findings in this study.

V. Risk Assessment and Exposures during Breast Feeding

Unlike pregnancy, exposures during breast feeding are intentional in that the mother has the choice of breast feeding or formula feeding her infant. Naturally, all mothers who plan to breast feed should avoid taking any medication that is not absolutely necessary. However, if a mother requires drug therapy after delivery, it is possible to devise a plan that minimizes the risk of drug exposure to the infant without depriving the mother and her infant of the benefits of breast feeding. If drug therapy is of short duration, it is usually unnecessary to terminate breast feeding while the mother takes the medication. Often the mother can reduce the amount of drug transferred to her infant by taking the medication right after breast feeding. However, in some cases it may be necessary to temporarily halt breast feeding while the mother undergoes treatment and resume breast feeding as soon as the therapy is no longer required.

When a mother must undergo chronic treatment, as for depression, psychosis, or epilepsy, the decision to continue breastfeeding during treatment can become extraordinarily difficult. Tables 1–3 clearly indicate that very little data has been published on the effects of psychotropic and antiepileptic drugs on the nursing infant. On the basis of small numbers of women, it appears that the infant is usually exposed to very small amounts of drug or chemical via the breast milk. However, there is a concern that chronic exposure of nursing infants to low levels of drugs or chemicals may be sufficient to interfere with normal development of the CNS. Lack of data from sound epidemiologic studies results in uncertainty on the part of the physician who must assess the risks of chronic drug exposure to the nursing infant. Because of this uncertainty in risk, many physicians recommend termination of breast feeding if their patient must undergo chronic drug therapy. Such recommendations often cause a great deal of turmoil for women who strongly desire to breast feed their infant. By discontinuing breast feeding, the risk of adverse effects to the infant from drug exposure is removed. However, at what cost to the infant? Failure to provide infants with breast milk, which is rich in nutrients and other protective factors, may introduce other potentially more serious risks.

If a woman decides to postpone treatment until the child is weaned she runs the risk of exacerbating her medical condition and being hospitalized, neither of which benefits the mother–infant pair. In one study of 25 clinically depressed women, five of them chose not to be treated with tricyclic antidepressants rather than give up breast feeding (Misri and Sivertz, 1991). All five women had exacerbation of their symptoms and two of them required hospitalization. No adverse effects were observed in any of the 20 breast-fed infants whose mothers underwent treatment with tricyclics. In another study, infants whose mothers experienced postpartum depression for 6 months and were not treated performed poorly on the Bayley Mental and Motor Scales when tested at 1 year of age and had lower percentiles on growth curves (Field, 1995).

Women whose diet or occupation puts them at an increased risk of having high levels of environmental contaminants in their breast milk (such as PCBs or mercury) must also decide whether the benefits of breast feeding outweigh the risks. As with pharmaceutical exposures, very little published data exist to determine whether the levels found in breast milk are unacceptably high levels for the nursing infant. Estimating the infant dose from breast milk concentrations of a particular chemical and comparing that dose to allowable daily intakes (ADIs) established by regulatory agencies provides some comparative information. However, it must be kept in mind that these regulatory levels were extrapolated from studies of animal toxicity and studies of accidental or occupational exposures in human adults (Poitrast *et al.*, 1988). Although ADIs have built-in safety factors of 100 to 1000, the safety of these levels for the human infant at a time when the CNS is still very vulnerable remains unclear.

It can be seen, then, that counseling a lactating woman who is exposed to potentially toxic substances is an important component of her postpartum care that requires careful assessment of the risks and the benefits of breast feeding in her particular situation. If chronic drug therapy is required and the mother chooses to breast feed, then the drug that is safest for the baby should be used. The dose of the drug and duration of treatment should be kept to the minimum amount necessary to achieve remission of the mother's symptoms (Friedman and Polifka, 1996; Wisner *et al.*, 1996). Breast-fed infants whose mothers are on medication or are exposed to toxic chemicals in their environment should be monitored for signs of toxicity, such as lethargy, irritability, rash, vomiting, and diarrhea. If these signs are evident in the nursing infant, then serum levels should be obtained in the infant, and, if necessary, breast feeding should be discontinued.

VI. Conclusion

Whereas a great deal is known about the health benefits of breast feeding for the developing infant, very little is known about the long-term effects of exposures to drugs and chemicals through the breast milk. In 1980, Wilson *et al.* reviewed the available pharmacokinetic data on drugs and chemicals consumed while breast feeding. They concluded that "data are lacking for most drugs and hence dosing via milk or risk to the infant remains speculative. . . . Few drugs are contraindicated in breast-feeding women, but supportive data for either proscriptions or permissive statements are often lacking" (pg. 1). In the years that have elapsed since Wilson *et al.* made these comments, very little epidemiologic or pharmacokinetic data have been added to the literature. In particular, the effects of exposure to breast milk contaminants on the neurodevelopment of children remains a neglected area of research. Risk assessment regarding breast milk exposure to drugs and chemicals is as speculative now as it was in 1980. More research is needed to clarify how exogenous substances are absorbed and metabolized by the nursing infant. Longitudinal follow-up on the neurodevelopment of nursed infants is also needed to determine if chronic exposure to low levels of breast milk contaminants has any long-term effects on the development of the infant. The knowledge accrued from such research will greatly enhance the ability of clinicians and their patients to assess the risks associated with lactational exposures.

References

Ahlborg, U. G., Brouwer, A., Fingerhut, M. A., *et al.* (1992). Impact of polychlorinated dibenzo-*p*-dioxins, dibenzofurans, and biphenyls on human and environmental health, with special emphasis on application of the toxic equivalency factor concept. *Eur. J. Pharmacol. Environ. Toxicol. Pharmacol. Sect.* **228,** 179–199.

Albers, J. M. C., Kreis, I. A., Liem, A. K. D., and van Zoonen, P. (1996). Factors that influence the level of contamination of human milk with poly-chlorinated organic compounds. *Arch. Environ. Contam. Toxicol.* **30,** 285–291.

Alexander, F. W. (1979). Sodium valproate and pregnancy. *Arch. Dis. Child.* **54,** 240.

Altshuler, L. L., Burt, V. K., McMullen, M., and Hendrick, V. (1995). Breastfeeding and sertraline: a 24-hour analysis. *J. Clin. Psychiatry* **56,** 243–245.

Amin-Zaki, L., Elhassani, S. B., Majeed, M. A., *et al.* (1974). Intrauterine methylmercury poisoning in Iraq. *Pediatrics* **54,** 587–595.

Amin-Zaki, L., Majeed, L. M., Elhassani, S., *et al.* (1979). Prenatal mercury poisoning, clinical observations over five years. *A.M.A. Dis. Child.* **133,** 172–177.

Amin-Zaki, L., Majeed, L. M., Greenwood, M. R., *et al.* (1981). Methylmercury poisoning in the Iraqi suckling infant: a longitudinal study over five years. *J. Appl. Toxicol.* **1,** 210–214.

Anonymous (1994a). Is breast feeding beneficial in the UK? Statement of the Standing Committee on Nutrition of the British Paediatric Association. *Arch. Dis. Child.* **71,** 376–380.

Anonymous (1994b). PCBs in breast milk. *Pediatrics* **94,** 122–123.

Anonymous (1996). Health effects of outdoor air pollution. *Am. J. Respir. Crit. Care Med.* **153,** 477–498.

Aschner, M., and Aschner, J. L. (1990). Mercury neurotoxicity: mechanisms of blood–brain barrier transport. *Neurosci. Biobehav. Rev.* **14,** 169–176.

Astley, S. J., and Little, R. E. (1990). Maternal marijuana use during lactation and infant development at one year. *Neurotoxicol. Teratol.* **12,** 161–168.

Atkinson, H. C., Begg, E. J., and Darlow, B. A. (1988). Drugs in human milk. Clinical pharmacokinetic considerations. *Clin. Pharmacokinet.* **14,** 217–240.

Bader, T. F., and Newman, K. (1980). Amitriptyline in human breast milk and the nursing infant's serum. *Am. J. Psychiatry* **137,** 855–856.

Baghurst, P. A., McMichael, A. J., Wigg, N. R., *et al.* (1992). Environmental exposure to lead and children's intelligence at the age of seven years: the port pirie cohort study. *N. Engl. J. Med.* **327,** 1279–1284.

Bakir, F., Damluji, S. F., Amin-Zaki, L., *et al.* (1973). Methylmercury poisoning in Iraq. *Science* **181,** 230-241.

Bardy, A. H., Granstrom, M.-L., and Hiilesmaa, V. K. (1982). Valproic acid and breast-feeding. In *Epilepsy, Pregnancy, and the Child.* D. Janz, L. Bossi, M. Dam, *et al.,* Eds. Raven Press, New York, pp. 359–360.

Becher, G., Skaare, J. L., Polder, A., and et al. (1995). PCDDs, PCDFs, and PCBs in human milk from different parts of Norway and Lithuania. *J. Toxicol. Environ. Health* **46,** 133–148.

Bellinger, D. (1994). Teratogen update: lead. *Teratology* **50,** 367–373.

Bellinger, D., Needleman, M. C., Bromfield, R., and Nimtz, M. (1984). A followup study of the academic attainment and classroom behavior of children with elevated dentine lead levels. *Biol. Trace El. Res.* **6,** 207–223.

Blacker, K. H., Weinstein, B. J., and Ellman, G. L. (1962). Mother's milk and chlorpromazine. *Am. J. Psychi.* **119,** 178–179.

Brandt, R. (1976). Passage of diazepam and desmethyldiazepam into breast milk. *Arzneimittelforsch.* **26,** 454–457.

Breyer-Pfaff, U., Nill, K., Entenmann, A., and Gaertner, H. J. (1995). Secretion of amitriptyline and metabolites into breast milk. *Am. J. Psychiatry* **152,** 812–813.

Brixen-Rasmussen, L., Halgrener, J., and Jorgensen, A. (1982). Amitriptyline and nortriptyline excretion in human breast milk. *Psychopharmacology* **76,** 94–95.

Burch, K. J., and Wells, B. G. (1992). Fluoxetine/norfluoxetine concentrations in human milk. *Pediatrics* **89,** 676–677.

Chaney, N. E., Franke, J., and Wadlington, W. B. (1988). Cocaine convulsions in a breast-feeding baby. *J. Pediatr.* **112,** 134–135.

Chaplin, S., Sanders, G. L., and Smith, J. M. (1982). Drug excretion in human breast milk. *Adv. Drug React. Ac. Pois. Rev.* **1,** 255–287.

Chasnoff, I. J., Lewis, D. E., and Squires, L. (1987). Cocaine intoxication in a breast-fed infant. *Pediatrics* **80,** 836–838.

Cole, A. P., and Hailey, D. M. (1975). Diazepam and active metabolite in breast milk and their transfer to the neonate. *Arch. Dis. Child.* **50,** 741–742.

Committee on Drugs, American Academy of Pediatrics (1994). The transfer of drugs and other chemicals into human milk. *Pediatrics* **93,** 137–105.

Cory-Slechta, D. A. (1996). Legacy of lead exposure: consequences for the central nervous system. *Otolaryngol. Head Neck Surg.* **114,** 224–226.

Cunningham, A. S., Jelliffe, D. B., and Jelliffe, E. F. P. (1991). Breastfeeding and health in the 1980s: a global epidemiologic review. *J. Pediatr.* **118,** 659–665.

Dabeka, R. W., and McKenzie, A. D. (1988). Lead and cadmium levels in commercial infant foods and dietary intake by infants 0–1 year old. *Food Addit. Contam.* **5,** 333–342.

Dabney, B. J.: Lead (REPROTEXT® Document). In "REPROTEXT® System" B. J. Dabney, Ed. MICROMEDEX, Inc., Englewood, Colorado (Edition expires 6/97).

Dahl, P., Lindstrom, G., Wiberg, K., and Rappe, C. (1995). Absorption of polychlorinated biphenyls, dibenzo-*p*-dioxins and dibenzofurans by breast-fed infants. *Chemosphere* **30,** 2297–2306.

Dahlstrom, A., Lundell, B., Curvall, M., *et al.* (1990). Nicotine and cotinine concentrations in the nursing mother and her infant. *Acta Paediatr. Scand.* **79,** 142.

de Andraca, I., and Uauy, R. (1995). Breastfeeding for optimal mental development. The α and the Ω in human milk. *World Rev. Nutr. Diet.* **78,** 1–27.

Dewailly, E., Ayotte, P., Laliberte, C., *et al.* (1996). Polychlorinated biphenyl (PCB) and dichlorodiphenyl dichloroethylene (DDE) concentrations in the breast milk of women in Quebec. *Am. J. Public Health* **86,** 1241–1246.

Dickinson, R. G., Harland, R. C., Lynn, R. K., and et al. (1979). Transmission of valproic acid (Depakene®) across the placenta: half-life of the drug in mother and baby. *J. Pediatr.* **94,** 832–835.

Duarte-Davidson, R., and Jones, K. C. (1994). Polychlorinated biphenyls (PCBs) in the UK population: estimated intake, exposure and body burden. *Sci. Total Environ.* **151,** 131–152.

Dusci, L. J., Good, S. M., Hall, R. W., and Ilett, K. F. (1990). Excretion of diazepam and its metabolites in human milk during withdrawal from combination high dose diazepam and oxazepam. *Br. J. Clin. Pharmacol.* **29,** 123–126.

Erickson, S. H., Smith, G. H., and Heidrich, F. (1979). Tricyclics and breast feeding. *Am. J. Psychiatry* **136,** 1483.

Erkkola, R., and Kanto, J. (1972). Diazepam and breast-feeding. *Lancet* **1,** 1235–1236.

Eschenhof, E., and Rieder, J. (1969). Antidepressivums amitriptylin in organisms untersvchungen uberdas schicksal des der ratte und des menscher. *Arzneimittelforsch.* **19,** 957.

Espir, M. L. E., Benton, P., Will, E., *et al.* (1976). Sodium valproate (Epilim)—some clinical and pharmacological aspects. In *Clinical and Pharmacological Aspects of Sodium Valproate (Epilim) in the Treatment of Epilepsy.* Legg, Ed. MCS Consultants, Tunbridge Wells, pp. 145–151.

Field, T. (1995). Infants of depressed mothers. *Inf. Behav. Dev.* **18,** 1–13.

Fisher, J. B., Edgren, B. E., Mammel, M. C., and Coleman, J. M. (1985). Neonatal apnea associated with maternal clonazepam therapy. A case report. *Obstet. Gynecol.* **66**(Suppl), 34S–35S.

Flores-Huerta, S., Hernandez-Montes, H., Argote, R. M., and Villalpando, S. (1992). Effects of ethanol consumption during pregnancy and lactation on the outcome and postnatal growth of the offspring. *Ann. Nutr. Metab.* **36,** 121–128.

Friedman, J. M., and Polifka, J. E. (1996). *The Effects of Drugs on the Fetus and Nursing Infant. A Handbook for Health Care Professionals.* The Johns Hopkins University Press, Baltimore, Md.

Froescher, W., Echelbaum, M., Niesen, M., *et al.* (1984). Carbamazepine levels in breast milk. *Ther. Drug Monit.* **6,** 266–271.

Funatsu, I., Yamashita, F., Ito, Y., *et al.* (1972). Polychlorbiphenyls (PCB)-induced fetopathy. *Kurame Med. J.* **19,** 43.

Gale, C. R., and Martyn, C. N. (1996). Breastfeeding, dummy use, and adult intelligence. *Lancet* **347,** 1072–1075.

Gallo, M. A., Scheuplein, R. J., and Van der Heijden, K. A., Eds. (1991). *Biological Basis for Risk Assessment of Dioxins and Related Compounds,* Banbury Report 35. Cold Spring Harbor Laboratory, Cold Spring Harbor, N.Y.

Gellert, R. J., and Wilson, C. (1979). Reproductive function in rats exposed prenatally to pesticides and polychlorinated biphenyls (PCB). *Environ. Res.* **18,** 437–443.

Gladen, B. C., and Rogan, W. J. (1991). Effects of perinatal polychlorinated biphenyls and dichlorodiphenyl dichloroethene on later development. *J. Pediatr.* **119,** 58–63.

Gladen, B. C., Rogan, W. J., Hardy, P., et al. (1988). Development after exposure to polychlorinated biphenyls and dichlorophenyl dichloroethene transplacentally and through human milk. *J. Pediatr.* **113,** 991–995.

Grandjean, P., Jorgensen, P. J., and Weihe, P. (1994). Human milk as a source of methylmercury exposure in infants. *Environ. Health Perspect.* **102,** 74–77.

Grandjean, P., Weihe, P., Needham, L. L., *et al.* (1995). Relation of a seafood diet to mercury, selenium, arsenic, and polychlorinated biphenyl and other organochlorine concentrations in human milk. *Environ. Res.* **71,** 29–38.

Hammond, P. B., and Dietrich, K. N. (1991). Lead exposure in early life: health consequences. *Rev. Environ. Toxicol.* **115,** 91–124.

Hanson, L. A., and Bergstrom, S. (1990). The link between infant mortality and birth rates—the importance of breastfeeding as a common factor. *Acta Paediatr. Scand.* **79,** 481–489.

Harada, M. (1976). Intrauterine poisoning: Clinical and epidemiological studies and significance of the problem. *Bull. Inst. Constit. Med. (Kumamoto University)* **25**(Suppl), 1–60.

Hardee, G. E., Steward, T., and Capomacchia, A. C. (1983). Tobacco smoke xenobiotic compound appearance in mother's milk after involuntary smoke exposures. I. Nicotine and cotinine. *Toxicol. Lett.* **15,** 109–112.

Higuchi, K., Ed. (1976). *PCB Poisoning and Pollution.* Academic Press, New York.

Hutzinger, O., Safe, S., and Zitko, V. (1974). *The Chemistry of PCB's.* CRC Press, Cleveland, Oh.

Isenberg, K. E. (1990). Excretion of fluoxetine in human breast milk. *J. Clin. Psychiatry* **51,** 169.

Jacobson, J. L. (1992). In *Final Report of the Workshop on Protocol Development for Determining Human Exposures to Toxic Chemicals, Body Burdens and Neurobehavioral and Developmental Assessment.* H. Daly, H. E. B. Humphrey, Health Committee, Science Advisory Board, International Joint Commission, Windsor, Ontario, p. 18.

Jacobson, J. L., and Jacobson, S. W. (1988). New methodologies for assessing the effects of prenatal toxic exposure on cognitive functioning in humans. In *Toxic Contaminants and Ecosystem Health: A Great Lakes Focus.* M. S. Evans, Ed. John Wiley and Sons, New York, pp. 373–387.

Jacobson, J. L., Jacobson, S. W., Fein, G. G., and et al. (1984). Prenatal exposure to an environmental toxin. A test of the multiple effects model. *Dev. Psychol.* **20,** 523–532.

Jacobson, J. L., Jacobson, S. W., and Humphrey, H. E. B. (1990a). Effects of exposure to PCBs and related compounds on growth and activity in children. *Neurotoxicol. Teratol.* **12,** 319–326.

Jacobson, J. L., Jacobson, S. W., and Humphrey, H. E. B. (1990b). Effects of *in utero* exposure to polychlorinated biphenyls and other contaminants on cognitive functioning in young children. *J. Pediatr.* **116,** 38–45.

Jacobson, S. W., Fein, G. G., Jacobson, J. L., *et al.* (1985). The effect of PCB exposure on visual recognition memory. *Child Dev.* **56,** 853–860.

Jensen, A. A. (1983). Chemical contaminants in human milk. *Residue Rev.* **89,** 1–128.

Johansen, H. R., Becher, G., Polder, A., and Skaare, J. U. (1994). Congener-specific determination of polychlorinated biphenyls and organochlorine pesticides in human milk from Norwegian mothers living in Oslo. *J. Toxicol. Environ. Health.* **42,** 157–171.

Kaneko, S., Sato, T., and Suzuki, K. (1979). The levels of anticonvulsants in breast milk. *Br. J. Clin. Pharmacol.* **7,** 624–627.

Kaneko, S., Suzuki, K., Sato, T., *et al.* (1981). The problems of anticonvulsant medication at the neonatal period: is breast feeding advisable? In *Epilepsy, Pregnancy and the Child.* D. Janz, L. Bossi, M. Dam, *et al.,* Eds., Raven Press, New York, pp. 343–347.

Kemp, J., Ilett, K. F., Booth, J., and Hackett, L. P. (1985). Excretion of doxepin and *N*-desmethyldoxepin in human milk. *Br. J. Clin. Pharmacol.* **20,** 497–499.

Kimbrough, R. D., and Jensen, A. A., Eds. (1989). *Halogenated Biphenyls, Terphenyls, Naphthalenes, Dibenzodioxins and Related Products,* 2nd ed. Elsevier, Amsterdam.

Knott, C., Reynolds, F., and Clayden, G. (1987). Infantile spasms on weaning from breast milk containing anticonvulsants. *Lancet* **2,** 272–273.

Kodama, H., and Ota, H. (1980). Transfer of polychlorinated biphenyls to infants from their mothers. *Arch. Environ. Health* **35,** 95–100.

Kok, T. H. H. G., Taitz, L. S., Bennett, M. J., and Holt, D. W. (1982). Drowsiness due to clemastine transmitted in breast milk. *Lancet* **1,** 914–915.

Koopman-Esseboom, C., Morse, D. C., Wesglas-Kuperus, N., *et al.* (1994). Effects of dioxins and polychlorinated biphenyls on thyroid hormone status of pregnant women and their infants. *Pediatr. Res.* **36,** 468–473.

Koopman-Esseboom, C., Weisglas-Kuperus, N., de Ridder, M. A. J., *et al.* (1996). Effects of polychlorinated biphenyl/dioxin exposure and feeding type on infants' mental and psychomotor development. *Pediatrics* **97,** 700–706.

Koup, J. R., Rose, J. Q., and Cohen, M. E. (1978). Ethosuximide pharmacokinetics in a pregnant patient and her newborn. *Epilepsia* **19,** 535–539.

Kuhnz, W., Jager-Roman, E., Deichl, A., *et al.* (1983). Carbamazepine and carbamazepine-10,11-epoxide during pregnancy and postnatal period in epileptic mothers and their nursed infants: pharmacokinetics and clinical effects. *Pediatr. Pharmacol.* **3,** 199–208.

Kuhnz, W., Koch, S., Jakob, S., *et al.* (1984). Ethosuximide in epileptic women during pregnancy and lactation period. Placental transfer, serum concentrations in nursed infants and clinical status. *Br. J. Clin. Pharmacol.* **18,** 671–677.

Lamm, S., Cole, B., Gly, K., and Ullman, W. (1973). Lead content of milk fed to infants 1971–1972. *N. Engl. J. Med.* **289,** 574–575.

Lanting, C. I., Fidler, V., Huisman, M., *et al.* (1994). Neurological differences between 9-year-old children fed breast milk or formula-milk as babies. *Lancet* **344,** 1319–1322.

Lawrence, R. A. (1994). *Breastfeeding. A Guide for the Medical Profession,* 4th ed. Mosby-Year Book, Inc., St Louis, Mo.

Lester, B. M., Cucca, J., Andreozzi, L., *et al.* (1993). Possible association between fluoxetine hydrochloride and colic in an infant. *J. Am. Acad. Child Adolesc. Psychiatry* **32,** 1253–1255.

Levitt, C. A., Kaczorowski, J., Hanvey, L., and et al. (1996). Breast-feeding policies and practices in Canadian hospitals providing maternity care. *Can. Med. Assoc. J.* **155,** 181–188.

Little, R. E., Anderson, K. W., Ervin, C. H., *et al.* (1989). Maternal alcohol use during breast-feeding and infant mental and motor development at one year. *N. Engl. J. Med.* **321,** 425–430.

Little, R. E., Lambert, M. D., and Worthington-Roberts, B. (1990). Drinking and smoking at 3 months postpartum by lactation history. *Paediatr. Perinat. Epidemiol.* **4,** 290–302.

Luck, W., and Nau, H. (1984). Exposure of the fetus, neonate, and nursed infant to nicotine and cotinine from maternal smoking. *N. Engl. J. Med.* **311,** 672.

Matheson, I., Pande, H., and Alertsen, A. R. (1985). Respiratory depression caused by *N*-desmethyldoxepin in breast milk. *Lancet* **2,** 1124.

Matheson, L., and Skjaeraasen, J. (1988). Milk concentrations of flupenthixol, nortriptyline and zuclopenthixol and between-breast differences in two patients. *Eur. J. Clin. Pharmacol.* **35,** 217–220.

McLachlan, M. S. (1993). Digestive tract absorption of polychlorinated dibenzo-p-dioxins, dibenzofurans, and biphenyls in a nursing infant. *Toxicol. Appl. Pharmacol.* **123,** 68–72.

Menkes, J. H. (1977). Early feeding history of children with learning disorders. *Dev. Med. Child. Neurol.* **19,** 169–171.

Merlob, P., Mor, N., and Litwin, A. (1992). Transient hepatic dysfunction in an infant of an epileptic mother treated with carbamazepine during pregnancy and breastfeeding. *Ann. Pharmacother.* **26,** 1563–1565.

Misri, S., and Sivertz, K. (1991). Tricyclic drugs in pregnancy and lactation: a preliminary report. *Int. J. Psychiatry Med.* **21,** 157–171.

Murakami, U. (1972). Organic mercury problem affecting intrauterine life. Proceedings of the International Symposium on the Effect of Prolonged Drug Usage on Fetal Development. In *Advances in Experimental Biology and Medicine.* M. A. Klingberg Ed., Vol. 27. Plenum Publishing Corp, New York, pp. 301–336.

Namihira, D., Saldivar, D., Pustilnik, L., *et al.* (1993). Lead in human blood and milk from nursing women living near a smelter in Mexico City. *J. Toxicol. Environ. Health* **38,** 225–232.

National Research Council (1980). Lead in the human environment. Committee on Lead in the Human Environment, Washington, DC.

Nau, H., Rating, D., Koch, S., *et al.* (1981). Valproic acid and its metabolites: placental transfer, neonatal pharmacokinetics, transfer via mother's milk and clinical status in neonates of epileptic mothers. *J. Pharmacol. Exp. Ther.* **219,** 768–776.

Neathery, M. W., and Miller, W. J. (1975). Metabolism and toxicity of cadmium, mercury, and lead in animals: a review. *J. Dairy Sci.* **58,** 1767–1781.

Needleman, H. L., Schell, A., Bellinger, D., *et al.* (1990). The long-term effects of exposure to low doses of lead in childhood: an 11-year follow-up report. *N. Engl. J. Med.* **322,** 83–88.

Negretti de Brätter, V. E., Brätter, P., Müller, J., *et al.* (1987). Ingestion of mercury during early infancy. In *Trace Element—Analytical Chemistry in Medicine and Biology.* P. Bratter, and P. Schramel, Eds., Vol. 4. Walter de Gruyter, Berlin, pp. 151–156.

Oskarsson, A., Palminger Hallén, I., and Sundberg, J. (1995). Exposure to toxic elements via breast milk. *Analyst* **120,** 765–770.

Oskarsson, A., Schütz, A., Skerfving, S., *et al.* (1996). Total and inorganic mercury in breast milk and blood in relation to fish consumption and amalgam fillings in lactating women. *Arch. Environ. Health* **51,** 234–241.

Paradise, L. J., Elster, B. A., and Tan, L. (1994). Evidence in infants with cleft palate that breast milk protects against otitis media. *Pediatrics* **94,** 853–860.

Patrick, M. J., Tilstone, W. J., and Reavey, P. (1972). Diazepam and breast-feeding. *Lancet* **1,** 542–543.

Perez-Reyes, M., and Wall, I. M. E. (1982). Presence of a Δ9-tetrahydrocannabinol in human milk. *N. Engl. J. Med.* **307,** 819–820.

Piazza, S. F., Haley, N. J., Clark, D. A., *et al.* (1987). Human milk contamination with nicotine and cotinine. *Pediatr. Res.* **21,** 401A.

Pirkle, J. L., Brody, D. J., Gunter, E. W., *et al.* (1994). The decline in blood lead levels in the United States. The National Health and Nutrition Examination Surveys (NHANES). *JAMA* **272,** 284–291.

Pittard, W. B. III, and O'Neal, W. Jr. (1986). Amitriptyline excretion in human milk. *J. Clin. Psychopharmacol.* **6,** 383–384.

Poitrast, B. J., Keller, W. C., and Elves, R. G. (1988). Estimation of chemical hazards in breast milk. *Aviat. Space Environ. Med.* **59(Suppl),** A87–A92.

Pons, G., Rey, E., and Matheson, I. (1994). Excretion of psychoactive drugs into breast milk. Pharmacokinetic principles and recommendations. *Clin. Pharmacokinet.* **27,** 270–289.

Pynnonen, S., Kanto, J., Sillanpaa, M., and Erkkola, R. (1977). Carbamazepine. Placental transport, tissue concentrations in foetus and newborn, and level in milk. *Acta Pharmacol. Toxicol.* **41,** 244–253.

Pynnonen, S., and Sillanpaa, M. (1975). Carbamazepine and mother's milk. *Lancet* **2,** 563.

Quinsey, P. M., Donohue, D. C., and Ahokas, J. T. (1995). Persistence of organochlorines in breast milk of women in Victoria, Australia. *Food Chem. Toxicol.* **33,** 49–56.

Rane, A., and Tunell, R. (1981). Ethosuximide in human milk and in plasma of a mother and her nursed infant. *Br. J. Clin. Pharmacol.* **12,** 855–858.

Riordan, J., and Auerbach, K. G. (1993). *Breastfeeding and Human Lactation.* Jones and Bartlett Publishers, Inc., Boston, MA.

Rogan, W. J., and Gladen, B. C. (1985). Study of human lactation for effects of environmental contaminants: The North Carolina Breast Milk and Formula Project and some other ideas. *Environ. Health. Perspect.* **60,** 215–221.

Rogan, W. J., and Gladen, B. C. (1991). PCBs, DDE, and child development at 18 and 24 months. *Ann. Epidemiol.* **1,** 407–413.

Rogan, W. J., Gladen, B. C., Hung, K. L., *et al.* (1988). Congenital poisoning by polychlorinated biphenyls and their contaminants in Taiwan. *Science* **241,** 334–336.

Rogan, W. J., Gladen, B. C., McKinney, J. D., *et al.* (1986a). Neonatal effects of transplacental exposure to PCBs and DDE. *J. Pediatr.* **109,** 335–341.

Rogan, W. J., Gladen, B. C., McKinney, J. D., *et al.* (1986b). Polychlorinated biphenyls (PCBs) and dichlorodiphenyl dichloroethene (DDE) in human milk: effects of maternal factors and previous lactation. *Am. J. Public Health* **76,** 172–177.

Rogan, W. J., Gladen, B. C., McKinney, J. D., *et al.* (1987). PCB's and DDE in human milk: Effects on growth, morbidity and duration of lactation. *Am. J. Public Health* **77,** 1294–1297.

Rogan, W. J., Gladen, B. C., and Wilcox, A. J. (1985). Potential reproductive and postnatal morbidity from exposure to polychlorinated biphenyls: epidemiologic considerations. *Environ. Health. Perspect.* **60,** 233–239.

Schardein, J. L. (1993). *Chemically Induced Birth Defects,* 2nd ed. Marcel Dekker, Inc., New York.

Schimmell, M. S., Katz, E. Z., Shaag, Y., *et al.* (1991). Toxic neonatal effects following maternal clomipramine therapy. *Clin. Toxicol.* **29,** 479–484.

Schou, M., and Amdisen, A. (1973). Lithium and pregnancy. III. Lithium ingestion by children breast-fed by women on lithium treatment. *Br. Med. J.* **2,** 138.

Schramel, P., Hasse, S., and Ovcar-Pavlu, J. (1988). Selenium, cadmium, lead, and mercury concentrations in human breast mik, in placenta, maternal blood, and the blood of the newborn. *Biol. Trace Elem. Res.* **15,** 111–124.

Schwartz, J., and Otto, D. (1987). Blood lead, hearing thresholds, and neurobehavioral development in children and youth. *Arch. Environ. Health* **42,** 152–160.

Scialli, A. R. (1992). *A Clinical Guide to Reproductive and Developmental Toxicology.* CRC Press, Inc., Boca Raton, FL.

Seegal, R. F., and Shain, W. (1992). Neurotoxicity of polychlorinated biphenyls: The role of orthosubstituted congeners in altering neurochemical function. In *The Vulnerable Brain and Environmental Risks,* 2nd ed. R. L. Isaacson, and K. F. Jensen, Eds., Plenum Press, New York, pp. 169–195.

Silva, P. A., Hughes, P., Williams, S., and Faed, J. M. (1986). Blood lead, intelligence, reading attainment, and behavior in eleven year old children in Dunedin, New Zealand. *J. Child Psychol. Psychiatry* **24,** 43–52.

Sim, M. A., and McNeil, J. J. (1992). Monitoring chemical exposure using breast milk: a methodological review. *Am. J. Epidemiol.* **136,** 1–11.

Skerfving, S. (1988). Mercury in women exposed to methylmercury through fish consumption, and in their newborn babies and breast milk. *Bull. Environ. Contam. Toxicol.* **41,** 475–482.

Sovner, R., and Orsulak, P. J. (1979). Excretion of imipramine and desipramine in human breast milk. *Am. J. Psychiatry* **136,** 451–452.

Spigset, O., Carleborg, L., Norstrom, A., and Sandlund, M. (1996). Paroxetine level in breast milk. *J. Clin. Psychiatry* **57,** 39.

Steen, B., Rane, A., Lonnerholm, G., *et al.* (1982). Phenytoin excretion in human breast milk and plasma levels in nursed infants. *Ther. Drug. Monit.* **4,** 331–334.

Steiner, E., Villen, T., Hallberg, M., and Rane, A. (1984). Amphetamine secretion in breast milk. *Eur. J. Clin. Pharmacol.* **27,** 123–124.

Stewart, R. B., Karas, B., and Springer, P. K. (1980). Haloperidol excretion in human milk. *Am. J. Psychiatry* **137,** 849–850.

Sykes, P. A., Quarrie, J., and Alexander, F. W. (1976). Lithium carbonate and breast-feeding. *Br. Med. J.* **2,** 1299.

Taddio, A., and Ito, S. (1994). Drug use during lactation. In *Maternal–Fetal Toxicology. A Clinician's Guide,* 2nd ed. G. Koren, Ed. Marcel Dekker, Inc., New York, pp. 133–219.

Taddio, A., Ito, S., and Koren, G. (1994). Excretion of fluoxetine and its metabolite in human breast milk. *Pediatr. Res.* **35,** 149A.

Tellier, L., and Aronson, R. A. (1994). Lead in breast milk: should mothers be routinely screened? *Wis. Med. J.* **93,** 257–258.

Tennes, K., Avitable, N., Blackard, C., *et al.* (1985). Marijuana: prenatal and postnatal exposure in the human. *Natl. Inst. Drug Abuse Res. Monogr. Ser.* **59,** 48–60.

Trundle, J. I., and Skellern, G. G. (1983). Gas chromatographic determination of nicotine in human breast milk. *J. Clin. Hosp. Pharm.* **8,** 289–293.

Tryphonas, H. (1995). Immunotoxicity of PCBs (Aroclors) in relation to Great Lakes. *Environ. Health Perspect.* **103(Suppl 9),** 35–46.

Uauy, R. D., Birch, D. G., Birch, E. E., and et al. (1990). Effect of dietary Ω-3 fatty acids on retinal function of very low birth weight neonates. *Pediatr. Res.* **28,** 485–492.

US Environmental Protection Agency (1986). Air Quality Criteria for Lead, Vol. VII and IX. US Environmental Protection Agency, Research Triangle Park, NC. EPA Publication No. EPA-600/8-83/028.

USP DI (1997). *USP DI (USP Dispensing Information), Volume 1. Drug Information for the Health Care Professional,* 17th ed. The US Pharmacopeial Convention, Rockville, MD.

von Unruh, G. E., Froescher, W., Hoffmann, F., and Niesen, M. (1984). Valproic acid in breast milk: how much is really there? *Ther. Drug. Monit.* **6,** 272–276.

Wesson, D. R., Camber, S., Harkey, M., and Smith, D. E. (1985). Diazepam and desmethyldiazepam in breast milk. *J. Psychoactive Drugs* **17,** 55–56.

Whalley, L. J., Blain, P. G., and Prime, J. K. (1981). Haloperidol secreted in breast milk. *Br. Med. J.* **282,** 1746–1747.

WHO/FAO (1989). Toxicological Evaluation of Certain Food Additives and Contaminants. WHO Food Additives Series: 24. Cambridge University Press, pp. 295–328.

Wiles, D. H., Orr, M. W., and Kolakowska, T. (1978). Chlorpromazine levels in plasma and milk of nursing mothers. *Br. J. Clin. Pharmacol.* **5,** 272–273.

Wilson, J. T., Brown, R. D., Cherek, D. R., *et al.* (1980). Drug excretion in human breast milk. Principles, pharmacokinetics and projected consequences. *Clin. Pharmacokinetics* **5,** 1–66.

Wilson, J. T., Brown, R. D., Hinson, J. L., *et al.* (1985). Pharmacokinetic pitfalls in the estimation of the breast milk/plasma ratio for drugs. *Ann. Rev. Pharmacol. Toxicol.* **25,** 667–689.

Winneke, G., Brockhous, A., Ewer U., *et al.* (1990). Results from the European multicenter study on lead neurotoxicity in children: implications for risk assessment. *Neurotoxicol. Teratol.* **2,** 553–559.

Wisner, K. L., and Perel, J. M. (1991). Serum nortriptyline levels in nursing mothers and their infants. *Am. J. Psychiatry* **148,** 1234–1236.

Wisner, K. L., Perel, J. M., and Findling, R. L. (1996). Antidepressant treatment during breast-feeding. *Am. J. Psychiatry* **153,** 1132–1137.

Wisner, K. L., Perel, J. M., and Foglia, J. P. (1995). Serum clomipramine and metabolite levels in four nursing mother–infant pairs. *J. Clin. Psychiatry* **56,** 17–20.

Wolff, M. (1983). Occupationally derived chemicals in breast milk. *Am. J. Ind. Med.* **4,** 259–281.

Wolff, M. (1993). Lactation. In *Occupational and Environmental Reproductive Hazards: A Guide for Clinicians.* M. Paul, Ed. Williams and Wilkins, Baltimore, MD., pp. 60–70.

Wong, K. C., and Hwang, M. Y. (1981). Children born to PCB poisoned mothers. *Clin. Med.* (*Taipei*) **7,** 83.

Woodward, A., Grgurinovich, N., and Ryan, P. (1986). Breast feeding and smoking hygiene. Major influences on cotinine in urine of smokers' infants. *J. Epidemiol. Community Health* **40,** 309–315.

World Health Organization (1977). Task force on environmental health criteria for lead. WHO, Geneva.

World Health Organization (1990). Environmental Health Criteria 101: Methylmercury. WHO, Geneva.

Yakushiji, T., Watanabe, I., Kuwabara, K., *et al.* (1984). Postnatal transfer of PCBs from exposed mothers to their babies: influence of breast-feeding. *Arch. Environ. Health* **39,** 368–375.

Yamaguchi, A., Yoshimura, T., and Kuratsune, M. (1971). A survey on pregnant women having consumed rice oil contaminated with chlorobiphenyls and their babies. *Fukuoka Igaku Zasshi* **62,** 112–117.

Ziegler, E. E., Edwards, B. B., Jensen, R. L., *et al.* (1978). Absorption and retention of lead by infants. *Pediatr. Res.* **12,** 29–34.

PART V

Behavioral Assessment

Teratology was originally conceived as the study of gross structural malformations that are observable at birth and can be attributed to some developmental disturbance. Congenital defects were known to the ancient Greeks and Romans, and the science of teratology is generally considered to have originated with Étienne Geoffroy Sainte-Hilaire in the 1820s (O'Rahilly and Müller, 1992). By comparison, the field of *behavioral teratology* is still in its infancy. This term was first used in print in the title of Werboff fand Gottlieb's classic 1963 article in the *Obstetrical and Gynecological Survey.* The purpose of their review was "to implicate yet another system which is susceptible to the teratogenic effects of drugs. That is, the behavior, or functional adaptation of the offspring to its environment" (p. 420). In the article, Werboff fand Gottlieb present the existing literature (which was brief at that time) on the behavioral effects of developmental drug and hormone administration in animals. The authors already recognized the importance of animal studies for investigating the behaviorally disruptive effects of developmental insults, in part because "in humans, it is an extremely difficult task to relate a behavioral deficit in a child or an adult back to his mother taking a drug during pregnancy" (p. 420).

Research in behavioral teratology accelerated over the next 10–15 years, and by the mid-to-late 1970s the field was reasonably well established as seen in reviews by Coyle and colleagues (1976) and by Vorhees and coworkers (1979). However, some criticisms began to be raised concerning the reliability and, particularly, the functional relevance of experimental findings in behavioral teratology. As stated by Cattabeni and Abbracchio in 1988, "these studies on animals have shown that nearly every chemical may reach the fetus and therefore influence its development . . . but there is no consensus on what behavioral tests to use in animal studies, on the significance and interpretation of results obtained with such sophisticated methodologies, and, most important, on how far to generalize from the animal studies to humans." Although Vorhees (1989) subsequently wrote a cogent reply to Cattabeni and Abbracchio, some of these issues remain important, such as the question of extrapolating from animal research to the human domain. It is noteworthy, therefore, that the chapters in Part V and the chapters that follow later in the volume illustrate not only an awareness of the extrapolation issue, but also the progress that has been made in this area.

The aim of this section is to familiarize readers with many of the basic methodologies used in contemporary behavioral teratology. Chapter 22, by Jerrold S. Meyer, introduces some basic considerations in behavioral assessment and then goes on to cover a variety of unconditioned behaviors and drug challenge methods applied to rodent studies. Although rodent learning and memory paradigms are not covered in detail, several references are provided for readers who are interested in this type of assessment. Chapter 23 by Merle G. Paule presents a number of important behavioral assessment methods for nonhuman primates. One recurring theme throughout both chapters is the use of test batteries to adequately characterize the behaviorally disruptive effects of putative developmental neurotoxicants. This provides much greater sensitivity than single measures for identifying neurotoxicant effects, and it also facilitates standardization of testing across different compounds and between different laboratories.

The material covered in this volume not only describes the results of past teratologic research, it also lays out the continuing challenge of detecting, validating, and mechanistically understanding the phenomena of developmental neurotoxicity. Although earlier chapters have presented some of the magnificent advances that have been made possible by the application of molecular genetic techniques, it would be wise to remember that changes in neuronal gene expression ultimately are manifested as behavior, and that functional (i.e., behavioral) deficits are the true *raison d'etre* of developmental neurotoxicology. For this reason, behavioral researchers must continue to work hand-in-hand with their cellular and molecular colleagues to elucidate the effects of developmental neurotoxicants at all levels of analysis.

Jerrold S. Meyer

References

Cattabeni, F., and Abbracchio, M. P. 1988. Behavioral teratology: an inappropriate term for some uninterpretable effects. *Trends Pharmacol. Sci.* **9,** 13–15.

Coyle, I., Wayner, M. J., and Singer, G. 1976. Behavioral teratogenesis: a critical evaluation. *Pharmacol. Biochem. Behav.* **4,** 191–200.

O'Rahilly, R., and Müller, F. 1992. *Human Embryology and Teratology.* Wiley-Liss, New York.

Vorhees, C. V. 1989. Behavioral teratology: what's not in a name: a reply to Cattabeni and Abbracchio. *Neurotoxicol. Teratol.* **11,** 325–327.

Vorhees, C. V., Brunner, R. L., and Butcher, R. E. 1979. Psychotropic drugs are behavioral teratogens. *Science* **205,** 1220–1225.

Werboff, J., and Gottlieb, J. S. 1963. Drugs in pregnancy: behavioral teratology. *Obstet. Gynecol. Survey* **18,** 420–423.

CHAPTER

22

Behavioral Assessment in Developmental Neurotoxicology

Approaches Involving Unconditioned Behaviors and Pharmacologic Challenges in Rodents

JERROLD S. MEYER
Department of Psychology
Neuroscience and Behavior Program
University of Massachusetts
Amherst, Massachusetts 01003

I. Introduction

The primary aim of this chapter is to describe and evaluate two broad kinds of test approaches used in behavioral assessment within developmental neurotoxicology. The approaches discussed are those involving unconditioned behavioral responses and also those that make use of drug challenges. We focus here on rodents, particularly rats and mice, which are the most commonly used animals in behavioral teratology. Readers inter-

* Abbreviations: 5-HT, serotonin; 5-HTP, 5-hydroxytryptophan; CBTS, Collaborative Behavioral Teratology Study; CD, coefficient of detection; CNS, central nervous system; CV, coefficient of variation; DA, dopamine; EEG, electroencephalogram; EMG, electromyogram; GABA, γ-aminobutyric acid; GD, gestational day; IUGR, intrauterine growth retardation; LQ, lordosis quotient; LTP, long-term potentiation; MAM, methylazoxymethanol; MAO, monoamine oxidase; NCTR, National Center for Toxicological Research; NE, norepinephrine; PD, postnatal day; PTZ, pentylenetetrazol; REM, rapid eye movement; THC, tetrahydrocannabinol.

ested in the assessment of learning and memory in these species are referred to reviews by Heise (1984), Miller and Eckerman (1986), Beninger (1989), and Spear *et al.* (1990).

There are distinct advantages as well as disadvantages to using rats or mice. Among the advantages are: (1) small size and relatively low purchase and maintenance costs; (2) ease of handling, treatment, and testing; (3) simplicity of breeding in the laboratory setting; (4) short gestation periods; (5) relatively large litter sizes, thereby providing many offspring for study; and (6) extensive background information concerning normal behavioral and neural development. The popularity of rats and mice as subjects in behavioral teratologic research attests to the strength of these advantages. Nevertheless, there are several important disadvantages that must be kept in mind. First, if one is interested in toxicant effects on higher cognitive functions, such functions are either lacking in rodents or at least are technically difficult to evaluate. Second, rats and mice differ substantially from both humans and nonhuman primates in their anatomic structure, fine motor skills, and dominant sensory modality. Primates are highly visual and can be readily trained on tasks that require grasping or fine manipulation, whereas rodents do not rely as heavily on the visual sense (indeed, albino strains have particularly poor vision) and, of course, they do not possess the hands and fingers of primates. These difference are disadvantageous to the extent that they limit the availability of rodent test procedures that are isomorphic to the procedures used in human behavioral testing. Third, the temporal characteristics of their central nervous system (CNS) development differ importantly from that of primates. For example, the birth of granule cells in the cerebellar cortex occurs postnatally instead of prenatally (Altman, 1969), and this is also true for the majority of granule cells in the dentate gyrus of the hippocampal formation (Schlessinger *et al.,* 1975). Another significant species difference concerns the timing of myelinogenesis, as myelin staining is not even detectable in the rat forebrain before approximately postnatal day (PD) 10 (Jacobson, 1963). Because of these and other differences, the first 1–2 weeks of postnatal life in rodents are sometimes considered roughly comparable to the third trimester of human brain development. This is at best a crude generalization, although it does serve to underscore the important fact that prenatal exposure of rats and mice is not developmentally equivalent to the full period of prenatal exposure in humans.

II. Basic Considerations in Behavioral Teratology

Ideally, certain methodologic principles should be followed in studies of behavioral teratology. Some of these principles are shown in Table 1 and are discussed briefly in the following paragraphs. It is important to note that most of these ideas are not original to the present author, but have been explicated by previous writers (e.g., Spear *et al.,* 1985; Adams, 1986; Vorhees, 1986; Abel, 1989; Stanton and Spear, 1990).

1. Multiple outcome measures should be taken that encompass diverse functional systems. It should be obvious that one cannot adequately characterize the potential behavioral teratogenic effects of a given compound without a diverse battery of tests that reflect a range of behavioral and/or physiologic functions. Some tests provide information about a fairly specific behavioral or physiologic process, whereas others (so-called apical tests) measure the integrated output of several functional subsystems. For example, Butcher (1976) has pointed out that successful acquisition and performance of even a simple maze task requires the integration of sensory, motor, motivation, and learning capacities.

2. Testing should be conducted across the period of postnatal development, from the neonatal period well into adulthood. Experience has shown that some functional effects of perinatal exposure to a compound may occur early in postnatal development and then disappear, whereas others may emerge as the nervous system matures (Spear, 1984). An example of the latter phenomenon was found by Spyker (1975) in studies of prenatal methylmercury exposure in mice. Some of the treated subjects who had previously appeared normal began to exhibit neurologic deficits, including tremor, ataxia, and coordination problems, when they reached 10–15 months of age.

3. Young pups do not behave simply like small adults, and therefore assessments should be appropriate for each age group studied. This principle extends to several features of the test situation, including the choice

TABLE 1 Some Methodologic Principles in the Behavioral Assessment of Developmental Neurotoxicity

1. Multiple outcome measures should be taken that encompass diverse functional systems.
2. Testing should be conducted across the period of postnatal development.
3. Assessments should be appropriate for each age group studied.
4. Test measures should be chosen that show reasonable comparability between nonhuman and human subjects within the limits of species differences in behavioral repertoire.
5. It may be necessary to "challenge" subjects to observe neurotoxic effects.
6. Attempts should be made to integrate functional results with other findings demonstrating neural abnormalities at the structural, electrophysiological, cellular, or molecular levels.
7. Although the neural and behavioral effects of developmental toxicant exposure are usually deleterious to the offspring, a positive outcome may occur in some instances.

of age-appropriate sensory demands, motor requirements, and reinforcers (Riley *et al.,* 1985). Moreover, certain behaviors (e.g., isolation-induced ultrasonic vocalizations, mouthing/suckling, wall climbing) only occur during a particular period of development and thus can be studied only during that time frame (Spear *et al.,* 1985).

4. The aim of developmental neurotoxicity testing in animals is to screen for possible neurobehavioral alterations in humans following early exposure to the same compound. Consequently, test measures should be chosen that show reasonable comparability between nonhuman and human subjects, within the limits of species differences in behavioral repertoire. Some examples include reflex development, sensory-evoked potentials, activity, sleep–wake cycles, auditory startle response, Pavlovian and operant learning, reproductive behavior, ingestive behavior, and aggressive behavior (for more information, see Stanton and Spear, 1990).

5. In some cases, the subjects' behavior under "baseline" conditions may show no effect of prior toxicant exposure, whereas behavioral deficits are manifested when the animals are appropriately challenged (Hannigan and Blanchard, 1988; Spear, 1996). This may be accomplished by stressing the subjects, by increasing task difficulty in learning studies, or by using drug challenges to perturb neurochemical functioning. The last part of this chapter discusses the use of pharmacologic challenges in behavioral teratology.

6. Attempts should be made to integrate functional results with other findings that demonstrate neural abnormalities at the structural, electrophysiologic, cellular, and molecular levels. Behavioral research holds a special position within developmental toxicology because it tells us how subjects may have become compromised in their ability to interact functionally with the physical and social environment. Nonetheless, good science revolves around mechanistic explanation, and it is therefore paramount that we pursue the neurobiologic substrates underlying these functional effects. Those behavioral researchers who do not have the training, facilities, or inclination to investigate neural mechanisms are encouraged to develop collaborations with other laboratories that do possess the necessary interest and expertise.

7. Although the neural and behavioral perturbations produced by developmental toxicant exposure are typically deleterious to the offspring, Cattabeni and Abbracchio (1988) have pointed out that this may not always be the case. For example, Little and Teyler (1996) reported that prenatal cocaine exposure unexpectedly led to an enhancement rather than a deficit in hippocampal long-term potentiation (LTP) in rabbits. Consequently, investigators must evaluate their subjects in an unbiased manner so that the outcome of chemical exposure is properly recognized and acknowledged even if it seems positive rather than negative.

III. Unconditioned Behaviors and Other Measures for Assessing Developmental Neurotoxicity

A. *Measures of Physical Growth*

Deficits in growth can serve as useful indicators of developmental toxicity, although Vorhees (1986, 1989) has pointed out that, at least for some substances, higher doses may be required to produce growth retardation than are needed for functional (i.e., behavioral) effects. Common growth measures include, but are not limited to, overall body weight and brain weight (either whole-brain or regional weights following dissection into specific areas of interest). If a toxicant causes intrauterine growth retardation (IUGR) when administered to a pregnant dam, this is reflected in reduced weight at birth of exposed compared to control offspring. Additional weight measures may then be taken at various postnatal intervals to determine whether the affected offspring show "catch-up" growth. Another possible pattern, which has been observed in some prenatal alcohol studies on mice (Middaugh and Boggan, 1991), involves no deficit in birth weight but a reduction in postnatal growth.

It is worth noting that slightly more sophisticated measures are available to investigators who wish to characterize changes in body or brain growth of their subjects. For example, group differences in body weight may reflect variation in skeletal growth (which, in turn, is regulated by growth hormone and other endocrine factors), lean body mass, and/or body fat. Measurement of body length (an index of skeletal growth) and composition (i.e., protein, lipid, and water) can be used to determine which of these components has been altered by a developmental manipulation. Likewise, if offspring brain weight has been affected, one may use brain DNA content, DNA concentration, and protein/DNA ratio to provide crude indices of altered cell number, packing density, and cell size respectively (for example, see Carlos *et al.,* 1992).

B. *Appearance of Physical Developmental Landmarks*

In addition to obtaining information on pre- and postnatal growth, investigators may also assess the time of appearance of standard landmarks of physical development. This is particularly useful when performing initial screening of a potential developmental toxicant (see later this chapter). Commonly measured landmarks include pinna unfolding, incisor eruption, eye opening, and (in females) vaginal opening. As previously discussed by Adams and Buelke-Sam (1981), a delay in physical maturation (e.g., day of eye opening) may be indicative of insult, but it does not necessarily reflect a deficit in CNS development.

C. Neural Excitability

Overall CNS excitability is governed by a complex interplay of excitatory and inhibitory synaptic transmission. These processes are mediated largely by the amino acid transmitters glutamate, γ-aminobutyric acid (GABA), and (especially in the spinal cord) glycine. Changes in CNS excitability produced by developmental neurotoxicants may be detected by assessing seizure susceptibility in offspring. Experimental seizures can be readily induced by several different techniques, including electroshock (Browning, 1987) or treatment with a convulsant drug such as pentylenetetrazol (PTZ) or strychnine (Faingold, 1987). For example, an early study by Pizzi and co-workers (1979) demonstrated that mice treated neonatally with monosodium glutamate exhibited enhanced susceptibility to PTZ-induced seizures in adulthood. Another approach is called *kindling,* which generally refers to the production of seizure activity by the repeated administration of an initially subconvulsant electrical stimulation to a discrete brain area (McNamara *et al.,* 1993). A pharmacologic variation of this phenomenon is cocaine kindling, in which subjects acquire a heightened susceptibility to cocaine-induced seizures when the drug is given subchronically in an appropriate regimen. Snyder-Keller and Keller (1995) found an increased rate of cocaine kindling in young male rats exposed to cocaine prenatally.

Because seizure susceptibility is not constant across postnatal development (Moshé *et al.,* 1993), age at testing is an important consideration when assessing toxicant-induced changes in CNS excitability. Technical constraints can also play a role in study design. For example, although kindling has been demonstrated in pups as young as PD 14–16 (Haas *et al.,* 1992), younger animals are difficult to study because of the need for placement of indwelling electrodes.

D. Sleep–Activity Cycle

Rats, mice, and other rodents exhibit characteristic circadian cycles of sleep and activity that are normally entrained to the light–dark cycle. Being nocturnal, rats and mice are most active during the dark phase of the cycle and spend most of their sleep time during the light phase. For example, rats sleep during approximately 80% of the light period and about 20% of the dark period (Borbely and Tobler, 1985). If one suspects that a particular toxicant might disrupt the development either of the principal circadian oscillator in the brain (i.e., the suprachiasmatic nucleus) or of the brainstem and diencephalic systems that regulate sleep and arousal states, then it may be of interest to study sleep–activity rhythms in exposed offspring.

1. Activity Rhythms

Depending on the information desired, there are several ways to investigate cycles of sleep and activity. For example, it is relatively simple to generate activity rhythms if the subjects are housed over a period of days or (preferably) weeks in an apparatus that continuously monitors locomotor or running (if a running wheel is present) behavior. Imposition of constant darkness causes circadian rhythms to "free-run," thereby permitting the determination of whether toxicant exposure has altered the periodicity of the circadian oscillator. Moreover, spectral analysis may be used to assess the amplitude of the activity rhythm under either standard light–dark conditions or under constant darkness. Developmentally, normal rat pups weaned at PD 14 and housed under a 12-hr light–dark cycle have already begun to exhibit a circadian activity rhythm at the time of weaning, and this rhythmicity gradually increases over the next several weeks (Honma and Honma, 1985).

2. Sleep Cycle

Rather than studying activity rhythms, an investigator may be interested in looking for possible treatment-induced changes in the sleep cycle. This is more technically demanding because it requires surgical implantation of indwelling electrodes and the measurement of electroencephalographic (EEG) activity. Like humans, rats exhibit intermittent periods of rapid-eye movement (REM) sleep characterized by a low-voltage, fast EEG. These REM epochs alternate with longer periods of non-REM sleep with a slower, higher voltage EEG. In rats, REM sleep constitutes roughly 15–20% of total sleep time (Borbely and Tobler, 1985). During the perinatal period, rat pups are asleep most of the time, but by PD 10 they have already begun to spend half of their time awake (Gramsbergen *et al.,* 1970).

E. Sensory Function

Several methodologic approaches can be used to assess the effects of toxicants on sensory function. These range from the elicitation of simple orienting responses and stimulus-induced reflexes, to complex psychophysical methods, and finally to the study of sensory evoked potentials. Each of these approaches are described here, although it should be mentioned that psychophysical and evoked potential techniques are generally considered too time- and labor-intensive to be used in primary screens for behavioral teratogenesis (see later this chapter). Hence, these techniques are most likely to be carried out in follow-up evaluation, after primary screening has already been completed.

1. Olfactory Orientation

Orienting responses can be elicited by almost every sensory modality. Such responses are highly variable, however, and they do not offer much information about sensory abnormalities unless the deficit is severe (i.e., the animal is nearly deaf or blind). One possible exception is an olfactory orientation test that was developed for use in young rats. Between PD 9 and 12, rat pups develop the ability to discriminate between home-cage versus clean bedding and to show approach behavior to the familiar stimulus (Gregory and Pfaff, 1971). By measuring the orientation of the subjects or their latency to reach the goal, one can obtain a simple index of olfactory functioning. It is important to note, however, that the latency measure must be used with caution because it can obviously be influenced by alterations not only in olfaction but also in motor activity.

2. Auditory Startle Reflex

A useful response for evaluating auditory system function is the auditory startle reflex. Startle reflexes refer to the rapid contraction of antagonistic muscles in the limbs, neck, and trunk that are elicited by sudden and strong auditory, tactile, or vestibular stimuli. These responses are thought to serve a protective function by preparing the organism for a potential physical blow or predatory attack. Automated systems to study auditory or tactile (air puff–induced) startle reflexes are available from several manufacturers, including San Diego Instruments (San Diego, CA) and Columbus Instruments (Columbus, OH) (Fig. 1). Auditory startle reflex latencies are quite short, as electromyographic (EMG) responses can be recorded within 5–10 ms following stimulus onset in rats. This finding indicates that the neural circuitry underlying the auditory startle reflex must be relatively simple, with only a few intervening synapses between the sensory receptors in the cochlea and the output motoneurons (Yeomans and Frankland, 1996).

Studies of developmental neurotoxicity typically use modifications of the basic auditory startle test procedure (Adams, 1986). For example, one may present the same auditory stimulus at a constant rate for a number of trials and examine the rate of response habitation. Another common approach involves the phenomenon of prepulse inhibition, which is an attenuation of the response by presentation of a weaker stimulus shortly before

FIGURE 1 One type of automated apparatus (SR-LAB) for testing acoustic startle reflexes in rodents. (Courtesy of San Diego Instruments, San Diego, CA.)

(e.g., 100 ms) the startle-eliciting stimulus. By varying the intensity of the prepulse stimulus and determining when response inhibition occurs, it is possible to determine auditory thresholds without the more laborious and time-consuming operant methods traditionally associated with sensory psychophysics. A third variation is fear-potentiated startle, which refers to enhancement of the startle response by prior presentation of a conditioned fear stimulus (for example, a light or tone that has been previously paired with foot shock). This procedure provides information concerning the efficacy of fear conditioning rather than information about the auditory system *per se.*

Although test batteries commonly assess auditory startle reflexes in adulthood, the basic response can be observed in rat pups as young as PD 12 or 13 (Bolles and Woods, 1964; Vorhees *et al.,* 1979b). However, some of the more complex phenomena associated with startle reflex testing may not emerge until later in development. For example, Hunt and colleagues (1994) found evidence for fear-potentiated startle in PD 23 but not PD 16 animals.

3. Psychophysical Methods

Psychophysics is the systematic study of sensory capacities by determining behavioral responses to physical changes in sensory stimuli. Although the theory and techniques of psychophysics were developed to investigate human sensory functions, "animal psychophysics" was subsequently established as a powerful tool for studying the sensory systems of animal subjects (see Stebbins, 1990, for a historic review). Psychophysical procedures usually require the use of operant learning tasks, but the topic is discussed here because the purpose of the task is to assess sensory function instead of learning or memory.

After subjects have initially been trained to perform an operant response such as a lever press for food or water reinforcement, a stimulus control element is introduced into the task. Animals may be asked whether they detect the presence of a sensory stimulus, or they may be required to discriminate between stimuli that vary along some chosen dimension. The simplest way to carry out such training is to use a single-response ("go/no-go") task in which, for example, lever pressing is reinforced in the presence but not the absence of the stimulus. Evans (1982) mentions certain disadvantages to this approach and points out that forced-choice methods are generally preferable, although such methods are admittedly more labor intensive and require more elaborate and costly equipment. The major advantages of forced-choice methods are that reinforcement is available on each trial and that response topography is the same for each subject choice (i.e., press one lever when you detect the stimulus and press the other lever when the stimulus is not detected). Several options are also available with respect to stimulus presentation procedures. Readers interested in more information on this topic are referred to the review by Rice (1994).

4. Sensory Evoked Potentials

If one has the appropriate training and equipment, then it may be desirable to screen for toxicant-induced sensory deficits using electrophysiologic rather than behavioral methods. This is most commonly accomplished by studying sensory evoked potentials, which are complex, time-locked electrical signals elicited by discrete sensory stimuli and recorded in cortical or subcortical structures associated with the transmission or processing of sensory information. Changes in the amplitude or latency of evoked potential waveforms can serve as a sensitive indicator of sensory system damage (Fox *et al.*, 1982; Boyes, 1992). For example, treatment of rat pups with lead resulted in abnormal visual evoked potentials throughout the preweaning period and into adulthood (Fox *et al.*, 1977). Two separate groups also found abnormalities in auditory evoked potentials in young rats exposed to cocaine prenatally (Church and Overbeck, 1990; Salamy *et al.*, 1992).

F. Motor Development

Postnatal development of reflexes and other aspects of motor function has been well characterized in rodents, particularly rats (Kallman, 1994). By assessing these behaviors at various ages, one can test for toxicant effects as indicated by a delayed onset of mature motor patterns, a persistence of immature behaviors, or a deficit in strength or coordination on a particular motor task.

1. Primitive Motor Behaviors

Two behaviors that are only present in young rat pups are the rooting reflex and a primitive type of locomotion that has been termed "pivoting." Rooting is a neonatal reflex that involves turning of the pup's head in the direction of a tactile stimulus applied to the snout. In newborns, this reflex is important for finding the mother's nipple in suckling; however, in normal animals the response wanes as the nervous system develops and other orienting mechanisms become functional. Pivoting refers to ambulation occurring mainly through the use of the forelimbs, upper torso, and head. This pattern seems to arise because the hindlimbs are not yet strong enough to support the animal's weight, and consequently the animal moves essentially by pivoting around its pelvis. Pivoting behavior was found to begin around PD 3, peak at PD 7, and nearly disappear by PD 15 as it is replaced by more adult-like ambulation (see later this chapter) (Altman and Sudarshan, 1975).

2. Other Motor Functions

Three motor functions commonly assessed in developmental test batteries are surface righting, midair righting, and negative geotaxis (Adams, 1986). Surface righting is the animal's ability to turn over from a supine position. Although this response may take up to a minute or more in newborn pups, the latency rapidly decreases across postnatal development. Subjects are typically tested daily beginning on PD 3 or 4 and ending on PD 7 or later, and an observer scores either the latency for each subject to complete the task or the number of subjects exhibiting successful righting within a specified time interval such as 15 sec. Midair righting is elicited by holding the animal upside down and dropping it onto a soft padded surface from a height of 60 cm. Altman and Sudarshan (1975) reported that in normal rats, midair head-turning begins to emerge on PD 13–14, and complete righting rapidly develops between PD 15 and 17. Finally, negative geotaxis is an orienting response in which an animal that has been positioned on an inclined surface with its head pointing downward will turn around 180° on the incline. In developmental neurotoxicology, subjects are typically tested from PD 6 to PD 9 or later for their latency to complete the 180° turn on a 25° inclined surface.

Last, three other motor functions that have been of interest to some investigators are mentioned. The first is forelimb grip strength, which can be assessed using several techniques. One method introduced by Tilson and Cabe (1978) involves the animal grasping a circular wire ring attached to a commercially available strain gauge and then pulling the subject away from the gauge until it releases its grip. An even simpler method is to induce the animal to suspend itself by grasping a thin wire, and then measuring the latency to fall off the wire. Bâ and Seri (1995) found an exponential increase in wire grasping time in rats tested between PD 10 and 25. The second function is motor coordination, which has classically been tested by means of the rotarod apparatus (Watzman and Barry, 1968). In this task, the subject is placed on a rotating rod or drum and must walk continuously to maintain its balance. One test approach is to increase the speed of rotation gradually and determine the point at which each subject falls off the rod. Alternatively, the speed is maintained at a constant rate and one records the percentage of subjects successfully completing a predetermined trial length (for example, 5 min). In either case, performance on the rotarod provides an overall index of motor coordination, vestibular control, and muscular strength. The third function is swimming, a complex motor pattern that has been well characterized developmentally (Schapiro *et al.,* 1970) and that has been used successfully in screening for a number of developmental toxicants (for example, Vorhees *et al.,* 1979b; Vorhees *et al.,* 1984; Rice and Millan, 1986). Assessment of swimming in adult animals typically involves testing subject endurance, and because rodents are generally good swimmers, task difficulty is often increased by some procedure such as adding a weight to the animal (Kallman, 1994).

G. Activity and Exploratory Behavior

1. Activity

One important feature of motor function not discussed in the previous section is locomotor activity. Activity can be a sensitive indicator of developmental toxicity, and hence measurement of this behavior is included in all screening batteries (see later this chapter). Locomotion can be divided into two elements: ambulation (horizontally directed movement) and rearing (vertically directed movement). It is generally desirable to assess both elements because they may be differentially affected by toxicant exposure. Despite its apparent simplicity, however, locomotor activity is considered an apical measure because it is influenced not only by the state of the animal's motor system but also by sensory and motivational factors. Furthermore, quite different results can be obtained depending on numerous procedural variables, including whether testing is carried out in the home cage or a separate testing chamber, the physical dimensions and complexity of the test environment, the method of motion detection, and whether subjects are habituated to the apparatus prior to testing (Reiter and MacPhail, 1982; Kallman, 1994). These considerations should be kept in mind when designing the activity component of a behavioral test battery.

Early studies of locomotor activity generally involved the use of running wheels, tilt- or jiggle-cages, or the open field test (Adams and Buelke-Sam, 1981). Each of these has its limitations, and significant criticisms have particularly been raised against the open field test as a valid and reliable measure of motor function (for example, see Walsh and Cummins, 1976). Indeed, behavior in the open field may be more indicative of exploratory and emotional tendencies than of motor activity *per se* (Maier *et al.,* 1988). Nevertheless, there are several advantages to this task that have led to its continued use in behavioral assessment. First, it is very inexpensive to set up, particularly in comparison to the commercially supplied, automated systems described later. Second, because the animal *must* be observed in the course of testing, the open field lends itself to the simultaneous measurement of other behaviors such as grooming and (in young animals) wall climbing (Spear *et al.,* 1985).

The most important methodologic advance in neurotoxicologic studies of motor activity has been the development of automated test systems. Although some of these systems involve video tracking methods, the most popular devices use infrared photobeam arrays. Such systems provide detailed quantitative output and can usually be set up to monitor many subjects at the same time in separate cages or other enclosures. Although some investigators have constructed their own homemade apparatus, excellent systems are available commercially from AccuScan (formerly Omnitech; Columbus, OH) and also the same firms mentioned earlier with respect to startle reflex testing (Fig. 2). Depending on the configuration, these commercial systems are typically capable of detecting ambulation, rearing, and in some cases "fine" movements such as those occurring in psychomotor stimulant-induced stereotypic behavior. Some systems continuously record the animal's location, thereby providing information about distance traveled, amount of time spent moving, pattern of movement, number of rotations, and so forth (Fitzgerald *et al.,* 1988). Despite the range of variation among various homemade and commercial systems, it is reassuring to note that qualitatively similar findings were obtained from six different laboratories using different equipment to test the acute effects of various compounds on motor activity (Crofton *et al.,* 1991).

Although many systems require placement of the animal into a specialized test chamber, others are designed for measuring activity in the home environment. This has the advantage of obviating the possible con-

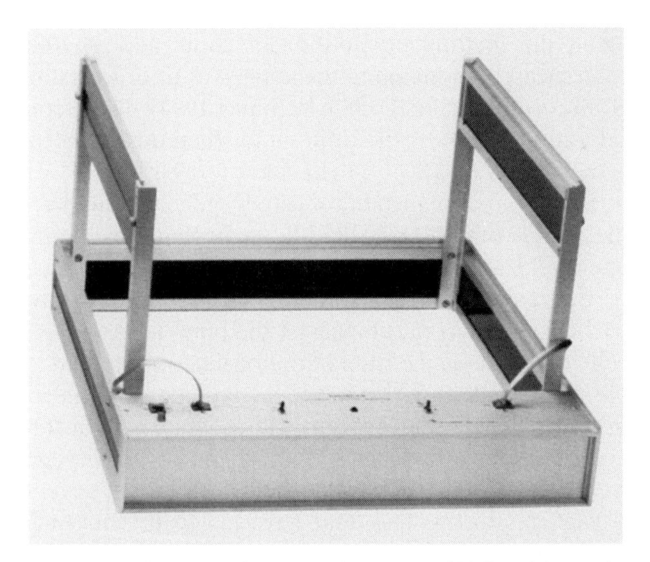

FIGURE 2 One type of automated apparatus (Digiscan) for testing locomotor activity in rodents. (Courtesy of AccuScan Instruments, Inc./dba Omnitech, Columbus, OH.)

founding effects of behavioral responses to novelty and also allows for around-the-clock monitoring if desired. In the simplest of such systems, the animals are maintained in standard clear plastic tubs that are placed within a photobeam matrix for purposes of testing. Alternatively, they may be housed in a figure eight residential maze system, which permits measurement of activity over long time periods (i.e., hours or days) in a complex environment (Reiter and MacPhail, 1982). The sensitivity of this approach has been demonstrated in numerous studies, such as the work by Balduini and co-workers (Balduini *et al.,* 1989, 1991) showing that the effects of prenatal treatment with the antimitotic alkylating agent methylazoxymethanol (MAM) on activity depend on the dose and the gestational age of exposure.

Activity can be assessed not only in adult animals, but also in pups. Up to PD 7–8 in mice (Fox, 1965) and PD 10 in rats (Westerga and Gramsbergen, 1990), pups ambulate primarily by pivoting and crawling. Thereafter, however, a rapid transition to walking occurs such that by PD 15 in rats, the animals have begun to display an adult-like gait (Westerga and Gramsbergen, 1990). Interestingly, there is also a large surge in the amount of locomotor activity between PD 10 and 15, followed by further changes that depend on how the behavior is measured (Campbell *et al.,* 1969; Altman and Sudarshan, 1975; Bâ and Seri, 1995). Given the rapid development of ambulation during the second postnatal week, testing of subjects around this time period may reveal whether prenatal or early postnatal toxicant exposure has caused a maturational delay.

2. Exploratory Behavior

Exploratory behavior can be defined as behavior directed toward acquiring information about the environment. Exploration is generally considered to be different than simple ambulation, although in practice it is often difficult to completely separate one from the other. Harro (1993) has reviewed some of the common methods used to assess exploratory behavior, including the open field, elevated plus-maze, and hole-board test (which involves the measurement of head-dipping into an array of holes placed in an open field apparatus; see File and Wardill, 1975). It is important to note that behavior in these test situations reflects several factors besides exploration, including ambulatory tendencies and fear of novelty (neophobia).

H. Ingestive Behaviors

Several studies have shown that exposure of adult animals to certain neurotoxic agents can alter the amount or pattern of food and water consumption (Evans, 1994). However, ingestive behavior has been a relatively neglected area of study in developmental neurotoxicology. At most, investigators have taken simple measures of daily food and water intake to augment data collected on postnatal weight gain. Ingestive behaviors are controlled, however, by complex neural mechanisms that receive and process many types of internal and external information. In the case of food intake and body weight regulation, for example, such information includes the current availability of oxidizable metabolic fuels, the sensory qualities of food (i.e., palatability), and other sensory cues that may have been conditioned to the reinforcing effects of food intake. Similar considerations apply to the regulation of drinking behavior and fluid balance. Moreover, there has been an explosion of knowledge concerning the genetic, neurochemical, and physiologic mechanisms that control eating behavior and body weight (see Rowland *et al.,* 1996). As these mechanisms are likely to be disrupted by developmental exposure to at least some neurotoxicants, it behooves investigators to begin examining more closely the complexities of ingestive behavior as an important component of secondary evaluation. Ingestive behavior can also be studied in young animals using simple procedures such as those described by Phifer and coworkers (1992).

I. Species- and Gender-Typical Behaviors

A final, broad area of assessment pertains to several types of behaviors that ethologists have generally found to be species and gender typical. These include grooming, reproductive behavior, aggressive behavior, and play. Some of the methods used to study these behaviors are presented here, along with selected examples of their use in developmental neurotoxicology.

1. Grooming

Celis and Torre (1993) have summarized the stereotyped pattern of grooming observed in rats. A grooming sequence typically begins with vibrational movements of the forelegs followed by licking of the forepaws, face and head washing, and finally grooming of the remainder of the body including the flanks, back, hindlimbs, and genitalia. Grooming bouts may also include scratching by one or both hindlimbs, or shaking of the body that is reminiscent of (though not as intense as) the "wet-dog shakes" that occur during withdrawal from some drugs of abuse.

In rodents, the incidence of grooming increases in conflict situations and following the presentation of novel stimuli. These findings led to the proposal that grooming behavior may be elicited or perhaps disinhibited at a particular level of arousal (Fentress, 1968a,b). Furthermore, many neuropeptides elicit "excessive grooming" (manifested as longer bouts of grooming) when injected directly into the brain (reviewed by Celis and Torre, 1993). Consequently, changes in the grooming behavior of prenatal drug–exposed subjects could be indicative of altered levels of arousal or changes in neuropeptide function. It is important to keep in mind, however, that grooming is incompatible with ambulation. Thus, if an experimental treatment leads to hyperlocomotion, grooming behavior may be reduced as secondary consequence of this effect.

One simple way to study grooming behavior is to observe or videotape subjects in an open field, and then to measure the frequency and overall duration of grooming bouts during a standard test period. Using this approach, we observed an enhancement of grooming in young male rats treated daily with Δ^9-tetrahydrocannabinol (THC) from PD 30 to 39, and then tested on PD 41 (Meyer, Benson, and Kunkle, unpublished observations, 1997). Alternatively, one may use the time-sampling procedure (occurrence of grooming or other behaviors of interest every 15 sec) described by Gispen and colleagues (1975).

2. Reproductive Behavior

Mating behaviors have been studied extensively in rodents, particularly rats. The state of female sexual receptivity, which is termed *behavioral estrus,* is induced by the secretion of estrogen and progesterone by the ovaries. When a male rat (or an experimenter) provides tactile stimulation of the flanks or anogenital region of a receptive female, she displays a characteristic spinal dorsoflexion accompanied by tail deflection. This is the classic lordosis posture, which is absolutely necessarily for penile insertion (i.e., intromission) and copulation by the male (Diakow, 1974; Pfaff *et al.,* 1978). Estrus females also display so-called proceptive behaviors such as "darting" and "hopping," which are stimulating to the male (Beach, 1976).

The central features of male rat sexual behavior include mounting (clasping the female's flanks followed by pelvic thrusting), intromission, and ejaculation. Male rats do not ejaculate each time they achieve intromission. Rather, ejaculation typically requires from 8 to 14 intromissions, each of which is separated by a pause, during which the male may groom itself.

A standard test for rodent sexual behavior involves placing the test subject in a Plexiglas chamber or other type of arena for 10–20 min with an appropriate stimulus animal of the opposite sex. Female subjects obviously must be in a state of estrus either occurring naturally in the intact, cycling animal (as determined by examination of vaginal smears) or induced artificially by ovariectomizing the animals and then treating them with an appropriate regimen of estrogen and progesterone. Stimulus animals have not been exposed to the neurotoxicant, and thus their behavior should not influence the outcome of the test. Testing is typically conducted under a red lamp during the dark phase of the animals' light–dark cycle. Common measures of female sexual behavior include lordosis quotient (LQ; percentage of mounting attempts that result in a lordosis response) and the frequency and intensity of proceptive behaviors. Common measures of male sexual behavior include latency to first mount, latency to first intromission, number of mounts to ejaculation, number of intromissions to ejaculation, latency from first intromission to ejaculation, and latency from ejaculation to next intromission (Abel, 1989).

Many screening batteries do not assess reproductive behavior at all, although some at least include simple measures of estrus cyclicity and fertility (for example, see Walker *et al.,* 1989). Yet many studies have shown that perinatal exposure to various drugs or other toxicants can alter the patterns of reproductive behaviors in adult offspring. One example comes from the work of Murphy and colleagues (1995), who showed that male offspring of dams given 5 mg/kg of THC twice daily from gestational day (GD) 14 through GD 19 later exhibited a significant increase in mount latency compared to controls, and furthermore none of the exposed subjects (compared to 50% of the controls) ejaculated during a 30-min test session. Female offspring were ovariectomized in adulthood and then treated with moderate doses of estrogen and progesterone that elicited at least some lordotic responses in 90% of the control females. In contrast, only 25% of the THC-exposed subjects exhibited any lordosis behavior, the latency to the first lordosis was increased, and the LQ was reduced, although not significantly (20% vs. 43% for the controls).

3. Aggressive Behaviors

Several investigators have emphasized the utility of classifying aggressive behaviors with respect to their social context, apparent function, and differing physiologic substrates. Among the most influential typologies of animal aggression have been the classic scheme of Moyer (1968) and the later proposal of Brain (1981), which differentiated between predatory attack, self-defensive behaviors, parental defensive behaviors, and social conflict. The term agonistic behavior is sometimes used to subsume both threatening/attack behaviors and submissive, defensive, and withdrawal behaviors that also occur in social encounters. Indeed, Grant (1963) and Grant and Macintosh (1963) found that agonistic interactions between pairs of male rats, mice, guinea pigs, or hamsters gave rise to a variety of postures and behaviors, including threats or attacks with biting of the opponent, aggressive postures, upright or sideways postures that possess both aggressive and defensive elements, aggressive grooming of the opponent, crouching, freezing, and fleeing.

Although a wide variety of aggression-eliciting situations have been introduced over the years, some of the older procedures are no longer used because they fail to meet current ethical standards of animal research (e.g., shock-elicited fighting or mouse-killing by rats). Among the paradigms still in common use for studying male aggression are 1) isolation-induced aggression—that is, housing a male animal in isolation for several weeks and then testing its response to an intruder; 2) resident-intruder aggression—that is, similar to that previously mentioned, except that the male is housed socially with one or more female conspecifics; and 3) brain stimulation–induced aggression—that is, aggressive behavior elicited by direct electrical stimulation of appropriate brain regions, usually in the hypothalamus (Koolhaas and Bohus, 1991; Olivier and Mos, 1992). Female rodents are usually studied with respect to maternal aggression (defense of the young). When assessing rodent aggressive behavior, some important considerations to keep in mind are choice of genetic strain (some are more aggressive than others), whether to study individual pairs of animals or the response of a social group to an intruder, whether to test repeated aggressive encounters in the same subjects (this can lead to animals becoming consistent "winners" or "losers," with marked behavioral and physiologic consequences in either case), and ensuring that physical harm is minimized during these agonistic interactions (Alleva, 1993).

Although agonistic behaviors are acutely affected by a number of different compounds, the influence of prenatal treatment on later aggressiveness has not been studied as extensively. One illustrative example is the report by Laviola *et al.* (1991) on the effects of prenatal oxazepam exposure on maternal aggression in mice. When tested for their responses to a male intruder on postpartum day 6, drug-exposed subjects exhibited substantially higher frequencies of fighting episodes, attacks, and offensive postures, and lower frequencies of submissive postures. These findings demonstrate the ability of this test paradigm to reveal a treatment-induced shift in agonistic tendencies from submissiveness towards increased aggressiveness.

4. Play Behavior

Play behavior in young mammals (including rodents) is thought to serve several potential functions, including the acquisition of social skills that are needed in later life. One such skill involves learning to recognize agonistic (i.e., threat or fear) displays and to perform the appropriate responses to these signals. Indeed, the behavioral precursors of aggression, dominance, and subordination are readily observable in the early play of many species.

Play-fighting in rats is much more prevalent in males than females. This activity begins around PD 17 and shows a large increase in frequency (in males) around the fourth week of life (Meaney and Stewart, 1981). Pouncing, wrestling, boxing, on-top posture (also called *pinning* when the bottom animal is on its back), and chasing are among the major behaviors that comprise play-fighting in rats.

As in the case of aggression, play has not received as much attention in the developmental toxicology literature as some of the other behaviors discussed earlier. Nevertheless, several researchers have recognized the importance of studying this and other social behaviors in the assessment of prenatal drug effects. In these experiments, play behavior is typically assessed in juvenile or periadolescent subjects that are observed in the presence of a stimulus animal (play partner) in a testing arena. Using this type of arrangement, Wood and coworkers (1994) found a significant reduction in pinning behavior in male and female rats (PD 28 to PD 36) that had been exposed to cocaine prenatally.

5. Isolation-Induced Ultrasonic Vocalizations

The final species-typical behavior considered here is the ultrasonic vocalizing that occurs when infant rodents are removed from the nest. Such vocalizations, which span a range of 30–50 kHz, appear to be a response to hypothermia rather than to social isolation (Blumberg *et al.,* 1992). When tested at ambient temperature, the rate of vocalizing of rat pups peaked at around PD 8 and fell to a low level by PD 14, the age at which the eyes have opened (Okon, 1971). Ultrasonic vocalizations are thought to serve a communicative function by attracting the mother and inducing her to return the pup to the nest (Allin and Banks, 1972). Blumberg and Alberts (1990, 1991) have proposed that rat pup ultrasonic vocalizations are an acoustic by-product of a respiratory maneuver known as "laryngeal braking," which involves the expulsion of air through a constricted larynx. This phenomenon occurs commonly in neonatal mammals in response to cold exposure, and it functions along with increased respiration rate to enhance pulmonary gas exchange.

Detection and measurement of ultrasonic vocalizations can be accomplished in a relatively simple manner using a commercially available "bat detector" (a device that transduces ultrasonic sounds to the audible range) (Fig. 3) and a tape recorder, or more sophisticated analyses can be performed with professional audio equipment (Winslow and Insel, 1991). Because of its simplicity, this behavior lends itself well to developmental neurotoxicity research (Cuomo *et al.,* 1987). However, obtaining useful results requires the usual care in selecting and controlling the experimental conditions. This is exemplified in the early work by Adams and her colleagues, which showed that the effects of maternal hypervitaminosis A or methylmercury treatment on the ultrasonic vocalizing of neonatal offspring depended on the bedding used in the test situation (clean vs. homecage) and the age of the subjects (Adams, 1982; Adams *et al.,* 1983).

FIGURE 3 PD 10 rat pup being tested for isolation-induced ultrasonic vocalizations. The microphone mounted above the animal is connected to a "bat detector" that was obtained from Ultra Sound Advice, London, UK.

IV. Behavioral Test Batteries in Developmental Neurotoxicology

Whereas some studies are designed to investigate the influence of a neurotoxicant on the development of one or just a few specific behavioral responses, there are times when a more comprehensive test battery is needed. This is particularly true when one is screening for possible neurotoxic effects of a previously untested compound. Test batteries must be constructed thoughtfully if useful results are to be obtained. As Vorhees (1985b, 1987) has pointed out, each test in the battery should possess the qualities of sensitivity, reliability, and validity. Conceptually, sensitivity refers to the ability of a given test to detect a specific type of dysfunction. It is defined quantitatively by statistical measures referred to as *coefficient of variation* (CV) and *coefficient of detection* (CD) (for more information, see Vorhees, 1985b, 1987). Reliability means that the test in question yields reproducible results both within the same laboratory (intralaboratory reliability) and between different laboratories (interlaboratory reliability). Finally, the overall validity of a test includes the concepts of construct validity and susceptibility (what biologic function or neural system is assessed by that test, and how closely the test results reflect damage to that system), criterion validity (whether the test meets the criterion of detecting the functional deficits induced by known developmental neurotoxicants [i.e., positive controls]), and cross-species predictive validity (ability of the test to correctly predict neurotoxicity in humans).

With respect to predictive validity, Vorhees (1989) pointed out that historically, most (if not all) neurobehavioral teratogens were recognized initially from human clinical findings rather than from animal screening. However, this is not necessarily because the available screening batteries lack sensitivity or validity. Indeed, a large workshop was conducted to assess qualitative and quantitative comparability of human and animal developmental neurotoxicity (Rees *et al.,* 1990). Although it is clear that improvements can be made in our current test batteries (see discussion later this chapter), the participating researchers generally concluded that animal testing serves its basic function of indicating areas of potential functional deficits in humans exposed to developmental neurotoxicants. The important point, therefore, is that new chemical compounds be tested before human exposure leads to adverse consequences for offspring.

An added challenge in the construction of screening batteries stems from the need to limit time and expense to manageable levels. Consequently, test batteries commonly include a number of apical measures that are relatively simple to carry out. Some types of learning tasks, however, may be relatively labor-intensive and time-consuming for the investigators.

Certain other features are commonly found in developmental neurotoxicity test batteries. For example, many batteries use a split-litter approach, in which various tests are conducted on different subsets of each litter. Moreover, it is critical that the data be analyzed using the appropriate statistical analyses. As littermates are generally more similar to each other than to animals in the same treatment group but in a different litter, good statistical practice requires that the litter be the unit of statistical analysis. The remainder of this section presents several rodent test batteries that have been developed for use in behavioral teratology screening.

A. *Test Batteries Designed for Rats*

1. Cincinnati Test Battery

TABLE 2 The Cincinnati Test Battery for Rats

Measure	Age at measurement (PD)
Physical growth	
Body weight	1, 4, 7, 14, 21, 30, 60
Developmental landmarks	
Upper and lower incisor eruption	8-completion
Eye opening	12-completion
Vaginal opening	30-completion
Sensorimotor function	
Surface righting	3–12
Negative geotaxis	6, 8, 10, 12
Swimming ontogeny	6–24 (even days only)
Pivoting locomotion	7, 9, 11
Olfactory orientation	9, 11, 13
Appearance of auditory startle response	10-completion (≤16)
Auditory/tactile startle (habituation and prepulse inhibition)	70, 80
Activity and exploration	
Figure eight activity monitor (or open field)	15–17, 40–42
Spontaneous alternation	40
Learning and memory	
M-maze	43–44
Biel (Cincinnati) water maze	50–55
Y-maze active avoidance	60–80
Passive avoidance	90–93

The Cincinnati Test Battery is an important screening instrument developed over a period of several years by Vorhees, Butcher, and Brunner at the University of Cincinnati (e.g., Vorhees, 1983; Vorhees *et al.,* 1979a,b, 1981; 1984). By 1985, the Cincinnati battery contained the items shown in Table 2 (Vorhees, 1985b). This battery provides extensive coverage of most areas of interest, and its sensitivity in detecting neurobehavioral teratologic effects has been demonstrated in many studies (Vorhees, 1983, 1985b; Vorhees *et al.,* 1979a,b, 1981, 1984). One notable feature of the Cincinnati Test Battery is its inclusion of swimming ontogeny and water maze tasks. As described earlier, swimming is a complex motor function that can be a sensitive indicator of toxicant-induced motor impairment. With regard to water mazes, the M-maze task is a brightness discrimination task based on a modified-T swimming maze, whereas the Biel (Cincinnati) maze task is a spatial learning task using a six-unit asymmetric multiple T-maze (Vorhees, 1983). Vorhees (1985b) has suggested that water mazes require less subject training and may, in some cases, be more sensitive to developmental neurotoxic effects than standard appetitive operant tasks.

2. Collaborative Behavioral Teratology Study Test Battery

By 1979, a number of researchers had recognized the need for a systematic interlaboratory evaluation of test sensitivity and reliability for screening potential developmental neurotoxicants. This concern led to the design of a new test battery by Adams, Buelke-Sam, Kimmel, and their colleagues at the National Center for Toxicological Research (NCTR). A large-scale study was set up (Collaborative Behavioral Teratology Study—CBTS) in which this battery would be tested by six different laboratories around the country, including NCTR itself and Vorhees' lab at Cincinnati. Two different test compounds were chosen for investigation: d-amphetamine sulfate (0.5 or 2.0 mg/kg on GD 12–15) and methylmercuric chloride (2.0 or 6.0 mg/kg on GD 6–9). Each study was carried out using four separate replications (four litters per treatment group in each replication) in order to provide a measure of intralaboratory reliability. The rationale, preliminary research, study design, results, and implications of CBTS were published in 1985 in a special issue of the journal *Neurobehavioral Toxicology and Teratology* (Adams *et al.,* 1985a,b,c; Buelke-Sam *et al.,* 1985; Kimmel and Buelke-Sam, 1985; Kimmel *et al.,* 1985; Nelson *et al.,* 1985). The scope of this endeavor is reflected by calculations showing that the project generated more than 1,000,000 data points and required approximately 44 man-years to complete (Buelke-Sam *et al.,* 1985).

The CBTS test battery is summarized in Table 3. It differs from the Cincinnati battery in several important respects, including less emphasis on sensorimotor function (although it should be noted that the Cincinnati battery did not include an adult assessment of auditory startle until the value of this measure was illustrated in CBTS testing) and the use of a visual discrimination operant task (including a discrimination reversal) instead of the water mazes and avoidance tasks used in the Cincinnati battery. Also noteworthy is the incorporation of a pharmacologic challenge test represented by an assessment of motor activity before and after administration of 0.5 mg/kg or (on a separate day) 1.0 mg/kg of d-amphetamine sulfate.

Due to space limitations, we summarize the results of the CBTS only briefly. In general, offspring of the untreated control dams exhibited similar parameters of physical development and behavior across different laboratories. This important finding indicates that the test measures used in the CBTS battery can yield reasonably comparable results from different investigators when the tests are administered in a standardized way. The test battery also exhibited adequate reliability. Although both intralaboratory variation (i.e., replication) and interlaboratory variation were statistically significant for most measures, such variation was generally modest. Litter was a significant factor for almost all measures, thus supporting the need to use litter as the unit of statistical analysis in neurobehavioral teratology. Most of the behavioral measures yielded moderate CVs (typically within the range of 18–40%) both within and between laboratories. Based on estimated CDs, a 5–25% change in outcome in the treated compared to the control offspring would be sufficient to be detected statistically. This is a reasonable range of values for this type of research. Behavioral performance of subjects was frequently influenced by prior testing. This fact must be taken into account when multiple behavioral measures are obtained on the same subjects. Gender differences occurred for some measures, such as locomotor activity. However, overall, female behavior was not more variable than that of males. There is no reason, therefore, that developmental neurotoxicity screening should not include both females and males. Finally, evaluation of the prenatal drug treatments revealed that exposure to the high dose of methylmercury led to increased auditory startle response amplitudes, particularly in females. In contrast, maternal amphetamine administration produced no reliable effects on offspring behavior.

TABLE 3 The Collaborative Behavioral Teratology Study Test Battery for Rats

Measure	Age at measurement (PD)
Physical growth	
Body weight	1, 7, 14, 21, 30, 60, 90, 110–120
Developmental landmarks	
Upper and lower incisor eruption	7-completion (≤13)
Eye opening	12-completion (≤16)
Testes descent	21-completion (≤26)
Vaginal opening	30-completion (≤40)
Sensorimotor function	
Negative geotaxis	7–10
Olfactory orientation	9–11
Auditory startle response (habituation)	18–19, 57–58
Activity and exploration	
Figure eight activity monitor (1-hr test)	21, 60
Figure eight activity monitor (23-hr test)	100–108
Learning and memory	
Visual discrimination operant task	75–89
Pharmacologic challenge	
Figure eight activity before and after D-amphetamine	120–131

The failure to observe an influence of prenatal amphetamine exposure on later behavior using the CBTS test battery could raise concerns about whether this battery is adequately sensitive for use as a routine screening instrument for developmental neurotoxicity. However, the statistical estimates of test sensitivity mentioned previously and the positive results with methylmercury raise the alternate possibility that the doses of amphetamine chosen for study were too low (and/or the dosing period was too brief) to produce functionally significant neurotoxic effects. Indeed, when Vorhees (1985a) used an abbreviated version of the Cincinnati Test Battery to investigate the effects of prenatal amphetamine in parallel with the CBTS, again no statistically significant effects were found, except for a small reduction in swimming direction scores on PD 6 and 10. These findings support the conclusion that the dosing regimen of amphetamine used in these experiments produces little or no impairment in the offspring, at least within the realms of growth, physical development, sensorimotor function, locomotor activity, and certain types of complex learning. Of course, no screening instrument can examine all types of learning. Moreover, neither the CBTS nor Cincinnati Test Battery includes tests of the species-typical behaviors discussed earlier. It may be wise for future test batteries to include some of these behaviors to insure that researchers are not overlooking neurotoxicant-induced impairments in various social or other kinds of species-typical behaviors.

3. European Collaborative Study

In Schwerzenbach, Switzerland, Jürg Elsner, along with colleagues in Basel and in Berlin, Germany, designed a study to evaluate the relative merits of a simplified manual test battery conducted on young animals versus complex automated test procedures conducted on adult subjects. Female rats were exposed to several different doses of methylmercury in their drinking water beginning 13 days before pairing with a stud male and continuing until offspring weaning at PD 21. Dosing, mating, and manual testing were all carried out in Basel. The manual test battery included 1-day assessments of pinna unfolding (PD 4), ear opening (PD 14), eye opening (PD 16), testes descent (PD 27), vaginal opening (PD 35), olfactory orientation (PD 14), righting reflex (PD 14), auditory startle reflex (PD 14), swimming (PD 14), pupillary reflex (PD 21), passive avoidance acquisition (PD 27), and passive avoidance retention (PD 34). Subsets of subjects were then transported to either Berlin or Schwerzenbach for adult testing. The automated procedures consisted of a visual discrimination reversal learning task based on a nose-poke response, a wheel-shaped activity monitor, and a lever-press task involving four training phases: 1) autoshaping, 2) discrete-trials responding, 3) a visual discrimination schedule, and 4) a spatial alternation schedule. The design, results, and interpretation of these experiments are reported in a series of papers published in *Neurobehavioral Toxicology and Teratology* (Elsner, 1986; Elsner *et al.,* 1986; Schreiner *et al.,* 1986; Suter and Schön, 1986).

The results of this interesting study indicated that both the manual test battery and the automated techniques yielded significant effects of prenatal methylmercury exposure, even at the lowest dose used (1.5 mg/l of drinking water). The authors emphasized that the automated operant procedures provided very precise information on the nature of the abnormal behavioral responses, including data suggesting possible attentional deficits in the treated offspring. They concluded that an ideal screening instrument should contain both manual testing of young subjects to assess the attainment of developmental landmarks and sensorimotor function, and also automated testing of acquisition and performance on various operant tasks.

B. Test Batteries Designed for Mice

Fox (1965) performed the first major investigation of sensorimotor/reflex development in mice. This classic study has served as the foundation for much of the subsequent work assessing the behavioral effects of developmental neurotoxicants in this species. Indeed, modified Fox batteries have revealed developmental delays following prenatal exposure to several different compounds, including ethanol (Zidell *et al.,* 1988), phencyclidine (Nicholas and Schreiber, 1983), and oxazepam (Alleva *et al.,* 1985).

The Fox battery has several limitations, however, such as an emphasis on early (preweaning) assessment and an absence of measures of either locomotor activity or learning. A number of mouse test batteries have therefore been constructed to remedy these problems. One example is the battery proposed by Kallman and Condie (1985), which still focuses on early development but adds a test of muscular coordination and a passive avoidance learning and retention task to the basic measures of sensorimotor function (i.e., righting reflex, forelimb placing, forepaw grasp, rooting reflex, cliff drop, auditory startle, and bar holding).

Interestingly, some of the more recent test batteries in mice have begun to include assessments of social behaviors. This can be seen in a series of studies by Pankaj and Brain (1991a,b,c) that examined the influence of prenatal exposure to benzodiazepine agonists, antagonists, and inverse agonists on early postnatal development and adult behavior. The test battery in this case included measures of agonistic and other behaviors in a resident–intruder paradigm. Different compounds produced gender-specific effects on aggressive and defensive behaviors, thus confirming the usefulness of including this type of measure in behavioral screening.

Perhaps the most ambitious mouse test batteries used to date are those of Petruzzi *et al.* (1995) and Rayburn *et al.* (1997). Petruzzi and co-workers examined the influence of gestational ozone exposure on offspring development. In addition to a modified Fox battery, the measures obtained in this study included social interaction tests at PD 23–25 and PD 43–45, assessment of locomotor activity at PD 57–60 using an automated apparatus, and performance in an 8-arm radial maze at PD 84–98. It is noteworthy that the social interaction tests yielded the most robust effects of the ozone treatment. The study of Rayburn and colleagues (1997) concerned the effects of prenatal exposure to the synthetic glucocorticoids betamethasone and dexamethasone. These investigators assessed offspring behavior on the following measures: negative geotaxis, separation-induced vocalizations, homing to the nest, auditory startle (both early and in adulthood), locomotor activity (both early and in adulthood), pain sensitivity, vision, wire coordination, social play, aggressive behavior (males only), elevated plus-maze (a test of exploratory behavior and anxiety), forced swim task (a "learned helplessness" paradigm), several different mazes to test spatial and other kinds of learning, and fertility. Although many of these measures were conducted in a much cruder fashion than, for example, in the CBTS or Cincinnati Test Battery (for example, auditory startle was carried out manually and just assessed the presence or absence of an ear and/or head twitch), the sheer breadth of this battery must be considered impressive.

In general, the mouse studies described illustrate the value of including social behaviors in behavioral screening, despite the added time and expense. It is hoped that the need for a more comprehensive assessment of all major categories of neurobehavioral function (including various species-typical as well as ingestive behaviors) will eventually lead to an expansion of standard rodent (including rat) test batteries to include these less traditional functions.

V. Pharmacologic Challenge Methods in Developmental Neurotoxicity Assessment

The term *pharmacologic challenge* refers to the procedure of testing the behavioral (or physiologic) reactivity of neurotoxicant-exposed and control offspring to one or more doses of a selected drug. There are at least three potential reasons for using this approach in assessing the effects of developmental neurotoxicants. As mentioned earlier, one reason is to "unmask" functional deficits that may not be apparent in the absence of the challenge treatment. A second purpose is to use a drug that is selective for a specific neurotransmitter system as a probe to determine whether that system has been influenced by toxicant exposure. Finally, offspring may be challenged with the same drug to which they were exposed prenatally, thereby determining whether sensitivity to the toxicant has been altered. Of course, these aims are not mutually exclusive, and one may use a given drug challenge for several purposes.

The proper use of pharmacologic challenges requires a careful consideration of drug choice, doses, age at testing, and other variables. Although it is most convenient to test subjects at only a single dose of the challenge agent, a more complete dose–response assessment greatly increases the possibility of detecting behavioral teratogenic effects. By repeatedly administering the drug challenge, one may also look for group differences in tolerance or sensitization (for example, see Hannigan and Pilati, 1991). Space limitations preclude a more detailed discussion of these issues; however, further information may be found in Hannigan and Blanchard (1988).

The present discussion focuses on the use of drugs to probe neurotransmitter system functioning. This type of challenge entails administering a drug that acts as a selective receptor agonist or antagonist, that evokes transmitter release, or that inhibits neurotransmitter synthesis, reuptake, or catabolism. In principle, one may use this approach to investigate the influence of developmental toxicant exposure on any neurotransmitter system for which appropriate pharmacologic tools are available. In practice, however, most of the research to date has focused on the monoaminergic neurotransmitters (i.e., dopamine [DA], norepinephrine [NE], and serotonin [5-HT]) because of their important behavioral functions, because many neurotoxicants (particularly drugs of abuse) interact significantly with these transmitters, and because of the availability of good pharmacologic agents (see Feldman *et al.,* 1997, for detailed information on the neuropharmacology of the monoaminergic systems).

A. *Dopaminergic Drug Challenges*

1. Amphetamine Challenge

Drug challenges in behavioral teratology frequently involve DA, as alterations in this neurotransmitter are thought to underlie some of the long-term effects of *in utero* exposure to many drugs (Middaugh and Zemp, 1985). The most commonly used DA-related drug is amphetamine, which is a potent DA releasing agent and also inhibits DA reuptake. Although amphetamine is not selective in its neurochemical actions (it also influences the noradrenergic and serotonergic systems), the acute effects of amphetamine administration are largely attributable to an enhancement of dopaminergic transmission (Feldman *et al.,* 1997).

The first reported use of an amphetamine challenge in behavioral teratology was that of Hughes and Sparber (1978), who examined the effects of prenatal methylmercury exposure on the behavior of rats in an autoshaped operant task. Although there were no group differences in operant responding under baseline conditions, significant differences did emerge following a challenge with 2.5 or 3.75 mg/kg of amphetamine. Other studies have used an amphetamine challenge to investigate the influence of prenatal ethanol on operant responding under a progressive-ratio reinforcement schedule in mice (Gentry *et al.,* 1995) and the influence of prenatal cocaine on acoustic startle responses in rats (Hughes *et al.,* 1996).

Amphetamine administration to adult rodents also leads to dose-dependent biphasic changes in locomotor activity (increased and then decreased) and an increase in focused stereotyped behaviors (Segal and Kuczenski, 1994). Hence, amphetamine-induced activity or stereotypy has been used to test for possible dopaminergic effects of many behavioral teratogens, such as ethanol (Blanchard *et al.,* 1987), cocaine (Simansky and Kachelries, 1996), haloperidol (Scalzo *et al.,* 1989), oxazepam (Alleva and Bignami, 1986), lead (Jason and Kellogg, 1981), tributyltin (Gårdlund *et al.,* 1991), and retinoic acid (Nolen, 1986). Young subjects likewise show behavioral activation to amphetamine, although the nature and intensity of the response varies considerably with age (Sobrian *et al.,* 1975; Lanier and Isaacson, 1977; Barrett *et al.,* 1982).

Amphetamine acts on the DA nerve terminals, but the resulting behavioral effects depend on the subsequent stimulation of postsynaptic DA receptors. From a screening standpoint, this is a useful feature of the amphetamine challenge because it should theoretically permit detection of prenatal drug-induced changes on either side of the synaptic complex. From the standpoint of understanding mechanism, however, this fact is clearly disadvantageous. Fortunately, this problem can be overcome to some extent by administering a direct agonist or antagonist for DA receptors, thereby probing the sensitivity of the combined receptor/signal transduction system.

2. Dopamine Agonists

The DA receptors are divided into two distinct families: D_1-like and D_2-like receptors (Seeman and Van Tol, 1994). The D_1-like family is composed of the D_1 and D_5 (alternatively D_{1A} and D_{1B}) subtypes, whereas the subtypes associated with the D_2-like family include D_2, D_3, and D_4. Most of the available dopaminergic agonists and antagonists are reasonably selective between receptor families, but not within their preferred family (although this situation has begun to improve with the ongoing development of new compounds).

Characteristic behavioral effects are produced in rodents by stimulation of D_1 or D_2 receptor agonists. Most studies of D_1-related behavioral effects have used the partial agonist SKF 38393, and systemic administration of this compound to adult rats and mice produces a robust enhancement of grooming behavior (Molloy and Waddington, 1987; Wachtel *et al.,* 1992). The effect of SKF 38393 on locomotor activity in adult animals is less clear, as several studies have found increased locomotor activity (e.g., Molloy and Waddington, 1987; Bruhwyler *et al.,* 1991), whereas Eilam and co-workers (1992) observed a decrease in locomotion following SKF 38393 treatment. Although there have also been conflicting reports concerning the effects of this compound on developing animals, Shieh and Walters (1996) found convincing evidence that SKF 38393 stimulates locomotor activity in 10- and 21-day-old rats previously habituated to the test apparatus.

The most commonly used D_2 agonists in behavioral teratology screening have been apomorphine and, to a lesser extent, quinpirole (LY 171555). Apomorphine is a mixed D_1/D_2 agonist that exerts its effects mainly through activation of D_2-like receptors, whereas quinpirole is generally characterized as a D_2/D_3 agonist with low affinity for D_1 receptors. We focus here mainly on apomorphine, although quinpirole shares many of apomorphine's functional properties. Low doses (typically 0.025 to 0.25 mg/kg) of apomorphine cause hypomotility (reduced locomotor activity) in rats and mice (Strömbom, 1975; Ljungberg and Ungerstedt, 1976), as well as increased yawning behavior in rats (Urba-Holmgren *et al.,* 1982). At higher doses (0.5 mg/kg and above), apomorphine administration is associated with hypermotility, stereotypy (including cage-climbing behavior), and hypothermia (Costall and Naylor, 1973; Riffee *et al.,* 1979; Protais *et al.,* 1984; Ogren and Fuxe, 1988). In a developmental study that tested subjects at ages ranging from PD 7 to PD 35, Shalaby and Spear (1980) found that high doses of apomorphine led to enhanced locomotor activity at all ages and increased wall climbing behavior at PD 7 and PD 14. However, hypomotility induced by low doses of apomorphine did not appear until PD 28.

The effects of low doses of apomorphine have generally been attributed to selective stimulation of presynaptic D_2 autoreceptors, with additional activation of postsynaptic, behaviorally stimulating D_2 receptors occurring at higher doses. Thus, the results of Shalaby and Spear (1980) were interpreted to indicate that DA autoreceptors may not become functional until the fourth or fifth week of life. Lin and Walters (1994), however, found evidence for hypomotility at PD 21 using several drugs thought to be more selective than apomorphine for D_2 autoreceptors. Moreover, some investigators have proposed that effects of low-dose apomorphine treatment are not due to autoreceptor activation, but rather the stimulation of a subpopulation of postsynaptic D_2-like receptors that inhibit rather than enhance locomotor activity (see Ståhle, 1992). It is possible that these are D_3 receptors, as apomorphine has a relatively high affinity for this receptor subtype (Sokoloff *et al.,* 1990), and administration of low doses of the partially selective D_3 agonist 7-OH-DPAT (7-OH-*N*-*N*-di-*n*-propyl-2-aminotetralin) decreases locomotion (Daly and Waddington, 1993), whereas the D_3 receptor antagonist U99194A stimulates locomotor activity (Waters *et al.,* 1993).

3. Dopamine Antagonists

Blockade of either D_1 or D_2 receptors leads to a suppression of spontaneous exploratory and locomotor behavior and the elicitation of a state known as *catalepsy* (Wanibuchi and Usuda, 1990). Catalepsy refers to a marked difficulty in initiating voluntary motor activity, usually demonstrated experimentally by showing that the subject does not change position when placed in an awkward or uncomfortable posture. Neuroleptic drugs like haloperidol produce catalepsy in developing as well as adult animals; however, the sensitivity to these drugs is lower around PD 15 than at either PD 10 or PD 20 (Meyer *et al.,* 1984). There is some evidence that this ontogenetic profile is due to developmental changes in the cholinergic system (Burt *et al.,* 1982).

There are several different methods for measuring catalepsy. Catalepsy in mice is studied by means of the ring immobility test, a procedure in which the animal is placed on a horizontal wire ring and its activity monitored for 5 min (Pertwee, 1972). Probably the most popular method for rats is the horizontal bar test. In this test, the subject is placed with its forepaws on an elevated wooden or metal bar, and the latency to remove its paws from the bar is recorded. Figure 4 shows a haloperidol-treated pup from a study in the author's laboratory that found a decreased cataleptic response in prenatal cocaine-exposed animals at PD 10 (Meyer *et al.,* 1994). Although catalepsy is a simple behavior to measure, best results are obtained by following certain experimental guidelines (Sanberg *et al.,* 1984; Ferré *et al.,* 1990). Furthermore, it is important to recognize that even though blockade of D_2 receptors is the proximal cause of neuroleptic-induced catalepsy (Sanberg, 1980), this behavior is modulated by other neurotransmitters such as acetylcholine (Burt *et al.,* 1982) and 5-HT (Wadenberg, 1996). Thus, changes in catalepsy associated with developmental neurotoxicant treatment do not necessarily reflect alterations in the dopaminergic system.

B. Noradrenergic Drug Challenges

1. Clonidine Challenge

The central adrenergic receptors are comprised of β_1 and β_2 receptors, and families of α_1 and α_2 receptors (i.e., α_{1A-D} and α_{2A-D}; Bylund *et al.,* 1994). The drug clonidine, which is a general α_2-adrenoceptor agonist, is frequently used as a challenge of the noradrenergic system. Very low doses of clonidine (e.g., 0.03–0.3 mg/kg) exert a sedative effect in adult rats and mice that is manifested by reduced ambulation and exploratory behavior (Strömbom, 1975; Ortmann *et al.,* 1982). Clonidine-induced sedation may be due to stimulation of α_2 autoreceptors on noradrenergic neurons of the locus coeruleus (De Sarro *et al.,* 1987; however, see also Nassif-Caudarella *et al.,* 1986). The sedative effects of clonidine are age dependent, in that rat pups at PD 7, 10, or 14 show behavioral activation as indicated by increased locomotion, wall climbing, and ultrasonic vocalizations (Reinstein and Isaacson, 1977; Hansen, 1993). By PD 20 or 21, a more adult-like hypoactivity is observed, which could be due to maturation of the α_2-autoreceptor system (Nomura *et al.,* 1980).

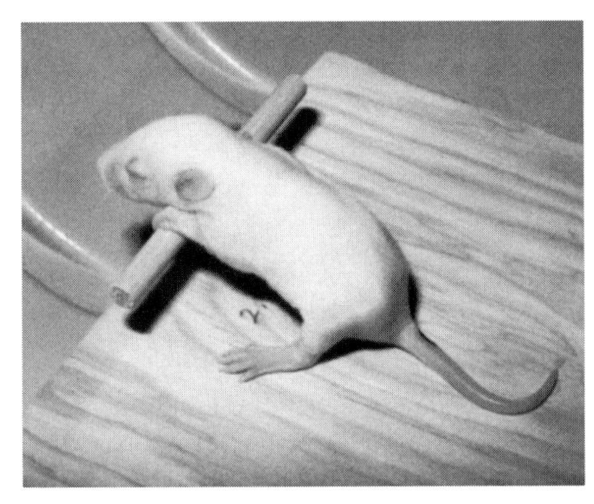

FIGURE 4 Measurement of haloperidol-induced catalepsy in a PD 10 rat pup using the horizontal bar test.

In contrast to the effects of low doses of clonidine, adult rats given a very high dose such as 40 or 50 mg/kg exhibit a behavioral syndrome consisting of tremor, ataxia, forelimb hyperextension and reciprocal forepaw treading, hindlimb abduction, tactile hyperreactivity, Straub tail, autonomic symptoms, and intermittent periods of catalepsy (Pranzatelli *et al.,* 1987). Some of these symptoms are similar to those found in the so-called serotonin syndrome (see next section). Mice treated with similar doses of clonidine become hyperaggressive if they are in a social setting (Ushijima *et al.,* 1984), but they engage in self-biting behavior if socially isolated (Katsuragi *et al.,* 1984). From a mechanistic standpoint, it is likely that the effects seen in both rats and mice are unrelated to α_2-adrenoceptor activation, but rather reflect other neurochemical actions of clonidine. In fact, the aggressive and self-mutilating responses in mice have been attributed to a blockade of central adenosine receptors (Katsuragi *et al.,* 1984; Ushijima *et al.,* 1984).

C. Serotonergic Drug Challenges

1. Serotonergic Agonists in Adult Animals

In early studies, challenge of the serotonergic system was often accomplished by administration of a 5-HT precursor such as L-tryptophan or 5-hydroxytryptophan (5-HTP), sometimes along with a monoamine oxidase (MAO) inhibitor or (in the case of 5-HTP) a peripherally acting decarboxylase inhibitor (to increase 5-HTP availability to the brain). The aim of these treatments was to enhance central 5-HT synthesis and synaptic overflow, thereby activating 5-HT–dependent behaviors. Although used occasionally still, this approach has largely been supplanted by the administration of agonists that are relatively selective for one or a few 5-HT receptor subtypes.

Quite a large number of different serotonergic receptors have been identified through a combination of pharmacologic and molecular biologic (i.e., cloning) techniques. These include (but are not limited to) the 5-HT_{1A}, 5-HT_{1B}, 5-HT_{1D}, 5-HT_{1E}, 5-HT_{1F}, 5-HT_{2A},

5-HT_{2B}, 5-HT_{2C}, 5-HT_3, and 5-HT_4 subtypes (for review, see Hoyer *et al.,* 1994). Some 5-HT_{1A} receptors function as somatodendritic autoreceptors, whereas synaptic (i.e., terminal) autoreceptors are of the 5-HT_{1B} or 5-HT_{1D} subtype. We focus on 5-HT_{1A} receptors and the 5-HT_2 receptor family, as these have received the most attention in behavioral challenge studies.

Many years ago, investigators found that treatment of adult rats with a 5-HT precursor and an MAO inhibitor elicited a striking behavioral syndrome characterized by reciprocal forepaw treading (sometimes also called *piano playing*), head weaving, tremor, hyperreactivity, hindlimb abduction and low body posture, Straub tail, and piloerection (reviewed by Green and Backus, 1990). Later work showed that another aspect of this syndrome is retraction of the lower lip, thus baring the lower incisors (Moore *et al.,* 1993). The serotonin syndrome is produced mainly by stimulation of postsynaptic 5-HT_{1A} receptors, as it can be elicited by the prototypical 5-HT_{1A} agonist 8-OH-DPAT (8-hydroxy-[2-di-*N*-propylamino]-tetralin), and the effect of 8-OH-DPAT is enhanced rather than diminished by prior depletion of 5-HT (Lucki and Wieland, 1990). However, investigators who might wish to use this syndrome as a behavioral screening paradigm for changes in 5-HT_{1A} receptor function should be aware that the receptors responsible for the syndrome are probably located in the brainstem and spinal cord rather than in the forebrain (Jacobs and Klemfuss, 1975).

At low doses (e.g., 0.1 mg/kg), 8-OH-DPAT does not elicit the serotonin syndrome, but rather reduces exploratory behavior (locomotor activity in a novel environment) and lowers core body temperature. Several studies suggest that these effects are due to activation of 5-HT_{1A} autoreceptors on the serotonergic neurons (Green and Backus, 1990). Another possible autoreceptor-mediated effect of 8-OH-DPAT is an enhancement of food intake in nondeprived rats (Lucki and Wieland, 1990).

Stimulation of 5-HT_2 receptors by agonists such as DOI (1-[2,5-dimethoxy-4-iodophenyl]-2-aminopropane) elicits a head twitch response in mice (Darmani *et al.,* 1990). In rats, 5-HT_2 receptor activation leads to a cluster of behaviors including head shakes, wet-dog shakes, and "skin jerks," which are periodic paraspinal muscle contractions along the animal's back (Pranzatelli, 1990). DOI administration also causes hyperthermia (Pranzatelli, 1990) and reduced food intake (Aulakh *et al.,* 1992), effects that are opposite those produced by low doses of 8-OH-DPAT.

2. Serotonergic Challenges in Developing Animals

Early studies of serotonergic challenges in developing animals typically used relatively nonselective agonists or antagonists because these were the only pharmacologic agents available at the time. For example, administration of the mixed agonist/antagonist quipazine to rat pups at PD 3–4 produced marked behavioral activation consisting of increased locomotion, wall climbing, forelimb paddling, hindlimb treading, and mouthing, the manifestation of an unusual position of the limbs (UPL–pup stationary with its paws up in the air and its elbows on the floor), and decreased twitching (Spear and Ristine, 1981; Ristine and Spear, 1985). Some of these behaviors resemble components of the serotonergic syndrome in adult animals, although important differences are also present. Suckling behavior may also be under serotonergic control in neonatal rodents, as treatment with the general serotonergic antagonist metergoline reduced infant rat suckling behavior at PD 3–4 and PD 7–8, but not at older ages (Spear and Ristine, 1982; Ristine and Spear, 1984).

Later studies have examined the influence of more selective serotonergic agonists such as 8-OH-DPAT and DOI in young rats. Preweanling (PD 17–18) animals exhibited a biphasic consummatory response to 8-OH-DPAT; that is, feeding behavior was stimulated by a low dose (0.03 mg/kg) but suppressed by a high dose (0.5 mg/kg) of the drug (Spear *et al.,* 1991). In contrast, neonates (PD 3–4) showed only the high dose effect, which suggests that 5-HT_{1A} autoreceptors (which are thought to underlie the stimulatory effect of 8-OH-DPAT on feeding) are not yet functional during the neonatal period. High doses of 8-OH-DPAT also produced behavioral activation in neonates (Kirstein and Spear, 1988), although similar treatment surprisingly failed to stimulate locomotor activity in preweanling animals (Frambes *et al.,* 1990). DOI administration to neonates increased mouthing behavior (opening and closing of the mouth), decreased probing (pushing against the floor or wall of the apparatus by the subject's snout), and elicited the same type of UPL seen following quipazine (Kirstein and Spear, 1988). Interestingly, no studies of DOI in young rats up to PD 20 have reported any incidence of head shaking, wet-dog shakes, or skin jerks (Jackson and Kitchen, 1989; Frambes *et al.,* 1990), suggesting that these responses to 5-HT_2 receptor stimulation may mature rather late in development.

VI. Summary

It is now well established that behavioral testing is a critical component of developmental neurotoxicity assessment. Current practice typically involves the examination of toxicant-exposed and control offspring for physical growth, attainment of physical developmental landmarks, sensory function, motor development, activity, and learning. Together, these measures encompass a broad range of basic processes, most of which can either be studied using rapid and simple manual techniques or have been adapted for automated assessment. We have seen, however, that a number of other neurobehavioral systems may also be useful for characterizing neurotoxicant effects. These include neural excitability, sleep–activity cycle, ingestive behavior, and a variety of species- and gender-typical behaviors such as grooming, reproductive behavior, aggressive behavior, play, and isolation-induced ultrasonic vocalizations.

Because of cost and time constraints, initial screening of potential developmental neurotoxicants cannot include all measures that might be of interest. Instead, initial screening is generally carried out using a battery of tests such as the Cincinnati Test Battery or the CBTS battery. It is important that the measures chosen for inclusion in a screening battery possess the requisite sensitivity, reliability, and validity. If the initial screening of a compound indicates signs of neurotoxicity, then further testing should be conducted using additional measures (including some involving social behaviors). Moreover, cellular, molecular, electrophysiologic, and anatomic studies should be carried out to determine the neural bases of the observed behavioral changes.

Finally, pharmacologic challenges can be extremely useful to unmask functional deficits that may not be apparent under baseline conditions and to probe the integrity of specific neurochemical systems. Such challenges are typically conducted using compounds that elicit robust and easily measured behavioral or physiologic responses. A number of monoaminergic drugs meet these criteria, including the mixed DA receptor agonist apomorphine, the D_1 agonist SKF 38393, the D_2/D_3 agonist quinpirole, the DA antagonist haloperidol, the α_2-adrenoceptor agonist clonidine, the 5-HT_{1A} agonist 8-OH-DPAT, and the 5-HT_2 agonist DOI. It is important to recognize, however, that pharmacologic challenges can be employed with any neurotransmitter system as long as there are appropriate agonists or antagonists that produce measurable functional effects.

References

Abel, E. L. (1989). *Behavioral Teratogenesis and Behavioral Mutagenesis: A Primer in Abnormal Development.* Plenum Press, New York.

Adams, J. (1982). Ultrasonic vocalizations as diagnostic tools in studies of developmental toxicity: an investigation of the effects of hypervitaminosis A. *Neurobehav. Toxicol. Teratol.* **4,** 299–304.

Adams, J. (1986). Methods in behavioral teratology. In *Handbook of Behavioral Teratology.* E. P. Riley and C. V. Vorhees, Eds. Plenum Press, New York, pp. 67–97.

Adams, J., and Buelke-Sam, J. (1981). Behavioral assessment of the postnatal animal: testing and methods development. In *Developmental Toxicology.* C. A. Kimmel and J. Buelke-Sam, Eds. Raven Press, New York, pp. 233–258.

Adams, J., Buelke-Sam, J., Kimmel, C. A., Nelson, C. J., and Miller, D. R. (1985a). Collaborative Behavioral Teratology Study: preliminary research. *Neurobehav. Toxicol. Teratol.* **7,** 555–578.

Adams, J., Buelke-Sam, J., Kimmel, C. A., Nelson, C. J., Reiter, L. W., Sobotka, T. J., Tilson, H. A., and Nelson, B. K. (1985b). Collaborative Behavioral Teratology Study: protocol design and testing procedures. *Neurobehav. Toxicol. Teratol.* **7,** 579–586.

Adams, J., Miller, D. R., and Nelson, C. J. (1983). Ultrasonic vocalizations as diagnostic tools in studies of developmental toxicity: an investigation of the effects of prenatal treatment with methylmercuric chloride. *Neurobehav. Toxicol. Teratol.* **5,** 29–34.

Adams, J., Oglesby, D. M., Ozemek, H., Rath, J., Kimmel, C. A., and Buelke-Sam, J. (1985c). Collaborative Behavioral Teratology Study: programmed data entry and automated test systems. *Neurobehav. Toxicol. Teratol.* **7,** 547–554.

Alleva, E. (1993). Assessment of aggressive behavior in rodents. In *Paradigms for the Measurement of Behavior. Methods in Neurosciences, Vol. 14.* P. M. Conn, Ed. Academic Press, San Diego, pp. 111–137.

Alleva, E., and Bignami, G. (1986). Prenatal benzodiazepine effects in mice: postnatal behavioral development, response to drug challenges, and adult discrimination learning. *Neurotoxicology* **7,** 303–317.

Alleva, E., Laviola, G., Tirelli, E., and Bignami, G. (1985). Short-, medium, and long-term effects of prenatal oxazepam on neurobehavioral development of mice. *Psychopharmacology* **87,** 434–441.

Allin, J. T., and Banks, E. M. (1972). Functional aspects of ultrasound production by infant albino rats (*Rattus norvegicus*). *Anim. Behav.* **20,** 175–185.

Altman, J. (1969). Autoradiographic and histological studies of postnatal neurogenesis. III. Dating the time of production and onset of differentiation of cerebellar microneurons in rats. *J. Comp. Neurol.* **136,** 269–294.

Altman, J., and Sudarshan, K. (1975). Postnatal development of locomotion in the laboratory rat. *Anim. Behav.* **23,** 896–920.

Aulakh, C. S., Hill, J. L., Yoney, H. T., and Murphy, D. L. (1992). Evidence for the involvement of 5-HT_{1C} and 5-HT_2 receptors in the food intake suppressant effects of 1-(2,5-dimethoxy-4-iodophenyl)-2-aminopropane (DOI). *Psychopharmacology* **109,** 444–448.

Bâ, A. and Seri, B. V. (1995). Psychomotor functions in developing rats: ontogenetic approach to structure–function relationships. *Neurosci. Biobehav. Rev.* **19,** 413–425.

Balduini, W., Elsner, J., Lombardelli, G., Peruzzi, G., and Cattabeni, F. (1991). Treatment with methylazoxymethanol at different gestational days: two-way shuttle box avoidance and residential maze activity in rat offspring. *NeuroToxicology* **12,** 677–686.

Balduini, W., Lombardelli, G., Peruzzi, G., Cattabeni, F., and Elsner, J. (1989). Nocturnal hyperactivity induced by prenatal methylazoxymethanol administration as measured in a computerized residential maze. *Neurotoxicol. Teratol.* **11,** 339–343.

Barrett, B. A., Caza, P., Spear, N. E., and Spear, L. P. (1982). Wall climbing, odors from the home nest and catecholaminergic activity in rat pups. *Physiol. Behav.* **29,** 501–507.

Beach, F. A. (1976). Sexual attractivity, proceptivity, and receptivity in female mammals. *Horm. Behav.* **7,** 105–138.

Blanchard, B. A., Hannigan, J. H., and Riley, E. P. (1987). Amphetamine-induced activity after fetal alcohol exposure and undernutrition in rats. *Neurotoxicol. Teratol.* **9,** 113–119.

Blumberg, M. S., and Alberts, J. R. (1990). Ultrasonic vocalizations by rat pups in the cold: an acoustic by-product of laryngeal braking? *Behav. Neurosci.* **104,** 808–817.

Blumberg, M. S., and Alberts, J. R. (1991). On the significance of similarities between ultrasonic vocalizations of infant and adult rats. *Neurosci. Biobehav. Rev.* **15,** 383–390.

Blumberg, M. S., Efimova, I. V., and Alberts, J. R. (1992). Ultrasonic vocalizations by rat pups: the primary importance of ambient temperature and the thermal significance of contact comfort. *Dev. Psychobiol.* **25,** 229–250.

Bolles, R. C., and Woods, P. J. (1964). The ontogeny of behaviour in the albino rat. *Anim. Behav.* **12,** 427–441.

Beninger, R. J. (1989). Methods for determining the effects of drugs on learning. In *Psychopharmacology. Neuromethods,* Vol. 13. A. A. Boulton, G. B. Baker, and A. J. Greenshaw, Eds. Humana Press, Clifton, pp. 623–685.

Borbely, A. A., and Tobler, I. (1985). Homeostatic and circadian principles in sleep regulation in the rat. In *Brain Mechanisms of Sleep.* D. J. McGinty, R. Drucker-Colín, A. Morrison, and P. L. Parmeggiani, Eds. Raven Press, New York, pp. 35–44.

Boyes, W. K. (1992). Testing visual system toxicity using evoked potential technology. In *The Vulnerable Brain and Environmental Risks, Volume I.: Malnutrition and Hazard Assessment.* R. L. Isaacson and K. F. Jensen, Eds. Plenum Press, New York, pp. 193–222.

Brain, P. F. (1981). Differentiating types of attack and defense in rodents. In *Multidisciplinary Approaches to Aggression Research.* P. F. Brain and D. Benton, Eds. Elsevier/North Holland, Amsterdam, pp. 53–77.

Browning, R. A. (1987). The role of neurotransmitters in electroshock seizure models. In *Neurotransmitters and Epilepsy.* P. C. Jobe and H. E. Laird II, Eds. Humana Press, Clifton, pp. 277–320.

Bruhwyler, J., Chleide, E., Liégeois, J. F., Delarge, J., and Mercier, M. (1991). Effects of specific dopaminergic agonists and antagonists in the open-field test. *Pharmacol. Biochem. Behav.* **39,** 367–371.

Buelke-Sam, J., Kimmel, C. A., Adams, J., Nelson, C. J., Vorhees, C. V., Wright, D. C., St. Omer, V., Korol, B. A., Butcher, R. E., Geyer, M. A., Holson, J. F., Kutscher, C. L., and Wayner, M. J. (1985). Collaborative Behavioral Teratology Study: results. *Neurobehav. Toxicol. Teratol.* **7,** 591–624.

Burt, D. K., Hungerford, S. M. Crowner, M. L., and Baez, L. A. (1982). Postnatal development of a cholinergic influence on neuroleptic-induced catalepsy. *Pharmacol. Biochem. Behav.* **16,** 533–540.

Butcher, R. E. (1976). Behavioral testing as a method for assessing risk. *Environ. Health Perspect.* **18,** 75–78.

Bylund, D. B., Eikenberg, D. C., Hieble, J. P., Langer, S. Z., Lefkowitz, R. J., Minneman, K. P., Molinoff, P. B., Ruffolo, R. R. Jr., and Trendelenburg, U. (1994). IV. International Union of Pharmacology Nomenclature of Adrenoceptors. *Pharmacol. Rev.* **46,** 121–136.

Campbell, B. A., Lytle, L. D., and Fibiger, H. C. (1969). Ontogeny of adrenergic arousal and cholinergic inhibitory mechanisms in the rat. *Science* **166,** 635–637.

Carlos, R. Q., Seidler, F. J., and Slotkin, T. A. (1992). Fetal dexamethasone exposure alters macromolecular characteristics of rat brain development: a critical period for regionally selective alterations. *Teratology* **46,** 45–59.

Cattabeni, F., and Abbracchio, M. P. (1988). Behavioral teratology: an inappropriate term for some uninterpretable effects. *Trends Pharmacol. Sci.* **9,** 13–15.

Celis, M. E., and Torre, E. (1993). Measurement of grooming behavior. In *Paradigms for the Measurement of Behavior. Methods in Neurosciences, Vol. 14.* P. M. Conn, Ed. Academic Press, San Diego, pp. 378–388.

Church, M. W., and Overbeck, G. W. (1990). Prenatal cocaine exposure in the Long-Evans rat: III. Developmental effects on the brainstem auditory-evoked potential. *Neurotoxicol. Teratol.* **12,** 345–351.

Costall, B., and Naylor, R. J. (1973). The role of telencephalic dopaminergic systems in the mediation of apomorphine-stereotyped behavior. *Eur. J. Pharmacol.* **24,** 8–24.

Crofton, K. M., Howard, J. L., Moser, V. C., Gill, M. W., Reiter, L. W., Tilson, H. A., and MacPhail, R. C. (1991). Interlaboratory comparison of motor activity experiments: implications for neurotoxicological assessments. *Neurotoxicol. Teratol.* **13,** 599–609.

Cuomo, V., De Salvia, M. A., Maselli, M. A., Santo, L., and Cagiano, R. (1987). Ultrasonic calling in rodents: a new experimental approach in behavioural toxicology. *Neurotoxicol. Teratol.* **9,** 157–160.

Daly, S. A., and Waddington, J. L. (1993). Behavioral effects of the putative D-3 dopamine receptor agonist 7-OH-DPAT in relation to other "D-2-like" agonists. *Neuropharmacology* **32,** 509–510.

Darmani, N. A., Martin, B. R., and Glennon, R. A. (1990). Withdrawal from chronic treatment with (±)-DOI causes supersensitivity to 5-HT_2 receptor-induced head-twitch behavior in mice. *Eur. J. Pharmacol.* **186,** 115–118.

De Sarro, G. B., Ascioti, C., Froio, F., Libri, V., and Nisticò, G. (1987). Evidence that locus coeruleus is the site where clonidine and drugs acting at alpha 1- and alpha 2-adrenoceptors affect sleep and arousal mechanisms. *Br. J. Pharmacol.* **90,** 675–685.

Diakow, C. (1974). Motion picture analysis of rat mating behavior. *J. Comp. Physiol. Psychol.* **88,** 704–712.

Eilam, D., Talangbayan, H., Canaran, G., and Szechtman, H. (1992). Dopaminergic control of locomotion, mouthing, snouth contact, and grooming: opposing roles of D_1 and D_2 receptors. *Psychopharmacology* **106,** 447–454.

Elsner, J. (1986). Testing strategies in behavioral teratology: III. Microanalysis of behavior. *Neurobehav. Toxicol. Teratol.* **8,** 573–584.

Elsner, J., Suter, K. E., Ulbrich, B., and Schreiner, G. (1986). Testing strategies in behavioral teratology: IV. Review and general conclusion. *Neurobehav. Toxicol. Teratol.* **8,** 585–590.

Evans, H. L. (1982). Assessment of vision in behavioral toxicology. In *Nervous System Toxicology.* C. L. Mitchell, Ed. Raven Press, New York, pp. 81–107.

Evans, H. L. (1994). Neurotoxicity expressed in naturally occurring behavior. In *Neurobehavioral Toxicity: Analysis and Interpretation.* B. Weiss and J. O'Donoghue, Eds. Raven Press, New York, pp. 111–135.

Faingold, C. L. (1987). Seizures induced by convulsant drugs. In *Neurotransmitters and Epilepsy.* P. C. Jobe and H. E. Laird II, Eds. Humana Press, Clifton, pp. 215–276.

Feldman, R. S., Meyer, J. S., and Quenzer, L. F. (1997). *Principles of Neuropsychopharmacology.* Sinauer Associates, Sunderland.

Fentress, J. C. (1968a). Interrupted ongoing behavior in two species of vole (*Microtus agrestis* and *Clethrionomys britannicus*). I. Response as a function of preceding activity and the context of an apparently 'irrelevant' motor pattern. *Anim. Behav.* **16,** 135–153.

Fentress, J. C. (1968b). Interrupted ongoing behavior in two species of vole (*Microtus agrestis* and *Clethrionomys brittanicus*). II. Extended analysis of motivational variables underlying fleeing and grooming behavior. *Anim. Behav.* **16,** 154–167.

Ferré, S., Guix, T., Prat, G., Jane, F., and Casas, M. (1990). Is experimental catalepsy properly measured? *Pharmacol. Biochem. Behav.* **35,** 753–757.

File, S. E., and Wardill, A. G. (1975). Validity of head-dipping as a measure of exploration in a modified hole-board. *Psychopharmacologia* **44,** 53–59.

Fitzgerald, R. E., Berres, M., and Schaeppi, U. (1988). Validation of a photobeam system for assessment of motor activity in rats. *Toxicology* **49,** 433–439.

Fox, D. A., Lewkowski, J. P., and Cooper, G. P. (1977). Acute and chronic effects of neonatal lead exposure on development of the visual evoked response in rats. *Toxicol. Appl. Pharmacol.* **40,** 449–461.

Fox, D. A., Lowndes, H. E., and Bierkamper, G. G. (1982). Electrophysiological techniques in neurotoxicology. In *Nervous System Toxicology*. C. L. Mitchell, Ed. Raven Press, New York, pp. 299–335.

Fox, W. M. (1965). Reflex-ontogeny and behavioural development of the mouse. *Anim. Behav.* **13,** 234–241.

Frambes, N. A., Kirstein, C. L., Moody, C. A., and Spear, L. P. (1990). 5-HT_{1A}, 5-HT_{1B}, and 5-HT_2 receptor agonists induce differential behavioral responses in preweanling rat pups. *Eur. J. Pharmacol.* **182,** 9–17.

Gårdlund, A. T., Archer, T., Danielsson, K., Danielsson, B., Fredriksson, A., Lindqvist, N. G., Lindström, H., and Luthman, J. (1991). Effects of prenatal exposure to tributyltin and trihexyltin on behaviour in rats. *Neurotoxicol. Teratol.* **13,** 99–105.

Gentry, G. D., Merritt, C. J., and Middaugh, L. D. (1995). Effects of prenatal maternal ethanol on male offspring progressive-ratio performance and response to amphetamine. *Neurotoxicol. Teratol.* **17,** 673–677.

Gispen, W. H., Wiegant, V. M., Greven, H. M., and de Wied, D. (1975). The induction of excessive grooming in the rat by intraventricular application of peptides derived from ACTH: structure-activity studies. *Life Sci.* **17,** 645–652.

Gramsbergen, A., Schwartze, P., and Prechtl, H. F. R. (1970). The postnatal development of behavioral states in the rat. *Dev. Psychobiol.* **3,** 267–280.

Grant, E. C. (1963). An analysis of the social behaviour of the male laboratory rat. *Behaviour* **21,** 260–281.

Grant, E. C., and Mackintosh, J. H. (1963). A comparison of the social postures of some common laboratory rodents. *Behaviour* **21,** 246–259.

Green, A. R., and Backus, L. I. (1990). Animal models of serotonin behavior. *Ann. N.Y. Acad. Sci.* **600,** 237–249.

Gregory, E. H., and Pfaff, D. W. (1971). Development of olfactory-guided behavior in infant rats. *Physiol. Behav.* **6,** 573–576.

Haas, K., Sperber, E. F., and Moshé, S. L. (1992). Kindling in developing animals: expression of severe seizures and enhanced development of bilateral foci. *Dev. Brain Res.* **68,** 140–143.

Hannigan, J. H., and Blanchard, B. A. (1988). Commentary: psychopharmacological assessment in neurobehavioral teratology. *Neurotoxicol. Teratol.* **10,** 143–145.

Hannigan, J. H., and Pilati, M. L. (1991). The effects of chronic postweaning amphetamine on rats exposed to alcohol *in utero:* weight gain and behavior. *Neurotoxicol. Teratol.* **13,** 649–656.

Hansen, S. (1993). Effect of clonidine on the responsiveness of infant rats to maternal stimuli. *Psychopharmacology* **111,** 78–84.

Harro, J. (1993). Measurement of exploratory behavior in rodents. In *Paradigms for the Measurement of Behavior. Methods in Neurosciences, Vol. 14.* P. M. Conn, Ed. Academic Press, San Diego, pp. 359–377.

Heise, G. A. (1984). Behavioral methods for measuring effects of drugs on learning and memory in animals. *Med. Res. Rev.* **4,** 535–558.

Honma, S., and Honma, K. (1985). Interaction between circadian and ultradian rhythms of spontaneous locomotor activity in rats during the early developmental period. In *Ultradian Rhythms in Physiology and Behavior. Experimental Brain Research Supplementum 12.* H. Schulz and P. Lavie, Eds. Springer-Verlag, Berlin, pp. 95–109.

Hoyer, D., Clarke, D. E., Fozard, J. R., Hartig, P. R., Martin, G. R., Myelcharane, E. J., Saxena, P. R., and Humphrey, P. P. A. (1994). VII. International Union of Pharmacology classification of receptors for 5-hydroxytryptamine (serotonin). *Pharmacol. Rev.* **46,** 157–203.

Hughes, H. E., Donohue, L. M., and Dow-Edwards, D. L. (1996). Prenatal cocaine exposure affects the acoustic startle response in adult rat. *Behav. Brain Res.* **75,** 83–90.

Hughes, J. A., and Sparber, S. B. (1978). d-Amphetamine unmasks postnatal consequences of exposure to methylmercury *in utero:* methods for studying behavioral teratogenesis. *Pharmacol. Biochem. Behav.* **8,** 365–375.

Hunt, P. S., Richardson, R., and Campbell, B. A. (1994). Delayed development of fear-potentiated startle in rats. *Behav. Neurosci.* **108,** 69–80.

Jackson, H. C., and Kitchen, I. (1989). Behavioural profiles of putative 5-hydroxytryptamine receptor agonists and antagonists in developing rats. *Neuropharmacology* **28,** 635–642.

Jacobs, B. L., and Klemfuss, H. (1975). Brainstem and spinal cord mediation of a serotonergic behavioral syndrome. *Brain Res.* **100,** 450–457.

Jacobson, S. (1963). Sequence of myelinization in the brain of the albino rat. A. Cerebral cortex, thalamus and related structures. *J. Comp. Neurol.* **121,** 5–29.

Jason, K. M., and Kellogg, C. K. (1981). Neonatal lead exposure: effects on development of behavior and striatal dopamine neurons. *Pharmacol. Biochem. Behav.* **15,** 641–649.

Kallman, M. J. (1994). Assessment of motoric effects. In *Developmental Neurotoxicology.* G. J. Harry, Ed., CRC Press, Boca Raton, pp. 103–122.

Kallman, M. J., and Condie, L. W. Jr. (1985). A test battery for screening behavioral teratogens in mice. *Neurobehav. Toxicol. Teratol.* **7,** 727–731.

Katsuragi, T., Ushijima, I., and Furukawa, T. (1984). The clonidine-induced self-injurious behavior of mice involves purinergic mechanisms. *Pharmacol. Biochem. Behav.* **20,** 943–946.

Kimmel, C. A., and Buelke-Sam, J. (1985). Collaborative Behavioral Teratology Study: background and overview. *Neurobehav. Toxicol. Teratol.* **7,** 541–545.

Kimmel, C. A., Buelke-Sam, J., and Adams, J. (1985). Collaborative Behavioral Teratology Study: implications, current applications and future directions. *Neurobehav. Toxicol. Teratol.* **7,** 669–673.

Kirstein, C. L., and Spear, L. P. (1988). 5-HT_{1A}, 5-HT_{1B}, and 5-HT_2 receptor agonists induce differential behavioral responses in neonatal rats. *Eur. J. Pharmacol.* **150,** 339–345.

Koolhaas, J. M., and Bohus, B. (1991). Animal models of human aggression. In *Animal Models in Psychiatry, II. Neuromethods, Vol. 19.* A. A. Boulton, G. B., Baker, and M. T. Martin-Iverson, Eds. Humana Press, Clifton, pp. 249–271.

Lanier, L. P., and Isaacson, R. L. (1977). Early developmental changes in the locomotor response to amphetamine and their relation to hippocampal function. *Brain Res.* **126,** 567–575.

Laviola, G., de Acetis, L., Bignami, G., and Alleva, E. (1991). Prenatal oxazepam enhances mouse maternal aggression in the offspring, without modifying acute chlordiazepoxide effects. *Neurotoxicol. Teratol.* **13,** 75–81.

Lin, M.-Y., and Walters, D. E. (1994). Dopamine D_2 autoreceptors in rats are behaviorally functional at 21 but not 10 days of age. *Psychopharmacology* **114,** 262–268.

Little, J. Z., and Teyler, T. J. (1996). Prenatal cocaine exposure leads to enhanced long-term potentiation in region CA1 of hippocampus. *Dev. Brain Res.* **92,** 117–119.

Ljungberg, T., and Ungerstedt, U. (1976). Automatic registration of behaviour related to dopamine and noradrenaline transmission. *Eur. J. Pharmacol.* **36,** 181–188.

Lucki, I., and Wieland, S. (1990). 5-Hydroxytrypamine$_{1A}$ receptors and behavioral responses. *Neuropsychopharmacology* **3,** 481–493.

Maier, S. E., Vandenhoff, P., and Crowne, D. P. (1988). Multivariate analysis of putative measures of activity, exploration, emotionality, and spatial behavior in the hooded rat (*Rattus norvegicus*). *J. Comp. Psychol.* **102,** 378–387.

McNamara, J. O., Bonhaus, D. W., and Shin, C. (1993). The kindling model of epilepsy. In *Epilepsy: Models, Mechanisms, and Concepts.*

P. A. Schwartzkroin, Ed. Cambridge University Press, Cambridge, pp. 27–47.

Meaney, M. J., and Stewart, J. (1981). A descriptive study of social development in the rat (*Rattus norvegicus*). *Anim. Behav.* **29,** 34–45.

Meyer, J. S., Robinson, P., and Todtenkopf, M. S. (1994). Prenatal cocaine treatment reduces haloperidol-induced catalepsy on postnatal day 10. *Neurotoxicol. Teratol.* **16,** 193–199.

Meyer, M. E., Smith, R. L., and Van Hartesveldt, C. (1984). Haloperidol differentially potentiates tonic immobility, the dorsal immobility response, and catalepsy in the developing rat. *Dev. Psychobiol.* **17,** 383–389.

Middaugh, L. D., and Boggan, W. O. (1991). Postnatal growth deficits in prenatal ethanol-exposed mice: characteristics and critical periods. *Alcohol Clin. Exp. Res.* **15,** 919–926.

Middaugh, L. D., and Zemp, J. W. (1985). Dopaminergic mediation of long-term behavioral effects of *in utero* drug exposure. *Neurobehav. Toxicol. Teratol.* **7,** 685–689.

Miller, D. B., and Eckerman, D. A. (1986). Learning and memory measures. In *Neurobehavioral Toxicology.* Z. Annau, Ed. Johns Hopkins University Press, Baltimore, pp. 94–149.

Molloy, A. G., and Waddington, J. L. (1987). Assessment of grooming and other behavioural responses to the D-1 dopamine receptor agonist SK & F 38393 and its R- and S-enantiomers in the intact adult rat. *Psychopharmacology* **92,** 164–168.

Moore, N. A., Rees, G., Sanger, G., and Perrett, L. (1993). 5-HT_{1A}-mediated lower lip reaction: effects of 5-HT_{1A} agonists and antagonists. *Pharmacol. Biochem. Behav.* **46,** 141–143.

Moshé, S. L., Stanton, P. K., and Sperber, E. F. (1993). Sensitivity of the immature central nervous system to epileptogenic stimuli. In *Epilepsy: Models, Mechanisms, and Concepts.* P. A. Schwartzkroin, Ed. Cambridge University Press, Cambridge, pp. 171–198.

Moyer, K. E. (1968). Kinds of aggression and their physiological basis. *Commun. Behav. Biol. A* **2,** 65–87.

Murphy, L. L., Gher, J., and Szary, A. (1995). Effects of prenatal exposure to delta-9-tetrahydrocannabinol on reproductive, endocrine and immune parameters of male and female rat offspring. *Endocrine* **3,** 875–879.

Nassif-Caudarella, S., Kempf, E., and Velley, L. (1986). Clonidine-induced sedation is not modified by single or combined neurochemical lesions of the locus coeruleus, the median and dorsal raphe nuclei. *Pharmacol. Biochem. Behav.* **25,** 1211–1216.

Nelson, C. J., Felton, R. P., Kimmel, C. A., Buelke-Sam, J., and Adams, J. (1985). Collaborative Behavioral Teratology Study: statistical approach. *Neurobehav. Toxicol. Teratol.* **7,** 587–590.

Nicholas, J. M., and Schreiber, E. C. (1983). Phencyclidine exposure and the developing mouse: behavioral teratological implications. *Teratology* **28,** 319–326.

Nolen, G. A. (1986). The effects of prenatal retinoic acid on the viability and behavior of the offspring. *Neurobehav. Toxicol. Teratol.* **8,** 643–654.

Nomura, Y., Oki, K., and Segawa, T. (1980). Pharmacological characterization of the central alpha-adrenoceptors which mediate clonidine-induced locomotor hypoactivity in the developing rat. *Naunyn-Schmiedeberg's Arch. Pharmacol.* **311,** 41–44.

Ogren, S. O., and Fuxe, K. (1988). Apomorphine and pergolide induce hypothermia by stimulation of dopamine D-2 receptors. *Acta Physiol. Scand.* **133,** 91–95.

Okon, E. E. (1971). The temperature relations of vocalization in infant Golden hamsters and Wistar rats. *J. Zool. Lond.,* **164,** 227–237.

Olivier, B., and Mos, J. (1992). Rodent models of aggressive behavior and serotonergic drugs. *Prog. Neuropsychopharmacol. Biol. Psychiatry* **16,** 847–870.

Ortmann, R., Mutter, M., and Delini-Stula, A. (1982). Effect of yohimbine and its diastereoisomers on clonidine-induced depression of exploration in the rat. *Eur. J. Pharmacol.* **77,** 335–337.

Pankaj, V., and Brain, P. F. (1991a). Effects of prenatal exposure to benzodiazepine-related drugs on early development and adult social behaviour in Swiss mice—I. Agonists. *Gen. Pharmacol.* **22,** 33–41.

Pankaj, V., and Brain, P. F. (1991b). Effects of prenatal exposure to benzodiazepine-related drugs on early development and adult social behaviour in Swiss mice—II. Antagonists. *Gen. Pharmacol.* **22,** 33–51.

Pankaj, V., and Brain, P. F. (1991c). Effects of prenatal exposure to benzodiazepine-related drugs on early development and adult social behaviour in Swiss mice—III. Inverse agonists. *Gen. Pharmacol.* **22,** 53–60.

Pertwee, R. G. (1972). The ring test: a quantitative method for assessing the 'cataleptic' effect of cannabis in mice. *Br. J. Pharmacol.* **46,** 753–763.

Petruzzi, S., Fiore, M., Dell'Omo, G., Bignami, G., and Alleva, E. (1995). Medium and long-term behavioral effects in mice of gestational exposure to ozone. *Neurotoxicol. Teratol.* **17,** 463–470.

Pfaff, D. W., Diakow, C., Montgomery, M., and Jenkins, F. A. (1978). X-ray cinematographic analysis of lordosis in female rats. *J. Comp. Physiol. Psychol.* **92,** 937–941.

Phifer, C. B., Denzinger, A., and Hall, W. G. (1992). The early presence of food-oriented appetitive behaviour in developing rats. *Dev. Psychobiol.* **24,** 453–461.

Pizzi, W. J., Unnerstall, J. R., and Barnhardt, J. E. (1979). Neonatal monosodium glutamate administration increases susceptibility to chemically-induced convulsions in adult mice. *Neurobehav. Toxicol. Teratol.* **1,** 169–173.

Pranzatelli, M. R. (1990). Evidence for involvement of 5-HT_2 and 5-HT_{1C} receptors in the behavioral effects of the 5-HT agonist 1-(2,5-dimethoxy-4-iodophenyl aminopropane)-2 (DOI). *Neurosci. Lett.* **115,** 74–80.

Pranzatelli, M. R., Schultz, L., and Snodgrass, S. R. (1987). High-dose clonidine motor syndrome: relationship to serotonin syndrome. *Behav. Brain Res.* **24,** 221–232.

Protais, P., Bonnet, J. J., Costentin, J., and Schwartz, J.-C. (1984). Rat climbing behaviour elicited by stimulation of cerebral dopamine receptors. *Naunyn-Schmiedeberg's Arch. Pharmacol.* **325,** 93–101.

Rayburn, W. F., Christenen, H. D., and Gonzalez, C. L. (1997). A placebo-controlled comparison between betamethasone and dexamethasone for fetal maturation: differences in neurobehavioral development of mice offspring. *Am. J. Obstet. Gynecol.* **176,** 842–851.

Rees, D. C., Francis, E. Z., and Kimmel, C. A. (1990). Scientific and regulatory issues relevant to assessing risk for developmental neurotoxicity: an overview. *Neurotoxicol. Teratol.* **12,** 175–181.

Reinstein, D. K., and Isaacson, R. L. (1977). Clonidine sensitivity in the developing rat. *Brain Res.* **135,** 378–382.

Reiter, L. W., and MacPhail, R. C. (1982). Factors influencing motor activity measurements in neurotoxicology. In *Nervous System Toxicology* C. L. Mitchell, Ed. Raven Press, New York, pp. 45–65.

Rice, D. C. (1994). Testing effects of toxicants on sensory system function by operant methodology. In *Neurobehavioral Toxicity: Analysis and Interpretation.* B. Weiss and J. O'Donoghue, Eds. Raven Press, New York, pp. 299–318.

Rice, S. A., and Millan, D. P. (1986). Validation of a developmental swimming test using Swiss Webster mice perinatally treated with methimazole. *Neurobehav. Toxicol. Teratol.* **8,** 69–75.

Riffee, W. H., Wilcox, R. E., and Smith, R. V. (1979). Stereotypic and hypothermic effects of apomorphine and N-n-propylnorapomorphine in mice. *Eur. J. Pharmacol.* **54,** 273–277.

Riley, E. P., Hannigan, J. H., and Balaz-Hannigan, M. A. (1985). Behavioral teratology as the study of early brain damage: considerations for the assessment of neonates. *Neurobehav. Toxicol. Teratol.* **7,** 635–638.

Ristine, L. A., and Spear, L. P. (1984). Effects of serotonergic and cholinergic antagonists on suckling behavior of neonatal, infant, and weanling rat pups. *Behav. Neural Biol.* **1,** 99–126.

Ristine, L. A., and Spear, L. P. (1985). Is there a "serotonergic syndrome" in neonatal rat pups? *Pharmacol. Biochem. Behav.* **22,** 265–269.

Rowland, N. E., Morien, N., and Li, B.-H. (1996). The physiology and brain mechanisms of feeding. *Nutrition* **12,** 626–639.

Salamy, A., Dark, K., Salfi, M., Shah, S., and Peeke, H. V. S. (1992). Perinatal cocaine exposure and functional brainstem development in the rat. *Brain Res.* **598,** 307–310.

Sanberg, P. R. (1980). Haloperidol-induced catalepsy is mediated by postsynaptic dopamine receptors. *Nature* **284,** 472–473.

Sanberg, P. R., Pevsner, J., and Coyle, J. T. (1984). Parametric influences on catalepsy. *Psychopharmacology* **82,** 406–408.

Scalzo, F. M., Ali, S. F., and Holson, R. R. (1989). Behavioral effects of prenatal haloperidol exposure. *Pharmacol. Biochem. Behav.* **34,** 727–731.

Schapiro, S., Salas, M., and Vukovich, K. (1970). Hormonal effects on ontogeny of swimming ability in the rat: assessment of central nervous system development. *Science* **168,** 147–151.

Schlessinger, A. R., Cowan, W. M., and Gottlieb, D. I. (1975). An autoradiographic study of the time of origin and the pattern of granule cell migration in the dentate gyrus of the rat. *J. Comp. Neurol.* **159,** 149–176.

Schreiner, G., Ulbrich, B., and Bass, R. (1986). Testing strategies in behavioral teratology: II. Discrimination learning. *Neurobehav. Toxicol. Teratol.* **8,** 567–572.

Seeman, P., and Van Tol, H. H. M. (1994). Dopamine receptor pharmacology. *Trends Pharmacol. Sci.* **15,** 264–270.

Segal, D. S., and Kuczenski, R. (1994). Behavioral pharmacology of amphetamine. In *Amphetamine and its Analogs.* A. K. Cho, Ed., Academic Press, San Diego, pp. 115–150.

Shalaby, I. A., and Spear, L. P. (1980). Psychopharmacological effects of low and high doses of apomorphine during ontogeny. *Eur. J. Pharmacol.* **67,** 451–459.

Shieh, G.-J., and Walters, D. E. (1996). Stimulating dopamine D_1 receptors increases the locomotor activity of developing rats. *Eur. J. Pharmacol.* **311,** 103–107.

Simansky, K. J., and Kachelries, W. J. (1996). Prenatal exposure to cocaine selectively disrupts motor responding to D-amphetamine in young and mature rabbits. *Neuropharmacology* **35,** 71–80.

Snyder-Keller, A. M., and Keller, R. W. Jr. (1995). Prenatal cocaine alters later sensitivity to cocaine-induced seizures. *Neurosci. Lett.* **191,** 149–152.

Sobrian, S. K., Weltman, M., and Pappas, B. A. (1975). Neonatal locomotor and long-term behavioral effects of d-amphetamine in the rat. *Dev. Psychobiol.* **8,** 241–250.

Sokoloff, P., Giros, B., Martres, M.-P., Bouthenet, M.-L., and Schwartz, J.-C. (1990). Molecular cloning and characterization of a novel dopamine receptor (D_3) as a target for neuroleptics. *Nature* **347,** 146–151.

Spear, L. P. (1984). Age at time of testing reconsidered in neurobehavioral teratological research. In *Neurobehavioral Teratology.* J. Yanai, Ed., Elsevier, Amsterdam, pp. 315–328.

Spear, L. P. (1996). Assessment of the effects of developmental toxicants: pharmacological and stress vulnerability of offspring. In *Behavioral Studies of Drug-Exposed Offspring: Methodological Issues in Human and Animal Research. NIDA Research Monograph 164.* C. L. Wetherington, V. L. Smeriglio, and L. P. Finnegan, Eds. National Institute on Drug Abuse, Rockville, MD, pp. 125–145.

Spear, L. P., Enters, E. K., and Linville, D. G. (1985). Age-specific behaviors as tools for examining teratogen-induced neural alterations. *Neurobehav. Toxicol. Teratol.* **7,** 691–695.

Spear, L. P., Frambes, N. A., and Goodwin, G. (1991). Low doses of the 5-HT_{1A} receptor agonist 8-OH-DPAT increase ingestive behavior in late preweanling and postweanling, but not neonatal rat pups. *Eur. J. Pharmacol.* **203,** 9–15.

Spear, N. E., Miller, J. S., and Jagielo, J. A. (1990). Animal learning and memory. *Annu. Rev. Psychol.* **41,** 169–211.

Spear, L. P., and Ristine, L. A. (1981). Quipazine-induced behavior in neonatal rat pups. *Pharmacol. Biochem. Behav.* **14,** 831–834.

Spear, L. P., and Ristine, L. A. (1982). Suckling behavior in neonatal rats: psychopharmacological investigations. *J. Comp. Physiol. Psychol.* **96,** 244–255.

Spyker, J. M. (1975). Behavioral teratology and toxicology. In *Behavioral Toxicology.* B. Weiss and F. B. Laties, Eds. Plenum Press, New York, pp. 311–344.

Ståhle, L. (1992). Do autoreceptors mediate dopamine agonist-induced yawning and suppression of exploration? A critical review. *Psychopharmacology* **106,** 1–13.

Stanton, M. E., and Spear, L. P. (1990). Workshop on the Qualitative and Quantitative Comparability of Human and Animal Developmental Neurotoxicity, Work Group I report: Comparability of measures of developmental neurotoxicity in humans and laboratory animals. *Neurotoxicol. Teratol.* **12,** 261–267.

Stebbins, W. C. (1990). Perception in animal behavior. In *Comparative Perception. Volume 1. Basic Mechanisms.* M. A. Berkley and W. C. Stebbins, Eds. Wiley & Sons, New York, pp. 1–26.

Strömbom, U. (1975). Effects of low doses of catecholamine receptor agonists on exploration in mice. *J. Neural Transm.* **37,** 229–235.

Suter, K. E., and Schön, H. (1986). Testing strategies in behavioral teratology: I. Testing battery approach. *Neurobehav. Toxicol. Teratol.* **8,** 561–566.

Tilson, H. A., and Cabe, P. A. (1978). Assessment of chemically-induced changes in the neuromuscular function of rats using a new recording grip meter. *Life Sci.* **23,** 1365–1370.

Urba-Holmgren, R., Holmgren, B., and Anias, J. (1982). Pre- and postsynaptic dopaminergic receptors involved in apomorphine induced yawning. *Acta Neurobiol. Exp.* **42,** 115–125.

Ushijima, I., Katsuragi, T., and Furukawa, T. (1984). Involvement of adenosine receptor activities in aggressive responses produced by clonidine in mice. *Psychopharmacology* **83,** 335–339.

Vorhees, C. V. (1983). Fetal anticonvulsant syndrome in rats: dose– and period–response relationships of prenatal diphenylhydantoin, trimethadione and phenobarbital exposure on the structural and functional development of the offspring. *J. Pharmacol. Exp. Ther.* **227,** 274–287.

Vorhees, C. V. (1985a). Behavioral effects of prenatal d-amphetamine in rat: a parallel trial to the collaborative behavioral teratology study. *Neurobehav. Toxicol. Teratol.* **7,** 709–716.

Vorhees, C. V. (1985b). Comparison of the Collaborative Behavioral Teratology Study and Cincinnati behavioral teratology test batteries. *Neurobehav. Toxicol. Teratol.* **7,** 625–633.

Vorhees, C. V. (1986). Principles of behavioral teratology. In *Handbook of Behavioral Teratology.* E. P. Riley and C. V. Vorhees, Eds. Plenum Press, New York, pp. 23–48.

Vorhees, C. V. (1987). Reliability, sensitivity and validity of behavioral indices of neurotoxicity. *Neurotoxicol. Teratol.* **9,** 445–464.

Vorhees, C. V. (1989). Concepts in teratology and developmental toxicology derived from animal research. *Ann. N.Y. Acad. Sci.* **562,** 31–41.

Vorhees, C. V., Brunner, R. L., and Butcher, R. E. (1979a). Psychotropic drugs as behavioral teratogens. *Science* **205,** 1220–1225.

Vorhees, C. V., Butcher, R. E., Brunner, R. L., and Sobotka, T. J. (1979b). A developmental test battery for neurobehavioral toxicity in rats: a preliminary analysis using monosodium glutamate, calcium carrageenan, and hydroxyurea. *Toxicol. Appl. Pharmacol.* **50,** 267–282.

Vorhees, C. V., Butcher, R. E., Brunner, R. L., and Wootten, V. (1984). Developmental toxicity and psychotoxicity of sodium nitrite in rats. *Chem. Toxicol.* **22,** 1–6.

Vorhees, C. V., Butcher, R. E., Brunner, R. L., Wootten, V., and Sobotka, T. J. (1981). Developmental neurobehavioral toxicity of butylated hydroxyanisol (BHA) in rats. *Neurobehav. Toxicol. Teratol.* **3,** 321–329.

Wachtel, S. R., Brooderson, R. J., and White, F. J. (1992). Parametric and pharmacological analyses of the enhanced grooming response elicited by the D_1 dopamine receptor agonist SKF 38393 in the rat. *Psychopharmacology* **109,** 41–48.

Wadenberg, M.-L. (1996). Serotonergic mechanisms in neuroleptic-induced catalepsy in the rat. *Neurosci. Biobehav. Rev.* **20,** 325–339.

Walker, R. F., Guerriero, F. J., and Wier, P. J. (1989). Design of a primary screen for developmental neurotoxicity. *Toxicol. Industrial Health* **5,** 231–245.

Walsh, R. N., and Cummins, R. A. (1976). The open-field test: a critical review. *Psychol. Bull.* **83,** 482–504.

Wanibuchi, F., and Usuda, S. (1990). Synergistic effects between D-1 and D-2 dopamine antagonists on catalepsy in rats. *Psychopharmacology* **102,** 339–342.

Waters, N., Svensson, K., Haadsma-Svensson, S. R., Smith, M. W., and Carlsson, A. (1993). The dopamine D3-receptor: a postsynaptic receptor inhibitory on rat locomotor activity. *J. Neural. Transm.* **94,** 11–19.

Watzman, N., and Barry, H. III. (1968). Drug effects on motor coordination. *Psychopharmacologia* **12,** 414–423.

Westerga, J., and Gramsbergen, A. (1990). The development of locomotion in the rat. *Dev. Brain Res.* **57,** 163–174.

Winslow, J. T., and Insel, T. R. (1991). The infant rat separation paradigm: a novel test for novel anxiolytics. *Trends Pharmacol. Sci.* **12,** 402–404.

Wood, R. D., Bannoura, M. D., and Johanson, I. B. (1994). Prenatal cocaine exposure: effects on play behavior in the juvenile rat. *Neurotoxicol. Teratol.* **16,** 139–144.

Yeomans, J. S., and Frankland, P. W. (1996). The acoustic startle reflex: neurons and connections. *Brain Res. Rev.* **21,** 301–314.

Zidell, R. H., Hatoum, N. S., and Thomas, P. T. (1988). Fetal alcohol effects: evidence of developmental impairment in the absence of immunotoxicity. *Fund. Appl. Toxicol.* **10,** 189–198.

CHAPTER

23

Assessment of Behavior in Primates

MERLE G. PAULE
Behavioral Toxicology Laboratory
Division of Neurotoxicology
National Center for Toxicological Research
Jefferson, Arkansas 72079

I. Introduction

Given the large number of compounds estimated to be neuroactive (Anger and Johnson, 1985) and, thus, potentially neurotoxic, it seems that a focused effort on the development of approaches for determining the effects of such chemicals on important aspects of brain function should continue with some urgency. Assessment of behavior in the context of neurotoxicology is carried out under the abject belief that behavior is an observable manifestation of nervous system function that can be studied, manipulated, and quantitated. Behavior, then, becomes a window through which we can see the nervous system at work. By carefully arranging the situations under which behaviors occur, it is possible to observe relatively specific aspects of nervous system function: learning, forgetting, counting, timing, nursing, socializing, discriminating (in the sensory modality of your choice), and so on. The list of specific functions that can be observed noninvasively in intact organisms is large, and therein lies some of the problem associated with this area of research: given limited resources, which lines of inquiry are most likely to provide data that are most relevant to the human condition? It is proposed here that there are now numerous examples of specific behaviors common to a variety of species, including humans, that can be routinely studied in the animal

*Abbreviations: BNBAS, Brazelton Neonatal Behavior Assessment Scale; CNS, central nervous system; ENNS, Early Neonatal Neurobehavioral Scale; NCTR, National Center for Toxicological Research; OTB, Operant Test Battery; PCB, polychlorinated biphenyl; THC, delta-9-tetrahydrodocannabinol.

laboratory. In many cases, the relevance of these measures to humans is clear: interesting and important behavioral endpoints appear to be plentiful. The proper choice of animal model is also of paramount concern (McMillan and Owens, 1995). Use of nonhuman primates in developmental neurotoxicology studies is important for a variety of reasons, not the least of which is that monkeys are often "the best model with which to characterize effects on nervous system development produced by toxicants" (Rice, 1987). Their pregnancies are characterized by a long gestational period, after which development to sexual maturity is also protracted. Thus, as in humans, a very long period of development (potential vulnerability?) is presented in these animals (Rice, 1987). In addition, nonhuman primates are particularly valuable when data from other species are equivocal, when humans are likely to be exposed to the agent of interest during pregnancy, and for studies of human-derived substances and/or other products that are inactive in nonprimates (Hendrickx and Pinkerd, 1990). (See also Evans, 1990, for a clear discussion of the validity, ethics, and importance to public health of using nonhuman primates as experimental subjects.)

A discussion of behavior assessments with respect to developmental neurotoxicology serves to focus attention on processes that generally occur early in the life of an organism, be they the timing of neurotoxic exposure or the period during which the behavioral assessment of the effects of a previous exposure is made. This chapter focuses on strategies and methods for assessing behaviors of individuals that are observable prior to the completion of puberty, and thus excludes aspects of sexual behavior. The cutoff here to limit behavioral observations to immature subjects is not meant to suggest that important alterations in the nervous system do not continue beyond puberty, but serves to focus on the topic at hand. Attributes of the anatomy and physiology of the rhesus monkey, specifically, at a variety of developmental stages can be found in Bourne (1975) and the developmental stages to be discussed here are those described by Golub and Gershwin (1984) and include: newborn (24 hr postnatal); neonate (1 day–1 month); infant (1–12 months); juvenile (12–24 months); and adolescent (2–4 years). Although the examples to be discussed here refer primarily to macaque species, the principles are applicable to most nonhuman, and usually human, primates. In many cases, concordance of assessments between species is possible because many of the techniques used in the animal laboratory are often adaptations of those used clinically for humans and, more recently, techniques developed primarily for laboratory animals are being used with human subjects (e.g., Overman *et al.,* 1992, 1996a,b; Glassman *et al.,* 1994; Paule, 1994, 1995). The maintenance of task continuity across species also allows for the quantitative determination of interspecies similarities and differences in important aspects of brain function and assists in the extrapolation of data from laboratory animals to humans. It should also be mentioned that in many cases, the types of behavioral assessments proposed for use in developing animals are also directly applicable to adult and aging animals as well. An earlier bibliographic work can be referenced for access to more in-depth, historic material concerning the behavioral assessments of nonhuman primates (Akins *et al.,* 1980), many of which can be found in Schrier *et al.* (1965).

II. The Animal Model

Because it is the job of the neurotoxicologist to make predictions—based (ideally) on laboratory animal data—about the kinds of nervous system malfunctions likely to occur in humans when exposed to a particular neuroactive agent, the use of animal models that approximate the human condition closely is the most expeditious way in which to generate the most relevant data: they serve to lessen the uncertainty in extrapolating findings from the animal laboratory to humans. Importantly, nonhuman primates are also capable of an incredibly rich behavioral repertoire, much of which overlaps with that of humans. For these reasons, nonhuman primates are used in the neurotoxicology laboratory in the hope that the information they provide proves invaluable in identifying adverse effects of compounds that affect the nervous systems of humans.

III. Assessment of Newborns and Neonates

Behavioral assessments at birth and throughout the period preceding weaning often require separation of the offspring from the dam. Thus, although these observations are clearly important for the early detection of exposure effect and subsequent monitoring of its progression or recession, they often necessitate immobilization of the dam and/or the rearing of offspring in a nursery setting. Nursery care, obviously, is labor intensive and requires appropriate facilities. In those cases in which offspring are left with their dams, assessments of maternal–infant and other social interactions are also possible and can be used to look for treatment effects. This discussion does not include methods for assessing social interactions between animals, although clearly those kinds of readily observable functions are extremely important: the effects of developmental exposure to chemicals on mother–infant interactions (e.g., Golub *et al.,* 1981b) and on interactions of offspring with other members of their social structure have often been studied, and the literature on such methods is large (e.g., Schlemmer *et al.,* 1976; Smith and Byrd, 1983; Burbacher *et al.,* 1990).

Within minutes of birth, observations of heart rate, respiratory effort, responsiveness (state or irritability), skeletomuscular status (muscle tone, flaccidity, flexion), and skin color are often used to develop numeric expressions (Apgar scores) of the human newborn's condition (Apgar, 1952; Apgar and James, 1962). These same observations can be made in nonhuman primates and adaptations of these procedures constitute part of a neonatal neurobehavioral test battery proposed by Golub (1990).

Relatively intense neonatal behavior assessment of humans occurs routinely in the clinic because of the need to determine whether developmental abnormalities are present during this first opportunity to observe brain function in offspring. A variety of test batteries have been developed for assessing multiple aspects of human neonatal and infant nervous system function: for example, the Brazelton Neonatal Behavior Assessment Scale (BNBAS) (Als *et al.,* 1977; Brazelton, 1978), the Early Neonatal Neurobehavioral Scale (ENNS) (Scanlon *et al.,* 1974), the Prechtl examination (Prechtl and Beintema, 1964), and the Graham test (Rosenblith, 1961). Similar test batteries have also been developed for use in nonhuman primates (e.g., Taylor *et al.,* 1980, Goodlin *et al.,* 1982; Golub and Gershwin, 1984; Champoux and Suomi, 1988; Golub, 1990). These batteries typically focus on levels of arousal, activity, and reflex assessments.

Golub (1990) has proposed that specific monkey neonatal neurobehavioral test batteries be used in safety testing protocols. It is suggested that these batteries include Apgar-type measurements (respiration rate, muzzle color, initial state, and observed and elicited muscle tone at several locations), assessments of reflex development [palmer grasp, righting, clasp–grasp, visual following, lipsmack-orient, sucking, rooting, snout, pupillary response to light, nystagmus response to rotation, glabellar tap elicited eye blink, Moro response (loss of support)], prone progression (ability to move forward when prone), sustained arousal after stimulation (buildup), and habituation (response to repeated pinprick of the heel).

Overnight rest–activity, sleep–wake patterns can be obtained by videotaping neonates and scoring the observations with respect to quiet or active sleeping, quiet or active waking, startles, limb extensions, and so forth. For comparison, normal macaque sleep–wake patterns have been described previously (Sackett *et al.,* 1979). In addition, information about tremors, spasms, and convulsions can be captured.

Neonatal motor and postural maturation, locomotion, rest patterns, and exploratory activities can be quantified via visual observation of spontaneous behavior that occurs during the first 2 weeks of life. Activities such as climbing up or down the side of an enclosure and the manner in which it is done can be used to categorize developmental stage and pattern. After the initial 2-week period of observation, assessments of visual exploration and scanning are additional measures of interest (Golub, 1990).

IV. Assessment of Developing Animals

Many of the behaviors that are assessed during the neonatal period (first month of life) can be assessed repeatedly as the animals mature (Golub, 1990). This is important because treatment effects may manifest as developmental delays that can be detected only with repeated measurements. Knowledge of when transition periods for specific behaviors exist can help to focus the observation periods and maximize the likelihood that an effect will be detectable (Golub, 1990).

Early perceptual–cognitive skills can also be assessed in monkeys in much the same way they are in humans by use of forced-choice looking procedures and novelty test paradigms (Grant-Webster *et al.,* 1990). Metrics of visual acuity can be obtained using forced-choice preferential looking procedures in which subjects are presented with a series of vertical black and white stripes that are paired with homogenous gray fields matched for brightness. Subjects fixate preferentially on the stripes. By altering stripe width and determining the smallest width fixated, one can estimate visual acuity. One week of maturation in the monkey appears to be equivalent to 1 month of development in the human, with achievement of adult acuity seen in some *Macaques* by 6 weeks of age and in some humans as early as 6 months of age (Grant-Webster *et al.,* 1990).

Normal human and monkey infants show significant preference for novel versus familiar objects. Novelty test paradigms have been used to study the development of memory because responding to novel stimuli implies that some aspects of the familiar stimuli have been "remembered" such that subjects no longer consider them novel. It has also been demonstrated in children that novelty preference at early ages correlates with subsequent IQ scores (Fagan and Singer, 1983; Fagan and Montie, 1988); nonhuman primate versions of apparatus for assessing visual exploration have also been described (Levin *et al.,* 1986). Grant-Webster *et al.* (1990) performed a series of studies in monkeys in which subjects were presented a variety of recognition problems. They recorded novelty preference scores, and it was demonstrated that at-risk monkey offspring (exposed to methylmercury, low birth weight, perinatal insult) did not show significant novelty preference. These studies were consistent with those in at-risk children [exposed to polychlorinated biphenyls (PCBs) *in utero*], in which it was shown that exposed infants spent significantly less time looking at novel stimuli than did controls. Importantly, these measures were sensitive to magnitude of exposure (dose) because the magnitude of the deficits noted corresponded to umbilical cord blood–PCB levels (Jacobson *et al.,* 1985).

In studies on the effects of gestational and perinatal maternal exposure to delta-9-tetrahydrocannabinol (THC), Golub *et al.* (1981a) reported that monkey offspring showed treatment-related alterations in visual attention. The results suggested that offspring exposed to THC during development failed to limit their responses to novel stimuli when tested at 1–2 years of age and would predict cognitive problems in humans similarly exposed. Indeed, Fried (1995) has reported long-term behavioral consequences of gestational exposure to marijuana smoke in humans, a finding that serves to validate the monkey model.

Given that relatively young nonhuman primates exhibit near-adult or adult levels of certain perceptual abilities, it then follows that they should be capable of exhibiting other aspects of central nervous system (CNS) function that involve complex integration. Nonverbal and nonsocial problem-solving abilities are common to a variety of species and can be modeled relatively easily in laboratory animals using operant behavioral techniques: By requiring subjects to make specific responses (operate something such as a lever, a button, a press-plate, a hidden food item) in the presence of specific stimuli (light, tone, food puzzle) in order to obtain reinforcers (food, water, money, etc.), tasks can be devised that require subjects to exhibit aspects of what are thought to be specific complex brain functions such as learning, visual discrimination, and so on. Such procedures are readily automatable and have been shown to be quite applicable for even infant monkeys (Rice, 1979; Rice and Gilbert, 1990) where aspects of learning and memory were assessable in very young animals (training began within the first week of life). Relatively simple tasks such as nonspatial visual discrimination (e.g., respond to the illuminated key, not the darkened one; respond to the red key and not the green one for access to infant formula) and reversal (instead of responding to the red key, which previously resulted in reinforcement, now respond to the green key) are readily trainable in young animals (Rice and Gilbert, 1990). In a series of studies in which young *Macaca fascicularis* were assessed using a variety of automated procedures, animals began training at about 1 month of age to perform a variable ratio task (the number of responses required to obtain a reinforcer varies from reinforcer to reinforcer, i.e., is unpredictable); subsequent tasks were added to the subjects' behavioral repertoire such that a battery of assessments was effected within the first year of life (Rice and Gilbert, 1990). In addition, performance of some of these and related tasks proved sensitive to developmental lead exposure (e.g., Bushnell and Bowman, 1979; Rice and Willes, 1979; Rice *et al.,* 1979; Levin and Bowman, 1983, 1986, 1989; Rice, 1985, 1988, 1992; Lilienthal *et al.,* 1986) and gestational exposure to methylmercury (Gilbert *et al.,* 1993). Although long-term developmental exposure to heavy metals clearly seems to have adverse impacts on subsequent CNS function in nonhuman primates, it has also been demonstrated that even short-term intrapartum exposures to obstetric analgesics can have an impact on subsequent behavior in offspring (Golub *et al.,* 1988; Golub and Donald, 1995).

V. Age-Dependent Sensitivity

An important issue in nonhuman primate (and human) studies concerns age-related alterations in sensitivity to agents that affect CNS function. In studies concerning the acute behavioral effects of amphetamine and cocaine in rhesus monkeys performing a battery of operant tasks (see later this chapter), tremendous age effects were noted. Adult subjects (>7 years old) were 10–30 times more sensitive to some of the behavioral effects of these agents than were juveniles (1.5 years old); adolescents were intermediate in sensitivity (Paule *et al.,* 1997). These data demonstrate clearly that developing animals are not just small adults, and it is of paramount importance to consider age at exposure when attempting to extrapolate findings. It may also be possible to determine the functional status of particular neurotransmitter systems by using such pharmacologic probes. For example, because both amphetamine and cocaine are thought to exert their behavioral effects via their interaction with the dopamine system, it would appear that in juvenile monkeys, the dopamine system is not fully functional, and that although it is more mature in adolescents, it may not reach adult status until after puberty. By choosing agents that target specific receptors and/or neurotransmitters in brain, it should be possible to ascertain functional development profiles for the system(s) of interest. Such observations in nonhuman primates may have direct applicability in the development of rational drug therapies for pediatric populations and in determining whether children are at greater or lesser risk for neurotoxicity from environmental exposures.

VI. Interspecies Comparisons Using an Automated Test Battery Approach

Hopefully, it is clear that many types of problem-solving tasks can be modeled in animals in exactly the same way they can be modeled in humans. This is important because problem-solving abilities represent rich and essential functional domains shared by both animals and humans. By invoking the concept of face validity, their relevance to humans seems clear. Use of identical behavioral measures in both laboratory subjects and humans is also important because it can provide laboratory animal test validity (Paule and Adams, 1994). Importantly, many of these types of tasks can be automated, and thus the labor required to collect data using these approaches can be minimized.

The National Center for Toxicological Research (NCTR) Operant Test Battery (OTB) has been used in both laboratory animals (Paule, 1990) and humans. The OTB (originally described in Schulze *et al.,* 1988) consists of five tasks, each of which is thought to allow assessment of a different aspect of complex brain function. Subjects interact with a behavioral panel on which are located a variety of manipulanda (levers, press-plates) and visual stimuli (colors, geometric forms). Different stimuli and response types are associated with each of the tasks. An unfortunate feature of these kinds of behavioral assessments in animals is that they can require substantial subject training, which may take weeks to months, depending on the task(s) of interest. Major advantages include their automatability, the ability to use training (acquisition) data as measures of insult, and the ability to sample behavior repeatedly over extended periods of time (years).

For the *motivation task,* a single response lever is presented to the subject, on which increasing amounts of "work" must be performed in order to obtain additional reinforcers (banana-flavored food pellets). In this task (also called a *progressive ratio* task because the ratio of work/reinforcer progresses throughout the session), the first reinforcer may cost only two lever presses, but the next reinforcer costs four, the next six, the next eight, and so on. Measurements such as response speed and number of lever presses made for the last reinforcer delivered (i.e., breakpoint) provide metrics of behavior believed to be related to a subject's motivation. Interestingly, in children performing this same task, no measure of responding in this task correlates with intelligence (Full Scale, Verbal or Performance IQ) (Paule *et al.,* 1998): children with Full Scale IQ scores of about 50 work just as hard to earn reinforcers (nickels in this case) as do children with IQs of more than 120. In studies looking at the effects of drugs on brain function, it is very important to be able to assess motivation in order to understand more fully how an agent may be affecting the behaviors of interest.

Often, the acute effects of drugs on other aspects of brain function in monkeys (e.g., time perception, learning, etc.) can be observed in the absence of any treatment-related effects on performance of this task (e.g., Schulze *et al.,* 1988, 1989a,b, 1992; Buffalo *et al.,* 1994). Conversely, it has been observed that performance in this task is sensitive to the effects of chronic administration of marijuana smoke, whereas the performance of other OTB tasks is relatively unaffected (Paule *et al.,* 1992). These findings demonstrate the detectability of an apparent "amotivational syndrome" in the adolescent monkey during chronic marijuana smoke exposure, and again serve to validate the monkey model because a similar syndrome has been described for adolescent humans.

For the *color and position discrimination task,* subjects are initially required to observe the central of three press-plates on which one of four colors (red, yellow, blue, or green) has been projected. The subject's task is to discriminate the color and respond to it (press it), after which it is immediately extinguished and the two side press-plates are illuminated white. If the center color had been either red or yellow, a response to the left (now white press-plate) results in reinforcer delivery; if the center color had been either blue or green, a response to the right (now white press-plate) results in reinforcer delivery. Measurement of choice accuracy and speed of responding to the initial color and to make either a correct or incorrect choice are obtained. In children subjects performing this same task, accuracy, response rate, and correct choice response latencies have all been shown to correlate significantly with Full Scale IQ (Paule *et al.,* 1998). In comparison to the performance of other OTB tasks by monkeys, color and position discrimination behavior is relatively insensitive to acute drug effects and is often the least sensitive of all. However, in a study on the acute behavioral effects of atropine in rhesus macaques, performance of this task was second only to the learning task in sensitivity, suggesting that atropine may have some preferential effects on color vision, as well as learning, in monkeys.

For the *time estimation task,* subjects must hold a response lever in the depressed position for some minimum amount of time (10 sec) but no longer than some maximum amount of time (14 sec). Thus, a 4-sec window of opportunity exists during which reinforcers can be obtained. Lever releases that are either too short or too long have no programmed consequences; subjects initiate the next trial by depressing the lever. Aspects of performance (accuracy and response rate) of this task by children are also correlated significantly with intelligence (Paule *et al.,* 1998). Performance of this task by monkeys is very sensitive to disruption by psychotropic drugs; generally, time estimation behavior is affected at doses lower than or equal to those affecting performance of the other OTB tasks (see, for example, Schulze *et al.,* 1989b; Buffalo *et al.,* 1993; Ferguson and Paule, 1993; Frederick *et al.,* 1995). This particular sensitivity to the acute effects of drugs may make timing behavior a good candidate for use in screening efforts because it is the most likely of all of the OTB behaviors to be affected by psychoactive agents.

The *short-term memory task* is a version of a type of procedure referred to as delayed matching-to-sample. Similar procedures have been used for many decades and, even early in the use of such paradigms, comparisons of task performance were made between monkeys and children (see, for example, Weinstein, 1940). Correct performance of this task requires that subjects observe a sample stimulus (in the OTB, this is a geometric form projected onto a press-plate) and then choose a matching stimulus from several choice stimuli presented some time later (e.g., 2–30 sec). This task is performed using the same press-plates described earlier for the color and position discrimination task. Initially, a sample stimulus appears on the central of three press-plates, and subjects respond to it (press it) to extinguish it immediately and start a delay period during which all three press-plates are darkened. After the delay, all three press-plates are illuminated with different geometric forms, one of which matches the sample stimulus for that trial. Reinforcers are delivered to subjects responding to the matching form. Choice accuracies and response latencies for each delay and those measures collapsed across delays (overall measures) are some of the variables collected. Plots of decreasing accuracy with increasing delays represent "forgetting" functions, the slopes of which represent metrics for rate of memory decay. Accuracy measures at 0 and very short delays are thought to be dependent on discrimination and encoding abilities. As delays increase, choice response latencies also increase and, particularly at the longer delays, these measures are thought to be associated with aspects of attention. Perhaps not surprisingly, most measures of performance of this task by children correlate significantly with their intelligence (Full Scale, Verbal and Performance (IQ) (Paule *et al.,* 1998). In earlier studies on the acute behavioral effects of THC and marijuana smoke in the monkey (Schulze *et al.,* 1988, 1989a), it was demonstrated that performance of this short-term memory task was quite sensitive to treatment effects, being the OTB task most sensitive to the effects of marijuana smoke and, for THC, second only in sensitivity to the time estimation task. In addition, the effects of these agents on short-term memory in the monkey model served to provide further validation of the approach because, in humans, the cannabinoids are well-known to adversely affect performance on short-term memory tasks.

For the *learning task* subjects are presented with four response levers, and over the course of each test session are to learn a specific sequence of lever presses that changes with each test session. Thus, subjects must repeatedly acquire new response sequences and the task used is called a *repeated acquisition task.* As used in the NCTR OTB, initially subjects are presented with a one-lever "sequence" (i.e., a single response to only the correct one of the four levers is required for reinforcer delivery). After demonstration that the one lever "sequence" has been mastered (some performance criterion has been met), the sequence length is incremented so that two correct responses must be made prior to reinforcer delivery. After demonstration that the two-lever sequence has been mastered, the sequence length is incremented again so that three correct responses must be made prior to reinforcer delivery, and so on up to six lever response sequences. Thus, this task is a form of incremental repeated acquisition and provides metrics of learning; accuracy and response rates are obtained for each level (lever sequence length) attained and measures of maximum sequence length mastered are recorded. Overall measures of accuracy and response rate are obtained by collapsing data across lever sequence length. Plots of errors made versus correct sequences completed provide learning curves for each level of task difficulty. Perhaps not surprisingly, sequence length mastered and accuracy and response rate for this task in children all correlate significantly with IQ measures (Paule *et al.,* 1998). This learning task is the most sensitive of the OTB tasks to disruption by the anticholinergic compound atropine; learning behavior is disrupted at doses that do not affect performance of the other OTB tasks (Schulze *et al.,* 1992).

The NCTR OTB has been applied in studies designed to examine the behavioral consequences to offspring of chronic cocaine administration during pregnancy in rhesus monkeys (Paule *et al.,* 1996; Morris *et al.,* 1996a,b, 1997). To date, no effects of cocaine exposure *in utero* have been detected using these procedures; all study animals, regardless of exposure history, learned to perform these tasks in similar fashion. In addition, the use of pharmacologic challenges to unmask subtle treatment effects (see Zenick, 1983; Walsh and Tilson, 1986) in these animals has failed to demonstrate any differences in the behavioral effects of a variety of psychotropic agents as a function of gestational cocaine exposure. In light of evidence suggesting that, in humans, gestational cocaine exposure alone has only very subtle, if any, behavioral effects in offspring, it might not be surprising that the functional domains represented by OTB task performance in the monkey are also not affected by such exposure.

VII. Future Directions

It seems clear that there are a host of behavioral assessments that can be used in both human and nonhuman primates, some of which are already being used in both the animal laboratory and the clinical setting. Continued exploration of how performance in one task relates to performance in others should help to determine which specific tasks are tapping into the same or similar functional domains, thus making interpretation of data obtained using one task more interpretable and/or extrapolatable to others. Determining the degree to which measures from existing behavioral tasks correlate with the more traditional pencil and paper neuropsychologic instruments is also valuable in helping to focus research on tasks that shed light on specific brain functions of interest. These kinds of comparisons are only possible using human versions of tasks that are applicable in the animal laboratory; human performance of both types of tasks must be compared to determine the relationship(s), if any, of one to the other. If, for example, performance by children of a fixed consecutive number task (press the right lever eight times before pressing the left lever to get a reinforcer) is shown to correlate with arithmetic capability as determined by some other standardized instrument, it could be reasoned that animal performance of the same task might be useful in exploring potential drug or toxicant effects on mathematic ability in humans.

It has been said that, "To overcome or reduce confounding in neurodevelopmental studies on human health and environment, efforts should be made to develop, apply and validate neurodevelopmental outcome measures which are less likely to be influenced by socio-hereditary factors" (Winneke, 1995). If we are able to develop behavioral tools that show little interspecies differences, the likelihood that they will be sensitive to cultural and other factors would seem to be small.

VIII. Conclusions

It is clear that many behavioral measures obtainable from monkeys (and other species) are relevant to human brain function as demonstrated by their correlation with traditional measures of intelligence in humans. Exactly the same behavioral endpoints can be obtained in nonhuman and human primates, and perhaps other animals; thus, analogous brain functions can be modeled across species. Toxicity data obtained from appropriate behavioral studies in developing nonhuman primate animal models should be extremely useful in the neurotoxic risk assessment process. Data from other animal models may also provide important information along these same lines, once the similarities and differences between the species are understood.

For exemplary purposes, the use of the automated system developed at the FDA's NCTR was highlighted; a battery of behavioral tasks that models specific aspects of complex brain function, including motivation, color and position discrimination, time estimation, short-term memory, and learning. Previous studies have indicated that each of the behaviors monitored in the NCTR OTB is representative of a different domain of complex brain function in that correlations of endpoints between tasks are usually nonexistent or minimal (Paule, 1990). Certainly the test battery approach is one that provides a variety of information and allows assessment of a variety of brain functions. Similar strategies have been proposed for assessing the effects of environmental exposures in children wherein many behavioral measures, in addition to questionnaire data, are highlighted (Krasnegor *et al.,* 1994). Specific batteries of assessments are suggested for specific ages.

Comparative data indicate that the performance of well-trained laboratory monkeys in the OTB is generally indistinguishable from that of children (Paule *et al.,* 1990). In addition, human performance in this same battery correlates significantly with measures of intelligence (IQ scores) (Paule *et al.,* 1998). As mentioned earlier, such correlations serve to highlight the relevance of these measures when obtained from appropriate animal models. OTB and IQ data obtained from children have been used to develop mathematic models for estimating the IQ of children based on OTB data alone (Hashemi *et al.,* 1994). Such approaches may someday provide a means for estimating the intelligence of our animal subjects, as well, and for determining the effects of neuroactive agents on intelligence.

Because laboratory animals other than monkeys can also perform tasks similar or identical to those in the NCTR OTB, such tools may provide the opportunity for extensive interspecies comparisons of cognitive processes, and provide the means for studying the effects of neuroactive (and thus potentially neurotoxic) agents using relevant endpoints in other animal models, when appropriate. Approaches similar to those used in the NCTR OTB are clearly valuable in providing important comparative data, and several different approaches are beginning to emerge. Tasks applicable to both animals and humans include those that use maze performance (e.g., Overman *et al.,* 1992, 1996b), joy-stick manipulation (Rumbaugh, *et al.,* 1989; Washburn and Rumbaugh,

1992a,b; Andrews, 1993; Andrews and Rosenblum, 1993, 1994a,b; Washburn and Hopkins, 1994; Lincoln *et al.,* 1995) and touch-screen technology (Roberts *et al.,* 1988; Bhatt and Wright, 1992; Gold *et al.,* 1994; Steckler and Sahgal, 1995). The task ahead involves continuing to determine which particular tasks provide information relevant to the human condition and, when that is known, applying such tasks in the animal laboratory to provide risk assessment information.

References

Akins, F. R., Mace, G. S., Hubbard, J. W., and Akins, D. L. Eds. (1980). *Behavioral Development of Nonhuman Primates: An Abstracted Bibliography* New York, IFI/Plenum.

Als, H., Tronick, E., Lester, B. M., and Brazelton, T. B. (1977). The Brazelton neonatal behavior assessment scale (BNBAS). *J. Abnorm. Child Psych.* **5**(3), 215–231.

Andrews, M. W. (1993). Video task paradigm extended to Saimiri. *Percept. Mot. Skills* **76**(1), 183–191.

Andrews, M. W., and Rosenblum, L. A. (1993). Live-social-video reward maintains joystick task performance in bonnet macaques [see comments]. *Percept. Mot. Skills* **77**(3 pt 1), 755–763.

Andrews, M. W., and Rosenblum, L. A. (1994a). Automated recording of individual performance and hand preference during joystick-task acquisition in group-living bonnet macaques (*Macaca radiata*). *J. Comp. Pyschol.* **108**(4), 358–362.

Andrews, M. W., and Rosenblum, L. A. (1994b). Relative efficacy of video versus food-pellet reward of joystick tasks [comment]. *Percept. Mot. Skills* **78**(2), 545–546.

Anger, W. K., and Johnson, B. L. (1985). Chemicals affecting behavior. In: J. L. O'Donoghue, ed., *Neurotoxicity of Industrial and Commercial Chemicals,* CRC Press, Boca Raton, Florida, pp. 52–148.

Apgar, V. (1952). A proposal for a new method of evaluation of the newborn infant. *Anesth. Analg.* **32,** 260–267.

Apgar, V., and James, L. S. (1962). Further observations on the newborn scoring system. *JAMA* **181,** 419–428.

Bhatt, R. S., and Wright, A. A. (1992). Concept learning by monkeys with video picture images and a touch screen. *J. Exp. Anal. Behav.* **57,** 219–225.

Bourne, G. H. (1975). Collected anatomical and physiological data from the rhesus monkey. In G. H. Bourne Ed., *The Rhesus Monkey.* Vol. I. Academic Press, New York, pp. 1–63.

Brazelton, T. B. (1978). The Brazelton neonatal behavior assessment scale: introduction. *Monographs Soc. Res. Child Dev.* **43**(5–6), 1–13.

Buffalo, E. A., Gillam, M. P., Allen, R. R., and Paule, M. G. (1993). Acute effects of caffeine on several operant behaviors in rhesus monkeys. *Pharmacol. Biochem. Behav.* **46**(3), 733–737.

Buffalo, E. A., Gillam, M. P., Allen, R. R., and Paule, M. G. (1994). Acute behavioral effects of MK-801 in rhesus monkeys: assessment using an operant test battery. *Pharmacol. Biochem. Behav.* **48**(4), 935–940.

Burbacher, T. M., Sackett, G. P., and Mottet, N. K. (1990). Methylmercury effects on the social behavior of *Macaca fascicularis* infants. *Neurotox. Teratol.* **12**(1), 65–71.

Bushnell, P. J., and Bowman, R. E. (1979). Reversal learning deficits in young monkeys exposed to lead. *Pharmacol. Biochem. Behav.* **10,** 733–742.

Champoux, M., and Suomi, S. J. (1988). Behavioral development of nursery-reared rhesus macaque (*Macaca mulatta*) neonates. *Infant Behav. Dev.* **11,** 363–367.

Evans, H. L. (1990). Nonhuman primates in behavioral toxicology: issues of validity, ethics and publich health. *Neurotoxicol. Teratol.* **12,** 531–536.

Fagan, J. F., and Montie, J. E. (1988). Behavioral assessment of cognitive well-being in the infant. In J. F. Kavanagh, Ed., *Understanding Mental Retardation: Research Accomplishments and New Frontiers.* Baltimore, Paul H. Brookes, pp. 207–221.

Fagan, J. F., and Singer, L. T. (1983). Infant recognition memory as a measure of intelligence. In L. P. Lipsitt, ed. *Advances in Infancy Research,* Vol. 2. Norwood, NJ, Ablex.

Ferguson, S. A., and Paule, M. G. (1993). Acute effects of pentobarbital in a monkey operant behavioral test battery. *Pharmacol. Biochem. Behav.* **45,** 107–116.

Frederick, D. L., Schulze, G. E., Gillam, M. P., and Paule, M. G. (1995). Acute effects of physostigmine on complex operant behavior in rhesus monkeys. *Pharmacol. Biochem. Behav.* **50**(4), 641–648.

Fried, P. A. (1995). Prenatal exposure to marihuana and tobacco during infancy, early and middle childhood: effects and an attempt at synthesis. *Arch. Toxicol. Suppl.* **17,** 233–260.

Gilbert, S. G., Burbacher, T. M., and Rice, D. C. (1993). Effects of *in-utero* methylmercury exposure on a spatial delayed alternation task in monkeys. *Toxicol. Pharmacol.* **123,** 130–136.

Glassman, R. B., Garvey, K. J., Elkins, K. M., Kasal, K. L., and Couillard, N. L. (1994). Spatial working memory score of humans in a large radial maze, similar to published score of rats, implies capacity close to the magical number 7 ± 2. *Brain Res. Bull.* **34**(2), 151–159.

Gold, L. H., Polis, I., Roberts, A. C., Robbins, T. W., and Koob, G. F. (1994). Cambridge neuropsychological test automated battery (CANTAB): application to neuropsychological assessment in rhesus monkeys. *NIDA Res. Monograph* **141,** 303.

Golub, M. S. (1990). Use of monkey neonatal neurobehavioral test batteries in safety testing protocols. *Neurotox. Teratol.* **12**(5), 537–541.

Golub, M. S., and Donald, J. M. (1995). Effect of intrapartum meperidine on behavior of 3- to 12-month-old infant rhesus monkeys. *Biol. Neonate* **67,** 140–148.

Golub, M. S., Eisele, J. H., and Donald, J. M. (1988). Obstetric analgesia and infant outcome in monkeys: neonatal measures after intrapartum exposure to meperidine or alfentanil. *Am. J. Obs. Gyn.* **158,** 1219–1225.

Golub, M. S., and Gershwin, M. E. (1984). Standardized neonatal assessment in the rhesus monkey. In P. W. Nathanielsz and J. T. Parer, eds., *Research in Perinatal Medicine.* Perinatology Press, New York, pp. 55–81.

Golub, M. S., Sassenrath, E. N., and Chapman, L. F. (1981a). Regulation of visual attention in offspring of female monkeys treated chronically with delta-9-tetrahydrocannabinol. *Dev. Psychobiol.* **14,** 507–512.

Golub, M. S., Sassenrath, E. N., and Chapman, L. F. (1981b). Mother–infant interaction in rhesus monkeys treated clinically with delta-9-tetrahydrocannabinol. *Child Dev.* **52**(1), 389–392.

Goodlin, B. L., Caffery, S. A., and Sackett, G. P. (1982). Neonate behavioral assessment of pigtail macaques (*Macaca nemestrina*). *Int. J. Primatol.* **3,** 286.

Grant-Webster, K. S., Gunderson, V. M., and Burbacher, T. M. (1990). Behavioral assessment of young nonhuman primates: perceptual–cognitive development. *Neurotoxicol. Teratol.* **12**(5), 543–546.

Hashemi, R. R., Pearce, B. A., Hinson, W. G., Paule, M. G., and Young, J. F. (1994). IQ estimation of monkeys based on human data using rough sets. *Proceed. 3rd Intl. Wrkshp. Rough Sets Soft Comp.* pp. 400–407.

Hendrickx, A. G., and Pinkerd, P. E. (1990). Nonhuman primates and teratological research. *J. Med. Primatol.* **19,** 81–108.

Jacobson, S. W., Fein, G. G., Jacobson, J. L., Scwartz, P. M., and Dowler, J. K. (1985). The effect of intrauterine PCB exposure on visual recognition memory. *Child Dev.* **56,** 853–869.

Krasnegor, N. A., Otto, D. A., Bernstein, J. H., Burke, R., Chappell, W., Eckerman, D. A., Needleman, H. L., Oakley, G., Rogan, W., Terracciano, G., and Hutchinson, L. (1994). Neurobehavioral test strategies for environmental exposures in pediatric populations. *Neurotoxicol. Teratol.* **16**(5), 499–509.

Levin, E. D., Boehm, K. M., Hagquist, W. W., and Bowman, R. E. (1986). A visual exploration apparatus for infant monkeys. *Am. J. Primatol.* **10,** 195–199.

Levin, E. D., and Bowman, R. E. (1983). The effect of pre- or postnatal lead exposure on Hamilton search task in monkeys. *Neurobehav. Toxicol. Teratol.* **5,** 391–394.

Levin, E. D., and Bowman, R. E. (1986). Comparative sensitivity of the Hamilton search task and delayed spatial alternation in detecting long-term effects of neonatal lead exposure in monkeys. *Neurobehav. Toxicol. Teratol.* **8,** 219–224.

Levin, E. D., and Bowman, R. E. (1989). Long-term effects of chronic postnatal lead exposure on delayed spatial alternation in monkeys. *Neurotoxicol. Teratol.* **10,** 505–510.

Lilienthal, H., Winneke, G., Brockhaus, A., and Malik, B. (1986). Pre- and postnatal lead exposure in monkeys: effects on activity and learning set formation. *Neurobehav. Toxicol. Teratol.* **8,** 265–272.

Lincoln, H. III, Andrews, M. W., and Rosenblum, L. A. (1995). Pigtail macaque performance on a challenging joystick task has important implications for enrichment and anxiety within a captive environment. *Lab. An. Sci.* **45**(3), 264–268.

McMillan, D. E., and Owens, S. M. (1995). Extrapolating scientific data from animals to humans in behavioral toxicology and behavioral pharmacology. In L. W. Chang, ed., *Neurotoxicology: Approaches and Methods,* Academic Press, Orlando, Florida, pp. 323–332.

Morris, P., Binienda, Z., Gillam, M. P., Harkey, M. R., Zhou, C., Henderson, G. L., and Paule, M. G. (1996a). The effect of chronic cocaine exposure during pregnancy on maternal and infant outcomes in the rhesus monkey. *Neurotoxicol. Teratol.* **18**(2), 147–154.

Morris, P., Binienda, Z., Gillam, M. P., Klein, J., McMartin, K., Koren, G., Duhart, H. M., Slikker, W. Jr., and Paule, M. G. (1997). The effect of chronic cocaine exposure throughout pregnancy on maternal and infant outcomes in the rhesus monkey. *Neurotoxicol. Teratol.* **19**(1), 47–57.

Morris, P., Gillam, M. P., Allen, R. R., and Paule, M. G. (1996b). The effects of chronic cocaine exposure during pregnancy on the acquisition of operant behaviors by rhesus monkey offspring. *Neurotox. Teratol.* **18**(2), 155–166.

Overman, W. H., Bachevalier, J., Schuhmann, E., and Ryan, R. (1996a). Cognitive gender differences in very young children parallel biologically based cognitive gender differences in monkeys. *Behav. Neurosci.* **110**(4), 673–684.

Overman, W. H., Carter, L., and Thompson, S. (1992). Development of place memory in children as measured in a dry Morris maze. *Soc. Neurosci. Abst.* **18,** 332.

Overman, W. H., Pate, B. J., Moore, K., and Peuster, A. (1996b). Ontogeny of place learning in children as measured in the radial arm maze, Morris search task, and open field task. *Behav. Neurosci.* **110**(6), 1205–1228.

Paule, M. G. (1990). Use of the NCTR operant test battery in nonhuman primates. *Neurotoxicol. Teratol.* **12,** 413–418.

Paule, M. G. (1994). Analysis of brain function using a battery of schedule-controlled operant behaviors. In B. Weiss and J. O'Donoghue, eds. *Neurobehavioral Toxicity: Analysis and Interpretation.* Raven Press, New York, pp. 331–338.

Paule, M. G. (1995). Approaches to utilizing aspects of cognitive function as indicators of neurotoxicity. In L. Chang and W. Slikker, Jr., eds., *Neurotoxicology: Approaches and Methods,* Academic Press, Orlando, Florida, pp. 301–308.

Paule, M. G. (1997). Age-related sensitivity to the acute behavioral effects of cocaine and amphetamine in monkeys. *Neurotox. Teratol.* **19**(3), 240–241.

Paule, M. G., and Adams, J. (1994). Interspecies comparison of the evaluation of cognitive developmental effects of neurotoxicants in primates. In L. W. Chang, ed., *Principles of Neurotoxicology,* Marcel Dekker, Inc., New York, pp. 713–731.

Paule, M. G., Allen, R. R., Bailey, J. R., Scallet, A. C., Ali, S. F., Brown, R. M., and Slikker, W. Jr. (1992). Chronic marijuana smoke exposure in the rhesus monkey. II: Effects on progressive ratio and conditioned position responding. *J. Pharmacol. Exp. Therap.* **260**(1), 210–222.

Paule, M. G., Chelonis, J. J., Buffalo, E. A., Blake, D. J., and Casey, P. H. (1998). Operant test battery performance in children: correlation with IQ. Submitted, *Neurotox. Teratol.*

Paule, M. G., Forrester, T. M., Maher, M. A., Cranmer, J. M., and Allen, R. R. (1990). Monkey versus human performance in the NCTR operant test battery. *Neurotoxicol. Teratol.* **12,** 503–507.

Paule, M. G., Gillam, M. P., Binienda, Z., and Morris, P. (1996). Chronic cocaine exposure throughout gestation in the rhesus monkey: pregnancy outcomes and offspring behavior. *Ann. N.Y. Acad. Sci.* **801,** 301–309.

Prechtl, H., and Beintema, D. (1964). *The Neurological Examination of the Full-Term Newborn Infant.* London, William Heinemann Medical Books, Ltd.

Rice, D. C. (1979). Operant conditioning of infant monkeys (*Macaca fascicularis*) to toxicity testing. *Neurobehav. Toxicol.* **1**(suppl. 1), 85–92.

Rice, D. C. (1985). Chronic low-level lead exposure from birth produces deficits in discrimination reversal in monkeys. *Toxicol. Appl. Pharmacol.* **79,** 201–210.

Rice, D. C. (1987). Primate research: relevance to human learning and development. *Dev. Pharmacol. Ther.* **10,** 314–327.

Rice, D. C. (1988). Schedule-controlled behavior in infant and juvenile monkeys exposed to lead from birth. *Neurotoxicology* **9,** 75–88.

Rice, D. C. (1992). Behavioral effects of lead in monkeys tested during infancy and adulthood. *Neurotoxicol. Teratol.* **14,** 235–245.

Rice, D. C., and Gilbert, S. G. (1990). Automated behavioral procedures for infant monkeys. *Neurotox. Teratol.* **12,** 429–439.

Rice, D. C., Gilbert, S. G. and Willes, R. F. (1979). Neonatal low-level lead exposure in monkeys (*Macaca fascicularis*): locomotor activity, schedule-controlled behavior, and the effects of amphetamine. *Toxicol. Appl. Pharmacol.* **51,** 503–513.

Rice, D. C., and Willes, R. F. (1979). Neonatal low-level lead exposure in monkeys (*Macaca fascicularis*) effects on two-choice non-spatial form discrimination. *J. Environ. Pathol. Toxicol.* **2,** 1195–1203.

Roberts, A. C., Robbins, T. W., and Everitt, B. J. (1988). The effects of intradimensional and extradimensional shifts on visual discrimination learning in humans and non-human primates. *Quart. J. Exper. Pscyhol.* **40B**(4), 321–341.

Rosenblith, J. F. (1961). The modified Graham behavior test for neonates: test–retest reliability, normative data, and hypotheses for future work. *Biol. Neonate* **3,** 174–192.

Rumbaugh, D. M., Richardson, W. K., Washburn, D. A., Savage-Rumbaugh, E. S., and Hopkins, W. D. (1989). Rhesus monkeys (*Macaca mulatta*), video tasks, and implications for stimulus–response spatial contiguity. *J. Comp. Psychol.* **103,** 32–38.

Sackett, G. P., Fahrenbruch, C. E., and Ruppenthal, G. C. (1979). Development of basic and physiological parameters and sleep–wakefulness patterns in normal and at-risk neonatal pigtail macaques (*Macaca nemestrina*). In G. C. Ruppenthal, ed. *Nursery Care of Nonhuman Primates.* New York, Plenum Press, pp. 125–144.

Scanlon, J. W., Brown, M. U., Weiss, J. B., and Alper, M. H. (1974). Neurobehavioral responses of newborn infants after maternal epidural anesthesia. *Anesthesiology* **40,** 121–128.

Schlemmer, R. F. Jr., Casper, R. C., Siemsen, F. K., Garver, D. L., and Davis, J. M. (1976). Behavioral changes in a juvenil primate social colony with chronic administration of d-amphetamine. *Pyschopharm. Commun.* **2**(1), 49–59.

Schrier, A. M., Harlow, H. F., and Stollnitz, F. (eds.) (1965). *Behavior of Nonhuman Primates (Modern Research Trends).* Vol. 1. New York, Academic Press.

Schulze, G. E., Gillam, M. P., and Paule, M. G. (1992). Effects of atropine on operant test battery performance in rhesus monkeys. *Life Sci.* **51**(7), 487–497.

Schulze, G. E., McMillan, D. E., Bailey, J. R., Scallet, A., Ali, S. F., Slikker, W. Jr., and Paule, M. G. (1988). Acute effects of delta-9-tetrahydrocannabinol in rhesus monkeys as measured by performance in a battery of complex operant tests. *J. Exper. Pharm. Ther.* **245,** 178–186.

Schulze, G. E., McMillan, D. E., Bailey, J. R., Scallet, A. C., Ali, S. F., Slikker, W. Jr., and Paule, M. G. (1989a). Acute effects of marijuana smoke on complex operant behavior in rhesus monkeys. *Life Sci.* **45**(6), 465–475.

Schulze, G. E., Slikker, W. Jr., and Paule, M. G. (1989b). Multiple behavioral effects of diazepam in rhesus monkeys. *Pharmacol. Biochem. Behav.* **34,** 29–35.

Smith, E. O., and Byrd, L. D. (1983). Studying the behavioral effects of drugs in group-living nonhuman primates. *Prog. Clin. Biol. Res.* **131,** 1–31.

Steckler, T., and Sahgal, A. (1995). Psychopharmalogical studies in rats responding at touch-sensitive devices. *Psychopharmacology* **118**(2), 226–229.

Taylor, E. M., Sutton, D., and Lindeman, R. C. (1980). Somatic reflex development in infant macaques. *J. Med. Primatol.* **9,** 205–210.

Walsh, T. J., and Tilson, H. A. (1986). The use of pharmacological challenges. In Z. Annau, ed. *Neurobehavioral Toxicology.* Baltimore, The Johns Hopkins University Press, pp. 244–267.

Washburn, D. A., and Hopkins, W. D. (1994). Videotape versus pellet-reward preferences in joystick tasks by macaques. *Percept. Mot. Skills* **78,** 48–50.

Washburn, D. A., and Rumbaugh, D. M. (1992a). Investigations of rhesus monkey video-task performance: evidence for enrichment. *Contemp. Topics* **31,** 6–10.

Washburn, D. A., and Rumbaugh, D. M. (1992b). Testing primates with joystick-based automated apparatus: lessons from the Language Research Center's computerized test system. *Behav. Res. Meth. Instr. Comp.* **24,** 157–164.

Weinstein, F. (1940). Matching-from-sample by rhesus monkeys and by children. *J. Comp. Psych.* **31,** 195–213.

Winneke, G. (1995). Endpoints of developmental neurotoxicity in environmentally exposed children. *Toxicol. Lett.* **77**(1–3), 127–136.

Zenick, H. (1983). Use of pharmacological challenges to disclose neurobehavioral deficits. *Fed. Proceed.* **42,** 3191–3195.

PART VI

Clinical Assessment and Epidemiology

Introduction

The chapters in Part VI serve as a bridge between the chapters that precede and the chapters that follow. The preceding chapters describe the anatomy, physiology, biochemistry, toxicology, and function of the developing central nervous system (CNS). Most of the information comes from studies in animal or *in vitro* models, in which experimental manipulations of various kinds have helped to unravel the complexities of nervous system development. Such experimental models are necessary to gain a deep understanding of the mechanisms that underlie neurotoxic events.

The chapters that follow discuss studies of the effects of various developmental neurotoxins and approaches to determining the risks related to such exposures. These investigations focus on humans and concern outcome much more than mechanism.

Part VI deals with methods employed in human studies of developmental neurotoxicity. The techniques discussed are used to establish that exposure to a particular agent does, in fact, produce adverse effects on CNS development, and to characterize such effects once they are known to exist. Although each chapter differs in its approach and content, there are several themes that continually reemerge. These include the following:

1. *Assessment of neurodevelopment in humans depends on the age and developmental stage of the subjects.* Development of the CNS in humans is a process that begins in embryogenesis and continues without interruption through fetal life, birth, and for many years thereafter. Evaluation of developmental neurotoxic effects must take into account the ongoing nature of this process. Early "outcomes" are not outcomes at all, but only markers on a continuing pathway.

2. *Neurodevelopmental evaluation in humans requires multiple and repeated assessments.* Although this is also true in studies of other mammals and especially in studies of other primates, the greater complexity of human cognitive and behavioral endpoints makes multiple and repeated measurement essential. Human infants simply do not have the developmental maturity to express some effects, so assessment of these functions is not possible

until later in life. Higher language and cognitive functions are two familiar examples.

3. *Neurodevelopmental toxins usually produce specific rather than global effects on cognitive and behavioral functions.* Comparison with experimental studies in laboratory animals may be especially useful in defining the particular effects that are most important in studies of human infants and children.

4. *Studies of the effects of developmental neurotoxins in humans are confounded by factors other than the exposure.* It is not possible to eliminate all potentially confounding factors, even in well-controlled studies. Moreover, confounding factors generally exhibit a greater effect as subjects exposed to developmental neurotoxins grow older. Effects seen in infants are often not, therefore, predictive of the findings in the same children at a later age.

5. *Studies that employ combinations of appropriate techniques are most informative.* This research is by its very nature interdisciplinary. Effective human studies require knowledge, sophistication, and scientific rigor in epidemiologic design and analysis, as well as in the choice, use, and interpretation of tests of neurodevelopmental function.

This section and the sections that follow demonstrate the practical importance of our rapidly increasing knowledge of CNS development, function, and malfunction. Many of the methods that are now being used in humans are based on information learned through studies in experimental models. This knowledge is being used to improve our understanding of the nature and range of developmental neurotoxicity in humans.

J. M. Friedman, MD, PhD

CHAPTER 24

Evaluation of the Human Newborn Infant

ANNELOES VAN BAAR
Academic Medical Center
Department of Neonatology
Amsterdam, The Netherlands
and
Saint Joseph Hospital
Department of Psychosocial Care
Veldhoven, The Netherlands

* Abbreviations: ABR, auditory brainstem response; APIB, Assessment for Premature Infant Behavior; CBF, cerebral blood flow; CD, cerebral Doppler; CFM, cerebral function monitoring; CHT, congenital hypothyroidism; CNS, central nervous system; CRIB, Clinical Risk Index for Babies; CT, computed tomography; EEG, electroencephalography; ELBW, extremely low birthweight; FAE, fetal alcohol effects; FAS, fetal alcohol syndrome; GM, general movements; IQ, intelligence quotient; LBW, low birthweight; MND, minor neurologic dysfunctions; MRI, magnetic resonance imaging; NBAS, neonatal behavioral scale; NBRS, Neurobiologic Risk Score; NICU, neonatal intensive care unit; NIRS, near infrared spectroscopy; NNNS, NICU Network Neurobehavioral Scale; OAE, otoacoustic emission; PDA, patent ductus arteriosus; PERI, Perinatal Risk Inventory; PET, positron emission tomography; PVH–IVH, periventricular–intraventricular hemorrhages; PVL, periventricular leucomalacia; sd, standard deviations; SEP, somatosensory evoked potentials; SGA, small for gestational age; SPECT, single photon emission computed tomography; VEP, visual evoked potentials; VLBW, very low birthweight; VSGA, very small for gestational age.

I. Introduction

A human newborn always evokes a feeling of amazement and excitement about the new life. The life ahead will be wonderful, but also demanding. The infant's own dispositions and capacities are important, as well as those of the parents and other caregivers, the circumstances the family lives in, and the personal history that a newborn already has and will experience further on.

Especially at birth, awareness of the time dimension as a condition in human development is apparent. Although birth marks the beginning of life independent of the mother, the newborn has already gone through quite a developmental process from conception onward, which created it's own personal history of physical growth and experiences and, sometimes already, developmental difficulties. An evaluation of the human newborn therefore reflects: 1) an outcome of fetal development affected by previous conditions; 2) current characteristics, functioning, and capacities, as well as 3) potential future development. The reasons why the newborn is evaluated may be guided specifically by one of these facets and thus influence the choice of measurements and the report of the results. The reasons why the instruments to assess the newborn's condition were developed also reflect differences in emphasis on one of these three facets. Some assessments mainly summarize previous ante-, peri-, and neonatal events, some focus on current functioning and some provide a diagnosis and potential concomitant sequelae for future development. Examiners must be aware of these three facets of evaluations of the human newborn and their own reasons for using such assessments.

Next to this time dimension, researchers or clinicians may also evaluate the outcome of an examination differently, depending on the importance for an individual child or for a certain group of newborns. All individuals are unique, because of their own mix of genes, the interaction with their caregivers, the family characteristics, and their own biologic as well as psychological history, which all evolve in time. This implies that studies on groups of children with some special characteristic in common can result in only general outlines concerning development. Results on groups of children are important and useful in guiding further theory and treatment, but allow only limited and imperfect individual prediction for each individual child.

The predictive value of an assessment also depends on the level of detail with which a process or characteristic can and needs to be studied. Does the outcome differentiate subtle individual differences or a global subdivision into normal or abnormal?

The purpose of this chapter is to present strengths and limitations of several frequently used or promising instruments for the evaluation of the human newborn infant in order to provide some background information when evaluating research results or guidelines when planning a new project. Next to assessments of neurobehavioral functioning, brain imaging techniques, neurophysiologic examinations, biochemical measures, neonatal risk scores and a number of important characteristics of newborns are discussed. These assessments were developed partly for neonates in critical conditions, notably preterm infants. As the evaluation of the newborn starts immediately after birth, this section begins with "a first impression."

II. A First Impression

The first evaluation of the human newborn is made right after birth, guided by expectations and circumstances evolving from the course of pregnancy and delivery. With a term pregnancy and a delivery without obstetric complications, a normal healthy newborn is expected. The first impressions reflect all three main facets of an evaluation: pregnancy outcome, current functioning, and a glimpse of the future.

A. Pregnancy Outcome

The main outcome measures of pregnancy are the newborn's gestational age and birth weight. A gestational age between 37 and 42 weeks is considered to be a complete term pregnancy. Prematurity indicates a gestational age of less than 37 weeks, and often further delineation is used, such as *extreme prematurity,* indicating a gestational age less than 32 weeks and *immaturity* less than 28 weeks of gestation. Preterm infants, especially the extreme premature and immature children, are at risk for developmental problems or delay in all domains: somatic (growth, health, and visual and auditory perception), neuromotor, cognitive and behavioral development (Escobar *et al.,* 1991; Hack *et al.,* 1993; Hille *et al.,* 1994; Kato Klebanov *et al.,* 1994; van Baar, 1996).

A birthweight higher than the tenth percentile on a standard population growth chart in relation to the infant's gestational age is generally considered appropriate. A birthweight below the tenth percentile is referred to as small for gestational age (SGA) and often further delineated, indicating a birthweight below the third percentile as very small for gestational age (VSGA).

Concerning birthweight without referring to appropriateness for gestational age, frequently used indications of subgroups are low birthweight (<2500 g = LBW), very low birth weight (<1500 g = VLBW), and extremely low birth weight (<1000 g = ELBW). A systematic trend in prevalence of cerebral palsy has been reported according to these subgroups of birthweights, with increasing numbers associated to lower birthweights (Pharoah *et al.,* 1996).

Head circumference at birth is another growth parameter that can be of prognostic value, especially in extreme cases. Dolk (1991) reported outcome with regard to intelligence quotient (IQ) at 7 years of age of 41 term infants with microcephalic headcircumference, that is, three standard deviations (sd) or more below the mean for age, sex, and race; 51% were mentally retarded (IQ < 71) and a further 17% had an IQ between 71–80. For 16 children, further pathology was evident, and all but one of these children were mentally retarded. In 39% of the children with head circumference 2 sd below the mean, delay in mental development (IQ < 80) was seen. Interestingly, socioeconomic status was not related to outcome in the microcephalic children, whereas it was associated in the normocephalic and in the group with head circumference 2 sd below the mean.

B. Apgar Score

With regard to current functioning of the newborn, Virginia Apgar (1953) summarized several important facets of immediate extrauterine adaptation. The healthy newborn responds to strong external stimuli, starts breathing, maintains an adequate circulation with a sufficient heart rate, and shows a normal color and good muscle tone. Apgar designed a scoring system to evaluate the newborn's condition systematically at 1, 5, and 10 min after birth. Ratings of five aspects of functioning (heart rate, respiratory effort, reflex irritability, muscle tone, and color) vary between 0, 1, and 2, with a total score of 10 indicating optimal performance. Usually the score, if not optimal at 1 min, increases over time, reflecting recovery of the newborn and improved adaptation to extrauterine life. Apgar scores are determined in virtually all term newborns in the United States and in many infants in other countries in the world (Swaiman, 1994). However precise quantitation of the Apgar score may vary considerably among observers (Volpe, 1995). The committee on fetus and newborn (1986) states that a low (0–3) Apgar score must not be equated solely with asphyxia. A low 1 min Apgar is a useful index that the child needs special attention, and the 5 min Apgar indicates effectiveness of resuscitation efforts. A 5 min Apgar score of 7–10 is considered normal. Correlation with mortality in first year and future adverse neurologic outcome such as cerebral palsy increases with low Apgar scores at 10 min or later (Volpe, 1995).

C. Major Physical Anomalies

The first visual observation of the newborn also includes a check for major physical anomalies; when found, further systematic evaluation and diagnosis is necessary.

III. Further Evaluations

After the first assessments, a further physical examination within several hours or days after birth is usually done, for example, as described by Aucott (1997), looking for major somatic disorders. The results of these assessments and knowledge or suspicions of specific risk conditions guide further examinations for diagnosis and care of the newborn. In case of, for instance, maternal alcohol abuse during pregnancy, signs of the fetal alcohol syndrome (FAS) or fetal alcohol effects (FAE) are looked for, like the characteristic craniofacial abnormalities like microcephaly, short palpebral fissures and poorly developed philtrum, thin upper lip, and flattening of the maxillary area (Spohr *et al.,* 1994). In case of prenatal drug exposure, occurrence of neonatal withdrawal symptoms, such as tremors, sleeping problems, high-pitched crying, hyperactivity, feeding and respiratory problems, yawning, sneezing and sweating, are checked and treated when necessary (Finnegan, 1980).

Usually a general neurologic or behavioral assessment is made in order to collect a systematic impression of the newborn's neurobehavioral functioning. Swaiman (1994) and Volpe (1995) describe a general neurologic examination consisting of elements assessing functioning of cranial nerves (such as olfaction, vision, functioning of the eyes, facial motility, tongue function, and taste), motor function, neonatal reflexes, and sensory response, next to cranial vault examination. Several standardized neonatal neurobehavioral assessment protocols have also been developed, with the more recently published relying to some extent on earlier schemes. Important assessments for full-term neonates are the neurologic examination of Prechtl (Prechtl and Beintema, 1964; Prechtl, 1977), and the Brazelton neonatal behavioral assessment scale (NBAS) (Brazelton, 1973, 1984; Brazelton and Nugent, 1995).

A. Neurologic Examination of the Full-Term Newborn Infant

In the second edition of this examination, Prechtl (1977) emphasizes that infants with suspected brain damage must receive a carefully standardized and detailed examination in order not to overlook those with impairment of the nervous system or worry parents of healthy infants unduly. Due to the complexity of the nervous system, Prechtl thinks that no summary of a few reflexes can assess the condition of the nervous system reliably. A screening test was, however, provided to identify those newborns in need of an extensive evaluation. Next to assessing state, the screening examination consists of six items, namely posture, eyes, spontaneous motor activity, resistance to passive movements, traction test, and Moro response, of which 26 characteristics are assessed. These were selected for discriminating from an obstetric optimal group from an obstetric at-risk group. The full neurologic examination consists of 42 items on which 145 characteristics have to be noted, and 14 summary and appraisal items with 32 characteristics.

A most important conceptual contribution has been that Prechtl and Beintema categorized different behavioral states of the newborn: (1) eyes closed, regular respiration; (2) eyes closed, irregular respiration; (3) eyes open, no gross movements; (4) eyes open, gross movements; (5) crying. These "states," or "levels of alertness," as Volpe (1995) says, constitute the basis for the potential neurologic responses to be observed. Another important concept has been the determination of optimal responses or conditions, which often are more easily agreed on, as opposed to normal and at-risk conditions and responses. The optimality score gives an impression of the functional integrity of the nervous system; it can be used in combination with a clinical diagnosis, if that can be made. A diagnosis of neurologically abnormal is based on overall syndromes such as hyperexcitability, apathetic reactions, hypotonia, hypertonia, or a hemisyndrome (asymmetry) (Prechtl, 1977).

The Prechtl examination was used in the Groningen Perinatal Project, which concerned a 3 year cohort of 3162 single-born infants (Touwen *et al.,* 1980). Touwen *et al.* (1980) used the optimality concept and concluded that it is a helpful complementary refinement for obstetric and neurologic data. The predictive value of the Prechtl examination could be assessed further as both abnormal and normal subgroups of these children have been followed up and compared at 1½, 4, 6, 9, 12, and 14 years of age (Bierman-van Eendeburg *et al.,* 1981; Hadders-Algra *et al.,* 1988a; Hadders-Algra and Touwen, 1989; Lunsing *et al.,* 1992; Soorani-Lunsing *et al.,* 1994). A diagnosis of neonatal neurologic abnormality was made in 160 (5%) neonates, and about 9% of them developed severe neurologic sequelae; no neurologic handicap was found in the follow-up of the normal neonates (Lunsing *et al.,* 1992). Neonatal neurologic findings were also found to be related to the occurrence of minor neurologic dysfunctions (MND) at 9 years of age (Hadders-Algra *et al.,* 1988b), and also to the increase in MND between 9 and 12 years of age (Lunsing *et al.,* 1992); with the onset of puberty at 14 years, MND was found to decrease (Soorani-Lunsing *et al.,* 1994). Note that the majority of the 160 neurologically abnormal neonates recovered; at 9 years, 55% were diagnosed as normal, 35% had MND, and around 10% were again diagnosed as abnormal (Hadders-Algra *et al.,* 1988b).

B. Neonatal Behavioral Assessment Scale

A very important difference between the NBAS and the neurologic examination of Prechtl is the view on the more active role that Brazelton (1995) attributes to the newborn in the examination. For example, Prechtl considers states to be involuntary modes of neural activity, whereas Brazelton appreciates the newborns' capacity to control levels of stimuli by the use of states of conciousness in adapting to the environment. The neurologic examination of Prechtl is designed to distinguish between normal and abnormal children, or between optimal findings and reduced optimality (Prechtl, 1982). The NBAS was developed to study individual differences and to yield a comprehensive profile of neonatal behavior, including competencies and strengths, as well as identifying areas of difficulty or deviation (Brazelton and Nugent, 1995). Brazelton added "drowsy" as the third separate state next to (1) "deep sleep," (2) "light sleep," (4) "quiet alert," (5) "active alert," and (6) "crying."

The NBAS consists of 28 behavioral items scored on a 9-point scale and 18 reflex items scored on a 4-point scale to assess neurologic status. In the second edition, seven items were supplemented to summarize the quality of the baby's responsiveness and the amount of input needed from the examiner to organize the newborn's reactions. The examination consists of different packages of items: habituation, motor–oral, truncal, vestibular, and social–interactive. Factor analysis of the NBAS items has yielded a social–interactive dimension, an irritable state dimension, and a motor dimension (Brazelton and Nugent, 1995). Most frequently, the scores are reduced into seven clusters: habituation, orientation, motor, range of state, regulation of state, autonomic stability, and reflexes (Lester *et al.,* 1982).

A key facet of the NBAS is that the examiner has to facilitate the performance and organizational skills of the baby, which increases the importance of the interaction during the examination. Therefore, training is needed (as for other examinations) to learn to perform the examination in a standardized manner, but also to learn to feel comfortable in handling the baby and elicit the infant's best performance. Sufficient interscorer reliability, with 1-point differences at the 9-point scales seen as permissable, should be accomplished in a training procedure (Brazelton and Nugent, 1995).

Test–retest reliability has been found for the NBAS with low to moderate correlations. More important is that this is considered a less appropriate concept for a detailed examination such as the NBAS in view of the rapid development and change that takes place and should take place in the neonatal period. Scoring patterns of change over the first month as emerging from the NBAS at days 3, 14, and 30 postpartum, so-called recovery curves, reflecting adjustment to the extrauterine environment after birth, may reveal in themselves an important characteristic of the infant's functioning and abilities (Lester, 1984). Brazelton and co-workers advocate repeated administration of the scale in order to capture the full range of the infant's behavior.

The same appreciation of rapid change in the newborn period, as well as the detailed observation of individual differences rather than a diagnosis such as "normal" or "abnormal," precludes high expectations concerning predictive validity. However, some relationships between NBAS clusters and later developmental tests, or measures of temperament and mother–infant interactive behavior, have been reported and are summarized by Brazelton and Nugent (1995).

Over the years it was realized in clinical practice that the NBAS provided a good opportunity for developing a relationship between parents and examiner in showing and interpreting the meaning of the baby's behavior and the implications for caregiving. When the NBAS is used in such clinical sessions, the parents are urged to share their feelings or concerns throughout the examination (Brazelton and Nugent, 1995). In this way, the NBAS examination becomes an intervention session. Several studies evaluating the effect of such intervention have been done with mixed results (see the review in Brazelton and Nugent, 1995). Again, the authors warn that hopes for effective intervention must not be too high for a single session with the NBAS. Wolke (1995) describes a systematic intervention approach, pointing out principles of caretaking, including administration of the NBAS. Another approach is described by Keefer (1995), who describes how items of the NBAS can be integrated into a routine physical examination. This allows the practitioner to enhance the psychologic and educational value of the examination by narrating observations to the parents and rate the infant's behavior roughly as optimal, average, worrisome, or abnormal. In the last two cases, respectively, repeated screening and further evaluations are indicated.

The NBAS has been used extensively for research purposes. Comparisons of the NBAS performance between groups of children at risk for developmental problems and reference groups have been made, see also the review of Brazelton and Nugent (1995). In our own study in Amsterdam, a group of infants exposed prenatally to heroin, methadone, and cocaine were examined twice with the NBAS after treatment for withdrawal symptoms and compared to a drug-free reference group. No group differences were found with the NBAS at 40 weeks gestational age, but at 44 weeks the full-term drug-exposed infants showed less ability in orientation, especially in the visual items (van Baar *et al.,* 1989).

Several studies using the NBAS report on dose–response relationships in infants exposed prenatally to cocaine reflecting higher neurobehavioral risk in infants with heavier exposure that show poorer motor capacities and state regulation, lesser automatic stability and greater excitability (Delaney-Black *et al.,* 1996; Martin *et al.,* 1996; Tronick *et al.,* 1996).

C. Infants with Perinatal Risk Factors

In view of he critical condition at birth and during their first weeks of life, it is not surprising that specific neonatal assessments have been developed for infants with perinatal risk factors, notably preterm infants. Next to assessments focussing at neurobehavioral functioning, methods for postnatal estimation of gestational age have been developed.

1. Gestational Age Estimation

When uncertainty exists concerning the newborn's gestational age, several postnatal assessments can be used to make an estimation. Most of these assessments are based on observation of physical maturity criteria and a neurologic examination (Aucott, 1997). Dubowitz *et al.* (1970) designed a gestational age chart for estimation ±2 weeks based on 11 external physical criteria (such as ear firmness and form and nipple formation) and 10 neurologic criteria (such as arm recoil and posture). For estimation of gestational age in black infants whose gestational age was less than 33 weeks, the assessment was found to be inaccurate with a tendency to overestimation (Spinnato *et al.,* 1984).

Ballard *et al.* (1979) made an abbreviation of the gestational age chart by Dubowitz *et al.* (1970), which they later (1991) adapted for use with extremely premature infants. Ballard *et al.* based their chart on seven physical criteria and six criteria for neuromuscular maturity. The chart can be used for infants between 20 and 44 weeks gestational age; an overestimation from 2 to 4 days was found for gestational ages younger than 37 weeks.

Accurate determination of gestational age is especially important for later follow-up studies of preterm infants. They are frequently examined at ages corrected for prematurity, that is, taking into account the number of weeks or days of birth before 40 weeks of gestation, and are then compared to peers of the same gestational (maturational) age, despite their longer postnatal experience. DiPietro and Allen (1991) point out the inconsistency between correcting for every week for gestational ages of less than 37 weeks, whereas no correction is applied for term infants with a gestational age varying between 37 and 42 weeks.

2. Neurobehavioral Examinations of the Preterm Infant

Dubowitz and Dubowitz (1981) elaborated their neurologic criteria used for the assessment of gestational age, into a more extensive neurologic examination (designing a clear and easy to use form). They based their examination on work of Saint-Anne Dargassies (1977), Prechtl (1977), and Brazelton (1973) and tried to make a practical tool for day-to-day work in a neonatal unit. The examination consists of 2 habituation items, 16 items to judge movement and tone, 6 reflex items, and 9 neurobehavioral items that are assessed on a 5-point scale. For every item, one of the six states of the infant as described by Brazelton (1973) is also recorded. The result of the examination is a qualitative description of the infant's reactions, which can be followed when the examination is repeated. The difficulty in using the scale is that there are no clear-cut criteria for normal or abnormal responses outlined in the manual. Use of the examination thus depends heavily on the examiner's experience and knowledge of neurologic development. Attempts to quantify the results have been reported (e.g., van Baar *et al.,* 1989). Dubowitz *et al.* (1984) reported on a total of 129 infants born at a gestational age of <35 weeks and found a good correlation between outcome at 1 year and the neurologic examination at 40 postmenstrual weeks, which was divided into normal, borderline (two abnormal signs or one abnormal sign and suboptimal head control), or abnormal (marked trunk hypotonia plus head lag and three abnormal signs). The presence of periventricular hemorrhage in 29% (37) of the children was not as good a predictor of developmental outcome at 1 year as the neonatal neurologic assessment.

Allen and Capute (1989) designed a neurologic examination for preterm infants at term age. This consists of assessment of posture, extremity, and axial tone, deep-tendon reflexes, pathologic and primitive reflexes, symmetry, oromotor function, cranial nerve function, auditory and visual responses, and behavior scored as normal or minor or major abnormality.

Of the 85 preterm infants scoring abnormal, 38% (32) developed cerebral palsy and 27% (23) developed minor neuromotor dysfunction at ages between 1 and 5 years. Of the 125 infants classified as normal, 81% (101) developed no abnormalities, 13% (16) had minor neuromotor dysfunction, and 6% (8) showed cerebral palsy.

3. General Movements

Another neurologic observation has been developed by Hadders-Algra and Prechtl (1992) that emphasizes spontaneous general movements (GM). In newborn infants, GM have a "writhing" quality, that is, a tight appearance, relatively slow speed, and limited amplitude. These writhing GMs are gradually broken down until 8–12 weeks into a "fidgety" quality, which is seen as a normal motor phenomenon of an ongoing flow of small and minute movements. A follow-up study was done on 16 infants (10 preterms and 6 with hypoxic–ischemic encephalopathy) with an abnormal quality of GM. Fourteen of these infants had some kind of neurologic dysfunction at 18 months: 7 had cerebral palsy, 6 showed minor neurologic dysfunctions, and 1 had severe mental retardation.

Quality of GM was found to be strongly related to neurodevelopmental outcome at 1½ years of age. Absence of the age-specific fidgety GM might be a special predictor of disability (Hadders-Algra *et al.,* 1997). Prechtl *et al.* (1997) report that in a sample of 130 children with 74% born preterm, only 3 children out of 70 with normal fidgety movements had an abnormal neurodevelopmental outcome, and only 3 of 60 with abnormal or no fidgety movements had normal outcome at age 2 years. All children with abnormal fidgety movements also had abnormal writhing movements in the neonatal period, but these were found to have lower specificity and thus less predictive value.

4. Assessment for Premature Infant Behavior (APIB)

Als *et al.* (1982) designed the Assessment for Premature Infant Behavior (APIB). It is based on the NBAS for full-term infants and consists of assessments of five behavioral systems: physiologic (e.g., respiratory patterns, skin color), motor (e.g., tonus, posture, movements), state (level of consciousness, with states also labeled as diffuse or well defined), attentional–interactive (alertness and ability to attend to social stimuli and inanimate objects), and regulatory (regulatory activity of the infant to maintain itself in a balanced, well-modulated state). In addition, ratings of the amount of facilitation by the examiner needed to bring out the infant's best performance is documented. The APIB is performed in six packages of maneuvers, reflecting the increasingly demanding input: I–sleep/distal; II–uncover–supine; III–low tactile; IV–high tactile–vestibular; VI–attention–interaction. Part of the exam consists of a number of reflexes, referred to here as *systematically elicited movements.* Detailed scoring at 9-point scales for all behaviors to be noted are provided. The APIB takes at least 30 min to be performed and is undoubtedly taxing to learn for potential examiners, who need training but also experience in handling preterm infants.

Als (1997) states that predictive and construct validity of the APIB has been established in a sample of 160 term and preterm children studied at 9 months of age. In addition, the APIB showed effect of an intervention that improved individualized care in the neonatal intensive care unit (NICU) (Als, 1997).

5. NICU Network Neurobehavioral Scale (NNNS)

The NICU Network Neurobehavioral Scale (NNNS) has been developed by Lester and Tronick (1994) as comprehensive assessment of both neurologic integrity and behavioral functioning, including withdrawal and stress symptoms. The NNNS is applicable for infants around 34–46 weeks gestational age who are medically stable and cared for in an open crib or isolette. It consists of (1) classic neurologic items that assess active and passive tone, primitive reflexes, and items reflecting CNS integrity; (2) behavioral items including state, sensory, and interactive responses; and (3) stress–abstinence items, divided into seven categories; physiologic, autonomic, CNS, skeletal, visual, gastrointestinal, and state. Basically, the NNNS is an elaboration of the NBAS, with more reflexes to be assessed and stress symptoms to be noted. Somewhat stricter guidelines are given for the NNNS in the performance of the exam, and for many items several categories have been added in order to allow documentation of all potential observations. Even head supports had monitors for heart rate, respiratory rate, and oxygen saturation are required next to the flashlight, rattle, bell, ball, and foot probe known from the NBAS.

Napiorkowsky *et al.* (1996) described several differences on the NNNS specifically related to muscle tone, motor performance, following during orientation, and signs of stress, between 20 cocaine-exposed infants in the first 2 days of life when compared with 20 drug-free infants and 17 infants whose mothers had taken alcohol regularly during pregnancy.

6. Combination of Examinations

Contrary to some of the extensive examinations described previously, Amiel-Tison (1996) argues that a short neurologic evaluation at term age allows sufficient subdivision between infants who need follow-up and infants who do not. The latter group is comprised of infants with normal ultrasound examinations (see later this chapter) and a normal clinical outcome according to evaluation (with the exclusion, however, of very preterm [<29 weeks] or ELBW infants and in the case of seizures observed in the NICU). The neurologic evaluation described can be done in several minutes, although it may take repeated tries and thus more time to get the infant to participate. The examination consists of measuring head circumference, examining the cranial sutures, visual pursuit of a black and white bull's eye, social interaction (which should be eager), sucking reflex, raise to sit, passive axial tone, movements of fingers and thumb, and autonomic control during examination.

IV. Imaging of the Brain

Next to the neurobehavioral assessments, other techniques have been developed to evaluate the human newborn. Imaging techniques are used to acquire information concerning the structure of the brain. These techniques, such as ultrasound scanning and magnetic resonance imaging (MRI), are used to study brain development and to look for potential brain injuries.

Hypoxic–ischemic encephalopathy is one such an important injury in the perinatal period in both full-term and preterm infants. A deficit in oxygen supply to the brain may result from hypoxemia, a diminished amount of oxygen in the blood supply, and ischemia, a diminished amount of blood perfusing the brain. This often occurs as a result of asphyxia, and impairment in the exchange of respiratory gases, oxygen, and carbon dioxide. In premature infants, hypoxic–ischemic encephalopathy is often accompanied by intraventricular hemorrhages (Volpe, 1995). Pathology looked for with imaging techniques consists of ventricular dilatation (indicative of white matter atrophy), periventricular–intraventricular hemorrhages (PVH–IVH) or periventricular leucomalacia (PVL), and the occurrence of echolucencies or cysts (Volpe, 1995).

A. Ultrasound Examination

Cranial ultrasonography is a technique that visualizes structures of the brain based on echoes of ultrasonic waves through the anterior fontanel (Fawer and Calame, 1991). The gestational age of the infants, the timing of the scan concerning postnatal age, and the pathology looked for are all important factors for the ultrasound diagnosis (Fawer and Calame, 1991). In addition, de Vries (1996) warns that the quality of the equipment used can have great impact, as localized cystic lesions could not be seen with 5 MHz transducers but appeared with 7.5 MHz transducers, indicating that technical improvements may critically affect further diagnosis. A major advantage of this imaging technique is that it is noninvasive and can be used at bedside, even in a NICU. This allows repeated assessments and close following of structural brain development.

Several classifications have been formulated, taking into account the duration, severity, and the site of the lesions seen by using ultrasonography.

Germinal matrix–intraventricular hemorrhage (Volpe, 1995):

- Grade I — Germinal matrix–intraventricular hemorrhage with no or minimal intraventricular hemorrhage (<10% of ventricular area on parasagittal view)
- Grade II — Intraventricular hemorrhage (10–50% of ventricular area on parasagittal view)
- Grade III — Intraventricular hemorrhage (>50% of ventricular area on parasagittal view)
- Separate notation — Periventricular echodensity (location and extent)

Ventricular dilatation (Levene, 1981):

- Grade I — Dilatation >97th percentile, but <4 mm above
- Grade II — Hydrocephalus, dilatation >97th percentile, >4 mm above

Leucomalacia (de Vries *et al.,* 1992):

- Grade I — Transient echodensities persisting for more than 7 days
- Grade II — Transient echodensities evolving into small localized fronto–parieto cysts
- Grade III — Periventricular densities evolving into extensive periventricular cystic lesions
- Grade IV — Densities extending into the deep white matter evolving into extensive cystic lesions

A clear prognostic relationship to later developmental problems such as cerebral palsy has been found with the detection of cysts and ventricular dilatation in preterm infants, as approximately 60–90% of these infants have neurologic deficits at follow-up (Volpe, 1995).

De Vries (1996) reports that several studies showed that 75% of children with hydrocephalus following IVH had cerebral palsy with mental retardation present in around half the cases. In 10% of children with low-grade leucomalacia, spastic diplegia was seen, and of those with a more serious grade III, most were found to develop cerebral palsy. In infants with cystic PVL, visual impairments and cognitive problems are more common than among infants with PVH–IVH (see review in de Vries, 1996).

Jongmans *et al.* (1993) found that duration of 'flares,' that is, echodensities seen by ultrasonography in periventricular white matter, in a group of preterm (<34 weeks) children at 6 years of age was related to motor development.

Prechtl *et al.* (1997) found that their observations of abnormal GMs better predicted neurodevelopmental outcome until 2 years of age than a subdivision into low or high risk based on ultrasound examinations.

B. Magnetic Resonance Imaging (MRI)

Another technique to visualize the anatomy of the brain is MRI. It employs radiofrequency radiation in the presence of carefully controlled magnetic fields in order to produce images of the body in any plane. It usually displays the distribution of hydrogen nuclei and parameters relating to their motion in water and lipids (Bydder, 1991). Christmann and Haddad (1991) found that MRI provides more information on anatomy and cerebral maturation, neuropathologic features, pathophysiologic events of ischemia and cerebral leucomalacia than ultrasound examinations or computed tomography (CT) scans. They conclude that MRI is a useful technique, complementary to ultrasound examination in diagnosing cerebral injury. An advantage of MRI is that it explores all regions of the central nervous system (CNS) without difficulty with specific signals that may provide a better definition of cerebral disorders. Volpe (1995) states that the possibility to visualize the myelination of brain development is probably its most prominent value. Major limitations are that MRI is not a bedside exam and children often need to be sedated, as perfect immobility is required (Christmann and Haddad, 1991). De Vries (1996) states that it is unclear as of yet if the more time consuming MRI will be of greater predictive value concerning neurodevelopmental outcome than repeated ultrasound examinations in preterm infants.

Rutherford *et al.* (1996) concluded that neonatal MRI of a group of 16 full-term infants with hypoxic–ischaemic encephalopathy within the first 4 weeks of birth, was related to neurologic optimality scores during follow-up and to later MRI, with at least one MRI made between 12 and 24 months of age. Six children who had normal neurologic and developmental outcome until this age had a neonatal MRI with patchy white matter abnormalities. Abnormal outcome was seen in 10 children who had a neonatal MRI with extensive white matter abnormalities, most severe in those with basal ganglia atrophy.

Stewart and Kirkbride (1996) found with 41 preterm (<33 weeks) children who had neonatal ultrasound examinations and an MRI at 14 years of age that the neonatal ultrasound examinations accurately identified serious permanent lesions and predicted disabilities at 1, 8, and 14 years of age. Ventricular dilatation in 56% (23) and atrophy of the corpus callosum in 41% (17)

were the commonest abnormalities noted in the MRI scans.

C. Other Imaging Techniques

Other imaging techniques being used with newborns are described here to provide some information on the most important characteristics and potential use.

1. Computed Tomography (CT)

Computed tomography (CT) is an imaging technique based on the use of ionizing radiation as in conventional X-ray (Volpe, 1995). CT scanning provides more definitive information than ultrasound scanning in the evaluation of most parenchymal disorders, hemorrhages, or other fluid collections in the subdural and subarachnoid spaces and most posterior fossa lesions (Volpe, 1995). The use of radiation and the requirement of the transport of the infant to the CT scanner are important disadvantages of this technique, and it is not used frequently in the preterm population (de Vries, 1996).

2. Cerebral Doppler (CD)

The cerebral Doppler effect consists of ultrasound waves emitted by a transducer that are reflected by moving red blood cells and their frequency is shifted proportional to the blood velocity (Messer, 1991). In neonates, the open anterior fontanelle is used as a window to study the intracranial vessels and, therefore, the cerebral circulation. Changes in cerebral blood flow (CBF) have been identified in several neonatal pathologic conditions (see Volpe, 1995), for instance, apnea, brain death, and seizures.

3. Near Infrared Spectroscopy (NIRS)

Near infrared spectroscopy (NIRS) is a technique based on two facts. First, light in the near-infrared zone can pass through the thin skin, bone, and other tissues of the infant. Second, the appropriate choice of near-infrared wave lengths allows interpretation of changes in light absorption that reflect oxygenation (Wyatt and Delpy, 1991). This method may provide crucial information on cerebral hemoglobin oxygen saturation, cerebral blood volume, CBF, cerebral oxygen delivery, cerebral venous oxygen saturation, and cerebral oxygen utilization (Volpe, 1995). No ionizing radiation is employed and the apparatus is portable and can be used in the NICU, which makes it a promising technique (Wyatt and Delpy, 1991).

4. Positron Emission Tomography (PET)

Positron emission tomography (PET) is based on the images resulting from the reactions within tissue to the emission of a positron from an isotope that comes from a nuclear bombardment. Physiologic and biochemical measurements can be made that are determinations of regional CBF, metabolism, and blood volume (Altman and Volpe, 1991).

5. Single Photon Emission Computed Tomography (SPECT)

Single photon emission computed tomography (SPECT) is a radionuclear technique of cerebral perfusion (Haddad *et al.,* 1991). SPECT has been used to define regional CBF, but needs further improvements for obtaining quantitative values (Volpe, 1995).

V. Neurophysiologic Assessments

Functioning of the newborn's brain can be evaluated by measuring electroencephalic activity. The polygraphy of these measurements must be assessed by skilled interpreters.

A. Electroencephalogram (EEG)

The electroencephalogram (EEG) reflects the spontaneous electroencephalic activity of the brain. It is measured by nine electrodes placed on the infant's scalp in a standard pattern (Watanabe, 1992). Whereas Volpe finds that active sleep, the predominant state of the newborn infant, is the best state to evaluate the EEG, Watanabe (1992) argues that recording in all states is essential for proper assessment of the neonatal EEG. Rapid evolutionary changes of EEG patterns reflect development of the brain in the 13 weeks prior to term age (see Volpe, 1995).

De Vries (1996) warns for disturbances when measuring in a NICU. Murdoch Eaton *et al.* (1992) describe several methods of continuous monitoring the EEG in an NICU. They conclude that it remains to be determined if the EEG's contribution to diagnosis and therapeutic monitoring actually leads to improvements in clinical management and outcome. Hellström *et al.* (1991) find cerebral function monitoring (CFM), which uses a filtered and selectively amplified EEG signal obtained from only one pair of biparietal electrodes, to be a useful technique in an NICU for detection of seizures that are distinguished by amplitude characteristics. As only one channel represents activity of the entire head and spikes and sharp waves cannot be detected, Murdoch Eaton *et al.* (1992) see the use of CFM in conjunction with other assessments and as indications when a repeat polygraphic EEG should be done.

Abnormal neonatal EEG patterns are usually described in terms of continuity of background or paroxysmal activity. Background activity represents the setting in which a normal or abnormal pattern appears and from which such pattern is distinguished. Paroxysmal activity denotes electrical phenomena with abrupt onset and termination and showing higher amplitude than the background. Abnormal background EEGs are described in terms of depression, hyperexcitability, continuous pattern without sleep cycles, topographic abnormality, transient patterns, dysmature pattern. Paroxysmal abnormalities are interictal (such as focal spikes or sharp waves), ictal (repeated spikes or sharp waves), or status (such as subclinical or clinical status epilepticus with various electrical discharges) (Watanabe, 1992).

Serial EEG recordings are considered a useful prognostic tool. Watanabe (1992) found that, for the preterm infant, abnormal EEG findings in the subacute and chronic stages were of more prognostic value than in the acute stage of brain insult. Neurologic outcome after markedly abnormal neonatal EEG was abnormal in 80%, mildly abnormal in 13%, and normal in 7%; after moderately abnormal neonatal EEG, 17% turned out definitively abnormal, 28% mildly abnormal, and 56% normal; and after normal neonatal EEG, 96% were found to show normal neurodevelopmental outcome and 4% were mildly abnormal.

B. Evoked Potentials

Evoked potentials are averaged electrical responses occurring in the EEG in response to an auditory, a visual, or a sensory stimulus (de Vries *et al.,* 1994). Evoked potentials provide information concerning the perceptive abilities measured and the general neurologic condition of the infant at the time of assessment.

1. Auditory Brainstem Responses (ABR)

Auditory brainstem responses (ABR) are measures of electrical events generated within the auditory brainstem pathway. The function of the middle ear and cochlea and the central auditory pathways through the brainstem are measured, allowing assessment of audiologic disorders (de Vries *et al.,* 1994). An ABR response is first recordable in most preterm infants by 30 weeks gestational age (Kennedy, 1992). As peripheral abnormalities are often transient in high-risk infants, it is advised to perform ABR around 40 weeks postmenstrual age. An automated auditory brainstem response method, the ALGO-1 Plus, has been developed for testing under NICU conditions (van Straaten *et al.,* 1996). Otoacoustic emission (OAE), energy produced by the cochlea as response to an auditory stimulus (Kennedy, 1992), may become more important than ABR as a general auditory screening (Watkin, 1996; de Vries, 1996).

Absent or poor ABRs are often associated with subsequent hearing loss, but follow-up testing and behavioral and audiologic examinations are indicated before diagnosis of hearing loss is made (de Vries *et al.,* 1994).

2. Visual Evoked Potentials (VEP)

Visual evoked potentials (VEPs) are electrical signals generated by the visual cortex in response to visual stimulation (de Vries *et al.,* 1994). VEPs are elicited by flashes or by pattern stimuli and recorded from occipital electrodes. Visual acuity can be estimated by patterned stimuli, but patterned VEPs are not always feasible in the assessment of neonates. In newborn infants with visual impairment, VEPs are unrecordable or only recordable to large patterns or flashes. Flash VEPs are used easily even in intensive care, and can be recorded in normal neonates born at 24 weeks gestational age and older (Taylor, 1992).

When measuring VEPs in preterm infants, both the gestational age of the infant at birth and its chronologic age since birth need to be taken into account, as maturation in the visual system is accelerated in normal preterm infants (Taylor, 1992). Eken *et al.* (1996) report that VEPs were severely abnormal but did not provide a more reliable diagnosis than fixation at acuity cards at term date and at 3 months of age in nine children of gestational ages of more than 35 weeks who developed cerebral visual impairment and were followed until 18 months of age. Extensive cystic leucomalacia (as seen with ultrasound examination) was found to be highly predictive of cerebral visual impairment, as well as severe mental and motor deficits. In the total cohort of 65 children in the study, Eken (1996) found a normal or only delayed flash VEP for six children who showed a low acuity at or below the tenth percentile or no response to the acuity cards at term date and 3 months, but no cerebral visual impairment at 18 months. The flash VEP thus accurately predicted that acuity was going to be within the normal range.

3. Somatosensory Evoked Potential (SEP)

The somatosensory evoked potential (SEP) is the sequence of voltage changes generated in the brain and the pathway from a peripheral nerve, following a transient stimulus, usually electrical, of a peripheral sensory nerve (Cooke, 1992). The median nerve as well as the posterior tibial nerve can be stimulated. Sensitive amplification and averaging techniques enable discrimination between the evoked response and other large and more random physiologic potentials with which the signal is mixed. The latency of the response is measured. SEPs are the most time-consuming evoked potentials to perform in a preterm infant (de Vries, 1996).

De Vries (1993) reports a high predictive value with regard to neurodevelopmental outcome of SEPs in full-term infants with postasphyxial encephalopathy. In an unselected group of infants of a postmenstrual age of less than 30 weeks, SEPs are often poorly recognizable, but further maturational processes, like myelinization and synaptogenesis, can be followed using SEPs (Smit *et al.,* 1997). As there is a close proximity to sensory and motor tracts, it was hoped that SEPs would predict motor problems. However, inconsistent results have been found (de Vries, 1996; Ekert *et al.,* 1997).

VI. Biochemical Measures

In several contexts, biochemical measures to evaluate potentially important facets of the newborn's condition are necessary.

A. Screening Programs

Neonatal screening programs are used in the Netherlands for instance, for early detection of congenital hypothyroidism (CHT), a condition that may seriously affect development if left untreated, or when treatment is started late (Kooistra *et al.,* 1994).

B. Specific Analyses

Examples of studies in which specific biochemical evaluations are needed are studies on thyroid functioning and further development. In preterm infants, low levels of thyroxine were found to be associated to developmental delay (den Ouden *et al.,* 1996). Thyroxine supplementation was found to prevent the period of hypothyroxinemia in extreme preterm infants, but it did not affect developmental (van Wassenaer *et al.,* 1997) or behavioral outcome (Briët *et al.,* 1997).

In the case of prenatal drug exposure, analyses of urine, meconium, or hair are done to assess some degree of actual exposure (Ostrea *et al.,* 1992; Sallee *et al.,* 1995).

C. Experimental Studies

An example of still experimental studies using biochemical measures is the work of Gunnar and coworkers (1995), who use cortisol levels as a measure of stress reactivity. They found that high newborn cortisol levels in response to pain stimuli, perhaps reflecting neurobehavioral organization, predicted higher positive affect and lower distress to limitations at 6 months of age.

VII. Perinatal Risk Scores

As perinatal complications are often not isolated phenomena but combinations of related problems, several scales were developed to summarize these and therefore improve prediction of developmental outcomes (Molfese and Thompson, 1985).

A. Neurobiologic Risk Score (NBRS)

In the design of the neurobiologic risk score (NBRS) (Brazy *et al.,* 1991) it was assumed that brain cell injury contributes to long-term developmental problems of preterm infants. The mechanisms that may result in brain cell injury were considered hypoxemia, insufficient blood flow, inadequate substrate for celmetabolism, and direct tissue damage. The items therefore included in the NBRS are: 1. blood pH; 2. ventilation; 3. infection; 4. convulsions; 5. hypoglycemia; 6. PVL; 7. IVH; 8. Apgar scores; 9. arterial oxygen pressure (PaO_2), 10. apnea with bradycardia, 11. hypotension, 12. patent ductus arteriosus (PDA), and 13. hyperbilirubinemia.

In a follow-up study of VLBW children good correlations were found with developmental tests. Prediction improved when only the first seven variables were used (Brazy *et al.,* 1991). A later report of this group (Goldstein *et al.,* 1995) further specified the type of acidosis and studied the interaction with hypoxemia and hypotension. They found that especially the metabolic component of acidosis was related to neurodevelopmental outcome at 2 years, as was hypotension independently.

B. Clinical Risk Index for Babies (CRIB)

The clinical risk for babies (CRIB) was developed to allow assessment of performance of different hospitals in neonatal intensive care (The International Neonatal Network, 1993). It consists of six items: birthweight, gestational age, congenital malformations, maximum base access in first 12 hr, and minimum as well as maximum appropriate fraction of inspired oxygen. It was found to be a robust index of initial neonatal risk, more accurate than birthweight alone. CRIB also had the same predictive value as the score for acute neonatal physiology (SNAP), which consists of 26 variables (Richardson *et al.,* 1993).

C. Perinatal Risk Inventory (PERI)

Scheiner and Sexton (1991) developed a perinatal risk inventory (PERI), which showed promise in identifying children with future developmental problems, in a sample of VLBW infants, extremely premature infants, or infants with neurologic signs and low Apgar scores. The PERI consists of 18 items with regard to neonatal illnesses and treatment needed, imaging results, EEG and gestational age, appropriateness of birthweight, Apgar scores, and head circumference measures. They found better prediction with the PERI than with an abbreviated neonatal neurobehavioral assessment scale and a family status index.

VIII. Discussion

The assessments for evaluation of the human newborn presented in this chapter concern different levels of information. The assessments reflect facets of outcome of previous development, current functioning, and future development, that is, risk for later developmental problems. The level of interest for a clinician or a researcher depends, of course, on the reasons why a newborn needs to be evaluated. A single assessment, however, will not provide a complete picture of the infant, and combinations of several assessments are needed to improve diagnosis and prediction. The use of a perinatal risk inventory summarizes the risk factors involved for an infant and allows comparison for infants treated in different hospitals.

Next to all neonatal evaluation assessments some identification of the family status is useful. Socioeconomic status has consistently been found to be related to developmental outcome, especially with regard to cognitive functioning, even when serious structural brain damage had been seen (e.g., Vohr *et al.,* 1992). Family status indices must be regarded as markers, because these reflect both ecologic environment and related risk or opportunities as well as genetic factors. Such markers must, of course, be seen in their own context. Parental unemployment, for instance, may have quite a different impact in a family in which the parents have no other major problems, thus allowing them to pay much attention to their children, than in a family in which the parents are addicted to drugs such as heroin and spend their time finding ways of obtaining that illegal substance.

For all neurobehavioral and neurophysiologic examinations of the newborn, it is important that the examinations are preferably done in a quiet, semidarkened, warm room, midway between feedings, in order to avoid having the infant mainly in either sleep or crying states (Brazelton and Nugent, 1995; Prechtl, 1977). The examiner should be aware of the fact that uncontrollable factors may affect the performance of the infant, despite adequate equipment, training, and testing. Much attention has been paid to interobserver agreement and training in neurobehavioral assessments. The same holds, however, for interpretation and measurement performance of all other assessments.

With regard to prediction, it is important to note that most newborns diagnosed as normal indeed turned out normal with later developmental assessments. Many newborns diagnosed as abnormal were also found to be normal with later assessments. The reports on GMs may show otherwise, but this new assessment needs further confirmation and study in different groups of infants than reported so far. Possibly the nervous system, which still needs to mature and develop extensively during the first 2 years, develops compensatory mechanisms or signs of dysfunction are not (yet) detectable at the time of follow-up (Prechtl, 1982). Furthermore, what kind of outcome (e.g., cerebral palsy, cognitive functioning, behavioral problems) is measured, and in what way (e.g., a continuous IQ score), or a qualitative judgment (in terms of normal or abnormal) is important too.

Although many important neurobehavioral, imaging, neurophysiologic, and perinatal risk assessments have been discussed in this chapter, this is certainly not a complete review. In addition to the standardized measures and characteristics as described in this chapter, experimental procedures or observations may be used to acquire detailed information, for example, on mother–infant interaction (van Baar, 1977) or on vagal tone as a reflection of the autonomic nervous system (Porges, 1992).

In general, a research project on human newborn infants will include information on gestational age, birthweight, appropriateness of birthweight for gestational age, Apgar scores, occurrence of physical anomalies, and an index of family status. Results of an assessment of neurobehavioral functioning will be needed as well. When brain injuries are suspected, some imaging examination and neurophysiologic functioning will be assessed. Sometimes specific biochemical measures of the newborn are necessary. Neonatal illness and invasive treatment needed may be summarized in a perinatal risk inventory in order to complete the evaluation of the human newborn infant.

Acknowledgment

Bert Smit, neonatologist, is gratefully acknowledged for his advice and comments on an earlier draft of this manuscript.

References

Allen, M. C. and Capute, A. J. (1989). Neonatal neurodevelopmental examination as a predictor of neuromotor outcome in premature infants. *Pediatrics,* **83,** 498–506.

Als, H. (1997). Neurobehavioral development of the preterm infant. In *Neonatal–Perinatal Medicine, Diseases of the Fetus and Infant,* 6th ed. Vol. I, A. A. Fanaroff and R. J. Martin, Eds. Mosby, St. Louis. pp. 964–989.

Als, H., Lester, B. M., Tronick, E. Z., and Brazelton, T. B. (1982). Toward a research instrument for the assessment of preterm infant behavior (APIB). *Theory and Research in Behavioral Pediatrics,* Vol. 1, H. E. Fitzgerald, B. M. Lester, and M. W. Yogman, Eds. New York, Plenum Press, pp. 85–132.

Altman, D. I., and Volpe, J. J. (1991). Positron emission tomography in the study of neonatal brain. In *Imaging Techniques of the CNS of the Neonates.* J. Haddad, D. Christmann, and J. Messer, Eds. Springer-Verlag, Berlin, pp. 171–183.

Amiel-Tison, C. (1996). Does neurological assessment still have a place in the NICU? *Acta Paediatrica Suppl.* **416,** 31–38.

Apgar, V. (1953). A proposal for a new method of evaluation of the newborn infant. *Curr. Res. Anesthes.* **32,** 260–267.

Aucott, S. W. (1997). Physical examination and care of the newborn. In *Neonatal–Perinatal Medicine, Diseases of the Fetus and Infant,* 6th ed. Vol. I, A. A. Fanaroff and R. J. Martin, Mosby, St. Louis. pp. 403–408.

Ballard, J. L., Khoury, J. C., Wedig, K., Wang, L., Eilers-Walsman, B. L., and Lipp, R. (1991). New Ballard score, expanded to include extremely premature infants. *J. Pediatr.* **119,** 417–423.

Ballard, J. L., Novak, K. K., and Driver, M. A. (1979). A simplified score for assessment of fetal maturation of newly born infants. *J. Pediatr.* **95,** 769–774.

Bierman-van Eendenburg, M. E. C., Jurgens-van der Zee, A. D., Olinga, A. A., Huisjes, H. H., and Touwen, B. C. L. (1981). Predictive value of Neonatal Neurological Examination: a follow-up study at 18 months. *Dev. Med. Child Neurol.* **23,** 296–305.

Brazelton, T. B. (1973). *Neonatal Behavioral Assessment Scale.* Clinics in Developmental Medicine No. 50. Philadelphia, Lippincott.

Brazelton, T. B. (1984). *Neonatal Behavioral Assessment Scale,* 2nd ed. Clinics in Developmental Medicine No. 88. Philadelphia, Lippincott.

Brazelton, T. B., and Nugent, J. K. (1995). *Neonatal Behavioral Assessment Scale,* 3rd ed. Clinics in Developmental Medicine No. 137. Mac Keith Press, London.

Brazy, J. E., Eckerman, C. O., Oehler, J. M., Goldstein, R. F., and O'Rand, A. M. (1991). *J. Pediatr.* **118,** 783–792.

Briët, J. M., van Wassenaer, A. G., van Baar, A. L., Dekker, F. W., and Kok, J. H. (1997). Evaluation of the effect of thyroxine supplementation in very preterm born infants on the behavioral outcome. Submitted.

Bydder, G. M. (1991). Principles of magnetic resonance imaging. In *Imaging Techniques of the CNS of the Neonates.* J. Haddad, D. Christmann, and J. Messer, Eds. Springer-Verlag, Berlin, pp. 1–17.

Christmann, D., and Haddad, J. (1991). Magnetic resonance imaging: application to the neonatal period. In *Imaging Techniques of the CNS of the Neonates.* J. Haddad, D. Christmann, and J. Messer, Eds. Springer-Verlag, Berlin, pp. 17–79.

Committee on fetus and newborn (1986). Use and abuse of the Apgar score. *Pediatrics,* **78,** 1148–1149.

Cooke, R. W. I. (1992). Somatosensory evoked potentials. In *The Neurophysiological Examination of the Newborn Infant.* J. A. Eyre, Ed. Mac Keith Press, London, 66–79.

de Vries, L. S. (1993). Somatosensory-evoked potentials in term neonates with postasphyxial encephalopathy. *Clin. Perinatol.* **20,** 463–482.

de Vries, L. S. (1996). Neurological assessment of the preterm infant. *Acta Paediatrica,* **85,** 765–771.

de Vries, L. S., Eken, P., and Dubowitz, L. M. S. (1992). The spectrum of leukomalacia using cranial ultrasound. *Behav. Brain Res.* **49,** 1–6.

de Vries, L. S., Pierrat, V., and Eken, P. (1994). The use of evoked potentials in neonatal intensive care unit. *J. Perinat. Med.* **22,** 547–555.

Delaney-Black, V., Covington, C., Ostrea, E., Romero, A., Baker, D., Tagle, M., Nordstrom-Klee, B., Silvestre, M. A., Angelilli, M. L., Hack, C., and Long, J. (1996). Prenatal cocaine and neonatal outcome: evaluation of dose–response relationship. *Pediatrics,* **98,** 735–740.

den Ouden, A. L., Kok, J. H., Verkerk, P. H., Brand, R., and Verloove-vanhorick, S. P. (1996). *Pediatr. Res.* **39,** 142–145.

DiPietro, J. A., and Allen, M. C. (1991). Estimation of gestational: implications for developmental research. *Child Dev.* **62,** 1184–1199.

Dolk, H. (1991). The predictive value of microcephaly during the first year of life for mental retardation at seven years. *Dev. Med. Child Neurol.* **33,** 974–963.

Dubowitz, L. M. S., Dubowitz, V., and Goldberg. (1970). Clinical assessment of gestational age. *J. Pediatr.* **77,** 1–10.

Dubowitz, L. M. S., and Dubowitz, V. (1981). *Neurological Assessment of the Preterm and Full Term Born Infant.* Clinics in Developmental Medicine No 79. London, Heinemann.

Dubowitz, L. M. S., Dubowitz, V., Palmer, P. G., Miller, G., Fawer, C.-L., and Levene, M. I. (1984). Correlation of neurologic assessment in the preterm newborn infant. *J. Pediatr.* **105,** 452–456.

Eken, P. (1996). Cerebral visual impairment in infants with haemorrhage–ischaemic lesions of the neonatal brain. Thesis, University of Utrecht.

Eken, P., de Vries, L. S., van Nieuwenhuizen, O., Schalij-Delfos, N. E., Reits, D., and Spekreijse, H. (1996). Early predictors of cerebral visual impairment in infants with cystic leukomalacia. *Neuropediatrics,* **27,** 16–25.

Ekert, P. G., Taylor, M. J., Keenan, N. K., Boulon, J. E., and Whyte, H. E. (1997). Early somatosensory evoked potentials in preterm infants: their prognostic utility. *Biol. Neonate,* **71,** 83–91.

Escobar, G. J., Littenberg, B., and Petitti, D. B. (1991). Outcome among surviving very low birthweight infants: a meta-analysis. *Arch. Dis. Child.* **66,** 204–211.

Fawer, C. L., and Calame, A. (1991). Ultrasound. In: *Imaging Techniques of the CNS of the Neonates.,* J. Haddad, D. Christmann, and J. Messer, Eds. Springer-Verlag, Berlin, pp. 79–107.

Finnegan, L. P., Ed. (1980). *Drug Dependence in Pregnancy: Clinical Management of Mother and Child.* Tunbridge Wells, Castle House Publications Ltd.

Goldstein, R. F., Thompson, R. J., Oehler, J. M., and Brazy, J. E. (1995). Influence of acidosis, hypoxemia and hypotension on neurodevelopmental outcome in very low birthweight infants. *Pediatrics,* **95,** 238–243.

Gunnar, M. R., Porter, F. L., Wolf, C. M., Rigusato, J., and Larson, M. C. (1995). Neonatal stress reactivity: predictions to later emotional temperament. *Child Dev.* **63,** 290–303.

Hack, M., Weissman, B., Breslau, N., Klein, N., Borawski-Clark, E., and Fanaroff, A. A. Health of very low birthweight children during their first eight years. (1993). *J. Pediatr.* **122,** 887–892.

Haddad, J., Constantinesco, A., and Brunot, B. (1991). Single photon emission tomography of the brain perfusion in neonates. In: *Imaging Techniques of the CNS of the Neonates.* J. Haddad, D. Christmann, and J. Messer, Eds. Springer-Verlag, Berlin, pp. 161–171.

Hadders-Algra, M., Huisjes, H. J., and Touwen, B. C. L. (1988a). Preterm or small-for-gestational-age infants. Neurological and behavioural development at the age of 6 years. *Europ. J. Pediatr.* **147,** 460–467.

Hadders-Algra, M., Huisjes, H. J., and Touwen, B. C. L. (1988b). Perinatal risk factors and minor neurological dysfunction; significance for behaviour and school achievement at nine years. *Dev. Med. Child Neurol.* **30,** 482–491.

Hadders-Algra, M., Klip-van den Nieuwendijk, A. W. J., Martijn, A., and van Eykern, L. A. (1997). Assessment of general movements:

toward a better understanding of a sensitive method to evaluate brain function in young infants. *Dev. Med. Child Neurol.* **39,** 89–99.

Hadders-Algra, M., and Prechtl, H. F. R. (1992). Developmental course of general movements in early infancy. I. Descriptive analysis of change in form. *Early Hum. Dev.* **28,** 201–213.

Hadders-Algra, M., and Touwen, B. C. L. (1989). The long term significance of neurological findings at toddler's age. *Pädiatrische Grensgebiete,* **2,** 93–99.

Hellström-Westas, L., Rosen, I., and Svenningsen, N. W. (1991). Cerebral function monitoring during the first week of life in extremely small low birthweight (ESLBW) infants. *Neuropediatrics,* **22,** 27–32.

Hille, E. T., Den Ouden, A. L., Bauer, L., van den Oudenrijn, C., Brand, R., and Verloove-Vanhorick, S. P. (1994). School performance at nine years of age in very premature and very low birthweight infants: perinatal risk factors and predictors at five years of age. *J. Pediatr.* **126,** 426–434.

Jongmans, M., Henderson, S., de Vries, L. S., and Dubowitz, L. M. S. (1993). Duration of periventricular densities in preterm infants and neurological outcome at 6 years of age. *Arch. Dis. Child.* **69,** 9–13.

Kato Klebanov, P., Brooks-Gunn, J., and McCormick, M. (1994). Classroom behavior of very low birthweight elementary school children. *Pediatrics,* **94,** 700–708.

Keefer, C. H. (1995). The combined physical and behavioral neonatal examination: a parent centered approach to pediatric care. In: T. B. Brazelton and J. K. Nugent, Eds. *Neonatal Behavioral Assessment Scale,* 3rd ed. Clinics in Developmental Medicine No. 137. Mac Keith Press, London.

Kennedy, C. R. (1992). The assessment of hearing and brainstem function. In *The Neurophysiological Examination of the Newborn Infant.* J. A. Eyre, Ed. Mac Keith Press, London, pp. 79–93.

Kooistra, L., Laane, C., Vulsma, T., Schellekens, J. M. H., van der Meere, J. J., and Kalverboer, A. F. (1994). Motor and cognitive development in children with congenital hypothyroidism: a long-term evaluation of the effects of neonatal treatment. *J. Pediatr.* **124,** 903–909.

Lester, B. M. (1984). Data analysis and prediction. In T. B. Brazelton, Ed. *Neonatal Behavioral Assessment Scale,* 2nd ed. Clinics in Developmental Medicine No. 88. Philadelphia, Lippincott.

Lester, B. M., Als, H., and Brazelton, T. B. (1982). Regional obstetrical anaesthesia and newborn behavior: a reanalysis toward synergestic effects. *Child Dev.* **53,** 687–692.

Lester, B. M., and Tronick, E. Z. (1994). NICU Network Neurobehavioral Scale (NNNS). Unpublished manuscript.

Levene, M. I. (1981). Measurement of the growth of the lateral ventricles in preterm infants with real-time ultrasound. *Arch. Dis. Child.* **56,** 900–904.

Lunsing, R. J., Hadders-Algra, M., Huisjes, H. J., and Touwen, B. C. L. (1992). Minor neurological dysfunction from birth to 12 years, I: increase during late school-age. *Dev. Med. Child Neurol.* **34,** 399–403.

Martin, J. C., Barr, H., Martin, D. C., and Streissguth, A. (1996). Neonatal neurobehavioral outcome following prenatal exposure to cocaine. *Neurotoxicol. Teratol.* **18,** 617–625.

Messer, J. (1991). Cerebral doppler in the neonate. In *Imaging Techniques of the CNS of the Neonates.* J. Haddad, D. Christmann, and J. Messer, Eds. Springer-Verlag, Berlin, pp. 107–117.

Molfese, V. J., and Thompson, B. (1985). Optimality versus complications: assessing predictive values of perinatal scales. *Child Dev.* **56,** 810–823.

Murdoch Eaton, D., Connell, J., Dubowitz, V., and Dubowitz, L. (1992). Monitoring of the electroencephalogram during intensive care. In *The Neurophysiological Examination of the Newborn Infant.* J. A. Eyre, Ed. Mac Keith Press, London, pp. 48–66.

Napiorkowski, B., Lester, B. M., Freier, C., Brunner, S., Dietz, L., Nadra, A., and Oh, W. (1996). Effects of *in utero* substance exposure on infant neurobehavior. *Pediatrics* **98,** 71–75.

Ostrea, E., Brady, M., Gause, S., Raymundo, A. L., and Stevens, M. (1992). Drug screening of newborns by meconium analysis: a large scale, prospective, epidemiologic study. *Pediatrics* **89,** 107–113.

Pharoah, P. O. D., Platt, M. J., and Cooke, T. (1996). The changing epidemiology of cerebral palsy. *Arch. Dis. Child.* **75,** F169–F173.

Porges, S. W. (1992). Vagal tone: a marker of stress vulnerability. *Pediatrics,* **90,** 498–504.

Prechtl, H. F. R., and Beintema, D. (1964). *The Neurological Examination of the Full-Term Newborn Infant.* 2nd ed. Clinics in Developmental Medicine No. 12. Philadelphia, Lippincott.

Prechtl, H. F. R. (1977). *The Neurological Examination of the Full-Term Newborn Infant.* 2nd ed. Clinics in Developmental Medicine No. 63. Philadelphia, Lippincott.

Prechtl, H. F. R. (1982). Assessment methods for the newborn infant, a critical evaluation. In *Psychobiology of the Human Newborn.* P. Stratton, Ed. Wiley & Sons, New York, pp. 21–53.

Prechtl, H. F. R., Einspieler, C., Cioni, G., Bos, A. F., Ferrari, F., and Sontheimer, D. (1997). An early marker for neurological deficits after perinatal brain lesions. *Lancet* **349,** 1361–1363.

Richardson, D. K., Gray, J. E., McCormick, M. C., Workman, K., and Goldman, D. A. (1993). Score for neonatal acute physiology: a psysiologic severity index for neonatal intensive care. *Pediatrics,* **91,** 617–623.

Rutherford, M., Pennock, J., Schwieso, J., Cowan, F., and Dubowitz, L. M. S. (1996). Hypoxic–ischaemic encephalopathy: early and late magnetic resonance imaging findings in relation to outcome. *Arch. Dis. Child.* **75,** F145–F151.

Saint-Anne Dargassies, S. (1977). *Neurological Development in Full Term and Premature Neonate.* Elsevier, Amsterdam.

Sallee, F. R., Katikaneni, L. P., McArthur, P. D., Ibrahim, H. M., Nesbitt, L., and Sethutaman, G. (1995). Head growth in cocaine exposed infants: relationship to neonate hair level. *Dev. Behav. Pediatr.* **16,** 77–81.

Scheiner, A. P., and Sexton, M. E. (1991). Prediction of developmental outcome using a perinatal risk inventory. *Pediatrics,* **88,** 1135–1143.

Smit, B. J., Kok, J. H., de Vries, L. S., Wassenaer, A. G., van Dekker, F. W., and Ongerboer de Visser, B. W. (1998). Somatosensory evoked potentials in very preterm infants in relation to L-thyroxine supplementation. *Pediatrics* (in press).

Soorani-Lunsing, R. J., Hadders-Algra, M., Huisjes, H. J., and Touwen, B. C. L. (1994). Neurobehavioral relationships after the onset of puberty. *Dev. Med. Child Neurol.* **36,** 334–343.

Spinnato, J. A., Sibai, M., Shaver, D. C., and Anderson, G. D. (1984). Inaccuracy of Dubowitz gestational age in low birthweight infants. *Obstet. Gyn.* **63,** 491–495.

Spohr, H. L., Willms, J., and Steinhausen, H. C. (1994). The fetal alcohol syndrome in adolescence. *Acta Paediatrica Suppl.* **404,** 27–31.

Stewart, A., and Kirkbride, V. (1996). Very preterm infants at fourteen years: relationship with neonatal ultrasound brain scans and neurodevelopmental status at one year. *Acta Paediatrica Suppl.* **416,** 44–47.

Swaiman, K. F. (1994). *Pediatric Neurology, Principles and Practice.* 2nd ed. Mosby, St. Louis.

Taylor, M. J. (1992). Visual evoked potentials. In *The Neurophysiological Examination of the Newborn Infant.* J. A. Eyre, Ed. Mac Keith Press, London, pp. 11–48.

The International Neonatal Network. (1993). The CRIB (clinical risk index for babies) score: a tool for assessing initial neonatal risk and comparing performance of neonatal intensive care units. *Lancet,* **342,** 193–198.

Touwen, B. C. L., Huisjes, H. H., Jurgens-van der Zee, A. D., Bierman-van Eendenburg, M. E. C., Smrkovsky, M., and Olinga, A. A. (1980). Obstetrical condition and neonatal neurological morbidity. An analysis with the help of the optimality concept. *Early Hum. Dev.* **4/3,** 207–228.

Tronick, E. Z., Frank, D. A., Cabral, H., Mirochnik, M., and Zuckerman, B. (1996). Late dose–response effects of prenatal cocaine exposure on newborn neurobehavioral performance. *Pediatrics,* **98,** 76–83.

van Baar, A. L. (1996). Cognitive development of extreme preterm infants and stimulation at home. *Infant Behav. Dev.* **19,** 792.

van Baar, A. L. (1997). Socio-emotional development of children of drug dependent mothers. I. Infancy and toddler's age. Submitted.

van Baar, A. L., Fleury, P., Soepatmi, S., Ultee, C. A., and Wesselman, P. J. M. (1989). Neonatal behaviour after drug dependent pregnancy. *Arch. Dis. Child.* **64,** 235–240.

van Straaten, H. L. M., Groote, M. E., and Oudesluys-Murphy, A. M. (1996). Evaluation of automated auditory brainstem response infant hearing screening method in at risk neonates. *Europ. J. Pediatr.* **155,** 702–705.

van Wassenaer, A. G., Kok, J. H., de Vijlder, J. J. M., Briët, J. M., Smit, B. J., Tamminga, P., van Baar, A. L., Dekker, F. W., Endert, E., and Vulsma, T. (1997). Thyroxine supplementation in infants of less than 30 weeks gestational age; a placebo controlled, randomized, double-blind trial. *New Engl. J. Med.* **336,** 21–26.

Vohr, B., Garcia-Coll, C., Flanagan, P., and Oh, W. (1992). Effects of intraventricular hemorrhage and socioeconomic status on perceptual, cognitive and neurologic status of low birth weight infants at 5 years of age. *J. Pediatr.* **121,** 280–285.

Volpe, J. J. (1995). *Neurology of the Newborn.* 3rd ed. W. B. Saunders Co., Philadelphia.

Watanabe, K. (1992). The neonatal electroencephalogram and sleep-cycle patterns. In *The Neurophysiological Examination of the Newborn Infant.* J. A. Eyre, Eds. Mac Keith Press, London, 11–48.

Watkin, P. M. (1996). Outcomes of neonatal screening for hearing loss by otoacoustic emission. *Arch. Dis. Child.* **75,** F158–168.

Wolke, D. (1995). Parents' perceptions as guides for conducting NBAS clinical sessions. In T. B. Brazelton and J. K. Nugent, Eds. *Neonatal Behavioral Assessment Scale,* 3rd ed. Clinics in Developmental Medicine No. 137. Mac Keith Press, London.

Wyatt, J. S., and Delpy, D. T. (1991). Near infrared spectroscopy. In *Imaging Techniques of the CNS of the Neonates.* J. Haddad, D. Christmann, and J. Messer, Eds. Springer-Verlag, Berlin, pp. 147–161.

CHAPTER

25

Developmental Evaluation of the Older Infant and Child

CLAIRE D. COLES
Department of Psychiatry and Behavioral Sciences
Marcus Institute at Emory University
Department of Pediatrics
Emory University School of Medicine
Atlanta, Georgia 30306

JULIE A. KABLE
Marcus Institute at Emory University
Department of Pediatrics
Emory University School of Medicine
Atlanta, Georgia 30306

We are not objects in space but processes in time.
(KELLY, 1955)

Development is a process beginning at conception, unfolding throughout the life span, determined not only by the organism's genetic potential and by all the many environmental events encountered, but also by the interactions of these events. The goal of neurodevelopmental evaluation is to focus not only on the behavior or physical status of the individual, which is a static event, but on development, which is a process and can only be measured dynamically. Because it is almost impossible to measure anything continuously, one challenge that emerges from this effort is to decide on the frequency of measurement. In practice, the "evaluation of development" is often a single event or series of events sampled from this process and generalized to

*Abbreviations: ADHD, attention deficit hyperactivity disorder; BSID, Bayley Scales of Infant Development; CNS, central nervous system; FAS, fetal alcohol syndrome; PCB, polychlorinated biphenyls; RR, relative risk; SES, socioeconomic status.

represent the whole. Often, a second challenge is to define the particular aspect(s) of development that are of interest. At worst, the results of the evaluation may be inadequate or unpredictive. At best, conducting an effective and meaningful evaluation can provide information that is not available by other means.

The particular purpose of this chapter is to discuss the measurement of the effects of teratogens, or potential teratogens, on the development of the infant and child. Teratogens are environmental agents, albeit encountered very early, and are among many other environmental factors that have an impact on the developing organism. There are special characteristics of these agents. Because the exposure occurs during gestation, when the process of development is particularly rapid, teratogens can have profound effects on outcome, particularly when exposure is during critical periods (Coles, 1994). However, unlike many other significant environmental and genetic factors, the period of exposure is necessarily limited and confined to the earliest stages so that there is no continuing exposure. As a result, to understand the impact of the teratogen it becomes important to distinguish between prenatal and postnatal factors and how these unfold over the life span.

I. Measurement of Development and Measurement of the Teratogenic Effects

In an ideal world, measuring these outcomes would be a simple task. The exposed offspring would be evaluated as soon as the exposure ceased (usually at birth, but, perhaps, earlier during gestation) in order to obtain an immediate understanding of the impact on the aspects of development that are of interest (e.g., physical features, growth, neurodevelopment, behavior). In theory, immediately after the exposure ends should be the ideal time for measurement for two reasons. First, because effects should be maximized immediately after exposure and before compensating development can occur, and second, because there has been no time for other moderating environmental factors to have had effects that may obscure evidence of the teratogenic impact.

In fact, measurement is frequently done in the neonatal period and is relatively meaningful in identifying dysmorphias and growth retardation [i.e., decrements in birthweight, head circumference, and length, as well as differences in ponderal index (Miller and Hassanein, 1971)]. However, measurement during this period is of limited usefulness in identifying enduring neurodevelopmental effects, which are often the outcomes of interest. This limitation results from a number of factors. First, there may be acute reaction of the exposure that affects behavior temporarily (see later this chapter), but that does not have long-term impact. Second, the behaviors of interest may not yet be present in the immature organism. Finally, development results from a combination of factors interacting over time. Only when the organism is very seriously damaged can the majority of developmental outcomes be predicted from the neonatal period (Bornstein and Sigman, 1986).

Those who study development in childhood have been very concerned with this lack of continuity between early and later development, and this stability problem has been a persistent issue in measurement (Bornstein and Krasnegor, 1989). Indeed, whereas continuities in various skills (e.g., motor skills) are usually found, measurement at any point during the life span may or may not be related to measurement at different points in time. In some sense, any measured outcome is valid for that point in time. Whether it is meaningful as a predictor or whether it represents the whole span of that individual's life are different questions. Because of these discontinuities, the problem of when and what to measure to understand development meaningfully is a difficult one that is answered differently in different circumstances.

For instance, it would be highly desirable to be able to assess infants, early in life, and predict those who will need therapies, special education, and other interventions. Unfortunately, it is a frequent complaint that infant tests, such as the widely used and well-standardized Bayley Scales of Infant Development (BSID) (Bayley, 1969, 1993), do not predict standard scores on tests of intelligence or cognitive development (Bornstein and Sigman, 1986). Although this phenomenon is frustrating and requires that children be followed closely and reassessed repeatedly to collect the desired information, an understanding of the nature of development makes it obvious that this situation is inevitable. In fact, it would be unreasonable to expect a test that measures the prelinguistic development and motor skills that can be observed in the first year of life to predict outcomes that rely on cognitive processes that do not develop until years later. These later skills (e.g., language, academic skills) develop in response not only to innate abilities (not yet mature during infancy), but also as a result of environmental inputs. Because of such typical patterns, effective measurement depends on a clear understanding of the process of development, the timing of the development of various skills, and the relationships between early events, environmental factors, and later developmental processes.

A second problem is that development is extraordinarily complex and that, as the individual grows older, more and more processes and behaviors are available for assessment. Deciding what and how to measure becomes increasingly difficult, particularly if resources are limited. For this reason, the type of evaluation (measurement) chosen must depend on the questions to be answered as well as the process (or outcome) to be measured. Such considerations also determine the appropriate research design. Cross-sectional assessment of affected individuals at a particular point in time may be useful in the examination of specific subsystems or processes. For instance, if the hypothesis is that prenatal nicotine exposure is associated with dyslexia, assessment of a cohort of school-age children using a battery of reading tests would be an effective strategy. In contrast, understanding the overall pattern of development or the incidence of adverse outcomes due to a particular exposure (e.g., lead, cocaine) may require longitudinal follow-up of a cohort identified in infancy as well as assessment across many aspects of development.

Finally, because the developmental process, the interaction of the biologic and the environmental, continues throughout life, the study of this process must take into account the vast number of potentially "confounding" or moderating factors that may have an impact on the effects being examined. For instance, the elements that make up socioeconomic status (SES) have a profound effect on cognitive and language development in infancy and early childhood (Sameroff *et al.,* 1987) and academic functioning during the school years (Brooks-Gunn and Furstenberg, 1986). These effects further reduce the occupational and economic opportunities available during adulthood. Assessment of a cohort of individuals without taking such a factor into account, in some manner, may result in inaccurate conclusions.

II. Assessment of the Developmental Effects of Teratogens

In planning the evaluation of a developmental effect, there are many considerations to be taken into account. Several are discussed here, including:

1. *Patterns of effect.* Classification of some anticipated impacts of the teratogen on the developmental process.
2. *Postnatal environmental factors.* A description of the role of potentially mediating or confounding factors that are either universal to the developmental process or specific to the particular exposure to be studied.
3. *Developmental outcome variables.* Guidelines for selection of the process(es) to be studied.

III. Some Possible Patterns of Effect

Some teratogenic exposures have devastating effects on the organism, causing fetal wastage, mortality, and significant physical malformations (Scialli, 1992). Others are more subtle and difficult to ascertain. In addition, some exposures affect one system, some another. The result is that there are potentially many different patterns of developmental response, requiring different approaches to measurement.

A. Transient or Acute Effects

Transient or acute effects are temporary and limited to a specific time period during which these effects on the individual can be assessed. Neonatal abstinence syndrome in human infants whose mothers were maintained on methadone (Hutchings and Fifer, 1986) as well as other central nervous system (CNS) depressants is an example of such an effect. The consequences of such passive addiction can profoundly influence the organism's behavior during a limited period of time but do not have clear and direct long-term adverse effects on developmental processes. Such effects, however, are important to document as they may have medical significance and can be costly to treat during their expression phase. In addition, the behavioral responses associated with withdrawal from the drug may produce secondary effects as a result of the adverse impact on the organism's biopsychosocial environment.

Another type of transitory effect may be the growth retardation shown in some offspring of women who abuse substances in pregnancy. Although these effects may be persistent under certain conditions (e.g., fetal alcohol syndrome), in others, children generally reach the average range for weight and height (Zuckerman and Bresnahan, 1991; Tennes *et. al.,* 1985; Lifschitz and Wilson, 1991).

B. Continuous Effects

Some effects are clearly continuous throughout development, suggesting continuity in the impact of the developmental processes throughout the life span. Physical and sensory impairments associated with exposure to teratogens are the best example of this category of outcomes. Effects of this type are relatively easy to observe at any point in time to obtain an estimate of the effect. Even these continuous effects, however, may differentially affect the child's development depending on the supports and stressors present in the environment. The deformity of a limb associated with prenatal exposure to thalidomide can be used as an example of an effect of this nature (Tuchman-Duplessis, 1975). The presence or absence of the limb deformity is easily documented at any point during the course of development, but the implications of the physical malformation may be variable. During infancy, the malformation may result in functional loss of the limb, and this loss may lead to delays or abnormalities in fine and gross motor development. Early indices of cognitive status (i.e., test scores) also may be affected adversely if the required behavioral response depends on motor functioning. Finally, a reduced ability to interact with the environment in certain ways may actually limit cognitive development (Robinson and Fieber, 1988). By school age, disabilities occasioned by the limb deformity may be compensated for by the use of other appendages and/or adaptive technologies so that the limitations of one component of the structural motor system do not adversely effect other areas of motor functioning or the assessment of developmental status and/or cognitive functioning. However, the child's social and emotional status may be aversely affected if there are negative reactions from peers, resulting in social isolation. Consequently, the impact of the continuous effect can vary as a result of different environmental conditions and adaptive demands. Documenting this continuous effect may require assessments across the life span to address the variable nature of the secondary effects of the primary teratogenic effect.

C. Exacerbation of Effects

Some outcomes seem to worsen over time as they adversely affect the individual's ability to interact with the environment, resulting in continued exacerbation of the problem. These snow-ball effects require continued assessments to ascertain the rate and trajectory of the negative impact on developmental processes. Neurologically based learning problems associated with prenatal exposure to teratogens fall into this category as the exposure influences the rate of acquisition of material from the environment. Because the organism's rate of acquisition of new information suffers a negative impact, the developmental "lag" in relation to the nonexposed cohort widens with time. During the first decade of life, children with fetal alcohol syndrome (FAS) show this pattern in their cognitive and adapative development (Stratton *et al.,* 1996). During the first year of life, standard scores on developmental tests are often in the average to low average range even among children who show the classic facial features, but as they get older and there are increasing demands from their environment, they begin to fall further and further behind age mates (Coles *et al.,* 1991). By adolescence, they have difficulty with the cognitive demands of both school and workplace and with adaptive function and social skills (Stressiguth *et al.,* 1996). Children exposed prenatally to polychlorinated biphenyls (PCBs) have also shown a similar trajectory (Jacobson and Jacobson, 1990). Animal models have also shown a similar pattern of development, suggesting that this pattern is not the result of environmental factors (Mactutus and Fechter, 1985).

D. Fading Effects

Fading effects appear to diminish over time. The decrease in effect size over time appears to be a function of the multifactorial relationship between the impact of the prenatal environment, the postnatal environment, and the genetic material of the organism. Neurobehavioral compensation or brain plasticity may contribute to a reduction in effect size when an intact area of the organism's nervous system takes over responsibility for the function of the damaged area. This effect is commonly referred to as a *recovery of function response* (Kolb and Wishaw, 1985).

The diminution of a teratogenic effect may also result from the impact of environmental influences on the developing nervous system. Riley, Hannigan, and Balaz-Hannigan (1985) advocated assessing CNS integrity early in development to identify potential teratogenic effects before these complicating compensations occur. This approach is useful but, because the behavioral repertoire of the immature nervous system is limited, there are equal limitations on the questions that researchers can ask of their subjects during this time. Nevertheless, documentation of these effects is important for the same reason that documenting transient or acute effects is important, as there may be costs associated with their presence. In addition, aberrant behaviors may influence others in the child's psychosocial environment, resulting in secondary behavioral dysfunctions.

Although documenting the presence of such effect(s) is valuable in itself, an even more productive approach is to adopt a developmental perspective toward the investigation of this type of phenomenon. This approach would involve both assessment of effects that fade over time as well as the mediating factors that control this diminution. Such information could suggest methods of intervention to facilitate the recovery of function in affected individuals.

E. Masked (or Sleeper) Effects

Masked (or sleeper) effects are another type of potential teratogenic effect on developmental processes. This type of effect is not displayed until a sufficient level of development is obtained to necessitate accessing the damaged areas of the neurobehavioral system. The assumption made is that the area in the CNS that supports the behavior being observed has been affected by the prenatal exposure and that, when the process of development finally requires its use, the limitations become evident. Because sufficient demands were not made on the system previously, it could not be observed until that time. However, it is also possible that these problems arise due to an interaction between a system that had previously been noted to be impaired and a particular set of environmental factors.

The most commonly identified effects of this nature involve the assessment of academic skills or learning disabilities. Children who were exposed prenatally to high levels of alcohol have typically demonstrated difficulties with graphomotor skills and mathematic problem-solving ability (Streissguth *et al.,* 1993). These difficulties cannot be documented until at least the preschool years and may be more accurately measured in school-age children. Specific effects on executive functioning are also more readily identified in older children (Denckla, 1996). Such effects are hard to anticipate and therefore measure, unless there is preliminary evidence for their presence or a comprehensive theory regarding the teratogen's impact on the CNS. In fact, most information that exists of this nature is the result of investigators adopting a test battery approach in a longitudinal design. For instance, investigators have followed cohorts of drug-exposed children over time and at each time point developed a comprehensive battery to assess the child's development across a variety of domains deemed important for successful functioning at that stage of life. Streissguth *et al.* (1993), pioneers in the investigation of prenatal alcohol effects, describe their approach to this problem comprehensively, including their rationale for development of a battery of tests designed to identify teratogenic effects at different age levels.

F. Variable Effects

Damage to a given neurologic or behavioral system may also result in effects that have a different behavioral manifestation during different periods of development as a result of changing roles within the neurologic system. These variable effects will only show continuity if the investigator has an adequate understanding of the neural basis of the behavioral systems that are being sampled at different points in development. An example will make this point more clear. Teratogenic effects associated with maternal smoking during pregnancy are suggestive of a variable expression of a continuous effect. During the course of early development, infants of smokers have demonstrated significant deficits in the response to nonverbal (Saxton, 1978; Jacobson *et al.,* 1984; Picone *et al.,* 1982; Fried and Makin, 1987; Fried and Watkinson, 1988; Kristjansson *et al.,* 1989; Kable, 1997) and verbal (Makin *et al.,* 1991; Fried *et al.,* 1992; McCartney *et al.,* 1994; Kable, 1997) auditory material. Older children of smokers have also demonstrated significantly lower scores on reading achievement tests (Hardy and Mellits, 1972; Davie *et al.,* 1972; Butler and Goldstein, 1973; Fergusson and Lloyd, 1991). Although these systems on the surface appear different, auditory processing deficits have been linked to reading difficulties as a result of disrupting the process by which phonologic processing at the level of spoken language influences the development of the phonologic coding of the written language (Pennington, 1991). Consequently, an underlying mechanism of a central auditory processing deficit may be contributing to expression of both of these findings.

G. Secondary Effects

Secondary effects are the result of an interaction between a primary effect and the environment. This interaction results in the expression of another effect, which may have a more adverse impact on development than the primary effect. A child who has asthma as a result of prenatal and postnatal exposure to tobacco smoke may also develop behavioral problems as a result of the impact asthma has on the child's life. The behavioral problems become secondary effects associated with the tobacco smoke. The most well-documented secondary effects are those associated with FAS. Children with FAS are at high risk for school disruption, incarceration, suicide, substance abuse, and early pregnancy (Streissguth *et al.,* 1996). These effects result not from the primary impact of the alcohol exposure, but are the secondary effects of the alcohol-damaged individual interacting with the demands of the environment. Presumably, if environmental changes could be implemented, then these effects could be alleviated.

IV. Postnatal Environmental Factors

Obviously, prenatal exposure is not the only factor that impacts the development of the organism (at least in surviving organisms), although it may be that, if the effect has been particularly devastating, the teratogenic exposure may be a major determinant of subsequent development. In many cases, however, it is a relatively minor influence (see Fried, Chapter 32). In fact, in many studies of prenatal drug and alcohol exposure, the amount of variance in development accounted for by the drug is small relative to other contributing factors (Florey, 1988).

In addition to the organism's genetic make-up (which is usually inferred), postnatal environment is the major influence on subsequent development and must therefore interact with any effects of the teratogen. In theory, the environment may influence the expression of teratogenic effect in a variety of ways. For instance, if the environment is positive and supportive, it may be a protective factor, minimizing the potential effects or postpone the expression of cognitive effects until later in childhood, when higher levels of cognitive functioning are required. Streissguth *et al.* (1993) suggest such a pattern in their analyses of the first 7 years of their data on moderate prenatal alcohol exposure. In contrast, in a negative or nonsupportive environment, effects may be exacerbated or potentiated. Wilson (1989) and Wilson and colleagues research (1979; 1981) on the children of heroin users suggest such effects. There are also many other possible relationships that can be hypothesized.

How environmental factors are treated in studies of development depends on the focus of the research, the availability of appropriate samples for study, the methods to be used, and a variety of other experimental considerations. Appropriate methodology depends on an understanding of the effects of the variables to be evaluated in the research. For instance, in designing the investigation, environmental factors can be classified as either distal or proximal, as a way of explaining their impact on the outcomes of interest. In addition, their relationship with the experimental variable can be either as a confounder or as an effect modifier.

A. *Proximal and Distal Variables*

Exploring the relationship between a teratogen and the postnatal environment meaningfully requires that aspects of the environment be defined adequately. The impact of the environment on the developmental processes that have been affected by teratogens were alluded to previously without environment being clearly defined. The environment makes reference to the context, both socially and physically, in which the organism interacts and evolves (Wachs and Gruen, 1982). The assessment of the environment may be done through both proximal and distal variables.

Examination of proximal variables involves sampling the actual characteristics of the organism's environment through observation or report. The features of the environment that are believed to sample the behaviors of interest directly are examined. Although there are an infinite number of possible proximal variables, some that have been examined as potentially affecting child outcomes are: caregiver–child interaction (variously defined), nutrition, quality of physical environment, medication, educational experience, and perinatal factors. Examination of each of these factors requires the measurement of the developing child's specific and direct experience. Proximal variables have the advantage of providing the experimenter with very detailed information regarding potential confounds for understanding individual variability in teratogenic response. Their disadvantage is that these measurements are often time-consuming or difficult to implement in the research protocol. One of the most commonly used indices of social and physical environment in teratology research is the Caldwell HOME Scale (Caldwell and Bradley, 1979; Elardo and Bradley, 1981). This instrument requires a visit to the family's home and an observation of the parent and child over an extended period of time.

Measurement of distal variables involves categorizations of environmental characteristics. Such commonly used variables as social class (SES), parental education, and ethnic group are distal variables. These features do not necessarily result directly in an effect on a developmental process, but serve to identify a pattern of behaviors and contingencies that may have an impact on development. These variables are typically easy to sample, but the role that these indices of environment play in influencing the impact of teratogenic agents may be diverse and difficult to interpret. The complex relationship between SES and FAS is a good example and is discussed later in this chapter to illustrate the difficulty in interpreting the effects of a distal variable.

B. *Confounding and Effect Modifying Variables*

Until recently, in most studies of the behavioral effects of teratogens, environmental factors have been treated as confounders that could bias or obscure the investigation into the variable of interest. As a result, attempts are made to control for these factors. For instance, gender, race, social class (SES), and similar demographic factors, when they are found to be correlated with outcomes, are often included as covariates in regression procedures. In other studies, attempts are made to control for these factors by means of the experimental design. This can be done in several ways. The potentially confounding factor can be used as a grouping variable. If there are concerns about potential sex differences in the effects of the exposure, gender can be included as an independent variable and males and females considered separately. Because there are often differences in response to a teratogen in males and females, this is frequently done in animal studies (Weinberg, 1992). Another method used is to include control or contrast groups who have similar values on the variable of concern. For example, if cognitive status is believed to be affected by the exposure but is not the focus of the investigation, it is necessary to include a nonexposed group with similar scores on measures of cognition to serve as a control for bias associated with ability level. If this is not done, outcomes may be attributed inappropriately. Unfortunately, such factors are often ignored, which results in difficulty in interpreting the meaning of outcomes and makes comparison across studies difficult.

Some factors may be not confounders but, rather, "effect modifiers" (Bellinger, 1997). Effect modifiers are factors that, at different levels, differentially moderate the outcomes of interest. In such cases, the response is not homogeneous across all "conditions." In addition to confusing the results of the investigation when not accounted for, this difference in response is itself of significance. In such situations, it is not appropriate to "control" for such an effect because to do so would be to lose valuable information or to misinterpret the results. The presence of an effect modifier may affect the choice of statistical procedure to be used because, in these situations, linear models, such as those implied by regression procedures, are not accurate reflections of the real world.

There are several classes of variables that are often effect modifiers, and this issue should be considered when evaluating developmental effects of gender, ethnic status, SES, parental education, and similar factors. In addition, there are variables specific to particular situations. This possibility should be evaluated, particularly when the data obtained are not consistent with theory or logic. In such situations, the possibility of effect modifiers should be kept in mind in interpreting data. To illustrate these problems, the relationship between the reported incidence of FAS and SES are used as an example of the problems in interpretation of the effects of environment.

V. Fetal Alcohol Syndrome (FAS) and Socioeconomic Status (SES)

The role that the distal variable of social class plays in the expression of a teratogenic effect is often poorly understood. This is the situation prevailing in attempts to understanding the reported incidence of FAS. Although FAS was described by Lemoine and colleagues in 1968, it was first brought to the attention of the English speaking world by a series of publications from Seattle (e.g., Jones *et al.,* 1973) that described a pattern of birth defects that included facial dysmorphia, growth retardation, and neurocognitive deficits. In the next two decades, FAS was reported in many different countries and ethnic groups. This disorder has been identified in Germany (Spohr and Steinhauser, 1984), Scotland (O'Beattie *et al.,* 1983), Scandinavia (Olegard *et al.,* 1979), Japan (Tanaka *et al.,* 1981), and Canada (Bray and Anderson, 1989). In the United States, FAS has been reported among Native Americans (May *et al.,* 1983), African-Americans (Sokol *et al.,* 1986), and Caucasians (Streissguth *et al.,* 1978). The widespread recognition of this disorder supports the idea that alcohol-related teratogenesis can occur in any individual who is exposed to sufficient alcohol prenatally. However, examination of these data also suggests that FAS is identified more frequently among individuals in lower economic groups than among the more economically advantaged (Abel, 1995). For instance, in the southeastern state of Georgia, the Centers for Disease Control and Prevention's (CDC) Atlanta Metropolitan Birth Defects Survey found that FAS was identified in 0.3 per 1000 births in an inner city hospital that served a predominantly African-American, low-income population, whereas a few miles away, in a suburban hospital, serving a predominantly white, middle class population, the rate was 0.003 per 1000 (CDC, *MMWR,* 1993).

Similar patterns have been reported elsewhere. Dehaene *et al.* (1981) found that women who gave birth to children with FAS had a consistent pattern of characteristics relative to the general population of French women. In addition to drinking alcohol heavily, smoking cigarettes, and drinking coffee, these women were older (>30 years), had an average of more than 5 children, were unmarried, and of lower social class. In a later study, these authors noted that FAS was found in 1 of 800 births in the general population, but in 1 among 90 births in a low SES sample (Crepin *et al.,* 1989). In a Scottish sample, O'Beattie *et al.* (1983) found that the distribution of children with FAS was skewed toward the lower SES classes, with 60% of their identified cases of FAS falling in the lowest of the five class designations, versus 7.7% in the general populations being so classified.

Other studies have been done among disadvantaged minority groups in the United States, identifying high rates of FAS among African-Americans relative to Caucasians (odds ratio: 7.6 [Sokol *et al.,* 1986]) as well as among Native Americans (May *et al.,* 1983). Bray and Anderson (1989) reviewed studies done among the native people in Canada and found a similar pattern. There was a much higher incidence of cases in these populations than in any other group.

In the only study that has looked specifically at the effect of socioeconomic factors on the occurrence of FAS, Bingol *et al.* (1987) compared the offspring of upper middle class white women who were chronic alcoholics with those of lower class black and Hispanic women who also had this diagnosis. Only 4.5% of the children of the upper class white women received the diagnosis of FAS, whereas 70.9% of those of the lower class minority women received that same diagnosis (a problem noted in this study is that ethnic group and SES are, truly, confounded.)

For the investigator interested in the effects of prenatal alcohol exposure, this skewing of the incidence of FAS into economically disadvantaged groups raises a number of questions about the nature and etiology of this disorder as well as about the criteria for diagnosis. Theoretically, the distal variable, social class, should not be an etiologic factor itself, but rather a marker for other factors or more proximal variables that may be associated with the incidence of FAS. Whereas there are many factors that may be related to the diagnosis of FAS, some that have been considered include the possibility that patterns of alcohol use differ as a function of SES, differences in genetic susceptibility to alcohol effects among disadvantaged ethnic groups, and lifestyle factors associated with poverty.

A second possibility is that there are some biases inherent in the process of identification of those with FAS, which may include: 1) bias in ascertainment, such that there is a higher probability that low SES individuals will be identified with this disorder; or 2) bias inherent in the criteria of FAS, such that low SES individuals will be affected differentially in terms of the criteria used for diagnosis. For example, examination of the relationships between standard scores on cognitive assessments (i.e., IQ) suggests that the neurobehavioral criteria used for diagnosis are significantly influenced by SES with the result that the probability of being identified as cognitively impaired is significantly greater in low SES groups. The relative risk (RR) of having an IQ in the range classified as mentally retarded (IQ <70) is seven times higher for individuals in the lower class versus those drawn from the general population (Coles, 1994). Because evidence of cognitive impairment is one of the three criteria for FAS, low SES individuals have a higher probability of being so identified. This difference in probability significantly affects the likelihood of being identified as meeting the criteria for FAS. In fact, when Bingol and colleagues' (1987) data from different ethnic and social class groups of alcoholic women is reanalyzed, controlling for the effects of SES on IQ, those in the higher SES group were now more likely to be identified as alcohol affected (Coles, 1994).

Finally, there may be multiple unmeasured factors associated with minority status or poverty that potentiate the teratogenic effects of prenatal exposure. Thus, in a given situation, it may be difficult to be sure whether SES is a confounding factor or an effect modifier in relation to a teratogen until extensive research has been done.

VI. Developmental Outcomes Evaluated

There are consistent patterns in the research on the effects of prenatal exposure. After the identification of a potential teratogen through case reports, case studies, and retrospective cohort studies, broad scale epidemiologic studies are undertaken to identify prevalence and animal models are used to specify elements of exposure (Scialli, 1992). Developmental studies take much longer to organize and carry out. They are often longitudinal in nature and expensive to conduct, and during the initial period of research, have a global approach to the assessment of outcomes.

The particular outcomes that are selected for evaluation depend on several elements. These include the observations made during the initial identification of the teratogen, as well as standard principles of teratology, which suggest that physiology, growth, and neurobehavior are proper areas for assessment of effects (Vorhees, 1986). However, the selection of specific outcomes and research designs to identify the impact of a teratogen on development has to be made within the context of the information available on the specific substance of interest, which, in turn, depends on the extensiveness of previous research in the area. Historically, general ability models of cognitive functioning have been used and are beneficial for identifying large and dramatic teratogenic effects such as FAS but have not proved as useful for understanding the subtle neurologic deficits associated with many common teratogens. However, when investigation is in an early stage and little information exists, the most appropriate approach may be to develop a global test battery to assess whether the substance has any identifiable impact on a number of developmental processes. Necessarily, such a battery is general in nature, sampling from a number of potentially significant domains. It usually includes measures of cognitive and motor development in infancy, and intellectual and academic functioning in childhood, as these are the most broad scale measures of competence available. Language, visual–motor skills, and adaptive capacity are very often measured as well. Because attention is often identified as deficit in individuals born to substance abusers and in those with neurologic damage, measurement of this construct is often attempted as well.

As more research is done and the information on a specific exposure expands, it becomes possible to refine the focus of research and explore more specific hypotheses regarding domains of development. Examination of the literature on several drugs of abuse (e.g., alcohol, nicotine) that have been studied extensively reveals these patterns in the direction of research over time.

At present, because of the accumulation of knowledge on alcohol and nicotine, general global research is not considered very productive by most scientists in these areas and interest has focused on the examination of more specific domains. There are many advantages in more focused research, including greater efficiency in data collection, lower costs, and the potential for practical application of the results. To illustrate the developmental measurement of specific processes, we discuss a specific area that has been the focus of interest—attention.

A. *Attention*

Probably because attention deficit hyperactivity disorder (ADHD) is so commonly diagnosed (Barkley, 1990), there is considerable interest in the possibility that prenatal exposure to a variety of toxins may produce a neurologic deficit that underlies this disorder. As a result, there has been considerable energy directed at measurement of this construct and at identification of behaviors in infancy that may be associated with later attention. However, even when this construct is the focus of research interest, selecting outcome variables is not a simple task. In fact, attention has proved to be one of the most multiply defined constructs from the field of psychology.

1. Defining Attention

Before attention can be used as an outcome measure in the study of the effects of teratogens and toxins on development, it is necessary to define this construct. Cognitive theorists have used the term in mathematical equations that predict the rate of learning. In such research, attention to the stimuli is important as a prerequisite to learning about the relationships between the stimuli and various responses in which the organism might engage (Sutherland and Mackintosh, 1971). Clinical psychologists and psychiatrists have measured children's attention skills through parent and teacher ratings of the children's ability to stay on task and remain appropriately seated in the classroom (Barkley, 1990). Developmentalists study attention by looking at children's behavioral and physiologic responses to the onset, offset, and alteration of stimulus presentations (Fagan, 1970; O'Connor, 1980; Rose and Wallace, 1985). Each discipline is involved in examining some aspect of attention. However, to gain an appropriate understanding of attentional mechanisms, one must accept that attention is a system with many components that influence the encoding of information from the environment by the organism. There are several models of attention that delineate subcomponents of attention. Ruff and Rothbart (1996) have identified a three component model, which included the following subcomponents: selectivity, state of engagement, and higher level control. Mirsky (1989) has proposed a four factor model, which includes the ability to focus, sustain, encode, and shift attention with each system supported by different brain regions.

2. Methods of Assessing Attention

Even if a model of attention is selected, it is still necessary to select a method for assessment of the varying elements of the model. Attention has been assessed through a variety of different behavioral systems often determined by the response repertoire available to the organism at the time of the assessment. Observable behavioral responses have included facial expression, duration of looking, head turning, reaction time, performance on a task, and eye gaze changes. Physiologic responses used to assess attention have included heart rate, respiratory sinus arrhythmia, adrenocortical activity, and neurotramsitter activity (Ruff and Rothbart, 1996). Each of these responses is designed to index to what extent the brain is actively interested and processing the information presented to it. Selection of the behavioral response to be assessed should be made by considering the component of attention of interest based on a multidimensional theory of the attentional system (i.e., Mirsky, 1989; Mirsky *et al.,* 1991; Ruff and Rothbart, 1996), the degree of noise in its measurement (validity, reliability), the age of the subject, and the stability of the response. Of particular importance when the focus is on the developmental process is whether a system can be measured in infancy and whether the outcomes of such measurement can be related to constructs in later childhood. The use of cognitive models that incorporate principles of neurodevelopmental brain maturation (Diamond, 1990; Segalowitz and Rose-Krasnor, 1992; Bell and Fox, 1994; Diamond *et al.,* 1994) offer an exciting new way of investigating neurodevelopmental sequelae associated with exposure to teratogens.

3. Research on Attention in Infancy and Later Development

As noted previously, continuity in cognitive status from infancy to later childhood has been debated (Bornstein and Krasnegor, 1989) and has received little empirical support using traditional sensorimotor assessments of cognitive functioning (Bornstein and Sigman, 1986), that is, global developmental tests such as the Bayley Scales (Bayley, 1969) that rely on behavioral response. The development of information processing tasks, however, has led to new information regarding the continuity of cognitive processes (Columbo, 1993). These tasks assess infant attention and changes in attention as a function of stimulus properties such as novelty, familiarity, and complexity. These tasks moderately predict performance on later intelligence tests, a rate that exceeds the predictive accuracy of other available measures of infant cognitive functioning (Bornstein and Sigman, 1986; Columbo, 1993). For example, visual attention in 7-month-old infants has been found to predict IQs of 5-year-olds (Rose *et al.,* 1989) and of 11-year-olds (Rose and Feldman, 1995). Orienting responses to auditory stimuli derived from heart rate data from 4-month-olds have been related significantly to cognitive ability at 18-months (O'Connor, 1980) and 5 years (O'Connor *et al.,* 1984).

These tasks have also been able to discriminate between groups known to have some impairment of cognitive functioning (Miranda and Frantz, 1974; Rose and Wallace, 1985) suggesting that they have concurrent validity as well as predictive validity. Tasks of visual recognition memory also have been applied to differentiating cognitive ability after prenatal exposure to PCBs in infants (Jacobson *et al.,* 1985), to moderate levels of alcohol in infants (Jacobson *et al.,* 1992, 1993), and to methylmercury in crab eating monkeys (Gunderson *et al.,* 1988). These tasks offer an alternative method of investigating auditory and visual information processing skills of infants who have been exposed to teratogens and remove the motor developmental confounds found in other infant measures, which may limit the conclusions that can be drawn from data using sensorimotor tasks of infant functioning.

4. Specificity in Prediction for Measures of Infant Attention

Long-term follow-up studies have permitted the exploration of the nature of the continuity found in these infant attentional measures. Responses to prolonged exposure (i.e., fixation duration) and response to novelty using a paired comparison procedure have been found to be moderate predictors of childhood intelligence from infancy (Bornstein and Krasnegor, 1989; Colombo, 1993; Rose and Feldman, 1995). Columbo (1993) has posited that measures of attention in infancy share in common an underlying general cognitive factor as well as specific cognitive factors. He elaborates on this by further hypothesizing that the responses to prolonged stimulation (i.e., fixation duration) are believed to reflect speed of encoding, whereas novelty preference has been theorized to be linked to memory function.

Available data from one long-term study, however, has only partially supported this theory. Rose and Feldman (1995) concluded in their follow-up of 11-year olds who were tested at 7 months and 12 months of age that the infancy attention measures correlated most consistently with perceptual speed at 11 years of age, and that this persisted after controlling for overall IQ at 11 years. Visual recognition memory and visual exposure time at 7 months but not 12 months was related to delayed memory after controlling for overall IQ, suggesting some stability in memory functioning but not the continuity found for speed of processing.

5. Assessments of Attention Throughout Childhood

The effects of exposure to teratogens on attention during the preschool and later childhood years have been measured by a number of studies that have adopted the clinical psychologist's definition of attention. Some of these have used the traditional checklist methods that are commonly used in clinical practice and rely on reports of behavior given by parents and teachers (Denson *et al.,* 1975; Landesman-Dwyer and Ragozin, 1981; Naeye and Peters, 1984). In other studies, computerized attention tasks (vigilance tasks or continuous performance tasks) have been used (i.e., Streissguth *et al.,* 1986; Kristjansson *et al.,* 1989; Nanson and Hiscock, 1990; Fried *et al.,* 1992). These studies have been useful in identifying problems within the attentional system, but do not specify the nature of the problem that is needed to understand the neurologic substrate of the teratogenic effect.

Coles *et al.* (1997) is an example of a study that systematically attempted to identify the nature of neurologic damage associated with prenatal alcohol exposure in a sample of children who were between 7 and 8 years of age using a model of attentional mechanisms. Using a general ability model (i.e., IQ tests), clinical groups of FAS and ADHD were indistinguishable, but when tested as a four-factor model of attention (Mirsky, 1989), demonstrated very different patterns of difficulties with the attentional mechanisms of focusing, encoding, shifting, and sustaining.

Presently a number of studies are underway that examine behaviors believed to represent aspects of attention and that appear to be affected by prenatal exposure. Executive functioning, which is believed to be related to frontal lobe functioning, is currently a focus of interest (see Fried, Chapter 32). Such studies permit a more refined understanding of the relationship of prenatal exposure to teratogens and their pertinent outcomes. Investigating the impact of a teratogenic exposure on developmental outcome, either of an individual or in a more general sense, requires a refined understanding of a number of elements. Failure to understand the normative process of development or the ways in which delays and aberrations are expressed can lead to hasty conclusions. Similarly, ignoring the impact of the environment or the standards of research can under- or overestimate the effects of the exposure (Coles, 1993). The first step of any such project is to identify the exposure of interest and then to investigate thoroughly the status of research in this area. Such an investigation should provide the information necessary to formulate the research plan. If it is an area about which there is little information, an exploratory screening approach may be most appropriate, using measures tapping global functioning. During the initial stages of the examination of a problem, an extremely narrow focus is likely to neglect to measure relevant elements, resulting either in missing effects or misrepresenting the pattern of effects. Similarly, a longitudinal design, although difficult and expensive, may be appropriate in the early stages of research, when the timing and pattern of effects is not yet known.

When a teratogen has been under investigation for some time, a more refined and specific approach is warranted. At present, it is unnecessary to re-establish that alcohol is teratogenic or that cigarette use is associated with lower birthweight. Instead, studies should be well controlled and targeted on specific hypotheses that enlarge the knowledge base.

VII. Summary

Examination of previous studies also provides information about the appropriate control procedures and aspects of development that reward study and in other ways guide the design of the investigation. Eventually, the research plan is determined by the questions to be answered. Such a well-thought out and developmentally appropriate approach to the examination of the effects of teratogens and toxins provide the basis for future clinical practice as well as public policy.

References

Abel, E. L. (1995). An update on incidence in fetal alcohol syndrome: FAS is not an equal opportunity birth defect. *Neurotoxicol. Teratol.* **17,** 427–443.

Barkley, R. A. (1990). *Attention Deficit Hyperactivity Disorder: A Handbook for Diagnosis and Treatment.* The Guilford Press, New York.

Bayley, N. (1969). *Manual for the Bayley Scales of Infant Development.* San Antonio: Psychological Corporation.

Bayley, N. (1993). *Bayley Scales of Infant Development, Second Edition, Manual.* San Antonio: Psychological Corporation.

Bell, M. A., and Fox, N. A. (1994). Brain development over the first year of life: relations between electroencephalographic frequency and coherence and cognitive and affective behaviors. In *Human Behavior and the Developing Brain.* (G. Dawson and K. W. Fischer Eds.), pp. 314–345. The Guilford Press, New York.

Bellinger, D. C. (1997). *Effect Modification in Epidemiologic Studies of Neurotoxicant Exposure and Child Development.* Paper presented at the Annual Meeting of the Neurobehavioral Teratology Society, Palm Beach, Florida.

Bingol, N., Schuster, C., Fuchs, M., Iosub, S., Turner, G., Stone, R. K., and Gromisch, D. S. (1987). The influence of socioeconomic factors on occurrence of fetal alcohol syndrome. *Advances in Alcoholism and Substance Abuse: Special Issue: Children of Alcoholics.* **6,** 105–118.

Bornstein, M. H., and Krasnegor, N. A. (1989). *Stability and Continuity in Mental Development: Behavioral and Biological Perspectives.* Hillsdale, Lawrence Erlbaum Associates, Publishers, New Jersey.

Bornstein, M. H., and Sigman, M. D. (1986). Continuity in mental development from infancy. *Child Dev.* **57,** 251–274.

Bray, D. L., and Anderson, P. D. (1989). Appraisal of the epidemiology of fetal alcohol syndrome among Canadian native people. *Canad. J. Pub. Heal.* **80,** 42–45.

Brooks-Gunn, J., and Furstenberg, F. F. Jr. (1986). The children of adolescent mothers: physical, academic, and psychological outcomes. *Devel. Rev.* **6,** 224–251.

Butler, N. R., and Goldstein, H. (1973). Smoking in pregnancy and subsequent child development. *Brit. Med. J.* **4,** 545–548.

Caldwell, B. M., and Bradley, R. H. (1979). *Home Observation for Measurement of the Environment.* University of Arkansas Press, Little Rock.

Centers for Disease Control and Prevention. (1993). Fetal alcohol syndrome—United States, 1979–1991. *MMWR,* **173,** 575.

Coles, C. D. (1993). Commentary: saying "goodbye" to the "crack baby." *Neurotoxicol. Teratol.* **15,** 290–292.

Coles, C. D. (1994). Critical periods for prenatal alcohol exposure: Evidence from animal and human studies. *Alco. Heal. Res. World,* **18(1),** 22–29.

Coles, C. D. (1994). *Comments on the Relationship Between Fetal Alcohol Syndrome (FAS) and Socioeconomic Status (SES).* Presented at International Conference on Prevention: FAS and Alcohol Related Birth Defects, Detroit, MI, November 9, 1994. Convention proceedings to be published as a monograph by the National Institute on Alcoholism and Alcohol Abuse.

Coles, C. D., Brown, R. T., Smith, I. E., Platzman, K. A., Erickson, S., and Falek, A. (1991). Effects of prenatal alcohol exposure at school age. I: Physical and cognitive development. *Neurotoxicol. Teratol.* **13,** 357–367.

Coles, C. D., Platzman, K. A., Raskind-Hood, C. L., Brown, R. T., Falek, A., and Smith, I. E. (1997). A comparison of children affected by prenatal alcohol exposure and attention deficit, hyperactivity disorder. *Alcohol. Clin. Exper. Res.* **21,** 150–161.

Colombo, J. (1993). Infant cognition: predicting later intellectual functioning. In *Individual Differences and Development Series, Vol. 5.* Sage Publications, Newbury Park.

Crepin, G. Dehaene, Ph., and Samaille, C. (1989). Aspects cliniques, evolutifs epidemiologiques de l'alcoolisme foetal: un fleau tourjours d'actualite. *Bulletin de L'Academe National du Medcine,* **173,** 575–582.

Davie, R., Butler, N. R., and Goldstein, H. (1972). *From Birth to Seven.* Longmans, London.

Dehaene, Ph., Crepin, G., Delahousse, G., Querleu, D., Walbaum, R., Titran, M., and Samaille-Villette, C. (1981). Aspects epidemiologique du syndrome d'alcoolisme foetal: 45 observations en 3 ans. *La Nouvelle Press Medicale,* **10,** 2639.

Denckla, M. D. (1996). A theory and model of executive function: a neuropsychological perspective. In *Attention, Memory, and Executive Function.* (G. R. Lyon and N. A. Krasnegor Eds.), pp. 263–278, Brookes Publishers, Baltimore.

Denson, R., Nanson, J. L., and McWatters, M. A. (1975). Hyperkinesis and maternal smoking. *Can. Psych. Assoc. J.* **20,** 183–187.

Diamond, A. (Ed.) (1990). The development and neural bases of higher cognitive functions. *Ann. NY Acad. Sci.* 608.

Diamond, A., Weker, J. F., and Lalonde, C. (1994). Toward understanding commonalties in the development of object search, detour navigation, categorization, and speech perception. In *Human Behavior and the Developing Brain.* (G. Dawson & K. W. Fischer Eds.), pp. 380–426, The Guilford Press, New York.

Elardo, R. and Bradley, R. (1981). The home observation for measurement of the environment (HOME): a review of research. *Dev. Rev.* **1,** 113–145.

Fagan, J. F. III. (1970). Memory in the infant. *J. Exper. Child Psych.* **9,** 217–226.

Fergusson, D. M., and Lloyd, M. (1991). Smoking during pregnancy and its effects on child cognitive ability from the ages of 8 to 12 years. *Paediatr. Perinat. Epidem.* **5,** 189–200.

Florey, C. du V. (1988). Weak associations in epidemiological research: some examples and their interpretation. *Internat. J. Epidem.* **17,** 950–954.

Fried, P. A., and Makin, J. E. (1987). Neonatal behavioral correlates of prenatal exposure to marihuana, cigarettes, and alcohol in a low risk population. *Neurotoxicol. Teratol.* **9,** 1–7.

Fried, P. A., and Watkinson, B. (1988). Twelve- and twenty-four-month neurobehavioral follow-up of children prenatally exposed to marihuana, cigarettes, and alcohol. *Neurotoxicol. Teratol.* **10,** 305–313.

Fried, P. A., Watkinson, B., and Gray, R. (1992). A follow-up study of attentional behavior in 6-year-old children exposed prenatally to marihuana, cigarettes, and alcohol. *Neurotoxicol. Teratol.* **14,** 299–311.

Gunderson, V. M., Grant-Webster, K. S., Burbacher, T. M., and Mottet, N. K. (1988). Visual recognition memory deficits in methylmercury-exposed *Mecaca fasicularis* infants. *Neurotoxicol. Teratol.* **10,** 373–379.

Hardy, J. B., and Mellits, E. D. (1972). Does maternal smoking during pregnancy have a long-term effect on the child. *Lancet, ii,* 1331–1336.

Hutchings, D. E., and Fifer, W. P. (1986). Neurobehavioral effects in human and animal offspring following prenatal exposure to methadone. In *Handbook of Behavioral Teratology.* (E. P. Riley and C. V. Vorhees Eds.), pp. 141–159, Plenum Press, New York.

Jacobson, S. W., Fein, G. G., Jacobson, J. L., Schwartz, P. M., and Dowler, J. K. (1984). Neonatal correlates of prenatal exposure to smoking, caffeine, and alcohol. *Infant Behav. Dev.* **7,** 253–265.

Jacobson, S. W., Fein, G. G., Jacobson, J. L., Schwartz, P. M., and Dowler, J. K. (1985). The effect of PCB exposure on visual recognition memory. *Child Dev.* **56,** 853–860.

Jacobson, J. L. and Jacobson, S. W. (1990). Methodological issues in human behavioral teratology. In *Advances in Infancy Research (Vol. 6).* (C. Rovee-Collier and L. P. Lipsett Eds.), pp. 111–148, Ablex, Norwood, New Jersey.

Jacobson, S. W., Jacobson, J. L., O'Neill, J. M., Padgett, R. J., Frankowski, J. J. and Bihun, J.T. (1992). Visual expectation and dimensions of infant information processing. *Child Dev.* **63,** 711–724.

Jacobson, S. W., Jacobson, J. L., Sokol, R. J., Martier, S. S., and Ager, J. W. (1993). Prenatal alcohol exposure and infant information processing ability. *Child Dev.* **64,** 1706–1721.

Jones, K. L., Smith, D. W., Ulleland, C. N., and Streissguth, A. P. (1973). Pattern of malformation in offspring of chronic alcoholic mothers. *Lancet,* **1,** 1267–1271.

Kable, J. A. (1997). *Auditory vs. General Information Processing Deficits in Infants of Mothers who Smoked during their Pregnancy.* Poster presented at the Society for Research in Child Development, Washington, DC.

Kelly, G. A. (1955). *The Psychology of Personal Constructs.* Norton, New York.

Kolb, B. and Whishaw, I. Q. (1985). Recovery of function. In *Fundamentals of Human Neuropsychology.* pp. 631–663, WH Freeman & Co., New York.

Kristjansson, E. A., Fried, P. A., and Watkinson, B. (1989). Maternal smoking during pregnancy affects children's vigilance performance. *Drug Alc. Dep.* **24,** 11–19.

Landesman-Dwyer, S. and Ragozin, A. S. (1981). Behavioral correlates of prenatal alcohol exposure: a four-year follow-up study. *Neurobehav. Toxicol. Teratol.* **3,** 187–193.

Lemoine, P., Harousseau, H., Bortyru, J. P. and Menuet, J. C. (1968). Les enfants de parents alcooliques: Anomalies observees a propos de 127 cas. *Ouset Medicine,* **25,** 476–482.

Lifschitz, M. H. and Wilson, G. S. (1991). Patterns of growth and development in narcotic-exposed children. In *Methodological Issues in Controlled Studies on Effects of Prenatal Exposure to Drug Abuse.* (M. M. Kilbey and K. Asghar Eds.), National Institute on Drug Abuse Research Monograph: **114,** 323–339.

Mactutus, C. F. and Fechter, L. D. (1985). Moderate carbon monoxide exposure produces persistent, and apparently permanent, memory deficits in rats. *Teratology,* **31,** 1–12.

Makin, J., Fried, P. A., and Watkinson, B. (1991). A comparison of active and passive smoking during pregnancy: long-term effects. *Neurotoxicol. Teratol.* **13,** 5–12.

May, P. A., Hymbaugh, K. J., Aase, J. M., and Samet, J. M. (1983). Epidemiology of fetal alcohol syndrome among American Indians of the southwest. *Soc. Biol.* **30,** 374–387.

McCartney, J. S., Fried, P. A., and Watkinson, B. (1994). Central auditory processing in school-age children prenatally exposed to cigarette smoke. *Neurotoxicol. Teratol.* **16,** 269–276.

Miller, H. C., and Hassanein, K. (1971). Diagnosis of impaired fetal growth in newborn infants. *Pediatrics,* **48,** 511–522.

Miranda, S. B., and Frantz, R. L. (1974). Recognition memory in Down's Syndrome and normal infants. *Child Dev.* **45,** 651–660.

Mirsky, A. F. (1989). The neuropsychology of attention: elements of a complex behavior. In *Integrated Theory and Practice in Clinical Neuropsychology.* (E. Peregman Ed.), Lawrence Erlbaum Associates, Hillsdale, New Jersey.

Mirsky, A. F., Anthony, B. J., Duncan, C. C., Ahern, M. B., and Kellam, S. G. (1991). Analysis of the elements of attention: a neuropsychological approach. *Neuropsychol. Rev.* **2,** 75–88.

Naeye, R. L., and Peters, E. C. (1984). Mental development of children whose mothers smoked during pregnancy. *Obstet. Gynecol.* **64,** 601–607.

Nanson, J. L., and Hiscock, M. (1990). Attention deficits in children exposed to alcohol prenatally. *Alcohol. Clin. Exper. Res.* **14,** 656–661.

O'Beattie, J., Day, R. E., Cockburn, F., and Garg, R. A. (1983). Alcohol and the fetus in the west of Scotland. *Brit. Med. J.* **287,** 17–20.

O'Connor, M. J. (1980). A comparison of preterm and full-term infants on auditory discrimination at four months and on Bayley Scales of Infant Development at eighteen months. *Child Dev.* **51,** 81–88.

O'Connor, M. J., Cohen, S., and Parmelee, A. H. (1984). Infant auditory discrimination in pre-term and full-term infants as a predictor of 5-year intelligence. *Dev. Psych.* **20,** 159–165.

Olegard, R., Sabel, K.-G., Aronsson, M., Sadin, B., Johansson, P. R., Carlsson, C., Kyllerman, M., Iversen, K., and Hrbek, A. (1979). Effects on the child of alcohol abuse during pregnancy: retrospective and prospective studies. *Acta Paediatrica Scandanavia. Supplement,* **275,** 112–121.

Pennington, B. F. (1991). Dyslexia and other developmental language disorders. In *Diagnosing Learning Disorders.* pp. 45–81, Guilford, New York.

Picone, T. A., Allen, L. H., Olsen, P. N., and Ferris, M. E. (1982). Pregnancy outcome in North American women: II. Effects of diet, cigarette smoking, stress, and weight gain on placentas, and on neonatal physical and behavioral characteristics. *Am. J. Clin. Nutrit.* **36,** 1214–1224.

Riley, E. P., Hannigan, J. H., and Balaz-Hannigan, M. A. (1985). *Behavioral Teratology as the Study of Early Brain Damage: Considerations for the Assessment of Neonates.* Conference proceedings of the National Center for Toxicology Research *et al.:* Design considerations in screening for behavioral teratogens: results of the collaborative behavioral teratology study. *Neurobehav. Toxicol. Teratol.* **7,** 635–638.

Robinson, C., and Fieber, N. (1988). Cognitive assessment of motorically impaired infants and preschoolers. In *Assessment of Young Developmentally Disabled Children.* (T. D. Wachs and R. Sheehan Eds.), pp. 127–161, Plenum Press, New York.

Rose, S. A., and Feldman, J. F. (1995). Prediction of IQ and specific cognitive abilities at 11 years from infancy measures. *Dev. Psych.* **31,** 685–696.

Rose, S. A., Feldman, J. F., Wallace, I. F., and McCarton, C. (1989). Infant visual attention: relation to birth status and developmental outcome during the first five years. *Dev. Psych.* **25,** 560–576.

Rose, S. A., and Wallace, I. F. (1985). Visual recognition memory: a predictor of later cognitive functioning in pre-terms. *Child Dev.* **56,** 843–852.

Ruff, H. A., and Rothbart, M. K. (1996). *Attention in Early Development: Themes and Variations.* Oxford University Press, New York.

Sameroff, A. J., Siefer, R., Barocas, R., Zax, M., and Greenspan, S. (1987). Intelligence quotient scores of 4-year-old children: socio-environmental risk factors. *Pediatrics* **79,** 343–350.

Saxton, D. W. (1978). The behavior of infants whose mothers smoke in pregnancy. *Early Hum. Dev.* **2,** 363–369.

Scialli, A. R. (1992). *A Clinical Guide to Reproductive and Developmental Toxicology.* CRC Press. Boca Raton, Florida.

Segalowitz, S. J. and Rose-Krasnor, L. (Eds.) (1992). The role of frontal lobe maturation in cognitive and social development. *Brain Cog.* **20,** 1–213.

Sokol, R. J., Ager, J., Martier, S. S., Debanne, S., Ernhart, C., Kuzma, J., and Miller, S. I. (1986). Significant determinations of susceptibility to alcohol teratogenicity. *Ann. NY Acad. Sci.* **47,** 87–102.

Spohr, H.-L., and Steinhauser, H. C. (1984). Clinical, psychopathological, and developmental aspects in children with the fetal alcohol syndrome (FAS). In *CIBA Foundation Symposium 105: Mechanisms of Alcohol Damage in Utero.* pp. 197–217. CIBA Foundation, Pitman Publishers, London.

Stratton, K., Howe, C., and Battaglia, F. (Eds.) (1996). *Fetal Alcohol Syndrome: Diagnosis, Epidemiology, Prevention, and Treatment.* National Academy Press, Washington, DC.

Streissguth, A. P., Barr, H. M., Kogan, J., and Bookstein, F. L. (1996). *Understanding the Occurrence of Secondary Disabilities in Clients with Fetal Alcohol Syndrome (FAS) and Fetal Alcohol Effects (FAE).* University of Washington Publication Services, Seattle, Washington.

Streissguth, A. P., Barr, H. M., Sampson, P. D., Parrish-Johnson, J. C., Kirchner, G. L., and Martin, D. C. (1986). Attention, distraction, reaction time, at age 7-years and prenatal alcohol exposure. *Neurotoxicol. Teratol.* **8,** 717–725.

Streissguth, A. P., Bookstein, F. L., Sampson, P. D., and Barr, H. M. (1993). *The Enduring Effects of Prenatal Alcohol Exposure on Child Development: Birth Through Seven Years. A Partial Least Squares Solution.* University of Michigan Press, Ann Arbor.

Streissguth, A. P., Herman, C. S., and Smith, D. W. (1978). Intelligence, behavior and dysmorphogenesis in the Fetal Alcohol Syndrome: a report on 20 patients. *J. Pediatr.* **92,** 363–367.

Sutherland, N. S., and Mackintosh, N. J. (1971). *Mechanisms of Animal Discrimination Learning.* Academic Press, New York.

Tanaka, H., Masataka, A., and Suzuki, N. (1981). The fetal alcohol syndrome in Japan. *Brain Dev.* **3,** 305–311.

Tennes, K., Avitable, N., Blackard, C., Boyles, C., Hassoun, B., Holmes, L., and Kreye, M. (1985). Marijuana: prenatal and postnatal exposure in the human. In *Consequences of Maternal Drug Abuse.* (T. M. Pinkert Ed.), pp. 48–60. National Institute in Drug Abuse Research Monograph: 59.

Tuchman-Duplessis, H. (1975). Drug effects on the fetus. *Monographs on Drugs (Vol. 2).* ADIS Press, Sydney, Australia.

Vorhees, C. V. (1986). Principles of behavioral teratology. In *Handbook of Behavioral Teratology.* (E. P. Riley and C. V. Vorhees Eds.), pp. 23–48. Plenum Press, New York.

Wachs, T. D. and Gruen, G. E. (1982). *Early Experience and Human Development.* Plenum Press, New York.

Weinberg, J. (1992). Prenatal ethanol effects: sex differences in offspring stress responsiveness. *Alcohol,* **9,** 219–223.

Wilson, G. S. (1989). Clinical studies of infants and children exposed prenatally to heroin. *Ann. NY Acad. Sci.* **562,** 183–194.

Wilson, G. S., Desmond, M. M., and Wait, R. B. (1981). Follow-up of methadone treated women and their infants: health, developmental, and social implications. *J. Pediatr.* **98,** 716–722.

Wilson, G. S., McCreary, R., Kean, J., and Baxter, C. (1979). The development of preschool children of heroin-addicted mothers: a controlled study. *Pediatrics,* **63,** 135–141.

Zuckerman, B., and Brenahan, K. (1991). Developmental and behavioral consequences of prenatal drug and alcohol exposure. *Pediatr. Clin. North Am.* **38,** 1387–1406.

CHAPTER

26

Behavioral Evaluation of the Older Infant and Child

PETER A. FRIED
Department of Psychology
Carleton University
Ottawa, Ontario, Canada K1S 5B6

Behavioral teratology gained recognition and credence only after years of considerable resistance from traditional teratologists with early attempts to draw attention to dysfunctions other than malformations being handicapped by a paucity of research (Vorhees 1986, 1989). A major milestone in broadening the scope of teratology was the identification in North American of the *fetal alcohol syndrome* (FAS) (Jones and Smith, 1973; Jones *et al.,* 1973). This syndrome emphasized that there were motor and mental developmental effects in addition to altered morphogenesis resulting from maternal alcoholism. In the years since the description of the FAS, there has been a burgeoning body of research examining prenatal exposure to licit and illicit drugs revealing a wide range of neurobehavioral consequences in the very young baby.

However, when the temporal focus of the outcome measures moves beyond early infancy, the number of published neurobehavioral teratologic findings is still relatively limited. The reasons run the gamut from logistic to interpretative issues. Logistically, enormous commitment is required from investigators, participating families, and funding agencies. On the interpretative side, difficulties arise when attempting to combine the fields of developmental psychology and neurotoxicity while factoring into this mix the extensive passage of

*Abbreviations: ADHD, attention deficit hyperactivity disorder; BSID, Bayley Scales of Infant Development; FAS, fetal alcohol syndrome; IQ, intelligence quotient; OPPS, Ottawa Prenatal Prospective Study; WISC, Wechsler Intelligence Scale for Children.

time between prenatal events and later behavioral measures (Kilbey and Asghar, 1991; Wetherington *et al.*, 1996). The methodologic issues have led to some resistance and uncertainty in the acceptance of longitudinal teratologic findings—not so much from a theoretic or traditional point of view as was the case in the early 1970s with behavioral teratology, but rather on the basis of practicality and/or pragmatism revolving around the very nature that makes us human—our complexity and the complexity of our environment.

This chapter has two themes. The first includes methodologic, conceptual, and interpretative issues that envelop the study of behavioral teratology when the outcomes are being gleaned from subjects beyond the infant and toddler stage. The general methods, advantages, and difficulties involved in longitudinal studies have been addressed in detail elsewhere (Mednick *et al.*, 1984). Within the sphere of behavioral teratology certain aspects bear elaboration and emphasis. The latter part of the chapter focuses, with examples, on the importance of the recognition of specific cognitive–behavioral dysfunctions that may not manifest themselves until the child is well beyond infancy.

I. Methodologic, Conceptual, and Interpretative Issues

A. *Introduction*

The principles that, either implicitly or explicitly, underlie most investigations of the potential long-term effects of *in utero* drug exposure in humans are extensions of the basic tenets of teratology and toxicology as set forth by Wilson (1977). These general principles have been expanded and somewhat reformulated recently by Vorhees and co-workers (Vorhees and Butcher, 1982; Vorhees, 1986, 1989) to encompass specifically behavioral teratology. In terms of the focus of the present chapter, there are a number of concepts arising from Vorhees' reformulation that have particular relevance for this chapter. These include a wider temporal window of vulnerability in the fetus, a greater sensitivity for potential behavioral as compared to morphologic injury, the need to search for the consequences of fetal exposure well beyond birth, and the recognition that the type and magnitude of a behavioral teratogenic effect is moderated by both prenatal and postnatal environmental factors.

The behavioral teratologic model described by Vorhees was derived primarily from animal research, and, as a consequence, the postnatal factors described were essentially limited to issues of maternal rearing. In human teratologic studies, the postnatal factors are considerably more complex, varied, and, unfortunately from an experimental point of view, much more difficult to ascertain, and/or quantify. Consideration of the postnatal environment must include social, psychologic, and demographic factors. Although these variations have an impact on and interact with maternal rearing, their role can potentially extend well beyond that one domain. As the child gets older, this postnatal environment has the likelihood of exerting a greater influence on the outcomes in the offspring. Furthermore, the environment for a significant number of drugs-using women and their children is unpredictable and unstable, with a host of uncontrollable confounding factors. The consequence of these realities is that the causal attribution of negative outcomes to prenatal drug exposure becomes substantially more problematic.

Nowhere is this issue better exemplified than with the current public and scientific concern about the potential long-term effects of prenatal exposure to cocaine. As described elsewhere (Lester *et al.*, 1996; see also Richardson and Day, Chapter 27), very little is known about the specific long-term aspects of development, if any, that can be attributed to this drug. In spite of this dearth of evidence, many communities have implemented special school-based programs predicted on expectations of universal and permanent long-term neurobehavioral damage. The basis for such programs appears to be a combination of media hyperbole and infant studies that frequently use research designs with inadequate consideration of maternal lifestyle habits and postnatal environmental factors. A consideration of the difficulty in separating fact from fiction with respect to imputing effects of prenatal cocaine is the subject of an excellent open-peer commentary (Hutchings, 1993).

Despite impediments to experimental control and the consequent difficulty in determining cause-and-effect relationships, research in potential longer term effects clearly is essential. Although not all methodologic difficulties can be overcome, the recognition of these difficulties and the implementation of appropriate strategies can serve to increase substantially the validity of the interpretation of the results. However, in the discussion of control issues that are described in different portions of this chapter, it should be kept in mind that, from a conceptual framework, a paradox can arise. In the world of the typical user, the very attempt to isolate the statistically unique contribution of the drug in question may obscure the reality of the drug's effect (Fried, 1993; 1996; Lester *et al.*, 1996); for example, the vast majority of cocaine users are polydrug users. It may be argued that the vulnerability of offspring will only be truly ascertained when the design of studies include rather than exclude the many facets that are common to most drug-using pregnant women. No matter which approach is taken, it is encumbent on the searcher to be cognizant of the multiple concomitant and interdependent non-drug-of-interest factors that potentially contribute to the outcomes in the maturing child.

The various topics discussed later in this chapter include some issues that are critical in the evaluation of behavior in children who are subjects in longitudinal, behavioral teratogenic investigations. To facilitate explanation here, several longitudinal studies are referred to with a fairly heavy reliance on the prospective study that the author initiated in 1978. This latter work—the Ottawa Prenatal Prospective Study (OPPS)—has, as its focus, the consequences of prenatal exposure to marijuana and cigarettes in a predominantly low-risk, middle-class sample. The general testing procedures that have and are being followed have been outlined elsewhere (e.g., Fried, 1992; 1995) and are not detailed here.

B. General Research Design Issues

Most research investigating longer term consequences of prenatal drug exposure uses prospective designs that allow for ongoing, systematic observations as the child matures. However, it should be noted that there is a decided place for the alternative retrospective approach. In fact, as pointed out by Hans (1996), much of what we know about the consequences in children beyond infancy who were born to drug-using parents comes from a number of older studies using a retrospective design. Today, this approach can be particularly useful in filling a particular niche. With its lower cost and less complex research design, the retrospective strategy can focus on children who have been highly exposed *in utero* to permit initial identification of particularly salient impairments can be made. Such observations can then be used to direct the content of neurobehavioral batteries in prospective studies with the objective of confirming the retrospective findings (Jacobson and Jacobson, 1996).

The prospective longitudinal protocol usually involves the recruiting of subjects prenatally or at birth and then following them for an extended period of time. For behavioral teratologic research, the key methodologic advantages of this approach compared to the retrospective design are, in general terms, an enhanced ability to determine the extent and timing of drug exposure during pregnancy, a more accurate evaluation of potential confounding variables, and the opportunity to examine the stability of the putative effects. However, the prospective longitudinal design places two methodologic and conceptual issues in the forefront of the evaluation and determination of prenatal drug effects. The first revolves around the development and changing competencies in the maturing child and the second is the interaction between the postnatal environmental and the prenatal drug exposure. These considerations repeatedly surface in the ensuing portions of this chapter.

The longitudinal paradigm has served as the prototype for a wide variety of systematic child developmental studies since the late 1920s (Mednick *et al.*, 1984). This design has allowed important behavioral teratogenic observations to be made since the late 1950s, even though the primary variable of interest of these early studies was not maternal drug use. For example, the Collaborative Perinatal Study (an overview of procedures and results is presented in Broman, 1984), initiated in 1959, followed 50,000 pregnancies for 7 years in order to identify fetal, interpartum, and neonatal events that affect children's development. This seminal work provided some pivotal behavioral findings related to drug exposure during pregnancy by means of a selective follow-up of offspring studied in a standardized, systematic fashion. For example, even though quantitative data on alcohol intake was not recorded, maternal alcoholism was mentioned in enough hospital records for the Collaborative Perinatal Study to provide one of the first cohorts for examining FAS (Jones *et al.,* 1974). In another substudy, because maternal smoking habits during pregnancy were ascertained, the very large sample size permitted a unique approach to determine how smoking during pregnancy affects children's mental development. At the age of 7 years, a comparison was made among siblings whose mothers smoked in one but not in the other pregnancy, thus controlling, to some degree, genetic factors and many child-raising practices (Naeye and Peters, 1984).

The research strategy as employed in the Collaborative Perinatal Study has a major interpretative advantage in that it allows the determination of the continuity and stability of findings over time. In drugs research, this case assist in the interpretation of the role of prenatal exposure with later behavioral observations. An example of this has evolved within the OPPS with the relationship of maternal use of cigarettes and auditory functioning in the offspring.

Prenatal cigarette exposure was associated with decreased auditory habituation in babies between 3 and 6 days of age (Fried and Makin, 1987). At 1 and 2 years of age (Fried and Watkinson, 1988), maternal cigarette smoking was related to lowered auditory-based behavioral measures and, at 36 months, a negative dose–response association was noted (Fried and Watkinson, 1990) on the verbal subscale of the McCarthy Test (McCarthy, 1972) as well as on the expressive language portion of the Reynell Scale (Reynell, 1977). This dose–response association between maternal smoking and different aspects of language continued at 4 (Fried and Watkinson, 1990), 5 and 6 years of age (Fried *et al.*, 1992a) and was also apparent when the children were tested between the ages of 9 to 12 (Fried *et al.*, 1997). In addition, speech and language measures were associated with maternal passive smoking when the children were tested between 6 and 9 years of age (Makin *et al.*, 1991).

On nonlanguage tests, auditory function was also associated with maternal cigarette smoking in the OPPS sample. Attention and impulsiveness in 4- to 7-year-old children were examined, using auditory and visual stimuli in different tasks (Kristjansson and Fried, 1989). Deficits in the ability to inhibit impulsive responding in the auditory mode were particularly related to smoking during pregnancy. At the ages of 6–11 years, the children were assessed using a central auditory processing task (Keith, 1986) that made perceptual rather than linguistic demands. Maternal smoking was linearly associated with poorer performance on this task (McCartney *et al.*, 1994) and, as with all the earlier observations, the association remained after adjusting for other drug use, demographic variables, and prenatal and postnatal passive smoke exposure.

In 9–12 year old children in the OPPS sample, a negative dose–response association was observed between smoking and reading (Fried *et al.*, 1997). Because of the consistent observations linking auditory functions and maternal cigarette use, a variety of reading tasks was examined in order to assess underlying phonologic (auditory) and orthographic (visual) processing. Phonologic abilities provide the critical link between the association of sounds and letters. It forms a key basis for word recognition and print decoding and plays a critical role in the development of reading (e.g., Siegel, 1993). Phonologic capabilities, however, are not the only strategy that discriminates between good and poor readers. An additional process is the ability to form orthographic representations that require visual memory skills (Cunningham and Stanovitch, 1990). This visual route of word recognition, which involves accessing the word stored in lexical memory, does not involve the phonologic processing of letters into sounds.

With the 9–12-year-old children, a negative dose–response relationship between reading competence and *in utero* cigarette exposure was apparent with passage comprehension, reading of real words and the reading of pseudowords. On those tasks in which phonologic versus orthographic analysis was possible, a series of intriguing observations was made. The children born to nonsmokers had higher phonologic processing scores. However, in situations where a word could be read by using either a phonologic or an orthographic approach, the children of smokers were more likely to chose the latter compared to the children of nonsmokers. Furthermore, the children of smokers, when employing an orthographic strategy, were successful in applying this approach in the reading task.

These findings that the phonologic rather than the orthographic sphere of reading was more affected by maternal smoking and that a visual rather than an auditory approach to word analysis was chosen more often by the offspring of smokers would have been very difficult to interpret without the longitudinal auditory findings gathered since birth. The continuity of the negative relationship between auditory–language outcome variables and maternal smoking increases the confidence in the interpretation that the deficits in phonologic aspects of reading are, at least to some degree, the direct consequences of *in utero* exposure to constituents of smoke. The possible mechanisms have been discussed elsewhere (McCartney *et al.*, 1994).

C. Caretaker Variables

Even with the consistent findings noted across all ages tested, the previously described proposed association between maternal smoking and auditory functioning is likely moderated by the reciprocal relationship between the offspring's behavior and the parent's interaction with the child. The behavior that develops between child and caretaker is by no means a unidirectional process; the interactive pattern depends on the characteristics and reactions of each member of the dyad, and this transactional state of affairs (Sameroff and Chandler, 1975) may exacerbate certain *in utero* drug effects. In the OPPS longitudinal findings cited previously, if the child (particularly during infancy) responds to a parent in an altered way because of a deficit in auditory functioning, the caregiver may alter behavior toward the child, including less verbal communication. In the development of language, the extent of parental interaction plays a pivotal role (Siegel, 1981; Tamis-LeMonda and Bornstein, 1989), and therefore a decreased level of verbal interaction with the infant may well potentiate the *in utero* consequences of maternal smoking in a continuous transactional cycle.

Infant outcomes following specific intrauterine drug exposures are covered in different chapters in this volume but, in general terms, behaviors frequently associated with *in utero* drug exposure include disturbed sleeping patterns, less responsiveness or overresponsiveness to external stimulation, difficulty in consolability, and excessive crying. These altered infant behaviors clearly are likely to influence caregiver–infant interaction and may well have an impact on outcomes assessed when the child is considerably beyond the infant stage. Procedures that assist in determining the extent that maternal behavior may either exacerbate or compensate the possible infant neurocognitive deficits ought to be part of the prospective longitudinal design. As summarized by Griffith and Freier (1992), the caregiver's role is to adapt to the information-processing abilities and needs of the infant. When this role is not carried out successfully, the infant is likely to experience situations that impede development.

Within normative populations, the parent–infant interactions have been shown to be of major importance in accounting for lateral emotional and cognitive outcomes (Sameroff, 1979). With prenatal drug use, the caretaker–child interaction takes on an even more paramount role in influencing future performance. Compared to the general population, there is the increased likelihood for the infant to be raised in a high-risk, less than optimal environment. The negative consequences of this situation becomes progressively more influential as the child gets older (Kagan *et al.,* 1978). The multiple placements that are such an integral part of the lives of many drug-exposed offspring are an unfortunate part of this equation. Some of these children go through as many as eight foster care placements in the first year of life (Beckwith *et al.,* 1994). This variable must not be overlooked in the consideration of outcomes as the child matures.

Although the focus of the literature being discussed is prenatal drug use, the postnatal drug habits of the primary caretaker can also affect the child's cognitive–behavioral development and ought to be part of the assessment in longitudinal work. If the caretaker is intoxicated, responsiveness to the child's physical and mental needs is likely to be impaired. Furthermore, the lifestyle associated with ongoing drug use is associated with environmental risk factors (e.g., abuse, poor nutrition, neglect, maternal personality disorders) that can affect markedly the development of the maturing child. One example of the impact that continued drug use has beyond the early infancy stage was reported by Rodning *et al.* (1991). Attachment behavior was assessed in 15-month old prenatally drug-exposed toddlers for whom the caretaking environment was evaluated at 3 and 9 months. One of the major findings was that, whereas every child whose mother continued to abuse drugs was insecurely attached, half of the children whose mothers were abstinent during the first 15 months formed secure attachment relationships.

D. Control and Confounders

Numerous reviews and reports have stressed the necessity for, and suggested strategies for, assessing and controlling a wide range of potentially confounding variables (e.g., Jacobson and Jacobson, 1996; Kilby and Ashgar, 1991; Wetherington *et al.,* 1996). The quasi-experimental nature of prospective human drug studies prevents the manipulation of the host of complex variables that influence the outcomes of interest. As described by Streissguth *et al.* (1989), behavioral teratologic studies do not propose that the focus teratogen is the best predictor of the outcome, rather that the drug continues to predict the outcome after appropriate statistical control for the other significant predictors. In most studies the potential confounders that are considered include pre- and postnatal demographic background variables, drugs other than the one under consideration, and, to varying degrees, aspects of the caretaker–child interaction.

Many of the behavioral teratologic longitudinal studies underway at this time, particularly those that have not oversampled heavy users, report relatively subtle or no cognitive effects of maternal drug use beyond the infancy stage once confounding variables have been controlled. This however, may reflect an overly conservative interpretation of the role of "confounding" variables. In prenatal drug studies, when statistical significance is found, the amount of variance accounted for by the drugs is relatively small and, given the statistical procedures used, this is not surprising. In most studies, the criterion for the inclusion of control variables in multivariate statistical analyses is that they be weakly related to the outcome in question (Jacobson and Jacobson, 1996). An effect arising from the drug in question is only inferred if the outcome variable remains significant with the maternal drug use after controlling for the potential confounders. In the OPPS work, the behavioral effects uniquely associated with maternal drug use (marijuana or tobacco) range from 1.5% to 5% after the variance due to potentially confounding factors is partialled out (Fried, 1992). Similarly, in a study in which social levels of alcohol were found to contribute significantly to mental and motor scores, the amount of variance accounted for by that drug was approximately 1% (Streissguth *et al.,* 1980), and in a longitudinal study of 7-year-old FAS children the unique variance attributable to maternal alcohol consumption was 3% (Coles, personal communication, 1997). However, nondrug lifestyle factors (including some of the caretaker variables discussed here) account for up to 35% of the cognitive outcome variability (Fried, 1996).

The question arises as to whether the unique contribution of the drugs in fact reflects the "accurate" relationship between the behavioral outcome variables and the prenatal drug exposure. Elsewhere (Fried and Watkinson, 1988) it has been proposed that it is more likely that the drug's real association with the offspring's behavior lies somewhere between the drug's unique contribution after controlling for potentially confounding variables and its relationship with no potential confounds considered. In the latter approach, variance attributable to prenatal drug exposure may be as high as 12% whereas, as stated before, after accounting for other predictors, the unique attribution may fall to 1%. What is typically reported is the conservative, unique effects of drugs with all of the shared variance being attributed to the confounding variable.

Related to the question discussed previously is the recognition of some covariates, not as potential confounders, but rather as either mediating or moderating variables (Jacobson and Jacobson, 1996). Mediating variables are those factors that may explain the underlying processes through which prenatal exposure exerts its impact on the outcome being assessed. Using an example given earlier in this chapter, prenatal drug use may be associated with postnatal drug use, which, in turn, is associated with particular forms of maternal caregiving (Rodning *et al.,* 1991). Within the OPPS, birthweight, length of gestation, and variables that assess the postnatal environment have frequently been examined in the context of mediating variables (Fried, 1992).

Moderating variables are those factors that are hypothesized to interact with prenatal drug exposure to result in an altered degree of either vulnerability or resilience. Examples of moderating variables include the timing of exposure (Richardson *et al.,* 1995), maternal age (O'Connell and Fried, 1991), and ethnicity (Day, 1994).

Emphasis has been placed on the necessity of controlling for a broad range of potential confounders because of the recognition of the very real risk of falsely attributing an observed effect solely to the drug being studied—increasing the chance of Type 1 error. This has certainly plagued some of the early work examining the prenatal consequences of cocaine. However, teratologic studies, particularly those that attempt to assess the long-term outcomes, must also be concerned with Type II statistical error—the failure to detect a real drug effect. The consequences of such an interpretative error—that prenatal exposure to the drug in question is innocuous—takes on implications and ramifications that do not exist in most nondrug longitudinal studies.

In studies that focus on long-term consequences of prenatal drug exposure, inappropriate overcontrolling for confounders is just one potential contributor to the failure to detect significant prenatal drug exposure. A further cause of a Type II error, and one that is particularly applicable in studies that assess exposed children over many years, is a diminishing sample size as the work proceeds. Not only can such attribution result in a loss of statistical power, but it also may well result in a systematic loss across different drug groups with the higher risk subjects being harder to retain. The critical roles of extensive outreach activities, strategies that involve the whole family, and the quality of the subjects' relationship with the research staff have been common themes for minimizing attribution in a number of ongoing longitudinal prospective projects (e.g., Fried, 1991; Howard, 1992; Streissguth and Giunta, 1992).

II. Cognitive Measures in Older Infants and Children

A. *Introduction*

In addition to the methodologic challenges described earlier, the researcher working with the maturing prenatally drug-exposed child has to be cognizant of the repercussions of the cognitive growth and development that is taking place. For the teratologic researcher, longitudinal studies provide the opportunity to challenge the child with situations and tasks that may unmask deficits not evident at earlier ages. The absence of early deficits may reflect the nervous system's functional plasticity. Such adaptability allows a degree of reorganization and recovery after fetal exposure that is sufficient to preclude the observation of deficits in tests that examine overall global abilities. Deficits may also not be observed in infancy but might be seen as the child matures because of the different competencies and the increased complexity of neurocognitive processes available with increasing age reflecting normative nervous system development. As the child gets older, challenges can be presented that assess these new, qualitatively different cognitive processes—particularly those that require the integration of fundamental underlying specific domains of function.

B. *Global Measures*

As described earlier in this chapter, most prospective studies examining behavioral teratogenic effects have evaluated offspring during infancy. The test used most frequently to assess infants between 12 and 30 months is the Bayley Scales of Infant Development (BSID) (Bayley, 1986). This test focuses on the progression through developmental milestones with a heavy reliance on sensorimotor dexterity but provides relatively little information about which aspects of cognitive function are involved.

As the child gets slightly older, most of the reports in the teratogenic literature use IQ tests such as the McCarthy Scales of Children's Abilities (McCarthy, 1972) and the Standford–Binet Intelligence Scale (Thorndike *et al.,* 1986). Such tests provide standardized norms and enable the researchers to examine prenatal drug effects in certain broad areas of functioning. A decrement in task performance on these tests would require further domain-specific assessments to clarify and elaborate the nature of the deficit. A further interpretative issues arises when prenatal drug exposures are found to be associated with small decrements in IQ. The functional significance of, for example, 5 points (Streissguth *et al.,* 1989), is not understood and may be reflecting an impairment within a specific domain of cognitive function.

Although these instruments, which are comprised of a variety of problem-solving tasks, do require attentional and motivational competency (Krasnegor *et al.,* 1994), the nature of these broad-based tests and the method of administration makes it exceedingly difficult to separate noncognitive from cognitive aspects. For example, these tests consist of a highly structured set of tasks that focus the child on one activity at a time, severely limiting the opportunity to investigate self-regulatory behavior. Furthermore, these standard neuropsychologic assessments provide a relatively limited evaluation of functions that go beyond development milestones, well-learned information, or established cognitive sets. Although the traditional tests have their place in the assessment of the longer term consequences of drug exposure—particularly in the determination of the effects of a previously unstudied substance or of a sample being assessed at an age that has not been the subject of extensive previous research—they may not be capable of identifying some of the subtle domains of function that discriminate between the drug- and nondrug-exposed offspring. Lester and Tronick (1994) suggested that disorders of regulation (e.g., attention, affect) in infants may be particularly vulnerable to prenatal drug exposure but are not adequately assessed by such infant tests as the BSID (Bayley, 1986). Limitations of tests of global mental and motor development have been noted and discussed in the context of investigations of a number of drugs, including cocaine (Chasnoff *et al.,* 1992; Lester *et al.,* 1996) and marijuana (Fried, 1996).

Beyond the early infancy stage, the limitations imposed by both the content and structured nature of the standard IQ tasks becomes more problematic in interpreting possible behavioral teratologic effects. Abilities involved in regulatory mechanisms, goal setting, initiating activities, planning, social exchange, self-monitoring, and metacognition (i.e., knowledge about one's own cognitive processes and using that awareness to enhance/modify thinking) are among the neurobehavioral capacities that may fail to be challenged on the widely used tests of general measures of cognitive–behavioral competency.

C. Nonglobal Measures

Researchers have been increasing the use of tests that examine specific functions and processes rather than milestones, accumulated knowledge, or general abilities. In contrast to the traditional global tests mentioned previously, these neuropsychologic tests typically take a much narrower focus, examining processes that contribute substantially to individual differences in performance and that underlie cognitive and behavioral competencies.

From a teratologic viewpoint, these tests hold considerable promise, as a number of reports indicate that they are sensitive to prenatal drug effects. With infants, this approach has emphasized the assessment of encoding and information-processing skills (Bornstein and Sigman, 1986). The Fagan visual recognition test (Fagan and Singer, 1983), used to assess visual recognition memory and visual discrimination, is an example. The procedure involves the infant being shown two identical photos followed by a novel photo being paired with the familiar one. The normative preference, as expressed by the duration of visual gaze for the novel stimulus, assesses the ability to recall the previously seen stimulus and to discriminate it from the novel one. This procedure has been used to identify *in utero* cocaine effects in 13-month-old infants (Jacobson *et al.,* 1996).

The use of domain-specific and process-specific neuropsychologic tests to assess children beyond the infancy stage in longitudinal teratologic studies has increased quite noticeably in the past few years, reflecting a convergence of two factors. First, many of the longitudinal studies have been underway for a sufficient length of time so that the children are now old enough to be assessed with standardized neuropsychologic instruments that can examine aspects of behavior and cognition not possible in the very young. The second factor involves a shift from the global intellectual outcomes as the primary dependent variables of interest to a focus on underlying processes and specific domains that may be affected by maternal drug use.

There is accumulating evidence that a general description of cognitive abilities may not be capable of identifying nuances in neurobehavior that may discriminate between drug- and nondrug-exposed children. For example, as summarized by Hans (1992), a comparison of studies that contrasted opioid-exposed an nonopioid-exposed children during early childhood revealed little differences on standardized tests of cognitive ability. However, based on clinical observations and examinations of social and adaptive functioning, children of the drug-using women had increased problems in the area of activity, self-control, and tasks that required focused attention.

In young children exposed prenatally to cocaine, the few studies that have followed the offspring beyond early infancy have failed to find differences between exposed and nonexposed children on global measures of cognitive development. In a 30-month follow-up study of term and near-term children born to women of low socioeconomic status (Hurt *et al.,* 1995), *in utero* cocaine exposure was not associated with measures from the BSID. Chasnoff *et al.* (1992), using the Bayley scales, also found no cocaine effect at 24 months in a longitudinally followed sample of polydrug/cocaine, polydrug/noncocaine, and a control group. In the same sample at 3 years of age (Azuma and Chasnoff, 1993; Griffith *et al.,* 1994), no statistically significant effects attributable to cocaine on overall performance using the Stanford–Binet Intelligence Scale were found, although the cocaine exposed group scored significantly lower on verbal reasoning. Richardson *et al.* (1996) reported that, among 6-year-old offspring of light to moderate cocaine users participating in two longitudinal studies, no association was found between the prenatal drug exposure and growth, academic achievement, or intellectual ability assessed with the Standford–Binet Intelligence Scale, but deficits were noted in sustained attention on a computerized vigilance task.

The findings of the OPPS research examining the longer term consequences of *in utero* exposure to marijuana also emphasize the value of going beyond global tests of cognitive attainment. In reports arising from this longitudinal study, there have been no associations between maternal use of marijuana during pregnancy and global mental scores at 2 (Fried and Watkinson, 1988) and 3 (Fried and Watkinson, 1990) years of age. At 4 years of age (Fried and Watkinson, 1990), a number of cognitive domains distinguished the children of heavy (daily) marijuana users from the remainder of the sample, after adjusting for confounding variables. In particular, memory and verbal outcome measures derived from the McCarthy Scales (McCarthy, 1972) and a measure of receptive language derived from the Peabody Picture Vocabulary Test (Dunn and Dunn, 1981) were negatively associated with prenatal heavy marijuana use. These 4-year-old observations are similar to those reported by Day *et al.* (1994), who, in a high-risk, cohort of 3-year-old offspring of marijuana users, observed impairments on the short-term memory, verbal and abstract–visual reasoning subscales of the Stanford–Binet Intelligence Test. Griffith *et al.* (1994) also noted that, although use during pregnancy was not predictive of an overall lowered IQ in 3-year-old offspring of polydrug users, maternal marijuana use was related to a pooer performance on abstract–visual reasoning.

At this time, published data that focuses on prenatal marijuana exposure and cognitive functioning in offspring 4 years of age and older is limited to the OPPS sample. In 5- and 6-year-old children, maternal use of marijuana was not predictive of deficits on global cognitive skills (Fried *et al.,* 1992a). However, at the age of 6 years, prenatal habits were associated with increased omission errors on a vigilance task, possibly reflecting a deficit in sustained attention (Fried *et al.,* 1992b). Furthermore, parental ratings of children at that age indicated greater problems, particularly in the area of inattention and conduct.

Thus, up to the age of 6 years, prenatal marijuana exposure does not appear to be associated with deficits in global intelligence, but there are specific behaviors that do differentiate the offspring from comparison children that are manifest in early childhood. These behaviors include lessened self-regulatory abilities (possibly presenting as behavioral problems) and lower scores on tasks of sustained attention, abstract–visual reasoning, and facets of language and memory.

Not only was the impact of prenatal marijuana exposure not observed in global intelligence measures, but also the onset of effects occurred at 3 years of age or older. As has been suggested by several workers (Fried and Watkinson 1990; Chasnoff *et al.,* 1992; Hans, 1996) the emergence of drug-related deficits in older drug-exposed children may be due to the subtle effects that prenatal drug exposure has on the complex cognitive behavior that develop throughout childhood. It is by the examination of measures that extend beyond general intellectual outcomes in longitudinal, teratologic research that impairments not in evidence during infancy may be shown to emerge as the child matures.

D. Executive Function

1. Description

There has been an upsurge of interest in the psychologic construct of executive function—a term that connotes a "top-down" mental control process (Denckla, 1996). Athough there is no agreement in terms of its precise operationalization or exactly what ought to be included within executive functions, several general principles are widely acknowledged. These are discussed later in the chapter and are related to their applicability in longitudinal, behavioral teratologic research.

The term *executive functions* is a useful shorthand for a set of control processes that are domain-general (as opposed to domain-specific) and are involved in future oriented behaviors. The concept refers to mental control processes that include interference control, judgment, planning, and decision-making—in essence, an anticipatory, goal-oriented preparedness to act (Denckla, 1994).

Key components in executive functions include goal formulation (volition), planning, and effective performance (Lezak, 1995). This requires thinking of the future by conceptualizing where one is now, where one would like to be, and how to get there, including both the implementation and the appropriate cessation of the sequences of behavior needed to achieve a goal or solve a problem. Effective performance involves the identification of subgoals, conjuring up different response options, the weighing of the options in order to decide on their relative merits, and concomitant self-monitoring and self-correction. If done successfully, the regulation and modification of behavior yields a successful and efficient outcome.

Executive functioning encompasses processes that overlap with many aspects of attentional capacities (Barkley, 1997), including the ability to inhibit prepotent but inappropriate response tendencies, the ability to attend to several stimuli at once or to be sufficiently flexible to alter the focus of attention, resistance to distraction and interference, the ability to sustain behavior, and, when provided with appropriate feedback, the ability not to perseverate.

Working memory is frequently described as being a critical construct in executive functioning (Goldman-Rakic, 1987; Pennington *et al.,* 1996). Working memory has a limited capacity and its role is prospective rather than retrospective. Its purpose is to guide the selection of appropriate responses, even when the stimulus that gave rise to that information is no longer there. The action selection is predicted on taking into account various constraints such as the current social environment, objectives, and information retrieved from long-term memory. The "working" part of working memory is the "manipulation of representation systems" and occurs during the delay between the stimulus and the response (Denckla, 1996, p. 266). Rules or constraints initially explicitly made by others and then internalized are often verbally mediated. Thus, that aspect of language that bridges the temporal gap between memory and response and assists in the regulation of action is a key aspect of working memory (Barkley, 1997).

From both an interpretative and an experimental point of view it is important to recognize executive function as an overarching domain. As such, this construct serves to organize and integrate specific cognitive and output processes over some time interval. This on-line, integrative set of capacities are, by their very nature, comprised of subordinate cognitive operations. Thus the assessment of executive functioning's planning and organizational capacities is always potentially confounded by the competence in the underlying specific domains that are to be mentally manipulated. To separate the content competency from the overarching executive functioning processes is a critical aspect in the examination of this construct and is discussed later in this chapter.

2. Anatomical Substrates

Clinical and empirical evidence indicate that executive functions are primarily subserved by the prefrontal region of the brain (e.g., Fuster, 1989; Dennis, 1991; Stuss and Benson, 1994; Lezak, 1995), although, as has been emphasized by many workers (e.g., Denckla, 1994; Tranel *et al.,*1994), not all executive functions are prefrontally mediated, nor are all prefrontal functions consistent with the construct of executive function. The prefrontal region is anatomically complex, with various subregions and with reciprocal connections to other regions of the frontal lobes, posterior association cortices, and subcortical structures (Fuster, 1989). These connections underlie the integrative nature of executive functioning and its involvement with subordinate cognitive processes. Most information about both prefrontal lobe functioning and executive functions in humans is derived from behavioral, neuropsychologic investigations of individuals with frontal lobe damage (Luria 1966; Stuss and Benson, 1984; Lezak, 1995). As Stuss (1992) emphasized, executive function relates to the psychologic concept of prefrontal system function, and thereby involves not only the anatomical maturation of the frontal lobes, but also the integrative demands on multiple brain regions that come into play when executive control functions are operational.

Researchers investigating the developmental course of executive functioning usually link this domain in children to the maturation of the prefrontal lobes. Morphologic development in the prefrontal cortex continues after birth and, based on studies of myelinization, is among the last areas to develop, with full maturization being reached around puberty (Yakovlev and Lecours, 1967). Different areas within the prefrontal lobes mature at different times (Goldman-Rakic, 1987; Grattan and Eslinger, 1991; Stuss, 1992) and, functionally, can be differentiated (Grattan and Eslinger, 1991; Fuster, 1989; Petrides *et al.,* 1993; Stuss and Benson, 1994; Lezak, 1995; Pennington and Ozonoff, 1996). It is, therefore, not surprising that executive functioning, which is often viewed as a marker of prefrontal functioning, is considered to be a multistage process, with various functions maturing at different times in different ways (Passler *et al.,* 1985; Dennis, 1991; Levin *et al.,* 1991; Welsh *et al.,* 1991). The intimate association between executive function and the prefrontal area has several important ramifications for long-term outcomes in behavioral teratologic research.

3. Stages of Development

Studies that have been undertaken to examine the developmental course of executive function have assessed children from nonclinical (Passler *et al.,* 1985; Becker *et al.,* 1987; Levin *et al.,* 1991; Welsh *et al.,* 1991) and clinical (Chelune *et al.,* 1986; Welsh *et al.,* 1990; Weyandt and Willis, 1994) samples. A mixture of adapted adult neuropsychologic frontal lobe measures (e.g., Passler *et al.,* 1985) and children's developmental measures (e.g., Levin *et al.,* 1991; Welsh *et al.,* 1991) have been used. In general, the results have supported and emphasized the multistage aspect of executive function in that performance has always been noted to vary depending on the age of the child and the nature of the task. Early reports, suggesting that executive functions do not appear until late childhood (e.g., 10 years—Golden, 1981), were based on children's performance assessed via adult neuropsychologic measures and likely reflect the instruments used rather than the capabilities and competencies of young children. When developmentally appropriate measures are used, precursors of executive function can be seen in infants and toddlers.

The ability to anticipate and plan as demonstrated in such tasks as objective permanence and objects search (Piaget, 1954) can be thought of as an early emergence of precursors of executive function. Similarly, the domain of self-control assessed by tasks that evaluate the infant's ability to inhibit a response to a desired object (Vaughn *et al.,* 1984) may be viewed as an early manifestation of executive function. Self-control behaviors have been observed in infants as young as 18 months, with the capacity and stability of this behavior increasing with age.

The use of the Fagan test, which, as described earlier, detected poorer recognition memory and information processing in the 13-month-old offspring of heavy cocaine users (Jacobson and Jacobson, 1996), is an example of an infant test that places demands on such aspects of executive function as attention and working memory. The results of this study demonstrate the sensitivity of executive functioning to prenatal drug exposure and the recognition that behaviors (or at least precursors thereof) traditionally thought only to be detectable in later childhood can be investigated in infants. Early identification enhances the opportunity to consider such issues as stability of observations, the role of environmental factors, the effect of intervention, and the predictive validity of early observations.

As the child matures, three ages have been identified (Passler *et al.,* 1985; Levin *et al.,* 1991; Welsh *et al.,* 1991, Stuss, 1992) in which major gains in performance in particular aspects of executive function have been noted. In children, the earliest appearing executive functions are simple organized strategic and planned behavior, which involve recognition memory, response flexibility, verbal regulation of motor responses, the ability to resist distractions, and the capacity to inhibit maladaptive responding (Luria, 1959; Passler *et al.,* 1985; Welsh *et al.,* 1991). Significant improvement is noted in these behaviors as the children research 6 years of age. By the time a child is 10 years old, considerable competence is evident in tasks that required hypothesis testing and impulse control, and between the ages of 10 and 12 years, adult competency is reported in tests of set maintenance, impulse control, and the inhibition of inappropriate perseverative responding (Passler *et al.,* 1985; Chelune and Thompson, 1987; Levin *et al.,* 1991; Welsh *et al.,* 1991). Complex planning skills, which often involve considerable demands on working memory, motor sequencing, and the ability to generate words, are areas of executive functioning that do not reach adult levels until adolescence (Levin *et al.,* 1991; Welsh *et al.,* 1991).

There is considerable congruence with the emergence of particular aspects of executive function and Piaget's (1954) description of the cognitive content of children at different ages. Increasing language facility during the preoperational stage (2–7 years) permits simple conceptualization and verbally mediated self-control. During Piaget's concrete operation stage (7–11 years), emphasis is placed on the development of the ability to perform logical analysis, an understanding of more complex cause-and-effect relations, and the ability to monitor the efficacy of one's own behavior. During the final stage of cognitive development, formal operations (12 years and older), essentially adult forms of logic and symbolic representation are used to formulate alternative hypotheses and to test these alternatives.

In essence, these developmental studies (plus the protracted, nonuniform rate of prefrontal lobe maturation) underscore the fact that executive functioning is not a uniform domain but consists of several components that have differential developmental trajectories, emerging and maturing in stagelike fashion throughout childhood. Principal component analysis also supports viewing executive function as being a domain of considerable range that is not of a singular nature, but rather encompasses a number of distinct cognitive processes. Both Welsh *et al.* (1991) and Levin *et al.* (1991) observed, in their batteries examining executive function in children, that the measures clustered into three different factors. Although the two batteries overlapped on approximately only half of the tasks there is a high degree of agreement between the studies. Both labelled one factor as planning, which included primarily a complex motor sequencing task. A second factor in both studies included perseveration in hypotheses-testing tasks and impulse-control variables. The third factor in the two investigations included verbal fluency. In both studies, the tasks that loaded on this last factor had the most protracted course of development.

4. Involvement of IQ and Effortfulness

Typically, injury to the frontal lobes during adulthood (Fuster, 1989) does not affect global intellectual abilities. Furthermore, lesions that occur during childhood do not affect the development of cognitive performance on standardized IQ tests (Grattan and Eslinger, 1991).

As discussed by several authors (e.g., Duncan *et al.,* 1995; Pennington and Ozonoff, 1996) frontal lobes may not play a major role in the maintenance of accumulated information ("crystallized intelligence") assessed on many of the subtests of widely used IQ measures. However, this region of the brain may be of key importance in that aspect of competence ("fluid intelligence") that involves the ability to reason logically and abstractly.

Consistent with the linkage between the prefrontal region of the brain and executive functioning are the findings that the cognitive control processes involved in executive function are also relatively independent of IQ-measured intelligence of children (Welsh and Pennington, 1988; Welsh *et al.,* 1991). This relative independence may be important in the understanding of some outcomes in behavioral teratologic research in which, as described earlier in this chapter, several drugs appear to have little or no impact on general IQ but do have an impact on other areas of cognitive functioning. Executive functioning may serve to distinguish between intelligence as a capacity to engage in adaptive goal-directed behavior from intelligence as measured by global performance on standard psychometric intelligence tests. This is not to say, however, that there is no shared variance between general IQ and executive function, particularly in subtests that involve "fluid" intelligence, including timed problem-solving tasks and tasks that do not involve familiar, highly ingrained information and well-learned cognitive sets (Denckla, 1994; Duncan *et al.,* 1995).

A further note concerning the relationship between developmental maturity and executive function is that, as a child matures, some tasks that required planning, judgment, and other aspects of executive function lose their effortfulness, becoming somewhat automatized, and move out of the realm of executive function (Denckla, 1996). Thus, in longitudinal work with children, it is inappropriate to attempt to use a particular battery of executive function tasks for all age levels. Furthermore, the level of general intelligence also has an impact on the issue of the ease of certain tasks and must be considered in the assessment of executive functioning. The underlying processes for a given task are not the same for same-aged children who vary considerably in general intelligence. High IQ may result in certain tasks being too easy to invoke executive functions (Denckla, 1996; Pennington *et al.,* 1996).

5. Tasks Used for Assessment

Several authors have provided detailed descriptions of a large array of research and clinical tasks designed to assess executive function and its development (Denckla, 1994; Tranel *et al.,* 1994; Pennington and Ozonoff, 1996). In order to present an overview of the approach taken to investigate this construct and to highlight some of the issues that arise with the instruments, a few of the most widely used tests with young children and the aspects of executive function they are purported to assess are described in the following paragraphs.

Virtually every researcher who examines the development of executive function in school-aged children includes the Wisconsin Card Sort Test (Grant and Berg, 1948; Heaton, 1981), although several reviews of frontal lobe functions have noted that performance is not necessarily impaired after frontal lobe damage (Mountain and Snow, 1993; Stuss *et al.,* 1994; Tranel *et al.,* 1994). This test requires the ability to infer correct sorting strategies, to use working memory, and to shift cognitive sets. The task involves the sorting of a deck of cards (manually or, more typically in recent studies, on a computer monitor) according to one of three stimulus dimensions (form, color, or number). The subject is required to form hypotheses regarding the relevant sorting rule, to test the principle chosen, and to modify the hypothesis in response to positive or negative feedback. After 10 consecutive correct sortings have been accomplished, the sorting rule is changed without warning. Several outcomes can be obtained, including the number of categories achieved and the number of perseverative responses (persistence in responding to a particular sorting rule in spite of negative feedback). In children younger than 8 years, a somewhat analagous task, the Category Test (Reed *et al.,* 1965) can be used to examine both the ability to deduce classification strategies and the ability to shift the principle when it is no longer effective.

Consistent with executive functioning and frontal lobe development, the Wisconsin test has consistently been found to be sensitive to developmental maturation (e.g., Chelune and Thompson, 1987; Levin *et al.,* 1991; Welsh *et al.,* 1991; Weyandt and Willis, 1994) with adult competency achieved at about 10–12 years of age (Chelune and Baer, 1986; Welsh *et al.,* 1991). Furthermore, perseveration scores have been reported to be unrelated to intelligence test scores (Welsh *et al.,* 1991). However, this assessment tool (as well as the Category Test) involves not only the executive functions of hypotheses testing, representational memory, and shifting of sets, but also multiple underlying processes including visual discrimination and visual–motor responding that are not executive functions (e.g., Pennington *et al.,* 1996). The interpretation that deficits in the performance of the Wisconsin task or the Category Test is a manifestation of deficits in executive function can only be made after an assessment of these underlying specific domains.

Disk transfer tasks are also a widely used means of examining aspects of executive function that involve problem solving, working memory, future planning, motor sequencing, and concept formation within particular constraints. The two most commonly reported tasks of this type are the Tower of Hanoi (Borys *et al.,* 1982) and the Tower of London (Shallice, 1982). In the former, the subject must duplicate the experimenter's sample of discs arranged in the form of a tower on one of three vertical pegs. The discs differ in size and the subject, beginning with a standard initial position of the discs on one peg, is asked to transfer the discs to the third peg following the constraints that a larger disc may not be placed on top of a smaller one and only one disc can be moved at a time. Tasks using three or four discs (thereby varying in difficulty) have been given to children of different ages. These tasks can be presented to preschoolers by means of a cover story (Klahr and Robinson, 1981) in which, for example, the pegs are referred to as "trees" and the discs (in the three-disc problem) as "daddy, mommy, and baby monkeys" who are jumping from tree to tree.

The Tower of London task is similar, to some extent, to the Tower of Hanoi in terms of the executive functions involved, but the constraints are different. Three colored beads, which can be placed on pegs of three different heights, make up the apparatus and the child is asked to copy the examiner's three-bead pattern in a specified number of moves. The easiest problem requires two moves, whereas the most difficult requires five.

The differences in constraint between the Tower of Hanoi and Tower of London disc transfer tasks may require different aspects of executive functioning. Goel and Grafman (1995) argue that the Tower of Hanoi, unlike the Tower of London, may not place as much demand on a planning-ahead component because moves can be corrected or undone. Rather, the Tower of Hanoi depends heavily on the awareness of subgoals in order to achieve the overall goal. In this task the subgoals entail, on occasion, a counterintuitive backward move requiring an inhibition of the prepotent response that would satisfy the global goal.

Both of these disc-transfer tasks show significant developmental trajectories as seen by the change in competencies over ages as the tasks are made more difficult. For example, Welsh *et al.* (1991) noted that on the three-ring, but not on the four-ring version, of the Tower of Hanoi, children achieved adult levels of performance by 12 years. The more difficult the task, the greater the demand on working memory, which improves in the maturing child.

In addition to the executive functions involved in the successful performance of these tasks, underlying competency in memory and visuospatial skills are also required. Thus, in order to conclude an observed impairment in disc transfer tasks is due to executive functioning and not to a dysfunction in the requisite content domains, appropriate assessments must be undertaken. An example of this can be found in the report by Shallice (1982), who demonstrated in frontal lobe patients deficits in the executive functioning component of the Tower of London task, whereas abilities in the visuospatial and general reasoning domains were not impaired. Similarly, Goel and Grafman (1995), using adult patients with prefrontal cortical lesions, observed impairments on a Tower of Hanoi task that were not due to a decline in memory or general intelligence.

A third group of tasks frequently reported in the assessment of executive function fall under the description of verbal fluency. Tests of verbal fluency assess the ability to generate a maximum number of responses appropriate to a given set of stimulus conditions and constraints. The goal of generating a maximum number of words within a time limit places demands on focused attention and rapid response generation. Furthermore, these tests are sensitive to perseverative tendencies as the tasks require the nonrepeating of words. Two types of tests are typically employed. In one, there is the constraint that all words must begin with a specific consonant (Spreen and Strauss, 1991; Welsh *et al.,* 1991, adapted from McCarthy, 1972), whereas in the semantic condition (Benton and Hamsher, 1976) the words must all describe a particular category (e.g., animals). Of the two variations of verbal fluency tasks, the one in which the words must begin with a specified consonant is the more abstract and provides less structure. Increased verbal fluency is consistently noted in both of these tasks with age, and adult equivalence is not reached prior to adolescence (Welsh *et al.,* 1991).

In the verbal fluency tasks, as with other assessments of executive functioning, the abilities in the underlying content domain—in this case general vocabulary—must be established. Such as assessment, in a nonclinical sample, was reported by Welsh *et al.* (1991). In this study, verbal fluency was not associated with verbal IQ, thus supporting the discriminant validity of the executive function task.

It is evident that in assessing the overlying integrative properties of executive function there is the necessity of determining possible deficits in the underlying domain-specific input and/or output (Denckla, 1996). As the objective of longitudinal behavioral teratologic research is the identification of putative prenatal drug effects, the need to be able to discriminate between the deficiencies in the underlying content versus the overarching executive functions takes on a particularly critical role.

6. Aspects of Executive Function May Distinguish among Different Prenatal Drug Exposures

In the OPPS, differential cognitive outcomes have consistently been found to be associated with prenatal cigarette compared to prenatal marijuana exposure. As described earlier in this chapter, in contrast to *in utero* exposure to cigarettes, prenatal marijuana use was not associated with deficits in global intelligence in offspring up to the age of 6 years, although a number of behaviors consistent with executive function did appear to differentiate the marijuana-exposed group from control children. In the OPPS, as part of a large assessment battery, cognitive performance was examined in 9–12 year olds exposed prenatally to marijuana and/or cigarettes (Fried *et al.,* in press). When this battery was originally created, the possible link between prenatal marijuana exposure and executive function in offspring was not recognized and was not part of the consideration in the choice of assessment instruments. Fortuitously, within its mix of global intelligence and specific cognitive measures, the battery did contain tasks that have been used traditionally to assess executive function or that assess behaviors thought to reflect components of executive function. In addition, the battery included tests that permitted the examination of the underlying content domains of those tests that were considered to assess executive functions.

Included in the tests was the Weschler Intelligence Scale for Children (WISC) (Weschler, 1991), a consonent oral fluency task (Spreen and Strauss, 1991), a measure of auditory working memory in which the children were read blocks of sentences, required to fill in the missing word at the end of each sentence in the block and then remember, within each block, the missing words (Siegel and Ryan, 1989); the Category Test (Reed *et al.,* 1965) to examine visual analysis and hypothesis testing, computerized assessment of impulsivity using a differential reinforcement of a low rate of responding schedule, sustained attention on a continuous performance task (Gordon and McClure, 1984), and a tactile performance task (Reitan and Davison, 1974). This last test required a blindfolded subject to place blocks of varying shapes into their proper place on a form board, first with the dominant hand, then with the nondominant hand, and finally with both hands, thus assessing working memory and nonvisual motor sequencing.

Competency in the basic content domains of the executive function tasks outlined previously was assessed by performances on the following tasks. The Pegboard Test (Reitan and Davison, 1974), which required the child to place keyhole-shaped metal pegs into matching holes as quickly as possible, was used as a measure of motor coordination, speed, and accuracy. Visual perception and motor proficiency was examined by use of the Developmental Test of Visual–Motor Integration (Beery, 1982), which required the child to copy geometric forms in an untimed fashion. Vocabulary levels were ascertained from subtests on the WISC.

Consistent with results obtained at earlier ages (Fried and Watkinson, 1988, 1990; Fried *et al.,* 1992a), a dose-dependent relationship was found between prenatal cigarette exposure and the global intelligence score and all the WISC subtests. Again, extending the auditory–prenatal cigarette association described earlier, the verbal subtests of the WISC test maximally discriminated among levels of *in utero* cigarette exposure. Furthermore, among the non-WISC tests, only the two auditory-based assessments (fluency and auditory working memory) were negatively associated with maternal cigarette use. These observations persisted after controlling for potential confounds, including secondhand smoke.

In contrast to the associations noted with prenatal cigarette use, exposure to marijuana *in utero* was not associated with either global intelligence or the verbal subtests. Only the Block Design and Picture Completion subtests of the WISC appeared to suffer a negative impact by prenatal marijuana use. The Block Design subtest, in which subjects are required to assemble blocks to form a design identical to one presented in a picture, makes demands on perceptual organization, spatial visualization, and abstract conceptualization. The Picture Completion subtest requires subjects to identify a missing portion of an incompletely drawn picture, and tests the ability to differentiate essential from nonessential details. The negative relationship between these two subtests and maternal marijuana use is consistent with the findings (Day *et al.,* 1994; Griffith *et al.,* 1994), described earlier in this chapter, of poorer abstract–visual reasoning in 3-year olds exposed *in utero* to marijuana.

The role of Full Scale IQ on the relationship between the Block Design and Picture Completion subtests and the two drugs in question was examined by using IQ as a covariate. With the smoking groups, when IQ was used as a covariate, the negative relationship between the two subtests and maternal smoking completely disappeared, suggesting that the Full Scale IQ mediated the relationship with the two subtests. With the marijuana groups, the influence of the overall IQ was quite different. When the Full Scale IQ was used as a covariate with the Block Design and Picture Completion subtests, the association with marijuana was only slightly altered, suggesting that the association of *in utero* marijuana exposure with Block Design and Picture Completion was, to a considerable degree, independent of general overall intelligence.

Of the six non-WISC tests thought to assess aspects of executive function, the two that were associated with prenatal marijuana were the Category Test and the impulse control test. The former, like the Block Design and the Picture Completion of the WISC, requires visual analysis and hypothesis testing. The impulse control test can be interpreted as an assessment of the inhibition of prepotent responses. No association between maternal marijuana use and the performance of the children on the content-specific motor and visual perception tasks was noted.

The impact of prenatal marijuana use on the convergence of the executive functions of visual analysis and response inhibition is strikingly similar to one of the three factors identified by Welsh *et al.* (1991) in her factor analytical approach to executive function described earlier in this chapter. This factor, labeled *hypothesis testing and impulse control,* was defined by tasks requiring visual analysis and inhibition of prepotent responses. Tests that loaded onto the other two factors of Welsh *et al.* (1991) were based on verbal fluency, motor sequencing, and working memory tasks. In the OPPS, these aspects of executive functioning were found not to be related to maternal marijuana use in the 9–12-year-old offspring.

These OPPS results serve to highlight two factors. The marijuana findings, in which only certain aspects of executive function appear associated with *in utero* exposure, are consistent with the anatomic and developmental literature cited earlier indicating that executive functioning is not a uniform composite domain. Furthermore, the differential neurocognitive associations of maternal cigarette use versus maternal marijuana use emphasize the value of assessing different dimensions of cognitive functioning.

7. Dysfunction in Social Behaviors

The Wisconsin, Disk, and Verbal Fluency tasks described earlier are examples of extensively used instruments that have been employed to examine such executive functions as hypothesis generation and testing under constraints, appropriate perseveration and flexibility in cognitive sets, working memory, motor sequencing, inhibition and interference control, and different aspects of attention. One sphere, however, that is not captured by these neuropsychologic and experimental tasks involves social competency. Social disability is one of the most distinctive characteristics of both adults and children with frontal lobe damage (Grattan and Eslinger, 1991), with executive function deficits in such areas as self-control, self-monitoring, identification of subgoals, and empathy playing a major role in the dysfunction. It is therefore consistent with the proposal of impaired executive functioning as a teratogenic outcome that several workers have identified altered behaviors in interactive and interpersonal situations in children exposed prenatally to drugs.

In an investigation of symbolic play in 13-month-old (Rodning *et al.,* 1989) and 24-month-old toddlers (Beckwith *et al.,* 1994), children exposed prenatally to cocaine, phencyclidine, heroin, and/or methadone were less likely than matched children to partake in representational play, and were more likely to exhibit disorganized, poorly controlled play, such as the scattering and throwing of toys. Strauss *et al.* (1979), assessing 5-year-old children exposed prenatally to opiates, observed no differences compared to a control group in IQ but, based on clinical ratings, the opiate-exposed children showed more task-irrelevant behavior. Wilson *et al.* (1979) reported that the parents of 3- to 6-year-old children exposed prenatally to opiates rated their offspring, in contrast to parents of nonexposed children, as having greater problems in self-adjustment, social adjustment, uncontrollable temper, impulsiveness, aggressiveness, and making and keeping friends. Very similar observations were noted in 6-year-old methadone-exposed children based on data gathered from teachers and mothers (Rosen and Johnston, 1986 cited in Hans, 1992). Griffith *et al.* (1994), in a follow-up study of 3-year-old children exposed prenatally to a variety of drugs, concluded that the blind examiner's clinical observations and judgments are of particular importance in the assessment of any subtle effects that prenatal drug exposure may have. Overall, the types of impairments noted within the broad domain of social discourse and interaction, as well as such developmental psychopathologies as attention deficit hyperactivity disorder (Pennington and Ozonoff, 1996), can be interpreted as deficits within the construct of components of executive function. Viewed in this manner and recognizing the importance of separating these higher order integrative capacities from underlying competencies may yield useful insights into this realm of *in utero* drug effects.

III. Summary and Conclusions

To understand the behavioral effects of prenatal drug exposure beyond the infant years, the longitudinal follow-up study is the main weapon in our research methodologic arsenal. The ability to examine particular outcomes in a repeated fashion during the maturation of the offspring, to assess the possible stability, attenuation, or potentiation of findings over the years, and to better identify potentially confounding factors are the keys to unraveling (in the sense of both identifying and understanding) the longer term consequences of prenatal drug exposure. However the longitudinal teratologic approach comes at a high cost—literally and figuratively. The level of commitment required on the part of the investigative team, the families and the funding agencies involved is enormous. From a methodologic perspective, this research strategy is "costly," as it is replete with severe limitations that are couched primarily in terms of experimental control (e.g., polydrug use, postnatal environment factors), resulting in consequent interpretative issues. However, by the very recognition of these issues and by implementing, within the longitudinal research design, appropriate outcome measurements and statistical procedures, these "costs" can be substantially, although certainly not entirely, reduced.

A failure to find robust drug effects on infant behavioral development tests must not be interpreted as making later evaluation unnecessary. There are a myriad of reasons why effects may not be present or detected during infancy, ranging from the limitations of the infant to the limitations of the tests employed. Assessments that purport to measure the mental control processes subsumed under the term *executive functions* (or its precursors) may provide a very useful approach in the longitudinal behavioral teratologic field. This overarching process, with its association with the late maturing prefrontal region, makes this construct of higher order cognitive–behavioral capacities of considerable interest and, potentially, of major interpretative significance. By taking advantage of the information available in the executive function literature, including the protracted, stagelike normative development from late infancy onwards, the clinical and neuropsychologic procedures used to measure various aspects, and the nonunitary nature of this construct, fruitful progress may be made into the identification and understanding of the putative effects of *in utero* drug exposure on the highest aspects of human cognition.

Acknowledgments

The OPPS work described has been and continues to be supported by grants from NIDA. I thank my long-time research associates B. Watkinson, R. Gray, and H. Linttell. The stimulating and fruitful discussions with S. Gmora about executive function are sincerely appreciated.

References

Azuma, S. D. and Chasnoff, I. J. (1993). Outcome of children prenatally exposed to cocaine and other drugs. *Pediat.* **92,** 396–402.

Barkley, R. A. (1997). Behavioral inhibition, sustained attention, and executive functions: constructing a unifying theory of ADHD. *Psychol. Bull.* **121,** 65–94.

Bayley, N. (1986). *Bayley Scales of Infant Development.* Psychological Corporation, New York.

Becker, M. G., Isaac, W. and Hynd, G. (1987). Neuropsychological development of non-verbal behaviors attributed to 'frontal lobe' functioning. *Dev. Neuropsychol.* **3,** 27–298.

Beckwith, L., Rodning, C., Norris, D., Phillipsen, L., Khandabi, P., and Howard, J. (1994). Spontaneous play in two-year-olds born to substance abusing mothers. *Infant Health J.* **15,** 189–201.

Beery, K. (1982). *Developmental Test of Visual-Motor Integration.* Modern Curriculum Press, Cleveland.

Benton, A. L. and Hamsher, K. (1976). *Multilingual Aphasia Examination.* AJA Associates, Iowa City.

Bernstein, J. H. (1994). Assessment of developmental toxicity: neuropsychological batteries. *Environ. Health Perspec.* **102,** 141–144.

Bornstein, M. H. and Sigman, M. D. (1986). Continuity in mental development from infancy. *Child Dev.* **57,** 251–274.

Borys, S. V., Spitz, H. H. and Dorans, B. A. (1982). Tower of Hanoi performance of retarded young adults and nonretarded children as a function of solution length and goal state. *J. Exp. Child Psychol.* **33,** 87–110.

Broman, S. (1984). The Collaborative Perinatal Project: an overview. In (1984) *Handbook of Longitudinal Research. Vol. 1. Birth and Childhood Cohorts.* S. A. Mednick, M. Harway, and K. M. Finello, Eds. Praeger, New York. pp. 185–215.

Chasnoff, I. J., Griffith, D. R., Freier, C. and Murray, J. (1992). Cocaine/polydrug use in pregnancy: two-year follow-up. *Pediat.* **89,** 284–289.

Chelune, G. J. and Baer, R. A. (1986). Developmental norms for the Wisconsin Card Sorting Task. *J. Clin. Exp. Neuropsychol.* **8,** 219–228.

Chelune, G. J., Ferguson, W., Koon, R. and Dickey, T. O. (1986). Frontal lobe disinhibition in attention deficit disorder. *Child Psychiatry Human Dev.* **16,** 221–234.

Chelune, G. J. and Thompson, L. L. (1987). Evaluation of the general sensitivity of the Wisconsin Card Sorting Test among younger and older children. *Dev. Neuropsychol.* **3,** 81–90.

Cunningham, A. E. and Stanovitch, K. E. (1990). Assessing print exposure and orthographic processing skill in children: a quick measure of reading experience. *J. Educ. Psychol.* **82,** 733–740.

Day, N. L., Richardson, G. A., Goldschmidt, L., Robles, N., Taylor, Stofer, D. S., Cornelius, M. D. and Geva, D. (1994). Effect of prenatal marijuana exposure on the cognitive development of offspring at age three. *Neurotoxicol. Teratol.* **16,** 169–175.

Denckla, M. B. (1994). Measurement of executive function. In *Frames of Reference for the Assessment of Learning Disabilities: New Views on Measurement Issues.* (S. R. Lyon, Ed. Paul H. Brookes, Baltimore. pp. 117–142.

Denckla, M. B. (1996). A theory and model of executive function. In *Attention, Memory, and Executive Function.* G. Lyon and N. A. Krasnegor, Eds. Paul H. Brookes, Baltimore. pp. 263–278.

Dennis, M. (1991). Frontal lobe function in childhood and adolescence: a heuristic for assessing attention regulation, executive control and the intentional states important for social discourse. *Dev. Neuropsychol.* **7,** 327–358.

Duncan, J., Burgess, P. and Emslie, H. (1995). Fluid intelligence after frontal lobe lesions. *Neuropsychologia* **33,** 261–268.

Dunn, L. M. and Dunn, L. M. (1981). *Peabody Picture Vocabulary Test—Revised.* American Guidance Service, Circle Pines.

Fagan, J. F. and Singer, L. T. (1983). Infant recognition memory as a measure of intelligence. In *Advances in Infancy Research.* L. P. Lipsitt, Ed. Vol. 2, Ablex, Norwood. pp. 31–72.

Fried, P. A. (1991). Who is it going to be? Subject selection issues in prenatal drug exposure research. In *Methodological Issues in Epidemiological, Prevention, and Treatment Research on the Effects of Prenatal Drug Exposure on Women and Children* Research Monograph 117 M. M. Kilbey and K. Asghar, Eds. U.S. Department of Health & Human Services, National Institute on Drug Abuse, Rockville. pp. 121–136.

Fried, P. A. (1992). Clinical implications for smoking: determining long-term teratogenicity. In *Maternal Substance Abuse and the Developing Nervous System* I. S. Zagon and T. A. Slotkin, Eds. Academic Press, San Diego. pp. 77–96.

Fried, P. A. (1993). Prenatal exposure to tobacco and marijuana: effects during pregnancy, infancy and early childhood. *Clin. Obstet. Gynecol.* **36,** 319–337.

Fried, P. A. (1995). The Ottawa Prenatal Prospective Study (OPPS): methodological issues and findings—it's easy to throw out the baby with the bath water. *Life Sciences* **56,** 2159–2168.

Fried, P. A. (1996). Behavioral outcomes in preschool and school-age children exposed prenatally to marijuana: a review and speculative interpretation. In *Behavioral Studies of Drug-Exposed Offspring: Methodological Issues in Human and Animal Research.* C. L. Wetherington, V. L. Smeriglio, and L. P. Finnegan, Eds. National Institute on Drug Abuse Research Monograph 164. DHHS Pub. No. 96-4105. Washington, DC. pp. 242–260.

Fried, P. A. and Makin, J. E. (1987). Neonatal behavioral correlates of prenatal exposure to marijuana, cigarettes and alcohol in a low-risk population. *Neurotoxicol. Teratol.* **9,** 1–7.

Fried, P. A., O'Connell, C. M., and Watkinson, B. (1992a). 60- and 72-month follow-up of children prenatally exposed to marijuana, cigarettes, and alcohol: cognitive and language assessment. *J. Dev. Behav. Pediatr.* **13,** 383–391.

Fried, P. A. and Watkinson, B. (1988). 12- and 24-month neurobehavioral follow-up of children prenatally exposed to marijuana, cigarettes and alcohol. *Neurotoxicol. Teratol.* **10,** 305–313.

Fried, P. A. and Watkinson, B. (1990). 36- and 48-month neurobehavioral follow-up of children prenatally exposed to marijuana, cigarettes and alcohol. *J. Dev. Behav. Pediatr.* **11,** 323–343.

Fried, P. A., Watkinson, B. and Gray, R. (1992b). A follow-up study of attentional behavior in 6-year-old children exposed prenatally to marijuana, cigarettes and alcohol. *Neurotoxicol. Teratol.* **14,** 299–311.

Fried, P. A., Watkinson, B. and Gray, R. Differential effects on cognitive functioning to 9 to 12-year-olds prenatally exposed to cigarettes and marijuana. *Neurotoxicol. Teratol.* in press.

Fried, P. A., Watkinson, B., and Siegel, L. S. (1997). Reading and language in 9–13 year olds prenatally exposed to cigarettes and marijuana. *Neurotoxicol. Teratol.* **19,** 171–183.

Fuster, J. M. (1989). *The Prefrontal Cortex: Anatomy, Physiology, and Neuropsychology of the Frontal Lobe.* 2nd ed. Raven Press, New York.

Goel, V. and Grafman, J. (1995). Are the frontal lobes implicated in "planning" functions? Interpreting data from the Tower of Hanoi. *Neuropsychologia,* **33,** 623–642.

Golden, C. J. (1981). The Luria–Nebraska Children's Battery: theory and formulation. In *Neuropsychological Assessment and the School-Age Child.* G. W. Hynd and J. E. Obrzut, Eds. Grune & Stratton, New York. pp. 277–302.

Goldman-Rakic, P. S. (1987). Circuitry of primate prefrontal cortex and regulation of behavior by representational memory. In *Handbook of Physiology. The Nervous System: Higher Functions of the Brain.* F. Plum, Ed. American Physiology Association, Bethesda. pp. 373–417.

Gordon, M., and McClure, D. F. (1984). *Gordon Diagnostic System: Interpretative Supplement.* Clinical Diagnostics, Golden.

Grant, D. S., and Berg, E. A. (1948). A behavioral analysis of degree of reinforcement and ease of shifting to new responses in a Weigl-type card sorting problem. *J. Exp. Psychol.* **38,** 404–411.

Grattan, L. M., and Eslinger, P. J. (1991). Frontal lobe damage in children and adults: a comparative review. *Dev. Neuropsychol.* **7,** 283–326.

Griffith, D. R., Azuma, S. D. and Chasnoff, I. J. (1994). Three-year outcome of children prenatally exposed to drugs. *J. Am. Acad. Child Adolesc. Psychiatry* **33,** 20–27.

Griffith, D. R. and Freier, C. (1992). Methodological issues in the assessment of the mother–child interactions of substance-abusing women and their children. In *Methodological Issues in Controlled Studies on Effects of Prenatal Exposure to Drugs of Abuse.* M. M. Kilbey and K. Asghar, Eds. National Institute on Drug Abuse Research Monograph 114. DHHS Pub. No. (ADM) 91-1837. Supt. of Docs., U.S. Govt. Print. Off. Washington, DC. pp. 228–247.

Hans, S. L. (1992). Maternal opioid drug use and child development. In *Maternal Substance Abuse and the Developing Nervous System.* I. S. Zagon and T. A. Slotkin, Eds. Academic Press, San Diego. pp. 177–213.

Hans, S. L. (1996). Prenatal drug exposure: behavioral functioning in late childhood and adolescence. In *Behavioral Studies of Drug-Exposed Offspring: Methodological Issues in Human and Animal Research.* C. L. Wetherington, V. L. Smeriglio and L. P. Finnegan, Eds. National Institute on Drug Abuse Research Monograph 164. DHHS Pub. No. 96-4105. Washington, DC. pp. 261–276.

Heaton, R. K. (1981). *Wisconsin Card Sorting Test Manual.* Psychological Assessment Resources, Odessa.

Howard, J. (1992). Subject recruitment and retention issues in longitudinal research involving substance-abusing families: a clinical services context. In *Methodological Issues in Epidemiological, Prevention, and Treatment Research on the Effects of Prenatal Drug Exposure on Women and Children.* Research Monograph 117 M. M. Kilbey and K. Asghar, Eds. U.S. Department of Health & Human Services, National Institute on Drug Abuse, Rockville. pp. 155–165.

Hurt, H., Brodsky, N. L., Betancourt, L., Braitman, L. E., Malmud, E. and Gianneta, J. (1995). Cocaine-exposed children: follow-up through 30 months. *J. Dev. Behav. Pediat.* **16,** 29–35.

Hutchings, D. E. (1993). The puzzle of cocaine's effects following maternal use during pregnancy: are there reconcilable differences? *Neurotoxicol. Teratol.* **15,** 281–286.

Jacobson, L. J. and Jacobson, S. W. (1996). Prospective, longitudinal assessment of developmental neurotoxicity. *Environ Health Perspect.* **104,** 275–283.

Jacobson, S. W., Jacobson, J. L., Sokol, R. J., Martier, S. S. and Chiodo, L. M. (1996). New evidence for neurobehavioral effects of *in utero* cocaine exposure. *J. Pediatr.* **129,** 581–590.

Jones, K. L. and Smith, D. W. (1973). Recognition of the fetal alcohol syndrome in early infancy. *Lancet* **2,** 999–1001.

Jones, K. L., Smith, D., Streissguth, A. P. and Myrianthopoulos, N. C. (1974). Outcome of offspring of chronic alcoholic women. *Lancet* **1,** 1076–1078.

Jones, K. L., Smith, D., Ulleland, C. N., and Streissguth, A. P. (1973). Pattern of malformation in offspring of chronic alcoholic mothers. *Lancet* **1,** 1267–1271.

Kagan, J., Lapidus, D. R., and Moore, M. (1978). Infant antecedents of cognitive functioning: a longitudinal study. *Child Dev.* **34,** 899–911.

Keith, R. W. (1986). *SCAN-A screening test for auditory processing disorders.* The Psychological Corporation, Harcourt, Brace Jovanovitch, San Antonio.

Kilbey, M. M. and Asghar, K. (Eds.) (1991). *Methodological Issues in Controlled Studies on Effects of Prenatal Exposure to Drugs of Abuse.* National Institute on Drug Abuse Research Monograph 114. DHHS Pub. No. (ADM) 91-1837. Supt. of Docs., U.S. Govt. Print. Off. Washington, DC.

Klahr, D. and Robinson, M. (1981). Formal assessment of problem-solving and planning processes in preschool children. *Cog. Psychol.* **13,** 113–148.

Krasnegor, N. A., Otto, D. A., Bernstein, J. H., Burke, R., Chappell, W., Eckerman, D. A., Needleman, H. L., Oakley, G., Rogan, W., Terracciano, G., and Hutchinson, L. (1994). Neurobehavioral test strategies for environmental exposures in pediatric populations. *Neurotoxicol. Teratol.* **16,** 499–509.

Kristjansson, B., and Fried, P. A. (1989). Maternal smoking during pregnancy affects children's vigilance performance. *Drug Alcohol Depend.* **24,** 11–29.

Lester, B. M., LaGasse, L., Freier, K. and Brunner, S. (1996). Studies of cocaine-exposed human infants. In *Behavioral Studies of Drug-Exposed Offspring: Methodological Issues in Human and Animal Research.* C. L. Wetherington, V. L. Smeriglio and L. P. Finnegan, Eds. National Institute on Drug Abuse Research Monograph 164. DHHS Pub. No. 96-4105. Washington, DC. pp. 175–210.

Lester, B. M. and Tronick, E. (1994). The effect of prenatal cocaine exposure and child outcome: lessons from the past. *Infant Ment. Health J.* **15,** 107–120.

Levin, H. S., Culhane, K. A., Hartman, J., Evankovich, K., Mattson, A. J., Harward, H., Ringholtz, G., Ewing-Cobbs, L. and Fletcher, J. M. (1991). Developmental changes in performance on tests of purported frontal lobe functioning. *Dev. Neuropsychol.* **7,** 377–395.

Lezak, M. D. (1995). *Neuropsychological Assessment.* 3rd ed. Oxford University Press, New York. pp. 650–685.

Lifschitz, M. H., Wilson, G. S., Smith, E. O., and Desmond, M. M. (1985). Factors affecting head growth and intellectual function in children of drug addicts. *Pediat.* **575,** 269–274.

Luria, A. R. (1959). The directive function of speech in development and dissolution. *Word* **15,** 309–325.

Luria, A. R. (1966). *Higher Cortical Functions in Man.* Basic, New York.

Makin, J., Fried, P. A., and Watkinson, B. (1991). A comparison of active and passive smoking during pregnancy: long-term effects. *Neurotoxicol. Teratol.* **13,** 5–12.

McCarthy, P. (1972). *McCarthy Scales of Children's Abilities.* The Psychological Corporation, New York.

McCartney, J. S., Fried, P. A., and Watkinson, B. (1994). Central auditory processing in school-age children prenatally exposed to cigarette smoke. *Neurotoxicol. Teratol.* **16,** 269–276.

Mednick, S. A., Harway, M., and Finello, K. M. (Eds.) (1984) *Handbook of Longitudinal Research. Vol 1. Birth and Childhood Cohorts.* Praeger, New York.

Mountain, M. A., and Snow, W. G. (1993). Wisconsin Card Sorting Test as a measure of frontal pathology: a review. *Clin. Neuropsychologist* **7,** 108–118.

Naeye, R. L. and Peters, E. C. (1984). Mental development of children whose mothers smoked during pregnancy. *Obstet. Gynecol.* **64,** 601–607.

O'Connell, C. M., and Fried, P. A. (1991). Prenatal exposure to cannabis: a preliminary report of postnatal consequences in school-age children. *Neurotoxicol. Teratol.* **13,** 631–639.

Passler, M. A., Isaac, W. and Hynd, G. W. (1985). Neuropsychological development of behavior attributed to frontal lobe functioning in children. *Dev. Neuropsychol.* **4,** 349–370.

Pennington, B. F., Bennetto, L., McAleer, O., and Roberts, R. J. Jr. (1996). In *Attention, Memory, and Executive Function.* G. Lyon and N. A. Krasnegor, Eds. Paul H. Brookes, Baltimore. pp. 327–348.

Pennington, B. F. and Ozonoff, S. (1996). Executive functions and developmental psychopathology. *J. Child Psychol. Psychiat.* **37,** 51–87.

Petrides, M., Alivisatos, B., Evans, A. C. and Meyer, E. (1993). Dissociation of human mid-dorsolateral from posterior dorsolateral frontal cortex in memory processing. *Proc. Natl. Acad. Sci. U.S.A.* **90,** 873–877.

Piaget, J. (1954). *The Construction of Reality in the Child.* Basic Books, New York.

Reed, H. B. C., Reitan, R. M., and Klove, H. (1965). Influence of cerebral lesions on psychological test performances of older children. *J. Consulting Psychol.* **29,** 247–251.

Reitan, R., and Davison, L. (1974). *Clinical Neuropsychology.* John Wiley & Sons, New York.

Reynell, J. (1977). *Reynell Developmental Language Scales-Revised.* NFER Publishing, Windsor, England.

Richardson, G. A., Conroy, M. L., and Day, N. L. (1996). Prenatal cocaine exposure: effects on the development of school-age children. *Neurotoxicol. Teratol.* **18,** 627–634.

Richardson, G. A., Day, N. L., and Goldschmidt, L. (1995). Prenatal alcohol, marijuana, and tobacco use: infant mental and motor development. *Neurotoxicol. Teratol.* **17,** 479–487.

Rodning, C., Beckworth, L., and Howard, J. (1989). Characteristics of attachment organization and play organization in prenatally drug-exposed toddlers. *Dev. Psychopathol.* **1,** 277–289.

Rodning, C., Beckworth, L. and Howard, J. (1991). Quality of attachment and home environments in children prenatally exposed to PCP and cocaine. *Dev. Psychopathol.* **3,** 351–366.

Sameroff, A. J. (1979). The etiology of cognitive competence: a systems perspective. In *Infants at Risk.* R. B. Kearsley and I. E. Sigel, Eds. Wiley, New York.

Sameroff, A. J. and Chandler, M. J. (1975). Reproductive risk and the continuum of caretaking causality. In *Review of Child Development Research.* F. Horowitz, E. M. Hetherington, S. Scarr-Salapatek, and G. Siegel, Eds. Vol. 4. University of Chicago Press, Chicago. pp. 187–243.

Shallice, T. (1982). Specific impairments of planning. *Philos. Trans. R. Soc. London: Series B: Biol. Sci.* **298,** 199–209.

Siegel, L. S. (1981). Infant tests as predictors of cognitive and language development at two years. *Child Develop.* **52,** 545–557.

Siegel, L. S. (1993). Phonological processing deficits as the basis of a reading disability. *Dev. Rev.* **13,** 246–257.

Siegel, L. S. and Ryan, E. B. (1989). The development of working memory in normally achieving and subtypes of learning disabled children. *Child Dev.* **60,** 973–980.

Spreen, O. and Strauss, E. A. (1991). *A Compendium of Neuropsychological Tests.* Oxford University Press, London.

Strauss, M. E., Lessen-Firestone, J. K. Chavez, C. J. and Stryker, J. C. (1979). Children of methadone-treated women at five years of age. *Pharmacol. Biochem. Behav.* **11,** 3–6.

Streissguth, A. P., Barr, H. M., Martin, D. C., and Herman, C. S. (1980). Effects of maternal alcohol, nicotine, and caffeine use during pregnancy on infant mental and motor development at eight months. *Alcohol. Clin. Exp.* **4,** 152–164.

Streissguth, A. P., Barr, H. M., Sampson, P. D., Darby, B. L., and Martin, D. C. (1989). IQ at age 4 in relation to maternal alcohol use and smoking during pregnancy. *Dev. Psychol.* **25,** 3–11.

Streissguth, A. P. and Giunta, C. T. (1992). Subject recruitment and retention for longitudinal research: practical considerations for a nonintervention model. In *Methodological Issues in Epidemiological, Prevention, and Treatment Research on the Effects of Prenatal Drug Exposure on Women and Children.* Research Monograph 117 M. M. Kilbey and K. Asghar, Eds. U.S. Department of Health & Human Services, National Institute on Drug Abuse, Rockville. pp. 137–154.

Stuss, D. T. (1992). Biological and psychological development of executive functions. *Brain Cog.* **20,** 8–23.

Stuss, D. T. and Benson, F. (1994). Neuropsychological studies of frontal lobes. *Psychol. Bull.* **95,** 3–28.

Tamis-LeMonda, C. S., and Bornstein, M. H. (1989). Habituation and maternal encouragement of attention in infants as predictors of toddler language, play and representational competence. *Child Dev.* **60,** 738–751.

Thorndike, R., Hagen, E., and Sattler, J. (1986). *The Standford–Binet Intelligence Scale—Fourth Edition.* Riverside Publishing, Chicago.

Tranel, D., Anderson, S. W. and Benton, A. (1994). Development of the concept of 'executive function' and its relationship to the frontal lobes. In *Handbook of Neuropsychology.* F. Boller and J. Grafman, Eds. Vol. 9, Elsevier Science, Amsterdam. pp. 125–148.

Vaughn, B. E., Kopp, C. B. and Krakow, J. B. (1984). The emergence and consolidation of self-control from eighteen to thirty months of age: normative trends and individual differences. *Child Dev.* **55,** 990–1004.

Vorhees, C. V. (1986). Principles of behavioral teratology. In *Handbook of Behavioral Teratology.* E. P. Riley and C. V. Vorhees, Eds. Plenum Press, New York. pp. 23–48.

Vorhees, C. V. (1989). Concepts in teratology and developmental toxicology derived from animal research. In *Prenatal Abuse of Licit and Illicit Drugs.* D. E. Hutchings, Ed. *Ann. N.Y. Acad. Sci.* **562,** 31–41.

Vorhees, C. V., and Butcher, R. E. (1982). Behavioral teratology. In *Developmental Toxicology.* K. Snell, Ed. Praeger Press, New York. pp. 249–298.

Weschler, D. (1991). *Weschler Intelligence Scale for Children,* 3rd ed. The Psychological Corporation, New York.

Welsh, M. C. and Pennington, B. F. (1988). Assessing frontal lobe functioning in children: views from developmental psychology. *Dev. Neuropsychol.* **4,** 199–230.

Welsh, M. C., Pennington, B. F., and Groisser, D. B. (1991). A normative-developmental study of executive function: a window on prefrontal function in children. *Dev. Neuropsychol.* **7,** 131–149.

Welsh, M. C., Pennington, B. F., Ozonoff, S., Rouse, B. and McCabe, E. (1990). Neuropsychology of early-treated PKU: specific executive function deficits. *Child Dev.* **61,** 1697–1713.

Wetherington, C. L., Smeriglio, V. L., and Finnegan, L. P. (Eds.) (1996). *Behavioral Studies of Drug-Exposed Offspring: Methodological Issues in Human and Animal Research.* National Institute on Drug Abuse Research Monograph 164. DHHS Pub. No. 96-4105. Washington, DC.

Weyandt, L. L., and Willis, W. G. (1994). Executive functions in school-aged children: potential efficacy of tasks in discriminating clinical groups. *Dev. Neuropsychol.* **10,** 27–38.

Wilson, J. G. (1977). Current status of teratology—general principles and mechanisms derived from animal studies. In *Handbook of Teratology, General Principles and Etiology.* J. G. Wilson and F. C. Fraser, Eds. Vol. 1. Plenum Press, New York. pp. 47–74.

Wilson, G. S., McCreary, R., Kean, J., and Baxter, J. C. (1979). The development of preschool children of heroin addicted mothers: a controlled study. *Pediat.* **63,** 135–141.

Yakovlev, P. I., and Lecours, A. R. (1967). The myelogenetic cycles of regional maturation of the brain. In *Regional Development of the Brain Early in Life.* A. Minkowski, Ed. Blackwell Scientific, Oxford. pp. 3–70.

CHAPTER

27

Epidemiologic Studies of the Effects of Prenatal Cocaine Exposure on Child Development and Behavior

GALE A. RICHARDSON
NANCY L. DAY
Western Psychiatric Institute and Clinic
University of Pittsburgh School of Medicine
Pittsburgh, Pennsylvania 15213

*Abbreviations: BNBAS, Brazelton Neonatal Behavioral Assessment Scale; BSID, Bayley Scales of Infant Development; CBCL, Child Behavior Checklist; CNS, central nervous system; EEG, electroencephalogram; HSQ, Home Screening Questionnaire; MDI, Mental Development Index; PDI, Psychomotor Development Index.

This chapter addresses some of the methodologic issues involved in studying the effects of prenatal cocaine exposure on child development, reviews what is known about the long-term effects of prenatal cocaine use, and describes an ongoing longitudinal study investigating the consequences of cocaine use during pregnancy.

I. Methodologic Issues in the Study of Prenatal Cocaine Use

A. *Case Selection and Assessment of Substance Use during Pregnancy*

The first decision a researcher must make is about study sample selection. It is difficult to find a sufficient number of pregnant women who use drugs or are addicted. Alcohol and other drug use, particularly heavy levels of use and/or addiction, are not common phenomena among pregnant women. In addition, women may hide their use because substance use during pregnancy is stigmatized and, in some states, there are legal sanctions against prenatal substance use. To circumvent these problems, many researchers have sampled from substance abuse treatment facilities. Although this is a cost-effective method of sampling, there are disadvantages. Most importantly, the factors affecting whether a woman enters treatment are, in themselves, risk factors for poor pregnancy outcome. The outcome data are affected by the presence of a number of risk factors in addition to the drug use itself, which makes it difficult to attribute causality.

Medical record reviews are also used to identify substance-using women, although this method also has major methodologic problems. Medical records may be incomplete, leading to missing data. In addition, strong biases affect who is questioned about drugs, which women answer honestly, and who is labeled in the medical record as a substance user. Therefore, ascertainment from medical record review is likely to yield a very biased sample of highly problematic substance users and to miss many women who either were not questioned or who chose not to report their use.

The method of assessment also affects the composition of the sample. For example, urinalysis provides an independent and reliable measure of use. However, because a positive urine screen indicates only use over a short time period (Vereby, 1987; Julien, 1995), urinalysis selectively identifies the recent and most likely heavier users. In addition, women who are selected for screening are often a biased group, identified on the basis of demographic as well as clinical indicators. Matera *et al.* (1990) compared women who had urine screens for cocaine with an unscreened group and found the screened women were more likely to be black and primiparous. These factors influence the outcome, independent of any drug effects.

Laboratory tests have other deficiencies. They cannot differentiate the mode of drug use—for example, crack versus cocaine use. A description of the pattern of use is not obtained from laboratory tests. Several researchers have shown that the pattern of use, for example, binge use, is an important predictor of the effects of prenatal substance exposure (Sampson *et al.,* 1989; Schenker *et al.,* 1990; Day *et al.,* 1991). Some laboratory tests, such as hair and meconium analyses, do not provide estimates of first trimester exposure, critical information to obtain when evaluating the teratogenic effects of a drug.

Interviewing can be an effective way of obtaining information regarding the pattern of substance use. Interviews allow the researcher to differentiate patterns of use and measure use over longer time periods. However, interview data are susceptible to memory problems and misreporting. Questions must be carefully developed and interviewers carefully selected and highly trained to ensure the accuracy of reporting. Questions should, at a minimum, elicit information on quantity and frequency to describe the pattern of use. Interviewers must be familiar with the terms women use for drugs and with the methods of drug administration. For example, Richardson and Day (1994) found that women were hesitant to use the word "crack," but they would describe how they were using the drug, which enabled a more accurate description of their drug use.

The length of the recall period affects the accuracy of reporting. Recall of first trimester use at delivery is less reliable than recall of first trimester use at the end of the first trimester. To obtain comparable data, women should be asked to report their use at standardized time periods. Using different recall periods within a sample yields noncomparable estimates of use. For example, some researchers assess use at the first prenatal visit, which can vary from the first trimester to the time of delivery. In this case, the women are asked to remember across different time periods, and for differing lengths of time.

With the exception of tobacco, women decrease their use of drugs during pregnancy, usually during the first trimester (Day *et al.,* 1994). This decrease in substance use makes measurement more difficult because the researcher is trying to measure a changing behavior. In addition, women often do not know they are pregnant for some time after conception and may not think of use during this time as use during pregnancy. This information must be elicited during the interview to obtain accurate patterns of first trimester use. A technique for enhancing early pregnancy recall has been presented (Day and Robles, 1989). Thus, each method of sample selection must be considered for its advantages, disadvantages, and impact on the nature of the sample and subsequent conclusions that can be drawn.

B. Selection of an Appropriate Comparison Group and Measurement of Covariates

Study designs lacking a control or comparison group cannot be used to determine causality (e.g., Davis *et al.,* 1992). Without a comparison group it is impossible to separate the effects of the substance under study from the other factors that correlate with the substance use. Therefore, although case studies may be useful in the early stages of research for hypothesis generation, they provide no information regarding cause-and-effect relationships and are inadequate for hypothesis testing.

It is also important to consider the characteristics of the comparison group to be selected. Women who use substances differ from women who do not use. The epidemiology of substance use varies from drug to drug, but in general, women who use drugs are more likely to be single and of lower socioeconomic status (Day *et al.,* 1993). Racial distribution varies depending on the substance. Thus, drug use is a marker for a number of things, including lifestyle and demographic status. These characteristics are risk factors for pregnancy outcome, independent of substance exposure. Because of this, when women who use drugs are compared to women who do not use drugs, there are significant differences in the outcomes of their offspring. By contrast, there are few differences when women who use cocaine are compared to women who use drugs other than cocaine. The effects previously attributed to substance exposure may have been markers for the environment or lifestyle (Richardson *et al.,* 1993). If the comparison group differs from the drug-exposed group in ways other than the drug of interest, effects of exposure cannot be attributed to that drug alone. It is crucial to measure these covariates with the same level of accuracy as the drug of interest. Otherwise, causality may be misattributed.

For example, an important covariate of cocaine use is prenatal care. A lack of prenatal care is associated with negative pregnancy outcomes (Merkatz and Thompson, 1990). Gestational age at the first prenatal visit and the total number of visits relative to the gestational age at delivery are usually used to measure prenatal care, although the content and quality of the care may be more important. Women who do not receive adequate prenatal care are more likely to be single, young, and African-American, to have less education and income, and to have inadequate nutrition and housing, characteristics noted previously as correlated with illicit drug use. Moreover, cocaine users are less likely to receive prenatal care. Effects of prenatal cocaine use have often been reported using samples of women who have had inadequate prenatal care. It is unclear whether the poor outcomes that are observed result from the drug exposure, sociodemographic characteristics, or inadequate prenatal care (Ahmed *et al.,* 1990; Merkatz and Thompson, 1990; Lutiger *et al.,* 1991; Richardson *et al.,* 1993).

An additional problem is that cocaine users are generally polydrug users. They have different patterns of drug use than women who do not use cocaine, and these patterns also affect infant outcome. Analyses must control for other substance use such as alcohol, tobacco, and marijuana when evaluating the effects of cocaine (Richardson *et al.,* 1993).

C. Follow-up Rates

An additional methodologic difficulty is that women who use drugs are difficult to follow-up. Neuspiel *et al.* (1991) reported follow-up rates at 2 months of 31% for the cocaine-exposed group and 52% for the comparison group. Azuma and Chasnoff (1993) had follow-up rates at 3 years of 44% and 50% of the original cohort for the drug-exposed and comparison groups, respectively. Thus, there can be both a high rate of subject loss and a greater rate of attrition in the exposed group. These factors limit the conclusions that can be drawn and may seriously bias the interpretation of the findings.

II. Long-Term Effects of Prenatal Cocaine Exposure

The long-term effects of prenatal cocaine exposure on child development have received considerable media attention (e.g, Rist, 1990; Elliott and Coker, 1991; Gregorchik, 1992) and have engendered much concern. However, researchers are concerned about the accuracy of these reports. Follow-up beyond the neonatal period of offspring exposed to cocaine has been reported by very few researchers. Many of the publications in the research literature are weak, for the reasons detailed previously, and the results are conflicting.

One outcome of interest is the growth of the offspring exposed to cocaine. Chasnoff *et al.* (1986) followed groups of opiate-exposed, nonopiate-exposed (including cocaine), and control infants at 3, 6, 12, 18, and 24 months. Nonopiate-exposed infants had smaller head circumferences than controls at 18 and 24 months, but not at 3, 6, or 12 months. In another study, Chasnoff *et al.* (1992) compared infants of women who were cocaine, alcohol, and marijuana users; alcohol and marijuana users; and nondrug-users. The follow-up phases where the same as in the previous study. The average head circumferences were smaller in the two drug-exposed groups compared with the nondrug-exposed infants at each follow-up point, although the two drug groups did not differ from each other. Although the cocaine-exposed group was shorter than the nondrug-exposed group at 24 months, there were no other weight or length differences at any other age. Jacobson *et al.* (1994) found no effects of prenatal cocaine exposure on weight, length, or head circumference at 6.5 months. At 13 months, cocaine exposure was associated with increased weight, but there was no effect on height or head circumference. Hurt *et al.* (1995) reported that cocaine-exposed infants had smaller weights and head circumferences at 6, 12, 18, 24, and 30 months, but there was no control for differences such as prenatal care and alcohol and tobacco use between the groups.

Researchers have also investigated the effects of prenatal cocaine exposure on infant and child development. Chasnoff *et al.* (1986) reported that, at 3 months, the nonopiate-exposed infants had lower Bayley Scales of Infant Development (BSID) motor scores and, at 6 months, lower BSID mental scores when compared with the control group. There were no BSID differences at 12 or 24 months. In the later study (Chasnoff *et al.,* 1992), at 6 months, the two drug-exposed groups had lower BSID mental and motor scores than the nondrug-exposed infants, but the two drug groups did not differ from each other. Jacobson *et al.* (1996) found no effect of cocaine exposure on the BSID at 13 months. They did, however, find that offspring of heavy cocaine users did less well on tests of visual recognition memory and processing speed at 13 months.

Singer *et al.* (1994) compared very low birth weight infants at 16 to 18 months of age. The cocaine-exposed infants had lower mental and motor BSID scores than the nondrug-exposed offspring, although other drug use was not controlled in the analysis. By contrast, Graham *et al.* (1992) did not find differences on the BSID at 20 months of age in a comparison of cocaine-exposed, marijuana-exposed, and nonexposed infants, and Hurt *et al.* (1995) found that cocaine-exposed infants did not differ from comparison infants on the BSID at 6, 12, 18, 24, or 30 months, nor in language development at 30 months (Hurt *et al.,* 1997).

At 3 years of age, children who were exposed prenatally to cocaine had significantly lower verbal reasoning scores on the Stanford–Binet Intelligence Scale than the nonuser group, although the cocaine-exposed children did not differ from the alcohol–marijuana-exposed children (Azuma and Chasnoff, 1993; Griffith *et al.,* 1994). Both drug-exposed groups had significantly lower scores on abstract–visual reasoning compared with the nonuser group, but they did not differ from each other. There were no differences among the groups on the composite score. Prenatal cocaine use did predict more externalizing problems on the Child Behavior Checklist (CBCL).

Bender *et al.* (1995) enrolled three groups of women in a study of the relationship between crack use and human immunodeficiency virus (HIV) prevalence. The first group used crack both pre- and postnatally; the second group used crack only during the postpartum period; and the third group did not use crack during either time period. At 4–6 years of age, the children who were exposed both pre- and postnatally had lower scores on tests of receptive language ability and visual–motor integration compared to the other two groups when confounding variables were controlled. In another report, Hawley *et al.* (1995) compared 3- to 5-year-old children of mothers in treatment for crack addiction with children in Head Start whose mothers had been asked not to volunteer if they had used drugs during pregnancy. Only 9% of the Head Start mothers participated. All mothers were interviewed at 3–5 years postpartum about their drug use during pregnancy. There were no differences between the groups in cognitive or language development, although the mothers in the treatment group rated their children on the CBCL as having more internalizing problems. There were no confounding variables controlled for in these analyses. In a report from our ongoing studies of prenatal alcohol and marijuana exposure, it was found that first trimester cocaine exposure was associated with more errors of omission, a measure of attention problems, on a computerized vigilance task at 6 years (Richardson *et al.,* 1996a).

Thus, the data on the long-term effects of cocaine exposure are limited. The studies that have controlled for confounding variables and have selected appropriate control groups, however, have reported somewhat consistent results: There are few effects that can be attributed directly to prenatal cocaine exposure. The few findings that have been reported reflect effects on the development of the central nervous system (CNS). These include early differences in visual recognition memory and processing speed, poorer performance on the Stanford–Binet, and reports of behavior and attention problems. The effects of prenatal cocaine exposure will be explored in more detail with findings from the Maternal Health Practices and Child Development Project.

III. A Longitudinal Study of Prenatal Cocaine Exposure: The Maternal Health Practices and Child Development Project

A. *Sample Selection and Study Design*

This program of research is funded by the National Institute on Drug Abuse and the National Institute on Alcohol Abuse and Alcoholism. The project is a collaboration between Western Psychiatric Institute and Clinic and Magee-Womens Hospital in Pittsburgh, Pennsylvania. The sample reported on in this chapter consists of women and infants participating in a longitudinal investigation of the effects of prenatal cocaine–crack use. Written consent was obtained for both mothers and infants according to guidelines established by the University of Pittsburgh's Institutional Review Board and by the Research Review and Human Experimentation Committee of Magee-Womens Hospital. A Confidentiality Certificate, obtained from the Department of Health and Human Services, assured participants that their responses could not be subpoenaed.

This is an ongoing prospective study of women who attended a prenatal clinic at Magee-Womens Hospital. These women were not in drug treatment during their pregnancies; rather, they are a representative group of women attending a prenatal clinic. Women were recruited for the study between March 1988, and December 1992. Women who were at least 18 years of age were initially interviewed when they came for prenatal care during their fourth or fifth prenatal month. We asked the women about their use of cocaine, crack, alcohol, marijuana, tobacco, and other drugs for the year prior to pregnancy and for the first trimester. All women who reported using any cocaine or crack during the first trimester were enrolled. After a woman who reported using the drug was enrolled, the next woman interviewed who reported no cocaine or crack use both during the year prior to pregnancy and the first trimester was also enrolled. Ninety percent of those women eligible to be interviewed consented to participate in the study. Medical chart reviews of a random sample of women who refused to participate indicated that only 5% had a history of drug use during the current pregnancy.

Of the women initially interviewed, 325 (18%) met the inclusion criteria and were enrolled into the study. Women selected into the study were interviewed at 7 months of pregnancy about their substance use during the second trimester and at 24 hr postdelivery, when they were asked about third trimester substance use. Demographic, lifestyle, and psychologic characteristics were also assessed at each phase. Information regarding pregnancy, labor, and delivery conditions was abstracted from the medical charts.

All infants underwent comprehensive physical examination, generally within 24–48 hr of delivery, by study nurse clinicians who were unaware of prenatal exposure status. The infants' length, head, and chest circumference were measured. Gestational age, as measured by a modification of the Dubowitz assessment (Ballard *et al.,* 1979), was also assessed. Birthweight and Apgar scores were transcribed from the medical records. All full-term (38–42 weeks), singleton infants underwent assessment with the Brazelton Neonatal Behavioral Assessment Scale (BNBAS) (Brazelton, 1984) twice during the postpartum period, at approximately days 2 and 3 postpartum. Information regarding pregnancy, labor, and delivery conditions was abstracted from the medical charts by study nurses after the physical examination was completed. Adequacy of prenatal care, as defined by Kessner *et al.* (1973), is based on gestational age at the first prenatal visit, the total number of prenatal visits, and the gestational age at delivery.

Women were interviewed and their infants were assessed again at 1, 3, and 7 years. The children received physical examinations and age-appropriate developmental assessments at each follow-up phase. We have excellent follow-up rates at both our 1-year and 3-year phases. We have a completion rate of 93% of the eligible subjects at 1 year and a 95% rate at 3 years. The 7-year phase is still in progress.

B. Sample Characteristics and Pattern of Cocaine Use

Women were, on average, 25 years old, 52% were Caucasian, and 48% were African-American. Women were of lower socioeconomic status, with a mean family income of $650 per month and an average educational level of 12 years. Twenty percent were married, and 41% worked and/or went to school during the first trimester. Twenty-four percent were primigravidous and 42% were primiparous. According to the definition by Kessner *et al.* (1973), 47% of the women had adequate prenatal care, 48% had an intermediate level of care, and 5% had inadequate prenatal care.

Fifty-four percent of the infants were male. The average gestational age was 39.7 weeks and mean Apgar scores were 7.9 and 8.9 at 1 and 5 min, respectively. The mean birth weight was 3254 grams, the mean length was 49.7 cm, and the mean head circumference was 34.5 cm.

The mean age of the children at the 3-year assessment was 38.3 months. Twenty-four percent of the children attended preschool or day care. Six percent were not in maternal custody. The mean Stanford–Binet Intelligence Scale (4th Ed.) composite score was 93. This average is within the range expected for low socioeconomic status samples and is comparable to that obtained in our other studies.

For each substance, women were asked about the usual, maximum, and minimum quantity and frequency of use. In addition, we used a technique to enhance the accuracy of reporting of first trimester substance use. Women often do not change their substance use patterns until after they have confirmed their pregnancy, which may not occur until late in the first trimester. Therefore, we asked women detailed information about the time periods from conception to recognition of the pregnancy, and from recognition to the diagnosis of the pregnancy in order to get substance use information for the first trimester. We also asked whether their use during these time periods was more like what they had reported for prior to pregnancy use or more like their reported first trimester use. This allowed calculation of substance use during each month of the first trimester and a weighted estimate of the average daily use for the first trimester. For second and third trimester use, women reported use over the entire trimester. These methods have been used in our previous research on prenatal alcohol and marijuana exposure (Day and Robles, 1989).

For descriptive purposes, occasional users were defined as women who used less than one line of cocaine per day, and frequent users were women who used one or more lines of cocaine per day or an equivalent amount of crack. During the year prior to pregnancy, 23% of the women were frequent users of cocaine. In the first trimester, 19% of the women were frequent users. The mean level of cocaine use for the women who used during the first trimester was approximately 14 lines/day. In the second and third trimesters, 5% and 6%, respectively, were frequent users, using one or more lines per day. During the second and third trimesters, the mean use for the users was 5 lines/day. The rate of frequent cocaine use at both the 1-year and 3-year follow-ups was 7%. The mean level of use for the women who used at the 1-year and 3-year phases was 9 lines/day and 8 lines/day, respectively.

We did not use laboratory tests of cocaine to identify the sample. The most commonly used test, urine screening, reflects use only over a short period of time (Vereby, 1987; Julien, 1995). In addition, laboratory measures cannot give an estimate of the pattern of use, information that is critical in evaluating the effect of a potential teratogen. Interviewing was the method of choice in order to collect detailed information regarding the timing and pattern of cocaine use, especially during the first trimester. We did, however, have access to medical record data for the women who were screened for clinical purposes. In fact, 100% of the women who had a positive urine screen for cocaine when tested by the hospital were identified by our study interview as users. By contrast, 79% of the women who had a negative screen admitted use on the interview. Thus, our interview detected more cocaine use than did the toxicology screen, demonstrating that drug use information can be obtained reliably when interviews are well-constructed and interviewers are carefully selected and trained.

C. Covariates of Cocaine Use

Women who were frequent users of cocaine during the first trimester differed from women who did not use. Frequent first trimester users of cocaine were older (27 vs. 24 years) and more likely to be African-American (64% vs. 43%) than nonusers. They were less often married (10% vs. 26%) and they had lower monthly family incomes than women who were nonusers during the first trimester ($459 vs. $720). Fewer of the frequent first trimester users had adequate prenatal care compared with the nonusers (29% vs. 54%).

Women who were frequent users of cocaine during the first trimester were significantly more likely than nonusers to use other substances. Eighty-five percent of the frequent cocaine users smoked tobacco, compared with 45% of the noncocaine users. The frequent first trimester cocaine users were also more likely to be alcohol users (88% vs. 56%), marijuana users (64% vs. 18%), and users of illicit drugs (15% vs. 3%), such as amphetamines and tranquilizers, during the first trimester than were the noncocaine users.

Women who used cocaine during the third trimester were more likely to be African-American (84% vs. 42%), to have lower family incomes ($443 vs. $680/month), and to have had more pregnancies (4.7 vs. 2.9), births (2.6 vs. 2.0), miscarriages (0.8 vs. 0.3), and abortions (1.3 vs. 0.6) than women who did not use cocaine in the third trimester. There were no significant differences in age, education, marital status, work–school status, weight gain during pregnancy, or labor or delivery conditions among the groups. Third trimester users had significantly fewer prenatal visits (7.5 vs. 10.4), more pregnancy conditions (4.0 vs. 2.3), and fewer received adequate prenatal care (28% vs. 49%) than women who did not use cocaine in the third trimester.

Women who used cocaine in the third trimester also used more of other substances than women who did not use cocaine in the third trimester. During the third trimester, the cocaine users smoked more cigarettes (10 vs. 7 cigarettes/day), drank more alcohol, (1.1 vs. 0.2 drinks/day), used more marijuana (0.1 vs. 0.04 joints/day), and other illicit drugs (9% vs. 1% users) than women who did not use cocaine in the third trimester.

Women who were frequent users of cocaine during the first trimester continued to be different at the 3-year follow-up. They were more likely to use cocaine (28% vs. 3%), marijuana (24% vs. 17%), and tobacco (76% vs. 45%) at the 3-year follow-up when compared with women who did not use cocaine during the first trimester. Although the proportion who used alcohol did not differ, women who were frequent users of cocaine during the first trimester were more likely to be heavier users of alcohol than noncocaine users. In addition, women who used cocaine at the 3-year follow-up provided less stimulating and organized home environments, as measured by the Home Screening Questionnaire (HSQ) (Frankenburg and Coons, 1986), than women who did not use currently. These findings highlight the importance of controlling for both the prenatal and current characteristics of women who use cocaine when investigating the effects of prenatal cocaine exposure.

D. Findings

Stepwise regressions were used for the outcome analyses. The effects of cocaine were evaluated for each trimester separately. Covariates were included in the regression model if they were significantly related to the outcome in a bivariate analysis. The following variables were included in all of the analyses: child gender and age at assessment, maternal race and age, and prenatal and current alcohol, marijuana, and tobacco use. The following additional variables were added to specific outcomes: maternal height for growth outcomes; environmental stimulation, as measured by the HSQ, and preschool attendance for the cognitive development outcomes; one or two of the socioeconomic status variables depending on the outcome (income, education, marital status, work status); and maternal depression for the analyses of temperament and behavior.

1. Birth

At birth, prior to controlling for confounding variables, the infants of frequent first trimester cocaine users were significantly smaller in weight, length, and head circumference than were infants of occasional and nonusers. This same pattern applied to second and third trimester use. However, when the covariates were entered into stepwise multiple regressions, there were no significant relationships between cocaine use during any trimester and infant birth weight, length, or head circumference (Richardson *et al.,* in preparation). These results are consistent with preliminary findings reported by Richardson and Day (1994). Cocaine use was also not associated with an increased risk of minor physical anomalies, low birth weight, or small-for-gestational age, after controlling for confounding variables. First trimester cocaine use was a significant predictor of reduced gestational age, with infants of frequent users being born approximately 1 week earlier than infants of women who did not use cocaine first trimester.

At birth, the development of the CNS in term infants was assessed with the BNBAS and electroencephalogram (EEG)–sleep studies. The Lester *et al.* (1982) clusters were the outcome variables for the BNBAS. Poorer autonomic stability was significantly predicted by cocaine exposure during each trimester of pregnancy. An increased number of abnormal reflexes was associated with cocaine exposure during the first and second trimesters, decreased scores on motor maturity were significantly associated with second and third trimester exposure, and decreased ability to regulate state was associated with third trimester exposure (Richardson *et al.,* 1996b). The offspring of women who used cocaine throughout pregnancy demonstrated the poorest motor performance and autonomic stability, and the highest number of abnormal reflexes, followed by those who used only first trimester, then by the abstainers. There was a significant effect of first trimester exposure even among the offspring of women who quit using cocaine after the first trimester, but the offspring of women who used throughout pregnancy seemed to be the most impaired.

Analyses of the EEG–sleep studies demonstrated that infants of women who used cocaine displayed less well-developed spectral correlations between homologous brain regions (Scher *et al.,* in preparation). These findings represent a different EEG–sleep pattern than that observed in noncocaine-exposed infants or in infants exposed prenatally to alcohol or marijuana (Scher *et al.,* 1988).

2. One Year

At 1 year, prenatal and current cocaine use were not significant predictors of weight, length, or head circumference in the regression analyses. At this time, the BSID mental (MDI) and motor (PDI) scales were used to assess development. There was a significant effect of second trimester cocaine use on the PDI (Richardson, *et al.,* 1995). Second trimester cocaine use significantly predicted lower motor scores. On average, offspring of women who used cocaine in the second trimester had an adjusted PDI score of 104, compared with 114 for offspring of women who were not second trimester users. There was no effect of first or third trimester use, and there was no effect of prenatal or current cocaine use on performance on the MDI.

The Bates Infant Characteristics Scale was used to assess temperament at 12 months. This instrument has four subscales: fussy/difficult, unadaptable, persistent, and unsociable. After controlling for confounding factors, the fussy/difficult subscale was significantly predicted by cocaine use during the first, second, and third trimesters. Unadaptability and excessive persistence were both predicted by first and second trimester cocaine use (Richardson *et al.,* 1994). Current cocaine use did not predict the mother's rating of the infant's temperament.

To provide a measure of the infant's behavior that was independent of the mother's report, we analyzed the examiner ratings of the BSID Infant Behavior Record. Examiners, who were blind to maternal substance use, rated infants of women who used cocaine first trimester as less responsive, less reactive to test materials, and as having shorter attention spans than infants of women who did not use cocaine in the first trimester.

3. Three Years

At 3 years, prenatal cocaine use was not a significant predictor of child weight or height. However, first trimester cocaine use was a significant predictor of 3-year head circumference. Children of women who used cocaine frequently during the first trimester had significantly smaller head circumferences in comparison to the offspring of women who did not use cocaine in the first trimester (adjusted mean = 49.7 vs. 50.2 cm).

First trimester cocaine use also significantly predicted the performance of the children on the Stanford–Binet. Women who used cocaine frequently in the first trimester had children with significantly lower composite (adjusted mean = 90 vs. 93) and short-term memory scores (adjusted mean = 97 vs. 100) than did children of women who did not use cocaine in the first trimester. The children of women who used cocaine in the first trimester and stopped had lower Stanford–Binet scores than children of women who never used, confirming the first trimester effects found in the regression analyses. Current cocaine use was not a significant predictor of Stanford–Binet scores.

At 3 years, as at 12 months, cocaine exposure in early pregnancy was associated with differences in child temperament. Children of women who used frequently first trimester were rated as more fussy and difficult and as more unadaptable than those of women who did not use first trimester. These effects were also significant when we assessed the group of women who used only during the first trimester and then stopped.

First trimester cocaine use was also related to behavior problems reported by the mother on the CBCL (Achenbach, 1992). Children who were exposed to cocaine in the first trimester were rated by their mothers as having more behavior problems overall and more internalizing problems. The women who used throughout pregnancy rated their children as having more internalizing, externalizing, and total behavior problems than did the children of mothers who stopped and the children of mothers who had never used. In a further analysis, a logistic regression was used to analyze the scores on the CBCL that were above the borderline clinical cutoff score (Achenbach, 1992). First and third trimester use were significant predictors of meeting clinical criteria for internalizing behaviors. Children of women who used cocaine frequently throughout pregnancy were significantly more likely to reach clinical criteria for internalizing behavior problems (71% scored above the borderline clinical cutoff) compared with offspring of women who used frequently in the first trimester and stopped (30%) and the offspring of women who never used cocaine (24%). Current cocaine use did not predict the mother's rating of the child's behavior problems.

As an independent corroboration of the child's behavior, we analyzed the examiner ratings of the child's behavior during the Stanford–Binet assessment. Examiners, unaware of maternal substance use, rated children of women who used cocaine during the first trimester as having shorter attention spans, less focused attention, more restlessness, and as making more attempts to distract the examiner than children of women who did not use cocaine.

Thus, the results from the Maternal Health Practices and Child Development Project demonstrate subtle, persistent effects of prenatal cocaine exposure on the development of the CNS. At birth, the exposed offspring demonstrated problems with motor maturity, state regulation, and abnormalities on the EEG–sleep studies. At 1 year of age, motor development was delayed and the children were noted by mothers and independent observers to be more fussy and difficult and to have a reduced attention span. At 3 years, the exposed children continued to be rated as fussy and difficult by their mothers, and observers again reported them as having decreased attention and more restlessness. They scored significantly lower on the Stanford–Binet Intelligence Scale, both on the composite score and the short-term memory subscale. Exposed offspring also had a higher rate of behavior problems. Each of these findings remained significant after controlling for other prenatal substance exposure and for current cocaine and other substance use.

IV. Conclusions and Future Directions

There is a consistent pattern of CNS effects across phases. At birth, cocaine exposure was related to decreased autonomic stability, an increased number of abnormal reflexes, difficulties in the regulation of state, and decreased motor maturity, as well as differences in sleep parameters. At the 1-year follow-up, effects were noted on the Bates Scale, as the children were reported by their mothers to be more fussy and difficult. These findings were confirmed at 3 years of age. Independent observations from the examiners also substantiated these findings, as at 3 years, Stanford–Binet examiners reported that the children who were exposed prenatally to cocaine had decreased attention span, more difficulty focusing, and were more restless during testing. At 3 years, there were cognitive deficits as measured by the Stanford–Binet composite and short-term memory scores. Prenatal cocaine exposure was also associated with increased behavior problems. In addition, at 3 years of age, we identified a significant effect of prenatal cocaine exposure during the first trimester on head circumference, a growth parameter, but one that is highly correlated with CNS functioning.

Therefore, we have consistently found that CNS deficits result from prenatal cocaine exposure. We found the same pattern in our earlier studies of prenatal alcohol and marijuana use, in which exposure during the first trimester significantly predicted a decrease in IQ and an increase in behavior problems. Other researchers have reported a similar pattern of effects. In the Seattle Longitudinal Study, exposure to alcohol during the period prior to pregnancy recognition was associated with decreased IQ scores (Streissguth *et al.,* 1989) and increased attentional errors and reaction time at 4 years (Streissguth *et al.,* 1984). These CNS effects were found in the absence of growth and morphologic changes (Streissguth, 1992). These findings are also consistent with Spear's (1995) review of the animal research literature, in which neurobehavioral effects of cocaine were found in the absence of growth or morphologic anomalies at lower dose exposures.

These CNS effects are subtle. They would not be detectable in any one child. However, they are important because they demonstrate that prenatal cocaine exposure has an effect on the development of the CNS. We know from our earlier work with alcohol and marijuana, as well as from the child development literature, that behavior problems, lower IQ, and attention deficits are risk factors for the development of additional problems as the children mature. Addressing some of these deficits at an early age may help avoid some of the future problems. It is important, at the same time, to realize that the "crack baby" is a negative and pejorative image. These children do not fit this picture and they do not deserve that label. They do, however, deserve and need acknowledgment and remediation for their problems that will be effective in both the short- and long-term.

Our next step is to study this cohort at 7 years of age. We will continue to assess the domains we have previously, such as growth, intelligence, temperament, and behavior. In addition, we have added a neuropsychologic assessment to the 7-year phase to obtain more detailed information regarding CNS functioning. We will be able to obtain a better estimate of CNS effects as the children develop and as more demands are made on their cognitive abilities.

Acknowledgment

This research was supported by the National Institute on Drug Abuse grants DA05460, DA06839, and DA08916 (G. Richardson, Principal Investigator).

References

Achenbach, T. (1992). *Manual for the Child Behavior Checklist/2-3 and 1992 Profile.* University of Vermont Department of Psychiatry, Burlington, VT.

Ahmed, F., McRae, J., and Ahmed, N. (1990). Factors associated with not receiving adequate prenatal care in an urban black population: program planning implications. *Soc. Work Health Care* **14,** 107–123.

Azuma, S. D. and Chasnoff, I. J. (1993). Outcome of children prenatally exposed to cocaine and other drugs: a path analysis of three-year data. *Pediatrics* **92,** 396–402.

Ballard, J. L., Novak, K. K., and Driver, B. A. (1979). A simplified score for assessment of fetal maturation of newly born infants. *J. Pediatr.* **95,** 769–774.

Bender, S. L., Word, C. O., DiClemente, R. J., Crittenden, M. R., Persaud, N. A., and Ponton, L. E. (1995). The developmental implications of prenatal and/or postnatal crack cocaine exposure in preschool children: a preliminary report. *J. Dev. Beh. Pediatr.* **16,** 418–424.

Brazelton, T. B. (1984). *Neonatal Behavioral Assessment Scale,* 2nd ed. Lippincott, Philadelphia, PA.

Chasnoff, I., Burns, K., Burns, W., and Schnoll, S. (1986). Prenatal drug exposure: effects on neonatal and infant growth and development. *Neurobehav. Toxicol. Teratol.* **8,** 357–362.

Chasnoff, I., Griffith, D., Freier, C., and Murray, J. (1992). Cocaine/polydrug use in pregnancy: two-year follow-up. *Pediatrics* **89,** 284–289.

Davis, E., Fennoy, I., Laraque, D., Kanem, N., Brown, G., and Mitchell, J. (1992). Autism and developmental abnormalities in children with perinatal cocaine exposure. *J. Nat. Med. Assoc.* **84,** 315–319.

Day, N., Cottreau, C., and Richardson, G. (1993). The epidemiology of alcohol, marijuana, and cocaine use among women of childbearing age and pregnant women. *Clin. Obstet. Gynecol.* **36,** 232–245.

Day, N., Goldschmidt, L., Robles, N., Richardson, G., Cornelius, M., Taylor, P., Geva, D., and Stoffer, D. (1991). Prenatal alcohol exposure and offspring growth at 18 months of age: the predictive validity of two measures of drinking. *Alcohol: Clin. Exper. Res.* **15,** 914–918.

Day, N. L., Richardson, G. A., Geva, D., and Robles, N. (1994). Alcohol, marijuana, and tobacco: effects of prenatal exposure on offspring growth and morphology at age six. *Alcohol: Clin. Exp. Res.* **18,** 786–794.

Day, N. L. and Robles, N. (1989). Methodological issues in the measurement of substance use. In *Prenatal Abuse of Licit and Illicit Drugs.* D. Hutchings, Ed., *Ann. NY Acad. Sci.* **562,** 8–13.

Elliott, K. T. and Coker, D. R. (1991). Crack babies: here they come, ready or not. *J. Instruct. Psych.* **18,** 60–64.

Frankenburg, W. and Coons, C. (1986). Home Screening Questionnaire: its validity in assessing home environment. *J. Pediatr.* **108,** 624–626.

Graham, K., Feigenbaum, A., Pastuszak, A., Nulman, I., Weksberg, R., Einarson, T., Goldberg, S., Ashby, S., and Koren, G. (1992). Pregnancy outcome and infant development following gestational cocaine use by social cocaine users in Toronto, Canada. *Clin. Investig. Med.* **15,** 384–394.

Gregorchik, L. A. (May, 1992). The cocaine-exposed children are here. *Phi Delta Kappan* **73,** 709–711.

Griffith, D. R., Azuma, S. D., and Chasnoff, I. J. (1994). Three-year outcome of children exposed prenatally to drugs. *J. Am. Acad. Child Adolesc. Psych.* **33,** 20–27.

Hawley, T. L., Halle, T. G., Drasin, R. E., and Thomas, N. G. (1995). Children of addicted mothers: effects of the "crack epidemic" on the caregiving environment and the development of preschoolers. *Amer. J. Orthopsychiatry* **65,** 364–379.

Hurt, H., Brodsky, N., Betancourt, L., Braitman, L., Malmud, E. and Giannetta, J. (1995). Cocaine-exposed children: follow-up through 30 months. *J. Dev. Behav. Pediatr.* **16,** 29–35.

Hurt, H., Malmud, E., Betancourt, L., Brodsky, N. L., and Giannetta, J. (1997). A prospective evaluation of early language development in children with *in utero* cocaine exposure and in control subjects. *J. Pediatr.* **130,** 310–312.

Jacobson, J., Jacobson, S., and Sokol, R. (1994). Effects of prenatal exposure to alcohol, smoking, and illicit drugs on postpartum somatic growth. *Alcohol: Clin. Exp. Res.* **18,** 317–323.

Jacobson, J., Jacobson, S., Sokol, R., Martier, S. S., and Chiodo, L. M. (1996). New evidence for neurobehavioral effects of in utero cocaine exposure. *J. Pediatr.* **129,** 581–590.

Julien, R. (1995). *A Primer of Drug Action,* 7th ed. W. H. Freeman, New York.

Kessner, D., Singer, J., Kalk, C., and Schlesinger, E. (1973). Infant death: an analysis by maternal risk and health care. *Contrasts in Health Status* (Vol. 1), Institute of Medicine, Washington, DC.

Lester, B. M., Als, H., and Brazelton, T. B. (1982). Regional obstetric anesthesia and newborn behavior: a reanalysis toward synergistic effects. *Child Dev.* **53,** 687–692.

Lutiger, B., Graham, K., Einarson, T. R., and Koren, G. (1991). Relationship between gestational cocaine use and pregnancy outcome: a meta-analysis. *Teratol.* **44,** 405–414.

Matera, C., Warren, W. B., Moomjy, M., Fink, D. J., and Fox, H. E. (1990). Prevalence of use of cocaine and other substances in an obstetric population. *Am. J. Obstet. Gynecol.* **163,** 797–801.

Merkatz, I. R., and Thompson, J. E., Eds. (1990). *New Perspectives on Prenatal Care.* Elsevier Science Publishing Company, New York.

Neuspiel, D. R., Hamel, S. C., Hochberg, E., Greene, J., and Campbell, D. (1991). Maternal cocaine use and infant behavior. *Neurotoxicol. Teratol.* **13,** 229–233.

Richardson, G. A., Conroy, M. L., and Day, N. L. (1996a). Prenatal cocaine exposure: Effects on the development of school-age children. *Neurotoxicol. Teratol.* **18,** 627-634.

Richardson, G. A. and Day, N. L. (1994). Detrimental effects of prenatal cocaine exposure: Illusion or reality? *J. Am. Acad. Child Adolesc. Psychiatr.* **33,** 28–34.

Richardson, G. A., Day, N. L., and Goldschmidt, L. (March 1995). A longitudinal study of prenatal cocaine exposure: infant development at 12 months. Paper presented at the *Society for Research in Child Development,* Indianapolis, IN.

Richardson, G. A., Day, N. L., and McGauhey, P. J. (1993). The impact of prenatal marijuana and cocaine use on the infant and child. *Clin. Obstet. Gynecol.* **36,** 302–318.

Richardson, G. A., Hamel, S. C., Day, N. L., and Goldschmidt, L. (September 1994). Maternal rating of temperament of cocaine-exposed infants. Paper presented at the *Society for Behavioral Pediatrics,* Minneapolis, MN.

Richardson, G. A., Hamel, S. C., Goldschmidt, L., and Day, N. L. (1996b). The effects of prenatal cocaine use on neonatal neurobehavioral status. *Neurotoxicol. Teratol.* **18,** 519–528.

Richardson, G. A., Hamel, S. C., Goldschmidt, L., and Day, N. L. (In preparation). Growth and morphology of infants prenatally exposed to cocaine: a comparison of prenatal care and no prenatal care samples.

Rist, M. C. (1990). The shadow children. *Am. School Board J.* **177,** 19–24.

Sampson, P. D., Streissguth, A. P., Barr, H. M., and Bookstein, F. L. (1989). Neurobehavioral effects of prenatal alcohol: Part II. Partial least squares analysis. *Neurotoxicol. Teratol.* **11,** 477–491.

Schenker, S., Becker, H. C., Randall, C. L., Phillips, D. K., Baskin, G. S., and Henderson, G. I. (1990). Fetal alcohol syndrome: current status of pathogenesis. *Alcohol: Clin. Exp. Res.* **14,** 635–647.

Scher, M. S., Richardson, G. A., Coble, P., Day, N. L., and Stoffer, D. (1988). The effects of prenatal alcohol and marijuana exposure: disturbances in neonatal sleep cycling and arousal. *Ped. Res.* **24,** 101–105.

Scher, M. S., Richardson, G. A., and Day, N. L. (In preparation). Effects of prenatal cocaine/crack exposure on EEG–sleep studies at birth and one year.

Singer, L., Yamashita, T., Hawkins, S., Cairns, D., Baley, J., and Kliegman, R. (1994). Increased incidence of intraventricular hemorrhage and developmental delay in cocaine-exposed, very low birth weight infants. *J. Pediatr.* **124,** 765–771.

Spear, L. (1995). Neurobehavioral consequences of gestational cocaine exposure: a comparative analysis. In *Advances in Infancy Research.* C. Rovee-Collier and L. Lipsitt, Eds. Ablex Publishing Co., Norwood, NJ. pp. 55–105.

Streissguth, A. P. (1992). Fetal alcohol syndrome and fetal alcohol effects: a clinical perspective of later developmental consequences. In *Maternal Substance Abuse and the Developing Nervous System* I. Zagon and T. Slotkin, Eds. Academic Press, San Diego. pp. 5–25.

Streissguth, A. P., Barr, H., Sampson, P., Darby, B., and Martin, D. (1989). IQ at age 4 in relation to maternal alcohol use and smoking during pregnancy. *Dev. Psych.* **25,** 3–11.

Streissguth, A. P., Martin, D., Barr, H., Sandman, B., Kirchner, G., and Darby, B. (1984). Intrauterine alcohol and nicotine exposure: attention and reaction time in 4-year-old children. *Dev. Psych.* **20,** 533–541.

Vereby, K. (1987). Cocaine abuse detection by laboratory methods. In *Cocaine—A Clinician's Handbook.* A. M. Washton, and M. S. Gold, Eds. Guilford Press, New York.

CHAPTER

28

Assessment of Case Reports and Clinical Series

J. M. FRIEDMAN
Department of Medical Genetics
University of British Columbia
Vancouver, British Columbia, Canada, V6H 3N1

I. Introduction

Epidemiologic studies with quantitative assessment of behavioral and cognitive function are valuable methods of demonstrating the effects of developmental toxins on traits that can be measured quantitatively. The strengths and limitations of these methods are discussed in the preceding chapters. This chapter focuses on case reports and clinical series, which are based on the recognition of patterns of anomalies.

Aase (1990) describes the process of pattern recognition colorfully with respect to the physical anomalies that constitute dysmorphic syndromes:

> Disparaging remarks have been made about the "Aunt Minnie" school of diagnosis, the implication being that simple familiarity, rather than rigorous analysis, is a weak foundation for a diagnostic conclusion. It is true that those who rely heavily on this technique run the risk of jumping to unsubstantiated conclusions and diagnostically mislabeling their patients. Most dysmorphic syndromes, however, show considerable variation even in their most characteristic features. The real value of pattern recognition, therefore, is the ability to identify Aunt Minnie even when she is wearing a wig or playing the bagpipes.

Formal epidemiologic studies have not proved to be very useful in the identification of patterns of anomalies. In other words, epidemiology is not very good at finding "Aunt Minnie" in a crowd of people. Fortu-

*Abbreviations: CNS, central nervous system; FAS, fetal alcohol syndrome; PCB, polychlorinated biphenyls.

nately, "Aunt Minnie" is easy to find by pattern recognition.

II. Recognition of Patterns of Anomalies

The first evidence that an exposure is teratogenic in humans usually comes from case reports and clinical series (Smithells, 1987). Case reports are simply descriptions of anecdotal observations. A group of cases reported together is called a *clinical series.* Such groups are usually assembled as convenience samples and may not be representative of the full range of a condition or exposure.

Case reports and clinical series are most useful when the cases are described clearly and in sufficient detail to permit other investigators to assess the patterns observed independently. The most valuable clinical series and case reports for evaluating teratogenicity are those that document the agent, dose, route, frequency, and time in pregnancy of the exposures fully, and also characterize the abnormalities in the affected child or children well. Information excluding other possible causes of the abnormalities and describing any potentially confounding concomitant exposures or conditions is also important.

Almost all exposures currently recognized as being teratogenic in humans have initially been identified in case reports and clinical series. This has been possible because teratogenic exposures typically produce qualitatively distinct patterns of congenital anomalies rather than single birth defects in otherwise normal children. Clinical series can be compelling when they demonstrate the occurrence of a highly characteristic pattern of anomalies in the children of women who experienced similar well-defined exposures at similar times in pregnancy. The association is especially convincing if both the pattern of anomalies and the exposure are rare in other circumstances.

Table 1 provides some examples of human teratogenic syndromes that have been recognized because of a characteristic pattern of congenital anomalies that occurred among the children of women who had similar unusual exposures during pregnancy. In some instances (e.g., with fetal alcohol syndrome, toluene embryopathy, rice oil disease, or Minimata disease), the exposures involved such high doses that maternal toxicity often occurred. In some of the other conditions listed, the exposures occurred with doses that produced a beneficial therapeutic effect in the mother without signs of maternal toxicity. This was the case with thalidomide embryopathy, isotretinoin embryopathy, and warfarin embryopathy, for example.

It is important to note that the syndromes listed in Table 1, even when fully expressed, are not characterized by any single distinctive feature. In fact, many of the component features are rather common and can have a variety of different causes. When these features occur together, however, they constitute a distinctive pattern of anomalies that is uncommon except in children born to mothers who had the indicated exposure during pregnancy.

Even though substantial variability may exist among individuals affected with a given syndrome, the overall pattern persists. This principle is well illustrated by Down syndrome, a condition that is caused by a chromosomal abnormality, not by a developmental toxin. Most affected people are instantly recognizable as having Down syndrome, both by experienced physicians and by members of the general public. Although the component anomalies of the Down syndrome phenotype vary greatly from patient to patient, the gestalt remains very similar regardless of age, gender, ethnic origin, or severity of mental retardation.

Serious neurodevelopmental abnormalities are an important component of all of the teratogenic syndromes listed in Table 1. In fact, there are very few teratogenic effects recognized in humans that do not involve either cognitive or behavioral abnormalities or both. Embryonic and fetal development of the central nervous system (CNS) is extremely complex in humans, so it is not surprising that most exposures that are capable of interfering with normal development of the embryo may also affect development of its CNS. Moreover, CNS development continues throughout gestation, so the potential for damage by teratogen exposure exists throughout pregnancy.

III. Limitations of Pattern Recognition

There are two different approaches to identifying patterns of congenital anomalies caused by maternal exposures during pregnancy. One approach involves studying the children of women who had a particular exposure during pregnancy. The corresponding epidemiologic method is the cohort study. Most human investigations of potential neurodevelopmental toxins employ cohort studies.

An alternative approach is identifying a common exposure among the mothers of children with a certain abnormality or pattern of abnormalities. The corresponding epidemiologic design is the case-control study. Case-control studies and cohort studies can be evaluated statistically and can provide quantitative estimates of the strength of an observed association. Neither case reports nor clinical series provide reliable quantitative

TABLE 1 Some Human Teratogenic Syndromes Recognized on the Basis of a Characteristic Pattern of Congenital Anomalies

Agent	Dose	Susceptible period of prenatal development	Characteristic pattern of congenital anomalies	Frequent neurodevelopmental abnormalities	References
Alcohol	Abuse (more than 90 ml of absolute ethanol per day)	Throughout pregnancy	Fetal alcohol syndrome: growth retardation, microcephaly, typical facial appearance, cardiac malformations	Developmental delay, neurobehavioral deficits, brain malformations	Jones *et al.*, 1973; Ginsburg *et al.*, 1991; Coles, 1992
Toluene	Abuse (by inhalation)	Unknown, but probably throughout pregnancy	Toluene embryopathy: microcephaly, growth deficiency, typical facial appearance, small fingernails	Developmental delay, attention deficit disorder	Hersh *et al.*, 1985; Arnold *et al.*, 1994
Thalidomide	Usual therapeutic dose	20–36 days after conception	Thalidomide embryopathy: characteristic limb reduction defects, microtia, ocular abnormalities, cardiac malformations	Facial nerve palsy, autism, developmental delay	Newman, 1986; Smithells and Newman, 1992
Anticonvulsant medications (trimethadione, phenytoin, valproic acid, carbamazepine)	Usual therapeutic dose	Malformations early in first trimester; other features may occur later in gestation	Fetal anticonvulsant syndrome: typical facial appearance, growth retardation, microcephaly, small fingernails	Developmental delay, spina bifida (with valproic acid or carbamazepine)	German *et al.*, 1970; DiLiberti, *et al.*, 1984; Hanson, 1986; Robert, 1988; Jones *et al.*, 1989
Aminopterin	Very large dose (used in attempt to induce abortion)	First trimester	Aminopterin embryopathy: growth retardation, abnormal calvarial ossification, typical facial appearance, abnormal auricles, cleft palate	Hydrocephalus, developmental delay	Thiersch, 1952; Warkany, 1978
Warfarin	Usual therapeutic dose	Skeletal features—latter half of first trimester; CNS anomalies—throughout pregnancy	Warfarin embryopathy: nasal hypoplasia, stippled epiphyses on radiographs, eye anomalies, growth retardation	Congenital anomalies of the CNS	Warkany, 1976; Hall *et al.*, 1980
Isotretinoin	Usual therapeutic dose	First trimester	Isotretinoin embryopathy: typical facial appearance, cleft palate, cardiac and great vessel malformations	CNS malformations, developmental delay	Rosa, 1983; Lammer *et al.*, 1985; Rosa *et al.*, 1986
Polychlorinated biphenyls (PCBs)	Ingestion of maternally toxic amounts	Unknown	Rice oil disease: growth retardation; discoloration of skin and mucous membranes; natal teeth—abnormal calvarial calcification	Behavioral abnormalities, developmental delay, hypotonia	Miller, 1985; Hsu *et al.*, 1985; Tilson *et al.*, 1990
Methylmercury	Ingestion of maternally toxic amounts	Throughout pregnancy	Minimata disease	Cerebral palsy, psychomotor retardation	Murikami, 1972; Amin-Zaki *et al.*, 1974

estimates of the risk of adverse outcome in children of women with a particular exposure during pregnancy.

At least 5% of all children have a serious congenital anomaly or mental retardation that becomes apparent within the first year of life (Baird *et al.,* 1988). The frequency of learning disabilities and behavioral disorders in childhood and adolescence is even greater. Many women continue to smoke cigarettes, drink alcohol, or use other "social" drugs while pregnant. Occupational exposures frequently continue during pregnancy as well. Prescription drugs are widely used in pregnancy, and pregnant women may take over-the-counter medication even more often. Coincidental occurrence of a potentially toxic exposure in a pregnant woman and congenital anomalies in her child is, therefore, very common.

Case reports and clinical series are important means of raising causal hypotheses, but most such hypotheses are incorrect. In other words, pattern recognition is a method with good sensitivity but poor specificity. Pattern recognition is subjective, and subjective impressions can be erroneous. This can account for claims that have been made but not independently substantiated for malformation syndromes associated with certain teratogenic exposures.

To avoid these pitfalls, observation of a pattern of anomalies in a few case reports or clinical series alone is not considered sufficient evidence to establish the teratogenicity of an exposure in humans (Brent, 1986; Scialli, 1992; Shepard, 1995). Concluding that a particular exposure may be teratogenic requires consistent findings from several studies. The conclusion is strengthened if studies of different design (e.g., clinical series, case-control studies, and cohort studies) all produce similar findings. Additional evidence may be provided by a mammalian model in which the teratogenic effect is reproduced under conditions of exposure that are analogous to those that occur in humans. A true teratogenic effect is also expected to make biologic sense. For example it does not make biologic sense for an exposure that does not result in systemic absorption of the agent by the mother to be teratogenic.

IV. Syndromes of Cognitive or Behavioral Abnormalities

Diagnosis of psychiatric disease is based largely on the recognition of qualitatively distinct patterns of abnormal behavior and cognition (American Psychiatric Association, 1994). The approach employed is quite similar to the one used to identify patterns of physical anomalies resulting from teratogenic exposures in case reports and clinical series.

Pattern recognition has been used only to a limited extent in behavioral teratology studies. This is due, at least in part, to the complexity and variability of many functional phenotypes. Phenotypic complexity and variability do not necessarily preclude the recognition of patterns, however. The physical phenotypes that characterize many teratogenic syndromes are also complex and highly variable. The possibility that some human neurodevelopmental toxins might be identified more easily on the basis of characteristic patterns of behavioral or cognitive abnormalities must be remembered.

Initial recognition of a pattern of anomalies usually occurs in patients with very striking features (Hanson, 1996). As the pattern becomes better known, the variability becomes apparent, and patients with milder manifestations may be recognized. Often there is overlap between mildly affected patients and normal individuals who have some of the same features on a familial basis or as a result of normal variation. Identifying mildly affected individuals is a particular problem with teratogenic syndromes because severity of the manifestations is expected to vary with the magnitude of the mother's exposure during pregnancy, and mild exposures are generally more common than severe ones.

An important example of this problem is provided by alcohol-related birth defects (Jones *et al.,* 1973; Ginsburg *et al.,* 1991; Coles, 1992). Classic fetal alcohol syndrome (FAS) occurs only in the children of women who abuse large amounts of alcohol chronically during pregnancy. However, neurodevelopmental abnormalities also are found with increased frequency among the children of women who drink smaller amounts during pregnancy. It is often difficult to distinguish adverse fetal effects of alcohol use from adverse effects of other exposures (e.g., cigarette smoking, or poor nutrition) and other factors (e.g., poverty or social deprivation) that often are found in women who drink heavily during pregnancy.

Although all of the teratogenic syndromes listed in Table 1 are frequently associated with neurodevelopmental abnormalities, similar functional abnormalities occur in children whose mothers were not exposed to these agents during pregnancy. It is usually impossible to attribute the neurodevelopmental abnormalities in an individual child to exposure of the mother to a potentially toxic agent during pregnancy, unless the child also has typical physical features of a teratogenic syndrome.

Characteristic patterns of behavioral or cognitive deficits have been recognized in many malformation syndromes that result from single gene or chromosomal abnormalities. In some instances, the pattern of behavioral or cognitive alterations is so distinctive that it may provide the strongest available clue to a diagnosis. A few examples of genetic malformation syndromes with characteristic behavioral or cognitive phenotypes are given in Table 2.

Although Table 2 does not include any teratogenic syndromes, there is no reason to doubt that processes altered by genetic defects can also be affected by teratogenic exposures. The characteristic behavioral phenotype of FAS provides the best known example of a teratogenic syndrome of neurobehavioral dysfunction. The pattern of attention deficits, poor judgment, and impulsivity in association with serious intellectual limitations is often more typical than the physical features in an adult with FAS (Streissguth, 1992; Spohr *et al.*, 1993, 1994; Niccols, 1994).

How can additional characteristic patterns of cognitive or behavioral abnormalities be recognized? One approach is to investigate children who have typical physical manifestations of a teratogenic embryopathy. Not all children born to women who experience a potentially teratogenic exposure during pregnancy actually are damaged. Differences in fetal manifestations may reflect differences in dose, route of exposure, timing or duration of exposure, or the effects of concomitant exposures or conditions. In addition, genetic differences in susceptibility (e.g., in the capacity of the mother or the fetus to inactivate a toxic chemical) are probably important in some instances. In any case, it is a reasonable presumption that children who have the most severe physical manifestations of teratogenic damage are also most likely to exhibit a related behavioral or cognitive phenotype, if it exists.

Several methods of identifying characteristic patterns of neurodevelopmental abnormalities are available. Interviews with parents and teachers can provide useful qualitative cognitive and behavioral profiles of affected children. This approach has resulted in the recognition of some of the characteristic functional patterns listed

TABLE 2 Some Human Multiple Congenital Anomaly Syndromes in Which Highly Characteristic Behavioral or Cognitive Abnormalities Occur

Syndrome	Etiology	Characteristic behavioral or cognitive feature	Other typical features	References
Lesch–Nyhan syndrome	Mendelian: X-linked recessive mutation of *HPRT1*	Compulsive self-mutilation and aggressiveness, dysarthric speech	Poor growth, developmental delay, abnormal tone, choreoform and athetoid movements	Christie *et al.*, 1982; Seegmiller, 1996
Fragile X syndrome	Irregular mendelian: X-linked recessive unstable expansion of trinucleotide repeat sequence in *FMR1*	Rapid, perseverative, and impulsive speech; extreme shyness, gaze aversion, and other autistic-like behaviors; hyperactivity	Mental retardation, long face, large mandible, large ears, large testes	Simko *et al.*, 1989
Prader–Willi syndrome	Usually microdeletion of paternal 15q11 (abnormal imprinting and other mechanisms occur less frequently)	Obsessive hyperphagia; strength in visual motor discrimination and other perceptual skills; weakness in sequential processing and short-term memory	Mental retardation, hypotonia in infancy, obesity in childhood, hypogonadism	Curfs *et al.*, 1991; Whitman and Greenswag, 1995
Angelman syndrome	Usually microdeletion of maternal 15q11 (abnormal imprinting and other mechanisms occur less frequently)	Laughter with minimal or inappropriate stimulus; sleep disorder; love of water and reflections	Severe mental retardation, complete or almost complete lack of speech, ataxia, seizures	Clayton-Smith, 1993
Williams syndrome	Usually microdeletion of 7q11 (other mechanisms also occur)	"Cocktail party affect"; lack of reserve or distancing behavior with strangers; hypersensitivity to sounds	Mental retardation, characteristic appearance, supravalvular aortic stenosis	Gosch and Pankau, 1994
Sotos syndrome	Unknown	Tantrums, destructive behavior, social withdrawal, poor coordination	Mental retardation, large size in childhood, large hands and feet	Rutter and Cole, 1991

in Table 2 (Christie *et al.,* 1982; Greenswag, 1987; Gosch and Panku, 1994). Formal developmental testing is very useful in quantitating these differences and can often be enhanced by concentrating the assessment on areas of particular concern in a given syndrome. Analyzing the data obtained with techniques of numeric taxonomy may also aid in recognizing important patterns of anomalies (Sneeth and Sokol, 1973; Abbot *et al,* 1985). Affected and control subjects can be evaluated in a comparative fashion (Chambers *et al.,* 1996; Nulman *et al.,* 1997), but it may not be possible to blind the assessment if the affected subjects have a characteristic physical phenotype and the control subjects do not.

Once a specific pattern of cognitive or behavioral abnormalities has been identified in patients with physical features of a teratogen embryopathy, one can look for the same pattern of functional alterations in children who do not have the embryopathy but whose mothers experienced similar exposures during pregnancy. Finding a similar pattern of cognitive and behavioral abnormalities provides convincing evidence of a neurodevelopmental effect.

Although the recognition of a possible pattern of behavioral or cognitive abnormalities in case reports or clinical series may not be sufficient by itself to establish the developmental toxicity of a particular exposure in humans, such recognition can be a very important first step. Incorporating qualitative functional assessment into the evaluation of potential developmental neurotoxins can complement the use of quantitative methods. Recognizing patterns of functional abnormalities is likely to enhance our ability to identify behavioral and cognitive teratogenic exposures in humans.

References

Aase, J. M. (1990). *Diagnostic Dysmorphology.* Plenum Medical Book Co., New York, pp. 268–269.

Abbott, L. A., Bisby, F. A., and Rogers, D. J. (1985). *Taxonomic Analysis in Biology: Computers, Models and Databases.* Columbia University Press, New York.

American Psychiatric Association (1994). Diagnostic and Statistical Manual of Mental Disorders: DSM-IV. Washington DC. American Psychiatric Association.

Amin-Zaki, L., Elhassani, S., Majeed, M. A., Clarkson, T. W., Doherty, R. A., and Greenwood, M. (1974). Intra-uterine methylmercury poisoning in Iraq. *Pediatrics* **54,** 587–595.

Arnold, G. L., Kirby, R. S., Langendoerfer, S., and Wilkin-Haug, L. (1994). Toluene embryopathy: clinical delineation and developmental followup. *Pediatrics* **93,** 216–220.

Baird, P. A., Anderson, T. W., Newcombe, H. B., and Lowry, R. B. (1988). Genetic disorders in children and young adults: a population study. *Am. J. Hum. Genet.* **42,** 677–693.

Brent, R. A. (1986). Evaluating the alleged teratogenicity of environmental agents. *Clin. Perinatol.* **13,** 609–603.

Chambers, C. A., Johnson, K. A., Dick, L. M., Felix, R. J., and Jones, K. L. (1996). Birth outcomes in pregnant women taking fluoxetine. *N. Engl. J. Med.* **335,** 1010–1015.

Christie, R., Bay, C., Kaufman, I. A., Bakay, B., Borden, M., and Nyhan, W. L. (1982). Lesch-Nyhan disease: clinical experience with nineteen patients. *Develop. Med. Child Neurol.* **24,** 293–306.

Clayton-Smith, J. (1993). Clinical research on Angelman syndrome in the United Kingdom: observations on 82 individuals. *Am. J. Med. Genet.* **46,** 12–15.

Coles, C. D. (1992). Prenatal alcohol exposure and human development. In Development of the Central Nervous System: Effects of Alcohol and Opiates. M. W. Miller, Ed. Wiley-Liss, New York, pp. 9–36.

Curfs, L. M. G., Wiegers, A. M., Sommers, J. R. M., Borghgraef, M., and Fryns, J. P. (1991). Strengths and weaknesses of youngsters with Prader–Willi syndrome. *Clin. Genet.* **40,** 430–44.

DiLiberti, J. H., Farndon, P. A., Dennis, N. R., and Curry, C. J. R. (1984). The fetal valproate syndrome. *Am. J. Med. Genet.* **19,** 473–481.

German, J., Kowal, A., and Ehlers, K. H. (1970). Trimethadione and human teratogenesis. *Teratology* **3,** 349–362.

Ginsburg, K. A., Blacker, C. M., Able, E. L., and Sokol, R. J. (1991). Fetal alcohol exposure and adverse pregnancy outcomes. *Contrib. Gynecol. Obstet.* **18,** 115–129.

Gosch, A., and Pankau, R. (1994). Social–emotional and behavioral adjustment in children with Williams–Bueren syndrome. *Amer. J. Med. Genet.* **53,** 335–339.

Greenswag, L. R. (1987). Adults with Prader–Willi syndrome: a survey of 232 cases. *Develop. Med. Child. Neurol.* **29,** 145–152.

Hall, J. G., Pauli, R. M., and Wilson, K. M. (1980). Maternal and fetal sequelae of anticoagulation during pregnancy. *Am. J. Med.* **68,** 122–140.

Hanson, J. W. (1986). Teratogen update: fetal hydantoin effects. *Teratology* **33,** 349–353.

Hanson, J. W. (1996). Human teratology. In D. L. Rimoin, J. M. Connor, R. E. Pyeritz, Eds. *Emery and Rimoin's Principles and Practice of Medical Genetics,* 3rd ed., Churchill Livingstone, New York, pp. 697–724.

Hersh, J. H., Podruch, P. E., Rogers, G., and Weisskopf, B. (1985). Toluene embryopathy. *J. Pediatr.* **106,** 922–927.

Hsu, S. T., Ma, C. I., Hsu, S. K., Wu, S. S., Hsu, N. H., Yeh, C. C., and Wu, S. B. (1985). Discovery and epidemiology of PCB poisoning in Taiwan: a four-year followup. *Environ. Health Perspect.* **59,** 5–10.

Jones, K. L., Lacro, R. V., Johnson, K. A., Adams, J. (1989). Pattern of malformations in the children of women treated with carbamazepine during pregnancy. *N. Engl. J. Med.* **320,** 1661–1666.

Jones, K. L., Smith, D. W., Ulleland, C. N., and Streissguth, A. P. (1973). Pattern of malformation in offspring of chronic alcoholic mothers. *Lancet* **1,** 1267–1271.

Lammer, E. J., Chen, D. T., Hoar, R. M., Agnish, N. D., Benke, P. J., Braun, J. T., Curry, C. J., Fernhoff, P. M., Grix, A. W., and Lott, I. T., et al. (1985). Retinoic acid embryopathy. *N. Engl. J. Med.* **313,** 837–841.

Miller, R. W. (1985). Congenital PCB poisoning: a reevaluation. *Environ. Health Perspect.* **60,** 211–214, 1985.

Murakami, U. (1972). Organic mercury problem affecting intrauterine life. In International Symposium on the Effect of Prolonged Drug Usage on Fetal Development. M. A. Klingberg, A. Abramovici, J. Chemke, *Adv. Exper. Med. Biol.* **27,** 301–336.

Newman, C. G. H. (1986). The thalidomide syndrome: risks of exposure and spectrum of malformations. *Clin. Perinatol.* **13,** 555–573.

Niccols, G. A. (1994). Fetal alcohol syndrome: Implications for psychologists. *Clin. Psychol. Rev.* **14,** 91–111.

Nulman, I., Rovet, J., Stewart, D. E., Wolpin, J., Gardner, H. A., Theis, J. G. W., Kulin, N., and Koren, G. (1997). Neurodevelopment of children exposed *in utero* to antidepressant drugs. *N. Engl. J. Med.* **336,** 258–262.

Robert, E. (1988). Valproic acid as a human teratogen. *Congen. Anom.* **28,** S71–80.

Rosa, F. W. (1983). Teratogenicity of isotretinoin. *Lancet* **2,** 513.

Rosa, F. W., Wilk, A. L., and Kelsey, F. O. (1986). Teratogen update: vitamin A congeners. *Teratology* **33,** 355–364.

Rutter, S. C., and Cole, T. R. (1991). Psychological characteristics of Sotos syndrome. *Develop. Med. Child Neurol.* **33,** 898–902.

Scialli, A. R. (1992). A Clinical Guide to Reproductive and Developmental Toxicity. CRC Press, Boca Raton, Florida, pp. 257–273.

Seegmiller, J. E. (1996). Purine and pyrimidine metabolism. In *Emery and Rimoin's Principles and Practice of Medical Genetics,* 3rd ed., D. L. Rimoin, J. M. Connor, R. E. Pyeritz, Eds. Churchill Livingstone, New York, pp. 1921–1944.

Shepard, T. H. (1995). Catalog of Teratogenic Agents, 8th ed. Johns Hopkins University Press, Baltimore, pp. xii–xvi.

Simko, A., Hornstein, L., Soukup, S., and Bagamery, N. (1989). Fragile X syndrome: recognition in young children. *Pediatrics* **83,** 547–552.

Smithells, R. W. (1987). The demonstration of teratogenic effects of drugs in humans. In *Drugs and Pregnancy: Human Teratogenesis and Related Problems,* 2nd ed., D. F. Hawkins, Ed. Churchill Livingstone, New York, pp. 27–36.

Smithells, R. W., and Newman, C. G. H. (1992). Recognition of thalidomide defects. *J. Med. Genet.* **29,** 716–723.

Sneath, P. H. A., Sokal, R. R. (1973). *Numerical Taxonomy: The Principles and Practice of Numerical Classification.* W. H. Freeman and Company, San Francisco..

Spohr, H.-L., Willms, J., and Steinhausen, H.-C. (1993). Prenatal alcohol exposure and long-term developmental consequences. *Lancet* **341,** 907–910.

Spohr, H.-L., Willms, J., and Steinhausen, H.-C. (1994). The fetal alcohol syndrome in adolescence. *Acta. Paeditr.* (Suppl) **404,** 19–26, 1994.

Streissguth, A. P. (1992). Fetal alcohol syndrome and fetal alcohol effects: a clinical perspective of later developmental consequences. In *Maternal Substance Abuse and the Developing Nervous System.* I. S. Zagon, T. A. Slotkin, Eds. Academic Press, San Diego, pp. 5–25.

Thiersch, J. B. (1952). Therapeutic abortions with a folic acid antagonist, 4-aminopteroylglutamic acid (4-amino PGA) administered by the oral route. *Am. J. Obstet. Gynecol.* **63,** 1298–1304.

Tilson, H. A., Jacobson, J. L., and Rogan, W. J. (1990). Polychlorinated biphenyls and the developing nervous system: cross-species comparisons. *Neurotoxicol. Teratol.* **12,** 239–248, 1990.

Warkany, J. (1976). Warfarin embryopathy. *Teratology* **14,** 205–210.

Warkany, J. (1978). Aminopterin and methotrexate: folic acid deficiency. *Teratology* **17,** 353–357.

Whitman, B. Y., and Greenswag, L. R. (1995). Phychological and behavioral management. In *Management of Prader–Willi Syndrome,* 2nd ed. L. R. Greenswag, and R. C. Alexander, Eds. Springer-Verlag, New York, pp. 125–141.

PART VII

Specific Neurotoxic Syndromes

One of the most frequent and important questions raised when considering the science of developmental neurotoxicology is, "What categories of chemicals would induce maldevelopment of the nervous system?" Part VII is devoted to presenting and discussing some of the more specific neurotoxic syndromes as a result of chemical exposure.

It has been claimed that more than one-third of the toxic chemicals recognized by the U.S. Environmental Protection Agency are known to be neurotoxic, and the list is increasing rapidly with the implementation of more sensitive screening and assessment techniques. However, all of these chemicals may be classified into two primary exposure groups: involuntary exposure chemicals, such as environmental and occupational (industrial) chemicals; and voluntary exposure chemicals, such as drugs of abuse and therapeutic drugs.

It is obvious that it is impossible to cover all chemicals with the potential to induce developmental neurotoxicity. In Part VII, some of the prime examples are selected to illustrate the general principles, developmental–toxic consequences, and toxicologic mechanisms of action. Specific comparisons are also given between animal models and human situations.

Heavy metals, especially mercury, cadmium, and lead, are perhaps the best known environmental toxicants that have strong influences on the development of the nervous system. As a result of the massive outbreak of methylmercury poisoning in Minamata Bay, Japan, in the 1950s, the general toxicologic impact of methylmercury on the human population is referred to as *Minamata disease.* Developing fetuses and newborns were affected, resulting in the designation of *fetal Minamata disease* (congenital methylmercury poisoning). In Chapter 29, Louis W. Chang and Grace Liejun Guo provide an overview on the history, clinical aspects, pathology, and biomolecular basis of this disease.

In Chapter 30, Lloyd Hastings and Marian L. Miller present a comprehensive account of the neurotoxicity of cadmium—its metabolism and neurotoxic consequences as well as current thoughts on the mechanisms of action of the toxic metal on the developing nervous system.

Perhaps the toxic metal receiving the most attention is lead. Special interest in lead neurotoxicity is not rested on its potential for inducing gross abnormalities in the nervous system, but for its suppressive effect on the intellectual development (IQ) in children exposed to this toxicant. In Chapter 31, Deborah C. Rice gives a detailed review on lead exposure to young children and the subsequent neurobehavioral consequences. Although there is an overwhelming (and therefore sometimes confusing) amount of literature in the area of neurobehavioral toxicology of lead, Dr. Rice presents a clear concept and convincing evidence on this issue. A summary of lead-induced behavioral deficits in animals is also presented, something it is hoped that readers will welcome.

Pesticides are perhaps the most widely used chemicals in the world, from ordinary households to farms and commercial orchards. Most, if not all, pesticides are considered toxic to humans. They can be viewed as occupational and environmental toxicants. Chapter 32 by Durisalah Desaiah is devoted to discussing the different classes of pesticides and their adverse effects on the developing nervous system. The mechanisms of action on the neurotransmitter systems, receptors, and signal transduction pathways are also discussed.

The two main categories of voluntary exposure chemicals are the drugs of abuse and the therapeutic drugs; that is, individuals are exposed to these chemicals voluntarily and intentionally. Among the drugs of abuse, it is those also known as the "social drugs"—alcohol and cigarettes (nicotine)—whose use is most widespread. Children born to mothers who are alcoholics or heavy cigarette smokers frequently encounter physical as well as intellectual deficits. These toxic syndromes are comprehensively addressed in two chapters. In Chapter 33 by Gideon Koren, Bonnie O'Hayan, Jonathan Gladstone, and Irena Nulman, both the clinical aspects and the possible mechanisms contributing to the teratogenic effects of alcohol are presented and discussed. Fetal alcohol syndrome (FAS) is a well-recognized developmental problem in the offspring of mothers who abuse alcohol during pregnancy. In Chapter 34 by Edward D. Levin and Theodore A. Slotkin, the epidemiologic evidence concerning neurologic deficits and neurobehavioral changes in the offspring of mothers who smoked cigarettes during pregnancy is reviewed. The cellular effects as well as neurobehavioral alterations as a result of maternal nicotine exposure are also presented and discussed.

In Chapter 35, Merle G. Paule discusses the investigation of maternal drug abuse, with specific emphasis on several drugs of abuse, including marijuana, opiates, phencyclidine, and cocaine. This subject is a dilemma for basic scientists and for clinicians alike, as establishing an animal model that simulates human situations is not always successful.

Although therapeutic drugs are used to benefit individuals taking the medications, at times there are the tragic consequences that the "medicines" become toxic to the fetus. Two prime examples are included in Part VII. In Chapter 36 by Jane Adams and R. Robert Holson, the neural and behavioral teratology of three retinoids—vitamin A, all-*trans* retinoic acid, and 13-*cis* retinoic acid—are reviewed. Comparative consequences in both animal models and humans are also presented and discussed. Although antiepileptic drugs, such as phenobarbital, phenytoin, and valproic acid, enjoyed successful results in controlling epilepsy; these drugs pose certain risks for pregnant patients. Chapter 37, by Deborah K. Hansen and R. Robert Holson, reviews the effects of these drugs on the development of the brain as well as on learning and behavior.

Autism is one of the most devastating neurologic deficits found in children. This syndrome is believed to be, by and large, genetic in origin. In Chapter 38, Patricia M. Rodier approaches this syndrome from the standpoint of a neuroteratologist and neurotoxicologist. Teratogens such as thalidomide (also a therapeutic drug) are examined as chemicals that could induce autism or autism-like syndromes, thus establishing a chemical model for the study of genetic disease.

Louis W. Chang

CHAPTER

29

Fetal Minamata Disease

Congenital Methylmercury Poisoning

LOUIS W. CHANG
Departments of Pathology, Pharmacology, and Toxicology
College of Medicine
University of Arkansas for Medical Sciences
Little Rock, Arkansas 72205

GRACE LIEJUN GUO
Department of Pharmacology and Toxicology
University of Kansas Medical Center
Kansas City, Kansas 66160

I. Introduction

In the mid-1950s, a peculiar disease with characteristic neurologic signs and symptoms: sensory disturbance, cerebellar ataxia, and constriction of visual field was described in Minamata Bay, Japan. This disease was later found to be related to methylmercury effluent from an acetaldehyde plant at the bay. The term "Minamata disease" has since then been equated with methylmercury poisoning (Study Group of Minamata Disease, Kumamoto University, 1968).

In 1958, Progressor Kitamura of Japan noticed that many infants from mothers of Minamata Bay born after 1955 developed characteristic neurologic symptoms, including mental retardation and cerebral pasly (Kitamura *et al.,* 1960). Later clinical and epidemiologic studies concluded that methylmercury in Minamata also induced congenital intoxication via transplacental transfer of mercury to the fetus (Harada, M., 1964, 1974a,b,c, 1976; Matsumoto and Takeuchi, 1965; Harada, Y., 1975). The term *congenital methylmercury poisoning* or *fetal Minamata disease* (FMD) was used to describe such medical phenomenon (Matsumoto and Takeuchi, 1965; Synder, 1971; Harada, Y., 1977; Marsh *et al.,* 1980).

In the 1970s a massive outbreak of methylmercury poisoning also occurred in Iraq, inducing both adult and infantile forms of clinical episodes (Bakir *et al.,* 1973;

* Abbreviations: aNCAM, adult neural cell adhesion molecule; CNS, central nervous system; eNCAM, embryonic neural cell adhesion molecule; FMD, fetal Minamata disease; GTP, glutathione triphosphate; MAP, microtubule-associated protein; NCAM, neural cell adhesion molecule.

Amin-Zaki *et al.,* 1974, 1976, 1981). The major difference between the Japanese episode and the Iraqi episode is the chronicity of exposure: the Japanese situation represented chronic methylmercury exposure (contaminated seafood), whereas the Iraqi condition was a more acute form (contaminated grains) of poisoning.

II. Human Episodes

Of the patients with FMD in Japan, 64% were living in families with members inflicted with severe Minamata disease (Harada, M., 1964). However, it is notable that mothers of children with FMD usually displayed only mild or no clinical symptoms of mercury poisoning (Harada, M., 1972, 1974a) when compared to other adults similarly exposed to mercury. Mercury analyses on hairs from both mothers and infants showed a significant elevation of mercury levels (Kitamura *et al.,* 1960; Harada, M., 1976). Analyses of umbilical cords from newborns inflicted with FMD also showed elevated mercury concentrations (Harada, Y., 1975).

Typical neurologic symptoms in FMD are summarized in Table 1. The most characteristic symptoms are mental retardation, dysarthria, cerebellar symptoms (such as ataxia), deformity of limbs, primitive reflexes (grasping, sucking, etc.), hypersalivation, and stunted body growth. The expression of clinical symptoms varied widely, ranging from mildly affected individuals to severe mental retardations and complete physical incapacitation (Harada, M., 1974b,c). The first signs of neurologic deficiency in the Minamata cases usually occurred in infants at an early age. Delayed movements, failure to follow visual stimuli, and uncoordinated sucking and swallowing were early signs of the disease. These signs were followed by persisting primitive reflexes and marked impairment of coordination. Although there was constriction of visual fields in the Minamata patients, no blindness was recorded. In contrast, 5 out of 15 patients examined in the Iraqi episode revealed blindness (Amin-Zaki *et al.,* 1976). The difference in clinical pictures between the Japanese and Iraqi episodes is most likely due to the differences in exposure conditions (chronic vs. acute).

Motor function deficits and mental retardation usually increase in severity with age, whereas other symptoms such as primitive reflexes and cerebellar symptoms might reduce with time.

TABLE 1 Prevalence of Symptoms in 22 Cases of Prenatal Methylmercury Intoxication in Minamata

Symptoms	Prevalence (%)
Mental disturbance	100
Ataxia	100
Impairment of gait	100
Disturbance in speech	100
Disturbance in chewing and swallowing	100
Brisk and increased tendon reflex	82
Pathologic reflexes	54
Involuntary movement	73
Salivation	77
Forced laughing	27

Visual fields and hearing of the victims were not examined. Data are based on a study of Harada (1968).

(Reprinted with permission from Chang, L. W. (1984). Developmental toxicology of methylmercury. In *Toxicology of the Newborn.* S. Kacew and M. J. Reasor, Eds. Elsevier Pub., Amsterdam, NY. pp. 176.)

III. Neuropathology of Fetal Minamata Disease

The main pathologic information of human congenital methylmercury poisoning was obtained from a few autopsy examinations in the Minamata episode (Matsumoto and Takeuchi, 1965) and in the Iraqi cases (Choi *et al.,* 1978).

1. Minamata Cases

Although there are many cases of FMD, autopsy investigations were few. Characteristic neuropathologic changes (Matsumoto and Takeuchi, 1965) may be summarized as follows:

1. Bilateral cerebral atrophy and hypoplasia with a decrease and disappearance of cortical nerve cells. Many cortical nerve cells also appeared to be malformed, reduced in size, or with shrunken neuritic processes.
2. Cerebellar atrophy and hypoplasia: a reduction of cerebellar granule cell layer.
3. Abnormality in neural cytoarchitecture with atopic and disoriented neurons. There were remaining matrix cells as well as intramedullary preservation of nerve cells, suggesting a disruption of neuronal maturation and migration.
4. Hypoplasia of corpus callosum.
5. Dysmyelination of white matter.
6. Hydrocephalus.

Microscopic examinations revealed that there was widespread neuronal involvement throughout the brain. The lesions distributions in the central nervous system (CNS) in fetal and infantile Minamata patients were different than those in the adult form, which showed marked anatomic selectivity of pathologic involvement

(Fig. 1). In FMD, disruption of cerebral and cerebellar cytoarchitecture was prominent. Many neurons appeared to be hypoplastic, ectopic, dysplastic, and disoriented (Figs. 2 and 3), strongly indicative of disrupted neuronal maturation, growth, and migration.

Extensive neuronal loss, particularly the granule cells in the cerebellum and the pyramidal neurons in the cerebrum, occurred throughout the brain (Fig. 4). Hypomyelination and dysmyelination of the white matter was also noted (Fig. 4).

2. Iraqi Cases

Grossly, the brains appeared to be smaller (hypoplastic), with abnormal gyral patterns. Heterotopic gray and hypoplastic basal ganglia and corpus callosum were also noticeable.

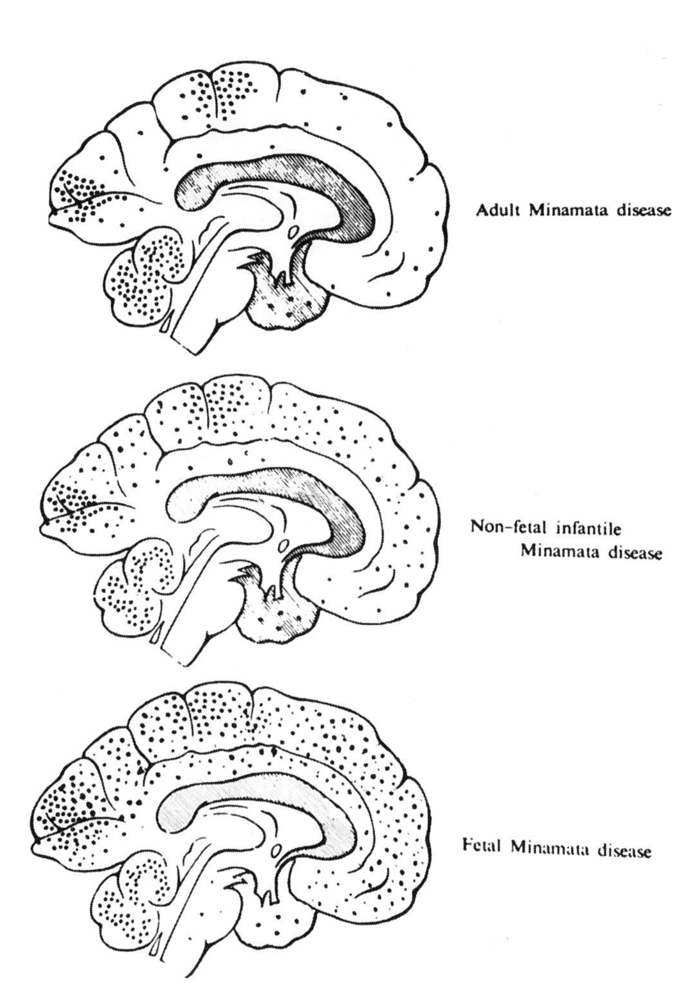

FIGURE 1 Comparison of lesion distribution and involvement in the brains of adult, infantile, and fetal Minamata disease. (Reprinted with permission from Chang, L. W. [1984]. Developmental toxicology of methylmercury. In *Toxicology of the Newborn.* S. Kacew and M. J. Reasor, Eds. Elsevier Pub., Amsterdam, NY. p. 178).

Microscopically, marked disruption of cytoarchitecture was evident, with nests of heterotopic and ectopic neurons, irregular aggregates of nerve cells, an undulating pattern of neuronal layers, neurons with hypoplastic neurites, and astrocytic proliferation. Unlike the Japanese cases, only mild neuronal necrosis was found in these patients.

Mercury analysis revealed that peak maternal hair–mercury levels could be correlated with the frequency of maternal symptoms during pregnancy and to neurologic effects in the infants (Marsh *et al.,* 1981, 1987; Cox *et al.,* 1989).

Furthermore, minimal clinical signs occurred in children whose mother's hair–mercury levels were between 68 and 180 ppm, and the second trimester of pregnancy appeared to pose greatest risk for exposure (Marsh *et al.,* 1981).

A. *Placental and Mammary Transfer*

Methylmercury was found to cross the placental barrier readily with a direct toxic effect on the fetus. It was noted that the blood–mercury of the affected infants was significantly higher than that of the mother (Amin-Zaki *et al.,* 1974). Kuhnert *et al.* (1981) compared the mercury levels in human maternal blood and cord blood and reported that there was 30% more methylmercury in the erythrocytes of fetal blood than in the erythrocytes of the maternal blood. Other study further indicated that transfer of mercury from maternal blood to fetal blood far exceeded the reverse (Reynolds and Pitkin, 1975; Hamada *et al.,* 1997). When the brain–mercury was compared, it was also found that the concentration of mercury in the fetal brain was at least twice that of their mothers (Amin-Zaki *et al.,* 1974). This information strongly endorsed the concept that the fetus serves as a "mercury trap" in pregnant mothers, resulting in higher tissue concentrations of mercury than in the maternal tissues.

Aside from placental transfer, the developing infant can also be exposed to mercury via the mother's milk (Amin-Zaki *et al.,* 1981; Grandjean *et al.,* 1994; Kacew, 1996). Amin-Zaki *et al.* (1974) reported that when the mother received substantial exposure to methylmercury, suckling infants may acquire blood–mercury levels in excess of 200 ppb, which is considered to be minimal toxic level for adults. Studies by Grandjean *et al.* (1994) also demonstrated that hair–mercury of infants increased with the length of the nursing period. Furthermore, increasing time interval from weaning to hair sampling did not decrease in mercury concentration, suggesting that a slow or absence of elimination of methylmercury during this time period (first year) of life.

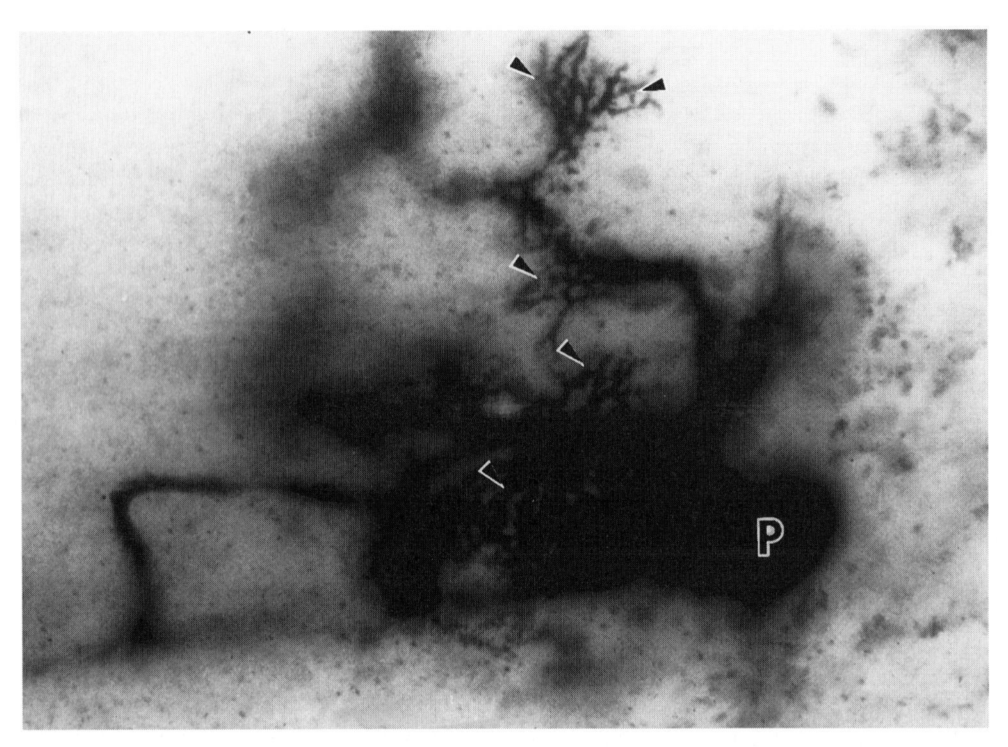

FIGURE 2 Purkinje neuron (P), cerebellum, fetal Minamata disease (human). Note the highly hypoplastic (stunted) dendritic processes (→) of the nerve cell. Rapid Golgi staining technique, × 400 (Reprinted with permission from Chang, L. W. [1984]. Developmental toxicology of methylmercury. In *Toxicology of the Newborn.* S. Kacew and M. J. Reasor, Eds. Elsevier Pub., Amsterdam, NY. p. 180).

B. Experimental Congenital Mercury Poisoning

Extensive research efforts have been made to elucidate the effects and mechanisms of mercury on the developing nervous system. Much of the basic information and concepts were presented in two detailed reviews by this author in the late 1970s and mid-1980s (Reuhl and Chang, 1979; Chang, 1984; Chang and Annau, 1984). An excellent review updating the current concepts in this area was also published by Massaro (1996). These reviews provided a large literature base covering information and concepts on developmental teratology, neuropathology, cytogenetic effects, neurobehavioral changes, and biochemical–molecular alterations in the CNS as a result of congenital mercury intoxication. Therefore, there is no need to rereview these subjects again here. Rather, in this chapter, emphasis is made on discussing one of the most characteristic effects of methylmercury on the developing brain: disruption of neuronal migration and neural cytoarchitecture. Attention is also devoted to the molecular impact of mercury on the cell adhesion molecules and on the cytoskeletal system of the nerve cells. Readers are encouraged to seek other important information from reviews published previously (Reuhl and Chang, 1979; Chang, 1984; Chang and Annau, 1984; Massaro, 1996).

C. Disruption of Neuronal Migration and Neural Cytoarchitecture

The cortical structures of the mammalian brain show a characteristic regularity in the geometric and spatial pattern of neuronal groups and layers. Early neurons are almost exclusively derived from the ventricular zone of the developing neural tube and then migrate to their destinated position in an "inside-out" manner; that is, the early arriving nerve cells form the deepest layer in the cortex, whereas the later arriving cells form the more superficial layers. Because the final neural functions depend heavily on the subsequent establishment of neural circuitry, proper positioning of the developing cortical neurons is essential for the establishment of a fully functional neural circuitry. As stated in an earlier section of this chapter, in both Japanese and Iraqi situations, the FMD brains showed severe developmental defects characterized primarily by dysplastic, disoriented, heterotopic, and ectopic neurons throughout the CNS.

With cinematography, Choi (1986) demonstrated that normal neuronal migration required interactions between the neuronal neuritic outgrowth and the astrocytes or another neuron. Exposure of organotypic cultures to methylmercury appeared to disrupt these inter-

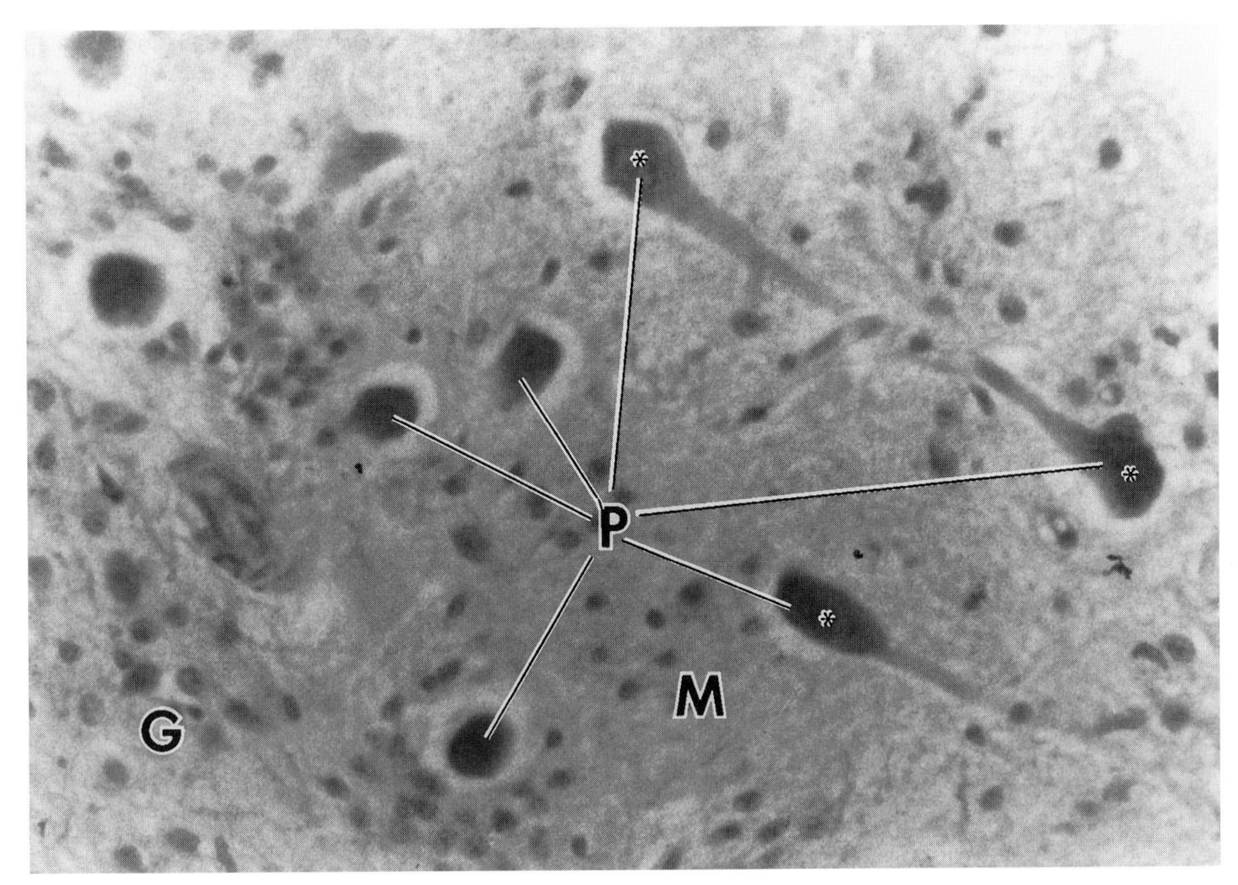

FIGURE 3 Cerebellar cortex, fetal Minamata disease (human). Note the ectopically located Purkinje cells (P) in the molecular layer (M) of the cerebellar cortex. Many of these Purkinje cells also showed a disoriented (horizontally oriented) position (*). G, granule cell layer, showing extensive cell loss. H&E, × 250 (Reprinted with permission from Chang, L. W. (1984). Developmental toxicology of methylmercury. In *Toxicology of the Newborn.* S. Kacew and M. J. Reasor, Eds. Elsevier Publ., Amsterdam, NY, p. 179).

actions, thus disrupting the neuronal movements (Choi, 1986). This observation spawned the concept that certain membrane changes induced by methylmercury would disrupt neuronal movements and, perhaps later, cell–cell contact establishments (synaptogenesis)

D. Biomolecular Basis of Neurotoxicity in FMD

Disruptions of neuronal migration and neural cytoarchitecture by methylmercury are closely related to two biomolecular changes: (1) alterations of neural cell adhesion molecules, and (2) disruption of neurocytoskeleton (microtubules).

1. Neural Cell Adhesion Molecule

Neural cell adhesion molecule (NCAM) is a surface glycoprotein and is a member of the immunoglobulin superfamily of molecules. There are three isoforms of NCAM, each of which is expressed differentially during development, undergoing posttranslational modification in phosphorylation and sulfation (Hoffman and Edelman, 1983). The embryonic NCAM (eNCAM) is the heavily sialylated isoform of NCAM. The high content of sialic acid renders a net negative charge in the membrane surface and also creates a repulsive force toward the negatively charged eNCAM in the adjacent cells. This situation inhibits membrane–membrane apposition and cell–cell adhesion as well as permits movement of cells related to each other during neuritic outgrowth and neuronal migration (Rutishauser and Jessel, 1988; Landmesser *et al.*, 1990). After the completion of neuronal migration, the eNCAM is then transformed to less sialylated adult form (aNCAM), which allows the cells to acquire tighter membrane–membrane intimacy and the establishment of synaptogenesis and stable cytoarchitecture.

The disruption of cell membrane integrity by methylmercury as observed by Choi (1986) would certainly disrupt the basic function of the associated eNCAM, thus effecting neuronal migration. Graff *et al.* (1994) further demonstrated that methylmercury, when given

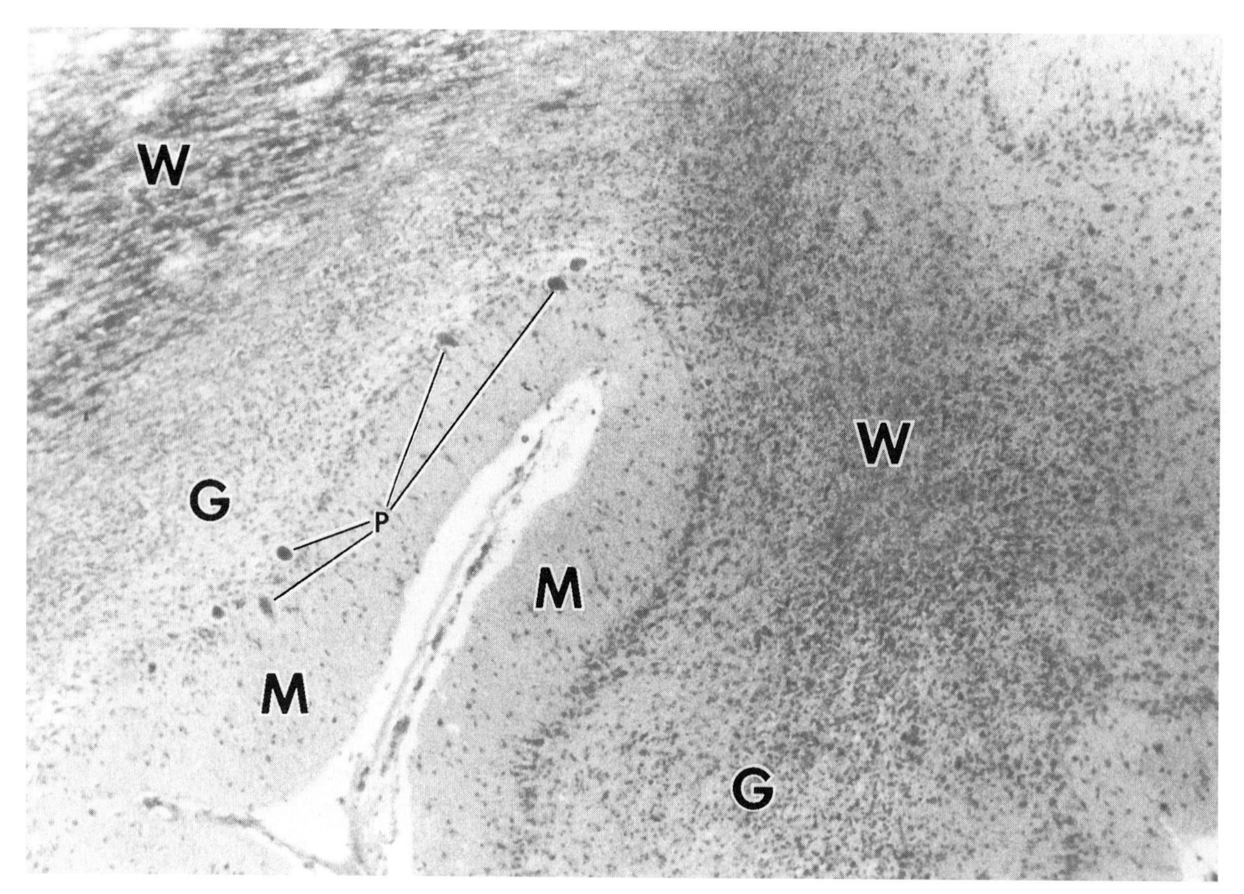

FIGURE 4 Cerebellar cortex, fetal Minamata disease (human). Note the generalized cell loss in the granule cell layer (G), hypoplastic and dysplastic Purkinje cells (P), and thinning of the white matter (W) with gliosis (hypo- and dysmyelination of the fiber tract, indicative by reduced LFB staining). M, molecular layer. LFB stain, × 100.

to neonatal rats, also disturbed posttranslational modifications of the NCAM molecule. There was a delayed conversion of the eNCAM to the poorly sialylated form of aNCAM (Lagunowich *et al.,* 1991). The prolonged retainment of high, negatively charged eNCAM during the developmental period would certainly also interfere with the establishment of synaptogenesis and proper cytoarchitecture (Regan, 1991; Reuhl and Dey, 1996). Indeed, reduced formation as well as malformation of synapses (dyssynaptogenesis) were reported by Chang *et al.* (1977) in rats exposed prenatally to methylmercury.

2. Neurocytoskeleton (Microtubules)

Microtubules (neurotubules) are another element that is important for cellular movements and kinetics. Microtubules are rod-shaped structures approximately 25 nm in diameter and are the largest of the three families (the other two being intermediate filaments and microfilaments) of the cytoskeleton. Associated with each of these cytoskeleton classes are families of proteins that modulate the stability and linkage of the cytoskeletal proteins with other cell structures. One of the more important cytoskeletal proteins is the microtubule-associated proteins (MAPs) (Matus, 1991). MAPs are important in the promotion of microtubule assembly (Nunez, 1986; Matus, 1991) in a stable fashion. The best described posttranslational modification of MAPs is phosphorylation. The phosphorylated form of MAP, especially MAP1B, is found in growing axons and in cross-bridges between microtubules (Matus, 1991), and this may be important in the development of the nervous system.

Metals, mercury included, have been reported to affect the cytoskeleton through interaction with microtubules (Cadrin *et al.,* 1988; Wasteneys *et al.,* 1988; Graff and Reuhl, 1990; Graff *et al.,* 1993; Graff and Reuhl, 1996). The regulation of microtubule assembly and stability may be disrupted when the following factors are affected: an adequate supply or synthesis of tubulin subunits and MAP; an adequate control of microtubule assembly under the influences of glutathione; *guanosine triphosphate* (GTP), posttranslational phosphorylation, and sulfhydryl accessibility. Methylmercury was found to influence all these factors (Chang and Verity, 1995; Chang, 1996; Verity, 1996).

It has been suggested that glutathione is an important regulator of tubulin sulfhydryl and could sequester toxic metals from cytoskeleton (Li and Chou, 1992). However, other investigations showed that depletion of intracellular glutathione by methylmercury had little influence on the disruption of microtubule assembly (Graff *et al.,* 1993).

It has been well established that free sulfhydryl groups are important for the polymerization of microtubules (Kuriyama and Sakai, 1974; Mellon and Rebhun, 1976). It is believed that methylmercury has high affinity for the sulfhydryl groups, leading to an inhibition on the polymerization of microtubules as well as collapse of assembled microtubules (Sager *et al.,* 1983; Miura *et al.,* 1984; Vogel *et al.,* 1985, 1988; Sager and Syversen, 1986; Graff and Reuhl, 1990; Graff *et al.,* 1993, 1997). Closer examination of cytoskeletal proteins following methylmercury exposure revealed a significant loss of polymerized microtubules, with perinuclear rearrangement and accumulation of micro- and intermediate filaments (Wasteneys *et al.,* 1988).

As posttranslationally modified α-tubulin increases during neuronal differentiation and maturation, microtubules become more resistant to methylmercury (Falconer *et al.,* 1994). It should be also pointed out that the primarily tyrosinated perikaryal microtubules are more vulnerable to methylmercury than are the highly acetylated microtubules in the neurites (Graff and Reuhl, 1995).

IV. Concluding Remarks

Both transplacental and mammary transfer are important routes of exposure to induce congenital or neonatal changes in the developing nervous system. The fetus acts as a "mercury trap," accumulating mercury even higher than the maternal tissue. The developing CNS is particularly sensitive and vulnerable to mercury. Neurologic deficits and pathologic changes in the brain were prominent findings in children born by mothers exposed to methylmercury during both the Japanese (Minamata) and Iraqi episodes of mercury poisoning. Among the various neuropathologic changes induced, perhaps the most characteristic abnormality is the induction of heterotopic (inappropriately located), dysplastic (inappropriately positioned), and hypoplastic (poor neuritic development) neurons in the CNS. This characteristic change of the CNS is believed to be the result of a disruption of neuronal migration during neural development.

The neuronal migratory disruption subsequently leads to abnormal neural cytoarchitecture, dyssynaptogenesis, and neural dysfunctions. Neuronal migratory disruption induced by methylmercury, mechanistically speaking, probably resulted from two main cellular alterations induced by methylmercury: (1) a disruption of cell membranes and their associated surface molecules, such as NCAM; and (2) a disruption of intracellular neurocytoskeletons, in particular the microtubule system (tubulin, MAPs, and stabilization of microtubules). Both NCAMs, microtubules, and MAPs are crucial factors in the normal development of the nervous system. Disruptions to these factors and systems by mercury leads to defective cell movements, neuronal migration, synaptogenesis, axonal transports, neuritic elongation and maintenance of neurites (axons and dendrites), and cytoarchitecture. Eventual neurologic dysfunction and behavioral changes would occur in later life.

Acknowledgment

The able assistance of Miss Mary A. B. Guo and Mrs. Marie Reese in the preparation of this manuscript is appreciated.

References

Amin-Zaki, L., Ehassani, S., Majeed, M. A., Clarkson, T. W., Doherty, R. A., Greenwood, M. R., and Giovanoli-Takubczak, T. (1976). Perinatal methylmercury poisoning in Iraq. *Am. J. Dis. Child* **130,** 1070–1076.

Amin-Zaki, L., El-Hassani, S. B., Majeed, M. A., Clarkson, T. W., Doherty, R. A., and Greenwood, K. R. (1974). Studies of infants postnatally exposed to methylmercury. *J. Pediatr.* **85,** 81–84.

Amin-Zaki, L., Majeed, M. A., Greenwood, M. R., El-Hassani, S. B., Clarkson, T. W., and Doherty, R. A. (1981). Methylmercury poisoning in the Iraqi suckling infant: a longitudinal study over five years. *J. Appl. Toxicol.* **1,** 210–214.

Bakir, F., Dambyi, S. F., Amin-Zaki, L., Murtadha, M., Khalida, A., Al-Rawi, N. Y., Tikriti, S., Dhahir, H. L., Clarkson, T. W., Smith, J. C., and Doherty, R. A. (1973). Methylmercury poisoning in Iraq. *Science* **181,** 230–241.

Cadrin, M., Wasteneys, G. O., Jones-Villeneuve, E. M. V., Brown, D. L. and Reuhl, K. R. (1988). Effects of methylmercury in retinoic acid–induced neuroectodermal derivatives of embryonal carcinoma cells. *Cell Biol. Toxicol.* **4,** 61–80.

Chang, L. W. (1984). Developmental toxicology of methylmercury. In *Toxicology and the Newborn.* S. Kacew and M. J. Reasor, Eds. Ch. 8. Elsevier Pub., Amsterdam, NY. pp. 173–197.

Chang, L. W. (1996). Mercury-related neurological syndromes and disorders. In *Mineral and Metal Neurotoxicology.* M. Yasui, M. J. Strong, K. Ota, and M. A. Verity, Eds. CRC Press, Boca Raton, FL. pp. 169–176.

Chang, L. W., and Annau, Z. (1984). Developmental neuropathology and behavioral teratology of methylmercury. In *Neurobehavioral Teratology.* J. Yanai, Ed. Elsevier Publ., Amsterdam, NY. pp. 405–432.

Chang, L. W., Reuhl, K. R., and Spyker, J. M. (1977). Ultrastructural study of the long-term effects of methylmercury on the nervous system after prenatal exposure. *Environ. Res.* **13,** 171–185.

Chang, L. W., and Verity, M. A. (1995). Mercury neurotoxicity: effects and mechanisms. In *Handbook of Neurotoxicology.* L. W. Chang and R. Dyer, Eds. Marcel Dekker, Inc., NY. pp. 31–60.

Choi, B. H. (1986). Methylmercury poisoning of the developing nervous system: I. Pattern of neuronal migration in the cerebral cortex. *Neurotox.* **7,** 591–600.

Choi, B. H., Lapham, L. W., Amin-Zaki, L., and Saleem, T. (1978). Abnormal neuronal migration, deranged cerebral cortical organization, and diffuse white matter osteocytosis of human fetal brain: a major effect of methylmercury poisoning *in utero. J. Neuropathol. Exp. Neurol.* **37,** 719–733.

Cox, C., Clarkson, T. W., Marsh, D. O., Amin-Zaki, L., Tikriti, S., and Myers, G. G. (1989). Dose–response analysis of infants prenatally exposed to methylmercury: an application of a single compartment model to single-strand hair analysis. *Environ. Res.* **49,** 318–332.

Falconer, M. M., Vallant, A., Reuhl, K. R., Lafferriere, N., and Brown, D. L. (1994). The molecular basis of microtubule stability in neurons. *Neurotox.* **15,** 109–122.

Graff, R. D., Falconer, M. M., Brown, D. L., and Reuhl, K. R. (1997). Altered sensitivity of posttranslationally modified microtubules to methylmercury in differentiating embryonal carcinoma-derived neurons. *Toxicol. Appl. Pharmacol.* **144,** 215–224.

Graff, R. D., Lagunowich, L. A., and Reuhl, K. R. (1992). Alterations in N-CAM expression by methylmercury. *Toxicologist* **12,** 312.

Graff, R. D., Philbert, M. A., Lowndes, H. Z., and Reuhl, K. R. (1993). The effects of glutathione depletion in methylmercury-induced microtubule disassembly in cultured embryonal carcinoma cells. *Toxicol. Appl. Pharmacol.* **120,** 20–28.

Graff, R. D., and Reuhl, K. R. (1990). Effects of repeated methylmercury exposure in interphase microtubules. *Toxicologist* **10,** 138 (abstract).

Graff, R. D., and Reuhl, K. R. (1996). Cytoskeletal toxicity of heavy metals. In *Toxicology of Metals.* L. W. Chang, Ed. CRC Press, Inc., Boca Raton, FL. pp. 639–658.

Grandjean, P., Jorgensen, P. J., and Weihe, P. (1994). Methylmercury from mother's milk: accumulation in infants. *Environ. Health Persp.* **102,** 74–77.

Hamada, R., Arimura, K., and Osame, M. (1997). Maternal–fetal mercury transport and fetal methylmercury poisoning. In *Metal Ions in Biological Systems* A. Sigel and H. Sigel, Eds. Vol. 34. Marcel Dekker, Inc., NY. pp. 405–420.

Harada, M. (1964). Neuropsychiatric disturbances due to organic mercury poisoning during the prenatal period. *Psychiat. Neurol. Jap.* **66,** 426–468.

Harada, M. (1972). Clinical studies on prolonged Minamata disease. *Psychiat. Neurol. Jap.* **74,** 668–678.

Harada, M. (1974a). Clinical and epidemiological studies on Minamata disease. *J. Constitut. Med.* **38,** 20–28.

Harada, M. (1974b). Mental deficiency due to methylmercury poisoning. *Brain Dev.* **6,** 378–387.

Harada, M. (1974c). Clinical studies on the "congenital" Minamata disease. *J. Constitut. Med.* **38,** 29–34.

Harada, M. (1976). Intrauterine poisoning: clinical and epidemiological studies and significance of the problem. *Bull. Inst. Constitut. Med. Kumamoto Univ.* **24,** 1–60.

Harada, Y. (1975). Organic mercury poisonings in children and fetus—infantile and congenital Minamata disease in Japan. *Kumanoto Domonkai Zattsushi* **58,** 18–33.

Harada, Y. (1977). Fetal methylmercury poisoning. In *Minamata Disease.* K. Tsubak and K. Inekayama, Eds. Ch. 3. Elsevier Science Publ., Amsterdam, NY. pp. 38–52.

Hoffman, S. and Edelman, G. M. (1983). Kinetics of hemophilic binding by embryonic and adult forms of the neural cell adhesion molecule. *Proc. Natl. Acad. Sci. U.S.A.* **89,** 5762–5766.

Kacew, S. (1996). Mammary heavy metal content: contribution of lactational exposure to toxicity in suckling infants. In *Toxicology of Metals.* L. W. Chang, Ed. CRC Press, Inc., Boca Raton, FL. pp. 1129–1137.

Kitamura, S., Kakita, T., and Kojoo, J. (1960). A supplement to the results of the epidemiological survey on Minamata disease. *J. Kumamoto Med. Soc.* **34,** 476–480.

Kuhnert, P. M., Kuhnert, B. R., and Erhard, P. (1981). Comparison of mercury levels in maternal blood, fetal cord blood, and placental tissues. *Am. J. Obstet. Gynecol.* **139,** 209–213.

Kuriyama, R., and Sakai, H. (1974). Role of tubulin-SH groups in polymerization to microtubules. *J. Biochem.* **76,** 651–654.

Lagunowich, L. A., Bhambhani, S., Graff, R. D., and Reuhl, K. R. (1991). Cell adhesion molecules in the cerebellum: targets for methylmercury toxicity. *Soc. Neurosci.* **17,** 115 (abstract).

Landmesser, L., Dahm, L., Tang, J., and Rutishauser, U. (1990). Polysialic acid as a regulator of intramuscular nerve branching during embryonic development. *Neuron* **4,** 655–667.

Li, W., and Chou, I. N. (1992). Effects of sodium arsenite in the cytoskeleton and cellular glutathione levels in cultured cells. *Toxicol. Appl. Pharmacol.* **114,** 132–139.

Marsh, D. O., Clarkson, T. W., Cox, C., Myers, G. J., Amin-Zaki, L., and Al-Tikriti, S. (1987). Fetal methylmercury poisoning, relationship between concentrations in single strand of maternal hair and child effects. *Arch. Neurol.* **44,** 1017–1022.

Marsh, D. O., Myers, G. J., and Clarkson, T. W. (1980). Fetal methylmercury poisoning clinical and toxicological data on 29 cases. *Ann. Neurol.* **7,** 348–353.

Marsh, D. O., Myers, G. J., and Clarkson, T. W. (1981). Dose–response relationship for human fetal exposure to methylmercury. *Clin. Toxicol.* **18,** 1311–1318.

Massaro, E. (1996). The development cytotoxicity of mercurials. In *Toxicology of Metals.* L. W. Chang, Ed. CRC Press, Inc., Boca Raton, FL. pp. 1047–1081.

Matsumoto, H., and Takeuchi, T. (1965). Fetal Minamata disease: a neurological study of two cases of intrauterine intoxication by a methylmercury compound. *J. Neuropath. Exp. Neurol.* **24,** 563–574.

Matus, A. (1991). Microtubule-associated proteins and neuronal morphogenesis. *J. Cell Sci. Suppl.* **15,** 61–67.

Mellon, M. G., and Rebhun, L. I. (1976). Sulfydryls and the *in vitro* polymerization of tubulin. *J. Cell Biol.* **70,** 226–238.

Miura, K., Inokawa, M., and Imura, M. (1984). Effects of methylmercury and some metal ions on microtubule networks in mouse glioma cells and *in vitro* tubulin polmerization. *Toxicol. Appl. Pharmacol.* **73,** 218–231.

Nunez, J. (1986). Differential expression of microtubule components during brain development. *Dev. Neurosci.* **8,** 125–141.

Regan, C. M. (1991). Regulation of neural cell adhesion molecule sialylation state. *Int. J. Biochem.* **23,** 513–523.

Reuhl, K. R., and Chang, L. W. (1979). Effects of methylmercury in the development of the nervous system—a review. *Neurotox.* **1,** 21–55.

Reuhl, K. R., and Dey, P. M. (1996). Cell adhesion molecules in metal neurotoxicity. In *Toxicology of Metals.* L. W. Chang, Ed. CRC Press, Inc., Boca Raton, FL. pp. 1097–1119.

Reynolds, W. A., and Pitkin, R. M. (1975). Transplacental passage of methylmercury and its uptake by primate fetal tissues. *Proc. Soc. Exp. Biol. Med.* **148,** 523–526.

Rutishauser, U., and Jessel, T. M. (1988). Cell adhesion molecules in vertebrate neural development. *Physiol. Rev.* **68,** 819–851.

Sager, P. R., Doherty, F. A., and Olmsted, J. B. (1983). Interaction of methylmercury with microtubules in cultured cells and *in vitro. Exp. Cell Res.* **146,** 127–137.

Sager, P. R., and Syversen, T. L. M. (1986). Disruption of microtubules by methylmercury. In *The Cytoskeleton: A Target for Toxic Agents.* T. W. Clarkson, P. R. Sager, and T. L. M. Syversen, Eds. Plenum Press, NY. pp. 97–116.

Study Group of Minamata Disease. (1968). *Minamata Disease.* Kumamoto Univ., Kumamoto City, Japan.

Synder, R. D. (1971). Congenital mercury poisoning. *N. Engl. J. Med.* **284,** 1014–1016.

Verity, M. A. (1996). Pathogenesis of methylmercury neurotoxicity. In *Mineral and Metal Neurotoxicology.* M. Yasui, M. J. Strong, K. Ota, and M. A. Verity, Eds. CRC Press, Boca Raton, FL. pp. 159–176.

Vogel, D. G., Margolis, R. L., and Mottet, N. K. (1985). The effects of methylmercury binding to microtubules. *Toxicol. Appl. Pharmacol.* **80,** 473–486.

Wasteneys, G. O., Cadrin, M., Reuhl, K. R. and Brown, D. L. (1988). The effects of methylmercury in the cytoskeleton of murine embryonal carcinoma cells. *Cell Biol. Toxicol.* **4,** 41–60.

CHAPTER

30

Developmental Neurotoxicity of Cadmium

LLOYD HASTINGS
MARIAN L. MILLER
Department of Environmental Health
College of Medicine
University of Cincinnati
Cincinnati, Ohio 45267

I. Introduction

Cadmium (Cd^{2+}) appears to be an especially potent neurotoxin in the developing nervous system, though this is not its only site of action. Although not studied nearly as extensively as some of the other heavy metals (e.g., lead, mercury), there is considerable evidence regarding the developmental neurotoxicity of cadmium. Prodan (1932) reviewed the history of Cd^{2+} and the early investigation of its general toxicity. In recent years, a great deal of research has been undertaken to examine the toxic effects of Cd^{2+}, resulting in a number of books and reviews [Foulkes, 1986; Friberg *et al.*, 1986; Agency for Toxic Substances and Disease Research (ATSDR),

*Abbreviations: 5HT, 5-hydroxytryptamine; ACh, acetylcholinesterase; ATSDR, Agency for Toxic Substances and Disease Research; CA, catecholamine; Ca^{2+}, calcium; CaM, calmodulin; Cd^{2+}, cadmium; CNS, central nervous system; DA, dopaminergic; EPA, (U.S.) Environmental Protection Agency; EPP, end plate potential; FR, fixed ratio; GI, gastrointestinal tract; GSH, glutathione; GSSG, oxidized glutathione; ig, intragastric; ip, intraperitoneal; iv, intravenous; MAO-A, monoamine oxidase A; MEPP, miniature end plate potential; MT, metallothioniens; NE, norepinephrine; Pb, lead; PN, postnatal day; PNS, peripheral nervous system; ppm, parts per million; sc, subcutaneous; SD, standard deviation; SEM, standard error of the mean; SOD, superoxide dismutase; WHO, World Health Organization.

1989; Robards and Worsfold, 1991; Waalkes *et al.,* 1992; World Health Organization (WHO), 1992a, 1992b]. Only a few, however, have specifically examined the toxic effects of Cd^{2+} on the nervous system (Tischner, 1980; Hastings, 1986; Babitch, 1988; Hastings, 1995). The focus of this review is restricted to the neurotoxicity resulting from Cd^{2+} exposure early in life.

Cadmium is a soft, silvery-white, pliable metal (atomic number 48, molecular weight 112.40), found in group IIb in the periodic table, along with zinc and mercury. In addition to a relatively low melting point (321°C) and boiling point (765°C), cadmium also has a relatively high vapor pressure (1 mm at 394°C), and its vapor is oxidized rapidly in air to form cadmium oxide. Cadmium has oxidation states of 0, +1, and +2, with +2 the most common. Cd^{2+} is used to denote all forms of cadmium discussed in this chapter unless specifically stated otherwise. An important factor determining the toxicity of most compounds is the species in which the element exists. For Cd^{2+}, however, this does not seem to be as critical as for other metals (e.g., mercury), and there is no evidence for any organocadmium compounds occurring in nature (WHO, 1992a). Other related topics, such as analytic methods for quantifying Cd^{2+} exposure, are discussed in greater detail by Robards and Worsfold (1991).

Cd^{2+} occurs naturally in the environment in mineral elements that are almost always associated with zinc and zinc–lead rich ores. In fact, most Cd^{2+} production occurs as a byproduct of zinc smelting. Although small quantities of Cd^{2+} occur naturally in the air, water, and soil, concentrations of any toxicologic significance are usually the result of anthropogenic activities. The largest source of airborne Cd^{2+} is the combustion of fossil fuels; other sources include mining and manufacturing operations, sludge-based and phosphate fertilizers, and incineration of municipal wastes (Robards and Worsfold, 1991). Approximately 3600 metric tons were used in manufacturing in the United States in 1985 (ATSDR, 1989), with the bulk of Cd^{2+} production used in metal plating processes, paint pigments, plastic stabilizers, and Ni-Cad batteries. In recent years, the use of Cd^{2+} in Ni-Cad batteries has increased tremendously. Most Cd^{2+} used in the United States is disposed of in landfills (ATSDR, 1989).

The primary route of Cd^{2+} exposure for most of the population is through food consumption. Cd^{2+} is bioaccumulated by many leafy plants, and the uptake of Cd^{2+} by crops grown on soil enriched by sludge application has been a matter of concern (Reddy and Dorn, 1985). Other foods, such as kidney and shellfish, also show elevated levels of Cd^{2+}. The average adult daily intake of Cd^{2+} has been estimated at 10–30 μg (Friberg *et al.,* 1986). Inhalation constitutes the second major source of Cd^{2+} exposure for the general population, with estimates ranging from 0.02 μg/day to 2 μg/day, depending on the degree of pollution in the surrounding area. The inhalation route becomes even more significant for persons who smoke, with Cd^{2+} in tobacco contributing up to 2 μg per pack smoked. The maternal blood supply is the major source of Cd^{2+} exposure for the fetus, although food consumption (through maternal milk or exogenous foods) provides the primary source of Cd^{2+} exposure in the neonate. Inhalation of secondhand cigarette smoke may also be a source of Cd exposure via the inhalation route for the neonate.

Route of exposure determines to a large extent the amount of Cd^{2+} absorption, with the chemical form being a much less important factor. In the adult, Cd^{2+} and most salts are only poorly absorbed from the gastrointestinal (GI) tract, with estimates ranging from 1% to 5% (ATSDR, 1989). In the neonate, however, absorption of Cd^{2+} by the GI tract has been shown to be much higher than in the adult, up to 55%(Clarkson *et al.,* 1985). Absorption from the lungs is much greater than from the gastrointestinal tract [ranging from 30% to 50% of the amount inhaled, with some estimates as high as 90% (Lee and Oberdörster, 1985)]. Exposure via this route is largely occupational, and therefore is of little concern during early development. The only exceptions might be for children who live near a zinc- or lead-smelting operation or who experience significant amounts of secondhand cigarette smoke. There is little information concerning absorption of Cd^{2+} and the resulting body burden of the metal under these conditions.

In the blood stream, Cd^{2+} is bound to red cells and serum albumin but is rapidly removed by the liver and kidney, with the kidney being the major site of storage. The kidney is considered the target organ for the toxic effects of Cd^{2+}; accumulation of Cd^{2+} greater than 200 μg/g tissue is associated with renal dysfunction (Piscator, 1986). Cd^{2+} is sequestered by the body without undergoing significant oxidation, reduction, or alkylation. Because Cd^{2+} binds readily to protein and nonprotein sulfhydryl groups, metallothioniens (MTs), low molecular weight proteins rich in cysteine, are thought to play an important role in the body's response to Cd^{2+}. MTs have a high affinity for Cd^{2+} and are induced by exposure to Cd^{2+} and other metals (Vallee, 1979). Most Cd^{2+} in the body is sequestered by MT, and this process is considered a major means of Cd^{2+} detoxification (Cherian and Goyer, 1978; Petering and Fowler, 1986). Excretion of Cd^{2+} occurs primarily in urine and is very slow, resulting in an extremely long biologic half-life of 25–30 years (Friberg *et al.,* 1986). Thus, chronic, low-level exposure can result in elevated Cd^{2+} body burdens.

The toxic effects of Cd^{2+} exposure on kidney, lung, and other organ systems have been studied extensively, primarily in mature organisms. The level of exposure required to elicit toxicity, the nature of the insult, and the mechanism(s) responsible for producing the toxicity have been elucidated. The neurotoxic properties of Cd^{2+}, however, are poorly understood. Although several neurotoxic effects have been attributed to Cd^{2+} exposure, the causal relationship between Cd^{2+} exposure in the developing organism and subsequent neurotoxicity is still largely unknown. This relationship is the topic of this review.

Discussion of the impact of Cd^{2+} exposure on developmental neurotoxicity is divided into three topics: the entrance of Cd^{2+} into the nervous system at different stages of development; the neurotoxicity resulting from early exposure to Cd^{2+}; and finally, possible mechanisms of neurotoxicity.

II. Entry of Cadmium into the Central Nervous System

A. *Cadmium Exposure during Gestation*

A fundamental question concerning the neurotoxicity of gestational exposure to Cd^{2+} is whether Cd^{2+} can cross the placental barrier. Results from early studies suggested that the placenta acted as a barrier to the passage of maternally administered Cd^{2+} (Berlin and Ullberg, 1963). It has since been shown that, whereas the placenta does act to restrict the entry of Cd^{2+} into the fetus, Cd^{2+} can cross the placental barrier if the dose is sufficiently elevated (Sonawane *et al.*, 1975). Maternal cadmium exposure during embryogenesis produces teratogenic effects including hydrocephalus, although exposure during the fetal period does not normally produce such gross malformations (Levin and Miller, 1980).

Most studies investigating Cd^{2+} toxicity have looked at whole body (fetal) or kidney Cd^{2+} levels as indicators of body burden rather than levels of Cd^{2+} in the brain. Although CNS pathology resulting from Cd^{2+} exposure has not necessarily been shown to be related to the concentration of Cd^{2+} in the brain (Nolan and Shaikh, 1986), ideally, gestational neurotoxicity resulting from Cd^{2+} exposure should be correlated with elevated levels of the metal in the fetal brain. However, studies employing a wide range of doses and varying exposure paradigms have consistently failed to find elevated levels of Cd^{2+} in the fetal brain (Table 1). Only a single study found an increase of Cd^{2+} in the fetal brain (Danielsson, 1984), and this required treatment of the pregnant dam with diethyl dithiocarbamate, a chelating agent. This may have enhanced the transport of Cd^{2+} across the blood–brain barrier, the Cd^{2+} having been chelated into a lipid-soluble complex.

Christley and Webster (1983) demonstrated that Cd^{2+} does not necessarily have to be present in tissue at detectable levels to produce toxicity. They found that cellular concentrations and cell damage after a teratogenic dose of Cd^{2+} were unrelated. They hypothesized that the ability of Cd^{2+} to inhibit many important enzymes (Vallee and Ulmer, 1972) could produce neuropathology at sites remote from the physical location of Cd^{2+}. Furthermore, most of the studies involving Cd^{2+} exposure during gestation usually demonstrated alterations in essential trace metals in the brain, even though there was no increase in brain Cd^{2+} levels. Alterations in levels of essential metals in the developing central

TABLE 1 Cadmium Levels in Fetal Brain

Species	Age	Detection method	Dose Cd (ppm)[a]	Brain Cd	Ref.
Mice	Fetal	^{109}Cd	0.0015	1.0 ± 0.4[b,c]	Webster (1998)
			0.24	0.9 ± 0.6	
			40.00	0.3 ± 0.1	
Rats	Fetal	AA[d]	0	0.021 ± 0.007[b,e]	Barański (1987)
			60	0.017 ± 0.009	
			180	0.022 ± 0.008	
Rats	Fetal	AA	0	0.12 ± 0.05[e,f]	Murthy *et al.* (1986)
			4.2	0.14 ± 0.06	
			8.4	0.18 ± 0.05	
Rats	Fetal	AA	0	<0.04 μg/g wet wt	Sowa and Steibert (1985)
			50	<0.04 μg/g wet wt	

[a]Exposure during gestation via the drinking water.
[b]Mean ± SD.
[c]ng/kg.
[d]Atomic absorption spectroscopy.
[e]μg/g wet weight (wet wt).
[f]Mean ± SEM.

nervous system (CNS) have well-known adverse consequences (Clarkson, *et al.,* 1985).

B. Cadmium Exposure in the Neonate

Unlike studies employing gestational exposure to Cd^{2+}, neonatal Cd^{2+} exposure does result in increased concentrations of the metal in the brain. In fact, the available data suggest that at a specified exposure level, more Cd^{2+} enters the brain of the developing organism than during any other stage of life, including adulthood. This has been substantiated by a number of studies (Table 2). The amount of Cd^{2+} entering the brain was three times as great when Cd^{2+} was administered at postnatal day (PN) 4 versus PN 21 or 70 (Wong and Klaassen, 1980). In a subsequent study, male rats 4 or 49 days of age were given seven injections of either 2 or 3 mg Cd^{2+}/kg over a 14-day period (Wong *et al.,* 1980). At both dose levels, the brains of the neonates contained more than 10 times the Cd^{2+} concentration of the adults. In rats exposed to Cd^{2+} on PN 0 versus PN 42, the brain concentrations were also 10 to 20 times greater at PN 0 (Valois and Webster, 1987a). In more recent studies, Gupta *et al.* (1993) found that gestational plus lactational exposure to rat dams via the drinking water (50 ppm) resulted in a twofold increase of Cd^{2+} in the brain. Choudhuri *et al.* (1996) reported that mice exposed to Cd^{2+} for 7 days postnatally had approximately four times the concentration of Cd^{2+} in their brains as compared to mice exposed as adults. However,

TABLE 2 Cadmium Levels in Neonatal Brain

Species	Age (PN)	Exposure/route	Detection method	Dose	Brain Cd	Ref.
Rat	1–42	Lactating dam or pup via water	AA[a]	5 ppm	<0.004 μg/g wet wt	Anderson *et al.* (1997)
	17–42	Pup via water			<0.004 μg/g wet wt	
Mice	7	ip	^{109}Cd	1 mg/kg	0.100 μg/g[b]	Choudhuri *et al.* (1996)
	14				0.058 μg/g	
	21				0.032 μg/g	
	30				0.025 μg/g	
	60				0.025 μg/g	
	96				0.022 μg/g	
Rat	7	Gestation/lactation via water	AA	0 ppm	0.030 μg/g[b]	Gupta *et al.* (1993)
	14				0.034 μg/g	
	21				0.039 μg/g	
	7			50 ppm	0.062 μg/g	
	14				0.065 μg/g	
	21				0.070 μg/g	
Mice	0	ip	^{109}Cd	84 μg/kg	0.9[b,c]	Valois and Webster (1987a)
	7				1.5	
	14				0.4	
	42				0.1	
	0			750 μg/kg	2.2	
	7				1.6	
	14				1.2	
	42				0.1	
Rats	4	iv, 2 hr postexposure	^{109}Cd	1 mg/kg	0.144 μg/g[b]	Wong and Klaasen (1982)
		iv, 1 hr postexposure			0.092 μg/g	
		iv, 2 days postexposure			0.080 μg/g	
		iv, 21 days postexposure			0.020 μg/g	
	70	iv, 2 hr postexposure			0.040 μg/g	
		iv, 1 hr postexposure			0.046 μg/g	
		iv, 2 days postexposure			0.049 μg/g	
		iv, 21 days postexposure			0.030 μg/g	
Rats	4	sc × 7, 12–16 hr postexposure	^{109}Cd	2 mg/kg	0.37[b,c]	Wong *et al.* (1980)
				3 mg/kg	0.30	
	49	sc × 7, 12–16 hr postexposure		2 mg/kg	0.025	
				3 mg/kg	0.025	

PN, postnatal days; ip, intraperitoneal; iv, intravenous; sc, subcutaneous.
[a]Atomic absorption spectroscopy.
[b]Extrapolated from graph.
[c]% of administered dose.

at a lower exposure level (5 ppm) during lactation, Andersson *et al.* (1997) found that Cd^{2+} levels in the brains of the offspring were below detection levels of their instruments. However, they were still able to detect alterations in serotonin levels (see page 551). These data, although showing some variability—especially when Cd^{2+} levels in the brain were measured using isotopic versus atomic absorption spectrophotometry—overall, show increased levels of Cd^{2+} in the brain after neonatal exposure. Thus, Cd^{2+} enters the CNS most freely right after birth, with entry decreasing until access is basically restricted around PN 21, concomitant with the shift of site of neural lesions from the CNS to the peripheral nervous system (PNS). The maturation of the blood–brain barrier is probably the single most important factor for the increased exclusion of Cd^{2+} from the brain. Just as important as whether Cd^{2+} can enter the CNS is the question of clearance of the metal from the brain. Newland *et al.* (1986) injected rat pups on PN 1 with 0, 1, 3, or 6 mg Cd/kg, but did not measure brain Cd^{2+} content until approximately PN 90. Cd^{2+} was still detectable in the brain in a dose-related fashion even after 3 months, showing that when Cd^{2+} enters the CNS, it appears to be rather persistent. Similar findings were reported by Brus *et al.* (1995), who exposed rat dams to 50 ppm $CdCl_2$ during gestation via the drinking water. Although Cd^{2+} levels were not measured at parturition, it can be assumed that they were negligible (Barański, 1987). However, brain Cd^{2+} levels measured at 6 weeks of age were highly elevated ($\approx$34 μg/g for exposed group vs 1 μg/g for controls).

Even though accumulation of Cd^{2+} in the brain is greatest during the neonatal period, transfer of the metal from mother to offspring by either the placenta or via milk is still highly restricted, especially at low levels of exposure (Whelton *et al.,* 1993). The Cd^{2+} that does gain entry into the CNS appears to remain for a long period of time, and is sufficiently toxic to alter certain aspects of the CNS. One of the body's initial responses to Cd^{2+} exposure is induction of MT, which, in other organ systems, has a protective function.

C. MT in the CNS

The MTs are low molecular weight, cysteine-rich, intracellular metalloproteins that have several putative physiologic roles, but the precise functions of which remain speculative. They are believed to be involved in the homeostasis of essential trace metals such as zinc and copper (Cousins, 1985), as well as detoxification of Cd^{2+} and other heavy metals (Cherian and Goyer, 1978; Petering and Fowler, 1986). Exposure to heavy metals (and other environmental factors such as stress) greatly induces the synthesis of MT (Waalkes and Klaassen, 1985). Cd^{2+} binds to MT with great affinity, and the subsequent storage of this complex in the kidney is considered to be a major route of detoxification (Webb, 1979). MTs are found in most tissues (Zelazowski and Piotrowski, 1974), but only a few studies have examined their occurrence and/or function in the CNS. Various animals, including rodents and primates, have detectable levels of MT in their brains (Waalkes and Klaassen, 1985; Gulati *et al.,* 1987b). Presumably, MTs in the CNS would perform as they do in other organs, protecting the tissue by sequestering and detoxifying the Cd^{2+}, although no solid experimental data exist to support this conjecture. The distribution of two isoforms of MT, MT-I and MT-II, both of which have been localized immunohistochemically in the nonneural cells of the brain, showed no apparent correlation with the known patterns of metal distribution in the brain (Young *et al.,* 1991). Furthermore, exposure to Cd^{2+} does not increase levels of MT in the brain (Ebadi, 1986; Onosaka *et al.,* 1984; Waalkes and Klaassen, 1985; Gulati *et al.,* 1987a), although exposure to stress does (Hidalgo *et al.,* 1990). Even exposure to high levels of Cd^{2+} (24 mg/kg, ip) or chronic exposure (5 mg/kg for 24 weeks) failed to increase levels of MT significantly in the brain (Onosaka and Cherian, 1981; Gulati *et al.,* 1987a). Although levels of MT did not increase in the brain, the Cd^{2+} that entered the brain was sequestered by MT (Gulati *et al.,* 1987a). The failure of Cd^{2+} to increase the levels of MT might be due to the fact that insufficient amounts of Cd^{2+} crossed the blood–brain barrier to trigger local production of MT. The question of whether Cd^{2+} exposure produced any neurotoxicity was not addressed in these studies.

Most of the preceding studies involved one or both isoforms of MT. The discovery of a third isoform of MT, MT-III (Palmiter *et al.,* 1992), may help to clarify the role of MT in Cd^{2+} neurotoxicity. MT-III is expressed only in the brain and only in neurons, unlike MT-I and -II, which are expressed in most organs, and in the brain are found in glia cells. Using Northern blot analysis to detect mRNA expression of MT isoforms, Choudhuri *et al.* (1996) found that MTs were not expressed in the brains of embryos or neonates before PN 14. *In situ* hybridization showed very high expression levels of MT in cerebellum, cerebrum, olfactory bulb, ependymal cell layer, and choroid plexus. Finally, there was an inverse correlation between amount of Cd^{2+} entering the CNS and expression of MT. During the period when entry of Cd^{2+} was greatest, the level of MT in the brain was at its lowest point. Consequently, MT would seem to play a minor role at most in protecting the CNS from the neurotoxicity resulting from early Cd^{2+} exposure.

The development of a transgenic knockout mouse in which MT-I and -II are inactive holds potential as a valuable tool for further exploring the relationship between MT and Cd^{2+} toxicity (Masters *et al.*, 1994). These mice were shown to be more susceptible to hepatic poisoning by Cd^{2+} than normal mice; neurotoxic effects were not assessed. However, in order to fully explore Cd^{2+} neurotoxicity and any protective role MT may play, the development of mice in which all three genes, including MT-III, are inactivated, will be required.

III. Neurotoxicity of Cadmium

A. *Neurotoxicity of Cd^{2+} Exposure during Gestation*

Because most studies involving gestational Cd^{2+} exposure failed to find Cd^{2+} in the brain, the assumption might be made that the neurotoxicity would not be observed in the offspring. However, most of the studies in which Cd^{2+} exposure was initiated during gestation have reported behavioral alterations in the offspring. One of the earliest studies to investigate the neurobehavioral toxicity resulting from gestational exposure to Cd^{2+} was by Hastings *et al.* (1978) (Table 3). Rat dams were exposed to 17.2 μg Cd^{2+}/ml (ppm) of drinking water for 90 days prior to mating and continuing through gestation. The long exposure period before mating was to allow the Cd^{2+} to reach a steady-state level as well as to permit full induction of MT. Although wheel-running activity was depressed in the offspring, acquisition of a discrimination task was comparable to controls. Cd^{2+} content (whole body) was not elevated in the offspring at birth, but there was a significant decrease in whole body iron. Thus, whereas alterations in behavior were observed, they most probably were not due to the direct effects of Cd^{2+} but instead to effects of Cd^{2+} on the placental transfer of nutrients, trace metals, or some other essential components. The inability to detect Cd^{2+} in small tissue samples (e.g., brain) limited the investigation of possible mechanisms of action. A confounding factor in this study—and in many of the remaining studies reviewed—was that birthweight was significantly lower for the Cd^{2+}-exposed group when compared to controls, although rate of growth was normal thereafter. The adverse effects of reduced birthweight on CNS development have been studied extensively (Jones and Crnic, 1986).

A related study (Cooper *et al.*, 1978) expanded the range of exposure to include both higher and lower levels [4.3, 8.8, 17.2 or 34.4 μg Cd^{2+}/ml (ppm)] as well as groups that received Cd^{2+} only during gestation. Exposure at the two highest levels resulted in reduced birthweight, reduced iron and copper levels (whole body), and reduced growth rates. Gestation-only exposure resulted in normal birthweight for the group receiving 17.2 ppm, whereas 34.4 ppm produced reduced birthweights. The high exposure groups (34.4 ppm, both continuous and gestation-only exposure) produced increased activity levels; the 17.2 ppm group showed decreased activity levels. Gestational exposure to Cd^{2+} at the two lowest levels was without effect in terms of activity levels or acquisition of a discrimination task. Thus, changes in activity levels were not observed in groups that did not also display a concomitant reduction in body weight, suggesting that if Cd^{2+} exposure does affect behavior adversely, it is probably via an indirect route.

Barański and co-workers (1983) investigated the neurotoxicity of gestational exposure to Cd^{2+} in a comprehensive series of studies. In the first study, they gavaged rat dams for 5 weeks prior to mating and throughout gestation (5 times/week at doses of 0.04, 0.4, or 4.0 mg Cd^{2+}/kg/day). Offspring were tested on measures of motor coordination and locomotor activity (two daily 5 min sessions in an automated activity cage). Under these exposure conditions, Cd^{2+} did not produce changes in fertility, embryo toxicity, or fetal lethality; nor were reductions in birthweight or growth observed in the offspring. Activity was reduced in only the two highest groups (0.4 and 4.0 mg Cd^{2+}/kg) for the males, although activity levels for females were reduced in all 3 exposure groups, demonstrating some gender bias. A separate measure of coordination was affected only in the two lower exposure groups for the males, but for females only the two higher groups were affected.

The results from these early studies illustrate two major problems associated with studies on Cd^{2+} neurotoxicity. The first is the lack of a readily available marker for level of internal exposure, such as the use of blood lead in the lead research literature. Cd^{2+} is cleared too rapidly from the blood to be a useful marker. Brain Cd^{2+} would be suitable, but often the levels (whole brain) do not change and the low concentrations involved require sophisticated methods for analysis. As a result, only a few studies have attempted to provide any indicator of internal exposure. Even when atomic absorption spectrophotometry with a graphite furnace is used, control values for levels of Cd^{2+} in the brain vary considerably (see Tables 1 and 2). The use of more sensitive techniques such as inductively coupled plasma–mass spectrophotometry (Hastings and Olson, 1993), which provides excellent sensitivity without many of the problems associated with atomic absorption, may help address this issue. Without such a marker and due to the many variations in exposure protocols, it is difficult to rigorously compare the results of the studies reported in the literature. The second problem is the

TABLE 3 Neurobehavioral Toxicity of Gestational Exposure to Cd

Dose	Gestation[a]	Route	Effects						Ref.
			Body weight	One-way avoidance	Activity	Reflex develop.	Motor coordination	Other behavior	
0.075 mg/kg	0–20	sc	→	↑	→	→		→ 2-way avoid	Pelletier and Satinder (1991)
0.225 mg/kg	0–20	sc	→	↑	→	→		→ 2-way avoid	
0.20 mg/kg	0–15	sc	Fetal toxicity	→	→	→	→	↑ aggression	Lehotzky *et al.* (1990)
0.62 mg/kg	0–15	sc	Fetal toxicity	↓	↓	→	↓	↑ aggression	
2.0 mg/kg	0–15	sc	Fetal toxicity	↓	↓	→	↓	↑ aggression	
0.71 mg/kg[b]	0–20	Water	→	→	↓				Ali *et al.* (1986)
1.2 mg/kg[c]	0–20	Water	↓	↓	↓	→			
60 ppm	1–20	Water	→	↓ F[d]	↓	→			Barański (1986)
0.02 mg/m^3	1–20 + 5 mo	Inhalation	→	↓ F	↑ F		→		Barański (1984)
0.16 mg/m^3	1–20 + 5 mo	Inhalation	↓	↓ F	↓ M		→		
0.04 mg/kg	1–20 + 5 wk	ig	→		→		↓ F		Barański *et al.* (1983)
0.4 mg/kg	1–20 + 5 wk	ig	→		↓		↓ F		
4.0 mg/kg	1–20 + 5 wk	ig	→		↓		↓ F		
4.3 ppm	0–20	Water	→		→			→ visual disc	Cooper *et al.* (1978)
8.6 ppm	0–20	Water	→		→			→	
17.2 ppm	0–20	Water	→		↓			→	
34.4 ppm	0–20	Water	↓		↑			→	
4.3 ppm	0–20 + 3 mo	Water	→		→			→	
8.6 ppm	0–20 + 3 mo	Water	→		→			→	
17.2 ppm	0–20 + 3 mo	Water	↓		→			→	
34.4 ppm	0–20 + 3 mo	Water	↓		→			→	
17.2 ppm	0–20 + 3 mo	Water	↓		↓			→ spatial disc	Hastings *et al.* (1978)

→, No change; ↑, increase; ↓, decrease.
[a]Days of gestation with Cd exposure, or gestation + postnatal exposure.
[b]4.2 ppm.
[c]8.4 ppm.
[d]F or M denotes sex.

fact that females appear to be more sensitive to Cd^{2+} exposure than males. These gender-related differences probably result from differences in absorption, distribution, and retention of Cd^{2+} (Murthy *et al.*, 1978). Except for a few studies that looked at the effects of Cd^{2+} on both sexes, most studies on the developmental neurotoxicity of Cd^{2+} use males almost exclusively in the protocols. Given the greater sensitivity of females to Cd^{2+} exposure, the applicability of data obtained from studies employing only males is open to question.

Because Cd^{2+} absorption is thought to be much higher via the inhalation route than through the oral route, Barański (1984) also investigated gestational Cd^{2+} exposure via inhalation. Rat dams were exposed to CdO aerosols (either 0, 0.02, or 0.16 mg Cd^{2+}/m^3) for 5 days a week, 5 hr daily, for 5 months and during gestation. The weights for all three groups were equal at birth, but the growth of pups was retarded in the highest exposure group. Once again, gender-related alterations in activity, avoidance behavior, and ambulation in an open field were observed. Employing much higher levels of exposure than those used in the initial studies, Barański (1985) intubated pregnant dams with daily doses of 2, 12, or 40 mg Cd^{2+}/kg from days 7–16 of pregnancy. Offspring in the high exposure group displayed congenital defects as well as increased body burdens of Cd^{2+} (brain Cd^{2+} was not measured). The middle exposure group displayed only lower birthweights; the birthweights of the 2 mg Cd^{2+}/kg group were comparable to controls. These more severe effects resulting from Cd^{2+} exposure only during gestation could be due to either the higher concentrations used, the lack of an acclimation period prior to gestation, or some combination of these factors.

Exposure to 60 ppm Cd^{2+} in the drinking water throughout gestation did not affect litter size, birthweight, or growth of the offspring (Barański, 1986). These results are in contrast to those of Cooper *et al.* (1978), who found gestational exposure to 34.4 ppm significantly reduced birthweight and growth. Even at these elevated exposure levels, brain Cd^{2+} levels at 2 weeks of age were not elevated, but by week 16, brain Cd^{2+} levels for both males and females were elevated. Physical and neuromuscular development were not impaired. Activity scores (measured for two daily 5 min periods at 10, 14, 18, and 22 weeks of age) were depressed at weeks 14 and 18 for females but only at week 14 for males. Acquisition of an avoidance task was impaired only in females. Because changes in brain copper and zinc concentrations were much more pronounced than changes in brain Cd^{2+}, Barański (1986) concluded that the alterations in behavior seen in adults were most probably the result of indirect effects of prenatal Cd^{2+} exposure. In this instance, neurotoxicity was observed without a concomitant reduction in body weight.

Whereas Barański's studies involving low levels of Cd^{2+} exposure employed gastric intubation, Ali *et al.* (1986) investigated the toxicity resulting from similar levels of exposure via the drinking water instead of a single daily bolus. Pregnant dams were exposed throughout gestation to either 4.2 or 8.4 ppm in the drinking water, resulting in an average daily exposure of about 0.7 and 1.2 mg Cd^{2+}/kg, respectively. A significant decrease in birthweight was observed at 8.4 ppm, whereas a significant decrease in growth was observed for both dose rates. These results are in contrast to Barański's findings of no effects on birthweights or growth from Cd^{2+} exposure at comparable levels via gastric intubation or at much higher levels via drinking water (60 ppm). The only differences seen by Ali *et al.* (1986) in morphologic development or reflex maturation were in cliff aversion and swimming behavior for both exposed groups. Offspring exposed to Cd^{2+} were hyperactive during neonatal development, but hypoactive at 60 days of age (activity was measured for 10 min sessions in an Actophotometers, Techno, India). A decrement in performance of a two-way avoidance task was observed in the 8.4 ppm group compared to controls at day 60 but not at day 90. Neurotoxicity was only observed when there were alterations in birthweight or growth (Table 3).

Long-term deficits in behavior were seen in offspring of dams exposed to a similar range of exposure levels (0.20, 0.62, or 2.0 mg $CdCl_2$ on days 7–15 of gestation), but via a subcutaneous (sc) route of administration (Lehotzky *et al.*, 1990). (The internal dose of Cd^{2+} was undoubtedly much higher, because the Cd^{2+} did not have to be absorbed through the gut.) The number of pups/litter was significantly reduced at the two higher doses of Cd^{2+}, although mean litter weights were not. Significant alterations were seen in motor coordination, open-field behavior, swim-stress test, pole-climbing avoidance (both acquisition and retention), and social interactions in these same groups. Doses sufficient to produce behavioral alterations were also toxic to the fetus.

Pelletier and Satinder (1991) examined the neurotoxic effects of gestational exposure to Cd^{2+} (either 0.075 or 0.225 mg Cd^{2+}/kg, sc) by employing the behavioral genetic teratology model (Satinder, 1985). The effects of Cd^{2+} on conditioned avoidance behavior was investigated using three strains of rats that were genetically different in terms of their ability to perform the avoidance task—that is, high, low, or normal rates of avoidance. No differences were observed in embryo toxicity, fetal mortality, birthweight, or reflex development, and no interactions were seen between Cd^{2+} exposure and genetic line. However, the "normal" group exposed to the high dose of Cd^{2+} demonstrated significantly better performance in a one-way avoidance task than nonexposed members of the strain, but no differences were seen in two-way or either-way responding. The facilitation of one-way avoidance behavior found in this study is in contrast to the results of Ali *et al.* (1986), Barański (1986), and Lehotzky *et al.* (1990), who found either impaired performance or no effect. Pelletier and Satinder (1991) interpreted the improved one-way avoidance performance to Cd^{2+}-induced hypernociception; however, no increase in sensitivity to electric shock was noted in the exposed animals.

In summary, exposure to Cd^{2+} during gestation was found to result in alterations in some aspect of behavior in all the studies reviewed. Usually these included reduction in activity or decreased acquisition of a one-way avoidance task. To characterize the results as a decrease in activity levels is an overgeneralization, however, because a number of ways of measuring activity were employed in the various studies, as were different durations of testing time. Moreover, conflicting results were obtained in some of the studies. Given the wide range in dose, duration, and route of exposure, the results are still fairly consistent. However, the question of whether the neurotoxicity resulting from gestational Cd^{2+} exposure is due to a direct action of Cd^{2+}, such as an alteration in carbohydrate metabolism (Chapatwala *et al.*, 1982), or inhibition of the synthesis of DNA and protein (Holt and Webb, 1986; Gupta *et al.*, 1993), or to an indirect action on the placental transport of essential metabolites or trace metals to the fetus (Hastings, 1978; Webster, 1978; Sowa and Steibert, 1985; Barański, 1987; Goyer, 1991) is unresolved. The evidence does suggest, however, that the fetal growth retardation often associated with gestational Cd^{2+} exposure is due to low iron and/or zinc levels (Webster, 1978; Kuhnert *et al.*, 1988; Sorell and Graziano, 1990) and this confounding factor must be kept in mind when interpreting the neurotoxicity that results from such exposure.

B. Neurotoxicity of Cd^{2+} in the Neonate

Although little Cd^{2+} enters the fetal brain, neurotoxicity can still result from such exposure. In contrast, entry of Cd^{2+} into the CNS is greatest during the neonatal period and, as a consequence, it is this stage of development that is the most susceptible to the neurotoxic effects of Cd^{2+}. During this period the blood–brain barrier is immature, and Cd^{2+} can enter into the CNS most freely. It is also during this stage of development that rapidly growing neurons are most susceptible to toxic insult (Jacobs, 1982).

The initial description of the neurotoxic properties of Cd^{2+} focused on the development of lesions in the sensory ganglia within 24 hr after exposure of adult rats to Cd^{2+} (Gabbiani *et al.*, 1967a). When neonates were exposed to Cd^{2+}, sensory ganglia were spared whereas hemorrhagic lesions were found in the cerebrum and cerebellum (Gabbiani *et al.*, 1967b; Webster and Valois, 1981), as well as the caudate–putamen and corpus callosum (Wong and Klaassen, 1982). The pathology in the CNS was initially characterized by hemorrhage, vacuolization of the capillary wall, thinning of the basement membrane, widening of intercellular junctions, and degeneration of the endothelium, suggesting that damage to the neural elements was secondary in nature (Nolan and Shaikh, 1986). The CNS was sensitive to Cd^{2+} exposure up to PN 20; exposure after PN 30 produced lesions only in the sensory ganglia of the PNS. This timeline of events suggests a critical period for entry of Cd^{2+} into the CNS that parallels maturation of the blood–brain barrier. Structurally immature capillaries may not be able to exclude Cd^{2+} from the brain, or alternatively, Cd^{2+} may gain access to the CNS as a result of the physiologic and biochemical peculiarities of the capillaries related to the metabolic needs of the rapidly growing brain (Webster and Valois, 1981). The latter theory is supported by data that identified that area of the brain undergoing the most vigorous postnatal growth as also the most sensitive to Cd^{2+}.

Lesions can also be produced in the CNS of weanling rats after chronic exposure via the drinking water (Murthy *et al.*, 1987). Both weanling (21 days) and adult rats were exposed to 100 ppm of Cd^{2+} for 120 days. Rats exposed as juveniles had degenerative changes in Purkinje cells of the cerebellar cortex, but not in the capillary endothelium. There were no CNS lesions in the adult exposed rats (sensory ganglia were not inspected). In this study using weanling rats, lesions in the CNS were observed although exposure was not initiated until after PN 20. In addition, the lesions occurred in neural tissue and not in the endothelial cells, as had been observed in other studies. With chronic exposure of the immature CNS, sufficient Cd^{2+} entered the brain to produce lesions without compromising the blood–brain barrier. When young but more mature rats (PN 35–42) were exposed chronically to $CdCl_2$ (10 ppm for 2 months, increased to 40 ppm for more than 18 months), lesions in the CNS were absent, but a frank peripheral neuropathy was observed at the end of exposure (Sato *et al.*, 1978). Presumably, in this instance, the level of exposure while the rats were still young was insufficient to cause any morphologic damage in the CNS. In terms of neuropathology, exposure of neonates to Cd^{2+} usually resulted in lesions in the CNS, whereas exposure of adults produced lesions only in the PNS. The primary toxic effect of Cd^{2+} appeared to involve the vasculature, with damage to the neural components secondary in nature.

The first study to examine the functional consequences of neonatal Cd^{2+} exposure was by Rastogi *et al.* (1977) (Table 4), who found that exposure to Cd^{2+} (0.1 or 1.0 mg/kg via gastric intubation) for the first 30 days of life resulted in hyperactivity. The lack of a dose-dependent relationship—that is, the increase in activity was the same at both levels of Cd^{2+} exposure, suggested that other factors or mechanisms were involved. However, the fact that only the higher dose significantly suppressed growth would imply that the differences observed in activity levels were not secondary to those associated with stunted growth.

Squibb and Squibb (1979) exposed neonates to Cd^{2+} (61, 122, or 244 ppm via the diet) and found activity levels to be depressed. One possible explanation for this

TABLE 4 Behavioral Toxicity after Neonatal Exposure to Cd

Species	Age (PN)	Exposure	Route	Body weight	Activity	Other behaviors affected	Ref.
Rat M/F[a]	5,6	0 mg/kg	sc	→	→	→	Holloway and Thor (1988a,b)
Rat	5,6	1 mg/kg	sc	→	→	→	
	5,6	2 mg/kg	sc	→	↑ M	Open field (PN 23–25); ↑ rough/tumble play M (PN 44); social memory M (PN 150)	
Rat	5,6	3 mg/kg	sc	↓			
	5,6	4 mg/kg	sc	↓			
Rat M	1	0 mg/kg	sc				Newland *et al.* (1986)
		1 mg/kg	sc				
		3 mg/kg	sc				
		6+ mg/kg	sc	↓ (PN 60)		Altered "transition" between FR25–FR75; *d*-amphetamine challenge; → (PN 70–102)	
Rat M/F	5	0 mg/kg	sc	→			Ruppert *et al.* (1985)
		1 mg/kg	sc	→			
		2 mg/kg	sc	→			
		4 mg/kg	sc	↓	↓/↑	First ↓, then ↑ in residential maze (PN 13–21)	
Rat M	6–20	1 mg/kg[b]	ig	→		↑ Rearing; ↓ open field (PN 50)	Smith *et al.* (1985)
	6	10 mg/kg[b]	ig	↓	→	→	
Rat M	5–15	0.25 mg/kg[b]	ig	↓ (2 wk)	→	↓ After high dose apomorphine (PN 50)	Smith *et al.* (1983)
Rat M	1	1–30 mg/kg[b]	sc			Dose related ↓ suckling (PN 2) and homing (PN 7)	Newland *et al.* (1983)
Rat M	6–16	0.25–7 mg/kg	ig	→	↑	↑ 0.25 mg only (20 min tilt cage) (PN 45); ↑ spatial discrimination task, reversal phase (4 and 7 mg) (PN 90)	Smith *et al.* (1982)
Rat M	4	2 mg/kg	sc	→	→		Wong and Klaassen (1982)
		4 mg/kg	sc	↓	↑	24 hr residential maze (PN 22)	
Mice	1	2 mg/kg[b]	sc	↓	↓	3 min, open field (PN 56); petechiae in CNS	Webster and Valois (1981)
	8	4 mg/kg[b]	sc	↓	↓	3 min, open field (PN 56)	
	15	6 mg/kg[b]	sc	→	→	→	
	22	8 mg/kg[b]	sc	→	→	→	
	21–66	25 ppm	Water	→		→	Corey-Slecta and Weiss (1981)
	21–66	50	Water	→		→	
	21–66	150	Water	↓		Taste aversion (in adults)	
	16–25	61–244 ppm	Diet	Diet[c]	↓	Wheel running (PN 16–25)	Squibb and Squibb (1979)
	0–30	0.1 mg/kg[b]	ig	→	↑	Selective Activity Meter 30 min (PN 30)	Rastogi *et al.* (1977)
	0–30	1 mg/kg[b]	ig	↓	↑	Selective Activity Meter 30 min (PN 30)	

→, no change; ↑, increase; ↓, decrease; PN, postnatal day; sc, subcutaneous; ig, intragastric; M, male; F, female.

[a]F or M denotes sex.

[b]$CdCl_2$.

[c]Body weight controlled by diet.

opposite finding was that the method used to assess activity in this study was quite different from that of the previous study. Rastogi *et al.* (1977) used an open-field type paradigm which is fairly passive in nature, whereas the Squibb and Squibb study used wheel running, a method that gives considerable feedback and was continued for a longer duration (9 days). However, as with the Rastogi *et al.* (1977) study, all three Cd^{2+} levels suppressed wheel-running activity to about the same degree. Body weight was controlled, so it was not known if Cd^{2+} suppressed growth under these conditions.

One of many confounding factors seen in toxicity studies is the reduction in growth rate accompanying exposure. The expression of overt toxicity (such as body weight loss or reduction in growth) is often present before neurotoxicity is observed. This makes it difficult to determine whether the neurotoxicity is due to the compound or to many numerous other factors that accompany loss of body weight or reduced growth. This distinction is critical when investigating the neurotoxic effects resulting from neonatal Cd^{2+} exposure (Table 4). An additional problem is encountered when exposure to the toxic agent is via the oral route. Cory-Slecta and Weiss (1981) investigated the effects of exposure to Cd^{2+} in the drinking water on growth in weanling rats and found that 150 ppm depressed growth, but 50 ppm or less did not. The rapidity with which Cd^{2+} affected body weight prompted them to examine the gustatory qualities of Cd^{2+} in solution. They found that Cd^{2+} in solution had a very averse taste and that the reduction in food and water intake alone could account for the sharp drop in body weight. As a result, studies that use the oral route of exposure must employ adequate controls for the effects associated with alterations in growth.

Given the fact that neonatal exposure to Cd^{2+} can cause frank neuropathology, several studies have examined the occurrence of lesions in the CNS in conjunction with alterations in behavior. Webster and Valois (1981) found that the earliest (and lowest) level of neonatal Cd^{2+} exposure resulted in the most severe lesions in the CNS, the greatest reduction in growth, and hypoactivity when tested on PN 56. Exposure to Cd^{2+} on PN 22 was without apparent effect. Their technique for measuring activity, however, was rather limited (a single 3 min period in the open field) and was confounded additionally by reduced growth in the exposed groups. In a second study that also looked at CNS lesions and behavior, Wong and Klaassen (1982) exposed 4-day-old rats to either 2 or 4 mg Cd^{2+} sc. Exposure to 4 mg resulted in an increase in activity on PN 22 as measured by a 24-hr period in a residential maze as well as a reduction in growth. Lesions in the cerebral cortex, cerebellum, and caudate–putamen were found in these rats, indications of the severity of the insult. Although there were slight differences in exposure protocols and the species tested were different, the physical symptoms of neurotoxicity were similar. Thus, the difference observed in the behavioral measures—hypo- versus hyperactivity—was most probably due to the different methods used to assess activity, rather than different underlying mechanisms of toxicity.

Both CNS neuropathology and behavior were evaluated in a study in which the experimental groups were formed based on severity of effect and not just the external exposure dose (Newland *et al.*, 1986). Newland *et al.* (1983) exposed rat pups to 6 mg Cd^{2+} on PN 1. Two groups resulted, one with and one without hydrocephalus. Hydrocephalus was linked with deficits in suckling behavior, in preference for home bedding, and in neurologic function; the behavior of the exposed group without hydrocephalus was comparable to controls. When Newland *et al.* (1986) tested the group of exposed rats free from signs of hydrocephalus as adults, along with rats that had been concurrently exposed to either 0, 1, or 3 mg/kg Cd^{2+} on PN 1, they found dose-related differences in transition from a fixed ratio (FR) 25 to a FR 75 schedule of reinforcement. Response output increased at 3 mg and decreased at 6 mg; challenge doses of D-amphetamine showed no interaction with exposure to Cd^{2+}. Thus, rats exposed to Cd^{2+} early in life that displayed no overt effects showed behavioral deficits when assessed later in life.

Changes in activity after exposure to a toxic compound has been extensively used as an indication of neurotoxicity (Maurissen and Mattsson, 1989) and even serves as a core component of the EPA's Guidelines for Neurotoxicity Screening Battery (EPA, 1991). Smith *et al.* (1982, 1983, 1985) looked specifically at the effects of early Cd^{2+} exposure on activity and possible underlying mechanisms of action. Rat pups exposed via gastric intubation to various concentrations of Cd^{2+} ranging from 0.25 to 7.0 mg/kg $CdCl_2$ on PD 5–15 showed no changes in growth (Smith *et al.*, 1982). The only group to show any changes in activity was the lowest exposure group, 0.25 mg/kg, which displayed hyperactivity (20 min session in a tilt cage, PN 45 or 46). Rats from the two highest exposure groups, 4 and 7 mg/kg, however, performed better than the controls on the reversal phase of a spatial discrimination task. Another study by Smith *et al.* (1983) using the single effective dose, 0.25 mg/kg, found no effect on activity levels. Exposure conditions were the same as before but activity was measured this time using a Stoelting Electronic Activity Meter (Stoelting Physiology Research Instruments, Chicago, IL) on PN 50. In this study there was, however, a drop in body weight 2 weeks after exposure had ceased. To examine any interaction of the dopamine system and Cd^{2+} on activity, the rats were challenged with varying doses of a dopamine agonist, apomorphine. Apomorphine at the highest dose was effective in altering activity, but only in rats categorized as having high baseline activity. The authors suggested that the failure to replicate their earlier findings could have been due to measuring different aspects of activity in the two studies. In the first study, the activity quantified was actually reactivity to a novel situation (tilt cage), whereas the second study measured spontaneous locomotor activity in the home cage. This caveat—to differentiate between spontaneous locomotor activity and reactivity to a novel environment—is certainly a valid point and has been addressed by others (Barnett and Cowan, 1976; Reiter, 1978). Comparisons between studies in terms of effect on "activity" can be made only when similar measuring techniques are employed.

Smith *et al.* (1985) subsequently conducted a study examining the effects of Cd^{2+} on eight categories of home cage behaviors over a 12-hr period on PN 50. The rat pups had been gavaged with either 1 mg/kg $CdCl_2$ on PN 6–20 or 10 mg/kg $CdCl_2$ on PN 6 only. The single high dose produced loss of weight, but did not result in any changes in behavior. The low dose, while not affecting growth, increased rearing and decreased open-field behavior. These studies demonstrate that the neurotoxic effects of Cd^{2+} are both subtle and complex and generalizations are not easily made.

The effects of Cd^{2+} exposure on activity was addressed specifically by Rupert *et al.* (1985). Zero, 1, 2, or 4 mg Cd^{2+}/kg was administered sc on PN 5. Only the highest dose (4 mg Cd^{2+}/kg) affected activity. Activity in a residential maze was depressed initially for the first 4 days, but increased during the last 2 days. Growth was also reduced in this group. Time of testing after exposure, as well as the duration of the test period itself, affect the nature of the response on the task. The effects of Cd^{2+} on activity shortly after administration were quite different when measured a week later. All these experimental factors contribute to the perplexing variations observed in activity levels after Cd^{2+}exposure. The challenge is to identify these factors that interact with Cd^{2+} exposure to produce adverse outcomes.

The effects of neonatal Cd^{2+} exposure on behavior were examined by Holloway and Thor (1988a,b), using slightly different assessment protocols. In the first study, male and female rat pups were exposed to 0, 1, 2, 3, or 4 mg Cd^{2+}/kg, sc on PN 5 or 6. Because of high mortality, rat pups exposed to 3 or 4 mg were eliminated from postweaning tests. This study was one of the few to evaluate both sexes. They found that in the 2 mg exposure group, only males displayed an increase in open-field activity (PN 23–25), along with an increase in rough and tumble play (PN 44). In the second study, the same male rats were tested again at approximately 150 days of age in a social recognition test. Cd^{2+} exposed rats failed to learn the identity of a strange rat as rapidly as controls, but there was no difference in activity levels (2 min open-field test) among the groups. Due to the fact that the olfactory bulb is a target organ for Cd^{2+} in the CNS (Arito *et al.*, 1981; Clark *et al.*, 1985; Sun *et al.*, 1995) and there is a close relationship between the olfactory system and the limbic system (Halasz, 1990), studies investigating the social–emotional effects of Cd^{2+} exposure would appear to be a promising area of research.

C. Studies on the Developmental Neurotoxicity of Cd^{2+} in Humans

Unlike the extensive experimental, clinical, and epidemiologic literature that exists investigating the neurotoxic effects of exposure to lead in humans (*e.g.*, Davis *et al.*, 1990), only a relatively few studies have been conducted that examine Cd^{2+} neurotoxicity in the human population. Furthermore, these studies all look at Cd^{2+} exposure in conjunction with exposure to some other metal, usually lead. Whereas a high correlation is known to exist between lead and Cd^{2+} concentrations in the blood (Kubota *et al.*, 1968), rarely is the occurrence of other metals studied or even measured in most lead studies. Even if behavioral abnormalities were found, it would be very difficult to assign a definitive role for Cd^{2+} in the etiology of these deficits. These studies, however, provide useful information on which to focus future studies of Cd^{2+} neurotoxicity in both animals and humans.

Phil and Parkes (1977) looked at Cd^{2+} content, along with 13 other metals, in the hair of normal or learning disabled children. The reliability of using metal content of hair as an indicator of either recent exposure or body burden, however, has been questioned (WHO, 1992a). Phil and Parkes (1977) found a significant relationship between elevated lead and Cd^{2+} content in the hair of learning disabled children in comparison to controls. The relative contribution of each metal to this finding could not be ascertained, however.

The relationship between hair Cd^{2+} and lead concentrations and different aspects of cognitive and motor functioning in a nonpreselected sample of school children was studied by Thatcher *et al.* (1982). Controlling for various demographic and socioeconomic variables, they found that Cd^{2+} had a significantly stronger effect on verbal IQ than lead and that lead had a more pronounced effect of performance IQ than Cd^{2+}. No evidence of synergistic effects between Cd^{2+} and lead was found. In this same cohort, analysis of sensory evoked potential function showed decreases in amplitude and increases in latency of peaks with increased metal concentration (Thatcher *et al.*, 1984). Lead was more strongly related to evoked potential measures recorded from central than from posterior areas, whereas Cd^{2+} was more strongly related to the latter. Difference in such spatial properties may possibly be related in some way to the selective association of the metals with deficits in either verbal or performance IQ.

To what extent subclinical levels of lead and Cd^{2+} influenced visual–perceptual functioning was investigated by Marlowe and co-workers (Marlowe *et al.*, 1983; Stellern *et al.*, 1983; Marlowe *et al.*, 1985). Hair Cd^{2+} and lead concentrations in mentally retarded children and children with borderline intelligence were higher than in controls (Marlowe *et al.*, 1983). Stellern *et al.* (1983) examined visual–perceptual performance by a group of elementary school students reported to have learning problems. Both Cd^{2+} and lead concentrations correlated significantly and negatively with visual–perceptual performance. Metal concentrations in the hair of a randomly selected sample of elementary school-age children were correlated with performance on the same visual–perceptual skills test (Marlowe *et al.*, 1985). Of the six metals examined, only aluminum and the interaction of aluminum with lead were related significantly to decreased visual–motor performance. Given the low levels of Cd^{2+} normally found in nonexposed individuals, the lack of a significant correlation between Cd^{2+} levels and performance scores was not surprising. This study further illustrates the need to examine exposure to toxic metals in general and not to just a single metal.

Level of Cd^{2+} in the hair is a measure of recent and/or current exposure. Bonithon-Kopp *et al.* (1986) obtained hair samples at birth from both mother and newborns to measure *in utero* exposure. The children were tested 6 years later on the McCarthy Scales of Childrens Abilities (McCarthy, 1972). A significant negative relationship between *in utero* lead and Cd^{2+} concentrations and the children's motor and perceptual abilities was found; however, there was no effect on memory or verbal skills. Significant correlations were found between newborn hair lead concentrations and the perceptual subscale, whereas for Cd^{2+} the correlations were significant for both perceptual and motor subscales—but only when using the maternal hair sample. Presumably, little Cd^{2+} crossed the placental barrier to affect the fetus directly, suggesting that the neurotoxic effects of *in utero* Cd^{2+} exposure most probably result from an indirect action of Cd^{2+} on placental blood flow or some similar mechanism (Tabacova *et al.*, 1994). Such a mode of action would be in agreement with the results from the animal literature.

Lewis *et al.* (1992) looked at *in utero* exposure but used levels of metals in amniotic fluid as the internal measure of exposure. Cd^{2+}, along with 6 other metals, was measured in the amniotic fluid. At 3 years of age, children were evaluated using the McCarthy Scales of Childrens Abilities. They formulated a toxic risk score derived from a composite value of all the metals measured that correlated negatively with cognitive skills and health status. In addition to the perceptual and motor subscales, the verbal subscale was also negatively correlated with the toxic risk score. However, the presence of Cd^{2+} was found to contribute very little to the overall results. The fact that little Cd^{2+} crossed the placental barrier into the amniotic fluid may explain this.

Roeleveld *et al.* (1990) reviewed the published literature on the developmental relationships between parental occupational exposure and neurotoxicity in the offspring. No data were available on the effects of prenatal or postnatal exposure to Cd^{2+} alone in humans, but some studies in which Cd^{2+} occurred in conjunction with other metals were reported. The main adverse effect of Cd^{2+} seemed to be reduced birthweight, which may compromise CNS development indirectly.

Finally, only a single study was found that linked Cd^{2+} exposure with any measure of neuropathology. An autopsy of a young child (2 years, 10 months of age) who died suddenly found marked cerebral edema in conjunction with elevated levels of cadmium in the brain as determined by X-ray microprobe analysis (Provias *et al.*, 1994). An environmental source of the Cd^{2+} exposure was not identified, however.

IV. Mechanisms of Cd^{2+} Neurotoxicity

In general, Cd^{2+} does not cross the placental barrier, but it can cross the blood–brain barrier and enter the CNS of neonates, resulting in both neuropathology and behavioral deficits. Like most toxic compounds, the adverse effects produced by Cd^{2+} occur widely and are mediated by a variety of mechanisms (Cooper *et al.*, 1984a,b,c; Foulkes, 1986). The following discussion on potential mechanisms of neurotoxicity reflects a similar multiplicity of modes of action. Furthermore, many of the studies investigating mechanisms of action have used adult animals, and it is quite clear that stage of development at which exposure occurs plays a major role in how Cd^{2+} affects the nervous system. Nevertheless, these adult studies, as well as the few that have investigated mechansims of action during early exposure, provide the extent of knowldege on Cd^{2+} neurotoxicity.

A. *Effects of Cd^{2+} on the Blood–Brain Barrier*

A review of the experimental data suggests two general conclusions concerning CNS neuropatholgoy and early exposure Cd^{2+}. First, Cd^{2+} gains entry into the

CNS due to an immature blood–brain barrier and, as the barrier matures, less of the metal enters the brain; and second, most of the resulting neuronal damage is secondary to vascular insult. Entry of Cd^{2+} into the brain is not a passive process, however; Cd^{2+} produces damage to the endothelial cells (Gabbiani *et al.,* 1974; Rohrer *et al.,* 1978; Webster and Valois, 1981) as well as the choroid plexus (Valois and Webster, 1987b, 1989). Thus, prolonged Cd^{2+} exposure in the adult may alter the blood–brain barrier in such a manner that the metal can gain access to the brain (Sun *et al.,* 1995). One mechanism by which this may occur is by the depletion of microvessel antioxidant substances in conjuction with increased lipid peroxidation (Shukla *et al.,* 1996). Once Cd^{2+} is in the brain, it may produce neuronal damage through similar oxidative mechanisms.

B. Cd^{2+} as a Ca^{2+} Blocker

Cd^{2+} is an exceptionally potent Ca^{2+} blocker. The release of a neurotransmitter from the presynaptic nerve terminal evoked by depolarization requires the influx of Ca^{2+} (Katz, 1969). Like many other polyvalent cations, Cd^{2+} blocks this influx causing a decrease in the end plate potential (EPP) at the neuromuscular junction (Forshaw, 1977; Satoh *et al.,* 1982; Cooper and Manalis, 1983; Guan *et al.,* 1987; Molgo *et al.,* 1989). Whereas Pb^{2+}, Co^{2+}, and other metals increase the spontaneous (not evoked) release of neurotransmitter, Cd^{2+} appears to be the only one which at low doses ($<100\ \mu M$) does not produce this dual effect (Cooper, 1984b). However, at higher doses (100–500 μM), Cd^{2+} solutions were found to increase spontaneous neurotransmitter release as indicated by an increase in miniature EPP (MEPPs). One hypothesis is that Cd^{2+} may be acting as either a partial blocker and/or an agonist at intracellular sites that normally bind Ca^{2+} (Guan *et al.,* 1987; Molgo *et al.,* 1989). Nerve terminals exposed to 1 m*M* Cd^{2+} for 3 hr in a Ca^{2+}-free medium can support transmitter release but not synaptic vesicle recycling, which presumably requires Ca^{2+} (Molgo *et al.,* 1989).

These studies that examined Cd^{2+} as a Ca^{2+} blocker were all conducted at the peripheral neuromuscular junction. In the CNS, where synaptic transmission in the olfactory cortex is thought to depend on glutamate receptors, Keran and Schoefield (1986) found Cd^{2+} to be just as potent a blocker of synaptic transmission in the CNS as the PNS. Cd^{2+} has also been shown to suppress adrenergic neurotransmission. A reduction in neurovascular transmission in the presence of Cd^{2+} was assumed to be mediated primarily by a reduction of transmitter release from the presynaptic terminal (Cooper and Steinbeg, 1977). A similar finding was reported by Williams *et al.* (1978) for Cd^{2+} concentrations greater than 0.25 μM, but at low levels (0.075-0.25 μM) Cd^{2+} enhanced neurotransmission. They suggested that this enhancement might be due to the inhibition of monoamine oxidase and catechol-*o*-methyl transferase—enzymes that metabolize catecholamines—by Cd^{2+} at low levels.

Although Cd^{2+} is well recognized as a Ca^{2+} blocker, its mode of interaction with Ca^{2+} is less well understood. Using rat brain microsomes, Shah and Pant (1991) found a dose-dependent inhibition of Ca^{2+}-ATPase activity by Cd^{2+} that resulted in a decrease in ATP-dependent Ca^{2+} uptake. Cd^{2+} had a similar effect on Ca^{2+} transport in rat intestinal epithelial cells (Verbost *et al.* 1987). Micromolar concentrations of external Cd^{2+} were only effective as a Ca^{2+} blocker when the membrane voltage was positive (Swandulla and Armstrong, 1989). Furthermore, the channel gate could close when the channel was still occupied by Cd^{2+}.

It has also been hypothesized that Cd^{2+} neurotoxicity may result from Cd^{2+} binding to calmodulin (CaM), and interfering with CaM's physiologic function (Suzuki *et al.,* 1985). CaM, a ubiquitous Ca^{2+}-binding protein, regulates many cellular processes by mediating the effects of Ca^{2+} (Cheung, 1980). CaM cannot distinguish between Cd^{2+} and Ca^{2+}, which have identical charges and similar ionic radii. When Cd^{2+} binds to CaM, the same conformational change occurs in CaM as when Ca^{2+} binds to it (Suzuki *et al.,* 1985). Sutoo *et al.* (1990) investigated whether Cd^{2+} could substitute for Ca^{2+} in binding with CaM and whether this interaction would disrupt normal physiologic functioning. They found that Cd^{2+} administration produced the same effect on dopamine synthesis as did Ca^{2+} administration and that this effect could be reversed by a CaM antagonist. Thus, one aspect of Cd^{2+} neurotoxicity may be manifested by the anomalous binding of Cd^{2+} to CaM, which in turn activates catecholamine (CA)-synthesizing enzymes in the brain at rates that may then result in malfunction. Cd^{2+}-exposed rats showed a significant decrease in CaM activity in the brain, and Vig *et al.* (1989) suggested that this resulted from Cd^{2+} binding to CaM, disrupting its normal cellular control by Ca^{2+}. Although provocative, the results from these studies mainly point out the need for additional research to more fully elucidate the relationship of Cd^{2+} with CaM and to understand what role perturbations in its normal function may play in the neurotoxicity resulting Cd^{2+} exposure.

C. Effects of Cd^{2+} Exposure on Neurotransmitters

Alterations in behavior resulting from exposure to Cd^{2+} exposure are presumably meditated by changes in neurotransmitter systems that subserve the behavior. Ribas-Ozonas *et al.* (1974) investigated this relationship and reported that intraventricular injections of Cd^{2+} resulted in increased levels of 5HIAA as well as increased activity. Cd^{2+} administered to juvenile rats resulted in decreased levels of 5HT and acetylcholinesterase (ACh), increased levels of DA, and no change in NE (Singhal *et al.,* 1976; Hrdina *et al.,* 1976) (Table 5). Cd^{2+} exposure in the neonate caused a similar increase in DA, but 5HT levels were also increased as was activity. It should be noted that in these various studies the level of neurotransmitter was measured in a variety of brain regions. Rarely was the change in neurotransmitter levels in the same direction in all regions examined. The data in Table 5 represent the predominant trends, although the exceptions occurring in individual areas may actually be just as important. The pattern of the overall fluctuations in neurotransmitters in individual brain regions may be the most meaningful measure (and the most difficult to interpret) in assessing the relationship between neurotransmitter systems and behavior. In addition, many of the studies failed to find a dose-dependent change in neurotransmitter levels, a feature common to many of the studies reviewed, regardless of the parameter being investigated. This makes it difficult to ascribe any cause-and-effect relationship with certainty. One confounding factor is that at higher Cd^{2+} exposure levels, kidney damage may occur. This would result in greater excretion of Cd^{2+}, which, in effect, would lower the actual level of exposure.

Given the potent effect of Cd^{2+} on ACh release at the neuromuscular junction, Hedlund *et al.* (1979) examined the effects of Cd^{2+} on muscarinic receptors in the brain. They found that exposure of rats to Cd^{2+} via drinking water resulted in a significant reduction of muscarinic binding sites in both the cortex and the striatum, although the level of Cd^{2+} present was similar throughout the brain. This is another example of the magnitude of the physiologic effect not correlating with tissue concentration of Cd^{2+}. Furthermore, exposure of the rats to 10 times the original concentration did not increase the extent of muscarinic inhibition. The absence of any type of dose-dependent relationship with effect makes understanding the mode of action of Cd^{2+} difficult.

Most *in vivo* exposures to Cd^{2+} have resulted in increased DA levels (Hrdina *et al.,* 1976; Singhal *et al.,* 1976; Rastogi *et al.,* 1977; Chandra *et al.,* 1985a; Murthy *et al.,* 1989). However, Cd^{2+} inhibited the uptake of DA into synaptosomes (Lai *et al.,* 1981; Hobson *et al.,* 1986) or reduced DA receptor density in striatal homogenates (Scheuhammer and Cherian, 1985). It has also been shown that Cd^{2+} inhibits membrane-bound Na^+-K^+-ATPase (Lai *et al.,* 1981; Magour *et al.,* 1981; Hobson *et al.,* 1986; Murthy *et al.,* 1989), but what role this inhibition plays in reducing DA uptake is still unresolved (Lai *et al.,* 1981; Hobson *et al.,* 1986).

Age of development at the time of exposure also affects how Cd^{2+} disturbs the various neurotransmitter systems. Murthy *et al.* (1986) exposed rats during gestation and found increased levels of 5HT in the offspring. Gupta *et al.* (1990) exposed both juvenile and adult rats to Cd^{2+} and found opposite effects depending on age—a decrease in 5HT in juveniles, an increase in adults. Andersson *et al.* (1997) found low-level exposure (5 ppm) to lactating dams resulted in increased levels of serotonin in the offspring. (Acetylcholine, dopamine, and noradrenalin were not affected, nor were there any signs of neuropathology.) Whereas the effects of Cd^{2+} on the dopaminergic (DA) system are fairly consistent, 5HT behavior shows much greater variability. This variability within similar brain regions and across studies may reflect differences in exposure protocols. The regional analysis of individual neurotransmitters showed variation also, increasing in some regions, decreasing in others, and not changing in still others (Rastogi *et al.,* 1977; Miele *et al.,* 1988; Das *et al.,* 1993). The absence of any correlation between Cd^{2+} content of a brain region and a corresponding increase (or decrease) in neurotransmitter level prompted Miele *et al.* (1988) to suggest that the observed change in monoamines in the brain may be due to an indirect effect of Cd^{2+}. They postulated that Cd^{2+} exposure acted as a stressor that increased adrenal corticosterone, which in turn precipitated the observed changes in NE, DA, and 5HT. Monoamine oxidase A (MAO-A), one of the key enzymes that regulates the metabolism of NE, DA, 5HT, and other biogenic amines has also been shown to be selectively inhibited by Cd^{2+} (Murthy *et al.,* 1989; Leung *et al.,* 1992).

In conclusion, Cd^{2+} exposure may well result in alterations in neurotransmitters within the nervous system. However, for the most part, no overall pattern emerges that describes the effect of Cd^{2+} on individual neurotransmitters or the changes in neurotransmitters with alterations in behavior. Only a few studies measured changes in neurotransmitter levels along with concomitant assessment of behavior. Furthermore, the variability in the results describing Cd^{2+} exposure and changes in neurotransmitters is equalled by the variability in the behavioral data.

D. Role of Oxidative Stress in Developmental Neurotoxicity of Cd

Cd^{2+} affects nervous system tissue in many other ways besides modifying neurotransmitter levels. These include inhibiting basal adenylate cyclase activity in homogenates of cerebellum, cerebrum, and brainstem (Ewers and Erbe, 1980), inhibiting methylation of phos-

TABLE 5 Effects of Cd Exposure on Brain Neurotransmitters

Species	Age	Dose Cd	Route	NE	DA	5HT	ACh	Other	Ref.
	0–42	5 ppm	Lactation, water			↓		↓ 5HIAA → neurotropins	Andersson *et al.* (1997)
Rat	Adult	3 mg/kg				↓	↓	→ ACh in cerebellum; ↑ ACh in cortex	Das *et al.* (1993)
	Forebrain nonsynaptic mitochondria							10 μM ↓ MAO-A → MAO-B; 100 μM ↓ both	Leung *et al.* (1992)
Rat	Juvenile	0.4 mg/kg	ip × 30 days			↓		↓ 5HIAA	Gupta *et al.* (1990)
Rat	Adult	0.4 mg/kg	ip × 30 days			↑		↑ 5HIAA	
Rat	Adult	100 ppm	Water	→	↑	↓		↓ MAO, ATPase	Murthy *et al.* (1989)
Rat	Adult	100 ppm	Diet		→	↓		↑ Lever pressing	Nation *et al.* (1989)
Rat	Juvenile	3 mg/kg	sc × 10 days	↑	↑	↑		Hypothalamus	Miele *et al.* (1988)
Rat	Juvenile	3 mg/kg	sc × 10 days	→	↓	↑		Striatum	
Rat	Juvenile	3 mg/kg	sc × 10 days	↓	↓	↓		Hippocampus	
Rat	Gestation	4.2 ppm	Water	→	→	↑			Murthy *et al.* (1986)
Rat	Gestation	8.4 ppm	Water	→	→	↑		↑ Striatal DA at 90 days	
Rat	Synaptosomes			↓	↓			Inhibited Na+/K+ ATPase	Hobson *et al.* (1986)
Rat	Adult	2 mg/kg	ip × 21 days	→	↑	↑			Chandra *et al.* (1985b)
Rat	Striatal homogenates							↓ Striatal D_2 receptor density	Scheuhammer and Cherian (1985)
Rat	Purified synaptosomes				↓			Inhibition of DA uptake	Lai *et al.* (1981)
Rat	Juvenile	1–10 mM $CdCl_2$	Water 8 days					↓ Muscarinic receptors in striatum and cortex	Hedlund *et al.* (1979)
	PN 0–30	0.1 mg/kg	ig	→	→	→		↑ 5HIAA; → MAO	Rastogi *et al.* (1977)
Rat	PN 0–30	1.0 mg/kg	ig	↑	↑	↑		↑ 5HIAA; → MAO	
Rat	Juvenile	0.25 mg/kg	ip 45 days	→	↑	↓	↓		Hrdina *et al.* (1976)
Rat	Juvenile	1 mg/kg	ip 45 days	→	↑	↓	↓		
Rat	Juvenile	1 mg/kg	ip 45 days	→	↑	↓	↓		Singhal *et al.* (1976)

→, No change; ↑, increase; ↓, decrease; ip, intraperitoneal; sc, subcutaneous; ig, intragastric.

pholipids in synaptosomal membranes (Wong and Lim, 1981), altering the membrane of phospholipid vesicles (Deleers *et al.*, 1986), and blocking of axonal transport (Gan *et al.*, 1986). Cd^{2+} exposure also decreases myelin-specific lipids in the brain, and it has been suggested that this alteration in lipid metabolism in early life causes or at least contributes to the neurotoxicity seen in early development (Gulati *et al.*, 1986, 1987a).

Finally, the neurotoxicity associated with Cd^{2+} exposure may result from the metal altering or impairing the defense mechanisms that protect against oxidative damage. Free radicals are constantly being produced by a number of processes within the tissue. Cd^{2+} exposure itself has been known for many years to increase the production of free radicals in tissues (Amoruso *et al.*, 1982). Free radicals attack cellular membranes, subcellular organelles, and induce lipid peroxidation (Sevanian, 1988). Subchronic exposure to Cd^{2+} has been shown to increase lipid peroxidation in the brain, resulting in a decrease in phospholipid content (Hussain *et al.*, 1985). Shukla *et al.* (1988a, 1988b) investigated the effects of Cd^{2+} on glutathione (GSH); GSH serves as an intracellular sulfhydryl buffer along with a variety of other physiologic functions (Orrenius *et al.*, 1983). Cd^{2+} lowered the concentration of reduced glutathione whereas the levels of oxidized glutathione (GSSG) increased in most brain regions except the hippocampus. Coadministration of the antioxidant vitamin E along with Cd^{2+} prevented this alteration in GSH status (Shukla *et al.*, 1988b). Administration of Cd^{2+} to rats decreased the activity of the free radical scavenging enzyme superoxide dismutase (SOD) in all brain regions except the hippocampus, whereas the concentration of lipid peroxides was increased in the same areas (Shukla *et al.*, 1987). Administration of vitamin E along with Cd^{2+} also reduced lipid peroxidation and increased SOD activity, again suggesting a protective effect for vitamin E (Shukla *et al.*, 1988c). Exposure to Cd^{2+} resulted in a persistent inhibition of glutathione peroxidase and catalase, enzymes that play an important role in the antioxidant defense mechanism, even after Cd^{2+} exposure had been terminated (Shukla *et al.*, 1989). Other measures indicative of free radical–induced lipid peroxidation, including membrane fluidity, intracellular calcium level, malonaldehyde level, phospholipids, and reduced glutathione, are altered by Cd^{2} exposure (Kumar *et al.*, 1996). Similar changes in the antioxidant defense system have been observed in the brains of rat pups exposed during gestation and lactation (Gupta *et al.*,1995). This increased production of free radicals in conjunction with the decrease in the defense mechanisms for handling oxidative damage could result in widespread toxicity.

V. Conclusions

Although there is ample data involving a wide variety of endpoints to suggest that Cd^{2+} produces neurotoxicity, the nature of the toxicity as well as the underlying mechanisms are still not well understood. It is possible to come to some general conclusions, however. The neurotoxic effects of prenatal Cd^{2+} exposure are probably due to an indirect action of Cd^{2+} on placental transfer of nutrients or essential metals. Conversely, the neurotoxic effects seen after neonatal Cd^{2+} exposure are probably related to the entry of Cd^{2+} into the nervous system, with or without cerebral vasculature damage—depending on level and duration of dose. Only during early development can Cd^{2+} cross the blood–brain barrier in sufficient quantity to produce neuropathology and/or functional deficits. A major confounding factor that makes interpretation of the data difficult is that, in order to obtain most neurotoxic effects, a dose sufficient to cause overt toxicity (e.g., body weight loss) is required.

However, when examined using *in vitro* procedures, Cd^{2+} has been shown to be a potent toxic compound capable of disrupting many neuronal processes. Cd^{2+} is a Ca^{2+} channel blocker and its ability to block neuronal transmission is well substantiated. It affects many cellular processes, either by blocking Ca^{2+} channels or by mimicking Ca^{2+}. Because Ca^{2+} is involved in so many different cellular processes, the number of opportunities for Cd^{2+} to produce its toxic effects are substantial. The real question, however, is which of these effects occur *in vivo.* Also needing further clarification is the role of MT in detoxifying Cd^{2+} once it enters into the brain.

Three issues must be addressed in future research: 1. a relevant and sensitive marker of internal exposure is needed; as well as 2. some uniform method of assessing neurobehavioral deficits; and 3. some measure of neural functioning (e.g., neurotransmitter level in specific brain regions). Detection of Cd^{2+} in specific brain regions is probably the best measure of internal exposure. As pointed out earlier, one analytic technique that shows extreme promise in this regard is inductively coupled plasma/mass spectrophotometry (Hastings and Olson, 1993). This technique is not only very sensitive, it can measure many different metals simultaneously. Cd^{2+} levels can be determined, as well as the effect of Cd^{2+} on all the essential and trace metals. Appropriate, well-defined neurobehavioral measures need to be chosen, a task that engenders continuous debate within the field of neurotoxicology. Although activity measures have been widely used, the type of activity evaluated has had little standardization. Consequently, exposure to Cd^{2+} has resulted in both increases and decreases in activity levels. Besides more defined measures of activity, tasks that focus on the emotional aspects of behavior may also be useful. Along with a sensitive measure of internal exposure and standardized assessment of behavior, some measure of neuronal functioning (e.g., neurotransmitter level) is required. All of these parameters have been examined in the studies reviewed, but usually only one or two in any one study. Comparisons between studies were hindered by the wide range in both exposure and assessment protocols. A comprehensive experimental design incorporating all these parameters is necessary to more fully elucidate the effects and mechanisms of Cd^{2+} neurotoxicity during early development.

Absent from this review of developmental Cd^{2+} neurotoxicity is a reference to any studies employing a molecular neurobiology approach, except for the work involving MT. In general, the field of neurotoxicity has been slow to incorporate the tools of molecular neurobiology. This may be due, in part, to the lack of definitive behavioral and/or neural endpoints that can be attributed to exposure to a toxic compound. As such endpoints are better identified, then the powerful tools of molecular neurobiology can be applied to aid in more fully understanding the mechanisms underlying neurotoxicity resulting from exposure to toxic agents.

References

Ali, M. M., Murthy, R. C., and Chandra, S. V. (1986). Developmental and longterm neurobehavioral toxicity of low level in-utero cadmium exposure in rats. *Neurobehav. Toxicol. Teratol.* **8,** 463–468.

Amoruso, M. A., Witz, G., and Goldstein, B. D. (1982). Enhancement of rat and human phagocyte superoxide anion radical production by cadmium. *Toxicol. Letters* **10,** 133–138.

Andersson, H., Petersson-Grawe, K., Lindqvist, E., Luthman, J., Oskarsson, A., and Olson, L. (1997). Low-level cadmium exposure of lactating rats causes alterations in brain serotonin levels in the offspring. *Neurotoxicol. Teratol.* **19,** 105–115.

Arito, H., Sudo, A., and Suzuki, Y. (1981). Aggressive behavior of the rat induced by repeated administration of cadmium. *Toxicol. Letters* **7,** 457–461.

ATSDR (Agency for Toxic Substances and Disease Registry) (1989). *Toxicological Profile for Cadmium.* U.S. Public Health Service.

Babitch, J. A. (1988). Cadmium Neurotoxicity. In *Metal Neurotoxicity* S. C. Bondy and K. N. Prasad, Eds. CRC Press, Boca Raton, pp. 141–166.

Barański, B. (1984). Behavioral alterations in offspring of female rats repeatedly exposed to cadmium oxide by inhalation. *Tox. Lett.* **22,** 53–61.

Barański, B. (1985). Effect of exposure of pregnant rats to cadmium on prenatal and postnatal development of the young. *J. Hyg. Epidemiol. Microbiol. Immunol.* **29,** 253–262.

Barański, B. (1986). Effect of maternal cadmium exposure on postnatal development and tissue cadmium, copper and zinc concentrations in rats. *Arch. Toxicol.* **58,** 255–260.

Barański, B. (1987). Effect of cadmium on prenatal development and on tissue cadmium, copper, and zinc concentrations in rats. *Environ. Res.* **42,** 54–62.

Barański, B., Stetkiewicz, I., Sitarek, K., and Szymczak, W. (1983). Effects of oral, subchronic cadmium administration on fertility, prenatal and postnatal progeny development in rats. *Arch. Toxicol.* **54,** 297–302.

Barnett, S. A., and Cowan, P. E. (1976). Activity, exploration, curiosity, and fear: An ethological study. *Interdisciplinary Sci. Rev.,* **1,** 43–62.

Berlin, M. and Ullberg, S. (1963). The fate of Cd^{109} in the mouse. *Arch. Environ. Health* **7,** 686–693.

Bonithon-Kopp, C., Huel, G., Moreau, T., and Wendling R. (1986). Prenatal exposure to lead and cadmium and psychomotor development of the child at 6 years. *Neurobehav. Toxicol. Teratol.* **8,** 307–310.

Brus, R., Kostrzewa, R. M., Felinska, W., Plech, A., Szkilnik, R., and Frydrych, J. (1995). Ethanol inhibits cadmium accumulation in brains of offspring of pregnant rats that consume cadmium. *Toxicol. Lett.* **76,** 57–62.

Chandra, S. V., Kalia, K., and Hussain, T. (1985a). Biogenic amines and some metals in brain of cadmium-exposed diabetic rats. *J. Appl. Toxicol.* **5,** 378–381.

Chandra, S. V., Murthy, R. C., and Ali, N. M. (1985b). Cadmium-induced behavioral changes in growing rats. *Ind. Health* **23,** 159–162.

Chapatwala, K. D., Baykin, M., Butts, A., and Rajanna, B. (1982). Effects of intraperitoneally injected cadmium on renal and hepatic gluconeogenic enzymes in rats. *Drug Chem. Toxicol.* **5,** 305–317.

Cherian, M. G. and Goyer, R. A. (1978). Metallothioneins and their role in the metabolism and toxicity of metals. *Life Sci.* **23**:1–10.

Cheung, W. Y. (1980). Calmodulin plays a pivotal role in cellular regulation. *Science* **207,** 19–27.

Choudhuri, S., Liu, W. L., Berman, N. E. J., and Klaassen, C. D. (1996). Cadmium accumulation and metallothionein expression in brain of mice at different stages of development. *Toxicol. Lett.* **84,** 127–133.

Christley, J. and Webster, W. S. (1983). Cadmium uptake and distribution in mouse embryos following maternal exposure during the organogenic period: A scintillation and autoradiographic study. *Teratology* **27,** 305–312.

Clark, D. E., Nation, J. R., Bourgeois, A. J., Hare, M. F., Baker, D. M., and Hinderberger, E. J. (1985). The regional distribution of cadmium in the brains of orally exposed adult rats. *Neurotoxicol.* **6,** 109–114.

Clarkson, T. W., Nordberg, G. F., Sager, P. R., Berlin, M., Friberg, L., Mattison, D. R., Miller, R. K., Mottet, N. K., Nelson, N., Parizek, J., Rodier, P. M. and Sandstead, H. (1985). An overview of the reproductive and developmental toxicity of metals. In: *Reproductive and Developmental Toxicity of Metals* T. W. Clarkson, G. F. Nordberg, and P. F. Sager, Eds. Plen Press: New York, pp. 1–26.

Cooper, G. P., Chourhury, H., Hastings, L., and Petering, H. G. (1978). Prenatal cadmium exposure: Effects on essential trace metals and behavior in rats. In: *Developmental toxicology of energy-related pollutants* D. D. Mahlum, M. R. Sikov, P. L. Hackett, and F. D. Andrew, Eds. Technical Information Center, U.S. Dept. Energy, Conf-771017, pp. 627–637.

Cooper, G. P. and Manalis, R. S. (1983). Influence of heavy metals on synaptic transmission: A review. *Neurotoxicol.* **4,** 69–84.

Cooper, G. P. and Manalis, R. S. (1984b). Cadmium: Effects on transmitter release at the frog neuromuscular junction. *European J. Pharmacol.* **99,** 251–256.

Cooper, G. P. and Steinberg, D. (1977). Effects of cadmium and lead on adrenergic neuromuscular transmission in the rabbit. *Am. J. Physiol.* **232,** C128–C131.

Cooper, G. P., Suszkiw, J. B., and Manalis, R. S. (1984a). Presynaptic effects of heavy metals. In *Cellular and Molecular Neurotoxicology* T. Narahashi, Ed. Raven Press, New York, pp. 1–21.

Cooper, G. P., Suszkiw, J. B., and Manalis, R. S. (1984c). Heavy metals: Effects on synaptic transmission. *Neurotoxicology* **5,** 247–266.

Cory-Slechta, D. A. and Weiss, B. (1981). Aversiveness of cadmium in solution. *Neurotoxicology* **2,** 711–724.

Cousins, R. J. (1985). Absorption, transport, and hepatic metabolism of copper and zinc: special reference to metallothionein and ceruloplasmin. *Physiol. Rev.* **65,** 238–309.

Danielsson, B. R. G. (1984). Placental transfer and fetal distribution of cadmium and mercury after treatment with dithiocarbamates. *Arch. Toxicol.* **55,** 161–167.

Das, K. P., Das, P. C., Dasgupta, S., and Dey, C. D. (1993). Serotonergic-cholinergic neurotransmitters' function in brain during cadmium exposure in protein restricted rat. *Biol. Trace Element Res.* **36,** 119–127.

Davis, J. M., Otto, D. A., Weil, D. E., and Grant, L. D. (1990). The comparative developmental neurotoxicity of lead in humans and animals. *Neurotoxicol. Teratol.* **12,** 215–229.

Deleers, M., Servais, J-P., and Wulfert, E. (1986). Neurotoxic cations induce membrane rigidification and membrane fusion at micromolar concentrations. *Biochemica et Biophysica Acta* **855,** 271–276.

Ebadi, M. (1986). Biochemical characterization of a metallothionein-like protein in rat brain. *Biol. Trace Elem. Res.* **11,** 101–116.

EPA (1991). Pesticide Assessment Guidelines—Subdivision F Hazard Evaluation: Human and Domestic Addendum 10—Neurotoxicity Series 81, 82, and 83. U.S. Environmental Protection Agency, Washington, D.C.

Ewers, U. and Erbe, R. (1980). Effects of lead, cadmium and mercury on brain adenylate cyclase. *Toxicol.* **16,** 227–237.

Forshaw, P. J. (1977). The inhibitory effect of cadmium on neuromuscular transmission in the rat. *European J. Pharmacol.* **42,** 371–377.

Foulkes, E. C. (Ed.) (1986). *Cadmium.* Springer-Verlag, Berlin.

Friberg, L., Elinder, C. G., Kjellström, T., and Nordberg, G. F., Eds. (1986). *Cadmium and Health. A Toxicological and Epidemiological Appraisal. Vol. II. Effects and Response.* CRC Press, Cleveland.

Gabbiani, G., Baic, D., and Deziel, C. (1967b). Toxicity of cadmium for the central nervous system. *Experimental Neurology* **18,** 154–160.

Gabbiani, G., Badonnel, M-C., Mathewson, S. M., Ryan, G. B. (1974). Acute cadmium intoxication. Early selective lesions of endothelial clefts. *Laboratory Invest.* **30,** 686–695.

Gabbiani, G., Gregory, A., and Baic, D. (1967a). Cadmium-induced selective lesions of sensory ganglia. *J. Neuropathol. Exp. Neurol.* **26,** 498–506.

Gan, S-D., Fan, M-M., and He, G-P. (1986). The role of microtubules in axoplasmic transport in vivo. *Brain Res.* 75–82.

Goyer, R. A. (1991). Transplacental transfer of cadmium and fetal effects. *Fund. & Appl. Toxicol.* **16,** 22–23.

Guan, Y. Y., Quastel, D. M., and Saint, D. A. (1987). Multiple actions of cadmium on transmitter release at the mouse neuromuscular junction. *Can. J. Physiol. Pharmacol.* **65,** 2131–2136.

Gulati, S., Gill, K. D., and Nath, R. (1986). Effect of cadmium on lipid composition of the weanling rat brain. *Acta Pharmacol. Toxicol. (Copenh)* **59,** 89–93.

Gulati, S., Gill, K. D., and Nath, R. (1987a). Effect of cadmium on lipid metabolism of brain: in vivo incorporation of labelled acetate into lipids. *Pharmacol. Toxicol.* **60,** 117–119.

Gulati, S., Paliwal, V. K., Sharma, M., Gill, K. D., and Nath, R. (1987b). Isolation and characterization of a metallothionein-like protein from monkey brain. *Toxicol.* **45,** 53–64.

Gupta, A., Gupta, A., Murthy, R. C., and Chandra, S. V. (1993). Neurochemical changes in developing rat brain after pre- and postnatal cadmium exposure. *Bull. Environ. Contam. Toxicol.* **51,** 12–17.

Gupta, A., Gupta, A., Murthy, R. C., Thakur, S. R., Dubey, M. P., and Chandra, S. V. (1990). Comparative neurotoxicity of cadmium in growing and adult rats after repeated administration. *Biochem. Int.* **21,** 97–105.

Gupta, A., Gupta, A., and Shukla, G. S. (1995). Development of brain free radical scavenging system and lipid peroxidation under the influence of gestational and lactational cadmium exposure. *Human. Experim. Toxicol.* **14,** 428–433.

Halasz, N. (1990). *The Vertebrate Olfactory System.* Akademiai Kiado, Budapest.

Hastings, L. (1986). Behavioral teratogenesis resulting from early cadmium exposure. In: *Handbook of Behavioral Teratology* (E. P. Riley and C. V. Vorhees, Eds.), Plenum Press, New York, pp. 321–333.

Hastings, L., Choudhury, H., Petering, H. G., and Cooper, G. P. (1978). Behavioral and biochemical effects of low-level prenatal cadmium exposure in rats. *Bull. Environm. Contam. Toxicol.* **20,** 96–101.

Hastings L. and Olson, L. (1993). Determination of metals in olfactory bulbs from Alzheimer's patients. *The Toxicologist* **13,** 129.

Hedlund, B., Gamarra, M. and Bartfai, T. (1979). Inhibition of striatal muscarinic receptors in vivo by cadmium. *Brain Res.* **168,** 216–218.

Hidalgo, J., Borras, M., Garvey, J. S., and Armario, A. (1990). Liver, brain, and heart metallothionein induction by stress. *J. Neurochem.* **55,** 651–654.

Hobson, M., Milhouse, M. and Rajanna, B. (1986). Effects of cadmium on the uptake of dopamine and norepinephrine in rat brain synaptosomes. *Bull. Environ. Contam. Toxicol.* **37,** 421–426.

Holloway, W. R., Jr., and Thor, D. H. (1988a). Cadmium exposure in infancy: effects on activity and social behaviors of juvenile rats. *Neurotoxicol. Teratol.* **10,** 135–142.

Holloway, W. R., Jr. and Thor, D. H. (1988b). Social memory deficits in adult male rats exposed to cadmium in infancy. *Neurotoxicol. Teratol.* **10,** 193–197.

Holt, D. and Webb, M. (1986). Comparison of some biochemical effects of teratogenic doses of mercuric mercury and cadmium in the pregnant rat. *Arch. Toxicol.* **58,** 243–248.

Hrdina, P. D., Peters, D. A. V., and Singhal, R. L. (1976). Effects of chronic exposure to cadmium, lead and mercury on brain biogenic amines in the rat. *Res. Commun. Chem. Pathol. Pharmacol.* **15,** 483–493.

Hussain, T., Ali, M. M., and Chandra, S. V. (1985). Effect of cadmium exposure on lipids, lipid peroxidation and metal distribution in rat brain regions. *Ind. Health* **23,** 199–205.

Jacobs, J. M. (1982). Vascular permeability and neurotoxicity. In: *Nervous System Toxicology.* C. L. Mitchell, Ed., Raven Press, New York, pp. 285–298.

Jones, A. P. and Crnic, L. S. (1986). Maternal mediation of the effects of malnutrition. In: *Handbook of Behavioral Teratology.* E. P. Ripley and C. V. Vorhees, Eds., Plenum Press, NY, pp. 409–425.

Katz, B. (1969). *The Release of Neural Transmitter Substances.* C. C. Thomas: Springfield, Ill.

Keran, Y. F. and Scholfield, C. N. (1986). Ca-channel blockers and the electrophysiology of synaptic transmission of the guinea-pig olfactory cortex. *Eur. J. Pharmacol.* **130,** 273–278.

Kubota, J. B., Lazar, S. and Losee, F. (1968). Copper, zinc, cadmium and lead in human blood from 19 locations in the United States. *Arch. Environ. Health* **16,** 788–793.

Kuhnert, D. R., Kuhnert, P. M., and Zarlingo, T. J. (1988). Association between placental cadmium and zinc and age and parity in pregnant women who smoke. *Obstet. Gynecol.* **71,** 67–70.

Kumar, R., Agarwal, A. K., and Seth, P. K. (1996). Oxidative stress-mediated neurotoxicity of cadmium. *Toxicol. Lett.* **89,** 65–69.

Lai, J. C., Liu, L., and Davidson, A. N. (1981). Differences in the inhibitory effect of Cd^{2+}, Mn^{2+} and Al^{3+} on the uptake of dopamine by synaptosomes from forebrain and from striatum of the rat. *Biochem. Pharmacol.* **30,** 3123–3125.

Lee, H. Y. and Oberdörster, G. (1985). Kinetics of intratracheally instilled $CdCl_2$ and Cd-thionein. *Toxicologist* **5,** 178.

Lehotzky, K., Ungvary, G., Polinak, D., and Kiss, A. (1990). Behavioral deficits due to prenatal exposure to cadmium chloride in CFY rat pups. *Neurotoxicol. Teratol.* **12,** 169–172.

Leung, T. K., Lim, L., and Lai, J. C. (1992). Differential effects of metal ions on type A and type B monoamine oxidase activities in rat brain and liver mitochondria. *Metab. Brain Dis.* **7,** 139–146.

Levin, A. A. and Miller, R. K. (1980). Fetal toxicity of cadmium in the rat: Maternal vs. fetal injections. *Teratology* **22,** 1–5.

Lewis, M., Worobey, J., Ramsay, D. S., and McCormack, M. K. (1992). Prenatal exposure to heavy metals: effect on childhood cognitive skills and health status. *Pediatrics* **89,** 1010–1015.

Margour, S., Kristof, V., Baumann, M., and Assmann, G. (1981). Effect of acute treatment with cadmium on ethanol anesthesia, body temperature, and synaptosomal Na^+-K^+-ATPase of rat brain. *Environ. Res.* **26,** 381–391.

Marlowe, M., Errera, J., and Jacobs, J. (1983). Increased lead and cadmium burdens among mentally retarded children and children with borderline intelligence. *Am. J. Men. Def.* **87,** 477–483.

Marlowe, M., Stellern, J., Errera, J., and Moon, C. (1985). Main and interaction effects of metal pollutants on visual-motor performance. *Arch. Environ. Health* **40,** 221–225.

Masters, B. A., Kelly, E. J., Quaife, C. J., Brinster, R. L., and Palmiter, R. D. (1994). Targeted disruption of metallothionein I and II genes increases sensitivity to cadmium. *Biochem.* **91,** 584–588.

Maurissen, J. P. J. and Mattsson, J. L. (1989). Critical assessment of motor activity as a screen for neurotoxicity. *Toxicol. Indust. Health* **5,** 195–201.

McCarthy, D. (1972). *McCarthy Scales of Childrens Abilities* (Manual), New York: Psychological Corp.

Miele, M., Desole, M. S., Demontis, P., Esposito, G., Congiu, A., and Anania, V. (1988). Neurochemical and behavioral effects of cadmium alone or associated with selenium in the rat. *Pharmacol. Res. Com.* **20,** 1063–1064.

Molgo, J., Pecot-Dechavassine, M., and Thesleff, S. (1989). Effects of cadmium on quantal transmitter release and ultrastructure of frog motor nerve endings. *J. Neural Transm.* **77,** 79–91.

Murthy, L. D., Rice, P., and Petering H. G. (1978). Sex differences with respect to the accumulation of oral cadmium in rats. In: *Trace Element Metabolism in Man and Animals,* vol. 3, R. Kirchgessner Ed. Technical University of Munich Press, Freising-Weihenstephen, West Germany, pp. 557–560.

Murthy, R. C., Ali, M. M., and Chandra, S. V. (1986). Effects of in-utero exposure to cadmium on the brain biogenic amine levels and tissue metal distribution in rats. *Ind. Health* **24,** 15–21.

Murthy, R. C., Saxena, D. K., Lal, B., and Chandra, S. V. (1989). Chronic cadmium-ethanol administration alters metal distribution and some biochemicals in rat brain. *Biochem. Int.* **19,** 135–143.

Murthy, R. C., Saxena, D. K., Sunderaraman, V., and Chandra, S. V. (1987). Cadmium induced ultrastructural changes in the cerebellum of weaned and adult rats. *Ind. Health* **25,** 159–162.

Nation, J. R., Frye, G. D., Von Stultz, J., and Bratton, G. R. (1989). Effects of combined lead and cadmium exposure: changes in schedule-controlled responding and in dopamine, serotonin, and their metabolites. *Behav. Neurosci.* **103,** 1108–1114.

Newland, M. C., Ng, W. W., Baggs, R. B., Gentry, G. D., Weiss, B., and Miller, R. K. (1986). Operant behavior in transition reflects neonatal exposure to cadmium. *Teratol.* **34,** 231–241.

Newland, M. C., Ng, W. W., Baggs, R. B., Miller, R. K., and Infurna, R. N. (1983). Acute behavioral toxicity of cadmium in neonatal rats. *Teratology* **27,** 65A–66A.

Nolan, C. V. and Shaikh, Z. A. (1986). The vascular endothelium as a target tissue in acute cadmium toxicity. *Life Sci.* **39,** 1403–1409.

Onosaka, S. and Cherian, M. G. (1981). The induced synthesis of metallothionein in various tissues of rat in response to metals. I. Effect of repeated injection of cadmium salts. *Toxicology* **22,** 91–101.

Onosaka, S., Tanaka, K., and Cherian, M. G. (1984). Effects of cadmium and zinc on tissue levels of metallothionein. *Environ. Health Perspec.* **54,** 67–72.

Orrenius, S., Ormstad, K., Thor, H., and Jewell, S. A. (1983). Turnover and functions of glutathione studied with isolated hepatic and renal cells. *Fed. Proc.* **42,** 3177–3188.

Palmiter, R. D., Findley, S. D., Whitmore, T. E., and Durnam, D. M. (1992) MT-III, a brain-specific member of the metallothionein gene family. *Biochem.* **89,** 6333–6337.

Pelletier, M. R. and Satinder, K. P. (1991). Low-level cadmium exposure increases one-way avoidance in juvenile rats. *Neurotoxicol. Teratol.* **13,** 657–662.

Petering, D. H. and Fowler, B. A. (1986). Discussion summary. Roles of metallothionein and related proteins in metal metabolism and toxicity: Problems and perspectives. *Environ. Health Perspec.* **65,** 217–224.

Phil, R. D. and Parkes, M. (1977). Hair element content in learning disabled children. *Science* **198,** 205–206.

Piscator, M. (1986). The nephropathy of chronic cadmium poisoning. In: *Cadmium* E. C. Foulkes, Ed, Springer-Verlag, Berlin. pp. 179–194.

Prodan, L. (1932). Cadmium poisoning: I. The history of cadmium poisoning and uses of cadmium. *J. Indus. Hyg.* **14,** 132–155.

Provias, J. P., Ackerley, C. A., Smity, C., Becker, L. E. (1994). Cadmium encephalopathy: a report with elemental analysis and pathological findings. *Acta Neuroopathol,* **88,** 583–586.

Rastogi, R. B., Merali, Z., and Singhal, R. L. (1977). Cadmium alters behaviour and the biosynthetic capacity for catecholamines and serotonin in neonatal rat brain. *J. Neurochem.* **28,** 789–794.

Reddy, C. S. and Dorn, C. R. (1985). Municipal sewage sludge application on Ohio farms: Estimation of cadmium intake. *Environ. Res.* **38,** 377–388.

Reiter, L. W. (1978). Use of activity measures in behavioral toxicology. *Environ. Health Perspect.* **26,** 9–20.

Ribas-Ozonas, B., Estomba, M. C., and Santos-Ruiz, A. (1974). Activation of serotonin and 5-hydroxyindoleacetic acid in brain structures after application of cadmium. In: *Trace Element Metabolism in Animals* W. G. Hoekstra, J. W. Suttie, H. E. Ganther, W. Mertz, Eds. University Park Press, Baltimore, pp. 476–478.

Robards, K. and Worsfold, P. (1991). Cadmium: toxicology and analysis. A review. *Analyst* **116,** 549–468.

Roeleveld, N., Zielhuis, G. A., and Gabreels, F. (1990). Occupational exposure and defects of the central nervous system in offspring: review. *Br. J. Ind. Med.* **47,** 580–588.

Rohrer, S. R., Shaw, S. M., and Lamar, C. H. (1978). Cadmium induced endothelial cell alterations in the fetal brain from prenatal exposure. *Acta Neuropathol.* **44,** 147–149.

Ruppert, P. H., Dean, K. F., and Reiter, L. W. (1985). Development of locomotor activity of rat pups exposed to heavy metals. *Toxicol. Appl. Pharmacol.* **78,** 69–77.

Satinder, K. P. (1985). Behavioral genetic teratology: An emerging research discipline. 8th ICLAS/CALAS Symposium. Verlag, New York, pp. 503–510.

Sato, K., Iwamasa, T., Tsuru, T., and Takeuchi, T. (1978). An ultrastructural study of chronic cadmium chloride-induced neuropathy. *Acta Neuropath.* **41,** 185–190.

Satoh, I., Asai, F., Itoh, K., Nishimura, M., and Urakawa, N. (1982). Mechanism of cadmium-induced blockade of neuromuscular transmission. *Europ. J. Pharmacol.* **77,** 251–257.

Scheuhammer, A. M. and Cherian, M. G. (1985). Effects of heavy metal cations, sulfhydryl reagents and other chemical agents on striatal D_2 dopamine receptors. *Biochem. Pharmacol.* **34,** 3405–3413.

Sevanian, A., ed. (1985). *Lipid Peroxidation in Biological Systems.* American Oil Chemist's Society, Champaign, IL.

Shah, J. and Pant, H. C. (1991). Effect of cadmium on Ca^{2+} transport in brain microsomes. *Brain Res.* **566,** 127–130.

Shukla, A., Shukla, G. S., and Srimal, R. C. (1996). Cadmium-induced alterations in blood-brain barrier permeability and its possible correlation with decreased microvessel antioxidant potential in rat. *Human Exp. Toxicol.* **15,** 400–405.

Shukla, G. S., Hussain, T., and Chandra, S. V. (1987). Possible role of regional superoxide dismutase activity and lipid peroxide levels in cadmium neurotoxicity: in vivo and in vitro studies in growing rats. *Life Sci.* **41,** 2215–2221.

Shukla, G. S., Hussain, T., and Chandra, S. V. (1988c). Protective effect of vitamin E on cadmium-induced alterations in lipofuscin and superoxide dismutase in rat brain regions. *Pharmacol. Toxicol.* **63,** 305–306.

Shukla, G. S., Hussain, T., Srivastava, R. S., and Chandra, S. V. (1989). Glutathione peroxidase and catalase in liver, kidney, testis and brain regions of rats following cadmium exposure and subsequent withdrawal. *Ind. Health* **27,** 59–69.

Shukla, G. S., Srivastava, R. S., and Chandra, S. V. (1988a). Glutathione status and cadmium neurotoxicity: studies in discrete brain regions of growing rats. *Fundam. Appl. Toxicol.* **11,** 229–235.

Shukla, G. S., Srivastava, R. S., and Chandra, S. V. (1988b). Prevention of cadmium-induced effects on regional glutathione status of rat brain by vitamin E. *J. Appl. Toxicol.* **8,** 355–359.

Singhal, R. L., Merali, Z., and Hrdina, P. D. (1976). Aspects of the biochemical toxicology of cadmium. *Federation Proc.* **35,** 75–80.

Smith, M. J., Garber, B., and Pihl, R. O. (1983). Altered behavior response to apomorphine in cadmium exposed rats. *Neurobehav. Toxicol. Teratol.* **5,** 161–165.

Smith, M. J., Phil, R. O., and Farrell, B. (1985). Longterm effects of early cadmium exposure on locomotor activity in the rat. *Neurobehav. Toxicol. Teratol.* **7,** 19–22.

Smith, M. J., Pihl, R. O., and Garber, B. (1982). Postnatal cadmium exposure and longtern behavior changes in the rat. *Neurobehav. Toxicol. Teratol.* **4,** 283–287.

Squibb, R. E., Jr. and Squibb, R. L. (1979). Effect of food toxicants on voluntary wheel running in rats. *J. Nutr.* **109,** 767–772.

Sonawane, B. R., Nordberg, M., Nordberg, G. F., and Lucier, G. W. (1975). Placental transfer of cadmium in rats: influence of dose and gestational age. *Environ. Health Perspect.* **12,** 97–102.

Sorell, T. L. and Graziano, J. H. (1990). Effect of oral cadmium exposure during pregnancy on maternal and fetal zinc metabolism in the rat. *Toxicol. Teratol.* **102,** 537–545.

Sowa, B. and Steibert, E. (1985). Effect of oral cadmium administration to female rats during pregnancy on zinc, copper, and iron content in placenta, foetal liver, kidney, intestine, and brain. *Arch. Toxicol.* **56,** 256–262.

Stellern, J., Marlowe, M., Cossairt, A., and Errera, J. (1983). Low lead and cadmium levels and childhood visual-perception development. *Perceptual & Motor Skills* **56,** 539–544.

Sun, T-J., Miller, M. L., and Hastings L. (1995). Effects of inhalation of cadmium on the rat olfactory system: behavior and morphology. *Neurotoxicol. Teratol.* **17,** 1–10.

Sutoo, D., Akiyama, K., and Imamiya, S. (1990). A mechanism of cadmium poisoning: the cross effect of calcium and cadmium in the calmodulin-dependent system. *Arch. Toxicol.* **64,** 161–164.

Suzuki, Y., Chao, S-H., Zysk, J. R., and Cheung, W. Y. (1985). Stimulation of calmodulin by cadmium ion. *Arch. Toxicol.* **57,** 205–211.

Swandulla, D. and Armstrong, C. M. (1989). Calcium channel block by cadmium in chicken sensory neurons. *Proc. Nat. Acad. Sci.* **86,** 1736–1740.

Tabacova, S., Little, R. E., Balabaeva, L., Pavlova, S., and Petrov, I. (1994). Complications of pregnancy in relation to maternal lipid peroxides, glutathione, and exposure to metals. *Reproduct. Toxicol.* **53,** 570–576.

Thatcher, R. W., Lester, M. L., McAlaster, R., and Horst, R. (1982). Effects of low levels of cadmium and lead on cognitive functioning in children. *Arch. Environ. Health* **37,** 159–166.

Thatcher, R. W., McAlaster, R., and Lester, M. L. (1984). Evoked potentials related to hair cadmium and lead in children. *ANYAS* **425,** 384–390.

Tischner, K. (1980). Cadmium. In: *Experimental and Clinical Neurotoxicology,* P. S. Spencer and H. H. Schaumburg, Eds. Williams & Wilkins, Baltimore, pp. 348–355.

Vallee, B. L. (1979). Metallothionein: historical review and perspectives. *Experientia* [Suppl] **34,** 19–40.

Vallee, B. L. and Ulmer, D. D. (1972). Biochemical effects of mercury, cadmium and lead. *Ann. Rev. Biochem.* **41,** 91–128.

Valois, A. A. and Webster, W. S. (1987a). Retention and distribution of cadmium in the mouse brain: an autoradiographic and gamma counting study. *Neurotoxicol.* **8,** 463–469.

Valois, A. A. and Webster, W. S. (1987b). The choroid plexus and cerebral vasculature as target sites for cadmium following acute exposure in neonatal and adult mice: An autoradiographic and gamma counting study. *Toxicol.* **46,** 43–55.

Valois, A. A. and Webster, W. S. (1989). The choroid plexus as a target site for cadmium toxicity following chronic exposure in the adult mouse: an ultrastructural study. *Toxicol.* **55,** 193–205.

Verbost, P. M., Senden, M. H. M. N., and van Os, C. H. (1987). Nanomolar concentrations of Cd^{2+} inhibit Ca^{2+} transport systems in plasma membranes and intracellular Ca^{2+} stores in intestinal epithelium. *Biochimica Biophys. Acta* **902,** 247–252.

Vig, P. J., Bhatia, M., Gill, K. D., and Nath, R. (1989). Cadmium inhibits brain calmodulin: in vitro and in vivo studies. *Bull. Environ. Contam. Toxicol.* **43,** 541–547.

Waalkes, M. P. and Klaassen, C. D. (1985). Concentration of metallothionein in major organs of rats after administration of various metals. *Fund. Appl. Toxicol.* **5,** 473–477.

Waalkes, M. P., Wahba, Z. Z., and Rodriguez, R. E. (1992). Cadmium. In: *Hazardous Materials Toxicology. Clinical Principles of Environmental Health,* J. B. Sullivan, Jr., and G. R. Krieger, Eds. Williams & Wilkins, Baltimore, pp. 845–852.

Webb, M. (1979). *The Metallothioneins: The Chemistry, Biochemistry, and Biology of Cadmium.* In: M. Webb Ed. Elsevier North-Holland, Biomedical Press, NY, pp. 195–266.

Webster, W. S. (1978). Cadmium-induced fetal growth retardation in the mouse. *Arch. Env. Health* **33,** 36–42.

Webster, W. S. (1988). Chronic cadmium exposure during pregnancy in the mouse: influence of exposure levels on fetal and maternal uptake. *J. Toxicol. Environ. Health* **24,** 183–192.

Webster, W. S. and Valois, A. A. (1981). The toxic effects of cadmium on the neonatal mouse CNS. *J. Neuropathol. Exp. Neurol.* **40,** 247–257.

Whelton, B. D., Toomey, J. M., Bhattacharyya, M. H. (1993). Cadmium-109 metabolism in mice. IV. Diet versus maternal stores as a source of cadmium transfer to mouse fetuses and pups during gestation and lactation. *J. Toxicol Environ. Health* **40,** 531–546.

WHO (World Health Organization) (1992a). *Cadmium.* Environmental Health Criteria 134, International Programme on Chemical Safety.

WHO (World Health Organization) (1992b). *Cadmium-Environmental Aspects.* Environmental Health Criteria 135, International Programme on Chemical Safety.

Williams, B. J., Laubach, D. J., Nechay, B. R., and Steinsland, O. S. (1978). The effects of cadmium on adrenergic neurotransmission *in vitro. Life Sci.* **23,** 1929–1934.

Wong, K-L., Cachia, R., and Klaassen, C. D. (1980). Comparison of the toxicity and tissue distribution of cadmium in newborn and adult rats after repeated administration. *Toxicol. Appl. Pharmacol.* **56,** 317–325.

Wong, K-L. and Klaassen, C. D. (1982). Tissue Distribution and retention of cadmium in rats during postnatal development: Minimal role of hepatic metallothionein. *Toxicol. Appl. Pharmacol.* **53,** 343–353.

Wong, K-L. and Klaassen, C. D. (1982). Neurotoxic effects of cadmium in young rats. *Toxicol. Appl. Pharmacol.* **63,** 330–337.

Wong, P. C. L. and Lim, L. (1981). The effects of aluminum, manganese and cadmium chloride on the methylation of phospholipids in the rat brain synaptosomal membrane. *Biochemical Pharmacol.* **30,** 1704–1705.

Young, J. K., Garvey, J. S., and Huang, P. C. (1991). Glial immunoreactivity for metallothionein in the rat brain. *GLIA* **4,** 602–610.

Zelazowski, A. S. and Piotrowski, J. K. (1974). The levels of metallothionein-like protein in animal tissues. *Experientia* **33,** 1624–1625.

CHAPTER

31

Developmental Lead Exposure
Neurobehavioral Consequences

DEBORAH C. RICE
Toxicology Research Division
Bureau of Chemical Safety, Food Directorate
Health Protection Branch, Health Canada
Ottawa, Ontario, Canada K1A 0L2

I. Introduction

Lead is undoubtedly the most studied developmental neurotoxicant; indeed, it is probably the most studied environmental toxicant for which developmental effects are the main concern. Concern over the potential health consequences of childhood lead exposure has generated a number of reviews (e.g., Grant and Davies, 1989; Needleman, 1992), including reports from United States and from international agencies (ATSDR, 1988; Mushak *et al.,* 1989; IPCS, 1995). Although lead is a common element in the earth's crust, its ubiquitous presence in bioavailable forms in the environment is due largely to the activities of humans (Lin-Fu, 1985). The addition of lead to gasoline in the 1920s resulted in a steep increase in the amount of lead emitted into the environment (Elias *et al.,* 1975) and, consequently, ubiquitous exposure of whole populations to increasing levels of lead (Fig. 1). At present, body burdens of people in industrialized societies are 2–3 orders of magnitude higher than natural background levels (NAS, 1980). The

*Abbreviations: ADHD, attention deficit hyperactivity disorder, BAER, brainstem auditory evoked potentials; CBCL, Child Behavior Checklist; DRL, differential reinforcement of low rate; FI, fixed interval; GCI, general cognitive index; IBR, [Bayley] Infant Behavioral Record; IPSC, International Programme on Chemical Safety; IQ, intelligence quotient; K-TEA, Kaufman Test of Educational Achievement; MDI, [Bayley] Mental Development Index; MSCA, McCarthy Scales of Children's Abilities; PDI, [Baylay] Psychomotor Development Index; SES, socioeconomic status; SID, [Bayley] Scales of Infant Development; WISC–R, Wechsler Intelligence Scales for Children–Revised.

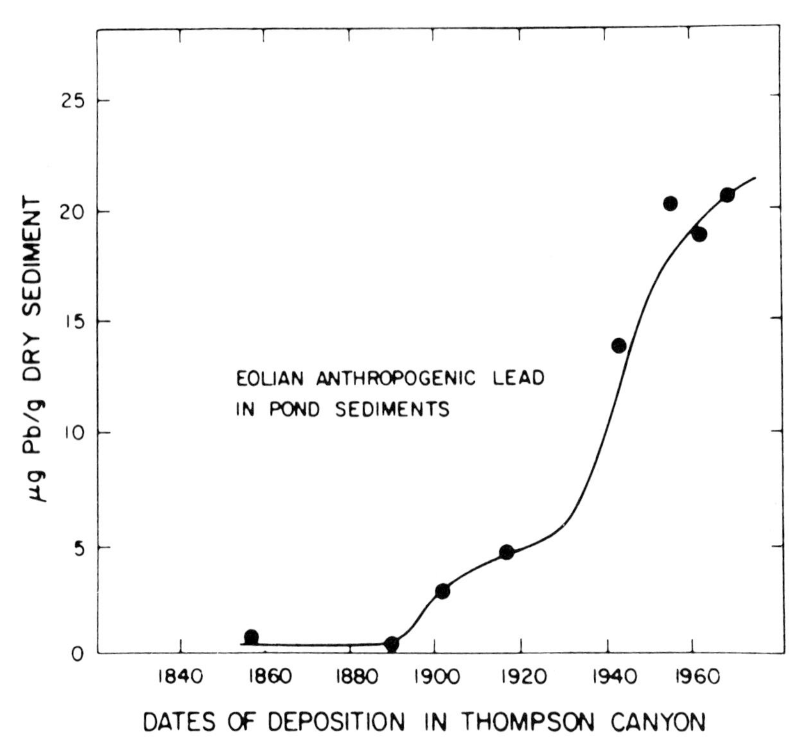

FIGURE 1 Lead levels in pond sediment in a terrestrial ecosystem located in an isolated canyon in a remote, uninhabited region, demonstrating the tremendous increase in atmospheric lead deposition in the 20th century. These data are consistent with other data in a number of media in areas around the world (e.g. Greenland snows), delineating worldwide lead pollution. [From Patterson, C. C. (1983). British mega exposures to industrial lead. In *Lead versus Health.* M. Rutter and R. Russell Jones, Eds. John Wiley & Sons, New York, pp. 17–32. Reproduced with permission].

average blood–lead level of young children in the United States and other industrialized countries has decreased dramatically from 15 μg/dl in the late 1970s to current levels of approximately 4 μg/dl. This dramatic reduction has been the result of actions undertaken by government and health officials in recognition of the serious threat to health imposed by elevated lead body burdens in infants and children. However, many children in the United States and elsewhere still have elevated body burdens of lead. Moreover, there is evidence that there is no threshold for lead-induced cognitive impairment at the range of blood–lead levels of present populations.

Lead poisoning in children was recognized before the turn of the 20th century (Turner, 1897; Gibson, 1892); by the mid-1920s, lead-based paint was recognized by American researchers as a serious and not uncommon source of illness in children (Lin-Fu, 1972, 1985; Rabin, 1989). In the 1940s it was recognized by astute physicians that children who had been treated for lead poisoning suffered permanent neuropsychologic damage (Byers and Lord, 1943), including poor school performance, impulsive behavior, short attention span, restlessness, and occasional neurologic signs. These observations were later replicated by other investigators (Thurston *et al.,* 1955; Jenkins and Mellins, 1957; Perlstein and Attala, 1966). In the early 1970s, deficits in IQ, fine motor performance, and behavioral disorders such as distractibility and constant need for attention were observed in children who had never exhibited overt signs of lead intoxication (de la Burdé and Choate, 1972; Lin-Fu, 1972). By the 1970s, concern arose in the United States and elsewhere that the many tons of lead being introduced into the environment every year by the use of leaded gasoline as well as other industrial processes were producing significant health effects, particularly in children.

II. Epidemiological Studies

A. *Assessment of Intelligence*

Virtually all studies on the behavioral effects of lead in children have included some measure of intelligence (IQ). These have sometimes differed between studies, and different instruments were necessarily used at different ages. One advantage of IQ tests is that they are standardized for the population, allowing straightforward comparisons between studies. The various subscales also provide global assessment of specific components of neuropsychologic function. The use of IQ as a measure of intellectual functioning has revealed lead-associated deficits even at low body burdens of lead. In addition, it has provided the opportunity to evaluate the combined evidence for intellectual impairment as a result of developmental lead exposure by using meta-analysis techniques.

1. Cross-Sectional Studies

In 1979, Needleman *et al.* studied more than 2,000 first and second graders using tooth lead in deciduous teeth as a marker of cumulative lead exposure. They reported a 4-point decrement in IQ measured by the Weschler Intelligence Scales for Children–Revised (WISC–R) between children with the highest and lowest decile tooth–lead levels after adjusting for confounders (724 μg/g vs. <8.7 μg/g, respectively) (Needleman *et al.*, 1985). Since the landmark study of Needleman *et al.*, other investigators have used tooth lead as a marker of cumulative lead exposure in cross-sectional studies. Winneke *et al.* (1983) studied 115 children in Stolberg, Germany. Children with tooth lead levels higher than 10 μg/g had covariate-adjusted IQs 4 points lower than those with average tooth lead levels of 4 μg/g. In a study in Southhampton, England, children with tooth lead levels higher than 8 μg/g had covariate-adjusted IQs 2 points less than children with tooth lead levels less than 2.5 μg/g (Smith *et al.*, 1983; Pocock *et al.*, 1987). Children in Christchurch, New Zealand, showed a small inverse correlation between dentine lead and IQ at age 8 or 9 years (Fergusson *et al.*, 1988 a,b) that became statistically insignificant after adjusting for confounders. In a study of 156 children in Denmark, children with circumpulpal dentine levels greater than 18.7 μg/g had a covariate-adjusted 6-point decrement on the WISC full-scale IQ and a 9-point decrement in verbal IQ when compared to children with dentine levels less than 5 μg/g. Average blood lead levels in the high-lead group were 5.7 μg/dl, and in the low-lead group were 3.7 μg/dl (Hansen *et al.*, 1989).

A number of cross-sectional studies have relied on blood lead as a marker of lead exposure. In a study of 114 6- to 7 year-old children in Nordenham, Germany, there was a small association between concurrent blood lead levels and measures of IQ (Winneke *et al.*, 1985). Yule *et al.* (1981) found a covariate-adjusted 7-point IQ difference between children with blood lead levels greater than 13 μg/dl compared with those with blood lead levels less than 12 μg/dl among 116 children age 6–12 years in Greenwich, England. In a study of low socioeconomic status (SES) children in the United States, significant associations were found between blood lead levels and IQ (Schroeder *et al.*, 1985; Hawk *et al.*, 1986). In a study of 501 school-age children in Edinburgh, Scotland, Fulton *et al.* (1987) reported a covariate-adjusted linear relationship between intellectual functioning and log blood lead concentration for blood lead values between about 5 and 25 μg/dl (mean, about 10 μg/dl), with no indication of a threshold for lead effect. Hatzakis *et al.* (1989) studied 533 children in Lavrion, Greece, with mean blood lead levels of 13.7 μg/dl (range, 7.4–63.9 μg/dl). A strong association was found between IQ and blood lead levels; there was a 9-point covariate-adjusted difference between children with blood lead levels higher than 45 μg/dl and those with blood lead levels less than 15 μg/dl. In a multicenter European study, 1,879 children 6–11 years old from eight institutions were studied (Winneke *et al.*, 1990). There was an inverse association between blood lead levels and performance on the WISC-R that was borderline statistically significant. A decrement of 3 points was calculated for an increase in blood lead from 5 to 20 μg/dl.

2. Prospective Studies

There are a number of prospective studies examining the developmental effects of environmental lead exposure in which a large body of data has been collected over a number of years. The mothers were recruited before the birth of their infants, and the infants were followed in a longitudinal manner. This design is stronger than a cross-sectional design, and these studies have provided convincing data regarding developmental deficits produced by low-level lead exposure.

In a study in Boston, 249 infants from middle- and upper middle-class homes were investigated (Bellinger *et al.*, 1987a). Performance on the Bayley Mental Development Index (MDI) at 6, 12, and 24 months of age was associated with cord but not postnatal blood lead levels. The difference between the high (mean, 14.6 μg/dl) and low (mean, 1.8 μg/dl) blood lead groups was 4–7 points. Assessment of these children at 57 months of age revealed that performance on the General Cognitive Index (GCI) of the McCarthy Scales of Children's Abilities (MSCA) was associated with blood lead levels at 24 months but not 57 months of age, after adjusting for possible confounders (Bellinger *et al.*, 1987b). Blood lead values averaged 6.8 μg/dl at 24 months and 6.4 μg/dl at 57 months. When these children were retested at 10 years of age, a 10 μg/dl increase in blood lead at 24 months was associated with a 5.8-point decline on full-scale IQ on the WISC–R (Fig. 2) and an 8.9-point decline on the Kaufman Test of Educational

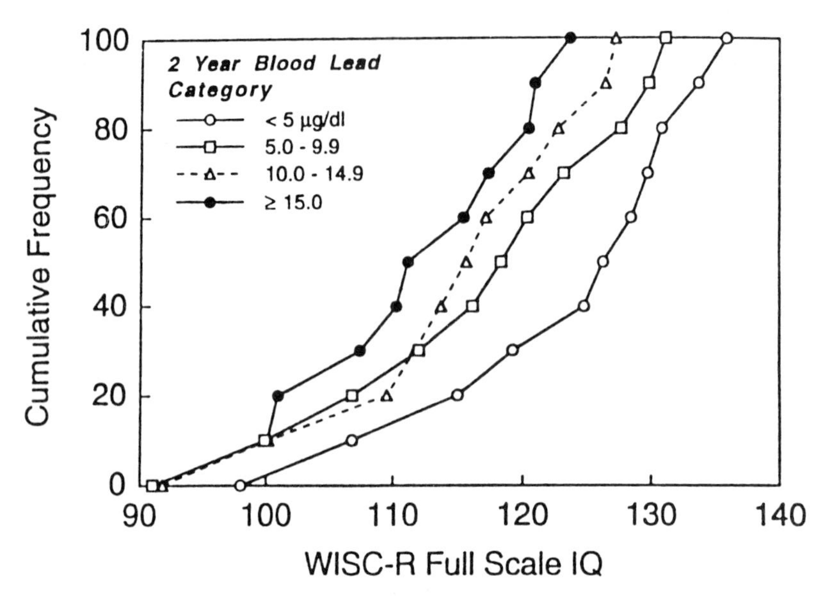

FIGURE 2 Cumulative frequency distribution of full-scale WISC–R IQ at age 10 for children in the Boston prospective study, stratified by blood lead categories at 2 years of age. [Reprinted from Bellinger, D. C. (1995). Interpreting the literature on lead and child development: the neglected role of the "Experimental System." *Neurotoxicol. Teratol.* **17,** 201–212, with permission of Elsevier Science.]

Achievement (K-TEA) (Bellinger *et al.,* 1992). Interestingly, blood lead level at 2 years of age was predictive of poorer performance on the K-TEA within IQ categories (Bellinger, 1995) (Fig. 3). This finding suggests that behavioral processes influenced by lead independent of effects on IQ are impairing school learning and performance.

In a prospective study of 305 infants from low-income areas in Cincinnati, Ohio, each log unit increment in blood lead (mean at birth = 4.5 μg/dl) was associated

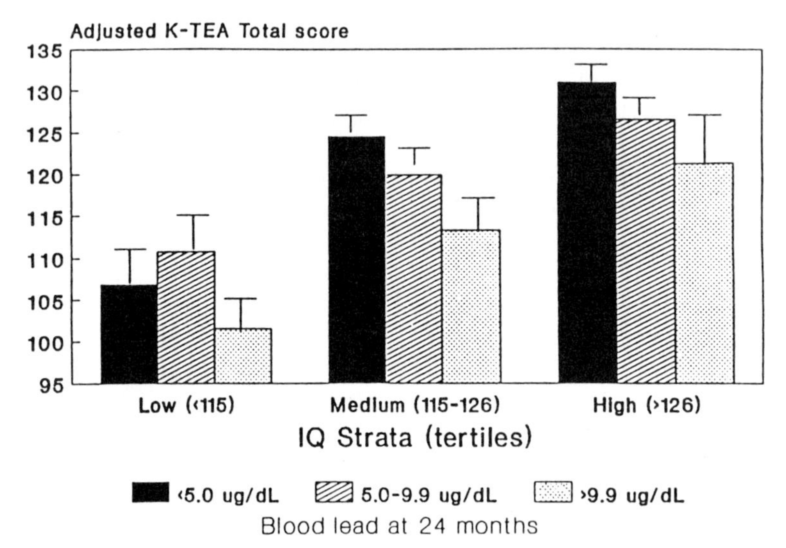

FIGURE 3 Adjusted Battery Composite scores of the Kaufman Test of Educational Achievement (Brief Form) at age 10 for children in the Boston prospective study stratified by full-scale IQ tertile and blood lead category at 2 years of age. Error bars represent 1 S.E. [Reprinted from Bellinger, D. C. (1995). Interpreting the literature on lead and child development: the neglected role of the "Experimental System." *Neurotoxicol. Teratol.* **17,** 201–212, with permission of Elsevier Science.]

with a covariate-adjusted reduction of 5.7 points on the Bayley MDI at 6 months (Dietrich *et al.,* 1987, 1989). The reduction was 8 points if the effect on gestational age and birthweight were included. At 1 year after birth, prenatal blood lead levels were negatively correlated with MDI, Bayley Psychomotor Development Index (PDI), and Bayley Infant Behavioral Record (IBR). The IBR revealed higher activity levels and more negative social–emotional responses. These effects appeared to have attenuated in this group of disadvantaged children by 4 years of age (Dietrich *et al.,* 1991). However, when these children were reassessed at 6.5 years of age, lifetime blood lead concentration in excess of 20 μg/dl was associated with a 7-point decrease in performance IQ on the WISC–R compared to children with blood lead levels less than 10 μg/dl (Dietrich *et al.,* 1993).

In a third prospective study in Port Pirie, South Australia, 537 children residing near a primary lead smelter were examined (McMichael *et al.,* 1986; Baghurst *et al.,* 1987; Wigg *et al.,* 1988). A decrease of 2 points on the MDI scale for every 10 μg/dl increase in blood lead level was observed at 24 months of age. Performance was found to be more related to postnatal rather than prenatal blood lead levels; however, no assessment, was performed before 2 years of age. In a follow-up assessment, previous blood lead levels were found to be inversely correlated with performance on the MSCA at 4 years of age (McMichael *et al.,* 1988). Subjects with blood lead levels of 30 μg/dl had a general cognitive score 7.2 points lower than children with blood lead levels of 10 μg/dl, with no evidence of a threshold for effect. These deficits persisted to 7 years of age, with a 4–5 point drop in IQ for blood lead levels between 10 and 30 μg/dl, using the integrated blood lead level across the 7 years as the measure of exposure (Baghurst *et al.,* 1992).

In a study of 541 children living in a smelter community in Yugoslavia, performance was compared to children living in a town with low lead exposure (average blood lead levels of 8.4 and 3.5μg/dl, respectively) (Wasserman *et al.,* 1992)). A 2.5-point deficit was observed on the Bayley MDI at 24 months of age. However, anemia was a significant covariate in this sample, being more strongly predictive of MDI scores than were lead levels. At 4 years of age, 332 of the original group were examined on the MSCA (Wasserman *et al.,* 1994). There was a covariate-adjusted inverse association with lead levels for all five of the GCI scales. Blood lead levels at 24–48 months of age were more predictive of GCI score than were cord blood lead concentrations. Children had received iron supplementation since 2 years of age, so that iron status was no longer an independent predictor of cognitive performance.

A study of 207 children in Sydney, Australia, failed to find any covariate-adjusted effects on the MSCA (Cooney *et al.,* 1989). Maternal and cord blood average lead levels were 9.1 and 8.1 μg/dl, respectively (range, 0–29 μg/dl). This middle class cohort had developmental scores that were approximately one standard deviation above the mean. Follow-up assessment of 175 children at 7 years of age also found no statistically significant association with blood lead history on the WISC-R (Cooney *et al.,* 1991).

3. Summary of Effects on IQ

The body of data on measures of intelligence tests has been referred to as inconsistent (Ernhart, 1992; Volpe *et al.,* 1992) on the basis of the fact that different subscales of IQ tests have been found to be affected in different studies, or in the same (prospective) study at different ages, or associated with lead body burden measured at different ages. The fact that the subscales of apparently greater sensitivity change between ages, or are different between studies, may be due to any of a number of factors. Assessment of intelligence in younger children has been performed using a number of scales, including the Bayley Scales of Infant Development (SID), particularly in very young children, and the MSCA. Most tests in older children used the WISC–R. It is generally recognized that early tests of intelligence such as the Bayley do not measure the same functions as tests used at school age such as the WISC–R, and have little predictive validity for individual children (Honzik, 1976). It would not be surprising, then, if there were little correlation between results of tests performed during infancy and tests performed on older children, either within or across studies. Despite this, prospective studies have been consistent in revealing lead-related deficits in IQ from infancy through at least early school age. In addition to the issue of comparability of the different instruments used for assessing intelligence, differences in concurrent or historic blood lead levels at the time of testing, as well as differences in pattern of blood lead levels over the lifetime, may contribute to differences in results: that is, different behavioral functions may have different sensitive periods (Rice, 1996).

It is critical to look at the epidemiologic evidence as a whole when assessing the strength of the evidence for decrement in IQ as a result of lead exposure in children. Meta-analyses by Gatsonis and Needleman (1992) and Schwartz (1994a) have discussed the details of the characteristics of the available studies, including control of covariates, sample characteristics (e.g., low vs. high SES), and the power of individual studies to detect a small effect of lead on IQ. Schwartz (1994a) found a 2.6-point decrement in IQ associated with an increase in blood lead from 10 to 20 μg/dl in a meta-analysis of

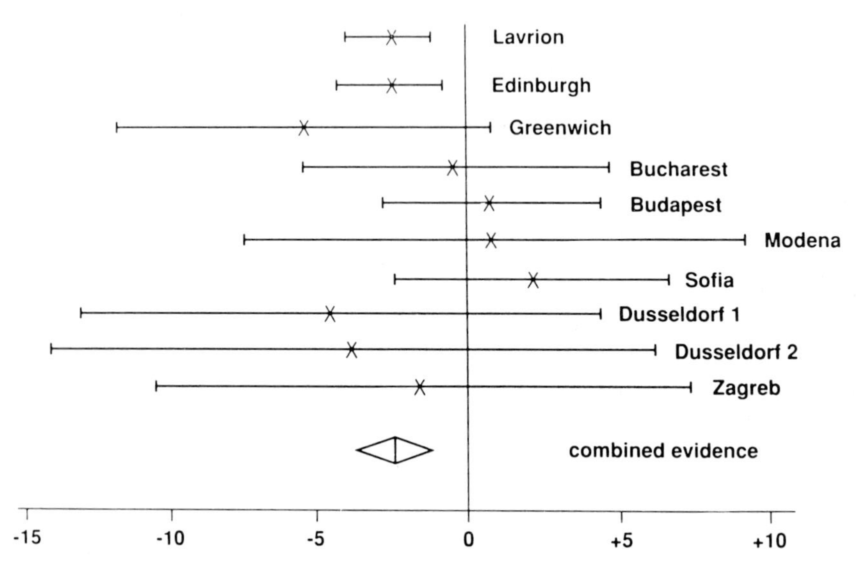

FIGURE 4 Estimated mean change in full-scale IQ with 95% confidence intervals for an increase in blood lead level from 10 to 20 μg/dl in the cross-sectional studies included in the IPCS meta-analysis. All but the top three studies were part of the European multicenter study. The meta-analysis estimates that full-scale IQ is reduced by 2.15 points, with 95% confidence intervals of −1.2 to −3.1 points ($p < 0.001$) [From IPCS. (1995) *Environmental Health Criteria 165: Inorganic Lead.* International Programme on Chemical Safety. World Health Organization, Geneva].

eight studies that met inclusion criteria, with no evidence of a threshold down to a blood lead level of 1 μg/dl. A series of meta-analyses was performed by the International Programme on Chemical Safety (IPCS, 1995). They used full- scale IQ as the outcome measure and performed separate analyses for the prospective and cross- sectional studies (Figs. 4 and 5). The effect estimate for each study was weighted by the inverse of its variance to account for differences in sample size and therefore the power of the study to detect an effect. Although for both analyses many of the 95% confidence intervals include zero, indicating lack of significance at the 0.05 level, the weighted common effect estimate was a decline of about 2–3 IQ points for an increase in blood lead concentration from 10 to 20 μg/dl.

B. Other Behavioral Endpoints

In addition to measures of intelligence such as IQ, studies have assessed performance and behavior using a number of other instruments. Investigators have used these measures presumably to identify endpoints that may be more sensitive to lead-induced impairment than a global measure such as IQ. In addition, in some instances endpoints have been chosen to try to elucidate the underlying behavioral processes responsible for the cognitive impairment observed as a result of increased body burden of lead.

Perhaps the ancillary assessment most often included, in both retrospective and prospective designs, is some version of the teachers' or parents' rating scale. In the cross-sectional study in Boston (Needleman *et al.*, 1979), teachers reported a dose-dependent impairment in functioning with increased dentine lead on such measures as

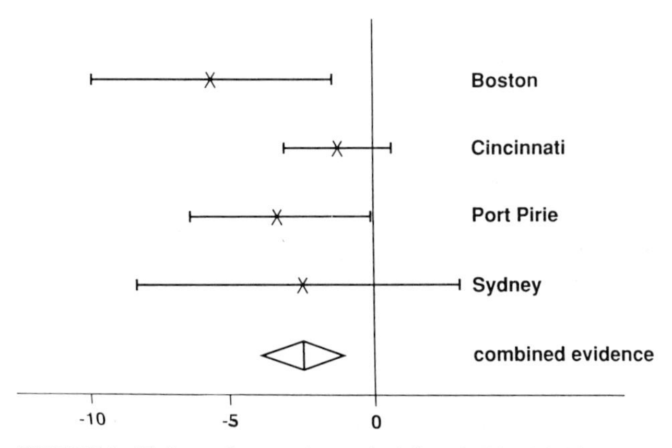

FIGURE 5 Estimated mean change in full-scale IQ with 95% confidence intervals for an increase in blood lead level from 10 to 20 μg/dl in the prospective studies included in the IPCS meta-analysis using blood lead values up to 3 years of age. The meta-analysis estimates that full-scale IQ is reduced by 2.6 points, with 95% confidence intervals of −1.2 to −4.0 points ($p < 0.001$) [From IPCS. (1995). *Environmental Health Criteria 165: Inorganic Lead.* International Programme on Chemical Safety. World Health Organization, Geneva].

distractible, not persistent, dependent, impulsive, easily frustrated, unable to follow simple and complex directions" (Fig. 6). These findings were replicated in a study using the same rating scale in a population of children in London (Yule *et al.,* 1984), which also reported a significant increase in hyperactivity, conduct problems, and inattentive–passive behavior on the Conners scale in these children. In a study of 247 first graders, Tuthill (1996) reported a robust relationship between hair lead and negative teacher ratings after adjusting for potential confounders on the same rating scale used by Needleman *et al.* (1979) and Yule *et al.* (1984). In addition, there was a strong relationship between physician-diagnosed attention deficit hyperactivity disorder (ADHD) and hair lead concentrations. In the study in children in New Zealand, Fergusson *et al.* (1988c) reported increased inattention and restlessness, short attention span, and increased distractibility as a function of dentine lead (mean, 6 μg/g) in children at 8 (n = 724) and 9 (n = 644) years of age. In a cross-sectional study in Mexico using blood as the measure of lead exposure, increased lead body burden was associated with decreased knowledge ability and socialization skills on a teachers' rating scale, as well as impaired performance on the WISC–R (Muñoz *et al.,* 1993). In the cross-sectional study in Scotland using blood lead as the independent variable, a dose-related increase on the aggressive–antisocial and hyperactive measures of the Rutter Scale were observed in a group of 6- to 9-year-old children with low overall blood lead levels (Thomson *et al.,* 1989). In a cross-sectional study in Dunedin, New Zealand, significant associations were found between blood lead levels and increased behavioral problems as assessed by both teachers and parents on the Rutter Behavioral Scale and increased scores on inattention and hyperactivity scales in the absence of changes in IQ (Silva *et al.,* 1988). In the Yugoslavia prospective study, blood lead levels were found to be associated with aggressive and destructive behavior, withdrawal, and sleep problems at 32 months as determined by maternal rating scales (Wasserman *et al.,* 1995). In the Boston prospective study, behavior was assessed using a teachers' rating scale when the children were 8 years old (Leviton *et al.,* 1993). Umbilical cord blood lead levels in girls were associated with an increased probability of being dependent and impersistant, whereas both umbilical and dentine lead levels were related to an inflexible and inappropriate approach to tasks. In boys, umbilical cord blood lead levels were associated with difficulty with simple directions and sequences of directions.

Several investigators have included measures of school performance in their assessments. In the cross-sectional study in Scotland (Fulton *et al.,* 1987), deficits in number skills and reading were identified in addition to deficits in IQ. In the New Zealand prospective study (Fergusson *et al.,* 1988a,b,c), robust deficits in school performance including reading, math, spelling, and handwriting, as assessed by the teachers, were present in the absence of deficits in IQ as measure by the WISC–R when the children were 9 years old. At 8–12 years of age, these children exhibited significant impairment in word recognition correlated with dentine levels at 6–7 years of age (Fergusson and Horwood, 1993). Yule *et al.* (1981) also found deficits in school performance such as spelling and reading in conjunction with decreased WISC–R scores in 6- to 12-year-old children. Girls, but not boys, in the Boston prospective study showed reading and spelling difficulties at 8 years of age related to dentine lead levels (Leviton *et al.,* 1993).

A number of studies have also included measures of visuomotor integration/attention/vigilance. The 1979 Boston cross-sectional study assessed performance on a simple reaction time task (Needleman *et al.,* 1979). Children with higher dentine lead levels had longer reaction times than children with lower dentine levels. This same paradigm was used in the London cohort (Hunter *et al.,* 1985), in which blood lead levels were the marker of lead exposure. Because the blood lead levels of some of the Boston children were known (Needleman, 1987), the two studies may be combined to reveal an orderly dose–effect relationship between increased reaction time and increased blood lead levels. Increased reaction time as a function of increased blood lead levels has also been replicated in the cohort of Greek children (Hatzakis *et al.,* 1987). This observation was pursued by Winneke's group in Germany by assessing vigilance and attention on a choice reaction time task using both visual and auditory stimuli. In the first study (Winneke *et al.,* 1983; Winneke and Kraemer, 1984), there was an indication of an effect on this device as a function of tooth lead in 9-year-old children in the absence of an effect on the German WISC–R. In a subsequent study in 6- to 9-year-old children, a robust effect was observed in both groups of children in two cities in Germany (Winneke *et al.,* 1989). In both studies there was a greater effect on false responses than on the number of failures to respond. These results have been replicated using the same device in a population of Greek children with higher blood lead levels (Hatzakis *et al.,* 1987). In the multicenter study involving 1,879 school-age children from eight European countries, there was a significant covariate-adjusted relationship between blood lead levels and poor performance on a test of constructional ability (the Bender Visual–Motor Gestalt Test) as well as on the serial choice reaction time task (Winneke *et al.,* 1990). In a subset of these children in Germany with very low average blood lead levels (<5 μg/dl), there was a negative association between lead levels and fine

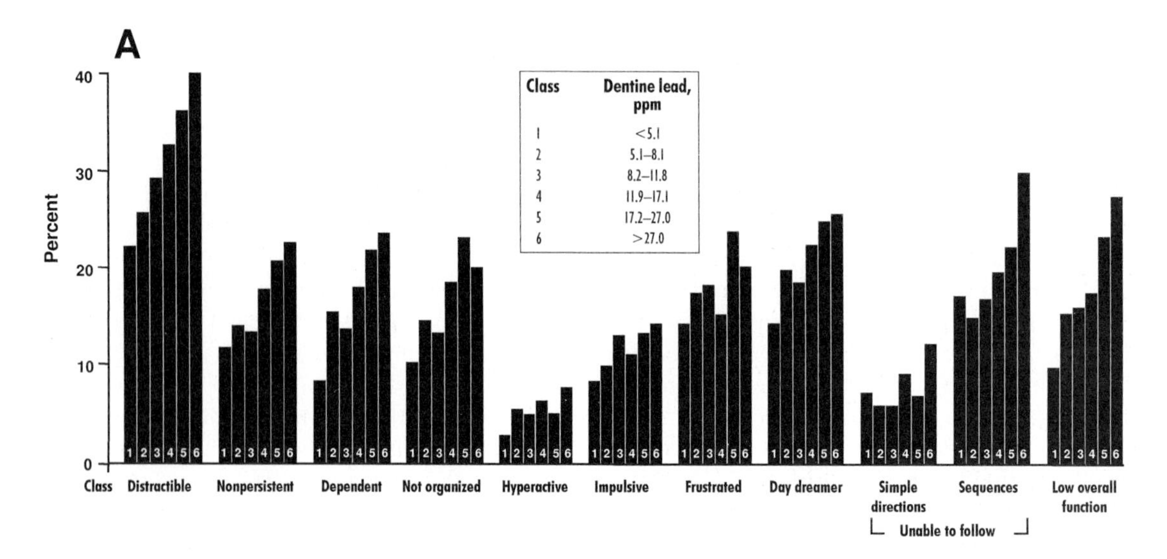
A
Percent
Class
Dentine lead, ppm
1 <5.1
2 5.1–8.1
3 8.2–11.8
4 11.9–17.1
5 17.2–27.0
6 >27.0
Distractible
Nonpersistent
Dependent
Not organized
Hyperactive
Impulsive
Frustrated
Day dreamer
Simple directions
Sequences
Low overall function
Unable to follow

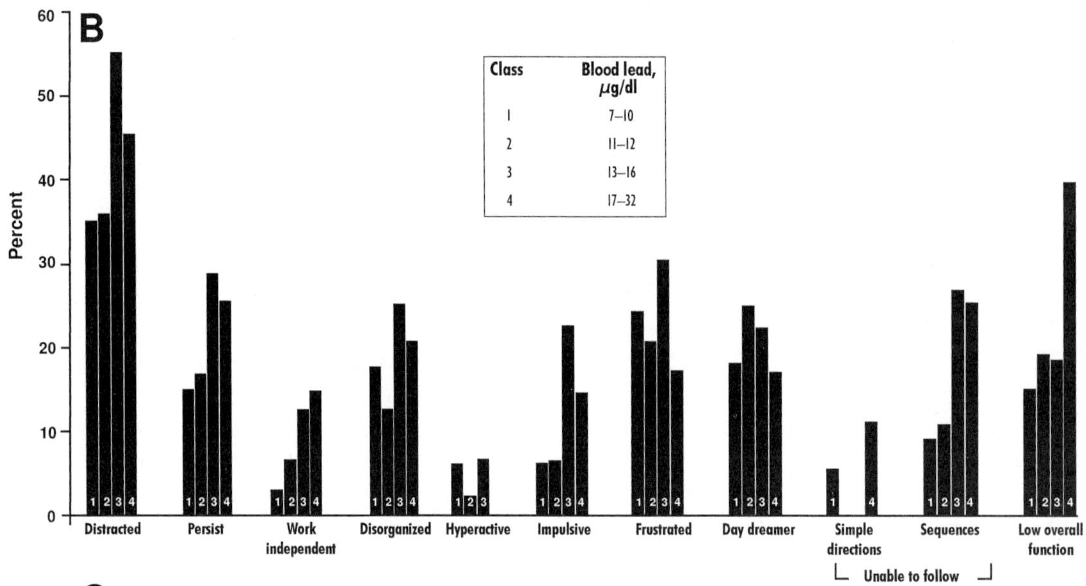
B
Percent
Class
Blood lead, μg/dl
1 7–10
2 11–12
3 13–16
4 17–32
Distracted
Persist
Work independent
Disorganized
Hyperactive
Impulsive
Frustrated
Day dreamer
Simple directions
Sequences
Low overall function
Unable to follow

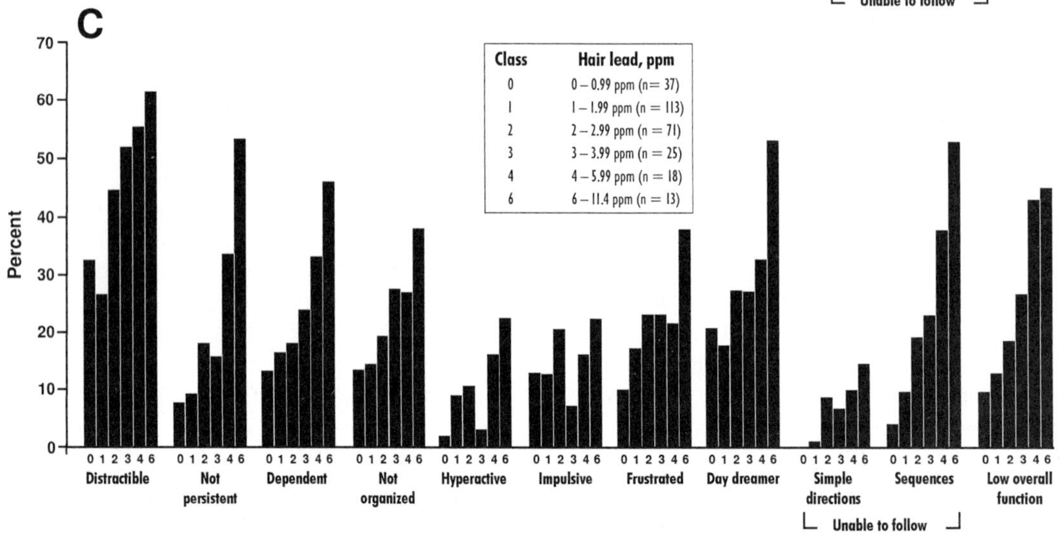
C
Percent
Class
Hair lead, ppm
0 0 – 0.99 ppm (n= 37)
1 1 – 1.99 ppm (n = 113)
2 2 – 2.99 ppm (n = 71)
3 3 – 3.99 ppm (n = 25)
4 4 – 5.99 ppm (n = 18)
6 6 – 11.4 ppm (n = 13)
Distractible
Not persistent
Dependent
Not organized
Hyperactive
Impulsive
Frustrated
Day dreamer
Simple directions
Sequences
Low overall function
Unable to follow

motor performance and pattern recognition (Winneke *et al.*, 1994). In the Port Pirie prospective study, evaluation of visuomotor integration revealed a covariate-adjusted inverse relationship related to both pre- and postnatal blood lead levels (Baghurst *et al.*, 1995). In a study in first graders in Denmark (Hansen *et al.*, 1989), attention was assessed using a continuous performance task. Performance was marginally associated with dentine lead levels; similar to the results from the German studies, false responses showed a greater correlation than failure to respond. While the child was performing the task, a trained psychologist assessed the child for on-task behavior. There was a significant correlation between lead body burden and off-task behavior. In the Boston prospective study, children were assessed at 10 years of age on the Wisconsin Card Sorting Test, a test of abstract thinking, sustained attention, and ability to change response strategy according to environmental requirements (Stiles and Bellinger, 1993). Children with higher recent blood lead levels performed more poorly, as well as perseverating in an old strategy after the "rules" of the test had changed. In a study designed to specifically address attentional processes, Bellinger *et al.* (1994a) studied a group of 79 19- and 20-year-olds. Dentine lead levels (average, 14 μg/g) were significantly associated with poorer performance on tests of ability to select and respond to critical information, as well as on the ability to shift focus adaptively as measured by the Wisconsin Card Sort Test. The authors concluded that, "[e]xecutive and self-regulation functions may be among the cognitive skills targeted by lead" (Bellinger *et al.*, 1994a, p. 98).

III. Long-Term Costs to Society

The decrease in IQ and increase in behavioral problems associated with increased lead body burden have significant monetary costs that must be borne by society. A global measure of failure that has been linked to increased lead burden is academic failure and the need for special education. In the Danish study, Lyngbye *et al.* (1990) found an increased need for special education, especially verbal, in first graders as a function of increased lead body burden. In a follow-up of the 1979 Boston cross-sectional study, Bellinger *et al.* (1984) assessed school performance in sixth graders as a function of their first-grade tooth lead levels. They found a tendency toward an increased need for remedial education and grade retention as a function of increased lead level.

The association between lead exposure and later problem behavior was examined prospectively in 1,782 8-year-old children in Boston (Bellinger *et al.*, 1994b). Tooth dentine levels (average, 3.4 ppm) but not cord blood levels (average, 6.8 μg/dl) were associated with total scores of problem behavior on the Teacher's Report Form of the Child Behavior Profile. Tooth lead levels were also associated with internalizing (consisting of scores for anxiety or withdrawal) and externalizing (consisting of inattentive, nervous–overactive, and aggressive scores). There was also a weak association between tooth lead levels and prevalence of "extreme" problem behavior scores. In a study of 2- to 5-year-old children (Sciarillo *et al.*, 1992), children with blood lead levels persistently higher than 15 μg/dl had higher total problem behavior scores than children with lower blood lead levels on the parent form of the Child Behavior Checklist (CBCL) (similar to the Child Behavior Profile used by Bellinger *et al.*, 1994b). Children with high blood lead levels were also three times more likely to be assigned problem behavior scores in the "clinical" range.

The consequences in older children of this poor social adjustment were explored in a retrospective cohort study of 301 boys that examined the association between bone lead levels and measures of social adjustment at 7 and again at 11 years of age (Needleman *et al.*, 1996). At 7 years of age, borderline associations after adjusting for covariates were observed between teachers' ratings and lead levels for aggression, delinquency, social problems, and externalizing behaviors on the CBCL. When children were 11 years old, parents of high-lead subjects reported significantly more somatic complaints, more delinquent and aggressive behavior, and higher internalizing and externalizing scores. Teachers' ratings at 11 years were associated with bone lead for a number of

FIGURE 6 Teachers' ratings of students on a forced-choice questionnaire. Proportion of negative comments within each group, as measured by (A) dentine lead levels, (B) blood lead, or (C) hair lead [A, from Needleman, H. L., Gunnoe, C., Leviton, A., Reed, R., Peresie, H., Maler, C., and Barrett, P. (1979). Deficits in psychologic and classroom performance of children with elevated dentine lead levels. *N. Engl. J. Med.* **300,** 689–695. Copyright 1979 Massachusetts Medical Society. All rights reserved; B, Reprinted with permission from Yule, W., Urbanowicz, M. A., Lansdown, R., and Millar, I. B. (1984). Teachers' ratings of children's behavior in relation to blood lead levels. *Br. J. Dev. Psych.* **2,** 295, © The British Psychological Society; and C, from Tuthill, R. W. (1996). Hair lead levels related to children's classroom attention-deficit disorder. *Arch. Environ. Health* **51,** 214–220. Reprinted with permission of the Helen Dwight Reid Educational Foundation. Published by Heldref Publications, 1319 18th St. N.W., Washington, D.C. 20036-1802. Copyright 1996.]

categories, including somatic complaints, anxious/depressed, social problems, attention problems, delinquent and aggressive behaviors, and internalizing and externalizing. High-lead subjects had higher scores in self-reports of delinquency at 11 years of age. High bone lead levels were associated with an increased risk of exceeding the clinical score for attention, aggression, and delinquency (Fig. 7), and high-lead subjects were more likely to obtain worse scores on all items of the CBCL during the 4-year period of observation.

The monetary cost of the ubiquitous exposure of fetuses and children to lead in industrialized societies has been calculated by Schwartz (1994b) in an estimation of the benefits of a 1 μg/dl reduction in the population mean blood lead concentration. He estimated medical costs associated with treatment of children with undue lead exposure, the increase in remedial education, and the costs associated with lower birth weight and reduced gestational age, among other factors (Table 1). The largest single cost is lost earnings as a result of decreased intellectual capability. In a later study using the powerful National Longitudinal Survey of Youth database to monetize the effect of decreased cognitive ability on earning capacity, the estimated loss of earnings was $7.59 billion per year for an increase in blood lead levels of 1 μg/dl in the U.S. population (Salkever, 1995). It is obvious that the "small" effect of lead on IQ is reflected in an enormous cost to society in lost potential and increased need for medical care and special education. Not included in this analysis was the disturbing possibility of increased criminality related to increased lead body burden.

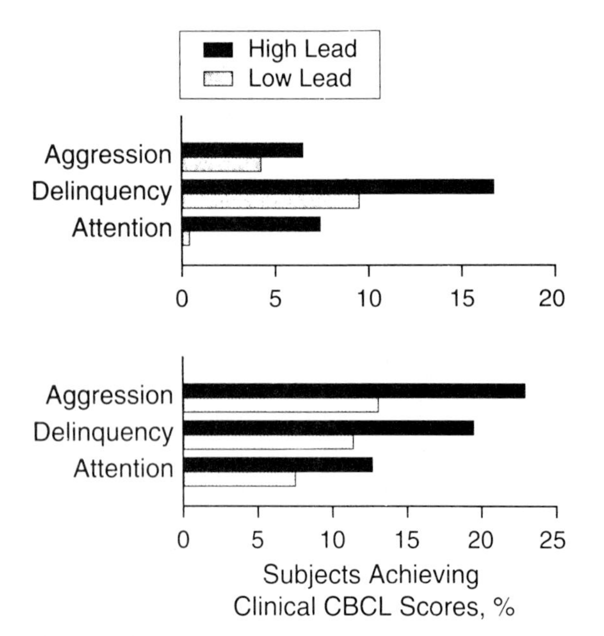

FIGURE 7 The association between bone lead concentration and clinical Child Behavior Checklist (CBCL) (>70) scores for aggression, delinquency and attention. High lead = above the median; low lead = below the median. Scores from parents (top) and teachers (bottom). [From Needleman, H. L., Riess, J. A., Tobin, M. J., Biesecker, G. E., and Greenhouse, J. B. (1996). Bone lead levels and delinquent behavior. *JAMA* **275,** 363–369].

TABLE 1 The Benefits per Year of a Reduction of 1 μg/dl Mean Blood Lead Concentration in the U.S. Population[a]

	Benefits, millions of U.S. dollars
Medical care	189
Compensatory education	481
Earnings	5060
Infant mortality	1140
Neonatal care	67
Total	6937

[a]Children only.
From Schwartz, J. (1994b). Societal benefits of reducing lead exposure. *Environ. Res.* **66,** 105–126.

IV. Animal Studies

Because of the particular vulnerability of children to lead toxicity, experimental investigations focused early on the effects of lead on the developing organism. Pentschew and Garro (1966) produced overt neurotoxicity, including encephalopathy, in suckling pups of rats exposed to a diet high in lead. Research in both the mouse (Silbergeld and Goldberg, 1973) and the rat (Sauerhoff and Michaelson, 1974) revealed increased locomotor activity as a result of high-dose developmental exposure.

Most of the research on the behavioral toxicity of developmental lead exposure in animal models has focused on characterization of the nature of the behavioral effects produced by lead, although there has also been an attempt to identify the behavioral processes underlying the observed lead-induced behavioral deficits (Rice, 1993). Whereas earlier studies used simple learning tasks, later research studied performance on tasks assessing complex learning and memory, sometimes taking cues from the results of epidemiologic studies in children. Lead research using animal models has revealed robust lead-induced impairment at low body burdens of lead on a wide range of behavioral tasks (Table 2). Effects have been observed at blood lead levels similar to those in children, with no indication of a no-observable-effect level.

It was recognized fairly early that the specific task parameters were an important determinant of the effects of lead on behavioral performance. In a study in rats exposed to lead prenatally, Winneke *et al.* (1977)

TABLE 2 Summary of Lead-Induced Behavioral Deficits in Animals

Task	Effect	Ref.
Visual discrimination	Increased errors on difficult task	Winneke *et al.,* 1977 Carson *et al.,* 1974
Discrimination reversal (spatial and nonspatial)	Increased errors	Bushnell and Bowman, 1979a,b Rice and Willes, 1979 Rice and Gilbert, 1990a Rice, 1985a, 1990 Gilbert and Rice, 1987
	Attention to irrelevant stimuli	Rice and Gilbert, 1990a Rice, 1985a, 1990 Gilbert and Rice, 1987
Spatial delayed alternation	Increased errors	Levin and Bowman, 1986 Rice and Karpinski, 1988 Rice and Gilbert, 1990b
	Perseveration	Rice and Karpinski, 1988 Rice and Gilbert, 1990b Cory-Slechta *et al.,* 1991
Hamilton Search Task	Increased errors	Levin and Bowman, 1986, 1983
Learning set	Retarded learning	Lilienthal *et al.,* 1986
Concurrent discrimination	Retarded learning Perseveration for position	Rice, 1992a
Repeated acquisition	Increased errors Perseveration for position	Cohn *et al.,* 1993
Delayed matching to sample (spatial and nonspatial)	Impaired at longer delays Perseveration for position	Rice, 1984
Fixed interval	Increase response rate	Rice, 1992b, 1988, 1985b Rice *et al.,* 1979 Cory-Slechta *et al.,* 1985, 1983 Cory-Slechta and Thompson, 1979 Angell and Weiss, 1982
	Altered temporal pattern of responding	Rice, 1992c
DRL	Increased rate of response	Rice, 1992c Alfano and Petit, 1981 Dietz *et al.,* 1978
	Retarded acquisition	Rice and Gilbert, 1985
Response duration	Decreased response duration Decreased response to external cue	Cory-Slechta *et al.,* 1981

found no effect on a visual discrimination problem that was easy for control rats (vertical vs. horizontal stripes), whereas lead-exposed rats were severely impaired on a difficult discrimination (bigger vs. smaller circle). In a study in lambs exposed pre- and postnatally (Carson *et al.,* 1974), six visual discrimination problems were presented sequentially. The first five were shape discriminations, whereas the sixth was a size-discrimination problem. Lead-exposed lambs were only impaired on the last problem, which was the most difficult problem for control lambs as measured by the number of days required to learn the task. This last problem represented a change in relevant stimulus dimension from form to size. These two studies provided a preview of two findings that would be observed consistently in later studies: difficult tasks are more sensitive to level-induced impairment than easier ones, as are studies in which there is a change in the relevant stimulus–response class.

The visual discrimination reversal task, which is a task used extensively in the elucidation of cognitive impairment produced by lead, is a test in which the formerly correct stimulus becomes the incorrect one, and vice versa. Typically, the subject is required to perform a series of such reversals. This allows the degree of improvement in performance across reversals to be assessed, which is indicative of how quickly the subject learns that the rules of the game change in a predictable pattern. Nonspatial (e.g., color, form) discrimination reversal performance has been found to be affected as a result of postnatal exposure in rhesus monkeys tested during infancy (Bushnell and Bowman, 1979a) and cynomolgus monkeys tested as juveniles (Rice and Willes,

1979). Monkeys with blood lead levels of 15 or 25 μg/dl during infancy and steady state levels of 11 or 13 μg/dl were impaired as juveniles on a series of nonspatial discrimination reversal tasks with irrelevant cues (Rice, 1985a). Analysis of the kinds of errors made by treated monkeys revealed that they were attending to irrelevant cues in systematic ways, suggesting that lead-treated monkeys were being distracted by these irrelevant cues to a greater degree than controls. In a study on possible sensitive periods for deleterious effects produced by lead, monkeys were exposed to lead either continuously from birth, during infancy only, or beginning after infancy (Rice and Gilbert, 1990a). Lead levels were about 30–35 μg/dl when monkeys were exposed to lead and given access to infant formula, and 19–22 μg/dl when dosed with lead after withdrawal of infant formula. When these monkeys were tested as juveniles on the series of nonspatial discrimination reversal problems, both the group dosed continuously from birth and the group dosed beginning after infancy made more errors and were more distracted by irrelevant cues than were controls. Interestingly, the group exposed only during infancy was unimpaired on these tasks, suggesting that the developmental period of lead exposure may be an important determinant of impairment.

Deficits on visual discrimination problems have also been observed in the absence of the requirement for reversal performance under some circumstances, such as high blood lead levels or increased task difficulty. Lilienthal *et al.* (1986) studied the effects of developmental lead exposure on learning set formation, in which a series of visual discrimination problems was learned sequentially. Rhesus monkeys exposed to lead *in utero* and continuing during infancy at doses sufficient to produce blood lead values up to 50 μg/dl in the lower dose group and 110 μg/dl in the high dose group displayed impaired improvement in performance across trials on any given problem, as well as impaired ability to learn successive problems more quickly as the experiment progressed. Such a deficit represents impairment in the ability to take advantage of previous exposure to a particular set of "rules," and is reminiscent of failure of lead-treated monkeys to improve as quickly as controls over a series of discrimination reversals. Concurrent discrimination performance was assessed in the group of monkeys described previously, in which the contribution of the developmental period of exposure to the behavioral toxicity of lead was explored by exposing them to lead continuously from birth, during infancy only, or beginning after infancy (Rice, 1992a). All three treated groups were impaired in their ability to learn a set of several problems concurrently. In addition, all three treated groups exhibited perseverative behavior, responding incorrectly more often than controls at the same position that had been responded on in the previous trial.

Performance on spatial tasks has also proved sensitive to lead-induced impairment. Spatial discrimination reversal tasks, analogous to the nonspatial discrimination reversal tasks described previously, consistently reveal lead-induced deficits. A subset of the monkeys in the Bushnell and Bowman (1979a) study, in which effects on both spatial and nonspatial discrimination reversal had been found during infancy, exhibited impairment on a series of spatial discrimination reversal tasks with irrelevant color cues at 4 years of age, despite the fact that lead exposure had ceased at 1 year of age and blood lead levels at the time of testing were at control levels (Bushnell and Bowman, 1979b). Monkeys with stable blood lead levels of 11 or 13 μg/dl, discussed previously, were impaired in the presence of but not in the absence of irrelevant stimuli on a series of spatial discrimination reversal tasks (Gilbert and Rice, 1987). As in the nonspatial discrimination reversal task, there was evidence that lead-exposed monkeys were attending to the irrelevant stimuli in systematic ways. Spatial discrimination reversal performance was also impaired in the group of monkeys in which the relevance of developmental period of exposure was assessed (Rice, 1990). Contrary to the result of the nonspatial discrimination reversal task, in which the group dosed only during infancy was unimpaired, all three dose groups were impaired to an equal degree, suggesting that spatial and nonspatial tasks may be affected differentially, depending on the development period of lead exposure.

The delayed spatial alternation task has also proved sensitive to lead-induced impairment in a number of studies. In this task, the subject is required to alternate responses between two positions, with no cues to signal which position is correct on any trial. Delays may be introduced between opportunities to respond in order to assess spatial memory. Rhesus monkeys exposed to lead from birth to 1 year of age, with peak blood levels as high as 300 μg/dl and levels of 90 μg/dl for the remainder of the first year of life, were markedly impaired on this task as adults (Levin and Bowman, 1986). In a study in cynomolgus monkeys, increasingly longer delays were introduced over successive sessions in adult monkeys from two studies: those with steady state blood lead levels of 11 or 13 μg/dl (Rice and Karpinski, 1988), as well as the groups in which potential sensitive periods were assessed (Rice and Gilbert, 1990b). All treated groups in both studies were impaired on the acquisition of the task because of indiscriminate responding on both buttons, as well as at longer delay values. In the study assessing sensitive periods, all three lead-exposed groups were impaired to an approximately equal degree, providing further evidence of a lack of sensitive period for lead-induced impairment on spatial tasks. Treated monkeys in both studies also displayed marked perseverative behavior, responding on the same position repeatedly. In a study in rats, improved performance on delayed alternation was observed in young and old animals, but not in rats exposed as adults (Cory-Slechta *et al.,* 1991). The training procedure consisted of many sessions of a cued alternation procedure: the rat had only to respond on the lever associated with a cue light as it alternated between positions from trial to trial. The authors interpreted the improved performance of the lead-treated groups as perseveration of alternation behavior as a result of the extensive training procedure. This explanation is consistent with the interpretation of the results of the monkey studies.

Assessment of spatial learning and memory on a repeated acquisition paradigm in rats exposed to lead beginning at weaning revealed a systematic pattern of perseveration for position that was principally responsible for the overall poorer performance of lead-treated subjects (Cohn *et al.,* 1993). Increased errors at longer delay values were also observed in monkeys on both the nonspatial and spatial versions of a delayed matching to sample paradigm (Rice, 1984). On the nonspatial matching task, lead-exposed monkeys perseverated on the position that had been responded on correctly on the previous trial, similar to the behavior of lead-exposed rats on the repeated acquisition task. Impaired performance has also been observed on another spatial memory task, the Hamilton Search Task, in monkeys exposed to lead *in utero* or postnatally (Levin and Bowman, 1983, 1986).

Intermittent schedules of reinforcement such as the fixed interval (FI) schedule have been used rather extensively in the study of the developmental effects of lead. Although this schedule requires the subject to make only one response at the end of a specified (uncued) interval, FI performance is typically characterized by a gradually accelerating rate of response terminating in reinforcement. Blood lead levels between about 10 and 30 μg/dl resulted in a dose-dependent increase in rate of response in both the rat and monkey (Cory-Slechta and Thompson, 1979; Rice *et al.,* 1979; Angell and Weiss, 1982; Cory-Slechta *et al.,* 1983, 1985; Rice, 1985b, 1988, 1992b). In general, other measures of FI performance were not affected. However, monkeys with blood lead levels of 40 μg/dl exhibited differences in the temporal distribution of responses across the interval compared to controls when tested as adults (Rice, 1992c). Effects of lead have also been examined on schedules assessing temporal discrimination. Cory-Slechta *et al.* (1981) assessed response duration performance, in which rats exposed to lead beginning at weaning were required to depress a lever for at least 3 sec. Lead-treated rats depressed the lever for a shorter time than did controls. Introduction of a tone signalling the 3-sec interval was effective in improving performance of control but not treated rats. On another test of temporal discrimination ability, the differential reinforcement of low rate (DRL) schedule, monkeys were required to space consecutive responses at least 30 sec apart in order to be reinforced (Rice, 1992c). Monkeys with steady state levels of 40 μg/dl exhibited a higher number of nonreinforced responses, lower number of reinforced responses, and shorter average time between responses over the course of the experiment than did control monkeys. Monkeys with steady state blood lead levels of 11 or 13 μg/dl learned the task at a slower rate, although they eventually were able to perform the DRL task in a way that was indistinguishable from controls (Rice and Gilbert, 1985). Increased rates of response (Alfano and Petit, 1981) and increased frequencies of responses emitted close together (short interresponse times) (Dietz *et al.,* 1978) have also been reported in rats performing on a DRL schedule. The increased rate of response on the FI may be considered to represent failure to inhibit inappropriate responding, because treated animals made more responses than controls without increasing reinforcement density; that is, their behavior was less efficient. The increase in rate of responding may also be considered to represent perseveration. The higher rate of response on the DRL schedule, which actually resulted in fewer reinforcements, clearly represents a failure to inhibit inappropriate responding, and may reflect the fact that lead-treated animals are less able to use internal cues for timing. This latter interpretation is also suggested by the poorer performance of lead-treated rats on the response duration task.

V. Effects on Sensory System Function

Although the effect of lead on cognitive function has been the focus of enormous effort, much less attention has been paid to lead's effects on sensory system function. Perhaps the most studied system with respect to developmental lead exposure is the auditory system. Changes in cortical evoked potentials (Otto *et al.,* 1981) and brainstem auditory evoked potentials (BAERs) (Otto, 1987; Robinson *et al.,* 1987) as a function of lead exposure have been observed in children. Effects on BAER latencies have been observed following both *in utero* and postnatal exposure in monkeys (Lilienthal *et al.,* 1990; Lasky *et al.,* 1995; Lilienthal and Winneke, 1996). A study of about 4,500 children with blood lead levels spanning from less than 5 μg/dl reported increased thresholds for pure tones at four frequencies using a psychophysical procedure, with no apparent threshold for the effect of lead (Schwartz and Otto, 1987). Thresholds for pure tone were determined in adult monkeys exposed to lead from birth, with concurrent blood lead levels of 60–140 μg/dl (Rice, 1997a). Three of six individuals were found to have elevated thresholds across most frequencies, with one monkey exhibiting severe impairment. Deficits have also been observed on tests of auditory processing, which may partially reflect deficits in auditory function *per se.* Children with higher dentine lead levels in the Needleman *et al.* (1979) study were impaired on the Seashore Rhythm test, in which the child was required to discriminate whether pairs of tones of various complexity were the same or different. Higher dentine lead was also associated with poor performance on the Sentence Repetition test, which required the repetition of sentences of increasing complexity. Dietrich *et al.* (1992) assessed auditory processing in a subset of children in the Cincinnati prospective study. Elevated blood lead levels were associated with a decreased ability to identify words when specific frequencies were filtered out (i.e., when information was missing). Differences in cortical potentials in response to speech sounds have also been reported in rhesus monkeys exposed to lead either pre- or postnatally (Molfese *et al.,* 1986; Morse *et al.,* 1987), suggesting altered auditory processing of complex stimuli.

There is evidence from industrial exposure in adults that lead produces deficits in visual function (Cavalleri *et al.,* 1982; Betta *et al.,* 1983; Williamson and Teo, 1986). Deficits were observed in low-luminance but not high-luminance spatial vision in rhesus monkeys with average blood lead levels of 85 μg/dl for the first year of life, and peak blood leads even higher early in life, whereas monkeys with blood lead levels of 55 μg/dl during the first year of life were not impaired (Bushnell *et al.,* 1977). In a study in monkeys with peak blood lead levels during infancy of 50 or 100 μg/dl and steady state levels of 30 or 40 μg/dl, described previously, spatial and temporal (motion) visual function was assessed during adulthood (Rice, unpublished). There was little evidence for impairment of spatial vision, although some individuals exhibited decreased low-frequency temporal vision. In a study on electrophysiologic correlates of lead exposure in rhesus monkeys, increased latencies and decreased amplitudes were observed in visual evoked potentials to light flashes as a result of developmental blood lead concentrations of approximately 50 or 100 μg/dl (Lilienthal *et al.,* 1988). There were no systematic effects on the electroretinogram. Changes in visual evoked potentials have also been observed in children as a function of lead body burden (Otto *et al.,* 1985).

In a study on the neuropathologic effects of lead on the visual system of monkeys (Reuhl *et al.,* 1989), morphometric analysis of various parts of areas V1 (primary visual cortex) and V2 revealed a decrease in neuronal volume density and decreased dendritic arborization in lead-exposed monkeys. No pathologic changes were observed in lateral geniculate nucleus or optic nerve. Blood lead levels during infancy were 90 preweaning and 50 μg/dl postweaning. However, blood levels increased to more than 200 μg/dl during a diet manipulation study when monkeys were 2.5 years old, 3 years before sacrifice for pathologic assessment.

VI. Possible Neuroanatomic Sites of Lead-Induced Damage

In utero exposure to lead at high doses in the rodent produces nonlocalized edema and lesions in the microvasculature (Pentschew and Garro, 1966; Michaelson and Sauerhoff, 1974), as well as decreased weight of neocortex, hippocampus, and cerebellum (Petit and Alfano, 1979; Petit *et al.,* 1983; Lorton and Anderson, 1986). Synaptogenesis may also be decreased in these same areas as a result of high-dose developmental lead exposure in rodents (Murray *et al.,* 1977; Petit and Le Boutillier, 1979; Bull *et al.,* 1983; Lorton and Anderson, 1986). *In utero* or neonatal exposure to high doses of lead in the monkey results in characteristic neuropathologic changes that may include edema, decreased cerebral weight, astrogliosis, and vascular damage (Tachon *et al.,* 1983; Lögdberg *et al.,* 1987, 1988), with vascular injury particularly evident in the apices of cortical gyri and in hippocampus following neonatal exposure (Zook *et al.,* 1980). Early studies in rats suggested that perinatal lead exposure results in the highest lead levels in the hippocampus, although other areas of the brain also contain significant concentrations of lead (Fjerdingstad *et al.,* 1974; Collins *et al.,* 1982). Although the evidence in humans is not well defined, lead probably does not concentrate preferentially in hippocampus (Okazaki *et al.,* 1963; Barry, 1975; Grandjean, 1978) and may preferentially accumulate in other areas of cortex in children suffering from lead poisoning (Niklowitz and Mandybur, 1975).

Initially, attention was focused on the hippocampus as a likely site of damage responsible for the observed behavioral effects of lead. The behavioral deficits observed in both rats (Petit and Alfano, 1979; Alfano and Petit, 1981) and monkeys (Bushnell and Bowman, 1979a) have been explained by inferring hippocampal damage, although the latter authors described inconsistencies between behavioral effects produced by lead and hippocampal damage (Bushnell and Bowman, 1979b). It has been suggested that damage to prefrontal cortex may be largely responsible for the behavioral impairment produced by lead (Rice and Karpinski, 1988; Levin *et al.,* 1992; Rice, 1993, 1997b). This hypothesis is based in part on the congruence between the tasks that are affected by lesions to specific brain areas (Fuster, 1984, 1989; Goldman-Rakic 1987a,b, 1988a,b, 1990; Markowitsch, 1988; Zola-Morgan and Squire, 1993; Murray, 1996) and developmental lead exposure: spatial delayed alternation, discrimination reversal, and delayed matching to sample. Perhaps more convincingly, lead exposure and lesions to prefrontal cortical areas apparently affect the same underlying behavioral processes in producing impaired performance. Both produce perseveration, increased distractibility, difficulty changing response strategy, lack of inhibitory control, and difficulty with the temporal ordering of behavior (Rice, 1997b). It is unlikely that damage to prefrontal cortex is solely responsible for the observed behavioral deficits, however.

In one study, the consequences of exposure to lead beginning at birth were examined on dendritic arborization, and neuronal and neuropil volume density in visual areas (Reuhl *et al.,* 1989) and limbic system (Reuhl and Rice, unpublished) in adult monkeys. No overt neuropathologic changes were observed by light microscopy in these monkeys. Neuropathologic assessment of limbic system was largely negative: no effects were observed in the hippocampus, amygdala, pyriform cortex, or inferior temporal cortex or gyrus. Decreased dendritic arborization was observed in superior temporal gyrus, which is involved in object discrimination and memory and auditory processing. Unfortunately, other cortical areas such as prefrontal cortex were not examined. Although these neuropathologic data are limited, they suggest that hippocampus and other limbic areas are not damaged as a result of chronic lead exposure even at relatively high blood lead levels.

VII. Conclusions

- There is compelling evidence for decreased intellectual function in children as measured by IQ associated with increased body burden of lead. Meta-analyses of both cross-sectional and longitudinal studies calculate a decrease of 2–3 points for an increase in blood lead concentrations from 10 to 20 μg/dl.
- Data from animal experiments support the evidence for impairment in overall functioning observed in children. Global impairment on a number of tasks of learning and short-term memory parallel the deficits observed on IQ in children.
- Other behavioral domains are also affected by lead in children, including impairment of attentional processes, perseverative behavior, impulsivity and hyperactivity, difficulty in changing response strategy, poor school performance, and problems in social adjustment.
- The behavioral processes affected by lead in animals mirror those identified in children. Increased distractibility, perseveration, failure to inhibit responses, and difficulty changing response strategy have been observed consistently in animals exposed to lead.
- The blood lead levels at which effects are observed is also consistent between the experimental and epidemiologic literature. There are a number of studies in children in which the mean blood lead level of the study population was 10 μg/dl or less, and there is no evidence for a threshold for lead-induced impairment at the range of blood lead levels in these studies. Schwartz (1994a) found no evidence for a threshold down to a blood lead concentration of 1 μg/dl. Monkeys with average blood lead levels about 10 μg/dl were impaired in comparison to monkeys with blood lead levels of 2–3 μg/dl. It is not surprising that there is no apparent threshold at current body burdens of lead, given that they are 2–3 orders of magnitude above natural background.
- There are enormous societal costs associated with the health consequences of lead exposure. Increased need for medical care and remedial education alone cost hundreds of millions of dollars a year to American society. The loss of earning power associated with the "small" estimated decrease in IQ as a result of lead exposure is estimated in the billions of U.S. dollars per year. Such monetary analyses, of course, do not address the decrease in quality of life as a consequence of the ubiquitous exposure of whole populations to lead.

References

Alfano, D. P., and Petit, T. L. (1981). Behavioral effects of postnatal lead exposure: Possible relationship to hippocampal dysfunction. *Behav. Neurol. Biol.* **32,** 319–333.

Angell, N. F. and Weiss, B. (1982). Operant behavior of rats exposed to lead before or after weaning. *Toxicol. Appl. Pharmacol.* **63,** 62–71.

ATSDR. (1988). *The Nature and Extent of Lead Poisoning in Children in the United States: A Report to Congress.* Agency for Toxic Substances and Disease Registry, Atlanta, Georgia.

Baghurst, P. A., McMichael, A. J., Tong, S., Wigg, N. R., Vimpani, G. V., and Robertson, E. F. (1995). Exposure to environmental lead and visual–motor integration at age 7 years: The Port Pirie cohort study. *Epidemiol.* **6,** 104–109.

Baghurst, P. A., McMichael, A. J., Wigg, N. R., Vimpani, G. V., Robertson, E. J., Roberts, R. J., and Tong, S.-B. (1992). Environmental exposure to lead and children's intelligence at the age of seven years. The Port Pirie cohort study. *New Engl. J. Med.* **327(18),** 1279–1284.

Baghurst, P. A., Robertson, E. F., McMichael, A. J., Vimpani, G. V., Wigg, N. R., and Roberts, R. J. (1987). The Port Pirie cohort study: lead effects on pregnancy outcome and early childhood development. *Neurotoxicol.* **8,** 395–402.

Barry, P. S. I. (1975). A comparison of concentrations of lead in human tissue. *Brit. J. Ind. Med.* **32,** 119–139.

Bellinger, D., Hu, H., Titlebaum, L., and Needleman, H.L. (1994a). Attentional correlates of dentine and bone lead levels in adolescents. *Arch. Environ. Med.* **49,** 98–105.

Bellinger, D., Leviton, A., Allred, E., and Rabinowitz, M. (1994b). Pre- and postnatal lead exposure and behavior problems in school-age children. *Environ. Res.* **66,** 12–30.

Bellinger, D., Leviton, A., Waternaux, C., Needleman, H. L., and Rabinowitz, M. (1987a). Longitudinal analysis of prenatal and postnatal lead exposure and early cognitive development. *New Engl. J. Med.* **316,** 1037–1043.

Bellinger, D., Needleman, H .L., Bromfield, R., and Mintz, M. (1984). A followup study of the academic attainment and classroom behavior of children with elevated dentine lead levels. *Biol. Trace Elem. Res.* **6,** 207–223.

Bellinger, D., Sloman, J., Leviton, A., Waternaux, C., Needleman, H., and Rabinowitz, M. (1987b). Low level lead exposure and child development: assessment at age 5 of a cohort followed from birth. In *Proceedings of the 6th International Conference on Heavy Metals in the Environment.* S. E. Lindberg and T. C. Hutchinson, Eds. CEP Consultants, Edinburgh, pp. 223–225.

Bellinger, D. C. (1995). Interpreting the literature on lead and child development: the neglected role of the "Experimental System." *Neurotoxicol. Teratol.* **17,** 201–212.

Bellinger, D. C., Stiles, K. M., and Needleman, H. L. (1992). Low-level lead exposure, intelligence and academic achievement: a long-term follow-up study. *Pediat.* **90,** 855–861.

Betta, A., DeSanta, A., Savonetto, C., and D'Andrea, F. (1983). Flicker fusion test and occupational toxicology: performance evaluation in workers exposed to lead and solvents. *Human Toxicol.* **2,** 83–90.

Bull, R. J., McCauley, P. T., Taylor, D. H., and Crofton, K. M. (1983). The effects of lead on the developing central nervous system of the rat. *Neurotoxicol.* **4,** 1–18.

Bushnell, P. J. and Bowman, R. E. (1979a). Reversal learning deficits in young monkeys exposed to lead. *Pharmacol. Biochem. Behav.* **10,** 733–742.

Bushnell, P. J. and Bowman, R. E. (1979b). Persistence of impaired reversal learning in young monkeys exposed to low levels of dietary lead. *J. Toxicol. Environ. Health* **5,** 1015–1023.

Bushnell, P. J., Bowman, R. H., Allen, J. R., and Marler, R. J. (1977). Scotopic vision deficits in young monkeys exposed to lead. *Sci.* **196,** 333–335.

Byers, R. K. and Lord, E. E. (1943). Late effects of lead poisoning on mental development. *Amer. J. Dis. Child.* **66,** 471–494.

Carson, T. L., Van Gelder, G. A., Karas, G. C., and Buck, W. B. (1974). Slowed learning in lambs prenatally exposed to lead. *Arch. Environ. Health* **29,** 154–156.

Cavalleri, A., Trimarshi, E., Gelmi, G., Barenffini, A., Minoia, C., Bisceldi, G., and Gallo, G. (1982). Effects of lead on the visual system of occupationally exposed subjects. *Scand. J. Work. Environ. Health* **8**(Suppl 1), 148–151.

Cohn, J., Cox, C., and Cory-Slechta, D. A. (1993). The effects of lead exposure on learning in a multiple repeated acquisition and performance schedule. *Neurotoxicol.* **14,** 329–346.

Collins, M. F., Hrdina, P. D., Whittle, E., and Singhal, R. L. (1982). Lead in blood and brain regions of rats chronically exposed to low doses of the metal. *Toxicol. Appl. Pharmacol.* **65,** 314–322.

Cooney, G. H., Bell, A., McBride, W., and Darter, C. (1989). Low-level exposure to lead: the Sydney Lead Study. *Dev. Med. Child Neurol.* **31,** 640–649.

Cooney, G. H., Bell, A., and Stavron, C. (1991). Low-level exposure to lead and neurobehavioral development: the Sydney study at seven years. In *Heavy Metals in the Environment,* CEC Consultants, Edinburgh, pp. 16–19.

Cory-Slechta, D. A., Bissen, S. T., Young, A. M., and Thompson, T. (1981). Chronic postweaning lead exposure and response duration performance. *Toxicol. Appl. Pharmacol.* **60,** 78–84.

Cory-Slechta, D. A., Pokora, M. J., and Widzowski, D. V. (1991). Behavioral manifestations of prolonged lead exposure initiated at different stages of the life cycle. II. Delayed spatial alternation. *Neurotoxicol.* **12,** 761–776.

Cory-Slechta, D. A. and Thompson, T. (1979). Behavioral toxicity of chronic postweaning lead exposure in the rat. *Toxicol. Appl. Pharmacol.* **47,** 151–159.

Cory-Slechta, D. A., Weiss, B., and Cox, C. (1983). Delayed behavioral toxicity of lead with increasing exposure concentration. *Toxicol. Appl. Pharmacol.* **71,** 342–352.

Cory-Slechta, D. A., Weiss, B., and Cox, C. (1985). Performance and exposure indices of rats exposed to low concentrations of lead. *Toxicol. Appl. Pharmacol.* **78,** 291–299.

de la Burdé, B. and Choate, M. S. (1972). Early asymptomatic lead exposure and development at school age. *J. Pediatr.* **87,** 638–642.

Dietrich, K. N., Berger, O. G., Succop, P. A., Hammond, P. B., and Bornschein, R. L. (1993). The developmental consequences of low to moderate prenatal and postnatal lead exposure: Intellectual attainment in the Cincinnati lead study cohort following school entry. *Neurotoxicol. Teratol.* **15,** 37–44.

Dietrich, K. N., Krafft, K. M., Bier, M., Berger, O., Succop, P. A., and Bornschein, R. L. (1989). Neurobehavioral effects of foetal lead exposure: the first year of life. In *Lead Exposure and Child Development: An International Assessment* M. A. Smith, L. D. Grant and A. I. Sors, Eds. Kluwer, Boston, pp. 320–331.

Dietrich, K. N., Krafft, K. M., Bornschein, R. L., Hammond, P. B., Berger, O., Succop, P. A. and Bier, M. (1987). Low-level lead exposure: effects on neurobehavioral development in early infancy. *Pediat.* **80,** 721–730.

Dietrich, K. N., Succop, P. A., Berger, O. G., Hammond, P. B. and Bornschein, R. L. (1991). Lead exposure and the cognitive development of urban preschool children: the Cincinnati lead study cohort at age 4 years. *Neurotoxicol. Teratol.* **13,** 203–211.

Dietrich, K. N., Succop, P. A., Berger, O. G. and Keith, R. W. (1992). Lead exposure and the central auditory processing abilities and cognitive development of urban children: the Cincinnati lead study cohort at age 5 years. *Neurotoxicol. Teratol.* **14,** 51–56.

Dietz, D. D., McMillan, D. E., Grant, L. D., and Kimmel, C. A. (1978). Effects of lead on temporally spaced responding in rats. *Drug Chem. Toxicol.* **1,** 401–419.

Elias, R., Hirao, Y., and Patterson, C. (1975). Impact of present levels of aerosol Pb concentrations on both natural ecosystems and humans. In *International Conference on Heavy Metals in the Environment, Vol. II,* pp. 257–271.

Ernhart, C. B. (1992). A critical review of low-level prenatal lead exposure in the human. 2. Effects on the developing child. *Reprod. Toxicol.* **6,** 21–40.

Fergusson, D. M., Fergusson, J. E., Horwood, L. J., and Kinzett, N. G. (1988a). A longitudinal study of dentine lead levels, intelligence, school performance and behavior. Part I. Dentine lead levels and exposure to environmental risk factors. *J. Child Psychol. Psychiat.* **29,** 781–792.

Fergusson, D. M., Fergusson, J. E., Horwood, L. J., and Kinzett, N. G. (1988b). A longitudinal study of dentine lead levels, intelligence, school performance and behavior. Part II. Dentine lead and cognitive ability. *J. Child Psychol. Psychiat.* **29,** 783–809.

Fergusson, D. M., Fergusson, J. E., Horwood, L. J., and Kinzett, N. G. (1988c). A longitudinal study of dentine lead levels, intelligence, school performance and behavior. Part III. Dentine lead levels and attention/activity. *J. Child Psychol. Psychiat.* **29,** 811–824.

Fergusson, D. M. and Horwood, L. J. (1993). The effects of lead levels on the growth of word recognition in middle childhood. *Intern. J. Epidemiol.* **22,** 891–897.

Fjerdingstad, E. J., Danscher, G., and Fjerdingstad, E. (1974). Hippocampus: selective concentration of lead in the normal rat brain. *Brain Res.* **80,** 350–354.

Fulton, M., Raab, G., Thomson, G., Laxen, D., Hunter, R., and Hepburn, W. (1987). Influence of blood lead on the ability and attainment of children in Edinburgh. *Lancet* **1(8544),** 1221–1226.

Fuster, J. M. (1984). Functional relationship between inferotemporal and prefrontal cortex. In *Cortical Integration* F. Reinoso-Suárez and C. Ajmone-Marsan, Eds. Raven Press, New York, pp. 341–352.

Fuster, J. M. (1989). *The Prefrontal Cortex.* Raven Press, New York.

Gatsonis, C. A., and Needleman, H. L. (1992). Recent epidemiological studies of low-level lead exposure and the IQ of children: A meta-analytic review. In *Human Lead Exposure* H. L. Needleman, Ed. CRC Press, Boca Raton, pp. 243–255.

Gibson, J. L. (1892). Notes on lead-poisoning as observed among children in Brisbane. *Proc. Intercolonial Med. Congress Austral.* **3,** 76–83.

Gilbert, S. G., and Rice, D. C. (1987). Low-level lifetime lead exposure produces behavioral toxicity (spatial discrimination reversal) in adult monkeys. *Toxicol. Appl. Pharmacol.* **91,** 484–490.

Goldman-Rakic, P. S. (1987a). Circuitry of primate prefrontal cortex and regulation of behavior by representational memory. In *Handbook of Physiology—The Nervous System V.* F. Plum and V. Mountcastle, Eds. American Physiological Society, Bethesda, Maryland, pp. 373–417.

Goldman-Rakic, P. S. (1987b). Development of cortical circuitry and cognitive function. *Child Devel.* **58,** 601–622.

Goldman-Rakic, P. S. (1988a). Changing concepts of cortical connectivity: parallel distributed cortical networks. In *Neurobiology of Neocortex* P. Rakic and W. Singer, Eds. John Wiley and Sons, New York, pp. 177–202.

Goldman-Rakic, P. S. (1988b). Topography of cognition: parallel distributed networks in primate association cortex. *Ann. Rev. Neurosci.* **11,** 137–156.

Goldman-Rakic, P. S. (1990). Cellular and circuit basis of working memory in prefrontal cortex of nonhuman primates. In *Progress in Brain Research,* H. B. M. Uylings, C. G. Van Eden, J. P. C. De Bruin, M. A. Corner, and M. G. P. Feenstra, Eds. Elsevier, New York.

Grandjean, P. (1978). Regional distribution of lead in human brains. *Toxicol. Lett.* **2,** 65–69.

Grant, L. D. and Davis, J. M. (1989). Effects of low-level lead exposure on paediatric neurobehavioral development: current findings and future directions. In *Lead Exposure and Child Development: An International Assessment.* M. A. Smith, L. D. Grant, and A. I. Sors, Eds. Kluwer, Boston, pp. 49–118.

Hansen, O. N., Trillingsgaard, A., Beese, I., Lyngbye, T., and Grandjean, P. (1989). A neuropsychological study of children with elevated dentine lead level: assessment of the effect of lead in different socio-economic groups. *Neurotoxicol. Teratol.* **11,** 205–214.

Hatzakis, A., Kokkevi, A., Katsouyanni, K., Maravelias, K., Salaminios, J. K., Kalandidi, A., Koutselinis, A., Stefanis, K., and Trichopoulis, D. (1987). Psychometric intelligence and attentional performance deficits in lead-exposed children. In *Proc. 6th Housernl. Conf. Heavy Metals in the Environment.* S. E. Lindberg and T. C. Hutchinson, Eds. CEP Consultants, Edinburgh, pp. 204–209.

Hatzakis, A., Kokkevi, A., Maravelias, C., Katsouyanni, K., Salaminlos, F., Kalandidi, A., Koutselinis, A., Stefanis, K., and Trichopoulos, D. (1989). Psychometric intelligence deficits in lead-exposed children. In *Lead Exposure and Child Development: An International Assessment.* M. A. Smith, L. D. Grant, and A. I. Sors, Eds. Kluwer, Boston, pp. 211–223.

Hawk, B. A., Schroeder, S. R., Robinson, G., Otto, D., Mushak, P., Kleinbaum, D., and Dawson, G. (1986). Relation of lead and social factors to IQ of low-SES children: a partial replication. *Am. J. Ment. Defic.* **91,** 178–183.

Honzik, M. (1976). Value and limitations of infant tests: an overview. In *The Origins of Intelligence: Infancy and Early Childhood* M. Lewis, Ed. Plenum Press, New York, pp. 59–95.

Hunter, J., Urbanowicz, M. A., Yule, W., and Lansdown, R. (1985). Automated testing of reaction time and its association with lead in children. *Int. Arch. Occup. Envir. Health* **57,** 27–34.

IPCS. (1995). *Environmental Health Criteria 165: Inorganic Lead.* International Programme on Chemical Safety. World Health Organization, Geneva.

Jenkins, C. D. and Mellins, R. B. (1957). Lead poisoning in children. *AMA Arch. Neurol. Psychiat.* **77,** 70–78.

Lasky, R. E., Maier, M. M., Snodgrass, E. B., Hecox, K. E., and Laughlin, N. K. (1995). The effects of lead on otoacoustic emissions and auditory evoked potentials in monkeys. *Neurotoxicol. Teratol.* **17,** 633–644.

Levin, E. D., and Bowman, R. E. (1986). Long-term lead effects on the Hamilton Search Task and delayed alternation in monkeys. *Neurobehav. Toxicol. Teratol.* **8,** 219–224.

Levin, E. and Bowman, R. (1983). The effect of pre- or postnatal lead exposure on Hamilton Search Task in monkeys. *Neurobehav. Toxicol. Teratol.* **5,** 391–394.

Levin, E. D., Schantz, S. L., and Bowman, R. E. (1992). Use of the lesion model for examining toxicant effects on cognitive behavior. *Neurotoxicol. Teratol.* **14,** 131–141.

Leviton, A., Bellinger, D., Allred, E. N., Rabinowitz, M., Needleman, H., and Schoenbaum, S. (1993). Pre- and postnatal low-level lead exposure and children's dysfunction in school. *Environ. Res.* **60,** 30–43.

Lilienthal, H., Lenaerts, C., Winneke, G,. and Hennekes, R. (1988). Alteration of the visual evoked potential and electroretinogram in lead-treated monkeys. *Neurotoxicol. Teratol.* **10,** 417–422.

Lilienthal, H. and Winneke, G. (1996). Lead effects on brain stem auditory evoked potential in monkeys during and after the treatment phase. *Neurotoxicol. Teratol.* **18,** 17–32.

Lilienthal, H., Winneke, G., Brockhaus, A., and Malik, B. (1986). Pre- and postnatal lead-exposure in monkeys: effects on activity and learning set formation. *Neurobehav. Toxicol. Teratol.* **8,** 265–272.

Lilienthal, H., Winneke, G., and Ewert, T. (1990). Effects of lead on neurophysiological and performance measures: animal and human data. *Environ. Health Perspec.* **89,** 21–25.

Lin-Fu, J. S. (1972). Undue absorption of lead among children—a new look at an old problem. *New Engl. J. Med.* **286,** 702–710.

Lin-Fu, J. S. (1985). Historical perspective on health effects of lead. In *Dietary and Environmental Lead: Human Health Effects* K. R. Mahaffey, Ed. Elsevier, New York, pp. 43–64.

Lögdberg, B., Berlin, M., and Schütz, A. (1987). Effects of lead exposure on pregnancy outcomes and the fetal brain of squirrel monkeys. *Scand. J. Work Environ. Health* **13,** 135–145.

Lögdberg, B., Brun, A., Berlin, M., and Schütz, A. (1988). Congenital lead encephalopathy in monkeys. *Acta Neuropathol.* **77,** 120–127.

Lorton, D. and Anderson, W. J. (1986). The effects of postnatal lead toxicity on the development of cerebellum in rats. *Neurobehav. Toxicol. Teratol.* **8,** 51–59.

Lyngbye, T., Hansen, O. L., Trillingsgaard, A., Beese, I., and Grandjean, P. (1990). Learning disabilities in children: significance of low-level lead exposure and confounding factors. *Acta Paediatr. Scand.* **79,** 352–360.

Markowitsch, H. J. (1988). Anatomical and functional organization of the primate prefrontal cortical system. *Comp. Prim. Biolog.* **4,** 99–153.

McMichael, A. J., Baghurst, P. A., Wigg, N. R., Vimpani, G. V., Robertson, E. F., and Roberts, R. J. (1988). Port Pirie Cohort Study: Environmental exposure to lead and children's abilities at the age of four years. *N. Engl. J. Med.* **319,** 468–475.

McMichael, A. J., Vimpani, G. V., Robertson, E. F., Baghurst, P. A., and Clark, P. D. (1986). The Port Pirie cohort study: Maternal blood lead and pregnancy outcome. *J. Epidemiol. Comm. Health* **40,** 18–25.

Michaelson, I. A. and Sauerhoff, M. W. (1974). Animal models of human disease: severe and mild lead encephalopathy in the neonatal rat. *Environ. Health Perspect.* **7,** 201–225.

Molfese, D. L., Laughlin, N. K, Morse, P. A., Linnville, S. E. Wetzel, W. F., and Erwin, R. J. (1986). Neuroelectrical correlates of categorical perception for place of articulation in normal and lead-treated rhesus monkeys. *J. Clin. Exper. Neuropsych.* **8,** 680–696.

Morse, P. A., Molfese, D., Laughlin, N. K., Linnville, S., and Wetzel, F. (1987). Categorical perception for voicing contrasts in normal and lead-treated rhesus monkeys: electrophysiological indices. *Brain Lang.* **30,** 63–80.

Muñoz, H., Romie, D., Palazuelos, E., Mancilla-Sanchez, T., Meneses-Gonzalez, J., and Hernandez-Avila, M. (1993). Blood lead level and neurobehavioral development among children living in Mexico City. *Arch. Environ. Health* **48(30),** 132–139.

Murray, E. A. (1996) (Ed). *Memory Systems in the Primate Brain.* Seminars in the Neurosciences. **8,** 1–67.

Murray, H. M., Gurule, M., and Zenick, H. (1977). Effects of lead exposure on the developing rat parietal cortex. In *Developmental Toxicity of Energy Related Pollutants, Proceedings of the 17th Annual Hanford Biology Symposium,* U.S. Department of Energy, Washington, D.C., pp. 520–535.

Mushak, P., Davis, J. M., Crocetti, A. J., and Grant, L. D. (1989). Review—Prenatal and postnatal effects of low-level lead exposure: integrated summary of a report to the U.S. Congress on childhood lead poisoning. *Environ. Res.* **50,** 11–36.

NAS. (1980). *Lead in the Human Environment.* National Academy of Sciences, Committee on Lead in the Human Environment. National Academy of Sciences, Washington.

Needleman, H. L. (Ed.) (1992) *Human Lead Exposure,* Boca Raton, CRC Press.

Needleman, H. L. (1987). Introduction: Biomarkers in neurodevelopmental toxicology. *Environ. Health Perspect.* **74,** 149–152.

Needleman, H. L., Geiger, S. K., and Frank, R. (1985). Lead and IQ scores: a reanalysis. *Science* **227,** 701–704.

Needleman, H. L., Gunnoe, C., Leviton, A., Reed, R., Peresie, H., Maher, C., and Barrett, P. (1979). Deficits in psychologic and classroom performance of children with elevated dentine lead levels. *N. Engl. J. Med.* **300,** 689–695.

Needleman, H. L., Riess, J. A., Tobin, M. J., Biesecker, G. E., and Greenhouse, J. B. (1996). Bone lead levels and delinquent behavior. *JAMA.* **275,** 363–369.

Niklowitz, W. J. and Mandybur, T. I. (1975). Neurofibrillary changes following childhood lead poisoning. *J. Neuropathol. Exp. Neurol.* **34,** 445–455.

Okazaki, H., Aronson, S. M., DiMaio, D. J., and Olvera, J. E. (1963). Acute lead encephalopathy of childhood, *Trans. Amer. Neurol. Assoc.* **88,** 245–250.

Otto, D., Baumann, S. B., Robinson, G. S., Schroeder, S. R., Kleinbaum, D. G., Barton, C. N., and Mushak, D. (1985). Auditory and visual evoked potentials in children with undue lead absorption. *Toxicologist* **5(1),** 81.

Otto, D. A. (1987). The assessment of neurotoxicity in children—electrophysiological methods. In *Toxic Substances and Mental Retardation. Neurobehav. Toxicol. Teratol.* S. R. Schroeder, Ed. Amer. Assoc. on Mental Deficiency, Washington, pp. 139–158.

Otto, D. A., Benignus, V. A., Muller, K. E., and Barton, C. N. (1981). Effects of age and body lead burden on CNS function in young children. I. Slow cortical potentials. *Electroencephalo. Clin. Neurophys.* **52,** 229–239.

Patterson, C. C. (1983). British mega exposures to industrial lead. In *Lead versus Health.* M. Rutter and R. Russell Jones, Eds. John Wiley & Sons, New York, pp. 17–32.

Pentschew, A., and Garro, F. (1966). Lead encephalomyelopathy of the suckling rat and its implications on the porphyrinopathic nervous diseases, with special reference to the permeability disorders of the nervous system's capillaries. *Acta Neuropathol.* (Berlin) **6,** 266–278.

Perlstein, M. A., and Attala, R. (1966). Neurological sequelae of plumbism in children. *Clin. Pediat.* **5,** 292–298.

Petit, T. L. and Alfano, D. P. (1979). Differential experience following developmental lead exposure: Effects on brain and behavior. *Pharmacol. Biochem. Behav.* **11,** 165–171.

Petit, T. L., Alfano, D. P., and Le Boutillier, J. C. (1983). Early lead exposure and the hippocampus: A review and recent advances. *Neurotoxicol.* **4,** 79–94.

Petit, T. L. and Le Boutillier, J. C. (1979). Effects of lead exposure during development on neocortical dendritic and synaptic structure. *Exper. Neurol.* **64,** 482–492.

Pocock, S., Ashby, D., and Smith, M. A. (1987). Lead exposure and children's intellectual performance. *Int. J. Epidemiol.* **16,** 57–67.

Rabin, R. (1989). Warnings unheeded: A history of child lead poisoning. *Public. Health* **79,** 1668–1674.

Reuhl, K. R., Rice, D. C., Gilbert, S. G., and Mallet, J. (1989). Effects of chronic developmental lead exposure on monkey neuroanatomy: Visual system. *Toxicol. Appl. Pharmacol.* **99,** 501–509.

Rice, D. C. (1984). Behavioral deficit (delayed matching to sample) in monkeys exposed from birth to low levels of lead. *Toxicol. Appl. Pharmacol.* **75,** 337–345.

Rice, D. C. (1985a). Chronic low-lead exposure from birth produces deficits in discrimination reversal in monkeys. *Toxicol. Appl. Pharmacol.* **77,** 201–210.

Rice, D. C. (1985b). Effect of lead on schedule-controlled behavior in monkeys. In *Behavioral Pharmacology: The Current Status.* L. S. Seiden and R. L. Balster, Eds. Alan R. Liss, New York, pp. 473–486.

Rice, D. C. (1988). Schedule-controlled behavior in infant and juvenile monkeys exposed to lead from birth. *Neurotoxicol.* **9,** 75–88.

Rice, D. C. (1990). Lead-induced behavioral impairment on a spatial discrimination reversal task in monkeys exposed during different periods of development. *Toxicol. Appl. Pharmacol.* **106,** 327–333.

Rice, D. C. (1992a). Effect of lead during different developmental periods in the monkey on concurrent discrimination performance. *Neurotoxicol.* **13,** 583–592.

Rice, D. C. (1992b). Lead-exposure during different developmental periods produces different effects in FI performance in monkeys tested as juveniles and adults. *Neurotoxicol.* **13,** 757–770.

Rice, D. C. (1992c). Behavioral effects of lead in monkeys tested during infancy and adulthood. *Neurotoxicol. Teratol.* **14,** 235–245.

Rice, D. C. (1993). Lead-induced changes in learning: Evidence for behavioral mechanisms from experimental animal studies. *Neurotoxicol.* **14,** 167–178.

Rice, D. C. (1996). Neurotoxicity of lead: Commonalities between experimental and epidemiological data. *Environ. Health Perspect.* **104**(Supp. 2), 337–351.

Rice, D. C. (1997a). Effects of lifetime lead exposure in monkeys on detection of pure tones. *Fundam. Appl. Toxicol.* **36,** 112–118.

Rice, D. C. (1997b). Anatomical substrates of behavioral impairment induced by developmental lead exposure in monkeys: Inferences from brain lesions. *Amer. Zoolog.* **37,** 409–425, 1997.

Rice, D. C., and Gilbert, S. G. (1990a). Sensitive periods for lead-induced behavioral impairment (nonspatial discrimination reversal) in monkeys. *Toxicol. Appl. Pharmacol.* **102,** 101–109.

Rice, D. C., and Gilbert, S. G. (1990b). Lack of sensitive period for lead-induced behavioral impairment on a spatial delayed alternation task in monkeys. *Toxicol. Appl. Pharmacol.* **103,** 364–373.

Rice, D. C., and Gilbert, S. G. (1985). Low lead exposure from birth produces behavioral toxicity (DRL) in monkeys. *Toxicol. Appl. Pharmacol.* **80,** 421–426.

Rice, D. C., Gilbert, S. G., and Willes, R. F. (1979). Neonatal low-level lead exposure in monkeys: Locomotor activity, schedule-controlled behavior, and the effects of amphetamine. *Toxicol. Appl. Pharmacol.* **51,** 503–513.

Rice, D. C., and Karpinski, K. F. (1988). Lifetime low-level lead exposure produces deficits in delayed alternation in adult monkeys. *Neurotoxicol. Teratol.* **10,** 207–214.

Rice, D. C., and Willes, R. F. (1979). Neonatal low-level lead exposure in monkeys (*Macaca fascicularis*): effect on two-choice nonspatial form discrimination. *J. Environ. Pathol. Toxicol.* **2,** 1195–1203.

Robinson, G. S., Keith, R. W., Bornschein, R. L., and Otto, D. A. (1987). Effects of environmental lead exposure on the developing auditory system as indexed by the brainstem auditory evoked potential and pure tone hearing evaluations in young children. In *Proc. 6th Intern. Conf. Heavy Metab. in the Environ.* (S. E. Lindberg and T. C. Hutchinson, Eds. CEP Consultants, Edinburgh, pp. 223–225.

Salkever, D. S. (1995). Updated estimates of earnings benefits from reduced exposure of children to environmental lead. *Environ. Res.* **70,** 1–6.

Sauerhoff, M. W. and Michaelson, I. A. (1974). Hyperactivity and brain catecholamines in lead-exposed developing rats. *Science* **182,** 1022–1024.

Schroeder, S. R., Hawk, B., Otto, D., Mushak, P., and Hicks, R. E. (1985). Separating the effects of lead and social factors on IQ. *Environ. Res.* **38,** 144–154.

Schwartz, J. (1994a). Low-level lead exposure and children's IQ: a meta-analysis and search for a threshold. *Environ. Res.* **65,** 42–55.

Schwartz, J. (1994b). Societal benefits of reducing lead exposure. *Environ. Res.* **66,** 105–124.

Schwartz, J. and Otto, D. (1987). Blood lead, hearing thresholds, and neurobehavioral development in children and youth. *Arch. Environ. Health.* **42,** 153–160.

Sciarillo, W., Alexander, A., and Farrell, K. (1992). Lead exposure and child behavior. *Am. J. Public Health* **82,** 1356–1360.

Silbergeld, E. K., and Goldberg, A. M. (1973). A lead-induced behavior disorder. *Life Sci.* **13,** 1275–1283.

Silva, P. A., Hughes, P., Williams, S., and Faed, J. M. (1988). Blood lead, intelligence, reading attainment, and behavior in eleven-year-old children in Dunedin, New Zealand. *J. Child Psychol. Psychiat.* **29,** 43–52.

Smith, M., Delves, H., Lansdown, R., Clayton, B., and Graham, P. (1983). The effects of lead exposure on urban children: the Institute of Child Health/Southhampton study. *Develop. Med. Child. Neurol.* **25**(Supp. 47).

Stiles, K. M. and Bellinger, D. C. (1993). Neuropsychological correlates of low-level lead exposure in school-age children: a prospective study. *Neurotoxicol. Teratol.* **15,** 27–35.

Tachon, P., Laschi, A., Briffaux, J. P., Brain, G., and Chambon, P. (1983). Lead poisoning in monkeys during pregnancy and lactation. *Sci. Total Environ.* **30,** 221–229.

Thomson, G., Raab, G., Hepburn, W., Hunter, R., Fulton, M. and Laxen, D. (1989). Blood-lead levels and children's behavior—Results from the Edinburgh lead study. *J. Child Psychol. Psychiat.* **30,** 515–528.

Thurston, D. L., Middelkamp, J. N., and Mason, E. (1955). The late effects of lead poisoning. *J. Pediat.* **47,** 413–423.

Turner, A. J. (1897). Lead poisoning among Queensland children. *Aust. Med. Gazette* **16,** 475–479.

Tuthill, R. W. (1996). Hair lead levels related to children's classroom attention-deficit disorder. *Arch. Environ. Health* **51,** 214–220.

Volpe, R. A., Cole, J. F., and Borelko, C. J. (1992). Analysis of prospective epidemiologic studies on the neurobehavioral effects of lead. *Environ. Geochem. Health* **14,** 133–140.

Wasserman, G., Jaramilo, B., Shrout, P., and Graziano, J. Lead exposure and child behavior problems at age 3 years. Paper presented at the 1995 Society for Research in Child Development, Indianapolis.

Wasserman, G. A., Graziano, J. H., Factor-Litvak, P., Pokovac, D., Morina, N., Musabegovic, A., Vrenezi, N., Capuni-Paracka, S., Lekic, V., Preteni-Redjepi, E., Hadzialjevic, S., Slavkovich, V., Kline, J., Shrout, P., and Stein, Z. (1994). Consequences of lead exposure and iron supplementation on childhood development at age 4 years. *Neurotoxicol. Teratol.* **16,** 233–244.

Wasserman, G. A., Graziano, J. H., Factor-Litvak, P., Popovac, D., Morina, N., Musabegovic, A., Vrenezi, N., Capuni-Paracka, S., Lekic, V., Preteni-Redjepi, E., Slavkovich, V., Kline, J., Shrout, P., and Stein, Z. (1992). Independent effects of lead exposure and iron deficiency anemia on developmental outcome at age 2 years. *J. Peds.* **121,** 695–703.

Wigg, N. R., Vimpani, C. V., McMichael, A. J., Baghurst, P. A., Robertson, E. F., and Roberts, R. J. (1988). Port Pirie cohort study: childhood blood lead and neuropsychological development at two years of age. *J. Epidemiol. Commun. Health* **42,** 213–219.

Williamson, A. M. and Teo, R. C. (1986). Neurobehavioral effects of occupational lead exposure. *Brit. J. Indust. Med.* **43,** 374–380.

Winneke, G., Altmann, L., Krämer, U., Turfield, M., Behler, R., Gutsmuths, F. J., and Mangold, M. (1994). Neurobehavioral and neurophysiological observations in six year old children with low lead levels in East and West Germany. *Neurotoxicol.* **15,** 705–714.

Winneke, G., Beginn, U., Ewert, T., Havestadt, C., Kraemer, U., Krause, C., Thron, H. L., and Wagner, H. M. (1985). Comparing the effects of perinatal and later childhood lead exposure on neuropsychologic outcome. *Environ. Res.* **38,** 155–167.

Winneke, G., Brockhaus, A., and Baltissen, R. (1977). Neurobehavioral and systemic effects of long-term blood lead elevation in rats. I. Discrimination learning and open-field behavior. *Arch. Toxicol.* **37,** 247–263.

Winneke, G., Brockhaus, A., Collet, W., and Kraemer, V. (1989). Modulation of lead-induced performance deficit in children by varying signal rate in a serial choice reaction task. *Neurotoxicol. Teratol.* **11,** 587–192.

Winneke, G., Brockhaus, A., Ewers, U., Kraemer, U., and Neuf, M. (1990). Results from the European multicenter study on lead neurotoxicity in children: implications for risk assessment. *Neurotoxicol. Teratol.* **12,** 553–559.

Winneke, G. and Kraemer, V. (1984). Neuropsychological effects of lead in children: Interaction with social background variables. *Neuropsychobiol.* **11,** 195–202.

Winneke, G., Kraemer, V., Brockhaus, A., Ewers, U., Kujanek, H., Lechner, H., and Janke, W. (1983). Neuropsychologic studies in children with elevated tooth-lead concentrations. II. Extended study. *Int. Arch. Occup. Envir. Health* **51,** 231–252.

Yule, W., Lansdown, R., Millar, I., and Urbanowicz, M. (1981). The relationship between blood lead concentration, intelligence, and attainment in a school population: a pilot study. *Dev. Med. Child. Neurol.* **23,** 567–576.

Yule, W., Urbanowicz, M. A., Lansdown, R., and Millar, I. B. (1984). Teachers' ratings of children's behavior in relation to blood lead levels. *Br. J. Dev. Psychol.* **2,** 295.

Zola-Morgan, S. and Squire, L. R. (1993). Neuroanatomy of memory. *Annu. Rev. Neurosci.* **16,** 547–563.

Zook, B. C., London, W. T., Wilpizeski, C. R., and Sever, J. L. (1980). Experimental lead paint poisoning in nonhuman primates III. Pathological findings. *J. Med. Primatol.* **9,** 343–360.

CHAPTER

32

Developmental Toxicity of Pesticides

DURISALA DESAIAH
University of Mississippi Medical Center
Jackson, Mississippi 39216

I. Introduction

Developmental toxicity, in general, includes the structural and functional aspects of an organ system resulting in death of the organism. Developmental toxicity also takes into consideration all of the abnormal development of the organism. In many organ systems the developmental abnormalities occur before birth, thus the teratogenic effects precedes, whereas in the central nervous system (CNS) there are several sensitive processes such as cell migration, synaptogenesis, and apoptosis that occur after birth and in the early years of life. The literature is vast in relation to teratology and developmental toxicity of a number of drugs, environmental chemicals, and so on, as well as of genetic alterations. The existing literature on the developmental toxicity of a number of agents in humans and in animals has been reviewed extensively (Schardein and Keller, 1989; Schardein, 1993; Rogers and Kavlock, 1996). The purpose of this chapter is to focus on the effects of specific agents such as pesticides on the developing nervous system. Even this subject is so large a single chapter is not enough to do justice. The reader is referred to other chapters in this book for detailed information on developmental neuroscience and toxicity.

Developmental neurotoxicity, for the purpose of this chapter, is any chemical agent that produces altered biochemical pathways leading to abnormal behavior in adult life as a result of exposure of the CNS during

* Abbreviations: 2,4,5-T,2,4,5-trichlorophenoxy acetic acid; 2,4-D,2,4-dichlorophenoxy acetic acid; ATP, adenosine triphosphate; ATPase, adenosine triphosphatase; CNS, central nervous system; DFP, diisopropylfluorophosphate; GABA, γ-aminobutyric acid; HCB, hexachlorocyclohexane; MACHAR, muscarinic acethylcholine receptor; MPP^+, 1-methyl-4 phenylpyridium ion; MPTP, 1-methyl-4-phenyl-1,2,3,6-tetrahydropyridine; PN, postnatal day; QNB, quiniclidinylbenzilate; SOD, superoxide dismutase; WHO, World Health Organization.

its development and maturation. The term *pesticides* encompasses a large number of chemicals that kill pests, including both animals and plants. These pesticides are classified, based on the specific organisms killed, as herbicides, insecticides, fungicides, and rodenticides. Each of these groups is further classified based on chemical structures such as organophosphates, organochlorines, botanicals, and others. Although there has been recorded history of use of the chemicals or extracts of plants for pest control during the 20th century, the actual development and use of the many synthetic chemicals started since the beginning of World War II. The pesticides are highly beneficial for humans because they kill the insects, weeds, and diseases of the crops, thus increasing the food production as well as the human health (Brooks, 1974; Matsumura, 1985; World Health Organization (WHO), 1990; Ecobichon, 1996). However, occupational, environmental, and accidental exposures in humans resulted in millions of poisonings and thousands of deaths worldwide (WHO, 1990). The rate of poisonings and death is much higher in the developing countries as compared to the Western countries (Coye, 1986; WHO, 1990; Weir *et al.,* 1992; Ecobichon, 1996). The compounds developed early on showed potent toxicity in mammalian species and residual toxicity in humans because of their amplification of environmental persistence. In view of their long-term adverse effects, the most used compounds, such as organochlorine pesticides, have been banned in the United States as well as in many developed countries. However, many of these organochlorine compounds and other more recently developed organophosphates and botanicals are being used in the developing countries. Environment does not have any boundaries and environmental pollution is a public health concern globally.

II. Effects on Adults

Many of the pesticide compounds are neurotoxic in adults. The neurobehavioural abnormalities resulted as a function of pesticide exposures show typical characteristics based on the structural similarities (Ecobichon, 1996; Moser, 1997). For example, organochlorine compounds produce hyperexcitability by inducing membrane depolarization, whereas the organophosphates cause typical cholinergic neurotoxicity. The mechanism of action to explain the neurotoxic symptoms varies with each compound, organochlorine or otherwise. [The reader is referred to Chapters 16 and 22 of *Cassarett & Doull's Toxicology* (Klassen, 1996) for detailed information on acute and chronic exposure in relation to different pesticides in humans and animals.] Mechanistic studies showed that the observed acute and chronic symptoms and death caused by organochlorine compounds such as DDT, lindane, endrin, Kepone, and so on, are due to their well-defined effects on nerve membrane depolarization by altering the ionic conductance (Narahashi, 1976; Joy, 1994), membrane bound enzymes (Cutkomp *et al.,* 1982; Desaiah, 1982; Matsumura, 1985), and a number of receptors and second messenger pathways (Eldefrawi *et al.,* 1985; Matsumura, 1985; Kodavanti *et al.,* 1988, 1989; Mehrotra *et al.,* 1988; Ecobichon, 1996). The organophosphate compounds, depending on the structure, are potent and irreversible inhibitors of acetylcholinesterase in the CNS (Aldridge and Johnson, 1971; Aldridge and Reiner, 1972; Ecobichon, 1979; Toia and Casida, 1979; Johnson, 1982). The buildup of the acetylcholine at the synapse is the cause for the observed symptoms and mortality. The carbamic acid esters are also known inhibitors of acetylcholinesterase, but the inhibition is reversible with time, thus these compounds are safer to humans in comparison to organophosphate compounds (Ecobichon, 1996). Data also suggest that the organophosphate compounds modulate other sites on the CNS besides the acetylcholinesterase (Richardson, 1995). Some of the organophosphates are known to produce delayed neurotoxicity (Abou-Donia, 1981; Johnson, 1982; Abou-Donia and Lapadula, 1990; Richardson, 1995: Ecobichon, 1996). A variety of other pesticides used to kill weeds, insects, and rodents are specific for the pest with a large safety margin to humans (Ecobichon, 1996). Because the purpose of this chapter is to review the literature on developmental neurotoxicity of pesticides, the information on adults is helpful in relating to the pesticide exposure during the development of the CNS.

III. Effects on the Developing Central Nervous System

There is an increased rate of birth defects in many countries, including the United States. These defects cannot be accounted for by genetic disorders alone. There are several drugs and environmental chemicals, such as pesticides, that have the potency to cause developmental neurotoxicity that reflects on adult behavior. In view of the large number of chemicals and drugs to which humans are exposed, it is highly unlikely to evaluate their potential as developmental neurotoxicants in humans. In these situations, one relies on the experimental data with animals to extrapolate the effects and to assess the risk to humans. Because of the intrinsic differences between the CNS of, for example, rats and humans, such extrapolation is difficult and often times results in false-negative signals. There are, however, several neuronal factors that determine the susceptibility to neurotoxic compounds in general, and to pesticides in particular.

A large body of evidence comes from rodent models to explain the susceptibility factors of the developing CNS to toxic substances. It has become a difficult task to compare the early CNS biologic events in animals to early development in humans in relation to toxic effects. It is imperative, therefore, to consider the developmental processes that are relatively similar in both humans and animals. Ideally, these processes should show similar responses to toxic compounds, and such is not the case all the time. The neurobiologic, neurochemical, and neurobehavioral processes that are intrinsic to the developing CNS play a role in determining the developmental neurotoxicity in adults. These processes are: neuronal proliferation; cell migration; synaptogeneses; apoptosis; receptors; neurotransmitters; and sensory, motor, and cognitive activities (Choi *et al.,* 1978; Bedi *et al.,* 1980; Rodier, 1980, 1994; Barks *et al.,* 1988; Olney, 1988; McDonald and Johnson, 1990; Court *et al.,* 1993; Christie *et al.,* 1995; Johnson, 1995; Institute for Environment and Health (IEH), 1996; Rodier *et al.,* 1996). The neurotransmitter systems such as monoaminergic, excitatory, and inhibitory pathways mediated by dopamine, glutamate, and γ-aminobutyric acid (GABA) are important in the development of neural networks. These systems are also implicated in behavioral aspects such as locomotor activity, which peaks before weaning in rats. The cholinergic system, however, develops slowly, peaking at 30 days postnatally and participates in attention behavior in the adult (Mabry and Campbell, 1977; Lauder *et al.,* 1986; Schambra *et al.,* 1989; Constantine-Paton *et al.,* 1990). Some of these neurotransmitter systems are also associated with the development of the human CNS and can be used for comparison, provided the time window of development is taken into consideration. For example, the first 3–4 weeks of life in rodents is comparable to third trimester of pregnancy through the first 2 years of life in humans.

The toxic substances of either environmental or pharmaceutical origin are shown to produce neurodevelopmental toxicity. The time of exposure during the development of the CNS is critical to explaining neurotoxicity in adult life. Many of the pesticides have been shown to effect the developing CNS adversely. These pesticides cross the different classifications, which include organochlorines, organophosphates, pyrethroids, and others. Exposure of neonates in the first 3–4 weeks of life to pesticides adversely affects adult behavior and growth. This is due to the fact that during this window of time there is increased axonal and dendritic outgrowth, synaptogenesis, and neuronal connections (Kolb and Whishaw, 1989). It is also established that animals during this period acquire sensory and motor skills (Bolles and Woods, 1964; Campbell *et al.,* 1969). Such a phenomenon is also true if the exposure to the pesticides occurs maternally. Several pesticides have been detected in breast milk of mothers around the world (Kimmel *et al.,* 1985).

Several organochlorine compounds have been shown to cause developmental neurotoxicity in rodents when exposed neonatally (Eriksson, 1992, 1997; Eriksson *et al.,* 1992). The cholinergic system was shown to be effected by the single dose of DDT given to rats at postnatal day (PN) 10 (Eriksson and Nordberg, 1986, 1990). The effects were region specific in the brain; cerebral cortex was highly effected, and no changes were observed in hippocampus. The 17-day-old mice showed an increased binding of ^{3}H-quiniclidinylbenzilate (QNB) suggesting increased muscarinic acethylcholine receptor (MACHAR) density. These changes were also evident in the adult life of these mice (Eriksson and Fredriksson, 1991). These neonatal changes on MACHAR were related to the functional disturbances as manifested in delayed locomotion, rearing, and total activity (Eriksson, 1997). Such a relationship of neonatal exposure to DDT and the functional aspects in the adult would serve as a marker for developmental neurotoxicity. Another example of the chlorinated compounds is chlordecone (Kepone), a well-characterized neurotoxicant in both animals and humans (Desaiah, 1982; Guzelian, 1982). Adult rodents, when given chlordecone, exhibited startle response, intensive tremors, and, eventually, death, probably due to respiratory failure (Tilson *et al.,* 1979; Desaiah, 1982; Guzelian, 1982). The adult animals treated with different doses of chlordecone showed abnormal neurobehavioral, neurochemical, and second messenger systems (Desaiah, 1982; Guzelian, 1982; Hong and Ali, 1982; Swanson and Woolley, 1982; Tilson and Mactutus, 1982; Komulainen and Bondy, 1987). Some of these abnormalities were also observed in humans exposed to chlordecone occupationally (Taylor, 1982). However, a few reports exist that chlordecone administered neonatally adversely effects the adult behavior in rodents, including learning deficits and tremors (Tilson *et al.,* 1982; Mactutus and Tilson, 1984; Mactutus *et al.,* 1984). Neonatally administered chlordecone on PN 4 showed changes in pituitary–adrenocortical function on days 77–78 (Rosecrans *et al.,* 1985). However, biochemical evaluation of neonatal treatment of chlordecone on adults is scanty. The data from my laboratories revealed that rat pups receiving a daily dose of 2.5 mg/kg up to PN 20 showed a decreased neuronal Na^{+}-K^{+} ATPase, calcium ATPase in synaptic membranes and F_1 ATPase in mitochondria (Jinna *et al.,* 1989). These changes, however, are reversible with time of withdrawal of the chlordecone treatment. The significant reduction of these enzymes could bring about changes in nerve membrane transport activity, impairment of calcium homeostasis, and biologic energy (ATP) deprivation (Desaiah, 1982; Koumulainen and Bondy, 1987). Early postnatal exposure of rats with Lindane (hexachlorocyclohexane), an insecticide, showed developmental behavioral changes in adults (Rivera *et al.,* 1990). Because Lindane is a lipophilic compound it can accumulate in the CNS and cause symptoms such as hyperactivity. A similar observation was made with rats exposed maternally to a fungicide, hexachlorobenzene (HCB) (Goldey and Taylor, 1992). These studies revealed that pups born to mothers exposed to HCB showed changes in acoustic startle response as a function of age. The pups at PN 23 showed a decreased acoustic startle response amplitude, and at PN 90 showed elevated acoustic startle response (Godley and Tyalor, 1992). Gestational and lactational exposure of a number of organochlorine compounds causes developmental neurotoxicity in adults. This is due mainly to the fact these compounds are highly lipophilic and can accumulate in the brain of the offspring through mother's milk and placental transfer. Children are particularly susceptible because the rate of intake of fluids such as milk and water is much higher than in adults. Children also play in dusty environments where some of these pesticides reside in soil.

The organophosphate group of pesticides, unlike organochlorine pesticides, do not persist in the environment, as they degrade rapidly (O'Brien, 1960; Heath, 1961; Ecobichon and Joy, 1994). The neurotoxicity in humans exposed occupationally and in experimental animals is well documented (Ecobichon, 1996). The organophosphate compounds are well-known specific inhibitors of acetylcholinesterase in the adult CNS (Aldridge and Reiner, 1972; Ecobichon, 1979). Diisopropylfluoro+phosphate (DFP) is one of the organophosphate compounds that produces delayed neurotoxicity in susceptible species (Abou-Donia, 1981; Johnson, 1982, 1990). Mice exposed neonatally on PN 10 with a single dose of 1.5 mg/kg showed a 50% reduction of acetylcholinesterase as well as increased motor behavior in adults (Eriksson, 1997). The density of MACHAR in these adult brains was also decreased (Ahlbom *et al.,* 1995). These studies suggest that neonatal exposure of some organophosphates modulate the neurochemical and neurobehavioral endpoints. Many other organophosphates such as fenthion, fenitrothion, and desbromoleptophos have shown irreversible effects on the gait in adult chicks exposed *in ovo* (Farage-Elawar and Francis, 1988). The gait imbalance was correlated with the decrease of acetylcholinesterase (Farage-Elawar and Francis, 1988). Chlorpyrifos, another organophosphate pesticide, showed developmental neurotoxicity. Rats treated with chlorpyrifos during gestation showed a significant reduction of acetylcholinesterase activity in maternal brain and, to a lesser extent, in brains at PN 3 (Chanda *et al.,* 1995; Chanda and Pope, 1996). Although there is a good correlation between the suppression of acetylcholinesterase and development of classic organophosphate-induced developmental neurotoxicity, there are other neurochemical alterations that would explain the cellular mechanisms involved in organophosphate neurotoxicity. Rats administered with a subtoxic dose of chlorpyrifos on PN 1 showed decreased DNA and protein synthesis in different areas of the brain even at PN 8 (Whitney *et al.,* 1995). Adenylylcyclase is highly critical for cell development through the cAMP mediated second messenger signaling cascade. Rats dosed with a subtoxic dose of chlorpyrifos on PN 1–4 or on PN 11–14 showed a decreased adenylylcyclase cascade in different brain regions as well as in the heart (Song *et al.,* 1997). These studies further demonstrated that chlorpyrifos modulates the G-proteins that link neurotransmitter and hormone receptors to adenylylcyclase and the expression of neurotransmitter receptors. These findings indicate that chlorpyrifos induces delayed neurotoxicity through the inhibition of adenylylcyclase cascade besides cholinergic pathways.

Pyrethroids and other botanicals also cause developmental neurotoxicity. Some of the synthetic pyrethroids share the same type of mechanism of action as DDT (Narahashi, 1985; Vijverberg and van den Bercken, 1990). Pyrethroids of type I produce prolonged depolarization of nerve membrane by blocking ionic channels, thus enhancing neurotransmitter release. Rats treated with bioallethrin on PN 10 showed neonatal and adult changes in MACHAR that were correlated with the hyperactive neurobehavioral changes such as spontaneous motor activity (Ahlboom *et al.,* 1995; Eriksson, 1997). Pyrethroids exhibit similar neurotoxicity as DDT, both in adults and neonates; hence, each might modulate the other's effects. Neonatal exposure of rats to DDT at low doses showed altered responses when challenged with bioallethrin in adults (Eriksson *et al.,* 1993; Johansson *et al.,* 1995; Eriksson, 1997).

Paraquat, a herbicide, is known to produce its major toxicity through inhibition of superoxide dismutase (SOD) (Sagar, 1987). One of its structural analogs, 1-methyl-4 phenylpyridium ion (MPP^+) is a metabolite of the well-known neurotoxicant MPTP (1-methyl-4-phenyl-1,2,3,6-tetrahydropyridine). MPTP, when converted to MPP^+, produces selective effects on the dopaminergic pathway, resulting in Parkinson-like disease in both animals and humans (Burns *et al.,* 1983; Ballard *et al.,* 1985). Neonatal exposure of mice with a single dose of paraquat (0.36 mg/kg body wt) or MPTP (0.30mg/kg/body wt) resulted in decreased levels of dopamine in 4-month-old mice (Fredriksson *et al.,* 1993). Because the dopaminergic pathway is known to mature rapidly during the postnatal development, exposure of environmental chemicals during this period could adversely effect the dopaminergic pathway in the adult. Other herbicide compounds of environmental significance are 2,4-dichlorophenoxy acetic acid (2,4-D) and 2,4,5-trichlorophenoxy acetic acid (2,4,5-T), both of which have shown neurotoxicity in adults. However, the developmental studies are rather scanty. When administered prenatally, a 1:1 mixture of 2,4-D and 2,4,5-T showed alterations in the ontogeny of dopamine, norepinephrine, and serotonin, with corresponding behavioral changes such as negative geotaxis, olfactory discrimination, and spontaneous locomotor activity (Mohammad and St. Omer, 1985, 1986; St. Omer and Mohammad, 1987). These findings were also reported in birds and other rodents (Sanderson and Rogers, 1981; Crampton and Rogers, 1983).

IV. Conclusions

In summary, this review identified several pesticides across different groups of chemicals as developmental neurotoxicants. These chemicals produce subtle changes in the developing CNS that are amplified in the adult behavior. The neurobehavioral changes in adults depend on the time of exposure during development and dose of the compound. There is a good correlation of the neurochemical changes with specific neurobehavioral modifications, and such changes are at times region specific in brain. There is, however, a paucity of information on the developmental neurotoxicity with many other pesticides, either that have been used or are currently being used, which makes it difficult to extrapolate the animal data to humans in risk assessment. Nevertheless, the pesticides discussed in this review are potent modulators of the developing brain.

Acknowledgments

The secretarial assistance of Mrs. Linda Clark is highly appreciated in the preparation of this manuscript.

References

Abou-Donia, M. B. (1981). Organophosphorus ester–induced delayed neurotoxicity. *Ann. Rev. Pharmacol. Toxicol.* **21,** 511–548.

Abou-Donia, M. B. and Lapadula, D. (1990). Mechanisms of organophosphorus ester–induced delayed neurotoxicity: type I and type II. *Ann. Rev. Pharmacol. Toxicol.* **30,** 405–440.

Ahlboom, J., Fredriksson, A., and Eriksson, P. (1995). Exposure to an organophosphate (DFP) during a defined period in neonatal life induces permanent changes in brain muscarinic receptors and behavior in mice. *Brain Res.* **677,** 13–19.

Aldridge, W. N. and Johson, M. K. (1971). Side effects of organophosphorus compounds; delayed neurotoxicity. *Bull WHO.* **44,** 259–263.

Aldridge, W. N. and Reiner, E. (1972). *Enzyme Inhibitors as Substrates.* North-Holland Elsevier, Amsterdam/New York.

Ballard, P. A., Tetrud, J. W., and Langston, J. W. (1985). Permanent human parkinsonism due to 1-methyl-4-phenyl-1,2,3,6-tetrahydropyridine (MPTP): seven cases. *Neurol.* **5,** 949–956.

Barks, J. D., Silverstein, F. S., Sims, K. Greenmeyer, J. T., and Johnson, M. V. (1988). Glutamate recognition sites in human fetal brain. *Neurosci. Lett.* **84,** 131–136.

Bedi, K. S., Thomas, Y. M., Davies, C. A., and Dobbing J. (1980). Synapse to neuron ratios of the frontal and cerebellar cortex of 30 day old and adult rats undernourished during early postnatal life. *J. Comp. Neurol.* **193,** 49–56.

Bolles, R. G. and Woods, P. J. (1964). The ontogeny of behavior in the albino rat. *Anim. Behav.* **12,** 427–441.

Brooks, G. T. (1974). *Chlorinated Insecticides. Technology and Application.* CRC, Cleveland, OH, **1,** 12–13.

Burns, R. S., Chiueh, C. C., Marky, S. P., Ebert, M. H., Jacobowitz, D. M., and Kopin, J. (1983). A primate model of parkinsonism: selective destruction of dopaminergic neurons in the pars compacta of the substantia nigra by *N*-methyl-4-phenyl-1,2,3,6-tetrahydropyridine. *Proc. Nat. Acad. Sci. USA* **80,** 4546–4550.

Campbell, B. A., Lytele, L. D., and Fibiger, H. C. (1969). Ontogeny of adrenergic arousal and cholinergic inhibitory mechanisms in the rat. *Science* **166,** 635–637.

Chanda, S. M., Harp, P., Liu, J., and Pope, C. N. (1995). Comparative developmental and maternal neurotoxicity following acute gestational exposure to chlorpyrifos in rats. *J. Toxicol. Env. Hlth.* **44,** 189–202.

Chanda, S. M. and Pope, C. N. (1996). Neurochemical and neurobehavioral effects of repeated gestational exposure to chlorpyrifos in maternal and developing rats. *Pharmacol. Biochem. Behav.* **53,** 771–776.

Choi, B. H., Lapham, L. W., Amin-Zake, L., and Saleem, T. (1978). Abnormal neuronal migration, deranged cerebral cortical organization and diffuse white matter astrocytosis of human fetal brain: a major effect of methyl mercury poisoning *in utero. J. Neuropath. Exper. Neurol.* **37,** 719–733.

Christie, D., Leiper, A. D., Chessells, J. M., and Vargha-Khadem, F. (1995). Intellectual performance after presymptomatic cranial radiopathy for leukemia: effects of age and sex. *Arch. Dis. Child.* **73,** 136–140.

Constantine-Paton, M., Cline, H. T., and Debshi, E. (1990). Patterned activity, synaptic convergence and the NMDA receptor in developing visual pathways. *Ann. Rev. Neurosci.* **13,** 129–154.

Court, J. A., Perry, E. K., Johnson, M., Piggott, M. A., Kervin, J. A., Perry, R. H., and Ince, P. G. (1993). Regional patterns of cholinergic and glutamate activity in the developing and aging human brain. *Dev. Brian Res.* **74,** 73–82.

Coye, M. J., Lowe, J. A., and Maddy, K. T. (1986). Biological monitoring of agricultural workers exposed to pesticides. 1. Cholinesterase activity determinators. *J. Occup. Med.* **28,** 619–627.

Crampton, M. A. and Rogers, L. J. (1983). Low doses of 2,4,5-richlorophenoxyacetic acid are behaviorally teratogenic to rats. *Experimentia* **39,** 891–892.

Cutkomp, L. K., Koch, R. B., and Desaiah, D. (1982). Inhibition of ATPases by chlorinated hydrocarbons. In *Insecticide Mode of Action.* J. R. Coats, Ed. Academic Press, San Diego, CA., pp. 45–69.

Desaiah, D. (1982). Biochemical mechanisms of chlordecone neurotoxicity: a review. *NeuroToxicol.* **3,** 103–110.

Ecobichon, D. J. (1979). Hydrolytic mechanisms of pesticide degradation. In *Advances in Pesticide Science. Biochemistry of Pests and*

Mode of Action of Pesticides and Formulation Chemistry. H. Geissbuhler Ed. Pergman, New York, pp. 516–524.

Ecobichon, D. J. (1996). Toxic effects of pesticides. In *Casaret & Doull's Toxicology, The Basic Science of Poisons* (5th edition). C. D. Klassen Ed. McGraw-Hill, New York, pp. 643–689.

Eicobichon, D. J. and Joy, R. M. (1994). *Pesticides and Neurological Diseases.* 2nd ed. CRC Press, Boca Raton, Florida.

Eldefrawi, M. E. S., Sherby, S. M., Abalis, I. M., and Eldefrawi, A. T. (1985). Interaction of pyrethroid and cyclodiene insecticides with nicotonic acetylcholine and GABA receptors. *NeuroToxicology* **6,** 47–62.

Eriksson, P. (1992). Neuroreceptor and behavioral effects of DDT and pyrethroids in immature and adult mammals. In *The Vulnerable Brain and Environmental Risks.* R. L. Isaccson and K. F. Jensen Eds. Plenum Press, New York, **2,** 235–251.

Eriksson, P. (1997). Developmental neurotoxicity of environmental agents in the neonate. *NeuroToxicology* **18,** 719–726.

Eriksson, P., Ahlbom, J., and Fredriksson, A. (1992). Exposure to DDT during a defined period in neonatal life induces permanent changes in brain muscarinic receptors and behavior in adult mice. *Brain Res.* **582,** 277–281.

Eriksson, P. and Fredriksson, A. (1991). Neurotoxic effects of two different pyrethroids, bioallethrin and deltamethrin on immature and adult mice: changes in behavioral and muscarinic receptor variables. *Toxicol. Appl. Pharmacol.* **108,** 78–85.

Eriksson, P., Johansson, U., Ahlbom, J., and Fredriksson, A. (1993). Neonatal exposure to DDT induced increased susceptibility to pyrethroid (bioallethrin) exposure at adult age—changes in cholinergic muscarinic receptor and behavioral variables. *Toxicol.* **77,** 21–30.

Eriksson, P. and Nordberg, A. (1986). The effects of DDT, DDOH-palmitic acid, and chlorinated paraffin on muscarinic receptors and the sodium-dependent chloline uptake in the central nervous system of immature mice. *Toxicol. Appl. Pharmacol.* **85,** 121–127.

Eriksson, P. and Nordberg, A. (1990). Effects of two pyrethroids, bioallethrin and deltamethrin, on subpopulations of muscarinic and nicotonic receptors in the neonatal mouse brain. *Toxicol. Appl. Parmacol.* **102,** 456–463.

Farage-Elawer, M. and Francis, B. M. (1988). Effects of fenthion, fenitrothion and desbromoleptophos on gait, acetylcholinesterase, and neurotoxic esterase in young chicks after *in ovo* exposure. *Toxicol.* **49,** 253–261.

Fredriksson, A., Fredriksson, M., and Eriksson, P. (1993). Neonatal exposure to paraquat on MPTP induces permanent changes in striatum dopamine and behavior in adult mice. *Toxicol. Apl. Pharmacol.* **122,** 258–264.

Goldey, E. S. and Taylor, D. H. (1992). Developmental neurotoxicity following premating maternal exposure to hexachlorobenzene in rats. *Neurotoxicol. Teratol.* **14,** 15–21.

Guzelian, P. S. (1982). Comparative toxicology of chlordeone (Kepon®) in humans and experimental animals. *Ann. Rev. Pharmacol. Toxicol.* **22,** 89–113.

Heath, D. F. (1961). *Organophosphorus Poisons. Anticholinesterase and Related Compounds.* Pergamon Press, London, UK.

Hong, J. S. and Ali, S. F. (1982). Chlordecone exposure in the neonate selectively alters brain and pituitary endorphin levels in prepubertal and adult rats. *NeuroToxicol.* **3,** 111–118.

Institute for Environment and Health (IEH). (1996). Perinatal Developmental Neurotoxicity. MRC, Leister, U.K.

Jinna, R. R., Uzodinma, J. E., and Desaiah, D. (1989). Age-related changes in rat brain ATPases during treatment with chlordecone. *J. Toxicol. Env. Hlth.* **27,** 199–208.

Johansson, U., Fredriksson, A., and Eriksson, P. (1995). Bioallethrin causes permanent changes in behavioral and cholinergic muscarinic receptor variables in adult mice exposed neonatally to DDT. *Eur. J. Pharmacol.* **293,** 159–166.

Johnson, M. H. (1995). The development of visual attention: a cognitive neuroscience perspective. In *The Cognitive Neurosciences,* M. S. Gazzaniga Ed. MIT Press, Cambridge, MA. pp. 735–747.

Johnson, M. K. (1982). The target for initiation of delayed neurotoxicity by organophosphorus esters: Biochemical studies and toxicological applications. In *Reviews of Biochemical Toxicology.* E. Hodgson, J. R. Bend, and R. M. Philpot, Eds. Elsevier, New York, **4,** 141–212.

Johnson, M. K. (1990). Organophosphates and delayed neuropathy—is NTE alive and well? *Toxicol. Appl. Pharmacol.* **102,** 385–399.

Joy, R. M. (1994). Chlorinated hydrocarbon insecticides, In D. J. Ecobichon, R. M. Joy Eds. *Pesticides and Neurological Diseases,* 2nd ed. CRC, Boca Raton, Florida, pp. 81–170.

Kimmel, C. A., Buelke-Sam, J., and Adams, J. (1985). Collaborative behavioral teratology study: implications, current applications and future directions. *Neurobehav. Toxicol. Teratol.* **7,** 669–673.

Klassen, C. D. (Ed.) (1996). *Casarett & Doull's Toxicology. The Basic Science of Poisons* (5th Edition), McGraw-Hill, New York.

Kodavanti, P. R. S., Mehrotra, B. D., Chetty, S. C., and Desaiah, D. (1988). Effect of selected insecticides on rat brain synaptosomal adenylate cyclase and phosphodiesterase. *J. Toxicol. Env. Hlth.* **25,** 207–215.

Kodavanti, P. R. S., Mehrotra, B. D., Chetty, S. C., and Desaiah, D. (1989). Inhibition of calmodulin activated adenylate cyclase in rat brain by selected insecticides. *NeuroToxicol.* **10,** 219–228.

Kolb, B. and Wishaw, I. Q. (1989). Plasticity in the neocortex: mechanisms underlying recovery from early brain damage. *Prog. Neurobiol.* **32,** 235–276.

Komulainen, H. and Bondy, S. C. (1987). Modulation of levels of free calcium within synaptosomes by organochlorine insecticides. *J. Pharmacol. Exp. Ther.* **241,** 575–581.

Lauder, J. M., Han, V. K. M., Henderson, P., Verdoorn, T., and Towle, A. C. (1986). Prenatal ontogeny of the GABAergic system in the rat brain: an immunocytochemical study. *Neuroscience* **19,** 465–493.

Mactutus, C. F. and Tilson, H. A. (1984). Neonatal chlordecone exposure impairs early learning and retention of active avoidance in the rat. *Neurobeh. Toxicol. Teratol.* **6,** 75–83.

Mactutus, C. F., Unger, K. L., and Tilson, H. A. (1984). Evaluation of neonatal chlordecone neurotoxicity during early development: initial characterization. *Neurobehav. Toxicol. Teratol.* **6,** 67–73.

Marby, P. D. and Campbell, B. A. (1977). Developmental Psychopharmacology. In *Handbook of Psychopharmacology.* Plenum Press, New York, **7,** 393–444.

Matsumura, F. (1985). *Toxicology of Insecticides.* Plenum, New York, pp. 122–128.

McDonald, J. W. and Johnston, M. V. (1990). Physiological and pathophysiological roles of excitatory amino acids during central nervous system development. *Brain Res. Rev.* **15,** 41–70.

Mehrotra, B. D., Reddy, R. S., and Desaiah, D. (1988). Effect of subchronic dieldrin treatment on calmodulin-regulated Ca^{2+} pump activity in rat brain. *J. Toxicol. Env. Hlth.* **25,** 461–469.

Mohammad, F. K. and St. Omer, V. E. V. (1985). Developing rat brain monoamine levels following *in utero* exposure to a mixture of 2,4-dichlorophenoxyacetic and 2,4,5, trichlorophenoxyacetic acids. *Toxicol. Lett.* **29,** 215–223.

Mohammad, F. K. and St. Omer, V. E. V. (1986). Behavioral and developmental effects in rats following *in utero* exposure to 2,4-D/2,4,5,T mixture. *Neurobehav. Toxicol. Teratol.* **8,** 551–560.

Moser, V. C., Becking, G. C., and Cuomo, V. et al. (1997). The IPCS collaborative study on neurobehavioral screening methods. V. Results of chemical testing. *NeuroToxicol.* **18,** 969–1056.

Narahashi, T. (1976). Effect of insecticides on nervous conduction and synaptic transmission. In C. F. Wilkinson Ed. *Insecticide Biochemistry and Physiology.* Plenum, New York, pp. 327–352.

Narahashi, T. (1985). Nerve membrane ionic channels as the primary target of pyrethroids. *NeuroToxicol.* **6,** 3–22.

Olney, J. W. (1988). Excitotoxic food additives: functional teratological aspects. *Progr. Brain Res.* **73,** 283–293.

O'Brien, R. D. (1960). *Toxic Phosphorus Esters. Chemistry, Metabolism and Biological Effects.* Academic Press. New York.

Resecrans, J. A., Squibb, R. E., Johson, J. H., Tilson, H. A., and Hong, J. S. (1985). Effects of neonatal chlordecone exposure on pituitary–adrenal function in adult Fischer-344 rats. *Neurobehav. Toxicol. Teratol.* **7,** 33–7.

Richardson, R. J. (1995). Assesment of the neurotoxic potential of chlorpyrifos relative to other organophosphorus compounds: a critical review of the literature. *J. Toxicol. Env. Hlth.* **44,** 135–165.

Rivera, S., Sanfelin, C., and Rodriguez, E. (1990). Behavioral changes induced in developing rats by an early postnatal exposure to lindane. *Neurotoxicol. Teratol.* **12,** 591–595.

Rodier, P. M. (198). Chronology of neuron development: animal studies and their clinical implications. *Dev. Med. Child Neurol.* **22,** 525–545.

Rodier, P. M. (1994). Comparative postnatal neurologic development. In *Prenatal Exposure to Toxicants.* H. L. Needlema, and D. Bellinger, Eds. Johns Hopkins University Press, Baltimore, pp. 3–23.

Rodier, P. M., Ingram, J. L., Tisdale, B., Nelson, S., and Romano, J. (1996). Embryological origin on autism: developmental anomalies of the cranial nerve motor nuclei. *J. Comp. Neurol.* **370,** 247–261.

Rogers, J. M. and Kavlock, R. J. (1996). Developmental toxicology. In *Casaret & Doull's Toxicology, The Basic Science of Poisons* (5th edition), C. D. Klassen, Ed. McGraw-Hill, New York, pp. 301–331.

Sagar, G. R. (1982). Uses and usefulness of paraquat. *Human Toxicol.* **6,** 7–11.

Sanderson, C. A. and Rogers, L. J. (1981). 2,4,5-trichlorophenoxyacetic acid causes behavioral effects in chickens at environmentally relevant doses. *Science* **211,** 593–594.

Schambra, U. B., Sulik, K. K., Petruz, P., and Lauder, J. M. (1989). Ontogeny of cholinergic neurons in the mouse forebrain. *J. Comp. Neurol.* **288,** 101–122.

Schardein, J. L. (1993). *Chemically Induced Birth Defects.* (2nd Ed.) Marcel Dekker, New York.

Schardein, J. L. and Keller, K. A. (1989). Potential human developmental toxicants and the role of animal testing in their identification and characterization. *CRC Crit. Rev. Toxicol.* **19,** 251–339.

Song, X., Seidler, F. J., Saleh, J. L., Zhang, J., Padilla, S., and Slotkin, T. A. (1997). Cellular mechanisms for developmental toxicity of chlorpyrifos: targeting the adenyl cyclase signaling cascade. *Toxicol. Appl. Pharamcol.* **145,** 158–174.

St. Omer, V. E. V. and Mohammad, F. K. (1987). Ontogeny of swimming behavior and brain catecholamine turnover in rats prenatally exposed to a mixture of 2,4-dichlorophenoxyacetic and 2,4,5-trichlorophenoxyacetic acids. *Neuropharmacol.* **26,** 1351–1358.

Swanson, K. L. and Woolley, D. W. (1982). Comparison of the neurotoxic effects of chlordecone and dieldrin in the rat. *NeuroToxicology* **3,** 81–102.

Taylor, J. R. (1982). Neurological manifestations in humans exposed to chlordecone and follow-up results. *NeuroToxicology* **3,** 9–16.

Tilson, H. A., Byrd, N., and Riley, M. (1979). Neurobehavioral effects of exposing rats to Kepone via the diet. *Environ. Hlth. Perspect.* **33,** 321.

Tilson, H. A. and Mactutus, C. F. (1982). Chlordecone neurotoxicity: a brief overview. *NeuroToxicology* **3,** 1–8.

Tilson, H. A., Squibb, R. E., and Burne, T. A. (1982). Neurobehavioral effects following a single dose of chlordecone (Kepone) administered neonatally to rats. *NeuroToxicology* **3,** 45–57.

Toia, R. F. and Casida, J. E. (1979). Phosphorylation. "Aging" and possible dealkylation reactions of saligenin cyclic phosphorus esters with α-chymotryprin. *Biochem. Pharmacol.* **28,** 211–216.

Vijverberg, H. P. M. and van den Bercken, J. (1990). Neurotoxicological effects and the mode of action of pyrethroid insecticides. *Crit. Rev. Toxicol.* **21,** 105–126.

Weir, S., Minton, N., and Murray, V. (1992). Organophosphate poisoning: the UK national poisons unit experience during 1984–1987. In *Clinical and Experimental Toxicology of Organophosphates and Carbamates.* B. Ballatyne and T. C. Barrs Eds. Butterworth-Heineman, Oxford, UK, pp. 463–470.

Whitney, K. D., Seidler, F. J., and Slotkin, T. A. (1995). Developmental neurotoxicity of chlorpyrifos: cellular mechanisms. *Toxicol. Appl. Pharmacol.* **134,** 53–62.

World Health Organization (WHO) (1990). Public Health Impact of Pesticides Used in Agriculture. WHO, Geneva.

CHAPTER

33

The Effects of Alcohol on The Fetal Brain

The Central Nervous System Tragedy

IRENA NULMAN
The Motherisk Program
Division of Clinical Pharmacology and Toxicology
Department of Pediatrics
Hospital for Sick Children
University of Toronto
Toronto, Ontario, Canada M5S 1A1

BONNIE O'HAYON
The Motherisk Program
Division of Clinical Pharmacology and Toxicology
Hospital for Sick Children
Faculty of Medicine
University of Toronto
Toronto, Ontario, Canada M5S 1A1

JONATHAN GLADSTONE
The Motherisk Program
Division of Clinical Pharmacology and Toxicology
Hospital for Sick Children
Faculty of Medicine
University of Toronto
Toronto, Ontario, Canada M5S 1A1

GIDEON KOREN
The Motherisk Program
Division of Clinical Pharmacology and Toxicology
Department of Pediatrics
The Research Institute
Hospital for Sick Children
University of Toronto
Toronto, Ontario, Canada M5S 1A1

* Abbreviations: ADH, alcohol dehydrogenase; ARBD, alcohol-related birth defects; ARND, alcohol-related neurodevelopmental disorder; ATP, adenosine triphosphate; BAC, blood-alcohol concentration; cAMP, (cyclic AMP) cyclic adenosine monophosphate; CNS, central nervous system; CT, computed tomography; FARA, fetal alcohol related abnormalities; FAS, fetal alcohol syndrome; IUGR, intrauterine growth retardation; LYP, long-term potentiation; MRI, magnetic resonance imaging; NAD, nicotinamide adenine dinucleotide; PET, positron emission tomography; SES, socioeconomic status

I. Introduction

Alcohol is a legal, socially acceptable drug the use of which is part of the lives of most American women. In a survey, 61.5% of American women reported consuming alcohol in the preceding year. Moreover, its use is highest among women of reproductive age—75% of women aged 18–34 reported alcohol use in the past year (National Institute on Drug Abuse, 1990), making it by far the most widely used human teratogen, with the number of consumers being several orders of magnitude larger than for any other teratogenic compound (Koren and Nulman, 1994).

The spectrum of alcohol's teratogenic effects spans a wide continuum that includes growth deficiency, central nervous dysfunction, craniofacial anomalies, and pathologic organ and skeletal conditions. Fetal alcohol syndrome (FAS) is the most severe expression of fetal alcohol related abnormalities (FARA) and may be seen in the offspring of women who drink heavily throughout pregnancy. Presently, alcohol is the most common preventable cause of birth defects and the leading cause of mental retardation ahead of Down syndrome and cerebral palsy

The time since the late 1970s has seen the accumulation of a large body of research into alcohol's teratogenicity. This chapter provides an overview of FARA, with focus on pharmacology of alcohol in the women and fetus, epidemiology, diagnostic issues, and risk factors.

II. Historical Perspective

Although the role of alcohol in causing human teratogenicity was not proved until the 1970s—the adverse effects of alcohol consumption during pregnancy have been noted throughout history. Several reviews of the history of alcohol and pregnancy are available (Warner Rosett, 1975; Rossett and Weiner, 1984; Abel, 1990). *Judges 13:7* states, "Behold, thou shalt conceive, and bear a son; and now drink no wine or strong drink." Plato is ascribed to having proclaimed that children should not be made in bodies saturated with drunkenness. Also, Aristotle is attributed to having stated that, "Foolish, drunken and harebrained women most often bring forth children like unto themselves, morose and languid." During the gin epidemic in 1726, the College of Physicians Report to the British Parliament described parental drinking as "a cause of weak, feeble and distempered children." In 1834, a report on drunkenness to the British House of Commons concluded that infants born to alcoholic mothers sometimes have a "starved, shriveled and imperfect look." The first scientific study on children of alcoholic mothers was reported in 1899 by Dr. William Sullivan, a Liverpool prison physician. He noted that maternal inebriety is unfavorable to normal development, and that jailed and detoxified women had improved birth outcomes.

Little attention was paid to the plausibility of alcohol's teratogenicity until the last few decades. An article by Lemoine *et al.*, in France in 1968, provided the first description in the medical literature of the effects of alcohol on the fetus. In was not until 1973, however, with the independent observation of Jones and Smith (1973) and by Jones *et al.* (1973), that a distinct dysmorphic syndrome associated with gestational alcoholism, FAS, was coined and recognized in the medical literature. Since the mid-1970s, thousands of articles have been published in the field of alcohol teratology.

III. Alcohol Pharmacology

Ethanol is a small molecule that is soluble in both water and lipids, enabling it to pass easily through cell membranes. It distributes throughout tissues and cells in proportion to their respective water contents. Therefore, highly vascular organs can equilibrate quickly with arterial blood and have an increased rate of alcohol distribution. During ethanol absorption, the brain, a highly vascular organ, achieves a higher ethanol concentration more rapidly than most other organs (Goldstein, 1983).

The metabolism of alcohol is almost entirely dependent on the liver (Lundsgaard, 1938), the only organ that contains the host of enzymes required to initiate the metabolic pathway. The first step in the metabolism of ethanol is its conversion to acetaldehyde, a process catalyzed by alcohol dehydrogenase (ADH).

NAD+ (NADH NAD+ (NADH
ethanol (acetaldehyde (acetate ($CO_2 + H_2O$
ADH ALDH

ADH is a general remover of hydrogen atoms from various compounds and is available in sufficient amounts to deal efficiently with ingested alcohol. ADH catalyzes the transfer of hydrogen atoms from ethanol to nicotinamide adenine dinucleotide (NAD), a cofactor for the reaction. Acetaldehyde is then oxidized to form acetate, which is later converted to carbon dioxide and water. For a given amount of ingested alcohol, women achieve higher blood alcohol levels due to smaller body size and higher fat content, resulting in a smaller volume of distribution for alcohol, and less gastric-ADH activity than in men, leading to less first-pass metabolism (Frezza *et al.,* 1990, Seitz *et al.,* 1993).

Gender differences in ethanol metabolism are also influenced by the sex hormones. Testosterone, for example, depresses the activity of hepatic ADH (Teschke and Wiese, 1982), resulting in a relative increase in ADH activity in the female liver before the age 50–53 (Maly and Sasse, 1991). There is also a delay in gastric emptying, which occurs during the luteal phase of a woman's menstrual cycle.

Among men and women with similar histories of alcohol abuse, more women incur liver injury than men (Morgan and Sherlock, 1977) and the typical female liver damage is more rapidly progressive than that of a male (Rankin, 1977). A similar pattern has been described for brain diminution, which occures in women after significantly shorter ethanol exposure (Mann *et al.,* 1992).

The actual blood–alcohol concentration (BAC) is dependent on the amount of ethanol ingested, gastrointestinal motility, vascularity of the mucous membranes, and the concentration and distribution of water in each organ. Because pregnancy may influence these factors, one may expect that the disposition of alcohol in pregnancy will be altered. For obvious ethical reasons, there are no controlled studies on the pharmacokinetics of alcohol in pregnant women. However, one can examine information from observational human and experimental animal studies.

There is a delay in the emptying time of a pregnant woman's stomach, which suggests a lower peak alcohol concentration maintained for a longer duration (Hinckers, 1978). In addition, pregnant women are in a state of water volume expansion, including water volume from the fetus, placenta, uterus, and amniotic fluid. Because alcohol is distributed according to the water content of body compartments, the pregnant women has a greater volume of distribution of alcohol, which, again, should contribute to her lower peak blood alcohol concentrations.

Pregnancy-induced alteration in volume of distribution may have implications for her fetus because the effect of alcohol on the fetus varies with the change in water concentration throughout pregnancy. In early pregnancy, the fetal water concentration tends to be higher (McCance and Widdowson, 1954), increasing alcohol's volume of distribution within the fetus. This is meaningful because ethanol crosses the placenta and rapidly achieves equilibrium between maternal and fetal fluid (Guerri and Sanchis, 1985).

It has been suggested that the altered water content in the mother and fetus may favor the passage of ethanol across the placenta into the fetus. Rat studies reveal an increasing gradient of ethanol concentration in fetal blood, amniotic fluid, and intragastric fetal content when compared to maternal blood (Traves *et al.,* 1995; Lopez-Tejero *et al.,* 1986, 1989a,b).

Limited or no ethanol-related ADH activity has been found in the placenta (Pares *et al.,* 1984), and the fetus itself has a low capacity to eliminate alcohol (Guerri and Sanchis, 1985). Ethanol concentration in the amniotic fluid at the end of pregnancy is associated with the contribution of fetal urine to the amniotic fluid (Hayashi *et al.,* 1991). Also, at the end of pregnancy the fetal sucking reflex develops, which may contribute to the high ethanol concentration found in the fetal intragastric fluid (Traves *et al.,* 1995).

Pregnant rats have been found to have significantly lower levels of ADH activity compared to virgin female rats, irrespective of ADH isoenzyme distribution, tolerance, or malnutrition status (Traves *et al.,* 1995). The combination of low ADH activity in the fetal liver (Sanchis and Guerri, 1986b) and marginal ethanol metabolism in the placenta, maternal metabolism is responsible for the elimination of ethanol from both the mother and the fetus.

Ethanol passes freely into breast milk and achieves almost the same concentration as it is in the woman's blood. Animal studies showed that chronic ethanol administration resulted in a loss of mammary cell polarization reduction in Goldi distyosomal elements, abnormalities in case in maturation and secretion, mammary gland, amino acid uptake, changes in pH and amount of lactose, lipoprotein lipase activity, and decrease in both absolute and relative mammary gland weight and protein content (Sanchis and Guerri 1986a; Vinas *et al.,* 1989). Alcohol interferes with prolactine secretion in lactating and nonlactating women (Volpi *et al.,* 1994).

Subramanian (1996) summarized that ethanol has a disruptive effect at the hypothalamic, pituitary, and mammary glands, all of which modulate mammary gland growth, initiation, and maintenance of lactation.

IV. Fetal Alcohol Syndrome

A. *Epidemiology and Economic Burden*

Many studies have been undertaken to estimate the occurrence of FAS in various North American communities. Results range from a incidence of 0 per 1,000 (Abel and Sokol, 1987, 1991) to a prevalence of 120 per 1,000 (Robinson *et al.,* 1987). Typically cited is an incidence for FAS of 0.5–3 cases per 1,000 births (Abel, 1995). In the Unites States, with an annual birth rate of 4 million, this figure can be translated into an estimated 2,000–12,000 children with FAS being born per year.

Abel reviewed the available epidemiologic studies on FAS and presented calculations for the incidence of FAS based on geographic, ethnic, and socioeconomic variables. He concluded that the mean incidence of FAS is estimated to be 0.97 cases per 1,000 live births in the general obstetric population. However, the incidence appears to vary both between and within countries. The general incidence of FAS was found to be more than 20 times higher in the United States (1.95 per 1,000) than in other countries (0.08 per 1,000). In addition, within the United States, studies conducted at locations characterized by low socioeconomic status (SES) and with African-American or Native American populations, found incidences of FAS 10 times higher (2.29 cases per 1,000) when compared to reports from sites with a predominantly middle/upper SES and Caucasian background (0.26 per 1,000).

Similar to other teratogens, the adverse effects of gestational alcohol exposure are not observed in all exposed individuals. FAS is generally seen only among children of women exposed to "heavy" drinking. "Heavy" drinking is an ambiguous term, variably defined by studies as consuming an average of 2 or more drinks per day, or 5–6 drinks on some occasions (binge), or a positive Michigan Alcoholism Screening Test score, or a clinical diagnosis of alcohol abuse (Abel and Sokol, 1987). Within this subpopulation of "heavy" drinkers, Abel's review of the literature (1995) found the incidence of FAS to be 4.3 per 100 live births. When reviewing the epidemiologic data, one must keep in perspective that children diagnosed with FAS represent only a small fraction of all the individuals affected with FARA.

Various attempts have been made to assess the economic costs of FAS (Harwood and Napolitano, 1985; Abel and Sokol, 1991; Bloss, 1994). Such estimates are inherently problematic due to uncertainties related to the prevalence of FAS and to the extent of lifetime health costs and other problems experienced. This ambiguity is well reflected in American cost estimates, which range from $75 million (Abel and Sokol, 1991) to $9.7 billion (Harwood and Napolitano, 1985). Regardless of the discrepancies, the costs of FAS and other FARA are extremely high for the individual, for the family, for the education and health care system, and for society at large.

B. Criteria for Definition and Diagnostic Problems

Jones and Smith (1973) coined the term *fetal alcohol syndrome (FAS)* to describe children they observed born to mothers who were chronic alcoholics throughout gestation. They noted that the affected children had a "similar pattern of craniofacial, limb and cardiovascular defects associated with prenatal-onset growth and developmental delay" (p. 1001) (Fig. 1). Since the time of their original definition, the criteria for diagnosis of FAS has been based on the presence of a triad of features: (1) pre- and/or postnatal growth retardation (weight, length and/or height $<$ 10th percentile), (2) CNS damage (signs of neurologic abnormality, developmental delay, or intellectual impairment), and (3) characteristic facial dysmorphology (microcephaly, micropthalmia, short palpebral fissures, poorly developed philtrum, thin upper lip, and flattened maxillary area) (Abel, 1990).

Numerous problems have come to light that contribute to the difficulty in making the diagnosis of FAS. First, research has demonstrated that in individuals exposed to alcohol prenatally, the presence and/or degree of abnormalities can vary considerably. An affected individual may have an IQ ranging from the normal to the severely mentally retarded range, and may have physical features that range from normal to obvious anomalies. Correspondingly, a child exposed to alcohol prenatally may present with normal growth and physical features, but with slight or substantial behavioral and cognitive abnormalities. Second, the phenotype of FAS varies with age and may alter or normalize as the child moves through childhood into adolescence and adulthood [Institute of Medicine (IOM), 1996a]. Third, many of the individual deficits seen in FAS are pathognomonic for fetal alcohol exposure and similar deficits may be seen in isolation or in combination without any history of prenatal alcohol exposure. Consequently, when the phenotype is "incomplete" or "atypical" or the clinician is inexperienced in this diagnosis, FAS may be misdiagnosed. Syndromes confused with FAS due to similar physical features may include Aarskog syndrome, Williams syndrome, Noonan's syndrome, Dubowitz syndrome, Bloom syndrome, fetal hydantoin syndrome, maternal phenylketonuria fetal effects, and fetal toluene syndrome. Other syndromes that may be confused for FAS due to similarities in the cognitive and behavioral profiles present include fragile X syndrome, velocardiofacial syndrome, Turner's syndrome, Opitz syndrome, and attention deficit hyperactivity disorder (IOM, 1996b).

Investigators have recognized that the birth defect known as FAS is only the "tip of the iceberg" of prenatal alcohol-related disabilities (Streissguth, 1996a). This viewpoint considers those with full expression of FAS as only a very small subsection of all the individuals affected physically, developmentally, behaviorally, and/or cognitively by fetal alcohol exposure. Although ter-

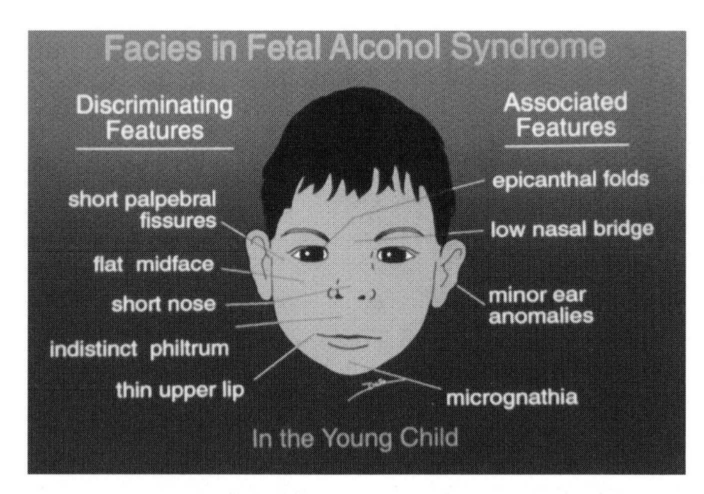

FIGURE 1 Facial morphologic abnormalities seen in a young child with fetal alcohol syndrome (FAS). (Reprinted with permission from the Project Cork Institute.)

minology such as *fetal alcohol effect, partial fetal alcohol effect,* and *alcohol-related birth defect* was originally developed to describe abnormalities found in animal studies, usage of these terms has appeared to clinically define those individuals with FARA without full expression of FAS. This has caused considerable confusion and various investigators believe the terms should be used only in the research and not in the clinical domain (Sokol and Clarren, 1989; Aase *et al.,* 1995).

Several attempts have been made to improve and clarify the criteria for diagnosis of affected individuals (Clarren and Smith, 1978; Rosett, 1980; Sokol and Clarren, 1989). The most recent and comprehensive advancement was undertaken by the Institute of Medicine of the National Academy of Sciences Committee to Study Fetal Alcohol Syndrome (IOM, 1996c). The committee attempted to resolve the issues confusing the clinical and research communities by delineating five diagnostic categories for FARA (Table 1). The first category contains those individuals with a diagnosis of FAS (triad of symptoms as previously mentioned) plus a confirmed history of maternal alcohol exposure (a pattern of excessive intake characterized by substantial, regular intake or heavy episodic drinking). The second group includes individuals with a diagnosis of FAS without a confirmed history of maternal alcohol exposure (accurate history not provided by birth mother, or inaccessible alcohol history for foster/adopted children). The third classification describes individuals as having partial FAS with confirmed maternal alcohol exposure. It includes those with confirmed alcohol exposure during gestation, some of the facial features of FAS, and evidence of growth, neurodevelopmental, behavioral, and/or cognitive abnormalities. Two additional categories were devised in which to place individuals with a history of maternal alcohol exposure plus evidence of conditions that have been noted in clinical or animal alcohol teratology research. Category four, alcohol-related birth defects (ARBD), designates those with physical anomalies, whereas category five, alcohol-related neurodevelopmental disorder (ARND), describes those with neurodevelopmental abnormalities and/or behavioral or cognitive abnormalities. A diagnosis of ARND represents a spectrum of FARA that may have as severe consequences for the patient as FAS.

C. Dysmorphology

As outlined previously, FARA is characterized by a triad of features: pre- and/or postnatal growth deficiency, visceral defects, specific craniofacial anomalies (Figs. 2, 3), and central nervous system (CNS) dysfunction. Although growth deficiency and craniofacial anomalies may be of considerable importance in making the clinical diagnosis of FARA, it is the structural and functional brain anomalies that ultimately are the most disabling to the individual. An overview of neuropathologic investigations using ultrasound (Ronen and Andrews, 1991), computed tomography (CT) (Goldstein and Arulanantham, 1978), magnetic resonance imaging (MRI) (Gabrielly *et al.,* 1990; Schaefer *et al.,* 1991; Mattson *et al.,* 1992; Coulter *et al.,* 1993; Robin and Zackai, 1994), positron emission tomography (PET) (Hannigan *et al.,* 1995), and autopsy (Clarren *et al.,* 1978; Clarren, 1979, 1981; Peiffer *et al.,* 1979; Wisniewski *et al.,* 1983) have clearly demonstrated the susceptibility of the brain to the teratogenic insults of alcohol. The affected brain typically is reduced in volume, (Figs. 4, 5) and certain regions (e.g., basal ganglia, corpus callosum, anterior vermis of cerebellum) show a disproportionately large

TABLE 1 Diagnostic for Fetal Alcohol Syndrome (FAS) and Alcohol-Related Effects (10 *M,* 1996)

Fetal Alcohol Syndrome

1. FAS with confirmed maternal alcohol exposure

A. Confirmed maternal alcohol exposure
B. Evidence of a characteristic pattern of facial anomalies that includes features such as short palpebral fissures and abnormalities in the premaxillary zone (*e.g.,* flat upper lip, flattened philtrum, and flat midface).
C. Evidence of growth retardation, as in at least one of the following:
—low birth weight for gestational age
—decelerating weight over time not due to nutrition
—disproportional low weight to height
D. Evidence of CNS neurodevelopment abnormalities, as in at least one of the following:
—decreased cranial size at birth
—structural brain abnormalities (*e.g.,* microcephaly, partial or complete agenesis of the corpus callosum, cerebellar hypoplasia)
—neurological hard or soft signs (as age appropriate), such as impaired fine motor skills, neurosensory hearing loss, poor tandem gait, poor eye-hand coordination

2. FAS without confirmed maternal alcohol exposure

B, C, and D as above

3. Partial FAS with confirmed maternal alcohol exposure

A. Confirmed maternal alcohol exposure
B. Evidence of some components of the pattern of characteristic facial anomalies
Either C or D or E
E. Evidence of a complex pattern of behavior or cognitive abnormalities that are inconsistent with developmental level and cannot be explained by familial background or environment alone, such as learning difficulties; deficits in school performance; poor impulse control; problems in social perception; deficits in higher level receptive and expressive language; poor capacity for abstraction or metacognition; specific deficits in mathematical skills; or problems in memory, attention, or judgment.

4. Alcohol-related birth defects (ARBD)

List of congenital anomalies, including malformations and dysplasias

Cardiac	Atrial septal defects	Aberrant great vessels
	Ventricular septal defects	Tetralogy of Fallot
Skeletal	Hypoplastic nails	Clinodactyly
	Shortened fifth digits	Pectus excavatum and carinatum
	Radioulnar synostosis	Klippel–Feil syndrome
	Flexion contractures	Hemivertebrae
	Camptodactyly	Scoliosis
Renal	Aplastic, dysplastic, hypoplastic kidneys	Ureteral duplications Hydronephrosis
	Horseshoe kidneys	
Ocular	Strabismus	Refractive problems secondary to small globes
	Retinal vascular anomalies	
Auditory	Conductive hearing loss	Neurosensory hearing loss
Other	Virtually every malformation has been described in some patients with FAS. The etiologic specificity of most of these anomalies to alcohol teratogenesis remains uncertain.	

5. Alcohol-related neurodevelopmental disorder (ARND)

Presence of:

A. Evidence of CNS neurodevelopmental abnormalities, as in any one of the following:
—decreased cranial size at birth
—structural brain abnormalities (*e.g.,* microcephaly, partial or complete agenesis of the corpus callosum, cerebellar hypoplasia)
—neurological hard or soft signs (as age appropriate), such as impaired fine motor skills, neurosensory hearing loss, poor tandem gait, poor eye-hand coordination

and/or:

B. Evidence of a complex pattern of behavior or cognitive abnormalities that are inconsistent with developmental level and cannot be explained by familial background or environment alone, such as learning difficulties; deficits in school performance; poor impulse control; problems in social perception; deficits in higher level receptive and expressive language; poor capacity for abstraction or metacognition; specific deficits in mathematical skills; or problems in memory, attention, or judgment.

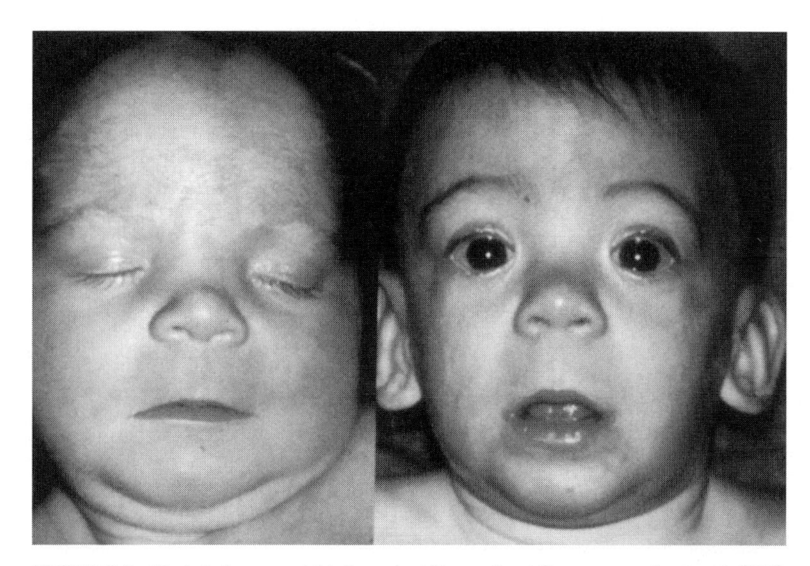

FIGURE 2 Facial changes at birth and at 8 month at the same patient with FAS. The child has microcephaly, low-set ears, widely spaced eyes, short palpebral fissure, short upturned nose, long and indistinct philtrum, thin upper lip, and micrognathia. [Printed with permission from Streissguth A. P., and Little, R. E. (1994). Unit 5: Alcohol Syndrome: Second Edition, of the Project Cork Institute Medical School Curriculum (slide lecture series) on *Biomedical Education: Alcohol Use and Its Medical Consequences,* produced by Dartmouth Medical School.]

decrease in size. Other abnormalities seen, including cerebral dysgenesis, enlarged ventricles, and abnormal neural/glial migration, indicates that alcohol-related brain anomalies in the fetus are structurally and functionally different than those seen in similarly developmentally disabled, functionally impaired, microcephalic children (Mattson *et al.,* 1994; Riley *et al.,* 1995). It is now apparent that a unique pattern of brain anomalies is attributable to prenatal alcohol exposure. Although growth deficiencies may normalize and craniofacial abnormalities may become less distinct with age, it is the brain injuries and their corresponding disabilities en-

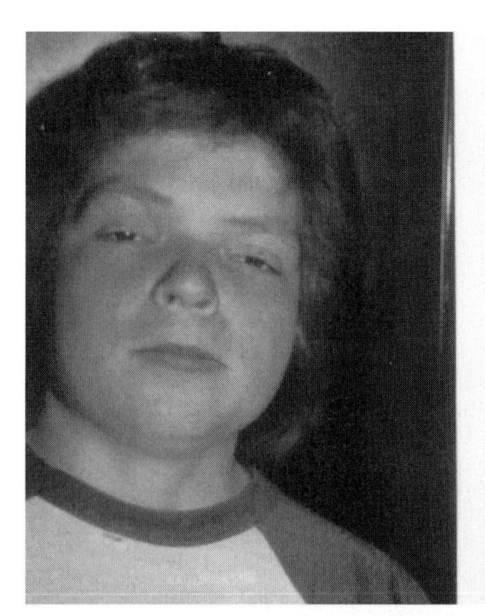

FIGURE 3 Face and body build in adult female with FAS. [Printed with permission from: Streissguth A. P., and Little, R. E. (1994). Unit 5: Alcohol Syndrome: Second Edition, of the Project Cork Institute Medical School Curriculum (slide lecture series) on *Biomedical Education: Alcohol Use and Its Medical Consequences,* produced by Dartmouth Medical School.]

FIGURE 4 Microcephaly and normal head circumference in children with FAS. [Printed with permission from: Streissguth A. P., and Little, R. E. (1994). Unit 5: Alcohol Syndrome: Second Edition, of the Project Cork Institute Medical School Curriculum (slide lecture series) on *Biomedical Education: Alcohol Use and Its Medical Consequences,* produced by Dartmouth Medical School.]

countered that are the hallmark of and the most unfortunate aspects of FARA.

Normal sensory input is essential for higher cortical functions performance. Therefore, early impairment of sensory pathways may contribute to abnormal intellectual development. Research into the developing visual system has shown that sensory impairment during the critical period of postnatal maturation interferes with normal development of visual perception, neural synapses, myelination, cell size, number, and organization (Globus and Scheibel, 1967; Wiesel, 1982).

A wide range of ophthalmologic abnormalities has been found to be associated with FAS, including microphthalmia, strabismus, visual impairment, short horizontal palpebral fissure, blepharoptosis, microcornia, corneal opacity, iris defects, cataracts, glaucoma, persistent hyaloid, and combinations of these abnormalities (Stromland, 1981, 1985; Chan *et al.,* 1991).

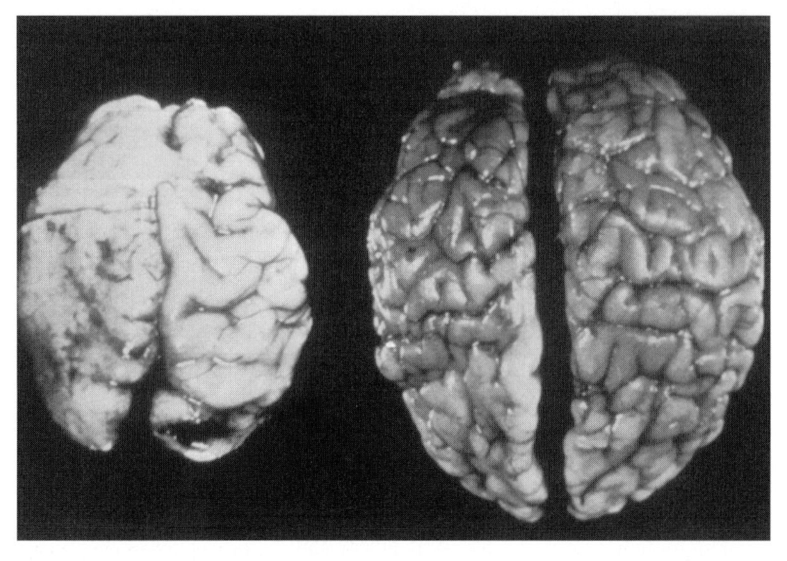

FIGURE 5 Two human brains. Normal brain (right) of an infant who died from causes other than FAS. (Left) Brain of an infant with FAS. [Reprinted with permission from the Project Cork Institute.]

Stromland and Hellstrom (1996) have suggested the eye to be a sensitive indicator of adverse effects of prenatal alcohol exposure. Optic nerve hypoplasia (the most common anomaly in their studied cohort) and a large number of other ophthalmologic signs were found to be persistent during 11 years of follow-up.

Similar observations have been made concerning sensory deprivation in the developing auditory system. FAS children were found to be at high risk for a variety of hearing disorders: delay in maturation of the auditory system, congenital sensorineural hearing loss, conductive hearing loss secondary to recurrent otitis media, and central hearing disorders (Church and Holloway, 1984). Hearing disorders are strongly associated with craniofacial anomalies, mental impairment, and ocular defects (Northern and Down, 1984).

Adequate hearing is necessary for proper speech, language, and intellectual development. A child with hearing loss is more likely to exhibit hyperactivity, distructability, and learning disabilities (Northern and Down, 1984). FAS is one of the most common causes of childhood hearing, speech and language disorders (Church and Holloway, 1984). However, by far, neurobehavioral sequellae of FARA are the most devastating in terms of the affected child to integrate into society appropriately.

In a prospective, longitudinal study focusing on the moderate amount of alcohol ingested by "social drinking," Streissguth *et al.* (1996b) looked at deficit in growth, morphology, and functional–teratogenic outcomes that were found in children with FAS. The findings of the study were in accordance with the experimental animal literature.

At that relatively moderate level of alcohol exposure, it is neurobehavioral function that displays the most reliably enduring and most lasting adverse effects of prenatal alcohol exposure. These neurobehavioral effects were present from birth through 14 years of age and were dependent on the level of alcohol consumed, generally with no threshold level. During the school-age years, the effects were associated more with binge-type drinking patterns. Drinking before pregnancy recognition was shown to have more adverse outcomes than drinking in midpregnancy, but the two were highly correlated. Slower information processing and attention problems were apparent throughout the 14 years, whereas learning difficulties were noted from ages 7 through 14 years. These effects were observed even in children without physical or facial features associated with FAS. To date, no method exists to quantify the brain damage caused by alcohol and its relation to dysfunctional behavior in an affected child. The behavior of children with FARA varies widely, and many other exposed offspring who do not exhibit the full bow of FAS diagnosis show neurobehavioral deficits that are as severe as those of FAS.

There was no statistical evidence of a "risk-free" drinking level or any threshold level for prenatal alcohol exposure in the dose–response analysis made by Streissguth *et al.* (1993). Attention, memory, speed, and reliability of information processing, arithmetic functioning, and phonologic processing were the most affected areas, and the neurobehavior abnormalities were not affected by birthweight and were "highly significant statistically despite the absence of postinfancy effects of prenatal alcohol on child weight and height" (p. 142).

V. Risk Factors in the Development of Alcohol-Related Abnormalities in the Fetus

The fact that not all children exposed to gestational alcohol consumption develop FARA and the reality that an increased incidence of FAS is associated with low SES and African-American and Native American communities suggest that other risk factors may be interacting with maternal alcohol exposure that further compound the risk to the fetus. Such risk factors may include poor maternal health, nutrition, and prenatal care; substance abuse; biologic susceptibility; and the dose, timing, and pattern of alcohol exposure. To complicate the research into such effects, risk factors tend to cluster in women who drink heavily.

Gladstone *et al.* (1997) showed that women in Ontario who had engaged in first-trimester binge alcohol consumption were significantly more likely to be single, young, to smoke cigarettes, to use cocaine, marijuana, and other illicit drugs.

Abel (1995) proposed two categoric types of maternal risk factors: permissive factors—predisposing behavioral, social, or environmental factors (e.g., alcohol consumption pattern, SES); and provocative factors—the biologic condition (e.g., high BAC, decreased antioxidant status) that increase fetal vulnerability to alcohol at the cellular level.

A. *Pattern of Drinking and Peak Blood Alcohol Concentration*

Numerous studies have been undertaken to elucidate the effects of mild, moderate, and heavy maternal alcohol consumption. Definitions of drinking patterns typically represent average daily, weekly, or monthly alcohol intake level. Unfortunately, the definitions of mild, moderate, and heavy vary considerably among studies, making generalizations and comparisons extremely difficult.

It is now evident that the pattern of drinking is a critical factor in predicting future teratogenic effects. Abel and Hannigan (1996) summarized this research by claiming that drinking patterns in terms of drinks or amount of absolute alcohol per week during pregnancy is not meaningful, given that blood alcohol level is a critical factor. There is a large body of evidence from rodent (Sulik *et al.,* 1981, 1984, 1986; Pierce and West, 1986; West *et al.,* 1990), primate (Clarren and Bowden, 1982; Clarren *et al.,* 1987, 1990, 1992; Clarren and Astley, 1992), and human models (Streissguth *et al.,* 1989a, 1989b, 1990, 1993, 1994a, 1994b), indicating that a single binge exposure or period of binge exposures is embryotoxic, fetotoxic, and sufficient to produce a wide spectrum of physical and neurodevelopmental deficits. Consistent findings have demonstrated that alcohol's teratogenic effects are related to maternal peak BAC obtained, and are not simply reflection of the total volume of alcohol consumed. Peak BAC is determined by both the dose and the rate of the alcohol consumption. Alcohol consumed in a binge pattern is capable of rapidly producing higher BACs that require a longer time for the fetus to clear than equal (or even greater) daily amounts that are consumed in a slower or more spread out manner. This observation of a "peak effect" is critical in setting a mechanistic framework for binge drinking, because it suggests that a critical threshold of BAC level may be required to induce neurotoxicity. Consequently, this may explain some of the large variability in the prevalence of FAS in the offspring of alcoholics. A very different pattern of BAC would be achieved by two different types of alcoholics: one who spreads 10 drinks throughout the day from morning to evening, and one who binges on 10 drinks in the evening (Clarren *et al.,* 1992; Gladstone *et al.,* 1996).

From the public health and prevention perspectives, there are several important points to consider regarding the deleterious effects of drinking in a binge pattern during pregnancy. It has been documented that most alcoholic women decrease their consumption of alcohol during pregnancy; nevertheless, when they do drink, they tend to binge (Little and Streissguth, 1978). Another area of concern is the high association between alcohol intoxication and unplanned or unprotected sexual activity (Wechsler and Isaac, 1992; Meilman, 1993; Wechsler *et al.,* 1994; Parker *et al.,* 1994). The fear is that women who engage in a binge style of alcohol consumption may increase their likelihood of unplanned pregnancy, and may then consequently unknowingly expose the fetus to continued binge until pregnancy is diagnosed. This is particularly alarming in the teenage population given their high rate of binge drinking (Johnston *et al.,* 1991; Kusserow, 1991), the rising rate of teenage pregnancy (National Centre for Health Statistics, 1990), and the tendency of adolescents to recognize their pregnancies later than adults (Cornelius *et al.,* 1994). These factors may combine to make the offspring of teenagers significantly more susceptible to FAS (Gladstone *et al.,* 1996).

B. *Effect of Timing on Fetal CNS Development*

Another key risk factor that has been shown to hold for both structural alterations and functional impairments is the timing of alcohol exposure. There are particular periods during gestation when the fetus and its CNS are most susceptible to the harmful effects of alcohol. Although, at present the exact temporal windows of fetal vulnerability are not known, there is some evidence of critical periods for a number of FAS expressions.

First trimester alcohol exposure is critical for organogenesis and for distinctive FAS facial dysmorphology (Sulik *et al.,* 1981, 1986). This is alarming because 50% of North American pregnancies are unplanned (Skrabanek, 1992), and a majority of women may not be aware of their pregnancy during the first 4–6 weeks of gestation. In light of this information, women who abuse alcohol should be advised of the risk of facial anomalies if alcohol exposure proceeds into the first trimester.

Alcohol exposure during gestation is strongly associated with intrauterine growth retardation (IUGR) and somatic growth defects (Ernhart *et al.,* 1985). IUGR is a term that describes fetuses or neonates that are small in weight, length, and head size for their gestational age. IUGR is not a transitory effect and growth to normal levels ("catch-up") is rare (Day *et al.,* 1991a,b). Smith (1986) reported that women who discontinued drinking before the end of the second trimester delivered babies that were not different in weight, length, and head circumference from unexposed infants. Microcephaly is associated with alcohol consumption throughout pregnancy (all three trimesters), this finding being supported by clinical and experimental evidence, and cessation of drinking after the second trimester may show improvement in head growth (Smith, *et al.,* 1986).

Alcohol exposure during the second and third trimesters produces neuronal loss (Miller and Potempa, 1990), altered neuronal circuitry (West and Hamre, 1985), and extensive gliosis (Goodlett *et al.,* 1993). The third trimester is an especially vulnerable period for the brain in response to alcohol insult (West *et al.,* 1986; Marcussen *et al.,* 1994), and a single exposure to high BACs produced a substantial loss of Purkinje cells (Goodlett *et al.,* 1990).

C. *Socioeconomic Status*

Low SES, a major permissive factor contributing to FAS, is an indicator for poor maternal nutrition, health, decreased access to prenatal care, and increased maternal stress. Each of these may independently and adversely affect pregnancy outcome (Abel and Hannigan, 1996). Although FAS occurs in all races, it is typically the poorer populations who have the largest numbers of FAS (Abel, 1995). It is likely that in children affected by FAS born to low SES families and lack of basic infant needs, including nutrition, stimulation, and education, act in a negative synergistic way to augment the damage caused by neurotoxicity.

D. Genetic Differences

Genetic variability in factors such as maternal rate of metabolism of alcohol, the rate of transport of nutrients across the placenta, and uterine blood flow may play a role in influencing the risk and the severity of alcohol-induced damage to the fetus. Genetic differences were found in neural sensitivity of the developing brain in such basic parameters as brain and cerebellar weight (Goodlett *et al.,* 1989).

Genetic factors influencing vulnerability to FAS may include differences in the various isoforms of ADH, although there is no convincing evidence for a racial predisposition to FARA based on enzymatic isoforms (Bosron *et al.,* 1980, 1983).

E. Polydrug Use

Women who abuse alcohol very commonly abuse other illicit drugs (Coles *et al.,* 1985; Kokotailo *et al.,* 1992; Gladstone *et al.,* 1997). Two examples for such synergistic effects may be forwarded: cocaine in combination with alcohol produces cocaethylene, an exceptionally active neurotoxic substance (Hearn *et al.,* 1991). Another example is the combination of maternal alcohol consumption and cigarette smoking, which increases the risk for low birth weight, microcephaly, and hearing difficulties (Wright *et al.,* 1983; Olsen *et al.,* 1991).

VI. Possible Mechanisms Contributing to the Teratogenic Effects of Alcohol

It is not possible at present to detail fully the mechanism with which ethanol exerts its effects during each stage of development. It has been postulated that alcohol-induced CNS abnormalities in the fetus are mediated through multiple mechanisms (West *et al.,* 1994). Prenatal alcohol exposure leads to disruption of most areas of brain development (West *et al.,* 1994).

A number of studies reported alterations in proliferation, neuronal migration, dendritic growth, conductivity among neurons, and even neuronal death (Miller 1986, 1993; Marcussen *et al.,* 1994).

Alcohol also has a number of detrimental effects on nonneuronal elements, and may affect indirectly brain function by altering the development of microvasculature in the cerebellum and hippocampus (Kelly *et al.,* 1990).

As ethanol crosses the placenta and the blood–brain barrier, it may affect the fetus directly or indirectly through secondary effects, leading, for example, to constriction of umbilical vessels and resulting in insufficient oxygen supply (Mukherjee and Hodgen 1982).

There is still concern whether alcohol *per se* or acetaldehyde is the teratogen associated with alcohol-induced brain damage in the fetus because heavy alcohol consumption typically results in high levels of both alcohol and acetaldehyde. Jones *et al.* (1991) postulated that inhibition of aldehyde dehydrogenase, the enzyme that converts acetaldehyde to acetate, exacerbates the effects of alcohol on the fetus by raising the levels of acetaldehyde.

Alcohol interferes with brain cell metabolism by decreasing protein synthesis and DNA methylation, mechanisms most commonly used to explain fetal growth retardation.

DNA methylation has been shown as one of the mechanisms involved in gene activity and function. Embryonic DNA is highly methylated, and alcohol-induced inhibition of fetal DNA methylation may be associated with the teratogenic effects of alcohol described in FAS (Garro *et al.,* 1991). Garro *et al.* (1991) also demonstrated that acetaldehyde is a 0^6-methylguanine transferase that has an important role in DNA repair activities. Espina *et al.* (1988) hypothesized that hypomethylation alters gene expression and may be responsible for the developmental abnormalities observed in FAS.

A. Free Radical Damage

An alternate explanation for alcohol damage has been free radical–induced mitochondrial damage. A study in fetal rat hepatocytes demonstrated that exposure to ethanol inhibits mitochondrial function, and thus creates oxidative stress in the hepatocyte and a subsequent drop in adenosine triphosphate (ATP) levels (Devi *et al.,* 1994). The excess amount of the intermediate of oxygen reduction, 0_2, and other short-lived reactive oxygenated free radicals were found to be associated with anomalies observed in FAS.

Free radicals are molecules with one or more unpaired electrons that are known to be highly unstable, reactive, and cytotoxic (De Groot and Littaner 1989; Nordmann *et al.,* 1992).

Radicals are normally scavenged by endogenous antioxidative enzymes; alterations of such balance by ethanol may increase oxidative stress (Nordmann *et al.,*

1992), which is highly damaging to cells. These disruptive mechanisms may affect proteins, lipids, chromosomes, and specific receptors.

Due to peroxidation, lipids may become free radicals themselves, and then may enhance chain reaction and membrane decomposition, manifested by changes in membrane fluidity and "leaking," levels of phospho-glyco-lipid compositions, and decrease in activity of calcium-, Na-, and K-ATPase, all of which have been shown in animal fetuses exposed to ethanol prenatally (Burmistrov *et al.,* 1991; Murdoch and Edwards, 1992; Arienti *et al.,* 1993). Gangliosides, which are actively involved in cellular migration, cell–cell interaction, neurite outgrowth, and other biologic processes, were found to be decreased in their activities after alcohol administration. Alcohol adversely interferes with interaction of water, membrane proteins, and gangliosides, leading to a new hydrogen bond and membrane fluidization. In animals, prenatal administration of gangliosides was found to antagonize some of the harmful effects of prenatal alcohol to the fetus (Klemm, 1990; Hungund and Mahadik, 1993). Concentration of gangliosides and of ganglioside-specific enzymes increase dramatically during brain development; therefore it has been suggested that prenatal alcohol exposure may affect the amount of gangliosides and their activity, thus contributing to the formation of FAS.

Alcohol-induced cellular damage due to oxygen radicals may also occur independently at ischemia or hypoxia. Neural crest cells, which are devoid of superoxide dismutase, an enzyme that catalyzes the conversion of 0_2 into H_20_2 and 0_2 and protects tissues against the deleterious effect of 0_2, are extremely sensitive to alcohol exposure (Davis *et al.,* 1990). This sensitivity may account for facial and visceral malformations because those structures derive from neural crest cell (Davis *et al.,* 1990).

The CNS may be more vulnerable than other organs to alcohol exposure because of its high dependence on uninterrupted blood flow, high content of polyunsaturated fatty acids, and relatively low levels of free radical scavenging enzymes and antioxidants (Abel, 1995).

Micronutrients such as iron, zinc, selenium, manganese, riboflavin, niacin and tryptophan (β-carotene and vitamins E and C) are needed for the free radical scavenging enzymatic defense mechanisms. They stabilize free radicals and inhibit their activities.

Alcohol abuse is associated with increase of cellular iron. The involvement of iron in free radical formation increases the potential for cellular damage, especially in the brain, where lipid peroxidation is very rapid (Nordmann *et al.,* 1992).

Zinc is known to be essential in the synthesis of protein, DNA, RNA, critical for cell duplication, and is a cofactor in enzymes involved in free radical defense mechanisms such as superoxide dismutase. Maternal alcohol intake can reduce fetal zinc levels and reduce the activity of superoxide dismutase. Zinc deficiency during development is teratogenic and, in combination with prenatal alcohol exposure, may interact synergistically to reduce fetal body and brain weight (Dreosti, 1993).

B. Acetaldehyde

Acetaldehyde production may also cause organic damage as it is an extremely reactive molecule and affects most tissues in the body. Although most of it is converted to acetate some enters the blood stream and reaches a plateau when both the alcohol dehydrogenase and cytochrome P450 systems are saturated. Acetaldehyde forms stable adducts with amino acid residues in proteins. These adducts are antigenic, giving rise to anti-adduct antibodies that can react with hepatocyte surface antigens causing the destruction of liver cells. It is important to note that alcoholics achieve significantly higher acetaldehyde plateaus than nonalcoholics, even when the same blood–alcohol level is attained. This is most likely due to the induction of the p450 system in alcoholics, resulting in a faster conversion of ethanol to acetaldehyde.

C. Prostaglandins

Alcohol causes both a direct and an indirect increase in prostaglandins in fetal tissues (Collier *et al.,* 1975). Increased prostaglandin levels lead to increased cyclic adenosine monophosphate (AMP) levels, which in turn may reduce the rate of cell division (Pastan *et al.,* 1975). In the brain, this process may interfere with stem cell division during neuronal proliferation. Following alcohol exposure in utero, decreased brain weight has been observed in fetuses with increased prostaglandin E and cAMP levels. Inhibiting prostaglandin production (by injecting acetylsalicylic acid prior to ethanol introduction) halves the number of defective offspring in animal models (Pennington, 1988).

D. Amino Acid Transport

An additional proposed mechanism links ethanol-induced growth retardation to fetal malnourishment secondary to ethanol's interference with placental essential amino acid transport (Lin, 1981). Whereas this could explain the fetal growth retardation commonly seen in FAS, when ethanol is administered directly to rat fetuses, thereby circumventing the placenta, malformation and growth retardation can still be documented (Brown *et al.,* 1979).

E. Fetal Hypoxia

There is evidence that alcohol-related impairments of essential amino acids (Fisher *et al.,* 1985), glucose (Snyder *et al.,* 1986) or vitamins and minerals (Schenker *et al.,* 1992) may be partly caused by impaired functioning of the oxygen-dependent Na-K ATPase membrane transport in the presence of hypoxia (Fisher *et al.,* 1986). Ethanol metabolism increases the liver's demands for oxygen to metabolize alcohol resulting in oxygen deprivation (Urgarte and Valenzuela, 1971). This relative hypoxia may be further intensified by ethanol-induced contraction of umbilical arteries and veins, and impaired oxygen unloading from hemoglobin by acidification of the blood (Yang *et al.,* 1986).

It has been suggested that episodes of acidosis and hypoxia are operative in impairing neurologic functioning of children with FAS. However, the theory of ethanol-induced hypoxia has been challenged, showing in pregnant ewes that maternal ethanol infusions actually produced a dose-dependent increase in uterine blood flow and fetal arterial oxygen pressure (Reynolds *et al.,* 1996).

F. Inhibition of Cell–Cell Adhesion

Another mechanism proposed for alcohol-induced damage is the sensitivity of L1-mediated neural cell adhesion molecules to ethanol. The L1 gene encodes for an essential cell membrane protein that helps the neuronal membranes stick to each other and to their extracellular matrix. Researchers have noted a startlingly similar picture of defects in people with FAS and in those with a rare genetic mutation in the cell adhesion molecule L1; mental retardation, hydrocephalus, and agenesis of the corpus callosum have all been observed in children with L1 mutations and with fetal alcohol syndrome (Ramanthan *et al.,* 1996). Subsequent rat studies have demonstrated that ethanol nearly completely inhibited the stickiness of cell adhesion molecules at a blood–alcohol level of 0.08%, a level defining legal intoxication in many U.S. states (Ramanthan *et al.,* 1996). The authors of this study concluded that, L1 plays a role in both neural development and learning. Ethanol inhibition of L1-mediated cell–cell interactions could contribute to FAS and ethanol-associated memory disorders.

G. Long-Term Potentiation

Finally, alcohol has been observed to disrupt long-term potentiation (LTP), a phenomenon that many neuroscientists believe is a prerequisite to memory and learning. LTP refers to long-lasting increases in the strength of synapses between neurons. In a study by Savage and Sutherland (Broun, 1996), pregnant rats were treated with alcohol equivalents of maternal consumption of 2–3 drinks per day. The adult offspring of the ethanol-treated rats were later tested on a series of maze tests. Interestingly, they performed equally well as controls on standard water-maze learning tests, but were strikingly worse when presented with a more difficult variation of the test. Whereas controls learned the new maze after just one trial, the offspring of alcohol-ingesting mothers required seven or eight trials to learn the new maze. The brains of the ethanol-exposed offspring were later examined and it was found that their neurons showed markedly reduced LTP in the hippocampus, an area essential for memory formation. It was also discovered that these neurons failed to release an important neurotransmitter when faced with changing stimuli, which is a sign that the neurons had lost the plasticity required for learning.

This last theory may prove very important in advancing knowledge regarding the mechanism of FARA because to date it proposes to explain how moderate drinking can induce teratogenesis. How much alcohol is considered damaging during pregnancy has never been definitively established because a distinct model of teratogenesis for moderate doses of alcohol is lacking.

VII. Primary and Secondary Disabilities

The majority of scientific assessments have been focused on studying preschool groups of children with gestational alcohol exposure. "FAS is not just a childhood disorder" (Streissguth *et al.,* 1991, p. 1967) and prenatal exposure to alcohol can cause a plethora of abnormalities and disabilities that have lifelong physical, mental, behavioral, and social consequences.

To address the long-term outcome of gestational alcohol, Streissguth *et al.* (1996c) define as primary disabilities those that reflect the FAS or ARND diagnosis. Secondary disabilities are those that an individual is not born with and could presumably be prevented through better understanding and appropriate intervention.

Primary disabilities associated with FAS or FAE (defined by this research group as some but not all the features of FAS) were examined in 473 individuals aged 3–51 years by presenting a wide spectrum of cognitive, behavioral, and language tests.

Patients with FAS (n = 178) had an average IQ of 79 and those with FAE (n = 295) had an average IQ of 90, but the adaptive behavior score was very low in both subgroups (61 and 67, respectively). The FAE individuals were presented with better cognitive abilities compared with those with FAS, but their behavioral functions, especially social adaptability, did not differ.

In this study, mental health problems were found in 90% of the sample. Disrupted school experience and trouble with the law was noted in 60% of assessed individuals. Confinement and inappropriate sexual behavior was noted in 50%, alcohol/drug problems was noted in 30%, and dependent living and problems with employment was noted in 80% of studied population.

Rates of secondary disabilities were nearly equal across the sexes. A diagnoses before 6 years of age was found to be a strong protective factor for all secondary disabilities (except mental health). Another protective factor was to be diagnosed as FAS rather than FAE. It is the combination of better cognition with poor social adaptability that leads individuals with FAE to disruptive school experience, inappropriate sexual behavior, drug problems, delinquency, dependent living, and unemployment. The determination of protective factors points on the importance of early diagnoses of FAS/FAE, and shows that patients with ARND abnormalities have more chances for serious secondary disabilities in their future.

Early diagnosis with proper intervention may change the appearance and course of the secondary disabilities as opposed to the primary disabilities, which most probably are not influenced by intervention.

VIII. Mental Health Problems

From the beginning of life, many FAS children start to develop medical complications, developmental delays, and psychiatric symptoms. Although hyperactivity and attention deficit are common problems both during preschool and school age (Iosub *et al.,* 1981; Majewski and Majewski, 1988), psychopathology is not restricted to these core symptoms. In the preschool period, eating disorders, enuresis, speech delay, and stereotypes also occur. Later, during early school age, problems such as speech delay and stereotypes are even more common, and problems such as anxiety or sleep disorders emerge. Moreover, there is an enormous variety of symptoms and high prevalence for any psychopathology. Steinhausen (1982, 1991a,b, 1996) showed that 63% of FAS children suffered from a single or (more commonly) more than one psychiatric disorder.

Mental health problems were found to be the most prevalent secondary disability recorded by Streissguth (1996c). Ninety percent (426) of 473 assessed patients presented with one or more psychiatric conditions. The most frequent mental health problems for children and adolescents in this group were attention deficit (61%), depression (50%), suicide threats (43%), and psychotic symptoms in 29%.

Evidence shows that severity of morphologic damage, psychopathology, and mental retardation tend to coincide in a subgroup of severely affected FAS children, who often come from extremely deprived backgrounds that include chronic maternal alcoholism and very often also paternal alcoholism. No study has yet tried to disentangle the effects of the teratogenic and environmental risk factors on the child's development, although the longitudinal studies have hinted that even a stimulating environment with sensitive parents or a good institution may not sufficiently compensate for prenatal damage due to alcohol exposure.

Although mental retardation and cognitive deficits are overrepresented among FAS children, mental functioning varies widely; many of these children function normally at school. It is interesting that, in contrast to other reports, Steinhausen's (1993) analysis shows no linear relation between degree of morphologic damage and intelligence. This relation may level off with increasing age or reflect the fact that dysmorphic features may be a crude measure of morphologic damage, especially of the brain.

Persisting mental impairment and psychiatric disorders cause serious problems for many FAS adolescents and young adults, and a considerable proportion of these patients remain dependent on support from others.

IX. Prenatal Alcohol Exposure and Attachment

At birth, there are signs of CNS dysfunction in infants born to mothers who report drinking large quantities of alcohol during pregnancy. These neonatal effects include irritability, autonomic instability, decreased sucking response, motor immaturity, slow habituation, low levels of arousal, distorted sleep patterns, and withdrawal symptoms (Streissguth *et al.,* 1983; Coles *et al.,* 1984, 1985). High-pitched crying, disturbed sleep, and feeding difficulties often follow withdrawal symptoms and may persist for days and weeks (Coles and Platzman, 1993). Behavioral difficulties may continue into the preschool period, with difficulties in cognitive functioning and sustained attention, emotional instability, increased activity level, rigidity, and irritability (Landesman-Dwyer *et al.,* 1981).

These neurobehavioral effects may have a significant impact on the mother–infant interaction and future attachment relationships (Meares *et al.,* 1982). Thus, the effects of alterations in infant behavior on infant attachment may be the most significant result of prenatal exposure to alcohol. O'Conner *et al.* (1992), using a causal modeling procedure and two alternative models, proposed that alcohol consumption following pregnancy was directly related to the mother's interaction with her child and resulted in a negative affective response in the child and in insecure attachment. One of the models tested the hypothesis that three independent and direct paths could be drawn between prenatal drinking and infant negative affect, maternal behavior, and attachment behavior, respectively. That model was based on the possibility that alcohol consumption affected mother and infant independently.

The results were that this group contained a high number of disorganized infants (32%) and that the mothers of these infants were the heaviest drinkers. Mothers who drank more had infants who displayed more negative affect in interaction, and expressed insecure attachment behavior. The mothers of these infants were less stimulating in the interaction process. Black and associates (1986) described children of alcoholics as ignoring, withdrawing, and avoiding conflict. These children were self-reliant and unable to trust other people when they needed help (Cork, 1979); they grew up perceiving adults as uncaring and insensitive. Research suggests a possible link between insecure attachment in infancy and subsequent child behavior problems (Lewis *et al.,* 1984; Erickson *et al.,* 1985; Crowell and Feldman, 1988), thus highlighting the need to examine pathways for later maladaptation. Maternal, emotional, and social aspects associated with the mother's drinking conspire to weaken the maternal–infant bond.

Future research should focus on studying the mechanisms of secure attachment and positive mother–child relationship because a better understanding of this complex process would lead to early intervention before the primary attachment relationship became disturbed.

X. Preventing Alcohol-Related Abnormalities in the Fetus

The recognition that prenatal alcohol exposure is associated with long-term physical, cognitive, behavioral, and social disabilities calls for cultural, sociological, medical, and public health interventions to prevent FARA.

Prevention of fetal alcohol effects, clearly the first line of defense against the effects of prenatal alcohol disorders, should be directed at different levels (IOM Committee to Study FAS, 1996d; Loebstein, 1997).

The primary level of prevention includes universal prevention, which strives to ensure that society as a group is aware of the hazardous consequences of drinking alcohol, particularly during pregnancy, and that abstinence before conception and throughout pregnancy is a prudent choice. Selective prevention intervention targets women who are at greater risk (i.e., women of reproductive age who drink more than occasionally). Whereas prevention and treatment of maternal alcohol abuse are extremely difficult and often unsuccessful, a more tangible measure of primary prevention of fetal alcohol exposure is effective contraception.

On the level of secondary prevention, when the diagnosis of heavy drinking in early pregnancy has been made, physicians should discuss the attendant fetal risk with the woman and her family in the same manner that other life-long risks are communicated. In some cases, women may choose to terminate the pregnancy. Although such decisions remain the responsibility of the woman, the physician has a major obligation to inform her accurately of fetal risk. It is important that everything possible be done to ensure discontinuation of drinking if the woman chooses to continue with the pregnancy, and to ensure successful follow-up after delivery.

On the level of tertiary prevention, the medical professionals should intervene as early as possible with FAS and ARND children in an attempt to prevent the development of secondary disabilities. Appropriate screening tools, including biomarkers of alcohol exposure, to identify drinking women and affected children should be implemented.

Training programs for physicians in the identification of FARA and development of centers with diagnostic capability, including neurodevelopmental testing, are essential for early diagnosis and future management of affected children in an attempt to maximize the child's postnatal development and long-term functioning. Alcohol-related fetal effects are the leading cause of mental delay and other forms of congenital brain injury. They can be prevented completely. Until appropriate policies are developed, there will be no lessening in the number of FARA.

XI. Conclusions

Alcohol is the most widely used human teratogen. Among all current substances of abuse, alcohol consumption during pregnancy poses by far the most serious problem. With alcohol being legally, culturally, and socially accepted, consumption is several orders of magnitude larger than that observed for any other known teratogenic compound. Since the mid-1980s, a significant body of scientific literature has shown that the most devastating consequences of alcohol exposure are its effect on the CNS, even when used in relatively moderate doses. Distinct patterns of brain damage include reduced volume of the diencephalus, vermis, basal ganglia, corpus callosum, and cerebellum, with a wide range of impairments at the molecular and biochemical levels resulting in behavioral effects and mental impairment, ranging from minor learning disabilities to mental retardation; hyperactivity; distractibility; reduced visual and auditory memory; poor judgment, adaptability, and social skills; hyperresponsiveness to stress; and somatosensory and auditory problems.

FAS and ARND are permanent manifestations of CNS dysfunction not found to be affected by time or environment. There is also no specific treatment for this disorder.

Despite considerable research in the field of alcohol teratology, the timing, specificity, and pathogenesis of alcohol teratogenicity remain uncertain. Similarly, the contributing role of risk factors such as socioeconomic variables, prenatal care, maternal health, genetic susceptibility, and concomitant exposures are not yet clear.

FAS is not a childhood disorder. There is a preventable long-term progression of the disorder into adulthood in which the maladaptive behaviors present a risk for a wide range of secondary disabilities. The costs of FARA are tremendously high for the individual, the family, for the education and health care systems, as well as for society. Primary prevention is the only treatment for the fetal alcohol induced CNS tragedy.

References

Aase, J. M., Jones, K. L., and Clarren, S. K. (1995). Do we need the term "FAE"? *Pediatrics* **95,** 428–430.

Abel, E. L. (1990). Historical background. In E. L. Abel, Ed. *Fetal Alcohol Syndrome.* Medical Economics Books. New Jersey, pp. 1–11.

Abel, E. L. (1995). An update on incidence of FAS: FAS is not an equal opportunity birth defect. *Neurotoxicol. Teratol.* **17,** 437–443.

Abel, E. L., and Hannigan, J. N. (1996). Risk factors pathogenesis. In: Spohr, H. L. and Steinhausen, H. *Alcohol, Pregnancy, and Developing Child.* Cambridge University Press. Cambridge. pp. 63–75.

Abel, E. L., Sokol, R. J. (1987). Incidence of fetal alcohol syndrome and economic impact of FAS-related anomalies. *Drug Alc. Depend.* **19,** 51–70.

Abel, E. L., and Sokol, R. J. (1991). A revised conservative estimate of the incidence of FAS and its economic impact. *Alc. Clin. Exp. Res.* **15,** 514–524.

Arienti, G., DiRenzo, G. C., Cosumi, E. V., Careim, E., and Corazzi, L., (1993). Rat brain microsome fluidity is modified by prenatal ethanol administration. *Neurochem. Res.* **18,** 335–338.

Black, C., Bucky, S. F., and Wilder-Padilla, S. (1986). The interpersonal and emotional consequences of being and adult child of an alcoholic. *Internat. J. Addict.* **21,** 213–231.

Bloss, G. (1994). The economic cost of FAS. *Alc. Health Res. World* **18,** 53–54.

Bosron, W. F., Li, T. K., and Vallee, B. L. (1980). New molecular forms of liver alcohol dehydrogenase: isolation and characterization of ADHI Indianapolis. *Proc. Natl. Acad. Sci. U.S.A.* **77,** 5784–5788.

Bosron, W. F., Li, T. K., and Vallee, B. L. (1983). Human liver alcohol dehydrogenase. ADH Indianapolis results from genetic polymorphism at the ADH2 gene locus. *Biocem. Genet.* **21,** 735–744.

Broun, S. (1996). New experiments underscore warnings on maternal drinking. *Science* **273,** 738–739.

Brown, N. A., Goulding, E. H., and Fabro, S. (1979). Ethanol embryotocitiy: direct effects on mammalian embryos *in vitro. Science* **206,** 573.

Burmistrov, S. O., Ketin, A. M., and Borodkin, Y. S. (1991). Changes in activity of antioxidative enzymes and lipid peroxidation levels in brain tissue of embryos exposed prenatally to ethanol. *Byull. Eksper. Biolog. Medits.* **112,** 606–607.

Chan, T., Bowell, R., O'Keefe, M., and Lanigan, B. (1991). Ocular manifestation in fetal alcohol syndrome. *Br. J. Ophthalmol.* **25,** 524–526.

Church, M. W., and Holloway, J. A. (1984). The effect of prenatal alcohol exposure on postnatal development of the brain stem auditory evoked potential in the rat. *Alc. Clin. Exp. Res.* **8,** 258–263.

Clarren, S. K. (1979). Neural tube defect and fetal alcohol syndrome. *J. Pediatr.* **95,** 328.

Clarren, S. K. (1981). Recognition of the fetal alcohol syndrome. *JAMA* **245,** 2436–2439.

Clarren, S. K., Alvord, E. C., Sumi, S. M., Streissguth, A. P., and Smith, D. W. (1978). Brain malformations related to prenatal exposure to ethanol. *J. Pediatr.* **92,** 64.

Clarren, S. K., and Astley, S. J. (1992). Pregnancy outcomes after weekly oral administration of ethanol during gestation in the pig-tailed macaque: comparing early gestational exposure to full gestational exposure. *Teratology* **45,** 1–9.

Clarren, S. K., Astley, S. J., and Bowden, D. M. (1988). Physical anomalies and developmental delays in nonhuman primate infants exposed to weekly doses of ethanol during gestation. *Teratology* **37,** 561–569.

Clarren, S. K., Astley, S. J., Bowden, D. M., Lai, H., Milam, A. H., Rudeen, P. K., and Shoemaker, W. J. (1990). Neuroanatomic and neurochemical abnormalities in non-human primate infants exposed to weekly doses of ethanol during gestation. *Clin. Exp. Res. Alc.* **14,** 675–683.

Clarren, S. K., Astley, S. J., Gunderson, V. M., and Spellman, D. (1992). Cognitive and behavioral deficits in nonhuman primates associated with very early embryonic binge exposures to ethanol. *J. Pediatr.* **121,** 789–796.

Clarren, S. K., and Bowden, D. M. (1982). Fetal alcohol syndrome: a new primate model for binge drinking and its relevance to human ethanol teratogenesis. *J. Pediatr.* **101,** 819–824.

Clarren, S. K., Bowden, D. M., and Astley, S. J. (1987). Pregnancy outcomes after weekly oral administration of ethanol during gestation in the pig-tailed macaque (*Macaca nemestrina*). *Teratology* **35,** 345–354.

Clarren, S. K., and Smith, D. W. (1978). The fetal alcohol syndrome. *N. Engl. J. Med.* **298,** 1063–1067.

Coles, C. D., and Platzman, K. A. (1993). Behavioral develpment in children prenatally exposed to drugs and alcohol. *Internat. J. Addict.* **28,** 1393–1433.

Coles, C. D., Smith, I. E., Fernhoff, P. M., and Falek, A. (1984). Neonatal alcohol withdrawal: characteristics in clinically normal, non-dysmorphic infants. *J. Pediatr.* **105,** 445–451.

Coles, C. D., Smith, I. E., Fernhoff, P. M., and Falek, A. (1985). Neonatal neurodevelopment characteristics as correlates of maternal alcohol use during gestation. *Alc. Clin. Exp. Res.* **9,** 454–460.

Collier, H. O. J., McDonald-Gibson, W. J., and Saeed, S. A. (1975). Letter: Stimulation of prostaglandin biosynthesis by capsaicin, ethanol and tyramine. *Lancet* **1,** 702.

Cork, R. M. (1979). The forgotten children: a study of children with alcoholic parents. *Alcoholism: Clin. Exper. Res.* **3,** 148–157.

Cornelius, M. D., Richardson, G. A., Day, N. L., Cornelius, J. R., Geva, D., and Taylor, P. N. (1994). A comparison of prenatal drinking in two recent samples of adolescents and adults. *J. Stud. Alcohol* **55,** 412–9.

Coulter, C. L., Leech, R. W., Schaefer, G. B., Scheithauer, B. W., and Brumback, R. A. (1993). Midline cerebral dysgenesis, dysfunction of the hypothalamic–pituitary axis, and fetal alcohol effects. *Arch. Neruol.* **50,** 771.

Crowell, H., and Feldman, S. S. (1988). Mothers' internal models of relationships and children's behavioral and developmental status; a study of mother–child interaction. *Child Dev.* **59,** 1273–1285.

Davis, W. L., Crawford, L. A., Cooper, O. J., Farmen, G. R., Thomas, D. L., Freeman, B. L., (1990). Ethanol induces the generation of reactive free radicals by neural crest cells *in vitro*. *J. Cranifac. Genet. Devel. Biol.* **10,** 277–293.

Day, N. L., Goldschmidt, L., Robles, N., Richardson, G., Cornellius, M., Taylor, P., Geva, D., and Stoffer, D. (1991a). Prenatal alcohol exposure and offspring growth at 18 months of age: the predictive validity of two measures of drinking. *Alc. Clin. Exp. Res.* **15,** 914.

Day, N. L., Rovles, N., Richardson, G., Geva, D., Taylor, P., and Scher, M. et al. (1991b). The effects of prenatal alcohol use on the growth of children at three years of age. *Alc. Clin. Exp. Res.* **15,** 67–71.

De Groot, M., and Littauer, A., (1989). Hypoxia, reactive oxygen and cell injury. *Free Rad. Biol. Med.* **6,** 541–551.

Devi, B. G., Henderson, G. I., Frosto, T. A., and Schenker, S. (1994). Effect of acute ethanol exposure on cultured fetal rat hepatocytes: relation to mitochondrial function. *Alc. Clin. Exp. Res.* **18,** 1436–1442.

Dreosti, I. E. (1993). Nutritional factors underlying the expression of the fetal alcohol syndrome. *N.Y. Acad. Sci.* **628,** 193–204.

Erickson, M. F., Sroufe, L. A., and Egeland, B. (1985). The relationship between quality of attachment and behavior problems in preschool in high-risk sample. *Mon. Soc. Res. Child Dev.* **50,** 147–166.

Ernhart, C. B., Wolf, A. W., Linn, P. L., Sokol, R. J., Kennard, M. J., and Filipovich, H. F. (1985). Alcohol-related birth defects: syndromal anomalies, intrauterine growth retardation, and neonatal behavioral assessment. *Alc. Clin. Exp. Res.* **9,** 447.

Espina, N., Lima, V., Lieber, C. S., and Garro, A. J. (1988). *In vitro* and *in vivo* inhibitory effects of ethanol and acetaldehyde on 06 methyl-guamine transferase. *Carcinogenesis* **9,** 761–766.

Fisher, S. E., Duffy, L., and Atkinson, M. (1986). Selective fetal malnutrition: effect of acute and chronic ethanol exposure upon rat placental Na,K-ATPase activity. *Alc. Clin. Exp. Res.* **10,** 150–153.

Fisher, S. E., Inselman, L. S., Duffy, L., Atkinson, M., Spencer, H., and Chang, B. (1985). Ethanol and fetal nutrition: effects of chronic ethanol exposure on rat placental growth and membrane-associated folic acid receptor binding activity. *J Pediatr Gastro. Nutr.* **4,** 645–649.

Frezza, M., di Padova, C., Pozzato, G., Terpin, M., Baraona, E., and Lieber, C. S. (1990). High blood alcohol levels in women: the role of decreased gastric alcohol dehydrogenase activity and first pass-metabolism. *N. Engl. J. Med.* **322,** 95–99.

Gabrielli, O., Salvolini, U., Coppa, G. V., Catassi, C., Rossi, R., Manac, A., Lanza, R., and Giorgi, P. L. (1990). Magnetic resonance imaging in the malformative syndromes with mental retardation. *Pediatr. Radiol.* **21,** 16.

Garro, A. J., McBeth, D. L., Lima, V., and Lieber, C. S. (1991). Ethanol consumption inhibits fetal DNA methylation in mice. Implications for the fetal alcohol syndrome. *Alc. Clin. Exp. Res.* **15,** 395–398.

Gladstone, J., Levy, M., Nulman, I., and Koren, G. (1997). Characteristics of pregnant women who engage in binge alcohol consumption. *Can. Med. Assoc. J.* **156,** 789–794.

Gladstone, J., Nulman, I., and Koren, G. (1996). Reproductive risks of binge drinking during pregnancy. *Repro. Toxicol.* **10,** 3–13.

Globus, A., and Scheibel, A. B. (1967). The effect of visual deprivationf on cortical neurons: a Goldy study. *Exp. Neurol.* **19,** 333–345.

Goldstein, D. B. (1983). *Pharmacology of Alcohol.* Oxford University Press, New York. p. 6.

Goldstein, G., and Arulanantham, K. (1978). Neural tube defects and tenal anomalies in a child with fetal alcohol syndrome. *J. Pediatr.* **93,** 636.

Goodlett, C. R., Leo, J. T., O'Callaghan, J. P., Mohoney, J. C., and West, J. R. (1993). Astrologistic induced by alcohol exposure during the brain growth spurt. *Develop. Brain Res.* **72,** 85–97.

Goodlett, C. R., Marcussen, B. L., and West, J. R. (1990). A single day of alcohol exposure during the brain growth spurt induces brain weight restriction and cerebellar Purkinje cell loss. *Alcohol* **7,** 107–114.

Goodlett, C. R., Nichols, J. M., and West, J. R. (1989). Genetic influence on alcohol-induced brain growth restriction: comparison of inbred strains of rats exposed to alcohol during the neonatal brain growth spurt. *Alc. Clin. Exp. Res.* **13,** 322.

Guerri, C., and Sanchis, R. (1985). Acetaldehyde and alcohol levels in pregnant rats and fetuses. *Alcohol* **2,** 267–270.

Hannigan, J. H., Martier, S. S., Chugani, H. T., and Sokol, R. J., (1995). Brain metabolilsm in children with fetal alcohol syndrome (FAS): A positron emission tomography study. *Alc. Clin. Exp. Res.* **19,** 53A.

Harwood, H. J., and Napolitano, D. M. (1985). Economic implications of the fetal alcohol syndrome. *Alc. Health Res. World* **10,** 38–43.

Hayashi, M., Shimazaki, Y., Kamata, S., Kakiichi, N., and Ikeda, M. (1991). Disposition of ethanol and acetaldehyde in maternal blood, fetal blood and amniotic fluid in near-term pregnant rats. *Bull. Environ. Contam. Toxicol.* **47,** 184–189.

Hearn, W. L., Flynn, D. D., and Hime, G. W., *et al.* (1991): A unique cocaine metabolite displays high affinity for the dopamine transporter. *J. Neurochem.* **56,** 698.

Hinckers, H. J. (1978). Characteristics of the physiology of alcohol during pregnancy: absorption of alcohol. *J. Perinat. Med.* **6,** 3.

Hungund, B. L., and Mahadik, S. P. (1993). Role of gangliosides in behavioural and biochemical action of alcohol: cell membrane structure and function. *Alc. Clin. Exp. Res.* **12,** 329–339.

Institute of Medicine (IOM) of the National Academy of Sciences Committe to Study Fetal Alcohol Syndrome. (1996a) Introduction. In K. Stratton, C. Howe, and F. Battaglia, Eds. *Fetal Alcohol Syndrome.* National Academy Press. pp. 17–32.

Institute of Medicine (IOM) of the National Academy of Sciences Committe to Study Fetal Alcohol Syndrome. (1996b) Issues and Research on Fetal Drug Effects. In D. Stratton, C. Howe, and F. Battaglia, Eds. *Fetal Alcohol Syndrome.* National Academy Press. pp. 33–51.

Institute of Medicine (IOM) of the National Academy of Sciences Committe to Study Fetal Alcohol Syndrome. (1996c). Diagnosis and Clinical Evaluation of Fetal Alcohol Syndrome. In K. Stratton, C. Howe, and F. Battaglia, Eds. *Fetal Alcohol Syndrome.* National Academy Press. pp. 63–81.

Institute of Medicine (IOM) of the National Academy of Sciences Committe to Study Fetal Alcohol Syndrome. (1996d) Prevention of Fetal Alcohol Syndrome. In K. Stratton, C. Howe, and F. Battaglia, Eds. *Fetal Alcohol Syndrome.* National Academy Press. pp. 116–142.

Iosub, S., Fuchs, M., Bingol, N., and Gromisch, D. C. (1981). Fetal alcohol syndrome revisited. *Pediatrics* **68,** 475–479.

Johnston, L. D., O'Malley, P. M., and Bachman, J. G. (1991). Drug use among American high school seniors, college students, and young adults, 1975–1990. II: College students and young adults. US Department of Health and Human Services. Rockville, MD.

Jones, K. L., Chambers, C. C., Johnson, K. A. (1991). The effect of disulfiram on the unborn baby. *Teratology* **43,** 438.

Jones, K. L., and Smith, D. W. (1973). Recognition of the fetal alcohol syndrome in early infancy. *Lancet* **2,** 999–1001.

Jones, K. L., Smith, D. W., Ulleland, C. H., and Streissguth, A. P. (1973). Pattern of malformation in offspring of chronic alcohol mothers. *Lancet* **1,** 1267–1271.

Kater, R. M. H., Tobon, F., and Iber, F. L. (1969). Increased rate of tolbutamide metabolism in alcoholic patients. *JAMA.* **207,** 363–365.

Kelly, S. J., Mahoney, J. C., and West, J. R. (1990). Changes in brain microvasculature resulting from early postnatal alcohol exposure. *Alcohol* **7,** 43–47.

Keppen, L. D., Pysher, T., and Rennert, O. M. (1985). Zinc defficiency acts as a co-teratogen with alcohol in fetal alcohol syndrome. *Pediatr. Res.* **19,** 944–947.

Klemm, W. R. (1990). Dehydration: a new alcohol theory. *Alcohol* **12,** 49–59.

Kokotailo, P. K., Adger, H. Jr., Duggan, A. K., Repke, J., and Joffe, A. (1992). Cigarette, alcohol, and other drug use by school age pregnant adolescents: prevalence, detection, and associated risk factors. *Pediatrics* **90,** 328–334.

Koren, G., and Nulman, I. (1994). Teratogenic drugs and chemicals in humans. In Koren G, Ed. *Maternal–Fetal Toxicology.* Marcel Dekker, New York, pp. 33–48.

Kusserow, R. P. (1991). Youth and alcohol: a national survey—drinking habits, access, attitudes and knowledge. US Deparment of Health and Human Services, Rockville, MD.

Landesman-Dwyer, S., Ragozin, A. S. and Little, R. (1981). Behavioral correlates of prenatal alcohl exposure: a four-year follow-up study. *Neurobehav. Toxicol. Teratol.* **3,** 187–193.

Lewis, N., Feiring, C., McGuffog, C., and Jaskir, J. (1984). Predicting psychopathology in six-year-olds from early social relations. *Child Dev.* **48,** 1277–1287.

Lin, G. W. J. (1981). Effect of ethanol feeding during pregnancy on placental transfer of alpha-amino isobutryic acid in the rat. *Life Sci.* **28,** 595–601.

Little, R. E., Streissguth, A. P. (1978). Drinking during pregnancy in alcoholic women. *Alc. Clin. Exp. Res.* **2,** 179–83.

Loebstein, R., Nulman, I., and Koren, G. (March/April, 1997). Fetal alcohol syndrome: an ongoing paediatric challenge. *Paediatr. Child Health 2* **2,** 1–4.

Lopez-Tejero, D., Arilla, E., Colas, B., Llobera, M., and Herrera, E. (1989a). Low intestinal lactase activity in offspring from ethanol-treated mothers. *Biol. Neonate* **55,** 204–213.

Lopez-Tejero, D., Ferrer, I., Llobera, M., Herrera, E. (1986). Effects of prenatal ethanol exposure on physical growth, sensory reflex maturation and brain development in the rat. *Neuropath. Appl. Neurobiol.* **12,** 251–260.

Lopez-Tejero, D., Llobera, M., and Herrera, E. (1989b). Permanent abnormal response to glucose load after prenatal ethanol exposure in rats. *Alcohol* **6,** 469–473.

Lundsgaard, E. (1938). Alcohol oxidation as a function of the liver. *C.R. Lab. Carlsberg Ser. Chim.* **22,** 333–337.

Majewski, F., and Majewski, B. (1988). Alcohol embryuopathy: symptoms, auxological data, frequency among the offspring, and pathogenesis. *Amsterdam, Excerpta Medica* 837–844.

Maly, I. P., Sasse, D. (1991). Intraacinar profiles of alcohol dehydrogenase and aldehyde dehydrogenase activities in human liver. *Gastroenterology* **101,** 1716–1723.

Mann, K., Batra, A., Gunthner, A., and Schroth, G. (1992). Do women develop alcoholic brain damage more readily than men? *Alc. Clin. Exp. Res.* **16,** 1052–1056.

Marcussen, B. L., Goodlett, C. R., Mahoney, J. C., and West, J. R. (1994). Alcohol-induced Purkinje cell loss during differentiation but not during neurogenesis. *Alcohol* **11,** 147–156.

Mattson, S. N., Riley, E. P., Jernigan, T. L., Ehlers, C. L., Delis, D. C., Jones, K. L., Stern, C., Johnson, K. A., Hesselink, J. R., and Bellugi, U. (1992). Fetal alcohol syndrome: a case report of neurophychological, MRI, and EEG assessment of two children. *Alc. Clin. Exp. Res.* **16,** 1001.

Mattson, S. N., Riley, E. P., Jernigan, T. L., Gracia, A., Kaneko, W. M., Ehlers, C. L., and Jones, K. L. (1994). A decrease in the size of the basal ganglia following prenatal alcohol exposure: a preliminary report, *Neurotoxicology* **16,** 283.

McCance, R. A., and Widdowson, E. W. (1954). Water metabolism. *Cold Spr. Harb. Symp. Quant. Biol.,* **19,** 155.

Meares, R., Penman, R., Milgrom-Friedman, J. and Baker, K. (1982). Some orgins of the 'difficult' child: the Brazelton scale and the mother's view of her newborn's character. *Brit. J. Med. Physc.* **55,** 77–86.

Meilman, P. C. (1993). Alcohol-induced sexual behavior on campus. *J. Am. Coll. Health* **42,** 27–31.

Miller, M. W. (1986). Effects of alcohol on the generation and migration of cerebral cortical neurons. *Science* **233,** 1308–1311.

Miller, M. W. (1993). Migration of cortical neurons altered by gestational exposure to alcohol. *Alc. Clin. Exper. Res.* **12,** 304–314.

Miller, M. W., and Potempa, G. (1990). Numbers of neurons and glia in mature rat somatosensory cortex: effects of prenatal exposure to alcohol. *J. Comp. Neurol.* **293,** 92–102.

Morgan, M. Y., and Sherlock, S. (1977). Sex-related differences among 100 patients with alcoholic liver disease. *Brit. Med. J.* **1,** 939–941.

Mukherjee, A. B., and Hodgen, G. D. (1982). Maternal ethanol exposure induced transient impairment of umbilical circulation and fetal hypoxia in monkeys. *Science* **218,** 700–701.

Murdoch, R. N., and Edwards, T. (1992). Alterations in the methylation of membrane phospholipids in the uterus and postimplantation embryo following exposiure to teratogenic doses of alcohol. *Biochem. Inter.* **28,** 1029–1037.

National Centre for Health Statistics. Advance report on final natality statistics, 1988. Monthly Vital Statistics report, Vol. 39, No. 4, Supplement, DHHS Publication No. (PHS) 90-1120, Hyattsville, MD: Department of Health and Human Services, Public Health Service; 1990.

Nordmann, R., Ribiere, C., and Rouach, H. (1992). Implication of free radical mechanisms in ethanol-induced cellular injury. *Free Rad. Biol. Med.* **12,** 219–239.

Northern, J. L., and Down, S. M. (1984). *Hearing in Children,* 3rd Ed. Williams & Wilkins, Baltimore.

O'Connor, M. H., Sigman, M., and Kasari, C. (1992). Attachment behavoir of infants exposed prenatally to alcohol: mediating effects of infant affect and mother–infant interaction. *Dev. Psychopath.* **4,** 243–256.

Olsen, J., Pereira, A., da C., and Olsen, S. F. (1991). Does maternal tobacco smoking modify the effect of alcohol on fetal growth? *Am. J. Pub. Health* **81,** 69–73.

Pares, X., Farres I., and Vallee, B. L. (1984). Organ specific alcohol metabolism. Placental X-ADH. *Biochem. Biophys. Res. Commun.* **119,** 1047–1055.

Parker, D. A., Harford, T. C., and Rosenstock, I. M. (1994). Alcohol, other drugs, and sexual risk-taking among young adults. *J. Subst. Abuse* **6,** 87–93.

Pastan, I. H., Johnson, G. S., and Anderson, W. B. (1975). Role of cyclic nucleotides in growth control. *Annu. Rev. Biochem.* **44,** 491–522.

Peiffer, J., Majewski, F., Fischbach, H., Bierich, J. R., and Volk, B. (1979). Alcohol embryo- and fetophathy: Neuropathology of 3 children and 3 fetuses. *J. Neurol. Sci.* **41,** 125.

Pennington, S. (1988). Ethanol-induced growth inhibition: the role of cAMP-dependent protein kinase. *Alc. Clin. Exp. Res.* **12,** 125–129.

Pierse, D. R., and West, J. R. (1986). Alcohol-induced microencephaly during the third trimester equivalent: relationship to dose and blood alcohol concentration. *Alcohol* **3,** 185–191.

Ramanthan, R., Wilkemeyer, M. F., Mittal, B., Perides, G., Charness, M. E. (1996). Alcohol inhibits cell–cell adhesion mediated by human L1. *J. Cell Biol.* **133,** 381–390.

Rankin, J. G. (1977). The natural history and management of the patient with alcoholic liver disease. In: MM Fisher and JG Rankin, Eds. *Alcohol and the Liver.* Vol. 3 of *Hepatology: Research and Clinical Issues.* Plenum Press, New York, 365–381.

Reynolds, J. D., Penning, D. H., Dexter, F., Atkins, B., Hardy, J., Poduska, D., and Brien, J. F. (1996). Ethanol increases uterine blood flow and fetal arterial blood oxygen tension in the near term pregnant ewe. *Alcohol* **13,** 251–6.

Riley, E. P., Mattson, S. N., Sowel, E. R., Jernigan, T. L., Sobel, D. F., and Jones, K. L. (1996). Abnormalities of corpus callosum in children prenatally exposed to alcohol. *Alc. Clin. Exp. Res.* **19,** 1198–1202.

Robin, N. H., and Zackai, E. H. (1994). Unusual craniofacial dusmorphia due to prenatal alcohol and cocaine exposure. *Teratology* **50,** 160.

Robinson, G. C., Conry, J. L., and Conry, R. F. (1987). Clinical profile and prevalence of fetal alcohol syndrome in an isolated community in British Columbia. *Can. Med. Assoc. J.* **137,** 203–207.

Ronen, G. M., and Andrews, W. L. (1991). Holoprosencephaly as a possible embryonic alcohol effect. *JAMA* **40,** 151.

Rosett, H. L. (1980). A clinical perspective of the fetal alcohol syndrome. *Alc. Clin. Exp. Res.* **4,** 119–122.

Rosett, H. L., and Weiner, L. (1984). *Alcohol and the Fetus.* New York, Oxford University Press, 21–27.

Sanchis, R., and Guerri, C. (1986a). Chronic ethanol intake in lactating rats: milk analysis. *Comp. Biochem. Physiol.* **85,** 107.

Sanchis, R., and Guerri, C. (1986b). Alcohol-metabolising enzymes in placenta and fetal liver: effect of chronic ethanol intake. *Alc. Clin. Exp. Res.* **10,** 39–44.

Schaefer, G. B., Shuman, R. M., Wilson, D. A., Saleeb, S., Domek D. B., Johnson, S. F., and Bodensteiner, J. B. (1991). Partial agenesis of the anterior corpus callosum: correlation between appearences, imaging, and neuropathology. *Pediatr. Neurol.* **7,** 39.

Schenker, S. J., Johnson, R. F., Mahuren, J. D., Henderson, G. I., Coburn, S. P. (1992). Human placental vitamin B6 (pyridoxal) transport: normal characteristics and effects of ethanol. *Am. J. Physiol.* **262,** R966–R974.

Scrabanek, P. (1992). Smoking and statistical overkill. *Lancet* **340,** 1208–1209.

Seitz, H. K., Egerer, G., and Simanowski, U. A., *et al.* (1993). Human gastric alcohol dehydrogenase activity: effect of age, sex and alcoholism. *Gut* **34,** 1433–1437.

Smith, D. E., Foundas, A., and Canale, J. (1986). Effects of perinatally administered alcohol on the development of the cerebeller granule cel. *Exp. Neurol.* **92,** 491–501.

Snyder, A. K., Singh, S. P., and Pullen, G. L. (1986). Ethanol-induced intrauterine growth retardation: correlation with placental glucose transfer. *Alc. Clin. Exp. Res.* **10,** 167–170.

Sokol, R. J., and Clarren, S. K. (1989). Guidelines for use of terminology describing the impact of prenatal alcohol on the offspring. *Alc. Clin. Exp. Res.* **13,** 597–598.

Steinhausen, A. P., Aase, J. M., Clarren, S. K., Randels, S. P., La Due, R. A. and Smith, D. F. (1991a). Fetal alcohol syndrome in adolescents and adults. *JAMA* **265,** 1961–1967.

Steinhausen, A. P., Randels, S. P., and Smith, D. F. (1991b). A test–retest study of intelligence in patients with fetal alcohol syndrome: implications for care. *J. Am. Acad. Child Adol. Psych.* **30,** 584–587.

Steinhausen, H., Nestler, V., and Huth, H. (1982). Psychopathology and mental functions in the offspring of alcoholic and epileptic mothers. *J. Am. Acad. Child Adol. Psych.* **21,** 268–273.

Steinhausen, H. C. (1996). Psychopathology and cognitive functioning in children with fetal alcohol syndrome. In H. Spohr and H. Steinhausen, Eds. *Alcohol, Pregnancy and the Developing Child.* Cambridge, Cambridge University Press, pp. 227–248.

Steinhausen, H.-C., Willms, J., and Spohr, H.-L. (1993). Long term psychopathological and cognitive outcome of children with fetal alcohol syndrome. *J. Am. Acad. Child Adol. Psych.* **32,** 990–994.

Streissguth, A. P., Aase I. M., Clarren, S. K., Randels, S. P., Robin, La Due, Smith, D. S. (1991). Fetal alcohol syndrome in adolescents and adults. *JAMA* **265,** 1961–1967.

Streissguth, A. P., Barr, H. M., Bookstein, F. L., and Sampson, P. D. (1993). *The Enduring Effects of Prenatal Alcohol Exposure on Child Development: Birth through 7 Years, a Partial Least Squares Solution.* Ann Arbor, MI: University of Michigan Press.

Streissguth, A. P., Barr, H. M., Kogan, J., and Bookstein, F. L. (1996c). *Understanding the Occurrence of Secondary Disabilities in Clients with Fetal Alcohol Syndrome and Fetal Alcohol Effects. Final Report.* Presented on Fetal Alcohol Syndrome Conference in Washington, DC. Sept. 1996.

Streissguth, A. P., Barr, H. M. and Martin, D. C. (1983). Maternal alcohol use and neonatal habituation assessed with the Brazelton scale. *Child Dev.* **54,** 1109–1118.

Streissguth, A. P., Barr, H. M., Olson, H. C., Sampson, P. D., Bookstein, F. L., and Burgess, D. M. (1994a). Drinking during pregnancy decreases word attack and arithmetic scores on standardized tests: adolescent data from a population-based prospective study. *Alc. Clin. Exp. Res.* **18,** 248–255.

Streissguth, A. P., Barr, H. M., and Sampson, P. D. (1990). Moderate prenatal alcohol exposure: effects on child IQ and learning problems at age 7 1/2 years. *Alc. Clin. Exp. Res.* **14,** 662–669.

Streissguth, A. P., Barr, H. M., Sampson, P. D., Bookstein, F. L., and Darby, B. L. (1989a). Neurobehavioral effects of prenatal alcohol. Part 1. Research strategy. *Neurotoxicol. Teratol.* **11,** 461–476.

Streissguth, A. P., Bookstein, F. L., and Barr, H. M. (1996b). A dose–response study of the enduring effects of prenatal alcohol exposure: birth to 14 years. In H. Spohn and H. Steinhausen, Eds. *Alcohol, Pregnancy and Developing Child.* Cambridge, Cambridge University Press. pp. 141–168.

Streissguth, A. P., Bookstein, F. L., Sampson, P. D., and Barr, H. M. (1989b). Neurobehavioral effects of prenatal alcohol. Part II. PLS analysis of neuropsychologic tests. *Neurotoxicol. Teratol.* **11,** 493–507.

Streissguth, A. P., Bookstein, F. L., Samspon, P. D., and Barr, H. M. (1996a). *The Enduring Effects of Prenatal Alcohol Exposure on Child Development.* Ann Arbor, University of Michigan Press, pp. 4.

Streissguth, A. P., Sampson, P. D., Olson, H. C., *et al.* (1994b). Maternal drinking during pregnancy: attention and short-term memory in 14-year old offspring—a longitudinal prospective study. *Alc. Clin. Exp. Res.* **18,** 202–218.

Stromland, K. (1981). Eyeground malformations in the fetal alcohol syndrome. *Neuropediatrics* **12,** 97–98.

Stromland, K. (1985). Ocular abnormalities in the fetal alcohol syndrome. *Acta Ophthalmol.* **63,** (Suppl 171) 1–50.

Stromland, R., and Hellstrom, A. (1996). Fetal alcohol syndrome—an opthalmological and socioeducational prospective study. *Pediatrics* **97,** 845–850.

Subramanian, M. G. (1996). Effects of ethanol on lactation. In E. L. Abel, Ed. *Fetal Alcohol Syndrome.* CRC Press, Boca Raton, FL. pp. 237–247.

Sulik, K. K., Johnston, M. C., Draft, P. A., Russell, W. E., and Dehart, D. B. (1986). Fetal alcohol syndrome and DiGeorge anomaly: critical ethanol exposure periods for craniofacial malformations as illustrated in an animal model. *Am. J. Med. Genet. Suppl.* **2,** 97–112.

Sulik, K. K., Johnston, M. C., and Webb, M. A. (1981). Fetal alcohol syndrome: embryogenesis in a mouse model. *Science.* **214,** 936–938.

Sulik, K. K., Lauder, J. M., Dehart, D. B. (1984). Brain malformations in prenatal mice following acute maternal ethanol administration. *Int. J. Dev. Neurosci.* **2,** 203–214.

Teschke, R., and Wiese, B. (1982). Sex-dependency of hepatic alcohol metabolizing enzymes. *J. Endocrinol. Invest.* **5,** 243–250.

Traves, C., Camps, L., and Lopez-Tejero, D. (1995). Liver alcohol dehydrogenase activity and ethanol levels during chronic ethanol intake in pregnant rats and their offspring. *Pharmacol. Biochem. Behav.* **52,** 93–99.

Ugarte, G., and Valenzuela, J. (1971). Mechanisms of liver and pancreas damage in man. In Y. Israel and J. Mardones, Eds. *Biological Basis of Alcoholism,* Wiley, New York, pp. 133–161.

Vinas, O., Vilaro, S., Herrera, E., and Remesar, X. (1989). Effects of chronic ethanol treatment on amino acid uptake and enzyme activities in the lactating rat mammary gland. *Life Sci.* **40,** 1745.

Volpi, R., Chiodera, P., Gramellini, D., Cigarini, C., Papadia, C., Caffarri, G., Rossi, G., and Coiro, V. (1994). Endogenous opioid medication of the inhibitory effect of ethanol on the prolactin response to breast stimulation in normal women. *Life Sci.* **54,** 739.

Warner, R. H. and Rosett, H. L. (1975). The effects of drinking on offspring: an historical survey of the American and British literature. *J. Stud. Alcohol* **36,** 395–420.

Weschler, H., Davenport, A., Dowdall, G., Moeykens, B., and Castillo, S. (1994). Health and behavioral consequences of binge drinking in college: a national survey of students at 140 campuses. *JAMA* **272,** 1672–1677.

Weshcler, H., and Isaac, N. (1992). "Binge" drinking at Massachusetts colleges: prevalence, drinking style, time trends, and associated problems. *JAMA* **267,** 2929–2931.

West, J. R., Goodlett, C. R., Bonthius, D. J., Hamre, K. M., and Marcussen, B. L. (1990). Cell population depletion associated with fetal alcohol brain damage: mechanisms of BAC-dependent cell loss. *Alc. Clin. Exp. Res.* **14,** 813–818.

West, J. R., and Hamre, K. M. (1985). Effects of alcohol exposure during different periods of development: changes in hippocampal mossy fibres. *Dev. Brain Res.* **17,** 280–284.

West, J. R., Hamre, K. M., and Cassel M. D. (1986). Effects of alcohol exposure during the third trimester equivilent on neuron number ni rat hippocampus and dentate gyrus. *Alc. Clin. Exp. Res.* **10,** 190–197.

West, J. R., Wei-Jung, A., Chen, and Pantazis, N. J., (1994). Fetal alcohol syndrome: the vulnerability of the developing brain and possible mechanisms of damage. *Metab. Brain Dis.* **9,** 291–322.

Wiesel, T. N. (1982). Postnatal development of the visual cortex and the influence of environment. *Nature* **299,** 583–592.

Wisniewski, K., Dambska, M., Sher, J. H., and Qazi, Q. (1983). A clinical neuropathology study of the fetal alcohol syndrome. *Neuropediatrics* **14,** 197.

Wright, J. T., Waterson, E. J., Barrison, I. G., Toplis, P. J., Lewis, I. G., Gordon, M. G., MacRae, K. D., Morris, N. F., and Murray-Lyon, I. M. (1983). Alcohol consumption, pregnancy, and low birth weight. *Lancet* **i,** 663–665.

Yang, H. Y., Shum, A. Y. C., Ng, H. T., and Chen, C. F. (1986). Effect of ethanol on human umbilical artery and vein *in vitro. Gyn. Obstet. Invest.* **21,** 131–135.

CHAPTER

34

Developmental Neurotoxicity of Nicotine

EDWARD D. LEVIN
THEODORE A. SLOTKIN
Departments of Psychiatry and Pharmacology
Duke University Medical Center
Durham, North Carolina 27710

I. Scope of Tobacco Use during Pregnancy

In the face of widespread dissemination of information concerning the adverse effects of smoking during pregnancy, nearly one third of all women of childbearing age continue to smoke (DiFranza and Lew, 1995). Despite self-reporting of high quit rates, it is now evident that most smokers continue throughout pregnancy, calling into question the effectiveness of antismoking publicity. Self-reported smoking at the beginning of pregnancy averages approximately 25% (Streissguth *et al.,* 1983; Fried *et al.,* 1984; Hickner *et al.,* 1984; Stewart and Dunkley, 1985; Kruse *et al.,* 1986; Aronson *et al.,* 1993; Bardy *et al.,* 1993; King *et al.,* 1993) and was originally thought to decline to as low as 18% by the third trimester (Fricker *et al.,* 1985; Kruse *et al.,* 1986; Sarvela and Ford,

*Abbreviations: ACh, acetylcholine; ADHD, attention deficit/hyperactivity disorder; α-BGT, α-bungarotoxin; ChAT, choline acetyltransferase; CNS, central nervous system; CO, carbon monoxide; HCN, hydrogen cyanide; IUGR, intrauterine growth retardation; ODC, ornithine decarboxylase; PN, postnatal [days]; SES, socioeconomic status; SIDS, sudden infant death syndrome.

1992). However, in a study, cotinine levels indicative of smoking were still detected in nearly 25% of 1,237 live births, more than a third of whose mothers had claimed to be nonsmokers (Bardy *et al.,* 1993). Similarly, in a large urban population, 41% of women smoking before pregnancy claimed to have quit during pregnancy (O'Campo *et al.,* 1992), but plasma measurements (maternal carboxyhemoglobin and thiocyanate) did not change or actually increased (Pley *et al.,* 1991). Thus, the risks associated with fetal exposure to nicotine and other constituents of tobacco smoke remain high despite continuing educational and medical intervention.

Rates of smoking during pregnancy are skewed across economic strata and ethnic groups in our society. Women in lower socioeconomic groups have nearly four times the rate of smoking than do their higher status counterparts (King *et al.,* 1993; Gazmararian *et al.,* 1996). Age is also a factor: considerably more adolescents smoke during pregnancy than do older women, and the majority of adolescents do not alter their smoking during successive pregnancies (Lodewijckx and De Groof, 1990). In some reports, young pregnant women smoked at higher rates even than young women in general (Tollestrup *et al.,* 1992). Studies restricted to at-risk populations find smoking rates at the beginning of pregnancy that are often greater than 50%, and these rates drop only slightly throughout pregnancy (Pley *et al.,* 1991; Wakefield and Jones, 1991; Latulippe *et al.,* 1992). One-fourth of these smokers even continue to smoke during the course of labor (Latulippe *et al.,* 1992).

However, some progress has been made among Hispanics and African-Americans. Hispanics are nearly three times more likely to quit during pregnancy than are non-Hispanic Caucasians, and Hispanics who continue to smoke show greater reductions in cigarette consumption (King *et al.,* 1993; Camilli *et al.,* 1994). The rate for African-American women who smoked during pregnancy has decreased from 37% in 1978 to less than 22% in 1990. A large part of this reduction is attributable to teenagers, whose rate of smoking during pregnancy declined from 36% to 7%, results that may be a significant contributor to the overall decrease in the incidence of low birthweight infants in this subgroup (Land and Stockbauer, 1993). When considered as a factor in the total population of smokers, however, these successes have to be weighed against the fact that unmarried white women are 40% more likely to smoke during pregnancy than are their nonpregnant counterparts (Williamson *et al.,* 1989).

It is evident that the tremendous publicity since the mid-1970s concerning the adverse effects of smoking during pregnancy has had some impact on at-risk subpopulations. However, at the same time, tobacco use during pregnancy remains quite high. Furthermore, the discrepancies between self-reported cigarette consumption and objective biologic measures call into question the degree to which inroads have actually been made into smoking during pregnancy, or whether we have merely increased the likelihood of deception. Given these issues, it is worthwhile to examine the estimated costs to society of maternal smoking.

II. Societal Cost of Maternal Smoking

A meta-analysis indicates that, in the United States, smoking during pregnancy accounts annually for an estimated 19,000–141,000 fetal deaths, 1,900–4,800 deaths during or immediately after parturition, and 2,200–4,200 deaths from sudden infant death syndrome (SIDS) (DiFranza and Lew, 1995). For SIDS, tobacco accounts for more cases than for all other abused substances, and given that the great majority of pregnant women who abuse other drugs also smoke cigarettes, the role of tobacco is most likely underappreciated (Haglund and Cnattingius, 1990; Kandall and Gaines, 1991). In addition to the obvious risk factors of decreased birthweight and premature birth that are indicated in the warning statement on cigarette packs, there is also increased infant mortality and an increased incidence of childhood brain tumors (U.S. Public Health Service, 1979; Picone *et al.,* 1982a,b; Preston-Martin *et al.,* 1982; Brooke *et al.,* 1989; Virji and Cottington, 1991; Cordier *et al.,* 1994). However, there are more subtle liabilities to maternal smoking that are probably even more widespread: neurobehavioral abnormalities that appear in childhood and adolescence and that are likely to be permanent (Rush and Callahan, 1989). Before turning to the neurobehavioral effects of smoking, it is useful to review the actual figures for developmental exposure and its known costs.

Maternal tobacco consumption is the major environmental risk factor for low birthweight (Morrison *et al.,* 1993). Mothers who self-reported smoking cigarettes during pregnancy were twice as likely to bear low birthweight infants than were nonsmokers (Aronson *et al.,* 1993). About 7% of total newborn hospital costs were due to the increased incidence of low birthweight associated with smoking, but this figure is undoubtedly low by at least half because of the inaccuracy of self-reportage. A more substantiated figure for fetal growth retardation caused by maternal smoking is a 2.6-fold increase when compared to women who claimed to have stopped smoking by the 18th week of gestation (Spinillo *et al.,* 1996). Intrauterine growth retardation (IUGR) is evident at birth but is no longer evident by 1 year of age (Schulte-Hobein *et al.,* 1992), although the resolution of growth deficits does not obviate the effects on neurobehavioral development (Rush and Callahan, 1989).

The attribution of fetal growth impairment to tobacco exposure is potentially confounded by the covariables of low socioeconomic status (SES) and poor prenatal care in the smoking population. It is therefore important to note that the relationship of smoking to reduced birthweight is dose-related: each μg/ml of cotinine in maternal serum results in a mean decrease of 1.3 g in birthweight and 0.06 mm in length (Bardy *et al.,* 1993). A reduction in the amount of smoking by an expectant mother attenuates the deficits (Li *et al.,* 1993). Smoking also overrides any genetic predisposition to having low birthweight infants (Wainright, 1983; Magnus *et al.,* 1985). Finally, animal models that incorporate fetal nicotine exposure without any of the covariables can reproduce the growth retarding effects of maternal smoking (Slotkin, 1992). It is thus evident that it is tobacco consumption itself that provides the major impetus for low birthweight infants in the smoking population.

Adverse effects of maternal smoking can continue into the postnatal period, as nicotine is readily passed through the milk. Breast-fed infants show urinary cotinine concentrations that are 10-fold higher than infants who are bottle fed, and actually have values in the range of adult smokers (Schulte-Hobein *et al.,* 1992). Although mothers who smoke tend to wean their babies earlier than do nonsmokers, infant cotinine levels are still higher than those of nonsmoking adult partners (Schulte-Hobein *et al.,* 1992), indicating that the newborn experiences high exposure levels when mothers continue to smoke after birth.

The continuation of the societal cost of maternal smoking is best exemplified by the high incidence of SIDS (Naeye *et al.,* 1976), the leading cause of death during the first postnatal year, causing thousands of deaths annually in the United States alone. SIDS is highly correlated both with smoking during pregnancy (Haglund and Cnattingius, 1990; Kandall and Gaines, 1991) and with postnatal exposure (Klonoff-Cohen *et al.,* 1995). It is now evident that the additional SIDS liability incurred with maternal smoking is most likely to represent perturbation of neural control of cardiorespiratory function (Milerad and Sundell, 1993). Smoking during pregnancy causes an increase in the rate of fetal breathing movements (Eriksen *et al.,* 1983), and after birth there is a dose-dependent increase in the incidence of obstructive apneas that can reach nearly three times that of babies born to nonsmokers (Kahn *et al.,* 1994). It has been suggested that, through hypoxia, carbon monoxide in tobacco smoke exerts a noxious effect on the development of central respiratory control mechanisms, which then remain particularly susceptible to causing SIDS (Hutter and Blair, 1996). However, animal studies suggest instead that nicotine itself is the agent that compromises the development and function of cardiorespiratory mechanisms (Krous *et al.,* 1981; Milerad and Sundell, 1993; Milerad *et al.,* 1994, 1995; Slotkin *et al.,* 1995, 1997b; Holgert *et al.,* 1995; Bamford *et al.,* 1996).

Kinney and co-workers (1993) characterized the distribution of nicotinic receptor binding sites in the developing human brainstem by quantitative tissue autoradiography and found high concentrations in tegmental nuclei during midgestation, higher than those later in development. Given the ability of nicotine to perturb cell development (Slotkin, 1992), this profile is consistent with targeting of brainstem areas by nicotine in causing SIDS (Krous *et al.,* 1981; Kinney *et al.,* 1993). More recent work has confirmed abnormalities of cholinergic target sites in the brainstem of infants that died of SIDS (Kinney *et al.,* 1995), results consonant with the effects of nicotine in animal models (Krous *et al.,* 1981; Slotkin *et al.,* 1987c; Navarro *et al.,* 1989a; McFarland *et al.,* 1991; Slotkin, 1992; Zahalka *et al.,* 1992, 1993c). Finally, fetal exposure of animals to nicotine can reproduce the cardiorespiratory collapse in response to hypoxia, the conditions thought to prevail in SIDS, and the operant mechanisms involve central and peripheral neural targets sensitive to nicotine (Slotkin *et al.,* 1995, 1997b). As the underlying causes of SIDS involve neural targeting, this review covers the issue of the role of developmental neurotoxicity of nicotine in causing the syndrome.

Despite all of these effects, it is important to note that smoking during pregnancy is not associated with gross malformations. A study of more than 67,000 pregnancies did not find any significant relationship between maternal smoking and congenital malformations, except for a slight trend toward higher incidence of neural tube defects (Evans *et al.,* 1979). In contrast, there is abundant evidence for adverse effects on neurobehavioral function. Although on the surface these effects may seem to be less important than is dysmorphogenesis, behavioral disruption and cognitive impairment can have dramatic effects on the quality and life course of an affected individual, and a profound impact on society in general.

III. Epidemiologic Associations of Neurobehavioral Deficits and Tobacco Use during Pregnancy

Although sequelae of maternal smoking are of obvious importance during gestational and neonatal development, it is equally important that there are long-term behavioral liabilities. The children of women who smoke are more likely to develop cognitive and learning deficits (Butler and Goldstein, 1973; Dunn and McBurney, 1977; Rantakallio, 1983; Naeye and Peters, 1984; Gueguen *et al.,* 1995), including attention deficit/hyperactivity disorder (ADHD) (Denson *et al.,* 1975; Fried, 1992), impaired attention and orientation (Landesman-Dwyer and Emanuel, 1979; Picone *et al.,* 1982a,b; Jacobson *et al.,* 1984; Streissguth *et al.,* 1984), and poor impulse control (Kristjansson *et al.,* 1989). Long-lasting deficits in cognitive function after maternal smoking have been seen in most studies (Hardy and Mellits, 1972; Lefkowitz, 1981). Fried (1989) found that maternal cigarette smoking was significantly correlated with impaired performance on the McCarthy Scale of Children's Abilities at 3 years of age. Streissguth and co-workers (1984) found impaired orientation and attentiveness in 4-year-old children. Kristjansson *et al.* (1989) found hyperactivity, impaired vigilance, and poor impulse control in 4- to 7-year olds. These long-lasting effects of developmental nicotine exposure may go undetected until they have an impact on school performance. A very long-term prospective study (Fogelman and Manor, 1988) found a highly significant decrement in school attainment by age 23 years, correlated with the initial effect on growth. By age 23 the growth effect was no longer evident, but there was still a strong effect on school performance. Thus, maternal smoking continues to exact a toll on the affected individuals and on society, far beyond the more well-known and publicized immediate perinatal consequences. It is worthwhile to discuss some of the details of these studies, as they point out not only the adverse consequences of smoking during pregnancy, but also the questions that have been raised about the actual causative agents in behavioral disruption.

Disruptive behavior in the offspring of women who smoke has been identified in epidemiologic studies (McGee and Stanton, 1994). However, the magnitude of effect is fairly small or not obviously related to dose and/or duration of smoking during pregnancy. In a New Zealand study, maternal smoking during pregnancy was associated with small but statistically detectable increases in rates of childhood problem behaviors, with children whose mothers smoked in excess of 20 cigarettes per day having mean scores that were 0.10 to 0.36 standard deviations higher than the scores of the offspring of nonsmokers (Fergusson *et al.,* 1993); smoking after pregnancy was not significantly associated with childhood problem behavior. In a large Finnish study, the incidence of delinquency by the age of 22 years more than doubled among the offspring of smokers. Again, no clear dose–response gradient was observed, and children of mothers who stopped smoking during the first trimester of the pregnancy had only a slightly smaller incidence of delinquency than children of mothers who continued smoking throughout pregnancy. These studies support the idea of a statistical connection between maternal smoking and problem behaviors, but fail to distinguish whether the net effect is due to fetal exposure to tobacco components, or is an epiphenomenon of a risk-taking lifestyle and an overall rebellious parental behavioral pattern that is likely to increase delinquency in children as an associative factor (Rantakallio *et al.,* 1992).

Maternal cigarette smoking also has been associated with cognitive deficits in the offspring. One study found that maternal smoking during pregnancy was associated with a 50% increase in the prevalence of idiopathic mental retardation, and children whose mothers smoked at least one pack a day (20 cigarettes) during pregnancy had more than a 75% increase (Drews *et al.,* 1996). It is more typical, however, to find more subtle cognitive deficits in the offspring of smokers. Impairment of visual and auditory orientation and habituation on the Brazelton and other tests has been seen in the infants of smoking mothers (Saxton, 1978; Picone *et al.,* 1982a; Fried and Makin, 1987; Fried *et al.,* 1987; Oyemade *et al.,* 1994), along with increased tremors and impaired neonatal reflex development (Picone *et al.,* 1982a; Fried and Makin, 1987; Fried *et al.,* 1987). Equally important, cognitive deficits have been found to persist. At 4–7 years of age, difficulties with attentional behavior were still noted in the offspring of women who smoked during pregnancy, as assessed by auditory and visual vigilance tasks. There was an increased activity level during the tasks and increased incidence of errors of commission in both auditory and visual tasks. This combination of results suggests that the deficits in attention may reflect impulsive responding and increased overall activity (Kristjansson *et al.,* 1989). In another study, long-term adverse behavioral consequences were seen after active and passive smoke exposure during pregnancy. Children of nonsmoking mothers performed better than did the children of smoking mothers on tests of speech and language skills, intelligence, visual–spatial abilities, and on the mother's rating of behavior. The performance of children of passive smokers was found, in most areas, to be in between those of the active smoking and nonsmoking groups (Makin *et al.,* 1991).

Fried and Watkinson (1988) and Fried and coworkers (1992a,b) also conducted long-term studies of the effects of maternal cigarette smoking on neurobehavioral outcome in the offspring. Maternal cigarette smoking during pregnancy was significantly associated with lower mental scores in their offspring and altered responses on auditory tests at 1 year. Although the effect was attenuated at 2 years (Fried and Watkinson, 1988),

3- and 4-year-old children still showed poorer language development and lower cognitive scores than the offspring of nonsmokers. At 5 and 6 years of age there was still a significant association between prenatal tobacco exposure and lower cognitive and receptive language scores (Fried *et al.,* 1992a,b). Importantly, in the 6-year-old children there was a demonstrable dose–response relationship between the magnitude of prenatal exposure and impulsive behavior as manifest with a response inhibition task and with increased errors of commission on a sustained vigilance task. Performance on a series of memory tasks, particularly those requiring verbal recall, was also impaired with maternal cigarette use (Fried *et al.,* 1992a,b).

The presence of a dose–response relationship in some epidemiologic studies demonstrating behavioral deficits in the offspring of women who smoked during pregnancy does not identify what components of smoking are the injurious factors. Indeed, in a study that evaluated neonatal hemoglobin levels and low birthweights, along with subsequent behavioral developmental factors such as hyperactivity, short attention span, and lower scores on spelling and reading tests, the higher frequency of abnormalities in the tobacco-exposed group was correlated with the physiologic variables, suggesting that fetal hypoxemia may contribute to the genesis of behavioral abnormalities (Naeye and Peters, 1984). Although in this study cognitive abnormalities were mild, with achievement test scores 2–4% lower than in nonexposed controls, the potential contributions of the various covariables of smoking reinforce the need for animal models that separate these factors from the role of nicotine as a neuroteratogen.

Finally, one essential question about the epidemiologic association of smoking during pregnancy and behavioral outcome is whether such exposure alters the propensity of the offspring themselves to take up smoking in later life. An initial study indicates that female offspring of smoking mothers are four times more likely to smoke as compared either to the offspring of nonsmokers or to children of women who did not smoke during pregnancy but did smoke after delivery (Kandel *et al.,* 1994). The importance of this effect is that it points to mechanisms involving nicotine itself: alteration of the function of nicotinic inputs that control activity of dopaminergic reward pathways could predispose the brain in a critical period of development, to the subsequent addictive influence of nicotine consumed more than a decade later in life. Again, whereas the eventual consequence of this type of putative effect is demonstrable in the offspring of smokers, the identification of specific mechanisms and the role of nicotine in adverse behavioral outcomes requires the use of appropriate animal models.

IV. Animal Models for Nicotine Effects on Neurobehavioral Development

Given the high societal cost of smoking during pregnancy in evoking perinatal mortality and morbidity, and adverse effects on behavioral development, why is it necessary to develop animal models of nicotine's effects? There are four basic reasons. First, cigarette smoking is highly associated with confounding variables that are known to play a role in adverse developmental outcomes, including low SES, poor nutrition and prenatal care, risk-taking behavior, and coabuse of other substances. If we are to concentrate on smoking cessation as an objective to prevent perinatal damage, then we need to be sure that it is smoking and not an associative variable or an interaction with that variable that is responsible. For example, genetic factors that make one more likely to smoke or to continue smoking during pregnancy, such as the presence of the D2A1 receptor trait (Comings *et al.,* 1996), may themselves be related to lower cognitive function in the offspring (Berman and Noble, 1995). Second, the identification of a specific component of tobacco smoke (*e.g.,* nicotine) as the injurious substance, would provide a mechanistic link to prove that the association of smoking with adverse developmental outcomes is not just a correlation, but rather has a distinct biologic basis. Third, identification of the mechanistic basis for the long-term adverse neurobehavioral effects of prenatal nicotine can help in the development of therapeutic strategies to reverse the deficits. Fourth, and perhaps most importantly, knowledge of whether nicotine is responsible for the effects of smoking on development would play an important role in determining whether and how to use nicotine delivery devices (patch, gum, inhaler) that are available for smoking cessation, so that these problems can be avoided in the first place.

A scheme of how cigarette smoking and its associated covariables contribute to perinatal risk and adverse behavioral outcomes is shown in Fig. 1. Nicotine, crossing the placenta, can act directly in the fetus to influence brain development or can alter general fetal development with secondary actions on the brain. Alternatively, nicotine can act on the maternal–fetal unit through its ability to produce uterine or placental vasoconstriction and a resultant hypoxia–ischemia. Cigarette use also involves coexposure to other elements of tobacco smoke, including carbon monoxide (CO) and hydrogen cyanide (HCN), both of which also interfere with oxygen delivery and use, and these additional factors may participate in the overall effects (Mactutus, 1989). Effects on maternal nutrition are also potential contributors, not only because of the anorexic effects of smoking,

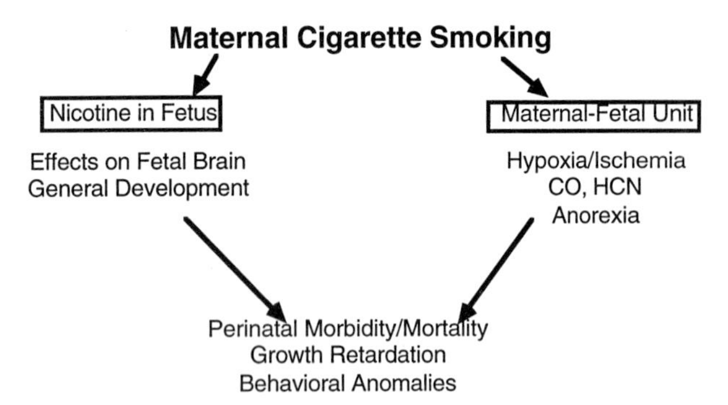

FIGURE 1 Variables contributing to adverse perinatal outcomes in maternal cigarette smoking.

but also because of the depletion of the antioxidant metabolites such as ascorbic acid and glutathion. Finally, the other lifestyle-related risk factors that are more prevalent in the smoking population may be primary contributors or may interact with direct effects of nicotine or effects of nicotine in the maternal–fetal unit. Despite the epidemiologic and statistical evidence available in the human population, it is impossible to prove beyond a doubt the causative relationship of smoking to adverse perinatal outcome without an appropriate animal model that separates these variables.

The most common modeling of the role of nicotine in the developmental effects of smoking pregnancy has, until the mid-1980s, been to inject nicotine into pregnant animals, usually rats. A bolus of nicotine produces high peak plasma levels as compared to human smoking (Fig. 2). Like smoking, nicotine injections produce ischemia and resultant tissue hypoxia (Martin and Becker, 1970, 1971; Cole *et al.,* 1972; Dow *et al.,* 1975; Slotkin *et al.,* 1986b, 1987a; Navarro *et al.,* 1988; McFarland *et al.,* 1991), effects that are exacerbated in smoking by the presence of CO, cyanide, and other chemical components of tobacco smoke. However, there are limitations to the utility with which the injection route can be applied to animal models of smoking. Most importantly, smoking involves multiple small doses of nicotine throughout the day, an impractical situation for animal administration. Instead, most investigations have involved relatively large doses of nicotine administered once or twice a day during pregnancy, typically 1.5–3 mg/kg (Martin and Becker, 1970, 1971; Nasrat *et al.,* 1986; Slotkin *et al.,* 1986b, 1987a). Although this delivers a net daily nicotine dose (corrected for metabolic differences between species) comparable to that in heavy smokers, it does so over a much more restricted period, producing far more intense short-term ischemia and hypoxia than would be experienced in humans. Even when an injection route is selected with a slowed rate of drug entry into the circulation (such as subcutaneous administration), these doses clearly produce blanching of the skin, cyanosis, and respiratory depression (McFarland *et al.,* 1991). As such, injection regimens may be viewed as examinations at the top of the dose–response curve, representative of a worst-case scenario of nicotine's fetal toxicity. Nevertheless, studies using injected nicotine have provided the pioneering work in this field, and significant information has been obtained that has led to the development of animal models more suitable for defining the effects of nicotine on development (Slotkin, 1992). Accordingly, the effects of nicotine injections are summarized here, although they do not prove the involvement of nicotine *per se* in adverse perinatal outcomes.

Nearly coincidentally with the first reports of the effects of maternal smoking on perinatal morbidity, mortality, and neurobehavioral development in humans in the mid-1970s (Cole *et al.,* 1972; Butler and Goldstein, 1973; Dow *et al.,* 1975; Naeye, 1978; Eriksson *et al.,* 1979; Meyer and Carr, 1987), Martin and Becker (1970, 1971) demonstrated that injection of nicotine into pregnant rats results in behavioral alterations in their offspring. Equally important, they pointed out that the effects are similar to those obtained with exposure to hypoxia alone, thus indicating a possible underlying mechanism relevant to both the animal model and human smoking, and opening the door to studies of the underlying biochemical and physiological variables. Because neural development in the neonatal rat corresponds to late gestational development in humans (Dobbing and Sands, 1979; Reinis and Goldman, 1980), some of these effects are elicited even with nicotine given postnatally: nicotine injected directly into newborn rats results in a pattern of noradrenergic hyperinnervation comparable to that seen in animals receiving drugs that lesion specific neuronal tracts within the central nervous system (CNS) (Jonsson and Hallman, 1980). Subsequently, it was found that a single episode of neonatal hypoxia is also sufficient to reproduce this effect (Slotkin *et al.,* 1986a).

Daily subcutaneous administration of nicotine to pregnant rats throughout gestation shares numerous features of cigarette smoking, with significant fetal resorption and IUGR accompanying the hypoxia–ischemia experienced with each injection (Nasrat *et al.,* 1986; Slotkin *et al.,* 1986b, 1987a,b; Navarro *et al.,* 1988). Measurement of sensitive biochemical markers of cell development confirms the presence of cell damage, interference with cell replication and growth, and deletion of neurons in favor of glia, a pattern typical of nonspecific injury, such as that elicited by exposure to heavy metals or hypoxia (Slotkin, 1992). At the synaptic level, a number of different transmitter pathways, notably catecholaminergic projections, display hyperactivity conse-

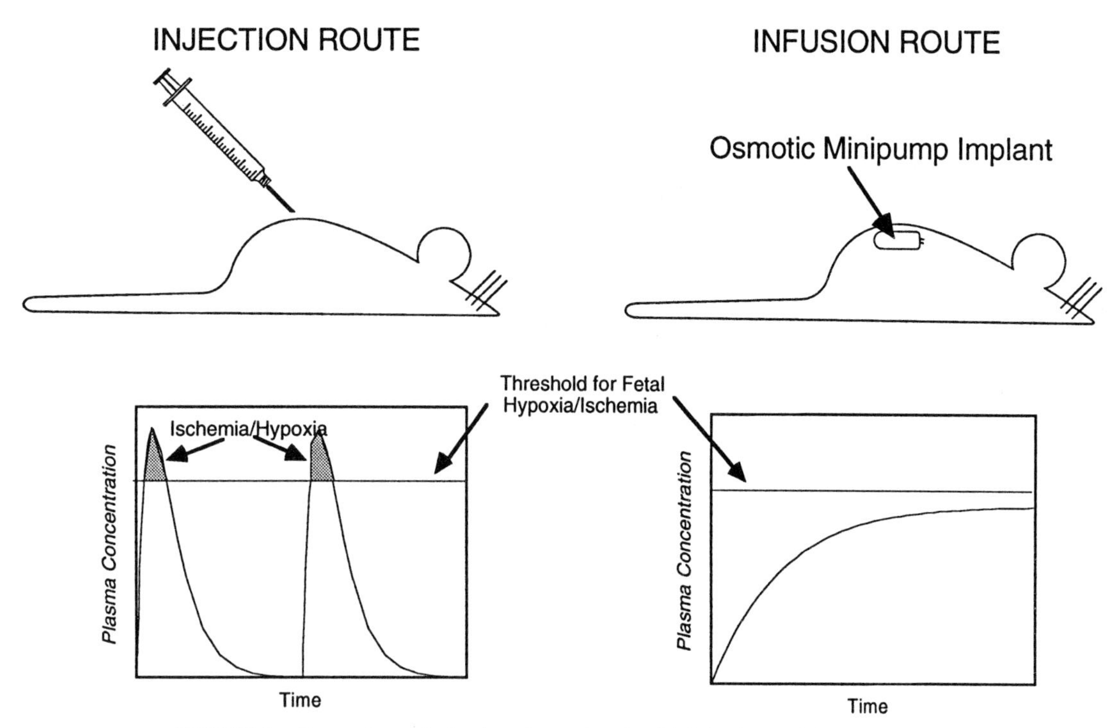

FIGURE 2 Comparison of nicotine plasma levels with injection and infusion routes.

quent to synaptic hyperproliferation in the damaged areas, again typical of early neuronal injury (Slotkin, 1992). Because hypoxia alone or exposure to other smoke products such as CO can elicit the same effects, it is unlikely that such damage is directly attributable to nicotine itself acting within the fetal brain (Fechter and Annau, 1977; Fechter *et al.,* 1986; Woody and Brewster, 1990; Slotkin, 1992). Hypoxia and ischemia are significant components of human cigarette smoking, and therefore these results are relevant to the issue of smoking and fetal brain damage. However, in order to determine whether nicotine acts on the fetal brain by itself, we must discard this wealth of information and turn instead to exposure models that do not involve acute hypoxic–ischemic episodes.

As shown in Fig. 2, continuous administration is an alternative animal model of nicotine exposure that isolates the effect of nicotine from acute effects on the maternal–fetal unit by achieving a steady-state drug level that does not surpass the threshold for compromised circulatory function. This was first attempted by the inclusion of nicotine in drinking water (Peters *et al.,* 1979; Peters, 1984). Although behavioral and neurotransmitter deficits were seen in the offspring, the authors also found that maternal weight gain was severely restricted by the treatment. Subsequently, other investigators have reported that nicotine in drinking water is extremely aversive and that animals reduce their fluid intake to hydropenic levels (Náquira and Arqueros, 1977; Murrin *et al.,* 1985, 1987). Indeed, a report by Zhu *et al.* (1996) found that continued inclusion of nicotine in drinking water throughout fetal and neonatal development results in severe maternal weight loss and growth inhibition of the offspring, likely as a result of reduced milk output from dehydration; consequently, development of brain regions that are largely devoid of nicotinic cholinergic receptors and that should be spared the effects of nicotine were compromised as much or more than regions enriched in nicotinic sites. Accordingly, studies with nicotine in drinking water have largely failed in the objective of designing an animal model that isolates the effects of nicotine on the fetus from other variables.

The advent of the implantable osmotic minipump has provided a solution to this problem (Fig. 2). Infusion avoids acute hypoxia–ischemia while enabling fetal exposures to be conducted at plasma levels of nicotine comparable to those in typical smokers (Murrin *et al.,* 1987; Lichtensteiger *et al.,* 1988; Fung, 1989; Slotkin, 1992). Because most studies have been conducted in the rat, it is important to determine what levels of exposure are appropriate in light of species differences in metabolism, pharmacokinetics, and pharmacodynamics of nicotine. An infused dose of 1.5–2 mg/kg/day produces plasma nicotine levels in rats similar to those of humans who smoke approximately 10–20 cigarettes per day (Murrin *et al.,* 1987; Lichtensteiger *et al.,* 1988; Fung, 1989). At this dose, fetal nicotine exposure causes neurochemical and behavioral changes in the offspring without evoking fetal resorption or growth retardation (Slotkin *et al.,* 1987b, 1997a; Navarro *et al.,* 1989b; Seidler *et al.,* 1992; Slotkin, 1992; Zahalka *et al.,* 1992, 1993c; Levin *et al.,* 1993, 1996b). Raising the dose to 4 mg/kg/day causes modest IUGR (Cutler *et al.,* 1996), which becomes increasingly severe as the dose is increased further to 6 mg/kg/day, which gives plasma levels in rats at the upper end of human exposures in heavy smokers (2–3 packs/day) (Lichtensteiger *et al.,* 1988). These higher dose levels reproduce some of the major epidemiologic features of cigarette smoking in man, including small reductions in maternal weight gain, increased rates of fetal resorption, and IUGR (Slotkin *et al.,* 1987b,c; Navarro *et al.,* 1988, 1989a,b, 1990b). Part of the discrepancy between plasma levels and perinatal effect resides in differential sensitivities of rats and humans to equivalent nicotine dose levels. That is, not only should a model incorporate the species differences in pharmacokinetics (requiring a higher dose in rats to achieve the same plasma levels as in humans), but also the pharmacodynamic differences that dictate the net pharmacologic effect of those doses. The rat is relatively insensitive to nicotine (Barnes and Eltherington, 1973), so that 6 mg/kg/day is appropriate for eliciting the electrophysiologic changes connoting nicotine's impact on CNS function (Lichtensteiger *et al.,* 1988).

Therefore, in discussing the developmental and behavioral effects of fetal nicotine exposure, it is important to evaluate both high- and low-dose paradigms, the former to identify all neurobehavioral parameters that are sensitive to perturbation by nicotine, and the latter to determine whether a given neural effect exhibits a lower threshold than does general developmental impairment.

The nicotine infusion model eliminates the problems of confounding variables such as injection stress, handling, hypodipsia, and episodic hypoxia–ischemia that are components of other administration paradigms; accordingly, this model is preferable for examination of the role of nicotine itself in neurobehavioral teratogenesis (Slotkin, 1992). Nevertheless, it may be possible to design even better models in which animals self-administer nicotine during pregnancy, to provide even a more realistic simulation of human exposure.

V. Effects of Prenatal Nicotine Exposure on Cellular Development

Before reviewing the evidence for behavioral teratogenesis by nicotine, it is important to understand the underlying basis for behavioral effects, namely how nicotine perturbs cellular development. In addition to causing fetal resorption and IUGR, infusions of high doses of nicotine (6 mg/kg/day) given to pregnant rats throughout gestation also evoke brain cell damage and reduction in brain cell numbers (Slotkin, 1992). One of the best examples of this effect is the elevation seen in ornithine decarboxylase (ODC), an enzymatic marker that is activated during the processes of cell damage and repair (Gillette and Mitchell, 1991; Vendrell *et al.,* 1991; Zawia and Harry, 1993; Kindy *et al.,* 1994) and during developmental delays (Slotkin and Thadani, 1980; Bell and Slotkin, 1986; Slotkin and Bartolome, 1986). Prenatal nicotine exposure produces a sustained increase in ODC throughout early postnatal life (Slotkin *et al.,* 1987c; Navarro *et al.,* 1989b), including the periods of neurogenesis, synaptogenesis, and gliogenesis (Fig. 3). The effect is present in regions that differ both in their maturational profiles (forebrain develops earlier than cerebellum) and in their nicotinic receptor concentrations (forebrain > cerebellum); the issue of receptor involvement is extremely important and is dealt with in greater detail in the next section. The elevation of ODC activity is associated with actual loss of brain cells, as indicated by shortfalls in the amount of DNA in each region (Slotkin *et al.,* 1987c; Navarro *et al.,* 1989b). As each brain cell has a single nucleus, the amount of DNA is an indicator of the number of cells (Winick and Noble, 1965). Prenatal nicotine exposure causes a progressive deficit in DNA during the brain growth spurt, and recovery of cell numbers only occurs after the closure of neurogenesis, implying that neurons have been replaced with glia (Slotkin *et al.,* 1987c).

In order for adverse effects on cell maturation to be translated into altered behavioral development, there must be eventual consequences for synaptic function. Given the existence of specific neurotransmitter receptors targeted by nicotine, namely nicotinic cholinergic receptors, the most obvious effects of nicotine exposure would be expected to be directed toward cholinergic function. Most (Sershen *et al.,* 1982; Hagino and Lee, 1985; Slotkin *et al.,* 1987b; van de Kamp and Collins, 1994) but not all (Fung and Lau, 1989) studies have found that chronic exposure of pregnant rats to nicotine increases the number of nicotinic receptor sites in the brains of the fetus and neonate, an effect that persists through the postnatal period of synaptogenesis before declining to control levels (Slotkin *et al.,* 1987b; van de Kamp and Collins, 1994). During the period of elevated receptor expression, nicotinic responses are also correspondingly supersensitive (Slotkin *et al.,* 1991). Although this effect is transient (Seidler *et al.,* 1992), the fact that hyperresponsiveness occurs during the synaptogenic period, when cholinergic activity "programs" structural and functional development of cholinergic

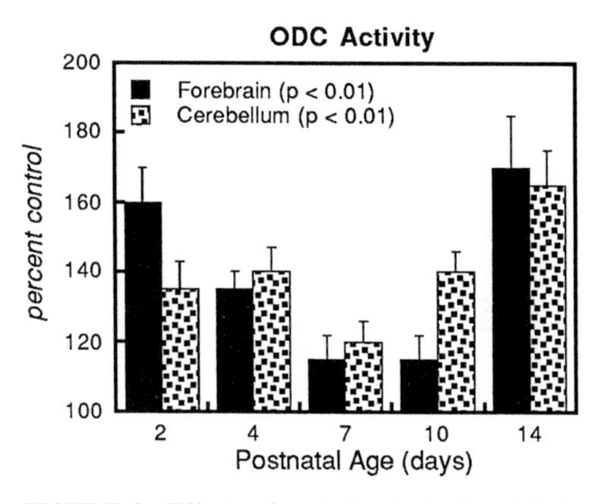

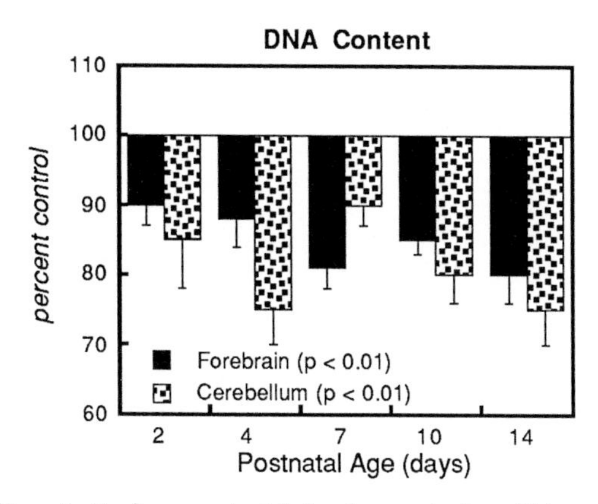

FIGURE 3 Effects of gestational nicotine infusion (6 mg/kg/day) on postnatal development of ornithine decarboxylase and DNA in rat brain regions.

target cells (Hohmann *et al.,* 1988; Navarro *et al.,* 1989a), indicates that prenatal nicotine exposure can sensitize cells such that the functional roles of subsequent trophic inputs are altered. This may produce a situation where abnormalities appear after a prolonged delay period in which neurobehavioral performance is normal, a *sine qua non* of behavioral teratogens.

Although most studies report up-regulation of nicotinic receptor expression in fetal brain during nicotine exposure, binding to some nicotinic receptor subclasses, notably that identified by α-bungarotoxin (α-BGT), seems to be relatively little affected (van de Kamp and Collins, 1994). Receptor subtypes show differential expression in developing brain regions and in particular, the α-BGT–sensitive site may play a role in neuritic outgrowth (Chan and Quik, 1993; Zoli *et al.,* 1995; Bina *et al.,* 1995; Broide *et al.,* 1996). Failure to desensitize and up-regulate these receptors could represent a major feature by which nicotine perturbs development. Research concerning the relative involvement of different nicotinic receptor subtypes in nicotine-induced neuroteratology is certainly warranted and may provide further insights into the mechanisms underlying the persistent neurobehavioral effects. Perhaps equally important, knowledge of the role of subtypes that mediate neurodevelopmental alterations might permit the design of nicotine substitution therapies that would spare fetal development.

In addition to effects exerted on nicotinic receptors and their linked trophic responses, prenatal nicotine exposure also compromises the development of cholinergic neuronal activity, as assessed with characteristic two biochemical markers (Navarro *et al.,* 1989a): choline acetyltransferase, which converts choline to acetylcholine (ACh), and the high affinity choline transporter that regulates ACh precursor availability. Choline acetyltransferase (ChAT) is a constitutive component of cholinergic nerve terminals and is required for neurotransmitter synthesis, but its activity is not the rate-limiting factor and the enzyme is unresponsive to nerve impulse activity (Cooper *et al.,* 1986). Instead, the relatively constant concentration of this enzyme within the nerve ending provides a reliable index of the number of cholinergic terminals and correlates well with the development of projections to target sites (Coyle, 1976; Coyle and Yamamura, 1976; Ross *et al.,* 1977; Zahalka *et al.,* 1992, 1993c). ACh synthesis is limited by the ability of the terminals to take up the precursor, choline, via the high affinity transporter that *is* responsive to nerve impulse activity and can change acutely over a course of minutes to hours to compensate for increased demand for transmitter release (Simon *et al.,* 1976; Klemm and Kuhar, 1979; Murrin, 1980). By combining these two factors to evaluate the ratio of choline uptake (or choline transporters) to ChAT activity, a measure of the net impulse activity per nerve terminal can be obtained. This index has already proved useful in evaluations of central cholinergic tone in developing brain and in neurodegenerative diseases such as Alzheimer's disease (Navarro *et al.,* 1989a; Slotkin *et al.,* 1990, 1994; Zahalka *et al.,* 1993a).

Applying the activity ratio method to developing rat forebrain (Fig. 4) indicates that cholinergic tone does not develop monotonically from low to high values, but rather displays a distinct peak centered around postnatal (PN) day 10 (Navarro *et al.,* 1989a). In animals exposed to nicotine prenatally, the postnatal peak of nerve activity is blunted. Using radiolabeled hemicholinium-3 to quantitate the number of high affinity choline transporter molecules also indicates initial neuronal hypoactivity in the postnatal period after prenatal nicotine exposure and, in some specific target regions such as the hippocampus, severe and persistent deficits eventually reemerge, a potential harbinger of behavioral teratogen-

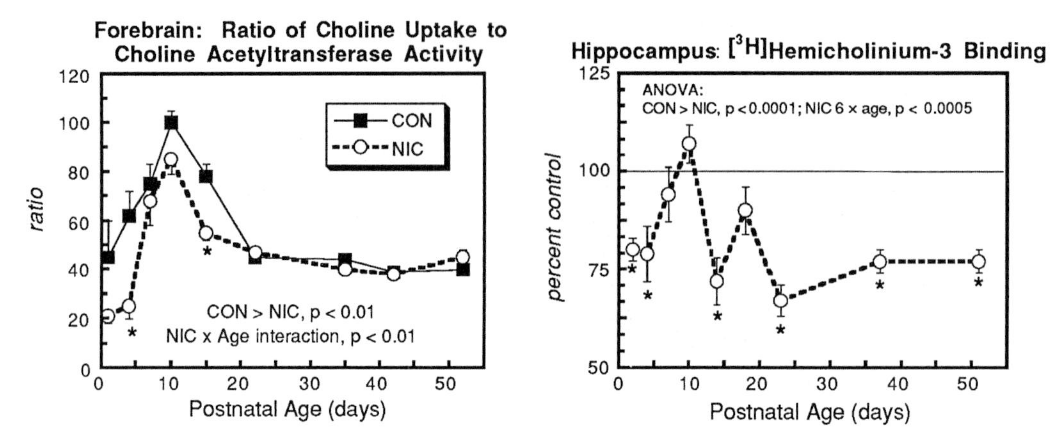

FIGURE 4 Effects of prenatal nicotine exposure on indices of cholinergic neuronal activity in forebrain and hippocampus.

esis (Zahalka *et al.,* 1992). Compounding the problem, cholinoceptive mechanisms eventually also become compromised, leading to even greater impairment of synaptic function (Zahalka *et al.,* 1993c).

Given the close anatomic association of cholinergic nicotinic mechanisms and catecholaminergic transmitter systems, it would be expected that prenatal nicotine exposure would also target these systems (Lichtensteiger *et al.,* 1988). Indeed, in the nicotine group, noradrenergic and dopaminergic projections show the same initial deficits in synaptic activity as seen for cholinergic neurons (Navarro *et al.,* 1988). Norepinephrine levels and turnover (an index of synaptic activity) are severely subnormal after birth in the nicotine cohort, and although apparent recovery occurs by 3 weeks of age, a subsequent and persistent deficit reemerges after puberty (Fig. 5). Synaptic reactivity to stimulation is also subnormal (Seidler *et al.,* 1992). At 30 days of age, before the reemergence of deficits in norepinephrine turnover, an acute challenge with nicotine fails to release any neurotransmitter in the prenatal nicotine group, whereas controls show a robust response.

Studies of cell development and synaptic function thus indicate conclusively that nicotine by itself, without participation of other components of tobacco or of the epiphenomena of maternal cigarette smoking, has an adverse effect on brain development. Late reemergence of synaptic dysfunction is paralleled by perturbed behaviors and neuroendocrine parameters linked to catecholaminergic pathways (Lichtensteiger and Schlumpf, 1985; Ribary and Lichtensteiger, 1989). Although it is tempting to attribute the functional deficits simply to decreased impulse activity, there is a logical inconsistency in such a conclusion. In the mature nervous system, long-term decreases in impulse activity are compensated by up-regulation of postsynaptic receptor sites juxtaposed to the nerve terminal, with the decrease in input and increased responsiveness offsetting each other to result in normal functional status. However, direct measurements of adrenergic receptor binding sites in

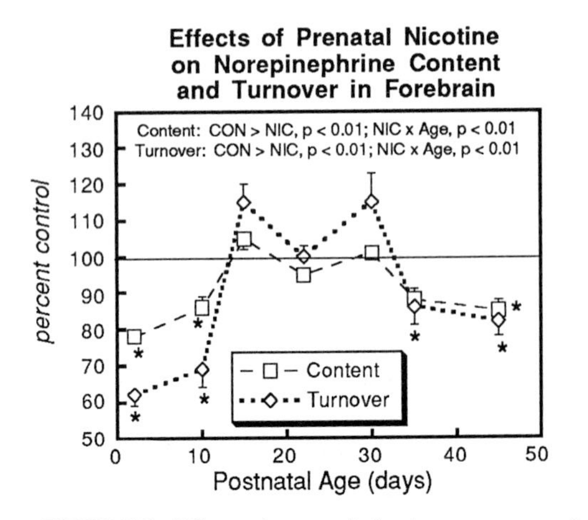

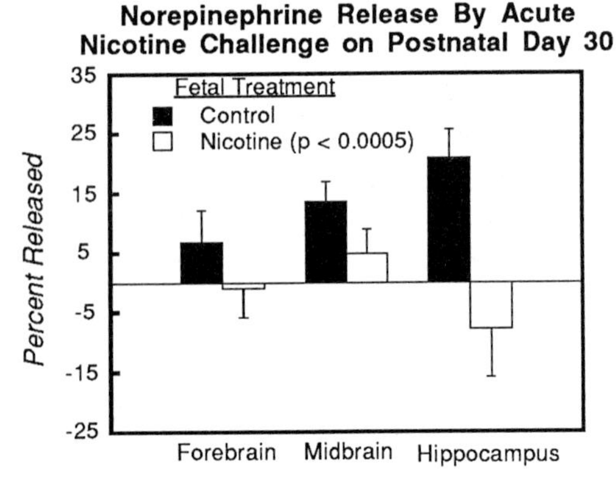

FIGURE 5 Effects of prenatal nicotine exposure on indices of noradrenergic neuronal activity and reactivity.

brain regions of animals exposed to maternal nicotine infusions reveal no upregulation of the β- or α_2-receptor subclasses, and only minor, transient increases in the α_1-receptor population (Navarro *et al.,* 1990b). Why, then, are these animals functionally subresponsive? Early in synaptic development, exposure of postsynaptic receptor sites to neurotransmitter provides information that enables the target cells to "program" their subsequent function. Accordingly, early denervation or receptor blockade produces permanent shortfalls in receptor-mediated responses rather than causing upregulation of receptor numbers and responses as would be the case with denervation in adulthood (Deskin *et al.,* 1981; Criswell *et al.,* 1989; Hou *et al.,* 1989a,b; Kostrzewa and Saleh, 1989; Kudlacz *et al.,* 1990a,b). Conversely, early exposure to neurotransmitter stimulation uniquely promotes the development of receptor linkages to postsynaptic function (Giannuzzi *et al.,* 1995; Zeiders *et al.,* 1997). The early postnatal deficits in noradrenergic activity caused by prenatal nicotine exposure would thereby reduce exposure of the sites to transmitter precisely during the critical period in which responses are being imprinted. Work in the peripheral nervous system has provided key information to illustrate this phenomenon. Over the first 3 weeks postnatally, cardiac β-adrenergic receptors increase in their number and linkage to heart rate control, coincidentally with a developmental surge in noradrenergic neuronal activity (Seidler and Slotkin, 1981; Slotkin, 1986). When this surge is blunted by prenatal nicotine exposure, the number of β-receptor sites develops more slowly (Fig. 6); the functional correlate is that heart-rate responses to a β-adrenergic agonist (isoproterenol) are subnormal, requiring six times the dose to achieve a 50 beat/min increase (Navarro *et al.,* 1990a). Although receptor numbers resolve to normal by young adulthood, the performance deficits do not: heart-rate responses elicited either with an adrenergic drug or physiologically through central stimulation of the nerve supply to the heart remain subnormal. Because neuronal input was reduced during the critical period of response programming, the nicotine-exposed animals never achieve proper reactivity to neuronal stimuli. The same factors operate to produce long-term deficits in cholinergic receptor signaling in the CNS: the blunting of cholinergic tone during synaptogenesis leads to impairment of cholinergic receptor development and of the linkage of these receptors to cellular responses (Zahalka *et al.,* 1992, 1993c). These studies thus provide not only a demonstration that nicotine itself is a neurochemical–neurobehavioral teratogen, but also provide evidence for the underlying cellular mechanisms by which behavioral patterns are programmed by early neuronal input, findings that may be of considerable importance in the general understanding of behavioral teratogenesis. The functional deficits in peripheral adrenergic systems are also discussed in terms of its impact on physiologic function later in this chapter.

One of the key questions in nicotine-induced neurobehavioral anomalies is whether the effects occur only when there is gross morphologic change or growth impairment. In obstetric practice, the criterion of low birthweight is considered to be one of the most critical risk factors; in the case of cigarettes, for example, the U.S. Surgeon General's warning to pregnant women points out the relationship of smoking to "fetal injury, premature birth, and low birth weight." Whereas in clinical studies, factors like growth impairment are difficult to separate from drug effects targeted to the nervous system, animal models are ideal for separation of these variables. It is important to note, therefore, that all of the nicotine-induced neurodevelopmental defects found

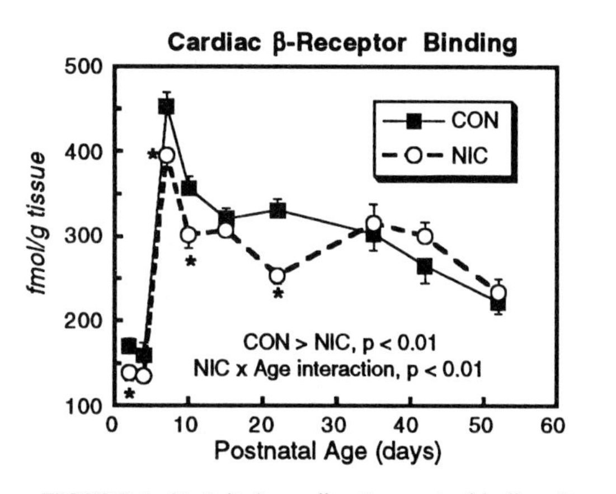

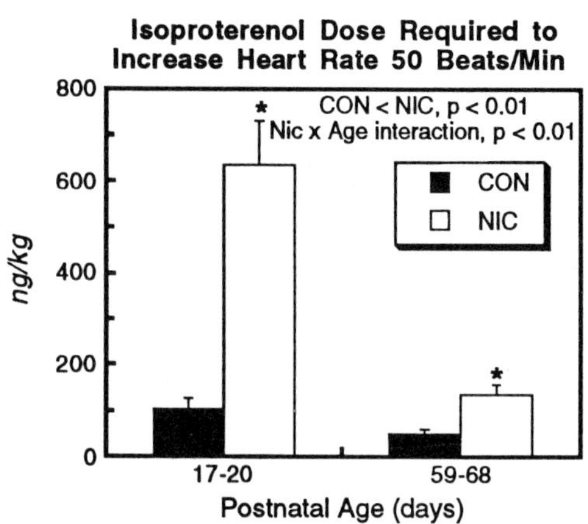

FIGURE 6 Deficits in cardiac β-receptor binding sites and in the physiologic response to receptor stimulation after prenatal nicotine exposure.

at high dose levels are also present when the dose of nicotine is reduced (2 mg/kg/day) to produce plasma levels comparable to those in moderate smoking (Navarro *et al.,* 1989b; Seidler *et al.,* 1992; Zahalka *et al.,* 1992, 1993c); under these conditions, drug effects on maternal weight gain, fetal resorption, litter size, and all standard morphologic variables are no longer present. Thus, doses of nicotine that are below the threshold for eliciting growth impairment are already fully capable of altering nervous system development. Comparable behavioral alterations at the lower dose of nicotine have also been reported (Lichtensteiger and Schlumpf, 1985; Ribary and Lichtensteiger, 1989; Levin *et al.,* 1993, 1996b; Cutler *et al.,* 1996).

These results indicate that, unlike standard teratogens or fetotoxins, which spare brain development relative to effects on the rest of the organism, nicotine targets the nervous system specifically. The important corollary is that, for nicotine (and by implication, for cigarette smoking), the use of a standard marker such as prematurity or birthweight is inadequate to assess whether there is a "safe" level of exposure. Equally vital, the question of why nicotine is an atypical fetotoxin–teratogen opens the door to mechanistic studies in a relatively new area in the study of birth defects: the targeting of a teratogen to a highly specific population of cell receptors.

VI. Targeting the Nicotinic Cholinergic Receptor

The lower dose threshold for adverse effects of fetal nicotine exposure on neuronal development as compared to growth impairment suggests that specific elements within the nervous system are targeted by nicotine. The most likely candidate is the nicotinic cholinergic receptor that transduces the cholinergic neuronal signal into cell membrane depolarization. The presence of nicotinic receptors in the fetal nervous system has been demonstrated biochemically by *in vitro* radioligand binding techniques and morphologically by *in situ* receptor autoradiography (Hagino and Lee, 1985; Larsson *et al.,* 1985; Lichtensteiger *et al.,* 1987; Slotkin *et al.,* 1987b; Cairns and Wonnacott, 1988). The course of development is caudal to rostral (Lichtensteiger *et al.,* 1988), so that nicotine targeting of these sites would progress from the brainstem in midgestation to the forebrain in later fetal development. However, the presence of receptor binding sites does not necessarily prove that the receptors are functionally connected to membrane depolarization, an event that presumably must occur before prenatal nicotine exposure can influence neuronal development. This can be determined by looking for the upregulation of receptor sites that accompanies desensitization from prolonged nicotine–induced depolarization (Schwartz *et al.,* 1982; Clarke *et al.,* 1985; Schwartz and Kellar, 1985). Thus, chronic nicotine administration in adult rats, where the receptors are definitively coupled to depolarization, results in increases in the number of nicotinic receptors, readily measured with radioligand binding. Similarly, up-regulation of receptors in fetal rat brain has been found, confirming that long-term cellular stimulation is occurring when nicotine is administered during gestation (Slotkin *et al.,* 1987b). Importantly, the same degree of receptor stimulation is obtained even with low-dose paradigms (Navarro *et al.,* 1989b), a finding that explains why neuronal effects are robust at a nicotine exposure level that does not inhibit growth.

However, why should cell depolarization by nicotine lead to abnormalities of nervous system development? The answer is that the ontogeny of synaptic transmission is a key element enabling developing cells to decide between replication and differentiation. A vast amount of literature exists that describes the role of neurotransmitters as trophic factors in developing cells. The ability of neurally active substances to alter membrane polarity in cells that contain specific receptors for these compounds has been demonstrated to control the expression of genes that are ordinarily turned on only when the cell enters differentiation (Patterson and Chun, 1977; Black, 1978, 1980; Mytilineou and Black, 1978). ACh, released by developing cholinergic neuronal projections, provides such trophic signals to selective target areas within the brain so that lesioning of these tracts leads to architectural disorganization (Hohmann *et al.,* 1988; Navarro *et al.,* 1989a). It is therefore possible to propose a model that explains both the exquisite sensitivity of the developing brain to prenatal nicotine exposure as well as the pattern of disrupted cell and synaptic development: fetal stimulation of nicotinic receptors by the drug results in premature onset of cell differentiation at the expense of replication, leading to shortfalls in cell number and aberrant patterns of synaptogenesis and synaptic activity. This model has been verified by studying the effects of acute nicotine administration on DNA synthesis in fetal and neonatal rat brain regions (McFarland *et al.,* 1991). Within 30 min of exposure to a single dose of nicotine, and persisting for several hours, DNA synthesis is reduced by as much as 50%, with the rank order of effect brainstem ≥ cerebral cortex > cerebellum, the same order as for regional nicotinic receptor concentrations (Fig. 7). The effect is restricted to replicating cells, as there are no alterations in factors reflecting synthesis of other macromolecules. Extensive studies demonstrate that the effect is, in fact, mediated directly via nicotinic receptors within the developing

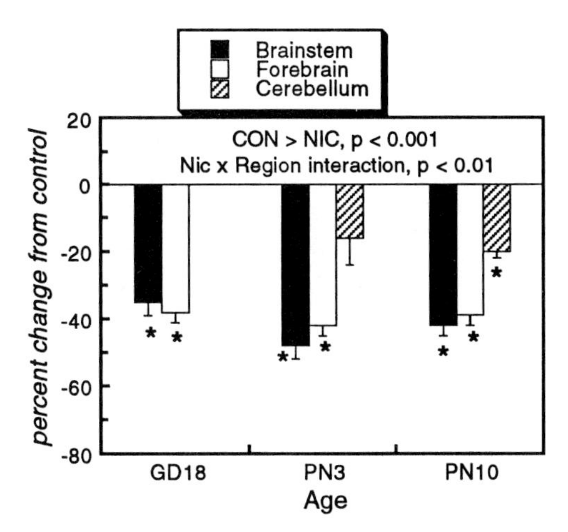

FIGURE 7 Acute inhibition of DNA synthesis in rat brain regions caused by a single dose of nicotine.

CNS: the deficit is still present when systemic ischemic–hypoxic effects of nicotine are eliminated, and inhibition of DNA synthesis can be evoked even by minute doses of nicotine introduced directly into the CNS. Finally, maternal nicotine administration produces the acute arrest of cell replication in fetal brain, thus indicating that the fetal nicotinic receptors are fully capable of eliciting the same developmental switchover to differentiation (McFarland *et al.,* 1991). One corollary of the targeting of nicotinic cholinergic receptors for interference with cell replication is that other perturbations that cause excessive cholinergic stimulation, including dietary choline supplementation or inhibition of ACh breakdown (cholinesterase inhibitors), evoke the same types of defects as do nicotine exposure (Bell and Slotkin, 1985, 1986; Bell *et al.,* 1986; Whitney *et al.,* 1995; Campbell *et al.,* 1997). These findings confirm conclusively that the selectivity and high sensitivity of the developing CNS to fetal nicotine exposure is a direct consequence of a specific, receptor-mediated process normally intended to control the timing of cell differeniation, but in this case elicited inappropriately and prematurely by nicotine.

If direct stimulation of nicotinic cholinergic receptors in the fetal brain is largely responsible for disrupted cell development and consequent effects on synaptic function and behavior, then the developmental profiles of receptors should define the critical sensitive period for nicotine's effects. Because these receptors begin to proliferate in midgestation (Lichtensteiger *et al.,* 1987, 1988; Naeff *et al.,* 1992), the earliest phases of fetal development should be relatively "silent" for adverse effects of nicotine. Indeed, examination of the same cell and synaptic markers that characterize the adverse effects of gestational nicotine exposure has confirmed that early gestational treatment has little or no effect (Slotkin *et al.,* 1993). Similarly, nicotine withdrawal does not represent a significant factor (Slotkin *et al.,* 1993), because it is nicotine itself that causes cell injury. These corollaries of receptor targeting have important implications for decisions about the management of smoking cessation in pregnant women, as is discussed at the end of this chapter.

Despite the clear-cut targeting of cell replication by nicotine, there are discrepancies in ascribing the entire spectrum of nicotine's effects to this mechanism. First, adverse effects are not strictly limited to brain regions that are usually considered to be enriched in cholinergic innervation (as based on the distribution in the mature brain), although these do tend to be more affected (Slotkin *et al.,* 1987a,b,c; Navarro *et al.,* 1989a,b; McFarland *et al.,* 1991). In part, these more widespread effects of nicotine may reflect transient expression of cholinergic phenotypes in regions that are not cholinergically enriched in maturity (Zahalka *et al.,* 1992, 1993a,b), as well as less regionally distinct nicotinic receptor expression or even transient overexpression during critical developmental periods (Slotkin *et al.,* 1987b; Court and Clementi, 1995; Court *et al.,* 1995). These qualifications do not explain, however, the fact that CNS cell loss actually shows greater deficits *after* the termination of nicotine exposure and beyond the period in which neurogenesis is occurring, a finding incompatible with actions targeted solely toward cell replication (Slotkin *et al.,* 1987c; Navarro *et al.,* 1989b). Indeed, the changes in synaptic activity and behavior typically show delayed postnatal abnormalities that appear only after a prolonged period of apparently normal function (Lichtensteiger and Schlumpf, 1985; Lichtensteiger *et al.,* 1987, 1988; Navarro *et al.,* 1988; Ribary and Lichtensteiger, 1989; von Ziegler *et al.,* 1991; Seidler *et al.,* 1992).

It is thus evident that fetal nicotine exposure also influences later events in the terminal differentiation or survival of neurons, such that functional alterations can appear after a considerable delay. As shown previously, increases in the activity of ODC in postnatal rat brain occur after prenatal nicotine (Slotkin *et al.,* 1987c; Navarro *et al.,* 1989b), implying that cell injury is continuing to occur even after the drug exposure ceases. One possibility, then, is that nicotine evokes changes in brain cell development that lead ultimately to cell death, either as a result of cryptic injury or of enhancement of the programmed dying-back of neurons (apoptosis). Studies have evaluated the effects of prenatal nicotine exposure on expression of the nuclear transcription factor, c-*fos* (Slotkin *et al.,* 1997a). Although c-*fos* is elevated briefly during acute cell stimulation (Curran *et al.,* 1990; Sheng and Greenberg, 1990; Edwards, 1994; Curran and Morgan, 1995), including that caused by nicotine (Koistinaho, 1991; Koistinaho *et al.,* 1993; Sharp *et al.,* 1993; Slotkin *et al.,* 1997a), prolonged constitutive expression of c-*fos* is highly unusual and is most closely related to apoptosis. Chronic c-*fos* expression by itself actually elicits cell death in even otherwise healthy cells (Smeyne *et al.,* 1993; Miao and Curran, 1994; Curran and Morgan, 1995; Preston *et al.,* 1996). It is therefore quite important that it was found that exposure of fetal rats to nicotine produces chronic elevations of c-*fos* expression that are detectable through late gestation and into the postnatal period, days after nicotine exposure ceases. The effect displayed the same critical period as had been shown previously for adverse effects of nicotine on cell numbers and synaptic function (Slotkin, 1992). These results strongly support the idea that, in addition to the immediate effects on cell replication, developmental nicotine exposure evokes delayed neurotoxicity by altering the program of neural cell differentiation to produce apoptosis.

Thus, there is conclusive evidence that nicotine itself is a neuroteratogen, targeting cell development of specific cells containing nicotinic cholinergic receptors. There are both immediate deleterious effects on cell replication as well as delayed neurotoxic effects reflecting subsequent cell loss from cryptic injury and/or apoptosis. The eventual deterioration of synaptic function in both cholinergic and catecholaminergic systems indicates the need to examine behavioral studies of both basal function in the affected regions as well as tests that challenge the performance limits.

VII. Behavioral Effects

Complementing epidemiologic studies that demonstrate adverse behavioral consequences of smoking during pregnancy, experimental studies in rodents have confirmed that prenatal nicotine exposure by itself can evoke adverse cognitive effects (Martin and Becker, 1971; Genedani *et al.,* 1983; Sorenson *et al.,* 1991; Yanai *et al.,* 1991, 1992) and hyperactivity (Richardson and Tizabi, 1994). However, these functional effects are not as robust as are the neurochemical defects described previously. This apparent discrepancy is not surprising in light of the redundancy in neural systems, sometimes referred to as *neural reserve.* Indeed, alternative neural pathways or adaptations can make up for major neuronal deficits that occur with perinatal injury (Lorber, 1980). This may underlie the fact that some studies have not seen deficits (Bertolini *et al.,* 1982; Sobrian *et al.,* 1995) and others have seen only subtle effects after pharmacologic or behavioral challenge (Cutler *et al.,* 1996; Levin *et al.,* 1996b). Challenges can serve not only to uncover nicotine-induced deficits for which there is compensation, they can also serve to help determine which neurobehavioral and neurochemical systems underlie the nicotine-induced effects and compensation. An approach using challenges thus increases our knowledge base concerning mechanisms of nicotine-induced neurobehavioral teratology and provides leads for the development of strategies to treat such deficits or to avoid them in the first place.

A. *Locomotor Activity*

Hyperactivity has been found in young rats after prenatal nicotine exposure. Studies with injected nicotine have identified sex selectivity, with males being more severely affected; in keeping with the idea of adaptive change that offsets nicotine-induced damage, the hyperactivity is less evident at older ages (Martin *et al.,* 1976; Martin and Martin, 1981). However, such studies invariably confound the effects of nicotine with hypoxia–ischemia because of the use of the injection route. It is therefore extremely important that locomotor hyperactivity has also been seen after prenatal nicotine exposure via low-dose minipump infusions throughout gestation (Fung, 1988). The specificity for locomotor activity was demonstrated by the absence of comparable effects on motor reflex development (Fung and Lau, 1989). Similarly, exposure to prenatal nicotine via infusions restricted to late gestation produced hyperactivity at about the same age and confirmed the male sex-selectivity of the effect (Lichtensteiger *et al.,* 1988; Schlumpf *et al.,* 1988). These results are in keeping with the neurochemical data that indicate a late gestational critical period for nicotine-induced CNS cell damage (Slotkin *et al.,* 1993).

If effects on locomotor activity are directly related to underlying neurochemical damage, then there should be a dependence of the behavioral change on synaptic mechanisms that are known to be affected by prenatal nicotine exposure. Unfortunately, these relationships have not been demonstrable consistently. Richardson and Tizabi (1994) used the nicotine infusion model and examined effects on male rats. Nicotine-induced hyperactivity was accompanied by increased dopamine concentrations in the substantia nigra and decreased dopamine levels in the ventral tegmental area and striatum, whereas no consistent changes were seen in frontal cortex or nucleus accumbens. There was also no correlation between alterations in dopamine levels, which were present throughout the nicotine group, and hyperactivity, which was displayed only by certain individuals. Similarly, Fung and Lau (1989) did not find nicotine-induced changes in locomotor activity to be related to alterations in striatal nicotinic or dopaminergic receptor binding, although, again, many of the changes in receptor binding were restricted to male rats. Interestingly, as with hyperkinesis in humans, systemic administration of amphetamine attenuated the hyperactivity of nicotine-exposed pups. Nevertheless, it is evident that the net effect of prenatal nicotine exposure on locomotor activity is a complex combination of the disruption of CNS development, superimposed on numerous adaptive and compensatory mechanisms that obscure the assignment of a one-to-one relationship between behavior and cell development.

B. Cognitive Function

In contrast to the dramatic alterations in neurochemical measures, prenatal nicotine administration has been found to cause relatively modest cognitive defects, again likely reflecting compensatory mechanisms such that functional impairment is sometimes not seen when parameters are limited to basal behavioral performance. With some tasks prenatal nicotine administration has been found to impair learning and memory persistently, whereas with other tasks no baseline deficits have been found. These disparities in the sensitivity to prenatal nicotine help to delineate the specificity for cue structure and motivational influences, but still leave open the possibility that functional deficits would be unmasked with pharmacologic or behavioral challenge to partially compensated for neural effects.

Spontaneous alternation is a task displaying persistent effects of prenatal nicotine. This tests the innate tendency of rodents and other animals to seek out novel locations, typically assessed as the alternation of consecutive arm entries in a T-maze. The alternation rate develops from near chance levels (50%) at weaning to adult levels of about 80–85% by adolescence (Douglas, 1975). Johns *et al.* (1982, 1993) injected pregnant guinea pigs with graded doses of nicotine and found impaired spontaneous alternation in adolescence and deficits in acquisition and reversal learning of brightness discrimination in adulthood. However, many of these effects could be attributable to hypoxic–ischemic insult from the injected nicotine, as similar results are obtained with hypoxic episodes alone (Martin and Becker, 1971). It is therefore important that we have found comparable effects using the low-dose (2 mg/kg/day) osmotic minipump paradigm that avoids hypoxia. In contrast to the normal developmental course, the nicotine-exposed rats showed an opposite pattern, with an initial acceleration in the development of alternation apparent from weaning until the onset of puberty, followed by a deterioration of alternation after puberty (Levin *et al.,* 1993). Thus, nicotine induces a precocious development of spontaneous alternation but ultimately impairs performance in adulthood. As noted from neurochemical studies, this pattern mimics that of the eventual deterioration of synaptic performance.

A prenatal nicotine-induced deficit was also seen in the study of rewarded T-maze spatial alternation (Levin *et al.,* 1996b), and here there was a sex-selective effect akin to that seen for locomotor activity. After low-dose prenatal nicotine infusion, male offspring showed a significant deficit at the 0 sec delay point, although normal performance was seen at longer delays (Fig. 8). This selective impairment at the shortest delay time did not appear to be one of memory retention, but rather may have been related to a persisting attentional impairment after prenatal nicotine exposure, a finding reminiscent of the types of attentional deficits seen in offspring of women who smoke. Sex selectivity in males has also been seen by other investigators using the nicotine injection route and evaluating learned avoidance (Genedani *et al.,* 1983) and appetitive operant tasks (Martin and Becker, 1971). Sex differences are not limited to males, however: Peters and Ngan (1982), using the injection route, found that females showed selective deficits in appetitive maze learning behavior as well as in shock-avoidance learning.

Other spatial tasks, such as the radial-arm maze and the Morris water maze, have shed light on the behavioral effects of prenatal nicotine exposure. Yanai and coworkers (1992) found that nicotine injected into pregnant mice during mid- to late gestation caused a significant deficit in radial-arm maze and water maze performance. Mice may be more sensitive than are rats to adverse effects of nicotine on these tasks, as there are only very modest and transient effects of low or

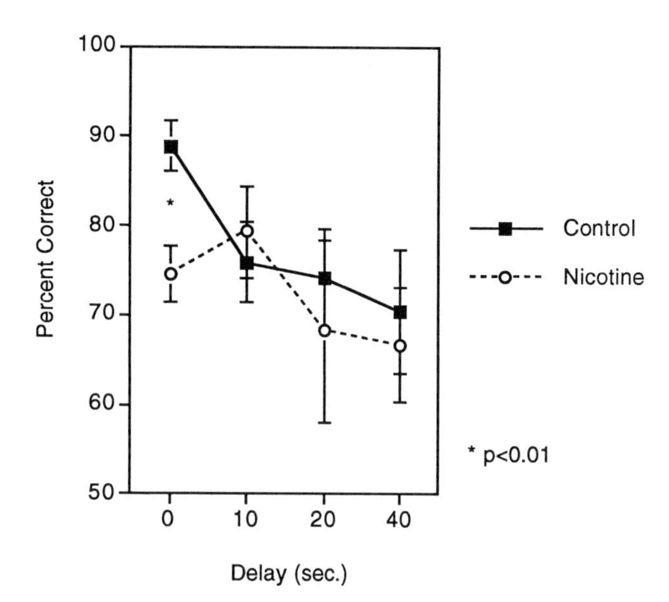

FIGURE 8 Selective prenatal nicotine-induced deficits of males in T-maze spatial alternation.

moderate nicotine infusion regimens on radial-arm maze acquisition in rats and no effects on learning in a delayed matching to position task in the water maze in either male or female rats (Levin *et al.,* 1993, 1996b; Cutler *et al.,* 1996). Alternatively, the effects seen in mice could represent hypoxia-induced deficits because of the use of the injection route. Nevertheless, mouse models could be useful for determining the relationship between nicotinic receptor dynamics and persistent cognitive effects of prenatal nicotine administration, as various mouse strains display quite different concentrations and regulatory dynamics of nicotinic receptors, along with differential behavioral responses to nicotine (Collins *et al.,* 1988; Collins and Marks, 1989; Marks *et al.,* 1989a,b; van de Kamp and Collins, 1994).

C. Response to Challenge

The relative resistance of rats to the adverse effects of prenatal nicotine exposure on radial-arm maze performance could also represent a more efficient compensation for underlying damage. If this is so, then further complication of the task should uncover the deficits. Indeed, it has been found that changing the cue structure by shifting the room in which rats are run in the maze adversely affects choice accuracy in all rats (demonstrating complication of the basic task), but in the group exposed to low-dose prenatal nicotine infusions, there is significantly greater impairment of choice accuracy (Fig. 9) (Levin *et al.,* 1996b). In another study, Sorenson *et al.* (1991) found that using a confinement procedure between choices in the radial-arm maze uncovered significant nicotine-induced deficits.

These results all indicate that the behavioral damage caused by prenatal nicotine exposure can be uncovered when performance is challenged as opposed simply to examining basal characteristics. Pharmacologic challenges can also uncover adverse effects of prenatal nicotine and, in addition, can help identify the specific neurotransmitter pathways involved in altered behavior. Again, using the radial-arm maze significant effects of nicotine exposure on the response to subsequent drug challenges were found (Levin *et al.,* 1993): male offspring were significantly more responsive to the amnestic effects of the nicotinic antagonist mecamylamine (Fig. 10). In adult rats, it was also found that a 3-week chronic course of nicotine infusion could result in hypersensitivity of the amnestic effects in the radial-arm maze of nicotinic antagonism with mecamylamine (Levin and Rose, 1990). As was found for neurochemical defects caused by prenatal exposure to nicotine, the behavioral results thus point to cholinergic systems as a specific target. In this case, if cholinergic systems are partially compromised in the nicotine group, they are more sensitive to impairment by a nicotinic antagonist. In addition, if partial compensation is occurring, then the maintenance of behavioral performance is more dependent on the ability to activate cholinergic pathways and again shows greater impairment when the effect of ACh is blocked by mecamylamine. It is most critical to note that the cholinergic deficit is not apparent unless the system is challenged with mecamylamine: basal performance (zero mecamylamine) is normal.

Concordance of neurochemical and behavioral findings also extends to the other major pathways targeted by prenatal nicotine exposure, namely catecholaminergic systems. Again, challenge with specific antagonists is required to uncover performance deficits. The β-adrenergic antagonist propranolol caused a significant dose-related impairment in choice accuracy in control

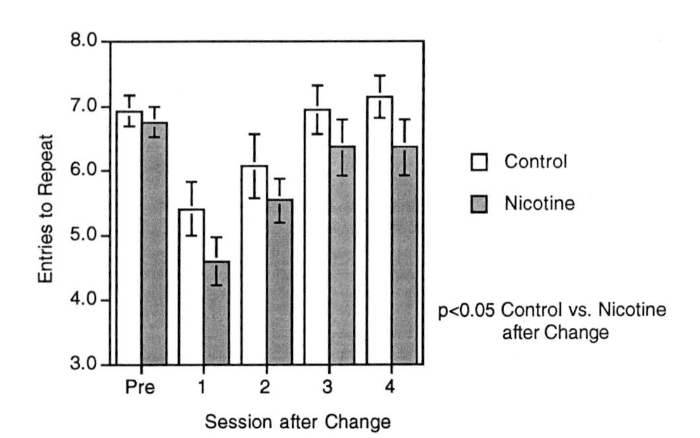

FIGURE 9 Environmental change uncovers prenatal nicotine-induced deficits in working memory performance accuracy as assessed in the radial-arm maze.

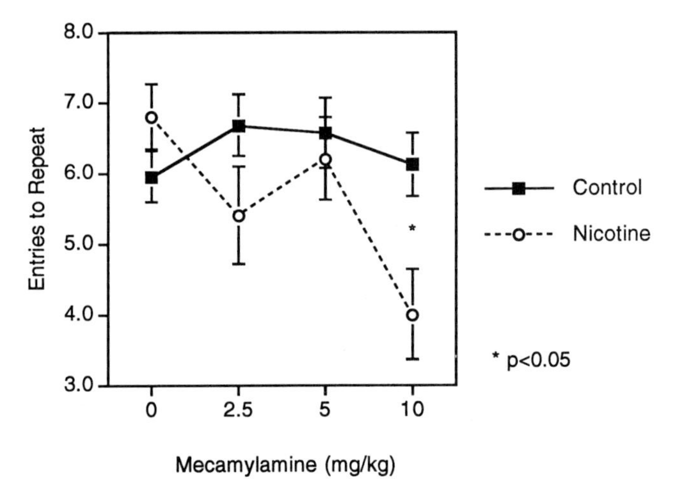

FIGURE 10 Adverse effects of prenatal nicotine exposure on behavioral performance is revealed by challenging the animals with the nicotinic antagonist mecamylamine.

rats (Fig. 11), indicating involvement of noradrenergic systems in radial-arm maze performance (Levin *et al.*, 1993). The nicotine group showed slightly poorer basal performance (albeit not statistically significantly different from controls) but, most importantly, showed no reactivity to propranolol whatsoever, indicating the noradrenergic pathways are not participating in maintenance of behavioral performance in the nicotine group. The involvement of noradrenergic systems is highly species- and strain-dependent, as, in a different rat strain, it was found that propranolol can improve performance (Cutler *et al.*, 1996). Nevertheless, the elimination of adrenergic components is still displayed, as this effect is also eliminated by prenatal nicotine administration.

For noradrenergic systems, the synaptic alterations caused by prenatal nicotine exposure indicated defects in impulse activity, and therefore one would predict that behavioral damage would not be limited to β-adrenergic components, but rather would be shared by multiple transduction pathways. Indeed, sex-specific behavioral effects directed toward α-noradrenergic responses have been found (Fig. 12). The α-agonist phenylpropanolamine caused a significant deficit in control females but not in females exposed prenatally to nicotine (Levin *et al.*, 1996b), once again indicating elimination of the involvement of a neural pathway mediating radial-arm maze behavior; the system is compensated to maintain normal behavior under basal conditions but is incapable of responding to challenge.

Prenatal nicotine exposure thus causes alterations in cognitive performance that may not be apparent until the system is challenged either by complicating the behavioral paradigm or by introducing pharmacologic agents that uncover the defects. The specific targeting of cholinergic and noradrenergic systems may have important consequences with regard to developmental

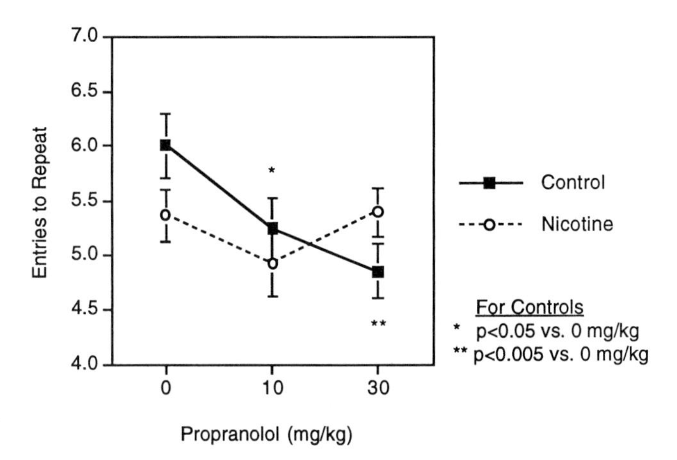

FIGURE 11 Loss of effect of the β-adrenergic antagonist propranolol in the radial-arm maze after prenatal nicotine exposure.

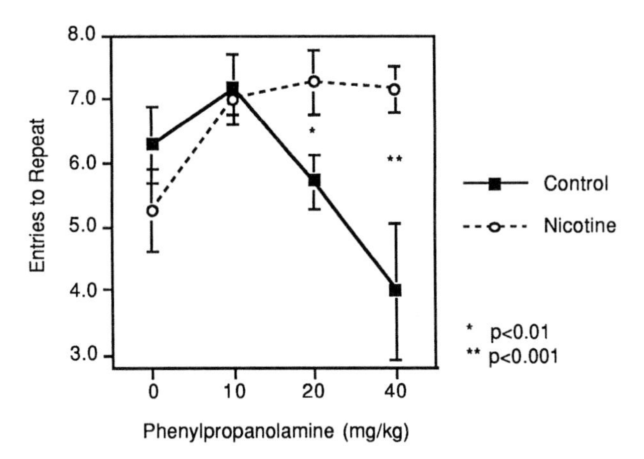

FIGURE 12 Loss of effect of the α-noradrenergic agonist phenylpropanolamine in female rats after prenatal nicotine exposure.

plasticity. Noradrenergic systems participate in the functional recovery from neonatal neural insult such as brain lesions and birth trauma. Lesions of the medial frontal cortex in adult rats cause working memory deficits (Kolb *et al.*, 1982, 1983; Kolb and Sutherland, 1992) that are not seen when the lesions are made in during infancy (Kolb and Sutherland, 1992; Sutherland *et al.*, 1982), a reflection of the ability of the developing brain to compensate for injury. Lesioning of noradrenergic or dopaminergic systems impairs the ability to recover cognitive performance after early postnatal lesions (Sutherland *et al.*, 1982; De Brabander *et al.*, 1992; Kolb and Sutherland, 1992). The effects of prenatal nicotine in attenuating the response of catecholaminergic systems may thus amplify the cellular deficits caused by nicotine exposure. To test this hypothesis, a study of the combined effects of prenatal nicotine exposure and neonatal medial frontal cortical lesions demonstrated that rats not exposed to nicotine prenatally did not show any discernible effect of the lesion with regard to T-maze spatial alternation performance. However, the combination of lesion with low-level prenatal nicotine infusions caused a pronounced impairment of choice accuracy.

Thus, in a variety of tasks, prenatal nicotine causes a persistent cognitive deficit that may be revealed only when behavior is challenged by more difficult tasks or by pharmacologic interventions. Altered drug responses can also help to determine the actual neurotransmitter systems involved in the persistent cognitive defects as well as uncovering the pathways that permit the compensation that can mask the deficits under basal conditions. The involvement of specific neural pathways that are also involved in developmental plasticity means that apparently subtoxic prenatal nicotine exposures can interact with birth trauma, hypoxic brain damage, or other nonspecific insults to produce a loss of compensatory mechanisms that ordinarily provide the plastic reserve of the developing brain. Prenatal nicotine exposure and neonatal neural trauma may individually have adverse effects, but together they are potentially devastating.

Cholinergic and noradrenergic systems play prominent roles not only in the developing brain, but also in mediating peripheral autonomic effects that are themselves critical during the perinatal period and particularly in maintaining physiologic competence during parturition (Lagercrantz and Slotkin, 1986; Slotkin and Seidler, 1988). Given that interference with the development of noradrenergic responsiveness is not limited to the brain but rather is shared by autonomic targets such as the heart (Fig. 6), it is important to examine the consequences of prenatal nicotine exposure on perinatal physiologic performance.

VIII. Physiologic Effects

The identification of nicotine as a neurobehavioral teratogen that acts by targeting nicotinic cholinergic receptors in the fetal nervous system opens up the issue of nicotinic receptor sites outside the CNS that may also be targeted. These include autonomic ganglia and the adrenal medulla (Rosenthal and Slotkin, 1977; Bareis and Slotkin, 1978; Hagino and Lee, 1985; Cairns and Wonnacott, 1988; Lichtensteiger *et al.*, 1988; Naeff *et al.*, 1992; Slotkin *et al.*, 1995). If development of peripheral sites is affected similarly, then the neurobehavioral teratology of nicotine will extend to the "behaviors" of the autonomic nervous system, including cardiovascular physiology. One example (Fig. 6) has been provided of a peripheral target in the form of adverse effects on cardiac β-adrenergic receptor development and adrenergic responsiveness (Navarro *et al.*, 1990a). This becomes an important issue in light of the profound increase in perinatal morbidity–mortality and in the incidence of SIDS in the offspring of smokers (Valdes-Dapena, 1981; Kinney and Filiano, 1988; Haglund and Cnattingius, 1990; Kandall and Gaines, 1991; DiFranza and Lew, 1995). Although the IUGR present with maternal smoking undoubtedly contributes to these risks (Morrison *et al.*, 1993), the fact that nicotine-induced neuroteratogenesis occurs even when growth impairment is absent raises the essential question of whether nicotine is responsible for physiologic deficits that compromise perinatal and neonatal survival over and above low birthweight.

Adverse effects of nicotine on neonatal cardiovascular and respiratory control have been postulated to underlie SIDS (Hunt and Brouillette, 1987; Stramba-Badiale *et al.*, 1992; Poets *et al.*, 1993), with the precipitating event assumed to be a period of hypoxia due to sleep apnea or airway obstruction. Similarly, during parturition, the fetus experiences episodes of intense hypoxia (Faber *et al.*, 1985; Lagercrantz and Slotkin, 1986). Survival during the hypoxic episodes involves physiologic mechanisms that are unique to the developing organism, including autonomous adrenal catecholamine secretion (i.e., without participation of neural reflexes) (Lagercrantz and Slotkin, 1986; Slotkin and Seidler, 1988), the presence of atypical adrenergic receptors in the myocardium (Drugge *et al.*, 1985; Lin *et al.*, 1992; Kauffman *et al.*, 1994), and a rapid transition in centrally mediated respiratory reflexes from the inhibitory hypoxic response that predominates in the fetus to an excitatory postnatal response (Schwieler, 1968; Henderson-Smart, 1981). In extending the nicotine animal models to studies of perinatal physiology, Slotkin *et al.* (1995) found that prenatal nicotine exposure leads to the complete loss of the adrenomedullary release of catecholamines in response to hypoxia in newborn rats, and consequently, elevated mortality during a hypoxic episode (Fig. 13). Within the first few minutes of introducing hypoxia, normal rats show a slight cardioacceleration followed by a moderate decrease in heart rate, whereas nicotine-exposed animals show no tachycardia and an immediate and profound drop in heart rate (Slotkin *et al.*, 1997b).

Deficiencies in cardiac β-receptor complement and in β-receptor–mediated cell stimulation (Fig. 6) undoubtedly contribute to the adverse effect of nicotine on cardiac responsiveness. However, the absence of an adrenomedullary response is particularly important because the fetal and neonatal heart do not possess functional sympathetic innervation, and are therefore completely dependent on circulating catecholamines as the source of adrenergic stimulation (Lagercrantz and Slotkin, 1986; Slotkin, 1986; Slotkin and Seidler, 1988). Why, then, does prenatal nicotine exposure lead to adrenomedullary incompetence? Direct depolarization of chromaffin cells in the nicotine-exposed group does evoke catecholamine release, so there is no inherent loss of the ability of these cells to produce or release catecholamines on stimulation. As splanchnic innervation is not functional in the fetus and neonate (Slotkin and Seidler, 1988), effects on centrally mediated reflexes can also be ruled out as contributory factors. In order to understand this deficiency, then, how the immature adrenal medulla responds to hypoxia must be explored.

There are two key features to adrenomedullary function in the fetus and neonate, illustrated in Fig. 14. First, although the tissue does not possess functional innervation, it nevertheless responds to specific stimuli, notably hypoxia, with a massive autonomous release of catecholamines (Lagercrantz and Slotkin, 1986; Slotkin and Seidler, 1988), far beyond the response that is

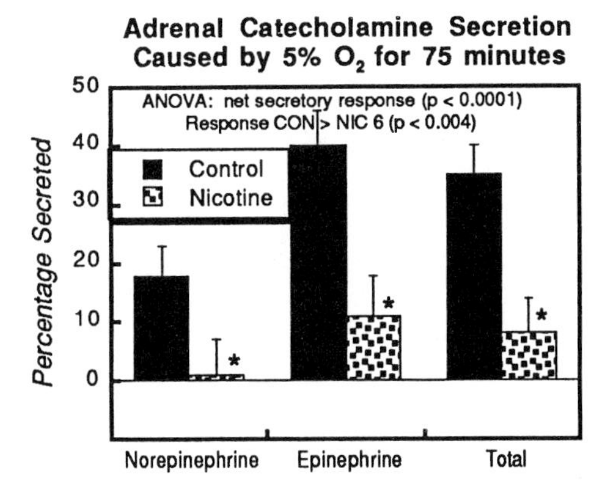

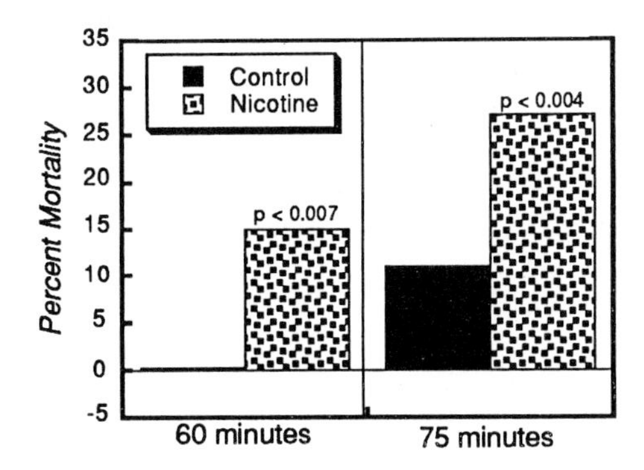

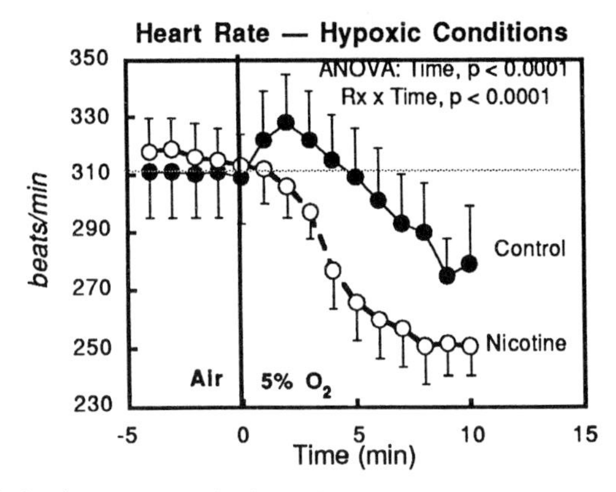

FIGURE 13 Prenatal nicotine exposure leads to the loss of adrenomedullary catecholamine secretion in response to hypoxia and a resultant increase in mortality during hypoxia. Cardiac responses are compromised.

achieved when the tissue is mature and innervated. The autonomous release mechanism is a characteristic only of the immature chromaffin cell and is lost completely when the cell undergoes terminal differentiation (Slotkin and Seidler, 1988; Thompson *et al.,* 1997). The second feature is that terminal cell differentiation and the loss of the autonomous response to hypoxia are caused by the ingrowth of the splanchnic nerve and stimulation of the nicotinic cholinergic receptors on the chromaffin cells (Slotkin and Seidler, 1988). Ordinarily, this is not a problem because the autonomous response is then replaced with a reflexly mediated response, so that catecholamine secretion occurs in response to hypoxia in both the preinnervation phase (autonomous response) and the postinnervation phase (reflex response). However, when chronic prenatal nicotine exposure occurs, the chromaffin cells are depolarized prematurely because they already possess functional nicotinic receptors (Rosenthal and Slotkin, 1977) and are thus forced into terminal differentiation at a point where innervation is not yet present. The fetus and neonate are then left with no response capability whatsoever during a hypoxic episode, because neither autonomous nor reflex responses occur. This is essentially the same basic mechanism as for CNS neurobehavioral teratogenesis by nicotine, namely, targeting of cholinergic cells for aberrant differentiation patterns because of premature and inappropriate stimulation by nicotine. When this occurs in peripheral sites, such as the chromaffin cells of the adrenal medulla, the relevant endpoint is physiologic collapse during episodes of hypoxia that ordinarily should be nonlethal.

In addition to compromising adrenomedullary mechanisms that have an impact ultimately on cardiac function, prenatal nicotine exposure also targets central noradrenergic function as a second defect that influences the response to perinatal hypoxia (Slotkin *et al.,* 1995). In normal neonates, hypoxia elicits little or no release of norepinephrine in brainstem areas that mediate respiratory inhibition, whereas substantial release occurs in nicotine-exposed animals. In the fetus, noradrenergic input inhibits breathing movements in response to hypoxia, a strategy that reduces muscle work and conserves energy (Rigatto, 1979; Blanco *et al.,* 1984; Dawes, 1984; Eden and Hanson, 1987; Bamford and Hawkins, 1990). Some of these characteristics persist into the neonatal

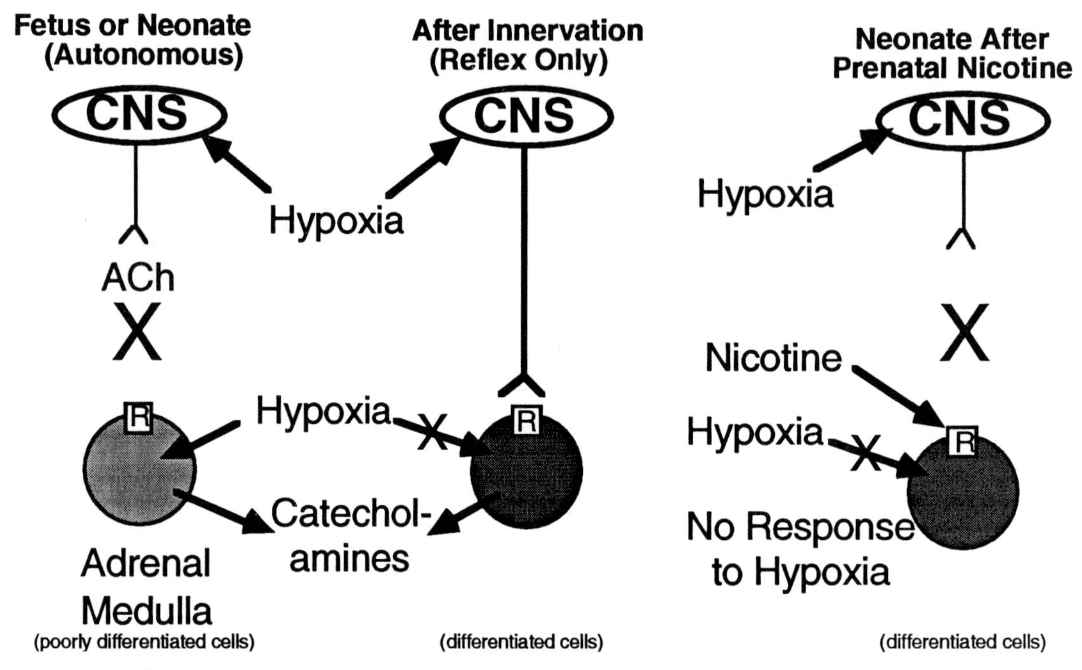

FIGURE 14 Development of the adrenal medulla and its response to hypoxia. In the fetus or neonate, the tissue responds autonomously to hypoxia so that catecholamines are released even though innervation is not functional. Ordinarily, the onset of innervation replaces the autonomous response with reflexly mediated release. With prenatal nicotine exposure, chronic, drug-induced depolarization of the chromaffin cells forces them into terminal differentiation prematurely so that autonomous responsiveness is lost before the onset of innervation, leading to an inability to respond to hypoxia.

period; unlike the adult's ability to maintain increased ventilation in response to hypoxia, the neonate responds in a biphasic manner with an immediate increase, followed by respiratory depression (Brady and Ceruti, 1966). The transition in patterns is a postnatal event, with apnea predominating with prematurity in humans as well as other species (Schwieler, 1968; Henderson-Smart, 1981). The phase in which apnea dominates in response to hypoxia is associated with central noradrenergic mechanisms (Bamford and Hawkins, 1990). Thus, the ability of nicotine to sensitize the CNS so that there is excessive release of norepinephrine during hypoxia would tend to elicit an inhibitory ventilatory response, contributing to respiratory collapse and death.

Accordingly, animal models of prenatal nicotine exposure have provided the first mechanistic proof of the epidemiologic link between smoking during pregnancy and perinatal morbidity and mortality associated with hypoxic birth trauma. In addition, however, these results share a number of key points with current hypotheses of the origin of SIDS that also implicate dysfunction of central respiratory control (Hunt and Brouillette, 1987) and/or of cardiovascular responses to hypoxia experienced during spontaneously occurring periods of apnea (Schulte *et al.,* 1982; Poets *et al.,* 1993; Southall *et al.,* 1993). Victims, siblings, and parents of victims of SIDS often demonstrate prolonged sleep apnea, excessive periodic breathing, impaired arousal during sleep, and diminished ventilatory sensitivity to hypoxia and hypercarbia (Valdes-Dapena, 1981; Pettigrew and Rahilly, 1985). Specifically, forebrain and brainstem respiratory centers and autonomic control of cardiac rhythm are suggested as the likely sites of dysfunction (Pettigrew and Rahilly, 1985; Quattrochi *et al.,* 1985; Stramba-Badiale *et al.,* 1992; Poets *et al.,* 1993; Meny *et al.,* 1994; Kinney *et al.,* 1995). The adrenomedullary, cardiac, and brainstem abnormalities found in nicotine-exposed fetuses and neonates may thus represent the reason for cardiorespiratory failure during an apneic episode or during a period of airway obstruction thought to precipitate death in SIDS. In this scenario, an infant exposed prenatally to nicotine would be at greater risk for cardiorespiratory failure during neonatal apneic episodes specifically because of: 1. premature loss of the autonomous adrenomedullary catecholamine secretory response to hypoxia that is essential to maintaining cardiovascular function; 2. impaired cardiac responsiveness to adrenergic stimulation because of reduced β-receptor numbers and deficient β-receptor coupling to cellular responses; and 3. excessive CNS norepinephrine release during hypoxia and persistence of predominantly inhibitory noradrenergic effects on respiration. These models thus provide physiologic mechanisms that may underlie SIDS while at the same time providing a mechanistic explanation for the association of maternal smoking with birth trauma and SIDS.

Because the fetal environment is hypoxic, similar mechanisms may be operating to produce the high incidence of fetal loss that accounts for tens of thousands of resorptions and spontaneous abortions annually in the United States as a result of smoking during pregnancy (DiFranza and Lew, 1995). Such effects are observed routinely in animal models of prenatal nicotine exposure even when using infusion routes that avoid acute hypoxic–ischemic episodes (Slotkin *et al.,* 1987c). In addition, there is every reason to suspect that the same factors may operate with early postnatal nicotine exposure, via nursing and secondhand smoke, because development of autonomic and central respiratory control is still not complete at that time (Milerad and Sundell, 1993; Bonham *et al.,* 1995; Holgert, 1995; Holgert *et al.,* 1995; Klonoff-Cohen *et al.,* 1995; Milerad *et al.,* 1995; Etzel *et al.,* 1997). It should also be pointed out that prenatal nicotine exposure is not the only way in which to achieve cardiorespiratory failure originating in hypoxia intolerance. Chronic treatment with glucocorticoids produces the same result, although by slightly different mechanisms (impaired adrenal response with addition of impaired cardiac conduction characteristics) (Kauffman *et al.,* 1994); as factors such as drug abuse and maternal stress elevate glucocorticoid levels, the endpoint of intolerance to hypoxia may explain interactions of these factors with smoking in producing parturitional morbidity and mortality as well as SIDS.

IX. Relevance of Animal Models to Human Nicotine Exposure

It is apparent that nicotine itself accounts for a major proportion of the neuroteratogenesis, behavioral abnormalities, and physiologic deficits associated with maternal cigarette smoking. These effects are superimposed on the known adverse effects of episodic fetal ischemia and hypoxia that are present in smoking but not in animal models that comprise exposure to nicotine alone. The most critical question for public health decisions is: What strategy should be followed in trying to get pregnant women to cease smoking, and more specifically, what role should be played by nicotine substitutes such as patches, gum, or inhalers? The current view is to intervene with nicotine substitutes where possible (Law and Tang, 1995). Certainly, the associated hypoxia and ischemia produced by smoking, both as a result of nicotine received in a bolus and the CO and cyanide found in cigarette smoke, contribute significantly to the liability for perinatal injury and to adverse effects on nervous system development. For these epiphenomena of smoking, the variable of IUGR may provide a partial index of the degree of sensitivity of the fetus. However, for the factors related to nicotine itself, the standard indices of safety are inappropriate because impaired nervous system development occurs below the threshold for growth suppression. Furthermore, if nicotine substitutes are chosen that deliver more nicotine than does smoking, the nicotine-related aspects of fetal nervous system damage may be as bad or worse; this is especially true when typical abuse patterns include smoking on top of the noncigarette nicotine source. With the advent of over-the-counter nicotine patches and the resultant decrease in physician input into smoking cessation counseling, superimposition of smoking and the nicotine patch may be expected to produce a worsening of perinatal morbidity, mortality, and neurobehavioral teratogenesis.

Although the only truly safe course is to discontinue smoking or nicotine exposure entirely during pregnancy, it is necessary to recognize that this is not always practicable, and that increased pressure on the pregnant smoker may simply increase the incidence of deception about the actual status of cigarette consumption. With regard to women who cannot or will not discontinue smoking at the beginning of pregnancy, animal models do permit some important decisions to be made about counseling and therapy. First, the finding that nicotine acts on developing neural tissue through specific nicotinic receptors that proliferate primarily in mid- to late gestation indicates that intervention with a nicotine substitute can be instituted in the first trimester, which, for standard teratogens, is considered to be the most sensitive period. Receptor targeting, however, indicates that the first trimester should be relatively safe as compared to later gestational exposures. During the first trimester, avoiding the covariables of hypoxia and ischemia that are components of cigarette smoking but not of nicotine substitutes should obviate many of the initial problems that result in spontaneous abortion, fetal resorption, or IUGR. Furthermore, this would allow sufficient time for counseling and careful explanation to have an impact on smoking cessation for the remainder of pregnancy. This proposed strategy is the opposite of that usually taken, which is to counsel first and then to institute nicotine substitution later.

The second consideration is that of dose and route of delivery of nicotine substitutes to be used in pregnancy. If nicotine itself causes neurobehavioral teratogenesis and increased risk of perinatal mortality and SIDS, then there should be more concern about the total dose delivered. Clearly, lower dose nicotine substitutes are preferable to high dose intervention, provided that this is tolerated well and does not precipitate a return to smoking. Furthermore, because the placenta protects the fetus to some extent by delaying the entry of nicotine into the fetal circulation, steady-state situations such as those achieved with the patch deliver relatively higher total doses to the fetus than episodic administration paradigms, such as with nicotine gum or inhalers. At the very least, pregnant women using nicotine patches should be instructed to remove the patch overnight to permit a decline in the fetal nicotine burden.

X. Conclusions

Nicotine is a neurobehavioral teratogen. Because it affects the basic processes of cell development, it targets not only developing cholinergic systems, but also many related neurotransmitter pathways, including central and peripheral catecholaminergic systems. In light of the critical role played by these systems in physiologic and behavioral development, the alterations caused by nicotine are likely to play important roles in perinatal morbidity and mortality associated with maternal smoking, as well as evoking persistent behavioral and learning deficits in the offspring. Long-term lowering of the response profile of a variety of neurotransmitter systems in nicotine-exposed offspring may have direct consequences in impaired synaptic function, and may set the stage for a predisposition to greater damage from other developmental insults.

It is clear that care must be taken before using nicotine replacement therapy in pregnancy, as any source of nicotine would still have demonstrable effects on neurodevelopment. Because nicotine acts through its specific receptors that arise in midgestation, exposure during early gestation is likely to be a window of opportunity for nicotine substitution in smoking cessation treatment. In any case, identification of nicotine as an injurious substance means that careful attention has to be given to the dose and pharmacokinetic characteristics with which substitution therapy is undertaken. In particular, one would predict that continual nicotine exposure via skin patches, which allow for complete equilibration of the fetus with the maternal circulation, would have more severe effects than does intermittent exposure via nicotine gum, nasal spray, or inhalers. This hypothesis needs to be tested rigorously in animal models before making definitive conclusions.

Knowledge of the mechanisms underlying the neurobehavioral deficits associated with prenatal nicotine exposure points the way to potential treatments to offset cognitive defects such as ADHD. Ironically, nicotine itself has been shown to diminish symptoms of ADHD effectively in adults (Levin *et al.,* 1996a). If pregnant women with cognitive dysfunctions smoke as a form of self-medication, their children may run a greater risk of cognitive impairment from the prenatal insult of nicotine exposure, on top of any inheritable characteristics. Maternal smoking may thus contribute to the apparent heritability of cognitive dysfunction, or may amplify genetic predisposition to dysfunction.

Nicotine exposure is probably the most widespread prenatal drug insult in the world. It continues largely unabated despite educational and medical interventions and is becoming of ever greater significance with the spread of smoking to the Third World countries and the disappearance of societal prohibitions on women's smoking. The exponential growth of smoking in women of childbearing age and its attendant consequences of persistent neurobehavioral abnormalities have tremendous impact on society.

References

Aronson, R. A., Uttech, S., and Soref, M. (1993). The effect of maternal cigarette smoking on low birth weight and preterm birth in Wisconsin, 1991. *Wis. Med. J.* **92,** 613–617.

Bamford, O. and Hawkins, R. L. (1990). Central effects of an α_2-adrenergic antagonist on fetal lambs: a possible mechanism for hypoxic apnea. *J. Dev. Physiol.* **13,** 353–358.

Bamford, O. S., Schuen, J. N., and Carroll, J. L. (1996). Effect of nicotine exposure on postnatal ventilatory responses to hypoxia and hypercapnia. *Respir. Physiol.* **106,** 1–11.

Bardy, A. H., Seppala, T., Lillsunde, P., Kataja, J. M., Koskela, P., Pikkarainen, J., and Hiilesmaa, V. K. (1993). Objectively measured tobacco exposure during pregnancy: neonatal effects and relation to maternal smoking. *Br. J. Obstet. Gynaecol.* **100,** 721–6.

Bareis, D. L. and Slotkin, T. A. (1978). Responses of heart ornithine decarboxylase and adrenal catecholamines to methadone and sympathetic stimulants in developing and adult rats. *J. Pharmacol. Exp. Ther.* **205,** 164–174.

Barnes, C. D., and Eltherington, L. G. (1973). *Drug Dosage in Laboratory Animals: A Handbook* (Revised Edition). University of California Press, Berkeley, CA.

Bell, J. M. and Slotkin, T. A. (1985). Perinatal dietary supplementation with a commercial soy lecithin preparation: effects on behavior and brain biochemistry in the developing rat. *Dev. Psychobiol.* **18,** 383–394.

Bell, J. M. and Slotkin, T. A. (1986). Polyamines as intermediates in developmental neurotoxic events. *Neurotoxicology* **7,** 147–160.

Bell, J. M., Whitmore, W. L., Barnes, G., Seidler, F. J., and Slotkin, T. A. (1986). Perinatal dietary exposure to soy lecithin: altered sensitivity to central cholinergic stimulation. *Int. J. Dev. Neurosci.* **4,** 497–501.

Berman, S. M. and Noble, E. P. (1995). Reduced visuospatial performance in children with the D2 dopamine receptor A1 allele. *Behav. Genet.* **25,** 45–58.

Bertolini, A., Bernardi, M., and Genedani, S. (1982). Effects of prenatal exposure to cigarette smoke and nicotine on pregnancy, offspring development and avoidance behavior in rats. *Neurobehav. Toxicol. Teratol.* **4,** 545–548.

Bina, K. G., Guzman, P., Broide, R. S., Leslie, F. M., Smith, M. A., and Odowd, D. K. (1995). Localization of alpha 7 nicotinic receptor subunit mRNA and alpha-bungarotoxin binding sites in developing mouse somatosensory thalamocortical system. *J. Comp. Neurol.* **363,** 321–332.

Black, I. B. (1978). Regulation of autonomic development. *Ann. Rev. Neurosci.* **1,** 183–214.

Black, I. B. (1980). Developmental regulation of neurotransmitter phenotype. *Curr. Top. Dev. Biol.* **15,** 27–40.

Blanco, C. E., Hanson, M. A., Johnson, P., and Rigatto, H. (1984). Breathing pattern of kittens during hypoxia. *J. Appl. Physiol.* **56,** 12–17.

Bonham, A. C., Kappagoda, C. T., Kott, K. S., and Joad, J. P. (1995). Exposing young guinea pigs to sidestream tobacco smoke decreases rapidly adapting receptor responsiveness. *J. Appl. Physiol.* **78,** 1412–1420.

Brady, J. P. and Ceruti, E. (1966). Chemoreceptor reflexes in the new-born infant: effects of varying degrees of hypoxia on heart rate and ventilation in a warm environment. *J. Physiol.* **184,** 631–645.

Broide, R. S., Robertson, R. T., and Leslie, F. M. (1996). Regulation of alpha(7) nicotinic acetylcholine receptors in the developing rat somatosensory cortex by thalamocortical afferents. *J. Neurosci.* **16,** 2956–2971.

Brooke, O. G., Anderson, H. R., Bland, J. M., Peacock, J. L., and Stewart, C. M. (1989). Effects on birth weight of smoking, alcohol, caffeine, socioeconomic factors, and psychosocial stress. *Br. Med. J.* **298,** 795–801.

Butler, N. R. and Goldstein, H. (1973). Smoking in pregnancy and subsequent child development. *Br. Med. J.* **4,** 573–575.

Cairns, N. J. and Wonnacott, S. (1988). [3H](-)nicotine binding sites in fetal human brain. *Brain Res.* **475,** 1–7.

Camilli, A. E., McElroy, L. F., and Reed, K. L. (1994). Smoking and pregnancy: a comparison of Mexican-American and non-Hispanic white women. *Obstet. Gynecol.* **84,** 1033–1037.

Campbell, C. G., Seidler, F. J., and Slotkin, T. A. (1997). Chlorpyrifos interferes with cell development in rat brain regions. *Brain Res. Bull.* **43,** 179–189.

Chan, J. and Quik, M. (1993). A role for the nicotinic alpha-bungarotoxin receptor in neurite outgrowth in PC12 cells. *Neuroscience* **56,** 441–451.

Clarke, P. B. S., Schwartz, R. D., Paul, S. M., Pert, C. B., and Pert, A. (1985). Nicotinic binding in rat brain: autoradiographic comparison of [^{3}H]acetylcholine, [^{3}H]nicotine, and [^{125}I]-α-bungarotoxin. *J. Neurosci.* **5,** 1307–1315.

Cole, P. V., Hawkins, L. H., and Roberts, D. (1972). Smoking during pregnancy and its effect on the fetus. *J. Obstet. Gynæcol. Brit. Commonw.* **79,** 782–787.

Collins, A. C. and Marks, M. J. (1989). Chronic nicotine exposure and brain nicotinic receptors: influence of genetic factors. In *Progress in Brain Research.* A. Nordberg, K. Fuxe, B. Holmstedt, and A. Sundwall, Ed. Elsevier Science Publishers B. V. (Biomedical Division), pp. 137–146.

Collins, A. C., Miner, L. L., and Marks, M. J. (1988). Genetic influences on acute responses to nicotine and nicotine tolerance in the mouse. *Pharmacol. Biochem. Beh.* **30,** 269–278.

Comings, D. E., Ferry, L., Bradshaw-Robinson, S., Burchette, R., Chiu, C., and Muhleman, D. (1996). The dopamine D2 receptor (DRD2) gene: a genetic risk factor in smoking. *Pharmacogenetics* **6,** 73–79.

Cooper, J. R., Bloom, F. E., and Roth, R. H. (1986). *The Biochemical Basis of Neuropharmacology,* 5th Ed. Oxford University Press, New York.

Cordier, S., Iglesias, M. J., Le Goaster, C., Guyot, M. M., Mandereau, L., and Hemon, D. (1994). Incidence and risk factors for childhood brain tumors in the Ile de France. *Int. J. Canc.* **59,** 776–782.

Court, J. and Clementi, F. (1995). Distribution of nicotinic subtypes in human brain. *Alz. Dis. Assoc. Dis.* **9,** 6–14.

Court, J. A., Perry, E. K., Spurden, D., Griffiths, M., Kerwin, J. M., Morris, C. M., Johnson, M., Oakley, A. E., Birdsall, N. J. M., Clementi, F., and Perry, R. H. (1995). The role of the cholinergic system in the development of the human cerebellum. *Dev. Brain Res.* **90,** 159–167.

Coyle, J. T. (1976). Neurochemical aspects of the development of the dopaminergic innervation to the striatum. In *Brain Dysfunction.* M. A. B. Brazier and F. Coceani, Ed. Raven Press, New York, pp. 25–39.

Coyle, J. T. and Yamamura, H. I. (1976). Neurochemical aspects of the ontogenesis of cholinergic neurons in the rat brain. *Brain Res.* **118,** 429–440.

Criswell, H., Mueller, R. A., and Breese, G. R. (1989). Priming of D_1-dopamine receptor responses: long-lasting behavioral supersensitivity to a D_1-dopamine agonist following repeated administration to neonatal 6-OHDA-lesioned rats. *J. Neurosci.* **9,** 125–133.

Curran, T., Abate, C., Cohen, D. R., Macgregor, P. F., Rauscher, F. J., Sonnenberg, J. L., Connor, J. A., and Morgan, J. I. (1990). Inducible proto-oncogene transcription factors: third messengers in the brain? *Cold Spr. Harb. Symp. Quant. Biol.* **55,** 225–234.

Curran, T. and Morgan, J. I. (1995). Fos: an immediate-early transcription factor in neurons. *J. Neurobiol.* **26,** 403–412.

Cutler, A. R., Wilkerson, A. E., Gingras, J. L., and Levin, E. D. (1996). Prenatal cocaine and/or nicotine exposure in rats: preliminary findings on long-term cognitive outcome and genital development at birth. *Neurotoxicol. Teratol.* **18,** 635–643.

Dawes, G. E. (1984). The central control of fetal breathing and skeletal muscle movements. *J. Physiol.* **346,** 1–18.

De Brabander, J. M., van Eden, C. G., de Bruin, J. P. C., and Feenstra, M. G. P. (1992). Activation of mesocortical dopaminergic system in the rat in response to neonatal medial prefrontal cortex lesions. Concurrence with functional sparing. *Brain Res.* **581,** 1–9.

Denson, R., Nanson, J. L., and McWatters, M. A. (1975). Hyperkinesis and maternal smoking. *Can. Psych. Assoc. J.* **20,** 183–187.

Deskin, R., Seidler, F. J., Whitmore, W. L., and Slotkin, T. A. (1981). Development of α-noradrenergic and dopaminergic receptor systems depends on maturation of their presynaptic nerve terminals in the rat brain. *J. Neurochem.* **36,** 1683–1690.

DiFranza, J. R. and Lew, R. A. (1995). Effect of maternal cigarette smoking on pregnancy complications and sudden infant death syndrome. *J. Fam. Pract.* **40,** 385–394.

Dobbing, J. and Sands, J. (1979). Comparative aspects of the brain growth spurt. *Early Hum. Dev.* **3,** 79–83.

Douglas, R. J. (1975). The development of hippocampal function: implications for theory and therapy. In *The Hippocampus, Vol. 2: Neurophysiology and Behavior.* R. L. Isaacson and K. H. Pribram, Ed. Plenum Press, New York, pp. 327–361.

Dow, T. G. B., Rooney, P. J., and Spence, M. (1975). Does anemia increase the risks to the fetus caused by smoking in pregnancy? *Brit. Med. J.* **4,** 253–254.

Drews, C. D., Murphy, C. C., Yeargin-Allsopp, M., and Decoufle, P. (1996). The relationship between idiopathic mental retardation and maternal smoking during pregnancy. *Pediatrics* **97,** 547–553.

Drugge, E. D., Rosen, M. R., and Robinson, R. B. (1985). Neuronal regulation of the development of the a-adrenergic chronotropic response in the rat heart. *Circ. Res.* **57,** 415–423.

Dunn, H. G. and McBurney, A. K. (1977). Cigarette smoking and the fetus and child. *Pediatrics* **60,** 772.

Eden, G. J. and Hanson, M. A. (1987). Maturation of the respiratory response to acute hypoxia in the newborn rat. *J. Physiol.* **392,** 1–9.

Edwards, D. R. (1994). Cell signalling and the control of gene transcription. *Trends Pharmacol. Sci.* **15,** 239–244.

Eriksen, P. S., Gennser, G., Lofgren, O., and Nilsson, K. (1983). Acute effects of maternal smoking on fetal breathing and movements. *Obstet. Gynecol.* **61,** 367–372.

Eriksson, M., Larsson, G., and Zetterstrom, R. (1979). Abuse of alcohol, drugs and tobacco during pregnancy: consequences for the child. *Paediatrician* **8,** 228–242.

Etzel, R. A., Balk, S. J., Bearer, C. F., Miller, M. D., Shea, K. M., Simon, P. R., Falk, H., Miller, R. W., Rogan, W., and Hendrick, J. G. (1997). Environmental tobacco smoke: a hazard to children. *Pediatrics* **99,** 639–642.

Evans, D. R., Newcombe, R. G., and Campbell, H. (1979). Maternal smoking habits and congenital malformations: a population study. *Br. Med. J.* **2,** 171–3.

Faber, J. J., Anderson, D. F., Morton, M. J., Parks, C. M., Pinson, C. W., Thornburg, K. L., and Willis, D. M. (1985). Birth, its physiology, and the problems it creates. In *The Physiological Development of the Fetus and Newborn.* C. T. Jones and P. W. Nathanielsz, Ed. Academic Press, London, pp. 371–380.

Fechter, L. D., and Annau, Z. (1977). Toxicity of mild prenatal carbon monoxide exposure. *Science* **197,** 680–682.

Fechter, L. D. Mactutus, C. F., and Storm, J. E. (1986). Carbon monoxide and brain development. *Neurotoxicology* **7,** 463–473.

Fergusson, D. M., Horwood, L. J., and Lynskey, M. T. (1993). Maternal smoking before and after pregnancy: effects on behavioral outcomes in middle childhood. *Pediatrics* **92,** 815–822.

Fogelman, K. R. and Manor, O. (1988). Smoking in pregnancy and development into early adulthood. *Br. J. Med.* **297,** 1233–1236.

Fricker, H. S., Burgi, W., Kaufmann, H., Bruppacher, R., Kipfer, H., and Gugler, E. (1985). The course of pregnancy in a representative Swiss sample (Aarau Pregnancy and Newborn Infant Study). II. Stimulants in pregnancy. *Schweiz. Med. Wochenschr.* **115,** 381–386.

Fried, P. A. (1989). Cigarettes and marijuana: are there measurable long-term neurobehavioral teratogenic effects? *NeuroToxicology* **10,** 577–584.

Fried, P. A. (1992). Clinical implications of smoking: determining long-term teratogenicity. In *Maternal Substance Abuse and the Developing Nervous System.* I. S. Zagon and T. A. Slotkin, Ed. Academic Press, New York, pp. 77–96.

Fried, P. A., Innes, K. S., and Barnes, M. V. (1984). Soft drug use prior to and during pregnancy: a comparison of samples over a four-year period. *Drug Alcohol. Depend.* **13,** 161–176.

Fried, P. A. and Makin, J. E. (1987). Neonatal behavioural correlates of prenatal exposure to marihuana, cigarettes and alcohol in a low risk population. *Neurotoxicol. Teratol.* **9,** 1–7.

Fried, P. A., O'Connell, C. M., and Watkinson, B. (1992a). 60- and 72-month follow-up of children prenatally exposed to marijuana, cigarettes, and alcohol: cognitive and language assessment. *J. Dev. Behav. Pediatr.* **13,** 383–391.

Fried, P. A. and Watkinson, B. (1988). 12- and 24-month neurobehavioural follow-up of children prenatally exposed to marihuana, cigarettes and alcohol. *Neurotoxicol. Teratol.* **10,** 305–313.

Fried, P. A., Watkinson, B., Dillon, R. F., and Dulberg, C. S. (1987). Neonatal neurological status in a low-risk population after prenatal exposure to cigarettes, marijuana, and alcohol. *J. Dev. Behav. Pediatr.* **8,** 318–326.

Fried, P. A., Watkinson, B., and Gray, R. (1992b). A follow-up study of attentional behavior in 6-year-old children exposed prenatally to marihuana, cigarettes, and alcohol. *Neurotoxicol. Teratol.* **14,** 299–311.

Fung, Y. K. (1988). Postnatal behavioural effects of maternal nicotine exposure in rats. *J. Pharm. Pharmacol.* **40,** 870–872.

Fung, Y. K. (1989). Postnatal effects of maternal nicotine exposure on the striatal dopaminergic system in rats. *J. Pharm. Pharmacol.* **41,** 576–578.

Fung, Y. K. and Lau, Y. S. (1989). Effects of prenatal nicotine exposure on rat striatal dopaminergic and nicotinic systems. *Pharmacol. Biochem. Behav.* **33,** 1–6.

Gazmararian, J. A., Adams, M. M., and Pamuk, E. R. (1996). Associations between measures of socioeconomic status and maternal health behavior. *Am. J. Prev. Med.* **12,** 108–115.

Genedani, S., Bernardi, M., and Bertolini, A. (1983). Sex-linked differences in avoidance learning in the offspring of rats treated with nicotine during pregnancy. *Psychopharmacology* 93–95.

Giannuzzi, C. E., Seidler, F. J., and Slotkin, T. A. (1995). β-Adrenoceptor control of cardiac adenylyl cyclase during development: agonist pretreatment in the neonate uniquely causes heterologous sensitization, not desensitization. *Brain Res.* **694,** 271–278.

Gillette, J. H. and Mitchell, J. L. A. (1991). Ornithine decarboxylase: a biochemical marker of repair in damaged tissue. *Life Sci.* **48,** 1501–1510.

Gueguen, C., Lagrue, G., and Janse-Marec, J. (1995). Effect of smoking on the fetus and the child during pregnancy. *J. Gynecol. Obstet. Biol. Reprod.* **24,** 853–859.

Hagino, N. and Lee, J. W. (1985). Effect of maternal nicotine on the development of sites for [^{3}H]nicotine binding in the fetal brain. *Int. J. Dev. Neurosci.* **3,** 567–571.

Haglund, B. and Cnattingius, S. (1990). Cigarette smoking as a risk factor for sudden infant death syndrome: a population-based study. *Am. J. Pub. Health* **80,** 29–32.

Hardy, J. B. and Mellits, E. D. (1972). Does maternal smoking during pregnancy have a long-term effect on the child? *Lancet* **2,** 1332–1336.

Henderson-Smart, D. L. (1981). The effect of gestational age on the incidence and duration of recurrent apnoea in newborn babies. *Aust. Paediatr. J.* **17,** 273–276.

Hickner, J., Westenberg, C., and Dittenbir, M. (1984). Effect of pregnancy on smoking behavior: a baseline study. *J. Fam. Pract.* **18,** 241–244.

Hohmann, C. F., Brooks, A. R., and Coyle, J. T. (1988). Neonatal lesions of the basal forebrain cholinergic neurons result in abnormal cortical development. *Dev. Brain Res.* **42,** 253–264.

Holgert, H. (1995). *Postnatal Development and Adaptation of Rat Adrenal Medulla and Peripheral Arterial Chemoreceptors.* PhD Thesis, Karolinska Institute, Stokholm, Sweden.

Holgert, H., Hokfelt, T., Hertzberg, T., and Lagercrantz, H. (1995). Functional and developmental studies of the peripheral arterial chemoreceptors in rat: effects of nicotine and possible relation to sudden infant death syndrome. *Proc. Natl. Acad. Sci. U.S.A.* **92,** 7575–7579.

Hou, Q.-C., Baker, F. E., Seidler, F. J., Bartolome, M., Bartolome, J., and Slotkin, T. A. (1989a). Role of sympathetic neurons in development of β-adrenergic control of ornithine decarboxylase activity in peripheral tissues: effects of neonatal 6-hydroxydopamine treatment. *J. Dev. Physiol.* **11,** 139–146.

Hou, Q.-C., Eylers, J. P., Lappi, S. E., Kavlock, R. J., and Slotkin, T. A. (1989b). Neonatal sympathectomy compromises development of responses of ornithine decarboxylase to hormonal stimulation in peripheral tissues. *J. Dev. Physiol.* **12,** 189–192.

Hunt, C. E. and Brouillette, R. T. (1987). Sudden infant death syndrome: 1987 perspective. *J. Pediatr.* **110,** 669–678.

Hutter, C. D. and Blair, M. E. (1996). Carbon monoxide—does fetal exposure cause sudden infant death syndrome? *Med. Hypoth.* **46,** 1–4.

Jacobson, S. W., Fein, G. G., Jacobson, J. J., Schwartz, P. M., and Dowler, J. K. (1984). Neonatal correlates of prenatal exposure to smoking, caffine, and alcohol. *Inf. Behav. Dev.* **7,** 253–265.

Johns, J. M., Louis, T. M., Becker, R. F., and Means, L. W. (1982). Behavioral effects of prenatal exposure to nicotine in guinea pigs. *Neurobehav. Toxicol. Teratol.* **4,** 365–369.

Johns, J. M., Walters, P. A., and Zimmerman, L. I. (1993). The effects of chronic prenatal exposure to nicotine on the behavior of guinea pigs (*Cavia porcellus*). *J. Gen. Psychol.* **120,** 49–63.

Jonsson, G. and Hallman, H. (1980). Effects of neonatal nicotine administration on the postnatal development of central noradrenalin neuron. *Acta. Physiol. Scand.* **479** (Supp), 25–26.

Kahn, A., Groswasser, J., Sottiaux, M., Kelmanson, I., Rebuffat, E., Franco, P., Dramaix, M., and Wayenberg, J. L. (1994). Prenatal exposure to cigarettes in infants with obstructive sleep apneas. *Pediatrics* **93,** 778–783.

Kandall, S. R. and Gaines, J. (1991). Maternal substance abuse and subsequent sudden infant death syndrome (SIDS) in offspring. *Neurotoxicol. Teratol.* **13,** 235–240.

Kandel, D. B., Wu, P., and Davies, M. (1994). Maternal smoking during pregnancy and smoking by adolescent daughters. *Am. J. Pub. Health* **84,** 1407–1413.

Kauffman, K. S., Seidler, F. J., and Slotkin, T. A. (1994). Prenatal dexamethasone exposure causes loss of neonatal hypoxia tolerance: cellular mechanisms. *Pediatr. Res.* **35,** 515–522.

Kindy, M. S., Hu, Y. G., and Dempsey, R. J. (1994). Blockade of ornithine decarboxylase enzyme protects against ischemic brain damage. *J. Cereb. Blood Flow Metab.* **14,** 1040–1045.

King, G., Barry, L., and Carter, D. L. (1993). Smoking prevalence among perinatal women: the role of socioeconomic status, race, and ethnicity. *Conn. Med.* **57,** 721–728.

Kinney, H. C. and Filiano, J. J. (1988). Brainstem research in sudden infant death syndrome. *Pediatrician* **15,** 240–250.

Kinney, H. C., Filiano, J. J., Sleeper, L. A., Mandell, F., Valdes-Dapena, M., and White, W. F. (1995). Decreased muscarinic receptor binding in the arcuate nucleus in sudden infant death syndrome. *Science* **269,** 1446–1450.

Kinney, H. C., O'Donnell, T. J., Kriger, P., and White, W. F. (1993). Early developmental changes in [^{3}H]nicotine binding in the human brainstem. *Neuroscience* **55,** 1127–1138.

Klemm, N. and Kuhar, M. J. (1979). Post-mortem changes in high affinity choline uptake. *J. Neurochem.* **32,** 1487–1494.

Klonoff-Cohen, H. S., Edelstein, S. L., Lefkowitz, E. S., Srinivasan, I. P., Kaegl, D., Chang, J. C., and Wiley, K. J. (1995). The effect of passive smoking and tobacco exposure through breast milk on sudden infant death syndrome. *JAMA* **273,** 795–798.

Koistinaho, J. (1991). Nicotine-induced Fos-like immunoreactivity in rat sympathetic ganglia and adrenal medulla. *Neurosci. Lett.* **128,** 47–51.

Koistinaho, J., Pelto-Huikko, M., Sagar, S. M., Dagerlind, Å., Roivainen, R., and Hökfelt, T. (1993). Differential expression of immediate early genes in the superior cervical ganglion after nicotine treatment. *Neuroscience* **56,** 729–739.

Kolb, B., Pittman, K., Sutherland, R. J., and Wishaw, I. Q. (1982). Dissociation of the contributions of the prefrontal cortex and dorsomedial thalamic nucleus to spatially guided behavior in the rat. *Behav. Brain Res.* **6,** 365–378.

Kolb, B. and Sutherland, R. J. (1992). Noradreneline depletion blocks behavioral sparing and alters cortical morphogenesis after frontal cortex damage in rats. *J. Neurosci.* **12,** 2321–2330.

Kolb, B., Sutherland, R. J., and Whishaw, I. Q. (1983). A comparison of the contributions of the frontal and parietal association cortex to spatial localization in rats. *Behav. Neurosci.* **97,** 13–27.

Kostrzewa, R. M. and Saleh, M. I. (1989). Impaired ontogeny of striatal dopamine D_1 and D_2 binding sites after postnatal treatment of rats with SCH-23390 and spiroperidol. *Dev. Brain Res.* **45,** 95–101.

Kristjansson, E. A., Fried, P. A., and Watkinson, B. (1989). Maternal smoking during pregnancy affects children's vigilance performance. *Drug Alc. Depend.* **24,** 11–19.

Krous, H. F., Campbell, G. A., Fowler, M. W., Catron, A. C., and Farber, J. P. (1981). Maternal nicotine administration and fetal brain stem damage: a rat model with implications for sudden infant death syndrome. *Am. J. Obstet. Gynecol.* **140,** 743–746.

Kruse, J., Le Fevre, M., and Zweig, S. (1986). Changes in smoking and alcohol consumption during pregnancy: a population-based study in a rural area. *Obstet. Gynecol.* **67,** 627–632.

Kudlacz, E. M., Navarro, H. A., Eylers, J. P., and Slotkin, T. A. (1990a). Adrenergic modulation of cardiac development in the rat: effects of prenatal exposure to propranolol via continuous maternal infusion. *J. Dev. Physiol.* **13,** 243–249.

Kudlacz, E. M., Navarro, H. A., Eylers, J. P., and Slotkin, T. A. (1990b). Prenatal exposure to propranolol via continuous maternal infusion: effects on physiological and biochemical processes mediated by beta adrenergic receptors in fetal and neonatal rat lung. *J. Pharmacol. Exp. Ther.* **252,** 42–50.

Lagercrantz, H. and Slotkin, T. A. (1986). The "stress" of being born. *Sci. Amer.* **254** (April), 100–107.

Land, G. H. and Stockbauer, J. W. (1993). Smoking and pregnancy outcome: trends among black teenage mothers in Missouri. *Am. J. Pub. Health* **83,** 1121–1124.

Landesman-Dwyer, S. and Emanuel, I. (1979). Smoking during pregnancy. *Teratol.* **19,** 119–126.

Larsson, C., Nordberg, A., Falkeborn, Y., and Lundberg, R.-Å. (1985). Regional [^{3}H]acetylcholine and [^{3}H]nicotine binding in developing mouse brain. *Int. J. Dev. Neurosci.* **3,** 667–671.

Latulippe, L. G., Marcoux, S., Fabia, J., Weber, J. P., and Tennina, S. (1992). Smoking during labour. *Can. J. Pub. Health* **83,** 184–187.

Law, M. and Tang, J. L. (1995). An analysis of the effectiveness of interventions intended to help people stop smoking. *Arch. Intern. Med.* **155,** 1933–1941.

Lefkowitz, M. M. (1981). Smoking during pregnancy: long-term effects on the offspring. *Develop. Psychol.* **17,** 192–194.

Levin, E. D., Briggs, S. J., Christopher, N. C., and Rose, J. E. (1993). Prenatal nicotine exposure and cognitive performance in rats. *Neurotoxicol. Teratol.* **15,** 251–260.

Levin, E. D., Conners, C. K., Sparrow, E., Hinton, S., Meck, W., Rose, J. E., Ernhardt, D., and March, J. (1996a). Nicotine effects on adults with attention-deficit/hyperactivity disorder. *Psychopharmacology* **123,** 55–63.

Levin, E. D. and Rose, J. E. (1990). Anticholinergic sensitivity following chronic nicotine administration as measured by radial-arm maze performance in rats. *Behav. Pharmacol.* **1,** 511–520.

Levin, E. D., Wilkerson, A., Jones, J. P., Christopher, N. C., and Briggs, S. J. (1996b). Prenatal nicotine effects on memory in rats: pharmacological and behavioral challenges. *Dev. Brain Res.* **97,** 207–215.

Li, C. Q., Windsor, R. A., Perkins, L., Goldenberg, R. L., and Lowe, J. B. (1993). The impact on infant birth weight and gestational age of cotinine-validated smoking reduction during pregnancy. *JAMA* **269,** 1519–1524.

Lichtensteiger, W., Ribary, U., Schlumpf, M., Odermatt, B., and Widmer, H. R. (1988). Prenatal adverse effects of nicotine on the developing brain. *Prog. Brain Res.* **73,** 137–157.

Lichtensteiger, W. and Schlumpf, M. (1985). Prenatal nicotine affects fetal testosterone and sexual dimorphism of saccharin preference. *Pharmacol. Biochem. Behav.* **23,** 439–444.

Lichtensteiger, W., Schlumpf, M., and Ribary, U. (1987). Pharmacological modifications of neuroendocrine ontogenesis. Development of receptors, nicotine and catecholamines. *Ann. Endocrinol.* **48,** 393–399.

Lin, W., Seidler, F. J., McCook, E. C., and Slotkin, T. A. (1992). Overexpression of α_2-adrenergic receptors in fetal rat heart: receptors in search of a function. *J. Dev. Physiol.* **17,** 183–187.

Lodewijckx, E. and De Groof, V. (1990). Smoking and alcohol consumption by Flemish pregnant women, 1966–83. *J. Biosoc. Sci.* **22,** 43–51.

Lorber, J. (1980). Is your brain really necessary? *Science* **210,** 1232–1234.

Mactutus, C. F. (1989). Developmental neurotoxicity of nicotine, carbon monoxide, and other tobacco smoke constituents. *Ann. NY Acad. Sci.* **562,** 105–122.

Magnus, P., Berg, K., Bjerkedal, T., and Nance, W. E. (1985). The heritability of smoking behaviour in pregnancy, and the birth weights of offspring of smoking-discordant twins. *Scand. J. Soc. Med.* **13,** 29–34.

Makin, J., Fried, P. A., and Watkinson, B. (1991). A comparison of active and passive smoking during pregnancy: long-term effects. *Neurotoxicol. Teratol.* **13,** 5–12.

Marks, M. J., Romm, E., Campbell, S. M., and Collins, A. C. (1989a). Variation of nicotinic binding sites among inbred strains. *Pharmacol. Biochem. Behav.* **33,** 679–689.

Marks, M. J., Stitzel, J. A., and Collins, A. C. (1989b). Genetic influences on nicotine responses. *Pharmacol. Biochem. Behav.* **33,** 667–668.

Martin, J. C. and Becker, R. F. (1970). The effects of nicotine administration *in utero* upon activity in the rat. *Psychon. Sci.* **19,** 59–60.

Martin, J. C. and Becker, R. F. (1971). The effects of maternal nicotine absorption or hypoxic episodes upon appetitive behavior of rat offspring. *Dev. Psychobiol.* **4,** 133–147.

Martin, J. C. and Martin, D. C. (1981). Voluntary activity in the aging rat as a function of maternal drug exposure. *Neurobehav. Toxicol. Teratol* **3,** 261–264.

Martin, J. C., Martin, D. C., Radow, B., and Sigman, G. (1976). Growth, development and activity in rat offspring following maternal drug exposure. *Exp. Aging Res.* **2,** 235–51.

McFarland, B. J., Seidler, F. J., and Slotkin, T. A. (1991). Inhibition of DNA synthesis in neonatal rat brain regions caused by acute nicotine administration. *Dev. Brain Res.* **58,** 223–229.

McGee, R. and Stanton, W. R. (1994). Smoking in pregnancy and child development to age 9 years. *J. Paediatr. Child Health* **30,** 263–268.

Meny, R. G., Carroll, J. L., Carbone, M. T., and Kelly, D. H. (1994). Cardiorespiratory recordings from infants dying suddenly and unexpectedly at home. *Pediatrics* **93,** 44–49.

Meyer, D. C. and Carr, L. A. (1987). The effects of perinatal exposure to nicotine on plasma LH levels in prepubertal rats. *Neurotoxicol. Teratol.* **9,** 95–98.

Miao, G. G. and Curran, T. (1994). Cell transformation by c-*fos* requires an extended period of expression and is independent of the cell cycle. *Mol. Cell Biol.* **14,** 4295–4310.

Milerad, J., Larsson, H., Lin, J., and Sundell, H. W. (1995). Nicotine attenuates the ventilatory response to hypoxia in the developing lamb. *Pediatr. Res.* **37,** 652–660.

Milerad, J., Rajs, J., and Gidlund, E. (1994). Nicotine and cotinine levels in pericardial fluid in victims of SIDS. *Acta. Paediat.* **83,** 59–62.

Milerad, J. and Sundell, H. (1993). Nicotine exposure and the risk of SIDS. *Acta. Paediatr. Suppl.* **389,** 70–72.

Morrison, J., Williams, G. M., Najman, J. M., Andersen, M. J., and Keeping, J. D. (1993). Birthweight below the tenth percentile: the relative and attributable risks of maternal tobacco consumption and other factors. *Environ. Health Perspect.* **3,** 275–277.

Murrin, L. C. (1980). High-affinity transport of choline in neuronal tissue. *Pharmacology* **21,** 132–140.

Murrin, L. C., Ferrer, J. R., Wanyun, Z., and Haley, N. J. (1987). Nicotine administration to rats: methodological considerations. *Life Sci.* **40,** 1699–1708.

Murrin, L. C., Ferrer, J. R., and Zeng, W. (1985). Nicotine administration during pregnancy and its effect on striatal development. *Neurosci. Abs.* **11,** 69.

Mytilineou, C. and Black, I. B. (1978). Development of adrenergic nerve terminals: the effects of decentralization. *Brain Res.* **158,** 259–268.

Naeff, B., Schlumpf, M., and Lichtensteiger, W. (1992). Prenatal and postnatal development of high-affinity [^{3}H]nicotine binding sites in rat brain regions—an autoradiographic study. *Dev. Brain Res.* **68,** 163–174.

Naeye, R. L. (1978). Effects of maternal cigarette smoking on the fetus and placenta. *Br. J. Obstet. Gynecol.* **85,** 732–737.

Naeye, R. L., Ladis, B., and Drage, J. S. (1976). Sudden infant death syndrome. A prospective study. *Am. J. Dis. Child.* **130,** 1207–1210.

Naeye, R. L. and Peters, E. C. (1984). Mental development of children whose mothers smoked during pregnancy. *Obstet. Gynecol.* **64,** 601–607.

Náquira, D. and Arqueros, L. (1977). Water intake in rats affected by addition of nicotine to drinking water. *Acta Physiol. Lat. Amer.* **28,** 73–76.

Nasrat, H. A., Al-Hachim, G. M., and Mahmood, F. A. (1986). Perinatal effects of nicotine. *Biol. Neonate* **49,** 8–14.

Navarro, H. A., Mills, E., Speidler, F. J., Baker, F. E., Lappi, S. E., Tayyeb, M. I., Spencer, J. R., and Slokin, T. A. (1990a). Prenatal nicotine exposure impairs beta-adrenergic function: persistent chronotropic subsensitivity despite recovery from deficits in receptor binding. *Brain Res. Bull.* **25,** 223–237.

Navarro, H. A., Seidler, F. J., Eylers, J. P., Baker, F. E., Dobbins, S. S., Lappi, S. E., and Slotkin, T. A. (1989a). Effects of prenatal nicotine exposure on development of central and peripheral cholinergic neurotransmitter systems. Evidence for cholinergic trophic influences in developing brain. *J. Pharmacol. Exp. Ther.* **251,** 894–900.

Navarro, H. A., Seidler, F. J., Schwartz, R. D., Baker, F. E., Dobbins, S. S., and Slotkin, T. A. (1989b). Prenatal exposure to nicotine impairs nervous system development at a dose which does not affect viability or growth. *Brain Res. Bull.* **23,** 187–192.

Navarro, H. A., Seidler, F. J., Whitmore, W. L., and Slotkin, T. A. (1988). Prenatal exposure to nicotine via maternal infusions: effects on development of catecholamine systems. *J. Pharmacol. Exp. Ther.* **244,** 940–944.

Navarro, H. A., Slotkin, T. A., Tayyeb, M. I., Lappi, S. E., and Seidler, F. J. (1990b). Effects of fetal nicotine exposure on development of adrenergic receptor binding in rat brain regions: selective changes in α_1-receptors. *Res. Comm. Sub. Ab.* **11,** 95–103.

O'Campo, P., Faden, R. R., Brown, H., and Gielen, A. C. (1992). The impact of pregnancy on women's prenatal and postpartum smoking behavior. *Am. J. Prev. Med.* **8,** 8–13.

Oyemade, U. J., Cole, O. J., Johnson, A. A., Knight, E. M., Westney, O. E., Laryea, H., Hill, G., Cannon, E., Fomufod, A., Westney, L. S., *et al.* (1994). Prenatal substance abuse and pregnancy outcomes among African American women. *J. Nutr.* **124,** 994S–999S.

Patterson, P. H. and Chun, L. L. Y. (1977). The induction of acetylcholine synthesis in primary cultures of dissociated rat sympathetic neurons. II. Developmental aspects. *Dev. Biol.* **60,** 473–481.

Peters, D. A. (1984). Prenatal nicotine exposure increases adrenergic receptor binding in the rat cerebral cortex. *Res. Commun. Chem. Pathol. Pharmacol.* **46,** 307–317.

Peters, D. A., Taub, H., and Tang, S. (1979). Postnatal effects of maternal nicotine exposure. *Neurobehav. Toxicol.* **1,** 221–225.

Peters, M. A. and Ngan, L. L. (1982). The effects of totigestational exposure to nicotine on pre- and postnatal development in the rat. *Arch. Int. Pharmacodyn. Ther.* **257,** 155–167.

Pettigrew, A. G. and Rahilly, P. M. (1985). Brainstem auditory evoked responses in infants at risk of sudden infant death. *Early Hum. Dev.* **11,** 99–111.

Picone, T. A., Allen, L. H., Olsen, P. N., and Ferris, M. E. (1982a). Pregnancy outcome in North American women. II. Effects of diet, cigarette smoking, stress, and weight gain on placentas, and on neonatal physical and behavioral characteristics. *Am. J. Clin. Nutr.* **36,** 1214–1224.

Picone, T. A., Allen, L. H., Schramm, M. M., and Olsen, P. N. (1982b). Pregnancy outcome in North American women. I. Effects of diet, cigarette smoking, and psychological stress on maternal weight gain. *Am. J. Clin. Nutr.* **36,** 1205–1213.

Pley, E. A., Wouters, E. J., Voorhorst, F. J., Stolte, S. B., Kurver, P. H., and de Jong, P. A. (1991). Assessment of tobacco-exposure during pregnancy; behavioural and biochemical changes. *Eur. J. Obstet. Gynecol. Reprod. Biol.* **40,** 197–201.

Poets, C. F., Stebbens, V. A., Samuels, M. P., and Southall, D. P. (1993). The relationship between bradycardia, apnea, and hypoxemia in preterm infants. *Pediatr. Res.* **34,** 144–147.

Preston, G. A., Lyon, T. T., Yin, Y. X., Lang, J. E., Solomon, G., Annab, L., Srinivasan, D. G., Alcorta, D. A., and Barrett, J. C. (1996). Induction of apoptosis by c-Fos protein. *Mol. Cell. Biol.* **16,** 211–218.

Preston-Martin, S., Yu, M. C., Benton, B., and Henderson, B. E. (1982). N-Nitroso compounds and childhood brain tumors: a case-control study. *Cancer Res* **42,** 5240–5245.

Quattrochi, J. J., McBride, P. T., and Yates, A. J. (1985). Brainstem immaturity in sudden infant death syndrome: a quantitative rapid Golgi study in dendritic spines in 95 infants. *Brain Res.* **325,** 39–48.

Rantakallio, P. (1983). A follow-up study up to the age of 14 of children whose mothers smoked during pregnancy. *Acta. Paediatr. Scand.* **72,** 747–753.

Rantakallio, P., Laara, E., Isohanni, M., and Moilanen, I. (1992). Maternal smoking during pregnancy and delinquency of the offspring: an association without causation? *Int. J. Epidemiol.* **21,** 1106–1113.

Reinis, S. and Goldman, J. M. (1980). *The Development of the Brain: Biological and Functional Perspectives.* Charles C. Thomas, Springfield, IL.

Ribary, U. and Lichtensteiger, W. (1989). Effects of acute and chronic prenatal nicotine treatment on central catecholamine systems of male and female rat fetuses and offspring. *J. Pharmacol. Exp. Ther.* **248,** 786–792.

Richardson, S. A. and Tizabi, Y. (1994). Hyperactivity in the offspring of nicotine-treated rats: role of the mesolimbic and nigrostriatal dopaminergic pathways. *Pharmacol. Biochem. Behav.* **47,** 331–337.

Rigatto, H. (1979). A critical analysis of the development of peripheral and central respiratory chemosensitivity during the neonatal period. In *Central Nervous Control Mechanisms in Breathing.* C. von Euler and H. Lagercrantz, Ed. Pergamon Press, New York, pp. 137–148.

Rosenthal, R. N. and Slotkin, T. A. (1977). Development of nicotinic responses in the rat adrenal medulla and long-term effects of neonatal nicotine administration. *Br. J. Pharmacol.* **60,** 59–64.

Ross, D., Johnson, M., and Burge, R. (1977). Development of cholinergic characteristics in adrenergic neurones is age dependent. *Nature* **267,** 536–639.

Rush, D. and Callahan, K. R. (1989). Exposure to passive cigarette smoking and child development. *Ann. NY Acad. Sci.* **562,** 74–100.

Sarvela, P. D. and Ford, T. D. (1992). Indicators of substance use among pregnant adolescents in the Mississippi Delta. *J. Sch. Health.* **62,** 175–179.

Saxton, D. W. (1978). The behaviour of infants whose mothers smoke in pregnancy. *Early Hum. Dev.* **2,** 363–369.

Schlumpf, M., Gahwiler, M., Ribary, U., and Lichtensteiger, W. (1988). A new device for monitoring early motor development: prenatal nicotine-induced changes. *Pharmacol. Biochem. Behav.* **30,** 199–203.

Schulte, F. J., Albani, M., Schnizer, H., and Bentele, K. (1982). Neuronal control of neonatal respiration—sleep apnea and the sudden infant death syndrome. *Neuropediatrics* **13** (Suppl.), 3–14.

Schulte-Hobein, B., Schwartz-Bickenbach, D., Abt, S., Plum, C., and Nau, H. (1992). Cigarette smoke exposure and development of infants throughout the first year of life: influence of passive smoking and nursing on cotinine levels in breast milk and infant's urine. *Acta. Paediatr.* **81,** 550–557.

Schwartz, R. D. and Kellar, K. J. (1985). *In vivo* regulation of [^{3}H]acetylcholine recognition sites in brain by nicotinic cholinergic drugs. *J. Neurochem.* **45,** 427–433.

Schwartz, R. D., McGee, R., and Kellar, K. J. (1982). Nicotinic cholinergic receptors labeled by [^{3}H]acetylcholine in rat brain. *Mol. Pharmacol.* **22,** 56–62.

Schwieler, G. H. (1968). Respiratory regulation during postnatal development in cats and rabbits and some of its morphological substrates. *Acta. Physiol. Scand.* **304** (Suppl.), 1–123.

Seidler, F. J., Levin, E. D., Lappi, S. E., and Slotkin, T. A. (1992). Fetal nicotine exposure ablates the ability of postnatal nicotine challenge to release norepinephrine from rat brain regions. *Dev. Brain Res.* **69,** 288–291.

Seidler, F. J. and Slotkin, T. A. (1981). Development of central control of norepinephrine turnover and release in the rat heart: responses to tyramine, 2-deoxyglucose and hydralazine. *Neuroscience* **6,** 2081–2086.

Sershen, H., Reith, M. E., Banay-Schwartz, M., and Lajtha, A. (1982). Effects of prenatal administration of nicotine on amino acid pools, protein metabolism, and nicotine binding in the brain. *Neurochem. Res.* **7,** 1515–1522.

Sharp, B. M., Beyer, H. S., McAllen, K. M., Hart, D., and Matta, S. G. (1993). Induction and desensitization of the c-Fos mRNA response to nicotine in rat brain. *Mol. Cell. Neurosci.* **4,** 199–208.

Sheng, M. and Greenberg, M. E. (1990). The regulation and function of c-*fos* and other immediate early genes in the nervous system. *Neuron* **4,** 477–485.

Simon, J. R., Atweh, S., and Kuhar, M. J. (1976). Sodium-dependent high affinity choline uptake: a regulatory step in the synthesis of acetylcholine. *J. Neurochem.* **26,** 909–922.

Slotkin, T. A. (1986). Endocrine control of synaptic development in the sympathetic nervous system: the cardiac-sympathetic axis. In *Developmental Neurobiology of the Autonomic Nervous System.* P. M. Gootman, Ed. Humana Press, Clifton, NJ, pp. 97–133.

Slotkin, T. A. (1992). Prenatal exposure to nicotine: what can we learn from animal models? In *Maternal Substance Abuse and the Developing Nervous System.* I. S. Zagon and T. A. Slotkin, Ed. Academic Press, New York, pp. 97–124.

Slotkin, T. A. and Bartolome, J. (1986). Role of ornithine decarboxylase and the polyamines in nervous system development: a review. *Brain Res. Bull.* **17,** 307–320.

Slotkin, T. A., Cho, H., and Whitmore, W. L. (1987a). Effects of prenatal nicotine exposure on neuronal development: selective actions on central and peripheral catecholaminergic pathways. *Brain Res. Bull.* **18,** 601–611.

Slotkin, T. A., Cowdery, T. S., Orband, L., Pachman, S., and Whitmore, W. L. (1986a). Effects of neonatal hypoxia on brain development in the rat: immediate and long-term biochemical alterations in discrete regions. *Brain Res.* **374,** 63–74.

Slotkin, T. A., Greer, N., Faust, J., Cho, H., and Seidler, F. J. (1986b). Effects of maternal nicotine injections on brain development in the rat: ornithine decarboxylase activity, nucleic acids and proteins in discrete brain regions. *Brain Res. Bull.* **17,** 41–50.

Slotkin, T. A., Lappi, S. E., Mccook, E. C., Lorber, B. A., and Seidler, F. J. (1995). Loss of neonatal hypoxia tolerance after prenatal nicotine exposure: implications for sudden infant death syndrome. *Brain Res. Bull.* **38,** 69–75.

Slotkin, T. A., Lappi, S. E., and Seidler, F. J. (1993). Impact of fetal nicotine exposure on development of rat brain regions: critical sensitive periods or effects of withdrawal? *Brain Res. Bull.* **31,** 319–328.

Slotkin, T. A., Lappi, S. E., Tayyeb, M. I., and Seidler, F. J. (1991). Chronic prenatal nicotine exposure sensitizes rat brain to acute postnatal nicotine challenge as assessed with ornithine decarboxylase. *Life Sci.* **49,** 665–670.

Slotkin, T. A., McCook, E. C., and Seidler, F. J. (1997a). Cryptic brain cell injury caused by fetal nicotine exposure is associated with persistent elevations of c-*fos* protooncogene expression. *Brain Res.* **750,** 180–188.

Slotkin, T. A., Nemeroff, C. B., Bissette, G., and Seidler, F. J. (1994). Overexpression of the high affinity choline transporter in cortical regions affected by Alzheimer's disease: evidence from rapid autopsy studies. *J. Clin. Invest.* **94,** 696–702.

Slotkin, T. A., Orband-Miller, L., and Queen, K. L. (1987b). Development of [3H]nicotine binding sites in brain regions of rats exposed to nicotine prenatally via maternal injections or infusions. *J. Pharmacol. Exp. Ther.* **242,** 232–237.

Slotkin, T. A., Orband-Miller, L., Queen, K. L., Whitmore, W. L., and Seidler, F. J. (1987c). Effects of prenatal nicotine exposure on biochemical development of rat brain regions: maternal drug infusions via osmotic minipumps. *J. Pharmacol. Exp. Ther.* **240,** 602–611.

Slotkin, T. A., Saleh, J. L., McCook, E. C., and Seidler, F. J. (1997b). Impaired cardiac function during postnatal hypoxia in rats exposed to nicotine prenatally: implications for perinatal morbidity and mortality, and for sudden infant death syndrome. *Teratology* **55,** 177–184.

Slotkin, T. A. and Seidler, F. J. (1988). Adrenomedullary catecholamine release in the fetus and newborn: secretory mechanisms and their role in stress and survival. *J. Dev. Physiol.* **10,** 1–16.

Slotkin, T. A., Seidler, F. J., Crain, B. J., Bell, J. M., Bissette, G., and Nemeroff, C. B. (1990). Regulatory changes in presynaptic cholinergic function assessed in rapid autopsy material from patients with Alzheimer disease: implications for etiology and therapy. *Proc. Natl. Acad. Sci. U.S.A.* **87,** 2452–2455.

Slotkin, T. A. and Thadani, P. V. (1980). Neurochemical teratology of drugs of abuse. In *Advances in the Study of Birth Defects, Vol. 4: Neural and Behavioural Teratology.* T. V. N. Persaud, Ed. MTP Press, Lancaster, England, pp. 199–234.

Smeyne, R. J., Vendrell, M., Hayward, M., Baker, S. J., Miao, G. G., Schilling, K., Robertson, L. M., Curran, T., and Morgan, J. I. (1993). Continuous c-*fos* expression precedes programmed cell death *in vivo. Nature* **363,** 166–169.

Sobrian, S. K., Ali, S. F., Slikker, W. J., and Holson, R. R. (1995). Interactive effects of prenatal cocaine and nicotine exposure on maternal toxicity, postnatal development and behavior in the rat. *Mol. Neurobiol.* **11,** 121–143.

Sorenson, C. A., Raskin, L. A., and Suh, Y. (1991). The effects of prenatal nicotine on radial-arm maze performance in rats. *Pharmacol. Biochem. Behav.* **40,** 991–993.

Southall, D. P., Noyes, J. P., Poets, C. F., and Samuels, M. P. (1993). Mechanisms for hypoxæmic episodes in infancy and early childhood. *Acta. Pædiat. Suppl.* **389,** 60–62.

Spinillo, A., Capuzzo, E., Piazzi, G., Baltaro, F., Iasci, A., and Nicola, S. (1996). Effect measures for behavioral factors adversely affecting fetal growth. *Am. J. Perinatol.* **13,** 119–123.

Stewart, J. P. and Dunkley, G. C. (1985). Smoking and health care patterns among pregnant women. *Can. Med. Assoc. J.* **133,** 989–994.

Stramba-Badiale, M., Lazzarotti, M., and Schwartz, P. J. (1992). Development of cardiac innervation, ventricular fibrillation, and sudden infant death syndrome. *Am. J. Physiol.* **263,** H1514–H1522.

Streissguth, A. P., Darby, B. L., Barr, H., Smith, J., and Martin, D. (1983). Comparison of drinking and smoking patterns during pregnancy over a six year interval. *Am. J. Obstet. Gynecol.* **145,** 716–723.

Streissguth, A. P., Martin, D. C., Barr, H. M., Sandman, B. M., Kirchner, G. L., and Darby, B. L. (1984). Intrauterine alcohol and nicotine exposure: attention and reaction time in 4-year-old children. *Dev. Psychol.* **20,** 533–541.

Sutherland, R. J., Kolb, B., Whishaw, I. Q., and Becker, J. B. (1982). Cortical noradrenaline depletion eliminates sparing of spatial learning after neonatal frontal cortex damage in the rat. *Neurosci. Lett.* **32,** 125–130.

Thompson, R. J., Jackson, A., and Nurse, C. A. (1997). Developmental loss of hypoxic chemosensitivity in rat adrenomedullary chromaffin cells. *J. Physiol.* **498,** 503–510.

Tollestrup, K., Frost, F. J., and Starzyk, P. (1992). Smoking prevalence of pregnant women compared to women in the general population of Washington State. *Am. J. Prev. Med.* **8,** 215–220.

U.S. Public Health Service (1979). *Smoking and Health: A Report of the Surgeon General.* US Government Printing Office, Washington, DC.

Valdes-Dapena, M. A. (1981). Sudden infant death syndrome: a review of the medical literature. *DHHS Publication HSA* **81-5271,** 1–18.

van de Kamp, J. L. and Collins, A. C. (1994). Prenatal nicotine alters nicotinic receptor development in the mouse brain. *Pharmacol. Biochem. Behav.* **47,** 889–900.

Vendrell, M., Zawia, N. H., Serratosa, J., and Bondy, S. C. (1991). c-*fos* and ornithine decarboxylase gene expression in brain as early markers of neurotoxicity. *Brain Res.* **544,** 291–296.

Virji, S. K. and Cottington, E. (1991). Risk factors associated with preterm deliveries among racial groups in a national sample of married mothers. *Am. J. Perinatol.* **8,** 347–353.

von Ziegler, N. I., Schlumpf, M., and Lichtensteiger, W. (1991). Prenatal nicotine exposure selectively affects perinatal forebrain aromatase activity and fetal adrenal function in male rats. *Dev. Brain Res.* **62,** 23–31.

Wainright, R. L. (1983). Change in observed birth weight associated with change in maternal cigarette smoking. *Am. J. Epidemiol.* **117,** 668–675.

Wakefield, M. A. and Jones, W. R. (1991). Cognitive and social influences on smoking behaviour during pregnancy. *Aust. NZ J. Obstet. Gynaecol.* **31,** 235–239.

Whitney, K. D., Seidler, F. J., and Slotkin, T. A. (1995). Developmental neurotoxicity of chlorpyrifos: cellular mechanisms. *Toxicol. Appl. Pharmacol.* **134,** 53–62.

Williamson, D. F., Serdula, M. K., Kendrick, J. S., and Binkin, N. J. (1989). Comparing the prevalence of smoking in pregnant and nonpregnant women, 1985 to 1986. *JAMA* **261,** 70–74.

Winick, M. and Noble, A. (1965). Quantitative changes in DNA, RNA and protein during prenatal and postnatal growth in the rat. *Dev. Biol.* **12,** 451–466.

Woody, R. C. and Brewster, M. A. (1990). Telencephalic dysgenesis associated with presumptive maternal carbon monoxide intoxication in the first trimester of pregnancy. *J. Toxicol. Clin. Toxicol.* **28,** 467–475.

Yanai, J., Pick, C. G., Rogel-Fuchs, Y., and Zahalka, E. A. (1991). Alteration in septohippocampal cholinergic receptors and related behavior after prenatal exposure to nicotine. In *Effects of Nicotine on Biological Systems.* F. Adlkofer and K. Thurau, Ed. Birkhäuser, Boston, pp. 401–406.

Yanai, J., Pick, C. G., Rogel-Fuchs, Y., and Zahalka, E. A. (1992). Alterations in hippocampal cholinergic receptors and hippocampal behaviors after early exposure to nicotine. *Brain Res. Bull.* **29,** 363–368.

Zahalka, E., Seidler, F. J., Lappi, S. E., Yanai, J., and Slotkin, T. A. (1993a). Differential development of cholinergic nerve terminal markers in rat brain regions: implications for nerve terminal density, impulse activity and specific gene expression. *Brain Res.* **601,** 221–229.

Zahalka, E. A., Seidler, F. J., Lappi, S. E., McCook, E. C., Yanai, J., and Slotkin, T. A. (1992). Deficits in development of central cholinergic pathways caused by fetal nicotine exposure: differential effects on choline acetyltransferase activity and [^{3}H]hemicholinium-3 binding. *Neurotoxicol. Teratol.* **14,** 375–382.

Zahalka, E. A., Seidler, F. J., and Slotkin, T. A. (1993b). Dexamethasone treatment *in utero* enhances neonatal cholinergic nerve terminal development in rat brain. *Res. Comm. Chem. Pathol. Pharmacol.* **81,** 191–198.

Zahalka, E. A., Seidler, F. J., Yanai, J., and Slotkin, T. A. (1993c). Fetal nicotine exposure alters ontogeny of M1-receptors and their link to G-proteins. *Neurotoxicol. Teratol.* **15,** 107–115.

Zawia, N. H. and Harry, G. J. (1993). Trimethyltin-induced c-fos expression: adolescent vs neonatal rat hippocampus. *Toxicol. Appl. Pharmacol.* **121,** 99–102.

Zeiders, J. L., Seidler, F. J., and Slotkin, T. A. (1997). Ontogeny of regulatory mechanisms for β-adrenoceptor control of rat cardiac adenylyl cyclase: targeting of G-proteins and the cyclase catalytic subunit. *J. Mol. Cell. Cardiol.* **29,** 603–615.

Zhu, J., Takita, M., Konishi, Y., Sudo, M., and Muramatsu, I. (1996). Chronic nicotine treatment delays the developmental increase in brain muscarinic receptors in rat neonate. *Brain Res.* **732,** 257–260.

Zoli, M., Lenovere, N., Hill, J. A., and Changeux, J. P. (1995). Developmental regulation of nicotinic ACh receptor subunit mRNAs in the rat central and peripheral nervous systems. *J. Neurosci.* **15,** 1912–1939.

CHAPTER

35

Maternal Drug Abuse and Adverse Effects on Neurobehavior of Offspring

MERLE G. PAULE
Behavioral Toxicology Laboratory
Division of Neurotoxicology
National Center for Toxicological Research
Jefferson, Arkansas 72079

I. Introduction

The nonmedical uses of abused substances represent serious social and medical problems throughout the world (e.g., DHHS 1991). Cocaine, for example, is such a potent reinforcer that in the laboratory setting, animals will self-administer it to the point of severe debilitation and even death (Johanson, 1984; Bozarth and Wise, 1985). In the United States, illicit use in general is widespread throughout various age and socioeconomic groups (Miller, 1983; Abelson and Miller, 1985; NIDA 1996). The medical consequences of inappropriate drug use or abuse are reflected in toxicities that occur in many organ systems (Gay, 1982: Loper, 1989), most notably the central nervous system (CNS) as evidenced by the well-known effects of abused drugs on brain and behavior (Benowitz, 1993; Prakash and Das, 1993).

An issue of primary concern is the prevalence of drug abuse in women of childbearing age (Miller, 1983; Abelson and Miller, 1985; NIDA, 1989) and particularly in pregnant women (Chasnoff *et al.,* 1985; Chasnoff, 1989; Day *et al.,* 1993). In this chapter the focus is on issues surrounding illicit drug abuse during pregnancy and the potential for such use to have adverse consequences for the developing fetus and offspring. Prenatal exposure to alcohol, tobacco smoke–nicotine, and antiepileptic drugs are not discussed here, as they are addressed specifically in other chapters. It should be remembered, however, that in the context of prenatal exposure, the basic pharmacodynamic principles that apply to alcohol, nicotine, and the anticonvulsants also

*Abbreviations: CNS, central nervous system; DHHS, Department of Health and Human Services; IUGR, intrauterine growth retardation; NBAS, Neonatal Behavioral Assessment Scale; NMDA, *N*-methyl-D-aspartate; PCP, phencyclidine; PN, postnatal day.

apply to the psychoactive compounds to be discussed in this chapter. Several reviews are available that discuss the broad issues of maternal drug use in depth (Chiriboga, 1993; Zuckerman and Bresnahan, 1991; Behnke and Eyler, 1993; Frank *et al.,* 1996) and should the reader desire more information on specific aspects of drug abuse during pregnancy, entire volumes have been devoted to the topic (e.g., Hutchings, 1989; Lewis and Bendersky, 1995). Problems associated with the study of illicit drug use in pregnant women are also discussed in this chapter, and additional resources on specific topics are presented. Finally, reasons for using animal models to garner information about processes associated with gestational drug exposure that may otherwise not be obtainable from human studies are presented.

II. Drug Access to the Developing Embryo–Fetus

Although the pharmacology of many drugs of abuse differ markedly, they all share certain properties. Notably, these include the ability to access and affect the CNS by crossing the blood–brain barrier (BBB). The placenta in many respects is similar to the BBB in its ability to limit the passage of chemicals from maternal blood into fetal tissue. Thus, it should not be surprising that drugs that are able to gain access to the brain can also gain access to the embryo or fetus through the placenta, and thus gain access to, *in utero,* developing tissue during virtually any period of development. Given that all abused substances alter nervous tissue function in some manner, it can be reasoned that if such alterations proved inappropriate or detrimental to ongoing developmental processes, then the potential for such changes subsequently to manifest as real, if not profound, functional changes in exposed offspring seems clear.

III. Observations for Specific Drugs

A. *Marijuana*

Marijuana is the illegal drug most commonly used during pregnancy (Day *et al.,* 1994a). The incidence of such use varies greatly depending on location, but ranges from a low of about 5% to a high of about 34% (Zuckerman and Bresnahan, 1991; Behnke and Eyler, 1993). The demonstration of cannabinoid receptors in brain (Devane *et al.,* 1988; Herkenham *et al.,* 1991; Howlett *et al.,* 1991; Johnson *et al.,* 1992) and brain constituents that bind to cannabinoid receptors (Devane *et al.,* 1992; Fride and Mechoulam, 1993; Smith *et al.,* 1994; Felder *et al.,* 1996) serves to highlight the potential for cannabinoids to target nervous system tissue. It is apparent from the literature that well-controlled studies on the effects of developmental exposure to marijuana in humans are rare at best. However, Fried (1995), Fried and co-workers (1992), and Day and colleagues (1994a,b) have for many years been following relatively large numbers of children exposed prenatally to marijuana. In general, their findings suggest that exposure to marijuana during pregnancy is related to subsequent adverse outcomes for mental and motor development and that age at time of assessment and time of gestational exposure is important (e.g., Richardson *et al.,* 1995). In addition, it has been reported that young maternal age may increase the offspring's risk of negative effects from prenatal marijuana exposure (Cornelius *et al.,* 1995), in that reduced gestational age was observed for children born to teenagers who smoked marijuana during pregnancy. This effect was observed to a greater degree in adolescent mothers than in adult mothers from the same clinic, even though the levels of prenatal marijuana exposure were lower in the adolescent mothers.

Assessment of marijuana-exposed offspring early in the postnatal period [postnatal days (PN) 9 and 30] has been reported by some investigators to be associated with negative outcomes, which include a higher incidence of tremors, increased startle reactions, and other behaviors thought to indicate alterations in state regulation of the nervous system (Fried, 1995). Other investigators have not replicated these findings (Tennes *et al.,* 1985). In subsequent work-ups of the population mentioned earlier, Fried (1995) reported no other marijuana-associated effects until the children were 4 years of age, when subtests of verbal ability and memory were significantly worse for marijuana-exposed offspring. At 1 year of age, subjects were assessed using a series of instruments to monitor a variety of nervous system functions. These included sensory perceptual ability, memory, problem solving, vocalization and the onset of words, aspects of object constancy, gross and fine motor movement, and temperament (Fried, 1995). No associations were noted between prenatal marijuana exposure and any outcomes, and these findings were consistent with the lack of effects reported by others (Tennes *et al.,* 1985; Astley and Little, 1990). Using the same assessment instruments at 24 months of age, no effects of prenatal marijuana exposure could be demonstrated and additional assessments of language abilities also proved negative. At 3 years of age, further assessments of language expression and comprehension as well as aspects of other verbal, perceptual, memory, motor, and general cognitive skills also proved negative for an effect of prenatal marijuana exposure (Fried, 1995). Interestingly, in another group of marijuana-exposed offspring considered to be at higher risk due to their lower social status, Day and colleagues (1994b) reported that although no effect of prenatal marijuana could be shown for global measures of intelligence (Stanford–Binet Intelligence Scale), there was a significant impact of exposure on short-term memory. Likewise, Fried had earlier demonstrated the lack of effects of marijuana exposure on global intellectual measures in children at age 4 years, whereas performance on subscales for verbal ability and memory were inferior to those for children whose mothers were not regular marijuana users. Reassessment of these same subjects at 5 and 6 years of age with the same instruments used in their 4-year assessments failed to uncover any significant effects of maternal marijuana use. Attempts to explain this apparent discrepancy have included the observation that positive environmental factors such as preschool and day care (Day *et al.,* 1994b) can serve to ameliorate the deficits in verbal and memory scores in 4- to 5-year-old children (Fried, 1995) associated with marijuana exposure *in utero.* Thus, those children adversely affected by *in utero* exposure to marijuana were able to "catch up" with their nonexposed counterparts when placed in presumably stimulating settings.

It was also postulated that general or global cognitive abilities may be insensitive to the adverse effects of developmental exposure to marijuana and that assessment of more specific characteristics such as sustained attention and impulse control might prove more appropriate (Fried, 1995). Thus, in subsequent assessments of the same subjects at 6 years of age, tasks thought to monitor aspects of those constructs were employed. In those studies, it was found that prenatal marijuana exposure was not associated with aspects of impulse control, but that it did predict poorer performance in the sustained attention task (Fried, 1995). In other assessments of 6- to 9-year olds (O'Connell and Fried, 1991), preliminary findings suggested that children of marijuana users had increased conduct problems, scored more poorly on visual memory and perception tasks, had language problems, and were more distractable. The authors point out that these behaviors are the same as those that appeared abnormal in these same children during earlier assessments (Fried *et al.,* 1992) and also in those subjects assessed by Day and colleagues (1994b) at age 3 years. A general consensus seems to be emerging that prenatal marijuana exposure causes real albeit subtle changes in the subsequent function of offspring: global measures do not demonstrate these effects, whereas specific aspects of brain function (i.e., short-term memory and attention) do. In addition, it seems that a nurturing environment can serve to ameliorate many of the negative effects associated with gestational exposure to marijuana.

B. Opiates

For purposes of discussion here, the term *opiates* is used to refer to those compounds both natural and synthetic that share morphine-like analgesic properties. Heroin and methadone are the most common opiates used for prolonged periods during pregnancy in humans. Acute exposures to opiate analgesics such as morphine, meperidine, and alfentanil occur routinely during labor, and data concerning the effects of such exposures should be considered pertinent to the issue of *in utero* opiate effects because all of these agents interact with the same family of subcellular receptors.

Estimates of opiate use among pregnant women range from about 1% to 2% to as high as 21% (Behnke and Eyler, 1993), with as many as 10,000 infants being born each year in the United States alone to women who used opiates during pregnancy (Hans, 1989). Human studies do not indicate that prenatal exposure to opiates results in any overt structural abnormalities in offspring (Finnegan, 1979; Kaltenbach and Finnegan, 1989). As is often the case for other drugs of abuse, many investigators have reported that offspring of opiate-using mothers are characterized by low birthweights (Lifschitz *et al.,* 1983), whereas other investigators do not find an effect of *in utero* opiate exposure on such parameters (Strauss *et al.,* 1976). In studies where low birthweights have been reported, offspring reach normal size at about 1½ years of age. In addition, in some studies where prenatal opiate exposure has been associated with smaller head circumference in offspring, the association did not hold up after analyses were controlled for several prenatal confounds (Lifschitz *et al.,* 1983). In a review by Chiriboga (1993), it was reported that low birthweight is a consistent finding for both heroin and methadone exposure *in utero,* and that intrauterine growth retardation (IUGR) is the primary cause of low birthweights in these infants. In addition, it was noted that infants exposed to opiates during gestation and reared in poor environments exhibited very delayed development, suggesting "the disproportionate adverse effect an impoverished environment has on the development" of these children (Chiriboga, 1993).

The most widely reported and robust finding in offspring of opiate-abusing mothers is a neonatal abstinence, or withdrawal, syndrome. In such cases, maternal opiate use results in the development of a passive drug addiction (physical dependence) in the fetus that is manifest by withdrawal symptoms in offspring during the neonatal period. Although withdrawal symptoms associated with *in utero* heroin exposure can persist for about 10 days postnatally, withdrawal symptoms associated with *in utero* methadone exposure can persist for up to 8 weeks (Brown and Zuckerman, 1991), and exposure to both heroin and methadone prenatally has been reported to cause a more prolonged and variable pattern of neurobehavioral abnormalities that lasts for up to 6 months (Hutchings *et al.,* 1993; Zuckerman and Bresnahan, 1991). Fortunately, there do not appear to be long-lasting functional consequences associated with either heroin or methadone exposure during gestation; after the withdrawal symptoms abate, infants appear normal (i.e., Bayley Scales of Mental Development, IQ) out to at least preschool (Kaltenbach and Finnegan, 1987, 1989; Hutchings *et al.,* 1993).

Animal models generally have not demonstrated that chronic opiate exposure during gestation causes congenital anomalies in offspring (see Lee and Chiang, 1985, for a review). Pregnancy outcomes tend to worsen in a dose-related fashion with opiate exposure during pregnancy and fetal growth is often stunted, but these effects are usually seen only at high doses. Of interest is the demonstration by Golub and colleagues (Golub *et al.,* 1988; Golub and Donald, 1995) of significant behavioral effects in rhesus monkeys caused by the intrapartum administration of the opiate analgesics meperidine or alfentanil. The significance of these findings lies in the observation that these effects were noted not only in the neonatal period after the drugs had been eliminated, but also, in the case of meperidine, out to 12 months of age. Both the pattern of development of spontaneous motor activity and performance of several cognitive tasks (discrimination–reversals, delayed spatial alternation, and continuous performance) were altered by acute exposure to meperidine during labor. Although these drugs are not heroin or methadone, these findings suggest a potential for opiates to affect the behavior of offspring for prolonged periods of time. It must also be pointed out here that these findings occurred after acute treatments during labor, not after chronic exposures throughout pregnancy, during which adaptive processes may intercede to attenuate such effects.

C. Phencyclidine

Phencyclidine (PCP, Angel Dust) is an antagonist at the *N*-methyl-D-aspartate (NMDA) excitatory amino acid receptor and it also binds to sigma opiate receptors and inhibits the uptake of catecholamines and serotonin (see Ali *et al.,* 1993; Lodge and Johnson, 1990). PCP remains a relatively widely used drug, particularly among the younger members of the population. In an earlier report on the incidence of PCP use in a population of more than 2,300 pregnant women, 0.8% reportedly used the drug during pregnancy, whereas more than 7% admitted prior use (Golden *et al.,* 1984). Unfortunately, there have been very few studies that have examined the developmental effects of PCP use during pregnancy in humans. In those that have emerged, PCP use during gestation has been reported to be associated with IUGR, increased meconium in amniotic fluid, precipitate labor, and drug withdrawal symptoms in the newborn (Tabor *et al.,* 1990) as well as small size at birth, symptoms of neonatal narcotic withdrawal, temperament and sleep problems (Wachsman *et al.,* 1989), and "abnormal neurobehavior in the newborn period" (Chasnoff *et al.,* 1986, p. 357). Other studies reported no effect on birthweight or head circumference, but increases in state lability, poor consolability, and sudden outbursts of agitation were noted (Chasnoff *et al.,* 1983). Problems inherent with human studies of this nature (not only for PCP but also for the other drugs to be mentioned), however, are many and often confound the interpretation of such observations. Such issues are discussed in more detail later in the chapter with respect to studies on the effects of prenatal cocaine exposure.

PCP is known to cross the placenta in both humans (Marble *et al.,* 1980; Ahmad, 1987; Knight et al., 1994) and animals (Cummings *et al.,* 1979; Nicholas and Schreiber, 1982; Ali *et al.,* 1989). PCP has been detected in the urine of newborns of PCP-using women up to 7 days postpartum (Marble *et al.,* 1980; Ahmad, 1987) and appears to concentrate in the fetus, as evidenced by brain (Ahmad *et al.,* 1987; Ali *et al.,* 1989) and plasma levels (Cummings *et al.,* 1979) that are higher in fetuses than in their maternal counterparts.

In animal studies designed to examine the effects of gestational exposure to PCP, surprisingly few effects have been noted at doses that do not also cause maternal weight loss. Only at very high doses (120 mg/kg) that are highly toxic to the dams has PCP been reported to increase physical malformations in mice (Marks *et al.,* 1980). In a study in which some maternal weight loss also accompanied gestational PCP treatment, rat offspring were shorter, lighter, and less viable than controls, and they exhibited delayed development of specific behaviors and developmental landmarks such as eyelid opening (Nabeshima *et al.,* 1987). In a very thorough neurochemical study of the effects of prenatal PCP exposure on the development of dopamine and NMDA systems in the rat, it was shown that such PCP treatment did not affect levels of dopamine, serotonin, or NMDA receptor complexes postnatally, even as much as 100 days postnatal. A moderate but transient effect on glutamate-evoked dopamine release was noted early in development (PND 8), but this effect was undetectable when assessed on PN 21 (Ali *et al.,* 1993). In rats, neither the dopamine nor the NMDA systems are very well developed during gestation and are quite immature, even at birth. Thus, because of their immaturity, these systems might be protected from the adverse effects of

PCP on dopamine and PCP receptors that have been noted to occur in adults (Quirion *et al.,* 1982).

In rodent studies, chronic treatment with relatively large doses of PCP has been shown to alter the seizure threshold to pentylenetetrazol-induced seizures in an age-dependent manner, either increasing, decreasing, or not changing subsequent CNS excitability depending on the specific postnatal period during which exposure occurred (Sircar *et al.,* 1994). In mice, developmental exposure to PCP has been associated with subsequent increases in the number of muscarinic cholinergic receptors and with reduced performance in both a radial arm and a water maze. The maze effects were dependent on both the maze type and the developmental period of drug administration (Yanai *et al.,* 1992). Others have shown that, in rats, chronic gestational exposure to PCP at quite low doses (0.5 mg/kg) compared with those generally used in other animal studies (5–20 mg/kg) results in increased sensitivity to later challenges with PCP (Howard and Takeda, 1990). Still others have shown that although exposure to PCP during gestation resulted in lower body weight at birth and an apparent embryolethal effect in males, it did not alter the number of live births, implantation sites, or sensitivity to subsequent apomorphine-induced wall climbing behavior (Fico *et al.,* 1990). In rat offspring of dams treated with PCP throughout the last 2 weeks of gestation, there were no differences in locomotor activity from birth to 30 days of age (Hutchings *et al.,* 1984). Whether developmental exposure to PCP results in any long-term effects appears to be dependent on the specific developmental events occurring at the time of exposure and the specific endpoint monitored for effect.

D. Cocaine

As with most drugs of abuse, cocaine use by humans during pregnancy has also been associated with deleterious effects on the mother–infant dyad, including: abruptio placentae; spontaneous abortion; reduced maternal weight gain during pregnancy; reduced length of pregnancy; congenital malformations; and reduced infant body weight, body length, and head circumference at birth (Bandstra and Burkett, 1991; Lindenberg *et al.,* 1991; Nair and Watson, 1991; Slutsker, 1992; Kain *et al.,* 1993;). Specific nervous system effects reported for infants have included hemorrhage, seizures, and poor performance on a number of perinatally administered rating scales for neurobehavioral development (Bandstra and Burkett, 1991; Lindenberg *et al.,* 1991; Slutsker, 1992). However, the effects of cocaine exposure during pregnancy on the residual behavioral function of offspring remain unclear (Hill and Tennyson, 1986; Lindenberg *et al.,* 1991; Slutsker, 1992). For example, using Apgar scores measured within minutes of birth (Apgar, 1952; Apgar and James, 1962), some investigators reported that cocaine exposure during pregnancy reduced these scores (Ryan *et al.,* 1987; Cohen *et al.,* 1991; Neuspiel *et al.,* 1991), whereas other investigators have reported normal scores (Chasnoff and Griffith, 1989; Chasnoff *et al.,* 1989; Eyler *et al.,* 1994). Similarly, using neonatal abstinence scores obtained within hours of birth (Finnegan, 1986), some investigators have reported increased scores associated with cocaine exposure during pregnancy (Oro and Dixon, 1987) whereas others have reported no effects (Chasnoff *et al.,* 1987; Ryan *et al.,* 1987). Several investigators have reported decreased scores on the Brazelton Neurobehavioral Assessment Scale when assessed within several days of birth (Chandler *et al.,* 1980) in cocaine exposed offspring (Chasnoff *et al.,* 1987, 1989; Chasnoff and Griffith 1989; Hume *et al.,* 1989), whereas others have reported normal scores (Neuspiel *et al.,* 1991; Richardson and Day, 1991). Dose-related effects have been reported in 3-week-old infants for assessments using the Neonatal Behavioral Assessment Scale (NBAS), whereas there were no effects in these same infants when assessed at 2–3 days of age (Tronick *et al.,* 1996).

In addition to the contradictory nature of the results from these studies, which have focussed on the assessment of behavioral function close to the time of birth, there have been very few studies concerning longer term behavioral consequences of cocaine exposure during pregnancy (Hill and Tennyson, 1986; Bandstra and Burkett, 1991; Lindenberg *et al.,* 1991). Regrettably, the findings from the few longer term studies that have been conducted are also unclear. For example, when the Brazelton Neurobehavioral Assessment Scale was administered repeatedly throughout the first month following birth, some investigators reported decreased scores (Coles *et al.,* 1992), whereas other investigators reported normal scores (Neuspiel *et al.,* 1991). Administration of the Movement Assessment Scale (Chandler *et al.,* 1980) at 4 months of age revealed significant effects of cocaine exposure during pregnancy (Schneider and Chasnoff 1992). However, repeated administration of the Denver Developmental Testing Scale (Frankenburg *et al.,* 1970) throughout the first 15 months of life failed to demonstrate any effect of cocaine exposure during pregnancy on subsequent infant development (Graham *et al.,* 1989). Furthermore, repeated assessments of the Bayley Scales of Infant Development (Bayley, 1969) disclosed an effect of gestational cocaine exposure following testing at 6 months of age, but there were no effects evident at earlier or later testing times (Chasnoff *et al.,* 1992). Finally, administration of the specialized Nursing Child Assessment Scale (Barnard *et al.,* 1983) between 2 and 4 months of age did not reveal an effect of cocaine exposure during pregnancy (Neuspiel *et al.,* 1991).

These apparent inconsistencies in the human literature have been attributed to difficulties in the techniques of measurement that preclude accurate determination of the type, dose, and pattern of cocaine use (Day and Robles 1989; Bandstra and Burkett, 1991; Lindenberg *et al.,* 1991; Nair and Watson, 1991; Slutsker, 1992) and to difficulties in controlling for a wide range of potentially confounding variables, such as other drug use, race, socioeconomic status, and level of prenatal care (Bandstra and Burkett, 1991; Lindendberg *et al.,* 1991; Slutsker, 1992). Such limitations can seriously undermine assessments concerning the long-term effects of gestational cocaine exposure on growth and development.

Although numerous reports in the rodent literature suggest that prenatal cocaine administration can lead to subsequent behavioral alterations in the offspring (e.g., Spear, 1995), and reports using a rabbit model demonstrate that cocaine has long-term consequences after *in utero* exposure (Romano *et al.,* 1995; Romano and Harvey, 1996a,b), a multitude of other reports suggest that it does not (Vorhees, 1995). Such observations have led others previously to posit that "little consensus exists concerning cocaine's developmental toxicity from these models" (Vorhees, 1995p. 90). Current observations suggest that, under certain circumstances and in certain animal models, cocaine can clearly be shown to affect aspects of nervous system chemistry, architecture, and function; thus, its toxicity can be demonstrated and explored using animal models. Such toxicity should then indicate where the most likely problem areas are to be found in humans, should toxicity manifest. An important issue now would seem to be assessing the risk of toxicity associated with cocaine exposure as experienced by humans.

Against this background, a long-term project was initiated in nonhuman primates investigating the effects of chronic cocaine exposure during pregnancy on maternal and infant outcomes. Of primary importance were assessments of several complex behaviors in the offspring as measures of the functional status of the brain. These animal studies were undertaken so that important procedural variables that are difficult, if not impossible, to manage in clinical studies—such as the exact magnitude and duration of cocaine exposure—could be known. To ensure the greatest relevance for the results, the project was conducted using the rhesus monkey as the animal model because, when compared to most other animals, nonhuman primates provide the closest approximation of the human condition with respect to pharmacology, physiology, metabolism, and perhaps most importantly, placentation. In addition, the use of the monkey provides opportunities to monitor very specific behaviors that can also be observed in humans. Initial findings from these studies have been published (Morris *et al.,* 1996a,b, 1997; Paule *et al.,* 1996) and are summarized here.

The importance of the use of appropriate animals models in laboratory studies can not be overemphasized: they provide important and—in many cases—preferred alternatives to human studies because critical features of experimental design can be strictly controlled (Dow-Edwards, 1989). Although generalization across species is always problematic, there is growing recognition of the fact that, in contrast to frequently used species such as the rat, the nonhuman primate represents a more appropriate experimental animal model for investigating human development (Kemnitz *et al.,* 1984). This view is based on the greater similarity between the two species with respect to pregnancy and fetal biology during the prenatal period (Kemnitz *et al.,* 1984) and with respect to anatomic, physiologic, and behavioral development during the postnatal period (Bourne, 1975; Golub and Gershwin, 1984). In addition, because many measures of neonatal assessment in the nonhuman primate are directly comparable to those evaluated clinically in the human infant, and because neurobehavioral evaluations are undergoing improved standardization, the nonhuman primate has become of increasing value as a model for early human development (Golub and Gershwin, 1984). Indeed, considerable effort has been devoted to the evaluation of nonhuman primate infants exposed to a variety of risk factors, including pharmacologic and toxicologic agents (Golub *et al.,* 1981; Levin and Bowman, 1983; Gilbert *et al.,* 1993).

Initial pharmacokinetic studies in the rhesus monkey model helped establish appropriate doses of cocaine for modeling known human exposures (Duhart *et al.,* 1993). In addition, transplacental pharmacokinetic studies in the rhesus monkey have allowed the determination of actual fetal exposures following maternal cocaine administration (Binienda *et al.,* 1993) and show that fetal exposure is about one-third that of the mother. Initial studies on the chronic administration of cocaine in the pregnant monkey began at about 4 weeks of gestation and dosing continued until term (about 23–24 weeks). Cocaine hydrochloride was given by intramuscular injection, 3 times per day, 5 days per week; dose groups included saline controls and cocaine groups receiving either 0.3 or 1.0 mg/kg per injection. These doses are equivalent to 0.0, 1.0, and 3.0 mg/kg/day, respectively, and approximate human doses of about 0, 70, and 210 mg/day for a 70-kg person; thus our high-dose group would roughly equate to the human use of about a quarter of a gram of cocaine hydrochloride per day. The 5-day per week dosing regimen was adopted to model the multiple day binging behavior often seen during human self-administration. In addition, previous pharmacokinetic studies showed that the plasma distribution profiles of cocaine in rhesus monkeys after intramuscular injections are very similar to those obtained in humans after intranasal instillation or snorting (Duhart *et al.,* 1993).

Although the effects of chronic cocaine exposure during pregnancy on maternal and infant outcomes in these first studies were essentially negative (Morris *et al.,* 1996a), the animal model proved viable and the data indicated a dose-related trend toward decreased birthweights and head circumferences, effects often reported in the clinical literature. In subsequent studies an escalating dosing procedure was used, in which the initial doses given were 1.0 mg/kg per injection (3 mg/kg/day, or the equivalent of approximately 210 mg/day for a 70-kg person) as in the earlier studies. After 1–2 weeks of dosing at this level, however, the dose was increased to 1.5 mg/kg per injection, and after another 1–2 weeks it was escalated again to 2.0 mg/kg per injection, and so on until term. It was necessary to gradually increase the administered doses because some animals would not eat at doses of as low as 1.0 mg/kg per injection, thus a period of time was allowed for animals to develop tolerance to the well-known appetite-suppressing effects of cocaine prior to dose escalation. Maternal and infant outcomes for pregnancies maintained under this exposure regimen also were not significantly different from those for control pregnancies: there were no significant effects of treatment on gestational length; maternal weight gain; or offspring birthweight, length, and head circumference.

The behavioral assessments of offspring from these studies involved monitoring the acquisition of performance of specific food-reinforced tasks designed to model aspects of learning, color and position discrimination, motivation, and short-term memory. These behaviors (detailed in Schulze *et al.,* 1988) were shown previously to be remarkably similar in monkeys and children (Paule *et al.,* 1990). Here again, the data analyzed to date indicate that the ability of offspring to learn to perform these tasks was not affected by gestational cocaine exposure: all subjects acquired these behaviors in like fashion (Morris *et al.,* 1996b).

These observations seemed remarkable because individual doses attained at term were as high as 8.5 mg/kg per injection, or 25.5 mg/kg/day, which would equate to daily exposures of more than 1.5 g of cocaine hydrochloride per day for a 70-kg person. In the absence of any obvious effects of such chronic cocaine exposure throughout most of pregnancy, the period of exposure was extended to include the whole of gestation rather than waiting until pregnancy was detected at about 3–4 weeks post fertilization. In these studies, females were exposed to escalating doses of cocaine for several months prior to mating, with the dose continuing to escalate after pregnancies were established. It is interesting and perhaps important to note that pregnancies could not be established if daily doses exceeded 9.0 mg/kg, given in 3 divided doses of 3.0 mg/kg per injection. These doses, which equate to the human use of about 630 mg (0.63 g) of cocaine hydrochloride per day for a 70-kg person, disrupted the menstrual cycles in these animals such that they did not conceive. It might also be true that women who routinely self-administer more than about 0.5 g of cocaine per day simply do not get pregnant, although no reports on this have emerged. If that is the case, it would seem reasonable to presume that the cocaine exposure levels experienced during most human pregnancies would be lower than this amount. This supposition presumes, of course, that drug-use patterns remain relatively constant prior to and during pregnancy, or at least do not escalate after impregnation, a situation not unlikely because most human studies actually report declining drug use as pregnancies progress.

For those animals that did become pregnant while receiving chronic cocaine administration, dose escalation continued after pregnancies were established as in the previous study. For this total gestational exposure cohort, the number of animals in each group was increased such that there were 10 subjects in both the treated group and the control group (ns of 3 were used previously). The results of these studies indicate that total gestational exposure under the escalating dosing regimen does result in significantly smaller offspring (decreased body weight and overall length) with smaller heads (decreased crown circumference, Morris *et al.,* 1997). It is unknown whether these effects resulted from the increased duration of exposure (i.e., exposure throughout the whole of pregnancy versus exposure beginning after pregnancies were detected at about 3–4 weeks of gestation) or whether the increase in sample size provided sufficient statistical power for demonstrating such effects (Morris *et al.,* 1997). The key objective of these studies remains the determination of whether prenatal exposure to cocaine is associated with long-lasting effects on the behavior of offspring, and these studies are, as of this writing, still underway. Implicit in this effort is the hypothesis that prenatal cocaine has effects on development that are detectable later as altered behavioral responses to pharmacologic challenges with drugs that affect specific systems within the CNS. There have been numerous reports in the rodent literature (reviewed in Spear, 1995) and in some rabbit studies (Murphy *et al.,* 1995; Romano *et al.,* 1995; Romano and Harvey, 1996a,b) that offer data to support this hypothesis: Heyser *et al.* (1992) reported that rats fail to show the usual cocaine-induced place preference if they were exposed to cocaine during gestation; rats also show reduced sensitivity to cocaine-induced wall climbing when cocaine is administered postnatally (Meyer *et al.*, 1992). In addition, the usual stimulation of locomotor activity associated with a postnatal cocaine challenge is attenuated following previous prenatal cocaine exposure (Sobrian *et al.*, 1990; Heyser *et al.*, 1992; Meyer *et al.*, 1992), as is the postnatal effect of amphetamine (Sobrian *et al.*, 1990; Hughes *et al.*, 1991; Meyer *et al.*, 1992; Simansky and Kachelries, 1996). The effect of cocaine to inhibit the reuptake of monoamines (including serotonin, dopamine, and norepinephrine) is well known (e.g., Benowitz, 1993; Dow-Edwards, 1995). It has also been suggested that cocaine may increase the release of these neurotransmitters from nerve terminals (Pitts and Marwah, 1987). In either case, the postsynaptic effects of these neurotransmitters are increased (Gold and Vereby, 1984). If these effects of cocaine were to occur during a sensitive or critical period of development, the increased monoaminergic activity could lead to increased synaptic reactivity (Whitaker-Azmitia, 1991) as part of the trophic influence of neurotransmitters on neural development (Mirmiran *et al.*, 1985). In fact, it has been reported by some investigators that prenatal cocaine exposure affects monoaminergic receptor development (Scalzo *et al.*, 1990; Henderson *et al.*, 1991; Seidler and Slotkin, 1992). Therefore, it is logical to monitor these kinds of developmental effects by administering drugs that have selective affinities for these types of receptors; if a change in receptor responsiveness occurs as a consequence of prenatal cocaine exposure, then such a change should be reflected in altered responsiveness to drugs with affinity for those receptors (Zenick, 1983; Walsh and Tilson, 1986).

In offspring from the original cohort of subjects [saline controls, low (1.0 mg/kg/day), high (3.0 mg/kg/day), and escalating (up to 25.5 mg/kg/day) doses of cocaine], dose–effect curves for the behavioral effects of cocaine were determined after they had learned how to perform the behavioral tasks mentioned earlier. Briefly, subjects were "challenged" with a variety of doses of cocaine given intravenously and the acute effects on the behaviors mentioned earlier were determined. The initial dose–effect curve was determined when subjects were approximately 1.5 years of age and a second determination was obtained in these same animals as adolescents when they were 3 years old. In addition, dose–response data were obtained for the same behavioral tasks in a set of adult (10–11 years old) male animals for comparison. The results indicated that young monkeys, in general, are much less sensitive (ca. 10-fold in juveniles and about 3-fold in adolescents) to the behaviorally disruptive effects of cocaine than are adults. There was no indication that cocaine exposure *in utero* affected the sensitivity of juvenile or adolescent offspring to the behaviorally disruptive effects of acute cocaine administration.

Thus, to date, it has been demonstrated in the monkey model that exposure to cocaine throughout the whole of gestation can produce offspring that are smaller in stature than control animals: decreased body weight, length, and head circumference. The doses used prior to impregnation (about the equivalent of 0.6 g of cocaine hydrochloride per day for a 70-kg human) appeared to approximate the maximal tolerated doses, because menstrual cycles were disrupted at higher doses. Escalating exposures throughout the entire pregnancy then resulted in the reported effects on offspring size. Under the behavioral assessment regimen currently employed (monitoring the acquisition and performance of several food-reinforced operant tasks designed to model motivation, color and position discrimination, learning and short-term memory, described in Morris *et al.*, 1996b), no effects of *in utero* exposure have yet been detected. However, the behavioral data for the offspring from the total gestational exposure group (i.e., the stunted babies) have yet to be analyzed.

Drug challenges of the offspring from the groups in which treatment began only after pregnancies were detected also have not revealed any effects of gestational drug exposure on subsequent responsivity to cocaine. However, because young monkeys are very insensitive to the behaviorally disruptive effects of cocaine when compared to adults (e.g., 10–30 times less sensitive; Paule, 1997), it may be necessary to challenge the offspring after adult sensitivity has been established to determine whether maturation of the systems responsive to cocaine has been altered by developmental cocaine exposure.

The findings of a lack of any behavioral effects, with or without cocaine challenge, seem remarkable in light of the findings that prenatal cocaine exposure alters the development of the rhesus monkey frontal cortex (Lidow, 1995) and affects development of dopamine neurons (Ronnekleiv and Naylor, 1995; Choi and Ronnekleiv, 1996) as well as expression of dynorphin and enkephalin mRNA in fetal rhesus monkey brain (Chai *et al.*, 1997). Whereas the daily doses (20 mg/kg) used in the Lidow study (1995) were in the same range as those attained near the end of pregnancies in studies by Morris *et al.* (1996a,b, 1997), the route of administration (oral) was different. Clearly, the kinetics of cocaine in blood are quite different after oral administration than after intramuscular administration and may influence the developmental effects of cocaine exposure. Ronnekleiv and Naylor (1995), however, used the same route of exposure (intramuscular) and very similar doses (3.0 mg/kg per injection, 4 injections per day, or 12 mg/kg/day) and found alterations in the biochemistry of dopamine neurons when fetuses were examined on day 60 of gestation. Exposures in that study were, however, 7 days/week rather than the 5 days/week used in the studies by Morris *et al.* (1996a,b, 1997), and doses were given four times per day rather than three. It remains to be seen whether the effects of cocaine on dopamine neurons reported by Ronnekleiv and Naylor (1995) persist in the offspring, and whether any functional alterations are associated with such changes if they do persist. As for the offspring produced in the laboratories of Morris *et al.* (1996a,b, 1997), challenges with several other pharmacologic probes of the dopaminergic and other neurotransmitter systems will be made as the animals age. Data from these studies should provide information about whether the functional status of these systems has been affected by gestational exposure to cocaine.

It would appear that, in all of the monkey studies discussed, the exposures used are very near the maximal tolerated doses, and thus likely present a near worse-case exposure scenario, at least for the monkey model. Under such exposures, offspring that are significantly smaller than control animals can be produced. Whether any functional consequences of *in utero* exposure to cocaine are demonstrable remains to be determined. It could be that the types of behaviors being monitored (i.e., the highly practiced operant tasks for food reinforcement) are insensitive to whatever lasting perturbations—if any—occur in the primate brain. Perhaps social behaviors or the classically conditioned responses noted to be affected in the rabbit model will turn out to be the kinds of behaviors sensitive to gestational cocaine exposure. It is known that in children, at least, performance of the operant behaviors used in our assessment battery (Paule 1990, 1995) is significantly correlated with IQ (Paule, 1994). It has also been shown by others that global measures of brain function, such as IQ, are often not sensitive to subtle developmental perturbations, whereas specific functional domains might be. Of importance now is to increase the behaviors monitored in the monkey model and to determine the relevance of the rabbit model, in which clear behavioral effects have been caused by gestational exposure to cocaine at doses that appear to be very relevant to human exposures.

References

Abelson, H. I., and Miller, J. P. (1985). A decade of trends in cocaine use in the household population. *National Institute Of Drug Abuse Research Monograph Series.* **61,** 35–49.

Ahmad, G. (1987). Abuse of phencyclidine (PCP): a laboratory experience. *Clin. Toxicol.* **25,** 341–346.

Ahmad, G., Halsall, L. C., and Bondy, S. C. (1987). Persistence of phencyclidine in fetal brain. *Brain Res.* **415**(1), 194–196.

Ali, S. F., Ahmad, G., Slikker, W. Jr., and Bondy, S. C. (1989). Effects of gestational exposure to phencyclidine: distribution and neurochemical alterations in maternal and fetal brain. *Neurotoxicology* **10**(3), 383–392.

Ali, S. F., Holson, R. R., Newport, G. D., Slikker, W. Jr., and Bowyer, J. F. (1993). Development of dopamine and *N*-methyl-D-aspartate systems in rat brain: the effect of prenatal phencyclidine exposure. *Brain Res.* **73**(1), 25–33.

Apgar, V. (1952). A proposal for a new method of evaluation of the newborn infant. *Anesth. Analg.* **32,** 260–267.

Apgar, V., and James, L. S. (1962). Further observations on the newborn scoring system. *JAMA* **181,** 419–428.

Astley, S., and Little, R. (1990). Maternal marijuana use during lactation and infant development at one year. *Neurotoxicol. Teratol.* **12,** 161–168.

Bandstra, E. S., and Burkett, G. (1991). Maternal–fetal and neonatal effects of in-utero cocaine exposure. *Sem. Perinat.* **15,** 288–301.

Barnard, K., Eyres, S., Lobo, M., and Snyder, C. (1983). An ecological paradigm for assessment and intervention. In *New Approaches to Developmental Screening of Infants.* T. B. Brazelton, and B. M. Lester, Eds. Elsevier, New York, pp. 199–218.

Bayley, N. (1969). Bayley Scales of Infant Development. New York: Psychological Corporation.

Behnke, M., and Eyler, F. D. (1993). The consequences of prenatal substance use for the developing fetus, newborn, and young child. *Int. J. Addict.* **28**(13), 1341–1391.

Benowitz, N. L. (1993). Clinical pharmacology and toxicology of cocaine. *Pharmacol. Toxicol.* **72,** 3–12.

Binienda, Z., Bailey, J. R., Duhart, H. M., Slikker, W. Jr., and Paule, M. G. (1993). Transplacental pharmacokinetics and maternal/fetal plasma concentrations of cocaine in pregnant macaques near term. *Drug Met. Disp.* **21,** 364–368.

Bourne, G. H. (1975). Collected anatomical and physiological data from the rhesus monkey. In, *The Rhesus Monkey,* G. H. Bourne Ed., Vol. I. Academic Press, New York, pp. 1–63.

Bozarth, M. A., and Wise, R. A. (1985). Toxicity associated with long-term intravenous heroin and cocaine self-administration in the rat. *JAMA* **245,** 81–83.

Brown, E., and Zuckerman, B. (1991). The infant of the drug abusing mother. *Pediatr. Ann.* **20**(10), 555–559.

Chai, L., Choi, W. S., and Ronnekleiv, O. K. (1997). Maternal cocaine treatment alters dynorphin and enkephalin mRNA expression in brains of fetal rhesus macaques. *J. Neurosci.* **17**(3), 1112–1121.

Chandler, L. S., Andrews, M. S., and Swanson, M. W. (1980). *Movement Assessment Of Infants.* Rolling Bay, Washington.

Chasnoff, I. J. (1989). Drug use and women: establishing a standard of care. *Ann. N.Y. Acad. Sci.* **562,** 208–210.

Chasnoff, I. J., Burns, W. J., Hatcher, R. P., and Burns, K. A. (1983). Phencyclidine: effects on the fetus and neonate. *Develop. Phamacol. Therap.* **6**(6), 404–408.

Chasnoff, I. J., Burns, W. J., Schnoll, S. H., and Burns, K. A. (1985). Cocaine use in pregnancy. *N. Engl. J. Med.* **313**(11), 666–669.

Chasnoff, I. J., Burns, K. A., Burns, W. J., and Schnoll. (1986). Prenatal drug exposure: effects on neonatal and infant growth and development. *Neurobehav. Toxicol. Teratol.* **8**(4), 357–362.

Chasnoff, I. J., Burns, K. A., and Burns, W. J. (1987). Cocaine use in pregnancy: perinatal morbidity and mortality. *Neurotoxicol. Teratol.* **9,** 291–293.

Chasnoff, I. J., and Griffith, D. R. (1989). Cocaine: clinical studies of pregnancy and the newborn. *Ann. NY Acad. Sci.* **562,** 260–266.

Chasnoff, I. J., Griffith, D. R., MacGregor, S., Dirkes, K., and Burns, K. A. (1989). Temporal patterns of cocaine use in pregnancy: perinatal outcome. *JAMA* **261,** 1741–1744.

Chasnoff, I. J., Griffith, D. R., Freier, C., and Murray, J. (1992). Cocaine/poly drug use in pregnancy: two-year followup. *Pediatrics* **89,** 284–289.

Chiriboga, C. A. (1993). Fetal effects. *Neurol. Clin.* **11**(3), 707–728.

Choi, W. S., and Ronnekleiv, O. K. (1996). Effects of *in utero* cocaine exposure on the expression of mRNAs encoding the dopamine transporter and the D1, D2 and D5 dopamine receptor subtypes in fetal rhesus monkey. *Brain Res. Dev. Brain Res.* **96**(1-2), 249–260.

Cohen, H. R., Green, J. R., and Crombleholme, W. R. (1991). Peripartum cocaine use: estimating risk of adverse pregnancy outcome. *Int. J. Gynecol. Obstet.* **35,** 51–54.

Coles, C. D., Platzman, K. A., Smith, I., James, M. E., and Falek, A. (1992). Effects of cocaine and alcohol use in pregnancy on neonatal growth and neurobehavioral status. *Neurotoxicol. Teratol.* **14,** 23–33.

Cornelius, M. D., Taylor, P. M., Geva, D., and Day, N. L. (1995). Prenatal tobacco and marijuana use among adolescents: effects on offspring gestational age, growth, and morphology. *Pediatrics* **95**(5), 738–743.

Cummings, A. J., Jones, H. M., and Cooper, J. E. (1979). Transplacental disposition of phencyclidine in the pig. *Xenobiotica* **9**(7), 447–452.

Day, N. L., and Robles, N. (1989). Methodological issues in the measurement of substance abuse. *Ann. NY Acad. Sci.* **562,** 8–13.

Day, N. L., Cottreau, C. M., and Richardson, G. A. (1993). The epidemiology of alcohol, marijuana and cocaine use among women of childbearing age and pregnant women. *Clin. Obst. Gyn.* **36,** 232–245.

Day, N. L., Richardson, G. A., Geva, D., and Robles, N. (1994a). Alcohol, marijuana, and tobacco: effects of prenatal exposure on offspring growth and morphology at age six. *Alc. Clin. Exp. Res.* **18**(4), 786–794.

Day, N. L., Richardson, G. A., Goldschmidt, L., Robles, N., Taylor, P. M., Stoffer, D. S., Cornelius, M. D., and Geva, D. (1994b). Effect of prenatal marijuana exposure on the cognitive development of offspring at age three. *Neurotox. Teratol.* **16**(2), 169–175.

Devane, W. A., Dysarz, F. A. III, Johnson, M. R., Melvin, L. S., and Howlett, A. C. (1988). Determination and characterization of a cannabinoid receptor in rat brain. *Molec. Pharmacol.* **34**(5), 605–613.

Devane, W. A., Hanus, L., Breuer, A., Pertwee, R. G., Stevenson, L. A., Griffin, G., Gibson, D., Mandelbaum, A., Etinger, A., and Mechoulam, R. (1992). Isolation and structure of a brain constituent that binds to the cannabinoid receptor. *Science* **258**(5090), 1946–1949.

Department of Health and Human Services (DHHS). (1991). *Public Health Report.* **106,** 59–68.

Dow-Edwards, D. L. (1989). Long-term neurochemical and neurobehavioral consequences of cocaine use during pregnancy. *Ann. NY Acad. Sci.* **562,** 280–289.

Dow-Edwards, D. L. (1995). Developmental toxicity of cocaine: mechanisms of action. In *Mothers, Babies, and Cocaine: The Role of Toxins in Development,* M. Lewis and M. Bendersky, Eds. Lawrence Erlbaum Associates, Hillsdale, New Jersey, pp 5–17.

Duhart, H. M., Fogle, C. M., Gillam, M. P., Bailey, J. R., Slikker, W. Jr., and Paule, M. G. (1993). Pharmacokinetics of cocaine in pregnant and nonpregnant rhesus monkeys. *Repr. Tox.* **7,** 429–437.

Eyler, F. D., Behnke, M., Conlon, M., Woods, N. S., and Frentzen, B. (1994). Prenatal cocaine use: a comparison of neonates matched on maternal risk factors. *Neurotoxicol. Teratol.* **16,** 81–87.

Felder, C. C., Nielsen, A., Briley, E. M., Palkovits, M., Priller, J., Axelrod, J., Nguyen D. N., Richardson, J. M., Riggin, R. M., Koppel, G. A., Paul, S. M., and Becker, G. W. (1996). Isolation and measurement of the endogenous cannabinoid receptor agonist, anandamide, in brain and peripheral tissues of human and rat. *FEBS Lett.* **393**(2–3), 231–235.

Fico, T. A., Banks, A. N., and Hutchings, D. E. (1990). Prenatal phencyclidine in rats: effects on apomorphine-induced climbing. *Pharmacol. Biochem. Behav.* **35**(1), 93–97.

Finnegan, L. P. (1979). Pathophysiological and behavioural effects of the transplacental transfer of narcotic drugs to the foetuses and neonates of narcotic-dependent mothers. *Bull. Narc.* **31**(3,4), 1–58.

Finnegan, L. P. (1986). Neonatal abstinence syndrome: assessment and pharmacotherapy. In *Neonatal Therapy: An Update.* F. F. Rubiltelli and B. Cranati, Eds. Elsevier Science Publishers, New York, pp. 122–126.

Frank, D. A., Bresnahan, K., and Zuckerman, B. S. (1996). Maternal cocaine use: impact on child health and development. *Curr. Prob. Ped.* **26**(2), 57–70.

Frankenburg, W. K., Dodd, J. B., and Fandal, A. (1970). *The Revised Denver Developmental Screening Test Manual.* University of Colorado Press, Denver.

Fride, E. and Mechoulam, R. (1993). Pharmacological activity of the cannabinoid receptor agonist, anandamide, a brain constituent. *Eur. J. Pharmacol.* **231**(2), 313–314.

Fried, P. A. (1995). Prenatal exposure to marijuana and tobacco during infancy, early and middle childhood: effects and an attempt at synthesis. *Arch. Toxicol. Suppl.* **17,** 233–260.

Fried, P. A., O'Connell, C. M., and Watkinson, B. (1992). 60- and 72-month follow up of children prenatally exposed to marijuana, cigarettes, and alcohol: cognitive and language assessment. *J. Dev. Behav. Ped.* **13**(6), 383–391.

Gay, R. (1982). Clinical management of acute and chronic cocaine poisoning. *Ann. Emerg. Med.* **11,** 562–571.

Gilbert, S. G., Burbacher, T. M., and Rice, D. C. (1993). Effects of *in-utero* methylmercury exposure on a spatial delayed alternation task in monkeys. *Toxicol. Pharmacol.* **123,** 130–136.

Gold M. S. and Vereby, K. (1984). The psychopharmacology of cocaine. *Psychiat. Ann.* **14,** 714–723.

Golden, N. L., Kuhnert, B. R., Sokol, R. J., Martier, S., and Bagby, B. S. (1984). Phencyclidine use during pregnancy. *Am. J. Obstet. Gyn.* **148**(3), 254–259.

Golub, M. S. and Donald, J. M. (1995). Effect of intrapartum meperidine on behavior of 3- to 12-month-old infant rhesus monkeys. *Biol. Neonate* **67,** 140–148.

Golub, M. S., Eisele, J. H., and Donald, J. M. (1988). Obstetric analgesia and infant outcome in monkeys: neonatal measures after intrapartum exposure to meperidine or alfentanil. *Am. J. Obstet. Gyn.* **158,** 1219–1225.

Golub, M. S. and Gershwin, M. E. (1984). Standardized neonatal assessment in the rhesus monkey. In *Research in Perinatal Medicine.* P. W. Nathanielsz and J. T. Parer, Eds. Perinatology Press, New York, pp. 55–81.

Golub, M. S., Sassenrath, E. N., and Chapman, L. F. (1981). Regulation of visual attention in offspring of female monkeys treated chronically with delta-9-tetrahydrocannabinol. *Dev. Psychobiol.* **14,** 507–512.

Graham, K., Dimitrakoudis, D., Pelligrini, E., and Koren, G. (1989). Pregnancy outcome following first trimester exposure to cocaine in social users in Toronto. *Vet. Hum. Tox.* **31,** 143–148.

Hans, S. L. (1989). Developmental consequences of prenatal exposure to methadone. *Ann. NY Acad. Sci.* **562,** 195–207.

Henderson, M. G., McConnaughey, M. M., and McMillen, B. A. (1991). Long-term consequences of prenatal exposure to cocaine or related drugs: effects on rat brain monoaminergic receptors. *Brain Res. Bull.* **26,** 941–945.

Herkenham, M., Lynn, A. B., Johnson, M. R., Melvin, L. S., de Costa, B. R., and Rice, K. C. (1991). Characterization and localization of cannabinoid receptors in rat brain: a quantitative *in vitro* autoradiographic study. *J. Neurosci.* **11**(2), 563–583.

Heyser, C. J., Miller, J. S., Spear, N. E., and Spear, L. P. (1992). Prenatal exposure to cocaine disrupts cocaine-induced conditioned place preference in rats. *Neurotox. Teratol.* **14,** 57–64.

Hill, R. M. and Tennyson, L. M. (1986). Maternal drug therapy: effect on fetal and neonatal growth and neurobehavior. *Neurotoxicol.* **7,** 121–140.

Howard, S. G. and Takeda, H. (1990). Effect of prenatal exposure to phencyclidine on the postnatal development of the cholinergic system in the rat. *Develop. Neurosci.* **12**(3), 204–209.

Howlett, A. C., Champion-Dorow, T. M., McMahon, L. L., and Westlake, T. M. (1991). The cannabinoid receptor: biochemical and cellular properties in neuroblastoma cells. *Pharmacol. Biochem. Behav.* **40**(3), 565–569.

Hughes, H. E., Pringle, G. F., Scribani, L. A., and Dow-Edwards, D. L. (1991). Cocaine treatment in neonatal rats affects the adult behavioral response to amphetamine. *Neurotox. Teratol.* **13,** 335–339.

Hume, R. F., O'Donnell, K. J., Stanger, C. L., Killam, A. P., and Gingras, J. L. (1989). *In utero* cocaine exposure: observations of fetal behavioral state may predict neonatal outcome. *Am. J. Obstet. Gyn.* **161,** 685–690.

Hutchings, D. E. (1989). *Prenatal Abuse of Licit and Illicit Drugs.* D. E. Hutchings, Ed. *Ann. NY Acad. Sci.* Vol. 562.

Hutchings, D. E., Bodnarenko, S. R., and Diaz-DeLeon, R. (1984). Phencyclidine during pregnancy in the rat: effects on locomotor activity in the offspring. *Pharmacol. Biochem. Behav.* **20**(2), 251–254.

Hutchings, D. E., Zmitrovich, A., Church, S., and Malowany, D. (1993). Methadone during pregnancy: the search for a valid animal model. *Ann. Ist. Super. Sanita.* **29**(3), 439–444.

Johanson, C. E. (1984). Assessment of the dependence potential of cocaine in animals, NIDA Research Monograph 50. Rockville, Maryland, *National Institute on Drug Abuse,* pp. 54–71.

Johnson, M. R., Rice, K. C., Howlett, A. C., Melvin, L. S., and Herkenham, M. (1992). The cannabinoid receptor—pharmacologic identification, anatomical localization and cloning. *NIDA Research Monograph* **119,** 86–89.

Kain, Z. N., Rimar, S., and Barash, P. G. (1993). Cocaine abuse in the parturient and effects on the fetus and neonate. *Anesth. Analges.* **77,** 835–845.

Kaltenbach, K. A. and Finnegan, L. P. (1987). Perinatal and developmental outcome of infants exposed to methadone *in- utero. Neurotox. Teratol.* **99**(4), 311–313.

Kaltenbach, K. A. and Finnegan, L. P. (1989). Prenatal narcotic exposure: perinatal and developmental effects. *Neurotoxicology* **10**(3), 597–604.

Kemnitz, J. W., Houser, W. D., Eisele, S. G., Engle, M. J., Perelman, R. H., and Farrell, P. M. (1984). Pregnancy and fetal development in the rhesus monkey. I maternal metabolism and fetal growth. In *Research in Perinatal Medicine.* P. W. Nathanielsz and J. T. Parer, Eds. Perinatology Press, New York, pp.1–27.

Knight, E. M., James, H., Edwards, C. H., Spurlock, B. G., Oyemade, U. J., Johnson, A. A., West W. L., Cole, O. J., Westney, L. S., Westney, O. E., *et al.* (1994). Relationships of serum illicit drug concentrations during pregnancy to maternal nutritional status. *J. Nutrit.* **124**(6 Suppl), 973S–980S.

Lee, C. C. and Chiang, C. N. (1985). Maternal–fetal transfer of abused substances: pharmacokinetic and pharmacodynamic data. In *Prenatal Drug Exposure: Kinetics and Dynamics.* C. N. Chiang and C. C. Lee, Eds. *NIDA Research Monograph* **60,** pp. 110–147.

Levin, E. D. and Bowman, R. E. (1983). The effect of pre- or postnatal lead exposure on Hamilton search task in monkeys. Neurobehav. *Toxicol. Teratol.* **5,** 391–394.

Lewis, M. and Bendersky, M. (1995). *Mothers, Babies and Cocaine: The Role of Toxins in Development.* M. Lewis and M. Bendersky, Eds. Lawrence Erlbaum Associates, Hillsdale, New Jersey.

Lidow, M. S. (1995). Prenatal cocaine exposure adversely affects development of the primate cerebral cortex. *Synapse* **21**(4), 332–341.

Lifschitz, M. H., Wilson, G. S., Smith, E. O., and Desmond, M. M. (1983). Fetal and postnatal growth of children born to narcotic-dependent women. *J. Pediatr.* **102**(5), 686–691.

Lindenberg, C. S., Alexander, E. M., Gendrop, S. C., Nencioli, M., and Williams, D. G. (1991). A review of the literature on cocaine abuse in pregnancy. *Nurs. Res.* **40,** 69–75.

Lodge, D. and Johnson, K. M. (1990). Noncompetitive excitatory amino acid receptor antagonists. *Trends Pharmacol. Sci.* **11,** 81–86.

Loper, K. A. (1989). Clinical toxicology of cocaine. *Med. Tox. Adv. Drug Expos.* **4,** 174–185.

Marble, R. D., Thomas, R. G., and Sterling, M. L. (1980). Screening for angel dust in newborns. *Pediatrics* **66,** 334.

Marks, T. A., Worthy, W. C., and Staples, R. E. (1980). Teratogenic potential of phencyclidine in the mouse. *Teratology* **21**(2), 541–546.

Meyer, J. S., Sherlock, J. D., and MacDonald N. R. (1992). Effects of prenatal cocaine on behavioral responses to a cocaine challenge on postnatal day 11. *Neurotox. Teratol.* **14,** 183–189.

Miller, J. P. (1983). National survey on drug abuse. Department of Health and Human Services. Publication No. (ADM) 83–1263.

Mirmiran, M., Brenner, E., Vander Gugten, J., and Swaab, D. F. (1985). Neurochemical and electrophysiological disturbances mediate developmental behavioral alterations produced by medicines. *Neurobehav. Toxicol. Teratol.* **7,** 677–683.

Morris, P., Binienda, Z., Gillam, M. P., Harkey, M. R., Zhou, C., Henderson, G. L., and Paule, M. G. (1996a). The effect of chronic cocaine exposure during pregnancy on maternal and infant outcomes in the rhesus monkey. *Neurotoxicol. Teratol.* **18**(*2*), 147–154.

Morris, P., Gillam, M. P., Allen, R. R., and Paule, M. G. (1996b). The effects of chronic cocaine exposure during pregnancy on the acquisition of operant behaviors by rhesus monkey offspring. *Neurotox. Teratol.* **18**(*2*), 155–166.

Morris, P., Binienda, Z., Gillam, M. P., Klein, J., McMartin, K., Koren, G., Duhart, H. M., Slikker, W. Jr., and Paule, M. G. (1997). The effect of chronic cocaine exposure throughout pregnancy on maternal and infant outcomes in the rhesus monkey. *Neurotoxicol. Teratol.* **19**(*1*), 47–57.

Murphy, E. H., Hammer, J. G., Schumann, M. D., Groce, M. Y., Wang, X. H., Jones, L., Romano, A. G., and Harvey, J. A. (1995). The rabbit as a model for studies of cocaine exposure *in utero. Lab. Ann. Sci.* **45**(2), 163–168.

Nabeshima, T., Yamaguchi, K., Hiramatsu, M., Ishikawa, K., Furukawa, H., and Kameyama, T. (1987). Effects of prenatal and perinatal administration of phencyclidine on the behavioral development of rat offspring. *Pharmacol. Biochem. Behav.* **28**(3), 411–418.

Nair, B. S. and Watson, R. R. (1991). Cocaine and the pregnant woman. *J. Repr. Med.* **36,** 862–867.

Neuspiel, D. R., Hamel, S. C., Hochberg, E., Greene, J., and Campbell, D. (1991). Maternal cocaine use and infant behavior. *Neurotoxicol. Teratol.* **13,** 229–233.

Nicholas, J. M. and Schreiber, E. C. (1982). Transfer of phencyclidine (PCP) and metabolites across the mouse placenta. *Fed. Proc.* **41,** 1713.

NIDA, National Institute on Drug Abuse. (1989). *National Household Survey on Drug Abuse: Population Estimates 1988.* Department of Health and Human Services. Publication No. (ADM) 893–1636.

National Institute on Drug Abuse, (NIDA). (1996). *National Pregnancy and Health Survey, Drug Abuse among Women Delivering Live Births: 1992.* National Institutes of Health Publication No. 96–3819.

O'Connell, C. M. and Fried, P. A. (1991). Prenatal exposure to cannabis: a preliminary report of postnatal consequences in school-age children. *Neurotoxicol. Teratol.* **13,** 631–639.

Oro, A. S. and Dixon, S. D. (1987). Perinatal cocaine and methamphetamine exposure: maternal and neonatal correlates. *J. Pediatr.* **111,** 571–578.

Paule, M. G. (1990). Use of the NCTR operant test battery in nonhuman primates. *Neurotoxicol. Teratol.* **12,** 413–418.

Paule, M. G. (1994). Analysis of brain function using a battery of schedule-controlled operant behaviors. In *Neurobehavioral Toxicity: Analysis And Interpretation.* B. Weiss and J. O'Donoghue, Eds. Raven, New York, pp. 331–338.

Paule, M. G. (1995). Approaches to utilizing aspects of cognitive function as Indicators of neurotoxicity. In *Neurotoxicology: Approaches and Methods.* L. Chang and W. Slikker, Jr., Eds. Academic Press, Orlando, Florida, pp. 301–308.

Paule, M. G. (1997). Age-related sensitivity to the acute behavioral effects of cocaine and amphetamine in monkeys. *Neurotox. Teratol.* **19**(3), 241–242.

Paule, M. G., Forrester, T. M., Maher, M. A., Cranmer, J. M., and Allen, R. R. (1990). Monkey versus human performance in the NCTR operant test battery. *Neurotoxicol. Teratol.* **12,** 503–507.

Paule, M. G., Gillam, M. P., Binienda, Z., and Morris, P. (1996). Chronic cocaine exposure throughout gestation in the rhesus monkey: pregnancy outcomes and offspring behavior. *Ann. NY Acad. Sci.* **801,** 301–309.

Pitts, D. K. and Marwah, J. (1987). Cocaine modulation of central monoaminergic neurotransmission. *Pharmacol. Biochem. Behav.* **26,** 453–461.

Prakash, A. and Das, G. (1993). Cocaine and the nervous system. *Int. J. Pharmacol. Ther. Toxicol.* **31,** 575–581.

Quirion, R., Bayhor, M. A., Zerbe, R. L., and Pert, C. B. (1982). Chronic phencyclidine treatment decreases phencyclidine and dopamine receptors in rat brain. *Pharmacol. Biochem. Behav.* **17,** 699–702.

Richardson, G. A. and Day, N. L. (1991). Maternal and neonatal effects of moderate cocaine use during pregnancy. *Neurotoxicol. Teratol.* **13,** 455–460.

Richardson, G. A., Day, N. L., and Goldschmidt, L. (1995). Prenatal alcohol, marijuana, and tobacco use: infant mental and motor development. *Neurotox. Teratol.* **17**(4), 479–487.

Romano, A. G. and Harvey, J. A. (1996a). Prenatal exposure to cocaine disrupts discrimination learning in adult rabbits. *Pharmacol. Biochem. Behav.* **53**(3), 617–621.

Romano, A. G. and Harvey, J. A. (1996b). Elicitation and modification of the rabbit's nictitating membrane reflex following prenatal exposure to cocaine. *Pharmacol. Biochem. Behav.* **53**(4), 857–862.

Romano, A. G., Kachelries, W. J., Simansky, K. J., and Harvey, J. A. (1995). Intrauterine exposure to cocaine produces a modality-specific acceleration of classical conditioning in adult rabbits. *Pharmacol. Biochem. Behav.* **52**(2), 415–420.

Ronnekleiv, O. K. and Naylor, B. R. (1995). Chronic cocaine exposure in the fetal rhesus monkey: consequences for early development of dopamine neurons. *J. Neurosci.* **15**(11), 7330–7343.

Ryan, L., Ehrlich, S., and Finnegan, L. (1987). Cocaine abuse in pregnancy: effects on the fetus and newborn. *Neurotoxicol. Teratol.* **9,** 295–299.

Scalzo, F. M., Ali, S. F., Frambes, N. A., and Spear, L. P. (1990). Weanling rats exposed prenatally to cocaine exhibit an increase in striatal D2 dopamine binding associated with an increase in ligand affinity. *Pharmacol. Biochem. Behav.* **37,** 371–373.

Schneider, J. N. and Chasnoff, I. J. (1992). Motor assessment of cocaine/polydrug exposed infants at 4 months of age. *Neurotoxicol. Teratol.* **14,** 97–101.

Schulze, G. E., McMillan, D. E., Bailey, J. R., Scallet, A., Ali, S. F., Slikker, W. Jr., and Paule, M. G. (1988). Acute effects of delta-9-tetrahydrocannabinol in rhesus monkeys as measured by performance in a battery of complex operant tests. *J. Exper. Pharm. Ther.* **245,** 178–186.

Seidler, F. J. and Slotkin, T. A. (1992). Fetal cocaine exposure causes persistent noradrenergic hyperactivity in rat brain regions: effects on neurotransmitter turnover and receptors. *J. Pharmacol. Exp. Ther.* **263**(2), 413–421.

Simansky, K. J. and Kachelries, W. J. (1996). Prenatal exposure to cocaine selectively disrupts motor responding to D-amphetamine in young and mature rabbits. *Neuropharmacology* **35**(1), 71–78.

Sircar, R., Veliskova, J., and Moshe, S. L. (1994). Chronic neonatal phencyclidine treatment produces age-related changes in pentylenetetrazol-induced seizures. *Brain Res.* **81**(2), 185–191.

Slutsker, L. (1992). Risks associated with cocaine use during pregnancy. *Obst. Gyn.* **79,** 778–789.

Smith, P. B., Compton, D. R., Welch, S. P., Razdan, R. K., Mechoulam, R., and Martin, B. R. (1994). The pharmacological activity of anandamide, a putative endogenous cannabinoid, in mice. *J. Pharmacol. Exp. Therap.* **270**(1), 219–227.

Sobrian, S. K., Burton, L. E., Robinson, N. L., Ashe, W. K., James, H., Stokes, D. L., and Turner, L. M. (1990). Neurobehavioral and immunological effects of prenatal cocaine exposure in rat. *Pharmacol. Biochem. Behav.* **35,** 617–629.

Spear, L. P. (1995). Alterations in cognitive function following prenatal cocaine exposure: studies in an animal model. In *Mothers, Babies, and Cocaine: The Role of Toxins in Development,* M. Lewis and M. Bendersky, Eds. Lawrence Erlbaum Associates, Hillsdale, New Jersey, pp. 207–227.

Strauss, M. E., Starr, R. H., Ostrea, E. M., Chavez, C. J., and Stryker, J. C. (1976). Behavioral concomitants of prenatal addiction to narcotics. *J. Pediatr.* **89**(5), 842–846.

Tabor, B. L., Smith-Wallace, T., and Yonekura, M. L. (1990). Perinatal outcome associated with PCP versus cocaine use. *Am. J. Drug Alc. Abuse* **16**(3–4), 337–348.

Tennes K., Avitable, N., Blackard, C., Boyles, C., Hassoun, B., Holmes, L., and Kreye, M. (1985). Marijuana: prenatal and postnatal exposure in the human. In *Current Research on the Consequences of Maternal Drug Abuse.* T. M. Pinkert, Ed. National Institute on Drug Abuse Research Monograph 59. DHHS Pub. No. (ADM)85–1400. US Govt Printing Office, Washington, DC. pp. 48–60.

Tronick, E. Z., Frank, D. A., Cabral, H., Mirochnick, M., and Zuckerman, B. (1996). Late dose–response effects of prenatal cocaine exposure on newborn neurobehavioral performance. *Pediatrics* **98**(1), 76–83.

Vorhees, C. V. (1995). A review of developmental exposure models for CNS stimulants: cocaine. In *Mothers, Babies, and Cocaine: The Role of Toxins in Development,* M. Lewis and M. Bendersky, Eds. Lawrence Erlbaum Associates, Hillsdale, New Jersey, pp. 71–94.

Wachsman, L., Schuetz, S., Chan, L. S., and Wingert, W. A. (1989). What happens to babies exposed to phencyclidine (PCP) *in utero*? *Am. J. Drug Alc. Abuse* **15**(1), 31–39.

Walsh, T. J. and Tilson, H. A. (1986). The use of pharmacological challenges. In *Neurobehavioral Toxicology.* Z. Annau, Ed. The Johns Hopkins University Press, Baltimore, pp. 244–267.

Whitaker-Azmitia, P. M. (1991). Role of serotonin and other neurotransmitter receptors in brain development: basis for developmental pharmacology. [Review]. *Pharmacol. Reviews* **43**(4), 553–561.

Yanai, J., Avraham, Y., Levy, S., Maslaton, J., Pick, C. G., Rogel-Fuchs, Y., and Zahalka, E. A. (1992). Alterations in septohippocampal cholinergic innervations and related behaviors after early exposure to heroin and phencyclidine. *Brain Res. Dev. Brain Res.* **69**(2), 207–214.

Zenick, H. (1983). Use of pharmacological challenges to disclose neurobehavioral deficits. *Fed. Proceed.* **42,** 3191–3195.

Zuckerman, B. and Bresnahan, K. (1991). Developmental and behavioral consequences of prenatal drug and alcohol exposure. *Ped. Clin. North Am.* **38**(6), 1387–1406.

CHAPTER

36

The Neurobehavioral Teratology of Vitamin A Analogs

JANE ADAMS
Department of Psychology
University of Massachusetts
Boston, Massachusetts 02125

R. ROBERT HOLSON
Division of Reproductive and Developmental Toxicology
Department of Health and Human Services
Food and Drug Administration
National Center for Toxicological Research
Jefferson, Arkansas 72079

I. Introduction

Current therapeutic use of oral retinoids (isotretinoin and etretinate) for the treatment of dermatologic disorders has refocused interest in the developmental neurotoxicity of vitamin A–related compounds. This chapter presents information on the neurobehavioral teratogenicity of three retinoids: vitamin A, all-*trans* retinoic acid (tretinoin), and 13-*cis* retinoic acid (isotretinoin). Animal and human data are reviewed in an effort to provide stage-specific and dose-dependent information on the effects of prenatal exposure to these agents (see earlier work by Adams, 1993a,b). Because all of the animal studies of the effects of prenatal exposure to retinoids on offspring behavior have been conducted in rats, our review is restricted to this animal model of behavioral and neural outcomes. In this review, we examine first the chemical and biologic relationships between different retinoids. We then discuss rat and human studies of neural malformations induced by embryonic exposure to each of the three compounds.

*Abbreviations: CNS, central nervous system; GD, gestational day; IU, international units; PN, postnatal day.

Third, we examine the postnatal functional characteristics in regard to survival, weight, and general behavioral effects. We then discuss specific behavioral alterations that have been reported. Finally, the consistency of the functional outcomes with known neuropathology is discussed.

II. The Chemical Relationships between Retinoids

Marcus and Coulton (1990) have described thoroughly the chemical and physiologic characteristics of compounds in the vitamin A family. Vitamin A refers generically to compounds possessing the biologic properties of retinol. It is lipid soluble and is available from dietary sources in mixtures that are about half retinol or retinol esters and half carotenoids. The term *retinoid* refers to retinol plus its natural derivatives (most of which are lipid soluble) and synthetic analogs (some of which are water soluble). More than 2000 compounds have been identified in this category, but substantive information regarding neurobehavioral teratologic activity is available for only 3 compounds: vitamin A (retinol), all-*trans* retinoic acid (tretinoin), and 13-*cis* retinoic acid (isotretinoin). Within the body, oxidative processes convert retinol into all-*trans* retinoic acid, and all-*trans* retinoic acid is further isomerized to create 13-*cis* retinoic acid (a bidirectional process). All-*trans* retinoic acid is the primary active form of vitamin A in many tissues. It is able to promote cellular differentiation and bone growth as well as retinol does, but it cannot support the role of retinol in visual or reproductive processes (Lammer and Armstrong, 1992). Different physiologic functions are mediated by different forms of the molecule, and all-*trans* retinoic acid can be as much as 10–100 times more potent than retinol in certain body systems (Marcus and Coulton, 1990).

All retinoids possess teratologic activity. The syndrome of malformations produced by retinoids includes skeletal, craniofacial (external ear malformations, palatal malformations), nervous system, and cardiac abnormalities. The quantitative relationships between pharmacokinetic characteristics of different retinoids and their teratogenic potency in mice has been examined elegantly by Kocchar and colleagues (Kochhar *et al.*, 1984; Kochhar and Penner, 1987; Kochhar *et al.*, 1988). This work has provided compelling evidence that all-*trans* retinoic acid and metabolites are the active agents of teratogenesis. Thus, the teratogenic activity of retinol and of 13-*cis* retinoic acid are believed to be mediated largely due to metabolic conversion to all-*trans* retinoic acid. Kochhar and colleagues have also examined the structure–activity relationships for the potency of these three compounds in producing skeletal and palatal malformations. Their work suggests that retinol is approximately one-quarter as potent as all-*trans* retinoic acid and isotretinoin is approximately one-eighth as potent as all-*trans* retinoic acid in producing skeletal and palatal malformations in mice. Adams (1993) examined these relative potency characteristics with regard to central nervous system (CNS) malformations and behavioral endpoints of teratogenesis reported in rat studies: generally, the relative potencies appeared to hold up for these endpoints.

In this chapter, the effects on CNS structure and function for each of these three retinoids is examined. Dosage levels of the different retinoid compounds can be compared more easily when vitamin A, which is normally expressed in international units (IU), is converted to retinol equivalents expressed in mg. The US Pharmacopeia suggests conversion methods where 1 IU of vitamin A (retinol) equals 0.33 micrograms of retinol. Thus, 10,000 IU of vitamin A would equal 3.3 mg retinol, and 100,000 IU would equal 33 mg retinol.

During the discussion of the teratogenicity of vitamin A, all-*trans* retinoic acid, and 13-*cis* retinoic acid, three tables are used heavily (Tables 1, 2, and 3). For simplicity, the data presented within these tables is divided into three categories. Table 1 presents teratologic effects on prenatal lethality (effects on measured resorptions or lethal effects inferred from reduced litter size) and CNS malformations. Table 2 presents postnatal outcome data derived from studies in which pups were reared at least through the weaning period. Table 3 describes specific behavioral alterations caused by prenatal exposure to the retinoid compounds. The reader should be aware that certain cited publications are single experiments in which similarly treated litters were allocated for the study of fetal versus postnatal endpoints; other publications represent work accomplished in several experiments. As a result, a complete one-to-one relationship does not always exist between all doses and stages of treatment and all outcomes examined. When an endpoint has not been examined or is not reported, dashes (—) are used to signify this study characteristic.

III. Retinoid Teratogenicity: Effects on the Central Nervous System

A. *Hypervitaminosis A*

1. Studies in Rats

Since the beginning of experimental teratology in the 1950s and 1960s, treatment with excess vitamin A has been used in laboratory investigations of gross malformations of the CNS (see review by Kalter, 1968). Early studies focused primarily on identifying periods of greatest vulnerability for the induction of specific gross malformations and on examining dose–response relationships. Given then-available techniques, these early studies of hypervitaminosis A focused on exencephaly as a measure of CNS teratogenesis.

Table 1 lists characteristics of selected teratology studies in rats. The early studies often expressed dose levels on a per-dam basis, so estimates of dosage levels per kg body weight have been made. In these estimates, an average rat weight of 0.25 kg was used. Gestational days (GD) were assigned using the morning of plug discovery as GD 0. Using this estimation and restricting our focus to the induction of exencephaly, it can be seen that doses at or above 120,000 IU vitamin A/kg administered on multiple days prior to GD 11 appear capable of producing this gross CNS malformation in rats. It should also be noted that these doses produce high levels of resorption. When treatment is restricted to a single day of administration, a larger dose of approximately 240,000 IU/kg produces abnormalities to a similar degree. Respectively, these doses are approximately

TABLE 1 Studies of Malformations of the Central Nervous System Induced by Embryonic Exposure to Retinoid Compounds

Compound	Species	Dose (a)	Route	Gestational days (GD)	Prenatal lethality	CNS malformations	Study
Vitamin A (aqueous)	Wistar rat	60,000 IU/kg	Gavage	2,3,4 . . . 16	—	Not produced	Cohlan, 1954
		100,000 IU/kg			—	Not produced	
		140,000 IU/kg			90% of pg.	52% exenceph.	Cohlan, 1953
Vitamin A (aqueous)	Wistar rat	120,000 IU/kg	Gavage	8–10	44% resorp.	26% exenceph.	Langman and Welch, 1966
		160,000 IU/kg			63% resorp.	19% exenceph.	
		200,000 IU/kg			70% resorp.	13% exenceph.	
		240,000 IU/kg			79% resorp.	19% exenceph.	
				>10	—	Not produced	
Vitamin A (oily)	Wistar rat	240,000 IU/kg	Oral	5–7	—	9% exenceph.	Giroud and Martinet, 1955
				8–10	—	53% exenceph.	
				10–12	—	5% exenceph.	
				>12	—	Not produced	
Vitamin A (oily)	Fischer 344 rat	100, 000 IU/kg	Gavage	8–10	Yes	Not examined	Vorhees, 1974
Vitamin A	Human	≥25,000 IU/day	Oral	Variable	—	Yes	Rosa *et al.*, 1986
Vitamin A	Human	≥10,000 IU/day	Oral	Variable	—	Yes	Rothman *et al.*, 1995
All-*trans* retinoic acid	Sprague Dawley	20 mg/kg	Gavage	7–9	Yes, high	—	Church and Tilak, 1996
		30 mg/kg			Yes, high	—	
All-*trans* retinoic acid	"Charles River Rats"	10 mg/kd(d)	Gavage	9	Increased resorption	Low incidence exencephaly	Nolen, 1972
All-*trans* retinoic acid	Sprague Dawley rat	5 mg/kg	Gavage	8–10	No	Not produced	Nolen, 1986
				11–13	No	Not produced	
				14–16	No	Not produced	
All-*trans* retinoic acid	Sprague Dawley (CD)	10 mg/kg	Gavage	8–10	Increased resorptions	Exencephaly	Holson *et al.*, 1998a,b
		2.5 mg/kg 10 mg/kg		11–13	No	No exencephaly; abnormal medullary nuclei (10 mg/kg)	
		12.5 mg/kg		14–16	No	No exencephaly	
13-*cis* retinoic acid	Rat	75 mg/kg	—	8–10	—	— (Craniofacial abnormalities)	Anonymous, 1991
13-*cis* retinoic acid	Wistar rat	150 mg/kg	Gavage	11–13	No	Not produced	Jensh *et al.*, 1990
		125 mg/kg			No	Not produced	
13-*cis* retinoic acid	Human	0.5–1.5 mg/kg	Oral	1–60 variable durations	Yes, 40%	Hindbrain most common	Lammer *et al.*, 1985

comparable to 40 mg and 80 mg of retinol/kg. A committee of the Teratology Society consensually agreed that a dose of 50 mg retinol/kg is the lowest teratogenic dose in the rat (Anonymous, 1987).

As can be seen in Table 1, the most vulnerable period for the production of exencephaly in the rat is between GD 8–10 (see review by Kalter, 1968; Geelen, 1979). Other malformations produced during this time period include meningoencephaloceles, meningoceles, eye defects, ear malformations, and cleft palate. It should be recognized that extensive work has been conducted in other species and that hypervitaminosis A also produces exencephaly and other malformations in mice and hamsters, although the period of greatest sensitivity for induction varies (in mice, GD 7–8, Murakami and Kameyama, 1965; in hamsters, GD 8, Shenefelt, 1972). One should also be aware that major malformations of other structures have a different time course. For example, Giroud and Martinet (1954) reported a 71% incidence of cleft palate in rats following exposure on GD 10–12, a time too late for the induction of neural tube closure defects.

2. Studies in Humans

An absence of experimental studies of retinol teratogenicity in humans precludes a reliable estimation of the active dose range; however, clinical reports of a malformation syndrome similar to what is seen in animal studies are noteworthy. Rosa *et al.* (1986) reviewed several human case reports and suggested that malformations were associated with supplemental dose levels of 25,000 IU vitamin A/day or more. It is important to recognize that these are supplemental doses per day (not per kg), and that these supplemental doses are superimposed on dietary intake levels. Among members of the U.S. population, typical dietary intake of vitamin A is estimated at approximately 7000–8000 IU/day. Thus, actual vitamin A levels represent the sum of approximately 32,000–33,000/day (Anonymous, 1987). In some of the human case reports, doses as high as 68,000 IU/day were estimated. As a result of these data, a committee of the Teratology Society has suggested that doses above 25,000 IU/day are teratogenic in humans, with respect to producing gross malformations (Anonymous, 1987).

Of concern is a report by Rothman and colleagues (1995) that argues that 10,000 IU/day of supplemental vitamin A may be teratogenic. Although this is lower than what has been suggested in clinical reports, it is derived from the first epidemiologic study to examine vitamin A teratogenicity systematically. Further work is necessary to better define the minimal teratogenic dose in humans.

B. All-trans Retinoic Acid

1. Studies in Rats

Similar to vitamin A, the sensitive period for the production of exencephaly following exposure to all-*trans* retinoic acid is between GD 8-10 (see Table 1). Within this exposure period, doses of 10 mg/kg all-*trans* retinoic acid are capable of producing this gross malformation.

As can be seen in Table 1, when administered between GD 11–13 this same dose produces no gross malformations or effects on prenatal lethality. However, a 10 mg/kg dose of all-*trans* retinoic acid on GD 11–13, does produce structural abnormalities in medullary nuclei. As discussed in Section IV of this chapter, GD 11–13 represent a particularly vulnerable period for more "subtle" effects on the brain that are incompatible with postnatal survival.

2. Studies in Humans

Human data are not available from experimental studies of all-*trans* retinoic acid exposure during pregnancy.

C. 13-cis Retinoic Acid (Isotretinoin)

1. Studies in Rats

Isotretinoin, 13-*cis* retinoic acid, is teratogenic in rodents only at levels much higher than those for all-*trans* retinoic acid (Kochhar *et al.,* 1984; Klug *et al.,* 1989). This difference in potency is believed due to the low isomerization rate of isotretinoin to all-*trans* retinoic acid, the proposed active agent of teratogenicity (Klug *et al.,* 1989), and the low level of placental transfer of 13-*cis* retinoic acid (Kochhar *et al.,* 1984). A critical examination of teratology data by members of the Teratology Society (Anonymous, 1987) led to the consensus that the lowest teratogenic dose of isotretinoin in the rat is 150/mg/kg/day administered on GD 8–10; this was later revised to 75 mg/kg/day (Anonymous, 1991).

Similar to all-*trans* retinoic acid, exposure to 13-*cis* retinoic acid on GD 11–13 also causes postnatal lethality in the absence of effects on prenatal survival or gross malformations (see Section IV of this chapter).

2. Studies in Humans

Although isotretinoin has the lowest potency of all retinoids previously discussed, it is the most problematic due to its human therapeutic use in the teratogenic dose range for the treatment of cystic acne. Lammer and colleagues (1985) conducted a study of prospectively ascertained exposed pregnancies and nonexposed, physician-matched controls. In the isotretinoin-exposed group, all women took therapeutic doses of 0.5–1.5 mg/kg for varying periods during the first 60 days of pregnancy. These women had a 40% spontaneous abortion rate, and liveborn infants had a 25% rate of major malformations. The cluster of malformations included craniofacial, cardiac, thymic, and nervous system abnormalities. As shown in Table 1, major malformations of the CNS were primarily in hindbrain structures. These included cerebellar hypoplasia, agenesis of the vermis, malformation-induced hydrocephalus, and abnormalities of the pons and medulla (Lammer *et al.,* 1985; Lammer and Armstrong, 1992). Abnormalities in the hippocampus and cortex have also been reported from autopsy data (Lammer and Armstrong, 1992).

IV. Postnatal Functional Alterations Resulting from Prenatal Exposure to Retinoids

A. *Hypervitaminosis A*

1. Studies in Rats

Investigations of the effects of prenatal hypervitaminosis A on postnatal functional development were conducted in rodents in the early 1970s and played a pivotal role in the establishment of "behavioral teratology" as a discipline (see historic reviews by Vorhees, 1986; Hutchings, 1983). These studies explored dose–response and period–response relationships for the expressed purpose of establishing principles governing the relationships between prenatal exposure, anatomic changes, and developmental insults defined by functional deficits. Studies of excess vitamin A exposure contributed to a sound foundation from which three primary tenets emerged. As stated by Vorhees (1986), these are: 1. Behavioral teratogenic effects are demonstrable at doses lower than doses causing malformations if the agent is capable of producing adverse behavioral effects; 2. the type and magnitude of the behavioral teratogenic effects depend on the dose of the agent reaching the developing nervous system; and 3. the type and magnitude of the behavioral teratogenic effects depend on the stage of development of the organism when exposed.

Table 2 presents the results of studies that investigated the effects of prenatal exposure to retinoid compounds on postnatal survival, offspring weight, and behavioral function. As can be seen in Table 2, postnatal studies of the effects of gestational exposure to vitamin A were primarily conducted using submalforming doses of less than 100,000 IU/kg (33 mg retinol/kg). Most of these studies reported no effects on postnatal survival, but effects on postnatal weight were occasionally present. Of primary importance is the fact that behavioral alterations were typically seen at these submalforming dose levels when treatments occurred on GD 8–10. The information in Table 2 makes it clear that doses less than 100,000 IU/kg when administered during this vulnerable period for gross CNS malformations produce reliable effects on postnatal behavior. As shown in Table 2, when vitamin A is administered at later stages of embryogenesis, behavioral alterations have also been reported. The research by Vorhees *et al.* (1978), Hutchings *et al.* (1973), and Hutchings and Gaston (1974) clearly illustrates that functional deficits can be produced by exposures restricted to late organogenesis.

2. Studies in Humans

Functional endpoints have not been addressed systematically in human studies. Case reports of excess vitamin A exposure describe developmental delay as a feature of children with malformations (Rosa *et al.,* 1986).

B. *All-trans Retinoic Acid*

1. Studies in Rats

Evaluation of Table 2 indicates that all-*trans* retinoic acid produces functional effects at lower dosage levels than are needed to produce gross malformations. Functional alterations also occur following exposures during later periods of embryogenesis. Most striking are the effects on postnatal survival that occur following exposures on GD 11–13 at dose levels that produced no effects on resorption or malformation. For example, treatment with 10 mg/kg all-*trans* retinoic acid on GD 11–13 does not alter prenatal survival or gross malformation rates, but it nevertheless produces 100% postnatal lethality (Holson *et al.,* 1998a,b). Nolen (1986) has also reported that 5 mg/kg administered during this stage produces a 90% postnatal death rate. In an effort to determine the lowest dose of all-*trans* retinoic acid that would be compatible with postnatal survival, Holson *et al.* (1998a) examined the effects of 10, 5, and 2.5 mg/kg doses given on GD 11–13. At the 10 mg/kg dose, all pups died prior to postnatal day (PN) 28; at 5 mg/kg, 80% of the pups died by PN 28; at 2.5 mg/kg, 38% died. This 2.5 mg/kg exposure, however, did produce significant effects on body weight, whole brain weight, and cerebellar weight. The same investigators have reported similar effects on cerebellar weight when higher doses (10 mg/kg; 12 mg/kg) are administered later in gestation (GD 14-16).

To understand the extreme effect on postnatal survival that occurs following exposure to all-*trans* retinoic

TABLE 2 Studies of Postnatal Outcome Following Prenatal Exposure to Retinoid Compounds

Compound	Species	Dose	Route/GD	Reduced survival	Offspring weight reduced	Behavioral alteration	Study
Vitamin A (oily)	Albino rat	600,000[a]	Oral/9	No	No	Yes	Malakhovskii, 1969
Vitamin A (oily)	Sprague Dawley	100,000[a]	Gavage/8–10	No	No	Yes	Butcher *et al.,* 1972
Vitamin A (oily)	Fischer 344 rat	40,000[a]	Gavage/8–10	No	No	Yes	Vorhees, 1974
		25,000[a]		No	No	Yes	
		10,000[a]		No	No	Yes	
Vitamin A (oily)	Sprague Dawley	80,000[a]	Gavage/5–7	No	No	No	Vorhees *et al.,* 1978
		80,000[a]	8–10	No	No	Yes	
		80,000[a]	11–13	No	Yes	Yes	
		80,000[a]	14–16	No	No	No	
		80,000[a]	17–19	No	No	Yes	
Vitamin A (aqueous)	Sprague Dawley	40,000[a] 80,000[a]	Gavage/8–10	No	No	Yes	Adams, 1982
Vitamin A (oily)	Sprague Dawley	80,000[a]	Gavage/11–15	No	No	Yes	Bruses *et al.,* 1991
Vitamin A (oily)	Wistar	240,000[a]	Gavage/13–14	No	Yes	Yes	Hutchings *et al.,* 1973
Vitamin A (oily)	Wistar	360,000[a]	Gavage/16–17	Yes	No	Yes	Hutchings and Gaston, 1974
Vitamin A (oily)	Sprague Dawley	40,000[a]	Gavage/7–20	No	Yes	Yes	Vorhees *et al.,* 1979
Vitamin A (oily)	Sprague Dawley	80,000[a] 160,000[a]	Gavage/6–20	No	No	Yes, both	Saillenfait and Vannier, 1988
Vitamin A (aqueous)	Sprague Dawley	25,000[a]	Gavage/6–19	No	No	No	Kutz *et al.,* 1989
		50,000[a]		No	No	No	
		100,000[a]		Yes	Yes	No	
All-*trans* retinoic acid	Sprague Dawley	10[b]	Gavage/7–9	Yes	Yes	Yes	Church and Tilak, 1996
All-*trans* retinoic acid	Sprague Dawley	5[b]	Gavage/8–10	No	No	Yes	Nolen, 1986
			14–16	No	Yes	Yes	
		5[b]	Gavage/11–13	Yes, 90%	No	Yes	
		5[b]	14–16	No	No	Yes	
		2.5[b]	11–13	No	Yes	Yes	
		2.5[b]	14–16	No	Yes	Yes	
		6[b]	Gavage/14–16	No	Yes	Yes	
		4[b]		No	No	Yes	
		2[b]		No	No	No	
All-*trans* retinoic acid	Sprague Dawley (CD)	10[b]	Gavage/8–10	No	No	—	Holson *et al.,* 1998a,b,c
		10[b]	11–13	Yes, 100%	Yes	Yes[c]	
		5[b]	11–13	Yes, 90%	Yes	—	
		2.5[b]	11–13	Yes, 38%	Yes	Yes[c]	
		10[b]	14–16	No	No	—	
		12[b]	14–16	No	No	—[d]	
All-*trans* retinoic acid	Rat	2.5[b]	Gavage/14–16	No	Yes	Yes	Nishimura, 1995
		5.0[b]		Yes	Yes	Yes	
13-*cis* retinoic acid	Wistar	150[b]	Gavage/11–13	Yes	—	—	Jensh *et al.,* 1990
		125[b]		Yes	—	—	
		100[b]		No	No	Yes	
		50[b]		No	No	Yes	
13-*cis* retinoic acid	Wistar	100[b]	Gavage/11–13	No	No	Yes	Jensh *et al.,* 1991
		50[b]		No	No	No	
13-*cis* retinoic acid	Human	0.5–1.5[b]	Oral/1–60 Variable duration	Yes	—	Yes	Adams and Lammer, 1991, 1995; Adams *et al.,* 1991

[a]IU/kg.
[b]mg/kg.
[c]Treatment with 10 mg/kg or 5 mg/kg on days 11-13 has also been shown to reduce whole brain weight, cerebellar weight, and the structure of the inferior olive and area postrema
[d]Treatment with 12 mg/kg on days 14-16 has also been shown to reduce cerebellar weight

acid on GD 11–13, Holson and colleagues (1998b) examined more subtle aspects of neuroanatomy as well as maternal and infant behavior. Rats exposed to 10 mg/kg all-*trans* retinoic acid during GD 11–13 were examined following cesarian section on GD 21. These fetuses were shown to have medullary abnormalities in the area postrema and the inferior olivary nucleus, despite normal gross anatomy of the brainstem. Such abnormalities may account for the pups' inability to nurse or initiate breathing effectively (Holson *et al.,* 1998b). These abnormalities are consistent with neuropathologic and functional characteristics of human infants exposed to isotretinoin during embryogenesis (Lammer and Armstrong, 1992).

2. Studies in Humans

Data are not available on human exposure to all-*trans* retinoic acid.

C. 13-cis Retinoic Acid

1. Studies in Rats

It has been suggested that 75 mg/kg/day is the lowest teratogenic dose of 13-*cis* retinoic acid when administered during the sensitive period (GD 8–10) to rats (Anonymous, 1991). Preliminary studies by Jensh and colleagues (1990; 1991) suggest that behavioral alterations are produced by doses of 50–100 mg/kg/day in the offspring of Wistar rats treated on GD 11–13. Thus, as with all-*trans* retinoic acid, a specific period of vulnerability around GD 11–13 is present for the effects of isotretinoin on behavior whereby functional effects occur at lower doses. As shown in Table 2, exposure levels of 125 mg 13-*cis* retinoic acid/kg or greater on GD 11–13 produce postnatal death in the grossly normal offspring of Wistar rats. This can be contrasted with the 2.5 mg/kg dose of all-*trans* retinoic acid, which increases postnatal mortality when administered during the same time period to Sprague Dawley rats. Clearly, a large difference in toxic potential exists for the administered doses of these compounds.

2. Studies in Humans

Investigations of the behavioral sequelae of human exposure to 13-*cis* retinoic acid are now in progress. Work by Adams and Lammer (1993;1995) represents a longitudinal follow-up of the prospective cases identified in studies first reported by Lammer and colleagues in 1985. Neuropsychologic evaluations of these children have been done at 5 years of age and are now ongoing in the children at 10–11 years of age. The results of data collected at 5 years of age have documented a striking degree of cognitive impairments in children born to women who took therapeutic doses (0.5–1.5 mg/kg/day) of this drug at some point and for varying durations during the first 60 days of pregnancy. At 5 years of age, 44 prospectively ascertained children exposed to isotretinoin and 40 physician matched controls were seen (Adams and Lammer, submitted). Striking impairments in general intellectual functioning were found at the age of 5 years on the Stanford–Binet IV. Among the 44 Accutane-exposed children, 43.1% were functioning in the subnormal range (13.6% functioning in the mentally retarded range; 29.5% in the range of borderline intelligence), and only 56.9% performed at average to above average levels. All of the Accutane-exposed children that performed in the mentally retarded range have major malformations. Of the 13 exposed children functioning in the borderline range of intelligence (full-scale scores between 70–85), 6 have major malformations. Only 1 of these 6 has a major CNS malformation. Two children with major malformations scored in the average to superior range of mental ability, neither having CNS abnormalities. Thus, these findings demonstrate that embryonic exposure to Accutane can produce subnormal general mental ability in the absence of CNS pathology which would be detected by current medical evaluations. Indeed, 47% of the children with subnormal mental ability have no major malformations. These data, therefore, suggest that individual variations in actual intrauterine exposure or in sensitivity to that exposure result in a spectrum of outcomes. However, because the women in the study took the medication at a relatively common therapeutic dosage level over varying time intervals, these data cannot be used to address direct dose-related effects on outcome, nor yet to identify sensitive periods within the first 60 days of pregnancy.

V. Specific Behavioral Alterations Associated with Prenatal Exposure to Retinoids

A. Hypervitaminosis A

1. Studies in Rats

As shown in Table 3, a variety of behaviors have been examined in rats exposed to Vitamin A during prenatal development. These diverse measures constitute assessments of reflex and motor development, activity levels, and learning abilities. Studies of excess vitamin A exposure have consistently shown effects on all categories of functioning. Likewise, effects have been seen following exposures during different segments of the embryonic period as well as during fetal development. Effects on learning have been reported at doses as low as 10,000 IU/kg in Fischer rats when exposures

TABLE 3 Specific Postnatal Behavioral Alterations Following Prenatal Exposure to Retinoids

Compound	Species	Dose	Route	GD	Behaviors evaluated[a]	Alterations present	Study
Vitamin A (oily)	Albino rat	600,000[b]	Oral	9	PW activity	Yes	Malakhovskii, 1969
					Active avoidance	Yes	
Vitamin A (oily)	Sprague Dawley	100,000[b]	Gavage	8–10	PW swim maze	Yes	Butcher *et al.*, 1972
Vitamin A (oily)	Fischer 344 rat	40,000[b] 25,000[b] 10,000[a]	Gavage	8–10	PW open field activity	No	Vorhees, 1974
					Shock avoidance	Yes, all groups	
Vitamin A (aqueous)	Sprague Dawley	40,000[b] 80,000[b]	Gavage	8–10	Negative geotaxis	No	Adams, 1982
					Ultrasonic Vocalization	Yes, 80,000	
Vitamin A (oily)	Sprague Dawley	80,000[b]	Gavage	5–7 8–10 11–13 14–16 17–19	PreW T-maze activity	Yes, 11–13	Vorhees *et al.*, 1978
					PW activity	Yes, 8–10	
					Biel maze	Yes, 8–10, 11–13, 17–19	
Vitamin A (oily)	Sprague Dawley	80,000[b]	Gavage	11–15	Reflex development	Yes	Bruses *et al.*, 1991
					T-maze learning	No	
					Conditioned turning	Yes	
Vitamin A (oily)	Wistar	240,000[b]	Gavage	13–14	Operant learning (auditory)	Yes	Hutchings *et al.*, 1973
Vitamin A (oily)	Wistar	360,000[b]	Gavage	16–17	Operant learning (auditory)	Yes	Hutchings and Gaston, 1974
Vitamin A (oily)	Sprague Dawley	80,000[b] 160,000[b]	Gavage	6–20	Motor development	Yes, both	Saillenfait and Vannier, 1988
					Open field	Yes, both	
					Rotorod	No	
					M water maze	No	
					Nocturnal activity	Yes, 160,000	
					Auditory startle	No	
Vitamin A (aqueous)	Sprague Dawley	25,000[b] 50,000[b] 100,000[b]	Gavage	6–19	Motor development	No	Kutz, *et al.*, 1989
					Open field activity	No	
					Operant learning	No	
Vitamin A (oily)	Sprague Dawley	40,000[b]	Gavage	7–20	Motor development	Yes	Vorhees *et al.*, 1979
					PW open field	Yes	
					Biel maze	Yes	
					Active avoidance	Yes	
					Passive avoidance	No	
					Rotorod	Yes	
All-*trans* retinoic acid	Sprague Dawley	10[c]	Gavage	7–9	Open field activity	Yes	Church and Tilak, 1996
All-*trans* retinoic acid	Sprague Dawley	5[c]	Gavage	8–10 14–16	Reflex development	No	Nolen, 1986
					Auditory startle	Yes, both	
					PW activity	Yes, both	
					M-maze	Yes, GD 14–16	
					PW open field	Yes, GD 8–10	
					Negative geotaxis	Yes, both	
		2.5[c] 2.5[c] 5[c] 5[c]	Gavage	11–13 14–16 11–13 14–16	Reflex development	Yes, all groups	
					Auditory startle	Yes, 2.5 on GD 11–13 and 5.0 on GD 14–16	
					PW activity	No	
					M-maze	Yes, both GD 14–16	
					PW open field	Yes, 5.0 on GD 14–16	
					Negative geotaxis	Yes, both GD 11–13	

TABLE 3 (*Continued*)

Compound	Species	Dose	Route	GD	Behaviors evaluated[a]	Alterations present	Study
All-*trans* retinoic acid	Sprague Dawley	4[c] 6[c]	Gavage	14–16	Reflex development	No	
					Negative geotaxis	Yes, 4 and 6 mg	
					Auditory startle	Yes, 4 and 6 mg	
					PW open field	Yes, 6 mg	
					M-maze	Yes, 6 mg	
					Active avoidance	Yes, 6 mg	
					Amphetamine challenge	Yes, 4 and 6 mg	
					Running wheel	Yes, 6 mg	
All-*trans* retinoic acid	Sprague Dawley (CD)	10[c]	Gavage	11–13	Nursing behavior	Yes	Holson *et al.*, 1988a,b,c
					Breathing	Yes	
		2.5[c]		11–13	Negative geotaxis	No	
					Auditory startle	No	
					Open field activity	No	
					Complex maze	No	
					Residential running wheel	Yes	
					Amphetamine challenge	Yes	
All-*trans* retinoic acid	Rat	2.5[c] 5.0[c]	Gavage	14–16	Motor development	Yes, 5.0	Nishimura, 1995
					Behavior in open field	Yes, both	
					Biel maze	No	
					Shuttle box learning	No	
13-*cis* retinoic acid	Wistar	100[c] 50[c]	Gavage	11–13	Developmental milestones	Yes, 100 mg	Jensh *et al.*, 1990
					Ultrasonic vocalization	Yes, 50 mg	
					PW open field	Yes, 100 mg	
					Activity wheel	Yes, 100 mg	
					Water T-maze	No	
					Active avoidance	Yes, 100 mg	
13-*cis* retinoic acid	Wistar	100[c] 50[c]	Gavage	11–13	Ultrasonic vocalization	Yes, both	Jensh *et al.*, 1991
13-*cis* retinoic acid	Human	0.5–1.5[c]	Oral	1–60 variable	General mental ability	Yes	Adams and Lammer, 1991, 1995
					Specific abilities	Yes	Adams *et al.*, 1991

[a]PW refers to evaluation at postweaning ages; PreW refers to evaluation at preweaning ages.
[b]IU/kg.
[c]mg/kg.

occur during GD 8–10 (Vorhees, 1974). Most behavioral studies, however, have been conducted in Sprague Dawley rats and at doses in the 40,000–100,000 IU range. Across these doses, all categories of behavior have been shown to be affected. Doses as low as 80,000 IU/kg appear able to effect learning in Sprague Dawley rats following administration at later stages of development (Vorhees *et al.,* 1978). It is important to note that significant deficits have been reported on measures of learning that require different forms of processing from the animal. In a study by Vorhees, Brunner, and Butcher (1979), rats were exposed to 40,000 IU/kg from GD 7 to GD 20. In this study, active avoidance learning and spatial learning in a complex water maze were affected; passive avoidance learning was not.

2. Studies in Humans

Data are not available on the specific behavioral characteristics of children exposed to excess vitamin A *in utero.*

B. All-*trans* Retinoic Acid

1. Studies in Rats

A variety of behaviors have also been examined in rats exposed to all-*trans* retinoic acid prenatally. In work by Nolen (1986), exposures to 5 mg/kg doses on GD 8–10 did not disrupt reflex development; however, at later periods (GD 11–13, GD 14–16) reflex development was delayed. Indeed, at the later treatment periods, reflex development was disrupted following doses of 2.5 mg/kg. In work by Holson and colleagues (1998c), however, 2.5 mg/kg administered on GD 11–13 did not disrupt negative geotaxis, the only index of reflex development used. However, Holson and colleagues (1998b)have seen severe effects on the early development of suckling reflexes and breathing patterns prior to the death of pups exposed to 10 mg/kg on GD 11–13.

Nolen (1986) reported effects on activity and on learning in rats exposed to 5 or 6 mg/kg on GD 14–16, but not on GD 11–13. Investigations of activity and learning by Holson and colleagues (1998c) have been restricted to animals exposed to 2.5 mg/kg all-*trans* retinoic acid on GD 11–13. In these rats, effects on running wheel activity and amphetamine-challenged open field activity were seen. However, the learning ability of water-deprived rats in a complex spatial maze (dry) was not affected. This contrasts with disrupted learning abilities measured under shock avoidance and water escape conditions (Nolen, 1986). Given that the treatment regimen used by Holson and colleagues is known to reduce whole brain weight, cerebellar weight, and to alter certain medullary nuclei, effects on learning would be anticipated. Demonstrating these effects may depend on alternative assessment techniques.

2. Studies in Humans

Effects of exposure to all-*trans* retinoic acid have not been directly investigated.

C. 13-*cis* Retinoic Acid

1. Studies in Rats

Jensh *et al.* (1991) have shown that Wistar rats exposed to 100 mg/kg of isotretinoin on GD 11–13 have alterations in reflex development, activity levels, and active avoidance learning.

2. Studies in Humans

Although 43% of the children exposed to isotretinoin during embryogenesis have subnormal intellectual ability, this figure does not capture the full spectrum of learning-related effects. Additional children exhibit learning disabilities, although their general mental ability is in the normal range. Evaluations were performed on the scores of the children across different neuropsychologic categories of functioning (assessed at age 5 years). This showed that lowest scores were seen on measures of visual–motor integration and visual–perceptual analysis (visual–spatial abilities), attention, and organizational abilities. Language-based abilities appeared to be a relative strength for the individual children. In order to quantify the number of children in the average to above average range of general mental ability who demonstrated this learning disability profile, performance on information and vocabulary subtests (language-based measures) was compared to performance on pattern analysis and bead memory (visual–spatial measures). Within-child differences of greater than 1.5 standard deviations (visual–perceptual scores <language-based scores by at least 1.5 standard deviations) were found in 16% of the isotretinoin-exposed children judged to be normal by psychometric criteria, versus only 7% of the control children who performed at average to above average levels of general mental ability. Thus, when all adverse cognitive outcomes are considered, it has been shown that 43% of the exposed children are identified as mentally retarded or of borderline intelligence versus only 9.8% of the age and demographically matched controls. An additional 16% of exposed cases demonstrate the significant learning disability profile, whereas only 7% of the control children met these defining criteria. Collectively, then, early embryonic exposure to isotretinoin is associated with a 59% incidence of intellectual deficits, versus an incidence of 17% in matched control children.

VI. Consistency of Behavioral Effects with Neuropathology

Comparing human and rat data on the nature of neural effects reported across the three retinoids, one sees that hindbrain neuropathology is consistent. Comparing behavioral effects also suggests certain similarities. Studies in rats exposed to vitamin A, all-*trans* retinoic acid, and 13-*cis* retinoic acid have shown effects on activity levels, on active avoidance learning, and on spatial learning in a complex water maze. Human studies of neurobehavioral effects of retinoids are restricted to studies of 13-*cis* retinoic acid. These studies suggest effects on general mental ability, and specific effects on attention and organizational ability and on visual–spatial processing. These functional manifestations of the disrupted neuroanatomy in rats and in humans are somewhat consistent. Effects in rats on activity levels and on altered inhibitory abilities during active avoidance learning may be generally conceived as similar to human effects on attention and organizational abilities.

This conception is plausible if one considers that these behaviors are mediated by similar circuitry involving hindbrain projections to frontal brain areas.

For both rats and humans, alterations in general learning ability as well as in spatial learning–processing abilities have been reported. It is important to recognize that many human mental retardation syndromes include hindbrain and craniofacial abnormalities as sequelae (see reviews by Adams, 1996; in press). This association between hindbrain abnormalities and mental retardation suggests that hindbrain abnormalities (cerebellar, in particular) may interfere with the ability to process and transmit information from hindbrain centers to other regions important in learning and spatial processing (such as frontal areas and parietal areas, respectively). Such "bottom-up" contributions to learning during development have been largely overlooked historically, but are now gaining increased attention (Akshoomoff and Courchesne, 1992; Daum and Ackerman, 1995).

This chapter demonstrates that not only are retinoids teratogenic in their ability to produce malformations, but they also produce a continuum of defects, including postnatal death and more subtle behavioral alterations. Effects on all of these endpoints are dose and stage dependent, and also vary with the relative teratogenic potency of each retinoid.

Acknowledgments

This work was partially funded by NIH/NICHD grant #HD29510 (Jane Adams, Principal Investigator). The authors also wish to thank Candyce McCarthy for assistance in preparation of the manuscirpt.

References

Adams, J. (1982). Ultrasonic vocalizations as diagnostic tools in studies of developmental toxicity: an investigation of the effects of hypervitaminosis A. *Neurobeh. Toxcicol. Teratol.* **4,** 299–304.

Adams, J. (1993a). Structure–activity and dose–response relationships in the neural and behavioral teratogenesis of retinoids. *Neurotoxicol. Teratol.* **15,** 193–202.

Adams, J. (1993b). Neural and behavioral pathology following prenatal exposure to retinoids. In *Retinoids in Clinical Practice.* G. Koren, Ed. Marcell-Dekker, New York, pp. 111–128.

Adams, J. (1996). Similarities in genetic mental retardation and neuroteratogenic syndromes. *Pharmacol. Biochem. Behav.* **55(4),** 683–690.

Adams, J. (in press). On neurodevelopmental disorders: perspectives from neurobehavioral teratology. In *Neurodevelopmental Disorders: Contributions to a New Framework from the Cognitive Neurosciences.* MIT Press, Boston.

Adams, J. and Lammer, E. J. (1991). Relationship between dysmorphology and neuropsychological function in children exposed to isotretinoin *in utero.* In *Functional Neuroteratology of Short Term Exposure to Drugs.* Fujii, T. and Boer. G. J., Ed. Teikyo University Press, Tokyo. pp. 159–170.

Adams, J. and Lammer, E. J. (1995). Human isotretinoin exposure: the teratogenesis of a syndrome of cognitive deficits. *Neurotoxicol. Teratol.* **17,** 386.

Adams, J., Lammer, E. J., and Holmes, L. B. (1991). A syndrome of cognitive dysfunctions following human embryonic exposure to isotretinoin. *Teratology* **43,** 497.

Akshoomoff, N. A., and Courchesne, E. (1992). A new role for the cerebellum in cognitive operations. *Behav. Neurosci.* **106,** 731–738.

Anonymous. (1987). Position paper by the Teratology Society: vitamin A during pregnancy. Public Affairs Committee of the Teratology Society. *Teratology* **35(2),** 267–269.

Anonymous. (1991). Recommendations for isotretinoin use in women of childbearing potential. The Public Affairs Committee and the Council of the Teratology Society. *Teratology* **44(1),** 1–6.

Bruses, J. L., Berninsone, P. M., Ojea, S. I., and Azcurra, J. M. (1991). The circling training rat model as a behavioral teratology test. *Pharm. Biochem. Behav.* **38,** 739–745.

Butcher, R. E., Brunner, R. L., Roth, T., and Kimmel, C. A. (1972). A learning impairment associated with maternal hypervitaminosis-A in rats. *Life Sci.* **2(1),** 141–145.

Church, M. W., and Tilak, J. P. (1996). Differential effects of prenatal cocaine and retinoic acid on activity level throughout day and night. *Pharm. Biochem. Behav.* **55,** 595–605.

Cohlan, S. Q. (1953). Excessive intake of vitamin A as a cause of congenital anomalies in the rat. *Science* **117,** 535–536.

Cohlan, S. W. (1954). Congential anomalies in the rat produced by excessive intake of vitamin A during pregnancy. *Pediatrics* **13,** 556–567.

Daum, I. and Ackerman, H. (1995). Cerebellar contributions to cognition. *Behav. Brain Res.* **67,** 201–210.

Durston, A. J., Timmermans, J. P. M., Hage, W. J., Kendriks, H. F. J., de Vries, N. J., Heideveid, M., Nieuwkoop, P. D. (1989). Retinoic acid causes an anteroposterior transformation in the developing central nervous system. *Nature* **340,** 140–144.

Eskessen, M. B., Jensh, R. P., Kochhar, D. M., Till, M. K. (1990). Postnatal behavioral sequelae of prenatal exposure to 13-cis retinoic acid. *Teratology* **41(5),** 621–622. Abstract.

Gal, I., Sharman, I. M., Pryse-Davies, J. (1972). Vitamin A in relation to human congenital malformations. *Adv Teratol* **5,** 143–154.

Gaston, J., Hutchings, D. E. (1974). The effects of vitamin A excess administered during the mid-fetal period on learning and development in rat offspring. *Dev Psychobiology* **7(3),** 225–233.

Geelen, J. A. (1979). Hypervitaminosis A induced teratogenesis. *CRC Crit. Rev. Toxicol.* **6,** 351–375.

Gibbon, J., Hutchings, D. E., and Kaufman, M. A. (1973). Maternal vitamin A excess during the early fetal period: effects on learning and development in the offspring. *Dev Psychobiology* **6(5),** 445–457.

Giroud, P. A. and Martinet, M. (1955). Hypervitaminoses A et anomalies chez le foetus de rat. *Rev. Intern. Vitaminol.* **26,** 10–18.

Holson, R. R., Gazzara, R. A., Ferguson, S. A., and Adams, J. (1997). A behavioral and neuroanatomical investigation of the lethality caused by gestational days 11–13 retinoic acid exposure. *Neurotoxicol. Teratol.* **19(5),** 347–353.

Holson, R. R., Gazzara, R. A., Ferguson, S. A., and Adams, J. (1997). Behavioral effects of low dose gestational day 11–13 retinoic acid exposure. *Neurotoxicol. Teratol.* **19(5),** 355–362.

Holson, R. R., Gazzara, R. A., Ferguson, S. A., Ali, S. F., LaBorde, J. B., and Adams, J. (1997). Gestational retinoic acid exposure: a sensitive period for effects on neonatal mortality and cerebellar development. *Neurotoxicol. Teratol.* **19(5),** 335–346.

Hutchings, D. E. (1983). Behavioral teratology: a new frontier in neurobehavioral research. In *Handbook of Experimental Pharmacology: Teratogenesis and Reproductive Toxicology,* Vol 65. Johnson E. D. and Hutchings D. M., Eds. Springer-Verlag, Berlin.

Hutchings, D. E. and Gaston, J. (1974). The effects of vitamin A excess administered during the mid-fetal period on learning and development in rat offspring. *Dev. Psychobiol.* **7(3),** 225–233.

Hutchings, D. E., Gibbon, J., and Kaufman, M. A. (1973). Maternal vitamin A excess during the early fetal period: effects on learning and development in the offspring. *Dev. Psychobiol.* **6(5),** 445–457.

Jensh, R. P., Kochhar, D. M., Till, M. K., and Eskessen, M. B. (1990). Postnatal behavioral sequelae of prenatal exposure to 13-*cis* retinoic acid. *Teratology* **41(5),** 621–622.

Jensh, R. P., Kochhar, D. M., Till, M. K., and Eskessen, M. B. (1991). Effects of prenatal exposure of isotretinoin (13-*cis* retinoic acid) on neonatal vocalization. *Teratology* **43(5),** 497.

Kalter, H. (1968). *Teratology of the Central Nervous System.* University of Chicago Press, Chicago, IL, Ch 3.

Klug, S., Kraft, J. C., Wildi, E., Merker, H. J., Persaud, T. V. N., Nau, H., and Neubert, D. (1989). Influence of 13-*cis* and all-*trans* retinoic acid on rat embryonic development *in vitro:* correlation with isomerisation and drug transfer to the embryo. *Arch. Toxicol.* **63,** 185–192.

Kochhar, D. M. and Penner, J. D. (1987). Developmental effects of isotretinoin and 4-oxo-isotretinoin: the role of metabolism in teratogenicity. *Teratology* **36,** 67–76.

Kochhar, D. M., Penner, J. D., and Satre, M. A. (1988). Derivation of retinoic acid metabolites from a teratogenic dose retinol (vitamin A) in mice. *Tox. Appl. Pharmaco.* **96,** 429–441.

Kochhar, D. M., Penner, J. D., and Tellone, C. I. (1984). Comparative teratogenic activities of two retinoids: effects on palate and limb development. *Teratogen. Carcinogen. Mutagen.* **4,** 377–387.

Kraft, J. C., Lofberg, B., Chahoud, I., Bochert, G., Nau, H. (1989). Teratogenicity and placental transfer of all-trans-, 13-cis, 4-Oxo-all-trans-, and 4-Oxo-13-cis-retinoic acid after administration of a low oral dose during organogenesis in mice. *Tox Applied Pharmaco* **100,** 162–176.

Kutz, S. A., Troise, N. J., Cimprich, R. E., Yearsley, S. M., and Rugen, P. J. (1989). Vitamin A acetate: a behavioral teratology study in rats. *Drug Chem. Toxicol.* **12,** 259–275.

Lammer, E. J. and Armstrong, D. L. (1992). Malformations of hindbrain structures among humans exposed to isotretinoin (13-*cis*-retinoic acid) during early embryogenesis. In *Retinoids in Normal Development and Teratogenesis.* G. Morriss-Kay, Ed. Oxford University Press: London, pp. 281–295.

Lammer, E. J., Chen, D. T., Hoar, R. M., Agnish, N. D., Benke, P. J., Braun, J. T., Curry, C. J., Ferhnoff, P. M., Grix, A. W., Lott, I. T., et al. (1985). Retinoic acid embryopathy. *N. Engl. J. Med.* **313(14),** 837–841.

Lammer, E. J. Hayes, A. M., Schunior, A., and Holmes, L. B. (1988). Unusually high risk for adverse outcomes of pregnancy following fetal isotretinoin exposure. *Am. J. Hum. Gen.* **43,** A58.

Langman, J. and Welch, G. W. (1966). Effect of vitamin A on development of the central nervous system. *J. Comp. Neural.* **128,** 1–16.

Lorente, C. A., Miller, S. A. (1978). The effect of hypervitaminosis A on rat palatal development. *Teratology* **18(2),** 277–284.

Madden, M., Ong, D. E., Chytil, F. (1990). Retinoid-binding protein distribution in the developing mammalian nervous system. *Development* **109,** 75–80.

Malakhovskii, V. G. (1969). Behavioral disturbances in rats receiving teratogenic agents antenatally. *Bull. Exp. Biol. Med.* **68(11),** 1230–1232.

Malakhovskii, V. G. (1971). Antenatal effect of pyrimethamine and vitamin A on behavior of the rat progeny. *Bull Exp Biol Med* **71(3),** 254–256.

Marcus, R. and Coulton, A. M. (1990). Fat soluble vitamins. In *The Pharmacological Basis of Therapeutics.* Goodman and Gilman, Ed. Pergamon Press, Elmsford, New York, pp.1553–1563.

Morriss, G. M. (1972). Morphogenesis of the malformations induced in rat embryos by maternal hypervitaminosis A. *J Anat* **113(2),** 241–250.

Murakami, U. and Kameyama, Y. (1965). Malformations of the mouse fetus caused by hypervitaminosis A of the mother during pregnancy. *Arch. Environ. Health* **10,** 732.

Nanda, R. (1974). Effect of vitamin A on the potentiality of rat palatal processes to fuse in vivo and in vitro. *Cleft Palate* **11,** 123–133.

Nanda, R., Romeo, D. (1977). Effect of intraamniotic administration of vitamin A on rat fetuses. *Teratology* **16,** 35–40.

Nishimura, T. (1995). Collaborative Work of 13th Behavioral Teratology Committee. Evaluation of the battery test as a behavioral teratology examination with retinoic acid. *Teratolgoy* **52,** 11B.

Nolen, G. A. (1972). The effects of various levels of dietary protein on retinoic acid–induced teratogenicity in rats. *Teratology* **5,** 143–152.

Nolen, G. A. (1986). The effects of prenatal retinoic acid on the viability and behavior of the offspring. *Neurobehav. Tox. Tera.* **8,** 643–654.

Pratt R. M., Goulding, E. H., Abbott, B. D. (1987). Retinoic acid inhibits migration of cranial neural crest cells in the cultured mouse embryo. *J Craniofacial Gen Dev Bio* **7,** 205–217.

Rosa, F. W. (1983). Teratogenicity of isotretinoin. *Lancet* **II.,** 513.

Rosa, F. W., Wilk, A. L., and Kelsey, F. O. (1986). Vitamin A congeners. *Teratology* **33,** 355–364.

Rothman, K. J., Moore, L. L., Singer, M. R., Nguyen, U. D. T., Mannino, S., and Milunsky, A. (1995). Teratogenicity of high vitamin A intake. *N. Engl. J. Med.* **333(21),** 1369–1373.

Saillenfait, A. M. and Vannier, B. (1988). Methodological proposal in behavioural teratology testing: assessment of propoxyphene, chlorpromazine, and vitamin A as positive controls. *Teratology* **37,** 185–199.

Shenefelt, R. E. (1972). Morphogenesis of malformations in hamsters caused by retinoic acid: relation to dose and stage at treatment. *Teratology* **5,** 103.

Snell, K. (1982). "Developmental Toxicology". New York: Praeger.

Stange, L., Carlstrom, K., Eriksson, M. (1978). Hypervitaminosis A in early human pregnancy and malformations of the central nervous system. *ACTA Obstet Gynecol Scand* **57,** 289–291.

Vorhees, C. V. (1974). Some behavioral effects of maternal hypervitaminosis A in rats. *Teratology* **10,** 269–274.

Vorhees, C. V. (1986). Principels of behavioral teratology. In *Handbook of Behavioral Teratology.* Riley, E. P. and Vorhees, C. B. Eds. New York, Plenum Press 23–48.

Vorhees, C. V., Brunner, R. L., and Butcher, R. E. (1979). Psychotropic drugs as behavioral teratogens. *Science* **205(21),** 1220–1225.

Vorhees, C. V., Brunner, R. L., McDaniel, C. R., and Butcher, R. E. (1978). The relationship of gestational age to vitamin A induced postnatal dysfunction. *Teratology* **17,** 271–276.

Willhite, C. C., Hill, R. M., Irving, D. W. (1986). Isotretinoin induced craniofacial malformations in humans and hamsters. *Journal of Craniofacial Genetics and Developmental Biology,* **2,** 193–209.

CHAPTER

37

Developmental Neurotoxicity of Antiepileptic Drugs

DEBORAH K. HANSEN
R. ROBERT HOLSON
Division of Reproductive and Developmental Toxicology
Department of Health and Human Services
Food and Drug Administration
National Center for Toxicological Research
Jefferson, Arkansas 72079

I. Introduction

The antiepileptic drugs (AEDs) phenobarbital, phenytoin, valproic acid (VPA), and carbamazepine have proved very useful in the treatment of various types of epilepsy. In addition, VPA in particular is increasingly being used in the treatment of various psychiatric conditions. One major concern with these drugs is their adverse effects during pregnancy. It has been estimated that more than 11,000 babies are born each year in the United States to epileptic women taking AEDs (Kelly, 1984). Most AEDs have been associated with teratogenic effects in offspring in animal models as well as in humans, and reports of long-term neurodevelopmental effects in humans have begun to appear.

This review focuses on the effects of these drugs on development of the brain and effects on learning and behavior. First the animal literature is discussed, followed by a summary of the data in humans. Most of the available literature, both human and animal, has focused on those AEDs which have received the most use, phenobarbital and phenytoin; therefore, this review focuses on those two compounds. The literature concerning VPA is increasing, and this drug will be reviewed as well.

II. General Toxicity

The barbiturates are highly lethal anesthetics with a marked ability to depress brainstem respiratory centers.

*Abbreviations: AED, antiepileptic drugs; CAR, conditional avoidance responding; GABA, γ-aminobutyric acid; GD, gestational day; IQ, intelligence quotient; MRI, magnetic resonance imaging; PN, postnatal day; sc, subcutaneous; VPA, valproic acid.

At therapeutic doses, sedation is a common effect, and this is true of phenytoin and VPA as well (Goodman and Gilman, 1985). Consequently, exposure to high doses of any of the drugs mentioned previously is likely to depress food intake acutely, resulting in weight loss, a primary marker of toxicity in dams and offspring. The question, then, is not whether such compounds are toxic, but rather, how toxic such compounds are at doses, routes, and exposure schedules common in animal research. Regrettably, as discussed later in this chapter, too few researchers routinely report toxicity or even exposure-related mortality of their treatment regimes. However, the scanty material available leaves little doubt that common treatment paradigms have high toxicity, often including fetal and maternal lethality. These effects are summarized in Table 1, together with maximal and minimal effective dose levels.

To begin with maternal toxicity, high doses of barbiturates have caused maternal lethality in some studies (Armitage, 1952; Becker et al., 1958), and in rats gestational exposure to barbiturates or phenytoin is frequently accompanied by lowered maternal weight gain (Vorhees, 1983; Martin *et al.,* 1985; Shapiro *et al.,* 1985; Vorhees and Minck, 1989; Weisenburger *et al.,* 1990; Pizzi and Jersey, 1992), but this is not invariable (Martin *et al.,* 1979; Pizzi *et al.,* 1996). The situation with mice is less certain; Yanai's laboratory has not reported substantial reductions in food intake (Iser-Strenger and Yanai, 1986) in dams given phenobarbital in their feed, although in a nearly identical paradigm, Kuprys and Tabakoff (1983) reported a 50% reduction in mean daily food intake. Effects of gestational barbiturate exposure on maternal weight gain in mice have followed the reports on food intake—little effect in Yanai's hands (Rogel-Fuchs *et al.,* 1994), but a large drop in weight gain in the Kuprys and Tabakoff (1983) experiment. There are too few gestational studies in phenytoin or valproate-exposed mice to allow any conclusions about maternal toxicity, and several rat studies have not seen maternal weight loss (Sobrian and Nandedkar, 1986; Vorhees, 1987a,b). Thus, one may anticipate reduced maternal food intake and weight gain at substantial dose levels with phenobarbital and phenytoin—this outcome may be less likely with VPA.

The observed level of maternal toxicity hardly seems to account for what appears to be very substantial mortality in rodents exposed gestationally or neonatally to common AEDs (Table 1). Mortality is enhanced after gestational exposure to barbiturates, in rats (Armitage, 1952; Vorhees, 1983, 1985; Martin *et al.,* 1985; Takagi *et al.,* 1986), guinea pigs (Becker *et al.,* 1958), gerbils (Chapman and Cutler, 1988), and mice (Zemp and Middaugh, 1975; Yanai and Iser, 1981). Doses that increased mortality ranged from 40 to 80 mg/kg, or possibly even less when barbiturates were administered in the diet (Yanai and Iser, 1981). Neonatal barbiturate exposure, too, is often lethal in rats (Schain and Watanabe, 1975) and mice (Yanai and Iser, 1981; Fishman *et al.,* 1989). Degree of lethality varies across studies, with a doubling of control levels at the common 50 mg/kg gestational phenobarbital dose in rats (11%, Vorhees, 1985), and mortality in mice at roughly the same dose ranging as high as 27% (Yanai and Iser, 1981). Similarly, neonatal mortality is substantially increased, twofold at 60 mg/kg in rats (Schain and Watanabe, 1975), and 45% at 50 mg/kg in mice (Fishman *et al.,* 1989).

Phenytoin also is quite lethal to exposed infants at doses commonly used in animal studies. Gestational phenytoin exposure in rats commonly increases mortality (Elmazar and Sullivan, 1981; Ikeda, 1982; Vorhees,

TABLE 1 General Toxicity[a]

Effect		Species	Antiepileptic drug (mg/kg)		
			Phenobarbital	Phenytoin	Valproic acid
Maternal toxicity		Rat	↑ (G,N)	↑ (G,N)	0(G),NA(N)
		Mouse	?	NA	NA
Offspring mortality		Rat	↑ (G,N)	↑ (G),NA(N)	↑ (G),NA(N)
		Mouse	↑ (G,N)	NA(G), ↑ (N)	0(G),NA(N)
Offspring body weight		Rat	?	↓ (G),NA(N)	?(G),NA(N)
		Mouse	?(G), ↓ (N)	NA(G),?(N)	0(G),NA(N)
Dose	Maximum	Rat	80(G),60(N)	1000(G),NA(N)	300(G),NA(N)
		Mouse	100(G),60(N)	NA(G),50(N)	300(G),NA(N)
	Minimum	Rat	5(G),10(N)	10(G),NA(N)	20(G),NA(N)
		Mouse	20(G),20(N)	NA(G),25(N)	NA

[a]Reported toxic effects of AED treatment by species and compound.

G, gestational exposure; N, neonatal exposure; NA, no information available; ↑, increased effect; ↓, reduced effect; 0, no effect.

1983; Shapiro *et al.,* 1985; Takagi *et al.,* 1986; Shapiro and Babalola, 1987; Vorhees and Minck, 1989; Zengel *et al.,* 1989; Weisenburger *et al.,* 1990; Minck *et al.,* 1991; Pizzi and Jersey, 1992), although, again, such effects are not universal (Gauron and Rowley, 1980; Cassidy *et al.,* 1984; Vorhees *et al.,* 1995). Mortality occurs predominately postparturition and is reported to be around one-third at a 100 mg/kg dose of the more bioactive phenytoin salt (Elmazar and Sullivan, 1981; Pizzi and Jersey, 1992), a mortality not achieved at less than 200 mg/kg of the phenytoin acid (Vorhees, 1983). We are unaware of reports concerning lethality of neonatal phenytoin exposure in rats, or gestational exposure in mice, but neonatal exposure in mice is, again, quite lethal (Fishman *et al.,* 1989; Ohmori *et al.,* 1992, 1997); as little as 35 mg/kg on postnatal (PN) 2–4 killed one-third of the exposed pups (Ohmori *et al.,* 1997).

Although data are sparse, it also seems that gestational VPA exposure can be lethal in monkeys (Mast *et al.,* 1986; Lockard *et al.,* 1983; Hendrickx *et al.,* 1988) as well as rats (Vorhees *et al.,* 1994), although in rats this may occur only at a high dose (300 mg/kg) (Sobrian and Nandedkar, 1986; Vorhees, 1987a; Vorhees *et al.,* 1991a, 1994).

Despite this lethality, developmental exposure to antiepileptics has surprisingly little effect on offspring body weight (Table 1). Gestational barbiturate exposure in rats evidently seldom depresses body weight (Harris and Case, 1979; Gauron and Rowley, 1980; Gupta *et al.,* 1980a; Martin *et al.,* 1985; Pizzi *et al.,* 1996), and when it does, the effect is small (Gupta *et al.,* 1980b; Martin *et al.,* 1979; Chapman and Cutler, 1988). Much the same may be said of the effects of neonatal exposure to barbiturates in rats (Schain and Watanabe, 1975; Harris and Case, 1979; Gupta and Yaffe, 1981; Patsalos and Wiggins, 1982), and of gestational exposure in mice, with the same lab sometimes seeing a modest effect (Middaugh *et al.,* 1975a; Yanai and Bergman, 1981; Yanai and Iser, 1981) and sometimes not (Middaugh *et al.,* 1981b; Yanai and Wanich, 1985). Only neonatal barbiturate exposure in the Yanai laboratory generally reduces body weight (Yanai and Tabakoff, 1980; Yanai and Bergman, 1981; Yanai and Iser, 1981; Bergman *et al.,* 1982; Yanai and Wanich, 1985; Yanai *et al.,* 1985; Iser-Strenger and Yanai, 1986; Rogel-Fuchs *et al.,* 1994), and even here the weight reduction seldom reaches 10%.

Finally, Table 1 also gives maximal and minimal doses reported to have an effect on any variable, by species and dose. It is clear that treatment effects are seen at dose levels well below those accompanied by substantial mortality. Certainly one direction that research in this field should and can move toward is lower, less toxic exposure levels. A second area of interest is the high neonatal lethality caused by *in utero* exposure to phenytoin or the barbiturates. It would be instructive to know why it is that so many of these pups die.

III. Effects of Antiepileptic Drugs on the Brain

Prior to considering what is known about AED effects on developing brain, it is important to review briefly an old and on the whole vexatious literature concerning phenytoin neurotoxicity in adult humans and animals. Beginning with the clinical literature, it has long been known that cerebellar signs are among the possible side effects of phenytoin therapy (Goodman and Gilman, 1985). Such symptoms usually resolve if treatment is discontinued, but in a smattering of cases that have come to autopsy [or to magnetic resonance imaging (MRI)], these disorders have been accompanied by clear histologic damage to the cerebellum, including loss of Purkinje cells (Hofmann, 1958; Kokenge *et al.,* 1965; Ghatak *et al.,* 1976; Breiden-Arendts and Gullotta, 1981; Botez *et al.,* 1985; Gessaga and Urich, 1985; Krause *et al.,* 1988; Masur *et al.,* 1989; Abe and Yagishita, 1991; Luef *et al.,* 1996; Kuruvilla and Bharucha, 1997). Of course, such reports are suspect, because these patients were all being treated for preexisting brain damage (epilepsy), and epileptics can have cerebellar damage without phenytoin exposure (Dam, 1972). However, a few reports of phenytoin overdose followed by crippling cerebellar damage in relatively uncomplicated cases exist (Masur *et al.,* 1989; Kuruvilla and Bharucha, 1997), although even here the literature is equivocal (Luef *et al.,* 1996).

Such multiply confounded clinical reports are often resolved by animal research, but in this instance results are as puzzling for animals as for humans. In a number of studies, prolonged exposure to high levels of phenytoin has produced damage in the cerebellum of rats (Kokenge *et al.,* 1965; Perez del Cerro and Snider, 1967; Bernocchi *et al.,* 1983), mice (Volk and Kirchgassner, 1985; Volk *et al.,* 1986), gerbils (Garzon *et al.,* 1983), and cats (Utterback *et al.,* 1958; Kokenge *et al.,* 1965). Notable in this research is the finding that cats are especially sensitive to phenytoin-induced cerebellar ataxia and damage (Kokenge *et al.,* 1965; Utterback *et al.,* 1958), and the claim that cerebellar axonal damage is caused by phenytoin-induced swelling of smooth endoplasmic reticulum caused by induction of P450s (Volk and Kirchgassner, 1985; Volk *et al.,* 1986). However, in some of the most careful work in this field, Dam (1970, 1972) and Dam and Nielsen (1970) could find no cerebellar damage in pigs, rats, or monkeys exposed chronically to high phenytoin doses. Other laboratories have also reported negative findings (Nikolarakis *et al.,* 1985;

Palm *et al.,* 1986; Dowson *et al.,* 1992). We are left, then, with the suspicion that under conditions as yet poorly defined, phenytoin dosages in the same range as those used in developmental studies are sometimes, but not always, neurotoxic to cerebellum and perhaps other brain regions.

This direct neurotoxicity has also been reported in cell culture studies. Because neuronal cell cultures typically use fetal cells, the finding of such neurotoxicity in cell cultures taken from fetal spinal cord (Bergey *et al.,* 1981; Serrano *et al.,* 1988) or cortex (Swaiman *et al.,* 1980; Sher *et al.,* 1985) suggests that AED neurotoxicity may also be seen in, and perhaps accentuated for, the developing brain.

Regardless of whether direct neurotoxicity is involved, there is certainly growing evidence that in laboratory animals AEDs cause brain damage when treatment occurs early in development. The crudest but most readily available evidence for such damage is provided by studies reporting effects on weight or neuronal constituents in whole brain (Table 2). These effects have been poorly studied in rats, with predictably mixed consequence. Thus prenatal barbiturate exposure can reduce rat whole brain weight (Patsalos and Wiggins, 1982), but such effects are not invariant (Martin *et al.,* 1979; Gupta *et al.,* 1980a). In a series of far better studies, Schain, Diaz and collaborators have demonstrated a clear barbiturate reduction of weight, DNA, RNA, protein and cholesterol in rats exposed neonatally to phenobarbital (Schain and Watanabe, 1975, 1976; Diaz *et al.,* 1977; Diaz and Schain, 1978). These studies are among the few to exclude nutritional factors by using appropriate controls. Similarly, a collaborative study in which 27 of 30 laboratories independently found reductions in rat brain weight after gestational phenytoin treatment (Tachibana *et al.,* 1996) leaves little doubt that such effects are real (see also Elmazar and Sullivan, 1981), despite repeated failures to obtain reduced brain weight in the Vorhees laboratory (Vorhees, 1985, 1987b; Minck *et al.,* 1989). Finally, in the rat, neonatal phenytoin can also reduce whole brain weight (Patsalos and Wiggins, 1982).

In short, phenytoin and barbiturates can reduce whole brain weight following either gestational or neonatal exposure, in mice and rats. At this time it is unclear whether VPA has a similar impact on the brain. There is the usual paucity of information on this question. Patsalos and Wiggins (1982) found phenobarbital and phenytoin but not valproate effects on brain weight; similarly, in cell cultures, valproate neurotoxicity seems, at best, modest (Swaiman *et al.,* 1980; Sher *et al.,* 1985). The only other report we have been able to locate did find reduced whole brain weight in neonatal mice immediately after several days' exposure to large amounts of VPA (Thurston *et al.,* 1981). As with so much else about valproate effects, resolution of this question awaits further research.

TABLE 2 Effects on Brain

Effect		Species	Antiepileptic drug		
			Phenobarbital	Phenytoin	Valproic acid
Whole brain	Weight	Rat	?(G) ↓ (N)	↓ (G,N)	0(G),NA(N)
		Mouse	?(G) ↓ (N)	NA(G), ↓ (N)	NA(G), ↓ (N)
	Cell number	Rat	NA(G), ↓ (N)	NA	NA
		Mouse	?(G),NA(N)	NA	NA
	Myelin	Rat	↓ (G,N)	↓ (G,N)	↓ (G),NA(N)
		Mouse	NA	NA	NA
Cerebellum	Weight	Rat	NA(G), ↓ (N)	?(G),NA(N)	NA
		Mouse	↓ (G,N)	NA	NA
	Cell number	Rat	NA(G),0(N)	NA	NA
		Mouse	↓ (G,N)	NA	NA
	Myelin	Rat	NA(G), ↓ (N)	NA	NA
		Mouse	NA	NA	NA
Hippocampus	Weight	Rat	NA	0(G),NA(N)	NA
		Mouse	0(G),?(N)	NA	NA
	Cell number	Rat	NA	0(G),NA(N)	NA
		Mouse	↓ (G,N)	NA	NA
Cortex cellularity		Rat	NA	0(G),NA(N)	NA
		Mouse	?(G), ↓ (N)	NA	NA

G, gestational exposure; N, neonatal exposure; NA, no information available; ↑, increased effect; ↓, reduced effect; 0, no effect.

Effects on whole brain weight will necessarily be mirrored in at least select subregions of the brain. One region in which such effects are seen is the cerebellum. Given the long-debated possibility of AED neurotoxicity in cerebellum, it is surprising that there do not appear to be many attempts to evaluate these effects following AED exposure in rats. Thus the impact of developmental exposure to phenytoin in rats has only been assessed in two studies using gestational exposure; of these, one reported weight reduction in the cerebellum (Elmazar and Sullivan, 1981), the other did not (Minck *et al.,* 1989). Still more inexplicably, we have been unable to locate any reports of the impact of gestational barbiturate exposure on rat cerebellum. Neonatal phenobarbital exposure in rats, however, has been well described, with good controls for nutritional effects. These studies found that as with whole brain, weight and cholesterol content of the cerebellum was reduced (Schain and Watanabe, 1975, 1976; Diaz *et al.,* 1977; Diaz and Schain, 1978). Unlike whole brain, RNA, DNA, and protein content were unaffected (Schain and Watanabe, 1975, 1976; Diaz *et al.,* 1977; Diaz and Schain, 1978).

In contrast to the rat literature, the Yanai laboratory has done an outstanding job of describing effects of barbiturates on the developing cerebellum in mice. After either gestational or neonatal phenobarbital exposure, these researchers find clear evidence of Purkinje cell death, reduced weight, reductions in the number of granule cells, and electron microscopic evidence of cell degeneration (Yanai *et al.,* 1979, 1985; Yanai and Tabakoff, 1980; Yanai and Bergman, 1981; Yanai and Iser, 1981; Bergman *et al.,* 1982; Fishman *et al.,* 1982, 1983, 1989; Kleinberger and Yanai, 1985; Yanai and Waknin, 1985). Two of these studies (Yanai and Waknin, 1985; Yanai *et al.,* 1985) reported similar effects of neonatal phenobarbital on three separate strains of mice. Like phenobarbital, neonatal phenytoin exposure can also damage the mouse cerebellum, reducing weight and Purkinje cell number (Fishman *et al.,* 1989; Ohmori *et al.,* 1992, 1997). It is also important to note that the Yanai group has not found any well-demarcated sensitive developmental period for these results. This again suggests a direct neurotoxic effect on cerebellar neurons rather than interference with cell division or programmed cell death. This is particularly true for the Purkinje cells; the neonatal exposure that can kill such cells occurs well after cell division is completed in these cerebellar neurons.

The hippocampus may be another target of AED effects on brain. Again, there is little evidence one way or the other from rat studies, with a single report of membrane order abnormalities following prenatal phenytoin (Vorhees *et al.,* 1990) or VPA treatment (Vorhees *et al.,* 1991a). There are also no studies looking at such effects following VPA exposure or phenytoin exposure in mice, so, again, the Yanai laboratory and barbiturate exposure is the source for most of the available data. As with cerebellum, gestational or neonatal barbiturates damage the mouse hippocampus, reducing both pyramid and granule cell number (Yanai *et al.,* 1979, 1981; Yanai and Tabakoff, 1980; Yanai and Bergman, 1981; Bergman *et al.,* 1982; Kleinberger and Yanai, 1985), evidently without always reducing overall hippocampal weight (Kleinberger and Yanai, 1985). Similar effects may also be seen in cortex and even the olfactory bulbs of mice treated neonatally with phenobarbital (Yanai and Bergman, 1981; Bergman *et al.,* 1982; Roselli-Austin and Yanai, 1989), although such effects seem more tenuous following *in utero* exposure to barbiturates (Yanai *et al.,* 1979, 1982; Yanai and Tabakoff, 1980). Because the entirety of rat forebrain is reduced in weight, DNA, RNA, protein, and cholesterol content following neonatal phenobarbital (Schain and Watanabe, 1975, 1976; Diaz *et al.,* 1977; Diaz and Schain, 1978), it is probable that similar effects will eventually be described in rat hippocampus and neocortex.

If there are large gaps in our understanding of AED effects on regional weight and cellularity, there are still larger deficiencies in knowledge of the impact of developmental AED exposure on neurotransmitters and their receptors. Starting with the monoamines, information is limited predominately to mice given barbiturates. Here Middaugh *et al.* (1981b) reported that, following prenatal barbiturates, whole brain levels of norepinephrine and dopamine were reduced, whereas serotonin content was unaltered. Yanai's group looked at more restricted brain regions, and reported reductions in dopamine in mouse hypothalamus but not striatum (Yanai *et al.,* 1985), plus alterations in striatal spiroperidol binding and apomorphine-induced climbing that depended on age at exposure (Iser-Strenger and Yanai, 1986). Direct effects on the norepinephrine system were also seen; cell density in the murine locus ceruleus was reduced, together with norepinephrine concentrations in the hippocampus (Yanai and Pick, 1987).

Because all AEDs seem to share an effect on the γ-aminobutyric acid (GABA) system, several investigators have examined developmental AED effects on that system with uncertain results. On the whole, the Yanai group has seen little barbiturate effect on this neurotransmitter or on GABA uptake sites or receptors (Fares *et al.,* 1990; Pick *et al.,* 1993). However, Vorhees (1985) found alterations in a variety of brain amino acids, including increased whole brain GABA, after prenatal phenytoin treatment in rats. Although more work is certainly needed in this area, preliminary findings do not suggest a focused impact of AEDs on the developing

GABA system. It is still too early to speculate about AED impact on other amino acid neurotransmitters.

Finally, Yanai and colleagues have reported a substantial barbiturate effect on acetylcholine neurotransmission in the hippocampus. After gestational or neonatal exposure, muscarinic receptor density in hippocampus increases, along with second messenger systems related to that receptor (Rogel-Fuchs *et al.*, 1992, 1994; Zahalka *et al.*, 1995; Abu-Roumi *et al.*, 1996). However, acetylcholine and choline acetyltransferase concentrations were normal (Kleinberger and Yanai, 1985; Rogel-Fuchs *et al.*, 1992). This seems to imply that septal acetylcholine input to the hippocampus could be reduced, resulting in a compensatory receptor upregulation. In support of this possibility, these researchers have shown that grafting fetal septal tissue containing cholinergic neurons into the hippocampus can protect against the behavioral deficits shown by mice subjected to developmental barbiturates (Rogel-Fuchs *et al.*, 1994).

To summarize, barbiturates, phenytoin, and perhaps VPA are neurotoxic in the developing brain. This neurotoxicity seems targeted to cerebellum, but hippocampus and other brain regions are also affected. Such neurotoxicity seems to lack narrow windows of vulnerability timed to developmental schedules of affected neurons (Bergman *et al.*, 1980), yet evidently such effects are more readily obtained in young than in adult animals (although this possibility requires direct experimental examination). To date, however, only the Israeli group has made any sustained effort to identify and examine such neurologic effects, and their attention has focused on barbiturates and mice. Certainly other laboratories involved in this research need to emulate the example set by Yanai and co-workers. This need is particularly acute for valproate, an increasingly popular antiepileptic, the effects of which on the developing brain are all but unknown.

IV. Effects of Antiepileptic Drugs on Behavior

Because many of the active investigators in this field have backgrounds in psychology, it is not remarkable that the behavioral effects of developmental AED exposure are better described than the biochemical or neuroanatomic effects. There is still much to be done in this area, however, before we can be at all confident that we have an adequate behavioral profile of AED-exposed rodents; more yet is required to link this behavioral profile to effects on specific brain regions or neurotransmitter systems.

As with the preceding topics, behavioral alterations in AED-exposed rodents are summarized in Table 3. That table, and this review, are organized by behavioral category, and we begin by noting that developmental AED exposure has a very general impact on spatial learning. Indeed, an impairment in maze learning was reported as far back as 1952, when Armitage tested rats given gestational barbiturates on the Maier maze in the first AED study of its kind. Subsequent experimenters have replicated this early report of maze learning deficits following gestational barbiturates (Becker *et al.*, 1964; Murai, 1966; Vorhees, 1983, 1985; Pizzi *et al.*, 1996). Curiously, the later reports have not found the marked deficits described by the earlier studies. Thus the Vorhees group found no effect of gestational phenobarbital

TABLE 3 Effects on Behavior

Effect	Species	Antiepileptic drug		
		Phenobarbital	Phenytoin	Valproic acid
Spatial learning	Rat	↓ (G,N)	↓ (G),NA(N)	↓ (G),NA(N)
	Mouse	↓ (G,N)	NA	NA
Activity	Rat	?(G,N)	↑ (G),NA(N)	?(G),NA(N)
	Mouse	↑ (G),?(N)	NA	NA
Coordination/motor	Rat	↓ (G),NA(N)	↓ (G),NA(N)	NA
	Mouse	NA	NA(G), ↓ (N)	NA
Operant learning	Rat	↓ (G,N)	NA	NA
	Mouse	↓ (G),NA(N)	NA	NA
Conditioned avoidance	Rat	↓ (G,N)	0(G),NA(N)	NA
	Mouse	NA	NA	NA
Seizure threshold	Rat	?(G),NA(N)	NA	↓ (G),NA(N)
	Mouse	↑ (G),0(N)	NA	NA

G, gestational exposure; N, neonatal exposure; NA, no information available; ↑, increased effect; ↓, reduced effect; 0, no effect.

on the T, Biel, or M mazes, and a problem with spontaneous alternation that was marginally significant (Vorhees, 1983, 1985). Likewise, in a study of primidone, a compound whose active metabolite is phenobarbital, Pizzi *et al.* (1996) reported only a modest deficit on the radial eight-arm maze, and then only in males and at high doses. In the only neonatal phenobarbital study of rat offspring, however, rats were given either 10 or 20 mg/kg phenobarbital over PN 5–45. This treatment regime produced a substantial deficit on the Lashley III maze, even at the 10 mg/kg dose (Fonseca *et al.,* 1976). Thus, although more needs to be done in this area, it seems likely that either neonatal or gestational phenobarbital can impair spatial learning in rats.

As with other aspects of phenobarbital functional teratology, this deficit has been explored most intensively in mice by Yanai and colleagues. These researchers describe a general decrement on spatial learning tasks, including radial eight-arm maze (Kleinberger and Yanai, 1985; Pick and Yanai, 1985; Pick *et al.,* 1987; Yanai and Pick, 1987, 1988; Yanai *et al.,* 1989b; Zahalka *et al.,* 1995), Morris water maze (Rogel-Fuchs *et al.,* 1992, 1994; Zahalka *et al.,* 1995), and spontaneous alternation (Pick and Yanai, 1984, did not find this deficit, but subsequent studies did—Kleinberger and Yanai, 1985; Zahalka *et al.,* 1995) in mice exposed to phenobarbital *in utero.* In their hands, neonatal phenobarbital treatment in mice has similar consequences, again seen on spontaneous alternation (Pick and Yanai, 1984; Kleinberger and Yanai, 1985), Morris maze (Rogel-Fuchs *et al.,* 1992, 1994), and radial eight-arm maze (Kleinberger and Yanai, 1985; Pick and Yanai, 1985; Pick *et al.,* 1987). Because this laboratory frequently compares gestational phenobarbital treatment [via the dam's food on gestational days (GD) 8–17) to neonatal treatment (sc injection on PN 2–22), within the same experiment, it is also possible to determine which period is the more sensitive for production of spatial learning deficits. These studies suggest that there is no overall pattern in the temporal locus of these spatial impairments. Not only are deficits produced at both exposure periods, but the magnitude of deficits at the two periods may also be test specific. Thus spontaneous alternation was more affected by neonatal than gestational phenobarbital exposure (Pick and Yanai, 1984; Kleinberger and Yanai, 1985), radial eight-arm maze deficits were more severe after gestational than neonatal phenobarbital exposure (Kleinberger and Yanai, 1985; Pick and Yanai, 1985; Pick *et al.,* 1987), and treatment at either developmental period produced impairments on Morris water maze of roughly equal magnitude (Rogel-Fuchs *et al.,* 1992, 1994). Similarly, in the only study of its kind in rats, Murai (1966) found that phenobarbital treatment on GD 4–7 produced maze learning problems only slightly greater than those seen after GD 16–19 exposure. Thus, as with phenobarbital-induced brain damage, these behavioral effects evidently can be produced by exposing the developing brain to phenobarbital at any point from early neurogenesis through the late neonatal period, if not later still.

It is conventional wisdom that spatial learning deficits in rodents are the result of hippocampal damage. As we have seen, there is reason to suspect such damage in phenobarbital-exposed rodents, and several experiments from the Yanai laboratory address this possibility directly. As already noted, Yanai has found that transplantation of septal acetylcholine-containing cells but not norepinephrine-containing locus ceruleus cells into the hippocampus protected against these spatial learning impairments (Yanai and Pick, 1987, 1988; Rogel-Fuchs *et al.,* 1994). 6-Hydroxydopamine–lesioning of the septal dopamine system was also partially protective, an effect that Yanai *et al.* (1989c) attributed to release of the septo-hippocampal system from dopaminergic inhibition. These experiments point toward the next generation of research—attribution of behavioral deficits to abnormalities in specific brain regions or neurotransmitter systems.

What Yanai has done to examine spatial learning deficits in phenobarbital-exposed mice is comparable to what Vorhees and colleagues have done for gestational phenytoin exposure in rats. These workers describe a broad range of behavioral deficits in their animals, deficits now replicated by other laboratories, including a collaborative study by 30 laboratories in Japan (Tachibana *et al.,* 1996). Chief among these effects is a substantial difficulty with spatial learning tasks, including spontaneous alternation (Vorhees, 1983), the Biel or Cincinnati water maze (Minck *et al.,* 1991; Tachibana *et al.,* 1996; Vorhees, 1983; 1987b; Vorhees and Minck, 1989), and the Morris and radial eight-arm mazes (Weisenburger *et al.,* 1990). Discussion of these deficits is deferred to the section on AED effects on activity measures. It is also important that Vorhees sees a similar impact on Biel maze and spontaneous alternation following gestational VPA exposure (Vorhees, 1987a; Vorhees *et al.,* 1994). Although to date there are no reports on the effects of valproate or phenytoin on spatial tasks in mice, or for that matter of neonatal phenytoin or valproate treatment on rat spatial behavior (Table 3), these findings taken as a whole strongly suggest that exposure to any of the common AEDs during development impairs performance on spatial learning tasks.

Turning to AED effects on activity (Table 3), the literature is quite weak on the results of barbiturate treatment. At this writing, it is impossible to know whether this is because there are no such effects or because real effects have not been identified. Thus gestational phenobarbital treatment in rats has variously been reported to reduce activity in male but not female offspring (Pizzi *et al.,* 1996), to increase activity in females but not males (Sobrian and Nandedkar, 1986), or to increase male activity following gestational exposure to sodium pentobarbital but not phenobarbital (Martin *et al.,* 1985). Results are equally confused for running wheel activity. Murai (1966) found no effect of gestational phenobarbital in rats, whereas Martin *et al.* (1985) observed increased activity in males exposed gestationally to 40 but not 80 mg/kg/day phenobarbital, with no effects after similar exposure to sodium pentobarbital. In rats, neonatal phenobarbital has been reported either to reduce female activity to male control levels without altering activity of treated males (Fonseca *et al.,* 1976), or to reduce male activity (Diaz and Schain, 1978).

Results are no clearer with mice. In a series of studies in mice exposed gestationally to phenobarbital, Middaugh and co-workers placed offspring in a large open field for a single 3-min session. They reported that at PN 21, activity of treated males rose to control female levels, whereas treated female activity did not change (Middaugh *et al.,* 1975a). At PN 75, the same 3-min activity measure gave roughly the same results (Midduagh *et al.,* 1981b), whereas by PN 160 the previously refractory treated females had also become "hyperactive" due to slowed adaptation over the 3-min exposure (Middaugh *et al.,* 1981a). To further confuse matters, Yanai's group found activity decreased in the open field after neonatal exposure to phenobarbital in mice, but this effect was seen in only one of two mouse strains (Yanai *et al.,* 1989a).

Thus activity effects of developmental exposure to barbiturates seem to be sex specific. However, it does not appear that these sex-specific responses necessarily involve effects on the sexual differentiation of behavior, because the animals' activity can become less like that of the opposite sex, which is to say, reduced in males and increased in females.

This failure to find consistent phenobarbital effects on activity is important precisely because such changes are prominent in rats after gestational phenytoin. A proportion but by no means all of such rats display a striking spontaneous circling behavior (Vorhees, 1983, 1987b,c; Minck *et al.,* 1989, 1991; Vorhees and Minck, 1989; Weisenburger *et al.,* 1990; Pizzi and Jersey, 1992). This gestational treatment also produces substantial hyperactivity in both sexes (Vorhees, 1983, 1987b; Vorhees and Minck, 1989; Minck *et al.,* 1991; Tachibana *et al.,* 1996), although some studies have reported that this hyperactivity was more pronounced in one sex, either males (Mullenix *et al.,* 1983; Pizzi and Jersey, 1992) or females (Vorhees *et al.,* 1992). Finally, these hyperactive rats are also prone to rear less (Vorhees, 1983; Tachibana *et al.,* 1996), perhaps due to the press of activity.

Circling, hyperactivity, and performance deficits are common in rats with midear infections. Why this is so is not known, but certainly this phenytoin syndrome in rats bears more than a passing resemblance to a midear problem. Moreover, maze deficits and hyperactivity were aggravated in circling rats, although noncirclers also displayed substantial problems (Vorhees and Minck, 1989; Weisenburger *et al.,* 1990; Minck *et al.,* 1991; Vorhees *et al.,* 1991b, 1992, 1995). Accordingly, Minck *et al.* (1989, 1991) examined the midear otoconia of circling rats exposed gestationally to phenytoin. Although some abnormalities were detected, they were not found in all circling rats, and hence did not appear to explain this syndrome. Vorhees and colleagues have accordingly excluded midear problems as the primary cause of circling, and circling as the primary cause of hyperactivity or maze deficits. Of course, this does not completely rule out other forms of midear problems, let alone central vestibular abnormalities, as a cause of the circling. It is also possible that noncircling rats with (less pronounced) maze deficits and hyperactivity simply have a less-pronounced form of an underlying vestibular disorder.

Turning to other treatments, species, and exposure periods, there are evidently no published reports of neonatal phenytoin effects on rat behavior, and likewise no reports of effects of gestational or neonatal phenytoin or VPA on activity in mice. There are three rather sketchy descriptions of activity in rats exposed gestationally to valproate. Sobrian and Nandedkar (1986) found that activity was increased in exposed females, unaltered in exposed males. In contrast, Vorhees has reported activity reductions in gestationally exposed rats of both sexes (Vorhees, 1987a; Vorhees *et al.,* 1994). Again, it is too early to draw any conclusions about effects of VPA, although it is unlikely that the Vorhees group could have missed anything resembling their phenytoin syndrome.

Despite known cerebellar deficits in phenobarbital-treated mice and scattered but persistent reports of cerebellar damage in phenobarbital-exposed adults (human and animal), there has been no concerted attempt to functionally evaluate the cerebellum after developmental AEDs. Indeed, the few reports that have been published are almost anecdotal in nature. In mice, Ohmori *et al.* (1992) have seen impaired gait and negative geotaxis in consequence of neonatal phenytoin exposure; similarly, gestational phenobarbital in guinea pigs produced severe motor problems described as "spasticity" (Becker *et al.,* 1958), whereas rats similarly treated had difficulty with an early form of the beam-balancing task (Becker *et al.,* 1964). Only in the case of rats exposed to phenytoin *in utero* has there been any real attempt to assess motor control. The results to date strongly suggest substantive deficits in motor coordination, especially early in life. In such animals, development of air righting is strikingly slowed (Elmazar and Sullivan, 1981; Vorhees, 1987b; Minck *et al.,* 1991; Vorhees *et al.,* 1994; Tachibana *et al.,* 1996; but see Cassidy *et al.,* 1984). Swimming development is also mildly impaired (Vorhees, 1983), and one study has described difficulties with rotorod and beam balancing (Elmazar and Sullivan, 1981). Of course, vestibular problems could easily account for the previously mentioned findings. To date, there are no published accounts of the effects of VPA on motor coordination.

Behavioral measures other than those previously mentioned have been used only occasionally in the study of developmental AED exposure. There are, for example, several studies suggestive of operant deficits in such animals. Martin *et al.* (1979) found that fixed ratio performance was impaired in rats treated gestationally with 40 but not 80 mg/kg/day phenobarbital, whereas, conversely, a schedule that rewarded a delayed response was affected only by the higher 80 mg/kg/day dose. In a study combining gestational and neonatal phenobarbital treatment, Harris and Case (1979) described an impairment on a fixed ratio 20 schedule. Mice also may have problems on operant tasks following gestational phenobarbital treatment. Middaugh and colleagues reported deficits on an fixed ratio 40 schedule, but again, this problem was prominent after 40 not 80 mg/kg/day (Middaugh *et al.,* 1975a,b). We are unaware of any reports of operant testing in rodents subjected to developmental phenytoin or VPA, but in light of the previously mentioned reports, this is an area that deserves closer scrutiny.

There are also a few studies that assessed conditional avoidance responding (CAR) in AED-treated rats but not mice. Martin *et al.* (1979, 1985) and Gauron and Rowley (1980) found this behavior to be impaired in rats given gestational phenobarbital, as did Harris and Case (1979) after phenobarbital exposure spanning gestation and the neonatal period. However, Elmazar and Sullivan (1981) found that two-way CAR was normal in the offspring of rat dams given phenytoin during pregnancy. As discussed later in this chapter, there are theoretical reasons for examining CAR after developmental AED treatment, and the studies mentioned previously certainly suggest that there might be problems, at least with the barbiturates.

One final paradigm exhausts the bulk of functional–behavioral assessments used to describe AED impact on development. This is seizure threshold, a not unreasonable choice for the evaluation of compounds with acute antiepileptic effects. Despite the face validity of such measures, this literature is extremely sparse, and even contradictory. Sobrian and Nandedkar (1986) found reductions in pentylenetetrazol-induced seizure thresholds after gestational phenobarbital in rats, whereas Murai (1966) saw increased thresholds for electrically induced seizures in such animals. In mice treated with phenobarbital gestationally but not neonatally, auditory seizures were enhanced (Yanai *et al.,* 1981). No reports of seizure susceptibility have been found for phenytoin-treated animals, but following prenatal VPA treatment in rats, Sobrian and Nandedkar (1986) noted that pentylenetetrazol-induced seizures were reduced in female but increased in male offspring. In contrast, in an abstract by Pizzi *et al.* (1988), a substantial reduction in pentylenetetrazol-induced seizures in adult rats exposed to valproate prenatally was reported. Conclusions are not possible at this stage, other than to point out that the entire issue deserves closer attention.

To summarize this section, developmental AED exposure causes substantial problems on standard spatial learning paradigms, and this is evidently true across species, exposure periods, and the three compounds reviewed here. There can also be no question that prenatal phenytoin treatment in rats produces a syndrome involving circling, hyperactivity, and motor problems, in addition to the mentioned spatial learning problems. At present, it is uncertain what causes these behavioral impairments, or even whether such effects have the same primary causes across compounds. Here there are several possibilities that can be investigated further with behavioral as well as neuroanatomic techniques.

For example, the ubiquitous spatial learning problems could be caused by damage to the hippocampus. Hippocampal lesions may be the most exhaustively described of all brain-damage syndromes in rodents, and such damage produces a clear profile of behavioral abnormalities. In addition to memory deficits on classic spatial learning tasks, hippocampal-lesioned animals have a dense passive avoidance deficit, they show lowered habituation of activity rates, increased operant response rates, and improved performance on two-way CAR. It would not be overly difficult to subject AED-exposed animals to such a test battery, with hippocampal-lesioned animals as positive controls. In addition, other laboratories should emulate the Yanai group by evaluating the hippocampus of AED-exposed rodents neurochemically and neuroanatomically. An attempt to replicate and extend their finding of increased QNB binding in hippocampi of phenobarbital-exposed mice might be a logical place to begin.

As to the gestational phenytoin syndrome in rats, this set of deficits strongly suggests abnormalities somewhere within the auditory–vestibular cerebellar system. Test batteries sensitive to cerebellar damage have been described, and could easily be applied to this question. Indeed, it is astonishing that only one laboratory has assessed rotorod performance in the early 1980s (Elmazar and Sullivan, 1981). Again, animals with developmental damage to the cerebellum would provide excellent positive controls. As with possible hippocampal deficits, a careful functional and neuroanatomic evaluation of the vestibular system and cerebellum is also in order, and certainly not just for animals exposed prenatally to phenytoin.

Finally, almost no work of any sort has been done with VPA (or, indeed, even with phenytoin in mice), and there is a dearth of rat studies using neonatal exposure to any of these compounds. Moreover, there are intriguing reports of deficits on CAR and operant behavior, and of poorly understood alterations in seizure susceptibility. These are just some of the areas that merit behavioral analysis, hopefully combined with more determined attacks on the neuroanatomic and neurochemical substrates of what is becoming a striking set of behavioral anomalies.

V. Developmental Toxicity Studies

A thorough review of the animal literature on developmental toxicity studies of these AEDs is beyond the scope of this review. The reader is referred to Finnell and Dansky (1991) for more information on this topic.

The earliest animal studies suggesting a teratogenic effect of AEDs concerned phenytoin. This drug produces abnormal development *in vivo* in a number of species, including the rat (Harbison and Becker, 1972; for others see Finnell and Dansky, 1991), the mouse (reviewed, Finnell and Dansky, 1991), and the cat (Khera, 1979).

Most of the experimental work has focused on the ability of phenytoin to produce orofacial clefting in mice, probably due to the reports of this anomaly in humans exposed to the drug. However, the drug has been reported to produce other defects in the mouse, including growth retardation, skeletal defects, hydronephrosis, open eye, hydrocephalus, and ectrodactyly (Harbison and Becker, 1969). Teratogenic effects in mice appeared to be a result of drug exposure and not the presence of a maternal seizure disorder (Finnell, 1981; Finnell and Chernoff, 1982). A genetic susceptibility to phenytoin-induced defects has been suggested by several investigators (Johnston *et al.,* 1979; Hansen and Hodes, 1983; Finnell and Chernoff, 1984).

The mechanism for phenytoin-induced embryotoxicity remains unknown despite a good deal of work in this area (reviewed, Hansen, 1991). An arene oxide produced by cytochrome P-450–mediated metabolism has been suggested to be the teratogenic agent (Martz *et al.,* 1977; Blake and Martz, 1980). Alternatively, Wells' group has suggested that cooxidation of phenytoin by prostaglandin H synthase may play a role in embryotoxicity (Kubow and Wells, 1986, 1989; Wells and Vo, 1989; Wells *et al.,* 1989).

Antagonism of folate has also been suggested to play a role in embryotoxicity (reviewed, Hansen, 1991). Supplementation with exogenous folate has been reported to decrease (Marsh and Fraser, 1973), to increase (Sullivan and McElhatton, 1975) or to have no effect (Marsh and Fraser, 1973; Mercier-Perot and Tuchmann-Duplessis, 1974) on the incidence of phenytoin-induced defects. Overall, there appears to be no definitive data to support a role for folate deficiency in phenytoin-induced embryotoxicity.

It has been suggested that phenytoin might disrupt normal palatal development via a glucocorticoid-mediated mechanism. Both compounds show a similar strain sensitivity (Biddle and Fraser, 1976, 1977; Hansen and Hodes, 1983). A teratogenic dose of phenytoin increases maternal glucocorticoid levels for a much longer period of time than does a nonteratogenic dose (Hansen *et al.,* 1988). There is conflicting evidence on the effect of maternal adrenalectomy on the incidence of clefting. At this time, the role of glucocorticoids in phenytoin-induced teratogenicity is unknown.

There is surprisingly little information available in the animal literature concerning the embryotoxicity of phenobarbital. Daily dietary administration of 50 or 150 mg of the compound/kg of chow to mice on GD 6–16 resulted in no difference in the incidence of dead or resorbed fetuses and only a slight increase in the frequency of fetuses with cleft palate (Sullivan and McElhatton, 1975). Primidone at much higher doses produced an increase in palatal defects but also produced maternal toxicity at the two highest doses. Primidone is metabolized in part to phenobarbital, so it is not clear if primidone or its phenobarbital metabolite was responsible for the produced embryotoxicity. Phenobarbital also increased the percentage of fetuses with cleft palate but had no effect on fetal weight or resorptions (Fritz *et al.,* 1976). Sullivan and McElhatton (1975) also observed a dose-related increase in cleft palate after phenobarbital or primidone exposure. Finnell *et al.* (1987) have reported strain differences in the response to phenobarbital similar to those observed after phenytoin treatment. Overall, it appears that in animal models, phenobarbital is less embryotoxic than phenytoin.

Valproic acid is teratogenic in a variety of species including rats (Ong *et al.,* 1983), rabbits (Petrere *et al.,* 1986), and nonhuman primates (Mast *et al.,* 1986; Hendrickx *et al.,* 1988), as well as humans (Winter *et al.,* 1987; Ardinger *et al.,* 1988). The neural tube in mice

(Eluma *et al.*, 1984; Paulson *et al.*, 1985) and humans (Bjerkedal *et al.*, 1982; Robert and Guibaud, 1982) seems to be especially susceptible to the effects of the drug. Strain differences in susceptibility have been reported (Finnell *et al.*, 1988) suggesting an underlying genetic predisposition.

Several possible mechanisms for VPA-induced embryotoxicity have been investigated. Although it has been suggested that valproate might cause a zinc deficiency, *in vitro* work has suggested that this is not the mechanism of embryotoxicity (Coakley and Brown, 1986). Valproic acid alters β-oxidation of fatty acids in adult animals and produces hepatotoxicity; it has been reported to decrease coenzyme A (Becker and Harris, 1983) and carnitine (Ohtani *et al.*, 1982; Laub *et al.*, 1986), both of which are required for fatty acid oxidation. However, the mechansims of hepatotoxicity and embryotoxicity appear to be different. Valproate does not alter embryonic coenzyme A levels (Brown *et al.*, 1985; Coakley *et al.*, 1986), and carnitine is not able to decrease valproate-induced embryotoxicity (Andrews *et al.*, 1996; Hansen, unpublished observations, 1997).

Most of the later work has involved effects of valproate on the vitamin folic acid. *In vivo* administration of the folate metabolite folinic acid was able to decrease the incidence of valproate-induced neural tube defects in mice (Trotz *et al.*, 1987; Wegner and Nau, 1991). However, this metabolite was not able to alter the incidence of neural tube defects in other studies either *in vivo* (Elmazar *et al.*, 1992; Hansen *et al.*, 1995) or *in vitro* (Hansen and Grafton, 1991; Hansen, 1993). At this point, the evidence is still not clear on the role of folic acid in VPA-induced embryotoxicity.

In summary, phenytoin, phenobarbital, and VPA are teratogenic in a number of animal species, producing a variety of developmental defects. Strain differences in susceptibility have been reported for all three AEDs, suggesting an underlying genetic predisposition.

VI. Human Studies

The majority of studies has indicated an increased risk for congenital malformations among epileptic women taking AEDs. This risk is typically two- to threefold that of the general population; however, the risk for specific malformations may be much higher (e.g., neural tube defects from exposure to VPA). Cardiovascular defects and orofacial clefts are the most commonly observed defects; abnormalities of the skeletal, nervous, urogenital, gastrointestinal, and respiratory systems have also been reported (reviewed in Dansky and Finnell, 1991).

Many of the studies on the complications of epilepsy during pregnancy have focused on major malformations and have seldom included neurobehavioral endpoints. Those few that have included such endpoints have only made these assessments at one time with little or no follow-up. Many have included only an examination of general intelligence. Some reviews on this subject include those of Adams *et al.* (1990), Dansky and Finnell (1991), Granstrom and Gaily (1992), and Dessens *et al.* (1994).

A complication in many of these studies is the issue of polytherapy. Many epileptics take more than one drug to control seizures. Epidemiologic studies have seldom had large enough numbers of cases to implicate a single AED. Therefore, in many of the studies reviewed, AEDs in general may be implicated, rather than a specific drug.

A number of reports have suggested that decreased head circumference is a result of prenatal AED exposure. Speidel and Meadow (1972), Nelson and Ellenberg (1982), Jager-Roman *et al.* (1982), Hill *et al.* (1982) and Majewski *et al.* (1981) observed microcephaly after exposure to various AEDs. Conflicting data were presented by Leavitt *et al.* (1992), who observed a nonsignificant decrease in head circumference among children exposed to various AEDs, as well as Granstrom (1982) and Latis *et al.* (1982), who observed no adverse effects in small groups of children born to epileptic mothers. In some of these studies, height and weight were also decreased by AED treatment, suggesting an overall growth-retarding effect of the drugs.

Phenytoin in particular was suggested to cause microcephaly by Hanson *et al.* (1976) and VanDyke *et al.* (1988), and carbamazepine was implicated by Hiilesmaa *et al.* (1981). Vert *et al.* (1982) reported that phenobarbital either alone or in combination decreased head circumference. Although Hiilesmaa *et al.* (1981) observed no catch-up growth in head circumference by 18 months of age, Hanson *et al.* (1976), using data from the Collaborative Perinatal Project of the National Institute of Neurological and Communicative Disorders and Stroke found that the decreased head circumference observed at birth in children exposed to phenytoin had resolved by 7 years of age.

Mild to moderate mental retardation has been observed in several studies, including those of Speidel and Meadow (1972), Hill *et al.* (1974, 1982), Majewski *et al.* (1981), Vert *et al.* (1982), and Kaneko (1991). The mental development index of the Bayley Scales of Infant Development, which examines emerging cognitive skills such as communication, was decreased by exposure to AEDs; Leavitt *et al.* (1992) observed no difference in the psychomotor development index. Nelson and Ellenberg (1982) observed that IQ scores below 70 were more common among the offspring exposed prenatally to AEDs, but this difference was not statistically significant.

Polytherapy has been suggested to produce more severe defects than monotherapy. The decrease in the mental development index of the Bayley scale was more significant with polytherapy than with monotherapy (Leavitt *et al.,* 1992). Losche *et al.* (1994) also observed that children exposed to multiple AEDs scored lower on tests of intelligence, verbal IQ, and motor skills than children exposed to monotherapy. Hill *et al.* (1982) also found that polytherapy resulted in lower developmental quotients than did monotherapy, particularly when compared to monotherapy with phenytoin. However, Gaily *et al.* (1988) found no differences in IQ scores at age 5.5 years when comparing children exposed to polytherapy or monotherapy.

Nelson and Ellenberg (1982) found that, generally, the children of women who had a seizure during pregnancy had lower IQ scores, but this difference was not significant. Losche *et al.* (1994) observed that offspring of women who had a high number of seizures during pregnancy scored lower on visual perception and psycholinguistic skills tests. Gaily *et al.* (1990) concluded that maternal seizures during pregnancy increased the risk of specific cognitive dysfunction.

The prospective study of Gaily *et al.* (1990) used a variety of tests to examine neurodevelopment at age 5.5 years. They found a significant decrease in nonverbal development using the block design subtest from the Wechsler Preschool and Primary Scale of Intelligence. They also observed a significant decrease in auditory memory. There were no differences in the verbal subtests of the Wechsler, in several auditory ability tests, in motor praxis tests, or in comprehension of linguistic structures. The authors concluded that more case than control children had a specific cognitive dysfunction, but 77% of prenatally exposed children had no dysfunction. High drug levels appeared to be a risk factor for cognitive dysfunction.

Gaily *et al.* (1990) found a significant difference in the maternal educational level between the case and the control children, and the maternal educational level also correlated with several of the test outcomes. Other authors have also suggested that a "nonoptimal pyschosocial environment" might contribute to the altered cognitive development of children born to epileptic mothers (Granstrom and Gaily, 1992), a confounding factor that has not always been taken into account in these studies.

Steinhausen *et al.* (1994) did a prospective study from birth to 6 years of age on a group of 73 children who had been exposed prenatally to various AEDs. They found no differences in mental or motor development between control and study children when they were assessed during the second year of life. However, when preschool-age children were assessed, there were a number of differences, with children exposed prenatally to AEDs showing lower scores on verbal and motor tests as well as on various intelligence tests. These decreases occurred only in the children exposed prenatally to AEDs, and they did not occur in children of epileptic mothers who had not taken AEDs or among children of epileptic fathers.

Phenytoin specifically has been reported to decrease IQ scores at 7 years of age (Hanson *et al.,* 1976). Van Dyke *et al.* (1988) also observed an increase in developmental delay or mental retardation and school or learning problems among children exposed prenatally to phenytoin. Scolnik *et al.* (1994) found that phenytoin decreased global IQ scores and the developmental language scale. However, Shapiro *et al.* (1976) found no difference in IQ scores at 4 years of age following exposure to phenytoin.

VanOverloop *et al.* (1992) examined 20 children 4–8 years of age who had been exposed prenatally to phenytoin with or without concurrent phenobarbital exposure. They observed decreases in intelligence, visual motor integration, and verbal comprehension test scores among these children when compared to age-matched controls. They also noted decreased spontaneous activity among these children and lower scores on the block design and maze subtests of the Wechsler.

Phenobarbital was reported to have no effect on neurologic or behavioral development (Van der Pol *et al.,* 1991) or on motor or mental development at 4 years of age (Shapiro *et al.,* 1976). Ardinger *et al.* (1988) did observe developmental delay or neurologic abnormality among children exposed prenatally to valproate either alone or in combination with other AEDs.

Carbamazepine either alone or in combination with phenobarbital has been reported to have no effect on neurologic or behavioral development (Van der Pol *et al.,* 1991). Scolnik *et al.* (1994) also observed no impairment in global IQ scores or on the developmental language scale following prenatal exposure to carbamazepine. However, Ornoy and Cohen (1996) found no difference in the Bayley psychomotor development index of children prenatally exposed to carbamazepine, but significant decreases on the mental development index of the Bayley Scales were observed.

VII. Conclusions

In conclusion, the three AEDs reviewed here clearly alter neurobehavioral development in animal models. AEDs in general, and phenytoin in particular, appear to alter neurobehavioral development in humans. Several studies have shown lower intelligence scores in children exposed prenatally to these drugs, and in several cases specific test scores were also decreased. It is interesting to note that phenytoin produced spatial learning problems in rodents and altered the performance of children exposed prenatally on block design, visual motor integration, and maze tests. This might imply that the visuospatial skills of these children are impaired.

There is much experimental work left to do in this area. As was pointed out, little is known about the regions of the brain that may be affected by these compounds. Even less is known about the effects of AEDs on neurotransmitters and their receptors. Finally, even though there is a large amount of behavioral data available, there is surprisingly little information on the effects of these drugs on motor development.

In particular, more work needs to be done on the effects of VPA and carbamazepine on the developing brain. With their increasing use in the treatment of various psychiatric conditions, it is important to determine if these drugs alter early brain development.

References

Abe, H. and Yagishita, S. (1991). Chronic phenytoin intoxication occurred below the toxic concentration in serum and its pathological findings [Japanese]. *No to Shinkei—Brain Nerve* **43,** 89–94.

Abu-Roumi, M., Newman, M. E., and Yanai, J. (1996). Inositol phosphate formation in mice prenatally exposed to drugs: relation to muscarinic receptors and postreceptor effects. *Brain Res. Bull.* **40,** 183–186.

Adams, J., Vorhees, C. V., and Middaugh, L. D. (1990). Developmental neurotoxicity of anticonvulsants: Human and animal evidence on phenytoin. *Neurotox. Teratol.* **12,** 203–214.

Andrews, J. E., Ebron-McCoy, M., and Kavlock, R. J. (1996). The effects of carnitine on the *in vitro* and *in vivo* developmental toxicity of valproic acid. *Teratology* **53,** 88–89.

Ardinger, H. H., Atkin, J. F., Blackston, R. D., Elsas, L. J., Clarren, S. K., Livingstone, S., Flannery, D. B., Pellock, J. M., Harrod, M. J., Lammer, E. J., Majewski, F., Schinzel, A., Toriello, H. V., and Hanson, J. W. (1988). Verification of the fetal valproate syndrome phenotype. *Am. J. Med. Genet.* **29,** 171–185.

Armitage, S. G. (1952). The effects of barbiturates on the behavior of rat offspring as measured in learning and reasoning situations. *J. Comp. Physiol. Psychol.* **45,** 146–152.

Becker, C.-M. and Harris, R. A. (1983). Influence of valproic acid on hepatic carbohydrate and lipid metabolism. *Arch. Biochem. Biophys.* **223,** 381–392.

Becker, R. F., Boneau, C. A., Shearin, C. A., and King, J. E. (1964). Behavioral alterations of young rats with a history of oversedation at birth. *Neurology* **14,** 510–520.

Becker, R. F., Flannagan, E., and King, J. E. (1958). The fate of offspring from mothers receiving sodium pentobarbital before delivery. *Neurology* **8,** 776–782.

Bergey, G. K., Swaiman, K. F., Schrier, B. K., Fitzgerald, S., and Nelson, P. G. (1981). Adverse effects of phenobarbital on morphological and biochemical development of fetal mouse spinal cord neurons in culture. *Ann. Neurol.* **9,** 584–589.

Bergman, A., Feigenbaum, J. J., and Yanai, J. (1982). Neuronal losses in mice following both prenatal and neonatal exposure to phenobarbital. *Acta Anat.* **114,** 185–192.

Bergman, A., Rosselli-Austin, L., Yedwab, G., and Yanai, J. (1980). Neuronal deficits in mice following phenobarbital exposure during various periods in fetal development. *Acta. Anat.* **108,** 370–373.

Bernocchi, G., Bottiroli, G., Cavanna, O., Arrigoni, E., Scelsi, R., and Manfredi, L. (1983). Fluorescence histochemical patterns of Purkinje cell layer in rat cerebellum after long-term phenytoin administration. *Bas. Appl. Histochem.* **27,** 45–53.

Biddle, F. G. and Fraser, F. C. (1976). Genetics of cortisone induced cleft palate in the mouse-embryonic and maternal effects. *Genetics* **84,** 743–754.

Biddle, F. G. and Fraser, F. C. (1977). Cortisone-induced cleft palate in the mouse. A search for the genetic control of the embryonic response trait. *Genetics* **85,** 289–302.

Bjerkedal, T., Czeizel, A., Goujard, J., Kallen, B., Mastroiacova, P., Nevin, N., Oakley, G. Jr., and Robert, E. (1982). Valproic acid and spina bifida. *Lancet* **ii,** 1096.

Blake, D. A., and Martz, F. (1980). Covalent binding of phenytoin metabolites in fetal tissue. In *Phenytoin-Induced Teratology and Gingival Pahtology.* T. M. Hassell, M. C. Johnston, and K. H. Dudley, Eds. Raven Press, New York, pp. 75–80.

Botez, M. I., Gravel, J., Attig, E., and Vezina, J. L. (1985). Reversible chronic cerebellar ataxia after phenytoin intoxication: possible role of cerebellum in cognitive thought. *Neurology* **35,** 1152–1157.

Breiden-Arends, C. and Gullotta, F. (1981). Diphenylhydantoin, epilepsy, cerebellar atrophy—histological and electron microscope examinations. (Author's transl). [German]. *Fortschritte der Neurologie–Psychiatrie* **49,** 406–414.

Brown, N. A., Farmer, P. B., and Coakley, M. (1985). Valproic acid teratogenicity: demonstration that the biochemical mechanism differs from that of valproate hepatotoxicity. *Biochem. Soc. Trans.* **13,** 75–77.

Cassidy, C. L., Rose, G. P., and Hend, R. W. (1984). Lack of behavioural teratogenic effect of phenytoin in rats. *Toxicol. Lett.* **22,** 39–46.

Chapman, J. B. and Cutler, M. G. (1988). Phenobarbitone: Adverse effects on reproductive performance and offspring development in the Mongolian gerbil (*Meriones unguiculatus*). *Psychopharmacology* **94,** 365–370.

Coakley, M. E. and Brown, N. A. (1986). Valproic acid teratogenicity in whole embryo culture is not prevented by zinc supplementation. *Biochem. Pharmacol.* **35,** 1052–1055.

Coakley, M. E., Rawlings, S. J., and Brown, N. A. (1986). Short-chain carboxylic acids, a new class of teratogens: studies of potential biochemical mechanisms. *Environ. Health Persp.* **70,** 105–111.

Dam, M. (1970). Number of Purkinje's cells after diphenylhydantoin intoxication in pigs. *Arch. Neurol.* **22,** 64–67.

Dam, M. (1972). The density and ultrastructure of the Purkinje cells following diphenylhydantoin treatment in animals and man. *Acta Neurol. Scand.* **48,** 13–51.

Dam, M. and Nielsen, M. (1970). Purkinje's cell density after diphenylhydantoin intoxication in rats. *Arch. Neurol.* **23,** 555–557.

Dansky, L. V. and Finnell, R. H. (1991). Parental epilepsy, anticonvulsant drugs, and reproductive outcome: epidemiologic and experimental findings spanning three decades; 2: Human studies. *Reprod. Toxicol.* **5,** 301–335.

Dessens, A. B., Boer, K., Koppe, J. G., van de Poll, N. E., and Cohen-Kettenis, P. T. (1994). Studies on long-lasting consequences of prenatal exposure to anticonvulsant drugs. *Acta Paediatr.* **Suppl. 404,** 54–64.

Diaz, J. and Schain, R. J. (1978). Phenobarbital: effects of long-term administration on behavior and brain of artificially reared rats. *Science* **199,** 90–91.

Diaz, J., Schain, R. J., and Bailey, B. G. (1977). Phenobarbital-induced brain growth retardation in artificially reared rat pups. *Biol. Neonate.* **32,** 77–82.

Dowson, J. H., Wilton-Cox, H., and James, N. T. (1992). Lipopigment in rat hippocampal and Purkinje neurones after chronic phenytoin administration. *J. Neurol. Sci.* **107,** 105–110.

Elmazar, M. M. A. and Sullivan, F. M. (1981). Effect of prenatal phenytoin administration on postnatal development of the rat: a behavioral teratology study. *Teratology* **24,** 115–124.

Elmazar, M. M. A., Thiel, R., and Nau, H. (1992). Effect of supplementation with folinic acid, vitamin B_6, and vitamin B_{12} on valproic acid–induced teratogenesis in mice. *Fund. Appl. Toxicol.* **18,** 389–394.

Eluma, F. O., Sucheston, M. E., Hayes, T. G., and Paulson, R. B. (1984). Teratogenic effects of dosage levels and time of administration of carbamazepine, sodium valproate, and diphenylhydantoin on craniofacial development in the CD-1 mouse fetus. *J. Craniofac. Genet. Devel. Biol.* **4,** 191–210.

Fares, F., Weizman, A., Pick, C. G., Yanai, J., and Gavish, M. (1990). Effect of prenatal and neonatal chronic exposure to phenobarbital on central and peripheral benzodiazepine receptors. *Brain Res.* **506,** 115–119.

Finnell, R. H. (1981). Phenytoin-induced teratogenesis: a mouse model. *Science* **211,** 483–484.

Finnell, R. H., Bennett, G. D., Karras, S. B., and Mohl, V. K. (1988). Common hierarchies of susceptibility to the induction of neural tube defects in mouse embryos by valproic acid and its 4-propyl-4-pentenoic acid metabolite. *Teratology* **38,** 313–320.

Finnell, R. H. and Chernoff, G. F. (1982). Mouse fetal hydantoin syndrome: effects of maternal seizures. *Epilspsia.* **23,** 423–429.

Finnell, R. H. and Chernoff, G. F. (1984). Variable patterns of malformation in the mouse fetal hydantoin syndrome. *Amer. J. Med. Genet.* **19,** 463–471.

Finnell, R. H. and Dansky, L. V. (1991). Parental epilepsy, anticonvulsant drugs, and reproductive outcome: epidemiologic and experimental findings spanning three decades; 1: Animal studies. *Reprod. Toxicol.* **5,** 281–299.

Finnell, R. H., Shields, H. E., Taylor, S. M., and Chernoff, G. F. (1987). Strain differences in phenobarbital-induced teratogenesis in mice. *Teratology* **35,** 177–185.

Fishman, R. H. B., Gaathon, A., and Yanai, J. (1982). Early barbiturate treatment eliminates peak serum thyroxine levels in neonatal mice and produces ultrastructural damage in the brains of adults. *Dev. Brain Res.* **5,** 202–205.

Fishman, R. H. B., Ornoy, A., and Yanai, J. (1983). Ultrastructural evidence of long-lasting cerebellar degeneration after early exposure to phenobarbital in mice. *Exp. Neurol.* **79,** 212–222.

Fishman, R. H., Ornoy, A., and Yanai, J. (1989). Correlated ultrastructural damage between cerebellum cells after early anticonvulsant treatment in mice. *Int. J. Devel. Neurosci.* **7,** 15–26.

Fonseca, N. M., Sell, A. B., and Carlini, E. A. (1976). Differential behavioral responses of male and female adult rats treated with five psychotropic drugs in the neonatal stage. *Psychopharmacologia* **46,** 263–268.

Fritz, H., Muller, D., and Hess, R. (1976). Comparative study of the teratogenicity of phenobarbitone, diphenylhydantoin, and carbamazepine in mice. *Toxicology* **6,** 322–330.

Gaily, E., Kantolas-Sorsa, E., and Granstrom, M.-L. (1988). Intelligence of children of epileptic mothers. *J. Pediatr.* **113,** 677–684.

Gaily, E., Kantolas-Sorsa, E., and Granstrom, M.-L. (1990). Specific cognitive dysfunction in children with epileptic mothers. *Devel. Med. Child Neurol.* **32,** 403–414.

Garzon, P., Lopez-Ortega, G., Sandoval-Rojas, S., Mora-Galindo, J., Bastidas-Ramirez, B.E., and Navarro-Ruiz, A. (1983). Phenytoin effects on tissues of Mongolian gerbils (*Meriones unguiculatus*). *Gen. Pharmacol.* **14,** 343–347.

Gauron, E. F. and Rowley, V. N. (1980). Critical periods for diphenylhydantoin and phenobarbital administration during gestation. *Psych. Report* **47,** 1163–1166.

Gessaga, E. C. and Urich, H. (1985). The cerebellum of epileptics. *Clin. Neuropath.* **4,** 238–245.

Ghatak, N. R., Santoso, R. A., and McKinney, W. M. (1976). Cerebellar degeneration following long-term phenytoin therapy. *Neurology* **26,** 818–820.

Goodman, L. S. and Gilman, A. G. (1985). *The Pharmacological Basis of Therapeutics,* 7th Ed., MacMillan, New York.

Granstrom, M.-L. (1982). Development of the children of epileptic mothers: preliminary results from the prospective Helsinki study. In *Epilepsy, Pregnancy, and the Child.* D. Janz, Ed. Raven Press, New York, pp. 403–408.

Granstrom, M.-L. and Gaily, E. (1992). Psychomotor development in children of mothers with epilepsy. *Neurology* **42 (Suppl. 5a),** 144–148.

Gupta, C., Shapiro, B. H., and Yaffe, S. J. (1980a). Reproductive dysfunction in male rats following prenatal exposure to phenobarbital. *Pediatr. Pharmacol.* **1,** 55–62.

Gupta, C., Sonawane, B. R., Yaffe, S. J., and Shapiro, B. H. (1980b). Phenobarbital exposure *in utero:* alterations in female reproductive function in rats. *Science* **208,** 508–510.

Gupta, C. and Yaffe, S. J. (1981). Reproductive dysfunction in female offspring after prenatal exposure to phenobarbital: critical period of action. *Pediatr. Res.* **15,** 1488–1491.

Hansen, D. K. (1991). The embryotoxicity of phenytoin: a review of possible mechanisms. *Proc. Soc. Exp. Biol. Med.* **197,** 361–368.

Hansen, D. K. (1993). *In vitro* effects of folate derivatives on valproate-induced neural tube defects in mouse and rat embryos. *Toxicol. In Vitro* **7,** 735–742.

Hansen, D. K. and Grafton, T. F. (1991). Lack of attenuation of valproic acid–induced effects by folinic acid in rat embryos *in vitro. Teratology* **43,** 575–582.

Hansen, D. K., Grafton, T. F., Dial, S. L., Gehring, T. A., and Siitonen, P. H. (1995). Effect of supplemental folic acid on valproic acid–induced embryotoxicity and tissue zinc level *in vivo. Teratology* **52,** 277–285.

Hansen, D. K. and Hodes, M. E. (1983). Comparative teratogenicity of phenytoin among several inbred strains of mice. *Teratology* **28,** 175–179.

Hansen, D. K., Holson, R. R., Sullivan, P. A., and Grafton, T. F. (1988). Alterations in maternal plasma corticosterone levels following treatment with phenytoin. *Toxicol. Appl. Pharmacol.* **96,** 24–32.

Hanson, J. W., Myrianthopoulos, N. C., Harvey, M. A. S., and Smith, D. W. (1976). Risks to the offspring of women treated with hydantoin anticonvulsants, with emphasis on the fetal hydantoin syndrome. *J. Pediatr.* **89,** 662–668.

Harbison, R. D. and Becker, B. A. (1969). Relation of dosage and time of administration of diphenylhydantoin to its teratogenic effect in mice. *Teratology* **2,** 305–311.

Harbison, R. D. and Becker, B. A. (1972). Diphenylhydantoin teratogenicity in rats. *Toxicol. Appl. Pharmacol.* **22,** 193–200.

Harris, R. A. and Case, J. (1979). Effects of maternal consumption of ethanol, barbital, or chlordiazepoxide on the behavior of the offspring. *Behav. Neural Biol.* **26,** 234–247.

Hendrickx, A. G., Nau, H., Binkerd, P., Rowland, J. M., Cukierski, M. J., and Cukierski, M. A. (1988). Valproic acid developmental toxicity and pharmacokinetics in the rhesus monkey: an interspecies comparison. *Teratology* **38,** 329–345.

Hiilesmaa, V. K., Teramo, K., Granstrom, M.-L., and Bardy, A. H. (1981). Fetal head growth retardation associated with maternal antiepileptic drugs. *Lancet* **2,** 165–167.

Hill, R. M., Verniaud, W. M., Horning, M. G., McCulley, L. B., and Morgan, N. F. (1974). Infants exposed *in utero* to antiepileptic drugs. *Am. J. Dis. Child.* **127,** 645–653.

Hill, R. M., Verniaud, W. M., Rettig, G. M., Tennyson, L. M., and Craig, J. P. (1982). Relationship between antiepileptic drug exposure of the infant and developmental potential. In *Epilepsy, Pregnancy, and the Child.* D. Janz, Ed. Raven Press, New York, pp. 409–417.

Hofmann, W. W. (1958). Cerebellar lesions after parenteral Dilantin administration. *Neurology* **8,** 210–214.

Ikeda, H. (1982). Pharmacological studies on the functional development of the central nervous system in first generation rats born to phenytoin-treated mothers. *Folia Pharmacolog. Japonica* **79,** 65–76.

Iser-Strenger, C. and Yanai, J. (1986). Alterations in mice dopamine receptor characteristics after early exposure to phenobarbital. *Brain Res.* **395,** 57–65.

Jager-Roman, E., Rating, D., Koch, S., Gopfert-Geyer, I., Jacob, S., and Helge, H. (1982). Somatic parameters, diseases, and psychomotor development in the offspring of epileptic parents. In *Epilepsy, Pregnancy, and the Child.* D. Janz, Ed. Raven Press, New York, pp. 425–432.

Johnston, M. C., Sulik, K. K., and Dudley, K. H. (1979). Genetic and metabolic studies of the differential sensitivity of A/J and C57BL/6J mice to phenytoin. *Teratology* **19,** 33A.

Kaneko, S. (1991). Antiepileptic drug therapy and reproductive consequences: functional and morphological effects. *Reprod. Toxicol.* **5,** 179–198.

Kelly, T. E. (1984). Teratogenicity of anticonvulsant drugs. I. Review of the literature. *Am. J. Med. Genet.* **19,** 413–434.

Khera, K. S. (1979). A teratogenicity study on hydroxyurea and diphenylhydantoin in cats. *Teratology* **20,** 447–452.

Kleinberger N. and Yanai, J. (1985). Early phenobarbital-induced alterations in hippocampal acetylcholinesterase activity and behavior. *Dev. Brain Res.* **22,** 113–123.

Kokenge, R., Kutt, H., and McDowell, F. (1965). Neurological sequelae following Dilantin overdose in a patient and in experimental animals. *Neurology* **15,** 823–829.

Krause, K., Bonjour, J., Berlit, P., Kynast, F., Schmidt-Gayk, H., and Schellenberg, B. (1988). Effect of long-term treatment with antiepileptic drugs on the vitamin status. *Drug-Nutrient Interact.* **5,** 317–343.

Kubow, S. and Wells, P. G. (1986). *In vitro* evidence for prostaglandin synthetase-catalyzed bioactivation of phenytoin to a free radical intermediate. *Pharmacologist* **28,** 195.

Kubow, S. and Wells, P. G. (1989). *In vitro* activation of phenytoin to a reactive free radical intermediate by prostaglandin synthetase, horseradish peroxidase, and thyroid peroxidase. *Mol. Pharmacol.* **35,** 504–511.

Kuprys, R. and Tabakoff, B. (1983). Prenatal phenobarbital treatment and temperature-controlling dopamine receptors. *Pharmacol. Biochem. Behav.* **18,** 401–406.

Kuruvilla, T. and Bharucha, N. E. (1997). Cerebellar atrophy after acute phenytoin intoxication. *Epilepsia* **38,** 500–502.

Latis, G. O., Battino, D., Boldi, B., Breschi, F., Ferraris, G., Moise, A., Molteni, B., and Simionato, L. (1982). Preliminary data of a neuropediatric follow-up of infants born to epileptic mothers. In *Epilepsy, Pregnancy, and the Child.* D. Janz, Ed. Raven Press, New York, pp. 419–423.

Laub, M. C., Paetzke-Brunner, I., and Jaeger, G. (1986). Serum carnitine during valproic acid therapy. *Epilepsia* **27,** 559–562.

Leavitt, A. M., Yerby, M. S., Robinson, N., Sells, C. J., and Erickson, D. M. (1992). Epilepsy in pregnancy: developmental outcome of offspring at 12 months. *Neurology* **42 (Suppl. 5),** 141–143.

Lockard, J. S., Levy, R. H., Koch, K. M., Maris, D. O., and Friel, P. N. (1983). A monkey model for status epilepticus: carbamazepine and valproate compared to three standard anticonvulsants. *Adv. Neurol.* **34,** 411–419.

Losche, G., Steinhausen, H.-C., Koch, S., and Helge, H. (1994). The psychological development of children of epileptic parents. II. The differential impact of intrauterine exposure to anticonvulsant drugs and further influential factors. *Acta Paediatr.* **83,** 961–966.

Luef, G., Burtscher, J., Kremser, C., Birbamer, G., Aichner, F., Bauer, G., and Felber, S. (1996). Magnetic resonance volumetry of the cerebellum in epileptic patients after phenytoin overdosages. *Eur. Neurol.* **36,** 273–277.

Majewski, F., Steger, M., Richter, B., Gill, J., and Rabe, F. (1981). The teratogenicity of hydantoins and barbiturates in humans, with consideration on the etiology of malformations and cerebral disturbances in the children of epileptic parents. *Biol. Res. Preg.* **2,** 37–45.

Marsh, L. and Fraser, F. C. (1973). Studies on Dilantin-induced cleft palate in mice. *Teratology* **7,** 23A.

Martin, J. C., Martin, D. C., Lemire, R., and Mackler, B. (1979). Effects of maternal absorption of phenobarbital upon rat offspring development and function. *Neurobehav. Toxicol.* **1,** 49–55.

Martin, J. C., Martin, D. C., Mackler, B., Grace, R., Shores, P., and Chao, S. (1985). Maternal barbiturate administration and offspring response to shock. *Psychopharmacology* **85,** 214–220.

Martz, F., Failinger, C. III, and Blake, D. A. (1977). Phenytoin teratogenesis: correlation between embryopathic effect and covalent binding of putative arene oxide metabolite in gestational tissue. *J. Pharmacol. Exp. Ther.* **203,** 231–239.

Mast, T. J., Cukierski, M. A., Nau, H., and Hendrickx, A. G. (1986). Predicting the human teratogenic potential of the anticonvulsant, valproic acid, from a non-human primate model. *Toxicology* **39,** 111–119.

Masur, H., Elger, C. E., Ludolph, A. C., and Galanshi, M. (1989). Cerebellar atrophy following acute intoxication with phenytoin. *Neurology* **39,** 432–433.

Mercier-Parot, L. and Tuchmann-Duplessis, H. (1974). The dysmorphogenic potential of phenytoin: experimental observations. *Drugs* **8,** 340–353.

Middaugh, L. D., Santos, C. A., and Zemp, J. W. (1975a). Effects of phenobarbital given to pregnant mice on behavior of mature offspring. *Develop. Psychobiol.* **8,** 305–313.

Middaugh, L. D., Santos, C. A., and Zemp, J. W. (1975b). Phenobarbital during pregnancy alters operant behavior of offspring in C57BL/6J mice. *Pharmacol. Biochem. Behav.* **3,** 1137–1139.

Middaugh, L. D., Simpson, L. W., Thomas, T. N., and Zemp, J. W. (1981a). Prenatal maternal phenobarbital increases reactivity and retards habituation of mature offspring to environmental stimuli. *Psychopharmacology* **74,** 349–352.

Middaugh, L. D., Thomas, T. N., Simpson, L. W., and Zemp, J. W. (1981b). Effects of prenatal maternal injections of phenobarbital on brain neurotransmitters and behavior of young C57 mice. *Neurobehav. Toxicol. Teratol.* **3,** 271–275.

Minck, D. R., Acuff-Smith, K. D., and Vorhees, C. V. (1991). Comparison of the behavioral teratogenic potential of phenytoin, mephenytoin, ethotoin, and hydantoin in rats. *Teratology* **43,** 279–293.

Minck, D. R., Erway, L. C., and Vorhees, C. V. (1989). Preliminary findings of a reduction of otoconia in the inner ear of adult rats prenatally exposed to phenytoin. *Neurotoxicol. Teratol.* **11,** 307–311.

Mullenix, P., Tassinari, M. S., and Keith, D. A. (1983). Behavioral outcome after prenatal exposure to phenytoin in rats. *Teratology* **27,** 149–157.

Murai, N. (1966). Effect of maternal medication during pregnancy upon behavioral development of offspring. *Tohoku J. Exp. Med.* **89,** 265–272.

Nelson, K. B. and Ellenberg, J. H. (1982). Maternal seizure disorder, outcome of pregnancy, and neurologic abnormalities in the children. *Neurology* **32,** 1247–1254.

Nikolarakis, K., Kittas, C., Papacharalampous, N., and Papageorgiou, K. (1985). Effect of phenytoin on Purkinje cells in the cerebellum. [German]. *Zentralblatt Fur Allgemeine Pathol. Patholog. Anat.* **130,** 229–232.

Ohmori, H., Kobayashi, T., and Yasuda, M. (1992). Neurotoxicity of phenytoin administered to newborn mice on developing cerebellum. *Neurotoxicol. Teratol.* **14**, 159–165.

Ohmori, H., Yamashita, K., Hatta, T., Yamasaki, S., Kawamura, M., Higashi, Y., Yata, N., and Yasuda, M. (1997). Effects of low-dose phenytoin administered to newborn mice on developing cerebellum. *Neurotoxicol. Teratol.* **19,** 205–211.

Ohtani, Y., Endo, F., and Matsuda, I. (1982). Carnitine deficiency and hyperammonemia associated with valproic acid therapy. *J. Pediatr.* **101,** 782–785.

Ong, L. L., Schardein, J. L., Petrere, J. A., Sakowski, R., Jordan, H., Humphrey, R. R., Fitzgerald, J. E., and de la Iglesia, F. (1983). Teratogenesis of calcium valproate in rats. *Fund. Appl. Toxicol.* **3,** 121–126.

Ornoy, A. and Cohen, E. (1996). Outcome of children born to epileptic mothers treated with carbamazepine during pregnancy. *Arch. Dis. Child.* **75,** 517–520.

Palm, R., Hallmans, G., and Wahlstrom, G. (1986). Effects of long-term phenytoin treatment on brain weight and zinc and copper metabolism in rats. *Neurochem. Pathol.* **5,** 87–106.

Patsalos, P. N. and Wiggins, R. C. (1982). Brain maturation following administration of phenobarbital, phenytoin, and sodium valproate to developing rats or to their dams: effects on synthesis of brain myelin and other subcellular membrane proteins. *J. Neurochem.* **39,** 915–923.

Paulson, R. B., Sucheston, M. E., Hayes, T. F., and Paulson, G. W. (1985). Teratogenic effects of valproate in the CD-1 mouse fetus. *Arch. Neurol.* **42,** 980–983.

Perez del Cerro, M. and Snider, R. S. (1967). Studies on Dilantin intoxication. *Neurology* **17,** 452–466.

Petrere, J. A., Anderson, J. A., Sakowski, R., Fitzgerald, J. E., and de la Iglesia, F. A. (1986). Teratogenesis of calcium valproate in rabbits. *Teratology* **34,** 263–269.

Pick, C. G., Statter, M., Schachar, D. B., Youdim, M. B., and Yanai, J. (1987). Normal zinc and iron concentrations in mice after early exposure to phenobarbital. *Int. J. Develop. Neurosci.* **5,** 391–398.

Pick, C. G., Weizman, A., Fres, F., Gavish, M., Kanner, B. I., and Yanai, J. (1993). Hippocampal gamma-aminobutyric acid and benzodiazepine receptors after early phenobarbital exposure. *Dev. Brain Res.* **74,** 111–116.

Pick, C. G. and Yanai, J. (1984). Long-term reduction in spontaneous alternations after early exposure to phenobarbital. *Int. J. Develop. Neurosci.* **2,** 223–228.

Pick, C. G. and Yanai, J. (1985). Long term reduction in eight arm maze performance after early exposure to phenobarbital. *Int. J. Develop. Neurosci.* **3,** 223–227.

Pizzi, W. J., Alexander, T. D., and Loftus, J. T. (1996). Developmental and behavioral effects of prenatal primidone exposure in the rat. *Pharmacol. Biochem. Behav.* **55,** 481–487.

Pizzi, W. J. and Jersey, R. M. (1992). Effects of prenatal diphenlhydantoin treatment on reproductive outcome, development, and behavior in rats. *Neurotox. Teratol.* **14,** 111–117.

Pizzi, W. J., Unnerstall, J. R., and Bart, S. (1988). Behavioral teratology of anticonvulsant drugs: phenytoin and valproic acid. *Teratology* **37,** 574.

Robert, E. and Guibaud, P. (1982). Maternal valproic acid and congenital neural tube defects. *Lancet* **ii,** 937.

Rogel-Fuchs, Y., Newman, M. E., Trombka, D., Zahalka, E. A., and Yanai, J. (1992). Hippocampal cholinergic alterations and related behavioral deficits after early exposure to phenobarbital. *Brain Res. Bull.* **29,** 1–6.

Rogel-Fuchs, Y., Zahalka, E. A., and Yanai, J. (1994). Reversal of early phenobarbital-induced cholinergic and related behavioral deficits by neuronal grafting. *Brain Res. Bull.* **33,** 273–279.

Rosselli-Austin, L. and Yanai, J. (1989). Neuromorphological changes in mouse olfactory bulb after neonatal exposure to phenobarbital. *Neurotox. Teratol.* **11,** 227–230.

Schain, R. J. and Watanabe, K. (1975). Effect of chronic phenobarbital administration upon brain growth of the infant rat. *Exp. Neurol.* **47,** 509–515.

Schain, R. J. and Watanabe, K. (1976). Origin of brain growth retardation in young rats treated with phenobarbital. *Exp. Neurol.* **50,** 806–809.

Scolnik, D., Nulman, I., Rovet, J., Gladstone, D., Czuchta, D., Gardner, H. A., Gladstone, R., Ashby, P., Weksberg, R., Einarson T., and Koren, G. (1994). Neurodevelopment of children exposed *in utero* to phenytoin and carbamazepine monotherapy. *JAMA* **271,** 767–770.

Serrano, E. E., Kunis, D. M., and Ransom, B. R. (1988). Effects of chronic phenobarbital exposure on cultured mouse spinal cord neurons. *Ann. Neurol.* **24,** 429–438.

Shapiro, B. H. and Babalola, G. O. (1987). Developmental profile of serum androgens and estrous cyclicity of male and female rats exposed, perinatally, to maternally administered phenytoin. *Toxicol. Lett.* **39,** 165–175.

Shapiro, B. H., Bardales, R. M., and Lech, G. M. (1985). Perinatal induction of hepatic aminopyrine *N*-demethylase by maternal exposure to phenytoin. *Pediatr. Pharmacol.* **5,** 201–207.

Shapiro, S., Slone, D., Hartz, S. C., Rosenberg, L., Siskind, V., Monson, R. P., Mitchell, A. A., Heinonen, O. P., Idanpaan-Heikkila, J., Haro, S., and Saxen, L. (1976). Anticonvulsants and parental epilepsy in the development of birth defects. *Lancet* **1,** 272–275.

Sher, P. K., Neale, E. A., Graubard, B. I., Habig, W. H., Fitzgerald, S. C., and Nelson, P. G. (1985). Differential neurochemical effect of chronic exposure of cerebral cortical cell culture to valproic acid, diazepam, or ethosuximide. *Pediatr. Neurol.* **4,** 232–237.

Sobrian, S. K. and Nandedkar, A. K. N. (1986). Prenatal antiepileptic drug exposure alters seizure susceptibility in rats. *Pharmacol. Biochem. Behav.* **24,** 1383–1391.

Speidel, B. D. and Meadow, S. R. (1972). Maternal epilepsy and abnormalities of the fetus and newborn. *Lancet* **2,** 839–843.

Steinhausen, H.-C., Losche, G., Koch S., and Helge, H. (1994). The psychological development of children of epileptic parents. I. Study design and comparative findings. *Acta Paediatr.* **83,** 955–960.

Sullivan, F. M. and McElhatton, P. R. (1975). Teratogenic activity of the antiepileptic drugs phenobarbital, phenytoin, and primidone in mice. *Toxicol. Appl. Pharmacol.* **34,** 271–282.

Swaiman, K. F., Schrier, B. K., Neale, E. A., and Nelson, P. G. (1980). Effects of chronic phenytoin and valproic acid exposure on fetal mouse cortical cultures. *Ann. Neurol.* **8,** 230.

Tachibana, T., Terada, Y., Fukunishi, K., and Tanimura, T. (1996). Estimated magnitude of behavioral effects of phenytoin in rats and its reproducibility: a collaborative behavioral teratology study in Japan. *Physiol. Behav.* **60,** 941–952.

Takagi, S., Alleva, F. R., Seth, P. K., and Balazs, T. (1986). Delayed development of reproductive functions and alteration of dopamine receptor binding in hypothalamus of rats exposed prenatally to phenytoin and phenobarbital. *Toxicol. Lett.* **34,** 107–113.

Thurston, J. H., Hauhart, R. E., Schulz, D. W., Naccarato, E. F., Dodson, W. E., and Carroll, J. E. (1981). Chronic valproate administration produced hepatic dysfunction and may delay brain maturation in infant mice. *Neurology* **31,** 1063–1069.

Trotz, M, Wegner, C., and Nau, H. (1987). Valproic acid–induced neural tube defects: reduction by folinic acid in the mouse. *Life Sci.* **41,** 103–110.

Utterback, R. A., Ojeman, R., and Malek, J. (1958). Parenchymatous cerebellar degeneration with Dilantin intoxication. *J. Neuropathol. Exp. Neurol.* **17,** 516–517.

Van der Pol, M. C., Hadders-Algra, M., Huisjes, H. J., and Touwen, B. C. L. (1991). Antiepileptic medication in pregnancy: late effects on the children's central nervous system development. *Am. J. Obstet. Gynecol.* **164,** 121–128.

Van Dyke, D. C., Hodge, S. E., Heide, F., and Hill, L. R. (1988). Family studies in fetal phenytoin exposure. *J. Pediatr.* **113,** 301–306.

VanOverloop, D., Schnell, R. R., Harvey, E. A., and Holmes, L. B. (1992). The effects of prenatal exposure to phenytoin and other anticonvulsants on intellectual function at 4 to 8 years of age. *Neurotox. Teratol.* **14,** 329–335.

Vert, P., Deblay, M. F., and Andre, M. (1982). Follow-up study on growth and neurologic development of children born to epileptic mothers. In *Epilepsy, Pregnancy, and the Child.* D. Janz, Ed. Raven Press, New York, pp. 433–436.

Volk, B. and Kirchgassner, N. (1985). Damage of Purkinje cell axons following chronic phenytoin administration: an animal model of distal axonopathy. *Acta Neuropathol.* **67,** 67–74.

Volk, B., Kirchgassner, N., and Detmar, M. (1986). Degeneratoin of granule cells following chronic phenytoin administration: an electron microscopic investigation of the mouse cerebellum. *Exp. Neurol.* **91,** 60–70.

Vorhees, C. V. (1983). Fetal anticonvulsant syndrome in rats: dose- and period-response relationships of prenatal diphenylhydantoin, trimethadione and phenobarbital exposure on the structural and functional development of the offspring. *J. Pharmacol. Exp. Ther.* **227,** 274–287.

Vorhees, C. V. (1985). Fetal anticonvulsant syndrome in rats: effects on postnatal behavior and brain amino acid content. *Neurobehav. Toxicol. Teratol.* **7,** 471–482.

Vorhees, C. V. (1987a). Behavioral teratogenicity of valproic acid: selective effects on behavior after prenatal exposure to rats. *Psychopharmacology* **92,** 173–179.

Vorhees, C. V. (1987b). Maze learning in rats: A comparison of performance in two water mazes in progeny prenatally exposed to different doses of phenytoin. *Neurotox. Teratol.* **9,** 235–241.

Vorhees, C. V. (1987c). Fetal hydantoin syndrome in rats: dose–effect relationships of prenatal phenytoin on postnatal development and behavior. *Teratology* **35,** 287–303.

Vorhees, C. V., Acuff-Smith, K. D., Minck, D. R., and Butcher, R. E. (1992). A method for measuring locomotor behavior in rodents: contact-sensitive computer-controlled video tracking activity assessment in rats. *Neurotox. Teratol.* **14,** 43–49.

Vorhees, C. V., Acuff-Smith, K. D., Moran, M. S., and Minck, D. R. (1994). A new method for evaluating air-righting reflex ontogeny in rats using prenatal exposure to phenytoin to demonstrate delayed development. *Neurotox. Teratol.* **16,** 563–573.

Vorhees, C. V., Acuff-Smith, K. D., Schilling, M. A., and Moran, M. S. (1995). Prenatal exposure to sodium phenytoin in rats induces complex maze learning deficits comparable to those induced by exposure to phenytoin acid at half the dose. *Neurotox. Teratol.* **17,** 627–632.

Vorhees, C. V. and Minck, D. R. (1989). Long-term effects of prenatal phenytoin exposure on offspring behavior in rats. *Neurotox. Teratol.* **11,** 295–305.

Vorhees, C. V., Rauch, S. L., and Hitzemann, R. J. (1990). Prenatal phenytoin exposure decreases neuronal membrane order in rat offspring hippocampus. *Int. J. Devl. Neuroscience* **8,** 283–288.

Vorhees, C. V., Rauch, S. L., and Hitzemann, R. J. (1991a). Prenatal valproic acid exposure decreases neuronal membrane order in rat offspring hippocampus and cortex. *Neurotox. Teratol.* **13,** 471–474.

Vorhees, C. V., Weisenburger, W. P., Acuff-Smith, K. D., and Minck, D. R. (1991b). An analysis of factors influencing complex water maze learning in rats: effects of task complexity, path order and escape assistance on performance following prenatal exposure to phenytoin. *Neurotox. Teratol.* **13,** 213–222.

Wegner, C. and Nau, H. (1991). Diurnal variation of folate concentrations in mouse embryo and plasma: the protective effect of folinic acid on valproic acid–induced teratogenicity is time dependent. *Reprod. Toxicol.* **5,** 465–471.

Weisenburger, W. P., Minck, D. R., Acuff-Smith, K. D., and Vorhees, C. V. (1990). Dose–response effects of prenatal phenytoin exposure in rats: effects on early locomotion, maze learning, and memory as a function of phenytoin-induced circling behavior. *Neurotox. Teratol.* **12,** 145–152.

Wells, P. G. and Vo, H. P. N. (1989). Effects of the tumor promoter 12-O-tetradecanolylphorbol-13-acetate on phenytoin-induced embryopathy in mice. *Toxicol. Appl. Pharmacol.* **97,** 398–405.

Wells, P. G., Zubovits, J. T., Wong, S. T., Molinari, L. M., and Ali, S. (1989). Modulation of phenytoin teratogenicity and embryonic covalent binding by acetylsalicylic acid, caffeic acid, and alpha-phenyl-*N-t*-butylnitrone: implications for bioactivation by prostaglandin synthetase. *Toxicol. Appl. Pharmacol.* **97,** 192–202.

Winter, R. M., Donnai, D., Burn, J., and Tucker, S. M. (1987). Fetal valproate syndrome: is there a recognisable phenotype? *J. Med. Genet.* **24,** 692–695.

Yanai, J. and Bergman, A. (1981). Neuronal deficits after neonatal exposure to phenobarbital. *Exp. Neurol.* **73,** 199–208.

Yanai, J., Bergman, A., and Feigenbaum, J. J. (1985). Genetic factors influencing neurosensitivity to early phenobarbital administration in mice. *Acta Anat.* **115,** 40–46.

Yanai, J., Bergman, A., Shafer, R., Yedwab, G., and Tabakoff, B. (1981). Audiogenic seizures and neuronal deficits following early exposure to barbiturate. *Dev. Neurosci.* **4,** 345–350.

Yanai, J., Fares, F., Gavish, M., Greenfeld, Z., Katz, Y., Marcovici, B. G., Pick, C. G., Rogel-Fuchs, Y., and Weizman, A. (1989b). Neural and behavioral alterations after early exposure to phenobarbital. *Neurotoxicology* **10,** 543–554.

Yanai, J., Guttman, R., and Stern, E. (1989a). Genotype-treatment interaction in response of mice to early barbiturate administration. *Biol. Neonate* **56,** 109–116.

Yanai, J. and Iser, C. (1981). Stereologic study of Purkinje cells in mice after early exposure to phenobarbital. *Exp. Neurol.* **74,** 707–716.

Yanai, J., Laxer, U., Pick, C. G., and Trombka, D. (1989c). Dopaminergic denervation reverses behavioral deficits induced by prenatal exposure to phenobarbital. *Dev. Brain Res.* **48,** 255–261.

Yanai, J. and Pick, C. G. (1987). Studies on noradrenergic alterations in relation to early phenobarbital-induced behavioral changes. *Int. J. Dev. Neurosci.* **5,** 337–344.

Yanai, J. and Pick, C. G. (1988). Neuron transplantation reverses phenobarbital-induced behavioral birth defects in mice. *Int. J. Dev. Neurosci.* **6,** 409–416.

Yanai, J., Rosselli-Austin, L., and Tabakoff, B. (1979). Neuronal deficits in mice following prenatal exposure to phenobarbital. *Exp. Neurol.* **64,** 237–244.

Yanai, J. and Tabakoff, B. (1980). Altered sensitivity to ethanol following prenatal exposure to barbiturate. *Psychopharmacology* **68,** 301–303.

Yanai, J. and Waknin, S. (1985). Comparison of the effects of barbiturate and ethanol given to neonates on the cerebellar morphology. *Acta Anat.* **123,** 145–147.

Yanai, J. and Wanich, A. (1985). Prenatal versus neonatal long-term effect of phenobarbital on mouse microsomal drug-oxidizing system. *Biol. Neonate* **48,** 269–273.

Yanai, J., Woolf, M., and Feigenbaum, J. J. (1982). Autoradiographic study of phenobarbital's effect on development of the central nervous system. *Exp. Neurol.* **78,** 437–449.

Zahalka, E. H., Rehavi, M., Newman, M. E., and Yanai, J. (1995). Alterations in hippocampal hemicholinium-3 binding and related behavioural and biochemical changes after prenatal phenobarbitone exposure. *Psychopharmacology* **122,** 44–50.

Zemp, J. W. and Middaugh, L. D. (1975). Some effects of prenatal exposure to D-amphetamine sulfate and phenobarbital on developmental neurochemistry and on behavior. *Addict. Dis. Int. J.* **2,** 307–331.

Zengel, A. E., Keith, D. A., and Tassinari, M. S. (1989). Prenatal exposure to phenytoin and its effect on postnatal growth and craniofacial proportion in the rat. *J. Craniofac. Genet. Devel. Biol.* **9,** 147–160.

CHAPTER

38

Neuroteratology of Autism

PATRICIA M. RODIER
Department of Obstetrics and Gynecology
University of Rochester
Rochester, New York 14642

I. Introduction

By definition, neuroteratology could be the study of all developmental injuries to the nervous system in both animals and humans. Historically, however, the field has come to be directed at studies of animals exposed to teratogens. Those who define themselves as neuroteratologists rarely study human subjects and almost never study brain injury from genetic causes. Psychiatrists, developmental pediatricians, and clinical psychologists do study brain damage in humans, and their investigations frequently focus on genetic syndromes rather than exposure to exogenous to agents. Animal teratologists are always under pressure to reveal mechanisms, which encourages them to study and restudy known teratogens, rather than searching for new hazards. Clinicians are more commonly expected to identify hazards by recognizing new congenital syndromes or providing more accurate descriptions of old ones. Thus, over the years, neuroteratologists have diverged from their clinical colleagues not only in their methods, but in what they choose to study. Much animal work is directed at understanding the very general deficits identified only as "developmental delay" and "learning disabilities" rather than specific, often genetic, human syndromes, such as William syndrome or Down syndrome. Con-

* Abbreviations: CN, cranial nerve; CNS, central nervous system; DZ, dizygotic; GD, gestational day; MRI, magnetic resonance imaging; MZ, monozygotic; NTD, neural tube defect; PCR, polymerase chain reaction; RA, retinoic acid; RARE, retinoic acid responsive elements; VPA, valproic acid.

versely, much human work is very far removed from any consideration of mechanisms. With some exceptions, the two research directions have moved in parallel with little contact.

II. Autism as a Congenital Disorder of the Nervous System

The behavioral symptoms of autism appear to reflect the most devastating of all congenital injuries to the central nervous system (CNS), and yet, they have never been studied by neuroteratologists. A parent of a child with autism once posed the question, "Why would those who study injuries to the developing brain choose to study minor defects with subtle behavioral effects, and ignore major defects with dramatic behavioral effects?" That is a good question for which neuroteratology has no answer. Apart from the historic trends of the field, some explanation can be found in the behavioral criteria by which autism and related disorders are diagnosed. They do not invite animal experiments. The diagnostic criteria for autism are shown in Table 1. The other "autism spectrum disorders"—pervasive developmental disorder and Asperger syndrome—share many of the same features.

Only a few of the diagnostic behaviors present possibilities for animal studies, and when one tries to reduce those to explicit experiments, difficulties become apparent. For example, stereotypies appear in mental retardation, blindness, schizophrenia, and obsessive–compulsive disorder (reviewed in Ridley, 1994) as well as autism. Thus, the demonstration of an increase in stereotypies would be difficult to interpret as indicative of an autism-like condition in an animal. Deficits in social interaction are seen after many brain injuries in animals, including temporal lobe lesions (Kluver and Bucy, 1939), neonatal lead exposure (Laughlin *et al.,* 1991), prenatal methylmercury exposure (Burbacher *et al.,* 1990), and maternal deprivation (Seay and Harlow, 1965). The fact that so many various teratogenic conditions produce similar social effects in animal studies suggests that the animal tasks commonly used do not discriminate different types of social deficit very well. This might be an area for development, but it would require the creation of animal tests related to the specific social abnormalities of autism. In rodents, at least, there may be no analogs of comfort-seeking, joint attention, turn-taking, and so forth, just as there are no obvious analogs for language. Much of our most specific behavioral information on deficits in autism relates to behaviors that probably are exclusive to humans, such as language (e.g., Ferrier *et al.,* 1991; Tager-Flusberg, 1992), associative pointing (Osterling and Dawson, 1994), and imitation (e.g., Smith and Bryson, 1994). With the possible exception of the links between stereotypies and dopaminergic and serotonergic function, there are no behaviors in the diagnostic criteria of autism with a known connection to a brain region or system. For example, although we know parts of the human CNS in which injury can eliminate expressive or receptive language

TABLE 1 Diagnostic Criteria for Autistic Disorder

Autistic disorder

A. A total of six (or more) items from (1), (2), and (3), with at least two from (1) one each from (2) and (3):
   (1) qualitative impairment in social interaction, as manifested by at least two of the following:
      (a) marked impairment in the use of multiple nonverbal behaviors such as eye-to-eye gaze, facial expression, body postures, and gestures to regulate social interaction
      (b) failure to develop peer relationships appropriate to developmental level
      (c) a lack of spontaneous seeking to share enjoyment, interests, or achievements with other people (e.g., by a lack of showing, bringing, or pointing out objects of interest)
      (d) lack of social or emotional reciprocity
   (2) qualitative impairments in communication as manifested by at least one of the following:
      (a) delay in, or total lack of, spoken language (not accompanied by an attempt to compensate through alternative modes of communication such as gesture or mime)
      (b) in individuals with adequate speech, marked impairment in the ability to initiate or sustain a conversation with others
      (c) stereotyped and repetitive use of language or idiosyncratic language
      (d) lack of varied, spontaneous make-believe play or social imitative play or social imitative play appropriate to developmental level
   (3) restricted, repetitive and stereotyped patterns of behavior, interests and activities, as manifested by:
      (a) encompassing preoccupation with one or more stereotyped and restricted patterns of interest that is abnormal in either intensity or focus
      (b) apparently inflexible adherence to specific, non-functional routines or rituals
      (c) stereotyped and repetitive motor mannerisms (e.g., hand or finger flapping or twisting, or complex whole-body movements)
      (d) persistent preoccupation with parts of objects

B. Delays or abnormal functioning in at least one of the following areas, with onset prior to age 3 years: (1) social interaction (2) language as used in social communication, or (3) symbolic or imaginative play.

C. The disturbance is not better accounted for by Rett's Disorder or Childhood Disintegrative Disorder.

Diagnostic and Statistical Manual of Mental Disorders–IV (DSM-IV)—American Psychiatric Association (1994). Washington, DC: American Psychiatric Association.

altogether, we have no information on CNS structures related to the kind of language abnormalities observed in autism (e.g., Frith and Happe, 1994). Similarly, we do not know what kinds of lesions would lead to failures of gestural communication or to lack of imaginative play.

Despite the lack of guidance for animal studies provided by much of the literature on symptoms of autism, there is clear evidence that this is an organic disorder that can be induced by some teratogens. At least two well-known teratologic causes of early brain damage have been recognized as factors in some cases of autism: intrauterine rubella infection (Chess, 1971) and postnatal herpes simplex infection (Delong *et al.,* 1981). There are no data to suggest that these infections are selectively destructive for specific brain regions, but they are known to target developing tissue, thus showing effects dependent on the stage of development when the infection occurred. These cases do not suggest an area of the brain for study, but they do emphasize the developmental nature of the injury.

The fact that the behavioral symptoms of autism can be caused by a number of brain-damaging genetic defects, such as phenylketonuria, fragile X syndrome, tuberous sclerosis, and Rett syndrome (Reiss *et al.,* 1986), reinforces the idea of the organic nature of the defect. Unfortunately, the brain damage in these conditions is not selective. It is so widespread and/or variable that little information on autism is likely to come from these cases. Furthermore, most cases of these diseases do not have symptoms of autism. The association of autism with Joubert's syndrome (Joubert *et al.,* 1969; Holroyd *et al.,* 1991) may be slightly more informative. First, it is thought that all cases of this rare genetic defect have autism, and second, the cranial nerve malfunctions and cerebellar pathology of the syndrome point to an injury centered in the brainstem.

Most anatomic studies of autism have been limited by the small number of brains available for histologic study and the lack of obvious hypotheses suggested by the symptoms. Most have found no differences in histology, although it has not been always clear what areas were examined, or what would have been classified as a difference. Studies in one lab have suggested increased packing density of cells in limbic system structures (Kemper and Bauman, 1992). Only one finding has been replicated in several labs. The Purkinje cells of the cerebellum are reduced in number in patients with autism (Bauman and Kemper, 1985; Ritvo *et al.,* 1986). Some studies indicate changes in other cell types: patchy reductions in the granule cells, low cell counts in the deep cerebellar nuclei, and possible changes in the configuration of the superior olive (Bauman and Kemper, 1994). The cerebellar results have not been consistent with regard to such data as locations of greatest injury, but they are the best documented differences observed. Interestingly, the cerebellar injuries are supported by imaging studies, which suggest that the gross form of the cerebellum is deviant in many cases of autism, with the posterior lobe of the vermis small in proportion to the anterior lobe (Courchesne *et al.,* 1994; Hashimoto *et al.,* 1995). Again, the results have not been completely consistent, but they have been reported by several labs. In addition, the cerebellar effects reported remind us that Joubert syndrome is characterized by agenesis of the cerebellar vermis and other brainstem defects. Thus, the magnetic resonance imaging (MRI) cerebellar findings and histologic cerebellar findings in cases of unknown cause agree with data from a genetic syndrome that brainstem injury may be related to autism.

It is natural that neurochemical studies have been applied to autism, because a number of other congenital syndromes are already known to result from neurotransmitter disturbances. However, the failure of the vast array of psychoactive drugs to affect the core symptoms of the disease suggests that a reasonable neurochemical hypothesis would have to invoke an alteration of function much more complex than a simple elevation or depression of function of a single transmitter. In fact, one major investigator has suggested that the most reasonable direction, after many years of negative results, would be to suspend work in this area of inquiry (Rutter, 1996). This is not to say that there is no hope of finding some neurochemical deviation eventually—almost any kind of injury would alter neurotransmitters locally, if not over large regions of the CNS—but it is fair to say that this approach has not provided any data to support the idea that studies of neurotransmitters are likely to explain the etiology of autism.

III. Thalidomide Exposure and Autism

In 1994, a teratologic study with key information about the origin of autism was published. Miller and Strömland had initiated a study of the effects of thalidomide exposure on eye motility in the population of thalidomide victims in Sweden. That prenatal exposure to thalidomide had cranial nerve effects had been known since the mid-1960s (d'Avignon and Barr, 1964) and the authors of the new study hoped to use this sample with a high incidence of rare forms of strabismus to learn more about them. In particular, the well-documented time table of somatic malformations associated with thalidomide offered the chance to determine when the embryo is susceptible to brainstem injuries leading to eye motility problems (Miller, 1991). To the surprise of the investigators, first one, then another, and another of their subjects was observed to have a syndrome of global behavioral anomalies. In the end, 5 of about 100 subjects were determined to have autism (Strömland *et al.,* 1994). Compared to a rate of approximately 1/1,000 in Sweden, this is a highly significant increase in the incidence of the disorder. However, the morphologic evaluation of all cases suggested an even greater effect: All the cases with autism had a pattern of physical anomalies that indicated exposure during a very narrow window of embryonic life, between gestational days (GD) 20–24. In fact, the members of the sample with evidence of such early exposure numbered only 15, making the rate of autism during the critical period 33%.

Like children with Joubert syndrome, the thalidomide-exposed cases of autism had malfunctions of several motor cranial nerves. The neurologic abnormalities observed in the five thalidomide autistic cases included the following: three cases of Duane syndrome, a failure of cranial nerve (CN) VI (abducens) to innervate the lateral rectus muscle of the eye with subsequent reinnervation of the muscle by CN III (oculomotor) (Hotchkiss *et al.,* 1980); four cases of Moebius syndrome, a failure of the CN VII (facial) to innervate the facial muscles, often associated with other cranial nerve symptoms (May, 1986); and two cases of abnormal lacrimation, a failure of the neurons of the superior salivatory nucleus (CN VII) to innervate the lacrimal apparatus with subsequent misinnervation of the structure by neurons that normally supply the submandibular glands (Ramsey and Taylor, 1980). One patient had gaze paresis, suggesting malfunction of CN III (oculomotor).

Each patient had hearing deficits and ear malformations. The auditory symptoms do not necessarily indicate an injury to CN VIII, which carries auditory and vestibular information, because there is no way to exclude the possibility of damage located at more central stations along the sensory pathways. In addition, thalidomide-induced malformations of the external ear are known to be associated with malformations of the middle and inner ear (d'Avignon and Barr, 1964). Thus, peripheral injury is a possibility in the thalidomide-induced auditory deficits.

Are the neurologic and physical features of the thalidomide cases representative of many cases of autism, or are these patients atypical? The literature on autistic cases of unknown etiology suggests that the features of the thalidomide cases are not rare. In a study of eye motility in autism, 7 of 34 cases had strabismus, and 31 of 34 cases had abnormal optokinetic nystagmus (Scharre and Creedon, 1992). In a review of studies of brainstem auditory responses in patients with autism, although hearing deficits were an exclusionary criterion in most studies, nonetheless, 35–50% of the cases tested had evidence of peripheral hearing deficits (Klin, 1993). Furthermore, victims of Moebius syndrome (congenital diplegia of the facial muscles and lateral rectus) have a high rate of autism—about 30% (Gillberg and Steffenberg, 1989; Strömland and Miller, 1997). Minor external terata are seen in a significant proportion of children with autism, the most common and most discriminating between autism and mental retardation being low-set, malformed, and, especially, posteriorly rotated ears (e.g., Walker, 1977; Rodier *et al.,* 1997). Thus, there is nothing new about the idea that people with autism have neurologic defects and malformations indicative of very early injury to the brainstem. What is new is the idea that these associations may be the critical in understanding the origin of the disease.

Obviously, the importance of the thalidomide study does not lie in its ability to tell us what causes autism. Very few cases could have been exposed to thalidomide, which was removed from the market long ago. Rather, this study establishes the principle that exogenous insults can cause autism, and that the critical period for insult is in the third and fourth week of intrauterine life. Furthermore, because almost no neurons have begun to form at this time, the critical period predicts that autism does not arise from direct injury to the forebrain or even to the cerebellum. These form too late to have been exposed to the teratogen (Bayer *et al.,* 1993). Instead, the initiating injury must be confined to the brainstem tegmentum. Secondary effects, of course, might follow from the initial injury, and malfunctions of areas undamaged by the initial insult might occur because of distortion of normal input from the damaged hindbrain.

Days 20–24 of gestation fall at the stage when the human neural tube is closing (Streeter, 1948). This is the period when the first neurons appear in all vertebrates, and in each species most are destined to form motor nuclei of cranial nerves (Taber-Pierce, 1973; Bayer *et al.,* 1993). Thus, the cranial nerve motor symptoms and the time of exposure in the thalidomide cases are consonant with the hypothesis that thalidomide interfered with neuron production for the cranial nerve motor nuclei. Whether it had this effect by a cytotoxic action or by disrupting the pattern formation of the rhombomeres from which the neurons arise is not known. We know nothing about the neuroanatomy of the thalidomide cases except what we can guess from their symptoms and the critical period. Indeed, the teratology literature contains not even one study of the effects of thalidomide on the CNS in humans or animals. Furthermore, because previous studies of neuroanatomy in autism typically have focused on the forebrain, there was no evidence at the time of the thalidomide study to answer the question of whether other people with autism have alterations of the cranial nerve nuclei.

IV. Studies Confirming the Results of the Thalidomide Study

Armed with the neuroanatomic predictions of the teratologic cases, it became possible to revisit the anatomic basis of autism with specific hypotheses. The incomplete brainstem of a well-documented adult with autism was examined and compared to control tissue with attention to the cranial nerve motor neurons (Rodier *et al.,* 1996). Grossly, the tissue appeared normal, but study of serial histologic sections indicated several anatomic abnormalities. The most obvious was the near absence of the facial nucleus. Counts of the number of facial neurons on one side of the brain were reduced by 95%. A small cluster of normal-looking neurons appeared in the most anterior part of the nucleus, but in more caudal sections there were no motor neurons at all. In addition, the fiber patterns characteristic of the nucleus were altered. That is, nuclear groups in the CNS are distinguished not only by a cluster of neurons, but also by the way that fiber pathways added later in development travel around nuclei, rather than through them. The way that axons "respect" the boundaries of a nucleus leaves nuclei relatively free of fibers. In contrast, there is often a concentration of fibers of passage on the borders of a nucleus. Histologically, the pattern of axons can be just as definitive as the pattern of cell bodies for recognizing a nucleus. If facial neurons in the autopsy tissue had died after the axons around them were in place, the outline of the nucleus would have remained visible, because fiber pathways do not move once established. Instead, the region that should have contained the facial nucleus was filled with fibers and showed no borders. This suggested that the neurons of the facial nucleus had never formed, or had been lost very soon after their production period. The superior olive, an auditory relay nucleus, was missing. The region where it should have appeared lacked both the fiber patterns and the neurons by which the superior olive is identified. The hypoglossal nucleus was present, but retained some structure characteristic of the embryo, and some of its motor cells appeared to have failed to differentiate.

Reconstruction of the area surrounding the pontomedullary junction revealed another striking abnormality: the distance from the trapezoid body to the inferior olive was decreased by several millimeters in the brain from a person with autism, with all the more caudal structures in normal relation to each other, but shifted orally. The missing nuclei had not been lost from the space they normally occupy. Rather, the space itself was missing. Both the superior olive and most of the facial nucleus are thought to arise from the fifth rhombomere (Chisaka *et al.,* 1992). Thus, it seems likely that this brain sustained an injury focused on that embryologic segment of the neural tube. This would explain both the absent nuclei and the shortening of the brainstem. The shortening is especially interesting because it provides the strongest possible evidence that this brain was injured during neural tube formation. It is impossible to imagine an injury later in development that would allow the later-forming tracts to pass through the missing region. The seamless union of the structures of the fourth and sixth rhombomeres indicates that they developed from the outset in the same position they hold in the mature brain.

One thing that makes the thalidomide study results surprising is that the critical period for autism induction is so early in gestation. There has long been a prevailing opinion in teratology that insults so early either kill the embryo or allow complete recovery. Thus, it would be useful to confirm experimentally in animals that teratogens can alter the cranial nerve nuclei as the neural tube closes, but leave the conceptus viable. Because thalidomide does not cause the same terata in rodents as in humans, this experiment had to be carried out with another teratogen. Valproic acid (VPA) was selected because the malformations it produces are similar to those of thalidomide (Binkerd *et al.,* 1988; Collins *et al.,* 1991). While the animal studies were underway, the first report of autism after human valproate exposure was published (Christianson *et al.,* 1994). Animals exposed to a teratogen during the critical period exhibit reductions in the number of cranial nerve motor neurons (Rodier *et al.,* 1996). They are distinguished from controls by shortening of the hindbrain in the region that forms from the fifth rhombomere, just as the autopsy brain was. Finally, preliminary data demonstrate that the animal model has secondary changes in the cerebellum like those reported in human cases of autism (Ingram *et al.,* 1996).

V. Genetics of Autism

Despite conclusive evidence that autism can be caused by teratogens, the majority of cases are thought to be genetic in origin. There is nothing contradictory in this statement, because a pattern of multiple etiologies is known to account for other disorders. For example, heart defects can be produced by many teratogens in the laboratory but have a major genetic component in humans (e.g., Boughman *et al.,* 1993). One mechanism by which such patterns arise has been demonstrated by studying neural tube defects (NTDs) in mice with mutations of the *Pax3* gene (Dempsey and Trasler, 1983). Mice with two mutant alleles have spontaneous open neural tubes, whereas heterozygotes and wild-type homozygotes rarely exhibit NTDs spontaneously. When a teratogenic exposure (in this case, retinoic acid) was imposed on these genetic backgrounds, there was an interaction between the external insult and the inherent developmental pattern decreed by the level of *Pax3* function. Thus, wild-type homozygotes exhibited only a 15% incidence of NTDs after retinoic acid exposure, but the same treatment led to a 50% rate of NTDs in heterozygotes.

In the case of autism, no specific mutations have been shown to be strongly associated with the disorder. Rather, it is family studies that convince us that autism is a genetic syndrome. It is worthwhile to consider these in some detail, for they are excellent models for the kinds of studies that might be useful in studying other congenital brain damage syndromes.

Investigators suspected that autism ran in families, but because individuals with the diagnosis rarely, if ever, have childen, it was impossible to follow the disorder through several generations of a large family. The rate in siblings, however, is clearly elevated. About 3–8% of siblings of cases have the same disorder (Jones and Szatmari, 1988; Smalley *et al.,* 1988; Ritvo *et al.,* 1989). If the incidence in the North American population is 1:750 (Bryson *et al.,* 1988), then the incidence in siblings is about 20–60 times as high as would be expected in the general population.

Twin studies are a traditional way to estimate the magnitude of the genetic contribution to a disease. If a condition is caused solely by genetic abnormality, then the concordance rate for the condition in monozygotic (MZ) twins should approach 100%. However, MZ twins share the same intrauterine environment as well as identical genetic backgrounds. Thus, we cannot say that even perfect concordance proves a genetic cause for a condition. One solution to this problem is to subtract the concordance rate observed in dizygotic (DZ) twins from that observed in MZ twins as a correction for environmental factors. The correction is not completely accurate, for DZ twins are genetically more similar than unrelated individuals, and teratologists know that even twins do not share identical environmental influences (for example, twins can be at slightly different developmental stages, so that an exposure affects one but not the other; twins are sometimes discordant for such other influences as infections). However, the difference between MZ and DZ concordance rates gives a conservative estimate of the role of genetic influences in a disorder. For autism, the earliest studies suggested that this difference was substantial; it equaled more than 30% (Folstein and Rutter, 1977). With larger and larger samples, the estimate has been altered to about 50% (Bailey, 1995), but the concordance rates for the full diagnosis do not tell the whole story.

The series of twin studies carried out in England have been distinguished by their attention to the fact that family members sometimes show partial expression of genetic disorders, rather than the complete array of symptoms. By screening twins for symptoms such as language delay, Medical Research Council (MRC) investigators in the U.K. have been able to show that many more cases of autism are genetic than can be predicted by concordance for diagnosis. The difference between MZ and DZ twins becomes about 80% when a broader definition of the disorder is used (Folstein and Rutter, 1977; Bailey *et al.,* 1995; Le Couteur *et al.,* 1996). Related studies have reported that parents, too, have some symptoms. Landa and colleagues (1991, 1992) have shown that pragmatic language oddities can be found in 30–40% of parents of children with autism, and that these rates are significantly higher than those in parents of children with other disabilities.

Investigators now suspect that almost all cases of autism are genetic in etiology. However, given the wide range of expression of symptoms in closely related individuals, with MZ twins varying from having identical symptoms to having one twin with a full diagnosis and the other twin with no detectable symptoms, understanding the penetrance of the genetic factors is a difficult task. To what extent is the variability of expression dependent on the genetic background on which the putative genetic factors are expressed, and to what extent do early exogenous insults play a role? One summary views the twin data as favoring an important role for the environment (Le Couteur *et al.,* 1996), on the basis that variation of behavioral symptoms of autism within twin pairs is as great as the variation between twin pairs. Thus, even after major genetic factors are identified, autism is likely to require extensive study by neuroteratologists.

In contrast to the productivity of the family studies that tell us autism has a genetic basis, attempts to determine what genes are involved have typically been negative. Linkage studies to determine the chromosomal location of genetic abnormalities in autism are difficult because of the infrequent availability of large families with many cases. Screening studies looking for linkage with restriction length polymorphisms at as many as 90 sites have found no strongly positive results (Spence *et al.,* 1985). Studies of markers for several candidate genes related to neurotransmitter function (Martineau *et al.,* 1994) and general growth factors (Herault *et al.,* 1994) have been negative. Inconsistent results were reported for the *EN2* homeogene (*EN2* is analogous to *invected* in *Drosophila* and is known to be required for normal cerebellar development in the mouse) on chromosome

2 (Petit *et al.,* 1995). A positive but minor association with autism has been documented for markers of the *c-Harvey-ras* gene (an oncogene thought to have a general role in development) on chromosome 11 (Herault *et al.,* 1995).

A positive relationship to autism has been reported for the short form polymorphism of the serotonin transporter gene enhancer (Cook *et al.,* 1997) using a transmission–disequilibrium test. The results are interesting but suggest that any contribution of this gene is small. That is, the significant effect is based on a switch from the 50–50 transmission rate expected if there is no linkage of a locus to a diagnosis to a transmission rate of 60–40. Furthermore, there has been no confirmation that the children positive for the short form of the enhancer have any change in serotonin levels. Previous studies of the same polymorphism in lymphoblasts indicate that cells with the short form polymorphism have reductions in expression and uptake of serotonin (Heils *et al.,* 1996; Lesch *et al.,* 1996), whereas studies of blood and urine levels of serotonin in cases of autism have suggested that some severely retarded patients with the disorder (and some with mental retardation alone) have elevated levels of serotonin (e.g., Schain and Freedman, 1961; Piven *et al.,* 1991). Brain levels of the transmitter have not been shown to differ between people with autism and controls (Ross *et al.,* 1985). Thus, it is difficult to know the meaning of the genetic finding.

In general, it is difficult to choose an efficient approach for the search for genes involved in autism without a biologically plausible hypothesis about the neurobiology of the disorder. The more we know about the behavioral and anatomic human phenotype, the more likely it is that such hypotheses can be developed.

VI. Embryology of Autism

Although it has been difficult to construct hypotheses about the neurobiology of autism based on its symptoms, it may be easier to construct hypotheses based on its embryology. To review from Sections III and IV: Autism must be initiated by a disturbance of development around the time of neural tube closure. This is the critical period deduced from the ear anomalies of the thalidomide cases. The cranial nerve malfunctions of those cases support the same time of injury, because this is the period when the motor cranial nerve nuclei are forming. Human autopsy material from a case of autism of unknown cause shows injury to the brainstem tegmentum, including shortening of the brainstem, that suggests the same time of injury. The loss of the facial nucleus in the histologic case is consonant with the thalidomide cases' malfunctions of the same cranial nerve. Experimental animal studies show that shortening and deficits of neurons in the cranial nerve nuclei can be induced by exposure to VPA around the time of neural tube closure. Preliminary data suggest that such injuries have secondary effects on the cerebellum, the structure most widely acknowledged to be abnormal in cases with autism. The wider literature supports the contention that the same neurologic malfunctions—such as facial diplegia, eye motility abnormalities, and hearing deficits—observed in teratologic cases are seen in many cases of autism. The literature on minor malformations in autism agrees that the initiating injury occurs as the neural tube is closing.

In fact, the teratologic origin of autism hardly seems a matter of hypothesis. It is hard to imagine a set of converging findings coming from more areas of investigation than those reviewed here, and all agree that some cases of autism arise from injuries to the closing neural tube. The challenge is whether this hypothesis can be extended to include cases in which the insult is endogenous, rather than exogenous, in origin.

To incorporate the genetic data, we need to consider whether there are genes active at the same stage of embryogenesis that have control over the phenotype that might be related to the injuries described in the teratologic cases. Fortunately, the genes active during neural tube closure have been among the best studied of all developmental genes. They are the genes active during formation and growth of the rhombomeres—the segments of the developing neural tube. They include the *Hox* genes, which are thought to be critical to the specification of the axial identity of cells of the embryo. Many *Hox* genes have important roles in the development of the rhombomeres and their products.

Suppose that mutations of the *Hox* genes can occur spontaneously, or be inherited, and that they lead to maldevelopment of the brainstem. In fact, there is a wealth of information from knockout mice that tells us what happens to the embryo when the normal functions of these genes are lacking. Suppose that the expression or function of some of the genes controlling development of the hindbrain can also be influenced by teratogens, such as thalidomide and VPA. If that is the case, then it would be perfectly reasonable that genetic and teratologic cases of autism might exhibit the same anatomic and neurologic defects. In fact, it is well-known that several of the *Hox* genes are sensitive to teratogens. Thalidomide and VPA have not been tested, but retinoic acid (RA) has been the subject of numerous investigations (e.g., Marshall *et al.,* 1992). RA alters the function of *Hox* genes by way of retinoic acid responsive elements (RAREs) that are part of the genes themselves. For example, *Hoxa1* and *Hoxb2* have RAREs (Conlon and Rossant, 1992; Langston and Gudas, 1992;

Langston *et al.,* 1997), and it is disturbance of these genes via their RAREs that is thought to be the cause of the brain and craniofacial symptoms observed after human and animal exposure to RA (e.g., Means and Gudas, 1995). If disturbance of any of these genes by teratogens or genetic engineering yields a phenotype like that of autism, then we can have a unifying hypothesis that might cover all cases of autism.

Let us consider first the embryologic changes induced by retinoids in humans. Adams and Lammer (1991) have reported a wide range of phenotypic effects, from normal appearance and mental function to major and minor malformations accompanied by severe retardation. Not all the cases were exposed early enough to put the *Hox* genes controlling the rhombomeres at risk, but, even so, the most common malformations among isotretinoin-exposed cases are ear malformations and facial asymmetries. There are neurologic anomalies, too: Moebius syndrome and eye motility problems. MRIs of the cases reveal cerebellar anomalies, including agenesis of the cerebellar vermis. There is no question that cases with severe retardation exhibit the highest rates of malformation, but neurologic symptoms and craniofacial anomalies are present in a few cases with IQs in the normal range. One of these has been diagnosed with Asperger syndrome (Adams, 1996, personal communication).

From animal studies it is known that the immediate effect of RA on the rhombomeres is to alter their identity (e.g., Marshall *et al.,* 1992). One result is that anterior segments take on the characteristics of those posterior to them, producing a shortening of the brainstem at the level of the anterior rhombomeres. This shortening lies anterior to that observed in the histology case and VPA model described in Section IV. Overall, the effects of retinoids on form and function are strikingly similar to those seen in the thalidomide cases and to the neuroanatomy described in human histologic and MRI studies of autism.

It was long believed that no mutations would ever be found in any of the *Hox* genes in any mammal. Because their role is so critical to embryonic development, it was thought that any functional alteration was likely to be fatal. This is both good and bad for the unifying hypothesis. On the surface it sounds ridiculous to propose that mutations do occur in genes in which none has ever been observed, despite intensive study. However, a gene in which variation is uncommon might be a good candidate for one in which abnormality could have strong phenotypic effects.

Reviewing the outcome from a number of different knockouts of early developmental genes, it appears that several result in characteristics that might relate them to autism. The most impressive in its correspondence to the phenotype of autism is the knockout of *Hoxa1*. The knockout studies indicate that heterozygotes for nonfunctional *Hoxa1* are similar to controls, but that homozygotes have many phenotypic effects, as described later. Homozygotes die within a few days after birth, presumably because their brainstem injuries interfere with eating and breathing.

Two slightly different knockout strains have been created (Lufkin *et al.,* 1991; Chisaka *et al.,* 1992). For simplicity, the effects of both are described without regard to whether they have been observed in both models or only one. The brainstem lacking *Hoxa1* function exhibits loss of the fifth rhombomere and its products along with parts of the fourth and sixth rhombomeres. The missing nuclei are the superior olive, the abducens, and all but the most anterior part of the facial nucleus. More caudal structures are shifted orally, so that the structures of the fourth and sixth rhombomeres are adjacent, and the brainstem is shortened. This shortening is at the same level where shortening occurred in the VPA model and in the human histologic case. *Hoxa1* knockouts have malformations of the inner, middle, and external ears. At every point for which there are data available, these mice match both the thalidomide cases and the histologic results (Carpenter *et al.,* 1993; Mark *et al.,* 1993). To strengthen the connection further, *Hoxa1* is expressed in the CNS as the neural tube is closing, and at no other time in life (Murphy and Hill, 1991). That is, its time of expression is an excellent match for the period when thalidomide induces autism. The developmental biology of *Hoxa1* suggests that mutations of this gene are ideal candidates for contributing to the phenotype of autism.

VII. Testing *HoxA1* as a Candidate Gene in Autism

Since 1995, the causes of three rare birth defects have been identified by groups who recognized a phenotype similar to that of a human disorder in mutant mice. They then proceeded to evaluate the candidate gene for mutations in large human kindreds with multiple cases of the defect. There is no single or simple method for asking whether a gene is normal, and these studies used different methods, but once a genetic abnormality is identified, it can be expected that other family members with the same congenital defect must share the same abnormality in the same gene. (Other families might have a different mutation leading to a similar anomaly.) Thus, once a mutation was documented, it was possible to test other affected and unaffected individuals for the same change in nucleic acid sequence to determine whether the mutation was the cause of the disorder in question. Muragaki and colleagues (1996) showed that affected family members from three kindreds with synpolydactyly shared abnormal expansions of a polyalanine repeat sequence in the gene *HoxD13.* The clue to the correct locus came from mice engineered to lack the function of *Hoxd13* (Dolle *et al.,* 1993). The animals had limb defects similar, but far from identical, to human synpolydactyly. Hand-foot-genital syndrome was shown to arise from a single base substitution in *HoxA13* (Mortlock and Innis, 1997). A naturally occurring mouse with a deletion in *Hox13* had previously been shown to have related malformations (Mortlock and Innis, 1997). Human holoprosencephaly has been related to an abnormal sequence in *Sonic hedgehog* (Roessler *et al.,* 1996), one of the classic early developmental genes long studied in mice and examined in knockouts (Chiang *et al.,* 1996). The *Sonic hedgehog* studies demonstrated that the gene is linked to midline defects of the brain and face, even though individuals with the mutation varied from expressing no detectable effect, to minor changes in dentition, to full holoprosencephaly. This is encouraging for studies of disorders like autism, in which family studies (Section V) have demonstrated that penetrance is variable. Another interesting feature of the *Sonic hedgehog* studies is that mice exhibited holoprosencephaly only when they were homozygous for the null mutation of the gene, but the affected human cases were heterozygotes. This suggests that humans may be more sensitive than mice to decreases in *Sonic hedgehog* function and, perhaps, the function of other early developmental genes as well.

The human sequence of the two exons of *HoxA1* has been published (Hong *et al.,* 1995). Thus, it was possible to develop polymerase chain reaction (PCR) primers to amplify the coding regions from genomic DNA extracted from human blood samples and to proceed to study those regions for deviations from the normal sequence (Ingram *et al.,* 1997). Instead of the optimal three- or four-generation families with multiple cases of autism, a small set of families with some evidence of a genetic etiology was collected. Each proband had autism, Asperger syndrome, or pervasive developmental disorder (often called the *autism spectrum disorders*). The second affected individual in the family had one of these or severe language delay, a characteristic clearly related to autism in twin studies (e.g., Le Couteur *et al.,* 1996). Furthermore, although many studies have concentrated on sibling pairs, these authors were convinced (because of the risk of similar teratogenic exposures to siblings) that affected parents, uncles, aunts, and cousins were actually better evidence of "familial" factors than were siblings. Thus, probands with affected family members from all these categories were included.

Two families whose external malformations suggested different etiologies were used for the initial sequencing of the exons. Affected individuals in one family have high IQs, simple ears, and brachydactyly type 4 of the thumbs. The second family has one son with autism and mental retardation and one son with Asperger syndrome and a normal IQ. Both children have ear malformations, including lobes attached to the skin of the neck, but not fully fused with the rest of the auricle. The parents have no obvious symptoms or malformations.

After reading and rereading of many nucleic acid sequences of the second exon from the two families, no deviation was seen in the sequence of any subject. However, the first exon sequence from some individuals in the second family was suspect. The pattern suggested that these individuals might be heterozygotes. Because no deviation in sequence had ever been observed in *Hoxa1* in any mammal, such an abnormality was of great interest, regardless of whether it could be related to autism.

Cloning the DNA of the family members separated the maternal and paternal alleles of each case and revealed that the brothers were heterozygous for the same single base substitution in the first exon of *Hoxa1.* One parent had the same anomaly of the gene, whereas the other had two normal alleles. The substitution was one that would alter the amino acid code for the protein product of the gene, changing a histidine to an arginine, and thus was likely to be functionally important.

Because the brothers had not only autism but ear anomalies, it seemed highly likely that their mutation was responsible for their symptoms, but the first step in proving that was to determine the rate at which the newly discovered polymorphism occurs in controls. Fortunately, the sequence anomaly coincided with a restriction site, which allowed testing of PCR products by digestion with a restriction enzyme. Thus, the expensive and time-consuming step of cloning was not needed to test for the anomaly. It was anticipated that the rate of occurrence might be very low, but this was not the case. In about 100 controls, 18 exhibited the same polymorphism. This result necessitated the collection of a larger set of familial cases of autism and related disorders to allow a test of whether the abnormal sequence occurs more frequently in subjects with these disorders than in controls. A preliminary test of association of the mutant allele with autism was highly significant. That is, in a sample of about 20 familial cases of autism spectrum disorders, the rate of the mutation approached 50%.

It is important to note that other mutations related to autism may exist in *HoxA1.* That is, the coding sequence was studied in only two families—one produced the polymorphism just described, and the other lacks this polymorphism. Either or both of these families could have abnormalities in the untranslated, untranscribed regions of the sequence of *HoxA1,* and that is also true for the larger set of probands examined for the first polymorphism. Alternatively, other loci may be involved. It has long been hypothesized that multiple genes are linked to autism (e.g., Pickles *et al.,* 1995), because the family data do not fit any single allele model (Jorde *et al.,* 1991). However, it is not clear how well models would fit if a broad diagnosis were employed in ascertaining affected cases. Given the low penetrance of the genetic factors in the disorder as deduced from family studies, genetic research using a broader definition is needed to settle this issue. In any case, the field now seems to be at a point where the genes linked to the disorder are likely to be identified quickly. Neuroteratologists who can determine how environmental factors interact with those genes are needed to solve the problems of the etiology of autism.

VIII. Neuroteratology and Developmental Disabilities

The intent of this chapter has been to review the area where neuroteratology intersects with the study of autism. Thus, it does not provide a sense of all the research in the field. Much of that work concentrates on descriptions of the human behaviors that distinguish people with autism. Readers interested in a broader view of autism research might begin by reading the report of the NIH State of the Science Conference, held in 1995 (*Journal of Autism and Developmental Disorders,* volume 26). It is an embarrassment that no neuroteratologists were speakers at this important event, but the reason is obvious: no one in the field had ever published a paper on autism.

Anyone who identifies him- or herself as a neuroteratologist would agree that developmental disabilities are among the greatest health problems facing our population. The specialties of the field should give its practitioners excellent qualifications for unraveling the mysteries of congenital disorders. A developmental perspective is important, and many of those who have worked on human developmental disabilities have training in postnatal development but not prenatal development. This is an area in which neuroteratologists can make a great contribution, but only if they keep themselves current on what is known about the structural and functional development of the nervous system, an area of explosive growth in recent years.

A frontal assault on develomental disabilities may require working collaboratively with other specialists—clinical psychologists, developmental biologists, epidemiologists, developmental pediatricians, neurologists, psychiatrists, and geneticists, to name only some of the experts who have interests complementary to those of neuroteratology. In such collaborations, it is important to understand that techniques can be learned quickly. There is no reason that a behaviorist cannot learn to perform other measures, whether they are biochemical or molecular or anatomic, with the support of someone who is expert. However, years of experience in reading the literature of a particular specialty are harder to share. To be productive, we need to develop relationships that allow us to see problems from other points of view. We need expert help to test our ideas, not just to advise us on techniques. In fact, the greatest advantage of reaching out to other fields is the intellectual excitement that results. What we already know may turn out to be a revelation to someone else. Communicating with scientists who know something different can give us ideas we never dreamed of.

We are making progress on autism, but what about other human syndromes? Neuroteratologists should be in the vanguard of those studying Down syndrome, or bipolar disease, or obsessive–compulsive disorder. We can do more, if we choose to broaden the range of problems we are willing to address.

References

Adams, J. (1996). Personal communication.

Adams, J. and Lammer, E. J. (1991). Relationship between dysmorphology and neuropsychological function in children exposed to isotretinoin "*in utero.*" In *Functional Neuroteratology of Short-Term Exposure to Drugs* (T. Fujii and G. J. Boer, Eds.), Teikyo University Press, Tokyo.

American Psychiatric Association. (1994). Diagnostic Criteria for Autistic Disorder. *Diagnostic and Statistical Manual of Mental Disorders–IV (DSM–IV).* Washington, DC, APA. pp. 70–71.

Bailey, A. LeCouteur, A. Gottesman, I. Bolton, P. Simonoff, E. Yuzda, E., and Rutter, M. (1995). Autism as a strongly genetic disorder: Evidence from a British twin study. *Psychol. Med.* **25,** 63–78.

Bauman, M. L., and Kemper, T. L. (1985). Histoanatomic observations of the brain in early infantile autism. *Neurology* **35,** 866–874.

Bauman, M. L., and Kemper, T. L. (1994). Neuroanatomic observations in autism. In *The Neurobiology of Autism.* M. L. Bauman and T. L. Kemper Eds. Johns Hopkins University Press, Baltimore, pp. 119–145.

Bayer, S. A., Altman, J., Russo, R. J., and Zhang, X. (1993). Timetables of neurogenesis in the human brain based on experimentally determined patterns in the rat. *Neurotoxicology* **14,** 83–144.

Binkerd, P. E., Rowland, J. M., Nau, H., and Hendrickx, A. G. (1988). Evaluation of valproic acid (VPA) developmental toxicity and pharmacokinetics in Sprague-Dawley rats. *Fund. Appl. Tox.* **11,** 485–493.

Boughman, J. A., Neill, C. A., Ferencz, C., and Loffredo, C. A. (1993). The genetics of congenital heart disease. In: *Perspectives in Pediatric Cardiology,* Vol. 4. *Epidemiology of Congenital Heart Disease: The Baltimore–Washington Infant Study 1981–1989.* Futura Publishing, Mount Kisco, NY, pp. 123–167.

Bryson, S. E., Clark, B. S., and Smith, I. M. (1988). First report of a Canadian epidemiological study of autistic syndromes. *J. Child Psychol. Psychiat.* **29,** 433–45.

Burbacher, T. M., Sackett, G. P., and Mottet, N. K. (1990). Methylmercury effects on the social behavior of *Maccaca fascicularis* infants. *Neurotoxicol. Teratol.* **12,** 65–71.

Carpenter, E. M. Goddard, J. M. Chisaka, O. and Manley, N. R. (1993). Loss of *HoxA1* (*Hox-1.6*) function results in the reorganization of the murine hindbrain. *Development* **118,** 1063–1075.

Chess, S. (1971). Autism in children with congenital rubella. *J. Autism Child. Schiz.* **1,** 33–47.

Chiang, C. et al. (1996). Cyclopia and defective axial patterning in mice lacking *Sonic hedgehog* gene function. *Nature* **383,** 407–413.

Chisaka, O., Musci, T. S., and Capecchi, M. R. (1992). Developmental defects of the ear, cranial nerves and hindbrain resulting from targeted disruption of the mouse homeobox gene *Hox-1.6. Nature* **355,** 516–520.

Christianson, A. L., Chesler, N., and Kromberg, J. G. R. (1994). Fetal valproate syndrome: clinical and neurodevelopmental features in two sibling pairs. *Devel. Med. Child Neurol.* **36,** 357–369.

Collins, M. D., Walling, K. M., Resnick, E., and Scott, W. J. (1991). The effect of administration time on malformations induced by three anticonvulsant agents in C57BL/6J mice with emphasis on forelimb ectrodactyly. *Teratology* **44,** 617–627.

Conlon, R. A., and Rossant, J. (1992). Exogenous retinoic acid rapidly induces anterior ectopic expression of murine *Hox*-2 genes *in vivo. Development*, **116,** 357–368.

Cook, E. H. Jr., Courchesne, R. Lord, C., Cox, N. J., Yan, S., Lincoln, A., Haas, R., Courchesne, E., and Leventhal, B. L. (1997). Evidence of linkage between the serotonin transporter, and autistic siorder. *Molec. Psychiat.* **2,** 247–250.

Courchesne, E., Townsend, J., and Saitch, O. (1994). The brain in infantile autism: posterior fossa structures are abnormal. *Neurology,* **44,** 214–228.

d'Avignon, M., and Barr, B. (1964). Ear abnormalities and cranial nerve palsies in thalidomide children. *Arch. Otolaryngol.* **80,** 136–140.

DeLong, G. R., Beau, S. C., and Brown, F. R. (1981). Acquired reversible autistic syndrome in acute encephalopathic illness in children. *Arch. Neurol.* **38,** 191–194.

Dempsey, E. E., and Trasler, D. G. (1983). Early morphological abnormalities in splotch mouse embryos and predisposition to gene- and retinoic acid-induced neural tube defects. *Teratology* **28,** 461–472.

Dolle, P., Diedrich, A., LeMeur, M., Schimmang, T., Schuhbaur, B., Chambon, P., and Duboule, D. (1993). Disruption of the *Hoxd-13* gene induces localized heterochrony leading to mice with neotenic limbs. *Cell,* **75,** 431–441.Ferrier, L. J., Bashir, A. S., Meryash, D. L., Johnston, J., and Wolff, P. (1991). Conversational skills of individuals with fragile-X syndrome: a comparison with autism and Down syndrome. *Dev. Med. Child Neurol.* **33,** 776–788.

Folstein, S. E., and Rutter, M. L. (1977). Infantile autism: a genetic study of 21 twin pairs. *J. Child Psycol. Psychiat.* **18,** 297–321.

Frith, U., and Happe, F. (1994). Language and communication in autistic disorders. *Phil. Trans. Roy. Soc.* (London) Series B **346,** 97–104.

Gillberg, C., and Steffenberg, S. (1989). Autistic behavior in Moebius syndrome. *Acta Pediatr. Scand.* **78,** 314–316.

Hashimoto, T., Tayama, M., Murakawa, K., Yoshimoto, T., Miyazaki, M., Harada, M., and Kuroda, Y. (1995). Development of the brainstem and cerebellum in autistic patients. *J. Aut. Dev. Disord.* **25,** 1–18.

Heils, A., Teufel, A., Petri, S., Stoeber, G., Riederer, P., Bengel, D., Lesch, K.-P. (1996). Allelic variation of human serotonin transporter gene expression. *J. Neurochem.* **66,** 2621–2624.

Herault, J. et al. (1994). Lack of association between three genetic markers of brain growth factors and infantile autism. *Bio. Psychiat.* **35,** 281–283.

Herault, J., Petit, E., Matineau, J., Perrot, A., Lenoir, P., Cherpi, C., Bartheleme, C., Sauvage, D., Mallet, J., Muh, J. P., and Lelord, G. (1995). Autism and genetics: clinical approach and association study with two markers of HRAS gene. *Am. J. Med Genet.* **60,** 276–281.

Holroyd, S., Reiss, A. L., and Bryan, R. N. (1991). Autistic features in Joubert's syndrome: a genetic disorder with agenesis of the cerebellar vermis. *Biol. Psychiat.* **29,** 287–294.

Hong, Y. S., et al. (1995). Stucture and function of the *HOX A1* human homeobox gene cDNA. *Gene* **159,** 209–214.

Hotchkiss, M. G., Miller, N. R., Clark, A. W., and Green, W. R. (1980). Bilateral Duane's retraction syndrome. *Arch. Ophthalmol.* **98,** 870–874.

Ingram, J. L., Croog, V. J., Tisdale, B., Rodier, P. M. (1996). Valproic acid treatment in rats reproduces the cerebellar anomalies associated with autism. *Teratology* **53,** 86.

Ingram, J. L., Stodgell, C. J., Kern, J. R., Hyman. S. L., Figlewicz, D. A., and Rodier, P. M. (1997). Mutations in *HoxA-1* are associated with some cases of autism. *Teratology* **55,** 50.

Jones, M. B., and Szatmari, P. (1988). Stoppage rules and genetic studies of autism. *J. Aut. Dev. Dis.* **18,** 31–40.

Jorde, L. B. et al. (1991). Complex segregation analysis of autism. *Am. J. Hum. Genet.* **49,** 932–938.

Joubert, M., Eisenring, J. J., Robb, J. P., and Andermann, F. (1969). Familial agenesis of the cerebellar vermis. *Neurology* **19,** 813–825.

Kemper, T. L., and Bauman, M. L. (1992). Neuropathology of infant autism. In H. Naruse and E. M. Ornitz, Eds. *Neurobiology of Infantile Autism.* Elsevier: Amsterdam, pp. 43–57.

Klin, A. (1993). Auditory brain stem responses in autism: brainstem dysfunction or peripheral hearing loss? *J. Aut. Dev. Dis.* **23,** 15–34.

Kluver, H., and Bucy, P. (1939). Preliminary analysis of functions of the temporal lobe in monkeys. *Arch. Neurol. Psychiat.* **42,** 979–1000.

Landa, R., Folstein, S. E., Isaacs, C. (1991). Spontaneous narrative—discourse performance of parents of autistic individuals. *J. Speech Hear. Res.* **34,** 1339–1345.

Landa, R., Piven, J., Wzorek, M., Gayle, J. O., Chase, G. A., and Folstein, S. E. (1992). Social language use in parents of autistic individuals. *Psycholog. Med.* **22,** 245–254.

Langston, A. W., and Gudas, L. J. (1992). Identification of a retinoic acid responsive enhancer 3′ of the murine homeobox gene *Hox-1.6. Mech. Dev.* **38,** 217–227.

Langston, A. W., Thompson, Jr., and Gudas, L. J. (1997). Retinoic acid responsive enhancers located 3′ of the *Hox A* and *Hox B* homeobox gene clusters: functional analysis. *J. Biol.Chem.* **272,** 2167–2175.

Laughlin, N. K., Bushnell, P. J., and Bowman, R. E. (1991). Lead exposure and diet: differential effects on social development in the rhesus monkey. *Neurotoxicol. Teratol.* **13,** 429–440.

Le Couteur, A., Bailey, A., Goode, S., Pickles, A., Robertson, S., Gottesman, I., and Rutter, M. (1996). A broader phenotype of autism: the clinical spectrum in twins. *J. Child. Psychol. Psychiat.* **37,** 785–801.

Lesch, K.-P., Bengel, D., Heils, A., Sabol, S. Z., Greenberg, B. D., Petri, S., Benjamin, J., Muller, C. R., Hamer, D. H., and Murphy, D. L. (1996). Association of anxiety-related traits with a polymorphism in the serotonin transporter gene regulatory region. *Science* **274,**1527–1531.

Lufkin, T., Dierich, A., LeMeur, M., Mark, M., and Chambon, P. (1991). Disruption of the *Hox 1.6* homeobox gene results in defects in a region corresponding to its rostral domain of expression. *Cell* **66,** 1105–1119.

Mark, M., Lufkin, T., Vonesch, J.-L., Ruberte, E., Olivo, J.-L., Dollé, P., Gorry, P., Lumsden, A., and Chambon, P. (1993). Two rhombomeres are altered in *Hoxa-1* mutant mice. *Development* **11,** 319–338.

Marshall, H., Nonchev, S., Sham, M. H., Muchamore, I., Lumsden, A., and Krumlauf, R. (1992). Retinoic acid alters the hindbrain *Hox* code and induces transformation of rhombomeres 2/3 into a 4/5 identity. *Nature* **360,** 737–741.

Martineau, J. et al. (1994). Catecholaminergic metabolism and autism. *Devel. Med. Child Neurol.* **36,** 688–697.

May, M. (1986). *The Facial Nerve.* New York: Thieme, pp. 404–408.

Means, A. L., and Gudas, L. J. (1995). The roles of retinoids in vertebrate development. *Ann. Rev. Biochem.* **64,** 201–233.

Miller, M. T. (1991). Thalidomide embryopathy: a model for the study of congenital incomitant horizontal strabismus. *Trans. Am. Ophthalmol. Soc.* **89,** 623–674.

Mortlock, D. P., and Innis, J. W. (1997). Mutation of *HoxA13* in hand-foot-genital syndrome. *Nat. Genet.* **15,** 179–180.

Muragaki, Y., Mundlos, S., Upton, J., and Olsen, B. R. (1996). Altered growth and branching patterns in synpolydactyly caused by mutations of *HoxD13. Science,* **272,** 548–550.

Murphy, P. and Hill, R. E. (1991). Expression of the mouse labial-like homeobox containing genes *Hox2.9* and *Hox1.6* during segmentation of the hindbrain. *Development* **111,** 61–74.

Osterling, J. and Dawson, G. (1994). Early recognition of children with autism: a study of first birthday home videotapes. *J. Aut. Dev. Dis.* **24,** 247–257.

Petit, E., Herault, J., Matineau, J., Perrot, A., Bartheleme, C., Hameury, L., Sauvage, D., Lelord, G., and Muh, J. P. (1995). Association study with two markers of a human homeogene in infantile autism. *J. Med. Genet.* **32,** 269–274.

Pickles, A., Bolton, P., MacDonald, H., Bailey, A., LeCouteur, A.,Sim, C.-H., Rutter, M. (1995). Latent-class analysis of recurrence risks for complex phenotypes with selection and measurement error: a twin and family history study of autism. *Am. J. Hum. Genet.* **57,** 717–726.

Piven, J., Tsai, G., Nehme, E., Coyle, J. T., Chase, and Folstein, S. E. (1991). Platelet serotonin, a possible marker for familial autism. *J. Aut. Dev. Dis.,* **21,** 51–59.

Ramsey, J., and Taylor, D. (1980). Congenital crocodile tears: a key to the aetiology of Duane's syndrome. *Br. J. Ophthalmol.* **64,** 518–522.

Reiss, A. L., Feinstein, C. and Rosenbaum, K. N. (1986). Autism and genetic disorders. *Schiz. Bull.* **12,** 724–738.

Ridley, R. M. (1994). The psychology of perseverative and stereotyped behaviour. *Prog. Neurobiol.* **44,** 221–231.

Ritvo, E. R., Freeman, B. J., Scheibel, A. B., Duong, T., Robinson, H., Guthrie, D., and Ritvo, A. (1986). Lower Purkinje cell counts in the cerebella of four autistic subjects: initial findings of the UCLA–NSAC autopsy research report. *Am. J. Psychiat.* **143,** 862–866.

Ritvo, E. R. et al. (1989). The UCLA–University of Utah epidemiologic survey of autism: recurrence risk estimates and genetic counseling. *Am. J. Psychiat.* **146,** 1032–1036.

Rodier, P. M., Bryson, S. E., and Welch, J. P. (1997). Minor malformations and physical measurements in autism: data from Nova Scotia. *Teratology* **55,** 319–325.

Rodier, P. M., Ingram, J. L, Tisdale, B., Nelson, S., and Romano, J. (1996). Embryological origin for autism: developmental anomalies of the cranial nerve motor nuclei. *J. Comp. Neurol.* **370,** 247–261.

Roessler, E. et al. (1996). Mutations in the human *Sonic hedgehog* gene cause holoprosencephaly. *Nature Genet.* **14,** 357–360.

Ross, D. L., Klyklyo, W. M., and Anderson, G. M. (1985). Cerebrospinal fluid indolamine and monoamine effects in fenfluramine treatment of autism. *Ann. Neurol.* **18,** 394.

Rutter, M. (1996). Autism research: prospects and priorities. *J. Aut. Dev. Disord.* **26,** 257–275.

Scharre, J. E., and Creedon, M. P. (1992). Assessment of visual function in autistic children. *Optom. Vis. Sci.* **69,** 433–439.

Schain, R. J., and Freedman, D. X. (1961). Studies on 5-hydroxyindole metabolism in autistic and other mentally retarded children. *J. Peds.,* **58,** 315–320.

Seay, B., and Harlow, H. F. (1965). Maternal separation in the rhesus monkey. *J. Nerv. Ment. Dis.* **140,** 434–441.

Smalley, S. L., Asarnow, R. F., and Spence, A. (1988). Autism and genetics: a decade of research. *Arch. Gen. Psychiat.* **45,** 953–961.

Smith, I. M., and Bryson, S. E. (1994). Imitation and action: a critical review. *Psycholog. Bull.* **116,** 259–273.

Spence, A. M., Ritvo, E. R., Marazita, M. L., Funderburk, S. J., Sparkes, R. S., and Freeman, B. J. (1985). Genetics of autism. *Behav. Genet.* **15,** 1–13.

Streeter, G. L. (1948). Developmental horizons in human embryos. *Contr. Embryol. Carneg. Instn.* **30,** 213–230.

Strömland, K., and Miller, M. T. (1997). Moebius syndrome revisited. *Teratology* **55,** 34.

Strömland, K., Nordin, V., Miller, M., Akerstrom, B., and Gillberg, C. (1994). Autism in thalidomide embryopathy: a population study. *Devel. Med. Child. Neurol.* **36,** 351–356.

Taber-Pierce, E. (1973). Time of origin of neurons in the brain stem of the mouse. *Prog. Brain Res.* **40,** 53–65.

Tager-Flusberg, H. (1992). Autistic children's talk about psychological states in the early acquisition of a theory of mind. *Child Devel.* **63,** 161–172.

Walker, H. A. (1977). Incidence of minor physical anomaly in autism. *J. Aut. Child. Schiz.* **7,** 165–176.

PART VIII

Risk Assessment

The world's focus on risk assessment has intensified since the late 1980s. Decisions concerning the production and use of chemicals generally are now considered in terms of not just risk, but a balance between risk and health, economic, or environmental benefit. These risk–benefit decisions, usually made by risk managers, require quantitative risk assessments (Slikker *et al.,* 1996). In order for the risk manager to make a decision in which the benefit is estimated in quantitative terms, the risk assessment side of the equation must also be quantitative. If a patient together with his or her physician is making the decision whether to use a medication with known efficacy (e.g., improves health conditions 9 out of 10 times), then the risk of taking such medication should also be stated in quantitative terms (e.g., adverse effects in 1 of 100 exposures at a given dose). A similar argument could be made for economic benefit (e.g., quantitative monetary benefit) versus a quantitative risk. Therefore, there is a need for risk assessment procedures for developmental neurotoxicants that are quantitative.

A solid science foundation is necessary for quantitative risk assessment of developmental neurotoxicants. Because health risk is the product of expose X effect, exposure data is essential. In the case of developmental assessments in which target tissues are separated from the administered dose by the maternal environment and the placental and blood–brain barriers, sophisticated approaches must be used to obtain target tissue concentrations. The generation of quantitative effect indices is likewise challenging because of the multidisciplinary nature of available endpoints. An often used assessment approach may include neurochemical, neurohistologic, neurophysiologic, and behavioral endpoints to define a neurotoxicologic profile of an agent (Slikker and Gaylor, 1995). In addition, these developmental neurotoxicity endpoints cannot be viewed in isolation but must be assessed in relation to general maternal toxicity. A quantitative assessment of developmental neurotoxicity risk is most relevant in the absence of potentially confounding maternal toxicity.

In Chapters 39, 40, 41, and 42, the essential elements of quantitative risk assessment for developmental neurotoxicants are addressed. In Chapter 39, Carole A. Kimmel summarizes currently used approaches and emphasizes the importance of a firm science foundation for developmental neurotoxicant risk assessment. In Chapter 40 by James Schardein, the selection and use of animal models for risk assessment are presented. The concordance of the outcome of animal studies to know human data on developmental neurotoxicants is the major focus of this chapter. The importance of determining target tissue concentrations of developmental neurotoxicants in relation to the risk assessment process is emphasized in the Chapter 41 by Kannan Krishnan and Melvin Andersen. The primary focus is on the development and use of physiologically based pharmacokinetics (PBPK) models and their application in the risk assessment process. A quantitative risk assessment of the developmental neurotoxicant methanol is exemplified in Chapter 42 by William Slikker, Jr. and David Gaylor. The use of both continuous and quantal dose–response data of relevant endpoints is emphasized. The currently used and two quantitative approaches are utilized to assess the same data sets of fetal weight and soft-tissue malformations so that the risk assessment methods can be compared directly.

William Slikker, Jr.

References

Slikker, W. Jr., Crump, K. S., Andersen, M. E., and Bellinger, D. 1996. Biologically based, quantitative risk assessment of neurotoxicants. *Fund. Appl. Toxicol.* **29,** 18–30.

Slikker, W. Jr., and Gaylor, D. W. 1996. Risk assessment strategies for neuroprotective agents. *Ann. N. Y. Acad. Sci.* **765,** 198–208.

CHAPTER

39

Current Approaches to Risk Assessment for Developmental Neurotoxicity

CAROLE A. KIMMEL
National Center for Environmental Assessment
U.S. Environmental Protection Agency
Washington, DC 20460

I. Introduction

Risk assessment for developmental neurotoxicity is an evolving area. The field of developmental neurotoxicity (originally called *behavioral teratology*) dates from the early 1960s (see Butcher, 1985; Vorhees, 1986; and Kimmel, 1988, for a historic perspective). The Food and Drug Administration (FDA) has had the authority to require testing of drugs and other chemicals for developmental neurotoxicity since the mid-1960s and early 1970s, but the actual number of studies required is few. Even with the publication of the international guidelines for pharmaceutical testing (FDA, 1994) that provide the FDA somewhat more flexibility for requiring developmental neurotoxicity testing (tests not specified), very few studies have been conducted. In 1991, the U.S. Environmental Protection Agency (EPA) published the first specific guidelines for developmental neurotoxicity testing for pesticides and industrial chemicals (EPA, 1991b). Despite the more specific nature of these guidelines, they also are required on an ad hoc basis when there is some indication from other data that developmental neurotoxicity may be a concern, and have been invoked only a few times. Whether passage of the Food Quality Protection Act in 1996 and the resulting emphasis on children's health protection and endocrine disruptors brings about more frequent testing remains to be seen.

*Abbreviations: BMD, benchmark dose; CNS, central nervous system; EPA, (United States) Environmental Protection Agency; FDA, (United States) Food and Drug Administration; LOAEL, lowest observed adverse effect level; MOE, margin of exposure; NOAEL, no observed adverse effect level; NRC, National Research Council; PCB, polychlorinated biphenyls; PNS, peripheral nervous system; RfC, reference concentration; RfD, reference dose; SES, socioeconomic status; UF, uncertainty factor.

As for risk assessment approaches, developmental neurotoxicity is considered in the EPA's risk assessment guidelines for both developmental toxicity and for neurotoxicity. Developmental neurotoxicity was originally covered in the Guidelines for Developmental Toxicity Risk Assessment (1991a). More recently, it has been expanded on somewhat in the Guidelines for Neurotoxicity Risk Assessment (1998a). The basic risk assessment process that has evolved within EPA for noncancer health effects applies to developmental neurotoxicity as to all of developmental toxicity.

II. General Risk Assessment Paradigm

In 1983, the National Research Council (NRC) of the National Academy of Sciences published its landmark document on risk assessment (NRC, 1993): "Risk Assessment in the Federal Government: Managing the Process." The NRC recommended that federal regulatory agencies establish inference guidelines to promote consistency and technical quality in risk assessment, and to ensure that the risk assessment process is maintained as a scientific effort separate from risk management. In 1984, the EPA began work on risk assessment guidelines for carcinogenicity, mutagenicity, suspect development toxicants, chemical mixtures, and exposure assessment. Following extensive internal and external peer review and public comments, these first five guidelines were published on September 24, 1986 (EPA, 1986a–e). Since 1986, two of the original guidelines were revised and reissued, suspect developmental toxicants (retitled developmental toxicity; EPA, 1991a) and exposure (EPA, 1992), and additional risk assessment guidelines have been published for reproductive toxicity (EPA, 1996b) and neurotoxicity (EPA, 1998a). The original carcinogenicity guidelines have also been revised and proposed for public comment (EPA, 1996a).

The process of risk assessment as described by the NRC has four components, the first two being hazard identification and dose–response assessment. These two parts of the process constitute the toxicologic evaluation. This evaluation is aimed at characterizing the sufficiency and strength of the available toxicity data and may indicate some level of confidence in the data. Dose–response modeling may be included, if the data are sufficient. Another component, exposure assessment, derives estimates of potential human exposure based on various environmental and/or occupational scenarios. These exposure estimates are used in conjunction with the toxicity evaluation to estimate potential human risk or to determine the exposure margin between minimally toxic levels in animal or human studies and estimated human exposure levels in particular situations. The final component of the process is risk characterization and involves the integration of data and determination of human risk. Risk characterization is used along with social, economic, engineering, and other factors in weighing alternative regulatory options and in making a regulatory decision; this constitutes the risk management process. The NRC's stated purpose in separating risk assessment (i.e., the scientific part of the process) from risk management was to allow the scientific data to be evaluated fully without bias from nonscientific influences.

The EPA approach to noncancer risk assessment has evolved somewhat from the original NRC paradigm through the development of the various risk assessment guidelines indicated previously. The most recent Guidelines for Reproductive Toxicity Risk Assessment (EPA, 1996b) and the Guidelines for Neurotoxicity Risk Assessment (EPA, 1998a) discuss hazard characterization as the first step in the risk assessment process. Hazard characterization is a broadening of the original hazard identification concept in that it does not simply involve identifying a hazard, but rather identifies whether an agent is a hazard in the context of dose, route, timing, and duration of exposure. The dose–response component then involves the quantitative modeling or estimation of the dose–response relationship, and calculation of a reference dose (RfD), reference concentration (RfC), or other standard. In part, the basis for this approach is the presumption that at low doses, there is a nonlinear dose–response relationship and possibly a threshold for most noncancer health effects, whereas there is an assumption of low-dose linearity for cancer. These assumptions are being challenged for both cancer and noncancer health effects, and there is a move towards harmonization of risk assessment approaches in the two areas—that is, basing the dose–response assessment on mode of action and using common approaches in which mode of action is assumed to be the same.

The next section focuses on risk assessment for developmental neurotoxicity within the context of developmental toxicity and neurotoxicity. This discussion is drawn from the U.S. EPA's Guidelines for Developmental Toxicity Risk Assessment (1991a) and Guidelines for Neurotoxicity Risk Assessment (1998a). These documents can be viewed as consensus documents in these fields because of the wide range of comments from individuals within and outside EPA in the development of these documents. Much of the information in the guidelines was developed as a result of workshops involving a wide range of participants who addressed specific topics and helped to clarify issues of data interpretation and analysis. Research and publication within the area of developmental neurotoxicity continue to grow, and it is expected that this field will continue to evolve and mature.

III. Developmental Neurotoxicity Risk Assessment

Developmental neurotoxicity is defined as an adverse change in the structure or function of the central nervous system (CNS) and/or the peripheral nervous system (PNS) following developmental exposure to a chemical, physical, or biologic agent. This definition is based on that of Tilson (1990) for neurotoxicity. Developmental exposure may occur prior to conception, prenatally, or postnatally to the time of sexual maturation. Adverse effects may be detected at any time in the lifespan of the organism. Thus, prenatal exposure may lead to delayed-onset effects (e.g., behavioral changes that are not apparent until a child reaches school age or adolescence) or latent effects (e.g., effects that become evident only after an environmental challenge or aging). Adverse effects may also be progressive (i.e., effects that continue to worsen after the exposure has ceased) or residual (i.e., effects that persist beyond a recovery period). Some developmental neurotoxic effects may be reversible and the nervous system is known for its reserve capacity. However, with exposure during development, reversibility of one type of adverse effect may not mean that there is no further effect, but the later developmental stages may also be affected. For example, delays in development of early reflexes that are eventually attained may portend delays or alterations in other systems that must develop at a particular time in order to mature properly. Thus, reversible developmental neurotoxic effects should be interpreted with caution and should not be assumed to indicate that no further effects will be seen.

A. Hazard Characterization

Hazard characterization involves identification of key studies for evaluation; description of the types of effects and the nature of effects (whether irreversible, reversible, latent, residual, progressive, or delayed-onset); other health endpoints of concern; the dose–response data available for evaluation; the route, level, timing, and duration of exposure; the strengths and weaknesses of the database; and the relevance of the database to humans. The characterization of developmental neurotoxic effects may be based on human or animal data or may be a combination of the two.

Data in humans are likely to result from accidental exposures (e.g., exposure to methylmercury in Japan and Iraq, exposure to lead in house paint). Data in animals may come from standard testing studies, but more likely from experimental studies in the literature conducted for the purpose of characterizing an observation in humans. As mentioned previously, the number of required standard testing studies available in animals is minimal, because studies are required only on a case-by-case basis by most regulatory agencies.

1. Human Data

Jacobson and Jacobson (1996) reviewed the prospective longitudinal assessment of developmental toxicity in children, and discussed design and methodologic issues. Tests such as the Bayley Scales of Infant Development and the McCarthy Test of Children's Abilities are used to assess child development, including general mental ability, using age-specific scales. These types of evaluations have been used to detect the effects of lead (Bellinger *et al.,* 1987, 1991; Dietrich *et al.,* 1987, 1991) and polychlorinated biphenyls (PCBs) (Jacobson *et al.,* 1990) on children at different ages, but must be controlled and evaluated carefully to be useful in a risk assessment. For example, confounding influences in the physical and social environment may be assessed indirectly using various demographic factors such as maternal age and intelligence, perinatal risk factors (e.g., neonatal asphyxia) or socioeconomic status (SES); or more directly using measures such as the Home Observation for Measurement of the Environment (HOME) Inventory (Caldwell and Bradley, 1979) or other scales that measure family interactions, stress, and cohesiveness (see Jacobson and Jacobson, 1996, for a more complete description).

Epidemiologic studies provide the most reliable data for assessing developmental neurotoxicity. These may be in the form of cohort studies in which groups are defined by their exposure and are then followed over time to determine outcome. Case-control studies are also commonly used, in which groups are defined by a selected health effect, and the exposure history is then examined to determine whether affected individuals are more frequently exposed than are controls. Ecological studies are sometimes done, in which the study groups are defined by a general characteristic (e.g., zip code, county of residence) and their exposure history is assumed. Cross-sectional studies may be conducted in which exposure and outcome are examined at the same time; these studies may not account for the possibility that exposure was not present at the critical time for the effect to be produced. Cohort and case-control studies provide the most reliable data for determining the effect of an agent on the developing nervous system. All epidemiology studies have strengths and limitations that must be considered in assessing the data. Adams and Selevan (1994) have reviewed in detail the general design characteristics of human studies for evaluating developmental toxicity.

2. Animal Data

Animal studies may be available on the developmental neurotoxic effects of specific chemicals and data often come from the literature because, as indicated previously, data from standard testing studies are rarely available. Studies from the literature may provide important data on developmental neurotoxicity and on pharmacokinetics, effects of structural analogs, as well as data from *in vitro* studies designed to address questions about mechanisms of action. Literature studies sometimes are limited in terms of the number of dose levels tested, numbers of animals, and appropriate controls. Thus, the use of such data in risk assessment must include a careful evaluation of the design and conduct of such studies.

Data from several standard testing guideline studies can be used to evaluate the potential effects of chemicals on developmental neurotoxicity. Effects on structural development of the nervous system are observed in the prenatal developmental toxicity study (EPA, 1998b). Some limited information also can be obtained on developmental neurotoxicity from the two-generation study (EPA, 1998c). Data are available on growth, viability, and overt signs of toxicity, including neurologic signs. In the latest revision of these test guidelines, observations on brain weight and sexual maturation have been added to provide information that can be used as a basis for further studies, including developmental neurotoxicity. The developmental neurotoxicity testing guidelines (EPA, 1991b) for pesticides and industrial chemicals provide the most specific data. Criteria for selecting agents to be tested have been developed and are based on a weight-of-evidence approach using criteria similar to those proposed by Levine and Butcher (1990). These include structural developmental defects in the nervous system; neuropathology and neurotoxicity in adults; effects on sexual maturation; and other developmental toxicity.

The developmental neurotoxicity testing guideline includes the observation of physical and functional (reflex) development, tests of motor activity, gross sensory function, learning and memory, and neuropathology, which includes brain weight and morphometric measurements of the brain. Because of its design, the developmental neurotoxicity testing protocol may be conducted as a separate study, concurrently with or as a follow-up to a prenatal developmental toxicity study, or be folded into a multigeneration study in the second generation. Thus, these studies include exposure at least through the period of organogenesis and into the early postnatal period, or exposure may be continuous over several generations. The evaluation of data from these studies has been discussed in more detail in the Guidelines for Neurotoxicity Risk Assessment (EPA, 1998a), which include guidance on interpretation of developmental neurotoxic effects. Evaluation of structural, neurophysiologic, neurochemical, and behavioral endpoints of neurotoxicity are discussed and guidance is given for their interpretation. In many cases, the interpretation of neurotoxicity endpoints measured in adults does not differ, whether the exposures are during development or in the adult, other than for delays in development of adult neurologic structure or function. However, there are particular issues of importance in the evaluation of developmental neurotoxicity studies. For example, in studies where exposure is to the mother either during gestation or lactation, there is little control over selection of animals for certain characteristics, because offspring representative of all litters must be tested. Thus, no preexposure baseline testing is possible in these animals, except when a pharmacologic or physical challenge may be used to uncover an effect. For example, a challenge with a psychomotor stimulant such as D-amphetamine may unmask latent developmental neurotoxicity (Hughes and Sparber, 1978; Adams and Buelke-Sam, 1981; Buelke-Sam *et al.,* 1985).

Direct extrapolation of developmental neurotoxicity to humans is limited in the same way as for other endpoints of developmental toxicity (i.e., lack of knowledge about underlying toxicologic mechanisms and their significance). In the evaluation of a limited number of agents known to cause developmental neurotoxic effects in humans, however, Adams (1986) concluded that these agents produce similar developmental neurotoxic effects in animals and humans. Strong support for this conclusion came from a subsequent workshop on the Qualitative and Quantitative Comparability of Human and Animal Developmental Neurotoxicity (Kimmel *et al.,* 1990). Several agents, including ionizing radiation, lead, methylmercury, anticonvulsants, alcohol, and PCBs were considered, and both the animal studies and the human data were evaluated by experts. All of these agents are known to cause human developmental neurotoxicity, and there was a high degree of correlation with the types of effects seen in the animal studies when compared on the basis of functional categories (Stanton and Spear, 1990). This conclusion lends strong support for the use of experimental animals in assessing the potential risks for developmental neurotoxicity in humans, one of the major assumptions underlying developmental toxicity risk assessment. Some support was found for comparability on a quantitative basis as well, with greater species similarity when target organ dose or systemic dose data were available (Tyl and Sette, 1990).

However, data were available to make quantitative evaluations for only a few of the agents reviewed.

Important design issues to be evaluated for developmental neurotoxicity studies are similar to those for standard developmental toxicity (e.g., a dose–response approach with the highest dose producing minimal overt maternal or perinatal toxicity, the number of litters large enough for adequate statistical power, randomization of animals to dose groups and test groups, and the litter generally considered as the statistical unit). In addition, the use of a replicate study design provides added confidence in the interpretation of the data.

Agents that produce developmental neurotoxicity at a dose that is not toxic to the maternal animal are of special concern. However, adverse developmental effects are often produced at doses that cause mild maternal toxicity (e.g., 10–20% reduction in weight gain during gestation and lactation). At doses causing moderate maternal toxicity (i.e., 20% or more reduction in weight gain during gestation and lactation), interpretation of developmental effects may be confounded. Current information is inadequate to assume that developmental effects at doses causing minimal maternal toxicity result only from maternal toxicity; rather, it may be that the mother and developing organism are equally sensitive to that dose level. Moreover, regardless of whether developmental effects are secondary to maternal toxicity, the maternal effects may be reversible, whereas the effects on the offspring may be permanent. These are important considerations for agents to which humans may be exposed at minimally toxic levels either voluntarily or involuntarily, because several agents (e.g., alcohol) are known to produce adverse developmental effects at minimally toxic doses in adults (Coles *et al.*, 1991).

Although interpretation of developmental neurotoxicity data may be limited, it is clear that functional effects should be evaluated in light of other toxicity data, including other forms of developmental toxicity (e.g., structural abnormalities, perinatal death, growth retardation). For example, alterations in motor performance may be due to a skeletal malformation rather than a nervous system change. Changes in learning tasks that require a visual cue might be influenced by structural abnormalities of the eye. The level of confidence that an agent produces an adverse effect may be as important as the type of change seen, and confidence may be increased by such factors as reproducibility of the effect, either in another study of the same function or by convergence of the data from tests that purport to measure similar functions. A dose–response relationship is an extremely important measure of a chemical's effect; in the case of developmental neurotoxicity, both monotonic and biphasic dose–response curves are likely, depending on the function being tested.

3. Dose–Response Evaluation

A critical part of the hazard characterization is the evaluation of the dose–response relationship. Human studies covering a range of exposures are rarely available and, therefore, animal data are typically used for estimating exposure levels likely to produce adverse effects in humans. Evidence for a dose–response relationship is an important criterion in establishing a developmental neurotoxic effect. Most agents causing developmental neurotoxicity in humans tend to produce these effects at doses lower than those causing maternal toxicity, although the dose levels are typically within a 10-fold factor of maternal toxic doses. The evaluation of dose–response relationships includes identifying effective dose levels as well as doses associated with no increase in adverse effects when compared with controls. The lack of a dose–response relationship in the data may suggest that the effect is not related to the putative toxic agent or that the study was not controlled appropriately. Much of the focus is on identifying the critical effects(s) observed at the lowest observed adverse effect level (LOAEL) associated with that effect and the no observed adverse effect level (NOAEL). The NOAEL is defined as the highest dose at which there is no statistically or biologically significant increase in the frequency of an adverse effect. Alternatively, the data may be modeled and a dose estimated that is associated with some low level of risk in the observable range, the benchmark dose (BMD). The NOAEL or BMD can then be used for calculating a reference dose if developmental neurotoxicity is the effect seen at the lowest exposure level. The reference dose for methylmercury is based on a BMD derived from developmental neurotoxicity data (EPA, 1998d).

In addition to these considerations, pharmacokinetic factors and other aspects that might influence comparisons with human exposure scenarios should be taken into account. For example, dose–response curves may exhibit not only monotonic but also U-shaped or inverted U-shaped functions (Davis and Svendsgaard, 1990). Such curves are hypothesized to reflect multiple mechanisms of action, the presence of homeostatic mechanisms, and/or activation of compensatory or protective mechanisms. As for other types of developmental toxicity, the severity of the developmental neurotoxic effect is likely to be much greater at high doses than at lower doses, including the LOAEL or BMD. Determining what is an adverse effect at low doses sometimes is difficult (e.g., cholinesterase inhibition), but it is such effects that are likely to be the focus of a risk assessment. Thus, the identification of a critical adverse effect often requires considerable professional judgment and should consider factors such as the biologic plausibility of the effect, the evidence of a dose–response continuum, and the likelihood for progression of the effect with continued exposure.

4. Characterization of the Database

Once the data in both humans and animals have been reviewed thoroughly, the database is characterized as being sufficient or insufficient to proceed further in the risk assessment process. This characterization of hazard is done within the context of exposure or dose, route, duration, and timing of exposure, relative to the developmental period during which exposure occurred. Characterization of the hazard in this manner allows for better determination of the strengths and weaknesses, as well as the uncertainties, of the data associated with a particular chemical. This does not, however, address the nature and magnitude of the human health risks, which are determined after the exposure assessment is completed and the final characterization of the risk is conducted. The scheme used in the EPA's guidelines defines two broad categories: sufficient and insufficient (Table 1). The sufficient category can be subdivided into sufficient human data and sufficient experimental animal/limited human data. These categories define the minimum evidence necessary to determine sufficiency, and the strength of evidence for a database increases with replication of the findings and with testing of additional species. All data, regardless of whether indicative of a hazard potential, are to be weighed in making a judgment about the strength of evidence available to support a complete risk assessment for developmental neurotoxicity. Information of pharmacokinetics and mechanisms or on more than one route of exposure may reduce uncertainties in determining relevancy to humans. More evidence (e.g., more studies, more species, more endpoints) is necessary to judge that an agent is unlikely to pose a hazard than that required to judge a potential hazard. This is because it is more difficult,

TABLE 1 Characterization of the Health-Related Database[a]

Sufficient evidence	The sufficient evidence category includes data that collectively provide enough information to judge whether a human developmental neurotoxic hazard could exist. This category may include both human and animal evidence.
Sufficient human evidence	This category includes data from which there is sufficient evidence from epidemiologic studies (e.g., case-control and cohort studies) that provide convincing evidence for the scientific community to judge that a causal relationship is or is not supported. A case series in conjunction with strong supporting evidence may also be used. Supporting animal data may or may not be available.
Sufficient animal evidence/limited human evidence	This category includes data from experimental animal studies and/or limited human data that provide convincing evidence for the scientific community to judge if the potential for developmental neurotoxicity exists. The minimum evidence needed to judge that a potential hazard exists would be data demonstrating an adverse developmental neurotoxic effect in a single, appropriate, well-executed study in a single experimental animal species. The minimum evidence needed to judge that a potential hazard does not exist would include data from appropriate, well-conducted laboratory animal studies in several species (at least two) that evaluated a variety of the potential manifestations of developmental neurotoxicity and showed no effects at doses that were minimally toxic to the dam. Information on pharmacokinetics, mechanisms, or known properties of the chemical class may also strengthen the evidence.
Insufficient Evidence	This category includes data from which there is less than the minimum evidence sufficient for identifying whether a developmental neurotoxic hazard exists, such as when no data are available on developmental neurotoxicity, as well as in the case of studies in animals or humans that have a limited study design (e.g., inadequate conduct or reporting of clinical endpoints, small numbers of experimental animals, inappropriate dose selection). Furthermore, data from a single species reported to have no adverse developmental effects, or databases limited to information on structure/activity relationships, short-term tests, pharmacokinetics, or toxicity of metabolic precursors would be considered insufficient evidence of developmental neurotoxicity. Such information, along with data on CNS malformations from the prenatal developmental toxicity study, brain weights, or alterations in sexual maturation from the two-generation study, or adult neurotoxicity, could be used to support the need for additional testing.

[a]Modified from EPA, 1991a, 1998a.

both biologically and statistically, to support a finding of no apparent adverse effect than to support a positive effect.

The insufficient evidence category may include situations in which no data are available, or data are available from studies with limited study design or of a type (e.g., some short-term studies) that are not yet considered sufficient to use for human risk assessment. Such data are often useful in recommending the need for additional studies. In some cases, a substantial database may exist in which no one study is considered sufficient, but the body of data may be judged as meeting the sufficient evidence category.

B. Dose–Response Analysis

A threshold or low-dose nonlinear response relationship is generally assumed for developmental and neurotoxic effects, unless a genotoxic mechanism is known to be at work in causing the effects. Because of the threshold assumption, extrapolation to low-dose levels using mathematical models typically is not done. Instead, the data on developmental neurotoxicity are evaluated along with the data on other types of toxicity, and effects seen at the lowest dose levels are used to calculate the chronic RfD, RfC, or other exposure standard. The RfD or RfC is an estimate of a daily exposure to the human population over a lifetime that is assumed to be without appreciable risk of deleterious effects. These values are calculated by dividing the NOAEL (or LOAEL, if a NOAEL is not available) or BMD by various uncertainty factors (UFs) to account for interspecies differences in susceptibility, intrahuman variability, LOAEL to NOAEL (when a NOAEL is not available), subchronic data to a chronic exposure scenario, and a database factor if the database is deficient in the types and quality of studies available. The default value for uncertainty factors is usually 10, but may vary depending on the quality and type of information available. For example, if pharmacokinetic data are available to provide a measure of internal dose, the interspecies factor may be reduced from 10 to 3. An RfD or RfC for developmental toxicity (RfD_{DT} or RfC_{DT}) can also be calculated that is based on shorter term exposure. The RfD_{DT} and RfC_{DT} were proposed in the Guidelines for Developmental Toxicity Risk Assessment (EPA, 1991a) as a way of specifying risks to children, because of the known importance of critical periods for exposure at various stages throughout the developmental period. A major disadvantage of the NOAEL/UF approach is that it does not give an estimate of risk at particular dose levels, so that when exposure occurs at levels above the RfD or RfC, there is no way to estimate how great the risk may be at the level.

The BMD may also be used to derive the RfD or RfC by fitting a mathematical model to the data and deriving a lower confidence limit on the effective dose associated with some defined level of effect (e.g., a 5% or 10% increase in response). Using this approach, the BMD for the most sensitive endpoint is selected as the point of departure in calculating the RfD or RfC and uncertainty factors are applied in a similar manner as for the NOAEL. The EPA is writing guidance for application of the BMD approach in risk assessment. One of the difficulties with much of the data on neurobehavioral effects is that the endpoints of interest are continuous measures. In this situation, the degree of change that is considered "adverse" must be defined in order to apply the models appropriately. Models for use with continuous data have been proposed and are being evaluated (e.g., Crump, 1984, 1995; Gaylor and Slikker, 1990; Glowa and MacPhail, 1995). Categorical regression analysis has also been proposed for neurotoxicity because it can evaluate different types of data and derive estimates for short-term exposure (Rees and Hattis, 1994). Decisions about the most appropriate approach require professional judgment, taking into account the biologic nature of the continuous effect variable and its distribution in the population under study. In the case of dose–response analysis of U-shaped functions, estimates of the NOAEL/LOAEL or BMD are typically taken from the lowest part of the dose–response curve associated with impaired function or adverse effect.

C. Exposure Assessment

The exposure assessment is done to estimate human exposures from various sources for comparison with the toxicity data in order to characterize risk. An important point to remember is that, based on the definition of developmental toxicity used in the EPA guidelines (EPA, 1991a), exposure of almost any segment of the population, except nonreproducing adults, should be considered important in relation to developmental toxicity. The Guidelines for Estimating Exposures (EPA, 1986e, 1992) discuss the various approaches that can be taken to estimating exposures through actual monitoring or by modeling various sources and exposure scenarios. Default values have been published (EPA, 1990) for use in estimating exposures (e.g., food and water consumption in adults and children, soil ingestion in children, respiration rates of children and adults). In some cases, behavior patterns of importance in children (e.g., crawling on the floor, pica, and food preferences or patterns) should be taken into account (NRC, 1993). This discussion does not attempt to provide definitive guidance for an exposure assessment, but focuses on those issues of specific importance to developmental toxicity risk assessment.

Several exposure conditions are unique for developmental toxicity. For example, during pregnancy, a woman is exposed directly and her developing conceptus indirectly, except in the case of some physical agents. Exposure of the conceptus to a chemical agent is dependent on maternal absorption, distribution, metabolism, and placental metabolism and transfer. Transit time in the conceptus also may depend on its ability to metabolize and/or excrete the chemical. In a few cases, a chemical agent may have its primary effect on the maternal system (Daston, 1994), and the effect on the conceptus does not depend on exposure to the agent and/or its metabolites, but on some factor induced in the mother. Infants may be exposed to agents via breast milk, particularly agents that are fat soluble and stored in fat. Data also suggest that lead may be mobilized along with calcium from bone stores in women during pregnancy and lactation, and be excreted in milk (Silbergeld *et al.*, 1988; Silbergeld, 1991; Watson *et al.,* 1993). Thus, exposure of the conceptus and child may not be the same as for the pregnant or lactating mother, and measurement of the agent in cord blood and in breast milk may give a better estimate of exposure.

Children may also be exposed directly via water, food, air, and soil. Dust and paint or anything that can be mouthed by a young child should also be considered a potential source of exposure. The period and duration of exposure should be related to stage of development. For example, exposures during pregnancy should be characterized as first, second or third trimester of pregnancy (or more specific periods, if possible), and in children during infancy, early, middle, and late childhood (specific ages, if possible), adolescence, etc. These stages of development may have different sensitivities to an agent, and exposure estimates should be characterized for as many as possible. In addition, exposure to either parent prior to conception should be considered as a possibility in the production of adverse developmental effects (Olshan and Mattison, 1994).

Exposures that can cause developmental toxicity may be for short or long durations. Experimental studies have confirmed for several agents that a single exposure is sufficient to cause developmental neurotoxicity—that is, repeated exposure is not a necessary prerequisite for adverse outcomes. Agents that show accumulation with repeated exposures or that have a long half-life result in greater exposures over time. The pattern of exposure is extremely important in what the predicted outcome might be, and it is usually difficult to predict from one pattern to another unless pharmacokinetic data are available to illuminate the differences. In the case of intermittent exposures, peak exposure levels as well as average exposure over the time period of exposure should be considered.

Unique to exposures during development is that there are often latent effects, except in cases of spontaneous abortions that occur during or shortly after exposure. For example, effects of exposure during pregnancy may not be manifest until after birth in the form of structural malformations, growth retardation, cancer, mental retardation, or other functional defects. In some cases, the results of exposure may not become apparent until long after the developmental period, including neurotoxic effects that are not evident until adulthood or until another factor (e.g., disease, other environmental challenges, aging) intervene (e.g., Barone *et al.,* 1995).

D. Risk Characterization

Risk characterization is the final step in the risk assessment process and involves integration of the toxicity evaluation and the exposure assessment. The risk characterization includes not only the conclusions of the hazard characterization, dose–response analysis, and the human exposure assessment, but also information on the quality of the data for both the toxicity evaluation and the exposure assessment, relevancy to humans, strengths and weaknesses in the data, assumptions made, scientific judgments, and the level of confidence in the determination of potential risk to humans. All of these factors are part of the weight-of-evidence determination.

This weight-of-evidence information is communicated to the risk manager along with the quantitative information on the range of effective exposure levels, dose–response modeling estimates (e.g., the BMD), appropriate uncertainty factors to use, and the RfD or RfC, for both chronic and developmental toxicity. When developmental neurotoxicity is a more sensitive endpoint than any other used in the calculation of the chronic RfD or RfC, information on appropriate use of the data in this context should also be communicated. Because low-dose extrapolation currently is not used for developmental or neurotoxicity risk assessment, specific risk estimates at doses below the NOAEL or BMD are not made, as is done for cancer or mutagenicity risk assessment. Thus, the RfD or RfC (and other descriptors of risk discussed later) give an estimate of dose that is assumed to be without appreciable risk above background, and is used principally because of the general default assumption of a threshold or nonlinear low dose–response relationship for noncancer health effects.

Three types of descriptors of human risk are especially useful and important. The first of these is related to interindividual variability—that is, the range of variability in population response to an agent, and the potential for highly sensitive or susceptible subpopulations. A default assumption is made that the most sensitive individual in the population will be no more than 10-fold more sensitive than the average individual; thus a default 10-fold uncertainty factor is often applied in calculating the RfD or RfC to account for this potential difference. When data are available on highly sensitive or susceptible subpopulations, the risk characterization can be done on them separately, or by using more accurate factors to account for the differences. When data are not available to indicate differential susceptibility between developmental stages and adulthood, all stages of development may be assumed to be highly sensitive or susceptible. Certain age subpopulations can sometimes be identified as more sensitive because of critical periods for exposure; for example, pregnant or lactating women, infants, children, adolescents. In general, not enough is understood about how to identify sensitive subpopulations without specific data on each agent, although it is known that factors such as nutrition, personal habits (e.g., smoking, alcohol consumption, illicit drug abuse), quality of life (e.g., socioeconomic factors), preexisting disease (e.g., diabetes), race, ethnic background, or other genetic factors may predispose some individuals to be more sensitive than others to the developmental effects of various agents.

The second descriptor of importance is for highly exposed individuals. These are individuals who are more highly exposed because of occupation, residential location, behavior, or other factors. For example, children are more likely than adults to be exposed to agents deposited in dust or soil either indoors or outside, both because of the time spent crawling or playing on the floor or ground and because of the mouthing behavior of young children. The inherent sensitivity of children may also vary with age, so that both sensitivity and timing of exposure must be considered in the risk characterization. If population data are absent, various scenarios may be assumed representing high-end exposures using upper percentile or judgment-based values. This approach must be used with caution, however, to avoid overestimation of exposure.

The third descriptor that is sometimes used to characterize risk is the margin of exposure (MOE). The MOE is the ratio of the NOAEL (or BMD) from the most appropriate or sensitive species to the estimated human exposure level from all potential sources. Considerations for the acceptability of the MOE are similar to those for the uncertainty factor applied to the NOAEL or BMD to calculate the RfD or RfC.

The risk characterization is communicated to the risk manager, who uses the results along with other technologic factors (e.g., exposure reduction measures) as well as social and economic considerations in reaching a regulatory decision. Depending on the statute involved and other considerations, risk management decisions are usually made on a case-by-case basis. This may result in different, but appropriate, regulation of an agent under different statutes.

In 1996, the Food Quality Protection Act indicated that an additional 10-fold safety factor should be applied to account for potential pre- and post-natal toxicity and completeness of data with respect to exposure and toxicity to infants and children. The basis for application of such a factor is being debated within the EPA. Given the limited amount of information on developmental neurotoxicity for most chemicals, it may be appropriate to include this factor when such data are not available, especially if there is information from other studies that these types of health effects might occur (see criteria for triggering developmental neurotoxicity studies discussed in Section IV.A, Hazard Characterization). This may be useful as a risk management tool when data are unavailable, or when limited data suggest cause for concern.

IV. Research Needs

Research to improve the risk assessment process for developmental neurotoxicity is needed in a number of areas. For example, research is needed to delineate the mechanisms of developmental neurotoxicity and pathogenesis, elucidate the functional modalities that may be altered, provide comparative pharmacokinetic data, develop improved animal models to examine the neurotoxic effects of exposure during all stages of development (premating, early postmating, later gestational periods, neonatal, early childhood, adolescent, peripubertal), determine the long-term consequences of developmental exposures, especially related to aging and late-onset diseases (e.g., dementia, Parkinsonism), and develop early neurological/behavioral endpoints that are predictive of neurotoxicity at later stages of development or in adulthood. In addition, research is needed to provide insight into the concept of threshold, develop approaches for improved mathematical modeling of developmental neurotoxic effects, improve procedures for examining the effects of agents given by various routes of exposure, determine the effects of recurrent exposures over prolonged periods of time, and address the synergistic or antagonistic effects of mixed exposures and neurotoxic response. Research such as this aids in the evaluation and interpretation of data on developmental neurotoxicity, and should provide methods to assess risk more precisely.

References

Adams, J. (1986). Clinical relevance of experimental behavioral teratology. *Neurotoxicology* **7,** 19–34.

Adams, J., and Buelke-Sam, J. (1981). Behavioral testing of the postnatal animal: testing and methods development. In *Developmental Toxicology.* C. A. Kimmel and J. Buelke-Sam, Eds. Raven Press, New York, pp. 233–238.

Adams, J., and Selevan, S. G. (1994). Issues and approaches in human developmental toxicology. In *Developmental Toxicology.* C. A. Kimmel and J. Buelke-Sam, Eds. Raven Press, New York, pp. 287–305.

Barone, S., Stanton, M. E., and Mundy, W. R. (1995). Neurotoxic effects of neonatal triethyl tin (TET) exposure are exacerbated with aging. *Neurobiol. Aging* **16,** 723–735.

Bellinger, D., Leviton, A., Waternaux, C., Needleman, H., and Rabinowitz, M. (1987). Longitudinal analyses of prenatal and postnatal lead exposure and early cognitive development. *N. Engl. J. Med.* **316,** 1037–1043.

Bellinger, D., Sloman, J., Leviton, A., Rabinowitz, M., Needleman, H. L., and Waternaux, C. (1991). Low-level lead exposure and children's cognitive function in the preschool years. *Pediatrics* **87,** 219–227.

Buelke-Sam, J., Kimmel, C. A., and Adams, J. (1985). Design considerations in screening for behavioral teratogens: results of the collaborative behavioral teratology study. *Neurobehav. Toxicol. Teratol.* **7,** 537–589.

Butcher, R. E. (1985). An historical perspective on behavioral teratology. *Neurobehav. Toxicol. Teratol.* **7,** 537–540.

Caldwell, B. M., and Bradley, R. H. (1979). *Home Observation for Measurement of the Environment.* University of Arkansas Press, Little Rock.

Coles, C. D., Brown, R. T., Smith, I. E., Platzman, K. A., Erickson, S., and Falek, A. (1991). Effects of prenatal alcohol exposure at school age. I. Physical and cognitive development. *Neurotoxicol. Teratol.* **13,** 357–367.

Crump, K. S. (1984). A new method for determining allowable daily intakes. *Fundam. Appl. Toxicol.* **4,** 854–871.

Crump, K. S. (1995). Calculation of benchmark doses from continuous data. *Risk Analysis* **15,** 79–89.

Davis, J. M., and Svendsgaard, D. J. (1990). U-shaped dose–response curves: their occurrence and implication for risk assessment. *J. Toxicol. Environ. Health* **30,** 71–83.

Dietrich, K. N., Krafft, K. M., Bornschein, R. L., Hammond, P. B., Berger, O., Succop, P., and Bier, M. (1987). Low level fetal lead exposure effect on neurobehavioral development in early infancy. *Pediatrics* **80,** 721–730.

Dietrich, K. N., Succop, P. A., Berger, O. G., Hammond, P. B., and Bornschein, R. L. (1991). Lead exposure and the cognitive development of urban preschool children: the Cincinnati lead study cohort at age 4 years. *Neurotoxicol. Teratol.* **13,** 203–211.

Food and Drug Administration (FDA). (1994). International Conference on Harmonization; guideline on detection of toxicity to reproduction for medicinal products; availability; notice. *Fed. Regist.* **59,** 48746–48752.

Gaylor, D. W., and Slikker, W. (1990). Risk assessment for neurotoxic effects. *Neurotoxicology* **11,** 211–218.

Glowa, J. R., and MacPhail, R. C. (1995). Qualitative approaches to risk assessment in neurotoxicology. In *Neurotoxicology: Approaches and Methods.* L. Chang, and W. Slikker, Eds. Academic Press, New York, pp. 777–787.

Hughes, J. A., and Sparber, S. B. (1978). D-Amphetamine unmasks postnatal consequences of exposure to methylmercury *in utero:* methods for studying behavioral teratogenesis. *Pharmacol. Biochem. Behav.* **8,** 365–375.

Jacobson, J. L., Jacobson, S. W., and Humphrey, H. E. B. (1990). Effects of exposure to PCBs and related compounds on growth and activity in children. *Neurotoxicol. Teratol.* **12,** 319–326.

Jacobson, J. L., and Jacobson, S. W. (1996). Prospective, longitudinal assessment of developmental neurotoxicity. *Environ. Health Perspect.* **104,** 275–283.

Kimmel, C. A. (1988). Current status of behavioral teratology: science and regulation. *CRC Crit. Rev. Toxicol.* **19,** 1–10.

Kimmel, C. A., Rees, D. C., and Francis, E. Z., Eds. (1990). Proceedings of the workshop on the qualitative and quantitative comparability of human and animal developmental neurotoxicity. *Neurotoxicol. Teratol.* **12,** 173–292.

Levine, T. E., and Butcher, R. E. (1990). Workshop on the qualitative and quantitative comparability of human and animal developmental neurotoxicity. Work Group IV report: Triggers for developmental neurotoxicity testing. *Neurotoxicol. Teratol.* **12,** 281–284.

National Research Council (NRC). (1983). *Risk Assessment in the Federal Government. Managing the Process.* National Academy of Sciences, Washington, DC.

National Research Council (NRC). (1993). *Presticides in the Diets of Infants and Children.* National Academy of Sciences, Washington, DC.

Olshan, A. F., and Mattison, D. R., Eds. (1994). *Male-Mediated Developmental Toxicity.* Plenum Press, New York.

Rees, D. C., and Hattis, D. (1994). Developing quantitative strategies for animal to human extrapolation. In *Principles and Methods of Toxicology.* A. W. Hayes, Ed. Raven Press, New York, pp. 275–315.

Silbergeld, E. K. (1991). Lead in bone: implications for toxicology during pregnancy and lactation. *Environ. Health Perspect.* **91,** 63–70.

Silbergeld, E. K., Schwartz, J., and Mahaffey, K. (1988). Lead and osteoporosis: mobilization of lead from bone in postmenopausal women. *Environ. Res.* **47,** 79–94.

Stanton, M. E., and Spear, L. P. (1990). Workshop on the qualitative and quantitative comparability of human and animal developmental neurotoxicity. Work Group I report: Comparability of measures of developmental neurotoxicity in humans and laboratory animals. *Neurotoxicol. Teratol.* **12,** 261–267.

Tilson, H. A. (1990). Neurotoxicology in the 1990s. *Neurotoxicol. Teratol.* **12,** 293–300.

Tyl, R. W., and Sette, W. F. (1990). Workshop on the qualitative and quantitative comparability of human and animal developmental neurotoxicity. Work Group III report: Weight of evidence and quantitative evaluation of developmental neurotoxicity data. *Neurotoxicol. Teratol.* **12,** 275–280.

U.S. Environmental Protection Agency (EPA). (1986a). Guidelines for carcinogen risk assessment. *Fed. Regist.* **51,** 33992–34003.

U.S. Environmental Protection Agency (EPA). (1986b). Guidelines for mutagenicity risk assessment. *Fed. Regist.* **51,** 34006–34012.

U.S. Environmental Protection Agency (EPA). (1986c). Guidelines for the health risk assessment of chemical mixtures. *Fed. Regist.* **51,** 34014–34025.

U.S. Environmental Protection Agency (EPA). (1986d). Guidelines for the health assessment of suspect developmental toxicants. *Fed. Regist.* **51,** 34028–34040.

U.S. Environmental Protection Agency (EPA). (1986e). Guidelines for estimating exposures. *Fed. Regist.* **51,** 34042–34054.

U.S. Environmental Protection Agency (EPA). (1991a). Guidelines for developmental toxicity risk assessment. *Fed. Regist.* **56,** 63798–63825.

U.S. Environmental Protection Agency (EPA). (1991b). Pesticide assessment guidelines, subdivision F. Hazard evaluation: human and domestic animals. Addendum 10: Neurotoxicity, series 81, 82, and 83. Office of Pesticides and Toxic Substances, Washington, DC. EPA 540/09-91-123. Available from : NTIS, Springfield, VA. PB91-154617.

U.S. Environmental Protection Agency (EPA). (1992). Guidelines for exposure assessment. *Fed. Regist.* **57,** 22888–22935.

U.S. Environmental Protection Agency (EPA). (1996a). Proposed guidelines for carcinogen risk assessment. *Fed. Regist.* **61,** 17960–18011.

U.S. Environmental Protection Agency (EPA). (1996b). Guidelines for reproductive toxicity risk assessment. *Fed. Regist.* **61,** 56274–56322.

U.S. Environmental Protection Agency (EPA). (1998a). Guidelines for neurotoxicity risk assessment. *Fed. Regist.* in press.

U.S. Environmental Protection Agency (EPA). (1998b). Draft health effects test guidelines OPPTS 870.3700 prenatal developmental toxicity study.

U.S. Environmental Protection Agency (EPA). (1998c). Draft health effects test guidelines OPPTS 870.3800 reproduction and fertility effects.

U.S. Environmental Protection Agency (EPA). (1998d). Integrated Risk Information System (IRIS). Online. National Center for Environmental Assessment, Washington, DC.

Watson, L. R., Gandley, R., and Silbergeld, E. K. (1993). Lead: interactions of pregnancy and lactation with lead exposure in rats. *Toxicologist* **13,** 349.

Vorhees, C. V. (1986). Origins of behavioral teratology. In *Handbook of Behavioral Teratology.* E. P. Riley and C. V. Vorhees, Eds. Plenum Press, New York, pp. 3–22.

CHAPTER

40

Animal/Human Concordance

JAMES L. SCHARDEIN
WIL Research Laboratories, Inc.
Ashland, Ohio 44805

*Abbreviations: CNS, central nervous system; CO, carbon monoxide; DES, diethylstilbestrol; FAS, fetal alcohol syndrome; FDA, [United States] Food and Drug Administration; FHS, fetal hydantoin syndrome; GD, gestational day; IUGR, intrauterine growth retardation; PCB, polychlorinated biphenyls; PCP, phencyclidine; VPA, valproic acid.

I. Introduction

Chemicals that cause damage to the nervous system during fetal development represent a unique challenge in hazard identification (Goldey *et al.,* 1994). On the one hand, one would expect toxic damage occurring during the fetal stage and to likely result in widespread effects. On the other hand, if greater plasticity is evidenced by younger animals, then tremendous recovery may be possible following gestational exposure, without damage.

Because humans are infinitely more sensitive than laboratory animals to known human developmental toxicants (Schardein *et al.,* 1985; Schardein and Keller, 1989), the importance of using a diversity of tests that evaluate different structural and functional domains in the hazard identification process for potential human developmental neurotoxicants is clearly evident. Thus, one outcome of these considerations has been the promulgation of developmental neurotoxicity guidelines, initially in 1986 for triethylene glycol ether,[1] followed formally with guidelines containing testing for developmental neurotoxicity in 1991.[2] How such testing translates into risk assessment was documented in 1994[3] and again in 1995.[4] Together with the ICH guidelines established in 1993[5] and the EPA OPPTS guidelines published in draft form most recently in 1996,[6] these testing protocols constitute the means by which chemicals and drugs are assessed for developmental and reproductive toxicity, including "behavioral" teratology, neurotoxicity and developmental neurotoxicity, as discussed in this chapter.

It was pointed out in the mid-1980s that more than 850 workplace chemicals could be classified as neurotoxic, and exposed human populations exceed 1 million for 65 of these chemicals (Anger, 1986). Another estimate of the problem is that the total number of neurotoxicants including drugs, food additives, cosmetics, and pesticides ranges from 1,500 to 5,000, and if environmental and commercial agents are included, these numbers may well be tripled (McMillan, 1987). Identifying such agents should clearly be a priority.

The term *concordance* means *agreement.* When we speak of the term in scientific jargon with respect to developmental toxicology, we think of developmental effects induced in animals and humans with the same agent and the same congenital malformation(s) induced in animals and humans with the same agent. Another description of this parlance is that there is mimicry between animals and humans with respect to these outcomes. Thus, in one notable example, the primate was concordant to the human when administered the teratogen thalidomide during gestation: the drug induced malformations and other classes of developmental toxicity in both. It also induced concordant malformations (limbs) in both; the effect in primates mimicked that in humans.

For development, it is assumed that the outcomes seen in specific experimental animal studies are not necessarily the same as those produced in humans. This assumption is made because of the possibility of species-specific differences in timing of exposure relative to critical periods of development, pharmacokinetics, developmental patterns, placentation, or modes of action.[7] For instance, tail defects produced in experimental animals may be more pertinent to potential human development than suspected, not because humans have no tails, but because they may signal the likelihood of defects of other organ systems. It should also be acknowledged that concordance between effects in an animal model and a human does not guarantee identification of potential human teratogens. However, concordance does impart special considerations to those agents identified initially on this basis. Take the case of thalidomide: concordance in an early rabbit study led to the expedient confirmation of the drug as a human teratogen. For many other agents studied experimentally, concordance or lack of it has not served as valuable a service. Historically, in fact, concordant effects of human teratogens in animals have usually been demonstrated after other effects confirmed the ability of the animal species to detect an agent's potential toxicity, however general. It is the purpose of this presentation to evaluate the property of concordance to developmental toxicity, with emphasis on developmental aspects of neurotoxicity.

II. Species Sensitivity

Several evaluations have been made of the biologic literature with respect to study outcomes as a measure of species sensitivity in ascertaining the extent to which animal models are predictors of potential human developmental toxicity or teratogenicity. The results of such evaluations would provide the degree of concordance between human and animal data.

There are obvious shortcomings in evaluating species sensitivity from this perspective. The animal data are derived from experimental studies in which the agents are deliberately administered at dose levels during gestation that are designed to elicit toxicity either to the mother or the fetus or to both. Humans, however, are

[1]Fed. Regist., 51: (94): 17890-17894, May 15, 1986.

[2]PB91-154617, EPA 540/09-91-123, 1991.

[3]Fed. Regist. 59(158): 42360-42404, August 17, 1994.

[4]Fed. Regist. 60(192): 52032-52056, October 4, 1995.

[5]*ICH Harmonized Tripartite Guideline: Detection of Toxicity to Reproduction for Medicinal Products,* June 24, 1993.

[6]EPA 712-C-94-207 (OPPTS 870.3700) and EPA 712-C-94-208 (OPPTS 870.3800).

[7]Fed. Regist. 61(22): 56278, 1996.

generally exposed only to low doses of the agent at varying time periods and frequencies. Thus, animal studies are much more likely to demonstrate developmental toxicity than human case studies. Nonetheless, these kinds of evaluations provide us with some useful information with respect to species usefulness (Table 1).

A. *To Human Teratogens*

The first evaluation made was an analysis by the U.S. Food and Drug Administration (U.S. FDA) as reported by Brown and Fabro (1983). They found that, of drug studies conducted on 38 drugs they identified as human teratogens, the mouse was most likely to yield a positive report; it did so 85% of the time. It was followed, in decreasing sensitivity, by the rat, rabbit, hamster, and monkey, the last at 30% sensitivity.

An evaluation along the same lines was made of the 15 recognized human teratogens at the time and the individual species responses to those agents (Schardein *et al.,* 1985). Although every agent known to be teratogenic in humans was also teratogenic in one or more laboratory species, there were different degrees of responsiveness to the agents by species. They ranked, in order, guinea pig, mouse (positive in 10/15), rat, primate, and rabbit. The last species was about equally sensitive as the hamster, dog, and pig. The cat and ferret were not seriously considered, because these two species were not tested in the same frequency as the others. The same holds true for the most sensitive species identified in this manner, the guinea pig (positive in 6). If it is discounted, the species identified from the human teratogen perspective as indicative of greatest concordance, were essentially the same:

$$\text{mouse} > \text{rat} > \text{rabbit} \approx \text{primate}$$

B. *To Human Nonteratogens*

The U.S. FDA identified some 165 drugs to which teratogenic effects had not been reported in humans; the rank order was somewhat different, as reported by Brown and Fabro (1983). The primate, at a rate of 80%, reacted negatively, followed by rabbit and rat. The mouse and hamster followed equally.

A nearly identical result from a similar evaluation was reported by Schardein *et al.* (1985). They examined data from more than 1,100 experimental animal studies in which agents not known to be teratogenic in humans were evaluated. The rabbit and primate almost equally were most often unreactive, at a rate greater than 70%, followed by the rat, then the mouse and dog equally. Thus, from the perspective of studies of agents thought not to be teratogenic in humans, the rank order for concordance from these evaluations was:

$$\text{primate} \approx \text{rabbit} > \text{rat} > \text{mouse}$$

C. *To Human Developmental Toxicants*

The final evaluation of species sensitivity to developmental toxicity is one compiled on some 51 known developmental toxicants in humans (Schardein and Keller, 1989). Overall, the ability of the most commonly used laboratory animals to correctly "flag" an agent as a human developmental toxicant with a positive response was, in order, rat, mouse, hamster, primate, and rabbit. The range of their responses was quite accurate, from 77% to 98%. When evaluated from the standpoint of concordance to human effects for the specific classes of developmental toxicity (growth retardation, death, and malformation), the order changed somewhat to mouse, rat, primate, rabbit, hamster. Ironically enough, this is the same rank order demonstrated earlier for concordant responses to known human teratogens. These latter data suggest further that a positive effect in any class in animal models could serve as the only indicator of potentially adverse effects in humans, concordant class or not.

III. Concordance to Human Teratogens

A tabular summary of animal models concordant to the 16 known groups of human teratogens is in Table

TABLE 1 Concordance: Animal and Human Teratogenicity Data

Category	Concordant species sensitivity	Sample size	Ref.
Human teratogens	Mouse > rat > rabbit > hamster > primate	38	Brown and Fabro, 1983
	Guinea pig > mouse > rat > primate > rabbit ≈ hamster ≈ dog ≈ pig	15	Schardein *et al.,* 1985
Human nonteratogens	Primate > rabbit > rat > mouse ≈ hamster	165	Brown and Fabro, 1983
	Rabbit ≈ primate > rat > mouse ≈ dog	1,135	Schardein *et al.,* 1985
Human developmental toxicants	Rat > mouse > hamster > primate > rabbit	51	Schardein and Keller, 1989
To classes of development toxicity	Mouse > rat > primate > rabbit > hamster		

2. Also tabulated are the animal species that did not produce concordant teratogenic effects to these agents when tested in the laboratory. The individual responses to these agents are summarized in subsequent sections of this chapter.

A. Alcohol

Alcohol was first described as a human teratogen in modern times by Jones and Smith (1973). With estimates as high as 68% of women reporting ingestion of alcohol during pregnancy (Hill *et al.*, 1977), literally thousands among the 3 million babies born each year in this country (Witti, 1978) will be afflicted with fetal alcohol syndrome (FAS).

The abnormalities most typically associated with alcohol teratogenicity are central nervous system (CNS) dysfunctions, growth deficiencies, a characteristic cluster of facial abnormalities, and variable major and minor malformations (Clarren and Smith, 1978).

Animals have been satisfactory models for studying FAS (Table 3). The craniofacial anomalies, characterized in the human by midfacial hypoplasia (lips, nose, jaws, eyes), has been described in beagle puppies (Ellis and Pick, 1980), C57BL/6J mouse fetuses (Sulik *et al.*, 1981), lambs (Potter *et al.*, 1981), and the pig-tailed monkey fetus (Clarren and Bowden, 1982). The concordance is striking in the case of the mouse fetus when compared to the human (Fig. 1).

Growth deficiency in the human as the result of alcohol exposure is manifested by reduction in both length

TABLE 3 Developmental Effects Induced by Alcohol

Human features[a]	Concordant species
Craniofacial anomalies	Mouse, sheep, dog, pig-tailed monkey
Growth deficiency	Mouse, rat, pig, dog, sheep
Neurological abnormalities/ developmental delay/ intellectual impairment	Rat, guinea pig, sheep, pig-tailed monkey

[a]Data from: Jones *et al.*, 1976; Clarren and Smith, 1978; Rosett, 1980.

TABLE 2 Animal Concordance to Putative Human Chemical Teratogens

Human teratogen	Markers	Animal models	
		Concordant	Nonconcordant
Alcohol	Craniofacial, neural, growth	Mouse, dog, guinea pig, rat, sheep, pig, pig-tailed monkey	Rabbit, ferret, cyno monkey, opossum
Aminoglycoside antibiotics	Inner ear	Rat, guinea pig	Mouse, rabbit, rhesus monkey
Androgens	Genital	Rat, rabbit, mouse, dog, cow, rhesus monkey, guinea pig, pig, mole, hamster, opossum, hedgehog, goat	Sheep
Anticoagulants	Skeletal	Rat	Mouse, rabbit
Anticonvulsants	Craniofacial, heart, growth, neural, CNS	Rat, mouse, hamster, rhesus monkey (variable) (see text)	Rabbit, dog, cat, gerbil, rhesus monkey
Antithyroid agents	Thyroid	Mouse, rat, guinea pig, rabbit	Cow, chinchilla
Chemotherapeutic agents	Polymorphic malformations, growth, death	Rat, pig, ferret, guinea pig; mouse, rabbit, dog, cat, sheep, rhesus monkey (all variable) (see text)	Hamster
Cocaine	CNS, heart, genitourinary, neural, growth, death	Mouse, rat	Rabbit
DES	Uterine–testicular	Mouse, rat	Hamster, ferret, rhesus monkey
Lithium	Heart	—	Mouse, rat
Mercury (organic)	Neural	Rat, mouse, guinea pig, cyno monkey, cat	Hamster, ferret, dog, rabbit
PCB	Pigment, skeletal, neural	Rat, mouse, rhesus monkey (incomplete) (see text)	Guinea pig, cow
Penicillamine	Connective tissue	Rat	Mouse, hamster
Retinoids	CNS, ear, heart, thymus	Mouse, rat, primates (3 species)	Ferret, hamster, rabbit
Tetracycline	Teeth	—	Rat, rabbit, dog, mouse, guinea pig
Thalidomide	Limbs	Rabbit, primates (8 species), armadillo (?)	Mouse, rat, dog, hamster, cat, pig, ferret, guinea pig

CNS, central nervous system; DES, diethylstilbestrol; PCB, polychlorinated biphenyls.

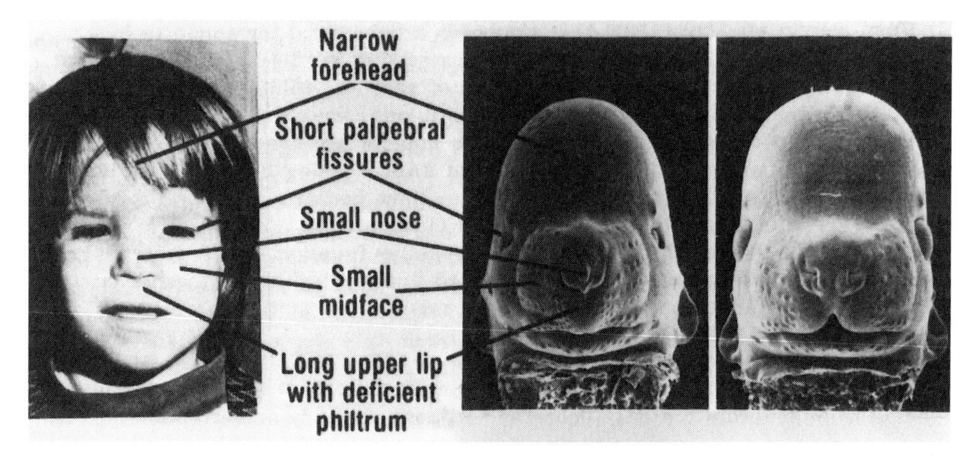

FIGURE 1 Fetal alcohol syndrome. Similarities in facies between human (left) and mouse fetus (center). A normal mouse fetus is also shown for comparison (right). (Courtesy Dr. K. Sulik, 1982).

and weight. Fetal body weight reduction has been recorded for fetuses of mice of several strains (Chernoff, 1977), rat (Sandor *et al.,* 1971), pig (Dexter and Tumbleson, 1980), and sheep (Potter *et al.,* 1981). It is not known whether reduced body length is concordant to the human, given the fact that this parameter is not usually assessed in animals.

Neurologic deficits and other CNS defects have also been observed in animal models whose mothers were treated with alcohol in pregnancy. Thus, rat pups of several strains have exhibited greater emotionality and significantly inferior learning abilities (Vincent, 1958), abolishment of reflexes (Skosyreva, 1973), microcephaly (Tze and Lee, 1975; Mankes *et al.,* 1982), and other effects on suckling behavior (Chen *et al.,* 1982). Even the subtle hypoplasia of the optic nerve has been observed in the Wistar rat fetus as well as in the human (Stromland and Pinazo-Duran, 1994). In the guinea pig, offspring from alcohol-treated mothers have shown evidenc of poor coordination and locomotion, sucking and feeding difficulties, as well as outright microscopic brain lesions (Papara-Nicholson and Telford, 1957). Lambs had reduced brain weights (Potter *et al.,* 1981), and pigtailed monkey offspring had neurologic findings similar to affected humans with the syndrome (Clarren and Bowden, 1982) as evidence of concordant CNS involvement. Several species, including the rabbit, ferret, opossum, and cyno monkey, have demonstrated teratogenic effects from maternal alcohol administration, but the effects were not concordant.

B. Aminoglycoside Antibiotics

About 60 cases of ototoxicity to the developing human fetus have been reported from treatment of the mother during pregnancy with any of several aminoglycoside antibiotics (Schardein, 1993). Drugs included are streptomycin, dihydrostreptomycin, kanamycin, and probably gentamicin. The ototoxicity ranges from minor hearing deficits and vestibular disturbances to outright high tone deafness, the effects attributable to toxicity of the eighth cranial nerve and subsequent cochlear damage.

Two species, the rat and guinea pig, have shown concordant effects to those of the human. In rats of several strains treated with streptomycin, dihydrostreptomycin, or kanamycin throughout gestation, there was damage to the organ of Corti and auditory impairment resulting in hearing deficits; the effects in some cases were not permanent (Fujimori and Imoi, 1957; Podvinec *et al.,* 1965; Onejeme and Khan, 1984). Guinea pigs administered kanamycin resulted in histologic changes in the inner ear of the offspring (Mesollela, 1963).

C. Androgens

Androgens and synthetic progestogens stimulate the action of the fetal testicular secretion on the urogenital sinus and external genitalia, but do not appreciably affect differentiation of the genital ducts or gonad proper (Grumbach and Ducharme, 1960). Thus, female offspring of women treated with these agents risk masculinization (pseudohermaphroditism). The lesion is characterized by clitoral hypertrophy with or without labioscrotal fusion (Wilkins *et al.,* 1958). There is usually a normal vulva, endoscopic evidence of a cervix, and a palpable, although usually infantile, uterus. The oviducts and ovaries are typically normal. These agents were identified as human teratogens in 1953 by Zander and Muller, and more than 250 human cases have been described to date (Schardein, 1993).

The effect in animals is comparable to that observed in humans. In virtually all species examined (some 13 species), maturation effects on gonadal structures are observed. In male offspring, the genitals are feminized, and in females, the genitals are masculinized as described earlier. The sheep is the solitary exception to concordance to the human effect (Keeler and Binns, 1968), and although dosages were sufficiently great to induce the effect, it may not have been looked for.

D. Anticoagulants

The coumarin anticoagulants, especially warfarin, have been identified as human teratogens, with about 60 cases reported to date. The characteristic embryopathy produced by these drugs when administered in the first trimester are malformations, developmental delay, low birthweight, and mental retardation, but the most consistent feature has been the skeletal malformations. These include nasal hypoplasia and bony abnormalities of the axial and appendicular skeleton, usually characterized as chondrodystrophy punctata (Shaul and Hall, 1977; Hall *et al.,* 1980).

The coumarin anticoagulants have been administered to laboratory animals, but no similar teratogenic effects to those reported in humans have been described with the exception of the rat. Although the abnormal chondrogenesis could not be replicated *in vitro* (Hallett and Holmes, 1979), studies with warfarin with added vitamin K_1 given orally at high doses to Sprague Dawley rats postnatally resulted in similar skeletal defects to those of the human (Howe and Webster, 1989, 1992, 1993). Both facial anomalies and skeletal stippling of limb bones have been demonstrated (Fig. 2).

E. Anticonvulsants

The anticonvulsant drugs considered or suspected as human teratogens fall into four groups, as follows:

1—barbiturates (phenobarbital, primidone)
2—hydantoins (phenytoin)
3—oxazolidinediones (trimethadione, paramethadione)
4—miscellaneous (carbamazepine, valproic acid)

As a group, these agents probably present a risk of 2- to 10-fold the normal rate for congenital malformations (Schardein, 1993). Collectively however, there appears to be several levels of hazard, the estimated risk ranging

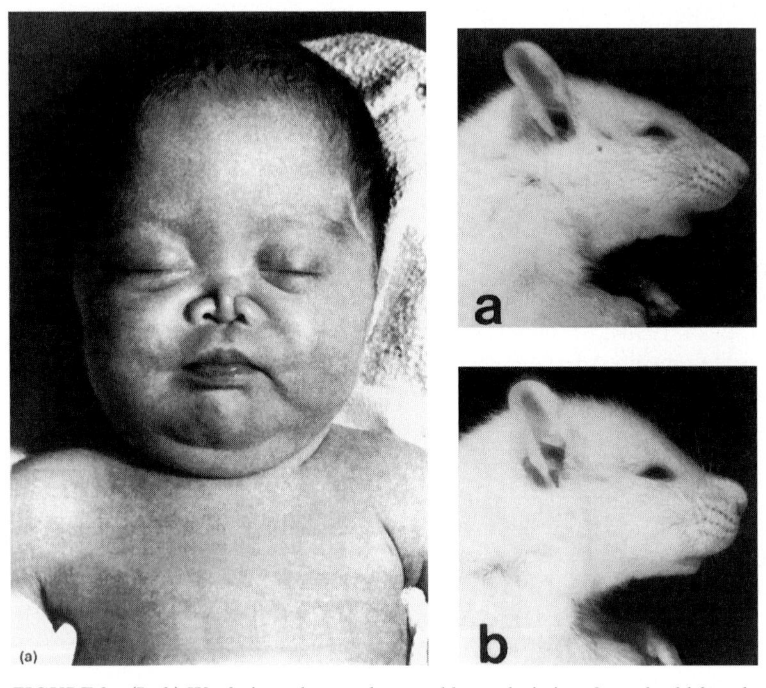

FIGURE 2 (Left) Warfarin embryopathy, nasal hypoplasia in a 3-week-old female infant. (From R. M. Pauli, J. D. Madden, K. J. Kranzler, W. Culpepper, and R. Port. Warfarin therapy initiated during pregnancy and phenotype chondrodysplasia punctata. *Journal of Pediatrics* 1976;88:506–508. Reprinted by permission of Mosby-Year Book, Inc.) (Right) Lateral view of a 3-week-old control male rat (a) and a 3-week warfarin-treated male (b). The warfarin-treated rat has a distinctive facial profile characterized by a short, slightly "upturned" and broader snout. (From The warfarin embryopathy: a rat model showing maxillonasal hypoplasia and other skeletal disturbances. A. M. Howe and W. S. Webster. *Teratology* 1992;46:379–390. Reprinted by permission of Wiley-Liss, Inc., a subsidiary of John Wiley & Sons, Inc.)

from about 75–80% of those evaluated for the oxazolidinediones and carbamazepine, 1–2% for valproic acid (VPA), to probably very low incidences for the barbiturates. Because of its widespread use, phenytoin probably accounts for the greatest number of cases of drug-induced malformation, perhaps as great as 5–10% in offspring of those whose mothers were treated in pregnancy. Each of the groups has its own pattern of developmental toxicity, and they are considered separately.

1. Barbiturates

Developmental toxicity expressed by the barbiturate anticonvulsant drugs primidone and/or phenobarbital in humans include as major effects pre- and postnatal growth deficiency, facial dysmorphia, cardiac defects, and neurologic effects, including mental retardation. Microcephaly is an associated feature.

Concordant data in laboratory animals is limited and no true model for the effects has been demonstrated (Table 4). Cardiac defects described in several mouse strains (Finnell *et al.,* 1987) and neural effects were reported in Sprague Dawley rats by another investigator (Vorhees, 1983).

TABLE 4 Developmental Effects Induced by Anticonvulsant Drugs

Major human features	Concordant species
Barbiturates[a]	
Facial dysmorphia	—
Growth deficiency	—
Cardiac defects	Mouse
Neural effects	Rat, mouse
Hydantoins[b]	
FHS: craniofacial anomalies, growth deficiency, neural effects	Mouse, rat
Oxazolidinediones[c]	
IUGR	Mouse
Facial dysmorphia	—
Heart disease	Mouse
Neural effects	Rat
Urogenital defects	—
Neonatal death	Rat
Carbamazepine[d]	
Craniofacial anomalies, growth deficiency, neural effects	—
Valproic acid[e]	
Facial dysmorphia	Rat, rhesus monkey
Neural tube defects	Hamster, mouse
Developmental delay	Rat
Postnatal growth retardation	Rat, rhesus monkey

[a]Data from: Seip, 1976; Rudd and Freedom, 1979; Myhre and Williams, 1981; Rating *et al.,* 1982.
[b]Data from: Hanson and Smith, 1975, 1977; Hanson *et al.,* 1976.
[c]Data from: Feldman *et al.,* 1977.
[d]Data from: Jones *et al.,* 1989.
[e]Data from: Dalens *et al.,* 1980; Robert, 1982; DiLiberti *et al.,* 1984; Ardinger *et al.,* 1988.

2. Hydantoins

Phenytoin induces a multisystem pattern of developmental and developmental neurotoxicity in infants of mothers treated with the drug during pregnancy (Adams *et al.,* 1990, Schardein, 1993). Characteristics of this fetal hydantoin syndrome (FHS) include a recognizable craniofacial phenotype; appendicular defects manifested by finger and nail hypoplasia, finger-like thumb and limb deformities; prenatal and postnatal growth deficiency; motor and mental deficiency; microcephaly; and cardiac, genitourinary, and skeletal anomalies as associated features.

Phenytoin is teratogenic in a number of species (Table 4), including the mouse, rat, rabbit, and primate; the hamster, dog, and cat have not demonstrated malformations (Schardein, 1993). However, the complete syndrome as observed in humans has only been reproduced in several strains of mice (Finnell, 1981). Systems anomalous in both human and mouse were growth deficiency; and ocular, neural, cardiac, renal, and skeletal defects. The syndrome was closely parallel in the two species, and correlated with maternal serum concentrations of drug. A less complete but acceptable concordance to the FHS syndrome has been reported for the Sprague Dawley CD strain rat (Lorente *et al.,* 1981). It was induced at very high doses of 1,000 mg/kg on gestation days (GD) 9, 11, and 13. The pattern of altered growth and development included growth retardation, craniofacial anomalies, distal phalangeal hypoplasia, and mental deficiency. Certain neurologic deficits postnatally, including delays in motor development and persistent impairment of locomotor function, were recorded in Wistar strain rats (Elmazar and Sullivan, 1981); other aspects of the FHS were not observed in this species. Other neurologic effects in Sprague Dawley strain rats were described by Vorhees (1983), including increased pivoting locomotion, delayed auditory startle and swimming development, increased maze errors, and increased rotational behavior, but again, other parameters of FHS were not evidenced in that study.

3. Oxazolidinediones

A characteristic phenotype common to treatment during pregnancy with either of two drugs of this type, paramethadione or trimethadione, has established these drugs as human teratogens. Comprising the fetal trimethadione syndrome, as it has come to be known, are a constellation of effects that include facial dysmorphism (including cleft lip and palate, high-arched palate, and malformed ears); congenital heart disease; neurologic effects including delayed mental development and speech impairment; and urogenital malformations. Intrauterine growth retardation (IUGR) and neonatal death have been associated with the syndrome of malformations as well.

In laboratory animals (Table 4), both drugs are teratogenic in the mouse, rat, and primate (Schardein, 1993). As with several other anticonvulsants, the mouse has proved to be closest concordant to the human, but it has not duplicated all features. At high doses of 1,045 mg/kg, IUGR, embryomortality, and congenital defects, especially those of the heart, were observed in CD-1 strain mice administered the drug on GD 8–10 (Brown *et al.,* 1979). The Sprague Dawley rat was also susceptible to some of the structural and functional effects of trimethadione administration. At doses of 250 mg/kg on GD 7–18, increased offspring mortality and several neurologic effects were observed, including increased ambulation, reduced spontaneous alternation frequency, and increased maze errors (Vorhees, 1983).

4. Carbamazepine

This is the most recent anticonvulsant recognized a probable human teratogen on the basis of a number of cases in which a number of recognizable clinical features were apparent. The fetal carbamazepine syndrome consists largely of craniofacial anomalies (upslanting palpebral fissures, short nose, long philtrum, epicanthal folds), pre- and postnatal growth deficiency, microcephaly, cardiac defects (especially ventricular septal defects), and hypoplastic fingernails (Schardein, 1993).

The drug is teratogenic in the mouse and rat (Schardein, 1993), but no concordance was apparent (Table 4).

5. Valproic Acid (VPA)

This anticonvulsant was recognized as a human teratogen a number of years ago on the basis of epidemiologic evidence. Clinically, there is facial dysmorphia, defects of the neural tube (especially spina bifida), developmental delay, and postnatal growth retardation (Schardein, 1993).

Several laboratory species induce acceptable degrees of concordance to specific features of the fetal valproate syndrome (Table 4). Two species demonstrate the neural tube defects. In the hamster (unknown strain), open neural tubes were reported in 9-day-old embryos following injection of 30 mg/100 g body weight 7 1/4 days after mating (Moffa *et al.,* 1984). The other species, the Han:NMRI strain mouse, was susceptible to induction of both spina bifida aperta (Fig. 3) and spina bifida occulata as in the human condition, from doses of 300–500 mg/kg given tid on GD 9 (Ehlers *et al.,* 1992). None of the other features of valproate were observed in either of these two species.

Craniofacial malformations have been induced in the Sprague Dawley rat that are similar to those of the fetal valproate syndrome (Binkerd *et al.,* 1988). Oral doses of 500 or 600 mg/kg on GD 8–17 produced low-set and posteriorly rotated ears, upturned nose, and dome-shaped cranium; growth retardation and underossification of axial and appendicular skeleton were reminiscent of the growth retardation and developmental delay observed in the human.

The rhesus monkey also exemplifies a model for animal concordance to certain features of the fetal valproate syndrome. Oral doses of 20–200 mg/kg (1–15 times the human therapeutic dose) on GD 21–50 induced a number of similar effects to those in the human (Hendrickx *et al.,* 1988a). Craniofacial malformations consisting of dome-shaped cranium, bulging forehead, and/or exophthalmia were observed (Fig. 3), and skeletal defects were the most significant findings. Fetal body weights, brain weights, and head circumference were reduced; embryo–fetal mortality was also observed. The effects in the monkey were dependent on pharmacokinetics and placental transfer characteristics in this species.

F. Antithyroid Agents

Antithyroid agents, particularly iodides, thiourea compounds, and imidazoles, induced goiter and hypothyroidism among offspring of mothers given excessive quantities during early and midpregnancy. Although not strictly a malformation, these agents have been characterized as human teratogens since the early part of the 20th century; some 140 cases have been reported in the literature (Schardein, 1993). The lesions induced by these drugs include goiter, hypothyroidism, and hyperplasia and hyperthyroidism may be present. Microscopically, there are variably sized thyroid follicles and little secretion; secondary compensatory hypertrophy is a common finding.

Every laboratory species tested, including mouse, rat, guinea pig, and rabbit, has been reported to possess concordant lesions to the human (Schardein, 1993).

G. Chemotherapeutic Agents

Of the cancer chemotherapeutic drugs, several antimetabolites and alkylating agents are considered human teratogens (Schardein, 1993). In the antimetabolite group, at least five drugs are placed in this category: aminopterin, methotrexate, azauridine, cytarabine, and fluorouracil. Of the alkylators, four drugs, busulfan, chlorambucil, cyclophosphamide, and mechlorethamine, are probable teratogens in humans. Together, the two groups account for some 35 cases of human

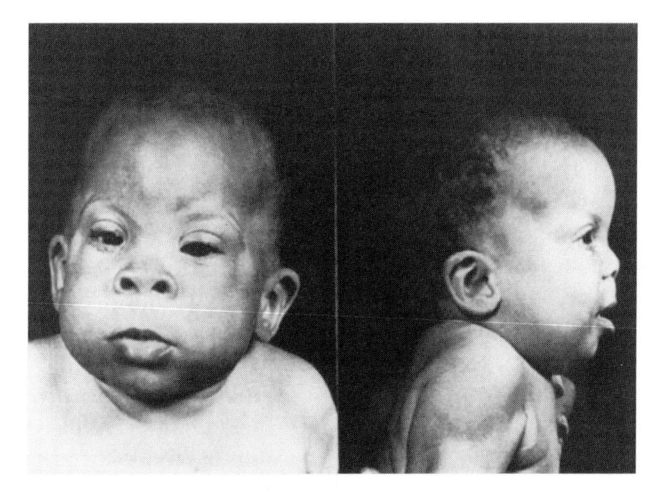

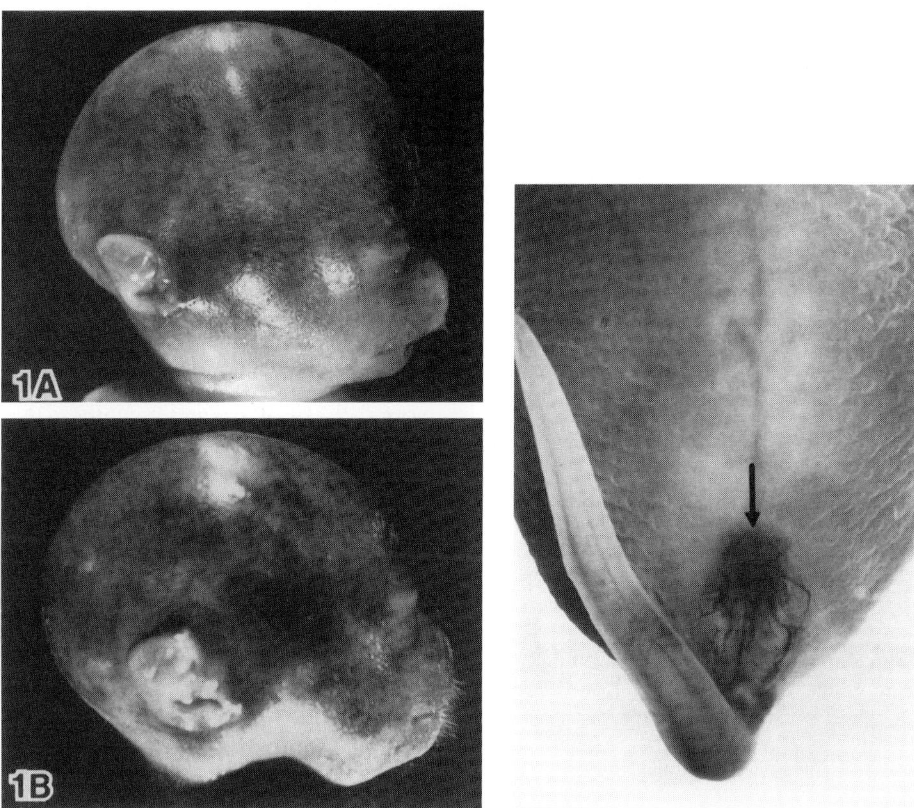

FIGURE 3 (Top) Fetal valproate syndrome. Note prominent metopic ridge, outer orbital ridge deficiency, and midface hypoplasia with short nose, broad nasal bridge, anteverted nostrils, low flat philtrum, and posterior angulation of the ears. (Verification of the fetal valproate syndrome phenotype. H. H. Ardinger, J. F. Atkin, R. D. Blackston, L. J. Elsas, S. K. Clarren, S. Livingstone, D. B. Flannery, J. M. Pellock, J. J. Harrod, E. J. Lammer, F. Majewski, A. Schinzel, H. V. Toriella, and J. W. Hanson. *Am. J. Med. Genet.* 1988;29:171–185. Copyright © 1988, John Wiley & Sons, Inc. Reprinted by permission of Wiley-Liss, Inc., a subsidiary of John Wiley & Sons, Inc.) Spina bifida is also a common feature. (Bottom 1B). Mandibular hypoplasia in fetus from valproate-treated rhesus monkey (200 mg/kg/day) on GD 21–50 (bottom 1A), control fetus. (Valproic acid developmental toxicity and pharmacokinetics in the rhesus monkey: an interspecies comparison. A. G. Hendricks, H. Nau, P. Binkerd, J. M. Rowland, J. R. Rowland, M. J. Cukierski, and M. A. Cukierski. *Teratology* 1988;38:329–345. Copyright © 1988, John Wiley & Sons, Inc. Reprinted by permission of Wiley-Liss, Inc., a subsidiary of John Wiley & Sons, Inc.) (Bottom right) Day 18 mouse fetus whose dam was exposed to 1,500 mg/kg sodium valproate on GD 9; the malformation (arrow) is spinal bifida aperta. (Valproic acid–induced spina bifida: a mouse model. K. Ehlers, H. Sturje, H.-J. Merker, and H. Nau. *Teratology* 1992;45:145–154. Copyright © 1988, John Wiley & Sons, Inc. Reprinted by permission of Wiley-Liss, Inc., a subsidiary of John Wiley & Sons, Inc.)

malformations. As might be expected from potent chemicals such as these with cytotoxic properties, their developmental toxicity profile is one of growth retardation, fetal–neonatal death, and congenital malformations. The last are polymorphic, representing numerous organ systems. Thus, human toxicity elicited with these drugs presents the whole gamut of developmental toxicity, except neural components; neurotoxicity has not been an associated feature.

Concordant malformations to those observed in humans have been recorded in a number of species and strains (Table 5). With rare exceptions, however, complete concordance to the extent of classifying the animal as a model has not occurred.

1. Antimetabolites

With aminopterin, the rat (Sansone and Zunin, 1954), sheep (James and Keeler, 1968), and dog and pig (Earl *et al.,* 1973) have all provided the gamut of developmental effects to include skull, limb, and jaw malformations; reduced fetal weight; and embryonic–fetal death. The findings in lambs closely parallel the human (Fig. 4). Similarly, with methotrexate, the rat (Wilson and Fradkin, 1967), rabbit (Jordan *et al.,* 1970), mouse (Skalko and Gold, 1974), rhesus monkey (Wilson, 1971), and cat (Khera, 1976) produced concordant developmental toxicity in those species. 6-Azauridine, less extensively studied than the other antimetabolites mentioned, induced concordant toxicity in the rat (Gutova *et al.,* 1971). The mouse (Nomura *et al.,* 1969) and rat (Chaube and Murphy, 1965) exhibited concordant developmental toxicity following cytarabine treatment to dams. With the remaining antimetabolite, 5-fluorouracil, the rat (Murphy, 1962), mouse (Dagg, 1960), guinea pig (Kromka and Hoar, 1973), and rhesus monkey (Wilson, 1971) had concordant findings to the human effects.

TABLE 5 Developmental Effects Induced by Chemotherapeutic Drugs

Major human features (agent and group)	Concordant species
Polymorphic malformations, growth retardation, death	
Antimetabolites	
Aminopterin	Rat, sheep, dog, pig
Methotrexate	Rat, rabbit, mouse, cat, rhesus monkey
Azauridine	Rat
Cytarabine	Mouse, rat
Fluorouracil	Mouse, rat, guinea pig, rhesus monkey
Alkylating agents	
Busulfan	Rat, mouse
Chlorambucil	Rat, mouse
Cyclophosphamide	Rat, mouse, rabbit, rhesus monkey
Mechlorethamine	Rat, mouse, ferret, rabbit

2. Alkylating Agents

Both the mouse (Pinto Machado, 1969) and rat (Weingarten *et al.,* 1971) had concordant developmental toxicity to that produced in the human with busulfan. The same two species were also concordant for chlorambucil (Murphy *et al.,* 1958; Tanimura *et al.,* 1965). With cyclophosphamide, the rat (Murphy, 1962), mouse (Hackenberger and Kreybig, 1965), rhesus monkey (Wilk *et al.,* 1978), and rabbit (Gerlinger and Clavert, 1964) have provided a similar pattern of developmental toxicity to that of the human. Mechlorethamine administered to the mouse (Thalhammer and Heller-Szollosy, 1955), rabbit (Gottschewski, 1964), rat (Haskin, 1948), and ferret (Mould *et al.,* 1973) resulted in concordance toxicity to the developing young.

H. Cocaine

The alkaloid cocaine is the most recent agent to be considered a human teratogen. Derived from the coca plant, its abuse as a street drug has accelerated to extraordinary levels since the mid-1980s. Unfortunately, some 2–3% of pregnant women in the United States are believed to use the drug during pregnancy (Lindenberg *et al.,* 1991). Thus, literally hundreds of babies of cocaine-addicted mothers are born with congenital malformations and other developmental toxicity.

The malformations characterizing cocaine-induced terata are CNS (micro- or hydrocephaly), cardiovascular, and genitourinary in type (Schardein, 1993). Some or all of these defects are thought to be due to disruptive vascular phenomena induced by the chemical's vasoconstrictive action (Chasnoff, 1991; Jones, 1991). Accompanying toxicity includes reduced birthweight–IUGR, spontaneous abortion, perinatal morbidity and mortality, and complications at birth. Impaired neonatal behavior and neurologic problems occur in significant frequency. Reported effects in this regard include abnormal brain wave patterns, short-term neurologic signs, depression of interactive behavior, and poor organizational responses to environmental stimuli (Schardein, 1993). Whether such neurologic findings translate into significant learning and behavioral problems remains to be seen.

Mice of two strains (DBA/2J and SWV) and the Sprague Dawley rat may be considered concordant species, although neither has shown all the markers observed in humans. Both species demonstrated the malformative effects of the CNS (Webster *et al.,* 1991) and

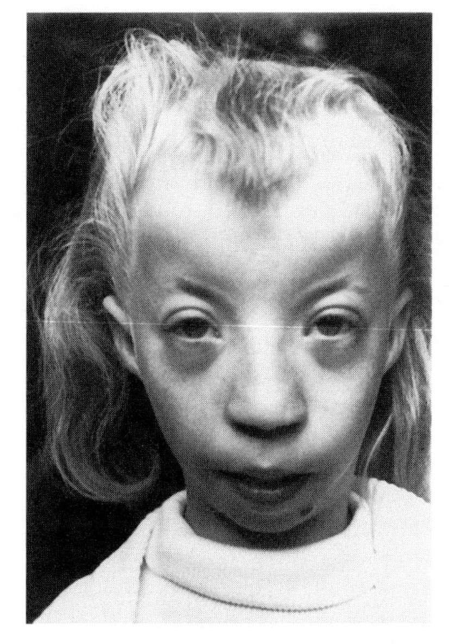

FIGURE 4 (Left) Child, age 9 from woman treated with aminopterin during pregnancy. (E. B. Shaw. Fetal damage due to maternal aminopterin ingestion—followup at age 9 years. *Am. J. Dis. Child* 1972;124:93–94. Reprinted with permission of the American Medical Association.) (Right) Lamb born to a ewe receiving aminopterin on GD 40–70. Note small ears and limb defects; the head characteristics are similar to those induced in humans. (Teratogenic effects on aminopterin in sheep. L. F. James and R. F. Keeler. *Teratology* 1968;1:407–412. Copyright © 1968, John Wiley & Sons, Inc. Reprinted by permission of Wiley-Liss, Inc., a subsidiary of John Wiley & Sons, Inc.)

heart and genitourinary systems (Finnell *et al.,* 1990). Both also evidenced increased fetal death and growth retardation (Fantel and MacPhail, 1980; Mahalik *et al.,* 1980). Neural toxicity was confined to rats of several strains and included increase in startle reaction and exploratory behavior (Foss and Riley, 1988), increased locomotor activity (Hutchings *et al.,* 1989), and other postnatal behavioral effects (Smith *et al.,* 1989).

I. Diethylstilbestrol

Diethylstilbestrol (DES) has been considered to be a human teratogen since the mid-1970s (Schardein, 1993). Rather than producing congenital malformations, however, the chemical induces latent effects on reproductive parameters. That is, it acts on the reproductive processes of daughters whose mothers were treated with the drug some years previously. The principal effect is on the vaginal epithelium, causing adenosis and, in extreme cases, clear cell adenocarcinoma. Associated with the vaginal lesions are irregular menses, reduced pregnancy rates, increased spontaneous abortion, and preterm delivery. More than 400 cases have been collected to date, and more recent findings suggest that there may well be adverse effects (testicular) on male offspring as well. No neural toxicity has been observed in either gender.

As far as the vaginal effect is concerned, several animal models exist. The first reported was the CD-1 strain mouse, which, after treatment with the drug for up to 18 months, was observed to exhibit both adenosis and adenocarcinoma (Newbold and McLachlan, 1982). This was confirmed in the same strain by others (Walker, 1983). Dose-related vaginal neoplasms at an incidence of 40–90 times higher than observed in humans were soon reported in Wistar rats (Miller *et al.,* 1982). Although the biologic processes between rodents and humans demonstrate some similarities, substantial differences exist, such as in tumor type induced, the relationship of treatment duration to induction, and so on; thus, the validity of the animal models has been questioned. Genital tract abnormalities have also been produced in the Syrian hamster (Gilloteaux *et al.,* 1982), ferret (Baggs and Miller, 1983) and rhesus monkey (Hendrickx *et al.,* 1988b), but no neoplasms have been recorded to date.

J. Lithium

Lithium carbonate, used therapeutically in treating manic-depressive psychosis, has been considered as a possible human teratogen, inducing congenital cardiac defects as a marker. Of the 25 cases or so documented, six have been termed Epstein's anomaly, in which the tricuspid valve is displaced into the right ventricle (Schardein, 1993). While the drug is teratogenic in rodents, it has not been in rabbits or primates, and there is no concordant species for this drug.

K. Methylmercury

Methyl (organic) mercury was first identified as a human teratogen in 1952 and later from environmental contamination in Japan (Schardein, 1993). Clinically, the victims presented symptoms of cerebral palsy (Murakami, 1970). Strabismus, blindness, speech disorders, and motor impairments (including ataxia, chorea, athetosis, ballismus-like movements, coarse tremors, myoclonus, and generalized convulsions) were associated features. Abnormal reflexes and mental retardation were common in these patients. Death occurred in about one-third of affected individuals. About 700 adult cases were verified. A concentration as low as 0.2 mg/g of mercury for a patient with neurologic symptoms was found in one study (Higgins, 1975). The affliction was not confined to adults. Neurologic deficits, microcephaly, and abnormal dentition were recorded in a total of about 40 infants born of the affected mothers. This effect was termed *fetal Minamata disease.* Pathologic changes have been well-described in the CNS (Matsumoto *et al.,* 1965). The fetal disease has since been documented in Sweden, the Soviet republics, the United States, and Iraq from mercury-contaminated foodstuffs. The chemical is one of the best known developmental neurotoxicants; a review of its properties in humans and animals has been published (Burbacher *et al.,* 1990).

Animal studies have been conducted on a number of mercury salts, and several have identified animal models bearing concordant developmental neural effects (Table 2). Mercury salts including chloride, sulfide, dicyandiamide, hydroxide, and dimethylmercury have all been implicated. In mice of at least three different strains, brain abnormalities, nonneural malformations and postnatal behavioral effects have been recorded (Murakami, 1972; Spyker and Smithberg, 1972), all concordant effects to the human. Abnormal swimming behavior has been produced in this species as well (Weiss and Doherty, 1975). Several strains of rats have also provided effects mimicking the human: brain lesions and fetal death (Tatetsu, 1968; Murakami, 1972), but in a developmental study designed to assess neurotoxic potential with dimethylmercury, only moderate evidence of nervous system damage was recorded in this species in contrast to the severe neurotoxicity demonstrated in the human (Goldey *et al.,* 1994). Similar effects were also reported for the cat (Moriyama, 1967; Khera, 1973) and guinea pig (Inouye and Kajiwara, 1988). Still another species, one in which studies have been performed for the purpose of comparison to the human condition, the cynomolgus monkey, has been reactive. In the first experiment, monkeys given 50 or 90 μg/kg methylmercury (hydroxide) produced microcephalic offspring of reduced size with functional changes of the CNS; there was also increased death as in the human (Burbacher *et al.,* 1984). The chemical given at various intervals in the reproductive cycle of the species produced varying effects, most notably reduction in viable offspring (Burbacher *et al.,* 1988). In the final study, the same investigators suppressed social interaction and increased nonsocial behavior in the offspring (Burbacher *et al.,* 1990). Methylmercury had developmental effects in several other species as well, but did not induce the neural component common to this developmental neurotoxicant.

L. Polychlorinated Biphenyls

The polychlorinated biphenyls (PCBs), a group of some 210 related chemicals, have been considered human teratogens since reports of food contamination in Japan in the mid-1970s (Schardein, 1993). One of these chemicals, Kaneclor 400, leaked in a manufacturing process for cooking oil, which was then ingested; some 1,500 adults became affected with this rice-oil cooking disease, or "Yusho" (Katsuki, 1969).

Toxicity was not limited to adults. Yusho was known to have affected nine pregnant women and their fetuses initially (Taki *et al.,* 1969), and four more later (Yamashita, 1977). A similar episode occurred in Taiwan in 1979 (Lan *et al.,* 1981), registering 2,000 more adult cases and eight fetal cases (Wong and Hwang, 1981). Eventually, a total of about 165 fetal cases were recorded. Congenital findings appear to be limited primarily to cola staining of the skin, due to increased melanin pigmentation in the epidermis (Kikuchi, 1984). Other findings include exophthalmus, abnormal dentition, edematous face, abnormal calcification of the skull, and IUGR. Examination of 7- to 9-year-old Japanese children indicated them to be apathetic, with "soft" neurologic signs (Harada, 1976) as evidence of developmental neurotoxicity. Other workers found lower intelligence scores among exposed children (Rogan *et al.,* 1988). In studies conducted in the United States on PCB-exposed women, in addition to clinically apparent developmental delay, neurodevelopmental effects were characterized primarily by motor deficits (Jacobson *et al.,* 1985; Rogan *et al.,* 1986; Gladen *et al.,* 1988; Jacobson and Jacobson, 1988).

Although not eliciting the pigmentation and the other structural developmental effects observed in humans, several species of laboratory animals have proved to be concordant models for the developmental neurotoxicity aspects of exposure (Table 2). With various PCBs, mice of several strains have demonstrated increased activity, increased latency in active avoidance testing, have shown neurologic signs, and "spinning" activity from exposure of dams to PCBs in the perinatal period (Chou *et al.,* 1979; Tilson *et al.,* 1979; Storm *et al.,* 1981). Several strains of rat pups have shown decreased activity and impaired neurologic development, swimming activity, active avoidance responses, and water maze acquisition (Shiota, 1976; Koja *et al.,* 1978; Overmann *et al.,* 1987; Pantaleoni *et al.,* 1988) following maternal exposure to PCBs. Rhesus monkey babies of mothers treated with PCBs evidenced increased activity and impaired learning (Bowman *et al.,* 1978, 1981; Mele *et al.,* 1986). A comparison of species effects in response to administration of PCBs has been made by Tilson *et al.* (1990), and a review of PCB developmental neurotoxicity in humans by Schantz (1996).

M. Penicillamine

The use of the drug D-penicillamine in the treatment of Wilson's disease has resulted in its identity as a human teratogen (Schardein, 1993). It has caused a generalized connective tissue defect, including lax skin, hyperflexibility of joints, vein fragility, varicosities, impaired wound healing, and inguinal hernias in nine human cases (Rosa, 1986). The effect is apparently reversible (Linares *et al.,* 1979) and is not accompanied by any other class of developmental or neural toxicity.

Penicillamine is teratogenic in mice, hamsters, and rats (Schardein, 1993). However, only in the Sprague Dawley rat has it produced connective tissue abnormalities concordant to that of the human (Table 2). A study in which the drug was fed at a concentration of 0.8% throughout gestation resulted in 21% of the fetuses being malformed, including the malformation cutis laxa, and increased resorption (Hurley *et al.,* 1982).

N. Retinoids

The retinoids constitute a diverse series of naturally occurring chemicals that are synthetic congeners of vitamin A. Of interest here in the context of human teratogens are three drugs—isotretinoin, tretinoin, and etretinate—which are considered responsible for more than 140 reported cases of congenital malformations (Schardein, 1993; Rosa *et al.,* 1994). All three drugs have therapeutic use in treating serious skin disorders.

Malformations characteristic of human exposure to these retinoids include ear, thymic, CNS, and cardiovascular anomalies (Lynberg *et al.,* 1990). Not all effects seen show mimicry between the three agents (Rosa, 1991). The embryopathy is often accompanied by growth retardation and spontaneous abortion.

All three retinoids discussed are potent teratogens in laboratory animals, inducing a variety of malformations that include as primary defects the CNS malformations observed in humans (Table 2). Among the model or concordant species are several strains of mice (Lofberg *et al.,* 1990; Padmanabhan *et al.,* 1990), rats (Aikawa *et al.,* 1982; Collins *et al.,* 1994), and mini-pig (Jorgensen, 1994). In addition, three species of primates: cynomolgous (Hummler *et al.,* 1990), rhesus (Wilson, 1971), and pig-tailed monkey (Newell-Morris *et al.,* 1980) are also concordant to the human condition. Doses required to elicit the pattern of defects range from about 12.5- to 250-fold the human therapeutic dose, with the primate (cyno monkey) most sensitive and the mouse least sensitive (Schardein, 1993). The high sensitivity of the primate as a model for retinoid toxicity is due to prolonged and pronounced exposures to both maternal and embryonic compartments by the parent compound and its 4-oxo metabolite, which is implicated as the proximate teratogenic agent in both this species and the human (Hummler *et al.,* 1994). The neurobehavioral aspects of the retinoids have been reviewed (Adams and Lammer, 1993).

O. Tetracyclines

The antibiotic tetracycline and its congeners deposit as fluorescent compounds in calcifying teeth and bones. Although not strictly a structural defect, the effect occurs prenatally up to 7 or 8 years of age and results in yellow to brown staining of the deciduous teeth that can be elicited fluorescently (Schardein, 1993). It is presumed that thousands of cases exist.

No animal model concordant for tetracycline effects is known.

P. Thalidomide

Thalidomide is, of course, the prototype teratogen identified in the human, and was responsible in the early 1960s of inducing characteristic hypoplastic and aplastic limb defects in perhaps as many as 7,700 babies, chiefly in West Germany, the United Kingdom, and Japan (Schardein, 1993). The United States was spared this misfortune through approval inaction by the FDA based on an incomplete drug submission by the German manufacturer and its American licensee. Coupled with the bilateral but asymmetric limb reduction defects perhaps best described as phocomelia, a number of other malformations were observed, especially of the ears, eyes, and viscera. However, it is the dysmelic arms and legs that characterize the induced anomaly.

Importantly, it was the concordant limb defects induced in the New Zealand rabbit (Somers, 1962) in the earliest published animal studies that confirmed the action of thalidomide on the developing fetus (Table 2), and although thalidomide has been shown to be teratogenic in some 15 species of experimental animals (the guinea pig being the sole exception), the induction of limb defects has been limited with any consistency to rabbits and primates (Fig. 5). In rabbits, at least six specific breeds as well as mixed, hybrid, crossbred, and common stock rabbits have demonstrated the characteristic limb defects at doses as low as 25 mg/kg/day during organogenesis, a level some 25-fold the effect level in the human (Schardein, 1993). Among primates, the cynomolgous, rhesus, stump-tailed, bonnet, Japanese, and green monkeys, the baboon and marmoset are concordant species, the bushbaby (*Galago* sp.) being the only primate species tested that was nonreactive. Doses of thalidomide in the range of 5–45 mg/kg/day (5- and 45-fold, respectively, the human effective dose) encompassing GD 18–48 in the multiple primate species were the experimental conditions necessary to induce the defect. It should be mentioned too, that a single phocomelic armadillo fetus has been recorded (Marin-Padilla and Benirschke, 1963), but the rarity of this species as an experimental model for development precludes the assumption that this event was, in fact, drug induced and therefore concordant.

IV. Concordance to Some Suspected Human Developmental Neurotoxicants

A. Carbon Monoxide

Carbon monoxide (CO, illuminating gas) is a known human neurotoxicant. This is not too surprising, given the fact that it is the major atmospheric pollutant in the United States and many other countries of the world, probably accounting for more than one-half of all airborne pollution (Schardein, 1993). Its toxicity resides in its ability to compete with oxygen in hemoglobin formation, thereby interfering with tissue oxygenation.

The pregnant woman, her fetus, and the newborn infant are all particularly vulnerable to low concentrations of CO (Longo, 1977). Of some 19 reported cases of maternal poisoning to the gas, 15 surviving infants subsequently developed neurologic sequelae of varying types of severity (Schardein, 1993). Even relatively low levels of CO that result in blood carboxyhemoglobin concentrations of 4–5% result in subtle alterations in

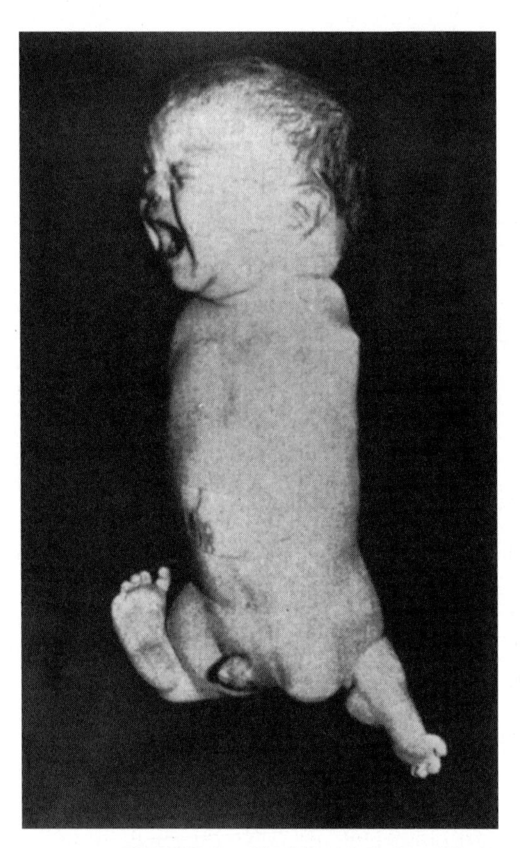

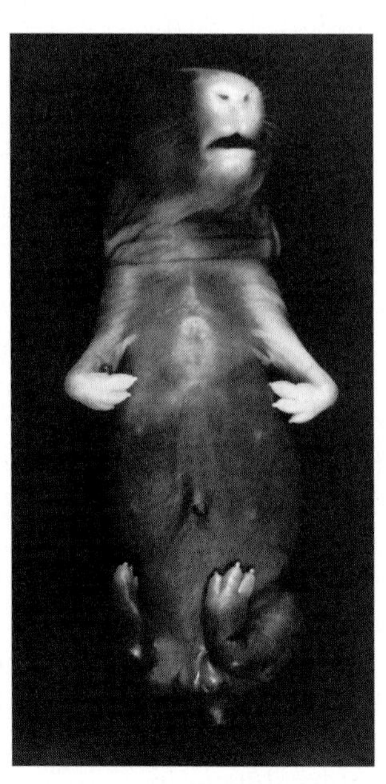

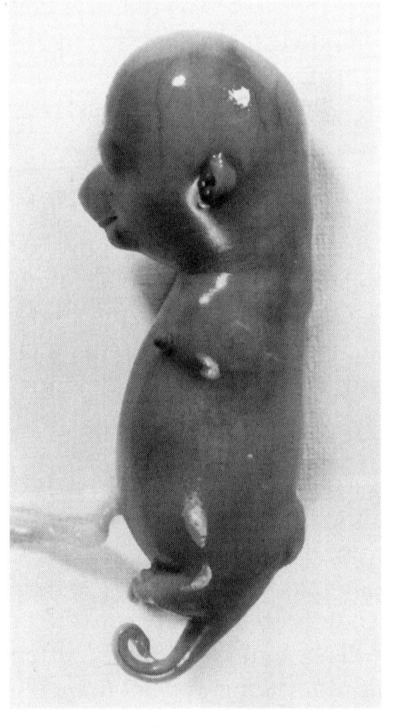

FIGURE 5 Thalidomide-induced limb malformations: (Left) Human. (H. B. Taussig. Thalidomide and phocomelia. *Pediatrics* 1962;30:654–659.) Pictures courtesy of Drs. W. Lenz and H. R. Wiedemann. (1962). *Pediatrics* **30,** 654–659. (Center) Bunny from doe receiving 150 mg/kg on GD 6–14 (Schardein, 1976, unpublished); (Right) Baboon fetus, 9 weeks of age from mother treated with unspecified amount of drug (Schardein, 1967, unpublished).

mental ability and performance of numerous functions in normal adults (Longo, 1977). In infants exposed *in utero,* neurologic findings have varied from subnormal mentality, retarded psychomotor development, "softening" of basal ganglia of brain, cerebral atrophy, microcephaly, and convulsions to frank mental retardation, cerebral palsy, and death.

Animals have not yet shown perfectly concordant neurotoxicity to the human (Table 6), but the reason probably lies in the fact that experimental exposures have not been replicated in the same manner at the same exposure levels as humans have been subjected to in environmental exposures. Rats exposed to 150 ppm inhalationally through pregnancy have exhibited behavioral effects (Mactutus and Fechter, 1984), mice exposed to 65 or 125 ppm CO by inhalation during organogenesis demonstrated aberrant righting reflex and negative geotaxis response (Singh, 1986), and rhesus monkeys exposed to CO produced fetuses with hemorrhagic necrosis of the cerebrum (Ginsberg and Myers, 1974).

B. Cigarette Smoking/Nicotine

Although not recognized as a human teratogen, cigarette smoking induces a number of adverse reproductive effects (Schardein, 1993). The principal chemical in tobacco smoke is the alkaloid nicotine. Included among the induced effects are a variety of neurologic impairments and behavioral alterations that have been reported among offspring of women smokers. Although such effects have been questioned by some (Hardy and Mellitis, 1972; Richardson *et al.,* 1989), some, but not all, of the neurotoxic effects reported include hyperkinesis (Denson *et al.,* 1975); less visual altertness and atypical sleep patterns (Landesman-Dwyer *et al.,* 1978); reduced vigilance performance (Kristjan *et al.,* 1989); subnormal performance on operant tasks such as head turning and suckling (Martin *et al.,* 1977); less satisfactory neurologic and intellectual maturity than normal to age 6½ years (Dunn *et al.,* 1977); lower scores of general and math ability at 8 months, reading comprehension at 9 months and significant impairment of reading ability at 7 years of age (Butler and Goldstein, 1973); effects on cognitive function at age 3 years (Sexton, 1990); and nonspecific behavioral changes (Pley *et al.,* 1991).

TABLE 6 Animal Concordance to Suspect or Known Human Developmental Neurotoxicants

Developmental neurotoxicant	Concordant animal models
Carbon monoxide	Rat, mouse, rhesus monkey
Cigarette smoke	—
Cadmium	Rat, mouse
Lead	Sheep, rat, squirrel monkey
Gasoline	—
Toluene	Rat
Heroin	—
Methadone	—
Phencyclidine	Mouse, rat

Laboratory animals have not been particularly good models for effects originating from smoking tobacco (Table 6). This is most likely due to the fact that replicate experimental procedures are difficult, if not possible, to achieve, given the wide number of exposure conditions necessary to duplicate effects. At any rate, neurotoxicity has not been demonstrated with smoking situations in laboratory animals. Effects on postnatal behavior were reported in rats given nicotine (Martin and Becker, 1970; Fung, 1988), but these could not be considered concordant effects to the type elicited in exposed human fetuses from chronic exposures to cigarette smoke.

C. Metals

Several metals have been identified as human developmental neurotoxicants.

1. Cadmium

Exposure to any of a variety of cadmium salts under certain conditions of exposure produced neurotoxicity in the developing human (Schardein, 1993). Indeed, there is a reported association between mental retardation and concentration of cadmium in the hair of school-age children (Marlowe *et al.,* 1983).

Developmental neurotoxicity has also been reported in both the mouse (Webster and Valois, 1981) and rat (Wong and Klaassen, 1982) (Table 6). The effects in both species on the incompletely developed CNS were elicited by treatment with cadmium early in postnatal life and consisted of activity alterations in open field or activity mazes.

2. Lead

Inorganic lead has been recognized as a poison for more than 1,000 years (Winter, 1979). Although the principal route of exposure is via food and beverages, it is the environmental sources (lead-based paint, automotive and industrial emissions, lead-glazed earthenware, and lead dust) that produce excessive toxicity (Klaassen *et al.,* 1986). Daily intake from all sources ranges from 0.1 mg to greater than 2 mg/day (NAS, 1972), although these values have probably decreased over time, given the banning of lead-based paint and use of unleaded fuel.

The chemical is clearly a reproductive toxicant in the widest sense of the term, with significant associations made between lead exposures and preterm birth or decreased gestational maturity, lower birthweight, reduced postnatal growth, and increased incidence of minor congenital anomalies as summarized by Dietrich (1991). High stillbirth and miscarriage rates have also been part of the toxicity profile (Scanlon, 1975). Of particular interest here are the early deficits in postnatal neurologic or neurobehavioral status attributed to lead. Developmental neurotoxicity by lead has been described in the human in several reports; the principal findings have been reduced intelligence (Needleman, 1985; Bellinger *et al.,* 1987) and mental retardation (Beattie *et al.,* 1975). Blood level concentrations of 30 μg/dl or greater are at risk to the pregnant mother (Swinyard *et al.,* 1983).

Developmental neurotoxicity has also been observed in offspring of dams treated with lead (Table 6). The comparative developmental neurotoxicity between humans and animals has been reviewed (Davis *et al.,* 1990). The greatest qualitative similarities were observed to involve relatively complex behavioral processes such as cognition and learning. In essence, some examples include slowed learning in lambs (Carson *et al.,* 1974) and rat pups (Brady *et al.,* 1975), reduced brain weight and brain lesions in the squirrel monkey (Logdberg *et al.,* 1988), and postnatal behavioral alterations in rats (Tesh and Pritchard, 1977). These are all acceptable evidences of concordance to human neurotoxicity induced with lead.

D. Solvents

1. Gasoline

Several reports have attested to a neurotoxic component along with structural anomalies following maternal exposure to gasoline; it was termed *fetal gasoline syndrome.* Hypertonia was the neurologic effect described in two offspring (Hunter *et al.,* 1979; Greenberg *et al.,* 1984).

No concordance to experimental animals has been demonstrated.

2. Toluene

Both occupational and exposures related to addictive abuse of toluene have led to a number of reports describing neurotoxicity in addition to developmental toxicity, as part of what has been termed *fetal solvents syndrome* (Schardein, 1993). Among the neurologic findings in offspring of exposed women have been mental disability (Toutant and Lippman, 1979); cerebellar dysfunction (Streicher *et al.,* 1981); delayed motor and cognitive development (Goodwin, 1988); and hypotonia, developmental delays, and CNS dysfunction (Hersh, 1989).

The rat would appear to represent a concordant species (Table 6). A persisting motor syndrome in this species has been described resembling to some extent at least, the syndrome seen in humans who are heavy abusers of toluene; particular reference was made to the wide-based ataxic gait of the animals (Pryor, 1991).

E. Opioids

Narcotics and drugs of abuse are difficult to assess with respect to toxicity, because their use is confounded by use of other agents, pattern of use, and so on. Nonetheless, it has been fairly well established that two of the agents in this category exhibit developmental neurotoxicity in offspring of maternal users. These are heroin (diacetylmorphine) and methadone.

1. Heroin

Although there were no distinctive neurologic problems demonstrated in human subjects exposed to heroin, Wilson (1989) reported weakness in the area of motor coordination and visual–motor perception function; "retardation" was much greater than in normal populations. Microcephaly was recorded in another report (Fulroth *et al.,* 1989).

There have been no good examples of animal concordance to these human effects, although CNS defects were observed in hamster fetuses of dams treated with the drug at very high doses (Geber and Schram, 1969).

2. Methadone

A report of neurobehavioral findings in 18-month-old children of women taking methadone may be predictive of later learning and behavioral problems (Rosen and Johnson, 1982). They found significant incidences of reduced head size; neurologic findings of tone discrepancies, developmental delays, and poor fine motor coordination; and significantly lower scores in mental and motor developmental indices. Other neurotoxic effects reported include decline in psychomotor performance (Strauss *et al.,* 1976); psychological problems (Wilson *et al.,* 1981); and poorer motor coordination, increased body tension, and delayed acquisition of motor milestones (Hans, 1989).

As with heroin, there has been no animal concordance. In fact, the drug did not affect several different operant tasks in behavioral studies conducted in rats (Hutchings *et al.,* 1979). CNS structural defects however, were produced in hamster embryos whose mothers were administered low doses of the drug (Geber and Schram, 1969).

F. Phencyclidine

Phencyclidine, a veterinary anesthetic used illicitly as a street drug ("PCP," "angel dust," etc.), has been documented as a possible neurotoxicant in humans. Abnormal behavior and spastic quadriparesis (Golden *et al.,* 1980), central nervous system defects including microcephaly (Strauss *et al.,* 1981; Michaud, 1982), and emotional problems (Chasnoff *et al.,* 1983) have been reported as evidence of this effect.

Both the mouse (Nicholas and Schreiber, 1983) and rat (Jordan *et al.,* 1979; Hutchings *et al.,* 1984; Nabeshima *et al.,* 1988) may be considered concordant to the human on the basis of responses to postnatal behavioral testing of pups whose dams received phencyclidine during pregnancy (Table 6).

V. Animal Developmental Neurotoxicants in Which Human Concordance Has Not Been Demonstrated

This presentation thus far has dealt with animal models for agents that affect human development. There are also agents that have shown developmental neurotoxicity in laboratory animals that have not (yet) demonstrated that property in humans (Table 7). This may be accounted for by a number of factors, the most important of which is the exposure level. Shown are representatives of a wide variety of chemical classes.

TABLE 7 Selected Animal Developmental Neurotoxicants (Behavioral Teratogens) That Have Not Been Confirmed in Humans

Agent(s)	Reactive species
Alcohols (methyl-, *t*-butyl-)	Rat
Azacytidine	Mouse
Carbon disulfide	Rat
Cycasin	Hamster, rat, ferret
Fungicides (carbendazim, maneb)	Rat, mouse
Hexachlorobenzene	Rat
3,3-Iminodipropionitrile	Rat
Insecticides	
(Chlordecone, chlordimeform, DDT, endrin, trichlorfon)	Rat, mouse, hamster, pig
Mecoprop Herbicide	Rat
Metals	
(Manganese, molybdenum, tellurium)	Mouse, sheep, rat
Neuroleptic drugs	
(Selected tranquilizers, antipsychotics, antidepressants, stimulants, sedatives)	Rat, mouse
Pesticides	
(Cyhexatin, ethylene dibromide)	Rabbit, mouse, rat
Polybrominated biphenyls	Rat
Potassium iodide	Rat
Substituted methylpyridines	Mouse

Several more compounds have been tested for developmental neurotoxicity in laboratory species because of suspicions that they might be active. They include the centrally active antidepressant drug fluoxetine (Vorhees *et al.,* 1994), the centrally active antipsychotic drug coded CI-943 (Henck *et al.,* 1995), the industrial chemical isopropanol (Bates *et al.,* 1994), and the plastic chemical acrylamide (Wise *et al.,* 1995). None has been developmentally neurotoxic.

Acknowledgment

The author gives special thanks to Mrs. Carmen Walthour for aid in the preparation of this manuscript.

References

Adams, J., and Lammer, E. J. (1993). Neurobehavioral teratology of isotretinoin. *Reprod. Toxicol.* **7,** 175–177.

Adams, J., Vorhees, C. V., and Middaugh, L. D. (1990). Developmental neurotoxicity of anticonvulsants: human and animal evidence on phenytoin. *Neurotoxicol. Teratol.* **12,** 203–214.

Aikawa, M., Sato, M., Noda, A., and Udaka, L. (1982). Toxicity study of etretinate. III. Reproductive segment 2 study in rats. *Yakuri to Chiryo,* **9,** 5095 passim 5143.

Anger, W. K. (1986). Workplace exposures, In Neurobehavioral Toxicology, A. Annau, Ed. Johns Hopkins Press, Baltimore, pp. 331–347.

Ardinger, H. H., Atkin, J. F., Blackston, R. D., Elsas, L. J., Clarren, S. K., Livingstone, S., Flannery, D. B., Pellock, J. M., Harrod, M. J., Lammer, E. J., Majewski, F., Schinzel, A., Toriella, H. V., and Hanson, J. W. (1988). Verification of the fetal valproate syndrome phenotype. *Am. J. Med. Genet.* **29,** 171–185.

Baggs, R. B., and Miller, R. K. (1983). Induction of congenital malformation by diethylstilbestrol in the ferret. *Teratology* **27,** 28A.

Bates, H. K., McKee, R. H., Bieler, G. S., Gardiner, T. H., Gill, M. W., Strother, D.E., and Masten, L. W. (1994). Developmental neurotoxicity evaluation of orally administered isopropanol in rats. *Fund. Appl. Toxicol.* **22,** 1152–158.

Beattie, A. D., Moore, M. R., Goldberg, A., Finlayson, M. J. W., Graham, J. F., Mackie, E. M., Main, J. C., McLaren, D. A., Murdock, R. M., and Stewart, G. T. (1975). Role of chronic low-level lead exposure in the aetiology of mental retardation. *Lancet* **1,** 589–592.

Bellinger, D., Leviton, A., Waternaux, C., Needleman, H., and Rabinowitz, M. (1987). Longitudinal analysis of prenatal lead exposure and early cognitive development. *N. Engl. J. Med.* **316,** 1037–1043.

Binkerd, P. E., Rowland, J. M., Nau, H., and Hendrickx, A. G. (1988). Evaluation of valproic acid (VPA) developmental toxicity and pharmacokinetics in Sprague-Dawley rats. *Fund. Appl. Toxicol.* **11,** 485–493.

Bowman, R. E., Heironimus, M. P., and Allen, J. R. (1978). Correlation of PCB body burden with behavioral toxicology in monkeys. *Pharmacol. Biochem. Behav.* **9,** 49–56.

Bowman, R. E., Heironimus, M. P., and Barsotti, D. A. (1981). Locomotor hyperactivity in PCB exposed rhesus monkeys. *Neurotoxicology* **2,** 251–268.

Brady, K., Herrara, Y., and Zenick, H. (1975). Influence of parental lead exposure on subsequent learning ability of offspring. *Pharm. Biochem. Behav.* **3,** 561–565.

Brown, N. A., and Fabro, S. (1983). The value of animal teratogenicity testing for predicting human risk. *Clin. Obst. Gynecol.* **26,** 467–477.

Brown, N. A., Shull, G., and Fabro, S. (1979). Assessment of the teratogenic potential of trimethadione in the CD-1 mouse. *Toxicol. Appl. Pharmacol.* **51,** 59–71.

Burbacher, T. R., Mohamed, M. K., and Mottett, N. K. (1988). Methylmercury effects on reproduction and offspring size at birth. *Reprod. Toxicol.* **1,** 267–278.

Burbacher, T. R., Monnett, C., Grant, K. S., and Mottet, N. K. (1984). Methylmercury exposure and reproductive dysfunction in the nonhuman primate. *Toxicol. Appl. Pharmacol.* **75,** 18–24.

Burbacher, T. R., Rodier, P. M., and Weiss, B. (1990). Methylmercury developmental neurotoxicity: a comparison of effects in humans and animals. *Neurotoxicol. Teratol.* **12,** 191–202.

Butler, N. R., and Goldstein, H. (1973). Smoking in pregnancy and subsequent child development. *Br. Med. J.* **4,** 573–575.

Carson, T. L., Van Gelder, G. A., Karas, G. L., and Buck, W. B. (1974). Slowed learning in lambs prenatally exposed to lead. *Arch. Environ. Health* **29,** 154–156.

Chasnoff, I. J. (1991). Cocaine and pregnancy: clinical and methodologic issues. *Clin. Perinatol.* **18,** 113–123.

Chasnoff, I. J., Burns, W. J., Hatcher, R. P., and Burns, K. A. (1983). Phencyclidine: Effects on the fetus and neonate. *Dev. Pharmacol. Ther.* **6,** 404–408.

Chaube, S. and Murphy, M. L. (1965). The teratogenic effects of cytosine arabinoside (CA) on the rat fetus. *Proc. Am. Assoc. Cancer Res.* **6,** 11.

Chen, J. S., Driscoll, C. D., and Riley, E. P. (1982). Ontogeny of suckling behavior in rats prenatally exposed to alcohol. *Teratology,* **26,** 145–153.

Chernoff, G. F. (1977). The fetal alcohol syndrome in mice: an animal model. *Teratology* **15,** 223–229.

Chou, S. M., Miike, T., Payne, W. M., and Davis, G. J. (1979). Neuropathology of "spinning syndrome" induced by prenatal intoxication with a PCB in mice. *Ann. NY Acad. Sci.* **320,** 373–395.

Clarren, S. K., and Bowden, D. M. (1982). A new primate model for binge drinking and its relevance to human ethanol teratogenesis. *Teratology* **25,** 35–36A.

Clarren, S. K., and Smith, D. W. (1978). The fetal alcohol syndrome. *N. Engl. J. Med.* **298,** 1063–1067.

Collins, M. D., Tzimos, G., Hummler, H., Burgin, H., and Nau, H. (1994). Comparative teratology and transplacental pharmacokinetics of all-*trans* retinoic acid, 13-*cis* retinoic acid, and retinyl palmitate following daily administration to rats. *Toxicol. Appl. Pharmacol.* **127,** 132–144.

Dagg, C. P. (1960). Sensitive stages for the production of developmental abnormalities in mice with 5-fluorouracil. *Am. J. Anat.* **106,** 89–96.

Dalens, B., Ranaud, E. J., and Gaulne, J. (1980). Teratogenicity of valproic acid. *J. Pediatr.* **97,** 332–333.

Davis, J. M., Otto, D. A., Weil, D. E., and Grant, L. D. (1990). The comparative developmental neurotoxicity of lead in humans and animals. *Neurotoxicol. Teratol.* **12,** 215–229.

Denson, R., Nanson, J. L., and McWalters, M. A. (1975). Hyperkinesis and maternal smoking. *Can. Psych. Assoc. J.* **20,** 183–187.

Deitrich, K. N. (1991). Human fetal lead exposure: Intrauterine growth, maturation,and postnatal neurobehavioral development. *Fund. Appl. Toxicol.***16,** 17–19.

Dexter, J. D., and Tumbleson, M. E. (1980). Fetal alcohol syndrome in Sinclair (S-1) miniature swine. *Teratology* **21,** 35–36A.

DiLiberti, J. H., Farndon, P. A., Dennis, N. R., and Curry, C. J. R. (1984). The fetal valproate syndrome. *Am. J. Med. Genet.,* **19,** 473–481.

Dunn, H. G., McBurney, A. K., Ingram, S., and Hunter, C. M. (1977). Maternal cigarette smoking during pregnancy and the child's subsequent development. II. Neurological and intellectual maturation to the age of 6½ years. *Can. J. Pub. Health* **68,** 43–50.

Earl, F. L., Miller, E., and Van Loon, E. J. (1973). Teratogenic research in beagle dogs and miniature swine. *Lab. Anim. Drug Test. Fifth Symp. Int. Comm. Lab. Anim.* Fischer, Stuttgart, pp. 233–247.

Ehlers, K., Sturje, H., Merker, H.-J., and Nau, H. (1992). Valproic acid–induced spina bifida: a mouse model. *Teratology* **45,** 145–154.

Ellis, F. W., and Pick, J. R. (1980). An animal model of the fetal alcohol syndrome in beagles. *Alcohol. Clin. Exp. Res.* **4,** 123–134.

Elmazar, M. M. A., and Sullivan, F. M. (1981). Effect of prenatal phenytoin administration on postnatal development of the rat: a behavioral teratology study. *Teratology* **24,** 115–124.

Fantel, A. G., and MacPhail, B. J. (1980). The teratogenicity of cocaine in rats and mice. *Teratology* **21,** 37A.

Feldman, G. L., Weaver, D. D., and Lovrien, E. W. (1977). The fetal trimethadione syndrome: report of an additional family and further delineation of this syndrome. *Am. J. Dis. Child.* **131,** 1389–1392.

Finnell, R. H. (1981). Phenytoin-induced teratogenesis: a mouse model. *Science* **211,** 483–484.

Finnell, R. H., Shields, H. E., Taylor, S. M., and Chernoff, G. F. (1987). Strain differences in phenobarbital-induced teratogenesis in mice. *Teratology* **35,** 177–185.

Finnell, R. H., Toloyan, S., van Waes, M., and Kalivas, P. W. (1990). Preliminary evidence for a cocaine-induced embryopathy in mice. *Toxicol. Appl. Pharmacol.* **103,** 228–237.

Foss, J. A., and Riley, E. P. (1988). Behavioral evaluation of animals exposed prenatally to cocaine. *Teratology* **37,** 517.

Fujimori, H., and Imoi, S. (1957). Studies on dihydrostreptomycin administered to the pregnant and transferred to their fetuses. *J. Jpn. Obstet. Gynecol. Soc.* **4,** 133–149.

Fulroth, R., Phillips, B., and Durand, D. J. (1989). Perinatal outcome of infants exposed to cocaine and/or heroin *in utero. Am. J. Dis. Child.* **143,** 905–910.

Fung, Y. K. (1988). Postnatal behavioral effects of maternal nicotine exposure in rats. *J. Pharm. Pha.* **40,** 870–872.

Geber, W. F., and Schram, L. C. (1969). Comparative teratogenicity of morphine, heroin, and methadone in the hamster. *Pharmacologist* **11,** 248.

Gerlinger, P., and Clavert, J. (1964). Action du cyclophosphamide injecte a des lapines gestantes sur les gonades embryonnaires. *C. R. Acad. Sci. [D](Paris),* **258,** 2899–2901.

Gilloteaux, J., Paul, R. J., and Steggles, A. W. (1982). Upper genital tract abnormalities in the Syrian hamster as a result of *in utero* exposure to diethylstilbestrol. I. Uterine cystadenomatous papilloma and hypoplasia. *Virchows Arch. [A]* **398,** 163–183.

Ginsberg, M. D., and Myers, R. E. (1974). Fetal brain damage following carbon monoxide intoxication: An experimental study. *Acta Obstet. Gynecol. Scand.* **53,** 309–317.

Gladen, B. C., Rogan, W. J., Hardy, P., Thulen, J., Tinglestad, J., and Tully, M. (1988). Development after exposure to polychlorinated biphenyls and dichlorodiphenyl dichloroethene transplacentally and through human milk. *J. Pediatr.* **13,** 991–995.

Golden, N. L., Sokol, R. J., and Rubin, I. L. (1980). Angel dust: possible effects on the fetus. *Pediatrics* **65,** 18–20.

Goldey, E. S., O'Callaghan, J. P., Stanton, M. E., Barone, S., and Crofton, K. M. (1994). Developmental neurotoxicity: evaluation of testing procedures with methylazoxymethanol and methyl mercury. *Fund. Appl. Toxicol.* **23,** 447–464.

Goodwin, T. M. (1988). Toluene abuse and renal tubular acidosis in pregnancy. *J. Obstet. Gynecol.* **71,** 715–718.

Gottschewski, G. H. M. (1964). Mammalian blastopathies due to drugs. *Nature* **201,** 1232–1233.

Greenberg, C. R., DeSa, J. D., Tenenbeim, M., and Evans, J. A. (1984). There is a fetal gasoline syndrome. *Proc. Greenwood Genet. Cent.* **3,** 107.

Grumbach, M. M., and Ducharme, J. R. (1960). The effects of androgens on fetal sexual development. Androgen-induced female pseudohermaphroditism. *Fertil. Steril.* **11,** 157–180.

Gutova, M., Elis, J., and Raskova, H. (1971). Teratogenic effect of 6-azauridine in rats. *Teratology* **4,** 287–294.

Hackenberger, I., and Kreybig, T. (1965). Vergleichende teratologische untersuchungen bei der maus und der ratte. *Arzneimittleforschung* **15,** 1456–1460.

Hall, J. G., Pauli, R. M., and Wilson, K. M. (1980). Maternal and fetal sequelae of anticoagulation during pregnancy. *Am. J. Med.* **68,** 122–140.

Hallett, J. J., and Holmes, L. B. (1979). An *in vitro* model for warfarin teratogenesis. *Pediatr. Res.* **13,** 486.

Hans, S. L. (1989). Developmental consequences of prenatal exposure to methadone. *Ann. NY Acad. Sci.* **562,** 195–207.

Hanson, J. W., Myrianthopoulos, N. C., Harvey, M. A. S., and Smith, D.W. (1976). Risks to the offspring of women treated with hydantoin anticonvulsants, with emphasis on the fetal hydantoin syndrome. *J. Pediatr.* **89,** 552–668.

Hanson, J. W., and Smith, D. W. (1975). The fetal hydantoin syndrome. *J. Pediatr.* **87,** 285–290.

Hanson, J. W., and Smith, D. W. (1977). Are hydantoins (phenytoins) human teratogens? *J. Pediatr.* **90,** 674–675.

Harada, M. (1976). Intrauterine poisoning. Clinical and epidemiological studies and significance of the problem. *Bull. Inst. Const. Med. Kumamoto Univ.* **25**(Suppl.): 1–60.

Hardy, J. B., and Mellitis, E. D. (1972). Does maternal smoking have a long-term effect on the child? *Lancet* **2,** 1332–1336.

Haskin, D. (1948). Some effects of nitrogen mustard on the development of external body form in the fetal rat. *Anat. Rec.* **102,** 493–511.

Henck, J. W., Petrere, J. A., and Anderson, J. A. (1995). Developmental neurotoxicity of CI-943: a novel antipsychotic. *Neurotoxicol. Teratol.* **17,** 13–24.

Hendrickx, A. G., Nau, H., Binkerd, P., Rowland, J. M., Rowland, J. R., Cukierski, M. J., and Cukierski, M. A. (1988a). Valproic acid developmental toxicity and pharmacokinetics in the rhesus monkey: an interspecies comparison. *Teratology* **38,** 329–345.

Hendrickx, A. G., Prahalada, S., and Binkerd, P. E. (1988b). Long-term evaluation of the diethylestilbestrol (DES) syndrome in adult female rhesus monkeys (*Macaca mulatta*). *Reprod. Toxicol.* **1,** 253–261.

Hersh, J. H. (1989). Toluene embryopathy: two new cases. *J. Med. Genet.* **25,** 333–337.

Higgins, I. T. T. (1975). Importance of epidemiological studies relating to hazards of food and environment. *Br. Med. Bull.* **31,** 230–235.

Hill, R. M., Craig, J. P., Chaney, M. D., Tennyson, L. M., and McCulley, L. B. (1977). Utilization of over-the-counter drugs during pregnancy. *Clin. Obstet. Gynecol.* **20,** 381–394.

Howe, A. M., and Webster, W. S. (1989). An animal model for warfarin teratogenicity. *Teratology* **40,** 259–260.

Howe, A. M., and Webster, W. S. (1992). The warfarin embryopathy: a rat model showing maxillonasal hypoplasia and other skeletal disturbances. *Teratology* **46,** 379–390.

Howe, A. M., and Webster, W. S. (1993). Warfarin embryopathy: a rat model showing nasal hypoplasia and "stippling." *Teratology* **48,** 185–186.

Hummler, H., Hendrickx, A. G., and Nau, H. (1994). Maternal toxicokinetics, metabolism, and embryo exposure following a teratogenic dosing regimen with 13-*cis*-retinoic acid (isotretinoin) in the Cynomolgous monkey. *Teratology* **50,** 184–193.

Hummler, H., Korte, R., and Hendrickx, A. G. (1990). Induction of malformations in the cynomolgous monkey with 13-*cis* retinoic acid. *Teratology* **42,** 263–272.

Hunter, A. G. W., Thompson, D., and Evans, J. A. (1979). Is there a fetal gasoline syndrome? *Teratology* **20,** 75–80.

Hurley, L. S., Keen, C. L., Lonnerdal, B., Mark-Savage, P., and Cohen, N. L. (1982). Reduction by copper supplementation of teratogenic effects of D-penicillamine and triethylene tetramine. *Teratology* **25,** 51A.

Hutchings, D. E., Bodnarenko, S. R., and Diaz-DeLeon, R. (1984). Phencyclidine during pregnancy in the rat: effects on locomotor activity in the offspring. *Pharmacol. Biochem. Behav.* **20,** 251–254.

Hutchings, D. E., Fico, T. A., and Dow-Edwards, D. L. (1989). Prenatal cocaine–maternal toxicity, fetal effects and locomotor activity in rat offspring. *Neurotoxicol. Teratol.* **11,** 65–69.

Hutchings, D. E., Towey, J. P., Gorinson, H. S., and Hunt, H. F. (1979). Methadone during pregnancy: assessment of behavioral effects in the rat offspring. *J. Pharmacol. Exp. Ther.* **208,** 106–112.

Inouye, M., and Kajiwara, Y. (1988). Developmental disturbances of the fetal brain in guinea pigs caused by methylmercury. *Arch. Toxicol.* **62,** 15–21.

Jacobson, J. L., and Jacobson, S. W. (1988). New methodologies for assessing the effects of prenatal toxic exposure on cognitive functioning in humans. In: *Toxic Contaminants and Ecosystem Health. A Great Lakes Focus.* M. Evans, ed. Wiley, pp. 373–388.

Jacobson, S. W., Fein, G. G., Jacobson, J. L., Schwartz, P. M., and Dowler, J. K. (1985). The effect of intrauterine PCB exposure on visual recognition memory. *Child. Develop.* **56,** 853–860.

James, L. F., and Keeler, R. F. (1968). Teratogenic effects of aminopterin in sheep. *Teratology* **1,** 407–412.

Jones, K. L. (1991). Developmental pathogenesis of defects associated with prenatal cocaine exposure: fetal vascular disruption. *Clin. Perinatol.* **18,** 139–146.

Jones, K. L., Lacro, R. V., Johnson, K. A., and Adams, J. (1989). Pattern of malformations in the children of women treated with carbamazepine during pregnancy. *N. Engl. J. Med.* **320,** 1661–1666.

Jones, K. L., and Smith, D. W. (1973). Recognition of the fetal alcohol syndrome in early infancy. *Lancet* **2,** 999–1001.

Jones, K. L., Smith, D. W., and Hanson, J. W. (1976). The fetal alcohol syndrome: Clinical delineation. *Ann. NY Acad. Sci.* **273,** 130–139.

Jordan, R. L., Terapane, J. F., and Schumacher, H. J. (1970). Studies on the teratogenicity of methotrexate in rabbits. *Teratology* **3,** 203.

Jordan, R. L., Young, T. R., Dinwiddie, S. H., and Harry, G. J. (1979). Phencyclidine-induced morphological and behavioral alterations in the neonatal rat. *Pharmacol. Biochem. Behav.* **11** (Suppl.): 39–45.

Jorgensen, K. D. (1994). Teratogenic activity of tretinoin in the Gottingen mini-pig. *Teratology* **50,** 26A–27A.

Katsuki, S. (1969). Foreward. Reports of the study for "Yusho" (chlorobiphenyls poisoning). *Fukuoka Acta Med.* **60,** 407.

Keeler, R. F., and Binns, W. (1968). Teratogenic compounds of *Veratrum californicum* (Durand). V. Comparison of cyclopian effects of steroidal alkaloids from the plant and structurally related compounds from other sources. *Teratology* **1,** 5–10.

Khera, K. S. (1973). Effects of methyl mercury in cats after pre- or post-natal treatment. *Teratology* **7,** A20.

Khera, K. S. (1976). Teratogenicity studies with methotrexate, aminopterin, and acetylsalicylic acid in domestic cats. *Teratology* **14,** 21–28.

Kikuchi, M. (1984). Autopsy of patients with Yusho. *Am. J. Ind. Med.* **5,** 19–30.

Klaassen, C. D., Amdur, M. O., and Doull, J. (Eds.) (1986). *Casarett and Doull's Toxicology. The Basic Science of Poisons.* 3rd Ed., Macmillan, New York.

Koja, T., Fujisaki, T., Shimizu, T., Kishita, C., and Fukuda, T. (1978). Changes of gross behavior with polychlorinated biphenyls (PCB) in immature rats. *Kagoshima Daigaka Igaka Zasshi* **30,** 377–381.

Kristjan, E. A., Fried, P. A., and Watkinson, B. (1989). Maternal smoking during pregnancy affects children's vigilance performance. *Drug Al. Dep.* **24,** 11–19.

Kromka, M., and Hoar, R. M. (1973). Use of guinea pigs in teratological investigations.*Teratology* **7,** A21–A22.

Lan, C. F., Chen, H. S., Shieh, L. L., and Chen, Y. H. (1981). An epidemiological study on polychlorinated biphenyls poisoning in Taichung area. *Clin. Med. (Taipei)* **7,** 96–100.

Landesman-Dwyer, S., Keller, L. S., and Streissguth, A. P. (1978). Naturalistic observations of newborn: effects of maternal alcohol intake. *Alcohol. Clin. Exp. Res.* **2,** 171–177.

Linares, A., Zarranz, J. J., Rodriguez-Alarcon, J., and Diaz-Perez, J. L. (1979). Reversible cutis laxa due to maternal D-penicillamine treatment. *Lancet* **2,** 43.

Lindenberg, C. A., Alexander, E. M., Gendrop, S. C., Nencioli, M., and Williams, D. G. (1991). A review of the literature on cocaine abuse in pregnancy. *Nurs. Res.* **40,** 69–75.

Lofberg, B., Chahoud, I., Bochert, G., and Nau, H. (1990). Teratogenicity of the 13-*cis* and all-*trans* isomers of the aromatic retinoid etretin: correlation to transplacental pharmacokinetics in mice during organogenesis after a single oral dose. *Teratology* **41,** 707–716.

Logdberg, B., Brun, A., Berlin, M., and Schutz, A. (1988). Congenital lead encephalopathy in monkeys. *Acta Neuropathol.* **77,** 120–127.

Longo, L. D. (1977). The biological effects of carbon monoxide on the pregnant woman, fetus, and newborn infant. *Am. J. Obstet. Gynecol.* **129,** 69–103.

Lorente, C. A., Tassinari, M. S., and Keith, D. A. (1981). The effects of phenytoin on rat development: an animal model system for fetal hydantoin syndrome. *Teratology* **24,** 169–180.

Lynberg, M. C., Khoury, M. J., Lammer, E. J., Waller, K. O., Cordero, J. F., and Erickson, J. D. (1990). Sensitivity, specificity, and positive predictive value of multiple malformations in isotretinoin embryopathy surveillance. *Teratology* **42,** 513–519.

Mactutus, C. F., and Fechter, L. D. (1984). Prenatal exposure to carbon monoxide: learning and memory deficits. *Science* **223,** 409–411.

Mahalik, M. P., Gautieri, R. F., and Mann, D. E. (1980). Teratogenic potential of cocaine hydrochloride in CF-1 mice. *J. Pharm. Sci.* **69,** 703–706.

Mankes, R. F., Rosenblum, I., Benitz, K.-F., LeFevre, R., and Abraham, R. (1982). Teratogenic and reproductive effects of ethanol in Long–Evans rats. *J. Toxicol. Environ. Health* **10,** 267–276.

Marin-Padilla, M., and Benirschke, K. (1963). Thalidomide induced alterations in the blastocyst and placenta of the armadillo, *Dasypus novemcinctus mexicanus,* including a choriocarcinoma. *Am. J. Pathol.* **43,** 999–1016.

Marlowe, M., Errera, J., and Jacobs, J. (1983). Increased lead and cadmium burdens among mentally retarded children and children with borderline intelligence. *Am. J. Ment. Defic.* **87,** 477–483.

Martin, J., Martin, D. C., Lund, C.A., and Streissguth, A.P. (1977). Maternal alcohol ingestion and cigarette smoking and their effects on newborn conditioning. *Alcohol. Clin. Exp. Res.* **1,** 243–247.

Martin, J. C., and Becker, R. F. (1970). The effects of nicotine administration *in utero* upon activity in the rat. *Psychon. Sci.* **19,** 59–60.

Matsumoto, H., Koya, G., and Takeuchi, T. (1965). Fetal Minamata disease. *J. Neuropathol. Exp. Neurol.* **24,** 563–574.

McMillan, D. E. (1987). Risk assessment for neurobehavioral toxicity. *Environ. Health Perspect.,* **76,** 155–161.

Mele, P. C., Bowman, R. E., and Levin, E. D. (1986). Behavioral evaluation of perinatal PCB exposure in rhesus monkeys: Fixed-interval performance and reinforcement–omission. *Neurobehav. Toxicol. Teratol.* **8,** 131–138.

Mesollela, C. (1963). Experimental studies on the toxic effect of kanamycin on the internal ear of the guinea pig during intrauterine life. *Arch. Ital. Laringol.* **71,** 37.

Michaud, J. (1982). Agenesis of the vermis with fusion of the cerebellar hemispheres, septooptic dysplasia and associated abnormalities. Report of a case. *Acta Neuropathol. (Berl.)* **56,** 161–166.

Miller, R. K., Baggs, R. B., Odoroff, C. L., and McKenzie, R. C. (1982). Transplacental carcinogenicity of diethylstilbestrol (DES): a Wistar rat model. *Teratology* **25,** 62A.

Moffa, A. M., White, J. A., Mackay, E. G., and Frias, J. L. (1984). Valproic acid, zinc and open neural tubes in 9-day old hamster embryos. *Teratology* **29,** 47A.

Moriyama, H. (1967). [A study on congenital Minamata disease. 1. Effects of organic mercury administration on pregnant animals, with reference to the mercury content in the maternal and fetal organs]. *J. Kumamoto Med. Soc.* **41,** 506–528.

Mould, G. P., Curry, S. H., and Beck, F. (1973). The ferret, a useful model for teratogenic study. *Naunyn Schmiedebergs Arch. Pharmacol.* **279** (Suppl.): R–18.

Murakami, U. (1970). Embryo-fetotoxic effect of some organic mercury compounds. *Nagoya Daigaku* **18,** 33–43.

Murakami, U. (1972). The effect of organic mercury on intrauterine life. In: *Drugs and Fetal Development.* M. A. Klingberg, A. Abramovici, and J. Chemke, eds. Plenum Press, New York, pp. 301–336.

Murphy, M. L. (1962). Teratogenic effects in rats of growth inhibiting chemicals, including studies on thalidomide. *Clin. Proc. Child. Hosp.* **18,** 307–322.

Murphy, M. L., Moro, A. D., and Lacon, C. (1958). Comparative effects of five polyfunctional alkylating agents on the rat fetus, with additional notes on the chick embryo. *Ann. NY Acad. Sci.* **68,** 762–782.

Myhre, S. A., and Williams, R. (1981). Teratogenic effects associated with maternal primidone therapy. *J. Pediatr.* **99,** 160–162.

Nabeshima, T., Hiramatsu, M., Yamaguchi, K., Kasugai, M., Ishizaki, K., Kawashima, K., Itoh, K., Ogawa, S., Katoh, A., and Furukawa, H. (1988). Effects of prenatal administration of phencyclidine on the learning and memory processes of rat offspring. *J. Pharmacob.* **11,** 816–823.

NAS (National Academy of Science) (1972). Committee on Medical and Biological Effects of Atmospheric Pollutants: Lead. Airborne lead in perspective, NAS, Washington, DC.

Needleman, H. L. (1985). The neurobehavioral effects of low level exposure to lead in childhood. *Int. J. Ment. Health* **14,** 64–77.

Newbold, R. R., and McLachlan, J. A. (1982). Vaginal adenosis and adenocarcinoma in mice exposed prenatally or neonatally to diethylstilbestrol. *Cancer Res.* **42,** 2003–2011.

Newell-Morris, L., Sirianni, J. E., Shepard, T. H., Fantel, A. G., and Moffett, B. C. (1980). Teratologic effects of retinoic acid in pigtail monkey (*Macaca nemestrina*). II. Craniofacial features. *Teratology* **22,** 87–101.

Nicholas, J. M., and Schreiber, E. C. (1983). Phencyclidine exposure and the developing mouse: behavioral teratological implications. *Teratology* **28,** 319–326.

Nomura, A., Watanabe, M., Yamagata, K., and Ohta, K. (1969). [Teratogenic effects of 1-β-D-arabinofuranosyl-cytosine (AC-1075) in mice and rats]. *Gendai No Rinsho* **3,** 758.

Onejeme, A., and Khan, K. M. (1984). Morphologic study of effects of kanamycin on the developing cochlea of the rat. *Teratology* **29,** 57–71.

Overmann, S. R., Kostas, J., Wilson, L. R., Shain, W., and Bush, B. (1987). Neurobehavioral and somatic effects of perinatal PCB exposure to rats. *Environ. Res.* **44,** 56–70.

Padmanabhan, R., Vaidya, H. R., and Abu-Alatta, A. A. F. (1990). Malformations of the ear induced by maternal exposure to retinoic acid in the mouse fetuses. *Teratology* **42,** 25A.

Pantaleoni, G., Fanini, D., Sponta, A. M., Palumbo, G., Giorgi, R., and Adams, P. M. (1988). Effects of maternal exposure to polychlorobiphenyls (PCBs) on F1 generation behavior in the rat. *Fund. Appl. Toxicol.* **11,** 440–449.

Papara-Nicholson, D., and Telford, I. R. (1957). Effects of alcohol on reproduction and fetal development in the guinea pig. *Anat. Rec.* **127,** 438–439.

Pauli, R. M., Madden, J. D., Kranzler, K. J., Culpepper, W., and Port, R. (1976). Warfarin therapy initiated during pregnancy and phenotype chondrodysplasia punctata. *J. Pediatr.* **88,** 506–508.

Pinto Machado, J. (1969). Inhibition de l'osteogenese provoquee par le busulfan chez l'embryon de souris. Demonstration par l'alizarine. *C. R. Soc. Biol. (Paris)* **163,** 1712–1715.

Pley, E. A. P., Wouters, E. J. M., Voorhors, F. J., Stolte, S. B., Jurver, P. H. J., and Dejong, P. A. (1991). Assessment of tobacco exposure during pregnancy—behavioral and biochemical changes. *Eur. J. Obstet. Gynecol.* **40,** 197–201.

Podvinec, S., Mihaljevic, B., Marcetic, A., and Simonovic, M. (1965). Schadigungen des fotalen Cortischen Organs durch Streptomycin. *Monatsschr. Ohrenheilkd.* **99,** 20–24.

Potter, B. J., Belling, G. B., Mano, M. T., and Hetzel, B. S. (1981). Teratogenic effects of ethanol in pregnant sheep: a model for the fetal alcohol syndrome. In: *Man, Drugs, Society—Current Perspective* (Proc. Pan-Pac. Conf. Drugs Alcohol, 1st) Australian Foundation Alcohol Drug Dependency, Canberra, pp. 303–306.

Pryor, G. T. (1991). A toluene-induced motor syndrome in rats resembling that seen in some human solvent abusers. *Neurotoxicol. Teratol.* **13,** 387–400.

Rating, D., Nau, H., Jager-Roman, E., Gopfert-Geyer, I., Koch, S., Beck-Mannagetta, G., Schmidt, D., and Helge, H. (1982). Teratogenic and pharmacokinetic studies of primidone during pregnancy and in the offspring of epileptic women. *Acta Paediatr. Scand.* **71,** 301–311.

Richardson, G. A., Day, N. L., and Taylor, P. M. (1989). The effect of prenatal alcohol, marijuana, and tobacco exposure on neonatal behavior. *Infant Behav. Dev.* **12,** 199–209.

Robert, E. (1982). Valproic acid and spina bifida: A preliminary report. *MMWR* **31,** 565–566.

Rogan, W. J., Gladen, B. C., Hung, K.-L., Koong, S.-L., Shia, L.-Y., Taylor, J. S., Wu, Y.-C. Yang, D., Rogan, N. B., and Hsu, C.-C. (1988). Congenital poisoning by polychlorinated biphenyls and their contaminants in Taiwan. *Science* **241,** 334–336.

Rogan, W. J., Gladen, B. C., McKinney, J. D., Carreras, N., Hardy, P., Thulen, J., Tinglestad, J., and Tully, M. (1986). Neonatal effects of transplacental exposure to PCBs and DDE. *J. Pediatr.* **109,** 335–341.

Rosa, F. (1991). Detecting human retinoid embryopathy. *Teratology* **43,** 419.

Rosa, F., Piazza-Hepp, T., and Goetsch, R. (1994). Holoprosencephaly with 1st trimester atopical tretinoin. *Teratology* **49,** 418–419.

Rosa, F. W. (1986). Teratogen update: penicillamine. *Teratology* **33,** 127–131.

Rosen, T. S., and Johnson, H. L. (1982). Methadone exposure: Effects on behavior in early infancy. *Pediatr. Pharmacol.* **2,** 192–196.

Rosett, H. L. (1980). A clinical perspective of the fetal alcohol syndrome. *Alcohol Clin. Exp. Res.* **4,** 119–122.

Rudd, N. L., and Freedom, R. M. (1979). A possible primidone embryopathy. *J. Pediatr.* **94,** 835–837.

Sandor, S., Amels, D., Cosma, M., Gherman, M., and Logofat, T. (1971). Action of ethanol on the prenatal development of albino rats (an attempt of multiphasic screening). *Rev. Roum. Embryol. Cytol.* **8,** 105–118.

Sansone, G., and Zunin, C. (1954). Embriopatie sperimentali da somministrazione diantifolici. *Acta Vitaminol. (Milano)* **8,** 73–79.

Scanlon, J. W. (1975). Dangers to the human fetus from certain heavy metals in the environment. *Rev. Environ. Health* **2,** 39–64.

Schantz, S. L., (1996). Developmental neurotoxicity of PCBs in humans: what do we know and where do we go from here? *Neurotoxicol. Teratol.* **18,** 217–227.

Schardein, J. L. (1993). *Chemically Induced Birth Defects* 2nd Ed, Marcel Dekker, New York.

Schardein, J. L., and Keller, K. A. (1989). Potential human developmental toxicants and the role of animal testing in their identification and characterization. *CRC Crit. Rev. Toxicol.* **19,** 251–339.

Schardein, J. L., Schwetz, B. A., and Kenel, M. F. (1985). Species sensitivites and prediction of teratogenic potential. *Environ. Health Perspect.* **61,** 55–67.

Seip, M. (1976). Growth retardation, dysmorphic facies and minor malformations following massive exposure to phenobarbitone *in utero. Acta Paediatr. Scand.* **65,** 617–621.

Sexton, M. (1990). Prenatal exposure to tobacco. 2. Effects on cognitive functioning at age 3. *Int. J. Epidemiol.* **19,** 72–77.

Shaul, W. L., and Hall, J. G. (1977). Multiple congenital anomalies associated with oral anticoagulants. *Am. J. Obstet. Gynecol.* **127,** 191–198.

Shaw, E. B. (1972). Fetal damage due to maternal aminopterin ingestion—followup at age 9 years. *Am. J. Dis. Child.* **124,** 93–94.

Shiota, K. (1976). Postnatal behavioral effects of prenatal treatment with PCBs (polychlorinated biphenyls) in rats. *Okajimas Fol. Anat. Jpn.* **53,** 105–114.

Singh, J. (1986). Early behavioral alterations in mice following prenatal carbon monoxide exposure. *Neurotoxicol.* **7,** 475–482.

Skalko, R. G., and Gold, M. P. (1974). Teratogenicity of methotrexate in mice. *Teratology* **9,** 159–164.

Skosyreva, A. M. (1973). [Effect of ethyl alcohol on the development of embryos at the organogenesis stage]. *Akush. Ginekol. (Mosk.)* **4,** 15–18.

Smith, R. F., Mattran, K. M., Kurkjian, M. F., and Kurtz, S. L. (1989). Alterations in offspring behavior induced by chronic prenatal cocaine dosing. *Neurotoxicol. Teratol.* **11,** 35–38.

Somers, G. F. (1962). Thalidomide and congenital abnormalities. *Lancet* **1,** 912–913.

Spyker, J. M., and Smithberg, M. (1972). Effects of methylmercury on prenatal development in mice. *Teratology* **5,** 181–190.

Storm, J. E., Hart, J. L., and Smith, R. F. (1981). Behavior of mice after pre- and postnatal exposure to Aroclor 1254. *Neurobehav. Toxicol. Teratol.* **3,** 5–9.

Strauss, A. A., Modanlou, H. D., and Bosu, S. K. (1981). Neonatal manifestations of maternal phencyclidine (PCP) abuse. *Pediatrics* **68,** 550–552.

Strauss, M. E., Starr, R. H., Ostrea, E. M., Chavez, C. J., and Stryker, J. C. (1976). Behavioral concomitants of prenatal addition to narcotics. *J. Pediatr.* **89,** 842–846.

Streicher, H. Z., Gabow, P. A., and Moss, A. H. (1981). Syndrome of toluene sniffing in adults. *Ann. Intern. Med.* **94,** 758–762.

Stromland, K., and Pinazo-Duran, M. D. (1994). Optic nerve hypoplasia: comparative effects in children and rats exposed to alcohol during pregnancy. *Teratology* **50,** 100–111.

Sulik, K. K., Johnston, M. C., and Webb, M. A. (1981). Fetal alcohol syndrome: embryogenesis in a mouse model. *Science* **214,** 936–938.

Swinyard, C. A., Sutton, D. B., and Saloum, L. M. (1983). Lead in the environment: experimental studies of lead toxicity in animals and their relevance to marginal lead toxicity in children. *Cong. Anom.* **23,** 29–60.

Taki, I., Hisanga, S., and Amagase, Y. (1969). Report on Yusho (chlorobiphenyls poisoning). Pregnant women and their fetuses. *Fukuoka Acta Med.* **60,** 471–474.

Tanimura, T., Yasuda, M., Kanno, Y., and Nishimura, H. (1965). [Comparison of teratogenic effects between single and repeated administration of clinical aspects]. *Kaibogaku Zasshi* **40,** 13.

Tatetsu, M. (1968). Experimental manifestation of "congenital Minamata disease." *Psychiat. Neurol. Jpn.* **70,** 162.

Taussig, H. B. (1962). Thalidomide and phocomelia. *Pediatrics* **30,** 654–659.

Tesh, J., and Pritchard, A. (1977). Lead and the neonate. *Teratology* **15,** 23A.

Thalhammer, O., and Heller-Szollosy, E. (1955). Exogene Bildungsfehler ("Missbildungen") durch Lostinjekion bei der graviden

Maus (Ein Beitrag zur Pathogenese von Bildungfehlen). *Z. Kinderheilk.* **76,** 351.

Tilson, H. A., Davis, G. J., McLachlan, J. A., and Lucier, G. W. (1979). The effects of polychlorinated biphenyls given prenatally on the neurobehavioral development of mice. *Environ. Res.* **18,** 466–474.

Tilson, H. A., Jacobson, J. L., and Rogan, W. J. (1990). Polychlorinated biphenyls and the developing nervous system: cross-species comparisons. *Neurotoxicol. Teratol.* **12,** 239–248.

Toutant, C., and Lippman, S. (1979). Fetal solvents syndrome. *Lancet* **1,** 1356.

Tze, W. J., and Lee, M. (1975). Adverse effects of maternal alcohol consumption on pregnancy and foetal growth in rats. *Nature* **257,** 479–480.

Vincent, N. M. (1958). The effects of prenatal alcoholism upon motivation, emotionality, and learning in the rat. *Am. Psychol.* **13,** 401.

Vorhees, C. V. (1983). Fetal anticonvulsant syndrome in rats: dose– and period–response relationships of prenatal diphenylhydantoin, trimethadione and phenobarbital exposure on the structural and functional development of the offspring. *J. Pharmacol. Exp. Ther.* **227,** 274–287.

Vorhees, C. V., Acuff-Smith, K. D., Schilling, M. A., Fisher, J. E., Moran, M. S., and Buelke-Sam, J. (1994). A developmental neurotoxicity evaluation of the effects of prenatal exposure to fluoxetine in rats. *Fund. Appl. Toxicol.* **23,** 194–205.

Walker, B. E. (1983). Complications of pregnancy in mice exposed prenatally to DES. *Teratology* **27,** 73–80.

Webster, W. S., Brown-Woodman, P. D. C., Lipson, A. H., and Ritchie, H. E. (1991). Fetal brain damage in the rat following prenatal exposure to cocaine. *Neurotoxicol. Teratol.* **13,** 621–626.

Webster, W. S., and Valois, A. A. (1981). The toxic effects of cadmium on the neonatal mouse CNS. *J. Neuropathol. Exp. Neurol.* **40,** 247–257.

Weingarten, P. L., Ream, J. R., and Pappas, A. M. (1971). Teratogenicity of myleran against musculoskeletal tissues in the rat. *Clin. Orthoped.* **75,** 236.

Weiss, B., and Doherty, R. A. (1975). Methyl mercury poisoning. *Teratology* **12,** 311–314.

Wilk, A. L., McClure, H. M., and Horigan, E. A. (1978). Induction of craniofacial malformations in the rhesus monkey with cyclophosphamide. *Teratology* **17,** 24A.

Wilkins, L., Jones, H. W., Holman, G. H., and Stempfel, R. S. (1958). Masculinization of female fetus associated with administration of oral and intramuscular progestins during gestation: nonadrenal pseudohermaphroditism. *J. Clin. Endocrinol. Metab.* **18,** 559–585.

Wilson, G. S. (1989). Clinical studies of infants and children exposed prenatally to heroin. *Ann. NY Acad. Sci.* **562,** 183–194.

Wilson, G. S., Desmond, M. M., and Wait, R. B. (1981). Followup of methadone-treated and untreated narcotic dependent women and their infants: Health, developmental and social implications. *J. Pediatr.* **98,** 716–722.

Wilson, J. G. (1971). Use of rhesus monkeys in teratological studies. *Fed. Proc.* **30,** 104–109.

Wilson, J. G., and Fradkin, R. (1967). Interrelations of mortality and malformations in rats. *Absts. Seventh Annu. Meet., Teratology Society* pp. 57–58.

Winter, R. (1979). *Cancer-Causing Agents. A Preventive Guide.* Crown Publishers, New York.

Wise, L. D., Gordon, L. R., Soper, K. A., Duchai, D. M., and Morrissey, R. E. (1995). Developmental neurotoxicity evaluation of acrylamide in Sprague–Dawley rats. *Neurotoxicol. Teratol.* **17,** 189–198.

Witti, F. P. (1978). Alcohol and birth defects. *FDA Consumer* **12,** 20–23.

Wong, K. C., and Hwang, M. Y. (1981). Children born to PCB poisoned mothers. *Clin. Med.* (*Taipei*) **7,** 83–87.

Wong, K.-L., and Klaassen, C. D. (1982). Neurotoxic effects of cadmium in young rats. *Toxicol. Appl. Pharmacol.* **63,** 330–337.

Yamashita, F. (1977). Clinical features of polychlorobiphenyls (PCB)-induced fetopathy. *Paediatrician* **6,** 20–27.

Zander, J., and Muller, H. A. (1953). Uber die Methylandrostendiolbehandlung wahrend einer Schwangerschaft. *Geburtschilfe Frauenheilkd.* **13,** 216–222.

CHAPTER

41

Physiologically Based Pharmacokinetic Models in the Risk Assessment of Developmental Neurotoxicants

KANNAN KRISHNAN
Group of Research on Human Toxicology
Faculty of Medicine
University of Montreal
Montreal, Quebec, Canada, H3C 3J7

MELVIN ANDERSEN
ICF Kaiser, Inc.
Research Triangle Park, North Carolina 27709

*Abbreviations: 2,4-D, 2,4-dichlorophenoxy acetic acid; AUCB, area under the blood concentration versus time curve; MBDE, mass balance differential equation; NOAEL, no observable adverse effect level; P_{bb}, brain : blood partition coefficient; PBPK, physiologically based pharmacokinetic; PC, partition coefficient; P_{cw}, cyclohexane : water partition coefficient; P_{ow}, *n*-octanol : water partition coefficient; P_{tb}, tissue : blood partition coefficient; RBC, red blood cells; REDC, relative effective dose to the child.

I. Introduction

Health risk assessment of developmental neurotoxicants is conducted in four steps, namely: 1. hazard identification; 2. exposure assessment; 3. dose–response assessment; and 4. risk characterization. The dose–response segment of the risk assessment process in this case focuses to establish the quantitative relationship between the exposure concentration or administered dose and the incidence of neurotoxic effects in developing human fetuses or infants. Frequently, the available data are limited to those collected in nonhuman mammalian species exposed to high doses, at different time frames of development, by routes and scenarios different from anticipated human exposures. Therefore, several extrapolations (e.g., test animal species to humans, high dose to low dose, route to route, one developmental stage to another) may be needed to establish the dose–response relationship in humans. One way of improving the scientific basis of the conventional extrapolation process is to approach it from the mechanistic perspective; that is, deal with the extrapolations of the components of the exposure–response continuum (i.e., exposure → tissue dose, tissue dose → tissue response). In other words, an understanding of the mechanistic determinants of the pharmacokinetic and pharmacodynamic processes at the quantitative level is fundamental for conducting scientifically sound extrapolations. Dose–response assessment procedures for developmental neurotoxicants using an appropriate measure of target tissue does (e.g., area under the brain concentration versus time curve, maximal brain concentration) are therefore likely to enhance the scientific basis and credibility of the process. For example, if the dose measures associated with a predetermined level of response or no-response in one species for a particular exposure scenario are known, then it is possible to determine the equivalent exposure concentration for other species, exposure scenarios, routes, and developmental stages, provided an appropriate pharmacokinetic model exists. Mechanistically based pharmacokinetic models are of use in this context.

Mechanistic models are those that are based on proven or plausible mechanisms responsible for the system behavior being modeled. The system in the present context refers to an organism and the behavior refers to pharmacokinetics. Physiologically based pharmacokinetic (PBPK) models are examples of mechanistic models used in toxicology. PBPK modeling refers to the development of validated, mathematical descriptions of absorption, distribution, metabolism, and excretion of chemicals in biota based on proven/plausible interrelationships among certain critical mechanistic determinants (Krishnan and Andersen, 1994). The mechanistic determinants are physiological characteristics of the animal, the physicochemical characteristics of the substance, and the rates of biochemical processes relating to the substance being modeled.

The development of PBPK models for volatile organics and gaseous anesthetics dates back to the research work of Haggard published in the 1920s (Haggard, 1924). Further developments in the PBPK modeling of volatile organics were contributed by Kety (1951), Mapleson (1963), Riggs (1970), and Fiserova-Bergerova (1975). In the biopharmaceutics arena, PBPK modeling traces back to Teorell's pioneering work in the 1930s (Teorell, 1937a,b). Beginning in early 1960s, scientists trained in chemical engineering also developed PBPK models of various drugs, particularly antineoplastics (Bischoff *et al.*, 1971; Dedrick *et al.*, 1972). Subsequently, the PBPK modeling approach has found extensive application in toxicology and risk assessment, enabling a better understanding of the pharmacokinetic mechanism of toxicity, efficient design of experiments, prediction of tissue dosimetry, and conduct of dose extrapolations (Ramsey and Andersen, 1984; Clewell and Andersen, 1987).

This chapter presents the basic principles and methods relating to the construction of PBPK models, and their application in the risk assessment of developmental neurotoxicants.

II. Steps in Physiologically Based Pharmacokinetic Modeling

The development of PBPK models is performed in four interconnected steps: model representation, model parameterization, model simulation, and model validation (Krishnan and Andersen, 1994). Model representation involves the development of conceptual and mathematical descriptions of the relevant compartments of the animal, as well as the exposure and clearance pathways of the chemical. Model parameterization involves obtaining the point estimates or a range of values for each of the parameters specified in the model equations. Model simulation involves obtaining simultaneous solution to the set of model equations using a numerical integration algorithm, simulation software, and a computer. Finally, the model validation phase involves comparison of *a priori* predictions of the model with relevant experimental data to refute, validate, or refine the model, and also characterization of the sensitivity, variability, and uncertainty associated with the model parameters and predictions.

A. *Model Representation*

1. Conceptual Representation

The conceptual PBPK model represents both the animal and the chemical (Fig. 1). The number and nature of model compartments are determined by the mechanisms and sites of absorption, distribution, metabolism, and excretion. Only the processes and tissues that critically influence or determine the pharmacokinetics of substances need to be considered. The whole purpose of modeling is to develop a simplified representation of the system without significant loss in the accuracy of system behavior. In order to predict the pharmacokinetics of chemicals in intact animals, a certain number of processes and tissues need to be considered. The basic physiologic processes include respiration, blood circulation, urinary and fecal excretion, and metabolism. Of

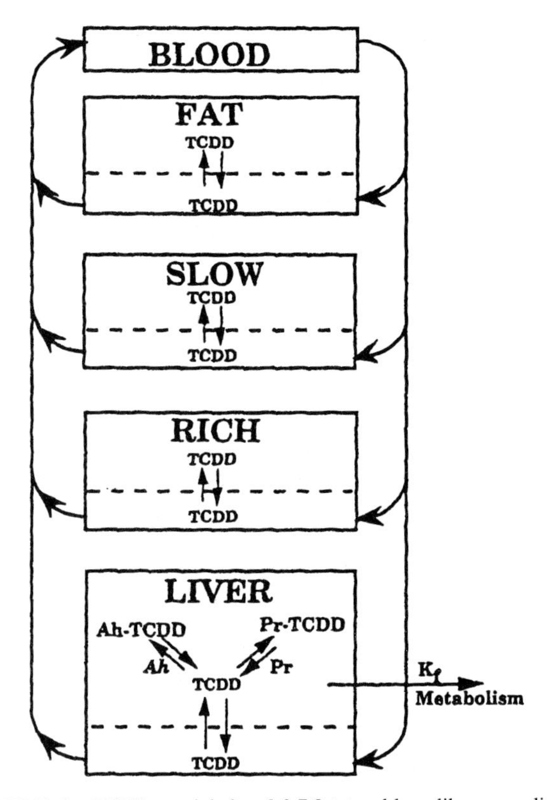

FIGURE 1 PBPK model for 2,3,7,8-tetrachlorodibenzo-*p*-dioxin (TCDD). Metabolism and protein binding are included in the liver. Each tissue compartment comprises of two subcompartments, namely, tissue blood and cellular matrix. Uptake after subcutaneous administration was described as a first order process with the chemical appearing in mixed venous blood. [Reproduced from Andersen *et al.* (1993), with permission of the Society for Risk Analysis.]

these, if a chemical is not volatile and is neither taken up nor eliminated via lungs to any significant extent, then the respiratory process does not have to be included in the model to obtain adequate simulations of the kinetics of that chemical. Similar arguments apply when the chemical being modeled is not excreted significantly via urine or feces, or metabolized in a specific organ (e.g., extrahepatic tissues). Blood circulation, however, is a basic process that is considered in all physiologic models. Even though the heart is not physically represented as a separate compartment, the blood circulation is expressed as a composite of blood flow to the various tissues in the body.

The tissues, individually or together, are represented as compartments in PBPK models. Target tissue is typically represented as a separate compartment. However, when the concentration versus time profiles are identical for several tissues, these tissues are lumped into a single compartment. In other words, when the critical determinants of tissue dosimetry do not vary among several tissues, each of them does not have to be represented as a separate compartment. This approach is illustrated by the frequent lumping of adipose tissues, slowly perfused tissues, and richly perfused tissues into composite compartments in PBPK models (Fig. 1). The adipose tissue or fat compartment in PBPK models represents all fat depots in the body, such as the perirenal, epididymal, and omental fat, and the slowly perfused tissues reflect the lumping of muscles and skin tissue. The richly perfused tissues compartment refers to the lumping of tissues such as the adrenals, kidney, thyroid, brain, lung, heart, testis, and the hepatoportal system. A tissue (e.g., brain) lumped within the richly perfused tissue compartment can be treated separately, if necessary (Krishnan *et al.*, 1992).

2. Mathematical Representation

The functions of each compartment (e.g., solubility, metabolism, binding) are described with mass balance differential equations (MBDEs) of the following kind:

$$\frac{dA_t}{dt} = CL_u C_a - CL_e C_{vt} - Cl_{met} C_{vt} \qquad (1)$$

where $\frac{dA_t}{dt}$ = rate of change in the amount of chemical in tissue (mg/hr); CL = clearance (L/hr), C = Concentration (mg/L), a = arterial, u = uptake, e = elimination, met = metabolic, and vt = venous blood leaving tissue.

The uptake clearance in Eq. [1] can be described according to Fick's law of simple diffusion, which states that the flux of a chemical is proportional to its concentration gradient:

$$\frac{dA_t}{dt} = K\Delta C \qquad (2)$$

where K = transfer constant (L/hr). This term, K, can be regarded as a mass transfer constant, cm/hr, times the area, A. Alternatively, it is estimated by a diffusivity, cm^2/hr, times area, cm^2, divided by a diffusion length, cm. In each case, it has units of clearance, flow/time.

The proportionality constant in Eq. [2] corresponds to the rate of chemical transfer from blood across tissue membrane barriers. The rate of change in the amount of chemical in cellular matrix (Fig. 2; *dAcm/dt*) is equal

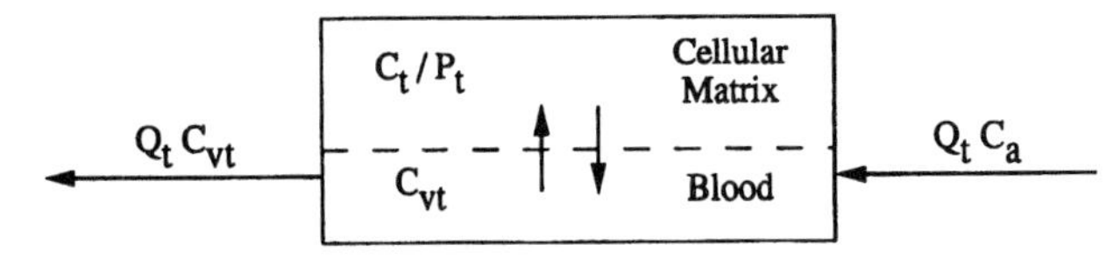

FIGURE 2 Schematic of a tissue compartment. Q_t is tissue blood flow rate, C_a is arterial blood concentration, C_{vt} is the concentration of the chemical in venous blood leaving the tissue, C_t is the tissue concentration and P_t is the tissue : blood partition coefficient. [Reproduced from Krishnan and Andersen (1994), with permission of Raven Press.]

to the product of the diffusional clearance (PA_t) and the concentration gradient:

$$\frac{dA_{cm}}{dt} = PA_t\left(C_a - \frac{C_t}{P_{tb}}\right) \quad (3)$$

where C_t = tissue concentration and P_{tb} = tissue : blood partition coefficient.

The rate of change in the tissue blood subcompartment (Fig. 2; dA_{tb}/dt); equals the sum of the retention from blood flow plus the net flux from cellular matrix:

$$\frac{dA_{tb}}{dt} = Q_t(C_a - C_{vt}) - PA_t(C_t/P_{tb} - C_{vt}) \quad (4)$$

If diffusion of chemical from tissue blood to cellular matrix is slow with respect to tissue blood flow, both Eqs. [3 and 4] are necessary. If the reverse is true, then a single equation is sufficient to describe the rate of change in the amount of chemical in the whole tissue mass (i.e., cellular matrix plus tissue blood):

$$\frac{dA_t}{dt} = Q_t(C_a - C_{vt}) \quad (5)$$

Some examples of mathematical descriptions used in PBPK models for simulating absorption, distribution, metabolism, and excretion of chemicals are provided in Table 1. More of the mathematical descriptions used in PBPK models are reviewed in Krishnan and Andersen (1994) and Hoang (1995). To solve these equations, the numeric values of several physiologic, physicochemical, and biochemical parameters are required.

TABLE 1 Examples of Mathematical Descriptions Used in PBPK Models of Inhaled Vapors

Item	Equation
Arterial blood concentration	$C_a = \frac{Q_p C_i + Q_c C_v}{Q_c + (Q_p/P_{ba})}$
Venous blood concentration	$C_v = \frac{\sum_{t=1}^{n} Q_t C_{vt}}{Q_c}$
Tissue mass-balance	$\frac{dA_t}{dt} = Q_t\,(C_a - C_{vt}) - RAM$
Amount in tissue	$A_t = \int_0^t \frac{dA_t}{dt}$
Concentration in tissue	$C_t = A_t/V_t$
Concentration in venous blood leaving tissue	$C_{vt} = C_t/P_{tb}$

A, C, P, Q and *V* represent amount, concentration, partition coefficient, blood or air flow rate, and volume. The subscripts *a, v, i, c, p, t, ba,* and *tb* refer to arterial, venous, inhaled, cardiac, pulmonary, tissue, blood : air, and tissue : blood. RAM is the rate of amount metabolized in the tissue.

B. Model Parameterization

1. Physiologic Parameters

Numeric values of physiologic parameters required for PBPK models can be obtained from biomedical literature. Compilations of the values of several physiologic parameters for developing and adult animals and humans have been published (e.g., Arms and Travis, 1988; Farris *et al.,* 1993; Brown *et al.,* 1997). Several researchers have also assembled such data in an effort to model the kinetics of substances in pregnant and lactating animals and humans (e.g., Fisher *et al.,* 1990; Luecke *et al.,* 1994, 1995; Gray, 1995). These data sources should be useful as such, or at least as a useful starting point for modeling the kinetics of developmental neurotoxicants.

2. Partition Coefficients

The blood : air and tissue : blood partition coefficients (PCs) represent the relative distribution of un-ionized chemicals between the two matrices at equilibrium. A number of *in vivo* and *in vitro* methods are available for the estimation of tissue : blood PCs (reviewed in Krishnan and Andersen, 1994; Knaak *et al.,* 1995). The most frequently used *in vivo* method involves collecting blood and tissue concentration data at steady state and finding the intersect of the best fit line with a unit slope drawn through the log–log plot of tissue versus blood concentration. The *in vitro* methods include vial equilibration, equilibrium dialysis, and ultrafiltration.

In search of an alternative for the *in vitro* systems to estimate the tissue : blood partition coefficients (P_{tb}), several authors have attempted to develop theoretic relationships between brain : blood partition coefficients (P_{bb}) and *n*-octanol : water partition coefficients (P_{ow}) (Fiserova-Bergerova and Diaz, 1986; Gargas *et al.,* 1989; Paterson and Mackay, 1989). These empiric relationships do not have any predictive power because species-specific or chemical-specific differences in mechanistic determinants of the tissue partitioning process are not taken into account. Kaliszan and Markuszewski (1996) performed a more thorough regression analysis to relate P_{bb} to not only P_{ow} of chemicals, but also to molecular bulkiness descriptors. Initially, the following regression relationships were proposed:

$$\log P_{bb} = 0.476 + 0.541 \log P_{ow} - 0.00794\, M_m \quad \text{(}H_2\text{ histamine antagonists)} \quad (6)$$

$$\log P_{bb} = -0.088 + 0.272 \log P_{ow} - 0.00116\, M_m \quad \text{(anesthetics and organic pollutants)} \quad (7)$$

where M_m = molecular mass.

An improved association in the case of H_2 histamine receptor antagonists was obtained by replacing log P_{ow} with log cyclohexane : water partition coefficient (log P_{cw}) in the regression equation:

$$\log P_{bb} = 1.296 + 0.309 \log P_{cw} - 0.0057\ M_m \quad (8)$$

The statistical quality of regression could further be improved when M_m in Eq. [8] was replaced with other bulkiness–related parameters—that is, polarizability, refractivity:

$$\log P_{bb} = 1.979 + 0.373 \log P_{cw} - 0.00275\ V_{wav} \quad (9)$$

where V_{wav} = water accessible volume.

Because the value of the second coefficient in Eqs. [6–9] is less than 1, it implies that the net polarity of the brain : blood system is lower than that of the octanol : water or cyclohexane : water systems. Based on these considerations, it is evident that equations that explicitly take into account the neutral lipid levels in tissues and blood would be more useful in predicting P_{bb}.

Poulin and Krishnan (1995a,b) proposed a tissue composition–based algorithm for predicting P_{tb} of chemicals that do not bind to macromolecules. The approach involves the calculation of tissue : water and blood : water PCs for a chemical, and the division of one by the other to provide estimates of tissue : blood PCs, as follows:

$$PC_{tb} = \frac{[P_{ow} \times F_{nt}] + [P_{ow} \times 0.3\ F_{pt}] + [0.7\ F_{pt}] + [F_{wt}]}{[P_{ow} \times F_{nb}] + [P_{ow} \times 0.3\ F_{pb}] + [0.7\ F_{pb}] + [F_{wb}]} \quad (10)$$

where P_{ow} = n-octanol or oil : water partition coefficient, F = volume fraction, nt = neutral lipids in tissues, nb = neutral lipids in blood, pt = phospholipids in tissues, pb = phospholipids in blood, wt = water in tissues, and wb = water in blood.

In Eq. [10], the numerator represents the calculation of the tissue : water PC, whereas the denominator estimates blood : water PC. $P_{o:w}$ is used as a surrogate for the neutral lipid : water partitioning, and phospholipids are described to behave as a mix of 70% water and 30% neutral lipid. The $P_{o:w}$ in Eq. [10] can be predicted, in turn, from molecular structure information. In other words, using the data on tissue and blood levels of lipids and water along with molecular structure information, the P_{tb}s can be predicted. This approach has been validated extensively (Poulin and Krishnan, 1996), and as such may be useful for predicting the PCs for constructing PBPK models for developmental neurotoxicants.

3. Biochemical Parameters

Biochemical parameters required for PBPK modeling include the rates of absorption, biotransformation, macromolecular binding, and excretion. Several *in vitro* and *in vivo* methods are potentially useful for estimating these rate constants (Krishnan and Andersen, 1994). The most common *in vivo* method involves the collection of blood kinetic data and its analysis with a PBPK model that has one or two unknowns. If these unknowns refer to specific biochemical parameters (e.g., maximal velocity for metabolism and Michaelis affinity constant), then their numeric values can be estimated by a curve fitting exercise. This approach has been used to estimate the rate constants for absorption, metabolism, and elimination of a number of chemicals (Krishnan and Andersen, 1994). Subcellular fractions, postmitochondrial preparations, isolated cells, tissue slices, and isolated perfused organs are potentially useful *in vitro* systems that may provide biochemical constants useful for PBPK modeling. However, the question of *in vitro–in vivo* extrapolation has not been resolved for all these experimental systems (Krishnan and Andersen, 1993; De Jongh and Blaauboer, 1996; Carlile *et al.,* 1997; Houston and Carlile, 1997). However, some works have indicated the possibility of scaling metabolic rate constants to intact animals, particularly from hepatocyte preparations (e.g., Kedderis and Held, 1996).

Until validated animal-replacement methods become available, one approach that can be used is to specify the theoretically plausible range of a specific biochemical parameter to obtain range of predictions of pharmacokinetic profiles of chemicals. This approach is applicable when the first-pass metabolism is not predominant. Accordingly, if the partition coefficients of a chemical have been predicted using molecular structure information, the hepatic metabolic clearance can be set equal to zero or to blood flow rate to simulate pharmacokinetics of inhaled vapors (Poulin and Krishnan, 1998a,b). This approach implies that the hepatic clearance can neither exceed the liver blood flow rate and not be lower than zero; therefore, the pharmacokinetic profiles or the "envelope" obtained with this approach will contain all experimental data corresponding to that scenario of exposure (Poulin and Krishnan, 1998b). The envelope defines not only the range within which the experimental values fall, but also the limits that are not exceeded, regardless of the extent and nature of inhibition or induction of hepatic metabolism. This approach may be used as a first-cut analysis before conducting detailed dermal or inhalation pharmacokinetic modeling studies.

C. Model Simulation

Model simulation is the system behavior predicted by solving the set of differential equations and algebraic equations representing the quantitative interrelationships among the model parameters under specific exposure conditions. In PBPK modeling, simulation has several advantages: to analyze events over a long period of time during complex scenarios; to determine the solutions for nonlinear descriptions employed in the model; and to reduce animal use in toxicokinetic studies where feasible (Rideout, 1991).

Simulation in the context of PBPK modeling is useful for predicting the amounts and kinetic profiles of chemicals in blood and tissues by solving the set of differential equations. Basically, the numeric solution methodology involves:

1. calculating the area under the dependent variable versus the independent variable curve for predetermined time intervals, and
2. adding the results of step 1 with the numeric value of the dependent variable at the beginning of the time interval.

The temporal change in a dependent variable $\left(\frac{dA_t}{dt}\right)$ should be added to its initial or starting value ($A_{t,o}$) such that the resulting value of the dependent variable at the end of each time interval can be obtained (Davis, 1992). The solution to MDBEs in PBPK models is of the following form:

$$A_{t1} = A_{t0} + \left(\frac{dA_t}{dt} * dt\right) \tag{11}$$

where *dt* is the integration interval or the predetermined length of time for which a solution is obtained.

The MBDEs for various tissues in the PBPK models need to be solved before calculations of C_t, C_{vt}, C_a, and C_v at any given time can be attempted. The numeric solution obtained using Eq. [11] (referred to as the *Euler algorithm*) will be reliable so long as *dt* is extremely small.

Other algorithms useful for solving MBDEs in PBPK models are Runge-Kutta, Gear and Adams–Moulton. When commercially available simulation software packages (e.g., ACSL®, SIMUSOLV®, SCoP®, STELLA®) are used, all that is needed is to specify the type of algorithm preferred for PBPK modeling. An alternative methodology, which does not require commercial differential equation solving software, has become available (Haddad *et al.,* 1996). Scientists who have used spreadsheets for organizing, transforming, and plotting experimental data can use this methodology for PBPK modeling. The methodology developed by Haddad *et al.* (1996) involves keying the PBPK model parameters and equations into the spreadsheets and solving the differential equations per Eq. [11].

During repeated, continuous exposures, however, the rate of change in the amount of chemical in tissues becomes zero. For steady-state situations then, the elaborate PBPK models are not required; the tissue and blood concentrations of chemicals can be computed using simpler algebraic expressions that use PCs and metabolic constants as the sole input parameters (e.g., Pelekis *et al.,* 1997).

D. Model Validation

The model validation process ensures the adequacy of the model representation of the system under study. The approaches commonly applied for testing the adequacy of computer simulation models can be classified into four categories: 1. inspection approach, 2. subjective assessment, 3. discrepancy measures, and 4. statistical tests. The testing of the degree of concordance between PBPK model simulations and experimental data has generally been conducted by "eye-balling" or visual inspection approach. This approach involves visual comparison of the plots of simulated data (usually continuous and represented by solid lines) with experimental values (usually discrete and represented by symbols) against a common independent variable (usually time). The rationale behind this approach is that, the greater the commonality between the simulated and experimental data, the greater our confidence will be in the model. The visual inspection approach to model validation continues to be used pending the validation of statistical tests and discrepancy measure tests appropriate for application to PBPK models. While this visual inspection approach lacks mathematical rigor, it has the advantage of requiring the human modeler to become better informed about the behavior of the model. Haddad *et al.* (1995) screened various statistical procedures (correlation, regression, confidence interval approach, lack of fit F test, univariate analysis of variance, and multivariate analysis of variance) for their potential usefulness in testing the degree of agreement of PBPK model simulations and experimental data obtained in intact animals. Lack of fit F test has been suggested as a useful and practical way of evaluating the adequacy of simulation models. Particularly, this simple procedure permits the consideration of multiple datasets (e.g., data for several endpoints collected at various time intervals) in conducting an evaluation of model validity. The multivariate analysis of variance probably represents the most appropriate test, with the variance for the simulation data permitting.

The application of an appropriate statistical test provides a means of evaluating whether model simulations are significantly different from experimental values. Regardless of the outcome of such statistical analysis, it is often necessary and useful to be able to represent, in a quantitative manner, the extent to which the model simulations differ from experimental data. In this context, Krishnan *et al.* (1995) have developed a simple index to represent the degree of closeness or discrepancy between *a priori* model predictions and experimental data used during the model validation phase. The approach involved the calculation of the root mean square of the error (representing the difference between the individual simulated and experimental values for each sampling point in a time course curve), and dividing them by the root mean square of the experimental values. The resulting numeric values of discrepancy measures for several datasets (each corresponding to an endpoint) obtained in a single experimental study are then combined on the basis of a weighting proportional to the number of data points contained in the dataset. Such consolidated discrepancy indices obtained from several experiments (e.g., exposure scenarios, doses, routes, species) are averaged to get an overall discrepancy index referred to as *PBPK index*. The application of this kind of a "quantitative" method should help remove the ambiguity in communicating the degree of concordance or discrepancy between PBPK model simulations and experimental data.

The use of a discrepancy measure test or statistical test to show that *a priori* predictions of a particular endpoint are in agreement with the experimental data alone is insufficient. These approaches are only useful for providing a quantitative measure of differences, but not any information on either the model robustness or the reliability of the model structure. In this context, it is important to verify the influence of variability, uncertainty, and sensitivity associated with model parameters. A number of approaches, most often using Monte Carlo analysis, and examples of their applications are available in the literature (Farrar *et al.,* 1989; Bois and Tozer, 1990; Hattis *et al.,* 1990; Hetrick *et al.,* 1991; Krewski *et al.,* 1995; Varkonyyi *et al.,* 1995).

Most PBPK modeling efforts have judged the adequacy–validity by comparing *a priori* predictions to experimental data on blood, plasma, or exhaled air concentrations. The fact that the model predictions of one endpoint—that is, a measurement endpoint (e.g., plasma concentration of parent chemical) are adequate does not mean that all other endpoints of toxicologic and risk assessment relevance—that is, assessment endpoints (e.g., concentration of a metabolite in brain) would necessarily be predicted with the same level of accuracy. When data on measurement endpoints rather than assessment endpoints are used for validating the model, it is essential to choose those measurement endpoints that have the same kind of sensitivity and response pattern as the assessment endpoint.

III. The Brain as a Separate Compartment in Physiologically Based Pharmacokinetic Models

The fact that certain tissues are represented as individual compartments in PBPK models relates to the necessity of doing so, either because they represent the site of action or their characteristic time constant (i.e., the product of tissue volume and tissue : blood partition coefficient divided by the rate of blood flow to tissue) is different from other groups of organs or tissues. Because the characteristic time constant for several chemicals in brain is somewhat similar to other well-perfused tissues, brain has routinely been considered within the "richly perfused tissues" compartment in PBPK models. Whenever the concentration profile of chemicals in brain is required, this tissue, representing about 1–2% of body volume, is separated out of the richly perfused tissues group, and represented as a single homogeneous compartment (e.g., Krishnan *et al.,* 1992).

Although the brain may be described as a separate compartment, the rest of the body may still be considered a composite of several tissue compartments (e.g., slowly perfused tissues, adipose tissues, liver, other richly perfused tissues). In cases in which neither the accumulation of chemicals in fatty tissues nor hepatic metabolism are major determinants of pharmacokinetics of chemicals a simpler model in which the body is considered as a single compartment may be used. Using this approach, Kim *et al.* (1994) simulated the pharmacokinetics and brain dosimetry of 2,4-dichlorophenoxy acetic acid (2,4-D) in rats and rabbits. The PBPK model, in this case, consisted of the brain and the rest of the body interconnected by systemic circulation. The chemical uptake by the brain was described as a diffusion-limited process because 2,4-D is not readily diffusible through the blood–brain barrier. The rate of change in the brain tissue (dA_b/dt) was described as follows:

$$\begin{aligned}\frac{dA_b}{dt} = &\text{(rate of 2,4-D entry into brain via plasma}\\ &- \text{rate of 2,4-D exit out of the tissue via brain plasma)}\\ &- \text{(rate of 2,4-D into brain via cerebrospinal fluid}\\ &- \text{rate of 2,4-D exit out of the tissue via cerebrospinal fluid)}\end{aligned} \quad (12)$$

When there is evidence of significant activity in the brain toward the metabolism of a chemical (Strobel *et al.,* 1997), a term to account for metabolic clearance should be added to brain MBDE. In most PBPK modeling efforts, the brain has been represented as a single homogeneous compartment. However, when the concentration profiles for discrete areas of the brain must be simulated, the single brain compartment can be replaced with a detailed description (Rapoport *et al.,* 1979). Kim *et al.* (1995) extended a PBPK model for 2,4-D to simulate regional dosimetries of this chemical within the brain—that is, in hypothalamus, caudate neucleus, forebrain, brainstem, cerebellum, and hippocampus (Fig. 3). This modeling strategy may be preferred to treating the brain as a single homogeneous compartment, given the fact that the toxic effects are likely to be associated

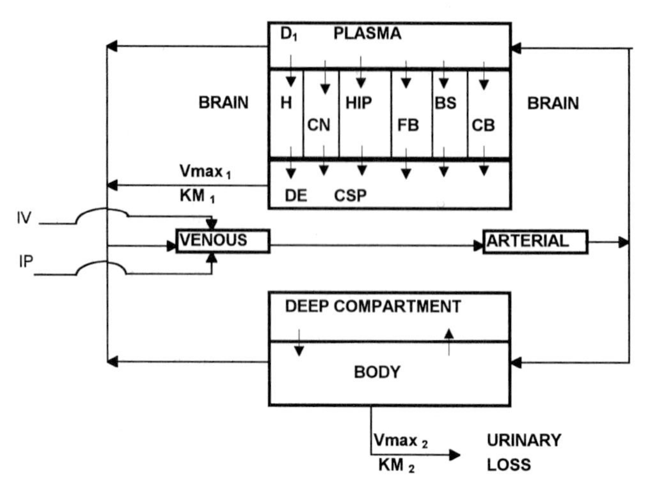

FIGURE 3 Schematic of a PBPK model for 2,4-dichlorophenoxy acetic acid (2,4-D). The brain compartment is a composite of hippothalamus (H), caudate nucleus (CN), hippocampus (HIP), forebrain (FB), brainstem (BS), and cerebellum (CB). IP and IV refer to intraperitoneal and intravenous, respectively. Vmax and KM are the maximal velocities and affinity constants for saturable clearance processes. [Reprinted from *Neurotoxicol. Teratol.*, Vol. 17, Kim *et al.*, Development of a physiologically based pharmacokinetic model for 2,4-dichlorophenoxyacetic acid dosimetry in discrete areas of the rabbit brain, pp. 111–120, Copyright 1995, with permission of Elsevier Science.]

with toxicant concentrations in specific regions of the brain (Bianchetti *et al.*, 1980).

PBPK models are useful in providing the concentration versus time profile of toxicants in target tissues such as the brain. For developmental neurotoxicants, the brain dosimetry of chemicals needs to be obtained in developing organisms for conducting scientifically sound dose–response assessments. Accordingly, tissue exposure, both during development inside and outside the mother (i.e., after parturition) should be considered. In other words, PBPK modeling efforts should account for both prenatal and postnatal exposures of developing organisms to neurotoxicants. The state of knowledge regarding the PBPK modeling of toxicants in developing organisms is reviewed in the following sections.

IV. Physiologically Based Pharmacokinetic Modeling of Fetal Exposure

Most of the PBPK models for neurotoxicants developed so far have not been adapted to simulate the disposition kinetics of such chemicals during periods of fetal growth and organ development. Some earlier modeling efforts have focused to simulate kinetics of chemicals on a particular day of gestation, most often during the latter periods of gestation (e.g., Olanoff and Anderson, 1980; Gabrielsson and Paalzow, 1983; Gabrielsson *et al.*, 1984, 1985; Gabrielsson and Growth, 1988). Such PBPK models require only one set of parameter estimates for both the mother and the fetus or embyro, corresponding to the day for which simulation is performed. Basically, in these models, in addition to the compartments representing the mother, a placental and a fetal compartment are included (Fig. 4). The weights of these latter compartments are expressed as a fraction of the mother's body weight. Following the inclusion of MBDEs for the placental and fetal compartments, the PBPK model simulates the profile of chemical kinetics in the fetal compartment for maternal exposures during that particular day of gestation.

The rate of change in the amount of chemical in the placental (dA_{pla}/dt) and fetal (dA_{fet}/dt) compartments can be described as follows (Fisher *et al.*, 1989):

$$\frac{dA_{pla}}{dt} = Q_{pla}(C_a - C_{pla}/P_{pla}) - dA_{fet}/dt \quad (13)$$

$$\frac{dA_{fet}}{dt} = Q_{fet}(C_{pla}/P_{pla} \cdot P_l/P - C_{fet}/P_{fet}) \quad (14)$$

where Q_{pla} = blood flow rate to placenta, Q_{fet} = blood flow rate to fetus, C_{pla} = concentration in placenta, C_{fet} = concentration in the fetus, P_{pla} = placenta : blood partition coefficient, P_{fet} = fetal tissues : blood partition coefficient, P_l = fetal blood : air partition coefficient, and P = maternal blood : air partition coefficient.

In Eqs. (13) and (14), chemical concentration in fetal drainage leaving the placenta is adjusted for difference

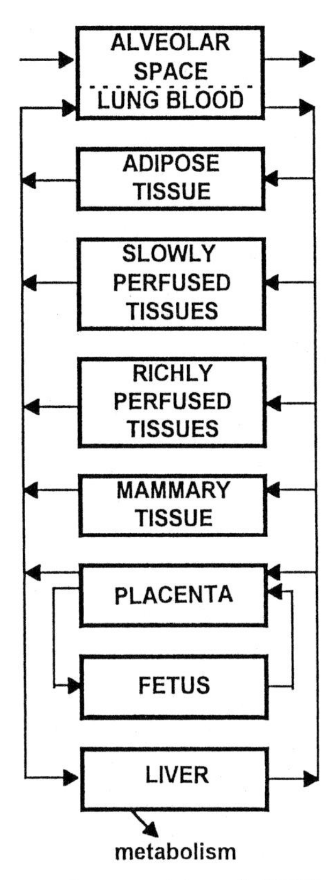

FIGURE 4 Conceptual representation of a PBPK model for simulating kinetics of volatile chemicals in a pregnant animal. [Adapted from Fisher *et al.* (1989).]

in chemical solubility between maternal and fetal blood. The inclusion of Eqs. (13) and (14) in a conventional adult PBPK model requires the numeric values of P_{pla}, P_{fet}, Q_{pla}, Q_{fet}, V_{pla}, *and* V_{fet} corresponding to the desired gestational day.

This PBPK modeling framework can be extended to simulate the kinetics of substances in growing fetus or embyro during the entire length of pregnancy (Fig. 5). This effort, however, requires the inclusion of a description of the temporal change in the numeric values of the parameters related to the various model compartments (i.e., maternal tissues, placenta, and fetus). The growth of the human embryo or fetus, for example, can be modeled using Verhulst logistic equation, a polynomial equation, or Gompertz equation (Wosilait *et al.,* 1992). Using the experimentally determined fetal weight changes, certain authors have developed simpler mathematical relationships (O'Flaherty *et al.,* 1992, 1995). These efforts have described the entire growth curve with a single smoothing equation, or with several regression equations, each of which describes a segment of the growth curve. This approach has been used to provide simulations of the concentration profiles of chemicals in the embryo or fetus at any time during the pregnancy, from conception to parturition (O'Flaherty *et al.,* 1992, 1995; Clarke *et al.,* 1993; Terry *et al.,* 1995; Ward *et al.,* 1997). Because the fetus is treated as a single homogeneous compartment, this model framework may not adequately simulate fetal tissue concentrations. Replacing the fetal compartment in the PBPK model shown in Fig. 5 with a network of several appropriate tissue compartments can facilitate the simulation of the tissue dosimetry of chemicals in the developing fetus (Fig. 6). The construction of such a submodel for the developing fetus requires the numeric values of physicochemical, biochemical, and physiologic parameters.

The estimation of partition coefficients can be accomplished either experimentally or using the validated animal-replacement techniques (Krishnan and Andersen, 1994; Poulin and Krishnan, 1996a,b,c). The physiologic parameters may be determined experimentally or obtained from the literature. For example, allometric equations to estimate the volumes of fetal organs as a function of body weight and gestational time are available in the literature (Luecke *et al.,* 1994, 1995; Gray, 1995). The simplest allometric approach involves calculating fetal organ weights (W_i) at any specific time of gestation as a function of fetal body weight (W_{fet}) as follows:

$$W_i = aW_{fet}^b \tag{15}$$

where a and b are constants determined by regression analysis. The values of a and b for predicting the weight of various fetal organs as a function of W_{fet} are listed in Table 2. The allometric equation for developing human brain has values of 0.1871 and 0.9585 for the coefficients a and b, respectively. It is important to realize that the brain weights of human embryo or fetus and adults are not adequately described by the general species allometric models (Luecke *et al.,* 1995).

The PBPK models of pregnancy are useful in simulating the pharmacokinetics and tissue dosimetry of developmental neurotoxicants in the mother and developing fetus, resulting from prenatal exposures (e.g., Gray, 1995; Kim *et al.,* 1996; Luecke *et al.,* 1997). Figure 7 presents an example of PBPK model-simulated concentrations of methylmercury in maternal and fetal tissues (Luecke *et al.,* 1997). These simulations underscore the usefulness of PBPK models in conducting extrapolation of fetal tissue concentrations from one developmental stage to another. When temporal change in the model parameters are accounted for, the PBPK model can be used to simulate the concentration profiles of chemicals on any particular day of gestation (Gray, 1995; Terry *et al.,* 1995; Luecke *et al.,* 1997). This aspect of PBPK modeling has important implications with regard to the risk assessment of developmental neurotoxicants, in that brain AUC or C_{max} values for any particular time frame can be obtained and correlated with observed responses.

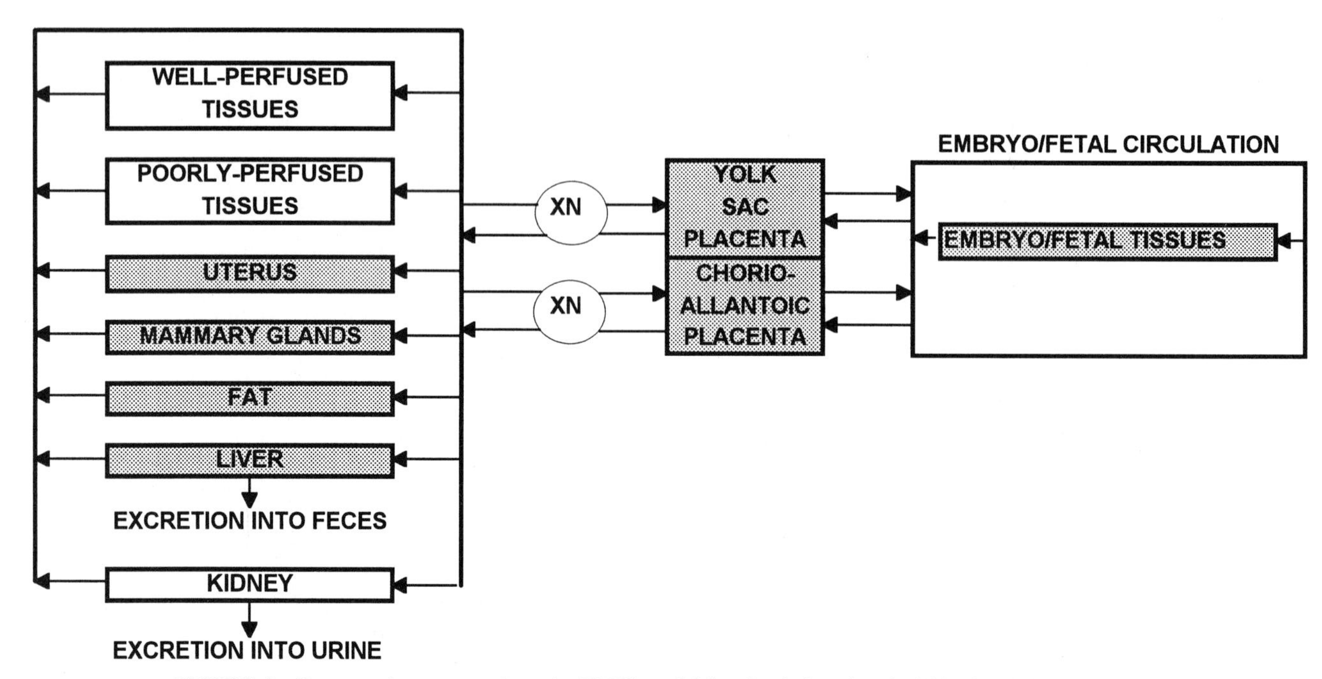

FIGURE 5 Conceptual representation of a PBPK model for simulating chemical kinetics during gestation. Shaded areas designate tissues that increase in relative volume during gestation. N is the number of fetuses. Even though the yolk sac and chorioallantoic placenta are shown in parallel, in fact a period of yolk sac dominance precedes the appearance and growth of chorioallantoic placenta later in gestation. [Reproduced from O'Flaherty *et al.* (1992), with permission of Academic Press.]

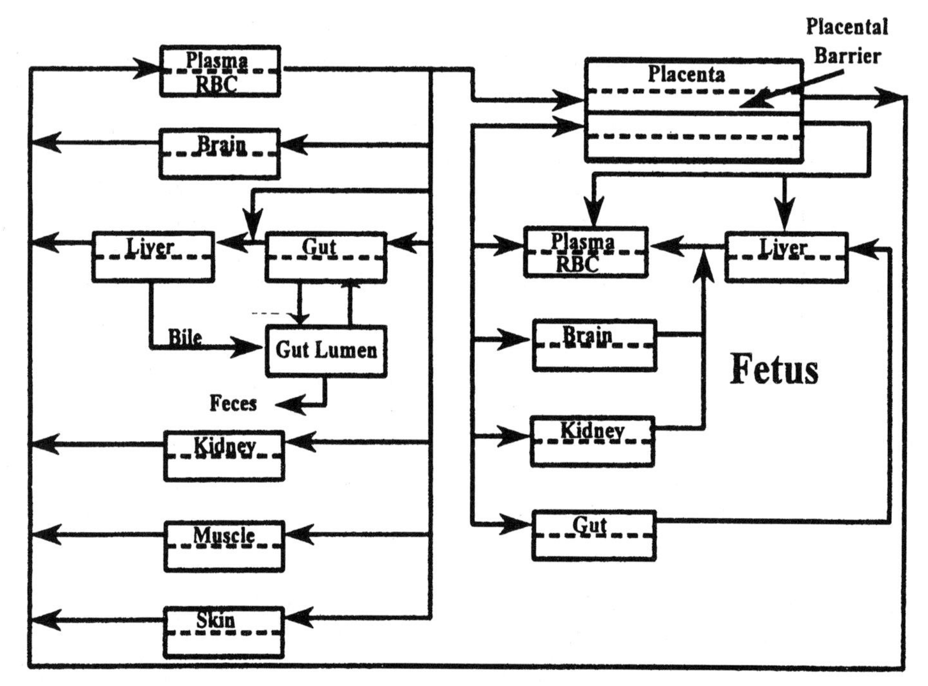

FIGURE 6 PBPK model for methylmercury in the pregnant rat and fetus. [Reproduced from Gray (1995), with permission of Academic Press.]

TABLE 2 Allometric Parameters for Fetal Organs and Tissues Subject to Change During Pregnancy[a]

Fetal organs/tissues	a	b
Adrenal	0.007467	0.8902
Bone	0.05169	0.9288
Bone marrow	0.01425	0.9943
Brain	0.1871	0.9585
Fat	0.1803	−0.9422
Heart	0.01012	0.9489
Kidney	0.004203	1.255
Liver	0.06050	0.9737
Lung	0.09351	1.552
Pancreas	0.1883	0.3854
Plasma	0.06796	0.9729
Skeletal muscle	0.02668	1.234
Spleen	0.0001302	1.204
Thymus	0.001218	1.093
Thyroid	0.0006470	1.023

The constants a and b are used in equation of the following kind $W_{tissue} = a\, W_{body}^{b}$. A third constant c (= 0.2332, −0.02127, −0.05945 or 0.02909, respectively) is used in case of fat, kidney, lung and spleen to accommodate growth rate differences in these organs and total weight of human embryo/fetus.

[a]Adapted from Luecke *et al.*, 1995.

Such analyses can provide insights regarding the pharmacokinetic mechanisms of toxicity, and a basis for conducting interspecies extrapolations of kinetically equivalent prenatal exposure levels.

V. Physiologically Based Pharmacokinetic Modeling of Lactational Exposure of Infants

After parturition, tissue dosimetry of neurotoxicants in the developing infants can be simulated using PBPK models in which the temporal changes in organ weights and other physiologic parameters are accounted for (Farris *et al.*, 1993). However, for situations in which breast feeding represents an important pathway of postnatal exposure of infants, physiologically based descriptions of the nursing mother and the nursed infant can be constructed and interconnected to simulate tissue dose of breast milk–driven chemicals in infants. Shelley *et al.* (1988) explored the application of PBPK models for simulating internal exposure of the nursing child due to maternal exposure to atmospheric contaminants in the occupational environment. The approach consisted of linking a three compartmental PBPK model for the mother with a two compartmental model representing the nursing child (Fig. 8). The infant exposure was described to be due exclusively to nursing by a mother exposed to contaminated air. The dose received by the child is then due solely to a direct transfer of bulk milk from the mother's mammary compartment to the child's gastrointestinal (GI) tract according to the nursing schedule and chemical concentration in mother's milk compartment. Transfer between the milk and the mammary tissues was described using two approaches. In the first approach, milk concentration (C_{mlk}) was calculated on the basis of continuous equilibration with mammary blood flow (Q_{mt}). The rate of change in the amount of chemical in the milk ($\frac{dA_{mk}}{dt}$) and mammary tissue ($\frac{dA_{mt}}{dt}$) compartments were then calculated as follows:

$$dA_{mt}/dt = Q_{mt}(C_a - C_{mlk}/P_{mlk}) - Q_{mlk}C_{mlk} \quad (16)$$

$$dA_{mk}/dt = Q_{mlk}C_{mlk} - Q_{skl}C_{mlk} \quad (17)$$

where Q_{mt} = rate of blood flow to mammary tissue, Q_{skl} = suckling rate of the child, Q_{mlk} = rate of milk production, and P_{mlk} = milk : blood partition coefficient of the chemical.

The second approach calculated C_{mlk} based on mammary tissue concentration at the time of milk production, treating milk and mammary tissue as a single compartment. When the milk and mammary tissue are combined, the MBDE for the combined volumes is:

$$dA_{mlk}/dt = Q_{mlk}(C_a - C_{mlk}/P_{mlk}) - Q_{skl}C_{mlk} \quad (18)$$

In the work of Shelley *et al.* (1988), GI absorption of a chemical in breast milk was considered to be instantaneous and its elimination from the infant body was described to be due solely to exhalation. Figure 9 presents an example of the predictions of this PBPK model of the area-under-the blood concentration versus time curve (AUCB) for mother and child during a 9-hr workshift and 3.5-hr nursing schedule. These simulation exercises allow the calculation of the effective dose to the child at a given point in time. The child's effective dose may be stated in terms of the fraction of the mother's effective dose. Accordingly, the relative effective dose to the child (REDC) is expressed by the following equation:

$$\text{REDC} = \frac{\text{AUCB(child)}}{\text{AUCB(mother)}} \quad (19)$$

Fisher *et al.* (1997) expanded this PBPK model structure to incorporate hepatic metabolism in the mother and a milk compartment that changes volume in response to milk reduction and suckling by the infant. The rate of change in the amount of chemical ingested by a nursing infant (R_{nurse}) was calculated as:

$$R_{nurse}\ (\text{mg/hr}) = C_{milk} \times V_{milk} \times \text{Nurse} \times S_{zone} \quad (20)$$

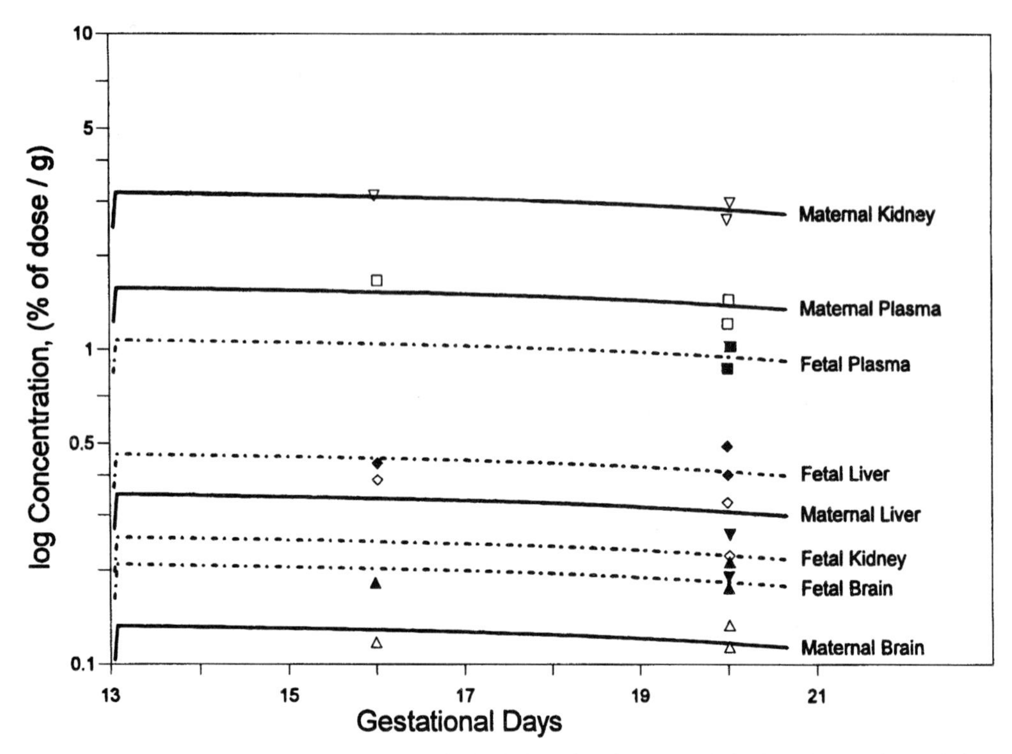

FIGURE 7 Comparison of rat PBPK model simulations (solid or dotted lines) with experimental data (symbols) on the concentration of methylmercury in maternal and fetal tissues and plasma. [Reprinted from *Comp. Meth. Prog. Biomed.,* Vol. 53, Leucke *et al.* A computer model and program for xenobiotic disposition during pregnancy, pp. 201–224, Copyright 1997, with permission from Elsevier Science.]

where V_{milk} = volume of milk currently in mammary tissue lumen (L), Nurse = infant nursing rate (hr^{-1}), and S_{zone} = 0 or 1 (numeric values corresponding to turning on and turning off, respectively, of nursing over a 24-hr period).

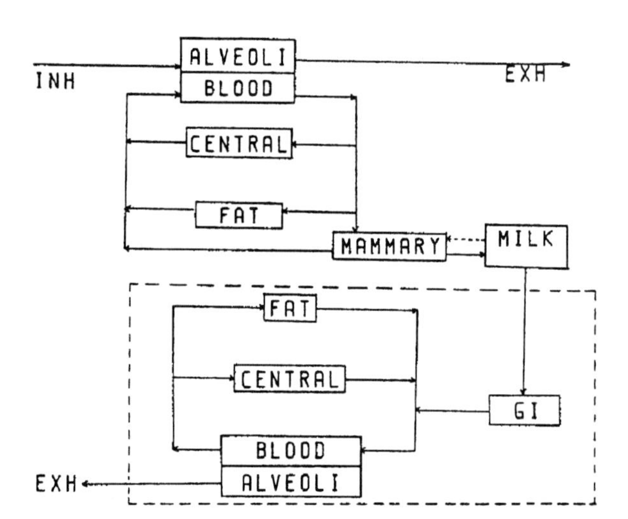

FIGURE 8 Conceptual model of a mother/child distribution system (INH = inhaled chemical, EXH = exhaled chemical, G = gastrointestinal). [Reprinted by permission of Elsevier Science from Shelley *et al. Applied Occupational and Environmental Hygiene,* Vol. 4, pp. 21–26, Copyright 1988 by Applied Industrial Hygiene, Inc.]

The rate of change in the volume of milk in mammary tissue lumen (R_{milk}) was described as:

$$R_{milk} = R_{prod} - R_{loss} \tag{21}$$

where R_{prod} = milk production rate (0.06 L/hr) and $R_{loss} = R_{nurse} \times V_{milk}$. Boundary conditions for V_{milk} were set at 0.01 L and 0.125 L. Before each nursing, V_{milk} was 0.125 L, and as a result of nursing, it diminished to a residual volume of 0.01 L.

In rodent studies, the lactation PBPK model framework proposed by Shelley *et al.* (1988) has been expanded to include hepatic metabolism in the mother and temporal changes in milk production, tissue volumes, and blood flow rates in both the mother and the pup (Fisher *et al.,* 1990). The resulting PBPK model describes the kinetics of chemicals in the pup following the ingestion of milk from mother exposed to contaminated air (Fig. 10) (Fisher *et al.,* 1990; Byczkowski *et al.,* 1994). To describe the dose received by the pup due to lactational transfer, two equations should be included in the adult PBPK model, one related to the milk compartment and the other to the pup suckling. The redistri-

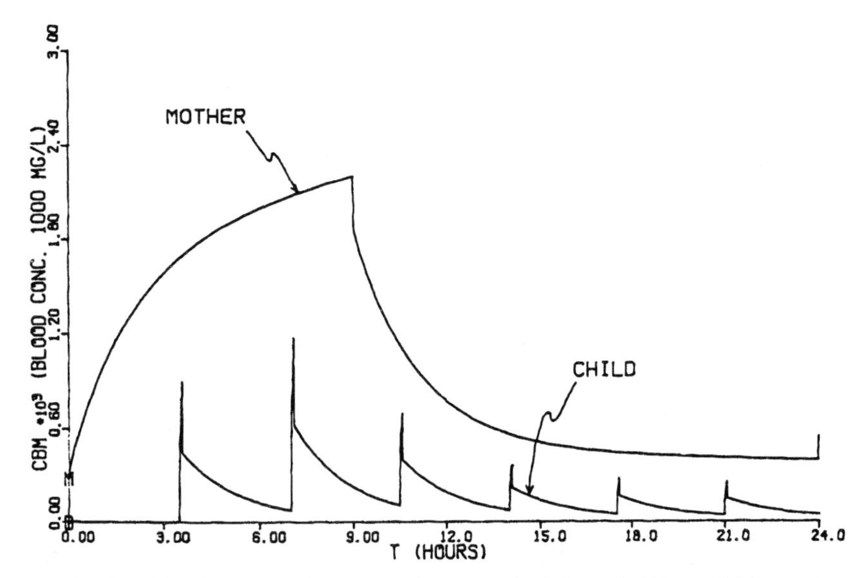

FIGURE 9 Blood concentration versus time curve for 9-hr workshift and 3.5-hr nursing schedule. [Reprinted by permission of Elsevier Science from A risk assessment approach for nursing infants exposed to volatile organics through the mother's occupational inhalation exposure, by Shelley *et al.*, *Applied Industrial Hygiene,* Vol. 4, pp. 21–26, Copyright 1988 by Applied Industrial Hygiene, Inc.]

bution of the ingested dose and tissue concentration profiles can then be simulated using a full-fledged PBPK description for the pup. While conducting simulations of tissue dosimetry, chemical input to pup PBPK model can accommodate exposures via contaminated food and air in addition to breast milk. Such multipathway exposure modeling can facilitate a scientifically sound characterization of risk to infants.

VI. Application of Physiologically Based Pharmacokinetic Modeling in the Risk Assessment of Developmental Neurotoxicants

The PBPK models are useful for simulating the AUCB, C_{max}, or average concentration of the parent

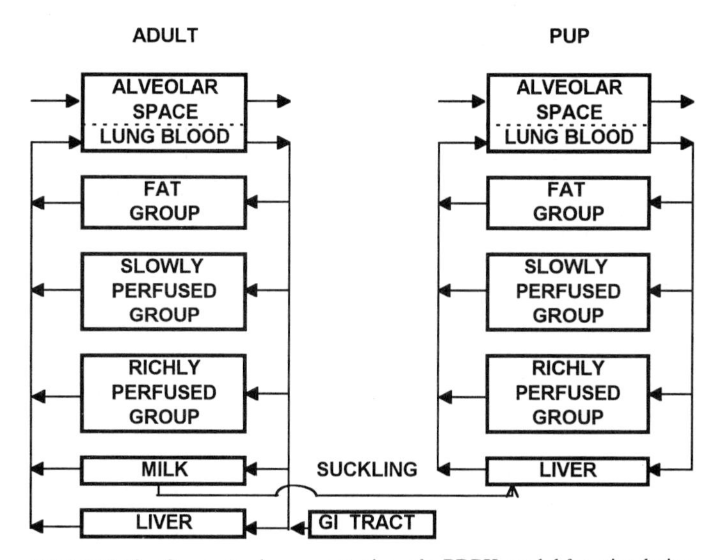

FIGURE 10 Conceptual representation of a PBPK model for stimulating the uptake and disposition of chemicals in the lactating rat and nursing neonate. [Adapted from Fisher *et al.* (1990).]

chemical or its metabolites in the various tissues as a function of dose administered. By relating these various dose measures to observed responses, appropriate dose surrogate(s) can be identified for risk assessment and extrapolation purposes. In the case of developmental neurotoxicants, for example, with the knowledge of brain AUC or C_{max} corresponding to no-observable adverse effect level (NOAEL) in one species, the PBPK modeling approach enables the evaluation of the equivalent dose in other species, for other exposure routes, exposure scenarios, and developmental stages (e.g., Kirshnan and Andersen, 1994; Welsch *et al.*, 1995). Such an approach can help reduce or eliminate the uncertainty factors related to pharmacokinetic differences between species, routes, and developmental stages (Clewell and Jannot, 1994; Welsh *et al.*, 1995). Even though PBPK models have been constructed for a number of developmental neurotoxicants [e.g., dioxins, polychlorinated biphenyls (PCBs), organochlorine pesticides, metals, organometallics], concrete examples of their application in risk assessment are only beginning to emerge.

The application of PBPK modeling to determine the reference dose of methylmercury has been demonstrated by Gearhart *et al.* (1995). These authors developed a PBPK model for methylmercury that consisted of an adult model having 11 compartments, representing both organ-specific and lumped tissues (Fig. 11). Methylmercury transport into and out of compartments was described as a flow-limited or perfusion-limited process. Flow-limited compartments are plasma, kidney, richly perfused tissues, slowly perfused tissues, brain blood, placenta, liver, and gut compartments. The concentration of methylmercury in each of the flow-limited tissue compartments is then simply a result of the arterial plasma concentration flowing to the compartment, tissue-plasma partition coefficients, and loss of methylmercury due to clearance processes. Transport to the three compartments of the adult model, namely red blood cells (RBCs), brain, and fetus, was described as being diffusion limited. There are four compartments in the model that are involved in methylmercury uptake or elimination: urine, hair, feces, and the intestinal lu-

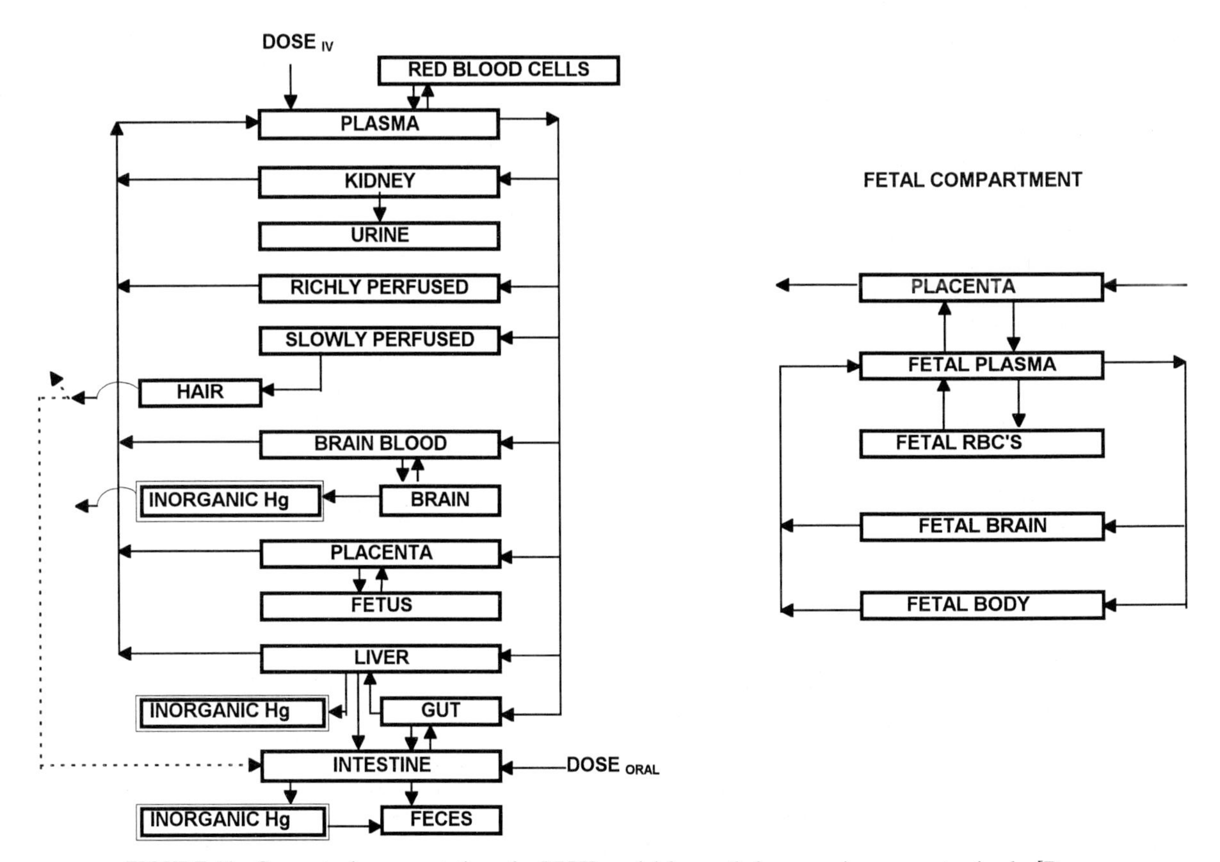

FIGURE 11 Conceptual representation of a PBPK model for methylmercury in pregnant animals. [Reproduced from Gearhart, J. M., Clewell, H. J. III, Crump, K. S., Shipp, A. M., and Silvers, A. (1995). Pharmacokinetic dose estimates of mercury in children and dose-response curves of performance tests in a large epidemiological study. *Water Air Soil Pollut.* **80,** 49–58, with kind permission of Kluwer Academic Publishers.]

men. The conversion of methylmercury to inorganic mercury by the gut flora, with subsequent excretion of inorganic mercury in feces, is also described. The demethylation of methylmercury in the tissues (except in the brain) is described as taking place in the liver, and the resulting inorganic mercury is assumed to accumulate in the kidney. In the brain, both demethylation and slowly reversible incorporation into tissue are described (Gearheart *et al.,* 1995).

The fetal submodel for methylmercury consists of four compartments that grow during the time of gestation (Fig. 11). These compartments correspond to fetal plasma, RBCs, brain, and the remaining fetal tissues. Fetal plasma flows to the brain and body, whereas the RBC compartment communicates with the plasma compartment in the same direction as in the adult model, as influenced by the diffusion parameter and RBC–plasma partition coefficient. This PBPK model described methylmercury pharmacokinetics coherently in the adult rat, monkey, and human, and predicted fetal levels of methylmercury from *in utero* exposure. Using this PBPK model, the results of a large epidemiologic study conducted in New Zealand were reanalyzed to evaluate the neurologic effects of prenatal methylmercury exposure in children. In this particular epidemiologic study, 6-year-old children whose mothers had been exposed to methylmercury via consumption of fish, were administered performance tests to ascertain academic attainment, language development, fine and gross motor coordination, intelligence, and social adjustment (Kjelstrom *et al.,* 1989). These responses were correlated with estimates of prenatal exposure based on average maternal hair concentrations during pregnancy.

Benchmark dose analysis coupled with this PBPK modeling effort indicated that fetal brain concentrations of methyl mercury associated with the NOAEL are on the order of 50 ppb, and result from a maternal dietary intake of methyl mercury in the range of 0.8 to 2.5 μg/kg/day. The reference dose determined using this mechanistic, quantitative modeling approach is about 3 to 8 times greater than the current US EPA value for methyl mercury (0.3 μg/kg/day).

With the current advances in the area of pharmacodynamic modeling, we are beginning to see quantitative models for neurotoxicity. These models are either empirical in nature (Dingemanse *et al.,* 1988), or mechanistically based and biologically motivated (e.g., Gaylor and Slikker, 1994; Leroux *et al.,* 1996). The PBPK model predictions of brain dosimetry can eventually be linked with mechanistically based models for neurotoxic effects, to simulate the dose-response curves more accurately. Such a scientifically-sound approach to the risk assessment of developmental neurotoxicants will not only enhance the credibility of the process but also provide a framework for integrating new mechanistic data as they evolve.

References

Andersen M. E., Mills J. J., Gargaas M. L., Kedderis L., Birnbaum L. S., Neubert D., and Greenlee W. F. (1993). Modeling receptor-mediated processes with dioxin: implications for pharmacokinetics and risk assessment. *Risk Anal.* **13,** 25–36.

Arms A. D., and Travis C. C. (1988). Reference physiological parameters in pharmacokinetic modeling. Office of Health and Environmental Assessment. U.S. EPA, Washington, DC. NTIS PB 88-196019.

Bianchetti G., Elghozi J. L., Gomeni R., Meyer P., and Morselli, P. L. (1980). Kinetics of distribution of dl-propranolol in various organs and discrete brain areas of the rat. *J. Pharmacol. Exp. Therapeut.* **2143,** 682–687.

Bischoff K. B., Dedrick R. L., Zakharo D. S., and Longstreth, J. A. (1971). Methotrexate pharmacokinetics. *J. Pharm. Sci.* **60,** 1128–1133.

Bois F. Y., and Tozer T. N. (1990). Precision and sensitivity of pharmacokinetic models for cancer risk assessment: tetrachloroethylene in mice, rats, and humans. *Toxicol. Appl. Pharmacol.* **102,** 300–315.

Brown R. P., Delp M. D., Lindstedt S. L., Rhomberg L. R., and Beliles, R. P. (1997). Physiological parameter value for physiologically based pharmacokinetic models. *Toxicol. Ind. Health* **13,** 407–484.

Byczkowski J. Z., Kinkead E. R., Leahy H. F., Randall, G. M., and Fisher, J. (1994). Computer simulation of the lactational transfert of tetrachloroethylene in rats using a physiologically based model. *Toxicol. Appl. Pharmacol.* **125,** 228–236.

Carlile D. J., Zomorodi K., and Houston J. B. (1997). Scaling factors to relate drug metabolic clearance in hepatic microsomes, isolated hepatocytes, and the intact liver. *Drug Metab. Disp.* **25,** 903–911.

Clarke D. O., Elswick B. A., Welsch F., and Conolly R. B. (1993). Pharmacokinetics of 2-methoxyethanol and 2-methoxyacetic acid in the pregnant mouse: a physiologically based mathematical model. *Toxicol. Appl. Pharmacol.* **121,** 239–252.

Clewell, H. J., and Andersen, M. E. (1987). Dose, species and route extrapolation using physiologically-based pharmacokinetic models. *Drink. Wat. Health* **8,** 159–184.

Clewell, H. J. III, and Jarnot, B. M. (1994). Incorporating of pharmacokinetics in noncancer risk assessment: example with chloropentafluorobenzene. *Risk Anal.* **14,** 265–276.

Davis, P. W. (1992). Differential equations for mathematics. *Science and Engineering.* Prentice-Hall, Englewood Cliffs, N.J.

De Jongh, J., and Blaauboer B. J. (1996). Simulation of toluene kinetics in the rat by a physiologically based pharmacokinetic model with application of biotransformation parameters derived independently *in vitro* and *in vivo. Fund. Appl. Toxicol.* **32,** 260–268.

Dedrick, R. L., Forrester D. D., and Ho, D. H. W. (1972). *In vivo–in vitro* correlation of drug metabolism: deamination of 1-B-D-arabinosyl cytosine. *Biochem. Pharmacol.* **21,** 1–16.

Dingemanse J., Danhof M., and Breimer D. D. (1988). Pharmacokinetic–pharmacodynamic modeling of CNS drug effects: an overview. *Pharmac. Ther.* **38,** 1–52.

Farrar, D., Allen, B., Crump, K., and Shipp, A. (1989). Evaluation of uncertainty in input parameters topharmacokinetic models and the resulting uncertainting in output. *Toxicol. Lett.* **49,** 371–385.

Farris, F. F., Dedrick, R. L., Allen, P. V., and Smith, J. C. (1993). Physiological model for the pharmacokinetics of methyl mercury in the growing rat. *Toxicol. Appl. Pharmacol.* **119,** 74–90.

Fentermacher, J. D., Patlak, C. S., and Blasberg, R. G. (1974). Transport of maternal between brain extracellular fluid, brain cells and blood. *Fed. Proc.* **33,** 2070–2074.

Fiserosa-Bergerova, V. (1975). Mathematical modeling of inhalation exposure. *J. Combust. Toxicol.* **32,** 201–210.

Fiserosa-Bergerova, V., and Diaz, M. L. (1986). Determination and prediction of tissue: gas partition coefficients. *Int. Arch. Occ. Env. Health* **58,** 75–87.

Fisher, J. W., Whittaker, T. A., Taylor, D. H., Clewell, H. J. III, and Andersen, M. (1989). Physiologically based pharmacokinetic modeling of the pregnant rat: a multiroute exposure model for trichloroethylene and its metabolite, trichloroacetic acid. *Toxicol. Appl. Pharamcol.* **99,** 395–414.

Fisher, J. W., Whittaker, T. A., Taylor, D. H., Clewell, H. J. III, and Andersen, M. E. (1990). Physiologically based pharmacokinetic modeling of the lactating rat and nursing pup: a multiroute exposure model for trichloroethylene and its metabolite, trichloroacetic acid. *Toxicol. Appl. Pharamcol.* **102,** 497–513.

Fisher, J., Mahle, D., Bankston, L., Greene, R., and Gearhart, J. (1997). Lactational transfer of volatile chemicals in breast milk. *Ind. Hyg. Ass. J.* **58,** 425–431.

Gabrielsson, J. L., and Paalzow, L. K. (1983). A physiological pharmacokinetic model for morphine disposition in the pregnant rat. *J. Pharmacokinet. Biopharm.* **11,** 147–163.

Gabrielsson, J. L., Paalzow, L. K., and Nordstorm, L. A. (1984). A physiologically based pharamcokinetic model for theophylline disposition in the pregnant and nonpregnant rat. *J. Pharmacokinet. Biopharm.* **12,** 149–165.

Gabrielsson, J. L., Johansson, P., Bondesson, U., and Paalzow, L. K. (1985). Analysis of methodone disposition in the pregnant rat by means of a physiological flow model. *J. Pharmacokinet. Biopharm.* **13,** 355–372.

Garielsson, J. L., and Growth, T. (1988). An extended physiological pharmacokinetic model of methadone disposition in the rat: validation and sensitivity analysis. *J. Pharmacokinet. Biopharm.* **16,** 183–201.

Gargas, M. L., Burgess, R. J., Voisard, D. E., Cason, G. H., and Andersen, M. E. (1989). Partition coefficient of low molecular weight volatile chemicals in various tissues and liquids. *Toxicol. Appl. Pharmacol.* **98,** 87–99.

Gaylor, D. W., and Slikker, W. Jr. (1994). Modeling for risk assessment of neurotoxic effects. *Risk Anal.* **14,** 333–338.

Gray, D. G. (1995). A physiologically based pharmacokinetic model for methyl mercury in the pregnant rat and fetus. *Toxicol. Appl. Pharmacol.* **132,** 91–102.

Gearhart, J. M., Clewell, H. J. III, Crump, K. S., Shipp, A. M., and Silvers, A. (1995). Pharmacokinetic dose estimates of mercury in children and dose–response curves of performance tests in a large epidemiological study. *Water Air Soil Pollut.* **80,** 49–58.

Haddad S., Gad S. C., Tardif R., and Krishnan K. (1995). Statistical approaches for the validation of physiologically-based pharmacokinetic models. *Fund. Appl. Toxicol.* **15,** 48 (Abstract 258).

Haddad S., Pelekis M., and Krishnan K. (1996). A methodology for solving physiologically based pharmacokinetic models without the use of simulation softwares. *Toxicol. Lett.* **85,** 113–126.

Haggard, H. W. (1924). The absorption, distribution and elimination of ethyl ether. Analysis of the mechanism of the absorption and elimination of such a gas or vapor as ethyl ether. *J. Biol. Chem.* **59,** 753–770.

Hattis, D., White, P., Marmorstein, L., and Koch, P. (1990). Uncertainty in pharmacokinetic modeling for perchloroethylene. I. Comparison of model structure, parameters, and predictions for low-dose metabolism rates for model drived by differents authors. *Risk Anal.* **10,** 449–458.

Hetrick, D. M., Jarabek, A. M., and Travis, C. C. (1991). Sensitivity analysis for physiologically based pharamcokinetic models. *J. Pharmacokin. Biopharm.* **19,** 1–20.

Hoang K. C. T. (1995). Physiologically based pharmacokinetic models: mathematical fundamentals and simulation implementations. *Toxicol. Lett.* **79,** 99–106.

Houston J. B., and Carlile D. J. (1997). Prediction of hepatic clearance from microsomes hepatocytes, and liver slices. *Drug Metab. Rev.* **29,** 891–922.

Kaliszan, R., and Markuszewski, M. (1996). Brain/blood distribution described by a combination of partition coefficient and molecular mass. *Int. J. Pharm.* **145,** 9–16.

Kedderis G. L., and Held S. D. (1996). Prediction of furan pharmacokinetics from hepatocyte studies: Comparison of bioactivation and hepatic dosimetry in rats, mice, and humans. *Toxicol. Appl. Pharmacol.* **140,** 124–130.

Kety, S. S. (1951). The theory and application of the exchange of inert gas at the lungs. *Pharmacol. Rev.* **3,** 1–41.

Kim, C. S., Gargas, M. L., and Andersen, M. E. (1994). Pharmacokinetic modeling of 2,4-dichlorophenoxyacetic acid (2,4-D) in rat and in rabbit brain following single dose administration. *Toxicol. Lett.* **74,** 189–201.

Kim, C. S., Slikker, W. Jr., Binienda, Z., Gargas, M. L., and Andersen, M. E. (1995). Developmental of a physiologically based pharmacokinetic model for 2,4-dichlorophenoxyacetic acid dosimetry in discrete areas of the rabbit brain. *Neurotoxicol. Teratol.* **17,** 111–120.

Kim, C. S., Binienda, Z., and Sandberg, J. A. (1996). Construction of a physiologically based pharmacokinetic model for 2,4-dichlorophenoxyacetic acid dosimetry in the developing rabbit brain. *Toxicol. Appl. Pharmacol.* **136,** 250–259.

Kjelstrom T., Kennedy P., Wallis S., Stewart A., Fribert L., Lind B., Wutherspoon T., and Mantell C. (1989). *Nat. Swedish Environ. Protec. Board Rep.* 3642.

Knaak J. B., Al-Bayati M. A., and Raabe O. G. (1995). Development of partition coefficients, V_{max} and K_m values, and allometric relationships. *Toxicol. Lett.* **79,** 87–98.

Krewski, D., Wang, Y., Bartlett, S., and Krishnan, K. (1995). Uncertainty, variability and sensitivity analysis in physiological pharmacokinetic models. *J. Biopharm. Stat.* **5,** 245–271.

Krishnan K., and Andersen M. E. (1993). *In vitro* toxicology and risk assessment. *Altern. Meth. Toxicol.* **9,** 185–203.

Krishnan, K. and Andersen, M. E. (1994). Physiologically based pharmacokinetic modeling in toxicology. *Principles and Methods of Toxicology.* Third Edition, Ed. A. Wallace Haves. Raven Press. Ltd., New York.

Krishnan K., Gargas M. L., Fennell T. R., and Andersen M. E. (1992). A physiologically-based description of ethylene oxide dosimetry in the rat. *Toxicol. Ind. Health* **8,** 121–140.

Krishnan K., Haddad S., and Pelekis M. (1995). A simple index for representing discrepancy between simulations of physiological pharmacokinetic models and experimental data. *Toxicol. Ind. Health* **11,** 413–321.

Leroux, B. G., Leisenring, W. M., Moolgavkar, S. H., and Faustman, E. M. (1996). A biologically-based dose–response model for developmental toxicity. *Risk Anal.* **16,** 449–458.

Luecke, R. H., Wosilait, W. D., Pearce, B. A., and Young, J. F. (1994). A physiologically based pharmacokinetic computer model for human pregnancy. *Teratology* **49,** 90–103.

Luecke, R. H., Wosilait, W. D., Pearce, B. A., and Young, J. F. (1995). Mathematical representation of organ growth in the human embyro/fetus. *Int. J. Bio-Med. Comp.* **39,** 337–347.

Luecke, R. H., Wosilait, W. D., Pearce, B. A., and Young, J. F. (1997). A computer model and program for xenobiotic disposition during pregnancy. *Comp. Meth. Prog. Biomed.* **53,** 201–224.

Mapelson, W. W. (1963). An electric analog for uptake and elimination in man. *Acta Pharmacol. Toxicol.* **18,** 197–204.

O'Flaherty, E. J., Scott, W., Schreiner, C., and Beliles, R. P. (1992). A physiologically based kinetic model of rat and mouse gestation: disposition of a weak acid. *Toxicol. Appl. Pharmacol.* **112,** 245–256.

O'Flaherty, E. J., Nau, H., McCandless, D., Beliles, R. P., Schreiner, C. M., and Scott, W. J. Jr. (1995). Physiologically based pharmacokinetics of methoxyacetic acid: dose-effect considerations in C57BL/6 mice. *Teratology* **52,** 78–89.

Olanoff, L. S., and Anderson, J. M. (1980). Controlled release of tetracycline-III: a physiological pharmacokinetic model of the pregnant rat. *J. Pharmacokinet. Biopharm.* **8,** 599–620.

Paterson, S., and Mackay, D. (1989). Correlation of tissue, blood and air partition coefficients of volatile organic chemicals. *Br. J. Ind. Med.* **46,** 321–328.

Pelekis, M., Poulin, P., and Krishnan, K. (1995). An approach for incorporating tissue composition data into physiological based pharmacokinetic models. *Toxicol. Ind. Health* **11,** 511–522.

Pelekis, M., Krewski, D., and Krishnan, K. (1997). Physiologically based algebraic expressions for predicting steady-state toxicokinetics of inhaled vapors. *Toxicol. Methods* **7,** 205–225.

Poulin, P., and Krishnan, K. (1995a). A biologically-based algorithm for predicting human tissue: blood partition coefficients of organic chemicals. *Hum. Exp. Toxicol.* **14,** 273–280.

Poulin, P., and Krishnan, K. (1995b). An algorithm for predicting tissue : blood partition coefficients of organic chemicals from *n*-octanol:water partition coefficient data. *J. Toxicol. Environ. Health* **46,** 117–129.

Poulin P., and Krishnan K. (1996a). A tissue composition-based algorithm for predicting tissue: air partition coefficients of organic chemicals. *Toxicol. Appl. Pharmacol.* **136,** 126–130.

Poulin P., and Krishnan K. (1996b). A mechanistic algorithm for predicting blood: air partition coefficients of organic chemicals with the consideration of reversible binding in hemoglobin. *Toxicol. Appl. Pharmacol.* **136,** 131–137.

Poulin, P., and Krishnan, K. (1996c). Molecular structure–based prediction of the partition coefficients of organic chemicals for physiological pharmacokinetic model. *Toxicol. Meth.* **6,** 117–137.

Poulin, P., and Krishnan, K. (1998a). A quantitative structure–toxicokinetic relationship model for highly metabolized chemicals. *Alternative to Laboratory Animals* **26,** 46–59.

Poulin, P., and Krishnan, K. (1998b). Molecular structure–based prediction of the toxicokinetics of inhaled vapors in humans. (in press)

Ramsey, J. C., and Andersen, M. E. (1984). A physiologically-based description of the inhalation pharmacokinetics of styrene in the rats and humans. *Toxicol. Appl. Pharmacol.* **73,** 159–175.

Rapoport, S. I., Ohno, K., and Pettigrew, K. D. (1979). Drug entry into the brain. *Brain Res.* **172,** 354–359.

Rideout, V. C. (1991). *Mathematical and Computer Modeling of Physiological Systems.* Prentice-Hall, New York.

Riggs, D. S. (1970). The mathematical approach to physiological problems: a critical treatise. MIT Press, Cambridge, MA.

Shelley M. L., Andersen M. E., and Fisher J. W. (1988). An inhalation distribution model for the lactating mother and nursing child. *Toxicol. Lett.* **43,** 23–29.

Slikker, W. Jr., Crump, K. S., Andersen, M. E., and Bellinger, D. (1996). Symposium overview. Biologically based, quantitative risk assessment of neurotoxicants *Fund. Appl. Toxicol.* **29,** 18–30.

Strobel H. W., Geng J., Kawashima H., and Wang H. (1997). Cytochrome P450-dependent biotransformation of drugs and other xenobiotic substrates in neutral tissue. *Drug. Metab. Rev.* **29,** 1079–1105.

Terry, K. K., Elswick, B. A., Welsch, F., and Conolly, R. B. (1995). Development of a physiologically based pharmacokinetic model describing 2-methoxy-acetic acid disposition in the pregnant mouse. *Toxicol. Appl. Pharmacol.* **132,** 103–114.

Theorell, T. (1937a). Kinetics of distribution of substances administered to the body. I. The extravascular modes of administration. *Arch. Int. Pharmacodyn.* **57,** 205–225.

Theorell, T. (1937b). Kinetics of distribution of substances administered to the body. II. The intravascular modes of administration. *Arch. Int. Pharmacodyn.* **57,** 226–240.

Varkonyyi, P., Bruckner, J. V., and Gallo, J. M. (1995). Effect of parameter variability on physiologically-based pharmacokinetic model predicting drug concentrations. *J. Pharm. Sci.* **84,** 381–384.

Ward, K. W., Blumenthal, G. M., Welsch, F., and Pollack, G. M. (1997). Development of a physiologically based pharmacokinetic model to describe the disposition of methanol in pregnant rats and mice. *Toxicol. Appl. Pharmacol.* **145,** 311–322.

Welsch, F., Blumenthal, G. M., and Conolly, R. B. (1995). Physiologically based pharmacokinetic models applicable to organogenesis: extrapolation between species and potential use in prenatal toxicity risk assessments. *Toxicol. Lett.* **82/83,** 539–547.

Wosilait, W. D., Luecke, R. H., and Yooung, J. F. (1992). A mathematical analysis of human embyronic and fetal growth data. *Growth Dev. Aging* **56,** 249–257.

CHAPTER

42

Quantitative Models of Risk Assessment for Developmental Neurotoxicants

WILLIAM SLIKKER, JR.
DAVID W. GAYLOR
Division of Neurotoxicology
National Center for Toxicological Research
Food and Drug Administration
Jefferson, Arkansas 72079

I. Introduction

Risk assessment procedures for developmental neurotoxicants are rarely applied even though there is firm evidence for human toxicity from this class of toxicant. The majority of the known human teratogens are reported to affect the craniofacial region and/or nervous system of the developing human, and there is ample evidence that neuroactive agents can cross the placenta and enter the developing nervous system (Slikker, 1994a,b). With examples at hand such as fetal alcohol syndrome, fetal Minamata disease, and developmental lead toxicity (Slikker, 1994a), there is reasonable need for procedures to assess the risk of developmental neurotoxicants. Probably due in part to the reliance of developmental neurotoxicology on both the principles of neurotoxicology and developmental toxicology, progress in achieving widely accepted risk-assessment procedures for developmental neurotoxicity has been slow. This area has received some attention in the United States Food and Drug Administration's (FDA's) Red Book II revision (Sobotka *et al.*, 1996) and the United States Environmental Protection Agency's (EPA's) Neurotoxicity Risk Assessment Guidelines (Boyes *et al.*, 1997).

In general, examples of the application of quantitative risk assessment procedures to developmental neurotoxicants are very few due to the meager amount of dose–response data available in the open literature. A series of reproductive toxicity data sets have been examined systematically with both currently used reference

* Abbreviations: BD, benchmark dosage; EPA, [United States] Environmental Protection Agency; FDA, [United States] Food and Drug Administration; GD, gestational day; LOAEL, lowest observable adverse effect level; NOAEL, no observable adverse effect level; PD, postnatal day.

dose procedures and the benchmark dose method (Kavlock *et al.*, 1995). Although the many comparisons completed generally supported the concept that similar outcomes are realized when currently used and benchmark methods are applied, the report did not concentrate on developmental neurotoxicity endpoints. The focus of this chapter, therefore, is to carefully compare developmental neurotoxicity data sets using the currently available no observed adverse effect level/lowest observable adverse effect level (NOAEL/LOAEL) approach to the benchmark approach and to the quantitative methods for continuous data developed by Gaylor and Slikker (1990).

The aliphatic alcohols, a subset of the volatile organic solvents, are widely used industrial solvents, cosolvents, and chemical intermediates (Rowe and McCollister, 1982). Estimates of exposure populations in the United States are large for these agents, ranging from 1 to 3 million for the top three of 13 solvents surveyed (Seta *et al.*, 1988). Among these organic solvents with large exposure populations is the shortest chain aliphatic alcohol, methanol. This agent is used primarily in the health services and electronic, business, and chemical industries. Workers exposed include assemblers, janitors, laboratory technicians and mechanical machine operators. Approximately 25% of this workforce is female, so that nearly a quarter of a million women a year may be exposed to methanol vapor in the workplace.

A series of systematic studies of the aliphatic alcohols with 1 to 10 carbons revealed that, although several of these agents produced fetal effects after prenatal inhalation exposure in the rat, only methanol and 1-propanol produced their effects in the absence of maternal toxicity (Nelson *et al.*, 1990). Because of this developmental toxicity data and because the female workforce potentially exposed to methanol is nearly 10 times larger than that exposed to 1-propanol, methanol was selected for further analysis using quantitative risk assessment procedures.

II. Risk Assessment of Methanol

Pregnant Sprague–Dawley rats (weighing 200–300 g at mating) were housed and exposed individually to 0, 5,000, or 10,000 ppm methanol from gestational day (GD) 1–19 or 20,000 ppm methanol from GD 7–15. Whole body exposures in Hinner-type exposure chambers were conducted 7 hr/day and maternal body weights were recorded weekly. On GD 20, pregnant females were weighed individually and euthanized by CO_2 asphyxiation. Live fetuses were removed surgically from the uterus, counted, weighed, sexed, and examined for external malformations. One-half of the fetuses was randomly selected for examination of skeletal malformations, and the other half of the fetuses was examined for visceral malformations and variations (Nelson *et al.*, 1985).

Methanol produces developmental toxicity in the rat including exencephaly, skeletal malformations, visceral malformations, and reduced fetal weight in the absence of detectable effects in the maternal animals (Nelson *et al.*, 1985).

III. Fetal Weight

The average fetal weights on GD 20 of pregnant dams exposed to methanol by inhalation in the study conducted by Nelson *et al.* (1985) are shown in Table 1. There are two sources of variations in fetal weights; variance among fetuses within a litter (σ_f^2) and the variance among the average fetal weights from different litters (σ_l^2). The overall variance in fetal weights is $\sigma^2 = (\sigma_f^2 + \sigma_l^2)$ with a standard of deviation of σ. From an analysis of variance, the standard deviation was estimated to be 0.3 g for both female and male fetuses. That is, the standard deviation was slightly less than 10% of the average fetal weights for females and males.

Typically, body weights can be described statistically by a normal (Gaussian) distribution. Various percentiles of the distribution can be estimated from the sample average and standard deviation. Obviously, low birthweights are undesirable and at some point reflect adverse development. In the absence of a specified fetal weight associated with adverse development, a low percentile of the control fetal weight distribution can be selected that may be undesirable and certainly abnormal (Gaylor and Slikker, 1990; Slikker and Gaylor, 1995).

For purposes of illustration in this chapter, the first percentile has been chosen as the cutoff for abnormally low fetal weight. Certainly, other percentiles could have been chosen. For a normal distribution, the first percen-

TABLE 1 Average Fetal Weights on GD 20 of Pregnant Sprague–Dawley Rats Exposed to Methanol by Inhalation

	Weight (g)	
Methanol concentration (ppm)	Female	Male
0 (Control)	3.15	3.34
5,000	3.19	3.30
10,000	2.93	3.12
20,000	2.76	2.82

From Nelson, B. K., Brightwell, W. S., Kahn, A., Burg, J. R., Weigel, W. W., and Goad, P. T. 1985. Teratological assessment of methanol and ethanol at high inhalation levels in rats. *Fund. Appl. Toxicol.* **5(4),** 727–736.

tile is 2.33 standard deviations below the average. The average fetal weight for females for the control group [derived from the three control groups in the experiments conducted by Nelson *et al.* (1985)] was 3.08 g. Hence, the first percentile is estimated as 3.08 − 2.33 (0.30) = 2.38 g for females, slightly more than 20% below the average weight. Males averaged 3.28 g, resulting in an abnormal cutoff of 3.28 − 2.33 (0.30) = 2.58 g.

The next step is to estimate the average fetal weight as a function of methanol concentration. These averages along with the standard deviation can be used to estimate the proportion of fetuses with abnormally low fetal weights for various concentrations of methanol. The logistic function is frequently used to model body/organ weight (W)

$$W = \frac{W_o}{(1 - a) + ae^{bC}}$$

where W_o is the weight of the controls, C is concentration of a chemical, and a and b are estimated from experimental dose–response data (see Rubinow, 1973).

In the study by Nelson *et al.* (1985), the dose–response curves were similar for both females and males, the only difference being the higher weights for males. For purposes of illustration, the dose–response results presented by Nelson *et al.* (1985) for both sexes were normalized to the percent of their respective controls. These data are plotted with the normalized logistic fetal weight curve in Fig. 1. Using the computer program PROC NLIN in SAS®, the estimated dose–response curve for females is

$$W = \frac{3.08}{0.93 + 0.07\, e^{.00006C}}$$

and for males

$$W = \frac{3.28}{0.93 + 0.07\, e^{.00006C}}$$

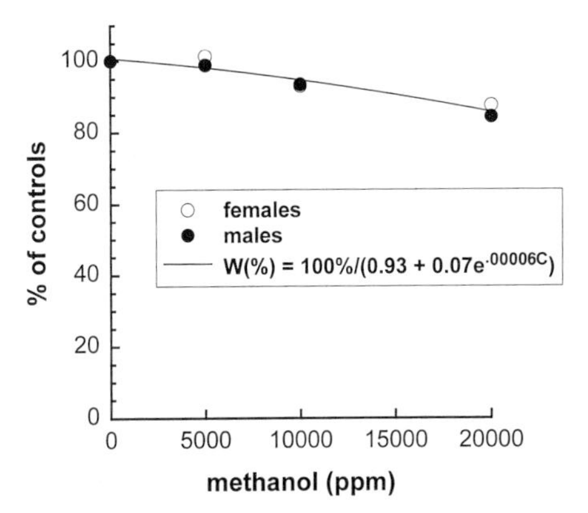

FIGURE 1 Fetal body weight normalized to controls (W%) versus methanol concentration (Males ●, Females ○). W(%) = 100% / ($0.93 + 0.07^{.00006C}$), where C is methanol concentration (ppm).

Using the average weights estimated by the logistic model and the standard deviation, the probability of abnormally low fetal weights can be estimated as a function of methanol concentration. For example, suppose we wish to estimate the concentration in which the proportion of abnormally low fetal weights is double—that is, from 1% to 2%. The 2% tile for a normal distribution is 2.05 standard deviations below the average. For females, the abnormal cutoff is 2.38 g. Two percent of the fetuses are estimated to be below this value when the average is equal to 2.38 + 2.05 (0.30) = 3.00 g. This is estimated to occur from logistic model at the concentration where

$$W = \frac{3.08}{0.93 + 0.07\, e^{.00006C}} = 3.00$$

Solving for C gives 5,400 ppm as the concentration in which the proportion of low fetal weights are estimated to double.

Because the average values and abnormal cutoff are increased by the same amount for males, the same concentration of 5,400 ppm is estimated to double the probability of low fetal weight in males. Concentrations with other levels of excess risk (additional probability that fetal weight is below the first percentile of control animals) are listed in Table 2.

The previous calculations assume that the standard deviation did not change with concentration. There is evidence in the experimental data that the standard deviation did increase at 20,000 ppm, but not at 10,000 ppm. The standard deviation in the 20,000 ppm group was 0.47. This poses an additional problem for calculations somewhat higher than 10,000 ppm. If the standard deviation is taken to be 0.47 at more than 10,000 ppm, the concentration corresponding to an excess risk of 10% needs to be recalculated. An excess risk of 10% occurs in which 11% of the fetuses have abnormally low weights. This occurs at 1.23 standard deviations below the average. For females, the abnormal cutoff is still 2.38 g based on the average and standard deviation of control animals. Eleven percent of the fetuses are estimated to be below this value when the

TABLE 2 Relationship between Concentrations of Methanol and Levels of Excess Risk[a]

Excess risk	Concentrations of methanol (ppm)
0.10	16,000
0.01	5,400
0.001	980

[a] Additional probability that fetal weight is below the first percentile of control animals.

average is equal to 2.38 + 1.23 (.47) = 2.96 g. For the logistic model, this occurs at a concentration *C*, where

$$W = 2.96 = \frac{3.08}{0.93 + 0.07\, e^{.00006C}}$$

Solving for *C* gives 7,800 ppm. Thus, the concentration producing an excess risk of 10% is between 7,800 and 16,000 ppm. If a relationship between the standard deviation and concentration could be estimated, a more precise estimate could be obtained.

Steps in the quantitative risk assessment of continuous data are:

1. Estimate the average and standard deviation of fetal weight for the control animals
2. Estimate a low percentile to identify abnormal values, from the statistical distribution of values, in this case the normal distribution
3. Establish a dose–response relationship to estimate average values as a function of dose (concentration)
4. Calculate the probabilities of values in the abnormal range from the estimated averages and standard deviation.

Being able to estimate risks provides a risk manager with additional information for making decisions. In the previous example, the NOAEL occurred at 5,000 ppm. If a typical safety factor of 10 were used for extrapolation from animals to humans and a safety factor of 10 were used to account for sensitive humans, an acceptable level for human exposure based on fetal weight might be set at 5,000/(10 × 10) = 50 ppm.

The concentration that doubles the proportion of abnormal fetal weights from 1% to 2% was estimated to be 5,400 ppm. Because this already accounts for sensitive individuals, an additional safety factor of 10 to account for greater sensitivity in humans than in rats gives an acceptable concentration of 5400/10 = 540 ppm. The risk at 980 ppm was estimated at 0.001. The risk at the NOAEL/100 = 50 ppm is estimated to be negligible. The ability to estimate risks associated with low exposures provides a risk assessor with more flexibility than simply the NOAEL/safety factor approach.

IV. Brain Malformations

The proportion of litters with one or more fetuses with malformations of the brain (hydrocephaly, exencephaly, or encephalocele) in the study by Nelson *et al.* (1985) was 0/45, 0/13, 2/15, and 4/15 at 0, 5,000, 10,000, and 20,000 ppm of methanol, respectively. Here, the brain malformations are considered adverse effects and the risk (proportions) are observed directly. Using the computer program GLOBAL 82 (Sci. Research Systems, Inc., Ruston, Louisiana, 1982), these data were fit by a polynomial–exponential model commonly used for quantal data. The proportion of abnormal litters (*P*) is estimated to be

$$P = 1 - e^{-(8.725 \times 10^{-10} \times C^2)}$$

where *C* is the concentration of methanol. Estimates of concentrations associated with various levels of risk are shown in Table 3 and Fig. 2.

The NOAEL was 5,000 ppm. If a typical safety factor of 10 were used for extrapolation from animals to humans and a safety factor of 10 were used to account for sensitive humans, an acceptable level for human exposure based on brain malformations might be set at 5,000/(10 × 10) = 50 ppm. From the model, the estimate of risk of brain malformations at 50 ppm is negligible—that is, for the safety factor approach, by definition, the NOAEL divided by the appropriate safety factors defines a dose at or below which the risk is negligible.

V. Discussion

Developmental toxicity studies of inhaled methanol conducted in the mouse yielded results comparable to the present rat studies. Rogers *et al.* (1993) reported that pregnant CD-1 mice exposed via inhalation at 1,000–15,000 ppm methanol for 7 hr/day on GD 6–15 exhibited reduced fetal weight gain at 10,000 ppm and higher and increased incidence of soft-tissue abnomalies at 5,000 ppm and higher. Similar abnormalities were observed by others in the mouse except that dosing was only 6 hr/day and significant effects were observed at 10,000 ppm and higher (Bolan *et al.*, 1993). A log–logistic dose–response model was applied to the exencephaly, cleft palate, or resorption data, and the benchmark dosage (BD) (the lower 95% confidence interval of the maximum likelihood estimates) was calculated to be 3,078 ppm. This BD for soft-tissue malformation or resorption in mice compares favorably with the excess risk of 0.01 in the present rat study (3,400 ppm methanol). It is argued by the authors that a quantitative approach has advantages over the NOAEL approach

TABLE 3 Relationship between Concentrations of Methanol and Levels of Excess Risk for Brain Malformations

Excess risk	Concentrations of methanol (ppm)
0.10	11,000
0.01	3,400
0.001	1,100

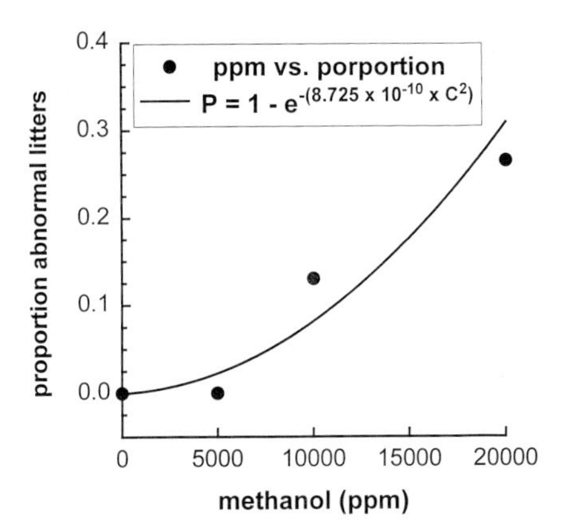

FIGURE 2 Proportion (P) of litters with brain malformations versus methanol concentration in ppm (C). $P = 1 - e^{-(8.725 \times 10^{-10} \times C^2)}$.

in that the NOAEL represents a statistical or often subjective no effect level that is dependent on study design and solely on a single point on the dose–response curve.

Developmental exposure to inhaled methanol has been studied in several species including the rat and mouse (Pollack and Brower, 1996). During inhalation exposures, the authors reported that the rate of methanol accumulation in the mouse was 2–3 times greater than in the rat, despite the fact that the mouse eliminates methanol twice as fast as the rat. The explanation offered by the authors is that the more rapid respiration rate and higher nasal cavity absorption of methanol in the mouse resulted in the higher methanol accumulation in the mouse. In both species, however, fetal methanol concentrations were approximately the same as in the mother.

Although it has been reported that neonatal rodent behavior is altered by prenatal methanol exposure, the lack of dose–response data for behavioral endpoints precludes these endpoints for use in quantitative risk assessments. In studies where an oral dose of 2% methanol in the drinking water on GD 15–17 or 17–19 was administered (average daily methanol intake of 2.5 gm/kg), several neonatal behavioral measures were altered as compared to vehicle control treated rats (Infurna and Weiss, 1986). Even though litter size, birthweight, and infant mortality did not differ between groups, methanol-exposed pups exhibited delayed onset of suckling on postnatal day (PD) 1 and extended nest locating time on PD 10. Although these data suggest that postnatal behavioral endpoints may be sensitive indicators of prenatal methanol exposure, the oral route of exposure makes comparison to inhalation studies problematic. In another study, (Stanton *et al.,* 1995) however, methanol exposure was via inhalation, and little evidence of effect was observed on a broad battery of tests beyond PD 1. Stanton *et al.* (1995) dosed Long–Evans rats with 15,000 ppm methanol or air for 7 hr/day on GD 7–19. Methanol exposure reduced maternal body weight transiently (4–7%) on GD 8–10 and pup body weight by 5% on PD 1, but no alteration in pup motor activity (PD 13–21), olfactory learning (PD 18), T-maze learning (PD 23–24), acoustic startle response (PD 24, 60), passive avoidance (PD 72), or visual evoked potentials (PD 160) were observed. This general lack of robust behavioral effects observed in offspring of methanol vapor–exposed pregnant dams is reinforced by the data of Weiss *et al.* (1996). The review committee concluded that of the many tasks conducted, only a few documented significant effects at a methanol dose of 4,500 ppm (6 hr/day) from GD 6 through PD 21 for both dams and pups (Weiss *et al.,* 1996). It remains to be determined if prenatal exposure to methanol vapor in the pregnant monkey results in significant dose–related postnatal behavior deficits in the neonatal or infant monkey offspring (Burbacher *et al.,* 1997).

VI. Summary

Even though biologically-based risk assessment guidelines for noncancer endpoints are gaining acceptance, quantitative and/or dose-response risk assessment models are not widely utilized. As new guidelines are implemented, dose-response data concerning critical, biologically-based biomarkers will be available as a basis of quantitative risk assessment. Strategies demonstrating the application of quantitative approaches need to be elaborated in order to evaluate their usefulness. A quantitative, dose-response risk assessment of the volatile organic compound methanol (ME) was completed in pregnant rats.

Methanol, of the 13 aliphatic alcohols examined, has substantial human exposure to women in the workplace and the ability, at high doses, to produce developmental neurotoxicity in the developing rodent without general maternal toxicity. Fetal weights for the various treatment groups (0, 5000, 10000 and 20000 ppm methanol) were measured: a) the average and standard deviation of fetal weight for the control animals were estimated, b) abnormal values were identified as the first percentile (2.33 standard deviations below the average) from the statistical distribution of the normally distributed control values, c) a dose-response relationship was established to estimate average values as a function of dose, and d) the probabilities of values in the abnormal fetal weight range were calculated from the estimated averages and their standard deviations. The fetal weight outcome indicated that excess risks of 0.10, 0.01 and 0.001 resulted from ME concentrations of 16,000, 5,400 and 980 ppm, respectively. For fetal brain abnormalities (i.e., hydrocephaly, exencephaly or encephalocele), excess risks of 0.1, 0.01 and 0.001 resulted from ME concentrations of 11,000, 3,400 and 1,100 ppm, respectively.

In conclusion, the use of continuous fetal weight data was as sensitive as fetal brain malformation data in estimating risk to methanol inhalation exposure in the developing rat. With a 10 fold adjustment for species differences, 980/10 = 98 and 1100/10 = 110 ppm, the quantitative, dose-response risk assessment approach provided estimations of risk (1 in 1,000) comparable to the NOAEL approach of 50 ppm. Both the NOAEL and quantitative, dose-response approaches resulted in doses comparable to but slightly lower than the existing permissible exposure limits (PEL) of 200 ppm. Advantages of the quantitative approach include the use of all the data defining the dose-response curve, the use of both continuous and quantal data, and the calculation of the risk resulting from exposure to a given dose.

References

Bolon, B., Dorman, D. C., Janszen, D., Morgan, K. T., and Welsch, F. (1993). Phase-specific developmental toxicity in mice following maternal methanol inhalation. *Fund. Appl. Toxicol.* **21,** 508–516.

Boyes, W. K., Dourson, M. L., Patterson, J., Tilson, H. A., Sette, W. F., MacPhail, R. C., Li, A. A., and O'Donoghue, J. L. (1997). EPA's neurotoxicity risk assessment guidelines. *Fund. Appl. Toxicol.* **40,** 175–184.

Burbacher, T., Grant, K. S., and Shen, D. (1997). *Reproductive and Offspring Developmental Effects of Inhaled Methanol in Nonhuman Primates.* Neurobehavioral Teratology Society, Abstract.

Gaylor, D. W., and Slikker, W. Jr. (1990). Risk assessment for neurotoxic effects. *Neurotoxicology* **11,** 211–218.

Infurna, R., and Weiss, B. (1986). Neonatal behavioral toxicity in rats following prenatal exposure to methanol. *Teratology* **33,** 259–265.

Kavlock, R. J., Allen, B. C., Faustman, E. M., and Kimmel, C. A. (1995). Dose–response assessments for developmental toxicity. IV. Benchmark doses for fetal weight changes. *Fund. Appl. Toxicol.* **26,** 211–222.

Nelson, B. K., Brightwell, W. S., Mackenzie, D. R., Kahn, A., Burg, J. R., Weigel, W. W., and Goad, P. T. (1985). Teratological assessment of methanol and ethanol at high inhalation levels in rats. *Fund. Appl. Toxicol.* **5(4),** 727–736.

Nelson, B. K., Brightwell, W. S., and Krieg, E. F. (1990). Developmental toxicology of industrial alcohols: a summary of 13 alcohols administered by inhalation to rats. *Toxicol. Indus. Health* **6,** 373–387.

Pollack, G. M., and Brouwer, K. L. R. (1996). Maternal–fetal pharmacokinetics of methanol. Health Effects Institute, Research Report No. 74, Topsfield, MA.

Rogers, J. M., Mole, M. L., Chernoff, N., Barbee, B. D., Turner, C. I., Logsdon, T. R., and Kavlock, R. J. (1993). The developmental toxicity of inhaled methanol in the CD-1 mouse, with quantitative dose–response modeling for estimation of benchmark doses. *Teratology* **47,** 175–188.

Rowe, V. K., and McCollister, S. B. (1982). Alcohols. In G. D. Clayton and F. E. Clayton, eds. *Patty's Industrial Hygiene and Toxicology,* Third Revised Edition, Volume 2C. Toxicology. John Wiley & Sons, New York, pp. 4527–4708.

Rubinow, S. I. (1973). *Mathematical Problems in the Biological Sciences.* Society for Industrial and Applied Mathematics, Philadelphia, PA.

Seta, J. A., Sundin, D. S., and Pedersen, D. H. (1988). *National Occupational Exposure Survey, Vol. 1. Survey Manual,* DHHS (NIOSH) Publication No. 88-106.

Slikker, W. Jr. (1994a). Principles of developmental neurotoxicology. *Neurotoxicology,* **15(1),** 11–16.

Slikker, W. Jr. (1994b). Placental transfer and pharmacokinetics of developmental neurotoxicants. In L. W. Chang, ed. *Principles of Neurotoxicology.* Marcel Dekker, Inc., New York, pp. 659–680.

Slikker, W. Jr., and Gaylor, D. (1995). Concepts on quantitative risk assessment of neurotoxicants. In L. W. Chang and W. Slikker, Jr., eds. *Neurotoxicology: Approaches and Methods* Academic Press, San Diego, pp. 771–776.

Sobotka, T. J., Ekelman, K. B., Slikker, W. Jr., Raffaele, K., and Hattan, D. G. (1996). Food and Drug Administration proposed guidelines for neurotoxicological testing of food chemicals. *Neurotoxicology,* **17(3–4),** 825–836.

Stanton, M. E., Crofton, K. M., Gray, L. E., Gordon, C. J., Boyes, W. K., Mole, M. L., Peele, D. B., and Bushnell, P. J. (1995). Assessment of offspring development and behavior following gestational exposure to inhaled methanol in the rat. *Fund. Appl. Toxicol.* **28,** 100–110.

Weiss, B., Stern, S., Soderholm, S. C., Cox, C., Sharma, A., Inglis, G. B., Preston, R., Balys, M. Reuhl, K. R., and Gelein, R. (1996). *Developmental Neurotoxicity of Methanol Exposure by Inhalation in Rats.* Health Effects Institute, Research Report No. 73.

INDEX

B

D

E

F

N

O

P

Q

R